W9-BFR-693

McGraw-Hill's

NATIONAL ELECTRICAL CODE® HANDBOOK

McGraw-Hill's

NATIONAL ELECTRICAL CODE® HANDBOOK

Twenty-second Edition

Based on the Current 1996
National Electrical Code®

by

J. F. McPartland
Senior Editorial Consultant
EC&M
Tenafly, New Jersey

and

Brian J. McPartland
Editor
Electrical Contractors
"Design and Installation Update"
Tappan, New York

McGRAW-HILL

New York San Francisco Washington, D.C. Auckland Bogotá
Caracas Lisbon London Madrid Mexico City Milan
Montreal New Delhi San Juan Singapore
Sydney Tokyo Toronto

"The Library of Congress has cataloged this
serial publication as follows:"

McGraw-Hill's national electrical code handbook.—16th ed- —
New York: McGraw-Hill, c1979-

v. : ill. ; 22 cm.

Editor: J. F. McPartland, 1979-
Based on 1978- ed. of: National Fire Protection Association. National electrical code.
Continues: NFPA handbook of the National electrical code.
ISSN 0277-6758 = McGraw-Hill's national electrical code handbook.

ISBN 0-07-045901-0

1. Electric engineering—Insurance requirements—Collected works. I. McPartland, Joseph F. II. National Fire Protection Association. National electrical code. III. Title: National electrical code handbook.

TK260.N2 621.319'24'0218 81-642618
AACR 2 MARC-S

Library of Congress [8311]

This 22nd edition is based on the 1996 NEC®.

Although every effort has been made to make the explanation of the Code *accurate, neither the Publisher nor the Author assumes any liability for damages that may result from the use of the Handbook.*

McGraw-Hill

A Division of The ***McGraw-Hill*** *Companies*

234567890 DOC/DOC 9010987

ISBN 0-07-045992-4

The sponsoring editor for this book was Harold B. Crawford, the editing supervisor was Margaret Lamb, and the production supervisor was Pamela A. Pelton. It was set in Melior by North Market Street Graphics.
Printed and bound by R. R. Donnelley & Sons.

This book is printed on acid-free paper.

Contents

Preface

This is a reference book of commentary, discussion, and analysis on the most commonly encountered rules of the 1996 National Electrical Code. Designed to be used in conjunction with the 1996 NE Code book published by the National Fire Protection Association, this Handbook presents thousands of illustrations—diagrams and photos—to supplement the detailed text in explaining and clarifying NEC regulations. Description of the background and rationale of specific Code rules is aimed at affording a broader, deeper, and readily developed understanding of the meaning and application of those rules. The style of presentation is conversational and intended to facilitate a quick, practical grasp of the ideas and concepts that are couched in the necessarily terse, stiff, quasi-legal language of the NEC document itself.

This Handbook follows the order of "articles" as presented in the NE Code book, starting with "Article 90" and proceeding through "Article 810" and "Chapter 9—Tables and Examples." The Code rules are referenced by "section" numbers (e.g., "250-45. Equipment Connected by Cord and Plug"). This format assures quick and easy correlation between NEC sections and the discussions and explanations of the rules involved. This companion reference to the NEC book expands on the rules and presents common interpretations that have been put on the many difficult and controversial Code requirements. A user of this Handbook should refer to the NEC book for the precise wording of a rule and then refer to the corresponding section number in this Handbook for a practical evaluation of the details.

Because many NEC rules do not present difficulty in understanding or interpretation, not all sections are referenced. But the vast majority of sections are covered, especially all sections that have proved troublesome or controversial. And particular emphasis is given to changes and additions that have been made in Code rules over recent editions of the NEC. Although this new edition of *McGraw-Hill's National Electrical Code® Handbook* does not contain the complete wording of the NE Code book, it does contain much greater analysis and interpretation than any other so-called Handbook contains, and it is more thoroughly illustrated than any other Code Handbook.

Today, the universal importance of the NE Code has been established by the federal government (OSHA and other safety-related departments), by state and local inspection agencies, and by all kinds of private companies and organizations. To meet the great need for information on the NEC, McGraw-Hill has been publishing a handbook on the National Electrical Code since 1932. Originally developed by Arthur L. Abbott in that year, the Handbook has been carried on in successive editions for each revision of the National Electrical Code.

One final point—words such as "workmanlike" are taken directly from the Code and are intended in a purely generic sense. Their use is in no way meant to deny the role women already play in the electrical industries or their importance to the field.

Joseph F. McPartland
Brian J. McPartland

Introduction to the National Electrical Code®

McGraw-Hill's National Electrical Code® Handbook is based on the 1996 Edition of the National Electrical Code as developed by the National Electrical Code Committee of the American National Standards Institute (ANSI), sponsored by the National Fire Protection Association (NFPA). The National Electrical Code is identified by the designation NFPA No. 70-1996. The NFPA adopted the 1996 Code at the NFPA Annual Meeting held in Washington, D.C., on May 25, 1995.

The National Electrical Code, as its name implies, is a nationally accepted guide to the safe installation of electrical wiring and equipment. The committee sponsoring its development includes all parties of interest having technical competence in the field, working together with the sole objective of safeguarding the public in its utilization of electricity. The procedures under which the Code is prepared provide for the orderly introduction of new developments and improvements in the art, with particular emphasis on safety from the standpoint of its end use. The rules of procedure under which the National Electrical Code Committee operates are published in each official edition of the Code and in separate pamphlet form so that all concerned may have full information and free access to the operating procedures of the sponsoring committee. The Code has been a big factor in the growth and wide acceptance of the use of electrical energy for light and power and for heat, radio, television, signaling, and other purposes from the date of its first appearance (1897) to the present.

The National Electrical Code is primarily designed for use by trained electrical people and is necessarily terse in its wording.

The sponsoring National Electrical Code Committee is composed of a Technical Correlating Committee and 20 Code-Making Panels, each responsible for one or more Articles in the Code. Each Panel is composed of experienced individuals representing balanced interests of all segments of the industry and the public concerned with the subject matter. The internal operations of the sponsoring committee are guided by a *Manual of Procedure for Code-Making Panels.* This Manual is published in pamphlet form, and copies are available from the NFPA, Batterymarch Park, Quincy, MA 02269.

The National Fire Protection Association also has organized an Electrical Section to provide the opportunity for NFPA members interested in electrical safety to become better informed and to contribute to the development of NFPA electrical standards. This new Handbook reflects the fact that the National Electrical Code was revised for the 1996 edition, requiring an updating of the previous Handbook which was based on the 1993 Edition of the Code. The established schedule of the National Electrical Code Committee contemplates a new edition of the National Electrical Code every three years. Provision is made under the rules of procedure for handling urgent emergency matters through a Tentative Interim Amendment Procedure. The Committee also has established rules for rendering Formal (sometimes called Official) Interpretations. Two general forms of findings for such Interpretations are recognized: (1) those making an interpretation of literal text and (2) those making an interpretation of the intent of the National Electrical Code when a particular rule was adopted. All Tentative Interim Amendments and Formal Interpretations are published by the NFPA as they are issued, and notices are sent to all interested trade papers in the electrical industry.

The National Electrical Code is purely advisory as far as the National Fire Protection Association is concerned but is very widely used as the basis of law and for legal regulatory purposes. The Code is administered by various local inspection agencies, whose decisions govern the actual application of the National Electrical Code to individual installations. Local inspectors are largely members of the International Association of Electrical Inspectors, 901 Waterfall Way, Suite 602, Richardson, TX 75080-7702. This organization, the National Electrical Manufacturers Association, the National Electrical Contractors Association, the Edison Electric Institute, the Underwriters' Laboratories, Inc., the International Brotherhood of Electrical Workers, governmental groups, and independent experts all contribute to the development and application of the National Electrical Code.

Brief History of the National Electrical Code®

The National Electrical Code was originally drawn in 1897 as a result of the united efforts of various insurance, electrical, architectural, and allied interests. The original Code was prepared by the National Conference on Standard Electrical Rules, composed of delegates from various interested national associations. Prior to this, acting on an 1881 resolution of the National Association of Fire Engineers' meeting in Richmond, Virginia, a basis for the first Code was suggested to cover such items as identification of the white wire, the use of single disconnect devices, and the use of insulated conduit.

In 1911, the National Conference of Standard Electrical Rules was disbanded, and since that year, the National Fire Protection Association (NFPA) has acted as sponsor of the National Electrical Code. Beginning with the 1920 edition, the National Electrical Code has been under the further auspices of the American National Standards Institute (and its predecessor organizations, United States of America Standards Institute, and the American Standards Association), with the NFPA continuing in its role as Administrative Sponsor. Since that date, the Committee has been identified as "ANSI Standards Committee C1" (formerly "USAS C1" or "ASA C1").

Major milestones in the continued updating of successive issues of the National Electrical Code since 1911 appeared in 1923, when the Code was rearranged and rewritten; in 1937, when it was editorially revised so that all the general rules would appear in the first chapters followed by supplementary rules in the following chapters; and in 1959, when it was editorially revised to incorporate a new numbering system under which each Section of each Article is identified by the Article Number preceding the Section Number.

For many years the National Electrical Code was published by the National Board of Fire Underwriters (now American Insurance Association), and this public service of the National Board helped immensely in bringing about the wide public acceptance which the Code now enjoys. It is recognized as the most widely adopted Code of standard practices in the U.S.A. The National Fire Protection Association first printed the document in pamphlet form in 1951 and has, since that year, supplied the Code for distribution to the public through its own office and through the American National Standards Institute. The National Electrical Code also appears in the National Fire Codes, issued annually by the National Fire Protection Association.

About the 1996 NE Code®

The trend for ever-increasing numbers of proposals for changes and adopted changes in successive editions of the NEC has not reversed itself. The 1996 NEC is based on a record number of proposals and comments that have resulted in literally hundreds of additions, deletions, and other modifications—both minor and major. There are completely new articles covering equipment and applications not previously covered by the Code. There are also new regulations and radical changes in old regulations that affect the widest possible range of everyday electrical design considerations and installation details.

Much of the analysis and discussion about the specifics related to the various additions, deletions, and modifications in the 1996 NEC are based the information available in two familiar documents that now have new titles. The publications formerly known as the "Technical Committee Report" (TCR) and the "Technical Committee Documentation" (TCD) have been rechristened the "Report on Proposals" (ROP) and the "Report on Comments" (ROC), respectively. From a practical standpoint, there is no substantive change. The renamed documents provide virtually the same information that their predecessors did. Both the 1995 ROP and the 1995 ROC for the 1996 NEC are available from the National Fire Protection Association, Batterymarch Park, Quincy, MA 02269, or by phoning, 1-(800) 344-3555. Those two documents are highly recommended references that will facilitate completion of the Herculean task that looms immediately ahead for every designer and installer.

Everyone involved in the layout, selection, estimation, specification, inspection, as well as installation, maintenance, replacement, etc., of electrical systems and equipment must make every effort to become as thoroughly versed in and completely familiar with the intimate details related to the individual change as is possible. And, this must be done as soon as possible.

Clearly, compliance with the NEC is more important than ever. As evidenced by the skyrocketing numbers of suits filed against *electrical* designers and installers continues. In addition, inspectors everywhere are more knowledgeable and competent and they are exercising more rigorous enforcement and generally tightening control over the performance of electrical work. Another factor is the Occupational Safety and Health Administration's *Design Safety Standard for Electrical Installations.* That standard, which borrowed heavily from the rules and regulations given in the NEC, is federal law and applies to all places of employment in general industry occupancies. Although the OSHA *Design-Safety Standard* is based heavily on the NEC, due to the relatively dynamic nature of the NEC, there will eventually be discrepancies. But, for those instances where a more recent edition of the Code permits something that is prohibited by the OSHA standard, OSHA officials have indicated that such an infraction—although still an infraction—will be viewed as what OSHA refers to as a "de minimus violation," which essentially boils down to no fine. Of course that is not always the case. "Listing" and "labeling" of products by third party testing facilities is *strongly recommended* by the NEC, and is made *mandatory* by Occupational Safety and Health Administration (OSHA) *Design Safety Standard for Electrical Systems.* The OSHA requirement for certification does, and always has, taken precedence over the less stringent position of the NEC regarding listing of equipment. To be certain as to whether or not OSHA must be followed instead of a more recent edition of the NEC—which will be the minority of times—one can write to Mr. Ray Donnelly, OSHA, Office of General Industry Compliance Assistance, 200 Constitution Ave, NW, Washington, DC 20210. The impact of the NEC—even on OSHA regulations, which are federal law—is a great indicator of the Code's far reaching effect.

The fact that the application of electrical energy for light, power, control, signaling, and voice/data communication, as well as for computer processing and computerized process-control continues to grow at a breakneck pace also demands greater attention to the Code. As the electrical percentage of the construction dollar continues upward, the high-profile and very visible nature of electrical usage demand closer, more penetrating concern for safety in electrical design and installation. In today's sealed buildings, with the entire interior environment dependent on the electrical supply, reliability and continuity of operation has become critical. Those realties demand not only a concern for eliminating shock and fire hazards, but also a concern for continuity of supply, which is essential for the safety of people, and, in today's business and industry, to protect data and processes, as well.

And, of course, one critical factor that, perhaps, emphasizes the importance of Code-expertise more than anything else, is the extremely competitive nature of construction and modernization projects, today. The restricted market and the overwhelming pressure to economize have caused some to employ extreme methods to achieve those ends without full attention to safety. The Code represents an effective, commendable, and, in many instances, legally binding standard that *must* be satisfied, which acts as a barrier to any compromises with basic electrical safety. It is a democratically developed consensus standard that the electrical industry has determined are the essential foundation for electrical design and installation; and compliance with the NEC will dictate a minimum dollar value for any project.

In this Handbook, the discussion delves into the letter and intent of Code rules. Read and study the material carefully. Talk it over with your associates; engage in as much discussion as possible. In particular, check out any questions or problems with your local inspection authorities. It is true that only time and discussion provide final answers on how some of the rules are to be interpreted. But now is the time to start. Do not delay. Use this Handbook to begin a regular, continuous, and enthusiastic program of updating yourself on this big new Code.

This Handbook's illustrated analysis of the 1996 NE Code is most effectively used by having your copy of the new Code book at hand and referring to each section as it is discussed. The commentary given here is intended to supplement and clarify the actual wording of the Code rules as given in the Code book itself.

Highlights of the 1996 NE Code® Changes

The following listed articles and sections indicate the major areas of change to the 1996 National Electrical Code. These articles and sections were amended or are completely new to the 1996 National Electrical Code.

Note: In the National Electrical Code itself, changes and additions of specific Code sections can be readily identified by a vertical line in the margin. A dot in the margin indicates deletions.

Article 90. Introduction

Sec. 90-2(a)(5) and (b)(5). Scope

Article 100. Definitions

- Bathrooms [relocated from 210-8(b)]
- Energized
- Live Parts
- Motor Control Center
- Nonlinear Load
- Raceway
- Service Conductors

Article 110. Requirements for Electrical Installations

Part A. General

Sec. 110-4. Voltages
Sec. 110-12(c). Integrity of Electrical Equipment and Connections
Sec. 110-14(c). Temperature Limitations
Sec. 110-16(a). Working Clearances and Exceptions 1 and 3
Sec. 110-16(e). Illumination
Sec. 110-16(f). Headroom
Sec. 110-17. Guarding of Live Parts

Part B. Over 600 Volts, Nominal

Sec. 110-31. Enclosure for Electrical Installations

Article 200. Use and Identification of Grounded Conductors

Sec. 200-6(d). Grounded Conductors of Different Systems
Sec. 200-10. Identification of Terminals: (b) Receptacles, Plugs, and Connectors

Article 210. Branch Circuits

Part A. General Provisions

Sec. 210-4(a). General (FPN) (new)
Sec. 210-6(c)(1), (2), and (4). 277 Volts to Ground
Sec. 210-7(d). Replacements
Sec. 210-8. Ground-Fault Circuit-Interrupter for Personal Protection: (a) Dwelling Units and (b) Other than Dwelling Units
Sec. 210-9. Circuits Derived from Autotransformers

Part B. Branch-Circuit Ratings

Sec. 210-21(b)(1). Receptacles, Exception Nos. 1 and 2
Sec. 210-22(c). Other Loads
Sec. 210-23. Permissible Loads: (a) 15- and 20-Ampere Branch Circuits, Exception
Sec. 210-24. Branch-Circuit Requirements—Summary
Sec. 210-25. Common Area Branch Circuits
Sec. 210-52(a). General Provisions
Sec. 210-52(b). Small Appliances
Sec. 210-52(c). Counter Tops
Sec. 210.52(d). Bathrooms
Sec. 210.52(e). Outdoor Outlets
Sec. 210-63. Heating, Air-Conditioning, and Refrigeration Equipment Outlet
Sec. 210-70. Lighting Outlets Required

Article 220. Branch-Circuit, Feeder, and Service Calculations

Sec. 220-3(a). Continuous and Noncontinuous Loads
Sec. 220-10(b). Continuous and Noncontinuous Loads
Sec. 220-17. Appliance Load—Dwelling Units
Sec. 220-35. Optional Calculations for Additional Loads to Existing Installations

Article 225. Outside Branch Circuits and Feeders

Sec. 225-8(b). Disconnect Required for Each, Exception No. 3
Sec. 225-8(d). Identification
Sec. 225-9. Overcurrent Protection
Sec. 225-19(d). Final Spans

Article 230. Services

Sec. 230-2(b). Identification
Sec. 230-9. Clearances from Building Openings
Sec. 230-24(b). Vertical Clearance from Ground
Sec. 230-28. Service Masts as Supports
Sec. 230-40. Number of Service-Entrance Conductor Sets
Sec. 230-43. Wiring Methods for 600 Volts, Nominal, or Less
Sec. 230-46. Unspliced Conductors
Sec. 230-95. Ground-Fault Protection of Equipment

Article 240. Overcurrent Protection

Sec. 240-3(b). Protection of Conductors
Sec. 240-20(b). Circuit Breaker as Overcurrent Device
Sec. 240-21. Location in Circuit

Sec. 240-24. Location in or on Premises
Sec. 240-85. Applications
Sec. 240-100. Feeders
Sec. 240-101. Branch Circuits

Article 250. Grounding

Sec. 250-24(a). Two or More Buildings or Structures Supplied from a Common Service
Sec. 250-32. Service Raceways and Enclosures
Sec. 250-33. Other Conductor Enclosures
Sec. 250-43. Fastened in Place or Connected by Permanent Wiring Methods (Fixed)—Specific: (g) Electric Signs
Sec. 250-76. Bonding for Over 250 Volts
Sec. 250-80. Bonding of Piping Systems and Exposed Structural Steel
Sec. 250-81. Grounding Electrode System
Sec. 250.91(b). Types of Equipment Grounding Conductors
Sec. 250-114. Continuity and Attachment of Equipment Grounding Conductors to Boxes

Article 300. Wiring Methods

Sec. 300-4(e). Cables and Raceways Installed in Shallow Grooves
Sec. 300-4(f). Insulated Fittings
Sec. 300-5(j). Ground Movement
Sec. 300-11. Securing and Supporting

Article 305. Temporary Wiring

Sec. 305-3(b). Approval
Sec. 305-6. Ground-Fault Protection for Personnel

Article 310. Conductors for General Wiring

Sec. 310-4. Conductors in Parallel Exception No. 4
Ampacity Tables 310-16 through 310-19, Note 3

Article 318. Cable Trays

Sec. 318-3. Uses Permitted. (a) Wiring Methods and (b) In Industrial Establishments
Sec. 318-6(a). Complete System
Sec. 318-6(j). Conduits and Cables Supported from Cable Trays
Sec. 318-9(e). Ventilated Channel Cable Trays

Article 331. Electrical Nonmetallic Tubing

Sec. 331-3. Uses Permitted
Sec. 331-11. Supports, Exception No. 1

Article 333. Armored Cable

Sec. 333-3. Uses Permitted
Sec. 333-7. Support, Exception No. 5
Sec. 333-9. Boxes and Fittings

Article 334. Metal-Clad Cable

Sec. 334-10(a). Support, Exception Nos. 1 and 2
Sec. 334-12. Fittings

Article 336. Nonmetallic-Sheathed Cable

Sec. 336-4. Uses Permitted
Sec. 336-18. Supports
Sec. 336-30. General: (3) Type NMS

Article 338. Service-Entrance Cable

Sec. 338-2. Uses Permitted as Service-Entrance Conductors
Sec. 338-6. Bends

Article 345. Intermediate Metal Conduit

Sec. 345-12. Supports, Exception Nos. 3 and 4
Sec. 345-14. Splices and Taps

Article 346. Rigid Metal Conduits

Sec. 346-3. Cinder Fill
Sec. 346-12. Supports, Exception Nos. 3 and 4

Article 347. Rigid Nonmetallic Conduit

Sec. 347-8. Supports
Sec. 347-9. Expansion Fittings
Sec. 347-12. Bushings

Article 348. Electical Metallic Tubing

Sec. 348-12. Supports, Exception

Article 350. Flexible Metal Conduit

Sec. 350-4. Uses Permitted
Sec. 350-5. Uses Not Permitted
Sec. 350-16. Bends
Sec. 350-18. Supports
Sec. 350-20. Fittings
Sec. 350-22. Trimming
Sec. 350-24. Splices and Taps

Article 351. Liquidtight Flexible Metal Conduit and Liquidtight Flexible Nonmetallic Conduit

Sec. 351-4(a). Permitted
Sec. 351-8. Supports
Sec. 351-27. Supports

Article 352. Surface Metal Raceways and Surface Nonmetallic Raceways

Sec. 352-5. Extension through Walls and Floors
Sec. 352-9. Gounding
Sec. 352-28. Extension through Walls and Floors
Sec. 352-29. Splices and Taps
Secs. 352-40 through 352-51. Strut-Type Channel Raceway

Article 362. Metal Wireways and Nonmetallic Wireways

Part A. Metal Wireways

Sec. 362-8. Supports
Sec. 362-9. Extension through Walls

Part B. Nonmetallic Wireways

Sec. 362-22. Supports
Sec. 362-23. Expansion Fittings

Article 364. Busways

Sec. 364-4(a). Uses Permitted, Exception
Sec. 364-8. Branches from Busways
Sec. 364-10. Rating of Overcurrent Protection—Feeders

Article 365. Cablebus

Sec. 365-1. Definition
Sec. 365-5. Overcurrent Protection

Article 370. Outlet, Device, Pull and Junction Boxes, Conduit Bodies and Fittings

Sec. 370-5. Short Radius Conduit Bodies
Sec. 370-16. Number of Conductors in Outlet, Device, and Junction Boxes, and Conduit Bodies
Sec. 370-18. Unused Openings
Sec. 370-23(b). Structural Mounting
Sec. 370-23(e). Exception Nos. 1 and 2
Sec. 370-23(g)(2). Conduit
Sec. 370-28(c). Covers

Article 380. Switches

Sec. 380-8(a). Location
Sec. 380-12. Grounding of Enclosures

Article 384. Switchboards and Panelboards

Sec. 384-1. Scope
Sec. 384-3(c). Used as Service Equipment, Exception
Sec. 384-4. Installation
Sec. 384-8. Clearances

Article 410. Lighting Fixtures, Lampholders, Lamps, and Receptacles

Sec. 410-1. Scope (FPN)
Sec. 410-2. Application of Other Articles
Sec. 410-30(c)(1). Electric-Discharge Fixtures, and Exception
Sec. 410-56(c). Isolated Ground Receptacles
Sec. 410-57(b), (c), and (d).
Sec. 410-58(b). Grounding-Pole Identification

Article 411. Lighting Systems Operating at 30 Volts or Less

Article 422. Appliances

Sec. 422-18(b). Ceiling Fans Exceeding 35 Lb (15.88 kg)

Article 424. Fixed Electric Space Heating Equipment

Sec. 424-19. Disconnecting Means

Article 430. Motors, Motor Circuits, and Controllers

Part A. General

Sec. 430-2. Adjustable Speed Drive Systems

Part B. Motor Circuit Conductors

Sec. 430-28. Feeder Traps

Part D. Motor Branch-Circuit Short-Circuit and Ground-Fault Protection

Sec. 430-52. Rating or Setting for Individual Motor Circuit

Part E. Motor Feeder Short-Circuit and Ground-Fault Protection

Sec. 430-62(b). Other Installations

Part G. Motor Controllers

Sec. 430-83(a). Horsepower Ratings at the Application Voltage

Part H. Motor Control Centers

Sec. 430-92 through Sec. 430-98

Part I. Disconnecting Means

Sec. 430-109. Type, Exception Nos. 1, 6, 7, and 8

Part M. Tables

Table 430-150. Full-Load Current Three-Phase Alternating-Current Motors
Table 430-151A. Conversion Table of Single-Phase Locked Rotor Currents for Selection of Disconnecting Means and Controllers as Determined from Horsepower and Voltages Rating
Table 430-151B. Conversion Table of Polyphase Design B, C, D, and E Maximum Locked-Rotor Currents for Selection of Disconnecting Means and Controllers as Determined from Horsepower and Voltage Rating and Design Letter
Table 430-152. Maximum Rating or Setting of Motor Branch-Circuit Short Circuit and Ground-Fault Protective Devices

Article 440. Air-Conditioning and Refrigerating Equipment

Sec. 440-3. Other Articles
Sec. 440-14. Location

Article 450. Transformers and Transformer Vaults

Sec. 450-3. Overcurrent Protection. (FPN No. 2)
Sec. 450-9. Ventilation (FPN No. 2)
Sec. 450-23. Less-Flammable Liquid-Insulated Transformers

Article 455. Phase Convertors

Sec. 455-2. Definitions

Article 500. Hazardous (Classified) Locations

Sec. 500-2. Location and General Requirements
Sec. 500-3. Special Precaution

Article 501. Class I Locations

Sec. 501-1. General, Exception No. 2
Sec. 501-4(a). Class I, Division 1, Exception Nos. 1, 2, and 3
Sec. 501-5. Sealing and Drainage
Sec. 501-5(d). Cable Seals, Class I, Division 1, Exception
Sec. 501-11. Flexible Cords, Class I, Divisions 1 and 2
Sec. 501-16(a). Bonding, Exception (FPN)

Article 502. Class II Locations

Sec. 502-4. Wiring Methods: (a) Class II, Division 1, Exception
Sec. 502-16(a). Bonding, Exception (FPN No. 2)
Sec. 502-16(b). Types of Equipment Grounding Conductors, Exception

Article 503. Class III Locations

Sec. 503-16(a). Bonding, Exception (FPN No. 2)
Sec. 503-16(b). Types of Equipment Grounding Conductors, Exception

Article 504. Intrinsically Safe Systems

Sec. 504-2. Definitions
Sec. 504-10(a). Control Drawing, Exception

Article 505. Class I, Zone 0, 1, and 2 Locations (New Article)

Article 511. Commercial Garages, Repair and Storage

Sec. 511-3(a). Up to a Level of 18 In. (457 mm) above the Floor, Exception
Sec. 511-4. Wiring and Equipment in Class I Locations
Sec. 511-6(c). Grounded and Grounding Conductors

Sec. 511-9. Electric Vehicle Charging
Sec. 511-16. Grounding

Article 514. Gasoline Dispensing and Service Stations

Sec. 514-16. Grounding

Article 516. Spray Application, Dipping, and Coating Processes

Sec. 516-2. Classification of Locations

Article 517. Health Care Facilities

Sec. 517-3. Definitions
Sec. 517-10. Exception No. 2
Sec. 517-13(a). Exception Nos. 1 and 3
Sec. 517-17(a). Feeders
Sec. 517-18(a). Patient Bed Location
Sec. 517-18(c). Pediatric Locations, Exception
Sec. 517-19(b). Patient Bed Location Receptacles
Sec. 517-21. Ground-Fault Circuit-Interrupter Protection for Personnel
Sec. 517-30. Essential Electrical Systems for Hospitals

Article 525. Carnivals, Circuses, Fairs, and Similar Events (New Article)

Article 550. Mobile Homes, Manufactured Homes, and Mobile Home Parks

Sec. 550-2. Definitions
Sec. 550-5. Power Supply

Article 625. Electric Vehicle Charging System Equipment

Article 680. Swimming Pools, Fountains, and Similar Installations

Sec. 680-4. Definitions
Sec. 680-12. Disconnecting Means
Sec. 680-28. Double Insulated Pool Pumps

Article 695. Fire Pumps

Article 700. Emergency Systems

Sec. 700-4. Tests and Maintenance
Sec. 700-9(c). Fire Protection

Article 725. Class 1, Class 2, and Class 3 Remote-Control, Signaling, and Power-Limited Circuits (Extensively Revised Throughout)

Article 727. Instrumentation Tray Cable Type ITC (New Article)

Article 760. Fire Alarm Systems (Revised)

Chapter 9. Tables

McGraw-Hill's

NATIONAL ELECTRICAL CODE® HANDBOOK

ARTICLE 90. INTRODUCTION

90-1. Purpose.

(a). This section clearly and simply describes the function of the NE Code in relation to electrical design and installation work. But it is important to understand that the NE Code is intended only to assure that electrical work is done safely—that is, to provide a system that is "essentially free from hazard."

The NE Code is recognized as a legal criterion of safe electrical design and installation. It is used in court litigation and by insurance companies as a basis for insuring buildings. Because the Code is such an important instrument of safe design, it must be thoroughly understood by all electrical designers. They must be familiar with all sections of the Code and should know the accepted interpretations which have been placed on many specific rulings of the Code. They should keep abreast of formal interpretations which are issued by the NE Code committee. They should know the intent of Code requirements—i.e., the spirit as well as the letter of each provision. They should keep informed on interim amendments to the Code. And, most important, they should keep this Code handbook handy and study it often.

(b). As stated, compliance with the provisions of the National Electrical Code can effectively minimize fire and accident hazards in any electrical design. The Code (throughout this manual, the word "Code" refers to the National Electrical Code) sets forth requirements, recommendations, and suggestions and constitutes a minimum standard for the framework of electrical design. As stated in its own introduction, the Code is concerned with the "practical safeguarding of persons and property from hazards arising from the use of electricity" for light, heat, power, radio, signaling, and other purposes.

Although the Code assures minimum safety provisions, actual design work must constantly consider safety as required by special types or conditions of electrical application. For example, effective provision of automatic protective devices and selection of control equipment for particular applications involve engineering skill of the designer, above routine adherence to Code requirements. Then, too, designers must know the physical characteristics—application advantages and limitations—of the many materials they use for enclosing, supporting, insulating, isolating, and, in general, protecting electrical equipment. The task of safe application based on skill and experience is particularly important in hazardous locations. Safety is not automatically made a characteristic of a system by simply observing codes. Safety must be designed into a system.

(c). The National Electrical Code contains provisions considered necessary for safety, but does not provide information of a design nature, other than for safety purposes, and should not be used to ensure adequate or efficient forms of installation. These features should be obtained from design manuals or through the services of a competent consulting engineer or electrical contractor.

90-2. Scope.

(a). The Code applies to all electrical work—indoors and outdoors—other than that work excluded by the rules of part **(b)** in this section.

By the addition of the word "equipment" in parts **(a)(2)** and **(a)(3)** of the 1993 Code, it is made clear that the NE Code applies to electrical circuits, systems, and components in their manner of installation as well as use.

Part **(a)(1)** makes clear that the NE Code also applies to "floating buildings" because the safety of Code compliance is required for all places where people are present. Coverage of floating buildings is contained in NEC Art. 553.

The scope of the NEC includes the installation of optical fiber cable, part **(a)(4).** As part of the high-technology revolution in industrial and commercial building operations, the use of light pulses transmitted along optical fiber cables has become an alternative method to electric pulses on metal conductors for control, signals, and communications. The technology of fiber optics has grown dramatically over recent years as a result of great strides in development of the fiber cables and associated equipment that converts electric pulses to light pulses and vice versa. For high rates of data transmission involved in data processing and computer control of machines and processes, optical fiber cables far outperform metallic conductors carrying electrical currents—all at a small fraction of the cost of metallic-conductor circuiting (Fig. 90-1).

NEC Art. 770, "Optical Fiber Cables," covers the use of such cables in association with conventional metallic-conductor circuits. Nonconductive optical fiber cables are permitted to be installed in the same raceway and enclosures as metallic-conductor circuits where the functions of the two different types of cables are associated with the same equipment, operation, or process.

According to Sec. 90-2(a)(5), certain utility-owned or -operated occupancies *must* be wired per the NEC. The wording in this section along with the companion rule of Sec. 90-5(b)(5) is intended to indicate those portions of a utility's electrical installations that are subject to the rules of the NEC. Basically stated, any utility occupancy that is *not* an "integral part" of a "generating plant, substation, or control center" must comply with the NEC in all respects. Clearly, any office space, storage area, garage, warehouse, or other nonpower-generating area of a building or structure is *not* an "integral part" of the gener-

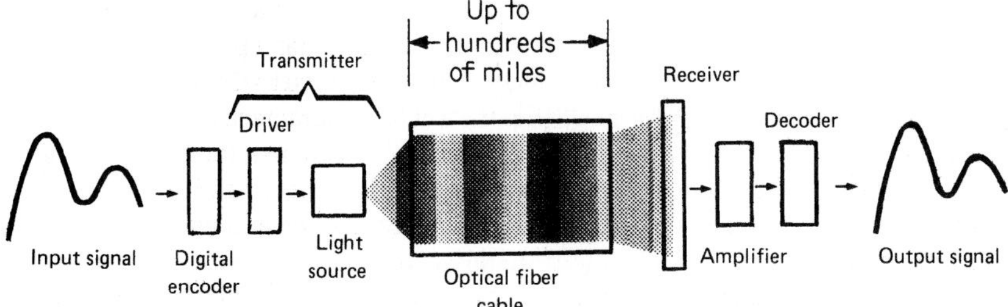

A telephone conversation is first transformed into an electrical signal. This input signal is scanned by a digital encoder, which reduces it to a series of "on's" and "off's." The driver, which activates the laser light source, transmits the digital "on's" as a pulse of light and the "off's" as the absence of a pulse. The light travels through the optical fiber cable to its destination, where it is received, amplified, and fed into a digital decoder. The decoder translates the pulses back into the original electrical signal.

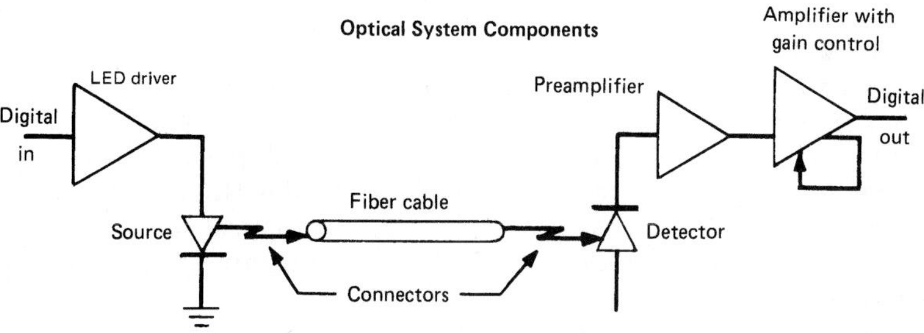

Fig. 90-1. The NEC covers the technology of fiber optics for communication and data transmission.

ation, transmission, or distribution of electrical energy; then it *is* covered by the NEC.

Specifically, the two rules were adopted to address an on-going problem where utilities were installing outdoor lighting, on private property, in accordance with the *National Electrical Safety Code* (NESC) rather than the National Electrical Code. The NESC is the standard used by utilities for their own installations. And it recognizes methods, techniques, and ratings not permitted by the NEC on the basis that the utilities will always use highly competent electricians who are thoroughly familiar with any hazard that might arise from such application. However, because the utilities would employ the NESC instead of the NEC, they could use smaller wire sizes and lower-rated equipment—in some cases, eliminating the disconnect all together, and the utility would always win any bid when competing with the electrical contractor on outdoor lighting jobs. While utility competition is not a problem itself, any installation in accordance with the NESC that is accessible to the general public *is.* Use of the NESC instead of the NEC at other than a "generating plant, substation, or control center" is contrary to the public well-being.

The primary concern with the wording accepted is the term *integral part*. It should be interpreted to mean "integral part of the process" (i.e., generation, transformation, or distribution of electric energy). It should *not* be taken to mean an "integral part" of the building or structure. Just because an office is *in* a generating plant, it shouldn't be exempt from the NEC. The same should apply to the cafeteria, bathrooms, and other areas within the plant that are not directly related to task of generating and delivering electrical energy.

It should be noted that equipment installed to perform associated functions, such as outdoor lighting at an outdoor substation, is intended to be considered as "integral" to the process and is therefore exempt from compliance with the NEC (Fig. 90-2). *But,* other loads, such as parking-lot lighting at a generating plant, should not be considered to be an integral part of the generation, transformation, or distribution of electrical energy and must comply with the NEC.

Fig. 90-2. Those buildings and structures that are directly related to the generation, transmission, or distribution of energy are intended to be excluded from compliance with the NEC. However, the rules covering this matter also indicate that functionally associated electrical equipment—such as the outdoor lighting for the utility-owned and -operated outdoor substation—are also exempt from the NEC. *But, only that equipment associated with the generation, transmission, or distribution of electrical power by the utility.*

(b). The rules of the Code do not apply to the electrical work described in **(1)** through **(5)**. The most common controversy that arises concerns exclusion of electrical work done by electric utilities (power companies).

In Sec. 90-2(b)(5), the rewrite of the rule to incorporate the old FPN is intended to limit utility exemption. This rewrite is the culmination of the work performed over a number of years by an electrical contractor group that was concerned for public safety where utilities were performing work—outdoor lighting, specifically—on private property for private companies using the

NESC instead of the NEC, which they felt was intended to cover such installations. And, although the intent was to prevent utilities from performing such installations *without* satisfying the NEC, the wording finally accepted appears to leave a gaping loophole, but does it?

The wording finally accepted recognizes non-NEC-complying utility installations on "private property by established rights such as easements." That permission could be misread to allow utilities to ignore NEC regulations where they are granted an easement on a piece of private property. When considered in conjunction with the remainder of the rule, Sec. 90-2(b)(5) really only allows the utility to deviate from NEC requirements where any such work—including outdoor lighting—is directly related to the generation, transmission, or distribution of electrical energy. An easement granted by an owner to the utility does *not* in itself exempt the utility from compliance with the NEC, unless it can demonstrate that the electrical installation is "associated" with the utility's mandate to deliver power. See Fig. 90-2.

This rule emphatically explains that not all electrical systems and equipment belonging to utilities are exempted from Code compliance. Electrical circuits and equipment in buildings or on premises that are used exclusively for the "generation, control, transformation, transmission, and distribution of electric energy" are considered as being safe because of the competence of the utility engineers and electricians who design and install such work. Code rules do not apply to such circuits and equipment—nor to any "communication" or "metering" installations of an electric utility. But, any conventional electrical systems for power, lighting, heating, and other applications within buildings or on structures belonging to utilities *must* comply with Code rules where such places are not "used exclusively by utilities" for the supply of electric power to the utilities' customers.

An example of the kind of utility-owned electrical circuits and equipment *covered* by Code rules would be the electrical installations in, say, an office building of the utility. But, in the Technical Committee Report for the 1987 NEC, the Code panel for Art. 90 stated that it is not the intent of this rule to have NEC regulations apply to "office buildings, warehouses, etc., that *are* an *integral* part of a utility generating plant, substation, or control center." NEC rules would not apply to any wiring or equipment in a utility generating plant, substation, or control center and would not apply to conventional lighting and power circuits in office areas, warehouses, maintenance shops, or any other areas of utility facilities used for the generation, transmission, or distribution of electric energy for the utility's customers. But NEC rules would apply to all electrical work in *other* buildings occupied by utilities—office buildings, warehouses, truck garages, repair shops, etc., that are not part of a generating plant or substation. And that opinion was reinforced by the actions of the Code-making panel (CMP) that sat for the 1996 NEC. (See Fig. 90-3.)

90-4. Enforcement. This is one of the most basic and most important of Code rules because it establishes the necessary conditions for use of the Code.

The NE Code stipulates that, when questions arise about the meaning or intent of any Code rule as it applies to a particular electrical installation, the electrical inspector having jurisdiction over the installation is the only one autho-

Fig. 90-3. *Circuits* and *equipment* of any utility company are exempt from the rule of the NEC when the particular installation is part of the utility's system for transmitting and distributing power to the utility's customers—*provided* that such an installation is accessible only to the utility's personnel and access is denied to others. Outdoor, fenced-in utility-controlled substations, transformer mat installations, utility pad-mount enclosures, and equipment isolated by elevation are typical utility areas to which the NEC does not apply. The same is also true of indoor, locked transformer vaults or electric rooms. (Sec. 90-2.) But, electrical equipment, circuits, and systems that are involved in supplying lighting, heating, motors, signals, communications, and other load devices that serve the needs of personnel in buildings or on premises owned (or leased) and operated by a utility are subject to NE Code rules, just like any other commercial or industrial building, provided that the buildings or areas are not integral parts of a generating plant or substation.

rized by the NE Code to make interpretations of the rules. The wording of Sec. 90-4 reserves that power to the local inspection authority along with the authority to approve equipment and materials and to grant the special permission for methods and techniques that might be considered alternatives to those Code rules that specifically mention such "special permission."

Up until the 1975 NE Code, Sec. 90-4 aimed only at giving the inspector the right to "interpret" Code rules. But the 1975 NE Code, for the first time, specifically gave the inspector the authority to "waive" specific requirements—that is, to disregard the wording and meaning of individual Code rules. However, use of this authority by the inspector was allowed (1) only in "industrial establishments and research and testing facilities" and (2) only where the inspector is satisfied that the safety objectives sought by the Code rule are achieved by the particular design and/or installation techniques that are used as alternatives to the specifics of the Code rule. The 1975 NE Code referred to the necessity for "establishing and maintaining effective safety and maintenance procedures" whenever a rule was "waived" or whenever "alternative methods" were accepted.

The present NE Code permits the electrical inspector to "waive specific requirements" or "permit alternate methods" in *any* type of electrical installation. The last sentence of Sec. 90-4 no longer limits waiver of Code rules to industrial establishments and research and testing facilities. In residential, commercial, and institutional electrical systems—as well as in industrial—inspectors may now accept design and/or installation methods that do not conform to a specific Code rule, provided they are satisfied that the safety objectives of the Code rule are achieved (Fig. 90-4).

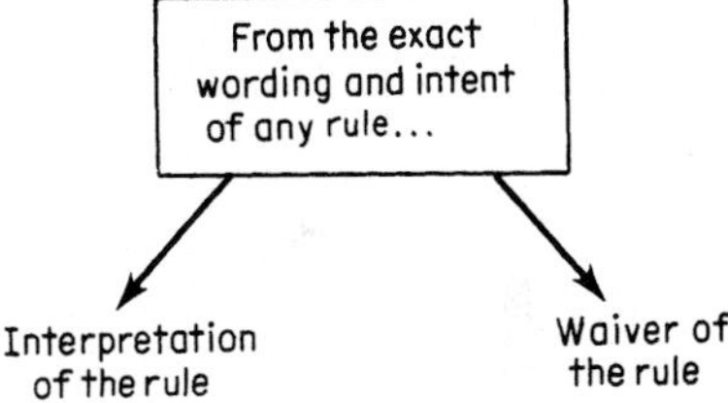

Fig. 90-4. *Inspector's authority* may be exercised either by enforcement of that individual's interpretation of a Code rule or by waiver of the Code rule when the inspector is satisfied that a specific non-Code-conforming method or technique satisfies the safety intent of the Code. (Sec. 90-4).

It should be noted that permission for any such variations from normal NE Code practice is granted only to the authority having jurisdiction in enforcing the Code. This permission is given to the electrical inspectors to permit them to recognize practices at variance with the Code, provided such practices are carefully controlled and under rigorously maintained conditions. In addition, such practices must be based on sound engineering principles and techniques which otherwise assure the safety of personnel and freedom from shock and fire hazard that the NE Code itself seeks to provide.

This recognition of practices at variance with the Code is provided only for special conditions and must not be interpreted as a general permission to engage in non-Code methods, techniques, or design procedures. In fact, it is likely that inspectors will exercise this authority only with reluctance and then with great care, because of the great responsibility this places on the inspector.

90-5. Mandatory Rules and Explanatory Material. Although the NE Code consists essentially of specific regulations on details of electrical design and installation, there is much explanatory material in the form of notes to rules. Compliance with the Code consists in satisfying all requirements and conditions that are stated by use of the word "shall." That word, anywhere in the Code, designates a mandatory rule. Failure to comply with any mandatory Code rule constitutes a "Code violation."

90-6. Formal Interpretations. Official interpretations of the National Electrical Code are based on specific sections of specific editions of the Code. In most cases, such official interpretations apply to the stated conditions on given installations. Accordingly, they would not necessarily apply to other situations that vary slightly from the statement on which the official interpretation was issued.

As official interpretations of each edition of the Code are issued, they are published in the NFPA Fire News and press releases are sent to interested trade papers.

All official interpretations issued on a specific Code edition are reviewed by the appropriate Code-making panel during the period when the specific Code edition is being revised. In reviewing an interpretation, a Code panel may agree with the interpretation findings and clarify the Code text to avoid further misunderstanding of intent, or the panel may reject the findings of the interpretation and alter the Code text to clarify the Code panel's intent. On the other hand, the Code panel may not recommend any change in the Code text because of the special conditions described in the official interpretation. For these reasons, the NFPA does not catalog official interpretations issued on previous editions of the Code. And in reviewing all previous interpretations, it can be stated that practically none of them would apply to the present edition of the Code because of revised Code wording that materially changes the intent.

If anyone feels that a past interpretation applies to the present text, they should submit it in the form of a proposed Code change when revisions for the next edition of the Code are being considered.

With the wide adoption of the Code throughout the country, the authority having jurisdiction has the prime responsibility of interpreting Code rules in its area and disagreements on the intent of particular Code rules in its area; and disagreements on the intent of particular Code rules should be resolved at the local level if at all possible. There is no guarantee that the authority having jurisdiction will accept the findings of an official interpretation.

90-7. Examination of Equipment for Safety. It is not the intent of the National Electrical Code to include the detailed requirements for internal wiring of electrical equipment. Such information is usually contained in individual standards for the equipment concerned.

The last sentence does not intend to take away the authority of the local inspector to examine and approve equipment, but rather to indicate that the requirements of the National Electrical Code do not generally apply to the internal construction of devices which have been listed by a nationally recognized electrical testing laboratory.

Although the specifics of Code rules on examination of equipment for safety are presented in Sec. 110-2 and Sec. 110-3, the general Code statement on this matter is made here in Sec. 90-7. Although the **Code** does place emphasis on the need for third-party certification of equipment by independent testing laboratories, it does not make a flat rule to that effect. However, the rules of OSHA are very rigid in insisting on product certification.

The clear effect of OSHA regulations is to *require* that "listed," "labeled," "accepted," and/or "certified" equipment be used whenever available. If any electrical system component is "of a kind" that *any* nationally recognized testing lab "accepts, certifies, lists, labels, or determines to be safe," then *that* component *must* be so designated in order to be acceptable for use under OSHA regulations. For instance, because liquidtight flexible metal conduit is a product that is listed and labeled by Underwriters Laboratories Inc. (UL), it is clearly and certainly a violation of the OSHA rule to use any nonlisted, nonlabeled *version* of liquidtight flexible metal conduit. A nonlisted, nonlabeled, noncertified component may be used *only* if it is "of a kind" that *no* nationally recognized lab covers. And even then, the nonrecognized component must be inspected or tested by another federal agency or by a state, municipal, or other local authority responsible for enforcing occupational safety provisions of the **NE Code**.

Every electrical designer and installer must exercise great care in evaluating any and all equipment and products used in electrical work to assure compliance with OSHA rules on certification by a nationally recognized testing lab. In many cases, "listed" components are combined to make up an assembled piece of equipment, but the entire assembly is not listed or labeled as an assembly and may constitute an assembly that is not safe for use. For instance, a listed circuit breaker and a listed magnetic starter may be combined in a listed electric cabinet, but that does not of itself constitute a listed "circuit-breaker combination motor starter." A listed "combination motor control unit" is an entire assembly that has been tested as an assembly and is labeled as such. It is critically important to be fully aware of the meaning and extent of the "listing" or "label" for every product. This is particularly true of combination products and custom-made equipment.

Codes and standards must be carefully interrelated and followed with care and precision. Modern work that fulfills these demands should be the objective of all electrical construction people.

Chapter One

ARTICLE 100. DEFINITIONS

Accessible: "(As applied to wiring methods.)"

Accessible: "(As applied to equipment.)"

Accessible, Readily: (Readily Accessible.)

Note that the first definition for "accessible" makes reference to "concealed" and "exposed." Because these words are critically important to applications of wiring methods and equipment, their definitions must be carefully studied and cross-referenced with one another, as well as related to Code rules using these words. Since discussions involving the definitions can become complex and murky, it should be noted that Sec. 90-4 gives the local inspector the authority to make the binding interpretation.

Note that there are definitions for words that apply to wiring methods and definitions that apply to equipment. Wiring methods are any of the NE Code-recognized techniques for running circuits between equipment. These include conductors in EMT (electric metallic tubing), rigid metal, or nonmetallic conduit; wireways; cable trays; underfloor raceways; all the cable assemblies covered in Arts. 330 to 340; busways; and the other hookup methods covered in Chap. 3. "Equipment" covers all the products that are connected or hooked by any of these recognized wiring methods.

For electrical applications above suspended ceilings with lift-out panels, it is necessary to distinguish between "equipment" and "wiring methods." It also is necessary to distinguish between ceiling spaces used for air handling and those not used as air-handling spaces. Typical applications of Code rules to both those categories are as follows for nonair-handling ceilings.

1. Wiring methods above lift-out ceiling panels are considered to be "exposed" because the definition of that word includes reference to "behind panels designed to allow access." A typical application of this definition is where wireways are installed in a hung ceiling. Section 362-2 says that wireways are permitted only for exposed work. Therefore, a wireway may be installed in the open and visible, or it may be installed above lift-out panels of a suspended ceiling.

   Any wiring methods above a ceiling made up of lift-out ceiling elements are, therefore, accessible, as shown in Fig. 100-1.

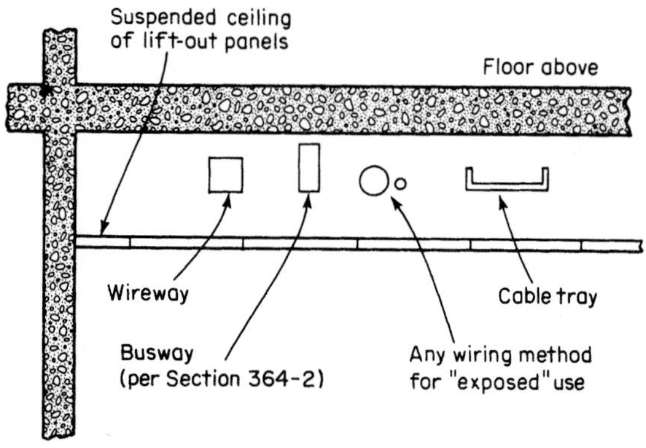

Fig. 100-1

2. The rules on installation of a busway used to be worded the same as that quoted above for a wireway. But now, Exception No. 1 to Sec. 364-4(a) permits busways above panels, if means of access are provided. It further limits such use to a totally enclosed, nonventilated busway, without plug-in switches or CBs on the busway—except for an individual fixture or "other" load—and only in ceiling space that is not used for air handling. Exception No. 2 to Sec. 364-4(a) permits a busway in an air-handling ceiling space when installed in accordance with Sec. 300-22(c)—that is, a totally enclosed, nonventilated, insulated busway that "has no provisions" to accommodate any plug-in fuse or CB.
3. Section 318-6(h) requires cable trays to be exposed and accessible. Note that the two words "exposed" and "accessible" must be taken "as applied to wiring methods. A cable tray may be used above a suspended, nonair-handling ceiling; and, if used with a wiring method permitted by Sec. 300-22(c), it may be used also above an air-handling ceiling.
4. In many of the articles on cables it is noted, under "Uses Permitted," that the cable may be used for both "exposed" and "concealed" work. Such cables may, therefore, be used above suspended ceilings.
5. Fuses and circuit breakers that provide overcurrent protection required by the NE Code are generally required by Sec. 240-24 to be "readily accessible"—that is, they must be capable of being reached quickly. Note that the

definition for "readily accessible" applies to **equipment** rather than **wiring methods** and is a different concept from that of the word "accessible." Fuses and/or CBs in a distribution panel or switchboard or motor control center are **not** readily accessible, for instance, if a bunch of crates piled on the floor block access and present an obstacle in getting to the fuses or CBs. They also are not readily accessible if it is necessary to get a portable ladder or stand on a chair or table to get at them.

Equipment is not "readily accessible" if conditions shown in Fig. 100-2 exist.

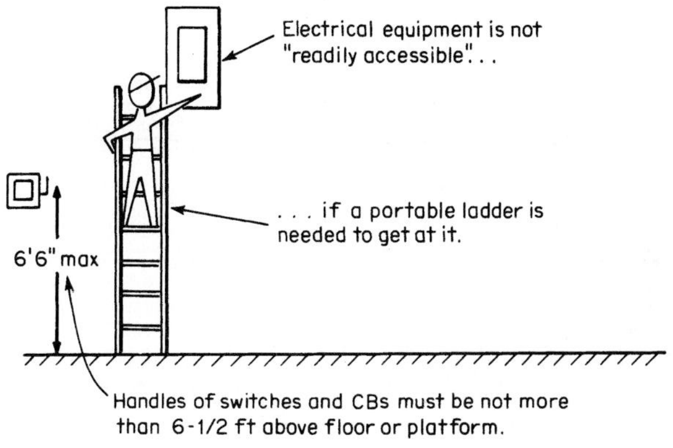
Electrical equipment is not "readily accessible"...
... if a portable ladder is needed to get at it.
6'6" max
Handles of switches and CBs must be not more than 6-1/2 ft above floor or platform.

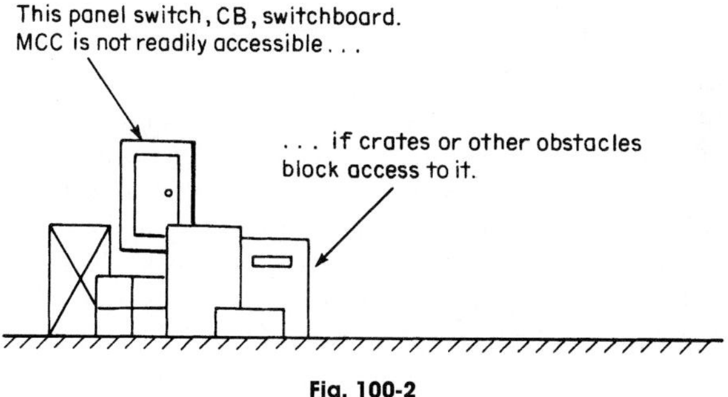
This panel switch, CB, switchboard. MCC is not readily accessible...
... if crates or other obstacles block access to it.

Fig. 100-2

Exception No. 1 in Sec. 240-24 permits overcurrent devices to be used high up on a busway where access to them could require use of a ladder, but not on a busway above a suspended ceiling (Sec. 364-4).

Supplementary overcurrent devices—over and above those required by the NE Code and which the Code defines as supplementary protection [Sec. 424-22(c) covering fuses or CBs permitted above a suspended ceiling for protection of electric duct heaters]—do not have to be readily accessible (Sec. 240-10 and Exception No. 2 in Sec. 240-24) and may be installed above a suspended ceiling. A CB- or fusible-type panelboard may be used in the ceiling space to satisfy Sec. 424-22(c).

Exception No. 3 of Sec. 240-24 permits service overcurrent protection to be sealed, locked, or otherwise not "readily accessible."

In general, aside from the cases noted above, fuses and CBs must **not** be used above a suspended ceiling because they would be not readily accessible. But equipment that is not required by a Code rule to be "accessible" or to be "readily accessible" may be mounted above a suspended ceiling, as indicated in Fig. 100-3.

If it is necessary to use a portable (but not a fixed) ladder to get to a switch or CB or transformer or other piece of equipment, then the equipment is not "readily accessible." The term "readily accessible" implies a need for performing

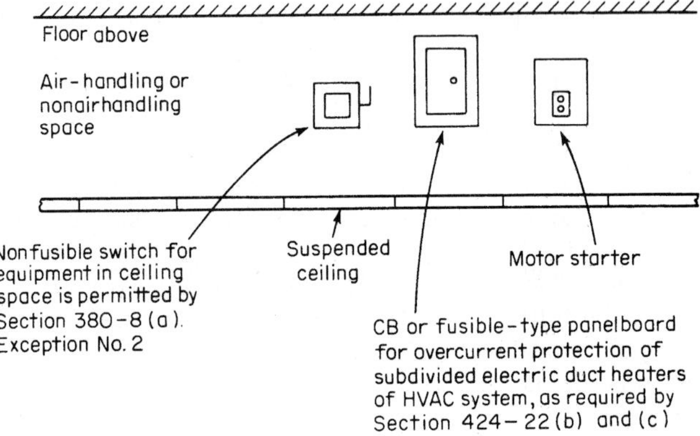

Fig. 100-3

promptly an indicated act, for example, to reach quickly a disconnecting switch or circuit breaker without the use of ladders, chairs, etc. The installation of such a switch or circuit breaker at a height above 6½ feet (1.98 m) from a standing level is not considered "readily accessible."

An analysis In judging whether electrical equipment is "readily accessible," it is first necessary to determine if the equipment is "accessible" in the National Electrical Code meaning of that word. If it can be determined that the equipment *is* accessible, then the equipment can be judged to be "readily accessible" if, when it is reached, it is not necessary to use a portable ladder or otherwise climb to get at it, and the equipment is not blocked by obstacles that have to be moved.

Any equipment that has to be "readily accessible" must be so only for "those to whom ready access is requisite"—which clearly and intentionally allows for making equipment *not* readily accessible to other than authorized persons, such as by providing a lock on the door, with the key possessed by or available to those who require ready access.

Because the definition of "readily accessible" contains a last phrase that says "See 'Accessible'," logic dictates that it is first necessary to satisfy the definition of the word "accessible." And the wording of that definition clearly establishes that there is no Code violation for "readily accessible" equipment to be behind a door under lock and key to make it accessible only to authorized persons.

Note that the definition for "accessible" does not say that a door to an electrical room is prohibited from being locked. In fact, the wording of the definition, by referring to "locked doors," actually presumes the existence and, therefore, the acceptability of "locked doors" in electrical systems. The only requirement implied by the wording is that locked doors, where used, must not "guard" against access—that is, the disposition of the key to the lock must be such that those requiring access to the room are not positively excluded.

To be "accessible," equipment must not be "guarded" by locked doors and must not be "*guarded*" by elevation. The critical word is "guarded." The definition is *not* intended to mean that equipment *cannot* be "behind" locked doors or that equipment *cannot* be mounted up high where it *can* be reached with a portable ladder. To make equipment "not accessible," a door lock or high mounting must be such that it positively "guards" against access. For instance, equipment mounted up high where conditions are such that it cannot be reached with a portable ladder is *not* "accessible." Equipment behind a locked door for which a key is not possessed by or available to persons who require access to the equipment is *not* "accessible." A common example of that latter condition occurs in multitenant buildings where a disconnect for the tenant of one occupancy unit is located behind the locked door of another tenant's occupancy unit from which the first tenant is effectively and legally excluded.

The definition of "accessible" has always been intended to recognize that equipment may be fully "accessible" even though installed behind a locked door or at an elevated height. Equipment that is high-mounted but can be reached with a ladder that is fixed in place or a portable ladder *is* "accessible" (although the equipment would not be "readily accessible" if a portable ladder had to be used to reach it). Similarly, it has always been within the meaning of the definition that equipment behind a locked door *is* "accessible" to anyone who possesses a key to the lock or to a person who is authorized to obtain and use the key to open the locked door. In such cases, conditions do *not* "guard" against access.

There is no rule in the NEC that prohibits use of a lock on the door of an electrical room. In fact, some Code rules *require* equipment to be in rooms with locked doors—while maintaining the condition of the room as "accessible" or "readily accessible."

Section 110-31 says, "Electrical installations in a vault, room, or closet or in an area surrounded by a wall, screen, or fence, access to which is *controlled by lock and key* or other approved means, *shall be considered to be accessible* to qualified persons only." [Italics supplied.]

In Art. 450 on transformers, where a transformer is used in a vault, two separate Code rules must be satisfied simultaneously. Section 450-13 says, "Transformers and transformer vaults *shall be readily accessible* to qualified personnel for inspection and maintenance." [Italics supplied.]

Section 450-43, covering rules on doors for the above-described vaults, which must be "readily accessible," says, "Vault doorways shall be protected as follows: **(c) Locks.** Doors shall be equipped with locks, and doors shall be kept locked, access being allowed only to qualified persons."

Those two rules combine to require that a transformer in a vault *must* be "readily accessible" *and* that the vault must be equipped with a locked door. Obviously, the "locked door" and "readily accessible" are not mutually exclusive and are completely compatible with each other.

Conditions of "accessibility" can be evaluated in accordance with the following:

CASE I—EQUIPMENT IN A LOCKED ROOM, LOCKED SPACE, OR LOCKED, FENCED-IN AREA

Condition A: Accessible, But *Not* Readily Accessible

This condition exists if the key to the locked door or gate is available to anyone requiring access to the equipment within—but, when the room, space, or area is entered, a portable ladder, chair, scaffold, hoist, or lift is needed to reach high-mounted equipment. Such a ladder or climbing or lifting device must be capable of reaching the equipment and be on the premises and available to anyone requiring access to the equipment—that is, the ladders or lifting or climbing devices must be permanently installed. Need for a *portable* ladder or other *portable* climbing or lifting device makes the equipment *not* "readily accessible" although it is "accessible."

Condition B: Accessible and Readily Accessible

This condition exists if the key to the locked door or gate is available to anyone requiring access to the equipment—and, when the room or space is entered, the equipment may be reached by a person standing on the floor or on a readily accessible platform—without need for a portable ladder or other climbing or lifting device and without need to move stored objects, materials, or other obstacles blocking access to the equipment. Any high-mounted equipment—suspended from overhead or located on an elevated platform or balcony—is also "readily accessible" if a permanently installed ladder or stairway provides for reaching the equipment.

Condition C: Not Accessible

(1) This condition exists if the key for the locked door or gate is not available to anyone requiring access because that person does not know who has the key and/or does not know how to contact the person with the key who will surrender the key or make its use available. This would typically be encountered in multiple-occupancy buildings like office buildings, apartment houses, shopping centers, etc., where, either as part of original design or as a result of alterations or modernization, a switch or circuit breaker for one tenant is within locked space of another tenant. In such a condition, "close approach" to the equipment by the first tenant is "guarded" by the locked door or gate. OR:

(2) This condition exists if the key for the locked door or gate is available to anyone requiring access—but the equipment is so high-mounted or so otherwise installed that available portable ladders or other climbing or lifting devices cannot reach the equipment. In such a condition, "close approach" to the equipment is "guarded" by "elevation."

CASE II—EQUIPMENT IN AN UNLOCKED ROOM, SPACE, OR FENCED-IN AREA

Condition A: Accessible But *Not* Readily Accessible

This condition exists if a portable ladder, chair, scaffold, hoist, or lift is needed to reach high-mounted equipment and such ladder or climbing or lifting device is capable of reaching the equipment and is on the premises and is available to anyone requiring access to the equipment.

Condition B: Accessible and Readily Accessible

This condition exists if the equipment may be reached by a person standing on the floor or on a readily accessible platform—without need for a portable ladder or other climbing or lifting device and without need to move stored objects, materials, or other obstacles blocking access to the equipment. Any high-mounted equipment—suspended from overhead or located on an elevated platform or balcony—is also "readily accessible" if a permanently installed ladder or stairway provides for reaching the equipment.

Condition C: Not Accessible

The equipment is so high-mounted or otherwise so installed that available portable ladders or other climbing or lifting devices cannot reach the equipment. In such a condition, "close approach" to the equipment is "guarded" by "elevation."

Ampacity:

"Ampacity" is the maximum amount of current in amperes that a conductor may carry under specific conditions of use. This definition provides a clear, logical description of what a conductor's ampacity is: "The current in amperes a conductor can carry continuously *under the conditions of use* without exceeding its temperature rating."

For instance, NEC Table 310-16 gives ampacities of various sizes of copper and aluminum conductors, with ampacities being higher for 75°C (THW) conductors and 90°C (THHN, XHHW, RHH) conductors than for 60°C (TW) conductors because the higher-temperature-rated conductors have insulations that are capable of withstanding the higher heat of greater I^2R. And in the table, the ampacities shown are for the "condition" that not more than three conductors are used in a raceway or cable and for the "condition" that the ambient temperature of air around the conductors is not over 30°C (86°F). As now covered by the revised wording of Note 8 to the Ampacity Tables of 0 to 2000 V in the 1993 NEC, if more than three current-carrying conductors are used in a raceway, *the ampacity of each conductor must be reduced as required by the note. And further, where the ambient temperature exceeds 30°C, the ampacity of each conductor must be reduced for elevated ambient temperature in accordance with the table of factors at the bottom of Table 310-16 (or Table 310-18).*

As stated in the Code-making panel's substantiation for revision of Note 8 to the Ampacity Tables, "Where *both* conditions (high ambients and four or more conductors in a raceway or cable) exist, *both deratings should be applied, one on the other.*"

Approved:

The electrical inspector having jurisdiction on any specific installation is the person who will decide what conductors and/or equipment are "approved." Although inspectors are not required to use "listing" or "labeling" by a national testing lab as the deciding factor in their approval of products, they invariably base their acceptance of products on listings by testing labs. Certainly the NE Code almost makes the same insistence as OSHA does that, whenever possible, acceptability must be based on some kind of listing or certification of a national lab. But on this matter, the OSHA law takes precedence—a "listed," "labeled," or otherwise "certified" product must *always* be used in preference to the same "kind" of product that is not recognized by a national testing lab (Fig. 100-4).

Fig. 100-4

Bathrooms:

A lot of the controversy that was generated by the question "What is a bathroom?" is eliminated by this definition of the word "bathroom." As used in the rule of Sec. 210-8, a "bathroom" is "an area" (which means it could be a room or a room plus another area) that contains first a "basin" (sometimes called a "sink") and then at least one more plumbing fixture—a toilet, a tub, and/or a shower. A small room with only a "basin" (a washroom) is not a "bathroom." Neither is a room that contains only a toilet and/or a tub or shower (Fig. 100-5). Fig. 100-6 shows application in hotel and motel bathrooms.

GFCI-protected receptacles are required . . .

THIS IS HOW A "BATHROOM" IS DEFINED

NOTE: If a room is not a bathroom according to the definition of Section 210-8(b), then the requirement of Section 210-52 for "at least one wall receptacle outlet . . . adjacent to the basin location" does not apply. If, however, a receptacle is installed in a room that is *not* a "bathroom"—such as the one above containing a basin *only*—GFCI protection is not required for the receptacle because it is not a bathroom receptacle.

TYPICAL BEDROOM SUITE IN ONE-FAMILY HOUSE OR APARTMENT UNIT

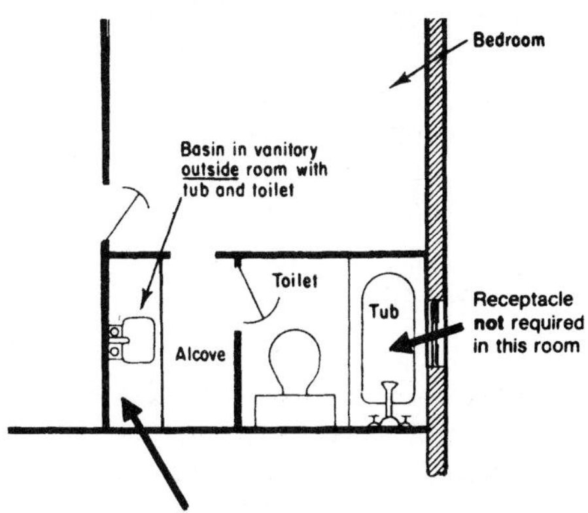

Although this **area** with basin is **outside** room with tub and toilet, the intent of Section 210-52 requires a receptacle at basin; and Section 210-8(a) requires that it be GFCI-protected.

NOTE: It is important to understand that the *Code* meaning of "bathroom" refers to the total "area" made up of the basin in the alcove *plus* the "room" that contains the tub and toilet. Although a receptacle is *not* required in the "room" with the tub and toilet, if one is installed in that room, it must be GFCI-protected because such a receptacle is technically "in the bathroom," just as the one at the basin location is "in the bathroom."

Fig. 100-5

. . . In a "bathroom" of a dwelling unit . . .

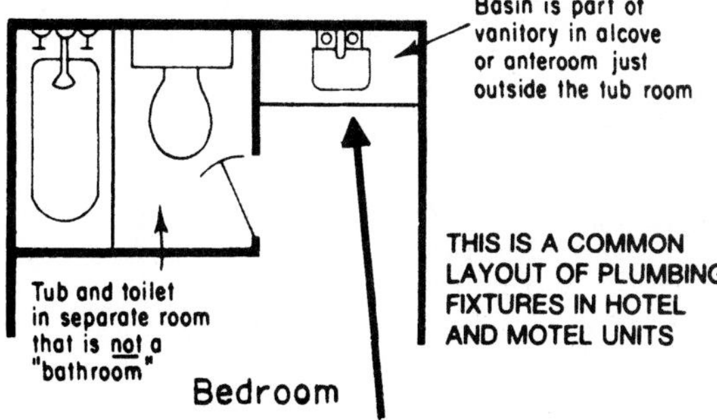

Guest rooms in hotels and motels are required by Section 210-60 to have the same receptacle outlets required by Section 210-52 for "dwelling units." The requirement for a wall receptacle outlet at the "basin location" applies to bathrooms; and the anteroom area with only a basin is, by definition and intent, part of the "bathroom." A receptacle at the basin would, therefore, be required. Section 210-8(b) applies to bathrooms, in hotels and motels — and GFCI protection is required for this receptacle.

. . . and in a "bathroom" of a hotel/motel guestroom.

Fig. 100-6

Bonding:

This definition covers a general concept that metal parts are conductively connected by a cable, wire, bolt, screw, or some other metallic connection of negligible impedance. The term is used frequently throughout the **Code** to imply that metal parts that are "bonded" together have no potential difference between them. Two common "bonding" techniques are bonding of switchboards and bonding of panelboards. When bonding is done by a short length of bare or insulated conductor, the conductor is referred to as a "bonding jumper." Examples of bonding are given in Figs. 100-7 and 100-8.

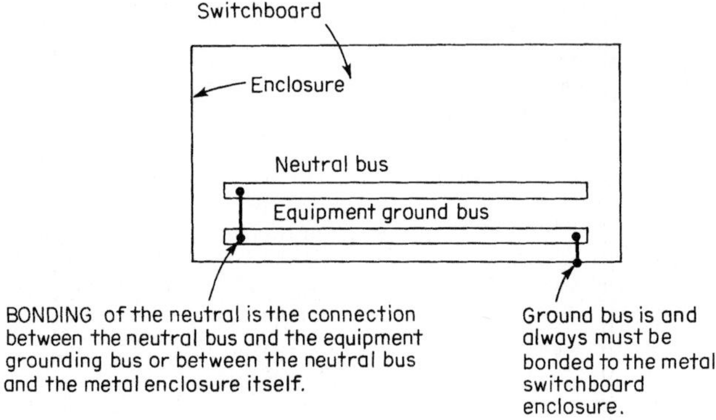
Switchboard
Enclosure
Neutral bus
Equipment ground bus
BONDING of the neutral is the connection between the neutral bus and the equipment grounding bus or between the neutral bus and the metal enclosure itself.
Ground bus is and always must be bonded to the metal switchboard enclosure.

Fig. 100-7

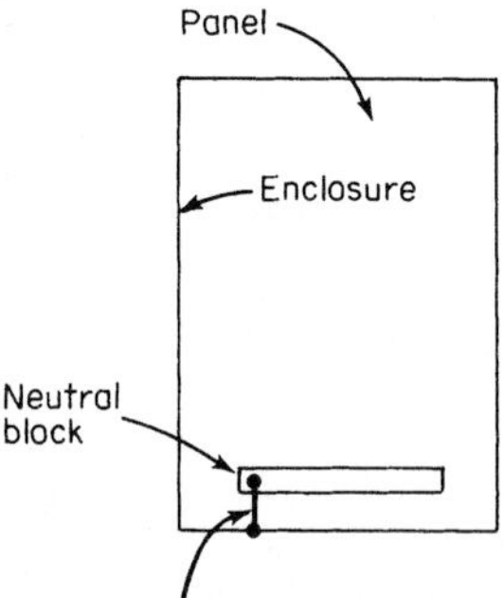
Panel
Enclosure
Neutral block

BONDING is the insertion of a bonding screw into the panel neutral block to connect the block to the panel enclosure, or it is use of a bonding jumper from the neutral block to an equipment grounding block that is connected to the enclosure.

NOTE: Bonding – the connection of the neutral terminal to the enclosure or to the ground terminal that is, itself, connected to the enclosure – might also be done in an individual switch or CB enclosure.

Fig. 100-8

Bonding Jumper:

This is any bare or insulated conductor used to provide bonding between metal parts in a system—such as between a metal switchboard enclosure and metal service conduits that stub-up under a service switchboard. The bonding jumper or jumpers provide for making an electrically conductive connection between the metal switchboard enclosure and the metal conduits—as required by NEC Sec. 250-71(a) and (b). An example is shown in Fig. 100-9.

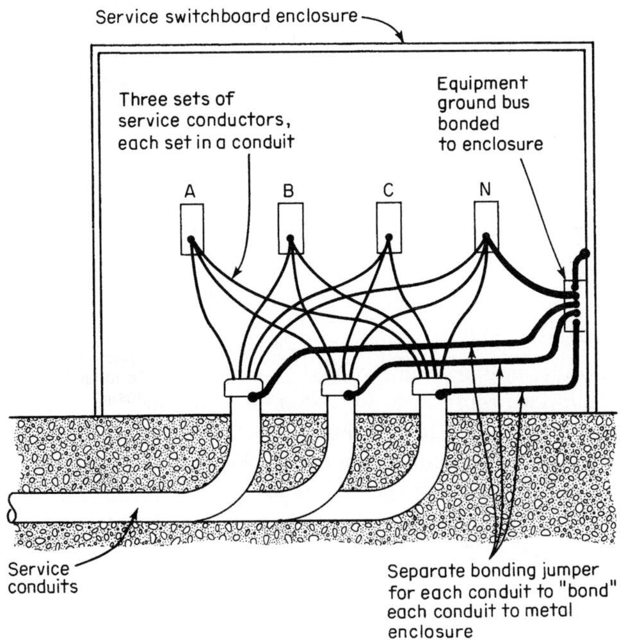

Fig. 100-9

Branch Circuit:

A branch circuit is that part of a wiring system extending beyond the final automatic overload protective device which is approved for use as branch-circuit protection. Thermal cutouts or motor overload devices are not branch-circuit protection. Neither are fuses in luminaires or in plug connections, where used for ballast protection or individual fixture protection. Such supplementary overcurrent protection is not a substitute for branch-circuit protection and does not establish the point of origin of the branch circuit. The extent of a branch circuit is illustrated in Fig. 100-10.

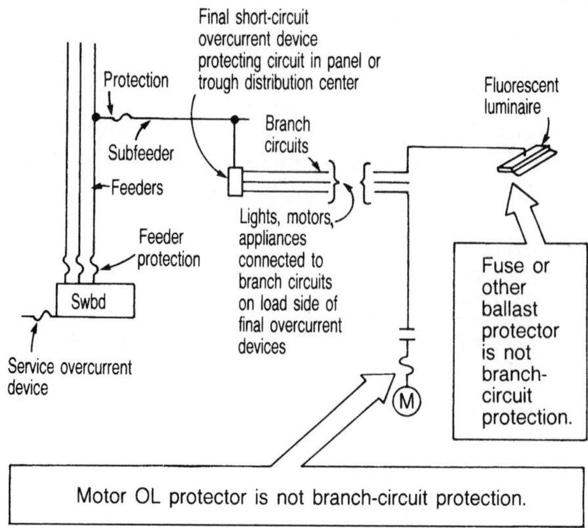

Fig. 100-10

In its simplest form, a branch circuit consists of two wires which carry current at a particular voltage from protective device to utilization device.

The branch circuit represents the last step in the transfer of power from the service or source of energy to utilization devices. First, the loads are circuited. Then the circuits are lumped on the feeders. Finally, the distribution system is connected to one or more sources of power.

Branch Circuit, Appliance:

Refer to Fig. 100-11.

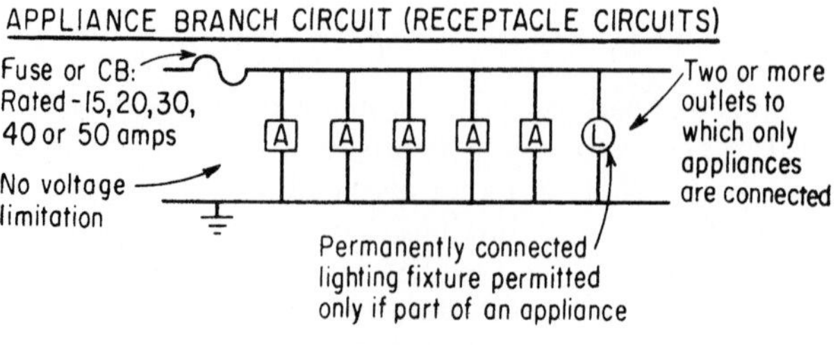

Fig. 100-11

Branch Circuit, General Purpose:

Refer to Fig. 100-12.

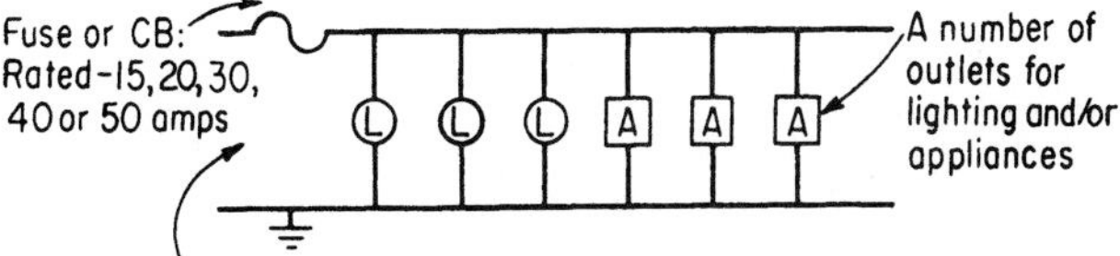

Circuit voltage shall not exceed 150 volts to ground for circuits supplying lampholders, fixtures or receptacles of standard 15-amp rating. For fluorescent, incandescent or mercury lighting under certain conditions, voltage to ground may be as high as 300 volts. In certain cases, voltage for electric discharge lighting may be up to 600 volts ungrounded.

Fig. 100-12

Branch Circuit, Individual:

Refer to Fig. 100-13.

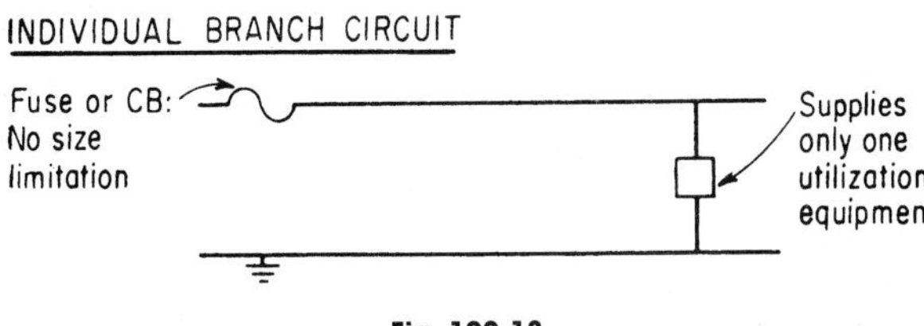

Fig. 100-13

Branch Circuit, Multiwire:

A multiwire branch circuit must be made up of a neutral or grounded conductor—as in corner-grounded delta systems—and at least two ungrounded or "hot" conductors. The most common multiwire circuits are shown in Fig. 100-14.

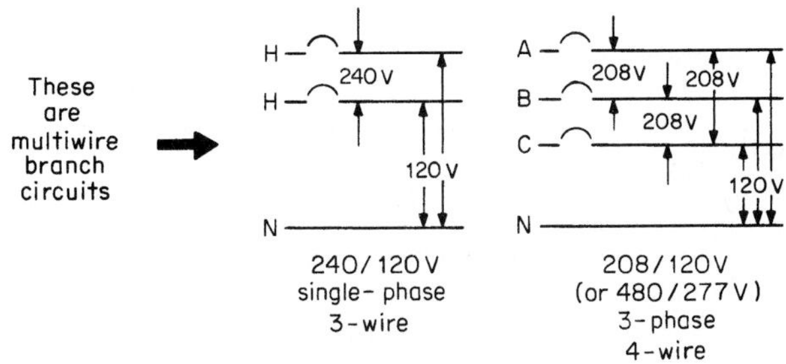

Fig. 100-14

A 3-wire, 3-phase circuit (without a neutral or grounded conductor) is not a "multiwire branch circuit," even though it does consist of "multi" wires, as shown in Fig. 100-15.

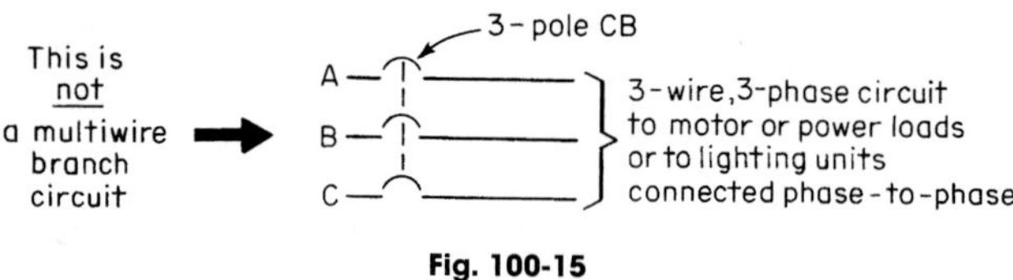

Fig. 100-15

Building:

Most areas have building codes to establish the requirements for buildings, and such codes should be used as a basis for deciding the use of the definition given in the National Electrical Code. The use of the term "fire walls" in this definition has resulted in differences of opinion among electrical inspectors and others. Since the definition of a fire wall may differ in each jurisdiction, the processing of an interpretation of a "fire wall" has been studiously avoided in the National Electrical Code because this is a function of building codes and not a responsibility of the National Electrical Code.

Cabinet:

The door of a cabinet is hinged to a trim covering wiring space, or gutter. The door of a cutout box is hinged directly to the side of the box. Cabinets usually contain panelboards; cutout boxes contain cutouts, switches, or miscellaneous apparatus.

Concealed:

Any electrical equipment that is closed in by structural surfaces is considered to be "concealed" as shown in Fig. 100-16.

Circuits run in an unfinished basement or an accessible attic are not "rendered inaccessible by the structure or finish of the building," and are therefore considered as exposed work rather than a concealed type of wiring.

Conduit Body:

An added sentence notes that FS and FD boxes—as well as larger cast or sheet metal boxes—are not considered to be "conduit bodies," as far as the NE Code is concerned. Although some manufacturers' literature refers to FS and FD boxes as conduit fittings, care must be used to distinguish between "conduit bodies" and "boxes" in specific Code rules. For instance, the next to last sentence of Sec. 370-16(c) prohibits splicing and use of devices in conduit bodies that are

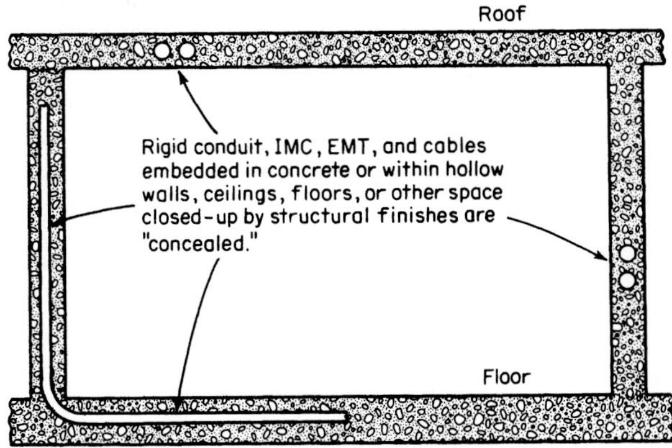

Fig. 100-16

not "durably and legibly" marked with their cubic inch capacity by the manufacturer. However, FS and FD boxes are not conduit bodies and may contain splices and/or house devices. Table 370-16(a) lists FS and FD boxes as "boxes." See Fig. 100-17.

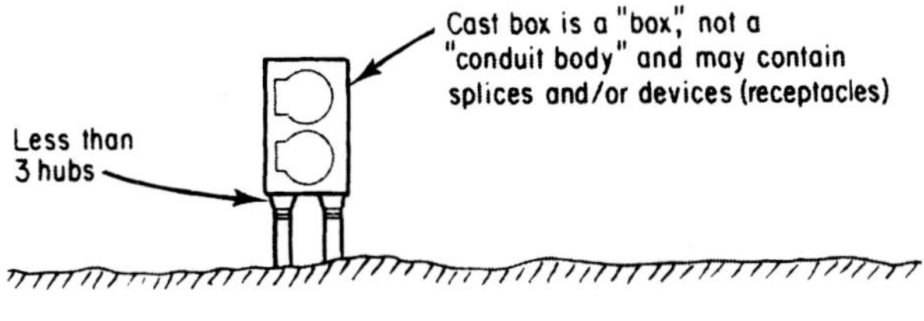

Fig. 100-17

Continuous Load:

Any condition in which the maximum load current in a circuit flows without interruption for a period of not less than 3 hr.

Demand Factor:

Two terms constantly used in electrical design are "demand factor" and "diversity factor." Because there is a very fine difference between the meanings for the words, the terms are often confused.

Demand factor is the ratio of the maximum demand of a system, or part of a system, to the total connected load on the system, or part of the system, under consideration. This factor is always less than unity.

Diversity factor is the ratio of the sum of the individual maximum demands of the various subdivisions of a system, or part of a system, to the maximum demand of the whole system, or part of the system, under consideration. This factor generally varies between 1.00 and 2.00.

Demand factors and diversity factors are used in design. For instance, the sum of the connected loads supplied by a feeder is multiplied by the demand factor to determine the load which the feed must be sized to serve. This load is termed the maximum demand of the feeder. The sum of the maximum demand loads for a number of subfeeders divided by the diversity factor for the subfeeders will give the maximum demand load to be supplied by the feeder from which the subfeeders are derived.

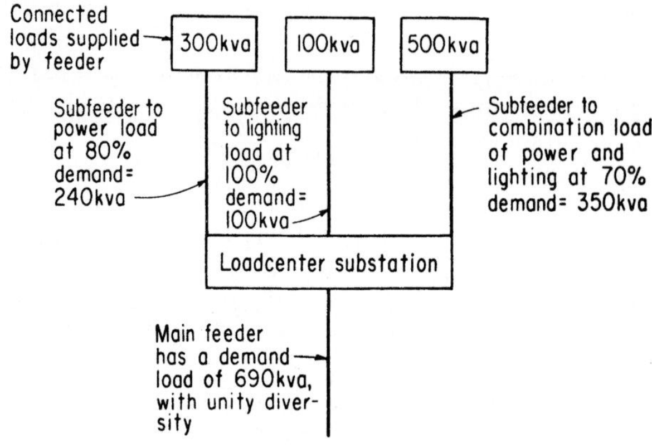

1. Sum of individual demands = 240 + 100 + 350 = 690 kva.
2. Sizing the substation at unity diversity, the required

$$\text{kva} = \frac{690}{1.00} = 690 \text{ kva.}$$

3. To meet this load, use a 750-kva substation.
4. If analysis dictates the use of a diversity factor of 1.4 , the

$$\text{required kva} = \frac{690}{1.40} = 492 \text{ kva.}$$

5. To meet this load, use a 500-kva substation.
6. Primary feeder to unit substation must have capacity to match the substation load.

Fig. 100-18

It is common and preferred practice in modern design to take unity as the diversity factor in main feeders to loadcenter substations to provide a measure of spare capacity. Main secondary feeders are also commonly sized on the full value of the sum of the demand loads of the subfeeders supplied.

From power distribution practice, however, basic diversity factors have been developed. These provide a general indication of the way in which main feeders can be reduced in capacity below the sum of the demands of the subfeeders they supply. On a radial feeder system, diversity of demands made by a number of transformers reduces the maximum load which the feeder must supply to some value less than the sum of the transformer loads. Typical application of demand and diversity factors for main feeders is shown in Fig. 100-18.

Device:

Switches, fuses, circuit breakers, controllers, receptacles, and lampholders are "devices."

Dwelling:

Dwelling unit. Because so many **Code** rules involve the words "dwelling" and "residential," there have been problems applying **Code** rules to the various types of "dwellings"—one-family houses, two-family houses, apartment houses, condominium units, dormitories, hotels, motels, etc. The present **NE Code** includes terminology to eliminate such problems and uses definitions of "dwelling" coordinated with the words used in specific **Code** rules.

A "dwelling unit" is now defined as "one or more rooms" used "as a housekeeping unit" and *must* contain space or areas specifically dedicated to "eating, living, and sleeping" and *must* have "permanent provisions for cooking and sanitation." A one-family house is a "dwelling unit." So is an apartment in an apartment house or a condominium unit. But, a guest room in a hotel or motel or a dormitory room or unit is not a "dwelling unit" if it does not contain "permanent provisions for cooking"—which must mean a built-in range or counter-mounted cooking unit (with or without an oven).

Any "dwelling unit" must include all the required elements shown in Fig. 100-19.

Exposed: "(As applied to wiring methods.)"

Wiring methods and equipment that are not permanently closed in by building surfaces or finishes are considered to be "exposed." See Fig. 100-20.

Feeder:

A "feeder" is a set of conductors which carry electric power from the service equipment (or from a transformer secondary, a battery bank, or a generator switchboard where power is generated on the premises) to the overcurrent protective devices for branch circuits supplying the various loads.

A feeder may originate at a main distribution center and feed one or more subdistribution centers, one or more branch-circuit distribution centers, one or more branch circuits (as in the case of plug-in busway or motor circuit taps to a feeder), or a combination of these. It may be a primary or secondary voltage circuit, but its function is always to deliver a block of power from one point to another point at which the power capacity is apportioned among a number of other circuits. In some systems, feeders may be carried from a main distribution switchboard to subdistribution switchboards or panelboards from which subfeeders originate to feed branch-circuit panels or motor branch circuits. In still

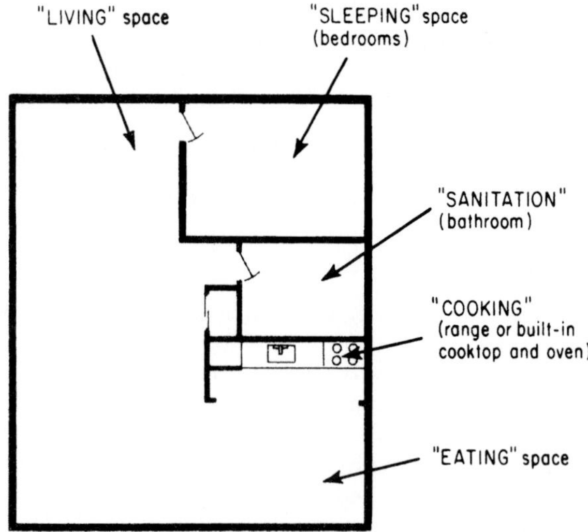

NOTE: Eating, living, and sleeping space could be one individual area, as in an efficiency apartment But the unit must contain a "bathroom," defined in Section 210-8(b) as "an area including a basin with one or more of the following: a toilet, a tub, or a shower." And the unit must contain permanent cooking equipment.

Fig. 100-19

other systems, either or both of the two foregoing feeder layouts may be incorporated with transformer substations to step the distribution voltage to utilization levels.

Grounded Conductor:

This is the conductor of an electrical system which is intentionally connected to a grounding electrode at the service of a premises, at a transformer secondary, or at a generator or other source of electric power. See Fig. 100-21. It is most commonly a neutral conductor of a system, but may be one of the phase legs—as in the case of a corner-grounded delta system.

Grounding one of the wires of the electrical system is done to limit the voltage upon the circuit which might otherwise occur through exposure to lightning or other voltages higher than that for which the circuit is designed. Another purpose in grounding one of the wires of the system is to limit the maximum voltage to ground under normal operating conditions. Also, a system which operates with one of its conductors intentionally grounded will provide for automatic opening of the circuit if an accidental or fault ground occurs on one of its ungrounded conductors.

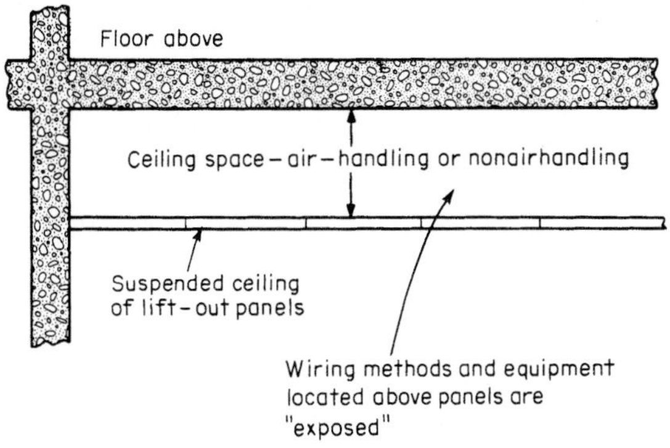

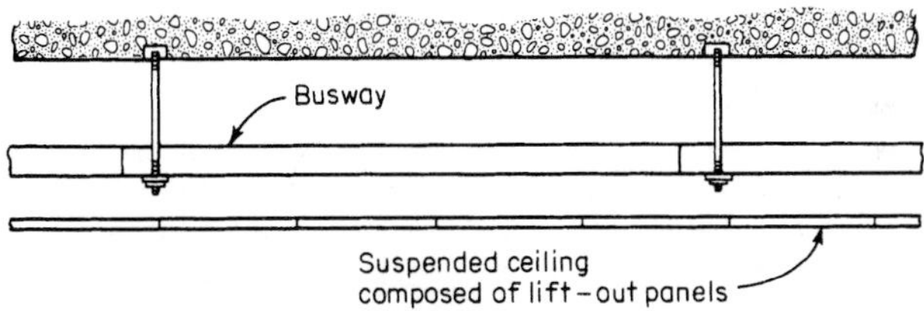

BUSWAY above this suspended ceiling is considered "exposed" as required by Section 364-4.

Fig. 100-20

Selection of the wiring system conductor to be grounded depends upon the type of system. In 3-wire, single-phase systems, the midpoint of the transformer winding—the point from which the system neutral is derived—is grounded. For grounded 3-phase wiring systems, the neutral point of the wye-connected transformer(s) or generator is usually the point connected to ground. In delta-connected transformer hookups, grounding of the system can be effected by grounding one of the three phase legs, by grounding a center-tap point on one of the transformer windings (as in the 3-phase, 4-wire "red-leg" delta system), or by using a special grounding transformer which establishes a neutral point of a wye connection which is grounded.

Grounding Conductor, Equipment:

The phrase "equipment grounding conductor" is used to describe any of the electrically conductive paths that tie together the noncurrent-carrying metal enclosures of electrical equipment in an electrical system. The term "equipment grounding conductor" includes bare or insulated conductors, metal raceways (rigid metal conduit, intermediate metal conduit, EMT), and metal cable

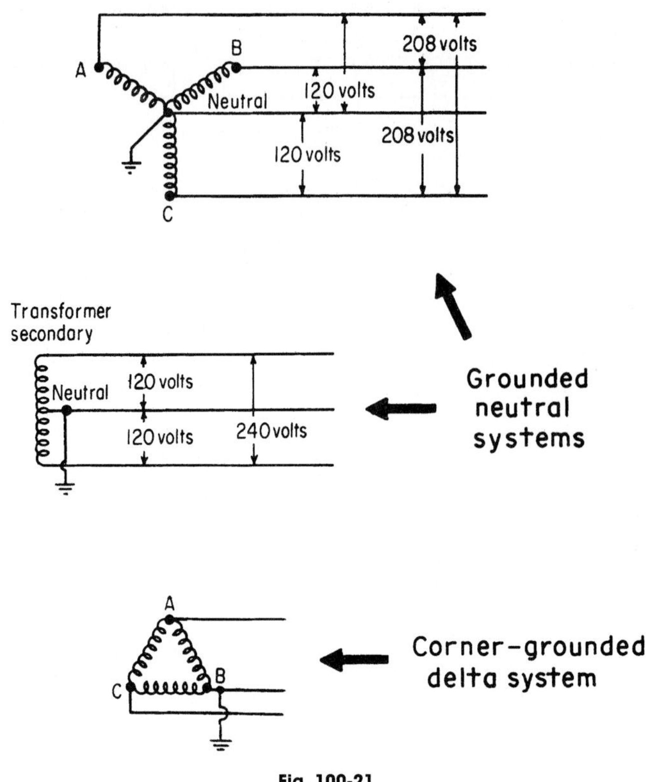

Fig. 100-21

jackets where the Code permits such metal raceways and cable enclosures to be used for equipment grounding—which is a basic Code-required concept as follows:

Equipment grounding is the grounding of all metal enclosures that contain electrical wires or equipment when an insulation failure in such enclosures might place a potential on the enclosures and constitute a shock or fire hazard. It is a permanent and continuous bonding together (i.e., connecting together) of all noncurrent-carrying metal enclosures and frames of electrical equipment—conduit, boxes, cabinets, housings of lighting fixtures and machines, frames of motors, etc.—and connection of this interconnected system of enclosures to the system grounding electrode. The interconnection of all metal enclosures must be made to provide a low-impedance path for fault-current flow along the enclosures in the event that one of the energized conductors within a metal enclosure should make contact with the enclosure and thereby energize it. This fault-current flow assures operation of overcurrent devices which will open a circuit in the event of a fault. By opening a faulted circuit, the system prevents dangerous voltages from being present on equipment enclosures which could be touched by personnel, with consequent electric shock to such personnel.

Simply stated, grounding of all metal enclosures of electric wires and equipment prevents any potential-above-ground on the enclosures. Such bonding together and grounding of all metal enclosures are required for both grounded electrical systems (those systems in which one of the circuit conductors is intentionally grounded) and ungrounded electrical systems (systems with none of the circuit wires intentionally grounded).

As shown in the sketch of Fig. 100-22, metal enclosures, metal raceways, and metal cable armors may serve as the "equipment grounding conductor." When nonmetallic raceways or cables are used, a bare or insulated conductor must be used within the raceway or cable to provide the interconnection of all metal enclosures. And the overall equipment grounding system must be connected to a grounding electrode at the service or power source (local transformer or a generator) and must also be connected at the service or source to the system-grounded conductor (such as a grounded neutral) if a grounded conductor is used.

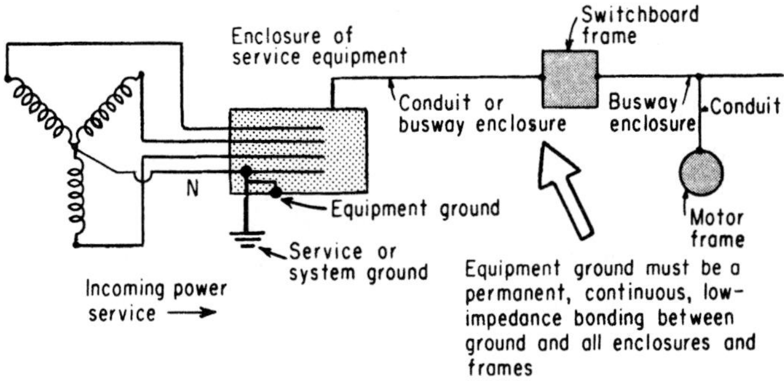

Fig. 100-22

Effective equipment grounding depends upon sure connection of all equipment grounding conductors of adequate ampacity. Equipment grounding is extremely important for grounded electrical systems to provide the automatic fault clearing, which is one of the important advantages of grounded electrical systems. A low-impedance path for fault current is necessary to permit enough current to flow to operate the fuses or circuit breakers protecting the circuit.

In a grounded electrical system with a high-impedance equipment ground return path (equipment grounding conductor path), if one of the phase conductors of the system (i.e., one of the ungrounded conductors of the wiring system) should accidentally come in contact with one of the metal enclosures in which the wires are run, not enough fault current would flow to operate the overcurrent devices. In such a case, the faulted circuit would not automatically open, and a dangerous voltage would be present on the conduit and other metal enclosures.

Grounding Electrode Conductor:

The conductor that runs from the bonded neutral block or busbar in service equipment to the system grounding electrode is clearly and specifically identified as the "grounding electrode conductor." See Fig. 100-23. It is also the conductor used to ground the bonded neutral of a transformer secondary or a generator.

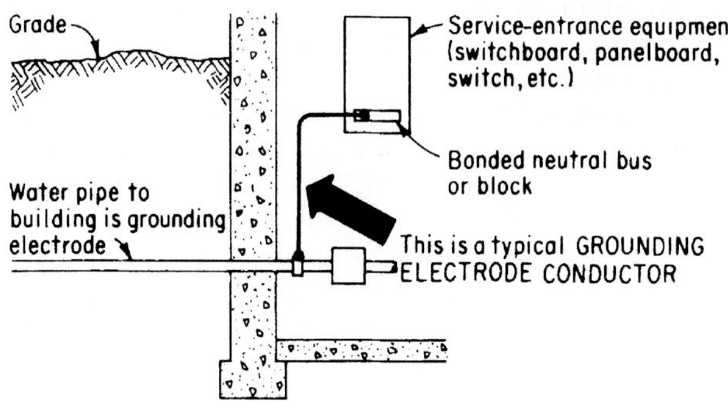

Fig. 100-23

Ground-Fault Circuit-Interrupter:

This revised definition makes clear that the device described is a GFCI (breaker or receptacle) of the type listed by UL and intended to eliminate shock hazards to people. Such devices must operate within a definite time from initiation of ground-fault current above the specified trip level (4 to 6 mA, as specified by UL). See Fig. 100-24.

Ground-Fault Protection of Equipment:

Although any type of ground-fault protection is aimed at protecting personnel using an electrical system, the so-called ground-fault protection required by

Fig. 100-24

NEC Secs. 215-10, 230-95, and 240-13 for 480Y/277-V disconnects rated 1,000 A or more, for example, is identified in Sec. 230-95 as "ground-fault protection of equipment." The so-called ground-fault circuit interrupter (GFCI), as described in the previous definition and required by Sec. 210-8 for residential receptacles and by other NEC rules, is essentially a "people protector" and is identified in Sec. 210-8 as "ground-fault protection for personnel." Because there are Code rules addressing these distinct functions—people protection versus equipment protection—this definition distinguishes between the two types of protection.

Identified:

It is the intent of the Code that this single word "identified" be substituted for the phrase "approved for the purpose" *with no change in concept from its definition* as it was given in previous editions of the NE Code. It is intended that this term permit code-making panels to include reference to function, use, environment, or specific purpose in the text of Code rules, for example, "identified for use in wet locations," "identified as suitable for Class I, Div. 1 locations," "identified as weatherproof."

Another purpose of the change from "approved for the purpose" to "identified" was to eliminate the implication that the local inspector (the authority having jurisdiction for Code enforcement) is the sole judge of a product's acceptability as a consequence of the definition for "approved." The term "identified" is intended to make clear that manufacturers and testing laboratories have responsibility to assist the inspector by their own indications of suitability of products for their intended uses.

Although the definition of "identified" does not specifically require that products be marked to designate specific application suitability, there is that suggestion; and such marking on a product, its label, or the box or wrapping in which the product is supplied would best comply with the definition of "identified."

In Sight From:

The phrase "in sight from" or "within sight from" or "within sight" means visible and not more than 50 ft (15.24 m) away. These phrases are used in many Code rules to establish installation location of one piece of equipment with respect to another. A typical example is the rule requiring that a motor-circuit disconnect means must be *in sight from* the controller for the motor [Sec. 430-102(a)]. This definition in Art. 100 gives a single meaning to the idea expressed by the phrases—not only that any piece of equipment that must be "in sight from" another piece of equipment must be visible, but also that the distance between the two pieces of equipment must not be over 50 ft (15.24 m). If, for example, a motor disconnect is 51 ft (15.48 m) away from the motor controller of the same circuit, it is *not* "within sight from" the controller even though it is actually and readily visible from the controller. In the interests of safety, it is arbitrarily defined that separation of more than 50 ft (15.24 m) diminishes visibility to an unacceptable level.

Interrupting Rating:

This definition covers both "interrupting ratings" for overcurrent devices (fuses and circuit breakers) and "interrupting ratings" for control devices (switches, relays, contactors, motor starters, etc.).

Labeled:

The label of a nationally recognized testing laboratory on a piece of electrical equipment is a sure and ready way to be assured that the equipment is properly made and will function safely when used in accordance with the application data and limitations established by the testing organization. Each label used on an electrical product gives the exact name of the type of equipment as it appears in the listing book of the testing organization.

Typical labels are shown in Fig. 100-25(*a*).

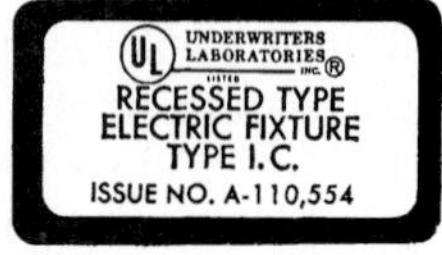

Fig. 100-25(*a*)

Underwriters Laboratories Inc., the largest nationally recognized testing laboratory covering the electrical field, describes its "Identification of Listed Products" as shown in Fig. 100-25(*b*).

It should be noted that the definitions for "labeled" and "listed" do not require that the testing laboratory be "nationally recognized." But OSHA rules *do* require such "labeling" or "listing" to be provided by a "nationally recognized" testing lab. Therefore, even though those NEC definitions acknowledge

The Listing Mark may appear in various forms as authorized by Underwriters Laboratories Inc. Typical forms which may be authorized are shown below:

UNDERWRITERS LABORATORIES INC. ®

UND. LAB. INC. ®

Underwriters
Lab. Inc. ®

Underwriters Laboratories Inc.®

LISTED

Listing Marks include one of the forms illustrated above, the word "Listed", and a control number assigned by UL. The product name as indicated in this Directory under each of the product categories is generally included as part of the Listing Mark text, but may be omitted when in UL's opinion, the use of the name is superfluous and the Listing Mark is directly and permanently applied to the product by stamping, molding, ink-stamping, silk screening or similar processes.

Separable Listing Marks (not part of a name plate and in the form of decals, stickers or labels) will always include the four elements: UL's name and/or symbol, the word "Listed", the product or category name, and a control number.

The complete four element Listing Mark will appear on the smallest unit container in which the product is packaged when the product is of such a size that only the symbol Ⓤ can be applied to the product or when the product size, shape, material or surface texture makes it impossible to apply any legible marking to the product.

LOOK FOR THE LISTING MARK

Fig. 100-25(*b*)

that a local inspector may accept the label or listing of a product by a testing organization that is qualified and capable even though it operates in a small area or section of the country and is not "nationally recognized," OSHA requirements may only be satisfied when "labeling" or "listing" is provided by a "nationally recognized" testing facility.

Listed:

As a result of broader, more intensive and vigorous enforcement of third-party certification of electrical system equipment and components, OSHA and the NE Code have made it necessary that all electrical construction people be fully aware of and informed about testing laboratories. The following organizations are widely known and recognized by governmental agencies for their independent product testing and certification activities. Each should be contacted directly for full information on available product listings and other data on standards and testing that *are* recognized by OSHA.

Underwriters Laboratories Inc.
333 Pfingsten Rd
Northbrook, IL 60062-2096
Other UL offices and testing stations:

1285 Walt Whitman Rd.
Melville, NY 11746-3081

1655 Scott Blvd.
Santa Clara, CA 95050-4169

12 Laboratory Dr.
Research Triangle Park, NC 27709-3995

Factory Mutual Engineering Corp.
1151 Boston-Providence Turnpike
Norwood, MA 02062

Electrical Testing Laboratories, Inc.
2 East End Ave.
New York, NY 10021

United States Testing Co., Inc. (UST)
1415 Park Avenue
Hoboken, NJ 07030

Canadian Standards Association (CSA)
178 Rexdale Blvd.
Rexdale, Ontario, Canada

Publications of nationally recognized testing laboratories may be obtained by writing to the various test labs at the above addresses.

Live Parts:

This definition indicates what is meant by that term as it is used throughout the Code. The inclusion of this term is not intended to indicate any change in application, such as where this term is used in relation to Conditions 1, 2, and 3 and the Table in Sec. 110-16(a). There is no intent to revise application of the work space rules which are based on the presence or absence of exposed "live parts." As applied to the workspace rules, equipment enclosures are still to be considered "exposed live parts" and treated as such for the purpose of establishing whether the clearance shown in Condition 1, 2, or 3 applies.

Nonlinear Load:

Those loads that cause distortion of the current waveform are defined as "nonlinear loads." A typical nonlinear load current and voltage waveform are shown in Fig. 100-26. As can be seen, while the voltage waveform [Fig. 100-26(*b*)] is a sinusoidal, 60-Hz, wave, the current waveform [Fig. 100-26(*a*)] is a series of pulses, with rapid rise and fall times, and does not follow the voltage waveform.

The FPN following this definition is *not* intended to be a complete list, but rather, just a few examples. There are many more such loads. The substantiation for inclusion of this FPN stated in part:

It has been known within the entertainment industry for some time that due to the independent single-phase phase-control techniques applied to three-phase, four-wire feeder, solid-state dimming can cause neutral currents in excess of the phase currents. This is in

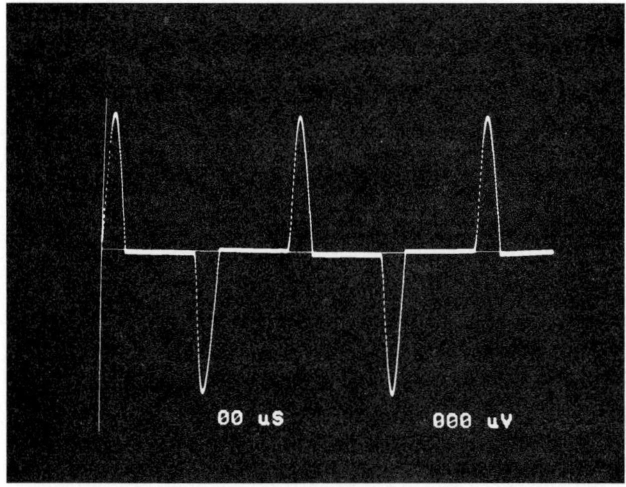

Fig. 100-26(*a*)

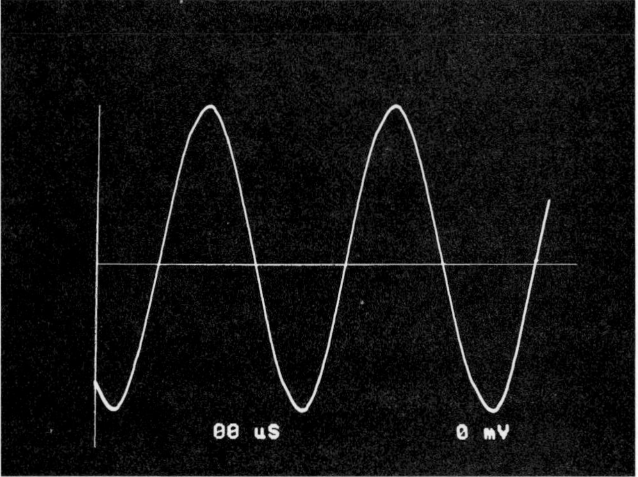

Fig. 100-26(*b*)

addition to the harmonics generated. This situation is dealt with in theaters in Secs. 520-27 and 520-51, etc. Dimming is also used in nontheatrical applications such as hotel lobbies, ballrooms, conference centers, etc. This effect must be taken into account wherever solid-state dimming is employed.

Plenum:

This definition is intended to clarify use of this word, which is referred to in Sec. 300-22(b) and other sections. A "plenum" is a compartment or chamber to which one or more air ducts are connected and which forms part of an air distribution system. This definition replaces the fine print note that was in Sec. 300-22(b) of the 1987 NEC. As now noted in the text of Sec. 300-22(b), a plenum is an enclosure "specifically fabricated to transport environmental air." The definition further clarifies that an air-handling space above a suspended ceiling or under a raised floor (such as in a computer room) is not a plenum, but is "other spaces" used for environmental air, as covered by Sec. 300-22(c).

Premises Wiring (System):

Published discussions of the Code panel's meaning of this phrase make clear the panel's intent that premises wiring includes all electrical wiring and equipment on the load side of the "service point," including any electrical work fed from a "separately derived system"—such as a transformer, generator, computer power distribution center, an uninterruptible power supply (UPS), or a battery bank. Premises wiring includes all electrical work installed on a premises. Specifically, it includes all circuits and equipment fed by the service or fed by a separately derived electrical source (transformer, generator, etc.). This makes clear that all circuiting on the load side of a so-called computer power center or computer distribution center [enclosed assembly of an isolating transformer and panelboard(s)] must satisfy all NEC rules on hookup and *grounding.* When a "computer power center" is supplied with factory-wired branch-circuit "whips" (lengths of flexible metal conduit or liquidtight flex—with installed conductors), the use of equipment grounding conductors and grounding connections at the power center and at receptacles fed by the circuits must satisfy all rules of Art. 250 (especially Sec. 250-26) on grounding, as well as the rules of Art. 645, Electronic Computer/Data Processing Systems.

Other sources, such as solar photovoltaic systems or storage batteries, also constitute "separately derived systems." All NEC rules applicable to premises wiring also pertain to the load side wiring of batteries and solar power systems.

Raceway:

Whenever this term is used in the Code, it must be understood that the meaning includes all the many enclosures used for running conductors between cabinets and housings of electrical distribution components—like panels, switches, motor starters, etc.—and housings of utilization equipment—like lighting fixtures, motors, heaters, etc. Any raceway must be an "enclosed" channel for conductors. Cable tray is a "support system" and not a "raceway." When the Code refers to "conduit," it means only those raceways containing the word "conduit" in their title. But "EMT" is not conduit. Table 1 of Chap. 9 in the back of the Code book refers to "Conduit and Tubing." The Code, thus, distinguishes between the two. "EMT" is tubing.

The FPN which appeared in the 1993 Code and in previous editions was incorporated into the definition itself. By doing so, the laundry list of equipment that is to be considered "raceway" is part of the Code and *not* just advisory; it has the force of law. Notice, too, that cablebus is no longer included.

Receptacle:

Each place where a plug cap may be inserted is a "receptacle," as shown in Fig. 100-27.

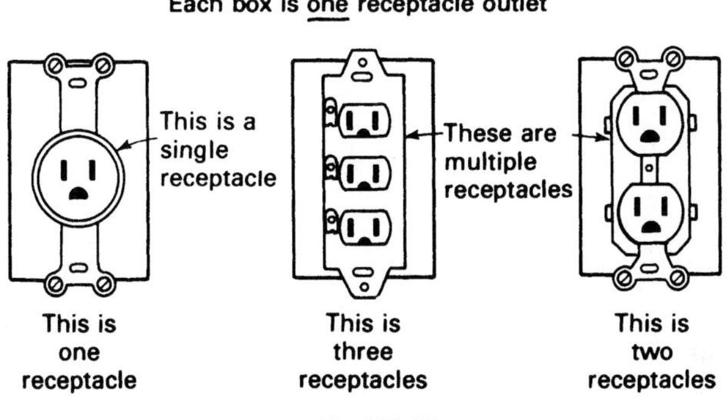

Fig. 100-27

Only a single receptacle can be served by an individual branch circuit. See Secs. 210-21(b) and 555-4.

Receptacle Outlet:

The "outlet" is the outlet box. But this definition must be carefully related to Sec. 220-3(c)(7) for calculating receptacle loads in other than dwelling occupancies. For purposes of calculating load, Sec. 220-3(c)(7) states that "For receptacle outlets, each single or multiple receptacle" must be taken as a load of 180 VA. Because a single, duplex, or triplex receptacle is a device on a single mounting strap, the rule requires that 180 VA must be counted for each strap, whether it supports one, two, or three receptacles.

Remote-Control Circuit:

The circuit that supplies energy to the operating coil of a relay, a magnetic contactor, or a magnetic motor starter is a "remote-control circuit" because that circuit controls the circuit that feeds through the contacts of the relay, contactor, or starter as shown in Fig. 100-28.

A control circuit as shown is any circuit which has as its load device the operating coil of a magnetic motor starter, a magnetic contactor, or a relay. Strictly speaking, it is a circuit which exercises control over one or more other circuits. And these other circuits controlled by the control circuit may themselves be control circuits, or they may be "load" circuits—carrying utilization current to a lighting, heating, power, or signal device. The sketch clarifies the distinction between control circuits and load circuits.

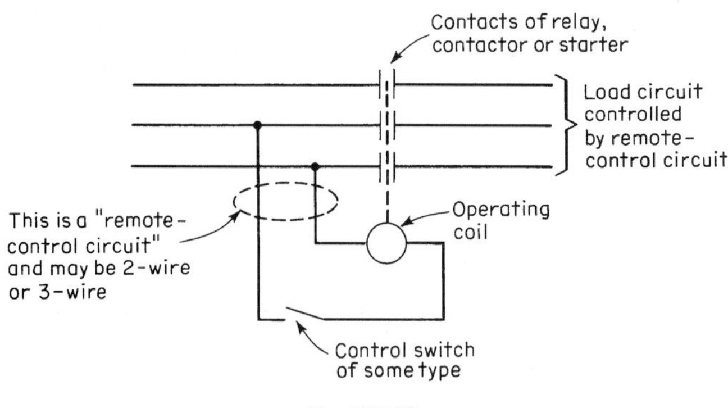

Fig. 100-28

The elements of a control circuit include all the equipment and devices concerned with the function of the circuit: conductors, raceways, contactor operating coils, source of energy supply to the circuit, overcurrent protective devices, and all switching devices which govern energization of the operating coil.

The NE Code covers application of remote-control circuits in Art. 725 and in Secs. 430-71 through 430-74.

Service:

The word "service" includes *all* the materials and equipment involved with the transfer of electric power from the utility distribution line to the electrical wiring system of the premises being supplied. Although service layouts vary widely, depending upon the voltage and amp rating, the type of premises being served, and the type of equipment selected to do the job, every service generally consists of "service-drop" conductors (for overhead service from a utility pole line) or "service-lateral" conductors (for an underground service from either an overhead or underground utility system)—plus metering equipment, some type of switch or circuit-breaker control, overcurrent protection, and related enclosures and hardware. A typical layout of "service" for a one-family house breaks down as in Fig. 100-29.

That part of the electrical system which directly connects to the utility-supply line is referred to as the "service entrance." Depending upon the type of utility line serving the house, there are two basic types of service entrances—an overhead and an underground service.

The **overhead service** has been the most commonly used type of service. In a typical example of this type, the utility supply line is run on wood poles along the street property line or back-lot line of the building, and a cable connection is made high overhead from the utility line to a bracket installed somewhere high up on the building. This wood pole line also carries the telephone lines, and the poles are generally called "telephone poles."

The aerial cable that runs from the overhead utility lines to the bracket on the outside wall of the building is called the "service drop." This cable is installed

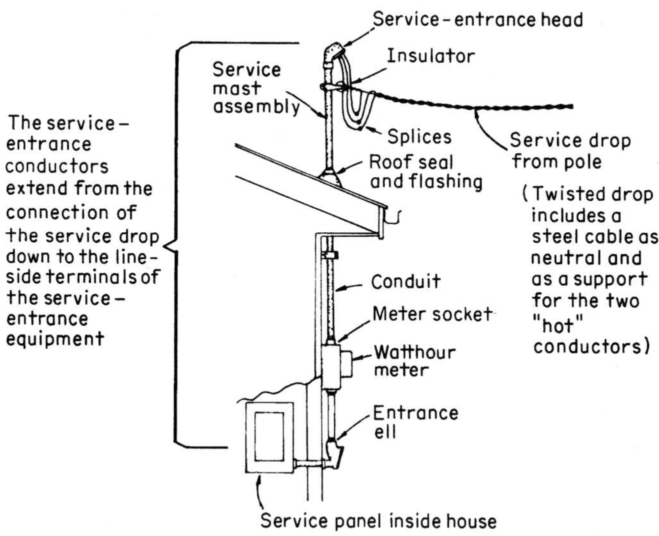

Fig. 100-29

by the utility-line worker. At the bracket which terminates the service drop, conductors are then spliced to the drop cable conductors to carry power down to the electric meter and into the building.

The **underground service** is one in which the conductors that run from the utility line to the building are carried underground. Such an underground run to a building may be tapped from either an overhead utility pole line or an underground utility distribution system. Although underground utility services tapped from a pole line at the property line have been used for many years to eliminate the unsightliness of overhead wires coming to a building, the use of underground service tapped from an underground utility system has only started to gain widespread usage in residential areas over recent years. This latter technique is called "URD"—which stands for *U*nderground *R*esidential *D*istribution.

As noted above, when a building is supplied by an overhead drop, an installation of conductors must be made on the outside of the building to pick up power from the drop conductors and carry it into the meter enclosure and service-entrance equipment (switch, CB, panelboard, or switchboard) for the building. On underground services, the supply conductors are also brought into the meter enclosure on the building and then are run into the service equipment installed, usually, within the building.

Service Conductors:

This is a general term that covers all the conductors used to connect the utility-supply circuit or transformer to the service equipment of the premises served. This term includes "service-drop" conductors, "service-lateral" (underground service) conductors, and "service-entrance" conductors. In an overhead distribution system, the service conductors begin at the line pole where connection is made. If a primary line is extended to transformers installed outdoors on private property, the service conductors to the building proper begin at the secondary terminals of the transformers.

Where the supply is from an underground distribution system, the service conductors begin at the point of connection to the underground street mains.

In every case the service conductors terminate at the service equipment.

Service Drop:

As the name implies, these are the conductors that "drop" from the overhead utility line and connect to the service-entrance conductors at their upper end on the building or structure supplied. See Fig. 100-30.

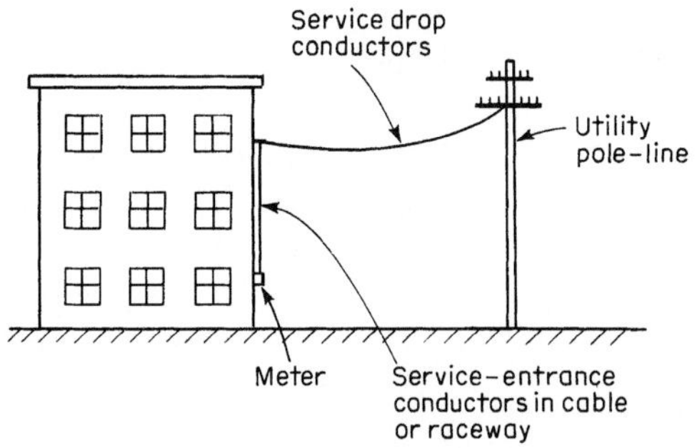

Fig. 100-30

Service Lateral:

This is the name given to a set of underground service conductors. A service lateral serves a function similar to that of a service drop as shown in Fig. 100-31.

Service Point:

"Service point" is the "point of connection between the facilities of the serving utility and the premises' wiring." All equipment on the load side of that point is subject to NE Code rules. Any equipment on the line side is the concern of the power company and is not regulated by the Code. This definition of "service point" must be construed as establishing that "service conductors" originate at that point. The whole matter of identifying the "service conductors" is covered by this definition.

The definition of "service point" does tell where the NE Code becomes applicable, and does pinpoint the origin of service conductors. And that is a critical task, because a corollary of that determination is identification of that equipment which is, technically, "service equipment" subject to all applicable NE Code rules on such equipment. Any conductors between the "service

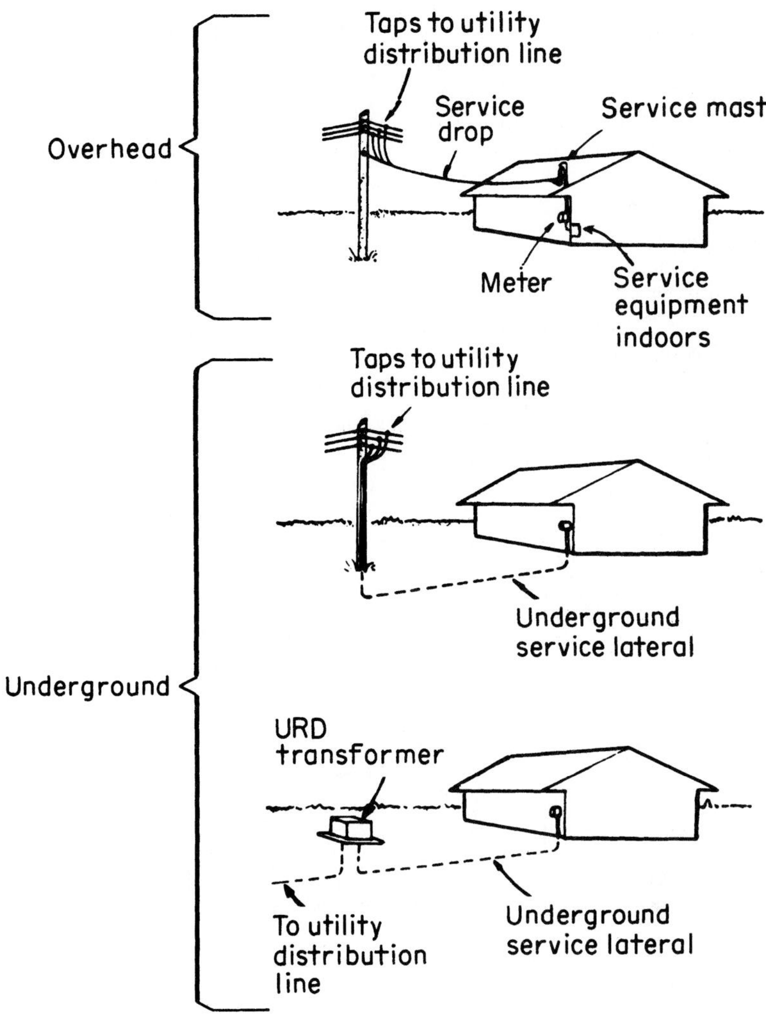

Fig. 100-31

point" of a particular installation and the service disconnect are identified as service conductors and subject to NE Code rules on service conductors (Fig. 100-32).

Solar Photovoltaic System:

This refers to the equipment involved in a particular application of the developing technology of solar energy conversion to electric power. This definition correlates to NEC Art. 690 covering design and installation of electrical systems for direct conversion of the sun's light into electric power.

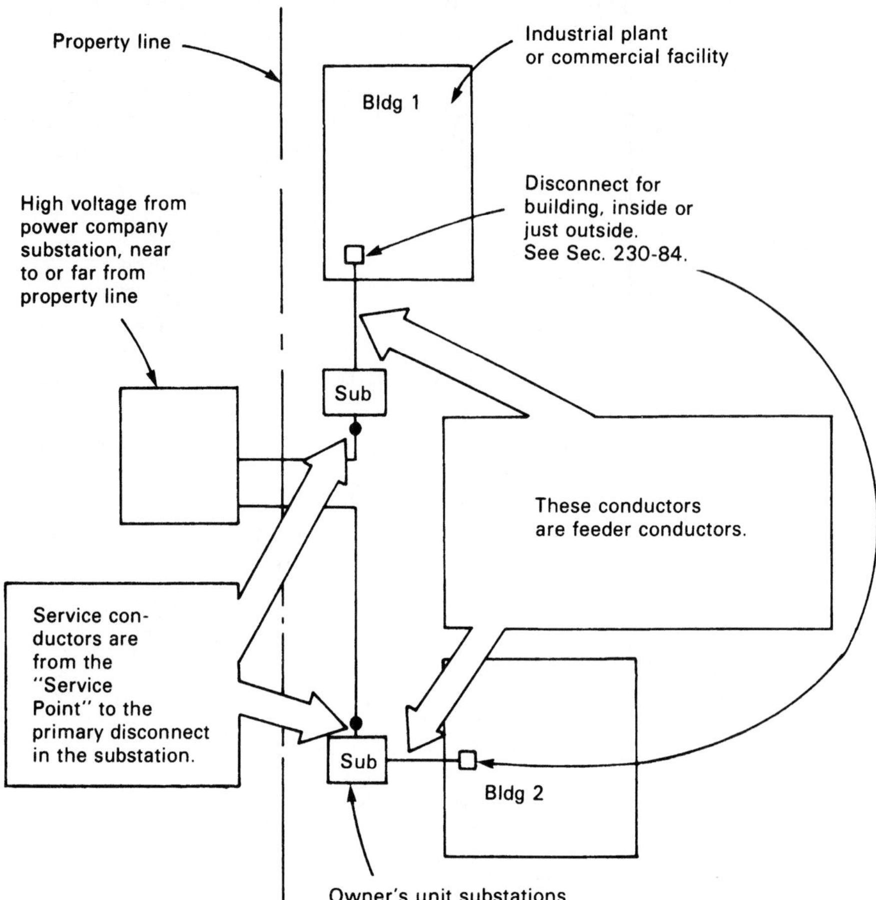

Fig. 100-32. NE Code rules apply on load side of "service point"—not from property line. (Sec. 230-200.)

Special Permission:

It *must* be carefully noted that any Code reference to "special permission" as a basis for accepting any electrical design or installation technique *requires* that such "permission" be in *written form.* Whenever the inspection authority gives "special permission" for an electrical condition that is at variance with Code rules or not covered fully by the rules, the authorization must be "written" and not simply verbal permission.

Switches:

Bypass isolation switch This is "a manually operated device" for bypassing the load current around a transfer switch to permit isolating the transfer switch for maintenance or repair without shutting down the system. The second paragraph of Sec. 700-6 permits that "means" be provided to bypass and isolate transfer equipment. This definition ties into that rule.

Transfer switch This is a switch for transferring load-conductor connections from one power source to another. And a note indicates that such a switch may be automatic or nonautomatic—but is subject to the applicable rules of Article 230 (Sec. 230-83), 517, 700, 701, and 702.

Voltage to Ground:

For a grounded electrical system, voltage to ground is the voltage that exists from any ungrounded circuit conductor to either the grounded circuit conductor (if one is used) or the grounded metal enclosures (conduit, boxes, panelboard cabinets, etc.) or other grounded metal, such as building steel. Examples are given in Fig. 100-33.

For an ungrounded electrical system, voltage to ground is *taken to be* equal to the maximum voltage that exists between any two conductors of the system. This is based on the reality that an accidental ground fault on one of the ungrounded conductors of the system places the other system conductors at a

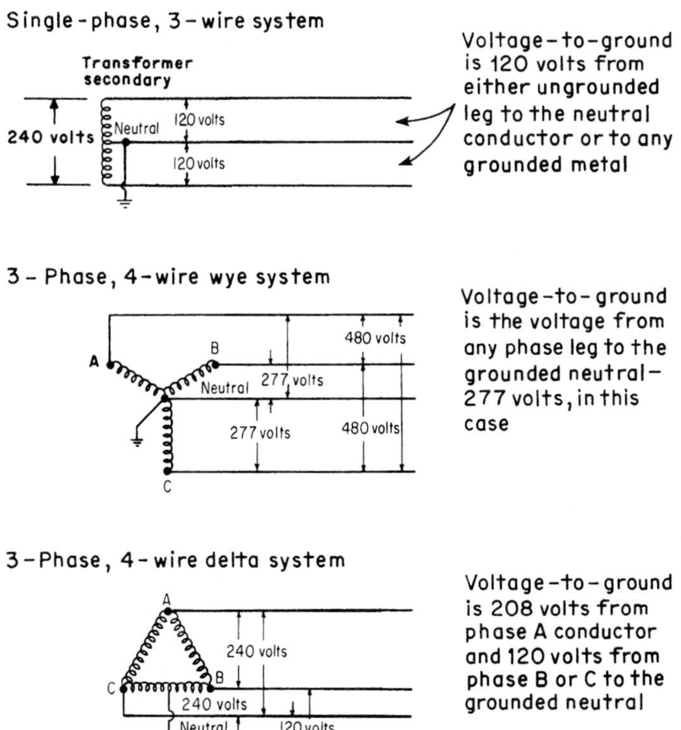

Fig. 100-33

voltage aboveground that is equal to the value of the voltage between conductors. Under such a ground-fault condition, the voltage to ground is the phase-to-phase voltage between the accidentally grounded conductor and any other phase leg of the system. On, say, a 480-V, 3-phase, 3-wire ungrounded delta system, voltage to ground is, therefore, 480 V, as shown in Fig. 100-34.

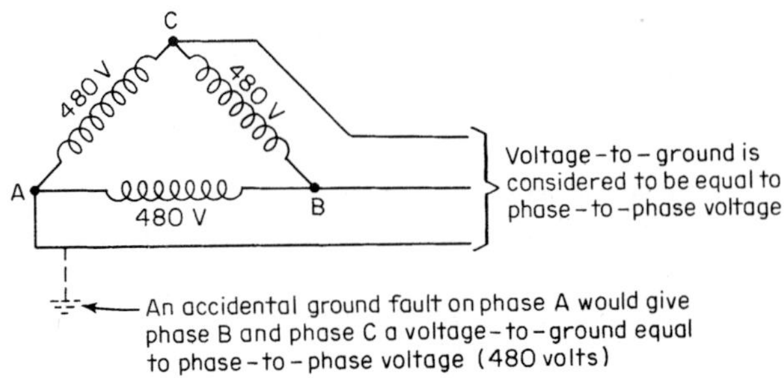

Fig. 100-34

In many Code rules, it is critically necessary to distinguish between references to "voltage" and to "voltage to ground." The Code also refers to "voltage between conductors," as in Sec. 210-6, to make very clear how rules must be observed.

ARTICLE 110. REQUIREMENTS FOR ELECTRICAL INSTALLATIONS

110-2. Approval. This section of the NE Code regulates the use of electrical products and equipment. The intent of the NEC is to place strong insistence on third-party certification of the essential safety of the equipment and component products used to assemble an electrical installation.

The use of custom-made equipment is also covered in OSHA rules. Every manufacturer of custom equipment must provide documentary safety-test data to the owner on whose work premises the custom equipment is installed. And it seems to be a reasonable conclusion from the whole rule itself that custom-equipment assemblies must make maximum use of "listed," "labeled," or "certified" components.

The word "approved," as used in this rule, must be taken to have the meaning given in the NE Code definition for that word. "Approved" is "acceptable to the authority having jurisdiction." The electrical inspector having jurisdiction on any specific installation is the final judge of what conductors and/or equipment are "approved." Although inspectors are not *required* to use "listing" or "labeling" by a national testing lab as the deciding factor in their approval of products, they invariably base their acceptance of products on listings by testing labs. The note in this section almost makes the same insistence as OSHA does that, whenever possible, acceptability must be based on some kind of listing or certification of a national lab.

110-3. Examination, Identification, Installation, and Use of Equipment. This section presents general rules on "Examination, Identification, Installation, and Use of Equipment." Part **(a)** lists eight factors that must be evaluated in determining acceptability of equipment for Code-recognized use.

The FPN to Sec. 110-3(a)(1) notes that, in addition to "listing" or "labeling" of a product by UL or another test lab to certify the conditions of its use, acceptability *may* be "identified by a description *marked on* or *provided with* a product to identify the suitability of the product for a specific purpose, environment, or application." This is a follow-through on the definition of the word "identified," as given in Art. 100. The requirement for identification of a product as specifically suited to a given use is repeated at many points throughout the Code.

Item **(3)**, an important consideration for electrical inspectors to include in their examination to determine suitability of equipment for safe and effective use, is "wire-bending and connection space." See Fig. 110-1. This factor is included because of increasing concern over inadequate gutter space at conductor terminal locations in enclosures for switches, CBs, and other control and protection equipment. This general mention of the need for sufficient conductor bending space is aimed at avoiding poor terminations and conductor damage that can result from excessively sharp conductor bends required by tight gutter spaces at terminals. Specific rules that cover this consideration are given in Sec. 373-6 on "Deflection of Conductors" at terminals or where entering or leaving cabinets or cutout boxes—covering gutter widths and wire-bending spaces.

Part **(b)** of this section is a critically important Code rule because it incorporates, as part of the NE Code itself, all the application regulations and limitations published by product-testing organizations, such as UL, Factory Mutual, ETL, etc. That rule clearly and certainly says, for instance, that any and every product listed in the UL *Electrical Construction Materials Directory* (Green Book) must be used exactly as described in the application data given with the listing in the book. Because the *Electrical Construction Materials Directory* and the other UL books of product listings, such as the *Hazardous Location Equipment Directory* (Red Book) and the *Electrical Appliance and Utilization Equipment Directory* (Orange Book), contain massive amounts of installation and application instruction, all those specific bits of application data become mandatory NE Code regulations as a result of the rule in Sec. 110-3(b). The data given in the UL listing books supplement and expand upon rules given in the NE Code. In fact, effective compliance with NE Code regulations can be assured only by careful study and observance of the limitations and conditions spelled out in the application instructions given in the UL listings books or similar instructions provided by other national testing labs.

In the preface to its *Electrical Construction Materials Directory* (the Green Book), UL points out certain basic conditions that apply to products listed.

1. In general, equipment listed is intended for use in ordinary locations, which may be dry, damp, or wet locations, as defined in the NE Code. All limitations on use specified in the NE Code and in the general information preceding the section on listing in the Green Book must be carefully

Fig. 110-1. Equipment must be evaluated for adequate gutter space to assure safe and effective bending of conductors at terminals. [Sec. 110-3(a)(3).]

observed. Equipment and products for use in hazardous locations are covered in the UL Red Book, *Hazardous Location Equipment Directory.*

2. Listed equipment has been investigated only for use indoors in dry locations, unless outdoor use is specifically permitted by NE Code rules, is indicated in the UL listing information, or is obvious from the designation of the equipment in the listing (such as "Swimming Pool Fixtures").
3. The amperage or wattage marking on power-consuming equipment is valid only when the equipment is supplied at its marked rated voltage. In general, current input to resistive loads increases in direct proportion to input voltage increase. Current input to an induction motor with a fixed load increases in direct proportion to decrease in input voltage.
4. All permanently connected equipment and appliances provided with terminals are intended for use with *copper* supply conductors. Terminals that are suitable for *either* copper or aluminum conductors are marked to indicate that. Such marking must be independent of any marking on terminal connectors and must be on a wiring diagram or other readily visible location. Permanently connected equipment and appliances with pigtail leads are intended for use with copper supply conductors. But aluminum supply wires may be used if they are spliced to the pigtails by splicing devices that are suitable for joining copper to aluminum.
5. A very important qualification is indicated for the temperature ratings of terminations. Although application data on minimum required tempera-

ture ratings of conductors connected to equipment terminals are given in **NE Code** Sec. 110-14(c), this matter is more comprehensively covered in the *UL General Information Directory.*

A basic general rule in the preface of the listing of "Equipment for Use in Ordinary Locations" in the UL *Electrical Construction Materials Directory* states that, for distribution and control equipment,

> Except as noted in the following paragraphs or in the information at the beginning of some product categories, the termination provisions are based on the use of 60°C ampacities for wire sizes No. 14-1 AWG and 75°C ampacities for wire sizes Nos. 1/0 AWG and larger, as specified in Table 310-16 of the **National Electrical Code.**
>
> Some distribution and control equipment is marked to indicate the required temperature rating of each field-installed conductor. If the equipment, normally intended for connection by wire sizes within the range 14-1 AWG, is marked "75°C only" or "60/75°C," it is intended that 75°C insulated wire may be used at full 75°C ampacity. Where the connection is made to a circuit breaker or switch within the equipment, such a circuit breaker or switch must also be marked for the temperature rating of the conductor.
>
> A 75°C conductor temperature marking on a circuit breaker or switch normally intended for wire sizes 14-1 AWG does not in itself indicate that 75°C insulated wire can be used unless (1) the circuit breaker or switch is used by itself, such as in a separate enclosure or (2) the equipment in which the circuit breaker or switch is installed is also so marked.
>
> "A 75 or 90°C conductor temperature marking on a terminal (e.g., AL7, CU7AL, AL7CU, or AL9, CU9AL, AL9CU) does not in itself indicate that 75 or 90°C insulated wire can be used unless the equipment in which the terminals are installed is marked for 75 or 90°C."
>
> Higher temperature rated conductors than specified may be used if the size is based on the above statements.

Application of these data to various types of equipment is repeated in many sections of the *Electrical Construction Materials Directory.* And, as the UL wording says, this temperature limitation on terminals applies to the terminals on all equipment—circuit breakers, switches, motor starters, contactors, etc.—except where some other specific condition is recognized in the general information preceding the product category. This is a vitally important matter, which has been widely disregarded in general practice.

When terminals are tested for suitability at 60 or 75°C, the use of 90°C conductors operating at their higher current ratings poses definite threat of damage to terminals on switches, breakers, etc. Many termination failures experienced in electrical equipment suggest overheating even where the load current did not exceed the current rating of the breaker or switch or other equipment.

When a 60°C-rated terminal is fed by a conductor operating at 90°C, there will be substantial heat conducted from the 90°C conductor metal to the 60°C-rated terminal; and, over a period of time, that can damage the termination—even though the load current does not exceed the equipment current rating and does not exceed the ampacity of the 90°C conductor. Whenever two metallic parts at different operating temperatures are tightly connected, the higher-temperature part (say, the 75 or 90°C wire) will give heat to the lower-temperature part (the 60°C terminal) and thereby raise its temperature over 60°C.

The basic UL limitation on termination temperature rating for all distribution and control equipment is shown in Fig. 110-2.

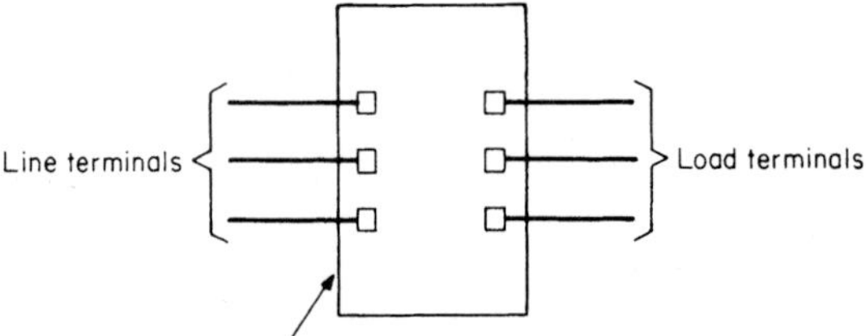

Panelboard, switch, transformer, motor starter, etc.

UNLESS A PIECE OF ELECTRICAL EQUIPMENT IS MARKED OTHERWISE circuit conductors connected to the terminals must not operate at more than the 60°C ampacity for conductor sizes No. 14 up to No. 1 and must not operate at more than a 75°C ampacity for conductor sizes No. 1/0 and larger (refer to *NEC* Tables 310-16 through -19).

FOR SWITCHES, CONTACTORS, ETC. WIRED WITH No. 14-1 AWG: Use TW wire (or use THW, THHN, RHH, XHHW, or other higher-temperature wire at the ampacity of the corresponding size of TW wire).

FOR SWITCHES, CONTACTORS, ETC. WIRED WITH No. 1/0 AWG OR LARGER: Use TW, THW, THWN or XHHW wire at their ampacities permitted up to 75°C (or use RHH, THHN, or other higher-temperature wire at the ampacity of the corresponding size of 75°C wire).

Fig. 110-2. [Secs. 110-3(b) and 110-14(c)]

The same UL Directory contains the following regulations on circuit-breaker terminals:

Circuit breakers and circuit-breaker enclosures as listed herein are for use with copper conductors unless marked to indicate which terminals are suitable for use with aluminum conductors. Such markings shall be independent of any marking on terminal connectors and shall be on a wiring diagram or other readily visible location.

1. Circuit-breaker enclosures are marked to indicate the temperature rating of all field-installed conductors.

2. Circuit breakers of continuous current rating of 125 A or less are marked as being suitable for 60°C only, 75°C only, or 60/75°C wire.

3. Circuit breakers rated 125 A or less and marked suitable for use with 75°C-rated wire are intended for field use with 75°C wire at full 75°C ampacity only when the circuit breaker is installed in a circuit-breaker enclosure or individually mounted in an industrial control panel with no other component next to it, unless the end-use equipment (panelboard, switchboard, service equipment, power outlet, etc.) is also marked suitable for use with 75°C wire.

4. A circuit breaker of continuous current rating of more than 125 A is suitable for use with 75°C wire.

A suitable marking is required in a circuit-breaker enclosure, whether or not terminals are mounted therein, if it is intended that the breaker to be mounted therein is to be used with aluminum wire.

Note: Where a terminal is marked with a torque value (lb·in. or lb·ft), it is mandatory to use a torque wrench or screwdriver when tightening terminals on circuit breakers. And Sec. 110-3(b) makes such torquing a mandatory NEC rule.

For any given size of conductor, the greater ampacity of a higher-temperature conductor is established by the ability of the conductor insulation to withstand the I^2R heat produced by the higher current flowing through the conductor. But it must not be assumed that the equipment to which that conductor is connected is also capable of withstanding the heat that will be thermally conducted from the metal of the conductor into the metal of the terminal to which the conductor is tightly connected.

Although this limitation on the operating temperature of terminals in equipment does reduce the advantage that higher-temperature conductors have over lower-temperature conductors, there are still many advantages to using the higher-temperature conductors because of their reduced cross-section areas that permit more economical raceway fills—either smaller conduit for a given number of conductors or more conductors in a given size of conduit. However, where higher-temperature conductors are used, they must be applied at the ampacities of corresponding sizes of 60 or 75°C conductors—as required.

In conductor sizes No. 14, No. 12, and No. 10, with their 15-, 20-, and 30-A load ratings, as set by the footnote to NE Code Table 310-16, there is no difference in load rating between the same size conductors of different temperature rating. A No. 12 TW copper conductor, for instance, has a 20-A basic rating; and a No. 12 THW (75°C), a No. 12 RHH (90°C), a No. 12 XHHW (75°C in wet or dry locations or 90°C in dry locations only), a No. 12 XHHW-2 (90°C in wet or dry locations), and a No. 12 THHN (90°C) are all rated at a basic loading value of 20 A. Even though the different types of conductors in sizes No. 14, No. 12, and No. 10 have the same load-current ratings, there is still an advantage in using the thin-wall-insulated, higher-temperature wires because they permit greater conduit fill.

In those conductor sizes where there are differences in ampacity between conductors of different temperature ratings (conductors No. 8 and larger), careful consideration must be given to a number of factors involved in selection of the correct and most effective conductor for any specific circuit application. Attention must certainly be paid to the above-described temperature limitation at equipment terminals. Selection of circuit conductors must also be based on load limitation required by NE Code Secs. 210-22(c), 220-3(a), and 220-10(b) for "continuous load"—that is, any case "where the maximum current is expected to continue for three hours or more," such as branch circuits and feeders for commercial and industrial lighting or similar loads that are left on all day or for periods over 3 hr. And the greater conduit fill of thin-wall-insulated, high-temperature conductors (offering use of smaller conduits) must also be factored into conductor selection, along with the impact of conductor load-current derating due to conduit fills over three current-carrying conductors. All these considerations are interrelated, and the most effective and most economical circuit makeup for any case can be determined only by thorough study and careful calculation. For typical details of this kind of analysis to relate all the applicable NE Code and UL rules, see Sec. 310-15 on ampacity of conductors.

110-4. Voltages. In all electrical systems there is a normal, predictable spread of voltage values over the impedances of the system equipment. It has been common practice to assign these basic levels to each nominal system voltage. The highest value of voltage is that at the service entrance or transformer secondary, such as 480Y/227 V. Then considering voltage drop due to impedance in the circuit conductors and equipment, a "nominal" midsystem voltage designation would be 460Y/265, and finally a "load" or "outlet" voltage is given as 440Y/254. Variations in "nominal" voltages have come about because of (1) differences in utility-supply voltages throughout the country, (2) varying transformer secondary voltages produced by different and often uncontrolled voltage drops in primary feeders, and (3) preferences of different engineers and other design authorities.

110-6. Conductor Sizes. In this country, the American Wire Gage (AWG) is the standard for copper wire and for aluminum wire used for electrical conductors. The American Wire Gage is the same as the Brown & Sharpe (B & S) gage. The largest gage size is No. 0000; above this size the sizes of wires and cables are stated in circular mils.

The circular mil is a unit used for measuring the cross-sectional area of the conductor, or the area of the end of a wire which has been cut square across. One circular mil (commonly abbreviated cmil) is the area of a circle 1/1,000 in. in diameter. The area of a circle 1 in. in diameter is 1,000,000 cmil; also, the area of a circle of this size is 0.7854 sq in.

To convert square inches to circular mils, multiply the square inches by 1,273,200.

To convert circular mils to square inches, divide the circular mils by 1,273,200 or multiply the circular mils by 0.7854 and divide by 1,000,000.

In interior wiring the gage sizes 14, 12, and 10 are usually solid wire; No. 8 and larger conductors in raceways are required to be stranded. (See Sec. 310-3.)

A cable (if not larger than 1,000,000 cmil) will have one of the following numbers of strands: 7, 19, 37, or 61. In order to make a cable of any standard size, in nearly every case the individual strands cannot be any regular gage number but must be some special odd size. For example, a No. 2/0 cable must have a total cross-sectional area of 133,100 cmil and is usually made up of 19 strands. No. 12 has an area of 6,530 cmil and No. 11, an area of 8,234 cmil; therefore each strand must be a special size between Nos. 12 and 11.

110-7. Insulation Integrity. Previous editions of the Code contained *recommended* values for testing insulation resistance. It was found that those values were incomplete and not sufficiently accurate for use in modern installations, and the recommendation was deleted from the Code. However, basic knowledge of insulation-resistance testing is important.

Measurements of insulation resistance can best be made with a megohmmeter insulation tester. As measured with such an instrument, insulation resistance is the resistance to the flow of direct current (usually at 500 or 1,000 V for systems of 600 V or less) through or over the surface of the insulation in electrical equipment. The results are in ohms or megohms, but insulation readings will be in the megohm range.

110-8. Wiring Methods. All Code-recognized wiring methods are covered in Chap. 3 of the NE Code—Arts. 300 through 384.

110-9. Interrupting Rating. Interrupting rating of electrical equipment is divided into two categories: current at fault levels and current at operating levels.

Equipment intended to clear fault currents must have interrupting rating equal to the maximum fault current that the circuit is capable of delivering at the line (not the load) terminals of the equipment. See Fig. 110-3. The internal impedance of the equipment itself may not be factored in to use the equipment at a point where the available fault current on its line side is greater than the rated, marked interrupting capacity of the equipment.

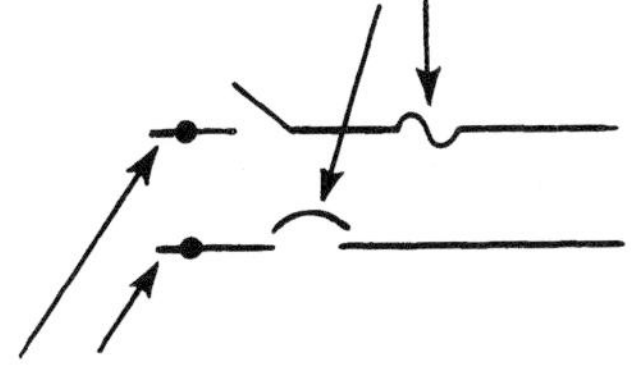

. . . must have an interrupting rating at least equal to the maximum fault current that the circuit could deliver into a short circuit on the *line side* of the device.

NOTE: That means that the fault current "available" at the line terminals of all fuses and circuit breakers *must be known* in order to assure that the device has a rating sufficient for the level of fault current.

Fig. 110-3. (Sec. 110-9.)

If overcurrent devices with a specific IC (interrupting capacity) rating are inserted at a point on a wiring system where the available short-circuit current exceeds the IC rating of the device, a resultant downstream solid short circuit between conductors or between one ungrounded conductor and ground (in grounded systems) could cause serious damage to life and property.

Since each electrical installation is different, the selection of overcurrent devices with proper IC ratings is not always a simple task. To begin with, the amount of available short-circuit current at the service equipment must be known. Such short-circuit current depends upon the capacity rating of the utility primary supply to the building, transformer impedances, and service conductor impedances. Most utilities will provide this information. But, be aware that Sec. 110-9 specifically requires all such calculations.

Downstream from the service equipment IC ratings of overcurrent devices may be reduced to lower than those at the service, depending on lengths and sizes of feeders, line impedances, and other factors. However, large motors and capacitors, while in operation, will feed additional current into a fault, and this must be considered when calculating short-circuit currents.

Manufacturers of overcurrent devices have excellent literature on figuring short-circuit currents, including graphs, charts, and one-line-diagram layout sheets to simplify the selection of proper overcurrent devices.

Equipment intended only for control of load or operating currents, such as contactors and unfused switches, must be rated for the current to be interrupted, but does not have to be rated for interrupting available fault level, as shown in Fig. 110-4.

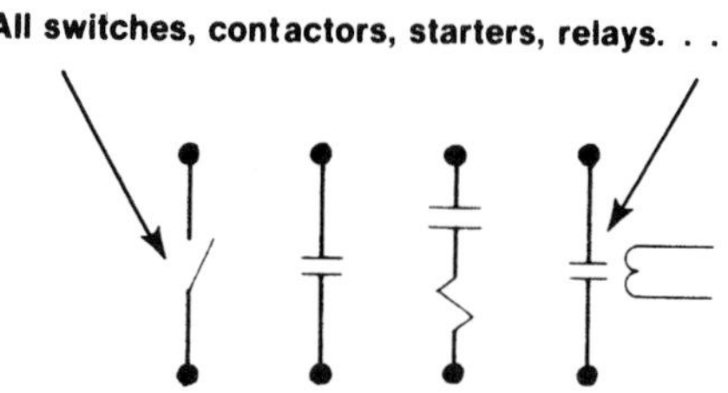

. . . must have an interrupting rating at least equal to "the current that must be interrupted"—which could be full-load current or, in the case of isolating or disconnect switches, some lesser value of operating current (such as transformer magnetizing current).

Fig. 110-4. (Sec. 110-9.)

110-10. Circuit Impedance and Other Characteristics. This section requires that all equipment be rated to withstand the level of fault current that is let through by the circuit protective device in the time it takes to operate—without "extensive damage" to any of the electrical components of the circuit as illustrated in Fig. 110-5.

The phrase "the component short-circuit withstand ratings" was added to this rule a few editions back. The intent of this addition is to require all circuit components that are subjected to ground faults or short-circuit faults to be capable of withstanding the thermal and magnetic stresses produced within them from the time a fault occurs until the circuit protective device (fuse or CB) opens to clear the fault, without extensive damage to the components.

A change was made in the 1996 Code, but, for the most part, this change was editorial in nature. However, the Code-making panel (CMP) responsible for Art. 110 has indicated that this section is not intended to establish a quantifiable amount of damage that is permissible under conditions of short circuit. The general requirement presented here is just that, a general rule. Specifics regarding what damage is or is not acceptable only depends on the product standard. For example, the permissible damage for a Type E (the so-called self-protected) motor starter is different from the requirements for other types of motor starters. The Type E unit must satisfy a more rigorous performance criterion than the others. Therefore, Sec. 110-10 would also essentially require a more rigorous performance criterion for the Type E starter but *not* for the others.

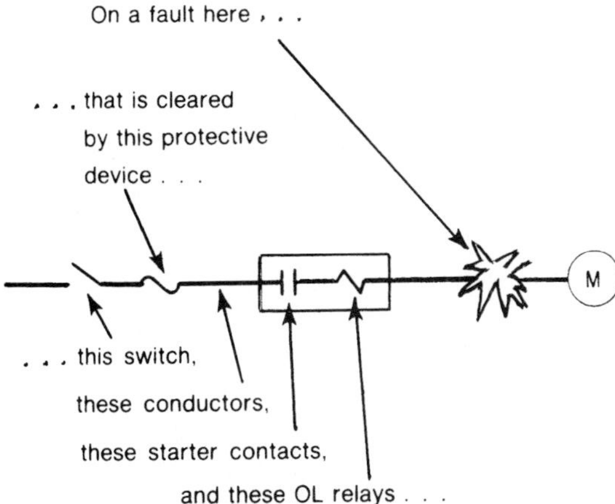
On a fault here . . .
. . . that is cleared by this protective device . . .
M
. . . this switch,
these conductors,
these starter contacts,
and these OL relays . . .
. . . must all be rated to safely withstand the energy of the fault let-through current

Fig. 110-5. (Sec. 110-10.)

110-11. Deteriorating Agents. Equipment must be "identified" for use in the presence of specific deteriorating agents, as shown on the typical nameplate in Fig. 110-6.

110-12. Mechanical Execution of Work. This statement has been the source of many conflicts because opinions differ as to what is a "neat and workmanlike manner."

The Code places the responsibility for determining what is acceptable and how it is applied in the particular jurisdiction on the authority having jurisdiction. This basis in most areas is the result of:

1. Competent knowledge and experience of installation methods
2. What has been the established practice by the qualified journeyman in the particular area
3. What has been taught in the trade schools having certified electrical training courses for apprentices and journeymen

Examples which generally would not be considered as "neat and workmanlike" include nonmetallic cables installed with kinks or twists; unsightly exposed runs; wiring improperly trained in enclosures; slack in cables between supports; flattened conduit bends; or improvised fittings, straps, or supports. See Fig. 110-7.

It has long been required in specific Code rules that unused openings in boxes and cabinets be closed by a plug or cap (see Secs. 370-18 and 373-4). The requirement for such plugging of open holes is also a general rule to provide fire-resistive integrity of all equipment—boxes, raceways, auxiliary gutters, cabinets, equipment cases or housings (Fig. 110-8).

As required by Sec. 110-12(b), in concrete manholes and pull boxes for underground electrical systems, safe and ready access demands proper train-

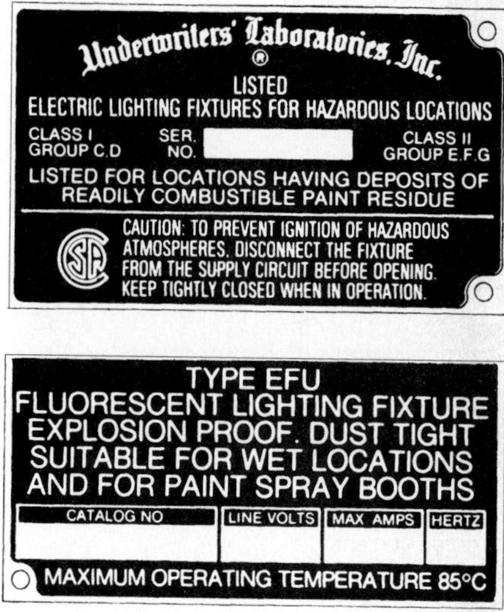

"Identified"—This kind of marking on a product makes it "recognizable as suitable for the specific purpose" as indicated on the labels.

Fig. 110-6.

ing and racking of conductors. This rule requires "racked" conductors in "subsurface" enclosures to provide adequate room for safe and easy movement of persons who have to do installation and maintenance work in such enclosures.

Part **(c)** presents a requirement for current-carrying parts—buswork, terminals, etc.—that is similar to the rule of Sec. 250-118. Both rules effectively prohibit conductive surfaces from being rendered nonconductive due to the introduction of paint, lacquer, or other substances. It should be noted that this rule is not intended to prohibit the use of "cleaners." Use of cleaning agents is recognized, but only those agents that do *not* contaminate conductive surfaces. Be certain that any type of cleaners used for maintenance purposes is suitable for the specific application.

Wording has been added to indicate that defective equipment may not be used. Although wording prohibiting the use of damaged or otherwise defective equipment may seem superfluous, apparently many installers were using or reusing damaged equipment. At complete odds with common sense, such practice puts those who use and maintain the system at risk and is now expressly forbidden. And although not specifically mentioned, any equipment that is damaged during the construction phase should be considered as covered by the rule of Sec. 110-12(c) and should be replaced.

Fig. 110-7. Stapling of BX to bottoms of joists and ragged drilling of joists add up to an unsightly installation that does not appear "workmanlike." (Sec. 110-12.)

Fig. 110-8. Unused openings in *any* electrical enclosure must be plugged or capped. Any punched knockout that will not be used *must* be closed, as at arrow.

110-14. Electrical Connections. Proper electrical connections at terminals and splices are absolutely essential to ensure a safe installation. Improper connections are the cause of most failures of wiring devices, equipment burndowns, and electrically oriented fires.

Terminals and splicing connectors must be "identified" for the *material* of the conductor or conductors used with them. Where in previous NEC editions this rule called for conductor terminal and splicing devices to be "suitable" for the material of the conductor (i.e., for aluminum or copper), the wording now requires that terminal and splicing devices must be "identified" for use with the material of the conductor. And devices that combine copper and aluminum conductors in direct contact with each other must also be "identified for the purpose and conditions of use."

Although the NEC definition of *identified* does not specifically require that products be marked to designate specific application suitability, there is that suggestion; and such marking on the product, its label, or the box or wrapping in which the product is supplied would best comply with the definition "identified," and such marking is widely done by manufacturers to satisfy requirements of UL. Specific rules of UL require terminals and splice devices to be marked to indicate suitability with copper or aluminum conductors. The term "identified" is intended to make clear that the manufacturers and testing laboratories have a responsibility to assist the inspector by their own indications of suitability of products for their intended uses.

In general, approved pressure-type wire splicing lugs or connectors bear no marking if approved for only copper wire. If approved for copper, copper-clad aluminum, and/or aluminum, they are marked "AL-CU"; and if approved for aluminum only, they are marked "AL." Devices listed by Underwriters Laboratories Inc. indicate the range or combination of wire sizes for which such devices have been listed. Terminals of 15- and 20-A receptacles not marked "CO/ALR" are for use with copper and copper-clad aluminum conductors only. Terminals marked "CO/ALR" are for use with aluminum, copper, and copper-clad aluminum conductors.

The vast majority of distribution equipment has always come from the manufacturer with mechanical set-screw-type lugs for connecting circuit conductors to the equipment terminals. Lugs on such equipment are commonly marked "AL-CU" or "CU-AL," indicating that the set-screw terminal is suitable for use with *either* copper or aluminum conductors. But, such marking on the lug itself is not sufficient evidence of suitability for use with aluminum conductors. UL requires that equipment with terminals that are found to be suitable for use with *either* copper or aluminum conductors must be marked to indicate such use on the label or wiring diagram of the equipment—completely independent of a marking like "AL-CU" on the lugs themselves. A typical safety switch, for instance, would have lugs marked "AL-CU," *but also must have* a notation on the label or nameplate of the switch that reads like this: "Lugs suitable for copper or aluminum conductors."

There are two possible ways to go when using aluminum conductors with distribution equipment that comes with mechanical set-screw terminals about which there may be some reservations on their effectiveness:

1. A termination device or copper pigtail may be put on the end of each aluminum conductor to provide an "end" that is suitable, tested, and proved for effective use in a set-screw-type lug. A number of manufacturers make "adapters" which are readily crimped onto the end of an aluminum conductor to "convert" the end to copper or an alloy that will not exhibit the creep and cold-flow disadvantages of aluminum in a set-screw termination. This is shown in Fig. 110-9.

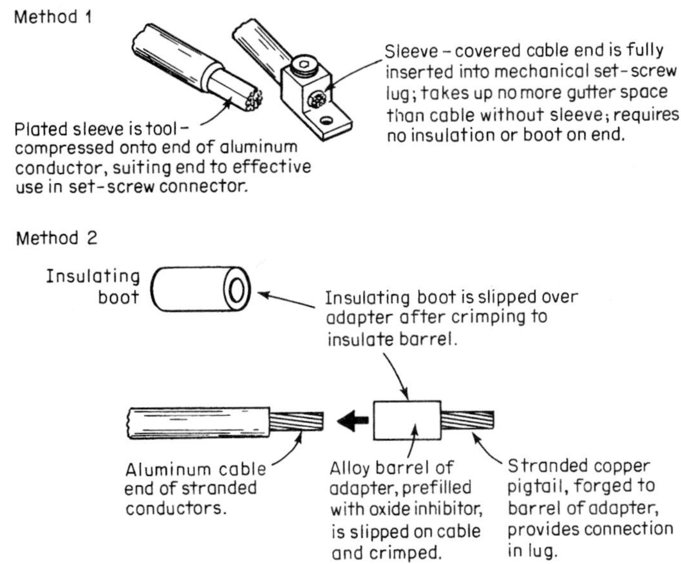
Method 1
Sleeve - covered cable end is fully inserted into mechanical set-screw lug; takes up no more gutter space than cable without sleeve; requires no insulation or boot on end.
Plated sleeve is tool-compressed onto end of aluminum conductor, suiting end to effective use in set-screw connector.
Method 2
Insulating boot
Insulating boot is slipped over adapter after crimping to insulate barrel.
Aluminum cable end of stranded conductors.
Alloy barrel of adapter, prefilled with oxide inhibitor, is slipped on cable and crimped.
Stranded copper pigtail, forged to barrel of adapter, provides connection in lug.

Fig. 110-9. (Sec. 110-14.)

2. Another way to attack the problem is to remove the set-screw lug from the switch or breaker or panel and replace the set-screw lug with a crimp-type lug designed to accept an aluminum conductor which is crimped in the barrel of the lug.

UL-listed equipment must be used in the condition as supplied by the manufacturer—in accordance with NE Code rules and any instructions covered in the UL listing in the *Electrical Construction Materials Directory* (the UL Green Book)—as required by NE Code Sec. 110-3(b). Unauthorized alteration or modification of equipment in the field voids the UL listing and can lead to very dangerous conditions. For this reason, any arbitrary or unspecified changing of terminal lugs on equipment is *not acceptable unless such field modification is recognized by UL* and spelled out very carefully in the manufacturers' literature and on the label of the equipment itself.

For instance, UL-listed authorization for field changing of terminals on a safety switch is described in manufacturers' catalog data and on the switch label itself. It is obvious that field replacement of set-screw lugs with compression-type lugs can be a risky matter if great care is not taken to assure that the size, mounting holes, bolts, and other characteristics of the compression lug line up with and are fully compatible for replacement of the lug that is removed. Careless or makeshift changing of lugs in the field has produced overheating, burning, and failures. To prevent junk-box assembly of replacement lugs, UL requires that any authorized field replacement data *must* indicate the specific lug to be used and *also must* indicate the tool to be used in making the crimps. Any crimp connection of a lug should always be done with the tool specified by the lug manufacturer. Otherwise, there is no assurance that the type of crimp produces a sound connection of the lug to the conductor.

Section 110-14(a), last sentence, prohibits use of more than one conductor in a terminal (see Fig. 110-10) *unless the terminal is approved for the purpose* (meaning approved for use with two or more conductors in the terminal).

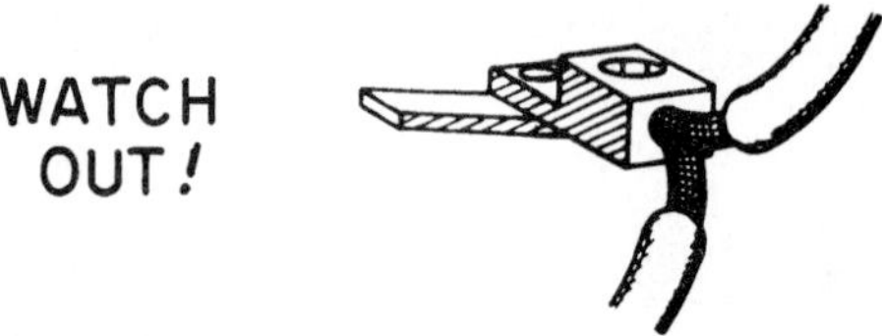

Fig. 110-10. [Sec. 110-14(a).]

Use of the word "identified" in the last sentence of Sec. 110-14(a) could be interpreted to require that terminals suited to use with two or more conductors must somehow be marked. For a long time terminals suited to and acceptable for use with aluminum conductors have been marked "AL-CU" or "CU-AL" right on the terminal. Twist-on or crimp-type splicing devices are "identified" both for use with aluminum wires and for the number and sizes of wires permitted in a single terminal—with the identification marked on the box in which the devices are packaged or marked on an enclosed sheet.

For set-screw and compression-type lugs used on equipment or for splicing or tapping-off, suitability for use with two or more conductors in a single barrel of a lug could be marked on the lug in the same way that such lugs are marked with the range of sizes of a single conductor that may be used, e.g., "No. 2 to No. 2/0." But the intent of the Code rule is that any single-barrel lug used with two or more conductors must be tested for such use (such as in accordance with UL 486B standard), and some indication must be made by the manufacturer that the lug is properly suited and rated for the number and sizes of conductors to be inserted into a single barrel. Again, the best and most effective way to identify a lug for such use is with marking right on the lug, as is done for "AL-CU." But the second sentence of Sec. 110-3(a)(1) also allows such identification to be "provided with" a product, as on the box or on an instruction sheet. See Fig. 110-11.

A fine-print note after the first paragraph of Sec. 110-14 calls attention to the fact that manufacturers are marking equipment, terminations, packing cartons, and/or catalog sheets with specific values of required tightening torques (pound-inches or pound-feet). Although the National Electrical Code does not presently contain explicit requirements on torquing, it is certainly true that the NEC objective of providing for the "practical safeguarding of persons and prop-

Fig. 110-11. A *terminal* with more than one conductor terminated in a single barrel (hole) of the lug (at arrows) must be "identified" (marked, listed, or otherwise tested and certified as suitable for such use).

erty from hazards arising from the use of electricity" cannot be fulfilled without diligent attention to so safety-related a consideration as proper connections. NEC Sec. 110-3(b), which requires all listed and labeled products to be used in accordance with any instructions included with the listing and labeling, can and should be construed to require that all electrical connections be torqued to values specified in UL standards. Although that puts the installer to the task of finding out appropriate torque values, many manufacturers are presently publishing "recommended" values in their catalogs and spec sheets. In the case of connector and lug manufacturers, such values are even printed on the boxes in which the devices are sold.

A proposal was submitted for the NE Code to make torquing of terminals mandatory. In response to that, the Code-making panel commented that "The proposal is adequately covered in Section 110-3(b)." That statement appears to say that torquing is, in fact, already a Code requirement, because Sec. 110-3(b) incorporates as Code rules all the requirements on torquing contained in UL standards. Torque is the amount of tightness of the screw or bolt in its threaded hole; that is, torque is the measure of the twisting movement that produces rotations around an axis. Such turning tightness is measured in terms of the force applied to the handle of the device that is rotating the screw or bolt and the distance from the axis of rotation to the point where the force is applied to the handle of the wrench or screwdriver:

$$\text{Torque (lb·ft)} = \text{force (lb)} \times \text{distance (ft)}$$
$$\text{Torque (lb·in.)} = \text{force (lb)} \times \text{distance (in.)}$$

Because there are 12 inches in a foot, a torque of "1 pound foot" is equal to "12 pound inches." Any value of "pound-feet" is converted to "pound-inches" by multiplying the value of "pound-feet" by 12. To convert from "pound-inches" to "pound-feet," the value of "pound-inches" is divided by 12.

Note: The expressions "pound-feet" and "pound-inches" are preferred to "foot-pounds" or "inch-pounds," although the expressions are used interchangeably.

Torque wrenches and torque screwdrivers are designed, calibrated, and marked to show the torque (or turning force) being exerted at any position of the turning screw or bolt. Figure 110-12 shows typical torque tools and their application.

Section 110-14(b) covers splice connectors and similar devices used to connect fixture wires to branch-circuit conductors and to splice circuit wires in junction boxes and other enclosures. Much valuable application information on such devices is given in the UL *Electrical Construction Materials Directory,* under the heading of "Wire Connectors and Soldering Lugs."

The new last sentence of Sec. 110-14(b) states that connectors or splices used with directly buried conductors must be *listed* for the application.

This new wording makes the use of listed connectors and splice kits mandatory where used directly buried. As indicated by the submitter of this proposal for a change, such equipment *is* listed, *is* commercially available, and *should be* used.

Part **(c)** of Sec. 110-14 reiterates the UL rules regarding temperature limitations of terminations, which are made mandatory by Sec. 110-3(b). It is worth noting that the information given by UL in the preface for "Equipment for Use in General Locations" is more detailed. The FPN following this section is intended to indicate that if information in a general or specific UL rule permits or requires different ratings and/or sizes, the UL rule MUST BE followed.

The new last sentence indicates acceptability of 90°C-rated wire where applied in accordance with the temperature limitations of the termination.

The inclusion of the UL data regarding temperature limitations in this section of the 1993 **Code** apparently caused a bit of confusion. Apparently, some inspectors and others were reading the new wording as prohibiting the use of conductors with a 90°C-rated insulation. In an effort to clarify that point, wording was added to specifically recognize that 90°C-rated conductors "shall be permitted to be used for derating purposes."

While the accepted wording clearly recognizes the use of 90°C-insulated conductors, it seems to limit their use to applications where derating is required, but it shouldn't. That is not what was intended.

90°C-insulated conductors may be used in virtually any application that 60°C- or 75°C-rated conductors may be used and in some that the lower-rated conductors cannot. But the ampacity of the 90°C-rated conductor must *never* be taken to be more than that permitted in the column that corresponds with the temperature rating of the terminations to which the conductor will be con-

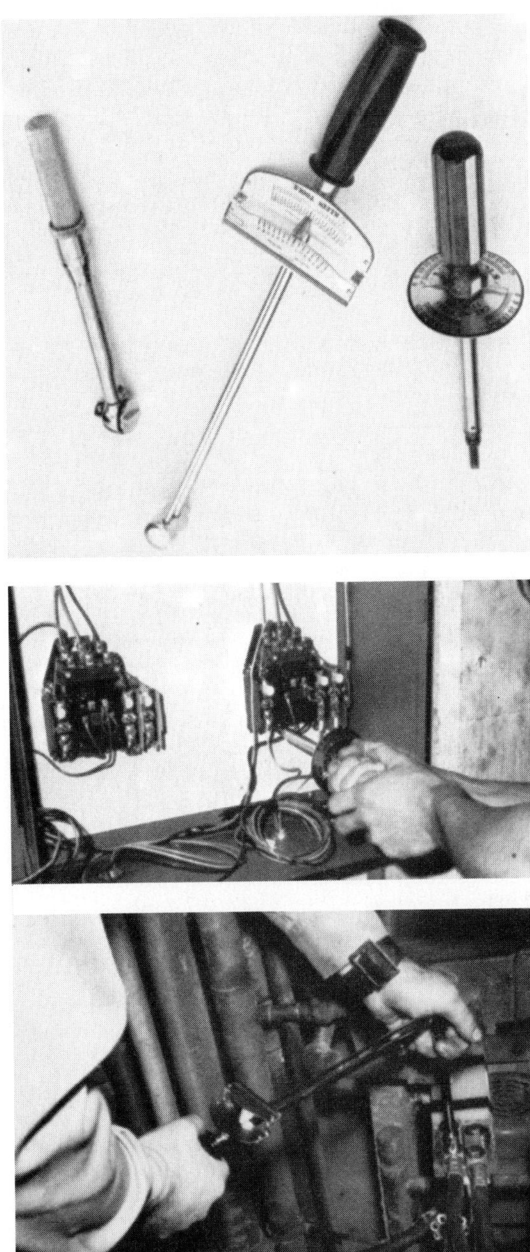

Fig. 110-12. Readily available torque tools are (at top, L–R): torque screwdriver, beam-type torque wrench, and ratchet-type torque wrench. These tools afford ready compliance with the implied requirement of the fine-print note in the **Code** rule. [Sec. 110-14.]

nected. And that applies to both ends of the conductor. For example, consider a No. 6 THHN copper conductor, which has a Table 310-16 ampacity of 55 A. But, if the No. 6 is supplied from a CB with, say, a 60/75°C-rating, it may be considered to be a 50-A wire, provided the equipment end is also rated 60/75°C or 75°C. But, if the equipment is rated at 60°C, the No. 6 THHN copper conductor may carry no more than the 60°C-ampacity (40 A) shown in Table 310-16 for a No. 6 copper wire. The wording in the last sentence is intended to indicate as much (Fig. 110-13).

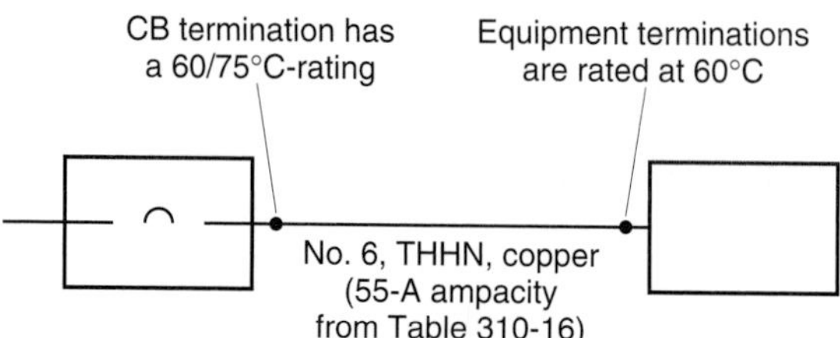
Sec. 110-14(c)
CB termination has a 60/75°C-rating
Equipment terminations are rated at 60°C
No. 6, THHN, copper (55-A ampacity from Table 310-16)
Although this conductor can safely carry 55 A, because the equipment terminations are rated for 60°C, the No. 6 must carry no more than the current shown in the 60°C-column in Table 310-16 for a No. 6 copper conductor, i.e., 40 A.

Fig. 110-13. The wording added refers to the use of 90°C-rated conductors for the purpose of derating. *But,* 90°C-insulated conductors may be used even where derating is not required, provided they are taken as having an ampacity not greater than the ampacity shown in the column from Table 310-16 that corresponds to the temperature rating of the terminations—at both ends—to which the conductors will be connected.

110-16. Working Space About Electric Equipment (600 Volts, Nominal, or Less). The basic rule of Sec. 110-16(a) calls for safe work clearances in accordance with Table 110-16(a). The intent of the second paragraph of Sec. 110-16(a), which calls for a work space at least 30 in. (762 mm) wide in front of electrical equipment, is to provide sufficient "elbow room" in front of such equipment as column-type panelboards and single enclosed switches [e.g., 12 in. (305 mm) wide] to permit the equipment to be operated or maintained under safer conditions (Fig. 110-14).

And, as required by the third sentence of the second paragraph of Sec. 110-16(a), clear work space in front of any enclosure for electrical equipment must be deep enough to allow doors, hinged panels, or covers on the enclosure to be opened to an angle of at least 90°. Any door or cover on a panelboard or cabinet that is obstructed from opening to at least a 90° position makes it difficult for any personnel to install, maintain, or inspect the equipment in the enclosure safely. Full opening provides safer access to the enclosure and minimizes potential hazards (Fig. 110-14).

In Code Table 110-16(a), the dimensions of working space in the direction of access to live electrical parts operating at 600 V or less must be carefully observed. All minimum clearances are 3 ft (914 mm). Section 110-16(d) requires

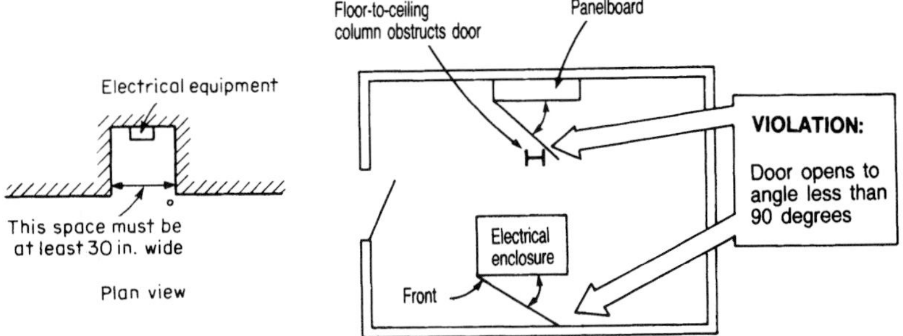

Fig. 110-14. Working space required in front of electrical equipment for side-to-side clearance and door opening. [Sec. 110-16(a).]

a working space at least 3 ft (914 mm) deep in front of switchboards or motor control centers that have live parts normally exposed on their front. The minimum of 3 ft was adopted for Code Table 110-16(a) to make all electrical equipment—panelboards, switches, breakers, starters, etc.—subject to the same 3-ft (914-mm) minimum to increase the level of safety and assure consistent, uniform spacing where anyone might be exposed to the hazard of working on any kind of live equipment. Application of Code Table 110-16(a) to the three "conditions" described in Sec. 110-16(a) is shown in the sketches making up this handbook's Table 110-1. Figure 110-15 shows a typical example of Condition 3.

Fig. 110-15. Condition 3 in Code Table 110-16(a) for the rule covered by Sec. 110-16(a) applies to the case of face-to-face enclosures, as shown here where two switchboards face each other. The distance indicated must be at least 3 or 4 ft depending on the voltage of enclosed parts. [Sec. 110-16(a).]

Exception No. 2 of Sec. 110-16(a) allows working clearance of less than the distances given in Table 110-16(a) for live parts that are operating at not over 30 V RMS, 42 V peak, or 60 V DC. The last phrase recognizes the safety of low-voltage circuits like the Class 2 control and power-limited circuits covered by

Art. 725. It is intended to allow less than the 3-ft (914-mm) minimum spacing of Table 110-16(a) for live parts of low-voltage control or power-limited circuits that *are integral parts* of switchboards or control centers that contain other bus and equipment rated at 208/120 V or 480/277 V, for which the table clearances do apply.

According to Sec. 110-16(a), Exception No. 1, minimum depth of work space behind equipment rated 600 V, and less, must be provided where access is needed when de-energized. The past few editions of the NEC have required a minimum depth of work space behind equipment rated over 600 V where access was required only when the equipment was de-energized. For equipment rated over 600 V requiring rear access only when *de-energized,* Sec. 110-34(a), Exception No. 1 mandates that the depth of work space must not be less than 30 in. (762 mm). For the 1996 NEC, the same rule applies to equipment rated 600 V or less (Fig. 110-16).

Exception No. 3, the new Exception in Sec. 110-16(a), permits smaller work space when replacing equipment at existing facilities provided procedures are established to ensure safety.

This exception is strange because it only has the effect of permitting existing violations to be brought into compliance by replacing the equipment. That is, the distance between two facing pieces of equipment may be less than required by Table 310-16 *provided* I am replacing equipment at an existing facility. For that to happen, the equipment that *is* installed would have to be in violation of Sec. 110-16(a) to begin with. This is also a strange rule because until replacement is made, any such installation would still be in violation of the Code. It can't be used for any new installation—even in an existing building—because the literal wording only recognizes such application where "equipment is being replaced." Therefore, from a practical purpose, this rule has very little, if any, application at all.

As an added safety measure, to prevent the case where personnel might be trapped in the working space around burning or arcing electrical equipment, the rule of Sec. 110-16(c) requires *two* "entrances" or directions of access to the working space around any equipment enclosure that contains "overcurrent devices, switching devices, or control devices," where such equipment is rated 1,200 A or more and is over 6 ft (1.83 m) wide. (Note that this rule is no longer limited just to "switchboards and control panels," as it was in the 1987 and previous Code editions.) Sec. 110-16(c) requires *two* "entrances" or directions of access to the working space around switchboards, motor-control centers, distribution centers, panelboard lineups, UPS cubicles, rectifier modules, substations, power conditioners, and any other equipment that is rated 1,200 A or more and is over 6 ft (1.83 m) wide.

At each end of the working space at such equipment, an entranceway or access route at least 24 in. (610 mm) wide must be provided. Because personnel have been trapped in work spaces by fire between them and the only route of exit from the space, rigid enforcement of this rule is likely. Certainly, design engineers should make two paths of exit a standard requirement in their drawings and specs. Although the rule does not *require* two doors into an electrical equipment room, it may be necessary to use two doors in order to obtain the

Table 110-1. Clearance Needed in Direction of Access to Live Parts in Enclosures for Switchboards, Panelboards, Switches, CBs, or Other Electrical Equipment—Plan Views [Sec. 110-16(a)]

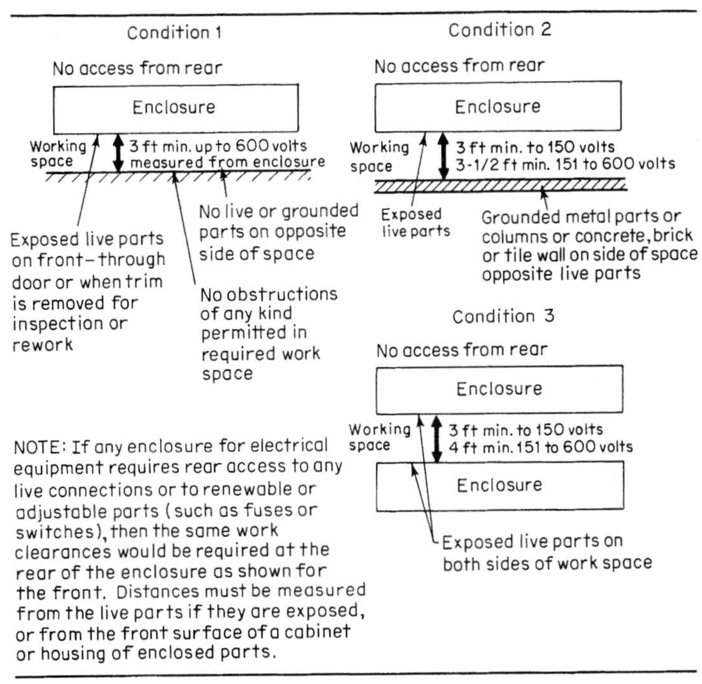

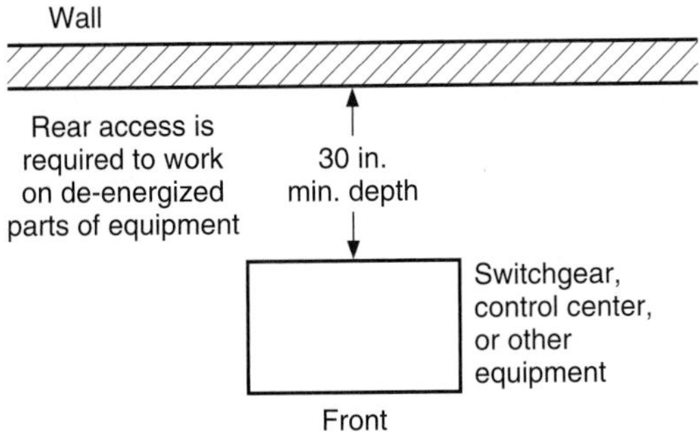

Fig. 110-16. Where access is needed, but only when the equipment is de-energized, the work space need only be 30 in. deep. This addition to Exception No. 1 in Sec. 110-16(a) applies *only* to those cases where access is needed only when the equipment is *de-energized.* As always, if access is needed at the rear for "examination, adjustment, servicing, or maintenance" when the equipment is energized, the depth, as well the other aspects of work space must satisfy the basic rule. And, if there is never a need to gain access to the rear of the equipment, there is no minimum depth required by the Code, but careful attention should be paid to any clearances required by the equipment manufacturer.

required two entrances to the required work space—especially where the switchboard or control panel is in tight quarters and does not afford a 24-in. (610 mm) wide path of exit at each end of the work space.

In Fig. 110-17, sketch "A" shows compliance with the **Code** rule—providing two areas for entering or leaving the defined dimensions of the work space. In that sketch, placing the switchboard with its front to the larger area of the room and/or other layouts would also satisfy the intent of the rule. It is only necessary to have an assured means of exit from the defined work space. If the space in front of the equipment is deeper than the required depth of work space, then a person could simply move back out of the work space at any point along the length of the equipment. That is the objective of Exception No. 2 to Sec. 110-16(c), which recognizes that where the space in front of equipment is twice the minimum depth of working space required by Table 110-16(a) for the voltage of the equipment and the conditions described, it is not necessary to have an entrance at each end of the space (Fig. 110-18). In such cases, a worker can move directly back out of the working space to avoid fire. For any case where the depth of space is not twice the depth value given in Table 110-16(a) for working space, an entranceway or access route at least 24 in. (6.10 mm) wide must be provided at each end of the working space in front of the equipment.

Sketch "B" in Fig. 110-17 shows the layout that must be avoided. With sufficient space available in the room, layout of any equipment rated 1,200 A or more with only one exit route from the required work space would be a clear violation of the rule. As shown in sketch "B," a door at the right end of the working space would eliminate the violation. *But,* if the depth *D* in sketch "B" is equal to or greater than twice the minimum required depth of working space from Table 110-16(a) for the voltage and "conditions" of installation, then a door at the right is *not* needed and the layout would *not* be a violation.

Exception No. 2 states that when the defined work space in front of an electrical switchboard or other equipment has an entranceway at only one end of the space, the edge of the entrance nearest the equipment must be at least 3, 3½, or 4 ft (914 mm, 1.07 m, or 1.22 m) away from the equipment—as designated in Table 110-16(a) for the voltage and conditions of installation of the particular equipment. This **Code** requirement requires careful determination in satisfying the precise wording of the rule. Figure 110-19 shows a few of the many possible applications that would be subject to the rule.

Section 110-16(d) requires lighting of work space at "service equipment, switchboards, panelboards, or motor control centers installed indoors." The basic rule and its exception are shown in Fig. 110-20.

The last sentence in Sec. 110-16(d) points out the **Code**-making panel's intent. That is, if an adjacent fixture provides adequate illumination, another fixture *is not* required.

It should be noted that although lighting is required for safety of personnel in work spaces, nothing specific is said about the kind of lighting (incandescent, fluorescent, mercury-vapor), no minimum footcandle level is set, and such details as the position and mounting of lighting equipment are omitted. All that is left to the designer and/or installer, with the inspector the final judge of acceptability.

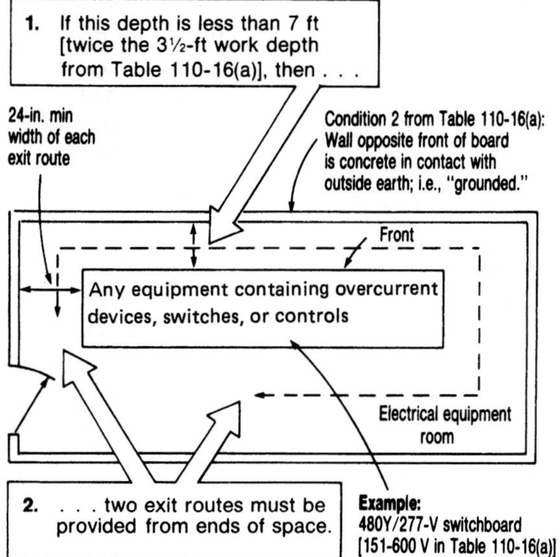
1. If this depth is less than 7 ft [twice the 3½-ft work depth from Table 110-16(a)], then . . .
24-in. min width of each exit route
Condition 2 from Table 110-16(a): Wall opposite front of board is concrete in contact with outside earth; i.e., "grounded."
Front
Any equipment containing overcurrent devices, switches, or controls
Electrical equipment room
2. . . . two exit routes must be provided from ends of space.
Example: 480Y/277-V switchboard [151-600 V in Table 110-16(a)]

(A) Complies

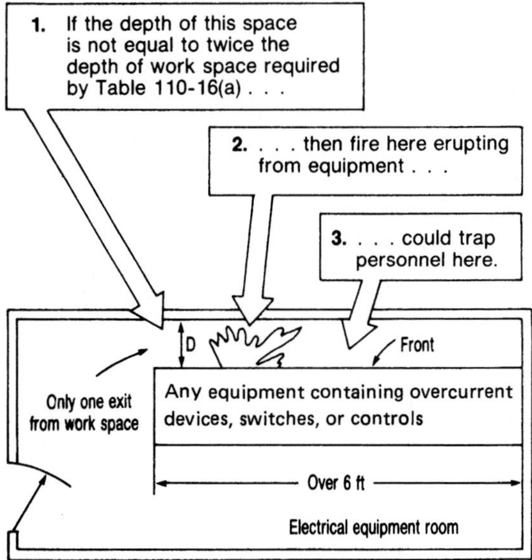
1. If the depth of this space is not equal to twice the depth of work space required by Table 110-16(a) . . .
2. . . . then fire here erupting from equipment . . .
3. . . . could trap personnel here.
D
Front
Only one exit from work space
Any equipment containing overcurrent devices, switches, or controls
Over 6 ft
Electrical equipment room

(B) Violation

Fig. 110-17. There must be two paths out of the workspace required in front of any equipment containing fuses, circuit breakers, motor starters, switches, and/or any other control or protective devices, where the equipment is rated 1,200 A or more and is over 6 ft (1.83 m) wide. [Sec. 110-16(c).]

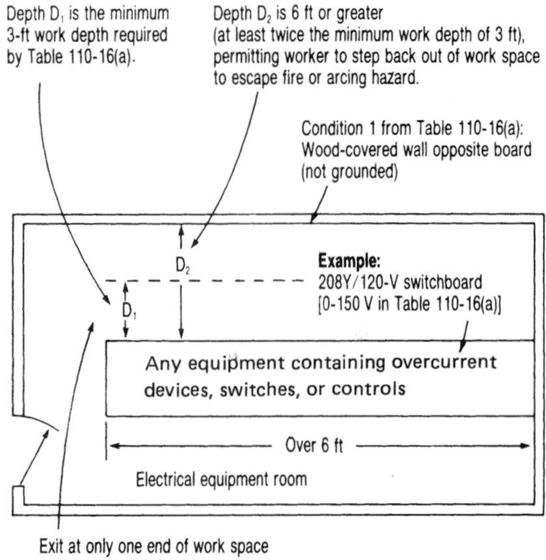
Depth D_1 is the minimum 3-ft work depth required by Table 110-16(a).
Depth D_2 is 6 ft or greater (at least twice the minimum work depth of 3 ft), permitting worker to step back out of work space to escape fire or arcing hazard.
Condition 1 from Table 110-16(a): Wood-covered wall opposite board (not grounded)
D_2
D_1
Example: 208Y/120-V switchboard [0-150 V in Table 110-16(a)]
Any equipment containing overcurrent devices, switches, or controls
Over 6 ft
Electrical equipment room
Exit at only one end of work space

Fig. 110-18. This satisfies Exception No. 2 to Sec. 110-16(c).

In Sec. 110-16(e), minimum headroom, which must extend to 6½ ft (1.98 m) or the height of the equipment, in working spaces is required around electrical equipment. The rule applies to "service equipment, switchboards, panelboards, or motor control centers." The Exception permits "service equipment or panelboards, in existing dwelling units, that do not exceed 200 amperes" to be installed with less than 6½ ft (1.98 m) of headroom—such as in crawl spaces under single-family houses. But the permission for reduced headroom of the equipment described in the Exception applies only in existing "dwelling units" that meet the definition of that phrase. In any space other than an existing dwelling unit, all indoor service equipment, switchboards, panelboards, or control centers must have headroom in any space around the equipment that is work space required by Sec. 110-16(a). And this headroom must be at least 6½ ft (1.98 m) high, but never less than the height of the equipment.

Details on lighting and headroom are shown in Fig. 110-20. But, in that sketch, it should be noted that the 6-ft 6-in. (1.98 m) headroom must be available for the entire length of the work space. It must be clearance from the floor up to the light fixture or to any other overhead obstruction—and not simply to the ceiling or bottom of the joists.

110-17. Guarding of Live Parts (600 Volts, Nominal, or Less). After the 1968 NEC, old Sec. 110-17(a)(3), accepting guard rails as suitable for guarding live parts, was deleted. It was felt that a guard rail is not proper or adequate protection in areas accessible to other than qualified persons.

Live parts of equipment should in general be protected from accidental contact by complete enclosure; i.e., the equipment should be "dead-front." Such construction is not practicable in some large control panels, and in such cases the apparatus should be isolated or guarded as required by these rules.

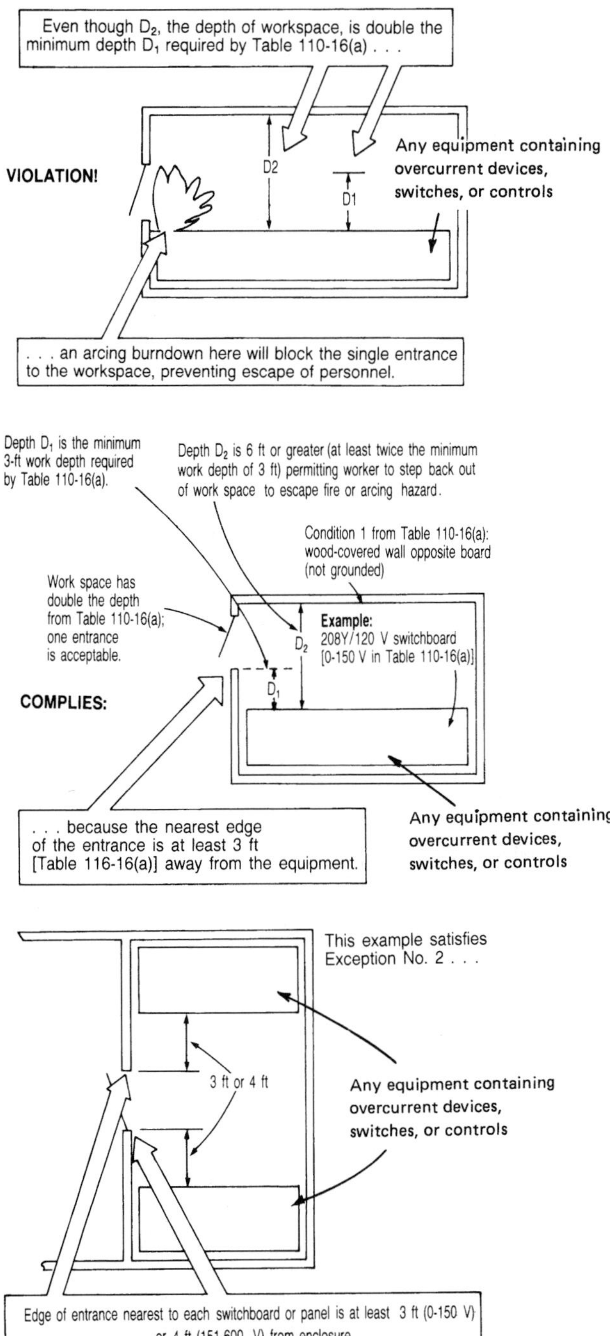

Even though D_2, the depth of workspace, is double the minimum depth D_1 required by Table 110-16(a) . . .
VIOLATION!
D2
D1
Any equipment containing overcurrent devices, switches, or controls
. . . an arcing burndown here will block the single entrance to the workspace, preventing escape of personnel.
Depth D_1 is the minimum 3-ft work depth required by Table 110-16(a).
Depth D_2 is 6 ft or greater (at least twice the minimum work depth of 3 ft) permitting worker to step back out of work space to escape fire or arcing hazard.
Condition 1 from Table 110-16(a): wood-covered wall opposite board (not grounded)
Work space has double the depth from Table 110-16(a); one entrance is acceptable.
Example: 208Y/120 V switchboard [0-150 V in Table 110-16(a)]
COMPLIES:
. . . because the nearest edge of the entrance is at least 3 ft [Table 116-16(a)] away from the equipment.
Any equipment containing overcurrent devices, switches, or controls
This example satisfies Exception No. 2 . . .
3 ft or 4 ft
Any equipment containing overcurrent devices, switches, or controls
Edge of entrance nearest to each switchboard or panel is at least 3 ft (0-150 V) or 4 ft (151-600 V) from enclosure

Fig. 110-19. Arcing burndown must not block route of exit. [Sec. 110-16(c).]

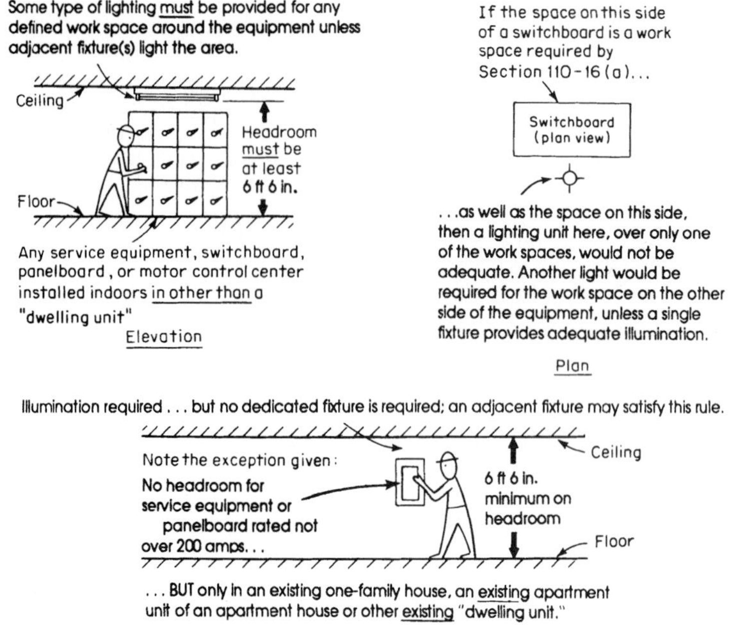

Some type of lighting must be provided for any defined work space around the equipment unless adjacent fixture(s) light the area.
Ceiling
Headroom must be at least 6 ft 6 in.
Floor
Any service equipment, switchboard, panelboard, or motor control center installed indoors in other than a "dwelling unit"
Elevation
If the space on this side of a switchboard is a work space required by Section 110-16 (a)...
Switchboard (plan view)
...as well as the space on this side, then a lighting unit here, over only one of the work spaces, would not be adequate. Another light would be required for the work space on the other side of the equipment, unless a single fixture provides adequate illumination.
Plan
Illumination required... but no dedicated fixture is required; an adjacent fixture may satisfy this rule.
Note the exception given:
No headroom for service equipment or panelboard rated not over 200 amps...
6 ft 6 in. minimum on headroom
Ceiling
Floor
... BUT only in an existing one-family house, an existing apartment unit of an apartment house or other existing "dwelling unit."

Fig. 110-20. Electrical equipment requires lighting and 6½-ft headroom at *all* work spaces around equipment. [Secs. 110-16(d) and (e).]

110-18. Arcing Parts. The same considerations apply here as in the case covered in Sec. 110-17. Full enclosure is preferable, but where this is not practicable, all combustible material must be kept well away from the equipment.

110-21. Marking. The marking required in Sec. 110-21 should be done in a manner which will allow inspectors to examine such marking without removing the equipment from a permanently installed position. It should be noted that the last sentence in Sec. 110-21 requires electrical equipment to have a marking durable enough to withstand the environment involved (such as equipment designed for wet or corrosive locations).

110-22. Identification of Disconnecting Means. As shown in Fig. 110-21, it is a mandatory Code rule that all disconnect devices (switches or CBs) for load devices and for circuits be clearly and permanently marked to show the purposes of the disconnects. This is a "must" and, under OSHA, it applies to all existing electrical systems, no matter how old, and also to all new, modernized, expanded, or altered electrical systems. This requirement for marking has been widely neglected in electrical systems in the past. Panelboard circuit directories must be fully and clearly filled out. And all such marking on equipment must be in painted lettering or other substantial identification.

Effective identification of all disconnect devices is a critically important safety matter. When a switch or CB has to be opened to de-energize a circuit quickly—as when a threat of injury to personnel dictates—it is absolutely necessary to identify quickly and positively the disconnect for the circuit or equip-

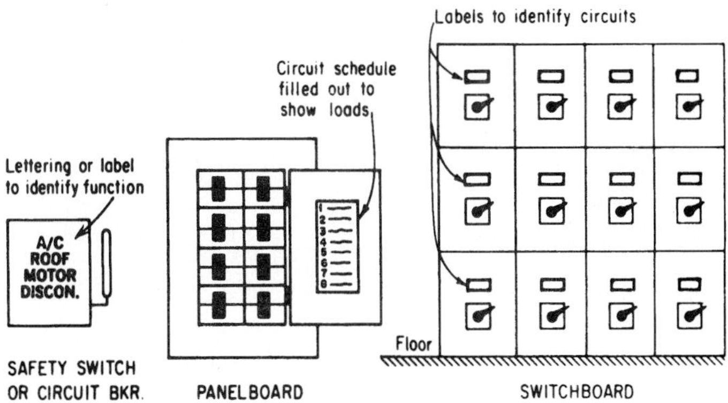

Fig. 110-21. All circuits and disconnects must be identified. OSHA regulations make NE Code Sec. 110-22 mandatory and retroactive for existing installations and for all new, expanded or modernized systems—applying to switches as well as circuit breakers. (Sec. 110-22.)

ment that constitutes the hazard to a person or property. Painted labeling or embossed identification plates affixed to enclosures would comply with the requirement that disconnects be "legibly marked" and that the "marking shall be of sufficient durability." Paste-on paper labels or marking with crayon or chalk could be rejected as not complying with the intent of this rule. Ideally, marking should tell exactly what piece of equipment is controlled by a disconnect (switch or CB) and should tell where the controlled equipment is located and how *it* may be identified. Figure 110-22 shows a case of this kind of identification as used in an industrial facility where all equipment is marked in two languages because personnel speak different languages. And that is an old installation, attesting to the long-standing recognition of this safety feature.

The second paragraph of Sec. 110-22 says that where circuit breakers or fused switches are used in a series combination with downstream devices that do not have an interrupting rating equal to the available short-circuit current but are dependent for safe operation on upstream protection that is rated for the short-circuit current, enclosure(s) for such "series rated" protective devices must be marked in the field "Caution—Series Rated System ______ A Available. Identified Replacement Component Required."

Manufacturers make available multimeter distribution equipment for multiple-occupancy buildings with equipment containing a main service protective device that has a short-circuit interrupting rating of some value (e.g., 65,000 A) that is connected in series with feeder and branch-circuit protective devices of considerably lower short-circuit interrupting ratings (say 22,000 or 10,000 A). Because all of the protective devices are physically very close together, the feeder and branch-circuit devices do not have to have a rated interrupting capability equal to the available short-circuit current at their points of installation. Although such application is a literal violation of NEC Sec. 110-9, which calls for all protective devices to be rated for the short-circuit current available at their supply terminals, "series rated" equipment takes advantage of the ability of the protective devices to operate in series (or in *cascade* as it is sometimes

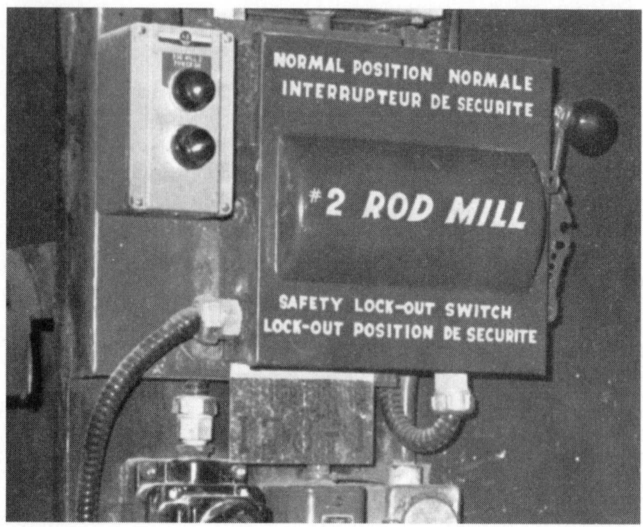

Fig. 110-22. Identification of disconnect switch and pushbutton stations is "legibly marked" in both English and French—and is of "sufficient durability to withstand the environment"—as required by the Code rule. (Sec. 110-22.)

called) with a fault current interruption on, say, a branch circuit being shared by the three series protective devices—the main, feeder, and branch circuit. Such operation can enable a properly rated main protective device to protect downstream protective devices that are not rated for the available fault. When manufacturers combine such series protective devices in available distribution equipment, they do so on the basis of careful testing to assure that all of the protective devices can operate without damage to themselves. Then UL tests such equipment to verify its safe and effective operation and will list such equipment as a "Series Rated System."

Because UL listing is based on use of specific models of protective devices to assure safe application, it is critically important that all maintenance on such equipment be based on the specific equipment. For that reason, this Code rule demands that the enclosure(s) for all such equipment be provided with "readily visible" markings to alert all personnel to the critical condition that must always be maintained to assure safety. Thus, all series-rated equipment enclosure(s) must be marked. An FPN, which called attention to a critically important and often overlooked requirement was accepted, then later rejected. It referred to Sec. 384-13.

The rule of Sec. 110-22 has long required that *every* disconnect be marked to indicate exactly what it controls. And that marking must be legible and sufficiently durable to withstand the environment to which it will be exposed. *And,* the rule of Sec. 384-13 requires that any modifications also be reflected in the circuit directory of panelboards. While most are aware of the requirement in Sec. 110-22, it seems as if very few pay any attention the Sec. 384-13. The FPN following Sec. 110-22 is intended to eliminate that.

It should be noted that the marking of disconnects is one of the few requirements that is made retroactive by OSHA. That is, regardless of when the disconnect was installed, or when a modification was performed, the purpose of every disconnect *must be* marked at the disconnect. If your facility or your customer's facility does not have such markings for each and every disconnect, every effort should be made to ensure that a program to provide such markings is initiated and completed. Failure to do so could result in heavy fines should you be subject to an inspection by OSHA (Fig. 110-22).

B. Over 600 V, Nominal

110-30. General. Figure 110-23 notes that high-voltage switches and circuit breakers must be marked to indicate the circuit or equipment controlled. This requirement arises because Sec. 110-30 says that high-voltage equipment must comply with preceding sections of Art. 110. Therefore, the rule of Sec. 110-22 calling for marking of all disconnecting means must be observed for high-voltage equipment as well as for equipment rated up to 600 V.

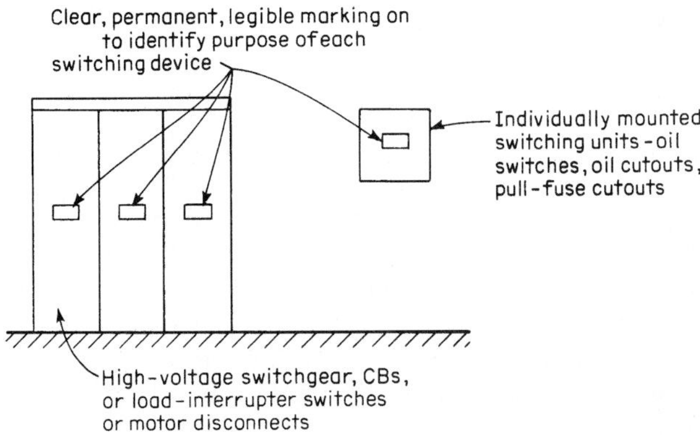

Fig. 110-23. High-voltage switches and breakers must be properly marked to indicate their function. (Sec. 110-30.)

110-31. Enclosure for Electrical Installations. Figure 110-24 illustrates the rules which cover installation of high-voltage equipment indoors in places accessible to unqualified persons. Installation must be in a locked vault or locked area, or equipment must be metal-enclosed.

For any equipment, rooms, or enclosures where the voltage exceeds 600 V, permanent and conspicuous warning signs reading DANGER—HIGH VOLTAGE, KEEP OUT *must* always be provided. It is a safety measure that alerts unfamiliar or unqualified persons who may for some reason gain access to a locked, high-voltage area.

Section 110-31(b) specifies that outdoor installations *with exposed live parts* must *not* provide access to unqualified persons. Elevation may be used to prevent access [Sec. 110-34(e)], or equipment may be enclosed by a wall, screen, or fence under lock and key, as shown in Fig. 110-25. Outdoor installations that

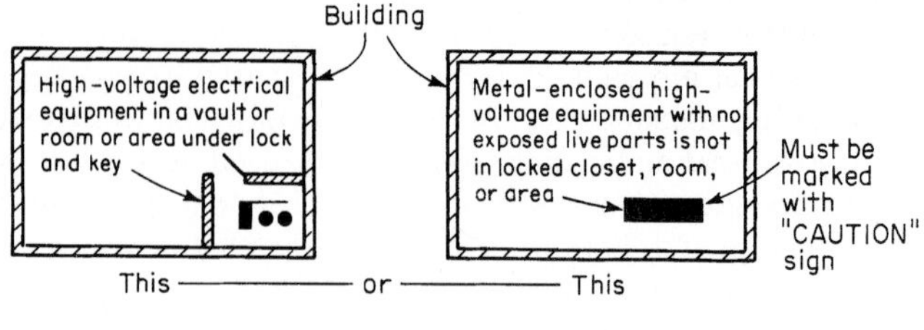

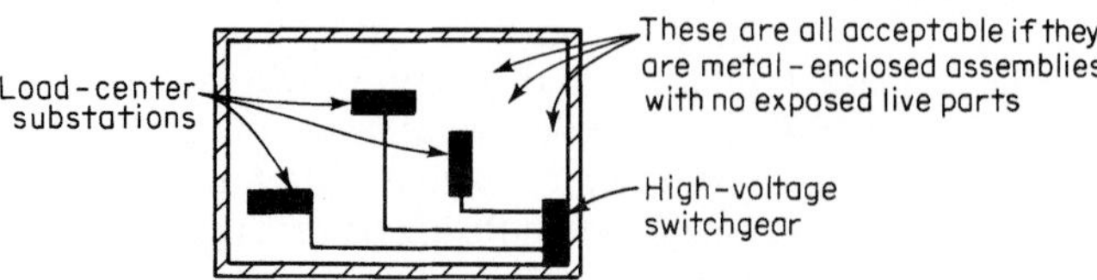

Fig. 110-24. NE Code rules on high-voltage equipment installations in buildings accessible to electrically unqualified persons. [Sec. 110-31(a).]

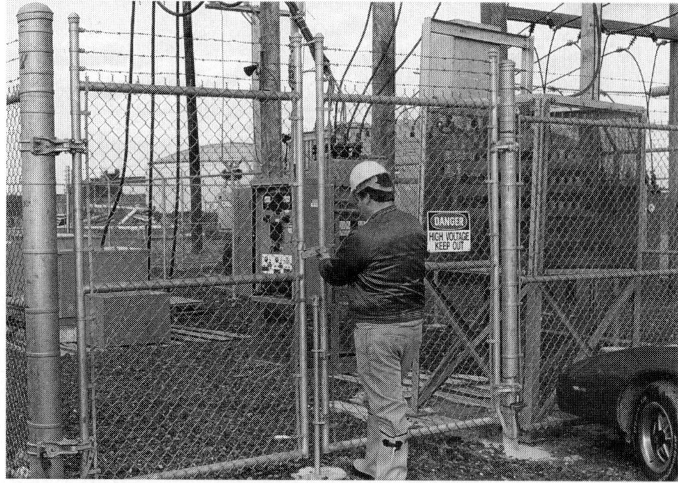

Fig. 110-25. High-voltage equipment enclosed by a wall, screen, or fence at least 7 ft (2.13 m) high with a lockable door or gate is considered as accessible only to qualified persons. [Sec. 110-31(b).]

are open to unqualified persons must comply with Art. 225, "Outside Branch Circuits and Feeders."

Where high-voltage equipment is installed in places accessible only to qualified persons, the rules of Secs. 110-34, 710-32, and 710-33 apply. In such areas, circuit conductors may be installed in conduit, in duct systems, in metal-clad cable, as bare wire, cable, and busbars, or as nonmetallic-sheathed cables or conductors as permitted by NE Code Secs. 710-3 through 710-6.

The last sentence of Sec. 110-31(c) recognizes a difference in safety concern between high-voltage equipment accessible to "unqualified persons"—who may not be qualified as electrical personnel but are adults who have the ability to recognize warning signs and the good sense to stay out of electrical enclosures—and "the general public," which includes children who cannot read and/or are not wary enough to stay out of unlocked enclosures (Fig. 110-26).

Fig. 110-26. *Metal-enclosed* high-voltage equipment accessible to the general public—such as pad-mount transformers or switchgear units installed outdoors or in indoor areas where the general public is not excluded—must have doors or hinged covers locked (arrow) if the bottom of the enclosure is less than 8 ft (2.44 m) above the ground or above the floor.

The rationale submitted with the proposal that led to this change in the **Code** rule noted:

> Where metal-enclosed equipment rated above 600 volts is accessible to the general public and located at an elevation less than 8 feet, the doors should be kept locked to prevent children and others who may be unaware of the contents of such enclosures from opening the doors.
>
> However, in a controlled environment where the equipment is marked with appropriate caution signs as required elsewhere in the **NE Code**, and only knowledgeable people have access to the equipment, the requirement to lock the doors on all metal-enclosed equipment rated above 600 volts and located less than 8 feet (2.44 m) above the floor does not contribute to safety and may place a burden on the safe operation of systems by delaying access to the equipment.

Note that Sec. 110-31(a)(1) does not require locking indoor metal-enclosed equipment that is accessible to unqualified persons, but such equipment is required to be marked with "WARNING" signs [Sec. 110-34(c)].

In Sec. 110-31(c) and the Exception following it, "bolted or screwed on" covers, as well as in-ground box covers over 100 lb (45.4 kg), are recognized as preventing access to the general public.

For equipment rated over 600 V, nominal, Sec. 110-31(a) requires that access be limited to qualified persons only, by installing such equipment within a "vault, room, closest, or in an area that is surrounded" by a fence, etc., with locks on the doors. In part **(c)** of Sec. 110-31, the **Code** recognizes additional methods for preventing unauthorized access to metal-enclosed equipment where it is *not* installed in a locked room or in a locked, fenced-in area.

For equipment rated over 600 V, Sec. 110-31(c) has required that the equipment *cover* or *door* be locked unless the enclosure is mounted with its bottom at least 8 ft off the ground. In that way, access to the general public is restricted and controlled. Now, in addition, the bolts or screws used to secure a cover or door may serve to satisfy the rule of Sec. 110-31(c), *provided* the enclosure is used only as a pull, splice, or junction box. Where accessible to the general public, an enclosure used for any other purpose must have its cover locked unless it is mounted with its bottom at least 8 ft (2.44 m) above the floor (Fig. 110-27).

Fig. 110-27. Access by the general public to any metal enclosure containing circuits or equipment rated over 600 V, nominal, must be prevented. New wording recognizes the bolts or screws on the covers of boxes used as pull, splice, and junction boxes as satisfying the requirement for preventing access. A new exception to this rule recognizes that covers weighing over 100 lb are inherently secured and do not require bolts or screws for the cover or door. *Remember,* the new permission given in the basic rule and the new exception is for pull, splice, and junction boxes, *only.*

A new exception was adopted to recognize that there is no need to secure the cover on an in-ground box that weighs at least 100 lb. This correlates with the rule of Sec. 370-72(e), which states that covers weighing 100 lb (45.4 kg) satisfy the basic requirement for securing covers given in this section.

110-32. Work Space about Equipment. Figure 110-28 points out the basic **Code** rule of Sec. 110-32 relating to working space around electrical equipment.

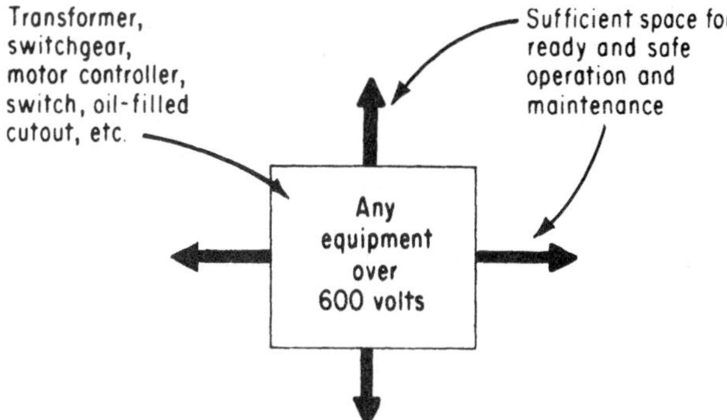

Fig. 110-28. This is the general rule for work space around any high-voltage equipment. (Sec. 110-32.)

Figure 110-29 shows the rules on adequate headroom and necessary illumination for safely working on high-voltage electrical equipment.

Figure 110-30 shows required side-to-side working space for adequate elbow room in front of high-voltage equipment.

110-33. Entrance and Access to Work Space. Entrances and access to working space around high-voltage equipment must comply with the rules shown in Fig. 110-31.

Exception No. 2 says that if the depth of space in front of a switchboard is at least twice the minimum required depth of working space from Table 110-

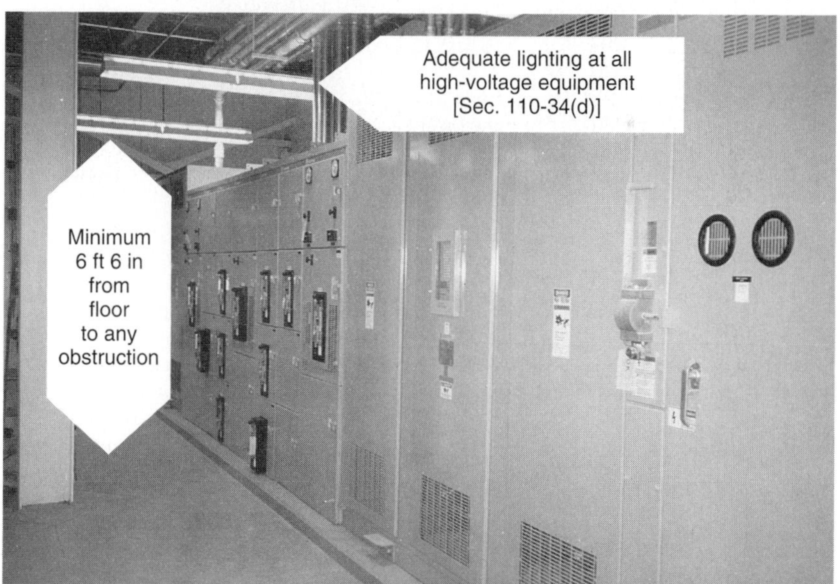

Fig. 110-29. Sufficient headroom and adequate lighting are essential to safe operation, maintenance, and repair of high-voltage equipment. (Sec. 110-32.)

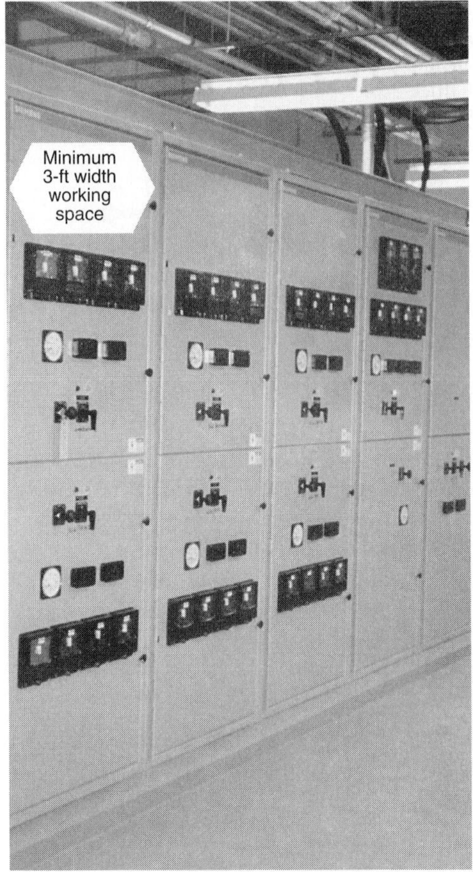
Minimum
3-ft width
working
space

Fig. 110-30. Working space in front of equipment must be at least 3 ft (914 mm) wide measured parallel to front surface of the enclosure. (Sec. 110-32.)

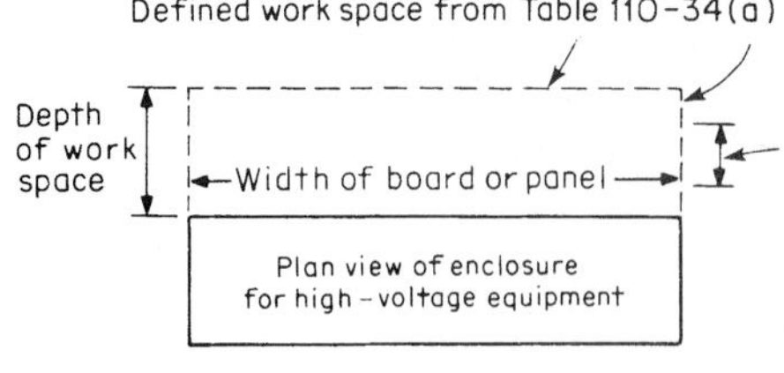
Defined work space from Table 110-34(a)
Depth
of work
space
Width of board or panel
Plan view of enclosure
for high-voltage equipment
One entrance 24 in. wide by 6-1/2 ft high must be provided to the defined work space. But, if the board or panel is over 6 ft wide, there must be one entrance at each end of of the work space, except where the depth of space is at least twice the value in Table 110-34 (a).

Fig. 110-31. Access to required work space around high-voltage equipment must be assured. (Sec. 110-33.)

34(a), any person in the working space would be capable of moving back out of the working space to escape any fire, arcing, or other hazardous condition. In such cases, there is no need for a path of exit at either end or at both ends of the working space. But where the depth of space is not equal to twice the minimum required depth of working space, there must be an exit path at each end of the working space in front of switchgear or control equipment enclosures that are wider than 6 ft (1.83 m). And what applies to the front of a switchboard also applies to working space at the rear of the board if rear access is required to work on energized parts.

The paragraph after Exception No. 2 specifies minimum clearance distance between high-voltage equipment and edge of entranceway to the defined work space in front of the equipment, where only one access route is provided. Based on Table 110-34(a)—which gives minimum depths of clear working space in front of equipment operating at over 600 V—the rule in this section presents the same type of requirement described for Sec. 110-16(c), Exception No. 2. Based on the particular voltage and the conditions of installation of the high-voltage switchgear, control panel, or other equipment enclosure, the nearest edge of an entranceway must be a prescribed distance from the equipment enclosure. Refer to the sketches given for Sec. 110-16(c).

110-34. Work Space and Guarding. Application of Code Table 110-34(a) to working space around high-voltage equipment is made in the same way as shown for Code Table 110-16(a)—except that the depths are greater to provide more room because of the higher voltages.

As shown in Fig. 110-32, a 30-in. (762-mm)-deep work space is required behind enclosed high-voltage equipment that requires rear access to "de-energized" parts. The Exception to Sec. 110-34(a)(3) notes that working space is not required behind dead-front equipment when there are no fuses, switches, other parts, or connections requiring rear access. But the rule adds that if rear access is necessary to permit work on "de-energized" parts of the enclosed assembly, the work space must be at least 30 in. (762 mm) deep. This is intended to prohibit cases where switchgear requiring rear access is installed too close to a wall behind it, and personnel have to work in cramped quarters to reach taps, splices, and terminations. However, it must be noted that this applies only where "de-energized" parts are accessible from the back of the equipment. If energized parts are accessible, then Condition 2 of Sec. 110-34(a) would exist, and the depth of working space would have to be anywhere from 4 to 10 ft (1.22 to 3.05 m) depending upon the voltage [see Table 110-34(a)].

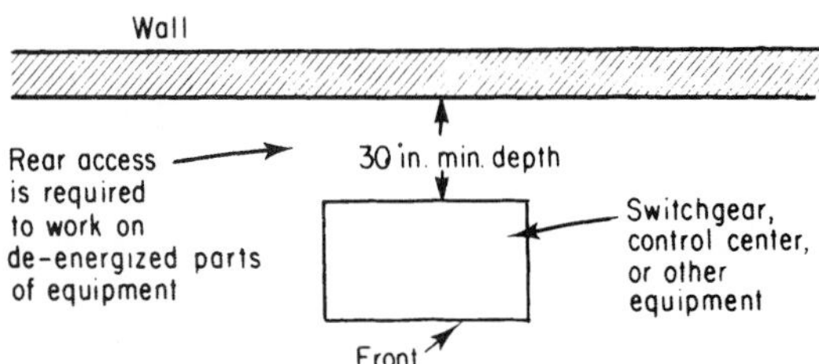

Fig. 110-32. Space for safe work on de-energized parts. [Sec. 110-34(a).]

Section 110-34(c) requires that the entrances to all buildings, rooms, or enclosures containing live parts or exposed conductors operating in excess of 600 V be kept locked, except where such entrances are under the observation of a qualified attendant at all times. The last paragraph in this section requires use of warning signs to deter unauthorized personnel.

The rule of Sec. 110-34(d) on lighting of high-voltage work space is shown in Fig. 110-29. Note that the rule calls for "adequate illumination," but does not specify a footcandle level or any other characteristics.

Figure 110-33 shows how "elevation" may be used to protect high-voltage live parts from unauthorized persons.

Fig. 110-33. Elevation may be used to isolate unguarded live parts from unqualified persons. [Sec. 110-34(e).]

110-40. Temperature Limitations at Terminations. Terminations for equipment supplied by conductors rated over 2,000 V may carry the 90°C ampacity values given in Tables 310-67 through 310-86.

The proposal to include this section pointed out that the ampacity values given in tables for conductors rated above 2,000 V were all 90°C-rated values. And, with the increased attention that has been focused on the coordination between conductor ampacity and temperature limitations of the equipment, some question had been raised regarding the use of the 90°C ampacity values in Tables 310-67 through 310-86 with equipment intended to be supplied by conductors rated over 2,000 V. The rule of Sec. 110-40 allows the conductors covered in Tables 310-67 through 310-86 to carry the full 90°C ampacity and be connected virtually without concern for the equipment terminations, unless otherwise marked to indicate such application is *not* permitted.

This rule was accepted based, in part, on information provided in the proposal regarding American National Standards Institute (ANSI) acceptance of the use of such conductors at their full 90°C ampacity where tested for such operation. The Code-making panel (CMP) added the qualifying statement "unless otherwise identified" to indicate that such application is permitted where equipment has been so tested. In fact, the wording accepted actually assumes that equipment intended to be supplied by conductors rated over 2,000 V—i.e., the conductors covered by Tables 310-67 through 310-86—*is* tested at the full 90°C ampacity. But, if the equipment is otherwise "identified," it must be used as indicated by the manufacturer.

It should be noted that although the rule is contained in part **B** of Art. 110, which covers equipment rated over 600 V, because the tables mentioned cover conductors rated over 2,000 V, it only applies to the terminations on equipment intended to be supplied by conductors rated over 2,000 V. The terminations on all other equipment supplied by conductors rated from 601 to 2,000 V must be coordinated with the ampacity value corresponding to the temperature rating of the terminations (Fig. 110-34).

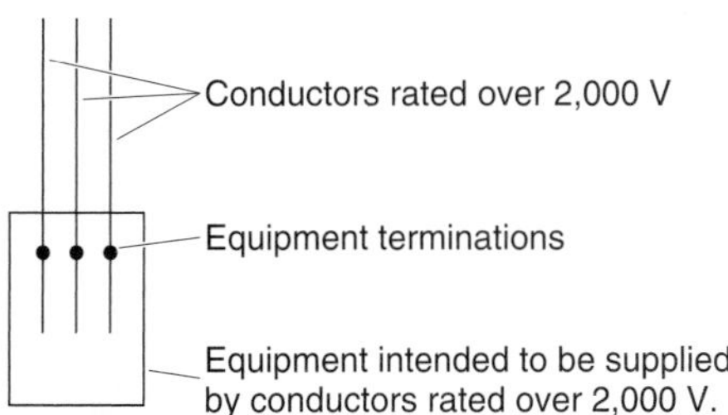
Conductors rated over 2,000 V
Equipment terminations
Equipment intended to be supplied
by conductors rated over 2,000 V.
Wording of Sec. 110-40 indicates terminations in such equipment **may be assumed** to be suitable for carrying the 90°C ampacity shown in Tables 310-67 to 310-86. If the terminations are not suited for connection to conductors at their full 90°C ampacity, the equipment must be marked by the manufacturer. **Remember, this only applies to conductors and equipment rated over 2,000 V.**

Fig. 110-34. When derating with conductors rated over 2,000 V, the temperature rating of the terminations may be assumed to be 90°C unless otherwise marked.

Chapter Two

ARTICLE 200. USE AND IDENTIFICATION OF GROUNDED CONDUCTORS

200-2. General. As indicated in the Exception, some circuits and systems may be operated without an intentionally grounded conductor—that is, without a grounded neutral or a grounded phase leg. Sections 250-7, 503-13(a), and 517-61(a)(1) *require* use of *ungrounded* circuits. Ungrounded circuits are required in anesthetizing locations where flammable anesthetics are used—which include hospital operating rooms, delivery rooms, emergency rooms, and any place where flammable anesthetics are administered. [See Sec. 517-61(a)(1).]

The rule in the last paragraph of this section requires that an insulated grounded-neutral conductor must have the same insulation voltage rating as the ungrounded conductors in all circuits rated up to 1,000 V—which means in all the commonly used 240/120-, 208/120-, and 480/277-V circuits. To correlate with Sec. 250-152 on minimum voltage rating of insulation on grounded neutrals of high-voltage systems, Secs. 250-152 and 200-2 state that where an insulated, solidly grounded neutral conductor is used with any circuit rated over 1,000 V—such as in 4,160/2,400- or 13,200/7,600-V solidly grounded neutral circuits—the neutral conductor does not have to have insulation rated for either phase-to-phase or phase-to-neutral voltage, but must have insulation rated for at least 600 V. See Sec. 250-152. [Of course, a bare, solidly grounded neutral conductor may be used in such circuits that constitute service-entrance conductors, are direct-buried portions of feeders, or are installed overhead, outdoors—as specified in Sec. 250-152. But when an insulated neutral is used, the above rule on 600-V rating applies.] Both Sec. 250-152 and Sec. 200-2 represent exceptions to Sec. 310-2 requiring conductors to be insulated.

200-6. Means of Identifying Grounded Conductors. The basic rule in part **(a)** requires that any grounded neutral conductor or other circuit conductor that is operated intentionally grounded must have a white or natural gray outer finish for the *entire length* of the conductor *if* the conductor is No. 6 size or smaller. See Fig. 200-1.

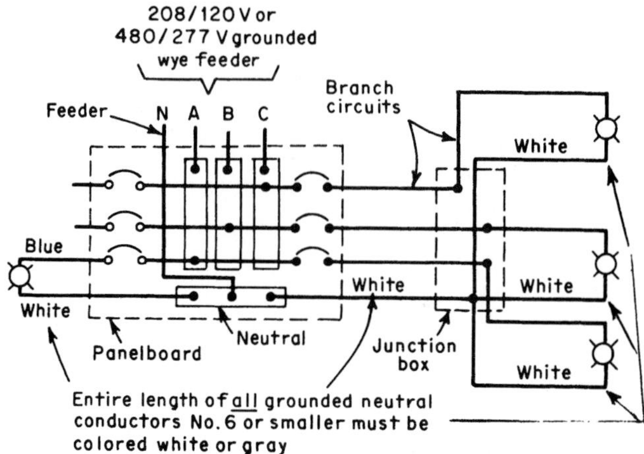

Fig. 200-1. Generally any grounded circuit conductor that is No. 6 size or smaller must have a continuous white or natural gray outer finish. [Sec. 200-6(a).]

Exempted by Exception No. 2. from the basic requirement for a white or gray neutral are multiconductor varnished-cloth insulated cables, fixture wires as covered by Sec. 402-8, branch-circuit neutrals with color tracer stripes (other than green) to distinguish among multiple neutrals of different systems in the same raceway or other enclosure, and neutrals of aerial cable—which may have a raised ridge on the exterior of the neutral to identify them.

Exception No. 3 to the basic rule that requires use of continuous white or natural gray color along the entire length of any insulated grounded conductor (such as grounded neutral) in sizes No. 6 or smaller permits use of a conductor of other colors (black, purple, yellow, etc.) for a grounded conductor in a *multiconductor cable* under certain conditions (see Fig. 200-2):

1. That such a conductor is used only where qualified persons supervise and do service or maintenance on the cable—such as in industrial and mining applications.
2. That every grounded conductor of color other than white or gray will be effectively and permanently identified at all terminations by distinctive white marking or other effective means applied at the time of installation.

This permission for such use of grounded conductors in multiconductor cable allows the practice in those industrial facilities where multiconductor cables are commonly used—although the rule does not limit the use to industrial occupancies.

An Exception identical to the one described above under Sec. 200-6(a) also follows the rule of Sec. 200-6(b) that requires any grounded conductor larger

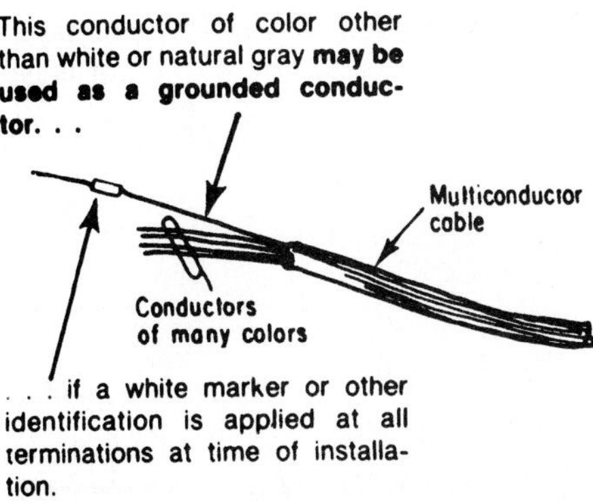
This conductor of color other than white or natural gray may be used as a grounded conductor. . .
Multiconductor cable
Conductors of many colors
. . . if a white marker or other identification is applied at all terminations at time of installation.
NOTE: This permission applies to No. 6 and smaller conductors —as well as to conductors larger than No. 6.

Fig. 200-2. [Sec. 200-6(a), Exception No. 3.]

than No. 6 to be either white or gray color for its whole length or to be marked with white identification (such as white tape) at all terminations at time of installation.

For conductors installed in raceways or in general-use cable assemblies such as NM cable or BX cable [NEC Type (AC)], continuous white or gray marking along the entire lengths of grounded conductors is not required for conductors larger than No. 6. Such grounded conductors may be of color other than white or gray provided that a "distinctive white marking"—such as white tape or paint—is applied to the conductor surface at all points of splice or termination to readily designate that it *is* a grounded conductor. See Fig. 200-3.

In the rule of part **(d)**, color coding must distinguish between grounded circuit conductors where branch circuits and/or feeders of different systems are in the same raceway. This rule assures that differentiation between neutral (grounded circuit) conductors of different wiring systems in the same raceway or other enclosure is provided for feeder circuits *as well as* branch circuits. Section 210-5(a) has long required such differentiation for branch circuits—requiring, for instance, that if a 208/120-V, 3-phase, 4-wire circuit is in the same conduit with a 480/277-V, 3-phase, 4-wire circuit, one neutral must be white or natural gray; but the neutral of the other circuit of a different voltage must be white with a color tracer stripe along the conductor length or must have some other means of distinguishing between the two neutrals of the different systems. The rule here makes the same requirement for differentiating neutrals when a conduit contains more than one feeder (or feeder and branch circuit) of different systems. (See Fig. 210-10.)

The Code-making panel has repeatedly indicated that white *or* gray may be used for one system's neutral, *but* the other system's neutral *must be* white with a color tracer *or* must be "reidentified" as recognized in either Sec. 200-6(a) or (b), whichever applies. Use of any means indicated by those sections—other than white—is intended to be recognized.

87

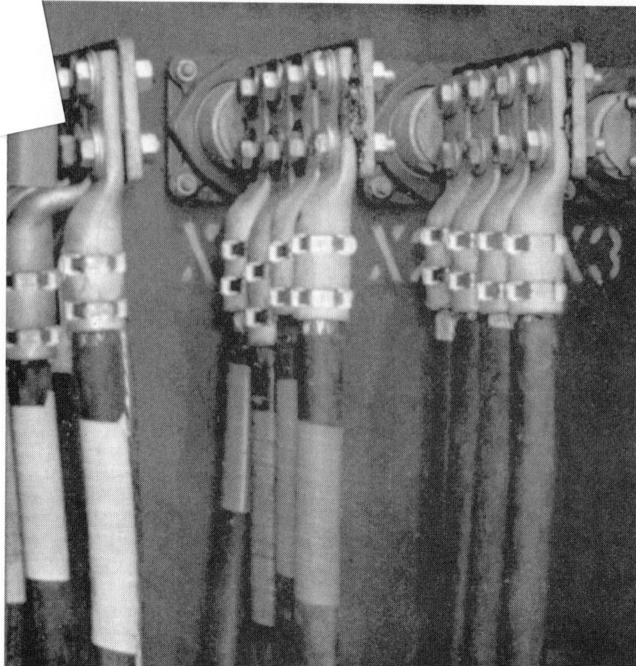

Fig. 200-3. Conductors of colors other than white or gray—in sizes larger than No. 6—may be used as grounded neutrals or grounded phase legs if marked white at all terminations—such as by white tape on the grounded feeder neutrals, at left. [Other color tapes are used on other circuit conductors to identify the three phases as A, B, and C—as required by Sec. 384-3(f).] [Sec. 200-6(b).]

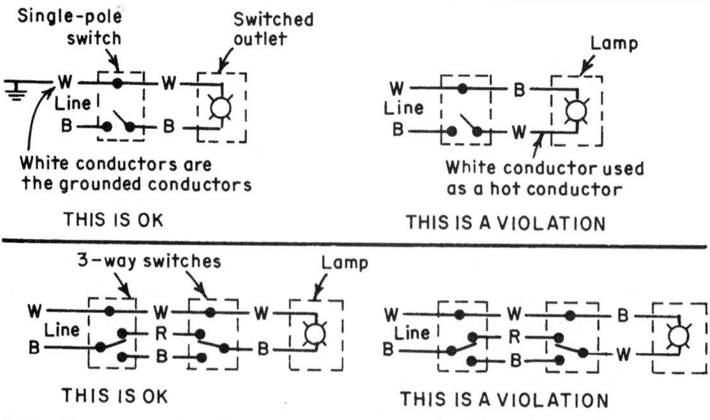

Fig. 200-4. A white or gray-colored conductor must normally be used *only* as a grounded conductor (the grounded circuit neutral or grounded phase leg of a delta system). (Sec. 200-7.)

200-7. Use of White or Natural Gray Color. Although the basic rule here limits conductors with white or gray outer covering to use only as grounded conductors [that is, as grounded neutral or grounded phase conductors (see Fig. 200-4)], a number of exceptions to the rule are noted:

Figure 200-5 shows a white-colored conductor used for an ungrounded phase conductor of a feeder to a panelboard. As shown in the left side of the panel bottom gutter, the white conductor has black tape wrapped around its end for a length of a few inches. This almost satisfies Exception No. 1 of Sec. 200-7, which permits a white conductor to be used for an ungrounded (a hot phase leg) conductor if the white is "permanently reidentified"—such as by wrapping with black or other color tape—to indicate clearly and effectively that the conductor is ungrounded. The wording of the Exception, however, might be interpreted to require that the black or other color tape be wrapped over the entire visible length of the conduit bushing.

Fig. 200-5. White conductor in lower left of panel gutter is used as an ungrounded phase conductor of a feeder, with black tape wrapped around the conductor end to "reidentify" the conductor as *not* a grounded conductor. (Sec. 200-7, Exception No. 1.)

Exception No. 2 indicates conditions under which a white conductor in a cable (such as BX or nonmetallic-sheathed cable) may be used for an ungrounded (hot-leg) conductor—*without need* for "reidentification" (such as painting or taping). When used as described, the white conductor is acceptable even though it is not a grounded conductor. Figure 200-6 shows examples of correct and incorrect hookups of switch loops where the hot supply is run first to the switched outlet, then to switches.

Exception No. 3 covers flexible cords for connecting appliances to a receptacle outlet. Exception No. 4 applies to circuits derived from the secondary side of transformers that step down to less than 50 V. Such circuits may use a white conductor.

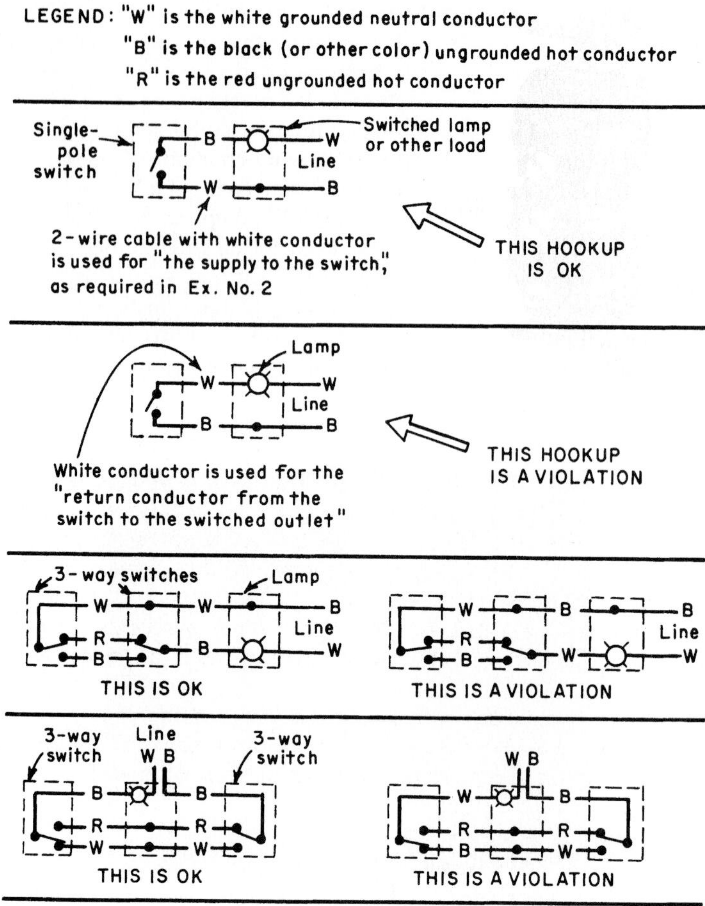

Fig. 200-6. For switch loops from load outlets with hot supply to the load outlet, white conductor in cable must be the "supply to the switch." (Sec. 200-7, Exception No. 2.)

200-10. Identification of Terminals. Part (b) permits a neutral terminal on a receptacle to be identified by the word "white" or the letter "W" marked on the receptacle as an alternative to identification of a neutral terminal by use of terminal parts (screw, etc.) that are "substantially white in color."

Marking of the word "white" or the letter "W" provides the required identification of the neutral terminal on receptacles that require white-colored plating on *all* terminals of a receptacle for purposes of corrosion resistance or for connection of aluminum conductors. Obviously, if all terminals are white-colored, color no longer serves to identify or distinguish the neutral as it does if the hot-conductor terminals are brass-colored. And as the rule is worded, the marking "white" or the letter "W" may be used to identify the neutral terminal on receptacles that have all brass-colored terminal screws. See Fig. 200-7.

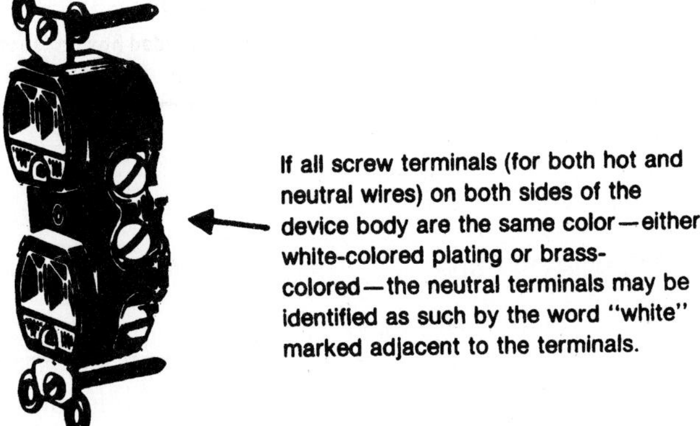

If all screw terminals (for both hot and neutral wires) on both sides of the device body are the same color—either white-colored plating or brass-colored—the neutral terminals may be identified as such by the word "white" marked adjacent to the terminals.

Fig. 200-7.

The third sentence of part **(b)** permits a push-in-type wire terminal to be identified as the neutral (grounded) conductor terminal either by marking the word "white" or the letter "W" on the receptacle body adjacent to the conductor entrance hole or by coloring the entrance hole white—as with a white-painted ring around the edge of the hole.

The Exception to part **(b)** makes clear that a 2-wire plug cap must have its neutral terminal identified if the plug is of the polarized type required by Sec. 410-42(a).

The rule of part **(c)** is shown in Fig. 200-8.

Part **(e)** of Sec. 200-10 requires that the neutral terminal of appliances be identified—to provide proper connection of field-installed wiring (either fixed wiring connection or attachment of a cord set).

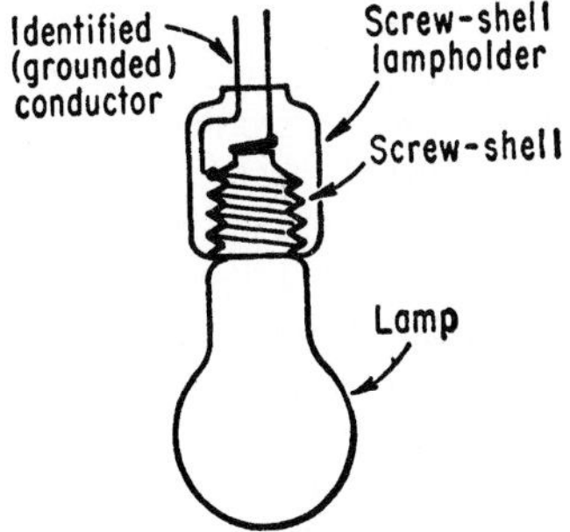

SCREW-SHELL LAMPHOLDERS are wired so that ungrounded conductor is connected to center terminal to reduce shock hazard. The identified (grounded) conductor is connected to the screw-shell.

Fig. 200-8. Screw-shell sockets must have the grounded wire (the neutral) connected to the screw-shell part. [Sec. 200-10(c).]

The rule applies to "appliances that have a single-pole switch or a single-pole overcurrent device in the line or any line-connected screw-shell lampholder" and requires simply that some "means" (instead of "marking") be provided to identify the neutral. As a result, use of white color instead of marking is clearly recognized for such neutral terminals of appliances.

ARTICLE 210. BRANCH CIRCUITS

210-1. Scope. Article 210 covers all branch circuits other than those "which supply only motor loads," which are covered in Art. 430. This section makes clear that the article covers branch circuits supplying lighting and/or appliance loads as well as branch circuits supplying any combination of those loads plus motor loads or motor-operated appliances. Where motors or motor-operated appliances are connected to branch circuits supplying lighting and/or appliance loads, the rules of *both* Art. 210 and Art. 430 apply. Article 430 alone applies to branch circuits that supply only motor loads.

210-2. Other Articles for Specific-Purpose Branch Circuits. Exceptions to the requirements for general-purpose branch circuits are indicated for "Closed-Loop and Programmed Power Distribution" as covered in NEC Art. 780. Design and hookup details for that type of branch-circuit distribution system are given in Code Art. 780 and represent special conditions which substantially differ from the usual Code rules on conventional branch-circuit wiring systems.

210-3. Rating. A branch circuit is rated according to the rating of the overcurrent device used to protect the circuit. A branch circuit with more than one outlet *must* normally be rated at 15, 20, 30, 40, or 50 A (Fig. 210-1). That is, the protective device must have one of those ratings for multioutlet circuits, and the conductors must meet the other size requirements of Art. 210. Under the definition for "receptacle" in NE Code Art. 100, it is indicated that a duplex convenience outlet (a duplex receptacle) is two receptacles and not one—even though there is only one box. Thus, a circuit that supplies only one duplex receptacle has "more than one outlet," that is, more than one point at which current is taken from the circuit to supply utilization equipment.

The Exception to this section gives limited permission to use multioutlet branch circuits rated over 50 A—but *only* to supply nonlighting loads and *only* in industrial places where maintenance and supervision assure that only qualified persons will service the installation. This Exception recognizes a real need in industrial plants where a single machine or piece of electrically operated equipment is going to be provided with its own dedicated branch circuit of adequate capacity—in effect, an individual branch circuit—but where such machine or equipment is required to be moved around and used at more than one location, requiring multiple points of outlet from the individual branch circuit to provide for connection of the machine or equipment at any one of its intended locations. *But at any time, only one machine or equipment will be connected to the branch circuit.*

BASIC RULE —

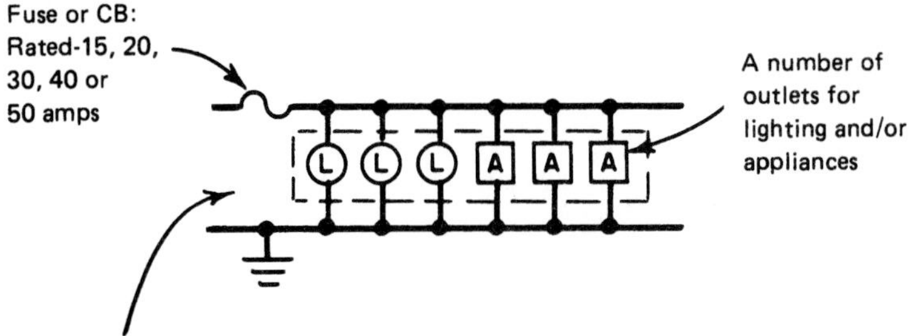
Fuse or CB:
Rated-15, 20,
30, 40 or
50 amps
A number of
outlets for
lighting and/or
appliances
L
A

Circuit voltage shall not exceed 150 volts to ground for circuits supplying lampholders, fixtures or receptacles of standard 15-amp rating. For incandescent or electric-discharge lighting under certain conditions, voltage to ground may be as high as 300 volts. In certain cases, voltage for electric discharge lighting may be up to 500 volts "between conductors" and may be an ungrounded circuit.

EXCEPTION —

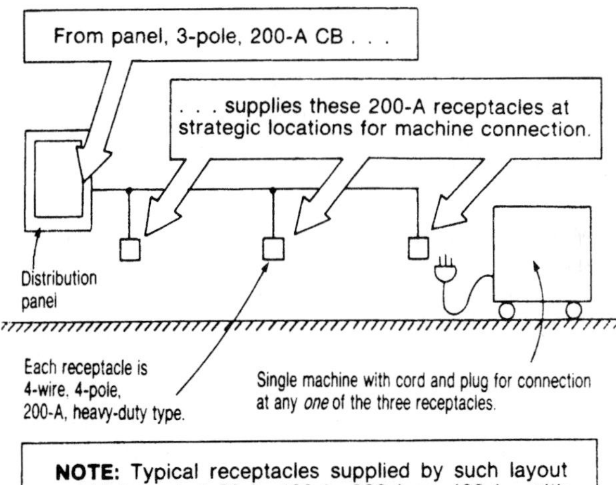
From panel, 3-pole, 200-A CB . . .
. . . supplies these 200-A receptacles at strategic locations for machine connection.
Distribution panel
Each receptacle is 4-wire, 4-pole, 200-A, heavy-duty type.
Single machine with cord and plug for connection at any *one* of the three receptacles.

NOTE: Typical receptacles supplied by such layout could be rated 60 A, 100 A, 200 A or 400 A—with their supply circuit of the same rating. Such hookups are common in jet airplane hangars for supplying cord-connected equipment used for servicing individual planes in their hangar bays.

Fig. 210-1. A multioutlet branch circuit must usually have a rating (of its overcurrent protective device) at one of the five values set by Sec. 210-3. (Sec. 210-3.)

It is important to note that it is the size of the overcurrent device which determines the rating of any circuit covered by Art. 210, even when the conductors used for the branch circuit have an ampere rating higher than that of the protective device. In a typical case, for example, a 20-A circuit breaker in a panelboard might be used to protect a branch circuit in which No. 10 conductors are used as the circuit wires. Although the load on the circuit does not exceed 20 A and No. 12 conductors would have sufficient current-carrying capacity to be used in the circuit, the No. 10 conductors with their rating of 30 A were selected to reduce the voltage drop in a long homerun. The rating of the circuit is 20 A because that is the size of the overcurrent device. The current rating of the wire does not enter into the ampere classification of the circuit.

Although multioutlet branch circuits are limited in rating to 15, 20, 30, 40, or 50 A, a branch circuit to a single-load outlet (for instance, a branch circuit to one machine or to one receptacle outlet) may have any ampere rating (Fig. 210-2). For instance, there could be a 200-A branch circuit to a special receptacle outlet or a 300-A branch circuit to a single machine.

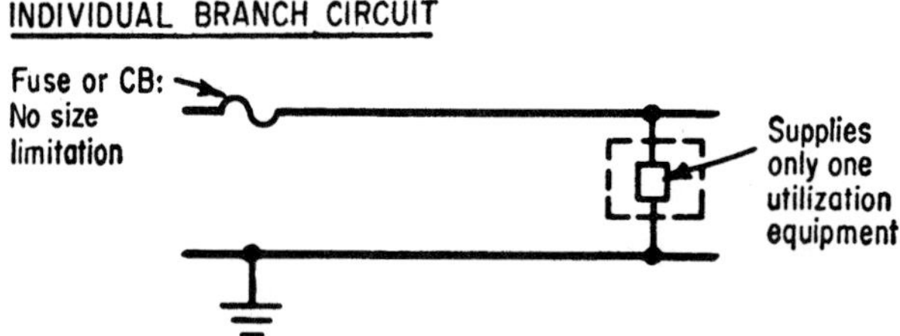

Fig. 210-2. A circuit to a single load device or equipment may have any rating. (Sec. 210-3.)

210-4. Multiwire Branch Circuits. A "branch circuit" as covered by Art. 210 may be a 2-wire circuit or may be a "multiwire" branch circuit. A "multiwire" branch circuit consists of two or more ungrounded conductors having a potential difference between them and an identified grounded conductor having equal potential difference between it and each of the ungrounded conductors and which is connected to the neutral conductor of the system. Thus, a 3-wire circuit consisting of two opposite-polarity ungrounded conductors and a neutral derived from a 3-wire, single-phase system or a 4-wire circuit consisting of three different phase conductors and a neutral of a 3-phase, 4-wire system is a *single* multiwire branch circuit. This is only one circuit, even though it involves two or three single-pole protective devices in the panelboard (Fig. 210-3). This is important, because other sections of the Code refer to conditions involving "one branch circuit" or "the single branch circuit." (See Secs. 250-24 and 410-31.)

The wording of part **(a)** of this section makes clear that a multiwire branch circuit may be considered to be either "a single circuit" or "multiple circuits." This coordinates with other Code rules that refer to multiwire circuits as well

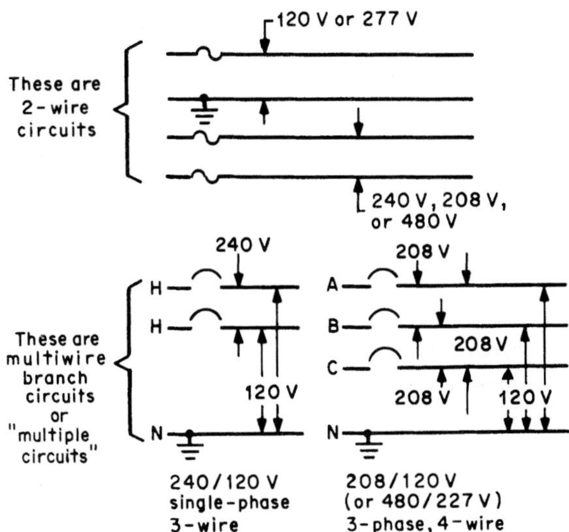

Fig. 210-3. Branch circuits may be 2-wire or multiwire type. (Sec. 210-4.)

as rules that call for two or more circuits. For instance, Sec. 220-4(b) requires that at least *two* 20-A small appliance branch circuits be provided for receptacle outlets in those areas specified in Sec. 210-52—that is, the kitchen, dining room, pantry, and breakfast room of a dwelling unit. The wording of this rule recognizes that a single 3-wire, single-phase 240/120-V circuit run to the receptacles in those rooms is equivalent to *two* 120-V circuits and satisfies the rule of Sec. 220-4(b).

In addition, a "multiwire" branch circuit is considered to be a single circuit of multiple-wire makeup. That will satisfy the rule in Sec. 410-31, which recognizes that a multiwire circuit is a single circuit when run through end-to-end connected lighting fixtures that are used as a raceway for the circuit conductors. Only *one circuit*—either a 2-wire circuit or a multiwire (3- or 4-wire) circuit—may be run through fixtures connected in a line.

The FPN following part **(a)** of Sec. 210-4 warns of the potential for "neutral overload" where line-to-neutral nonlinear loads are supplied. This results from the additive harmonics that will be carried by the neutral in multiwire branch circuits. In some cases, where the load to be supplied consists of, or is expected to consist of, so-called nonlinear loads that are connected line-to-neutral, it may be necessary to use an oversized neutral (up to two sizes larger), or each phase conductor could be run with an individual full-size neutral. Either way, a derating of 80% would be required for the number of conductors (see Note 10 to Tables 310-16 through 310-19).

Part **(b)** is discussed after part **(c)**.

The basic rule of part **(c)** of this section states that multiwire branch circuits (such as 240/120-V, 3-wire, single-phase and 3-phase, 4-wire circuits at

THIS IS THE BASIC RULE

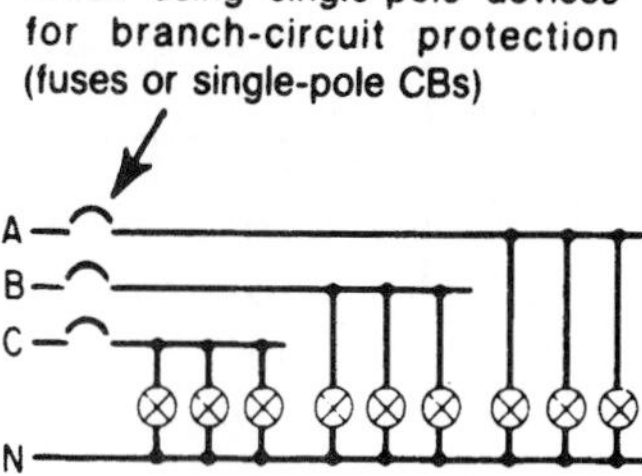

. . . multiwire branch circuits shall supply only line-to-neutral connected loads.

Fig. 210-4. With single-pole protection only line-to-neutral loads may be fed. (Sec. 210-4.)

Ex. No.1 A multiwire branch circuit may supply a single utilization equipment with line-to-line and line-to-neutral voltage using single-pole switching devices in branch-circuit protection.

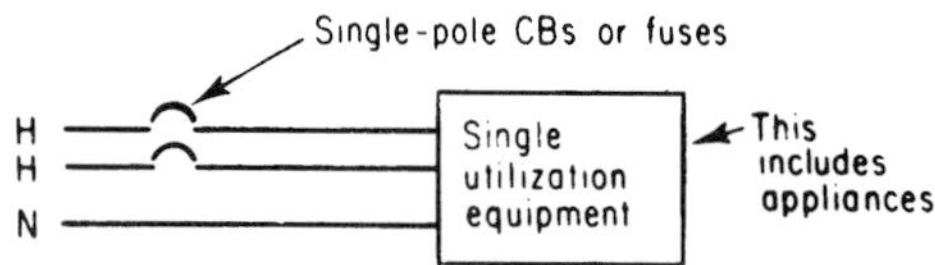

Ex. No. 2 If a multipole CB is used, loads may be connected line-to-line and/or line-to-neutral.

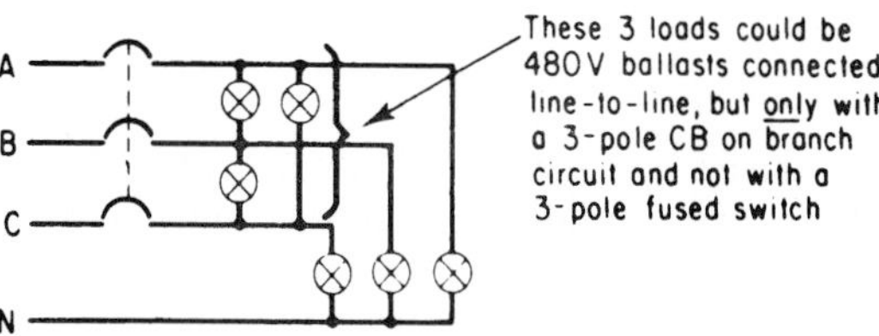

Fig. 210-5. Line-to-line loads may only be connected on multiwire circuits that conform to the Exceptions given. (Sec. 210-4.)

208/120 V or 480/277 V) may be used only with loads connected from a hot or phase leg to the neutral conductor (Fig. 210-4). The two exceptions to that rule are shown in Fig. 210-5.

Exception No. 1 permits use of single-pole protective devices for an individual circuit to "only one utilization equipment"—in which the load may be connected line-to-line as well as line-to-neutral. "Utilization equipment," as defined in Art. 100, is "equipment which utilizes electric energy for mechanical, chemical, heating, lighting, or similar purposes." The definition of "appliance," in Art. 100, notes that an appliance is "*utilization equipment,* generally other than industrial, normally built in standardized sizes or types . . . such as clothes washing, air conditioning, food mixing, deep frying, etc." Because of those definitions, the wording of Exception No. 1 opens its application to commercial and industrial equipment as well as residential.

Exception No. 2 permits a multiwire branch-circuit to supply line-to-line connected loads only when it is protected by a multipole CB. The intent of Exception No. 2 is that line-to-line connected loads may be used (other than in Exception No. 1) *only* where the poles of the *circuit protective device* operate together, or simultaneously. A multipole CB satisfies the rule, but a fused multipole switch would not comply because the hot circuit conductors are *not* "opened simultaneously by the branch-circuit overcurrent device." This rule requiring a multipole CB for any circuit that supplies line-to-line connected loads as well as line-to-neutral loads was put in the **Code** to prevent equipment loss under the conditions shown in Fig. 210-6. Use of a 2-pole CB in the sketch would cause opening of both hot legs on any fault and prevent the condition shown.

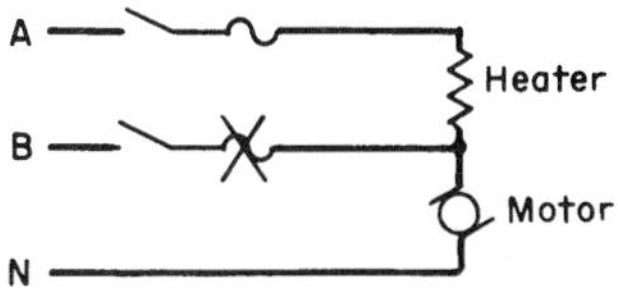

Should fuse B open, the heater and motor would be in series on 115 volts, and the motor could burn out if not properly protected.

Fig. 210-6. Single-pole protection can expose equipment to damage. (Sec. 210-4.)

At the end of part **(c)**, a fine-print note calls attention to Sec. 300-13(b), which requires maintaining the continuity of the grounded neutral wire in a multiwire branch circuit by pigtailing the neutral to the neutral terminal of a receptacle. Exception No. 2 of Sec. 210-4(c) and Sec. 300-13(b) are both aimed at the same safety objective—to prevent damage to electrical equipment that can result when two loads of unequal impedances are series-connected from hot leg to hot leg as a result of opening the neutral of an energized multiwire branch circuit or are series-connected from hot leg to neutral. Section 300-13(b) prohibits dependency upon device terminals (such as internally connected screw terminals of duplex receptacles) for the splicing of neutral conductors in multiwire (3- or 4-wire) circuits. *Grounded neutral wires* must not depend on device connection (such as the break-off tab between duplex-receptacle screw terminals) for continuity. White wires can be spliced together, with a pigtail to the neutral terminal on the receptacle. If the receptacle is removed, the neutral will not be opened.

This rule is intended to prevent the establishment of unbalanced voltages should a neutral conductor be opened *first* when a receptacle or similar device is replaced on energized circuits. In such cases, the line-to-neutral connections downstream from this point (farther from the point of supply) could result in a considerably higher-than-normal voltage on one part of a multiwire circuit and damage equipment, because of the "open" neutral, if the downstream line-to-neutral loads are appreciably unbalanced. Refer to the description given in Sec. 300-13 of this book.

Some electrical inspectors have applied an interpretation to the rule of part **(c)** and its Exception No. 2 to require a multipole CB on multiwire circuits with line-to-line loads to provide safety to maintenance personnel or any persons working on or replacing lighting fixtures or receptacle outlets connected line-to-line in existing systems. This has been a widely discussed and very controversial problem in **Code** interpretation over recent years. Shock hazard may exist in replacing or maintaining any piece of electrical equipment where *only one* of *two* hot supply conductors has been opened. If only one of the two single-pole CBs or single-pole switches controlling a 240-V circuit or a 240/120-V, 3-wire circuit is shut off, the load will be de-energized and electrical workers may presume that they will not contact any hot terminals of equipment supplied and then be surprised by electric shock from the other hot leg.

Part **(b)** of this section requires a 2-pole (simultaneously operating) circuit breaker or switch to be used at the panelboard (supply) end of a 240/120-V, single-phase, 3-wire circuit that supplies one or more split-wired duplex receptacles or combination wiring devices *in a dwelling unit* (a one-family house, an apartment in an apartment house, a condominium unit, or any other occupancy that conforms to the definition of "dwelling unit" given in Art. 100). In addition to covering split-wired receptacles, this rule applies the same concept of safety to cover duplex switches and switch-receptacle combinations, as found in the "combination line" of wiring devices that have either two receptacles, two switches, or a switch and receptacle on a single mounting strap (yoke), with a break-off fin for isolating the devices—as well as night-lights, pilot-light lamps, or other visual or audible indicators which are incorporated into common-yoke assemblies of wiring devices.

Figure 210-7 shows that a 2-pole CB, two single-pole CBs with a handle tie that enables them to be used as a 2-pole disconnect, or a 2-pole switch ahead of branch-circuit fuse protection will satisfy the requirement that both hot legs must be interrupted when the disconnect means is opened to de-energize a multiwire circuit to a split-wired receptacle. This **Code** rule provides the greater safety of disconnecting both hot conductors simultaneously to prevent shock hazard in replacing or maintaining any piece of electrical equipment where *only one* of *two* hot supply conductors has been opened.

Note: Circuits supplying split-wired receptacles in commercial, industrial, and institutional occupancies (places that are not "dwelling units") may use single-pole CBs, plug fuses without switches, or single-pole switches ahead of fuses.

It should also be noted that although a 2-pole switch ahead of fuses may satisfy as the simultaneous disconnect required ahead of split-wired receptacles, such a switch does *not* satisfy as the simultaneous multipole "branch-circuit protective device" that is required by Exception No. 2 of Sec. 210-4 when a

ANY SPLIT-WIRED WIRING DEVICE IN A DWELLING UNIT MUST HAVE MULTIPOLE, SIMULTANEOUS DISCONNECT MEANS AT CIRCUIT ORIGIN. . .

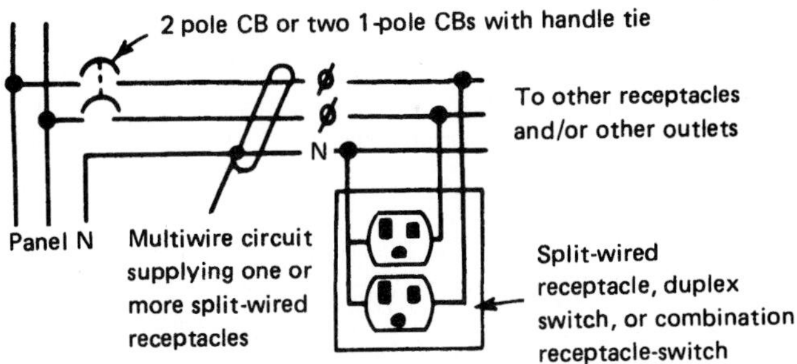

. . . OR THIS MAY BE DONE . . .

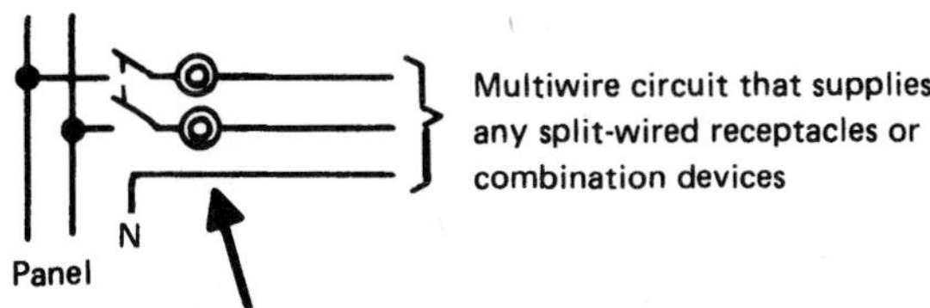

2-pole switch ahead of plug fuses will satisfy as disconnect on circuit to one or more split-wired receptacles, provided there are no phase-to-phase-connected loads on the circuit.

. . . BUT THIS WOULD BE A VIOLATION!

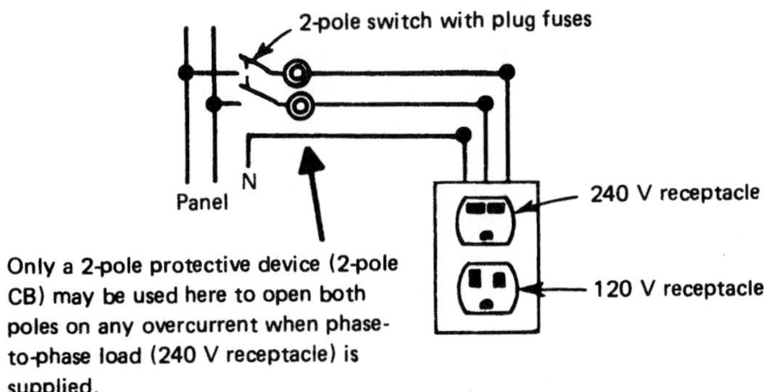

Only a 2-pole protective device (2-pole CB) may be used here to open both poles on any overcurrent when phase-to-phase load (240 V receptacle) is supplied.

Fig. 210-7

multiwire circuit supplies any loads connected phase-to-phase. In such a case, a 2-pole CB *must* be used because fuses are single-pole devices and do not assure simultaneous opening of all hot legs on overcurrent or ground fault.

It should be noted that the threat of motor burnout shown in the diagram of Fig. 210-6 can exist just as readily where the 230-V resistance device and the 115-V motor are fed from a dual-voltage (240-V, 120-V) duplex receptacle as where loads are fixed wired. As shown in Fig. 210-8, the rule of Sec. 210-4 does clearly call for a 2-pole CB (and not single-pole CBs or fuses) for a circuit supplying a dual-voltage receptacle. In such a case, a line-to-line load and a line-to-neutral load could be connected and subjected to the condition shown in Fig. 210-6.

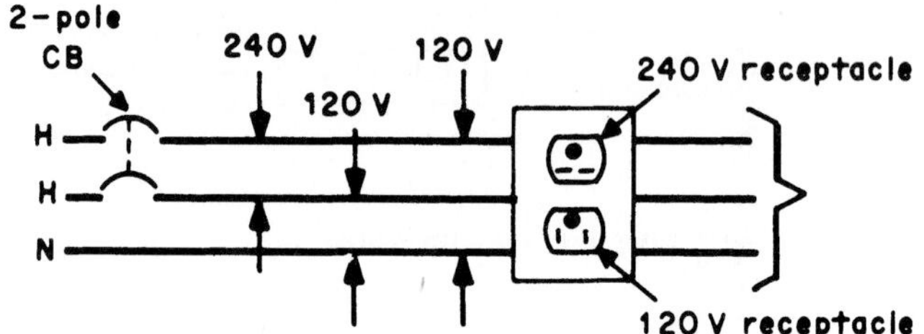

Fig. 210-8. A dual-voltage receptacle requires a 2-pole CB on its circuit. (Sec. 210-4.)

In part **(d)**, an important rule for multiwire branch circuits requires some means of identification of each of the hot (ungrounded) conductors of branch circuits in a building that contains wiring systems operating at two or more different voltage levels.

This **Code** rule restores the need to identify phase legs of circuits where more than one voltage system is used for multiwire branch circuits in a building. For instance, a building that utilizes both 208Y/120-V circuits and 480Y/277-V circuits must have separate and distinct color coding of the hot legs of the two voltage systems—*or* must have some means other than color coding, such as tagging, marking tape (color or numbers), or some other identification that will satisfy the inspecting agency. And this new rule further states that the "means of identification must be permanently posted at each branch-circuit panelboard"—to tell how the individual phases in each of the different voltage systems are identified (Fig. 210-9).

The wording of the new rule requires that the "means of identification" must distinguish between all conductors "by phase and by system." But, if a building uses only one voltage system—such as 208Y/120 V or 240/120 V single phase—then, no identification is required for the circuit phase (the "hot" or ungrounded) legs. And where a building utilizes two or more voltage systems, the separate, individual identification of ungrounded conductors must be done whether the circuits of the different voltages are run in the same or separate raceways.

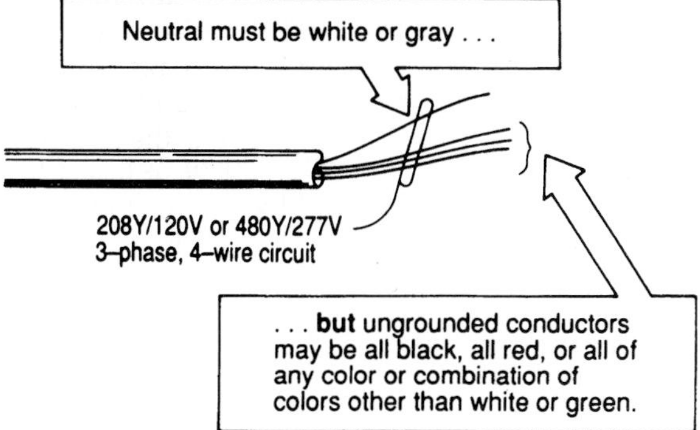
WHEN BUILDING CONTAINS ONLY
ONE SYSTEM VOLTAGE FOR CIRCUITS:
Neutral must be white or gray . . .
208Y/120V or 480Y/277V
3–phase, 4–wire circuit
. . . but ungrounded conductors may be all black, all red, or all of any color or combination of colors other than white or green.

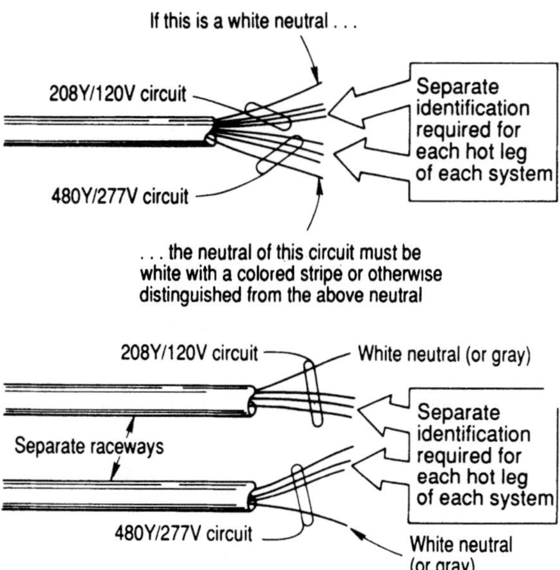
IF THERE ARE TWO SYSTEM VOLTAGES:
If this is a white neutral . . .
208Y/120V circuit
480Y/277V circuit
Separate identification required for each hot leg of each system
. . . the neutral of this circuit must be white with a colored stripe or otherwise distinguished from the above neutral
208Y/120V circuit
White neutral (or gray)
Separate raceways
Separate identification required for each hot leg of each system
480Y/277V circuit
White neutral (or gray)

Fig. 210-9. Separate identification of ungrounded conductors is required only if a building utilizes more than one nominal voltage system. Neutrals must be color-distinguished if circuits of two voltage systems are used in the same raceway (Sec. 210-5), but *not* if different voltage systems are run in separate raceways. [Sec. 210-4(d).]

210-5. Color Code for Branch Circuits. Code rules on color coding of conductors (Sec. 210-5) apply only to branch-circuit conductors and do not directly require color coding of feeder conductors. However, NE Code Sec. 384-3(f) does require identification of phase legs of feeders to panelboards, switchboards, etc.—and that requires some technique for marking the phase legs. Many design engineers do insist on color coding of feeder conductors to afford effective balancing of loads on the different phase legs.

Color coding of branch-circuit conductors is divided into three categories:

Grounded conductor: The grounded conductor of a branch circuit (the neutral of a wye system or a grounded phase of a delta) must be identified by a continuous white or natural gray color, for the entire length of conductors No. 6 or smaller. Where wires of different systems (such as 208/120 and 480/277) are installed in the same raceway, box, or other enclosure, the neutral or grounded wire of one system must be white or gray; and the neutral of the other system must be white with a color stripe. If there are three or more systems in the same raceway or enclosure, the additional neutrals must be white with colored tracers other than green. The point is that neutrals of different systems must be distinguished from each other when they are in the same enclosure [Code Sec. 210-5(a) and Fig. 210-10].

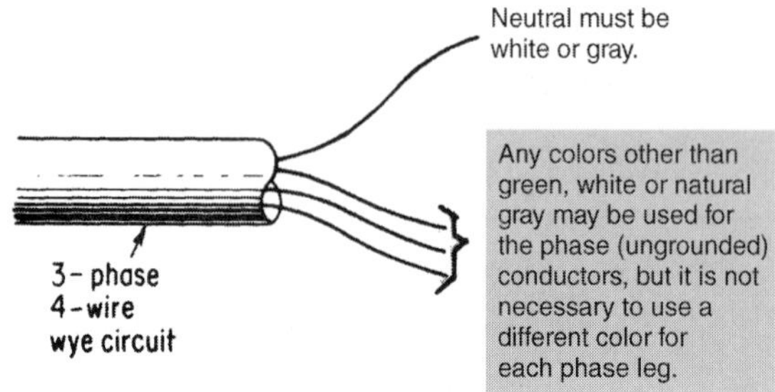
WHEN THERE IS ONLY ONE SYSTEM VOLTAGE:
Neutral must be white or gray.
Any colors other than green, white or natural gray may be used for the phase (ungrounded) conductors, but it is not necessary to use a different color for each phase leg.
3-phase 4-wire wye circuit

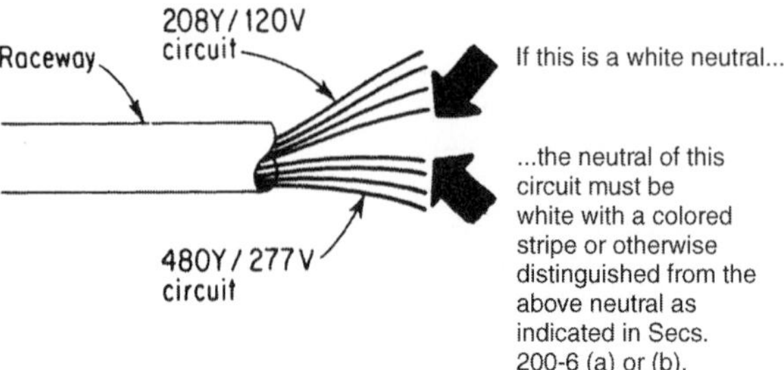
BUT, Neutrals of different systems must be distinguished
Raceway
208Y/120V circuit
If this is a white neutral...
...the neutral of this circuit must be white with a colored stripe or otherwise distinguished from the above neutral as indicated in Secs. 200-6 (a) or (b).
480Y/277V circuit

Fig. 210-10. Grounded circuit conductors (neutrals) must have color identification. [Sec. 210-5(a).]

As already noted, Exceptions to Sec. 200-6 modify the basic rule that requires use of continuous white or natural gray color along the entire length of any insulated grounded conductor (such as a grounded neutral) in size No. 6 or smaller. Likewise, Exception No. 2 to Sec. 210-5(a) also permits use of a conductor of other colors (black, purple, yellow, etc.) for a grounded conductor in a *multiconductor* cable under the conditions given in Exception No. 3 of Sec. 200-6(a).

That permission for such use of grounded conductors in multiconductor cable allows the practice in commercial and industrial facilities where multiconductor cables are commonly used.

Hot conductors: The NE Code requires that individual hot conductors of a multiwire circuit be identified where a building has more than one nominal voltage system. [See Sec. 210-4(d).]

Grounding conductor: An equipment grounding conductor of a branch circuit (if one is used) must be color-coded green or green with one or more yellow stripes—or the conductor may be bare [Sec. 210-5(b)].

However, Exception No. 1 refers to Sec. 250-57(b), which says that an equipment grounding conductor larger than No. 6 may be other than a green-insulated conductor or a green-with-yellow-stripe conductor. Section 250-57 permits an equipment grounding conductor with insulation that is black, blue, or any other color—provided that one of the three techniques specified in Sec. 250-57 is used to identify this conductor as an equipment grounding conductor.

The first technique consists of stripping the insulation from the insulated conductor for the entire length of the conductor appearing within a junction box, panel enclosure, switch enclosure, or any other enclosure. With the insulation stripped from the conductor, the conductor then appears as a bare conductor, which is recognized by the Code for grounding purposes. A second technique which is acceptable is to paint the exposed insulation green for its entire length within the enclosure. If, say, a black insulated conductor is used in a conduit coming into a panelboard, the length of the black conductor within the panelboard can be painted green to identify this as an equipment grounding conductor.

The third acceptable method is to mark the exposed insulation with green-colored tape or green-colored adhesive labels. Figure 210-12 summarizes the rules on identification of equipment grounding conductors. Green-colored conductors must not be used for any purpose other than equipment grounding.

Color coding of circuit conductors (or some other method of identifying them), as required by Sec. 210-4(d), is a wiring consideration that deserves the close, careful, complete attention of all electrical people. By providing ready identification of the two or three phase legs and neutrals in wiring systems, color coding is the easiest and surest way of balancing loads among the phase legs, thereby providing full, safe, effective use of total circuit capacities. In circuits where color coding is missing or not effectively applied, loads or phases get unbalanced, many conductors are either badly underloaded or excessively loaded, and breakers or fuses often are increased in size to eliminate tripping due to overload on only one phase leg. Modern electrical usage—for reasons of safety and energy conservation, as well as full, economic application of system

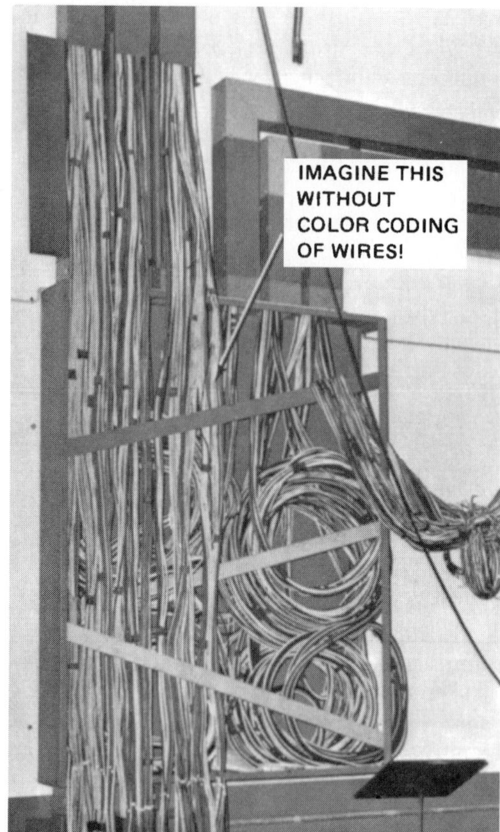

Fig. 210-11. Although not Code required, color identification of branch-circuit phase legs is needed for safe and effective work on grouped circuits. (Sec. 210-5.)

equipment and materials—demands the many real benefits that color coding can provide.

For the greater period of its existence, the NE Code required a very clear, rigid color coding of branch circuits for good and obvious safety reasons. Color coding of hot legs to provide load balancing is a safety matter. Section 220-4(d) requires balancing of loads from branch-circuit hot legs to neutral. Section 220-22 bases sizing of feeder neutrals on clear knowledge of load balance in order to determine "maximum unbalance." And mandatory differentiation of voltage levels is in the safety interests of electricians and others maintaining or working on electrical circuits, to warn of different levels of hazard.

Because 99 percent of electrical systems involve no more than two voltage configurations for circuits up to 600 V, and because there has been great standardization in circuit voltage levels, there should be industrywide standardization on circuit conductor identifications. A clear, simple set of rules could

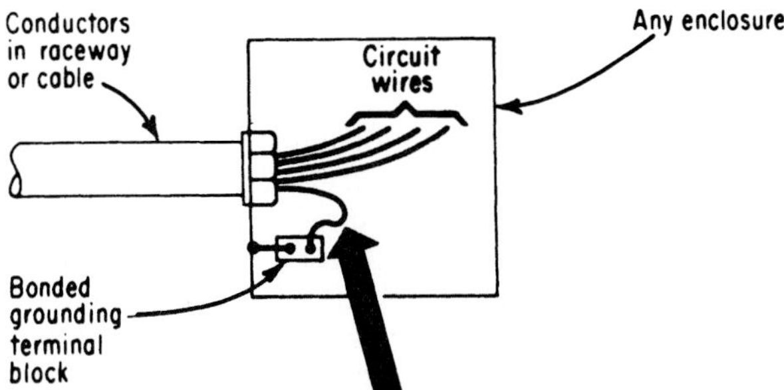

Equipment grounding conductor may be bare, or covered to show a green color or green with one or more yellow stripes.

Exception No.1

But, for a grounding conductor larger than No. 6, an insulated conductor of other than green color or green with yellow stripes may be used provided *one* of the following steps is taken:

1. Stripping the insulation from an insulated conductor of another color (say, black) for the entire length that is exposed in the box or other enclosure— so that the conductor appears as a bare conductor.
2. Painting the exposed insulation green for its entire length within the enclosure.
3. Wrapping the entire length of exposed insulation with green-colored tape or green-colored adhesive labels.

Fig. 210-12. Equipment grounding conductor for branch circuit. [Sec. 210-5(b).]

cover the preponderant majority of installations, with exceptions made for the relatively small number of cases where unusual conditions exist and the local inspector may authorize other techniques. Color coding should follow some basic pattern—such as the following:

- 120-V, 2-wire circuit: grounded neutral—white; ungrounded leg—black.
- 240/120-V, 3-wire, single-phase circuit: grounded neutral—white; one hot leg—black; the other hot leg—red.
- 208Y/120-V, 3-phase, 4-wire: grounded neutral—white; one hot leg—black; one hot leg—red; one hot leg—blue.
- 240-V, delta, 3-phase, 3-wire: one hot leg—black; one hot leg—red; one hot leg—blue.
- 240/120-V, 3-phase, 4-wire, high-leg delta: grounded neutral—white; high leg (208-V to neutral)—orange; one hot leg—black; one hot leg—blue.
- 480Y/277-V, 3-phase, 4-wire: grounded neutral—gray, or where run in the same raceway with circuits from another system, white with a color tracer; one hot leg—brown; one hot leg—orange; one hot leg—yellow.
- 480-V, delta, 3-phase, 3-wire: one hot leg—brown; one hot leg—orange; one hot leg—yellow.

By making color coding a set of simple, specific color designations, standardization will assure all the safety and operating advantages of color coding to all electrical systems. Particularly today, with all electrical systems being subjected to an unprecedented amount of alterations and additions because of continuing development and expansion in electrical usage, conductor identification is a regular safety need over the entire life of the system. (Fig. 210-11.)

A second step in clarification and expansion of color coding would require color coding for all circuits—multiwire branch circuits, branch circuits without a neutral or other grounded conductor, feeders, and even for service conductors. As now indicated in Sec. 384-3(f) of the NE Code, correct, effective loading of all circuits—to get full capacity of all phases and to prevent unknown unbalances with the attendant chance of oversizing of protection and overloading of conductors—depends upon ready identification of all conductors at all points in a system.

Of course, there are alternatives to "color" identification throughout the length of conductors. Color differentiation is almost worthless for color-blind electricians. And it can be argued that color identification of conductors poses problems because electrical work is commonly done in darkened areas where color perception is reduced even for those with good eyesight. The NE Code already recognizes white tape or paint over the conductor insulation end at terminals to identify neutrals where conductors are larger than No. 6 (Sec. 200-6). Number markings spaced along the length of a conductor on the insulation (1, 2, 3, etc.)—particularly, say, white numerals on black insulation—might prove very effective for identifying and differentiating conductors. Or the letters "A," "B," and "C" could be used to designate specific phases. Or a combination of color and markings could be used. But some kind of conductor identification is essential to safe, effective hookup of the ever-expanding array of conductors used throughout buildings and systems today. And the method used for identifying ungrounded circuit conductors must be posted at each branch-circuit panelboard to comply with requirements of Sec. 210-4(d). Although not required by Sec. 210-5(a), the method used to distinguish the grounded (neutral) conductors for the different systems should also be included with that information required for the ungrounded (phase) conductors.

210-6. Branch Circuit Voltage Limitations. Voltage limitations for branch circuits are presented in Sec. 210-6. In general, branch circuits serving lampholders, fixtures, or receptacles of the standard 15-A or less rating are limited to operation at a maximum voltage rating of 120 V.

Part **(a)** applies specifically to dwelling units—one-family houses, apartment units in multifamily dwellings, and condominium and co-op units—and to guest rooms in hotels and motels and similar residential occupancies. In such occupancies, any lighting fixture or any receptacle for plug-connected loads rated up to 1,440 VA or for motor loads of less than ¼ hp must be supplied at not over 120 V between conductors.

Note: The 120-V supply to the above-type loads may be derived from (1) a 120-V, 2-wire branch circuit, (2) a 240/120-V, 3-wire branch circuit, or (3) a 208/120-V, 3-phase, 4-wire branch circuit. Appliances rated 1,440 VA or more (ranges, dryers, water heaters, etc.) may be supplied by 240/120-V circuits in accordance with Sec. 210-6(c)(5).

Part **(b)** permits a circuit with not over 120 V between conductors to supply medium-base screw-shell lampholders, ballasts for fluorescent or HID lighting fixtures, and plug-connected or hard-wired appliances—in any type of building or on any premises (Fig. 210-13).

Part **(c)** applies to circuits with over 120 V between conductors (208, 240, 277, or 480 V) but not over 277 V (nominal) to ground. This is shown in Fig. 210-14, where all of the circuits are "circuits exceeding 120 V, nominal, between conductors and not exceeding 277 V, nominal, to ground." Circuits of any of those voltages are permitted to supply incandescent lighting fixtures with mogul-base screw-shell lampholders, ballasts for electric-discharge lighting fixtures or plug-connected or hard-wired appliances, or other utilization equipment. It is important to note that this section no longer contains the requirement for a minimum 8-ft (2.44-m) mounting height for incandescent or electric-discharge fixtures with mogul-base screw-shell lampholders used on 480/277-V systems.

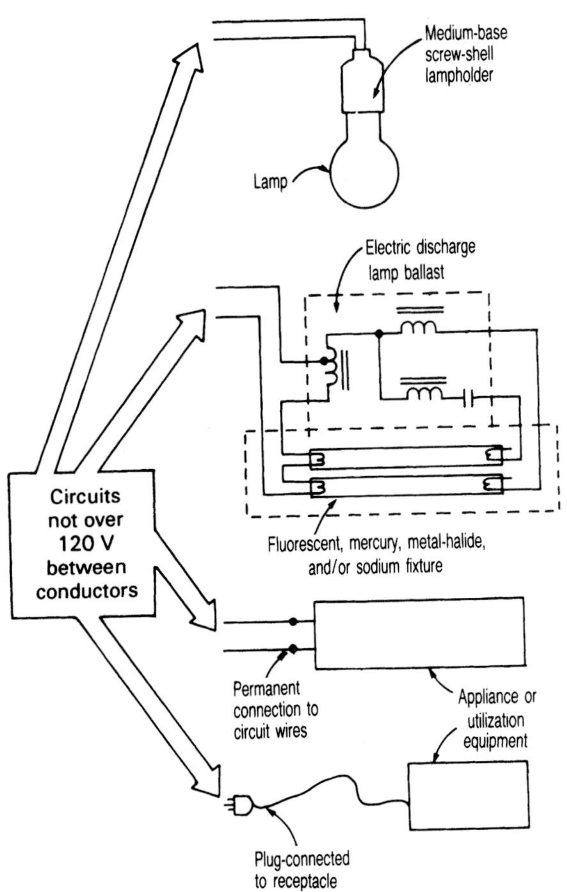

Fig. 210-13. In any occupancy, 120-V circuits may supply these loads. [Sec. 210-6(b)]

A UL-listed electric-discharge lighting fixture rated at 277 V nominal may be equipped with a medium-base screw-shell lampholder and does not require a mogul-base screw-shell. The use of the medium-base lampholder, however, is limited to "listed electric-discharge fixtures." For 277-V incandescent fixtures, Sec. 210-6(c)(3) continues the requirement that such fixtures be equipped with "mogul-based screw-shell lampholders."

Fluorescent, mercury-vapor, metal-halide, high-pressure sodium, low-pressure sodium, and/or incandescent fixtures may be supplied by 480/277-V, grounded-wye circuits—with loads connected phase-to-neutral and/or phase-to-phase. Such circuits operate at 277 V to ground even, say, when 480-V ballasts are connected phase-to-phase on such circuits. Or lighting could be supplied by 240-V delta systems—either ungrounded or with one of the phase legs grounded, because such systems operate at not more than 277 V to ground.

Use of incandescent lighting at over 150 V to ground is accepted by the **NE Code** in commercial, institutional, and industrial buildings and premises.

Although 277-V incandescent lamps are available with medium screw bases, and fixtures for them are available with medium screw-shell lampholders, use of such equipment violates the **NE Code** rule in Sec. 210-6(c)(1). And Sec. 210-6(b)(1) limits medium base, screw-shell lampholders to use only where connected to circuit wires with not more than 120 V between the wires—such as 2-wire 120-V circuits.

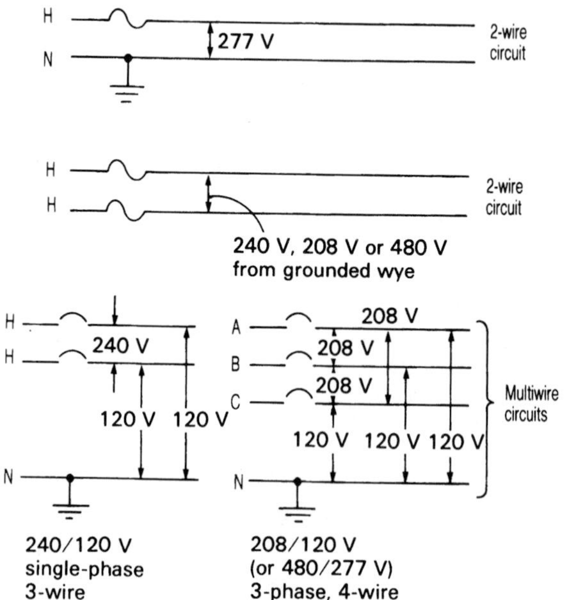

Fig. 210-14. These circuits may supply incandescent lighting with mogul-base screw-shell lampholders for over 120 V between conductors, electric-discharge ballasts, and cord-connected or permanently wired appliances or utilization equipment. [Sec. 210-6(c).]

Section 210-6(c) permits installations on 480/277-V, 3-phase, 4-wire wye systems—with equipment connected from phase-to-phase (480-V circuits) or connected phase-to-neutral (277-V circuits). In either case, the voltage to ground is only 277 V. In any such application, it is important that the neutral point of the 480/277-V wye be grounded to limit the voltage aboveground to 277 V. If the neutral were not grounded and the system operated ungrounded, the voltage to ground, according to the Code, would be 480 V (see definition "voltage to ground," Art. 100), and lighting equipment could be used on such circuits only for outdoor applications as specified under Sec. 210-6(d) (discussed later).

On a neutral-grounded 480/277-V system, incandescent, fluorescent, mercury-vapor, metal-halide, high-pressure sodium, and low-pressure sodium equipment can be connected from phase-to-neutral on the 277-V circuits. If fluorescent or mercury-vapor fixtures are to be connected phase-to-phase, some Code authorities contend that autotransformer-type ballasts cannot be used when these ballasts raise the voltage to more than 300 V, because, they contend, the NE Code calls for connection to a circuit made up of a grounded wire and a hot wire. (See Sec. 410-78.) On phase-to-phase connection these ballasts would require use of 2-winding, electrically isolating ballast transformers. The wording of Sec. 410-78 does, however, lend itself to interpretation that it is only necessary for the *supply system* to the ballast to *be grounded*—thus permitting the two hot legs of a 480-V circuit to supply an autotransformer because the hot legs are derived from a neutral-grounded "system." But Sec. 210-9 can become a complicating factor. Use of a 2-winding (isolating) ballast is clearly acceptable and avoids all confusion.

Section 210-6(d) of the NE Code permits fluorescent and/or mercury-vapor units to be installed on circuits rated over 277 V (nominal) to ground and up to 600 V between conductors—but only where the lamps are mounted in permanently installed fixtures on poles or similar structures for the illumination of areas such as highways, bridges, athletic fields, parking lots, at a height not less than 22 ft, or on other structures such as tunnels at a height not less than 18 ft (5.49 m). (See Fig. 210-15.) Part **(d)** covers use of lighting fixtures on 480-V ungrounded circuits—such as fed from a 480-V delta-connected or wye-connected ungrounded transformer secondary.

This permission for use of fluorescent and mercury units under the conditions described is based on phase-to-phase voltage rather than on phase-to-ground voltage. This rule has the effect of permitting the use of 240- or 480-V ungrounded circuits for the lighting applications described. But as described above, autotransformer-type ballasts may not be permitted on an ungrounded system if they raise the voltage to more than 300 V [Sec. 410-78]. In such cases, ballasts with 2-winding transformation would have to be used.

Certain electric railway applications utilize higher circuit voltages. Infrared lamp industrial heating applications may be used on higher circuit voltages as allowed in Sec. 422-15(c) of the Code.

Caution: The concept of maximum voltage not over "120 V . . . *between* conductors," as stated in Sec. 210-6(a), has caused considerable discussion and controversy in the past when applied to split-wired receptacles and duplex receptacles of two voltage levels. It can be argued that split-wired general-

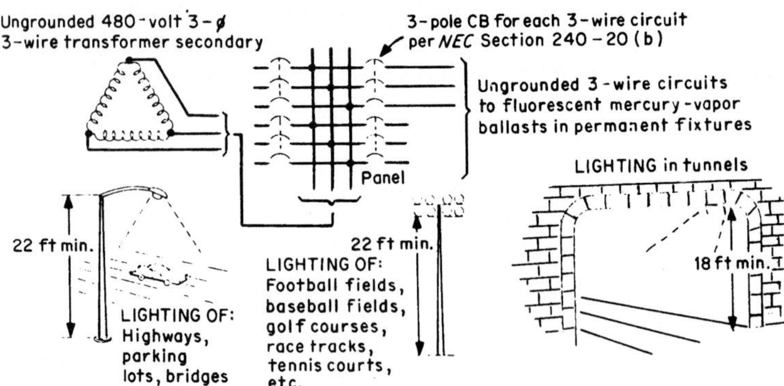

Fig. 210-15. Ungrounded circuits, at up to 600 V between conductors, may supply lighting only as shown. [Sec. 210-6(d).]

purpose duplex receptacles are not acceptable in dwelling units and in hotel and motel guest rooms because they are supplied by conductors with *more* than 120 V between them—that is, 240 V on the 3-wire, single-phase, 120/240-V circuit so commonly used in residences. The two hot legs connect to the brass-colored terminals on the receptacle, with the shorting tab broken off, and the voltage between those conductors *does* exceed 120 V. The same condition applies when a 120/240-V duplex receptacle is used—the 240-V receptacle is fed by conductors with more than 150 V between them. But, the Code-making panel ruled that the use of duplex receptacles connected on 240/120-V, 3-wire, single-phase circuits in dwelling units and guest rooms in hotels, motels, dormitories, etc., is *not* prohibited. See Fig. 210-16.

Objection to split-wired receptacles in dwelling units has a sound basis: Two appliances connected by 2-wire cords to a split-wired receptacle in a kitchen do present a real potential hazard. With, say, a coffee maker and a toaster plugged into a split-wired duplex, if the hot wire in each appliance should contact the metal enclosure of the appliance, there would be 240 V between the two appliance enclosures. The user would be exposed to the extremely danger-

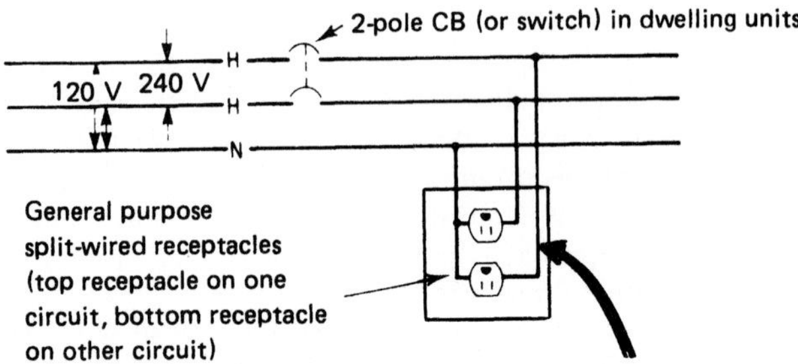

Fig. 210-16. Split-wired receptacles *are* permitted in residential occupancies ("dwelling units") and in all other types of occupancies (commercial, institutional, industrial, etc.).

ous chance of touching each appliance with a different hand—putting 240 V across the person, from hand-to-hand, through the heart path (Fig. 210-17). Use of nonsplit-wired receptacles on the usual spacing of up to 4 ft (1.22 m) does tend to separate appliances on different hot legs.

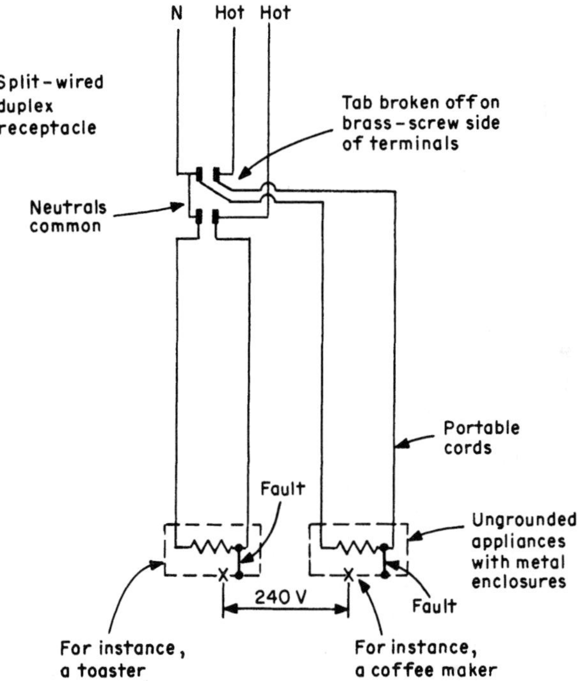

Fig. 210-17. This can be dangerous.

Use of split-wired receptacles and other receptacles with more than 120 V between terminals is, of course, completely acceptable in any commercial, institutional, or industrial location. And it is also perfectly acceptable at any time to use a split-wired receptacle for switch control of one plug-in point leaving the other hot all the time (Fig. 210-18).

Section 210-6(c) (5) clearly permits "permanently connected utilization equipment" to be supplied by a circuit with voltage between conductors in excess of 120 V, and permission *is* intended for the use of 277-V heaters in dwelling units, as used in high-rise apartment buildings and similar large buildings that may be served at 480/277 V.

210-7. Receptacles and Cord Connectors. Section 210-7 states that "receptacles installed on 15- and 20-A branch circuits shall be of the grounding type." This requires that on all new installations a grounding-type device must be used (Fig. 210-19). This rule came into effect in 1968. Prior to that, ungrounded devices were permitted on branch circuits. If a residence or office is equipped

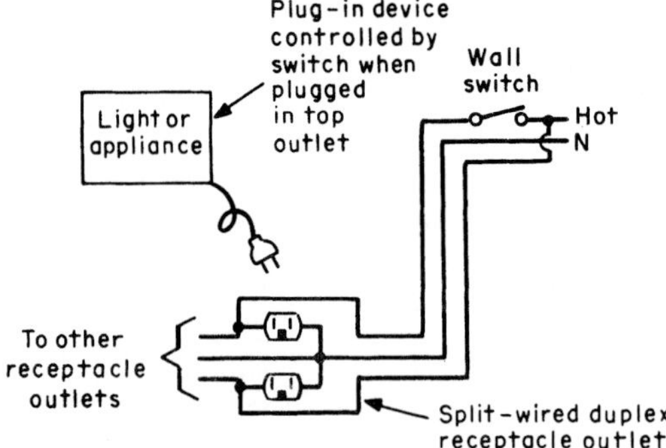

Fig. 210-18. Split-wiring of receptacles to control one of the receptacles may be done from the same hot leg of a 2-wire circuit or with separate hot legs of a 3-wire, 240/120-V circuit. (Sec. 210-6.)

with these older devices, it is not necessary to replace them as long as they are serviceable.

In all cases where a grounding-type receptacle is installed, it shall be grounded. For nonmetallic-sheathed cable the grounding conductor is run with the branch-circuit conductors. The armor of Type AC metal-clad cable, the sheath of ALS cable, and certain metallic raceways are acceptable as grounding means (Fig. 210-20).

The purpose here is to make certain that grounding-type receptacles are used and grounded where a ground is available, and nongrounding-type receptacles are used where grounding is impractical so that no one will be deceived as to the availability of a grounding means for appliances.

The rule of Sec. 210-7(b), requiring grounding of the ground terminal of receptacles and cord connectors, has two important Exceptions. The first permits receptacles mounted on portable or vehicle-mounted generators, in accordance with Sec. 250-6, to have their ground terminals left ungrounded, when the generator frame itself is not grounded (Fig. 210-21). But where such receptacles are mounted on portable or vehicle-mounted generators, the grounding terminal of the receptacle *must* be bonded to the generator frame [Sec. 250-6(a) (2)].

Exception No. 2 to this section exempts replacement receptacles from the basic rule because in some cases a grounding means may not be available where replacing tab-slot (nongrounding) type receptacles [see discussion under part **(d)** of this section].

Part **(c)** indicates that grounding of receptacles and cord connectors must be provided by connection to the circuit equipment grounding conductor. For correlation purposes, a note after Sec. 210-7(c) refers to the use of "quiet grounding" for electrical noise reduction for sensitive equipment, as covered in Sec. 250-74, Exception No. 4.

The rewrite of Sec. 210-7(d) clarifies the application and yet another field marking.

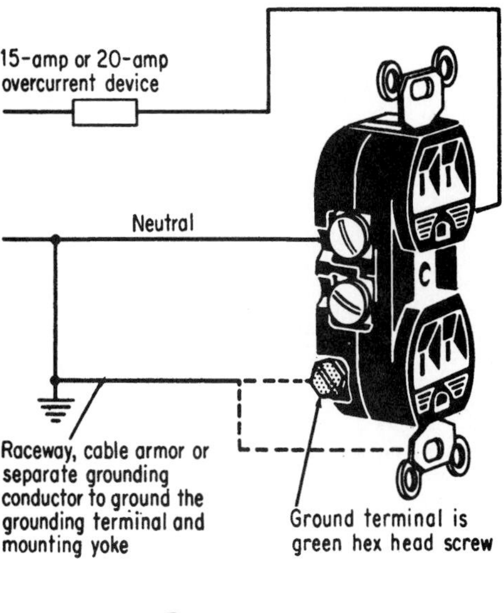
15-amp or 20-amp
overcurrent device
Neutral
Raceway, cable armor or
separate grounding
conductor to ground the
grounding terminal and
mounting yoke
Ground terminal is
green hex head screw

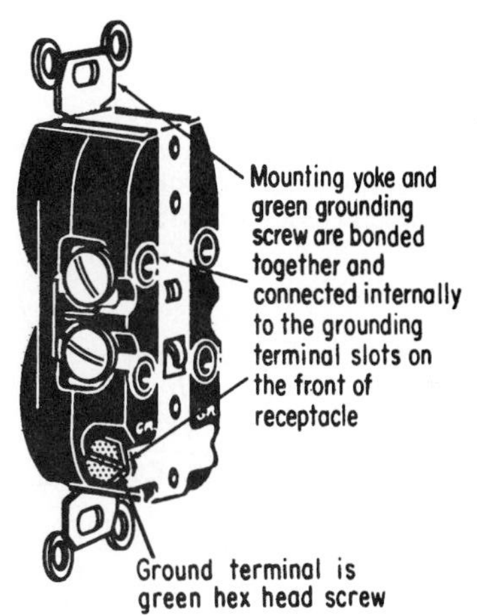
Mounting yoke and
green grounding
screw are bonded
together and
connected internally
to the grounding
terminal slots on
the front of
receptacle
Ground terminal is
green hex head screw

Fig. 210-19. Any receptacle on a 15- or 20-A branch circuit must be a grounding type. (Sec. 210-7.)

The basic requirement of Sec. 210-7(d) calls for replacement of defective receptacles in accordance with the requirements indicated in parts **(d)** (1), (2), or (3).

Sec. 210-7(d)(1) generally requires any two-slot, nongrounding receptacle that becomes defective to be replaced with a grounding-type receptacle if there is a "grounding means" in the outlet box. The grounding means may be the equipment grounding conductor generally required to be run with the circuit conductors by Sec. 210-7(c) to ground receptacles. Or, it may be an equipment ground installed in accordance with the Sec. 250-50(b), Exception. In either

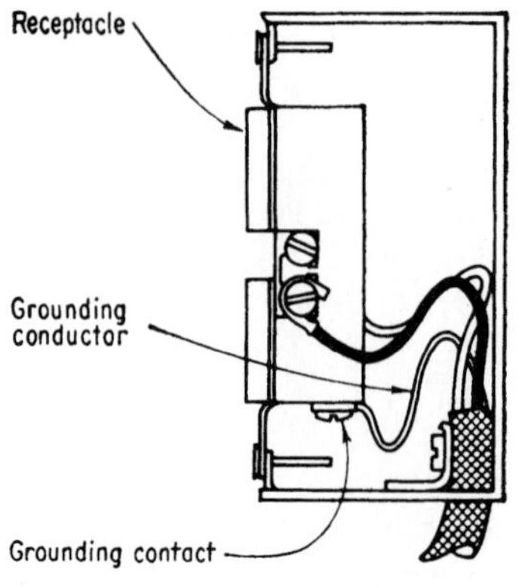
Receptacle
Grounding conductor
Grounding contact

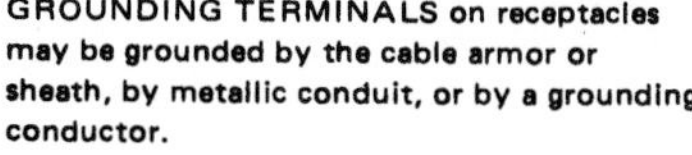

GROUNDING TERMINALS on receptacles may be grounded by the cable armor or sheath, by metallic conduit, or by a grounding conductor.

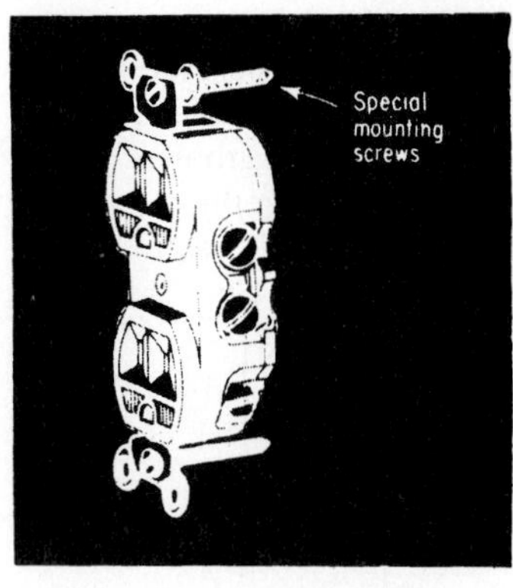
Special mounting screws

SELF–GROUNDING RECEPTACLE includes mounting screw equipped with pressure clip to assure tight contact between screws and device yoke, eliminating the need for a grounding jumper from green hex-head screw to metal box.

Fig. 210-20. The grounded terminal must be grounded in an approved manner. [Sec. 210-7(c).]

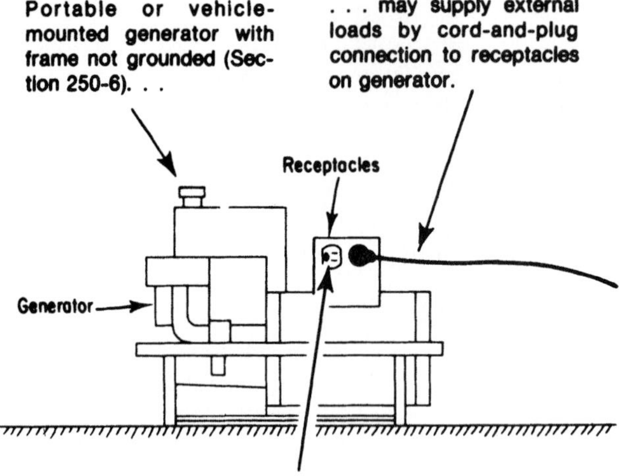
Portable or vehicle-mounted generator with frame not grounded (Section 250-6). . .
. . . may supply external loads by cord-and-plug connection to receptacles on generator.
Receptacles
Generator
Receptacle grounding terminals do not have to be grounded but *must* be bonded to generator frame.

Fig. 210-21. Grounding is not required for generator-mounted receptacles. [Sec. 210-7(b).]

case, a grounding-type receptacle must be used as a replacement for the defective nongrounding type. And, the green ground screw on the receptacle must be connected to the available grounding means. It should be noted that it is *not* necessary to install a ground in accordance with Sec. 250-50, Exception, but if one *chooses* to do so, the use of a grounding-type receptacle becomes mandatory, unless a GFCI is required. And, in either case, the ground installed in accordance with the Sec. 250-50 Exception must be connected to the ground screw on the receptacle.

Next Sec. 210-7(d)(2) requires that where any receptacle is to be replaced, if the **Code** now requires a receptacle in that location to be GFCI-protected, a GFCI receptacle must be provided. That requirement has the effect of making all GFCI rules retroactive where replacement is made. That is, normally there is no requirement to apply a new **Code** to an existing building and bring that building into compliance with the new **Code**. But, now, in an existing installation, in, say, a bathroom of a commercial office building, if a receptacle should need replacement, Sec. 210-8(b)(1) in conjunction with Sec. 210-7(d)(2) would require the replacement receptacle to be GFCI protected.

Part **(3)** of Sec. 210-7(d) covers those instances where a replacement must be made at a receptacle enclosure that does not contain a grounding means. In that situation the **Code** offers three options for replacement. It should be noted that there is no evidence in any available documentation to suggest that this section is presenting these options in an order of preference. Any of the three indicated replacement alternatives is equally acceptable.

The first option is to replace the existing two-slot (nongrounding) receptacle with another nongrounding-type receptacle. Although acceptable, such replacement may not be the most desirable where the receptacle has to supply grounded equipment. The use of a "cheater" or plug adapter with such equipment or the possibility of the user breaking off the ground prong on the grounding-type plug are equally unpalatable. Use of the alternative presented in either part **b** or **c** would be better.

The rule of part **b** recognizes a GFCI receptacle as replacement of a nongrounding-type receptacle where there is no ground in the box. Such a receptacle will provide automatic opening of the supply under fault conditions without connection to an equipment grounding conductor. This is accomplished through the internal electronic circuitry of the GFCI device. Given the much lower threshold (5 mA ± 1 mA) at which the GFCI receptacle operates versus the instantaneous trip for a 15-A CB, the GFCI offers a much better level of shock protection.

The next sentence requires a GFCI receptacle to be installed as a replacement for a nongrounding receptacle in a box where a ground is not available to be marked "No Equipment Ground." This requirement is intended to alert anyone using that outlet that, even though the GFCI receptacle has a U-shaped ground slot, there is no connection to the grounding system. This is important where supplying electronic equipment, such as computers, electronic cash registers, etc., because an equipment grounding conductor is critical for proper operation of such equipment. The field marking, "No Equipment Ground" will serve to prevent a user from being fooled into thinking that there *is* a ground connection.

If there is no ground in the box, Sec. 210-7(d)(3)c permits replacement of a nongrounding-type receptacle with a GFCI-protected grounding-type receptacle. And, in addition to the marking "No Equipment Ground," the receptacle must be marked, "GFCI Protected." The warning about the absence of an equipment ground is required for the same reason it is required at the GFCI receptacle. And the "GFCI Protected" marking is provided to help service electricians in the future. A lot of time could be wasted if the GFCI device protecting the grounding-type receptacle tripped, leaving the protected receptacle de-energized and the service electrician was unaware of the upstream GFCI device. By providing such a field marking, any service electrician will be immediately advised of the possibility that the upstream GFCI device—receptacle or breaker—may have de-energized the circuit.

The only possible combination not addressed is where a two-slot (nongrounding) receptacle is used as a replacement at a location that is now required to be GFCI-protected. There is no intent to prohibit the use of a nongrounding-type receptacle and a GFCI breaker to satisfy the need for replacement and GFCI protection. Such application is not mentioned because the CMP didn't feel that anyone would use that combination, given the costs associated with such application. They felt most would just use a GFCI receptacle. Because the use of a nongrounding receptacle and GFCI breaker is not specifically mentioned, some inspectors may not recognize their use. But, if it is recognized by the local inspection authority, it seems as if the field marking "GFCI Protected" should be provided, while the other field marking "No Equipment Ground" may be omitted (Fig. 210-22). This **Code** rule has long recognized that it could be misleading and therefore dangerous to replace a nongrounding receptacle with a grounding receptacle without grounding the green hex-head screw (or other ground terminal) of the receptacle or providing line-side GFCI protection.

210-8. Ground-Fault Circuit-Interrupter Protection for Personnel. Part **(a)** of Sec. 210-8 of the **NE Code** is headed "Dwelling Units." The very clear and detailed definition of those words, as given in Art. 100 of the **NE Code**, indicates that all the ground-fault circuit interruption rules apply to:

- All one-family houses
- Each dwelling unit in a two-family house
- Each apartment in an apartment house
- Each dwelling unit in a condominium

GFCI protection is required by Sec. 210-8 for all 125-V, single-phase, 15- and 20-A receptacles installed in *bathrooms* of dwelling units [part **(a)(1)**] and hotel or motel units [part **(b)**] and in garages of dwelling units only (Fig. 210-24). The requirement for GFCI protection in "garages" is included because home owners do use outdoor appliances (lawn mowers, hedge trimmers, etc.) plugged into garage receptacles. Such receptacles require GFCI protection for the same reason as "outdoor" receptacles. In either place, GFCI protection may be provided by a GFCI circuit breaker that protects the whole circuit and any receptacles connected to it, or the receptacle may be a GFCI type that incorporates the components that give it the necessary tripping capability on low-level ground faults.

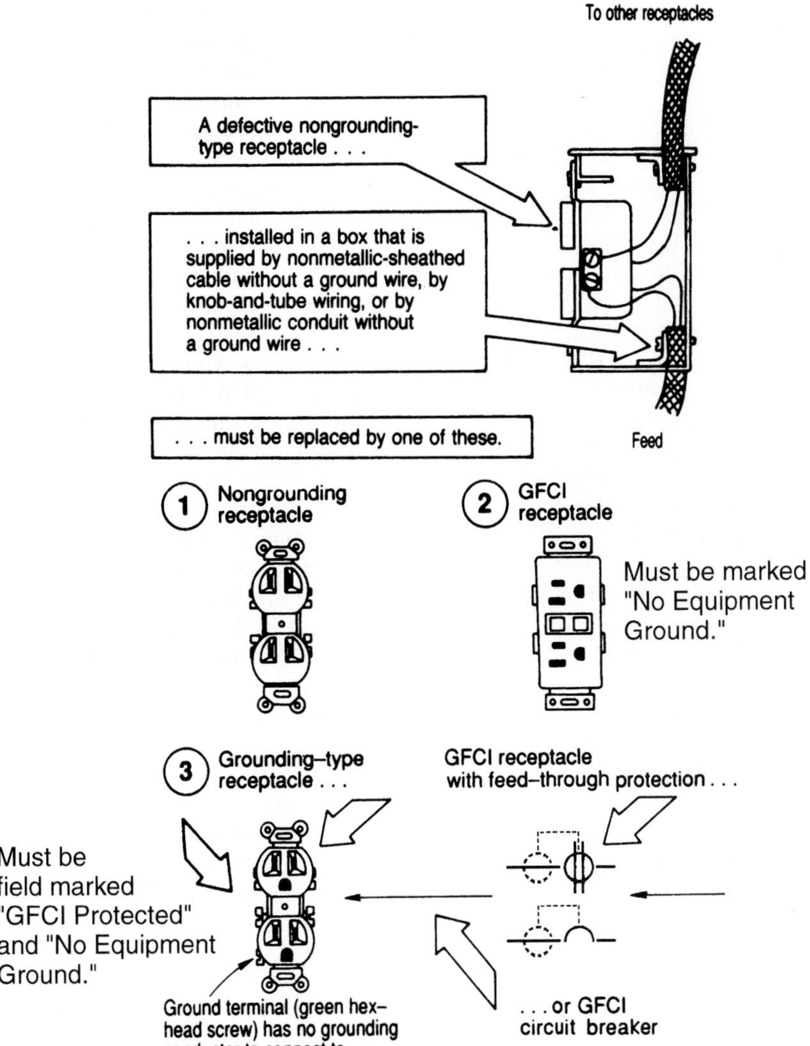
To other receptacles
A defective nongrounding-type receptacle . . .
. . . installed in a box that is supplied by nonmetallic-sheathed cable without a ground wire, by knob-and-tube wiring, or by nonmetallic conduit without a ground wire . . .
Feed
. . . must be replaced by one of these.
1 Nongrounding receptacle
2 GFCI receptacle
Must be marked "No Equipment Ground."
3 Grounding–type receptacle . . .
GFCI receptacle with feed–through protection . . .
Must be field marked "GFCI Protected" and "No Equipment Ground."
Ground terminal (green hex–head screw) has no grounding conductor to connect to.
. . . or GFCI circuit breaker

Fig. 210-22. A nongrounding-type receptacle, a GFCI-type receptacle, or GFCI protection, must always be used when replacing a nongrounding receptacle in any case where the box does not contain an equipment grounding conductor. [Sec. 210-7(d).]

As noted above, GFCI protection is required by Sec. 210-8(b) in bathrooms of all occupancies. This includes commercial office buildings, industrial facilities, schools, dormitories, theaters—bathrooms in ALL non-dwelling occupancies. The rule here extends the same protection of GFCI breakers and receptacles to bathrooms in all nondwelling-type occupancies as for receptacles in bathrooms of dwelling units.

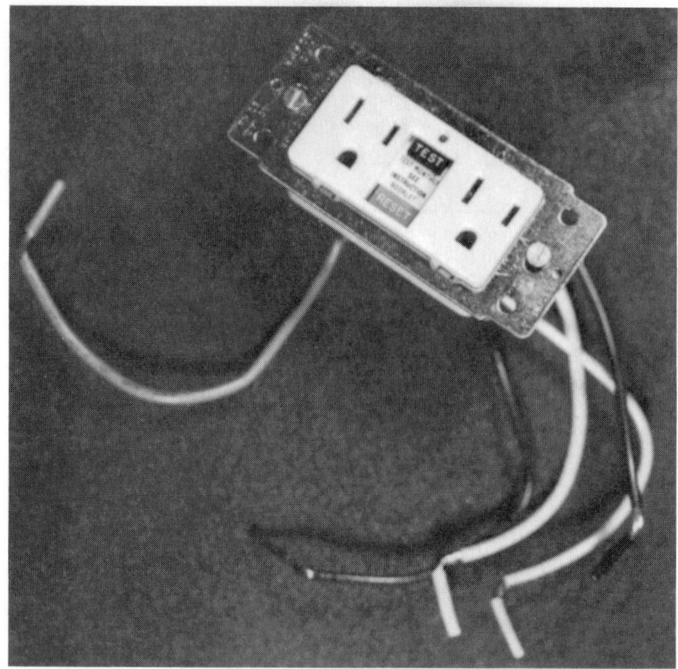

Fig. 210-23. A GFCI receptacle may replace existing nongrounding device in box without an equipment grounding means and may feed downstream receptacles. [Sec. 210-7(d).]

The rule of Sec. 210-8(a)(2) requiring GFCI protection in garages applies to both attached garages and detached (or separate) garages associated with "dwelling units"—such as one-family houses or multifamily houses where each unit has its own garage. Section 210-52 requires at least one receptacle in an attached garage and in a detached garage if electric power is run out to the garage.

Part **(a)(2)** of Sec. 210-8 says that 15- and 20-A receptacles in tool huts, workshops, storage sheds, and other "unfinished accessory buildings" at dwellings must be GFCI-protected. In addition to requiring GFCI protection for receptacles installed in a garage at a dwelling unit, other outbuildings, such as tool sheds and the like, must have GFCI protection for all 15- and 20-A, 125-V receptacles that are at "grade level." Curiously enough, this CMP eliminated the definition for "direct grade level access" in part **(a)(4),** but uses a version of that phrase to identify those receptacles covered by this rule. From available documentation it can be determined that the qualifier "grade level portions" is supposed to indicate that the rule covers receptacles that are accessible from the ground or first floor of such a structure.

It should be noted that this rule in no way requires a receptacle to be installed in such a building. But, where a 15- or 20-A, 125-V receptacle *is* installed in a location that is accessible from the ground floor of "unfinished accessory buildings used for storage or work areas," it must be GFCI-protected. The same exceptions that apply to garages apply to these structures.

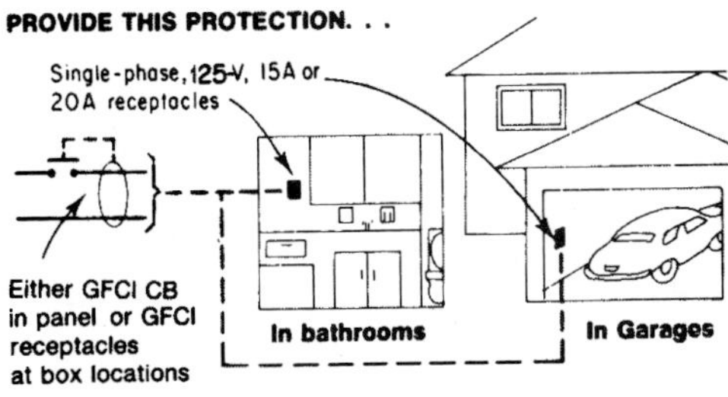

If a receptacle is installed to supply a freezer in a garage, it does *not* have to be GFCI protected . . .
. . . BUT AT LEAST ONE OTHER RECEPTACLE must be installed in the garage for using hand-held electric tools or appliances, and it MUST be GFCI-protected.

CEILING-MOUNTED RECEPTACLE for plugging in the power cord from an electric garage-door operator is "not readily accessible," and Exception No. 1, therefore, permits it to be non-GFCI protected.

Fig. 210-24. GFCI protection is required for receptacles in garages as well as in bathrooms. [Sec. 210-8(a)(2).]

Although the basic rule of part **(a)(2)** requires *all* 125-V, single-phase, 15- or 20-A receptacles installed in garages or sheds, etc., to have ground-fault circuit-interrupter protection, Exception No. 1 excludes a ceiling-mounted receptacle that is used solely to supply a cord-connected garage-door operator. And, as worded, the Exception excludes *any* receptacles that "are not readily accessible"—that is, any receptacle that requires use of a portable ladder or a chair to get up to it.

From Exception No. 2 to part **(a)(2)**, a receptacle in garages or accessory buildings for "dedicated appliances"—those that are put in place and not normally moved because of their weight and size—are excluded from the need for GFCI because they will not normally be used to supply hand-held cord-connected appliances (lawn mowers, hedge trimmers, etc.) that are used outdoors.

Note: Any receptacle in a garage that is excluded from the requirement for GFCI protection, as noted in the two Exceptions of this section, may not be considered as the receptacle that is required by Sec. 210-52(g) to be installed in an attached or detached garage with electric power. Thus if a non-GFCI-protected receptacle is installed in a garage at the ceiling for connection of a door operator or if a non-GFCI-protected receptacle is installed in a garage for a freezer, at least one additional receptacle must be installed in the garage to satisfy Sec. 210-52(g) for general use of cord-connected appliances, and such a receptacle *must* be installed not over 5½ ft (1.67 m) above the floor and *must* have GFCI protection (either in the panel CB or in the receptacle itself).

Application of the two Exceptions of Sec. 210-8(a)(2) may prove troublesome, because receptacle outlets are most commonly installed in a garage during construction of a house and before it is known what appliances might be used. Under such conditions, GFCI receptacles would be required because "dedicated space" appliances are not in place. Then if, say, a freezer is later installed in the garage, the GFCI receptacle could be replaced with a non-GFCI device. But such replacement would not be acceptable for a receptacle that is the only one in the garage, because such a receptacle is required by Sec. 210-52(g) and is not subject to Exception No. 2 of Sec. 210-8(a).

Part **(a)(3)** of Sec. 210-8, on outdoor receptacles, requires GFCI protection of all 125-V, single-phase, 15- and 20-A receptacles installed "outdoors" at dwelling units. Because hotels, motels, and dormitories are not "dwelling units" in the meaning of the **Code** definition, outdoor receptacles at such buildings do not require GFCI protection. The rule specifies that such protection of outdoor receptacles is required for *all* receptacles outdoors at dwellings (Fig. 210-25). The phrase "direct grade level access" has been deleted from part **(a)(3)**. Because the qualifier "grade level access" has been deleted, apartment units constructed above ground level would need GFCI protection of receptacles installed outdoors on balconies. Likewise, GFCI protection would be required for any outdoor receptacle installed on a porch or other raised part of even a one-family house even though there is no "grade-level access" to the receptacle, as in the examples of Fig. 210-25.

The **Code** has long required GFCI protection for outdoor receptacles at dwelling units, but only those that had "direct grade level access." That qualifier, as well as its definition, has been eliminated from the 1996 **Code**, which makes GFCI protection mandatory for *all* outdoor receptacles at dwelling units, regardless of mounting height. Previously exempted locations—such as on balconies and decks that were greater than 6½ ft (1.98 m) above grade—are now covered by this rule and must have GFCI protection.

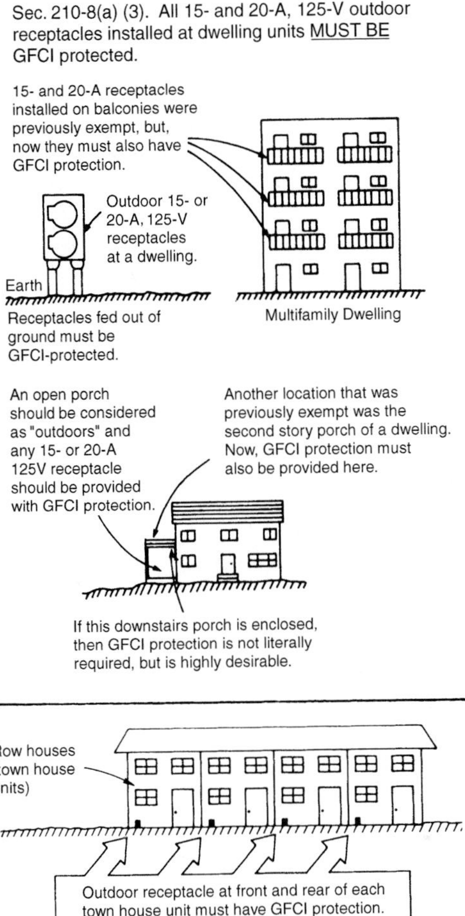

Sec. 210-8(a) (3). All 15- and 20-A, 125-V outdoor receptacles installed at dwelling units MUST BE GFCI protected.
15- and 20-A receptacles installed on balconies were previously exempt, but, now they must also have GFCI protection.
Outdoor 15- or 20-A, 125-V receptacles at a dwelling.
Earth
Receptacles fed out of ground must be GFCI-protected.
Multifamily Dwelling
An open porch should be considered as "outdoors" and any 15- or 20-A 125V receptacle should be provided with GFCI protection.
Another location that was previously exempt was the second story porch of a dwelling. Now, GFCI protection must also be provided here.
If this downstairs porch is enclosed, then GFCI protection is not literally required, but is highly desirable.
Row houses (town house units)
Outdoor receptacle at front and rear of each town house unit must have GFCI protection.
In short, at dwellings ALL 15- and 20-A, 125-V receptacles installed outdoors must be GFCI protected.

Fig. 210-25. For dwelling units, *all* outdoor receptacles require GFCI protection. [Sec. 210-8(a)(3).]

The only Exception to Sec. 210-8(a)(3) is for 15- and 20-A, 125-V receptacles that are installed to supply snow melting and deicing equipment in accordance with Art. 426. Such a receptacle does *not* require GFCI protection provided it is installed on a dedicated circuit and in an inaccessible location. Under those circumstances to supply deicing and snow melting equipment only, GFCI protection may be omitted (Fig. 210-25).

According to the rule of Sec. 210-8(a)(4) and (5), all 125-V, single-phase, 15- and 20-A receptacles installed in crawl spaces at or below grade and/or in unfinished basements must be GFCI-protected. This is intended to apply only to those basements or portions thereof that are unfinished (not habitable). Sec.

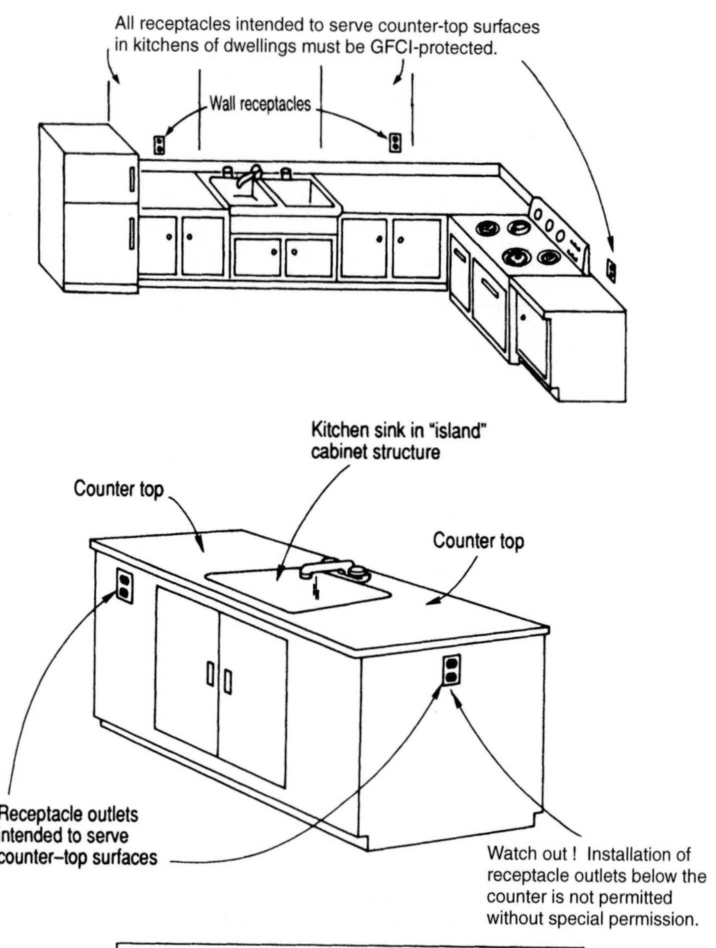
All receptacles intended to serve counter-top surfaces in kitchens of dwellings must be GFCI-protected.
Wall receptacles
Kitchen sink in "island" cabinet structure
Counter top
Counter top
Receptacle outlets intended to serve counter–top surfaces
Watch out ! Installation of receptacle outlets below the counter is not permitted without special permission.
NOTE: These same receptacles would require GFCI protection if the "island" contained a range top, not a sink.

Fig. 210-26. GFCI protection must be provided for receptacles in kitchen. Receptacles in face of island cabinet structure in kitchen, if permitted, must be GFCI-protected. [Sec. 210-8(a)(6).]

210-52(g) requires that at least one receptacle outlet must be installed in the basement of a one-family dwelling, in addition to any installed for laundry equipment. This rule applies to basements of all one-family houses but not to apartment houses, hotels, motels, dormitories, and the like.

Sec. 210-8(a)(5) Exception No. 1 permits receptacles that are not readily accessible to be installed without GFCI protection. And Exception No. 2 allows the use of a single receptacle for one appliance or a duplex receptacle for two appliances without GFCI protection.

According to part **(a)(6)**, GFCI protection is required for all 125-V, single-phase, 15- or 20-A receptacles installed in any kitchen sink or within 6 ft (1.83 m) wet bar sink "to serve the counter-top surfaces." This will provide GFCI-protected receptacles for appliances used on counter tops in kitchens or bars in dwelling units. This would include any receptacles installed in the vertical surfaces of a kitchen "island" that includes counter-top surfaces with or without additional hardware such as a range, grill, or even a sink. Because so many kitchen appliances are equipped with only 2-wire cords (toasters, coffee makers, electric fry pans, etc.), their metal frames are not grounded and are subject to being energized by internal insulation failure, making them shock and electrocution hazards. Use of such appliances close to any grounded metal—the range, a cooktop, a sink—creates the strong possibility that a person might touch the energized frame of such an appliance and at the same time make contact with a faucet or other grounded part—thereby exposing the person to shock hazard. Use of GFCI receptacles within arm's reach of the sink [6 ft (1.83 m) to either side of the sink] will protect personnel by opening the circuit under conditions of dangerous fault current flow through the person's body (Fig. 210-26).

Part **(a)(7)** requires that 15- and 20-A, 125-V counter-top receptacles installed within 6 ft (1.83 m) of wet bar sink must be GFCI protected.

Although the requirement for GFCI protection of *kitchen* counter-top receptacles is no longer based on their distance from the kitchen sink, the 6-ft (1.83 m) limitation is still the determining factor with wet bar counter-top receptacles. Any 15- or 20-A receptacles installed to serve countertops that are within 6 ft (1.83 m) from the outside edge of a wet bar sink must be provided with GFCI protection.

Although the rule of Sec. 210-8(a)(6) and (7) only applies to counter-top receptacles in kitchens and within 6 ft (1.83 m) of wet bar sinks, voluntary application in commercial kitchens and at wet bars in taverns and the like would certainly enhance safety in such locations and should be considered.

210-9. Circuits Derived from Autotransformers. The top of Fig. 210-27 shows how a 110-V system for lighting may be derived from a 220-V system by means of an autotransformer. The 220-V system either may be single phase or may be one leg of a 3-phase system. That hookup complies with the basic rule. In the case illustrated the "supplied" system has a grounded wire solidly connected to a grounded wire of the "supplying" system: 220-V single-phase system with one conductor grounded.

Autotransformers are commonly used to supply reduced voltage for starting induction motors.

Exception No. 1 permits the use of an autotransformer in existing installations for an individual branch circuit without connection to a similar identified grounded conductor where transforming from 208 to 240 V or vice versa (see Fig. 210-27). Typical applications are with cooking equipment, heaters, motors, and air-conditioning equipment. For such applications transformers are commonly used. This has been a long-established practice in the field of voltage ranges where a hazard is not considered to exist.

Buck or boost transformers are designed for use on single- or 3-phase circuits to supply 12/24-V or 16/32-V secondaries with a 120/240-V primary. When connected as

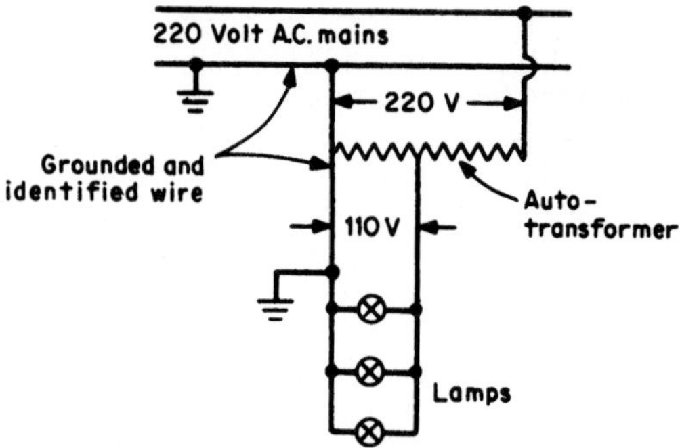

Autotransformer used to derive a 2-wire 110-V system for lighting from a 220-V power system. (Sec. 210-9.)

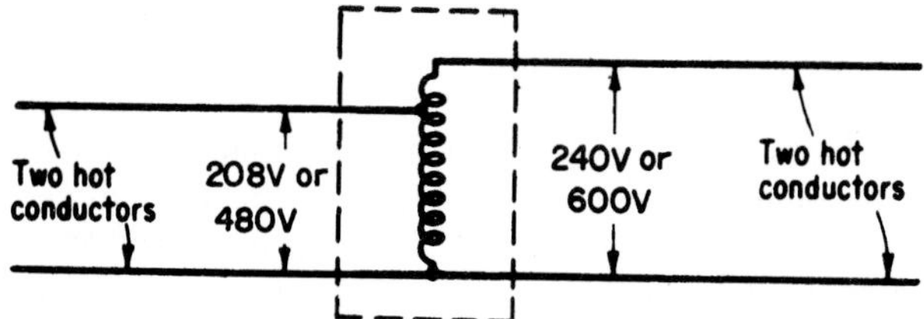

Fig. 210-27. Autotransformers with and without grounded conductors are recognized. (Sec. 210-9, Exceptions No. 1 and No. 2.)

autotransformers the kVA load they will handle is large in comparison with their physical size and relative cost.

Exception No. 2 permits 480- to 600-V or 600- to 480-V autotransformers without connection to grounded conductor—but only for industrial occupancies.

210-10. Ungrounded Conductors Tapped from Grounded Systems. This section permits use of 2-wire branch circuits tapped from the outside conductors of systems, where the neutral is grounded on 3-wire DC or single-phase, 4-wire, 3-phase, and 5-wire 2-phase systems.

Figure 210-28 illustrates the use of unidentified 2-wire branch circuits to supply small motors, the circuits being tapped from the outside conductors of a 3-wire DC or single-phase system and a 4-wire 3-phase wye system.

All poles of the disconnecting means used for branch circuits supplying permanently connected appliances must be operated at the same time. This requirement applies where the circuit is supplied by either circuit breakers or switches.

In the case of fuses and switches, when a fuse blows in one pole, the other pole may not necessarily open, and the requirement to "manually switch together" involves only the manual operation of the switch. Similarly, when a pair of circuit breakers is connected with handle ties, an overload on one of the conductors with the return circuit through the neutral may open only one of the

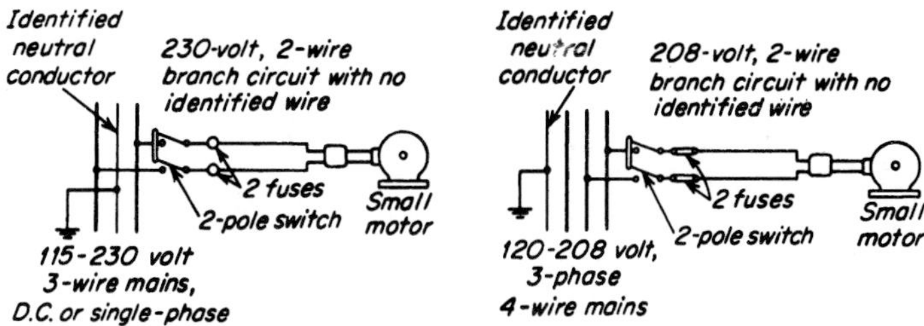

Fig. 210-28. Tapping circuits of ungrounded conductors from the hot legs of grounded systems. (Sec. 210-10.)

circuit breakers; but the manual operation of the pair when used as a disconnecting means will open both poles. The words "manually switch together" should be considered as "operating at the same time," i.e., during the same operating interval, and apply to the equipment used as a disconnecting means and not as an overcurrent protective device.

Circuit breakers with handle ties are, therefore, considered as providing the disconnection required by this section. The requirement to "manually switch together" can be achieved by a "master handle" or "handle tie" since the operation is intended to be effected by manual operation. The intent was not to require a common trip for the switching device but to require that it have the ability to disconnect ungrounded conductors by one movement of the hand. For service disconnecting means see Sec. 230-71.

210-19. Conductors—Minimum Ampacity and Size. In past NEC editions, the basic rule of this section has said—and *still does* say—that the conductors of a branch circuit must have an ampacity that is not less than the maximum current load that the circuit will supply. Obviously, that is a simple and straightforward rule to assure that the conductors are not operated overloaded. But up to the 1981 NEC, the rule further required that branch-circuit conductors have "an ampacity of not less than the rating of the branch circuit" (Fig. 210-29). Because Sec. 210-3 clearly notes that the amp rating of a multioutlet circuit (typical lighting and appliance branch circuits) is set by the rating of the circuit protective device, the conductors of the circuit were, therefore, always required to have an ampacity of not less than the rating of the fuses or circuit-breaker poles that protect them. And there were no exceptions to that rule.

The wording of this rule now requires the circuit conductors to have an ampacity not less than "the rating of the branch circuit" *only* for a multioutlet branch circuit that supplies receptacles for cord- and plug-connected loads. The concept here is that plug receptacles provide for random, indeterminate loading of the circuit; and, by matching conductor ampacity to the amp rating of the circuit fuse or CB, overloading of the conductors can be avoided. But for multioutlet branch circuits that supply fixed outlets—such as lighting fixture outlets or hard-wired connections to electric heaters or other appliances—it is acceptable to have a condition where the conductor ampacity is adequate for the load current but, where there is no standard rating of protective device that corresponds to the conductor ampacity, the circuit fuse or CB rating is the next

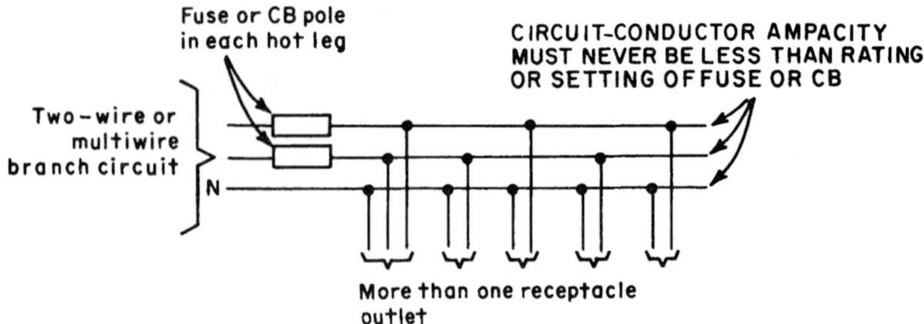

Fig. 210-29. This is the basic rule for any multioutlet branch circuit supplying one or more receptacles. [Sec. 210-19(a).]

higher standard rating of protective device above the ampacity value of the conductor.

For multioutlet branch circuits (rated at 15, 20, 30, 40, or 50 A), the ampacities of conductors usually correspond to standard ratings of protective devices when there is only one circuit in a cable or conduit. Standard rated protective devices of 15, 20, 30, 40, or 50 A can be readily applied to conductors that have corresponding ampacities from Tables 310-16 through 310-19 and their footnotes—i.e., 15 A for No. 14, 20 A for No. 12, 30 A for No. 10, and 40 A for No. 8, with 55-A rated No. 6 used for a 50-A circuit. But when circuits are combined in a single conduit so that more than three current-carrying conductors are involved, the ampacity derating factors of Note 8 to Table 310-16 often result in reduced ampacity values that do not correspond to standard fuse or CB ratings. It is to such cases that the rule of Sec. 210-19(a) may be applied.

For instance, assume that two 3-phase, 4-wire multioutlet circuits supplying fluorescent lighting are run in a single conduit. Two questions arise: (1) How much load current may be put on the conductors? and (2) What is the maximum rating of overcurrent protection that may be used for each of the six hot legs?

The eight wires in the single conduit (six phases and two neutrals) must be taken as eight conductors when applying Note 8 of Table 310-16 because the neutrals to electric-discharge lighting carry harmonic currents and must be counted as current-carrying conductors [Note 10(c) of Table 310-16]. Note 8 then shows that the No. 14 wires must have their ampacity reduced to 70 percent (for 7 to 9 wires) of the 20-A ampacity given in Table 310-16 for No. 14 TW. With the eight No. 14 wires in the one conduit, then, each has an ampacity of 0.7 × 20, or 14 A. Because Sec. 210-19(a) requires circuit wires to have an ampacity at least equal to the rating of the circuit fuse or CB if the circuit is supplying receptacles, use of a 15-A fuse or 15-A circuit breaker would *not* be acceptable in such a case because the 14-A ampacity of each wire *is* less than "the rating of the branch circuit" (15 A). *But* because the circuits here are supplying fixed lighting outlets, as stated in the original assumption, Sec. 210-19(a) would accept the 15-A protection on wires with 14-A ampacity. In such a case, it is not only necessary that the design load current on each phase must not exceed 14 A, but if the lighting load *is* continuous (operating steadily for three or more hours), the load on each 15-A CB or fuse must not exceed 0.8 × 15, or 12 A [as required by Sec. 210-22(c) and Sec. 384-16(c)].

Refer to the discussion of Sec. 210-22(c) and the discussion of ampacity and derating under Sec. 310-15.

In part **(b)**, the rule also calls for the same approach to sizing conductors for branch circuits to household electric ranges, wall-mounted ovens, counter-mounted cooking units, and other household cooking appliances (Fig. 210-30).

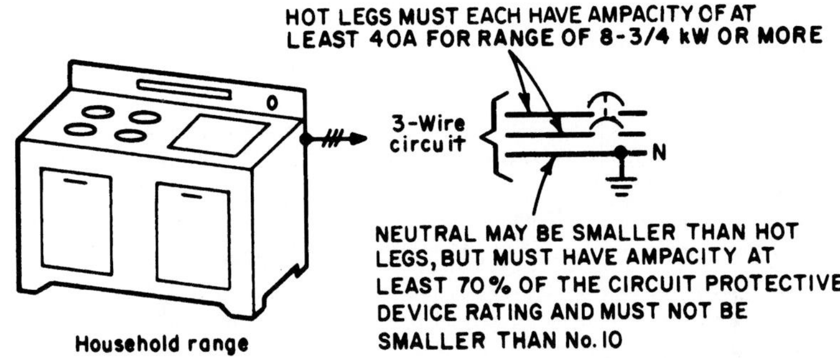
HOT LEGS MUST EACH HAVE AMPACITY OF AT LEAST 40A FOR RANGE OF 8-3/4 kW OR MORE
3-Wire circuit
N
NEUTRAL MAY BE SMALLER THAN HOT LEGS, BUT MUST HAVE AMPACITY AT LEAST 70% OF THE CIRCUIT PROTECTIVE DEVICE RATING AND MUST NOT BE SMALLER THAN No. 10
Household range

Fig. 210-30. Sizing circuit conductors for household electric range. [Sec. 210-19(b).]

The maximum demand for a range of 12-kW rating or less is sized from NEC Table 220-19 as a load of 8 kW. And 8,000 W divided by 230 V is approximately 35 A. Therefore, No. 8 conductors with an ampacity of 40 A may be used for the range branch circuit.

On modern ranges the heating elements of surface units are controlled by five-heat unit switches. The surface-unit heating elements will not draw current from the neutral unless the unit switch is in one of the low-heating positions. This is also true to a greater degree as far as the oven-heating elements are concerned, so the maximum current in the neutral of the range circuit seldom exceeds 20 A. Because of that condition, Exception No. 2 permits a smaller-size neutral than the ungrounded conductors, but not smaller than No. 10.

A reduced-size neutral for a branch circuit to a range, wall-mounted oven, or cook-top must have ampacity of not less than 70 percent of the circuit rating, which is determined by the current rating or setting of the branch-circuit protective device. This is a change from previous wording that required a reduced neutral to have an ampacity of at least 70 percent of "the ampacity of the ungrounded conductors." Under that wording, a 40-A circuit (rating of protective device) made up of No. 8 TW wires for the hot legs could use a No. 10 TW neutral—because its 30-A ampacity is at least 70 percent of the 40-A ampacity of a No. 8 TW hot leg (0.7 × 40 = 28 A). But if No. 8 THHN (55-A ampacity) is used for the hot legs with the same 40-A protected circuit, the neutral ampacity would have to be at least 70 percent of 55 A (0.7 × 55 = 38.5 A) and a No. 10 TW (30 A) or a No. 10 THW (35 A) could not be used. The new wording bases neutral size at 70 percent of the protective-device rating (0.7 × 40 A = 28 A), thereby permitting any of the No. 10 wires to be used, and does not penalize use of higher-temperature wires (THHN) for the hot legs.

Exception No. 1 permits taps from electric cooking circuits (Fig. 210-31). Because Exception No. 1 says that taps on a 50-A circuit must have an ampacity of at least 20 A, No. 14 conductors—which have an ampacity of 20 A in Table 310-16—may be used.

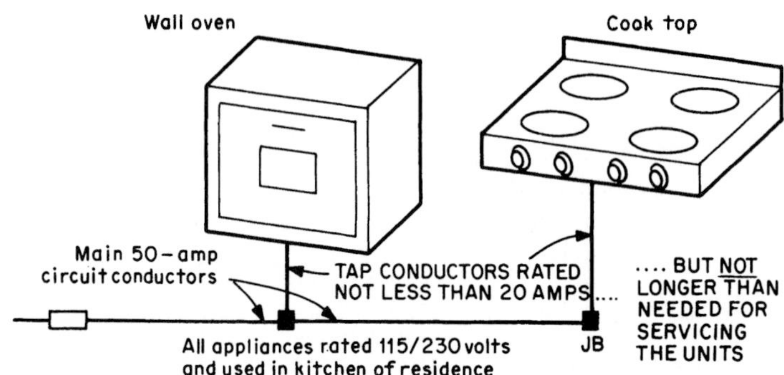

Fig. 210-31. Tap conductors may be smaller than wires of cooking circuit. [Sec. 210-19(b), Exception No. 1.]

Exception No. 1 applies to a 50-A branch circuit run to a counter-mounted electric cooking unit and wall-mounted electric oven. The tap to each unit must be as short as possible and should be made in a junction box immediately adjacent to each unit. The words "no longer than necessary for servicing the appliance" mean that it should be necessary only to move the unit to one side in order that the splices in the junction box be accessible.

Section 210-19(c) sets No. 14 as the smallest size of general-purpose circuit conductors. But tap conductors of smaller sizes are permitted as explained in Exceptions No. 1 and No. 2 (Fig. 210-32). No. 14 wire, not longer than 18 in. (457 mm), may be used to supply an outlet unless the circuit is a 40- or 50-A branch circuit, in which event the minimum size of the tap conductor must be No. 12.

The wording of Sec. 210-19(c), Exception No. 1, specifically excludes receptacles from being installed as indicated here because they are not tested for such use. That is, when tested for listing, receptacles are not evaluated using 18-in. (457-mm) taps of the size specified in Table 210-24 and protected as indicated by Sec. 210-19, Exception No. 1. As a result, receptacles have been prohibited from being supplied by tap conductors, as is permitted by this exception for other loads.

It should be noted that the supply of receptacle outlets as indicated by Sec. 210-19(c), Exception No. 1, is also at odds with the Canadian Electrical Code. Therefore, excluding receptacle outlets from this application also serves the organized attempt to harmonize standards with Canada.

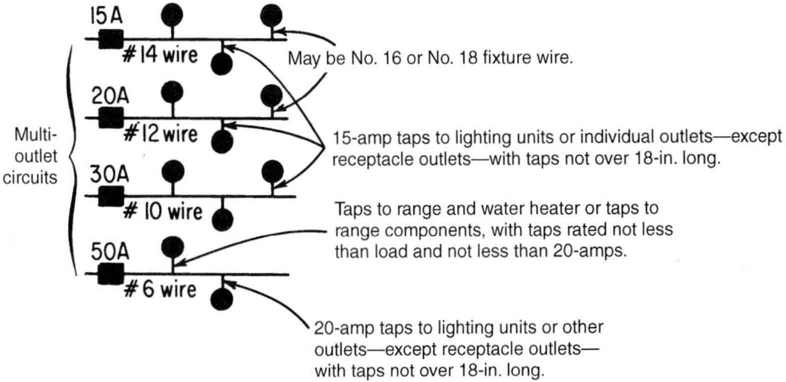

Fig. 210-32. Tap conductors may be smaller than circuit wires. [Sec. 210-19(c), Exception Nos. 1 and 2.]

210-20. Overcurrent Protection. According to the basic Code rule of this section, the rating or setting of an overcurrent device in any branch circuit must not exceed the current-carrying capacity of the circuit conductor or may be the next higher value of overcurrent device where conductor ampacity does not match the rating of a standard fuse or CB. Section 240-3 applies to the rating of overcurrent protection. Figure 210-33 shows the basic rules that apply to use of overcurrent protection for branch circuits. (Section 240-2 designates other Code articles that present data and regulations on overcurrent protection for branch circuits to specific types of equipment.)

Branch-circuit taps—as covered in Secs. 210-19 and 210-20—are considered protected by the branch-circuit overcurrent devices, even where the rating or setting of the protective device is greater than the amp rating of the tap conductors, fixture wires, or cords.

When only three No. 12 TW or THW conductors of a branch circuit are in a conduit, each has a load-current rating of 20 A (see the footnote to Table 310-16, Art. 310) and must be protected by a fuse or CB rated not over 20 A. This satisfies Sec. 210-19, which requires branch-circuit conductors to have an ampacity not less than the rating of the branch circuit—and Sec. 210-3 notes that the rating of a branch circuit is established by the rating of the protective device. It also satisfies Sec. 210-20, which says:

> Branch-circuit conductors . . . shall be protected by overcurrent protective devices having a rating or setting . . . *not exceeding that specified in Section 240-3* for conductors. . . .

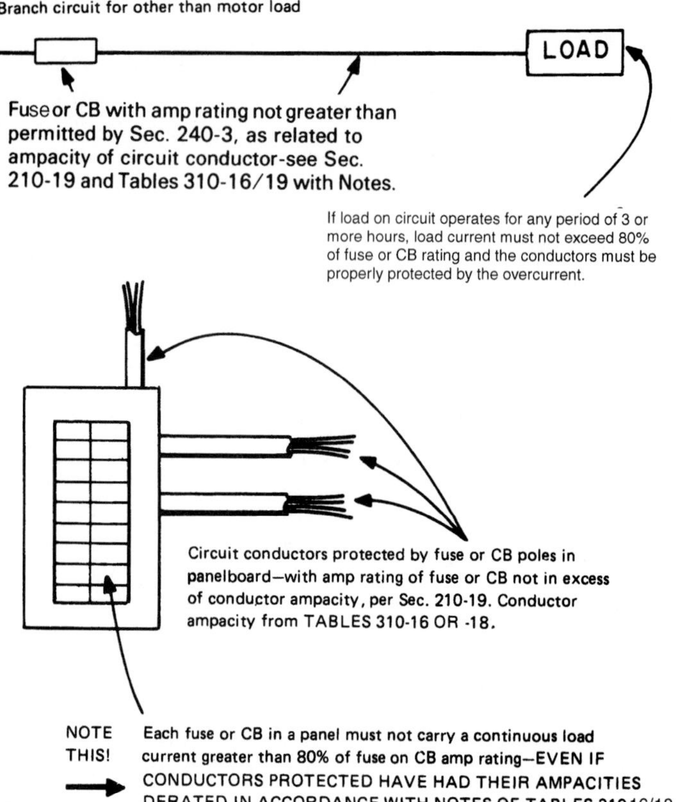
Branch circuit for other than motor load
LOAD
Fuse or CB with amp rating not greater than permitted by Sec. 240-3, as related to ampacity of circuit conductor-see Sec. 210-19 and Tables 310-16/19 with Notes.
If load on circuit operates for any period of 3 or more hours, load current must not exceed 80% of fuse or CB rating and the conductors must be properly protected by the overcurrent.
Circuit conductors protected by fuse or CB poles in panelboard—with amp rating of fuse or CB not in excess of conductor ampacity, per Sec. 210-19. Conductor ampacity from TABLES 310-16 OR -18.
NOTE THIS!
Each fuse or CB in a panel must not carry a continuous load current greater than 80% of fuse on CB amp rating—EVEN IF CONDUCTORS PROTECTED HAVE HAD THEIR AMPACITIES DERATED IN ACCORDANCE WITH NOTES OF TABLES 310-16/18

Fig. 210-33. Branch-circuit protection involves a number of rules (Sec. 210-20.)

The basic rule of Section 240-3 says:

> Conductors . . . shall be protected against overcurrent *in accordance with their ampacities* as specified in Section 310-15.

That rule says that conductors are required to be protected at a current value indicated by the table and its accompanying notes, such as Note 8, which reduces ampacities from the table values.

In Table 310-16, which applies to conductors in raceways and in cables and covers the majority of conductors used in electrical systems for power and light, the ampacities for sizes No. 14, No. 12, and No. 10 are particularly significant because copper conductors of those sizes are involved in the vast majority of branch circuits in modern electrical systems. No. 14 has an ampacity of 20, No. 12 has an ampacity of 25, and No. 10 has an ampacity of 30. The typical impact of that on circuit makeup and loading is as follows:

No. 12 TW or THW copper is shown to have an ampacity of 25; and based on the general UL requirement that equipment terminals be limited to use with conductors loaded not in excess of 60°C ampacities for wires up to No. 1 AWG, No. 12 THHN or XHHW copper conductors must also be treated as having a 25-A continuous rating (the ampacity of 60°C No. 12) and *not* 30 A, as shown in Table 310-16. *But,* the footnote to Table 310-16 limits all No. 12 copper to not over 20-A load current by requiring that they be protected at not more than 20 A.

The ampacity of 25 A for No. 12 TW and THW copper wires interacts with Note 8 to Tables 310-16 to 310-19 where there are, say, six No. 12 TW current-carrying wires for the phase legs of two 3-phase, 4-wire branch circuits in one conduit supplying, say, receptacle loads. In such a case, the two neutrals of the branch circuits do not count in applying Note 8, and only the six phase legs are counted to determine how much all conductors must have their ampacities *de-rated* to the "Percent of Values in Tables"—as stated at the top of the table in Note 8. In the case described here, that literally means that each No. 12 phase leg may be used at an ampacity of 0.8 × 25, or 20 A. And the footnote to Table 310-16 would require use of a fuse or CB rated not over 20 A to protect each No. 12 phase leg. Each No. 12 would be protected at the current value that represents the maximum I^2R heat input that the conductor insulation can withstand. In that example, the derated ampacity of the No. 12 conductors (20 A) is not in excess of "the rating of the branch circuit"—that is, the 20-A rating of the fuse or CB protecting the circuit. Thus, Sec. 210-19(a) is completely and readily satisfied because the ampacity and protective device rating came out to the same value. Thus the circuits described could be used for supplying receptacles and/or fixed-load outlets. The only other possible qualification is that Sec. 384-16(c) would require the load current on each of the phase legs to be further limited to no more than 80 percent of the 20-A rating of the overcurrent device—that is, to 16 A—if the load current is "continuous" (operates steadily for 3 hr or more), a condition not likely for receptacle-fed loads. Refer to further analysis in Sec. 310-15.

210-21. Outlet Devices. Specific limitations are placed on outlet devices for branch circuits: Lampholders must not have a rating lower than the load to be served; and lampholders connected to circuits rated over 20 A must be heavy-duty type (that is, rated at least 660 W if it is an "admedium" type and at least 750 W for other types). Because fluorescent lampholders are not of the heavy-duty type, this excludes the use of fluorescent luminaires on 30-, 40-, and 50-A circuits. The intent is to limit the rating of lighting branch circuits supplying fluorescent fixtures to 20 A. The ballast is connected to the branch circuit rather than the lamp, but by controlling the lampholder rating, a 20-A limit is established for the ballast circuit. Most lampholders manufactured and intended for use with electric-discharge lighting for illumination purposes are rated less than 750 W and are not classified as heavy-duty lampholders. If the luminaires are individually protected, such as by a fuse in the cord plug of a luminaire cord connected to, say, a 50-A trolley or plug-in busway, some inspectors have permitted use of fluorescent luminaires on 30-, 40-, and 50-A circuits. But such protection in the cord plug or in the luminaire is supplementary (Sec. 240-10), and branch-circuit protection of 30-, 40-, or 50-A rating would still exclude use of fluorescent fixtures according to Sec. 210-21(a).

Section 210-21(b) contains two paragraphs of importance. Part **(b)(1)** reads: "A **single** receptacle installed on an individual branch circuit shall have an ampere rating of not less than that of the branch circuit." Since the branch-circuit overcurrent device determines the branch-circuit rating (or classification), a single receptacle (not a duplex receptacle) supplied by an individual branch circuit cannot have a rating *less than* the branch-circuit overcurrent device, as shown in Fig. 210-34.

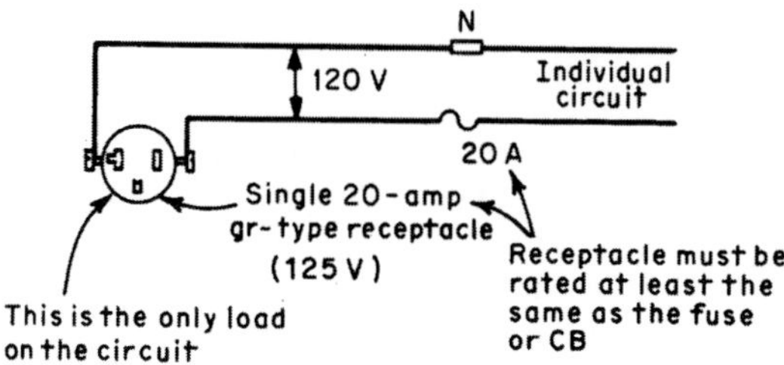

Fig. 210-34. Receptacle amp rating must *not* be less than circuit protection rating for an individual circuit. [Sec. 210-21(b).]

Receptacles must have ratings at least equal to the load. On circuits having two or more outlets, receptacles shall be rated as follows:

- On 15-A circuits—not over 15-A rating
- On 20-A circuits—15- or 20-A rating
- On 30-A circuits—30-A rating
- On 40-A circuits—40- or 50-A rating
- On 50-A circuits—50-A rating

For multioutlet branch circuits rated over 50 A, as permitted under the limited conditions described in the discussion on the Exception to Sec. 210-3, every receptacle must have a rating not less than the branch-circuit rating.

210-22. Maximum Loads. Part **(c)** of this section says that a branch circuit supplying a continuous current load must have a rating (the rating of the circuit fuse or CB) not less than 125 percent of the continuous-current load. The concept is the same as that described in Sec. 220-3(a). The idea of the rule is that 125 percent of a total continuous-load current gives a circuit rating such that the continuous-load current does "not exceed 80 percent of the rating of the branch circuit." One is the reciprocal of the other.

Sec. 210-22(c) says that (for loads other than motor loads) the rating of a branch circuit must be at least equal to 1.25 times the continuous-load current when the load will constitute a continuous load—such as store lighting and similar loads that are on for periods of 3 or more hours.

Because multioutlet branch circuits, such as lighting circuits, are rated in accordance with the rating or setting of the circuit overcurrent protective device, this rule has the effect of saying that the rating of the protective device must equal at least 1.25 times the continuous-load current (Fig. 210-35). Where a circuit also supplies some noncontinuous load (not "on" for periods of 3 hr or more) in addition to continuous-load current, the branch circuit protective device must have a rating not less than the noncontinuous-load current plus 1.25 times the continuous-load current.

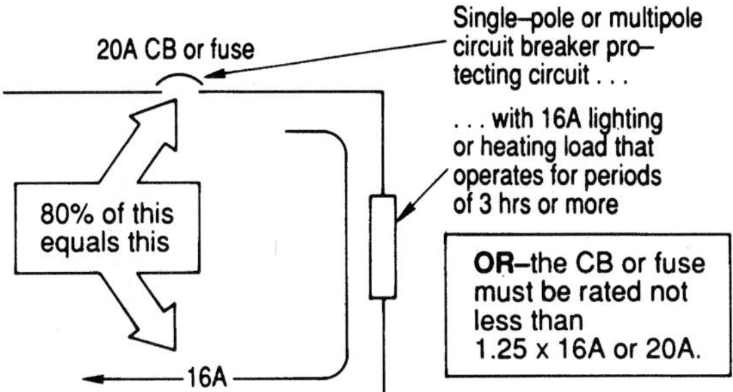

Fig. 210-35. Branch-circuit protective device must be rated not less than 125 percent of the continuous load current. [Sec. 210-22(c).]

Section 210-22(c) states that calculation of conductor ampacity, as well as the minimum rating for non-100-percent-rated protective devices, must be correlated with the nature of the load—that is, continuous or noncontinuous. In the 1993 and other recent editions of the NEC, there was a move afoot to make it clear that where the load to be supplied was continuous—that is, was energized for 3 hr or more—the *overcurrent (OC) device,* not the *conductor,* must be rated for the noncontinuous load plus 125 percent of the continuous load. By definition, when one determines a conductor's ampacity, that value *is* a continuous value. However, with OC devices, unless marked otherwise, the continuous load may not exceed 80 percent of the device's rating. To that end, the wording of this section, as well as the wording in Secs. 220-3(a) and 220-10(c), was modified to indicate that the rating of the OC device must not be less than 125 percent of the continuous plus the noncontinuous load. And, all references to sizing the conductor on a similar basis were eliminated. With the 1996 Code, all of that is out the window.

The rule of Sec. 210-22(c) now states that the OC device must never be rated less than the sum of the noncontinuous load plus 125 percent of the noncontinuous load, unless, of course, the OC device and its assembly are listed to carry a continuous load equal to their long-time rating. That part remains as it was in the past few editions of the Code. But, evaluating the minimum conductor size's minimum ampacity is a bit more involved.

Basically, there are two tests for establishing minimum conductor ampacity where the load is continuous or for a combination of continuous and noncontinuous load. First, the conductor size selected must have table ampacity that is at least equal to the noncontinuous load plus 125 percent of the continuous load prior to derating. Second, if derating is required, the conductor size selected must be properly protected by the OC device—selected as indicated above—after the required derating is applied.

The requirements spelled out here for branch circuits are repeated—word for word—in Secs. 220-3(a) and 220-10(b), which cover calculations for branch circuits and feeders, respectively. Application—in all its possible combinations—is best illustrated by analyzing a feeder supplying 100 A of discharge lighting, so even though the following example discusses a feeder, the same rules apply to analyzing minimum required ampacity for branch-circuit conductors.

question If a 3-phase, 4-wire circuit is supplying 100 A of electric discharge lighting in a commercial facility, what is (1) the minimum rating for a non-100-percent-rated OC device, and (2) the minimum conductor size permitted to serve this load?

answer (part 1) The minimum rating of OC device must be:

$$\begin{aligned}\text{Minimum OC Device Rating} &= \text{noncontinuous} + (1.25 \times \text{continuous})\\ &= 0 + (1.25 \times 100\text{ A})\\ &= 0 + 125\text{ A}\\ \text{Minimum OC Device Rating} &= 125\text{ A}\end{aligned}$$

answer (part 2) The minimum size conductor must be:

The conductor must have an ampacity that is equal to, or greater than, the noncontinuous load plus 125 percent of the continuous *before* any derating. From Table 310-16, we see that a No. 2 copper THHN conductor has an ampacity of 130 A. So, it passes the first test of acceptability. Next, it must be properly protected by the 125-A OC device after any required derating is applied. Here, an 80 percent derating is required for number of conductors because the load is electric discharge lighting, which is a nonlinear load, and, therefore, Note 10 requires the neutral to be counted as a fourth current-carrying conductor. According to the Note 8 table, the 130-A value shown in Table 310-16 must be derated to 80 percent (4 to 6 conductors) of 130 A Table 310-16 value.

$$\begin{aligned}\text{Conductor Ampacity for a THHN No. 2, copper} &= 130\text{ A} \times 0.80\\ \text{Conductor Ampacity for a THHN No. 2, copper} &= 104\text{ A}\end{aligned}$$

Now, after derating, the ampacity of the THHN No. 2 copper conductor is 104 A. But, a conductor with that ampacity is *not* properly protected by a 125-A-rated OC device. Therefore, the next conductor size, a THHN, No. 1 copper must be considered.

A quick analysis shows that: (1) the prederated table value passes the first test (150 A from Table 310-16), and (2) if derated to 80 percent of the table value, it passes the second test because it is properly protected by a 125-A-rated OC device (150 A × 0.80 = 120 A).

It should be noted that if, say, a 70 percent derating is required, the THHN No. 1 copper conductor's ampacity will be 105 A (150 A × 0.70), and the next larger conductor size, that is a THHN No. 1/0, copper, must be evaluated (Fig. 210-36).

In addition to all that, the rule of Sec. 110-14(c) and UL requirements regarding termination temperature limitations must be observed. For the No. 1 conductor derated at 80 percent, the final ampacity is 120 A, which is less than 75°C ampacity of 130, but more than the 60°C ampacity of 110 A shown in Table 310-16. But, even if the terminals on the OC device and its enclosure are not marked to indicate that use of conductors at the full 75°C ampacity is permitted, there is no problem here because the maximum current will be no more than 100 A, as given.

Minimum rating of overcurrent device = noncontinuous + (1.25 × continuous)
= 0 + (1.25 × 100 A)
= 125 A

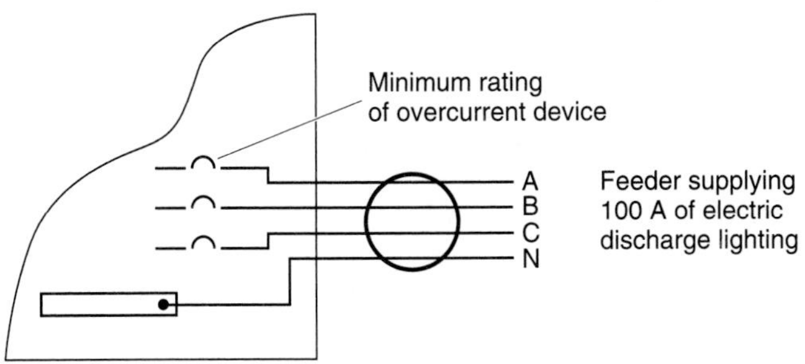

Minimum conductor size must pass two tests of acceptability: (1) pre-derating ampacity equal to, or greater than, the sum of noncontinuous + (1.25 × continuous) loads. (2) Ampacity must be such that conductor is protected per Sec. 240-3 after any derating.

(*a*)

Determine device rating (Noncontinuous + 125% of continuous) → Go to → Table 310-16 select conductor size → Go to → Equal to, or greater than, device rating? → Yes → Is derating required? → No → Minimum conductor size established

Equal to, or greater than, device rating? → No → Table 310-16 select conductor size

Is derating required? → Yes → Properly protected per Sec. 240-3 after derating? → Yes → Minimum conductor size established

Properly protected per Sec. 240-3 after derating? → No → Table 310-16 select conductor size

(*b*)

Fig. 210-36. The rules for sizing conductors where supplying continuous loads has been modified as indicated above (*a*). The "Flow Chart" (*b*) shows the decision making process in simplified form. First, establish the device rating—noncontinuous load plus 1.25 times the continuous load. Next find a conductor with a table ampacity equal to, or greater than, the device rating. If no derating is required, simply use that size conductor. Where derating is required, if the conductor's derated ampacity is properly protected by the OC device selected, use that size conductor. If not, go back to the ampacity tables and begin again.

Although the above-mentioned limitation applies only to loads other than motor loads, Sec. 384-16(c) says that "the total load on any overcurrent device located in a panelboard shall not exceed 80 percent of its rating where in normal operation the load will continue for 3 hours or more," which is the same rule stated reciprocally.

NOTE: In both of the above cases, neither the 125 percent of continuous load nor the 80 percent load limitation applies "where the assembly including the overcurrent device is approved for continuous duty at 100 percent of its rating."

It is very important to understand that the UL and Code rules calling for load limitation to 80 percent of the rating of the protective device are based on the inability of the protective device itself to handle continuous load without overheating. And, if a protective device has been designed, tested, and "listed" (such as by UL) for continuous operation at its full-load rating, then there is no requirement that the load current be limited to 80 percent of the breaker or fuse rating (or that the breaker or fuse rating must be at least 125 percent of continuous load). The 80 percent continuous limitation is not at all based on or related to conductor ampacity—which is a separate, independent determination.

Section 384-16(c) requires any CB or fuse in a panelboard to have its load limited to 80 percent; and only one exception is made for such protective devices in a panel: A continuous load of 100 percent is permitted only when the protective device assembly (the CB unit or fuses in a switch) is approved for 100 percent continuous duty. (And there are no such devices rated less than 600 A. UL has a hard and fast rule that any breaker not marked for 100 percent of continuous load must have its continuous load limited to 80 percent of its rating.)

210-23. Permissible Loads. A single branch circuit to one outlet or load may serve any load and is unrestricted as to amp rating. Circuits with more than one outlet are subject to NE Code limitations on use as follows: (The word "appliance" stands for any type of utilization equipment.)

1. Branch circuits rated 15 and 20 A may serve lighting units and/or appliances. The rating of any one cord- and plug-connected appliance shall not exceed 80 percent of the branch-circuit rating. Appliances fastened in place may be connected to a circuit serving lighting units and/or plug-connected appliances, provided the total rating of the fixed appliances fastened in place does not exceed 50 percent of the circuit rating (Fig. 210-37). *Example:* 50 percent of a 15-A branch circuit = 7.5 A. A room-air-conditioning unit fastened in place, with a rating not in excess of 7.5 A, may be installed on a 15-A circuit having two or more outlets. Such units may not be installed on one of the small appliance branch circuits required in Sec. 220-4(b).

   However, modern design provides separate circuits for individual fixed appliances. In commercial and industrial buildings, separate circuits should be provided for lighting and separate circuits for receptacles.

2. Branch circuits rated 30 A may serve fixed lighting units (with heavy-duty-type lampholders) in other than dwelling units or appliances in any occupancy. Any individual cord- and plug-connected appliance which draws more than 24 A may not be connected to this type of circuit (Fig. 210-38).

   Because an *individual* branch circuit—that is, a branch circuit supplying a single outlet or load—may be rated at any ampere value, it is important to note that the omission of recognition of a 25-A *multioutlet* branch circuit does not affect the full acceptability of a 25-A *individual* branch

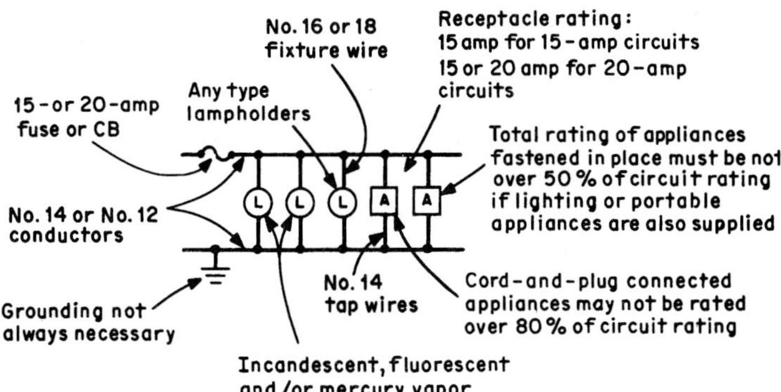

Fig. 210-37. General-purpose branch circuits—15 or 20 A. [Sec. 210-23(a).]

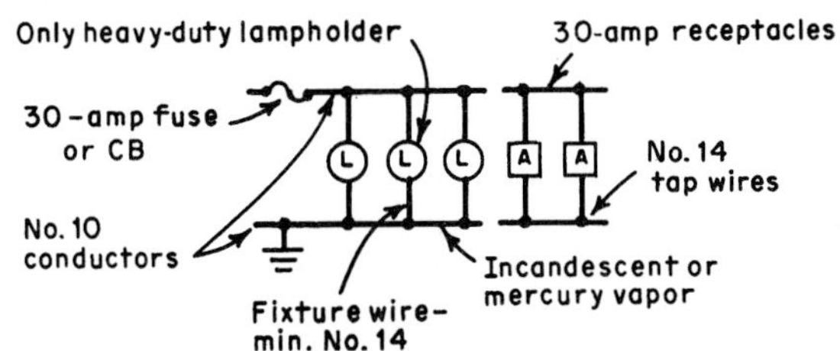

Fig. 210-38. Multioutlet 30-A circuits. [Sec. 210-23(b).]

circuit supplying a single outlet. A typical application of such a circuit would be use of No. 10 TW aluminum conductors (rated at 25 A in Table 310-16), protected by 25-A fuses or circuit breaker, supplying, say, a 4,500-W water heater at 230 V. The water heater is a load of 4,500 ÷ 230, or 19.6 A—which is taken as a 20-A load. Then, because Sec. 422-14(b) designates water heaters as continuous loads (in tank capacities up to 120 gal), the 20-A load current multiplied by 125 percent equals 25 A, and satisfies Sec. 422-4(a), Exception No. 2, on the required minimum branch-circuit rating. The 25-A rating of the circuit overcurrent device also satisfies Sec. 422-27(e), which says that the overcurrent protection must not exceed 150 percent of the ampere rating of the water heater.

No. 10 aluminum, with a 60°C ampacity of 25 A, may be used instead of No. 12 copper (rated 20 A). But the need for and the application possibilities of a 25-A *multioutlet* branch circuit have always been extremely limited. Such a circuit has never been considered suitable to supply lighting loads in dwelling units (where aluminum branch-circuit conductors have been primarily used). But for heavy-current appliances (16 to 20 A) realistic loading dictates use of an *individual* branch circuit, which *may* be rated at 25 A.

3. Branch circuits rated 40 and 50 A may serve fixed lighting units (with heavy-duty lampholders) or infrared heating units in other than dwelling

units or cooking appliances in any occupancy (Fig. 210-39). It should be noted that a 40- or 50-A circuit may be used to supply any kind of load equipment—such as a dryer or a water heater—where the circuit is an individual circuit to a single appliance. The conditions shown in that figure apply only where more than one outlet is supplied by the circuit. Figure 210-40 shows the combination of loads.

4. A multioutlet branch circuit rated over 50 A—as permitted by Sec. 210-3—is limited to use only for supplying industrial utilization equipment (machines, welders, etc.) and may *not* supply lighting outlets.

Except as permitted in Sec. 660-4 for portable, mobile, and transportable medical x-ray equipment, branch circuits having two or more outlets may supply only the loads specified in each of the above categories. It should be noted that any other circuit is not permitted to have more than one outlet and would be an individual branch circuit.

Application of those rules—and other Code rules that refer to "dwelling unit"—must take into consideration the NE Code definition for that phrase. A "dwelling unit" is defined as "one or more rooms" used "as a housekeeping unit" and must contain space or areas specifically dedicated to "eating, living, and sleeping" and must have "permanent provisions for cooking and sanitation." A one-family house is a "dwelling unit." So is an apartment in an apartment house or a condominium unit. But, a guest room in a hotel or motel or a dormitory room or unit is not a "dwelling unit" if it does not contain "permanent provisions for cooking"—which must mean a built-in range or counter-mounted cooking unit (with or without an oven).

It should be noted that the requirement calling for heavy-duty type lampholders for lighting units on 30-, 40-, and 50-A multioutlet branch circuits excludes the use of fluorescent lighting on these circuits because lampholders are not rated "heavy-duty" in accordance with Sec. 210-21(a) (Fig. 210-41). Mercury-vapor units with mogul lampholders may be used on these circuits provided tap conductor requirements are satisfied.

As indicated, multioutlet branch circuits for lighting are limited to a maximum loading of 50 A. Individual branch circuits may supply any loads. Excepting motors, this means that an individual piece of equipment may be supplied by a branch circuit which has sufficient carrying capacity in its conductors, is protected against current in excess of the capacity of the conductors, and supplies only the single outlet for the load device.

Fixed outdoor electric snow-melting and de-icing installations may be supplied by any of the above-described branch circuits. (See Sec. 426-4 in Art. 426, "Fixed Outdoor Electric De-Icing and Snow-Melting Equipment.")

210-24. Branch-Circuit Requirements—Summary. Table 210-24 summarizes the requirements for the size of conductors where two or more outlets are supplied. The asterisk note also indicates that these ampacities are for copper conductors where derating is not required. Where more than three conductors are contained in a raceway or a cable, Note 8 to Tables 310-16 through 310-19 specifies the load-current derating factors to apply for the number of conductors involved. A 20-A branch circuit is required to have conductors which have an

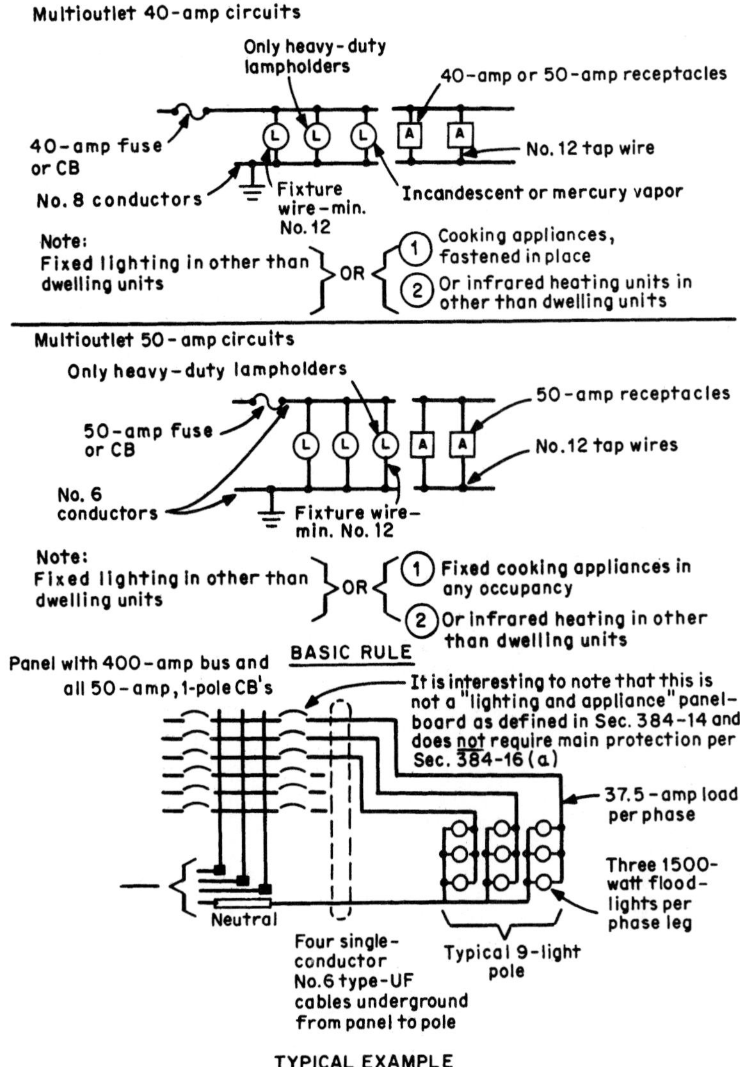

Fig. 210-39. Larger circuits. [Sec. 210-23(c).]

ampacity of 20 A and also must have the overcurrent protection rated 20 A where the branch circuit supplies two or more outlets. Refer to the detailed discussion of conductor ampacity and load-current limiting under Sec. 310-15, where Table 310-16 and its notes are explained.

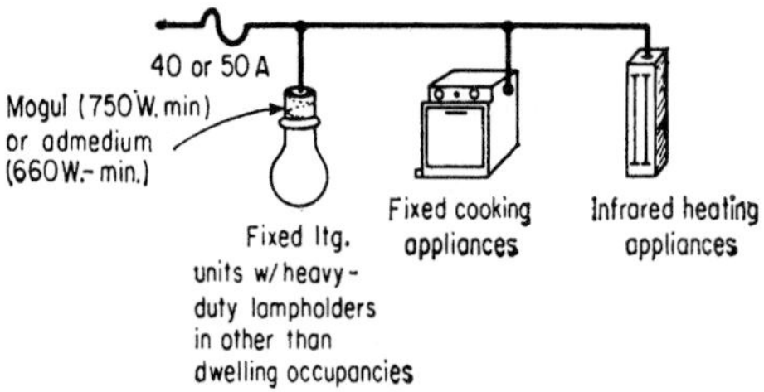

Fig. 210-40. Only specified loads may be used for multioutlet circuit. [Sec. 210-23(c).]

210-25. Common Area Branch Circuits. The first part of this rule states that branch circuits within a dwelling unit may not supply any other dwelling or its associated loads. This is a basic safety concern. In the past, there have been cases where the supply of loads in adjacent dwellings has resulted in injury and death where people mistakenly thought everything was electrically isolated when it was not. As a result, supply of any loads other than those "within that dwelling unit or loads associated only with that dwelling unit" has long been prohibited.

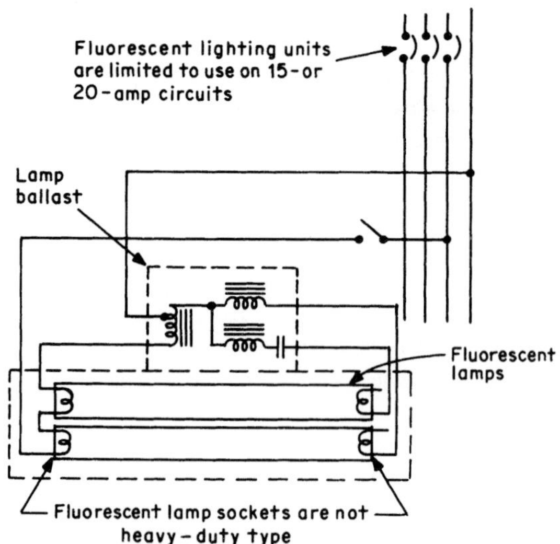

Fig. 210-41. Watch out for this limitation on fluorescent equipment. (Sec. 210-23.)

It should be noted that a common area panel is required in virtually every two- and multifamily dwelling. The explosion of local ordinances regarding interconnected smoke detectors in such occupancies, as well as the growth of the so-called common area and the vast array of equipment that may be supplied in such an area, today, has assured us that a common area panel must be provided. Indeed, in some of the more expensive complexes, the common load may be equal to, or greater than, the combined load of all of the dwellings. Remember that loads such as lighting for the parking lot, landscape, hallways, stairways, walkways, and entrance ways, as well as fountain pumps, sprinkler systems, etc.—in short any common area load—must be supplied from this common area panel.

The second sentence addresses installation of the common area panel at two- and multifamily dwellings. Basically stated, a separate panel to supply common area loads must be provided and it must be supplied directly from the service conductors, have its own meter, be suitable for use as service equipment, or be supplied from a disconnect that is, etc. That statement is based on the change in wording that now prohibits supplying the common area panel from "equipment that supplies an individual dwelling unit." Clearly, if a meter supplying any individual unit was also used to monitor usage on the common area panel, the literal wording of Sec. 210-25 would be violated because that "equipment" (the meter) supplies "an individual dwelling unit."

Curiously, the literal wording would also be satisfied if the whole building were on a single meter. In such a case, the common area panel would be supplied from "equipment" that supplies *many* dwellings, not "an individual" unit. But, even then, the common area panel would have to be supplied directly from the single meter and satisfy other rules (e.g., be suitable as service equipment, etc.) as necessary. In no case may the common area panel be supplied from a panel in another dwelling or, as it now states, from any equipment that supplies a single unit (Fig. 210-42).

210-50. General. Part **(b)** simply requires that wherever it is known that cord- and plug-connected equipment is going to be used, receptacle outlets must be installed. That is a general rule that applies to any electrical system in any type of occupancy or premises.

210-52. Dwelling Unit Receptacle Outlets. This section sets forth a whole list of rules requiring specific installations of receptacle outlets in all "dwelling units"—i.e., one-family houses, apartments in apartment houses, and other places that conform to the definition of "dwelling unit." As indicated, receptacle outlets on fixed spacing must be installed in every room of a dwelling unit except the bathroom. The **Code** rule lists the specific rooms that are covered by the rule requiring receptacles spaced no greater than 12 ft (3.66 m) apart in any continuous length of wall.

In part **(a)**, the required receptacles must be spaced around the designated rooms and any "similar room or area of dwelling units." The wording of this section assures that receptacles are provided—the correct number with the indicated spacing—in those unidentified areas so commonly used today in residential architectural design, such as *greatrooms* and other big areas that combine living, dining, and/or recreation areas.

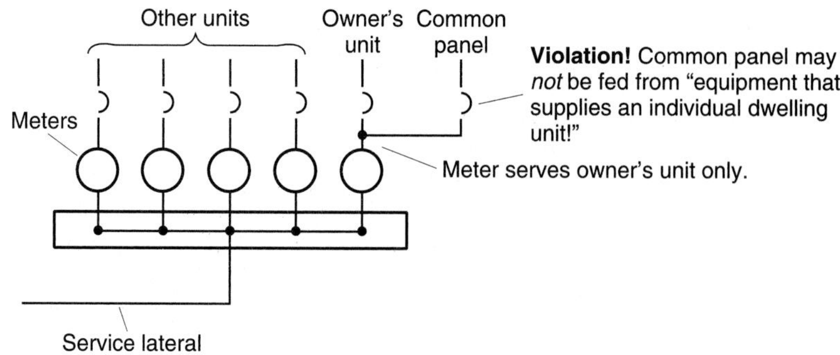
Multifamily
occupancy
Other units
Owner's unit
Common panel
Violation! Common panel may not be fed from "equipment that supplies an individual dwelling unit!"
Meters
Meter serves owner's unit only.
Service lateral

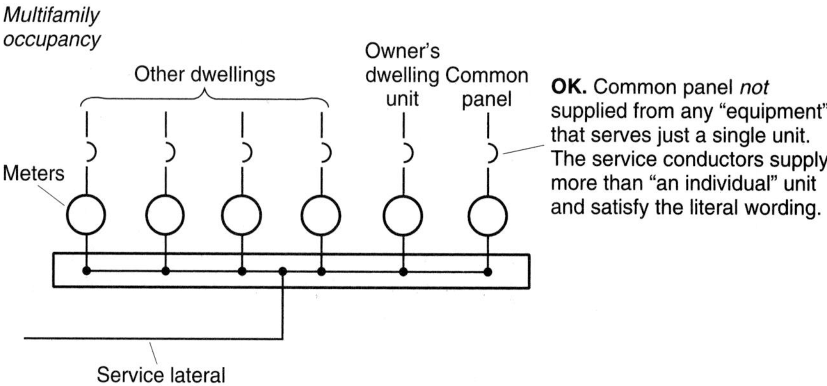
Multifamily
occupancy
Other dwellings
Owner's dwelling unit
Common panel
OK. Common panel not supplied from any "equipment" that serves just a single unit. The service conductors supply more than "an individual" unit and satisfy the literal wording.
Meters
Service lateral

Fig. 210-42. Rewording of this rule has answered a number of questions regarding its application. The rule clearly prohibits the supply of the common area panel from any individual unit's "equipment." The term "equipment" is defined in Art. 100 and includes virtually every part of an electrical installation. As indicated by the literal wording, the common area panel must be supplied from a point in the system that serves more than a single unit. In the diagram above, that would be the service lateral because beginning with the taps to the individual meters, the "equipment" is serving one unit. And the common area panel may not be supplied from such equipment.

As shown in Fig. 210-43, general-purpose convenience receptacles, usually of the duplex type, must be laid out around the perimeters of living room, bedrooms, and all the other rooms. Spacing of receptacle outlets should be such that no point along the floor line of an unbroken wall is more than 6 ft (1.83 m) from a receptacle outlet. Care should be taken to provide receptacle outlets in smaller sections of wall space segregated by doors, fireplaces, bookcases, or windows. Although Sec. 210-52(h) calls for one receptacle outlet for each dwelling-unit hallway that is 10 ft (3.05 m) or more in length, this section does not specify location or require more than a single receptacle outlet. However, good design practice would dictate that a convenience receptacle should be

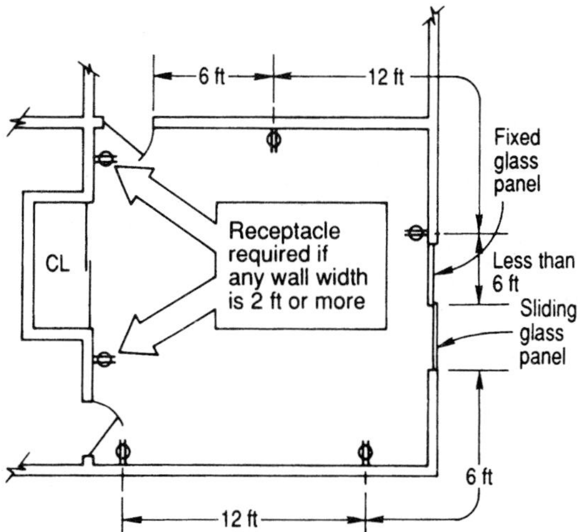

Fig. 210-43. From any point along wall, at floor line, a receptacle must be not more than 6 ft away. Required receptacle spacing considers a fixed glass panel as wall space and a sliding panel as a doorway. [Sec. 210-52(a).]

provided for each 10 ft (3.05 m) of hall length. And they should be located as close as possible to the middle of the hall.

In determining the location of a receptacle outlet, the measurement is to be made along the floor line of the wall and is to continue around corners of the room, but is not to extend across doorways, archways, fireplaces, passageways, or other space unsuitable for having a flexible cord extended across it. The location of outlets for special appliances within 6 ft (1.83 m) of the appliance [Sec. 210-50(c)] does not affect the spacing of general-use convenience outlets but merely adds a requirement for special-use outlets.

Figure 210-44 shows two wall sections 9 ft (2.74 m) and 3 ft (914 mm) wide extending from the same corner of the room. The receptacle shown located in the wider section of the wall will permit the plugging in of a lamp or appliance located within 6 ft (1.83 m) of either side of the receptacle. The same rule would apply to the other wall shown.

Receptacle outlets shall be provided for all wall space within the room except individual isolated sections which are less than 2 ft (610 mm) in width. For example, a wall space 23 in. (584 mm) wide and located between two doors would not need a receptacle outlet.

In measuring receptacle spacing for exterior walls of rooms, the fixed section of a sliding glass door assembly is considered to be "wall space" and the sliding glass panel is considered to be a doorway. In previous NEC editions the entire width of a sliding glass door assembly—both the fixed and movable panels—was required to be treated as wall space in laying out receptacles "so that no point along the floor line in any wall space is more than 6 feet (1.83 m)" from a receptacle outlet. The wording takes any fixed glass panel to be a continuation

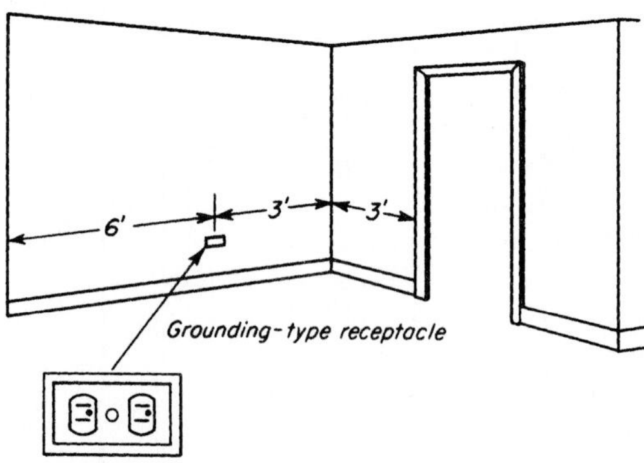

Fig. 210-44. Location of the receptacle as shown will permit the plugging in of a lamp or appliance located 6 ft on either side of the receptacle. [Sec. 210-52(a).]

of the wall space adjoining it, but the sliding glass panel is taken to be the same as any other doorway (such as with hinged doors) (Fig. 210-43).

The last sentence of the first paragraph of part **(a)** requires fixed room dividers and railings to be considered in spacing receptacles. This is illustrated by the sketch of Fig. 210-45. In effect, the two side faces of the room divider provide additional wall space, and a table lamp placed as shown would be more than 6 ft (1.83 m) from both receptacles A and B. Also, even though no place on the wall is more than 6 ft (1.83 m) from either A or B, a lamp or other appliance placed at a point such as C would be more than 6 ft (1.83 m) from B and out of reach from A because of the divider. This rule would ensure placement of a receptacle in the wall on both sides of the divider or in the divider itself if its construction so permitted.

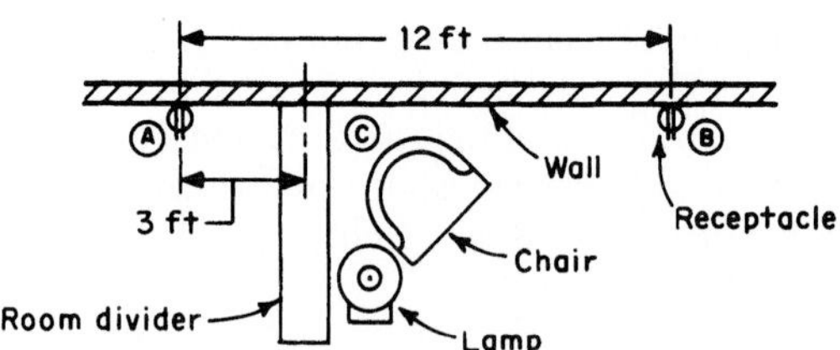

Fig. 210-45. Fixed room dividers must be counted as wall space requiring receptacles. [Sec. 210-52(a).]

Recessed or surface-mounted floor receptacles must be within 18 in. (457 mm) of the wall to qualify as one of the "required" receptacle outlets in a dwelling. The previous wording used in this next to last paragraph indicated floor-mounted receptacles were not considered to satisfy the rule of Sec. 210-52(a) unless they were "located close to the wall." The use of a specific dimension, regardless of its arbitrary nature, is much more desirable than the relative term "close." The use of nonspecific, relative, and subjectively interpreted terms—such as "close" or "large"—opens the door for conflict and makes applying or enforcing a given rule that much more difficult.

The use of either surface-mounted or recessed receptacle outlets has grown since "railings" were required to be counted as "wall space" by the 1993 **Code**. Now, where floor-mounted receptacle outlets are provided—either surface-mounted or recessed—to serve as a required receptacle outlet in a dwelling for any so-called wall space, such an outlet must be no more than 18 in. (457 mm) from the wall (Fig. 210-46).

As noted in the last paragraph of Sec. 210-52(a), any receptacle that is an integral part of a lighting fixture or an appliance or a cabinet may not be used to satisfy the specific receptacle requirements of the section. For instance, a receptacle in a medicine cabinet or lighting fixture may not serve as the required bathroom receptacle. And a receptacle in a post light may not serve as the required outdoor receptacle for a one-family dwelling.

In spacing receptacle outlets so that no floor point along the wall space of the rooms designated by Sec. 210-52(a) is more than 6 ft (1.83 m) from a receptacle, a receptacle that is part of an appliance must not generally be counted as one of the required spaced receptacles. However, the Exception at the end of part **(a)** states that a receptacle that is "factory installed" in a "permanently installed electric baseboard heater" (not a portable heater) may be counted as one of the required spaced receptacles for the wall space occupied by the heater. Or a receptacle "provided as a separate assembly by the manufacturer" may also be counted as a required spaced receptacle. But, such receptacles must not be connected to the circuit that supplies the electric heater. Such a receptacle must be connected to another circuit.

Because of the increasing popularity of low-density electric baseboard heaters, their lengths are frequently so long [up to 14 ft (4.27 m)] that required maximum spacing of receptacles places receptacles above heaters and produces the undesirable and dangerous condition where cord sets to lamps, radios, TVs, etc., will droop over the heater and might droop into the heated-air outlet. And UL rules prohibit use of receptacles above certain electric baseboard heaters for that reason. Receptacles in heaters can afford the required spaced receptacle units without mounting any above heater units. They satisfy the UL concern and also the preceding note near the end of Sec. 210-52(a) that calls for the need "to minimize the use of cords across doorways, fireplaces, and similar openings"—and the heated-air outlet along a baseboard heater is a "similar opening" that must be guarded (Fig. 210-47).

A fine-print note at the end of part **(a)** points out that the UL instructions for baseboard heaters (marked on the heater) may prohibit the use of receptacles above the heater because cords plugged into the receptacle are exposed to heat damage if they drape into the convection channel of the heater and contact the energized heating element.

A rewrite of Sec. 210-52(b) serves to clarify application and prohibits one long-time practice.

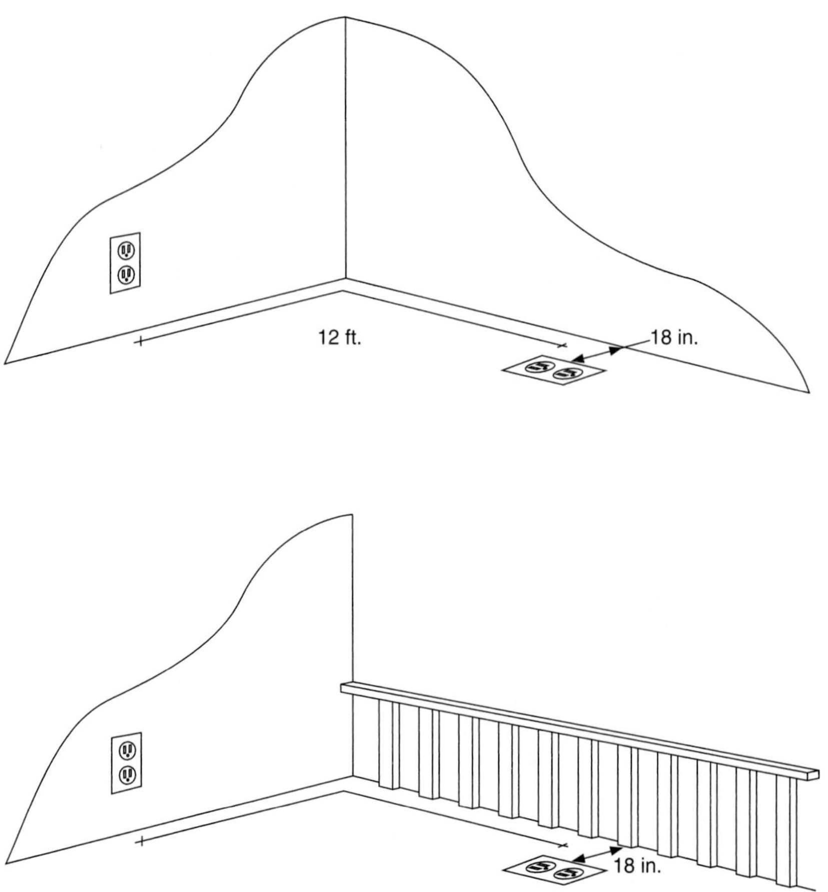

Fig. 210-46. Any receptacle outlet that is intended to serve as one of the required outlets in a dwelling must be no more than 18 in. (457 mm) from the "wall space."

Part **(b)(1)** of this section requires two, or more, 20-A branch circuits to supply all receptacle outlets required by Sec. 210-52(a) and (c) in the kitchen, etc. And the new part **(b)(2)** states that no other outlets may be supplied from those small appliance branch circuits. Those two requirements had both been contained in part **(b)(1)** of the 1993 and previous Codes. However, because the two rules were combined in a single paragraph, it was not always easy to determine to which part a given exception applied.

The basic rule of Sec. 210-52(b)(1) states that those receptacles required every 12 ft (3.66 m) [Sec. 210-52(a)], those that serve countertop space [Sec. 210-52(c)], and the refrigerator receptacle in the kitchen, dining room, pantry, etc., must be supplied by one of the two, or more, 20-A small appliance branch circuits.

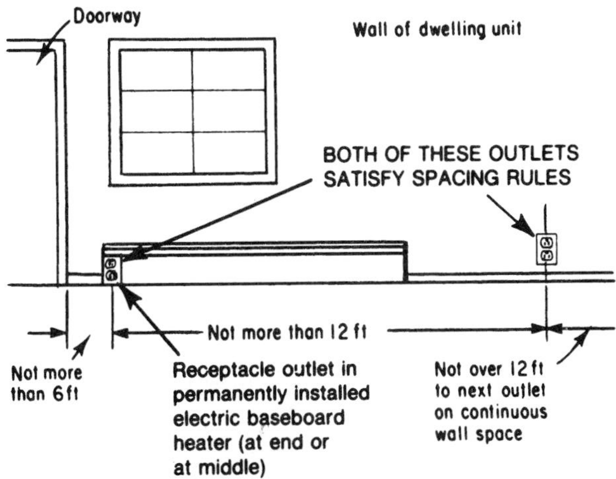

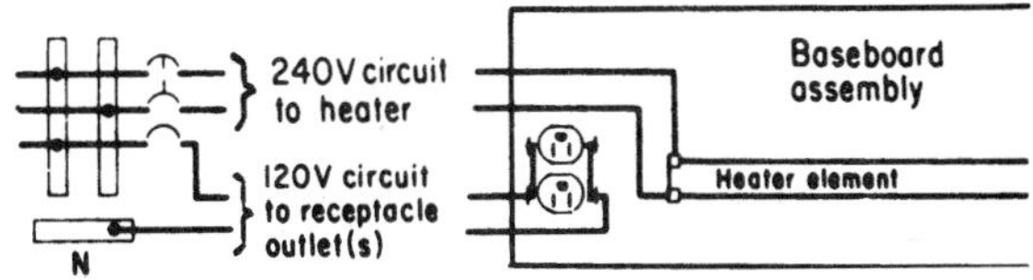

Fig. 210-47. Receptacles in baseboard heaters may serve as "required" receptacles. [Sec. 210-52(a), Exception.]

The wording used here must be carefully examined. Because the new wording only specifically permits the refrigerator receptacle and those receptacles required by Sec. 210-52(a) and (c) on the small appliance branch circuits, the installation of any other receptacles on the small appliance branch circuits is effectively prohibited. Any receptacle installed for specific equipment such as dishwashers, garbage disposals, trash compactors, etc.—which are *not* required by part **(a)** or **(c)**—must be supplied from a 15- or 20-A general-purpose branch circuit.

The exceptions to part **(b)(1)** are exceptions to the rule that *all required* receptacle outlets must be supplied from the two, or more, 20-A branch circuits. The first exception recognizes the use of a switched receptacle supplied from a general-purpose branch circuit where such a receptacle is provided instead of a lighting outlet in accordance with Exception No. 1 to Sec. 210-70(a). That rule specifically excludes kitchens from employing a switched receptacle instead of a lighting outlet, but, in those other rooms and areas identified in the basic rule, a wall-switched receptacle outlet supplied from a general-purpose branch circuit may count as a required receptacle (Fig. 210-48).

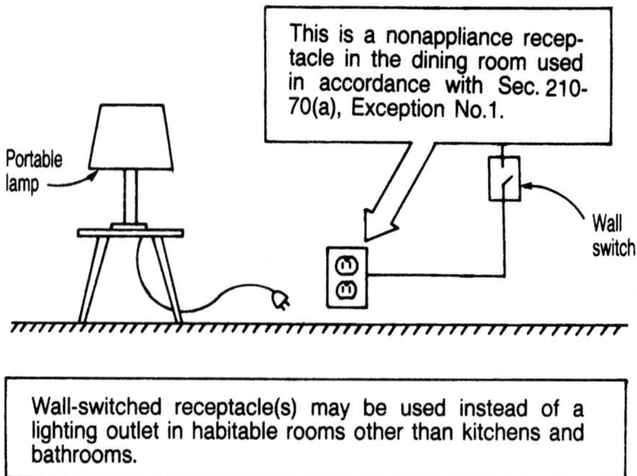
This is a nonappliance receptacle in the dining room used in accordance with Sec. 210-70(a), Exception No.1.
Portable lamp
Wall switch
Wall-switched receptacle(s) may be used instead of a lighting outlet in habitable rooms other than kitchens and bathrooms.

Fig. 210-48. For those rooms and areas identified by Sec. 210-52(b)(1), other than the kitchen, a wall switch-controlled receptacle may be supplied from a general-purpose branch circuit *and* serve as one of the required receptacles.

In Sec. 210-52(b)(1), Exception No. 2, the Code recognizes the supply of the required receptacle for a refrigerator from an individual 15-A branch circuit. Many refrigerators installed in dwellings are rated at 12 A and could be supplied from a 15-A circuit. Rather than mandate the use of 20-A-rated circuit for those cases where a 15-A circuit is adequate, it is permissible to use a 15-A-rated circuit, provided the supply to the refrigerator receptacle is a dedicated branch circuit—i.e. *no* other outlets supplied.

It should be noted that it is no longer permissible to supply an outdoor receptacle from the small appliance branch circuit. This was previously recognized in Sec. 210-52(b)(1), Exception No. 2 and served to limit the number of GFCIs needed at a dwelling. That is, because grade-level-accessible outdoor receptacles were required to have GFCI protection, the Code permitted supplying the outdoor receptacle using the feed-through capability of the GFCIs installed in the kitchen rather than require an additional GFCI device, which provided for economy. Now, however, supplying an outdoor receptacle from the small appliance branch circuit is prohibited (Fig. 210-49).

Section 210-52(b)(2) states that only those outlets identified in part **(b)(1)**—and no other outlets—may be installed on the two, or more, small appliance branch circuits. Outlets for lighting and hard-wired appliances, as well as "unrequired" receptacles for equipment, must be installed on 15- or 20-A general-purpose circuits.

The first exception to Sec. 210-52(b)(2) allows a clock hanger receptacle to be installed on a small appliance branch circuit, or it may be supplied from a general-purpose circuit. The second recognizes a receptacle provided for control power or clock, fan, or light in a gas-fired cooking unit. Note that only a *receptacle* outlet is permitted. Any hard-wired connection for such auxiliary functions on a gas-fired unit must be supplied from a general-purpose branch circuit and *not* from the small appliance branch circuits (Fig. 210-50).

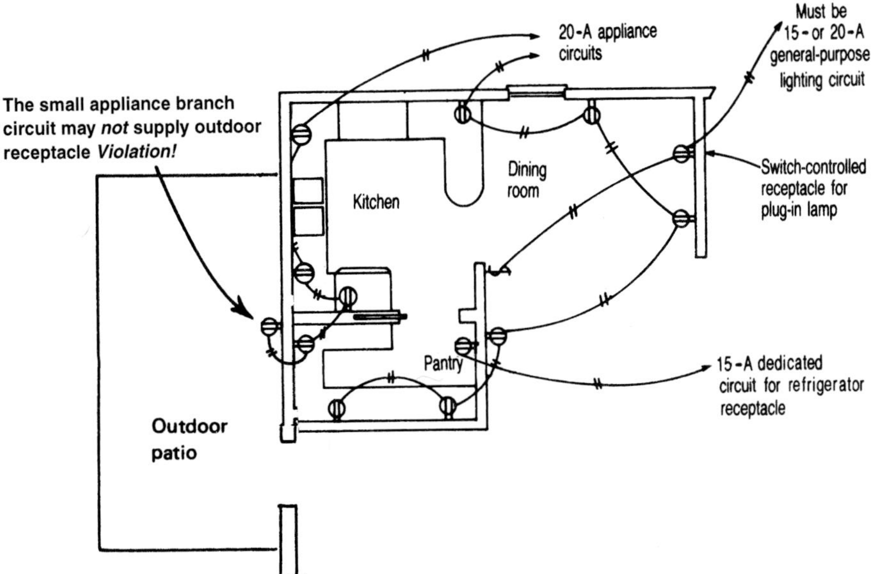

Fig. 210-49. Summary of Sec. 210-52(b)(1) and its two exceptions. Although primarily editorial in nature, the rewrite of this section eliminated a widely used Exception that permitted supply of an outdoor receptacle from one of the small appliance branch circuits. As indicated above, supply of an outdoor receptacle from any of the two, or more, 20-A small appliance branch circuits is prohibited.

Section 210-52(c) was revised primarily to clarify application, but it also presents new requirements and restrictions regarding installation of counter-top receptacles in kitchens and dining rooms.

This section is broken into five subparts—(c)(1) through (5). The first four subparts identify those counter spaces in the kitchen and dining room that must be provided with receptacle outlets and indicate the number required, while the last subpart indicates where the receptacle outlet must be installed.

In part **(c)(1)** the NEC puts forth the spacing requirements for receptacle outlets installed at counter spaces along the wall. Basically stated, each wall counter space that is 12 in. (305 mm) or wider must have at least one receptacle outlet to supply cord- and plug-connected loads. The receptacles must be placed so that no point along the wall line is more than 24 in. (610 mm) from an outlet. That translates into one outlet every 4 ft (1.22 m). It should be noted that the term "measured horizontally" is intended to recognize application as shown in Fig. 210-51. There is no need to measure "around the corner" in that case.

As given in subpart **(c)(2),** each free standing (island) counter top that measures 24 in. (610 mm) or more by 12 in. (305 mm) or more must be provided with one receptacle outlet. The same dimensions apply to peninsular counter tops, which is a counter top that extends from another counter or a wall. The dimensions are to be measured from "the connecting edge," which is an imaginary line at the end of the peninsula where it attaches to the other counter (Fig. 210-52). If the area to the right of the so-called connecting edge in Fig. 210-52 measures

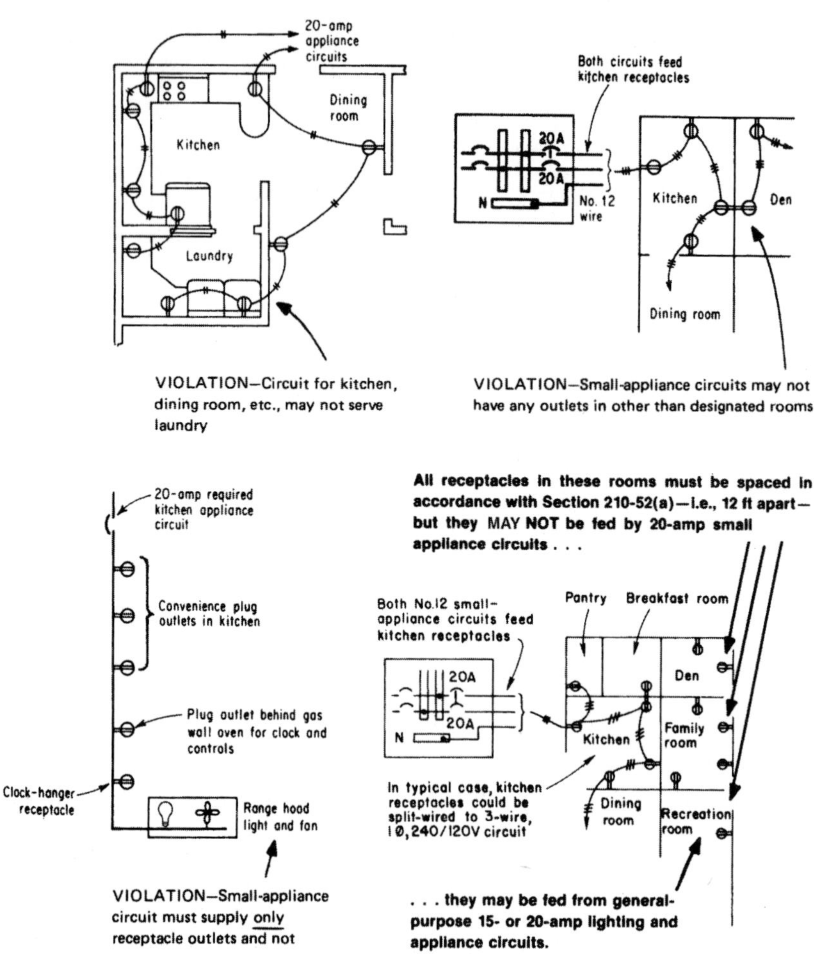

Fig. 210-50. A summary of Sec. 210-52(b)(2) and its two Exceptions. A clock-hanger receptacle and/or a receptacle for the supply of auxiliary equipment on a gas-fired range, oven, or cook-top may also be supplied from the two, or more, small appliance branch circuits.

24 in. (610 mm) (or more) by 12 in. (305 mm) (or more), at least one receptacle outlet must be provided. In no case is more than one receptacle outlet *required* at either an island or peninsula counter space, although more may be desirable. If additional receptacles are provided, they must be supplied from one of the 20-A small appliance branch circuits. And, whether the outlet is required or desired, it must be installed as indicated by the last subpart of this section.

Subpart **(c)(4)** gives the long-time rule regarding pieces of counter tops that are separated by cook-tops, sinks, etc. As indicated, each such piece must be treated as an individual counter. And, if the dimensions are as described in parts **(1)**, **(2)**, or **(3)**, as applicable, of this section, at least one receptacle outlet must be provided.

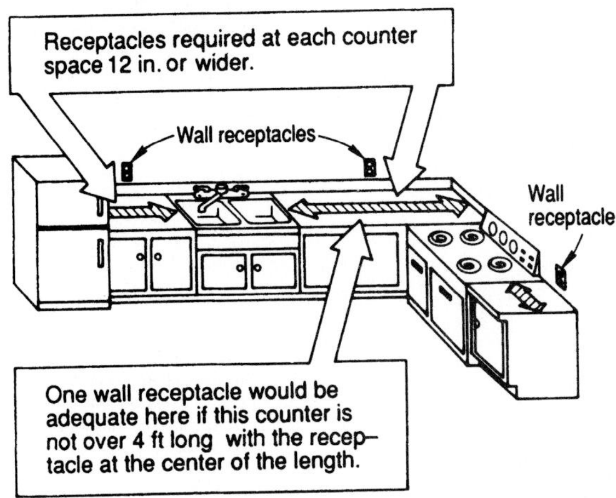
Any point along the wall line of each length of counter top must *not* be over 24 in., measured horizontally, from a receptacle outlet.
Receptacles required at each counter space 12 in. or wider.
Wall receptacles
Wall receptacle
One wall receptacle would be adequate here if this counter is not over 4 ft long with the receptacle at the center of the length.

Fig. 210-51. The term "measured horizontally" can essentially be translated as "when you are facing the counter." There is no need to measure around the corner here because that would effectively measure the area twice. If the stove were *not* there and the counter continued around the corner, as in the case of a peninsula counter, the measurement should be continued from the "connecting edge," which here would be the imaginary line where the stove meets the wall counter.

The requirements in Sec. 210-52(c)(5) mandate where the required receptacle outlets may be installed. That is, a given receptacle outlet may not be counted as one of the required outlets unless it is installed on top of, or not more than 18 in. (457 mm) above, the counter it is intended to serve. Additionally, no receptacle may be installed face-up in a counter top. That is a "make sense" proposition inasmuch as a receptacle installed face-up would eventually become a "drain" for soup, milk, water, or whatever else is eventually spilled on the counter. Only the so-called tombstone or doghouse enclosures would be acceptable for surface mounting. And, any outlet located above the counter must be no more than 18 in. (457 mm) above. Although not entirely clear, it is assumed that the 18 in. must be measured from the counter surface, *not* the top of the backsplash.

Note that the basic rule generally requires the outlet to be mounted above, or on top of, the counter. The basic rule does *not* recognize installation of an outlet *below* the counter space. However, where the counter does not extend more than 6 in. (152 mm) beyond "its support base," the exception to Sec. 210-52(c)(5) permits installation of a receptacle outlet below the counter, but *only* where the local inspector authorizes such installation. In addition to accommodate the physically impaired, this exception addresses those instances where locating the outlet on, or above, the counter is not possible. In addition to either of those

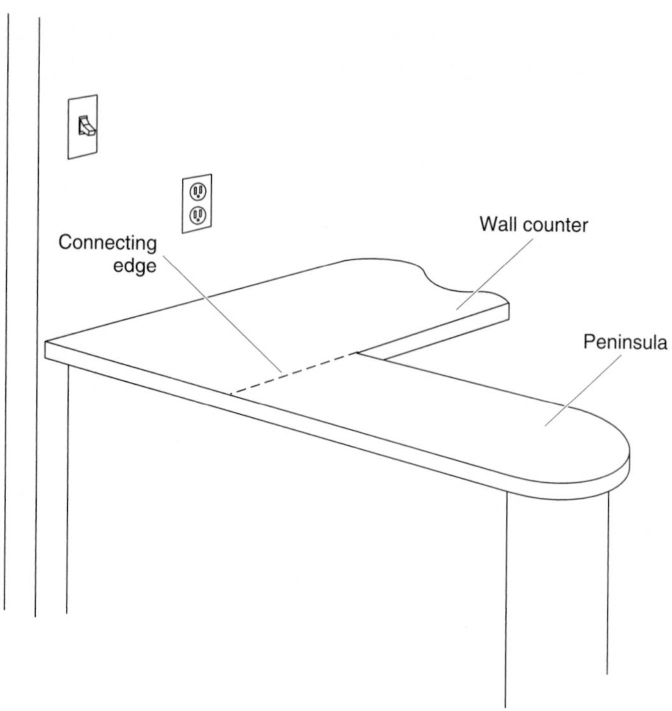

Fig. 210-52. As covered in the last sentence of part **(c)(3)**, the area to be considered as "peninsula" counter begins at the imaginary line as shown above. If that area has a long dimension of 24 in. (610 mm), or more, by 12 in. (305 mm), or more, at least one receptacle outlet is required to serve that counter space in kitchens and dining rooms at dwellings.

cases, the local electrical inspector is permitted to authorize any below-the-counter installation deemed appropriate. But, in any case, a receptacle outlet that is mounted more than 12 in. (305 mm) below the counter, or one rendered inaccessible by a appliance, does not qualify as one of the outlets required by part **(c)**. And, be aware that any below-the-counter installations must be authorized in writing and that document must be retained (Fig. 210-53).

Part **(d)** requires the installation of at least one wall receptacle outlet adjacent to each wash basin location in bathrooms of dwelling units—and Sec. 210-60 requires the same receptacle in bathrooms of hotel and motel guest rooms. The **Code** now requires a dedicated circuit for bathroom receptacles installed in dwellings. In every bathroom, at least one receptacle outlet must be installed at each basin and any such outlet(s) must be supplied from a dedicated 20-A branch circuit (Fig. 210-54).

Part **(e)** requires that at least one outdoor receptacle "accessible at grade level and not more than 6′ 6″ (1.98 m) above grade" must be installed at the front and back for every one-family house ("a one-family dwelling") and grade-level accessible unit in a two-family dwelling. The definition of "one-family dwel-

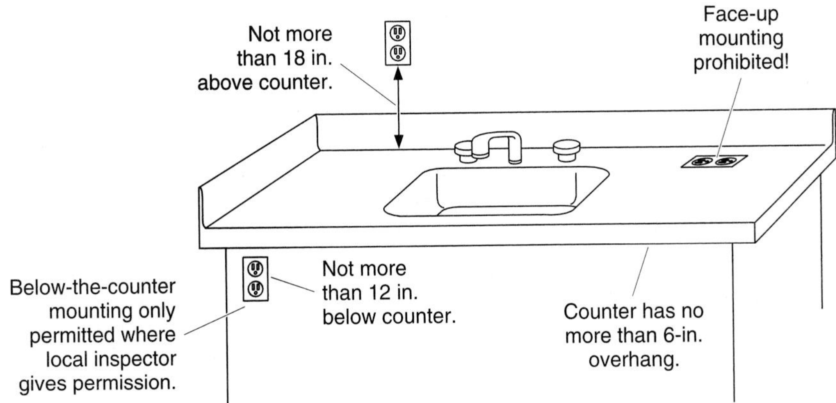
Not more
than 18 in.
above counter.
Face-up
mounting
prohibited!
Below-the-counter
mounting only
permitted where
local inspector
gives permission.
Not more
than 12 in.
below counter.
Counter has no
more than 6-in.
overhang.

Inaccessible receptacles.

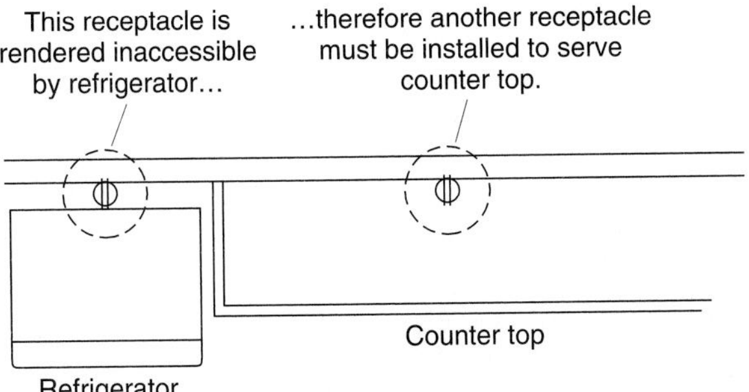
This receptacle is
rendered inaccessible
by refrigerator...
...therefore another receptacle
must be installed to serve
counter top.
Counter top
Refrigerator

Receptacle located behind an appliance, making the receptacle inaccessible, does not count as one of the required "counter-top" receptacles.

Fig. 210-53. This new section indicates where required outlets intended to serve counter space in dwellings must be installed. Remember that below-the-counter mounting is only permitted where written permission from the local inspector is received.

ling" (Art. 100) makes clear that an outdoor receptacle is not required for outdoor balconies of apartment units, motels, hotels, or other units in multiple-occupancy buildings.

Part **(e)** also requires that townhouse-type one-family dwellings be provided with one GFCI-protected outdoor receptacle outlet at the front of each dwelling and one at the rear of each dwelling. This rule is aimed at providing adequacy in the availability of outdoor receptacles for one-family dwelling units. (Fig. 210-55).

The use of outdoor appliances at two-family houses has been judged to be as common as at one-family houses, and the need for outdoor receptacles to eliminate use of extension cords from within the house is recognized by the rule calling for outdoor receptacles for two-family houses.

Sec. 210-52(d). Receptacle outlets installed in bathrooms in dwellings must be supplied from dedicated—no other outlets—20-A branch circuit.

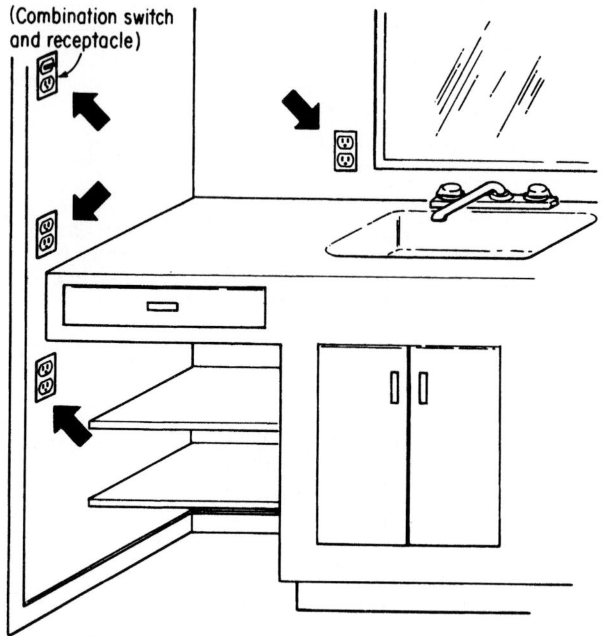

LOCATION of receptacle will vary, depending upon available wall space. Arrows show several possibilities. A receptacle in a medicine cabinet or in the bathroom lighting fixture does not satisfy this rule.

Fig. 210-54. A dedicated 20-A branch circuit must be provided to supply required receptacle outlets installed "adjacent" to bathroom sinks, as well as any other receptacle outlets installed in the bathroom. The outlet installed below the counter space here may, or may not, be viewed as satisfying the relative term "adjacent." But, the restriction against below-the-counter mounting in kitchens and dining rooms of dwellings is not repeated for bathrooms.

For a two-family dwelling, the rule requires a separate outdoor receptacle outlet for each dwelling unit in a two-family house where each dwelling unit is an upstairs-and-downstairs unit—that is, each unit has living space (i.e., kitchen and living room) located "at grade level." And, as with a one-family dwelling, the receptacle outlets could contain a single, duplex, or triplex receptacle—installed on the outside wall or fed underground. [Note that a receptacle in a post light does not qualify as the required outdoor receptacle, because a receptacle "that is part of any lighting fixture" is excluded by the last paragraph of Sec. 210-52(a).] The clear intent of the rule, however, is *not* to require outdoor receptacles for a dwelling unit that is totally on the second floor of a two-family house, with only its entrance door on the first floor, providing access to the stairway.

In a multiple-occupancy building—such as adjacent up-and-down duplex units in "townhouses"—if adjacent units are separated by fire-rated walls,

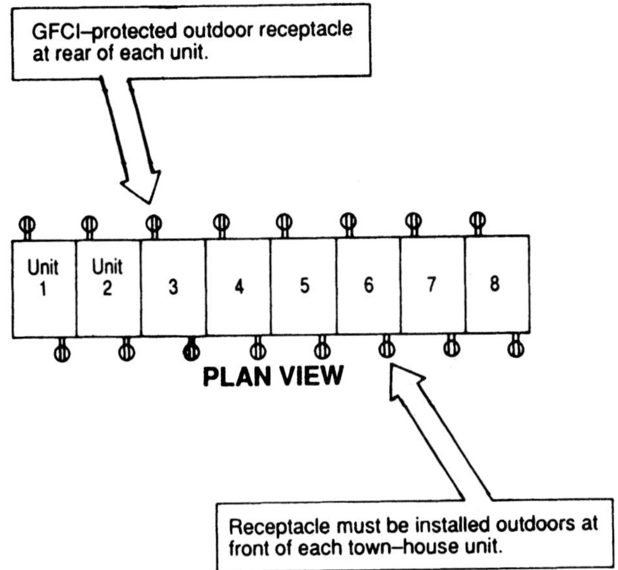

Fig. 210-55. Front and rear-receptacle outlets are required outdoors for town-house-type one-family dwellings. [Sec. 210-52(e).]

each unit is considered to be a separate building and each is, by Code definition, a "one-family dwelling," even though the appearance of a continuous structure might make it seem like a multifamily dwelling or apartment house. Each such unit is, therefore, required to have at least two GFCI-protected outdoor receptacles.

Notice that although "grade level" accessibility was eliminated as a qualifier for GFCI protection of outdoor receptacles at dwelling, the outdoor receptacles required by part **(e)** must be "accessible" at grade level and no more than 6 ft 6 in. (1.98 m) above grade.

Part **(f)** requires that at least one receptacle—single or duplex or triplex—must be installed for the laundry of a dwelling unit. Such a receptacle and any other receptacles for special appliances must be placed within 6 ft (1.83 m) of the intended location of the appliance. And part **(g)** requires a receptacle outlet in a basement in addition to any receptacle outlet(s) that may be provided as the required receptacle(s) to serve a laundry area in the basement. One receptacle in the basement at the laundry area located there may *not* serve as *both* the required "laundry" receptacle and the required "basement" receptacle. A separate receptacle has to be provided for each requirement to satisfy the Code rules.

Section 210-52(g) requires that at least one receptacle (other than for the laundry) must be installed in the basement of a one-family dwelling, one in an attached garage, and one in a detached garage *if* power is run to the detached garage. This rule calls for at least one receptacle outlet in the basement of a one-family house, in addition to any required for a basement laundry (Fig. 210-56). It calls for at least one receptacle in an attached garage of a one-family house.

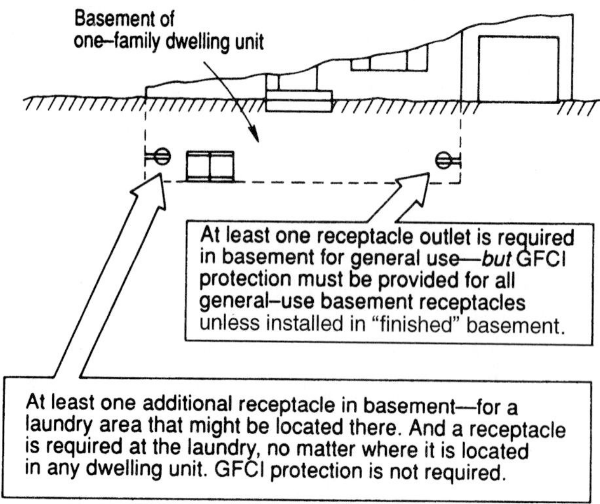

Basement of
one–family dwelling unit
At least one receptacle outlet is required
in basement for general use—*but* GFCI
protection must be provided for all
general–use basement receptacles
unless installed in "finished" basement.
At least one additional receptacle in basement—for a
laundry area that might be located there. And a receptacle
is required at the laundry, no matter where it is located
in any dwelling unit. GFCI protection is not required.

Fig. 210-56. Only one basement receptacle is required (in addition to any for the laundry), *but all* general-purpose receptacles in *unfinished* basements must be GFCI protected. [Sec. 210-52(g).]

But for a detached garage of a one-family house, the rule simply requires that one receptacle outlet must be installed in the detached garage *if*—for some reason other than the NEC—electric power is run to the garage, such as where the owner might desire it or some local code might require it (Fig. 210-57). The rule itself does *not* require that electric power be run to a detached garage to supply a receptacle there.

If the required "basement" receptacle is installed in an "unfinished" basement—that is a basement that has *not* been converted to, or constructed as, a recreation room, bedroom, den, etc.—such a receptacle would be required to be provided with GFCI protection [Sec. 210-8(a)(5)]. And, that same rule requires that any additional receptacles in an unfinished basement be GFCI-protected. In addition, *all* receptacles installed in a dwelling-unit garage (attached or detached) must have GFCI protection, as required by Sec. 210-8(a)(2).

With the wording of the rules of Sec. 210-52(f) and Sec. 210-8(a)(5), it would be acceptable for a one-family dwelling to have one basement receptacle with GFCI protection if the basement is unfinished, but any other receptacles that are optionally installed must also have GFCI protection. The one or more receptacles provided for a laundry area in the basement are *excluded* from need for GFCI protection by Exception No. 2 of Sec. 210-8(a)(5).

In part **(h)**, a receptacle outlet is required in any dwelling-unit hallway that is 10 ft (3.05 m) or more in length. This provides for connection of plug-in appliances that are commonly used in halls—lamps, vacuum cleaners, etc. The length of a hall is measured along its centerline.

Figure 210-58 shows required receptacles for a one-family dwelling.

210-60. Guest Rooms. The number of receptacles in a guest room of a hotel or motel must be determined by the every-12-ft (3.66-m) rule of Sec. 210-52(a) but

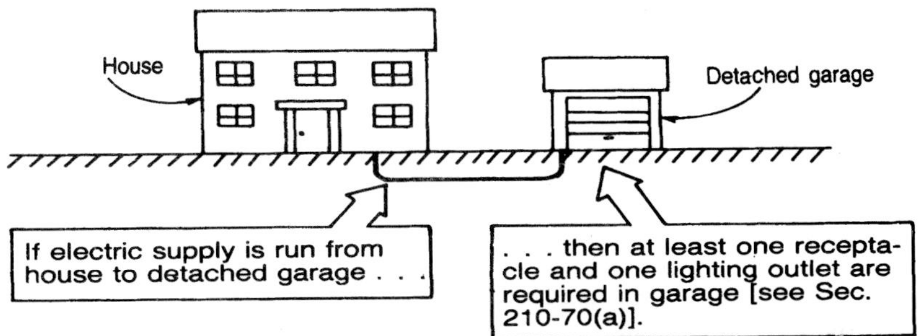
House
Detached garage
If electric supply is run from house to detached garage . . .
. . . then at least one receptacle and one lighting outlet are required in garage [see Sec. 210-70(a)].

Fig. 210-57. Detached garage may be required to have a receptacle and lighting outlet. [Sec. 210-52(g)]

may be located where convenient for the furniture layout, exempted from the rule that "no point along the floor line in any wall space is more than 6 ft (1.83 m) . . . from an outlet." The intent of the rule is that the *number* of receptacles must satisfy Sec. 210-52(a) but *spacing* of the receptacles is exempted from the every-12-ft (3.66-m) rule. In such cases, the spacing requirements of not more than 12 ft (3.66 m) between receptacles, etc., do not have to be observed.

210-62. Show Windows. The rule here calls for one receptacle in a show window for each 12 ft (3.66 m) of length (measured horizontally) to accommodate portable window signs and other electrified displays (Fig. 210-59).

210-63. Heating, Air-Conditioning, and Refrigeration Equipment Outlet. A general-purpose 125-V receptacle outlet must be installed within 25 ft (7.62 m) of heating, air-conditioning, refrigeration equipment on rooftops *and* in attics and crawl spaces (Fig. 210-60).

This rule provides a readily accessible outlet for connecting 120-V tools and/or test equipment that might be required for the maintenance or servicing of mechanical equipment in attics and crawl spaces as well as rooftops. Each such receptacle must be on the same level and within 25 ft (7.62 m) of the heating, refrigeration, and air-conditioning equipment. This receptacle must not be fed from the load side of the disconnecting means for the mechanical equipment. And it must be GFCI-protected to satisfy Sec. 210-8(b)(2). Only rooftop units on one- and two-family dwelling units are excluded from this requirement.

210-70. Lighting Outlets Required. The basic rule of part **(a)** requires "at least one wall switch-controlled lighting outlet" in rooms, halls, stairways, attached garages, "detached garages with electric power," and at outdoor entrances. This rule requires a wall switch-controlled lighting outlet in every *attached* garage of a dwelling unit (such as a one-family house). But, for a *detached* garage of a dwelling unit, a switch-controlled lighting outlet is required *only* if the garage is provided with electric power—whether the provision of power is done as an optional choice or is required by a local code. Note that the NEC rule here does not itself require running power to the detached garage for the lighting outlet, but simply says that the lighting outlet must be provided *if* power is run to the garage.

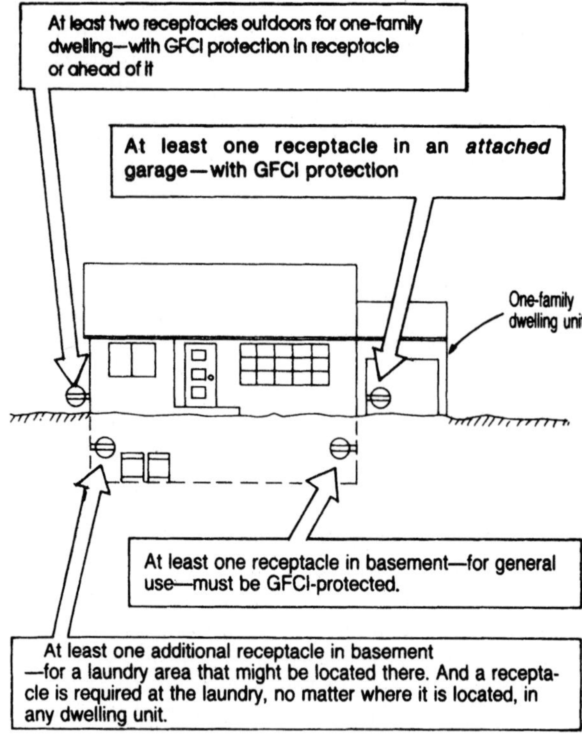
At least two receptacles outdoors for one-family dwelling—with GFCI protection in receptacle or ahead of it
At least one receptacle in an *attached* garage—with GFCI protection
One-family dwelling unit
At least one receptacle in basement—for general use—must be GFCI-protected.
At least one additional receptacle in basement —for a laundry area that might be located there. And a receptacle is required at the laundry, no matter where it is located, in any dwelling unit.

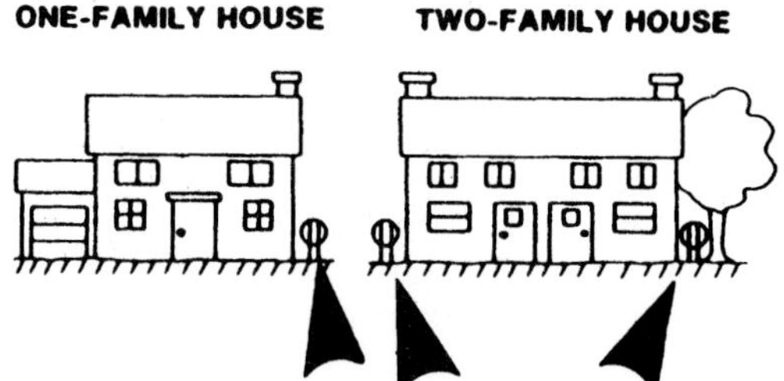
ONE-FAMILY HOUSE
TWO-FAMILY HOUSE
At least two receptacles must be installed out-doors—one at the front and one at the back—for a one-family dwelling and for each dwelling unit of a two-family dwelling at grade level—with GFCI protection in or ahead of each receptacle.

Fig. 210-58. These specific receptacles are required for dwelling occupancies. [Sec. 210-52(e), (f), and (g).]

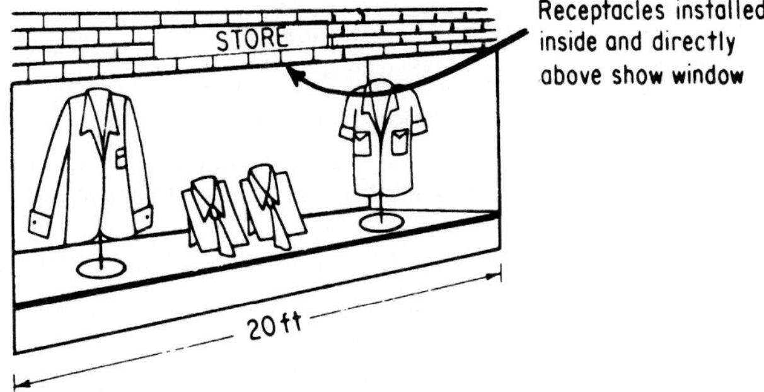

For a 20-ft-long store show window, a minimum of two receptacles must be installed, one for each 12 linear ft or major fraction thereof of show window length.

Fig. 210-59. Receptacles are required for show windows in stores or other buildings. [Sec. 210-62.]

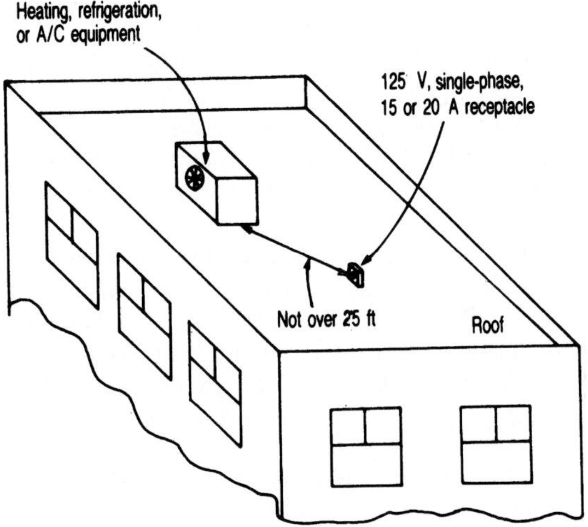

Fig. 210-60. Maintenance receptacle outlet required for rooftop mechanical equipment as well as for such equipment in attics and crawl spaces. [Sec. 210-63.]

The word “bathrooms” is in the basic rule because various building codes do not include bathrooms under their definition of “habitable rooms.” So the word “bathroom” was needed to assure that the rule covered bathrooms. The rule does not stipulate that the required “lighting outlets” must be ceiling lighting outlets; they also may be wall-mounted lighting outlets (Fig. 210-61).

A note clarifies that “a vehicle door in an attached garage is *not* considered as an outdoor entrance.” This makes it clear that the **Code** does not require such a light outlet at any garage door that is provided as a vehicle entrance because the lights of the car provide adequate illumination when such a door is being used during darkness. But the wording of this note does suggest that a rear or side door that is provided for personnel entry to an attached garage would be "considered as an outdoor entrance" because the note excludes only "vehicle" doors. Such personnel entrances from outdoors to the garage would seem to require a wall-switched lighting outlet.

At least one lighting outlet must be installed in every attic, underfloor space, utility room, and basement if it is used for storage or if it contains equipment requiring servicing. In such cases, the lighting outlet must be controlled by a wall switch at the entrance to the space. A lamp socket controlled by a pull chain or a canopy switch cannot be used. And each such required lighting outlet must be installed "at or near the equipment requiring servicing."

The last paragraph of part **(a)**, just before the Exceptions, states that lighting outlets for indoor stairways are required and must be controlled by a wall switch at each floor level connected by a stairway of six or more steps. This rule has the effect of requiring 3-way switching for control of the lighting outlet illuminating such stairways.

Three exceptions are given to the basic requirements. Exception No. 1 notes that in rooms other than kitchens and bathrooms, a wall switch-controlled receptacle outlet may be used instead of a wall switch-controlled lighting outlet. The receptacle outlet can serve to supply a portable lamp, which would give the necessary lighting for the room. Exception No. 2 states that "in hallways, stairways, and at outdoor entrances remote, central, or automatic control of lighting shall be permitted." This latter recognition appears to accept remote, central, or automatic control as an alternative to the wall switch control mentioned in the basic rules.

Part **(a)**, Exception No. 3, indicates two conditions under which the use of an occupancy sensor is permitted to control any of the required lighting outlets designated in the basic rule: (1) where used in addition to the required wall switch and (2) where the sensor is equipped with manual override and is mounted at the "customary" switch location. Notice that the literal wording only permits such control for "lighting outlets." Therefore, even though it's not entirely clear, it must be assumed that occupancy sensor control of a receptacle outlet—installed in accordance with the first exception to this section—is *not* permitted.

For lighting outlets—other than those required by part **(c)** of Sec. 210-70—in commercial, industrial, or institutional buildings, occupancy sensors may be used for lighting control with, or without, a conventional wall switch. There is also no requirement for manual override of occupancy sensors installed in other than dwellings. And, the actual sensor location is determined as indicated by the manufacturer—which may, or may not, be the customary wall switch location. Both of those details are design considerations. The requirements defined by this exception apply only to control of required lighting outlets at dwellings.

Where using any occupancy sensor for control of lighting, the use of a sensor that fails in the "on" position would be preferable to one that fails "off" or one that fails "as is." Such a fail-safe feature on any sensor is not required but is

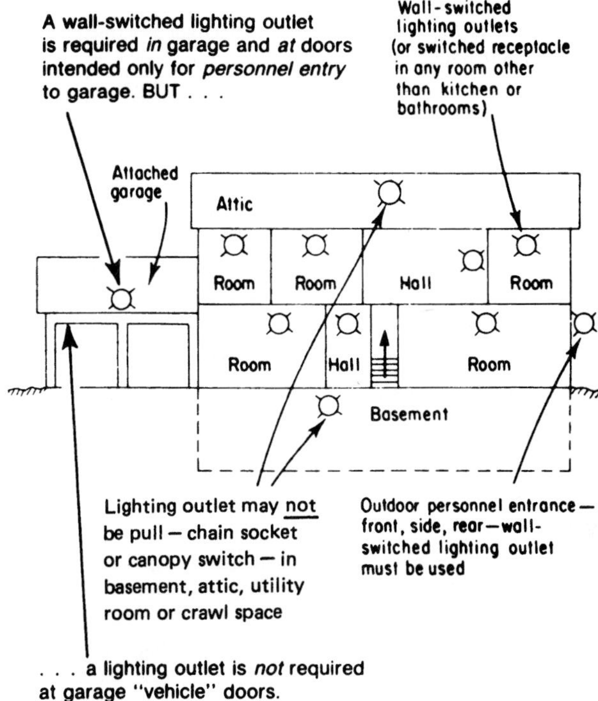
A wall-switched lighting outlet is required *in* garage and *at* doors intended only for *personnel entry* to garage. BUT . . .
Wall-switched lighting outlets (or switched receptacle in any room other than kitchen or bathrooms)
Attached garage
Attic
Room
Room
Hall
Room
Room
Hall
Room
Basement
Lighting outlet may not be pull – chain socket or canopy switch – in basement, attic, utility room or crawl space
Outdoor personnel entrance — front, side, rear — wall-switched lighting outlet must be used
. . . a lighting outlet is *not* required at garage "vehicle" doors.

EXCEPTION:

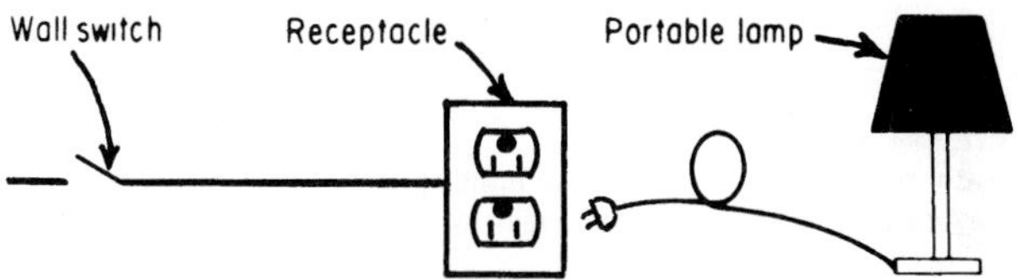
Wall switch
Receptacle
Portable lamp

Wall-switched receptacle(s) may be used instead of a lighting outlet in habitable rooms other than kitchens and bathrooms.

Fig. 210-61. Lighting outlets required in dwelling units. (Sec. 210-70.)

preferable because control of the lighting outlet can be provided from the conventional wall switch or manual override until such time as the sensor can be replaced (Fig. 210-62).

But note carefully that every kitchen and every bathroom must have at least one wall switch-controlled lighting outlet (Fig. 210-63).

Section 210-70(b) notes that "at least one wall switch controlled lighting outlet or wall switch controlled receptacle shall be installed in guest rooms in hotels, motels, or similar occupancies."

Part **(c)** requires that a wall switch-controlled lighting outlet must be provided in attics or underfloor spaces housing heating, A/C, and/or refrigeration

Sec. 210-70(a), Exception No. 3
Required outlets may be controlled by occupancy sensors.
Such control may be provided by either:

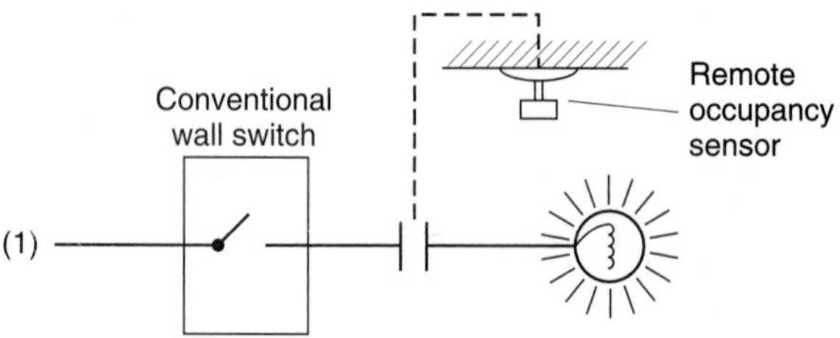

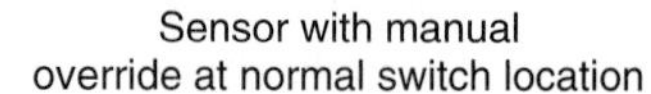

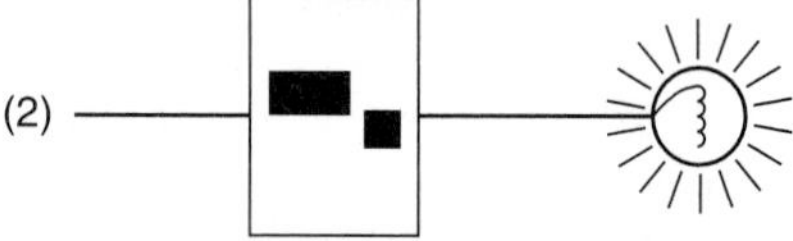

It is *not* clear if control of a receptacle outlet installed in accordance with Exception No. 1 to Sec. 210-70(a) is permitted.

Fig. 210-62. Occupancy sensor control for lighting outlets in dwellings must be as shown above. Either a sensor and a conventional wall switch or a sensor with a manual override installed at the "customary" wall-switch location must be provided. Remember, the rule here only applies to dwellings.

Kitchen and all bathrooms—
Each must have at least one lighting outlet that is **wall-switch-controlled** (not pullchain or switch in fixture or canopy)

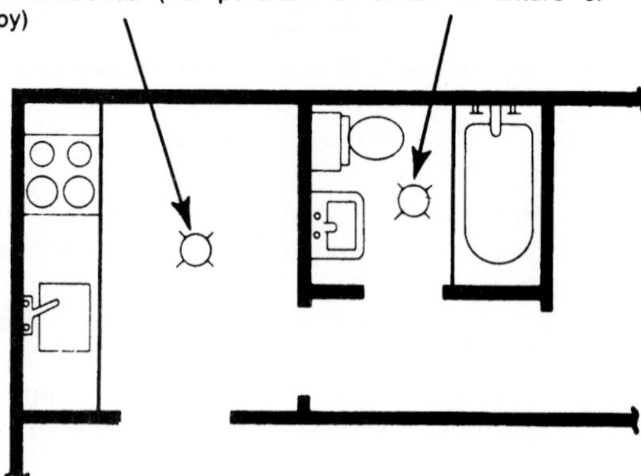

Fig. 210-63. Switch-controlled lighting outlet in kitchen and bathroom. [Sec. 210-70(a).]

equipment—*in other than dwelling units.* The lighting outlet must be located at or near the equipment to provide effective illumination. And the control wall switch must be installed at the point of entry to the space.

ARTICLE 215. FEEDERS

215-1. Scope. "Feeders" are the conductors which carry electric power from the service equipment (or generator switchboard, where power is generated on the premises) to the overcurrent protective devices for branch circuits supplying the various loads. "Subfeeders" originate at a distribution center other than the service equipment or generator switchboard and supply one or more other distribution panelboards, branch-circuit panelboards, or branch circuits. Code rules on feeders apply also to all subfeeders (Fig. 215-1).

For the given circuit voltage, feeders and subfeeders must be capable of carrying the amount of current required by the load, plus any current which may be required in the future. Selection of the size of a feeder depends upon the size and nature of the known load computed from branch-circuit data, the anticipated future load requirements, and voltage drop.

Article 215 deals with the determination of the minimum sizes of feeder conductors necessary for safety. Overloading of conductors may result in insulation breakdowns due to overheating; overheating of switches, busbars, and terminals; the blowing of fuses and consequent overfusing; excessive voltage drop; and excessive copper losses. Thus the overloading will in many cases create a fire risk and is sure to result in very unsatisfactory service.

The actual maximum load on a feeder depends upon the total load connected to the feeder and the demand factor. If at certain times the entire connected load is in operation, the demand factor is 100 percent; i.e., the maximum load, or maximum demand, is equal to the total connected load. If the heaviest load ever carried is only one-half the total connected load, the demand factor is 50 percent.

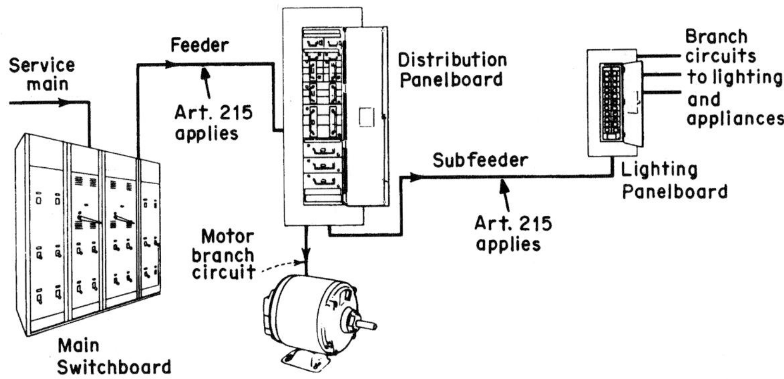

Fig. 215-1. Article 215 applies only to those circuits that conform to the NEC definition of "feeder." (Sec. 215-1.)

215-2. Minimum Rating and Size. There are two steps in the process of predetermining the maximum load that a feeder will be required to carry: first, a reasonable estimate must be made of the probable connected load; and, second, a reasonable value for the demand factor must be assumed. From a survey of a large number of buildings, the average connected loads and demand factors have been ascertained for lighting and small appliance loads in buildings of the more common classes of occupancy, and these data are presented in Sec. 220-3 and part **B** of Art. 220 as minimum requirements.

The load is specified in terms of voltamperes per square foot for certain occupancies. These loads are here referred to as standard loads, because they are minimum standards established by the **Code** in order to assure that the feeders and branch circuits will have sufficient carrying capacity for safety.

In this section, the last two sentences of the first paragraph note that it is never necessary for feeder conductors to be larger than the service-entrance conductors (assuming use of the same conductor material and the same insulation). In particular, this is aimed at those cases where the size of service-entrance conductors for a dwelling unit is selected in accordance with the higher-than-normal ampacities permitted by Note 3 to **Code** Table 310-16 for services to residential occupancies. If a set of service conductors for an individual dwelling unit are brought in to a single service disconnect (a single fused switch or circuit breaker) and load and the service conductors are sized for the increased ampacity value permitted by Note 3 to Table 310-16, diversity on the load-side feeder conductors gives them the same reduced heat-loading that enables the service conductors to be assigned the higher ampacity. This rule simply extends the permission of Note 3 to those feeders and is applicable for any such feeder for a dwelling unit (a one-family house or an apartment in a two-family or multifamily dwelling, such as an apartment house) or for mobile-home feed (Fig. 215-2). See the discussion on Note 3 of Table 310-16 in Art. 310.

Part **(a)** specifies that the feeder wires must never have an ampacity of less than 30 A when the feeder supplies at least the number of circuits as shown in Fig. 215-3.

As shown in Fig. 215-4, the rule of part **(b)** of this section requires that the ampacity of feeder conductors must be at least equal to that of the service conductors where the total service current is carried by the feeder conductors. In the case shown, No. 4 TW aluminum is taken as equivalent to No. 6 TW copper and has the same ampacity (55 A).

A note at the end of Sec. 215-2 comments on voltage drop in feeders. It should be carefully noted that the **NEC** does not establish any mandatory rules on voltage drop for either branch circuits or feeders. The references to 3 and 5 percent voltage drops are purely advisory—i.e., recommended maximum values of voltage drop. The **Code** does not consider excessive voltage drop to be unsafe.

The voltage-drop note suggests not more than 3 percent for feeders supplying power, heating, or lighting loads. It also provides for a maximum drop of 5 percent for the conductors between the service-entrance equipment and the connected load. If the feeders have an actual voltage drop of 3 percent, then only 2 percent is left for the branch circuits. If a lower voltage drop is obtained in the feeder, then the branch circuit has more voltage drop available, provided that the total drop does not exceed 5 percent. For any one load, the total voltage drop is made up of the voltage drop in the one or more feeders plus the voltage drop in the branch circuit supplying that load.

Again, however, values stated in the FPN on voltage drop are recommended values and are not intended to be enforced as a requirement.

Voltage drop must always be carefully considered in sizing feeder conductors, and calculations should be made for peak load conditions. For maximum efficiency, the size of feeder conductors should be such that voltage drop up to

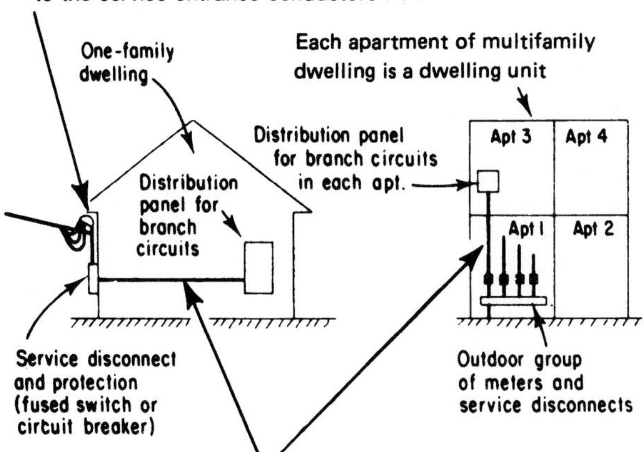

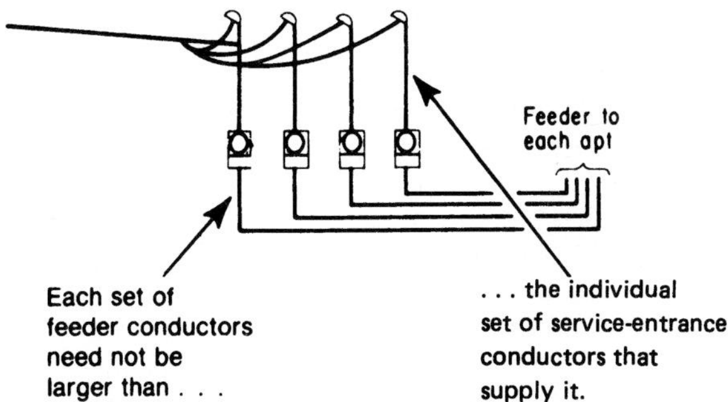

Fig. 215-2. Feeder conductors need not be larger than service-entrance conductors when higher ampacity of Note 3, Table 310-16, is used. (Sec. 215-2.)

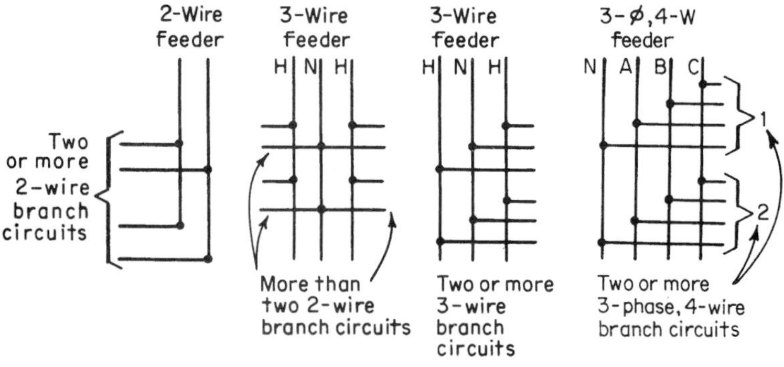

EXAMPLE

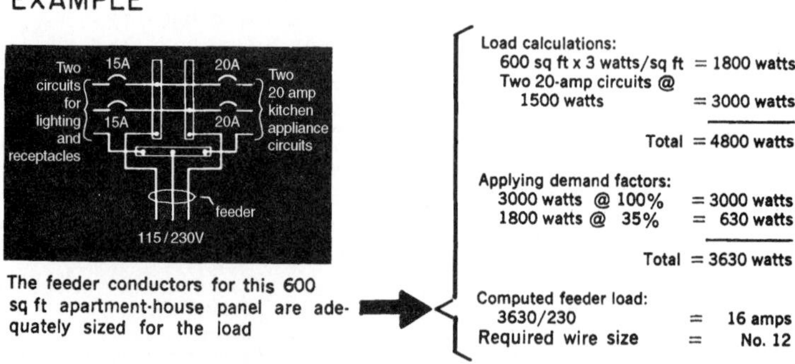

BUT, THE FEEDER CONDUCTORS MUST NOT BE SMALLER THAN No. 10 FROM SEC. 215-2 (a) (3).

Fig. 215-3. Feeder must have an ampacity of at least 30 A in these cases. Conductors must not be smaller than No. 10 TW copper or No. 8 TW aluminum. [Sec. 215-2(a).]

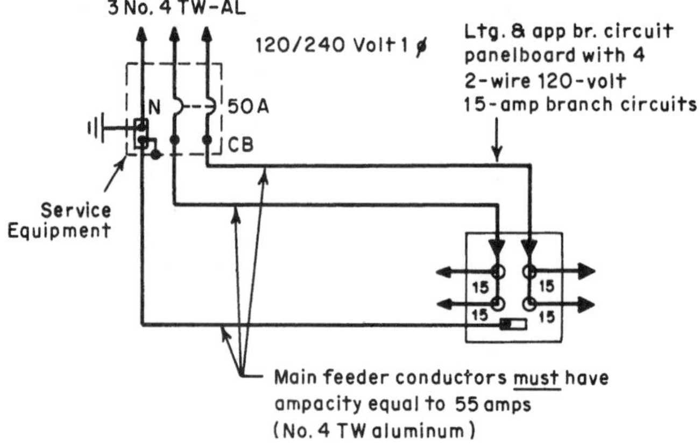

Fig. 215-4. Feeder conductors must not have ampacity less than service conductors. [Sec. 215-2(b).]

the branch-circuit panelboards or point of branch-circuit origin is not more than 1 percent for lighting loads or combined lighting, heating, and power loads and not more than 2 percent for power or heating loads. Local codes may impose lower limits of voltage drop. Voltage-drop limitations are shown in Fig. 215-5 for NEC levels and better levels of drop, as follows:

1. For combinations of lighting and power loads on feeders and branch circuits, use the voltage-drop percentages for lighting load (at left in Fig. 215-5).
2. The word *feeder* here refers to the overall run of conductors carrying power from the source to the point of final branch-circuit distribution, including feeders, subfeeders, sub-subfeeders, etc.
3. The voltage-drop percentages are based on nominal circuit voltage at the source of each voltage level. Indicated limitations should be observed for each voltage level in the distribution system.

There are many cases in which the above-mentioned limits of voltage drop (1 percent for lighting feeders, etc.) should be relaxed in the interests of reducing the prohibitive costs of conductors and conduits required by such low drops. In many installations a 5 percent drop in feeders is not critical or unsafe—such as in apartment houses.

Voltage-drop tables and slide calculators are available from a good number of electrical equipment manufacturers. Voltage-drop calculations vary according to the actual circuit parameters, e.g., AC or DC, single- or multiphase, power factor, circuit impedance, line reactance, types of enclosures (nonmetallic or metallic), length and size of conductors, and conductor material (copper, copper-clad aluminum, or aluminum).

Calculations of voltage drop in any set of feeders can be made in accordance with the formulas given in electrical design literature, such as those shown in Fig. 215-6. From this calculation, it can be determined if the conductor size initially selected to handle the load will be adequate to maintain voltage drop

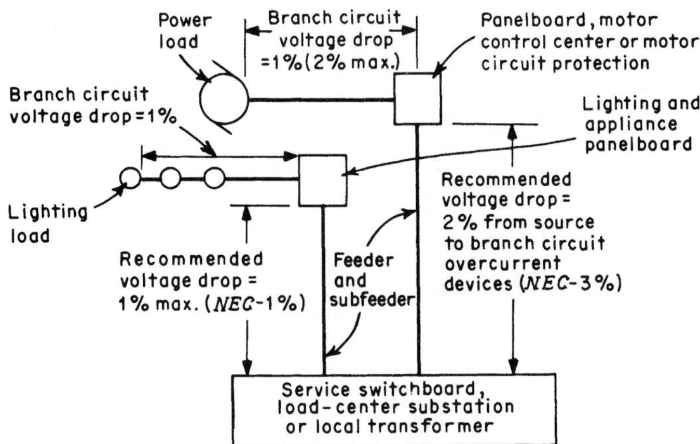

Fig. 215-5. Recommended basic limitations on voltage drop. (Sec. 215-2, FPN.)

Two-wire, single-phase circuits (inductance negligible)

$$V = \frac{2k \times L \times I}{d^2} = 2R \times L \times I$$

$$d^2 = \frac{2k \times L \times I}{V}$$

V = drop in circuit voltage (volts)
R = resistance per ft of conductor (ohms/ft)
I = current in conductor (amperes)

Three-wire, single-phase circuits (inductance negligible)

$$V = \frac{2k \times L \times I}{d^2}$$

V = drop between outside conductors (volts)
I = current in more-heavily loaded outside conductor (amps)

Three-wire, three-phase circuits (inductance negligible)

$$V = \frac{2k \times L \times I}{d^2} \times 0.866$$

V = voltage drop of 3-phase circuit

Four-wire, three-phase balanced circuits (inductance negligible)

Lighting loads
Voltage drop between one outside conductor and neutral equals one-half of drop calculated by formula above for 2-wire circuits.

Motor loads
Voltage drop between any two outside conductors equals 0.866 times the drop determined by formula above for two-wire circuits.

In the above formulas:

L = one-way length of circuit (ft)
d^2 = cross-section area of conductor (circular mils)
k = resistivity of conductor metal (cir mil-ohms/ft)
= 12 for circuits loaded to more than 50% of allowable circuit capacity
= 11 for circuits loaded less than 50%
= 18 for aluminum or copper-clad aluminum conductors

Example: **230-V two-wire heating circuit. Load is 24 A. Circuit size is No. 10 AWG copper, and the one-way circuit length is 200 ft.**

$$VD = \frac{24 \times 200 \times 24}{10{,}380} = \frac{115{,}200}{10{,}380} = 11$$

An 11-V drop on a 230-V circuit is about a 5 percent drop (11/230 = 0.0478). No. 8 AWG copper conductors would be needed to reduce the voltage drop to 3 percent on the branch circuit and allow 2 percent more on the feeder.

Fig. 215-6. Calculating voltage drop in feeder circuits. (Sec. 215-2, FPN.)

within given limits. If it is not, the size of the conductors must be increased (or other steps taken where conductor reactance is not negligible) until the voltage drop is within prescribed limits. Many such graphs and tabulated data on voltage drop are available in handbooks and from manufacturers. Figure 215-7 shows an example of excessive voltage drop—over 10 percent in the feeder.

215-4. Feeders with Common Neutral. A frequently discussed Code requirement is that of Sec. 215-4, covering the use of a common neutral with more than one set of feeders. This section says that a common neutral feeder may be used for two or three sets of 3-wire feeders or two sets of 4-wire feeders. It further requires that all conductors of feeder circuits employing a common neutral feeder must be within the same enclosure when the enclosure or raceway containing them is metal.

A common neutral is a single neutral conductor used as the neutral for more than one set of feeder conductors. It must have current-carrying capacity equal to the sum of the neutral conductor capacities if an individual neutral conductor were used with each feeder set. Figure 215-8 shows a typical example of a common neutral, used for three-feeder circuits. A common neutral may be used only with feeders. It may never be used with branch circuits. A single neutral of a multiwire branch circuit is not a "common neutral." It is the neutral of only a single circuit even though the circuit may consist of 3 or 4 wires. A feeder common neutral is used with more than one feeder circuit.

215-7. Ungrounded Conductors Tapped from Grounded Systems. Refer to Sec. 210-10 for a discussion that applies as well to feeder circuits as to branch circuits.

215-8. Means of Identifying Conductor with the Higher Voltage to Ground. The wording of this section recognizes orange as the preferred color of the high leg of a 4-wire delta supply without disturbing current practices in various local

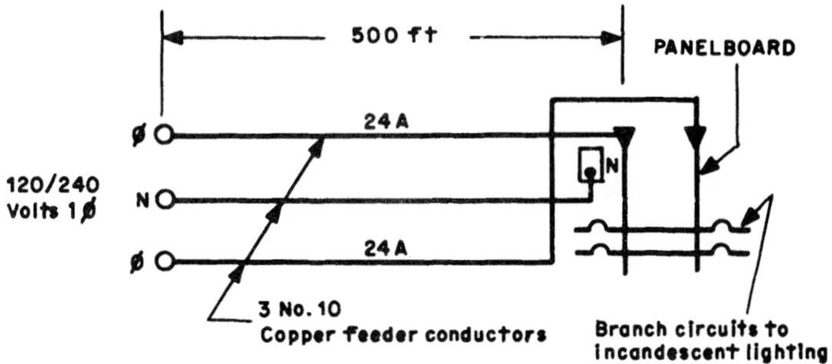

1. No. 10 copper conductor has a resistance of 1.018 ohms per 1000 ft (Table 8, Chapter 9).
2. The two 500-ft lengths of circuit conductors total 1000 ft and have a resistance of 1.018 ohms.
3. Voltage Drop = load current x conductor resistance
   = 24 amps x 1.018 ohms = **24.43 volts**
4. $\frac{24.43}{240}$ = 10.2% VOLTAGE DROP–*NEC* SUGGESTS MAX. 3%

Fig. 215-7. Feeder voltage drop should be checked. [Sec. 215-2, FPN.]

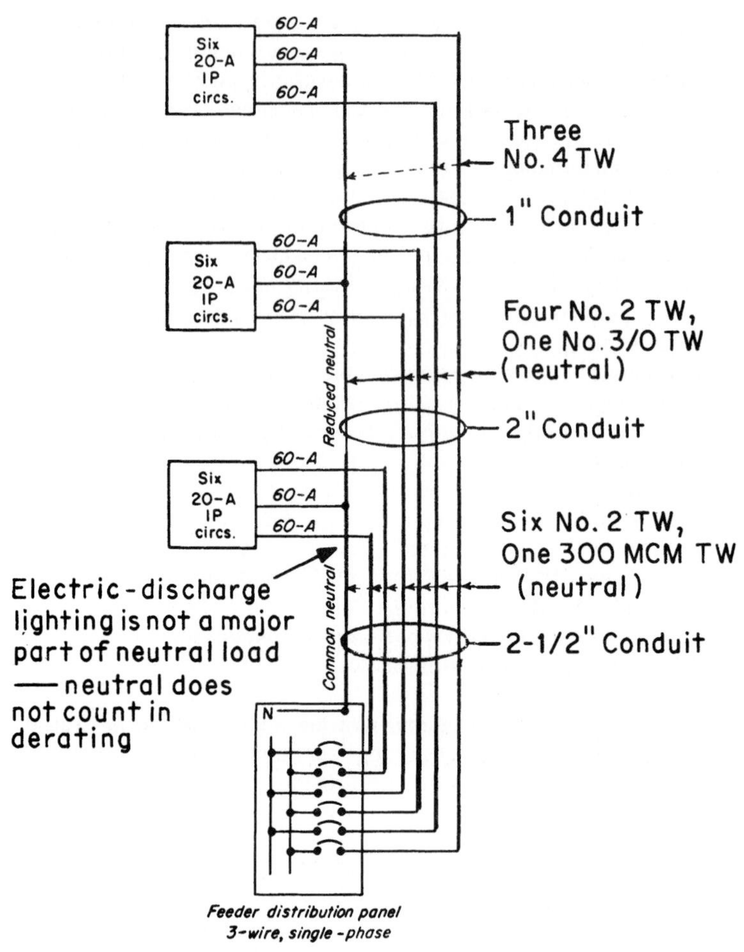

Fig. 215-8. Example of three feeder circuits using a single, "common neutral"—with neutral size reduced as permitted. (Sec. 215-4.)

areas where other colors (such as red, yellow, or blue) or other means of identification are required by electric utility regulations or by local code (Fig. 215-9).

Note that identification of the phase leg with 208 V to ground is required only at those points in the system where the neutral is present—such as in panelboards, motor-control centers, and other enclosures where circuits are connected. The purpose of this is to warn that 208 V, not 120 V, exist from the high leg to the neutral. Such indication minimizes the chance that a 208-V circuit might be accidentally or unwittingly connected to 120-V loads, such as lamps or appliances of 120-V operating coils in motor starters. Such connection would burn out 120-V equipment and presents a hazard to personnel.

215-9. Ground-Fault Circuit-Interrupter Protection for Personnel. A ground-fault circuit-interrupter may be located in the feeder and protect all branch circuits connected to that feeder. In such cases, the provisions of Sec. 210-8 and Art. 305 on temporary wiring will be satisfied and additional *downstream* ground-fault protection on the individual branch circuits would not be required. It should be mentioned, however, that downstream ground-fault protection is more desirable than ground-fault protection in the feeder because less equipment will be de-energized when the ground-fault circuit-interrupter opens the supply in response to a line-to-ground fault.

As shown in Fig. 215-10, if a ground-fault protector is installed in the feeder to a panel for branch circuits to outdoor residential receptacles, this protector will satisfy the NEC as the ground-fault protection required by Sec. 210-8 for such outdoor receptacles.

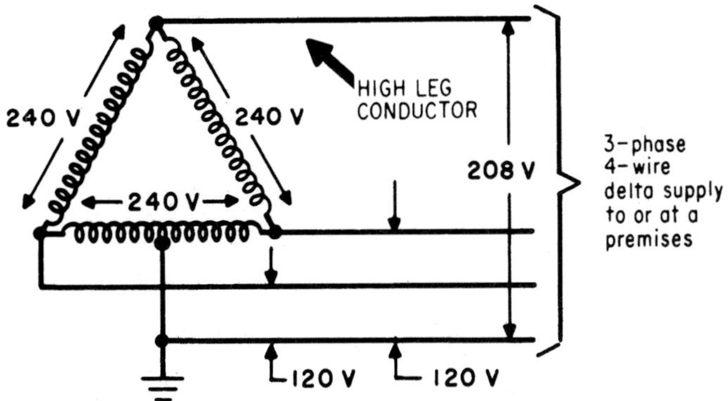

HIGH-LEG CONDUCTOR may be orange in color or may be some other color—such as red or yellow—as long as the color or tagging or other identification clearly distinguishes this as the one with higher voltage to ground at any connection point where the neutral is present.

Fig. 215-9. Identifying the high leg of 4-wire delta circuits. (Sec. 215-8.)

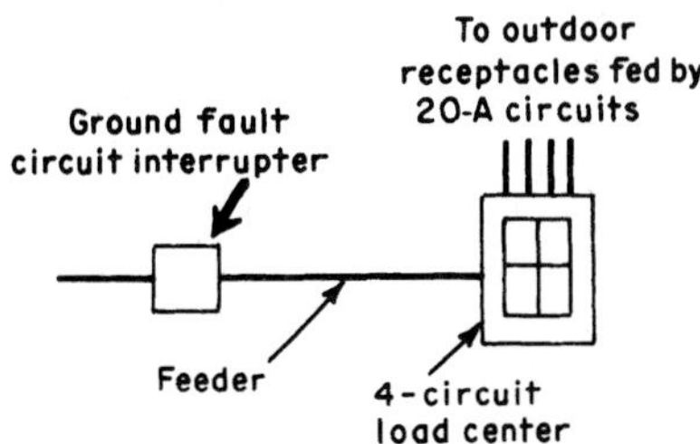

Fig. 215-10. GFCI in feeder does satisfy as protection for branch circuits. (Sec. 215-9.)

215-10. Ground-Fault Protection of Equipment. This section mandates equipment ground-fault protection for every feeder disconnect switch or circuit used on a 480Y/277-V, 3-phase, 4-wire feeder where the disconnect is rated 1,000 A or more, as shown in Fig. 215-11. This is a very significant Code requirement for ground-fault protection of the same type that has long been required by Sec. 230-95 for every *service* disconnect rated 1,000 A or more on a 480Y/277-V service.

An exception notes that feeder ground-fault protection is not required on a feeder disconnect if equipment ground-fault protection is provided on the supply (line) side of the feeder disconnect.

The substantiation submitted as the basis for the addition of this new rule stated as follows:

> Substantiation: The need for ground-fault equipment protection for 1000 amp or larger 277/480 grounded system is recognized and required when the service equipment is 277/480 volts. This proposal will require the same needed protection when the service equipment is not 277/480 volts. Past proposals attempted to require these feeders be treated as services in order to achieve this protection, but treating a feeder like a service created many other concerns. This proposal only addresses the feeder equipment ground-fault protection needs when it is not provided in the service equipment.

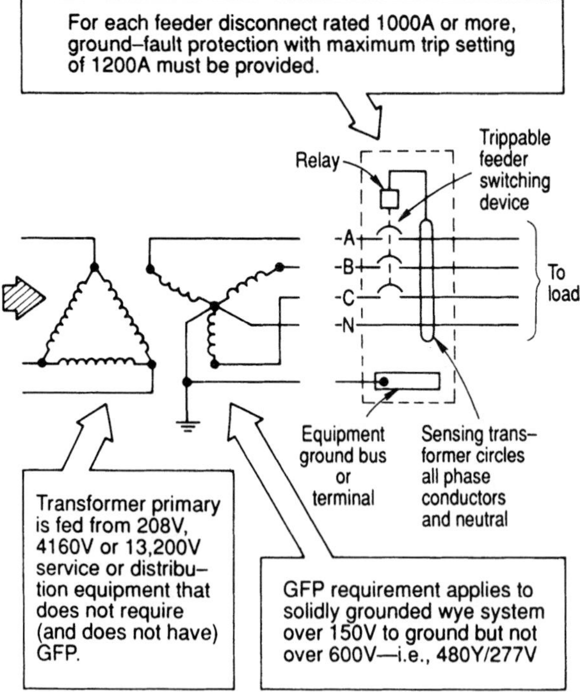

Fig. 215-11. A 480Y/277-V feeder disconnect rated 1,000 A or more must have ground-fault protection (GFP) if there is not GFP on its supply side. (Sec. 215-10.)

As noted, this rule calls for this type of feeder ground-fault protection when ground-fault protection is not provided on the supply side of the feeder disconnect, such as where a building has a high-voltage service (say, 13,200 V) or has, say, a 208Y/120-V service with a load-side transformer stepping-up the voltage to 480Y/277 V—because a service at either one of these voltages (e.g., 13.2 kV and 208 Y/120 V) is not required by Sec. 230-95 to have GFP.

215-11. Circuits Derived from Autotransformers. New to the 1996 NEC, this section recognizes application of autotransformers for supplying a feeder to a panelboard or group of overcurrent devices. This is the same permission given for branch circuits (see Sec. 210-9).

ARTICLE 220. BRANCH-CIRCUIT, FEEDER, AND SERVICE CALCULATIONS

220-1. Scope. All the calculations and design procedures covered by Art. 220 involve mathematical manipulation of units of voltage, current, resistance, and other measures of electrical conditions or characteristics.

220-2. Voltages. NE Code references to voltages vary considerably. The Code contains references to 120, 125, 115/230, 120/240, and 120/208 V. Standard voltages to be used for the calculations that have to be made to observe the rules of Art. 220 are 120, 120/240, 208Y/120, 240, 480Y/277, 480, 600Y/347, and 600 V. But use of lower voltage values (115, 230, 440, etc.) as denominators in calculations would not be a Code violation because the higher current values that result would assure Code compliance because of greater capacity in circuit wires and other equipment.

In all electrical systems there is a normal, predictable spread of voltage values over the impedances of the system equipment. It has been common practice to assign these basic levels to each nominal system voltage. The highest value of voltage is that at the service entrance or transformer secondary, such as 480Y/277 V. Then considering a voltage drop due to impedance in the circuit conductors and equipment, a "nominal" midsystem voltage designation would be 460Y/265. Variations in "nominal" voltages have come about because of (1) differences in utility-supply voltages throughout the country, (2) varying transformer secondary voltages produced by different and often uncontrolled voltage drops in primary feeders, and (3) preferences of different engineers and other design authorities. The reference to 600Y/347V is one of many additions that were made to the 1996 NEC that attempt to harmonize the NEC with the Canadian Electrical Code.

Because the NE Code is produced by contributors from all over the nation and of varying technical experiences, it is understandable that diversity of designations would creep in. As with many other things, we just have to live with problems until we solve them.

To standardize calculations, part **B** of Chap. 9 also specifies that nominal voltages of 120, 120/240, 240, and 208/120 V are to be used in computing the ampere load on a conductor. [Dividing these voltages into the watts load will produce lower current values than would 230 and 115 V; thus use of the lower voltage values results in larger (safer) conductor sizes.] *All* branch-circuit, feeder, and service conductor calculations made at those lower voltage levels would obviously satisfy the NEC minimum requirements on conductor sizes.

In some places, the NE Code adopts 115 V as the basic operating voltage of equipment designed for operation at 110 to 125 V. That is indicated in Tables 430-148 to 430-151. References are made to "rated motor voltages" of 115, 230, 460, 575, and 2,300 V—all values over 115 are integral multiples of 115. The last note in Tables 430-149 and 430-150 indicates that motors of those voltage ratings are applicable on systems rated 110–120, 220–240, 440–480, and 550–600 V. Although the motors can operate satisfactorily within those ranges, it is better to design circuits to deliver rated voltage. These Code voltage designations for motors are consistent with the trend over recent years for manufacturers to rate equipment for corresponding values of voltage.

Where calculations result in values involving a fraction of an ampere, the fraction may be dropped if it is less than 0.5. A value such as 20.7 A should be continued to be used as 20.7 or rounded off as the next higher whole number, in this case 21 A. Again, this is on the safe side. There are occasions, however, when current values must be added together. In such cases, it is on the safe side to retain fractions less than 0.5, since several fractions added together can result in the next whole ampere.

220-3. Computation of Branch Circuits. Part **(a)** is essentially the same as the requirement of Sec. 210-22(c). (Refer to that discussion.)

Article 220 gives the basic rules on calculations of loads for branch circuits and feeders. This note warns that Sec. 600-6(c) is another rule on this subject and requires that the 20-A branch circuit that must be supplied for a sign on the outside of every commercial occupancy must be taken as a computed load of 1,200 VA.

The task of calculating a branch-circuit load and then determining the size of circuit conductors required to feed that load is common to all electrical system calculations. Although it may seem to be a simple matter (and it usually is), there are many conditions which make the problem confusing (and sometimes controversial) because of the NE Code rules that must be observed.

The requirements for loading and sizing of branch circuits are covered in Art. 210 and in Sec. 220-3. In general, the following basic points must be considered:

- The ampacity of branch-circuit conductors must not be less than the maximum load to be served [Sec. 210-19(a)].
- The ampacity of branch-circuit conductors must generally not be less than the rating of the branch circuit. Section 210-19(a) requires that the conductors of a branch circuit that supplies any receptacle outlets must have an ampacity not less than the rating of the branch circuit, which rating is determined by the rating or setting of the overcurrent device protecting the circuit.
- The rating of a branch circuit is established by the rating or setting of its OC (overcurrent) protective device (Sec. 210-3).

- The normal, maximum, continuous ampacities of conductors in cables or raceways are given in Tables 310-16 to 310-19 for both copper and aluminum.
- These normal ampacities of conductors may have to be derated where there are more than three conductors in a cable or raceway (Note 8 to Tables 310-16 through 310-19).
- The current permitted to be carried by the branch-circuit protective device (fuse or CB) and conductors may have to be reduced if the load is continuous [Sec. 210-22(c) and Sec. 220-3(a)].

Section 210-20 says that the rating or setting of the branch-circuit overcurrent protective device is not to exceed that specified in Sec. 240-3 for conductors. Section 240-3 says that conductors shall be protected against overcurrent in accordance with their ampacities; but part **(b)** of that rule allows that where the ampacity of the conductor does not correspond with the standard ampere rating of a fuse or a circuit breaker, the next higher standard device rating shall be permitted if this rating does not exceed 800 A and if the wire being protected is not part of a branch circuit supplying receptacles for plug-connected portable tools, appliances, or other plug-in loads. In selecting the size of the branch-circuit overcurrent device, the rule of *both* Secs. 210-19(a) and 210-20 must be satisfied, because all **Code** rules bearing on a particular detail must always be observed.

Section 210-19(a) does not permit *any* case where branch-circuit conductors supplying one or more receptacle outlets would have an ampacity of less than the ampere rating of the circuit protective device. Section 210-19(a) thereby correlates to and repeats the first phrase in part **(b)** of Sec. 240-3 to prohibit using a protective device of "the next higher standard rating" on branch circuits to receptacles. However, a branch circuit that supplies only hard-wired outlets—such as lighting outlets or outlets to fixed electric heaters—may have its overcurrent protection selected as the next higher standard rating of protective device above the ampacity of a conductor when the conductor ampacity does not correspond to a standard fuse or CB rating—as permitted by part **(b)** of Sec. 240-3.

Section 220-3(a) must be evaluated against all those background data from Art. 210. Although determination of ampacities from Tables 310-16 through 310-19 yields the maximum allowable *continuous* current ratings of conductors, there are **Code** rules that limit the load that may be carried continuously (3 hr or more) to no more than 80 percent of the rating of the circuit protective device. Section 210-22(c) says that the total load on a branch circuit must not be more than the sum of noncontinuous load *plus* 125 percent of the continuous load. Although the rating of the branch circuit is set by the ampere rating or setting of the overcurrent device protecting the circuit, the conductors of a branch circuit supplying any receptacles may not have ampacity (either normal or derated) of less than the rating of the protective device [Section 210-19(a)]. As a repetition of the rule of Sec. 210-22(c), Sec. 220-3(a) requires a branch-circuit protective device to be rated not "less than the noncontinuous load plus 125 percent of the continuous load."

This wording is the reciprocal way of saying what Sec. 384-16(c) says—that the continuous load of a circuit must not exceed 80 percent of the rating of the branch-circuit protective device. (From Sec. 210-3, the "rating" of a branch circuit is determined by the "ampere" rating or setting of the overcurrent device.") The wording of both Sec. 210-22(c) and Sec. 220-3(a) states the need to limit heating effect. As shown in Fig. 220-1, although circuit has a total load current of 20 A, the loading satisfies the wording that the continuous load shall not exceed 80 percent of the rating of the branch-circuit overcurrent protective device (0.8 × 20 = 16 A). But, according to the rule of Sec. 220-3(a), with 4 A of noncontinuous load, the above circuit could carry only that amount of continuous load which, when multiplied by 1.25, would equal 16 A. Then, 16/1.25 = 12.8 A. Thus the maximum continuous load that would be permitted in addition to the 4-A noncontinuous load is 12.8 A. The branch-circuit rating (20 A) is "not less than the noncontinuous load [4 A] plus 125 percent of the continuous load" (1.25 × 12.8 A = 16 A).

[Section 220-10(b) also has the effect of limiting a continuous load to not more than 80 percent of the rating of any feeder CB or fuse protection that is not UL-listed for continuous loading to 100 percent of its rating.]

Although the rules of Secs. 210-22(c), 220-10(b), and 220-3(a) are aimed at limiting the load on the circuit protective device, the conductor's ampacity also must be based on the nature of the load. Just as is required for the overcurrent device, the conductor's ampacity must not be less than the noncontinuous load plus 125 percent of the continuous load, *except* where derating—either for number of conductors, Note 8, or elevated ambient temperature, Ambient Tem-

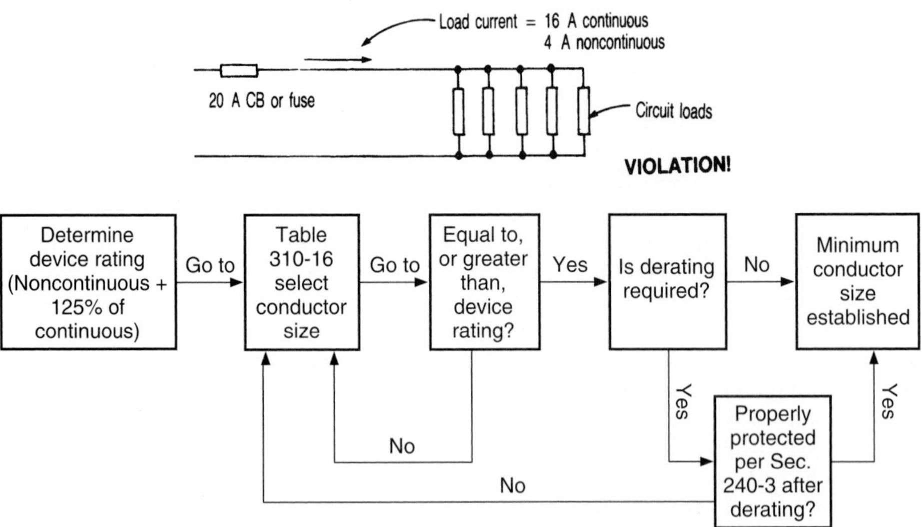

Fig. 220-1. Continuous load does not exceed 80 percent of the circuit rating (20 A); but the 20-A CB rating is *less* than "the noncontinuous load plus 125 percent of the continuous load" and violates Sec. 220-3(a). Flow chart for establishing minimum OC device rating and conductor size where supplying "continuous" load. [Sec. 210-22(c), 220-3(a), 220-10(b)]

perature Correction Factors at the bottom of Tables 310-16 through 310-19—is needed. In those cases, the conductor's ampacity must be less than the sum of noncontinuous plus 125 percent of the continuous load *before* any derating is applied. *And,* after derating *is* applied, the conductor's ampacity must be such that the overcurrent device protects the conductor as required or permitted by Sec. 240-3.

The continuous-current limitation, as set forth in those NE Code sections, is not established because the conductors cannot carry 100 percent of their rated current continuously. The conductors still have the same ampacity—the same maximum allowable continuous current rating. Likewise, a fused switch or circuit breaker can, itself, withstand the heat produced within it by 100 percent of its current rating. The 80 percent limit is set because of the following:

1. Conductors of any circuit must connect to the terminals of the fusible switch or circuit breaker that provides disconnect and protection for a branch circuit or feeder.
2. Current flow through a circuit produces heating in the fusible switch or circuit breaker as well as in the conductors.
3. The heat produced in the switch or CB does not generally harm the switch or CB itself, but that heat is readily conducted into the end lengths of conductors that are attached to the terminals.
4. Although the conductors can take the heat input from 100 percent of their own current rating, the extra heat conducted into the conductor from the switch or CB adds to the heat load on the conductors adjacent to the terminations.
5. For a continuous load, excessive heat will be produced in the conductor insulation if the conductor is already carrying its full rated current; and that can cause damage to the conductor insulation.

The effect of the rule of Sec. 220-3(a) is that any CB or fuse for a branch circuit supplying a total continuous load must have its load current limited to 80 percent, and only one exception is made for such protective devices—a continuous load of 100 percent of the fuse or CB rating is permitted *only* when the protective device assembly (fuse in switch or CB) is approved for 100 percent continuous duty. (And there are no such devices rated less than 225 A, so the exceptions referring to 100 percent rated devices do not apply to any branch circuits of less than 225 A. In addition, UL has a hard and fast rule that *any* breaker *not* marked for 100 percent continuous load must have its load limited to 80 percent of its rating.)

Code Table 220-3(b) lists certain occupancies (types of buildings) for which a minimum general lighting load is specified in voltamperes per square foot. In each type of building, there must be adequate branch-circuit capacity to handle the total load that is represented by the product of voltamperes per square foot times the square-foot area of the building. For instance, if one floor of an office building is 40,000 sq ft (3,720 m^2) in area, that floor must have a total branch-circuit capacity of 40,000 times 3½ VA/sq ft [Code Table 220-3(b)] for general lighting. Note that the total load to be used in calculating required circuit capacity must never be taken at less than the indicated voltamperes per square foot times square feet for those occupancies listed. Of course, if the branch-circuit load for lighting is determined from a lighting layout of specific fixtures of known voltamp rating, the load value must meet the previous voltamperes-per-square-foot minimum; and if the load from a known lighting layout is greater, then the greater voltamp value must be taken as the required branch-circuit capacity.

Note that the bottom of Table 220-3(b) requires a minimum general lighting load of ½ VA/sq ft to cover branch-circuit and feeder capacity for halls, corridors, closets, and all stairways. Likewise, an additional ¼ VA/sq ft must be provided for storage areas.

As indicated in Sec. 220-3(b), when the load is determined on a voltamperes-per-square-foot basis, open porches, garages, unfinished basements, and unused areas are not counted as part of the area. Area calculation is made using the *outside* dimensions of the "building, apartment, or other area involved."

When fluorescent or mercury-vapor lighting is used on branch circuits, the presence of the inductive effect of the ballast or transformer creates a power factor consideration. Determination of the load in such cases must be based on the total of the voltampere rating of the units and not on the wattage of the lamps.

Based on extensive analysis of load densities for general lighting in office buildings, Table 220-3(b) requires a minimum unit load of only 3½ VA/sq ft—rather than the previous unit value of 5—for "office buildings" and for "banks."

A double-asterisk note at the bottom of the table requires the addition of another 1 VA/sq ft to the 3½ value to cover the loading added by general-purpose receptacles in those cases where the actual number of receptacles is not known at the time feeder and branch-circuit capacities are being calculated. In such cases, a unit load of 4½ VA/sq ft must be used, and the calculation based on that figure will yield minimum feeder and branch-circuit capacity for both general lighting and all general-purpose receptacles that may later be installed.

Of course, where the actual number of general-purpose receptacles is known, the general lighting load is taken at 3½ VA/sq ft for branch-circuit and feeder capacity, and each strap or yoke containing a single, duplex, or triplex receptacle is taken as a load of 180 VA to get the total required branch-circuit capacity, with the demand factors of Table 220-13 applied to get the minimum required feeder capacity for receptacle loads.

Part **(c)** covers rules on providing branch-circuit capacity for loads other than general lighting and designates specific amounts of load that must be allowed for each outlet. This rule establishes the minimum loads that must be allowed in computing the minimum required branch-circuit capacity for general-use receptacles and "outlets not used for general illumination." Item **(3)** requires that the actual voltampere rating of a recessed lighting fixture be taken as the amount of load that must be included in branch-circuit capacity. This permits local and/or decorative recessed lighting fixtures to be taken at their actual load value rather than having them be taken as "other outlets," which would require a load allowance of "180 voltamperes per outlet"—even if each such fixture were lamped at, say, 25 W. Or, in the case where a recessed fixture contained a 300-W lamp, allowance of only 180 VA would be inadequate.

Similarly, track lighting, sign, and outline lighting must also be considered separately. Such lighting is *not* part of the general lighting load and therefore must be accounted for as indicated in the specific sections that cover those types of equipment.

Receptacle Outlets

The last sentence of Sec. 220-3(c)(7) calls for "each single or each multiple receptacle *on one strap*" to be taken as a load of "not less than 180 voltamperes"—in commercial, institutional, and industrial occupancies. The rule requires that every general-purpose, single or duplex or triplex convenience receptacle outlet in nonresidential occupancies be taken as a load of 180 VA, and that amount of circuit capacity must be provided for each such outlet (Fig. 220-2). Code intent is that each individual device strap—whether it holds one, two, or three receptacles—is a load of 180 VA. This rule makes clear that branch-circuit and feeder capacity must be provided for receptacles in nonresidential occupancies in accordance with loads calculated at 180 VA per receptacle strap.

If a 15-A, 115-V circuit is used to supply *only* receptacle outlets, then the maximum number of general-purpose receptacle outlets that may be fed by that circuit is

$$15 \text{ A} \times 115 \text{ V} \div 180 \text{ VA or 9 receptacle outlets}$$

For a 20-A, 115-V circuit, the maximum number of general-purpose receptacle outlets is

$$20 \text{ A} \times 115 \text{ V} \div 180 \text{ VA or 12 receptacle outlets}$$

See Fig. 220-3.

Note: In these calculations, the actual results work out to be 9.58 receptacles on a 15-A circuit and 12.77 on a 20-A circuit. Some inspectors round off these values to the nearest integral numbers and permit 10 receptacle outlets on a 15-A circuit and 13 on a 20-A circuit.

Although the Code gives the above-described data on maximum permitted number of receptacle outlets in commercial, industrial, institutional, and other nonresidential installations, there are no such limitations on the number of receptacle outlets on residential branch circuits. There are reasons for this approach.

In Sec. 210-52, the Code specifies where and when receptacle outlets are required on branch circuits. Note that there are no specific requirements for receptacle outlets in commercial, industrial, and institutional installations other than for store show windows in Sec. 210-62 and roof A/C equipment in Sec. 210-63. There is the general rule that receptacles do have to be installed where flexible cords are used. In nonresidential buildings, if flexible cords are not used, there is no *requirement* for receptacle outlets. They have to be installed only where they are needed, and the number and spacing of receptacles are completely up to the designer. But because the Code takes the position that receptacles in nonresidential buildings only have to be installed where

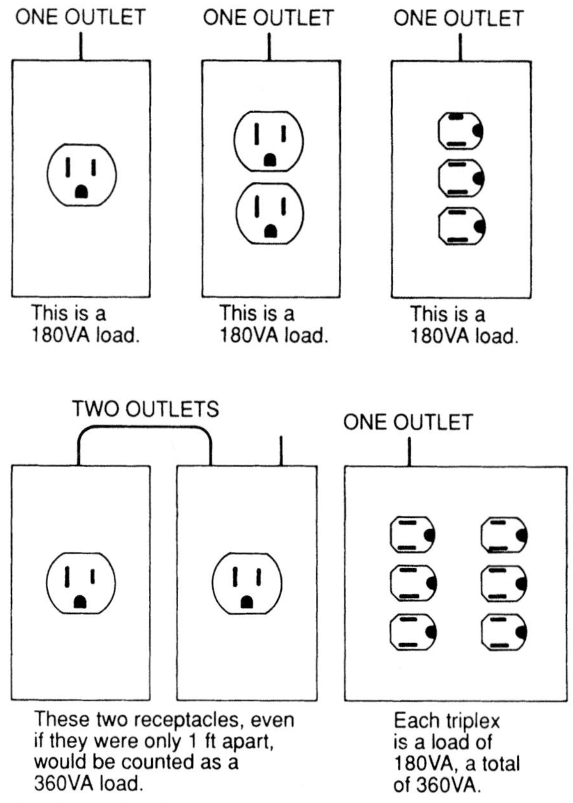

Fig. 220-2. Classification of single, duplex, and triplex receptacles. (Sec. 220-3.)

needed for connection of specific flexible cords and caps, it demands that where such receptacles are installed, each must be taken as a load of 180 VA.

A different approach is used for receptacles in dwelling-type occupancies. The Code simply assumes that cord-connected appliances will always be used in all residential buildings and requires general-purpose receptacle outlets of the number and spacing indicated in Secs. 210-52 and 210-60. These rules cover one-family houses, apartments in multifamily houses, guest rooms in hotels and motels, living quarters in dormitories, etc. But because so many receptacle outlets are required in such occupancies and because use of plug-connected loads is intermittent and has great diversity of load values and operating cycles, the Code notes at the bottom of Table 220-3(b) that the loads connected to such receptacles are adequately served by the branch-circuit capacity required by Sec. 220-4, and no additional load calculations are required for such outlets.

In dwelling occupancies, it is necessary to first calculate the total "general lighting load" from Sec. 220-3(b) and Table 220-3(b) (at 3 VA/sq ft for dwellings or 2 VA/sq ft for hotels and motels, including apartment houses without provi-

15A, 115V CIRCUIT—Maximum of 9 receptacle outlets

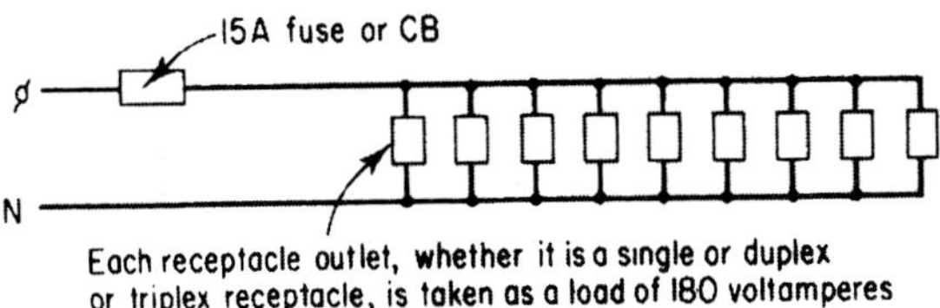
15A fuse or CB
ø
N
Each receptacle outlet, whether it is a single or duplex or triplex receptacle, is taken as a load of 180 voltamperes

20A, 115V CIRCUIT—Maximum of 12 receptacle outlets

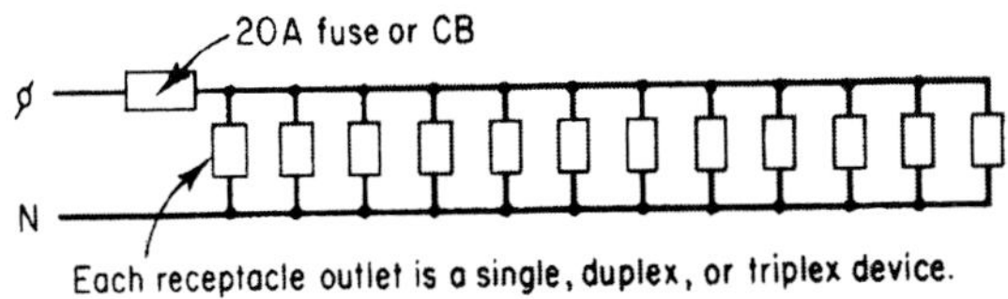
20A fuse or CB
ø
N
Each receptacle outlet is a single, duplex, or triplex device.

Fig. 220-3. Number of receptacles per circuit, nonresidential occupancy. [Sec. 220-3(c)(6).]

sions for cooking by tenants) and then provide the minimum required number and rating of 15-A and/or 20-A general-purpose branch circuits to handle that load as covered in Sec. 220-4(a). As long as that basic circuit capacity is provided, any number of lighting outlets may be connected to any general-purpose branch circuit, up to the rating of the branch circuit if loads are known. The lighting outlets should be evenly distributed among all the circuits. Although residential lamp wattages cannot be anticipated, the Code method covers fairly heavy loading.

When the above Code rules on circuits and outlets for general lighting in dwelling units, guest rooms of hotels and motels, and similar occupancies are satisfied, general-purpose convenience receptacle outlets may be connected on circuits supplying lighting outlets; or receptacles only may be connected on one or more of the required branch circuits; or additional circuits (over and above those required by Code) may be used to supply the receptacles. But no matter how general-purpose receptacle outlets are circuited, *any number* of general-purpose receptacle outlets may be connected on a residential branch circuit—with or without lighting outlets on the same circuit.

And when small-appliance branch circuits are provided in accordance with the requirements of Sec. 220-4(b), *any number* of small-appliance receptacle outlets may be connected on the 20-A small-appliance circuits—*but only* receptacle outlets may be connected to these circuits and only in the specified rooms.

Section 210-52(a) applies to spacing of receptacles connected on the 20-A small-appliance circuits, as well as spacing of general-purpose receptacle outlets. That section, therefore, establishes the *minimum* number of receptacles that must be installed for greater convenience of use.

Exception No. 1 to Sec. 220-3(c) requires branch-circuit capacity to be calculated for multioutlet assemblies (prewired surface metal raceway with plug outlets spaced along its length). Exception No. 1 says that each 1-ft length of such strip must be taken as a 180-VA load when the strip is used where a number of appliances are likely to be used simultaneously. For instance, in the case of industrial applications on assembly lines involving frequent, simultaneous use of plugged-in tools, the loading of 180 VA/ft must be used. (A loading of 180 VA for each 5-ft (1.52 m) section may be used in commercial or institutional applications of multioutlet assemblies when use of plug-in tools or appliances is not heavy.) Figure 220-4 shows an example of such load calculation.

Exception No. 3 permits branch-circuit capacity for the outlets required by Sec. 210-62 to be calculated as shown in Fig. 220-5—instead of using the load-per-outlet value from part **(c)**.

As noted in Exception No. 5, in calculating the size of branch-circuit, feeder, or service conductors, a load of 5,000 VA may be used for a household electric dryer when the actual dryer rating is not known. This is an Exception to Sec. 220-3(c) (1), which specifies that the "ampere rating of appliance or load served" shall be taken as the branch-circuit load for an outlet for a specific appliance. And where more than one dryer is involved, the demand factors of Table 220-18 may be used.

220-4. Branch Circuits Required. After following the rules of Sec. 220-3 to assure that adequate branch-circuit capacity is available for the various types of load that might be connected to such circuits, Sec. 220-4(a) requires that the minimum required number of branch circuits be determined from the total computed load, as computed from Sec. 220-3, and from the load rating of the branch circuits used.

For example, a 15-A, 115-V, 2-wire branch circuit has a load rating of 15 A times 115 V, or 1,725 VA. If the load is resistive, like incandescent lighting or electric heaters, that capacity is 1,725 W. If the total load of lighting, say, that was computed from Sec. 220-3 were 3,450 VA, then exactly two 15-A, 115-V, 2-wire branch circuits would be adequate to handle the load, provided that the load on the circuit is not a "continuous" load (one that operates steadily for 3 hr or more). Because Sec. 220-3(a) requires that branch circuits supplying a continuous load be loaded to not more than 80 percent of the branch-circuit rating, if the above load of 3,450 VA was a continuous load, it could not be supplied by *two* 15-A, 115-V circuits loaded to full capacity. A continuous load of 3,450 VA could be fed by *three* 15-A, 115-V circuits—divided among the three circuits in such a way that no circuit has a load of over 15 A times 115 V times 80 percent, or 1,380 VA. If 20-A, 115-V circuits are used, because each such circuit has a continuous load rating of 20 times 115 times 80 percent, or 1,840 VA, the total load of 3,450 VA can be divided between two 20-A, 115-V circuits. (A value of "120 V" could be used instead of "115 V.")

example Given the required unit load of 3 VA/sq ft for dwelling units [Table 220-3(b)], the **Code**-minimum number of 20-A, 120-V branch circuits required to supply general lighting and general-purpose receptacles (not small appliance receptacles in kitchen, dining room, etc.) in a 2,200-sq-ft-one-family house is three circuits. Each such 20-A circuit has a capacity of 2,400 VA. The required total circuit capacity is 2,200 times 3 VA/sq ft, or 6,600 VA. Then dividing 6,600 by 2,400 equals 2.75. Thus, at least three such circuits would be needed.

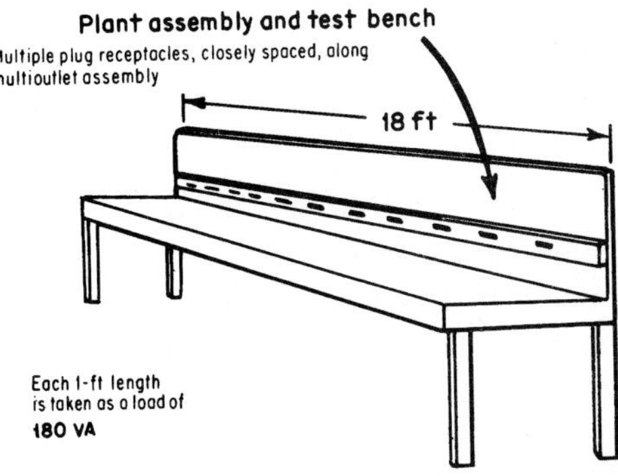
Plant assembly and test bench
Multiple plug receptacles, closely spaced, along multioutlet assembly
18 ft
Each 1-ft length is taken as a load of 180 VA
Load allowed for this bench = 18 X 180 VA = 3240 VA

Fig. 220-4. Calculating required branch-circuit capacity for multioutlet assembly. (Sec. 220-3(c), Exception No. 1.)

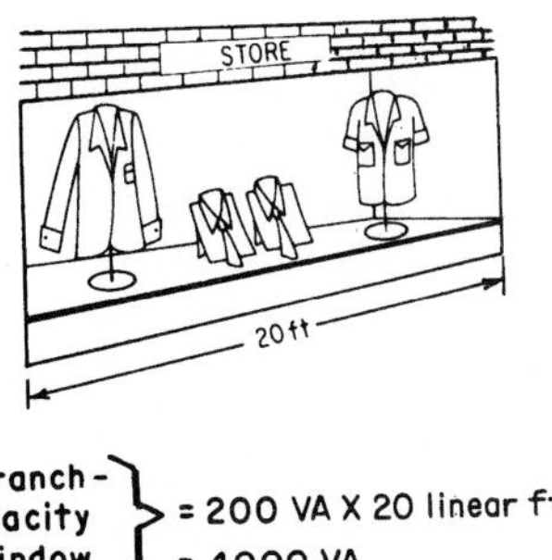
STORE
20 ft
Required branch-circuit capacity for show window = 200 VA X 20 linear ft = 4000 VA

Fig. 220-5. Alternate method for calculating show-window circuit capacity. (Sec. 220-3(c), Exception No. 3.)

example In Sec. 220-3(b), the NE Code requires a minimum unit load of 3 VA/sq ft for general lighting in a school, as shown in Table 220-3(b). For the school in this example, *minimum capacity for general lighting* would be

$$1{,}500 \text{ sq ft} \times 3 \text{ VA/sq ft} \qquad \text{or } 4{,}500 \text{ VA}$$

By using 115-V circuits, when the total load capacity of branch circuits for general lighting is known, it is a simple matter to determine how many lighting circuits are needed. By dividing the total load by 115 V (using "120 V" would yield a lower current), the total current capacity of circuits is determined:

$$\frac{4{,}500 \text{ VA}}{115 \text{ V}} = 39.1 \text{ A}$$

But, because the circuits will be supplying continuous lighting loads (over 3 hr), it is necessary to multiply that value by 1.25 in order to keep the load on any circuit to not more than 80 percent of the circuit rating. Then, using either 15- or 20-A, 2-wire, 115-V circuits (and dropping the fraction of an ampere) gives

$$\frac{1.25 \times 39 \text{ A}}{15 \text{ A}} = 3.25$$

which means four 15-A circuits, or

$$\frac{1.25 \times 39 \text{ A}}{20 \text{ A}} = 2.43$$

which means three 20-A circuits. And then each circuit must be loaded without exceeding the 80 percent maximum on any circuit.

Part **(b)** of Sec. 220-4 requires that two or more 20-A branch circuits be provided to supply all the receptacle outlets required by Sec. 210-52(b) in the kitchen, pantry, dining room, breakfast room, and any similar area of any dwelling unit—one-family houses, apartments, and motel and hotel suites with cooking facilities or serving pantries. That means that at least one 3-wire, 20-A, 240/120- or 208/120-V circuit shall be provided to serve only receptacles for the small-appliance load in the kitchen, pantry, dining room, and breakfast room of any dwelling unit. Of course, two 2-wire, 20-A, 120-V circuits are equivalent to the 3-wire circuit and could be used. If a 3-wire, 240/120-V circuit is used to provide the required two-circuit capacity for small appliances, the 3-wire circuit can be split-wired to receptacle outlets in these areas.

Part **(c)** of Sec. 220-4 requires that at least one 20-A branch circuit be provided for the one or more laundry receptacles installed, as required by Sec. 210-52(f), at the laundry location in a dwelling unit. Further, the last sentence of part **(c)** prohibits use of the laundry circuit for supplying outlets that are not for laundry equipment and because laundry outlets are required by Sec. 210-50(c) to be within 6 ft. (1.83 m) of the intended location of the appliance, it would seem that any receptacle outlet more than 6 ft. (1.83 m) from laundry equipment could not be connected to the required 20-A laundry circuit (Fig. 220-6).

Part **(d)** of Sec. 220-4 makes clear that a feeder to a branch-circuit panelboard and the main busbars in the panelboard must have a minimum ampacity to serve the *calculated* total load of lighting, appliances, motors, and other loads supplied. And the amount of feeder and panel ampacity required for the general lighting load must not be less than the amp value determined from the circuit voltage and the total voltamperes resulting from the minimum unit load from Table 220-3(b) (voltamperes per square foot) times the area of the occupancy supplied by the feeder—even if the actual connected load is less than the calculated load determined on the voltamperes-per-square-foot basis. (Of course, if the connected load is greater than that calculated on the voltamperes-per-square-foot basis, the greater value of load must be used in determining the number of branch circuits, the panelboard capacity, and the feeder capacity).

It should be carefully noted that the first sentence of Sec. 220-4(d) states, "Where the load is computed on a voltamperes-per-square-foot (0.93 sq m) basis, the *wiring system* up to and including the branch-circuit panelboard(s) shall be provided to serve not less than the calculated load." Use of the phrase "wiring system up to and including" requires that a feeder must have capacity for the total minimum branch-circuit load determined from square-foot area times the minimum unit load [voltamperes per square foot from Table 220-

3(b)]. And the phrase clearly requires that amount of capacity to be allowed in every part of the distribution system supplying the load. The required capacity would, for instance, be required in a subfeeder to the panel, in the main feeder from which the subfeeder is tapped, and in the service conductors supplying the whole system.

Actually, reference to "wiring system" in the wording of Sec. 220-4(d) presents a requirement that goes beyond the heading, "Branch Circuits Required," of Sec. 220-4 and, in fact, constitutes a requirement on *feeder* capacity that supplements the rule of the second sentence of Sec. 220-10(a). This requires a feeder to be sized to have enough capacity for "the computed load"—as determined by part **A** of this article (which means, as computed in accordance with Sec. 220-3).

A second part of Sec. 220-4(d) affects the required minimum number of branch circuits. Although the feeder and panelboard must have a minimum

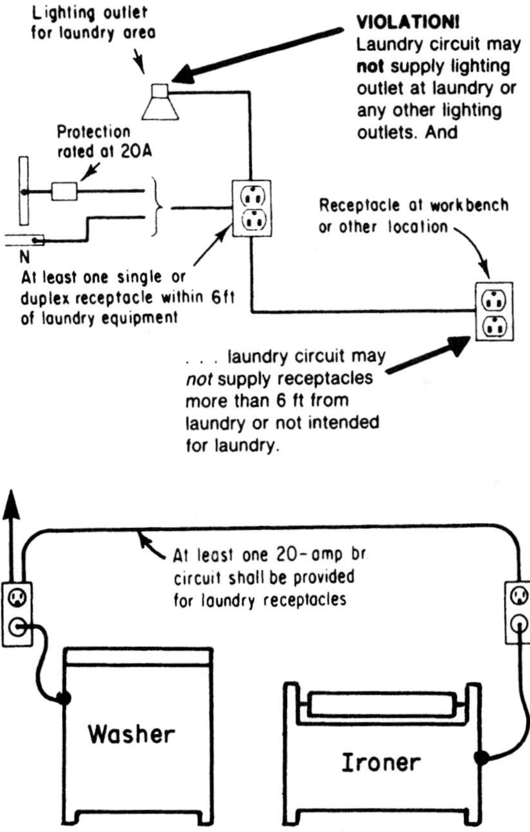

Fig. 220-6. *No* "other outlets" are permitted on 20-A circuit required for laundry receptacle(s). [Sec. 220-4(c).]

ampacity for the *calculated* load, it is only necessary to install the number of branch-circuit overcurrent devices and circuits required to handle the actual connected load in those cases where it is less than the calculated load. The last sentence of Sec. 220-4(d) is clearly an exception to the basic rule of the first sentence of Sec. 220-4(a), which says that "The minimum number of branch circuits *shall* be determined from the *total computed* load. . . ." Instead of having to supply *that minimum* number of branch circuits, it is necessary to have only the number of branch circuits required for the actual total "connected load."

example For an office area of 200 × 200 ft, a 3-phase, 4-wire, 460/265-V feeder and branch-circuit panelboard must be selected to supply 277-V HID lighting that will operate continuously (3 hr or more). The actual continuous connected load of all the lighting fixtures is 92 kVA. What is the minimum size of feeder conductors and panelboard rating that must be used to satisfy Sec. 220-4?

200 ft × 200 ft = 40,000 sq ft
40,000 sq ft @ minimum of 3.5 VA/sq ft = 140,000 VA

The minimum computed load for the feeder for the lighting is

140,000 VA ÷ [(480)(1.732)] = 168 A per phase

The actual connected lighting load for the area, calculated from the lighting design, is

92,000 VA ÷ [(480)(1.732)] = 111 A per phase

Sizing of the feeder and panelboard must be based on 168 A, *not* 111 A, to satisfy Sec. 220-4(d).

The next step is to correlate the rules of Sec. 220-4(a) and (d) with those of Sec. 220-10. Section 220-10(a) requires a feeder to be sized for the "computed load" as determined by part **A** [Sec. 220-3(b)]. The feeder to the continuous calculated load of 168 A must have an ampacity at least equal to that load, and the feeder protective device must be sized at 125 percent of the continuous load of 168 A, when using a CB or fused switch that is not UL-listed for continuous operation at 100 percent of rating, as required by Sec. 220-10(b).

168 × 1.25 = 210 A [Sec. 220-10(b)]

1. Assuming use of a non-100 percent rated protective device, the overcurrent device must be rated not less than 1.25 × 168 A, or 210 A—which calls for a standard 225-A circuit breaker or fuses (the standard rating above 210-A).
2. Although feeder conductors with an ampacity of 210 A would be adequate for the load, they would not be properly protected (Sec. 240-3) by 225-A device after derating. The feeder must have an ampacity that is not less than 210 A (168 A × 1.25) before derating *but* must also be properly protected by the 225-A rated device *after* derating.
3. Using Table 310-16, the smallest size of feeder conductor that would be protected by 225-A protection after 80 percent derating for number of conductors is a No. 4/0 THHN or XHHW copper, with an ampacity of 260 A before derating (260 A × .8 = 208 A).
4. Because the UL requires that conductors larger than No. 1 AWG must be used at no more than their 75°C ampacities to limit heat rise in equipment terminals, the selected No. 4/0 THHN or XHHW copper conductor must not operate at more than 230 A—which is the table value of ampacity for a 75°C No. 4/0 copper conductor. And the load current of 168 A is well within that 230 A maximum.

Thus, all requirements of Sec. 220-10(b) and UL are satisfied.

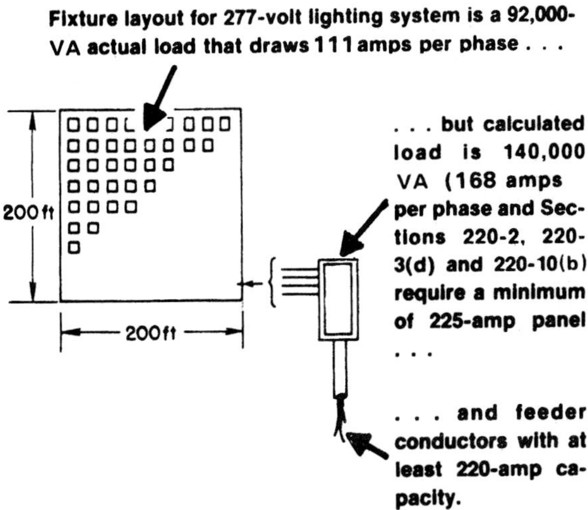
Fixture layout for 277-volt lighting system is a 92,000-VA actual load that draws 111 amps per phase . . .
200ft
200ft
. . . but calculated load is 140,000 VA (168 amps per phase and Sections 220-2, 220-3(d) and 220-10(b) require a minimum of 225-amp panel . . .
. . . and feeder conductors with at least 220-amp capacity.

Section 384-13 requires the panelboard here to have a rating not less than the minimum required ampacity of the feeder conductors—which, in this case, means the panel must have a busbar rating not less than 168 A. A 225-A panelboard (i.e., the next standard rating of panelboard above the minimum calculated value of load current—168 A) is therefore required, even though it might seem that a 125-A panel would be adequate for the actual load current of 111 A.

The number of branch-circuit protective devices required in the panel (the number of branch circuits) is based on the size of branch circuits used and their capacity related to connected load. If, say, all circuits are to be 20-A, 277-V phase-to-neutral, each pole may be loaded no more than 16 A because Sec. 220-3(a) requires the load to be limited to 80 percent of the 20-A protection rating. With the 111 A of connected load per phase, a single-circuit load of 16 A calls for a minimum of 111 ÷ 16, or 8 poles per phase leg. Thus a 225-A panelboard with 24 breaker poles would satisfy the rule of Sec. 220-4(d).

220.10. General.

Calculating Feeder Load

The key to accurate determination of required feeder conductor capacity in amperes is effective calculation of the total load to be supplied by the feeder. Feeders and subfeeders are sized to provide sufficient power to the circuits they supply. For the given circuit voltage, they must be capable of carrying the amount of current required by the load, plus any current which may be required in the future. The size of a feeder depends upon known load, future load, and voltage drop.

The minimum load capacity which must be provided in any feeder or subfeeder can be determined by considering NE Code requirements on feeder load. As presented in Sec. 220-10, these rules establish the minimum load capacity to be provided for all types of loads.

Part **(a)** of Sec. 220-10 requires feeder conductors to have ampacity at least equal to the sum of loads on the feeder, as determined from Sec. 220-3. Then part **(b)** rules on the rating of any feeder protective device.

If an overcurrent protective device for feeder conductors is not UL-listed for continuous operation at 100 percent of its rating, the load on the device must not exceed the noncontinuous load plus 125 percent of the continuous load. The first paragraph of part **(b)** applies to feeder overcurrent devices—circuit breakers and fuses in switch assemblies—and requires that the rating of any such protective device must generally never be less than the amount of noncontinuous load of the circuit (that amount of current that will not be flowing for 3 hr or longer) plus 125 percent of the amount of load current that will be continuous (flowing steadily for 3 hr or longer) (Fig. 220-7).

For any given load to be supplied by a feeder, after the minimum rating of the overcurrent device is determined from the above calculation (noncontinuous plus 125 percent of continuous), then a suitable size of feeder conductor must be selected. For each ungrounded leg of the feeder (the so-called phase legs of the circuit), the conductor ampacity must be at least equal to the amount of noncontinuous current plus the amount of continuous current, from the NEC tables of ampacity (Tables 310-16 through 310-19 and their accompanying notes).

Although the rules of Secs. 210-22(c), 220-10(b), and 220-3(a) are aimed at limiting the load on the circuit protective device, the conductor's ampacity also must be based on the nature of the load. Just as is required for the overcurrent device, the conductor's ampacity must not be less than the noncontinuous load plus 125 percent of the continuous load, *except* where derating—either for number of conductors, Note 8, or elevated ambient temperature, Ambient Temperature Correction Factors at the bottom of Tables 310-16 through 310-19—is needed. In those cases, the conductor's ampacity must be less than the sum of noncontinuous plus 125 percent of the continuous load *before* any derating is applied. *And,* after derating *is* applied, the conductor's ampacity must be such that the overcurrent device protects the conductor as required or permitted by Sec. 240-3.

Note that the conductor size increase described above applies only to the ungrounded or phase conductors because they are the ones that must be properly protected by the rating of the protective device. A neutral or grounded conductor of a feeder does not have to be increased; its size must simply have ampacity sufficient for the neutral load as determined from Sec. 220-22.

The Exception for Sec. 220-10(b) notes that a circuit breaker or fused switch that is UL-listed for continuous operation at 100 percent of its rating may be loaded right up to a current equal to the device rating. Feeder ungrounded conductors must be selected to have ampacity equal to the noncontinuous load plus the continuous load. The neutral conductor is sized in accordance with Sec. 220-22, which permits reduction of neutral size for feeders loaded over 200 A that do not supply electric-discharge lighting, data processing equipment, or other "nonlinear loads" that generate high levels of harmonic currents in the neutral.

Fuses for feeder protection The rating of a fuse is taken as 100 percent of rated nameplate current when enclosed by a switch or panel housing. But, because of the heat generated by many fuses, the maximum continuous load permitted on a fused switch is restricted by a number of NEMA, UL, and NE Code rules to 80

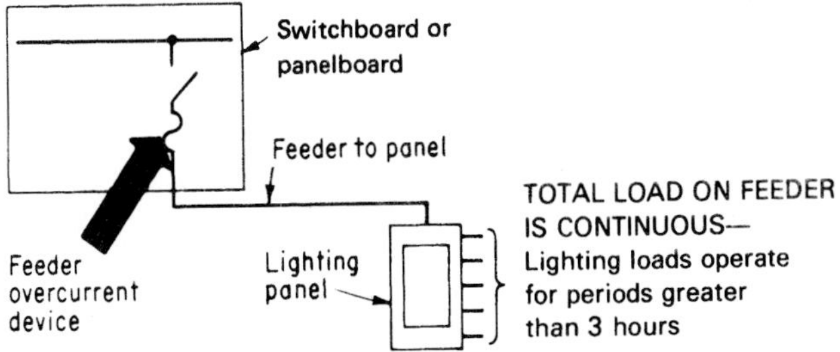

FEEDER OVERCURRENT DEVICE must be rated not less than 125% of the continuous load *and* the feeder conductors must be sized so they have an ampacity such that they are properly protected by the rating of the feeder CB or fuses, as required by Sec. 240-3. Another way of saying that is "the continuous load must not exceed 80% of the rating of the protection."

EXAMPLE: For this feeder, with conductors rated at 380 A, the maximum continuous load permitted for a conventional fused switch is 80% of the 400-A fuse rating [400X0.8 = 320A].

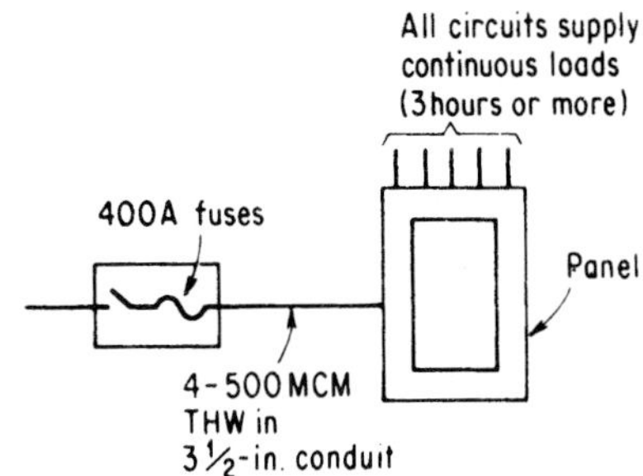

Rating of fuses must be at least equal to 125% times the continuous load and the 400-A rating is proper protection for conductors with an ampacity of 380 A.

Fig. 220-7. Feeders must generally be loaded to no more than 80 percent for a continuous load. [Sec. 220-10(b).]

percent of the rating of the fuses. Limitation of circuit-load current to *no more than* 80 percent of the current rating of fuses in equipment is done to protect the switch or other piece of equipment from the heat produced in the fuse element—and also to protect attached circuit wires from excessive heating close to the terminals. The fuse itself can actually carry 100 percent of its current rating continuously without damage to itself, but its heat is conducted into the adjacent wiring and switch components.

NEMA standards require that a fused, enclosed switch be marked, as part of the electrical rating, "Continuous Load Current Not to Exceed 80 Percent of the Rating of Fuses Employed in Other Than Motor Circuits" (Fig. 220-8). That derating compensates for the extra heat produced by continuous operation. Motor circuits are excluded from that rule, but a motor circuit is required by the **NE Code** to have conductors rated at least 125 percent of the motor full-load current—which, in effect, limits the load current to 80 percent of the conductor ampacity and limits the load on the fuses rated to protect those conductors. But, the UL *Electrical Construction Materials Directory* does recognize fused bolted-pressure switches and high-pressure butt-contact switches for use at 100 percent of their rating on circuits with available fault currents of 100,000, 150,000, or 200,000 rms symmetrical A—as marked (Fig. 220-9). (See "Fused Power Circuit Devices" in that UL publication.)

Manual and electrically operated switches designed to be used with Class L current-limiting fuses rated 601 to 4,000 A, 600 V AC are listed by UL as "Fused Power Circuit Devices." This category covers bolted-pressure-contact switches and high-pressure, butt-type-contact switches suitable for use as feeder devices or service switches if marked "Suitable for Use As Service Equipment." Such devices "have been investigated for use at *100 percent of their rating* on circuits having available fault currents of 100,000, 150,000 or 200,000 rms symmetrical amperes" as marked.

CB for feeder protection The nominal or theoretical continuous-current rating of a CB generally is taken to be the same as its trip setting—the value of current at which the breaker will open, either instantaneously or after some intentional time delay. But, as described above for fuses, the real continuous-current rating of a CB—the value of current that it can safely and properly carry for periods of 3 hr or more—frequently is reduced to 80 percent of the nameplate value by codes and standards rules.

The UL *Electrical Construction Materials Directory* contains a clear, simple rule in the instructions under "Circuit Breakers, Molded-Case." It says:

> Unless otherwise marked, circuit breakers should not be loaded to exceed 80 percent of their current rating, where in normal operation the load will continue for three or more hours.

A load that continues for 3 hr or more is a *continuous* load. If a breaker is marked for *continuous* operation, it may be loaded to 100 percent of its rating and operate continuously.

There are some CBs available for continuous operation at 100 percent of their current rating, but they must be used in the mounting and enclosure arrangements established by UL for 100 percent rating. Molded-case CBs of the 100 percent continuous type are made in ratings from 225 A up. Information on use of 100 percent rated breakers is given on their nameplates.

Figure 220-10 shows two examples of CB nameplate data for two types of UL-listed 2,000-A, molded-case CBs that are specifically tested and listed for continuous operation at 100 percent of their 2,000-A rating—*but* only under the conditions described on the nameplate. These two typical nameplates clearly indicate that ventilation may or may not be required. Because most switchboards have fairly large interior volumes, the "minimum enclosure" dimensions shown on these nameplates (45 by 38 by 20 in.) (1,143 by 965 by 503 mm) usually are readily achieved. *But,* special UL tests must be performed if these dimensions are *not* met. Where busbar extensions and lugs are connected to the

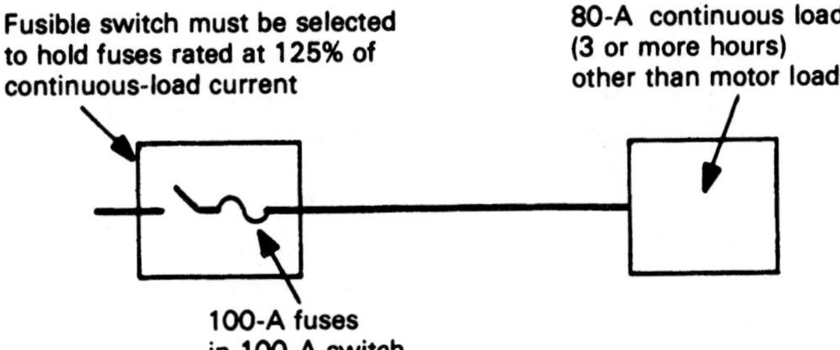

Circuit conductors must be sized so that they have ampacity of at least 100-A protection before derating and be properly protected by the next higher standard rating of protective device above ampacity of conductors.

Fig. 220-8. For branch circuit or feeder, fuses in enclosed switch must be limited for continuous duty. [Sec. 220-10(b).]

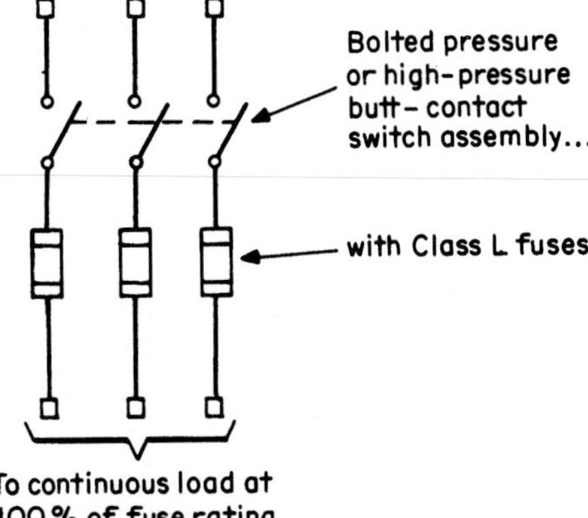

Fig. 220-9. Some fused switches may be used at 100 percent rating for continuous load. [Sec. 220-10(b).]

CB within the switchboard, the caution about copper conductors does not apply, and aluminum conductors may be used.

If the ventilation pattern of a switchboard does not meet the ventilation pattern and the required enclosure size specified on the nameplate, the CB must be applied at 80 percent rating. Switchboard manufacturers have UL tests conducted with a CB installed in a specific enclosure, and the enclosure may receive a listing for 100 percent rated operation even though the ventilation pattern or overall enclosure size may not meet the specifications. In cases where the breaker nameplate specifications are not met by the switchboard, the customer would have to request a letter from the manufacturer certifying that a 100 percent rated listing has been received. Otherwise, the breaker must be applied at 80 percent.

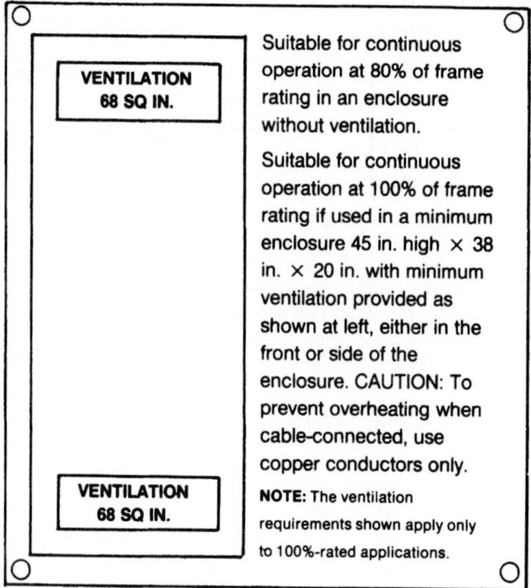

EXAMPLE 1
VENTILATION
68 SQ IN.
Suitable for continuous operation at 80% of frame rating in an enclosure without ventilation.
Suitable for continuous operation at 100% of frame rating if used in a minimum enclosure 45 in. high × 38 in. × 20 in. with minimum ventilation provided as shown at left, either in the front or side of the enclosure. CAUTION: To prevent overheating when cable-connected, use copper conductors only.
VENTILATION
68 SQ IN.
NOTE: The ventilation requirements shown apply only to 100%-rated applications.

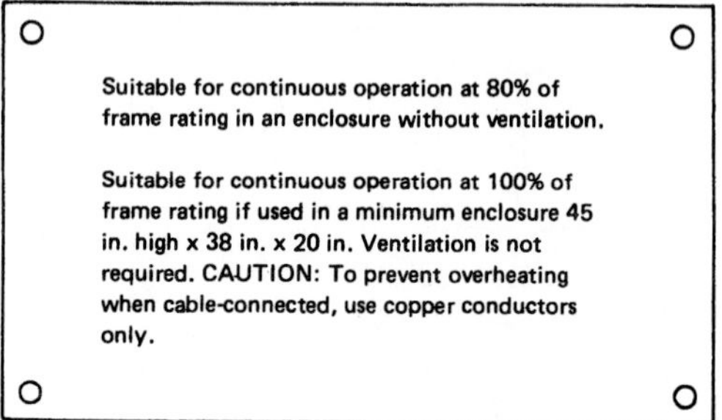

EXAMPLE 2
Suitable for continuous operation at 80% of frame rating in an enclosure without ventilation.
Suitable for continuous operation at 100% of frame rating if used in a minimum enclosure 45 in. high x 38 in. x 20 in. Ventilation is not required. CAUTION: To prevent overheating when cable-connected, use copper conductors only.

Fig. 220-10. Nameplates from CBs rated for 100 percent continuous loading. [Sec. 220-10(b).]

To realize savings with devices listed by UL at 100 percent of their continuous-current rating, use must be made of a CB manufacturer's data sheet to determine the types and ampere ratings of breakers available that are 100 percent rated, along with the frame sizes, approved enclosure sizes, and the ventilation patterns required by UL, if any. According to UL Standard 489, paragraph 33.37A, "a circuit breaker having a frame size less than 600 amperes shall not be marked suitable for continuous operation at 100 percent of rating." Of course, trip units with lower ampere ratings may be installed in the 600-A frame.

It is essential to check the instructions given in the UL listing to determine **if** and under what conditions a CB (or a fuse in a switch) is rated for continuous operation at 100 percent of its current rating.

220.11. General Lighting. For general illumination, a feeder must have capacity to carry the total load of lighting branch circuits determined as part of the lighting design and not less than a minimum branch-circuit load determined on a voltamperes-per-square-foot basis from the table in Sec. 220-2(b).

Demand factor permits sizing of a feeder according to the amount of load which operates simultaneously.

Demand factor is the ratio of the maximum amount of load that will be operating at any one time on a feeder to the total connected load on the feeder under consideration. This factor is frequently less than 1. The sum of the connected loads supplied by a feeder is multiplied by the demand factor to determine the load which the feeder must be sized to serve. This load is termed the maximum demand of the feeder:

$$\text{Maximum demand load} = \text{connected load} \times \text{demand factor}$$

Tables of demand and diversity factors have been developed from experience with various types of load concentrations and various layouts of feeders and subfeeders supplying such loads. Table 220-11 of the NE Code presents common demand factors for feeders to general lighting loads in various types of buildings (Fig. 220-11).

The demand factors given in Table 220-11 may be applied to the total branch-circuit load to get required feeder capacity for lighting (but must not be used in calculating branch-circuit capacity). Note that a feeder may have capacity of less than 100 percent of the total branch-circuit load for only the types of buildings designated in Table 220-11, that is, for dwelling units, hospitals, hotels, motels, and storage warehouses. In all other types of occupancies, it is assumed that *all* general lighting will be operating at the same time, and each feeder in those occupancies must have capacity (ampacity) for 100 percent of the voltamperes of branch-circuit load of general lighting that the feeder supplies.

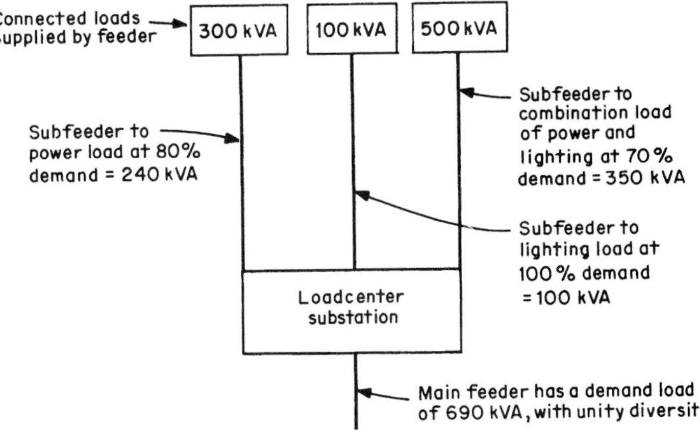

Fig. 220-11. How demand factors are applied to connected loads. (Sec. 220-11.)

example If a warehouse feeder fed a total branch-circuit load of 20,000 VA of general lighting, the minimum capacity in that feeder to supply that load must be equal to 12,500 VA plus 50 percent times (20,000 − 12,500) VA. That works out to be 12,500 plus 0.5 × 7,500 or 16,250 VA.

But, the note to Table 220-11 warns against using any value less than 100 percent of branch-circuit load for sizing any feeder that supplies loads that will all be energized at the same time.

220.12. Show-Window Lighting. If show-window lighting is supplied by a feeder, capacity must be included in the feeder to handle 200 VA per linear foot of show-window length. Because that is the same loading as given in Sec. 220-3(c), Exception No. 3, it works out to be a 100 percent demand for the entire branch-circuit load of show-window lighting.

220-13. Receptacle Loads—Nondwelling Units. This rule permits two possible approaches in determining the required feeder ampacity to supply receptacle loads in "other than dwelling units," where a load of 180 VA of feeder capacity must be provided for all general-purpose 15- and 20-A receptacle outlets. (In dwelling units and in guest rooms of hotels and motels, no feeder capacity is required for 15- or 20-A general-purpose receptacle outlets. Such load is considered sufficiently covered by the load capacity provided for general lighting.) But in other than dwelling units, where a load of 180 VA of feeder capacity must be provided for all general-purpose 15- and 20-A receptacle outlets, a *demand factor* may be applied to the total calculated receptacle load as follows. Wording of this rule makes clear that *either* Table 220-11 or Table 220-13 may be used to apply demand factors to the total load of 180-VA receptacle loads when calculating required ampacity of a feeder supplying receptacle loads connected on branch circuits.

In other than dwelling units, the branch-circuit load for receptacle outlets, for which 180 VA was allowed per outlet, may be added to the general lighting load and may be reduced by the demand factors in Table 220-11. That is the basic rule of Sec. 220-13 and, in effect, requires any feeder to have capacity for the total number of receptacles it feeds and requires that capacity to be equal to 180 VA (per single or multiple receptacle) times the total number of receptacles (straps)—with a reduction from 100 percent of that value permitted only for the occupancies listed in Table 220-11.

Because the demand factor of Table 220-11 is shown as 100 percent for "All Other" types of occupancies, the basic rule of Sec. 220-13 as it appeared prior to the 1978 NE Code required a feeder to have ampacity for a load equal to 180 VA times the number of general-purpose receptacle outlets that the feeder supplied. That is no longer required. Recognizing that there is great diversity in use of receptacles in office buildings, stores, schools, and all the other occupancies that come under "All Others" in Table 220-11, Sec. 220-13 contains a table to permit reduction of feeder capacity for receptacle loads on feeders. Those demand factors apply to *any* "nondwelling" occupancy.

The amount of feeder capacity for a typical case where a feeder, say, supplies panelboards that serve a total of 500 receptacles is shown in Fig. 220-11.

Although the calculation of Fig. 220-12 cannot always be taken as realistically related to usage of receptacles, it is realistic relief from the 100 percent demand factor, which presumed that all receptacles were supplying 180-VA loads simultaneously.

220-14. Motors. Any feeder that supplies a motor load or a combination load (motors plus lighting and/or other electrical loads) must satisfy the indicated NEC sections of Art. 430. Feeder capacity for motor loads is usually taken at 125 percent of the full-load current rating of the largest motor supplied, plus the sum of the full-load currents of the other motors supplied.

220-15. Fixed Electric Space Heating. Capacity required in a feeder to supply fixed electrical space heating equipment is determined on the basis of a load equal to the total connected load of heaters on all branch circuits served from the feeder. Under conditions of intermittent operation or where all units cannot operate at the same time, permission may be granted for use of less than a 100 percent demand factor in sizing the feeder. Sections 220-30, 220-31, and 220-32 permit alternate calculations of electric heat load for feeders or service-entrance conductors (which constitute a service feeder) in dwelling units. But reduction of the feeder capacity to less than 100 percent of connected load must be authorized by the local electrical inspector. Feeder load current for heating must not be less than the rating of the largest heating branch-circuit fed.

220.16. Small Appliance and Laundry Loads—Dwelling Unit. For a feeder or service conductors in a single-family dwelling, in an individual apartment of a multifamily dwelling with provisions for cooking by tenants, or in a hotel or motel suite with cooking facilities or a serving pantry, at least 1,500 VA of load

Take the total number of general-purpose receptacle outlets fed by a given feeder. . .

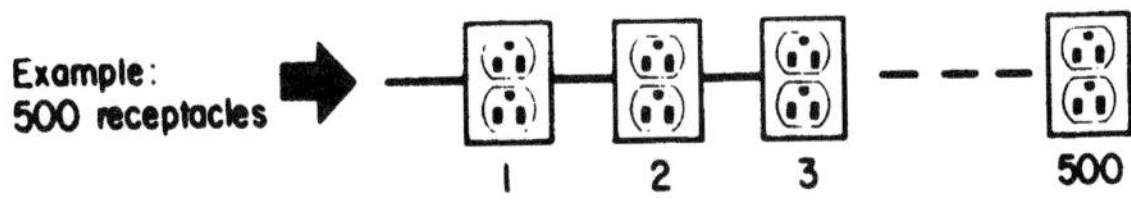

. . . multiply the total by 180 voltamperes [required load of Section 220-2(c) (6) for each receptacle]. . .

500 × 180 VA = 90,000 VA

Then apply the demand factors from Table 220-13:

First 10 kVA or less @ 100% demand = 10,000 VA
Remainder over 10 kVA @ 50% demand
= (90,000 − 10,000) × 50%
= 80,000 × 0.5 = 40,000 VA
Minimum demand-load total = 50,000 VA

Therefore, the feeder must have a capacity of 50 kVA for the total receptacle load. Required minimum ampacity for that load is then determined from the voltage and phase-makeup (single- or 3-phase) of the feeder.

Fig. 220-12. Table 220-13 permits demand factor in calculating feeder demand load for general-purpose receptacles. (Sec. 220-13.)

must be provided for each 2-wire, 20-A small appliance circuit (to handle the small appliance load in kitchen, pantry, and dining areas). The total small appliance load determined in this way may be added to the general lighting load, and the resulting total load may be reduced by the demand factors given in Table 220-11.

A feeder load of at least 1,500 VA must be added for each 2-wire, 20-A laundry circuit installed as required by Sec. 220-4(c). And that load may also be added to the general lighting load and subjected to the demand factors in Table 220-11.

220-17. Appliance Load—Dwelling Unit(s). For fixed appliances (fastened in place) other than ranges, clothes dryers, air-conditioning equipment, and space heating equipment, feeder capacity in dwelling occupancies must be provided for the sum of these loads; but, if there are at least four such fixed appliances, the total load of four or more such appliances may be reduced by a demand factor of 75 percent (**NE Code** Sec. 220-17). Wording of this rule makes clear that a "fixed appliance" is one that is "fastened in place."

As an example of application of this **Code** provision, consider the following calculation of feeder capacity for fixed appliances in a single-family house. The calculation is made to determine how much capacity must be provided in the service-entrance conductors (the service feeder):

Water heater	2,500 W	230 V =	11.0 A
Kitchen disposal	½ hp	115 V = 6.5 A + 25% =	8.1 A
Furnace motor	¼ hp	115 V =	4.6 A
Attic fan...........	¼ hp	115 V = 4.6 A	0.0 A
Water pump....	½ hp	230 V =	3.7 A
Load in amperes on each ungrounded leg of feeder =			27.4 A

To comply with Sec. 430-24, 25 percent is added to the full-load current of the ½ hp, 115-V motor because it is the highest-rated motor in the group. Since it is assumed that the load on the 115/230-V feeder will be balanced and each of the ¼-hp motors will be connected to different ungrounded conductors, only one is counted in the above calculation. Except for the 115-V motors, all the other appliance loads are connected to both ungrounded conductors and are automatically balanced. Since there are four or more fixed appliances in addition to a range, clothes dryer, etc., a demand factor of 75 percent may be applied to the total load of these appliances. Seventy-five percent of 27.4 = 20.5 A, which is the current to be added to that computed for the lighting and other loads to determine the total current to be carried by the ungrounded (outside) service-entrance conductors.

The above demand factor may be applied to similar loads in two-family or multifamily dwellings.

220.18. Electric Clothes Dryers—Dwelling Unit(s). This rule prescribes a minimum demand of 5 kVA for 120/240-V electric clothes dryers in determining branch-circuit and feeder sizes. Note that this rule applies only to "household" electric clothes dryers, and not to commercial applications. This rule is helpful because the ratings of electric clothes dryers are not usually known in the planning stages when feeder calculations must be determined (Fig. 220-13).

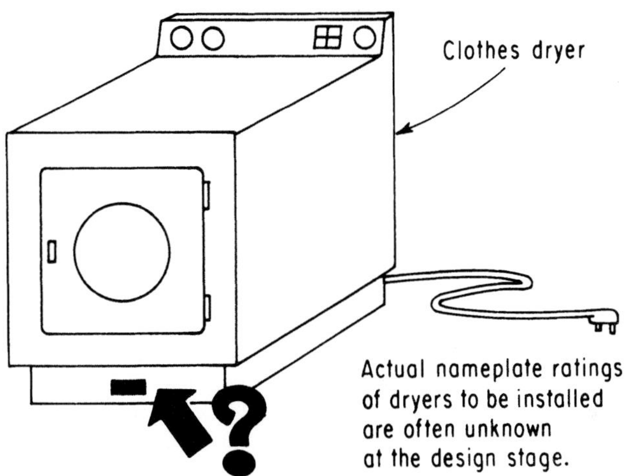
Clothes dryer
Actual nameplate ratings
of dryers to be installed
are often unknown
at the design stage.

Fig. 220-13. Feeder load of 5 kVA per dryer must be provided if actual load is not known. (Sec. 220-18.)

When sizing a feeder for one or more electric clothes dryers, a load of 5,000 VA or the nameplate rating, whichever is larger, shall be included for each dryer—subject to the demand factors of Table 220-18 when the feeder supplies a number of clothes dryers, as in an apartment house.

220-19. Electric Ranges and Other Cooking Appliances—Dwelling Unit(s). Feeder capacity must be allowed for household electric cooking appliances rated over 1¾ kW, in accordance with Table 220-19 of the Code. Feeder demand loads for a number of cooking appliances on a feeder may be obtained from Table 220-19.

Note 4 to Table 220-19 permits sizing of a branch circuit to supply a single electric range, a wall-mounted oven, or a counter-mounted cooking unit in accordance with that table. That table is also used in sizing a feeder (or service conductors) that supplies one or more electric ranges or cooking units. Note that Sec. 220-19 and Table 220-19 apply only to such cooking appliances in a "dwelling unit" and do not cover commercial or institutional applications.

Figure 220-14 shows a typical NEC calculation of the minimum demand load to be used in sizing the branch circuit to the range. The same value of demand load is also used in sizing a feeder (or service conductors) from which the range circuit is fed. Calculation is as follows:

A branch circuit for the 12-kW range is selected in accordance with Note 4 of Table 220-19, which says that the branch-circuit load for a range may be selected from the table itself. Under the heading "Number of Appliances," read across from "1." The maximum demand to be used in sizing the range circuit

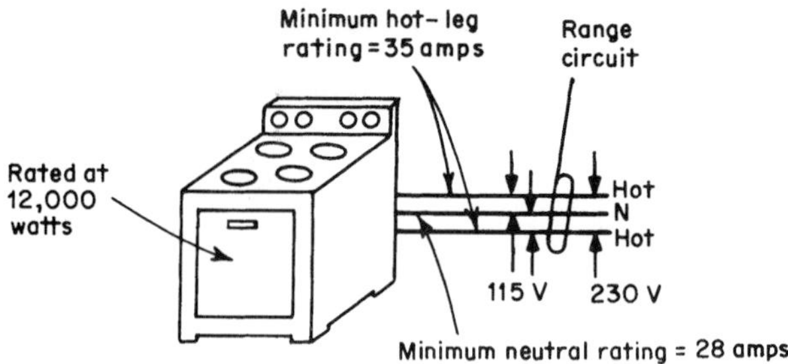

Fig. 220-14. Minimum amp rating of branch-circuit conductors for a 12-kW range. (Sec. 220-19.)

for a 12-kW range is shown under the heading "Maximum Demand" to be not less than 8 kW. The minimum rating of the range-circuit ungrounded conductors will be

$$\frac{8{,}000 \text{ W}}{230 \text{ V}} = 34.78 \text{ or } 35 \text{ A}$$

NE Code Table 310-16 shows that the minimum size of copper conductors that may be used is No. 8 (TW—40 A, THW—45 A, XHHW or THHN—50 A). No. 8 is also designated in Sec. 210-19(b) as the minimum size of conductor for any range rated 8¾ kW or more.

The overload protection for this circuit of No. 8 TW conductors would be *40-A fuses or a 40-A circuit breaker.* If THW, THHN, or XHHW wires are used for the circuit, they must be taken as having an ampacity of not more than 40 A and protected at that value. That requirement follows from the UL rule that conductors up to No. 1 AWG size must be used at the 60°C ampacity for the size of conductor, regardless of the actual temperature rating of the insulation—which may be 75°C or 90°C. The ampacity used must be that of TW wire of the given size.

Although the two hot legs of the 230/115-V, 3-wire circuit must be not smaller than No. 8, Exception No. 2 to Sec. 210-19(b) permits the neutral conductor to be smaller, but it specifies that it must have an ampacity not less than 70 percent of the rating of the branch-circuit CB or fuse and may never be smaller than No. 10.

For the range circuit in this example, the neutral may be rated

$$70\% \times 40 \text{ A (rating of branch-circuit protection)} = 28 \text{ A}$$

This calls for a No. 10 neutral.

Figure 220-15 shows a more involved calculation for a range rated over 12 kW. Figure 220-16 shows two units that total 12 kW and are taken at a demand load of 8 kW, as if they were a single range. Figure 220-17 shows another calculation for separate cooking units on one circuit. And a feeder that would be used to supply any of the cooking installations shown in Figs. 220-14 through 220-17 would have to include capacity equal to the demand load used in sizing the branch circuit.

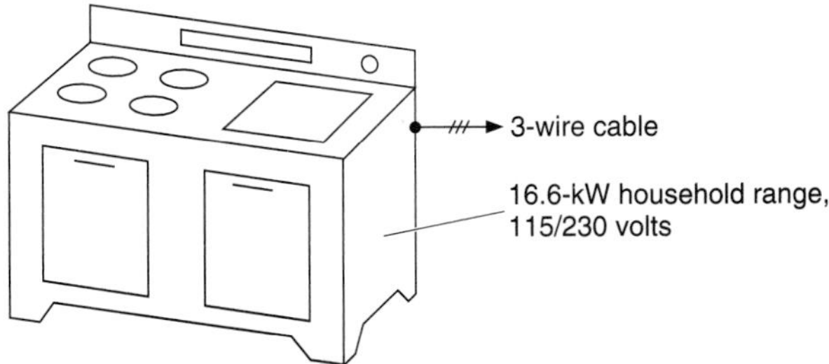

Refer to *NE Code* Table 220-19.
1. Column A applies to ranges rated not over 12 kW, but this range is rated 16.6 kW.
2. Note 1, below the Table, tells how to use the Table for ranges over 12 kW and up to 27 kW. For such ranges, the maximum demand in Column A must be increased by 5% for each additional kW of rating (or major fraction) above 12 kW.
3. This 16.6-kW range exceeds 12 kW by 4.6 kW.
4. 5% of the demand in Column A for a single range is 400 watts (8000 watts × 0.05).
5. The maximum demand for this 16.6-kW range must be increased above 8 kW by 2000 watts:
   400 watts (5% of Column A) × 5 (4 kW + 1 for the remaining 0.6 kW)
6. The required branch circuit must be sized, therefore, for a total demand load of 8000 watts + 2000 watts = 10,000 watts
7. Required size of branch circuit—

$$\text{Amp rating} = \frac{10{,}000 \text{ w}}{230 \text{ v}} = 43 \text{ amps}$$

USING 60C CONDUCTORS, AS REQUIRED BY UL, THE UNGROUNDED BRANCH CIRCUIT CONDUCTORS WOULD CONSIST OF NO. 6 TW CONDUCTORS.

Fig. 220-15. Sizing a branch circuit for a household range over 12 kW. (Sec. 220-19.)

A feeder supplying more than one range (rated not over 12 kW) must have ampacity sufficient for the maximum demand load given in Table 220-19 for the number of ranges fed. For instance, a feeder to 10 such ranges would have to have ampacity for a load of 25 kW.

Other Calculations on Electric Cooking Appliances

The following "roundup" points out step-by-step methods of wiring the various types of household electric cooking equipment (ranges, counter-mounted cooking units, and wall-mounted ovens) according to the NEC.

Tap Conductors

Section 210-19(b), Exception No. 2, gives permission to reduce the size of the neutral conductor of a 3-wire range branch circuit to 70 percent of the rating of the CB or fuses protecting the branch circuit. However, this rule does not apply to smaller taps connected to a 50-A circuit—where the smaller taps (none less than 20-A ratings) must all be the same size. Further, it does not apply when individual branch circuits supply each wall- or counter-mounted cooking unit and all circuit conductors are of the same size and less than No. 10.

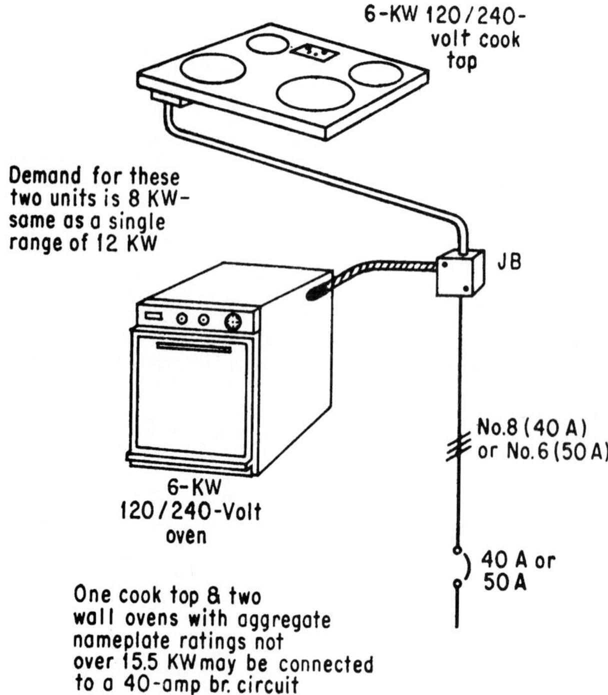

Fig. 220-16. Two units treated as a single-range load. (Sec. 220-19.)

Section 210-19(b), Exception No. 1, permits tap conductors, rated not less than 20 A, to be connected to 50-A branch circuits that supply ranges, wall-mounted ovens, and counter-mounted cooking units. These taps cannot be any longer than necessary for servicing. Figure 220-18 illustrates the application of this rule.

In Sec. 210-19(b), Exception No. 1, the wording "no longer than necessary for servicing" encourages the location of circuit junction boxes as close as possible to each cooking and oven unit connected to 50-A circuits. A number of counter-mounted cooking units have integral supply leads about 36 in. (914 mm) long, and some ovens come with supply conduit and wire in lengths of 48 to 54 in. (1.22 to 1.37 m). Therefore, a box should be installed close enough to connect these leads.

Feeder and Circuit Calculations

Section 220-19 permits the use of Table 220-19 for calculating the feeder load for ranges and other cooking appliances that are individually rated more than 1¾ kW.

Note 4 of the table reads: "The branch-circuit load for one wall-mounted oven or one counter-mounted cooking unit shall be the nameplate rating of the

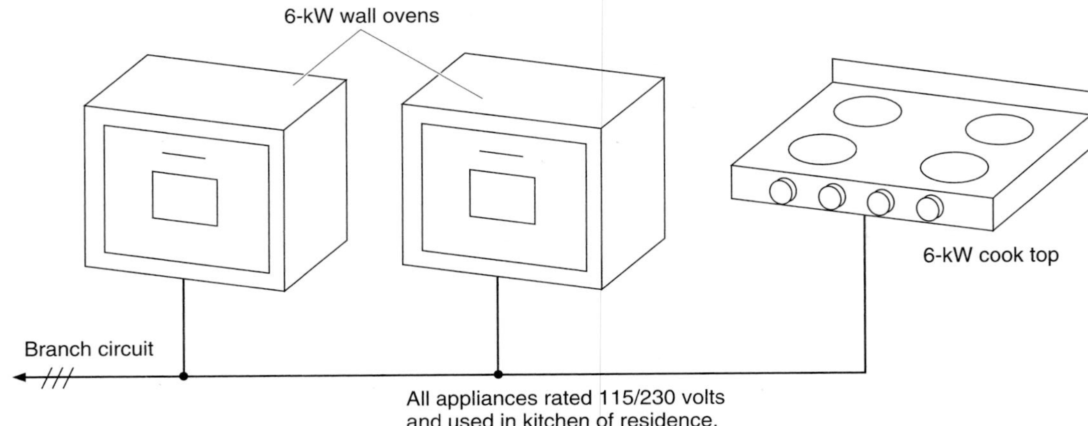

1. Note 4 of Table 220-19 says that the branch-circuit load for a counter-mounted cooking unit and not more than two wall-mounted ovens, all supplied from a single branch circuit and located in the same room, shall be computed by adding the nameplate ratings of the individual appliances and treating this total as a single range.

2. Therefore, the three appliances shown may be considered to be a single range of 18-kW rating (6 kW + 6 kW + 6 kW).

3. From Note 1 of Table 220-19, such a range exceeds 12 kW by 6 kW and the 8-kW demand of Column A must be increased by 400 watts (5% of 8000 watts) for each of the 6 additional kilowatts above 12 kW.

4. Thus, the branch-circuit demand load is—

8000 WATTS + (6 × 400 WATTS) = 10,400 WATTS

Fig. 220-17. Determining branch-circuit load for separate cooking appliances on a single circuit. (Sec. 220-19.)

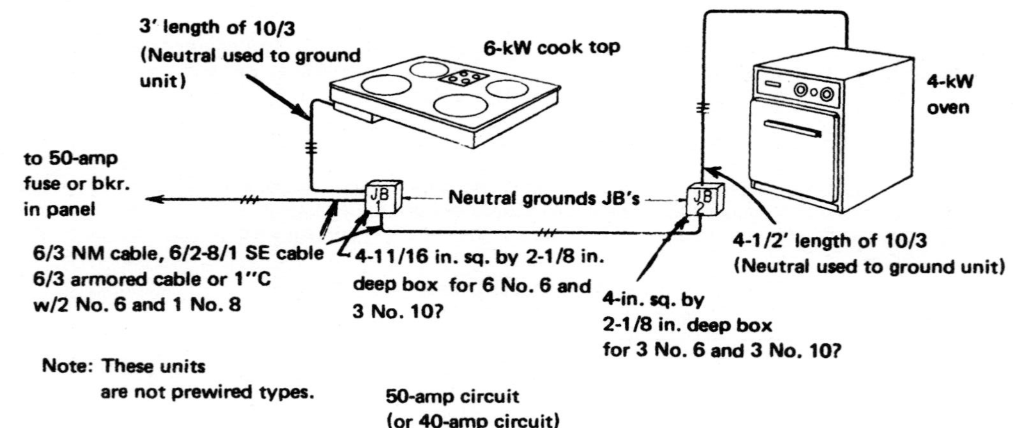

NEC rules permit a 50-amp circuit to supply cook tops and ovens. Typical arrangement shows such a circuit. Junction box sizes are computed from Table 370-16(a) and Table 370-16(b) for No. 6 conductor combinations. Taps to each unit are No. 10 to permit the use of the neutral as an equipment ground. Using the neutral to ground the junction boxes is permitted by Sec. 250-60.

Fig. 220-18. One branch circuit to cooking units. (Sec. 220-19.)

appliance." Figure 220-19 shows a separate branch circuit to each cooking unit, as permitted.

Common sense dictates that there is no difference in demand factor between a single range of 12 kW and a wall-mounted oven and surface-mounted cooking unit totaling 12 kW. This is explained in the last sentence of Note 4 of Table 220-19. The mere division of a complete range into two or more units does not change the demand factor. Therefore, the most direct and accurate method of computing the branch-circuit and feeder calculations for wall-mounted ovens and surface-mounted cooking units within each occupancy is to total the kilowatt ratings of these appliances and treat this total kilowatt rating as a single range of the same rating. For example, a particular dwelling has an 8-kW, 4-burner, surface-mounted cooking unit and a 4-kW wall-mounted oven. This is a total of 12 kW, and the maximum permissible demand given in Column A of Table 220-19 for a single 12-kW range is 8 kW.

Similarly, it follows that if the ratings of a 2-burner, counter-mounted cooking unit and a wall-mounted oven are each 3.5 kW, the total of the two would be 7 kW—the same total as a small 7-kW range. Because the 7-kW load is less than 8¾ kW, Note 3 of Table 220-19 permits Column C of Table 220-19 to be used in lieu of Column A. The demand load is 5.6 kW (7 kW times 0.80). Range or total cooking and oven unit ratings less than 8¾ kW are more likely to be found in small apartment units of multifamily dwellings than in single-family dwellings.

Because the demand loads in Column A of Table 220-19 apply to ranges not exceeding 12 kW, they also apply to wall-mounted ovens and counter-mounted cooking units within each individual occupancy by totaling their aggregate nameplate kilowatt ratings. Then if the total rating exceeds 12 kW, Note 1 to the table should be used as if the units were a single range of equal rating. For example, assume that the total rating of a counter-mounted cooking unit and two wall-mounted ovens is 16 kW in a dwelling unit. The maximum demand for a single 12-kW range is given as 8 kW in Column A. Note 1 requires that the maximum demand in Column A be increased 5 percent for each additional kilowatt or major fraction thereof that exceeds 12 kW. In this case 16 kW exceeds 12 kW by 4 kW. Therefore, 5 percent times 4 equals 20 percent, and 20 percent of 8 kW is 1.6 kW. The maximum feeder and branch-circuit demand is then 9.6 kW (8 kW plus 1.6 kW). A 9,600-W load would draw over 40 A at 230 V, thereby requiring a circuit rated over 40 A.

For the range or cooking unit demand factors in a multifamily dwelling, say a 12-unit apartment building, a specific calculation must be made, as follows:

1. Each apartment has a 6-kW counter-mounted cooking unit and a 4-kW wall-mounted oven. And each apartment is served by a separate feeder from a main switchboard. The maximum cooking demand in each apartment feeder should be computed in the same manner as previously described for single-family dwellings. Since the total rating of cooking and oven units in each apartment is 10 kW (6 kW plus 4 kW), Column A of Table 220-19 for one appliance should apply. Thus, the maximum cooking demand load on each feeder is 8 kW.

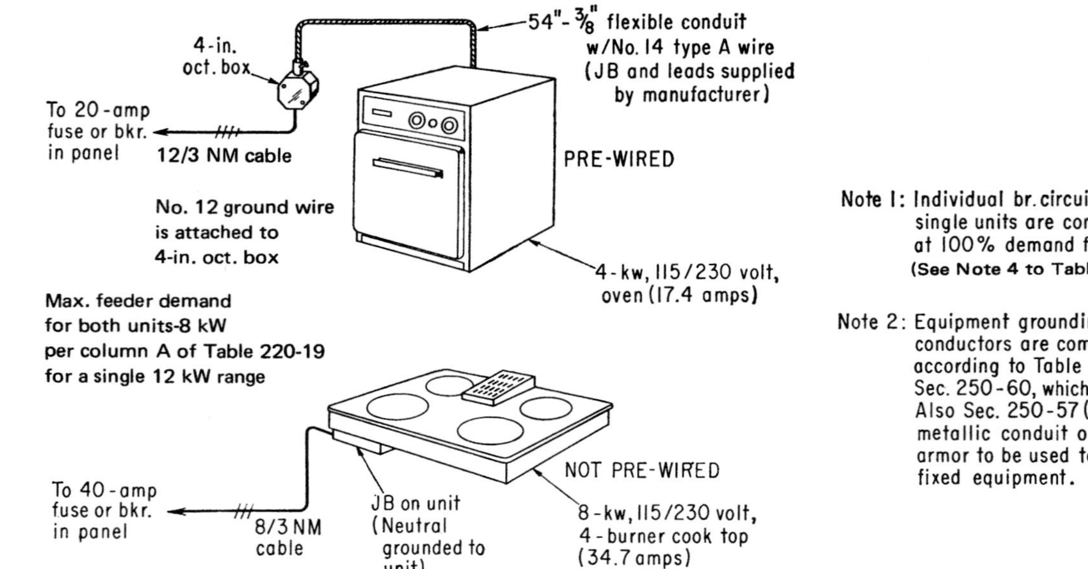

An 8-kW cook top is supplied by an individual No. 8 (40-amp) branch circuit, and a No. 12 (20-amp) branch circuit supplies a 4-kW oven. Such circuits are calculated on the basis of the nameplate rating of the appliance. In most instances individual branch circuits cost less than 50-amp, multi-outlet circuits for cooking and oven units.

Fig. 220-19. Separate branch circuit to cooking units. (Sec. 220-19.)

2. In figuring the size of the main service feeder, Column A should be used for 12 appliances. Thus, the demand would be 27 kW.

As an alternate calculation, assume that each of the 12 apartments has a 4-kW counter-mounted cooking unit and 4-kW wall-mounted oven. This would total 8 kW per apartment. In this case Column C of Table 220-19 can be used to determine the cooking load in each separate feeder. By applying Column C on the basis of a single 8-kW range, the maximum demand is 6.4 kW (8 kW times 0.80). Therefore, 6.4 kW is the cooking load to be included in the calculation of each feeder. Notice that this is 1.6 kW less than the previous example where cooking and oven units, totaling 10 kW, had a demand load of 8 kW. And this is logical, because smaller units should produce a smaller total kilowatt demand.

On the other hand, it is better to use Column A instead of Column C for computing the main service feeder capacity for twelve 8-kW cooking loads. The reason for this is that Column C is inaccurate where more than five 8-kW ranges (or combinations) and more than twelve 7-kW ranges (or combinations) are to be used. In these instances, calculations made on the basis of Column C result in a demand load greater than that of Column A for the same number of ranges. As an example, twelve 8-kW ranges have a demand load of 30.72 kW (12 times 8 kW times 0.32) in applying Column C, but only a demand load of 27 kW in Column A. And in Column A the 27 kW is based on twelve 12-kW ranges. This discrepancy dictates use of Column C only on the limited basis previously outlined.

Branch-Circuit Wiring

Where individual branch circuits supply each counter-mounted cooking unit and wall-mounted oven, there appears to be no particular problem. Figure 220-19 gives the details for wiring units on individual branch circuits.

Figure 220-18 shows an example of how typical counter-mounted cooking units and wall-mounted ovens are connected to a 50-A branch circuit.

Several manufacturers of cooking units provide an attached flexible metal conduit with supply leads and a floating 4-in. (101.2 mm) octagon box as a part of each unit. These units are commonly called "prewired types." With this arrangement, an electrician does not have to make any supply connections in the appliance. Where such units are connected to a 50-A circuit, the 4-in. (101.2 mm) octagon box is removed, and the flexible conduit is connected to a larger circuit junction box, which contains the No. 6 circuit conductors.

On the other hand, some manufacturers do not furnish supply leads with their cooking units. As a result, the electrical contractor must supply the tap conductors to these units from the 50-A circuit junction box. See Fig. 220-18. In this case, connections must be made in the appliance as well as in the junction box. Figure 220-20 shows a single branch circuit supplying the same units as shown in Fig. 220-18.

40-A Circuits

The NEC does recognize a 40-A circuit for two or more outlets, as noted in Sec. 210-23(c). Because a No. 8 (40-A) circuit can supply a single range rated not over

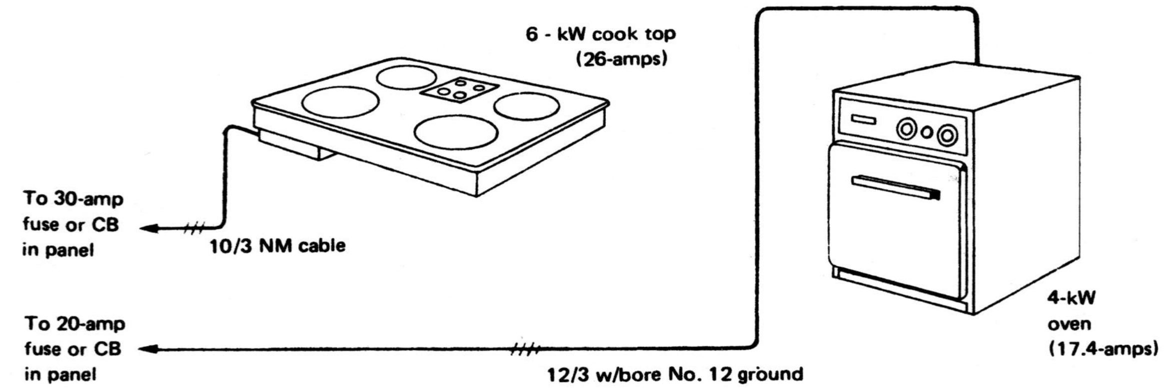
6 - kW cook top
(26-amps)
To 30-amp
fuse or CB
in panel
10/3 NM cable
To 20-amp
fuse or CB
in panel
12/3 w/bore No. 12 ground
4-kW
oven
(17.4-amps)
Same size and type of units as in Fig. 220-18,
but wired on individual circuits
Individual branch circuits supply the same units that appear in Fig. 220-18. With this arrangement, smaller branch circuits supply each unit with no JBs required. Although two additional fuse or CB poles are required in a panelboard, overall labor/material costs are less than the 50-amp circuit shown in Fig. 220-18. However, one disadvantage to individual circuits is that smaller size circuits will not handle larger units, which may be installed at a later date.

Fig. 220-20. Separate circuits have advantages. (Sec. 220-19.)

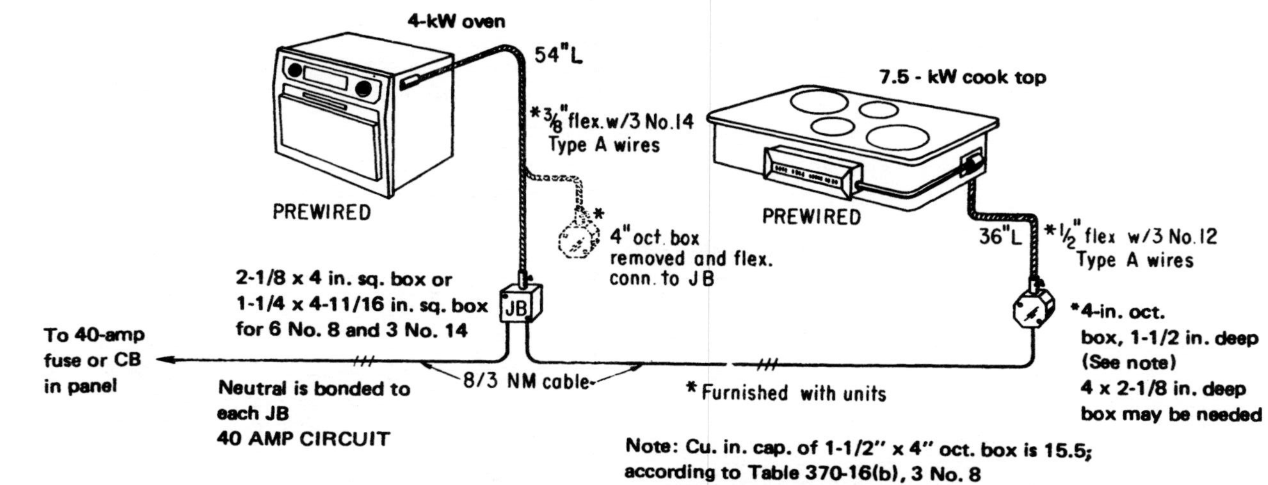

The *NEC* permits 40-amp circuits in lieu of 50-amp circuits where the aggregate nameplate rating of cook tops and ovens is less than 15.5 kW. Most ranges are less than 15.5 kW and so are most combinations of cook tops and ovens.

Fig. 220-21. A single 40-A circuit may supply units. (Sec. 220-19.)

15.4 kW, it can also supply counter- and wall-mounted units not exceeding the same total of 15.4 kW. The rating of 15.4 kW is determined as the maximum rating of equipment that may be supplied by a 40-A branch circuit, which has a capacity of 9,200 W (40 A × 230 V). From Note 1 to Table 220-19, a 15.4-kW load would require a demand capacity equal to 8,000 W plus [(15.4 – 12) × 0.05 × 8,000] = 8,000 W plus 3 × 0.05 × 8,000 = 8,000 plus 1,200 = 9,200 W.

Figure 220-21 shows an arrangement of a No. 8 (40-A) branch circuit supplying one 7.5-kW cooking unit and one 4-kW oven. Or individual branch circuits may be run to the units.

220-20. Kitchen Equipment—Other than Dwelling Unit(s). Commercial electric cooking loads must comply with Sec. 220-20 and its table of feeder demand factors for commercial electric cooking equipment—including dishwasher booster heaters, water heaters, and other kitchen equipment. Space-heating, ventilating, and/or air-conditioning equipment are excluded from the phrase "other kitchen equipment."

At one time, the **Code** did not recognize demand factors for such equipment. **Code** Table 220-20 is the result of extensive research on the part of electric utilities. The demand factors given in Table 220-20 may be applied to *all* equipment (except the excluded heating, ventilating, and air-conditioning loads) that is *either* thermostatically controlled *or* is used only on an intermittent basis. Continuously operating loads, such as infrared heat lamps used for food warming, would be taken at 100 percent demand and not counted in the "Number of Units" that are subject to the demand factors of Table 220-20.

The rule says that the minimum load to be used in sizing a feeder to commercial kitchen equipment must not be less than the sum of the largest two kitchen equipment loads. If the feeder load determined by using Table 220-20 on the total number of appliances that are controlled or intermittent and then adding the sum of load ratings of continuous loads like heat lamps is less than the sum of load ratings of the two largest load units—then the minimum feeder load must be taken as the sum of the two largest load units.

example Find the minimum demand load to be used in sizing a feeder supplying a 20-kW quick-recovery water heater, a 5-kW fryer, and four continuously operating 250-W food-warmer infrared lamps—with a 208Y/120V, 3-phase, 4-wire supply.

Although the water heater, the fryer, and the four lamps are a total of 1 + 1 + 4, or 6, unit loads, the 250-W lamps may not be counted in using Table 220-20 because they are continuous loads. For the water heater and the fryer, Table 220-20 indicates that a 100 percent demand must be used where the "Number of Units of Equipment" is 2. The feeder minimum load must then be taken as

$$\begin{aligned}\text{Water heater @ 100\%}\quad & 20\ \text{kW}\\ \text{Fryer @ 100\%}\quad & \underline{5\ \text{kW}} = 25\ \text{kW}\\ \text{Four 250-W lamps @ 100\%}\quad & 1\ \text{kW} + 25\ \text{kW} = 26\ \text{kW}\end{aligned}$$

Then, the feeder must be sized for a minimum current load of

$$\frac{26 \times 1{,}000}{208 \times 1.732} = 72\ \text{A}$$

The two largest equipment loads are the water heater and the dryer:

$$20 \text{ kW} + 5 \text{ kW} = 25 \text{ kW}$$

and they draw

$$\frac{25 \times 1{,}000}{208 \times 1.732} = 69 \text{ A}$$

Therefore, the 72-A demand load calculated from Table 220-20 satisfies the last sentence of the rule because that value is "not less than" the sum of the largest two kitchen equipment loads. The feeder must be sized to have at least 72 A of capacity (a minimum of No. 3 TW, THW, THHN, RHH, etc.).

Figure 220-22 shows another example of reduced sizing for a feeder to kitchen appliances.

220-21. Noncoincident Loads. When dissimilar loads (such as space heating and air cooling in a building) are supplied by the same feeder, the smaller of the two loads may be omitted from the total capacity required for the feeder if it is unlikely that the two loads will operate at the same time.

220-22. Feeder Neutral Load. This section covers requirements for sizing the neutral conductor in a feeder, that is, determining the required amp rating of the neutral conductor. The basic rule of this section says that the minimum required ampacity of a neutral conductor must be at least equal to the "feeder neutral load"—which is the "maximum unbalance" of the feeder load.

"The maximum unbalanced load shall be the maximum net computed load between the neutral and any one ungrounded conductor. . . ." In a 3-wire, 120/240-V, single-phase feeder, the neutral must have a current-carrying capacity at least equal to the current drawn by the total 120-V load connected between the more heavily loaded hot leg and the neutral. As shown in Fig. 220-23, under unbalanced conditions, with one hot leg fully loaded to 60 A and the other leg open, the neutral would carry 60 A and must have the same rating as the loaded hot leg. Thus No. 6 THW hot legs would require No. 6 THW neutral (copper).

It should be noted that straight 240-V loads, connected between the two hot legs, do not place any load on the neutral. As a result, the neutral conductor of such a feeder must be sized to make up a 2-wire, 120-V circuit with the more heavily loaded hot leg. Actually, the 120-V circuit loads on such a feeder would be considered as balanced on both sides of the neutral. The neutral, then, would be the same size as each of the hot legs if only 120-V loads were supplied by the feeder. If 240-V loads also were supplied, the hot legs would be sized for the total load; but the neutral would be sized for only the total 120-V load connected between one hot leg and the neutral, as shown in Fig. 220-24.

But, there are qualifications on the basic rule of Sec. 220-22, as follows:

1. When a feeder supplies household electric ranges, wall-mounted ovens, counter-mounted cooking units, and/or electric dryers, the neutral conductor may be smaller than the hot conductors but must have a carrying capacity at least equal to 70 percent of the current capacity required in the

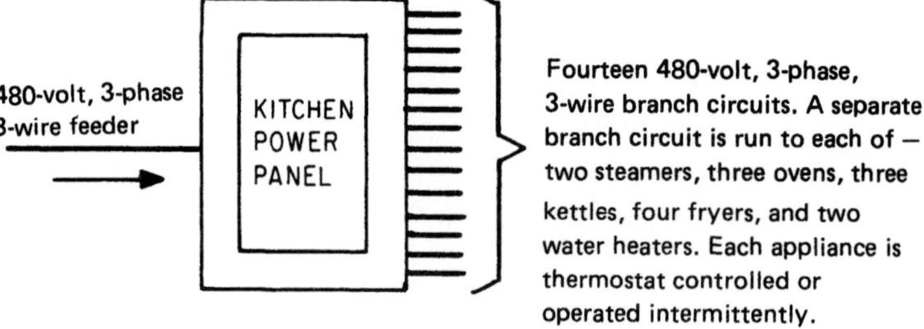
480-volt, 3-phase
3-wire feeder
KITCHEN
POWER
PANEL
Fourteen 480-volt, 3-phase, 3-wire branch circuits. A separate branch circuit is run to each of – two steamers, three ovens, three kettles, four fryers, and two water heaters. Each appliance is thermostat controlled or operated intermittently.

Kitchen panel supplies fourteen 480-volt, 3-phase, 3-wire branch circuits.
A separate branch circuit is run to each of—
Two steamers,
Three ovens,
Three kettles,
Four fryers, and
Two water heaters
The 14 appliances make up a total connected load of 303.3 kVA

QUESTION:
Is a full-capacity feeder (303.3 kVA/480 X 1.73 = 366 amps) required here? Or can a demand factor be applied?

ANSWER:
Although it is possible that all of the appliances might operate simultaneously, it is not expected that they will all be operating at full connected load. Table 220-20 of the *NE Code* does permit use of a demand factor on a feeder for commercial electric cooking equipment (including dishwasher, booster heaters, water heaters and other kitchen equipment). As shown in the Table, for six or more units, a demand factor of 65% can be applied to the feeder sizing:

366 amps × 0.65 = 238 amps

The feeder must have a least that much capacity, and that amp rating must be at least equal to or greater than the sum of the amp ratings of the two largest load appliances served. Capacity must be included in the building service entrance conductors for this load.

Fig. 220-22. Demand factor for commercial-kitchen feeder. (Sec. 220-20.)

ungrounded conductors to handle the load (i.e., 70 percent of the load on the ungrounded conductors). Table 220-19 gives the demand loads to be used in sizing feeders which supply electric ranges and other cooking appliances. Table 220-18 gives demand factors for sizing the ungrounded circuit conductors for feeders to electric dryers. The 70 percent demand

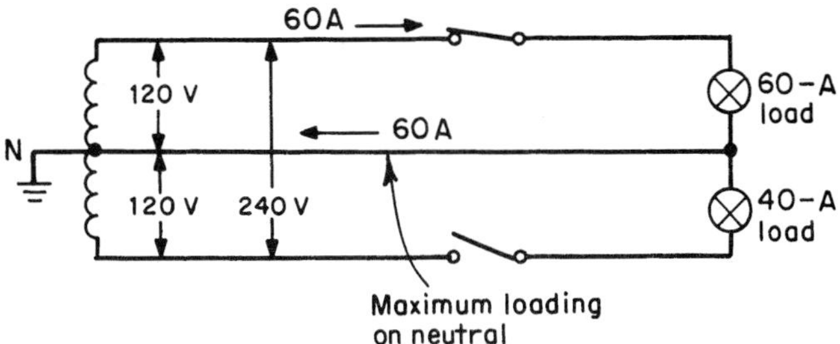
60A
120 V
60-A load
60 A
N
120 V
240 V
40-A load
Maximum loading on neutral

Fig. 220-23. Neutral must be sized the same as hot leg with heavier load. (Sec. 220-22.)

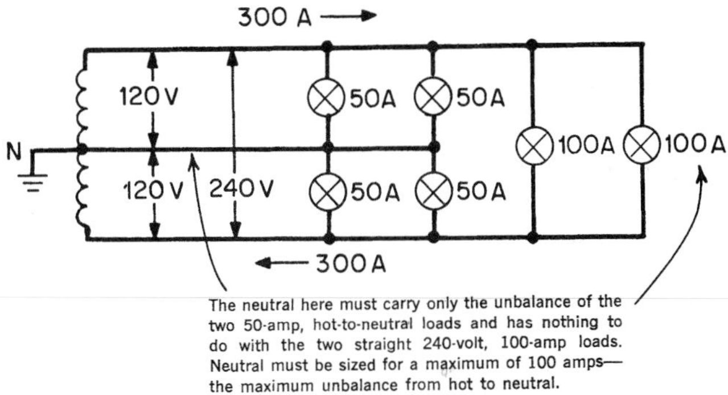
300 A
120 V
50A
50A
N
100A
100 A
120 V
240 V
50A
50A
300 A
The neutral here must carry only the unbalance of the two 50-amp, hot-to-neutral loads and has nothing to do with the two straight 240-volt, 100-amp loads. Neutral must be sized for a maximum of 100 amps— the maximum unbalance from hot to neutral.

Fig. 220-24. Neutral sizing is not related to phase-to-phase loads. (Sec. 220-22.)

factor may be applied to the minimum required size of a feeder phase (or hot) leg in order to determine the minimum permitted size of neutral, as shown in Fig. 220-25.

2. For feeders of three or more conductors—3-wire, DC; 3-wire, single-phase; and 4-wire, 3-phase—a further demand factor of 70 percent may be applied to that portion of the unbalanced load in excess of 200 A. That is, in a feeder supplying only 120-V loads evenly divided between each ungrounded conductor and the neutral, the neutral conductor must be the same size as each ungrounded conductor up to 200-A capacity, but may be reduced from the size of the ungrounded conductors for loads above 200 A by adding to the 200 A only 70 percent of the amount of load current above 200 A in computing the size of the neutral. It should be noted that this 70 percent demand factor is applicable to the unbalanced load in excess of 200 A and not simply to the total load, which in many cases may include 240-V loads on 120/240-V, 3-wire, single-phase feeders or 3-phase loads or phase-to-phase connected loads on 3-phase feeders. Figure 220-26 shows an example of neutral reduction as permitted by Sec. 220-22.

WATCH OUT!

The size of a feeder neutral conductor may *not* be based on less than the current load on the feeder phase legs when the load consists of electric-discharge

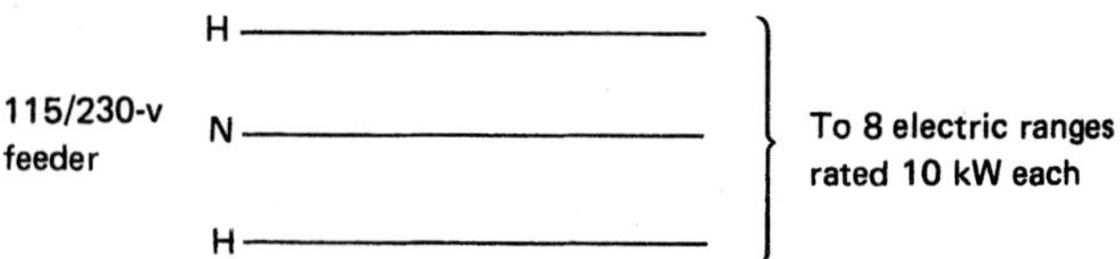

From Table 220-19—DEMAND LOAD for 8 10-kW ranges = 23 kW

LOAD ON EACH UNGROUNDED LEG = $\frac{23{,}000\text{ W}}{230\text{ V}}$ = 100 amps (e. g., No. 1TW)

Required minimum Neutral capacity = 70% X 100 amps = 70 amps (e.g., No. 4TW)

Fig. 220-25. Sizing the neutral of a feeder to electric ranges. (Sec. 220-22.)

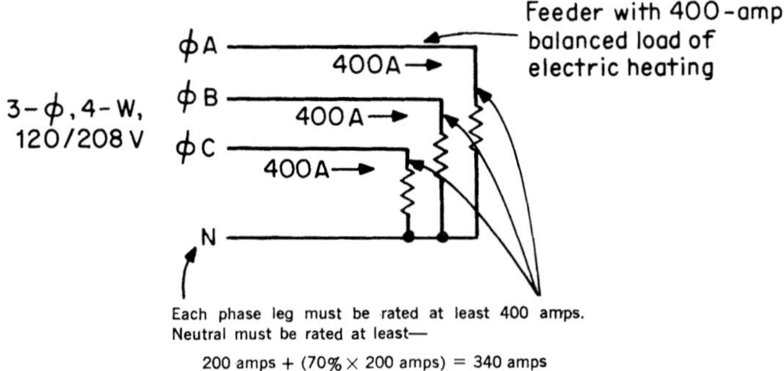

Fig. 220-26. Neutral may be smaller than hot-leg conductors on feeders over 200 A. (Sec. 220-22.)

lighting, data processing equipment, or similar equipment. The foregoing reduction of the neutral to 200 A plus 70 percent of the current over 200 A does not apply when all or most of the load on the feeder consists of electric-discharge lighting, electronic data processing equipment, and similar electromagnetic or solid-state equipment. In a feeder supplying ballasts for electric-discharge lamps and/or computer equipment, there must not be a reduction of the neutral capacity for that part of the load which consists of discharge light sources, such as fluorescent mercury-vapor or other HID lamps. For feeders supplying only electric-discharge lighting or computers, the neutral conductor must be the same size as the phase conductors no matter how big the total load may be (Fig. 220-27). Full-sizing of the neutral of such feeders is required because, in a balanced circuit supplying ballasts or computer loads, neutral current approximating the phase current is produced by third (and other odd-order) harmonics developed by the ballasts. For large electric-discharge lighting or computer loads, this factor affects sizing of neutrals all the way back to the service. It also affects rating of conductors in conduit because such a feeder circuit consists of *four* current-carrying wires, which requires application of an 80 percent reduc-

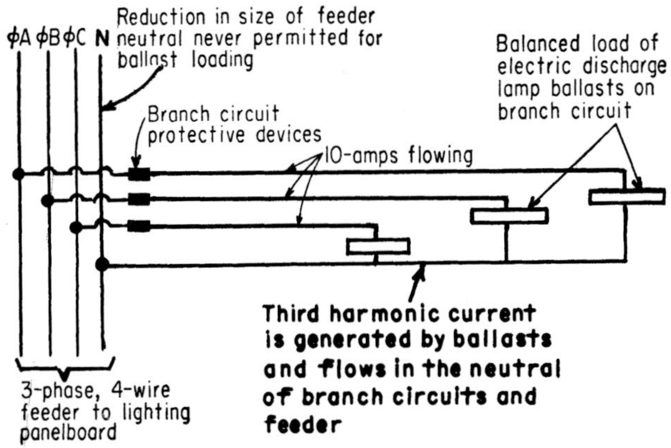

EXAMPLE:

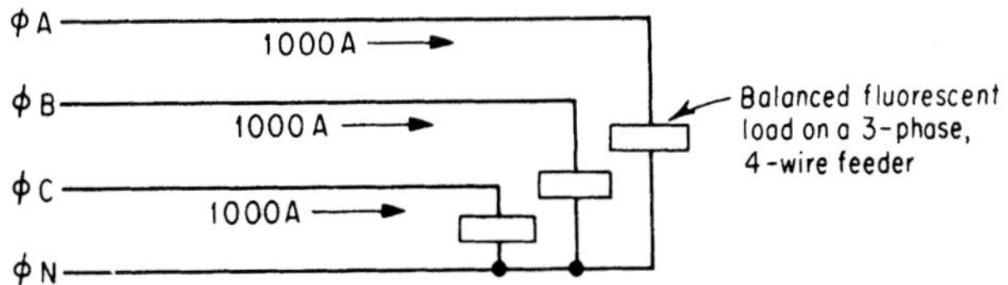

There must be no reduction in amp rating of this neutral. It must have 1000-amp rating like the phase conductors.

Fig. 220-27. Full-size neutral for feeders to ballast loads or computers. (Sec. 220-22.)

tion factor. [See Note 8 and Note 10(c) of the "Notes to Tables 310-16 through 310-19" in the NE Code.]

In the case of a feeder supplying, say, 200 A of fluorescent lighting and 200 A of incandescent, there can be no reduction of the neutral below the required 400-A capacity of the phase legs, because the 200 A of fluorescent lighting load cannot be used in any way to take advantage of the 70 percent demand factor on that part of the load in excess of 200 A.

It should be noted that the Code wording in Sec. 220-22 prohibits reduction in the size of the neutral when electric-discharge lighting and/or computers are used, even if the feeder supplying the electric-discharge lighting load over 200 A happens to be a 120/240-V, 3-wire, single-phase feeder. In such a feeder, however, the third harmonic currents in the hot legs are 180° out of phase with each other and, therefore, would not be additive in the neutral as they are in a 3-phase, 4-wire circuit. In the 3-phase, 4-wire circuit, the third harmonic components of the phase currents are in phase with each other and add together in the neutral instead of canceling out. Figure 220-28 shows a 120/240-V circuit.

Figure 220-29 shows a number of circuit conditions involving the rules on sizing a feeder neutral.

220-30. Optional Calculation—Dwelling Unit. This section sets forth an optional method of calculating service demand load for a residence. This method may be used instead of the standard method as follows:

1. Only for a one-family residence or an apartment in a multifamily dwelling, or other "dwelling unit"
2. Served by a 120/240-V or 120/208-V 3-wire, 100-A or larger service or feeder
3. Where the total load of the dwelling unit is supplied by one set of service-entrance or feeder conductors

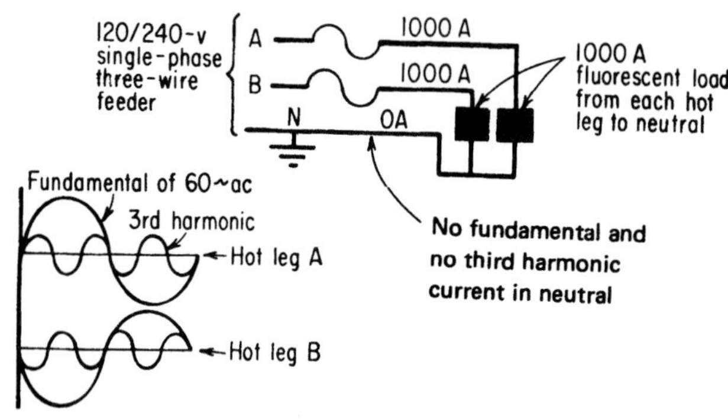

As shown, both the fundamental and harmonic currents are 180° out of phase and both cancel in the neutral. Under balanced conditions, the neutral current is zero. But the literal wording of Sec. 220-22 says there can be no reduction in neutral capacity when fluorescent lighting is supplied. As a result, there should be no use of the 70% factor for current over 200 amps as there would be for incandescent loading. Neutral here must be rated for 1000 amps.

Fig. 220-28. No reduction of neutral capacity even with zero neutral current. (Sec. 220-22.)

This method recognizes the greater diversity attainable in large capacity installations. It therefore permits a smaller size of service-entrance conductors for such installations than would be permitted by using the load calculations of Sec. 220-10 through Sec. 220-22.

In making this calculation, the heating load or the air-conditioning load may be disregarded as a "noncoincident load," where it is unlikely that two dissimilar loads (such as heating and air conditioning) will be operated simultaneously. In previous NEC editions, 100 percent of the air-conditioning load was compared with 100 percent of the total connected load of four or more separately controlled electric space heating units, and the smaller of the two loads was not counted in the calculation. In the present NEC, 100 percent of the air-conditioning load is compared with only *40 percent* of the total connected load

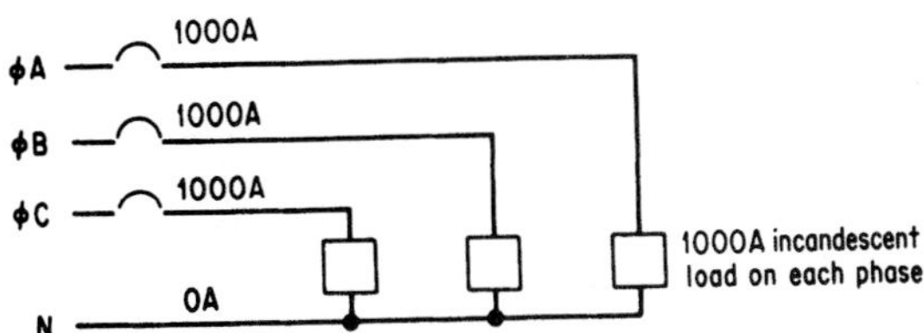

1. Incandescent lighting only

Serving an incandescent load, each phase conductor must be rated for 1000 amps. But neutral only has to be rated for 200 amps plus (70% x 800 amps) or 200 + 560 = 760 amps.

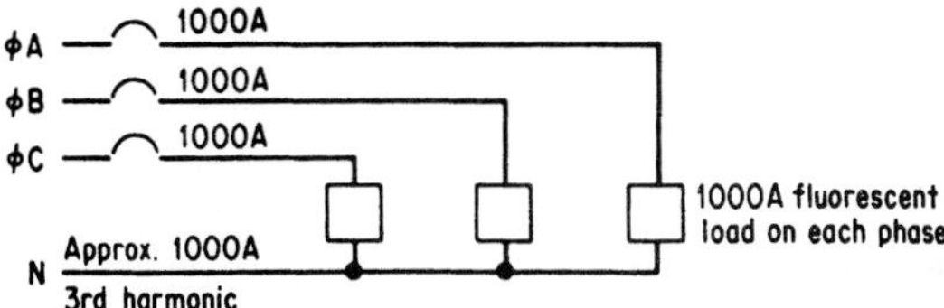

2. Electric discharge lighting only

Because load is electric discharge lighting, there can be no reduction in the size of the neutral. Neutral must be rated for 1000 amps, the same as the phase conductors, because the third harmonic currents of the phase legs add together in the neutral. This applies also when the load is mercury-vapor or other metallic-vapor lighting.

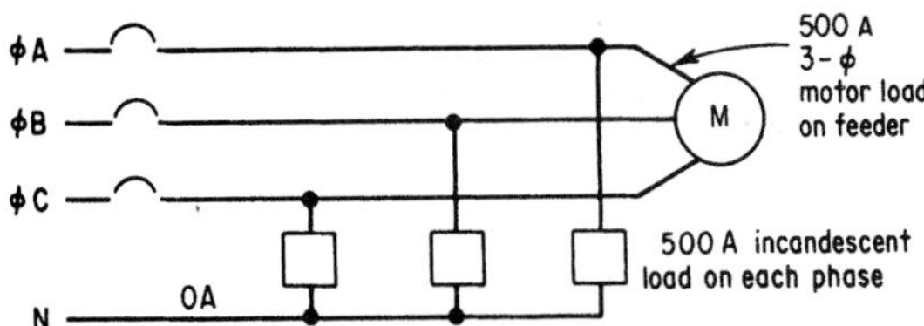

3. Incandescent plus motor load

Although 1000 amps flow on each phase leg, only 500 amps is related to the neutral. Neutral, then, is sized for 200 amps plus (70% x 300 amps) or 200 + 210 = 410 amps. The amount of current taken for 3-phase motors cannot be "unbalanced load" and no capacity has to be provided for this in the neutral.

Fig. 220-29. Sizing the feeder neutral for different conditions of loading. (Sec. 220-22.)

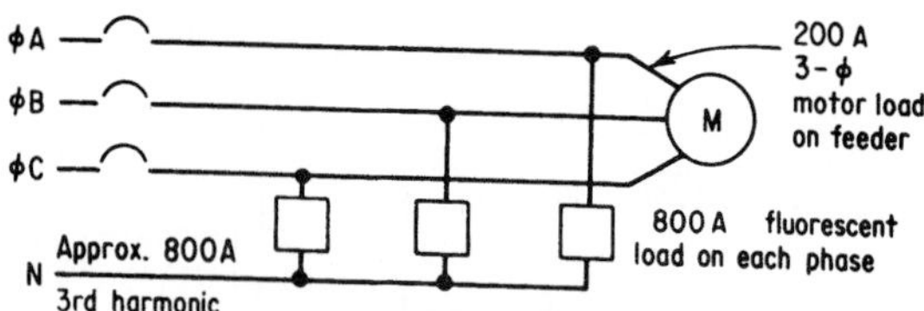

4. Electric discharge lighting plus motor load
Here, again, the only possible load that could flow on the neutral is the 800 amps flowing over each phase to the fluorescent lighting. But because it is fluorescent lighting there can be no reduction of neutral capacity below the 800-amp value on each phase. The 70% factor for that current above 200 amps DOES NOT APPLY in such cases.

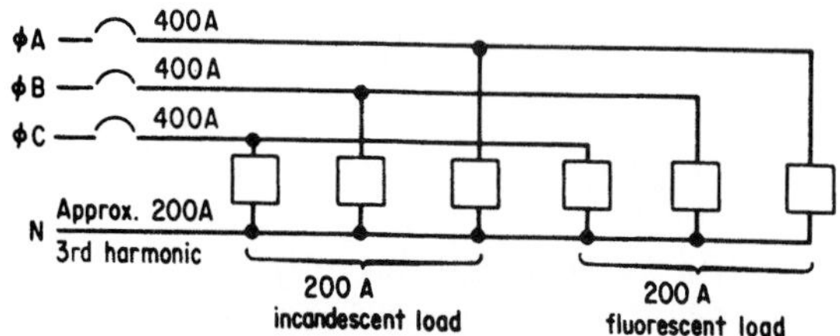

5. Incandescent plus electric discharge lighting
Each phase leg carries a total of 400 amps to supply the incandescent load plus the fluorescent load. But because there can be no reduction of neutral capacity for the fluorescent and because the incandescent load is not over 200 amps, the neutral must be sized for the maximum possible unbalance, which is 400 amps.

Fig. 220-29. (*Continued*)

of four or more electric space heaters, and the lower value is omitted from the calculation.

For instance, if a dwelling unit had 3 kW of air-conditioning load and 6 kW of electric heaters, the electric heating load of 6 kW was formerly used in the calculation because it is greater than the 3-kW air-conditioning load—even though the heating load was then subjected to the diversity factor of 40 percent in the calculation itself. *Now,* with these same loads, the 3-kW air-conditioning load is greater than 40 percent of the 6-kW heating load ($0.40 \times 6 = 2.4$ kW) and the calculation will be made using the 3-kW air-conditioning load at 100 percent demand, and the electric heating load will be disregarded.

example A typical application of the data and table of Sec. 220-30, in calculating the minimum required size of service conductors, is as follows:

A 1,500-sq ft (139.5 m^2) house (excluding unoccupied basement, unfinished attic, and open porches) contains the following specific electric appliances:

12-kW range
2.5-kW water heater
1.2-kW dishwasher

9 kW of electric heat (in five rooms)
5-kW clothes dryer
6-A, 230-V AC unit

When using the optional method, if a house has air conditioning as well as electric heating, there is recognition in Sec. 220-21 that if "it is unlikely that two dissimilar loads will be in use simultaneously," it is permissible to omit the smaller of the two in calculating required capacity in feeder or in service-entrance conductors. In Sec. 220-30, that concept is spelled out in the table to require adding only the larger of either the total air-conditioning load or 40 percent of the connected load of *four or more* separately controlled electric space heating units. For the residence considered here, these loads would be as follows:

$$\text{Air conditioning} = 6\text{ A} \times 230\text{ V} = 1.38\text{ kVA}$$
$$40\%\text{ of heating (five separate units)} = 9\text{ kW} \times 0.4 = 3.6\text{ kW (3600 VA)}$$

Because 3.6 kW is greater than 1.38 kVA, it is permissible to omit the air-conditioning load and provide a capacity of 3.6 kW in the service or feeder conductors to cover *both* the heating and air-conditioning loads.

The "other loads" must be totaled up in accordance with Sec. 220-30:

	Voltamperes
1. 1,500 VA for each of two small appliance circuits (2-wire, 20-A) required by Sec. 220-4(b)(1)	3,000
Laundry branch circuit (3-wire, 20-A)	1,500
2. 3 VA/sq ft of floor area for general lighting and general-use receptacles (3 × 1,500 sq ft)	4,500
3. Nameplate rating of fixed appliances:	
Range	12,000
Water heater	2,500
Dishwasher	1,200
Clothes dryer	5,000
Total	29,700

In reference to Table 220-30, load categories 1, 3, and 4 are not applicable here: "Air conditioning" has already been excluded as a load because 40 percent of the heating load is greater. There is no "central" electric space heating; and there are *not* "less than four" separately controlled electric space heating units.

The total load of 29,700 VA, as summed up above, includes "all other load," as referred to in **Code** Table 220-30. *Then:*

1. Take 40% of the 9000-W heating load	3,600
2. Take 10 kVA of "all other load" at 100% demand	10,000
3. Take the "remainder of other load" at 40% demand factor: (29,700 − 10,000) × 40% = 19,700 × 0.4	7,880
Total demand	21,480

Using 230- and 115-V values of voltage rather than 240- and 120-V values, ampacities may then be calculated. At 230 V, single phase, the *ampacity of each service hot leg would then have to be*

$$\frac{21{,}480\text{ W}}{230\text{ V}} = 89.5\text{ or }90\text{ A}$$

But, Sec. 230-42(b) (2) requires a minimum conductor rating when demand load is 10 kW (10 kVA) or more:

Minimum service conductor required = 100 A

Then the neutral service-entrance conductor is calculated in accordance with Sec. 220-22, based on Sec. 220-10. All 230-V loads have no relation to required neutral capacity. The water heater and electric space heating units operate at 230 V, 2-wire and have no neutrals. By considering only those loads served by a circuit with a neutral conductor and determining their maximum unbalance, the minimum required size of neutral conductor can be determined.

When a 3-wire, 230/115-V circuit serves a total load that is balanced from each hot leg to neutral—that is, half the total load is connected from one hot leg to neutral and the other half of total load from the other hot leg to neutral—the condition of maximum unbalance occurs when all the load fed by one hot leg is operating and all the load fed by the other hot leg is off. Under that condition, the neutral current and hot-leg current are equal to half the total load watts divided by 115 V (half the volts between hot legs). But that current is exactly the same as the current that results from dividing the *total* load (connected hot leg to hot leg) by 230 V (which is twice the voltage from hot leg to neutral). Because of this relationship, it is easy to determine neutral-current load by simply calculating hot-leg current load—total load from hot leg to hot leg divided by 230 V.

In the example here, the neutral-current load is determined from the following steps that sum up the components of the neutral load:

	Voltamperes
1. Take 1,500 sq ft at 3 VA/sq ft [Table 220-3(b)]	4,500
2. Add three small appliance circuits (two kitchen, one laundry) at 1,500 VA each (Sec. 220-16)	4,500
Total lighting and small appliance load	9,000
3. Take 3,000 VA of that value at 100% demand factor (Secs. 220-11 & 220-16; Table 220-11)	3,000
4. Take the balance of the load (9,000 – 3,000) at 35% demand factor: 6,000 VA × 0.35	2,100
Total of 3 and 4	5,100

Assuming an even balance of this load on the two hot legs, the neutral load under maximum unbalance will be the same as the total load (5,100 VA) divided by 230 V (Fig. 220-30):

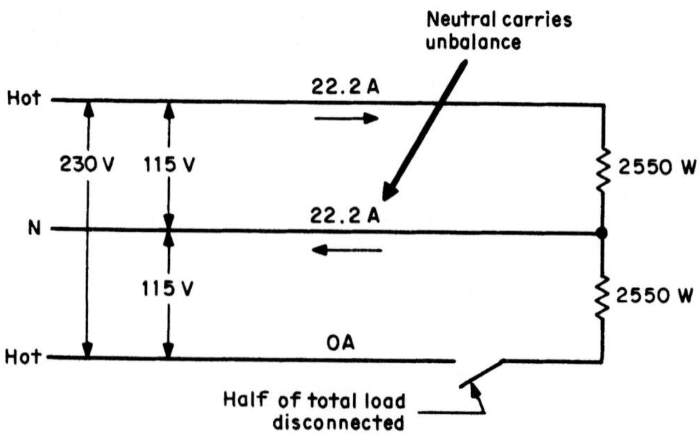

Fig. 220-30. Neutral current for lighting and receptacles. (Sec. 220-30.)

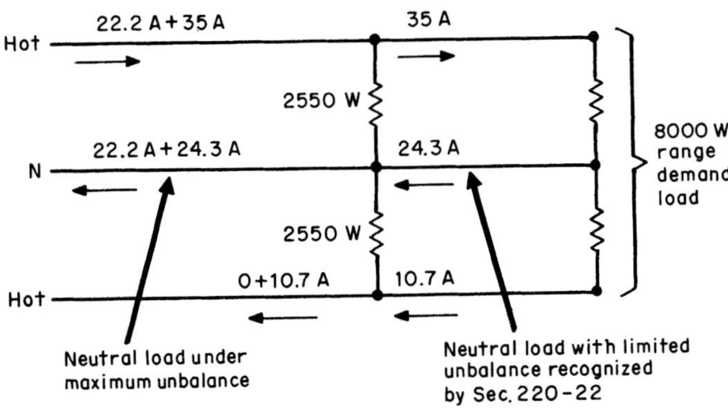

Fig. 220-31. Neutral for lighting, receptacles, and range. (Sec. 220-30.)

$$\frac{5{,}100 \text{ VA}}{230 \text{ V}} = 22.17 \text{ A}$$

And the neutral unbalanced current for the range load can be taken as equal to the 8,000-W range demand load multiplied by the 70 percent demand factor permitted by Sec. 220-22 and then divided by 230 V (Fig. 220-31):

$$\frac{8{,}000 \times 0.7}{230} = \frac{5{,}600}{230} = 24.34 \text{ A}$$

Then, the neutral-current load that is added by the 115-V, 1,200-W dishwasher must be added (Fig. 220-32):

The clothes dryer contributes neutral load due to the 115-V motor, its controls, and a light. As allowed in Sec. 220-22, the neutral load of the dryer may be taken at 70 percent of the load on the ungrounded hot legs. Then (5,000 VA × 0.7) ÷ 230 V = 15.2 A.

The minimum required neutral capacity is, therefore,

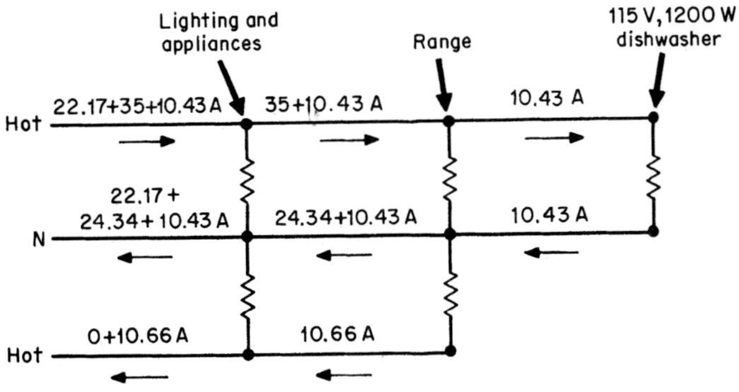

Fig. 220-32. Neutral current for all but dryer. (Sec. 220-30.)

	22.17 A
	24.34 A
	10.43 A
	15.2 A
Total:	72 A (Fig. 220-32)

From **Code** Table 310-16, the neutral minimum for 72 A would be:

No. 3 copper TW, THW, THHN or XHHW
No. 2 aluminum TW, THW, THHN or XHHW

And the 75 or 90°C conductors must be used at the ampacity of 60°C conductors, as required by UL instructions in the UL's *Electrical Construction Materials Directory.*

Note: The above calculation of the minimum required capacity of the neutral conductor differs from the calculation and results shown in Example No. 2(a) in Chap. 9 in the **NE Code** book. In the book, the 1,200-W dishwasher load is added as a 230-V load to the range load and general lighting and receptacle load. To include a 115-V load as a 230-V load (and then divide the total by 230 V, as shown) does not accurately represent the neutral load that the 115-V, 1,200-W dishwasher will produce. In fact, it yields exactly half the neutral load that the dishwasher represents. The optional calculation method of Sec. 220-30 does indicate in part **(3)** that fixed appliances be added at nameplate load and does not differentiate between 115-V devices and 230-V devices. It simply totals all load and then applies the 100 percent and 40 percent demand factors as indicated. That method clearly is based on well-founded data about diversity of loads and is aimed at determining a reasonable size of the service hot legs. But, calculation of the feeder neutral in accordance with Sec. 220-22 is aimed at determining the *maximum unbalanced current* to which the service neutral might be subjected.

Although the difference is small between the **NE Code** book value of 64 A (or 67 A if 15,400 VA is divided by 230 V) and the value of 72 A determined here, precise calculation should be made to assure real adequacy in conductor ampacities. The difference actually changes the required minimum size of neutral conductor from No. 4 up to No. 3 for copper. A load like a dishwasher, which draws current for a considerable period of time and is not just a few-minute device like a toaster, should be factored into the calculation with an eye toward adequate capacity of conductors.

220-31. Optional Calculation for Additional Loads in Existing Dwelling Unit. This covers an optional calculation for additional loads in an existing dwelling unit that contains a 120/240- or 208/120-V, 3-wire service of any current rating. The method of calculation is similar to that in Table 220-30.

The purpose of this section is to permit the *maximum load* to be applied to an *existing* service without the necessity of increasing the size of the service. The calculations are based on numerous load surveys and tests made by local utilities throughout the country. This optional method would seem to be particularly advantageous when smaller loads such as window air conditioners or bathroom heaters are to be installed in a dwelling with, say, an existing 60-A service, as follows:

If there is an existing electric range, say, 12 kW (and no electric water heater), it would not be possible to add any load of substantial rating. Based on the formula 13,800 VA (230 V × 60 A) = 8,000 + 0.4 (X – 8,000), the total "gross load" that can be connected to an existing 115/230-V, 60-A service would be X = 22,500 VA. Actually, it can be greater if a value of 240 V is used [240 V × 60 A = 14,400 VA].

example Thus, an existing 1,000-sq ft dwelling with a 12-kW electric range, two 20-A appliance circuits, a 750-W furnace circuit, and a 60-A service would have a gross load of:

	Voltamperes
1,000 sq ft × 3 VA/sq ft	3,000
Two 20-A appliance circuits @ 1,500 VA each	3,000
One electric range @	12,000
Furnace circit @	750
Gross voltamperes	18,750

Since the *maximum* permitted gross load is 22,500 VA, an appliance not exceeding *3,750* VA could be added to this existing 60-A service. However, the tabulation at the end of this section lists air-conditioning equipment, central space heating, and less than four separately controlled space heating units at 100 percent demands; and if the appliance to be added is one of these, then it would be limited to *1,500 VA:*

From the 18,750-VA gross load we have 8,000 VA @ 100 percent demand + [10,750 VA (18,750 – 8,000) × 0.40] or 12,300 VA. Then, 13,800 VA (60 A × 230 V) – 12,300 VA = 1,500 VA for an appliance listed at *100 percent demand.*

220-33. Optional Calculation—Two Dwelling Units. This section provides an optional calculation for sizing a feeder to "two dwelling units." It notes that if calculation of such a feeder according to the basic long method of calculating given in part **B** of Art. 220 exceeds the minimum load ampacity permitted by Sec. 220-32 for three identical dwelling units, then the *lesser* of the two loads may be used. This rule was added to eliminate the obvious illogic of requiring a greater feeder ampacity for two dwelling units than for the three units of the same load makeup. Now optional calculations provide for a feeder to one dwelling unit, two dwelling units, or three or more dwelling units.

220-34. Optional Method—Schools. The optional calculation for feeders and service-entrance conductors for a school makes clear that feeders "within the building or structure" must be calculated in accordance with the standard long calculation procedure established by part **B** of Art. 220. *But* the ampacity of any individual feeder does not have to be greater than the minimum required ampacity for the whole building, regardless of the calculation result from part **B.**

The last sentence in this section excludes portable classroom buildings from the optional calculation method to prevent the possibility that the demand factors of Table 220-34 would result in a feeder or service of lower ampacity than the connected load. Such portable classrooms have air-conditioning loads that are not adequately covered by using a watts-per-square-foot calculation with the small area of such classrooms.

220-35. Optional Calculations for Additional Loads to Existing Installations. Because of the universal practice of adding more loads to feeders and services in all kinds of existing premises, this calculation procedure is given in the Code. To determine how much more load may be added to a feeder or set of service-entrance conductors, at least one year's accumulation of measured maximum-demand data must be available. Then, the required spare capacity may be calculated as follows:

$$\text{Additional load capacity} = \text{ampacity of feeder or service conductors} - [(1.25 \times \text{existing demand kVA} \times 1{,}000) \div \text{circuit voltage}]$$

where "circuit voltage" is the phase-to-phase value for single-phase circuits and 1.732 times the phase-to-phase value for 3-phase circuits.

A third required condition is that the feeder or service conductors be protected against overcurrent, in accordance with applicable **Code** rules on such protection.

ARTICLE 225. OUTSIDE BRANCH CIRCUITS AND FEEDERS

225-4. Conductor Covering. The wiring method known as "open wiring" is recognized in Art. 225 as suitable for overhead use outdoors—"run between buildings, other structures or poles" (Fig. 225-1). This is derived from Secs. 225-1, 225-4, 225-14, 225-18, and 225-19. Section 225-4 requires open wiring to be insulated or covered if it comes within 10 ft (3.05 m) of any building or other structure, which it must do if it attaches to the building or structure. Insulated conductors have a dielectric covering that prevents conductive contact with the conductor when it is energized. Covered conductors—such as braided, weatherproof conductors (sometimes referred to as TBWP)—have a certain mechanical protection for the conductor but are not rated as having insulation, and thus there is no protection against conductive contact with the energized conductor.

Because Sec. 225-4 says that conductors in "*cables*" (except Type MI) must be of the rubber or thermoplastic type, a number of questions arise.

1. What kind of "*cable*" does the **Code** recognize for overhead spans between buildings, structures, and/or poles?
2. May an overhead circuit from one building to another or from lighting fixture to lighting fixture on poles use service-entrance cable, UF cable, or Type NM or NMC nonmetallic-sheathed cable?

The **Code** covers specific types of cables in turn (Arts. 330 through 340), but only in Art. 321 does the **Code** refer to use for outdoor overhead applications. And no other mention is made in the **Code** of cable suitable for overhead, outdoor spans.

The **Code** refers to overhead cable assemblies in Art. 342, "Nonmetallic Extensions." This includes a nonmetallic cable assembly with an integral sup-

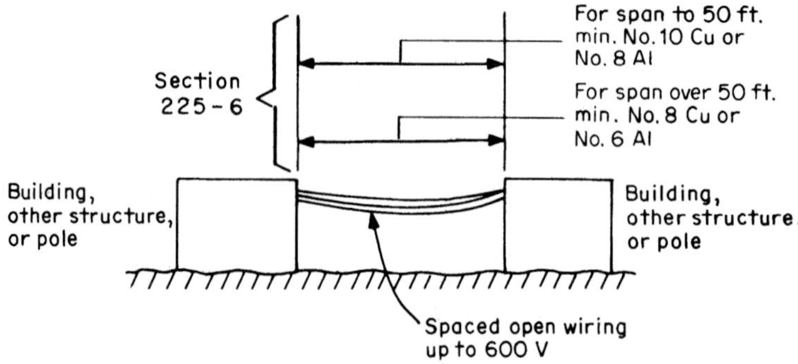

Fig. 225-1. Open wiring is OK for overhead circuits. (Sec. 225-4.)

porting messenger cable (within the assembly) for "aerial" applications. But this aerial cable is suitable only for limited use indoors, in industrial plants that are dry locations. This cable assembly is listed as an "aerial cable" by UL, but it is the *only* aerial cable listed by UL. But again, *it may not be used outdoors.*

Use of service-entrance cable between buildings, structures, and/or poles is not supported by Art. 338. No mention is made of overhead use; and Sec. 338-4 requires that "unarmored" SE cable be supported as required by Sec. 336-18, which says that the cable must be "secured in place at intervals not exceeding 4½ feet (1.37 m)." No exceptions are given. It could be argued that such support could be made to a messenger cable to which the SE cable is secured, and such practice is recognized in Sec. 321-3(a)(3).

Where Sec. 338-3(b) refers to SE cable as "a feeder to supply other buildings," it clearly refers only to "type SE" cable with a bare neutral and an overall outer jacket. (Type USE cable is recognized by UL and the **Code** for use as a feeder or branch circuit underground where all conductors are insulated.) The UL listing on type SE cable recognizes it only for aboveground installation. That certainly means on building surfaces or in raceway, but there is no mention of aerial or overhead use.

Use of Type MI, MC, or UF cable for outdoor, overhead circuits is supported by Sec. 321-3. There are no exceptions given to the support requirements in Sec. 336-18 that would let NM or NMC be used aerially, and such cables are not recognized by Sec. 321-3 for use as "messenger supported wiring."

Service-Drop Cable

The **NE Code** has Art. 321, "Messenger Supported Wiring," which covers use of "service-drop" cable, but the UL has no listing for or reference to such cable. The **NE Code** does make reference to it; and its use for aerial circuits between buildings, structures, and/or poles is particularly dictated (Fig. 225-2). Experience with this cable is very extensive and highly satisfactory. It is an engineered product specifically designed and used for outdoor, overhead circuiting.

NE Code Secs. 230-21 through 230-29 cover use of service-drop cable for overhead service conductors. Because the general rules of Art. 225 on outside branch circuits and feeders do make frequent references to other sections of Art. 230, it is logical to equate cables for overhead branch circuits and feeders to cables for overhead services. Although the rules of Art. 321 refer to a variety of messenger-supported cable assemblies, for outdoor circuits, use of service-drop cable is the best choice—because such cable is covered by the application rules of Secs. 230-21 through 230-29. Other types of available aerial cable assemblies, although not listed by UL, might satisfy some inspection agencies. But, in these times of OSHA emphasis on codes and standards, use of service-drop cable has the strongest sanction.

One important consideration in the use of service-drop cable as a branch circuit or feeder is the general **Code** prohibition against use of bare circuit conductors. Section 310-2 requires conductors to be insulated. An exception notes that bare conductors may be used where "specifically permitted." Bare *equip-*

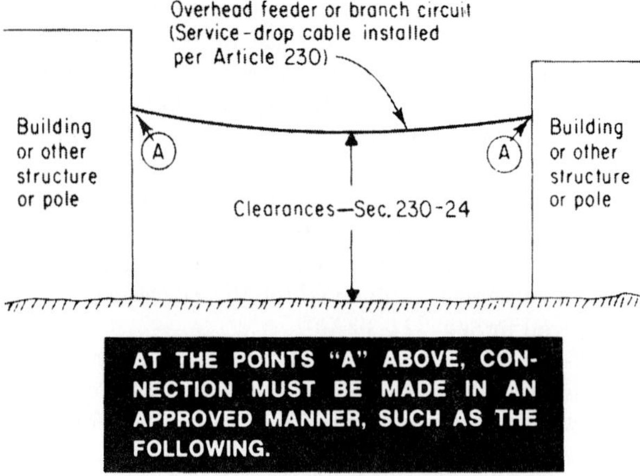

AT THE POINTS "A" ABOVE, CONNECTION MUST BE MADE IN AN APPROVED MANNER, SUCH AS THE FOLLOWING.

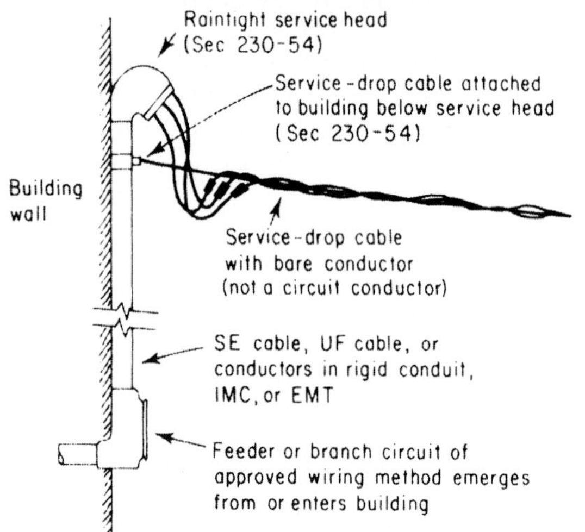

Fig. 225-2. Aerial cable for overhead circuits. (Sec. 225-4.)

ment grounding conductors are permitted in Sec. 250-91(b). A bare conductor for SE cable is permitted in Sec. 338-1(c). Bare neutrals are permitted for service-entrance conductors in Sec. 230-41, for underground service-entrance (service lateral) conductors in Sec. 230-30, and for *service-drop conductors* in Sec. 230-22, Exception *when used as service conductors.* When service-drop cable is used as a feeder or branch circuit, however, there is no permission for use of a bare circuit conductor—although it may be acceptable to use the bare conductor of the service-drop cable as an equipment grounding conductor. And where service-drop cable is used as a feeder from one building to another, it would seem that a bare neutral could be acceptable as a grounded neutral conductor—as permitted in the last sentence of Sec. 338-3(b), first paragraph. (If service-drop cable is used as a feeder to another building and the bare conductor is *not* used as an equipment grounding conductor, then a grounding electrode *must* be installed at the other building in accordance with Sec. 250-24.)

When service-drop cable is used between buildings, the method for leaving one building and entering another *must* satisfy Secs. 230-43, 230-52, and 230-54. This is required in Sec. 225-11.

The Exception to Sec. 225-4 excludes equipment grounding conductors and grounded circuit conductors from the rules on conductor covering. This Exception permits equipment grounding conductor *and* grounded circuit conductors (neutrals) to be bare or simply covered (but not insulated) as permitted by other Code rules.

Because the matter of outdoor, overhead circuiting is complex, check with local inspection agencies on required methods. As NE Code Sec. 90-4 says, the local inspector has the responsibility for making interpretations of the rules.

225-6. Minimum Size of Conductor. Open wiring must be of the minimum sizes indicated in Sec. 225-6 for the various lengths of spans indicated.

Section 225-6 gives a definition of *festoon lighting* as "a string of outdoor lights suspended between two points more than 15 feet (4.57 m) apart" (Fig. 225-3). Such lighting is used at carnivals, displays, used-car lots, etc. Such application of lighting is limited because it has a generally poor appearance and does not enhance commercial activities.

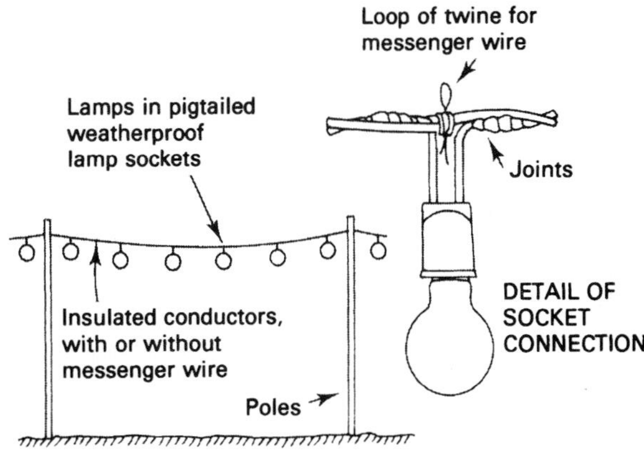

Fig. 225-3. Festoon lighting is permitted outdoors. (Sec. 225-6.)

Overhead conductors for festoon lighting must not be smaller than No. 12; and where any span is over 40 ft (12.2 m) (Sec. 225-13), the conductors must be supported by a messenger wire, which itself must be properly secured to strain insulators. But the rules on festoon lighting do not apply to overhead circuits between buildings, structures, and/or poles.

225-7. Lighting Equipment Installed Outdoors. Part **(b)** permits a common neutral for outdoor branch circuits—something not permitted for indoor branch circuits (a neutral of a 3-phase, 4-wire circuit is *not* a common neutral). For two

208Y/120-V multiwire circuits consisting of six ungrounded conductors (two from each phase) and a single neutral (serving both circuits) feeding a bank of floodlights on a pole, if the maximum calculated load on any one circuit is 12 A and the maximum calculated load on any one phase is 24 A, the ungrounded circuit conductors may be No. 14, but the neutral must be at least No. 10. This rule clearly states the need to size a common neutral for the *maximum* (most heavily loaded) phase leg made up by multiple conductors connected to any one phase and supplying loads connected phase-to-neutral.

Part **(c)** covers use of 480/277-V systems for supplying incandescent and electric-discharge lighting fixtures. A minimum mounting height of 8 ft (2.44 m) is no longer required for lighting fixtures installed outdoors on buildings, structures, or poles on industrial or commercial premises and fed by 480/277-V circuits. This section rules that such fixtures must be not less than 3 ft (914 mm) from "windows, platforms, fire escapes, and the like." And the wording of this section is not just limited to fixtures mounted on buildings, structures, or poles but could apply to fixtures mounted at ground level for lighting of signs, building facades, and other decorative or ornamental lighting.

225.8. Disconnection. For a group of buildings under single management, disconnect means must be provided for each building, as in Fig. 225-4. The rule of part **(b)** requires that the conductors supplying each building in the group be provided with a means for disconnecting all ungrounded conductors from the supply.

For large-capacity, multibuilding industrial premises with a single owner, Exception No. 1 permits use of the feeder switch in the main building as the only disconnect for each feeder to an outlying building, provided the switches in the main building are accessible to the occupants of the outlying building.

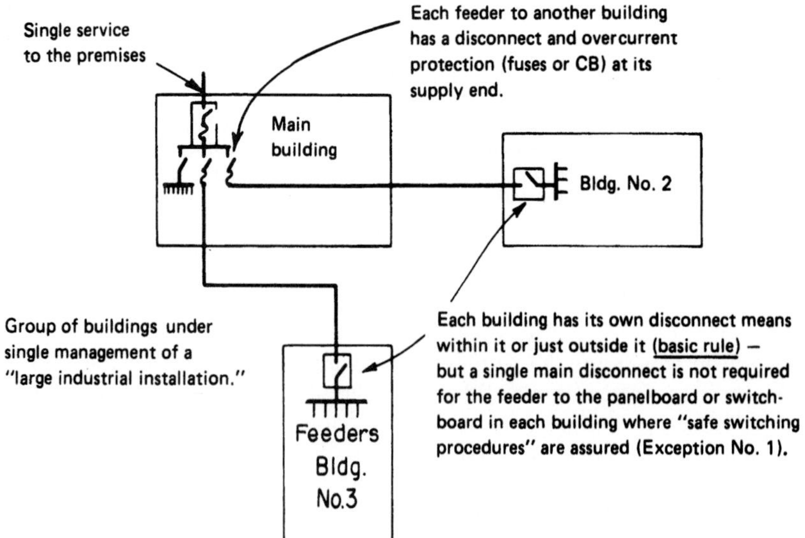
Single service
to the premises
Each feeder to another building
has a disconnect and overcurrent
protection (fuses or CB) at its
supply end.
Main
building
Bldg. No. 2
Group of buildings under
single management of a
"large industrial installation."
Each building has its own disconnect means
within it or just outside it (basic rule) —
but a single main disconnect is not required
for the feeder to the panelboard or switchboard in each building where "safe switching
procedures" are assured (Exception No. 1).
Feeders
Bldg.
No.3

Fig. 225-4. Each building must have its own disconnect means. (Sec. 225.8.)

Where a disconnect is to be used at each outbuilding, the effect of the last sentence of Sec. 225-8(b) (requiring location of the disconnect in accordance with Sec. 230-71 and 230-72) is to clearly require that the disconnect for any outbuilding be located within or just outside each building "nearest the point of entrance of the service-entrance conductors." However, many inspection and engineering authorities prefer a readily accessible feeder disconnect *within* each outlying building, regardless of the distance from the main building.

For any "integrated electrical system" as defined and regulated by Art. 685, Exception No. 2 suspends the basic rule calling for a feeder disconnect at each building.

Exception No. 3 eliminates the need for individual disconnects for individual lighting standards. The literal wording calls for a disconnect at each "structure." The addition of this Exception indicates the CMP's intent, which is to permit one disconnect for a number of lighting poles.

Part **(c)** requires that the disconnect for each building of a multibuilding layout be recognized for service use—usually that means that the disconnect means for each building must be listed by UL, or another national test lab, as suitable for service equipment. The Exception waives that requirement for wiring device switches (snap switches) used for ON-OFF control of lighting or other loads under the conditions noted.

225.10. Wiring on Buildings. Compared with the 1984 and previous NEC editions, additional wiring methods are recognized for installation on the outdoor surfaces of buildings. Rigid, nonmetallic conduit may be used for outside wiring on buildings, as well as the other raceway and cable methods covered in this section. For a long time, rigid PVC was not permitted for such application.

225.14. Open-Conductor Spacings. Open wiring runs must have a minimum spacing between individual conductors (as noted in Sec. 225-14) in accordance with Table 230-51(c), which gives the spacing of the insulator supports on a building surface and the clearance between individual conductors on the building or run in spans (Fig. 225-5).

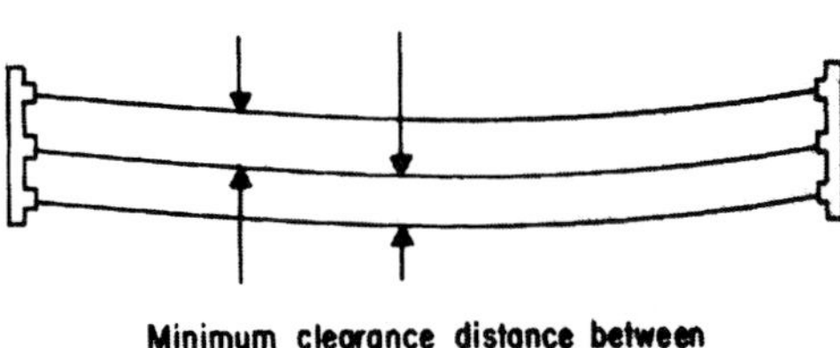
Minimum clearance distance between
conductors is given in Table 230-51 (c)

Fig. 225-5. Spacing of open-wiring conductors. (Sec. 225-14.)

It should be noted that Sec. 225-14 and Table 230-51(c) require that the *minimum spacing* between individual conductors in spans run overhead be 3 in. (76 mm) for circuits up to 300 V (such as 120, 120/240, 120/208, and 240 V).

For circuits up to 600 V, such as 480 Δ and 480/277 V, the *minimum spacing* between individual conductors must be at least 6 in. (152 mm).

225-18. Clearance from Ground. Open conductors must be protected from contact by persons by keeping them high enough aboveground or above other positions where people might be standing. And they must not present an obstruction to vehicle passage or other activities below the lines (Fig. 225-6).

Section 225-18 applies only to "open conductors" and gives the conditions under which clearances must be 10, 12, 15, or 18 ft (3.05, 3.66, 4.57, or 5.49 m)—for conductors that make up either a branch circuit or a feeder [not service-drop conductors, which are subject to Sec. 230-24(b)]. Although the wording of the third, fourth, and fifth paragraphs is the same as that of those referring to corresponding clearances in Sec. 230-24(b), application of Sec. 225-18 is limited to "open conductors"—and Art. 225 gives no clue as to minimum clearances for triplex or quadruplex cables commonly used for outdoor overhead branch circuits and feeders. This does pose a serious problem in Code application, especially in relation to Art. 321, which recognizes both indoor and outdoor use of a variety of messenger supported assemblies (not "open conductors") for overhead branch circuits or feeders. Article 321 gives no clearances from ground for such circuits, but Sec. 321-2 does make such circuits subject to "applicable provisions of Article 225"—which, as noted, gives clearances only for "open conductors."

As Secs. 225-18 and 230-24(b) stand, "open conductors" for an overhead *branch circuit or feeder* require only a 10-ft (3.05 m) clearance from ground for circuits up to 150 V to ground; *but* "open conductors" for a service drop up to 150 V must have a clearance of not less than 12 ft (3.66 m) from ground.

Open conductors are strung from one building to another, for circuits not over 150 volts to ground.

Open conductors, not over 600 V

Not less than 10 ft

10 ft min.

Elevated platform

Pedestrian walkway

Individual open conductors supported by poles pass over both a residential driveway and a road bearing truck traffic:

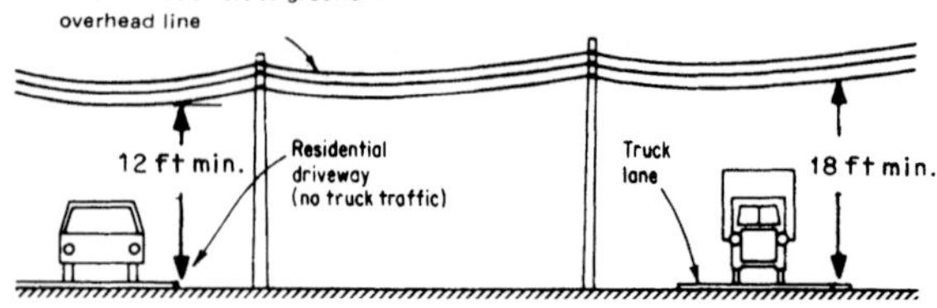

Fig. 225-6. Conductor clearance from ground. (Sec. 225-18.)

The rules of this section agree with the clearances and conditions set forth in the *NESC* (*National Electrical Safety Code*) for open conductors outdoors. The distances given for clearance from ground must conform to maximum voltage at which certain heights are permitted.

225-19. Clearances from Buildings for Conductors of Not Over 600 Volts, Nominal. The basic minimum required clearance for outdoor conductors running above a roof is 8-ft (2.44-m) vertical clearance from the roof surface.

The wording of this section is substantially different from that of the 1987 and previous editions of the NEC and completely changes the concept involved. The rule no longer differentiates between "conductors not fully insulated" and "fully insulated conductors"—as was done in the 1987 Sec. 225-19(a). The basic ideas behind the rules are as follows:

1. Any branch-circuit or feeder conductors—whether insulated, simply covered, or bare—must have a clearance of at least 8 ft (2.44 m) vertically from a roof surface over which they pass. And that clearance must be maintained not less than 3 ft (914 mm) from the edge of the roof in all directions. (The use of fully insulated conductors with only a 3-ft (914-mm) vertical or diagonal clearance above a roof is *no longer* allowed.)
2. A roof that is subject to "pedestrian or vehicular traffic" must have conductor clearances "in accordance with the clearance requirements of Sec. 225-18." This is an extremely obscure rule that is impossible to apply with comprehension. It might be interpreted to require 10 or 15-ft (3.05 or 4.57-m) clearance for a roof subject to pedestrian traffic and 18 ft (5.49 m) for a roof subject to vehicular traffic. This rule is made difficult by the reference to Sec. 225-18 because that section applies only to "open conductors" and does not at all cover the several types of insulated (and sometimes jacketed) cable assemblies that are far more commonly used than "open conductors."

In parts **(b)** and **(c)**, overhead conductor clearance from signs, chimneys, antennas, and other nonbuilding or nonbridge structures must be at least 3 ft (914 mm)—vertically, horizontally, or diagonally. The minimum required clearance here was reduced from a 5-ft (1.52-m) requirement in the 1987 NEC.

As indicated in Fig. 225-7, Exception No. 2 to Sec. 225-19(a) may apply to circuits that are operated at 300 V or less.

225-22. Raceways on Exterior Surfaces of Buildings. Condensation of moisture is very likely to take place in conduit or tubing located outdoors. The conduit or tubing should be considered suitably drained when it is installed so that any moisture condensing inside the raceway or entering from the outside cannot accumulate in the raceway or fittings. This requires that the raceway shall be installed without "pockets," that long runs shall not be truly horizontal but shall always be pitched, and that fittings at low points be provided with drainage openings.

In order to be raintight, all conduit fittings must be made up wrench-tight. Couplings and connectors used with electrical metallic tubing shall be of the raintight type. See Sec. 348-8.

225-24. Outdoor Lampholders. This section applies particularly to lampholders used in festoons. Where "pigtail" lampholders are used, the splices should be

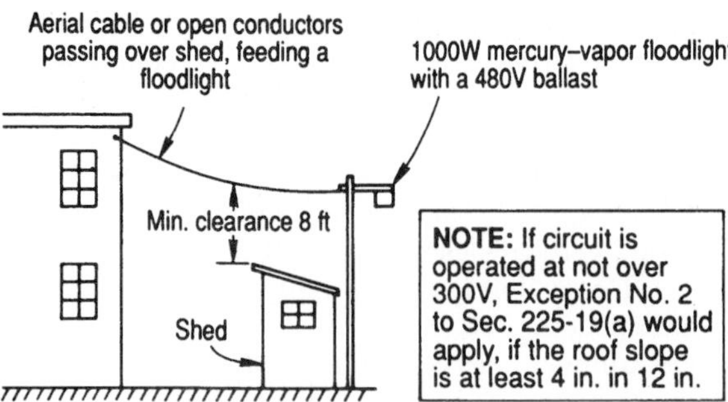
Aerial cable or open conductors passing over shed, feeding a floodlight
1000W mercury–vapor floodlight with a 480V ballast
Min. clearance 8 ft
Shed
NOTE: If circuit is operated at not over 300V, Exception No. 2 to Sec. 225-19(a) would apply, if the roof slope is at least 4 in. in 12 in.

Fig. 225-7. Conductors—whether or not they are fully insulated for the circuit voltage—must have at least 8-ft (2.44-m) vertical clearance above a roof over which they pass. (Sec. 225-19.)

staggered (made a distance apart) in order to avoid the possibility of short circuits, in case the taping for any reason should become ineffective.

According to the UL Standard for Edison-Base Lampholders, "pin-type" terminals shall be employed only in lampholders for temporary lighting or decorations, signs, or specifically approved applications.

225-25. Location of Outdoor Lamps. In some types of outdoor lighting it would be difficult to keep all electrical equipment above the lamps, and hence a disconnecting means may be required. A disconnecting means should be provided for the equipment on each individual pole, tower, or other structure if the conditions are such that lamp replacements may be necessary while the lighting system is in use. It may be assumed that grounded metal conduit or tubing extending below the lamps would not constitute a condition requiring that a disconnecting means must be provided.

225-26. Vegetation. Coming to grips with a long-standing controversy, outdoor conductors and equipment may not be supported from or mounted on trees. Trees or any other "vegetation" must not be used "for support of overhead conductor spans." Wording of this rule removed the words "or other electric equipment." The effect is to permit outdoor lighting fixtures to be mounted on trees and to be supplied by an approved wiring method—conductors in a raceway or Type UF cable—attached to the surface of the tree. The Code Panel defeated several proposals to have a new exception added to this section to specifically recognize use of lighting fixtures and their supply wiring directly on trees (Fig. 225-8).

The one exception to this rule recognizes that temporary wiring may be run on trees in accordance with Article 305.

ARTICLE 230. SERVICES

230-2. Number of Services. For any building, the service consists of the conductors and equipment used to deliver electric energy from the utility supply

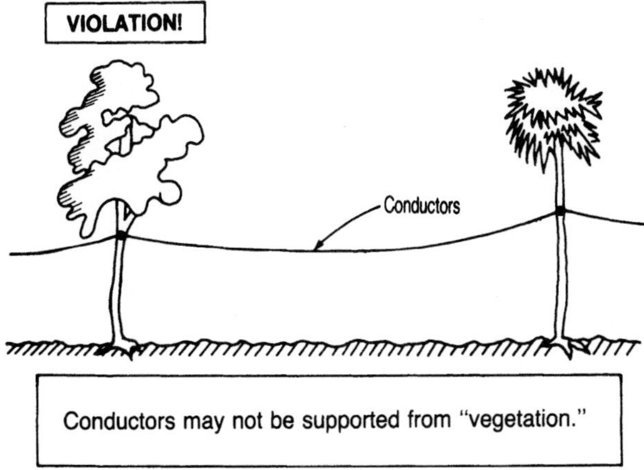

Fig. 225-8. The rule of Sec. 225-26 prohibits wiring installed on trees.

lines to the interior distribution system. Service may be made to a building either overhead or underground, from a utility pole line or from an underground transformer vault.

The first sentence of this rule requires that a building or structure be supplied by "only one service." Because "service" is defined in Art. 100 as "The *conductors* and equipment for delivering energy from the electricity supply system to the wiring system of the premises supplied," use of one "service" corresponds to use of one "service drop" or one "service lateral." Thus, the basic rule of this section requires that a building or other structure be fed by only one service drop (overhead service) or by only one set of service lateral conductors (underground service). As shown in Fig. 230-1, a building with only one service drop to it satisfies the basic rule even when more than one set of service-entrance conductors are tapped from the single drop (or from a single lateral circuit).

The seven Exceptions to that basic rule cover cases where two or more service drops or laterals may supply a single building or structure.

The second paragraph of this section introduces a requirement that applies to any installation where more than one service is permitted by the Code to supply one building. It requires a "permanent plaque or directory" to be mounted or placed "at each service drop or lateral or at each service-equipment location" to advise personnel that there are other services to the premises and to tell where such other services are and what building parts they supply.

Exceptions No. 1 and No. 2 permit a separate drop or lateral for supply to a fire pump and/or to emergency electrical systems, such as emergency lighting or exit lights.

Exception No. 2 to the basic rule that a building "shall be supplied by only one service" recognizes use of an additional power supply to a building from any "parallel power production systems." This would permit a building to be fed by a solar photovoltaic, wind, or other electric power source—in addition

Fig. 230-1. One set of service-drop conductors supply building from utility line (coming from upper left) and two sets of SE conductors are tapped through separate metering CTs. (Sec. 230-2.)

to a utility service—just as an emergency or standby power source is also permitted (Fig. 230-2).

Exception No. 3 recognizes another situation in which more than one service (i.e., more than one service drop or lateral) may be used. By "special permission" of the inspection authority, more than one service may be used for a multitenant building when there is no single space that would make service equipment available to all tenants.

Two or more services to one building are permitted when the total demand load of all the feeders is more than 2,000 A, up to 600 V, where a single-phase

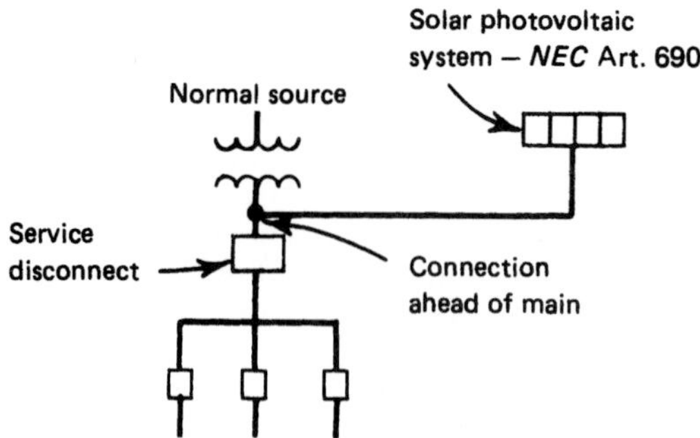

Fig. 230-2. Electric power generated by a solar voltaic assembly or by a wind-driven generator may be used as a source of power in "parallel" with the normal service. (Sec. 230-2.)

service needs more than one drop, or by special permission (Fig. 230-3). Exception No. 4 relates capacity to permitted services. Where requirements exceed 2,000 A, two or more sets of service conductors may be installed. Below this value, special permission is required to install more than one set. The term "capacity requirements" appears to apply to the total calculated load for sizing service-entrance conductors and service equipment for a given installation.

Cases of separate light and power services to a single building and separate services to water heaters for purposes of different rate schedules are also exceptions to the general rule of single service. And if a single building is so large that

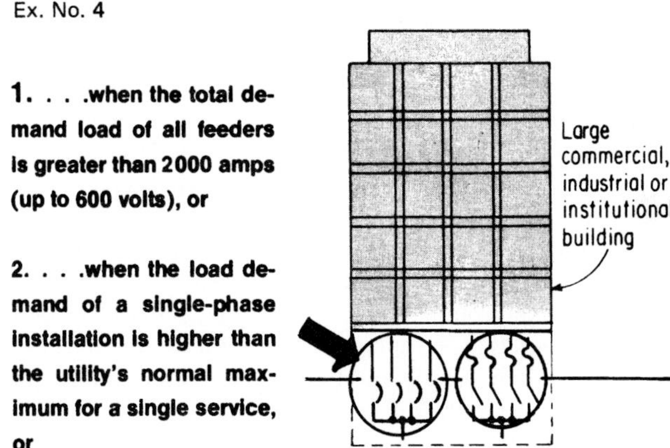
Ex. No. 4
1. . . .when the total demand load of all feeders is greater than 2000 amps (up to 600 volts), or
2. . . .when the load demand of a single-phase installation is higher than the utility's normal maximum for a single service, or
Large commercial, industrial or institutional building

3. . . .when special permission is obtained from the inspection authority.

NOTE: "Two or more services" means two or more service drops or service laterals—not sets of service-entrance conductors tapped from one drop or lateral.

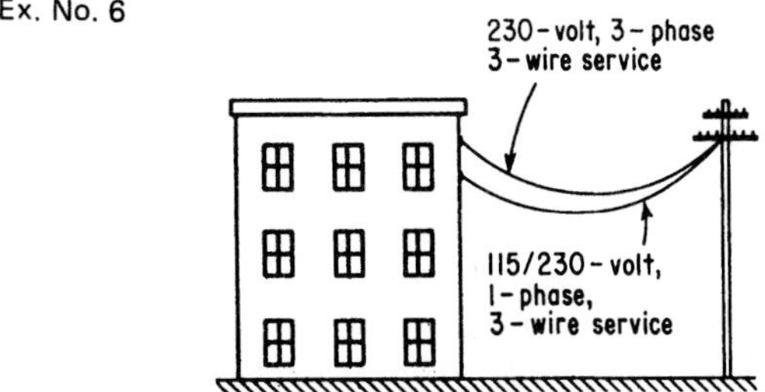
Ex. No. 6
230-volt, 3-phase
3-wire service
115/230-volt,
1-phase,
3-wire service

Fig. 230-3. Exceptions to Sec. 230-2 permit two or more services under certain conditions. (Sec. 230-2.)

one service cannot handle the load, special permission can be given for additional services.

Exception No. 5 requires special permission to install more than one service to buildings of large area. Examples of large-area buildings are high-rise buildings, shopping centers, and major industrial plants. In granting special permission the authority having jurisdiction must examine the availability of utility supplies for a given building, load concentrations within the building, and the ability of the utility to supply more than one service. Any of the special-permission clauses in the Exceptions in Sec. 230-2 require close cooperation and consultation between the authority having jurisdiction and the serving utility.

Exception No. 6 is illustrated at the bottom of Fig. 230-3.

Exception No. 7 to the basic rule requiring that any "building or other structure" be supplied by "only one service" adds an important qualification of that rule as it applies only to Sec. 230-40, Exception No. 2, covering service-entrance layouts where two to six service disconnects are to be fed from one drop or lateral and are installed in separate individual enclosures at one location, with each disconnect supplying a separate load. As described in Sec. 230-40, Exception No. 2, such a service equipment layout may have a separate set of service-entrance conductors run to "*each or several*" of the two to six enclosures. Exception No. 7 notes that where a separate set of underground conductors of size 1/0 or larger is run to each or several of the two to six service disconnects, the several sets of underground conductors are considered to be one service (that is, one service lateral) even though they are run as separate circuits, that is, connected together at their supply end (at the transformer on the pole or in the pad-mount enclosure or vault) *but not* connected together at their load ends. The several sets of conductors are taken to be "one service" in the meaning of Sec. 230-2, although they actually function as separate circuits (Fig. 230-4).

Although Sec. 230-40, Exception No. 2, applies to "service-entrance conductors" and service equipment layouts fed by *either* a "service drop" (overhead service) or a "service lateral" (underground service), Exception No. 7 is addressed specifically and only to service "lateral" conductors (as indicated by the word "underground") because of the need for clarification based on the **Code** definitions of "service drop," "service lateral," "service-entrance conductors, overhead system," and "service-entrance conductors, underground system." (Refer to these definitions in the **Code** book to clearly understand the intent of Exception No. 7 and its relation to Sec. 230-40, Exception No. 2.)

The matter involves these separate but related considerations:

1. Because a "service lateral" may (and usually does) run directly from a transformer on a pole or in a pad-mount enclosure to gutter taps where short tap conductors feed the terminals of the service disconnects, most layouts of that type literally do not have any "service-entrance conductors" that would be subject to the application permitted by Sec. 230-40, Exception No. 2—other than the short lengths of tap conductors in the gutter or box where splices are made to the lateral conductors.
2. Because Sec. 230-40, Exception No. 2, refers only to sets of "service-entrance conductors" as being acceptable for individual supply circuits tapped from *one* drop or lateral to feed the separate service disconnects,

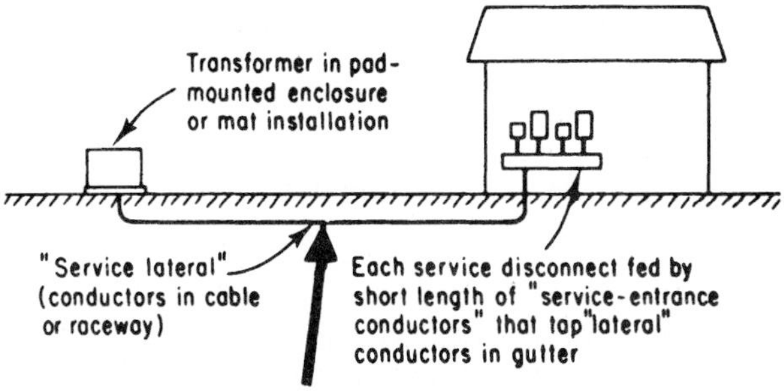
The 1975 NE Code had this limitation on service laterals (and this is still acceptable)—
Transformer in pad-mounted enclosure or mat installation
"Service lateral" (conductors in cable or raceway)
Each service disconnect fed by short length of "service-entrance conductors" that tap "lateral" conductors in gutter
"Service lateral" conductors are not "service entrance" conductors and were, therefore, not applicable to the subdivision permission of Section 230-40, Exception No. 2. The requirement of Section 230-2 for one set of service lateral conductors demanded one circuit of single-conductor or parallel-conductor makeup.

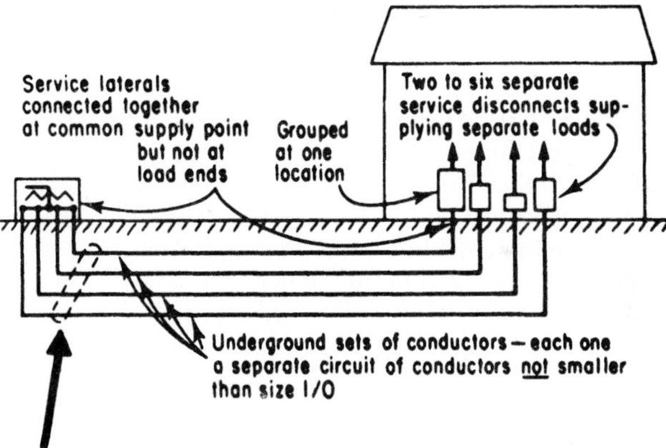
Now, Exception No. 7 considers this type of hookup to be one set of service lateral conductors —
Service laterals connected together at common supply point but not at load ends
Grouped at one location
Two to six separate service disconnects supplying separate loads
Underground sets of conductors—each one a separate circuit of conductors not smaller than size 1/0
This is one service lateral, in the meaning of the basic rule of Section 230-2.

Fig. 230-4. "One" service lateral may be made up of several circuits. (Sec. 230-2.)

that rule clearly does not apply to "service lateral" conductors which by definition are not "service-entrance conductors." So there is no permission in Sec. 230-40, Exception No 2, to split up "service lateral" capacity. And the basic rule of Sec. 230-2 has the clear, direct requirement that a building or structure be supplied through only *one* lateral for any underground service. That is, either a service lateral must be a single circuit of one set of conductors, or if circuit capacity requires multiple conductors per phase leg, the lateral must be made up of sets of conductors in parallel—connected together at *both* the supply and load ends—in order to constitute a single circuit (that is, one lateral).

3. Exception No. 7 permits "laterals" to be subdivided into separate, nonparallel sets of conductors in the way that Sec. 230-40, Exception No. 2, permits such use for "service-entrance conductors"—*but only* for conductors of 1/0 and larger and *only* where each separate set of lateral conductors (each separate lateral circuit) supplies *one* or *several* of the two to six service disconnects.

Exception No. 7 recognizes the importance of subdividing the total service capacity among a number of sets of smaller conductors rather than a single parallel circuit (that is, a number of sets of conductors connected together at *both* their *supply and load* ends). The single parallel circuit would have much lower impedance and would, therefore, require a higher short-circuit interrupting rating in the service equipment. The higher impedance of each separate set of lateral conductors (not connected together at their load ends) would limit short-circuit current and reduce short-circuit duty at the service equipment, permitting lower IC (interrupting capacity)-rated equipment and reducing the destructive capability of any faults at the service equipment.

Part **(b)** of Sec. 230-2 mandates the use of plaques or directories where more than one source of supply is provided to a single building or structure. This directory must be placed at *each* service. So, if there are two services, there should be two plaques at *each* service location. The directory (or directories) must identify all feeders and branch circuits supplied from that service.

230-3. One Building or Other Structure Not to Be Supplied Through Another. The service conductors supplying each building or structure shall not *pass through the inside* of another building, unless they are in a raceway encased by 2 in. (50.8 mm) of concrete or masonry (Fig. 230-5). Section 230-6 points out that conductors in a raceway enclosed within 2 in. (50.8 mm) of concrete or masonry are considered to be "outside" the building even when they are run within the building.

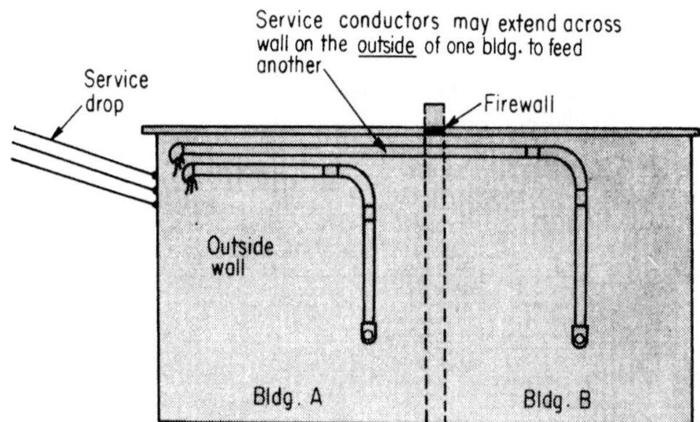

Fig. 230-5. This is not a violation of the basic rule of Sec. 230-3. (Sec. 230-3.)

A building as defined in Art. 100 is a "structure which stands alone or which is cut off from adjoining structures by fire walls with all openings therein protected by approved fire doors." A building divided into four units by such fire walls may be supplied by four separate service drops, but a similar building without the fire walls may be supplied by only one service drop, except as permitted in Sec. 230-2.

A commercial building may be a single building but may be occupied by two or more tenants whose quarters are separate, in which case it might be undesirable to supply the building through one service drop. Under these conditions special permission may be given to install more than one service drop.

230-6. Conductors Considered Outside of Building. Conductors in conduit or duct enclosed by concrete or brick not less than 2 in. (50.8 mm) thick are considered to be outside the building, even though they are actually run within the building. Figure 230-6 shows how a service conduit was encased within a building so that the conductors are considered as entering the building right at the service protection and disconnect where the conductors emerge from the concrete, to satisfy the rule of Sec. 230-70(a), which requires the service disconnect to be as close as possible to the point where the SE conductors enter the building. Figure 230-7 shows an actual case of this application, where forms were hung around the service conduit and then filled with concrete to form the required case.

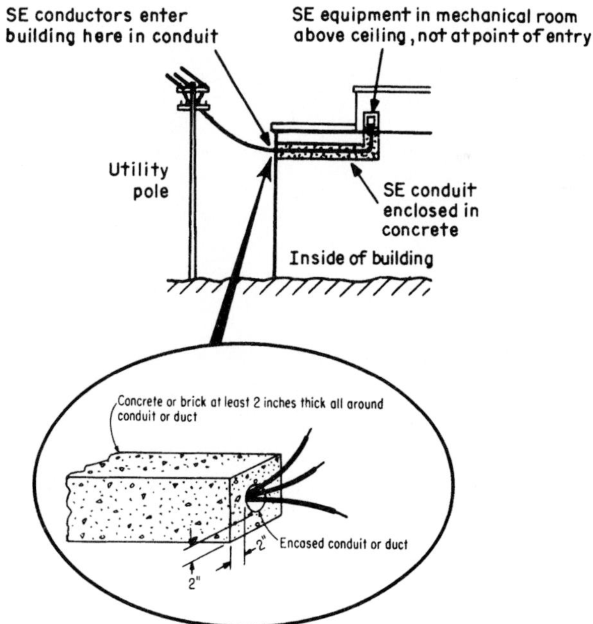

Fig. 230-6. "Service raceways" in concrete are considered "outside" a building. (Sec. 230-6).

Fig. 230-7. Top photo shows service conduit carried above suspended ceiling, without the SE disconnect located at the point of entry. When conduit was concrete-encased, the service conductors then "enter" the building at the SE disconnect—where they emerge from the concrete. Service conduit enters building at lower left and turns up into SE disconnect (right) in roof electrical room.

230-7. Other Conductors in Raceway or Cable. Although the basic rule permits only service-entrance conductors to be used in a service raceway or service cable, exceptions do recognize the use of grounding conductors in a service raceway or cable and also permit conductors for a time switch if overcurrent protection is provided for the conductors, as shown in Fig. 230-8.

230-8. Raceway Seal. Figure 230-9 indicates that Sec. 300-5(g) may apply to underground service conduits. Where service raceways are required to be sealed—as where they enter a building from underground—the sealing compound used must be marked on its container or elsewhere as suitable for safe and effective use with the particular cable insulation, with the type of shielding used, and with any other components it contacts. Some sealants attack certain insulations, semiconducting shielding layers, etc.

230-9. Clearance from Building Openings. Any service drop conductors—open wiring or multiplex drop cable—must have the 3-ft (914-mm) clearance from windows, doors, porches, etc., to prevent mechanical damage to and accidental contact with service conductors (Figs. 230-10 and 230-11). The clearances

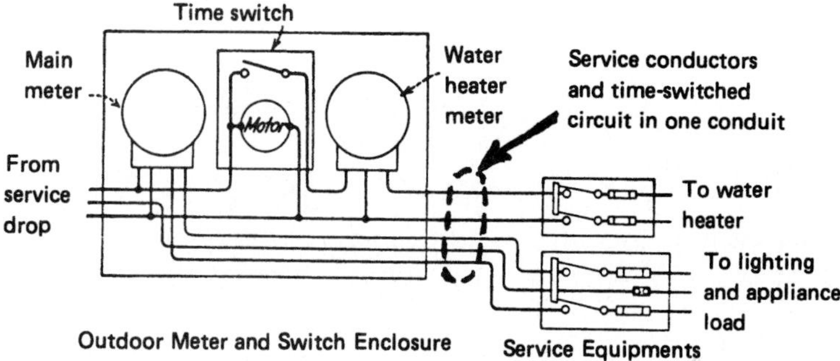

Fig. 230-8. A time switch with its control circuit connected on the supply side of the service equipment. (Sec. 230-7.)

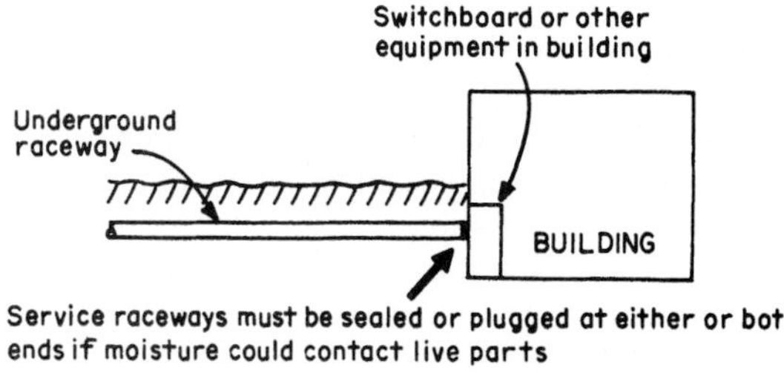

Fig. 230-9. Service raceways may have to be sealed. (Sec. 230-8.)

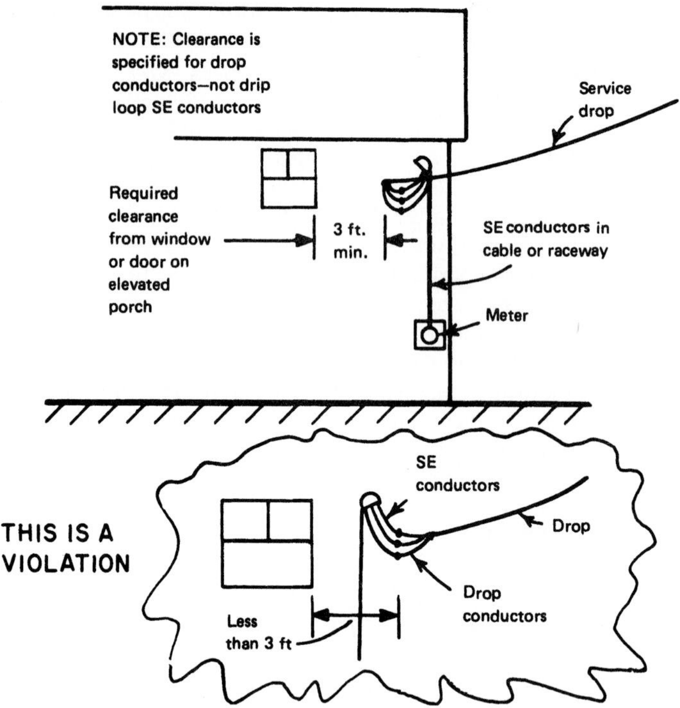

Fig. 230-10. Drop conductors must have clearance from building openings. (Sec. 230-9.)

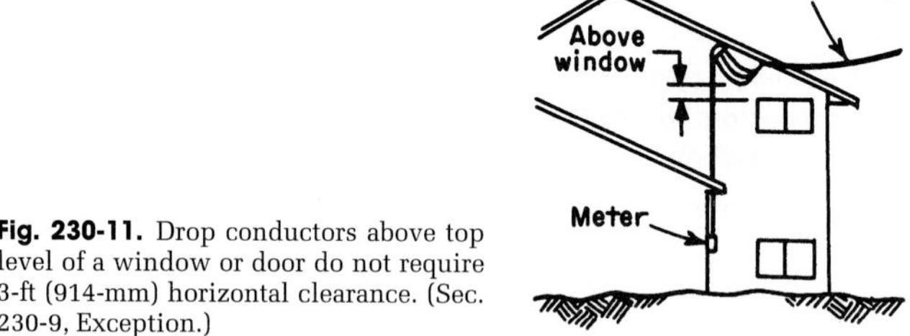

Fig. 230-11. Drop conductors above top level of a window or door do not require 3-ft (914-mm) horizontal clearance. (Sec. 230-9, Exception.)

required in Secs. 230-24, 230-26, and 230-29 are based on safety-to-life considerations in that wires are required to be kept a reasonable distance from people who stand, reach, walk, or drive under service-drop conductors. As the Exception notes, conductors that run above the top level of a window do not have to be 3 ft (914 mm) away from the window.

The third paragraph of this section says that service-drop or service-entrance conductors must not be mounted on or secured to a building wall directly beneath an elevated opening through which supplies or materials are moved into and out of the building. Such installations of conductors—say, beneath a high door to a barn loft—would obstruct access to the opening and present a hazard to personnel (Fig. 230-12).

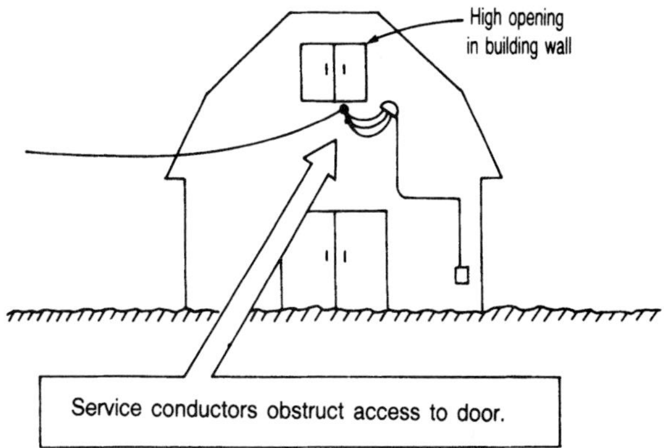

Fig. 230-12. This violates the rule of the last paragraph of Sec. 230-9.

230-22. Insulation or Covering. The use of "covered"—not "insulated"—wire, such as TBWP (triple-braid weatherproof wire), has resulted in 14 accidents. Because of nine electrocutions and four hospitalizations, the use of *only* insulated wire for ungrounded conductors is required by this rule.

The only overhead service conductor that is permitted to be bare is a grounded conductor of a multiconductor cable. The grounded neutral of open wiring must be insulated to the same voltage level as the ungrounded conductors.

230-24. Clearances. There are three exceptions to the basic rule of part **(a)** that service-drop conductors must have at least an 8-ft (2.44-m) vertical clearance from the highest point of roofs over which they pass.

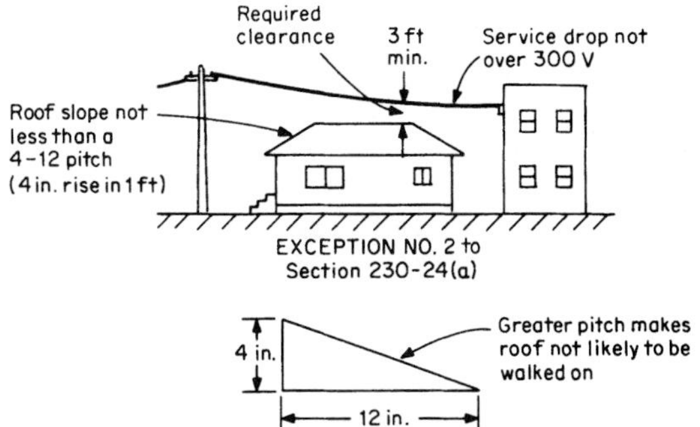

Fig. 230-13. Service-drop conductors may have less than 8-ft (2.44-m) roof clearance. (Sec. 230-24.)

Exception No. 1 to the basic rule calling for 8-ft (2.44-m) clearance of service-drop conductors above a roof requires that clearance above a flat roof subject to pedestrian traffic or used for auto and/or truck traffic must observe the heights for clearance of drop conductors from the ground as given in part **(b)** of Sec. 230-24.

The intent of Exception No. 2 is that where the roof has a slope greater than 4 in. (102 mm) in 12 in. (305 mm), it is considered difficult to walk upon, and the height of conductors could then be less than 8 ft (2.44 m) from the highest point over which they pass but in no case less than 3 ft (914 mm) except as permitted in Exception No. 3. Figure 230-13 shows the rule. Figure 230-14 shows the conditions permitted by Exception No. 3.

Part **(b)** covers service-drop clearance to ground, as shown in Fig. 230-15. The four dimensions of clearance from ground—10, 12, 15, and 18 ft (3.05, 3.66, 4.57, and 5.49 m)—are qualified by voltage levels and, for the 10-ft (3.05-m) mounting height, by the phrase "only for service-drop cables." These NE Code rules are in agreement with the *National Electrical Safety Code.* Where mast-type service risers are provided, the clearances in Sec. 230-24(b) will have to be considered by the installer.

230-28. Service Masts as Supports. Figure 230-16 illustrates this rule.

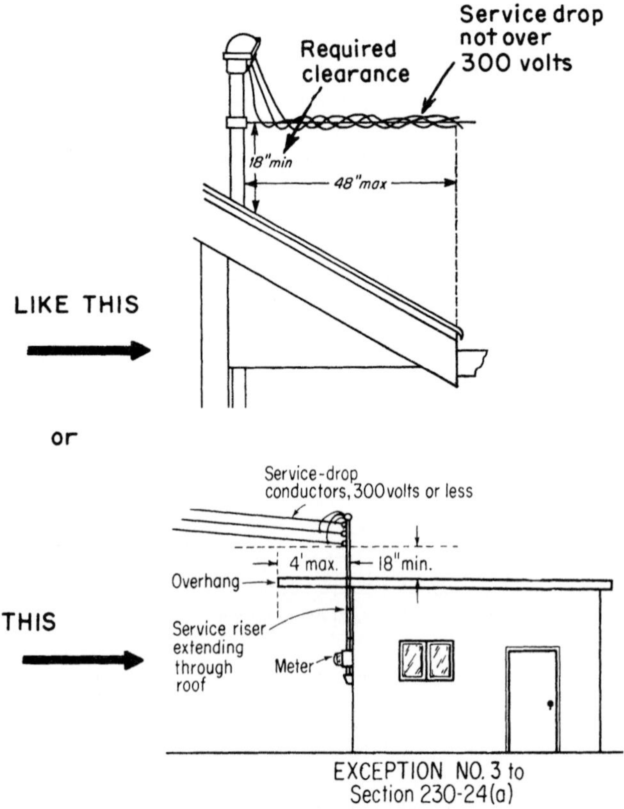

Fig. 230-14. Reduced clearance for service drop. (Sec. 230-24.)

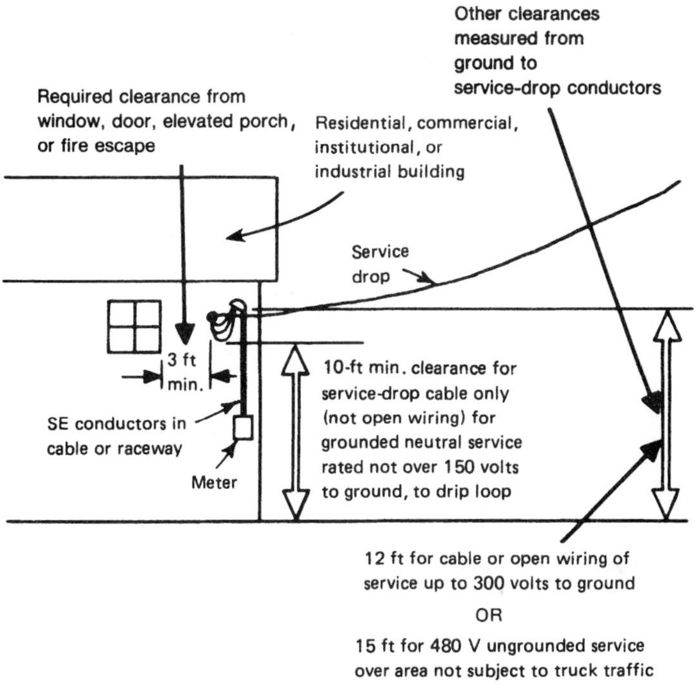

Fig. 230-15. Service-drop clearance to ground. (Sec. 230-24.)

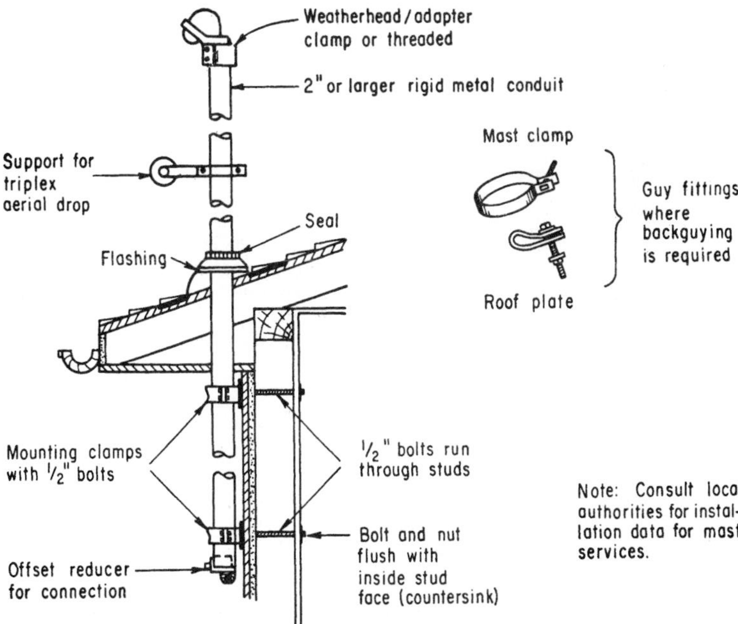

Fig. 230-16. Service mast must provide adequate support for connecting drop conductors. (Sec. 230-28.)

230-30. Insulation. The Exceptions to Secs. 230-30 and 230-41(a) clarify the use of aluminum, copper-clad aluminum, and bare copper conductors used as grounded conductors in service laterals and service-entrance conductors (Fig. 230-17).

For service lateral conductors (underground service), an individual grounded conductor (such as a grounded neutral) of *aluminum* or *copper-clad aluminum* without insulation or covering may *not* be used in a raceway underground. A bare *copper* neutral may be used—in a raceway, in a cable assembly, or even directly buried in soil where local experience establishes that soil conditions do not attack copper.

The way this rule was worded in the 1975 NE Code, an individual *bare* aluminum neutral (along with individual insulated aluminum phase legs) of an underground service lateral appeared to be permitted if the circuit was installed

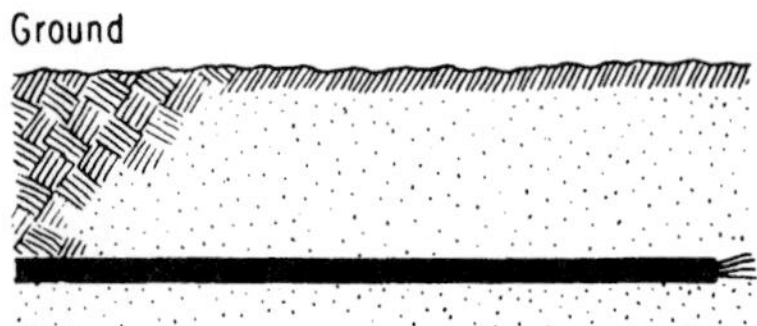

INSULATED PHASE CONDUCTORS and a bare copper neutral for an underground service lateral in buried raceway. Note: A bare aluminum or copper-clad aluminum neutral could be used here when part of a moisture- and fungus-resistant cable.

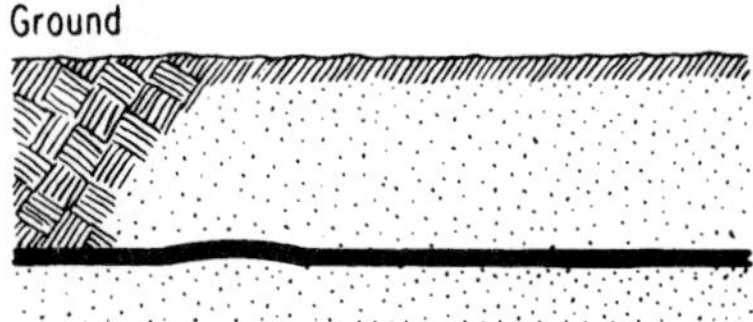

BARE COPPER NEUTRAL in a direct-buried cable assembly with moisture- and fungus-resistant outer covering. Note: A bare aluminum or copper-clad aluminum could be used like this, but it must be within the same type of cable assembly.

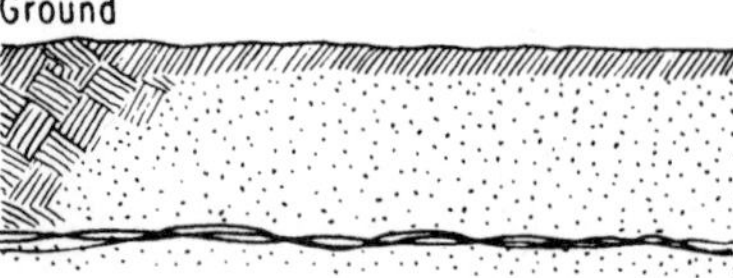

TYPE USE PHASE CONDUCTORS and a bare copper neutral directly buried where soil conditions are suitable for the bare copper.

Fig. 230-17. Sections 230-30 and 230-41 permit bare neutrals for service conductors. (Secs. 230-30 and 230-41.)

in conduit or other raceway, and that interpretation was common. The wording of part **(d)** of the Exception permits an aluminum grounded conductor of an underground service lateral to be without individual insulation or covering "when part of a cable assembly identified for underground use" where the cable is directly buried or run in a raceway. Of course, a lateral made up of individual insulated phase legs and an *insulated* neutral is acceptable in underground conduit or raceway (Fig. 230-18).

230-32. Protection Against Damage. Underground service lateral conductors—whether directly buried cables, conductors in metal conduit, conductors in nonmetallic conduit, or conductors in EMT—must comply with Sec. 300-5 for protection against physical damage (Fig. 230-19).

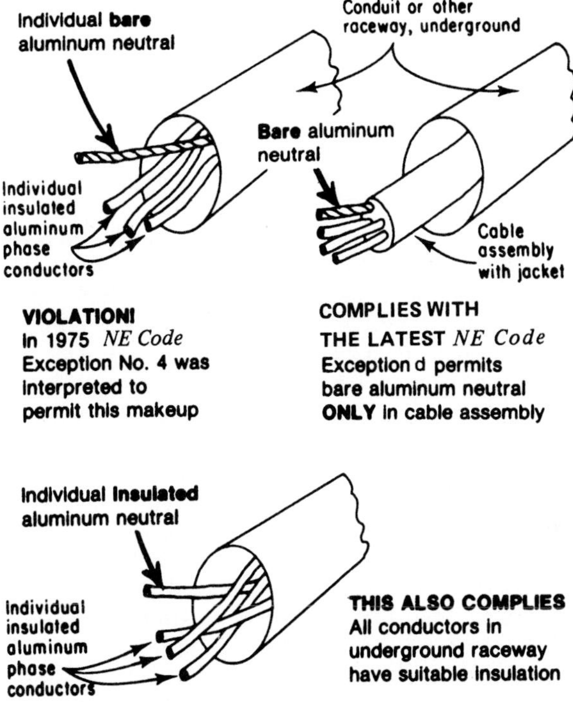

Fig. 230-18. Underground bare aluminum grounded leg must always be in a cable assembly. (Sec. 230-30.)

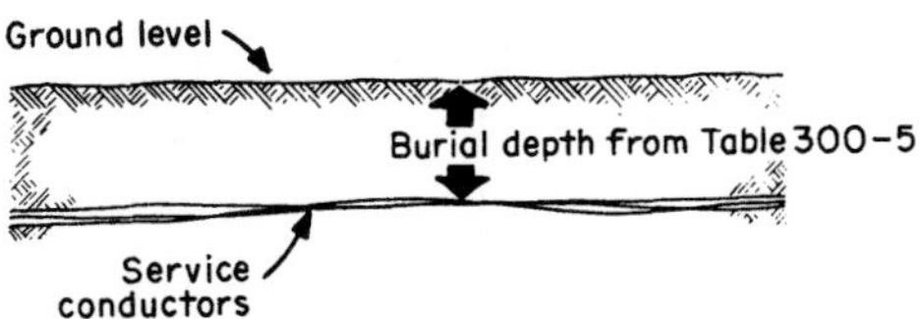

Fig. 230-19. Protecting underground service conductors. (Sec. 230-32.)

230-40. Number of Service-Entrance Conductor Sets. As a logical follow-up to the basic rule of Sec. 230-2, which requires that a single building or structure must be supplied "by only one service" (that is, only one service drop or lateral), this rule calls for only one set of SE conductors to be supplied by each service drop or lateral that is permitted for a building. Exception No. 1 covers a multiple-occupancy building (a two-family or multifamily building, a multitenant office building, or a store building, etc.). In such cases, a set of SE conductors for each occupancy or for groups of occupancies is permitted to be tapped from a single drop or lateral (Fig. 230-20).

When a multiple-occupancy building has a separate set of SE conductors run to each occupancy, in order to comply with Sec. 230-70(a), the conductors should either be run on the outside of the building to each occupancy or, if run inside the building, be encased in 2 in. (50.8 mm) of concrete or masonry in accordance with Sec. 230-6. In either case the service equipment should be located "nearest to the entrance of the conductors inside the building," and each occupant would have "access to his disconnecting means."

Any desired number of sets of service-entrance conductors may be tapped from the service drop or lateral, or two or more subsets of service-entrance conductors may be tapped from a single set of main service conductors, as shown for the multiple-occupancy building in Fig. 230-20.

Exception No. 2 permits two to six disconnecting means to be supplied from a single service drop or lateral where each disconnect supplies a separate load (Fig. 230-21). Exception No 2, recognizes the use of, say, six 400-A sets of service-entrance conductors to a single-occupancy or multiple-occupancy building in lieu of a single main 2,500-A service. It recognizes the use of up to six subdivided loads extending from a single drop or lateral in a *single-occupancy* as well as multiple-occupancy building. Where single metering is required, doughnut-type CTs could be installed at the service drop.

The real importance of this rule is to eliminate the need for "paralleling" conductors of large-capacity services, as widely required by inspection authorities to satisfy previous editions of the NEC (Fig. 230-21). This same approach could be used in subdividing services into smaller load blocks to avoid the use of the equipment ground-fault circuit protection required by Sec. 230-95.

This rule can also facilitate expansion of an existing service. Where less than six sets of service-entrance conductors were used initially, one or more additional sets can be installed subsequently without completely replacing the original service. Of course, metering considerations will affect the layout.

But, the two to six disconnects (circuit breakers or fused switches) must be installed close together at one location and not spread out in a building.

230-41. Insulation of Service-Entrance Conductors. Except for use of a bare neutral, as permitted, all service-entrance conductors must be insulated and may not simply be "covered"—as discussed under Sec. 230-22. The same wording is used in part *d* of the Exception in Sec. 230-41 as described above for Sec. 230-30. In this section, the reference is to "service-entrance conductors" instead of "service lateral conductors." But, again, a *bare individual* aluminum or copper-clad aluminum grounded conductor (grounded neutral or grounded phase leg) may be used in a raceway or a cable assembly or for direct burial where "identified" for direct burial.

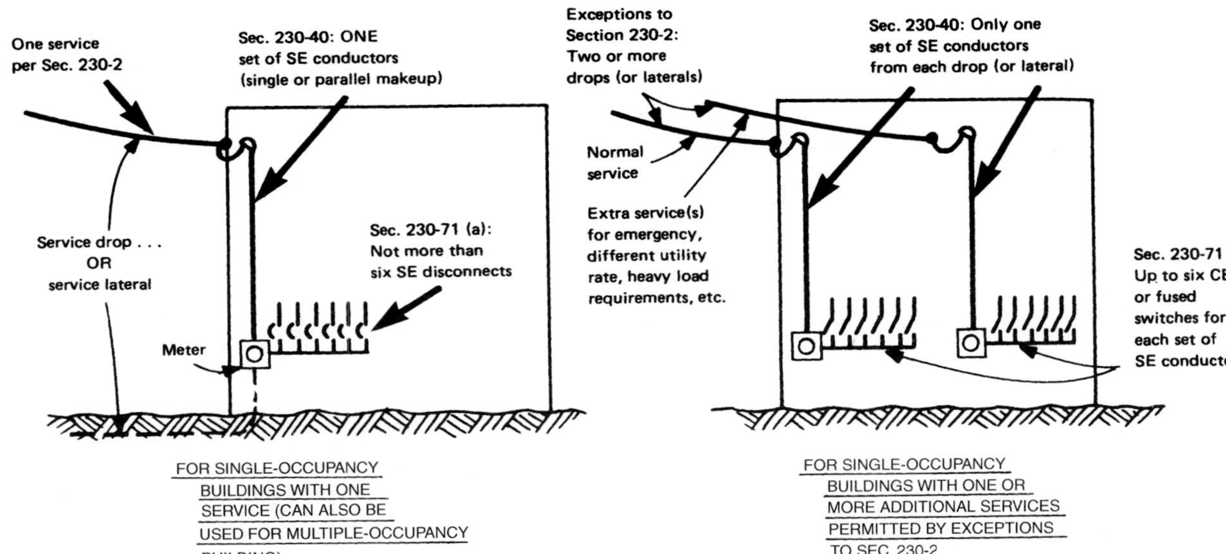

Fig. 230-20. Service layouts must simultaneously satisfy Secs. 230-2, 230-40, 230-71, and all other NEC rules that are applicable. (Sec. 230-40.)

Sec. 230-2:
One service

Utility pole

Sec. 230-40, Ex. No. 2:
A separate set of
SE conductors may be
run to each of not more
than six SE disconnects
in separate enclosures

Metering from
point of drop
connection

Meter

Two to six service
disconnects, each fed
by a separate set of
SE conductors

FOR SINGLE OCCUPANCY BUILDINGS
SUCH AS FACTORIES, SCHOOLS AND STORES
OR FOR MULTIPLE-OCCUPANCY BUILDINGS

Sec. 230-40, Ex. No. 1: Separate sets of SE
conductors tapped from one drop (or lateral)
to feed each of any number of occupancy units

Service drop is carried along top of building with service entrance
conductors for each occupancy tapping the drop through a service
head fitting. Or service can be made underground to a splice box or
gutter.

All service conductors
run on outside of building.

Sec. 230-2:
One service
drop. . . .
OR
. . . lateral

Meter

6 7 8 9 10
Apt.1 2 3 4 5

Building
of any
number
of floors

First floor

Single common metering shown.
Individual metering could be used.

Service entrance equipment
in each occupancy may consist
of up to six switches or CBs.

FOR MULTIPLE-OCCUPANCY BUILDING
(SEPARATE TENANTS IN APARTMENTS,
OFFICES, STORES, ETC.)

Fig. 230-20. *(Continued)*

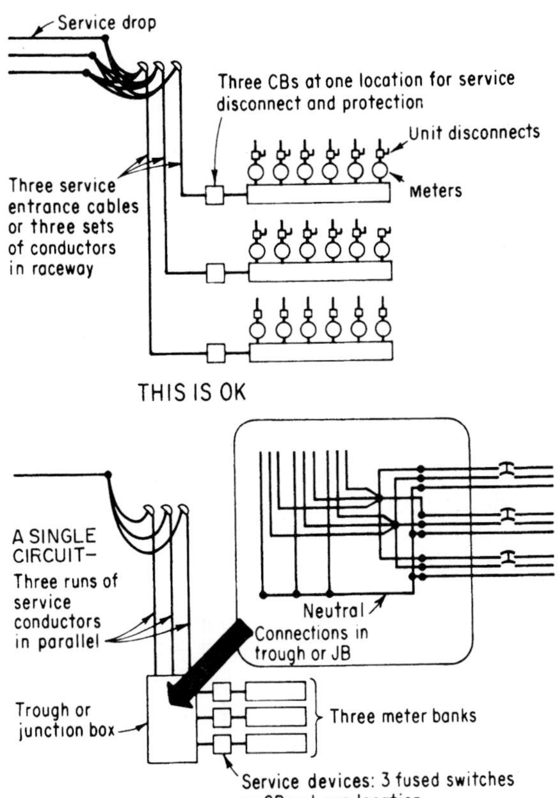

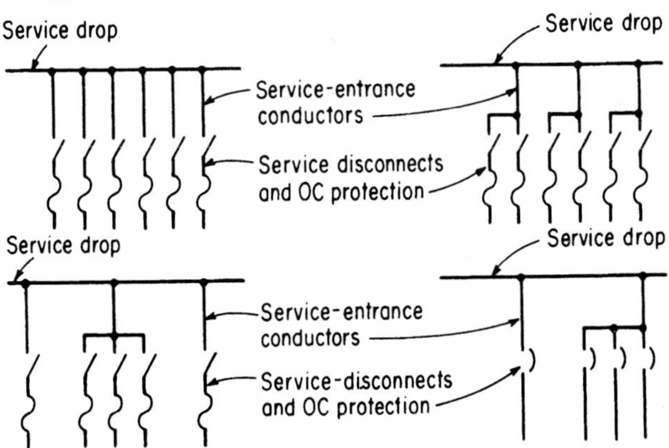

From two to six separate sets of service-entrance conductors may be supplied by a single service drop for either single- or multiple-occupancy buildings. Disconnects can be of same or different ratings, and each set of service-entrance conductors can be installed using any approved wiring method.

Fig. 230-21. Tapping sets of service-entrance conductors from one drop (or lateral). (Sec. 230-40.)

Aluminum SE cable with a bare neutral may be used aboveground as SE conductors. *But,* an aluminum SE cable with a bare neutral may be used underground *only* if it is "identified" for underground use in a raceway or directly buried. Conventional-style SE-U aluminum SE cable with a bare neutral is not "identified" for use underground but may be used, as the first sentence of Sec. 230-40 describes, as "service-entrance conductors entering or on the exterior of buildings or other structures." In "SE-U," the "U" stands for "unarmored" not "underground."

230-42. Size and Rating. Sizing of service-entrance conductors involves the same type of step-by-step procedure as set forth for sizing feeders covered in Art. 220. A set of service-entrance conductors is sized just as if it were a feeder. In general, the service-entrance conductors must have a minimum ampacity—current-carrying capacity—selected in accordance with the ampacity tables and rules of Sec. 310-15, sufficient to handle the total lighting and power load as calculated in accordance with Article 220. Where the Code gives demand factors to use or allows the use of acceptable demand factors based on sound engineering determination of less than 100 percent demand requirement, the lighting and power loads may be modified.

According to the Exception of Sec. 230-42(a), the "maximum allowable current" of busways used as service entrance conductors must be taken to be the amp value for which the busway has been listed or labeled. This is an exception to the basic rule that requires the ampacity of service-entrance conductors to be "determined from Section 310-15"—which does not give ampacities of busways.

From the analysis and calculations given in the feeder circuit section, a total power and lighting load can be developed to use in sizing service-entrance conductors. Of course, where separate power and lighting services are used, the sizing procedure should be divided into two separate procedures.

When a total load has been established for the service-entrance conductors, the required current-carrying capacity is easily determined by dividing the total load in kilovoltamperes (or kilowatts with proper correction for power factor or the load) by the voltage of the service.

From the required current rating of conductors, the required size of conductors is determined. Sizing of the service neutral is the same as for feeders. Although suitably insulated conductors must be used for the phase conductors of service-entrance feeders, the NE Code does permit use of bare grounded conductors (such as neutrals) under the conditions covered in Secs. 230-30 and 230-41.

An extremely important element of service design is that of fault consideration. Service busway and other service conductor arrangements must be sized and designed to assure safe application with the service disconnect and protection. That is, service conductors must be capable of withstanding the let-through thermal and magnetic stresses on a fault.

After calculating the required circuits for all the loads in the electrical system, the next step is to determine the minimum required size of service-entrance conductors to supply the entire connected load. The NE Code procedure for sizing SE conductors is the same as for sizing feeder conductors for the entire load—as set forth in Sec. 220-10. Basically, the service "feeder" capacity must be not less than the sum of the loads on the branch circuits for the different applications.

The *general lighting load* is subject to demand factors from Table 220-11, which takes into account the fact that simultaneous operation of all branch-circuit loads, or even a large part of them, is highly unlikely. Thus, service or feeder capacity does not have to equal the connected load. The other provisions of Art. 220 are then factored in.

Part **(b)** of Sec. 230-42 makes a 100-A service conductor ampacity a mandatory minimum if the system supplied is a one-family dwelling with more than five 2-wire branch circuits (or the equivalent of that for multiwire circuits) or if a one-family dwelling has an initial computed load of 10,000 W. Now that three 20-A small appliance branch circuits are required in a single-family dwelling, the average new home will need a 100-A, 3-wire service, because even without *electric* cooking, heating, drying, or water heating appliances, more than five 2-wire branch circuits will be installed.

230-43. Wiring Methods for 600 Volts, Nominal, or Less. The list of acceptable wiring methods for running service-entrance conductors does include flexible metal conduit (Greenfield) and liquidtight flexible metal conduit, but limits use of such raceways to a maximum length of 6 ft (1.83 m) and an equipment grounding conductor must be run with it. Although such raceways were prohibited under previous NEC editions, effectively bonded flexible metal conduit and liquidtight flexible metal conduit in a length not over 6 ft (1.83 m) may be used as a raceway for service-entrance conductors (Fig. 230-22). A length of flex

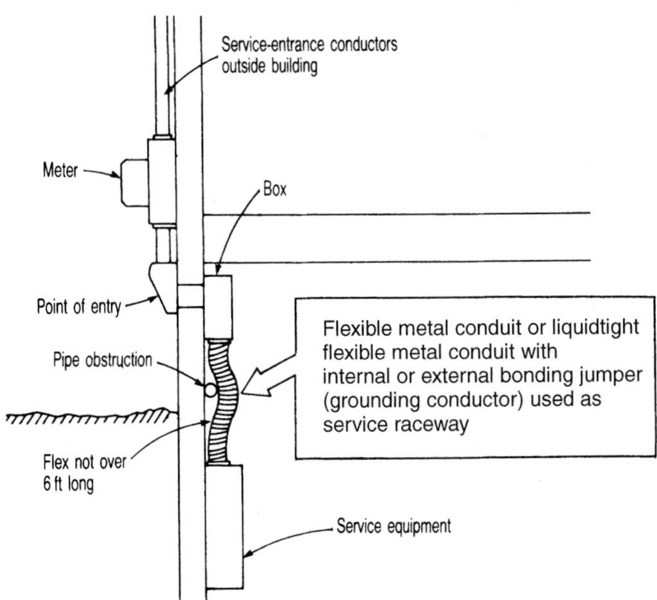

Fig. 230-22. These two flexible conduits may be used for service raceway. [Sec. 230-43.]

or liquidtight flex not longer than 6 ft (1.83 m) may be used as a service raceway provided an equipment grounding conductor sized from Table 250-94 (and with a cross-section area at least 12½ percent of the csa of the largest service phase conductor for conductors larger than 1,100 MCM copper or 1,750 MCM aluminum) is used. This rule recognizes that the flexibility of such raceway is often needed or desirable in routing service-entrance conductors around obstructions in the path of connections between metering equipment and service-entrance switchboards, panelboards, or similar enclosures. The required equipment grounding conductor may be installed either inside or outside the flex, using acceptable fittings and termination techniques for the grounding conductor.

It should be noted that *liquidtight* flexible metal conduit is recognized as an acceptable service raceway. And, liquidtight flexible *nonmetallic* conduit—of any length—may be used as a service raceway containing service-entrance conductors. Use of liquidtight flexible nonmetallic conduit for a service raceway must satisfy all of the rules of Sec. 351-22 to Sec. 351-30. Such flex may not be used in lengths over 6 ft (1.83 m) long and an equipment grounding conductor will normally be required.

230-46. Unspliced Conductors. For Exception No. 3, an underground service conduit usually terminates at the inside of the building wall unless the building has no basement. A metal conduit or a service cable may terminate at this point or may be run directly to the service equipment. From the terminal box, the conductors are run to the service equipment in rigid metal conduit or electrical metallic tubing or in an auxiliary gutter and may terminate at any suitable point behind the switchboard.

Figure 230-23 shows the conditions of Exception No. 3 and Exception No. 4. The sketch on the right shows a form of construction sometimes employed where the inside meter of an existing installation is removed and an outdoor meter is installed. New service-entrance conductors are connected to the service drop and are carried down to a meter fitting in raceway or cable. From the meter, the outside service conductors return in the raceway or cable and are spliced to the old service-entrance conductors. These splices are permitted by Exception No. 4.

Exception No. 5 recognizes that, where a busway is used for service-entrance conductors, the sections must be connected together, and such connections are exempted from the rule that SE conductors must not be spliced.

Lastly, Exception No. 6 recognizes splicing of damaged "existing" service conductors. For existing service conductors, only, if a dig-in or other accident causes the service conductors to be damaged or severed, this Exception permits splicing instead of replacing the service conductors, provided a "listed" splice kit is used. Additionally, where the local inspector grants "special permission," this Exception allows "extension" of existing service conductors. Remember "special permission" by definition (Art. 100) *must be* in writing. Make certain that the inspector gives *written* permission.

230-50. Protection of Open Conductors and Cables Against Damage—Aboveground. In part **(a)**, exposed service-entrance cable that is attached to a building "near sidewalks (and) walkways" must be protected against physical damage by

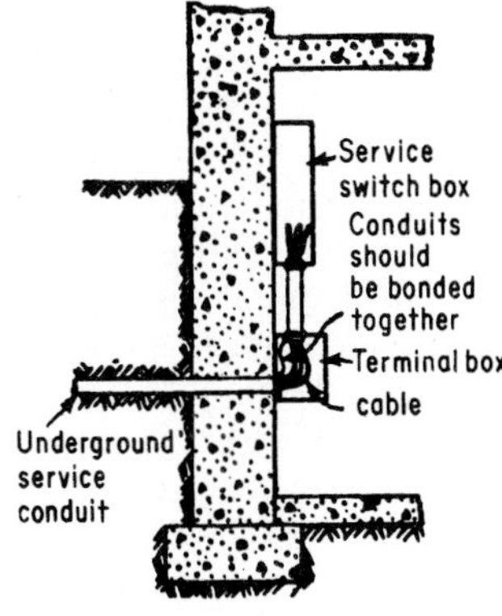

Exception No. 3 Terminal box used at the end of an underground service conduit.

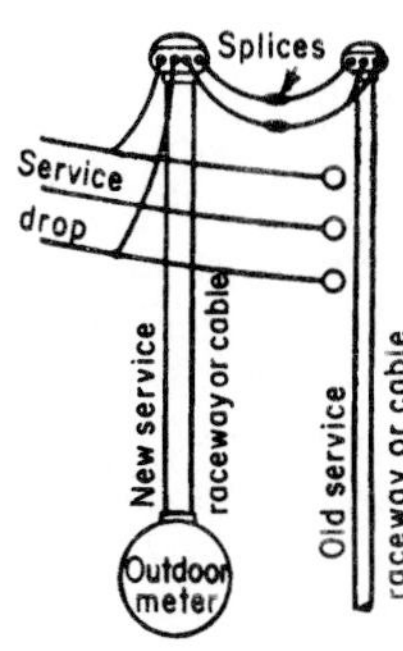

Exception No. 4 Splices in service-entrance conductors where an outdoor meter is installed in place of an indoor meter for an existing installation.

Fig. 230-23. Permitted splices in service-entrance conductors. (Sec. 230-46.)

sleeving with rigid metal conduit, IMC, rigid PVC conduit, EMT, or some similar protection—just as it would be if adjacent to a "driveway." This is an important protection rule that has a great number of applications (Fig. 230-24).

Part **(b)** Exception allows use of types MI and MC cables for service entrance or service lateral applications, without need for mounting at least 10 ft (3.05 m) above grade—provided they are not exposed to damage or are protected.

230-51. Mounting Supports. Service-entrance cable must be clamped to building surface by straps at intervals not over 30 in. (762 mm). The spacing of 30 in.

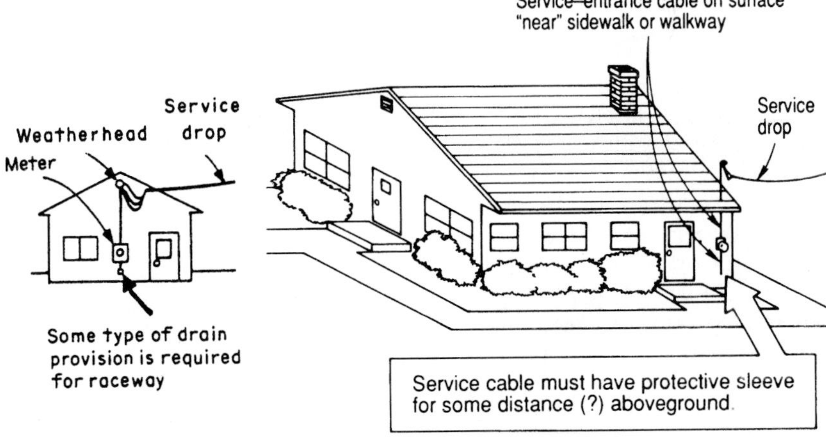

Fig. 230-24. Outdoor service raceway must be raintight *and* drained and SE cable must be protected. (Secs. 230-50 and 230-53.)

(762 mm) replaces the old maximum interval of 4½ ft (1.37 m). Closer spacing of the cable clamps will make a more secure, neater installation. And the cable must still be clamped within 12 in. (305 mm) of the service weather head and within 12 in. (305 mm) of cable connection to a raceway or enclosure.

230-53. Raceways to Drain. Service-entrance conductors in EMT or rigid conduit must be made raintight, using raintight raceway fittings, and must be equipped with a drain hole in the service ell at the bottom of the run or must be otherwise provided with a means of draining off condensation (Fig. 230-24).

230-54. Overhead Service Locations. When rigid metal conduit, IMC, or EMT is used for a service, the raceway must be provided with a service head (or weather head). Figure 230-25 shows details of service-head installation. As covered in part **(b)**, service cable may be installed without a servicehead provided it is bent to form a "gooseneck;" then either tape and print or tape the end with a "self-sealing water-resistant thermoplastic: That is, where no service head is used at the upper end of a service cable, the cable should be bent over so that the individual conductors leaving the cable will extend in a downward direction, and the end of the cable should be carefully taped and painted or sealed with water-resistant tape to exclude moisture.

Part **(c)** of this section requires that service heads be located above the service-drop attachment. Although this arrangement alone will not always prevent water from entering service raceways and equipment, such an arrangement will solve most of the water-entrance problems. An exception to this rule permits a service head to be located not more than 24 in. (610 mm) from the service-drop termination where it is found that it is impractical for the service head to be located above the service-drop termination. In such cases a *mechanical connec-*

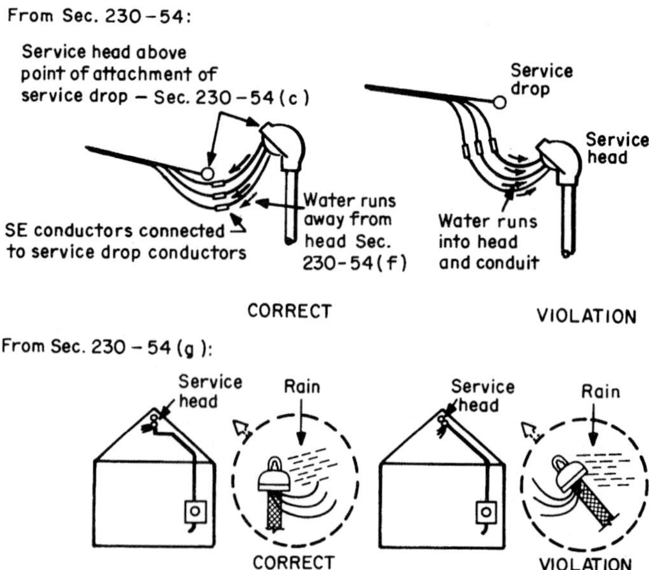

Fig. 230-25. Location of service head minimizes entrance of rain. (Sec. 230-54.)

tor is required at the lowest point in the drip loop to prevent siphoning. This Exception will permit the Code-enforcing authority to handle hardship cases that may occur.

The intent of part **(g)** is to require use of connections or conductor arrangements, both at the pole and at the service, so that water will not enter connections and siphon under head pressure into service raceways or equipment.

230-56. Service Conductor with the Higher Voltage-to-Ground. This Code rule repeats the requirement of Sec. 215-8 that the "high" leg (the 208-V-to-ground leg) of a 240/120-V 3-phase, 4-wire delta system must be identified by marking to distinguish it from the other hot legs, which are only 120 V to ground. One method permitted is color-coding the so-called high-leg orange. The rule recognizes "other means," but any such "identification" must be provided "at each termination or junction point." Clearly, the use of an overall orange-colored insulation will most easily satisfy this rule. See Sec. 215-8.

230-70. General. Part **(a)** covers the place of installation of a service disconnect. The disconnecting means required for every set of service-entrance conductors must be located at a readily accessible point anywhere outside, or inside, nearest to the point at which the service conductors enter the building (Fig. 230-26). The service disconnect switch (or circuit breaker) is generally placed on the inside of the building as near as possible to the point at which the conductors come in. And part **(b)** requires lettering or a sign on the disconnect(s) to identify it (them) as "Service Disconnect."

Although the Code does not set any maximum distance from the point of conductor entry to the service disconnect, various inspection agencies set maximum limits on this distance. For instance, service cable may not run within the building more than 18 in. (457 mm) from its point of entry to the point at which it enters the disconnect. Or service conductors in conduit must enter the disconnect within 10 ft (3.05 m) of the point of entry. Or, as one agency requires, the disconnect must be within 10 ft (3.05 m) of the point of entry, but overcurrent protection must be provided for the conductors right at the point at which they emerge from the wall into the building. The concern is to minimize the very real and proven potential hazard of having unprotected service conductors within the building. Faults in such unprotected service conductors must burn themselves clear and such application has caused fires and fatalities.

Switches or circuit breakers used for service-entrance disconnecting means must be approved for use as service equipment. This rule is meant to require that the switch be listed and labeled by the UL as suitable for service entrance. Check manufacturers' catalogs on this.

230-71. Maximum Number of Disconnects. Service-entrance conductors must be equipped with a readily accessible means of disconnecting the conductors from their source of supply. As stated in part **(a)**, the disconnect means for each service and each set of SE conductors permitted by Sec. 230-2 and 230-40 Exception No. 1, respectively, may consist of not more than six switches or six circuit breakers, in a common enclosure or grouped individual enclosures, located either outside the building wall, or inside, as close as possible to the point at which the conductors enter the building. Figure 230-27 shows the basic application of that rule to a single set of SE conductors.

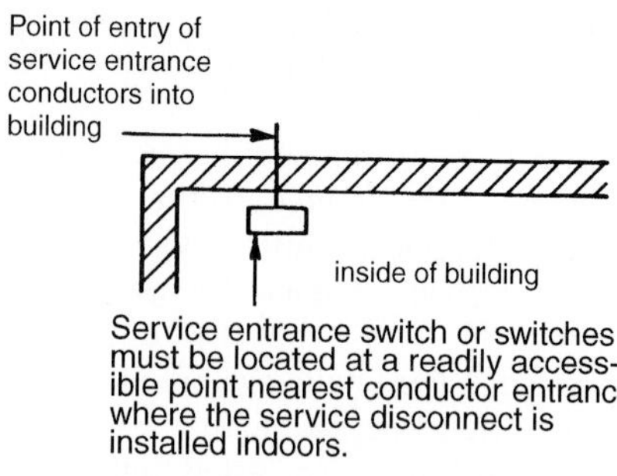
Point of entry of service entrance conductors into building
inside of building
Service entrance switch or switches must be located at a readily accessible point nearest conductor entrance where the service disconnect is installed indoors.
GENERAL CONCEPT

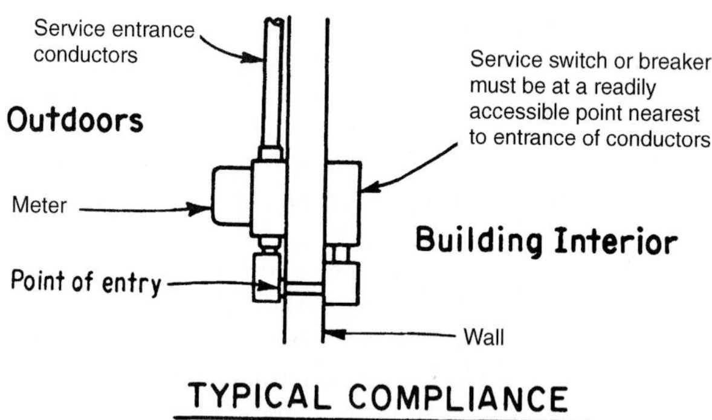
Service entrance conductors
Outdoors
Service switch or breaker must be at a readily accessible point nearest to entrance of conductors
Meter
Building Interior
Point of entry
Wall
TYPICAL COMPLIANCE

Fig. 230-26. Service disconnect must open current for any conductors within building. (Sec. 230-70.)

The Exception to part **(a)** says that when control power for a ground-fault protection system is tapped from the line side of the service disconnect means, the disconnect for the control power circuit is not counted as one of the six permitted disconnects for a service. A ground-fault-protected switch or circuit breaker supplying power to the building electrical system *counts* as one of the six permitted disconnects. But a disconnect supplying only the control-circuit power for a ground-fault protection system, installed as part of the listed equipment, does not count as one of the six service disconnects.

The rule of this section correlates "number of disconnects" with Secs. 230-2 and 230-40, which permit a separate set of SE conductors to be run to each occupancy (or group of occupancies) in a multiple-occupancy building, as follows:

Section 230-2 permits more than one "service" to a building—that is, more than one service drop or lateral—under the conditions set forth in the Exceptions. As set forth in the first sentence of Sec. 230-40 each such "service" must supply *only* one set of SE conductors in a building that is a single-occupancy (one-tenant) building, and each set of SE conductors may supply up to six SE disconnects grouped together at one location—in the same panel or switch-

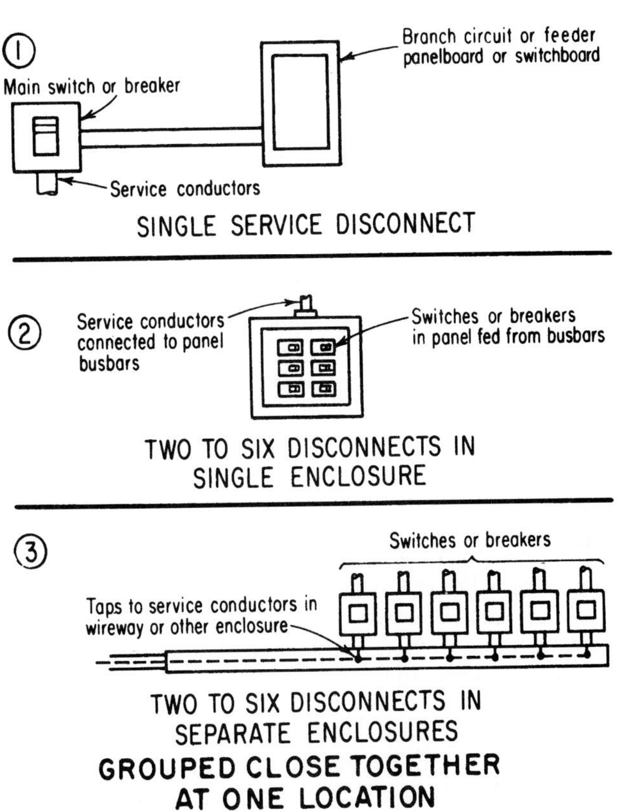

Fig. 230-27. The three basic ways to provide service disconnect means. (Sec. 230-71.)

board or in grouped individual enclosures. If the grouped disconnects for one set of SE conductors are not at the same location as the grouped disconnects for one or more other sets of SE conductors, for those situations described and permitted in the Exceptions to Sec. 230-2, then a "plaque or directory" must be placed at each service-disconnect grouping to tell where the other group (or groups) of disconnects are located and what loads each group of disconnects serves.

Exception No. 1 to Sec. 230-40 says that a single service drop or lateral may supply *more than one set* of SE conductors for a multiple-occupancy building. Then at the load end of each of the sets of SE conductors, in an individual occupancy or adjacent to a group of occupancy units (apartments, office, stores), up to six SE disconnects may be supplied by each set of SE conductors.

The first sentence of part **(a)** to Sec. 230-71 ties directly into Sec. 230-40, Exception No. 1. It is the intent of this basic rule that, where a multiple-occupancy building is provided with more than one set of SE conductors tapped from a drop or lateral, each set of those SE conductors may have up to six switches or circuit breakers to serve as the service disconnect means for that set of SE conductors. The rule does recognize that six disconnects for each set of SE conductors at a multiple-occupancy building with, say, 10 sets of SE conductors tapped from a drop or lateral does result in a total of 6 × 10, or 60, disconnect devices for completely isolating the building's electrical system from the utility supply. Sec. 230-72(b) also recognizes use of up to six disconnects for each of the "separate" services for fire pumps, emergency lighting, etc., which are recognized in Sec. 230-2 as being separate services for specific purposes.

Although the basic rule of Sec. 230-40 specifies that only one set of SE conductors may be tapped from a single drop for a building with single occupancy, Exception No. 2 to Sec. 230-40 recognizes that a separate set of SE conductors may be run from a single service drop or lateral to each of up to six service disconnects mounted in separate enclosures at one location, constituting the disconnect means for a single service to a single-occupancy building.

For any type of occupancy, a power panel (not a lighting and appliance panel, as described in Sec. 384-14) containing up to six switches or circuit breakers may be used as service disconnect. A lighting and appliance panel used as service equipment for renovation of an existing service in an individual residential occupancy (but not for new installations) may have up to six main breakers or fused switches. However, a lighting and appliance panel used as service equipment for new buildings of any type must have not more than two main devices—with the sum of their ratings not greater than the panel bus rating. See Sec. 384-16(a).

The first sentence of Sec. 230-71(a) and that of Sec. 230-72(a) note that from one to six switches (or circuit breakers) may serve as the service disconnecting means for each class of service for a building. For example, if a *single-occupancy* building has a 3-phase service and a separate single-phase service, each such service may have up to six disconnects (Fig. 230-28). Where the two sets of service equipment are not located adjacent to each other, a plaque or directory must be installed at each service-equipment location indicating

AS PERMITTED BY EXCEPTIONS TO SEC. 230-2
Lighting service— 120/240 V
Power service— 440 V 3-phase
Each class of service may consist of 1 to 6 fused switches or CBs in a common enclosure, or a group of separate enclosures, grouped together at a common location

Fig. 230-28. Each separate service may have up to six disconnect devices. (Sec. 230-71.)

where the other service equipment is—as required by the second paragraph of Sec. 230-2.

Part **(b)** notes that single-pole switches or circuit breakers equipped with handle ties may be used in groups as single disconnects for multiwire circuits, simultaneously providing overcurrent protection for the service (Fig. 230-29). Multipole switches and circuit breakers may also be used as single disconnects. The requirements of the Code are satisfied if all the service-entrance conductors can be disconnected with no more than six operations of the hand—regardless of whether each hand motion operates a single-pole unit, a multipole unit, or a group of single-pole units with "handle ties" or a "master handle" controlled by a single hand motion. Of course, a single main device for service disconnect and overcurrent protection—such as a main CB or fused switch—gives better protection to the service conductors.

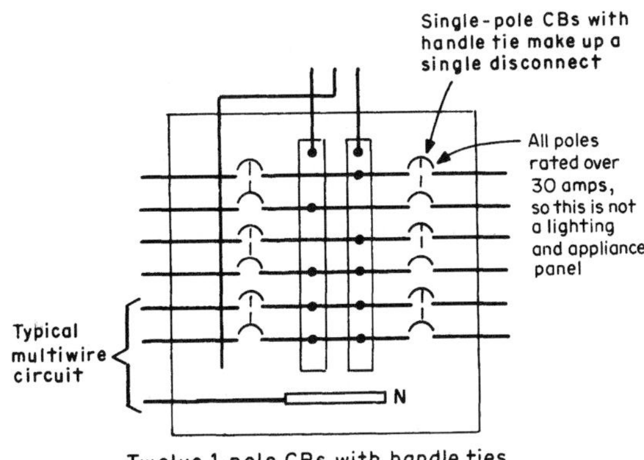

Fig. 230-29. This arrangement constitutes six disconnects. (Sec. 230-71.)

The FPN to this section refers to Sec. 384-16(a), which requires a higher degree of overcurrent protection for *lighting and appliance* branch-circuit panelboards. Each such panelboard must be individually protected on the supply side by not more than two main circuit breakers or two sets of fuses having a combined rating not greater than that of the panelboard. Exception No. 2 to Sec. 384-16(a) eliminates the need for individual protection for a lighting and appliance branch-circuit panelboard where such panelboards are used as service-entrance equipment for renovation of an existing installation (but *not* for new jobs) in an *individual residential occupancy.* Examples of these provisions are shown in illustrations in Sec. 384-16. It should be noted that these rules concern only a lighting and appliance branch-circuit panelboard, which is defined in Sec. 384-14 as a panelboard having more than 10 percent of its overcurrent devices rated 30 A or less, for which neutral connections are provided. Panelboards other than that type which are used as service equipment can still follow the basic six-switch rule.

The reference to Sec. 430-95 is intended to point out the limitations associated with installations where the service equipment is within a motor control center. For such installations, the rule of Sec. 430-95 mandates the use of a single main disconnect.

230-72. Grouping of Disconnects. The basic rule of part **(a)** requires that for a service disconnect arrangement of more than one disconnect—such as where two to six disconnect switches or CBs are used, as permitted by Sec. 230-71(a)—all the disconnects making up the service equipment "for each service" must be grouped and not spread out at different locations. The basic idea is that anyone operating the two to six disconnects must be able to do it while standing at one location. Service conductors must be able to be readily disconnected from all loads at one place. And each of the individual disconnects must have lettering or a sign to tell what load it supplies (Fig. 230-30).

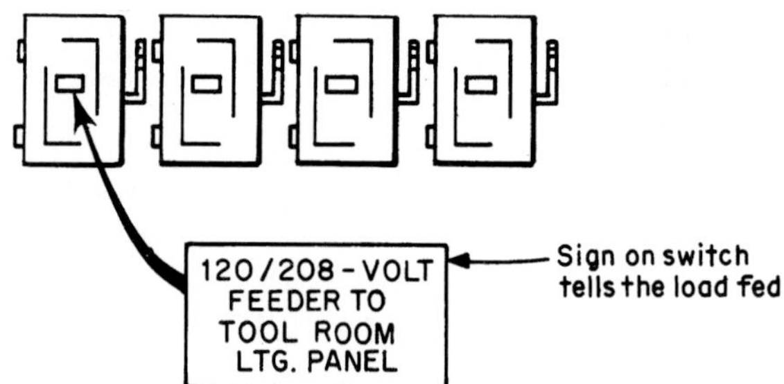

Fig. 230-30. Two to six disconnect switches or CBs must be grouped and *identified.* (Sec. 230-72.)

This rule makes clear that the two to six service disconnects that are permitted by Sec. 230-71(a) for *each* "service" or for "*each* set of SE conductors" at a multiple-occupancy building must be grouped. But, where permitted by the Exceptions to Sec. 230-2, the individual groups of two to six breakers or switches do not have to be together, and if they are not together, a sign at each location must tell where the other service disconnects are. (See Sec. 230-2.) Each grouping of two to six disconnects may be within a unit occupancy—such as an apartment—of the building.

The special or emergency-service equipments permitted by Sec. 230-2 do not have to be grouped with the regular service equipment. It should also be noted that Sec. 700-12(d) requires emergency services to be widely separated from the other services, to prevent failure of both due to a single fault.

The Exception to part **(a)** *permits* (*Note:* it permits, it does not *require*) one of the two to six service disconnects to be located remote from the other disconnecting means that are grouped in accordance with the basic rule—PROVIDED THAT *the remote disconnect is used only to supply a water pump that is also intended to provide fire protection.* In a residence or other building that gets its water supply from a well, a spring, or a lake, the use of a remote disconnect for the water pump will afford improved reliability of the water supply for fire suppression in the event that fire or other faults disable the normal service equipment. And it will distinguish the water-pump disconnect from the other normal service disconnects, minimizing the chance that firefighters will unknowingly open the pump circuit when they routinely open service disconnects during a fire. This Exception ties into the rule of Sec. 230-72(b), which *requires* (not simply permits) remote installation of a fire-pump disconnect switch that is permitted to be tapped ahead of the one to six switches or CBs that constitute the normal service disconnecting means (see Sec. 230-82, Exception No. 5). The Exception provides remote installation of a *normal service disconnect* when it is used for the same purpose (water pump used for fire fighting) as the *emergency service disconnect* (fire pump) covered in Sec. 230-72(b). In both cases, remote installation of the pump disconnect isolates the critically important pump circuit from interruption or shutdown due to fire, arcing-fault burndown, or any other fault that might knock out the main (normal) service disconnects.

A wide variety of layouts can be made to satisfy the Code *permission* for remote installation of a disconnect switch or CB service as a *normal* service disconnect (one of a maximum of six) supplying a water pump. Figure 230-31 shows three typical arrangements that would basically provide the isolated fire-pump disconnect.

Part **(b)**, as noted above, makes it mandatory to install emergency disconnect devices where they would not be disabled or affected by any fault or violent electrical failure in the normal service equipment (Fig. 230-32). Figure 230-33 shows a service disconnect for emergency and exit lighting installed very close to the normal service switchboard. An equipment burndown or fire near the main switchboard might knock out the emergency circuit. And the tap for the switch, which is made in the switchboard ahead of the service main, is particularly susceptible to being opened by *an arcing failure in the board*. The switch should be 10 or 15 ft (3.05 or 4.57 m) away from the board. And because the switchboard is fed from an outdoor transformer-mat layout directly outside the building, the tap to the safety switch would have greater reliability if it was made from the transformer secondary terminals rather than from the switchboard service terminals. Although the rule sets no specific distance of separation, remote locating of emergency disconnects is a mandatory Code rule.

In part **(b)**, the phrase "permitted by Section 230-2" makes clear that each separate service permitted for fire pumps or for emergency service may be equipped with up to six disconnects in the same way as the normal service—or any service—may have up to six SE disconnects. And the disconnect or disconnects for a fire-pump or emergency service must be remote from the normal service disconnects, as shown in Fig. 230-32.

Part **(c)** applies to applications of service disconnect for multiple-occupancy buildings—such as apartment houses, condominiums, town houses, office buildings, and shopping centers. Part **(c)** requires that the disconnect means for each occupant in a multiple-occupancy building be accessible to each occu-

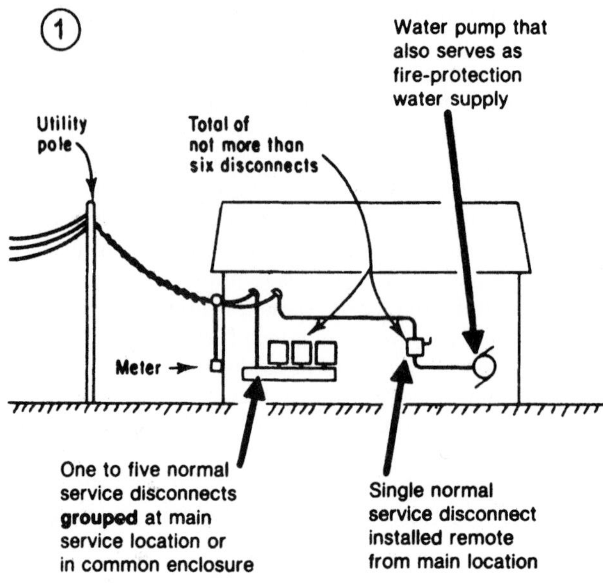

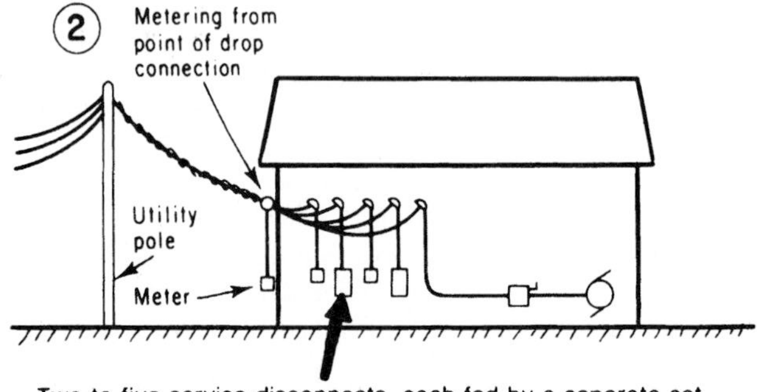

Fig. 230-31. Rule *permits* remote installation of one of two-to-six service disconnects to protect fire-pump circuits (typical layouts). (Sec. 230-72.)

pant. For instance, for the occupant of an apartment in an apartment house, the disconnect means for de-energizing the circuits in the apartment must be in the apartment (such as a panel), in an accessible place in the hall, or in a place in the basement or outdoors where it can be reached.

As covered by the Exception to part **(c)**, the access for each occupant as required by paragraph **(c)** would be modified where the building was under the

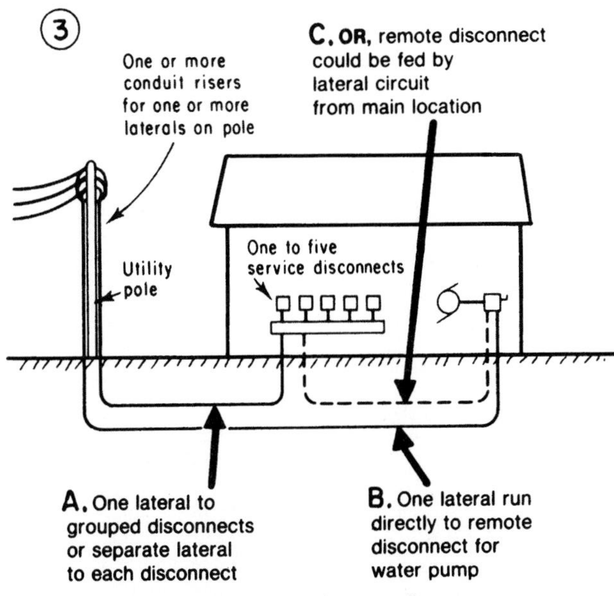
3
One or more conduit risers for one or more laterals on pole
C. OR, remote disconnect could be fed by lateral circuit from main location
Utility pole
One to five service disconnects
A. One lateral to grouped disconnects or separate lateral to each disconnect
B. One lateral run directly to remote disconnect for water pump

Fig. 230-31. (*Continued*)

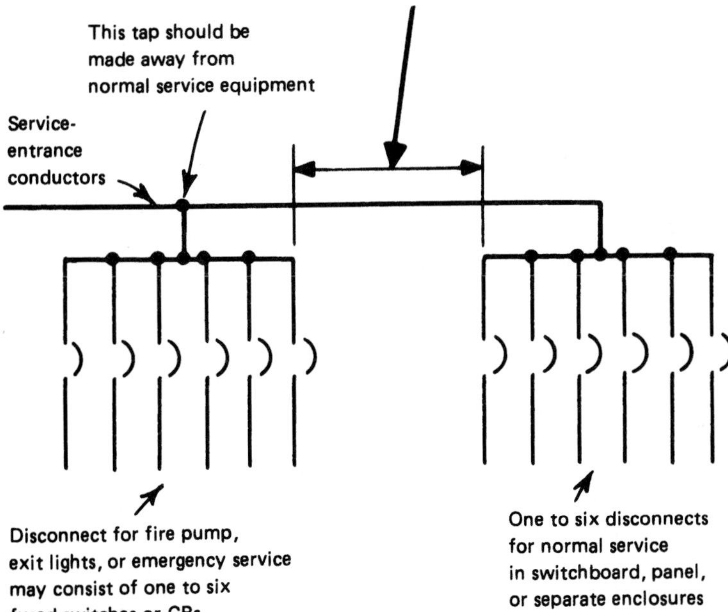
THIS DISTANCE SHOULD ISOLATE EMERGENCY DISCONNECT(S) FROM FAULTS IN NORMAL SERVICE EQUIPMENT.
This tap should be made away from normal service equipment
Service-entrance conductors
Disconnect for fire pump, exit lights, or emergency service may consist of one to six fused switches or CBs
One to six disconnects for normal service in switchboard, panel, or separate enclosures

Fig. 230-32. Emergency service disconnects must be isolated from faults in normal SE equipment. (Sec. 230-72.)

Fig. 230-33. Emergency disconnect close to service switchboard and fed by tap from it could readily be disabled by fault in board. (Sec. 230-72.)

management of a building superintendent or the equivalent and where electrical service and maintenance were furnished. In such a case, the disconnect means for more than one occupancy may be accessible only to authorized personnel.

Multiple-occupancy buildings having individual occupancy above the second floor were required by past **NEC** editions to have service equipment grouped in a common accessible place, with the disconnect means consisting of not more than six switches (or six CBs), as in Fig. 230-34. Although specific provisions requiring that application are no longer in the **NE Code**, it was required under previous **Code** editions. Exception No. 1 to Sec. 230-40 permits a separate set of SE conductors to be run to each tenant unit (e.g., apartment) in *any* multiple-occupancy building, with the service disconnect for each unit located within the unit—and no limitation is placed on the height or number of floors in the building.

230-75. Disconnection of Grounded Conductor. In this section the other means for disconnecting the grounded conductor from the interior wiring may be a screw or bolted lug on the neutral terminal block. The grounded conductor must not be run straight through the service equipment enclosure with no means of disconnection.

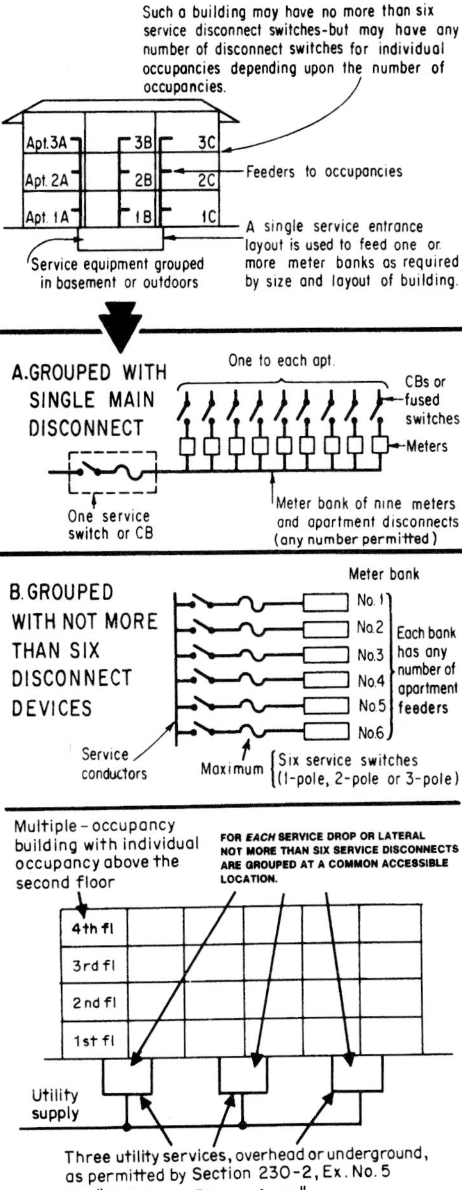

Fig. 230-34. These rules on grouping of service disconnects are no longer in the NEC but represent good and acceptable practice that has been followed widely. (Sec. 230-72.)

230-76. Manually or Power Operable. Any switch or CB used for service disconnect must be manually operable. In addition to manual operation, the switch may have provision for electrical operation—such as for remote control of the switch, provided it can be manually operated to the open or OFF position.

Code wording clearly indicates that an electrically operated breaker with a mechanical trip button which will open the breaker even if the supply power is dead is suitable for use as a service disconnect. The manually operated trip button assures that the breaker "can be opened by hand." To provide manual closing of electrically operated circuit breakers, manufacturers provide emergency manual handles as standard accessories. Thus such breaker mechanisms can be both closed and opened manually if operating power is not available.

Local requirements on the use of electrically operated service disconnects should be considered in selecting such devices.

230-78. Externally Operable. If a switch can be opened and closed without exposing the operator to contact with live parts, it is an externally operable switch, even though access to the switch handle requires opening the door of a cabinet. The Exception pertains to electrically operated switches and circuit breakers, and it explains that such switches or CBs are required to be externally operable only to the *open* position, and not necessarily to the *closed* position (Fig. 230-35).

230-79. Rating of Disconnect. Aside from the limited conditions covered in parts **(a)** and **(b)**, this section requires that service equipment (in general) shall have a rating not less than 60 A, applicable to both fusible and CB equipment. Part **(c)** requires 100-A minimum rating of a single switch or CB used in the service disconnect for any "one-family dwelling" with an *initial* load of 10 kVA or more or where the *initial* installation contains more than *five 2-wire* branch circuits. It

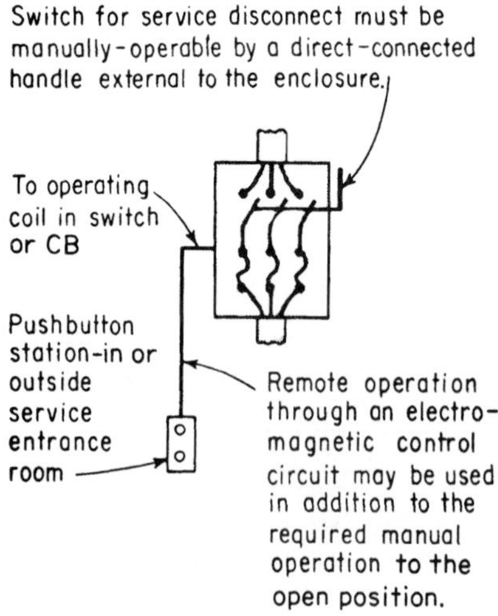
Switch for service disconnect must be manually-operable by a direct-connected handle external to the enclosure.
To operating coil in switch or CB
Pushbutton station-in or outside service entrance room
Remote operation through an electromagnetic control circuit may be used in addition to the required manual operation to the open position.
Note: Some local codes require both manual and electrical means of operation.

Fig. 230-35. Manual operation of any service switch is required. (Sec. 230-78.)

should be noted that the rule applies to one-family houses only, because of the definition of "one-family dwelling" as given in Art. 100. It does not apply to apartments or similar dwelling units that are in two-family or multifamily dwellings.

If the demand on a total connected load, as calculated from Sec. 220-10 through Sec. 220-21 or any of the applicable optional calculations permitted by part **C** of Art. 220, is 10 kVA or more, a 100-A service disconnect, as well as 100-A rated service-entrance conductors [Sec. 230-42(b)(2)], must be used. Any one-family house with an electric range rated 8¾ kW must always have a 100-A rated disconnect (or service equipment) because such a range is a demand load of 8 kW and the two required 20-A kitchen appliance circuits come to a demand load of 3,000 W [Sec. 220-16(a)] at 100 percent from Table 220-11—and the 8 kW plus 3 kW exceeds the 10-kW (or 10 kVA) level at which a minimum 100-A rated service is required.

If a 100-A service is used, the demand load may be as high as 23 kVA. By using the optional service calculations of Table 220-30, a 23-kVA demand load is obtained from a connected load of 42.5 kVA. This shows the effect of diversity on large-capacity installations.

230-80. Combined Rating of Disconnects. Figure 230-36 shows an application of this rule, based on determining what rating of a single disconnect would be required *if* a single disconnect were used instead of multiple ones. It should be noted that the sum of ratings above 400 A does comply with the rule of this section and with Exception No. 3 of Sec. 230-90(a) even though the 400-A service-entrance conductors could be heavily overloaded. Exception No. 3 exempts this type of layout from the need to protect the conductors at their rated ampacity, as required in the basic rule of Sec. 230-90. The Code assumes that the 400-A rating of the service-entrance conductors was carefully calculated from Art. 220 to be adequate for the maximum sum of the demand loads fed by the five disconnects shown in the layout.

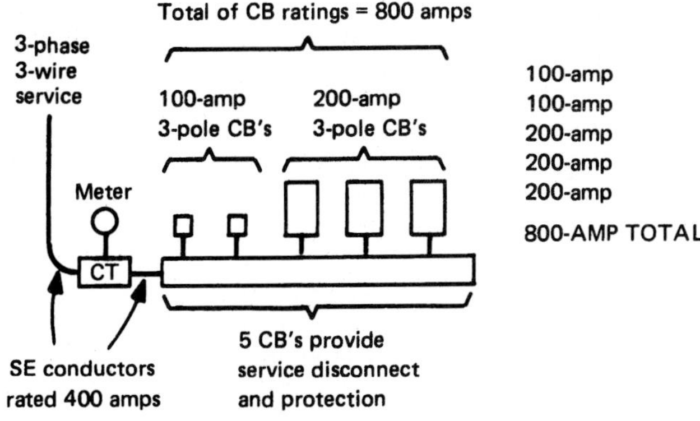

From Art. 220, calculation of demand load indicated that a single disconnect for this service must be rated at least 400 amps. The rating of multiple disconnects must total at least that value.

Fig. 230-36. Multiple disconnects must have their sum of ratings at least equal to the minimum rating of a single disconnect. (Sec. 230-80.)

230-82. Equipment Connected to the Supply Side of Service Disconnect. Cable limiters, fuses or CBs away from the building, high-impedance shunt circuits (such as potential coils of meters, etc.), supply conductors for time switches, surge-protective capacitors, instrument transformers, lightning arresters and circuits for emergency systems, fire-pump equipment, and fire and sprinkler alarms may be connected on the supply side of the disconnecting means. Emergency-lighting circuits, surge-protective capacitors, and fire alarm and other protective signaling circuits, when placed ahead of the regular service disconnecting means, must have separate disconnects and overcurrent protection.

Exception No. 1 to the rule prohibiting equipment connections on the line side of the service disconnect permits "cable limiters or other current-limiting devices" to be so connected.

Cable limiters are used to provide protection for individual conductors that are used in parallel (in multiple) to make up one phase leg of a high-capacity circuit, such as service conductors. A cable limiter is a cable connection device that contains a fusible element rated to protect the conductor to which it is connected.

As indicated in Exception No. 3, meters can be connected on the supply side of the service disconnecting means and overcurrent protective devices if the meters are connected to service not in excess of 600 V where the grounded conductor bonds the meter cases and enclosures to the grounding electrode.

As permitted by Exception No. 6, an electric power production source that is auxiliary or supplemental to the normal utility service to a premises may be connected to the supply (incoming) side of the normal service disconnecting means. This Exception to the basic rule that, "Equipment shall not be connected to the supply side of the service disconnecting means," permits connection of a solar photovoltaic system into the electrical supply for a building or other premises, to operate as a parallel power supply.

Exception No. 8 recognizes that control power for a ground-fault protection system may be tapped from the supply side of the service disconnect means. Where a control circuit for a ground-fault system is tapped ahead of the service main and "installed as part of listed equipment," suitable overcurrent protection and a disconnect must be provided for the control-power circuit.

230-83. Transfer Equipment. The rule on transfer switch disconnection from one source before connection to another source has an exception to apply to "parallel power production systems," as permitted by Exception No. 2 of Sec. 230-2. Exception No. 1 permits two or more sources to be connected in parallel through transfer switches. Either manual or automatic transfer means may be provided (Fig. 230-37).

230-90. (Overcurrent Protection) Where Required. The intent in paragraph **(a)** is to assure that the overcurrent protection required in the service-entrance equipment protects the service-entrance conductors from "overload." It is obvious that these overcurrent devices cannot provide "fault" protection for the service-entrance conductors if the fault occurs in the service-entrance conductors, but can protect them from overload where so selected as to have proper rating. Conductors on the load side of the service equipment are considered as feeders or branch circuits and are required by the Code to be protected as described in Arts. 210, 215, and 240.

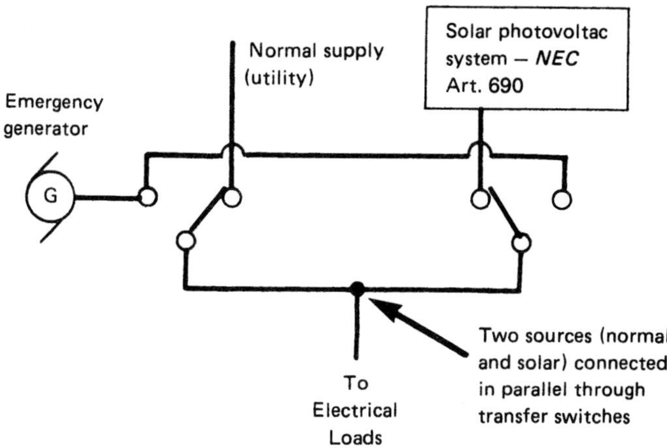

Fig. 230-37. This is an exception to the basic rule that two sources cannot be connected simultaneously to loads. (Sec. 230-83.)

Each ungrounded service-entrance conductor must be protected by an overcurrent device in series with the conductor (Fig. 230-38). The overcurrent device must have a rating or setting not higher than the allowable current capacity of the conductor, with the Exceptions noted.

The rule of Exception No. 1 says that if the service supplies one motor in addition to other load (such as lighting and heating), the overcurrent device may be rated or set in accordance with the required protection for a branch circuit supplying the one motor (Sec. 430-52) plus the other load, as shown in Fig.

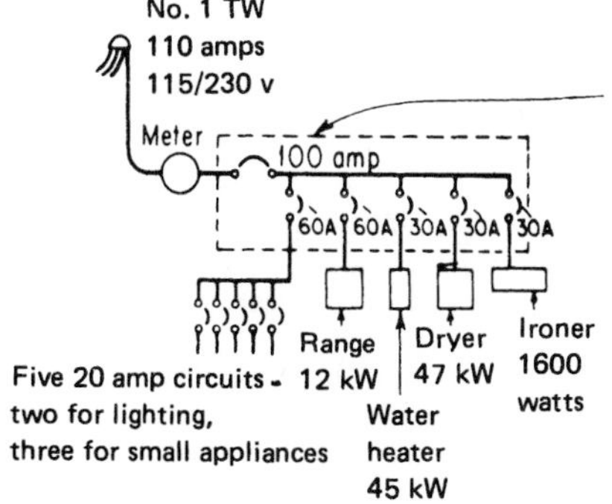

Service equipment with one overall 100-amp main disconnect and fuse. Current through service conductors limited to 100 amperes. Without a main disconnect and overcurrent device, current is not limited and current over 110 amps could flow. Sum of protective devices is 210 amps per hot leg.

Note: Service-entrance conductors must be selected with adequate ampacity for the calculated service demand load, from Secs. 220-10 through 220-21, or any applicable optional calculation covered in Part C of Art. 220.

Fig. 230-38. Single main service protection must not exceed conductor ampacity (or may be next higher rated device above conductor ampacity). (Sec. 230-90.)

230-39. Use of 175-A fuses where the calculation calls for 170-A conforms to Exception No. 2 of Sec. 230-90—next higher standard rating of fuse (Sec. 240-6). For motor branch circuits and feeders, Arts. 220 and 430 permit the use of overcurrent devices having ratings or settings higher than the capacities of the conductors. Article 230 makes similar provisions for services where the service supplies a motor load or a combination load of both motors and other loads.

If the service supplies two or more motors as well as other load, then the overcurrent protection must be rated in accordance with the required protection for a feeder supplying several motors plus the other load (Sec. 430-63). Or if the service supplies only a multimotor load (with no other load fed), then Sec. 430-62 sets the maximum permitted rating of overcurrent protection.

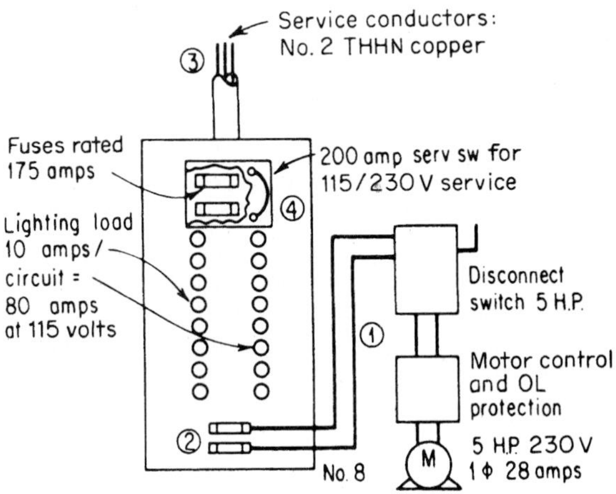

1. Size of motor branch circuit conductors: 125% x 28 amps equals 35 amps. This requires No. 8's.

2. Size of motor branch circuit fuses: 300% x 28 amps equals 84 amps. This requires maximum fuse size of 90 amps. Smaller fuses, such as time-delay type, may be used.

3. Size of service entrance conductors must be adequate for a load of 125% x 28 amps plus 100 amps (continuous lighting load of 80 amps x 1.25) or 128 amps.

4. Size of main fuses: 90 amps (from 2 above) plus 100 amps (80 amps x 1.25) equals 190 amps. This requires maximum fuse size of 200 amps. Again, smaller fuses may and should be used where possible to improve the overload protection on the circuit conductors.

Fig. 230-39. Service protection for lighting plus motor load. (Sec. 230-90.)

Exception No. 3. Not more than six CBs or six sets of fuses may serve as overcurrent protection for the service-entrance conductors even though the sum of the ratings of the overcurrent devices is in excess of the ampacity of the service conductors supplying the devices—as illustrated in Fig. 230-40. The grouping of single-pole CBs as multipole devices, as permitted for disconnect means, may also apply to overcurrent protection. And a set of fuses is all the fuses required to protect the ungrounded service-entrance conductors.

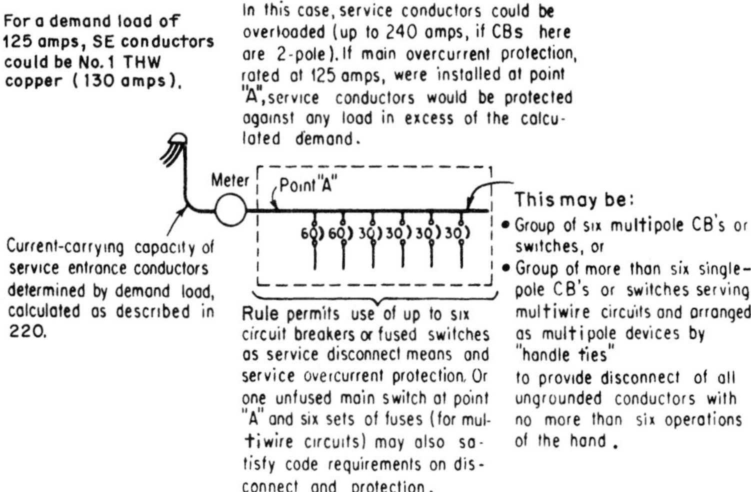

For a demand load of 125 amps, SE conductors could be No. 1 THW copper (130 amps).
In this case, service conductors could be overloaded (up to 240 amps, if CBs here are 2-pole). If main overcurrent protection, rated at 125 amps, were installed at point "A", service conductors would be protected against any load in excess of the calculated demand.
Meter
Point "A"
60 60 30 30 30 30
Current-carrying capacity of service entrance conductors determined by demand load, calculated as described in 220.
Rule permits use of up to six circuit breakers or fused switches as service disconnect means and service overcurrent protection. Or one unfused main switch at point "A" and six sets of fuses (for multiwire circuits) may also satisfy code requirements on disconnect and protection.
This may be:
• Group of six multipole CB's or switches, or
• Group of more than six single-pole CB's or switches serving multiwire circuits and arranged as multipole devices by "handle ties" to provide disconnect of all ungrounded conductors with no more than six operations of the hand.

Fig. 230-40. With six subdivisions of protection, conductors could be overloaded. (Sec. 230-90.)

This Exception ties into Sec. 230-80. Service conductors are sized for the *total* maximum demand load—applying permitted demand factors from Art. 220. Then each of the two to six feeders fed by the SE conductors is also sized from Art. 220 based on the load fed by each feeder. When those feeders are given overcurrent protection in accordance with their ampacities, it is frequently found that the sum of those overcurrent devices is greater than the ampacity of the SE conductors which were sized by applying the applicable demand factors to the total connected load of all the feeders. Exception No. 3 recognizes that possibility as acceptable even though it departs from the rule in the first sentence of Sec. 230-90(a). The assumption is that if calculation of demand load for the SE conductors is correctly made, there will be no overloading of those conductors because the diversity of feeder loads (some loads "on," some "off") will be adequate to limit load on the SE conductors.

Assume that the load of a building computed in accordance with Art. 220 is 255 A. Under Sec. 240-3(b), 300-A fuses or a 300-A CB may be considered as the proper-size overcurrent protection for service conductors rated between 255 and 300 A if a single service disconnect is used.

If the load is separated in such a manner that six 70-A CBs could be used instead of a single service disconnect means, total rating of the CBs would be greater than the ampacity of the service-entrance conductors. And that would be acceptable.

Exception No. 4 to Sec. 230-90(a) is shown in Fig. 230-41 and is intended to prevent opening of the fire-pump circuit on any overload up to and including stalling or even seizing of the pump motor. Because the conductors are "outside the building," operating overload is no hazard; and, under fire conditions, the pump must have no prohibition on its operation. It is better to lose the motor than attempt to protect it against overload when it is needed.

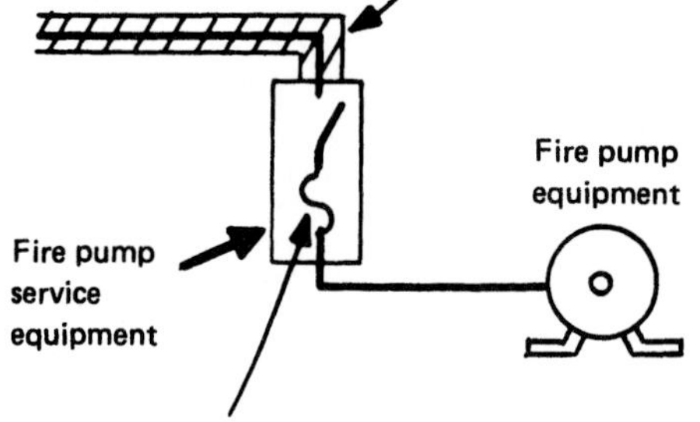
If the service conductors to the fire-pump room enter the fire-pump service equipment directly from the outside or if they are encased in 2-in.-thick concrete . . .
. . . they are judged to be "outside of the building," and . . .
Fire pump equipment
Fire pump service equipment
. . . the overcurrent protective device (fuses or CB) must be rated or set to carry the motor locked-rotor current indefinitely.

Fig. 230-41. (Sec. 230-90.)

230-95. Ground-Fault Protection of Equipment. Fuses and CBs, applied as described in the previous section on "Overcurrent Protection," are sized to protect conductors in accordance with their current-carrying capacities. The function of a fuse or CB is to open the circuit if current exceeds the rating of the protective device. This excessive current might be caused by operating overload, by a ground fault, or by a short circuit. Thus, a 1,000-A fuse will blow if current in excess of that value flows over the circuit. It will blow early on heavy overcurrent and later on low overcurrents. But it will blow, and the circuit and equipment will be protected against the damage of the overcurrent. But, there is another type of fault condition which is very common in grounded systems and will not be cleared by conventional overcurrent devices. That is the phase-to-ground fault (usually arcing) which has a current value less than the rating of the overcurrent device.

On any high-capacity feeder, a line-to-ground fault (i.e., a fault from a phase conductor to a conduit, to a junction box, or to some other metallic equipment enclosure) can, and frequently does, draw current of a value less than the rating or setting of the circuit protective device. For instance, a 500-A ground fault on a 2,000-A protective device which has only a 1,200-A load will not be cleared by the device. If such a fault is a "bolted" line-to-ground fault, a highly unlikely fault, there will be a certain amount of heat generated by the I^2R effect of the current; but this will usually not be dangerous, and such fault current will merely register as additional operating load, with wasted energy (wattage) in the system. But, bolted phase-to-ground faults are very rare. The usual phase-to-ground fault exists as an intermittent or arcing fault, and an arcing fault of the same current rating as the essentially harmless bolted fault can be fantastically destructive because of the intense heat of the arc.

Of course, any ground-fault current (bolted or arcing) above the rating or setting of the circuit protective device will normally be cleared by the device. In such cases, bolted-fault currents will be eliminated. But, even where the protective device eventually operates, in the case of a heavy ground-fault current which adds to the normal circuit load current to produce a total current in excess of the rating of the normal circuit protective device (fuse or CB), the time delay of the device may be minutes or even hours—more than enough time for the arcing-fault current to burn out conduit and enclosures, acting just like a torch, and even propagating flame to create a fire hazard.

In spite of the growth of effective and skilled application of conventional overcurrent protective devices, the problem of ground faults persists and even grows with expanding electrical usage. In the interests of safety, definitive engineering design must account for protection against such faults. Phase overcurrent protective devices are normally limited in their effectiveness because (1) they must have a time delay and a setting somewhat higher than full load to ride through normal inrushes, and (2) they are unable to distinguish between normal currents and low-magnitude fault currents which may be less than full-load currents.

Dangerous temperatures and magnetic forces are proportional to current for overloads and short circuits; therefore, overcurrent protective devices usually are adequate to protect against such faults. However, the temperatures of arcing faults are, generally, independent of current magnitude; and arcs of great and extensive destructive capability can be sustained by currents not exceeding the overcurrent device settings. Other means of protection are therefore necessary. A ground-detection device which "sees" only ground-fault current can be coupled to an automatic switching device to open all three phases when a line-to-ground fault exists on the circuit.

Section 230-95 requires ground-fault protection equipment to be provided for each service *disconnecting means* rated 1,000 A or more in a solidly grounded-wye electrical service that operates with its ungrounded legs at more than 150 V to ground. Note that this applies to the rating of the disconnect, not to the rating of the overcurrent devices or to the capacity of the service-entrance conductors.

The wording of the first sentence of this section makes clear that service GFP (ground-fault protection) is required under specific conditions: only for grounded-wye systems that have voltage over 150 V to ground and less than

600 V phase-to-phase. In effect, that means the rule applies only to 480/277-V grounded-wye and *not* to 120/208-V systems or any other commonly used systems (Fig. 230-42). Recent recognition of the 600Y/340-V distribution systems—used in Canada—would subject any system so-rated to the rule of Sec. 230-95. And, each disconnect rated 1,000 A, or more, must be provided with equipment GFP. GFP is *not* required on any systems operating over 600 V phase-to-phase.

In a typical GFP hookup as shown in Fig. 230-43, part **(a)** of the section specifies that a ground-fault current of 1,200 A or more must cause the disconnect to open all ungrounded conductors. Thus the maximum GF pick-up setting permitted is 1,200 A, although it may be set lower.

With a GFP system, at the service entrance a ground fault anywhere in the system is immediately sensed in the ground-relay system, but its action to open the circuit usually is delayed to allow some normal overcurrent device near the point of fault to open if it can. As a practical procedure, such time delay is designed to be only a few cycles or seconds, depending on the voltage of the circuit, the time-current characteristics of the overcurrent devices in the system, and the location of the ground-fault relay in the distribution system. Should any of the conventional short-circuit overcurrent protective devices fail to operate in the time predetermined to clear the circuit, and if the fault continues, the ground-fault protective relays will open the circuit. This provides added overcurrent protection not available by any other means.

The rule requiring GFP for any service disconnect rated 1,000 A or more (on 480/277-V or 600/347-V services) specifies a maximum *time delay of 1 sec for ground-fault currents of 3,000 A or more* (Fig. 230-44).

The maximum permitted setting of a service GFP hookup is 1,200 A, but the time-current trip characteristic of the relay must assure opening of the disconnect in not more than 1 sec for any ground-fault current of 3,000 A or more. This change in the **Code** was made to establish a specific level of protection in GFP equipment by setting a maximum limit on i^2t of fault energy.

The reasoning behind this change was explained as follows:

The amount of damage done by an arcing fault is directly proportional to the time it is allowed to burn. Commercially available GFP systems can easily meet the 1-sec limit. Some users are requesting time delays up to 60 sec so all downstream overcurrent devices can have plenty of time to trip thermally before the GFP on the main disconnect trips. However, an arcing fault lasting 60 sec can virtually destroy a service equipment installation. Coordination with downstream overcurrent devices can and should be achieved by adding GFP on feeder circuits where needed. The **Code** should require a reasonable time limit for GFP. Now, 3,000 A is 250 percent of 1,200 A, and 250 percent of setting is a calibrating point specified in ANSI 37.17. Specifying a maximum time delay starting at this current value will allow either flat or inverse time-delay characteristics for ground-fault relays with approximately the same level of protection.

Selective coordination between GFP and conventional protective devices (fuses and CBs) on service and feeder circuits is now a very clear and specific task as a result of rewording of Sec. 230-95(a) that calls for a maximum time delay of 1 sec at any ground-fault current value of 3,000 A or more.

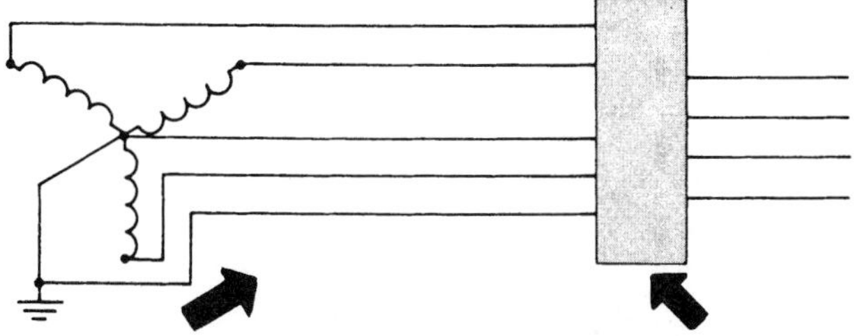

. . . applies to solidly grounded wye services over 150 volts to ground but not over 600 volts phase-to-phase — i.e., 480Y/277 V or 600/347 V.
For *each* service disconnect rated 1000 amps or more, ground-fault protection with maximum trip setting of 1200 amps must be provided.

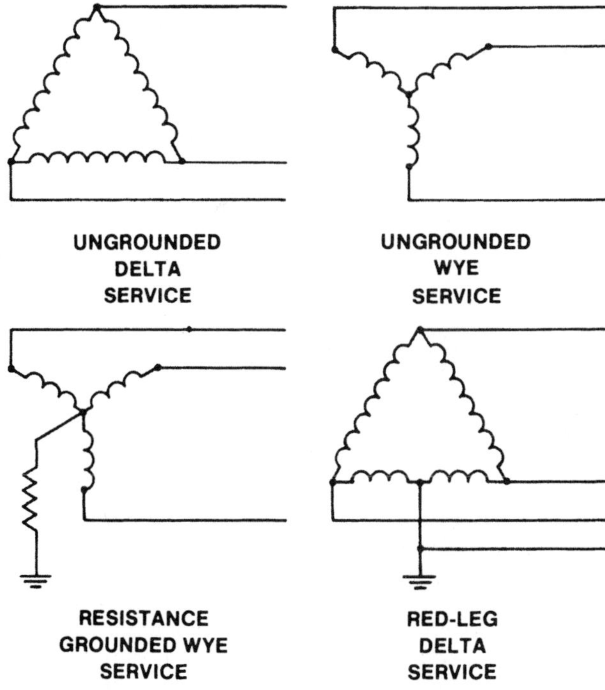

Fig. 230-42. Service ground-fault protection is mandatory. (Sec. 230-95.)

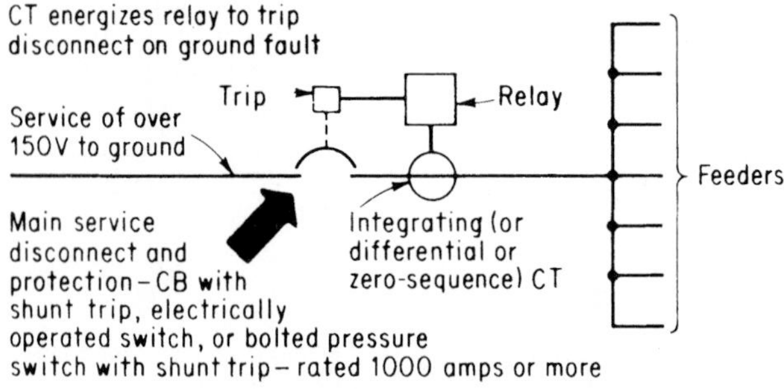
CT energizes relay to trip disconnect on ground fault
Trip
Relay
Service of over 150V to ground
Feeders
Main service disconnect and protection – CB with shunt trip, electrically operated switch, or bolted pressure switch with shunt trip – rated 1000 amps or more
Integrating (or differential or zero-sequence) CT

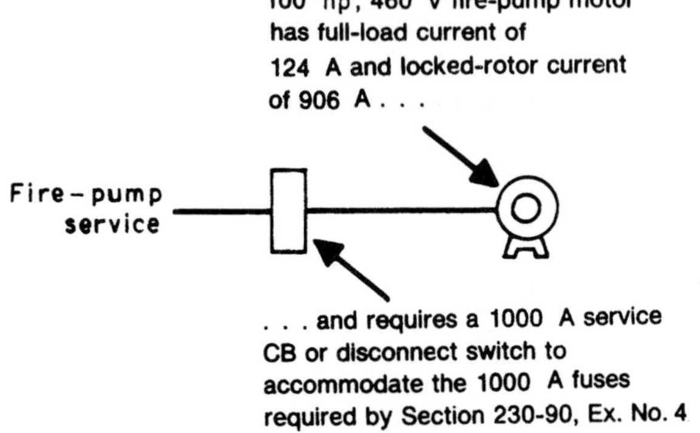
100 hp, 460 V fire-pump motor has full-load current of 124 A and locked-rotor current of 906 A . . .
Fire – pump service
. . . and requires a 1000 A service CB or disconnect switch to accommodate the 1000 A fuses required by Section 230-90, Ex. No. 4
GROUND-FAULT PROTECTION MUST NOT BE USED!

Fig. 230-43. GFP is required for each disconnect rated 1,000 A or more, but not for a fire-pump disconnect. (Sec. 230-95.)

For applying the rule of Sec. 230-95, the rating of any service disconnect means shall be determined as shown in Fig. 230-45.

Because the rule on required service GFP applies to the rating of each service disconnect, there are many instances where GFP would be required if a single service main disconnect is used but *not* if the service subdivision option of Sec. 230-71(a) is taken, as shown in Fig. 230-46.

By Exception No. 1 to Sec. 230-95, continuous industrial process operations are exempted from the GFP rules of parts **(a)**, **(b)**, and **(c)** where the electrical system is under the supervision of qualified persons who will effect orderly shutdown of the system and thereby avoid hazards, greater than ground fault itself, that would result from the nonorderly, automatic interruption that GFP would produce in the supply to such critical continuous operations. Exception No. 1 excludes GFP requirements where a nonorderly shutdown will introduce additional or increased hazards. The idea behind that is to provide maximum protection against service outage of such industrial processes. With highly trained personnel at such locations, design and maintenance of the electrical

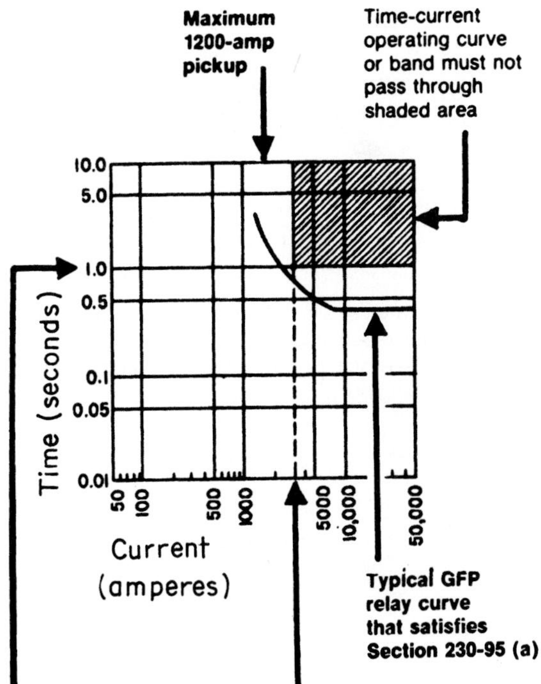

Fig. 230-44. The rule specifies maximum energy let-through for GFP operation. [Sec. 230-95(a).]

system can often accomplish safety objectives more readily without GFP on the service. Electrical design can account for any danger to personnel resulting from loss of process power versus damage to electrical equipment.

Exception No. 2 excludes fire-pump service disconnects from the basic rule that requires ground-fault protection on any service disconnect rated 1,000 A or more on a grounded-wye 600/347V or 480/277-V system.

Because fire pumps are required by Sec. 230-90, Exception No. 4, to have overcurrent protection devices large enough to permit locked-rotor current of the pump motor to flow without interruption, larger fire pumps (100 hp and more) would have disconnects rated 1,000 A or more. Without Exception No. 2, those fire-pump disconnects would be subject to the basic rule and would have to be equipped with ground-fault protection. But GFP on any fire pump is objectionable on the same basis that Sec. 230-90, Exception No. 4, wants nothing less than protection rated for locked rotor. The intent is to give the pump motor every chance to operate when it functions during a fire, to prevent opening of the motor circuit or any overload up to and including stalling or seizing of the shaft or bearings. For the same reason, Sec. 430-31 exempts fire

FUSED SWITCH (bolted pressure switch, service protector, etc.)

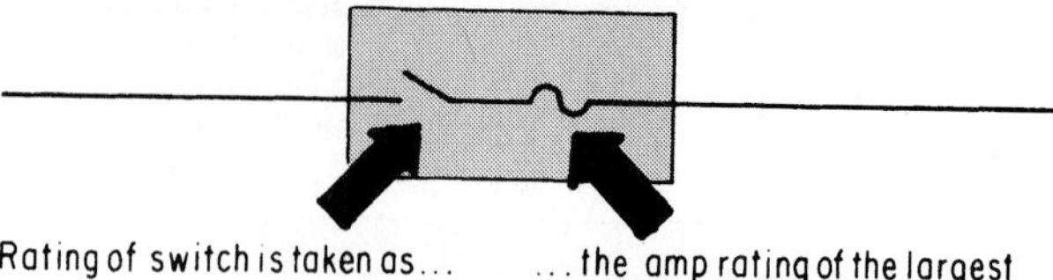

Rating of switch is taken as... ...the amp rating of the largest fuse that can be installed in the switch fuseholders.

EXAMPLE

If 900-amp fuses are used in this service switch, ground-fault protection would be required, because the switch can take fuses rated 1200 amps—which is above the 1000-amp level at which GFP becomes mandatory.

CIRCUIT BREAKER

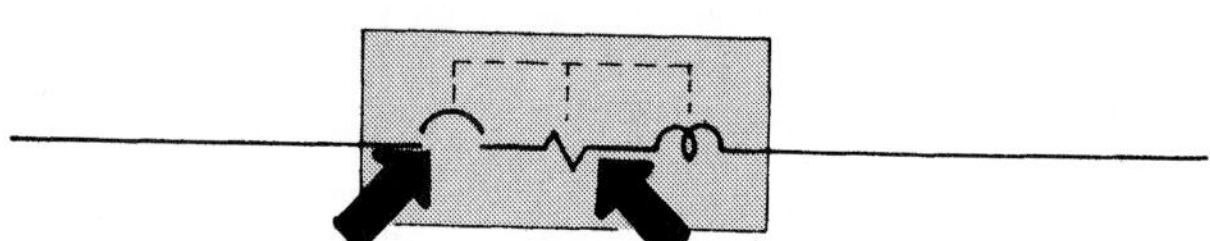

Rating of breaker is taken as... ...the maximum continuous current rating (pickup of long time-delay) for which the trip device in the breaker is set or can be adjusted.

Example: GFP would be required for a service CB with, say, an 800-amp trip setting if the CB had a trip device that can be adjusted to 1000 amps or more.

Fig. 230-45. Determining rating of service disconnect for GFP rule. (Sec. 230-95.)

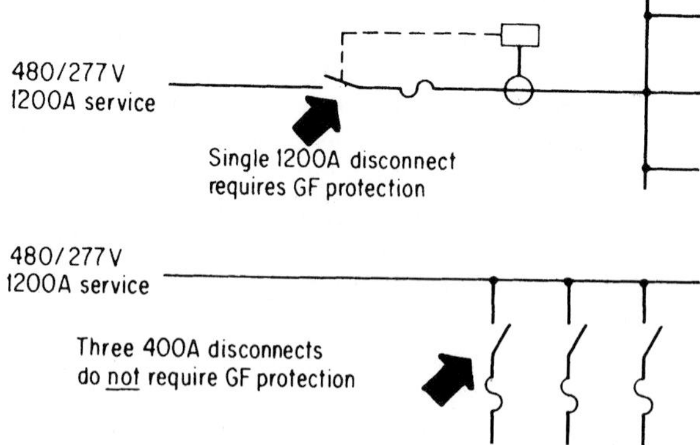
480/277V
1200A service
Single 1200A disconnect
requires GF protection
480/277V
1200A service
Three 400A disconnects
do not require GF protection

Fig. 230-46. Subdivision option on disconnects affects GFP rule. (Sec. 230-95.)

pumps from the need for overload protection, and Sec. 430-72, Exception No. 4, requires overcurrent protection to be omitted from the control circuit of a starter for a fire pump.

And it should be noted that Exception No. 2 in Sec. 230-95 says that ground-fault protection "*shall not apply*" to fire-pump motors—making omission of GFP *mandatory.*

Important considerations are given in fine-print notes in this section. Obviously, the selection of ground-fault equipment for a given installation merits a detailed study. The option of subdividing services discussed under *six service entrances from one lateral* (Sec. 230-2, Exception No. 7) should be evaluated. A 4,000-A service, for example, could be divided using five 800-A disconnecting means, and in such cases GFP would not be required.

One very important note in Sec. 230-95(b) warns about potential desensitizing of ground-fault sensing hookups when an emergency generator and transfer switch are provided in conjunction with the normal service to a building. The note applies to those cases where a solid neutral connection from the normal service is made to the neutral of the generator through a 3-pole transfer switch. With the neutral grounded at the normal service and the neutral bonded to the generator frame, ground-fault current on the load side of the transfer switch can return over two paths, one of which will escape detection by the GFP sensor, as shown in Fig. 230-47. Such a hookup can also cause nuisance tripping of the GFP due to normal neutral current. Under normal (nonfaulted) conditions, neutral current due to normal load unbalance on the phase legs can divide at common neutral connection in transfer switch, with some current flowing toward the generator and returning to the service main on the conduit—indicating falsely that a ground fault exists and causing nuisance tripping of GFP. The note points out that "means or devices" (such as a 4-pole, neutral-switched transfer switch) "may be needed" to assure proper, effective operation of the GFP hookup (Fig. 230-48). VERY IMPORTANT!

Because of so many reports of improper and/or unsafe operation (or failure to operate) of ground-fault protective hookups, part **(c)** of Sec. 230-95 *requires* (a mandatory rule) that *every* GFP hookup be "performance tested when first installed." And the testing MUST *be done on the job site!* Factory testing of a GFP system does not satisfy this **Code** rule. This rule requires that such testing be done according to instructions . . . provided with the equipment." A written record must be made of the test and must be available to the inspection authority.

Figure 230-49 shows two basic types of GFP hookup used at service entrances.

230-200. General (Services Exceeding 600 Volts, Nominal). The rules on high-voltage services given in the provisions of Art. 230 apply only to equipment on the load side of the "service-point." Because there has been so much controversy over identifying what is and what is not "service" equipment in the many complicated layouts of outdoor high-voltage circuits and transformers, the definition in Art. 100 provides clarification. In any particular installation, identification of that point can be made by the utility company and design personnel. The definition clarifies that the property line is not the determinant as to where **NE Code** rules must begin to be applied. This is particularly important in cases

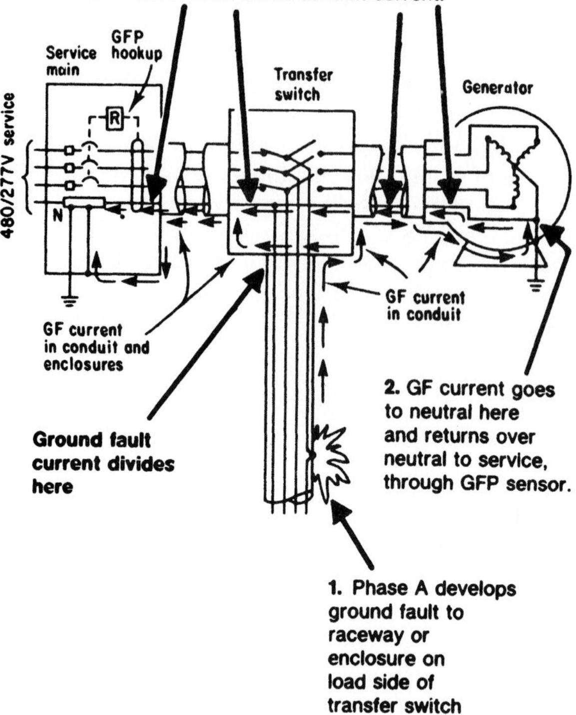

Fig. 230-47. Improper operation of GFP equipment can result from emergency system transfer switch. (Sec. 230-95.)

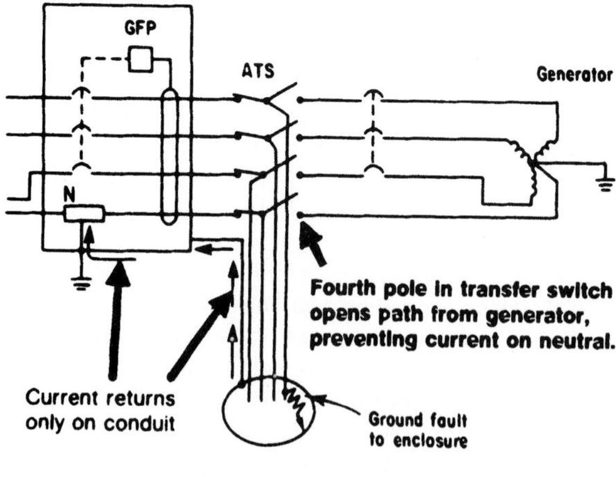

Fig. 230-48. Four-pole transfer switch is one way to avoid desensitizing GFP. (Sec. 230-95.)

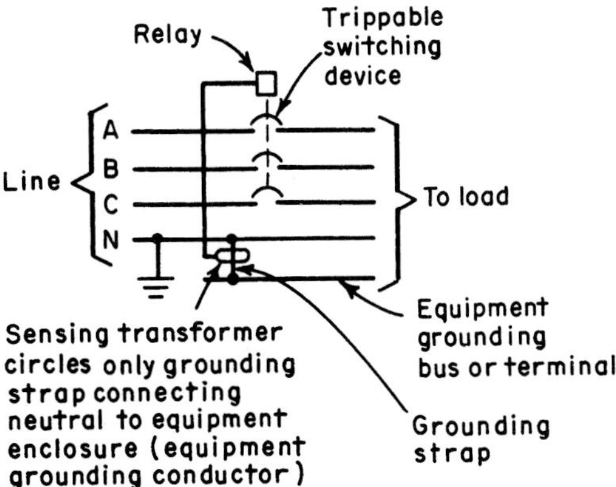

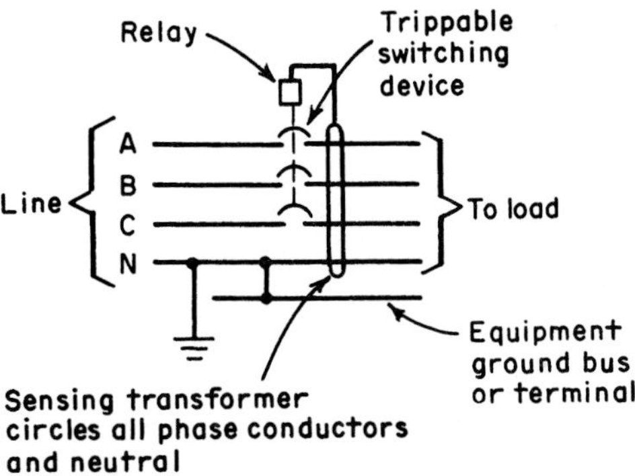

Fig. 230-49. Types of ground-fault detection that may be selected for use at services. (Sec. 230-95.)

of multibuilding industrial complexes where the utility has distribution circuits on the property. See "Service Point" in Art. 100.

Section 230-200 says that "service conductors and equipment used on circuits exceeding 600 volts" must comply with *all* the rules in Art. 230 (including any "applicable provisions" that cover services up to 600 V). And Sec. 230-10 says that for services up to 600 V, the "service conductors" are those conductors—whether on the primary or secondary of a step-down transformer or transformers—that carry current from the "service point" (where the utility connects to the customer's wiring) to the service disconnecting means for a building or structure. See "Service Point" in Art. 100. All conductors between the defined points—"service point" and "service disconnecting means"—must comply with all requirements for service conductors, whether above or below 600 V.

Section 230-205 says the service disconnect means must be located in accordance with Sec. 230-70. Section 230-70 calls for the "service disconnecting means" to be either outside or inside the building or structure as close as possible to where the service conductors enter. Those rules identify the service disconnect means and apply to the conductors entering a building or structure. All conductors between the "service point" and the "service disconnecting means" must be treated as "service conductors," regardless of voltage. Conductors operating at over 600 V must satisfy the rules of part **H** of Art. 230 in addition to satisfying all rules of Art. 230 up to part **H.**

Design and layout of any "service" are critically related to safety, adequacy, economics, and effective use of the whole system. It is absolutely essential that we know clearly and surely what circuits and equipment of any electrical system constitute the "service" and what parts of the system are not involved in the "service." For instance, in a system with utility feed at 13.2 kV and step-down to 480/277 V, the mandatory application of Sec. 230-95 requiring GFP hinges on establishing whether the "service" is on the primary or secondary side of the transformers. If the secondary is the service, where the step-down transformer belongs to the utility and the "service point" is on its secondary, we have a mandatory need for GFP and none of the **Code** rules on service would apply to any of the 13.2-kV circuits—regardless of their length or location. If the transformer belongs to the customer and the "service point" is on the primary side, the primary is the service, Sec. 230-95 does not require GFP on services over 600 V phase-to-phase, all the primary circuit and equipment must comply with all of Art. 230, and the secondary circuits are feeders and do not have to comply with any of the service regulations.

The whole problem involved here is complex and requires careful, individual study to see clearly the many interrelated considerations. Let us look at a few important things to note about **Code** definitions as given in Art. 100:

1. "Service conductors" run to the *service equipment* of the *premises* supplied. Note that they run to "premises" and are not required to run to a "building." The **Code** does not define the word *premises,* but a typical dictionary definition is "a tract of land, including its buildings." But for many years, the **Code** rule of what was Sec. 230-201 in the 1990 **Code** (now Art. 100) did refer to "service conductors to the building." Although that phrase no longer is used in the rule, the wording does clearly aim at establishing the service conductors to the "building or other structure served" (Fig. 230-50).
2. "Service equipment" *usually* consists of "a circuit breaker or switch and fuses, and their accessories, located near the point of entrance of *supply* conductors to a building or other structure, or an otherwise defined area." Note that the service equipment is the means of cutoff of the supply, and the service conductors may enter "a building" or "other structure" or a "defined area." But, again, a service does not necessarily have to be to "a building." It could be to such a "structure" as an outdoor switchgear or unit substation enclosure.

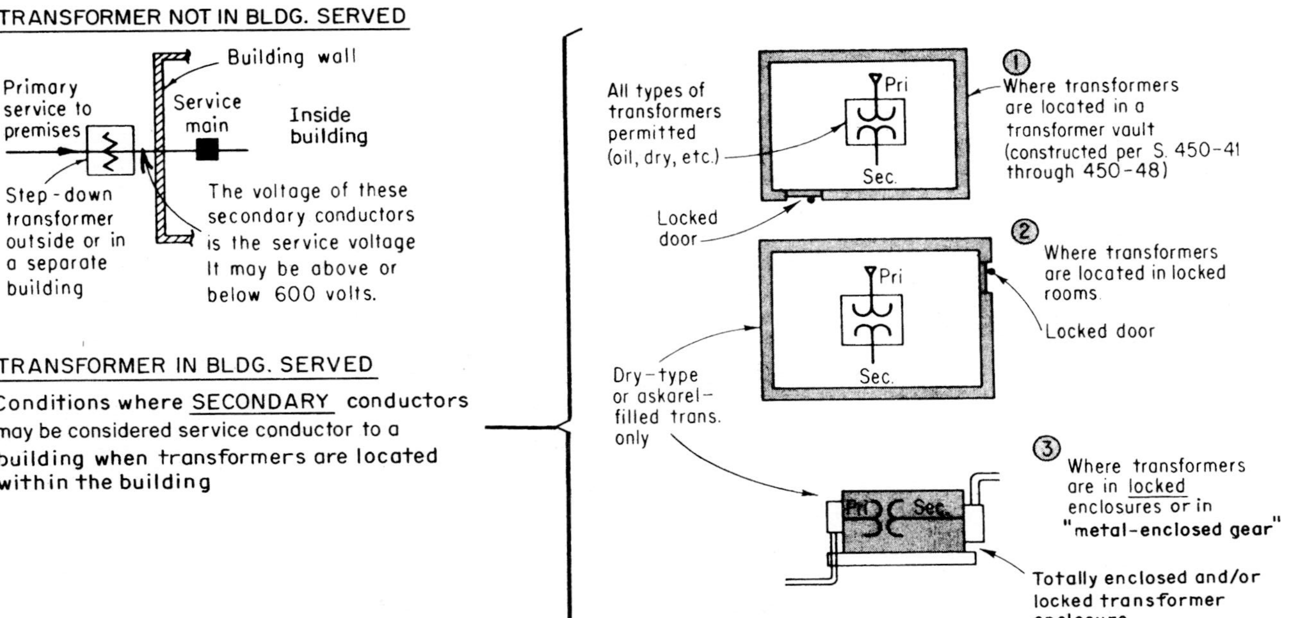
TRANSFORMER NOT IN BLDG. SERVED
Building wall
Primary service to premises
Service main
Inside building
Step-down transformer outside or in a separate building
The voltage of these secondary conductors is the service voltage It may be above or below 600 volts.
TRANSFORMER IN BLDG. SERVED
Conditions where SECONDARY conductors may be considered service conductor to a building when transformers are located within the building
All types of transformers permitted (oil, dry, etc.)
Pri
Sec.
① Where transformers are located in a transformer vault (constructed per S. 450-41 through 450-48)
Locked door
② Where transformers are located in locked rooms.
Locked door
Dry-type or askarel-filled trans. only
③ Where transformers are in locked enclosures or in "metal-enclosed gear"
Totally enclosed and/or locked transformer enclosure

Fig. 230-50. Where the transformer belongs to the utility, the "service point" is on its secondary and the secondary conductors are the service conductors to the building or structure. (Sec. 230-202 and Art. 100.)

The wording in Art. 100 bases identification of "service conductors" as extending from the "service point." Because of the definition of "service point," it is essential to determine whether the transformers belong to the power company or the property owner.

If a utility-owned transformer that handles the electrical load for a building is in a locked room or locked enclosure (accessible only to qualified persons) in the building and is fed, say, by an underground high-voltage (over 600 V) utility line from outdoors, the secondary conductors from the transformer would be the "service conductors" to the building. *And* the switching and control devices (up to six CBs on fused switches) on the secondary would constitute the "service equipment" for the building. Under such a condition, if any of the secondary section "service disconnects" were rated 1,000 A or more, at 480/277-V grounded wye, they would have to comply with Sec. 230-95, requiring GFP for the service disconnects.

However, if the utility made primary feed to a transformer or unit substation belonging to the owner, then the primary conductors would be the service conductors and the primary switch or CB would be the "service disconnect." In that case, no GFP would be needed on the "service disconnect" because Sec. 230-95 applies only up to 600 V, and there is no requirement for GFP on high-voltage services (Fig. 230-51). Also, in that case, there would be no need for GFP on the secondary section disconnects, because they would not be "service disconnects"—and those are the same disconnects that might be subject to Sec. 230-95 if the transformer belonged to the utility. However, Secs. 215-10 or 240-13 may require such protection for these secondary section disconnects. (See also Fig. 230-52.)

230-202. Service-Entrance Conductors. This section specifies the minimum conductor size, that is, No. 6 in a raceway and No. 8 in a multiconductor cable. Additionally, it indicates that only those wiring methods given in Sec. 710-4 may be used. That section gives the wiring methods that are acceptable for use as service-entrance conductors where it has been established that primary conductors (over 600 V) are the service conductors or where the secondary conductors are the service conductors and operate at more than 600 V. The basic conduits that may be used are rigid metal conduit, intermediate metal conduit, and rigid nonmetallic conduit. In addition, cable tray, cable bus, or "other identified" raceways or even type MC may be used. Note, too, that bare conductors, bare buswork, or open runs of type MV are permitted as indicated. And the NEC no longer requires concrete encasement of the nonmetallic conduit.

Paragraph **(a)** of Sec. 710-4 points out that *cable tray* systems are also acceptable for high-voltage services, provided that the cables used in the tray are "identified as service-entrance conductors." Section 318-3(a) recognizes "*multiconductor* service-entrance cable" for use in tray, for cables rated up to 600 V. High-voltage (over 600 V) service-entrance cables may be used if the cables are "identified"—which, in today's strict usage, virtually means that such cable must be listed by a nationally recognized test lab (UL, etc.) as suitable for the purpose. Article 338 on "Service-Entrance Cable" does not refer to any high-voltage cable for service-entrance use. Details of this section are shown in Fig. 230-53.

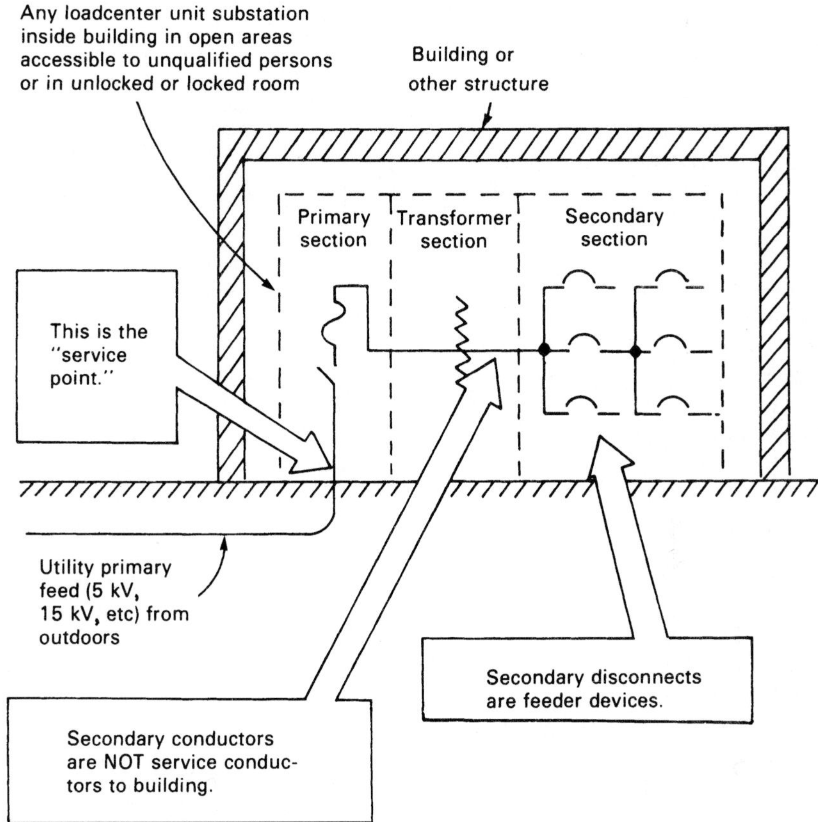
Any loadcenter unit substation inside building in open areas accessible to unqualified persons or in unlocked or locked room
Building or other structure
Primary section
Transformer section
Secondary section
This is the "service point."
Utility primary feed (5 kV, 15 kV, etc) from outdoors
Secondary disconnects are feeder devices.
Secondary conductors are NOT service conductors to building.

Fig. 230-51. The primary is the "service" for *any* indoor transformer belonging to the owner and fed by utility line. (Sec. 230-200 and Sec. 230-10.)

230-203. Warning Signs. Any warning sign for use where unauthorized persons might contact live parts must always include the phrase "KEEP OUT" immediately after the phrase "DANGER HIGH VOLTAGE." Court cases involving electrical accidents have established the importance of a sign giving a command "KEEP OUT" after advising of the potential hazard "DANGER HIGH VOLTAGE." This is an important factor in making such a sign act as an effective deterrent. And such wording should always be used on signs of this type, whether the sign is required by Code rule [see also Sec. 110-17(c) and Sec. 110-34(c)] or not required.

230-204. Isolating Switches. An air-break isolating switch must be used between an oil switch or an air, oil, vacuum, or sulfur hexachloride CB and the supply conductors, unless removable truck panels or metal-enclosed units are used providing disconnect of all live parts in the removed position. This line-side disconnect assures safety to personnel in maintenance (Fig. 230-54). Part **(d)** requires a grounding connection for an isolating switch, as in Fig. 230-55.

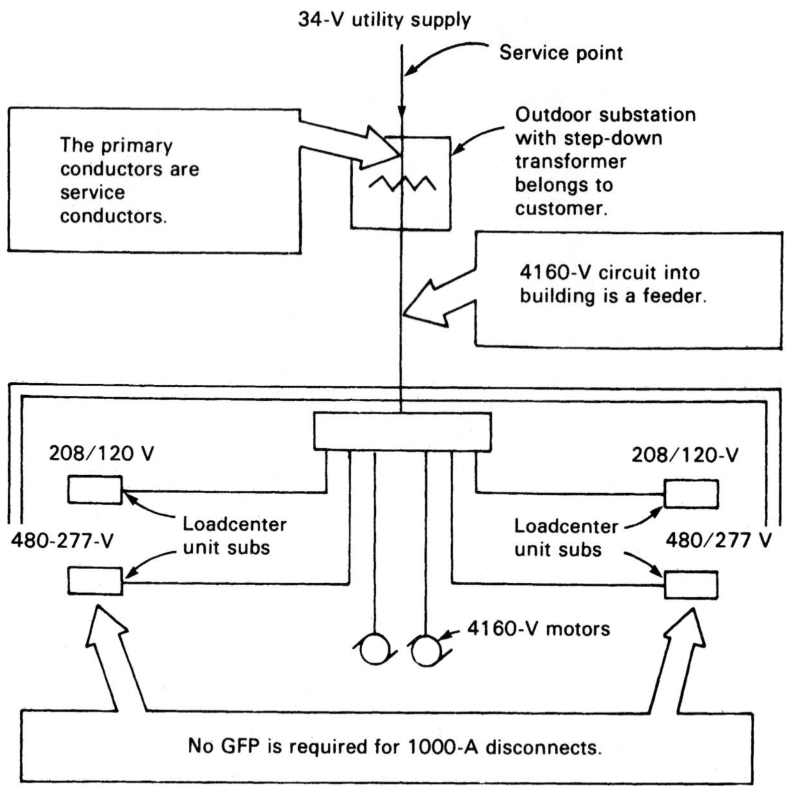

Fig. 230-52. The primary circuit must be taken as the "service conductors" where the "service point" is on the primary side of an outdoor transformer.

230-205. Disconnecting Means. In part **(a)**, the basic rule requires a high-voltage service disconnect means to be located "outside or inside nearest the point of entrance of the service conductors" into the building or structure being supplied—as for 600-V equipment in Sec. 230-70.

Part **(b)**, covering the electrical fault characteristics, requires that the service disconnect be *capable of closing,* safely and effectively, on a fault equal to or greater than the maximum short-circuit current that is available at the line terminals of the disconnect. The last sentence notes that where fuses are used within the disconnect or in conjunction with it, the fuse characteristics may contribute to fault-closing rating of the disconnect. The idea behind this rule is to assure that the disconnect switch may be safely closed on a level of fault that can be safely interrupted by the fuse.

230-208. Protection Requirements. Service conductors operating at voltages over 600 V must have a short-circuit (not overload) device in each ungrounded conductor, installed either (1) on load side of service disconnect, or (2) as an integral part of the service disconnect.

All devices must be able to detect and interrupt all values of current in excess of their rating or trip setting, which must be as shown in Fig. 230-56.

Sec. 710-4(a) HIGH-VOLTAGE SERVICE CONDUCTORS FOR LOCATIONS ACCESSIBLE TO OTHER THAN QUALIFIED PERSONS

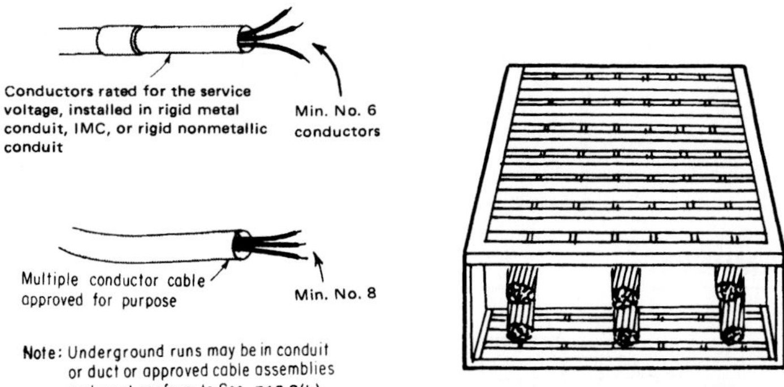

Sec. 230-212 SERVICE CONDUCTORS OPERATING AT MORE THAN 15 kV

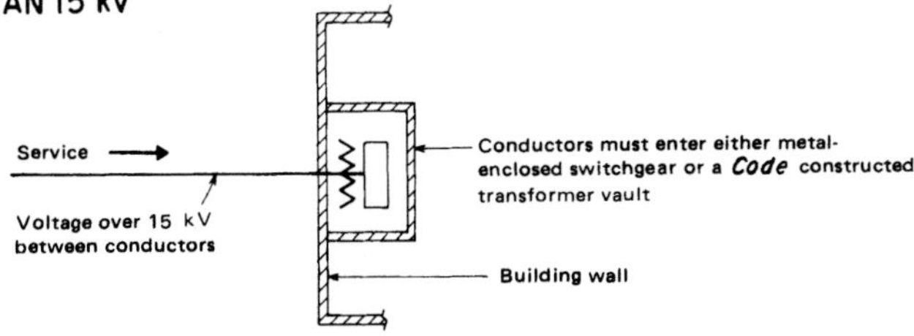

Sec. 710-8 POTHEAD ON SERVICE CONDUCTORS

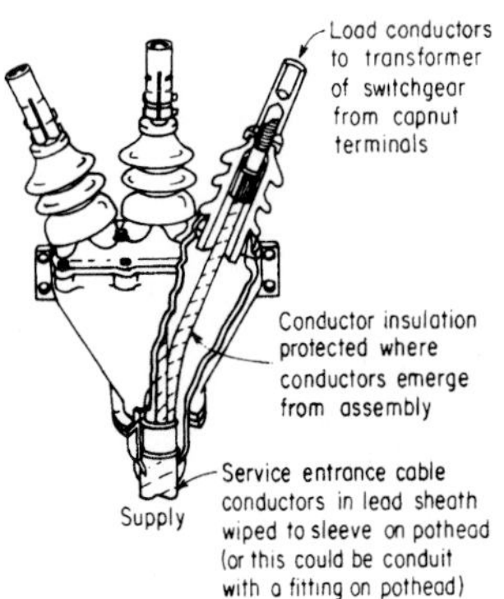

REMEMBER, SEC. 230-6 ALSO APPLIES TO HIGH-VOLTAGE CONDUCTORS (OVER 600 V NOMINAL) AS GIVEN IN SEC. 230-200, THEREFORE CONDUCTORS ENCLOSED IN MASONRY ARE CONSIDERED AS INSTALLED OUTSIDE THE BUILDING

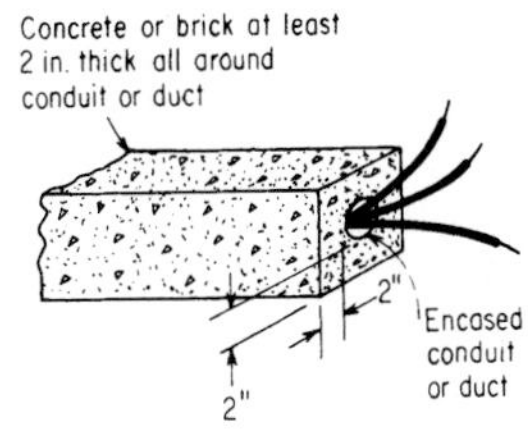

Fig. 230-53. Provisions for service conductors rated over 600 V (refer to subpart letter identification of rules). (Sec. 230-202.)

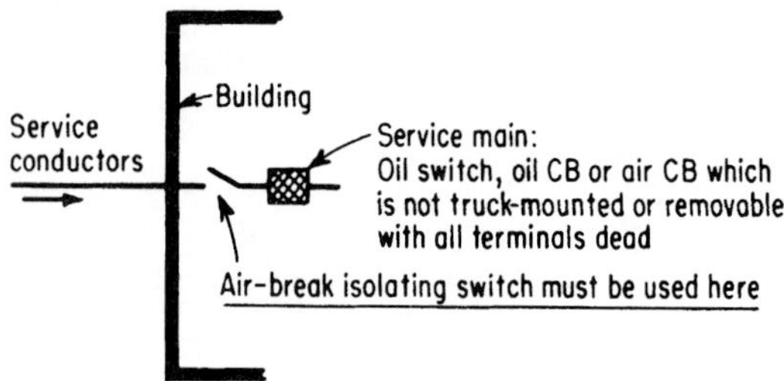

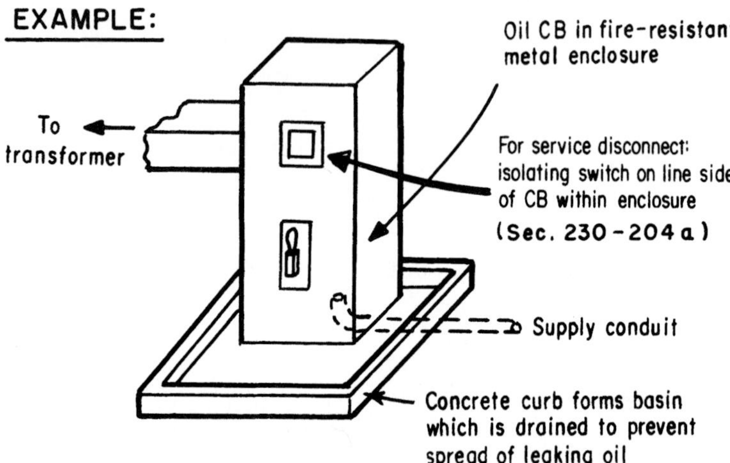

Fig. 230-54. Isolating switch may be needed to kill line terminals of service disconnect. (Sec. 230-204.)

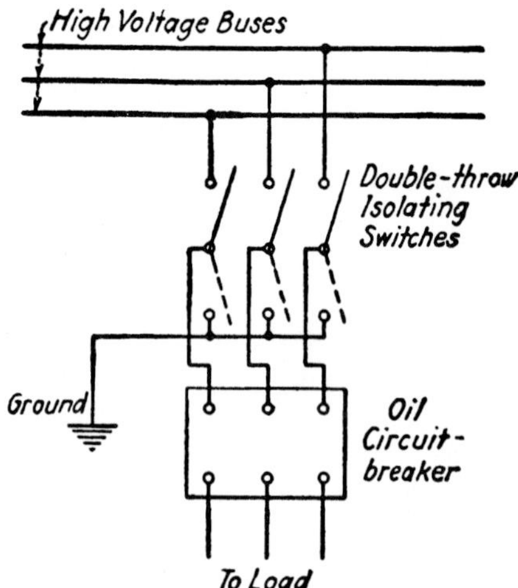

Fig. 230-55. One method for grounding the load side of an open isolating switch. (Sec. 230-204.)

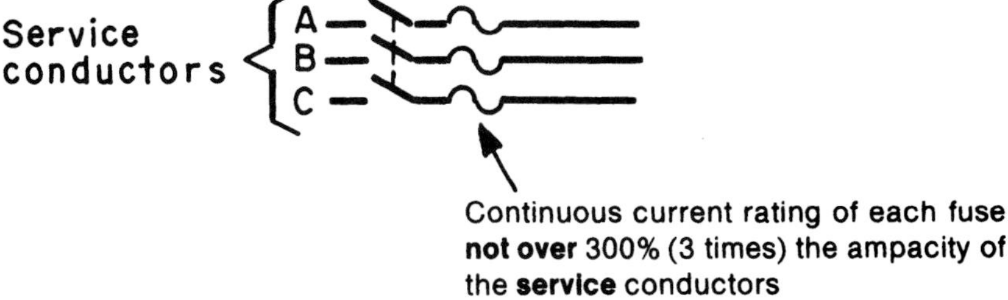

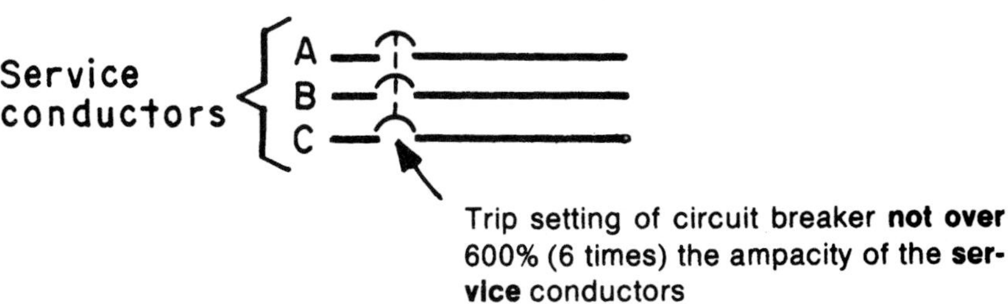

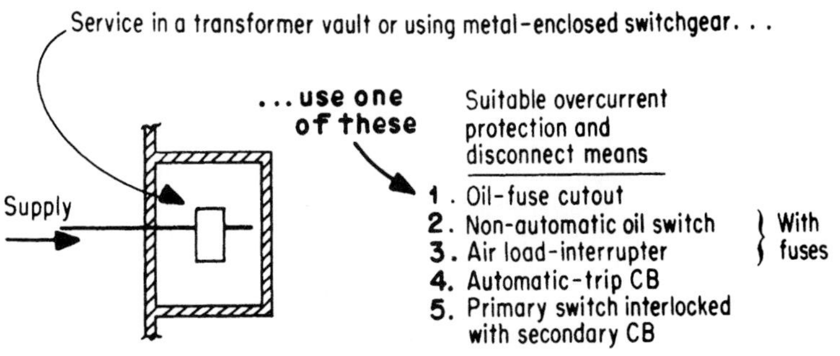

Fig. 230-56. Maximum permitted rating or setting of high-voltage overcurrent protection for service. (Sec. 230-208.)

The difference between 300 percent for fuses and 600 percent for CBs is explained as follows:

The American National Standards Institute (ANSI) publishes standards for power fuses. The continuous-current ratings of power fuses are given with the letter "E" following the number of continuous amps—for instance, 65E or 200E or 400E. The letter "E" indicates that the fuse has a melting time-current characteristic in accordance with the standard for E-rated fuses:

> The melting time-current characteristics of fuse units, refill units, and links for power fuses shall be as follows:
>
> (1) The current-responsive element with ratings 100 amperes or below shall melt in 300 seconds at an rms current within the range of 200 or 240 percent of the continuous current rating of the fuse unit, refill unit, or fuse link.
>
> (2) The current-responsive element with ratings above 100 amperes shall melt in 600 seconds at an rms current within the range of 220 to 264 percent of the continuous current rating of the fuse unit, refill unit, or fuse link.
>
> (3) The melting time-current characteristic of a power fuse at any current higher than the 200 to 240 or 264 percent specified in (1) or (2) above shall be shown by each manufacturer's published time current curves, since the current-responsive element is a distinctive feature of each manufacturer.
>
> (4) For any given melting time, the maximum steady-state rms current shall not exceed the minimum by more than 20 percent.

The fact that E-rated fuses are given melting times at 200 percent or more of their continuous-current rating explains why NE Code Secs. 230-208 and 240-100 set 300 percent of conductor ampacity as the maximum fuse rating but permit CBs up to 600 percent. In effect, the 300 percent for fuses times 2 (200 percent) becomes 600 percent—the same as for CBs.

Part **(b)** of this section permits overcurrent protection for services over 600 V to be loaded up to 100 percent of its rating even on continuous loads (operating for periods of 3 hr or more).

ARTICLE 240. OVERCURRENT PROTECTION

240-1. Scope. For any electrical system, required current-carrying capacities are determined for the various circuits—feeders, subfeeders, and branch circuits. Then these required capacities are converted into standard circuit conductors which have sufficient current-carrying capacities based on the size of the conductors, the type of insulation on the conductors, the ambient temperature at the place of installation, the number of conductors in each conduit, the type and continuity of load, and judicious determination of spare capacity to meet future load growth. Or if busway, armored cable, or other cable assemblies are to be used, similar considerations go into selection of conductors with required current-carrying capacities. In any case, the next step is to provide overcurrent protection for each and every circuit:

The overcurrent device for conductors or equipment must automatically open the circuit it protects if the current flowing in that circuit reaches a value which will cause an excessive or dangerous temperature in the conductor or conductor installation.

Overcurrent protection for conductors must also be rated for safe operation at the level of fault current obtainable at the point of their application. Every fuse and circuit breaker for short-circuit protection must be applied in such a way that the fault current produced by a bolted short circuit on its lead terminals will not damage or destroy the device. Specifically this requires that a short-circuit overcurrent device have a proven interrupting capacity at least equal to the current which the electrical system can deliver into a short on its line terminals. That is, the calculation for the short-circuit interrupting rating must *not* include the impedance of the device itself. That impedance may only be applied to the calculation for the next device downstream.

But safe application of a protective device does not stop with adequate interrupting capacity for its own use at the point of installation in the system. The speed of operation of the device must then be analyzed in relation to the thermal and magnetic energy which the device permits to flow in the faulted circuit. A very important consideration is the provision of conductor size to meet the potential heating load of short-circuit currents in cables. With expanded use of circuit-breaker overcurrent protection, coordination of protection from loads back to the source has introduced time delays in operation of overcurrent devices. Cables in such systems must be able to withstand any impressed short-circuit currents for the durations of overcurrent delay. For example, a motor circuit to a 100-hp motor might be required to carry as much as 15,000 A for a number of seconds. To limit damage to the cable due to heating effect, a much larger size conductor than necessary for the load current alone may be required.

A device may be able to break a given short-circuit current without damaging itself in the operation; but in the time it takes to open the faulted circuit, enough energy may get through to damage or destroy other equipment in series with the fault. This other equipment might be a cable or busway or a switch or motor controller—any circuit component which simply cannot withstand the few cycles of short-circuit current which flows in the period of time between initiation of the fault and interruption of the current flow.

The FPN following Sec. 240-1 often raises questions about the approved use of conductors and overcurrent protection to withstand faults.

example Assume a panelboard with 20-A breakers rated 10,000 A IC (interrupting capacity) and No. 12 copper branch-circuit wiring. Available fault current at the point of breaker application is 8,000 A. The short-circuit withstand capability of a No. 12 copper conductor with plastic or polyethylene insulation rated 60°C would be approximately 3,000 A of fault or short-circuit current for one cycle.

question: Assuming that the CB (circuit breaker) will take at least one cycle to operate, would use of the conductor where exposed to 8,000 A violate Sec. 110-9 or 110-10? These sections state that overcurrent protection for conductors and equipment is provided for the purpose of opening the electrical circuit if the current reaches a value which will cause an excessive or dangerous temperature in the conductor or conductor insulation. The 8,000-A available fault current would seem to call for use of conductors with that rating of short-circuit withstand. This could mean that branch-circuit wiring from all 20-A CBs in this panelboard must be *No. 6 copper* (the next larger size suitable for an 8,000-A fault current).

answer: As noted in UL Standard 489, a CB is required to operate safely in a circuit where the available fault current is up to the short-circuit current value for which the breaker is rated. The CB must clear the fault without damage to the insulation of conductors of proper size for the rating of the CB. A UL-listed, 20-A breaker is, therefore, tested and rated to be used with 20-A rated wire (say, No. 12 THW) and will protect the wire in accordance with Sec. 240-3 when applied at a point in a circuit where the short-circuit current available does not exceed the value for which the breaker is rated. This is also true of a 15-A breaker on No. 14 (15-A) wire, for a 30-A breaker on No. 10 (30-A) wire, and all wire sizes.

UL 489 states:

> A circuit breaker shall perform successfully when operated under conditions as described in paragraphs 21.2 and 21.3. There shall be no electrical or mechanical breakdown of the device, and the fuse that is indicated in paragraph 12.16 shall not have cleared. Cotton indicators as described in paragraphs 21.4 and 21.6 shall not be ignited. There shall be no damage to the insulation on conductors used to wire the device. After the final operation, the circuit breaker shall have continuity in the closed position at rated voltage.

240-3. Protection of Conductors. Aside from flexible cords and fixture wires, conductors for all other circuits must conform to the rules of Sec. 240-3.

Clearly, the rule wants overcurrent devices to prevent conductors from being subjected to currents in excess of the ampacity values for which the conductors are rated by Tables 310-16 through 310-19 *with all notes.*

The wording mentions Sec. 310-15, which includes all notes to Tables 310-16 through 310-19. That is important because it points out that when conductors have their ampacities derated because of conduit fill (Note 8 to the tables) or because of elevated ambient temperature, the conductors must be protected at the *derated* ampacities and *not* at the values given in the tables.

Specifically, the general rule is that the device must be rated to protect conductors in accordance with their safe allowable current-carrying capacities. Of course, there will be cases where standard ampere ratings and settings of overcurrent devices will not correspond with conductor capacities. In such cases, part **(b)** permits the next larger standard size of overcurrent device to be used where the rating of the protective device is 800 A or less, unless the circuit in question is a multioutlet receptacle circuit for cord- and plug-connected portable loads, in which case the next smaller standard size *must* be used. Therefore, a basic guide to effective selection of the amp rating of overcurrent devices for any feeder or service application is given in various subsections [**(a)** through **(m)**].

For example, if a circuit conductor of, say, 500 kcmil THW copper (not more than three in a conduit at not over 86°F ambient) satisfies design requirements and NE Code rules for a particular load current not in excess of the conductor's table ampacity of 380 A, then the conductor *may* be protected by a 400-A rated fuse or CB.

Section 240-6, which gives the "Standard Ampere Ratings" of protective devices to correspond to the word "standard" in part **(b)**, shows devices rated at 350 and 400 A, but none at 380 A. In such a case, the NE Code accepts a 400-A rated device as "the next higher standard device rating" above the conductor ampacity of 380 A.

But, such a 400-A device would permit load increase above the 380 A that is the safe maximum limit for the conductor. Better conductor protection could be achieved by using a 350-A-rated device, which will prevent such overload.

For application of fuses and CBs, parts **(b)** and **(c)** have this effect:

1. If the ampacity of a conductor does not correspond to the rating of a standard-size fuse, the next *larger* rating of fuse may be used only where that rating is 800 A or less. Over 800 A, the next *smaller* fuse must be used, as covered in part **(c).** For any circuit over 800 A, Sec. 240-3(c) prohibits the use of "the next higher standard" rating of protective device (fuse or CB) when the ampacity of the circuit conductors does not correspond with a standard ampere rating of fuse or CB. The rating of the protection may not exceed the conductor ampacity. Although it would be acceptable to use a protective device of the next lower standard rating (from Sec. 240-6) below the conductor ampacity, there are many times when greater use of the conductor ampacity may be made by using a fuse or CB of rating lower than the conductor ampacity but not as low as the next lower standard rating. Listed fuses and CBs are made with ratings between the standard values shown in Sec. 240-6.

   For example, if the ampacity of conductors for a feeder circuit is calculated to be 1,540 A, Sec. 240-3(c) does not permit protecting such a conductor by using the next higher standard rating above 1,540–1,600 A. The next lower standard rating of fuse or CB shown in Sec. 240-6 is 1,200 A. Such protection could be used, but that would sacrifice 340 A (1,540 minus 1,200) of conductor ampacity. Because listed 1,500-A protective devices are available and would provide for effective use of almost all the conductor's 1,540-A capacity, this rule specifically recognizes such an application as safe and sound practice. Such application is specifically recognized by the fine print Note at the end of Sec. 240-6.

   In general, Sec. 240-6 is not intended to require that all fuses or CBs be of the standard ratings shown. Intermediate values of protective device ratings may be used, provided all **Code** rules on protection—especially the basic first sentence of Sec. 240-3, which requires conductors to be protected at their ampacities—are satisfied (Fig. 240-1).
2. A nonadjustable-trip breaker (one without overload trip adjustment above its rating—although it may have adjustable short-circuit trip) must be rated in accordance with the current-carrying capacity of the conductors it protects—except that the next higher standard rating of CB may be used if the ampacity of the conductor does not correspond to a standard unit rating. In such a case, the next higher standard setting may be used only where the rating is 800 A or less. An example of such application is shown in Fig. 240-1, where a nonadjustable CB with a rating of 1,200 A is used to protect the conductors of a feeder circuit which are rated at 1,140 A. As shown there, use of that size CB to protect a circuit rated at 1,140 A (3×380 A = 1,140 A) clearly violates Sec. 240-3(c) because the CB is the next higher rating above the ampacity of the conductors—on a circuit rated over 800 A. With a feeder circuit as shown (three 500 kcmil THW, each rated at 380 A), the CB must *not* be rated over 1,140 A. A standard 1,000-

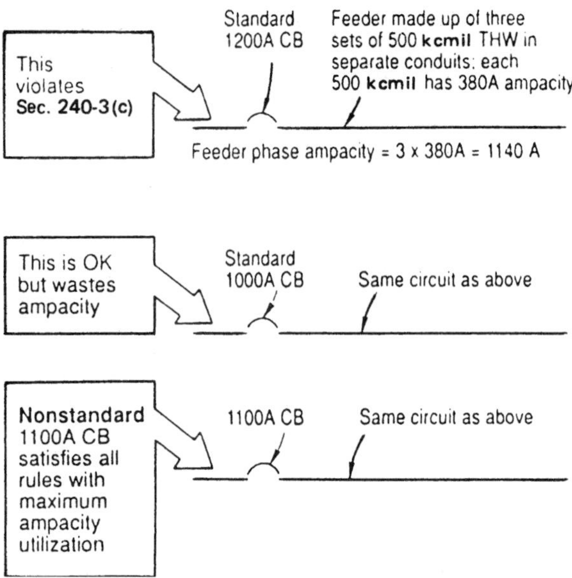

Fig. 240-1. Protection in accordance with Sec. 240-3(c) may use standard or nonstandard rated fuses or circuit breakers. (Secs. 240-3 and 240-6.)

A CB would satisfy the Code rule—being the *next lower* rated protective device from Sec. 240-6. Or a 1,100-A fuse could be used. Of course, if 500 kcmil THHN or XHHW conductors are used instead of THW conductors, then each 500 is rated at 430 A, three per phase would give the circuit an ampacity of 1,290 A (3 × 430), and the 1,200-A CB would satisfy the basic rule in Sec. 240-3(c).

It should be noted, however, that Sec. 240-3(b) requires that the rating of overcurrent protection must *never* exceed the ampacity of circuit conductors supplying one or more receptacle outlets on a branch circuit with more than one outlet. This wording in Sec. 240-3(b) coordinates with the rules described under Sec. 210-19(a) on conductor ampacity. The effect of that rule is to require that the rating of the overcurrent protection must not exceed the Code-table ampacity (NEC Table 310-16) or the derated ampacity dictated by Note 8 to the tables for any conductor of a multioutlet branch circuit supplying any receptacles for cord- and plug-connected portable loads. If a standard rating of fuse or CB does not match the ampacity (or derated ampacity) of such a circuit, the next lower standard rating of protective device must be used. *But,* where branch-circuit conductors of an individual circuit to a single load or a multioutlet circuit supply *only* fixed connected (hard-wired) loads—such as lighting outlets or permanently connected appliances—the next larger standard rating of protective device *may* be used in those cases where the ampacity (or derated ampacity) of the conductor does not correspond to a standard rating of protective device—but, again, that is permitted only up to 800 A, above which the next lower rating of fuse or CB must be used, as described under Sec. 210-19(a).

Section 240-3(f) refers the matter of protecting motor-control circuits to Art. 430 on motors.

Section 240-3(l) applies to the protection of the remote-control circuit that energizes the operating coil of a magnetic contactor, as distinguished from a magnetic motor starter (Fig. 240-2). Although it is true that a magnetic starter is a magnetic contactor with the addition of running overload relays, part **(l)** refers only to the coil circuit of any magnetic contactor but not to protection of the coil circuit of a magnetic starter.

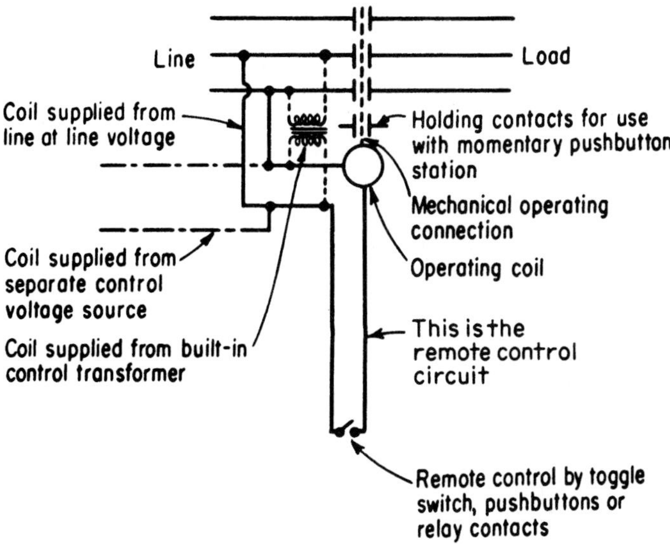

Fig. 240-2. Coil-circuit wires of magnetic contactor must be protected as required by Sec. 725-23. (Sec. 240-3.)

Section 725-23 covers control wires for magnetic contactors used for control of lighting or heating loads, but not motor loads. Section 430-72 covers that requirement for motor-control circuits. In Fig. 240-3, the remote-control conductors may be considered properly protected by the branch-circuit overcurrent devices (A) if these devices are rated or set at not more than 300 percent of (3 times) the current rating of the control conductors. If the branch-circuit overcurrent devices were rated or set at more than 300 percent of the rating of the control conductors, the control conductors would have to be protected by a separate protective device located at the point (B) where the conductor to be protected receives its supply. (See Sec. 725-23, Exception No. 3.)

Section 240-3(i) permits the secondary circuit from a transformer to be protected by means of fuses or a CB in the primary circuit to the transformer—*if* the transformer has no more than a 2-wire primary circuit *and* a 2-wire secondary.

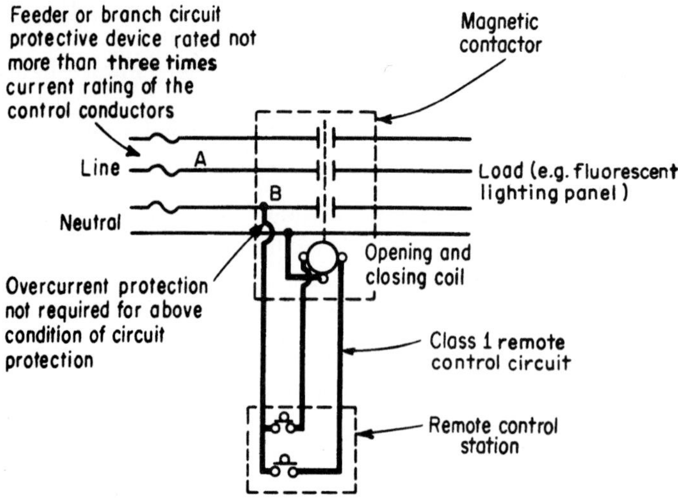

For instance, 30-amp fuses at A would be adequate protection if No. 14 wire, rated at 15 amps, is used for the remote-control circuit because 30 amps is less than 3 X 15 amps. If fuses at A were over 45 amps, then 15-amp protection would be required at B for No. 14 wire.

Fig. 240-3. Protecting a remote-control circuit in accordance with Sec. 725-23. (Sec. 240-3.)

As shown in Fig. 240-4, by using the 2-to-1 primary-to-secondary turns ratio of the transformer, 20-A primary protection will protect against any secondary current in excess of 40 A—thereby protecting, say, secondary No. 8 TW wires rated at 40 A. As the wording of the rule states, the protection on the primary (20 A) must not exceed the value of the secondary conductor ampacity (40 A) multiplied by the secondary-to-primary transformer voltage ratio (120 ÷ 240 = 0.5). Thus, 40 A × 0.5 = 20 A. But it should be carefully noted that the rating of the primary protection must comply with the rules of Sec. 450-3(i)(b)(1).

Section 240-3(i) clearly and emphatically states that the secondary conductors from a transformer may *not* be protected by overcurrent protection on the primary side of the transformer—*except* for a transformer with a *2-wire* sec-

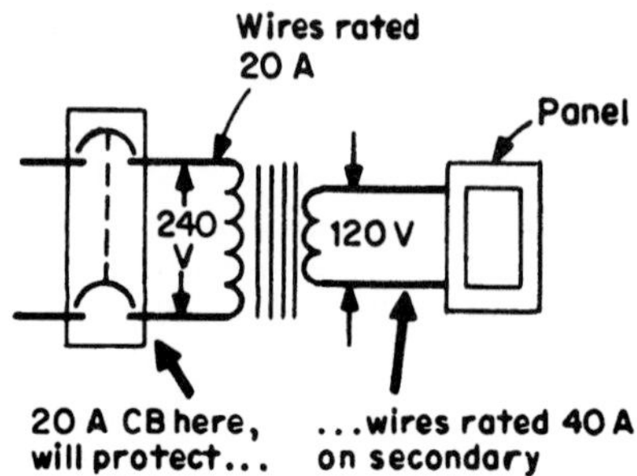

Fig. 240-4. Primary fuses or CB may protect secondary circuit for 2-wire to 2-wire transformer. (Sec. 240-3.)

ondary. That has long been the intent of the Code, but much discussion and controversy have regularly concentrated on this matter because the Code has not previously had the simple prohibition against secondary protection by a primary CB or set of fuses. The whole issue of transformers and overcurrent protection is now firmly established as follows.

The basic way to provide overcurrent protection for a dry-type transformer rated 600 V or less (with a rated primary current of more than 9 A) is to use fuses or CBs rated at not more than 125 percent of the transformer primary full-load current (TPFLC) to protect *both* the transformer and the circuit conductors that supply the transformer primary. [This is presented in Sec. 450-3(b)(1).] These circuit conductors must have an ampacity of not less than the rating of the overcurrent protection or must have an ampacity such that the overcurrent protective device is "the next higher standard device rating" above the conductor ampacity, as described in part **(b)** of Sec. 240-3. *But the primary circuit protection is not acceptable as suitable protection for the secondary conductors of a transformer with more than 2 wires on its secondary*—even if the secondary conductors have an ampacity equal to the ampacity of the primary conductors times the primary-to-secondary voltage ratio. Figure 240-5 covers these points.

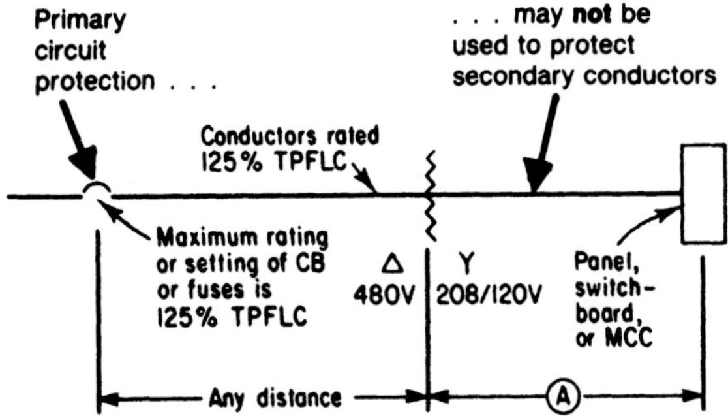

Distance "A" from transformer to first protection on the secondary side is limited to 10 or 25 ft, subject to the requirements of parts (b), (d) and (j) of Section 240-21. If overcurrent protection is placed at the transformer secondary connection to protect secondary conductors, the circuit can run any distance to the panel.

Fig. 240-5. Part **(b)** *clearly* resolves long-standing controversy. (Sec. 240-3.)

When primary devices are used for protection of 3- and 4-wire transformer secondaries, it is possible that an unbalanced load may greatly exceed the secondary conductor ampacity, which was selected assuming balanced conditions. As shown in Fig. 240-6, if the primary CB is set at 20 A, it will protect the primary No. 12 wires, which are rated for 20 A. Under conditions of full load of 20 A in the primary and a balanced secondary loading (left), the secondary current is 40 A, and the primary CB will protect the secondary No. 8 TW wires at their 40-A rating. But unbalance (right) can permit overloading of the secondary conductors without an increase in the primary current. Thus, the primary CB will not clear a 100 percent overload on the secondary wires.

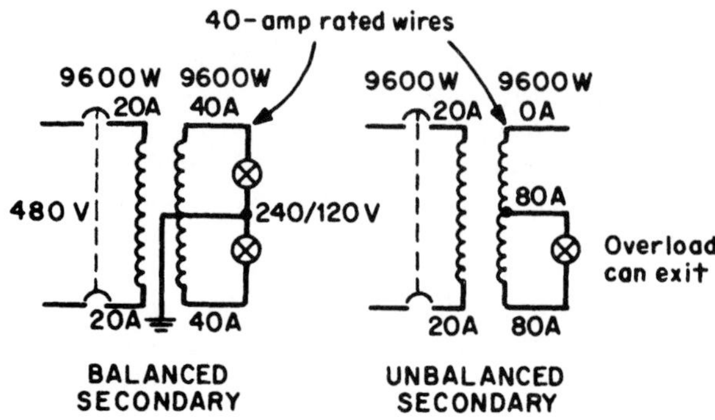

Fig. 240-6. Why primary protection may not do the job for 3-wire or 4-wire secondary 40-A rated wires. (Sec. 240-3.)

However, part **(i)** recognizes such primary protection of the secondary conductors of single-phase, 2-wire to 2-wire transformers if the primary overcurrent protection complies with Sec. 450-3 (e.g., not over 125 percent of rated primary current) and does not exceed the value determined by multiplying the secondary conductor ampacity by the secondary-to-primary transformer voltage ratio. In such applications, the lengths of the primary or secondary conductors are not limited.

The basic rule of Sec. 240-3(a) represents a basic concept in Code application. When conductors supply a load to which loss of power would create a hazard, this rule states it is not necessary to provide "overload protection" for such conductors, *but* "short-circuit protection" *must* be provided. By "overload protection," this means "protection at the conductors' ampacity"—i.e., protection that would prevent overload (Fig. 240-7).

Several points should be noted about this rule.

1. This requirement is reserved only for applications where circuit opening on "overload" would be more objectionable than the overload itself, "such as in a material handling magnet circuit." In that example mentioned in the rule, loss of power to such a magnet while it is lifting a heavy load of steel would cause the steel to fall and would certainly be a serious hazard to personnel working below or near the lifting magnet. To minimize the hazard created by such power loss, the circuit to it *need not* be protected at the conductor ampacity. A higher value of protection may be used—letting the circuit sustain an overload rather than opening on it and dropping the steel. Because such lifting operations are usually short-time, intermittent tasks, occasional overload is far less a safety concern than the dropping of the magnet's load.

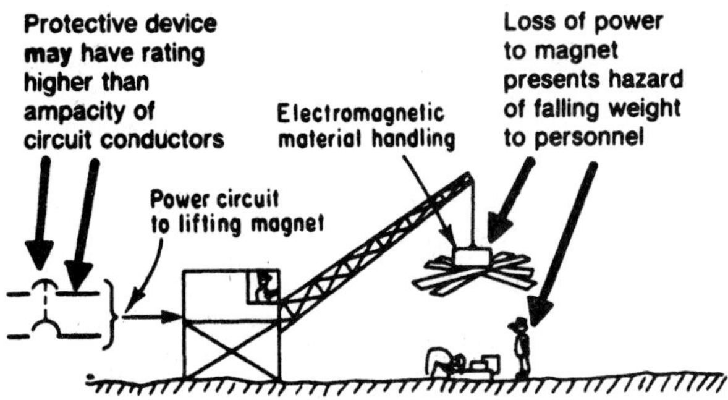
Protective device
may have rating
higher than
ampacity of
circuit conductors
Electromagnetic
material handling
Loss of power
to magnet
presents hazard
of falling weight
to personnel
Power circuit
to lifting magnet

Fig. 240-7. If "overload protection" creates a hazard, it *may* be eliminated. (Sec. 240-3.)

2. The rule to eliminate *only* "overload protection" is not limited to a lifting magnet circuit, which is mentioned simply as an example. Other electrical applications that present a similar concern for "hazard" would be equally open to use of this rule.
3. Although Sec. 240-3(a) *requires* elimination of overload protection and requires short-circuit protection, it gives no guidance on selecting the actual rating of protection that must be used. For such circuits, fuses or a CB rated, say, 200 to 400 percent of the full-load operating current would give freedom from overload opening. Of course, the protective device ought to be selected with as low a rating as would be compatible with the operating characteristics of the electrical load. And it must have sufficient interrupting capacity for the circuit's available short-circuit current.
4. Finally, it should be noted that this is *not* a mandatory rule but a *permissible* application. It says ". . . overload protection *shall not be required* . . ."; it does *not* say that overload protection "shall *not be used.*" Overload protection *may be used,* or it *may be eliminated.* Obviously, careful study should always go into application of this requirement.

240-4. Protection of Flexible Cords and Fixture Wires. The basic rules of the first paragraph are that

1. *All flexible cords and extension cords* must be protected at the ampacity given for each size and type of cord or cable in NEC Tables 400-5(A) and 400-5(B). "Flexible cords" includes "tinsel cord"—No. 27 AWG wires in a cord that is attached directly or by a special plug to a portable appliance rated not over 50 W.
2. *All fixture wires* must be protected in accordance with their ampacities, as given in Table 402-5.
3. The required protection may be provided by use of supplementary overcurrent protective devices (usually fuses), instead of having branch-circuit protection rated at the low values involved.

Then the basic rules are modified by an Exception applying to each of the above rules:

Exception No. 1 applies only to a flexible cord or a tinsel cord (not an "extension cord") that is "approved for and used with a specific *listed* (by UL or other recognized test lab) appliance or portable lamp." Such a cord, under the conditions stated, is not required to be protected at its ampacity from NEC Table 400-5 where it is

- Tinsel cord or No. 18 cord or larger, connected to a branch circuit rated not over 20 A
- No. 16 cord or larger, connected to a branch circuit rated not over 30 A
- Cord with 20 A or greater ampacity, connected on a circuit rated up to 50 A

Note that "extension cords" are *not* covered by Exception No. 1 because they are *not* "approved for and used with a specific listed appliance." As a result, No. 16 and No. 18 extension cords *must* be fused at their ampacity values (7, 10, 13 A, etc., from Table 400-5). *But,* Exception No. 3 of this section says that a "listed" extension cord set that contains No. 16 wire does not require protection at its amp rating and is considered to be protected by a 20-A branch-circuit fuse or CB. Extension cords with larger wire may also be used in any length on a 20-A branch circuit.

Exception No. 2 gives the conditions under which fixture wire does not have to be protected at the ampacity value given in Table 402-5 for its particular size *if* the fixture wire is any one of the following:

- No. 18 wire, not over 50 ft long, connected to a branch circuit rated not over 20 A
- No. 16 wire, not over 100 ft long, connected to a branch circuit rated not over 20 A
- No. 14 or larger wire, of any length, connected to a branch circuit rated not over 30 A
- No. 12 or larger wire, of any length, connected to a branch circuit rated not over 50 A

From those rules, No. 16 or No. 18 fixture wire may be connected on any 20-A branch circuit provided the "run length" (the length of any one of the wires used in the raceway) is not more than 50 ft—such as for 4- to 6-ft fixture whips [Sec. 410-67(c)]. *But,* for remote-control circuits run in a raceway from a magnetic motor starter or contactor to a remote pushbutton station or other pilot-control device, Secs. 430-72(b) and 725-12 require that a No. 18 wire be protected at not over 7 A and a No. 16 wire at not over 10 A—where fixture wires are used for remote-control circuit wiring, as permitted by Sec. 725-16(a) and (b).

The top sketch in Fig. 240-8 shows use of No. 18 or No. 16 fixture wire without need for separate protection when tapped from a 20-A branch circuit supplying recessed fixtures where higher temperature wire is required by Sec. 410-67(c), where, for instance, a fixture calls for 150°C wire to its hot terminal box. Of course, use of fixture wire must also satisfy all other Code rules that apply—as in Art. 400 on cords, Art. 402 on fixture wires, and Sec. 725-16 on fixture wires for control circuits.

In Exception No. 1, the tabulation shows minimum sizes of flexible cords that may be used on circuits rated at 20, 30, 40, and 50 A. A similar tabulation in Exception No. 2 for fixture wire eliminates cross-referencing Sec. 240-4 to amp ratings of tap conductors given in Exception No. 1 of Sec. 210-19(c).

According to Exception No. 3, any listed extension cord set using No. 16 gage wire is considered protected by 20-A branch-circuit protection (bottom of Fig.

Exception No. 2 —

Typical use of fixture wire: 4-to-6-ft length of flex for fixture whip in ceiling, containing two No. 18 Type AF wires (for 6-A fixture load, see Section 402-5) or two No. 16 Type AF wires (for 8-A fixture load). Section 240-4 permits No. 16 and No. 18 fixture wire to be protected at 20 A.

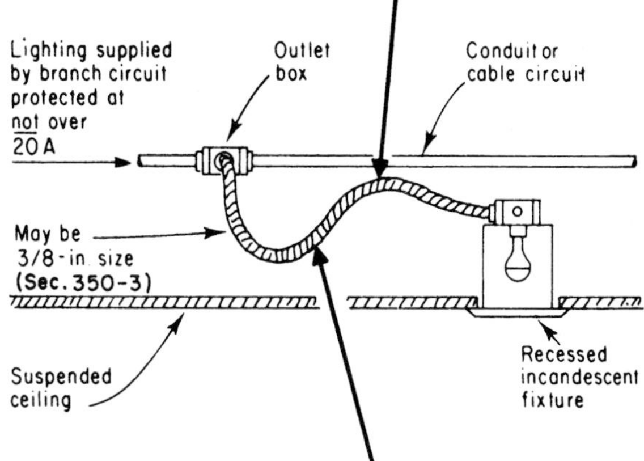

IMPORTANT!! Flex is equipment grounding conductor because AF wires in flex are tapped from circuit protected at not over 20-A as permitted in Section 250-91(b), Exception No. 1.

Exception No. 3 —

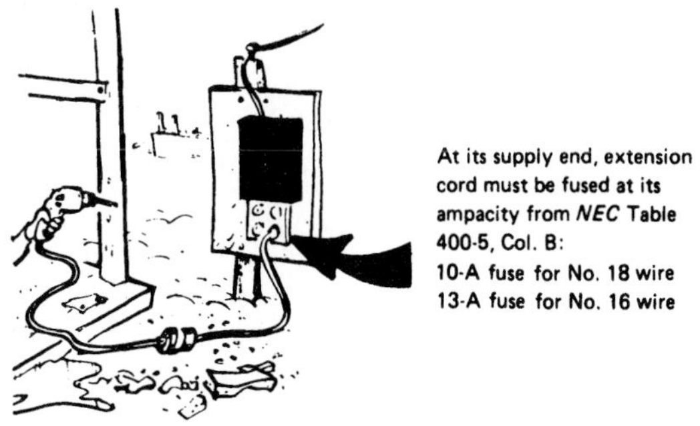

At its supply end, extension cord must be fused at its ampacity from *NEC* Table 400-5, Col. B:
10-A fuse for No. 18 wire
13-A fuse for No. 16 wire

NOTE: Exception No. 3 permits a "listed extension cord set that contains No. 16 wire to be connected to a receptacle on a branch circuit protected at 15 or 20 A — without need for separate protection for the cord set.

Fig. 240-8. Separate rules cover fixture wires and extension cords. (Sec. 240-4.)

240-8). The basic rule of this section requires that "extension cords" be protected against overcurrent in accordance with their ampacities as specified in Table 400-5. Based on that rule, manufactured extension cord sets with No. 18 wire must incorporate overcurrent protection in the plug cap of the set, with a rating of not over 10 A for the No. 18 wire, from Column B of Table 400-5. When such an extension cord is plugged into a receptacle, if an overload (above 10 A) is placed on the cord, the protection will open the circuit.

Although this section in the 1984 NEC required a similar protection (rated at 13 A) for No. 16 extension cord sets over 25 ft long, a revision of Exception No. 3 permits a No. 16 extension cord set of any length to be used without any special protection for the cord set—provided it is plugged into a receptacle or a branch circuit rated not over 20 A.

240-6. Standard Ampere Ratings. This is a listing of the "standard ampere ratings" of fuses and CBs for purposes of Code application. However, an important qualification is made by the fine-print note (FPN) at the end of this section. Although this NEC section designates "STANDARD ampere ratings" for fuses and circuit breakers, UL-listed fuses and circuit breakers of other intermediate ratings are available and may be used if their ratings satisfy Code rules on protection. For instance, Sec. 240-6 shows standard rated fuses at 1,200 A, then 1,600 A. But if a circuit was found to have an ampacity of, say, 1,530 A and, because Sec. 240-3(c) says such a circuit may not be protected by 1,600-A fuses, it is not necessary to drop down to 1,200-A fuses (the next lower standard size). This FPN fully intends to recognize use of 1,500-A fuses—which would satisfy the basic rule of Sec. 240-3(c) for protection rated over 800 A. (Fig. 240-1.)

The Exception to part **(a)** of Sec. 240-6 designates specific "additional standard ratings" of *fuses* at 1, 3, 6, 10, and 601 A. These values apply *only* to fuses and *not* to CBs. The 601-A rating gives Code recognition to use of Class L fuses rated less than 700 A. The reasoning of the Code panel was:

> An examination of fuse manufacturers' catalogs will show that 601 amperes is a commonly listed current rating for the Class L nontime-delay fuse. Section 430-52 (Exception No. 2d) also lists this current rating as a break point in application rules.
>
> Without a 601 ampere rating, the smallest standard fuse which can be used in Class L fuse clips is rated 700 amperes. Since the intent of Table 430-152 and Section 430-52 is to encourage closer short-circuit protection, it seems prudent to encourage availability and use of 601-ampere fuses in combination motor controllers having Class L fuse clips.
>
> Because ratings of inverse time circuit breakers are not related to fuse clip size, a distinction between 600 and 601 amperes in circuit breakers would serve no useful purpose. Hence, inverse-time circuit breaker ratings are listed separately. Such separation also facilitates recognition of other fuse ratings as standard.

The smaller sizes of fuses (1, 3, 6, and 10 A) listed as "standard ratings" provide more effective short-circuit and ground-fault protection for motor circuits—in accordance with Sec. 430-52, Sec. 430-40, and UL requirements for protecting the overload relays in controllers for very small motors. The Code panel reasoning was as follows:

> Fuses rated less than 15 amperes are often required to provide short circuit and ground-fault protection for motor branch circuits in accordance with Section 430-52.
>
> Tests indicate that fuses rated 1, 3, 6 and 10 amperes can provide the intended protection in motor branch circuits for motors having full load currents less than 3.75

> amperes (3.75 × 400% = 15). These ratings are also those most commonly shown on control manufacturers' overload relay tables. Overload relay elements for very small full load motor currents have such a high resistance that a bolted fault at the controller load terminals produces a short-circuit current of less than 15 amperes, regardless of the available current at the line terminals. An overcurrent protective device rated or set for 15 amperes is unable to offer the short circuit or ground fault protection required by Section 110-10 in such circuits.
>
> An examination of fuse manufacturers' catalogs will show that fuses with these ratings are commercially available. Having these ampere ratings established as standard should improve product availability at the user level and result in better overcurrent protection.
>
> Since inverse time circuit breakers are not readily available in the sizes added, it seems appropriate to list them separately.

Listing of those smaller fuse ratings has a significant effect on use of several small motors (fractional and small-integral-horsepower sizes) on a single branch circuit as described under Sec. 430-53(b).

Section 240-6(b) states that if a circuit breaker has external means for changing its continuous-current rating (the value of current above which the inverse-time overload—or long-time delay—trip mechanism would be activated), the breaker must be considered to be a protective device of the maximum continuous current (or overload trip rating) for which it might be set. This type of CB adjustment is available on molded-case, insulated-case, and air power circuit breakers. As a result of that rule and Sec. 240-3, the circuit conductors connected to the load terminals of such a circuit breaker must be of sufficient ampacity as to be properly protected by the maximum current value to which the adjustable trip might be set. That means that the CB rating must not exceed the ampacity of the circuit conductors, except that where the ampacity of the conductor does not correspond to a standard rating of CB, the next higher standard rating of CB may be used, up to 800 A (Fig. 240-9).

Prior to the 1987 edition, the NEC did not require that a circuit breaker with adjustable or changeable trip rating must have load-circuit conductors of an ampacity at least equal to the highest trip rating at which the breaker might be used. Conductors of an ampacity less than the highest possible trip rating could be used provided that the actual trip setting being used did protect the conductor in accordance with its ampacity, as required in NEC Sec. 240-3. Since the 1990 edition, such application may be made only in accordance with the Exception to this rule, which says that an adjustable-trip circuit breaker may be used as a protective device of a rating lower than its maximum setting and used to protect conductors of a corresponding ampacity in accordance with Sec. 240-3(b) *if* the trip-adjustment is

1. Located behind a removable and sealable cover, or
2. Part of a circuit breaker which is itself located behind bolted equipment enclosure doors accessible only to qualified persons or part of a circuit breaker that is locked behind doors (such as in a room) accessible only to qualified persons

Although this rule and its Exception permit use of conductors with ampacity lower than the maximum possible trip setting of a CB under the conditions given, this does not apply to fusible switches, and it is never necessary for a fusible switch to have its connected load-circuit conductors of ampacity equal

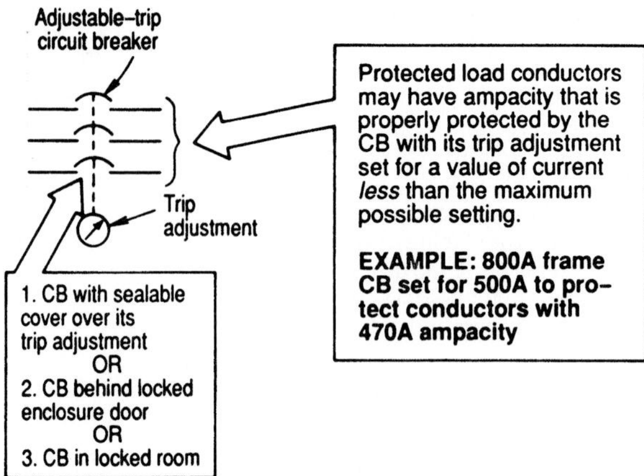

Adjustable-trip circuit breaker
Trip adjustment
Protected load conductors may have ampacity that is properly protected by the CB with its trip adjustment set for a value of current *less* than the maximum possible setting.
EXAMPLE: 800A frame CB set for 500A to protect conductors with 470A ampacity
1. CB with sealable cover over its trip adjustment
OR
2. CB behind locked enclosure door
OR
3. CB in locked room

Fig. 240-9. An adjustable-trip circuit breaker that has access to its trip-adjustment limited only to qualified persons may be taken to have a rating less than the maximum value to which the continuous rating (the long-time or over-load adjustment) might be set. (Sec. 240-6.)

to the maximum rating of a fuse that might be installed in the switch—provided that the actual rating of the fuse used in the switch does protect the conductor at its ampacity.

240-8. Fuses or Circuit Breakers in Parallel. The basic rule *prohibits* the use of parallel fuses, which at one time was acceptable when fused switches had ratings above 600 A. However, fused switches and *single fuses* (such as Class L) are now readily available in sizes up to 6,000 A. Moreover, this rule prohibits the use of CBs in parallel unless they are tested and approved as a single unit. At one time, this Code rule did not mention CBs. However, it is acceptable to factory-assemble CBs or fuses in parallel and have them tested and approved as a unit.

The Exception recognizes fuses or CBs in parallel where "factory assembled" and "listed as a unit." Such units are used to increase the rating of overcurrent protection in marine, over-the-road, off-road, commercial, and industrial installations. Use of other than listed units that are manufactured as units is a clear and direct violation.

240-10. Supplementary Overcurrent Protection. Supplementary overcurrent protection is commonly used in lighting fixtures, heating circuits, appliances, or other utilization equipment to provide individual protection for specific components within the equipment itself. Such protection is not branch-circuit protection and *the* NE Code *does not require supplemental overcurrent protective devices to be readily accessible.* Typical applications of supplemental overcurrent protection are fuses installed in fluorescent fixtures and cooking or heating equipment where the devices are sized to provide lower overcurrent protection than that of the branch circuit supplying such equipment. This is discussed under Secs. 424-19 and 424-22 on electric space heating equipment.

In application, Sec. 240-10 is frequently tied into the rule of Sec. 240-24. An example of this relationship is explained in the following comments of a consulting engineer:

> It has, for several years, been a common design practice in our office to specify the installation of fusible disconnect switches at motors located above suspended ceilings. As permitted by NEC Sec. 424, fuses and/or circuit breakers may be installed above suspended ceilings for the protection of fixed electric space heating equipment. But we specify fusible disconnects for air-handling equipment, with and without integral electric strip heaters and for electric duct heaters.
>
> Possibly the most frequently occurring condition is that of small fan and cooling units that have fractional horsepower (1/12 to 1/4 hp), 120-V, single-phase motors. We have routinely specified the installation of switch-and-plug fuse-combination units as the disconnecting means for the motor. We also routinely specify fusible disconnects for larger 3-phase fan-and-coil units mounted above suspended ceilings.

Past editions of the Code prohibited the fuses within the disconnects described in the comment from being considered as the required branch-circuit, short-circuit, and ground-fault protection because they are *not* readily accessible. As a result, many resorted to envoking the permission given in Sec. 240-10 as covering the fuse within the disconnect mounted in the overhead and additional overcurrent protection was provided to serve as the required protection. Now, Exception No. 4 to Sec. 240-24(a) allows overcurrent protection to be mounted in a nonreadily accessible location but only where adjacent to the equipment it *supplies.*

240-12. Electrical System Coordination. This rule applies to any electrical installation where hazard to personnel would result from disorderly shutdown of electrical equipment under fault conditions. The purpose of this rule is to permit elimination of "overload" protection—i.e., protection of conductors at their ampacities—and to eliminate unknown or random relation between operating time of overcurrent devices connected in series.

[As worded in previous NEC editions, this rule permitted elimination of overload protection for circuit conductors only for "industrial locations." Now, it may be applied in any electrical system (industrial, commercial, institutional, or residential).]

The section recognizes two requirements, both of which must be fulfilled to perform the task of "orderly shutdown."

One is selective coordination of the time-current characteristics of the short-circuit protective devices in series from the service to any load—so that, automatically, any fault will actuate only the short-circuit protective device closest to the fault on the line side of the fault, thereby minimizing the extent of electrical outage due to a fault.

The other technique that must also be included if *overload* protection is eliminated is "overload indication based on monitoring systems or devices." A note to this section gives brief descriptions of both requirements and establishes only a generalized understanding of "overload indication." Effective application of this rule depends upon careful design and coordination with inspection authorities.

It should be noted, however, that it says that the technique of eliminating overload protection to afford orderly shutdown "shall be permitted"—but does *not require* such application. Although it could be argued that the wording implies a mandatory rule, consultation with electrical inspection authorities on this matter is advisable because of the safety implications in nonorderly shutdown due to overload.

240-13. Ground-Fault Protection of Equipment. Equipment ground-fault protection—of the type required for 480Y/227-V service disconnects—is now required for each disconnect rated 1,000 A or more that serves as a main disconnect for a building or structure. Like Sec. 215-10, this section expands the application of protection against destructive arcing burndowns of electrical equipment. The intent is to equip a main building disconnect with GFP whether the disconnect is technically a service disconnect or a building disconnect on the load side of service equipment located elsewhere. This was specifically devised to cover those cases where a building or structure is supplied by a 480Y/277-V feeder from another building or from outdoor service equipment. Because the main disconnect (or disconnects) for such a building serves essentially the same function as a service disconnect, this requirement makes such disconnects subject to all of the rules of Sec. 230-95, covering GFP for services (Fig. 240-10).

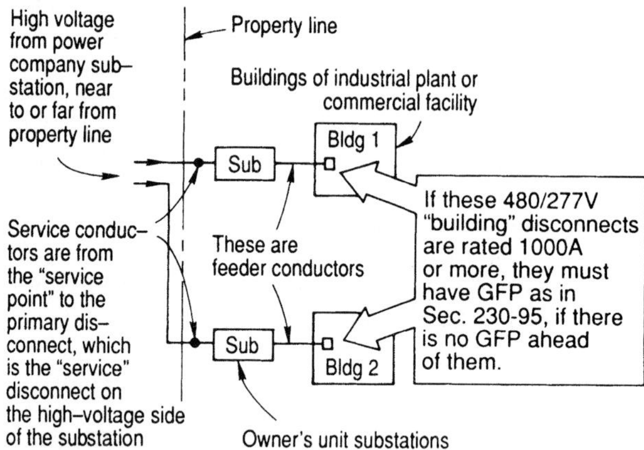

Fig. 240-10. Ground-fault protection is required for the feeder disconnect for each building—either at the building or at the substation secondary. (Sec. 240-13.)

The last part of this section is intended to clarify that the rule applies to the rating of *individual* disconnects and *not* to the sum of disconnects. Where an individual disconnect is rated 1000 A, or more, GFP protection must be provided.

The first exception here excludes the need for such GFP from disconnects for critical processes where automatic shutdown would introduce additional or different hazards. And as with service GFP, the requirement does not apply to fire-pump disconnects.

Exception No. 3 also suspends the need for GFP on a building or structure disconnect *if* such protection is provided on the upstream (line) side—either service or feeder disconnect GFP—of the building disconnect. The rule also stipulates that there must not be any desensitizing of the ground-fault protection because of neutral and/or grounding electrode connections that return current to the neutral on the load side of the service ground-fault sensing hookup.

240-20. Ungrounded Conductors. A fuse or circuit breaker must be connected in series with each ungrounded circuit conductor—usually at the supply end of the conductor. A current transformer and relay that actuates contacts of a CB is considered to be an overcurrent trip unit, like a fuse or a direct-acting CB (Fig. 240-11).

Although part **(b)** basically requires a CB to open all ungrounded conductors of a circuit simultaneously, parts **(1)**, **(2)**, and **(3)** cover acceptable uses of a number of single-pole CBs instead of multipole CBs.

The basic rule on use of single-pole versus multipole CBs is covered in this section.

Circuit breakers must open simultaneously all ungrounded conductors of circuits they protect; i.e., they must be multipole CB units, except that individual single-pole CBs may be used for protection of each ungrounded conductor of certain types of circuits, including 3-wire single-phase circuits, or lighting or appliance branch circuits connected to 4-wire, 3-phase systems provided that such lighting or appliance circuits are supplied from a grounded-neutral system and the loads are connected line-to-neutral. Figure 240-12 covers the rules.

Note: Two single-pole circuit breakers may not be used on "ungrounded 2-wire circuits"—such as 208-, 240-, or 480-V single-phase, 2-wire circuits. A 2-pole CB must be used if protection is provided by CBs. Use of single-pole CBs with handle ties but not common-trip is not allowed. This rule is intended to assure that a ground fault will trip open both conductors of an ungrounded 2-wire circuit derived from a grounded system. However, use of fuses for protection of such a circuit is permitted even though it will present the same chance of a fault condition as shown in Fig. 240-12.

Although 1-pole CBs may be used, as noted, it is better practice to use multipole CBs for circuits to individual load devices which are supplied by two or more ungrounded conductors. It is never wrong to use a multipole CB; but, based on the rules given here and in Sec. 210-4, it may be a violation to use single-pole CB units. A 3-pole CB must always be used for a 3-phase, 3-wire circuit supplying phase-to-phase loads fed from an ungrounded delta system, such as 480-V outdoor lighting for a parking lot, as permitted by Sec. 210-6(b).

Refer also to Sec. 210-4 for limitation on use of single-pole protective devices with line-to-neutral loads. And Sec. 110-3(b) requires that use of single-pole CBs be related to UL rules as described in Fig. 240-13.

Part **(c)** of this section excludes "closed-loop power distribution systems" from the need for fuse or circuit-breaker protection. In such systems (covered by NEC Art. 780), "listed devices" that provide equivalent protection may be used instead.

240-21. Location in Circuit. The basic rule of this section is shown in Fig. 240-14.

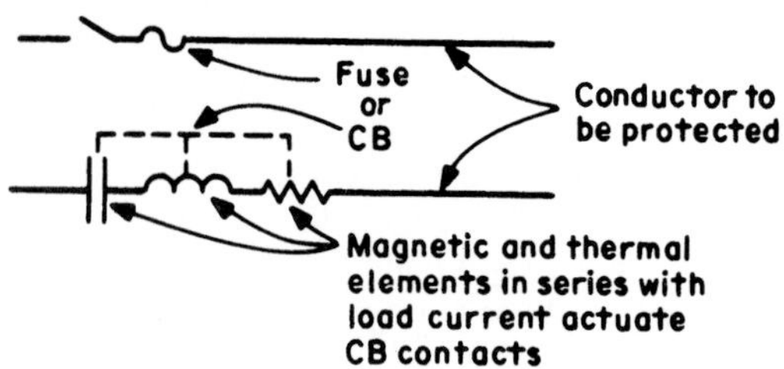

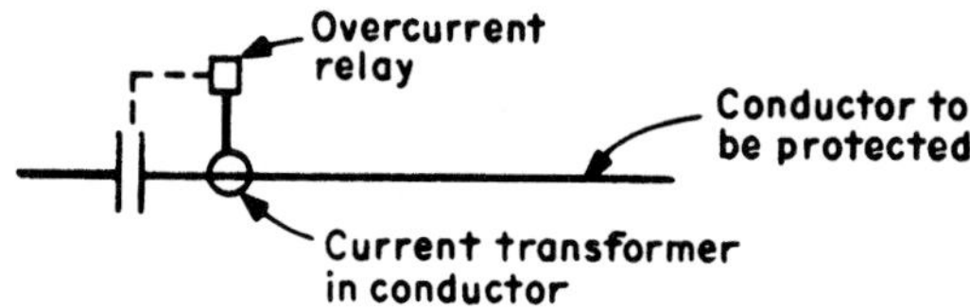

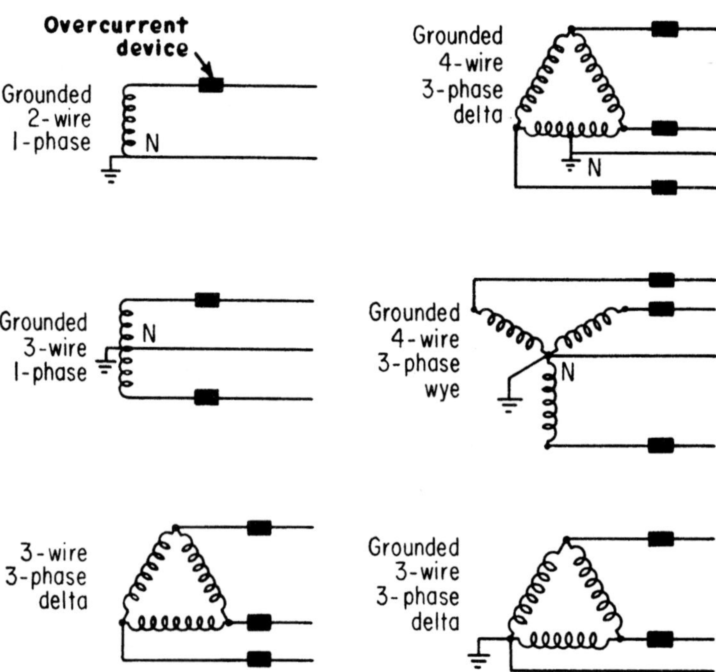

Fig. 240-11. A fuse or overcurrent trip unit must be connected in series with each ungrounded conductor. (Sec. 240-20.)

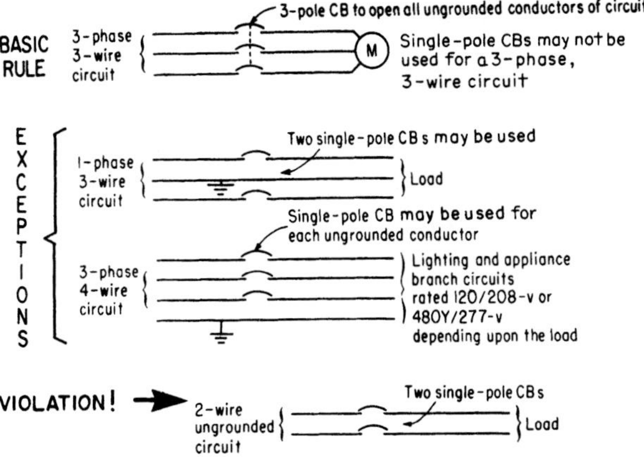

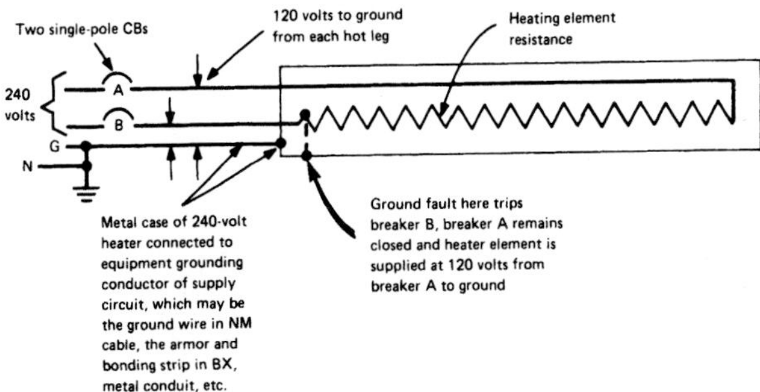

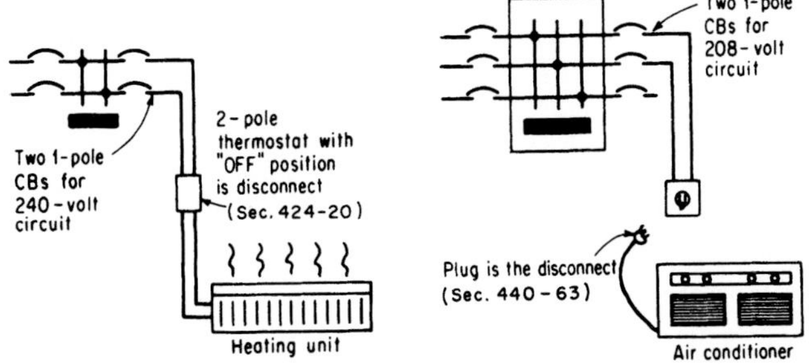

Fig. 240-12. Single-pole vs. multipole breakers. (Sec. 240-20.)

1. "Single-pole CBs rated 120 volts ac are suitable for use in a single-phase multiwire circuit where the neutral is connected to the load."

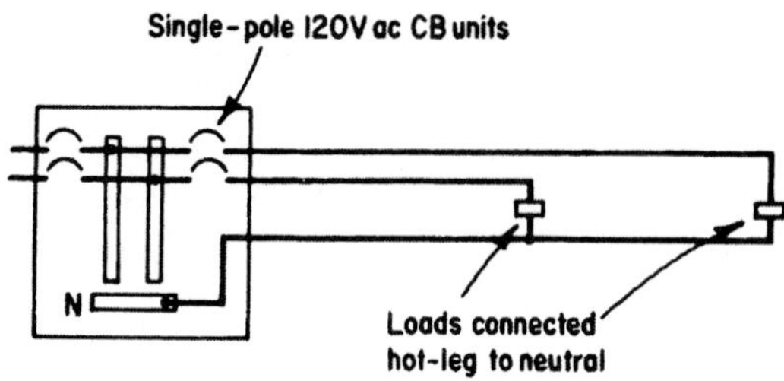

2. "Single-pole circuit breakers rated 120/240 volts ac are suitable for use in a single-phase multiwire circuit *with or without* the neutral connected to the load."

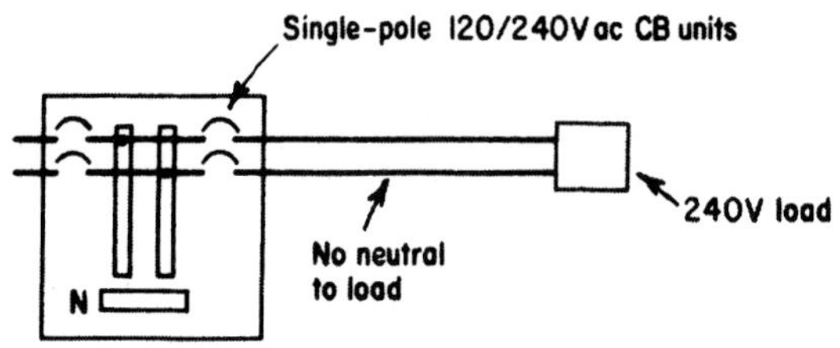

Fig. 240-13. NE Code rules must be correlated with the above UL requirements. (Sec. 240-20.)

Although basic Code requirements dictate the use of an overcurrent device at the point at which a conductor received its supply, subparts (b) through (n) effectively present exceptions to this rule in the case of taps to feeders. That is, to meet the practical demands of field application, certain lengths of unprotected conductors may be used to tap energy from protected feeder conductors.

These "exceptions" to the rule for protecting conductors at their points of supply are made in the case of 10-, 25-, and 100-ft (3.05-, 7.62-, and 30.5-m) taps from a feeder, as described in Sec. 240-21, parts **(b)**, **(d)**, and **(m)**. Application of the tap rules should be made carefully to effectively minimize any sacrifice in safety. The taps are permitted without overcurrent protective devices at the point of supply.

Section 240-21(b) says that unprotected taps not over 10 ft (3.05 m) long (Fig. 240-15) may be made from feeders or transformer secondaries provided:

BASIC RULE

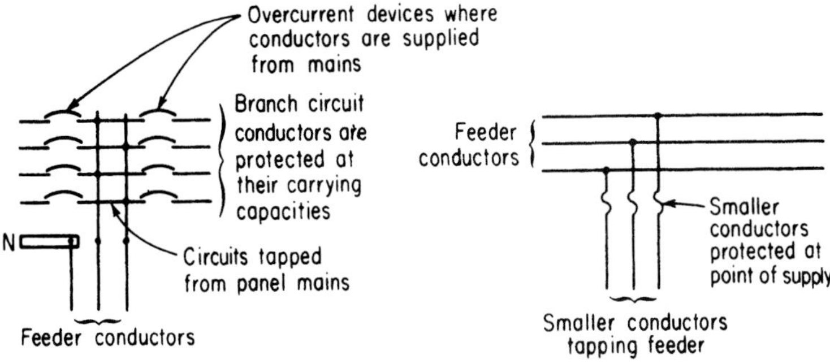

PROTECTING TWO SIZES OF CONDUCTORS

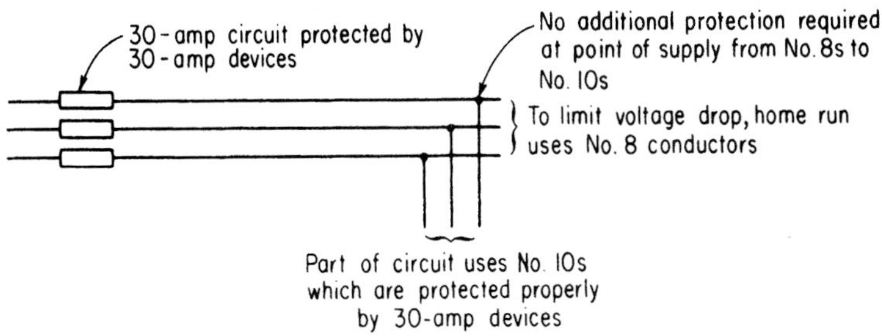

Fig. 240-14. Conductors must be protected at their supply ends. (Sec. 240-21.)

1. The smaller conductors have a current rating that is not less than the combined computed loads of the circuits supplied by the tap conductors and must have ampacity of—

   Not less than the rating of the "device" supplied by the tap conductors, or

   Not less than the rating of the overcurrent device (fuses or CB) that might be installed at the termination of the tap conductors.

   Important Limitation: For any 10-ft (3.05-m) unprotected feeder tap installed in the field, the rule limits its connection to a feeder that has protection rated *not* more than 1,000 percent of (10 times) the ampacity of the tap conductor where the tap conductors do not remain within the enclosure or vault in which the tap is made. Under the rule, unprotected No. 14 tap conductors are not permitted to tap a feeder any larger than 1,000 percent of the 20-A ampacity of No. 14 copper conductors—which would limit such a tap for use with a maximum feeder protective device of not over 10 × 20 A, or 200 A.

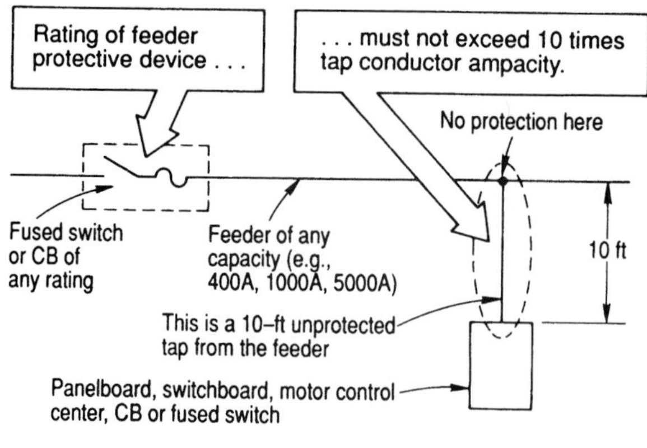
Rating of feeder protective device . . .
. . . must not exceed 10 times tap conductor ampacity.
No protection here
Fused switch or CB of any rating
Feeder of any capacity (e.g., 400A, 1000A, 5000A)
10 ft
This is a 10–ft unprotected tap from the feeder
Panelboard, switchboard, motor control center, CB or fused switch

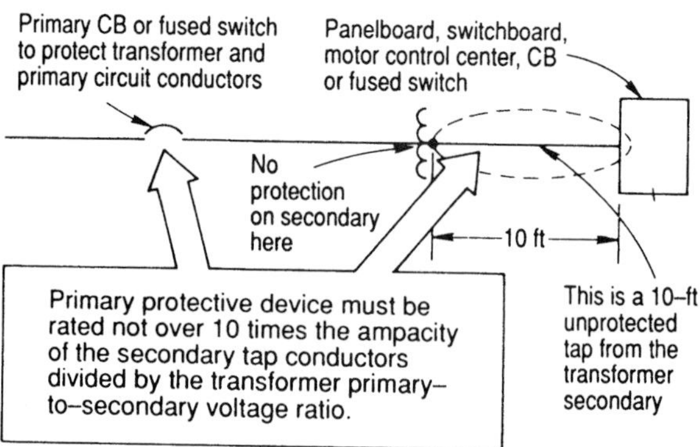
Primary CB or fused switch to protect transformer and primary circuit conductors
Panelboard, switchboard, motor control center, CB or fused switch
No protection on secondary here
10 ft
Primary protective device must be rated not over 10 times the ampacity of the secondary tap conductors divided by the transformer primary–to–secondary voltage ratio.
This is a 10–ft unprotected tap from the transformer secondary

EXAMPLE: If the above transformer is stepping 480V single phase down to 240/120V single phase, and the 10–ft tap conductors on the transformer secondary have an ampacity of 100A, the primary feeder CB must be rated not more than 10 times 100A divided by two (480/240), or 500A.

Fig. 240-15. Ten-foot taps may be made from a feeder or a transformer secondary. (Sec. 240-21.)

2. The tap does not extend beyond the switchboard, panelboard, disconnect, or control device which it supplies.
3. The tap conductors are enclosed in conduit, EMT, metal gutter, or other approved raceway when not a part of the switchboard or panelboard.

Section 240-21(b) specifically recognizes that a 10-ft (3.05-m) tap may be made from a transformer secondary in the same way it has always been permitted from a feeder. In either case, the tap conductors must not be over 10 ft (3.05

m) long and must have ampacity not less than the amp rating of the switchboard, panelboard, disconnect, or control device—or the tap conductors may be terminated in an overcurrent protective device rated not more than the ampacity of the tap conductors. In the case of an unprotected tap from a transformer secondary, the ampacity of the 10-ft (3.05-m) tap conductors would have to be related through the transformer voltage ratio to the size of the transformer primary protective device—which in such a case would be "the device on the line side of the tap conductors."

Taps not over 25 ft (7.62 m) long (Fig. 240-16) may be made from feeders, as noted in part **(c)** of Sec. 240-21, provided:

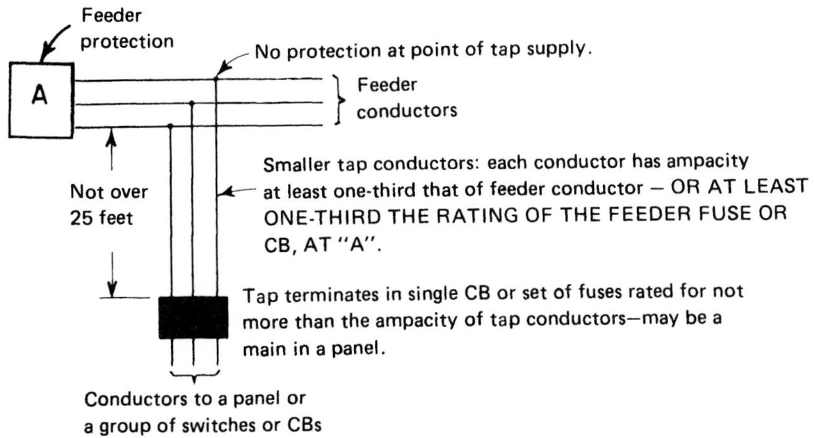

Fig. 240-16. Sizing feeder taps not over 25 ft (7.62 m) long. (Sec. 240-21.)

1. The smaller conductors have a current rating at least one-third that of the feeder overcurrent device rating or of the conductors from which they are tapped.
2. The tap conductors are suitably protected from mechanical damage. In previous **Code** editions, the 25-ft (7.62-m) feeder tap without overcurrent protection at its supply end simply had to be "suitably protected from physical damage"—which could accept use of cable for such a tap. Now, the rule requires such tap conductors to be "enclosed in a raceway"—just as has always been required for 10-ft (3.05-m) tap conductors.
3. The tap is terminated in a single CB or set of fuses which will limit the load on the tap to the ampacity of the tap conductors.

Examples of Taps

Figure 240-17 shows use of a 10-ft (3.05-m) feeder tap to supply a single motor branch circuit. The conduit feeder may be a horizontal run or a vertical run, such as a riser. If the tap conductors are of such size that they have a current rating at least one-third that of the feeder conductors (or protection rating) from which they are tapped, they could be run a distance of 25 ft (7.62 m) without protection at the point of tap-off from the feeder because they would comply with the rules of Sec. 240-21(c) which permit a 25-ft (7.62-m) tap if the conductors terminate in a single protective device rated not more than the conductor ampacity. Section 364-12 generally requires that any busway used as a feeder must have overcurrent protection on the busway for any subfeeder or branch circuit tapped from the busway. The use of a cable-tap box on busway without overcurrent protection (as shown in the conduit installation of Fig. 240-17) would usually be a violation. But, Exception No. 1 to Sec. 364-12 clearly eliminates such protection where making taps. Refer to Secs. 240-24 and 364-12.

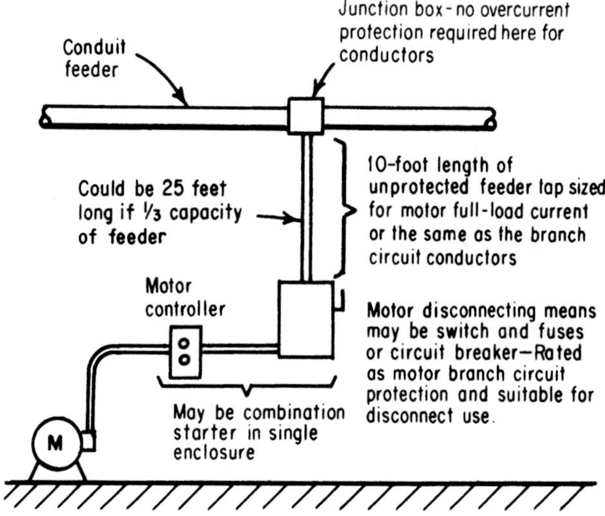

Fig. 240-17. A 10-ft (3.05-m) tap for a single motor circuit. (Sec. 240-21.)

A common application of the 10-ft (3.05-m) tap exception is the supply of panelboards from conduit feeders or busways, as shown in Fig. 240-18. The case shows an interesting requirement that arises from Sec. 384-16, which requires that lighting and appliance panelboards be protected on their supply side by overcurrent protection rated not more than the rating of the panelboard busbars. If the feeder is a busway, the protection must be placed (a requirement of Sec. 364-12) at the point of tap on the busway. In that case a 100-A CB or fused switch on the busway would provide the required protection of the panel, and the panel would not require a main in it. But, if the feeder circuit is in conduit, the 100-A panel protection would have to be in the panel or just ahead of it. It could not be at the junction box on the conduit because that would make it not readily accessible and therefore a violation of Sec. 240-24.

With a conduit feeder, a fused-switch or CB main in the panelboard could be rated up to the 100-A main rating. Some inspectors might accept installation of the overcurrent protection in the overhead because of the permission in Sec. 240-24(a), Exception No. 4, which now allows an OC device to be accessible by portable means where "adjacent to . . . equipment" it supplies. However, *most* do *not* believe such application is covered by Exception No. 4.

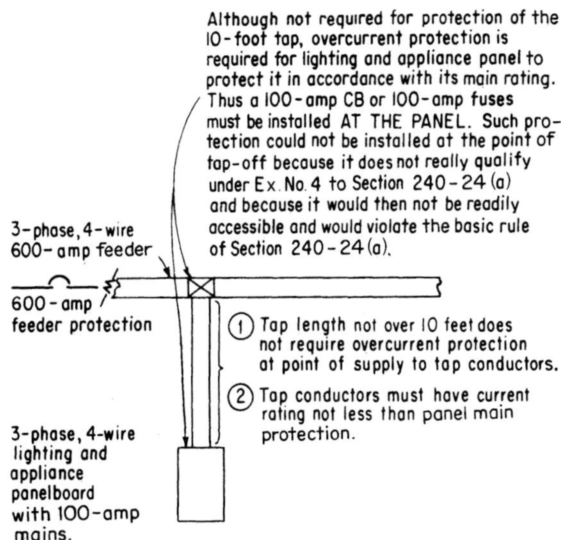

Although not required for protection of the 10-foot tap, overcurrent protection is required for lighting and appliance panel to protect it in accordance with its main rating. Thus a 100-amp CB or 100-amp fuses must be installed AT THE PANEL. Such protection could not be installed at the point of tap-off because it does not really qualify under Ex. No. 4 to Section 240-24 (a) and because it would then not be readily accessible and would violate the basic rule of Section 240-24 (a).
3-phase, 4-wire 600-amp feeder
600-amp feeder protection
1 Tap length not over 10 feet does not require overcurrent protection at point of supply to tap conductors.
2 Tap conductors must have current rating not less than panel main protection.
3-phase, 4-wire lighting and appliance panelboard with 100-amp mains.

Fig. 240-18. A 10-ft (3.05-m) tap to lighting panel with unprotected conductors. (Sec. 240-21.)

For transformer applications, typical 10- and 25-ft (3.05- and 7.62-m) tap considerations are shown in Fig. 240-19.

Figure 240-20 shows application of part **(d)** of Sec. 240-21 in conjunction with the rule of Sec. 450-3(b) (2), covering transformer protection. As shown in Example 1, the 100-A main protection in the panel is sufficient protection for the transformer and the primary and secondary conductors when these conditions are met:

1. Tap conductors have ampacity at least one-third that of the 125-A feeder conductors.
2. Secondary conductors are rated at least one-third the ampacity of the 125-A feeder conductors, based on the primary-to-secondary transformer ratio.
3. Total tap is not over 25 ft (7.62 m), primary plus secondary.
4. All conductors are in conduit.
5. Secondary conductors terminate in the 100-A main protection that limits secondary load to the ampacity of the secondary conductor and simultaneously provides the protection required by the lighting panel and is not rated over 125 percent of transformer secondary current.

10-FT TAP

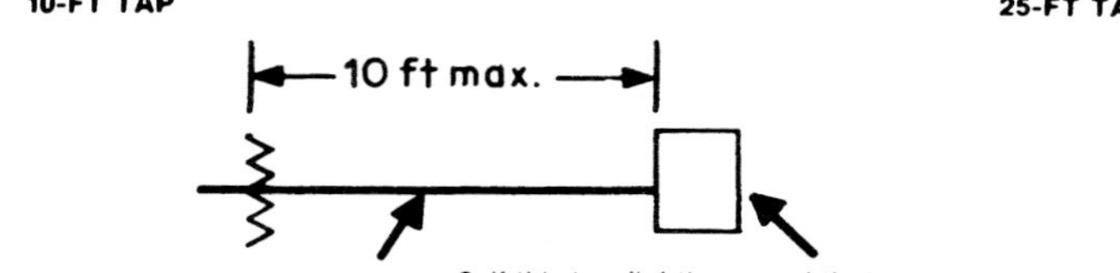

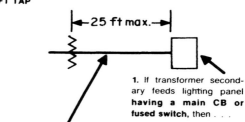

1. A 10-ft tap may be made from transformer secondary to a panel, switchboard, MCC, etc.

2. If this is a lighting panel that requires main protection, a fused switch or CB must be installed as a main protective device in the panel or just ahead of it, at the end of the 10-ft tap.

3. If the panel is *not* a lighting panel (such as a panel with 240-volt or 480-volt heating circuits or other makeup that does not make it a lighting panel as specified in Section 384-14), then main protection is not required at all, and the 10-ft tap conductors terminate in main lugs of the panel switchboard, or other equipment.

25-FT TAP

25 ft max.

1. If transformer secondary feeds lighting panel **having a main CB or fused switch,** then . . .

2. . . . secondary tap conductors from transformer may be 25 ft long, as permitted by Section 240-21(d), but *only* where the tap terminates in a single CB or set of fuses.

3. Or, a 25-ft tap may be made from a transformer to a CB or fused switch in an individual enclosure or serving as a main in a switchboard or MCC.

NOTE: From a single transformer secondary of adequate capacity, more than one set of 10-ft tap conductors may be run to more than one panel or other distribution equipment.

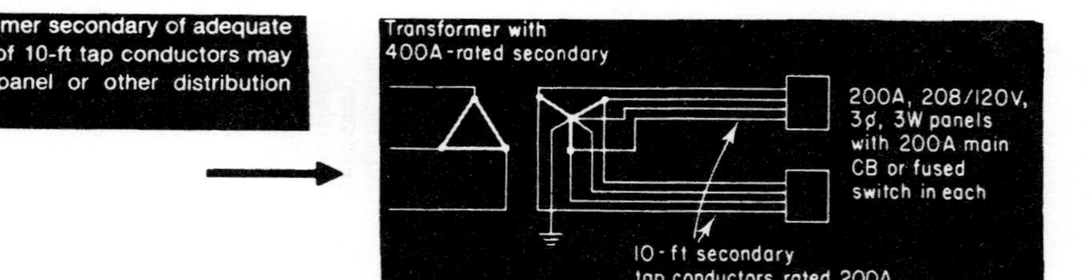

Fig. 240-19. Taps from transformer secondaries. (Sec. 240-21.)

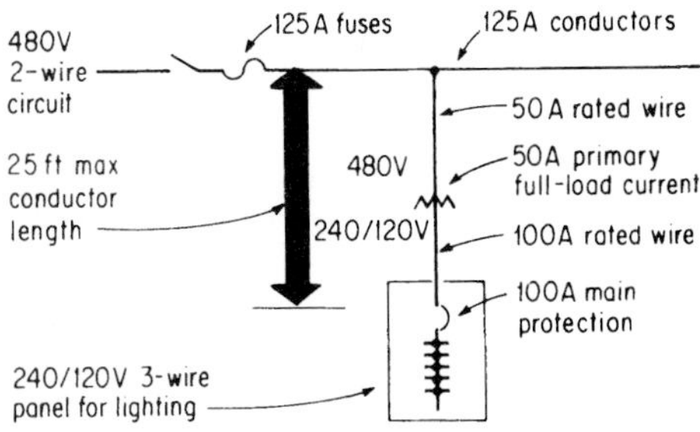

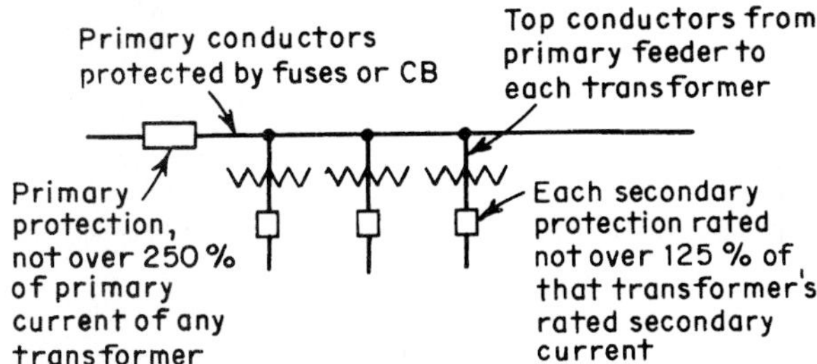

Fig. 240-20. Feeder tap of primary-plus-secondary not over 25 ft (7.62 m) long. (Sec. 240-21.)

6. Primary feeder protection is not over 250 percent of transformer rated primary current, as recognized by Sec. 450-3(b) (2), and the 100-A main breaker in the panel satisfies as the required "overcurrent device on the secondary side rated or set at not more than 125 percent of the rated secondary current of the transformer."

In Example 2, each set of tap conductors from the primary feeder to each transformer may be the same size as primary feeder conductors *or* may be smaller than primary conductors if sized in accordance with Sec. 240-21(d)—which permits a 25-ft (7.62-m) tap from a primary feeder to be made up of both primary and secondary tap conductors. The 25-ft (7.62-m) tap may have any part of its length on the primary or secondary but must not be longer than 25 ft (7.62 m) and must terminate in a single CB or set of fuses.

Figure 240-21 shows another example of Secs. 240-21(d) and 450-3(b) (2). Because the primary wires tapped to each transformer from the main 100-A feeder are also rated 100 A and are therefore protected by the 100-A feeder protection, all the primary circuit to each transformer is excluded from the allowable 25 ft (7.62 m) of tap to the secondary main protective device. The 100-A protection in each panel is not over 125 percent of the rated transformer secondary current. It, therefore, provides the transformer protection required by Sec. 450-3(b) (2). The same device also protects each panel at its main busbar rating of 100 A.

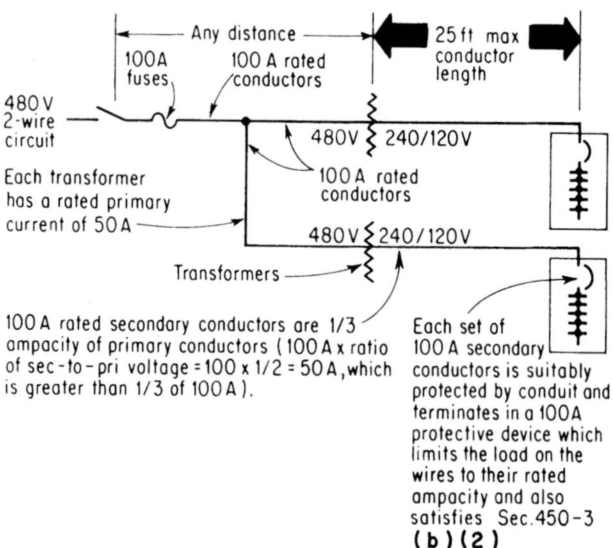

Fig. 240-21. Sizing a 25-ft (7.62-m) tap and transformer protection. (Sec. 240-21.)

Figure 240-22 compares the two different 25-ft (7.62-m) tap techniques covered by part **(c)** and part **(d)**.

As shown in Fig. 240-23, Sec. 240-21(i) gives permission for unprotected taps to be made from generator terminals to the first overcurrent device it supplies—such as in the fusible switch or circuit breakers used for control and protection of the circuit that the generator supplies. As the rule is worded, no maximum length is specified for the generator tap conductors. But because the tap conductors terminate in a single circuit breaker or set of fuses rated or set for the tap-conductor ampacity, tap conductors up to 25 ft (7.62 m) long would comply with the basic concept given in Sec. 240-21(c) for 25-ft (7.62-m) taps. And Sec. 445-5, which is referenced, requires the tap conductors to have an ampacity of at least 115 percent of the generator nameplate current rating.

Part **(e)** is another departure from the rule that conductors must be provided with overcurrent protection at their supply ends, where they receive current from other larger conductors or from a transformer. Sec. 240-21(e) permits a longer length than the 10-ft unprotected tap of part **(b)** and the 25-ft (7.62-m) tap of part **(c)** and **(d)**. Under specified conditions that are similar to the requirements of the 25-ft-tap exception, an unprotected tap up to 100 ft (30.5 m) in length may be used in "high-bay manufacturing buildings" that are over 35 ft (10.7 m) high *at the walls*—but only "where conditions of maintenance and supervision assure that only qualified persons will service the system." Obviously, that last phrase can lead to some very subjective and individualistic determinations by the authorities enforcing the Code. And the phrase "35 ft (10.7 m) high at the walls" means that this Exception cannot be applied where the height is over 35 ft (10.7 m) at the peak of a triangular or curved roof section but less than 35 ft (10.7 m) at the walls.

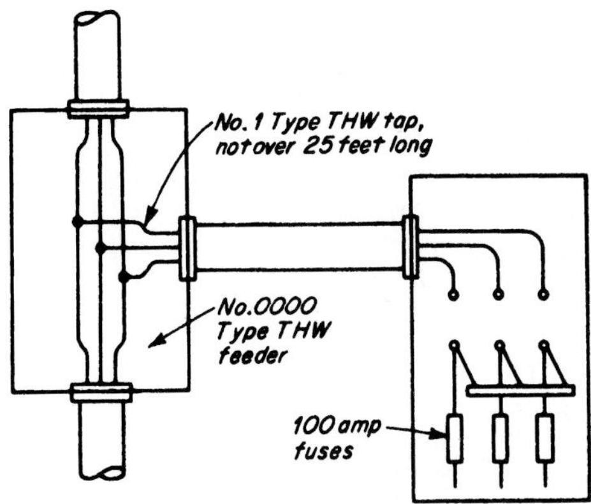

25-ft tap — Sec. 240-21 (c)

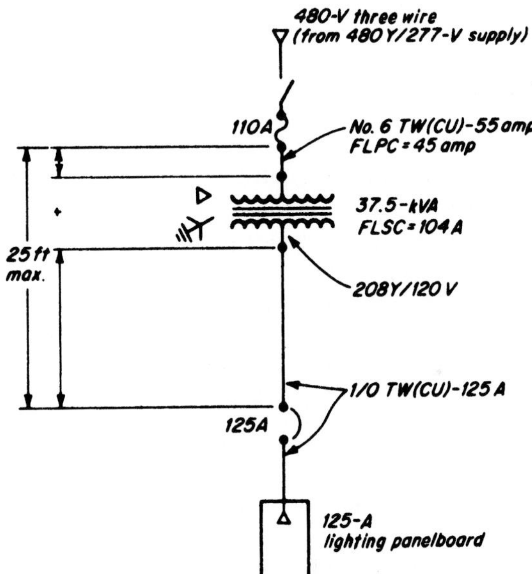

25-ft tap — Sec. 240-21 (d)

Fig. 240-22. Examples show difference between the two types of 25-ft (7.62-m) taps covered by parts **(c)** and **(d)**. (Sec. 240-21.)

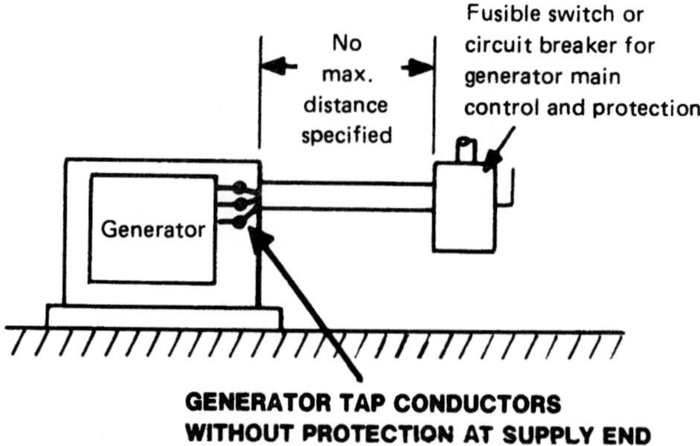

Fig. 240-23. Unprotected tap may be made from a generator's output terminals to the first overcurrent device. (Sec. 240-21(i).)

The 100-ft (30.5-m) tap exception must meet specific conditions:

1. From the point at which the tap is made to a larger feeder, the tap run must not have more than 25 ft (7.62 m) of its length run horizontally, and the sum of horizontal run and vertical run must not exceed 100 ft (30.5 m). Figure 240-24 shows some of the almost limitless configurations of tap layout that would fall within the dimension limitations.
2. The tap conductors must have an ampacity equal to at least one-third of the rating of the overcurrent device protecting the larger feeder conductors from which the tap is made.
3. The tap conductors must terminate in a circuit breaker or fused switch, where the rating of overcurrent protection is not greater than the tap-conductor ampacity.
4. The tap conductors must be protected from physical damage and must be installed in metal or nonmetallic raceway.
5. There must be no splices in the total length of each of the conductors of the tap.
6. The tap conductors must not be smaller than No. 6 copper or No. 4 aluminum.
7. The tap conductors must not pass through walls, floors, or ceilings.
8. The point at which the tap conductors connect to the feeder conductors must be at least 30 ft (9.14 m) above the floor of the building.

As shown in Fig. 240-24, the tap conductors from a feeder protected at 1,200 A are rated at not less than one-third the protection rating, or 400 A. Although 500 kcmil THW copper is rated at 380 A, that value does not satisfy the minimum requirement for 400 A. But if 500 kcmil THHN or XHHW copper, with an ampacity of 430 A, were used for the tap conductors, the rule would be satisfied. However, in such a case, those conductors would have to be used as if their ampacity were 380 A for the purpose of load calculation because of the general UL rule of 75°C conductor terminations for connecting to equipment rated over 100 A—such as the panelboard, switch, motor-control center or

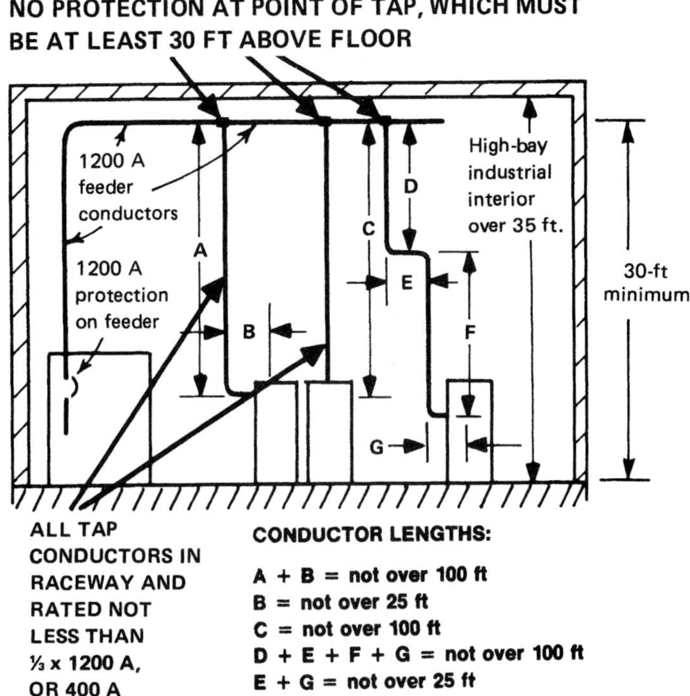

Fig. 240-24. Unprotected taps up to 100 ft long may be used in "high-bay manufacturing buildings."

other equipment fed by the taps. And the conductors for the main feeder being tapped could be rated less than the 1,200 A shown in the sketch if the 1,200-A protection on the feeder was selected in accordance with Secs. 430-62 or 430-63 for supplying a motor load or motor and lighting load. In such cases, the overcurrent protection may be rated considerably higher than the feeder conductor ampacity. But the tap conductors must have ampacity at least equal to one-third the *feeder protection rating.*

Section 240-21(j) applies exclusively to industrial electrical systems. Conductors up to 25 ft (6.72 m) long may be tapped from a transformer secondary without overcurrent protection at their supply end and without need for a single circuit breaker or set of fuses at their load end. Normally, a transformer secondary tap over 10 ft (3.05 m) long and up to 25 ft (6.72 m) long must comply with the rules of Sec. 240-21(c) and (d)—which call for such a transformer secondary tap to be made with conductors that require no overcurrent protection at their supply end but are required to terminate at their load end in a single CB or single set of fuses with a setting or rating not over the conductor ampacity. However, Sec. 240-21(j) permits a 10- to 25-ft (3.05- to 6.72-m) tap from a transformer secondary without termination in a single main overcurrent device—*but* it limits the application to "Industrial Installations." The tap conductor ampacity must be at least equal to the transformer's secondary current rating and must be at least equal to the sum of the ratings of overcurrent devices supplied by the tap conductors. The conductors could come into main-lugs-only of a power panel if the conductor ampacity is at least equal to the sum of the ratings of the protective devices supplied by the busbars in the panel. Or the tap could be made to an auxiliary gutter from which a number of individually enclosed circuit breakers or fused switches are fed from the tap—provided that the ampacity of the tap conductors is at least equal to the sum of the ratings of protective devices supplied.

An example of the application of 240-21(j) is shown at the top of Fig. 240-25. If that panel contains eight 100-A circuit breakers (or eight switches fused at 100 A), then the 25-ft (6.72-m) tap conductors must have an ampacity of at least 8 × 100 A, or 800 A. In addition, the tap conductor ampacity must be not less than the secondary current rating of the transformer. The layout of a similar application at an auxiliary gutter is shown at the bottom of Fig. 240-25.

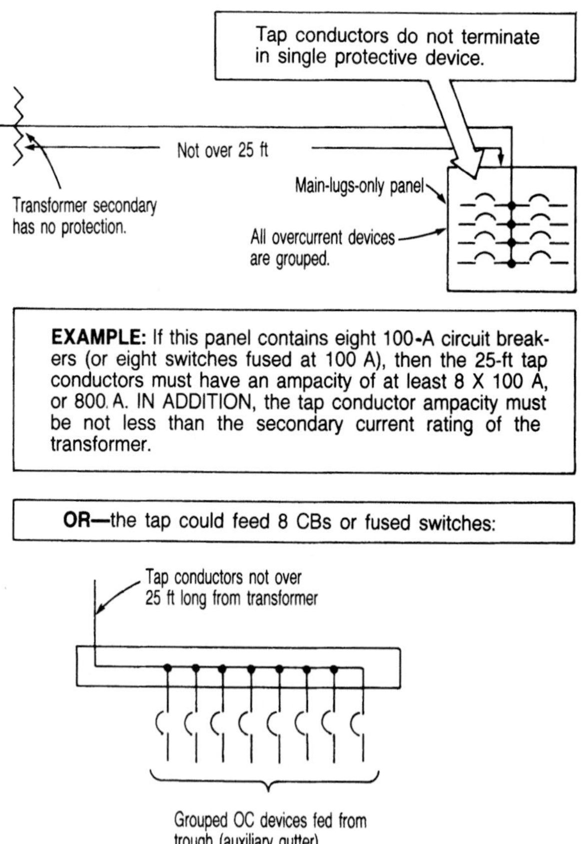

Fig. 240-25. These tap applications are permitted for transformer secondaries only in "industrial" electrical systems.

The rule of part **(m)** allows outdoor feeder taps and unprotected secondary conductors from outdoor transformers to run for any distance *outdoors.* Physical protection for the conductors must be provided and they must terminate in a single CB or set of fuses. The CB or set of fuses must be part of, or adjacent to, the disconnect, which may be installed anywhere outdoors or indoors as close as possible to the point of conductor entry. Part **(3)** emphasizes that such unprotected conductors must not be run within any building or structure. As is the case with service conductors, these tap conductors must be terminated at an OC device as soon as they enter.

240-22. Grounded Conductors. The basic rule prohibits use of a fuse or CB in any conductor that is intentionally grounded—such as a grounded neutral or a grounded phase leg of a delta system. Figure 240-26 shows the two exceptions to that rule and a clear violation of the basic rule.

240-23. Change in Size of Grounded Conductor. In effect, this recognizes the fact that if the neutral is the same size as the ungrounded conductor, it will be protected wherever the ungrounded conductor is protected. One of the most obvious places where this is encountered is in a distribution center where a small grounded conductor may be connected directly to a large grounded feeder conductor.

240-24. Location in or on Premises. According to part **(a)**, overcurrent devices must be readily accessible. And in accordance with the definition of "readily accessible" in Art. 100, they must be "capable of being reached quickly for operation, renewal, or inspections, without requiring those to whom ready access is requisite to climb over or remove obstacles or to resort to portable ladders, chairs, etc." (Fig. 240-27).

Although the Code gives no maximum heights at which overcurrent protective devices are considered readily accessible, some guidance can be obtained from Sec. 380-8, which provides detailed requirements for location of switches and CBs. This section states that switches and CBs shall be so installed that the center of the grip of the operating handle, when in its highest position, will not be more than 6 ft 7 in. (2.0 m) above the floor or working platform.

Exception No. 1 covers any case where an overcurrent device is used in a busway plug-in unit to tap a branch circuit from the busway. Section 364-12 requires that such devices consist of an externally operable CB or an externally operable fusible switch. These devices must be capable of being operated from the floor by means of ropes, chains, or sticks. Exception No. 2 refers to Sec. 240-10, which states that where supplementary overcurrent protection is used, such as for lighting fixtures, appliances, or internal circuits or components of equipment, this supplementary protection is not required to be readily accessible. An example of this would be an overcurrent device mounted in the cord plug of a fixed or semifixed luminaire supplied from a trolley busway or mounted on a luminaire that is plugged directly into a busway. Exception No. 3 acknowledges that Sec. 230-92 permits service overcurrent protection to be sealed, locked, or otherwise made not readily accessible. Figure 240-28 shows these details.

Section 240-24 clarifies the use of plug-in overcurrent protective devices on busways for protection of circuits tapped from the busway. After making the general rule that overcurrent protective devices must be readily accessible

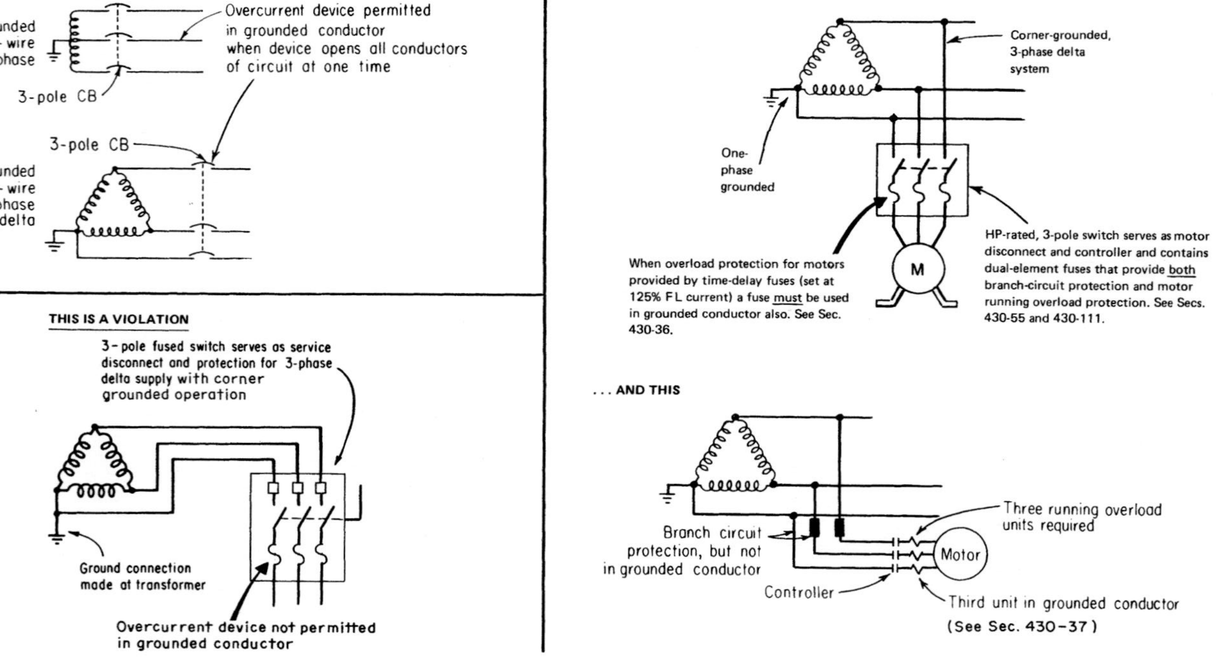

EXCEPTION NO. 1 PERMITS THIS
Grounded 3-wire 1-phase
3-pole CB
Overcurrent device permitted in grounded conductor when device opens all conductors of circuit at one time
3-pole CB
Grounded 3-wire 3-phase delta
THIS IS A VIOLATION
3-pole fused switch serves as service disconnect and protection for 3-phase delta supply with corner grounded operation
Ground connection made at transformer
Overcurrent device not permitted in grounded conductor
EXCEPTION NO. 2 RECOGNIZES THIS . . .
Corner-grounded, 3-phase delta system
One-phase grounded
M
When overload protection for motors provided by time-delay fuses (set at 125% FL current) a fuse must be used in grounded conductor also. See Sec. 430-36.
HP-rated, 3-pole switch serves as motor disconnect and controller and contains dual-element fuses that provide both branch-circuit protection and motor running overload protection. See Secs. 430-55 and 430-111.
. . . AND THIS
Branch circuit protection, but not in grounded conductor
Controller
Motor
Three running overload units required
Third unit in grounded conductor
(See Sec. 430-37)

Fig. 240-26. Overcurrent protection in grounded conductor. (Sec. 240-22.)

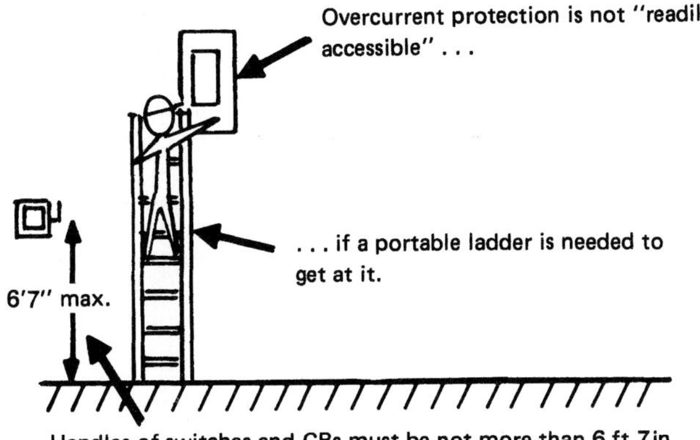
Overcurrent protection is not "readily accessible" . . .
. . . if a portable ladder is needed to get at it.
6'7" max.
Handles of switches and CBs must be not more than 6 ft 7 in. above floor or platform (Sec. 380-8).

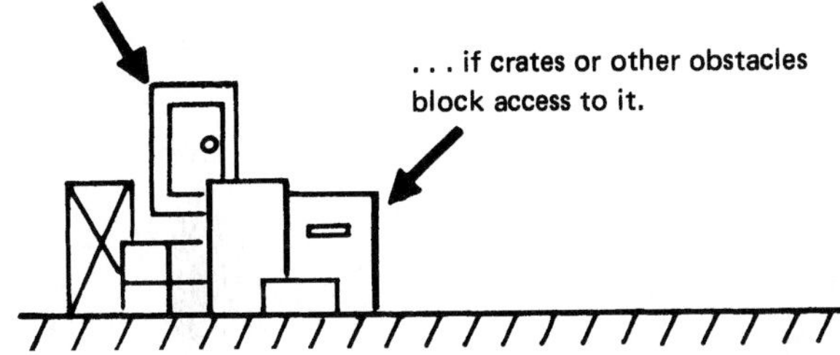
Overcurrent device in a panel, switch, CB, switchboard, MCC is not readily accessible . . .
. . . if crates or other obstacles block access to it.

Fig. 240-27. Overcurrent devices must be "readily accessible." (Sec. 240-24.)

(capable of being reached without stepping on a chair or table or resorting to a portable ladder), Exception No. 1 notes that it is not only *permissible* to use busway protective devices up on the busway—it is *required* by Sec. 364-12. Such devices on high-mounted busways are not "readily accessible" (not within reach of a person standing on the floor). The wording of Sec. 364-12 makes clear that this requirement for overcurrent protection in the device on the busway applies to subfeeders tapped from the busway as well as branch circuits tapped from the busway.

The rule of Exception No. 4 recognizes the installation of an OC device in an inaccessible location where mounted adjacent to "motors, appliances, or other equipment. . . ." The term "equipment" is defined in Art. 100. That definition seems to give broad permission for application of this Exception. It seems that locating OC devices for conductor protection would not be permitted. Clearly, for motors, appliances, and transformers, the OC device that *supplies* such "equipment" may be mounted in an inaccessible location. The rules of NE Code Secs. 240-24, 364-12, and 380-8 must be correlated with each other to assure effective Code compliance.

Part (a)

FEEDER TAP FROM BUSWAY

Plug-in connection for tapping off feeder or sub-feeder must contain overcurrent protection, which does not have to be within reach of person standing on floor

Busway feeder

Panelboard, switchboard, motor control center, or trough with two or more branch circuits tapped off

BRANCH-CIRCUIT TAP FROM BUSWAY

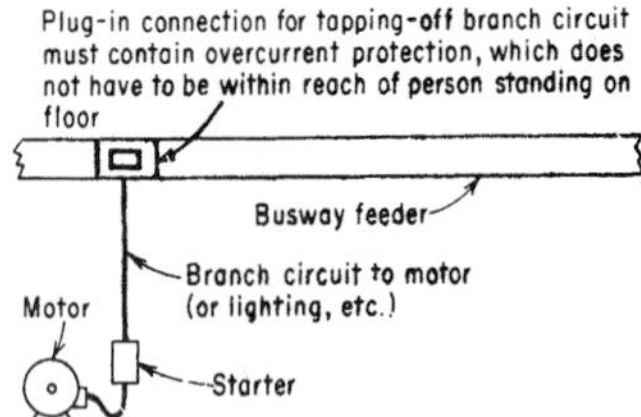

LUMINAIRE FED BY CORD FROM BUSWAY

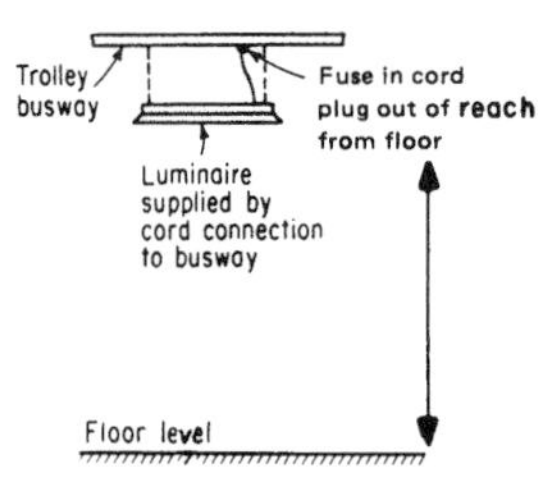

FOR SERVICE

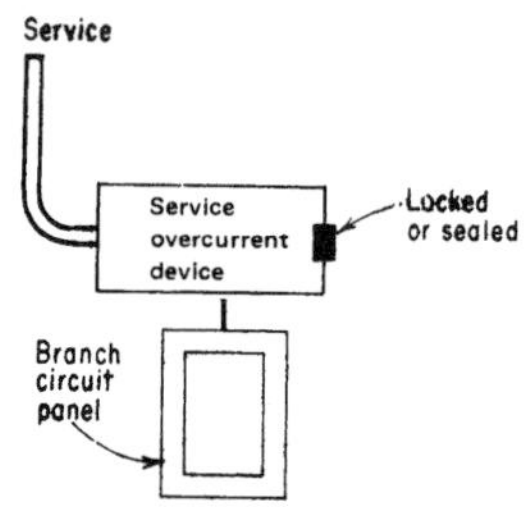

Part (b)

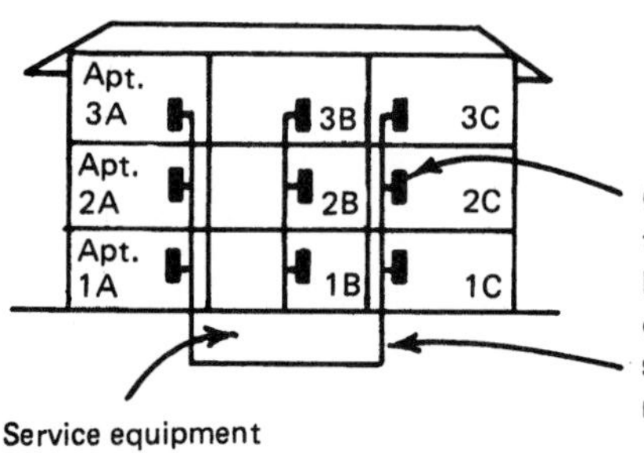

Fig. 240-28. Fuses or CBs that are permitted to be *not* readily accessible. (Sec. 240-24.)

Part **(b)** applies to apartment houses and other multiple-occupancy buildings, as described in Fig. 240-28.

In addition, it is important to note that parts **(c)** and **(d)** of Sec. 240-24 require that overcurrent devices be located where they will not be exposed to physical damage or in the vicinity of easily ignitible material. Panelboards, fused switches, and circuit breakers may *not* be installed in *clothes closets* in any type of occupancy—residential, commercial, institutional, or industrial. But they may be installed in other closets that do not have easily ignitible materials within them—provided that the working clearances of Sec. 110-16 [30-in. (762 mm) wide work space in front of the equipment, 6¼-ft (1.9-m) headroom, illumination, etc.] are observed and the work space is "not used for storage," as required by Sec. 110-16(b).

Section 240-24(e) flatly prohibits what was a somewhat common practice for dwellings. In certain areas of the nation, overcurrent protective devices were located in areas such as kitchens and bathrooms. Although it is still permissible to locate the overcurrent protective devices in the kitchen, the rule of part (e) now forbids location of the overcurrent devices within the bathroom.

240-33. Vertical Position. Figure 240-29 shows the basic requirements of Secs. 240-30, 240-32, and 240-33.

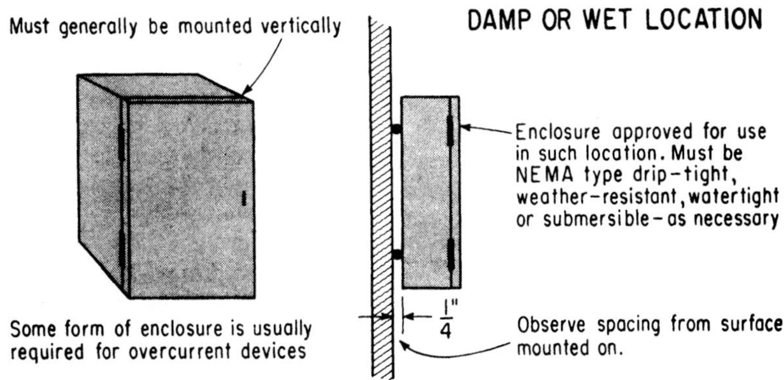

Fig. 240-29. Enclosures for overcurrent protection. (Sec. 240-30.)

240-40. Disconnecting Means for Fuses. The basic rules are shown in Fig. 240-30. Exception No. 2 is illustrated in Fig. 240-31.

240-50. General (Plug Fuses). Plug fuses must not be used in circuits of more than 125 V between conductors, but they may be used in grounded-neutral systems where the circuits have more than 125 V between ungrounded conductors but not more than 150 V between any ungrounded conductor and ground (Fig. 240-32). And the screw-shell of plug fuseholders must be connected to the load side of the circuit.

240-51. Edison-Base Fuses.

240-52. Edison-Base Fuseholders.

240-53. Type S Fuses.

240-54. Type S Fuses, Adapters, and Fuseholders. Rated up to 30 A, plug fuses are Edison-base or Type S. Section 240-51(b) limits the use of Edison-base fuses to replacements of existing fuses of this type. Type S plug fuses are required by Sec. 240-53 for all new plug-fuse installations. Type S plug fuses must be used in Type S fuseholders or in Edison-base fuseholders with a Type S adapter inserted, so that a Type S fuse of one ampere classification cannot be replaced with a higher-amp rated fuse (Fig. 240-33). Type S fuses, fuseholders, and

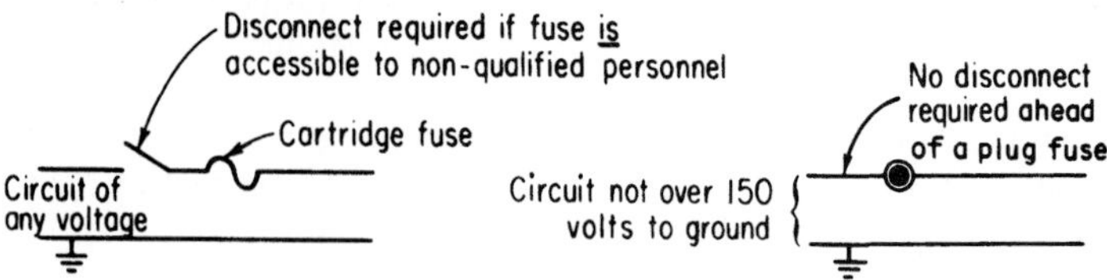

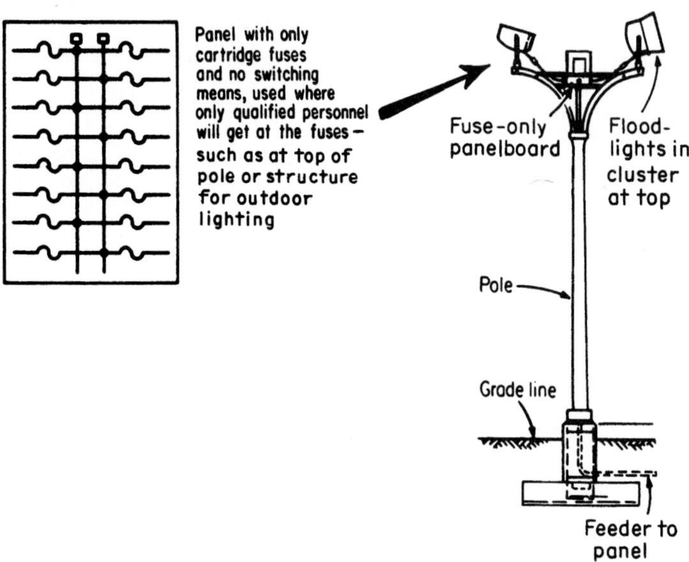

Fig. 240-30. Disconnect means for fuses. (Sec. 240-40.)

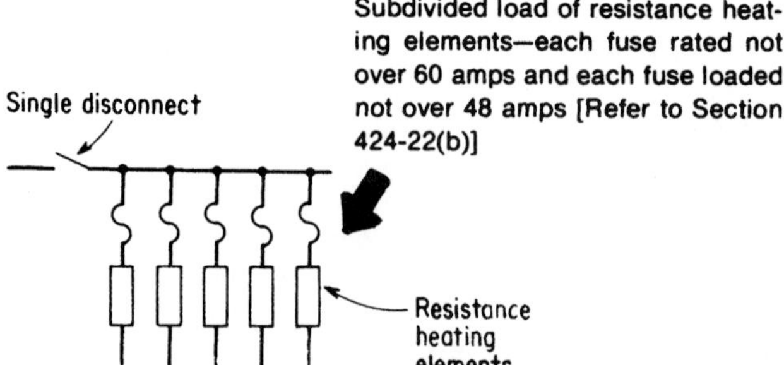

Fig. 240-31. Single disconnect for one set of fuses is permitted for electric space heating with subdivided resistance-type heating elements. (Sec. 240-40.)

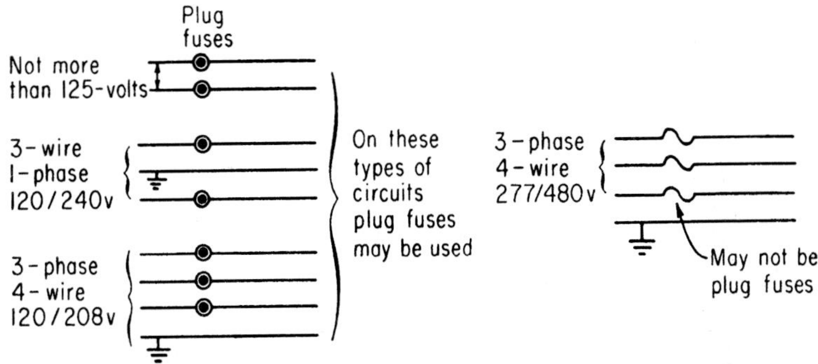

Fig. 240-32. Using plug fuses. (Sec. 240-50.)

adapters are rated for three classifications based on amp rating and are noninterchangeable from one classification to another. The classifications are 0–15, 16–20, and 21–30 A. The 0- to 15-A fuseholders or adapters must not be able to take any fuse rated over 20 A. The purpose of this rule is to prevent overfusing of 15- and 20-A circuits.

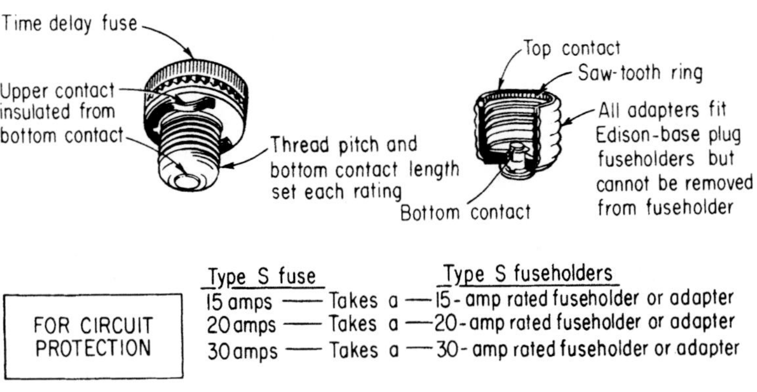

Fig. 240-33. Type S plug fuse. (Sec. 240-53.)

240-60. General (Cartridge Fuses). The last sentence of part **(b)** must always be carefully observed. It is concerned with an extremely important matter:

The installation of current-limiting fuses demands extreme care in the selection of the fuse clips to be used. Because current-limiting fuses have an additional protective feature (that of current limitation, i.e., extremely fast operation to prevent the flow of the extremely high currents which many modern circuits can produce into a ground fault or short circuit) as compared to noncurrent-limiting fuses, some condition of the mounting arrangement for current-limiting fuses must prevent replacement of the current-limiting fuses by noncurrent-limiting. This is necessary to maintain safety in applications where, for example, the busbars of a switchboard or motor control center are braced in accordance with the maximum let-through current of current-limiting fuses which protect the busbars, but would be exposed to a much higher potential value of fault let-through current if noncurrent-limiting fuses were used to replace the current-limiting fuses. The possibility of higher current flow than that for which the busbars are braced is created by the lack of current limitation in the noncurrent-limiting fuses.

Section 240-60(b) takes the above matter into consideration when it rules that "fuseholders for current-limiting fuses shall not permit insertion of fuses that are not current limiting." To afford compliance with the Code and to obtain the necessary safety of installation, fuse manufacturers provide current-limiting fuses with special ferrules or knife blades for insertion only in special fuse clips. Such special ferrules and blades do permit the insertion of current-limiting fuses into standard NEC fuse clips, to cover those cases where current-limiting fuses (with their higher type of protection) might be used to replace noncurrent-limiting fuses. But the special rejection-type fuseholders will not accept noncurrent-limiting fuses—thereby assuring replacement only with current-limiting fuses.

The very real problem of Code compliance and safety is created by the fact that many fuses with standard ferrules and knife-blade terminals are of the current-limiting type and are made in the same construction and dimensions as corresponding sizes of noncurrent-limiting fuses, for use in standard fuseholders. Such current-limiting fuses are not marked "current limiting" but may be used to obtain limitation of energy let-through. Replacement of them by standard nonlimiting fuses could be hazardous.

Class J and L fuses Both the Class J (0–600 A, 600 V AC) and Class L (601–6,000 A, 600 V AC) fuses are current-limiting, high-interrupting-capacity types. The interrupting ratings are 100,000 or 200,000 rms symmetrical amperes, and the designated rating is marked on the label of each Class J or L fuse. Class J and L fuses are also marked "current limiting," as required in part **(c)** of Sec. 240-60.

Class J fuse dimensions are different from those for standard Class H cartridge fuses of the same voltage rating and ampere classification. As such, they will require special fuseholders that will not accept noncurrent-limiting fuses. This arrangement complies with the last sentence of NEC Sec. 240-60(b).

Class K fuses These are subdivided into Classes K-1, K-5, and K-9. Class K fuses have the same dimensions as Class H (standard NE Code) fuses and are interchangeable with them. Classes K-1, K-5, and K-9 fuses have different degrees of current limitation but are not permitted to be labeled "current limiting" because physical characteristics permit these fuses to be interchanged with noncurrent-limiting types. Use of these fuses, for instance, to protect equipment busbars that are braced to withstand 40,000 A of fault current at a point where, say, 60,000 A of current would be available if noncurrent-limiting fuses were used is a clear violation of the last sentence of part **(b)**. As shown in Fig. 240-34, because such fuses can be replaced with nonlimiting fuses, the equipment bus structure would be exposed to dangerous failure. Classes R and T have been developed to provide current limitation and prevent interchangeability with noncurrent-limiting types.

Fuseholders in main switch of motor
control center require Class K-1 fuses
for current limitation to protect busbars.
Fuseholders furnished permit replacement of K-1
fuses with non-current-limiting type.

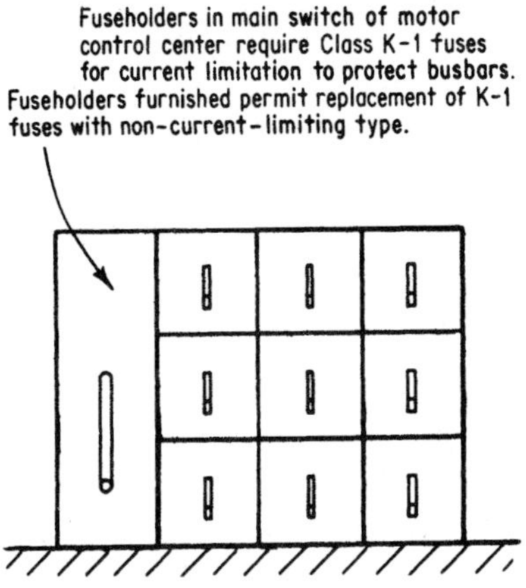

Fig. 240-34. Current-limiting fuseholders must be rejection type. (Sec. 240-60.)

Class R fuses These fuses are made in two designations: RK1 and RK5. UL data are as follows:

Fuses marked "Class RK1" or "Class RK5" are high-interrupting-capacity types and *are* marked "current limiting." Although these fuses will fit into standard fuseholders that take Class H and Class K fuses, special rejection-type fuseholders designed for Class RK1 and RK5 fuses will not accept Class H and Class K fuses. In that way, circuits and equipment protected in accordance with the characteristics of RK1 or RK5 fuses cannot have that protection reduced by the insertion of other fuses of a lower protective level.

Other UL application data that affect selection of various types of fuses are as follows:

Fuses designated as Class CC (0–20 A, 600 V AC) are high-interrupting-capacity types and are marked "current limiting." They are not interchangeable with fuses of higher voltage or interrupting rating or lower current rating.

Class G fuses (0–60 A, 300 V AC) are high-interrupting-capacity types and are marked "current limiting." They are not interchangeable with other fuses mentioned above and below.

Fuses designated as Class T (0–600 A, 250 and 600 V AC) are high-interrupting-capacity types and are marked "current limiting." They are not interchangeable with other fuses mentioned above.

Part **(c)** requires use of fuses to conform to the marking on them. Fuses that are intended to be used for current limitation must be marked "current limiting."

Class K-1, K-5, and K-9 fuses are marked, in addition to their regular voltage and current ratings, with an interrupting rating of 200,000, 100,000, or 50,000 A (rms symmetrical). (See Fig. 240-35.)

Class CC, RK1, RK5, J, L, and T fuses are marked, in addition to their regular voltage and current ratings, with an interrupting rating of 200,000 A (rms symmetrical).

Although it is not required by the Code, manufacturers are in a position to provide fuses that are advertised and marked indicating they have "time-delay"

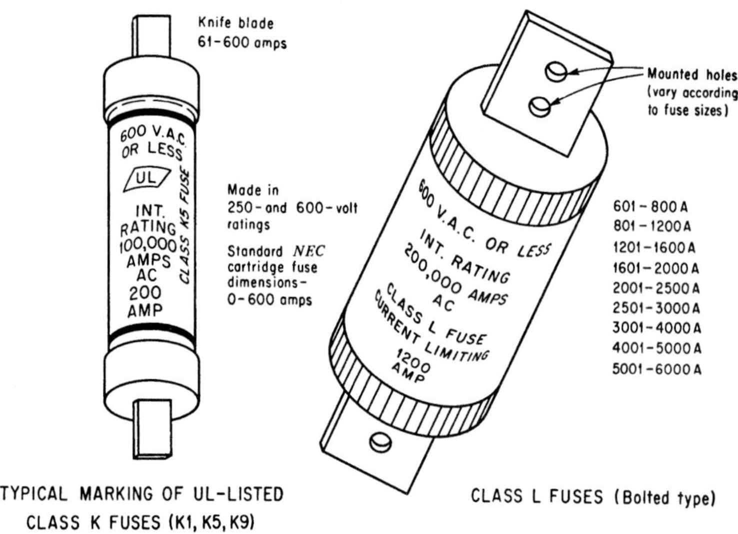

Fig. 240-35. Fuses must be applied in accordance with marked ratings. (Sec. 240-60.)

characteristics. In the case of Class CC, Class G, Class H, Class K, and Class RK fuses, time-delay characteristics of fuses (minimum blowing time) have been investigated. Class G or CC fuses, which can carry 200 percent of rated current for 12 sec or more, and Class H, Class K, or Class RK fuses, which can carry 500 percent of rated current for 10 sec or more, may be marked with "D," "time delay," or some equivalent designation. Class L fuses are permitted to be marked "time delay" but have not been evaluated for such performance. Class J and T fuses are not permitted to be marked "time delay."

240-61. Classification. This section notes that any fuse may be used at its voltage rating or at any voltage below its voltage rating.

240-80. Method of Operation (Circuit Breakers). This rule requiring trip-free manual operation of circuit breakers ties in with that in Sec. 230-76, although this rule requires manual operation to *both* the closed and the open positions of the CB. According to the Exception, a power-operated circuit breaker used as a service disconnect means must be capable of being opened by hand but does not have to be capable of being closed by hand. The general rule of Sec. 240-80 requires circuit breakers to be "capable of being *closed and opened* by manual operation." That rule also says that if a CB is electrically or pneumatically operated, it must also provide for manual operation. This Exception recognizes that Sec. 230-76(2) only requires a power-operated CB to be capable of being *opened* by hand (Fig. 240-36).

240-81. Indicating. This rule requires the up position to be the ON position for any CB. *All* circuit breakers—not just those "on switchboards or in panelboards"—must be ON in the up position and OFF in the down position if their handles operate vertically rather than rotationally or horizontally. This is an

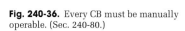

Electrically operated power CB for service disconnect may be tripped manually and provides for manual closing of its contacts by opening cover to get at handle inside.

Fig. 240-36. Every CB must be manually operable. (Sec. 240-80.)

expansion of the rule that previously applied only to circuit breakers on switchboards or in panelboards. This brings the rule into agreement with that of the second paragraph of Sec. 380-7—which makes the identical requirement for *all* circuit breakers *and* switches in individual enclosures. Switches and circuit breakers in individual enclosures must be marked to clearly show ON and OFF positions and vertically operated switches and CBs must be ON when in the up position (Fig. 240-37).

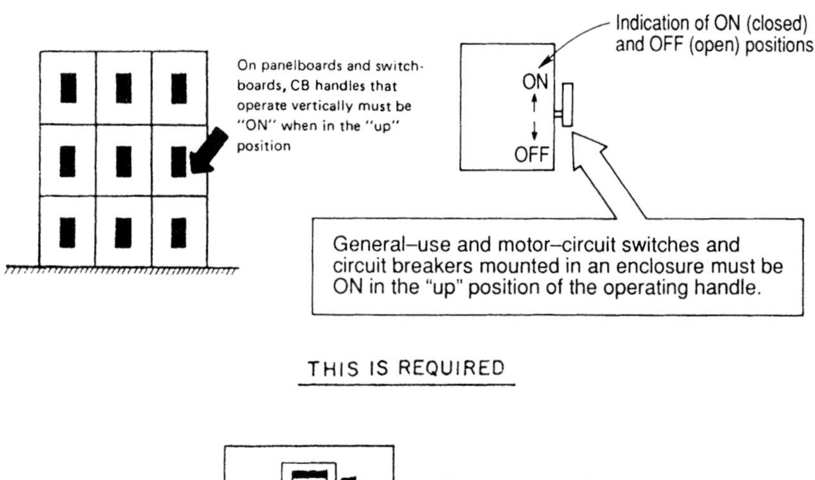

On panelboards and switchboards, CB handles that operate vertically must be "ON" when in the "up" position
Indication of ON (closed) and OFF (open) positions
ON
OFF
General–use and motor–circuit switches and circuit breakers mounted in an enclosure must be ON in the "up" position of the operating handle.
THIS IS REQUIRED

Panel main CB is "ON" when handle is in the down position

THIS IS A VIOLATION!

Fig. 240-37. Handle position of CB in any kind of enclosure must be "ON" in the up position. (Sec. 240-81.)

240-83. Marking. Part **(a)** requires that the marking of a CB's ampere rating must be durable and visible after installation. That marking is permitted to be made visible by removing the trim or cover of the CB.

The last paragraph of part **(c)** says that enclosures containing series-rated protective devices must be marked with the "additional series combination interrupting rating" of the enclosed devices (Fig. 240-38).

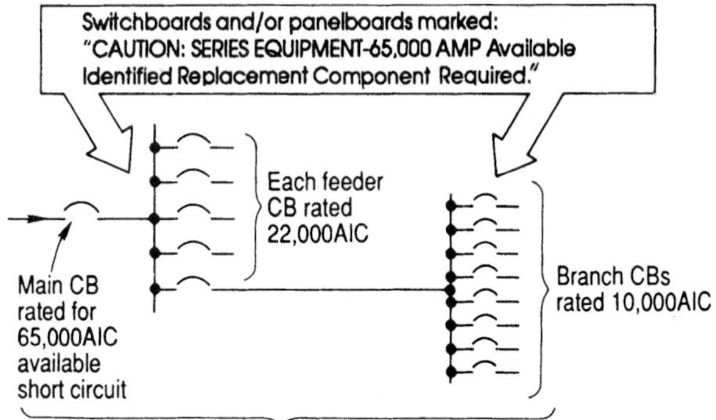
Switchboards and/or panelboards marked:
"CAUTION: SERIES EQUIPMENT-65,000 AMP Available
Identified Replacement Component Required."
Each feeder
CB rated
22,000AIC
Main CB
rated for
65,000AIC
available
short circuit
Branch CBs
rated 10,000AIC
All breakers are close together but have been tested and
listed as safe for use even though feeder and branch CBs do
not have interrupting capacity for the available fault current.

Fig. 240-38. An "additional series combination interrupting rating" must be "marked" on equipment. [Sec. 240-83(c).]

An enclosure containing circuit breakers used at a point in a system with higher available short-circuit current than the breakers' IC rating, as part of a UL-listed series rated system, with an upstream CB of adequate IC must have marking like "Caution—Series Rated System _____ A Available. Identified Replacement Component Required." This rule is in recognition of the safety and effectiveness of tested and UL-listed applications of circuit breakers in series to achieve more economical use of breakers with lower than normal interrupting rating. The requirement calls for marking of "end-use equipment, such as switchboards and panelboards" to advise maintenance and operating personnel of the elevated interrupting rating of the enclosed protective devices that are used as a part of a UL-listed series rated system. By marking end-use equipment, maintenance and operating personnel are alerted that the IC rating of the enclosed devices is greater than that which is marked on the devices because they are used as part of a listed series rated system. This rule in Sec. 240-83(c) corresponds to a similar concern addressed by the wording of Sec. 110-22 of the 1993 NEC.

Part **(d)** of this section requires that any CB used to switch 120- or 277-V fluorescent lighting be approved for the purpose and be marked "SWD" (Fig. 240-39).

In commercial and industrial electrical systems, ON-OFF control of lighting is commonly done by the breakers in the lighting panel, eliminating any local wiring-device switches. UL states that "circuit breakers marked SWD are suit-

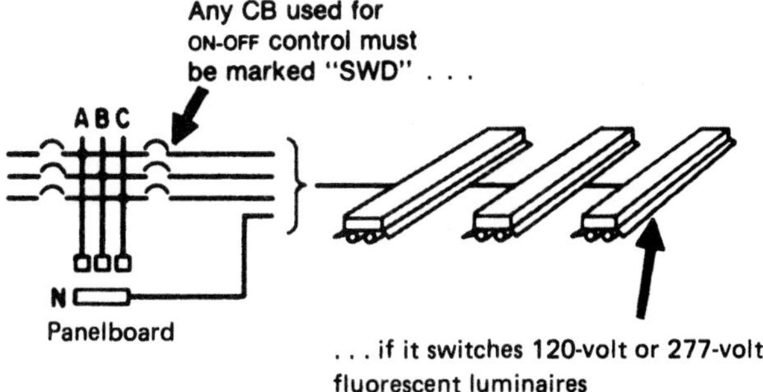

Fig. 240-39. Circuit breakers used for switching lights must be "SWD" type. [Sec. 240-83(d).]

able for switching 120- or 277-V fluorescent lighting on a regular basis." Such listing indicates that any CB used for regular switching of 120- or 277-V fluorescents must be marked SWD and that breakers *not* so marked are *not* suitable for panel switching of lighting. Both 277- and 120-V CBs are required to be designated as suitable for regular or frequent switching of lighting of any kind. Such CBs are available with the "SWD" marking and are suitable for switching duty because they are ruggedly constructed.

Under this rule, only those breakers bearing the designation "SWD" (switching duty) may be used as snap switches for lighting control. Type "SWD" breakers have been tested and found suitable for the greater frequency of ON-OFF operations required for switching duty than for strictly overcurrent protection, in which the breaker is used only for generally infrequent disconnect for circuit repair or maintenance.

The rule of part **(e)** requires specific voltage markings on circuit breakers.

240-85. Applications. This section repeats UL data regarding interpretation of voltage markings. The wording explains circuit-breaker voltage markings in terms of the device's suitability for grounded and ungrounded systems. Designation of only a phase-to-phase rating—such as "480 V"—indicates suitability for grounded or ungrounded systems. But voltage designations showing a phase-to-neutral voltage by "slash" markings—like 480Y/277 V or 120/240 V—indicate that such circuit breakers are limited exclusively to use in grounded neutral electrical systems.

The first part of Sec. 240-85 calls attention to the marking that identifies a two-pole breaker's suitability for use on corner-grounded systems. Two-pole devices marked 240 V or 480 V must be further identified by a marking "1ϕ-3ϕ" to be used on corner-grounded delta systems.

240-100. Feeders (Over 600 Volts, Nominal). This section presents rules on overcurrent protection for high-voltage (over 600 V) feeder conductors. It requires short-circuit protection of adequate interrupting capacity for its point of use. Although the rule calls for "short-circuit" protection, it does *not* require that conductors be protected in accordance with their rated ampacities. Refer to Sec. 230-208, which is referenced in this section (Fig. 240-40).

TYPE OF DEVICE

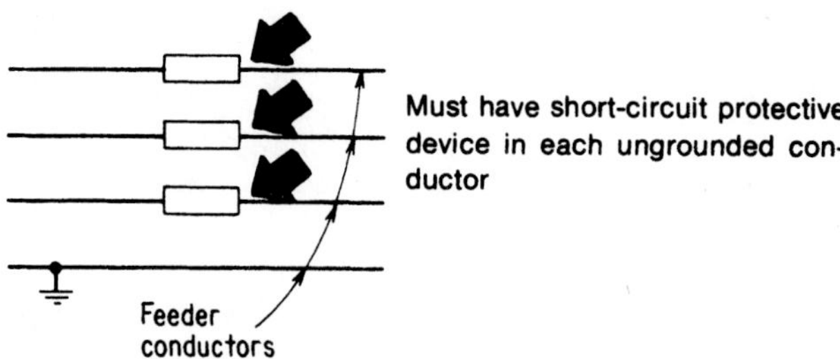
Must have short-circuit protective device in each ungrounded conductor
Feeder conductors

REQUIRED MINIMUM RATING

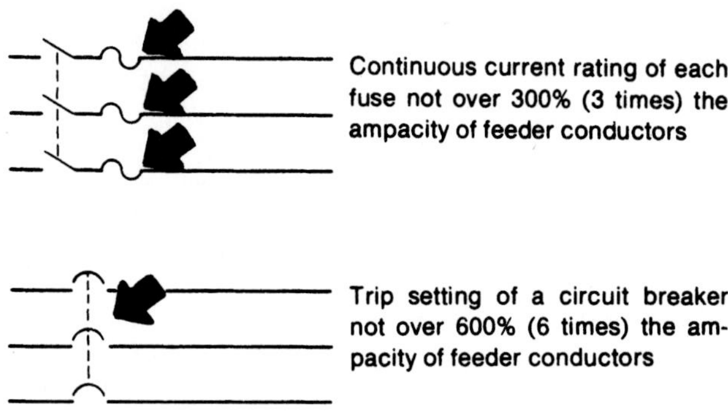
Continuous current rating of each fuse not over 300% (3 times) the ampacity of feeder conductors
Trip setting of a circuit breaker not over 600% (6 times) the ampacity of feeder conductors

Fig. 240-40. Overcurrent protection of high-voltage (over 600 V) feeder conductors. (Sec. 240-100.)

A high-voltage feeder:

1. *Must* have a "short-circuit protective device in *each* ungrounded conductor." OR—
2. It may be protected by a CB equipped with overcurrent relays and current transformers in *only two phases* and arranged as described in Secs. 710-20 and 710-21, which cover overcurrent protection of high-voltage service conductors.

The requirement on maximum value of high-voltage overcurrent protection is as follows:

A FUSE must be rated in continuous amps at *not* more than THREE TIMES the ampacity of the circuit conductor.

A CIRCUIT BREAKER must have a long-time trip element rated *not* more than SIX TIMES the ampacity of the feeder conductor.

As explained under Sec. 230-208, E-rated fuses used for circuits over 600 V can carry 200 percent of their rated current continuously. Therefore, THREE TIMES the current rating of such a fuse is the *same value* of protection as SIX TIMES the current rating of a CB of the same amp rating as the fuse. For instance, three times the 100-A rating of an E-rated fuse is actually three times 200 A, or 600 A—which is the same as six times the 100-A rating of a CB.

240-101. Branch Circuits. This applies to high-voltage branch circuits, which are invariably branch circuits for high-voltage motors and as such must also satisfy Sec. 430-125 covering overload and fault-current protection for high-voltage motor circuits. A high-voltage branch circuit must be protected by a short-circuit protective device in *each* ungrounded conductor *or* it may be protected by a CB with relays in *only two* phase legs, as described in Secs. 710-20 and 710-21.

ARTICLE 250. GROUNDING

250-1. Scope. One of the most important, but least understood, considerations in design of electrical systems is that of grounding. The word *grounding* comes from the fact that the technique itself involves making a low-resistance connection to the earth or to ground. For any given piece of equipment or circuit, this connection may be a direct wire connection to the grounding electrode which is buried in the earth; or it may be a connection to some other conductive metallic element (such as conduit or switchboard enclosure) which is connected to a grounding electrode.

The purpose of grounding is to provide protection of personnel, equipment, and circuits by eliminating the possibility of dangerous or excessive voltages.

There are two distinct considerations in grounding for electrical systems: grounding of one of the conductors of the wiring system, and grounding of all metal enclosures which contain electrical wires or equipment when an insulation failure in such enclosures might place a potential on the enclosures and constitute a shock or fire hazard. The types of grounding are:

1. *Wiring system ground.* This consists of grounding one of the wires of the electrical system, such as the neutral, to limit the voltage upon the circuit which might otherwise occur through exposure to lightning or other voltages higher than that for which the circuit is designed. Another purpose in grounding one of the wires of the system is to limit the maximum voltage to ground under normal operating conditions. Also, a system which operates with one of its conductors intentionally grounded will provide for automatic opening of the circuit if an accidental or fault ground occurs on one of its ungrounded conductors (Fig. 250-1).
2. *Equipment ground.* This is a permanent and continuous bonding together (i.e., connecting together) of all noncurrent-carrying metal parts of equipment enclosures—conduit, boxes, cabinets, housings, frames of motors, and lighting fixtures—and connection of this interconnected system of enclosures to the system grounding electrode (Fig. 250-2). The interconnection of all metal enclosures must be made to provide a low-impedance path for fault-current flow along the enclosures to assure operation of overcurrent devices which will open a circuit in the event of a fault. By

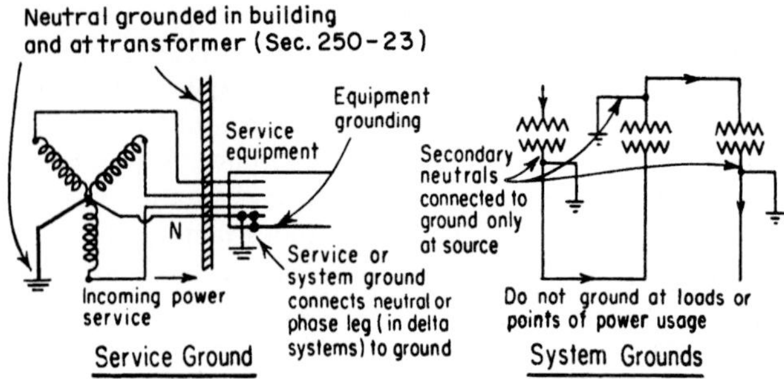

Fig. 250-1. Operating a system with one circuit conductor grounded. (Sec. 250-1.)

opening a faulted circuit, the system prevents dangerous voltages from being present on equipment enclosures which could be touched by personnel, with consequent electric shock to such personnel.

Simply stated, grounding of all metal enclosures of electric wires and equipment prevents any potential-above-ground on the enclosures. Such bonding

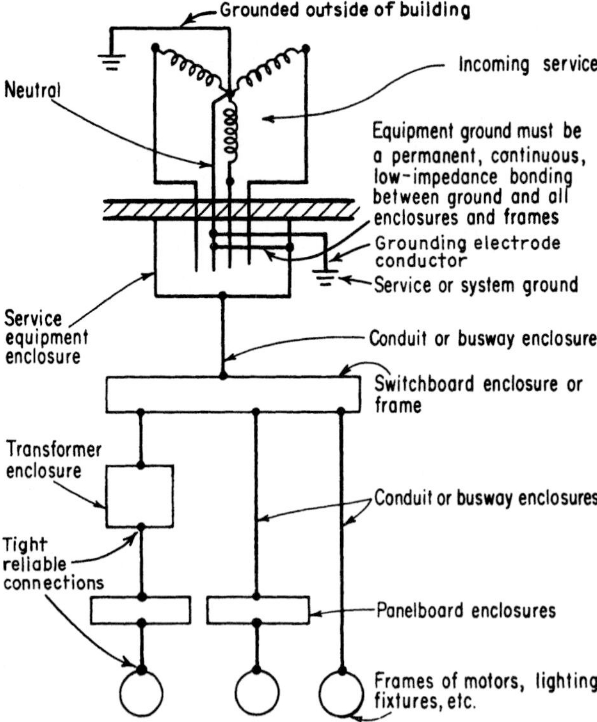

Fig. 250-2. Equipment grounding is interconnection of metal enclosures of equipment and their connection to ground. (Sec. 250-1.)

together and grounding of all metal enclosures are required for both grounded electrical systems (those systems in which one of the circuit conductors is intentionally grounded) and ungrounded electrical systems (systems with none of the circuit wires intentionally grounded).

But effective equipment grounding is extremely important for grounded electrical systems to provide the automatic fault clearing which is one of the important advantages of grounded electrical systems. A low-impedance path for fault current is necessary to permit enough current to flow to operate the fuses or CB protecting the circuit.

In a grounded electrical system with a high-impedance equipment ground-return path, if one of the phase conductors of the system (i.e., one of the ungrounded conductors of the wiring system) should accidentally come in contact with one of the metal enclosures in which the wires are run, it might produce a condition where not enough fault current would flow to operate the overcurrent devices. In such a case, the faulted circuit would not automatically open, and a dangerous voltage would be present on the conduit and other metal enclosures. This voltage presents a shock hazard and a fire hazard due to possible arcing or sparking from the energized conduit to some grounded pipe or other piece of grounded metal.

250-5. Alternating-Current Circuits and Systems to Be Grounded. Part **(a)** does recognize use of ungrounded circuits or systems when operating at less than 50 V. But grounding of circuits under 50 V is required, as shown in Fig. 250-3.

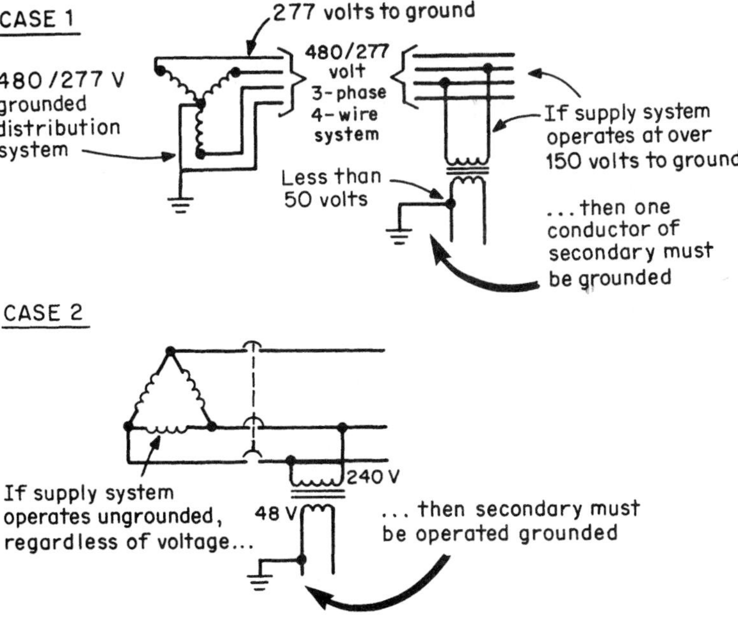

Fig. 250-3. Circuits under 50 V may have to be grounded. (Sec. 250-5.)

According to part **(b)(1)** of this rule, all alternating-current wiring systems from 50 to 1,000 V *must* be grounded if they can be so grounded that the maximum voltage to ground does not exceed 150 V. This rule makes it *mandatory* that the following systems or circuits operate with one conductor grounded:

1. 120-V, 2-wire systems or circuits must have one of their wires grounded.
2. 240/120-V, 3-wire, single-phase systems or circuits must have their neutral conductor grounded.
3. 208/120-V, 3-phase, 4-wire, wye-connected systems or circuits must be operated with the neutral conductor grounded.
4. Where the grounded conductor is uninsulated as permitted for service drop, service lateral, and service entrance conductors (Exceptions to Secs. 230-22, 230-30, 230-41).

In all the foregoing systems or circuits, the neutrals must be grounded because *the maximum voltage to ground does not exceed 150 V* from any other conductor of the system when the neutral conductor is grounded.

In parts **(2)** and **(3)** of this section, all systems of any voltage up to 1,000 V must operate with the neutral conductor solidly grounded whenever any loads are connected phase-to-neutral, so that the neutral carries load current. *All* 3-phase, 4-wire wye-connected systems and all 3-phase, 4-wire delta systems (the so-called "red-leg" systems) must operate with the neutral conductor solidly grounded if they are used as a circuit conductor. That means that the neutral conductor of a 240/120-V, 3-phase, 4-wire system (with the neutral taken from the midpoint of one phase) must be grounded. It is also mandatory that 480Y/277-V, 3-phase, 4-wire interior wiring systems have the neutral grounded if the neutral is to be used as a circuit conductor—such as for 277-V lighting. And if 480-V autotransformer-type fluorescent or mercury-vapor ballasts are to be supplied from 480/277-V systems, then the neutral conductor will have to be grounded at the voltage source to conform to Sec. 410-78, even though the neutral is not used as a circuit conductor. Of course, it should be noted that 480/277-V systems are usually operated with the neutral grounded to obtain automatic fault-clearing of a grounded system (Fig. 250-4).

Although the NE Code does not require grounding of electrical systems in which the voltage to ground would exceed 150 V, it does recommend that ground-fault detectors be used with ungrounded systems which operate at more than 150 and less than 1,000 V. Such detectors indicate when an accidental ground fault develops on one of the phase legs of ungrounded systems. Then the indicated ground fault can be removed during downtime of the industrial operation—i.e., when the production machinery is not running.

Many industrial plants prefer to use an ungrounded system with ground-fault detectors instead of a grounded system. With a grounded system, the occurrence of a ground fault is supposed to draw enough current to operate the overcurrent device protecting the circuit. But such fault-clearing opens the circuit—which may be a branch circuit supplying a motor or other power load or may be a feeder which supplies a number of power loads; and many industrial plants object to the loss of production caused by downtime. They would rather use the ungrounded system and have the system kept operative with a single ground fault and clear the fault when the production machinery is not in use.

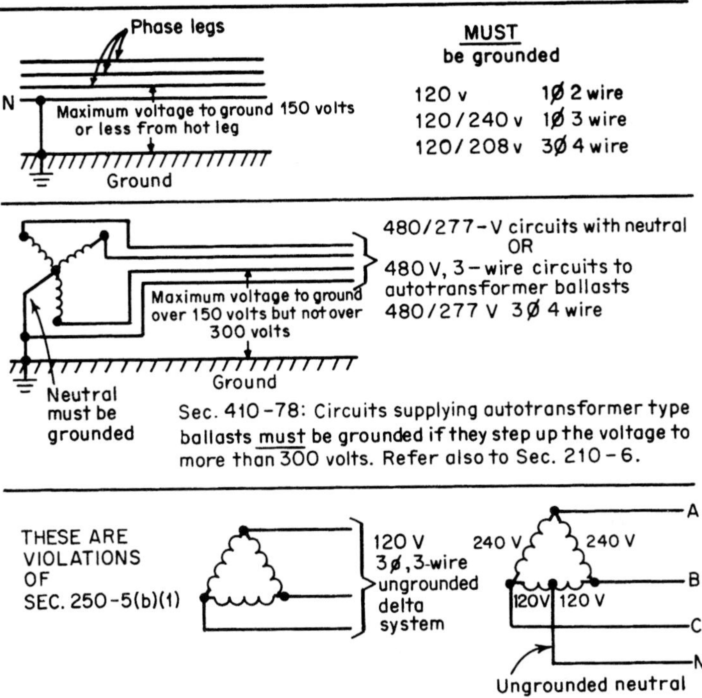

Fig. 250-4. Some systems or circuits must be grounded. (Sec. 250-5.)

In some plants, the cost of downtime of production machines can run to thousands of dollars per minute. In other plants, interruption of critical process is extremely costly.

The difference between a grounded and an ungrounded system is that a single ground fault will automatically cause opening of the circuit in a grounded system, but will not interrupt operations in an ungrounded system. However, the presence of a single ground fault on an ungrounded system exposes the system to the very destructive possibilities of a phase-to-phase short if another ground fault should simultaneously develop on a different phase leg of the system (Fig. 250-5).

Grounded neutral systems are generally recommended for high-voltage (over 600 V) distribution. Although ungrounded systems do not undergo a power outage with only one-phase ground faults, the time and money spent in tracing faults indicated by ground detectors and other disadvantages of ungrounded systems have favored use of grounded neutral systems. Grounded systems are more economical in operation and maintenance. In such a system, if a fault occurs, it is isolated immediately and automatically.

Grounded neutral systems have many other advantages. The elimination of multiple faults caused by undetected restriking grounds greatly increases service reliability. The lower voltage to ground which results from grounding the

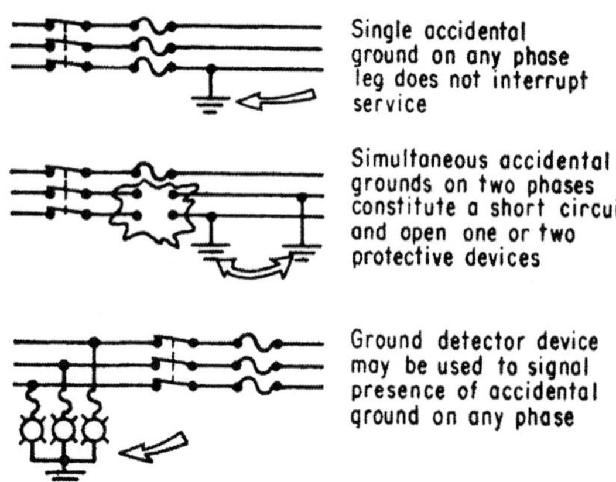
UNGROUNDED SYSTEMS
Single accidental ground on any phase leg does not interrupt service
Simultaneous accidental grounds on two phases constitute a short circuit and open one or two protective devices
Ground detector device may be used to signal presence of accidental ground on any phase

Fig. 250-5. Characteristics of ungrounded systems. (Sec. 250-5.)

neutral offers greater safety for personnel and requires lower equipment voltage ratings. And on high-voltage (above 600 V) systems, residual relays can be used to detect ground faults before they become phase-to-phase faults which have substantial destructive ability.

Exception No. 3 to the basic rule requiring grounding of one conductor of AC electrical systems recognizes use of *ungrounded* control circuits derived from transformers. According to the rules of part **(b)** of this section, any 120-V, 2-wire circuit *must* normally have one of its conductors grounded; the neutral conductor of any 240/120-V, 3-wire, single-phase circuit *must* be grounded; and the neutral of a 208/120-V, 3-phase, 4-wire circuit *must* be grounded. Those requirements have often caused difficulty when applied to control circuits derived from the secondary of a control transformer that supplies power to the operating coils of motor starters, contactors, and relays. For instance, there are cases where a ground fault on the hot leg of a grounded control circuit can cause a hazard to personnel by actuating the control circuit fuse or CB and shutting down an industrial process in a sudden, unexpected, nonorderly way. A metal-casting facility is an example of an installation where sudden shutdown due to a ground fault in the hot leg of a grounded control circuit could be objectionable. Because designers often wish to operate such 120-V control circuits ungrounded, Exception No. 3 of Sec. 250-5(b) permits ungrounded control circuits under certain specified conditions.

A 120-V control circuit may be operated ungrounded when all the following exist:

1. The circuit is derived from a transformer that has a primary rating less than 1,000 V.
2. Whether in a commercial, institutional, or industrial facility, supervision will assure that only persons qualified in electrical work will maintain and service the control circuits.

3. There is a need for preventing circuit opening on a ground fault—i.e., continuity of power is required for safety or for operating reliability.
4. Some type of ground detector is used on the ungrounded system to alert personnel to the presence of any ground fault, enabling them to clear the ground fault in normal downtime of the system (Fig. 250-6).

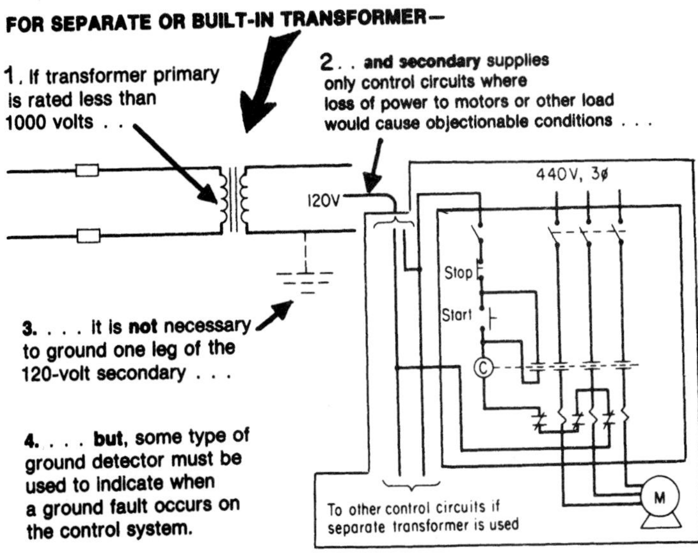

FOR SEPARATE OR BUILT-IN TRANSFORMER—
1. If transformer primary is rated less than 1000 volts . .
2. . . and secondary supplies only control circuits where loss of power to motors or other load would cause objectionable conditions . . .
3. . . . it is not necessary to ground one leg of the 120-volt secondary . . .
4. . . . but, some type of ground detector must be used to indicate when a ground fault occurs on the control system.
120V
440V, 3ø
Stop
Start
C
M
To other control circuits if separate transformer is used

Fig. 250-6. Ungrounded 120-V circuits may be used for controls. (Sec. 250-5.)

Although no mention is made of secondary voltage in this Code rule, this Exception permitting ungrounded control circuits is primarily significant only for 120-V control circuits. The NE Code has long permitted 240- and 480-V control circuits to be operated ungrounded. Application of this Exception can be made for any 120-V control circuit derived from a control transformer in an individual motor starter or for a separate control transformer that supplies control power for a number of motor starters or magnetic contactors. Of course, the Exception could also be used to permit ungrounded 277-V control circuits under the same conditions.

Exception No. 4 excludes from the need for grounded operation the circuits in flammable anesthetizing locations, as covered by Sec. 517-160(a)(1) or for electrolytic cells as covered in Art. 668.

A very important permission is given in Exception No. 5: Three-phase AC systems rated 480 to 1,000 V may be operated with a high-impedance connection to ground. As an alternative to solidly grounded or ungrounded operation of 3-phase AC systems rated 480 V (or up to 1,000 V), such systems may be high-impedance grounded. Such operation may be used *only* where qualified persons will service the installation, where continuity of power is required, where ground detectors are installed on the system, and where there are no line-neutral (277-V) loads being supplied. Exception No. 5 represents the first NEC recognition of impedance grounding as an alternative to solid grounding or nongrounding of electrical systems. Other sections of Art. 250 coordinate full recognition of high-impedance grounding of electrical systems.

As described in the title of this Exception, the high impedance used to ground the neutral point of the system is usually a resistor, which limits the ground-fault current to a low value. Such application is readily made to a 480/277-V, 3-phase, 3-wire system, with the neutral point connected to ground through the grounding resistor (Fig. 250-7).

As covered by the rule of part **(c)**, any AC system of 1,000 V or more must be grounded if it supplies portable equipment. Otherwise, such systems do not have to be grounded, although they *may* be grounded.

Part **(d)** of Sec. 250-5 has special meaning on grounding requirements for emergency generators used in electrical systems. It is best studied in steps:

1. The wording here presents the basic rule that covers grounding of "separately derived systems"—which has always been understood to refer to generator output circuits and transformer secondary circuits because such systems are "derived" separately from other wiring systems and have no conductor connected to the other systems.
2. For a separately derived system, if the voltage and hookup require grounding as specified in Sec. 250-5(b), then such systems have to be grounded and bonded as described in Sec. 250-26.
3. With respect to 2-winding transformers (i.e., single-phase or polyphase transformers that are *not* autotransformers and have *only* magnetic coupling from the primary to the secondary), there is no question that the secondary circuits are "separately derived," and grounding must always be done as required by Secs. 250-5(b) and 250-26.
4. But, when the rule of Sec. 250-5(d), is applied to 208/120- or 480/277-V generators used for emergency power in the event of an outage of the normal electric utility service, care must be taken with such a generator that is tied into the automatic transfer switch that is also fed from the normal service equipment. With a solidly connected neutral conductor running from the service equipment, through the transfer switch, to the generator neutral terminal, and with the neutral bonded and grounded at both the normal service and the generator, objectionable currents can flow. Section 250-21 prohibits grounding connections that produce objectionable flow of current over grounding conductors or grounding paths. The two grounding connections—at the service and at the generator—can produce objectionable current flow under both normal and fault conditions:

   Under normal conditions, neutral current of the connected load in the building has two paths of current flow from the common neutral point in the transfer switch back to the service neutral terminal. One path, of course, is over the neutral conductor from the transfer switch to the service equipment. The other path is over the neutral conductor from the transfer switch to the generator, at which point the current can flow back to the service equipment over the grounding conductor (the conduit and enclosure interconnections) that runs between the service equipment and the generator.

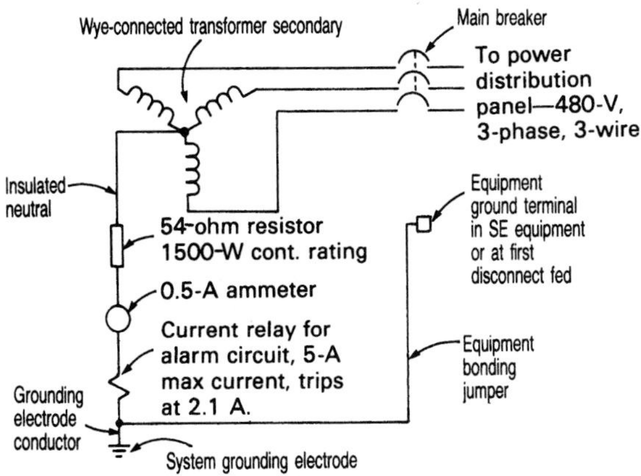

1—Diagram of resistance-grounded system shows how neutral of main transformer(s) is grounded through a resistor, ammeter and current relay. Ground fault on any phase causes current to flow from fault through ground at transformer, through current relay, ammeter and resistor back to transformer neutral. Resistor limits fault current; current relay, which trips at any current level above 2.1 amps, initiates alarm.

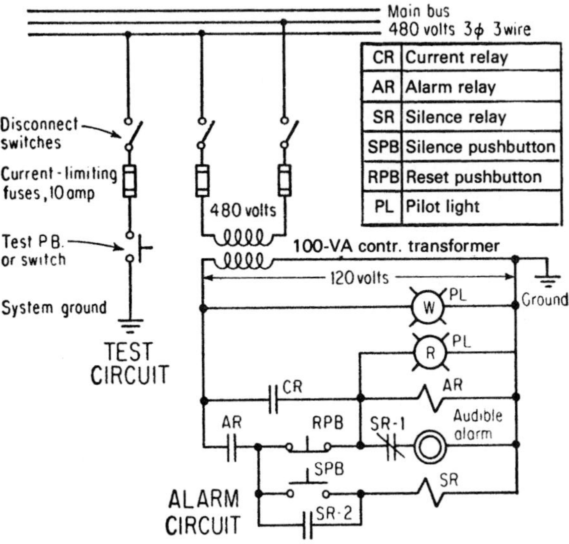

2—Test circuit (on left of diagram) allows application of temporary ground to system to test alarm function. In alarm circuit, contacts CR close when current relay, which is connected in neutral-conductor ground-circuit at main transformer, senses a ground fault (see Fig. 1). Operation of contacts CR initiates audible and visual alarm.

Fig. 250-7. This is a typical application of resistance-grounded system operation. [Sec. 250-5(b), Exception No. 5.]

Under ground-fault conditions, a similar double path for current flow can cause desensitizing of ground-fault protective equipment—as discussed and shown under Sec. 230-95, when a 3-pole transfer switch is used.

5. Elimination of the desensitizing of service GF protection can be accomplished by the use of a 4-pole transfer switch that prevents a solid neutral connection from the service equipment to the generator.
6. In the present NE Code, the grounding requirements of Sec. 250-5(d) apply to a generator *only* where the generator "has no direct electrical connection, including a solidly grounded circuit conductor" to the normal service. The rule would apply to a generator that fed its load without any tie-in through a transfer switch to *any* other system. The rule would not apply if a generator *does have* a solidly connected neutral from it to the service through a 3-pole, solid-neutral transfer switch (Fig. 250-8).

The first FPN (fine-print note) after Sec. 250-5(d) specifically identifies an on-site generator (emergency or standby) as "not a separately derived system" if the neutral conductor from the generator is connected solidly through a terminal lug in a transfer switch to the neutral conductor from the normal (usually, the power company) service to the premises. Therefore, the generator neutral point does not have to be bonded to the frame and connected to a grounding electrode.

The second FPN cautions that the neutral conductor from a generator to a transfer switch must be sized at least equal to 12½ percent of the cross-section area of the largest associated phase conductor (Sec. 445-5) to assure adequate conductivity (low impedance) for fault current that might return over that neutral when the generator is supplying the premises load, the neutral of both the generator and the normal service are connected solidly through the transfer switch (making the generator *not* a separately derived system), and the generator neutral is not bonded to the generator case and grounded at the generator. Under such a set of conditions, fault current from a ground fault in the premises wiring system would have to return to the point at the normal service equipment where the equipment grounding conductor (service equipment enclosure, metal conduits, etc.) is bonded to the service neutral. Only from that point can the fault current return over the neutral conductors, through the transfer switch to the neutral point of the generator winding. Section 445-5 requires that such a generator neutral must satisfy Sec. 250-23(b)—which says that a neutral that might function as an equipment grounding conductor must have a cross-section area at least equal to 12½ percent of the csa of the largest phase conductor of the generator circuit to the transfer switch (Fig. 250-9).

The effect of the rule of Sec. 250-5(d) on transfer switches is as follows:

3-pole transfer switch If a solid neutral connection is made from the service neutral, through the transfer switch, to the generator neutral, then bonding and grounding of the neutral at the generator are *not* required because the neutral is already bonded and grounded at the service equipment. And if bonding and grounding were done at the generator, it could be considered a violation of Sec. 250-21(a) and would have to be corrected by Sec. 250-21(b) (Fig. 250-8).

4-pole transfer switch Because there is no direct electrical connection of either the hot legs or the neutral between the service and the generator, the gen-

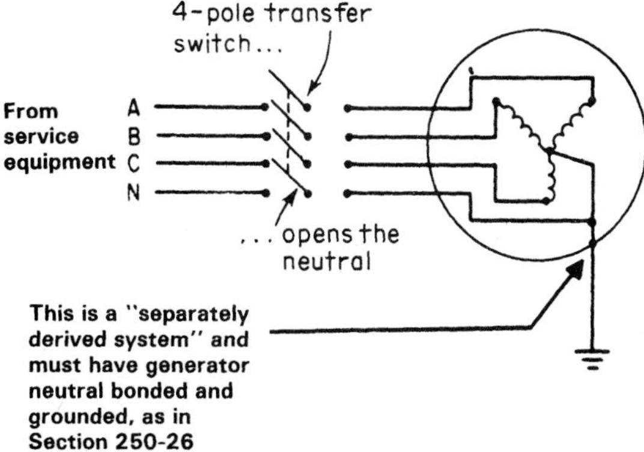

Generator neutral *must* be bonded when neutral is opened.

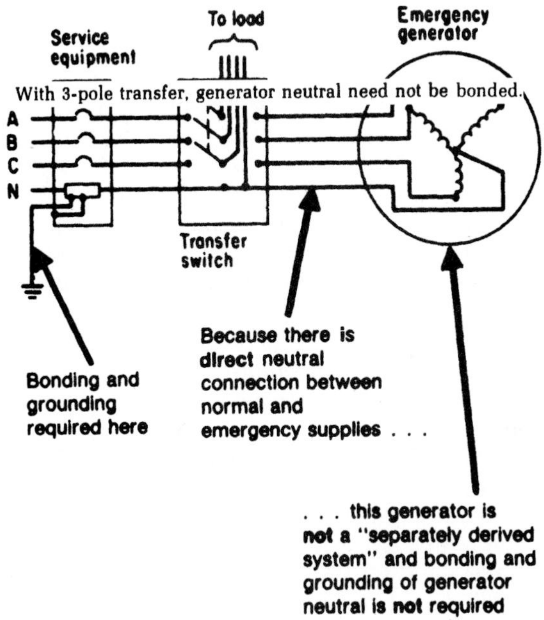

With 3-pole transfer, generator neutral need not be bonded.

Fig. 250-8. These are the choices on bonding and grounding of generator neutral. [Sec. 250-5(d).]

erator in such a hookup is a "separately derived system" and must be grounded and bonded to the generator case at the generator (Fig. 250-8).

It should be noted that the 4-pole transfer switch and other neutral-switching techniques came into use to eliminate problems of GFP desensitizing that were caused by use of a 3-pole transfer switch *when the neutral of the generator was*

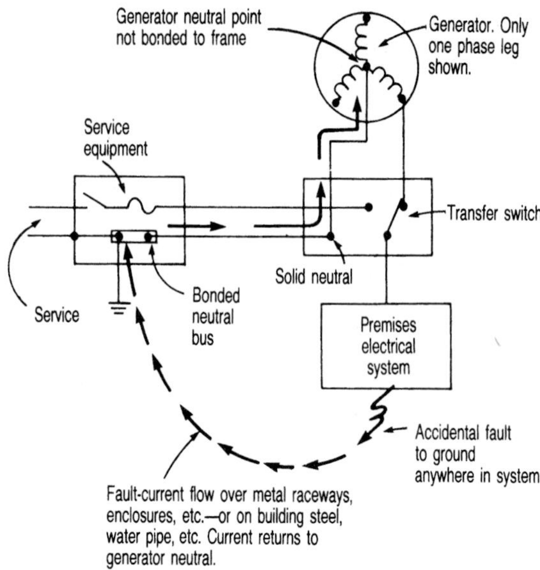

Fig. 250-9. Neutral conductor from service equipment to generator neutral point must be sized at least equal to 12½ percent of the cross-section area of the generator phase leg. [Sec. 250-5(d).]

bonded to the generator housing. By eliminating that bonding requirement for emergency generators in Sec. 250-5(d) it was the **Code** intent to make possible use of 3-pole transfer switches without disruption of service GFP. But that has not resulted and the neutral-switching concept has prevailed.

Although the rule in Sec. 250-5(d) permits use of an ungrounded and non-bonded generator neutral in conjunction with a 3-pole transfer switch, such application has been found to produce other conditions of undesirable current flow, resulting in other forms of desensitizing of service GFP—such as desensitizing a zero-sequence sensor used for GFP on the generator output. In such a hookup, with the system being supplied by the generator and the normal service open, ground-fault current returning over a metal raceway to the metal case of the transfer switch will flow to the bond point between the neutral and equipment ground at the normal service equipment and then return to the generator over the solid neutral, through the zero-sequence sensor. As a result, the use of a 4-pole transfer switch or some other technique that opens the neutral is the only effective way to avoid GFP desensitizing. Ground-fault protection is not compatible with a solid neutral tie between the service and an emergency generator—with or without its neutral bonded.

Refer to the discussion under Sec. 250-23(a) on the relationship between GFP desensitizing and the point of connection of the grounding electrode conductor.

250-6. Portable and Vehicle-Mounted Generators. Part **(a)** rules that the frame of a portable generator does not have to be grounded if the generator supplies only equipment mounted on the generator and/or plug-connected equipment through receptacles mounted on the generator, provided that the noncurrent-carrying metal parts of equipment and the equipment grounding conductor terminals are bonded to the generator frame. (See Fig. 250-10.)

A clarification in part **(a)** points out that, where a portable generator is used with its frame *not* grounded, the frame is permitted to act as the grounding electrode for any cord-connected tools or appliances plugged into the generator's receptacles (Fig. 250-10). This assures that tools and appliances that are required by Sec. 250-45 to be grounded do satisfy the Code when plugged into a receptacle on the ungrounded frame of a portable generator. It should also be noted that part **(c)** of this section requires the neutral conductor of the generator output to be bonded to the frame of the generator when it is not used as an emergency source connected to a transfer switch.

Part **(b)** notes that the frame of a vehicle-mounted generator may be bonded to the vehicle frame, which then serves as the grounding electrode—but only when the generator supplies only equipment mounted on the vehicle and/or cord- and plug-connected equipment through receptacles on the vehicle or generator. When the frame of a vehicle is used as the grounding electrode for a generator mounted on the vehicle, grounding terminals of receptacles on the generator must be bonded to the generator frame, which must be bonded to the vehicle frame.

If either a portable or vehicle-mounted generator supplies a fixed wiring system external to the generator assembly, it must then be grounded as required for any separately derived system (as, for instance, a transformer secondary), as covered in Sec. 250-26.

The wording of part **(c)** brings application of portable and vehicle-mounted generators into compliance with the concept described above in Sec. 250-5(d) on grounding and bonding of the generator neutral conductor. A generator neutral *must be* bonded to the generator frame when the generator is a truly separately derived source, such as the sole source of power to the loads it feeds, and is *not* tied into a transfer switch as part of a *normal emergency* hookup for feeding the load normally from the utility service and from the generator on an emergency or standby basis (Fig. 250-11). A note to this section refers to Sec. 250-5(d) and makes that rule applicable to grounding and bonding of portable generators that supply a fixed wiring system on a premises. In such a case, bonding of the neutral to the generator frame is not required if there is a solid neutral connection from the utility service, through a transfer switch to the generator, as shown in the bottom sketch of Fig. 250-11.

250-21. Objectionable Current over Grounding Conductors. Although parts **(a)** and **(b)** of this section permit "arrangement" and "alterations" of electrical systems to prevent and/or eliminate objectionable flow of currents over "grounding conductors or grounding paths," part **(d)** specifically prohibits any exemptions from NEC rules on grounding for "electronic equipment" and states that "currents that introduce noise or data errors" in electronic data-processing and computer equipment are not "objectionable" currents that allow modification of grounding rules. This paragraph emphasizes the Code's intent that electronic data-processing equipment must have its input and output circuits in full compliance with all NEC rules on neutral grounding, equipment grounding,

1. . . . the generator supplies equipment mounted on generator and/or cord-and-plug-connected equipment through receptacles on the generator, and

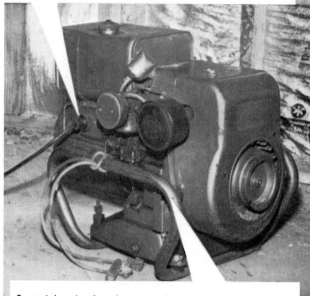

2. metal parts of equipment and equipment grounding terminals of receptacles are bonded to generator frame.

Grounding of portable generator frame is not required if these conditions exist. (Sec. 250-6).

When frame of portable generator is left ungrounded as permitted in Section 250-6 . . .

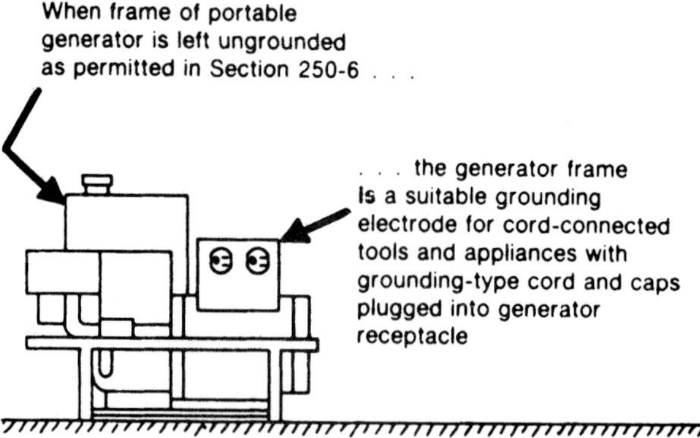

. . . the generator frame
Is a suitable grounding
electrode for cord-connected
tools and appliances with
grounding-type cord and caps
plugged into generator
receptacle

"Ungrounded" generator frame is acceptable grounding electrode. (Sec. 250-6).

Fig. 250-10. Grounding details for a portable generator.

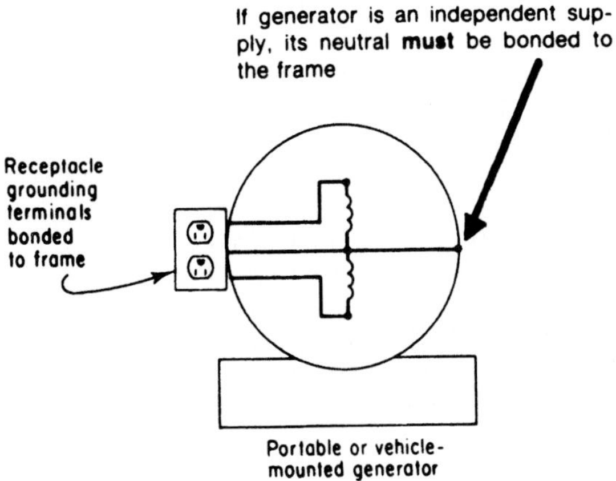

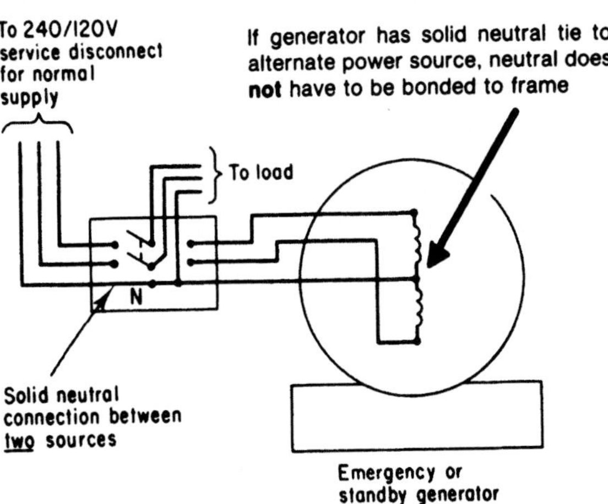

Fig. 250-11. Generator neutral may be required to be grounded. (Sec. 250-6.)

and bonding and grounding of neutral and ground terminal buses. Sec. 250-21(b) does offer alternative methods for correcting "objectionable current over grounding conductors" but part **(d)** specifically states that such modifications or alternative methods are not applicable to the on-site wiring for electronic or data-processing equipment if the only purpose is to eliminate "noise or data errors" in the electronic equipment. This paragraph amplifies the revised wording of "Premises Wiring" as given in Art. 100.

250-22. Point of Connection for Direct-Current Systems. On a 2- or a 3-wire DC distribution system, a neutral that is required to be grounded must be grounded at the supply station only.

As noted in the Exception, an on-site supply for a DC system must have a required grounding connection made at either the source of the DC supply or at the first disconnect or overcurrent device supplied. Because the basic rule says a DC source (from outside a premises) must have a required grounding connection made at "one or more supply stations" and *not* at "any point on premises wiring," an on-site DC source would be prohibited from having a grounding connection that might be required. This Exception resolves that basic problem by referring to a "DC system source . . . located on the premises."

250-23. Grounding Service-Supplied Alternating-Current Systems. As noted in part **(a)**, when a premises is supplied by an electrical system that has to be operated with one conductor grounded—either because it is required by the Code (e.g., 240/120-V, single phase) or because it is desired by the system designer (e.g., 240-V, 3-phase, corner grounded)—a connection to the grounding electrode must be made at the service entrance (Fig. 250-12). That is, the neutral conductor or other conductor to be grounded must be connected at the service equipment to a conductor which runs to a grounding electrode. The conductor that runs to the grounding electrode is called the "grounding electrode conductor"—an official definition in the NE Code.

The Code rule of the second sentence of Sec. 250-23(a) says that the connection of the grounding electrode conductor to the system conductor which is to be grounded must be made "at any accessible point from the *load end* of the service drop or service lateral" to the service disconnecting means. This means that the grounding electrode conductor (which runs to building steel and/or water pipe or driven ground rod) must be connected to the system neutral or other system wire to be grounded either in the enclosure for the service disconnect or in some enclosure on the supply side of the service disconnect.

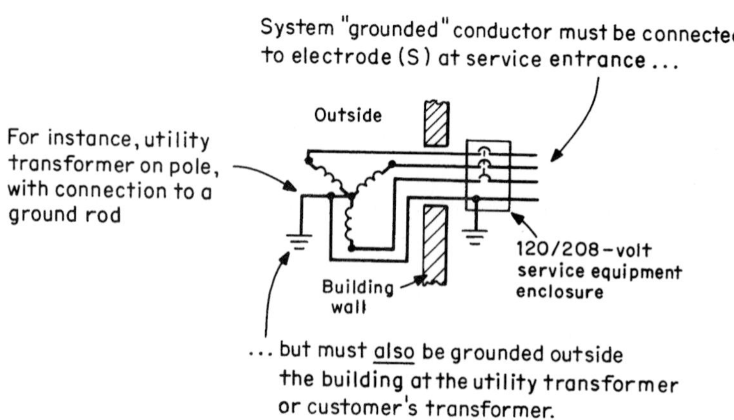

Fig. 250-12. Grounded interior systems must have *two* grounding points (Sec. 250-23.)

Such connection may be made, for instance, in the main service switch or CB or in a service panelboard or switchboard. Or, the grounding electrode conductor may be connected to the system grounded conductor in a gutter, CT cabinet, or meter housing on the supply side of the service disconnect (Fig. 250-13). The utility company should be checked on grounding connections in meter sockets or other metering equipment.

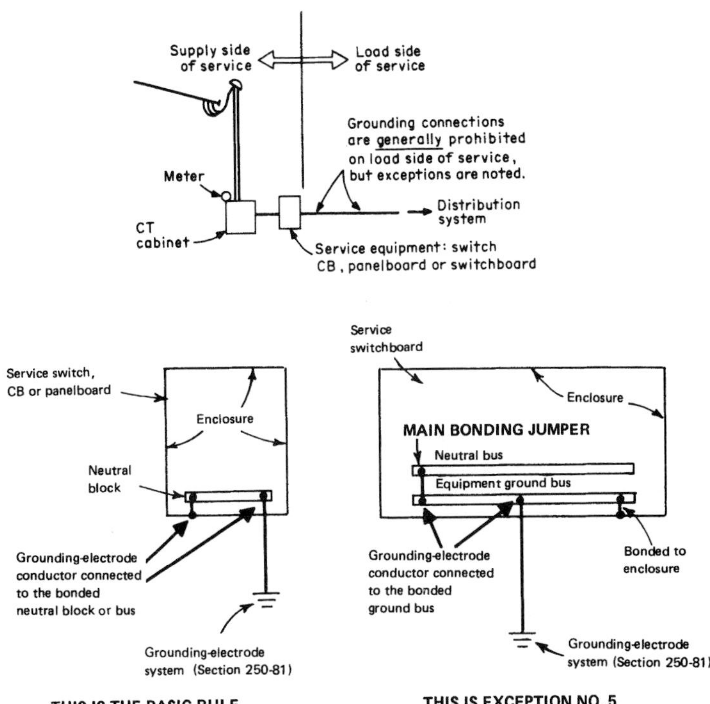

Fig. 250-13. Grounding connection must be made in SE equipment or on its line side. (Sec. 250-23.)

In addition to the grounding connection for the grounded system conductor at the point of service entrance to the premises, it is further required that another grounding connection be made to the same grounded conductor at the transformer which supplies the system. This means, for example, that a grounded service to a building must have the grounded neutral connected to a grounding electrode at the utility transformer on the pole, away from the building, as well as having the neutral grounded to a water pipe and/or other suitable electrode at the building, as shown in Fig. 250-12. And in the case of a building served from an outdoor transformer pad or mat installation, the conductor which is grounded in the building must also be grounded at the transformer pad or mat, per Sec. 250-23(a).

This section is concerned with the grounding of a utility-fed AC electric supply circuit that has one of its conductors operated intentionally grounded—such as any 240/120-V single-phase system; 208/120-V, 3-phase, 4-wire system; or 480/277-V, 3-phase, 4-wire system—whether required to have its neutral conductor grounded by Sec. 250-5(b) or not. The basic concept being conveyed is as follows:

1. For such grounding connection, a grounding electrode conductor must be connected to the grounded conductor (the neutral conductor) anywhere from the "load end of the service drop or lateral" to the *neutral block* or *bus* within the enclosure for the service disconnect—which includes a meter socket, a CT cabinet, or an auxiliary gutter or other enclosure ahead of the service disconnect, panelboard, or switchboard.
2. The grounding electrode conductor that is connected to the grounded neutral (or grounded phase leg) must be run to a "grounding electrode system," as specified in Secs. 250-81 and/or 250-83.
3. Exception No. 1 notes that the required grounding electrode connection for a local step-down transformer or a generator is treated somewhat differently from that for a utility-fed service.
4. Exception No. 5 permits the grounding electrode conductor to be connected to the equipment grounding bus in the service-disconnect enclosure—instead of to the neutral block or bus—for instance, where such connection is considered necessary to prevent desensitizing of a service GFP hookup that senses fault current by a CT-type sensor on the ground strap between the neutral bus and the ground bus. See Fig. 250-13. However, in any particular installation, the choice between connecting to the neutral bus or to the ground bus will depend on the number and types of grounding electrodes, the presence or absence of grounded building structural steel, bonding between electrical raceways and other metal piping on the load side of the service equipment, and the number and locations of bonding connections. The grounding-electrode-conductor may be connected to either the neutral bus or terminal lug or the ground bus or block in any system that has a conductor or a busbar bonding the neutral bus or terminal to the equipment grounding block or bus. Where the neutral is bonded to the enclosure simply by a bonding screw, the grounding electrode conductor *must* be connected to the neutral in all cases, because screw-bonding is not suited to passing high lightning currents to earth.

One of the most important and widely discussed regulations of the entire Code revolves around this matter of making a grounding connection to the system grounded neutral or grounded phase wire. The Code says, "Grounding connections shall not be made on the load side of the service disconnecting means." Once a neutral or other circuit conductor is connected to a grounding electrode at the service equipment, the general rule is that the neutral or other grounded leg must be insulated from all equipment enclosures or any other grounded parts on the load side of the service. That is, bonding of subpanels (or any other connection between the neutral or other grounded conductor and equipment enclosures) is prohibited by the NE Code.

There are some exceptions to that rule, but they are few and are very specific:

1. In a system, even though it is on the load side of the service, when voltage is stepped down by a transformer, a grounding connection *must* be made to the secondary neutral to satisfy Secs. 250-5(b) and 250-26.
2. When a circuit is run from one building to another, it may be necessary or simply permissible to connect the system "grounded" conductor to a grounding electrode at the other building—as covered by Sec. 250-24.
3. Section 250-61 permits frames of ranges, wall ovens, counter-top cook units, and clothes dryers to be "grounded" by connection to the grounded neutral of their supply circuit (Sec. 250-60).

The **Code** makes it a violation to bond the neutral block in a panelboard to the panel enclosure in other than a service panel. In a panelboard used as service equipment, the neutral block (terminal block) is bonded to the panel cabinet by the bonding screw provided. And such bonding is required to tie the grounded conductor to the interconnected system of metal enclosures for the system (i.e., service-equipment enclosures, conduits, busway, boxes, panel cabinets, etc.). It is this connection which provides for flow of fault current and operation of the overcurrent device (fuse or breaker) when a ground fault occurs. But, there must not be any connection between the grounded system conductor and the grounded metal enclosure system at any point on the load side of the service equipment, because such connection would constitute connection of the grounded system conductor to a grounding electrode (through the enclosure and raceway system to the water pipe or driven ground rod). Such connections, like bonding of subpanels, can be dangerous, as shown in Fig. 250-14.

This rule on not connecting the grounded system wire to a grounding electrode on the load side of the service disconnect must not be confused with the rule of Sec. 250-60, which permits the grounded system conductor to be used for grounding the frames of electric ranges, wall ovens, counter-mounted cooking units, and electric clothes dryers, but only for *existing* branch circuits. The connection referred to in Sec. 250-60 is that of an ungrounded metal enclosure to the grounded conductor for the purpose of grounding the enclosure.

A very important qualification in the third sentence of Sec. 250-23(a) on grounding connections for AC systems eliminates confusion and controversy about connection of the grounding electrode conductor when a building is fed by service conductors from a meter on a yard pole.

The basic rule of this section says that the grounding electrode conductor required for grounding both the grounded service conductor (usually a grounded neutral) and the metal enclosure of the service equipment *must be* connected to the grounded service conductor *within* or on the *supply side* of the service disconnect. But the rule further requires that the connection of the grounding electrode conductor *must be connected between the load end of the service drop or lateral and the service equipment.*

As a result of that requirement, if a service is fed to a building from a meter enclosure on a pole or other structure some distance away, as is commonly done on farm properties, and an overhead or underground run of service conductors is made to the service disconnect in the building, the grounding elec-

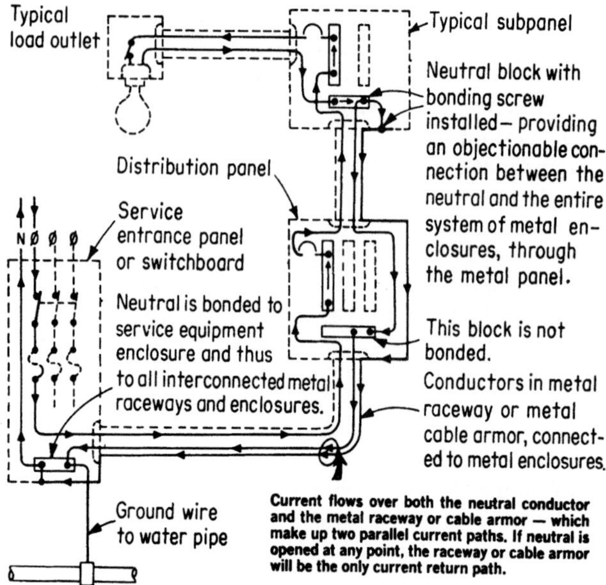
1. THIS CONDITION WILL EXIST....AND...
Typical load outlet
Typical subpanel
Neutral block with bonding screw installed– providing an objectionable connection between the neutral and the entire system of metal enclosures, through the metal panel.
Distribution panel
N Ø Ø Ø
Service entrance panel or switchboard
Neutral is bonded to service equipment enclosure and thus to all interconnected metal raceways and enclosures.
This block is not bonded.
Conductors in metal raceway or metal cable armor, connected to metal enclosures.
Ground wire to water pipe
Current flows over both the neutral conductor and the metal raceway or cable armor — which make up two parallel current paths. If neutral is opened at any point, the raceway or cable armor will be the only current return path.

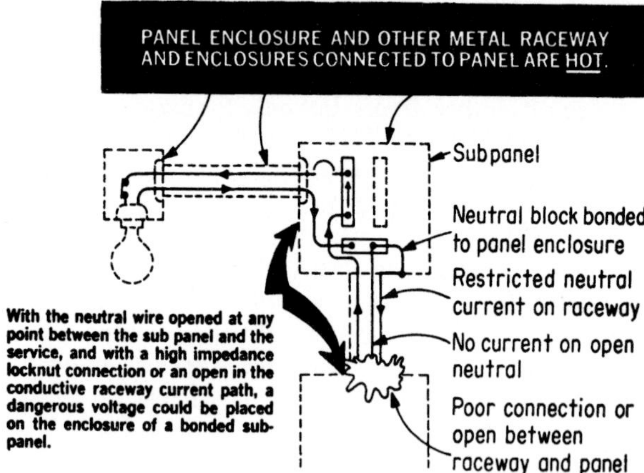
2. THIS HAZARD COULD DEVELOP
PANEL ENCLOSURE AND OTHER METAL RACEWAY AND ENCLOSURES CONNECTED TO PANEL ARE HOT.
Subpanel
Neutral block bonded to panel enclosure
Restricted neutral current on raceway
No current on open neutral
Poor connection or open between raceway and panel
With the neutral wire opened at any point between the sub panel and the service, and with a high impedance locknut connection or an open in the conductive raceway current path, a dangerous voltage could be placed on the enclosure of a bonded subpanel.

Fig. 250-14. NEC prohibits bonding of subpanels because of these reasons. (Sec. 250-23.)

trode conductor will not satisfy the Code if it is connected to the neutral in the meter enclosure but must be connected at the *load end* of the underground or overhead service conductors. And, the connection should preferably be made *within* the service-disconnect enclosure.

This rule on grounding connection is shown in Fig. 250-15. If, instead of an underground lateral, an overhead run were made to the building from the pole, the overhead line would be a "service drop." The rule of Sec. 250-23(a) would likewise require the grounding connection at the load end of the service drop. If a fused switch or CB is installed as service disconnect and protection at the load side of the meter on the pole, then that would establish the service at that point, and the grounding electrode connection to the bonded neutral terminal would be required at that point. The circuit from that point to the building would be a feeder and not service conductors. But, electrical safety and effective operation would require that an equipment grounding conductor be run with the feeder circuit conductors for grounding the interconnected system of conduits and metal equipment enclosures along with metal piping systems and building steel within the building. Or, if an equipment grounding conductor is not in the circuit from the pole to the building, the neutral could be bonded to the main disconnect enclosure in the building and a grounding electrode connection made at that point also. Either technique complies with the concepts of Sec. 250-24(a) and its Exception No. 2.

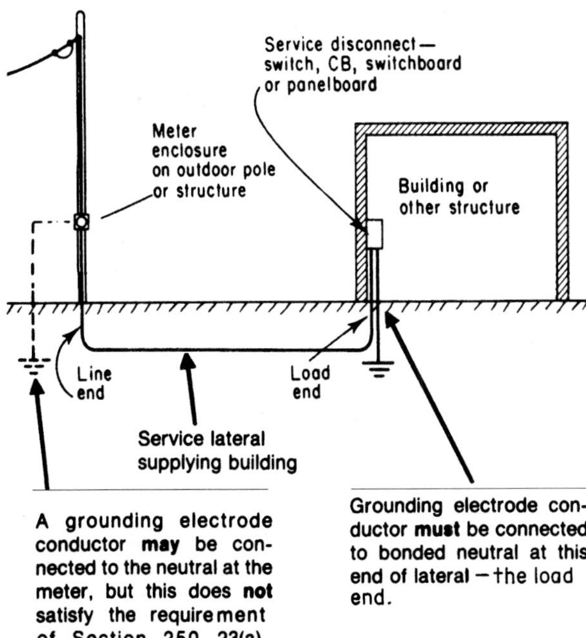
Service disconnect—
switch, CB, switchboard
or panelboard
Meter
enclosure
on outdoor pole
or structure
Building or
other structure
Line
end
Load
end
Service lateral
supplying building
A grounding electrode conductor **may** be connected to the neutral at the meter, but this does **not** satisfy the requirement of Section 250- 23(a).
Grounding electrode conductor **must** be connected to bonded neutral at this end of lateral—the load end.

Fig. 250-15. Connection to grounded conductor at load end of lateral or drop. (Sec. 250-23.)

There is an important Exception to the rule that each and every service for a grounded AC system have a grounding electrode conductor connected to the grounded system conductor anywhere on the supply side of the service disconnecting means (preferably within the service-equipment enclosure) and that the grounding electrode conductor be run to a grounding electrode at the service. Because controversy has arisen in the past about how many grounding electrode conductors have to be run for a dual-feed (double-ended) service with a secondary tie, Exception No. 4 recognizes the use of a single grounding electrode conductor connection for such dual services. It says that the single grounding electrode connection may be made to the "tie point of the grounded circuit conductors from each power source." The explanation on this Code permission was made by NEMA, the sponsor of the rule, as follows:

> Unless center neutral point grounding and the omission of all other secondary grounding is permitted, the selective ground-fault protection schemes now available for dual power source systems with secondary ties will not work. Dual power source systems are utilized for maximum service continuity. Without selectivity, both sources would be shut down by any ground fault. This proposal permits selectivity so that one source can remain operative for half the load, after a ground fault on the other half of the system.

Figure 250-16 shows two cases involving the concept of single grounding point on a dual-fed service:

In Case 1, if the double-ended unit substation is in a locked room in a building it serves or consists of metal-enclosed gear or a locked enclosure for each transformer, the secondary circuit from each transformer is a "service" to the building. The question then arises, "Does there have to be a separate grounding electrode conductor run from each secondary service to a grounding electrode?"

In Case 2, if each of the two transformers is located outdoors, in a separate building from the one they serve, in a transformer vault in the building they serve, or in a locked room or enclosure and accessible to qualified persons only or in metal-enclosed gear—then the secondary circuit from each transformer constitutes a service to the building. Again, is a separate grounding connection required for each service?

In both cases, a single grounding electrode connection may serve both services, as shown at the bottom of Fig. 250-16.

The Code rule in Exception No. 4 refers to "services that are dual fed (double ended) in a *common enclosure or grouped together.*" The phrase "common enclosure" can readily cover use of a double-ended loadcenter unit substation in a single, common enclosure. But the phrase "grouped together" can lend itself to many interpretations and has caused difficulties. For instance, if each of two separate services was a single-ended unit substation, do both the unit substations have to be in the same room or within the same fenced area outdoors? How far apart may they be and still be considered "grouped together"? As shown in Case 2 of Fig. 250-16, if separate transformers and switchboards are used instead of unit subsubstations, may one of the transformers and its switchboard be installed at the opposite end of the building from the other one? The Code does not answer those questions, but it seems clear that the wording does suggest that both of the services must be physically close and at least in

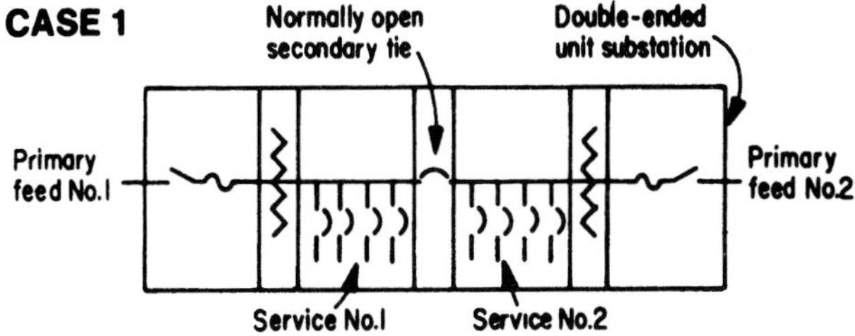

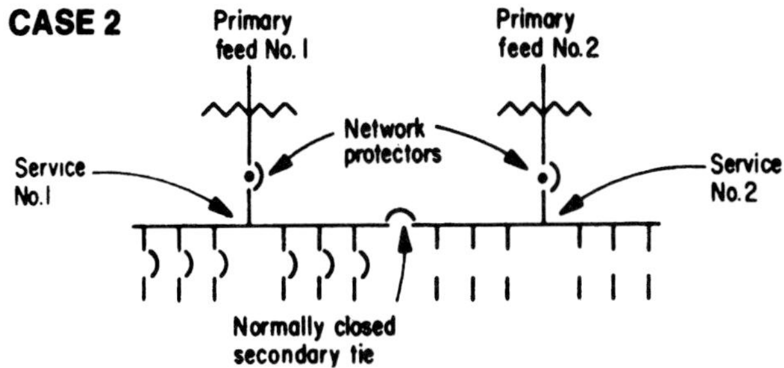

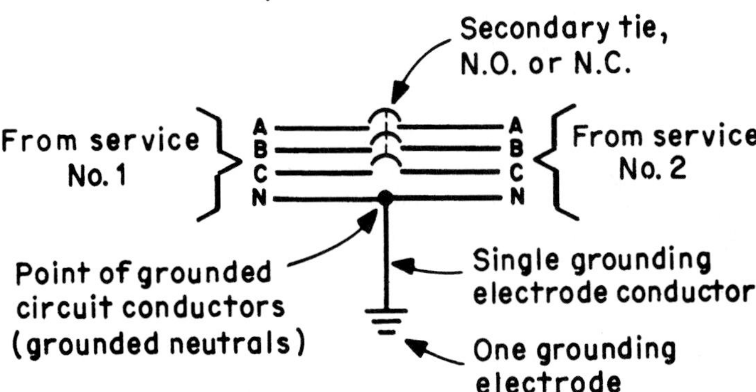

Fig. 250-16. *One* grounding connection permitted for a double-end service. (Sec. 250-23.)

the same room or vault or fenced area. That understanding has always been applied to other **Code** rules calling for "grouping"—such as for switches and CBs in Sec. 380-8 and for service disconnects in Sec. 230-72(a).

Part **(b)** requires that whenever a service is derived from a grounded neutral system, the grounded neutral conductor must be brought into the service-entrance equipment, even if the grounded conductor is not needed for the load supplied by the service. A service of less than 1,000 V that is grounded outdoors at the service transformer (pad mount, mat, or unit substation) must have the grounded conductor run to "each service disconnecting means" and bonded to the separate enclosure for "each" service disconnect. If two to six normal service disconnects [as permitted by Sec. 230-71(a)] are installed in separate enclosures (or even additional disconnect switches or circuit breakers for emergency, fire pump, etc.), the grounded circuit conductor must be run to a bonded neutral terminal in *each* of the separate disconnect enclosures fed from the service conductors. Exception No. 3 to this rule clarifies that if multiple service disconnect switches or circuit breakers are installed within "an assembly listed for use as service equipment"—such as in a service panelboard, switchboard, or multimeter distribution assembly—only a single grounded (neutral) conductor has to be run to the single, common assembly enclosure and bonded to it.

Running the grounded conductor to each individual service disconnect enclosure is required to provide a low-impedance ground-fault current return path to the neutral to assure operation of the overcurrent device, for safety to personnel and property. (See Fig. 250-17.) In such cases, the neutral functions strictly as an equipment grounding conductor, to provide a closed circuit back to the transformer for automatic circuit opening in the event of a phase-to-ground fault anywhere on the load side of the service equipment. Only one phase leg is shown in these diagrams to simplify the concept. The other two phase legs have the same relation to the neutral.

The same requirements apply to installation of separate power and light services derived from a common 3-phase, 4-wire, grounded "red-leg" delta system. The neutral from the center-tapped transformer winding must be brought into the 3-phase power service equipment as well as into the lighting service, even though the neutral will not be used for power loads. This is shown in Fig. 250-18 and is also required by Sec. 250-23(b), which states that such an unused neutral must be at least equal to the required minimum size of grounding electrode conductor specified in Table 250-94 for the size of phase conductors. In addition, if the phase legs associated with that neutral are larger than 1,100 kcmil, the grounded neutral must not be smaller than 12½ percent of the area of the largest phase conductor, which means 12½ percent of the total csa of conductors per phase when parallel conductors are used.

In any system where the neutral *is* required on the load side of the service—such as where 208Y/120-V or 480Y/277-V, 3-phase, 4-wire distribution is to be made on the premises—the neutral from the supply transformer to the service equipment is needed to provide for neutral current flow under conditions of load unbalance on the phase legs of the premises distribution system. But, even in a premises where all distribution on the load side of the service is to be solely 3-phase, 3-wire (such as 480-V, 3-phase, 3-wire distribution) and the neutral conductor is not required in the premises system, this **Code** rule says that the neutral must still run from the supply transformer to the service equipment.

The final part of the rule covers cases where the service phase conductors are paralleled, with two or more conductors in parallel per phase leg and neutral, and requires that the size of the grounded neutral must be calculated on the equivalent area for parallel conductors. If a calculated size of neutral (at least

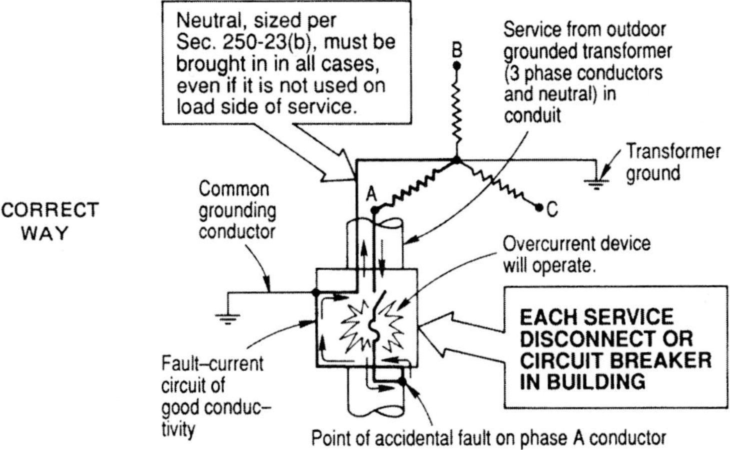

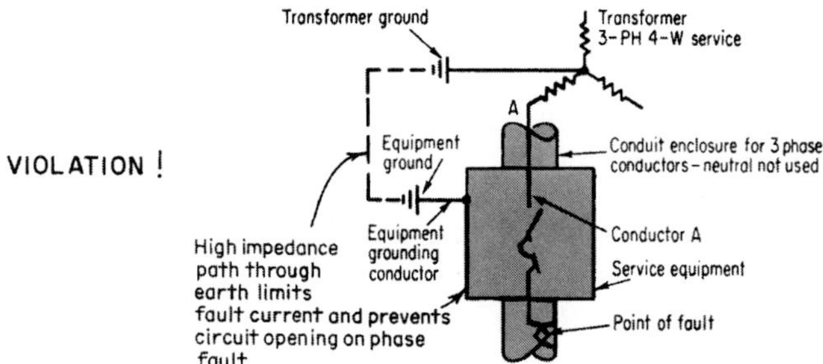

Fig. 250-17. Clearing of ground faults on the load side of any service disconnect depends upon fault-current return over a grounded circuit conductor (usually a neutral) brought into each and every enclosure for service disconnect switch or CB. (Sec. 250-23.)

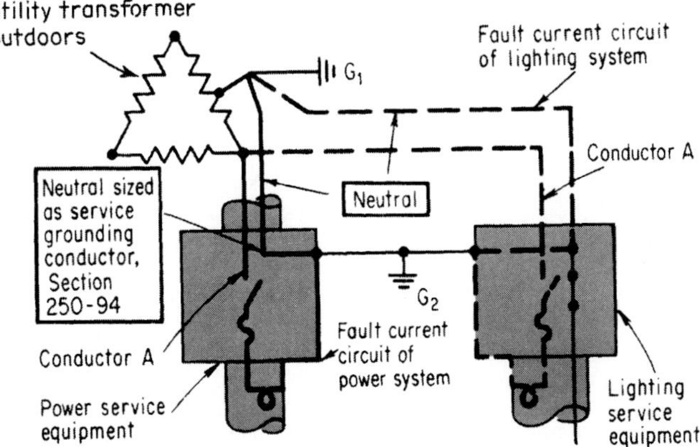

Fig. 250-18. Neutral must be brought in to each service equipment and bonded to enclosure. (Sec. 250-23.)

12½ percent of the phase-leg cross section) is to be divided among two or more conduits, and if dividing the calculated size by the number of conduits being used calls for a neutral conductor smaller than 1/0 in each conduit, the FPN calls attention to Sec. 310-4, which gives No. 1/0 as the minimum size of conductor that may be used in parallel in multiple conduits. For that reason, each neutral would have to be at least a No. 1/0, even though the calculated size might be, say, No. 1 or No. 2 or some other size smaller than No. 1/0. But, the Code rule does permit subdividing the required minimum 12½ percent grounded (neutral) conductor size by the total number of conduits used in a parallel run, thereby permitting a multiple makeup using a smaller neutral in each pipe.

As shown in Figure 250-19, the minimum required size for the grounded neutral conductor run from the supply transformer to the service is based on the size of the service phase conductors. In this case, the overall size of the service phase conductors is 4 × 500 kcmil per phase leg, or 2,000 kcmil. Because that is larger than 1,100 kcmil, it is not permitted to simply use Table 250-94 in sizing the neutral. Instead, 2,000 kcmil must be multiplied by 12½ percent. Then 2,000 kcmil × 0.125 equals 250 kcmil—the minimum permitted size of the neutral conductor run from the transformer to the service equipment. It is Code intent to permit the required 250-kcmil-sized neutral to be divided by the number of conduits. From NEC Table 8 in Chapter 9, it can be seen that four No. 2 conductors, each with a cross-section area of 66,360 circular mils, would approximate the area of one 250 kcmil (250,000 circular mils divided by 4 = 62,500 circular mils). But, because No. 1/0 is the smallest conductor that is permitted by Sec. 310-4 to be used in parallel for a circuit of this type, it would be necessary to use a No. 1/0 copper conductor in each of the four conduits, along

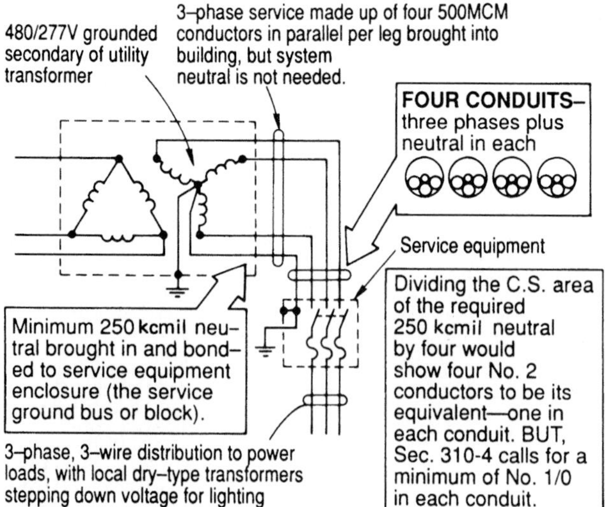

Fig. 250-19. Grounded service conductor must *always* be brought in. (Sec. 250-23.)

with the phase legs. It should be noted that Exception No. 4 to Sec. 310-4 would not apply unless the paralleled grounded conductors are being replaced at an "existing installation." It would be a literal violation to parallel No. 2s at a new installation.

250-24. Two or More Buildings or Structures Supplied from a Common Service. In Sec. 250-23(a), bonding of a panel neutral block (or the neutral bus or terminal in a switchboard, switch, or circuit breaker) to the enclosure is required in service equipment. Exception No. 2 of that section permits bonding of the neutral conductor on the load side of the service equipment in those cases where a panelboard (or switchboard, switch, etc.) is used to supply circuits in a building and the panel is fed from another building. This is covered in Sec. 250-24 which says that, where two or more buildings are supplied from a common service to a main building, a grounding electrode at each other building shall be connected to the AC system grounded conductor on the supply side of the building disconnecting means of a grounded system as shown in Fig. 250-20 or connected to the metal enclosure of the building disconnecting means of an ungrounded system. Those are the basic rules covered in parts **(a)** and **(b)** of this section. *But* Exception No. 1 to part **(a)** and Exception No. 1 to part **(b)** note that a grounding electrode at a separate building supplied by a feeder or branch circuit is not required where only one branch circuit is supplied and there is no noncurrent-carrying equipment in the building that requires grounding. An

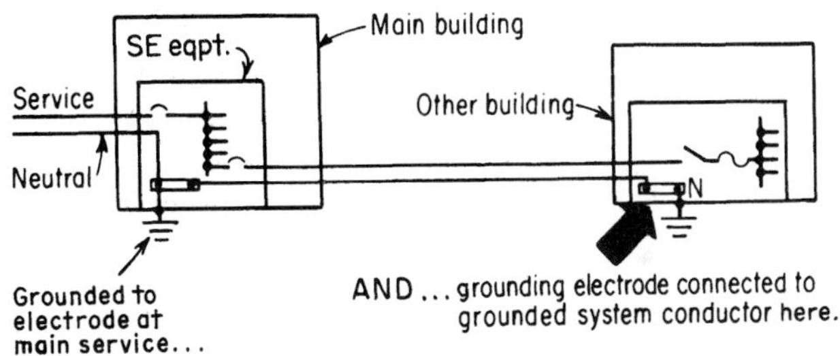

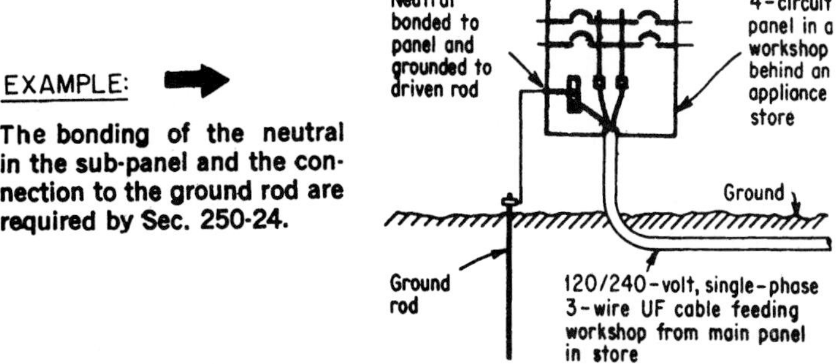

Fig. 250-20. Grounded conductor (e.g., a neutral) must be grounded at each building. (Sec. 250-24.)

example would be a small residential garage with a single lighting outlet or switch with no *metal* boxes, faceplates, or lighting fixtures within 8 ft (2.44 m) vertically or 5 ft (1.52 m) horizontally from a grounded condition.

Exception No. 2 to part **(a)** states that the grounded circuit conductor of a feeder to a separate building does not have to be bonded and grounded to a grounding electrode if an equipment grounding conductor is run with the circuit conductors for grounding any noncurrent-carrying equipment, water piping, or building metal frames in the separate building. (See Fig. 250-21.) And, as shown at the bottom of that illustration, the need for a grounding electrode at the outbuilding is eliminated because the words "equipment grounding conductor" as used in Exception No. 2 of Sec. 250-24(a) are understood to include metal "conduit" as indicated by Sec. 250-91(b), which recognizes an "equipment grounding conductor . . . enclosing the circuit conductors." If the separate building has an approved grounding electrode and/or interior metallic piping system, the equipment grounding conductor shall be bonded to the electrode and/or piping system. However, if the separate building does *not* have a grounding electrode—that is, does not have 10 ft (3.05 m) or more of underground metal water pipe, does not have grounded structural steel, and does not have any of the other electrodes recognized by Sec. 250-81—then at least one grounding electrode must be installed. That would most likely be a *made* electrode—such as a driven ground rod—and it must be bonded to the equipment ground terminal or equipment grounding bus in the enclosure of the panel, switchboard, circuit breaker, or switch in which the feeder terminates (Fig. 250-21).

When a grounding electrode connection is made to a grounded system conductor (usually a neutral) at a building that is fed from another building, the necessity for bonding the neutral block in such a subpanel is based on Secs. 250-24 and 250-54. The latter section says, "Where an ac system is connected to a grounding electrode in or at a building as specified in Sections 250-23 and 250-24, the same electrode shall be used to ground conductor enclosures and equipment in or on that building." Although the **Code** permits and even requires bonding at both ends, if the feeder circuit is in conduit, neutral current flows on the conduit because it is electrically in parallel with the neutral conductor, being bonded to it at both ends.

At one time, the rule of Exception No. 2 of part **(a)** of this section required bonding and grounding of the neutral conductor at any outbuilding in which livestock was housed. That bonding and grounding was required even if an equipment grounding conductor was run from the main building to the livestock building. That rigid rule has been removed from the **Code**, and it is no longer mandatory that the neutral must always be bonded and connected to a grounding electrode at an outbuilding housing livestock.

As was required in past **NEC** editions, the neutral had to be connected to a grounding electrode and bonded to the disconnect enclosure (and, as a result, bonded to the entire equipment grounding system of interconnected metal raceways and housings in the livestock building being fed). When a neutral is bonded in that way, without connection to a grounding electrode at the livestock building, any flow of normal unbalanced load current on the neutral produces a voltage drop on the neutral from the main disconnect in the livestock

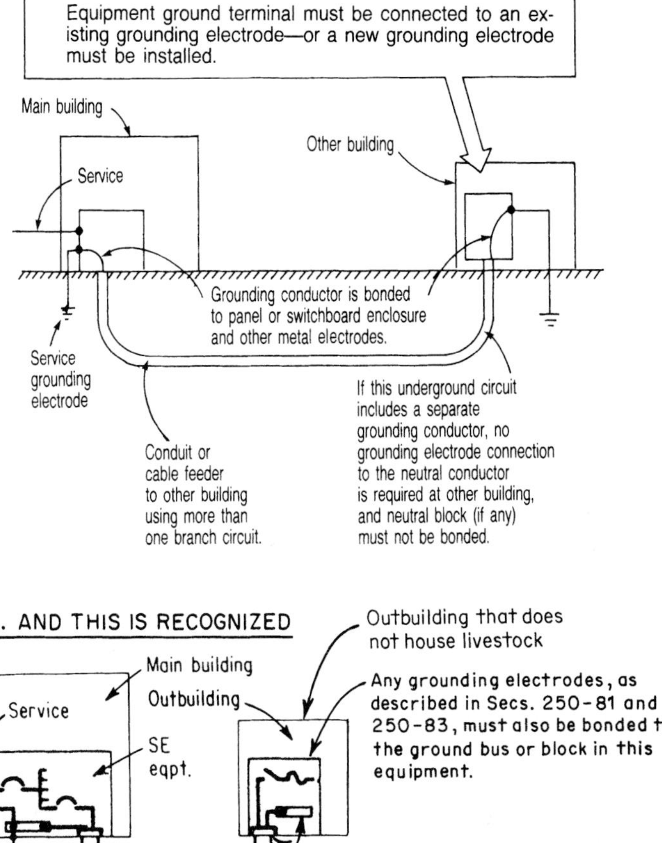
Equipment ground terminal must be connected to an existing grounding electrode—or a new grounding electrode must be installed.
Main building
Other building
Service
Grounding conductor is bonded to panel or switchboard enclosure and other metal electrodes.
Service grounding electrode
If this underground circuit includes a separate grounding conductor, no grounding electrode connection to the neutral conductor is required at other building, and neutral block (if any) must not be bonded.
Conduit or cable feeder to other building using more than one branch circuit.
. . . AND THIS IS RECOGNIZED
Outbuilding that does not house livestock
Main building
Outbuilding
Service
SE eqpt.
Any grounding electrodes, as described in Secs. 250-81 and 250-83, must also be bonded to the ground bus or block in this equipment.
Conduit is connected to metal enclosures at both ends by locknuts and bushings or by bonding jumpers.
Neutral
Metal conduit feeder to outbuilding consists of phase legs and neutral—but no separate grounding conductor.
Neutral block in panel or switchboard not bonded or grounded here

Fig. 250-21. Grounding connection at outbuildings may be eliminated. (Sec. 250-24.)

building back to the ground reference of the grounding electrode at the service of the main building. That voltage drop then appears as a potential to ground from equipment housings in the livestock building and can have an adverse effect on livestock. The reasoning behind elimination of the mandatory rule on neutral bonding in a livestock building was given as follows:

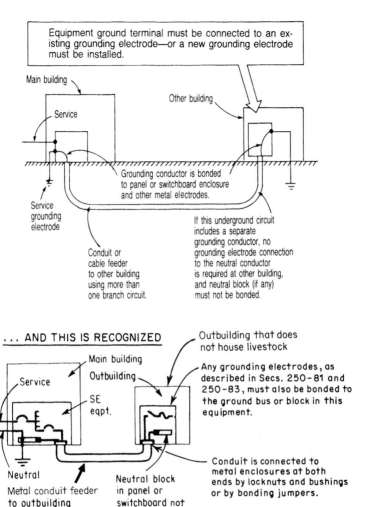

When livestock are housed in Building No. 2, and when a separate equipment grounding conductor is run from Building No. 1 to Building No. 2 for the purpose of grounding all metal equipment and parts, it shall be permissible to isolate the grounding conductor from the neutral in Building No. 2 if neutral-to-earth voltages cause distress to the confined livestock.

Substantiation: Neutral-to-earth voltages (stray voltages) are caused by many factors. One of the primary causes is voltage drop on secondary circuits due to circuit imbalance and long circuits. Second, voltage drops are imposed on neutral busbars in a building service entrance, which are in turn transmitted to metal grounding conductors, conduit, or panel grounding to a metallic water system. Livestock, particularly dairy animals and swine, are very sensitive to AC voltages that can occur when part of their body makes contact with the described metal equipment and part of their body is in contact with true earth. AC voltages over 1 V are known to cause dairy animals to go out of milk production and to inhibit the growth rate of swine.

(Note attached paper on stray voltage problems, page 11, item 7.) You will note that an alternate solution to resolving stray voltage problems is to isolate the neutral from the grounding conductors in the barn panel and run a separate fourth wire either back to the transformer or metering location (main farm service entrance).

By adding the proposed exception to NEC Sec. 250-24, it would make it legal or permissible to run the fourth wire and isolate the neutral and grounding wires at the panel serving building No. 2, which is assumed to be a grounded system having a grounding electrode connected to the neutral.

The last sentence of Exception No. 2 to part **(a)** does require that the equipment grounding conductor that is run with the feeder to the disconnect enclosure in a livestock building be an "insulated or covered copper" conductor and may not simply consist of metal conduit enclosing the feeder conductors. For such a feeder, the equipment grounding conductor must be insulated or covered copper *only* for an underground circuit or for any part of the circuit that is run underground. The purpose of that rule is to assure a more reliable equipment grounding path for such buildings to prevent potentials on equipment that would threaten livestock because of their great sensitivity to even very low voltages—as described earlier. An equipment grounding conductor run underground—directly buried or in a raceway—must be insulated or covered copper to prevent corrosion. *But* an overhead feeder to a barn building may be aluminum or copper multiplex cable with a bare messenger wire used as the equipment grounding conductor. Figure 250-22 shows the considerations in the Exception: (3) applies to any outbuilding; (1) and (2) apply only to a separate building housing livestock.

Figure 250-23 shows another condition in which a grounding electrode connection must be made at the *other* building, as specified in the basic rule of Sec. 250-24(b). For an ungrounded system, when, as shown in the sketch, an equipment grounding conductor is *not* run to the outbuilding, then a grounding electrode conductor must be run from the ground bus or terminal in the outbuilding disconnect to a suitable grounding electrode which *must be* provided.

According to Sec. 250-24(b), Exception No. 2, a circuit of an ungrounded system *does not* require a grounding electrode connection to the equipment ground terminal at an outbuilding disconnect *only* where the outbuilding uses not more than one branch circuit, where an equipment grounding conductor is run with the circuit conductors to the outbuilding, and where there are no existing electrodes at the outbuilding. If the ungrounded circuit to the outbuilding is a feeder supplying more than one branch circuit, then connection must be made from the

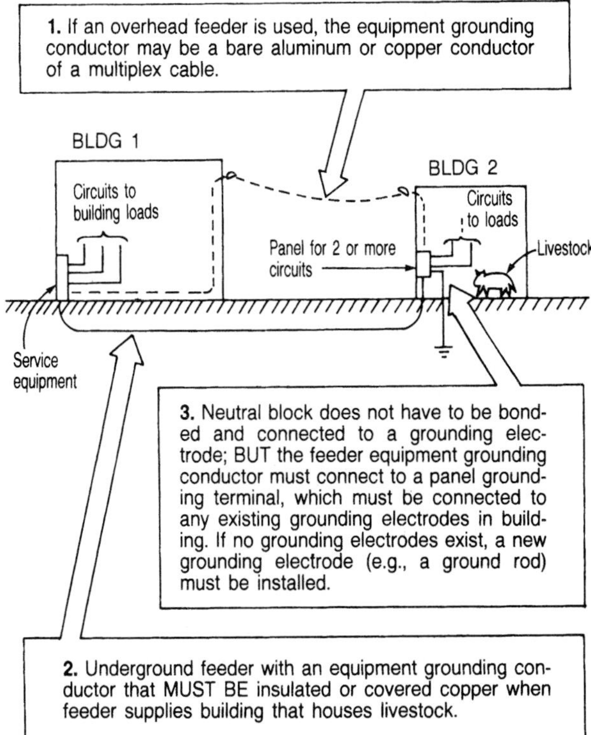

1. If an overhead feeder is used, the equipment grounding conductor may be a bare aluminum or copper conductor of a multiplex cable.
BLDG 1
Circuits to building loads
BLDG 2
Circuits to loads
Panel for 2 or more circuits
Livestock
Service equipment
3. Neutral block does not have to be bonded and connected to a grounding electrode; BUT the feeder equipment grounding conductor must connect to a panel grounding terminal, which must be connected to any existing grounding electrodes in building. If no grounding electrodes exist, a new grounding electrode (e.g., a ground rod) must be installed.
2. Underground feeder with an equipment grounding conductor that MUST BE insulated or covered copper when feeder supplies building that houses livestock.

Fig. 250-22. Building for livestock must be carefully grounded. [Sec. 250-24(a), Exception No. 2.]

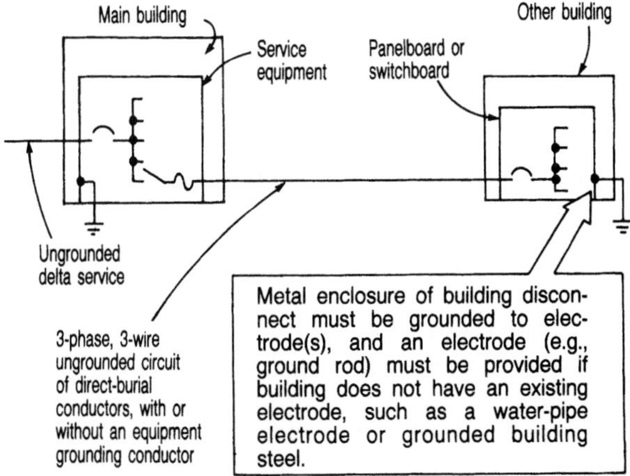

Fig. 250-23. Grounding connection for an ungrounded supply to outbuilding. (Sec. 250-24.)

equipment grounding terminal in the outbuilding disconnect to either an existing ground electrode at the building (water pipe, building steel, etc.) or to a new ground electrode installed for the purpose. And such a connection to a ground electrode must be made even if an equipment grounding conductor is run with the feeder conductors to the outbuilding. If the building houses livestock, that portion of the equipment grounding conductor that is run underground to the outbuilding must be insulated or covered copper.

In Fig. 250-23, if the 3-phase, 3-wire, ungrounded feeder circuit to the outbuilding had been run with a separate equipment grounding conductor which effectively connected the metal enclosure of the disconnect in the outbuilding to the grounding electrode conductor in the SE equipment of the main building, a connection to a grounding electrode would not be required provided that conditions a., b., c., and d. of Exception No. 2 are satisfied. But if all those conditions are not satisfied and a grounding electrode connection would be required at the outbuilding, then the equipment grounding conductor run to the outbuilding would have to be bonded to any grounding electrodes that were "existing" at that building—such as an underground metal water service pipe and/or a grounded metal frame of the building. All grounding electrodes that *exist* at the outbuilding must be bonded to the ground bus or terminal in the disconnect at the outbuilding, whether or not an equipment grounding conductor is run with the circuit conductors from the main building.

Part **(c)** covers design of the grounding arrangement for a feeder from one building to another building when the main disconnect for the feeder is at a remote location from the building being supplied—such as in the other building where the feeder originates. The rule prohibits grounding and bonding of a feeder to a building from another building *if* the disconnect for the building being fed is located in the building where the feeder originates. Part **(c)** correlates the grounding concepts of Sec. 250-24 with the disconnect requirements of Sec. 225-8(b), Exceptions No. 1 and No. 2. This is most easily understood by considering the following sequence of **Code** rules involved:

1. First, in Sec. 250-23(a), bonding of a panel neutral block (or the neutral bus or terminal in a switchboard, switch, or circuit breaker) to the enclosure is required in service equipment. Exception No. 2 of that section permits bonding of the neutral conductor on the load side of the service equipment in those cases where a panelboard (or switchboard, switch, etc.) is used to supply circuits in a building and the panel or switchboard in that building is supplied by a feeder from another building. This is covered in Sec. 250-24(a), which says that where two or more buildings are supplied from a common service to a main building, a grounding electrode at each other building shall be connected to the AC system grounded conductor on the supply side of the building disconnect means of a grounded system and the neutral must be bonded to the metal enclosure of the disconnect (the enclosure of the panelboard or switchboard or switch or circuit breaker) in the building being supplied. The basic rule calls for grounding and bonding within the building disconnect for the building being fed. That is, the feeder to the second building is treated like a service to the building.

2. But, Exception No. 2 of Sec. 250-24(a) gives the *option* of treating the feeder to a building from another building exactly like a feeder that did not leave the main building—that is, running an equipment grounding conductor and *not* grounding and bonding the feeder neutral in the building disconnect. Exception No. 2 to part **(a)** of Sec. 250-24 states that the grounded circuit conductor of a feeder to a separate building does not have to be bonded and grounded to a grounding electrode if an equipment grounding conductor is run with the circuit conductors for grounding any noncurrent-carrying equipment, water piping, or building metal frames in the separate building. If the separate building has an approved grounding electrode and/or interior metallic piping system, the equipment grounding conductor shall be bonded to the electrode and/or piping system. However, if the separate building does *not* have a grounding electrode—that is, does not have 10 ft (3.05 m) or more of underground metal water pipe, does not have grounded structural steel, and does not have any other electrodes recognized by Sec. 250-81—then at least one grounding electrode must be installed. That would most likely be a *made* electrode—such as a driven ground rod—and it must be bonded to the equipment ground terminal or equipment grounding bus in the enclosure of the panelboard, switchboard, circuit breaker, or switch in which the feeder terminates.
3. Although the use of an equipment grounding conductor and elimination of the need for a bonded connection to a grounding electrode, as described in (2) above, is given as an optional alternative to (1) above, part **(c)** of Sec. 250-24 makes elimination of a bonded connection to an electrode a *mandatory* method when the disconnect for the feeder is in the other building where the feeder originates, as covered by Sec. 225-8 (Fig. 250-24).
4. The basic rule of Sec. 225-8(b) says that for a group of buildings under single management, disconnect means must be provided for each building. This rule requires that the conductors supplying each building in the group be provided with a means for disconnecting all ungrounded conductors from the supply. And the rule specifically permits the disconnect for each building or structure to be the same kind as permitted for a service disconnect—that is, up to six switches or CBs, as covered in Sec. 230-71.

For large-capacity multibuilding industrial premises with a single owner, Exception No. 1 of Section 225-8(b) permits use of the feeder switch in the main building as the only disconnect for a feeder to an outlying building, provided the disconnect in the main building is accessible to the occupants of the outlying building. Now, when that option of Exception No. 1 of Sec. 225-8(b) is selected, eliminating a main disconnect for the feeder within the outbuilding being supplied, part **(c)** of Sec. 250-24 requires the following conditions to be met:

1. The grounded (neutral) conductor of the feeder *must not* be connected to a grounding electrode at the outbuilding and the grounded conductors must *not* be bonded to the panelboard or switchboard enclosure in the outbuilding.

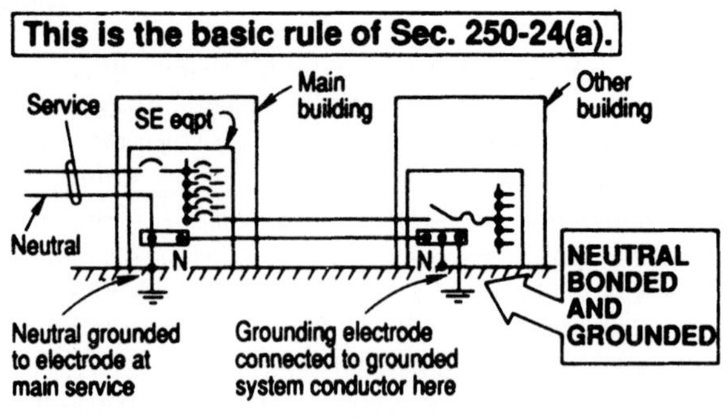
This is the basic rule of Sec. 250-24(a).
Service
SE eqpt
Main building
Other building
Neutral
N
N
NEUTRAL BONDED AND GROUNDED
Neutral grounded to electrode at main service
Grounding electrode connected to grounded system conductor here

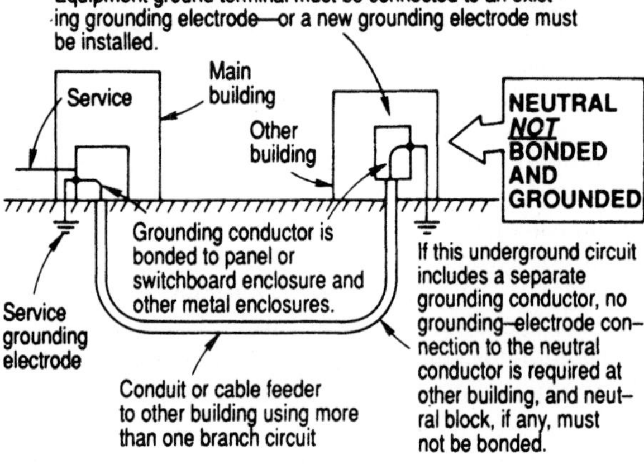
This optional alternative becomes mandatory . . .
Equipment ground terminal must be connected to an existing grounding electrode—or a new grounding electrode must be installed.
Main building
Service
Other building
NEUTRAL NOT BONDED AND GROUNDED
Grounding conductor is bonded to panel or switchboard enclosure and other metal enclosures.
Service grounding electrode
If this underground circuit includes a separate grounding conductor, no grounding-electrode connection to the neutral conductor is required at other building, and neutral block, if any, must not be bonded.
Conduit or cable feeder to other building using more than one branch circuit

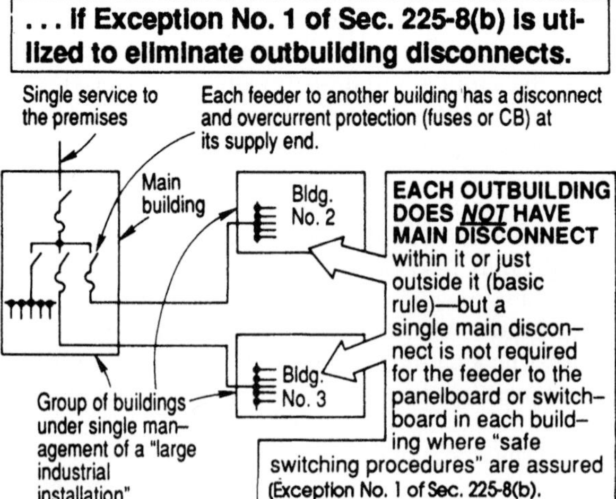
. . . If Exception No. 1 of Sec. 225-8(b) is utilized to eliminate outbuilding disconnects.
Single service to the premises
Each feeder to another building has a disconnect and overcurrent protection (fuses or CB) at its supply end.
Main building
Bldg. No. 2
Bldg. No. 3
EACH OUTBUILDING DOES NOT HAVE MAIN DISCONNECT within it or just outside it (basic rule)—but a single main disconnect is not required for the feeder to the panelboard or switchboard in each building where "safe switching procedures" are assured (Exception No. 1 of Sec. 225-8(b).
Group of buildings under single management of a "large industrial installation"

Fig. 250-24. Part (c) correlates grounding for multiple buildings under a single management with rule of Sec. 225.8(b) on location of feeder disconnect for an outbuilding. [Sec. 250-24(c).]

2. An equipment grounding conductor *must* be run with the feeder circuit conductors to the outbuilding to ground metal equipment enclosures, metal conduit, and other noncurrent-carrying metal parts of electrical equipment in the outbuilding. Any interior metal piping and building and structural metal frames in the outbuilding *must* be bonded to the equipment grounding conductor. In addition, any grounding electrodes at the outbuilding must be bonded to the feeder equipment grounding conductor. And if the outbuilding contains no grounding electrodes as covered in Sec. 250-81, a made electrode (a ground rod, etc.) must be installed and bonded to the equipment grounding conductor of the feeder.
3. The *bond* between the feeder equipment grounding conductor and any grounding electrodes at the outbuilding *must* be made in a junction box "located immediately inside or outside the second building or structure."

As follow-up to part **(c)**, the rule of part **(d)** says the conductor used to bond the feeder equipment grounding conductor to the one or more existing or new grounding electrodes at an outbuilding must be sized from Table 250-95, based on the rating of the overcurrent device protecting the ungrounded conductors of the feeder. This is a rule to clarify the sizing of the bonding conductor to building electrodes as required by Exception No. 2 of Sec. 250-24(a) and part **(2)** of Sec. 250-24(c). In effect, the wording of this rule in part **(d)** calls for the bonding conductor to be the same size as the equipment grounding conductor of the feeder, based on Table 250-95. And the same size of conductor would be used to connect to the interior metal piping systems and structural metal frame of the outbuilding.

Exception No. 1 notes that the grounding conductor would never have to be larger than the circuit ungrounded conductors. And Exception No. 2 says the grounding conductor would not have to be larger than No. 6 copper or No. 4 aluminum where it connects to a driven ground rod or other made electrode.

250-25. Conductor to Be Grounded—Alternating-Current Systems. Selection of the wiring system conductor to be grounded depends upon the type of system. In 3-wire, single-phase systems, the midpoint of the transformer winding—the point from which the system neutral is derived—is grounded. For grounded 3-phase wiring systems, the neutral point of the wye-connected transformer(s) or generator is the point connected to ground. In delta-connected transformer hookups, grounding of the system can be effected by grounding one of the three phase legs, by grounding a center-tap point on one of the transformer windings (as in the 3-phase, 4-wire "red-leg" delta system), or by using a special grounding transformer which establishes a neutral point of a wye connection which is grounded.

250-26. Grounding Separately Derived Alternating-Current Systems. A separately derived AC wiring system is a source derived from an on-site generator (emergency or standby), a battery-inverter, or the secondary supply of a transformer. Any such AC supplies required to be grounded by Sec. 250-5 must comply with Sec. 250-26:

1. Any system which operates at over 50 V but not more than 150 V to ground must be grounded [Sec. 250-5(b)].
2. This requires the grounding of generator windings and secondaries of transformers serving 208/120-V, 3-phase or 240/120-V, single-phase cir-

cuits for lighting and appliance outlets and receptacles, at loadcenters throughout a building, as shown for the very common application of dry-type transformers in Fig. 250-25.

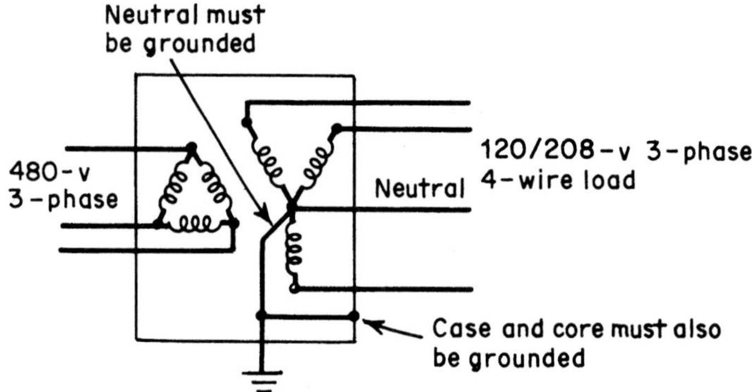

Fig. 250-25. Grounding is required for "separately derived" systems. (Sec. 250-26.)

3. All Code rules applying to both system and equipment grounding must be satisfied in such installations.

Referring to Fig. 250-26, the steps involved in satisfying the Code rules are as follows:

Step 1—Sec. 250-26(a)

A bonding jumper must be installed between the transformer secondary neutral terminal and the metal case of the transformer. The size of this bonding conductor is based on Sec. 250-79(c) and is selected from Table 250-94 of the Code—based on the size of the transformer secondary phase conductors and selected to be the same size as a required grounding electrode conductor. For cases where the transformer secondary circuit is larger than 1,100 kcmil copper or 1,750 kcmil aluminum per phase leg, the bonding jumper must be not less than 12½ percent of the cross-section area of the secondary phase leg.

example Assume this is a 75-kVA transformer with a 120/208-V, 3-phase, 4-wire secondary. Such a unit would have a full-load secondary current of

$$75{,}000 \div (208 \times 1.732) \text{ or } 209 \text{ A}$$

If we use No. 4/0 THW copper conductors for the secondary phase legs (with a 230-A rating), we would then select the size of the required bonding jumper from Table 250-94 as if we had 4/0 service conductors. The table shows that 4/0 copper service conductors require a minimum of No. 2 copper or No. 1/0 aluminum for a grounding electrode conductor. And the bonding jumper would have to be either of those two sizes.

If the transformer was a 500-kVA unit with a 120/208-V secondary, its rated secondary current would be

$$\frac{500 \times 1{,}000}{1.732 \times 208} = 1{,}388 \text{ A}$$

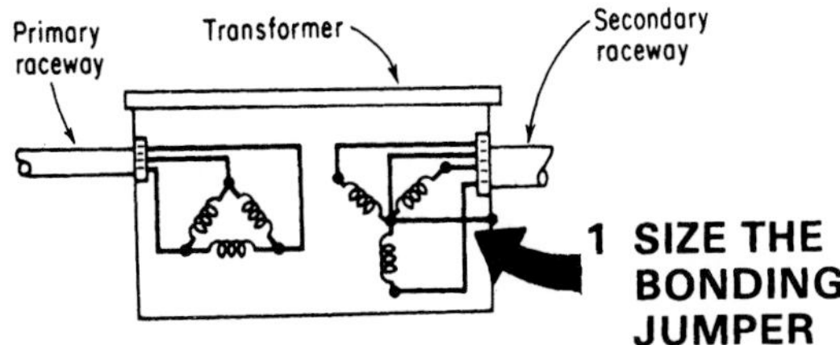

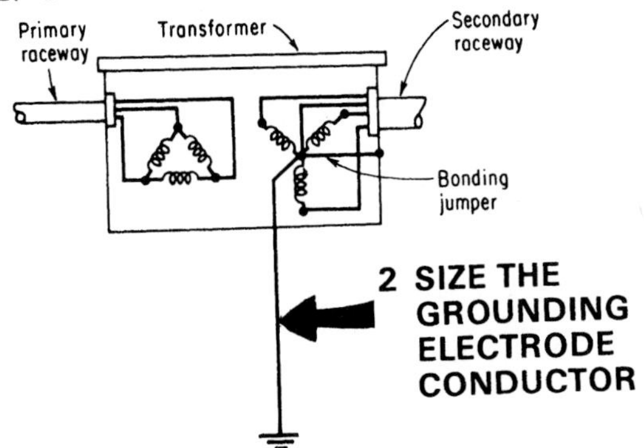

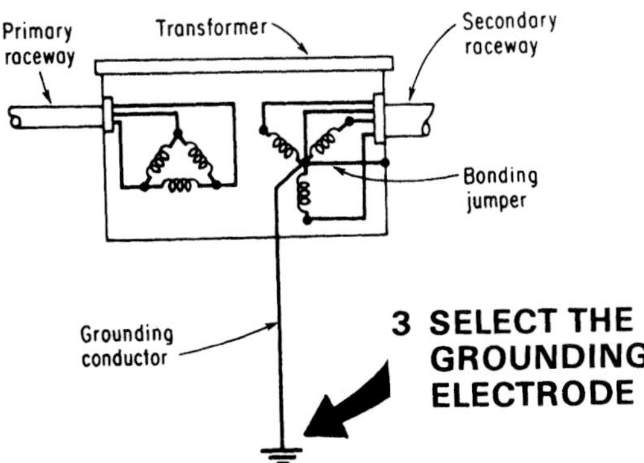

Fig. 250-26. Grounding a transformer secondary. (Sec. 250-26.)

Using, say, THW aluminum conductors, the size of each secondary phase leg would be four 700 kcmil aluminum conductors in parallel (each 700 kcmil THW aluminum is rated at 375 A, four are 4 × 375 or 1,500 A, which suits the 1,388-A load).

Then, because 4 × 700 kcmil equals 2,800 kcmil per phase leg and is in excess of 1,750 kcmil, Sec. 250-79(c) requires the bonding jumper from the case to the neutral terminal to be at least equal to 12½ percent × 2,800 kcmil (0.125 × 2,800) or 350 kcmil aluminum.

Step 2—Sec. 250-26(b)

A grounding electrode conductor must be installed from the transformer secondary neutral terminal to a suitable grounding electrode. This grounding conductor is sized the same as the required bonding jumper in Step 1. That is, this grounding electrode conductor is sized from Table 250-94 as if it is a grounding electrode conductor for a service with service-entrance conductors equal in size to the phase conductors used on the transformer secondary side. But, this grounding electrode conductor does *not* have to be larger than 3/0 copper or 250 kcmil aluminum when the transformer secondary circuit is over 1,100 kcmil copper or 1,750 kcmil aluminum.

example For the 75-kVA transformer in Step 1, the grounding electrode conductor must be not smaller than the required minimum size shown in Table 250-94 for 4/0 phase legs, which makes it the same size as the bonding jumper—i.e., No. 2 copper or No. 1/0 aluminum. But, for the 500-kVA transformer, the grounding electrode conductor is sized directly from Table 250-94—which requires a 3/0 copper or 250 kcmil aluminum where the phase legs are over 1,100 kcmil copper or 1,750 kcmil aluminum.

The last sentence of Sec. 250-26(a) and (b) permits the bonding and grounding connections to be made either right at the transformer or generator or at the first disconnect or overcurrent device fed from the transformer or generator, as in Fig. 250-27.

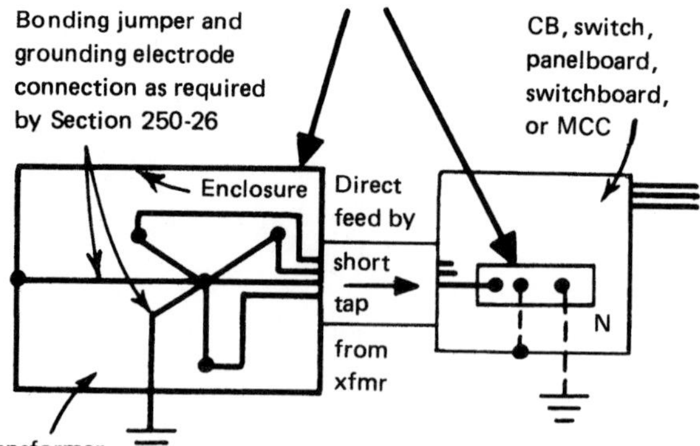

Fig. 250-27. Transformer secondary bonding and grounding must be "at the source" or at a secondary disconnect or protective device. (Sec. 250-26.)

The last sentence of Sec. 250-26(a) and (b), however, does say that the bonding and grounding "shall be made at the *source*"—which appears to indicate right at the transformer or generator, and *not* any other point—where the transformer supplies a system that "has no disconnecting means or overcurrent device." A local transformer that "has no disconnect or overcurrent devices" on its secondary is one that supplies only a single circuit and has overcurrent protection on its primary, such as a control transformer to supply motor starter coils. But in such applications, the transformer does not supply a separately derived "system" in the usual sense that a "system" consists of more than one circuit. It is true, however, that for a transformer supplying only one circuit, any required bonding and grounding of a secondary grounded conductor—such as the grounded leg of a two-wire 120-V control circuit—would normally have to be done right at the transformer.

Another interpretation that might be put on that last phrase is that it refers to a hookup in which the transformer secondary feeds *main lugs only* in a panelboard, switchboard, or motor control center. In such cases, the absence of a main CB or fused switch would mean that there is no disconnect means or overcurrent device for the overall "system" fed by the transformer, even though there are disconnects and overcurrent protection for the individual "circuits" that make up the "system." That interpretation would require bonding and grounding of the secondary right *at the source* (the transformer itself).

In both part **(a)** and part **(b)** of this section, Exception No. 2 exempts high-impedance grounded transformer secondaries or generator outputs from the need to provide direct (solid) bonding and grounding-electrode connections of the neutral. This simply states an exception to each part that is necessary to operate a high-impedance grounded system.

Step 3—Sec. 250-26(c)

The grounding electrode conductor, installed and sized as in Step 2, must be properly connected to a grounding electrode which must be "as near as practicable to and preferably in the same area as the grounding conduction connection to the system." That is, the grounding electrode must be as near as possible to the transformer itself. In order of preference, the grounding electrode must be:

1. The nearest available structural steel of the building, provided it is established that such building steel is effectively grounded.
2. The nearest available metal water pipe, provided it is effectively grounded. Although Section 250-112 clarifies the term "effectively grounded" by noting that the grounding connection to a grounding electrode must "assure a permanent and effective ground," it should be noted that the rule of Sec. 250-81(a) prohibits metallic water piping located more than 5 ft (1.52 m) from the "point of entrance to the building" from serving as a grounding electrode or grounding electrode conductor (Fig. 250-28). Thus, even though any metallic piping in the vicinity of a separately derived system must be bonded to the grounded conductor, a nearby cold water pipe cannot be used as a grounding electrode.

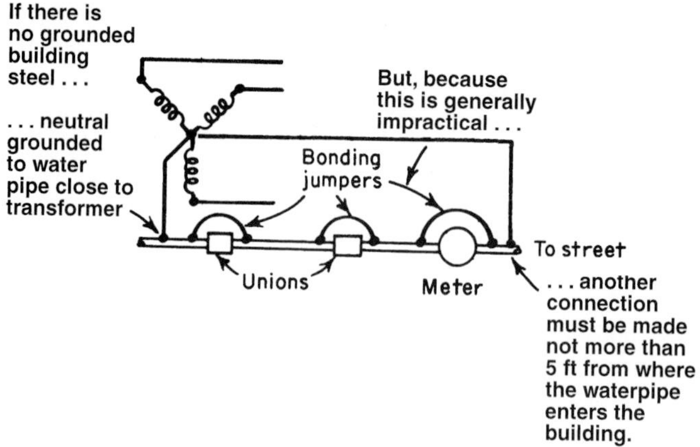
If there is no grounded building steel . . .
. . . neutral grounded to water pipe close to transformer
But, because this is generally impractical . . .
Bonding jumpers
To street
Unions
Meter
. . . another connection must be made not more than 5 ft from where the waterpipe enters the building.

Fig. 250-28. Sec. 250-112 defines "effectively grounded" water-pipe electrode. (Sec. 250-26.)

If there is no effectively grounded building steel, a connection to the water pipe must be made within 5 ft (1.52 m) from where the pipe enters the building. Even though the rule of Sec. 250-80(a) *mandates* bonding of the water piping in the vacinity of the separately derived system, Sec. 250-81 prohibits using the water piping beyond 5 ft (1.52 m) from the entry point as a grounding electrode or as a grounding electrode conductor. Therefore, two connections from the grounded (neutral) conductor of the separately derived system must be made to the water piping—one locally and one at the point of entry. The FPN following part **(c)** is intended to call this to the reader's attention. The size of such bonding jumpers must be at least the same size as the grounding conductor from the transformer to the water pipe and other electrodes. Of course, the water piping system must satisfy Sec. 250-80. There must be at least 10 ft (3.05 m) of the metal water piping buried in earth outside the building for the water-pipe system to qualify as a grounding electrode. There must always be a connection between an interior metal water piping system and the service-entrance grounded conductor (the neutral of the system which feeds the primary of any transformers in the building). That grounding connection for the neutral or other system grounded conductor must be made at the service. And where a metallic water piping system in a building is fed from a nonmetallic underground water system or has less than 10 ft (3.05 m) of metal pipe underground, the service neutral or other service grounded conductor must have a connection to a ground rod or other electrode in addition to the connection to the interior metal water piping system. Refer to Secs. 250-80 and 250-81. Where building steel or a metal water pipe is not available for grounding of local dry-type units, other electrodes may be used, based on Secs. 250-81 or 250-83.

Figure 250-29 shows techniques of transformer grounding that have been used in the past but are no longer acceptable along with an example of "case grounding" that *is* specifically recognized by the Exceptions to Sec. 250-26(a) and (b).

NOT PERMITTED FOR POWER TRANSFORMERS

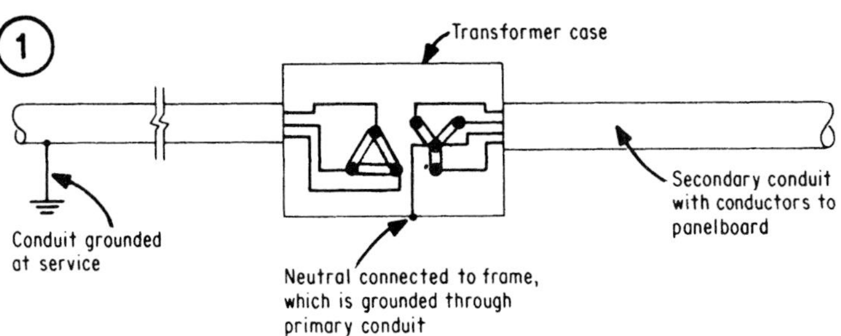

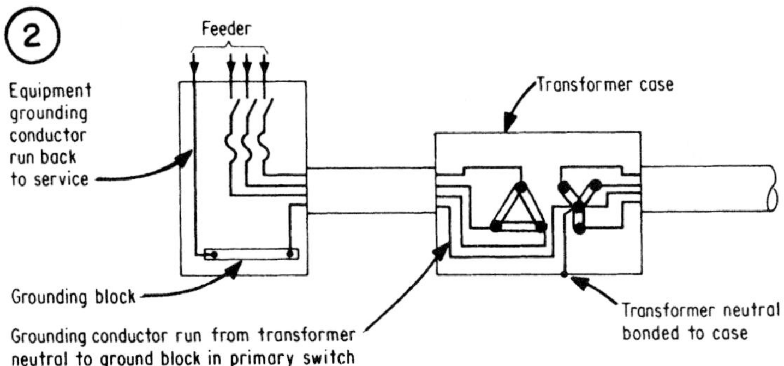

THIS IS PERMITTED

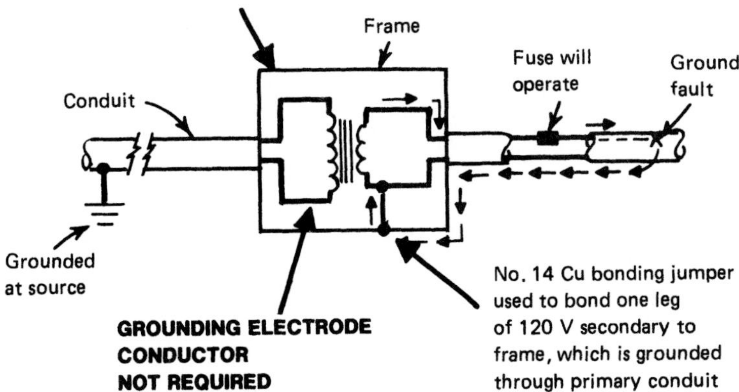

Fig. 250-29. Code rules regulate specific hookups for grounding transformer secondaries. (Sec. 250-26.)

Exception No. 1 to Sec. 250-26(b) exempts small transformers for control, signal, or power-limited circuits from the basic requirement for a grounding electrode conductor run from the bonded secondary grounded conductor (such as a neutral) to a grounding electrode (nearby building steel or a water pipe). Exception No. 1 to both parts **(a)** and **(b)** in this section apply to transformers used to derive control circuits, signal circuits, or power-limited circuits, such as circuits to damper motors in air-conditioning systems. A Class 1, Class 2, or Class 3 remote-control or signaling transformer that is rated not over 1,000 VA simply has to have a grounded secondary conductor bonded to the metal case of the transformer, and no grounding electrode conductor is needed, *provided that* the metal transformer case itself is properly grounded by grounded metal raceway which supplies its primary or by means of a suitable (Sec. 250-57) equipment grounding conductor that ties the case back to the grounding electrode for the primary system, as indicated at the bottom of Fig. 250-29. Exception No. 1 to Sec. 250-26(a) permits use of a No. 14 copper conductor to bond the grounded leg of the transformer secondary to the transformer frame, leaving the supply conduit to the transformer to provide the path to ground back to the main service ground, but depending on the connection between neutral and frame to provide effective return for clearing faults, as shown. Grounding of transformer housings must be made by connection to grounded cable armor or metal raceway or by use of a grounding conductor run with circuit conductor (either a bare conductor or a conductor with green-colored covering).

Because the rule on bonding jumpers for the secondary neutral point of a transformer refers to Sec. 250-79(c) and therefore ties into Table 250-94, the smallest size that may be used is No. 8 copper, as shown in that table. But for small transformers—such as those used for Class 1, Class 2, or Class 3 remote-control or signaling circuits—that large a bonding jumper is not necessary and is not suited to termination provisions. For that reason, Exception No. 1 to Sec. 250-26(a) and (b) permits the bonding jumper for such transformers rated not over 1,000 VA to be smaller than No. 8. The jumper simply has to be at least the same size as the secondary phase legs of the transformer and in no case smaller than No. 14 copper or No. 12 aluminum.

250-27. High-Impedance Grounded Neutral System Connections. This complete section covers grounding connections for a high-impedance grounded neutral system. When high-impedance grounding is used for a 480- to 1,000-V system in accordance with the conditions of Exception No. 5 of Sec. 250-5(b), the grounding connections must be made in compliance with the rules of this section, as follows:

1. The grounding impedance (usually a resistor) must be connected between the system neutral point and the grounding electrode conductor. The neutral point may be that of a wye transformer connection, or a neutral point may be derived from a 480-V delta system by using a zigzag grounding autotransformer.
2. The neutral conductor from the neutral point to the grounding impedance must be fully insulated because it is operating at a substantial voltage above ground.
3. The system neutral must not be connected to ground except through the impedance unit.

4. The neutral conductor from the neutral point to the grounding impedance may be installed in a separate raceway.
5. The equipment bonding jumper (the connection between the system equipment grounding conductors and the grounded end of the grounding impedance) must be an unspliced conductor run from the first disconnect or overcurrent device of the system to the ground end of the impedance device.
6. The grounding electrode conductor must be connected anywhere from the ground end of the impedance to the equipment ground bus or terminal in the service equipment or the first disconnect means.

250-33. Other Conductor Enclosures and Raceways. The basic rule requires all enclosures and raceways to be grounded. But, Exception No. 1 permits the installation of short runs as extensions from existing open wiring, knob-and-tube work, or nonmetallic-sheathed cable without grounding where there is little likelihood of an accidental connection to ground or of a person touching both the conduit, raceway, or armor and any grounded metal or other grounded surface at the same time. Additionally, the Exception permits "short sections" of enclosures and raceways used as sleeving or to otherwise support cables. The exact length of such ungrounded metallic enclosures and raceways is *not* given. That determination will ultimately be made by the local electrical inspector.

250-42. Equipment Fastened in Place or Connected by Permanent Wiring Methods (Fixed). The word "fixed" as applied to equipment requiring grounding now applies to "equipment fastened in place or connected by permanent wiring," as shown in Fig. 250-30. And that usage is consistently followed in other Code sections. As noted in Exception No. 4, enclosures for listed equipment—such

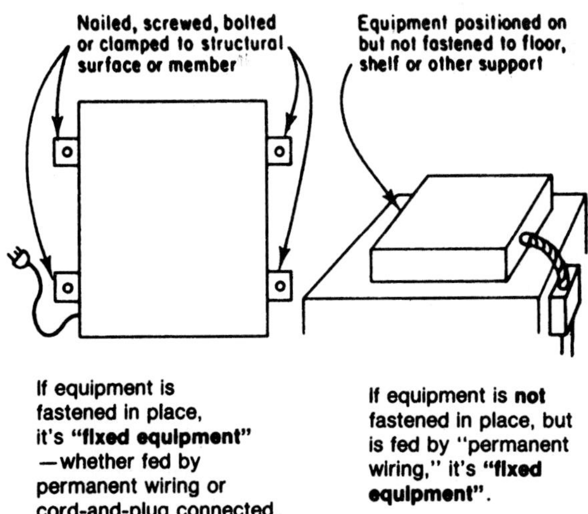
Nailed, screwed, bolted or clamped to structural surface or member
Equipment positioned on but not fastened to floor, shelf or other support
If equipment is fastened in place, it's "fixed equipment" —whether fed by permanent wiring or cord-and-plug connected.
If equipment is not fastened in place, but is fed by "permanent wiring," it's "fixed equipment".

Fig. 250-30. "Fixed" equipment is now clearly and readily identified for grounding rules. (Sec. 250-42.)

as data processing equipment and listed office equipment—operating at over 150 V to ground do not have to be grounded if protected by a system of double insulation or its equivalent.

250-43. Fastened in Place or Connected by Permanent Wiring Methods (Fixed)—Specific. This section presents what amounts to an incomplete, yet binding, roster of equipment that must be grounded. The various subparts identify those items that are *specifically* required to be grounded. Of particular note is that part **(g)** requires that *all* electric signs and associated equipment must have exposed, noncurrent-carrying metal parts grounded.

As required by part **(k)**, motor-operated water pumps, including the submersible type, must have their metal frames grounded. This Code rule clarifies an issue that was a subject of controversy. It means that a circuit down to a submersible pump in a well or cistern must include a conductor to ground the pump's metal frame even though the frame is not accessible or exposed to contact by persons.

Notice, too, that Section 250-43(l) makes grounding of metal well-casings mandatory for ALL types of occupancies. The grounding of metal well-casings was required for "Agricultural Buildings" in Sec. 547-8(d). In recognition of the fact that the shock hazard exists where a metal well-casing is used—regardless of the type of occupancy or installation—this section requires grounding ALL metal well-casings used with "submersible pumps." However, that section was eliminated in the 1996 NEC because it was viewed as redundant. Remember, according to the rule of Sec. 90-3, any requirements put forth in Chapters 1–4 apply to those special occupancies covered in Chapter 5, unless otherwise modified by the rules in Chapter 5. Therefore, extension of this requirement to other occupancies by inclusion in Sec. 250-43 made the wording in Sec. 547-8(d) superfluous and that section was deleted in the 1996 Code.

250-45. Equipment Connected by Cord and Plug. Figure 250-31 shows cord-connected loads that must either be operated grounded or be double-insulated. Except when supplied through an isolating transformer as permitted in paragraph **(d)** of this section, the frames of portable tools should be grounded by means of an equipment grounding conductor in the cord or cable through which the motor is supplied. Portable hand lamps used inside boilers or metal tanks should preferably be supplied through isolating transformers having a secondary voltage of 50 V or less, with the secondary ungrounded. Code-recognized double-insulated tools and appliances may be used in all types of occupancies other than hazardous locations, in lieu of required grounding.

OSHA regulations have made NE Code Sec. 250-45 retroactive, requiring grounded operation of cord- and plug-connected appliances in all existing as well as new installations. Check on local ruling on that matter.

250-46. Spacing from Lightning Rods. Lightning discharges with their steep wave fronts build up tremendous voltages to metal near the lightning rods, so the 6-ft (1.83-m) separation or bonding is required to prevent flashover with its attendant hazard.

250-50. Equipment Grounding Conductor Connections. Part **(a)** requires that the equipment grounding conductor at a service—such as the ground bus or terminal in the service-equipment enclosure, or the enclosure itself—must be con-

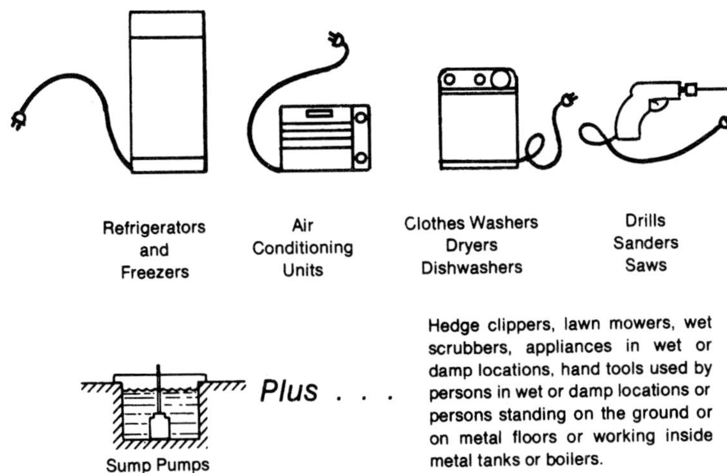
Refrigerators
and
Freezers
Air
Conditioning
Units
Clothes Washers
Dryers
Dishwashers
Drills
Sanders
Saws
Sump Pumps
Plus . . .
Hedge clippers, lawn mowers, wet scrubbers, appliances in wet or damp locations, hand tools used by persons in wet or damp locations or persons standing on the ground or on metal floors or working inside metal tanks or boilers.
NOTE: USE OF DOUBLE INSULATION ON TOOLS OR APPLIANCES ELIMINATES NEED FOR GROUNDING.

Fig. 250-31. Grounding cord and plug cap are required for shock protection. (Sec. 250-45.)

nected to the system *grounded* conductor (the neutral or grounded phase leg). The equipment ground and the neutral or other grounded leg must be bonded together and it must be done on the supply side of the service disconnecting means—which means either *within* or *ahead* of the enclosure for the service equipment (Fig. 250-32).

Part **(b)** requires the ground bus or the enclosure to be simply bonded to the grounding electrode conductor within or ahead of the service disconnect for an ungrounded system.

As shown at the top of Fig. 250-33, some switchboard sections or interiors include neutral busbars factory-bonded to the switchboard enclosure and are marked "suitable for use only as service equipment." They may not be used as subdistribution switchboards—i.e., they may not be used on the load side of the service except where used, with the inspector's permission, as the first disconnecting means fed by a transformer secondary or a generator and where the bonded neutral satisfies Sec. 250-26(a) for a separately derived system.

As illustrated at the bottom of Fig. 250-33, the Exception for parts **(a)** and **(b)** describes an accepted technique for using grounding-type receptacles for replacement of existing nongrounding devices or for circuit extensions—on wiring systems that do not include an equipment grounding conductor.

250-51. Effective Grounding Path. This section sets forth basic rules on the effectiveness of grounding. In effect, this rule defines the phrase "effective grounding path" and establishes *mandatory* requirements on the quality and quantity of conditions in any and every grounding circuit. The three parts of the section "shall" be satisfied.

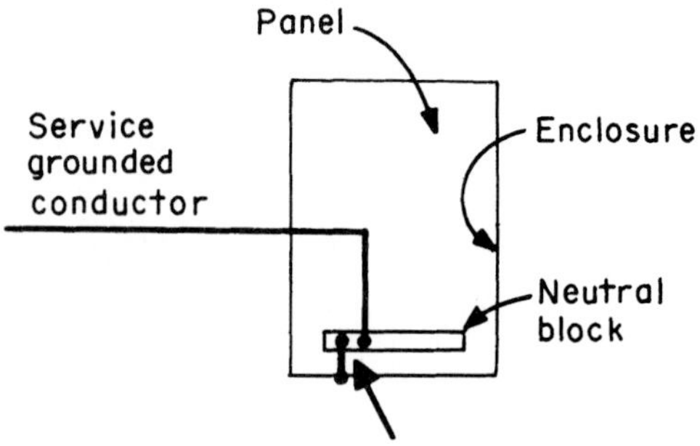

Bonding is the insertion of a bonding screw into the panel neutral block to connect the block to the panel enclosure, or it is use of a bonding jumper from the neutral block to an equipment grounding block that is connected to the enclosure.

NOTE: Bonding—the connection of the neutral terminal to the enclosure or to the ground terminal that is, itself, connected to the enclosure—might also be done in an individual switch or CB enclosure.

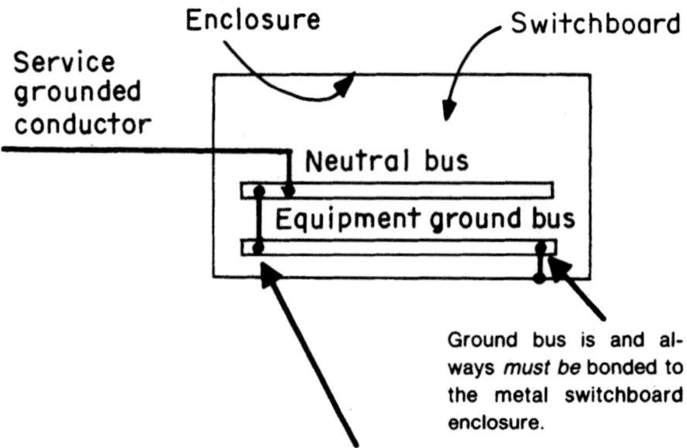

Ground bus is and always *must be* bonded to the metal switchboard enclosure.

Bonding of the neutral is the connection between the neutral bus and the equipment grounding bus or between the neutral bus and the metal enclosure itself.

Fig. 250-32. Equipment ground must be "bonded" to grounded conductor at the service equipment. (Sec. 250-50.)

Because each of the three required characteristics of grounding paths set forth in **(1)**, **(2)**, and **(3)** is a real and important factor of safety, it is logical and desirable that compliance with the Code hinges on carefully establishing each separate condition:

(1). That every ground path is "permanent and continuous" can be established by the installer by proper mounting, coupling, and terminating of the conductor or raceway intended to serve as the grounding conductor [as permit-

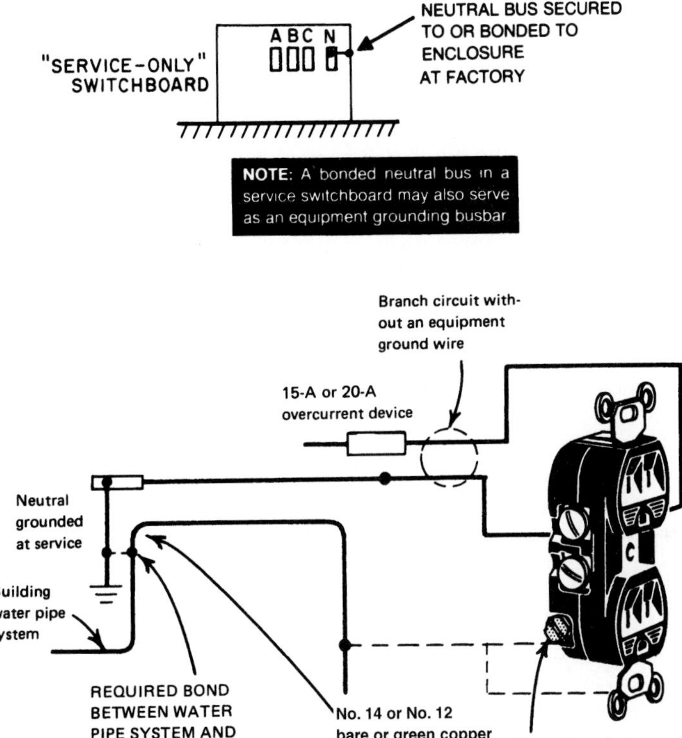

Fig. 250-33. These are bonding and grounding details covered by Sec. 250-50. (Sec. 250-50.)

ted by Sec. 250-91(a) and (b)]. And the condition can be visually checked by the electrical inspector, the design engineer, and/or any other authority concerned. There is nothing in the wording of that part **(1)** or in other Code rules to demand that any kind of actual test be made to verify the condition. However, an inspector could insist that the word "continuous" refers to a path of current and that only a continuity test with a meter or light or bell could positively assure that the path is "continuous."

(2). That every grounding conductor has "capacity to conduct safely any fault current likely to be imposed on it" can be established by falling back on those other Code rules [Secs. 250-93, 250-94, 250-95, 250-23(b), 250-26, 250-79, 680-25(a)(d)(e), etc.] that specifically establish a minimum required size of grounding conductor. Certainly, it is reasonable to conclude that adequate sizing of grounding conductors in accordance with those rules provides adequate "capacity to conduct safely . . . etc."

(3). But when we come to this part of Sec. 250-51, questions arise as to the intent of the rule; and much logic supports the argument that testing is essential to Code compliance. Certainly, on a grounded system, when a phase-to-ground fault occurs, the impedance of the fault-current path over the raceways or equipment grounding conductors must be low enough to cause enough current to flow to operate the fuse or CB nearest the fault on its line side (Fig. 250-34). To *know* for sure that impedance of any and every grounding conductor is "sufficiently low to limit . . . etc." requires that the actual value of impedance be measured; and such measurement not only involves use of testing equipment but also demands a broad and deep knowledge of the often sophisticated technology of testing in circuits operating on alternating current where inductance and capacitance are operative factors. In short, what is "sufficiently low impedance" and what does "facilitate the operation of the circuit protective devices" mean? And if testing *is* done, is it necessary to test "every" equipment grounding conductor? Are all these possible interpretations of the intent of the Code rule unrealistic? Is all of this beyond the capability of personnel and out of the economic reach of customers who have to pay for it? It all comes down to the hard question: *Just how much is safety worth?*

After all these questions and speculations, we are still left with the fact that Sec. 250-51(3) *is* a mandatory rule that can be fully satisfied only by testing. It is unreasonable to assume that the Code-making panel never intended this rule to be enforced. And the concept behind the rule as well as the logic of testing being essential to compliance are theoretical ideals with which every electrical person would agree.

The last sentence in this section prohibits the use of current flow through the earth as the sole equipment grounding conductor because earth impedance is too high and restricts fault-current flow, as shown at the bottom of Fig. 250-34. This rule is also in Sec. 250-91(c). Inspectors as well as computer CATV installers have often overlooked this very important Code rule, and the requirement needs to be more obviously emphasized.

250-53. Grounding Path to Grounding Electrode at Services. Section 250-53(a) requires all the bonded components—the service-equipment enclosure, the grounded neutral or grounded phase leg, and any equipment grounding conductors that come into the service enclosure—to be connected to a *common* grounding electrode (Sec. 250-54) by the single grounding electrode conductor. A common grounding electrode conductor shall be run from the common point so obtained to the grounding electrode as required by Code Secs. 250-53 and 250-54 (Fig. 250-35). Connection of the system neutral to the switchboard frame or ground bus within the switchboard provides the lowest impedance for the equipment ground return to the neutral. The main bonding jumper that bonds the service enclosure and equipment grounding conductors [which may be either conductors or conduit, EMT, etc., as permitted by Sec. 250-91(b)] to the grounded conductor of the system is required by part **(b)** of this section to be installed within the service equipment or within a service conductor enclosure on the line side of the service. This is the bonding connection required by Sec. 250-50(a) (Fig. 250-36). And it should be noted that in a service panel, equipment grounding conductors for load-side circuits may be connected to

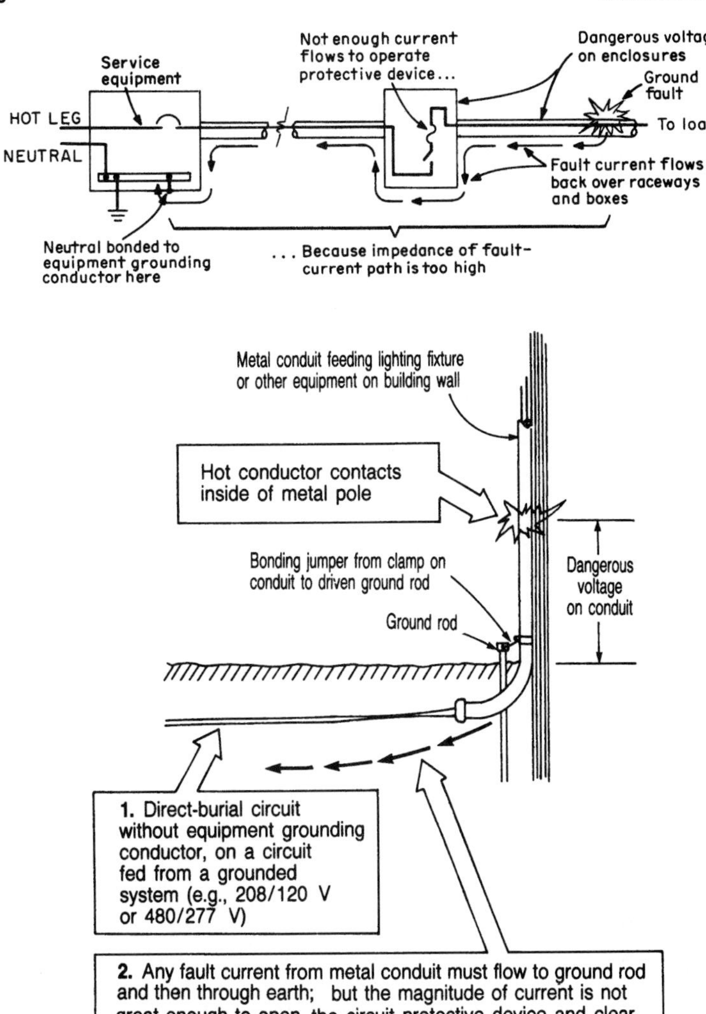

Fig. 250-34. These are violations of the basic concept of effective grounding. (Sec. 250-51.)

the neutral block, and there is no need for an equipment grounding terminal bar or block.

If a grounding conductor were used to ground the neutral to the water pipe or other grounding electrode and a separate grounding conductor were used to ground the switchboard frame and housing to the water pipe or other electrode, without the neutral and the frame being connected in the switchboard, the

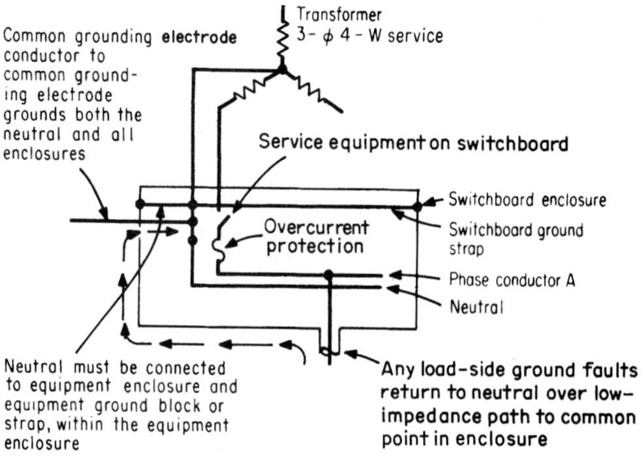

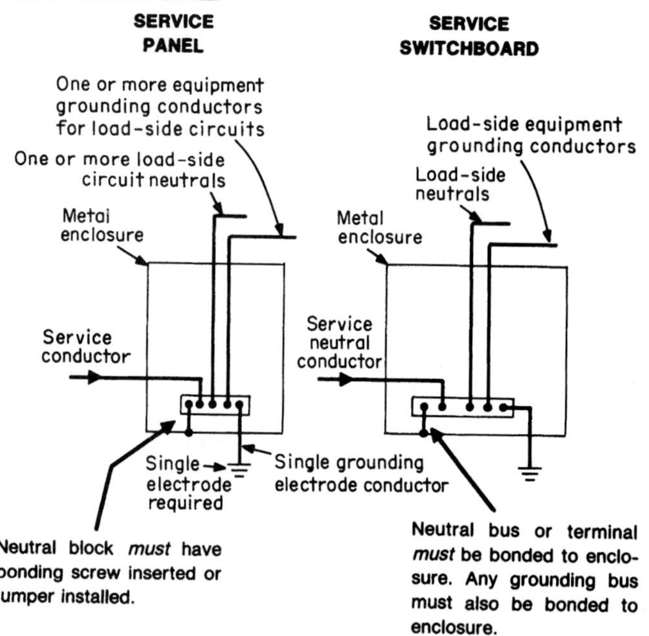

Fig. 250-35. Common grounding electrode conductor for service and equipment ground. (Sec. 250-53.)

length and impedance of the ground path would be increased. The proven hazard is that the impedance of the fault-current path can limit fault current to a level too low to operate the overcurrent devices "protecting" the faulted circuit.

Note that a number of grounding electrodes that are bonded together, as required by Sec. 250-81, are considered to be *one* grounding electrode.

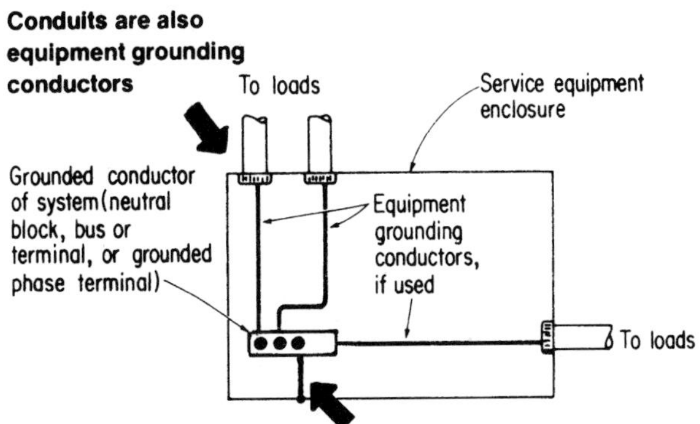

Main bonding jumper (wire, bus, screw or similar conductor) must be *within* service equipment or service conductor enclosure

Fig. 250-36. Main bonding jumper must be within SE enclosure. (Sec. 250-53.)

250-54. Common Grounding Electrode. The same electrode(s) that is used to ground the neutral or other grounded conductor of an AC system must also be used for grounding the entire system of interconnected raceways, boxes, and enclosures. The single, common grounding electrode conductor required by Sec. 250-53 connects to the single grounding electrode and thereby grounds the bonded point of the system and equipment grounds.

In any building housing livestock, all piping systems, metal stanchions, drinking troughs, and other metalwork with which animals might come in contact should be bonded together and to the grounding electrode used to ground the wiring system in the building. See Sec. 250-81.

250-57. Equipment Fastened in Place or Connected by Permanent Wiring Methods (Fixed)—Grounding. This section requires that metal equipment enclosures, boxes, and cabinets to be grounded must be grounded by metal cable armor or by the metal raceway that supplies such enclosures (rigid metal conduit, intermediate metal conduit, EMT, flex or liquidtight flex as permitted by Sec. 350-5 or 351-9), *or* by an equipment grounding conductor, such as where the equipment is fed by rigid nonmetallic conduit. Refer to Sec. 250-91(b). *But* in Sec. 250-57(b), the rule explicitly requires that *when* a separate equipment grounding conductor (i.e., other than the metal raceway or metal cable armor) is used for alternating-current circuits, it *must* be contained *within* the same raceway, cable, or cord or otherwise run with the circuit conductors (Fig. 250-37). External grounding of equipment enclosures or frames or housings is a violation for AC equipment. It is not acceptable, for instance, to feed an AC motor with a nonmetallic conduit or cable, without a grounding conductor in the conduit or cable, and then provide grounding of the metal frame by a grounding conductor connected to the metal frame and run to building steel or to a grounding-grid conductor. An equipment grounding conductor *must always* be run with the circuit conductors.

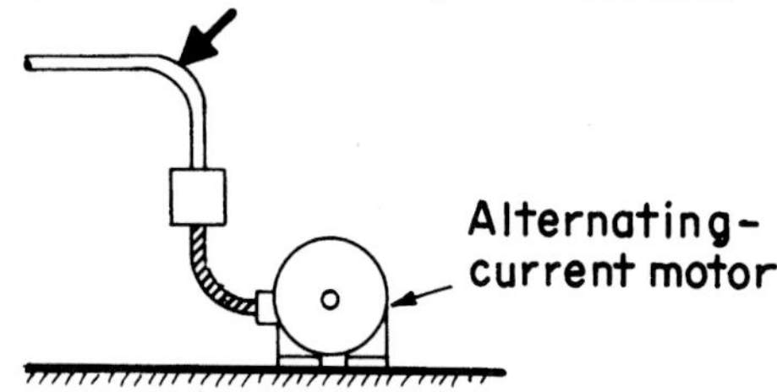
If equipment grounding conductor (other than raceway) is used to ground motor, it must be run in raceway with circuit wires
Alternating-current motor

Fig. 250-37. Equipment grounding conductor must be in raceway or cable with circuit conductors for AC equipment. (Sec. 250-57.)

The rule in Sec. 250-57(b) which insists on keeping an equipment grounding conductor physically close to AC circuit supply conductors is a logical follow-up to the rules of Sec. 250-51 which call for minimum impedance in grounding current paths to provide most effective clearing of ground faults. When an equipment grounding conductor is kept physically close to any circuit conductor that would be supplying the fault current (that is, the grounding conductor is in the "same raceway, cable, or cord or otherwise run with the circuit conductors"), the impedance of the fault circuit has minimum inductive reactance and minimum AC resistance because of mutual cancellation of the magnetic fields around the conductors and the reduced skin effect. Under such condition of "sufficiently low impedance," the meaning of Sec. 250-51(c) is best fulfilled—voltage to ground is limited to the greatest extent, the fault current is higher because of minimized impedance, the circuit overcurrent device will operate at a faster point in its time-current characteristic to assure maximum fault-clearing speed, and the entire effect will be to "facilitate the operation of the circuit protective devices in the circuit."

Exception No. 2 of Sec. 250-57(b) *excludes* DC circuits from the need to keep the grounding conductor close to the circuit conductors. Because there are no alternating magnetic fields around DC conductors, there is no inductive reactance or skin effect in DC circuits. The only impedance to current flow in a DC circuit is resistance—which will be the same for a DC ground-fault path whether or not the equipment grounding conductor is placed physically close to the circuit conductors which would supply the fault current in the event of a ground fault. External equipment grounding—by connection to grounded building steel or to an external ground grid—is, therefore, OK for DC equipment, provided the external grounding path is effectively tied back to the grounded conductor of the DC system.

The arrangement shown in Fig. 250-38 violates the basic rule of Sec. 250-57(b) because the lighting fixture, which must be grounded to satisfy Sec. 250-42, is not grounded in accordance with Secs. 250-57 and 250-91(b) or by an equipment grounding conductor contained within the cord, as noted in Sec. 250-57(b).

Note that Sec. 250-91(b) refers very clearly to an "equipment grounding conductor run with or enclosing the circuit conductors." Except for DC circuits [Sec. 250-57(b), Exception No. 2] and for isolated, ungrounded power sources

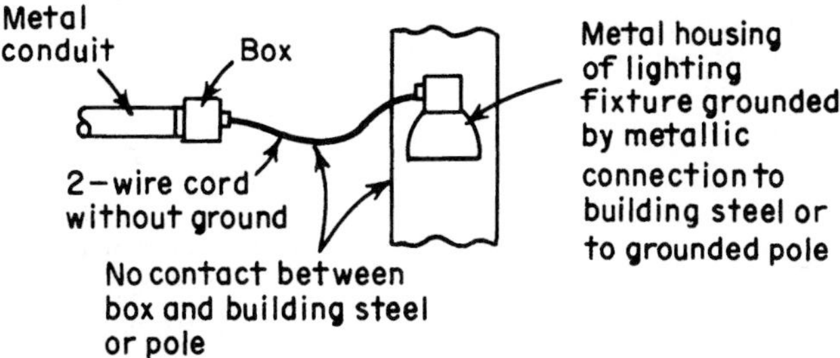

Fig. 250-38. Supply to AC equipment must include equipment grounding conductor. (Sec. 250-57.)

[Sec. 517-19(f) and (g)], an equipment grounding conductor of any type must not be run separately from the circuit conductors. The engineering reason for keeping the ground return path and the phase legs in close proximity (that is, in the same raceway) is to minimize the impedance of the fault circuit by placing conductors so their magnetic fields mutually cancel each other, keeping inductive reactance down, and allowing sufficient current to flow to "facilitate the operation of the circuit protective devices," as required by Sec. 250-51.

The hookup in Fig. 250-38 also violates the rule of the last sentence in Sec. 250-58(a), which prohibits use of building steel as the equipment grounding conductor for AC equipment. And the rules of Sec. 250-58 often have to be considered in relation to the rules of Sec. 250-57.

Note: CARE MUST BE TAKEN TO DISTINGUISH BETWEEN AN "EQUIPMENT *GROUNDING CONDUCTOR*" AS COVERED BY SEC. 250-57 AND AN "EQUIPMENT *BONDING JUMPER*" AS COVERED BY SEC. 250-79(e). A "*BONDING JUMPER*" MAY BE USED EXTERNAL TO EQUIPMENT BUT IT MUST NOT BE OVER 6 FT (1.83 m) LONG.

Exception No. 1 recognizes conductors of colors other than green for use as equipment grounding conductors if the conductor is stripped for its exposed length within an enclosure, so it appears bare, or if green coloring, green tape, or green label is used on the conductor at the termination. As shown in Fig. 250-39, the phase legs may or may not be required to be "identified by phase and system" [see Sec. 210-4(d)]. If color coding is used, the phase legs may be any color other than white, gray, or green. The neutral may be white or gray or any other color than green if it is larger than No. 6 and if white tape, marking, or paint is applied to the neutral near its terminations. The grounding conductor may be green or may be any insulated conductor of any color if all insulation is stripped off for the exposed length. Alternatives to stripping the black insulated conductor used for equipment ground include (1) coloring the exposed insulation green or (2) marking the exposed insulation with green tape or green adhesive labels.

Exception No. 4 permits specific on-the-job identification of an insulated conductor used as an equipment grounding conductor in a multiconductor cable. Such a conductor, regardless of size, may be identified in the same man-

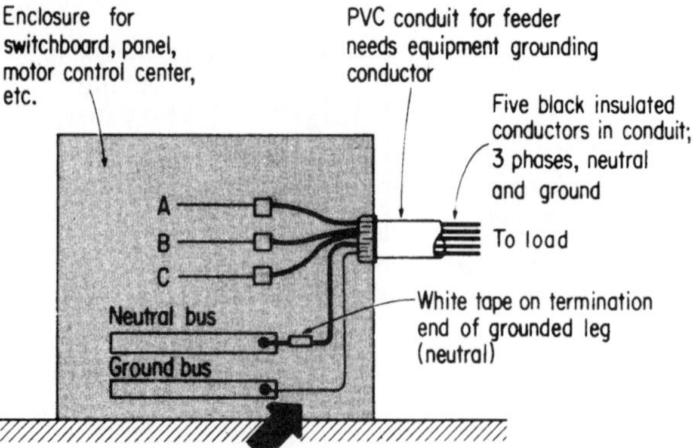

Fig. 250-39. Equipment grounding conductor larger than No. 6 may be a stripped conductor of any color covering. (Sec. 250-57.)

ner permitted by Exception No. 1 of that section for conductors larger than No. 6 used in raceway. The conductor may be stripped bare or colored green to indicate that it is a grounding conductor. But such usage is recognized only for commercial-, institutional- and industrial-type systems under the conditions given in the first two lines of the Exception.

250-58. Equipment Considered Effectively Grounded. This rule clarifies the way in which structural metal may be used as an equipment grounding conductor, consistent with the rule of Sec. 250-57(b) requiring a grounding conductor to be kept physically close to the conductors of any AC circuit for which the grounding conductor provides the fault return path.

Part **(a)** notes that if a piece of electrical equipment is attached and electrically conductive to a metal rack or structure supporting the equipment, the metal enclosure of the equipment is considered suitably grounded by connection to the metal rack PROVIDED THAT the metal rack itself is effectively grounded by metal raceway enclosing the circuit conductors supplying the equipment or by an equipment grounding conductor run with the circuit supplying the equipment. An example of such application is shown in Fig. 250-40. Although this example shows grounding of lighting fixtures to a rack, the Code rule recognizes any "electric equipment" when this basic grounding concept is observed. It is important to note that if a ground fault developed in equipment so grounded (as at point A), the fault current would take the path indicated by the small arrows. In such case, although the fault-current path through the steel rack is not close to the hot conductor in the flexible cord that is feeding the fault—as normally required by Sec. 250-57(b)—the distance of the external ground path is not great, from the fixture to the panel enclosure or box. Because such a short external ground path produces only a relatively slight increase in ground-path impedance, Sec. 250-58(a) permits it. The permission for external bonding of flexible metal conduit and liquidtight flex in Sec. 250-79(e) is based on the same acceptance of only slight increase of overall impedance of the ground path.

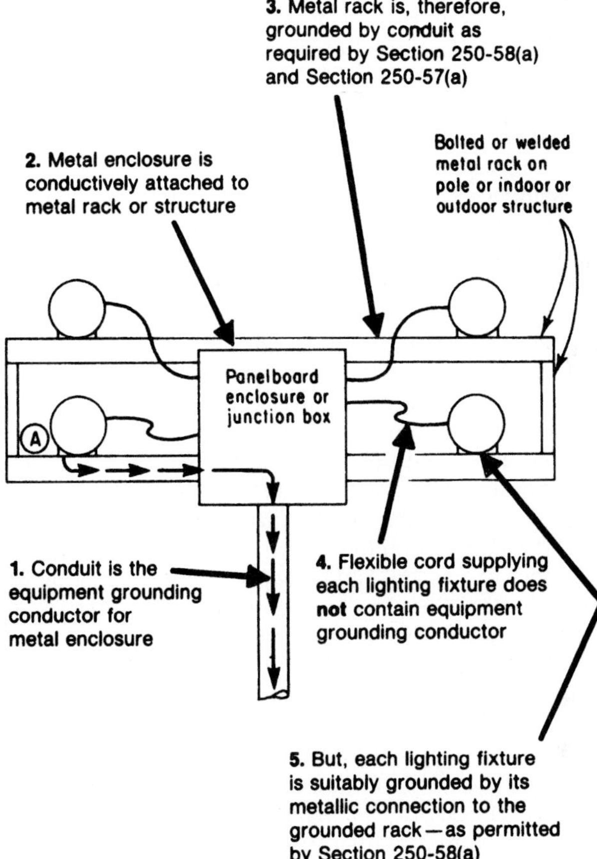
3. Metal rack is, therefore, grounded by conduit as required by Section 250-58(a) and Section 250-57(a)
2. Metal enclosure is conductively attached to metal rack or structure
Bolted or welded metal rack on pole or indoor or outdoor structure
Panelboard enclosure or junction box
A
1. Conduit is the equipment grounding conductor for metal enclosure
4. Flexible cord supplying each lighting fixture does not contain equipment grounding conductor
5. But, each lighting fixture is suitably grounded by its metallic connection to the grounded rack—as permitted by Section 250-58(a)

Fig. 250-40. This use of metal rack as equipment ground is permitted. (Sec. 250-58.)

The second sentence of Sec. 250-58(a) clearly prohibits using structural building steel as an equipment grounding conductor for equipment mounted on or fastened to the building steel—IF THE SUPPLY CIRCUIT TO THE EQUIPMENT OPERATES ON ALTERNATING CURRENT. BUT, structural building steel that is effectively grounded and bonded to the grounded circuit conductor of a DC supply system may be used as the equipment grounding conductor for the metal enclosure of DC-operated equipment that is conductively attached to the building steel.

It is important to understand the basis for the **Code** rules of Sec. 250-57(b) and Sec. 250-91(b) and their relation to the concept of Sec. 250-58(a):

Note that Sec. 250-91(b) refers very clearly to an "equipment grounding conductor run with or enclosing the circuit conductors." Except for DC circuits [Sec. 250-57(b), Exception No. 2] and for isolated, ungrounded power sources [(Sec. 517-19(f) and (g)], an equipment grounding conductor of any type must not be run separately from the circuit conductors. Keeping the ground return path and the phase legs in close proximity (that is, in the same raceway) minimizes the impedance of the fault circuit by placing conductors so their magnetic fields mutually cancel each other, keeping inductive reactance down and allowing sufficient current to flow to "facilitate the operation of the circuit protective devices," as required by Sec. 250-51.

The second sentence of Sec. 250-58(a) applies the above concept of ground-fault impedance to the metal frame of a building and prohibits its use as an equipment grounding conductor for AC equipment enclosures. As shown in Fig. 250-41, use of building steel as a grounding conductor provides a long fault return path of very high impedance because the path is separated from the feeder circuit hot legs—thereby violating Sec. 250-51(c). Ground-fault current returning over building steel to the point where the building steel is bonded to the AC system neutral (or other grounded) conductor is separated from the circuit conductor that is providing the fault current. Impedance is, therefore, elevated and the optimum conditions required by Sec. 250-51 are not present, so that the grounding cannot be counted on to "facilitate the operation" of the fuse or CB protecting the faulted circuit. The current may not be high enough to provide fast and certain clearing of the fault.

The first sentence of Sec. 250-58(a) accepts a limited variation from the basic concept of keeping circuit hot legs and equipment grounding conductors physically close to each other. When equipment is grounded by connection to a "metal rack or structure" that is specifically provided to support the equipment and *is* grounded, the separation between the circuit hot legs and the rack, which serves as the equipment grounding conductor, exists only for a very short length that will not significantly raise the overall impedance of the ground-fault path. Figure 250-42 shows another application of that type, similar to the one shown in Fig. 250-40. Although this shows a 2-wire cord as being acceptable, use of a 3-wire cord (two circuit wires and an equipment grounding wire) is better practice, at very slight cost increase.

Aside from the limited applications shown in Figs. 250-40 and 250-42, *required* equipment grounding must always keep the equipment grounding conductor alongside the circuit conductor for grounded AC systems. Of course, as long as required grounding techniques are observed, there is no objection to additional connection of equipment frames and housings to building steel or to grounding grids to provide potentials to ground. But the external grounding path is not suitable for clearing AC equipment ground faults.

250-59. Cord- and Plug-Connected Equipment. The proper method of grounding portable equipment is through an extra conductor in the supply cord. Then if the attachment plug and receptacle comply with the requirements of Sec. 250-59, the grounding connection will be completed when the plug is inserted in the receptacle.

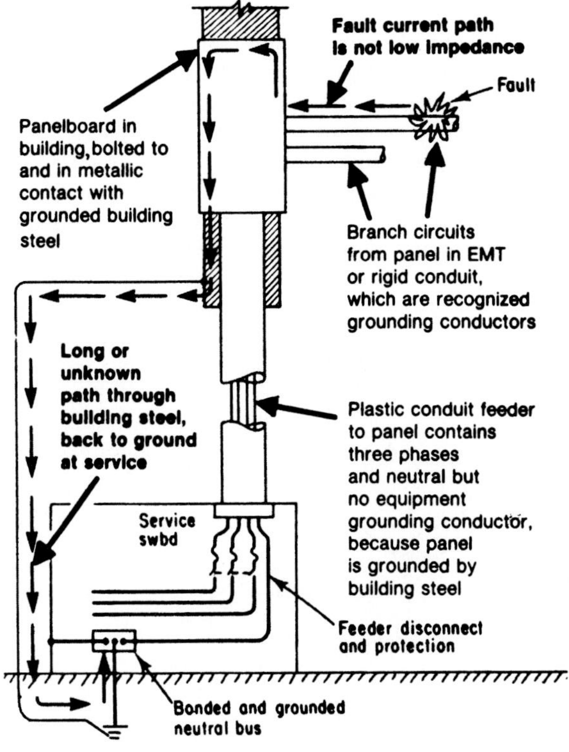
Fault current path
is not low impedance
Fault
Panelboard in
building, bolted to
and in metallic
contact with
grounded building
steel
Branch circuits
from panel in EMT
or rigid conduit,
which are recognized
grounding conductors
Long or
unknown
path through
building steel,
back to ground
at service
Plastic conduit feeder
to panel contains
three phases
and neutral but
no equipment
grounding conductor,
because panel
is grounded by
building steel
Service
swbd
Feeder disconnect
and protection
Bonded and grounded
neutral bus
THIS LAYOUT IS A VIOLATION

Fig. 250-41. Building metal frame is not an acceptable grounding conductor for AC equipment. (Sec. 250-58.)

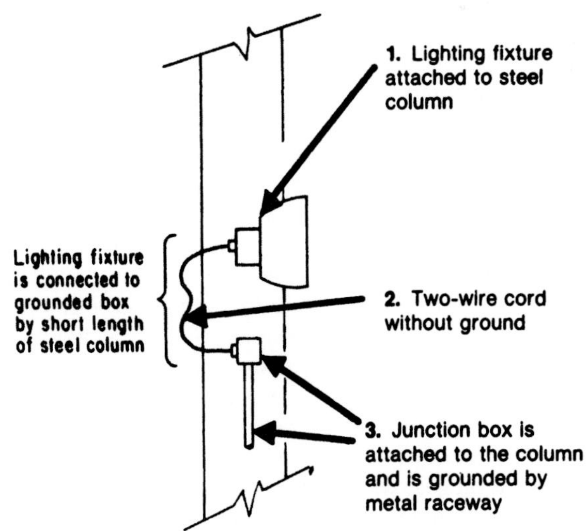
1. Lighting fixture
attached to steel
column
Lighting fixture
is connected to
grounded box
by short length
of steel column
2. Two-wire cord
without ground
3. Junction box is
attached to the column
and is grounded by
metal raceway

Fig. 250-42. This satisfies basic rule of Sec. 250-58(a). [Sec. 250-58(a).]

A grounding-type receptacle and an attachment plug should be used where it is desired to provide for grounding the frames of small portable appliances. The receptacle will receive standard two-pole attachment plugs, so grounding is optional with the user. The grounding contacts in the receptacle are electrically connected to the supporting yoke so that when the box is surface-mounted, the connection to ground is provided by a direct metal-to-metal contact between the device yoke and the box. For a recessed box a grounding jumper must be used on the receptacle or a self-grounding receptacle must be used. See Secs. 250-74 and 250-114.

Figure 250-43 shows a grounding-type attachment plug with a movable, self-restoring grounding member—as covered in the Exceptions of this section. Although recognized, availability of such a device is questionable.

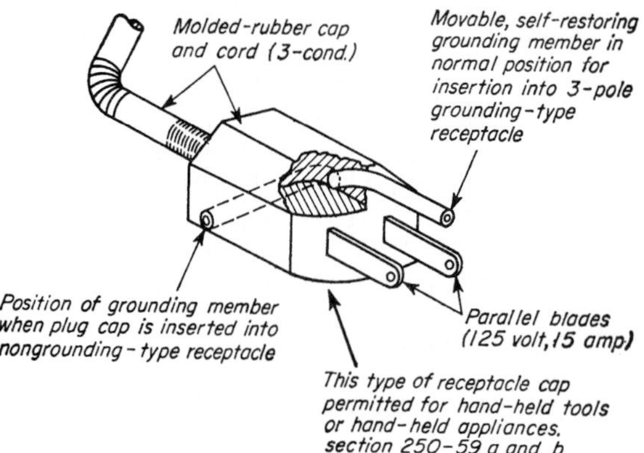

Fig. 250-43. This type of plug cap is permitted on cords for tools and appliances. (Sec. 250-59.)

250-60. Frames of Ranges and Clothes Dryers. Prior to the 1996 edition of the NEC, the frame of an electric range, wall-mounted oven, or counter-mounted cooking unit could be grounded by direct connection to the grounded circuit conductor (the grounded neutral) and thus could be supplied by a 3-wire cord set and range receptacle irrespective of whether or not the conductor to the receptacle contains a separate grounding conductor.

The 1996 NEC prohibits such application except on "existing branch circuits." That wording doesn't permit grounding of ranges or dryers with the neutral, unless the circuit itself—not the occupancy—is an existing circuit. For *all* new circuits and new construction, the neutral may *not* be used as a grounding conductor.

Where permitted to be so grounded parts **(a)** and **(b)** clarify the use of a No. 10 or larger grounded neutral conductor of a *120/208-V* circuit for grounding the frames of electric ranges, wall-mounted ovens, counter-mounted units, or clothes dryers. This method is acceptable whether the 3-wire supply is 120/208 or 120/240 V. However, a provision, applicable to both 3-wire supply voltages, does require that when using service-entrance cable having an uninsulated neutral conductor, the branch circuit must originate at the service-entrance equipment. The purpose of this provision is to prevent the uninsulated neutral from coming in contact with a panelboard supplied by a feeder and a separate grounding conductor (in the case of nonmetallic-sheathed cable). This would place the neutral in parallel with the grounding conductor, or with feeder *metal* raceways or cables if they were used. Insulated neutrals in such situations will prevent this (Fig. 250-44).

This permission is limited to existing circuits only!
Conditions when grounded neutral conductor (No. 10 or larger) may be used to ground metal frames of specified appliances.
Branch circuit (e.g.-SE cable)
Service
Service equipment panel
Electric ranges, wall-mounted ovens, counter-mounted cooking units or clothes dryers
Where SE cable with an uninsulated neutral conductor is used the branch circuit must originate at the service equipment
115/230- or 120/208-volt 3-wire rating

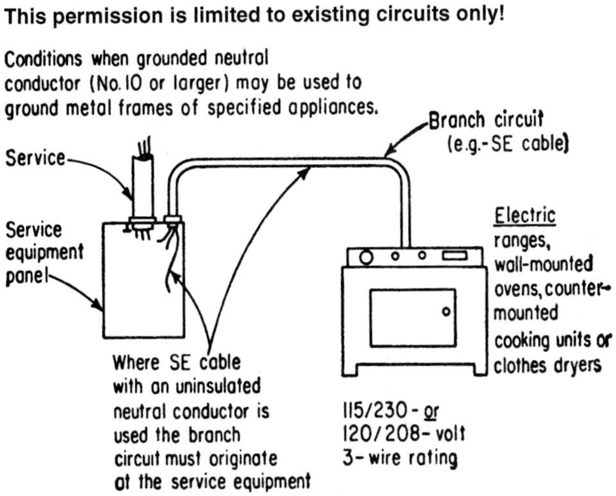

Fig. 250-44. Ranges and dryers may be grounded to the circuit neutral. (Sec. 250-60.)

Wording of the rule that permits frames of ranges and clothes dryers to be grounded by connection to the grounded neutral conductor of their supply circuits also permits the same method of grounding of "outlet or junction boxes" serving such appliances. The rule permits grounding of an outlet or junction box, as well as cooking unit or dryer, by the circuit grounded neutral (Fig. 250-45). That practice has been common for many years but has raised questions about the suitability of the neutral for such grounding. Now, the revised rule makes clear that such grounding of the box is acceptable. Figure 250-46 shows other details of such application. Without this permission to ground the metal box to the grounded neutral, it would be necessary to run a 4-wire supply cable to the box, with one of the wires serving as an equipment grounding conductor sized from Table 250-95.

Important: As shown in the asterisk note under Fig. 250-45, if a nonmetallic-sheathed cable is used, say, to supply a wall oven or cook-top, such cable is required by part **(c)** of Sec. 250-60 to have an *insulated* neutral. It would be a violation, for instance, to use a 10/2 NM cable with a bare No. 10 grounding

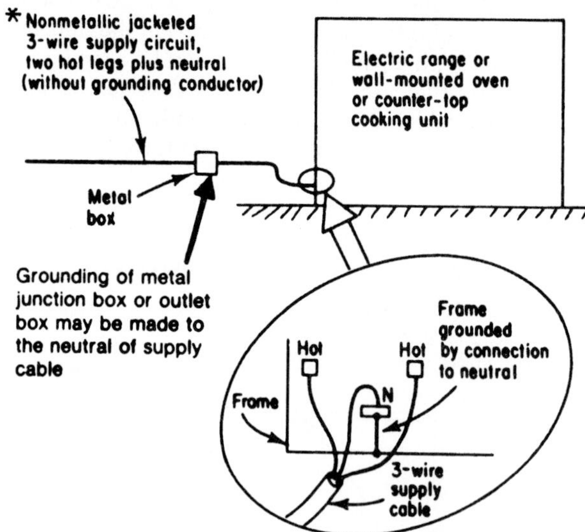
This permission is limited to existing circuits only!
* Nonmetallic jacketed 3-wire supply circuit, two hot legs plus neutral (without grounding conductor)
Electric range or wall-mounted oven or counter-top cooking unit
Metal box
Grounding of metal junction box or outlet box may be made to the neutral of supply cable
Hot
Hot
Frame grounded by connection to neutral
N
Frame
3-wire supply cable
* Service cable or NM or NMC cable. But NM or NMC cable must have an insulated neutral.

Fig. 250-45. Neutral may be used to ground boxes as well as appliances. (Sec. 250-60.)

conductor to supply a cooking appliance—connecting the two insulated No. 10 wires to the hot terminals and using the bare No. 10 as a neutral conductor to ground the appliance. An uninsulated grounded neutral may be used only when part of a service-entrance cable.

250-61. Use of Grounded Circuit Conductor for Grounding Equipment. Part **(a)** permits connection between a grounded neutral (or grounded phase leg) and equipment enclosures, for the purpose of grounding the enclosures to the grounded circuit conductor. The grounded conductor (usually the neutral) of a circuit may be used to ground metal equipment enclosures and raceways on the supply side of the service disconnect or the supply side of the first disconnect fed from a separately derived transformer secondary or generator output or on the supply side of a main disconnect for a separate building. The wording here includes the supply side of a separately derived system as a place where metal equipment parts or enclosures may be grounded by connection to the grounded circuit conductor (usually a neutral). It is important to note that, in the meaning of the code [as covered in Sec. 250-26(a) and in Sec. 250-23(a)], the phrase "on the supply side of the disconnecting means" includes connection within the enclosure of the disconnecting means.

Figure 250-47 shows such applications. At A, the grounded service neutral is bonded to the meter housing by means of the bonded neutral terminal lug in the

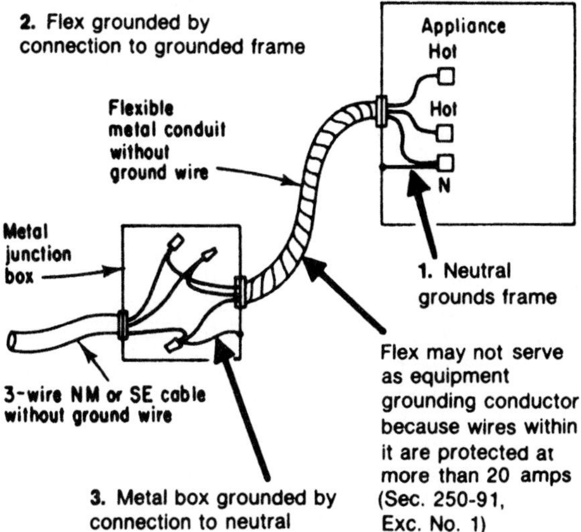

Watch Out! **Although generally permitted in the past, the 1996 Code prohibits such practice, except for *existing circuits!***

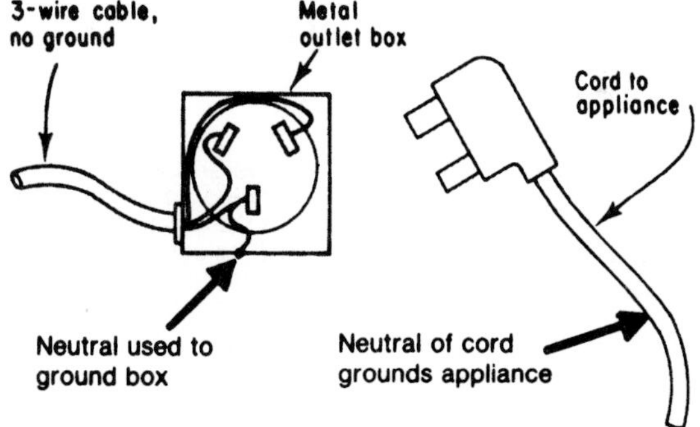

Fig. 250-46. These techniques may be used to ground boxes in circuit. (Sec. 250-60.)

socket—and the housing is thereby grounded by this connection to the grounded neutral, which itself is grounded at the service equipment as well as at the utility transformer secondary supplying the service. At B, the service equipment enclosure is grounded by connection (bonding) to the grounded neutral—which itself is grounded at the meter socket and at the supply transformer. These same types of grounding connections may be made for CT cabinets, auxiliary gutters, and other enclosures on the line side of the service-entrance disconnect means, including the enclosure for the service disconnect. In some areas, the utilities and inspection departments will not permit the arrangement shown in Fig. 250-47 because the connecting lug in the meter housing is not always accessible for inspection and testing purposes. At C, equipment is grounded to the neutral on the line (supply) side of the first disconnect fed from a step-down transformer (a separately derived system).

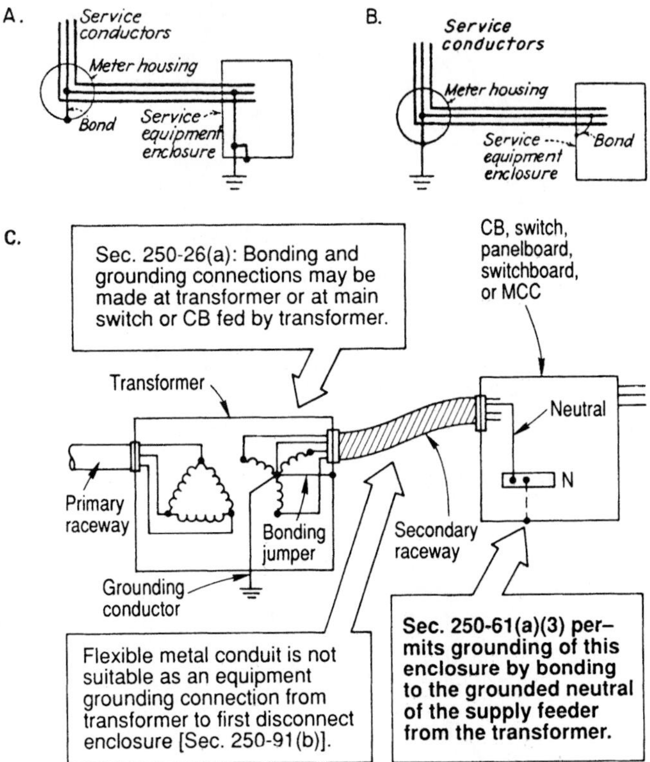
A.
Service conductors
Meter housing
Bond
Service equipment enclosure
B.
Service conductors
Meter housing
Service equipment enclosure
Bond
C.
Sec. 250-26(a): Bonding and grounding connections may be made at transformer or at main switch or CB fed by transformer.
CB, switch, panelboard, switchboard, or MCC
Transformer
Neutral
N
Primary raceway
Bonding jumper
Secondary raceway
Grounding conductor
Flexible metal conduit is not suitable as an equipment grounding connection from transformer to first disconnect enclosure [Sec. 250-91(b)].
Sec. 250-61(a)(3) permits grounding of this enclosure by bonding to the grounded neutral of the supply feeder from the transformer.

Fig. 250-47. Using grounded circuit conductor to ground equipment housings on line side of service or separately derived system. (Sec. 250-61.)

Aside from the permission given in the five exceptions to the rule of part **(b)** of this section, the wording of part **(b)** prohibits connection between a grounded neutral and equipment enclosures on the load side of the service. The wording supports the prohibition in Sec. 250-23 of grounding connections. So aside from the few specific exceptions mentioned, bonding between any system grounded conductor, neutral or phase leg, and equipment enclosures is prohibited on the load side of the service (Fig. 250-48). The use of a neutral to ground panelboard or other equipment (other than specified in the Exceptions) on the load side of service equipment would be extremely hazardous if the neutral became loosened or disconnected. In such cases any line-to-neutral load would energize all metal components connected to the neutral, creating a dangerous potential above ground. Hence, the prohibition of such a practice. This is fully described in Fig. 250-15.

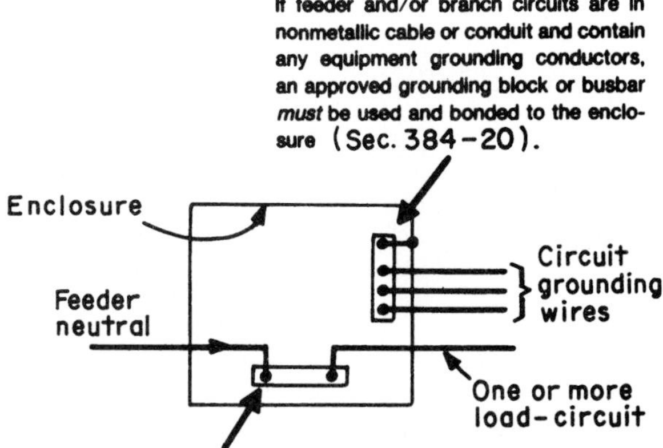

Fig. 250-48. Panel, switchboard, CB, and switch on load side of service within a single building. (Sec. 250-61.)

Although this rule of the Code prohibits neutral bonding on the load side of the service, Secs. 250-50(a) and 250-53(b) clearly require such bonding at the service entrance. And the exceptions to prohibiting load-side neutral bonding to enclosures are few and very specific:

- Exception No. 1 of Sec. 250-61(b) permits frames of ranges, wall ovens, counter-top cook units, and clothes dryers to be "grounded" by connection to the grounded neutral of their supply circuit, but only for *existing* circuits (Sec. 250-60).
- When a circuit is run from one building to another, it may be necessary, simply permissible, or expressly prohibited to connect the system "grounded" conductor to a grounding electrode at the other building—as covered by Sec. 250-24 and Exception No. 2 of Sec. 250-61(b).
- Exception No. 3 to Sec. 250-61(b) permits grounding of meter enclosures to the grounded circuit conductor (generally, the grounded neutral) on the *load side* of the service disconnect if the meter enclosures are located near the service disconnect, the service is not equipped with ground-fault protection, and the neutral is not less than the minimum required by Sec. 250-95, based on the rating of the service overcurrent device. There is no definition for the

word "near," but it can be taken to mean in the same room or general area. This rule applies, of course, to multioccupancy buildings (apartments, office buildings, etc.) with individual tenant metering (Fig. 250-49).

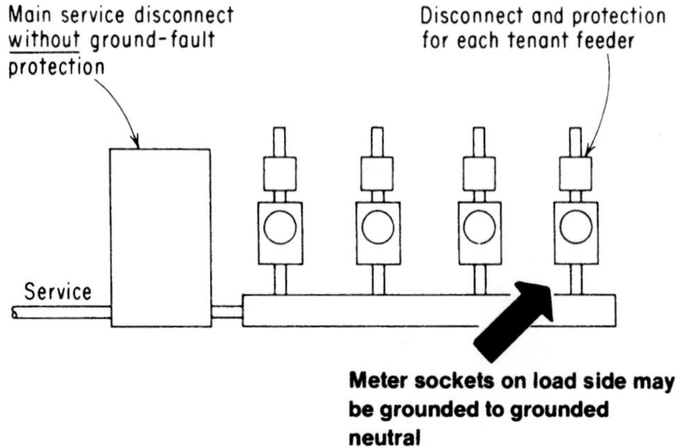

Fig. 250-49. Grounding meter enclosures to grounded conductor on *load side* of service disconnect. (Sec. 250-61.)

If a meter bank is on the upper floor of a building, as in a high-rise apartment house, or otherwise away from service disconnect, such meter enclosures would not meet the rule that they must be "near" the service disconnect. In such cases, the enclosures must not be grounded to the neutral. And if the service has ground-fault protection, meter enclosures on the load side must not be connected to the neutral, even if they are "near" the service disconnect.

Exception No. 4 of this section refers to Secs. 710-72 and 710-74, covering use of neutral conductor and grounding of electrode-type boilers.

250-70. General (Bonding). One of the most interesting and controversial phases of electrical work involves the grounding and bonding of secondary-voltage service-entrance equipment. Modern practice in such work varies according to local interpretations of Code requirements and specifications of design engineers. In all cases, however, the basic intent is to provide an installation which is essentially in compliance with National Electrical Code rules on the subject, using practical methods for achieving objectives.

In order to ensure electrical continuity of the grounding circuit, bonding (special precautions to ensure a permanent, low-resistance connection) is required at all conduit connections in the service equipment and where any nonconductive coating exists which might impair such continuity. This includes bonding at connections between service raceways, service cable armor, all service-equipment enclosures containing service-entrance conductors, including meter fittings, boxes, and the like.

The need for effective grounding and bonding of service equipment arises from the electrical characteristics of utility-supply circuits. In the common arrangement, service conductors are run to a building and the service overcurrent protection is placed near the point of entry of the conductors into the building, at the load end of the conductors. With such a layout, the service conductors are not properly protected against ground faults or shorts occurring on the supply side of the service overcurrent protection. Generally, the only protection for the service conductors is on the primary side of the utility's distribution transformer. By providing "bonded" connections (connecting with special care to reliable conductivity), any short circuit in the service-drop or service-entrance conductors is given the greatest chance of burning itself clear—because there is not effective overcurrent protection ahead of those conductors to provide opening of the circuit on such heavy fault currents. And for any contact between an energized service conductor and a grounded service raceway, fittings, or enclosures, bonding provides discharge of the fault current to the system grounding electrode—and again burning the fault clear. This condition of services is shown in Fig. 250-50.

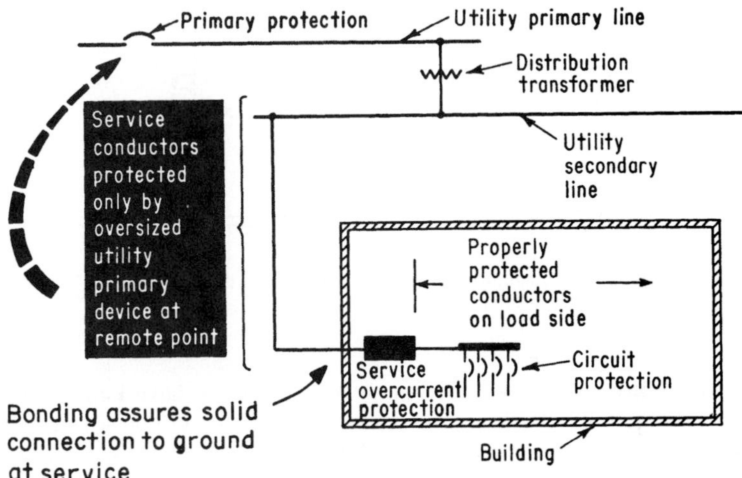

Fig. 250-50. Service bonding must assure burn-clear on shorts and grounds in service conductors. (Sec. 250-70.)

250-71. (Bonding) Service Equipment. Because of the requirement set forth in Sec. 250-70, all enclosures for service conductors must be grounded to prevent a potential aboveground on the enclosures as a result of fault—which would be a very definite hazard—and to facilitate operation of overcurrent devices anywhere on the supply side of the service conductors. However, because of the distant location of the protection and the normal impedance of supply cables, it is important that any fault to an enclosure of a hot service conductor of a grounded electrical system find a firm, continuous, low-impedance path to ground to assure sufficient current flow to operate the primary protective device or to burn the fault clear quickly. This means that all enclosures containing the service conductors—service raceway, cable armor, boxes, fittings, cabinets—must be effectively bonded together; that is, they must have low impedance through themselves and must be securely connected to each other to assure a continuous path of sufficient conductivity to the conductor which makes the connection to ground (Fig. 250-51).

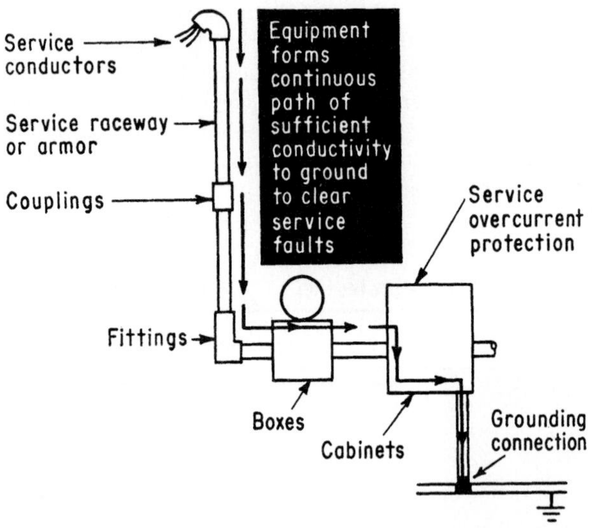

Fig. 250-51. Bonding assures low-impedance path through all service conductor enclosures. (Sec. 250-71.)

The spirit of the **Code** and good engineering practice have long recognized that the conductivity of any equipment ground path should be at least equivalent to 25 percent of the conductivity of any phase conductor with which the ground path will act as a circuit conductor on a ground fault. Or, to put it another way, making the relationship without reference to insulation or temperature rise, the impedance of the ground path must not be greater than four times the impedance of any phase conductor with which it is associated.

In ungrounded electrical systems, the same careful attention should be paid to the matter of bonding together the noncurrent-carrying metal parts of all enclosures containing service conductors. Such a low-impedance ground path will quickly and surely ground any hot conductor which might accidentally become common with the enclosure system.

Specific **NE Code** requirements on grounding and bonding are as follows:

1. Section 230-63 requires that service raceways, metal sheath of service cables, metering enclosures, and cabinets for service disconnect and protection be grounded. An exception to this rule is made in the case of certain lead-sheathed cable services as covered in Sec. 250-55. And, the **Code**

requires that flexible metal conduit or liquidtight flexible metal conduit used in a run of service raceway must be bonded around (Fig. 250-52). Section 230-43 states that rule on flex and lists the *only* types of raceway that may enclose service-entrance conductors.

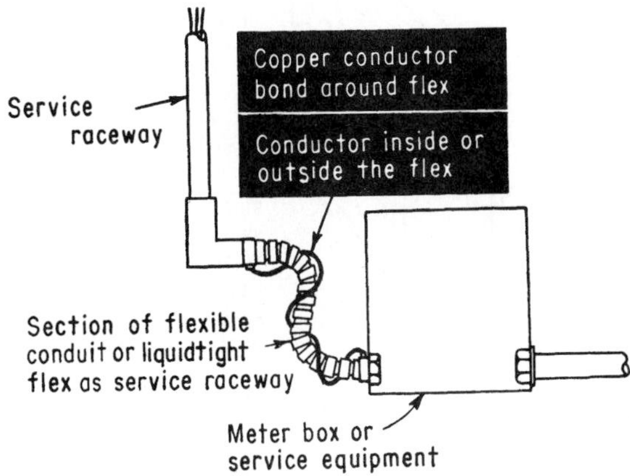

Fig. 250-52. Flex may be used as a service raceway, with a jumper. (Sec. 250-71.)

2. Section 250-32 also requires that service raceways and service cable sheaths or armoring—when of metal—be grounded.
3. Section 250-71 sets forth the service equipment which must be bonded—that is, the equipment for which the continuity of the grounding path must be specifically assured by using specific connecting devices or techniques. As indicated in Fig. 250-53, this equipment includes (1) service raceway, cable trays, cable sheath, and cable armor; (2) all service-equipment enclosures containing service-entrance conductors, including meter fittings, boxes, etc., interposed in the service raceway or armor; and (3) any conduit or armor that encloses a grounding electrode conductor that runs to and is connected to the grounding electrode or "system" of electrodes, as described under Sec. 250-81.

Part **(3)** of the rule to Sec. 250-71(a) on bonded connections for all interconnected service equipment makes *clear* that bonded terminations must be used at ends—and at all boxes, enclosures, etc., in the run—of conduit (rigid metal conduit, IMC or EMT) or cable armor that encloses a grounding electrode conductor. That means that connection of conduit or cable armor must be made using a bonding locknut or bonding bushing (with a bonding jumper around unpunched concentric or eccentric rings left in any sheet metal knockout) or must be connected to a threaded hub or boss. Such connections must comply with the techniques covered in Sec. 250-72.

The first sentence of Sec. 250-92(b) bears on the same matter and requires a metal enclosure for a grounding electrode conductor to be electrically in paral-

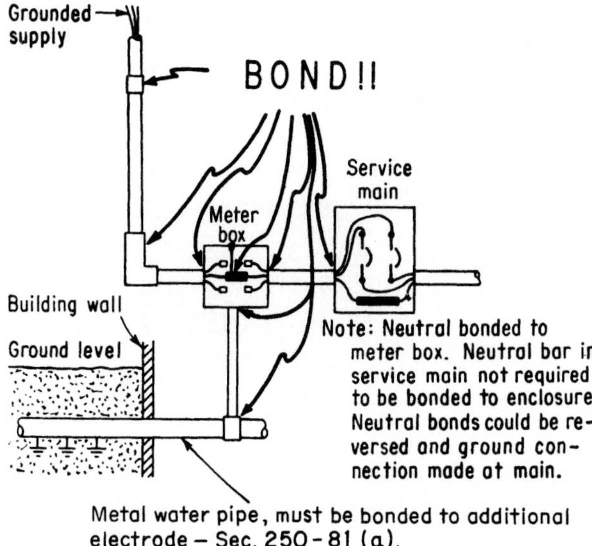

Fig. 250-53. "Bonding" consists of using prescribed fittings and/or methods for connecting components enclosing SE conductors. (Sec. 250-71.)

lel with the grounding electrode conductor. As a result, a metal conduit or EMT enclosing a grounding electrode conductor—whether or not such conduit or EMT is mechanical protection required by Sec. 250-92(a)—forms *part* of the grounding electrode conductor. In fact, such conduit or EMT has a much lower impedance than its enclosed grounding electrode conductor (due to the relation of magnetic fields) and is, therefore, even more important than the enclosed conductor in providing an effective path for current to the grounding electrode.

Assuring the continuity of raceway or armor for a grounding conductor does reduce the impedance of the ground path compared with what the impedance would be if the raceway or armor had poor connections or even opens. Effective bonding of the raceway or armor minimizes the DC resistance of the ground path and reduces the overall impedance which includes the choke action due to presence of magnetic material (steel conduit or armor), the increased inductive reactance of the circuit. Because of that, Sec. 250-71(a)(3) requires "bonding" connection of such raceway to any service enclosure, as shown in Fig. 250-54. Refer to Sec. 250-92(b).

Section 250-71(b) calls for a ready, effective "intersystem" bonding and grounding of different systems, such as communications (telephone), lightning rod systems, and CATV systems at the service equipment for *all* buildings, not just dwellings. The rule requires that there be an "accessible means external to enclosures" for bonding metal enclosures of, say, telephone equipment to metal enclosures of electrical system components to reduce voltage differences

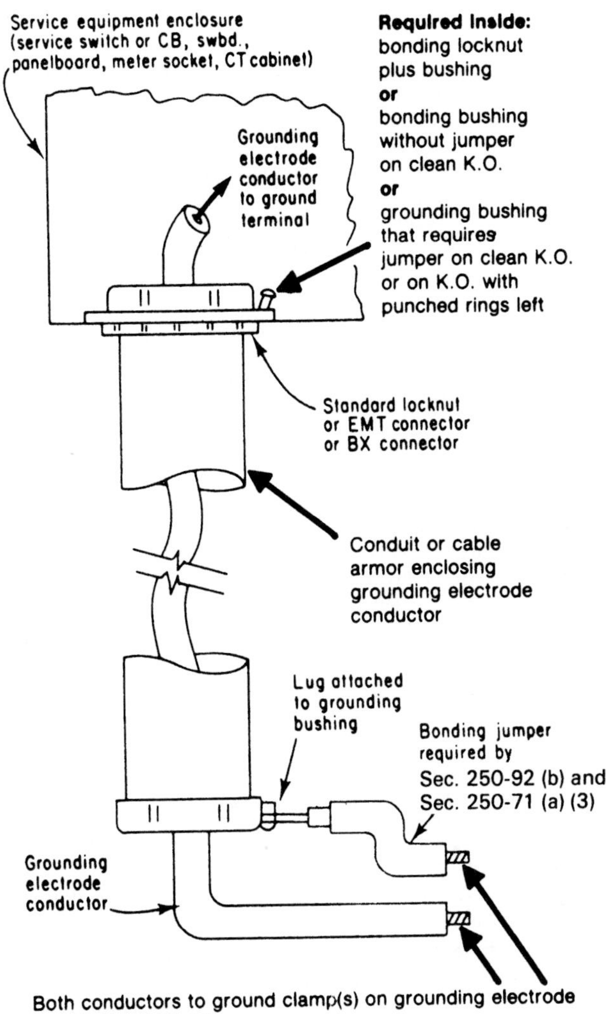

Fig. 250-54. Grounding-conductor enclosure must be "bonded" at both ends. (Sec. 250-71.)

between such metal enclosures as a result of lightning or power contacts. This rule was placed in the NE Code because Sec. 800-40(d) requires bonding interconnection between a building's power grounding electrode system and the "protector ground" (grounding electrode conductor) of telephone and other communications systems, and because making that bonded interconnection has become more difficult. Sections 810-21(j) and 820-40(d) also require such grounding interconnections.

The proposal for this **Code** addition included the following commentary:

In the past, the bond between communications and power systems was usually achieved by connecting the communications protector grounds to an interior water pipe. Where the power was grounded to a ground rod, the bond was connected to the power grounding-electrode conductor or to metallic service conduit, which were usually accessible. With growing use of plastic water pipe, the tendency for service equipment to be installed in finished areas where the grounding electrode conductor is often concealed, and the use of plastic entrance conduit, communications installers no longer have an easily identifiable point for connecting bonds or grounds.

Where lightning or external power fault currents flow in protective grounding systems, there can be dangerous potential differences between the equipment of those systems. Even with the required common or bonded electrodes, lightning currents flowing in noncommon portions of the grounding system result in significant potential differences as a result of inductive voltage drop in the noncommon conductor. If a current flows through a noncommon grounding conductor 10 ft (3.05 m) long, there can be an inductive voltage drop as high as 4000 volts. If that noncommon conductor is either the power grounding-electrode conductor or the communications-protector grounding conductor, the voltage will appear between communication-equipment and power-equipment enclosures. The best technical solution to minimizing that voltage is with a short bond between the service equipment and the communications-protector ground terminal. The conductor to the grounding electrode is then common, and the voltage drop in it does not result in a potential difference between systems.

An externally accessible point for intersystem bonding should be provided at the electrical service if accessible metallic service-entrance conduit is not present or if the grounding-electrode conductor is not accessible. This point could be in the form of a connector, tapped hole, external stud, a combination connector-SE cable clamp, or some other approved means located at the meter base or service equipment enclosure.

The first FPN at the end of this section describes a No. 6 copper "pigtail" that can be made available.

A definite shock hazard can arise if a *common* grounding electrode conductor is *not* used to ground *both* the bonded service neutral *and* the communications protector. The problem can be solved by simply bonding the ground terminal of the protector to a grounded enclosure of the service equipment (the service panelboard enclosure or the meter socket) and not using the separate telephone grounding electrode conductor.

250-72. Method of Bonding Service Equipment. Section 250-71 is very specific in listing the many types of equipment that require bonding connections, but the actual "how to" is often hazy. For virtually every individual situation where a bonding connection must be made, there is available a variety of products on the market which present the installer with a choice of different methods.

This section sets forth the specific means which may be used to connect service-conductor enclosures together to satisfy the bonding requirements of Sec. 250-71. These means include:

1. Bonding equipment to the grounded service conductor by means of suitable lugs, pressure connectors, clamps, or other approved means—except that soldered connections must not be used. Section 250-61 permits grounding of meter housings and service equipment to the grounded service conductor on the supply side of the service disconnecting means.
2. Threaded couplings in rigid metal conduit or IMC (intermediate metal conduit) runs and threaded bosses on enclosures to which rigid metal conduit or IMC connects.

3. Threadless couplings and connectors made up tight for rigid metal conduit, IMC, or electrical metallic tubing.
4. Bonding jumpers to securely connected metallic parts. Bonding jumpers must be used around concentric or eccentric knockouts which are punched or otherwise formed in such a manner that would impair the electrical current flow through the reduced cross section of metal that bridges between the enclosure wall and punched ring of the KO (knockout). And the bonding jumpers must be sized from Sec. 250-79(c).
5. Other devices (not standard locknuts and bushings) approved for the purpose.

Based on those briefly worded **Code** requirements, modern practice follows more or less standard methods.

Where rigid conduit is the service raceway, threaded or threadless couplings are used to couple sections of conduit together. Conduit connection to a meter socket may be made by connecting a threaded conduit end to a threaded hub or boss on the socket housing, where the housing is so constructed; by a locknut and bonding bushing; by a locknut outside with a bonding wedge or bonding locknut and a standard metal or completely insulating bushing inside; or sometimes by a locknut and standard bushing where the socket enclosure is bonded to the grounded service conductor. Conduit connections to KOs in sheet metal enclosures can be made with a bonding locknut (Fig. 250-55), a bonding wedge, or a bonding bushing where no KO rings remain around the opening through which the conduit enters and, where the box is listed for such use, even where the KOs have *not* all been removed. But, generally, where a KO ring does remain around the conduit entry hole, a bonding bushing or wedge with a jumper wire must be used to assure a path of continuity from the conduit to the enclosure. Figure 250-56 summarizes the various acceptable techniques. It should be noted that the use of the common locknut and bushing type of connection is not allowed. Neither is the use of double locknuts—one inside, one outside—and a bushing, although that is permitted on the load side of the service equipment. The special methods set forth in Sec. 250-72 are designed to prevent poor connections or loosening of connections due to vibration. This minimizes the possibility of arcing and consequent damage which might result when a service conductor faults to the grounded equipment.

Similar provisions are used to assure continuity of the ground path when EMT is the service raceway or when armored cable is used. EMT is coupled or connected by threadless devices—compression-type, indenter-type, or setscrew type, using raintight type outdoors. Although a threadless box connector is suitable to provide bonded connection of the connector to the metal raceway (rigid metal, IMC, EMT), it is also necessary to provide a "bonded" connection between the connector and the metal enclosure. A threadless box connector on the end of EMT used as service raceway provides satisfactory bonding of the EMT to the connector, but the last sentence of this rule says that a standard locknut or a standard bushing connected to the threaded end of the connector does not provide the required bonding of the connector to the metal service equipment to which the connector is connected. On the end of the connector, a bonding locknut or bonding bushings with or without jumpers must be used if the knockout is clean (all rings punched out or clean knockout punched on the

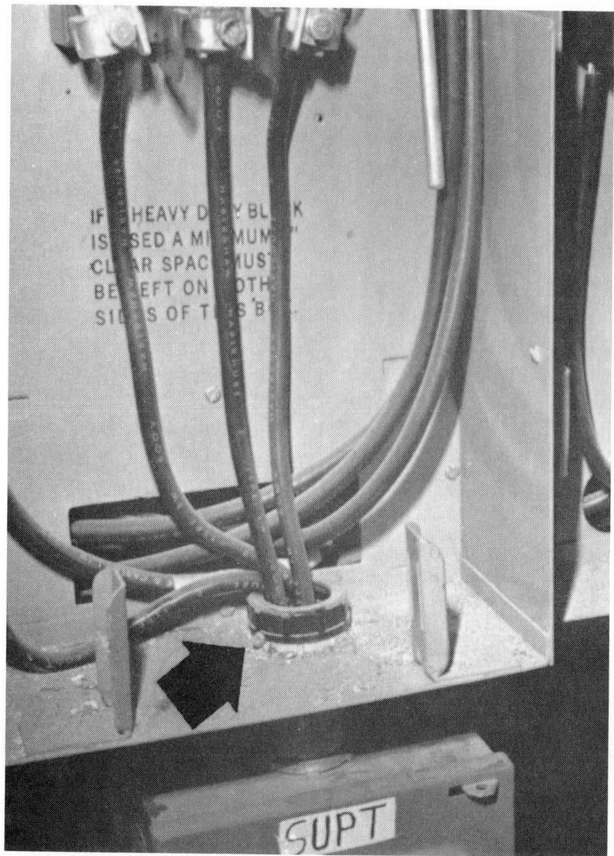

Fig. 250-55. Bonding locknut is a recognized method for bonding a service conduit nipple to a meter socket, when the KO is clean (no rings left in enclosure wall) or is cut on the job. With plastic bushing permitted, this is the most economical of the several methods for making a bonded conduit termination. (Sec. 250-72.)

job). If concentric or eccentric rings are left, a grounding locknut with a jumper, a grounding bushing with a jumper, or a grounding wedge with a jumper must be used to provide bonded connection around the perforated knockout. And fittings used with service cable armor must assure the same degree of continuity of ground path.

The use of bonding bushings, bonding wedges, and bonding locknuts is recognized without reference to types of raceways or types of connectors used with the raceways or cable armor. As a result, common sense and experience have molded modern field practice in making raceway and armored service cable connection to service cabinets. The top of Fig. 250-57 shows how a bonding wedge is used on existing connections at services or for raceway connec-

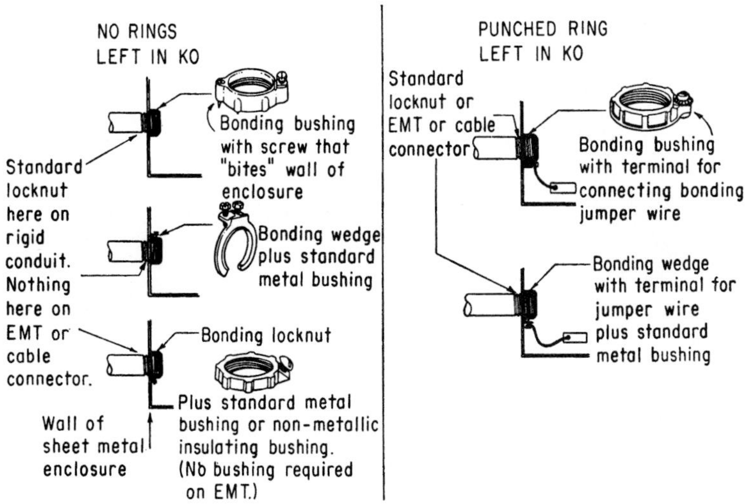

Fig. 250-56. Methods for "bonding" wiring methods to sheet metal enclosures. (Sec. 250-72.)

tions on the load side of the service—such as required by Sec. 501-16(b) for Class I hazardous locations. A bonding bushing, with provision for connecting a bonding jumper, is the common method for new service installations where some of the concentric or eccentric "doughnuts" (knockout rings) are left in the wall of the enclosure, therefore requiring a bonding jumper. Great care must be taken to assure that each and every type of bushing, locknut, or other fitting is used in the way for which it is intended to best perform the bonding function.

Figure 250-58 shows detailed application of the above rules to typical meter-socket installations. Meter-enclosure bonding techniques are shown in Fig. 250-59. Bonding details for current-transformer installations are shown in Fig. 250-60. Those illustrations are intended to portray typical field practice aimed at satisfying the various Code rules.

250-74. Connecting Receptacle Grounding Terminal to Box. The first paragraph requires that a jumper be used when the outlet box is installed in the wall (Fig. 250-61). Because boxes installed in walls are very seldom found to be perfectly flush with the wall, direct contact between device screws and yokes and boxes is seldom achieved. Screws and yokes currently in use were designed solely for the support of devices rather than as part of the grounding circuit.

Although the general rule states that a flush-type box, installed in a wall for a receptacle outlet, does require a bonding jumper from a grounded box to the receptacle grounding terminal, Exception No. 1 pertains to surface-mounted boxes and eliminates the need for a separate bonding jumper between a surface-mounted box and the receptacle grounding terminal under the conditions described. Although the Exception generally exempts surface-mounted boxes from the need for a bonding jumper from the box to the ground terminal of a receptacle installed in the box—because there is solid contact between the

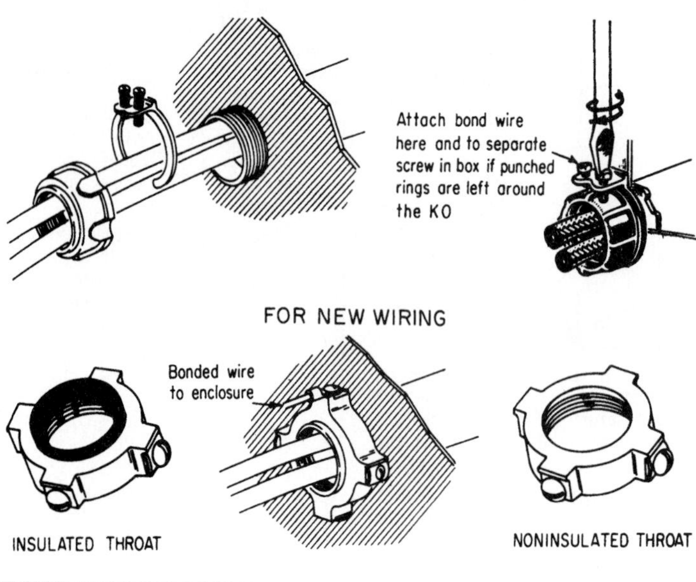

ALWAYS NEEDS JUMPER

NO JUMPER ON CLEAN KO

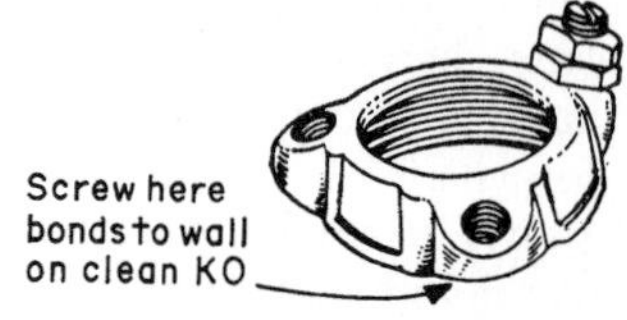

Bonding bushing with lug for jumper wire – may be used with jumper for clean KO or with rings left in wall of enclosure where *not* listed for connection with KOs in place.

Bonding bushing with screw that "bites" into enclosure wall may be used without a jumper on a clean KO or with a jumper when KO rings are left in wall or, where the box is listed for use *with* KOs remaining.

Fig. 250-57. Bonding bushings and similar fittings must be used in their intended manners. (Sec. 250-72.)

receptacles grounded mounting yoke and the ears on the box when installed—that is not applicable to a receptacle mounted in a raised box cover. A receptacle mounted in a raised box cover is connected to the cover by only a single screw, and that has been judged inadequate for grounding. A bonding jumper must be used on such a receptacle. (See Fig. 250-62.)

Figure 250-63 illustrates a grounding device which is intended to provide the electrical grounding continuity between the receptacle yoke and the box on

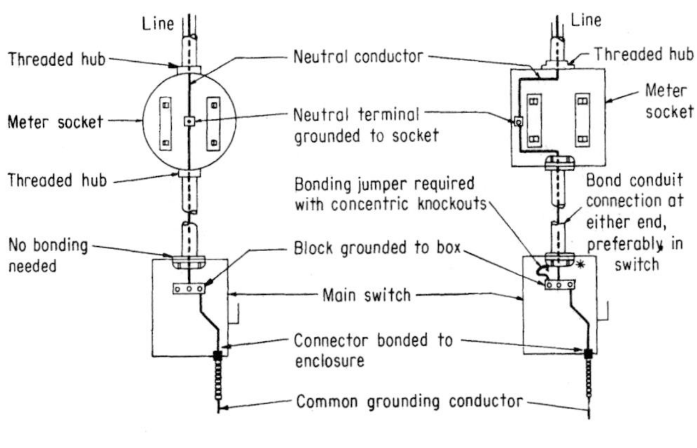

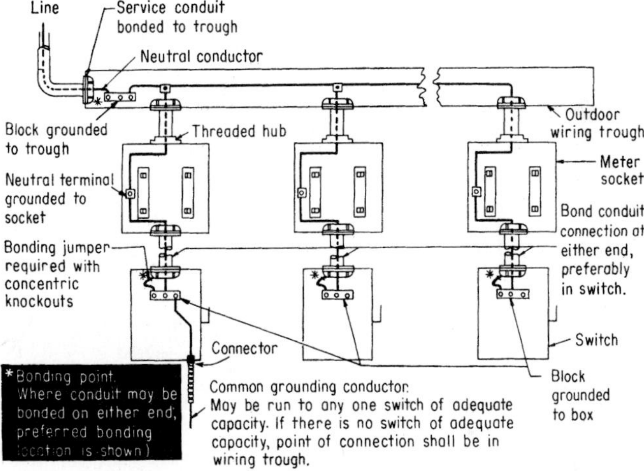

Fig. 250-58. Typical meter socket applications. (Sec. 250-72.)

which it is mounted and serves the dual purpose of both a mounting screw and a means of providing electrical grounding continuity in lieu of the required bonding jumper. As shown in the sketch, special wire springs and four-lobed machine screws are part of a receptacle design for use without a bonding jumper to box. This complies with Sec. 250-74, Exception No. 2.

Exception No. 3 permits nonself-grounding receptacles without an equipment grounding jumper to be used in floor boxes which are designed for and listed as providing proper continuity between the box and the receptacle mounting yoke.

Exception No. 4 allows the use of a receptacle with an isolated grounding terminal (no connection between the receptacle grounding terminal and the yoke). Sensitive electronic equipment that is grounded normally through the building ground is often adversely affected by pickup of transient signals which cause an imbalance in the delicate circuits. This is particularly true with highly intri-

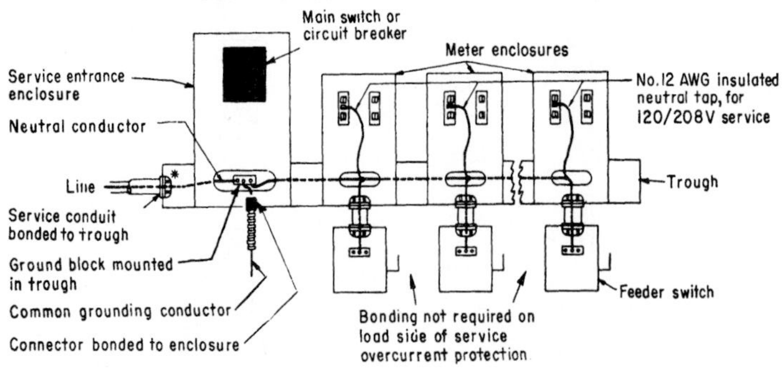

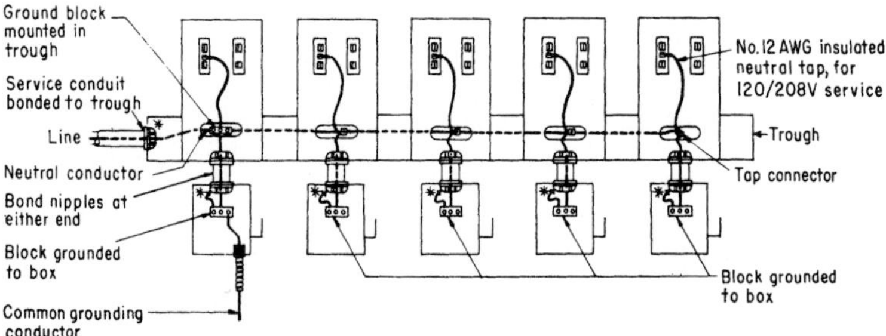

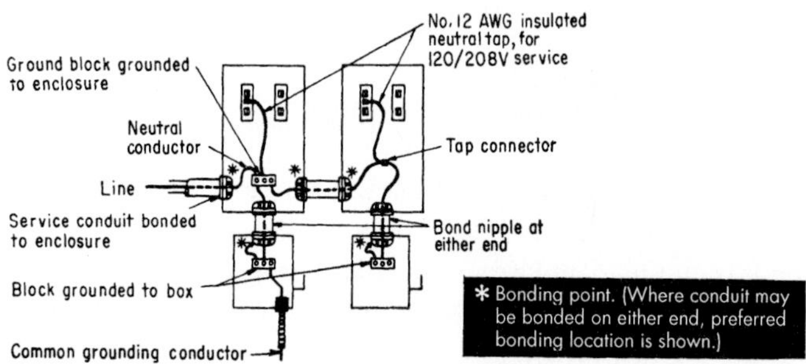

Fig. 250-59. Typical meter-enclosure installations (120/208- or 120/240-V services). (Sec. 250-72.)

cate medical and communications equipment, which often pick up unwanted currents, even of very low magnitude.

The use of an isolated grounding receptacle allows a "pure" path to be established back to the system grounding terminal, in the service disconnecting means, without terminating in any other intervening panelboard. In Fig. 250-64, a cutaway of an isolated grounding receptacle shows the insulation between the grounding screw and the yoke (top), and the hookup of the insulated grounding conductor to the common neutral-equipment-ground point of the electrical system (bottom).

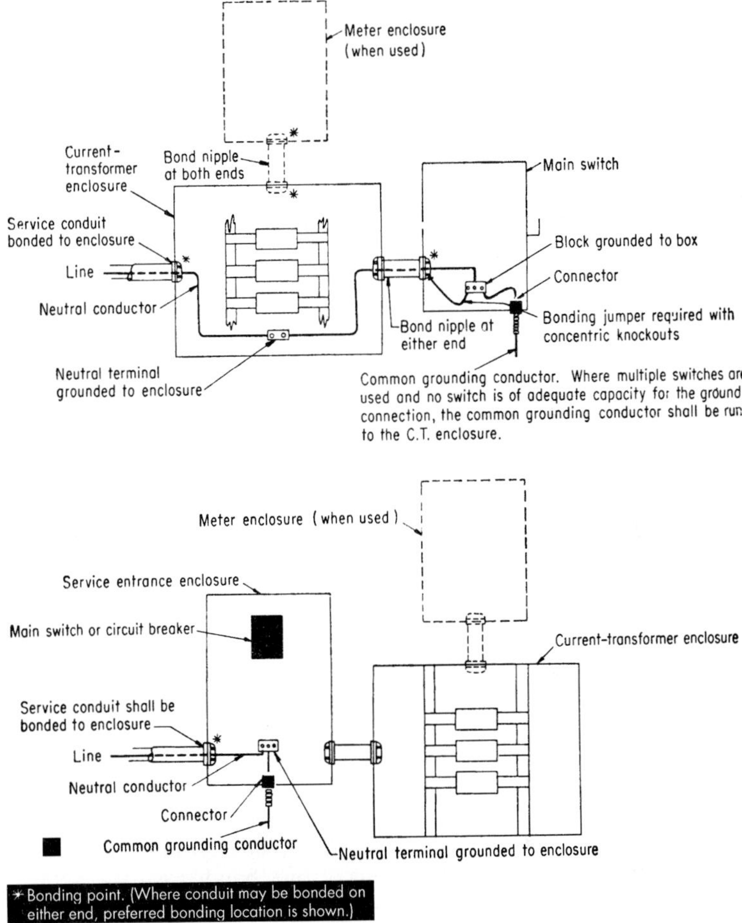

Fig. 250-60. Bonding at CT cabinets. (Sec. 250-72.)

The last sentence of Exception No. 4 permits an equipment grounding conductor from the insulated (quiet) ground terminal of a receptacle to be run, unbroken, all the way back to the ground terminal bus that is bonded to the neutral at the service equipment or at the secondary of a step-down transformer—but in no case may the isolated ground extend beyond the building in which it is used. That is, it must be bonded to a ground bus within the building it is run in, even if the "service equipment" is located in another building. Or the equipment grounding conductor may be connected to any ground bus in an intermediate panelboard fed from the service or transformer. *But,* the important point is to be sure the insulated ground terminal of the receptacle does tie into the equipment ground system that is bonded to the neutral.

This Exception must be observed very carefully to avoid violations that have been commonly encountered in the application of branch circuits to computer

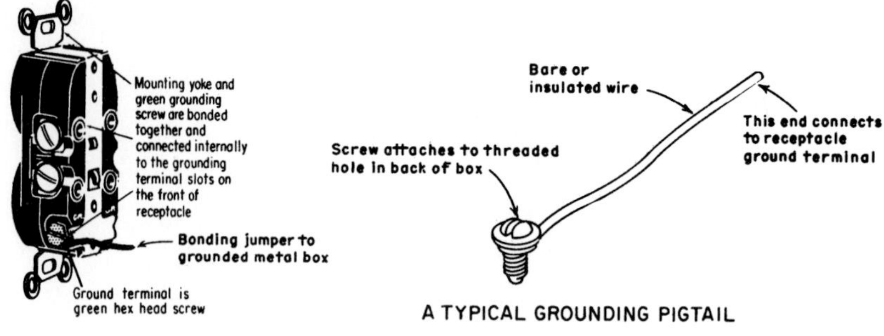

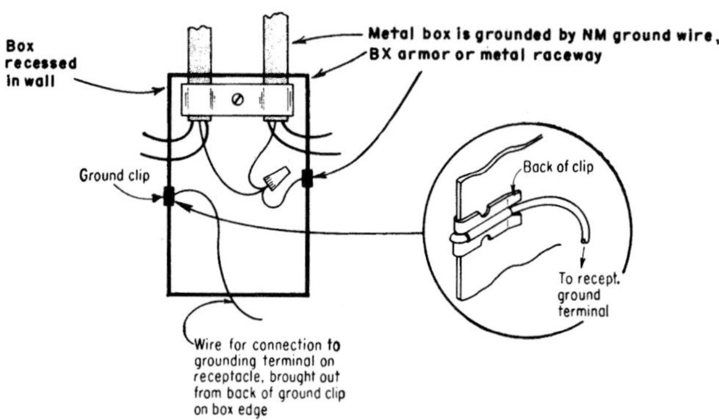

Fig. 250-61. Bonding jumper connects receptacle ground to grounded box. (Sec. 250-74.)

equipment—where manufacturers of computers and so-called computer power centers specified connection of "Quiet" receptacle ground terminals to a grounding electrode that is independent of (not bonded to) the neutral and bonded equipment ground bus of the electrical system. This practice developed to eliminate computer operating problems that were attributed to "electrical noise." Such isolation of the receptacle ground terminal does not provide an effective return path for fault-current flow and, therefore, constitutes a hazard.

Any receptacle grounding terminal (the green hex-head screw)—whether it is the common type with the mounting yoke or the type insulated from the yoke—must be connected back to the point at which the system neutral is bonded to the equipment grounding terminal and to the grounding electrode. That common (bonded) point may be at the service equipment (where there is no voltage step-down from the service to the receptacle), or the common neutral-equipment-ground point may be at a panelboard fed from a step-down transformer (as used in computer power centers).

When an isolated ground connection is made for the receptacle ground terminal, the box containing the receptacle must be grounded by the raceway supplying it and/or by another equipment grounding conductor run with the circuit wires. And those grounding conductors must tie into the same neutral-

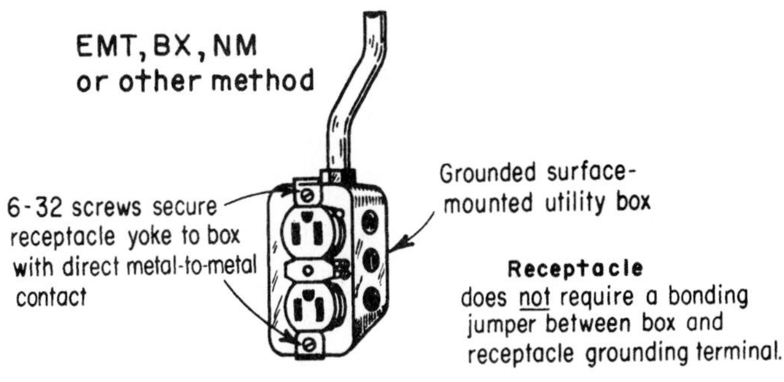
EMT, BX, NM
or other method
6-32 screws secure
receptacle yoke to box
with direct metal-to-metal
contact
Grounded surface-
mounted utility box
Receptacle
does not require a bonding
jumper between box and
receptacle grounding terminal.

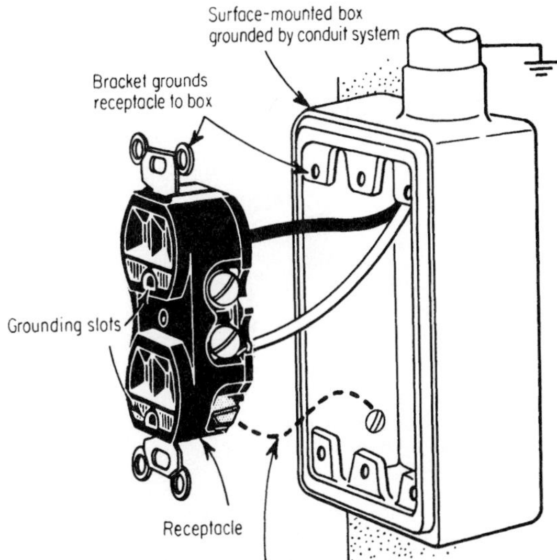
Surface-mounted box
grounded by conduit system
Bracket grounds
receptacle to box
Grounding slots
Receptacle
A jumper wire to connect the grounding-screw terminal to the grounded box
is not required with a surface-mounted box, but is required when receptacle
is used in a recessed box

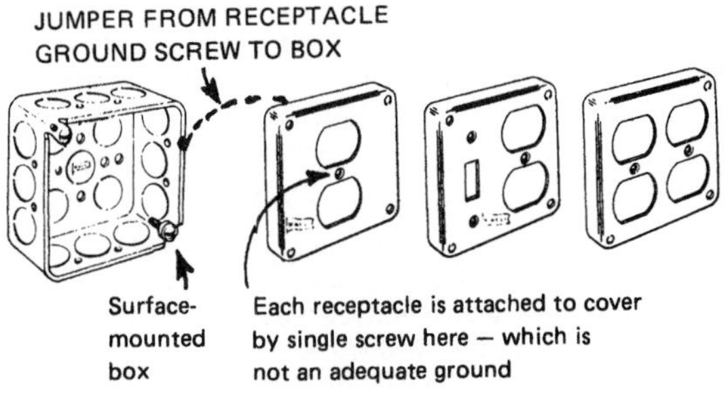
JUMPER FROM RECEPTACLE
GROUND SCREW TO BOX
Surface-
mounted
box
Each receptacle is attached to cover
by single screw here — which is
not an adequate ground

RECEPTACLE IN RAISED COVER REQUIRES BONDING JUMPER (cast covers for FS and FD boxes and other covers may contain a receptacle without a bonding jumper if they are "listed" as suitable for grounding.)

Fig. 250-62. Typical applications where a surface box does and does not need a receptacle bonding jumper. (Sec. 250-74.)

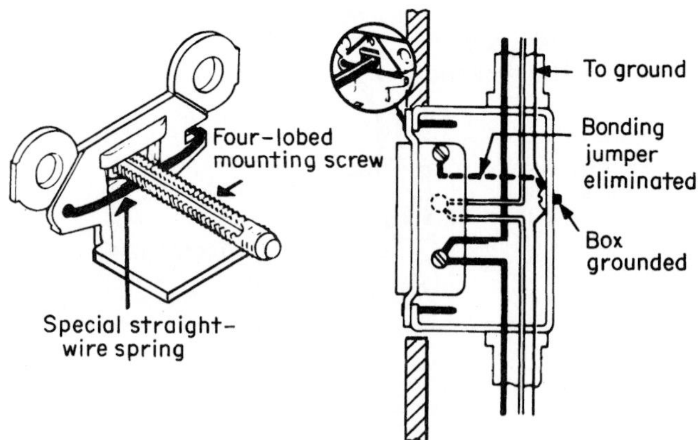

Fig. 250-63. Self-grounding screws ground receptacle in recessed box without bonding jumper. (Sec. 250-74.)

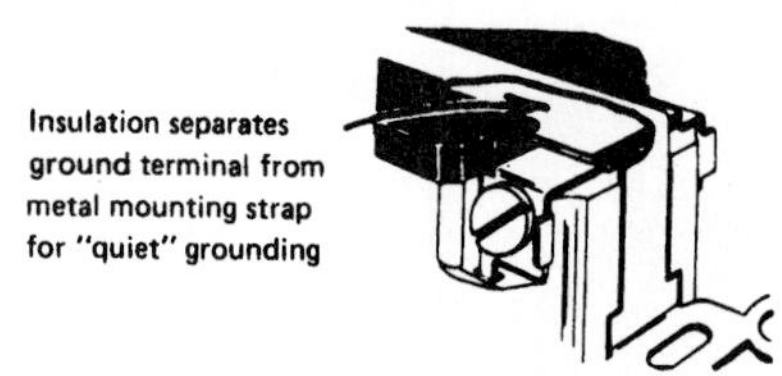

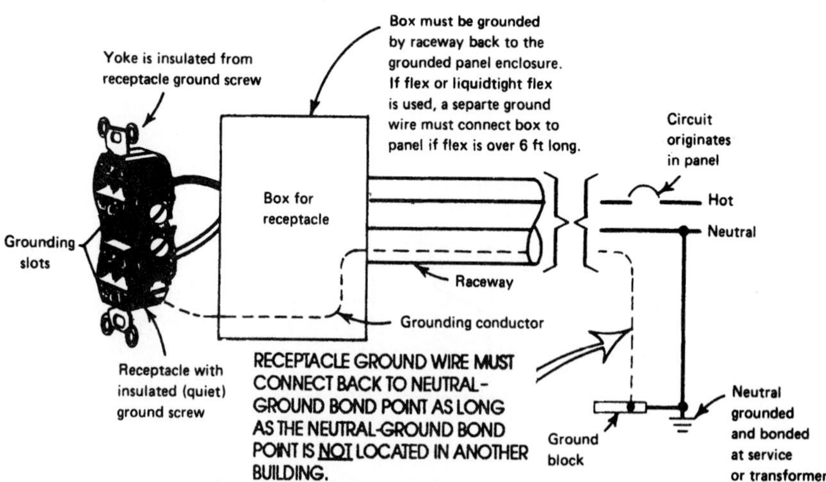

Fig. 250-64. Receptacles with isolated ground terminal are used with "clean" or "quiet" ground. (Sec. 250-74.)

equipment-ground point to which the receptacle isolated ground terminal is connected.

See comments that follow Sec. 384-27, Exception.

250-75. Bonding Other Enclosures. Metal raceways, cable sheaths, equipment frames and enclosures, and all other metal noncurrent-carrying parts must be carefully interconnected with Code-recognized fittings and methods to assure a low-impedance equipment grounding path for fault current—whether or not an equipment grounding conductor (a ground wire) is run within the raceway and connected enclosures. The interconnected system of metal raceways and enclosures must itself form a Code-conforming equipment grounding path—even if a "supplementary equipment grounding conductor" is used within the metal-enclosure grounding system (Fig. 250-65).

The Exception to this basic rule recognized that to reduce electromagnetic noise or interference on a grounding circuit, an insulating "spacer or fitting"

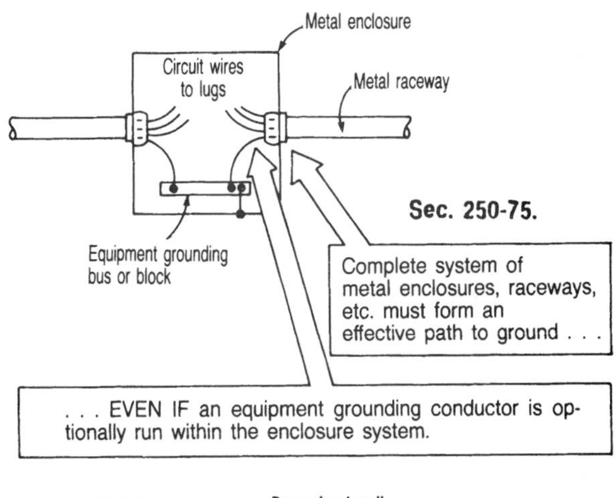

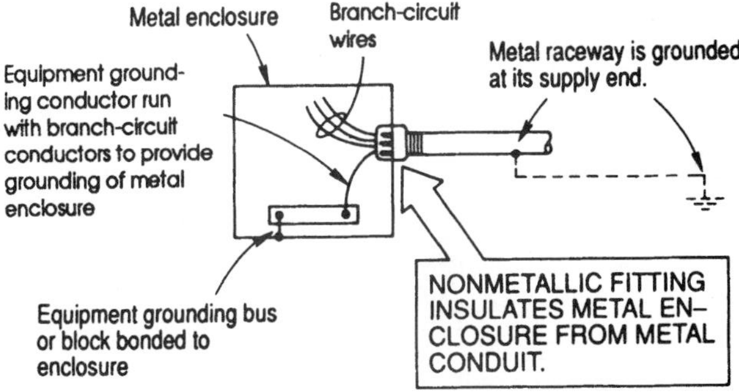

Fig. 250-65. Interconnected metal enclosures (boxes, raceways, cabinets, housings, etc.) must form a continuous equipment grounding path, even if a separate equipment grounding wire is run within the metal enclosure system, except as shown in bottom part of above, to eliminate "noise" on the grounding circuit. [Sec. 250-75.]

may be used to interrupt the electrical continuity of a metallic raceway system used to enclose the branch-circuit conductors at the point of connection to the metal enclosure of a single piece of equipment.

This exception permits interrupting the current path between a metal equipment enclosure and the metal conduit that supplies the enclosure—*but only* if the metal conduit is grounded at its supply end *and* an equipment grounding conductor is run through the conduit into the metal enclosure and is connected to an equipment grounding terminal of the enclosure, to provide safety grounding of the metal enclosure. Provisions for an equipment ground reference separate from the metallic raceway system is covered by Sec. 250-74, Exception No. 4 for electronic equipment that is cord- and plug-connected. This Exception covers a separate equipment ground reference for hard-wired sensitive electronic equipment.

250-76. Bonding for Over 250 Volts. Single locknut-and-bushing terminations are permitted for 120/240-V and 120/208-V systems. Any 480/277-V grounded system, 480-V ungrounded system, or higher must generally use double locknut-and-bushing terminals on clean knockouts of sheet metal enclosures (no concentric rings in wall) for rigid metal conduit and IMC (Fig. 250-66).

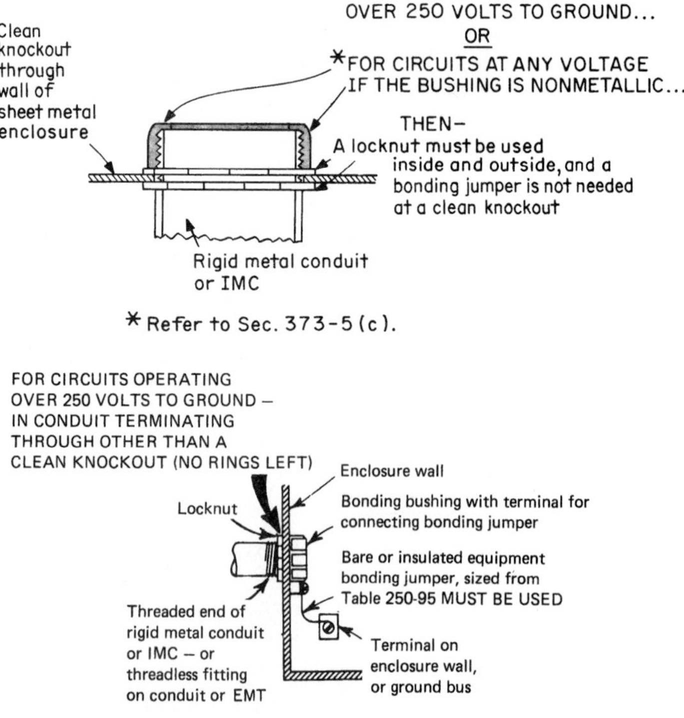

Fig. 250-66. For circuits over 250 volts to ground, a bonding jumper may be needed at conduit termination. [Sec. 250-76.]

Where good electrical continuity is desired on installations of rigid metal conduit or IMC, two locknuts should always be provided on clean knockouts (no rings left) of sheet metal enclosures so that the metal of the box can be solidly clamped between the locknuts, one being on the outside and one on the inside. The reason for not relying on the bushing in place of the inside locknut is that both the conduit and the box may be secured in place and if the conduit is placed so that it extends into the box to a greater distance than the thickness of the bushing, the bushing will not make contact with the inside surface of the box. But that possible weakness in the single-locknut termination does not exclude it from use on systems up to 250 V to ground.

The Exception to the main rule here has the effect of requiring that a bonding jumper must be used around any "oversized, concentric, or eccentric knockouts" in enclosures for circuits over 250 V to ground that are run in a metal raceway or cable unless the enclosure or fittings have been investigated and listed for use without a bonding jumper. Many such enclosures and fittings are so listed and are readily avaible. Clearly, the use of that equipment will serve to reduce labor costs. But, generally for such circuits, a bonding jumper must be used at any conduit or cable termination in other than a clean, unimpaired opening in an enclosure (Fig. 250-66). In any case where all the punched rings (the "doughnuts") are not removed and the box is *not* listed for use without a bonding jumper, or, where all the rings are removed but a reducing washer is used to accept a smaller size of conduit, a bonding jumper must be installed from a suitable ground terminal in the enclosure to a lug on the bushing or locknut of the termination of any conduit or cable containing conductors operating at over 250 V to ground. Such circuits include 480/277-V circuits (grounded or ungrounded); 480-, 550-, 600/347 V, and 600-V circuits; and higher-voltage circuits. A bonding jumper is not needed for terminations of conduit that carry such circuits through KOs that are punched on the job to accept the corresponding size of conduit. *But,* double locknuts (one inside and one outside the enclosure) must be used on threaded conduit ends, or suitable threadless connectors or other fittings must be used on rigid or flexible conduit, EMT, or cable.

250-77. Bonding Loosely Jointed Metal Raceways. Provision must be made for possible expansion and contraction in concrete slabs due to temperature changes by installing expansion joints in long runs of raceways run through slabs. See Sec. 300-7(b). Because such expansion joints are loosely jointed to permit back-and-forth movement to handle changes in gap between butting slabs, bonding jumpers must be used for equipment grounding continuity (Fig. 250-67). Expansion fittings may be selected as vibration dampers and deflection mediums as well as to provide for movement between building sections or for expansion and contraction due to temperature changes in long conduit runs. The fitting diagrammed in Fig. 250-67 provides for movement from the normal in all directions plus 30° deflection, is available up to 4-in (101.5 mm) in diameter, and may be installed in concrete.

250-78. Bonding in Hazardous (Classified) Locations. All raceway terminations in hazardous locations must be made by one of the techniques shown in Fig. 250-56 for service raceways. And as required by Sec. 501-16(a) such bonding tech-

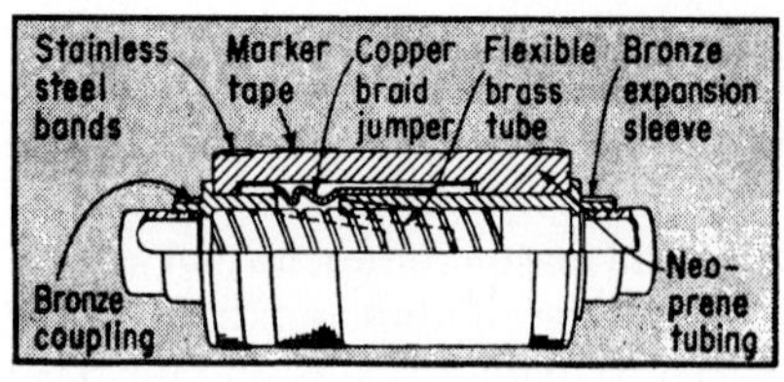

Fig. 250-67. Conduit expansion fitting includes bonding jumper for ground continuity. (Sec. 250-77.)

niques must be used in "all intervening raceways, fittings, boxes, enclosures, etc., between hazardous areas and the point of grounding for service equipment." Refer to Secs. 501-16(a), 502-16(a), and 503-16 (Fig. 250-68).

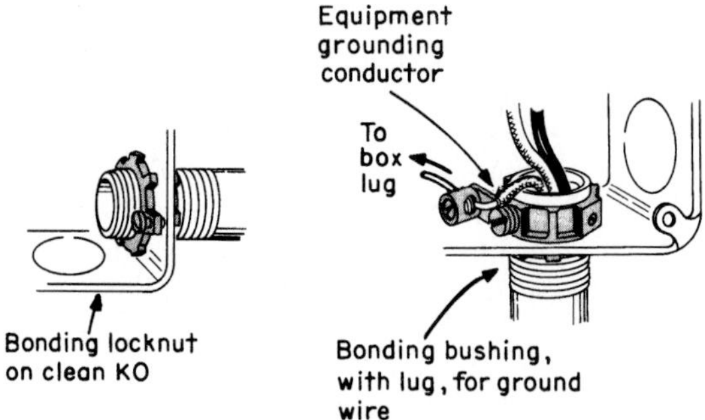

NOTE: Connection of threaded rigid metal conduit or IMC to a threaded boss or hub is considered to be a bonded conduit termination.

Fig. 250-68. Bonded raceway terminations must be used at sheet metal KOs in hazardous areas. (Sec. 250-78.)

250-79. Main and Equipment Bonding Jumpers. Part **(a)** calls for use of copper or other "corrosion-resistant" conductor material—which does include aluminum and copper-clad aluminum. Part **(c)** demands use of connectors, lugs, and other fittings that have been designed, tested, and listed for the particular application.

Part **(d)** covers sizing of any bonding jumper within the service equipment enclosure or on the line or supply side of that enclosure. Refer to the definition of "Bonding Jumper, Main" in Art. 100.

Figure 250-69 shows examples of sizing bonding jumpers in accordance with the first sentence of part **(d)** of this section.

At A, the bonding bushing and jumper are used to comply with part **(d)** of Sec. 250-72. Referring to Table 250-94, with 500-kcmil copper as the "largest service-entrance conductor," the minimum permitted size of grounding electrode conductor is No. 1/0 copper (or No. 3/0 aluminum). That therefore is the minimum permitted size of the required bonding jumper.

At B, with each service phase leg made up of two 500-kcmil copper conductors in parallel, the left-hand column heading in Table 250-94 refers to the "equiva-

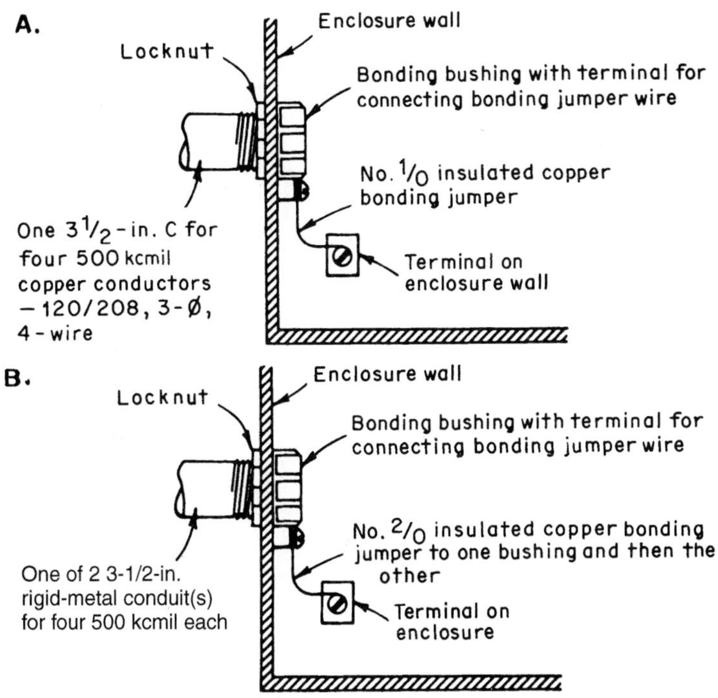

Fig. 250-69. Examples of the basic sizing of service bonding jumpers. (Sec. 250-79.)

lent for parallel conductors." As a result, the phase leg is taken at 2 × 500 or 1,000 kcmil, which is the physical equivalent of the makeup. Then Table 250-94 requires a minimum bonding jumper of No. 2/0 copper (or No. 4/0 aluminum).

Figure 250-70 shows an example of sizing a service bonding jumper in accordance with the second sentence of part **(d)** of this section. In this sketch, the jumper between the neutral bus and the equipment ground bus is defined by the NE Code as a "main bonding jumper" and the minimum required size of this jumper for this installation is determined by calculating the size of one service phase leg. With three 500 kcmil per phase, that works out to 1,500-kcmil copper per phase. Because that value is in excess of 1,100-kcmil copper, as noted in the Code rule, the minimum size of the main bonding jumper must equal at least 12½ percent of the phase leg cross-section area. Then,

$$12\tfrac{1}{2}\% \times 1{,}500 \text{ kcmil}$$
$$= 0.125 \times 1{,}500 = 187.5 \text{ kcmil}$$

Referring to Table 8 in Chap. 9 in the back of the Code book, the smallest conductor with at least that cross-section area (csa) is No. 4/0 with a csa of 211,600

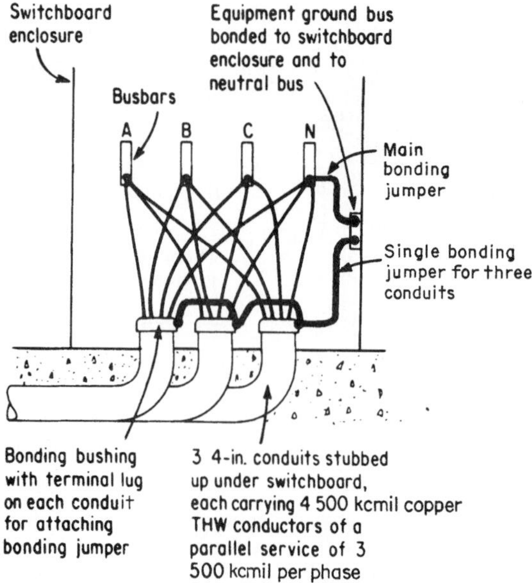

Fig. 250-70. Sizing main bonding jumper and other jumpers at service equipment. (Sec. 250-79.)

cmil, or 211.6 kcmil. Note that a No. 3/0 has a csa of only 167.8 kcmil. Thus No. 4/0 copper with any type of insulation would satisfy the Code.

The jumper shown in Fig. 250-70 running from one conduit bushing to the other and then to the equipment ground bus is defined by the NE Code as an "equipment bonding jumper." It is sized the same as a main bonding jumper (above). In this case, therefore, the equipment bonding jumper would have to be not smaller than No. 4/0 copper. And with the calculation that uses the 12½ percent value, if the jumper conductor is to be aluminum instead of copper, a calculation must be made as described below.

In the sketch of Fig. 250-70, if each of the three 4-in. (101.5 mm) conduits has a separate bonding jumper connecting each one individually to the equipment ground bus, the next to last sentence of part **(d)** may be applied to an individual bonding jumper for each separate conduit (Fig. 250-71). The size of a separate bonding jumper for each conduit in a parallel service must be not less than the size of the grounding electrode conductor for a service of the size of the phase conductor used in each conduit. Referring to Table 250-94, a 500-kcmil copper service calls for at least a No. 1/0 grounding electrode conductor. Therefore, the bonding jumper run from the bushing lug on each conduit to the ground bus must be at least a No. 1/0 copper (or 3/0 aluminum).

The third sentence of part **(d)** *requires* separate bonding jumpers when the service is made up of multiple conduits and the equipment bonding jumper is run within each raceway (such as plastic pipe) for grounding service enclosures. According to the third sentence of part **(d)**, when service-entrance con-

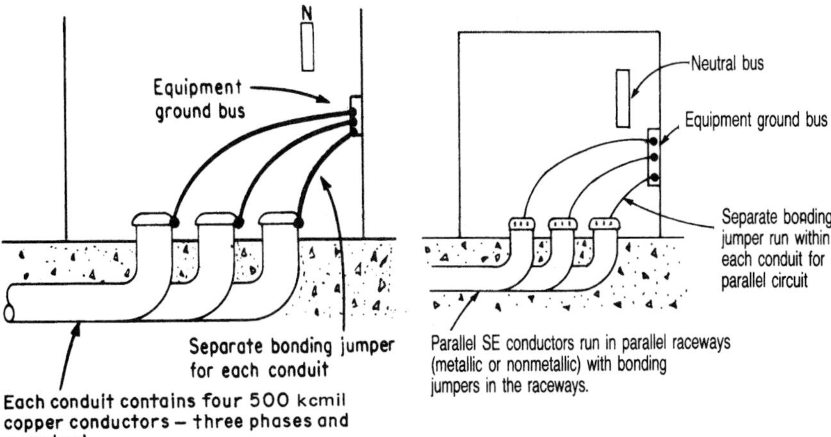

Fig. 250-71. An individual bonding jumper may be used for each conduit (left) and *must* be used as shown at right. [Sec. 250-79(d).]

ductors are paralleled in two or more raceways, an equipment bonding jumper that is routed within the raceways must also be run in parallel, one in each raceway, as at the right in Fig. 250-71. This clarifies application of nonmetallic service raceway where parallel conduits are used for parallel service-entrance conductors. As worded, the rule applies to both nonmetallic and metallic conduits where the bonding jumper is run within the raceways rather than from lugs on bonding bushings on the conduit ends. But for metallic conduits stubbed-up under service equipment, if the conduit ends are to be bonded to the service equipment enclosure by jumpers from lugs on the conduit bushings, either a single large common bonding jumper may be used—from one lug, to another lug, to another, etc., and then to the ground bus—or an individual bonding jumper (of smaller size from Table 250-94, based on the size of conductors in each conduit) may be run from each bushing lug to the ground bus.

The second sentence of part **(d)** sets minimum sizes of copper *and* aluminum service-entrance conductors above which a service bonding jumper must have a cross-section area "not less than 12½ percent of the area of the largest phase conductor." And the rule states that if the service conductors and the bonding jumper are of different material (i.e., service conductors are copper, say, and the jumper is aluminum), the minimum size of the jumper shall be based on the assumed use of phase conductors of the same material as the jumper and with an ampacity equivalent to that of the installed phase conductors (Fig. 250-72).

The last sentence in part **(d)** covers the sizing of a bonding jumper that is used to bond a raceway that contains a grounding electrode conductor. Such a raceway is required to provide mechanical protection for a grounding electrode conductor smaller than No. 6, as noted in Sec. 250-92(a). And protection is frequently provided for larger grounding electrode conductors.

At service equipment, a bonding jumper for a raceway containing a grounding electrode conductor only has to be at least the same size as the required

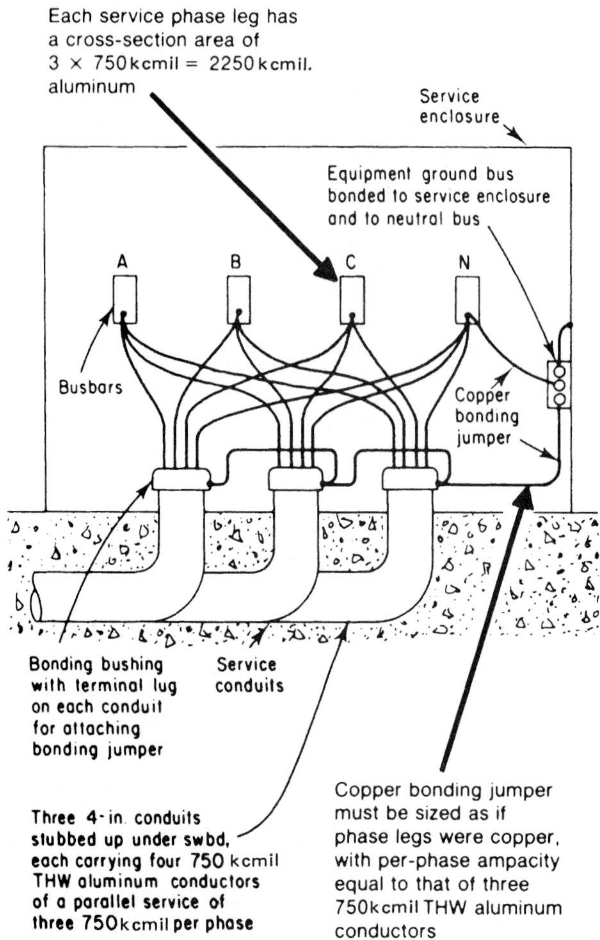

Fig. 250-72. Sizing a copper bonding jumper for aluminum service conductors. [Sec. 250-79(c).]

grounding electrode conductor, as shown in Fig. 250-73. The last sentence makes clear that the bonding jumper for a grounding electrode conductor conduit does *not* have to be sized at 12½ percent of the cross-section area of the largest phase conductor of the service—as required by the foregoing text of Sec. 250-79(d), when the largest service phase conductor is larger than 1100-kcmil (MCM) copper or 1,750-kcmil aluminum. That requirement for 12½ percent of the service phase size applies only to bonding jumpers that are used with conduits containing ungrounded service phase conductors, but not with raceway sleeves for grounding electrode conductors.

Part **(b)** of Sec. 250-92 covers details on the use of "metal enclosures" (such as conduit or EMT sleeves) for grounding electrode conductors. The grounding

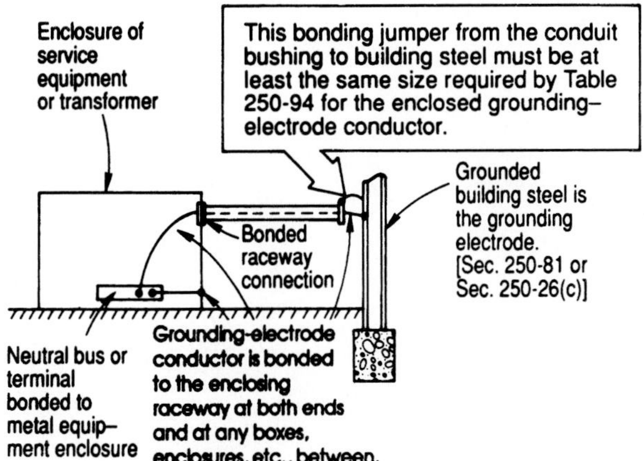
Enclosure of service equipment or transformer
This bonding jumper from the conduit bushing to building steel must be at least the same size required by Table 250-94 for the enclosed grounding–electrode conductor.
Grounded building steel is the grounding electrode. [Sec. 250-81 or Sec. 250-26(c)]
Bonded raceway connection
Grounding-electrode conductor is bonded to the enclosing raceway at both ends and at any boxes, enclosures, etc., between.
Neutral bus or terminal bonded to metal equip– ment enclosure

Fig. 250-73. A typical example of the rule on bonding jumpers for raceways enclosing grounding electrode conductors. [Sec. 250-79(d).]

conductor must be connected to its protective conduit at both ends so that any current that might flow over the conductor will also have the conduit as a parallel path. The regulation presented is actually a performance description of the rule of Sec. 250-71(a)(3), which specifically and simply requires that any conduit or armor enclosing a grounding electrode conductor be electrically parallel with the conductor. Bonded connections at both ends of an enclosing raceway must be used for any grounding electrode arrangement at a service and for grounding of a separately derived system, such as a generator or transformer secondary. Any bonding jumper for that application simply has to be at least the same size as required for the grounding electrode conductor run inside the conduit.

Part **(e)** requires a bonding jumper on the load side of the service to be sized as if it were an equipment grounding conductor for the largest circuit with which it is used. And sizing would have to be done from Table 250-95, as follows.

Figure 250-74 shows a floor trench in the switchboard room of a large hotel. The conductors are feeder conductors carried from circuit breakers in the main switchboard (just visible in upper right corner of photo) to feeder conduits going out at left, through the concrete wall of the trench, and under the slab floor to the various distribution panels throughout the building. Because the conduits themselves are not metallically connected to the metal switchboard enclosure, bonding must be provided from the conduits to the switchboard ground bus to assure electrical continuity and conductivity as required by NE Code Secs. 250-33, 250-42(e), 250-51, and 250-57.

1. The single, common, continuous bonding conductor that bonds all the conduits to the switchboard must be sized in accordance with NE Code Table 250-95, based on the highest rating of CB or fuses protecting any one of the total number of circuits run in all the conduits.

Fig. 250-74. Conduits in trench carry feeder conductors from switchboard at right (arrow) out to various panels and control centers. A single, common bonding jumper—run continuously from bushing to bushing—may be used to bond all conduits to the switchboard ground bus. [Sec. 250-79(e).]

2. Sizing of the single, common bonding jumper would be based on the highest rating of overcurrent protection for any one of the circuits run in the group of conduits. For instance, some of the circuits could be 400-A circuits made up of 500 kcmils in individual 3-in. (76-mm) conduits, and others could be parallel-circuit makeups in multiple conduits—such as 800-A circuits, with two conduits per circuit, and 1,200-A circuits, with three conduits. If, for instance, the highest-rated feeder in the group was protected by a 2,000-A circuit breaker, then the single, common bonding jumper for all the conduits would have to be 250-kcmil copper or 400-kcmil aluminum—determined readily from Table 250-95, by simply going down the left column to the value of "2,000" and then reading across. The single conductor is run through a lug on each of the conduit bushings and then to the switchboard ground bus.

In the case shown in Fig. 250-74, however, because the bonding jumper from the conduit ends to the switchboard is much longer than a jumper would be if the conduits stubbed-up under the switchboard, better engineering design might dictate that a separate equipment grounding *conductor* (rather than a "jumper") be used for each individual circuit in the group. If one of the conduits is a 3-in. (76-mm) conduit carrying three 500-kcmil conductors from a 400-A CB in the switchboard, the minimum acceptable size of bonding jumper (or equipment grounding conductor) from a grounding bushing on the conduit end to the switchboard ground bus would be No. 3 copper or No. 1 aluminum or copper-clad aluminum, as shown opposite the value of 400 A in the left column of Table 250-95. If another two of the 3-in. (76-mm) conduits are used for a feeder consisting of two parallel sets of three 500-kcmil conductors (each set of three 500-kcmil conductors in a separate conduit) for a circuit protected at 800 A, a single bonding jumper could be used, run from one grounding bushing to the other grounding bushing and then to the switchboard ground bus. This single bonding jumper would have to be a minimum No. 1/0 copper, from NE Code Table 250-95 on the basis of the 800-A rating of the feeder overcurrent protective device.

With such a long run for a jumper, as shown in Fig. 250-74, Code rules could be interpreted to require that the bonding jumper be subject to the rules of Sec. 250-95; that is, use of a bonding jumper must conform to the requirements for equipment grounding conductors. As a result, bonding of conduits for a parallel circuit makeup would have to comply with the second sentence in Sec. 250-95, which requires equipment grounding conductors to be run in parallel "where conductors are run in parallel in multiple raceways. . . ." That would then be taken to require that bonding jumpers *also* must be run in parallel for multiple-conduit circuits. And that concept is supported by the next-to-last sentence of part **(d)** of Sec. 250-79. In the case of the 800-A circuit above, instead of a single No. 1/0 copper jumper from one bushing lug to the other bushing lug and then to the ground bus, it would be necessary to use a separate No. 1/0 copper from each bushing to the ground bus, so the jumpers are run in parallel—as required for equipment grounding conductors. Figure 250-75 shows the two possible arrangements. The wording of Secs. 250-79(e) and 250-95 can be used to support either method. Bonding jumpers on the load side of service equipment are sized and routed the same as equipment grounding conductors because such bonding jumpers and equipment grounding conductors serve identical functions. And note that Sec. 250-95 requires the equipment grounding conductor for each of the conduits for a parallel circuit to be the full size determined from the circuit rating. In the case here, a No. 1/0 copper for each conduit is required, based on the 800-A rating of the feeder protective device.

An Exception notes that an equipment bonding jumper never has to be larger than the circuit conductors within a conduit being bonded.

Part **(f)** of Sec. 250-79 follows the thinking that was described in Sec. 250-58(a) for external grounding of equipment attached to a properly grounded metal rack or structure. A short length of flexible metal conduit, liquidtight flex, or any other raceway may, if the raceway itself is not acceptable as a grounding conductor, be provided with grounding by a "bonding jumper" (note: *not* an "equipment grounding conductor") run *either* inside or *outside* the raceway or enclosure PROVIDED THAT the *length* of the *equipment bonding jumper* is *not more* than 6 ft (1.83 m) and the jumper is routed with the raceway or enclosure.

Where an equipment bonding conductor is installed within a raceway, it must comply with all the Code rules on identification of equipment grounding conductors. A bonding jumper installed in flexible metal conduit or liquidtight flex serves essentially the same function as an equipment grounding conductor. For that reason, a bonding jumper should comply with the identification rules

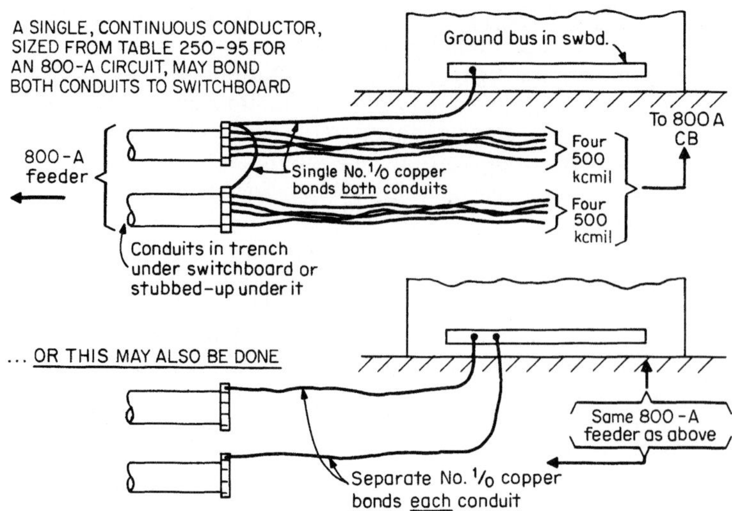

Fig. 250-75. One jumper may be used to bond two or more conduits on the load side of the service. [Sec. 250-79(e).]

of Sec. 310-12(b)—on the use of bare, green-insulated or green-taped conductors for equipment grounding.

Note that this application has limited use for the conditions specified and is a special variation from the concept of Sec. 250-57(b), which requires grounding conductors run inside raceways. Its big application is for external bonding of short lengths of liquidtight or standard flex, under those conditions where the particular type of flex itself is not suitable for providing the grounding continuity required by Secs. 350-5 and 351-9. Refer also to Sec. 250-91(b), Exceptions Nos. 1 and 2.

The top of Fig. 250-76 shows how an external bonding jumper may be used with standard flexible metallic conduit (so-called Greenfield). If the length of the flex is not over 6 ft (1.83 m), but the conductors run within the flex are protected at more than 20 A, a bonding jumper *must* be used either inside or outside the flex. An outside jumper must comply as shown. For a length of flex not over 6 ft (1.83 m), containing conductors that are protected at not more than 20 A and used with conduit termination fittings that are approved for grounding, a bonding jumper is *not* required—as covered in Sec. 250-91(b), Exception No. 1.

The bottom of Fig. 250-76 shows use of an external bonding jumper with liquidtight flexible metallic conduit. If liquidtight flex is not over 6 ft (1.83 m) long *but* is larger than 1¼-in. (31.75-mm) trade size, a bonding jumper must be used, installed *either* inside or outside the liquidtight. An outside jumper must comply as shown. If a length of liquidtight flex larger than 1¼ in. (31.75 mm) is short enough to permit an external bonding jumper that is not more than 6 ft (1.83 m) long between external grounding-type connectors at the ends of the flex, an external bonding jumper may be used. BUT WATCH OUT! The rule says the *jumper,* not the flex, must not exceed 6 ft (1.83 m) in length *AND* the

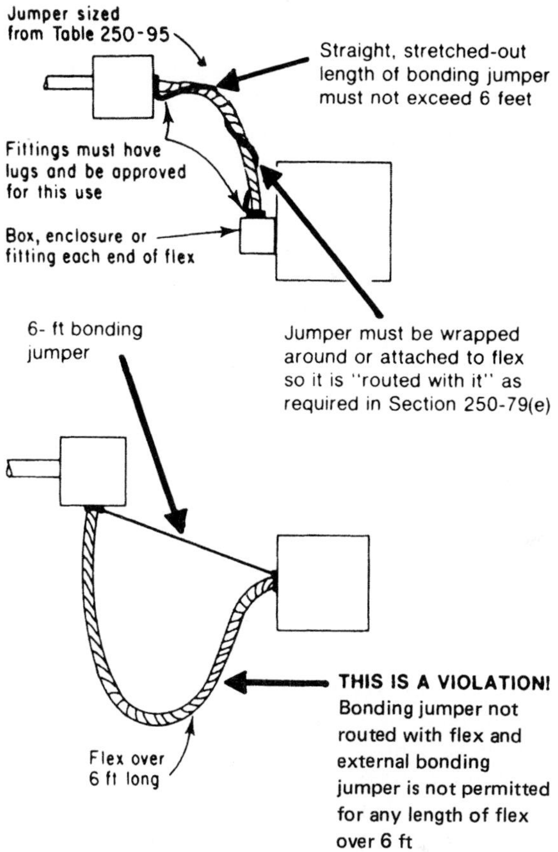
Jumper sized from Table 250-95
Straight, stretched-out length of bonding jumper must not exceed 6 feet
Fittings must have lugs and be approved for this use
Box, enclosure or fitting each end of flex
6- ft bonding jumper
Jumper must be wrapped around or attached to flex so it is "routed with it" as required in Section 250-79(e)
THIS IS A VIOLATION! Bonding jumper not routed with flex and external bonding jumper is not permitted for any length of flex over 6 ft
Flex over 6 ft long

Bonding jumper required for flex may be outside the flex.

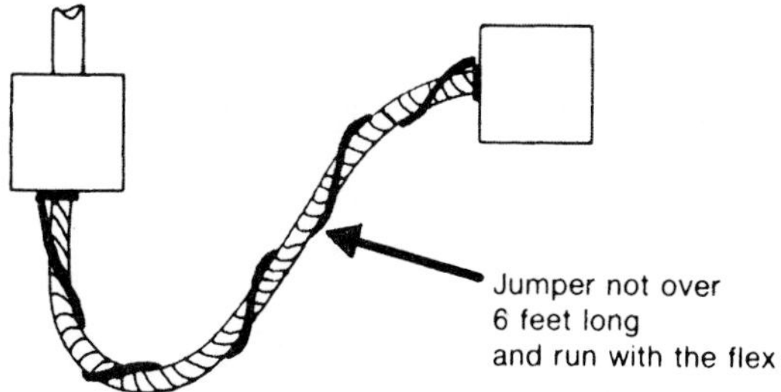
Jumper not over 6 feet long and run with the flex

Liquidtight flex larger than 1¼ in. size must have bonding jumper — inside or outside — in any length up to 6 ft.

Fig. 250-76. Bonding jumper rules for standard flex and liquidtight flex. [Sec. 250-79(f).]

jumper "shall be routed with the raceway"—that is, run along the flex surface and not separated from the flex.

250-80. Bonding of Piping Systems and Exposed Structural Steel. This section on bonding of piping systems in buildings is divided into two parts—metal *water* piping and *other* metal piping. This section is a rather elaborate sequence of phrases that may be understood in several ways. Of course, the basic concept is to ground any metal pipes that would present a hazard if energized by an electrical circuit.

Part **(a)** requires any "interior metal water piping system" to be bonded to the service-equipment enclosure, the grounded conductor (usually, a neutral) at the service, the grounding electrode conductor, *or* the one or more grounding electrodes used. All points of attachment of bonding jumpers for metal water-piping systems must be accessible. Only the connections (and not the entire length) of water-pipe bonding jumpers are required to be accessible for inspection. This rule applies where the metal water piping system does not have 10 ft (3.05 m) of metal pipe buried in the earth and is, therefore, not a grounding electrode. In such cases, though, this rule makes clear that the water piping system must be bonded to the service grounding arrangement. And the bonding jumper used to connect the interior water piping to, say, the grounded neutral bus or terminal (or to the ground bus or terminal) must be sized from Table 250-94 based on the size of the service conductors. The jumper is sized from that table because that is the table that would have been used *if* the water piping had 10 ft (3.05 m) buried under the ground, making it suitable as a grounding electrode. Note that the "bonding jumper" is sized from Table 250-94 [and not from Sec. 250-79(c)], which means it never has to be larger than No. 3/0 copper or 250-kcmil aluminum. Refer to the illustrations for Sec. 250-81, which also cover bonding of water piping.

The Exception to part **(a)** permits "isolated" metal water piping to be bonded to the main electrical enclosure (panelboard or switchboard) in each unit of a multitenant building—such as in each apartment of an apartment house, each store of a shopping center, or each office unit of a multitenant office building. See the top of Fig. 250-77. This Exception is intended to provide a realistic and effective way to bond interior metal water piping to the electrical grounding system in multitenant buildings where the metal water piping in each tenant's unit is fed from a main water distribution system of nonmetallic piping and is isolated from the metal water piping in other units. In apartment houses, multistore buildings, etc., it would be difficult, costly, and ineffective to use long bonding jumpers to tie the isolated piping in all the units back to the equipment grounding point of the building's service equipment—as required by the basic rule of Sec. 250-80(a). The objective of the basic rule is better achieved in such cases by simply bonding the isolated water piping in each occupancy to the equipment ground bus of the panelboard or switchboard serving the occupancy. The bonding jumper must be sized from Table 250-95 (*not* 250-94)—based on the largest rating of protective device for any circuit within the occupancy.

Following the Exception, this section puts forth the bonding requirements for separately derived systems. Basically stated, the grounded conductor must be

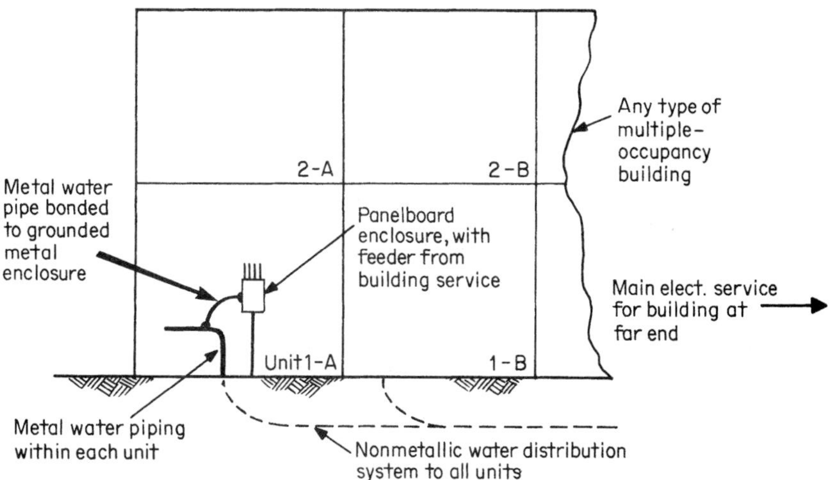

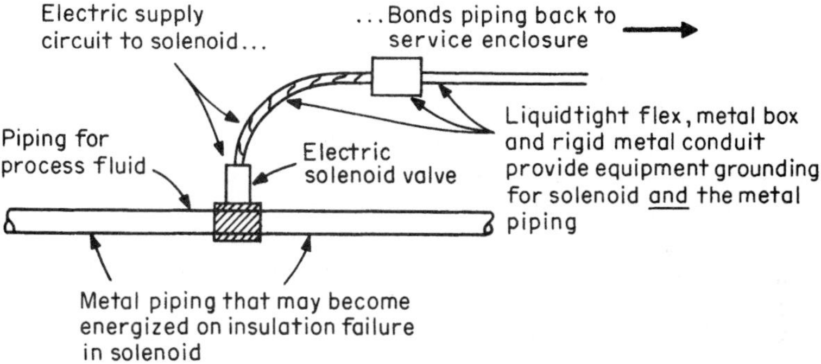

Fig. 250-77. Certain techniques are permitted as alternatives to the basic rules on grounding of metal piping systems. [Sec. 250-80.]

bonded to the interior metal water piping system within the area served by the separately dervived system. This is in addition to the bonding connection required between the grounded conductor and the building steel, or where no building steel is available, the connection to the water pipe is within 5 ft (1.52 m) from its point of entry. This local connection ensures that the metal water piping within the area supplied by the separately derived system is at the same potential with respect to ground as the metal enclosures and raceways associated with the separately derived system.

Part **(b)** requires a bonding connection from "other" (than water) metal piping systems—such as process liquids or fluids—that "may become energized" to the grounded neutral, the service ground terminal, the grounding electrode conductor, or the grounding electrodes. *But,* for these *other* piping systems the bonding jumper is sized from Table 250-95, using the rating of the overcurrent device of the circuit that *may* energize the piping.

Understanding and application of part **(b)** of Sec. 250-80 hinge on the reference to metal piping "which may become energized." What does that phrase mean? Is it not true that any metal piping "may" become energized? Or does the rule mean to apply only to metal piping that conductively connects to metal enclosures of electrical equipment—such as pump motors, solenoid valves, pressure switches, etc.—in which electrical insulation failure would put a potential on the metal piping system? It does seem that the latter case is what the Code rule means. And that concept is supported by the second paragraph of part **(b)**, which notes that where a particular circuit poses the threat of energization to a piping system, the equipment grounding conductor for that circuit (which could be the conduit or other raceway enclosing the circuit) may be used as the means of bonding the piping back to the service ground point. That has the effect of saying, for instance, that the equipment grounding conductor for a circuit to a solenoid valve in a pipe may also ground the piping (see bottom of Fig. 250-77).

Part **(c)**, which is new in the 1996 Code, now requires exposed building steel that is *not* effectively grounded to be bonded to the grounding electrode system. The connection may be made at the service equipment, to the grounded conductor, to the grounding electrode conductor where the grounding electrode conductor is "of sufficient size," or to any of the electrodes used.

The reference to "sufficient size" for connection to the grounding electrode conductor is intended to indicate that such a conductor must be sized per Sec. 250-94 based on the size of the largest service phase conductor. But, where, say, No. 6 copper is used for connection to a driven ground rod—as permitted in part **(a)** to the Exception—the bonding jumper from the nongrounded building must *not* be connected to that No. 6, unless it also satisfies the basic rule in Sec. 250-94. The same concept applies to parts **(b)** and **(c)** of the Exception to Sec. 250-94.

250-81. Grounding Electrode System. The rules in this section cover the grounding electrode arrangement required at the service entrance of a premise or in a building or other structure fed from a service in another building or other structure, as covered in Sec. 250-24. But, the rules do not apply to grounding of a separately derived system, such as a local step-down transformer, which is covered by part **(c)** of Sec. 250-26 (Fig. 250-78).

Section 250-81 calls for a "grounding electrode system" instead of simply a "grounding electrode" as required by previous NE Code editions. Up to the 1978 NEC, the "water-pipe" electrode was the premier electrode for service grounding, and "other electrodes" or "made electrodes" were acceptable *only* "where a water system (electrode) . . . is not available." If a metal water pipe to a building had at least 10 ft (3.05 m) of its length buried in the ground, that *had* to be used as the grounding electrode and no other electrode was required. The underground water pipe was the preferred electrode, the *best* electrode.

In the present NEC, of all the electrodes previously and still recognized by the NEC, the water pipe is the least acceptable electrode and is the only one that

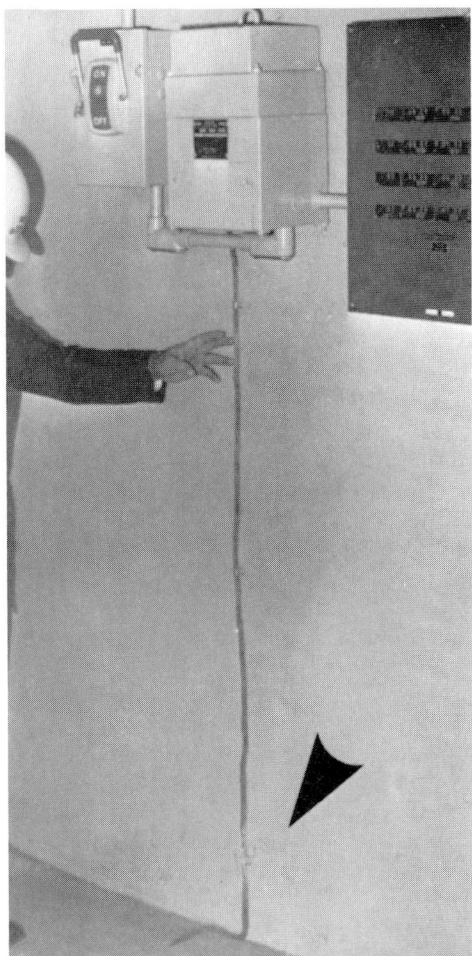

Fig. 250-78. Grounding electrode conductor from the bonded secondary neutral of this local transformer was connected to grounded building steel before concrete floor was poured. This installation is not covered by the rules of Sec. 250-81, but is covered by Sec. 250-26 and complies with those rules. (Sec. 250-81.)

may never be used by itself as the *sole* electrode. It must always be supplemented by at least one "additional" grounding electrode (Fig. 250-79). Any one of the other grounding electrodes recognized by the NEC *is* acceptable as the *sole* grounding electrode, by itself.

Take a typical water supply of 12-in. (3.05 mm)-diameter metal pipe running, say, 400 ft (122 m) underground to a building with a 4,000-A service. From Sec. 250-81(a), that water pipe, connected by a 3/0 copper conductor to the bonded

Fig. 250-79. Connection to an underground metal water-supply pipe is never adequate grounding for electric service equipment. (Sec. 250-81.)

service-equipment neutral, *must* be used as a grounding electrode but may *not* serve as the only grounding electrode. It must be supplemented by one of the other electrodes from Sec. 250-81 or 250-83. So the installation can be made acceptable by, say, running a No. 6 copper grounding electrode conductor from the bonded service neutral to an 8-ft (2.44-m), ½-in. (12.7 mm)-diameter ground rod. Although that seems like using a mouse to help an elephant pull a load, it is the literal requirement of Sec. 250-81. And if the same building did not have 10 ft (3.05 m) of metal water pipe in the ground, the 8-ft (2.44-m) ground rod would be entirely acceptable as the *only* electrode.

Although the last sentence in the first paragraph of Sec. 250-81 calls for the grounding electrode conductor to be "unspliced," Exception No. 1 to the basic rule says that the rule requiring "unspliced" grounding electrode conductor is relaxed to permit exothermic welded connections or listed irreversible pressure connectors to "extend" the grounding electrode conductor. This exception provides for the very common need for adding to the length of grounding electrode conductors because of unforeseen changes or accidental damage to such conductors.

The basic rule of Sec. 250-81 requires that all or any of the electrodes specified in **(a)**, **(b)**, **(c)**, and **(d)**, if they are available on the premises, must be bonded together to form a "grounding electrode system."

(a) If there is at least a 10-ft (3.05-m) length of underground metal water pipe, connection of a grounding electrode conductor must be made to the water pipe at a point less than 5 ft (1.52 m) from where the water piping enters the building.

(b) If, in addition, the building has a metal frame that is "effectively grounded," the frame must be bonded to the water pipe—or vice versa, because the rules do not spell out where actual connections are to be made for grounding electrode conductors and bonding jumpers. In the fine-print note, a definition is given for the phrase "effectively grounded" as applied to the metal frame of a building. To be "effectively grounded," the metal frame must be connected to earth by a low-impedance ground connection that will prevent "buildup of voltages" that could present a hazard to equipment or persons. Structural building steel that is bolted into a concrete footing or foundation in the earth would satisfy that definition.

(c) Then, if there is at least a total of 20 ft (6.1 m) of one or more ½-in. (12.7-mm)-diameter steel reinforcing bars or rods embedded in the concrete footing or foundation, a bonding connection must be made from one of the other electrodes to one of the rebars—and obviously that has to be done before concrete is poured for the footing or foundation.

When two or more grounding electrodes of the types described in Sec. 250-81 are to be combined into a "grounding electrode system," the size of the bonding jumper between pairs of electrodes must not be smaller than the size of grounding electrode conductor indicated in Table 250-94 for the particular size of the largest phase leg of the service feeder.

The "unspliced" grounding electrode conductor at the service may be connected to whichever one of the interbonded electrodes that provides the most convenient and effective point of connection.

The last sentence of this paragraph in Sec. 250-81 says that the grounding electrode conductor must be sized as if it were the grounding electrode conductor for whichever one of the interbonded electrodes that requires the largest grounding electrode conductor. If, for instance, a grounding electrode system consists of a metal underground water-pipe electrode supplemented by a driven ground rod, the grounding electrode conductor to the water pipe would have to be sized from Table 250-94; and on, say, a 2,000-A service, it would have to be a No. 3/0 copper or 250-kcmil aluminum, connected to the water-pipe electrode, which would require that size of grounding electrode conductor. A bonding jumper from the bonded grounding terminal or bus in the SE equipment to the driven ground rod would not have to be larger than a No. 6 copper or No. 4 aluminum grounding electrode conductor, just as it would be if the ground rod is used by itself as a grounding electrode. A bonding jumper between the water-pipe electrode and the ground rod would also have to be that size. There is negligible benefit in running larger than a No. 6 copper or No. 4 aluminum to a "made electrode," such as a ground rod, because the rod itself is the limiting resistance to earth.

In parts **(c)** and **(d)** of Sec. 250-81, the "20 feet (6.1-m) of bare copper conductor" referred to is going to be "available" at a building only if either 20-ft (6.1-m) arrangement has been specified by the electrical designer, because they are clearly and only grounding electrodes; whereas the other electrodes in parts

(a), **(b)**, and the rebars in **(c)** are specified by other than the electrical designer and may be "available."

If a building has all or some of the electrodes described, the above applications are mandatory. If it has none, then any one of the electrodes described in Sec. 250-83 may be used for service grounding.

In the last sentence of Sec. 250-81(a), an electrode (such as a driven ground rod) that supplements an underground water-pipe electrode may be "bonded" to any one of several points in the service arrangement. It may be "bonded" to (1) the grounding electrode conductor or (2) the grounded service conductor (grounded neutral), such as by connection to the neutral block or bus in the service panel or switchboard or in a CT cabinet, meter socket, or other enclosure on the supply side of the service disconnect or (3) grounded metal service raceway or (4) any grounded metal enclosure that is part of the service or (5) interior metal water piping at any convenient point. Part **(a)** calls for a bonding jumper around any water meter within a building and any place where piping on both sides of the meter is required to be grounded.

The second paragraph makes very clear that a ground rod or other "made" electrode that is used to supplement a water-pipe electrode does not require any larger than a No. 6 copper (or No. 4 aluminum) conductor for a bonding jumper that is the only connection from the ground rod to the grounding-electrode conductor, to the bonded neutral block or bus in the service equipment, to any grounded service enclosure or raceway, or to interior metal water piping.

The several requirements set by this section and the conditions established for application of the rules can be best understood by considering a step-by-step approach in making the necessary provisions for typical installations. To restate the above general description of Sec. 250-81:

First, take the case of a building fed by an underground water piping system with at least a 10-ft (3.05-m) length of *metal* water pipe buried in the earth ahead of the point at which the metal pipe enters the building (Fig. 250-80). Such buried pipe is a grounding electrode, and connection must be made to the underground pipe by a grounding electrode conductor sized from Table 250-94 and run from the grounding point in the service equipment. But now, a number of other factors must be accounted for, as follows:

1. EVEN THOUGH THE WATER PIPE IS A SUITABLE GROUNDING ELECTRODE, SEC. 250-81(a) REQUIRES THAT AT LEAST ONE MORE GROUNDING ELECTRODE MUST BE PROVIDED AND MUST BE BONDED TO THE WATER-PIPE ELECTRODE. A water pipe, by itself, is not an adequate grounding electrode and must be supplemented by at least one other electrode to provide a "grounding electrode system."
2. The additional electrode may be:
   - The metal frame of the building provided the frame is effectively grounded (embedded in earth and/or in buried concrete). Figure 250-80 shows an example of the metal frame electrode supplementing the water pipe.
   - OR, a concrete-encased electrode within and near the bottom of a concrete foundation or footing in direct contact with earth. The electrode

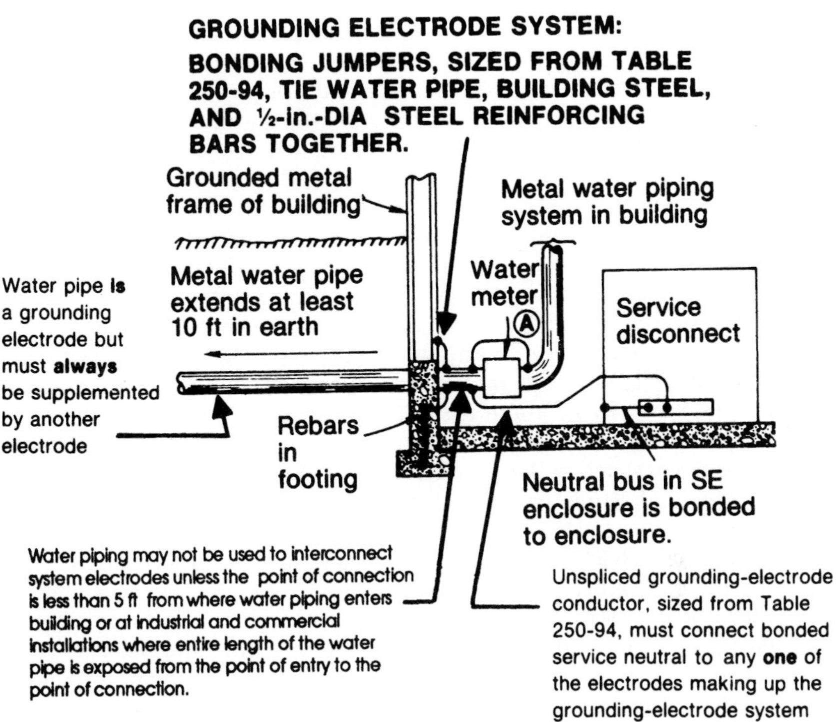
GROUNDING ELECTRODE SYSTEM:
BONDING JUMPERS, SIZED FROM TABLE 250-94, TIE WATER PIPE, BUILDING STEEL, AND ½-in.-DIA STEEL REINFORCING BARS TOGETHER.
Grounded metal frame of building
Metal water piping system in building
Water pipe is a grounding electrode but must always be supplemented by another electrode
Metal water pipe extends at least 10 ft in earth
Water meter
A
Service disconnect
Rebars in footing
Neutral bus in SE enclosure is bonded to enclosure.
Water piping may not be used to interconnect system electrodes unless the point of connection is less than 5 ft from where water piping enters building or at industrial and commercial installations where entire length of the water pipe is exposed from the point of entry to the point of connection.
Unspliced grounding-electrode conductor, sized from Table 250-94, must connect bonded service neutral to any one of the electrodes making up the grounding-electrode system
NOTE: At point "A", a bonding jumper must be used around the water meter and must be not smaller than the grounding-electrode conductor.

Fig. 250-80. Metal building frame and reinforcing bars must be used as an electrode if present. (Sec. 250-81.)

must consist of at least 20 ft (6.1 m) of one or more steel reinforcing bars or rods of not less than ½-in. (12.7-mm) diameter, or it must consist of at least 20 ft (6.1 m) of bare solid copper conductor not smaller than No. 4 AWG. If the building footing or foundation shown in Fig. 250-80 did contain such steel reinforcing, connection of a bonding conductor would have to be made to the steel and the conductor would have to be brought out for connection to the water pipe, the building steel, or the bonded service neutral.

- OR, a "ground ring encircling the building or structure," buried directly in the earth at least 2½-ft (762 mm) down. The ground ring must be "at least 20 ft (6.1 m)" of bare No. 2 or larger copper conductor. (In most cases, the conductor will have to be considerably longer than 20 ft (6.1 m) in order to "encircle" the building or structure.)
- OR, underground bare metal gas piping or other metal underground piping or tanks.

- OR, a buried 8-ft (2.44-m) ground rod or pipe or a plate electrode. An example of bonding of a supplemental grounding electrode is shown in Fig. 250-81. And an interesting point about the bonding of the ground rod to the neutral bus in the SE enclosure is the question of the size of the bonding jumper. Because the first paragraph of Sec. 250-81 requires that all the electrodes that make up a grounding electrode system be "bonded together" by a jumper sized from Table 250-94, and because the last sentence of Sec. 250-81(a) requires the supplemental electrode to be "bonded" to the service neutral or other specified point, it might appear that the conductor shown connecting the ground rod to the neutral bus in the sketch is a "bonding jumper" and must be the same size as the grounding electrode conductor (from Table 250-94). But, the second paragraph of Sec. 250-81(a) describes that conductor as a "bonding jumper," and it is not described as a "grounding electrode conductor." Therefore, a No. 6 copper or a No. 4 aluminum is the maximum size required for that connection—as stated in the second paragraph.

  In Sec. 250-81(c), the encased grounding electrode may be 20 ft (6.1 m) of bare No. 4 stranded copper conductor.

  Sections 250-81 and 250-83 list the acceptable electrodes and describe installation requirements.

3. As worded in Sec. 250-81, any of the four types of grounding electrodes mentioned there [**(a)**, **(b)**, **(c)**, and **(d)**] must be bonded together IF THEY ARE PRESENT. Note that the rule does not state that any of those electrodes must be provided. But if any or all of them are present, they must be bonded together to form a "grounding electrode system," sizing such bonding jumpers from Sec. 250-79(c). And where a water-pipe electrode, as described, is present, any one of the three electrodes in Sec. 250-81 may be used as the required "additional electrode."

In the case of a building fed by a nonmetallic underground piping system or one where there is *not* 10 ft (3.05 m) of metal pipe underground, the water piping system is *not* a grounding electrode—HOWEVER, THE INTERIOR METAL WATER PIPING SYSTEM AND ANY UNGROUNDED EXPOSED INTERIOR STRUCTURAL STEEL MEMBERS MUST BE BONDED TO THE SERVICE GROUNDING, as described above under Sec. 250-80(a). But when the water pipe is not a grounding electrode, another type of electrode must be provided to accomplish the service grounding. Any one of the other three electrodes of Sec. 250-81 [**(b)**, **(c)**, or **(d)**] may be used as the required grounding electrode. For instance, if the metal frame of the building is effectively grounded, a grounding electrode conductor, sized from Table 250-94 and run from the bonded service neutral or ground terminal to the building frame, may satisfy the **Code**, as shown in Fig. 250-82. Where none of the four electrodes described in Sec. 250-81 is present, one of the electrodes from Sec. 250-83 MUST BE USED, such as a ground rod as shown in Fig. 250-83. Note that any type of electrode other than a water-pipe electrode MAY BE USED BY ITSELF AS THE SOLE ELECTRODE.

As shown in Fig. 250-83, the literal wording of Sec. 250-94, Exception No. 1, would permit use of a No. 6 copper or No. 4 aluminum grounding electrode

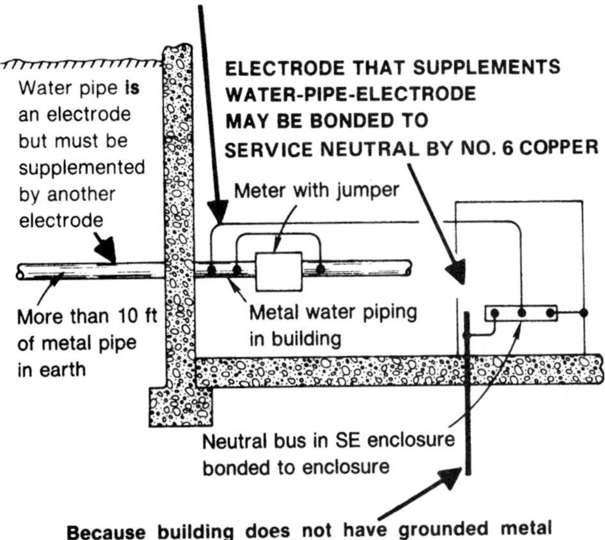
Grounding electrode conductor sized from Table 250-94 must connect bonded service neutral to clamp on water-pipe electrode, on either side of water meter, with same size bonding jumper around water meter.
ELECTRODE THAT SUPPLEMENTS WATER-PIPE-ELECTRODE MAY BE BONDED TO SERVICE NEUTRAL BY NO. 6 COPPER
Water pipe is an electrode but must be supplemented by another electrode
Meter with jumper
More than 10 ft of metal pipe in earth
Metal water piping in building
Neutral bus in SE enclosure bonded to enclosure
Because building does not have grounded metal frame or either of the electrodes described in Sections 250-81(c) and (d), the next-to-last sentence of Section 250-81(a) requires that one of the electrodes of Section 250-83 be used to supplement the water-pipe electrode—such as a driven ground rod.

Fig. 250-81. Supplementing water-pipe electrode in building without metal frame. (Sec. 250-81.)

conductor. It is consistent with long-time Code practice, based on tests, that the ground rod shown in Fig. 250-81 be connected to the neutral terminal or to the water pipe by a conductor that is not required to be larger than No. 6 copper or No. 4 aluminum.

The four illustrations shown for this section are only typical examples of the many specific ways in which the new rules on grounding electrodes may be applied.

A very important sentence of Sec. 250-81(a) says that "continuity of the grounding path or the bonding connection to interior piping shall not rely on water meters." The intent of that rule is that a bonding jumper always MUST BE USED around a water meter. This has been added because of the chance of loss of grounding if the water meter is removed or replaced with a nonmetallic water meter. The bonding jumper around a meter must be sized in accordance with Table 250-94. Although the Code rule does not specify that the bonding jumper around a meter be sized from that table, the reference to "bonding

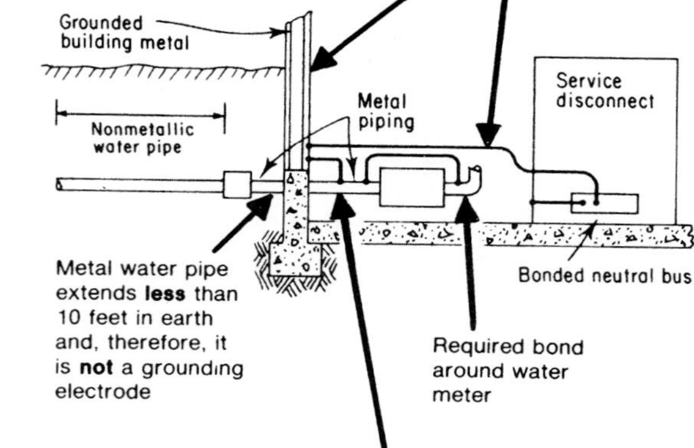
Section 250-81(b) recognizes the grounded metal building frame as a suitable grounding electrode by itself, connected to bonded service neutral by conductor sized from Table 250-94.
Grounded building metal
Nonmetallic water pipe
Metal piping
Service disconnect
Bonded neutral bus
Metal water pipe extends **less** than 10 feet in earth and, therefore, it is **not** a grounding electrode
Required bond around water meter
Section 250-80(a) requires interior metal water piping to be bonded to service grounding, such as by bonding to the grounding electrode (the building frame in this case), with jumper sized from Table 250-94.
NOTE: Rebars in foundation could also serve as the only electrode if building did not have metal frame.

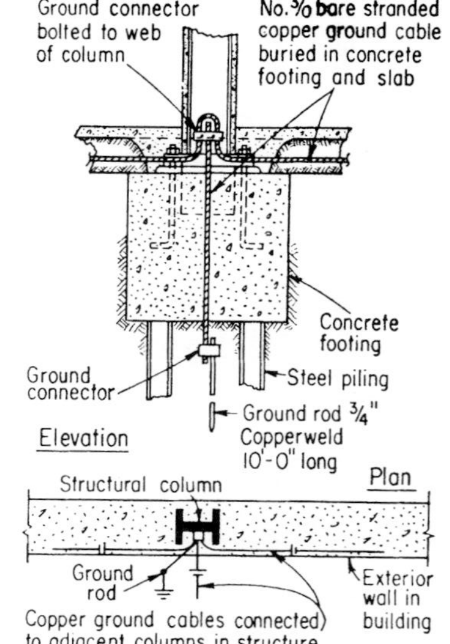
Ground connector bolted to web of column
No. 3/0 bare stranded copper ground cable buried in concrete footing and slab
Concrete footing
Steel piling
Ground connector
Ground rod 3/4" Copperweld 10'-0" long
Elevation
Plan
Structural column
Ground rod
Exterior wall in building
Copper ground cables connected to adjacent columns in structure
ONE METHOD of grounding building structural members to ground cable system.

Fig. 250-82. Building metal frame may be sole grounding electrode. (Sec. 250-81.)

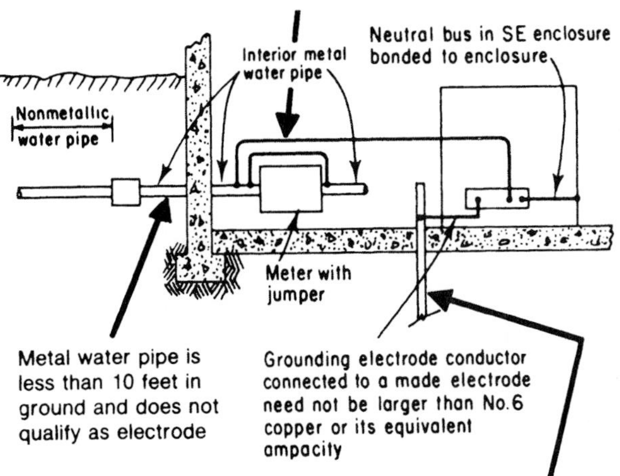
Section 250-80(a) requires interior metal water piping to be bonded to service grounding, such as by bonding to grounded neutral bus, using bonding jumper sized from Table 250-94.
Interior metal water pipe
Neutral bus in SE enclosure bonded to enclosure
Nonmetallic water pipe
Meter with jumper
Metal water pipe is less than 10 feet in ground and does not qualify as electrode
Grounding electrode conductor connected to a made electrode need not be larger than No. 6 copper or its equivalent ampacity
Because building does not have a grounded metal frame or either of the electrodes described in Section 250-81(c) and (d), Section 250-83 requires that at least one of the electrodes described in that Section be used—such as a ground rod.

Fig. 250-83. Ground rod may be only electrode in building without any of the electrodes in Sec. 250-81. (Sec. 250-81.)

jumper . . . in accordance with Sec. 250-94," as stated in the first paragraph of Sec. 250-81, would logically apply to the water-meter bond.

The concrete-encased electrode that is described in part **(c)** of Sec. 250-81, known as the "Ufer system," has particular merit in new construction where the bare copper conductor or steel reinforcing bar or rod can be readily installed in a foundation or footing form before concrete is poured. Installations of this type using a bare copper conductor have been installed as far back as 1940, and tests have proved this system to be highly effective.

The intent of "bottom of a concrete foundation" is to completely encase the electrode within the concrete, in the footing near the bottom. The footing shall be in direct contact with the earth, which means that dry gravel or polyethylene sheets between the footing and the earth are not permitted (Fig. 250-84).

It may be advisable to provide additional corrosion protection in the form of plastic tubing or sheath at the point where the grounding electrode leaves the concrete foundation.

For concrete-encased steel reinforcing bar or rod systems used as a grounding electrode in underground footings or foundations, welded-type connections (metal-fusing methods) may be used for connections that are encased in concrete. Compression or other type mechanical connectors may also be used.

250-83. Made and Other Electrodes. This section covers grounding electrodes that may be used if none of the electrodes of Sec. 250-81 is available. Or one of

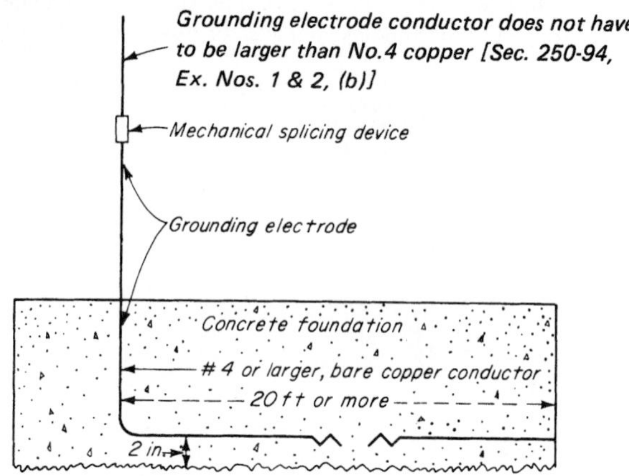
BARE COPPER CONDUCTOR... OR
Grounding electrode conductor does not have to be larger than No.4 copper [Sec. 250-94, Ex. Nos. 1 & 2, (b)]
Mechanical splicing device
Grounding electrode
Concrete foundation
4 or larger, bare copper conductor
20 ft or more
2 in.

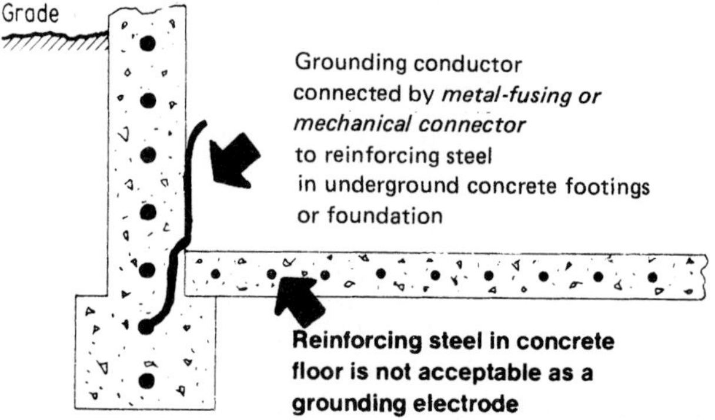
... 1/2 IN. DIAMETER REBARS OR RODS
Grade
Grounding conductor connected by metal-fusing or mechanical connector to reinforcing steel in underground concrete footings or foundation
Reinforcing steel in concrete floor is not acceptable as a grounding electrode

Fig. 250-84. The "Ufer" grounding electrode is concrete-encased, and the grounding electrode conductor does not have to be larger than No. 4 copper in either case. (Sec. 250-81.)

the electrodes of Sec. 250-83 may be used as the "additional electrode" required by Sec. 250-81(a) to supplement a water-pipe electrode.

Part **(a)** warns that a metal underground gas piping system must *never* be used as a grounding electrode. A metal underground gas piping system has been flatly disallowed as an acceptable grounding electrode because gas utility companies reject such practice and such use is in conflict with other industry standards.

As a general rule, if a water piping system or other approved electrode is not available, a driven rod or pipe is used as the electrode (Fig. 250-85). A rod or pipe driven into the ground does not always provide as low a ground resistance as is desirable, particularly where the soil becomes very dry. In some cases where several buildings are supplied, grounding at each building reduces the

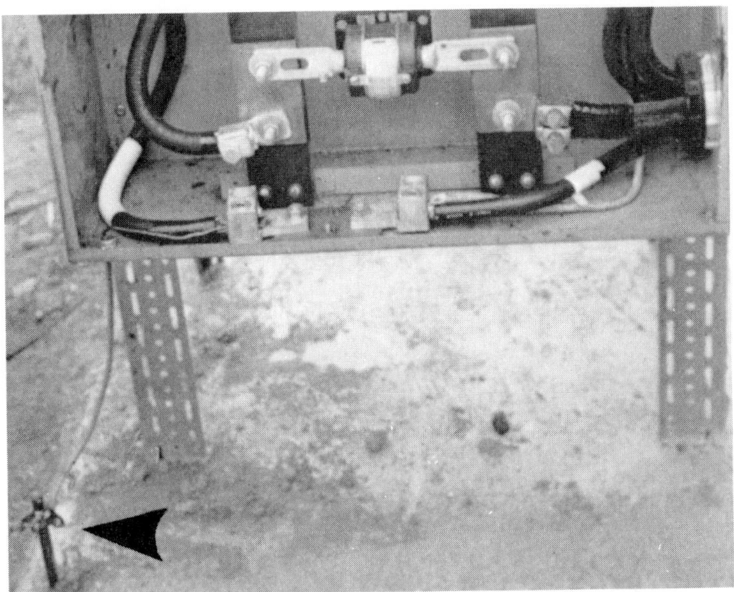

Fig. 250-85. A driven ground rod must have at least 8 ft (2.44 m) of its length buried in the ground, and if the end of the rod is aboveground (arrow), both the rod and its grounding-electrode-conductor attachment must be protected against physical damage.

ground resistance. (See Sec. 250-24.) Part **(e)** of Sec. 250-83 prohibits use of an aluminum grounding electrode.

Where it is necessary to bury more than one pipe or rod in order to lower the resistance to ground, they should be placed at least 6 ft (1.83 m) apart. If they were placed closer together, there would be little improvement.

Where two driven or buried electrodes are used for grounding two different systems that should be kept entirely separate from one another, such as a grounding electrode of a wiring system for light and power and a grounding electrode for a lightning rod, care must be taken to guard against the conditions of low resistance between the two electrodes and high resistance from each electrode to ground. If two driven rods or pipes are located 6 ft (1.83 m) apart, the resistance between the two is sufficiently high and cannot be greatly increased by increasing the spacing. The rule of this section requires at least 6 ft (1.83 m) of spacing between electrodes serving different systems.

The basic rule of Sec. 250-83(c)(3) calls for a ground rod to be driven straight down into the earth, with at least 8 ft (2.44 m) of its length in the ground (in contact with soil). If rock bottom is hit before the rod is 8 ft (2.44 m) into the earth, it is permissible to drive it into the ground at an angle—not over 45° from the vertical—to have at least 8 ft (2.44 m) of its length in the ground. However, if rock bottom is so shallow that it is not possible to get 8 ft (2.44 m) of the rod in the earth at a 45° angle, then it is necessary to dig a 2½-ft (762-mm)-deep trench and lay the rod horizontally in the trench. Figure 250-86 shows these rules.

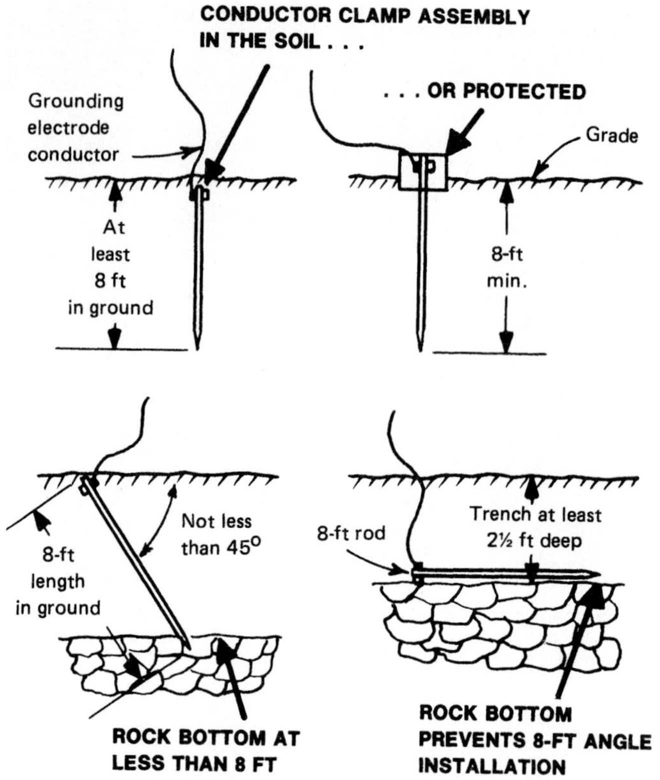

Fig. 250-86. In all cases, a ground rod must have at least 8 ft (2.44 m) of its length in contact with the soil.

A second requirement calls for the upper end of the rod to be flush with or below ground level—*unless* the aboveground end and the conductor clamp are protected either by locating it in a place where damage is unlikely or by using some kind of metal, wood, or plastic box or enclosure over the end (Sec. 250-117).

This two-part rule was added to the Code because it had become common practice to use an 8-ft (2.44-m) ground rod driven, say, 6½ ft (1.98 m) into the ground with the grounding electrode conductor clamped to the top of the rod and run over to the building. Not only is the connection subject to damage or disconnection by lawnmowers or vehicles, but also the length of unprotected, unsupported conductor from the rod to the building is a tripping hazard. The rule says—bury everything or protect it!

Of course, the buried conductor-clamp assembly that is flush with or below grade must be resistant to rusting or corrosion that might affect its integrity, as required by Sec. 250-115.

In addition to ground rods, plate electrodes are another form of "made" (manufactured) electrodes. Such electrodes are listed in the UL *Electrical Con-*

struction Materials Directory under the heading "Grounding and Bonding Equipment"—which also covers bonding devices, ground clamps, grounding and bonding bushings, ground rods, armored grounding wire, protector grounding wire, grounding wedges, ground clips for securing the ground wire to an outlet box, water-meter shunts, and similar equipment. Only listed devices are acceptable for use. And listed equipment is suitable only for use with copper, unless it is marked "AL" and "CU."

250-84. Resistance of Made Electrodes. This section on the resistance to earth of made electrodes clarifies Code intent and eliminates a cause of frequent controversy. The rule says that if a single made electrode (rod, pipe, or plate) shows a resistance to ground of over 25 ohms, *one* additional made electrode must be used in parallel, but there is then no need to make any measurement or add more electrodes or be further concerned about the resistance to ground. In previous Code editions, wording of this rule implied that additional electrodes had to be used in parallel with the first one until a resistance of 25 ohms or less was obtained. Now, as soon as the second electrode is added, it does not matter what the resistance to ground reads, and there is no need for more electrodes (Fig. 250-87).

The last sentence of Sec. 250-84 requires at least a 6-ft (1.83-m) spacing between any pair of made electrodes (ground rods, pipes, and/or plates), where more than one ground rod, pipe, or plate is connected to a single grounding electrode conductor, in any case where the resistance of a single grounding electrode is over 25 ohms to ground. And a note points out that even greater

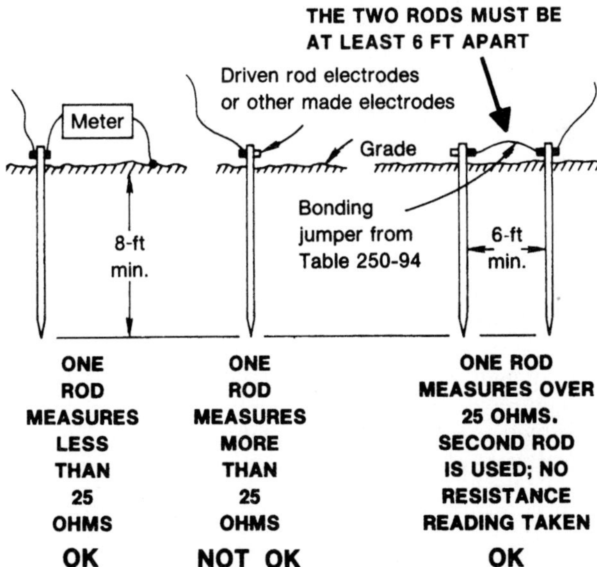

Fig. 250-87. Earth resistance of ground rod must be considered. (Sec. 250-84.)

spacing is better for rods longer than 8 ft (2.44 m). Separation of rods reduces the combined resistance to ground.

Insofar as made electrodes are concerned, there is a wide variation of resistance to be expected, and the present requirements of the National Electrical Code concerning the use of such electrodes do not provide for a system that is in any way comparable to that which can be expected where a good underground metallic piping can be utilized.

It is recognized that some types of soil may create a high rate of corrosion and will result in a need for periodic replacement of grounding electrodes. It should also be noted that the intimate contact of two dissimilar metals, such as iron and copper, when subjected to wet conditions can result in electrolytic corrosion.

Under abnormal conditions, when a cross occurs between a high-tension conductor and one of the conductors of the low-tension secondaries, the electrode may be called upon to conduct a heavy current into the earth. The voltage drop in the ground connection, including the conductor leading to the electrode and the earth immediately around the electrode, will be equal to the current multiplied by the resistance. This results in a difference of potential between the grounded conductor of the wiring system and the ground. It is therefore important that the resistance be as low as practicable.

Where made electrodes are used for grounding interior wiring systems, resistance tests should be conducted on a sufficient number of electrodes to determine the conditions prevailing in each locality. The tests should be repeated several times a year to determine whether the conditions have changed because of corrosion of the electrodes or drying out of the soil.

Figure 250-88 shows a ground tester being used for measuring the ground resistance of a driven electrode. Two auxiliary rod or pipe electrodes are driven to a depth of 1 or 2 ft (305 to 610 mm), the distances A and B in the figure being 50 ft (15.24 m) or more. Connections are made as shown between the tester and the electrodes; then the crank is turned to generate the necessary current, and the pointer on the instrument indicates the resistance to earth of the electrode being tested. In place of the two driven electrodes, a water piping system, if available, may be used as the reference ground, in which case terminals P and C are to be connected to the water pipe.

But, as noted above, where two made electrodes are used, it is not necessary to take a resistance reading, which is required in the case of fulfilling the requirement of 25 ohms to ground for one made electrode.

250-86. Use of Lightning Rods. This rule requires an individual "grounding electrode system" for grounding of the grounded circuit conductor (e.g., the neutral) and the equipment enclosures of electrical systems, and it prohibits use of the lightning ground electrode system for grounding the electrical system. Although the rule does *not* generally prohibit or require bonding between different grounding electrode systems (such as for lightning and for electric systems), it does note that such bonding is sometimes required by rules in the sections listed. And the note calls attention to the advantage of such bonding. There have been cases where fires and shocks have been caused by a potential difference between separate ground electrodes and the neutral of AC electrical circuits.

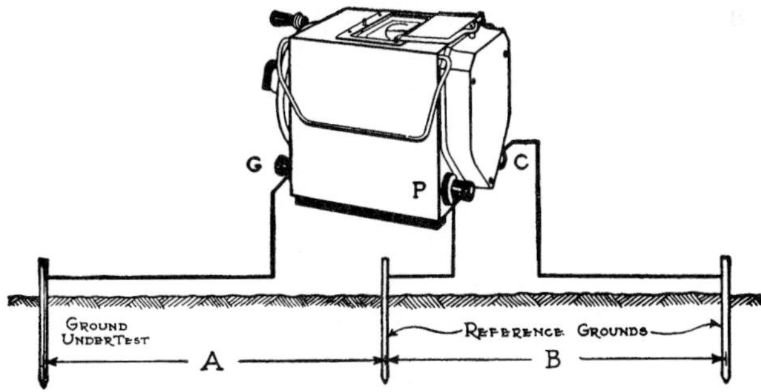

Fig. 250-88. Ground-resistance testing must be done with the proper instrument and in strict accordance with the manufacturer's instructions. (Sec. 250-84.)

250-91. Material (Grounding Conductors). Figure 250-89 shows the typical use of copper, aluminum, or copper-clad aluminum conductor to connect the bonded neutral and equipment ground terminal of service equipment to each of the one or more grounding electrodes used at a service. Controversy has been common on the permitted color of an insulated (or covered) grounding electrode conductor. Section 200-7 generally prohibits use of white or natural gray color for any conductor other than a "grounded conductor"—such as the grounded neutral or phase leg, as described in the definition of "grounded conductor." Green color is reserved for equipment grounding conductors, although there is no **Code** rule clearly prohibiting a green grounding electrode conductor. Refer to Sec. 310-12(b).

Exception No. 2 of Sec. 250-91(a) covers the installation of the grounding electrode conductor for a service layout consisting of two to six service disconnects in separate enclosures with a separate set of service-entrance conductors run down to each disconnect. Such an arrangement is covered in Sec. 230-40,

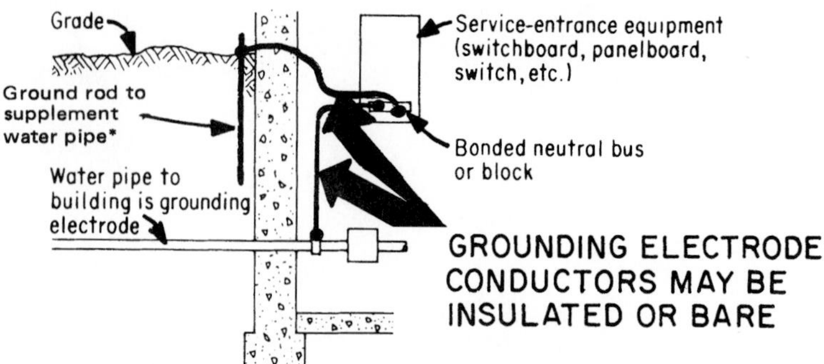

Fig. 250-89. An insulated grounding electrode conductor may be any color other than white, natural gray, or green. (Sec. 250-91.)

Exception No. 2. Previous Code editions required (and it still satisfies the Code to use) either a separate grounding electrode conductor run from each enclosure to the grounding electrode or a single, unspliced conductor looped from enclosure to enclosure. And a single grounding electrode conductor used to ground all the service disconnects has to be without splice [the last line in Sec. 250-91(a)]—run from one enclosure to the other and then to the water pipe or other grounding electrode. The specific rule of Exception No. 2 says that when a service disconnect consists of two to six switches or CBs in separate enclosures, a tap from each enclosure may be made to a single, grounding electrode conductor that *must* be without splice or joint. For such an arrangement (shown at the bottom left in Fig. 250-90), the last sentence of Exception No. 2 clarifies that there must be no break in the single, common grounding electrode conductor used to connect the taps to ground two or more service disconnects in individual enclosures.

The wording of Exception No. 2 is clear on the sizing of the main, common grounding electrode conductor. Wording of the exception requires the main grounding electrode conductor to be sized from Table 250-94, which has a footnote that covers multiple sets of SE conductors. The main grounding electrode conductor must be sized for the sum of the cross-section areas of the total number of conductors connected to one hot leg of the service drop. All considerations of function, adequacy, and safety would be satisfied by running the single, common, unspliced grounding electrode conductor from the disconnect for the largest of the several sets of service-entrance conductors directly to the grounding electrode. That single, common conductor could be sized directly from Table 250-94, based on the size of SE conductors as determined from the note under Table 250-94. Each other service-entrance disconnect would then have a tap conductor that taps to the unspliced common grounding electrode conductor.

The wording of the rule makes clear that the size of the grounding electrode tap to each separate enclosure may be determined from Table 250-94 on the basis of the largest service hot leg serving each enclosure, as shown in Fig. 250-

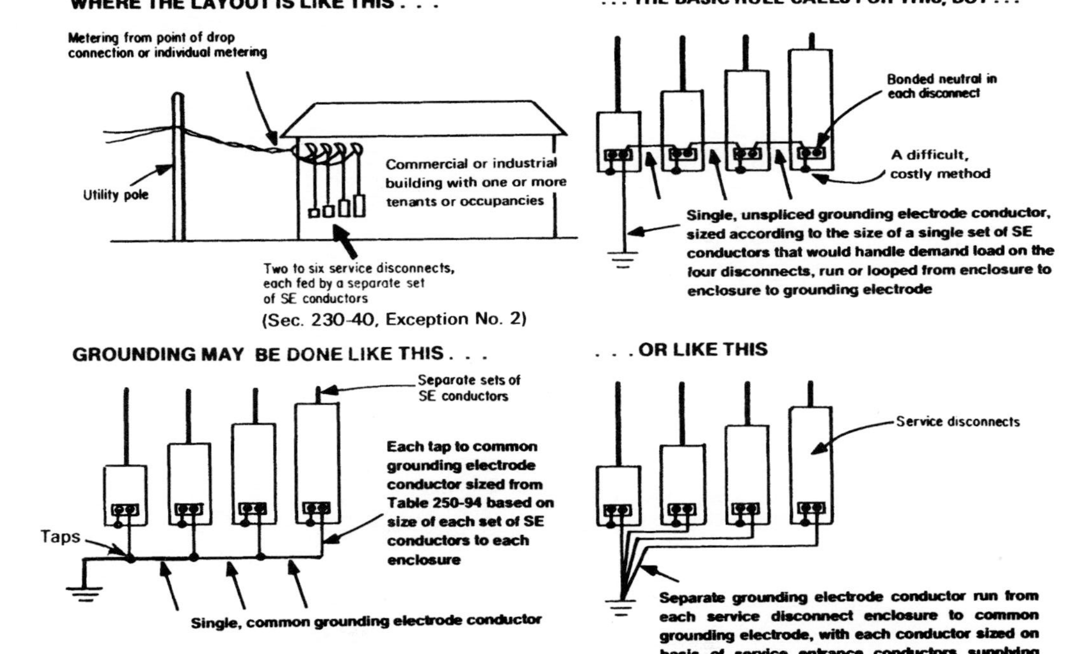

WHERE THE LAYOUT IS LIKE THIS . . .
Metering from point of drop connection or individual metering
Utility pole
Commercial or industrial building with one or more tenants or occupancies
Two to six service disconnects, each fed by a separate set of SE conductors
(Sec. 230-40, Exception No. 2)
. . . THE BASIC RULE CALLS FOR THIS, BUT . . .
Bonded neutral in each disconnect
A difficult, costly method
Single, unspliced grounding electrode conductor, sized according to the size of a single set of SE conductors that would handle demand load on the four disconnects, run or looped from enclosure to enclosure to grounding electrode
GROUNDING MAY BE DONE LIKE THIS . . .
Separate sets of SE conductors
Each tap to common grounding electrode conductor sized from Table 250-94 based on size of each set of SE conductors to each enclosure
Taps
Single, common grounding electrode conductor
. . . OR LIKE THIS
Service disconnects
Separate grounding electrode conductor run from each service disconnect enclosure to common grounding electrode, with each conductor sized on basis of service entrance conductors supplying each enclosure

Fig. 250-90. Grounding electrode conductor may be tapped for multiple service disconnects. (Sec. 250-91.)

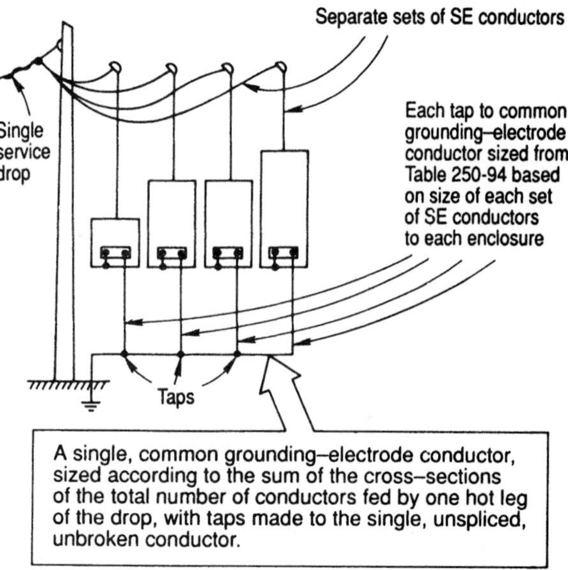
Separate sets of SE conductors
Single service drop
Each tap to common grounding–electrode conductor sized from Table 250-94 based on size of each set of SE conductors to each enclosure
Taps
A single, common grounding–electrode conductor, sized according to the sum of the cross–sections of the total number of conductors fed by one hot leg of the drop, with taps made to the single, unspliced, unbroken conductor.

Fig. 250-91. Rule covers sizing main and taps of grounding electrode conductor at multiple-disconnect services. A single common grounding electrode conductor must be "without splice or joint," with taps made to the grounding electrode conductor. (Sec. 250-91.)

91. Although that illustration shows an overhead service to the layout, the two to six service disconnects could be fed by individual sets of underground conductors, making up a "single" service lateral as permitted by Exception No. 7 to Sec. 230-2.

Part **(b)** describes the various types of conductors and metallic cables or raceways that are considered suitable for use as equipment grounding conductors. And the Code recognizes cable tray as an equipment grounding conductor as permitted by Art. 318.

Exception No. 1 recognizes flexible metal conduit, flexible metallic tubing, (see Art. 349) or liquidtight flexible metal conduit with termination fittings UL-listed for use as a grounding means (without a separate equipment grounding wire) if the total length of the flex is not over 6 ft (1.83 m) and the contained circuit conductors are protected by overcurrent devices rated at 20 A or less.

Standard flexible metal conduit (also known as "Greenfield") must be listed by UL although not *all* are listed as suitable for grounding in themselves. However, Sec. 350-14 of the NE Code as well as Exception No. 1 in part **(b)** permits flex to be used without any supplemental grounding conductor when any length of flex in a ground return path is not over 6 ft (1.83 m) and the conductors contained in the flex are protected by overcurrent devices rated not over 20 A and the *fittings* are listed as suitable for grounding (Fig. 250-92). Use of standard flex with the permission given in Sec. 250-79(e) for either internal or external bonding must be as follows:

1. When conductors within a length of flex up to 6 ft (1.83 m) are protected at more than 20 A, equipment grounding may not be provided by the flex, but a separate conductor must be used for grounding. If a length of flex is short enough to permit a bonding jumper not over 6 ft (1.83 m) long to be

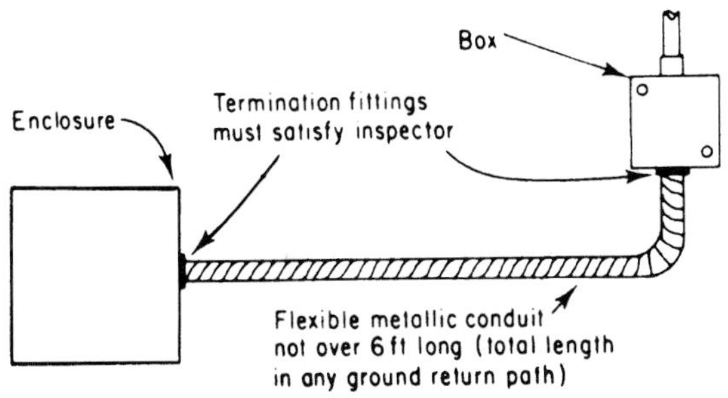

Flex not over 6 ft long is suitable as a grounding means (without a separate ground wire) if the conductors in it are protected by OC devices rated not more than 20 amps.

Fig. 250-92. Standard flex is limited in use without an equipment ground wire. (Sec. 250-91.)

run between external grounding-type connectors at the flex ends, while keeping the jumper *along* the flex, such an external jumper may be used where equipment grounding is required—as for a short length of flex with circuit conductors in it protected at more than 20 A. Of course, such short lengths of flex may also be "bonded" by a bonding jumper inside the flex, instead of external. Refer to Sec. 250-79(e).
2. Any length of standard flex that would require a bonding jumper longer than 6 ft (1.83 m) may not use an external jumper. In the **Code** sense, when the length of such a grounding conductor exceeds 6 ft (1.83 m), it is *not* a BONDING JUMPER BUT *IS* AN EQUIPMENT GROUNDING CONDUCTOR AND MUST BE RUN ONLY *INSIDE* THE FLEX, AS REQUIRED BY SEC. 250-57(b). Combining UL data with the rule of Sec. 250-79(e) and the Exception to Sec. 350-14 *every* length of flex that is over 6 ft (1.83 m) must contain an equipment grounding conductor run *only* inside the flex (Fig. 250-93).

In part *a* of Exception No. 1 it should be noted that exemption from the need for an equipment grounding conductor applies only to flex where there is not over 6 ft (1.83 m) of "total length in any ground return path." That means that from any branch-circuit load device—lighting fixture, motor, etc.—all the way back to the service ground, the total permitted length of flex without a ground wire is 6 ft (1.83 m). In the total circuit run from the service to any outlet, there could be one 6-ft (1.83-m) length of flex or two 3-ft (914-mm) lengths or three 2-ft (610-mm) lengths or a 4-ft (1.22-m) and a 2-ft (610-mm) length—where the

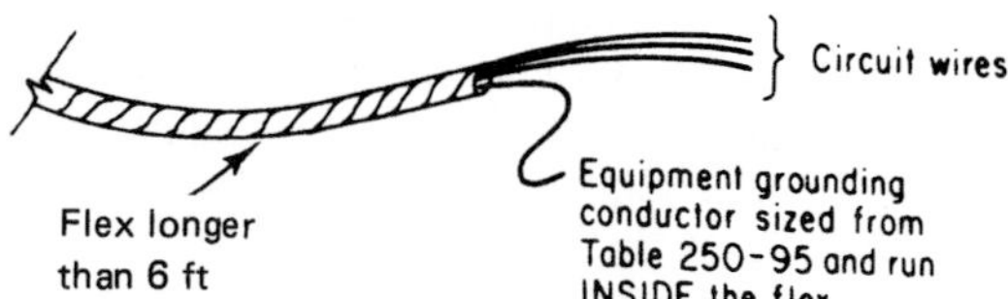

Fig. 250-93. Internal equipment grounding is required for any flex over 6 ft (1.83 m) long. (Sec. 250-91.)

flex lengths are in series as equipment ground return paths. In any circuit run—feeder to subfeeder to branch circuit—any length of flex that would make the total series length over 6 ft (1.83 m) would have to use an internal or external bonding jumper, regardless of any other factors.

In all cases, sizing of bonding jumpers for all flex applications is made according to Sec. 250-79(d), which requires the same minimum size for bonding jumpers as is required for equipment grounding conductors. In either case, the size of the conductor is selected from Table 250-95, based on the maximum rating of the overcurrent devices protecting the circuit conductors that are within the flex.

In part *b* of Exception No. 1, the term "listed for grounding," as applied to termination fittings, will require the authority having jurisdiction to verify the grounding capabilities of fittings by requiring only "listed" fittings to be used with these short conduit lengths. See also Secs. 350-14 and 351-9.

Exception No. 2 presents conditions under which *liquidtight* flexible metal conduit may be used without need for a separate equipment grounding conductor:

1. Both Exception No. 2 and the UL's *Electrical Construction Materials Directory* (the Green Book) note that any listed liquidtight flex in 1¼-in. (31.75-mm) and smaller trade size, in a length not over 6 ft (1.83 m), may be satisfactory as a grounding means through the metal core of the flex, without need of a bonding jumper (or equipment grounding conductor) either internal or external (Fig. 250-94). Liquidtight flex in 1¼-in. (31.75-mm) and smaller trade size may be used without a bonding jumper inside *or* outside provided that the "total length" of that flex "in any ground return path" is not over 6 ft (1.83 m). Thus, two or more separate 6-ft (1.83-m) lengths installed in a raceway run would not be acceptable with the bonding jumper omitted from all of them. In such cases, one 6-ft (1.83-m) length or more than one length that does not total over 6 ft (1.83 m) may be used with a bonding jumper, but any additional lengths 6 ft (1.83 m) or less in the same raceway run must have an internal or external bonding jumper sized from Table 250-95.

   The required conditions for use of liquidtight flex without need of a separate equipment bonding jumper (or equipment grounding conductor) are as follows:

   Where terminated in fittings investigated for grounding and where installed with not more than 6 ft (1.83 m) (total length) in any ground return path, liquidtight flexible metal conduit in the ⅜- and ½-in. trade

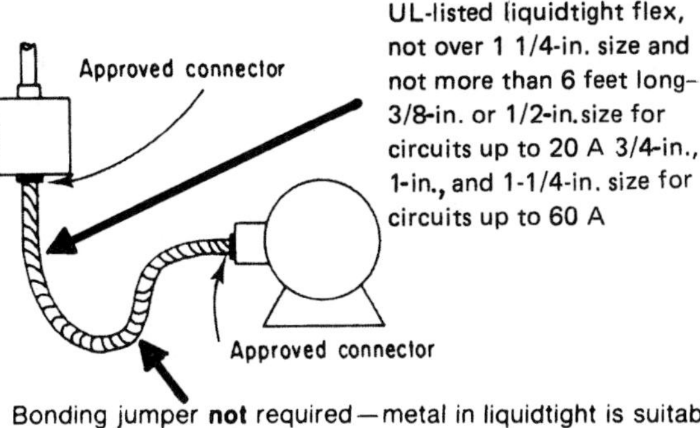

Fig. 250-94. Liquidtight flex may be used with a separate ground wire. (Sec. 250-91.)

sizes is suitable for grounding where used on circuits rated 20 A or less, and the ¾-, 1-, and 1¼-in. trade sizes are suitable for grounding where used on circuits rated 60 A or less. See the category "Conduit Fittings" (DWTT) with respect to fittings suitable as a grounding means.

The following are not considered to be suitable as a grounding means:

a. The 1½-in. and larger trade sizes
b. The ⅜- and ½-in. trade sizes where used on circuits rated higher than 20 A, or where the total length in the ground return path is greater than 6 ft (1.83 m)
c. The ¾-, 1-, and 1¼-in. trade sizes where used on circuits rated higher than 60 A, or where the total length in the ground return path is greater than 6 ft (1.83 m)

Although UL gives the same grounding recognition to their "listed" liquidtight flex, this Code rule covers liquidtight that the UL does not list, such as high-temperature type.

2. For liquidtight flex over 1¼ in., UL does not list any as suitable for equipment grounding, thereby requiring use of a separate equipment grounding conductor installed in *any* length of the flex, as required by Code. If a length of liquidtight flex larger than 1¼ in. is short enough to permit an external bonding jumper not more than 6 ft (1.83 m) long between external grounding-type connectors at the ends of the flex, an external bonding jumper may be used. BUT WATCH OUT! The rule says the *jumper,* not the flex, must not exceed 6 ft (1.83 m) in length AND the jumper "shall be routed with the raceway"—that is, run along the flex surface and not separated from the flex.
3. If any length of flex is *over* 6 ft (1.83 m), then the flex is not a suitable grounding conductor, regardless of the trade size of the flex, whether it is larger or smaller than 1¼ in. In such cases, an *equipment grounding conductor* (not a "bonding jumper"—the phrase reserved for short lengths) must be used to provide grounding continuity and IT MUST BE RUN

INSIDE THE FLEX, NOT EXTERNAL TO IT, IN ACCORDANCE WITH SEC. 250-57(b).

Part **(c)** of Sec. 250-91 is an extremely important rule that has particular impact on the use of electrical equipment outdoors. The first part of the rule accepts the use of "supplementary grounding electrodes"—such as a ground rod—to "augment" the equipment grounding conductor; *BUT* an equipment grounding conductor must always be used where needed and the connection of outdoor metal electrical enclosures to a ground rod is never a satisfactory alternative to the use of an equipment grounding conductor because use of just ground-rod grounding would have the earth as "the sole equipment grounding conductor" and that is expressly prohibited by the last clause of part **(c)**.

This whole matter of earth ground usually comes up as follows:

question: When direct-burial or nonmetallic-conduit circuits are run underground to supply lighting fixtures or other equipment mounted on metal standards or poles or fed by metal conduit run up a pole or building wall, is it necessary to run an equipment grounding conductor to ground the metal standard or pole or conduit if a ground rod has been driven for the same purpose?

answer: Yes. An equipment grounding conductor is necessary to provide low impedance for ground-fault current return to assure fast, effective operation of the circuit protective device when the circuit is derived from a grounded electrical system (such as 240/120 V, single-phase or 208Y/120 or 480Y/277 V, 3-phase). Low impedance of a grounding path "to facilitate the operation of the circuit protective devices in the circuit" is clearly and specifically required by **NE Code** Sec. 250-51. When a ground rod is used to ground an outdoor metal standard or pole or outdoor metal conduit and no other grounding connection is used, ground-fault current must attempt to return to the grounded system neutral by flowing through the earth. Such an earth return path has impedance that is too high, limiting the current to such a low value that the circuit protective device does not operate. In that case, a conductor that has faulted (made conductive contact) to a metal standard, pole, or conduit will put a dangerous voltage on the metal—exposing persons to shock or electrocution hazard as long as the fault exists. The basic concept of this problem—and **Code** violation—is revealed in Fig. 250-95.

The same undesirable condition would exist where a direct-burial or nonmetallic-conduit circuit feeds up through a metal conduit outdoors. As shown in Fig. 250-96, fault current would have to return through the high impedance of the earth.

The hazard of arrangements using only a ground rod as shown arises from the chance that a person might make contact with the energized metal standard or conduit and have a high enough voltage across the person to produce a dangerous current flow through the person's body. The actual current flow through the body will depend upon the contact resistances and the body resistance in conjunction with the voltage gradient (potential difference) imposed across the body. As shown in Fig. 250-97, a person contacting an energized metal pole can complete a circuit to earth or to some other pipe or metal that is grounded back to the system neutral.

It is important to note that part **(c)** of Sec. 250-91 applies to these situations where outside metal standards or conduits are grounded by means of a ground rod at the standard or conduit. The installations shown in Figs. 250-95, 250-96, and 250-97 are in violation of the last clause of Sec. 250-91(c) as well as Sec. 250-51(c). Figure 250-98 is another example of a violation of the rule.

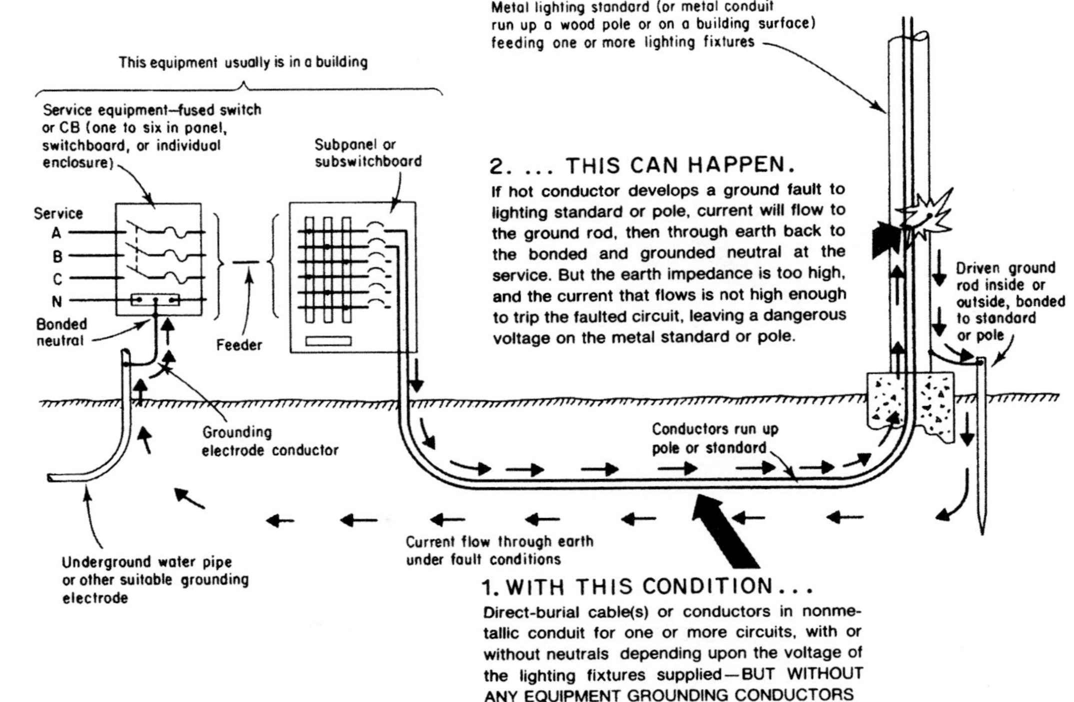

This equipment usually is in a building
Service equipment—fused switch or CB (one to six in panel, switchboard, or individual enclosure)
Service
A
B
C
N
Bonded neutral
Feeder
Subpanel or subswitchboard
Metal lighting standard (or metal conduit run up a wood pole or on a building surface) feeding one or more lighting fixtures
2. ... THIS CAN HAPPEN.
If hot conductor develops a ground fault to lighting standard or pole, current will flow to the ground rod, then through earth back to the bonded and grounded neutral at the service. But the earth impedance is too high, and the current that flows is not high enough to trip the faulted circuit, leaving a dangerous voltage on the metal standard or pole.
Driven ground rod inside or outside, bonded to standard or pole
Grounding electrode conductor
Conductors run up pole or standard
Underground water pipe or other suitable grounding electrode
Current flow through earth under fault conditions
1. WITH THIS CONDITION...
Direct-burial cable(s) or conductors in nonmetallic conduit for one or more circuits, with or without neutrals depending upon the voltage of the lighting fixtures supplied—BUT WITHOUT ANY EQUIPMENT GROUNDING CONDUCTORS

Fig. 250-95. This Code violation produces a dangerous condition at a metal standard. (Sec. 250-91.)

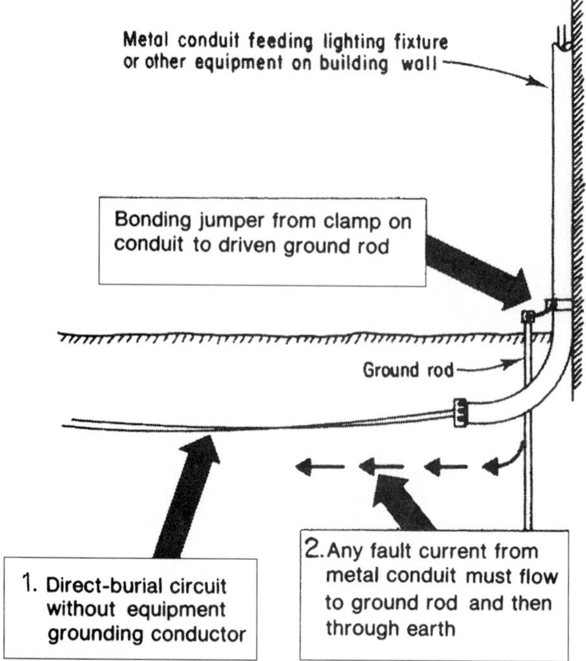

Fig. 250-96. This is also a dangerous condition for conduit up a pole or building. (Sec. 250-91.)

Compliance with the letter and spirit of those Code sections—and other sections on equipment grounding—will result from use of an equipment grounding conductor run with the circuit conductors—either closely placed in the same trench with direct-burial conductors (Type UF or Type USE) or pulled into nonmetallic conduit with the circuit conductors. Such arrangement is also dictated by Sec. 250-57(b), which requires that an equipment grounding conductor *must* be within the same raceway, cable, or cord or otherwise run with the circuit conductors.

Effective grounding methods that are in accordance with NE Code rules are shown in Figs. 250-99 and 250-100.

[Although NE Code Art. 338 on "Service-Entrance Cable" does not say that Type USE cable may be used as a feeder or branch circuit (on the load side of service equipment), the UL listing on Type USE cable says it "is suitable for all of the underground uses for which Type UF cable is permitted by the National Electrical Code."]

If an underground circuit to a metal standard or pole is run in metal conduit, a separate equipment grounding conductor is not needed in the conduit if the conduit end within the standard is bonded to the standard by a bonding jumper (Fig. 250-101). Section 250-91(b) recognizes metal conduits as suitable equipment grounding conductors in themselves.

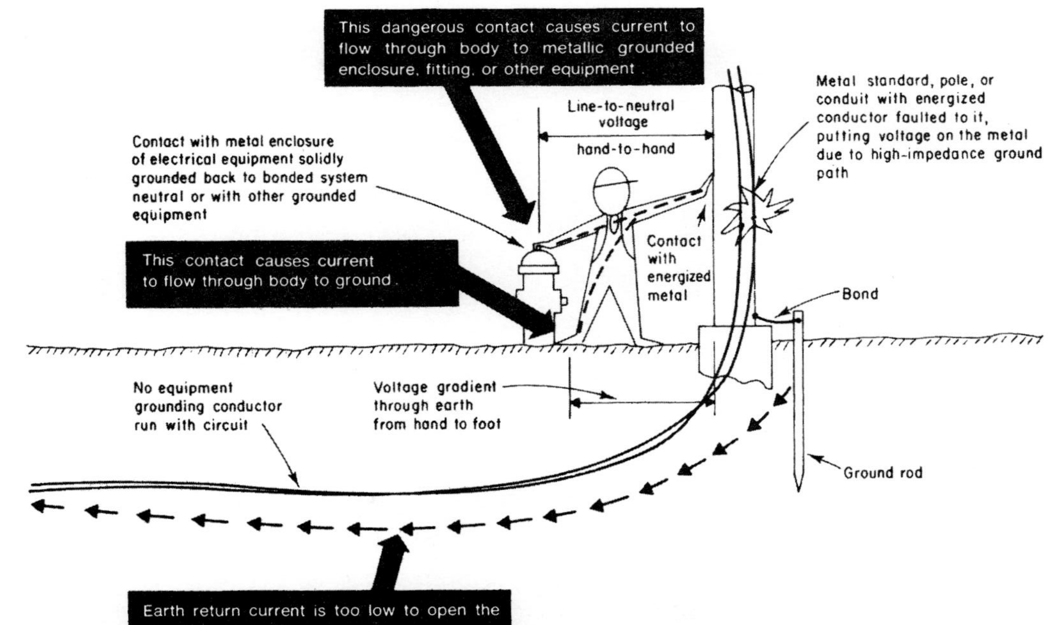
This dangerous contact causes current to flow through body to metallic grounded enclosure, fitting, or other equipment.
Line-to-neutral voltage
hand-to-hand
Metal standard, pole, or conduit with energized conductor faulted to it, putting voltage on the metal due to high-impedance ground path
Contact with metal enclosure of electrical equipment solidly grounded back to bonded system neutral or with other grounded equipment
This contact causes current to flow through body to ground.
Contact with energized metal
Bond
No equipment grounding conductor run with circuit
Voltage gradient through earth from hand to foot
Ground rod
Earth return current is too low to open the circuit fuses or CB.

Fig. 250-97. Ineffective grounding creates shock hazards. (Sec. 250-91.)

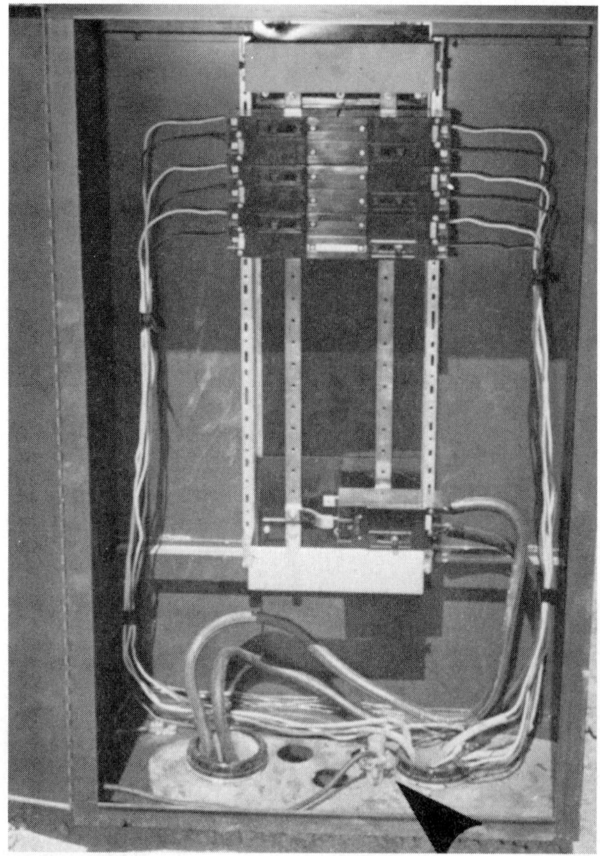

Fig. 250-98. Driven ground rod (arrow) has conductor run to it from a large lug at the left rear of the enclosure. All of the equipment grounding conductors from UF 480-V circuits to pole lights are connected to that lug. But the ground rod and earth path are the sole return path for fault currents. The two larger conductors make up a 480-V underground USE circuit, without the neutral or an equipment grounding conductor brought to the panel. [Sec. 250-91(c).]

250-92. Installation. Part **(a)**, in its first paragraph, covers physical protection for a grounding electrode conductor, as shown in Fig. 250-102.

A common technique for protecting bare or insulated grounding conductors (one which grounds the wiring system and equipment cases) makes use of a metal conduit sleeve, run open or installed in concrete. In all such cases, part **(b)** of this section covers details on the use of "metal enclosures" for grounding electrode conductors. The grounding conductor must be connected to its protective conduit at both ends so that any current which might flow over the conductor will also have the conduit as a parallel path. The regulation presented is actually a performance description of the rule of Sec. 250-71(a)(3) which specifically and simply requires that any conduit or armor enclosing a grounding

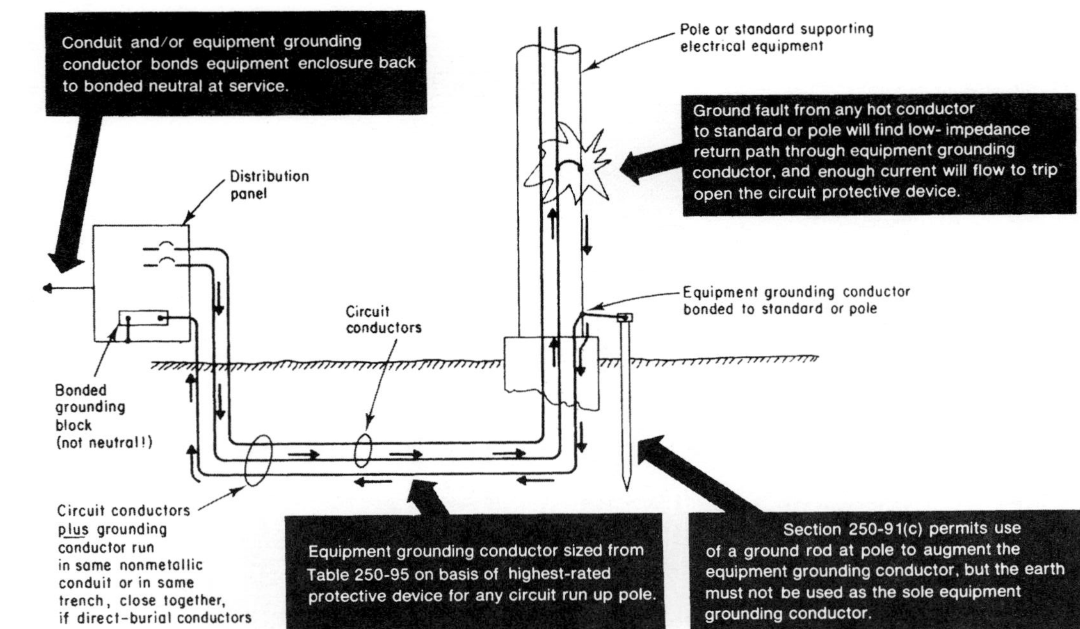
Conduit and/or equipment grounding conductor bonds equipment enclosure back to bonded neutral at service.
Distribution panel
Bonded grounding block (not neutral!)
Circuit conductors
Circuit conductors plus grounding conductor run in same nonmetallic conduit or in same trench, close together, if direct-burial conductors
Pole or standard supporting electrical equipment
Ground fault from any hot conductor to standard or pole will find low- impedance return path through equipment grounding conductor, and enough current will flow to trip open the circuit protective device.
Equipment grounding conductor bonded to standard or pole
Equipment grounding conductor sized from Table 250-95 on basis of highest-rated protective device for any circuit run up pole.
Section 250-91(c) permits use of a ground rod at pole to augment the equipment grounding conductor, but the earth must not be used as the sole equipment grounding conductor.

Fig. 250-99. Equipment grounding conductor assures effective fault-clearing. (Sec. 250-91.)

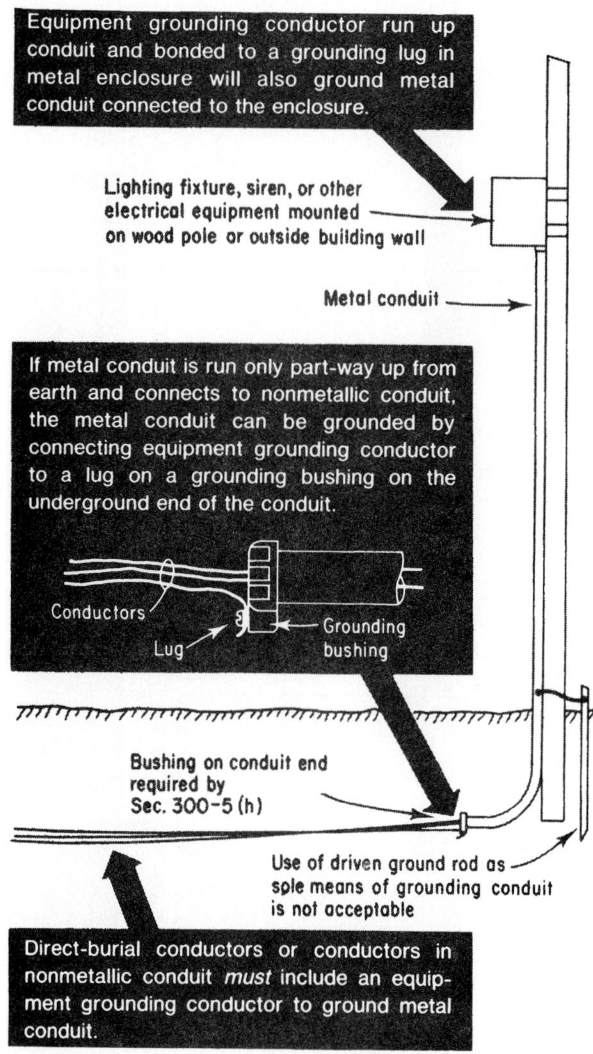
Equipment grounding conductor run up conduit and bonded to a grounding lug in metal enclosure will also ground metal conduit connected to the enclosure.
Lighting fixture, siren, or other electrical equipment mounted on wood pole or outside building wall
Metal conduit
If metal conduit is run only part-way up from earth and connects to nonmetallic conduit, the metal conduit can be grounded by connecting equipment grounding conductor to a lug on a grounding bushing on the underground end of the conduit.
Conductors
Lug
Grounding bushing
Bushing on conduit end required by Sec. 300-5 (h)
Use of driven ground rod as sole means of grounding conduit is not acceptable
Direct-burial conductors or conductors in nonmetallic conduit *must* include an equipment grounding conductor to ground metal conduit.

Fig. 250-100. Watch out for grounding details like these! (Sec. 250-91.)

electrode conductor be bonded to a service enclosure at one end and to the grounding electrode at the other end. The concept involved is to assure that any metal raceway (or other enclosure) containing a grounding electrode conductor is electrically in parallel with the conductor, as shown in Fig. 250-103. Bonded connection at both ends of an enclosing raceway must be used for any grounding electrode arrangement at a service and for grounding of a separately derived system, such as a generator or transformer secondary, as shown in Fig. 250-104. And in that illustration, where the transformer secondary phase leg is larger

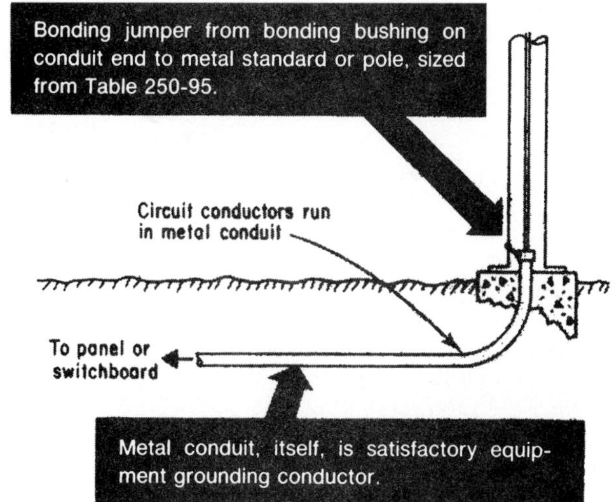

Fig. 250-101. Underground metal conduit to metal standard provides Code-acceptable ground-fault return path. (Sec. 250-91.)

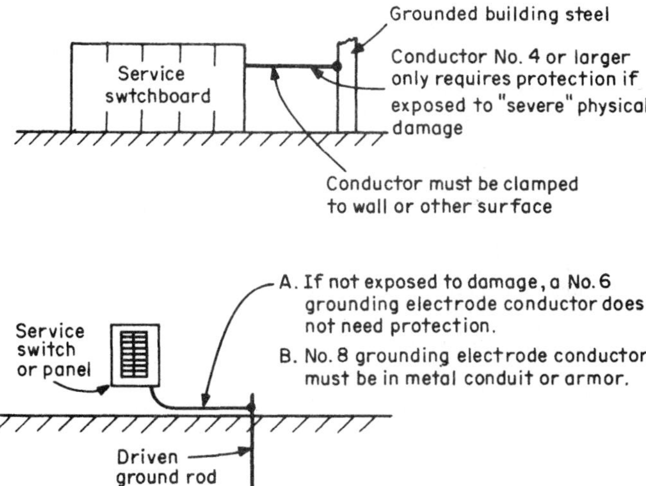

Fig. 250-102. Protection for grounding electrode conductor. (Sec. 250-92.)

than 1,100 kcmil copper or 1,750 kcmil aluminum, the bonding jumper from the conduit bushing to the case will, invariably, have to be larger than the grounding electrode conductor. Table 250-94 shows that a grounding electrode conductor never has to be larger than No. 3/0 copper or 250 kcmil whereas Sec. 250-79(c) requires a bonding jumper to have a cross-section area of at least 12½

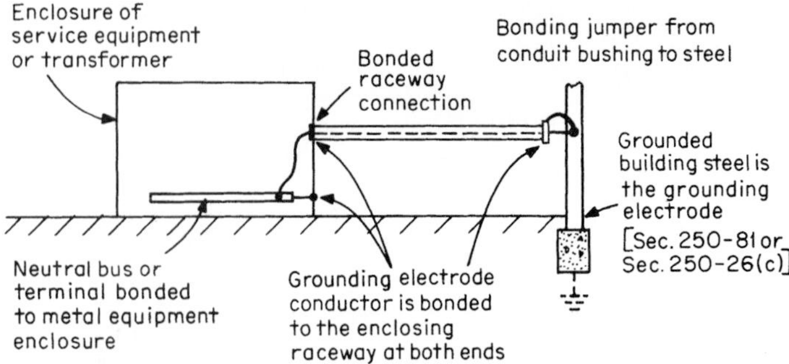

Fig. 250-103. Grounding electrode conductor must be electrically in parallel with enclosing raceway and other enclosures. (Sec. 250-92.)

percent of the cross-section area of the largest one of the transformer secondary phase legs (or service-entrance phase legs). (Usually, all the phase legs are the same size, so the rule can be read as "12½ percent of *one* of the phase legs.") Refer to Sec. 250-79(c).

The necessity for making a grounding electrode conductor electrically in parallel with its protective conduit applies to all applications of such conductors. If the protective conduit in any such case was arranged so that the conductor and conduit were not acting as parallel conductors—such as the conduit would be in Fig. 250-104 if there were no bonding jumper from the conduit bushing to the conductor lug—the presence of magnetic metal conduit (steel) would serve to greatly increase the inductive reactance of the grounding conductor to limit any flow of current to ground. The steel conduit would act as the core of a "choke" to restrict current flow. Figure 250-105 shows the "skin effect" of current flow over a conduit in parallel with a conductor run through it, resulting in a condition which makes it critically necessary to keep the conduit connected electrically in parallel with an enclosed grounding electrode conductor. The condition is as follows.

The presence of the steel conduit acts as an iron core to greatly increase the inductive reactance of the conductor. This choke action raises the impedance of the conductor to such a level that only 3 A flows through the conductor and the balance of 97 A flows through the conduit. This division of current between the conduit and the conductor points up the importance of assuring tight couplings and connectors throughout every conduit system and for every metal raceway system and metal cable jacketing. In particular, this stresses the need for bonding both ends of any raceway used to protect a grounding conductor run to a water pipe or other grounding electrode. Such conduit protection must be securely connected to the ground electrode and to the equipment enclosure in which the grounding conductor originates. If such conduit is left open, lightning and other electric discharges to earth through the grounding conductor will find a high-impedance path. The importance of high conductivity in the conduit system is also important for effective equipment grounding, even when a specific equipment grounding conductor is used in the conduit.

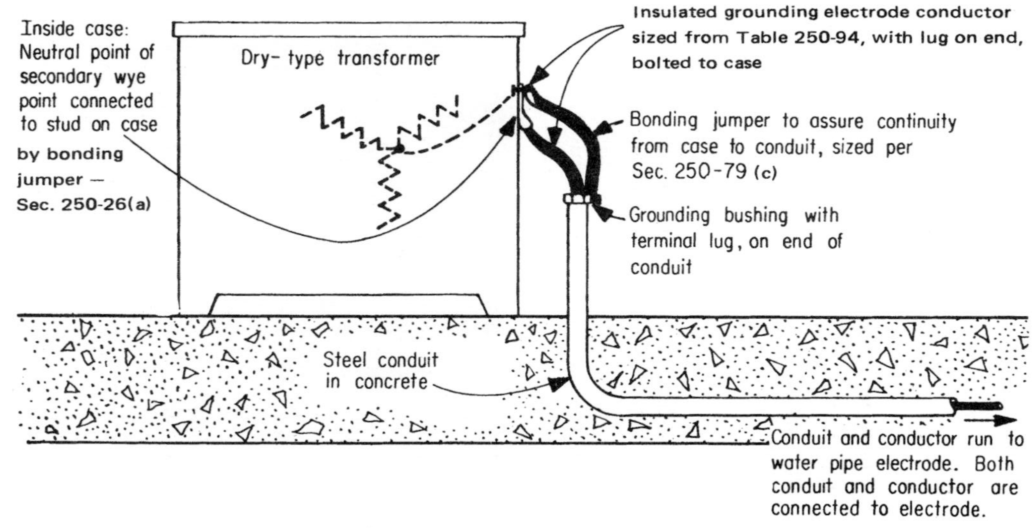

Fig. 250-104. Protective metal conduit on grounding conductor must always be electrically in parallel with conductor. (Sec. 250-92.)

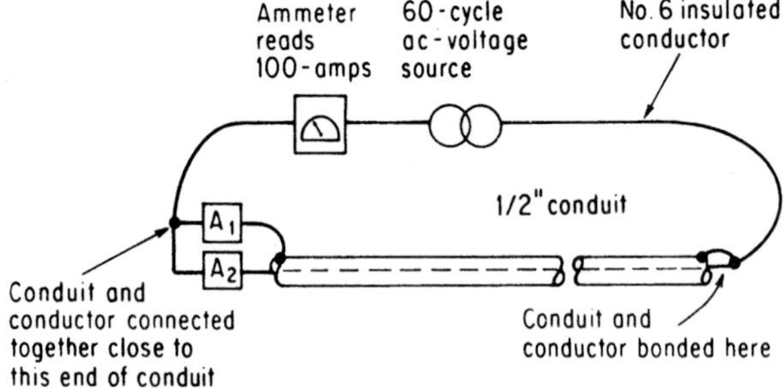

Ammeter A_1 (indicates amount of current in conduit) = 97 amps

Ammeter A_2 (indicates amount of current in conductor) = 3 amps

Fig. 250-105. Enclosing conduit is more important than the enclosed grounding electrode conductor. (Sec. 250-92.)

Figure 250-105 clearly shows the conduit itself to be a more important conductor than the actual conductor. Figure 250-106 is a clear example of a violation of this rule. Figure 250-107 shows two other examples of violations. Of course, nonmagnetic conduit—such as aluminum—would have a different effect, but it should also be in parallel because it would not be a low-reluctance

Fig. 250-106. Grounding electrode conductors are run in conduit from their connections to an equipment grounding bus in an electrical room to the point where they connect to the grounding electrodes. Without a bonding jumper from each conduit to the ground bus, this is a clear VIOLATION of the second paragraph rule of Sec. 250-92(b). (Sec. 250-92.)

core for the magnetic field around the enclosed conductor. The inductive reactance of the conductor would not be elevated and the magnetic choking would be minimized. However, the **Code** rules on bonding [Sec. 250-71] and parallel connection of conduit containing a grounding electrode conductor apply to *all metal* raceways. Aluminum conduit must, therefore, be connected the same as steel conduit or EMT.

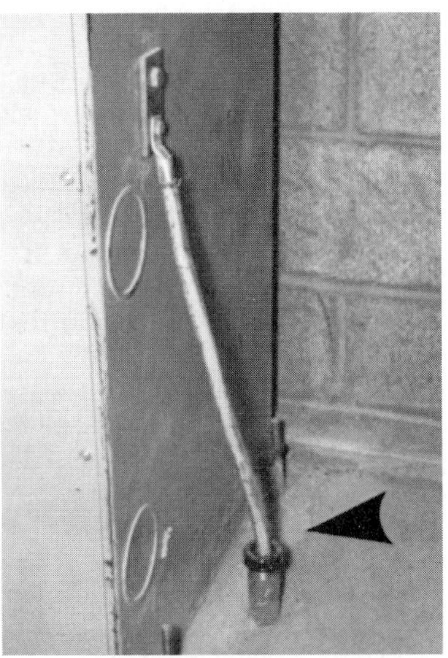

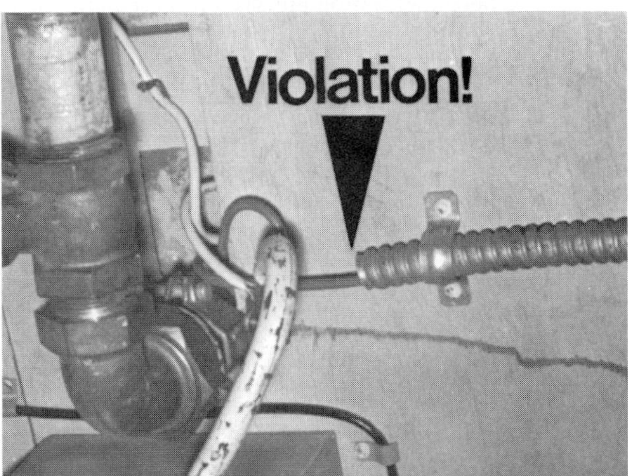

Fig. 250-107. Two examples of very clear violations of the rule that requires enclosing metal raceways (rigid metal conduit at top and flex at right) to be bonded at both ends to a grounding electrode conductor within the raceway. [Secs. 250-71 and 250-92(b).]

PVC conduit may be used to protect No. 4 or smaller grounding electrode conductors used in accordance with this section. Use of nonmetallic raceways for enclosing grounding electrode conductors will reduce the impedance below that of the same conductor in a steel raceway. The grounding electrode conductor will perform its function whether enclosed or not, the principal function of the enclosure being to protect the conductor from physical damage. Rigid nonmetallic conduit will satisfy this function.

The second paragraph of Sec. 250-92(a) requires special care for aluminum grounding electrode conductors. If such a conductor is to be run, say, along a concrete wall, it would have to be clamped to a wooden "running board" that is nailed to the wall—to keep it from "direct contact with masonry." And outdoors, such a conductor must be kept at least 18 in. (457 mm) above the ground. But, there has been controversy about these rules. Although the wording of this paragraph literally requires those installation methods for *all* aluminum and copper-clad aluminum grounding electrode conductors—whether bare or insulated, installed within or without raceway—some inspectors have taken the rules as applying only to *bare* conductors (Fig. 250-108). Because Sec. 230-30 and the NEC, in general, recognize insulated aluminum conductors in conduit underground and directly buried in soil, it is argued that *insulated* aluminum conductor may be run in direct contact with masonry and within 18 in. (457 mm) of earth—even directly buried in the soil if it is Type UF or USE, or within conduit or other raceway underground (Fig. 250-109). Of course, all such application must not have exposed (bare) aluminum, even short lengths, such as at terminations, that would violate the above rules. That means that terminations must be made at least 18 in. (457 mm) aboveground.

Part **(c)** of Sec. 250-92 covers installation of "equipment grounding conductors" as distinguished from "grounding electrode conductors." (Refer to "Definitions," Art. 100.) Where the equipment grounding conductor is a bare or insulated conductor (a cable) that is run by itself, instead of a metal raceway or

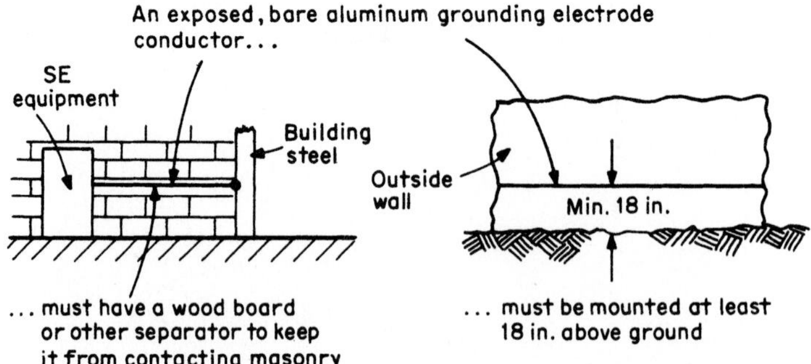
An exposed, bare aluminum grounding electrode conductor...
SE equipment
Building steel
Outside wall
Min. 18 in.
... must have a wood board or other separator to keep it from contacting masonry
... must be mounted at least 18 in. above ground
NOTE: An insulated aluminum conductor or a bare conductor installed in conduit may not be subject to these limitations, depending upon the inspector's interpretation.

Fig. 250-108. Does this controversial rule apply only to *bare* aluminum conductors? (Sec. 250-92.)

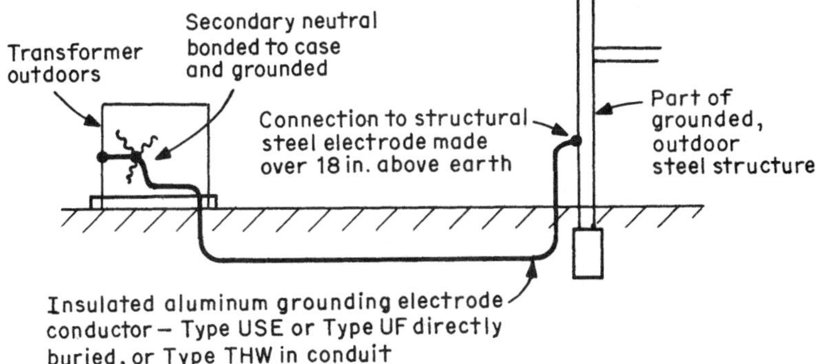

Fig. 250-109. This has been accepted but does violate literal Code wording. (Sec. 250-92.)

armor, all the rules of part **(a)**, as described above, must be satisfied. A separately run equipment ground wire is a basic exception to the normal Code practice of keeping the ground wire and circuit wires close together. When so run, it must follow the rules for a grounding electrode conductor, which is normally run by itself, either exposed or in raceway. If a steel conduit or tubing is used for mechanical protection of the grounding conductor, it needs to be bonded to the grounding conductor where it enters and where it leaves the protecting steel conduit in order to keep the impedance of the grounding circuit at an acceptable level, as described in part **(a)** above.

250-93. Size of Direct-Current Grounding Electrode Conductor. Figure 250-110 is a diagram of a balancer set used with a 2-wire 230-V generator to supply a 3-wire system as referred to in part **(a)** of this section.

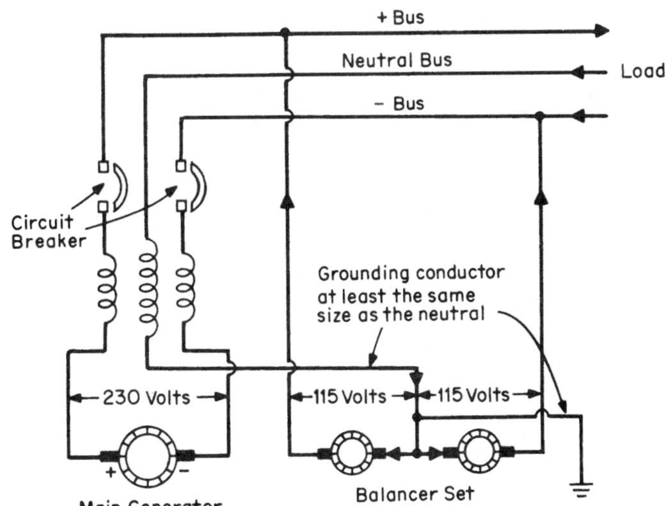

Fig. 250-110. Sizing a DC system grounding conductor. (Sec. 250-93.)

250-94. Size of Alternating-Current Grounding Electrode Conductor. For copper wire, a minimum size of No. 8 is specified in order to provide sufficient carrying capacity to ensure an effective ground and sufficient mechanical strength to be permanent. Where one of the service conductors is a grounded conductor, the same grounding electrode conductor is used for grounding both the system and the equipment. Where the service is from an ungrounded 3-phase power system, a grounding electrode conductor of the size given in Table 250-94 is required at the service.

If the sizes of service-entrance conductors for an AC system are known, the minimum acceptable size of grounding electrode conductor can be determined from NE Code Table 250-94. Where the service consists of only one conductor for each hot leg or phase, selection of the minimum permitted size of grounding electrode conductor is a relatively simple, straightforward task. If the largest phase leg is, say, a 500-kcmil copper THW, Table 250-94 shows No. 1/0 copper or No. 3/0 aluminum (reading across from "Over 350 kcmil thru 600 kcmil") as the minimum size of a grounding electrode conductor.

But, use of the table for services with multiple conductors per phase leg (e.g., four 500 kcmil for each of three phase legs of a service) is more involved.

The heading over the left-hand columns of this table is "Size of Largest Service-Entrance Conductor or Equivalent for Parallel Conductors." To make proper use of this table, the meaning of the word "equivalent" must be clearly understood. "Equivalent" means that parallel conductors per phase are to be converted to a single conductor per phase that has a cross-section area of its conductor material at least equal to the sum of the cross-section areas of the conductor materials of the two or more parallel conductors per phase. (The cross-section area of the insulation must be excluded.)

For instance, two parallel 500-kcmil copper RHH conductors in separate conduits would be equivalent to a single conductor with a cross-section area of 500 + 500, or 1,000 kcmil. From Table 250-94, the minimum size of grounding electrode conductor required is shown to be No. 2/0 copper or No. 4/0 aluminum—opposite the left column entry, "Over 600 kcmil thru 1,100 kcmil." Note that use of this table is based solely on the size of the conductor material itself, regardless of the type of insulation. No reference is made at all to the kind of insulation.

Figure 250-111 shows a typical case where a grounding electrode conductor must be sized for a multiple-conductor service. A 208/120-V, 3-phase, 4-wire service is made up of two sets of parallel copper conductors of the sizes shown in the sketch. The minimum size of grounding electrode conductor which may be used with these service-entrance conductors is determined by first adding together the physical size of the two No. 2/0 conductors which make up each phase leg of the service:

1. From NE Code Table 8 in Chap. 9 in the back of the Code book, which gives physical dimensions of the conductor material itself (excluding insulation cross-section area), each of the phase conductors has a cross-section area (csa) of 133,100 kcmil. Two such conductors per phase have a total csa of 266,200 kcmil.
2. The same table shows that the single conductor which has a csa at least equal to the total csa of the two conductors per phase is a 300 kcmil size of conductor. That conductor size is then located in the left-hand column

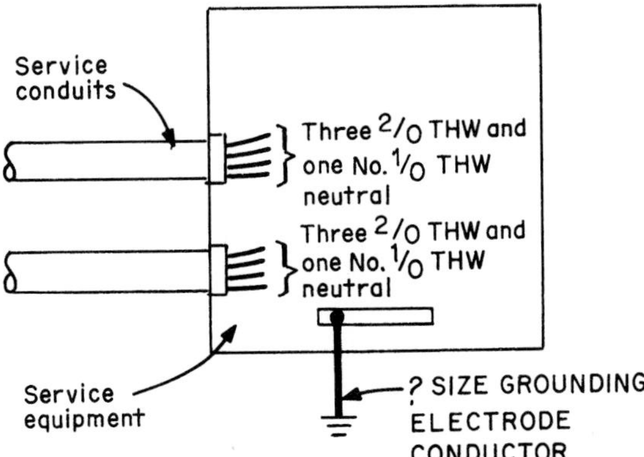

Fig. 250-111. Typical task of sizing the conductor to the grounding electrode. (Sec. 250-94.)

of Table 250-94 to determine the minimum size of grounding electrode conductor, which turns out to be No. 2 copper or No. 1/0 aluminum or copper-clad aluminum.

Figure 250-112 shows another example of conductor sizing, as follows:

1. The grounding electrode conductor A connects to the street side of the water meter of a metallic water supply to a building. The metallic pipe extends 30 ft (9.14 m) underground outside the building.
2. Because the underground metallic water piping is at least 10 ft (3.05 m) long, the underground piping system is a grounding electrode and must be used as such.
3. Based on the size of the service-entrance conductors (5 × 500 kcmil = 2,500 kcmil per phase leg), the minimum size of grounding electrode con-

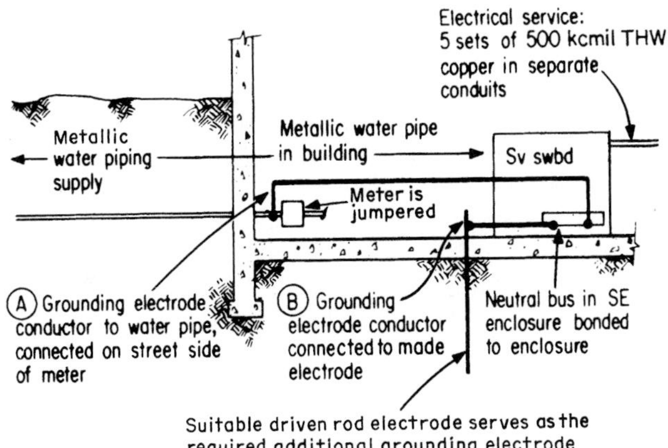

Fig. 250-112. Two different sizes of grounding electrode conductors are required for installations like this. (Sec. 250-94.)

ductor to the water pipe is No. 3/0 copper or 250-kcmil aluminum or copper-clad aluminum.

4. The connection to the ground rod at B satisfies the rule of Sec. 250-81(a) requiring a water-pipe electrode to be supplemented by another electrode.
5. But, the minimum size of grounding electrode conductor B required between the neutral bus and the made electrode is No. 6 copper or No. 4 aluminum, as covered by part **(a)** of the Exception in this section. Although the **Code** does not require the conductor to a made electrode to be larger than No. 6, regardless of the ampacity of the service phases, a larger size of conductor is commonly used for mechanical strength, to protect it against breaking or damage. As discussed under Sec. 250-81(a) and shown in Fig. 250-81, the conductor at B in Fig. 250-112 can be considered to be a bonding jumper, as covered by the last paragraph of Sec. 250-81(a), which also says that the conductor to the ground rod need not be larger than No. 6 copper or No. 4 aluminum.

Parts **(b)** and **(c)** of the Exception makes clear that a grounding electrode conductor does not have to be larger than a conductor-type electrode to which it connects. Section 250-81(c) recognizes a "concrete-encased" electrode—which must be at least 20 ft (6.1 m) of one or more ½-in. (12.7-mm)-diameter steel reinforcing bars or rods in the concrete or at least 20 ft (6.1 m) of bare No. 4 copper conductor (or a larger conductor), concrete-encased in the footing or foundation of a building or structure. Section 250-81(d) recognizes a "ground-ring" electrode made up of at least 20 ft (6.1 m) of No. 2 bare copper conductor (or larger), buried directly in the earth at a depth of at least 2½ ft (762 mm). Because each of those electrodes is described under Sec. 250-81, they are *not* "made electrodes," which are described under Sec. 250-83. As electrodes from Sec. 250-81, such electrodes would normally be subject to the basic rule of Sec. 250-94, which calls for connection to any such electrode by a grounding electrode conductor sized from Table 250-94—requiring up to No. 3/0 copper for use on high-capacity services. *But,* that is not required, as explained in these paragraphs of the two Exceptions.

Parts **(b)** and **(c)** recognize that there is no reason to use a grounding electrode conductor that is larger than a conductor-electrode to which it connects. The grounding electrode conductor need not be larger than No. 4 copper for a No. 4 concrete-encased electrode and need not be larger than No. 2 copper if it connects to a ground-ring electrode—as in parts **(c)** of the Exception. Where Table 250-94 would permit a grounding electrode conductor smaller than No. 4 or No. 2 (based on size of service conductors), the smaller conductor may be used—but the electrode itself must not be smaller than No. 4 or No. 2. See Fig. 250-84.

The first note under Table 250-94 correlates to Sec. 230-40, Exception No. 2, and Sec. 250-91(a), Exception No. 2, as follows:

When two to six service disconnects in separate enclosures are used at a service, with a separate set of SE conductors run to each disconnect, the size of a single common grounding electrode conductor must be based on the largest sum of the cross sections of the same phase leg of each of the several sets of SE conductors. When using multiple service disconnects in separate enclosures, with a set of SE conductors run to each from the drop or lateral (Sec. 230-40, Exception No. 2) and using a single, common grounding electrode conductor, either run continuous and unspliced from one disconnect to another and then to the grounding electrode, or with taps from each disconnect to a common grounding electrode conductor run to the electrode—as in Sec. 250-91(a), Exception No. 2, this note is used to determine the size of the grounding electrode conductor from Table 250-94. The "equivalent area" of the size of SE conductors is the largest sum of the cross-section areas of one ungrounded leg of each of the several sets of SE conductors.

250-95. Size of Equipment Grounding Conductors. When an individual equipment grounding conductor is used in a raceway—either in a nonmetallic raceway, as required by Sec. 347-4, or in a metal raceway where such a conductor is used for grounding reliability even though Sec. 250-91(b) accepts metal raceways as a suitable grounding conductor—the grounding conductor must have a minimum size as shown in Table 250-95. The minimum acceptable size of an equipment grounding conductor is based on the rating of the overcurrent device (fuse or CB) protecting the circuit, run in the same raceway, for which the equipment grounding conductor is intended to provide a path of ground-fault current return (Fig. 250-113). Each size of grounding conductor in the table is adequate to carry enough current to blow the fuse or trip the CB of the rating indicated beside it in the left-hand column. In Fig. 250-113, if the fuses are rated at 60 A, Table 250-95 shows that the grounding electrode conductor used with that circuit must be at least a No. 10 copper or a No. 8 aluminum or copper-clad aluminum.

Whenever an equipment grounding conductor is used for a circuit that consists of only one conductor for each hot leg (or phase leg), the grounding conductor is sized simply and directly from Table 250-95, as described. When a circuit is made up of parallel conductors per phase, say an 800-A circuit with two conductors per phase, an equipment grounding conductor is also sized in the same way and would, in that case, have to be at least a No. 1/0 copper or No. 3/0 aluminum. *But,* if such a circuit is made up using two conduits—that is,

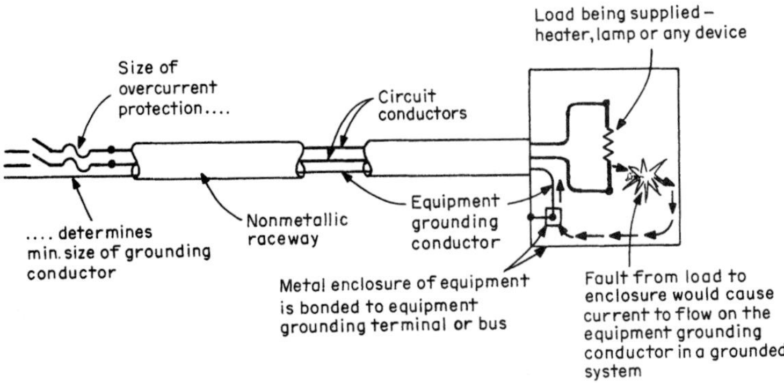
Load being supplied – heater, lamp or any device
Size of overcurrent protection....
Circuit conductors
.... determines min. size of grounding conductor
Nonmetallic raceway
Equipment grounding conductor
Metal enclosure of equipment is bonded to equipment grounding terminal or bus
Fault from load to enclosure would cause current to flow on the equipment grounding conductor in a grounded system

Fig. 250-113. Size of grounding conductor must carry enough current to operate circuit overcurrent device. (Sec. 250-95.)

three phase legs and a neutral in each conduit—Sec. 250-95 requires that an individual grounding conductor be run in each of the conduits *and* each of the two grounding conductors must be at least No. 1/0 copper or No. 3/0 aluminum (Fig. 250-114). Another example is shown in Fig. 250-115, where a 1,200-A protective device on a parallel circuit calls for No. 3/0 copper or 250-kcmil aluminum grounding conductor.

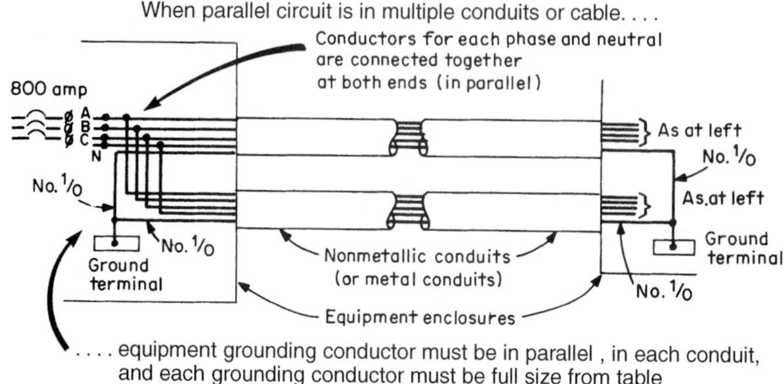
When parallel circuit is in multiple conduits or cable. . . .
Conductors for each phase and neutral are connected together at both ends (in parallel)
800 amp
A
B
C
N
No. 1/0
No. 1/0
Ground terminal
Nonmetallic conduits (or metal conduits)
Equipment enclosures
As at left
No. 1/0
As at left
Ground terminal
No. 1/0
. . . . equipment grounding conductor must be in parallel , in each conduit, and each grounding conductor must be full size from table

Fig. 250-114. Grounding conductor must be used in each conduit for parallel conductor circuits. (Sec. 250-95.)

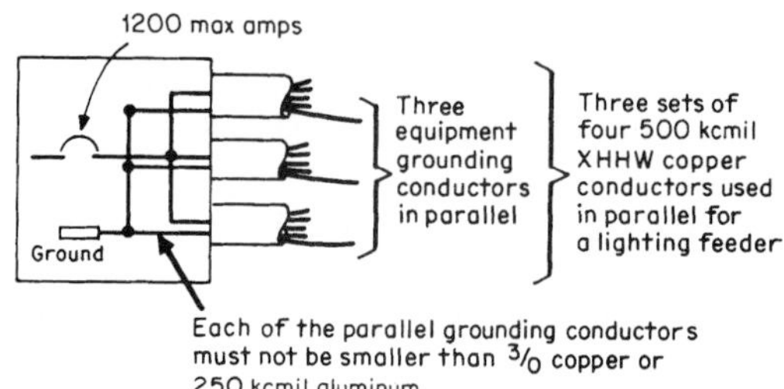
1200 max amps
Three equipment grounding conductors in parallel
Three sets of four 500 kcmil XHHW copper conductors used in parallel for a lighting feeder
Ground
Each of the parallel grounding conductors must not be smaller than 3/0 copper or 250 kcmil aluminum

Fig. 250-115. Using equipment grounding conductors in parallel. (Sec. 250-95.)

[Note in that example that each 500-kcmil XHHW circuit conductor has an ampacity of 430 A (Table 310-16), and three per phase gives a circuit ampacity of $3 \times 430 = 1{,}290$ A. Use of a 1,200-A protective device satisfies the basic rule of Sec. 240-3, protecting each phase leg within its ampacity. Because the load on the circuit is continuous (over 3 hr), the circuit is loaded to not over 960 A—satisfying Sec. 220-10(b), which requires a continuous load to be limited to no more than 80 percent of the circuit protection rating. Each circuit conductor is actually made up of three 500-kcmil XHHW, with a total-per-phase ampacity of $3 \times 430 = 1{,}290$ A. But load is limited to $0.8 \times 1{,}200 = 960$ A per phase. Each 500 kcmil is then carrying $960 \div 3 = 320$ A. Because that value is less than 380 A, which is the ampacity of a 500-kcmil THW copper, the use of XHHW conductors *does* comply with the UL requirement that size 1/0 and larger conductors connected to the equipment be rated at not over 75°C (such as THW) or, if 90°C conductors are used (such as XHHW), they must be used at no more than the ampacity of 75°C conductors of the same size. However, some authorities object to that usage on the grounds that the use of 1,200-A protection would not be acceptable to Sec. 240-3(c), if 75°C (THW) conductors were used, with an ampacity of only 3 × 380 A, or 1,140 A, per phase, because the rating is over the 800-A level and it is not permitted to go to the next larger size of protective device. Therefore, they note, the XHHW conductors are not actually being used as 75°C conductors; the load current could later be increased above the 75°C ampacity; and the application might be taken to violate the letter and intent of the UL rule, thereby violating Sec. 110-3(b) of the NEC.]

The fifth paragraph of Sec. 250-95 covers a similar concern for unnecessarily oversizing equipment grounding conductors. Because the minimum acceptable size of an equipment grounding conductor is based on the rating of the overcurrent protective device (fuse or CB) protecting the circuit for which the equipment grounding conductor is intended to provide a path of ground-fault return, a problem arises when a motor circuit is protected by a magnetic-only (a so-called instantaneous) circuit breaker. Because Sec. 430-52 and Table 430-152 permit an instantaneous-trip CB with a setting of 800 percent of (8 times) the motor full-load running current—and even up to 1,700 percent for an instantaneous CB or MSCP (motor short-circuit protector), if needed to handle motor inrush current—use of those high values of current rating permitted in Table 430-152 would result in excessively large equipment grounding conductors. Because such large sizing is unreasonable and not necessary, the rule says when sizing an equipment grounding conductor from Table 250-95 for a circuit protected by an instantaneous-only circuit breaker or by an MSCP, the rating of the motor running overload device must be used in the left-hand column of Table 250-95 [Fig. 250-116]—but the equipment grounding conductor may never be smaller than No. 14.

Exception No. 2 states that the equipment grounding conductor *need not* be larger than the circuit conductors. The main application for this Exception is for motor circuits where short-circuit protective devices are usually considerably larger than the motor branch-circuit conductor ampacity, as permitted in Sec. 430-52 (up to 1,300 percent) to permit starting of a motor without opening on inrush current. In such cases, literal use of Table 250-95 could result in grounding conductors larger than the circuit conductors.

Exception No. 3 points out that metal raceways and cable armor are recognized as equipment grounding conductors; Table 290-95 does not apply to them.

Figure 250-116 shows details of a controversy that often arises about Secs. 250-95 and 250-57(a). When two or more circuits are used in the same conduit, it is logical to conclude that a single equipment grounding conductor within the conduit may serve as the required grounding conductor for each circuit if it satisfies Table 250-95 for the circuit with the highest rated overcurrent protection. The common contention is that if a single metal conduit is adequate as the equipment grounding conductor for all the contained circuits, a single ground-

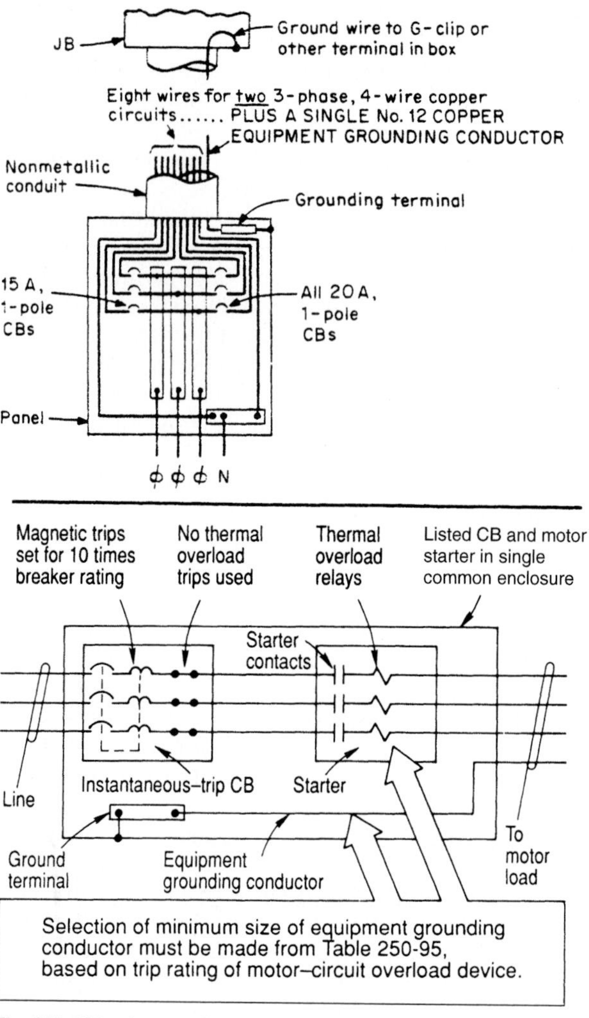

Fig. 250-116. These applications are covered by the fourth and fifth paragraphs of Sec. 250-95. (Sec. 250-95.)

ing conductor can serve the same purpose when installed in a nonmetallic conduit that connects two metal enclosures (such as a panel and a home-run junction box) where both circuits are within both enclosures. As shown, a No. 12 copper conductor satisfies Table 250-95 as an equipment grounding conductor for the circuit protected at 20 A. The same No. 12 also may serve for the circuit protected at 15 A, for which a grounding conductor must not be smaller than No. 14 copper. Such application is specifically permitted by the fourth paragraph of Sec. 250-95, just before the Exceptions. Although this will have primary application with PVC conduit where an equipment grounding conductor is required, it may also apply to circuits in EMT, IMC, or rigid metal conduit when an equipment grounding conductor is run with the circuit conductors to supplement the metal raceway as an equipment grounding return path.

250-112. To Grounding Electrode. The rule requires that the connection of a grounding electrode conductor to the grounding electrode "shall be accessible" (Fig. 250-117). [Section 250-80(a) also requires that any clamp for a bonding jumper to interior metal water piping must be accessible.] Inspectors want to be able to see and/or be able to get at any connection to a grounding electrode. But because there are electrodes permitted in Secs. 250-81 and 250-83 that would require underground or concrete-encased connections, an Exception was added to the basic rule to permit inaccessible connections in such cases (Fig. 250-118). Electrode connections that are *not* encased or buried—such as where they are made to exposed parts of electrodes that are encased, driven, or buried—*must* be accessible. This section now places the burden on the installer to make such connections accessible wherever possible.

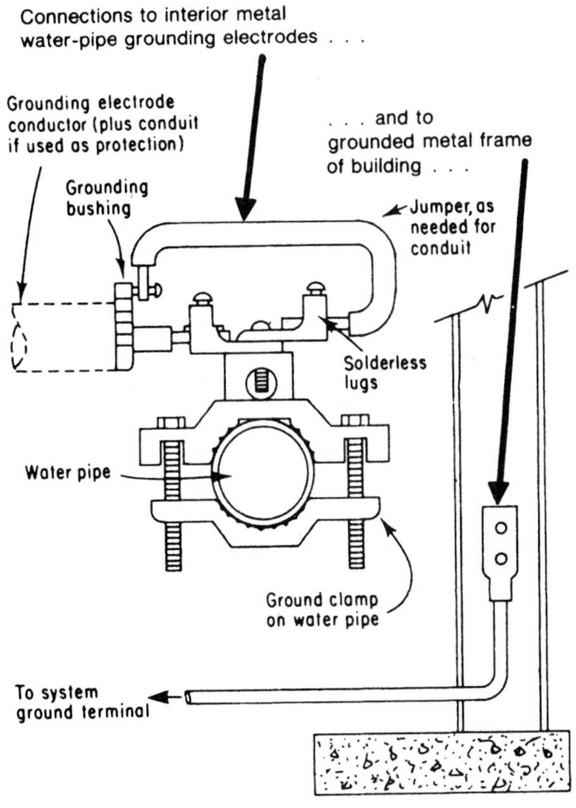

Fig. 250-117. Whenever possible, connections to grounding electrodes must be "accessible." (Sec. 250-112.)

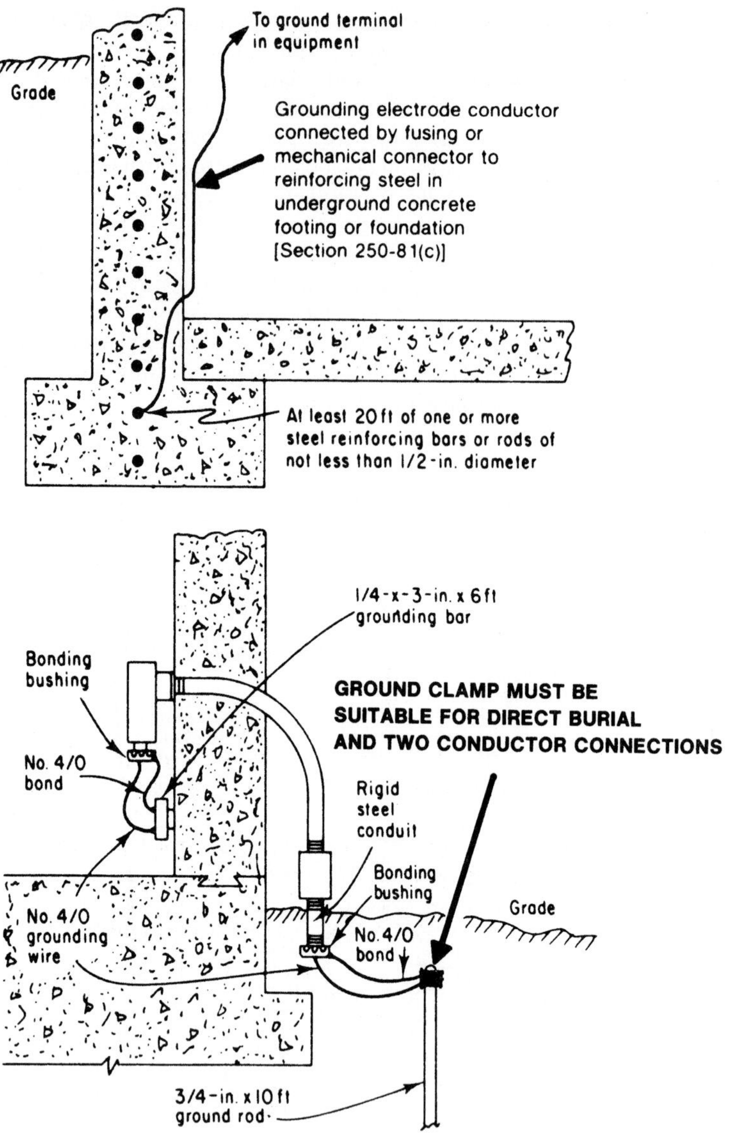

Fig. 250-118. Encased and buried electrode connections are permitted by Exception to basic rule. (Sec. 250-112.)

The second sentence of this section requires assured connection to a metal piping system electrode, as shown in Fig. 250-119. In a typical case of grounding for a local transformer within a building, Sec. 250-26(b) notes that grounding of the secondary neutral may be made to the water pipe but only within 5 ft (1.52 m) from its point of entry in the building. But Sec. 250-112 requires that bonding jumpers be used to assure continuity of the ground path back to the underground pipe for that portion permitted to serve as a grounding electrode by Sec. 250-81 wherever the piping may contain insulating sections or is liable to become disconnected. Bonding jumpers around unions, valves, water meters, and other points where a water-piping-system electrode might be opened must have enough slack to permit removal of the part. Hazard is created

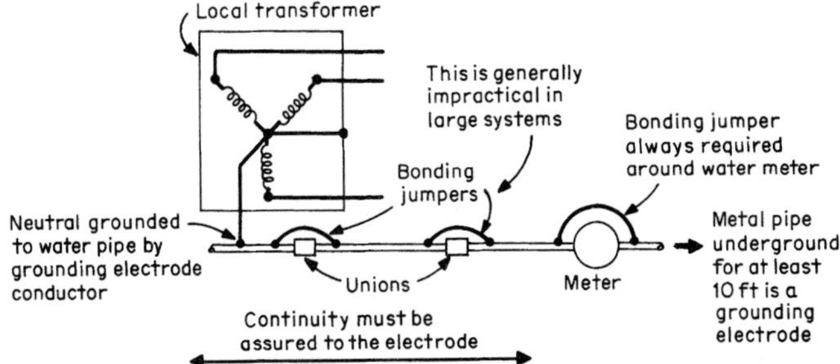

Fig. 250-119. Although required, bonding of metal piping can pose problems. Unless this is an "industrial" installation with qualified maintenance people and the water pipe is completely exposed and visible for the entire distance from its point of entry to the grounding electrode conductor connection, such connection from the grounded conductor must be made no further than 5 ft (1.52 m) from the pipe's entry point. (Sec. 250-112.)

when bonding jumpers are so short that they have to be removed to remove the equipment they jumper. Dangerous conditions have been reported about this matter. Bonding jumpers must be long enough to assure grounding integrity along piping systems under any conditions of maintenance or repair.

250-113. To Conductors and Equipment. There are many grounding and bonding fittings on the market which can be used to properly attach the grounding conductors. The one selected should satisfy the following principles:

1. It must be UL listed.
2. It should be rugged, strong, and well plated so that it will fasten and stay tight.
3. It must fasten mechanically.
4. It must have capacity for a large enough ground wire.
5. It must be compatible with the metals used in the system. For example: Aluminum conductors should not be connected with copper connectors. Compatible aluminum connectors are available for such requirements.

In addition to clamps, exothermic welding is recognized as a suitable method for connecting grounding conductors and bonding jumpers to conductors, equipment, and grounding electrodes, along with listed clamps, etc. This clarifies long-standing controversy and assures the acceptability of exothermic welding of connections.

250-114. Continuity and Attachment of Equipment Grounding Conductors to Boxes. The basic rule requires all ground wires in boxes to be solidly connected together. Then part **(a)** states that where a grounding conductor enters a metal outlet box it must be connected to the box by means of a grounding screw (used for no other purpose) or by an approved grounding device (such as the popular spring-steel grounding clip). Where several grounding conductors enter the same box they must be properly joined together and a final connection made to the grounding screw or grounding clip. In part **(b)**, covering nonmetallic boxes, grounding conductors must be attached to any metal fitting or wiring device required to be grounded.

From this rule, grounding conductors in any metal box must be connected to each other and to the box itself. Figure 250-120 shows a method of connecting ground wires in a box to satisfy the letter of Sec. 250-114. Note that the two ground wires are solidly connected to each other by means of a crimped-on spade tongue terminal, with one of the ground wires (arrow) cut long enough so that it is bent back out of the crimp lug to provide connection to the green hex-head screw on a receptacle outlet (if required by Sec. 250-74). The spade lug is secured firmly under a screw head, bonding the lug to the box. Of course, the specific connections could be made in other ways. For instance, the ground wires could be connected to each other by twist-on splicing devices; and connection of the ground wires to the box could be made by simply wrapping a single wire under the screw head or by connecting a wire from the splice connector to an approved grounding clip on the edge of the box (Fig. 250-121).

In all the drawings here, connection to the box is made either by use of a screw in a threaded hole in the side or back of the box or by an approved

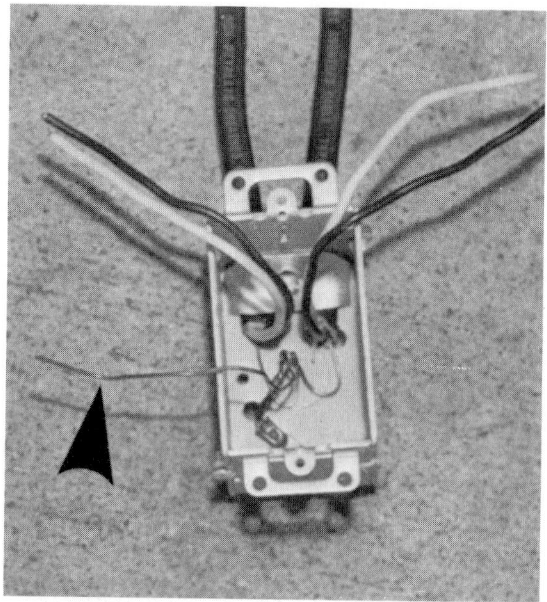

Fig. 250-120. Both ground wires are solidly bonded together in the crimped barrel of the spade lug, which is screwed to back of metal box. (Sec. 250-114.)

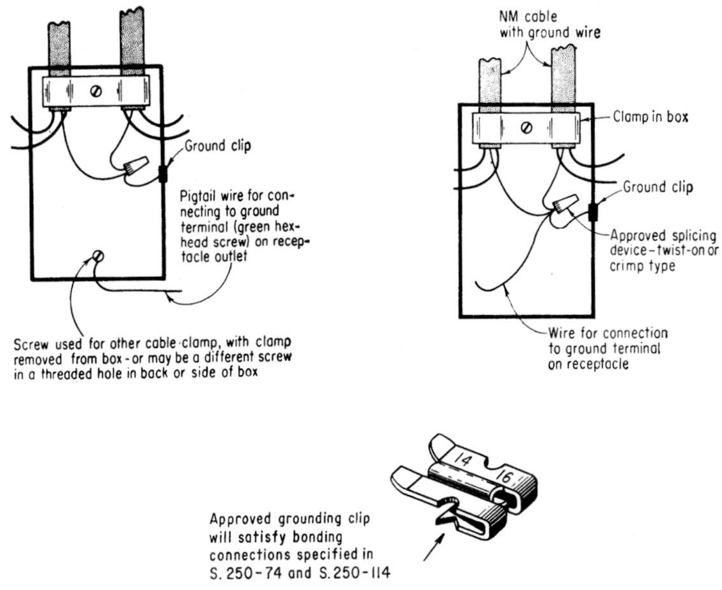

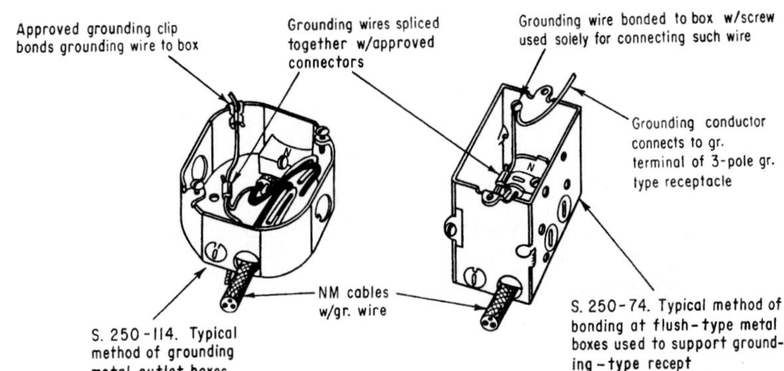

Fig. 250-121. All these techniques bond the ground wires together and to the box. (Sec. 250-114.)

ground clip device which tightly wedges a ground wire to the edge of the box wall, as shown in Fig. 250-122. Preassembled pigtail wires with attached screws are available for connecting either a receptacle or the system ground wire to the box.

Figure 250-123 shows connection of two cable ground wires by means of two grounding clips on the box edges (arrow). In the past, such use has been disallowed by some inspection authorities because the ground wires are not actually connected to each other but are connected only through the box. The clear wording of the Code rule here ("all conductors . . . joined within or *to* the box"), however, permits such practice.

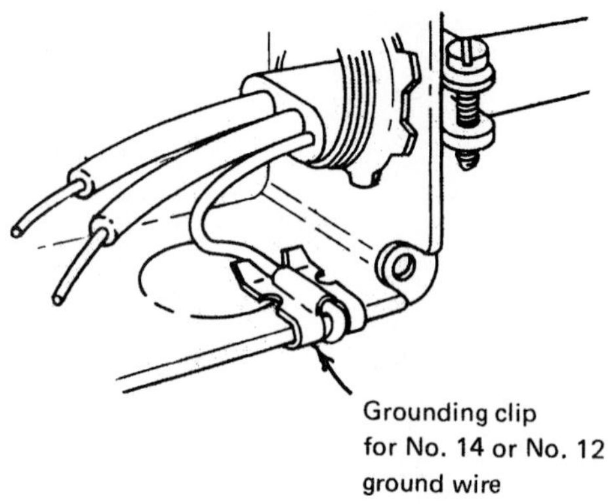

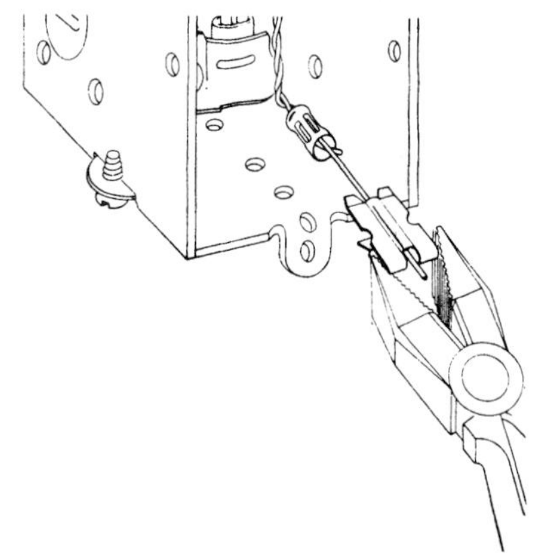

Installation of grounding clip.

Fig. 250-122. Ground clip is "an approved grounding device" of Sec. 250-114(a). (Sec. 250-114.)

Figure 250-124 shows another method that has been objected to as clear violation of NE Code Sec. 250-114(a) which requires that a screw used for connection of grounding conductors to a box "shall be used for no other purpose." Use of this screw, simultaneously, to hold the clamp is for "other purpose" than grounding. Objection is not generally made to use of the clamp screw for ground connection when, in cases where the clamp is not in use, the clamp is removed and the screw serves only the one purpose—to ground the grounding wires.

The Exception to this rule eliminates the need for connecting an isolated grounding conductor to all other grounding conductors.

250-115. Connection to Electrodes. Because Sec. 250-83(c)(3) requires *buried* or protected connections of grounding electrode conductors to ground rods, the third sentence of this section requires that a buried ground clamp be of such material and construction that it has been designed, tested, and marked for use

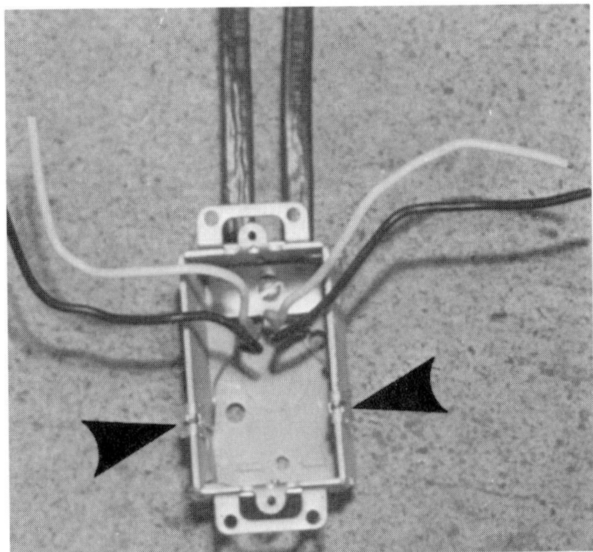

Fig. 250-123. Each ground wire is connected to the metal box by a ground clip (one on each side at arrows). The first sentence of Sec. 250-114 permits ground wires to be "spliced or joined . . . to the box" by use of ground clips or ground screw terminals in the box. (Sec. 250-114.)

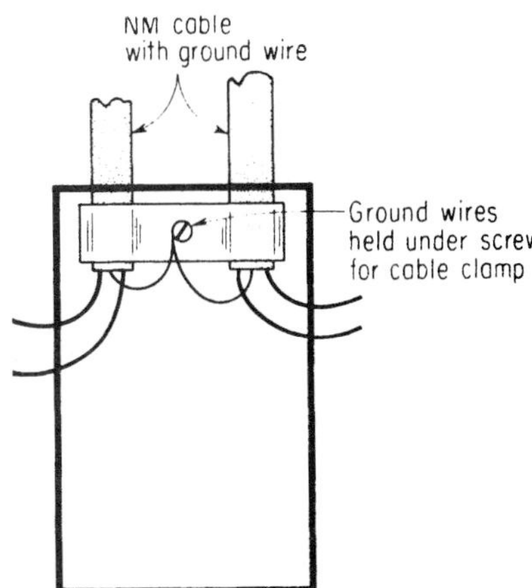

Fig. 250-124. This clearly violates Sec. 250-114(a). (Sec. 250-114.)

directly in the earth. And any clamp that is used with two or more conductors must be designed, tested, and marked for the number and types of conductors that may be used with it. This is shown in the bottom drawing of Fig. 250-118.

250-152. Solidly Grounded Neutral Systems. Figure 250-125 shows the details of this set of rules. This section does permit a neutral conductor of a solidly grounded "Y" system to have insulation rated at only 600 V, instead of requiring insulation rated for the high voltage (over 1,000 V). It also points out that a bare copper neutral may be used in such systems for service-entrance conductors or for direct buried feeders, and bare copper or copper-clad aluminum may be used for overhead sections of outdoor circuits.

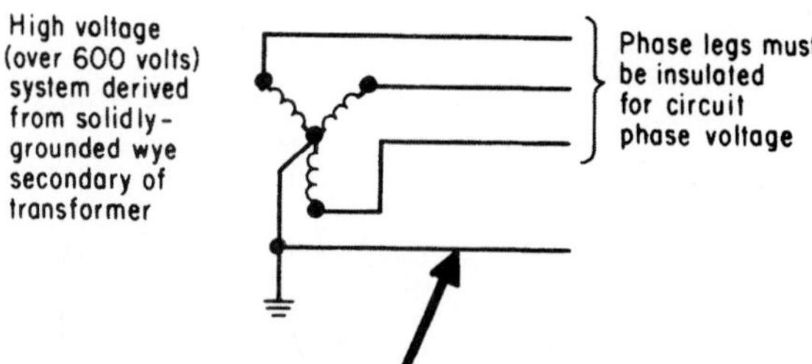
High voltage
(over 600 volts)
system derived
from solidly-
grounded wye
secondary of
transformer
Phase legs must
be insulated
for circuit
phase voltage
Solidly grounded neutral conductor must have insulation rated for at least 600 volts, although a bare copper neutral may be used for SE conductors or for direct-buried feeders, and bare copper or aluminum may be used for overhead parts of outdoor circuits.

Fig. 250-125. Neutral of high-voltage system generally must be insulated for 600 V. (Sec. 250-152.)

ARTICLE 280. SURGE ARRESTERS

280-1. Scope. This article is no longer entitled "Lightning Arresters," as it was up to the 1978 NE Code. It is now entitled "Surge Arresters," and a complete editorial rewrite has been done on the content of the article. Article 280 is the result of the work of a subcommittee of the Code-making panel and is aimed at updating the scope and details of Code rules. Installation, connection, and grounding requirements are covered for surge arresters installed on premises wiring systems, with completely new renumbering of the various sections.

In previous editions of the Code, use of lightning arresters was *required* for "industrial stations" where thunderstorms are frequent and lightning protection is needed. Article 280 no longer makes that requirement and is designed to be applied to any installations where surge arresters might be installed.

This basic requirement used to say that surge arresters are mandatory only "where thunderstorms are frequent"—but they are *not* needed if some other type of lightning protection is provided. And the mandatory need used to apply only to "industrial stations"—such as a generating station or substation serving principally a single industrial plant or factory, as distinguished from a station serving several customers of a public utility power company (Fig. 280-1). This Article no longer makes use of surge arresters mandatory, as previously required by Sec. 280-10, which has been deleted from the Code. Figure 280-2 shows a lightning arrester used on one of several high-voltage circuits serving the heavy electrical needs of a modern sports stadium.

280-3. Number Required. A double-throw switch which disconnects the outside circuits from the station generator and connects these circuits to ground

Fig. 280-1. Surge (lightning) arresters in an electric substation serving an industrial plant are commonly used in areas where lightning is a problem. (Sec. 280-1.)

Fig. 280-2. Lightning arrester (arrow) is a typical "surge arrester" and, where used, one arrester must be connected to each ungrounded circuit conductor—such as shown here for a 2,400-V grounded circuit supplying a transformer for stepping voltage down to supply lightning at this athletic stadium. (Sec. 280-3.)

would satisfy the condition for a single set of arresters for a station bus, as covered in the second sentence of this section.

280-4. Surge Arrester Selection. Figure 280-3 shows the position of a choke coil where it is used as a lightning-protection accessory to an arrester.

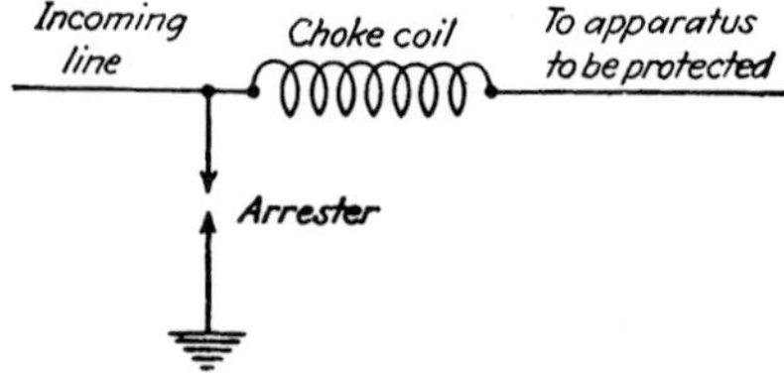

Fig. 280-3. Using a choke coil as an accessory to an arrester. (Sec. 280-4.)

In Sec. 280-4(b), ratings of surge arresters are covered by the basic rule that applies to silicon-carbide-type surge arresters with a fine-print note pointing up the difference in voltage rating of metal-oxide varistor (MOV) type arresters. This addresses the high-technology operating nature of the metal-oxide surge arrester (MOSA) as applied to premises wiring systems. The concern is to make an effective distinction between gapped silicon-carbide arresters, widely used in the past, and the newer metal-oxide block arresters. Manufacturers' application data on rating and other characteristics and the minimum duty-cycle voltage rating of an arrester for a particular method of system grounding must be observed carefully.

280-12. Routing of Surge Arrester Connections. This rule is particularly important because bends and turns enormously increase the impedance to lightning discharges and therefore tend to nullify the effectiveness of a grounding conductor.

280-24. Circuits of 1 kV and Over—Interconnections. These rules are aimed at assuring more effective lightning protection of transformers. Lightning protection of a transformer cannot be provided by a primary arrester that is connected only to a separate electrode. Common grounding of gaps or other devices must be used to limit voltage stresses between windings and from windings to case.

280-25. Grounding. This section refers to Secs. 250-131 and 250-132, which cover connection of lightning arresters. The second sentence covers the need to keep grounding conductors electrically in parallel with their enclosing metal raceway—as discussed under Secs. 250-71(a)(3) and 250-92(b). For instance, assume that a lightning arrester is installed at the service head on a conduit service riser, with the grounding conductor run inside the service conduit, bonded to the meter socket at the grounding lug, then run through a hole in the meter socket to the grounding electrode without a metal enclosure from the drilled hole to the electrode. In such a hookup, this rule requires the grounding conductor to be bonded to the conduit at the service head (Fig. 280-4). Ordinarily the meter enclosure has a threaded hub, which would mean the conduit would be in good electrical contact with the meter enclosure and would be bonded at the meter socket end. However, Sec. 280-25 requires that the grounding conductor, if in a metallic enclosure, be bonded at both ends. Therefore, bonding at the service head is necessary.

The reason given for putting that rule in the NEC was explained as follows:

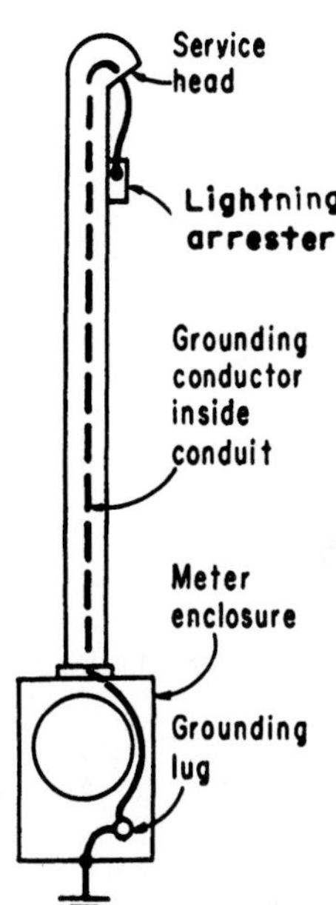

Fig. 280-4. Arrester grounding conductor must be bonded to both ends of enclosing metal raceway (or other enclosure). (Sec. 280-25.)

When conducting lightning currents, the impedance of a lightning arrester grounding conductor is materially increased if run through a metallic enclosure, especially if of magnetic material. The voltage drop in this impedance may be sufficient to cause arcing to the enclosure, and in any event it reduces the effectiveness of the lightning arrester. Bonding of the conductor to both ends of the enclosure is necessary to eliminate this detrimental effect where metallic enclosures are used.

Chapter Three

ARTICLE 300. WIRING METHODS

300-1. Scope. The Exceptions in part **(a)** of this section indicate clearly that not all the general requirements in Art. 300 apply to remote-control circuits, to signal circuits, to low-energy circuits, to fire protective signaling circuits, and to communications systems. Only those sections of Art. 300 that are referenced in Arts. 504, 725, 760, 770, 800, and 810 apply to the types of circuits covered by those articles. In effect, not all the regulations on wiring for general-purpose power and light circuits apply to the specialized circuits covered by Arts. 504, 725, 760, 770, 800, and 810.

300-3. Conductors. Part **(a)** requires that single conductors described in Table 310-13 must be used *only* as part of one of the wiring methods covered in Chapter 3.

Part **(b)** requires that all conductors of the same circuit—including the neutral and all equipment grounding conductors—must be run in the same raceway, cable tray, trench, cable, or cord. Exception No. 3 to that rule notes those sections of the NEC where an equipment grounding conductor may be run separately from the other conductors of the circuit or where other departures from the basic rule are recognized—such as isolated phase conductors in plastic conduit.

In part **(c) (1),** the words "cable or raceway" after the words "wiring enclosure" clearly indicate that it is the intent of the Code that circuits of different voltage up to 600 V may occupy the same wiring enclosure (cabinet, box, housing), cable, or raceway provided all the conductors are insulated for the maximum voltage of any circuit in the enclosure, cable, or raceway. It is the intent of the Code panel to indicate clearly that, for instance, motor power conductors and motor control conductors be permitted in the same conduit. In the past, there has been a long-standing controversy about the use of control-circuit conductors in the same conduit with power leads to motors.

For a long time it was argued that Sec. 300-3(e) in the 1975 Code required a separate raceway for each motor when the control conductors were run in the raceway with the power conductors (Fig. 300-1). This was supported by Sec. 725-26, which also indicates that Class 1 control conductors were permitted to be run *only in the raceway* for the power conductors which the control conductors actually control.

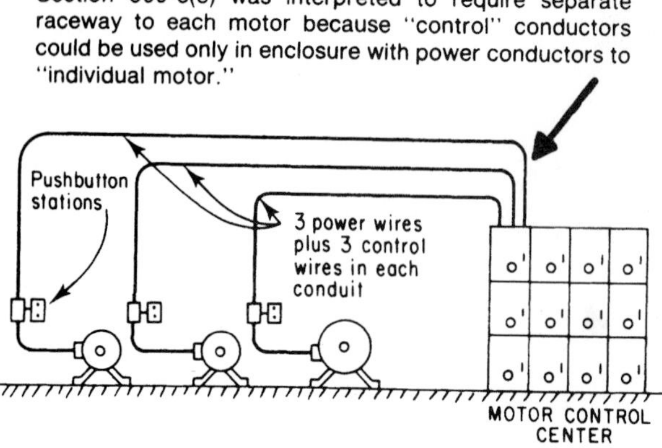
Section 300-3(e) was interpreted to require separate raceway to each motor because "control" conductors could be used only in enclosure with power conductors to "individual motor."
Pushbutton stations
3 power wires plus 3 control wires in each conduit
MOTOR CONTROL CENTER

Fig. 300-1. The 1975 NE Code limited mixing of control and power wires in raceways. (Sec. 300-3.)

The wording of Sec. 300-3(c) (1) recognizes the use of power and control wires in a single raceway to supply more than one motor, but such usage must be made to conform to the last sentence of Sec. 725-26. The two NE Code rules of Sec. 300-3(c) (1) and Sec. 725-26 must be put together carefully.

A common raceway, as shown in Fig. 300-2, may be used *only* where the two or more motors are required to be operated together in order to serve their load function. Many industrial and commercial installations have machines, manufacturing operations, or processes which are based on use of a number of motors driving various parts or stages of the task. In such cases, either all motors operate or none do. Use of all control wires and power wires in the same raceway does not produce a situation where a fault in one motor circuit could disable another circuit to a motor that might otherwise be kept operating.

But when a common raceway is used for power and control wires to separate, independent motors, a fault in one circuit could knock out all the others that do not have to shut down when one goes out. With motor circuits so closely associated with vital, important functions like elevators, fans, pumps, etc., in modern buildings, it is a safety matter to separate such circuits and minimize outage due to any fault in a single circuit. For safety's sake, the Code, in effect, says, "Do not put all your eggs in one basket." But the objectionable loss of more than one motor on a single fault does not apply where all motors must be shut down when any one is stopped—as in multimotor machines and processes.

For those cases where each motor is serving a separate, independent load—with no interconnection of their control circuits and no mechanical interlock-

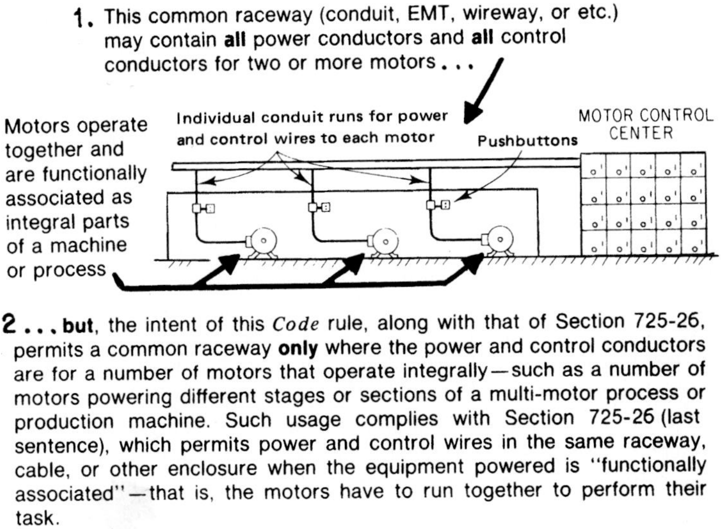

1. This common raceway (conduit, EMT, wireway, or etc.) may contain all power conductors and all control conductors for two or more motors . . .
Motors operate together and are functionally associated as integral parts of a machine or process
Individual conduit runs for power and control wires to each motor
Pushbuttons
MOTOR CONTROL CENTER
2 . . . but, the intent of this Code rule, along with that of Section 725-26, permits a common raceway only where the power and control conductors are for a number of motors that operate integrally—such as a number of motors powering different stages or sections of a multi-motor process or production machine. Such usage complies with Section 725-26 (last sentence), which permits power and control wires in the same raceway, cable, or other enclosure when the equipment powered is "functionally associated"—that is, the motors have to run together to perform their task.

Fig. 300-2. Mixing of power and control wires in common raceway is still limited. (Sec. 300-3.)

ing of their driven loads—the use of a separate raceway for each motor is required by the last sentence of Sec. 725-26, *but only* when control wires are carried in the raceways (Fig. 300-3). For the three motors shown, it would be acceptable to run the power conductors for all the motors in a single raceway and all the control circuit wires in another raceway. Such a hookup would not violate Sec. 725-26, although deratings would have to be made and there is the definite chance of loss of more than one motor on a fault in only one of the circuits in either the power raceway or the control raceway.

Separate conduit or raceway is required for power and control wires to each motor . . .
Pushbutton stations
3 power wires plus 3 control wires in each conduit
MOTOR CONTROL CENTER
. . . When these are individual motors that do not operate together as parts of a machine or process—that is, each motor has a separate, independent load that may operate by itself.

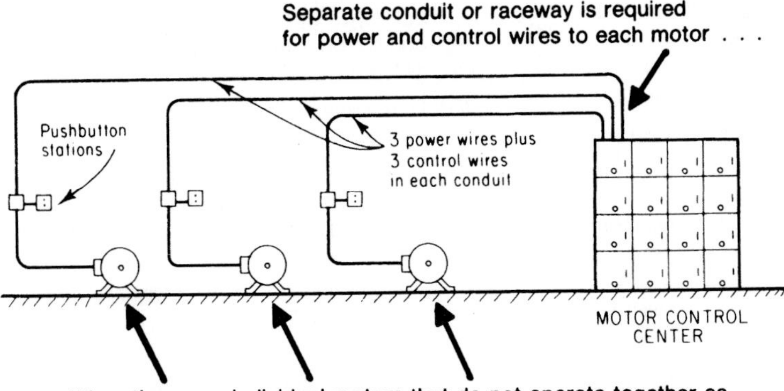

Fig. 300-3. Section 725-26 prohibits intermixing of power and starter coil-circuit wires when motors are not "functionally associated." (Sec. 300-3.)

The last sentence in part **(c)(1)** is intended to recognize nonshielded conductors—rated at 8 kV—occupying the same enclosure, raceway, or cable, regardless of the actual voltage carried (e.g., 2.4 and 6.9 kV). This is a practice that should be avoided because surface voltage of such cables is *not* zero. And some feel that the normal voltage discharge and leakage current will be increased where a difference of potential exists between two unshielded conductors. It is theorized that the result of the increased discharge and leakage will be premature insulation failure. Although the assertion of premature failure has not been determined empirically, there is no data to suggest otherwise. Any problem can be avoided by simply using separate raceway systems.

Part **(c) (2)** of Sec. 300-3 states that conductors operating at more than 600 V *must not* occupy the same equipment wiring enclosure, cable, or raceway with conductors of 600 V or less. But, the rule lists three exceptions to paragraph **(c) (2)** [not to paragraph **(c) (1)**]. Exception No. 3 is intended to apply to enclosures, not raceways, such as used for high-voltage motor starters, permitting the high-voltage conductors operating at over 600 V to occupy the same controller housing as the control conductors operating at less than 600 V (Fig. 300-4). In

SECTION 300-3(c) (2) PROHIBITS THIS—

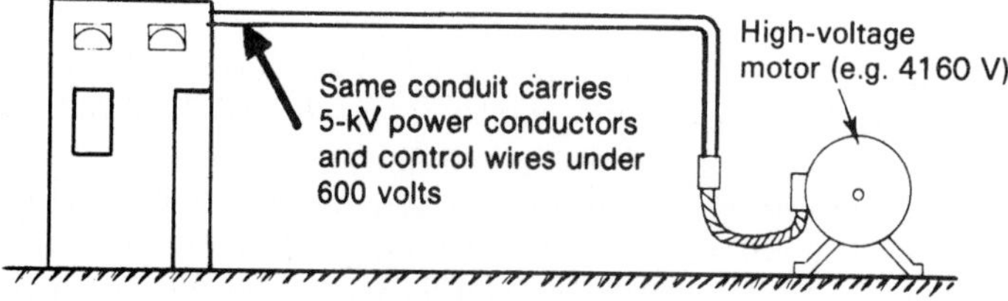

BUT, EXCEPTION NO. 3 PERMITS THIS—

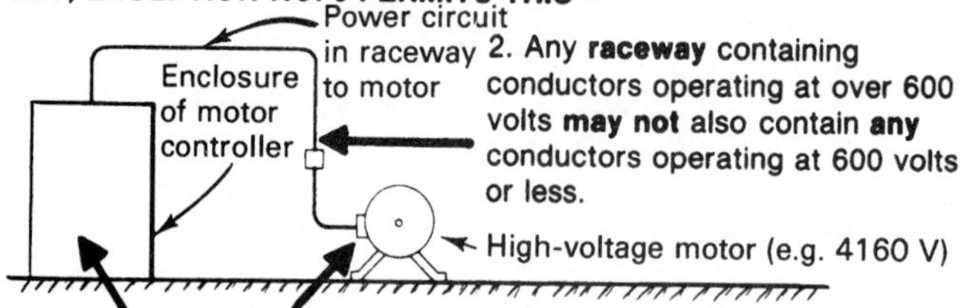

1. For any **individual** motor or starter—excitation, control, relay and/or ammeter conductors operating at 600 volts or less **may** occupy the same **starter or motor enclosure** as the conductors operating at over 600 volts.

Fig. 300-4. Control wires for high-voltage starters may be used in the starter enclosure, but not in *raceway* with power conductors. (Sec. 300-3.)

addition, Exception No. 1 of Sec. 300-32 specifically recognizes use of high- and low-voltage conductors in the same enclosure of "motors, switchgear and control assemblies, and similar equipment."

300-4. Protection Against Physical Damage. Part **(a)** gives the rules on protection required for cables and raceways run through wood framing members, as shown in Fig. 300-5. Where the edge of a hole in a wood member is in. (31.8 mm) from the nearest edge of the member, a 1/16-in. (1.5 steel plate must be used to protect any cable or flexible conduit a nails or screws. The same protection is required for any cable or flex duit laid in a notch in the wood. *But,* rigid metal conduit, EMT, IMC, and PVC conduits do not require such protection.

Clearance must be provided from the edge of a hole in a wood member to the edge of the wood member. Where the NE Code used to require a minimum of 1½ in. (38 mm) from the edge of a cable hole in a stud to the edge of the stud, the present NE Code requires only 1¼ in. (31.8 mm). This permits realistic compliance when drilling holes in studs that are 3½ in. (89 mm) deep. It also was taken into consideration that the nails commonly used to attach wall surfaces to studs were of such length that the 1¼-in. (31.8-mm) clearance to the edge of the cable hole afforded entirely adequate protection against possible penetration of the cable by the nail.

Figure 300-6 shows typical application of cable through drilled studs, with holes at centers and adequate clearance to edge of stud. Figure 300-7 shows an objectionable example of a drilled hole, violating the rule of part (2) of this section, which warns against "weakening the building structure." Figure 300-8 shows an acceptable way of protecting cables run through holes in wood members.

In part **(b)**, the rules on installations through metal framing members apply to nonmetallic-sheathed cable and to ENT (electrical nonmetallic tubing). Part **(1)** of Sec. 300-4(b) applies to NM cable run through slots or holes in metal framing members and requires that such holes must *always* be provided with bushings or grommets installed in the openings *before* the cable is pulled. But that requirement on protection by bushings or grommets in the holes does *not* apply to ENT run through holes.

Part **(2)** applies to both NM cable and ENT and requires that the cable or tubing be protected by a steel sleeve, a steel plate, or a clip when run through metal framing members in any case where nails or screws might be driven into the cable or tubing.

Part **(c)** requires cables and raceways above lift-out ceiling panels to be supported as they are required to be when installed in the open. They may not be treated as if they were being run through closed-in building spaces or fished through hollow spaces of masonry block.

Part **(d)** requires cables and raceways run along (parallel with) framing members (studs, joists, rafters) to have at least a 1¼-in. (31.8-mm) clearance from the nearest edge of the member; otherwise the cable or raceway must be protected against nail or screw penetrations by a steel plate or sleeve at least 1/16 in. (1.59 mm) thick.

Part **(d)** was added because of many persistent reports of nail and screw penetrations of both metallic and nonmetallic cables and raceway. The rule applies to both exposed and concealed locations. Exception No. 1 excludes intermediate metal conduit (IMC), rigid metal conduit, rigid nonmetallic conduit, and electrical metallic tubing (EMT) from the rule. The rule will apply to Romex (Type NM), BX (Type AC), flexible metal conduit, ENT (electrical nonmetallic

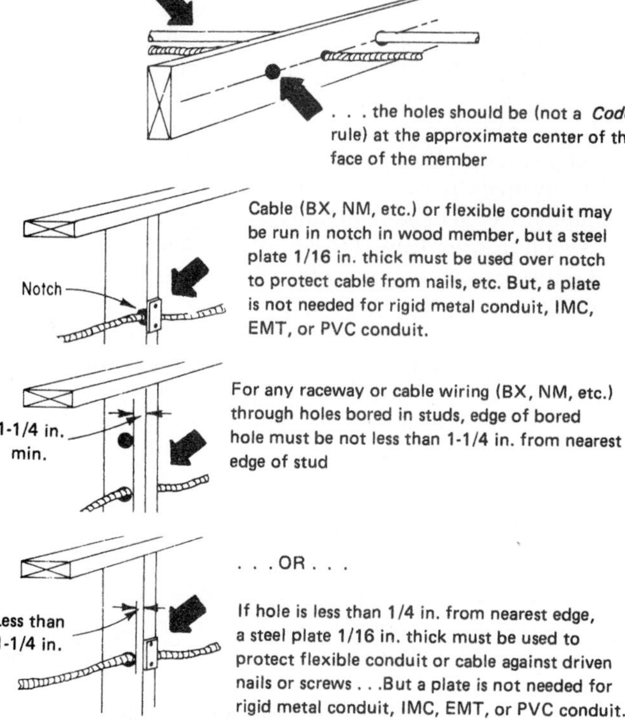
If BX, NM cable, or raceway wiring (rigid conduit, EMT, etc.) is used through holes bored in joists, rafters or similar wood members. . .
. . . the holes should be (not a *Code* rule) at the approximate center of the face of the member
Notch
Cable (BX, NM, etc.) or flexible conduit may be run in notch in wood member, but a steel plate 1/16 in. thick must be used over notch to protect cable from nails, etc. But, a plate is not needed for rigid metal conduit, IMC, EMT, or PVC conduit.
1-1/4 in. min.
For any raceway or cable wiring (BX, NM, etc.) through holes bored in studs, edge of bored hole must be not less than 1-1/4 in. from nearest edge of stud
. . . OR . . .
Less than 1-1/4 in.
If hole is less than 1/4 in. from nearest edge, a steel plate 1/16 in. thick must be used to protect flexible conduit or cable against driven nails or screws . . .But a plate is not needed for rigid metal conduit, IMC, EMT, or PVC conduit.

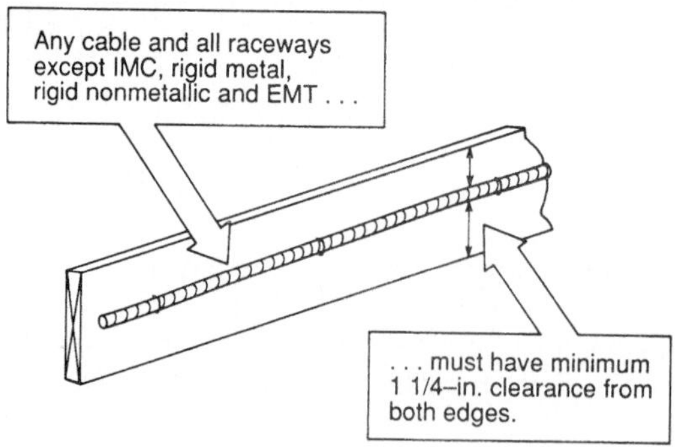
Any cable and all raceways except IMC, rigid metal, rigid nonmetallic and EMT . . .
. . . must have minimum 1 1/4–in. clearance from both edges.
NOTE: If clearance is less than 1 1/4 in., a 1/16–in. thick steel plate or sleeve must be used to protect cable or raceway.

Wiring on structural members must have clearance for protection against nails.

Fig. 300-5. Holes in wood framing must not weaken structure or expose cable to nail puncture. (Sec. 300-4.)

Drilled holes at the center of the face of a joist do not reduce the structural strength of the joist.

Signal and alarm wiring is run through the same stud holes as the NM cables. The NEC does not prohibit use of more than one cable through a single hole.

Fig. 300-6. Holes or notches in joists and studs must not weaken the structure of a building. (Sec. 300-4.)

Fig. 300-7. Excessive drilling of structural wood members can result in dangerous notching (arrow) that weakens the structure, violating Sec. 300-4(a) (2). (Sec. 300-4.)

tubing), Type MC cable, and all other cables and raceways, except those excluded (Fig. 300-5, bottom).

Exception No. 2 excludes from this rule concealed work in finished buildings and finished panels in prefab buildings—where cables may be fished. Exception No. 3 excludes mobile homes and recreational vehicles.

Part **(e)** presents protection requirements where cables are run in "shallow grooves." This rule calls for the same 1⁄16-in (1.59-mm) thick steel kick plate, sleeving, or the equivalent. Alternately, one may locate the cable with at least 1¼ in. (31.8 mm) of "free space" for the cable's entire concealed length. The "free space" is to be measured from the top of the groove to the top of the cable, *not* to the bottom of the groove. This rule is aimed at applications with exposed beam construction where grooves or channels are cut in the beams for supply conductors to lighting fixtures or ceiling fans.

300-5. Underground Installations. This section is a comprehensive set of rules on installation of underground circuits for circuits up to 600 V. [Higher-voltage circuits must satisfy Sec. 710-4(b).] Table 300-5 in the **Code** book establishes *minimum* burial depths for specific conditions of use. Fig. 300-9 shows burial

Fig. 300-8. Steel plates are attached to wood structure member to protect cable from penetration by nail or screw driven into finished wall, where the edge of the cable hole is less than 1¼ in. (31.8 mm) from the edge of the wooden member. (Sec. 300-4.)

requirements for rigid metal conduit and IMC. Figure 300-10 shows the basic depth requirements for the various wiring methods.

Because Table 300-5 does not specifically mention EMT, it could be taken to indicate that the **NEC** does not recognize EMT for underground use. But Sec. 348-1 (in condition No. 3) does recognize EMT for direct earth burial, and so does UL, with this stipulation: "In general, electrical metallic tubing in contact with soil requires supplementary corrosion protection." Note that such protection is not always mandatory. The UL note means to indicate that EMT may be buried without a protective coating (like asphalt paint) where local experience verifies that soil conditions do not attack and corrode the EMT.

Figure 300-11 shows modifications of basic burial depths. If a 2-in. (50.8-mm) -thick or thicker concrete pad is used in the trench over an underground circuit other than rigid metal conduit or IMC, the basic burial depth in Table 300-5 may be reduced by 6 in. (152 mm). It should be noted that intent here is for the concrete pad to be *in* the trench, right over the cable or raceway. The wording must be taken to mean that it may not be a walk or other concrete at grade level. And the burial depth may not be reduced by more than 6 in. (152 mm) no matter how thick the concrete pad is. This rule is at odds with Exception No. 2 to Table 710-4(b) where burial depth for high-voltage circuits may be reduced "6 in. (152 mm) for each 2 in. (50.8 mm) of concrete" or equivalent protection in the trench over the wiring method (other than rigid metal or IMC).

Note that, in Table 300-5, rigid metal conduit or IMC that is buried in the ground must have at least a 6-in. (152-mm) thick cover of earth or earth-plus-

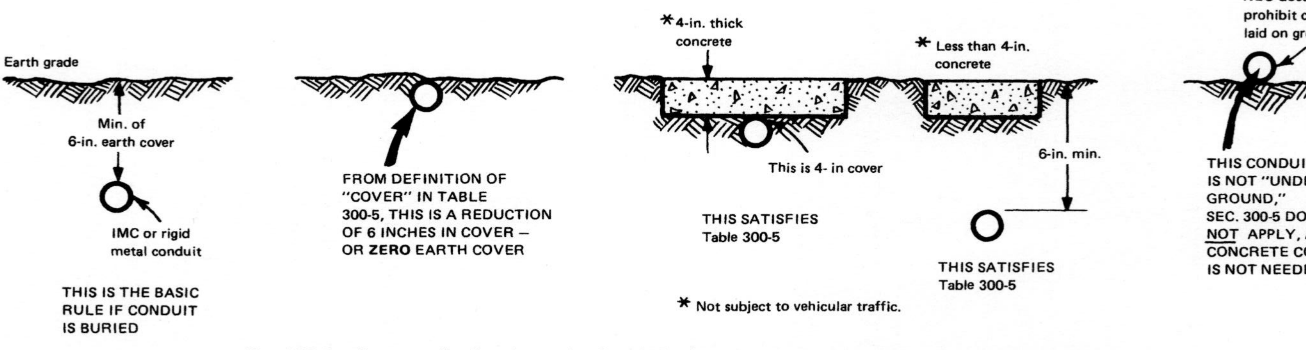
Earth grade
Min. of
6-in. earth cover
IMC or rigid
metal conduit
THIS IS THE BASIC
RULE IF CONDUIT
IS BURIED
FROM DEFINITION OF
"COVER" IN TABLE
300-5, THIS IS A REDUCTION
OF 6 INCHES IN COVER –
OR ZERO EARTH COVER
*4-in. thick
concrete
This is 4- in cover
THIS SATISFIES
Table 300-5
* Less than 4-in.
concrete
6-in. min.
THIS SATISFIES
Table 300-5
* Not subject to vehicular traffic.
NEC does not
prohibit conduit
laid on ground
THIS CONDUIT
IS NOT "UNDER –
GROUND,"
SEC. 300-5 DOES
NOT APPLY, AND
CONCRETE COVER
IS NOT NEEDED

Fig. 300-9. These are the details involved with the use of rigid metal conduit and IMC underground.

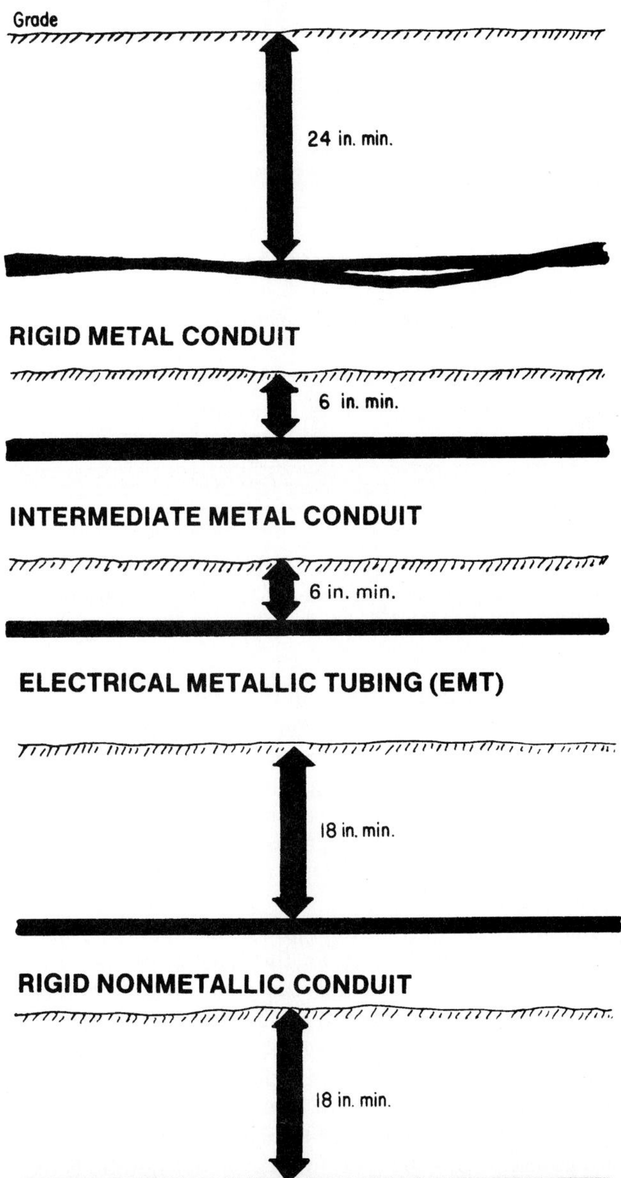

Fig. 300-10. These are the *basic* burial depths, *but* variations are recognized in Table 300-5 for certain conditions. (Sec. 300-5.)

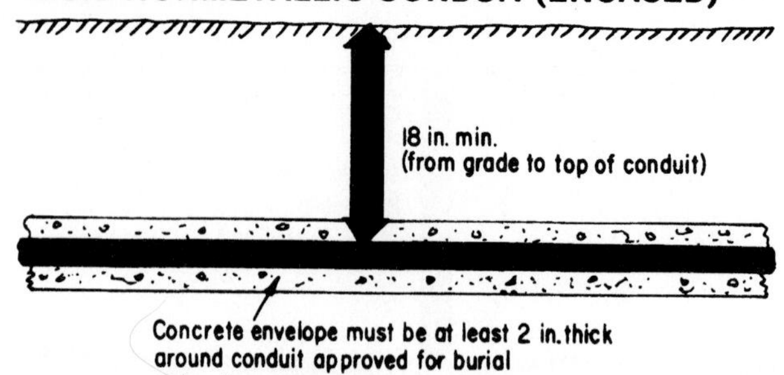
RIGID NONMETALLIC CONDUIT (ENCASED)
18 in. min.
(from grade to top of conduit)
Concrete envelope must be at least 2 in. thick
around conduit approved for burial
only when encased

Fig. 300-10. (*Continued.*)

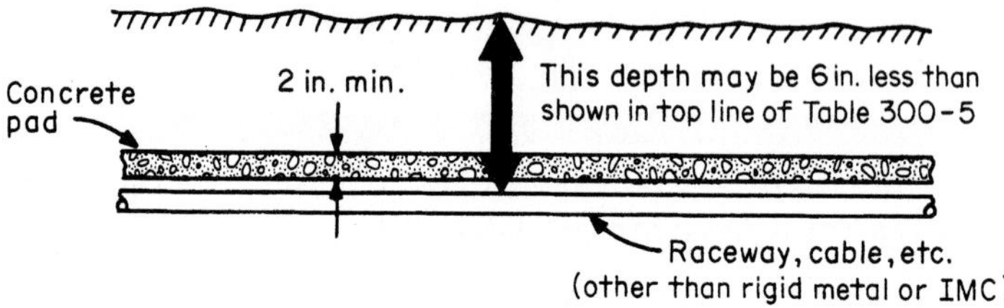
Concrete
pad
2 in. min.
This depth may be 6 in. less than
shown in top line of Table 300-5
Raceway, cable, etc.
(other than rigid metal or IMC)

Fig. 300-11. Concrete pad "in trench" permits only a 6-in. (152-mm) reduction of burial depth for circuits up to 600 V. (Sec. 300-5.)

concrete—even if it has a 2-in. (50.8-mm) thick concrete pad over it. But rigid metal conduit, IMC, or other raceways may be installed directly under a 4-in-or-thicker exterior slab that is not subject to vehicle traffic—without any need for earth cover. Given the fact that rigid metal conduit or IMC may be laid directly on the ground (which supports it for its entire length) and would not necessarily require any concrete cover, there is no reason why it cannot be laid on the ground or flush with the ground and covered with at least 4 in. (102 mm) of concrete. (See Fig. 300-9.)

Section 300-5 only applies to "Underground Installations" and is not applicable if the conduit is laid directly on the ground. No **Code** rule prohibits conduit laid on the ground, provided the conduit is "securely fastened in place" (Sec. 346-12) and is not exposed to physical damage—such as vehicular traffic—and many such installations have been made for years. But, when conduit is installed in the ground, there is serious concern about damage due to digging in the ground, which Sec. 300-5 addresses.

As shown in Fig. 300-12, Table 300-5 recognizes that raceways run under concrete slabs at least 4 in. (102 mm) thick or under buildings have sufficient protection against digging and are not required to be subject to the burial-depth requirements given in the top line of Table 300-5. Where raceways are so installed, the rule requires that the slab extend at least 6 in. (152 mm) beyond the underground raceway, as follows:

1. Any direct burial cable run under a building must be installed in a raceway, as required by Sec. 300-5(c), and the raceway may be installed in the

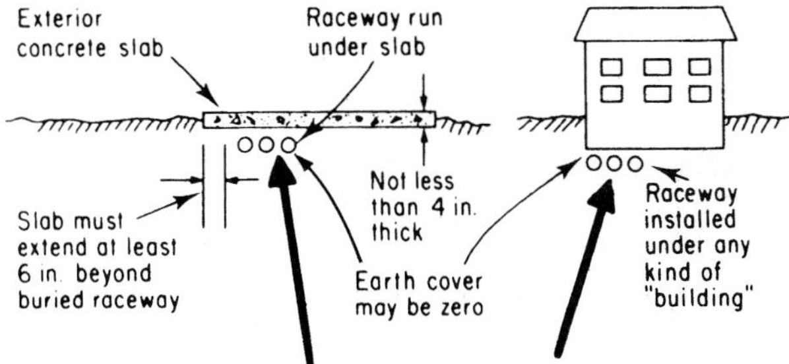

Fig. 300-12. Table 300-5 eliminates burial-depth requirements for direct buried "raceways" under specified conditions. (Sec. 300-5.)

earth, immediately under the bottom of the building—without any earth cover.

2. Any direct buried cable under a slab at least 4 in. (102 mm) thick (and not subject to vehicles) is subject to the 18-in. (457-mm) minimum burial-depth requirement of Table 300-5.

Figure 300-13 shows the mandatory 24-in. (610-mm) burial depth given in Table 300-5 for *any* wiring methods buried under public or private roads,

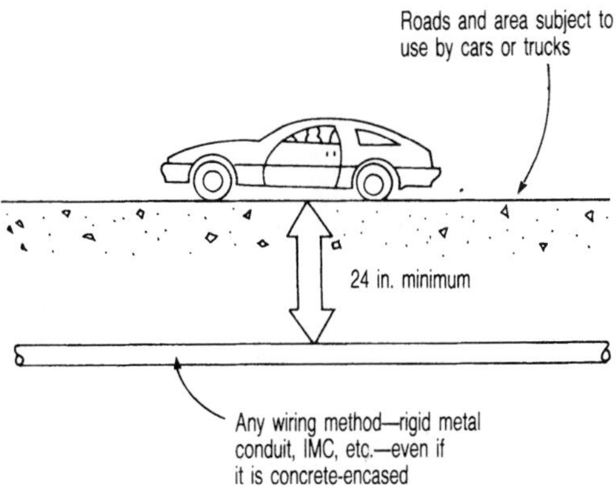

Fig. 300-13. All wiring methods must be at least 2 ft (610 mm) under vehicular traffic. (Sec. 300-5.)

alleys, driveways, parking lots, or other areas subject to car and truck traffic. A minimum earth cover of 24 in. (610 mm) is required for any underground cable or raceway wiring that is installed under vehicle traffic, regardless of concrete encasement or any other protective measure. This requires the minimum 2-ft (610 mm) earth cover for wiring under the designated areas—including driveways and parking areas of private residences. The minimum earth cover for cables and raceways under driveways and parking areas for one- and two-family dwellings is only 18 in. (457 mm).

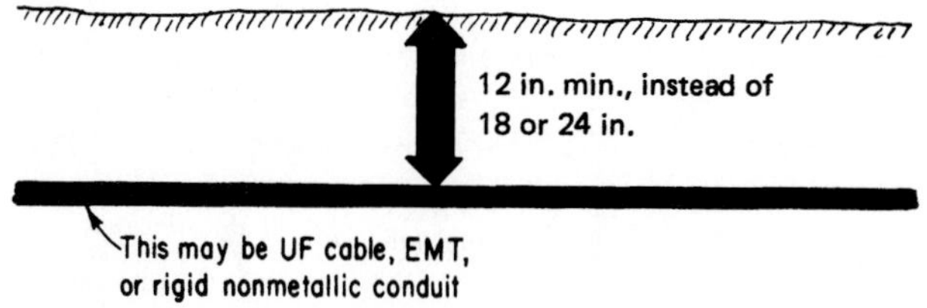

Fig. 300-14. This is OK only for a residential branch circuit rated not over 20 A and protected by a GFCI circuit breaker. (Sec. 300-5.)

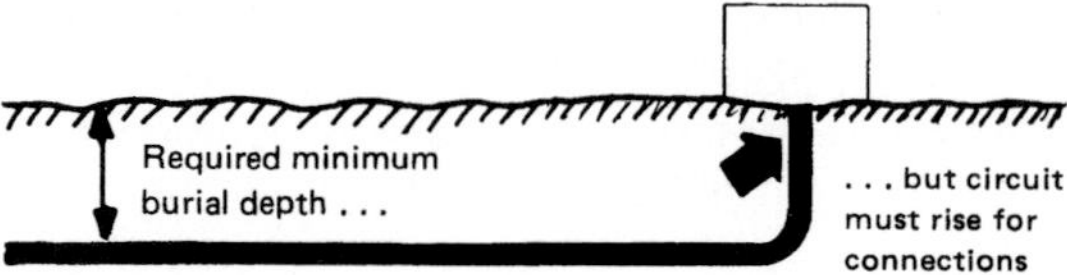

Of course, lesser depths than shown in Table 300-5 are permitted where cable or conductors in raceway come up to terminations or splices in boxes or equipment. NOTE 3.

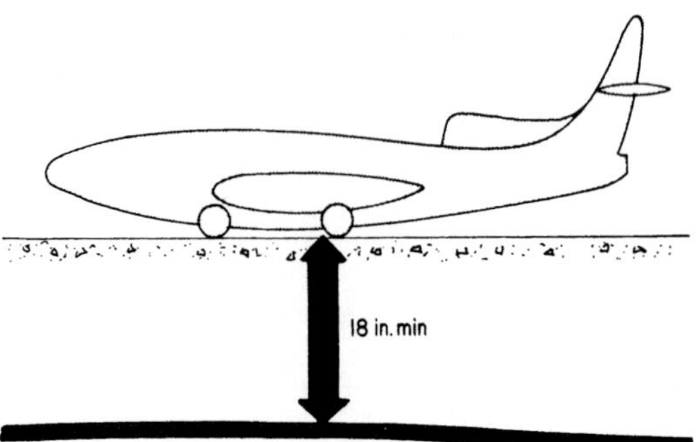

Any cables or raceways may be buried not less than 18 in. deep under airport runways and adjacent defined areas where trespass is prohibited.

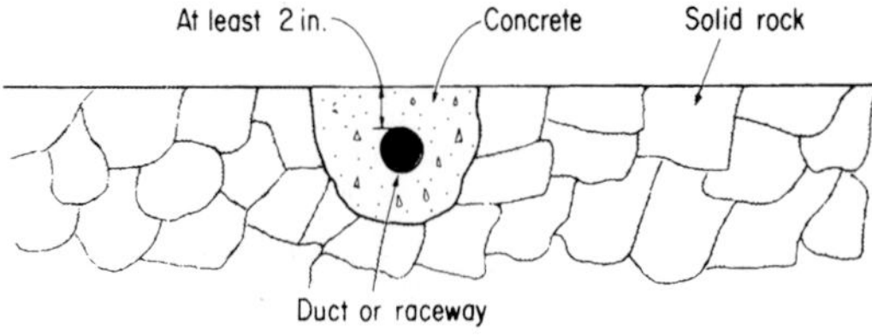

Duct and raceway installed in solid rock may be buried if concrete at least 2 in. thick covers the raceway and extends down to the rock surface.

Fig. 300-15. These applications are also covered in the table burial depths. (Sec. 300-5.)

Table 300-5 (second vertical column from right) gives limited use of lesser burial depth for the residential circuits described, as shown in Fig. 300-14. Any GFCI-protected residential "*branch circuit*" not over 120 V and protected at 20 A or less may be buried only 12 in. (305 mm) below grade, instead of, say, 24 in. (610 mm) as required for Type UF cable for any nonresidential use or for a residential "feeder."

Figure 300-15 shows three other special conditions for burial depth.

Table 300-5 (vertical column at right) recognizes reduced burial depth for low-voltage landscape lighting circuits and supply circuits to lawn sprinkler and irrigation valves, as shown in Fig. 300-16. This recognizes the reduced hazards and safety considerations for circuits operating at not more than 30 V.

Part **(b)** of Sec. 300-5 requires NEC grounding and bonding for all underground circuits and equipment. For instance, metal conduit and metal sheath or electrostatic shielding must be effectively grounded at all terminations by connection to grounded metal enclosure, by bonding jumper, etc., to limit voltage to ground and facilitate operation of overcurrent protective devices, as shown in Fig. 300-17.

Figure 300-18 shows the rule of part **(c)**.

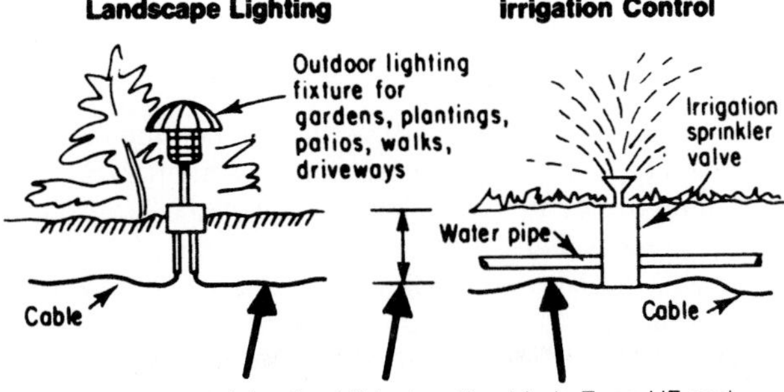

Fig. 300-16. Reduced burial depth for low-voltage landscape lighting and lawn-sprinkler controls. (Sec. 300-5.)

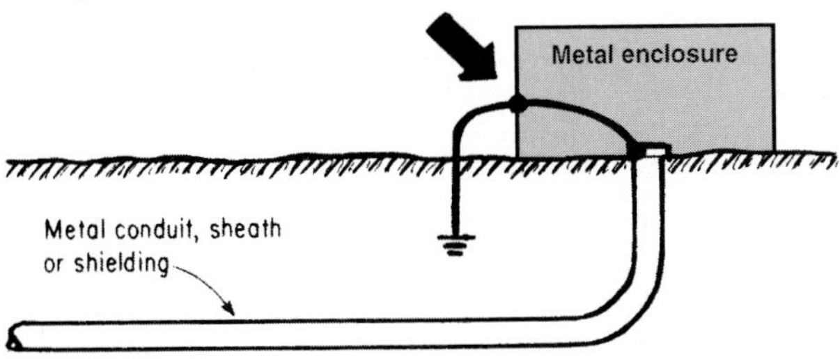

Fig. 300-17. Metal of wiring system must be grounded at *all* terminations. (Sec. 300-5.)

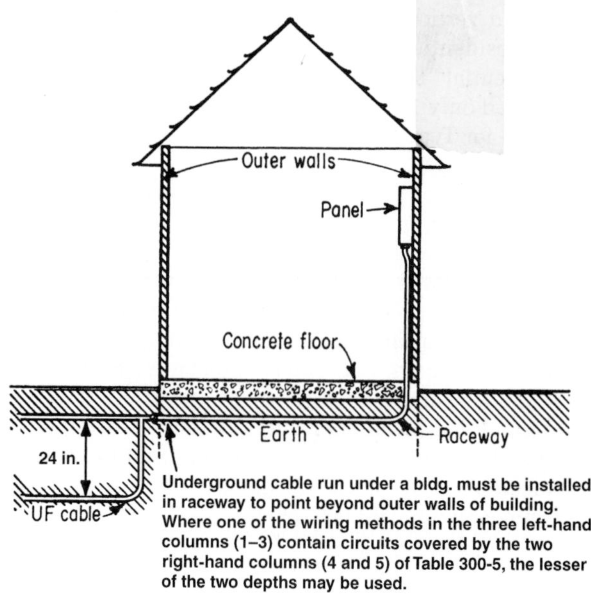
Outer walls
Panel
Concrete floor
Earth
Raceway
24 in.
UF cable
Underground cable run under a bldg. must be installed in raceway to point beyond outer walls of building. Where one of the wiring methods in the three left-hand columns (1–3) contain circuits covered by the two right-hand columns (4 and 5) of Table 300-5, the lesser of the two depths may be used.

Fig. 300-18. Burial of cable in earth is not permitted under a building. (Sec. 300-5.)

As shown at the top of Fig. 300-19, direct buried conductors or cables coming up a pole or on a building from underground installation must be protected from the minimum required burial distance below grade (from Table 300-5 but never required to be more than 18 in. (457 mm) into the ground) to at least 8 ft (2.44 m) above grade, as required by part **(d)** of this section. Where exposed to physical damage, raceways on buildings and raceways on poles must be rigid conduit, IMC, PVC Schedule 80, or equivalent, and the raceway or other enclosure for underground conductors must extend from below the ground line up to 8 ft (2.44 m) above finished grade. If a raceway on a building or on a pole is not subject to physical damage, EMT or Schedule 40 PVC may be used instead of other raceways.

Figures 300-20 and 300-21 show other rules of Sec. 300-5. Note that part **(f)** specifically requires that backfilled trenches must contain any necessary protection for raceways or cables buried in the trench. It specifies that sand or suitable running boards of wood or concrete or other protection must be afforded in those cases where backfill consists of heavy stones or sharp objects that otherwise would present the possibility of damage to the cable or raceway.

Part **(i)** of this section requires that an underground circuit made up of single-conductor cables for direct burial must have all conductors of the circuit run in the same trench. That rule raises the question: When an underground direct burial circuit is made up of conductors in multiple, must all the conductors be installed in the same trench? And if they are, is derating required for more than three current-carrying conductors in a trench, just as it would be for more than three conductors in a single raceway? The answer to both questions is "yes."

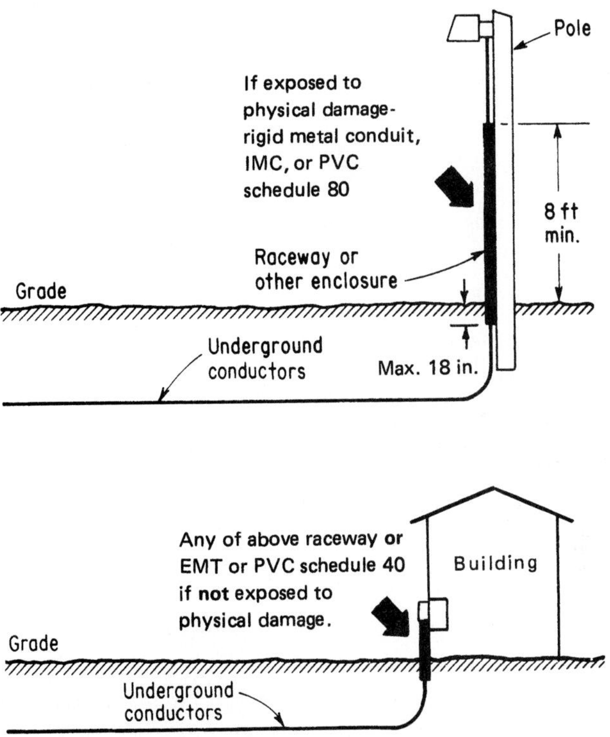

Fig. 300-19. Conductors from underground must be protected. (Sec. 300-5.)

The wording of the rule in part **(i)** clearly indicates that *all* the conductors making up a direct burial circuit of single conductors in parallel must be run in the same trench and *must* be "in close proximity." The wording of Sec. 300-5(i) also requires that all conductors of a circuit be run in the same raceway if a raceway is used (with building wire suitable for wet locations, such as THW). But, Exception No. 1 permits parallel conductor makeup in multiple raceways—with each raceway containing all hot, grounded, and (if used) grounding conductors of the circuit.

When multiple-conductor makeup of a circuit is installed with all the parallel-circuit conductors in the same trench, it is necessary to observe the rule of the second paragraph of Note 8 to Tables 310-16 to 310-19 of the NE Code. It says:

> Where single conductors . . . are stacked or bundled longer than 24 in. (610 mm) without maintaining spacing and are not installed in raceways, the ampacity of each conductor *shall* be reduced as shown in the above table.

That means that the same deratings must be made as when more than three conductors are used in a single conduit—as explained in Note 8. Certainly, direct burial single conductors are covered by that requirement because Table 310-16 specifically covers direct burial conductors.

Those Code rules often make for tricky and troublesome applications. For instance, as shown in Fig. 300-22, an underground circuit of Type USE insu-

Part (e)

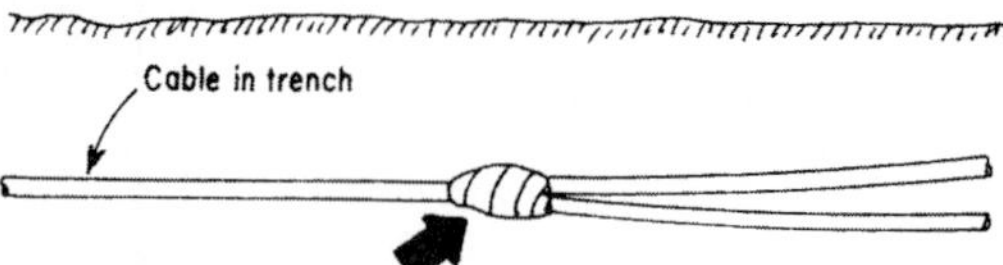

Splices or taps are permitted in trench without a box—but only if listed methods and materials are used.

Part (f)

Backfill of heavy rocks or sharp or corrosive materials must not be used if it may cause damage or prevent adaquate compaction of ground.

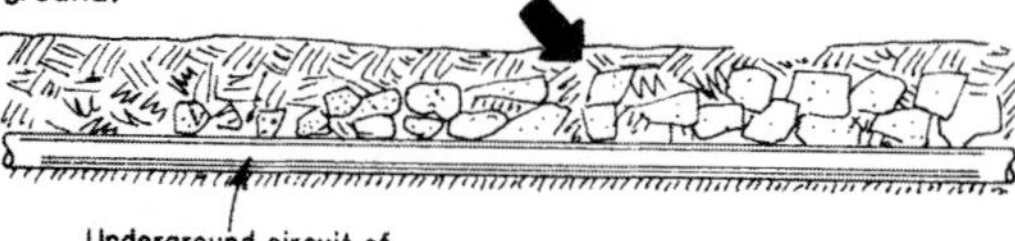

Part (g)

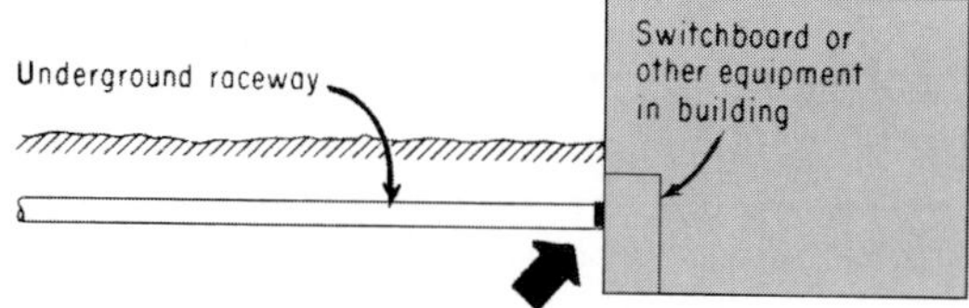

Conduits or other raceways must be sealed or plugged at either or both ends if moisture could contact live parts

Part (h)

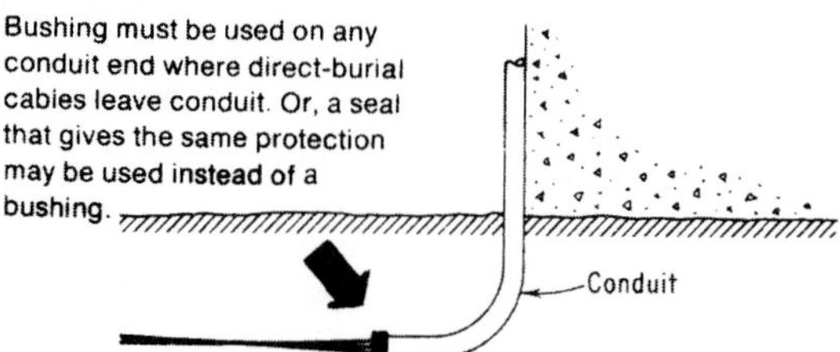

Bushing must be used on any conduit end where direct-burial cables leave conduit. Or, a seal that gives the same protection may be used instead of a bushing.
Conduit

Fig. 300-20. Underground wiring must satisfy these requirements. (Sec. 300-5.)

lated aluminum conductors might be used for a 3-wire, single-phase service to a multifamily dwelling. Because that is a residential service, Note 3 to Tables 310-16 to 310-19 may be observed to gain a higher-than-normal ampacity for the conductors.

Assuming that a 400-A conductor ampacity is indicated by the calculated demand load from Art. 220, each phase leg of the service feeder must have an

Fig. 300-21. For direct burial underground conductors, a box must be used at splice points, with conductors brought up in sweep ells and the box properly grounded—*unless listed* materials are used to make directly buried splices in the conductors in accordance with Sec. 110-14(b). (Sec. 300-5.)

ampacity of 400 A. Refer to Note 3 of the ampacity tables: a No. 4/0 THW aluminum has an ampacity of 200 A. Two such conductors per hot leg and two for the neutral would give the required 400-A capacity for the service.

But how should the parallel circuit be run?

All the circuit conductors *must* be run in close proximity in the same trench, as required by Sec. 300-5(i). That means all six USE conductors are in the same trench; and because the neutrals do not count as current-carrying conductors, the derating of these "bundled" conductors must be to 80 percent of the 200-A ampacity—as required for four conductors in the table of Note 8. With each 4/0 THW aluminum now derated to 160 A (0.8 × 200), the ampacity of each hot leg is only 320 A (2 × 160).

The rule requiring all conductors to be in the same trench makes the circuit of two 4/0 THW aluminum per leg inadequate. In referring to Table 310-16, it now becomes necessary to pick a larger size of THW aluminum—such that, derated to 80 percent, two of them will provide the required 400-A rating. A 350-kcmil THW aluminum has a normal rating of 250 A. Derated to 80 percent (250 × 0.8), it has the needed ampacity of 200 A, so that two of them in parallel per hot leg and neutral will have the ampacity of 400 A.

If the two parallel sets of conductors could have been run in separate trenches, the 4/0 THW aluminum conductors would have met the need.

Exception No. 1 of part **(i)** permits an underground circuit to be made up in parallel in two or more raceways—*without need for derating.* But, in such cases, each raceway must contain one of each of the phase legs, a neutral (if used) and an equipment grounding conductor (if used). With "A-B-C-N" in each raceway of a multiple group, that would be the same type of multiple-conduit-parallel-conductor makeup as required and commonly used for above-ground circuits.

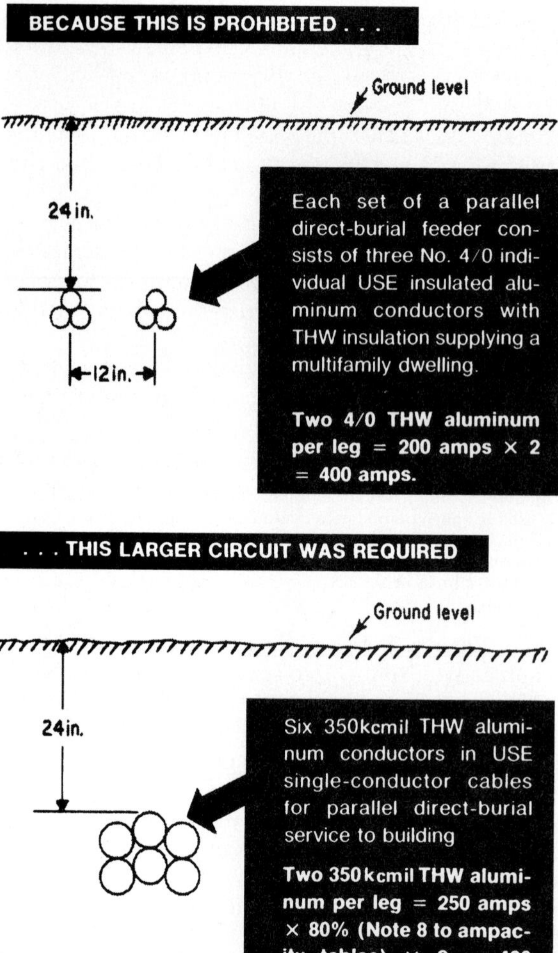

Fig. 300-22. Literal application of Code rules often imposes stiff requirements. (Sec. 300-5.)

Exception No. 2 permits "isolated-phase" makeup of underground circuits in multiple conduits—all phase "A" conductors in one conduit, all phase "B" conductors in a second conduit, all phase "C" conductors in a third conduit, all neutrals in a fourth conduit—with an equipment grounding conductor (or conductors, if needed) installed in a fifth conduit or installed in each of the three conduits carrying the phase conductors. *But,* that makeup is permitted only where the conduits are nonmetallic and are "in close proximity" to each other. See Sec. 300-20.

Part **(j)**, which is new in the 1996 Code, now makes it mandatory that frost heave and settling be accommodated. In those areas of the country that experience nature's cyclical freeze and thaw, the concern for frost damage to buried raceways and cables has generally been addressed either by installation below the frost line or expansion fittings and direct buried cables with slack. This was done to ensure that the raceway remains intact and its contained circuits remain operational. The Code now mandates the implementation of those materials and methods that will ensure the underground raceways and cables are *not* damaged by frost heave.

300-6. Protection Against Corrosion. These are general regulations that are repeated in more detail in the various articles covering raceways and enclosures. The last sentence in part **(a)** allows organic coatings to be applied to metallic boxes or cabinets to prevent corrosion when used outdoors, in lieu of the standard "4-dip" zinc galvanizing method.

Part **(b)** is a general rule that is best understood when related to the specific recommendations given in the UL Green Book for the various types of raceways. See Arts. 345, 346, and 348 for such data.

Figure 300-23 shows the *right* and *wrong* ways of installing equipment in indoor wet locations—as covered in part **(c)** of this section.

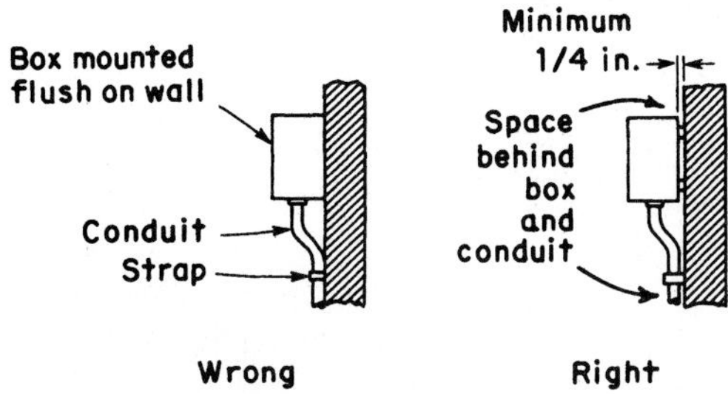

Fig. 300-23. Water or moisture must not be trapped in contact with metal. (Sec. 300-6.)

300-7. Raceways Exposed to Different Temperatures. Part **(a)** requires protection against moisture accumulation. If air is allowed to circulate from the warmer to the colder section of the raceway, moisture in the warm air will condense in the cold section of the raceway. This can usually be eliminated by sealing the raceway just outside the cold rooms so as to prevent the circulation of air. Sealing may be accomplished by stuffing a suitable compound in the end of the pipe (Fig. 300-24).

The FPN directs the reader to data regarding the use of expansion fittings for PVC conduit and provides guidance applying that data to IMC, EMT, and rigid steel conduit.

300-8. Installation of Conductors with Other Systems. Any raceway or cable tray that contains electric conductors must not contain "any pipe, tube, or equal for steam, water, air, gas, drainage or any service other than electrical."

300-10. Electrical Continuity of Metal Raceways and Enclosures. This is the basic rule requiring a permanent and continuous bonding together (i.e., connecting together) of all noncurrent-carrying metal parts of equipment enclosures—conduit, boxes, cabinets, enclosures, housings, frame of motors, and lighting

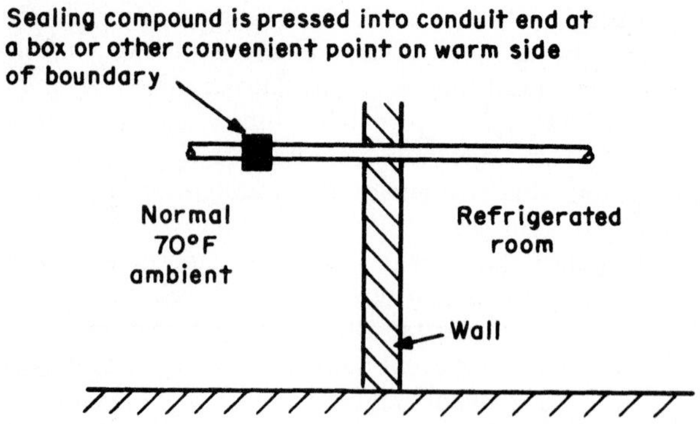

Sealing compound is pressed into conduit end at a box or other convenient point on warm side of boundary

Normal 70°F ambient

Refrigerated room

Wall

Fig. 300-24. Sealing protects against moisture accumulation in raceway. (Sec. 300-7.)

fixtures—and connection of this interconnected system of enclosures to the system grounding electrode at the service or transformer (Fig. 300-25). The interconnection of all metal enclosures must be made to connect all metal to the grounding electrode and to provide a low-impedance path for fault-current flow along the enclosures to assure operation of overcurrent devices which will

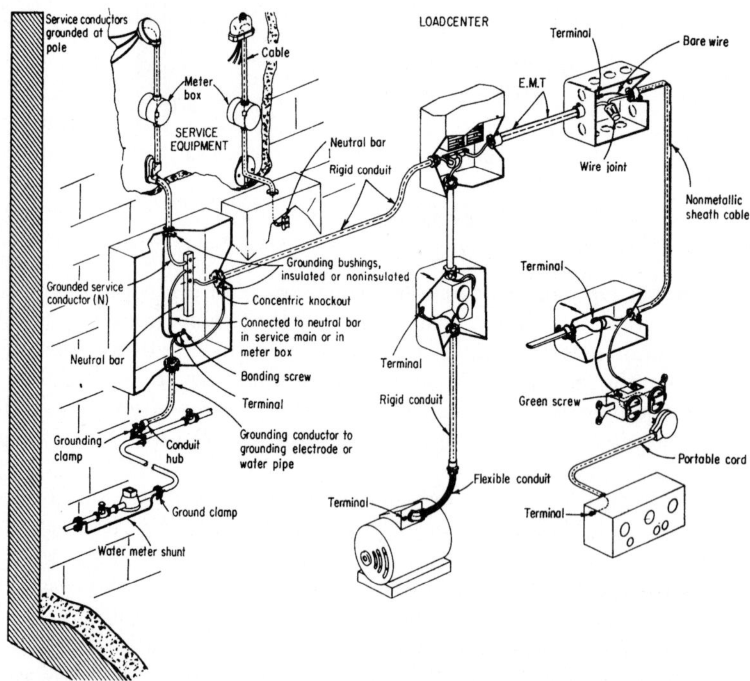

Fig. 300-25. All metal enclosures must be interconnected to form "a continuous electric conductor." (Sec. 300-10.)

open a circuit in the event of a fault. By opening a faulted circuit, the system prevents dangerous voltages from being present on equipment enclosures which could be touched by personnel, with consequent electric shock to such personnel.

Simply stated, this interconnection of all metal enclosures of electric wires and equipment prevents any potential aboveground on the enclosures. Such bonding together and grounding of all metal enclosures are required for both grounded electrical systems (those systems in which one of the circuit conductors is intentionally grounded) and ungrounded electrical systems (systems with none of the circuit wires intentionally grounded).

But effective equipment interconnection and grounding are extremely important for grounded electrical systems to provide the automatic fault-clearing which is one of the important advantages of grounded electrical systems. A low-impedance path for fault current is necessary to permit enough current to flow to operate the fuses or CB protecting the circuit.

300-11. Securing and Supporting. The basic rule in part **(a)** **(1)** and **(2)** does *not* permit "branch-circuit wiring" to be supported by the suspended ceiling support wires—regardless of whether or not the ceiling is fire-rated or nonfire-rated. The wording used in the Exceptions following each part gives limited acceptance to attaching wiring methods to the ceiling support wires of a hung ceiling that is essentially listed for such support.

This rule severely limits the use of ceiling support wires as a means of support for "wiring." As indicated by the Exception that follows both parts **(1)** and **(2)**, *only* ceilings *listed* for the support of "wiring and equipment" may do so. The support of wiring from additional support wires—that is, ones that are installed just to support wiring—is not addressed in the wording of this rule but may be acceptable. Check with the local inspection authority for guidance.

In part **(b)**, raceways are prohibited from being used as a means of support for cables or nonelectrical equipment. Telephone or other communication, signal, or control cables must not be fastened to electrical conduits—such as by plastic straps or any other means.

Although raceways must not be used as a means of support for other raceways, cables, or nonelectric equipment, Exception No. 1 permits large conduits with hanger bars or fittings intended to support smaller raceways. Exception No. 2 permits such applications as tying Class 2 thermostat cable to a conduit carrying power-supply conductors for electrically controlled heating and air-conditioning equipment that is controlled by the Class 2 wires. And, Exception No. 3 applies to those instances where raceways are permitted to support boxes or conduit bodies (Sec. 370-23) or fixtures [Sec. 410-16(f)].

300-13. Mechanical and Electrical Continuity—Conductors. Part **(b)** prohibits dependency upon device terminals (such as internally connected screw terminals of duplex receptacles) for the splicing of neutral conductors in multiwire (3-wire or 4-wire) circuits. *Grounded neutral wires* must not depend on device connection (such as the break-off tab between duplex receptacle screw terminals) for continuity. White wires can be spliced together with a pigtail to neutral terminal on receptacle. If receptacle is removed, neutral will not be opened (Fig. 300-26).

Do it this way...

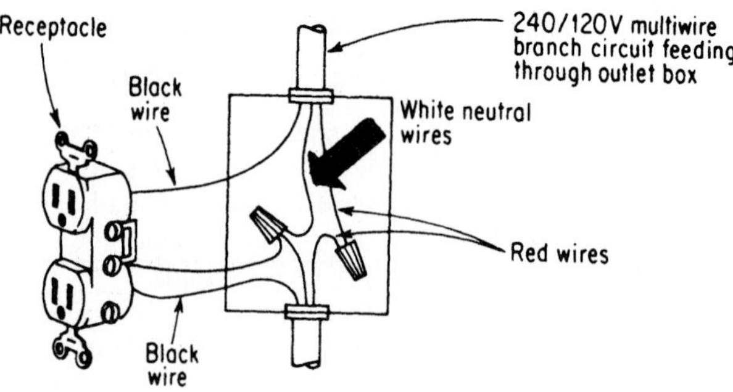

...or this way

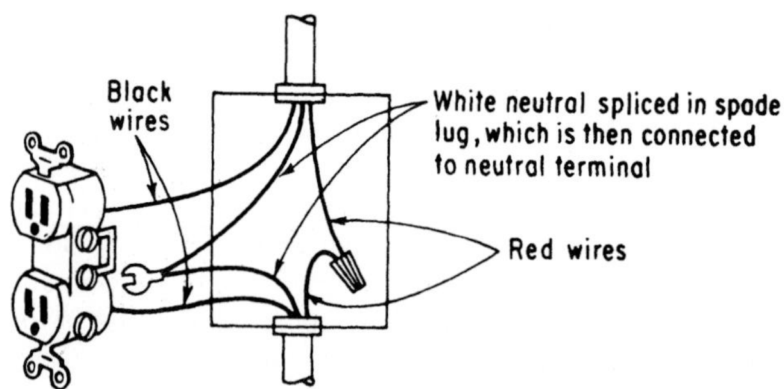

Fig. 300-26. Neutrals of multiwire circuit must *not* be spliced at receptacle terminals. (Sec. 300-13.)

This rule is to prevent the establishment of unbalanced voltages should a neutral conductor be opened *first* when a receptacle or similar device is replaced on energized circuits. In such cases, the line-to-neutral connections downstream from this point (farther from the point of supply) could result in a considerably higher-than-normal voltage on one part of a multiwire circuit and damage equipment, because of the "open" neutral, if the downstream line-to-neutral loads are appreciably unbalanced.

Note that this paragraph does not apply to 2-wire circuits or circuits which do not have a grounded conductor. This rule applies only where multiwire circuits feed receptacles or lampholders. This would most commonly be a 3-wire 240/120-V or a 3- or 4-wire 208/120-V or even a 480/277-V branch circuit.

The reason for the pigtailing requirement is to prevent the neutral conductor from being broken and creating downstream hazards. The problem lies in the inclination of electricians to work on hot circuits. Assume that a duplex receptacle on a 240/120-V 3-wire circuit becomes defective, and the first thing the electrician does, working hot, is to disconnect the neutral wires from the receptacle. Downstream, 2.4- and 12-A loads have been operating (plugged into addi-

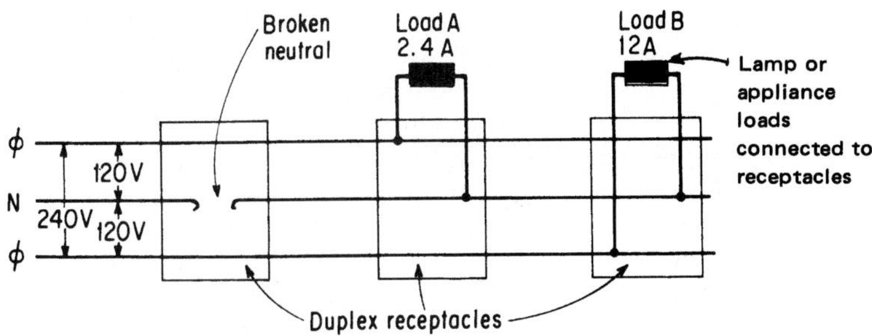

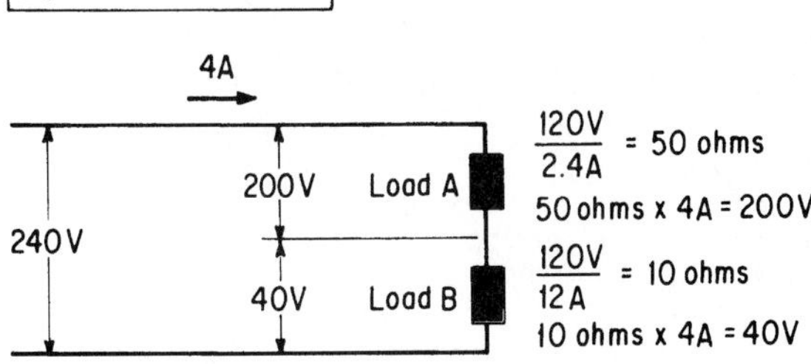

Fig. 300-27. Splicing neutrals on receptacle screws causes "open" in neutral if receptacle is removed. (Sec. 300-13.)

tional receptacles on the multiwire circuit), each connected to a different hot leg. When the neutral is broken by the electrician upstream, normal operation of the loads reverts to the condition shown in Fig. 300-27. The two loads are now in series across 240 V. As shown, load A now has 200 V impressed across it. It could run extremely hot and burn out. Load B now has only 40 V across it; if it is a motor-operated device, the low voltage could cause the motor to burn up. Both could cause injuries.

Also, in disconnecting the neutral, the electrician could get a 120-V shock if both the disconnected neutral conductor going downstream and the box were touched—not unlikely, since the neutral is usually considered to be dead—that is, at ground potential.

300.14. Length of Free Conductors at Outlets, Junctions, and Switch Points. The rule here applies only to the length of the conductor at its end. The Exception covers wires running through the box. Wires looping through the box and intended for connection to outlets at the box need have only sufficient slack that any connections can be made easily.

300.15. Boxes, Conduit Bodies, or Fittings—Where Required. Part **(a)** permits *either* a "box" or a "conduit body" to be used at splice points or connection points in *raceway* systems. Type T or Type L conduit bodies actually become a part of the conduit or tubing and should not contain more conductors than permitted for the raceway. Conduit bodies must not contain splices, taps, or devices unless they comply with the rules of Sec. 370-16(c). For conductors No. 4 or larger see Sec. 370-28(a). Use of boxes and fittings for splicing, for connections to switches or outlet wiring devices, or for pulling must conform to the many detailed rules of Art. 370. Refer to those rules for further discussion.

Exception No. 2 permits splices to be made within lighting fixture wiring compartments where the branch-circuit wires are spliced to fixture or ballast wires.

Part **(b)** accepts *only* a box for splices and connections to devices when the wiring system is "*cable*" instead of raceway (Fig. 300-28).

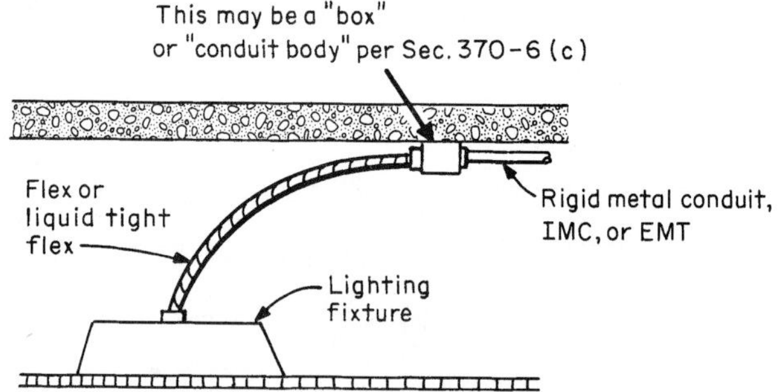

But with type AC cable (BX) Type NM cable, or any other cable, a "box" must be used at all splice and outlet points, and even where cable connects to raceway

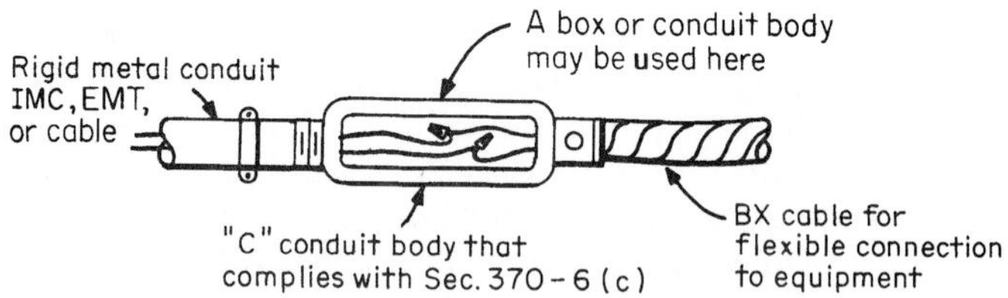

Fig. 300-28. Raceways and cables may use boxes or conduit bodies at conductor splice points. (Sec. 300-15.)

Exception No. 4 of this section recognizes use of wiring devices that have "integral enclosures." These are the so-called boxless devices made and acceptable for use in nonmetallic-sheathed cable systems (Type NM). Such listed devices do not require a separate box at each outlet because the construction of the device forms an integral box in itself.

Exception No. 5 recognizes the use of manufactured metallic wiring systems—"prefab" or "modular wiring" systems used for distribution of lighting and communications in the space above suspended ceilings. Such UL-listed systems are covered in Art. 604 and are designed with their own components for connections of their cable "whips." Therefore, they are exempted from the usual rule on boxes at splices, junctions, etc.

Exception No. 6 permits use of a "conduit body" at splice, pull, or connection points in cable systems. Of course, where conduit bodies are used in cable runs, they will have to be supported by straps on their hubs or by some other means. When used with rigid metal conduit or IMC, the conduit itself is adequate support for the conduit body if the conduit is clamped within 3 ft (914 mm) on two or more sides of the conduit body.

Exception No. 8 recognizes transition from Type AC cable to raceway without the need for a box, provided that no splice or termination is made in the conductors. This permits the common practice of changing from, say, BX to EMT for a run down a wall, with the armor stripped from a long length of the BX and the exposed wires run in the EMT. A suitable fitting made for connecting BX to EMT must be used (Fig. 300-29).

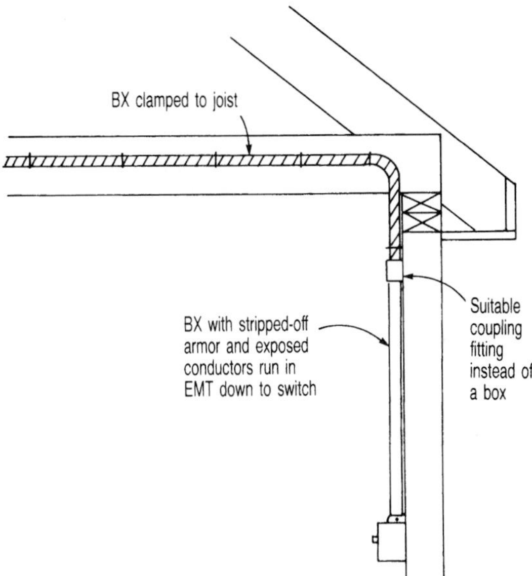

Fig. 300-29. This type of no-box connection for cable-to-raceway change is permitted by Exception No. 9. [Sec. 300-15(b).]

Exception No. 9 correlates with the rules of Secs. 300-5(e) and 110-14(b), which permit splicing of underground direct burial cables and conductors using listed splice kits without a box.

Part **(c)** of Sec. 300-15 was proposed to address the almost universal misuse of Type NM (Romex) connectors with other cables and even cords. The wording as it appears in the **Code**, however, requires that *any* fitting or connector must be "designed and listed" for the wiring method used. Be aware that, while it will be up to the manufacturers of this equipment to obtain the listing for these products, it is the designer-installer's responsibility to specify or use ONLY those fittings and connectors specifically listed for the application.

300-16. Raceway or Cable to Open or Concealed Wiring. Where the wires are run in conduit, tubing, metal raceway, or armored cable and are brought out for connection to open wiring or concealed knob-and-tube work, a fitting such as is shown in Fig. 300-30 may be used.

Where the terminal fitting is an accessible outlet box, the installation may be made as shown in Fig. 300-30.

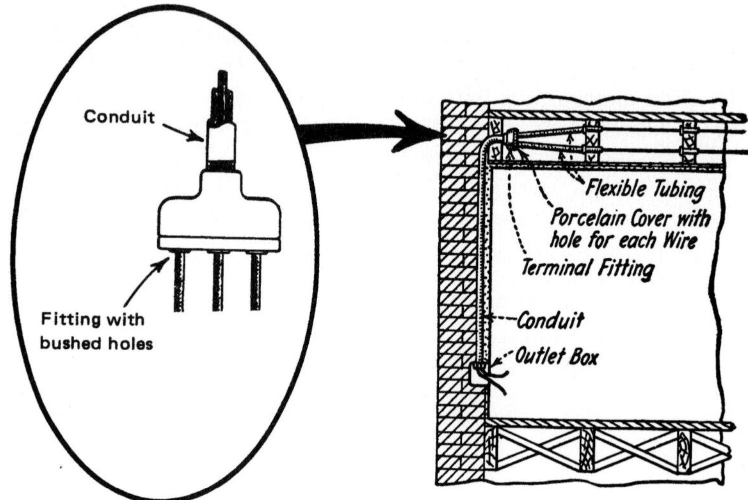

A terminal fitting satisfies the rule on transition from raceway to knob and tube.

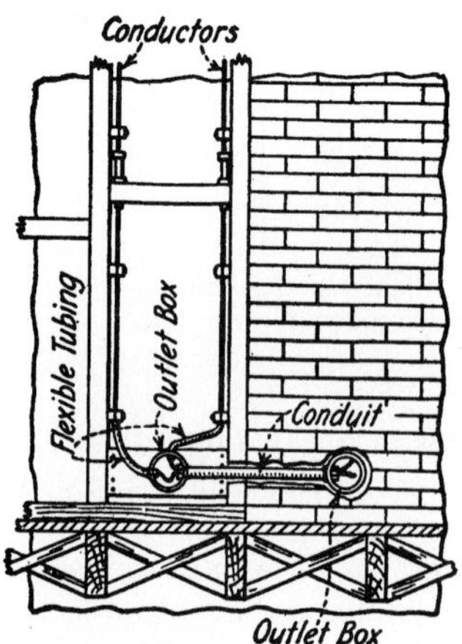

An outlet box may be used where raceway connects to open wiring or concealed knob and tube.

Fig. 300-30. These are techniques for connecting conduit to open wiring. (Sec. 300-16.)

Fig. 300-31. Conductors shown here have been pulled into the conduit before boxes and continuation of the raceway system were installed to supply underground circuits to outdoor building lighting. Under previous Code editions, this was a violation of Sec. 300-18. That section and its rules were removed after the 1981 NEC.

Figure 300-31 shows wires pulled into an incomplete raceway system. That used to be a Code violation, but the rule prohibiting it was removed.

300-18. Raceway Installations. This section restores a general Code concept that was in the 1981 and previous Codes (it was Sec. 300-18), requiring raceways to be installed as a complete system, with associated boxes and other enclosures, before pulling conductors into the raceway and box system. This rule is reinserted in the NEC because of reports of damage to conductors being pulled into incomplete raceway systems. The exceptions to the rule are intended to permit wiring of motors and fixture whips after the basic raceway system has been wired, as well as covering prewired assemblies.

300-19. Supporting Conductors in Vertical Raceways. Long vertical runs of conductors should not be supported by the terminal to which they are connected. Supports as shown in Fig. 300-32 may be used to comply with Sec. 300-19(a). (See also Figs. 300-33 and 300-34.)

300-20. Induced Currents in Metal Enclosures or Metal Raceways. By keeping all conductors of an AC circuit close together—in a raceway or a box or other enclosure—the magnetic fields around the conductors tend to oppose or cancel each other, thereby minimizing the inductive reactance of the circuit and also minimizing the amount of magnetic flux that can cause heat due to hysteresis loss (magnetic friction) in steel or iron and due to the I^2R losses of currents that are induced in adjacent metal. The rule of this section calls for always running a neutral conductor with the phase legs of an AC circuit to minimize such induction heating. The equipment grounding conductor must also be run close to the circuit conductors to achieve the reduction in inductive reactance and

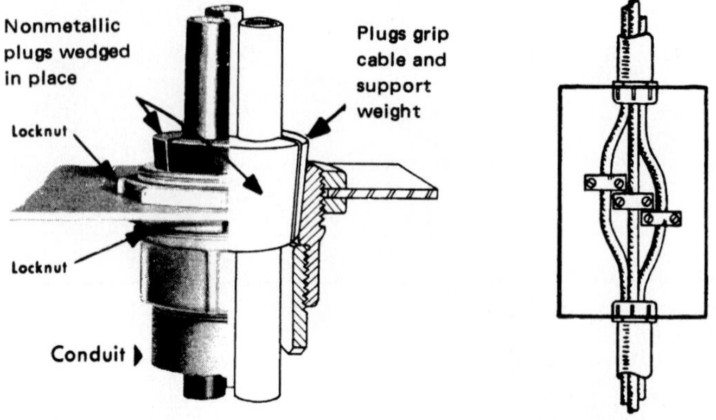

Conductor-support bushing screwed on end of conduit at a cabinet, pull box, or conductor-support box. (*Russell & Stoll.*)

Conductor-support box with single-wire cleats to clamp conductors.

Fig. 300-32. Some type of support must carry the weight of conductors in long risers. (Sec. 300-19.)

minimize the impedance of the fault-current return path when a fault does occur—thereby assuring the fastest possible operation of the protective device (fuse or CB) in the circuit. The reference to "all" grounding conductors in the basic rule is aimed at the so-called isolated grounding conductors, which must also be run with the circuit conductors from the outlet to the panel where that circuit originates. Where the isolated ground continues through that panel, it should be run with the feeder that supplies that panel, although *that* is not entirely clear (Fig. 300-35).

When an AC circuit is arranged in such a way that the individual conductors are not physically close for mutual cancellation of their field flux, it is particularly important to take precautions where a single conductor passes through a hole in any magnetic material—like a steel enclosure surface. The presence of the magnetic material forms a closed (circular) magnetic core that raises the flux density of the magnetic field around the conductor (that is, it greatly strengthens the magnetic field). Under such conditions, there can be substantial heating in the enclosure due to hysteresis (friction produced by the alternating reversals of the magnetic domains in the steel) and due to currents induced in the steel by the strong magnetic field. To minimize those effects, the second paragraph of Sec. 300-20 requires special treatment, such as that shown in Fig. 300-36. Or a rigid, nonmetallic board (fiberglass, plastic, etc.) should be used for the enclosure wall that the conductors pass through.

300-21. Spread of Fire or Products of Combustion. Application of this section to all kinds of building constructions is a very broad and expanding controversy in modern electrical work, in particular because of the phrase "substantially increased." The rule here requires that electrical installations shall be made to substantially protect the integrity of rated fire walls, fire-resistant or fire-

Fig. 300-33. Bore-hole cable, with steel wire armor, is permitted by the Exception to be supported only at the top of very high risers because the steel armor supports the length of the cable when the steel wires are properly clamped in the support ring of the type of fitting shown here. (Sec. 300-19.)

stopped walls, partitions, ceilings, and floors. Electrical installations must be so made that the possible spread of fire through hollow spaces, vertical shafts, and ventilating or air-handling ducts will be reduced to a minimum. These rules require close cooperation with building officials to avoid destruction of fire ratings when electrical installations extend through such areas.

Floor Penetrations

Certainly, the electrical industry has come to agree that poke-through wiring—that technique in which floor outlets in commercial buildings are wired through holes in concrete slab floors—is an acceptable wiring method if use is made of UL-listed poke-through fittings that have been tested and found to pre-

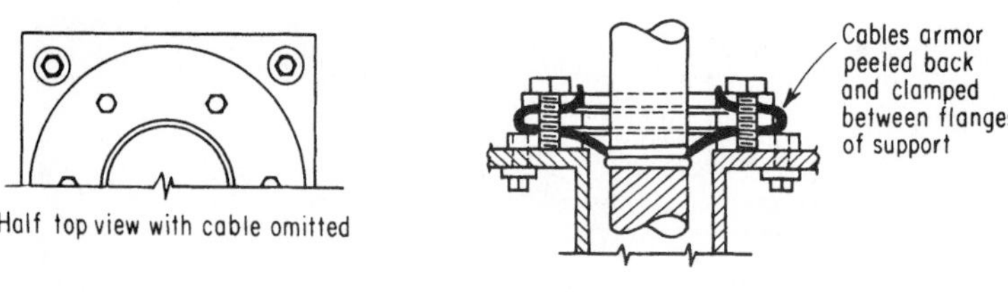
Half top view with cable omitted
Cables armor peeled back and clamped between flanges of support
TYPE FS CABLE SUPPORTS

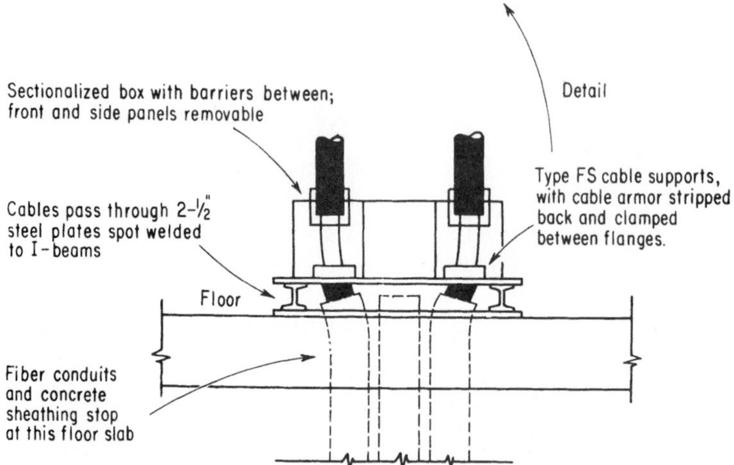
Detail
Sectionalized box with barriers between; front and side panels removable
Type FS cable supports, with cable armor stripped back and clamped between flanges.
Cables pass through 2-½" steel plates spot welded to I-beams
Floor
Fiber conduits and concrete sheathing stop at this floor slab

Fig. 300-34. Separate strands of cable armor are snubbed between flanges of support fitting at top of run. Partitioned enclosure protects unarmored sections of cable. (Sec. 300-19.)

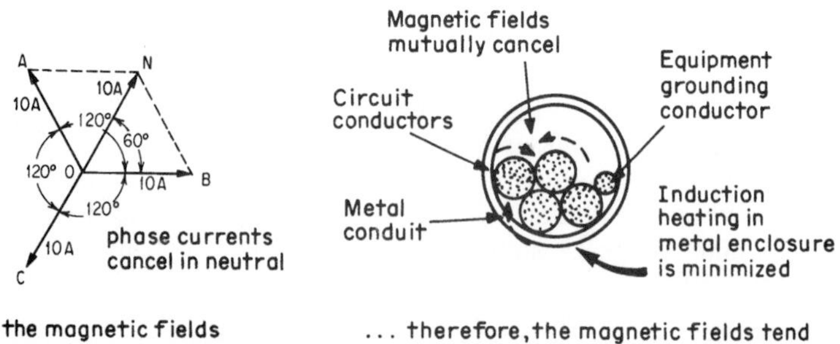
In a typical 3-phase circuit —
A
N
10A
10A
120°
60°
120° O
10A
B
120°
10A
C
phase currents cancel in neutral
Magnetic fields mutually cancel
Equipment grounding conductor
Circuit conductors
Metal conduit
Induction heating in metal enclosure is minimized
the magnetic fields around the conductors have the phase displacements of the currents . . .
. . . therefore, the magnetic fields tend to cancel each other (like the neutral current becomes zero under balanced loading) if the conductors are close together

Fig. 300-35. Close placement of AC conductors minimizes magnetic fields and induction. (Sec. 300-20.)

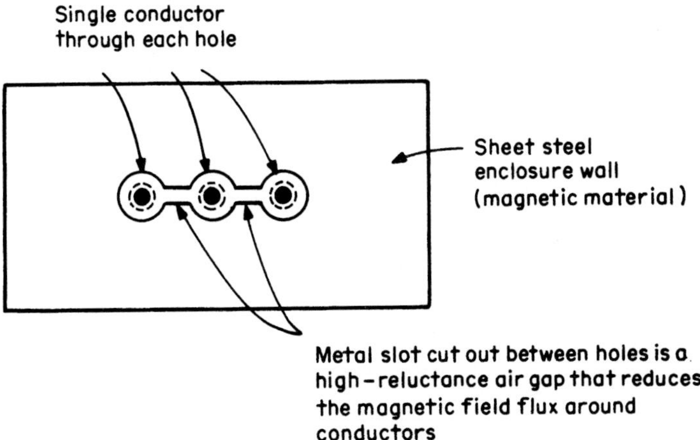
Single conductor through each hole
Sheet steel enclosure wall (magnetic material)
Metal slot cut out between holes is a high-reluctance air gap that reduces the magnetic field flux around conductors

Fig. 300-36. Induction heating is reduced by opening the magnetic core. (Sec. 300-20.)

serve the fire rating of the concrete floor. Throughout the country, the use of poke-through wiring continues to be a popular and very effective method of wiring floor outlets in office areas and other commercial and industrial locations. Holes are cut or drilled in concrete floors at the desired locations of floor outlets, and floor box assemblies are installed and wired from the ceiling space of the floor below. The method permits installation of each and every floor box at the precise location that best serves the layout of desks and other office equipment.

The wiring of each floor outlet at a poke-through location may be done basically in either of two ways—by some job-fabricated assembly of pipe nipples and boxes *or* by means of a manufactured through-floor assembly (Fig. 300-37) made expressly for the purpose and tested and listed by a nationally recognized testing lab, such as UL.

May either of the methods be used? A clear regulation of the Occupational Safety and Health Administration appears to rule decisively on this question. In the *Occupational Safety and Health Standards,* Subpart S—Electrical, paragraph 1910.303 only permits equipment that has been approved. And in paragraph 1910.339, Definitions, "approved" means "acceptable," and under "acceptable" OSHA clearly and flatly *demands* that an installation or equipment determined to be safe by a nationally recognized testing lab must *always* be used in preference to any equipment *not* certified by a testing lab. Thus, if a UL-listed poke-through fitting is available, then the use of any nonlisted, homemade assembly—which has not been determined to be safe—appears to be clearly not acceptable to OSHA and could be construed as a violation of **NE Code** Sec. 110-2, which calls for all equipment to be "approved."

Section 300-21 also applies to cable and/or conduit penetrations of fire-rated walls, floors, or ceilings without altering the fire rating of the structural surface (Figs. 300-38 and 300-39).

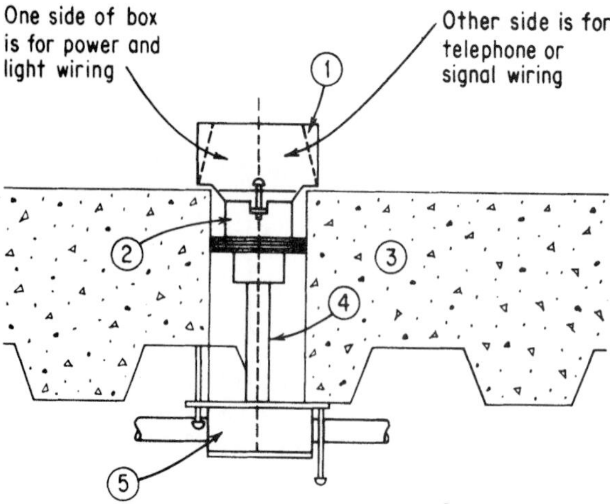
One side of box
is for power and
light wiring
Other side is for
telephone or
signal wiring
1
2
3
4
5

Numbered components include (1) combination floor service box, (2) fire-rated center coupling, (3) concrete slab, (4) barriered extension, and (5) barriered junction box.

UL-listed assembly is fire-rated for thickest concrete slab.

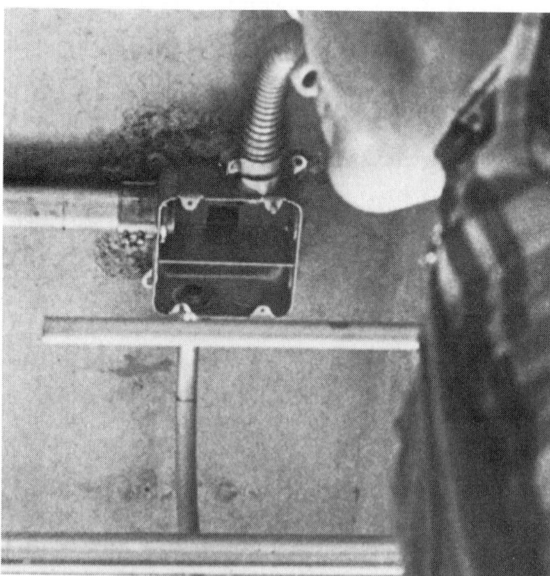

Bottom end of UL-listed poke-through fitting consists of a partitioned box for 120-V circuit and telephone circuit run through vertical channels of fitting into dual-service floor outlet box on top of slab.

Fig. 300-37. Several manufacturers make UL-listed poke-through assemblies. (Sec. 300-21.)

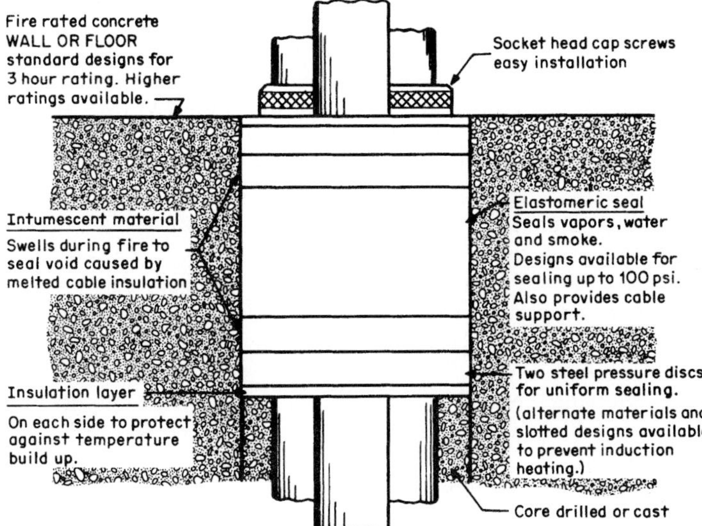
Fire rated concrete
WALL OR FLOOR
standard designs for
3 hour rating. Higher
ratings available.
Socket head cap screws
easy installation
Intumescent material
Swells during fire to
seal void caused by
melted cable insulation
Elastomeric seal
Seals vapors, water
and smoke.
Designs available for
sealing up to 100 psi.
Also provides cable
support.
Insulation layer
On each side to protect
against temperature
build up.
Two steel pressure discs
for uniform sealing.
(alternate materials and
slotted designs available
to prevent induction
heating.)
Core drilled or cast

Fig. 300-38. Fire-stop fitting for passing cables or conduit through a fire-rated wall, floor slab, or similar concrete surface, without altering the fire rating of the surface. (Sec. 300-21.)

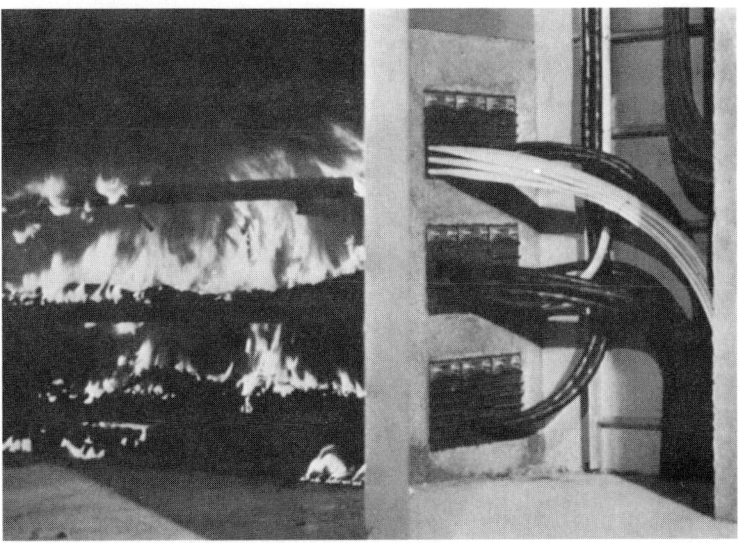

Fig. 300-39. Another type of device to provide for passing cable and/or conduit through fire-rated building surfaces without altering the conditions of fire resistance. (Sec. 300-21.)

Ceiling Penetrations

Another similar concern covered by this section is the installation of lay-in lighting fixtures in a fire-rated suspended ceiling. Suspended ceilings are usually evaluated only for their esthetic and acoustic value, but they also serve as fire-protective membranes for the floor above. Although concrete floor structures have various fire-resistance ratings by themselves (depending on the concrete thickness and aggregate used), some assemblies require some type of protective cover. When this is the case, the ceiling is tested in combination with the floor-slab structure for which the rating is desired.

Such a ceiling properly serves its function of fire protection until an installer cuts holes in it, such as for recessed-lighting fixtures or for air diffusers or grilles. Because of that, the acceptability of the overall ceiling system must be carefully determined.

The FPN following the basic rule provides some guidance, but the following presents a comprehensive step-by-step procedure for using the publications vaguely referenced in the FPN.

First, check the Underwriters Laboratories' *Electrical Construction Materials Directory* (commonly called the UL Green Book), which notes that recessed fixtures that have been shown to provide a degree of fire resistance with the floor, roof, or ceiling assemblies with *which they have been tested* are labeled as follows: "Recessed-type electric fixture classified for fire resistance; fire-resistance classification floor and ceiling Design No. — —."

Next, find the design referred to in the UL *Fire Resistance Index.* This booklet follows the format of the *Electrical Construction Materials Directory.* Refer to a design of the required fire rating and be sure that the fixtures are listed for use with that design.

Designers must specify the particular UL design that suits their requirement, note this in the specifications, and be certain that the lighting fixtures are fire-rated in accordance. But it is advisable for the electrical contractor to investigate the ceiling design for possible fire rating in all cases and to receive from the designer written confirmation of the exact nature and value of the rating, if one exists.

In the UL *Building Materials Directory,* various fire-rated assemblies are listed by "design number" and by "rating time." A companion publication, the *Fire Resistance Index,* contains detailed cross-section drawings of the assemblies, with all critical dimensions shown. Each pertinent element is usually flagged with an identifying number. Keyed to the number are clarifying statements listing additional critical limitations (such as the size and number of penetrations in the ceiling).

The top installation in Fig. 300-40 was tested and given a 1½-hr rating. No protective material was used between the fixture and the floor slab above. A somewhat better rating could have been obtained had protection been provided over the fixture.

At the bottom of Fig. 300-40 is a fixture with protection. When this construction was tested, failure occurred after 2 hr 48 min, and it received a 2-hr time rating. Even with this type of protection, the UL listing will limit the area occu-

Recessed fixture without protective covering (1½-hr rating)

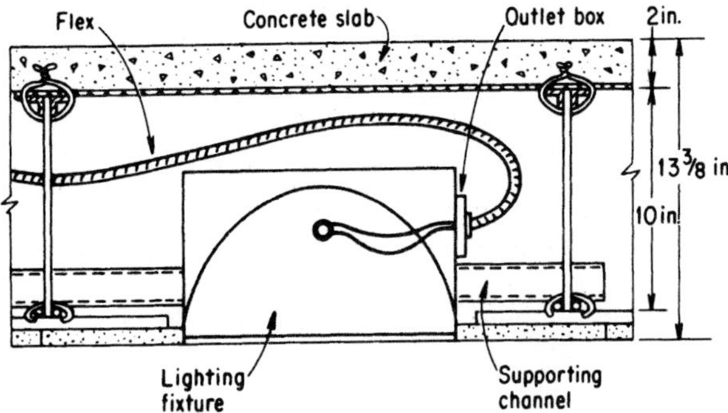

Recessed fixture with box board shell (2-hr rating)

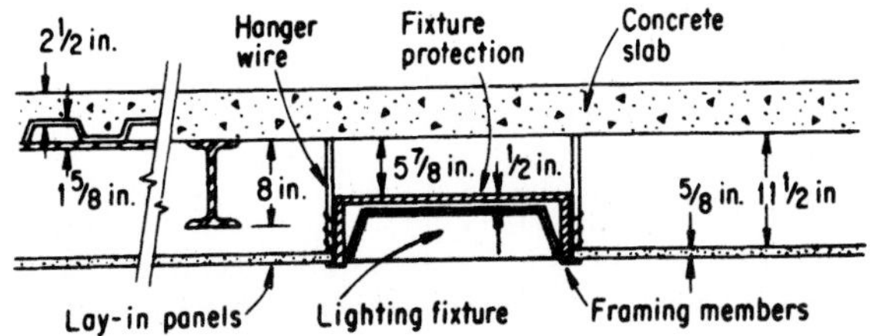

Fig. 300-40. The complete assembly of concrete slab plus fixture and ceiling gets a fire rating. (Sec. 300-21.)

pied by fixtures to 25 percent of the total ceiling area. (But a coffered ceiling may contain 100 percent lighted vaulted modules.)

Other Penetrations

Plasterboard (gypsum board) panels used so commonly for interior wall construction in modern buildings are fire-rated. UL and other labs make tests and assign fire ratings (in hours) to wall assemblies or constructions that make use of plasterboard. For instance, a wall made up of wood or metal studs with a single course of ⅝-in. (15.9-mm) plasterboard on each side of the studs would be assigned a 1-hr fire rating. A wall with two courses of ½- or ⅝-in. (12.7- or 15.9-mm) plasterboard on each side of the studs would be a 2-hr wall (Fig. 300-41). The assigned fire ratings are based on the thickness and number of courses of plasterboard. And the fire rating is for the wall assembly *without any penetrations into the wall.*

Because of the fire rating assigned to the assembly, any wall so constructed is fire-rated. The wall may be between rooms or between a room and a corridor or stairwell. And no distinction is made between an interior wall of an apartment,

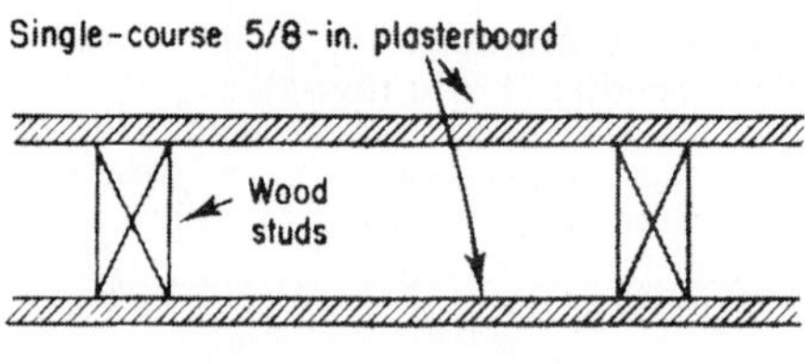

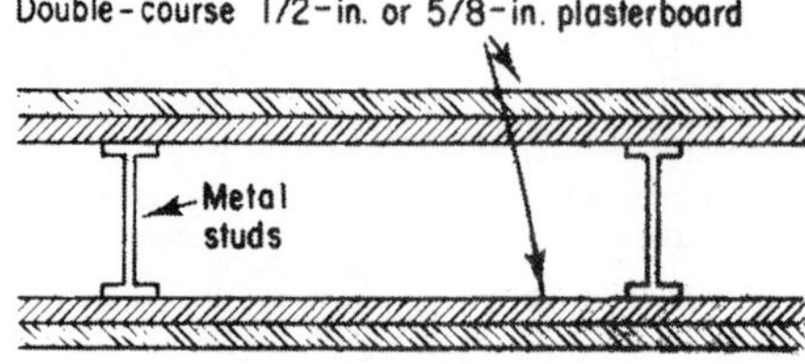

Note: These are only typical assemblies. Carefully determine fire-rating time for specfic walls.

Fig. 300-41. Wall assemblies using plasterboard are fire-rated by UL and others. (Sec. 300-21.)

say, and a wall that separates one apartment from another. All wall assemblies using plasterboard are fire-rated and immediately raise concern over violation of Sec. 300-21 if any electrical equipment is recessed in the wall.

When any electrical equipment is installed as a penetration of a wall, there is the immediate question, Does this substantially increase the possible spread of fire or the spread of products of combustion (smoke and/or heated air)? Building inspectors and electrical inspectors have generally permitted installation of wall switches, thermostats, dimmers, and receptacles in boxes recessed in plasterboard walls. In single-family houses, the entire interior is not considered to be compartmented. It is assumed that individual rooms or areas are not normally closed off from each other and that fire or smoke spread would not be affected at all by those penetrations. The consensus has been that such small openings cut in the plasterboard do not violate the letter or intent of Sec. 300-21, although there is no specific Code rule that exempts any wall from the concern of Sec. 300-21.

In apartment houses, office buildings, and other multioccupancy buildings, however, inspectors could logically question use of wiring devices installed in common walls between apartments or between an apartment and a corridor or stairwell. Such walls are assumed to be between interior spaces that are normally closed off from each other by the main doors to the individual apartments. Fire and/or smoke spread, which is normally restricted by the closed doors, might be considered *substantially increased* by any penetrations of those fire-rated walls (Fig. 300-42). Although switch and receptacle

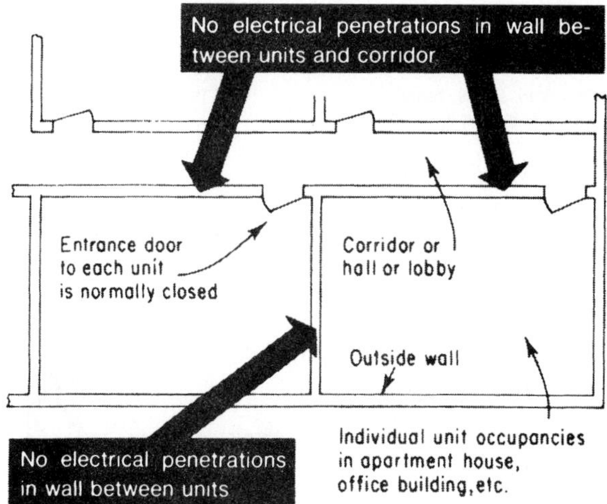

Fig. 300-42. Walls separating closed-off spaces must have maintained fire rating. (Sec. 300-21.)

boxes are usually accepted, use of a panelboard in a common wall has been rejected.

For larger electrical equipment, such as panelboards, the same general analysis would apply. For interior walls of private houses or individual unit occupancies in apartment houses, hotels, dormitories, office buildings, and the like, a panel installed in a wall between two rooms or spaces that are normally *not* closed off from each other cannot "substantially" contribute to greater fire and/or smoke spread. *But,* panelboards and similar large equipment should normally not be installed in fire-rated walls between spaces that *are* closed off from each other by doors that are normally closed.

When it is necessary to install a panelboard or other large equipment in a wall between areas that are normally closed off from each other, a boxed recess in the wall should be constructed of the fire-rated plasterboard to maintain the fire rating of the wall (Fig. 300-43). This is also common practice for installing recessed enclosures for fire extinguishers in corridor walls and medicine cabinets mounted in walls between apartment units.

Another technique that has been used to maintain fire rating where a panelboard is installed in a wall between individual apartments is to glue pieces of plasterboard to the top, bottom, sides, and back of the recessed panel. In one particular job, this was done as a corrective measure where panelboards had first been installed in such walls without attention to maintaining the fire rating of the wall. But the use of plasterboard directly affixed to the panelboard surfaces could be considered an unauthorized modification of the panel that voids UL listing because of improper application.

300-22. Wiring in Ducts, Plenums, and Other Air-Handling Spaces. Part (**a**) of this section applies only to wiring in the types of ducts described.

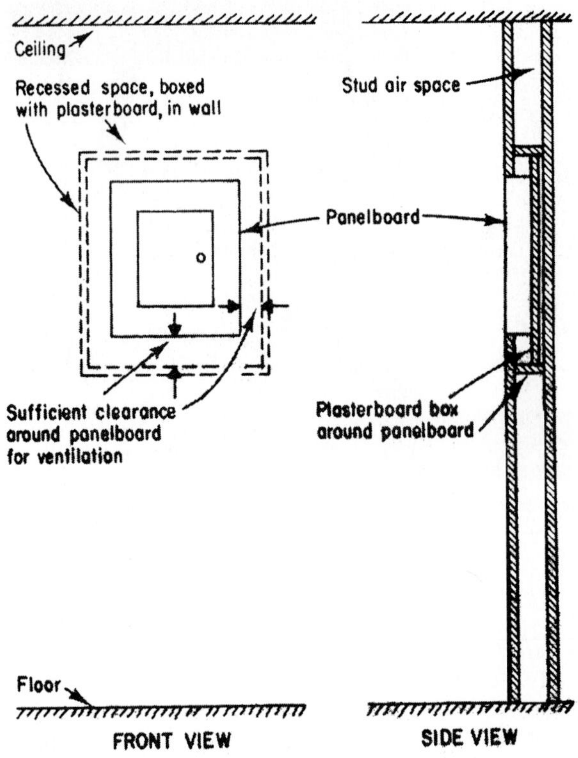

Fig. 300-43. Boxing of large-area penetrations has been required in fire walls. (Sec. 300-21.)

Part **(b)** covers use of wiring methods and equipment within "ducts or plenums"—which are channels or chambers intended and used only for supply or return of conditioned air. Such "ducts or plenums" are sheet metal or other types of enclosures which are provided expressly for air handling and must be distinguished from "Other Space Used for Environmental Air"—such as the space between a suspended ceiling and the floor slab above it. Space of that type is covered by Part **(c)** of this section. The space between a raised floor (Fig. 300-44) and the slab below the raised floor is also covered by Part **(c)**, unless the air-handling raised floor is within a computer room. Part **(d)** states that an air-handling raised floor used for data processing circuits must comply with Art. 645. One NFPA Standard defines a duct system as "a continuous passageway for the transmission of air which, in addition to ducts, may include duct fittings, dampers, plenums, fans, and accessory air handling equipment." The word *duct* is not defined, but a plenum is defined as "an air compartment or chamber to which one or more ducts are connected and which forms part of an air distribution system."

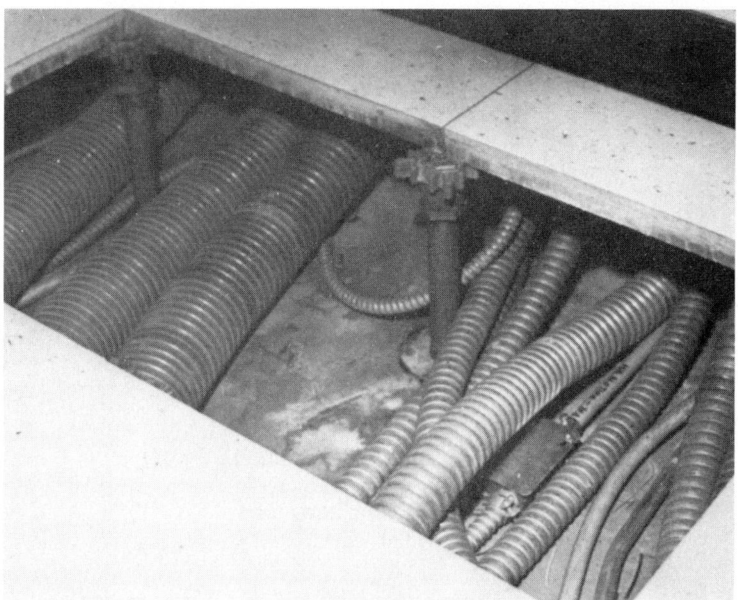

Fig. 300-44. Space under a raised floor, which is commonly used for circuits to data processing equipment and provides for passage of conditioned air to the room and to the equipment is covered in part **(d)**. (Sec. 300-22.)

Parts **(b)** and **(c)** of this section clearly limit acceptable wiring methods *ONLY* to the ones described. Part **(c)** permits use of totally enclosed, nonventilated, insulated busway in an air-handling ceiling space *provided* it is a nonplug-in-type busway that cannot accommodate plug-in switches or breakers. This one specific busway wiring method was added for hung-ceiling space used for environmental air. Surface metal raceway or wireway with metal covers or solid-bottom metal cable tray with solid metal covers may be used in air-handling ceiling space provided that the raceway is accessible, such as above lift-out panels. The air-handling space under a raised floor in a data-processing location is covered by Art. 645 on electronic computer and data processing equipment.

Figure 300-45 shows wiring methods for use in an air-handling ceiling space.

The Code panel has made clear that they generally oppose nonmetallic wiring methods in ducts and plenums and in air-handling ceilings, except for nonmetallic cable assemblies that are specifically listed for such use. It is also the intent of the Code that cables with an outer nonmetallic jacket should not be permitted in ducts or plenums. Although the jacket material, usually PVC, would not propagate a fire, it would contribute to the smoke and provide additional flammable material in the air duct. The last paragraph of part **(c)** permits use of nonmetallic equipment enclosures and wiring that are specifically UL-listed or classified for use in air-handling ceiling spaces.

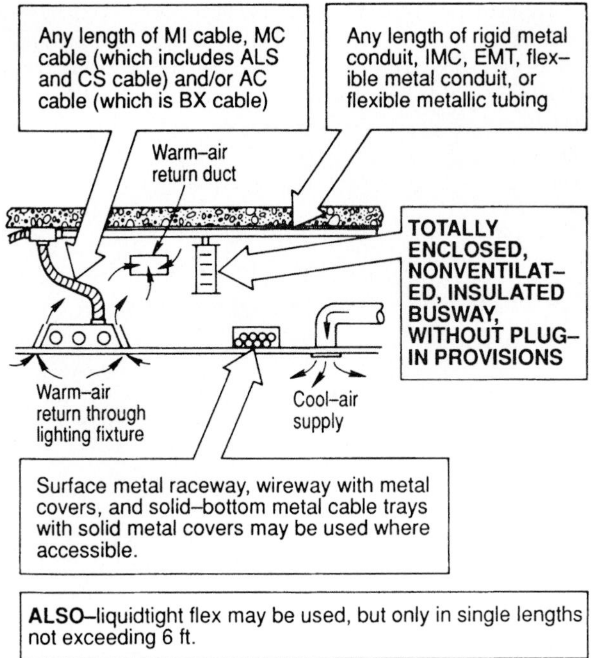

Fig. 300-45. Any wiring method other than these is a violation in air-handling space. (Sec. 300-22.)

In effect, the rules in parts **(b)** and **(c)** exclude from use in all air-handling spaces any wiring that is not metal-jacketed or metal-enclosed, to minimize the creation of toxic fumes due to burning plastic under fire conditions. Section 800-53(a) basically requires telephone, intercom, and other communications circuits to be wired with Type CMP cable or other types installed in compliance with Sec. 300-22 when such circuits are used in ducts or plenums or air-handling ceilings.

Wiring in air-handling space under raised floors in computer centers must use the wiring methods described in Sec. 645-5(d). Ventilation in the raised-floor space must be used only for the data processing area and the data processing equipment.

Although the rules of Sec. 300-22(c) apply only to air-handling spaces above suspended ceilings and beneath raised floors in other than computer rooms, such spaces are also subject to the general rules that apply to nonair-handling spaces. For a thorough understanding of this complex matter refer to these definitions in Art. 100: "accessible," "concealed," "exposed," and "readily accessible."

Because all those words or phrases are used in the Code and are critically important to applications of wiring methods and equipment, their definitions must be carefully studied and cross-referenced with each other, as well as related to Code rules using those words or phrases. Many common controversies about Code rules revolve around those words and phrases and interpretation of the definitions. Refer to the discussion on "suspended ceilings" given under the definition for "accessible" in Art. 100 of this handbook. In addition to that information, other rules relate to use above a suspended ceiling as follows:

1. Generally, all switches and CBs must be located so they may be operated from a readily accessible place, and the distance from the floor or platform up to the center of the handle in its highest position must not be over 6 ft 7 in. (2.0 m) (Sec. 380-8). Exception No. 2 of that rule does permit switches to be installed at high locations that are not readily accessible, even above suspended ceilings, *but only unfused switches,* because use of a fused switch would violate Sec. 240-24 on ready accessibility of overcurrent devices (the fuses in the switch) generally prohibits mounting of overcurrent devices in an inaccessible location. Limited permission for inaccessible mounting is given where the overcurrent device is mounted adjacent to a piece of equipment it supplies (see Sec. 240-24, Exception No. 4). For motors, however, Sec. 430-102 requires a motor disconnect switch to be in sight from the motor controller location. And Sec. 430-107 says one disconnecting means shall be readily accessible. That means *not* above a suspended ceiling, where it would *not* be readily accessible.
2. Section 430-102(b) permits a motor to be out of sight from the location of its controller, and there is no rule requiring that motor controllers be readily accessible. Motor controllers may be installed above suspended ceilings.
3. Section 450-13 requires transformers to be installed so they *are* readily accessible, but certain exceptions are made. Exception No. 1 permits dry-type transformers rated 600 V or less to be located "in the *open* on walls, columns, or structures"—without the need to be readily accessible. And Exception No. 2 permits dry-type transformers up to 600 V, 50 kVA, to be installed in "fire-resistant hollow spaces of buildings not permanently closed in by structure," provided the transformer is designed to have adequate ventilation for such installation. Refer to Sec. 450-13.

Air-Handling Ceilings

All the foregoing rules also apply to wiring and equipment installed above suspended ceilings in space used for air-conditioning purposes. But, in addition to those rules, the broad and detailed rules of Sec. 300-22(c) cover electrical installations in spaces above suspended ceilings when the space is used to handle environmental air. This section makes two basic determinations:

1. It lists all the wiring methods that are permitted in air-handling ceilings (which also may be used in nonair-handling ceiling spaces) and gives conditions and limitations for such use. This is a straightforward materials list which needs little or no interpretation (Fig. 300-45).
2. Section 300-22(c) further comments on other "electric equipment" that is permitted in such spaces. That refers to switches, starters, motors, etc. The basic condition that must be satisfied is that the wiring materials and

other construction of the equipment must be suitable for the expected ambient temperature to which they will be subjected.

Application of that Code permission on use of "equipment" calls for substantial interpretation. The designer and/or installer must check carefully with equipment manufacturers and with inspection agencies to determine what is acceptable in the air-handling space above a suspended ceiling. Practice in the field varies widely on this rule, and Code interpretation has proved difficult.

Exception No. 2 of this section recognizes the installation of motors and control equipment in air-handling ducts where such equipment has been specifically approved for the purpose. Equipment of this type is UL-listed and may be found in the *Electrical Appliance and Utilization Equipment List* under the heading "Heating and Ventilating Equipment."

Exception No. 3 is intended to exclude from the requirements those areas which may be occupied by people. Hallways and habitable rooms are being used today as portions of air-return systems, and while they have air of a heating or cooling system passing through them, the prime purpose of these spaces is obviously not air handling.

Exception No. 4 permits modular wiring systems to be used in air-handling spaces *provided* that the wiring system consists of metallic-jacketed cable assemblies and there is *not* a plastic outside sheath over the metal. (See Fig. 300-46.)

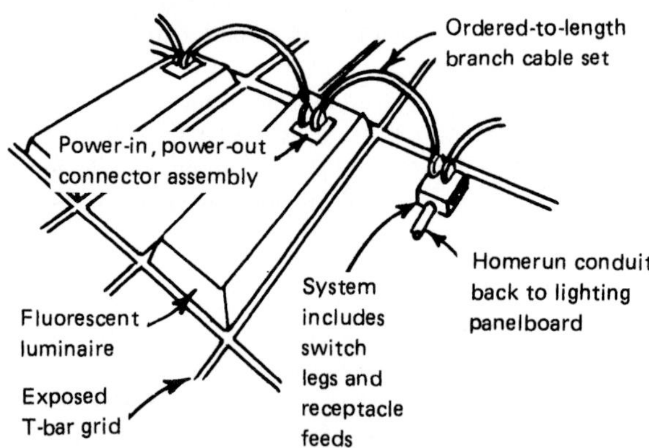

Fig. 300-46. Modular wiring systems, as recognized by Article 604, are permitted to be used in air-handling ceiling spaces.

Exception No. 5 permits Type NM cable to "pass through" a closed-in joist or stud space that is used for cold-air return, as shown in Fig. 300-47. This is allowed because NM cable is suitable to be used under the temperature and moisture conditions in such spaces, as used in "dwelling units," to which the Exception is limited.

Fig. 300-47. A joist space through which Type NM cable passes "perpendicular to the long dimension" of the space may be closed in to form a duct-like space for the cold-air return of a hot-air heating system—but only in a "dwelling unit."

ARTICLE 305. TEMPORARY WIRING

305-1. Scope. Although a temporary electrical system does not have to be made up with the detail and relative permanence that characterizes a so-called *permanent* wiring system, the specific rules of this article cover the only permissible ways in which a temporary wiring system may differ from a permanent system. Aside from the given permissions for variation from rules on permanent wiring, all temporary systems are required to comply in all other respects with Code rules covering permanent wiring (Fig. 305-1).

305-3. Time Constraints. In part **(a)**, the words "maintenance" and "repair" indicate that the less rigorous methods of temporary wiring may be used and that all rules on temporary wiring must be observed wherever maintenance or repair work is in process. This expands the applicability of temporary wiring beyond new construction, remodeling work, or demolition.

Part **(b)** recognizes use of temporary wiring for seasonal or holiday displays and decorations, as shown in Fig. 305-2.

Part **(c)** of this section permits temporary wiring to be used for other than simple construction work. Such wiring, as covered in this article, may also be used during emergency conditions or for testing, experiments, or development activities. As the proposal for this Code rule noted:

> Were it not permissible to use temporary wiring methods for testing purposes, it would be impossible to check, before placing in service, many electrical installations. Likewise, emergency conditions would remain without electric power and lighting until permanent installation could be made.

Fig. 305-1. Temporary wiring is not an "anything goes" condition and must comply with standard **Code** rules to prevent a rat-nest condition which can pose hazard to life and property. (Sec. 305-1.)

However, part **(d)** of this section is aimed at assuring that the equipment and circuits installed under this article are really "temporary" and not a back door to low-quality permanent wiring systems.

305-4. General. Although part **(a)** requires a temporary *service* to satisfy all the rules of Art. 230, part **(b)** recognizes use of temporary *feeders* that are single-conductor building wire or single-conductor "cable assemblies" used as open wiring (Fig. 305-3), multiconductor cable assemblies (Type NM, UF, etc.), or multiconductor cord or cable of the type covered by Art. 400 for hard usage or extra-hard usage "Flexible Cords and Cables"—which are not acceptable for use as feeder or branch-circuit conductors of permanent wiring systems. Section 400-8 specifically prohibits use of such cords and cables "as a substitute for the fixed wiring of a structure." As shown in Fig. 305-4, prewired portable cables with plug and socket assemblies are available for power risers in conjunction with GFCI-protected branch-circuit centers, or cable can be run horizontally on a single floor to suit needs. GFCI breakers may be used in temporary panelboards interconnected with cable and feeding standard receptacles in portable boxes, as shown in Fig. 305-5.

Section 305-4(c) requires temporary branch circuits to consist of single conductors in open wiring, multiconductor cable assemblies (Types NM and UF)

Fig. 305-2. Temporary wiring techniques are permitted for 90 days for such "experimental" work as energy demand analysis. (Sec. 305-3.)

or cords or cables covered in Table 400-4, provided that they originate in a panelboard *or* "an approved power outlet," which is one of the manufactured assemblies made for jobsite temporary wiring. As shown in Fig. 305-6, the temporary branch circuits for receptacle outlets may be part of a manufactured temporary system, which consists of cable harnesses and power centers (or outlets). Several variations of protection may be provided by such portable receptacle boxes, as shown in Fig. 305-6. Box 1 may have GFCI protection for its own receptacles without providing downstream protection. Box 2 may have the same protection as box 1 and in addition have GFCI protection for its 50-A outlet, thus providing protection for box 3. With this arrangement, box 1 will sense the ground fault from the worker at upper left and will trip, allowing boxes 2 and 3 to continue to provide power. Or, all three boxes could receive GFCI protection from a permanently mounted loadcenter feeding the 50-A receptacle outlet at upper left. In this case, the ground fault shown would interrupt the power to all boxes.

Section 305-6 makes it clear that only receptacles used under temporary job conditions require GFCI protection. The implication is that the nonmetallic-sheathed cable runs and pigtail connections traditionally associated with tem-

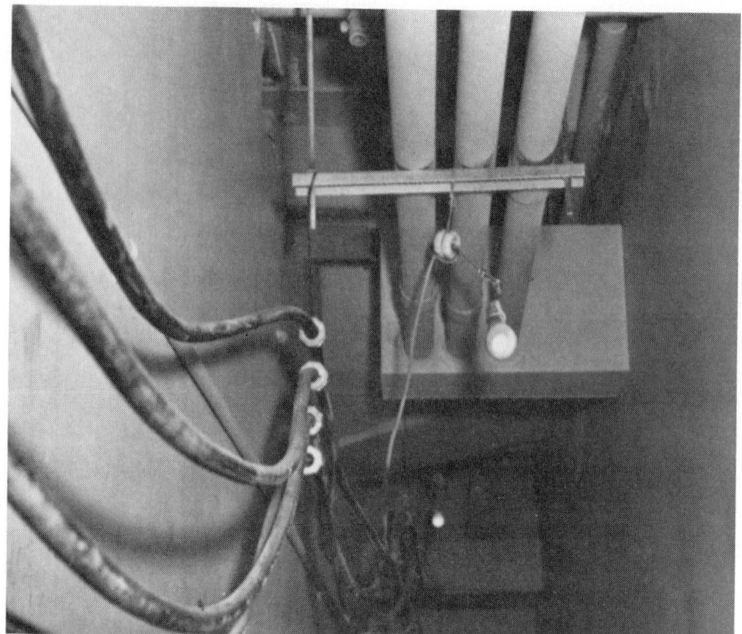

Fig. 305-3. Temporary feeders operating at not over 150 V to ground and where not subject to physical damage may be run as open conductors supported by insulators spaced not over 10 ft (3.05 m) apart. (Sec. 305-4.)

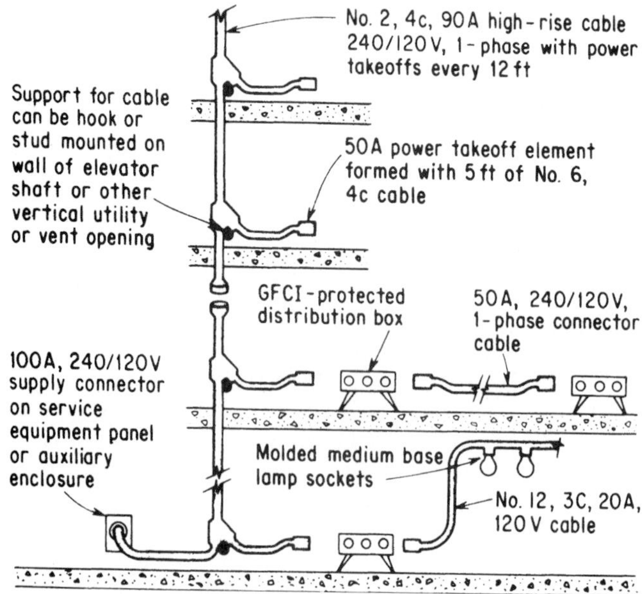

Fig. 305-4. Temporary feeders may be cord assemblies made especially for such use. (Sec. 305-4.)

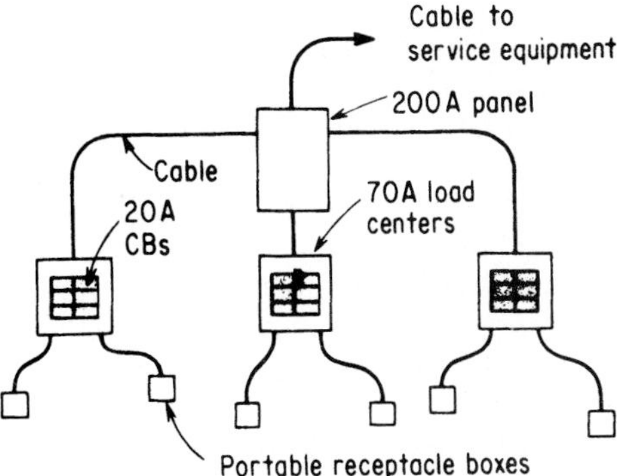

Fig. 305-5. Distribution for temporary power may utilize cable or raceway feeders. (Sec. 305-4.)

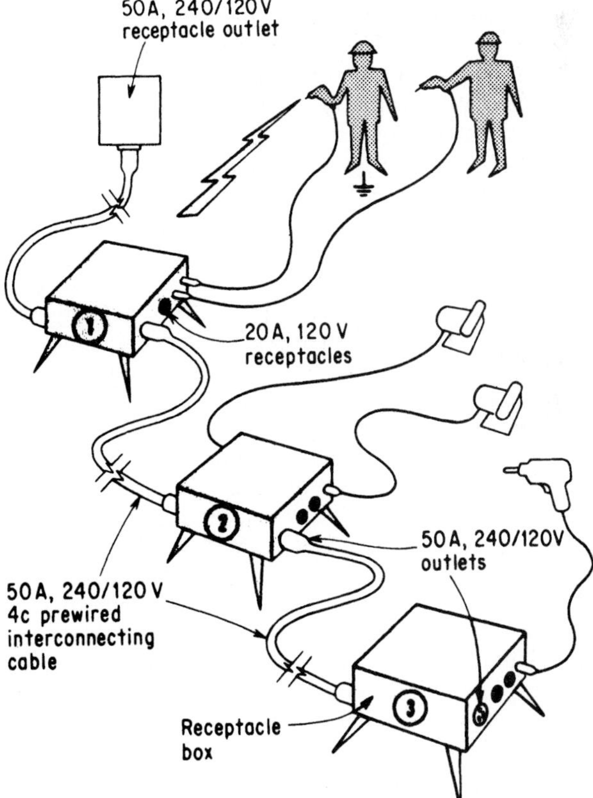

Fig. 305-6. Temporary branch circuits may be part of a manufactured system. (Sec. 305-4.)

porary power on the jobsite would not win awards for neatness and safety, but that once the permanent feeders and panelboards are in place and energized, the shock hazard is considerably reduced.

However, as long as portable tools are being used in damp locations in close proximity with grounded building steel and other conductive surfaces, the possibility of shock exists from faulty equipment whether it is energized from temporary or permanent circuits.

Standard panelboards used for temporary power on the jobsite may be fitted with GFCI circuit breakers for the protection of entire circuits, in accordance with the rules of Sec. 305-6. However, the many varieties of portable power distribution centers and modules have been developed with integral GFCI breakers protecting single-phase, 15- and 20-A, 120-V circuits. Other circuits (higher amperage, higher voltage, and 3-phase) are not required by the **NE Code** to have GFCI protection, and these usually are protected by standard overcurrent devices. A variety of cord sets are also available for use with GFCI-protected plug-in units to supply temporary lighting and receptacle outlets.

While a manufactured system of cable harnesses and power-outlet centers costs more than nonmetallic-sheathed cable runs and pigtail sockets, it is completely recoverable; and its cost can be written off over several jobs. From then on, with the exception of costs for setup and removal, storage, and transportation, much of the temporary power charges included in bids could be profit.

In previous **Code** editions, part **(c)** of this section required temporary wiring circuits to be "fastened at ceiling height every 10 feet (3.05 m)." But now, if such circuits operate at not over 150 V to ground and are not subject to physical damage, the fourth sentence in this paragraph permits open-wiring temporary branch circuits to be run at any height "supported on insulators at intervals of not more than 10 feet (3.05 m)." Open wiring must not be laid on the floor or ground.

In the interest of greater safety, part **(d)** prohibits use of both receptacles and lighting on the same temporary branch circuit on construction sites. The purpose is to provide complete separation of the lighting so that operation of an overcurrent device or a GFCI due to fault or overload of cord-connected tools will not simultaneously disconnect lighting (Fig. 305-7).

According to part **(e)**, every multiwire branch circuit must have a disconnect means that *simultaneously* opens all ungrounded wires of the temporary circuit. At the power outlet or panelboard supplying any temporary multiwire branch circuit (two hot legs and neutral or three hot legs and neutral), a multipole disconnect means must be used. Either a 2-pole or a 3-pole switch or circuit breaker would satisfy the rule; or single-pole switches of single-pole CBs may be used with "approved" handle ties to permit the single-pole devices to operate together (simultaneously) for each multiwire circuit, as shown for the multiwire lighting circuit in Fig. 305-7.

Part **(f)** requires lamps for general lighting on temporary wiring systems to be "protected from accidental contact or breakage." Protection must be provided by a suitable fixture or lampholder with a guard (Fig. 305-8). OSHA rules also require use of a suitable metal or plastic guard on each lamp. As shown in Fig. 305-9, commercial lighting strings provide illumination where required. Splice

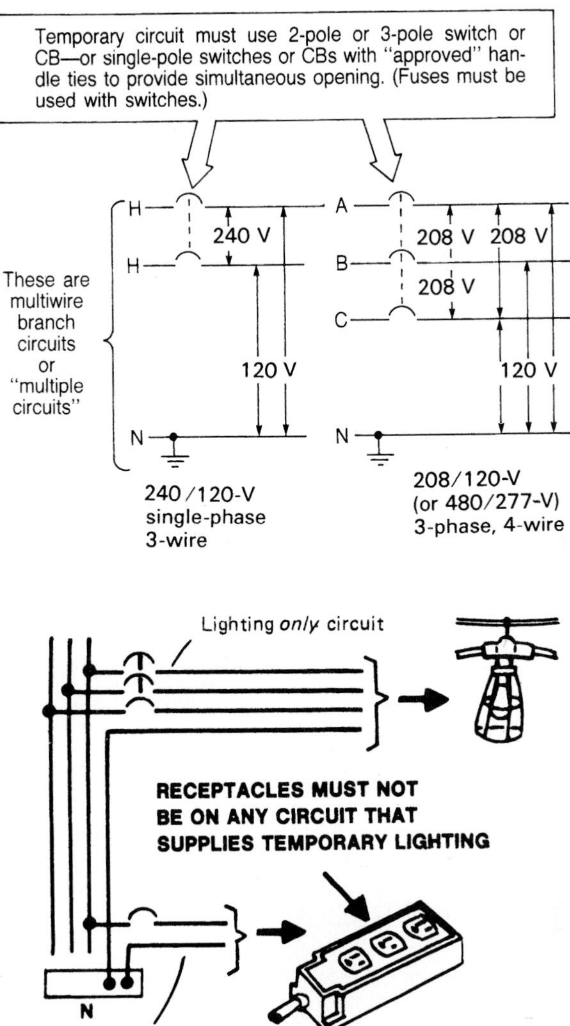
Temporary circuit must use 2-pole or 3-pole switch or CB—or single-pole switches or CBs with "approved" handle ties to provide simultaneous opening. (Fuses must be used with switches.)
These are multiwire branch circuits or "multiple circuits"
H
H
N
240 V
120 V
240/120-V single-phase 3-wire
A
B
C
N
208 V
208 V
208 V
120 V
208/120-V (or 480/277-V) 3-phase, 4-wire
Lighting only circuit
RECEPTACLES MUST NOT BE ON ANY CIRCUIT THAT SUPPLIES TEMPORARY LIGHTING
N
Receptacles only circuit

Fig. 305-7. This rule prevents loss of lighting when a defective, high-leakage, or overloaded Code-connected tool or appliance opens the branch-circuit protection of a circuit supplying one or more receptacles. (Sec. 305-4.)

enclosure is equipped with integral support means, and a variety of lamp-guard styles provide protection for lamp bulbs.

Part **(f)** requires grounding of metal lamp sockets. The high exposure to shock hazard on construction sites makes use of ungrounded metal-shell sockets extremely hazardous. When they are used, the shell *must* be grounded by a conductor run with the temporary circuit.

Fig. 305-8. A lampholder with a guard is proper protection for a lamp at any height in a temporary wiring system (above). Unguarded lamps at any height constitute a Code violation (right). (Sec. 305-4.)

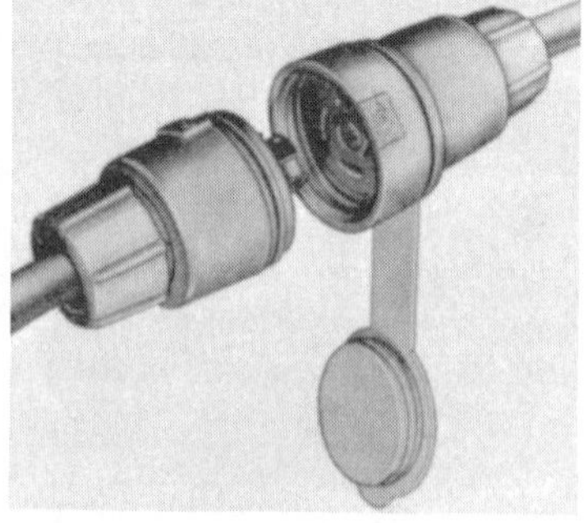

Special watertight plugs and connectors provide insurance against nuisance tripping caused by weather conditions on construction sites.

Fig. 305-9. Temporary lighting strings of cable and sockets are available from manufacturers. (Sec. 305-4.)

In part **(g)**, splices or tap-offs are permitted to be made in temporary wiring circuits of cord or cable without the use of a junction box or other enclosure at the point of splice or tap (Fig. 305-10). But this new permission applies only to nonmetallic cords and cables. A box, conduit body, or terminal fitting must be used when a change is made from a cord or cable circuit to a raceway system or to a metal-clad or metal-sheathed cable.

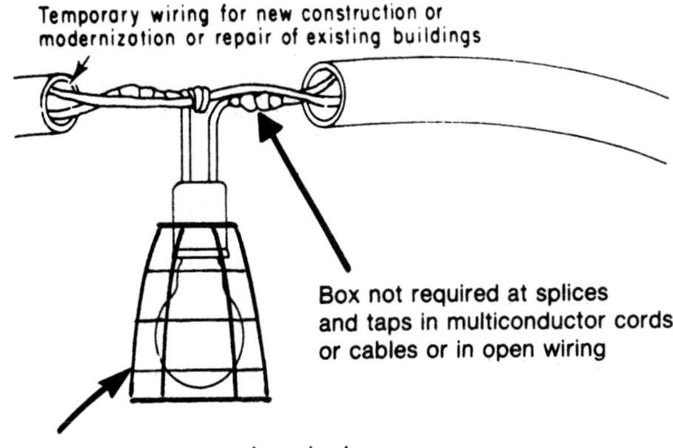
Temporary wiring for new construction or
modernization or repair of existing buildings
Box not required at splices
and taps in multiconductor cords
or cables or in open wiring
NEC requires a guard on the lamp.

Fig. 305-10. Splices may be used without boxes for cord and cable runs on construction sites. (Sec. 305-4.)

Regulations in part **(h)** require protection of flexible cords and cables from damage due to pinching, abrasion, cutting, or other abuse.

Part **(i)** calls for the use of proper fittings to secure cables that enter enclosures containing receptacles and/or switches.

305.6. Ground-Fault Protection for Personnel. This section covers the rules that concern GFCI protection for all "125-V, single-phase, 15- and 20-A receptacle outlets" on construction sites. (Note that there are no requirements for GFCI protection of 240-V receptacles, 3-phase receptacles, or receptacles rated over 20 A.)

The basic rule of part **(a)** of this section says that ground-fault circuit interrupters (of the type listed for use in temporary power applications—not GFCI circuit breakers or GFCI receptacles listed for permanent use in dwellings or at pools) must be used to provide personnel protection for all receptacles of the designated rating—that "are not part of the permanent wiring of the building or structure" (Fig. 305-11). In addition, the second sentence mandates GFCI protection for all permanently installed receptacles—it's presumed that this applies only to 15- and 20-A, 125-V receptacles, but this is not clear—that are used for temporary power (Fig. 305-12).

But one phrase in the Code rule significantly qualifies the *need* for GFCI protection of the designated receptacle outlets:

GFCI PROTECTION IS REQUIRED *ONLY* FOR THOSE RECEPTACLES THAT "ARE *IN USE* BY PERSONNEL."

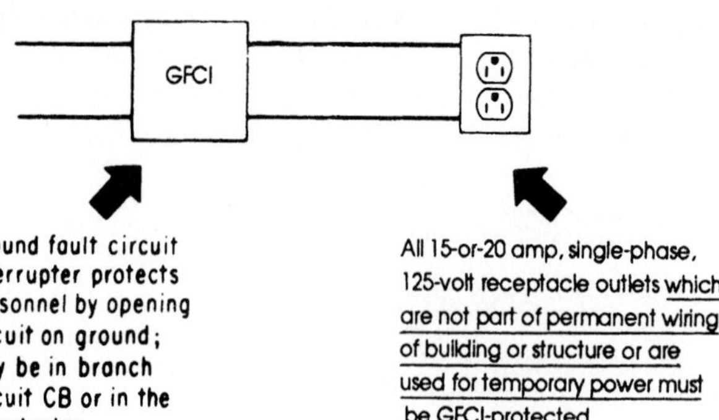
BASIC RULE
GFCI
Ground fault circuit interrupter protects personnel by opening circuit on ground; may be in branch circuit CB or in the receptacles
All 15-or-20 amp, single-phase, 125-volt receptacle outlets which are not part of permanent wiring of building or structure or are used for temporary power must be GFCI-protected

Fig. 305-11. GFCI protection on construction sites for receptacles in use. (Sec. 305-6.)

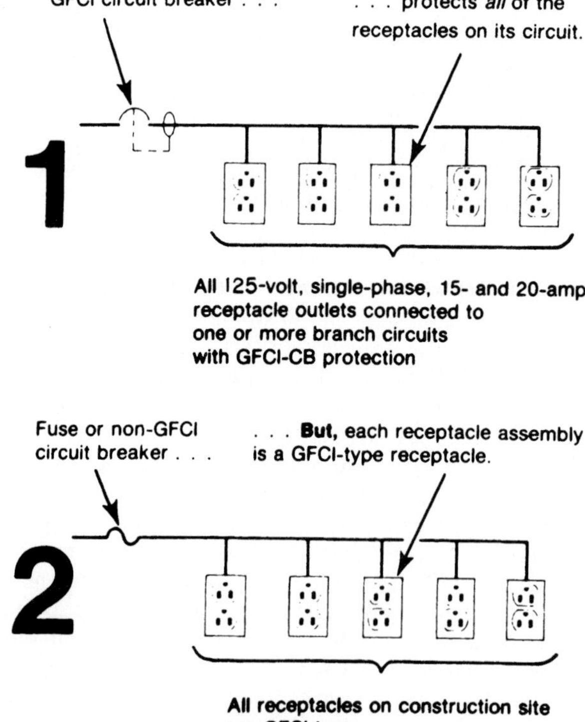
GFCI circuit breaker . . .
. . . protects *all* of the receptacles on its circuit.
1
All 125-volt, single-phase, 15- and 20-amp receptacle outlets connected to one or more branch circuits with GFCI-CB protection
Fuse or non-GFCI circuit breaker . . .
. . . **But,** each receptacle assembly is a GFCI-type receptacle.
2
All receptacles on construction site are GFCI type.

Fig. 305-12. Two ways to satisfy the basic rule on personnel shock protection at *temporary* receptacles on construction sites. (Sec. 305-6.)

That phrase clearly limits required GFCI protection to receptacles that are actually being used at any particular time. Receptacles *not* in use do not have to be GFCI-protected. This means that *portable* GFCI protectors may be used at only those outlets being used (Fig. 305-13). There is no need to use GFCI breakers in the panel to protect "*all*" receptacles or to use all GFCI-type receptacles. This seems to seriously confuse the task of electrical inspection: If all cord-

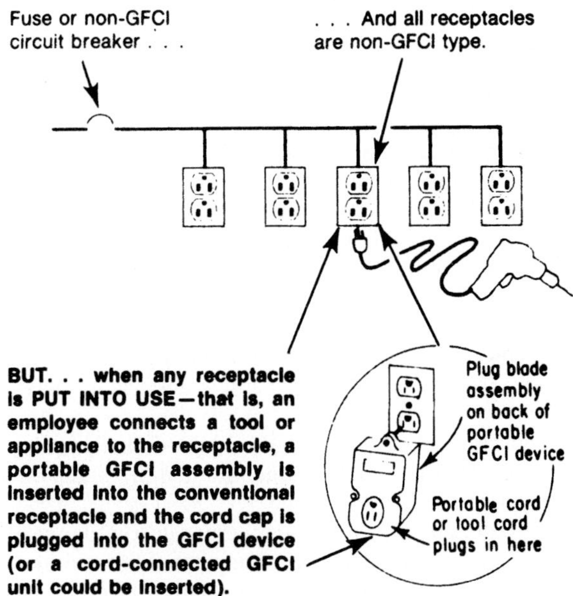
Fuse or non-GFCI circuit breaker . . .
. . . And all receptacles are non-GFCI type.
BUT. . . when any receptacle is PUT INTO USE—that is, an employee connects a tool or appliance to the receptacle, a portable GFCI assembly is inserted into the conventional receptacle and the cord cap is plugged into the GFCI device (or a cord-connected GFCI unit could be inserted).
Plug blade assembly on back of portable GFCI device
Portable cord or tool cord plugs in here

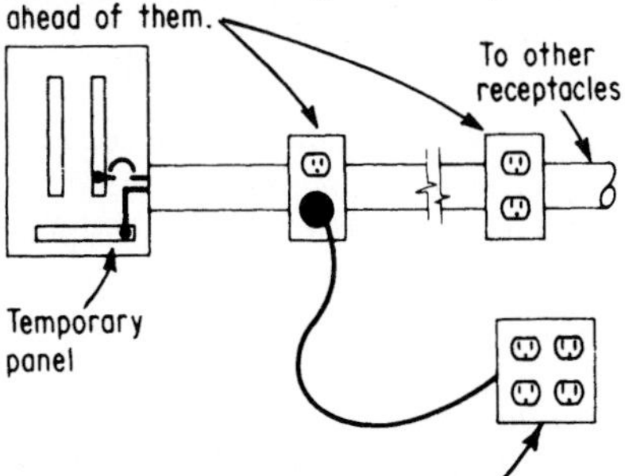
These 15A receptacles are fed by a temporary branch circuit without ground-fault protection ahead of them.
To other receptacles
Temporary panel
Wherever personnel are using cord-connected tools they plug in this portable ground-fault circuit interrupter having protected receptacles on its face for connection of the tools.

Fig. 305-13. Portable GFCI devices may be used to satisfy GFCI rule. (Sec. 305-6.)

connected tools and appliances are unplugged from receptacles when the inspector comes on the job, then *none* of the receptacles is "in use" and none of them has to have GFCI protection and there is no Code violation.

A new requirement given in part **(b)** for the 1996 Code extends the GFCI requirements to *all* receptacles "not covered in part **(a)**." That includes 3-phase and phase-to-phase receptacles of any current value permanently or temporarily installed. Alternately the Assured Equipment Grounding Program explained in the remainder of this section may be used, or cord sets with integral GFCI protection may be used.

Still another option for avoiding use of GFCI protection on construction sites is given in part **(b)** of this section. GFCI protection of receptacles *may be omitted* totally if a "written procedure" is established to assure testing and maintenance of "equipment grounding conductors" for receptacles, cord sets, and cord- and plug-connected tools and appliances used on the construction site (Fig. 305-14). In effect, the NE Code accepts such an equipment grounding conductor program as a measure that provides safety that is equivalent to the safety afforded by GFCI protection. GFCI protection is not required if all the following conditions are satisfied:

1. The inspection authority having jurisdiction over a construction site must approve a written procedure for an equipment grounding program.
2. The program must be enforced by a single designated person at the construction site.
3. "Electrical continuity" tests must be conducted on all equipment grounding conductors and their connections. The requirements on making such tests are vague, but they do call for:
   a. Testing of fixed receptacles where there is any evidence of damage.
   b. Testing of extension cords before they are first used and again where there is evidence of damage or after repairs have been made on such cords.
   c. Testing of all tools, appliances, and other equipment that connect by cord and plug before they are first used on a construction site, again any time there is any evidence of damage, after any repair, and at least every 3 months.

Obviously, those rules are very general and could be satisfied in either a rigorous, detailed manner or a fast, simple way that barely meets the qualitative criteria. The electrical contractor who has responsibility for the temporary wiring on any job site is the one to develop, write, and supervise the assured equipment grounding program, where that option is chosen as an alternative to use of GFCI protection. This whole NE Code approach to use of either GFCI or an "assured equipment grounding program" directly parallels the new OSHA approach to the matter of receptacle protection on construction sites.

It should be noted that the Assured Equipment Grounding Program described in part **(b)** is one of the most frequently cited violations during OSHA inspections. Implementation is a bureaucratic nightmare and is rarely successfully executed.

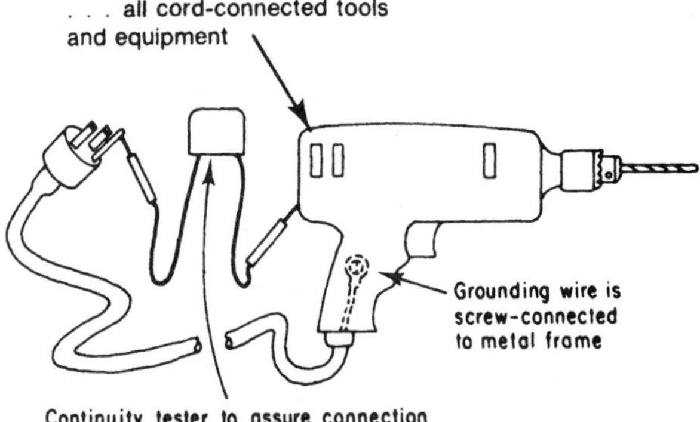
A written procedure must cover testing of...
. . . all cord-connected tools and equipment
Grounding wire is screw-connected to metal frame
Continuity tester to assure connection of equipment grounding conductor

. . . and all receptacles, cord sets, and extension cords.
Continuity tester or ohmmeter to check connections and assured grounding continuity
15- and 20-amp locking plugs and receptacles are also covered under the assured grounding program.

Fig. 305-14. Assured grounding program eliminates the need for GFCI. (Sec. 305-6.)

ARTICLE 310. CONDUCTORS FOR GENERAL WIRING

310-2. Conductors. Although conductors are generally required by this rule to be insulated for the phase-to-phase voltage between any pair of conductors, bare conductors may be used for equipment grounding conductors, for bonding jumpers, for grounding electrode conductors, and for grounded neutral conductors (Secs. 230-22, 230-30, 230-41, 250-57, 250-60, 250-91, and 338-3).

The application shown in Fig. 310-1 is a commonly encountered violation of Sec. 310-2 because it involves an unauthorized use of a bare conductor. Section 250-60 permits grounding of ranges, cook-tops, and ovens to the neutral conductor *only* where "the grounded conductor (the neutral) is insulated" or is a bare neutral of an SE cable.

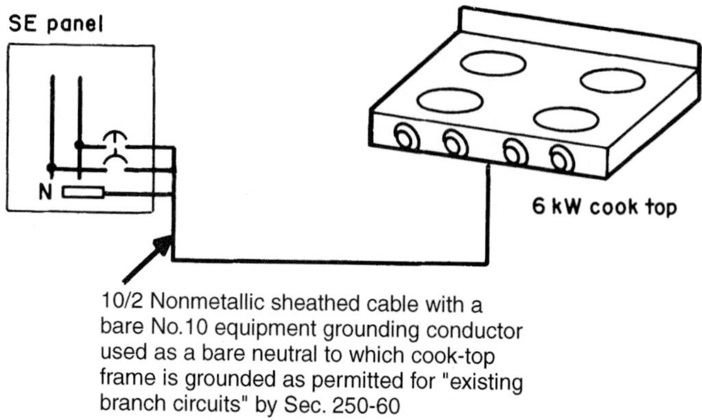
SE panel
N
6 kW cook top
10/2 Nonmetallic sheathed cable with a bare No.10 equipment grounding conductor used as a bare neutral to which cook-top frame is grounded as permitted for "existing branch circuits" by Sec. 250-60

Fig. 310-1. This is a controversial application that violates Secs. 310-2 and 336-2. (Sec. 310-2.)

Section 310-2 states that "conductors shall be insulated," except when covered or bare conductors (see definition in Art. 100) are specifically approved in this Code. As noted above, several sections in Art. 250 state that grounding conductors may be insulated or bare. Article 230 cites several instances when a grounded conductor may be uninsulated or bare.

Section 338-3(b) permits use of Type SE cable without individual insulation on the grounded circuit conductor to be used as a branch circuit for a range, a wall oven, a cook-top, or a clothes dryer if such a cable originates at the service equipment panel.

Nonmetallic-sheathed cable Types NM and NMC are covered in Art. 336 and do not enjoy the same status as Type SE cable does in Sec. 338-3(b). To the contrary, Sec. 336-30 states: "In addition to the insulated conductors, the [NM, NMC, or NMS] cable shall be permitted to have an insulated or bare conductor for equipment grounding purposes *only*." Use of the bare conductor as a neutral in addition to a grounding conductor would be a violation of Sec. 336-25.

The same evaluation would apply to UF cable, because Sec. 339-3(a)(4) requires UF cable to comply with the provisions of Art. 336 when used for interior wiring as a nonmetallic-sheathed cable. The bare grounding conductor in a Type NM, NMC, NMS, or UF cable cannot be used as a neutral conductor.

Although the basic rule of this section requires conductors to be insulated, a note refers to Sec. 250-152 on the use of solidly grounded neutral conductors in high-voltage systems. As an exception to the general rule that conductors must be insulated, Sec. 250-152 does permit a neutral conductor of a solidly grounded "Y" system to have insulation rated at only 600 V (Fig. 310-2). It also points out that a bare copper neutral may be used for service-entrance conductors or for direct buried feeders, and bare copper or copper-clad aluminum may be used for overhead sections of outdoor circuits.

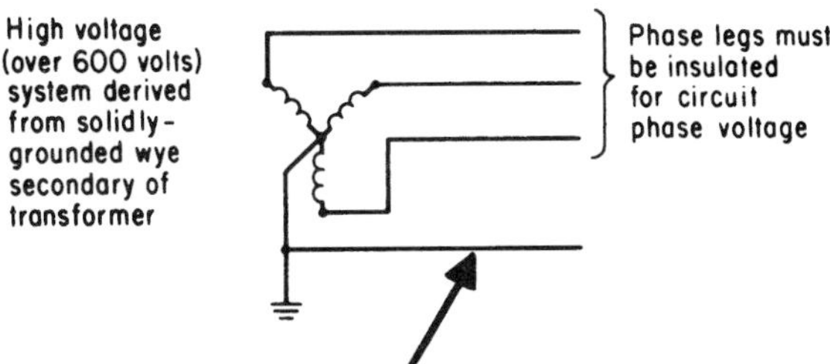
High voltage (over 600 volts) system derived from solidly-grounded wye secondary of transformer
Phase legs must be insulated for circuit phase voltage
Solidly grounded neutral conductor must have insulation rated for at least 600 volts, although a bare copper neutral may be used for SE conductors or for direct-buried feeders, and bare copper or aluminum may be used for overhead parts of outdoor circuits.

Fig. 310-2. A note refers to neutral conductors of solidly grounded high-voltage systems (Sec. 250-152). (Sec. 310-2.)

Of course, for such high-voltage systems, the phase legs—the ungrounded conductors—must be insulated for the circuit phase voltage. It is interesting, however, that there is no specific Code rule that requires insulation of any circuit to be rated for phase-to-phase voltage. Code rules do not distinguish between phase-to-phase voltage and phase-to-neutral voltage on grounded systems, with respect to insulation. Thus, the use of circuit conductors with insulation rated only for phase-to-neutral voltage would not constitute a violation of any specific Code rule, and such practice is used on high-voltage systems.

310-3. Stranded Conductors. No. 8 and larger conductors must be stranded when they are installed in conduit, EMT, or any other "raceway." The use of an insulated or stranded No. 8 copper conductor is required for the equipment bonding conductor required by Sec. 680-20(b)(1). But only a solid No. 8 copper conductor is required by Sec. 680-22(b) at swimming pools for bonding together noncurrent-carrying metal parts of pool equipment—metal ladder, diving board stands, pump motor frame, lighting fixtures in wet niches, etc.

310-4. Conductors in Parallel. The requirements of Sec. 310-4 for conductors in parallel recognize copper, copper-clad aluminum, and aluminum conductors in sizes 1/0 and larger. Also, this section makes it clear that the rules for paralleling conductors apply to grounding conductors (except for sizing which is accomplished in accordance with Sec. 250-95) when they are used with conductors in multiple.

Conductors that are permitted to be used in parallel (in multiple) include "phase" conductors, "neutral" conductors, and "grounded circuit" conductors. In the places where this section describes parallel makeup of circuits, a "grounded circuit conductor" is identified along with "phase" and "neutral" conductors to extend the same permission for paralleling to grounded legs of corner-grounded delta systems.

This section recognizes the use of conductors in sizes 1/0 and larger for use in parallel under the conditions stated, to allow a practical means of installing large-capacity feeders and services. Paralleling of conductors relies on a number of factors to ensure equal division of current, and thus all these factors must be satisfied in order to ensure that none of the individual conductors will become overloaded.

When conductors are used in parallel, *all* the conductors making up *each phase, neutral, or grounded* circuit conductor must satisfy the five conditions of the second paragraph in this section. Those characteristics—same length, same conductor material (copper or aluminum), same size, same insulation, and same terminating device—apply only to the paralleled conductors making up each phase or neutral of a parallel-makeup circuit. All the conductors of any phase or the neutral must satisfy the rule, but phase "A" conductors (all of which must be the same length, same size, etc.) may be different in length, material, size, etc., from the conductors making up phase "B" or phase "C" or the neutral. But, all phase "B" conductors must be the same length, same size, etc.; phase "C" conductors must all be the same; and neutral conductors must all be alike (Fig. 310-3). As the last sentence in the fine-print note explains, it is not the intent of this Code rule to require that conductors of one phase be the same as those of another phase or of the neutral. The only concern for safe operation of a parallel-makeup circuit is that all the conductors in parallel per phase leg (neutral, or grounded conductor) will evenly divide the load current and thereby prevent overloading of any one of the conductors. Of course, the realities of material purchase and application and good design practice will dictate that *all* the conductors of all phases and neutral will use the same conductor material, will have the same insulation, will have as nearly the same length as possible to prevent voltage drop from causing objectionable voltage unbalance on the phases, and will be terminated in the same way. The size of conductors may vary from phase to phase or in the neutral, depending upon load currents.

Figure 310-4 shows two examples of parallel-conductor circuit makeup. The photo at bottom shows six conductors used per phase and neutral to obtain 2,000-A capacity per phase, which simply could not be done without parallel conductors per phase leg. Note that a fusible limiter lug is used to terminate each individual conductor. Although limiter lugs are required by the NEC only as used in Sec. 450-6(a) (3), they may be used to protect each conductor of any parallel circuit against current in excess of the ampacity of the particular size of conductor. The CB or fuses on such circuits are rated much higher than the ampacity of each conductor.

Where large currents are involved, it is particularly important that the separate phase conductors be located close together to avoid excessive voltage drop and ensure equal division of current. It is also essential that each phase and the neutral, and grounding wires, if any, be run in each conduit even where the conduit is of nonmetallic material.

The sentence just before the FPN in this section calls for the same type of raceway or enclosure for conductors in parallel in separate raceways or cables.

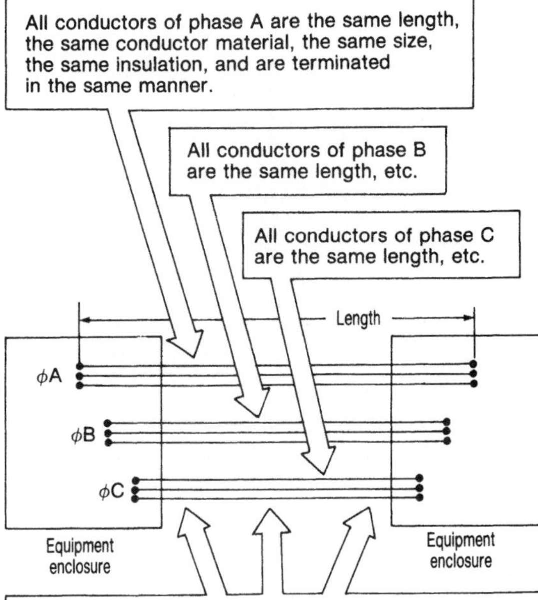
All conductors of phase A are the same length, the same conductor material, the same size, the same insulation, and are terminated in the same manner.
All conductors of phase B are the same length, etc.
All conductors of phase C are the same length, etc.
Length
ϕA
ϕB
ϕC
Equipment enclosure
Equipment enclosure

BUT, the length, conductor material, size, insulation, and terminating devices of the conductors of phase A are not the same as those of phase B or phase C, and phase B is not the same as phase C.

NOTE: This shows three conductors per phase. All nine conductors may be used in a single conduit with their ampacities derated to 70% of the value shown in Table 310-16. Or, three conduits may be used, with a phase A, B and C conductor in each, and no derating would be required.

Fig. 310-3. This is the basic rule on conductors used for parallel circuit makeup. (Sec. 310-4.)

The impedance of the circuit in a nonferrous raceway will be different from the same circuit in a ferrous raceway or enclosure. See Sec. 300-20.

From the **Code** tables of current-carrying capacities of various sizes of conductors, it can be seen that small conductor sizes carry more current per circular mil of cross section than do large conductors. This results from rating conductor capacity according to temperature rise. The larger a cable, the less is the radiating surface per circular mil of cross section. Loss due to "skin effect" (apparent higher resistance of conductors to alternating current than to direct current) is also higher in the larger conductor sizes. And larger conductors cost more per ampere than smaller conductors.

All the foregoing factors point to the advisability of using a number of smaller conductors in multiple to get a particular carrying capacity, rather than using a

Multiple conductors (two in parallel for each phase leg) are used for normal and emergency feeder through this automatic transfer switch.

Six conductors in parallel make up each phase leg and the neutral of this feeder. Fusible limiter lug on each conductor, although not required by **Code** on other than transformer tie circuits, is sized for the conductor to protect against division of current among the six conductors that would put excessive current on any conductor.

Fig. 310-4. These are examples of circuit makeup using conductors in parallel. (Sec. 310-4.)

single conductor of that capacity. In many cases, multiple conductors for feeders provide distinct operating advantages and are more economical than the equivalent-capacity single-conductor makeup of a feeder. But, it should be noted, the reduced overall cross section of conductor resulting from multiple conductors instead of a single conductor per leg produces higher resistance and greater voltage drop than the same length as a single conductor per leg. Voltage drop may be a limitation.

Figure 310-5 shows a typical application of copper conductors in multiple, with the advantages of such use. Where more than three conductors are in-

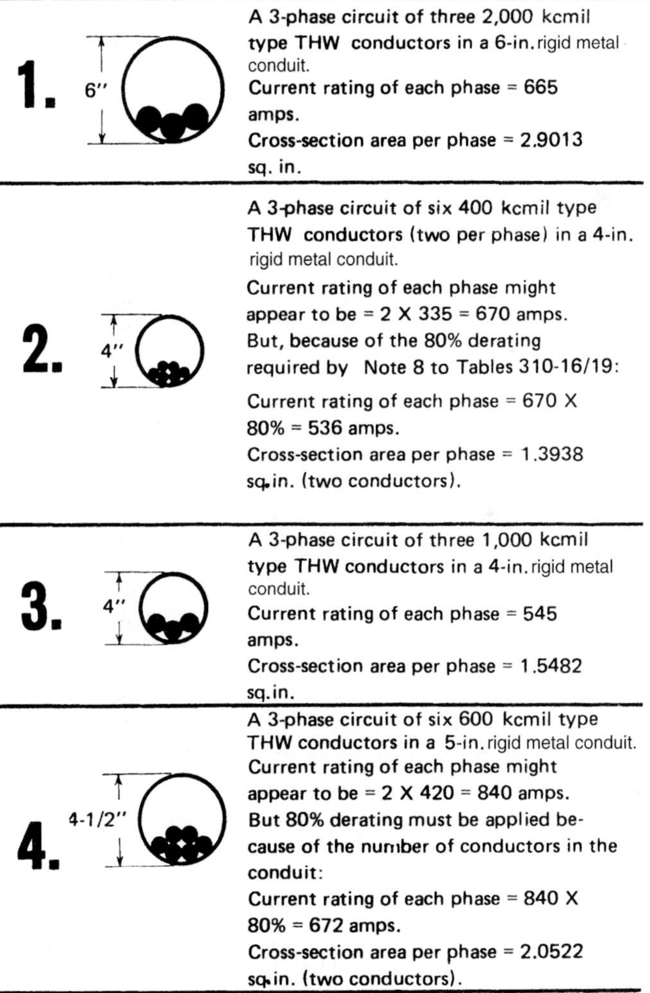

1.
6"
A 3-phase circuit of three 2,000 kcmil type THW conductors in a 6-in. rigid metal conduit.
Current rating of each phase = 665 amps.
Cross-section area per phase = 2.9013 sq. in.
2.
4"
A 3-phase circuit of six 400 kcmil type THW conductors (two per phase) in a 4-in. rigid metal conduit.
Current rating of each phase might appear to be = 2 X 335 = 670 amps.
But, because of the 80% derating required by Note 8 to Tables 310-16/19:
Current rating of each phase = 670 X 80% = 536 amps.
Cross-section area per phase = 1.3938 sq. in. (two conductors).
3.
4"
A 3-phase circuit of three 1,000 kcmil type THW conductors in a 4-in. rigid metal conduit.
Current rating of each phase = 545 amps.
Cross-section area per phase = 1.5482 sq. in.
4.
4-1/2"
A 3-phase circuit of six 600 kcmil type THW conductors in a 5-in. rigid metal conduit.
Current rating of each phase might appear to be = 2 X 420 = 840 amps.
But 80% derating must be applied because of the number of conductors in the conduit:
Current rating of each phase = 840 X 80% = 672 amps.
Cross-section area per phase = 2.0522 sq. in. (two conductors).

Fig. 310-5. The above circuit makeups represent typical considerations in the application of multiple-conductor circuits. (Sec. 310-4.)

stalled in a single conduit, the ampacity of each conductor must be derated from the ampacity value shown in NEC Table 310-16. The four circuit makeups show:

1. Without ampacity derating because there are more than three conductors in the conduit, circuit 2 would be equivalent to circuit 1.
2. A circuit of six 400 kcmils can be made equivalent in ampacity to a circuit of three 2,000 kcmils by dividing the 400s between two conduits [3 conductors/3-in. (76-mm) rigid metal conduit]. If three different phases are used in each of two 3-in. (76-mm) conduits for this circuit, the multiple circuit would not require ampacity derating to 80 percent, and its 670-A rating would exceed the 665-A rating of circuit 1.
3. Circuit 2 is almost equivalent to circuit 3 in ampacity.
4. Circuit 4 is equivalent to circuit 1 in ampacity, but uses less conductor copper and a smaller conduit. And the advantages are obtained even with the ampacity derating for conduit fill.

Except where the conductor size is governed by conditions of voltage drop, it is seldom economical to use conductors of sizes larger than 1,000 kcmil, because above this size the increase in ampacity is very small in proportion to the increase in the size of the conductor. Thus, for a 50 percent increase in the conductor size, i.e., from 1,000,000 to 1,500,000 cmil, the ampacity of a Type THW conductor increases only 80 A, or less than 15 percent, and for an increase in size from 1,000,000 to 2,000,000 cmil, a 100 percent increase, the ampacity increases only 120 A, or about 20 percent. In any case where single conductors larger than 500,000 cmil would be required, it is worthwhile to compute the total installation cost using single conductors and the cost using two (or more) conductors in parallel.

The next-to-last paragraph of Sec. 310-4 warns that when multiple conductors are used per circuit phase leg, they may require more space at equipment terminals to bend and install the conductors. Refer to Sec. 373-6.

Figure 310-6 shows an interesting application of parallel conductors. A 1,200-A riser is made up of three conduits, each carrying three phases and a neutral. At the basement switchboard, the 1,200-A circuit of three conductors per phase plus three conductors for the neutral originates in a bolted-pressure switch with a 1,200-A fuse in each of the three phase poles. Because the total of 12 conductors make up a *single* 3-phase, 4-wire circuit, a 400-A, 3-phase, 4-wire tap-off must tap all the conductors in the junction box at top. That is, the three phase A legs (one from each conduit) must be skinned and bugged together and then the phase A tap made from that common point to one of the lugs on the 400-A CB. Phases B and C must be treated the same way—as well as the neutral. The method shown in the photo was selected by the installer on the basis that the conductors in the right-hand conduit are tapped on this floor, the center-conduit conductors are tapped to a 400-A CB on the floor above, and the left-conduit conductors are tapped to a 400-A CB on the floor above that. But such a hookup can produce excessive current on some of the 500 kcmils. Because it does not have the parallel conductors of equal length at points of load-tap, the currents will not divide equally, and this is a violation of the second paragraph of Sec. 310-4, which calls for parallel conductors to "be the same length."

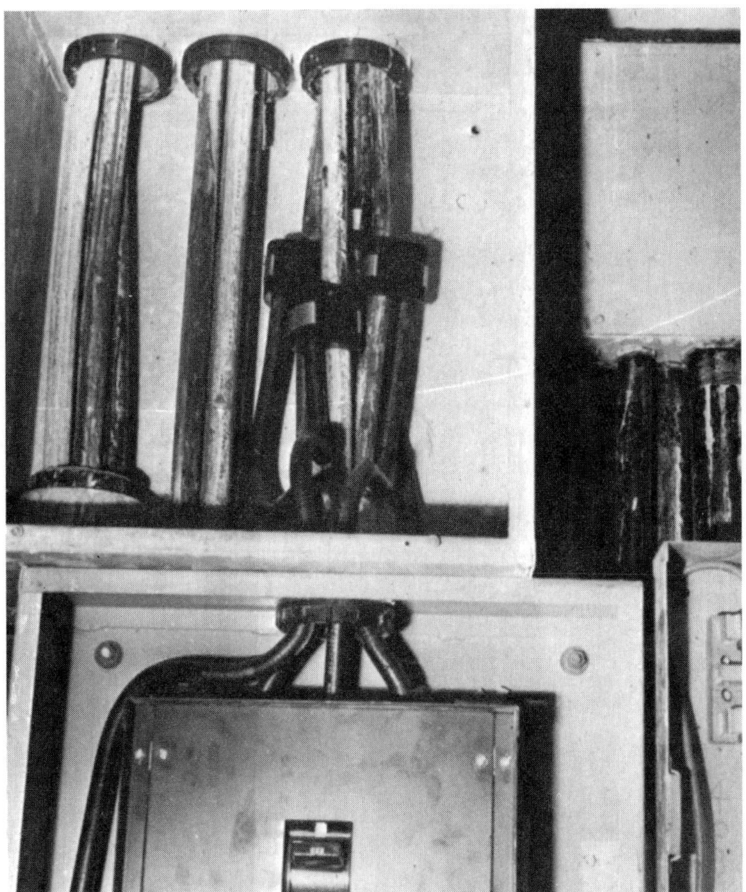

Fig. 310-6. A 1,200-A circuit of three sets of four 500-kcmil conductors (top) is tapped by a single set of 500 kcmils to a 400-A CB (bottom) that feeds an adjacent meter center in an apartment house. This was ruled a violation because the tap must be made from all the conductors of the 1,200-A circuit. (*Note:* The conduits feeding the splice box at top are behind the CB enclosure at bottom.) (Sec. 310-4.)

Exception No. 1 of this section clearly indicates long-time Code acceptance of paralleling conductors smaller than No. 1/0 for use in traveling cables of elevators, dumbwaiters, and similar equipment.

Exception No. 2 of Sec. 310-4 permits parallel-circuit makeup using conductors smaller than 1/0—*but all the conditions given must be observed.*

This Exception permits use of smaller conductors in parallel for circuit applications where it is necessary to reduce conductor capacitance effect or to reduce voltage drop over long circuit runs. As it was argued in the proposal for this Exception—

> If a No. 14 conductor, for example, is adequate to carry some load of not more than the 15-amp rating of the wire, there can be no reduction in safety by using two No. 14

wires per circuit leg to reduce voltage drop to acceptable limits—with a 15-amp fuse or CB pole protecting each pair of No. 14s making up each leg of the circuit.

Where conductors are used in parallel in accordance with this Exception, the rule requires that *all* the conductors be installed in the same raceway or cable. And that will dictate application of the last sentence of Sec. 310-4: "Conductors installed in parallel *shall* comply with the provisions of Art. 310 Note 8, Notes to Ampacity Tables of 0 to 2000 Volts." Thus a single-phase, 2-wire control circuit made up of two No. 14s for each of the two legs of the circuit would have to be considered as four conductors in a conduit, and the "ampacity" of each No. 14 would be reduced to 80 percent of the value shown in Table 310-16. If TW wires are used for the circuit described, the ampacity of each is no longer the value of 20 A, as shown in Table 310-16. With four of them in a conduit, each would have an ampacity of 0.8 × 20, or 16 A. Then using a 15-A fuse or CB pole for each pair of No. 14s would properly protect the conductors and would also comply with the "dagger" footnote of Table 310-16, which says that No. 14 must not have overcurrent protection greater than 15 A. (See Fig. 310-7.)

Parallel makeup with smaller than 1/0 conductors

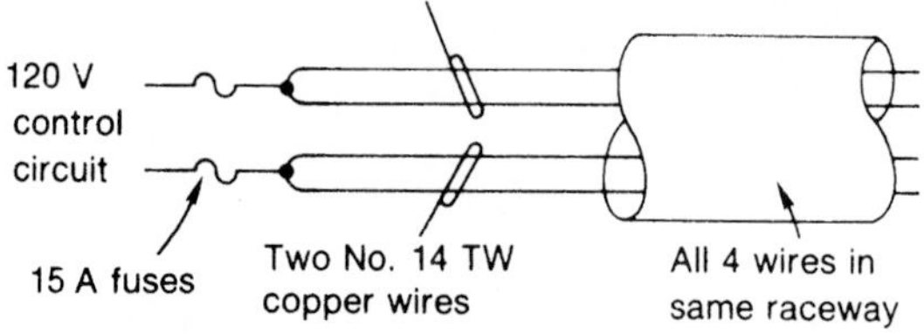

From Table 310-16, No. 14 TW copper has ampacity of 20.

"Ampacity reduction" from Note 8 to table: 20 amps × 0.8 = 16 amps.

15-AMP FUSES PROPERLY PROTECT THE NO. 14s AND SATISFY FOOTNOTE TO TABLE 310-16.

Fig. 310-7. Overcurrent protection must be rated not in excess of the ampacity of one conductor when conductors smaller than No. 1/0 are used in parallel. (Sec. 310-4.)

Exception No. 3 permits circuits operating at frequencies of 360 Hz or higher to use conductors smaller than 1/0 in parallel. Exception No. 3 permits parallel use of conductors smaller than 1/0 for circuits operating at 360 Hz or higher frequencies, provided that all the wires are in the same conduit, the ampacity of each wire is adequate to carry the entire current that is divided among the parallel wires, and the rating of the circuit protective device does not exceed the ampacity of any one of the wires. Such use of small conductors in parallel is very effective in reducing inductive reactance and "skin effect" in high-frequency circuits. Interweaving of the multiple wires per phase and neutral produces greater mutual cancellation of the magnetic fields around the wires and thereby lowers inductance and skin effect. Typical application of such usage is made for the 400-Hz circuits that are standard in the aerospace and aircraft industry.

Exception No. 4 now recognizes the use of parallel conductors in sizes down to No. 2 where used as a neutral, but only in an existing installation. This is a good idea where necessary to accommodate additive harmonics on the neutral of multiwire circuits. The use of two No. 2s provides about 25 percent *less* cross-sectional area, at the same time providing 25 percent *more* surface area than a 3/0. This serves to reduce the heating caused by "skin effect" because the "skin" area has been increased.

The limitation to "existing installation" is somewhat puzzling since this technique would be useful in new construction. It seems as if the need for paralleling must first be demonstrated through measurements. In any event paralleling grounded conductors smaller than 1/0 may only be done at existing installations.

310-6. Shielding. The effect of this **Code** rule is to require all conductors operating over 2 kV to be shielded, *unless* the conductor is UL-listed for operation unshielded at voltages above 2 kV. Because 2,300-V delta (which is over 2 kV) is the lowest general-purpose, high-voltage circuit in use today, unlisted conductors *must* be shielded for such circuits and any other voltages above that—such as 4,160/2,300-V, 3-phase, 4-wire wye (grounded or ungrounded neutral). *But note this*—UL does list 5-kV unshielded conductors for use in accordance with Sec. 310-6, Table 310-63, and other **Code** rules (Fig. 310-8).

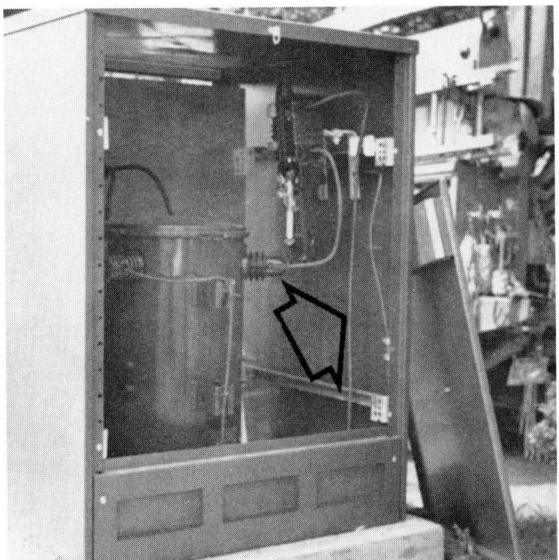

Fig. 310-8. A nonshielded conductor (arrow) is permitted for use on a 2,300-V circuit (phase-to-neutral), as shown here, *only* if the conductor is listed by UL or another national test lab and approved for use without electrostatic shielding. (Sec. 310-6.)

UL also lists shielded polyethylene insulated conductors up to 35 kV. And, in accordance with NE Code Table 310-64, UL has been listing Type RHH insulated conductors (rubber or cross-linked polyethylene insulation) with electrostatic shielding for operation up to 5 kV.

In addition to applicable NE Code, Insulated Power Cable Engineers Association (IPCEA), and UL data on use of cable shielding, manufacturers' data should be consulted to determine the need for shielding on the various types and constructions of available cables.

Shielding of high-voltage cables protects the conductor assembly against surface discharge or burning (due to corona discharge in ionized air) which can be destructive to the insulation and jacketing. It does this by confining and distributing stress in the insulation and eliminating charging current drain to intermittent grounds. It also prevents ionization of any tiny air spaces at the surface of the insulation by confining electrical stress to the insulation. Shielding, which is required by this Code rule to be effectively grounded, increases safety to human life by eliminating the shock hazard presented by the external surface of unshielded cables. By preventing electrical discharges from cable surfaces to ground, shielding also reduces fire or explosive hazards and minimizes any radio interference high-voltage circuits might cause.

Electrostatic shielding of cables makes use of both nonmetallic and metallic materials. As shown in accompanying sketches of typical cable assemblies, semiconductive tapes or extruded coverings of semiconductive materials are combined with metal shielding to perform the shielding function. Metallic shielding may be done with:

1. A copper shielding tape wrapped over a semiconducting shielding of nonmetallic tape that is applied over the conductor insulation (Fig. 310-9)

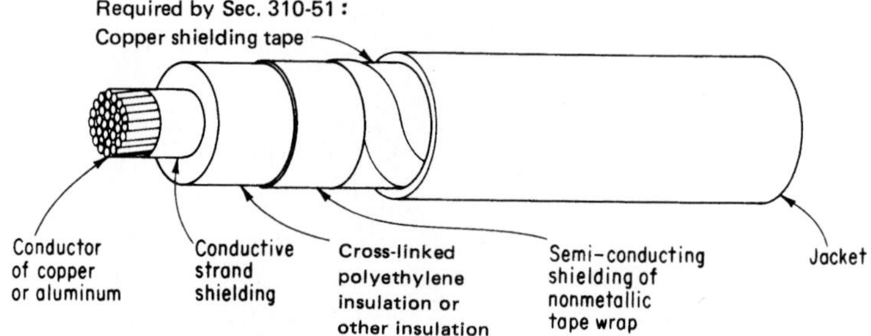

Fig. 310-9. A flat copper tape spiraled over the insulation is an electrostatic shield. (Sec. 310-6.)

2. A concentric wrapping of bare wires over a semiconducting, nonmetallic jacket over the conductor insulation (Fig. 310-10)
3. Bare wires embedded in the semiconducting, nonmetallic jacket that is applied over the insulation (Fig. 310-11)
4. A metal sheath over the conductor insulation, as with lead-jacketed cable

For many years, high-voltage shielded power cables for indoor distribution circuits rated from 5 to 15 kV were of the type using copper tape shielding and

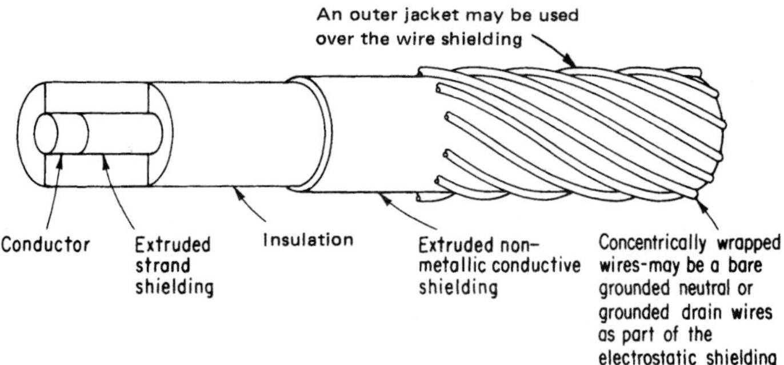

Fig. 310-10. Wires, instead of metal tape, are also used for electrostatic shielding (URD and UD type). (Sec. 310-6.)

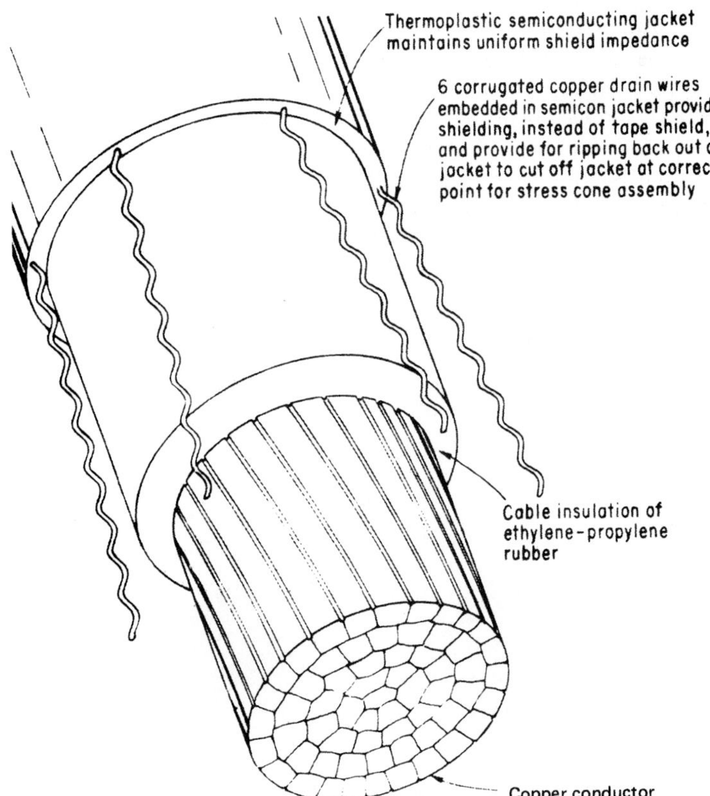

Fig. 310-11. Wires embedded in semiconducting jacket form another type of shielding. (Sec. 310-6.)

an outer overall jacket. But in recent years, cables shielded by concentric-wrapped bare wires have also come into widespread use—particularly for underground outdoor systems up to 15 kV. These latter cables are the ones commonly used for underground residential distribution (called "URD"). Such a conductor is shown in Fig. 310-10.

In addition to use for URD (directly buried with the concentric-wire shield serving as the neutral or second conductor of the circuit), concentric-wire-shielded cables are also available for indoor power circuits, such as in conduit, with a nonmetallic outer jacket over the concentric wires. Such cable assemblies are commonly called "drain-wire-shielded" cable rather than "concentric-neutral" cable because the bare wires are used only as part of the electrostatic shielding and not also as a neutral. Smaller-gage wires are used where they serve only for shielding and not as a neutral.

Figure 310-11 shows drain-wire-shielded high-voltage cable with electrostatic shielding by means of drain wires *embedded* in a semiconducting jacket over the conductor insulation. This type of drain-wire-shielded conductor is designed to be used for high-voltage circuits in conduit or duct for commercial and industrial distribution as an alternative to tape-shielded cables. For the same conductor size, this type of embedded drain-wire-shielded cable has a smaller outside diameter and lighter weight than a conventional tape-shielded cable. For the drain-wire cable the assembly difference reduces installation labor, permits reduced bending radius for tight conditions and easier pulling in conduit, and affords faster terminations (with stress cones) and splices. An extremely important result of the smaller overall cross-section area (csa) of the drain-wire-shielded cable is the chance to use smaller conduits—with lower material and labor costs—when conduit is filled to 40 percent of its csa based on the actual cable csa, as covered by Note 5 to the tables in Chap. 9 of the NE Code.

Another consideration in conductor assemblies is that of strand shielding. As shown in Fig. 310-12, a semiconducting material is tape-wrapped or extruded onto the conductor strands and prevents voids between the insulation and the strands, thereby reducing possibilities of corona cutting on the inside of the insulation.

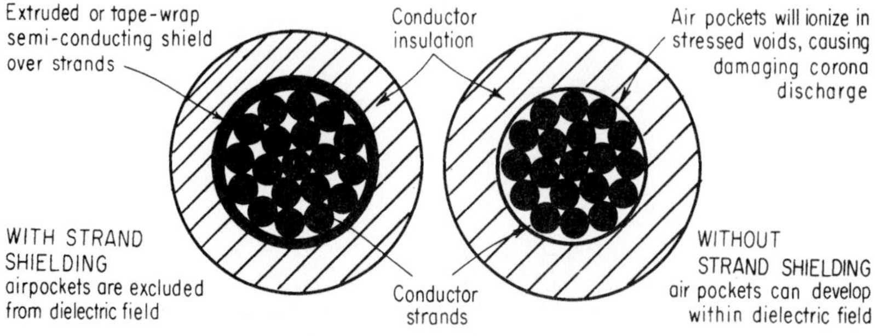
Extruded or tape-wrap semi-conducting shield over strands
Conductor insulation
Air pockets will ionize in stressed voids, causing damaging corona discharge
WITH STRAND SHIELDING
airpockets are excluded from dielectric field
Conductor strands
WITHOUT STRAND SHIELDING
air pockets can develop within dielectric field

Fig. 310-12. Strand shielding is part of the overall electrostatic shielding system on the conductor. (Sec. 310-6.)

Refer to Sec. 710-6 on terminating and grounding shielded conductors.

310-7. Direct Burial Conductors. The second sentence in this section appears to require shielding only on direct burial cables rated above 2,000 V, although a rule in Sec. 710-4(b) says without reference to voltage that "Nonshielded cables shall be installed" in conduit, which, in effect, prohibits direct burial of any nonshielded cables rated over 600 V (see Fig. 310-13). Correlation between this

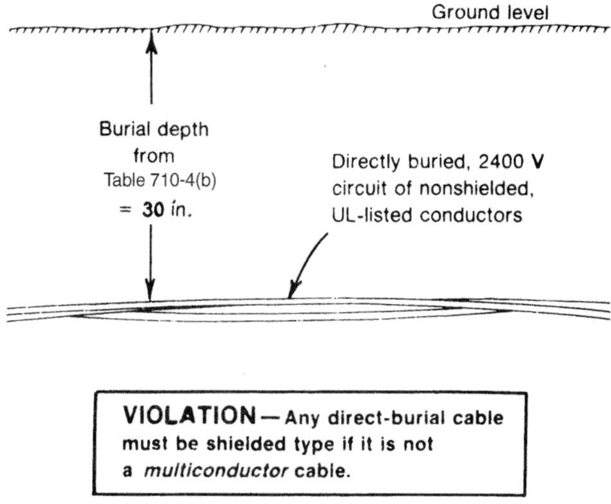

Fig. 310-13. This application is covered by both Sec. 310-7 and Sec. 710-4(b). (Sec. 310-7.)

section and Sec. 710-4 must be carefully made, because the Exception to Sec. 310-7 recognizes "nonshielded multiconductor cables" for direct burial in ratings up to 5 kV provided that the cable has an overall metallic sheath or armor.

This section also requires direct burial high-voltage cables to be "identified for such use," which, in effect, means listed by a test organization or designated by the inspection agency as suitable for direct burial.

Code rules on underground use of conductors rated up to 600 V are given in Sec. 310-8.

310-8. Wet Locations. Any conductor used in a "wet location" (refer to the definition under "location" in Art. 100) *must* be one of the designated types—each of which has the letter "W" in its marking to indicate suitability to *wet* locations. Any conduit run underground is assumed to be subject to water infiltration and is, therefore, a *wet location,* requiring use of only the listed conductor types within the raceway.

Figure 310-14 shows a clear violation of the last sentence of Sec. 310-8. In the photo, conductors marked RHW are run, from the junction box below the magnetic contactor, directly buried in the ground. Although Type RHW is suitable for wet locations, it is not approved for direct burial. If, however, the conductors were of the type that is marked "RHW-USE"—that is, it is listed and recognized

Fig. 310-14. Bundle of conductors (arrow) are Type RHW individual building conductors that would be suitable for installation in conduit underground but are not marked "USE"; and their use here, run directly buried to outdoor lighting poles, constitutes a violation of the last sentence of Sec. 310-8. (Sec. 310-8.)

as *both* a single-conductor RHW and a single-conductor Type USE (underground service entrance) cable—then such conductors would satisfy this section.

Where Sec. 310-7 refers specifically to "direct burial conductors," requires that they be "identified," and is directed at high-voltage cables, the rule in this section (which applies to conductors rated up to 600 V) simply says "conductors" for direct burial must be "listed"—which means certified by some kind of test lab—and not just "identified"—which means evaluated by other than the manufacturer but not necessarily "listed."

UF cable is acceptable for direct earth burial. Although Sec. 338-1(b) says USE cable is OK for "underground use," it does not say it is OK for direct burial. But the UL *Electrical Construction Materials Directory* (Green Book) notes that listed USE cable is recognized for "burial directly in the earth" (Fig. 310-15).

Besides UF and USE, then, what other cables can be directly buried? Section 334-3(5) does recognize MC cable for direct burial "when identified for such use." Sections 330-3(5) and (9) recognize MI cable for direct burial. Note that Sec. 330-3(6) permits MI in "fill" below grade.

For burial-depth requirements on directly buried cables, refer to Sec. 300-5, Table 300-5, and part **(i)** of that section. Cables approved for direct earth burial must be installed a minimum of 24 in. (610 mm) below grade, as given in Table 300-5, or at least 30 in. (762 mm) below grade for high-voltage cables as covered in Table 710-8(b), with its Exceptions.

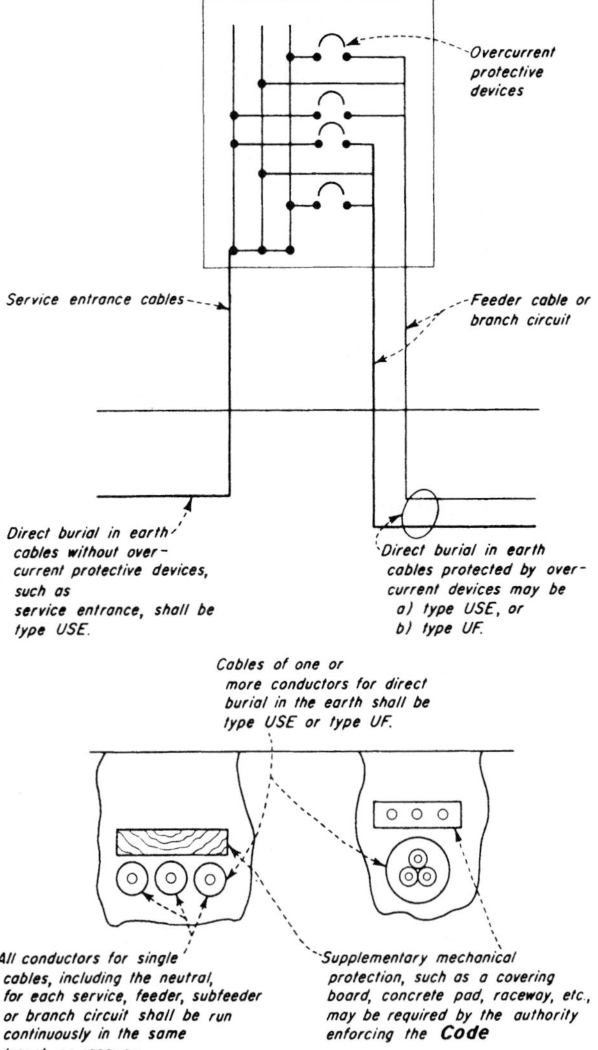

Fig. 310-15. Types USE and UF cables are designated by the letter "U" for underground use. (Sec. 310-8.)

Direct burial conductors should be trench-laid without crossovers; slightly "snaked" to allow for possible earth settlement, movement, or heaving due to frost action; and have cushions and covers of sand or screened fill to protect conductors against sharp objects in trenches or backfill. Figure 310-16 shows some recommended details on installing direct burial cables. Moreover, when conductors are routed beneath roadways or railroads, they should be additionally protected by conduits. And, to guard against damage which might occur

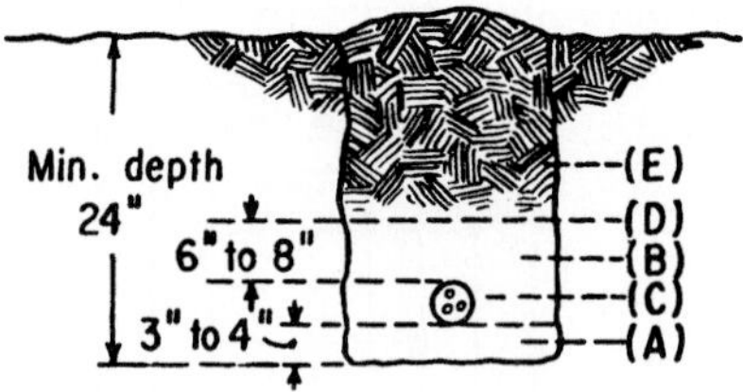

A—Soft bed of sand or screened fill.

B—Blanket of sand or screened fill 6 in. to 8 in. above top of cable.

C—Cable "snaked" slightly in trench for slack when earth settles. Keep single-conductor cables uniformly apart about 6 in. in trench. Avoid cable crossovers. Keep cable below frost line.

D—Add protective slab (creosoted plank, etc.) on sand fill in areas where future digging might occur. Enclose cable in pipe or conduit under highways or rail tracks.

E—Normal backfill.

Fig. 310-16. This satisfies the intent of Sec. 300-5(f). (Sec. 310-8.)

during future digging, conductors in soft fill should be covered by concrete slabs or treated planks.

Where prewired cable-in-conduit is being buried, it also should be slightly snaked, although it is unnecessary to provide sand beds or screen the backfill. Inasmuch as these complete conductor-raceway assemblies can be delivered on reels in specified factory-cut lengths, installation is simplified and expedited.

310-9. Corrosive Conditions. Figure 310-17 shows how conductors are marked to indicate that they are gasoline- and oil-resistant, such as Type THHN-THWN, for use in gasoline stations and similar places.

Fig. 310-17. Typical marking indicates suitability of conductors for use under unusual environmental conditions. (Sec. 310-9.)

310-10. Temperature Limitation of Conductors. This requirement is extremely important and is the basis of safe operation of insulated conductors. As shown in Table 310-13, conductors have various ratings—60°C, 75°C, 90°C, etc.

Since Tables 310-16 through 310-19 are based on an assumed ambient (surrounding) temperature of 30°C (86°F) (or 40°C), conductor ampacities are based on the ambient temperature plus the heat (I^2R) produced by the conductor (wire) while carrying current. Therefore, the type of insulation used on the conductor determines the maximum permitted conductor ampacity.

example A No. 3/0 THW copper conductor for use in a raceway has an ampacity of 200 according to Table 310-16. In a 30°C ambient the conductor is subjected to this temperature when it carries *no* current. Since a THW-insulated conductor is rated at 75°C, this leaves 45°C (75 minus 30) for increased temperature due to current flow. If the ambient temperature exceeds 30°C, the conductor maximum load-current rating must be reduced proportionally (see "Correction Factors" at the bottom of Table 310-16) so that the total temperature (ambient plus conductor temperature rise due to current flow) will not exceed the temperature rating of the conductor insulation (60°C, 75°C, etc.). For the same reason, conductor maximum load-current ratings must be reduced below the ampacity values where more than three conductors are contained in a raceway or cable (see Note 8 to Tables 310-16 through 310-19).

While it can be shown that smaller conductors, such as Nos. 14 and 12 60°C-insulated conductors, will not reach 60°C at their assigned ampacities (Table 310-16) in a 30°C ambient, ampacities beyond those listed in Tables 310-16 through 310-19 would create excessive voltage drop (*IR* drop) and would not be compatible with some termination devices.

Although conductor ampacities increase with the rating of conductor insulation, it should be noted that some terminations are designed only for 60° or 75°C maximum temperatures (ambient plus current). Accordingly, the higher-rated ampacities for conductors of 90°C, 110°C, etc., cannot be utilized unless the terminations have comparable ratings or where derating of such higher-amp conductors brings load current down to the allowable ampacities for 60° or 75°C conductors of the same size.

To find the temperature in degrees Fahrenheit (F) where the temperature is given in degrees Celsius (C), apply the formula

$$\text{Degrees F} = \tfrac{9}{5} \times \text{degrees C} + 32$$

Thus, the maximum operating temperature for Type T insulation is 60°C, $\tfrac{9}{5} \times 60° = 108°$. And $108° + 32° = 140°$, which is the same temperature on the Fahrenheit scale as 60 degrees on the Celsius scale.

Reversing the process, where the temperature is given in degrees F, gives

$$\text{Degrees C} = (\text{degrees F} - 32) \times \tfrac{5}{9}$$

The maximum operating temperature for Type THW insulation is 167°F.

$$167° - 32° = 135°$$

$135° \times \tfrac{5}{9} = 75°$, the corresponding temperature in degrees C.

Watch out when conductors are used in locations with elevated ambient temperatures—in boiler rooms, near furnaces, etc. All load ratings are based on a given ambient—such as 30°C, 86°F for conductors covered by **NE Code** Table 310-16. It is up to the designer and/or installer to make the necessary deratings *required* by the "Correction Factors" given with those tables. Equipment deterioration and ultimate thermal failure are the price of carelessness. Moisture or excessive dampness that may degrade aluminum terminations can also result in high-resistance terminations with resultant heating that damages or destroys equipment and conductors.

310-12. Conductor Identification. For part **(a)**, refer to the discussion given for Secs. 200-6 and 200-7. For part **(b)** refer to Exception Nos. 1 and 3 in Sec. 250-57(b). Section 310-12(a), Exception No. 4, now recognizes the use in multiconductor cables of a *grounded* conductor that is not white throughout its entire length provided that only qualified persons will service the installation. The rule requires that such grounded conductors be identified by white marking at their termination at the time of installation.

Similarly, a *grounding* conductor in a multiconductor cable may be identified at each end and at every point where the conductor is accessible by stripping the insulation from the entire exposed length or by coloring the exposed insulation green or by marking with green tape or green adhesive labels [Sec. 310-12(b), Exception No. 2].

310-13. Conductor Constructions and Applications. Table 310-13 presents application and construction data on the wide range of 600-V insulated, individual conductors recognized by the NE Code, with the appropriate letter designation used to identify each type of insulated conductor. Figure 310-18 shows a typi-

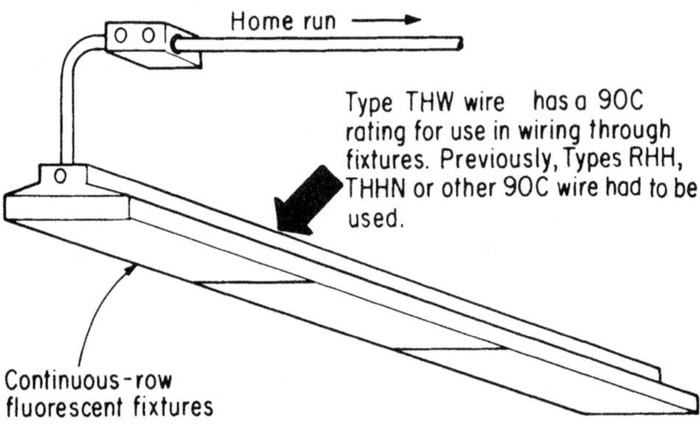
Home run
Type THW wire has a 90C rating for use in wiring through fixtures. Previously, Types RHH, THHN or other 90C wire had to be used.
Continuous-row fluorescent fixtures

Fig. 310-18. THW wire has the 90°C rating required of conductors within 3 in. (76 mm) of a ballast (Sec. 410-31). (Sec. 310-13.)

cal detail on application, as covered for Type THW conductor in NEC Table 310-13. Type THW wire has a special application provision for electric-discharge lighting, which makes THW the answer for installers needing a 90°C conductor for wiring end-to-end fixtures in compliance with Sec. 410-31.

Important data that should be noted in Table 310-13 are as follows:

1. The designation for "thousand circular mils" is "kcmil," which has been substituted for the long-time designation "MCM" in this table and throughout the NEC.
2. Type MI (mineral insulated) cable may have either a copper or an alloy steel sheath.

3. Type RHW-2 is a conductor insulation that is moisture- and heat-resistant rubber with a 90°C rating, for use in dry and wet locations.
4. Type XHHW-2 is a moisture- and heat-resistant cross-linked synthetic polymer with a 90°C rating, for use in dry and wet locations.
5. The suffix "LS" designates a conductor insulation to be "low smoke" producing and flame retardant. For example, Type THHN/LS is a THHN conductor with a limited smoke-producing characteristic.
6. Type THHW is a moisture- and heat-resistant insulation, rated at 75°C for wet locations and 90°C for dry locations. This is similar to THWN and THHN without the outer nylon covering but with thicker insulation.
7. All insulations using asbestos—A, AA, AI, AIA, AVA, etc.—have been deleted from Table 310-13 because they are no longer made.

Conductors for high-voltage circuits (over 2,000 V) must satisfy the specifications of Tables 310-61 through 310-84.

Conductors intended for 600-V (and up to 2,000-V) general wiring under the requirements of the National Electrical Code are required to be one of the recognized types listed in Code Table 310-13 and not smaller than No. 14 AWG. The National Electrical Code does not contain detailed requirements for insulated conductors since these are covered in separate standards such as those of Underwriters Laboratories Inc.

"Dry locations" in this case would mean for "general use" in dry locations.

Table 310-13 permits maximum operating temperatures of 90°C (194°F) in dry and damp (but *not* "wet") locations for Types FEP, FEPB, RHH, XHHW, and THHN wire; but the load-current ratings for Nos. 14, 12, and 10 copper conductors and Nos. 12 and 10 aluminum conductors are limited to those permitted by the maximum overcurrent protection ratings given in the footnote to Table 310-16. One reason is the inability of 15-, 20-, and 30-A CBs to protect these sized conductors against damage under short-circuit conditions. The other reason is that the wiring devices which are commonly connected by these sizes of conductors are not suitable for conditions encountered at higher current loadings.

Terminals of 15- and 20-A receptacles not marked "CO/ALR" are for use with copper and copper-clad aluminum conductors only. Terminals marked "CO/ALR" are for use with aluminum, copper, and copper-clad aluminum conductors. Screwless pressure terminal connectors of the conductor push-in type are for use only with copper and copper-clad aluminum conductors.

Terminals of receptacles rated 30 A and above not marked "AL-CU" are for use with copper conductors only. Terminals of receptacles rated 30 A and above marked "AL-CU" are for use with aluminum, copper, and copper-clad aluminum conductors.

The conductor material known as copper-clad aluminum is made from a metallurgical materials system by using a core of aluminum with a bonded outer skin of copper. There is 10 percent copper by volume (the outer skin) and 26.8 percent by weight. Terminations for copper-clad aluminum conductors should be marked "AL-CU" except where listings by Underwriters Laboratories indicate otherwise.

310-15. Ampacities. Sec. 310-15 states that ampacities of conductors may be determined by *either* of two methods. The first method is described in part **(a)** and is the old, tested, and familiar method of the NEC, based on Tables 310-16 through 310-19. The second permitted method is covered in part **(b)** of the NEC and is the complex, confusing, incomplete, and defective procedure that was presented in the 1987 NEC as the basic method, based on an elaborate formula given in part **(b)**.

The NEC ampacity determination procedure using the formula is permitted as an optional alternative "under engineering supervision." All of the ampacity tables based on the formula are in App. B in the back of the Code book, where information on the formula method and its related ampacity tables is introduced with the sentence, "This appendix is **not** part of the requirements of this Code, but is included for information purposes only." Thus, the 1987 NEC ampacity method is given as a nonmandatory, optional alternative to the old standby method.

In part **(a)**, a fine-print note (FPN) points out that "Tables 310-16 through 310-19 are application tables that are for use in determining conductor size on loads calculated in accordance with Article 220." Inasmuch as the NEC itself *requires* that Art. 220 be used at all times in calculating loads, the ampacity-determination method of part **(a)** is completely adequate for all conductor sizing in accordance with all Code rules.

It should be noted that even though conductors must now have an ampacity that is at least equal to the noncontinuous load plus 125 percent of the continuous load, before derating, as now required by Secs. 210-22(c), 220-3(a), and 220-10(b), the method for determining conductor ampacity remains the same. As indicated, to satisfy the new minimum sizing requirements, the conductor ampacity must have a table value at least equal to the rating of the overcurrent device before derating *and* must be properly protected as well as be adequate for the actual load after derating at the actual temperature rating permitted the overcurrent device's terminations.

The basic ampacity determination procedure of part **(a)** is to use Tables 310-16 through 310-19, with their 11 notes. The most commonly used table will be Table 310-16, which covers all 60, 75, and 90°C insulated copper, aluminum, and copper-clad aluminum conductors used in *any* raceway or *any* cable, either indoors or outdoors—above ground or underground—and any cable or conductor directly buried in the earth.

Tables 310-16 through 310-19 and accompanying Notes 1 through 11 provide the maximum continuous ampacities for copper, aluminum, and copper-clad aluminum conductors. Table 310-16 covers conductors rated up to 2,000 V where not more than three conductors are installed in raceway or cable or are directly buried in the earth—based on an ambient of 30°C.

Table 310-17 covers both copper conductors and aluminum or copper-clad aluminum conductors up to 2,000 V where conductors are used as single conductors in free air based on an ambient of 30°C.

Tables 310-18 and 310-19 apply to conductors rated 150 to 250°C, used either in raceway or cable or as single conductors in free air, based on an ambient of 40°C. Care must be taken in using these tables and in noting references to them throughout the Code.

All of these tables and their notes are intended to cover any condition of application that might be encountered. This overall method will indicate the "ampacity" of any of the conductor types for any condition of use. NEC Table 310-16, for instance, specifies ampacities for conductors where not more than three current-carrying conductors are contained in a single raceway or cable or directly buried in the earth provided that the ambient temperature is not in excess of 30°C (86°F). For higher ambient temperatures, the ampacity must be derated in accordance with the "Ampacity Correction Factors" at the bottom of each table. For applications of *more* that three current-carrying conductors in a conduit or cable, the ampacity must be derated in accordance with Note 8 to Table 310-16. And where *both* elevated ambient (above 30°C) and more than three conductors in a conduit or cable are present, both deratings must be made—one on top of the other. First, an ampacity must be developed from the appropriate table—correcting for the elevated ambient—then the derating factor given in Note 8 for the number of current-carrying conductors contained within the raceway or cable must be applied to determine conductor ampacity.

Table 310-16 gives ampacities under the two conditions described: when the raceway or cable containing the conductors is operating in an ambient temperature not over 30°C (86°F) and when there are not more than three current-carrying conductors in the raceway or cable. Under those conditions, the ampacity value shown in the table corresponds to the thermal limit of each particular insulation. But in any case, where either or both of the two conditions are exceeded, the ampacity of a conductor is *not* the value shown in the table and must be reduced from that value, using the given derating factors. And then protection must be based upon or provided at the reduced ampacity to ensure that the temperature limit of the insulation is not exceeded.

It should be clearly understood that any reduced ampacity, required because of a higher ambient and/or conductor bundling (Note 8), has the same meaning as the value shown in the table; each represents a current value above which excessive heating would occur under the particular conditions. And if there are two conditions that lead to excessive heating, then a greater reduction in current is required than if only one such condition exists.

Using the Ampacity Tables

An important step in the design of circuits is selection of the type of conductor to be used—TW, THW, THWN, RHH, THHN, XHHW, etc. The various types of conductors are covered in Art. 310 of the NE Code, and the ampacities of conductors with the different insulations and temperature ratings are given in Tables 310-16 through 310-19 for the varying conditions of use—in a raceway, in open air, at normal or higher-than-normal ambient temperatures. Conductors must be used in accordance with all the data in those tables and notes.

In selecting the type and temperature rating of wire for circuits, consideration must be given to a very important UL qualification indicated for the temperature ratings of equipment terminations. Although application data on minimum required temperature ratings of conductors connected to equipment terminals are not given in the NE Code, they nevertheless become part of the mandatory regulations of the Code because of Code Sec. 110-3(b). This section incorporates the instructions in UL and other listing books as part of the Code itself. It reads as follows:

> Listed or labeled equipment shall be used, installed, or both, in accordance with any instructions included in the listing or labeling.

A basic rule in the UL *Electrical Construction Materials Directory* states:

> **Distribution and Control Equipment Terminations** – Most terminals are suitable for use only with copper wire. Where aluminum or copper-clad aluminum wire can or shall be used, (some crimp terminals may be Listed only for aluminum wire) there is marking to indicate this. Such marking is required to be independent of any marking on terminal connectors, such as on a wiring diagram or other visible location. The marking may be in an abbreviated form such as "AL-CU".
>
> Except as noted in the following paragraphs or in the information at the beginning of some product categories, the termination provisions are based on the use of 60C ampacities for wire sizes No. 14-1 AWG, and 75C ampacities for wire sizes Nos. 1/0 AWG and larger, as specified in Table 310-16 of the National Electrical Code.
>
> Some distribution and control equipment is marked to indicate the required temperature rating of each field-installed conductor. If the equipment, normally intended for connection by wire sizes within the range 14-1 AWG, is marked "75C only" or "60/75C", it is intended that 75C insulated wire may be used at full 75C ampacity. Where the connection is made to a circuit breaker or switch within the equipment, such a circuit breaker or switch must also be marked for the temperature rating of the conductor.
>
> A 75C conductor temperature marking on a circuit breaker or switch normally intended for wire sizes 14-1 AWG does not in itself indicate that 75C insulated wire can be used unless (1) the circuit breaker or switch is used by itself, such as in a separate enclosure, or (2) the equipment in which the circuit breaker or switch is installed is also so marked.
>
> "A 75 or 90C conductor temperature marking on a terminal (e.g. AL7, CU7AL, AL7CU or AL9, CU9AL, AL9CU) does not in itself indicate that 75 or 90C insulated wire can be used unless the equipment in which the terminals are installed is marked for 75 or 90C."
>
> Higher temperature rated conductors than specified may be used if the size is based on the above statements.

This temperature limitation on terminals applies to the terminals on all equipment—circuit breakers, switches, motor starters, contactors, etc.—except where some other specific condition is recognized in the general information preceding the product category. Figure 310-19 illustrates this vitally important matter, which has been widely disregarded in general practice. When terminals are tested for suitability at 60 or 75°C, the use of 90°C conductors operating at their higher current ratings poses definite threat of heat damage to switches, breakers, etc. Many termination failures in equipment suggest overheating even where the load current did not exceed the current rating of the breaker, switch, or other equipment.

When a 60°C-rated terminal is fed by a conductor operating at 90°C, there will be substantial heat conducted from the 90°C conductor metal to the 60°C-rated terminal; and, over a period of time, that can damage the termination—even though the load current does not exceed the equipment current rating and does not exceed the ampacity of the 90°C conductor. Whenever two metallic parts at different operating temperatures are tightly connected together, the higher-

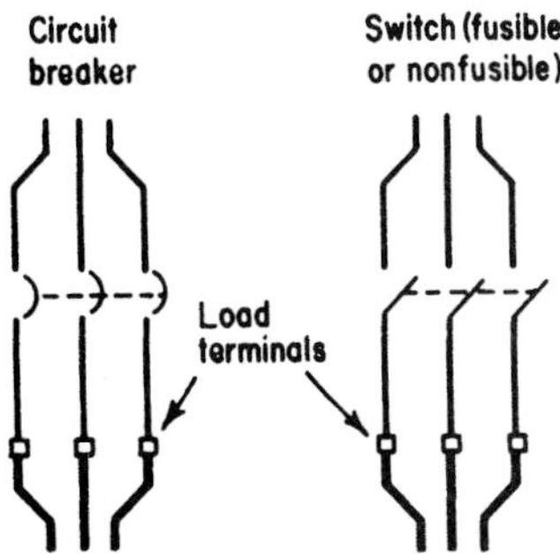

Unless a circuit breaker or switch is marked otherwise, circuit conductors connected to the terminals must not operate at more than a 60C ampacity for conductors in sizes No. 14 to No. 1 AWG and must not operate at more than a 75C ampacity for conductors in sizes No. 1/0 AWG and larger [refer to *NE Code* Tables 310-16 through -19].

IN GENERAL —
FOR CBs, SWITCHES, CONTACTORS, ETC. RATED 125 AMPS OR LESS Use TW wire (or use THW, THHN, RHH, XHHW or other higher-temperature wire at the ampacity of the corresponding size of TW wire).

FOR CBs, SWITCHES, CONTACTORS, ETC. RATED OVER 125 AMPS—Use TW wire at its 60 C ampacities or use THW, THWN or XHHW wire at ampacities permitted up to 75C (or use RHH, THHN or other higher-temperature wire at the ampacity of the corresponding size of 75C wire)

Fig. 310-19. UL specifies maximum temperature rating for conductors connecting to equipment terminals. (Sec. 310-15.)

temperature part (say 75 or 90°C wire) will give heat to the lower-temperature part (the 60°C terminal) and thereby raise its temperature over 60°C.

For any given size of conductor, the greater ampacity of a higher-temperature conductor is established by the ability of the conductor insulation to withstand the I^2R heat produced by the higher current flowing through the conductor. But it must not be assumed that the equipment to which that conductor is connected also is capable of withstanding the heat that will be thermally conducted from the metal of the conductor into the metal of the terminal to which the conductor is tightly connected.

Although this limitation on the operating temperature of terminals in equipment does somewhat reduce the advantage that higher-temperature conductors have over lower-temperature conductors, there are still many advantages to using the higher-temperature wires.

NEC Ampacities

As described in the definition of "ampacity" (in Art. 100), the ampacity of a conductor is the amount of current, in amperes, that the conductor can carry continuously under specified conditions of use without developing a temperature in excess of the value that represents the maximum temperature that the conductor insulation can withstand. For any particular application, one of the four NEC ampacity tables must be consulted—depending upon the particular wiring method (wires in raceway, cable, individual insulated wires, etc.) and depending upon the manner of installation (in free air, directly buried in the ground, in raceway underground, as messenger supported wiring, etc.). From the table that corresponds to the specific wiring method and conditions of use, the basic ampacity of any size and insulation of copper or aluminum conductor can be determined. And then any required adjustments can be made to the ampacity value, as needed for ambient temperature and number of conductors in a raceway or cable. NEC Table 310-16, for instance, specifies ampacities for conductors where not more than three conductors are contained in a single raceway or cable or earth provided that the ambient temperature is not in excess of 30°C (86°F). For higher ambient temperatures, the ampacity must be derated in accordance with the "Ampacity Correction Factors" at the bottom of each table. For applications of *more* than three conductors in a conduit or cable, the ampacity of conductors must be derated in accordance with Note 8 to Table 310-16. And where *both* elevated ambient (above 30°C) and more than three conductors in a conduit or cable are present, *both* deratings must be made—one on top of the other.

The NEC approach to setting ampacities of conductors is aimed at designating that level of current that will cause the conductor to reach its thermal limit—the current that the conductor can carry safely *but* above which the temperature rating of the conductor insulation would be exceeded and the conductor exposed to thermal degradation or damage to its insulation. This concept of ampacity is verified in the first sentence of the FPN to Sec. 240-1, where the wording has been virtually unchanged for 30 years and says:

> Overcurrent protection for conductors and equipment is provided to open the circuit if the current reaches a value that will cause an excessive or dangerous temperature in conductors or conductor insulation.

For purposes of consistent analysis, the discussions of ampacity here will be based on the first of the ampacity tables, Table 310-16. We will assume that the specific application corresponds to the conditions to which Table 310-16 applies—i.e., not more than three single insulated conductors (0 to 2,000 V) in a raceway or Type AC, NM, NMC, or SE cable. The particular considerations given to ampacity value for Table 310-16 can be made for any of the other tables.

NEC Table 310-16 gives ampacities under two conditions: when the raceway or cable containing the conductors is operating in an ambient temperature not over 30°C (86°F) and when there are not more than three current-carrying conductors in the raceway or cable. Under those conditions, the ampacities shown correspond to the thermal limit of each particular insulation. But in any case where either or both of the two conditions are exceeded, the ampacity of the conductors of a circuit must be reduced (and protection must be based upon or provided at the reduced ampacity!) to ensure that the temperature limit of the insulation is not exceeded:

1. If the ambient temperature is above 30°C, the ampacity must be reduced in accordance with the correction factors given with Table 310-16.
2. If more than three current-carrying conductors are used in a single cable or raceway, the conductors tend to be bundled in such a way that their heat-dissipating capability is reduced, and excessive heating will occur at the ampacities shown in the table. As a result, Note 8 to Tables 310-16 through 310-19 requires the reduction of ampacity, and conductors have to be protected at the reduced ampacity.

It should be clearly understood that any reduced ampacity, required because of a higher ambient and/or conductor bundling, has the same meaning as the value shown in the table: Each represents a current value above which excessive heating would occur under the particular conditions. And if there are two conditions that lead to excessive heating, then a greater reduction in current is required than if only one such condition exists.

In conductor size Nos. 14, 12, and 10, Table 310-16 clearly indicates that 90°C-rated conductors do, in fact, have higher ampacities than those given for the corresponding sizes of 60°C and 75°C conductors. As shown in Fig. 310-20, No. 12

1996 Edition

Table 310-16

Size	Temperature Rating of Conductor		
	60°C (140°F)	75°C (167°F)	90°C (194°F)
AWG kcmil	TYPES TW†, UF†	TYPES FEPW†, RH†, RHW†, THHW†, THW†, THWN†, XHHW† USE†, ZW†	TYPES TA, TBS, SA SIS, FEP†, FEPB†, MI RHH†, RHW-2, THHN†, THHW†, THW-2, THWN-2, USE-2, XHH, XHHW† XHHW-2, ZW-2
	COPPER		
18	....	....	14
16	....	....	18
14	20†	20†	25†
12	25†	25†	30†
10	30	35†	40†
8	40	50	55

Fig. 310-20. NE Code table shows higher ampacities for 90°C branch-circuit wires (Nos. 14, 12, and 10). (Sec. 310-15.)

TW and No. 12 THW copper conductors are both assigned an ampacity of 25 A under the basic application conditions of the table. *But,* a No. 12 THHN, RHH, or XHHW (dry location) has an ampacity of *30* A. However, the footnote to Table 310-16 (shown in Fig. 310-21) requires that "overcurrent protection" for No. 14,

†Unless otherwise specifically permitted elsewhere in this Code, the overcurrent protection for conductor types marked with an obelisk (†) shall not exceed 15 amperes for No. 14, 20 amperes for No. 12, and 30 amperes for No. 10 copper; or 15 amperes for No. 12 and 25 amperes for No. 10 aluminum and copper-clad aluminum after any correction factors for ambient temperature and number of conductors have been applied.

Fig. 310-21. Note below Table 310-16 radically alters conductor applications for Nos. 14, 12, and 10. (Sec. 310-15.)

No. 12, and No. 10 copper conductors be taken as 15, 20, and 30 A, respectively, regardless of the type and temperature rating of the insulation on the conductors. And the footnote says that these limitations on overcurrent protection apply after any correction factors for ambient temperature and/or number of conductors have been applied. When applied to selection of branch-circuit wires in cases where conductor ampacity derating is required by Note 8 of Tables 310-16 through 310-19 for conduit fill (over three wires in a raceway), the footnote to Table 310-16 affords advantageous use of the 90°C wires for branch-circuit makeup. The reason is that, as stated in Note 8, the derating of ampacity is based on taking a percentage of the actual ampacity value shown in the table, and the ampacity values for 90°C conductors are higher than those for 60 and 75°C conductors.

Application of Note 8 to Table 310-16 depends upon how many current-carrying conductors are in a raceway. A true neutral conductor (a neutral carrying current only under conditions of unbalanced loading on the phase conductors) is not counted as a current-carrying conductor. If a 208Y/120-V circuit or a 480Y/277-V circuit is made up of three phase legs and a true neutral in a conduit, the circuit is counted as only three conductors in the conduit, and derating for conduit fill, as described in Note 8 of Tables 310-16 through 310-19, is not necessary. But neutrals for circuits with these voltage ratings must be counted as current-carrying conductors if the major portion of the load consists of electric-discharge lighting, data processing equipment, or similar equipment [Note 10(c) of the tables]. Thus, if the circuit supplies fluorescent, mercury, or metal-halide lamps, the neutral is counted as the fourth current-carrying conductor because it carries third harmonic current, which approximates the phase-leg current, under balanced loading. Any such 4-wire circuit must have its load current derated to 80 percent of the ampacity given in Table 310-16, as required by Note 8 to those tables.

As shown in Fig. 310-22, the makeup of a branch circuit consists of selecting the correct size of wire for the particular load current (based on number of wires in the raceway, ambient temperature, and ampacity deratings) and then relating the rating of the overcurrent protective device to all the conditions.

In Table 310-16, which applies to conductors in raceways and in cables and covers the vast majority of conductors used in electrical systems for power and light, the ampacities for sizes No. 14, No. 12, and No. 10 are particularly significant because copper conductors of those sizes are involved in most branch circuits in modern electrical systems. Number 14 has an ampacity of 20, No. 12 has an ampacity of 25, and No. 10 has an ampacity of 30. The typical impact of that on circuit makeup and loading is as follows:

1. Number 12 TW or THW copper is shown to have an ampacity of 25; and based on the general UL requirement that equipment terminals be limited to use with 60°C conductors in sizes up to No. 1 AWG, No. 12 THHN or XHHW

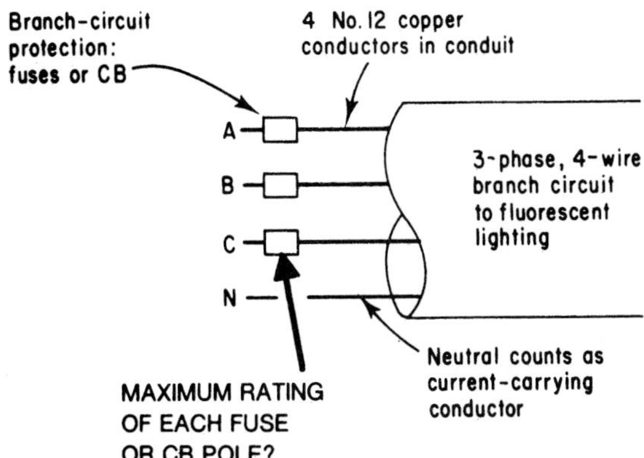

Fig. 310-22. Loading and protection of branch-circuit wires must also account for derating of wire "ampacity." (Sec. 310-15.)

copper conductors must also be treated as having a 25-A continuous rating. *But,* the footnote to Table 310-16 limits all No. 12 copper wires to a maximum load of 20 A by requiring that they be protected at not more than 20 A.

2. The ampacity of 25 A for No. 12 TW and THW copper wires interacts with Note 8 to Tables 310-16 through 310-19 where there are, say, six No. 12 TW current-carrying wires for the phase legs of two 3-phase, 4-wire branch circuits in one conduit supplying, say, receptacle loads. In such a case, the two neutrals of the branch circuits do not count in applying Note 8, and only each of the six phase legs must have its ampacity derated to the "Percent of Values in Tables as Adjusted for Ambient Temperature if Necessary" as stated at the upper right of the table in Note 8. In the case described here, that literally means that each No. 12 phase leg may be used at a derated ampacity of 0.8 × 25, or 20 A. And the footnote to Table 310-16 would require use of a fuse or CB rated not over 20 A to protect each No. 12 phase leg. Each No. 12 would then be protected at its new ampacity that represents the maximum I^2R heat input that the conductor insulation can withstand. The only other possible qualification is that Sec. 384-16(c) would require the load current on each of the phase legs to be further limited to no more than 80 percent of the 20-A rating of the overcurrent device—that is, 16 A—if the load current is "continuous" (operates steadily for 3 hr or more), a condition not likely for receptacle-fed loads.
3. If two 3-phase, 4-wire branch circuits of No. 12 TW or THW copper conductors are installed in a single conduit or EMT run and supply, say, fluorescent or other electric-discharge lighting, the two neutrals of the branch circuits would carry harmonic current even under balanced conditions and would have to be counted, along with the six phase legs, as current-carrying wires for ampacity derating in accordance with Note 8. In such a case, as the table of Note 8 shows, "7 through 9" conductors must have ampacity derated to 70 percent of the 25-A value shown in Table 310-16 (0.7 × 25 = 17.5 A), which gives each conductor an ampacity of 17.5 A. If 20-A overcurrent protection is used for each No. 12 phase leg

and the load is limited to no more than 17.5 or 16 A (0.8 × 20 A) for a continuous load, the application would satisfy Note 8 to Table 310-16. And Sec. 210-19(a) permits the use of 20-A protection as the next higher standard rating of protective device above the conductor ampacity of 17.5 A, provided that the circuit supplies only hard-wired outlets, such as lighting fixtures, and does not supply any receptacle outlets for "cord- and plug-connected portable loads" that could permit overloading of the conductors above 17.5 A.

Use of 20-A protection on conductors with an ampacity of 17.5 A is recognized only for fixed circuit loading (like lighting fixture outlets) and not for the variable loading that general-purpose convenience receptacles permit, because any increase in load current over 17.5 A would produce excessive heat input to the eight bundled No. 12 conductors in that conduit and would damage and ultimately break down the conductor insulation. Of course, the question arises: Over the operating life of the electrical system, how can the addition of excessive current be prevented? The practical, realistic answer is: It can't! It would be better to use 15-A protection on the No. 12 wires or use 90°C rated conductors.

4. No. 12 THHN or XHHW conductors—with their 90°C rating and consequently greater resistance to thermal damage—could be used for the two 3-phase, 4-wire, 20-A circuits to the electric-discharge lighting load, would satisfy all **Code** rules, and would not be subject to insulation damage. With eight current-carrying wires in the conduit, the 70 percent ampacity derating required by Note 8 would be applied to the 30-A value shown in Table 310-16 as the ampacity of No. 12 THHN, RHH, or XHHW (dry locations). Then because 0.7 × 30 = 21 A, the maximum of 20-A protection required by the footnote to Table 310-16 would ensure that the conductors were never subjected to excessive current and its damaging heat. And if the original loading on the conductors is set at 16 A [the 80 percent load limiting of Sec. 384-16(c)] for continuous operation of the lighting, any subsequent increase in load even up to the full 20-A capacity would not reach the 21-A maximum ampacity set by Note 8.

Figure 310-23 summarizes the applications described in 3 and 4 above.

Advantage of 90°C Wires

If the four circuit wires in Fig. 310-22 are 90°C-rated conductors—such as THHN, RHH, or XHHW—the loading and protection of the circuit must be related to required ampacity derating as shown in Fig. 310-24. The application is based on these considerations:

1. As described in Figs. 310-20 and 310-21, each No. 12 THHN has an ampacity of 30 A from Table 310-16, but the footnote to that table limits the overcurrent protection on any No. 12 THHN to not more than 20 A.
2. Because the neutral of the 3-phase, 4-wire circuit must be counted as a current-carrying wire, there are four conductors in the conduit—thereby requiring that each conductor have its ampacity derated to 80 percent of its table-value ampacity, as required by Note 8 of Table 310-16. Each No. 12 then has a new (derated) ampacity of 0.8 × 30, or 24 A.

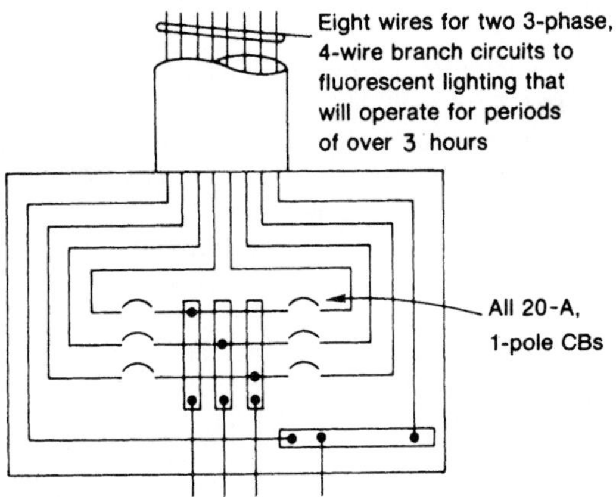

CASE 1
With TW or THW No. 12 copper conductors:
No. 12 ampacity = 25 amps, from Table 310-16
From Note 8, derating = 0.7 × 25 = 17.5 amps
From Sec. 384-16(c), max. load = 0.8 × 20 = 16 amps

CASE 2
With THHN, XHHW or RHH No. 12 copper conductors:
No. 12 ampacity = 30 amps, from Table 310-16
From Note 8, derating = 0.7 × 30 = 21 amps
From Sec. 384-16(c), max. load = 0.8 × 20 = 16 amps

IN CASE 1, CONDUCTORS ARE NOT PROTECTED IN ACCORDANCE WITH THE 17.5-A AMPACITY, IN CASE 2, THEY ARE PROTECTED AGAINST EXCESSIVE LOAD CURRENT.

Fig. 310-23. Conductors with 90°C insulation eliminate the possibility of conductor damage due to overload.

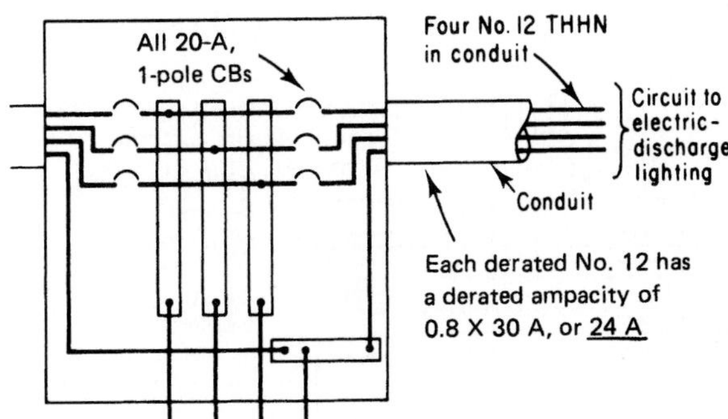

NOTE: TW or THW wires would have to be derated from 20 to 16 A.

Fig. 310-24. Derating of 90°C branch-circuit wires (Nos. 14, 12, and 10) is based on higher ampacities. (Sec. 310-15.)

3. By using a 20-A, single-pole protective device (fuse, single-pole CB, or one pole of a 3-pole CB), which is the maximum protection permitted by the Table 310-16 footnote, each No. 12 THHN easily complies with the Sec. 240-3 requirement that the branch-circuit wire have overcurrent protection in accordance with (usually not greater than) the conductor ampacity (24 A).
4. If the lighting load on the circuit is noncontinuous—that is, does *not* operate for any period of 3 hr or more—the circuit may be loaded up to its 20-A maximum rating.
5. If the load fed is continuous—full-load current flow for 3 hr or more—the load on the circuit must be limited to 80 percent of rating of each 20-A fuse or CB pole, as required by Sec. 384-16(c). Then 16 A is the maximum load. UL rules state that "unless otherwise marked, circuit breakers should not be loaded to exceed 80% of their current rating, where in normal operation the load will continue for 3 hours or more."

One question commonly asked relates to the typical ampere and voltage ratings of "100% Continuous-Rated" overcurrent devices that are available. UL presents the following data as guidance in their "Molded Case Circuit Breaker" Marking Guide:

> 35. **100 Percent Continuous Rated**—Unless otherwise marked for continuous use at 100 percent of its current rating, a circuit breaker is intended for use at no more than 80 percent of its rated current where in normal operation the load will continue for three hours or more. A breaker with a frame size of 250 A or more, or a multi-pole breaker of any current rating and rated greater than 250 V, may be marked to indicate it is suitable for continuous use at 100 percent of its current rating. The marking is, "Suitable for continuous operation at 100 percent of rating only if used in a circuit breaker enclosure Type ____ or in a cubicle space by ____ by ____ inches" or an equivalent statement. This type of breaker may also be marked to indicate it is to be used with wire sized for a 75°C conductor with 90°C insulation and used with 90°C wire connectors.

Figure 310-25 shows the use of two 3-phase, 4-wire circuits of THHN conductors in a single conduit. The 90°C wires offer distinct advantages (substantial economies) over use of either 60°C (TW) or 75°C (THW) wires for the same application, as follows.

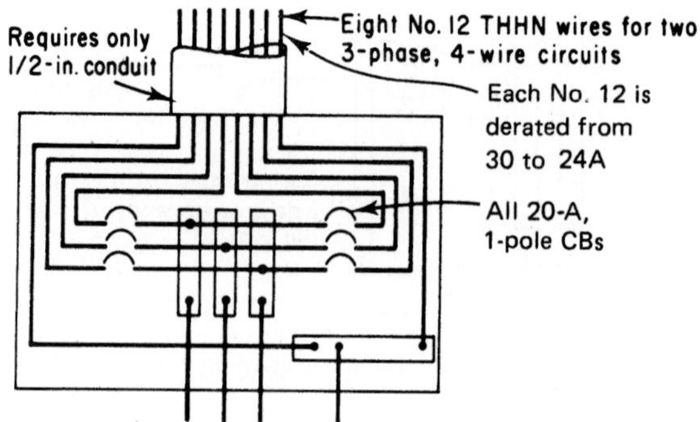

Fig. 310-25. The 90°C conductors can take derating without losing full circuit load-current rating. (Sec. 310-15.)

Resistive load If the circuit shown feeds only incandescent lighting or other resistive loads (or electric-discharge lighting does not make up "a major portion of the load"), then Note 10(c) of Tables 310-16 through 310-19 does not require the neutral conductor to be counted as a current-carrying conductor. In such cases, circuit makeup and loading could follow these considerations:

1. With the six phase legs as current-carrying wires in the conduit, Note 8 requires that the ampacity of each No. 12 be derated from its basic table value of 30 A to 80 percent of that value—or 24 A.
2. Then each No. 12 is properly protected by a 20-A CB or fuse—satisfying Sec. 210-20 and the footnote to Table 310-16.
3. If the circuit load is not continuous, each phase leg may be loaded to 20 A.
4. If the load is continuous, a maximum of 16 A (80 percent) must be observed to satisfy Sec. 384-16(c).
5. If the total load is made up of both continuous and noncontinuous loads, the sum of noncontinuous load *plus* 125 percent of the continuous load must not exceed the rating of the branch circuit, in this case, 20 A [see Sec. 220-3(a)].

Electric-discharge load If the two circuits of Fig. 310-25 supplied electric-discharge lighting (fluorescent, mercury-vapor, metal-halide, high-pressure sodium, or low-pressure sodium), makeup and loading would have to be as shown in Fig. 310-23, case 2. If that same makeup of circuits supplied noncontinuous loads, the circuit conductors could be loaded right up to 20 A per pole.

In Fig. 310-23, case 2, the only difference between such circuit makeups using RHH conductors and the ones using THHN is the need for ¾-in. (19 mm) conduit instead of ½-in. (12.7 mm) conduit because of the larger cross-section area of RHH and XHHW (see Tables 3A and 3B, Chap. 9, NE Code).

Feeder Applications

Applying the UL temperature limitation to selection of feeder conductors is generally similar to the procedure described above for selection of branch-circuit conductors.

Refer to Fig. 310-26:

1. Because the load on the feeder is continuous, the 100-A, 3-pole CB must have its load current limited so the 100-A protective-device rating is not less than 125 percent times the continuous current, as required by Sec. 220-10(b). Then, 100 ÷ 1.25 = 80 A. The 76-A load is, therefore, OK. (A CB or fused switch may be loaded continuously to 100 percent of the CB or fuse amp rating only when the assembly is UL-listed for such use.)
2. The CB load terminals are recognized by UL for use with 60°C conductors or higher-temperature conductors loaded not over the 60°C ampacity of the given size of conductor.
3. The feeder phase conductors must have an ampacity that is at least equal to the rating of the device—before any derating—and such that they are protected by the 100-A protective device in accordance with Sec. 240-3. That means they must have an ampacity of 100 A or have a lower value of ampacity for which the 100-A protection rating is the next higher standard rating of protective device above the conductor ampacity, when conduc-

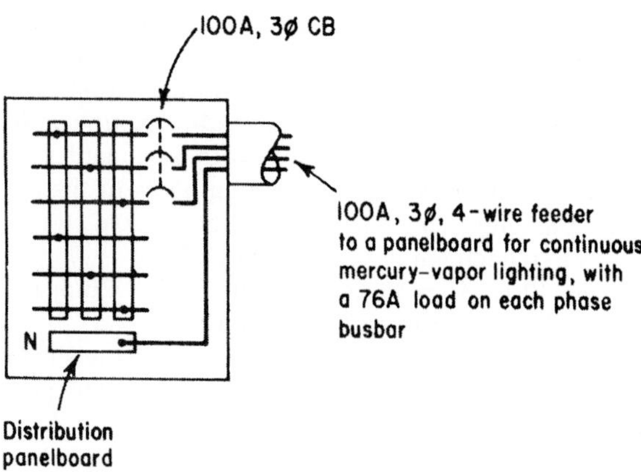

Fig. 310-26. Feeder conductors for up to 100-A equipment must use 60°C ampacity. (Sec. 310-15.)

tor ampacity does not correspond to a standard rating of protective device. The feeder neutral is not subject to this limitation because the neutral does not connect to the terminal of a switch, CB, starter, etc.—the devices for which heating would be a problem under continuous load. NEC Sec. 220-22 is the basic rule that covers sizing of a feeder neutral. Section 220-10(b) covers sizing of feeder phase conductors.

4. If 60°C copper conductors are used for this feeder, reference must be made to the second column of Table 310-16. This feeder supplies electric-discharge lighting; therefore, Note 10(c) of the table requires that the feeder neutral be counted as a current-carrying conductor because of the harmonic currents present in the neutral. Then, because there are four current-carrying conductors in the conduit, the ampacity of each conductor must be derated to 80 percent of its value in column 2 of Table 310-16, as required by Note 8 to Table 310-16. After the conductor is derated to 80 percent, it must have an ampacity such that it is properly protected by the 100 A protection. Because standard ratings of protective devices are 90 and 100 A (from Sec. 240-6), the derated ampacity of the required conductors must not be less than 91 A—which is the lowest ampacity value that may be protected by a 100-A protective device in accordance with Sec. 240-3(b), which permits the next higher standard rating of protective device above the conductor ampacity. From Table 310-16, a No. 1/0 TW conductor is rated at 125 A when only three conductors are used in a conduit. With four conductors in a conduit, the 125-A rating is reduced to 80 percent (0.8 × 125 A), or 100 A, which is properly protected by the 100-A CB.

Continuous load on a feeder must be limited to no more than 80 percent of any CB or fused switch that is not UL-listed and marked for continuous loading to 100 percent of its rating—without any relationship to conductor derating. When conductor ampacity is derated because more than three conductors are used in a raceway, the conductors must be protected at the *derated* ampacity. Then—*in addition*—if the CB or fused switch that provides the protection is not UL-listed and marked for 100 percent continuous load, continuous circuit loading must not exceed 80 percent of the rating of the CB or fuses.

The feeder circuit of four No. 1/0 TW conductors, rated at 100 A, would require a minimum of 2-in. (50.8-mm) rigid metal conduit. [*Note:* A reduced size of neutral could be used, because the need to upsize hot conductors to be properly protected by the 125 percent protective device required by Sec. 220-10(b) does not apply to the neutral. The neutral could be a No. 2 TW conductor, which has an ampacity of 76 A after derating of its ampacity from 95 A—0.8 × 95 A = 76 A, which is adequate for the load current under conditions of maximum unbalance of phase-to-neutral current.]

5. If 75°C conductors are used for this feeder circuit instead of 60°C conductors, the calculations would be different. With THW copper conductors, Table 310-16 shows that No. 1 conductors, rated at 130 A for not more than three current-carrying conductors in a conduit, would have an ampacity of 0.8 × 130 A, or 104 A, when four are used in the conduit and derated. The 104-A conductor ampacity would be properly protected by the 100-A CB that was shown to satisfy Sec. 220-10(b).

   Although UL listing and testing of the CB are based on the use of 60°C conductors, the use of No. 1 75°C THW conductors is acceptable because the terminals of the breaker in this case would not be loaded to more than the amp rating of a 60°C conductor of the same size. A No. 1 60°C TW conductor is rated at 110 A when not more than three current-carrying conductors are used in a conduit. When four conductors are in one conduit, the 60°C No. 1 wires are derated to 80 percent of 110 A, or 88 A. Because that value is greater than the load of 76 A on each CB terminal, the CB terminals are not loaded in excess of the 88-A allowable ampacity of 60°C No. 1 conductors, and the UL limitation is satisfied.

   Four No. 1 THW conductors, rated at 104 A, would require a minimum of 1½-in. (38-mm) rigid metal conduit. Or, four No. 1 RHH, THHN, or XHHW conductors could be used in 1½-in. (38-mm) rigid metal conduit.

6. If 90°C THHN (or XHHW or RHH) conductors are used for this feeder (in a dry location), No. 2 copper conductors could be used. From Table 310-16, No. 2 THHN with a basic ampacity of 130 A would be derated to 0.8 × 130 A, or 104 A—which would be properly protected within the conductor's ampacity by the 100-A CB. The ampacity of a 60°C No. 2 conductor is 95 A normally and derated to 80 percent is 0.8 × 95 A, or 76 A—which gives the conductors the same rating as the load. Under such a condition the load current is not in excess of the 76-A allowable ampacity of a 60°C No. 2 conductor, and the UL limitation is satisfied.

   Four No. 2 THHN, XHHW, or RHH conductors, rated exactly at 104 A, which is above the required minimum rating of 76 A, would require a minimum of 1¼-in. (31.8 mm) conduit (in dry locations only).

   Note: Of course, voltage drop in the feeder will vary with the different-size conductors and must be accounted for.

Figure 310-27 shows an example where feeder conductors could be used at up to the 75°C ampacity. In this case, because the load on the feeder is not electric-discharge lighting or data processing equipment, the neutral does *not* count as a current-carrying conductor. As a result, there are only three current-

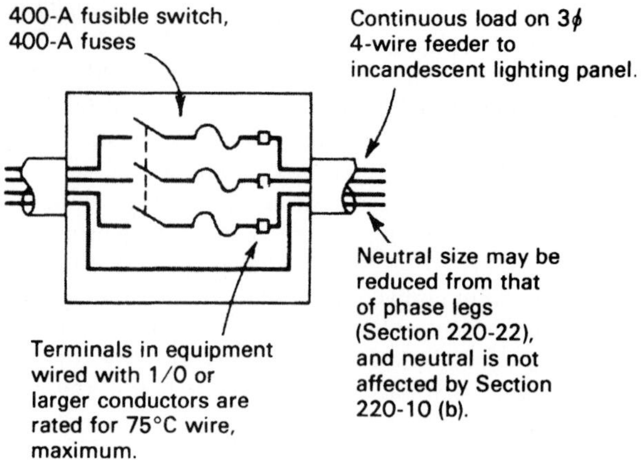
400-A fusible switch,
400-A fuses
Continuous load on 3ϕ
4-wire feeder to
incandescent lighting panel.
Neutral size may be
reduced from that
of phase legs
(Section 220-22),
and neutral is not
affected by Section
220-10 (b).
Terminals in equipment
wired with 1/0 or
larger conductors are
rated for 75°C wire,
maximum.

Fig. 310-27. Equipment wired with conductors No. 1/0 or larger may use the 75°C ampacity. (Sec. 310-15.)

carrying conductors in the conduit, and derating according to Note 8 is not required. The details are as follows:

1. Because the load on this feeder is continuous, Sec. 220-10(b) limits the load current to not more than 80 percent of the rating of the fuses. Immediately, we know that the feeder load may not exceed 80 percent of the 400-A fuse rating: $0.8 \times 400 = 320$ A.
2. If 75°C conductors are used, because they are permitted by UL test conditions, Table 310-16 shows that 700-kcmil THW aluminum conductors rated at 375 A could be used and suitably protected by the 400-A fuses in accordance with Sec. 240-3(b), because 400 A is the next higher standard fuse rating (from Sec. 240-6) above the 375-A rating of the conductors. In that case the maximum continuous feeder load of 320 A is well within the 375-A ampacity of the conductors.
3. If 90°C conductors are used, they must be used at no more than the ampacity of a 75°C conductor of the same size as the 90°C conductor. Table 310-16 shows that 600-kcmil XHHW aluminum conductors (in a dry location) have a 385-A rating, and the 400-A fuses constitute acceptable protection for those conductors in accordance with Sec. 240-3(b). The load on the feeder phase legs (320 A) is well within the 385-A ampacity of the 600-kcmil conductors. To check the suitability of the 90°C conductors, the ampere rating of a 75°C 600-kcmil aluminum conductor is given in Table 310-16 as 340 A. Because the load of 320 A is not in excess of the 340-A rating of a 75°C conductor, the UL limitation on maximum rating of conductor termination is satisfied.
4. The smaller conduit size required for the reduced size of higher-temperature conductors is a labor and material advantage.

Notes to Tables 310-16 through 310-19

Note 3 This note has been in the NE Code for a long time and permits use of certain conductors at ampacity values higher than those shown for the conductors in Table 310-16. For instance, a No. 2/0 THW copper conductor may be used at an ampacity of 200 A instead of at 175 A, as shown in Table 310-16. This permission has been given by the NE Code in recognition of the reality that residential service conductors are supplying loads of great diversity and of short operating periods or cycles, so that the conductors almost never see full demand load approaching their ampacity and certainly not for continuous operation (3 hr or more).

The higher allowable ampacity ratings for 120/240-V, 3-wire—*not* 2-phase and neutral from a 208Y/120-V system—single-phase dwelling services may also be utilized for feeder conductors, and Type USE is one of the types of conductors given the higher ampacities. The ampere rating of services that are subject to the higher ampacity values was increased from 200 to 400 A.

Those higher conductor ampacities permitted by Note 3 are applicable only for service conductors or feeder conductors used in "dwelling units," thereby limiting the application to one-family houses and individual apartments in apartment houses, condominiums, and the like—because only such units conform to the Code definition of "dwelling unit," as given in Art. 100. The wording excludes use of Note 3 in relation to service-entrance conductors to a whole building, such as a multifamily dwelling, as shown in Fig. 310-28. And in that drawing, prior to the 1990 Code, the wording did prohibit use of the higher ampacities of conductors if each of the SE runs to the individual apartments were a feeder instead of service-entrance conductors, as each run would be if a disconnect and protective device were used at each meter location. But reason dictates that the higher ampacities should be allowed in either case, even though each feeder does not carry "the total current supplied by that service." The phrase "total current supplied by that service" was removed in the 1990 NEC, which has the effect of permitting the use of the increased ampacities for feeders to individual "dwelling units" in a multifamily dwelling. And, although they are prohibited by the literal limitation to "dwelling units," it would be reasonable to permit the higher ampacities for service-entrance conductors to the whole building. Refer to Sec. 215-2(b) of the Code.

Prior to the 1990 NEC, in Note 3, it was clearly indicated that the higher allowable ampacities for residential occupancies using 3-wire, single-phase services also may be applied to 3-wire, single-phase feeders in those cases where the feeder conductors from the service equipment to a subpanel or other distribution point carry the total current supplied by the service conductors. Inclusion of the phrase "that supply the total load to a dwelling unit" in the 1993 Code would not recognize use of the increased ampacities given in Note 3 for a feeder as shown in the one-family house drawing of Fig. 310-28, if, say, swimming pool circuits or other loads are supplied from the service, but are not carried by the feeder. Now, the revision in the 1996 NEC of that portion in Note 3 which now reads "the main power feeder to a dwelling unit," recognizes application of Note 3 to just that type of feeder.

The last sentence of Note 3 permits the neutral conductor of these 3-wire services and 3-wire feeders to be two sizes smaller than the hot conductors because the neutral carries only the unbalanced current of the hot legs and is not at all involved with 2-wire, 240-V loads. That is how the neutral in SE cable is sized.

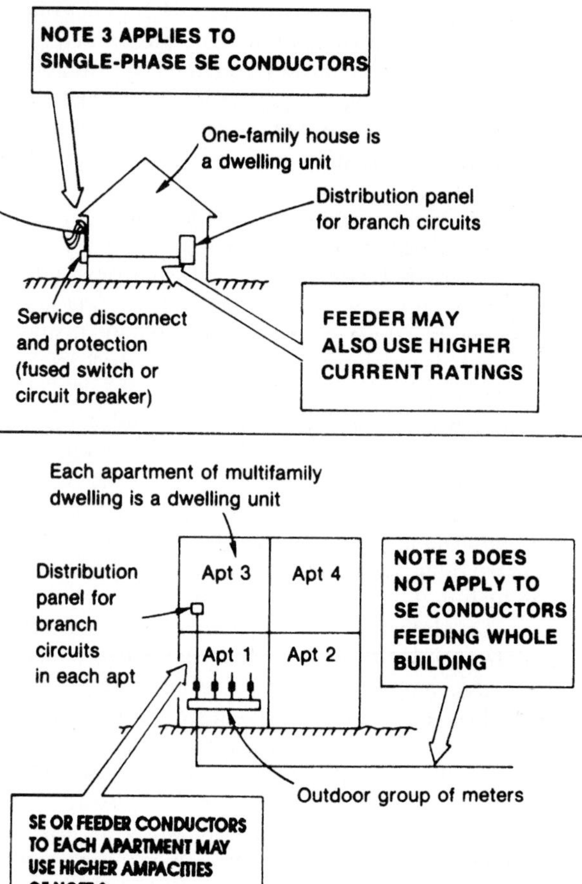

Fig. 310-28. Higher ampacities may be used for service conductors to "dwelling units."

Note 5 This note provides that, if an uninsulated conductor is used with insulated conductors in a raceway or cable, its size shall be the size that would be required for a conductor having the same insulation as the insulated conductors and having the required ampacity (Fig. 310-29).

example Two No. 6 Type THW conductors and one bare No. 6 conductor in a raceway or cable. The ampacity of the bare conductor would be 65 A.

If the insulated conductor were Type TW, the ampacity of the bare conductor would be 55 A.

Note 8(a) *Where more than three current-carrying conductors* are used in a raceway or cable, their current-carrying capacities must be reduced to compensate for the proximity heating effect and reduced heat dissipation due to reduced ventilation of the individual conductors that are bunched or form an enclosed group of closely placed conductors. Where the number of conductors

Bare conductor has the capacity of a conductor of its size that has...
... the same insulation as used on the insulated conductors run with the bare conductor.

Fig. 310-29. How to figure ampacity of a bare conductor, where permitted. (Sec. 310-15.)

in a raceway or cable exceeds three, the ampacity of each conductor shall be reduced as indicated in the table of Note 8.

If, for instance, four No. 8 THW copper conductors are used in a conduit, the *ampacity* of each No. 8 is reduced from the 50-A value shown in the table to 80 percent of that value. In such a case, each No. 8 then has a *new* reduced *ampacity* of 0.8 × 50 A, or 40 A. And, from Sec. 240-3, "Conductors shall be protected in accordance with their *ampacities*." Thus, 40-A-rated fuses or CB poles would be required for overcurrent protection—as the general rule.

The application of those No. 8 conductors and their protection rating is based on the general concept behind the **NE Code** tables of maximum allowable current-carrying capacities (called "ampacities"). The **NE Code** tables of ampacities of insulated conductors installed in a raceway or cable have always set the maximum continuous current that a given size of conductor can carry continuously (for 3 hr or longer) without exceeding the temperature limitation of the insulation on the conductor, that is, the current above which the insulation would be damaged. But, because the overcurrent devices were tested with conductors sized at 125 percent of the continuous current *plus* the noncontinuous, the ampacity—the continuous current—must be increased where supplying a continuous load [see Sec. 220-3(a)].

This concept has always been verified in the FPN to Sec. 240-1, where the wording has been virtually identical for over 30 years and says, "Overcurrent protection for conductors and equipment is provided to open the circuit if the current reaches a value that will cause an excessive or dangerous temperature in conductors or conductor insulation." To correspond with that objective, Sec. 240-3 says, "Conductors, other than flexible cords and fixture wires, shall be protected against overcurrent in accordance with their ampacities as specified in Section 310-15."

Table 310-16, for instance, gives ampacities under the conditions that the raceway or cable containing the conductors is operating in an ambient not over 30°C (86°F) and that there are *not more* than three current-carrying conductors in the raceway or cable. Under those conditions, the ampacities shown correspond to the thermal limit of the particular insulations. But if either of the two conditions is exceeded, ampacities have to be reduced to keep heat from exceeding the temperature limits of the insulation:

1. If ambient is above 30°C, the *ampacity* must be reduced in accordance with the correction factors given at the bottom of Table 310-16.

2. If more than three current-carrying conductors are used in a single cable or raceway, the conductors tend to be bundled in such a way that their heat-dissipating capability is reduced and excessive heating would occur at the ampacities shown in the table. As a result, Note 8 requires *reduction of ampacity,* and conductors must be protected at the reduced ampacity.

(*Note:* It should be clearly understood that any reduced ampacity—required for higher ambient and/or conductor bundling—has the same meaning as the value shown in a table: Each represents a current value above which excessive heating would occur under the particular conditions. And if there are two conditions that reduce heat dissipation, then more reduction of current is required than for one condition of reduced dissipation.)

Note 8 Requires Derating of "Ampacity"

Note 8 to Table 310-16 says, "Where the number of conductors in a raceway or cable exceeds three, the ampacities shall be reduced as shown in the following table." And that table has a heading on the right to require that any ampacity derating for elevated ambient temperature must be made in addition to the one for number of conductors. If, for instance, four No. 8 THHN current-carrying copper conductors are used in a conduit, the ampacity of each No. 8 is reduced from the 55-A value shown in Table 310-16 to 80 percent of that value. Each No. 8 then has a new (reduced) ampacity of 0.8 × 55 A, or 44 A. Then, if a derating factor must be applied because the conductors are in a conduit where the ambient temperature is, say, 40°C instead of 30°C, the factor of 0.91 (36–40°C) from the bottom of Table 310-16 must be applied to the 44-A current value to determine the final value of ampacity for the conductors (44 × 0.91 = 40 A). Moreover, Sec. 240-3 of the NEC states, "Conductors, other than flexible cords and fixture wires, shall be protected against overcurrent in accordance with their ampacities as specified in Section 310-15." Thus, fuses or CB poles rated at 40 A would be required. The ampacity of the conductors is changed and the conductors must be protected in accordance with the derated ampacity value and not in accordance with the tabulated value.

Because conductor "ampacity" is reduced when more than three conductors are used in a conduit, the overcurrent protection for each phase leg of a parallel makeup in a single conduit would generally have to be rated at not more than the sum of the *derated* ampacities of the number of conductors used per phase leg. That would satisfy Sec. 240-3, which requires conductors to be protected at their ampacities. Because ampacity is reduced in accordance with the percentage factors given in Note 8 for more than three conductors in a single conduit, that derating dictates the use of multiple conduits for parallel-makeup circuits to avoid the penalty of loss of ampacity.

Figure 310-30 shows examples of circuit makeups based on the unsafe concept of load limitation instead of ampacity derating, as applied to overcurrent rating and conductor ampacity—which is a Code violation, because the conductors are *not* protected in accordance with their ampacities.

Figure 310-31 shows a condition of bunched or bundled Type NM cables where they come together at a panelboard location. The paragraph following

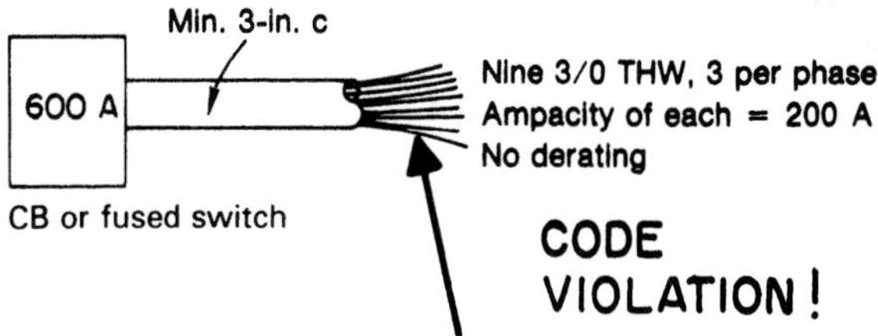

Note 8, max. conductor ampacity per phase
= 0.7 X 600 = 420 A, continuous or noncontinuous

Possible current in excess of conductor thermal limit
= 600 − 420 = 180 A

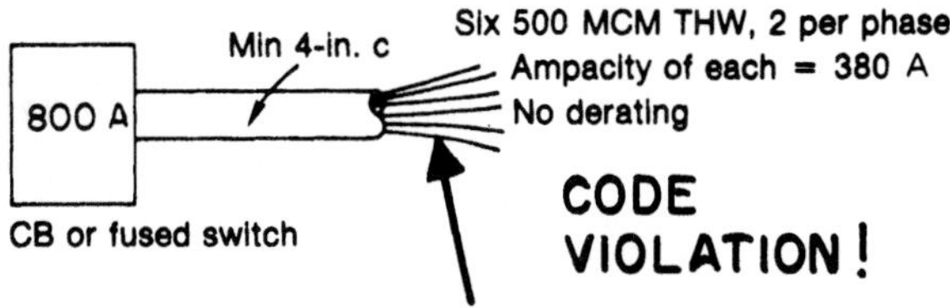

Note 8, max. conductor ampacity per phase
= 0.8 X 760 = 608 A, continuous or noncontinuous

Possible current in excess of conductor thermal limit
= 800 − 608 = 192 A

Fig. 310-30. Parallel-conductor makeup must not be used in single conduits without ampacity reduction, even if load current is limited to the conductor ampacity as shown here.

the table of Note 8 requires conductors in bundled cables to have their load currents reduced from the ampacity values shown in Table 310-16. The ampacity derating is required for conductor stacks or bundles that are longer than 2 ft (24 in., or 610 mm). For shorter bundles, derating is not required. And Exception No. 3 excludes the need for derating groups of (four or more) conductors installed in nipples not over 24 in. (610 mm) long.

Exception No. 4 of Note 8 says that underground conductors that are brought up above ground in a protective raceway [Sec. 300-5(d)] do not require derating if not more than four conductors are used and if the protective conduit has a length not over 10 ft (3.05 m) "above grade." The total length of raceway may exceed 10 ft (3.05 m). The phrase "above grade" clearly limits the length of protective conduit that may contain conductors *without* derating in accordance with Note 8(a). The 10-ft (3.05 m) length covers the length of 8 ft (2.44 m) above grade but not the 1½ ft (7.62 mm) into the earth given by Sec. 300-5(d) on conductors emerging from underground.

Note 8 does not apply to conductors in wireways and auxiliary gutters, as covered in Secs. 362-1 and 374-6. Wireways or auxiliary gutters may contain up to 30 conductors at any cross section [excluding signal circuits and control con-

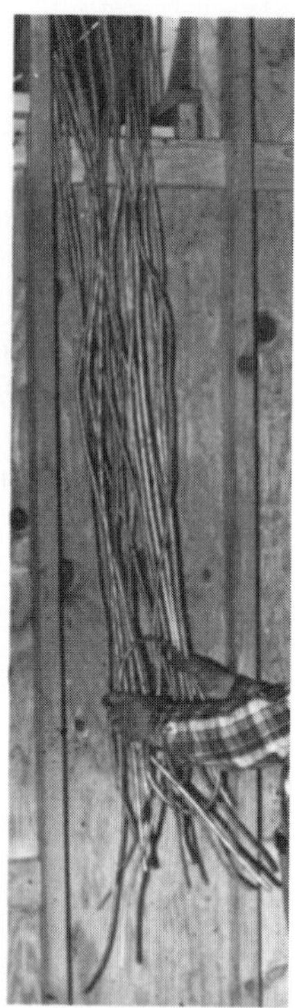

Fig. 310-31. If a large number of multiconductor cables are bundled together in a stud space, capacity derating in accordance with Note 8 would be required. If the individual cables are spaced apart and stapled, then the conductors in the cables may be loaded up to their rated ampacity values from Table 310-16. (Sec. 310-15.)

ductors used for starting duty only between a motor and its starter in auxiliary gutters (Sec. 374-5)]. The total cross-sectional area of the group of conductors must not be greater than 20 percent of the interior cross-sectional area of the wireway or gutter. And load-limiting factors for more than three conductors do not apply to wireway the way they do to wires in conduit. However, if the derating factors from Note 8 of the NE Code Tables 310-16 through 310-19 are used, there is no limit to the number of wires permitted in a wireway or an auxiliary gutter. But, the sum of the cross-sectional areas of all contained conductors at any cross section of the wireway must not exceed 20 percent of the cross-sectional area of the wireway or auxiliary gutter. More than 30 conductors may be used under those conditions.

Note 10 In the determination of conduit size, neutral conductors must be included in the total number of conductors because they occupy space as well as phase conductors. A completely separate consideration, however, is the relation of neutral conductors to the number of conductors, which determines whether ampacity derating must be applied to conductors in a conduit, as follows.

Neutral conductors which carry only unbalanced current from phase conductors (as in the case of normally balanced 3-wire, single-phase or 4-wire, 3-phase circuits supplying resistive loads) are not counted in determining ampacity derating of conductors on the basis of the number in a conduit, as described. A neutral conductor used with two phase legs of a 4-wire, 3-phase system to make up a 3-wire feeder is not a true neutral in the sense of carrying only current unbalance. Such a neutral carries the same current as the other two conductors under balanced load conditions and must be counted as a phase conductor when more than three conductors in conduit are derated.

Because the neutral of a 3-phase, 4-wire wye branch circuit or feeder to a load of fluorescent, metal-halide, mercury, or sodium lamp lighting or to electronic data processing equipment—the so-called information technology equipment—or any other nonlinear load will carry harmonic current even under balanced loading on the phases (refer to Sec. 220-22), such a neutral is not a true noncurrent-carrying conductor and must be counted as a phase wire when the number of conductors to arrive at an ampacity derating factor is determined for more than three conductors in a conduit. As a result, all the conductors of a 3-phase, 4-wire branch circuit or feeder to a fluorescent load would have an ampacity of only 80 percent of their nominal ampacity from Table 310-16 or other ampacity table. Because the 80 percent is a derating of ampacity, the conductors must be protected at the derated "ampacity" value.

Figure 310-32 shows four basic conditions of neutral loading and the need for counting the neutral conductor in loading a circuit to fluorescent or mercury ballasts, as follows:

Case 1—With balanced loads of equal power factor, there is no neutral current, and consequently no heating contributed by the neutral conductor. For purposes of heat derating according to the **Code**, this circuit produces the heating effect of only three conductors.

Case 2—With two phases loaded and the third unloaded, the neutral carries the same as the phases, but there is still the heating effect of only three conductors.

Case 3—With two phases fully loaded and the third phase partially loaded, the neutral carries the difference in current between the full phase value and the partial phase value, so that again there is the heating effect of only three full-load phases.

Case 4—With a balanced load of fluorescent ballasts, third harmonic current generation causes a neutral current approximating phase current, and there will be the heating effect of four conductors. Such a neutral conductor must be counted with the phase conductors when the load-current limitation due to conduit occupancy is determined, as required in part (**c**) of Note 10.

Although Note 10 exempts only neutral conductors from those conductors which must be counted in determining load-limiting factors for more than three conductors in a raceway or cable (per Note 8), similar exemption should be allowed for one of the "travelers" in a 3-way (or 3- and 4-way) switch circuit. As shown in Fig. 310-33, only one of the two conductors is a current-carrying conductor at any one time; therefore, the other should not be counted for load-limitation purposes where such switch legs are run in conduit or EMT along with other circuit conductors.

Note 11 This note makes it clear that an equipment grounding conductor or bonding conductor, which under normal conditions is carrying no current, does not have to be counted in determining ampacity derating of conductors when more than three conductors are used in a raceway or cable. As a result, equipment grounding and bonding conductors do not have to be factored into the calculation of required ampacity derating specified in Note 8(a).

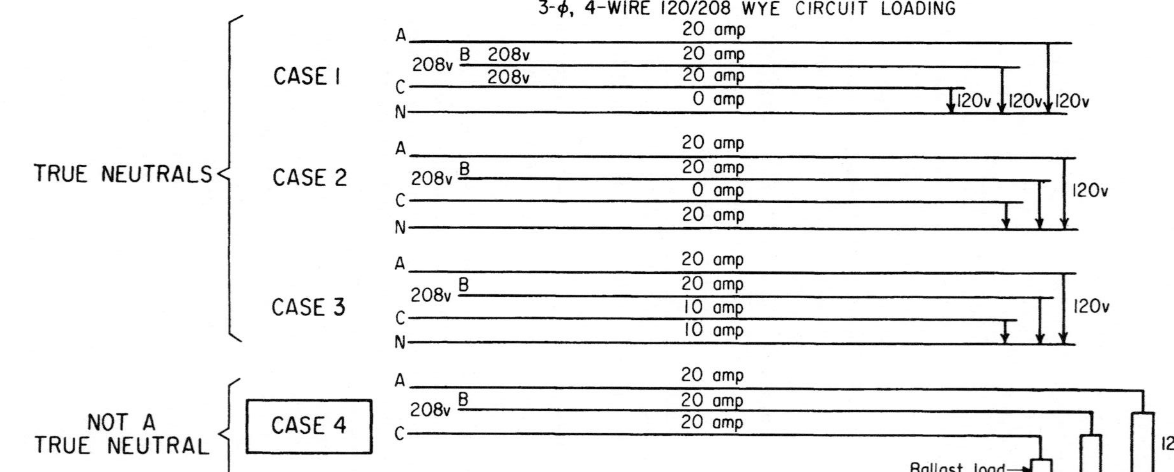

Fig. 310-32. All neutrals count for conduit fill, but only "true neutrals" do not count in determining ampacity derating for number of conductors in a raceway or cable [Note 10(a)]. (Sec. 310-15.)

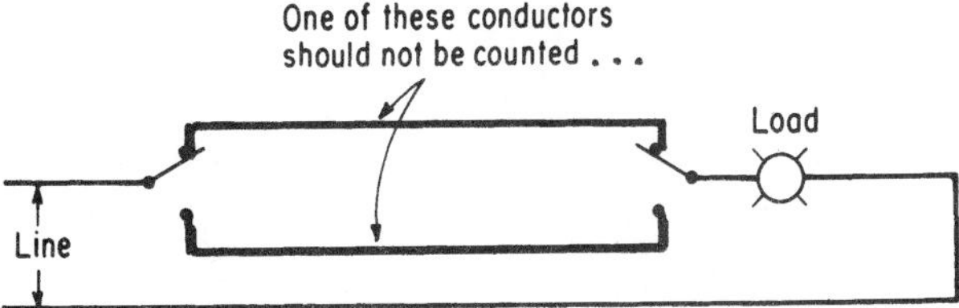
FOR DERATING PURPOSES —
One of these conductors should not be counted . . .
Load
Line
. . . where these wires are in conduit or EMT with other wires.

Fig. 310-33. The 3-wire run in conduit between 3-way switches contains only two current-carrying conductors. (Sec. 310-15.)

Ampacity of conductors over 600 V, nominal Since the 1975 edition, the NEC has added a vast amount of information and data for conductors rated over 600 V, up to 35 kV. For instance, Tables 310-67 through 310-84 give maximum continuous ampacities for copper and aluminum solid dielectric insulated conductors rated from 2,001 to 35,000 V.

ARTICLE 318. CABLE TRAYS

318-2. Definition. Cable trays are open, raceway-like support assemblies made of metal or suitable nonmetallic material and are widely used for supporting and routing circuits in many types of buildings. Troughs of metal mesh construction provide a sturdy, flexible system for supporting feeder cables, particularly where routing of the runs is devious or where provision for change or modification in circuiting is important. Ladder-type cable trays are used for supporting interlocked-armor cable feeders in many installations (Fig. 318-1). Where past Code editions treated a cable tray simply as a support system for cables, in the same category as a clamp or hanger, the Code today recognizes a cable tray as a conductor support method, somewhat like a raceway, under prescribed conditions, and an integral part of a Code-approved wiring method. However, cable trays are not listed under the Code definition of "raceway" in Art. 100. Any "raceway" must be an "enclosed" channel for conductors. A cable tray is a "support system" and not a "raceway."

318.3. Uses Permitted. The NEC recognizes a cable tray as a support for wiring methods that may be used without a tray (metal-clad cable, conductors in EMT, IMC, or rigid conduit, etc.), and a cable tray may be used in either commercial, industrial, or institutional buildings or premises (Fig. 318-2). Where cables are available in both single-conductor and multiconductor types—such as SE (service entrance) and UF cable—only the multiconductor type may be used in a tray. However, Sec. 318-3(b) permits use of single-conductor building wires in a tray. Single-conductor cables for use in a tray must be 1/0 or larger, listed for use in a tray, and "marked on the surface" as suitable for tray applications. In previous NEC editions, single-conductor cable in a tray had to be 250 kcmil or

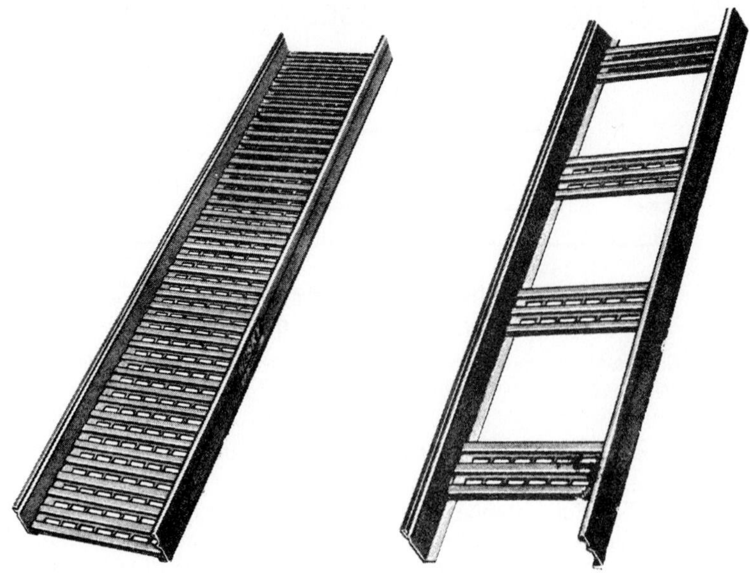
Trough-type (or expanded-metal-type) tray
Ladder-type tray

Fig. 318-1. Two basic types of cable tray. (Sec. 318-2.)

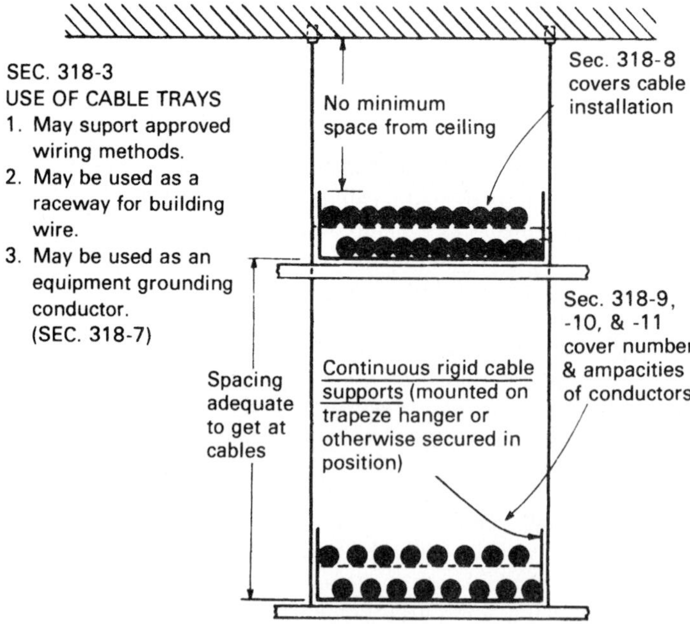
SEC. 318-3
USE OF CABLE TRAYS
1. May suport approved wiring methods.
2. May be used as a raceway for building wire.
3. May be used as an equipment grounding conductor. (SEC. 318-7)
No minimum space from ceiling
Sec. 318-8 covers cable installation
Sec. 318-9, -10, & -11 cover number & ampacities of conductors
Spacing adequate to get at cables
Continuous rigid cable supports (mounted on trapeze hanger or otherwise secured in position)

Fig. 318-2. Cable-tray use is subject to many specific rules in Art. 318. (Sec. 318-3.)

larger. Sizes 1/0 through 4/0 single-conductor cables may now be used but must be used in a ladder-type tray with rungs spaced not over 9 in. (229 mm) apart or in a ventilated trough cable tray. Sizes 250 kcmil and larger may be used in any kind of tray. This rule states that such use of building wire is permitted in industrial establishments only, where conditions of maintenance and supervision assure that only competent individuals will service the installed cable tray system. This applies to ladder-type trays, ventilated troughs, or 4-in. (102-mm) ventilated channel-type cable trays.

Single-conductor cables used in a cable tray must be a type specifically "listed for use in cable trays." This is a qualification on the rule that was in previous codes permitting use of single-conductor building wire (RHH, USE, THW, MV) in a cable tray. The wording permits any choice of conductor types that may be used, simply requiring that any type must be listed. It adds thin-wall-insulated cables, like THHN or XHHW, to the other types mentioned. Present UL standards make reference to cables designated "for CT (cable tray) use" or "for use in cable trays"—which is marked on the outside of the cable jacket. Such cables are subjected to a "vertical tray flame test," as used for Type TC tray cable and other cables. Only cables so tested and marked "VW-1" would be recognized for use in a cable tray.

Part **(c)** specifically recognizes use of the metal length of a cable tray as an equipment grounding conductor for the circuit(s) in the tray—in both commercial, industrial, and institutional premises where qualified maintenance personnel are available to assure the integrity of the grounding path. Section 318-3(d) specifically uses the word "only" when referring to cable types that are permitted to be used in cable trays in hazardous locations. In previous Code editions, wording was more open-ended and permitted specific cables without limiting use to only such cables.

As covered in part **(e)**, nonmetallic cable tray may be used in corrosive locations. This permits use of nonmetallic tray—such as fiberglass tray—in industrial or other areas where severe corrosive atmospheres would attack a metal tray. Such a tray is also permitted where "voltage isolation" is required.

318.4. Uses Not Permitted. Cable tray may be used in air-handling ceiling space *but* only to support the wiring methods permitted in such space by Sec. 300-22(c). This recognizes cable trays simply as supports for raceways or cables permitted in hung ceilings used for air conditioning. Section 318-6(h) requires cable trays to be exposed and accessible. Note that the two words "exposed" and "accessible" must be taken "as applied to wiring methods." Cable trays may be used above a suspended, nonair-handling ceiling with any of the wiring methods covered by Sec. 318-3. If used with a wiring method permitted by Sec. 300-22(c), cable trays may be used above an air-handling ceiling.

318.6. Installation. Part **(a)** makes clear that cable trays *must* be used as a complete system—that is, straight sections, angle sections, offsets, saddles, etc.—to form a cable support system that is continuous and grounded as required by Sec. 318-7(a). Cable trays must not be installed with separate, unconnected sections used at spaced positions to support the cable. Manufactured fittings or field-bent sections of tray may be used for changes in direction or elevation.

In the 1971 NEC, part **(c)** of Sec. 318-4 on Installation (now Sec. 318-6) read as follows:

Fig. 318-3. This was clearly a violation of Sec. 318-4(c) in the 1971 NEC because the tray does not connect to the transformer enclosures. The tray continuity required by Sec. 318-6(a) of the present NEC does seem to be violated by lack of connection to the enclosures. (Sec. 318-6.)

(c) Continuous rigid cable supports shall be mechanically connected to any enclosure or raceway into which the cables contained in the continuous rigid cable support extend or terminate.

That wording clearly made a violation of the kind of hookup shown in Fig. 318-3, where the tray does not connect to the transformer enclosures—and is not bonded by jumpers to those enclosures. However, an FPN added in the 1993 Code—which has been incorporated in the basic rule in Sec. 318-6(a) of the 1996 Code—indicates that use of "discontinuous segments" is intended to be acceptable—*But,* ground-continuity between the enclosure and the cable tray must be assured. Refer also to Sec. 318-7.

Section 318-6(e) notes that any multiconductor cables rated 600 V or less may be used in the same cable tray. Section 318-6(f) points out that although cables rated over 600 V must not be installed in the same cable tray with cables rated 600 V or less, there are two exceptions to that rule. High-voltage cables and low-voltage cables may be used in the same tray if a solid, noncombustible, fixed barrier is installed in the tray to separate high-voltage cables from low-voltage cables. Where the high-voltage (over 600 V) cables are Type MC, it is not necessary to have a barrier in the cable tray, and MC cables operating above 600 V may be used in the same tray with MC cables operating less than 600 V or with nonmetallic-jacketed cables operating at not over 600 V. But for high-

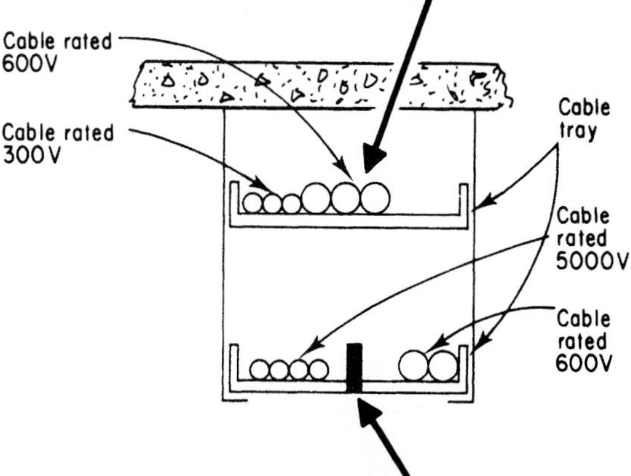
Multiconductor cables rated up to 600 volts may be used in the same tray, even when voltage ratings differ.
Cable rated 600V
Cable rated 300V
Cable tray
Cable rated 5000V
Cable rated 600V
A solid, noncombustible, fixed barrier must be used in tray to separate high-voltage and low-voltage cables with nonmetallic jackets — but barrier is not needed in tray if the high-voltage cables are Type MC.

Fig. 318-4. Cables of different voltage ratings may be used in the same tray. (Sec. 318-6.)

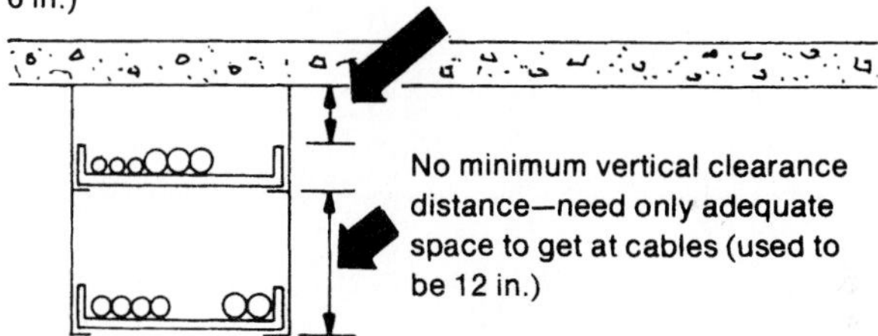
No minimum vertical clearance distance from tray top to ceiling, beam or other obstruction (used to be 6 in.)
No minimum vertical clearance distance—need only adequate space to get at cables (used to be 12 in.)

Fig. 318-5. Tray spacing must simply be adequate for cable installation and maintenance. (Sec. 318-6.)

voltage cables other than Type MC, a barrier must be used in the tray to separate high-voltage from low-voltage cables (Fig. 318-4).

Figure 318-5 shows the rule of part **(i)**.

318.7. Grounding. Part **(a)** requires cable trays to be grounded, just as conduit or other metal enclosures for conductors must be grounded. That rule combined with the last two sentences in part **(a)** of Sec. 318-6 makes cable trays comply with the Code concept that metal raceways constitute an equipment grounding conductor to carry fault currents back to the bonded neutral at a service, at a transformer secondary, or at a generator. Part **(b)** of Sec. 318-7 permits steel or aluminum cable tray to serve as an equipment grounding conductor for the circuits in the tray in much the same way as conduit or EMT may serve as the equipment grounding conductor—a return path for fault current—for the circuit conductors they contain, under the conditions specified in part **(b)**. Even though metallic cable trays are permitted to act as an equipment grounding conductor in the same way that a metallic conduit, tubing, or cable sheath might be, it should be noted that a cable tray is *not* a "raceway" as defined in Art. 100. Therefore, other Code rules that apply to "raceways" (e.g., Sec. 200-7 on grounded conductor identification) *do not* apply to cable trays.

Note that paragraph **(4)** under part **(b)** requires all tray system components, as well as "connected raceways," to be bonded together—either by the bolting means provided with the tray sections or fittings or by bonding jumpers, as shown in Fig. 318-6.

318.8. Cable Installation. Although splices are generally limited to use in conductor enclosures with covers and are prohibited in the various conduits, part **(a)** permits splicing of conductors in cable trays.

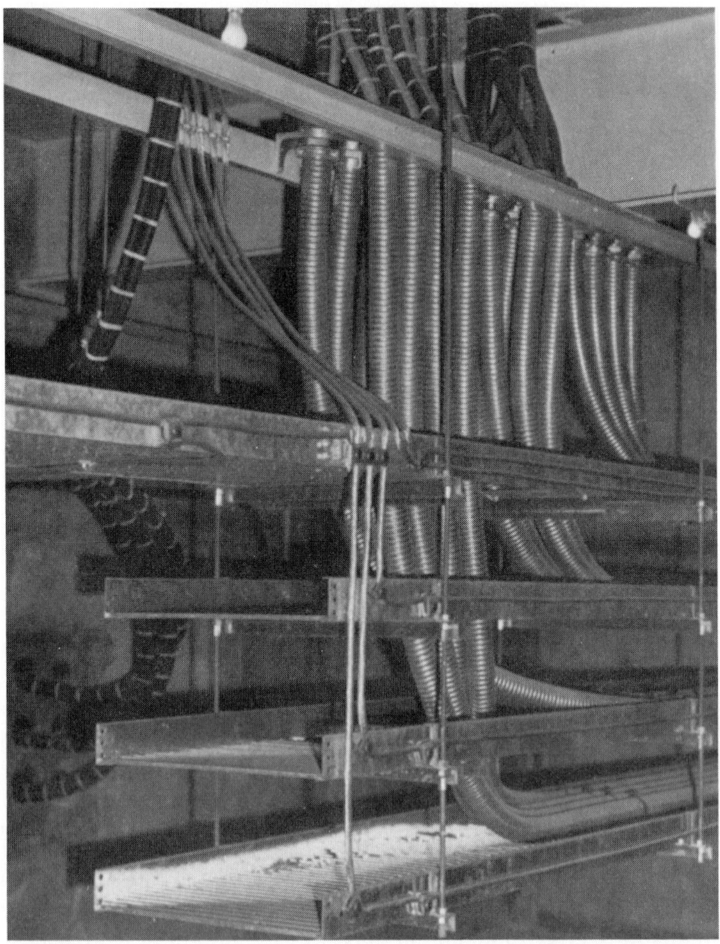

Fig. 318-6. Bare equipment bonding jumpers tie all tray runs together, with jumpers carried up to the equipment grounding bus in the switchboard above. (Sec. 318-7.)

Part **(c)** permits cables or conductors to drop out of a tray in conduits or tubing that have protective bushings and are clamped to the tray side rail by cable-tray conduit clamps to provide the bonded connections required by Sec. 318-7(b) (4).

Figure 318-7 shows how single-conductor cables must be grouped to satisfy part **(d)** of this section for a 1,200-A circuit made up of three 500-kcmil copper XHHW conductors per phase and three for the neutral. By distributing the phases and neutral among three groups of four and alternating positions, more effective cancellation of magnetic fluxes results from the more symmetrical placement—thereby tending to balance current by balancing inductive reactance of the overall 1,200-A circuit.

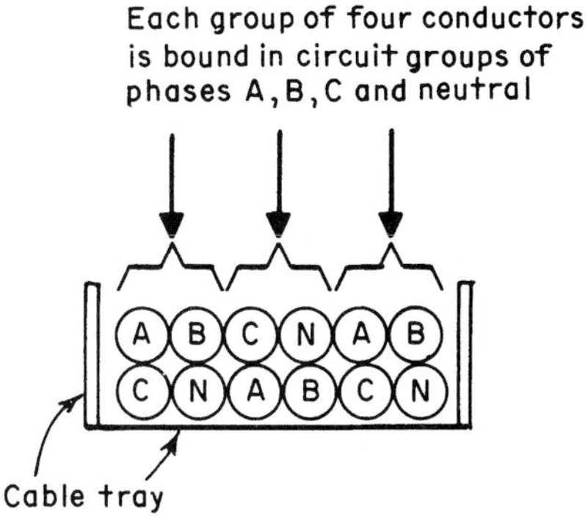

Fig. 318-7. A parallel 1,200-A circuit must have conductors grouped for reduced reactance and effective current balance. (Sec. 318-8.)

Part **(e)** prohibits stacking of single conductors Nos 1/0–4/0 in ladder or ventilated trough cable trays, unless done in accordance with Sec. 318-11(b)(4).

318.9. Number of Multiconductor Cables, Rated 2000 Volts or Less, in Cable Trays. These rules apply to multiconductor cables rated 2,000 V or less. For cables rated 2,001 V or higher, the number permitted in a cable tray is now covered in Sec. 318-12.

Section 318-9 is broken down into parts **(a)**, **(b)**, **(c)**, **(d)**, and **(e)**, each part covering a different condition of use. Section 318-9(a) applies to ladder or ventilated trough cable trays containing multiconductor power or lighting cables or any mixture of multiconductor power, lighting, control, and signal cables.

Section 318-9(a) has three subdivisions:

1. Where all the multiconductor cables are made up of conductors No. 4/0 or larger, the sum of the outside diameters of all the multiconductor cables in the tray must not be greater than the cable tray width, and the cables *must* be placed side by side in the tray in a single layer as shown in Fig. 318-8.
2. Where all the multiconductor cables in the tray are made up of conductors smaller than No. 4/0, the sum of the cross-sectional areas of all cables *must not exceed* the maximum allowable cable fill area in column 1 of Table 318-9 for the particular width of cable tray being used. The table shows, for instance, that if an 18-in. (457-mm)-wide ladder or ventilated

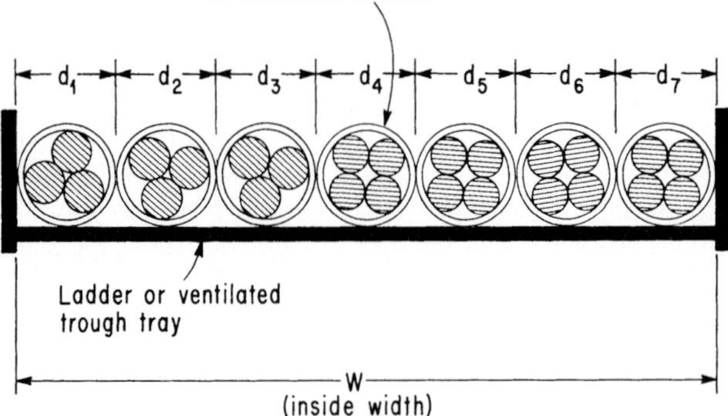

1. Cable-tray width (W) = at least $d_1 + d_2 + d_3 + d_4 + d_5 + d_6 + d_7$ in.
2. All cables *must* lie flat, side by side, in one layer.

Fig. 318-8. No. 4/0 and larger multiconductor cables *must* be in a *single* layer. (Sec. 318-9.)

trough cable tray is used with multiconductor cables smaller than No. 4/0, column 1 sets 21 sq in. (13,545 sq mm) as the maximum value for the sum of the overall cross-section areas of all the cables permitted in that tray, as in Fig. 318-9.

3. Where a tray contains one or more multiconductor cables No. 4/0 or larger along with one or more multiconductor cables smaller than No. 4/0, there are two steps in determining the maximum fill of the tray.

   First, the sum of the outside cross-section areas of all the cables smaller than No. 4/0 must not be greater than the maximum permitted fill area resulting from the computation in column 2 of Table 318-9 for the particular cable tray width. Then, the multiconductor cables that are No. 4/0 or larger must be installed in a single layer, and no other cables may be placed on top of them (Fig. 318-10). Note that the available cross-section area of a tray which can properly accommodate cables smaller than No. 4/0 installed in a tray along with No. 4/0 or larger cables is, in effect, equal to the allowable fill area from column 1 for each width of tray minus 1.2 times the sum of the outside diameters of the No. 4/0 or larger cables.

   Another way to look at this is to consider that, for any cable tray, the sum of the cross-section areas of cables smaller than No. 4/0, when added to 1.2 times the sum of the diameters of cables No. 4/0 or larger, must not exceed the value given in column 1 of Table 318-9 for a particular cable tray width.

For the installation shown in Fig. 318-10, assume that the sum of the cross-section areas of the seven cables smaller than No. 4/0 is 16 sq in. (10,320 sq mm) and assume that the diameters of the four No. 4/0 or larger cables are 3 in. (76 mm), 3.5 in. (89 mm), 4 in., (102 mm), and 4 in. (102 mm). The abbreviation "Sd" in column 2 of Table 318-9 represents "sum of the diameters" of No. 4/0

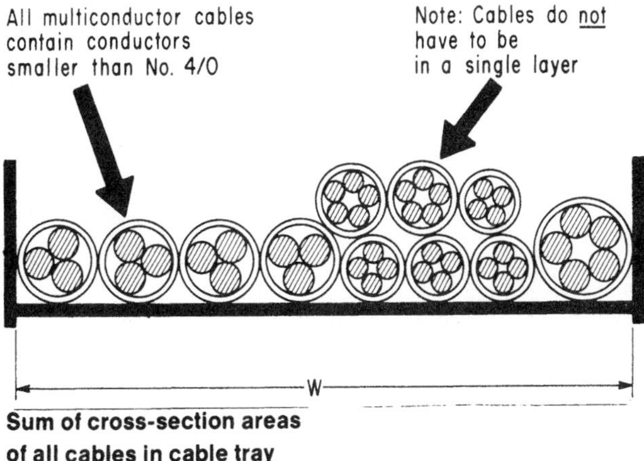
All multiconductor cables
contain conductors
smaller than No. 4/0
Note: Cables do not
have to be
in a single layer
W

Sum of cross-section areas of all cables in cable tray
= not less than cable fill in sq in. given in column 1 of Table 318-9 for the particular cable-tray width involved.

Note: Cross-section area (in sq in.) of each cable can be obtained from cable manufacturers' catalogs or spec sheets. If, in the case shown here, the sum of the cross-section areas of the 10 cables in the tray came to, say, 26 sq in., the smallest permissible width of cable tray would be 24 in., as shown in column 1 of Table 318-9. An 18-in.-wide cable tray would be good only for a sum of cable areas up to 21 sq in.

Fig. 318-9. Smaller than No. 4/0 cables may be stacked in tray. (Sec. 318-9.)

and larger cables installed in the same tray with cables smaller than No. 4/0. In the example here, then, Sd is equal to 3 + 3.5 + 4 + 4 = 14.5, and 1.2 × 14.5 = 17.4. Then we add the 16-sq-in. total of the cables smaller than No. 4/0 to the 17.4 and get 17.4 + 16 = 33.4. Note that this sum is over the limit of 28 sq in., which is the maximum permitted fill given in column 1 for a 24-in. (610-mm) wide cable tray. And column 1 shows that a 30-in. (762-mm) wide tray (with 35-sq-in. fill capacity) would be required for the 33.4 sq in. determined from the calculation of column 2, Table 318-9.

Section 318-9(b) covers use of multiconductor control and/or signal cables (not power and/or lighting cables) in ladder or ventilated trough with a usable inside depth of 6 in. (152 mm) or less. For such cables in ladder or ventilated trough cable tray, the sum of the cross-section areas of all cables at any cross section of the tray *must not* exceed 50 percent of the interior cross-section area of the cable tray. And it's important to note that a depth of 6 in. (152 mm) must be used in computing the allowable interior cross-section area of any tray that has a usable inside depth of more than 6 in. (152 mm) (Fig. 318-11).

Section 318-9(c) applies to solid-bottom cable trays with multiconductor power or lighting cables or mixtures of power, lighting, control, and signal cables. The maximum number of cables must be observed, as noted.

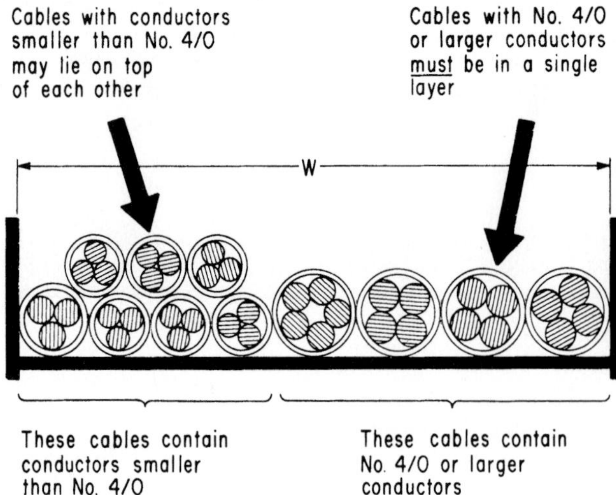
Cables with conductors smaller than No. 4/0 may lie on top of each other
Cables with No. 4/0 or larger conductors must be in a single layer
W
These cables contain conductors smaller than No. 4/0
These cables contain No. 4/0 or larger conductors

Inside width (W) of tray must not be less than that required by Table 318-9, based on the calculation indicated in column 2 of the table.

Fig. 318-10. Large and small cables have a more complex tray-fill formula. (Sec. 318-9.)

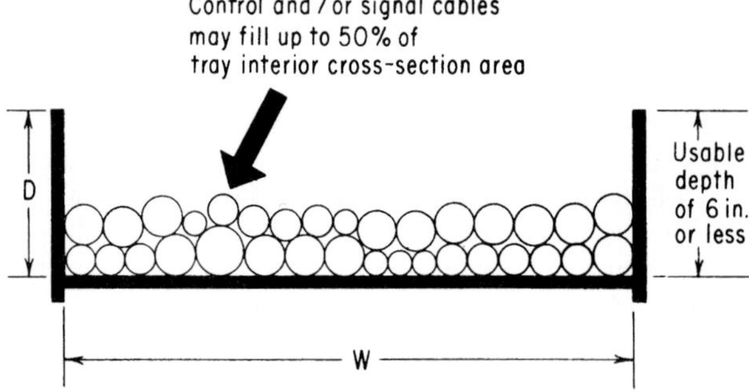
Control and / or signal cables may fill up to 50% of tray interior cross-section area
D
Usable depth of 6 in. or less
W

Permissible fill:

If the usable depth (D) in the above drawing is 6 in. or less, the sum of the cross-section areas of all contained cables must be not more than ½ × D × W. If the tray has a depth of more than 6 in., the value of (D) must be taken as 6 in. for computing tray fill.

Fig. 318-11. Tray fill for multiconductor control and/or signal cables is readily determined. (Sec. 318-9.)

318.10. Number of Single Conductor Cables, Rated 2,000 Volts or Less, in Cable Trays. This section covers the maximum permitted number of single-conductor cables in a cable tray and stipulates that the conductors must be evenly distributed on the cable tray. This section differentiates between (a) ladder or ventilated trough tray and (b) ventilated channel-type cable trays.

In Ladder or Ventilated Trough Tray

1. Where all cables are 1,000 kcmil or larger, the sum of the diameters of all single-conductor cables must not be greater than the cable tray width, as shown in Fig. 318-12. That means the cable tray width must be at least equal to the sum of the diameters of the individual cables.

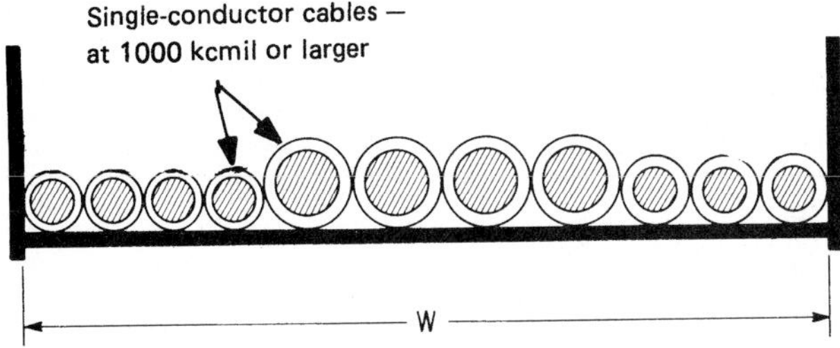

Tray must be wide enough to hold all cables side-by-side, as shown.

Fig. 318-12. For large cables, tray width must at least equal sum of cable diameters. (Sec. 318-10.)

2. Where all cables are from 250 up to and including 1,000 kcmil, the sum of the cross-section areas of all cables must not be greater than the maximum allowable cable fill areas in square inches, as shown in column 1 of Table 318-10 for the particular cable tray width.

example 1 Assume a number of cables, all smaller than 1,000 kcmil, have a total csa of 11 sq in. (7,095 sq mm) Column 1 of Table 318-10 shows that a fill of 11 sq in. (7,095 sq mm) is greater than that allowed for 6-in.-wide tray (6.5 sq in.) but less than the maximum fill of 13 sq in. permitted for 12-in.-wide tray. Thus, 12-in.-wide tray would be acceptable.

example 2 Assume four 4-wire sets of single-conductor, 500-kcmil RHH cables are used as power feeder conductors in a cable tray. Table 5 in Chap. 9 of the NE Code shows that the overall csa of each 500-kcmil RHH conductor (without outer covering) is 0.8316 sq in. The total area of 16 such conductors would be 16 × 0.8316, or 13.3 sq in. From Table 318-10, column 1, 13.3 sq in. is just over the maximum permissible fill for 12-in.-wide tray, but it is well below the maximum fill of 19.5 sq in. permitted for 18-in.-wide tray. Thus, 18-in.-wide tray is acceptable.

3. Where 1,000-kcmil or larger single-conductor cables are installed in the same tray with single-conductor cables smaller than 1,000 kcmil, the fill must not exceed the maximum fill determined by the calculation indicated in column 2 of Table 318-10—in a manner similar to the calculations indicated above for multiconductor cables.

example If nine 750-kcmil THW conductors are in a tray with six 1,000-kcmil THW conductors, the required minimum width (W) of the tray would be determined as follows:

1. The sum of the csa of the nine 750-kcmil conductors (those smaller than 1,000 kcmil) is equal to 9 × 1.2252 sq in. (from column 5, Table 5, Chap. 9, **NE Code**), or 11.03 sq in.
2. Each 1,000-kcmil THW conductor has an outside diameter of 1.404 in. The sum of the diameters of the 1,000-kcmil conductors is, then, 6 × 1.404, or 8.424 in.
3. Column 2 of Table 318-10 says, in effect, that to determine the minimum required width of cable tray it is necessary to add 11.03 sq in. (from 1 above) to 1.1 × 8.424 (from 2 above) and use the total to check against column 1 of Table 318-10 to get the tray width:

   $$11.03 + (1.1 \times 8.424) = 11.03 + 9.27 = 20.3 \text{ sq in.}$$

   From column 1, Table 318-10, the fill of 20.3 sq in. is greater than the 19.5 sq in. permitted for 18-in.-wide tray. But, this fill is within the permitted fill of 26 sq in. for 24-in.-wide tray. The 24-in.-wide tray is, therefore, the minimum size tray that is acceptable.
4. Where *any* cables in the tray are sizes 1/0 through 4/0, then all cables *must* be installed in a single layer. And the sum of the single-conductor cable diameters must not exceed the cable tray width as required for "Multiconductor Cables Rated 2000 V, Nominal, or Less" as covered in Sec. 318-9(a)(1).

In 4-in.-Wide Channel-Type Tray

Where single-conductor cables are installed in 4-in. (102-mm)-wide, ventilated channel-type trays, the sum of the diameters of all single conductors must not exceed the inside width of the channel.

318-11. Ampacity of Cables Rated 2000 Volts or Less in Cable Trays

Multiconductor Cables

When cable assemblies of more than one conductor are installed as required by Sec. 318-9, each conductor in any of the cables will have an ampacity as given in Table 310-16 or 310-18. Those are the standard tables of ampacities for cables with not more than three current-carrying conductors within the cable (excluding neutral conductors that carry current only during load unbalance on the phases). The ampacity of any conductor in a cable is based on the size of the conductor and the type of insulation on the conductor, as shown in Tables 310-16 and 310-18. For cables not installed in a cable tray, if a cable contains more than three current-carrying conductors, derating of the conductor ampacities must be made in accordance with Note 8 to Tables 310-16 through 310-19. But the last sentence of Sec. 318-11(a) flatly exempts cables in a tray from Note 8.

Exception No. 1 to the above determination of conductor ampacities is made in the case of any cable tray with more than 6 ft (1.83 m) of continuous, solid, unventilated covers. In such cases, the conductors in the cable have an ampacity of not more than 95 percent of the ampacities given in Table 310-16 or 310-18.

Exception No. 2 applies to a single layer of multiconductor cables with maintained spacing between cables, installed in uncovered tray.

Single-Conductor Cables

The ampacity of any single-conductor cable or single conductors twisted together is determined as follows:

600 kcmil and larger—Where installed in accordance with Sec. 318-10, the ampacity of any 600-kcmil or larger single-conductor cable in an uncovered tray is *not more* than 75 percent of the ampacity given for the size and insulation of conductor in Table 310-17 or for the size and insulation of conductor in Tables 310-17 and 310-19. Note that this means 75 percent of the free-air ampacity of the conductor. And if more than 6 ft (1.83 m) of the tray is continuously covered with a solid, unventilated cover, the ampacities for 600-kcmil and larger conductors must not exceed 70 percent of the ampacity value in Tables 310-17 and 310-19.

No. 1/0 through 500 kcmil—For any single-conductor cable in this range, installed in accordance with Sec. 318-10 in uncovered tray, its ampacity is not more than 65 percent of the ampacity value shown in Table 310-17 or 310-19. And if any such cables in this range are used in a tray that is continuously covered for more than 6 ft (1.83 m) with a solid, unventilated cover, the ampacities must not exceed 60 percent of the ampacity values in Tables 310-17 and 310-19.

Where No. 1/0 and larger single-conductor cables are installed in a single layer in an uncovered cable tray with a maintained spacing of not less than one cable diameter between individual conductors, the ampacities of such conductors are equal to the free-air ampacities given in Tables 310-17 and 310-19, as shown in Fig. 318-13.

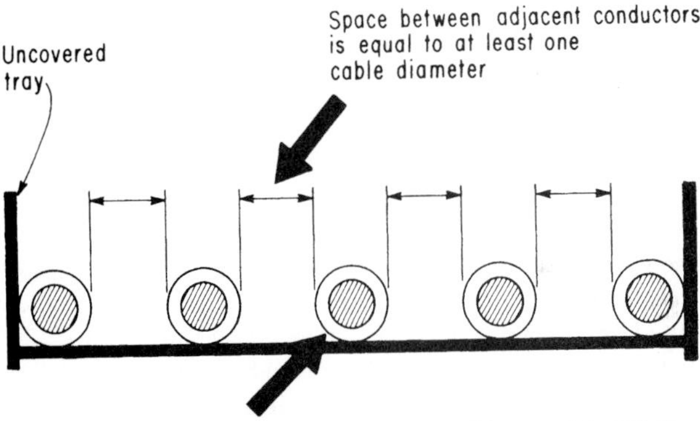

Fig. 318-13. With spacing, cables in tray may operate at free-air ampacity. (Sec. 318-11.)

318-12. Number of Type MV and Type MC Cables (2001 Volts or Over) in Cable Trays. This section applies only to high-voltage circuits in a tray. Type MV cable is a high-voltage cable now covered by new Art. 326. Type MC cable is the metal-clad cable operating above 2,000 V—a cable assembly long known as interlocked armor cable. [Type MC or other armored cable (e.g., ALS or CS) operating at voltages up to 2,000 V must conform to Secs. 318-10 and 318-11 on number and ampacities of cables when used in tray.]

Type MV and Type MC high-voltage cables must conform to the tray fill shown in Fig. 318-14.

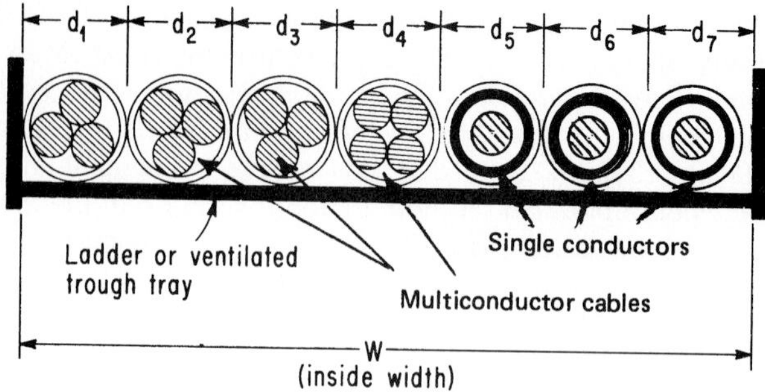

1. Cable-tray width (W) = at least $d_1 + d_2 + d_3 + d_4 + d_5 + d_6 + d_7$ in.
2. All cables *must* lie flat, side by side, in one layer.

Fig. 318-14. Tray must be wide enough for all high-voltage cables in a single layer. (Sec. 318-12.)

318-13. Ampacity of Type MV and Type MC Cables (2001 Volts or Over) in Cable Trays. This section covers the ampacities of MV and MC cables operating above 2,000 V in cable trays—both single-conductor and multiconductor. Exception No. 2 recognizes the improved heat dissipation afforded by spacing of the cables and allows use of the free-air ampacity tables in loading multiconductor cables. The spacing of "one cable diameter" is also recognized for single conductors in Sec. 318-13(b)(3).

ARTICLE 320. OPEN WIRING ON INSULATORS

320-1. Definition. Conductors for open wiring may be any of the general-use types listed in Table 310-13 for "dry" locations and "dry and wet" locations such as THW, XHHW, THHN, etc.

The conductors are secured to and supported by insulators of porcelain, glass, or other composition materials. In modern wiring practice open wiring is used for high-tension work in transformer vaults and substations. It is very commonly used for temporary work and is used for runs of heavy conductors for feeders and power circuits, as in manholes and trenches under or adjacent to switchboards, to facilitate the routing of large numbers of circuits fed into conduits.

320-3. Uses Permitted. This section limits open wiring on insulators to industrial or agricultural establishments, up to 600 V. Section 320-15 spells out such installations in unfinished attics and roof spaces.

320-6. Conductor Supports. Methods of dead-ending open cable runs are shown in Fig. 320-1.

Where heavy AC feeders are run as open wiring, the reactance of the circuit is reduced and hence the voltage drop is reduced by using a small spacing between the conductors. Up to a distance of 15 ft (4.57 m) between supports the 2½-in. (64-mm) spacing may be used if spacers are clamped to the conductors

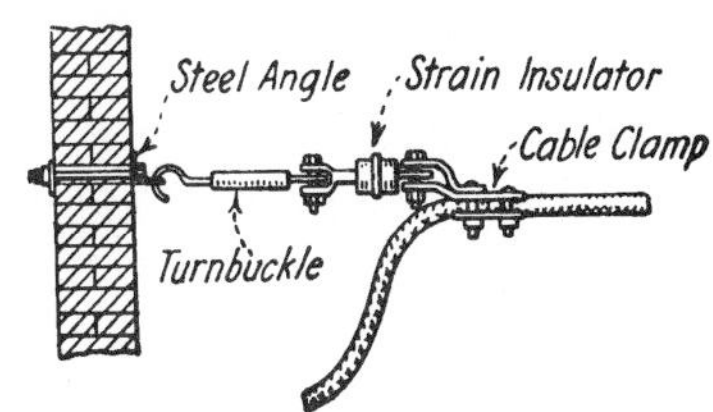

Method of dead-ending heavy conductors used in open wiring.

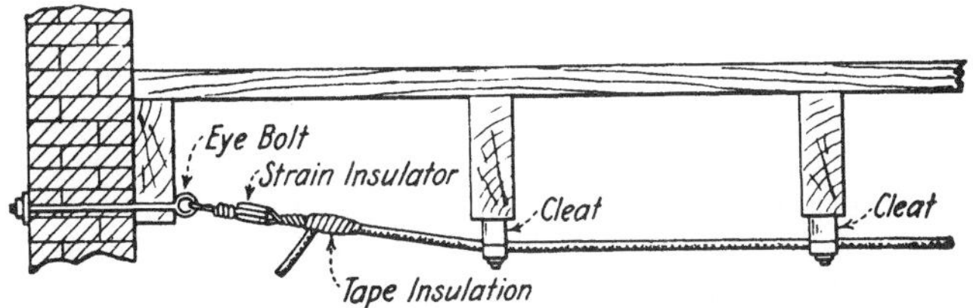

Method of dead-ending heavy conductors used in open wiring.

Fig. 320-1. Proven methods must be used for dead-ending open wiring. (Sec. 320-6.)

at intervals not exceeding 4½ ft (1.37 m). A spacer consists of the three porcelain pieces of the same form as used in the support, with a metal clamping ring.

In Exception No. 2, reference to "mill construction" is generally understood to mean the type of building in which the floors are supported on wooden beams spaced about 14 to 16 ft (4.27 to 4.88 m) apart. Wires not smaller than No. 8 may safely span such a distance where the ceilings are high and the space is free from obstructions.

320-7. Mounting of Conductor Supports. Figure 320-2 illustrates mounting of knobs and cleats for the support of No. 14, No. 12, and No. 10 conductors. For conductors of larger size, solid knobs with tie wires or single-wire cleats should be used.

320-12. Clearance from Piping, Exposed Conductors, etc. The additional insulation on the wire, referred to in this rule, is to prevent the wire from coming in contact with the adjacent pipe or other metal.

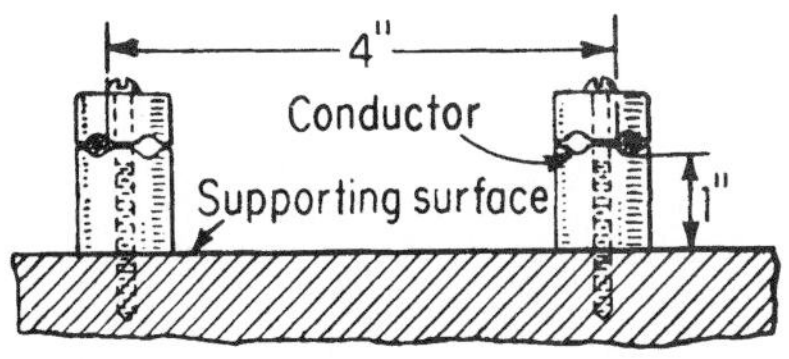

Split knobs for supporting small wires used in open wiring or concealed knob-and-tube work.

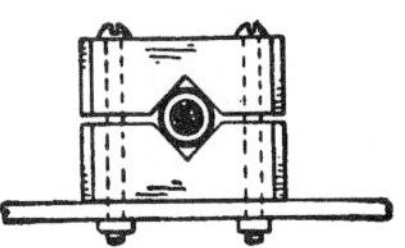

Single-wire cleat for supporting large conductors used in open wiring.

Fig. 320-2. Proper wiring support devices must be correctly mounted. (Sec. 320-7.)

ARTICLE 321. MESSENGER SUPPORTED WIRING

321-1. Definition. This article covers a wiring system that has long been manufactured and widely used in industrial installations. The basic construction of the wiring method has been used for many years as service-drop cable for utility supply to all kinds of commercial and residential properties.

From long-time application, messenger supported wiring is actually an old standard method, even though it has not been covered by the NE Code up until recent years. In Sec. 225-6, the Exception refers to "supported by messenger wire," and that phrase has been in the Code for over 30 years. Figure 321-1 shows an example of triplex service-drop cable used for supplying floodlights at an outdoor athletic field. So coverage of this type of wiring is important—especially for the vast amounts of outdoor use where messenger-cable wiring offers so many advantages over open wiring. But, messenger supported wiring is recognized for both indoor and outdoor branch circuits and/or feeders. Refer to the discussion under Sec. 225-4 for outdoor use of messenger supported wiring.

321-3. Uses Permitted. Messenger support is permitted for a wide range of cables and conductors—for use in commercial and industrial applications. Part **(b)** covers ordinary building wires supported on a messenger and recognizes use of messenger supported wiring in "industrial establishments *only*" with "any of the conductor types given in Table 310-13 or Table 310-62." Tables 310-13 and 310-62 include single-conductor Types MV, RHH, RHW, and THW and also accept all the other single-conductor types, such as THHN, XHHW, TW. All such application is recognized either indoors or outdoors—provided that any conductors exposed to the weather are "listed for use in wet locations" and are "sunlight-resistant" if exposed to direct rays of the sun.

321-5. Ampacity. In any specific application, the ampacity of any conductor "shall be determined by Sec. 310-15"—which covers ampacity tables for all the various conductors in the various wiring methods. The choice of table to be used depends upon the type of cable or conductor assembly being supported by the messenger. Some may opt to use ampacity tables for a "single conductor in free air" (Table 310-17), which is not applicable to, say, triplexed single conductors (bundled together) but is actually applicable only to single conductors that are "isolated" individually in air.

ARTICLE 324. CONCEALED KNOB-AND-TUBE WIRING

324-3. Uses Permitted. Note that this wiring method is restricted to use only for extensions of existing installations and is not Code-acceptable as a general-purpose wiring method for new electrical work. Under the conditions specified in **(1)** and **(2)**, concealed knob-and-tube wiring may be used only if special permission is granted by the local inspection authority having jurisdiction as noted in the second sentence of Sec. 90-4.

324-5. Conductors. Conductors for concealed knob-and-tube work may be any of the general-use types listed in Table 310-13 for "dry" locations and "dry and wet" locations such as TW, THW, XHHW, RHH, etc.

Fig. 321-1. Messenger-supported cable, used here to supply pole-mounted floodlights, may be constructed in a number of different assemblies, such as this service-drop cable with an ACSR messenger cabled with insulated conductors.

324-11. Unfinished Attics and Roof Spaces. Where wires are run on knobs or through tubes in a closed-in and inaccessible attic or roof space, the wiring is concealed knob-and-tube work; but if the attic or roof space is accessible, the wiring must conform to Art. 320 as open wiring on insulators. Both cases are covered by the foregoing rules.

Where the wiring is installed at any time after the building is completed, in a roof space having less than 3-ft (914-mm) headroom at any point, the wires may be run on knobs across the faces of the joists, studs, or rafters or through or on the sides of the joists, studs, or rafters. Such a space would not be used for storage purposes, and the wiring installed may be considered as concealed knob-and-tube work.

An attic or roof space is considered accessible if it can be reached by means of a stairway or a permanent ladder. In any such attic or roof space wires run through the floor joists where there is no floor must be protected by a running board and wires run through the studs or rafters must be protected by a running board if within 7 ft (2.13 m) from the floor or floor joists.

ARTICLE 325. INTEGRATED GAS SPACER CABLE TYPE IGS

325.1. Definition. Type IGS cable is a "factory assembly of one or more conductors, each individually insulated and enclosed in a loose-fit nonmetallic flexible conduit." The cable is for underground use—including direct earth burial—for service conductors, feeders, or branch circuits. The introduction of this cable to the NEC was recommended on the following basis:

> Underground cable costs are increasing at a high rate. A need exists for lower material costs and reduced cost for installation. Failures on underground cables are increasing, particularly direct-burial types.
>
> The new cable system overcomes all the above problems. The new cable system has the advantage of low first cost for materials and low installation cost. It eliminates the need for field pulling of cables into conduits and eliminates the cost of assembly of conduit in the field. The new system may be directly buried, plowed in, or bored in for further savings. It is a cable and conduit system.
>
> A tough natural-gas-approved pipe is used as the conduit. When it is pressurized, it will withstand much abuse. The gas pressure keeps out moisture and serves to monitor the cable for damage by insects or mechanical damage that can lead to future failure. The gas pressure can even be attached to an alarm to sound a loss of pressure or to trip a CB for hazardous locations. However, a loss of pressure in the cable will not cause it to fail. Even on dig-ins, the gas serves to warn the digger. The gas prevents combustion and burning on cable failure. The SF_6 gas is nontoxic, odorless, tasteless, and will not support combustion. It acts to put out a fire.
>
> UL has tested a 3/C 250 MCM [kcmil] Type IGS-EC 600-V cable in 2-in. conduit. The UL test at zero gauge pressure shows a breakdown voltage between conductors of 14,000 V after numerous short-circuit, breakdown, and humidity tests. When the cable is single conductor, the breakdown voltage is even higher on loss of pressure, as the polyethylene pipe or conduit provides additional insulating value.
>
> An award-winning installation was made in 1979 at 5 kV. Three installations of 3/C 250 MCM [kcmil] Type IGS-EC cable have been made for residential underground ser-

vice entrances in Oakland, California. The first was made in May of 1979 and all have been successful.

ARTICLE 326. MEDIUM VOLTAGE CABLE TYPE MV

326-1. Definition. This is a very limited definition of a Code designation—Type MV. The description of this cable type is amplified in the *Electrical Construction Materials Directory* of the Underwriters Laboratories, as follows:

Medium-Voltage Cable (PITY)

Medium-voltage cables are rated 5000 to 35,000 volts.

They are single or multiconductor, aluminum or copper, with solid extruded dielectric insulation and may have an extruded jacket, metallic covering or combination of both over the single conductors or over the assembled conductors in a multi-conductor power cable.

All insulated conductors rated higher than 8000 volts have electrostatic shielding. Cables rated 5000 or 8000 volts may be shielded or nonshielded.

Nonshielded cables are intended for use where conditions of maintenance and supervision ensure that only competent individuals service and have access to the installation.

Shielded cables are marked "MV-90" and are suitable for use in wet or dry locations at 90 C.

Nonshielded cables are marked either "MV-90" indicating suitability for use in wet or dry locations at 90 C maximum or "MV-90" Dry Locations Only" indicating suitability for use only in dry locations at 90 C maximum.

Cables marked "oil resistant I" or "oil resistant II" are suitable for exposure to mineral oil at 60 C or 75 C, respectively.

Cables marked "sunlight resistant" may be exposed to the direct rays of the sun.

Cables intended for installation in cable trays in accordance with Article 318 of the National Electrical Code are marked "for CT Use" or "for use in cable trays."

Cables with aluminum conductors are marked with the word "aluminum" or the letters "AL."

Cables are marked with their conductor size, voltage rating and insulation level (100 percent or 133 percent).

The basic standard used to investigate products in this category is UL1072, "Medium-Voltage Power Cables."

The Listing Mark of Underwriters Laboratories Inc. on the product is the only method provided by UL to identify products manufactured under its Listing and Follow-Up Service. The Listing Mark for these products includes the name and/or symbol of Underwriters Laboratories Inc. (as illustrated in the Introduction of this Directory) together with the word "LISTED," a control number and the following product name: "Medium-Voltage Cable."

Medium Voltage Cable, Classified in Accordance with UL1072, with Metric Conductor Sizes (PIVW)

This category covers medium voltage cables rated 2001 to 35,000 volts and in conductor sizes 10 through 500 sq mm.

The cables are single or multiconductor, aluminum or copper, with solid extruded dielectric insulation. An extruded jacket, metallic covering, or combination of both may be provided over single conductors or over the assembled conductors in a multiconductor power cable.

All insulated conductors rated 8001 volts and higher have electrostatic shielding. Cables rated 2001 to 8000 volts may be shielded or nonshielded.

Nonshielded cables are intended for use where conditions of maintenance and supervision ensure that only competent individuals service and have access to the installation.

Cables marked MV-75, MV-85, or MV-90 are suitable for use in wet or dry locations at 75 C, 85 C, or 90 C, respectively.

Cables which are suitable for use in dry locations only are so marked. Cables marked "oil resistant I" or "oil resistant II" are suitable for exposure to mineral oil at 60 C or 75 C, respectively.

Cables marked "sunlight resistant" may be exposed to the direct rays of the sun.

Cables intended for installation in cable trays are marked "For CT Use" or "For Use In Cable Trays".

Cables with aluminum conductors are marked with the word "Aluminum" or the letters "AL."

Cables are marked with conductor size in sq mm, voltage rating and insulation level (100 percent or 133 percent).

The basic standard used to investigate products in this category is UL1072, "Medium-Voltage Power Cables."

Look for Classification Marking on Product

The Classification Marking of Underwriters Laboratories Inc. on the product, the attached tag, the reel or the smallest unit container in which the product is packaged in the only method provided by UL to identify these products manufactured under its Classification and Follow-Up Service.

The Classification Marking for these products shall only be the name of Underwriters Laboratories Inc. together with the word "Classified," a control number, and the product name: "Medium Voltage Cable."

MEDIUM VOLTAGE CABLE
CLASSIFIED BY UNDERWRITERS LABORATORIES INC®
IN ACCORDANCE WITH UL1072, WITH METRIC
CONDUCTOR SIZES

326-3. Uses Permitted. Because the **Code** has an article and cable designation (Type MV) for cables operating above 2,000 V up to 35,000 V, it may be expected that electrical inspection authorities will insist that all cables in that voltage range must be Type MV to satisfy the **NE Code.**

Great care should be exercised in determining the attitude of local inspection authorities toward the meaning of this article. In particular, the relationship of Sec. 110-8 to Art. 326 should be determined. Section 110-8 states that "only wiring methods recognized as suitable are included in this **Code.**" The question to be answered is: Will electrical inspection agencies require all high-voltage conductors to be Type MV? Or will inspection agencies accept high-voltage conductors not specifically designated Type MV? In other words, because the **Code** has an accepted type of high-voltage cable, will it be permissible to use high-voltage cables that are not of this accepted type? For circuits in common use up to 600 V, Sec. 110-8 has consistently been interpreted to require that *any* conductor or cable must be one of the types specifically designated in the **Code**—Table 310-13 or elsewhere in Arts. 300 to 365. That is, conductors must be Type TW, THW, or one of the other designated types, and cable must be Type AC, NM, MI, MC, or other designated cable. It would be a **Code** violation to use any non-**Code**-designated wire or cable for systems up to 2,000 V. It would, therefore, seem to be similarly contrary to **Code** to use a non-**Code**-designated cable for higher-voltage circuits inasmuch as there is a **Code**-designated type (Type MV) for such applications.

Refer to Table 310-61 on Type MV conductors and to Sec. 318-13 for use of Type MV cables in tray.

ARTICLE 328. FLAT CONDUCTOR CABLE TYPE FCC

328-1. Scope. This article covers design and installation regulations on a branch-circuit wiring system that supplies floor outlets in office areas and other commercial and institutional interiors. (See Fig. 328-1.) The method may be used for new buildings or for modernization or expansion in existing interiors. FCC wiring may be used on any hard, sound, smooth floor surface—concrete, wood, ceramic, etc. The great flexibility and ease of installation of this surface-mounted flat-cable wiring system meet the need that arises from the fact that the average floor power outlet in an office area is relocated every two years.

Fig. 328-1. Flat conductor cable (FCC) supplies terminal base for floor-outlet pedestal at exact location required for desk in office area. FCC cable is taped in position over an insulating bedding tape and then covered with a flat steel tape (not yet installed here) to protect the three conductors (hot leg, neutral, and equipment grounding conductor) in the flat cable. Carpet squares are used to cover the finished cable runs.

Undercarpet wiring to floor outlets eliminates any need for core drilling of concrete floors—avoiding noise, water dripping, falling debris, and disruption of normal activities in an office area. Alterations or additions to Type FCC circuit runs are neat, clean, and simple and may be done during office working hours—not requiring the overtime labor rates incurred by floor drilling, which must be done at night or on weekends. The FCC method eliminates use of conduit or cable, along with the need to fish conductors.

Type FCC wiring offers versatile supply to floor outlets for power and communication—at any locations on the floor. The flat cable is inconspicuous under the carpet squares. Elimination of floor penetrations maintains the fire integrity of the floor, as required by Sec. 300-21.

A typical system might use separate flat-cable circuit layouts for 120-V power to floor-pedestal receptacles, telephone circuits, and data communications lines for CRT displays and computer units. For 120-V power, the flat cable contains three flat, color-coded (black, white, and green) No. 12 copper conductors for 20-A circuits—one hot conductor, one neutral, and one equipment grounding conductor. Telephone circuits use flat, 3-pair, No. 26 gauge conductors. And data connection circuits use flat RG62A/U coaxial cable that is only 0.09 in. (2.25 mm) high.

328-2. Definitions. The various components of a type FCC system are described here. Figure 328-2 shows typical components of an FCC system.

328-10. Coverings. Floor carpet squares used for covering Type FCC wiring must not be larger than 36 by 36 in. (914 by 914 mm). This rule eliminates questions that arose about the possibility of using "squares" of broadloom carpet that covered a room floor wall to wall.

In making an undercarpet installation, usual thinking would dictate installation of the cable layout first and then placement of the floor covering of carpet squares over the entire area. But some installers have found it easier and less expensive to first cover the entire floor area with the self-adhesive carpet squares and then plan the circuit layouts to keep the cable runs along the centerlines of carpet squares and away from the edges of the squares. After the layout is determined, it is a simple matter to lift only those carpet squares along the route of each run, install the cable and pedestal bases, and replace the self-stick carpet squares to restore the overall floor covering. That approach has proved effective and keeps carpet cutting to the middle of any square.

328-17. Crossings. Not more than two FCC cable runs may be crossed over each other at any one point. To prevent lumping under the floor carpets, this rule permits no more than two Type FCC cables to be crossed over each other at a single point. This applies to FCC power cable and FCC communications and data cables.

ARTICLE 330. MINERAL-INSULATED, METAL-SHEATHED CABLE

330-1. Definition. The data from the UL Green Book expand on the definition (Fig. 330-1) and cover application notes as follows.

Mineral-insulated metal-sheathed cable is labeled in a single-conductor construction from No. 16 AWG through No. 4/0 AWG, two- and three-conductor from No. 16 AWG through No. 4 AWG, four-conductor from No. 16 AWG through No. 6 AWG, and seven-conductor Nos. 16, 14, 12, and 10 AWG. The exterior sheath may be of copper or alloy steel.

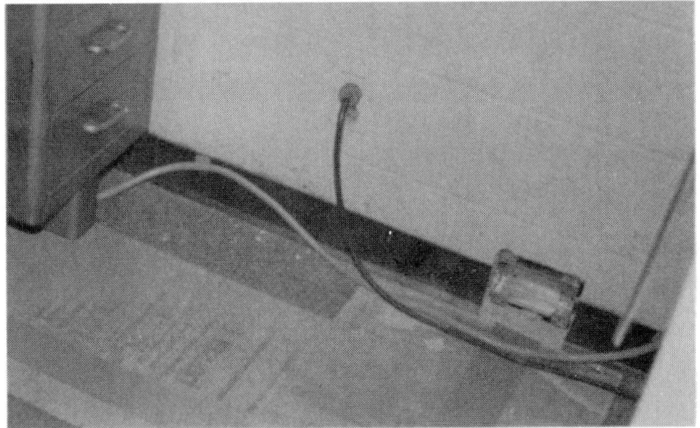

(*a*)

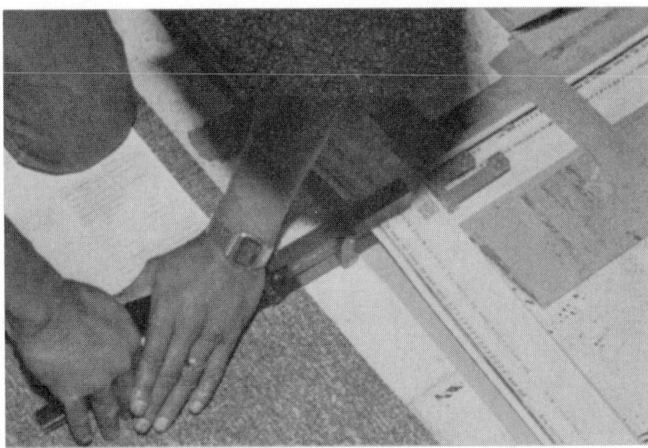

(*b*)

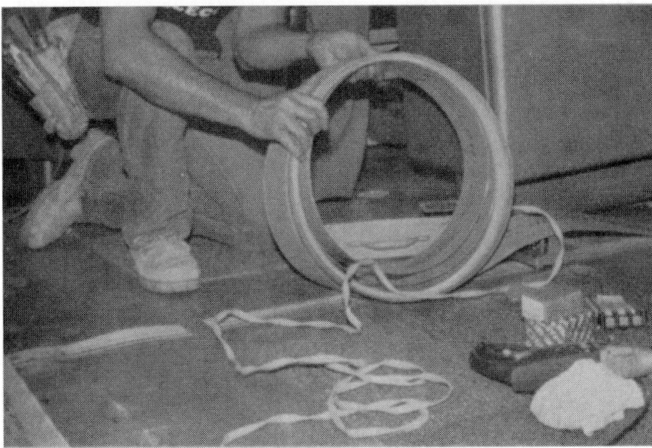

(*c*)

Fig. 328-2. Typical components of a Type FCC system: (*a*) Bottom shield in place. (*b*) Connecting the conductor. (*c*) Coil of top shield.

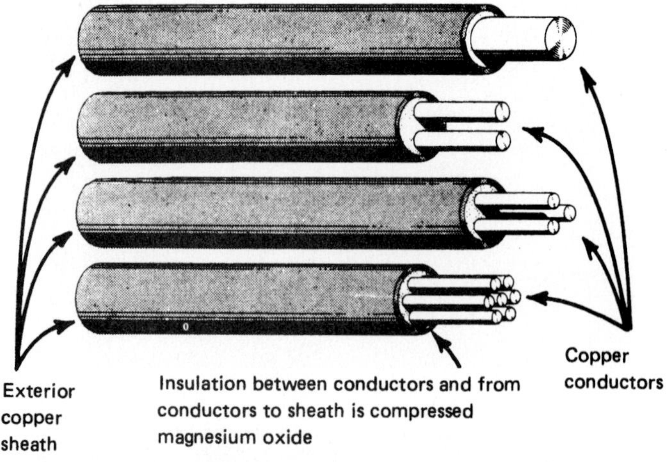

Fig. 330-1. Type MI is a single- or multiconductor cable that requires special termination. (Sec. 330-1.)

The standard length in which any size is furnished depends on the final diameter of the cable. The smallest cable, 1/C No. 16 AWG, has a diameter of 0.216 in. and can be furnished in lengths of approximately 1,900 ft. Cables of larger diameter have proportionally shorter lengths. The cable is shipped in paper-wrapped coils ranging in diameter from 3 to 5 ft (914 mm to 1.52 m)

The original intent behind development of this cable was to provide a wiring material which would be completely noncombustible, thus eliminating the fire hazards resulting from faults or excessive overloads on electrical circuits. To accomplish this, it is constructed entirely of inorganic materials. The conductors, sheath, and protective armor are of metal. The insulation is highly compressed magnesium oxide, which is extremely stable at high temperatures (fusion temperature of 2,800°C).

330-3. Uses Permitted. This section describes the general use of mineral-insulated metal-sheathed cable, designated Type MI. Briefly, it basically includes general use as services, feeders, and branch circuits in exposed and concealed work, in dry and wet locations, for underplaster extensions and embedded in plaster, masonry, concrete, or fill, for underground runs, or where exposed to weather, continuous moisture, oil, or other conditions not having a deteriorating effect on the metallic sheath (Fig. 330-2). The maximum permissible operating temperature for general use is 85°C (determined by present standard terminations). The cable itself, however, is recognized for 250°C in special applications. Permissible current ratings will be those given in Table 310-16 (or Table B-310-3 in App. B in the back of the NEC book). Type MI cable in its many sizes and constructions is suitable for all power and control circuits up to 600 V.

There is no question that MI cable can be used "in underground runs" as indicated in Sec. 330-3(10). But there is a question as to the meaning of "in underground runs." This question arises because of the wording in Sec. 310-7. Sections 310-7 and 310-8(b) state that cable suitable for direct burial in the

Fig. 330-2. Type MI is recognized for an extremely broad range of applications—for any kind of circuit, indoors or outdoors, wet or dry, and even in hazardous locations, as where MI motor branch circuits supply pumps in areas subject to flammable gases or vapors. (Sec. 330-3.)

earth must be of a type specifically approved for the purpose ("identified" and "listed" for such use). That would require that the local inspector be satisfied with direct burial of MI cable and that UL listing recognize such use.

Although the copper sheath of MI cable has good resistance to corrosion, acid soils may be harmful to the copper sheath. Direct earth burial in alkaline and neutral soils would generally be expected to create no problems, but in any direct burial application MI cable with an outer plastic or neoprene jacket would assure effective application and provide compliance with the phrase "protected against physical damage and corrosive conditions" in part **(10)** of Sec. 330-3. Such jacketed MI cables are available, and have been successfully used in direct burial applications.

The fact remains, however, that the Code is not clear on this subject, and local rulings may vary on this subject.

330-14. Fittings. Connections of Type MI cable must be carefully made in accordance with UL and manufacturers' application data to assure effective operation (Fig. 330-3).

330-15. Terminal Seals. This rule is applied in conjunction with that of Sec. 330-14 to assure *both* sealing of the cable end and means for connecting to enclosures (Fig. 330-4).

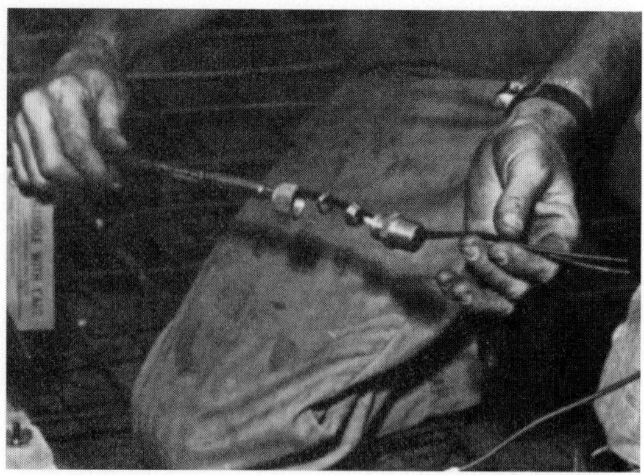

Fig. 330-3. Termination fitting for Type MI cable must be an approved connector, with its component parts assembled in proper sequence. (Sec. 330-14.)

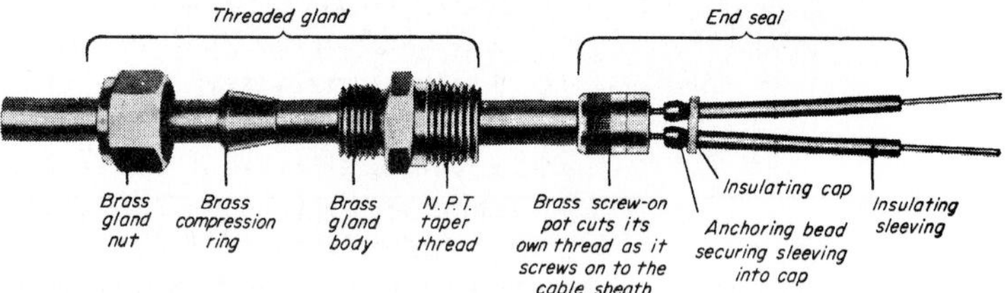

This typical fitting is approved for MI termination in hazardous locations, in accordance with Sec. 501-4.

Fig. 330-4. MI cable termination must provide end sealing and connection means. (Sec. 330-15.)

ARTICLE 331. ELECTRICAL NONMETALLIC TUBING

331-1. Definition. One type of plastic raceway defined in the Code (Fig. 331-1) is "ENT," electrical nonmetallic tubing, which is "a pliable corrugated raceway of circular cross section with integral or associated couplings, connectors, and fittings listed for the installation of electrical conductors. It is composed of a material that is resistant to moisture [and] chemical atmospheres and is flame retardant." ENT can be bent by hand, when being installed, to establish direction and lengths of runs.

331-3. Uses Permitted. ENT is permitted to be used as a general-purpose, "flexible"-type conduit in any type of occupancy (Fig. 331-2). Electrical nonmetallic tubing (ENT) is *not limited* to use in buildings up to three stories high. But, where the building does *not* exceed three stories above grade, ENT may be used in "exposed" locations. Where concealed, ENT may be used in a building of any height (Fig. 331-3)—subject to conditions given in Sec. 331-3 and Sec. 331-4. ENT may be used:

Fig. 331-1. ENT is a pliable, bendable plastic raceway for general-purpose use for feeders and branch circuits.

Fig. 331-2. ENT may be used in residential and nonresidential buildings. (Sec. 331-3.)

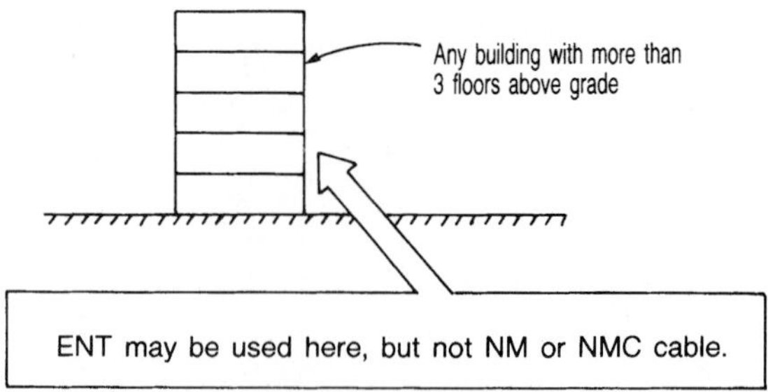
Any building with more than
3 floors above grade
ENT may be used here, but not NM or NMC cable.

Fig. 331-3. ENT (electrical nonmetallic tubing) may be used in any type of building, of any height. (Sec. 331-3.)

1. Concealed in walls, floors, and ceilings that provide a thermal barrier with at least a 15-min fire rating from listings of fire-rated assemblies. The rule permits ENT above a ½-in. (12.7 mm) Sheetrock ceiling.

   As previously indicated, ENT may be used *exposed* in a building that is not over three floors above grade. This is limited permission that recognizes ENT for exposed use under the same limitations that Sec. 336-5(a) places on use of Romex in buildings. And the paragraph immediately following part (**1**) in Sec 336-5(a) indicates that where there is a definition of the "first floor" Romex is accepted in four-story buildings where the first floor is used totally for vehicle parking, storage, or "similar use," where the space is not used for human occupancy or human habitation. ENT may be used *exposed* in such a four-story building.
2. In severe corrosive locations where suited to resist the particular atmosphere (but not "exposed").
3. In concealed, dry, and damp locations not prohibited by Sec. 331-4.
4. Above suspended ceilings with at least a 15-min fire rating.
5. Embedded in poured concrete with fittings that are listed or otherwise identified for that use.

Note that there is no *general* permission in this section to use ENT *exposed.*

331-4 Uses Not Permitted. ENT may not be used in *exposed* locations, except above suspended ceilings of 15-min fire-rated material in buildings of any height above grade. This section excludes ENT from hazardous locations—except for intrinsically safe circuits per Art. 504—from supporting fixtures or equipment, from use where the ambient temperature exceeds that for which the ENT is rated, from direct burial, and from exposed use, with exceptions as noted.

ENT must not be used in places of assembly, theaters, and similar locations unless the installation satisfies the rules of Arts. 518 and 520. This is intended to warn against improper use of ENT, as reported for installations in these types of locations.

331-5. Size. ENT is Code-recognized in ½ to 2-in. (12.7- to 50.8-mm) trade sizes. A full line of plastic couplings, box connectors, and fittings is available, which are attached to the ENT by mechanical method or cement adhesive (Fig. 331-4).

331-10. Bends—Number in One Run. ENT runs between "pull points"—boxes, enclosures, and conduit bodies—must not contain more than the equivalent of

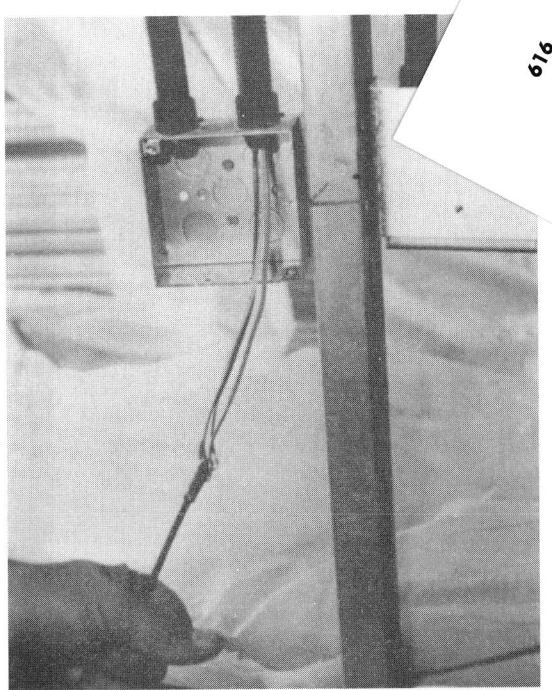

Fig. 331-4. Available in ½-, ¾-, and 1-in. (12.7, 19, and 25.4 mm) sizes, ENT has a full line of couplings and box connectors. (Sec. 331-5.)

four quarter bends (360°). This rule makes it clear that the practice of making 360° of bends between boxes and "fittings"—such as between "couplings," which comes under the definition for the word "fittings" as given in Art. 100 and previously used in this rule—is prohibited.

ARTICLE 333. ARMORED CABLE

333-1. Definition. This section identifies Type AC cable, which is the cable assembly long used and known as BX cable. All the regulations on use of Type AC cable are given in the sections of this **Code** article. Type AC armored cable, the commonly used BX cable, is covered by an article of its own and is separated on application and **Code** enforcement from the use of metal-clad cables, which are covered in Art. 334.

Type AC cable (BX) is listed and labeled by UL as "Armored Cable" in the *Electrical Construction Materials Directory.* The assembly contains the conductors within a jacket made of a spiral wrap of steel with interlocking of the edges of the strip (Fig. 333-1).

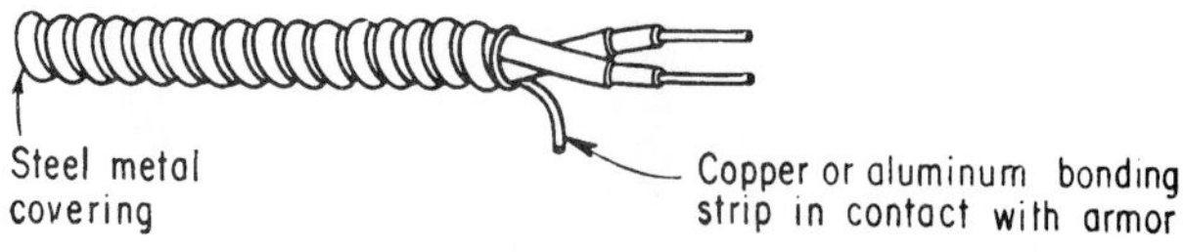

Fig. 333-1. Type AC cable contains insulated conductors plus bonding conductor under the armor. (Sec. 333-1.)

Armored cable assemblies of 2, 3, 4, or more conductors in sizes No. 14 AWG to No. 1 AWG—such conductors may even incorporate an optical fibercable—conform to the standards of the Underwriters Laboratories. These standards cover multiple-conductor armored cables for use in accordance with the National Electrical Code, in wiring systems of 600 V or less, at temperatures of 60, 70, or 90°C depending upon conductor insulation.

Armored cables of other types which do not come under these UL standards are listed by UL as "Metal-Clad Cable, Type MC" and are covered by Art. 334. One type of MC cable is commonly called "interlocked armor cable."

333-3. Uses Permitted. Type AC armored cable is familiarly used in all types of electrical systems for power and light branch circuits and feeders. Figure 333-2

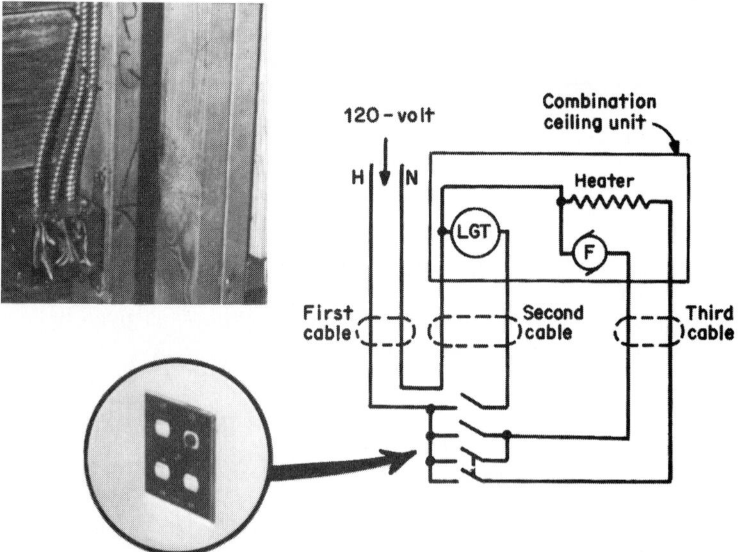

Fig. 333-2. Cable runs of 12/2 BX are used at junction box (above) which was then equipped with switches and pilot light (right) for light-heat-fan unit. Use of two 12/2 cables, with neutral in only one cable, is a violation of the concept covered in Sec. 300-20. A 12/4 cable could serve for all switch legs and satisfy the Code rule. (Sec. 333-3.)

shows use of three runs of 12/2 BX for the supply and two switch legs to a combination light-heat-fan unit in a bathroom. One 12/2 is the supply and the other cables control the appliance as shown in the wiring diagram. But the use of two 12/2 cables for the switch legs violates Sec. 300-20 because the neutral is not kept with all the conductors it serves. As a result, induction heating could be produced.

As noted in the first sentence of Sec. 333-3, Type AC cable may be used in cable trays *if* "identified for such use." To correlate to Sec. 318-(a)(1)—which recognizes Type AC cable in a tray—this rule basically calls for some marking or label attached to the cable to show that it is listed for "CT use" or some other designation that conveys the same idea (to assure that the cable has been tested and found acceptable—such as in accordance with the UL vertical tray flame test).

Type AC is also used for signal and control circuit work. It is particularly effective for running loudspeaker circuits in public address systems and other sound systems where the flexibility of the cable lends itself to ready installation on new construction or rewiring jobs and the armor provides much needed mechanical protection. Armored cable is also especially effective for wiring Class 2 control circuits—such as low-voltage relay switching circuits—where mechanical protection for the conductors and flexibility of installation are required.

Section 333-4 covers use of Type AC cable in hazardous locations, as covered in the Exceptions to Secs. 501-4(b) and 502-4(b), as well as the basic rule in 504-20. Type AC cable may be used for wiring of intrinsically safe equipment, such as instruments or signals in which the electric circuit is not capable of releasing enough energy under any fault condition to cause ignition of the hazardous atmosphere (Fig. 333-3). This same permission is also given for Type

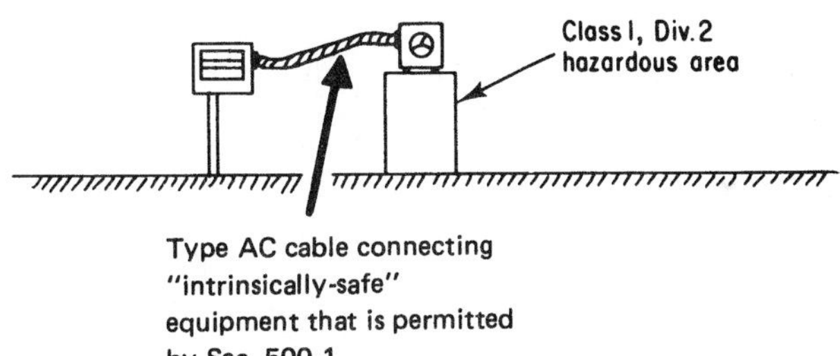

Fig. 333-3. Part **(b)(3)** recognizes BX cable for limited use in hazardous locations. (Sec. 333-6.)

NM cable, Type NMC cable, Type NMS cable, Type UF cable, rigid nonmetallic conduit, surface raceways, multioutlet assembly, underfloor raceways, cellular metal floor raceways, and wireways in their respective articles.

333-7. Support. Armored cable must be secured by approved staples, straps, or similar fittings, as shown in Fig. 333-4.

In exposed work, both as a precaution against physical damage and to ensure a workmanlike appearance, fastenings should be spaced not more than 24 to 30 in. (610 to 762 mm) apart. In concealed work in new buildings, the cable must be supported at intervals of not over 4½ ft (1.37 m) for Type AC to keep it out of the way of possible injury by mechanics of other trades. In either exposed work

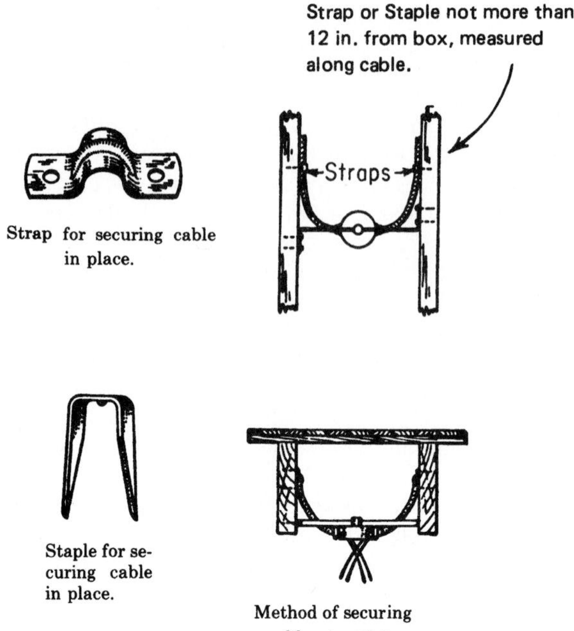

Fig. 333-4. BX must be clamped every 4½ ft (1.37 mm) and within 12 in. (305 mm) of terminations. (Sec. 333-7.)

or concealed work, the cable should be securely fastened in place within 1 ft (305 mm) of each outlet box or fitting so that there will be no tendency for the cable to pull away from the box connector.

Exception No. 2 of Sec. 333-7 limits Type AC cable to not over a 2-ft (610-mm) unclamped length for flexibility where such a cable feeds motorized equipment (such as a fan or unit heater) or connects to any enclosure or equipment where the flexibility of the BX length will isolate and suppress vibrations. Section 350-4, Exception No. 2, and Sec. 351-8, Exception No. 2, recognize flexible metal conduit and liquidtight flexible metal conduit for an unsecured length of up to 3 ft (914 mm) at any termination where flexibility is needed. Similarly, Exception No. 3 of this section permits lengths of BX up to 6 ft (1.83 m) long to be used without any staples, clamps, or other support where used in a hung ceiling as a lighting fixture whip or similar whip to other equipment (Fig. 333-5). This permits use of BX in the same manner as permitted for flexible metal conduit (Greenfield) or liquidtight flexible metal conduit in lengths from 4 to 6 ft (1.22 to 1.83 mm) as a connection from a circuit outlet box to a recessed lighting fixture [Sec. 410-67(c)]. This use of unclamped BX is an exception to the basic rule that it be clamped every 4½ ft (1.37 m) and within 22 in. (559.2 mm) of any outlet box or fitting. Exception No. 3 is an expansion on Exception No. 2 that permits up to a 2-ft (610-mm) length of BX to be used without clamping "at terminals where flexibility is necessary."

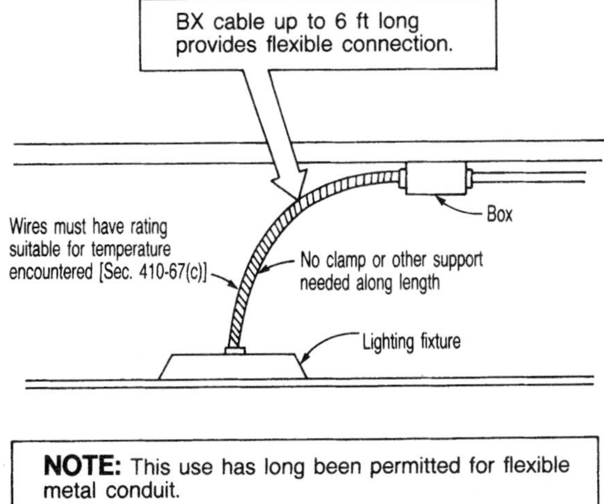
BX cable up to 6 ft long provides flexible connection.
Box
Wires must have rating suitable for temperature encountered [Sec. 410-67(c)]
No clamp or other support needed along length
Lighting fixture
NOTE: This use has long been permitted for flexible metal conduit.

Fig. 333-5. Armored cable (Type AC) may be used for 4- to 6-ft (1.22- to 1.83-mm) fixture whips, without supports, in an "accessible ceiling." (Sec. 333-7.)

Note that the requirements on clamping or securing of BX and flexible metal conduits must be observed for applications in suspended-ceiling spaces, whether for air handling, as covered in Sec. 300-22(c), or nonair handling.

The wording of Exception No. 5 to Sec. 333-7 recognizes that in an "other than vertical run," which means for horizontal runs, the hole in the framing member is to be considered as satisfying the support requirements given in this section. Such recognition is contingent upon the cable being secured with a listed clamping device within 12 in. (305 mm) of any splice or termination point. This ends a long-time dispute. Because the Code does not allow conductors to be supported by a hung ceiling or its support wires, many inspectors required Type AC—as well as other cables and raceways—to be otherwise secured at the distances spelled out in the basic rule. Now, it is clear that where run horizontally through framing members, no additional securing or support is needed for Type AC.

333-9. Boxes and Fittings. Note that a termination fitting—that is, a box connector—must be used at every end of Type AC cable entering an enclosure or a box (Fig. 333-6) unless the box has an approved built-in clamp to hold the cable armor, provide for the bonding of the armor to the metal box, and protect the wires in the cable from abrasion.

A standard type of box connector for securing the cable to knockouts or other openings in outlet boxes and cabinets is shown in Fig. 333-7. A plastic bushing, as shown, must be inserted between the armor and the conductors. The fiber bushing, which can be seen through slots in the connector after installation, prevents the sharp edges of the armor from cutting into the insulation on the conductors and so grounding the copper wire.

Fig. 333-6. Connectors for BX entering a panelboard cabinet or other enclosure must use approved fittings—some type of single-connector or duplex type (as shown, with two cables terminated at each connector through a single KO). (Sec. 333-9.)

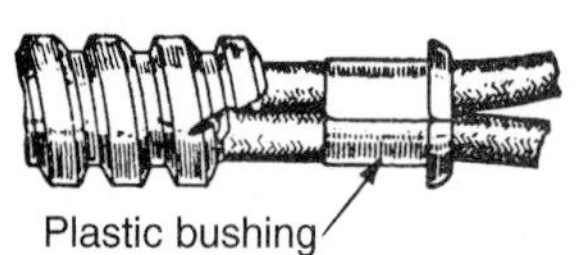

Plastic bushing to protect the conductors in Type AC cable from the sharp edges of the armor.

Box connector for Type AC cable

Fig. 333-7. Every BX termination must be equipped with a protective bushing and a box connector or clamp built into the box. (Sec. 333-9.)

The box shown in Fig. 333-8 is equipped with clamps to secure Type AC cables, making it unnecessary to use separate box connectors. The other box shown is similar but has the cable clamps outside, thus permitting one more conductor in the box. See Sec. 370-16(a)(1).

As covered in the last sentence of Sec. 333-9, "a box, fitting or conduit body"—such as a "C" conduit body—must be used where Type AC cable is connected to another wiring method. Figure 333-9 shows a typical application of this technique, which is also recognized in Exception No. 9 of Sec. 300-15(b).

333-11. Exposed Work. Exception No. 1 refers to a length not over 24 in. (610 mm) to a lighting fixture, a motor, or a range where some flexibility is necessary, as noted in Sec. 333-7, Exception No. 2.

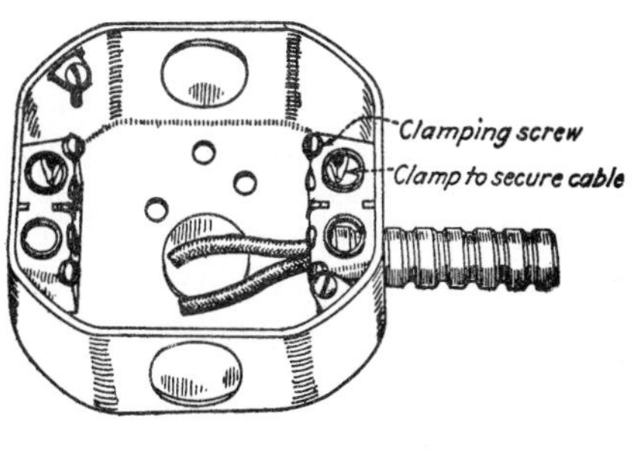
Clamping screw
Clamp to secure cable

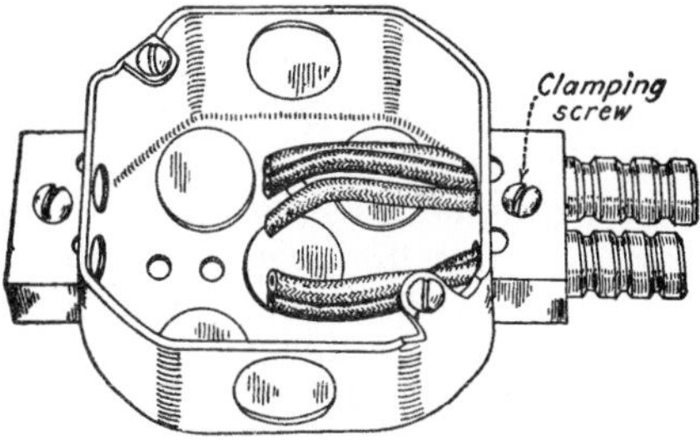
Clamping screw

Fig. 333-8. A box connector fitting is not required if box includes cable clamps for Type AC cable. (Sec. 333-9.)

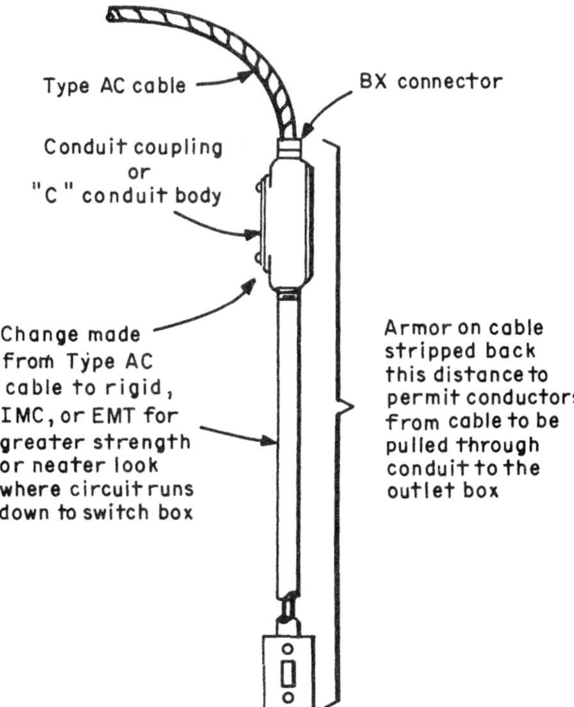
Type AC cable
BX connector
Conduit coupling or "C" conduit body
Change made from Type AC cable to rigid, IMC, or EMT for greater strength or neater look where circuit runs down to switch box
Armor on cable stripped back this distance to permit conductors from cable to be pulled through conduit to the outlet box

Fig. 333-9. This connection of BX to conduit or EMT is specifically recognized. (Sec. 333-9.)

Exception No. 3 permits up to a 6-ft (1.83-m) length of Type AC cable for a fixture whip as an Exception to the rule requiring AC cable to "closely follow" the building surface. This is related to Sec. 333-7, Exception No. 3, which permits 6-ft (1.83 m) fixture whips without clamping of the cable.

333-19. Construction. Note that Type AC cable is recognized for branch circuits and feeders, *but not* for service-entrance conductors, which *must* be one of the cables or wiring methods specified in Sec. 230-43. Type MC (metal-clad) cable, such as interlocked armor cable or the other cables covered in Art. 334, is recognized by Sec. 230-43 for use as service-entrance conductors.

Because the armor of Type AC cable is recognized as an equipment grounding conductor by Sec. 250-91(b)(6), its effectiveness must be assured by using an "internal bonding strip," or conductor, under the armor and shorting the turns of the steel jacket. The ohmic resistance of finished armor, including the bonding conductor that is required to be furnished as a part of all except lead-covered armored cable, must be within values specified by UL and checked during manufacturing. The bonding conductor run within the armor of the cable assembly is required by the UL standard.

Because the function of the bonding conductor in Type AC cable is simply to short adjacent turns of the spiral-wrapped armor, there is no need to make any connection of the bonding conductor at cable ends in enclosures or equipment. The conductor may simply be cut off at the armor end.

Construction of armored cable must permit ready insertion of an insulating bushing or equivalent protection between the conductors and the armor at each termination of the armor—such as the so-called "red head."

333-20. Conductors. UL data on conductors used within Type AC cables refer to the marking on the cable as follows:

> ACT—Indicates an armored cable employing conductors having thermoplastic insulation.
> AC—Indicates an armored cable employing conductors having thermosetting insulation.
> **Suffixes on Type letters:**
> No suffix—Indicates 60°C rating.
> B—Indicates 90°C rating with ampacity of 60°C rated conductors.
> H—Indicates 75°C rating.
> HH—Indicates 90°C rating.
> Aluminum—armored cable is Type ACTHH with Type THHN conductors in sizes 14-1 AWG copper only.

As required by the Exception in this section, armored cable (BX) installed within thermal insulation must have 90°C-rated conductors (Types THHN, RHH, XHHW), but the ampacity must be taken as that of 60°C-rated conductors. This requirement recognizes that the heat rise on conductors operating with reduced heat-dissipating ability (such as surrounded by fiberglass or similar thermal insulation) requires that the conductors have a 90°C-rated insulation. That temperature might be reached even with the wires carrying only 60°C ampacities. Although the wires must have 90°C insulation, they must not be loaded over those ampacity values permitted for TW (60°C), as shown in Table 310-16.

ARTICLE 334. METAL-CLAD CABLE

334-1. Definition. This article covers "Metal-Clad Cable," as listed by UL under that heading in the *Electrical Construction Materials Directory* (the Green Book). This section defines the type of cable assemblies covered by this article (Fig. 334-1). The definition for metal-clad cable—"a factory assembly of one or

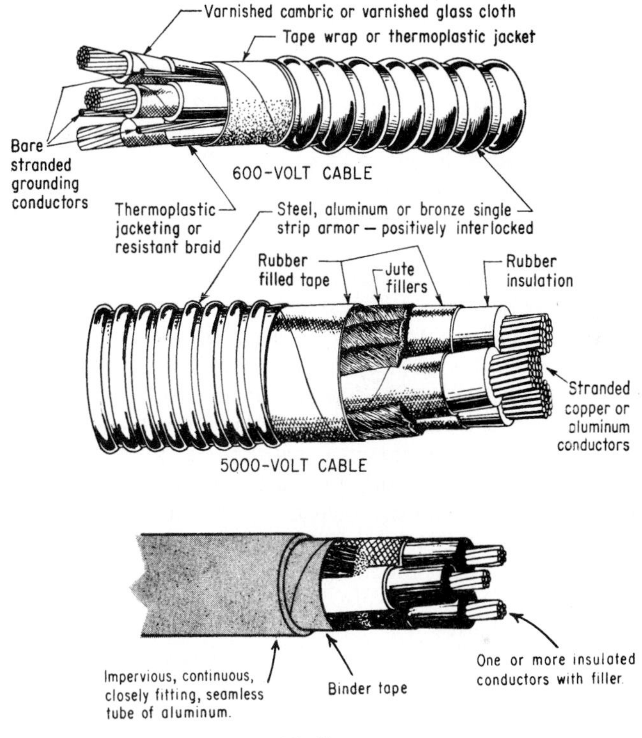

Fig. 334-1. These are some of the constructions in which Type MC cable is available. (Sec. 334-1.)

more conductors, each individually insulated and enclosed in a metallic sheath of interlocking tape, or a smooth or corrugated tube"—also covers Type ALS and Type CS cables. All metal-clad cables—Type MC, Type ALS, and Type CS—are covered in the single Art. 334.

Aluminum-sheathed (ALS) cable has insulated conductors with color-coded coverings, cable fillers, and overall wrap of Mylar tape—all in an impervious, continuous, closely fitting, seamless tube of aluminum. It may be used for both exposed and concealed work in dry or wet locations, with approved fittings. CS cable is very similar with a copper exterior sheath instead of aluminum.

Because the rules of these three cable types have been compiled into a single article, use of any one of the Type MC cables must be evaluated against the specific rules that now generally apply to all such cables. The Code no longer contains the designations Type ALS and Type CS. They are included now along with interlocked armored cable as Type MC cables.

One type of Type MC cable that has been used for many years under the name "interlocked-armor cable" is the heavy-duty, industrial feeder type of armored cable that is similar in appearance to but really different from standard BX armored cable, as covered in Art. 333. MC cable is a different, heavier-duty assembly than BX (Type AC), and great care must be taken to carefully distinguish between the design and installation regulations that apply to each of the cable types. This is particularly important now that the NE Code recognizes Type MC cable in the size range from No. 14 and larger. Now, because both cable assemblies are available in sizes No. 14 up to No. 1, armored cable must be carefully distinguished as either Type AC or Type MC. As clearly shown in NE Code Art. 334, Code rules are different for the two types of cable. And so are UL regulations as indicated in the Green Book. Always check the label on the cable.

Type MC is rated by UL for use up to 2,000 V. Cable rated 5,000 to 35,000 V is "listed" as Type MV. Type MC cable is recognized in three basic armor designs: (1) interlocked metal tape, (2) corrugated tube, and (3) smooth metallic sheath.

334-3. Uses Permitted. Although this section clearly lists all the permitted applications of any of the various forms of Type MC cable, care must be taken to distinguish between the different constructions, based on the Code rules (Fig. 334-2). For a long time, the interlocked-armor Type MC and the corrugated sheath Type MC have been designated by UL as "intended for aboveground use." But part **(5)** of this section recognizes Type MC cable as suitable for direct burial in the earth, when it is identified for such use.

The Exception to Sec. 334-4, warns against direct burial of Type MC cable. The fine distinction between acceptable and unacceptable use of directly buried Type MC cable will generally require discussion with local inspection authorities to assure clear understanding of the contrasting phrases in Secs. 334-3 and 334-4.

334-4. Uses Not Permitted. Type MC cables are permitted by Sec. 334-3 to be used exposed or concealed in dry or wet locations. But such cable must not be subjected to destructive, corrosive conditions—such as direct burial in the earth, in concrete, or exposed to cinder fills, strong chlorides, caustic alkalis, or vapors of chlorine or of hydrochloric acids, unless protected by materials suitable for the condition. But the Exception to this section says that Type MC cable may be used for underground runs where suitably protected against physical damage and corrosive conditions.

334-10. Installation. Figure 334-3 shows the maximum permitted spacing of supports for the larger-sized Type MC cable. The interlocked-armor Type MC has commonly been used on cable tray, as permitted in part **(b)** of this section (Fig. 334-4). In addition to the 6-ft (1.83-mm) interval, smaller cables—those with four or less conductors smaller than No. 8—must also be secured within 12 in (305 mm) of "each box, cabinet, or fitting."

334-11. Bending Radius. Figure 334-5 shows the bending-radius rules for the "smooth sheath" Type MC cables. Cable with interlocked or corrugated armor

Fig. 334-2. ALS (aluminum-sheathed) Type MC cable was used for extensive power and light wiring in refrigerated rooms and storage areas of a store. The ALS was surface-mounted (exposed) on clamps in this damp location. Because the cable assembly is a tight grouping of conductors within the sheath, there would be no passage of warm air from adjacent nonrefrigerated areas through the cable which crosses the boundaries between the areas. It was therefore not necessary to seal the cables to satisfy Sec. 300-7(a). (Sec. 334-3.)

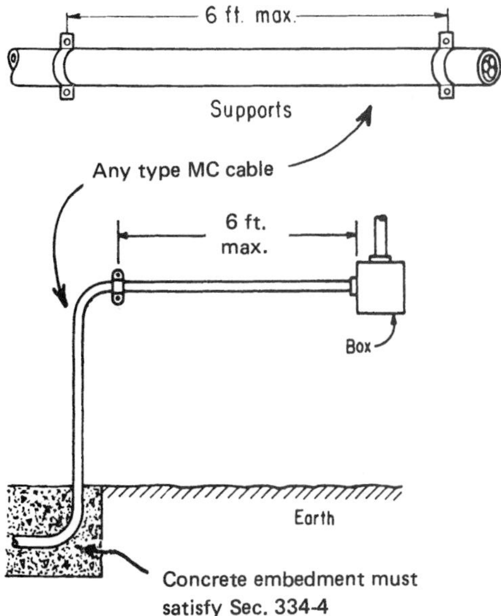

Fig. 334-3. Surface mounting of Type MC cable must be secured. (Sec. 334-10.)

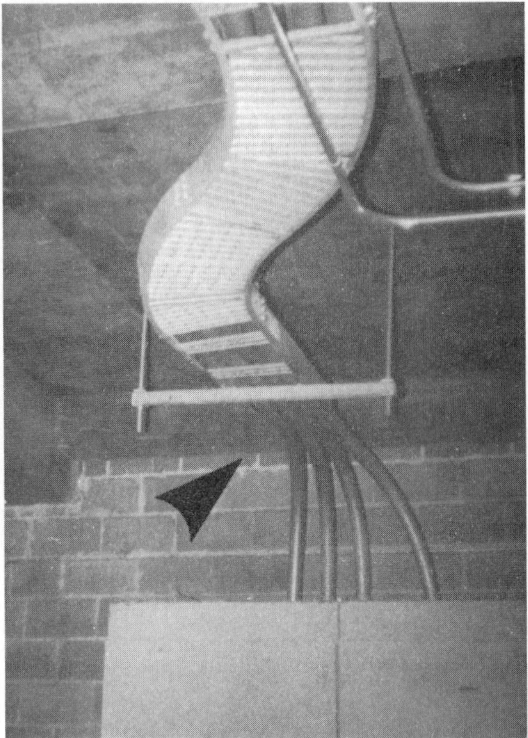

Fig. 334-4. Any Type MC cable is recognized for use in cable tray, and the interlocked-armor version has been widely used in tray, as shown here. (Sec. 334-10.)

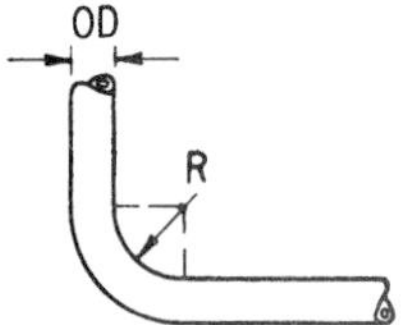

Bends: Radius (R) shall not be less than:
(a) 10 times OD for cables with OD 3/4 in. or less.
(b) 12 times OD for cables with OD over 3/4 in. but not over 1½ in.
(c) 15 times OD for cables with OD over 1½ in.

Fig. 334-5. Minimum radius values prevent excessively sharp, destructive bending of ALS or CS cable. (Sec. 334-11.)

must have a bending radius not less than 7 times the outside diameter of the cable armor. To conform to IPCEA rules on bending radius for shielded conductors in MC cable, the minimum value must be either 12 times the diameter of one of the conductors within the cable or 7 times the diameter of the MC cable itself, whichever is greater.

334-12. Fittings. Only approved, UL-listed connectors and fittings are permitted to be used with any Type MC cable. Such fittings are listed in the UL Green Book under "Metal-Clad Cable Connectors." Figure 334-6 shows typical approved

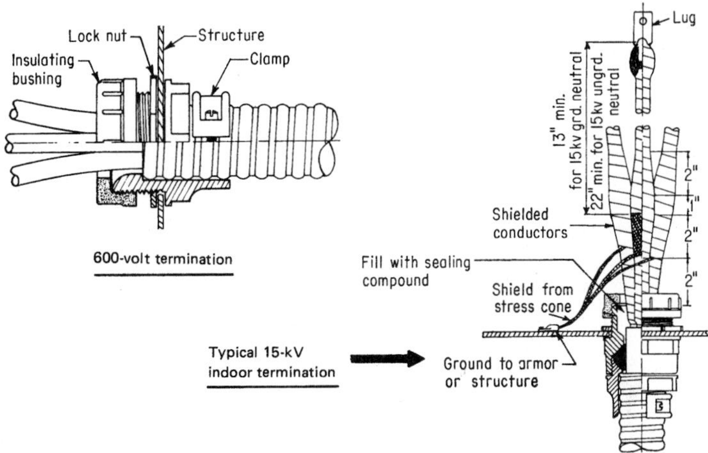

Fig. 334-6. Terminations for interlocked-armor cables must be approved devices, correctly installed. (Sec. 334-12.)

connectors for interlocked-armor Type MC cable. As shown at left, 600-V terminations for interlocked-armor cable to switchgear or other enclosures in dry locations can be made with connectors, a locknut, and a bushing in the typical basic assembly shown. In damp locations, compound-filled or other protective terminations may be desired. High-voltage connectors (5 through 35 kV) are generally filled with sealing compound and individual conductors terminated in a suitable manner, depending upon whether the conductors are shielded or not. Or the IA cable may terminate in a pothead for positively sealed and insulated terminations indoors or outdoors.

ARTICLE 336. NONMETALLIC-SHEATHED CABLE

336-1. Scope. This section makes clear that nonmetallic-sheath cable must be installed *and* manufactured as required here.

Refer to Sec. 310-2 of this handbook for a discussion on the use of only insulated conductors as circuit conductors and the prohibition on using the bare grounding conductor in NM cable as both a neutral and equipment grounding conductor for a circuit to a cooking appliance (Sec. 250-60).

336-2. Definition. Nonmetallic-sheathed cable is one of the most widely used cables for branch circuits and feeders in residential and commercial systems (Fig. 336-1). Such cable is commonly and generally called "Romex" by electri-

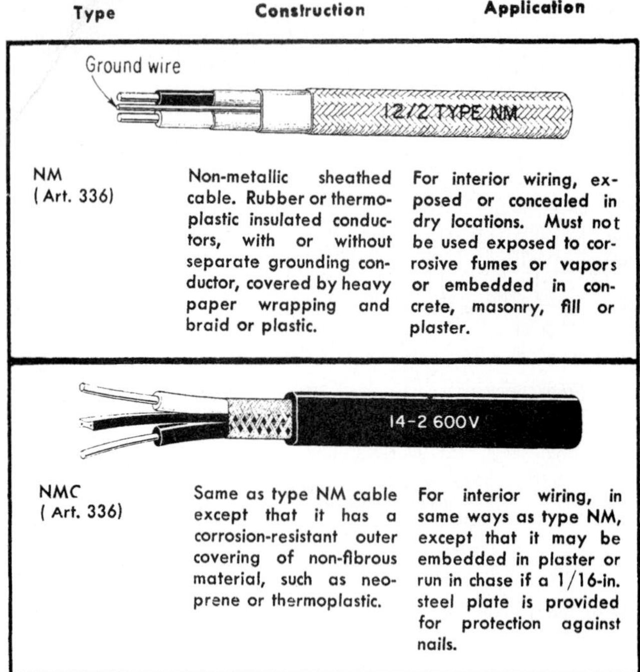

Type	Construction	Application
NM (Art. 336)	Non-metallic sheathed cable. Rubber or thermoplastic insulated conductors, with or without separate grounding conductor, covered by heavy paper wrapping and braid or plastic.	For interior wiring, exposed or concealed in dry locations. Must not be used exposed to corrosive fumes or vapors or embedded in concrete, masonry, fill or plaster.
NMC (Art. 336)	Same as type NM cable except that it has a corrosion-resistant outer covering of non-fibrous material, such as neoprene or thermoplastic.	For interior wiring, in same ways as type NM, except that it may be embedded in plaster or run in chase if a 1/16-in. steel plate is provided for protection against nails.

Fig. 336-1. Two of the three separate types of nonmetallic-sheathed cable are shown here. (Sec. 336-30.)

cal construction people, even though the word *Romex* is a registered trade name of the General Cable Corp. Industry usage has made the trade name a generic title so that nonmetallic-sheathed cable made by any manufacturer might be called Romex. This generic usage of a trade name also applies to the term *BX,* which is commonly used to describe any standard armored cable, made by any manufacturer—even though the term *BX* is a registered trade name of General Electric Co. Type NM cable has an overall covering of fibrous or plastic material which is flame-retardant and moisture-resistant. Type NMC is similar, but the overall covering is also fungus-resistant and corrosion-resistant. The letter "C" indicates that it is corrosion-resistant.

336-4. Uses Permitted. This type of wiring may be used either for exposed or for concealed wiring (Fig. 336-2) in any kind of building or structure.

1. NM cable may be used in one-family dwellings.
2. NM cable may be used in two-family dwellings.
3. NM cable may be used in multifamily dwellings.
4. NM cable may be used in "other structures.

But in any case, the dwelling, building, or structure must not have over three floors above grade. [See definition for building or structure with "three floors above grade" in Sec. 336-5(a)(1).]

Fig. 336-2. Although NM cable is most widely used for branch circuits, the larger sizes (No. 8 and up) are commonly used for feeders, as run here from apartment disconnects to tenant panelboards. (Sec. 336-4.)

Type NM or NMC cable must be "identified" for use in cable trays. This requirement essentially calls for UL listing and marking on the cable to make it "recognized" as suitable for installation in cable trays.

Although NM cable is limited to use in "normally dry locations," NMC—the corrosion-resistant type—is permitted in "dry, damp, moist, or corrosive locations." Because it has been widely used in barns and other animals' quarters where the atmosphere is damp and corrosive (due to animal vapors), NMC cable is sometimes referred to as "barn wiring."

Part **(b)** says that Type NMC that is run in a shallow chase in masonry, concrete, or adobe and covered over must be protected against nail or screw penetration by a 1/16-in (1.59-mm) (minimum) steel plate.

Part **(c)** specifies those permissible uses for Type NMS, which is a hybrid cable containing power, signaling, communication—voice/data/visual—and even optical fiber cable. This cable is the multifunction so-called smart house cable that provides for security, convenience, energy-saving capabilities, control, network access, cable TV, etc., in a dwelling.

Type NMS may be used for both concealed and exposed installations in "normally dry locations." And fishing of Type NMS is also permitted where excess moisture does not collect.

The reference to Art. 780 indicates that where such cable is used to wire the so-called smart-house, all requirements that apply to a commercial or industrial distribution control system also apply to the system Type NMS interconnects.

336-5. Uses Not Permitted. The first sentence of this section limits use of Type NM and Type NMC cables to any building that does not have more than three floors above grade (Fig. 336-3).

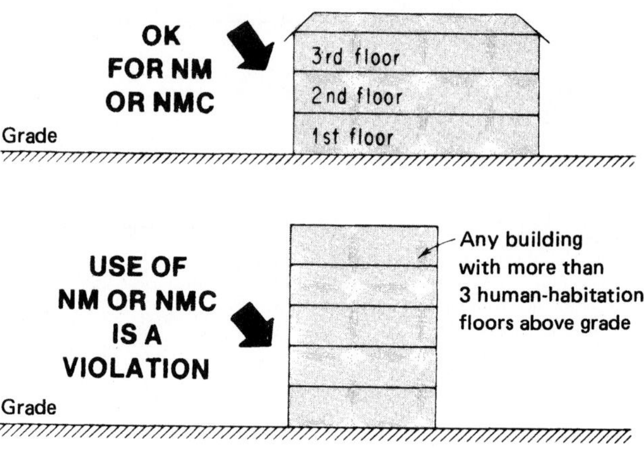
Type NM or type NMC may be used in any building with not more than 3 floors above grade.
OK FOR NM OR NMC
3rd floor
2nd floor
1st floor
Grade
USE OF NM OR NMC IS A VIOLATION
Any building with more than 3 human-habitation floors above grade
Grade
Note: NM or NMC may not be used at all in this building — not even on the first three floors.

Fig. 336-3. Nonmetallic-sheathed cable is limited in application. [Sec. 336-5(a).]

The Code rule limiting use of Types NM, NMS, and NMC cables to buildings not exceeding three floors above grade has produced difficulties in interpretation. The problem arises when buildings are built on hillsides or sloping grades, where the building will have three floors above grade on the uphill side and four floors above grade on the downhill side. The question then is: Is this a three-story building or a four-story building, and is use of Type NM cable permitted?

To clarify the issue, the second paragraph of Sec. 336-5(a) defines the first floor of a building and attempts to establish a basis for applying the Code rule. Whether or not a particular building will be considered as a three- or four-story building when installed on sloping grade depends upon the definition given for the first floor of a building.

The Code spells out that the first floor is that floor that has 50 percent or more of its exterior wall surface area level with or above grade. We must presume that the rule refers to the total wall surface area on all four sides of that floor of the building. And it is not clear how wall "surface area" can be "level with grade." If the bottom floor has over 50 percent of its wall area above adjacent finished grade, then the bottom floor is the first floor. In this rule, the wording is awkward and must be carefully related to building construction and exterior grading in determining whether or not Romex cable may be used in the building. If that bottom floor is the first floor in a building with four floors of dwelling space, then the building is a four-story building and use of Type NM or Type NMC cable is prohibited. But, a bulldozer or backhoe could be used to alter the actual steepness of the grade to create the required conditions that would exclude the bottom floor as the first floor of the building and would thereby permit use of nonmetallic-sheathed cable (see Fig. 336-4).

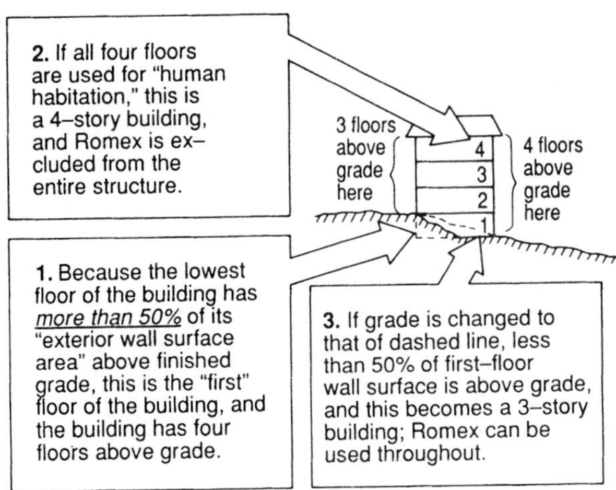

Fig. 336-4. Definition of "first floor" clarifies use of NM cable in buildings "not exceeding three floors above grade." [Sec. 336-4(a).]

But, the last sentence of this rule says that a three-story building wired in Romex is permitted "one additional level" (four in all) if the first floor level is totally used for auto parking or storage (Fig. 336-5).

The wording of this rule makes clear that NM, NMS, or NMC may be used only in a building that is not more than three stories high—and may not be used in any mid- or high-rise buildings of any type, as described above.

NM cable may not be embedded in masonry as noted in part **(b)(2)**, and the word "adobe" prohibits the use of nonmetallic-sheathed cable embedded in the chases in adobe brick, commonly used in the southwest part of the United

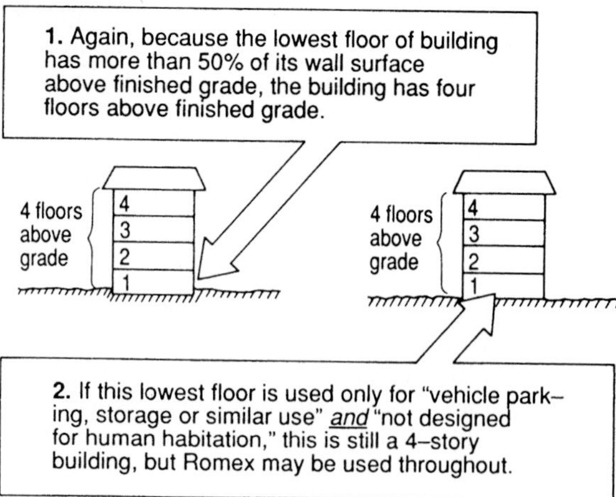

1. Again, because the lowest floor of building has more than 50% of its wall surface above finished grade, the building has four floors above finished grade.
4 floors above grade
4
3
2
1
4 floors above grade
4
3
2
1
2. If this lowest floor is used only for "vehicle parking, storage or similar use" and "not designed for human habitation," this is still a 4-story building, but Romex may be used throughout.

Fig. 336-5. A first floor that is *totally* garage or storage area may have three floors above it, and the whole building may be wired with NM or NMC cable. [Sec. 336-5(a).]

States. Adobe is a sun-dried brick material used for building construction. It is brittle and boxes embedded in it tend to become loose. This prohibits the cable from being run in the adobe or in the chases between the adobe bricks.

Although part **(b)** prohibits NM and NMS cable embedded in plaster or other construction materials and prohibits it run in chases between bricks, stones, etc., this section does *not* prohibit use of NMC cable for plaster embedment or in chases, and Sec. 336-3(b)(3) specifically permits it. The 1971 NEC had another sentence to the rule shown in part **(b)**, and it allowed use of NMC as follows:

> Where embedded in plaster or run in a shallow chase in masonry walls and covered with plaster within 2 inches of the finished surface, it [NMC cable] shall be protected against damage from nails by a cover of corrosion-resistant coated steel at least 1/16 inch in thickness and 3/4 inch wide in the chase or under the final surface finish.

That sentence permitted use of NMC (but not NM) under plaster, as shown in Fig. 336-6; however it was removed from Sec. 336-3(b) in the 1975 NEC and does not appear in the present NEC. Omission of that sentence raises the question: Is NMC permitted to be embedded in plaster as covered by the old rule described?

There is no definite answer to that question and the matter must be decided by the local inspector having jurisdiction. No data were made available to explain deletion of the old Code rule described. Because the present NEC rule prohibits NM and NMS embedded in plaster, but does not prohibit NMC in plaster, it can be argued that such use of NMC is Code-acceptable. But it would be necessary to protect the cable against the possibility of being damaged by driven nails—such as nails used to hang pictures or add construction elements on the wall. Sufficient protection against nail puncture of the cable is provided by a cover of corrosion-resistant coated steel of at least 1/16 in. (1.59 mm) thick-

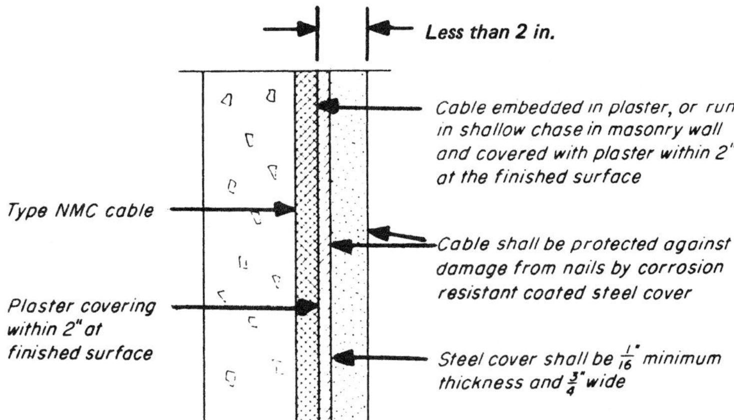

Fig. 336-6. This was permitted by previous editions of the **Code** and may still be acceptable. [Sec. 336-4(b)(3).]

ness and ¾ in. (19 mm) width. Such metal protection must be run for the entire length of the cable where it is less than 2 in. (50.8 mm) below the finished surface. The metal strip protection may be run in the chase or under the plaster finish. But, it must be carefully noted that both NM and NMC are prohibited by Sec. 336-5(a)(8) from embedment in cement, concrete, or aggregate—which is distinguished from plaster.

Section 336-3 correlates use of NM cable to the rule of the second paragraph of Note 8 to Tables 310-16 to 310-19, which says:

> Where single conductors or multiconductor cables are stacked or bundled longer than 24 in. (610 mm) without maintaining spacing and are not installed in raceways, the ampacity of each conductor shall be reduced as shown in the above table.

Bundled NM, NMS, or NMC cables will require ampacity derating in accordance with Note 8 to Table 310-16 when the whole bundle is tightly packed, thereby losing the ability of the inside cables to dissipate the heat generated in them. An example of this is shown in Fig. 310-31. This is true of NM cables as well as any other cables. And the derating percentage from the table in Note 8 must be based on the total number of insulated conductors in the group. For instance, fourteen 3-wire cables would have to be ampacity derated to 60 percent of the conductor ampacity ($14 \times 3 = 42$ conductors, at 60 percent, from Note 8).

If the ampacity derating required by Note 8 is not observed, stacking or bundling of groups of NM, NMS, or NMC cable runs can result in dangerous overheating of conductors and terminations.

336-6. Exposed Work—General. Figure 336-7 shows the details described in parts **(a)** and **(b)** of this section. The rules of this section tie into the rules of part **(c)**, covering use in unfinished basements, which are really places of "exposed work."

As covered in part **(c)** to Sec. 336-6, cables containing Nos. 14, 12, or 10 conductors must be run through holes drilled through joists. When running parallel to joists, any cable must be stapled to the wide, vertical face of a joist and

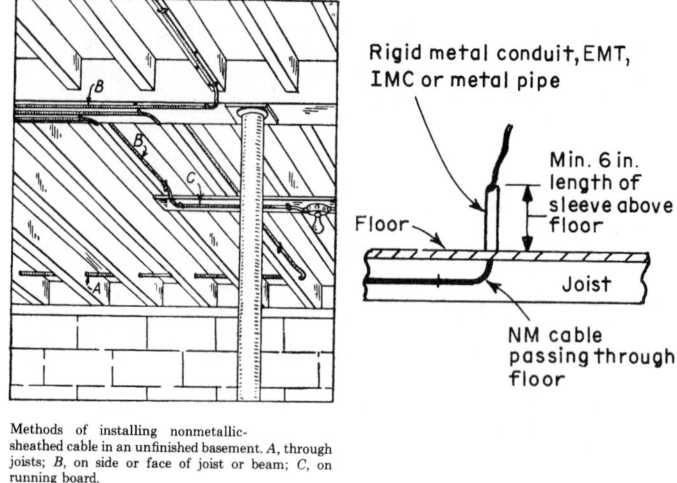

Methods of installing nonmetallic-sheathed cable in an unfinished basement. *A*, through joists; *B*, on side or face of joist or beam; *C*, on running board.

Fig. 336-7. This applies to unfinished basements and other exposed applications. (Sec. 336-6.)

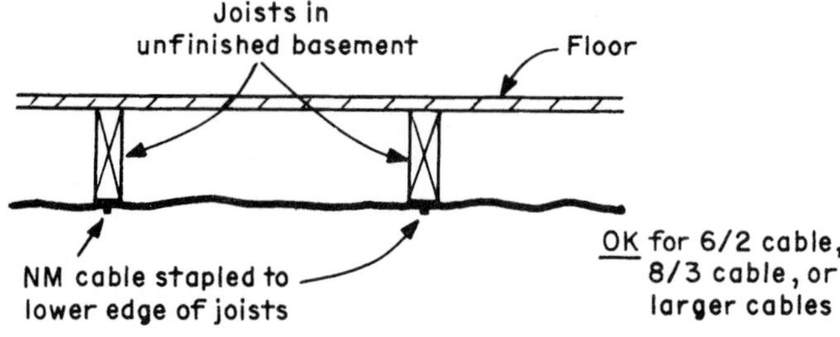

Fig. 336-8. Only large cables may be stapled to bottom edge of floor joists. [Sec. 336-6(c).]

never to the bottom edge. But, as shown in Fig. 336-8, larger cables may be attached to the bottom of joists when run at an angle to the joists.

336-18. Supports. Figure 336-9 shows support requirements for NM or NMC cable. Figure 336-10 shows a violation. In concealed work the cable should if possible be so installed that it will be out of reach of nails. Care should be taken to avoid wherever possible the parts of a wall where the trim will be nailed in place, e.g., door and window casings, baseboards, and picture moldings. See Sec. 300-4.

Connectors listed for use with Type NM, NMS, or NMC cable (nonmetallic-sheathed cable) are also suitable for use with flexible cord or service-entrance

Fig. 336-9. NM or NMC cables must be stapled every 4½ ft (1.37 m) where attached to the surfaces of studs, joints, and other wood structural members. It is not necessary to use staples or straps on runs that are supported by the drilled holes through which the cable is pulled. But there must be a staple within 12 in. (305 mm) of every box or enclosure in which the cable terminates. (Sec. 336-18.)

cable *if* such additional use is indicated on the device or carton. Connectors listed under the classifications "Armored Cable Connectors" and "Conduit Fittings" may be used with nonmetallic-sheathed cable when that is specifically indicated on the device or carton. Connectors for NM, NMS, or NMC cable are also suitable for use on Type UF cable (underground feeder and branch-circuit cable—NE Code Art. 339) in dry locations, unless otherwise indicated on the carton. Each connector covered in the listing is recognized for connecting only one cable or cord—unless it is a duplex connector for connecting two cables or if the carton is marked to indicate use with more than one cable or cord.

336-20. Boxes of Insulating Material. By using nonmetallic outlet and switch boxes a completely "nonmetallic" wiring system is provided. Such a system has economic advantage and other advantages in locations where corrosive vapors are present. See Sec. 370-3.

336-21. Devices of Insulating Material. Note this use of switch and outlet devices without boxes is limited to exposed cable systems and for rewiring in existing buildings. This reference must not be confused with that of Exception No. 2 in Sec. 336-18, which refers to approved wiring devices that incorporate their own wiring boxes, so they are devices "without a *separate* outlet box" and not devices "without boxes."

336-26 Conductors. The last paragraph before the FPN requires that NM, NMS, and NMC cables always have their conductors applied to the ampacity of Type TW wire—that is, the 60°C ampacity from Table 310-16. However, the insulation on the conductors must be rated at 90°C.

Fig. 336-10. Absence of stapling of the NM cables within 12 in. (305 mm) of entry into the panelboards is a clear violation of Sec. 336-15. (Sec. 336-18.)

ARTICLE 338. SERVICE-ENTRANCE CABLE

338-1. Definition. The Code contains no specifications for the construction of this cable; it is left to Underwriters Laboratories Inc. to determine what types of cable should be approved for this purpose. The types listed by the Laboratories at the present time conform to the following data:

> This Listing covers Service Entrance Cable designated Type SE and Type USE intended for use in accordance with Article 338 of the **National Electrical Code**, NFPA70.
>
> Service entrance cable, rated 600 volts, is Listed in sizes No. 12 AWG and larger for copper, and No. 10 AWG and larger for aluminum or copper-clad aluminum. Type SE cable contains Type RHW, XHHW, or THWN conductors. Type USE cable contains insulation equivalent to RHW or XHHW. If the type designation for the conductors is marked on the outside surface of the cable, the temperature rating of the cable corresponds to the rating of the individual conductors. When this marking does not appear, the temperature rating of the cable is 75 C. Type USE-2 contains insulation equivalent to RHW-2 or XHHW-2 and is rated 90 C wet or dry.

The cables are designated as follows:

Type SE—Cable for aboveground installation. The outer jacket or finish of Type SE is suitable for use where exposed to sun.

Types USE and USE-2—Cable for underground installation including burial directly in the earth. Cable in sizes No. 4/0 AWG and smaller and having all conductors insulated is suitable for all of the underground uses for which Type UF cable is permitted by the **National Electrical Code**. Types USE and USE-2 are not suitable for use in premises or aboveground except to terminate at the service equipment or metering equipment.

Submersible Water Pump Cable—Indicates a multi-conductor cable in which 2, 3, or 4 single-conductor Type USE or USE-2 cables are provided in a flat or twisted assembly. The cable is Listed in sizes 12 AWG to 4/0 AWG incl., copper, and 10 AWG to 4/0 AWG incl., aluminum or copper-clad aluminum. The cable is tag marked "For use within the well casing for wiring deep-well water pumps where the cable is not subject to repetitive handling caused by frequent servicing of the pump units." The insulation may also be surface-marked "Pump Cable." The cable may be directly buried in the earth in conjunction with this use.

Cables which are acceptable for installation in cable trays are so marked.

Many single-conductor cables are dual-rated (Type USE, RHW, or RHH) and may be used in raceways, for either service conductors or feeders and branch circuits.

Based upon tests which have been made involving the maximum heating that can be produced, an uninsulated conductor employed in a service cable assembly is considered to have the same current-carrying capacity as the insulated conductors even though it may be smaller in size.

Figure 338-1 shows two basic styles of service-entrance cable for aboveground use. The one without an armor over the conductors is referred to as "Type SE Style U"—the letter "U" standing for "unarmored." That cable is sometimes designated as "Type SEU." The cable assembly with the armor is designated "Type ASE" cable, with the "A" standing for "armored."

Figure 338-2 shows another type of SE cable, known as "Style SER"—the letter "R" standing for "round." In a typical assembly of that cable, three conductors insulated with Type XHHW cross link polyethylene are cabled together with fillers and one bare ground conductor with a tape over them and gray PVC overall jacket. For use aboveground in buildings, it is suitable for operation at 90°C in dry locations or 75°C in wet locations.

The three insulated conductors—a black, a red, and a blue—are used as the phase legs of the service and the bare conductor is used as the neutral.

Figure 338-3 shows Type USE cable for underground (including direct earth burial) applications of service or other circuits. Type USE may consist of one, two, or three conductors, Type RHW insulated wire with neoprene jacket suitable for operation in wet or dry locations at a maximum temperature of 75°C. It is for underground service entrance for direct earth burial, conduit, duct, or aerial applications.

Depending upon whether USE cable is used for service entrance, for a feeder, or for a branch circuit, burial depth must conform to Sec. 300-5 and its many specific rules on direct burial cable.

338-2. Uses Permitted as Service-Entrance Conductors. As would be expected, where used as service conductors, Type SE must comply with Art. 230, "Ser-

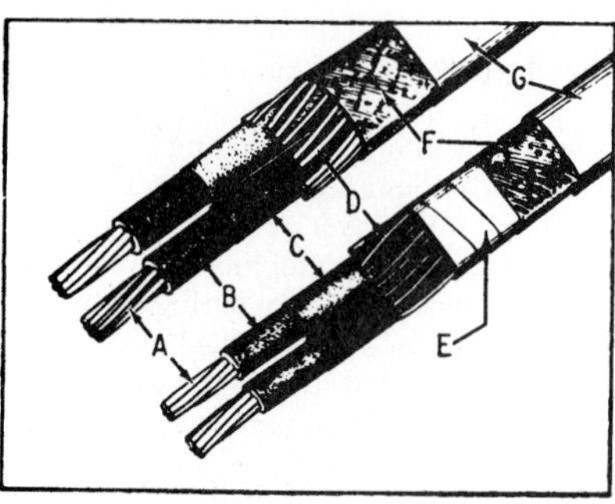
G
F
D
C
B
A
E

SERVICE ENTRANCE cables may consist of either copper or aluminum phase conductors (A) covered by heat resistant insulation (B) and moisture-resistant braid or tape (C) color coded for circuit identification, while basic assembly is enclosed by concentric neutral (D). Unarmored Type SE Style U is covered by variety of tapes (F) and outer braid (G) such as glass and cotton impregnated with moisture resistant and flame retardant finish labelled with pertinent data. Armored Style A additionally contains flat steel armor (E) as protection against physical abuse.

Fig. 338-1. Two types of aboveground SE cable. (Sec. 338-1.)

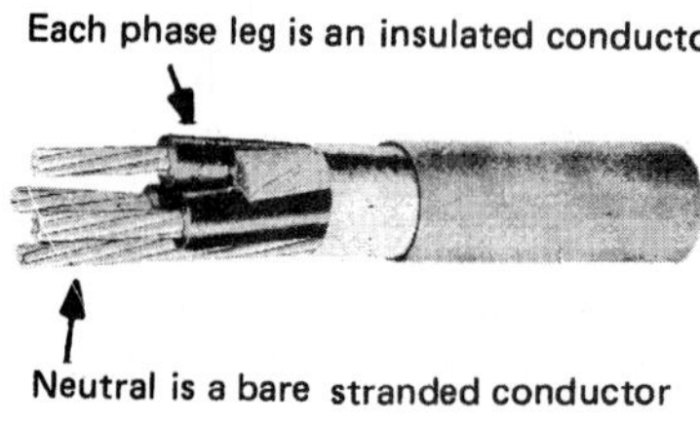
Each phase leg is an insulated conductor
Neutral is a bare stranded conductor

Fig. 338-2. Style SER cable contains individual conductors and no concentric neutral. (Sec. 338-1.)

Fig. 338-3. Type USE cable may be multiconductor or single conductor cable. (Sec. 338-1.)

vices." The wording in the second paragraph specifically permits Type USE to "emerge" from a trench and be run aboveground to terminate in meters or service equipment.

338-3. Uses Permitted as Branch Circuits or Feeders. Part **(a)** recognizes use of service-entrance cable for branch circuits and feeders within buildings or structures provided that all circuit conductors, including the neutral of the circuit, are insulated. Such use must conform to Art. 336 on installation methods—the same as those for Type NM cable. See Sec. 338-3(b).

Part **(b)** covers permitted uses of service-entrance cable that contains a bare conductor for the neutral but limits such application to 120/240- or 120/208-V systems. When an SE cable has an outer nonmetallic covering over the enclosed bare neutral, this **Code** rule permits the use of SE cable for circuits supplying ranges, wall-mounted ovens, and counter-mounted cooking units (Fig. 338-4). And in such cases, the bare conductor may be used as the neutral of the branch circuit as well as the equipment grounding conductor (see Sec. 250-60). However, price differences between SE cable and NM cable generally determine that SE cable will probably be used only where part of a circuit run is outdoors, or where 75°C supply conductors must be used to connect appliances. SE cables, in sizes 8/3 and smaller, generally cost more than corresponding sizes of NM cables. And even though 6/2, 8/1 SE cable costs slightly less than 6/3 NM cable, additional labor costs usually more than offset the total installation cost in favor of the 6/3 NM cable.

SE cable is also permitted to be used as a feeder from one building to another building, with the bare conductor used as a grounded neutral. Or an SE cable with a bare neutral may be used as a feeder within a building, if the bare neutral is used *only* as the equipment grounding conductor and one of the insulated conductors within the cable is used as the neutral of the feeder. (See Fig. 338-5.)

Part **(c)** requires that SE cable used to supply appliances not be subject to conductor temperatures in excess of the temperature specified for the insulation involved. The insulated conductors of SE cables are either 60 or 75°C, and if they are rated at 75°C, such marking will appear on the outer sheath. A cooking unit or oven that requires 75°C supply conductors would be an application for the use of SE cables, rated at 75°C. However, a review of UL listings for cooking units and ovens indicates that most such units do not require supply conductor ratings to exceed 60°C. The details in Fig. 338-4 show a method of connecting cooking units where the supply conductors are required to be 75 or 90°C.

338-4. Interior Installation Methods. Type SE cable used for interior branch circuits or feeders must satisfy all of the general wiring rules of Art. 300. This section requires the "installation" of unarmored SE cable (which is the usual type of SE cable) to satisfy Art. 336 on nonmetallic-sheathed cable (Type NM). All the rules of Art. 336 that cover *how* cable is "installed" must be satisfied. But the wording of Sec. 338-4 has caused difficulty.

Because SE cable must be installed in accordance with Art. 336, the relation between Sec. 338-4 and Sec. 336-5 raises the question:

Does the **NE Code** permit the use of SER (Service Entrance, Round) cable for feeders in a structure more than three floors above grade?

Inspectors have ruled that Sec. 338-4 ("installed in accordance with the provisions of Article 336") would bring SE cable under the provision of Sec. 336-4(a) limiting the use of Type SER cable to structures not over three floors above grade.

The chairperson of the **Code** panel says that it is the intent of the **Code** panel that this rule, in referring to Art. 336, means to limit SE-cable branch circuits or

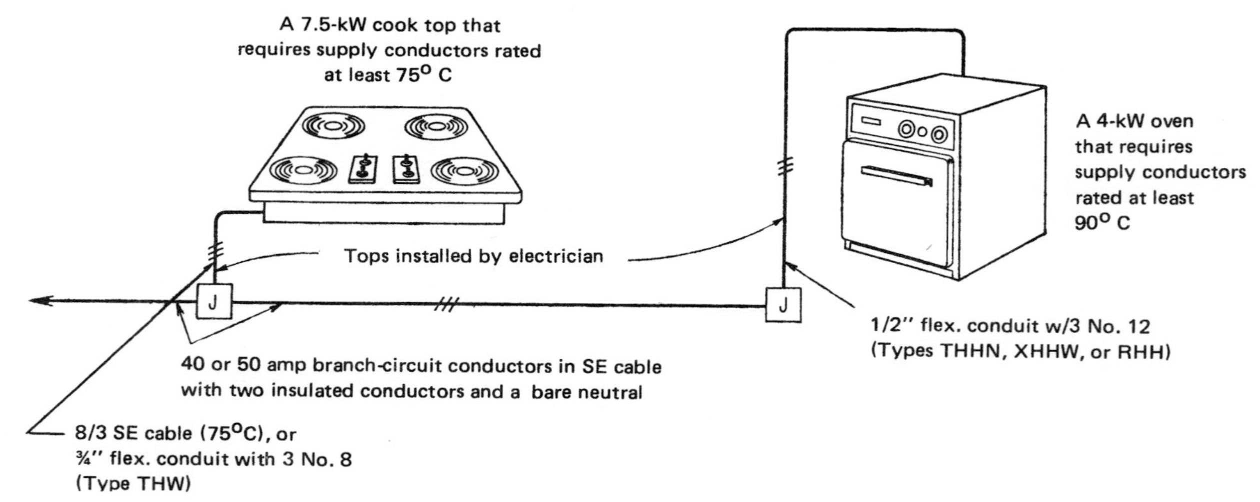

Fig. 338-4. Although previously permitted, SE cable with bare neutral may be used for branch circuit to range or other cooking units, *but only* on "existing branch circuits." (Sec. 338-3.)

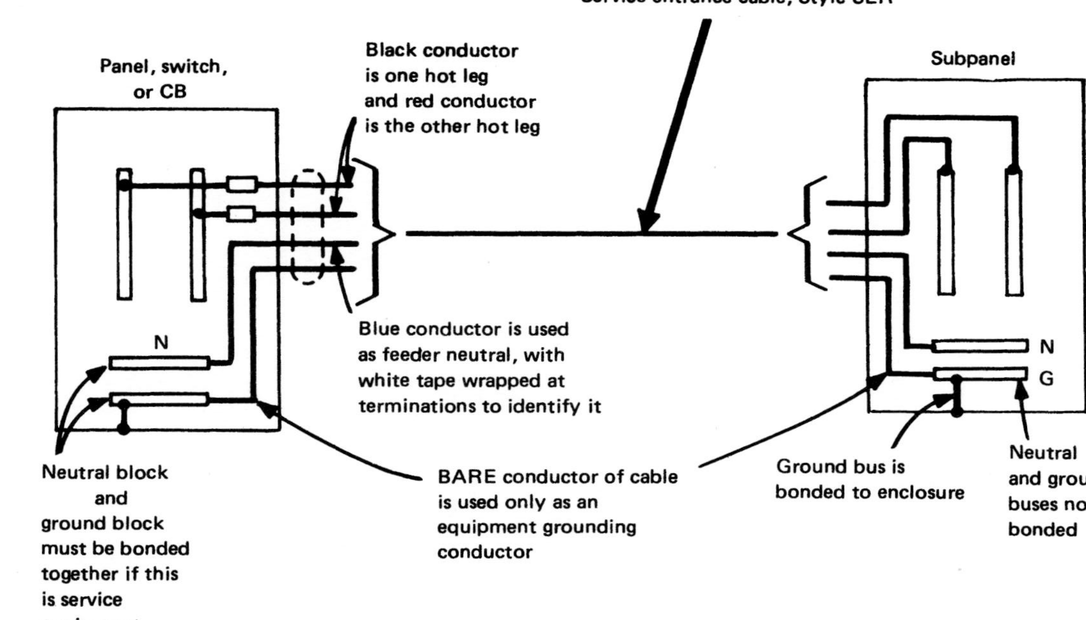
Service entrance cable, Style SER
Panel, switch, or CB
Black conductor is one hot leg and red conductor is the other hot leg
Subpanel
N
Blue conductor is used as feeder neutral, with white tape wrapped at terminations to identify it
N
G
Neutral block and ground block must be bonded together if this is service equipment
BARE conductor of cable is used only as an equipment grounding conductor
Ground bus is bonded to enclosure
Neutral and ground buses not bonded

Fig. 338-5. Typical application of SE cable with a bare neutral for use as a feeder within a building. (Sec. 338-4.)

feeders in the same way that Sec. 336-4 limits use of NM or NMC cable. That is, use of NM, NMC, or SE cable is limited to dwellings or other buildings—where the structure is not more than three floors above grade. As a result, the use of SE cable as a feeder, as shown in Fig. 338-5, would be a violation in any building with more than three floors above grade, such as in a high-rise apartment building.

ARTICLE 339. UNDERGROUND FEEDER AND BRANCH-CIRCUIT CABLE

339-1. Description and Marking. Figure 339-1 shows a violation of the Code rule that a bare conductor in a UF cable is for grounding purposes only.

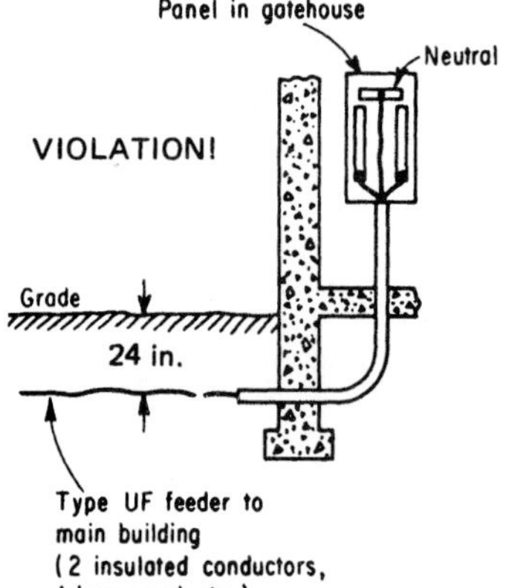

Fig. 339-1. Bare conductor in UF cable may not be used as a neutral. (Sec. 339-1.)

339-2. Other Articles. Figures 339-2 and 339-3 show details on compliance of UF cable with Sec. 300-5. Where UF comes up out of the ground, it must be protected for 8 ft (2.44 m) up on a pole and as described in Sec. 300-5(d).

339-3. Use. The rules of part **(a)** are shown in Fig. 339-4 and must be correlated to the rules of Sec. 300-5 on direct burial cables. The rule of **(2)** in part **(a)** corresponds to that of Sec. 300-5(i). If multiple conductors are used per phase and neutral to make up a high-current circuit, this rule requires all conductors to be run in the same trench or raceway and therefore subject to the derating factors of Note 8 to Tables 310-16 to 310-31. Refer to the paragraph right after the table in Note 8, in the NEC. Also see discussion under Sec. 300-5(i).

UF cable may be used underground, including direct burial in the earth, as feeder or branch-circuit cable when provided with overcurrent protection not in excess of the rated ampacity of the individual conductors. If single-conductor cables are installed, all cables of the feeder circuit, subfeeder, or

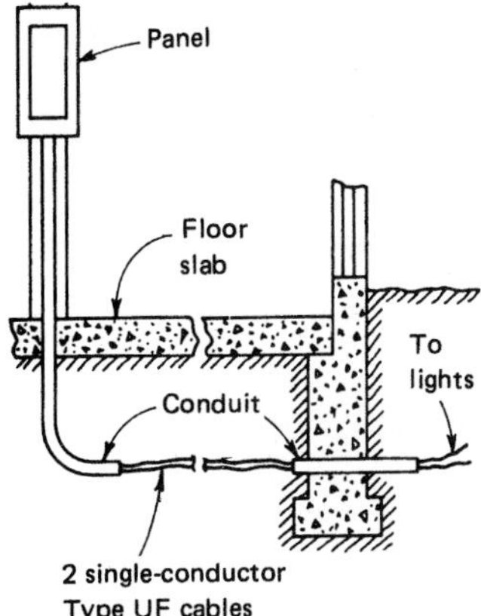

VIOLATION for cable to run under any building if not totally in raceway [Sec. 300-5 (c)]

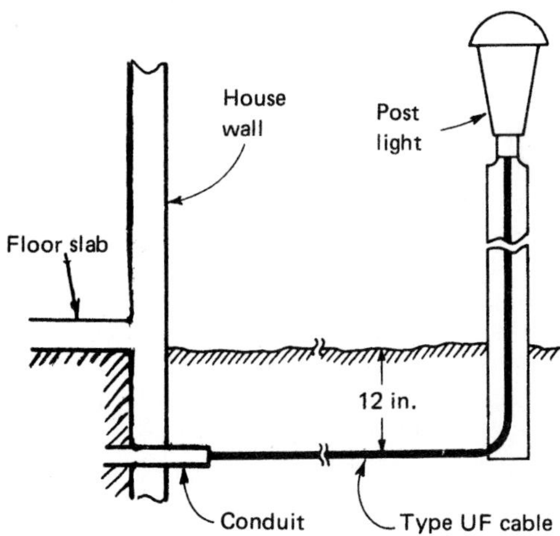

ACCEPTABLE for burial at only 12-in. depth when used for residential branch-circuit rated not more than 20 A at 120 V or less where GFCI protection is also provided [Table 300-5]

Fig. 339-2. UF cable must conform to Sec. 300-5 on direct-burial cables. (Sec. 339-2.)

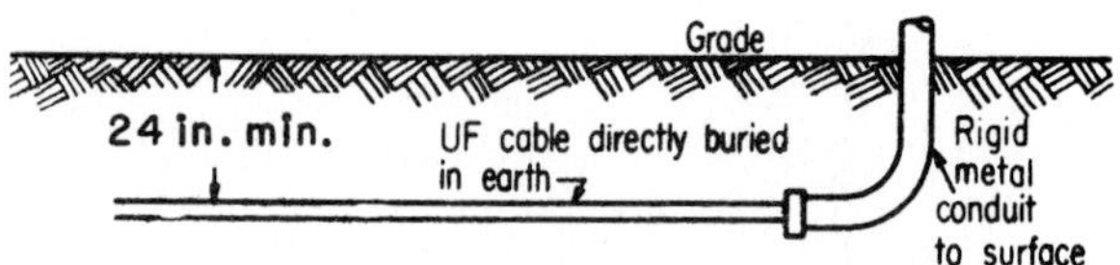

① UF CABLE WITHOUT SUPPLEMENTAL PROTECTION

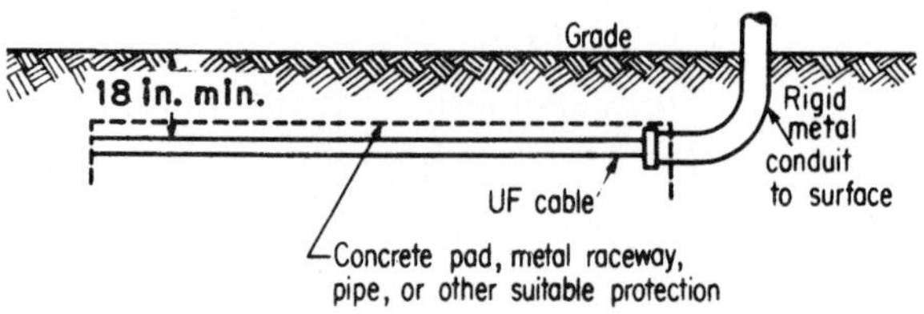

② UF CABLE WITH SUPPLEMENTAL PROTECTIVE COVERINGS

Fig. 339-3. The second qualifier under "Location of Wiring Method or Circuit" in Table 300-5 permits a 6-in. reduction of UF burial depth. (Sec. 339-2.)

branch circuit, including the neutral cable, must be run together in close proximity in the same trench or raceway. It may be necessary in some installations to provide additional mechanical protection, such as a covering board, concrete pad, raceway, etc., when required by the authority enforcing the **Code**. Multiple-conductor Type UF cable (but not single-conductor Type UF cables) may also be used for interior wiring when used in the same way as Type NM cable, complying with the provisions of Art. 336 of the **Code**. And UF may be used in wet locations.

The effect of the wording in part **(a)(4)**, where UF cable is used for interior wiring, is to require that its conductors must be rated at 90°C, with loading based on 60°C ampacity. This rule is a follow-up to the requirement that UF for interior wiring must satisfy the rules of Art. 336 on nonmetallic-sheathed cable (see Sec. 336-26, second paragraph).

As noted in part **(b)(8)**, single-conductor Type UF cable embedded in poured cement, concrete, or aggregate may be used for nonheating leads of fixed electric space heating cables, as covered in Secs. 424-43 and 426-23.

Application data of the UL are as follows:

> Cables suitable for exposure to direct rays of the sun are indicated by tag marking and marking on the surface of the cable with the designation "Sunlight Resistant."
>
> This cable may be terminated by using nonmetallic sheathed cable connectors (see Nonmetallic Sheathed Cable Connectors).
>
> If single conductor Type UF cable is terminated with a fitting not specifically recognized for use with single conductor cable, special care should be taken to assure it is properly secured and not subject to damage.

Only multiconductor Type UF cable may be used in cable tray, in accordance with Art. 318.

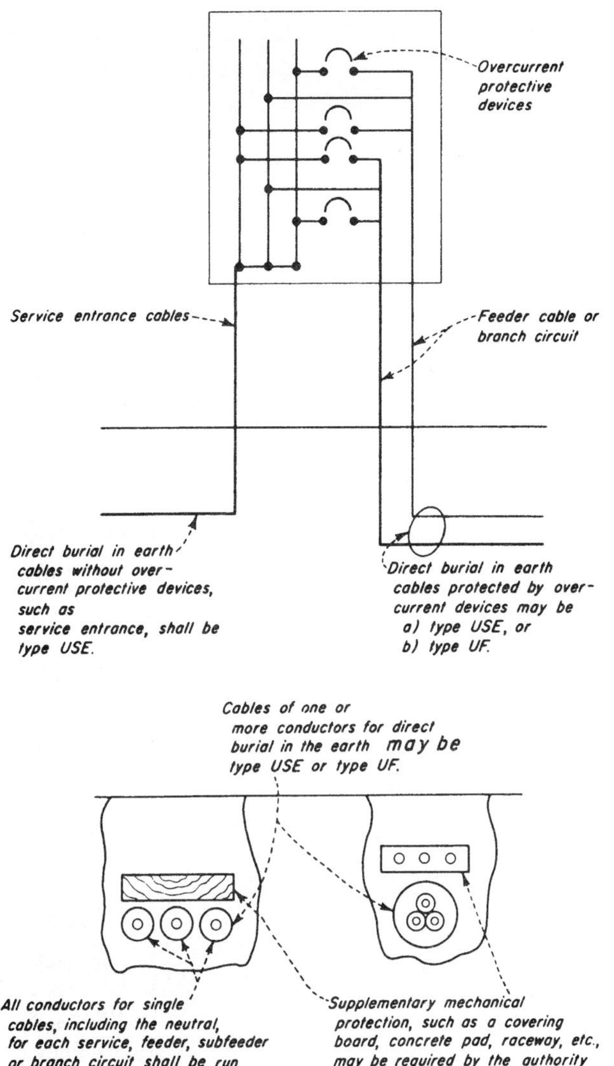

Fig. 339-4. UF cable may be used only as feeders or branch circuits. (Sec. 339-3.)

ARTICLE 340. POWER AND CONTROL TRAY CABLE

340-1. Definition. This article covers the use of a nonmetallic-sheathed power and control cable, designated Type TC cable (T and C are the initials for "Tray Cable"), which may be used in cable trays, in raceways, or where supported by a messenger wire outdoors (Fig. 340-1).

Fig. 340-1. This typical 3-conductor tray cable contains bare equipment grounding conductors. (Sec. 340-1.)

340-4. Use Permitted. Type TC tray cable is limited to use in industrial establishments where maintenance and supervision assure that only competent individuals will work on the cables.

NE Code Sec. 318-2 recognizes the use of Type TC power and control tray cable installed in a cable tray. Although specs on the construction and application of Type TC cable are covered in NE Code Secs. 340-1 through 340-3, great care must be used in relating those Code rules to the "Power and Control Tray Cable" in the UL Green Book. Type TC cable is recognized under Sec. 318-3(a)(9).

UL data on "Power and Control Tray Cable" include the following:

Power and Control Tray Cable (QPOR)

Type TC power and control tray cable is intended for use in accordance with Article 340 of the National Electrical Code. The cable consists of two or more insulated conductors, with or without one or more grounding conductors, with or without one or more optical fiber members and covered with a nonmetallic jacket. A single grounding conductor may be insulated or bare and may be sectioned. Any additional grounding conductor is fully insulated and has a distinctive surface marking. The cables are rated 600 or 2,000 volts.

The cable is Listed in conductor sizes No. 18 AWG to 1,000 kcmil copper or No. 12 to 1,000 kcmil aluminum or copper-clad aluminum. Conductor sizes within a cable may be mixed.

Cables with copper-clad aluminum conductors are surfaced printed "AL (CU-CLAD)" or "Cu-clad Al."

Cables with aluminum conductors are surface printed "AL."

For termination information, see Guide AALZ information.

If the type designation of the conductors is marked on the outside surface of the cable, the temperature rating of the cable corresponds to the rating of the individual conductors. When this marking does not appear, the temperature rating of the cable is 60 C unless otherwise marked on the surface of the cable.

Fittings for use with these cables are Listed by Underwriters Laboratories Inc. under the Outlet Bushings, Nonmetallic-Sheathed Cable Connectors, or Service Entrance Cable Fittings categories. Cables which have been investigated for use where exposed to direct rays of the sun are marked "sunlight resistant."

Cables investigated for direct burial in the earth are so identified.

Cables surface marked "Oil Resistant I" or "Oil Res I" are suitable for exposure to mineral oil at 60C. Cables suitable for exposure to mineral oil at 75C are surface marked "Oil Resistant II" or "Oil Res II."

Cables that comply with the Limited Smoke Test requirements specified in UL 1685, Standard Vertical-Tray Fire-Propagation and Smoke-Release Test for Electrical and Optical-Fiber Cables, are surface marked with a suffix "-LS."

Cables containing optical fiber members are identified with suffix "OF."

Regarding cable seals outlined in Article 501 of the National Electrical Code, Type TC Cable has a sheath which is considered to be gas/vapor tight but the cable has not been investigated for transmission of gases or vapors through its core.

The basic standard used to investigate products in this category is UL1277, "Electrical Power and Control Tray Cables With Optional Optical-Fiber Members."

The Listing Mark of Underwriters Laboratories Inc. on the product is the only method provided by UL to identify products manufactured under its Listing and Follow-Up Service. The Listing Mark for these products includes the name and/or symbol of Underwriters Laboratories Inc. (as illustrated in the Introduction of this Directory) together with the word "Listed," a control number, and a product name. A Power and Control Tray Cable which contains copper or copper-clad aluminum conductors has the product name, "Power and Control Tray Cable Type TC"; a Power and Control Tray Cable which contains aluminum conductors has the product name, "Aluminum Power and Control Tray Cable Type TC."

Note that this cable appears to be for cable tray only, but Type TC is recognized by Sec. 340-1 for use in raceway or with messenger support, in addition to use in tray.

340-5. Uses Not Permitted. Although (4) of this section has the effect of prohibiting the use of Type TC tray cable directly buried in the earth, the rule is modified by the phrase "unless identified for such use." The result of this wording is to permit Type TC cable directly buried in the earth where the cable is marked or otherwise approved for the purpose by the local electrical inspector. This permission for direct burial was added because the cable assembly was designed to withstand such application and because Type TC cable has been successfully and effectively used directly for years in many installations (Fig. 340-2 with burial conforming to Sec. 300-5). Such cable is listed for direct earth burial by UL, and the performance record has been excellent.

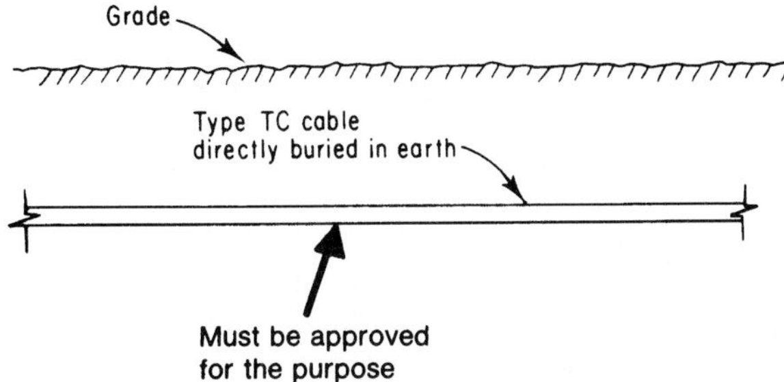

Fig. 340-2. Type TC (power and control tray cable) is recognized for direct earth burial. (Sec. 340-5.)

The effect of Exception No. 5 is to eliminate the need for a raceway to enclose or support the Type TC cable when it leaves a cable tray for connection to equipment. Such installation is *only* permitted where the Type TC cable has met the "crush" requirements applied to Type MC cable and the total length does *not* exceed 50 ft (15.24 m). Additionally, "qualified persons" must maintain the system, which only really excludes residences, and ground continuity must also be assured by a grounding conductor within the Type TC cable. But, if all those conditions are satisfied, Type TC may be run as open wiring from the tray to the equipment.

ARTICLE 342. NONMETALLIC EXTENSIONS

A nonmetallic extension is an assembly of two conductors without a metallic envelope, designed specially for a 15- or 20-A branch circuit as an extension from an existing outlet. Surface extensions are limited to residences and offices. Aerial extensions are limited to industrial purposes where it has been determined that the nature of the occupancy would require such wiring for connecting equipment.

ARTICLE 345. INTERMEDIATE METAL CONDUIT

This article covers a conduit with wall thickness less than rigid metal conduit but greater than that of EMT. Called "IMC," this intermediate metal conduit uses the same threading method and standard fittings for rigid metal conduit and has the same general application rules as rigid metal conduit. Intermediate metal conduit actually is a lightweight rigid steel conduit which requires about 25 percent less steel than heavy-wall rigid conduit. Acceptance into the **Code** was based on a UL fact-finding report which showed through research and comparative tests that IMC performs as well as rigid steel conduit in many cases and surpasses rigid aluminum and EMT in most cases.

Notice that IMC must be "listed." This is clearly spelled out in Sec. 345-1, "Definition." Although not literally required prior to the 1996 **Code**, only "listed" IMC is now recognized.

IMC may be used in any application for which rigid metal conduit is recognized by the **NEC**, including use in all classes and divisions of hazardous locations as covered in Secs. 501-4, 502-4, and 503-3. Its thinner wall makes it lighter and less expensive than standard rigid metal conduit, but it has physical properties that give it outstanding strength. The lighter weight facilitates handling and installation at lower labor units than rigid metal conduit. Because it has the same outside diameter as rigid metal conduit of the same trade size, it has greater interior cross-section area (Fig. 345-1). In the past this extra space was not recognized by the **NEC** to permit the use of more conductors than can be used in the same size of rigid metal conduit. However, with the elimination of Tables 2, 3A, 3B, and 3C, as well as the revisions of Tables 4 and 5 to more correctly reflect the interior area of raceways and the dimensions of conductors, the **Code** does permit greater fill in IMC (see Tables C4 and C4A in Appendix C).

345-3. Uses Permitted. The data of the UL supplement the requirements of part **(a)** on use of IMC, as follows:

> Listing of Intermediate Ferrous Metal Conduit includes standard 10 ft. lengths of straight conduit, with a coupling, special length either shorter or longer, with or without a coupling for specific applications or uses, elbows, bends, and nipples in trade sizes ½ to 4 in. incl. for installation in accordance with Article 345 of the **National Electrical Code.**
>
> Fittings for use with unthreaded intermediate ferrous metal conduit are listed under conduit fittings (Guide DWTT) and are suitable only for the type of conduit indicated by the marking on the carton.

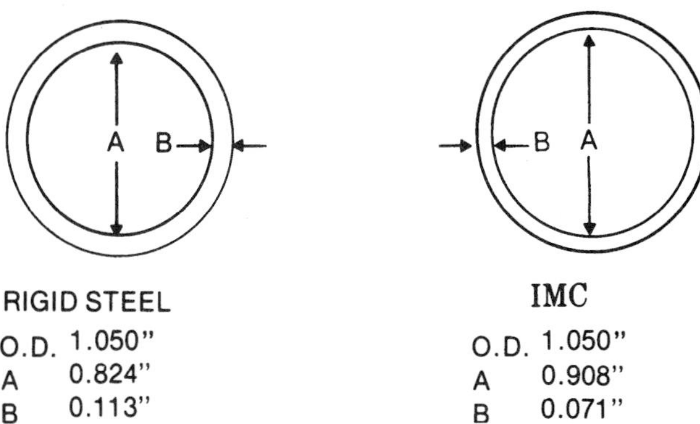

Fig. 345-1. Typical comparison between rigid and IMC shows interior space difference. (Sec. 345-1.)

Galvanized intermediate steel conduit installed in concrete does not require supplementary corrosion protection.

Galvanized intermediate steel conduit installed in contact with soil does not generally require supplementary corrosion protection.

In the absence of specific local experience, soils producing severe corrosive effects are generally characterized by low resistivity less than 2000 ohm-centimeters.

Wherever ferrous metal conduit runs directly from concrete encasement to soil burial, severe corrosive effects are likely to occur on the metal in contact with the soil.

Although literature on IMC refers to Type I and Type II IMC because of slight differences in dimensions due to manufacturing methods, the NEC considers IMC to be a single type of product and the rules of Art. 345 apply to all IMC.

Note that the wording in the UL data above includes the word "generally" in stating that IMC does not need additional protective material applied to the conduit when used in soil. That is intended to indicate that local soil conditions (acid versus alkaline) may require protection of the conduit against corrosion. And the UL note about corrosion of conduit running from concrete to soil must be observed. Refer to comments under Sec. 346-1 covering these conditions.

In part **(a)**, wording of the rule is significantly modified by the Exception, which specifically permits use of aluminum fittings and enclosures with steel intermediate metal conduit (Fig. 345-2). This same Exception is also given in Art. 346 on rigid metal conduit and Art. 348 on electrical metallic tubing. Tests have established that aluminum fittings and enclosures create no difficulty when used with steel raceways. The Exception is intended to counteract the implication of that phrase that cautions against use of dissimilar metals in a raceway system to guard against galvanic action. This section prohibits the use of dissimilar metals, "where practicable." This phrase is used frequently in the Code; in effect, it is saying, "You *shall* do it, if you can, or if the inspector thinks you can." By using this phrase, the Code recognizes that the contractor may not always be able to comply.

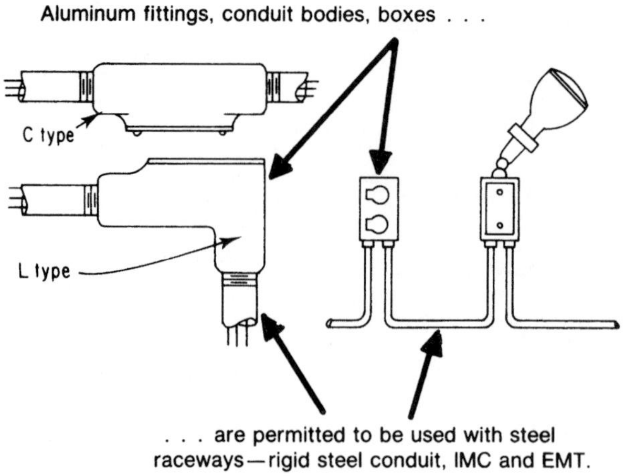

Aluminum fittings, conduit bodies, boxes . . .

C type

L type

. . . are permitted to be used with steel raceways—rigid steel conduit, IMC and EMT.

Fig. 345-2. NEC warning against use of dissimilar metals does not apply to this. (Sec. 345-3.)

In part **(b)**, wording of the rule intends to make clear that the galvanizing or zinc coating on the IMC does give it the measure of protection required when used in concrete or when directly buried in the earth. The last phrase, "judged suitable for the condition," refers to the need to comply with UL regulations such as those contained in UL's *Electrical Construction Materials Directory,* advising how and when steel raceways and other metal raceways may be used in concrete or directly buried in earth.

The UL data point out that there are soils where some difficulties may be encountered, and there are other soil conditions that present no problem to the use of steel or other metal raceways. The phrase "judged suitable for the condition" implies that a correlation was made between the soil conditions or the concrete conditions at the place of installation and the particular raceway to be used. This means that it is up to the designers and/or installers to satisfy themselves as to the suitability of any raceway for use in concrete or for use in particular soil conditions at a given geographic location. Of course, all such determinations would have to be cleared with the electrical inspection authority to be consistent with the meaning of Code enforcement.

For use of IMC in or under cinder fill, part **(c)** gives the limiting conditions. See Sec. 346-3.

345-7. Number of Conductors in Conduit. The rules on conduit fill are the same for IMC, rigid metal conduit, EMT, flexible metal conduit, flexible metallic tubing, and liquidtight flexible metallic tubing—for conduits ½ in. (12.7 mm) size and larger although different tables are used. Refer to Sec. 346-6.

345-12. Supports. The basic rule on clamping IMC is simple and straightforward (Fig. 345-3). Spacing may be increased to a maximum of 5 ft (1.52 m) where necessary because no structural member is available. But, the distance must not be extended, except as permitted in the exceptions. The exceptions allowing wider spacing of supports are the same as those covered in Sec. 346-12 for rigid metal conduit.

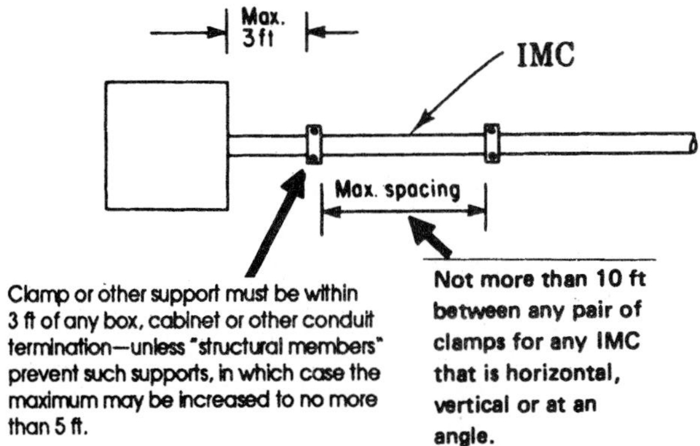

Fig. 345-3. All runs of IMC must be clamped in this way. (Sec. 345-12.)

Spacing between supports for IMC [greater than every 10 ft (3.05 m)] is the same as the spacing allowed for rigid metal conduit. The exceptions recognize the essential equality between the strengths of IMC and rigid metal conduit.

ARTICLE 346. RIGID METAL CONDUIT

346-1. Use. UL data on rigid metal conduit are similar to that on IMC and supplement the rules of this section, as follows:

> Galvanized rigid steel conduit installed in concrete does not require supplementary corrosion protection.
>
> Galvanized rigid steel conduit installed in contact with soil does not generally require supplementary corrosion protection.
>
> In the absence of specific local experience, soils producing severe corrosive effects are generally characterized by low resistivity less than 2000 ohm-centimeters.
>
> Wherever ferrous metal conduit runs directly from concrete encasement to soil burial, severe corrosive effects are likely to occur on the metal in contact with the soil.
>
> Supplementary nonmetallic coatings presently used have not been investigated for resistance to corrosion.
>
> Supplementary nonmetallic coatings of greater than 0.010-in. thickness applied over the metallic protective coatings are investigated with respect to flame propagation and detrimental effects to the basic corrosion protection provided by the protective coatings.

For rigid aluminum conduit, the UL application notes state:

> Aluminum conduit used in concrete or in contact with soil requires supplementary corrosion protection.
>
> Supplementary nonmetallic coatings presently used have not been recognized for resistance to corrosion.

For direct earth burial of rigid conduit and IMC, the UL notes must be carefully studied and observed:

1. Galvanized rigid steel conduit and galvanized intermediate steel conduit directly buried in soil do not *generally* require supplementary corrosion

protection. The use of the word "generally" in the UL instructions indicated that it is still the responsibility of the designer and/or installer to use supplementary protection where certain soils are known to produce corrosion of such conduits. Where corrosion of underground galvanized conduit is known to be a problem, a protective jacketing or a field-applied coating of asphalt paint or equivalent material must be used on the conduit. *But,* UL notes on "Supplementary nonmetallic coatings" must be observed for resistance to corrosion.

2. Aluminum conduit used directly buried in soil requires supplementary corrosion protection. But again, it is completely the task and responsibility of the designer and/or installer to select an effective protection coating for the aluminum conduit, because UL says "supplementary nonmetallic coatings presently used have not been *recognized* for resistance to corrosion." That could also be interpreted as a direct prohibition on the use of aluminum conduit directly buried.

The UL notes must also be observed in all use of metal conduits in concrete, as follows:

1. Galvanized rigid steel conduit and galvanized intermediate steel conduit installed in concrete *do not require* supplementary corrosion protection. See Sec. 348-1 on EMT.
2. Aluminum conduit installed in concrete *definitely requires* supplementary corrosion protection, but the supplementary protective coatings presently used "have not been recognized for resistance to corrosion."
3. *Watch out for this!* UL warns, "Wherever ferrous metal conduit runs directly from concrete encasement to soil burial, severe corrosive effects are likely to occur on the metal in contact with the soil." Supplementary protective coating on conduit at the crossing line can eliminate the conditions shown in Fig. 346-1.

346-3. Cinder Fill. Cinders usually contain sulfur, and if there is much moisture, sulfuric acid is formed, which attacks steel conduit. A cinder fill outdoors should be considered as "subject to permanent moisture." In such a place conduit runs should be provided corrosion protection as described, encased in 2 in. (50.8 mm) of concrete, or buried in the ground at least 18 in. (457 mm) below the fill. This would not apply if cinders were not present.

346-6. Number of Conductors in Conduit. The basic NE Code rule on the maximum number of conductors which may be pulled into rigid metal conduit, rigid nonmetallic conduit, intermediate metal conduit, electrical metallic tubing, flexible metal conduit, and liquidtight flexible metal conduit is contained in the single sentence of this section.

The number of conductors permitted in a particular size of conduit or tubing is covered in Chap. 9 of the Code in Tables C1 through C12A in Appendix C for conductors all of the same size used for either new work or rewiring. Tables 4 and 5 cover combinations of conductors of different sizes when used for new work or rewiring. For nonlead-covered conductors, three or more to a conduit, the sum of the cross-sectional areas of the individual conductors must not exceed 40 percent of the interior cross-section area (csa) of the conduit or tubing for new work or for rewiring existing conduit or tubing (Fig. 346-2). Note 4

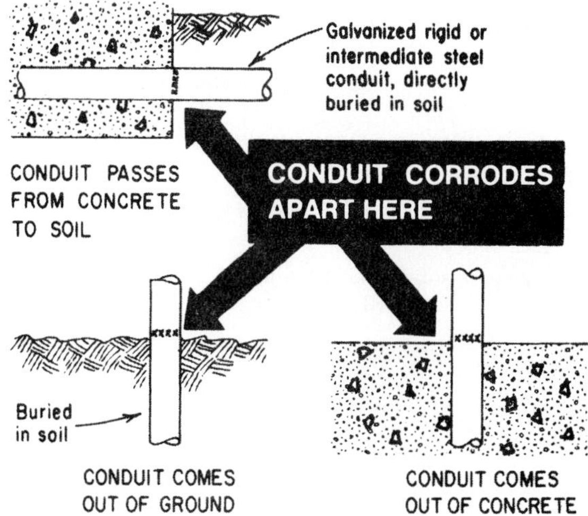

Fig. 346-1. Protective coating on section of conduit can prevent this corrosion problem. (Sec. 346-1.)

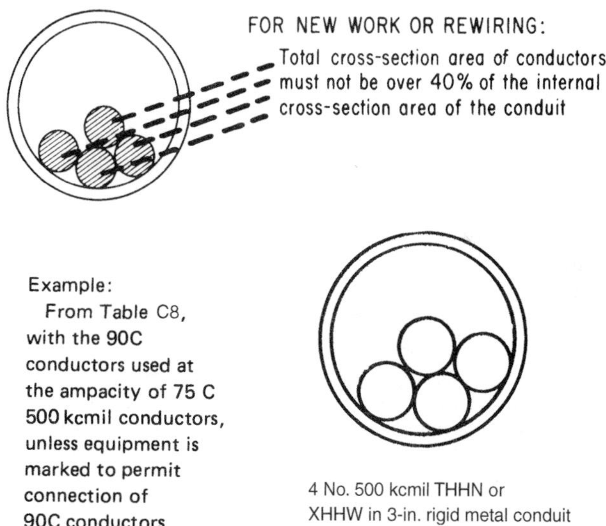

Fig. 346-2. For three or more conductors the sum of their areas must not exceed 40 percent of the conduit area. (Sec. 346-6.)

preceding all the tables in Chap. 9, in the back of the Code book, permits a 60 percent fill of conduit nipples not over 24 in. (610 mm) long and no derating of ampacities is needed.

When all conductors in a conduit or tubing are the same size, Tables C8 and C8A in Appendix C give the maximum allowable fill for conductors depending on conductor type up to 2,000 kcmil, for ½- to 6-in. (12.7 to 152 mm) rigid metal conduit.

question What is the minimum size of rigid metal conduit required for six No. 10 THHN wires?

answer Table 8C, Chap. 9, shows that six No. 10 THHN wires may be pulled into a ½-in. (12.7 m) rigid metal conduit.

question What size conduit is the minimum for use with four No. 6 RHH conductors with outer covering?

answer Table C8, Chap. 9, shows that a 1¼-in. (31.8 mm) minimum conduit size must be used for from four to six No. 6 RHH conductors.

question What is the minimum size conduit required for four No. 500-kcmil XHHW conductors?

answer Table C8 shows that 3-in. conduit may contain four 500-kcmil XHHW (or THHN) conductors.

When all the conductors in a conduit or tubing are not the same size, the minimum required size of conduit or tubing must be calculated. Table 1, Chap. 9, says that conduit containing three or more conductors of any type except lead-covered, for new work or rewiring, may be filled to 40 percent of the conduit csa. Note 6 to this table refers to Tables 4 through 8, Chap. 9, for dimensions of conductors, conduit, and tubing to be used in calculating conduit fill for combinations of conductors of different sizes.

example What size rigid metal conduit is the minimum required for enclosing six No. 10 THHN, three No. 4 RHH (without outer covering), and two No. 12 TW conductors (Fig. 346-3)?

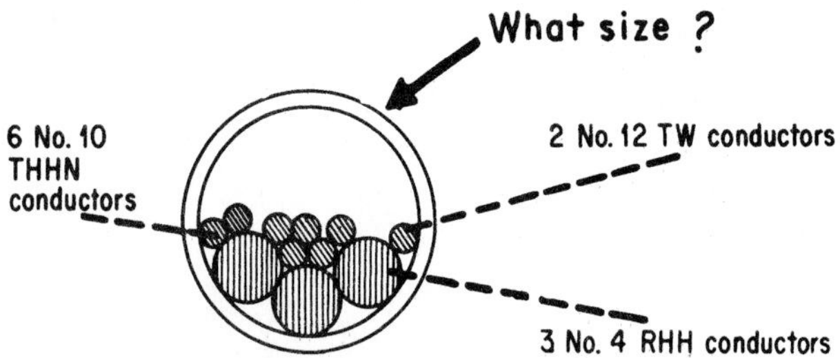

Fig. 346-3. Minimum permitted conduit size must be calculated when conductors are not all the same size. (Sec. 346-6.)

Cross-section areas of conductors:
From Table 5, Chap. 9:

No. 10 THHN	0.0211 sq in.
No. 4 RHH	0.0973 sq in.
No. 12 TW	0.0181 sq in.

Note: RHH without outer covering has same dimensions as THW.

Total area occupied by conductors:

6 No. 10 THHN	$6 \times 0.0211 = 0.1266$ sq in.
3 No. 4 RHH	$3 \times 0.0973 = 0.2919$ sq in.
2 No. 12 TW	$2 \times 0.0181 = 0.0968$ sq in.
Total area occupied by conductors	0.5153 sq in.

Referring to Table 4, Chap. 9:

The fourth column from the left gives the amount of square inch area that is 40 percent of the csa of the sizes of conduit given in the first column at left. The 40 percent column shows that 0.355 sq in. is 40 percent fill of a 1-in. (25.4-mm) conduit, and 0.610 sq in. is 40 percent fill of a 1¼-in. (31.8-mm) conduit. Therefore, a 1-in. (25.4-mm) conduit would be too small and—

A 1¼-in. (31.8-mm) rigid metal conduit is the smallest that may be used for these 11 conductors.

Example: What is the minimum size of conduit for four No. 4/0 TW and four No. 4/0 XHHW conductors?

From Table 5, a No. 4/0 TW has a csa of 0.3718 sq in. Four of these come to 4 × 0.3718 or 1.4872 sq in.

From Table 5 we find that four No. 4/0 XHHW have a csa of 1.2788 sq in.

$$1.4872 + 1.2788 = 2.766 \text{ sq in.}$$

From Table 4, 40 percent of the csa of 3-in. (76-mm) conduit is 3.000 sq in. A 2½-in. (64-mm) conduit would be too small.

Therefore—

A 3-in. (76-mm) conduit must be used.

Figure 346-4 shows how a conduit nipple is excluded from the normal 40 percent limitation on conduit fill. In this typical example, the nipple between

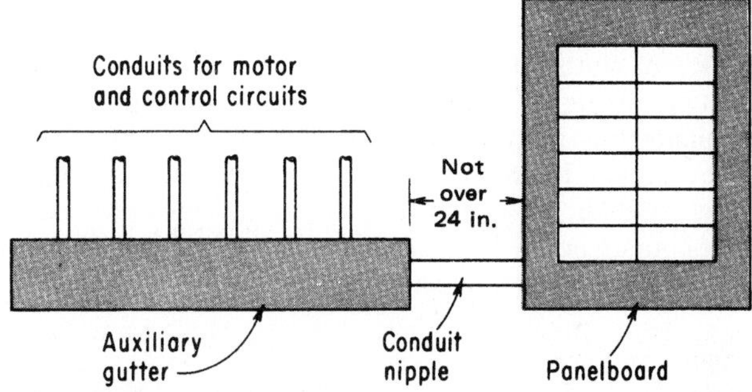

Fig. 346-4. Conduit nipples may be filled to 60 percent of csa and no derating is required. (Sec. 346-6.)

a panelboard and an auxiliary gutter contains 12 No. 10 TW wires, 6 No. 14 THHN wires, 3 No. 8 THW wires, and 2 No. 2 RHH wires (without outer covering). The minimum trade size of nipple that can be used in this case is 1¼ in. (31.8 mm). [Nipple may be filled to 60 percent of its csa if it is not over 24 in. (610 mm) long. Area of conductors = 12 × 0.0243 sq in. (csa of each No. 10 TW) plus 6 × 0.0097 sq in. (each No. 14 THHN) plus 3 × 0.0556 sq in. (each No. 8 THW) plus 2 × 0.1333, or a total of 0.7832 sq in., which is 60 percent of 1.30534 sq in. **NE Code** Table 4, Chap. 9, shows that a 1¼-in. (31.8-mm) nipple is the smallest that can be used. Sixty percent of the csa of 1¼-in. (31.8-mm) nipple = 0.6 × 1.562 or 0.9372 sq in.; 60 percent of the csa of 1-in. (25.4 mm) nipple = 0.6 × 0.888 or 0.532 sq in.] And the conductors do *not* have to be derated in accordance with Note 8 of Tables 310-16 through 310-19. If the nipple had been 25 in. (635.4 mm) long, calculation at 40 percent fill would have called for a 1½-in. (3.8-mm) size and all conductors would have had to be derated per Note 8.

THWN and THHN are the smallest-diameter building wires. The greatly reduced insulation wall on Type THWN or THHN gives these thin-insulated conductors greater conduit fill than TW, THW, or RHH. Type XHHW wire has the same conduit fills from No. 4 through 500 kcmil. And the nylon jacket on THWN and THHN has an extremely low coefficient of friction. THWN is a 75°C rated wire for general circuit use in dry or wet locations. THHN is a 90°C rated wire for dry locations only.

Although the same procedure applies, the tables in Appendix C and the various parts of Table 4 must be correlated with the type of raceway to be used. This is a major departure from past **Codes**, but provides for more realistic fill. Remember, that spare fill capacity may be desirable in certain applications—such as long underground runs to outbuildings. The **Code** permits fill to 40 percent but no more. If a raceway is filled to the 40 percent maximum permitted in Chap. 9, a new raceway will be required if additional circuits are desired at a later date.

To fill conduit to the **Code** maximum allowance is frequently difficult or impossible from the mechanical standpoint of pulling the conductors into the conduit, because of twisting and bending of the conductors within the conduit. Bigger-than-minimum conduit should generally be used to provide some measure of spare capacity for load growth; and in many cases, the conduit to be used should be upsized considerably to allow future installation of some larger anticipated size of conductors.

346-8. Bushings. As with IMC, rigid metal conduit always requires a bushing on the conduit end using locknuts and bushing for connection to knockouts in sheet metal enclosures (Fig. 346-5). But simply because a conduit can be secured to a sheet metal KO with two locknuts [one inside and one outside—as required by Sec. 250-76, Exception (b)], it does not mean the bushing may be eliminated. Of course, no bushing is needed where the conduit threads into a hub or boss on a fitting or an enclosure.

Fig. 346-5. Conduit terminations, other than threaded connections to threaded fittings or enclosure hubs, must be provided with bushings for protection of the conductors. (Sec. 346-8.)

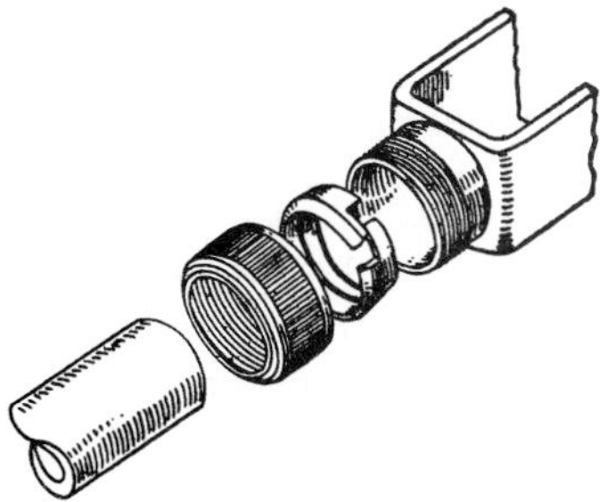

Fig. 346-6. Threadless connectors may be used on unthreaded end of conduit. (Sec. 346-9.)

346-9. Couplings and Connectors. Figure 346-6 shows a threadless connection of rigid metal conduit to the hub on a fitting. It is effective both mechanically and electrically if any nonconducting coating is removed from the conduit.

A running thread is considered mechanically weak and has poor electrical conductivity.

Where two lengths of conduit must be coupled together but it is impossible to screw both lengths into an ordinary coupling, the Erickson coupling or a swivel-coupling may be used (Fig. 346-7). They make a rigid joint which is both mechanically and electrically effective. Also, bolted split couplings are available.

It is not intended that conduit threads be treated with paint or other materials in order to assure watertightness. It is assumed that the conductors are approved for the locations and that the prime purpose of the conduit is for protection from physical damage and easy withdrawal of conductors for replace-

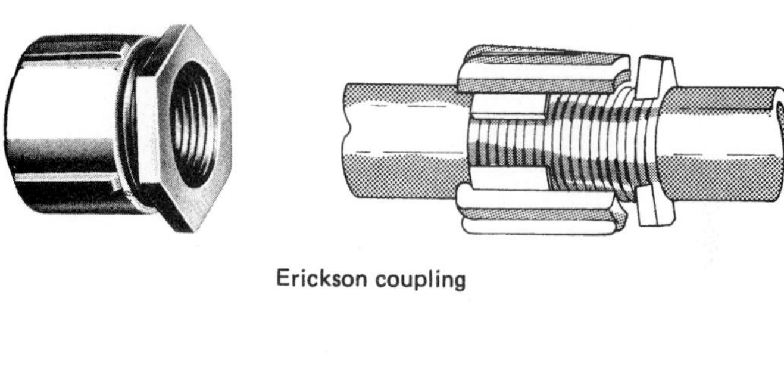

Erickson coupling

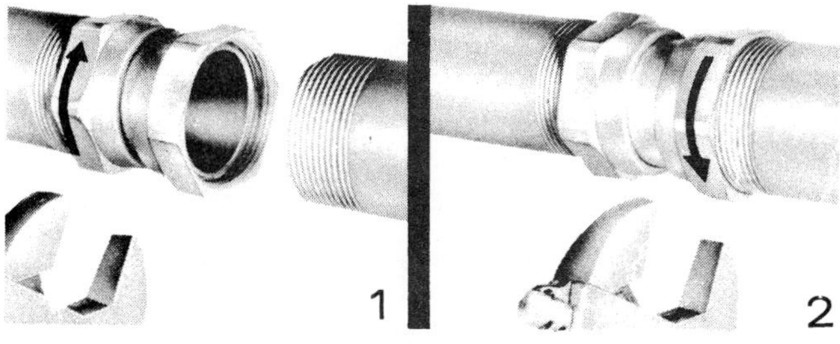

Swivel coupling

Fig. 346-7. Fittings provide for coupling conduits where conduits cannot be rotated (turned). (Sec. 346-9.)

ment. There are available pipe-joint compounds that seal against water without interrupting electrical conductivity.

346-10. Bends—How Made. The basic rule here recognizes bend radii in accordance with Table 346-10. And the Exception to that rule recognizes the bend radii in Table 346-10, Exception.

Table 346-10 gives minimum bending radii for bends in rigid metal conduit, IMC, or EMT using any approved bending equipment and methods. (See Fig. 346-8.) However, the Exception to this rule permits sharper bends (i.e., smaller

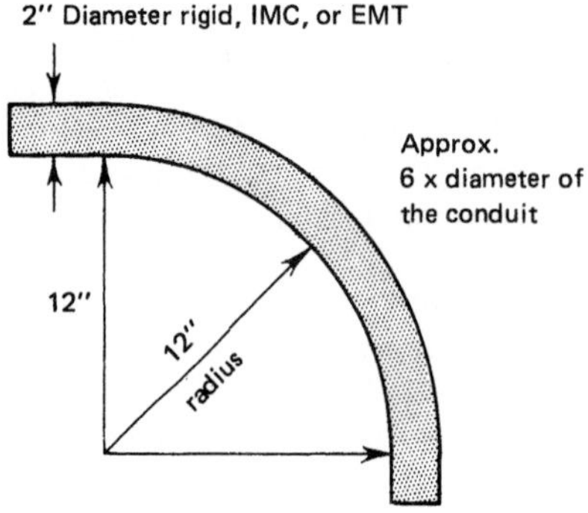

For conduit containing conductors without lead sheath

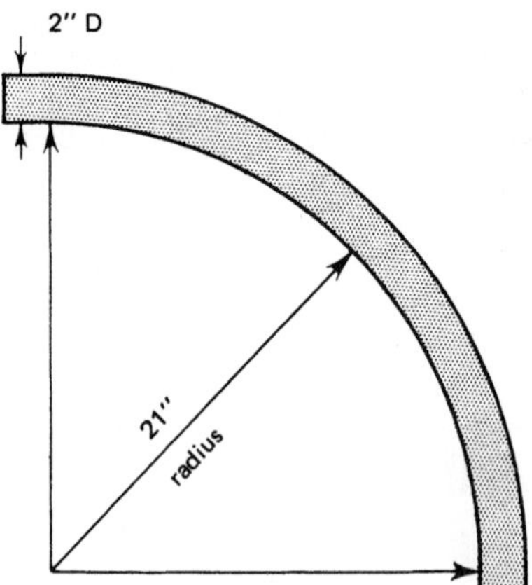

For conduit containing lead-sheathed conductors

NOTE: From Table 346-10 Exception, a bending radius of not less 9½ in. may be used for a one-shot bend on 2-in. rigid, IMC, or EMT if the conductors to be installed do not have a lead sheath.

Fig. 346-8. Minimum bending radii are specified to protect conductors from damage during pull-in. (Sec. 346-10.)

bending radii) if a one-shot bending machine is used in making a bend for which the machine and its accessories are designed. The minimum radii for one-shot bends are given in Table 346-10 Exception. All bending radii apply to any amount of bend—i.e., 45°, 90°, etc.

346-11. Bends—Number in One Run. There must be not more than the equivalent of four quarter bends (360°) between any two "pull points"—conduit bodies and boxes, as shown in Fig. 346-9. In previous Codes, the 360° of bends was

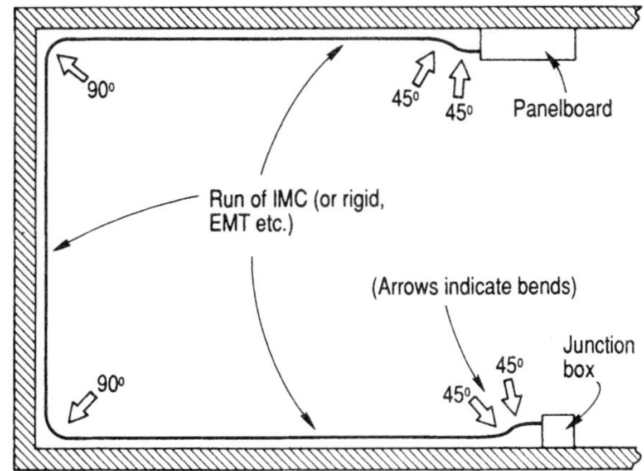

Fig. 346-9. Rigid metal conduit—like all other types of conduits—is limited to not over 360° of bends between "pull points," such as the panelboard and junction box shown here. (Sec. 346-11.)

permitted between boxes and "fittings" and even between "fitting and fitting." Because the word "fitting" is defined in Art. 100 and the term does include conduit couplings, bushings, etc., there could be very many bends in an overall run, totaling far more than 360° if the equivalent of four quarter bends could be made between each pair of conduit couplings. The present wording limits the 360° of bends to conduit runs between "pull points"—such as between switchboards and panelboards, between housings, boxes, and conduit bodies—all of which are "pull points."

The same concept of the number of bends permitted is given in all of the NEC Articles on raceways—ENT, EMT, IMC, rigid metal conduit, rigid nonmetallic conduit, etc.

ARTICLE 347. RIGID NONMETALLIC CONDUIT

347-1. Description. Nonmetallic conduit wiring systems include a wide assortment of products (Fig. 347-1).

All approved rigid nonmetallic conduits are suitable for underground installations. Some types are approved for direct burial in the earth while other types must be encased in concrete for underground applications. The non-

Fig. 347-1. Rigid nonmetallic conduit systems are made up of a wide variety of components—conduit, fittings, elbows, nipples, couplings, boxes, straps. (Sec. 347-1.)

metallic conduits include fiber conduit, asbestos-cement conduit, soapstone, rigid polyvinyl chloride conduit, polyethylene conduit, and styrene conduit. Of these, medium-density polyethylene conduit and styrene conduit are not UL-listed. High-density polyethylene conduit and the others *are* UL-listed. The listed and labeled conduits differ widely in weight, cost, and physical characteristics, but each has certain application advantages.

The only nonmetallic conduit approved for use aboveground at the present time is rigid polyvinyl chloride (PVC Schedule 40, or Schedule 80) (Fig. 347-2). Since not all PVC conduits are suitable for use aboveground, the UL label in each conduit length will indicate if the conduit is suitable for such use. For use of Schedule 80, see Fig. 347-3 and Secs. 300-5(d) and 710-4(b)(1).

UL application data are detailed and divide "Rigid Nonmetallic Conduit" into three categories with specific instructions on each category, as follows:

Rigid Nonmetallic Fiberglass Conduit (DZKT)

Nonmetallic Conduit is intended for use as described in the National Electrical Code.

This Listing covers fiberglass reinforced epoxy conduit intended for installation in accordance with Articles 347 and 710 of the National Electrical Code.

Fiberglass reinforced epoxy conduit may be used with wires rated 90 C or less.

Conduit marked "Rigid Nonmetallic Conduit Underground (Fiberglass Reinforced Epoxy Conduit)" is suitable for underground use only—for direct burial, with or without being encased in concrete.

Conduit marked "Rigid Nonmetallic Conduit (Fiberglass Reinforced Epoxy Conduit)" is suitable for use aboveground, underground and for direct burial with or without encasement in concrete. This conduit is suitable for exposed work where not subject to physical damage.

Fig. 347-2. PVC conduit is the only rigid nonmetallic conduit that may be used aboveground. And when enclosing conductors run up a pole (shown here feeding a floodlight at top), the PVC conduit must be Schedule 80 PVC conduit if it is exposed to physical damage, such as possible impact by trucks or cars. If the conduit is not so exposed, it may be Schedule 40 PVC conduit. See Sec. 300-5(d). (Sec. 347-1.)

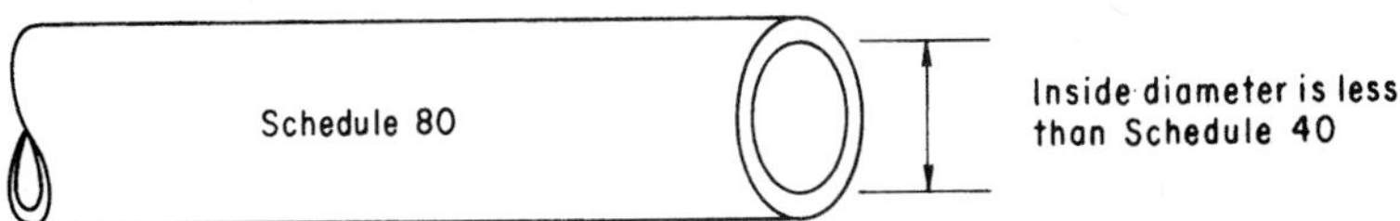

For conductor fill to 40% of the cross-section area, refer to data on wire-fill capacity marked on conduit surface.

Fig. 347-3. Extra-heavy-wall PVC conduit must have conductor fill limited to its reduced csa. (Sec. 347-1.)

Where the conduit emerges from underground installation, the wiring method shall be of a type recognized by the **National Electrical Code** for the purpose.

Fiberglass reinforced epoxy conduit is Listed in sizes ½ to 6 in. incl. Listing includes straight conduit, elbows and bends, unless otherwise noted.

Fiberglass reinforced epoxy conduit, elbows and bends (including fittings) which have been investigated for direct exposure to reagents are identified by the designation "Reagent Resistant" and are marked to indicate the specific reagents.

Fiberglass reinforced epoxy conduit is designed for connection to couplings, fittings, and boxes by use of a suitable epoxy-type cement or drive-on bell and spigot. Instructions supplied by the epoxy-type cement manufacturer describe the method of assembly and precautions to be followed.

Fittings for fiberglass reinforced epoxy conduit are Listed under CONDUIT FITTINGS (DWTT). For underground conduit other than fiberglass reinforced epoxy, see RIGID NONMETALLIC UNDERGROUND, PLASTIC CONDUIT (EAZX). For aboveground conduit other than fiberglass reinforced epoxy see RIGID NONMETALLIC SCHEDULE 40 AND SCHEDULE 80 PVC CONDUIT (DZYR).

The basic Standards used to investigate products in this category are UL 651, "Schedule 40 and Schedule 80 PVC Conduit, and UL 651A, "Type EB and Type A PVC Conduit and HDPE Conduit".

Rigid Nonmetallic, Schedule 40 and Schedule 80 PVC Conduit (DZYR)

This listing covers Rigid Nonmetallic PVC Conduit (Schedule 40 and Schedule 80) intended for installation in accordance with Articles 347 and 710 of The **National Electrical Code**. It is suitable for use aboveground, underground and for direct burial without encasement in concrete.

Unless marked for higher temperature, rigid nonmetallic conduit is intended for use with wires rated 75 C or less including where it is encased in concrete within buildings and where ambient temperature is 50 C or less. Where encased in concrete in trenches outside of buildings it is suitable for use with wires rated 90 C or less.

Schedule 40 conduit is suitable for exposed work where not subject to physical damage.

The marking "Schedule 80 PVC" identifies conduit suitable for use where exposed to physical damage, see Section 347-3(c) of The **National Electrical Code.**

Nonmetallic plastic conduit is listed in sizes ½ to 6 in. incl. Listing includes straight conduit, elbows and bends.

For additional Listings of Rigid Nonmetallic Conduit suitable for underground use, see the categories of Rigid Nonmetallic Fiberglass Conduit (DZKT) and Rigid Nonmetallic Underground Conduit, Plastic (EAZX).

Schedule 80 rigid PVC conduit has a reduced cross-sectional area available for wiring space. The actual cross-sectional area and the need for reference to **National Electrical Code** Chapter 9 Table 1 for wire fill capacity are prominently marked on the conduit surface.

Listed PVC conduit is inherently resistant to atmosphere containing common industrial corrosive agents and will also withstand vapors or mist of caustic, pickling acids, plating bath and hydrofluoric and chromic acids.

PVC conduit, elbows and bends (including couplings) which have been investigated for direct exposure to other reagents, may be identified by the designation "Reagent Resistant" printed on the surface of the product. Such special uses are described as follows:

PVC conduit, elbows and bends. Where exposed to the following reagents at 60 C or less: Acetic, Nitric (25 C only) acids in concentrations not exceeding ½ normal; hydrochloric acid in concentrations not exceeding 30 percent; sulfuric acid in concentrations not exceeding 10 normal; sulfuric acid in concentrations not exceeding 80 percent (25 C only); concentrated or dilute ammonium hydroxide; sodium hydroxide solutions in concentrations not exceeding 50 percent; saturated or dilute sodium chloride solution; cottonseed oil, or ASTM No. 3 petroleum oil.

PVC conduit is designed for connection to couplings, fittings and boxes by the use of a suitable solvent-type cement. Instructions supplied by the solvent-type cement manufacturer describe the method of assembly and precautions to be followed.

Fittings for rigid nonmetallic conduit are listed under Conduit Fittings, DWTT.

The basic standard used to investigate products in this category is UL651, "Schedule 40 and 80 Rigid PVC Conduit."

Rigid Nonmetallic Underground Conduit, Plastic (EAZX)

This listing covers plastic types of nonmetallic conduit, for use only when installed underground as raceway for installation of wires and cables in accordance with Articles 347 and 710 of the **National Electrical Code**. This conduit may be: (1) polyvinyl chloride (PVC) Type A or Type EB; or (2) high-density polyethylene (HDPE) Schedule 40. The various conduit types differ in their inside and outside diameters. For underground conduit of other than the plastic type, see Rigid Nonmetallic Fiberglass Conduit (DZKT).

The conduit is designed for use in underground work under the following conditions, as indicated on the Listing Mark, (1) when laid with its entire length in concrete (Type A), (2) when laid with its entire length in concrete in outdoor trenches (Type EB) and (3) direct burial with or without being encased in concrete (HDPE Schedule 40). The conduit is intended for use in ambient temperatures of 50 C or less and, unless marked otherwise, Type A and HDPE Schedule 40 conduit are intended for use with wires rated 75 C or less. Type EB conduit, Type A conduit encased in concrete in trenches outside of buildings, may be used with wires rated 90 C or less. HDPE Schedule 40 conduit, when directly buried or encased in concrete may be used with wire rated 90 C or less.

Where conduits emerge from underground installation the wiring method shall be of a type recognized by the **National Electrical Code** for the purpose.

Plastic underground conduit is Listed in sizes ½ to 6-in. incl. Listing includes straight conduit, elbows and bends unless otherwise noted.

PVC conduit is designed for joining with PVC couplings by the use of a solvent-type cement. HDPE conduit is designed for joining by threaded couplings, drive-on couplings, or a butt fusing process. Instructions supplied by the solvent-type cement manufacturer describe the method of assembly and precautions to be followed.

The basic standard used to investigate products in this category is UL651A, "Type EB and A PVC Conduit and HDPE Conduit."

The Listing Mark of Underwriters Laboratories Inc. on the product is the only method provided by UL to identify products manufactured under its Listing and Follow-Up Service. The Listing Mark for these products includes the name and/or symbol of Underwriters Laboratories Inc. (as illustrated in the Introduction of this Directory) together with the word "LISTED," a control number, and one of the following product names as appropriate: "Rigid Nonmetallic Conduit Underground (High Density Polyethylene, Schedule 40)"; "Rigid Nonmetallic Conduit Underground For Concrete Encasement Only (Type A)"; Rigid Nonmetallic Conduit Underground For Concrete Encasement In Outdoor Trenches Only. Not For Use in Ceilings, Floors or Walls (Type EB)."

Note: As a result of the wording and intent of NEC Sec. 110-3(b), all the above application data constitute mandatory rules of the NEC itself—subject to the same enforcement as any other NEC rules.

347-2. Uses Permitted. This section applies to use of the conduit for circuits operating at any voltage (up to 600 V and at higher voltages). The rules make rigid nonmetallic conduit a general-purpose raceway for interior and exterior wiring, concealed or exposed in wood or masonry construction—under the conditions stated. Only PVC is acceptable as a rigid nonmetallic conduit for in-building use (aboveground).

Aboveground applications of rigid nonmetallic conduit must be Schedule 40 or Schedule 80 PVC conduit, which is the only nonmetallic conduit listed for use aboveground.

Rigid nonmetallic conduit may be used aboveground to carry high-voltage circuits without need for encasing the conduit in concrete. That permission is also given in Sec. 710-4(a). Aboveground use is permitted indoors and outdoors.

Part **(g)** covers underground applications of all the types of rigid nonmetallic conduit—for circuits up to 600 V, as regulated by Sec. 300-5; and for circuits over 600 V, as covered by Sec. 710-4(b) (Fig. 347-4). Directly buried nonmetallic conduit carrying high-voltage conductors does not have to be concrete-encased if it is a type approved for use without concrete encasement. If concrete encasement is required, it will be indicated on the UL label and in the listing.

Fig. 347-4. All UL-listed rigid nonmetallic conduits are acceptable for use underground. PVC Schedule 40 and Schedule 80 and Type II fiber conduits do not require concrete encasement. Other types must observe UL and NEC rules on concrete encasement. (Sec. 347-2.)

Figure 347-5 shows both underground and aboveground application. Referring to the circled numbers: (1) The burial depth must be at least 18 in. (457 mm) for any circuit up to 600 V. The buried conduit may be Schedule 40 or Schedule 80 (either without concrete encasement) or Type A or Type EB (both require concrete encasement). Refer to Secs. 347-1 and 300-5. (2) The concrete encasement where the conduit comes up from its 18-in. (457-mm) depth was required at one

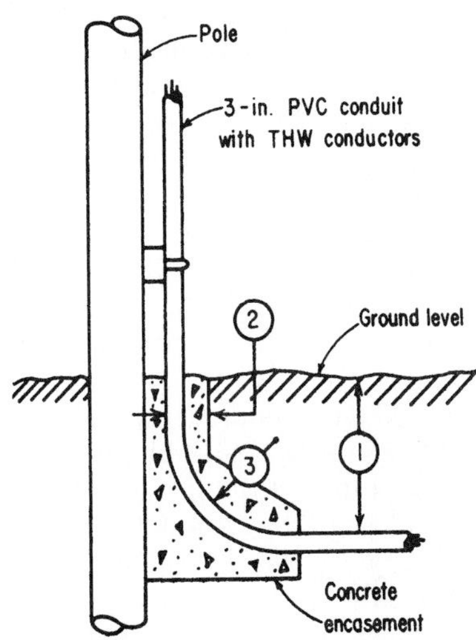

Fig. 347-5. Schedule 80 PVC conduit may run up pole from earth to aboveground use. (Sec. 347-2.)

time by the NEC, but is no longer required. (See Sec. 300-5.) (3) The radius of the bend must comply with Table 346-10 [minimum 18 in. (457 mm)]. The conduit aboveground, on a pole or on a building wall, must be Schedule 80 if the conduit is exposed to impact by cars or trucks or to other physical damage. If the conduit is not exposed to damage, it may be Schedule 40.

In many cases where nonmetallic conduit is used to enclose conductors suitable for direct burial in the earth, inspectors and engineering authorities have accepted use of any type of conduit—PVC, polyethylene, styrene, etc.—without concrete encasement and without considering application of Code rules to the conduit. The reasoning is that because the cables are suitable for direct burial in the earth, the conduit itself is not required at all and its use is above and beyond Code rules. But temperature considerations are real and related to effective, long-time operation of an installation. Temperature effects must not be disregarded in any conduit-conductor application.

347-3. Uses Not Permitted. It should be noted that nonmetallic conduit is not permitted in ducts, plenums, and other air-handling spaces. See Sec. 300-21 and the comments following Sec. 300-22.

Figure 347-6 shows a difference in application rules between rigid nonmetallic conduit and metal conduit with respect to supporting equipment. The Exception to part **(b)** allows limited use of rigid nonmetallic conduit for support of nonmetallic conduit bodies that do *not* contain devices or fixtures.

Parts **(d)** and **(e)** require care in use of the conduits so that they are not exposed to damaging temperatures. In using nonmetallic conduits care must be taken to assure temperature compatibility between the conduit and the conductors used in it. For instance, a conduit that has a 75°C temperature rating at which it might melt and/or deform must not be used with conductors which

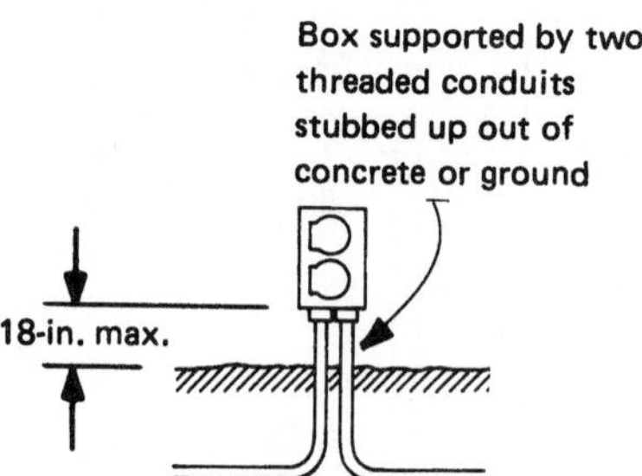

Fig. 347-6. This is O.K. for rigid metal conduit but not for rigid nonmetallic conduit. (Sec. 347-3.)

have a 90°C temperature rating and which will be loaded so they are operating at their top temperature limit. There is available PVC rigid conduit listed by UL and marked to indicate its suitability for use with all 90°C-rated conductors, thereby suiting the conduit to use with 90°C-rated conductors. The UL data described in Sec. 347-1 give the acceptable ambient temperatures and conductor temperature ratings that correlate to these NEC rules. Conductors with 90°C insulation may be used at the higher ampacities of that temperature rating only when the conduit is concrete encased (Fig. 347-7).

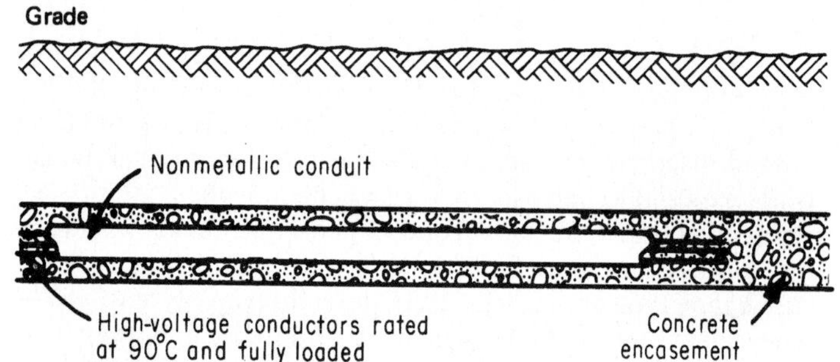

Fig. 347-7. UL data indicates that this violates Sec. 347-3(e). (Sec. 347-3.)

347-4. Other Articles. When equipment grounding is required for metal enclosures of equipment used with rigid nonmetallic conduit, an equipment grounding conductor must be provided. Such a conductor *must* be installed in the conduit along with the circuit conductors (Fig. 347-8). Refer to Secs. 250-57, 250-58, and 250-45.

347-5. Trimming. See Fig. 347-9.

347-8. Supports. In this section, Table 347-8, giving the maximum distance between supports for rigid nonmetallic conduit, permits greater spacing than previous NEC editions. For each size of rigid nonmetallic conduit, a single maximum spacing between supports, in feet, is given for all temperature ratings of conductors used in rigid nonmetallic conduit raceways (Fig. 347-10).

The Exception here is similar to exceptions added to Code articles covering other raceways and cables. This wording specifically recognizes holes in framing members as providing support for rigid nonmetallic conduit.

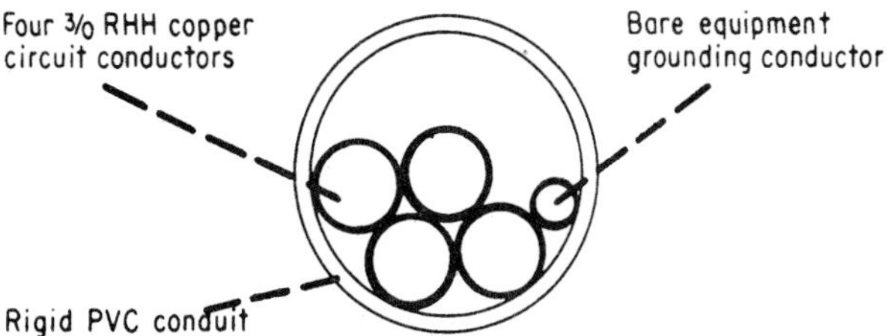

Equipment grounding conductor is sized from Table 250-95, based on the rating of the fuses or CB protecting the circuit conductor in the conduit. With 200-amp protection for the 3/0 conductors here, the equipment grounding conductor must be at least a No. 6 copper or No. 4 aluminum.

Fig. 347-8. Equipment grounding conductor must be used "within" the rigid nonmetallic conduit. (Sec. 347-4.)

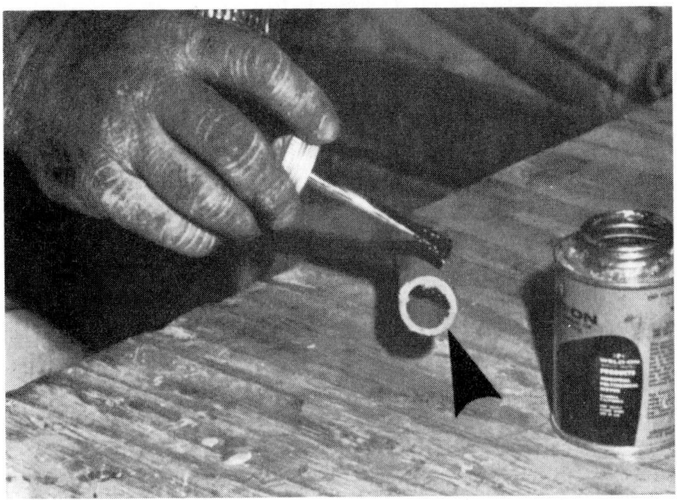

Fig. 347-9. PVC conduit is designed for connection to couplings and enclosures by an approved cement, but leaving rough edges in the conduit end is a clear violation of Sec. 347-5. (Sec. 347-5.)

347-9. Expansion Fittings. Where conduits are subject to constantly changing temperatures and the runs are long, expansion and contraction of PVC conduit must be considered. In such instances an expansion coupling should be installed near the fixed end of the run to take up any expansion or contraction that may occur. Available expansion couplings have a normal expansion range of 6 in. (152 mm). The coefficient of linear expansion of PVC conduit can be obtained from manufacturers' data.

Expansion couplings are normally used where conduits are exposed. In underground or slab applications such couplings are seldom used because expansion and contraction can be controlled by *bowing* the conduit slightly.

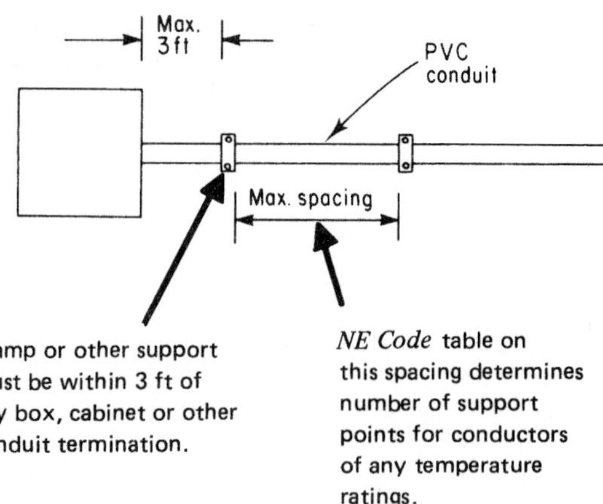

Fig. 347-10. Support rules on nonmetallic conduit are simple and direct. (Sec. 347-8.)

However, the rule of Sec. 300-5(j) now mandates that frost heave be addressed in underground installations. The FPN indicates methods—expansion joints—that may be used to satisfy the rule. Conduits left exposed for an extended period of time without expansion fittings during widely variable temperature conditions should be examined to see if contraction has occurred.

347-11. Number of Conductors. Refer to Sec. 346-6.

347-13. Bends—How Made. Refer to Sec. 346-10.

347-14. Bends—Number in One Run. Refer to Sec. 346-11.

ARTICLE 348. ELECTRICAL METALLIC TUBING

348-1. Use. As is the case with raceways and cables, the NEC only recognizes the use of "listed" EMT. EMT is a general-purpose raceway of the same nature as rigid metal conduit and IMC. Although rigid metal conduit and IMC afford maximum protection for conductors under all installation conditions, in many instances it is permissible, feasible, and more economical to use EMT to enclose circuit wiring rated 600 V or less. Because EMT is lighter than conduit, however, and is less rugged in construction and connection details, the NE Code restricts its use (Art. 348) to locations (either exposed or concealed) where it will not be subjected to severe physical damage or (unless suitably protected) to corrosive agents.

EMT distribution systems are constructed by combining wide assortments of related fittings and boxes. Connection is simplified by employing threadless components that include compression, indentation, and set-screw types.

Some questions have been raised about the acceptability of EMT directly buried in soil. The last paragraph of Sec. 348-1 gives EMT exactly the same recognition for direct burial that Sec. 346-1(c) gives to rigid steel conduit. The wording of both paragraphs is identical, and it certainly seems clear that if galvanizing is enough corrosion protection for rigid steel conduit, it must provide equivalent protection for EMT in direct burial. In the UL listing on "Electrical Metallic Tubing," a note says that "galvanized steel electrical metallic tubing in a concrete slab below grade level *may* require supplementary corrosion protection." (That word "may" leaves the decision up to the designer and/or installer.) But note that the rule carefully refers to "*galvanized* steel" EMT and not just to "steel" EMT.

The next note says, "In general, galvanized steel electrical metallic tubing in contact with soil requires supplementary corrosion protection." That sentence certainly admits that there are locations where soil conditions are such that supplementary corrosion protection is not required. Such an interpretation of UL intent would seem to be consistent with the equal approval that Secs. 345-3(b), 346-1(c), and 348-1 (last paragraph) give to direct burial of galvanized IMC rigid steel conduit and to galvanized electrical metallic tubing. The Code makes no distinction, where direct burial is concerned, between rigid conduit and EMT on the basis of their different wall thicknesses and structural strength differences. Nor does the Code distinguish between those two raceways with regard to corrosion protection.

It is reasonable to conclude that the phrase "judged suitable for the condition" means that past experience and local soil conditions should be considered in determining the acceptability of direct burial of EMT as well as IMC and rigid conduit, with appropriate attention given to additional protection against corrosion if necessary. Of course, the ruling of the local electrical inspector should be sought and followed.

Where corrosion protection has been provided and deemed suitable for the conditions, EMT burial depths must meet the minimum cover requirements of NE Code Sec. 300-5(a) and Table 300-5. But, because EMT is *not* specifically mentioned in Sec. 300-5(a) or Table 300-5, the minimum cover required must be determined in consultation with the local inspector, except that a 12-in. (305-mm) depth is permissible for residential branch circuits rated 120 V or less, provided with overcurrent protection of not more than 20 A, and provided the circuit also has GFCI protection and 6 in. (152 mm) for low-voltage (less than 30 V) control circuits for irrigation or lighting.

As noted in Sec. 345-3(a), Exception, and Sec. 346-1(b), Exception, permission is given for use of aluminum fittings and enclosures with steel electrical metallic tubing.

348-5. Size. The whole concern and discussion regarding the differences of actual cross-sectional area between the various raceways has been rendered moot. That is, in recognition of the differences between actual csa from one conduit or tubing, the tables in Chap. 9 covering csa—Table 4—and the tables covering maximum number of conductors of all the same size and insulation within a given size of raceway—now Tables C1 through C12A—have been completely rearranged and revised. The procedure remains the same, but the permitted fill is raceway-specific. And conductor dimensions have been corrected.

348-6. Number of Conductors in Tubing. Conductor fill for EMT is the same as described under Sec. 346-6 for rigid metal conduit.

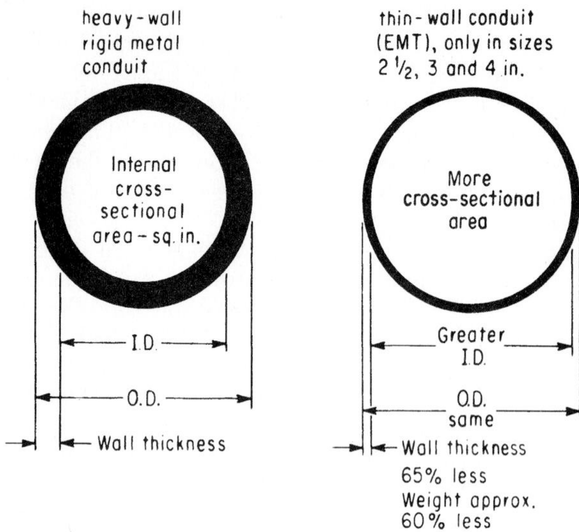

Trade size rigid and EMT	Inches outer dia. (O.D.) EMT and rigid	Wall thickness in. Rigid	Wall thickness in. EMT	Inside cross-sectional area sq. in. Rigid	Inside cross-sectional area sq. in. EMT	More C.S.A. % for EMT
2½	2.875	0.203	0.072	4.79	5.85	22%
3	3.500	0.216	0.072	7.38	8.84	19%
4	4.500	0.237	0.083	12.72	14.75	16%

Fig. 348-1. Larger sizes of EMT have same outside diameters as rigid and IMC. (Sec. 348-5.)

348-7. Threads. Here, the rules clarify Code intent. Threading of electrical metallic tubing is prohibited, but integral couplings used on EMT shall be permitted to be factory threaded. Such equipment has been used successfully in the past and has been found satisfactory. The revised Code rule recognizes such use. But it should be noted that this applies to EMT using *integral threaded fittings,* that is, fittings which are part of the EMT itself.

348-8. Couplings and Connectors. Couplings of the raintight type are required wherever electrical metallic tubing is used on the exteriors of buildings. (See Secs. 225-22 and 230-53.)

Sec. 370-17 requires that conductors entering a box, cabinet, or fitting be protected from abrasion. The end of an EMT connector projecting inside a box, cabinet, or fitting must have smooth, well-rounded edges so that the covering of the wire will not be abraded while the wire is being pulled in. Where ungrounded conductors of size No. 4 or larger enter a raceway in a cabinet, the EMT connector must have an insulated throat (insulation set around the edge of the connector opening) to protect the conductors. See Sec. 373-6(c). For conductors smaller than No. 4, an EMT connector does *not* have to be the insulated-throat type. Using THW conductors, a circuit of No. 4 conductors (a 2- or 3-wire circuit) requires a 1-in. (25.4-mm) size EMT (Table C1, Appendix C, NEC). Therefore, for THW or TW wire, there is no requirement for insulated-throat EMT connectors in ½- and ¾-in. (12.7- and 19-mm) sizes. A circuit, say, of three No. 1 THW wires would call for 1¼-in (31.8-mm) EMT, which would require use of insulated-throat connectors—or noninsulated-throat connector with a nonmetallic bushing on the connector end. In the larger sizes, the economics on the makeups can be significantly different. A 4-in. (102-mm) insulated-throat EMT connector might cost $18, whereas a noninsulated-throat connector in that size might cost $10 and $2 for a 4-in. (102-mm) plastic bushing (Fig. 348-2).

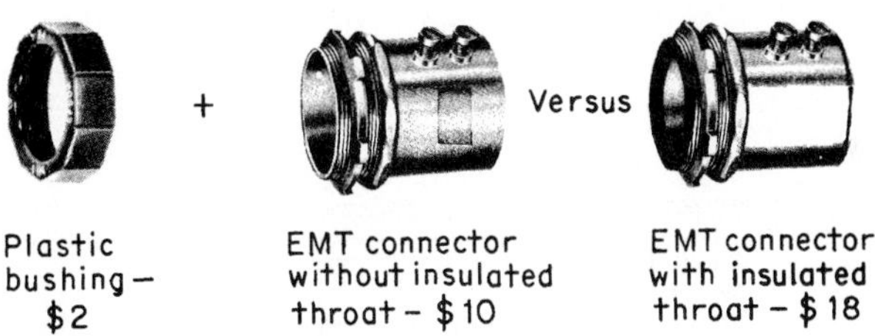

Fig. 348-2. Different-cost makeups for 4-in. EMT satisfy Code rules on EMT termination. (Sec. 348-8.)

When an EMT connector is used—either with or without an insulated throat to satisfy Sec. 373-6(c)—there is no requirement in Art. 348 that a bushing be used on the connector end. Note, however, that a bushing is required for rigid metal conduit and for IMC as covered in Secs. 346-8 and 345-15.

348-9. Bends—How Made. Refer to Sec. 346-10.

348-10. Bends—Number in One Run. Figure 348-3 shows EMT run from a panelboard to a junction box (JB) along the wall—with exactly a total of 360° of bend (from the panel: 45°, 45°, 90°, 90°, 45°, 45°).

348-12. Supports. Figure 348-4 shows this rule applied to an EMT layout. As stated in the basic rule of this section, EMT must be supported every 10 ft (3.05 m) and within 3 ft (914 mm) of each "outlet box, junction box, device box, cabinet, conduit body, or other tubing terminations." Prior to the 1993 NEC, this section referred to "each outlet box, junction box, cabinet and fitting." If the word "fitting" is taken to include couplings, then a strap must be used within 3 ft (914 mm) of each coupling. The definition of "fitting," given in Art. 100, includes locknuts and bushings. That wording was changed to provide a "laundry list" of enclosures that are covered by this rule. The intent was to clarify that supports are *not* required within 3 ft (914 mm) of EMT couplings.

As permitted by Exception No. 1, clamps on unbroken lengths of EMT may be placed up to 5 ft (1.52 m) from each termination at an outlet box or fitting where

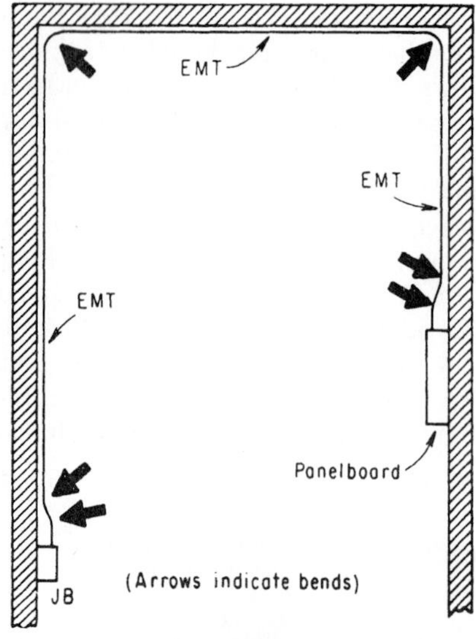

Fig. 348-3. EMT, like other conduit runs, is limited to not over 360° of bends between raceway ends. (Sec. 348-10.)

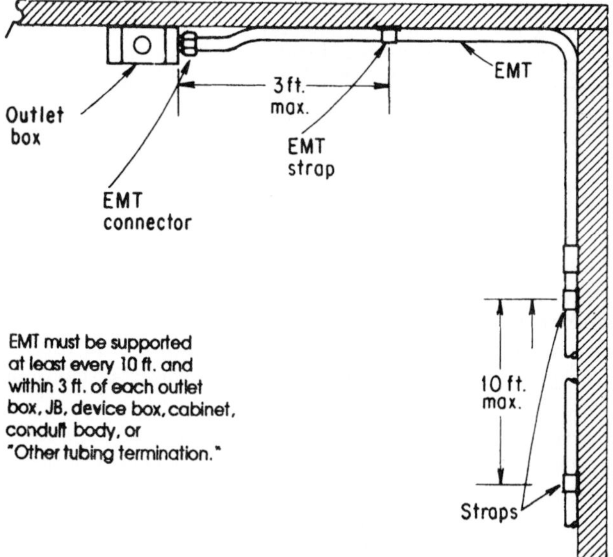

Fig. 348-4. EMT must be clamped within 3 ft (914 mm) of every enclosure or "fitting." (Sec. 348-12.)

structural support members do not *readily* permit support within 3 ft (914 mm). This Exception allows the first clamp to be up to 5 ft (1.52 m) from a termination of EMT at an outlet box.

The last paragraph following the two Exceptions makes clear that no additional means of support or securing are needed where framing provides support

for horizontal runs at least every 10 ft (3.05 m). Additio
secured within 3 ft (914 mm) of every termination.

ARTICLE 349. FLEXIBLE METALLIC TUBING

349-1. Scope. The first section of this article defines
rule indicates that flexible metallic tubing is intende
ject to physical damage" and gives use above suspended ceili
ple. Although this wording does not limit its use to air-handling ceilings, it does raise some questions for electrical inspectors with respect to accepting flexible metallic tubing as a general-purpose raceway.

The meaning of the phrase, "not subject to physical damage," is not clear. When the proposal was made to add flexible metallic tubing to the Code as a suitable raceway, it was indicated that it had been designed for certain specific applications and not for general use. It was specifically intended for use as the fixture whip on recessed fixtures where high-temperature wire is run from the branch-circuit junction box to the hot wiring compartment in lighting fixtures, an application long filled by flexible metallic conduit (Fig. 349-1). Although

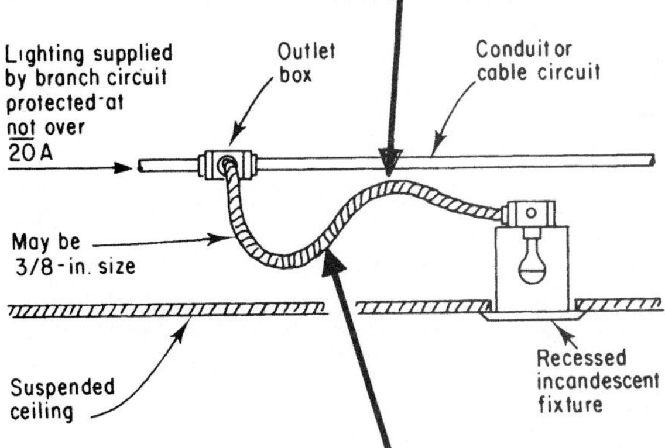
Typical use of flexible metallic tubing: 4-to-6-ft length for fixture whip in ceiling, containing two No. 18 Type AF wires (for 6-amp fixture load, see Section 402-5, Exceptions)
or two No. 16 Type AF wires (for 8-amp fixture load). Section 240-4 permits No. 16 and No. 18 fixture wire to be protected at 20 amps.
Lighting supplied by branch circuit protected at not over 20 A
Outlet box
Conduit or cable circuit
May be 3/8-in. size
Suspended ceiling
Recessed incandescent fixture
IMPORTANT!! Flex tubing is equipment grounding conductor because fittings are listed for grounding, and the AF wires in flex tubing are tapped from circuit protected at not over 20 amps, as permitted in Section 250-91(b), Exception No. 1.

Fig. 349-1. Flexible metallic tubing has limited application. (Sec. 349-1.)

shows Type AF wire, any fixture wire from Table 402-3 may be used d it has a temperature rating sufficient for the marking in the fixture supplied.

-3. Uses Permitted. This section limits the use of flexible metallic tubing to ranch circuits. In addition, branch-circuit conductors can only be installed in "dry locations"—either concealed or accessible—with systems rated no more than 1,000 V.

349-4. Uses Not Permitted. Part (6) limits use of flexible metallic tubing to lengths not over 6 ft (1.83 m) long. That limitation has the effect of ruling out flexible metallic tubing as a general-purpose raceway and limiting its use to short interconnections so commonly made with flexible metal conduit or liquidtight flexible metal conduit. But it does not appear that flexible metallic tubing, in spite of its resistance to moisture or liquid penetration, is an alternative to the use of liquidtight flexible metal conduit in wet locations.

ARTICLE 350. FLEXIBLE METAL CONDUIT

350-5. Uses Not Permitted. UL data supplement the **Code** data on use of standard flexible metal conduit—known also as "Greenfield" or simply "flex." The UL data note:

> These listings include flexible aluminum and steel conduit in trade sizes 5⁄16 to 4 in. incl. and flexible aluminum and steel conduit Type RW (reduced wall), in trade sizes from 3⁄8 to 3 in. incl., for installation in accordance with Article 350 of the **National Electrical Code.**
>
> Flexible metal conduit (steel or aluminum) shall not be used underground (directly buried or in duct which is buried) or embedded in poured concrete or aggregate, or in direct contact with earth or where subjected to corrosive conditions. In addition, flexible aluminum conduit shall not be installed in direct contact with masonry in damp locations.
>
> Flexible metal conduit no longer than six ft and containing circuit conductors protected by overcurrent devices rated at 20 amperes or less is suitable as a grounding means.
>
> Flexible metal conduit longer than six ft has not been judged to be suitable as a grounding means.
>
> See the category Conduit Fittings (DWTT) with respect to fittings suitable as a grounding means.
>
> To prevent possible damage to flexible aluminum conduit, and flexible aluminum and steel conduit Type RW, care must be exercised when installing connectors employing direct bearing set screws.
>
> Flexible aluminum conduit is marked at intervals of not more than one ft with the letters "AL".
>
> Flexible aluminum conduit Type RW is marked at intervals of not more than one ft with the letters "AL" and "RW".
>
> Flexible steel conduit Type RW is marked at intervals of not more than one ft with the letters "RW".

Where Sec. 350-5 prohibits use of flex "in wet locations, unless conductors are of . . . types approved for the specific conditions," interpretation has raised difficulty. Any conductor with a "W" designation—such as THW or XHHW—is recognized by Sec. 310-7 for use in wet locations. From the location" (under "Location" in Art. 100), any indoor plac spray or splashing or outdoors exposed to weather must u wire types—such as outdoor service-entrance conductors ta drop and run in conduit down the outside of a building. arises: May flex be used in such wet locations if the cond "W" type (say, THW)? It would seem that the parallel betwee tions would permit use of flex with THW wire in a wet location. However, there would still be concern for water getting inside the flex and running into enclosures or equipment. Because of that, the rule requires that flexible metal conduit used in wet locations (such as exposed to weather outdoors) must be installed to prevent water from entering the raceway and thereby entering other raceways or enclosures to which the flex is connected.

Note that the words in this rule require that "installation is such that water is *not likely* to enter other raceways or enclosures. . . ." This used to present somewhat of a conflict with Sec. 225-22, which requires that raceways exposed to outdoor weather "shall be made raintight." That is an absolute requirement, whereas the phrase "not likely" allows some possibility of water entry. But, flex is permitted to be exposed to weather outdoors as indicated by the Exception to Sec. 225-22.

350-10. Size. Part **(a)(2)** to this rule permits ⅜-in. flexible metal conduit to be used in lengths up to 6 ft (1.83 m) for connections to lighting fixtures. This provides correlation with Sec. 410-67(c), which includes 4 to 6 ft (1.22 to 1.83 m) of metal raceway for connecting recessed fixtures (generally the nonwired types). Figure 350-1 shows such application, and it is permissible to use No. 16 or No. 18 150°C fixture wire as shown in Fig. 349-1 for flex tubing.

Part **(a)(5)** permits ⅜-in. flex if it is "part of a listed assembly," which assumes it is supplied as part of UL-listed equipment, in lengths up to 25 ft (7.62 m) to connect "wired fixtures."

Part **(a)(3)** permits flex in ⅜-in. size to be used for the cable assemblies of modular wiring systems in hung ceilings [so-called "manufactured wiring systems" covered by Sec. 604-6(a)] in lengths over 6 ft (1.83 m) long. This is directed specifically to ceiling modular wiring. And the equipment grounding conductor run in such flex wiring assemblies may be either bare or insulated [see Sec. 604-6(a)(2)] (Fig. 350-1).

350-12. Number of Conductors. This section specifies that Table 1 of NEC Chap. 9 must be used in determining the maximum permitted number of conductors in ½- through 4-in. flex. Flexible metal conduit is permitted the same conductor fill as other types of conduit and tubing. The number of conductors permitted in ⅜-in. flex is given in Table 350-14.

350-14. Grounding. As shown in the UL data under Sec. 350-2, flex in any length over 6 ft (1.83 m) is not suitable as an equipment grounding conductor and an equipment grounding conductor must be used within the flex to ground metal enclosures fed by the flex. Exception No. 1 permits flex as an equipment grounding conductor *only* under the given conditions—which would be the same as shown in Fig. 349-1 for flex tubing. Refer to Sec. 250-91(b) and to the discussion of grounding and bonding in Sec. 250-79.

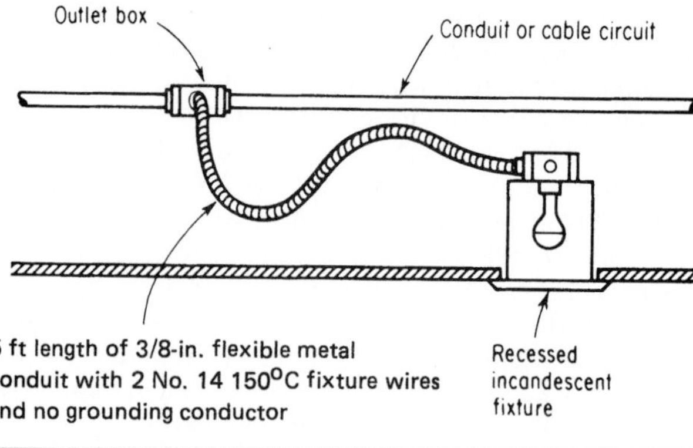

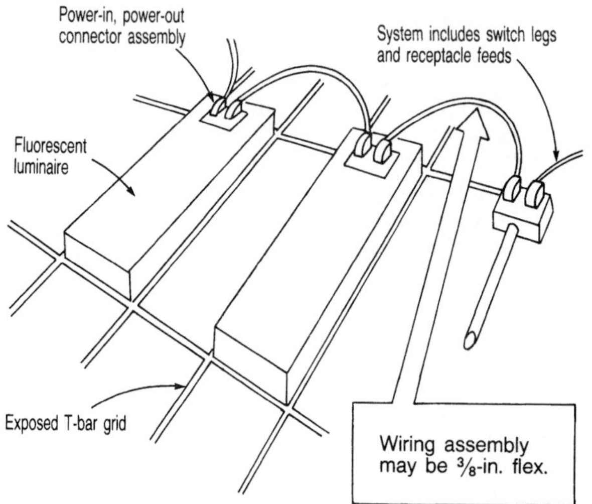

complete system in a ceiling space consisting of prewired cable lengths with snap-in connectors for fixtures and switches, may use ⅜-in. flexible metal conduit.

Fig. 350-1. Flex of ⅜-in. size may be used for fixture "whip." (Sec. 350-10.)

The second sentence of this section notes that an equipment bonding jumper used with flexible metal conduit may be installed inside the conduit or outside the conduit when installed in accordance with the limitations of Sec. 250-79. The Exception to this rule makes clear that flexible metal conduit when used as an equipment grounding conductor in itself is permitted only where a length of not over 6 ft (1.83 m) is inserted in any ground return path. The wording indicates that the total length of flex in any ground return path must not exceed 6 ft (1.83 m). That is, it may be a single 6-ft (1.83-m) length. Or, it may be two 3-ft (914-mm) lengths, three 2-ft (610-mm) lengths, or any total equivalent of 6 ft (183 m). If the total length of flex in any ground return path exceeds 6 ft (183 m), the rule appears to require an equipment grounding conductor to be run within or outside any length of flex beyond the permitted 6 ft (183 m) that is acceptable as a ground return path in itself.

It should be noted that the Exception to this section is not applicable to the use of flex in a hazardous location. The rules in Sec. 501-16(b) and Sec. 502-16(b) simply require bonding for flex, with only a very narrow Exception given in Sec. 502-16(b) (Fig. 350-2).

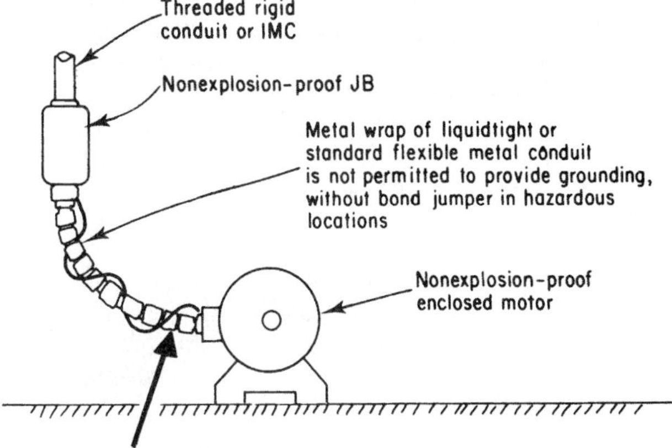
Where bonding of standard or liquidtight flex is flatly required, as in Class I, Div. 2 and Class II, Div. 2 locations . . .
Threaded rigid conduit or IMC
Nonexplosion-proof JB
Metal wrap of liquidtight or standard flexible metal conduit is not permitted to provide grounding, without bond jumper in hazardous locations
Nonexplosion-proof enclosed motor
. . .an internal or external bonding jumper must be used at all times, for any size and any length of the flex and must conform to Section 250-79(e) as noted in Section 501-16(b) and Section 502-16(b).

Fig. 350-2. Flex must always be bonded in Class I and Class II hazardous locations. (Sec. 350-14.)

The last sentence says that an equipment grounding conductor (or jumper) must *always* be installed for a length of metal flex that is used to supply equipment "where flexibility is required," such as equipment that is not fixed in place. That wording actually modifies the Exception, which describes the conditions under which a 6-ft (1.83-m) or shorter length of metal flex (Greenfield) may be used for grounding through the metal of its own assembly, without need for a bonding wire. Because experience has indicated many instances of loss of ground connection through the flex metal due to repeated movement of a flex whip connected to equipment that vibrates or flex supplying movable equipment, the last sentence requires use of an equipment bonding jumper, either inside or outside the flex in all cases where vibrating or movable equipment is supplied—for assured safety of grounding continuity. The rule applies to those lengths of 3 ft (914 mm) or less that are permitted by Sec. 350-18, Exception No. 2, because "flexibility is necessary."

350-16. Bends. Figure 350-3 shows the details of this section.

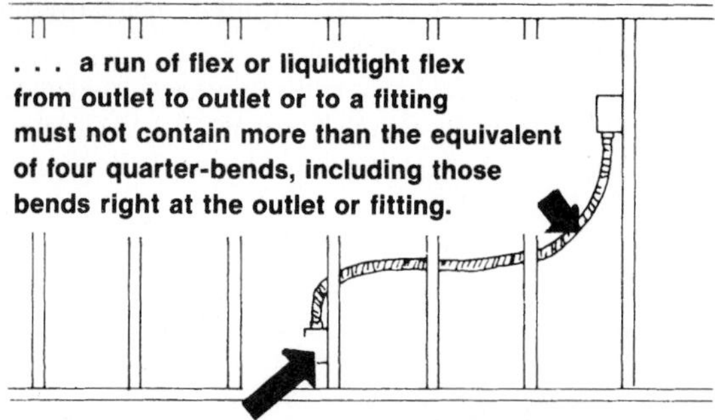

FOR EXPOSED OR CONCEALED WORK . . .
. . . a run of flex or liquidtight flex from outlet to outlet or to a fitting must not contain more than the equivalent of four quarter-bends, including those bends right at the outlet or fitting.
Angle connectors for flex connection to enclosures must not be used for *concealed* flex installations. Straight connectors are OK.

Fig. 350-3. Concealed or exposed flex must not have too many bends that could damage wires on pull-in. (Sec. 350-16.)

In this section and in Sec. 351-10, the limitation to no more than a total of 360° of bends between outlets applies to both exposed and concealed applications of standard metal flex and liquidtight metal flex.

Without restriction on the maximum number of bends in exposed and concealed work, bends could result in damage to conductors in a run with an excessive number of bends or could encourage installation of conductors prior to conduit installation, with conduit then installed as a cable system. A limit on number of bends for exposed and concealed work conforms with the requirements for other raceway systems, such as Secs. 345-11, 346-11, 347-14, and 348-10.

350-18. Supports. Straps or other means of securing the conduit in place should be spaced much closer together [every 4½ ft (1.37 m) and within 12 in. (305 mm) of each end] for flexible conduit than is necessary for rigid conduit. Every bend should be rigidly secured so that it will not be deformed when the wires are being pulled in, thus causing the wires to bind.

Figure 350-4 shows use of unclamped lengths of flex, as permitted by Exception No. 2. Figure 350-5 shows another example. Exception No. 3 is illustrated in Fig. 350-1.

ARTICLE 351. LIQUIDTIGHT FLEXIBLE METAL CONDUIT AND LIQUIDTIGHT FLEXIBLE NONMETALLIC CONDUIT

351-1. Scope. This article is divided into two parts. Part **A** covers *metal* liquidtight flex, and part **B** contains seven sections on *nonmetallic* liquidtight flex. Liquidtight metal flex (often called "Sealtite" as a generic term in industry usage, although that word is the registered trade name of the liquidtight flex made by Anaconda Metal Hose Division) is similar in construction to the common type of flexible metal conduit, but is covered with an outer sheath of thermoplastic material (Fig. 351-1).

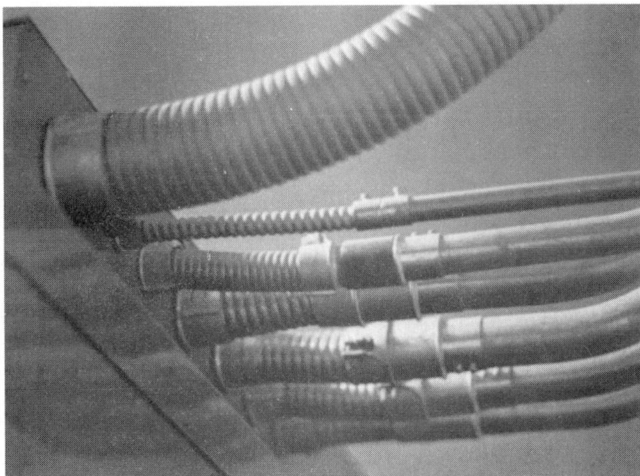

Fig. 350-4. Lengths of flex not over 3 ft (914 mm) long may be used without clamps or straps where the flex is used at terminals to provide flexibility for vibration isolation or for alignment of connections to knockouts. (Sec. 350-18.)

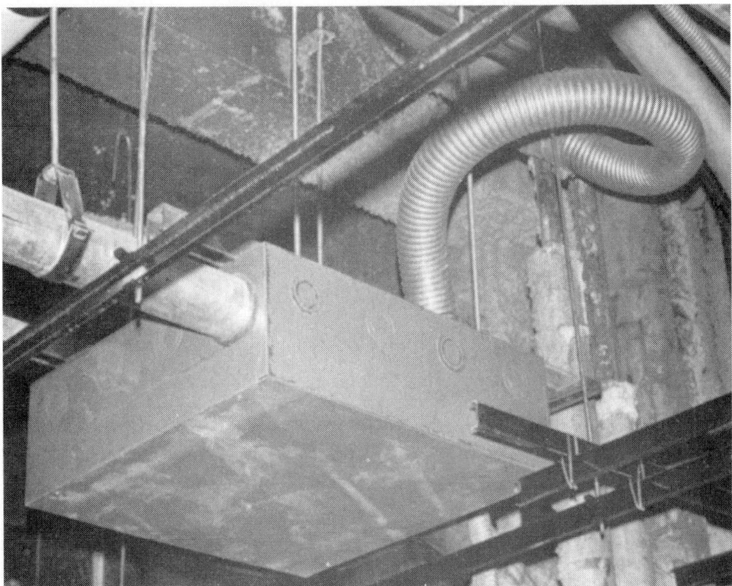

Fig. 350-5. A length of flex not over 3 ft (914 mm) long connects conduit to pull box in modernization job, providing the flexibility to feed from fixed conduit to box. (Sec. 350-18.)

Fig. 351-1. Plastic jacket on liquidtight flex suits it to outdoor use exposed to rain or indoor locations where water or other liquids or vapors must be excluded from the raceway and associated enclosures. In lengths under 6 ft (1.83 m), UL-listed metal liquidtight flex does not require a bonding jumper. (Sec. 351-1.)

351-4. Use. UL data on liquidtight metal flex say:

> Flexible liquid-tight conduit is intended for use in wet locations or where exposed to mineral oil, both at a maximum temperature of 60 degrees C and installed in accordance with the National Electrical Code. It is not intended for direct burial or where exposed to gasoline or similar light petroleum solvents unless so marked on the product.

That rule of UL has an effect on the conductors used in the flex. Because UL-listed liquidtight flexible metal conduit is intended for use at a maximum temperature of 60°C, conductors used in liquidtight flex must be 60°C-rated Type TW; or, if higher-temperature-rated conductors are used (THW, RHH, THHN, XHHW), they must be used at the 60°C ampacities of NE Code Table 310-16.

Although UL data limit its listed liquidtight flex to use at a maximum of 60°C, there are applications requiring higher-temperature-rated flex for foundries, near boilers, and in other hot places. Even though high-temperature flex is not UL-listed, it is generally consistent with NEC rules [Sec. 110-3(a)(5)] to use the higher-temperature flex when an application would exceed the 60°C rating of UL-listed flex. High-temperature flex is *not* a product listed by any test lab, and its use is, therefore, not contrary to Sec. 110-2. It would violate Secs. 351-4(b)(2) and 110-3(b) to use the listed 60°C flex in any way in which the loading on its contained conductors and the given ambient combined to produce a temperature over 60°C in the plastic jacket of the flex. Such application was recognized in past Codes, but the 1996 edition presents a complicating factor. In Sec. 351-4(a), the wording was changed to indicate that "listed liquidtight flexible metal conduit" is permitted. That wording would seem to eliminate the use of the so-called high-temperature liquidtight flex because it is *not listed.* Check with the local AHJ to see if the high-temperature liquidtight flex *is* permitted. Or—where ambient temperatures are too high, another raceway or cable—one suited to the environment—should be used. Refer to temperature correction factors in Table 310-16.

As noted in the UL data quoted above, liquidtight flexible metal conduit is permitted for use directly buried in the earth if it is "so marked on the product." The rule in Sec. 351-4(a) extends Code recognition to direct burial of liquidtight flexible metal conduit if it is "listed and marked" for such use. Based on many years of such application, liquidtight metal flex is recognized for direct burial, but any such use is permitted only for liquidtight flex that is "listed" by UL, or some other test lab, and is "marked" to indicate suitability for direct burial, to assure the installer and inspector of Code compliance. In the past, successful applications have been made in the earth and in concrete. Standard flexible metal conduit is prohibited from being used "underground or embedded in poured concrete or aggregate" [Sec. 350-2(6)]. But that prohibition is not placed on liquidtight metal flex.

351-5. Size. Refer to Sec. 350-18. Figure 351-2 satisfies Exception No. 2 of Sec. 350-18 if the No. 12 wires are stranded, as required in Sec. 430-145(b). Table 350-3 accepts four No. 12 THHN in ⅜-in. Greenfield or liquidtight.

351-6. Number of Conductors. Refer to Sec. 346-6 for ½-in. to 4-in. (12.7- to 102-mm) sizes, and to Table 350-12 for ⅜-in. size, see Tables C5 and C5A in Appendix C.

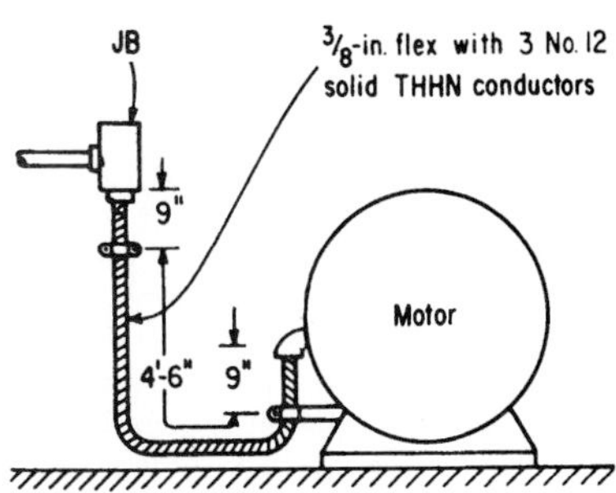

NOTE: The 6 ft length of liquidtight is suitable as an equipment grounding conductor, without bonding.

Fig. 351-2. Both standard flexible metal conduit and liquidtight may be used here. (Sec. 351-5.)

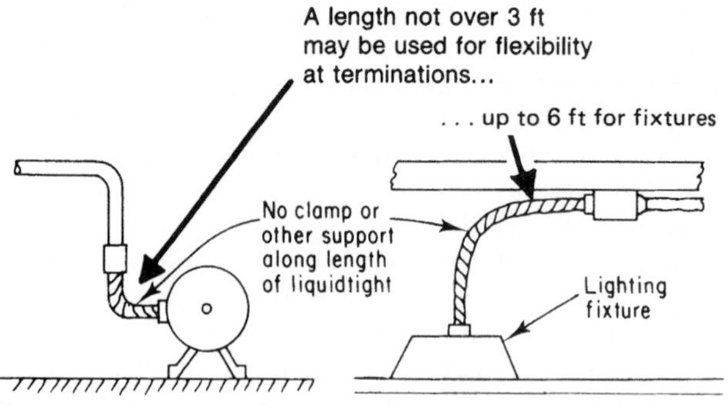
A length not over 3 ft
may be used for flexibility
at terminations...
. . . up to 6 ft for fixtures
No clamp or
other support
along length
of liquidtight
Lighting
fixture

Fig. 351-3. Unsupported length of liquidtight flex is O.K. at terminations. (Sec. 351-8.)

351-8. Supports. As shown in Fig. 351-3, Exception No. 2 permits a length of liquidtight flexible metal conduit not over 3 ft (914 mm) long to be used at terminals where flexibility is required without any need for clamping or strapping. Obviously, the use of flex requires this permission for short lengths without support.

In Exception No. 3, liquidtight metal flex is specifically recognized for a 4- to 6-ft (1.22- to 1.83-m) fixture "whip," without clamping of the flex. This covers a practice that has long been common. Either standard or liquidtight metal flex may be used to carry supply conductors to lighting fixtures—such as required by Sec. 410-67(c), where high-temperature wires must be run to a fixture terminal box.

351-9. Grounding. According to the basic rule, where flexible metal conduit and fittings have not been specifically approved as a grounding means, a separate grounding conductor (insulated or bare) shall be run inside the conduit [or outside, for lengths not over 6 ft (1.83 m)] and bonded at each box or similar equipment to which the conduit is connected.

The first sentence of this rule recognizes the metal in liquidtight flex as an equipment grounding conductor, as listed by UL:

> Where terminated in fittings investigated for grounding and where installed with not more than 6 feet (total length) in any ground return path, liquid-tight flexible metal conduit in the ⅜ and ½ in. trade sizes is suitable for grounding where used on circuits rated 20 amperes or less, and the ¾, 1, and 1¼ inch trade sizes are suitable for grounding where used on circuits rated 60 amperes or less. See the category "Conduit Fittings" (DWTT) with respect to fittings suitable as a grounding means.
>
> The following are not considered to be suitable as a grounding means:
>
> 1) The 1½ inch and larger trade sizes,
>
> 2) The ⅜ and ½ inch trade sizes where used on circuits rated higher than 20 amperes, or where the total length in the ground return path is greater than 6 feet,
>
> 3) The ¾, 1, and 1¼ inch trade sizes where used on circuits rated higher than 60 amperes, or where the total length in the ground return path is greater than 6 feet.

When a bonding jumper is required—such as for a length of the flex that is not over 6 ft (1.83 m) long but is over 1¼-in. (31.8-mm) size—the second sentence permits internal or external bonding of liquidtight flex as covered previously for standard flex and spelled out under Sec. 250-79. But for any size of flex run over 6 ft (1.83 m), *only* an internal equipment grounding conductor will satisfy this section and Sec. 250-91(b). The wording of Exception No. 1 focuses on a maximum total length of 6 ft (1.83 m) in any equipment "ground return path" where the liquidtight flex itself is used as the equipment grounding conductor. UL applications data in the *Electrical Construction Materials Directory* (the Green Book) make that same limitation to a "total length" of 6 ft (1.83 m).

As required by the last sentence following Exception 2, an equipment grounding conductor (or jumper) must *always* be installed for a length of liquidtight metal flex used to supply equipment that is not fixed in one place or location.

Thus Exception No. 2 actually modifies Exception No. 1, which states the conditions under which a 6-ft (1.83-m) or shorter length of liquidtight metal flex may be used for grounding through the metal of its own assembly, without need for a bonding wire. [Refer to Sec. 250-91(b), Exception No. 2.] Because experience has indicated many instances of loss of ground connection through the flex metal due to repeated movement of a flex whip supplying movable equipment, Exception No. 2 requires use of an equipment bonding jumper, either inside or outside the flex, in all cases where movable equipment is supplied—for assured safety of grounding continuity. The rule applies to those lengths of 3 ft (914 mm) or less that are permitted to be installed without clamping of the flex—as permitted by Sec. 351-8, Exception No. 2, because "flexibility is necessary."

351-10. Bends—Number in One Run. Figure 350-3 shows this rule.

351-22. Definition. Part **B** of Art. 351 covers liquidtight flexible nonmetallic conduit, which may be used for indoor or outdoor applications in residential, commercial, and industrial applications. It must not be used in any individual length over 6 ft (1.83 m) long (unless special approval is given because the "required degree of flexibility" necessitates a longer length) and is limited to a maximum of 2-in. (50.8-mm) size. Grounding requires a conductor within or outside the flex.

351-23. Use. Liquidtight flexible nonmetallic conduit may be used exposed or concealed and also may be used for direct burial in earth if "listed and marked for the purpose." This extends similar permission to liquidtight flexible *nonmetallic* conduit that was given for liquidtight flexible *metallic* conduit in the 1987 NEC. And Sec. 351-23(a) recognizes this nonmetallic flex for "concealed" as well as exposed locations.

351-24. Size. Although ½-in. trade size is the smallest recognized size of liquidtight flexible nonmetallic conduit for general use, Exception No. 1 notes that ⅜-in. liquidtight flexible metal conduit may be used for motor leads. This was added to coordinate with Sec. 430-145(b) for motors with detached junction boxes. The other two Exceptions allow 6-ft (1.83-m) lengths where "part of a listed assembly to supply lighting fixtures" and for electric signs.

ARTICLE 352. SURFACE METAL RACEWAYS AND SURFACE NONMETALLIC RACEWAYS

352-1. Use. At one time, this article was titled "Surface *Metal* Raceways." The article now includes both metallic and *nonmetallic* surface raceways (Fig. 352-1).

Right: Typical use of small metal surface raceway for extensions from existing receptacle outlets.

Below: Shallow switch or receptacle box for surface raceway.

Fig. 352-1. Surface raceway has become popular for new works as well as for modernization. (Sec. 352-1.)

352-2. Other Articles. In every type of wiring having a metal enclosure around the conductors, it is important that the metal be mechanically continuous in order to provide protection for the conductors and that the metal form a continuous electrical conductor of low impedance from the last outlet on the run to the cabinet or cutout box. A path to ground is thus provided through the box or cabinet, in case any conductor comes in contact with the metal enclosure, an outlet box, or any other fitting. See Sec. 250-91(b).

352-3. Size of Conductors. Manufacturers of metal surface raceways provide illustrations and details on wire sizes and conductor fill for their various types of raceway. It is important to refer to their specification and application data.

352-4. Number of Conductors In Raceways. The rules of conductor fill may now be applied to surface metal raceway in very much the same way as standard wireway. This rule applies wireway conductor fill and ampacity determination to any surface metal raceway that is over 4 sq in (2,580 sq mm) in cross section. As with wireway, if there are not more than 30 conductors in the raceway and they do not fill the cross-section area to more than 20 percent of its value, the conductors may be used without any conductor ampacity derating from Note 8 of Tables 310-16 through 310-19 (Fig. 352-2).

352-6. Combination Raceways. Metal surface raceways may contain separated systems as shown in Fig. 352-3.

ARTICLE 353. MULTIOUTLET ASSEMBLY

353-1. Other Articles. UL data are as follows:

> This Listing covers metal raceways with factory installed conductors and attachment plug receptacles without provision for field installation of additional conductors except where the product is marked to indicate the number, type, and size of additional conductors which may be field installed. Also covered are nonmetallic raceways with factory installed conductors and attachment plug receptacles either factory installed or separately Listed as Multioutlet Assembly Fittings for field installation.

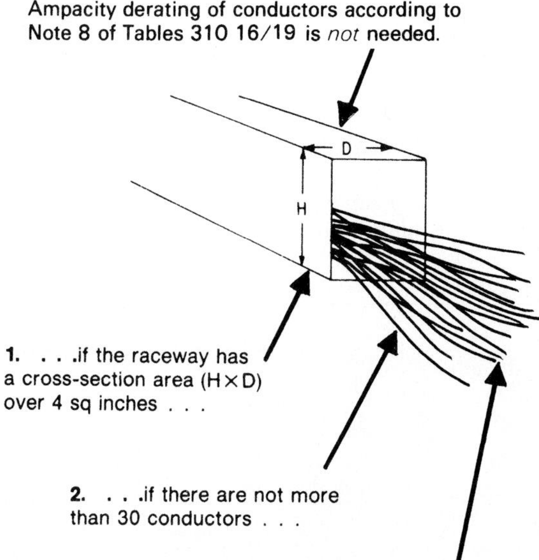
Ampacity derating of conductors according to Note 8 of Tables 310 16/19 is *not* needed.
D
H
1. . . .if the raceway has a cross-section area (H×D) over 4 sq inches . . .
2. . . .if there are not more than 30 conductors . . .
3. . . .and the sum of conductor cross-section areas does not exceed 20% of the interior cross-section area of the raceway.

Fig. 352-2. NEC rule permits conductor fill of metal surface raceway *without* ampacity derating of wires. (Sec. 352-4.)

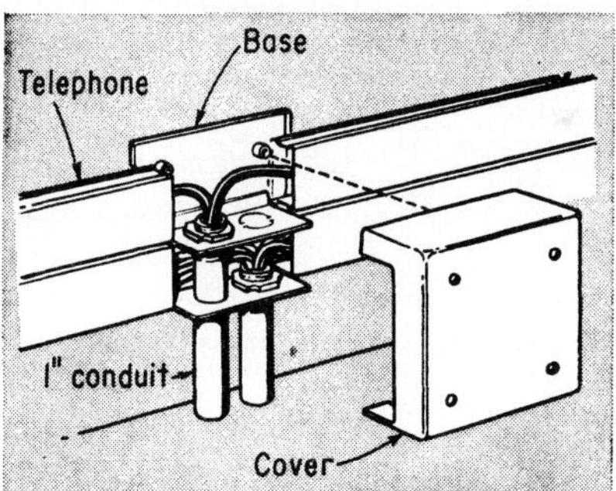
Base
Telephone
1" conduit
Cover

Fig. 352-3. For separating high and low potentials, combination raceway or tiered separate raceways may be used with barriered box assembly. (Sec. 352-6.)

Multioutlet Assemblies are for installation in accordance with Article 353 of the National Electrical Code.

353-2. Use. These assemblies are intended for surface mounting except that the metal type may be surrounded by the building finish or recessed so long as the front is not covered. The nonmetallic type may be recessed in baseboards.

In calculating the load for branch circuits supplying multioutlet assembly, see Sec. 220-3(c), Exception No. 1.

ARTICLE 354. UNDERFLOOR RACEWAYS

354-2. Use. Underfloor raceway was developed to provide a practical means of bringing conductors for lighting, power, and signaling systems to office desks and tables (Fig. 354-1). It is also used in large retail stores, making it possible to secure connections for display-case lighting at any desired location.

Fig. 354-1. Underfloor raceway system, with spaced grouping of three ducts (one for power, one for telephone, one for signals), is covered with concrete after installation on first slab pour. (Sec. 354-2.)

This wiring method makes it possible to place a desk or table in any location and it will always be over, or very near to, a duct line. The wiring method for lighting and power between cabinets and the raceway junction boxes may be conduit, underfloor raceway, wall elbows, and cabinet connectors.

354-3. Covering. The intent in paragraphs **(a)** and **(b)** is to provide a sufficient amount of concrete over the ducts to prevent cracks in a cement, tile, or similar floor finish. Figure 354-2 shows a violation. Two 1½- by 4½ in. (38- by 114-mm) underfloor raceways with 1-in. (25.4-mm)-high inserts are spaced ¾ in. (19 mm) apart by adjustable-height supports resting directly on a base floor-slab, as shown. After raceways are aligned, leveled, and secured, concrete fill is poured

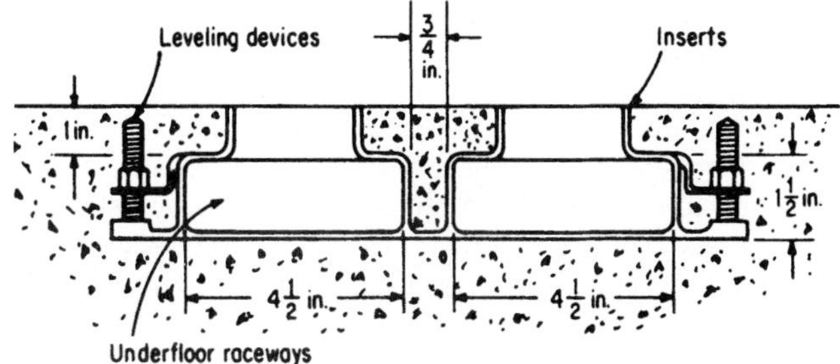

Fig. 354-2. The 1-in. (25.4-mm) cover is inadequate for raceways less than an inch apart. (Sec. 354-3.)

level with insert tops. But spacing between raceways must be at least 1 in. (25.4 mm); otherwise the concrete cover must be 1½ in. (38 mm) deep.

354-6. Splices and Taps. This section has a second paragraph that recognizes "loop wiring" where "unbroken" wires extend from underfloor raceways to terminals of attached receptacles, and then back into the raceway to other outlets. For purposes of this Code rule *only,* the "loop" connection method is not considered a splice or tap (Fig. 354-3).

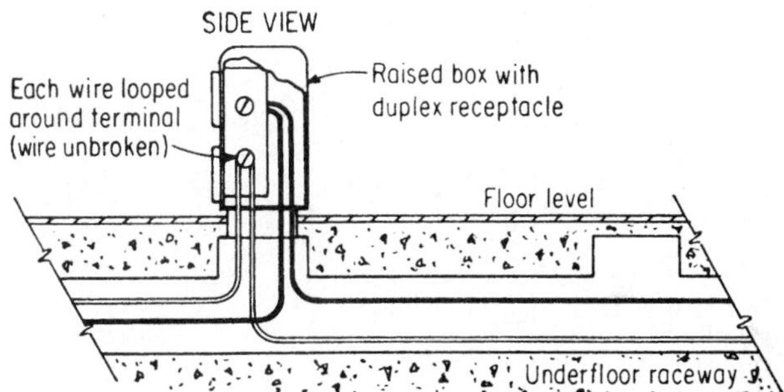

Fig. 354-3. "Loop" method permitted at outlets supplied from underfloor raceways. (Sec. 354-6.)

As noted in the Exception, splices and taps may be made in trench-type flush raceway with an accessible removable cover. The removable cover of the trench duct must be accessible after installation, and the splices and taps must not fill the raceway to more than 75 percent of its cross-section area.

ARTICLE 356. CELLULAR METAL FLOOR RACEWAYS

356-1. Definitions. This is a type of floor construction designed for use in steel frame buildings in which the members supporting the floor between the beams consist of sheet steel rolled into shapes which are so combined as to form cells, or closed passageways, extending across the building. The cells are of various shapes and sizes, depending upon the structural strength required.

The cellular members of this type of floor construction form raceways. A cross-sectional view of one type of cellular metal floor is shown in Fig. 356-1.

356-2. Uses Not Permitted. Connections to the ducts are made by means of *headers* extending across the cells. A header connects only to those cells which are to be used as raceways for conductors. Two or three separate headers, connecting to different sets of cells, may be used for different systems, for example, for light and power, signaling systems, and public telephones.

Figure 356-2 shows the cells, or ducts, with header ducts in place. By means of a special elbow fitting the header is extended up to a cabinet or distribution

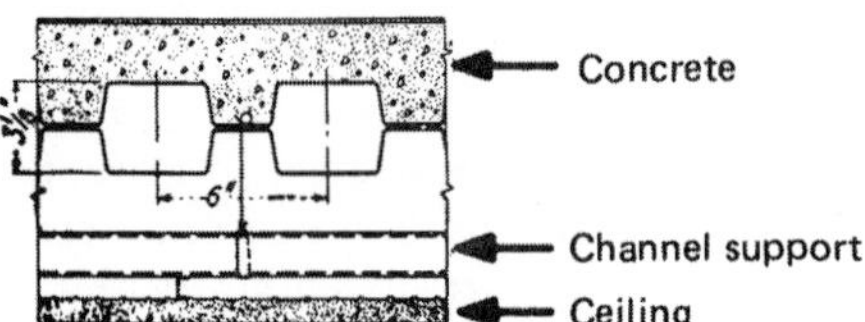

Fig. 356-1. Cross section of one type of cellular method floor construction. (Sec. 356-1.)

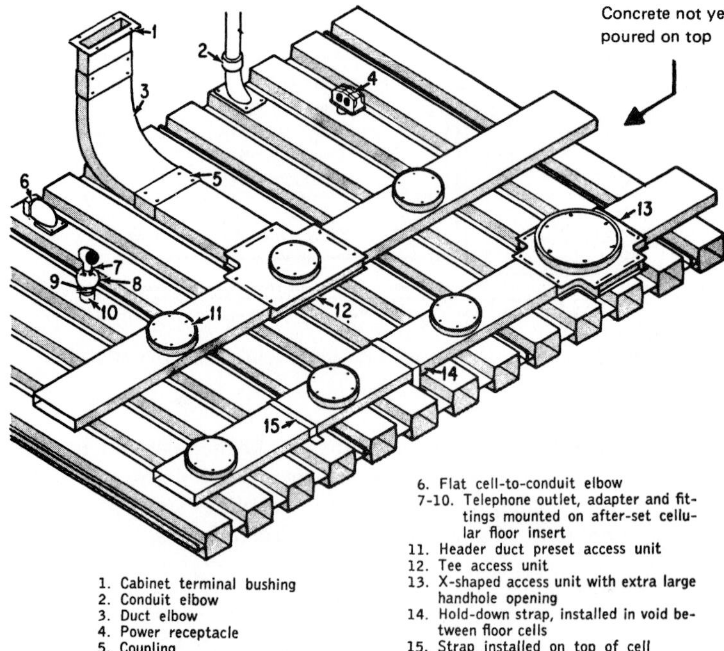

1. Cabinet terminal bushing
2. Conduit elbow
3. Duct elbow
4. Power receptacle
5. Coupling
6. Flat cell-to-conduit elbow
7-10. Telephone outlet, adapter and fittings mounted on after-set cellular floor insert
11. Header duct preset access unit
12. Tee access unit
13. X-shaped access unit with extra large handhole opening
14. Hold-down strap, installed in void between floor cells
15. Strap installed on top of cell

CELLULAR STEEL FLOOR contains unlimited number of channels for enclosing and isolating various electrical services. Wiring is routed from distribution panels to floor outlets through header ducts as shown.

Fig. 356-2. Components for electrical usage in cellular metal floor must be properly applied. (Sec. 356-2.)

center on a wall or column. A junction box or access fitting is provided at each point where the header crosses a cell to which it connects.

356-6. Splices and Taps. See Sec. 354-6.

356-8. Markers. The markers used with this system consist of special flat-head brass screws, screwed into the upper side of the cells and with their heads flush with the floor finish.

356-9. Junction Boxes. The fittings with round covers shown in Fig. 356-2 are termed *access fittings* by the manufacturer but actually serve as junction boxes. Where additional junction boxes are needed, a similar fitting of larger size is provided which may be attached to a cell at any point.

356-10. Inserts. The construction of an insert is shown in Fig. 356-3. A 1⅝-in. (41.2 mm)-diameter hole is cut in the top of the cell with a special tool. The lower end of the insert is provided with coarse threads of such form that the insert can be screwed into the hole in the cell, thus forming a substantial mechanical and electrical connection.

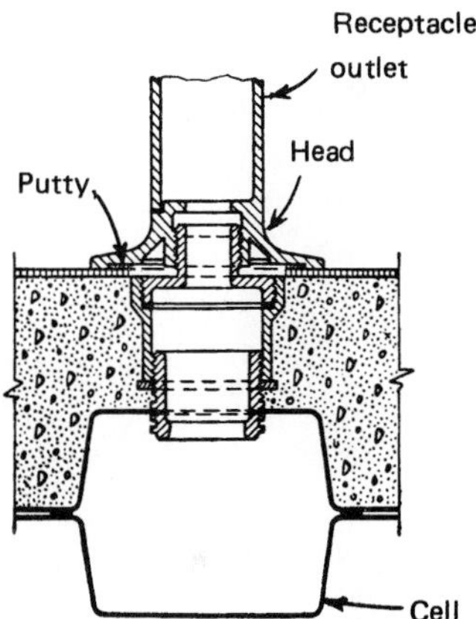

Fig. 356-3. Typical insert for connecting from cell to floor outlet assembly. (Sec. 356-10.)

ARTICLE 358. CELLULAR CONCRETE FLOOR RACEWAYS

358-1. Scope. The term *precast cellular concrete floor* refers to a type of floor construction designed for use in steel frame, concrete frame, and wall bearing construction, in which the monolithically precast reinforced concrete floor members form the structural floor and are supported by beams or bearing walls. The floor members are precast with hollow voids which form smooth round cells. The cells are of various sizes depending on the size of floor member used.

The cells form raceways which by means of suitable fittings can be adapted for use as underfloor raceways. A precast cellular concrete floor is fire resistant and requires no additional fireproofing.

358-5. Header. Connections to the cells are made by means of *headers* secured to the precast concrete floor, extending from cabinets and across the cells. A header connects only those cells which are used as raceways for conductors. Two or three separate headers, connected to different sets of cells, may be used for different systems, for example, for light and power, signaling, and telephones.

Figure 358-1 shows three headers installed, each header connecting a cabinet with separate groups of cells. Special elbows extend the header to the cabinet.

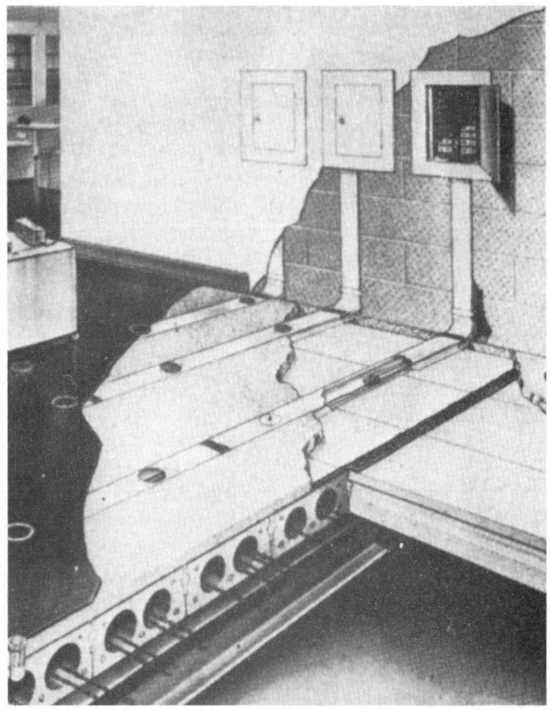

Fig. 358-1. Headers, flush with finished concrete pour, carry wiring to cells. (Sec. 358-5.)

358-7. Junction Boxes. Figure 358-2 shows how a JB must be arranged where a header connects to a cell.

358-8. Markers. Markers used with this system are special flat-head brass screws which are installed level with the finished floor. One type of marker marks the location of an access point between a header and a spare cell reserved for, but not connected to, the header. A junction box can be installed at the point located by the marker if the spare cell is needed in the future. The screw for this type marker is installed in the center of a special knockout provided in the top of the header at the access point. The second type of marker is installed over the center of cells at various points on the floor to locate and identify the cells below. Screws with specially designed heads identify the type of service in the cell.

358-9. Inserts. A 1⅞-in (47.5 mm)-diameter hole is cut through the floor and into the center of a cell with a concrete drill bit. A plug is driven into the hole

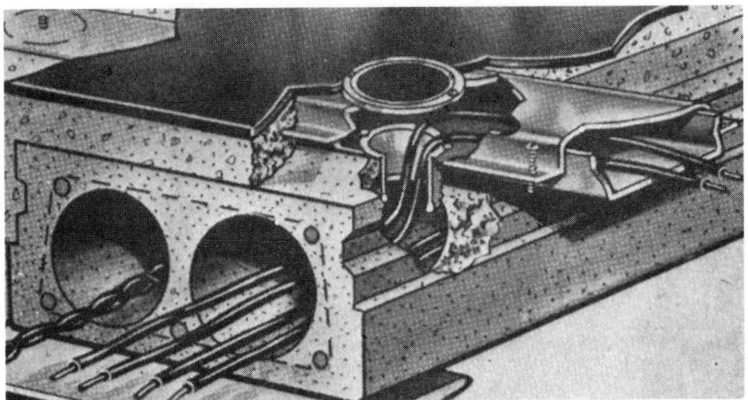

Fig. 358-2. Junction box is used to provide conductor installation from header to cell. (Sec. 358-7.)

and a nipple is screwed into the plug. The nipple is designed to receive an outlet with a duplex electrical receptacle or an outlet designed for a telephone or signal system.

ARTICLE 362. METAL WIREWAYS AND NONMETALLIC WIREWAYS

362-1. Definition. Metal wireways are sheet-metal troughs in which conductors are laid in place after the wireway has been installed as a complete system. Wireway is available in standard lengths of 1, 2, 3, 4, 5, and 10 ft (305, 610, and 914 mm and 1.22, 1.52, and 3.05 m), so runs of any exact number of feet can be made up without cutting the duct. The cover may be a hinged or removable type. Unlike auxiliary gutters, wireways represent a type of wiring, because they are used to carry conductors between points located considerable distances apart.

The purpose of a wireway is to provide a flexible system of wiring in which the circuits can be changed to meet changing conditions, and one of its principal uses is for exposed work in industrial plants. Wireways are also used to carry control wires from the control board to remotely controlled stage switchboard equipment. A wireway is approved for any voltage not exceeding 600 V between conductors or 600 V to ground. An installation of wireway is shown in Fig. 362-1.

362-2. Use. Figure 362-1 shows a typical Code-approved application of 4- by 4-in. (102- by 102-mm) wireway in an exposed location.

362-5. Number of Conductors. Wireways may contain up to 30 "current-carrying" conductors at any cross section (signal circuits and control conductors used for starting duty only between a motor and its starter are not "current-carrying" conductors). The total cross-sectional area of the group of

Fig. 362-1. Wireway in industrial plant—installed exposed, as required by Sec. 362-2 provides highly flexible wiring system that provides easy changes in the number, sizes, and routing of circuit conductors for machines and controls. Hinged covers swing down for ready access. Section 326-6 permits splicing and tapping in wireway. Section 362-11 covers use of conduit for taking circuits out of wireway. (Sec. 362-2.)

conductors must not be greater than 20 percent of the interior cross-sectional area of the wireway or gutter. And ampacity derating factors for more than three conductors do not apply to wireway the way they do to wires in conduit. However, if the derating factors from Note 8 of NE Code Tables 310-16 through 310-19 are used, there is no limit to the number of current-carrying wires permitted in a wireway or an auxiliary gutter. But, the sum of the cross-section areas of all contained conductors at any cross section of the wireway must not exceed 20 percent of the cross-section area of the wireway or auxiliary gutter. More than 30 conductors may be used under these conditions.

Exception No. 3 says that wireway used for circuit conductors for an elevator or escalator may be filled with any number of wires, occupying up to 50 percent of the interior cross section of the wireway, and no derating has to be made for fill.

The second sentence of the first paragraph has the effect of saying that any number of signal and/or motor control wires (even over 30) may be used in wireway provided the sum of their cross-section areas does not exceed 20 percent of wireway csa. And ampacity derating of those conductors is not required.

Figure 362-2 shows examples of wireway fill calculations. The example at the bottom shows a case where power and lighting wires (which *are* current-carrying wires) are mixed with signal wires. Because there are not over 30 power and light wires, no derating of conductor ampacities is needed. If, say, 31 power and light wires were in the wireway, then the power and light conductors would be subject to derating. If all 49 conductors were signal and/or con-

Basic rule

1. Any number of current-carrying conductors up to a maximum of 30, without derating.

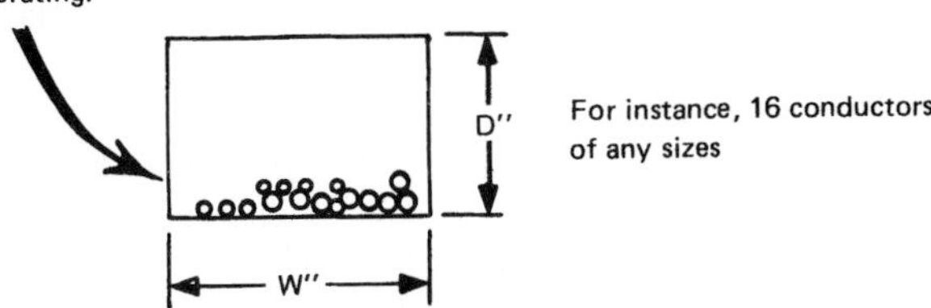
D''
W''
For instance, 16 conductors
of any sizes

2. The sum of the cross-section areas of all the conductors (from table 5 in Chap. 9 of the *NEC*) must not be more than 20% X W'' X D''

Note: Signal and motor control wires are not considered to be current-carrying wires. Any number of such wires are permitted to fill up 20% of wireway cross-section area.

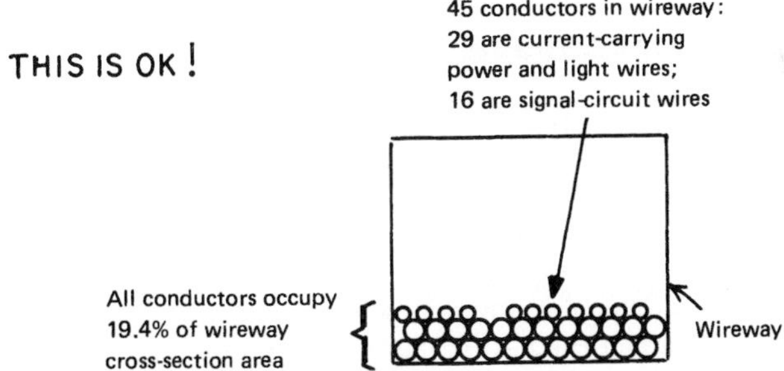
THIS IS OK!
45 conductors in wireway:
29 are current-carrying
power and light wires;
16 are signal-circuit wires
All conductors occupy
19.4% of wireway
cross-section area
Wireway

Fig. 362-2. Wireway fill and need for derating must be carefully evaluated. (Sec. 362-5.)

trol wires, no derating would be required. But, in all cases, wireway fill must not be over 20 percent.

362-6. Deflected Insulated Conductors. Deflected conductors in wireways must observe the rules on adequate enclosure space given in Sec. 373-6. This section is based on the following:

> Although wireways don't contain terminals or supplement spaces with terminals, pull boxes and conduit bodies don't either. This rule borrows language from both 374-9(d) and 370-28(a)(2), Exception, in an attempt to produce a consistent approach in the Code. Although in some cases the deflected conductors travel long distances in the wireway and are therefore easily inserted, in other cases the conductors are deflected again within inches of the first entry. The result is even more stress on the insulation than if they were entering a conduit body.

362-7. Splices and Taps. The conductors should be reasonably accessible so that any circuit can be replaced with conductors of a different size if necessary and so that taps can readily be made to supply motors or other equipment. Accessibility is ensured by limiting the number of conductors and the space they occupy as provided in Secs. 362-5 and 362-6.

362-8. Supports. Wireway must be supported every 5 ft (1.52 m). Wireway lengths over 5 ft (1.52 m) must be supported at *each end* or *joint, unless* listed for other support. In no case should the distance between supports for wireway exceed 10 ft (3.05 m).

362-11. Extensions from Wireways. Knockouts are provided in wireways so that circuits can be run to motors or other apparatus at any point. Conduits connect to such knockouts, as shown in Fig. 362-1.

Sections of wireways are joined to one another by means of flanges which are bolted together, thus providing rigid mechanical connection and electrical continuity. Fittings with bolted flanges are provided for elbows, tees, and crosses and for connections to cabinets. See Sec. 250-91(b).

Part **B** covers nonmetallic wireways. As given in Sec. 362-19, the maximum conductor fill is determined in the same manner as for metal wireways, but support requirements are slightly different. Sec. 362-22(a) requires support at 3-ft (914-mm) intervals, instead of 5-ft (1.52-m) ones.

ARTICLE 363. FLAT CABLE ASSEMBLIES

363-1. Definition. Type FC cable is a flat assembly with three or four parallel No. 10 special stranded copper conductors. The assembly is installed in an approved U-channel surface metal raceway with one side open. Then tap devices can be inserted anywhere along the run. Connections from tap devices to the flat cable assembly are made by "pin-type" contacts when the tap devices are fastened in place. The pin-type contacts penetrate the insulation of the cable assembly and contact the multistranded conductors in a matched phase sequence (phase 1 to neutral, phase 2 to neutral, and phase 3 to neutral).

Covers are required when the installation is less than 8 ft (2.44 m) from the floor. The maximum branch-circuit rating is 30 A.

Figure 363-1 shows the basic components of this wiring method.

363-3. Uses Permitted. Figure 363-2 shows a Type FC installation supplying lighting fixtures. As shown in the details, one tap device provides for circuit tap-off to splice to cord wires in the junction box; and the other device is simply a fitting to support the fixture from the lips of the channel.

ARTICLE 364. BUSWAYS

364-2. Definition. Busways consist of metal enclosures containing insulator-supported busbars. Varieties are so extensive that possibilities for 600-V distribution purposes are practically unlimited. Busways are available for either indoor or outdoor use as point-to-point feeders or as plug-in takeoff routes for power. Progressive improvements in busway designs have enhanced their electrical and mechanical characteristics, reduced their physical size, and simplified the methods used to connect and support them. These developments in turn have reduced installation labor to the extent that busways are most favor-

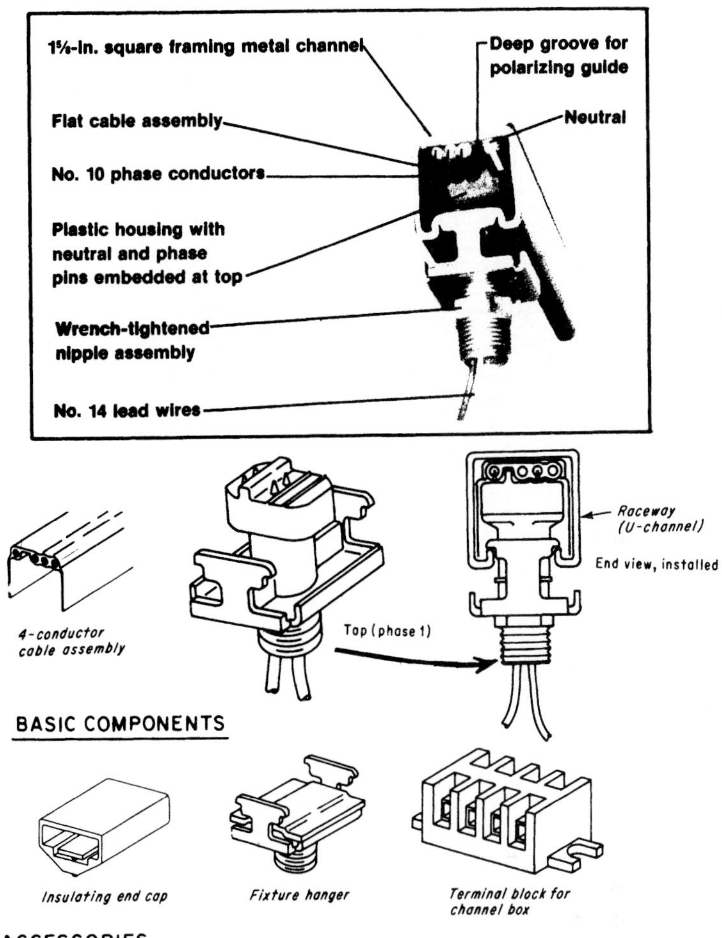

Fig. 363-1. Type FC wiring system uses cable in channel, with tap devices to loads. (Sec. 363-1.)

ably considered when it is required to move large blocks of power to loadcenters (via low-impedance feeder busway), to distribute current to closely spaced power utilization points (via plug-in busways), or to energize rows of lighting fixtures or power tools (via trolley busways).

Busways classed as indoor low-reactance assemblies can be obtained in small incremental steps up to 6,500 A for copper busbars and 5,000 A for aluminum. Enclosed outdoor busways are similarly rated. In the plug-in category, special assemblies are available up to 5,000 A, although normal 600-V AC requirements generally are satisfied by standard busways in the 225-to-1,000-A range. Where power requirements are limited, small compact busways are available with ratings from 250 down to 20 A.

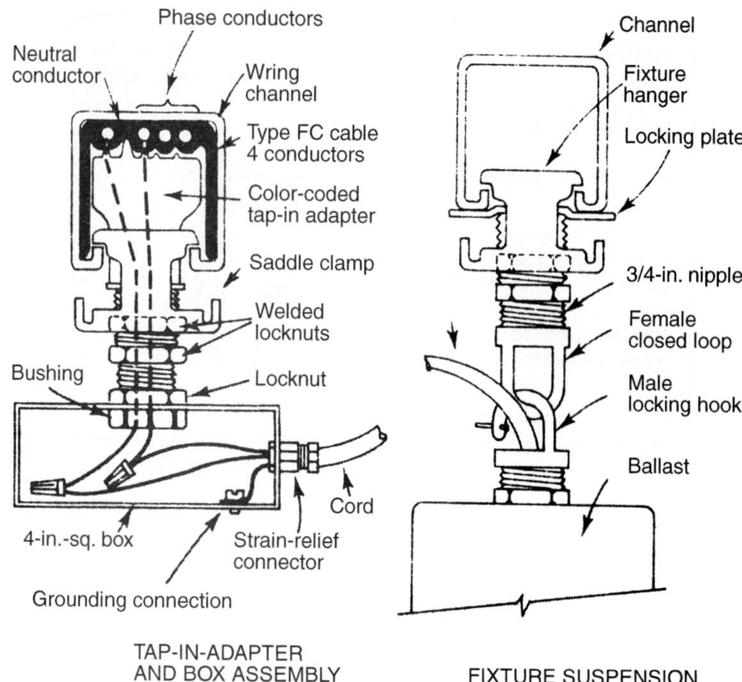

Fig. 363-2. Limited application of Type FC cable system includes use as branch-circuit wiring method to supply luminaires. (Sec. 363-3.)

Plug-in and clamp-on devices include fused and nonfusible switches and plug-in CBs rated up to about 800 A. Other plug-in devices include ground detectors, temperature indicators, capacitors, and transformers designed to mount directly on the busway.

364-4. Use. Figure 364-1 shows the most common way in which busways are installed—in the open.

Wiring methods above lift-out ceiling panels are considered to be "exposed"—because the definition of that word includes reference to "behind panels designed to allow access." This section calls for busways to be "located in the open and visible" but does permit busways above lift-out panels of a suspended ceiling, if means of access are provided. It limits such to totally enclosed, nonventilated busways, without plug-in switches or CBs on the busway and only in ceiling space that is not used for air handling. Figure 364-2 shows how other Code rules tie into this section. For instance, fuses and CBs that provide overcurrent protection required by the NE Code must generally be readily accessible—that is, they must be capable of being reached quickly (Sec. 240-24). Fuses and/or CBs are not readily accessible if it is necessary to get a portable ladder or stand on a chair or table to get at them. However, Exception No. 1 in Sec. 240-24 permits overcurrent devices to be used high up on a busway (as *required* by Sec. 364-12) where access to them could require use of

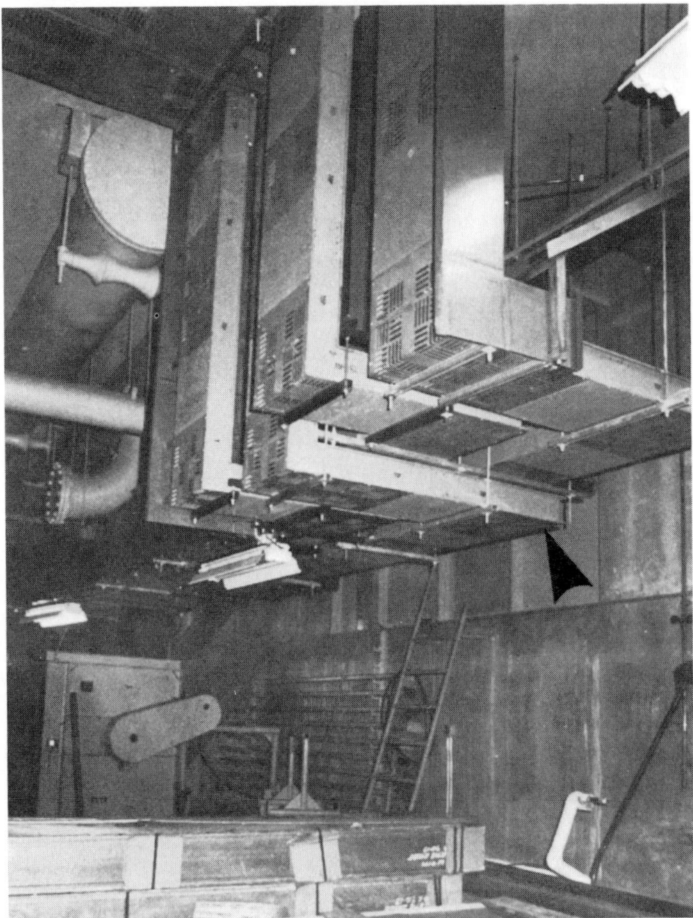

Fig. 364-1. Ventilated-type (with open grills for ventilation) busways may be used only "in the open" and must be "visible." Only the totally enclosed, nonventilating type may be used above a suspended ceiling. (Sec. 364-4.)

a ladder. But, as noted in part *b* of the Exception, totally enclosed nonventilated busways with no provisions for plug-in connections may be used in an air-handling space above a suspended ceiling. This Exception correlates to the permission for such busway use in spaces used for environmental air—as stated in Sec. 300-22(c).

Other data limiting application of busways are contained in the UL regulations on listed busways—all of which information becomes mandatory Code rules because of NEC Sec. 110-3(b). Such UL data are as follows:

Busways may carry various markings to indicate the intended use for which the busway was investigated and listed. Busways intended to supply and support industrial and commercial lighting fixtures is marked "Lighting Busway."

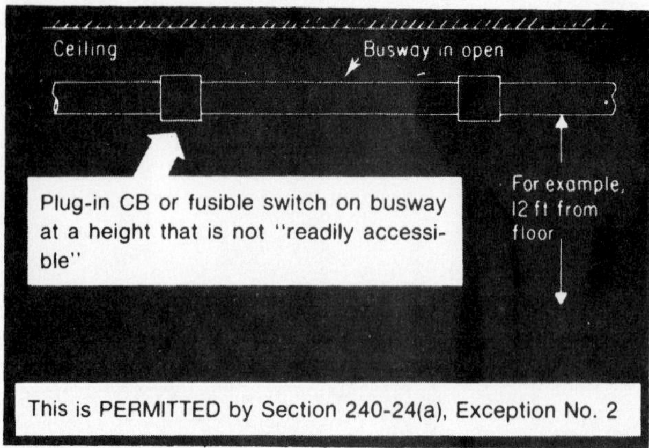

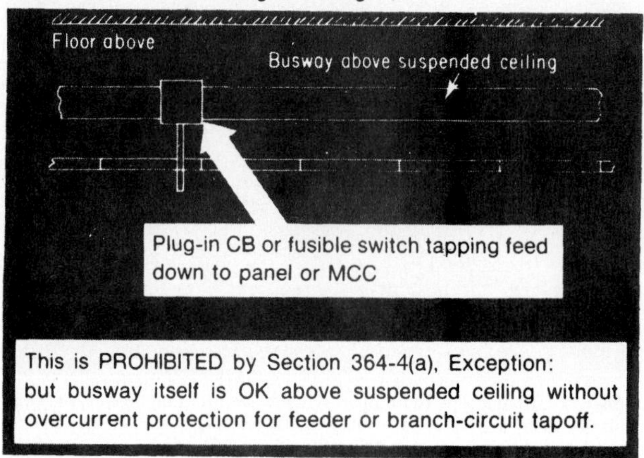

Fig. 364-2. Use of busways involves NEC rules on accessibility of overcurrent devices. (Sec. 364-4.)

Busways with sliding or other continuously movable means for tapping-off current to load circuits are marked "Trolley Busway." And, if the same busway is also acceptable for supporting and feeding lighting fixtures, it will *also* be marked "Lighting Busway." If a busway is designed to accept plug-in devices at any point along its length and is intended for general use, it is marked "Continuous Plug-in Busway." Busways marked "Lighting Busway" and protected by overcurrent devices rated in excess of 20 A are intended for use only with fixtures having heavy-duty lampholders—unless each fixture is equipped with additional overcurrent protection to protect the lampholders. This rule correlates with NE Code Secs. 210-23(b) and (c), which require only heavy-duty lampholders on branch circuits rated 30, 40, or 50 A. A "heavy-duty lampholder" is defined in Sec. 210-21(a) as one having "a rating of not less than 660 watts if of the admedium type and not less than 750 watts if of any other type." Medium-base lampholders—the ordinary 120-V incandescent lampholder—and all fluorescent lampholders are not "heavy-duty" lampholders and are acceptable on lighting busways fed by branch-circuit fuses or breakers rated not over 20 A. But, 50-A lighting busways, for instance, may be used as branch circuits to supply fixtures with incandescent, mercury-vapor, or other electric-discharge lamps with mogul-base, screw-shell lampholders or other lampholders rated "heavy-duty." In such applications, there is no need for additional overcurrent protection in each fixture. In such a case, the busway is used as a branch circuit in accordance with Sec. 364-12.

However, if fluorescent lighting fixtures are fed by a 50-A lighting busway and each fixture is individually fused at a few amps to protect the nonheavy-duty lampholders with the fuse in each fixture or its cord plug, that is permitted in the UL application information as well as by NE Code Sec. 364-12, Exception Nos. 2 or 3. In that case, the lighting busway is a feeder and each fixture tap is a "branch circuit." Note that in such a case the fuse in the plug or in the fixture is not "supplementary overcurrent protection," as described in Sec. 240-10—in spite of the conflict between Secs. 240-10 and 364-12, Exception No. 2. Figure 364-3 shows how those rules are applied. Note that the details involved are related to Sec. 210-21(a)—which prohibits fluorescent lampholders (nonheavy-duty) on circuits rated over 20 A—and Sec. 364-12, Exception No. 2—which identifies the fuse in the cord plug or in the fixture as "branch-circuit overcurrent device."

364-5. Support. As shown in Fig. 364-4, busway risers may be supported by a variety of spring-loaded hangers, wall brackets, or channel arrangements where busways pierce floor slabs or are supported on masonry walls or columns. As shown in Fig. 364-5, spring mounts for vertical busways may be located at successive floor-slab levels or, as indicated, supported by wall brackets located at intermediate elevations. Springs provide floating cradles for absorbing transient vibrations or physical shocks. Fire-resistant material is packed into space between the busway and the edges of slab-piercing throat.

364-6. Through Walls and Floors. Figure 364-6 shows a violation of this section, which requires a busway to be totally enclosed within the floor slab and for 6 ft (1.83 m) above it.

364-8. Branches from Busways. Figure 364-7 shows feeds into and out of a busway. The rule here requires that a cord connecting to a plug-in switch or CB on a busway must be supported by a "tension take-up device on the cord."

364-10. Rating of Overcurrent Protection—Feeders. The rated ampacity of a busway is fixed by the allowable temperature rise of the conductors. The ampacity can be determined in the field only by reference to the nameplate.

364-11. Reduction in Ampacity Size of Busway. Overcurrent protection—either a fused-switch or CB—is usually required in each busway subfeeder tapping power from a busway feeder of higher ampacity, protected at the higher ampacity. This is necessary to protect the lower current-carrying capacity of the subfeeder and should be placed at the point at which the subfeeder connects into the feeder. However, the Exception to this section provides that overcurrent protection may be omitted where busways are reduced in size, if the smaller

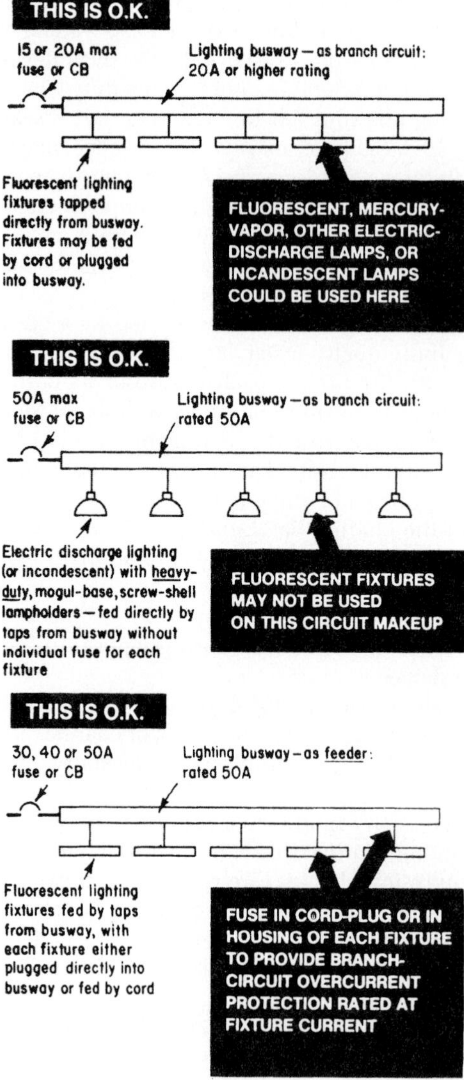

Fig. 364-3. These applications involve UL data and several Code sections. (Sec. 364-4.)

busway does not extend more than 50 ft (15.24 m) and has a current rating at least equal to one-third the rating or setting of the overcurrent device protecting the main busway feeder (Figs. 364-8 and 364-9), but *only* at an "industrial establishment." For all other installations, the basic rule for overcurrent protection at the point where the busway is reduced in size must be satisfied.

Where the smaller busway is kept within the limits specified, the hazards involved at industrial installations are very slight and the additional cost of

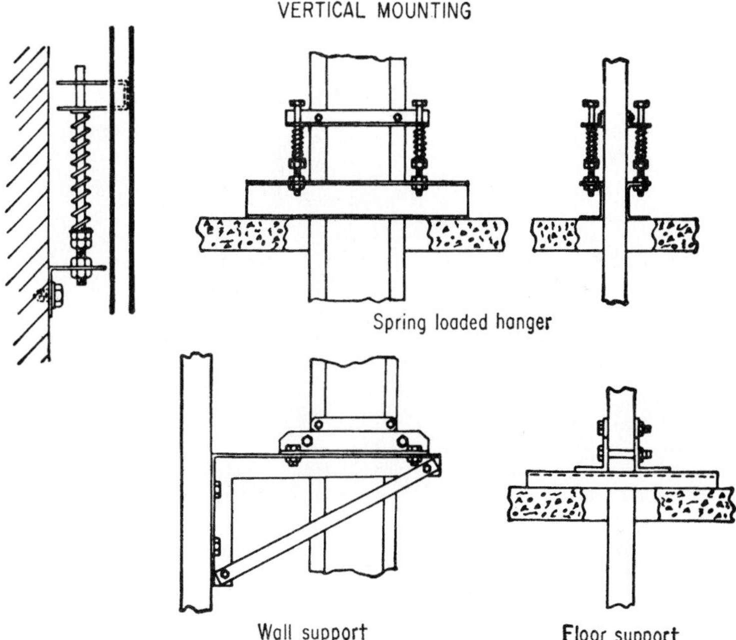

Fig. 364-4. Vertical busways runs should be supported at least every 16 ft (4.88 m). (Sec. 364-5.)

Fig. 364-5. Opening for a busway riser through a slab must be closed off, as required by Sec. 300-21. (Sec. 364-5.)

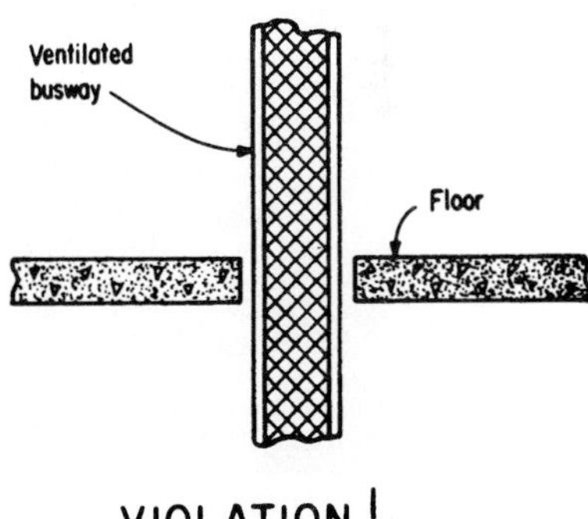

Fig. 364-6. Ventilated busways may not be used through a floor slab and for 6 ft (1.83 m) above the floor. (Sec. 364-6.)

Circuits fed from busway may be run in any conventional wiring method—such as EMT or rigid conduit (left) or as "suitable cord," such as "bvus-drop" cable down to machines. And cable-tap-boxes may be used (arrow at right) to connect feeder conductors that supply power to busway.

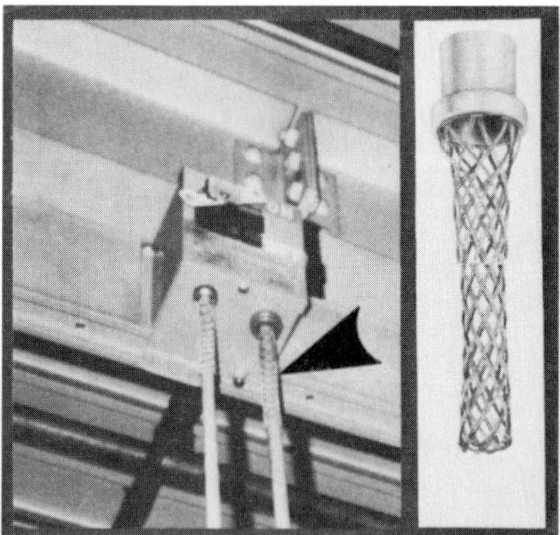

SOME TYPE of tension-relief device must be used on bus-drop cable or other suitable cord where it connects to a plug-in switch or CB on busway. Photo shows strain relief connector with mesh grip (arrow) on cord to bus-tap CB, which is equipped with hook-eye lever mechanism to provide operation of the CB by a hookstick from floor level—as required by Sec. 380-8(a), Ex. No. 1 and Sec. 364-12.

Fig. 364-7. Wiring details at busway connections must be carefully observed. [Sec. 364-8.]

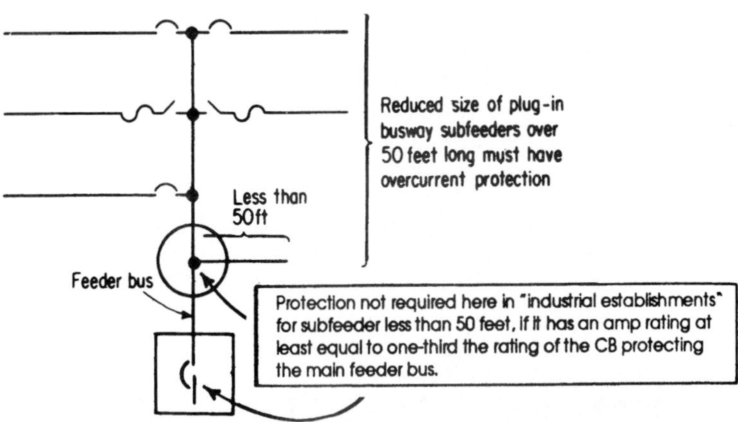

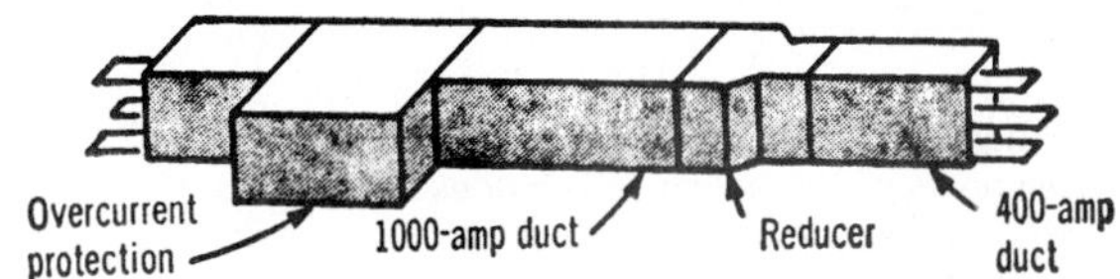

Fig. 364-8. A busway subfeeder may sometimes be used without protection. (Sec. 364-11.)

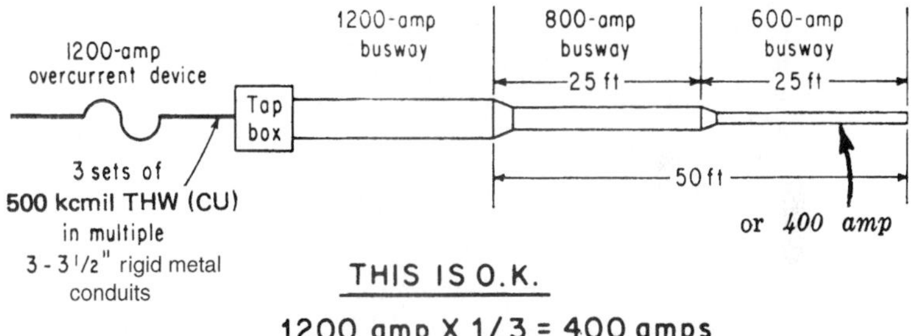

Fig. 364-9. Total length of a reduced busway is not over 50 ft (15.24 m). (Sec. 364-11.)

providing overcurrent protection at the point where the size is changed is not considered as being warranted.

364-12. Feeder or Branch Circuits. The rules of this section are interrelated with those of Secs. 240-24 and 380-8. The basic rule of this section makes it clear that branch circuits or subfeeders tapped from busway must have overcurrent protection on the busway at the point of tap. And if they are out of reach from the floor, all fused switches and CBs must be provided with some means for a person to operate the handle of the device from the floor (hookstick, chain operator, rope-pull operator, etc.).

Although no definition is given for "out of reach" from the floor, the wording of Sec. 380-8(a) can logically be taken to indicate that a switch or CB *is* "out of reach" if the center of its operating handle, when in its highest position, is more than 6½ ft (1.98 m) above the floor or platform on which the operator would be standing. Thus, a busway over 6½ ft (1.98 m) above the floor would require some means (hookstick, etc.) to operate the handles of any switches or CBs on the busway.

Figure 364-10 relates the rules of Sec. 240-24, Exception No. 1 to Sec. 364-12—with the rule of Sec. 240-24 *permitting* overcurrent devices to be "not readily accessible" when used up on a high-mounted busway and Sec. 364-12 *requiring* such protection to be mounted on the busway. To get at overcurrent protection in either case, personnel might have to use a portable ladder or chair or some other climbing technique. Again, 6 ft 7 in. (2.0 m) could be taken as the height above which the overcurrent protection is not "readily accessible"—or the height above which the Code considers that some type of climbing technique (ladder, chair, etc.) may be needed by some persons to reach the protective device.

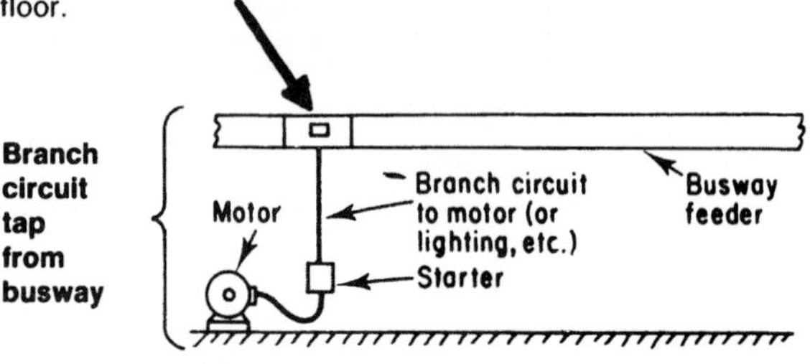
NE Code Section 364-12—Plug-in connection for tapping-off branch circuit shall contain overcurrent protection, which does not have to be within reach of person standing on floor.
Branch circuit tap from busway
Motor
Branch circuit to motor (or lighting, etc.)
Starter
Busway feeder

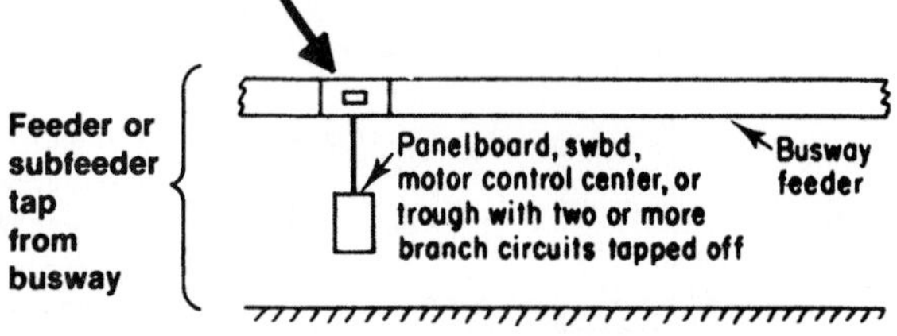
Plug-in connection for tapping-off a feeder or subfeeder shall contain overcurrent protection, which does not have to be within reach of person standing on floor.
Feeder or subfeeder tap from busway
Panelboard, swbd, motor control center, or trough with two or more branch circuits tapped off
Busway feeder

Fig. 364-10. Protection must always be used on busway for these taps—regardless of busway mounting height. (Sec. 364-12.)

Then, where the plug-in switch or CB on the busway is "out of reach" from the floor [that is, over 6½ ft (1.98 m) above the floor], provision must be made for operating such switches or CBs from the floor, as shown in Fig. 364-11. The plug-in switch or CB unit must be able to be operated by a hookstick or chain or rope operator if the unit is mounted out of reach up on a busway. Section 380-8 says all busway switches and CBs must be operable from the floor. Refer to Sec. 380-8. Figure 364-12 shows a typical application of hookstick-operated disconnects.

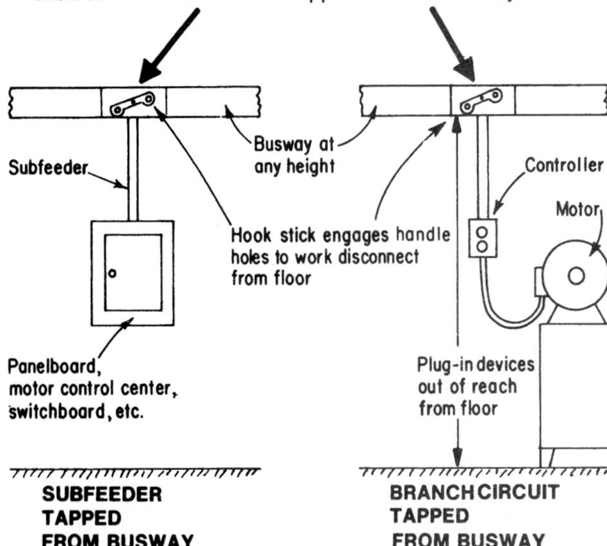
Plug-in device must be externally operable circuit breaker or fused switch that provides overcurrent protection for the subfeeder or branch circuit tapped from the busway.
Busway at any height
Subfeeder
Controller
Motor
Hook stick engages handle holes to work disconnect from floor
Panelboard, motor control center, switchboard, etc.
Plug-in devices out of reach from floor
SUBFEEDER TAPPED FROM BUSWAY
BRANCH CIRCUIT TAPPED FROM BUSWAY

Fig. 364-11. Busway plug-in devices must be operable from the floor or platform where operator stands. (Sec. 364-12.)

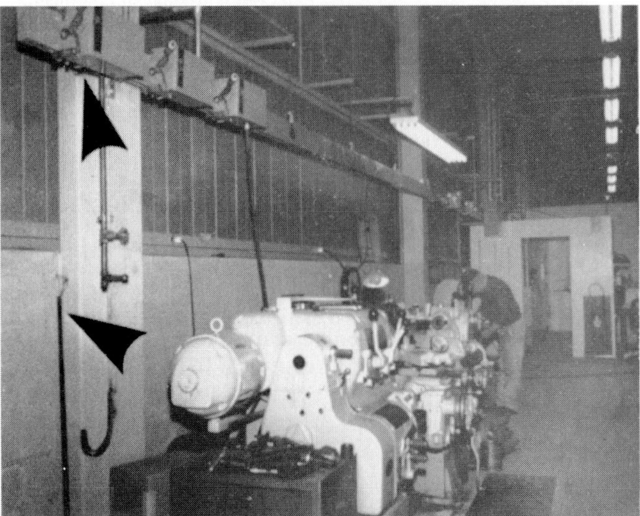

Fig. 364-12. Disconnects mounted up on the busway (top arrow) are out of reach from the floor but do have hook-eye lever operators to provide operation by person standing in front of machines. Although the NEC does not literally require ready availability of a hookstick, it is certainly the intent of the Code that one be handy (lower arrow). (Sec. 364-12.)

Figure 364-13 shows an application that has caused controversy because Sec. 364-12 says that any busway used as a feeder must have overcurrent protection on the busway for any subfeeder or branch circuit tapped from the busway. Therefore, use of a cable-tap box on busway without overcurrent protection could be ruled a Code violation. It can be argued that the installation shown—a 10- or 25-ft (3.05- or 7.62-m) tap without overcurrent protection on the busway—is covered by Exception No. 1 of that section, which recognizes taps as permitted in Sec. 240-21—including 10- and 25-ft (3.05- and 7.62-m) taps. But, as is now clearly spelled out in Exception No. 1 to Sec. 364-12, busways may be tapped as would any feeder. It is to be viewed simply as a conductor. Then, the rules given in the referenced parts of Sec 240-21 must be satisfied. Therefore, the application shown in Fig. 364-13 does satisfy the code.

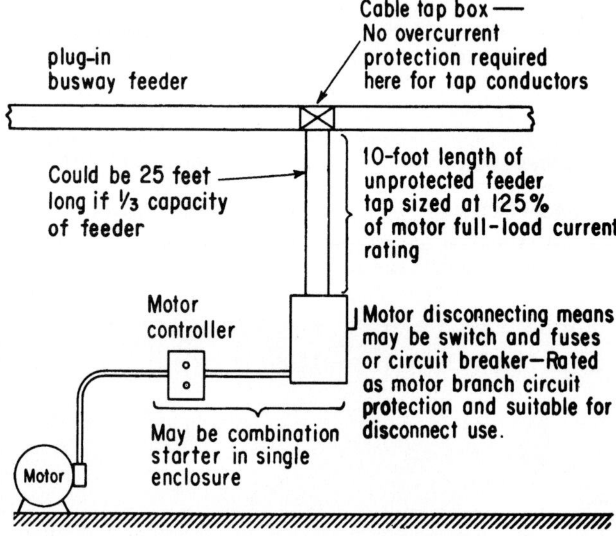

Fig. 364-13. This use of unprotected tap from busway conflicts with Sec. 364-12. (Sec. 364-12.)

364-13. Rating of Overcurrent Protection—Branch Circuits. Refer to data on busways on lighting branch circuits in Sec. 364-4 and Fig. 364-3.

ARTICLE 365. CABLEBUS

Cablebus is an approved assembly of insulated conductors mounted in "spaced" relationship in a *ventilated* metal protective supporting structure including fittings and conductor terminations. In general, cablebus is assembled at the point of installation from components furnished by the manufacturer.

Field-assembly details are shown in Fig. 365-1. First, the cablebus framework is installed in a manner similar to continuous rigid cable support systems. Next, insulated conductors are pulled into the cablebus framework. Then the conductors are supported on special insulating blocks at specified intervals. And finally, a removable (ventilated) top is attached to the framework.

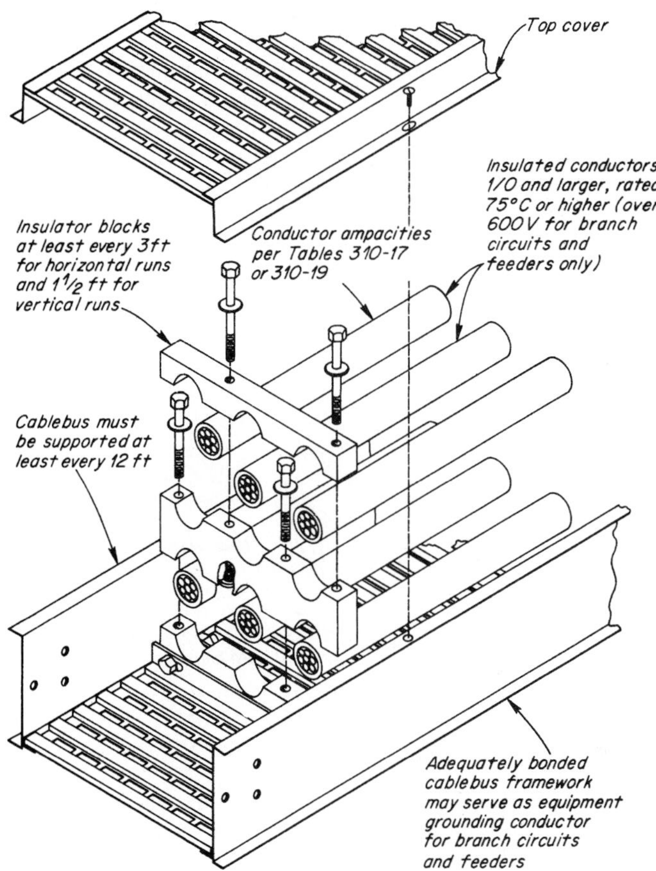

Fig. 365-1. Cablebus systems are field assembled from manufactured components. (Sec. 365-1.)

ARTICLE 370. OUTLET, DEVICE, PULL AND JUNCTION BOXES, CONDUIT BODIES AND FITTINGS

370-1. Scope. This rule makes clear that Art. 370 regulates use of conduit bodies when they are used for splicing, tapping, or pulling conductors. And this article does refer specifically to conduit bodies, to more effectively distinguish rules covering “boxes,” “conduit bodies,” and “fittings.” The rules of Art. 370

must be evaluated in accordance with the definitions given in Art. 100 for "conduit body" and "fitting." Capped elbows and SE elbows are "fittings" and are not "conduit bodies" and must not contain splices, taps, or devices. The pieces of equipment described in Art. 370 tie into Sec. 300-15, "Boxes or fittings—where required."

370-2. Round Boxes. The purpose of this rule is to require the use of rectangular or octagonal metal boxes having, at each knockout or opening, a flat bearing surface for the locknut or bushing or connector device to seat against a flat surface. But, round outlet boxes may be used with nonmetallic-sheathed cable because the cable is brought into the box through a knockout, without the use of a box connector to secure the cable to the box. However, the Exception to Sec. 370-17(c) permits only boxes of "nominal size 2¼ in. by 4 in." the so-called single gang boxes to be used without securing the NM or NMC cable to the box itself—as long as it is stapled to the stud or joist within 8 in. (203 mm) of the box. Because "round" boxes are *not* "single gang" boxes, it appears that all such round outlet boxes must be equipped with cable clamps to satisfy the Exception to Sec. 370-17(c) (Fig. 370-1). Shallow metal boxes with internal clamps for NM cable are acceptable as round boxes.

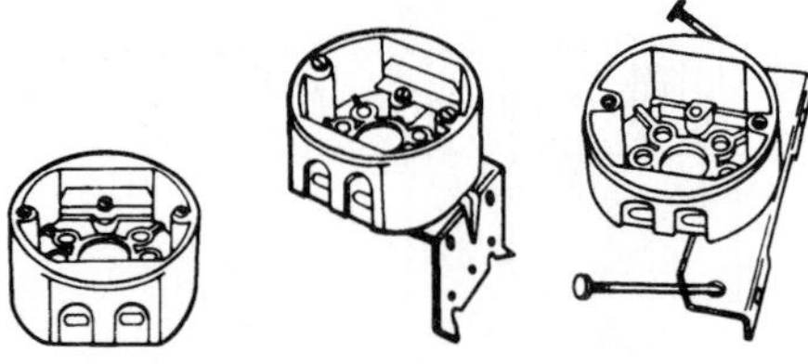

Round nonmetallic outlet boxes may be used only with NM cable

Fig. 370-1. Round boxes may be used only for connecting cables with internal clamps—such as NM or BX cable. (Sec. 370-2.)

370-3. Nonmetallic Boxes. Growth in the application of nonmetallic boxes over past years is the basis for the two Exceptions to this section, which regulate the conditions under which nonmetallic boxes may be used with metal raceways or metal-sheathed cable. The need and popularity of these boxes developed out of industrial applications where corrosive environments dictated their use to resist the ravages of various punishing atmospheres. In many applications it is desirable to use nonmetallic boxes along with plastic-coated metal conduits for a total corrosion-resistant system. Such application is recognized by the Code in the Exceptions of this section, although a limitation is placed requiring "internal" or "integral" bonding means in such boxes (Fig. 370-2).

According to the basic rule in the first sentence of this rule, nonmetallic boxes are permitted to be used only with open wiring on insulators, concealed knob-and-tube wiring, nonmetallic sheathed cable, electrical nonmetallic tubing, and rigid nonmetallic conduit (any "nonmetallic raceways"). Exception

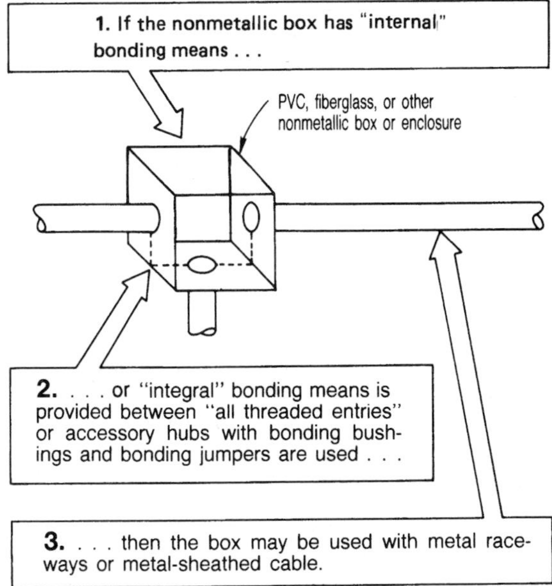
1. If the nonmetallic box has "internal"
bonding means . . .
PVC, fiberglass, or other
nonmetallic box or enclosure
2. . . . or "integral" bonding means is provided between "all threaded entries" or accessory hubs with bonding bushings and bonding jumpers are used . . .
3. . . . then the box may be used with metal raceways or metal-sheathed cable.

Fig. 370-2. Nonmetallic boxes are recognized for use with metal raceways and metal-sheathed cable. (Sec. 370-3.)

No. 1 requires internal bonding means in such boxes used with metal cable or raceways. The permission used to apply only to nonmetallic boxes sufficiently large—that is, over 100 cu in. (1,640 cu mm). Now, *any* size of box—PVC boxes, fiberglass boxes, or other nonmetallic boxes or enclosures—may be used with metal raceways or metal-sheathed cable. Exception No. 2 requires that "integral" bonding means between all "threaded" raceway and cable entries must be provided in the box for all metal conduits or metal-jacketed cables. That is, the grounding continuity from each raceway entry to each other raceway entry must be provided by using metal hubs and bonding bushings with lugs for the bonding jumper. And, as further specified in Exception No. 2, there must be a "provision for attaching an equipment grounding jumper inside the box." That requirement assures the safety of effective equipment grounding where metal raceway systems are used with nonmetallic boxes.

As worded, the manufacturer of the nonmetallic box must provide the necessary bonding means for all boxes with "threaded entries" for raceway and/or cable. The wording of this **Code** rule does recognize the use of accessory hubs with ground lugs to achieve equipment grounding continuity through such boxes. Such application does provide the required grounding continuity and is available for that purpose from the box manufacturers.

370-4. Metal Boxes. With a metal box in contact with metal walls or ceilings covered with metal, or with metal lath or with conductive thermal insulation, a stray current may flow to ground through an unknown path if a "hot" wire should accidentally become grounded on the box. To prevent this, the box must be effectively grounded by means of a separate grounding conductor.

370-15. Damp, Wet, or Hazardous (Classified) Locations. "Weatherproof" is defined as meaning "so constructed or protected that exposure to the weather will not interfere with its successful operation." A box or fitting may be considered weatherproof when so made and installed that it will exclude rain and snow. Such a box or fitting need not necessarily be sealed against the entrance of moisture.

The left part of Fig. 370-3 shows a fitting which is considered as weatherproof because the openings for the conductors are so placed that rain or snow

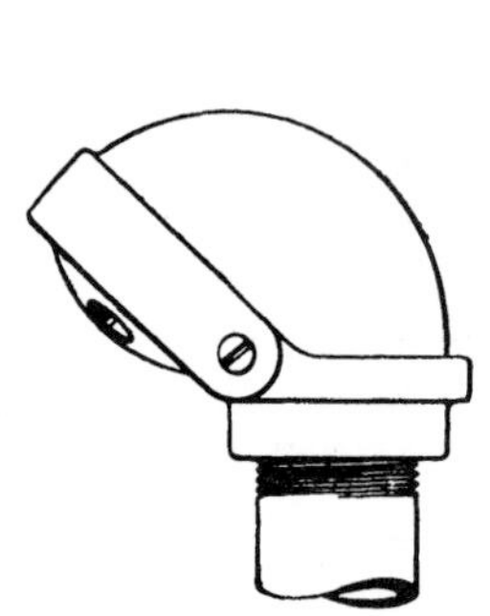

Fitting for use at the outer end of a service conduit.

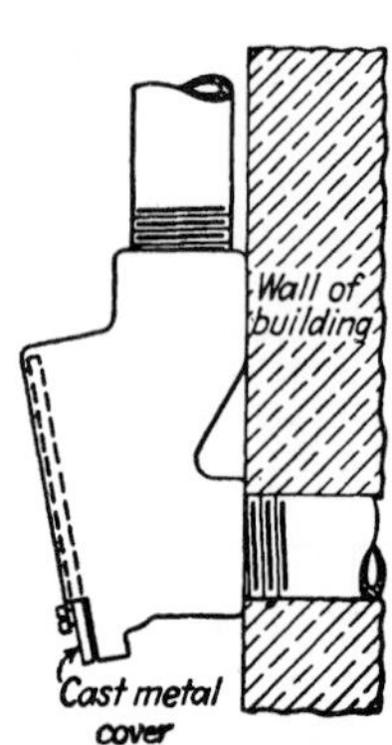

Type LB conduit fittting used where a service conduit passes through a building wall. See Sec. 370-18.

Fig. 370-3. Fittings must be suited to use in wet locations. (Sec. 370-15.)

cannot enter the fitting. On the right, it shows a fitting made weatherproof by means of a metal cover that slides under flanges on the face of the fitting, and, as required by Sec. 230-53, an opening is provided through which any moisture condensing in the conduit can drain out.

See the definitions of "wet locations" and "damp locations" in Art. 100.

Weatherproof boxes are for use in "wet locations" as defined by the NE Code. In "damp locations," boxes must be "located or equipped" to prevent water from entering or accumulating in the box. Boxes with threaded conduit hubs will normally prevent water from entering except for condensation within the box or connected conduit.

Caution: Extreme care must be exercised in correlating UL and NE Code rules on the use of boxes and enclosures in damp or wet locations because of uncertainties about their definitions. UL requires a weatherproof box for wet locations (such as outdoors exposed to rain). Weatherproof, in the NE Code definition, means only that it must be constructed or protected so that exposure to weather will not interfere with operation of contained equipment and does *not* mean that entry of water must be excluded. NE Code Sec. 370-15 on use of outlet boxes requires that boxes in *either damp or wet* locations must be "placed or equipped" to prevent entry of any moisture; or, if water does enter, the box must be drained so that water will not accumulate within the box. In damp locations (but not in wet locations), UL requires boxes to be "located or equipped" to prevent entry of water into the box.

From the above, it could be argued that a UL weatherproof box, which is intended for wet locations, does not satisfy NE Code Sec. 370-15 because that section requires exclusion of moisture from all boxes in wet (and damp) locations, and weatherproof boxes do not necessarily exclude moisture. It also could be contended that UL rules requiring exclusion of water from boxes in damp locations are more strict than the rule on use of weatherproof boxes in wet locations.

However, the last sentence of NE Code Sec. 370-15 says that boxes in wet locations "shall be approved for the purpose." Because UL says "weatherproof boxes are intended for use in wet locations," it seems clear that such usage is approved.

370-16. Number of Conductors in Outlet, Device, and Junction Boxes, and Conduit Bodies. Note that motor terminal housings are excluded from the rules on box conductor fill. And where any box or conduit body contains No. 4 or larger conductors, all the requirements of Sec. 370-28 on pull boxes must be satisfied. Refer to Sec. 370-28 for applications of conduit bodies as pull boxes.

Selection of any outlet or junction box for use in any electrical circuit work must take into consideration the maximum number of wires permitted in the box by Sec. 370-16. Safe electrical practice demands that wires *not* be jammed into boxes because of the possibility of nicks or other damage to insulation—posing the threat of grounds or shorts.

This section is broken down into three subparts. Part **(a)** establishes the volume of a box. Part **(b)** describes the method for determining how much volume is used by the various conductors, devices, etc. Part **(c)** applies to conduit bodies, only.

As stated in part **(a)(1)** of this section, Table 370-16(a) shows the maximum number of wires permitted in the *standard* metal boxes listed in that table. But that table applies only where all wires in a given box are all of the same size, i.e., all No. 14 or all No. 12, etc. Table 370-16(b) is provided for sizing a box where all the wires in the box are not the same size, by using so much cubic-inch space for each size of wire.

Table 370-16(a) includes the maximum number of No. 18 and No. 16 conductors that may be used in various sizes of boxes, and Table 370-16(b) gives the required box space for those sizes of conductors. Because of the extensive use of No. 18 and No. 16 wires for fixture wires and for control, signal, and communications circuits, these data are needed to assure safe box fill for modern electrical systems.

As stated in part **(a)(2)**, all other boxes—nonstandard, nonmetallic, or those metal boxes covered by part **(a)(1)** that are stamped with their cubic-inch capacity by the manufacturer—must consider their volume to be that which is stamped on them. And, as stated in the first paragraph of Sec. 370-16, the value of volume [part **(a)**] must never be less than the fill [part **(b)**].

Part **(a)(2)** of Sec. 370-16 covers boxes—metal and nonmetallic—that are not listed in Table 370-16(a) and conduit bodies with provision for more than two conduit entries (cross and T conduit bodies). And the basic way of determining correct wire fill is to count wires in accordance with the intent of Sec. 370-16(b) and then calculate the required volume of the box or conduit body by totaling up the volumes for the various wires from Table 370-16(b).

Part **(a)(2)** covers wire fill for metal boxes, up to 100 cu in. (1,640 cu mm) volume, that are not listed in Table 370-16(a) and for nonmetallic outlet and junction boxes. Although Code rules have long regulated the maximum number of conductors permitted in metal wiring boxes [such as given in Table 370-16(a)], there was no regulation on the use of conductors in nonmetallic device boxes up to the 1978 NEC. Now Sec. 370-16(a)(2) requires that *both* metal boxes not listed in Table 370-16(a) and nonmetallic boxes be durably and legibly marked by their manufacturer with their cubic inch capacities to permit calculation of the maximum number of wires that the Code will permit in the box. Calculation of the conductor fill for these boxes will be based on the marked box volume and the method of counting conductors set forth in Sec. 370-16(b). The conductor volume will be taken at the values given in Table 370-16(b), and deductions of space as required for wiring devices or for clamps must be made in accordance with the rules of Sec. 370-16(b). This requirement for marking of both metal and nonmetallic boxes arises from the wording of Sec. 370-16(a)(2), which refers to boxes other than those described in Table 370-16(a) and to nonmetallic boxes.

As shown in Fig. 370-4, a nonmetallic box for a switch has two 14/2 NM cables, each with a No. 14 ground. The wire count is four No. 14 insulated wires, plus two for the switch to be installed, and one for the two ground wires. That is a total of seven No. 14 wires. From Table 370-16(b), at least 2 cu in. of box volume must be allowed for each No. 14. This box must, therefore, be

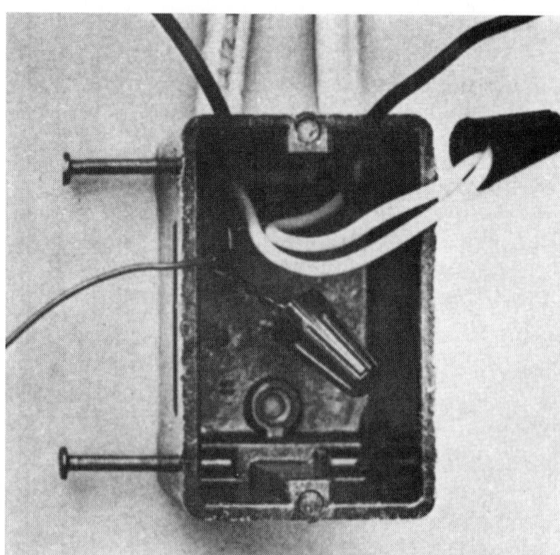

Fig. 370-4. Every nonmetallic box must be "durably and legibly marked by the manufacturer" with its cubic-inch capacity to permit calculation of number of wires permitted in the box—using Table 370-16(b) and the additions of wire space required to satisfy Sec. 370-16(a)(1). (Sec. 370-16.)

marked to show that it has a capacity of at least 7 × 2, or 14, cu in. (As shown, the ground wires are connected by a twist-on connector, with one end of the wire brought out to connect to a ground screw on the switch mounting yoke. Such a technique is required to provide grounding of a metal switchplate that is used on an outlet within reach of water faucets or other grounded objects. Refer to Secs. 250-42(e) and 410-56(d).

Part **(b)** of Sec. 370-16 describes the detailed way of counting wires in a box and subtracting from the permitted number of wires shown in Table 370-16(a) where cable clamps, fittings, or devices like switches or receptacles take up box space.

Important details of the wire-counting procedure of part **(b)** are as follows:

1. From the wording, it is clear that no matter how many ground wires come into a box, whether they are ground wires in NM cable or ground wires run in metal or nonmetallic raceways, a deduction of only one conductor must be made from the number of wires shown in Table 370-16(a) (Fig. 370-5). Or, as will be shown in later examples, one or more ground wires

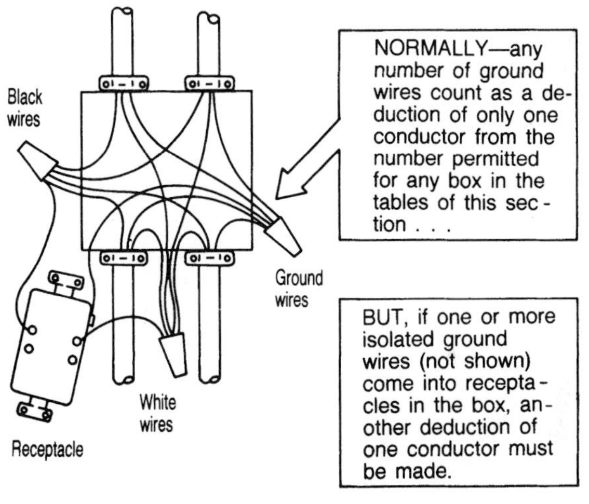

Black wires
NORMALLY—any number of ground wires count as a deduction of only one conductor from the number permitted for any box in the tables of this section . . .
Ground wires
BUT, if one or more isolated ground wires (not shown) come into receptacles in the box, another deduction of one conductor must be made.
White wires
Receptacle

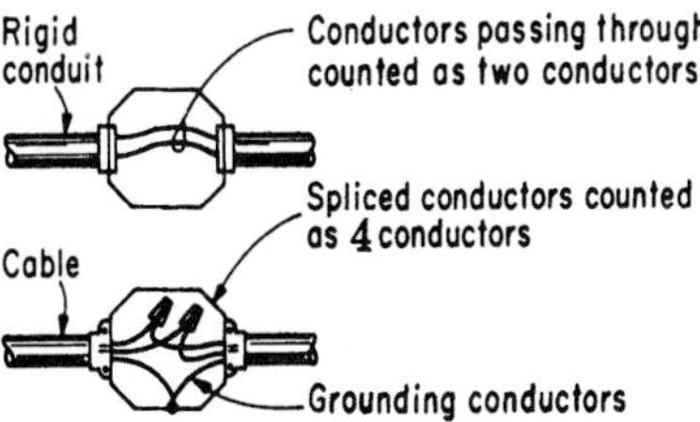

Any wire passing through counts as one, as follows:

Fig. 370-5. Count all ground wires as *one* wire (or two wires if isolated-ground wires are also used) of the largest size of ground wire in the box. (Sec. 370-16.)

in a box must be counted as a single wire of the size of the largest ground wire in the box. Any wire running unbroken through a box counts as one wire. Each wire coming into a splice device (crimp or twist-on type) is counted as one wire. And each wire coming into the box and connecting to a wiring device terminal is *one* wire.

When a number of "isolated-ground" equipment grounding conductors for receptacles come into a box along with conventional equipment grounding wires, each type of equipment ground wires must be counted as one conductor for purposes of wire count when determining the maximum number of wires permitted in a box. When a number of isolated-ground receptacles are used in a box (as for computer wiring), all the isolated-ground conductors count as a deduction of one from the number of wires given in Table 370-16(a) as permitted for the particular size of box. *And then,* another deduction of one conductor must be made for any other equipment grounding conductors (*not* isolated-ground wires).

2. Regarding the deduction of a wire from the Code-given number, Table 370-16(a), for fixture studs, cable clamps, and hickeys, does this apply to the above-mentioned items collectively regardless of number and combination, or does it apply to each item individually, such as clamps—minus one, studs—minus one, etc?

   Answer: It is the intent of parts **(b)(2)** and **(3)** to clarify that a deduction of one must be made from the number in the table for each *type* of device used in a box. A deduction of one must be made if the box contains cable clamps—whether one clamp or two clamps, a deduction of only one has to be made. A deduction of one must be made if the box contains a fixture stud. A deduction of one must be made if the box contains a hickey. Thus a box containing two clamps but no fixture studs or hickeys would have a deduction of one from the table number of wires for the clamps. If a box contained one clamp and one fixture stud, a deduction of two would be made because there are two *types* of devices in the box. Then, as given in part **(b)(4)**, in addition to the deductions for clamps, hickeys, and/or studs, a deduction of two conductors must be made for each mounting strap that supports a receptacle, switch, or combination device. In the 1993 and previous NEC editions, a deduction of only *one* conductor had to be made for each wiring device mounting strap (or yoke) installed in the box.

3. Must unused cable clamps be removed from a box? And if clamps are not used at all in a box, must they be removed to permit removal of the one-wire reduction?

   Answer: Unused cable clamps may be removed to gain space or fill in the box, or they may be left in the box if adequate space is available without the removal of the clamp or clamps. If one clamp is left, the one-wire deduction must be made. If no clamps are used at all in a box, such as where the cable is attached to the box by box connectors, the one-wire deduction is not made.

4. Is the short jumper installed between the grounding screw on a grounding-type receptacle and the box in which the receptacle is contained officially

classified as a *bonding jumper?* And is this conductor counted when the box wire count is taken?

Answer: The jumper is classed as a *bonding jumper.* Section 250-74 uses the wording "bonding jumper" in the section pertaining to this subject. This conductor is not counted because it does not leave the box. The last sentence of Sec. 370-16(b)(1) covers that point.

The first sentence of Sec. 370-16(a) requires that ganged boxes be treated as a single box of volume equal to the sum of the volumes of the sections that are connected together to form the larger box. An example of wire counting and correct wire fill for ganged boxes is included in the following examples. *Note:* In the examples given here, the same rules apply to wires in boxes for any wiring method—conduit, EMT, BX, NM.

Examples of Box Wire Fill

The top example in Fig. 370-6 shows how deductions must be made from the maximum permitted wires in a box containing cable clamps and a fixture stud. The example at the bottom shows a nonmetallic-sheathed cable with three No. 14 copper conductors supplying a 15-A duplex receptacle (one ungrounded conductor, one grounded conductor, and one "bare" grounding conductor).

After supplying the receptacle, these conductors are extended to other outlets and the conductor count would be as follows:

Circuit conductors	4
Grounding conductors	1
For internal cable clamps	1
For receptacle	2
Total	8

The No. 14 conductor column of Table 376-16(a) indicates that a device box not less than 3 by 2 by 3½ in. (76 by 50.8 by 89 mm) is required. Where a square box with plaster ring is used, a minimum of 4- by 1¼-in. (102- by 31.8-mm) size is required.

Table 370-16(a) includes the most popular types of metal "trade-size" boxes used with wires No. 14 to No. 6. Cubic-inch capacities are listed for each box shown in the table. According to paragraph **(a)(2)**, boxes other than those shown in Table 370-16(a) are required to be marked with the cubic-inch content so wire combinations can be readily computed.

Figure 370-7 shows another example with the counting data in the caption. The wire fill in this case violates the limit set by Sec. 370-16(a).

Figure 370-8 shows an example of wire-fill calculation for a number of ganged sections of sectional boxes. The photo shows a four-gang assembly of 3- by 2- by 3½-in. (76- by 50.8- by 89-mm) box sections with six 14/2 NM cables, each with a No. 14 ground wire and one 14/3 NM cable with a No. 14 ground. The feed to the box is 14/3 cable (at right side), with its black wire supplying the receptacle which will be installed in the right-hand section. The red wire serves as feed to three combination devices—one in each of the other sections—each device consisting of two switches on a single strap. When finished, the

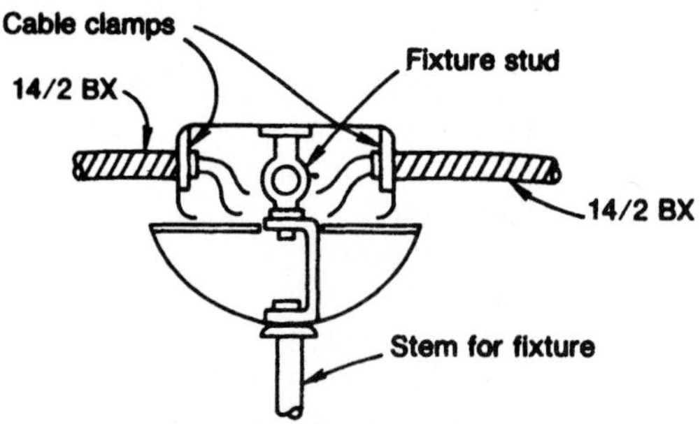

FOR A 4 × 1¼-IN. OCTAGONAL BOX:	
From Table 370-16(a)	6 No. 14 wires
Minus one for the two cable clamps and minus one for the fixture stud	2 No. 14 wires
MAX. NUMBER PERMITTED	4 No. 14 wires

NOTE: If NM cable were used, another deduction of one for the two ground wires would make use of this box a violation.

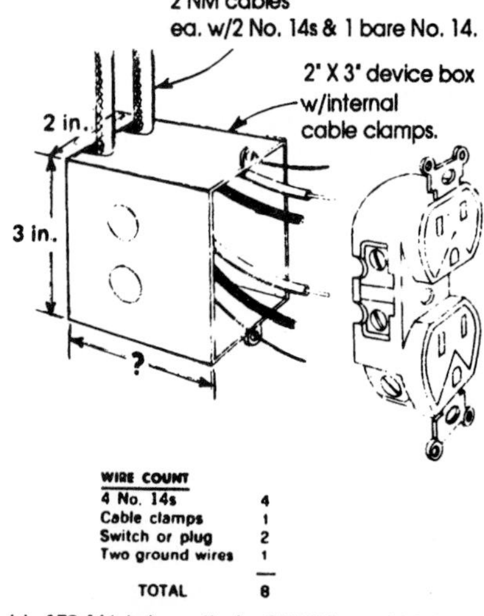

WIRE COUNT	
4 No. 14s	4
Cable clamps	1
Switch or plug	2
Two ground wires	1
TOTAL	8

Table 370-16(a) shows that a 2" X 3" box which is suitable for use with 8 No. 14 wires must be 3 1/2" deep.

Fig. 370-6. Correct wire count determines proper minimum size of outlet box. (Sec. 370-16.)

four-gang box will contain a total of six switches and one duplex receptacle. Each of the 14/2 cables will feed a switched load. All the white neutrals are spliced together and the seven bare No. 14 ground wires are spliced together, with one bare wire brought out to the receptacle ground terminal and one to the ground clip on the bottom of the left-hand section. The four-gang assembly is

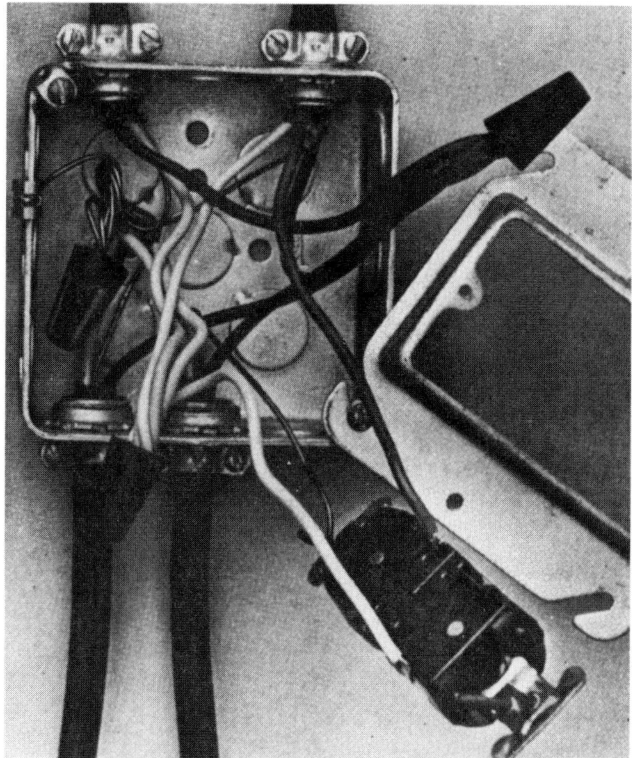

Fig. 370-7. THIS IS A CODE VIOLATION! A 4- × 4- × 1½-in. (102- × 102- × 38-mm) square metal box, generally referred to as a "1900" box, has four NM cables coming into it. At upper right is a 14/3 cable with No. 14 ground. The other three cables are 14/2 NM, each with a No. 14 ground. The red wire of the 14/3 cable feeds the receptacle to be installed in the one-gang plaster ring. The black wire of the 14/3 feeds the black wires of the three 14/2 cables. All the whites are spliced together, with one brought out to the receptacle, as required by Sec. 300-13(b). All the ground wires are spliced together, with one brought out to the grounding terminal on the receptacle and one brought out to the ground clip on the left side of the box. The wire count is as follows: nine No. 14 insulated wires, plus one for all of the ground wires and two for the receptacle. That is a total of 12 No. 14s. Note that box connectors are used instead of clamps and there is, therefore, no addition of one conductor for clamps. But Table 370-16(a) shows that a 4- × 4- × 1½-in. (102- × 102- × 38-mm) square box may contain only 10 No. 14 wires. Some think that this is *not* a violation. They say that because the area provided by the cover has not been considered. But, unless the cubic-inch capacity is marked on the cover, it may *not* be considered. Therefore, this is a violation. (Sec. 370-16.)

taken as a box of volume equal to 4 times the volume of one 3- by 2- by 3½-in. (76- by 50.8- by 89-mm) box. From Table 370-16(a), that volume is 18 cu in. for each sectional box. Then for the four-gang assembly, the volume of the resultant box is 4 × 18, or 72, cu in. Then wire fill for the four-gang assembly may be 4 times that permitted for the basic single gang box used in the assembly. Because a 3- by 2- by 3½-in (76- by 50.8- by 89-mm) box is shown in Table 370-16(a) to

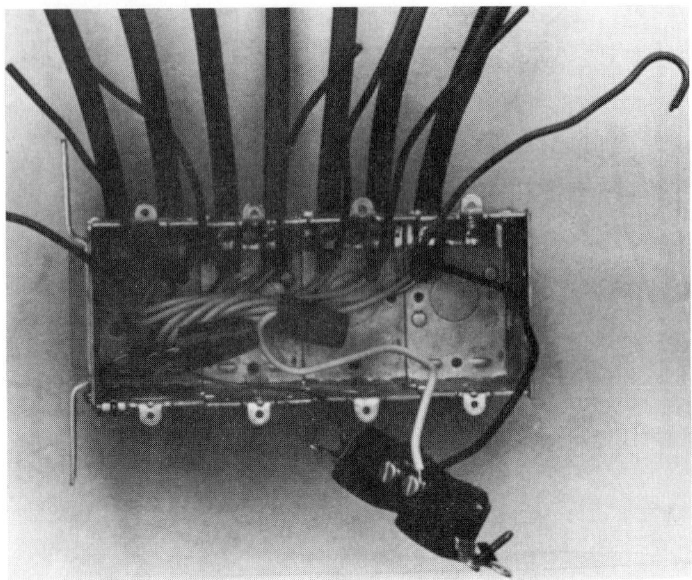

Fig. 370-8. Calculation of the proper minimum box size for the number of conductors used in gaged boxes must follow Sec. 370-16(a)(1), taking the assembly as a single box of the sum of the volumes of the ganged sections and filling it to the sum of the conductor count. (Sec. 370-16.)

have a permitted fill of 9 No. 14 wires, the four-gang assembly may contain 4 × 9, or 36, No. 14 wires—with deductions made as required by Sec. 370-16(b).

Deduct one wire for all the clamps; deduct one No. 14 for all the bare equipment ground wires; and deduct two No. 14s for each "strap containing one or more devices," which calls for a deduction of eight because there are four device "straps" (one for each of the three combination switches and one for the receptacle). The total deductions come to 1 + 1 + 8, or 10.

Deducting 10 from 36 gives a permitted fill of 26 No. 14 insulated circuit wires. In the arrangement shown, there are 6 cables with 2 insulated wires and 1 with 3 insulated wires, for a total of 15 insulated No. 14 wires. Because that is well within the maximum permitted fill of 26 No. 14 wires, such an arrangement satisfies Sec. 370-16.

The alternative method of counting wires and determining proper box size would be as follows:

1. There are 15 No. 14 insulated circuit wires.
2. Add one wire for all the cable clamps.
3. Add one wire for all the No. 14 ground wires.
4. Add two wires for each of the four device straps.

The total of the wire count is: 15 + 1 + 1 + 8, or 25, No. 14 wires.

Then dividing that among the four box sections gives six-plus wires per section—which is taken as seven No. 14 wires per section. Referring to Table 370-16(a), it will be noted that a 3- by 2- by 2½-in. (76- by 50.8- by 64-mm) box may contain six No. 14 wires. This calculation, therefore, establishes that the four-gang assembly could not be made up of 3- by 2- by 2½-in. (76- by 50.8- by 64-mm) boxes but would require 3- by 2- by 2¾-in. (76- by 50.8- by 79.8-mm) boxes—with a permitted fill of seven No. 14 wires per section, to accommodate the seven No. 14 wires per section.

The **Code** wire-counting method in Sec. 370-16(b) applies to Sec. 370-16(a)(2)—which applies where all the wires in a box are not the same size. As shown in Table 370-16(b), each No. 14 wire in a box must be allowed at least 2 cu in. of free space within the box. In the alternative calculation above, with a total of 25 No. 14 wires determined as the overall count, part **(b)** of Sec. 370-16 would require the box to have a minimum volume of 2 × 25, or 50, cu in. Each 3- by 2- by 2½-in. (76- by 50.8- by 64-mm) box has a volume of 12.5 [Table 370-16(a)]—for a total of 4 × 12.5, or 50, cu in. volume of the four-gang assembly. That volume satisfies the conductor volume and would permit use of 3- × 2- × 2½-in. (76- by 50.8- by 64-mm) boxes. But, the rule of Sec. 370-16(a) would require at least 3- × 2- × 2¾-in. (76- by 50.8- by 79.8-mm) boxes. Use of 3- × 2- × 2¾-in. (76- by 50.8- by 79.8-mm) boxes would give more room and provide easier and safer installation.

When different sizes of wires are used in a box, Table 370-16(b) must be used in establishing adequate box size. Using the same method of counting conductors as described in Sec. 370-16(b), the volume of cubic inches shown in Table 370-16(b) must be allowed for each wire depending upon its size. Where two or more ground wires of different sizes come into a box, they must all be counted as a single wire of the largest size used.

When deductions are made from the number of wires permitted in a box [Table 370-16(a)], as when devices, fixture studs, etc. are in the box, the deductions must "be based on the largest conductors entering the box" in any case where the conductors are of different sizes.

Figure 370-9 shows a calculation with different wire sizes in a box. When conduit or EMT is used, there are no internal box clamps and, therefore, no addition for clamps. In this example, the metal raceway is the equipment grounding conductor—so no addition has to be made for one or more ground wires. And the red wire is counted as one wire because it is run through the box without splice or tap. As shown in the wire count under the sketch, the way to account for the space taken up by the wiring devices is to take each one as two wires of the same size as the largest wire attached to the device—i.e., No. 12—as required in the end of the first sentence of part **(a)(2)**. Note that the neutral pigtail required by Sec. 300-13(b) is excluded from the wire count as it would be under Sec. 370-16(a)(1).

From Table 370-16(b) each No. 12 must be provided with 2.25 cu in.—a total of 7 × 2.25, or 15.75 cu in. for the No. 12s. Then each No. 14 is taken at 2 cu in.—a total of 4 × 2, or 8 cu in. for both. Adding the two resultant volumes—15.75 plus 8—gives a minimum required box volume of 23.75 cu in. From Table 370-16(a), a 4- by 4-in. square box 2⅛ in. deep, with 30.3 cu in. interior volume, would satisfy this application.

For the many kinds of tricky control and power wire hookups so commonly encountered today—such as shown in Fig. 370-10—care must be taken to count all sizes of wires and make the proper volume provisions of Table 370-16(b).

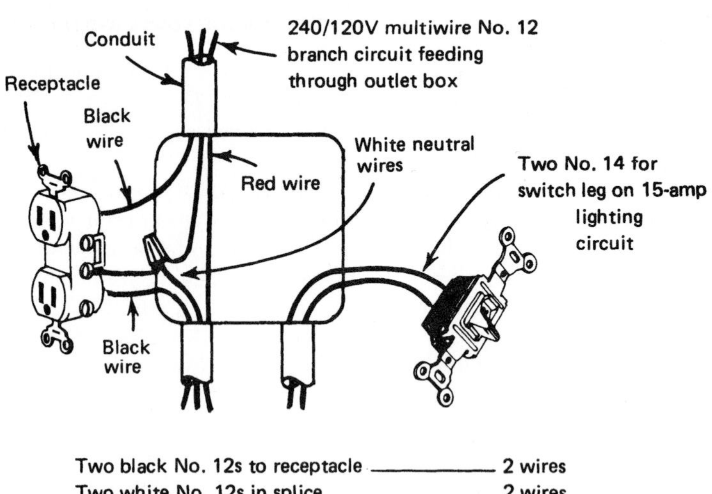

Two black No. 12s to receptacle	2 wires
Two white No. 12s in splice	2 wires
Pigtail to neutral terminal does not count	
Red wire running directly through box	1 wire
One receptacle strap	2 wires
One switch strap	2 wires
Total No. 12s =	7 wires
Two black No. 14s to switch	2 wires
Total No. 14s =	4 wires

Fig. 370-9. When wires are different sizes, volumes from Table 370-16(b) must be used. (Sec. 370-16.)

Fig. 370-10. Many boxes contain several sizes of wires—some running through, some spliced, and some connected to wiring devices. Calculation of minimum acceptable box size must be carefully made. The combination switch and receptacle here is on a single mounting strap, which is taken as two wires of the size of wires connected to it. (Sec. 370-16.)

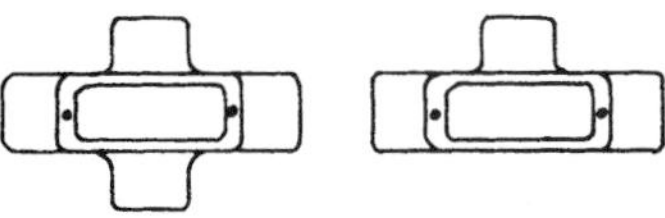

All such bodies must be durably and legibly marked by manufacturer with their cubic-inch capacities.

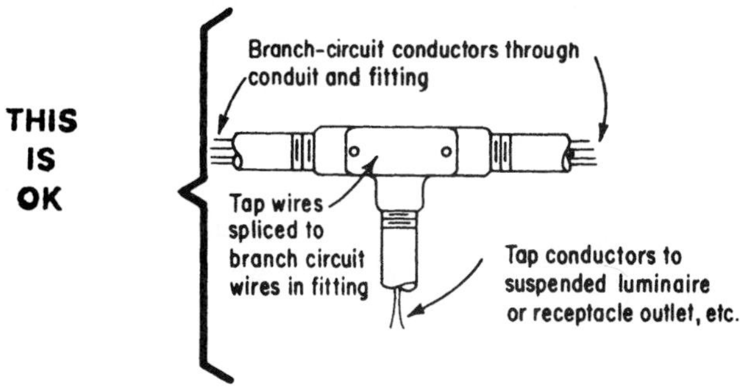

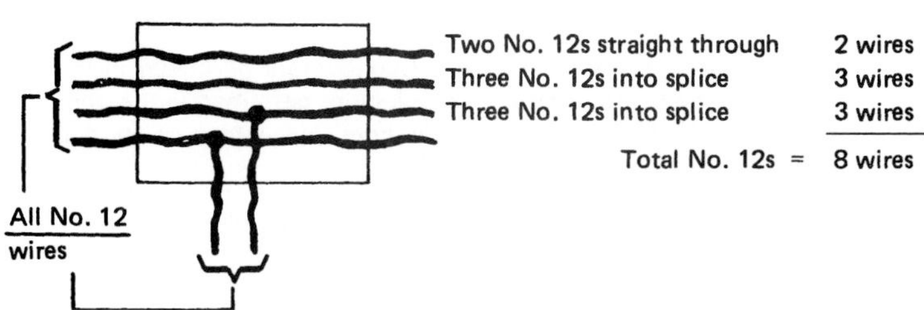

Fig. 370-11. Conduit bodies no longer must have "more than two entries" for conduit bodies to contain splices or taps. (Sec. 370-16.)

Table 370-16(a) gives the maximum number of wires permitted in boxes. But the last sentence of the first paragraph of Sec. 370-16(a)(2) does indicate that boxes may contain more wires if their internal volumes are marked and are greater than shown in Table 370-16(a).

Because the volumes in the table are minimums, most manufacturers continue to mark their products with the actual volume. This in many cases is considerably greater than the volumes shown in the table. The last sentence of Sec. 370-16**(a)(2)** says that boxes that are marked to show a cubic-inch capacity greater than the minimums in the table may have conductor fill calculated in accordance with their actual volume, using the volume per conductor given in Table 370-16(a)(2).

Conduit bodies must be marked with their cubic-inch capacity, and conductor fill is determined on the basis of Table 370-16(c). Such conduit bodies may contain splices or taps. An example of such application is shown in Fig. 370-11. Each of the eight No. 12 wires that are "counted" as shown at bottom must be provided with at least 2.25 cu in., from Table 370-6(b). The conduit body must, therefore, be marked to show a capacity of not less than 8×2.25, or 18, cu in.

For each No. 6 conductor used in the boxes or conduit bodies covered by Sec. 370-16(c), there must be at least 5 cu in. of box volume and a minimum space at least 1½ in. (38 mm) wide where any No. 6 is bent in a box or fitting.

Part **(c)** of Sec. 370-16 contains a number of provisions which must be carefully evaluated. Figure 370-12 shows the first rule. For instance, in that drawing, if a conduit body is connected to ½-in. (12.7-mm) conduit, the conduit and the conduit body may contain seven No. 12 TW wires—as indicated in Table C8, Appendix C, for rigid metal conduit—and the conduit body must have a csa at least equal to 2×0.3 sq in. (the csa of ½-in. conduit), or 0.6 sq in. That is really a matter for the fitting manufacturers to observe.

Cross-section area of conduit body must be . . .
Type C
conduit body
. . . at least twice the cross-section area of largest conduit connected to it . . .
No. 6 or
smaller
conductors
Type L conduit body
(LB, LR, LF, etc.)
No. 6 or
smaller
conductors
. . . and the maximum number of conductors permitted in the conduit body is the number of conductors permitted in the conduit connected to the conduit body, from *Code* tables on conduit fill (Chapter 9 and Appendex C).

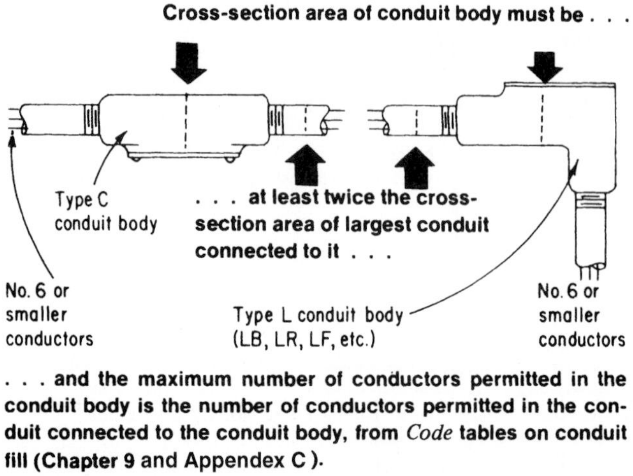

Fig. 370-12. For No. 6 and smaller conductors, conduit body must have a csa twice that of largest conduit. (Sec. 370-16.)

The second paragraph of part **(c)** covers the details shown in Fig. 370-13. The rule requires that where fittings are used as shown in the drawing, they must be supported in a rigid and secure manner. Because Sec. 370-23 establishes the correct methods for supporting of boxes and fittings, it must be observed, and that section refers to support by "conduits"—which seems to exclude such use on EMT because the NEC distinguishes between "conduit" and "tubing" (EMT), as in the headline for Table 1 of Chap. 9 in the back of the Code book. Figure 370-14 shows typical applications of those conduit bodies for splicing.

370-17. Conductors Entering Boxes, Conduit Bodies, or Fittings. Part **(b)** requires cables or raceways to be secured to *all metal* outlet boxes, conduit bodies, or fittings—such as by threaded connection, connector devices, or internal box clamps.

The first sentence of part **(c)** requires that a nonmetallic box must have a temperature rating at least equal to the lowest-temperature-rated conductor enter-

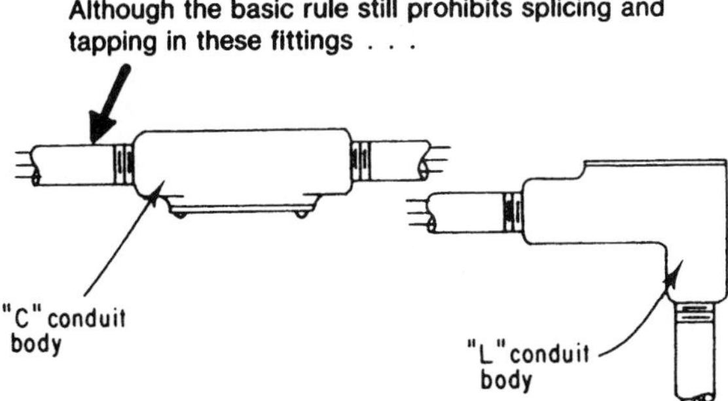
Although the basic rule still prohibits splicing and tapping in these fittings . . .
"C" conduit body
"L" conduit body
. . .Permission is now given to splice in such fittings, if Section 370-16(b) is satisfied; that is, if—

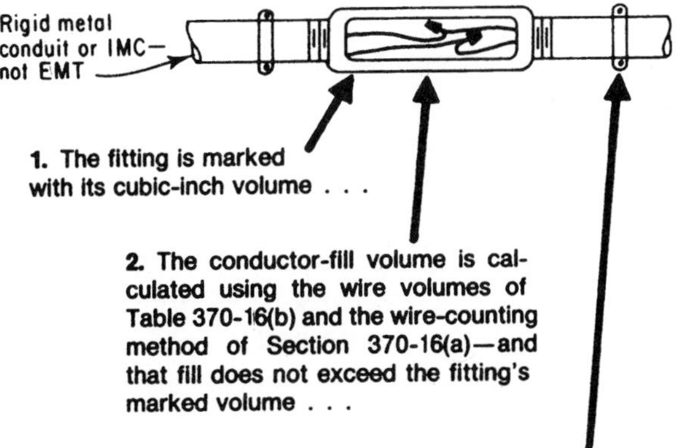
Rigid metal conduit or IMC—not EMT
1. The fitting is marked with its cubic-inch volume . . .
2. The conductor-fill volume is calculated using the wire volumes of Table 370-16(b) and the wire-counting method of Section 370-16(a)—and that fill does not exceed the fitting's marked volume . . .
3. And the fitting is "supported in a rigid and secure manner"—such as by the "conduit," if the conduit is clamped on each side of the fitting as described in the next-to-last paragraph of Section 370-23.

Fig. 370-13. Splices may be made in "C" and "L" conduit bodies—if the conditions shown in this illustration are satisfied. (Sec. 370-16.)

ing the box. This rule assumes that the lowest-temperature-rated conductor in a box must be suited to the temperature in the box. The box would, therefore, be properly applied if it has a temperature rating at least equal to that of the lowest-rated conductor.

Part **(c)** also requires that, where nonmetallic-sheathed cable is connected to nonmetallic boxes, the cable must enter through a knockout (KO) opening provided for nonmetallic-sheathed cable, and not through a hole made at any point on the box. At least ¼ in. (6.35 mm) of the cable sheath must be brought inside the box.

Another very important limitation in this Code section applies to the need for clamping nonmetallic-sheathed cable at a KO where the cable enters anything other than a single gang box. The Code has always accepted the use of non-

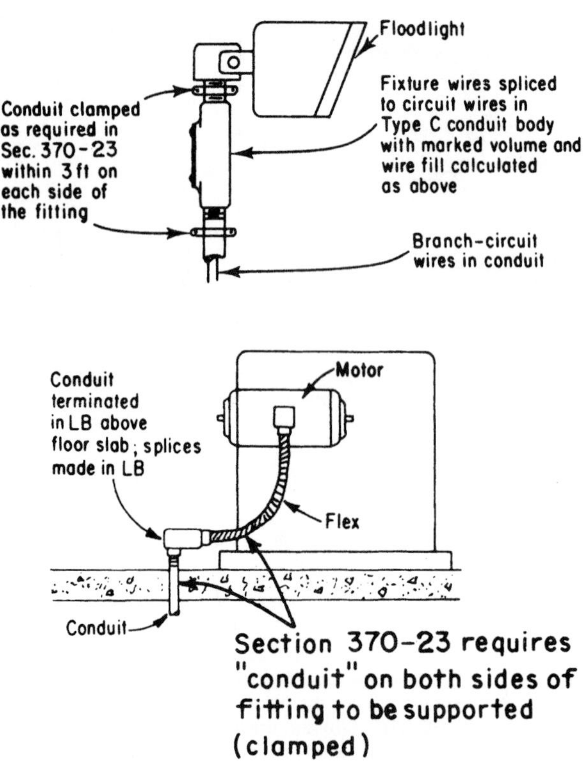
Floodlight
Fixture wires spliced to circuit wires in Type C conduit body with marked volume and wire fill calculated as above
Conduit clamped as required in Sec. 370-23 within 3 ft on each side of the fitting
Branch-circuit wires in conduit
Conduit terminated in LB above floor slab; splices made in LB
Motor
Flex
Conduit
Section 370-23 requires "conduit" on both sides of fitting to be supported (clamped)

Fig. 370-14. Splicing in "C" or "L" conduit bodies is common practice. (Sec. 370-16.)

metallic-sheathed cable without box clamps or any type of connector where the cable is stapled within 8 in. (203 mm) of the box. The cable is then brought into the box through an NM cable KO on the box, without any kind of a connector at the KO or any clamps in the box (Fig. 370-15). But the intention of the Exception to part **(c)** is that boxes or enclosures other than single gang boxes must be provided with a clamp or connector to secure nonmetallic-sheathed cable to such boxes (Fig. 370-16). *Only single gang nonmetallic boxes may be used without a cable clamp at the box KOs.* Where the Code permits elimination of a cable clamp if the cable is clamped to the stud within 8 in. (203 mm) of the box, the rule specifies that the 8-in. (203-mm) length be measured *along the cable* and not simply from the point of the cable strap to the box edge itself.

When used with open wiring on insulators, knob-and-tube work, or nonmetallic-sheathed cable, nonmetallic boxes have the advantage that an accidental contact between a "hot" wire and the box will not create a hazard.

370-19. Boxes Enclosing Flush Devices. A through-the-wall box is a box which is manufactured to be installed in a partition wall so that a receptacle or switch may be attached to either side; therefore, it is not necessary to use two standard boxes one facing each side and connected by a jumper.

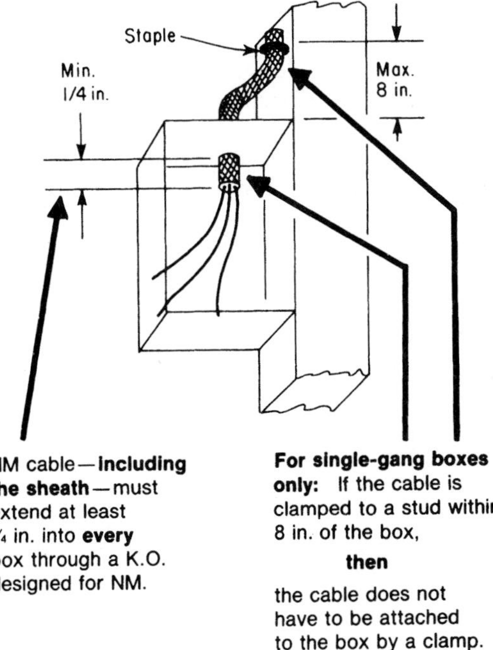
Staple
Min. 1/4 in.
Max. 8 in.
NM cable—**including the sheath**—must extend at least ¼ in. into **every** box through a K.O. designed for NM.
For single-gang boxes only: If the cable is clamped to a stud within 8 in. of the box,
then
the cable does not have to be attached to the box by a clamp.

Fig. 370-15. NM cable does not have to be clamped to single-gang boxes. (Sec. 370-17.)

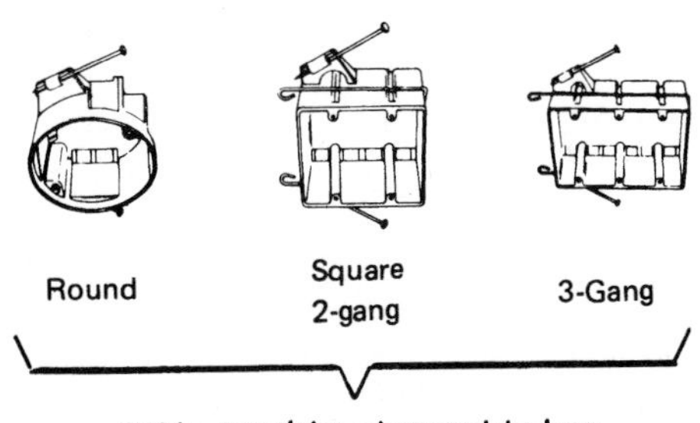
Round
Square 2-gang
3-Gang
Cable *must* be clamped to box

Fig. 370-16. NM cable must be clamped to all nonmetallic boxes that are *not* "single gang boxes." (Sec. 370-17.)

From the literal wording of Sec. 370-19, it could be interpreted to prohibit the use of such boxes. If a single device were used on only one side, then the other side would have to be closed by a cover of a thickness as required in Secs. 370-40(b) and 370-41—generally 14 gage. If a device is installed on both sides and the requirements of Sec. 410-56(d) regarding faceplates are followed, the boxes could be considered to comply with the Code because the walls and backs of the boxes would be enclosed.

If the screws used for attaching the receptacles and switches to boxes were used also for the mounting of boxes, a poor mechanical job would result, since the boxes would be insecurely held whenever the devices were not installed and the screws loosened for adjustment of the device position. Hence the prohibition.

370-21. Repairing Plaster and Drywall or Plasterboard. The purpose of Secs. 370-20 and 370-21 is to prevent openings around the edge of the box through which fire could be readily communicated to combustible material in the wall or ceiling. For this reason inspection authorities do not allow square or hexagonal boxes in ceilings, finished with Sheetrock, without the use of "mud rings."

370-22. Exposed Surface Extensions. The extension should be made as illustrated in Fig. 370-17. The extension ring is secured to the original box by two screws passing through ears attached to the box.

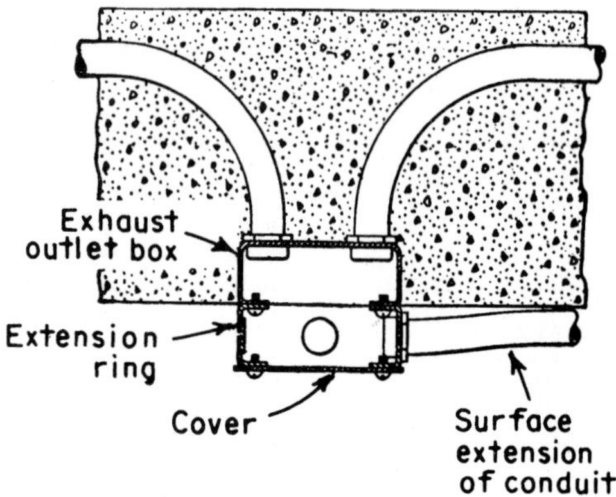

Fig. 370-17. Extension ring must be secured to box for surface extension (Sec. 370-22.)

As noted in the Exception, a surface extension may be made from the *cover* of a concealed box, *if* the cover mounting design is secure, the extension wiring method is flexible, and grounding does not depend upon connection between the box and the cover. This Exception to the basic rule, which requires use of a box or extension ring over the concealed box, permits a method that provides a high degree of reliability.

370-23. Supports. The Code rule strictly requires all boxes, conduit bodies, and fittings to be fastened in their installed position—and the various paragraphs of this section cover different conditions of box support for commonly encountered enclosure applications. The one widely accepted exception to that rule—although actually not recognized by the Code—is the so-called throw-away or floating box, which is a junction box used to connect flexible metal conduit from a recessed fixture to flex or BX branch-circuit wiring, in accordance with Sec. 410-67(c) (Fig. 370-18). In such cases, the connection of the fixture "whip" [the 4 to 6 ft (1.22 to 1.83 m) of flex with high-temperature wires, e.g., 150°C Type AF] is made to the branch-circuit junction box which hangs down

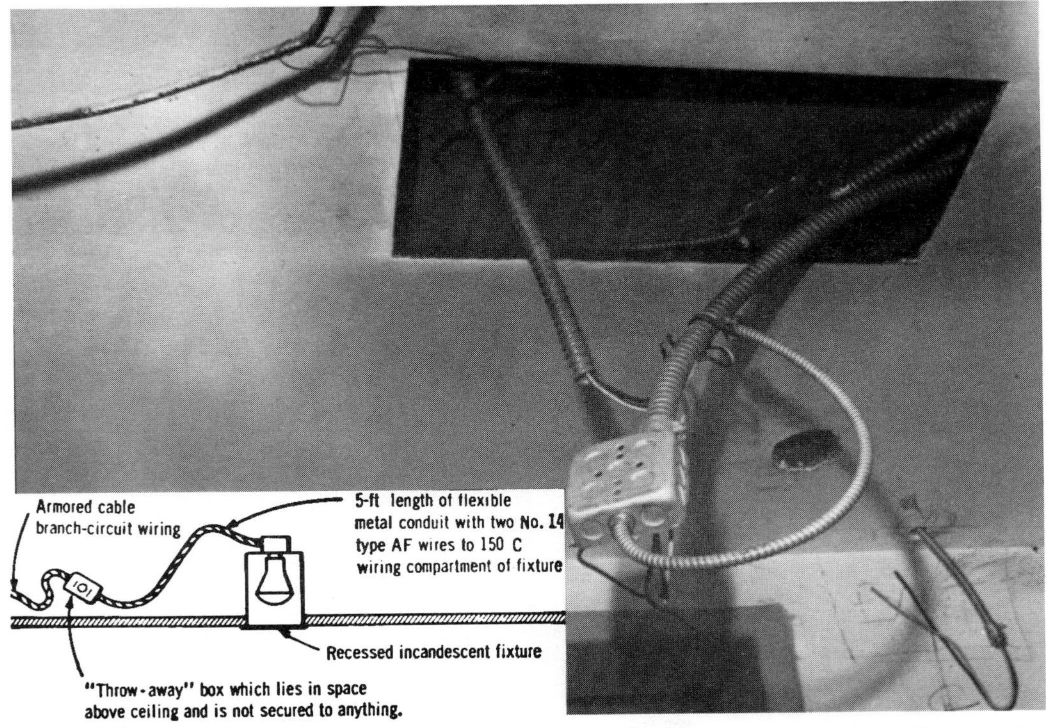

Fig. 370-18. Fixture supply flex is tapped out of junction box fed by flex or BX branch-circuit wiring in ceiling space. Box is later "thrown-away," unattached, into ceiling space. (Sec. 370-23.)

through the ceiling opening, and then the junction box is pushed back out of the way in the ceiling space and the fixture raised into position. But with suspended ceilings of lift-out panels, there is no need to leave such a loose box in the ceiling space, because connection can be made to a fixed box before the ceiling tiles are laid in place.

Figure 370-19 shows the rule of part **(b)(1)** of this section.

An outlet box built into a concrete ceiling, as shown in Fig. 370-20, seldom needs any special support. At such an outlet, if it is intended for a fixture of great weight to be safely hung on an ordinary ⅜-in. (9.53 mm) fixture stud, a special fixture support consisting of a threaded pipe or rod is required, such as is shown in Fig. 370-21.

In a tile arch floor (Fig. 370-21) a large opening must be cut through the tile to receive the conduit and outlet box.

The requirement of metal or wood supports for boxes applies to concealed work in walls and floors of wood-frame construction and other types of construction having open spaces in which the wiring is installed. In walls or floors of concrete, brick, or tile where conduit and boxes are solidly built into the wall or floor material, special box supports are not usually necessary.

As covered in part **(c)**, in an existing building, boxes may be flush mounted on plaster or any other ceiling or wall finish. Where no structural members are available for support, boxes may be affixed with approved anchors or clamps. Figure 370-22 illustrates that. For cutting metal boxes into existing walls, "Madison Holdits" are used to clamp the box tightly in position in the opening. Actually, the local inspector can determine acceptable methods of securing "cut-in" device boxes because this provision provides appreciable latitude for such decisions.

And according to part **(c)**, framing members of suspended ceiling systems may be used to support boxes if the framing members are rigidly supported and securely fastened to each other and to the building structure.

Figure 370-23 shows box-support methods that are covered by part **(d)** of Sec. 370-23 for boxes not over 100 cu in. (1,640 cu mm) in volume that do not contain devices or support fixtures. The rule there applies to "conduit" [rigid metal conduit and IMC, but not EMT or PVC conduit, Sec. 347-3(b)] used to support boxes—as for overhead conduit runs. A box may be supported by two properly clamped conduit runs (rigid metal conduit or IMC) that are threaded into entries on the box or into field-installed hubs attached to the box. This permission to use separate, field-installed hubs offers a relaxation of the previous demand for direct threaded connection to the box. Where locknut and bushing connections are used instead of threaded conduit connection to a threaded hub or similar connection, the box must be independently fastened in place. Figure 370-24 shows an installation that is most likely a violation of part **(d)**. However, it may be acceptable because an Exception to part **(d)** permits "conduit or electrical metallic tubing" to support conduit bodies that are not larger than the largest trade size of conduit or EMT used.

The rules of part **(e)** of Sec. 370-23 are shown in Fig. 370-25. The rule recognizes the support of elevated threaded-hub junction boxes [not over 100 cu in. (1,640 cu mm)] by conduits emerging from a floor or concrete or the earth, such

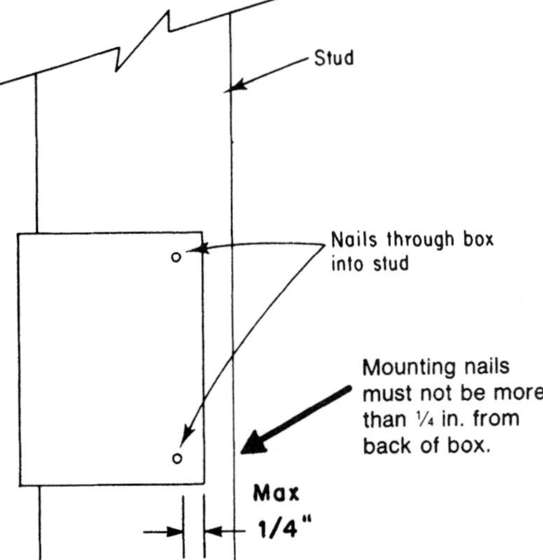

Fig. 370-19. Box-mounting nails must not obstruct box interior space. (Sec. 370-23.)

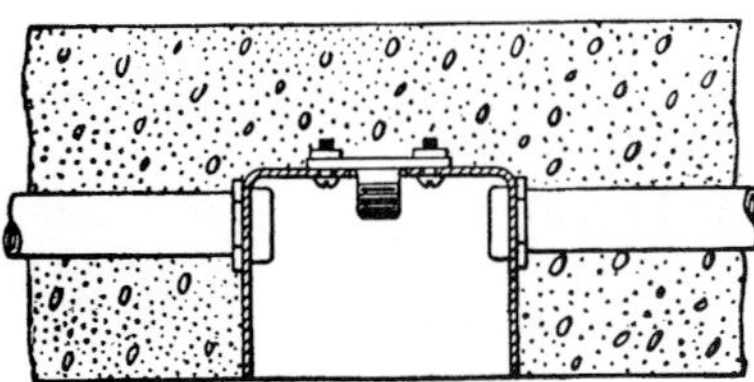

Fig. 370-20. Box in concrete is securely supported. (Sec. 370-23.)

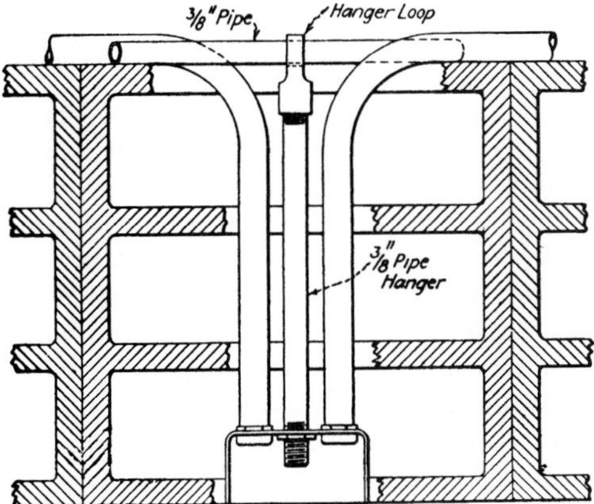

Fig. 370-21. Box in tile arch ceiling requires pipe-hanger support if very heavy lighting fixture is to be attached to the box stud. (Sec. 370-23.)

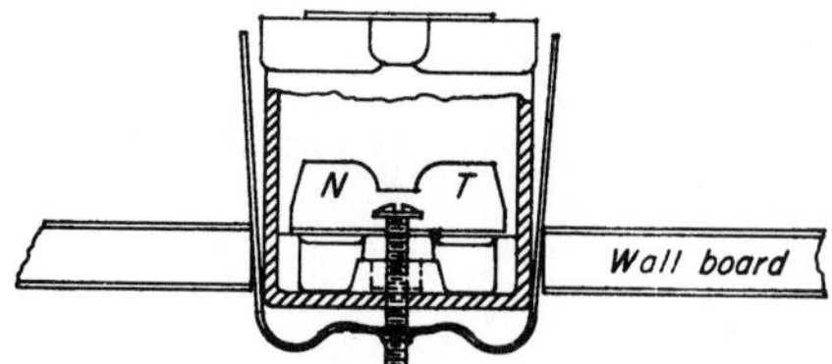

1. Inserting box and bracket through wall board.

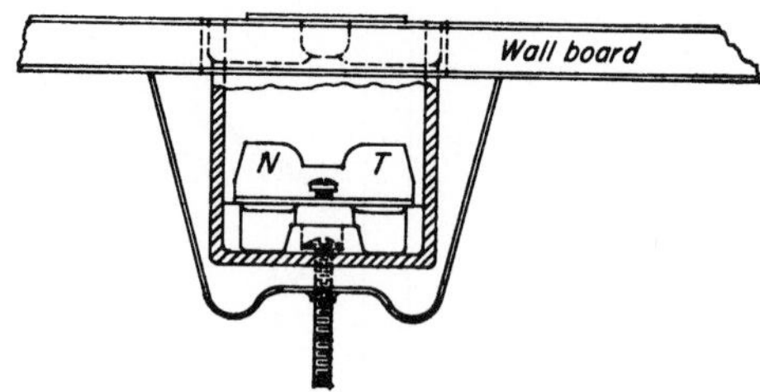

2. Box anchored to wall.

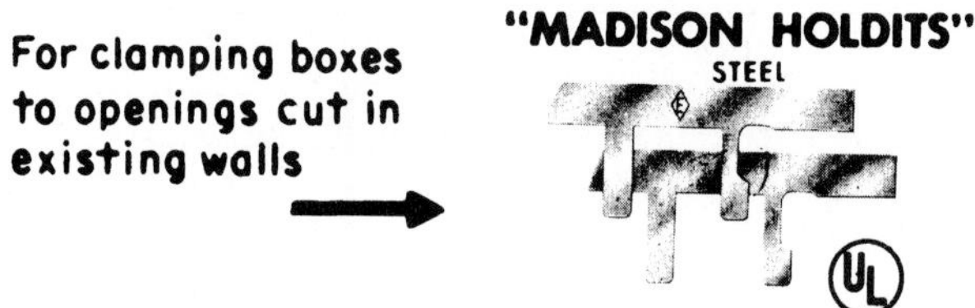

Fig. 370-22. Part (c) of Sec. 370-23 refers to these types of clamping devices (Sec. 370-23.)

as those used near swimming pools, patios, or shrubbery. Support by a single conduit is not recognized. Figure 370-26 shows several installations that are in violation of these rules.

The intent of part **(e)** is that a lighting fixture may be supported on a box that is itself supported by two or more rigid metal or intermediate metal conduits threaded into the box or into hubs that are field-installed to the box. The wording of this rule permits the two or more conduits to be threaded into either threaded entries in the box itself or into field-installed hubs that are "identified for the purpose" and properly connected to the box. And the wording permits the box to "support fixtures" or contain devices, receptacles, or switches.

Although not clear in the past, the wording of part **(e)** now does make it clear, this rule accepts only *metal* conduits—rigid metal and IMC—as the support. Nonmetallic conduit is prohibited from supporting boxes or any equipment by Sec. 347-3(b).

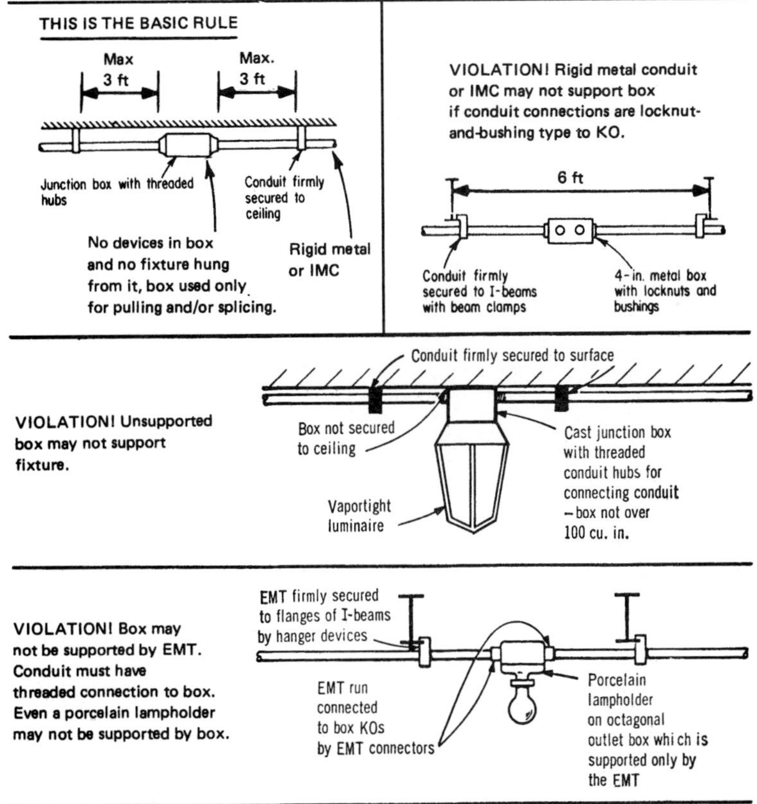
THIS IS THE BASIC RULE
Max 3 ft
Max. 3 ft
Junction box with threaded hubs
Conduit firmly secured to ceiling
No devices in box and no fixture hung from it, box used only for pulling and/or splicing.
Rigid metal or IMC
VIOLATION! Rigid metal conduit or IMC may not support box if conduit connections are locknut-and-bushing type to KO.
6 ft
Conduit firmly secured to I-beams with beam clamps
4-in. metal box with locknuts and bushings
Conduit firmly secured to surface
VIOLATION! Unsupported box may not support fixture.
Box not secured to ceiling
Cast junction box with threaded conduit hubs for connecting conduit – box not over 100 cu. in.
Vaportight luminaire
VIOLATION! Box may not be supported by EMT. Conduit must have threaded connection to box. Even a porcelain lampholder may not be supported by box.
EMT firmly secured to flanges of I-beams by hanger devices
EMT run connected to box KOs by EMT connectors
Porcelain lampholder on octagonal outlet box which is supported only by the EMT

Fig. 370-23. Box may be supported by "conduit" that is clamped, but box must not contain or support anything. (Sec. 370-23.)

Part **(g)** permits a pendant box (such as one containing a START-STOP button) to be supported from a multiconductor cable, using, say, a strain-relief connector threaded into the hub on the box, or some other satisfactory protection for the conductors.

370-24. Depth of Outlet Boxes. Sufficient space should be provided inside the box so that the wires do not have to be jammed together or against the box, and the box should provide enough of an enclosure so that in case of trouble, burning insulation cannot readily ignite flammable material outside the box.

370-25. Covers and Canopies. This rule requires every outlet box to be covered up—by a cover plate, a fixture canopy, or a faceplate, which has the openings for a receptacle, snap switch, or other device installed in a box.

Part **(a)** requires all metal faceplates to be grounded as required by Sec. 250-42. Because metal faceplates are *exposed* conductive parts, they must satisfy Sec. 250-42(a), which, in effect, says that ungrounded metal faceplates shall not be installed within 8 ft (2.44 m) vertically or 5 ft (1.52 m) horizontally of ground or grounded objects—laundry tubs, bathtubs, shower baths, plumb-

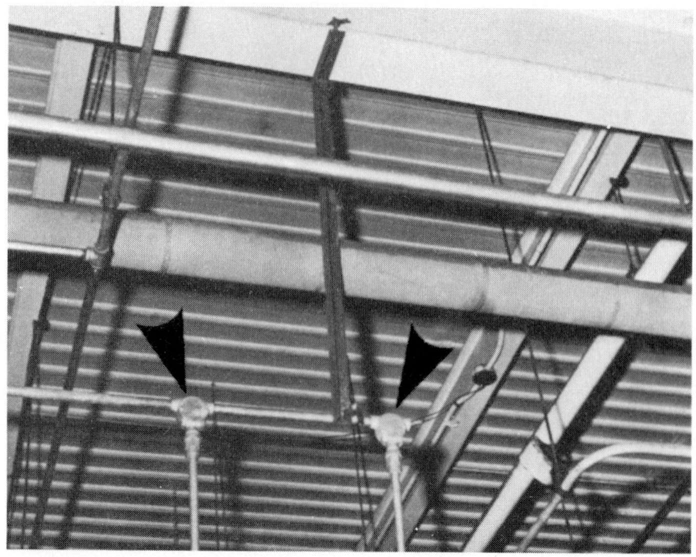

Fig. 370-24. "T" conduit bodies are supported by the rigid conduit that connects to their threaded hubs, but the conduit body at right does not have the conduit supported on two sides of the conduit body. Note angle iron brace to conduit from I-beam flange. (Sec. 370-23.)

ing fixtures, steam pipes, radiators, or other grounded surfaces—that are subject to contact by persons.

And part **(b)** of Sec. 250-42 requires metal faceplates to be grounded in wet or damp locations (which have been judged to include bathrooms)—unless the faceplates are isolated from contact.

A metal faceplate, if not grounded, may become "alive" by reason of contact of the ungrounded circuit wire with the plate or switch box, and a hazard is thus created if the plate is within reach from any conductive object. The hazard still exists, however, if a plate of insulating material is attached by means of metal screws with exposed metal heads. Insulated screws and metal screws with insulated heads are available.

When a metal faceplate is attached to a switch or receptacle in a grounded metal box, it is thereby grounded and complies with Sec. 410-21, which covers grounding of lighting fixtures and faceplates. This is true for a faceplate on a switch or a receptacle because the faceplate attaches to the metal mounting strap of the device and that strap is connected to the ears on the grounded metal box by the mounting screws and, sometimes, additionally by a bonding jumper used for grounding the receptacle ground terminal and strap. In a nonmetallic box fed by NM cable, a receptacle mounting strap is grounded by connection of the cable ground wire to the ground terminal (green hex-head screw)—thereby grounding a metal faceplate attached to the receptacle strap. But in the case of a snap switch in a nonmetallic box, the switch must be equipped with a ground terminal on its strap to permit connection of the NM cable ground wire to the strap—thereby assuring grounding of a metal faceplate attached to the strap of

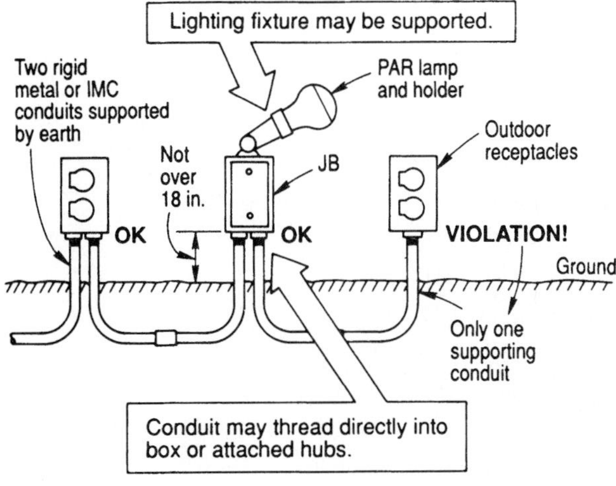

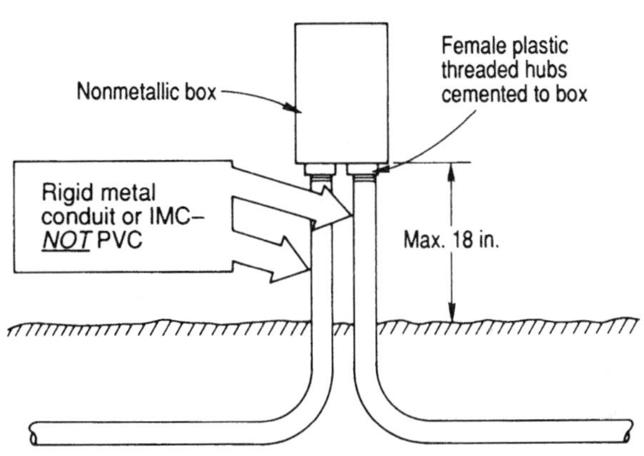

Fig. 370-25. Boxes fed out of the ground or a concrete floor, patio, or walk must observe these rules. (Sec. 370-23.)

the switch. Figure 370-27 shows such a switch used to ensure effective grounding of metal wall plates when nonmetallic switch boxes are used.

It should be noted that Secs. 370-25(a) and 250-42(a) combine to require grounding of metal faceplates that are "within 8 ft (2.44 m) vertically or 5 ft (1.52 mm) horizontally . . ." etc. But, Sec. 410-56(d), which is referenced after Sec. 370-25(a), flatly requires grounding of *any* metal faceplate if it is attached to a wiring device in a box fed by a wiring system that contains an equipment grounding means—regardless of any distances to grounded objects. In Fig. 370-27, when a metal faceplate is to be used for a switch in a nonmetallic box,

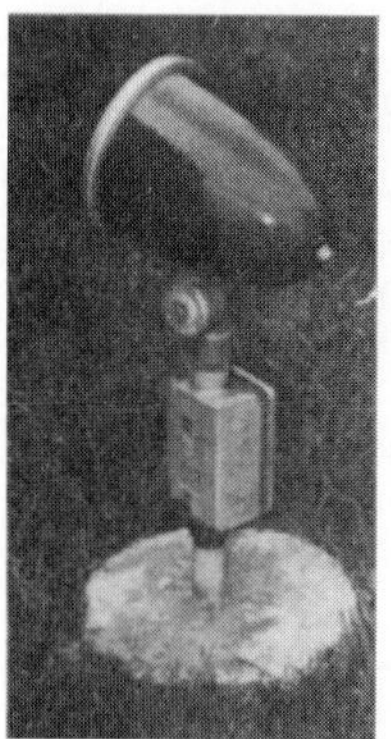

Fig. 370-26. A *single* rigid metal conduit may not support a box, even with concrete fill in the ground (left). Box may not be supported on EMT, even with several connections used (center). Method at right is a violation on three counts: EMT, not "conduit," supports the box; only one hub on box is connected; box is more than 18 in. above ground. (Sec. 370-23.)

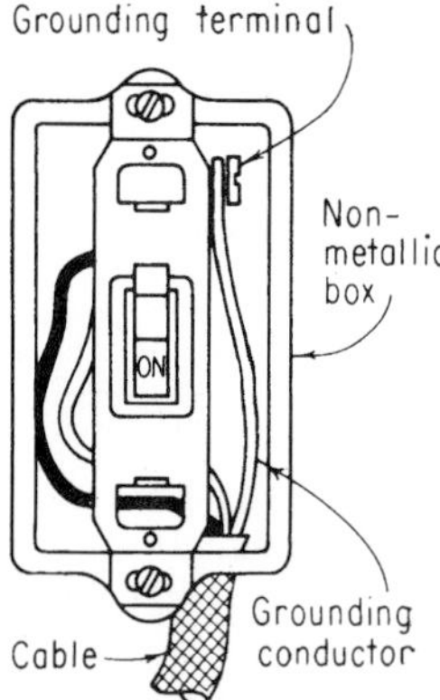

Fig. 370-27. Grounding switch must be used for metal faceplate on any non-metallic box. (Sec. 370-25.)

Sec. 370-25(a) requires the plate to be grounded if it is "within 8 ft (2.44 m) . . ." etc. But, if the faceplate is *not* in a wet or damp location and is *not* "within 8 ft (2.44 m) . . ." etc., it does *not* have to be grounded to satisfy Sec. 370-25. *However,* if NM with a ground wire is used, then Sec. 410-18(a) requires use of that type of switch to ground the metal faceplate, no matter where it is used. But, if NM cable without a ground wire is used, Sec. 410-18(b) would make it mandatory to use a faceplate of insulating material—thereby prohibiting the metal faceplate. So to use a metal faceplate on any nonmetallic box, the NM cable must have a ground wire and the type of switch shown must be used.

In part **(b)** of Sec. 370-25, if the ceiling or wall finish is of combustible material, the canopy and box must form a complete enclosure. The chief purpose of this rule is to require that no open space be left between the canopy and the edge of the box where the finish is wood or other combustible material. Where the wall or ceiling finish is plaster, the requirement does not apply, since plaster is not classed as a combustible material; however, the plaster must be continuous up to the box, leaving no opening around the box.

370-27. Outlet Boxes. Part **(b)** requires floor boxes to be completely suitable for the particular way in which they are used. Adjustable floor boxes and associated service receptacles can be installed in every type of floor construction. A metal cap keeps the assembly clean during pouring of concrete slabs. After the concrete has cured, this cap can then be removed and discarded and floor plates and service fittings added.

As noted in part **(c)**, a ceiling paddle fan must not be supported from a ceiling outlet box—unless the box is UL-listed as suitable as the sole support means for a fan (Fig. 370-28). The vibration of ceiling fans places severe

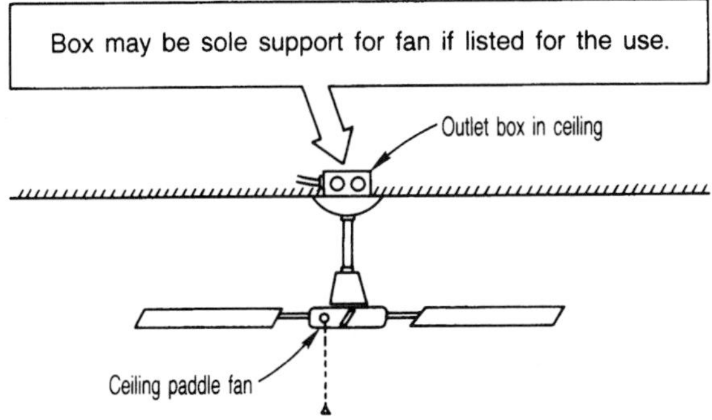

Fig. 370-28. Support for ceiling fans must be suited to the dynamic loading of the vibrating action. [Sec. 370-27(c).]

dynamic loads on the screw attachment points of boxes. But boxes designed and listed for this application pose no safety problems. (See Sec. 422-18.)

370-28. Pull and Junction Boxes. As noted in Sec. 370-16, conduit bodies must be sized the same as pull boxes when they contain No. 4 or larger conductors.

For raceways containing conductors of No. 4 or larger size, the NE Code specifies certain minimum dimensions for a pull or junction box installed in a raceway run. These rules also apply to pull and junction boxes in cable runs—but instead of using the cable diameter, the minimum trade size raceway required for the number and size of conductors in the cable must be used in the calculations. Basically there are two types of pulls—straight pulls and angle pulls. Figure 370-29 covers straight pulls. Figure 370-30 covers angle pulls. In all the cases shown in those illustrations, the depth of the box only has to be sufficient to permit installation of the locknuts and bushings on the largest conduit. And the spacing between adjacent conduit entries is also determined by the diameters of locknuts and bushings—to provide proper installation. Depth is the dimension not shown in the sketches.

EXAMPLES:

1.

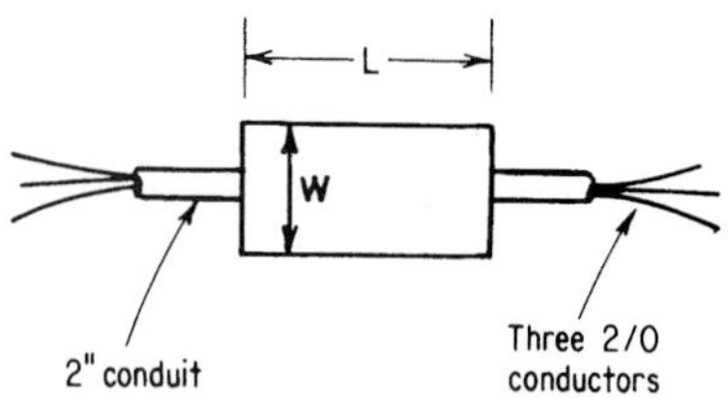

L = 8 × 2 in. = 16 in. minimum
W = Whatever width is necessary to provide proper installation of the conduit locknuts and bushings within the enclosure.

2.

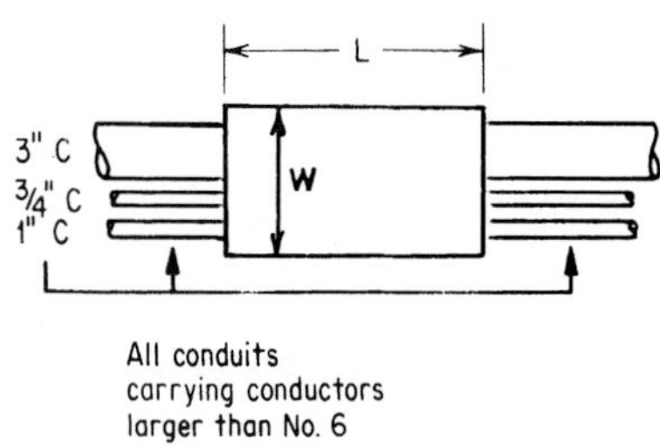

The 3-in. conduit is the largest.
Therefore—
L = 8 × 3 in. = 24 in. minimum
W = Width necessary for conduit locknuts and bushings.

Fig. 370-29. In straight pulls, the length of the box must be not less than 8 times the trade diameter of the largest raceway. (Sec. 370-28.)

According to the rule of part **(a)(2)** in sizing a pull or junction box for an angle or "U" pull, if a box wall has more than one row of conduits, "each row shall be calculated separately and the single row that provides the maximum distance shall be used." Consider the following:

A pull box has two rows of conduits entering one side (or wall) of the box for a right-angle pull. What is the minimum required inside distance from the wall with the two rows of conduit entries to the opposite wall of the box?

Row 1: One 2½-in. (64-m) and one 1-in. (25.4 mm) conduit.

Row 2: One ½-in. (12.7-mm), two 1¼-in. (31.8-mm), one 1½-in. (38-mm), and two ¾-in. (19-mm) conduits.

EXAMPLE:

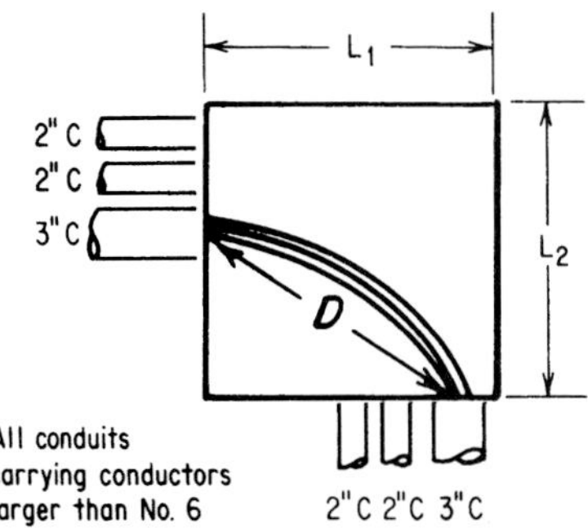

The 3-in. conduit is the largest.
Therefore—

L_1 = 6 × 3 in. + (2 in. + 2 in.) = 22 in. min.
L_2 = 6 × 3 in. + (2 in. + 2 in.) = 22 in. min.
D = 6 × 3 in. = 18 in., minimum distance between raceway entries enclosing the same conductors

Fig. 370-30. Box size must be calculated for angle pulls. For boxes in which the conductors are pulled at an angle or in a "U," the distance between each raceway entry inside the box and the opposite wall of the box must not be less than 6 times the trade diameter of the largest raceway in a row. And the distance must be increased for additional raceway entries by the amount of the maximum sum of the diameters of all other raceway entries in the same row on the same wall of the box. The distance between raceway entries enclosing the same conductors must not be less than 6 times the trade diameter of the larger raceway. (Sec. 370-28.)

Interpretation of 1987 NEC Sec. 370-18(a)(2):

1. 6 × diameter of "largest raceway" entering box wall:

$$6 \times 2\tfrac{1}{2} \text{ in.} = 15 \text{ in.}$$

2. Add "the maximum sum of diameters of all other raceway entries in any one row on the same wall of the box."

Row 2 will give "the maximum sum":

$$\tfrac{1}{2} + (2 \times 1\tfrac{1}{4}) + 1\tfrac{1}{2} + (2 \times \tfrac{3}{4}) = 6 \text{ in.}$$

3. Adding the two results: 15 in. + 6 in. = 21 in.

That is the minimum size box dimension to the wall opposite the wall where the conduits enter.

Calculating with Revised Rule of Present NEC:

Calculating each row "separately" and taking the box dimension from the row that gives the maximum distance.

Row 1: 6 × largest raceway (2½ in.) + other entries (1 in.) = 16 in.

Row 2: 6 × the largest raceway (1½ in.) + other entries [½ in. + (2 × 1¼ in.) + (2 × ¾ in.)] = 9 in. + ½ in. + 4 in. = 13½ in.

Result: The minimum box dimension must be the 16 in. dimension from run No. 1 which "provides the maximum distance" calculated.

Figure 370-31 shows a more complicated conduit and pull box arrangement, which requires more extensive calculation of the minimum permitted size. In

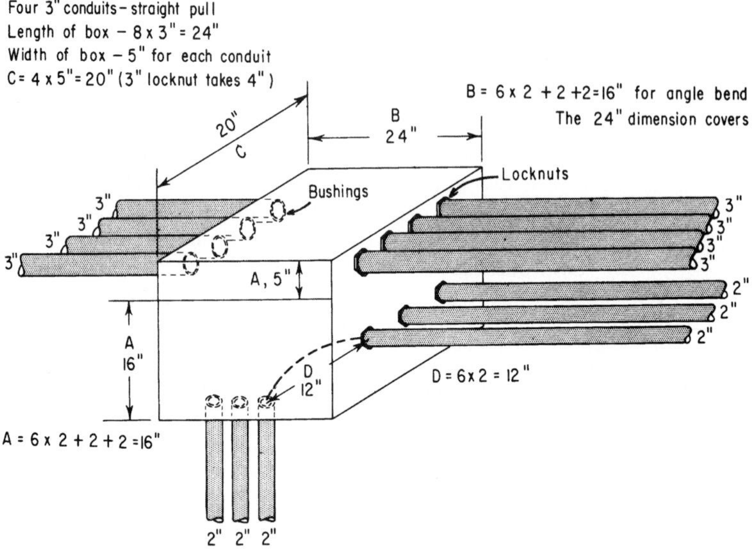

Fig. 370-31. A number of calculations are involved when angle and straight pulls are made in different directions and different planes. (Sec. 370-28.)

this particular layout shown, the upper 3-in. (76-mm) conduits running straight through the box represent a problem separate from the 2-in. (50.8-mm) conduit angle pulls. In this case the 3-in. (76-mm) conduit establishes the box length in excess of that required for the 2-in. (50.8-mm) conduit. After computing the 3-in. (76-mm) requirements, the box size was calculated for the angle pull involving the 2-in. (50.8 mm) conduit.

Subparagraph **(3)** of Sec. 370-28(a) permits smaller pull or junction boxes where such boxes have been approved for and marked with the maximum number and size of conductors and the conduit fills are *less* than the maximum permitted in Table 1, Chap. 9. This rule provides guidelines for boxes which have been widely used for years, but which have been smaller than the sizes normally required in subparagraphs **(1)** and **(2)**. These smaller pull boxes must be listed by UL under this rule.

There are many instances where an installation is made in which raceways and conductors are not matched so as to utilize maximum conduit fill as permitted by the **Code**. An example would be a 2-in. (50.8-mm) conduit with six No. 4 THHN conductors. The **Code** would permit up to 16 conductors depending upon the type of insulation. It was felt that in such installations provisions should be made for the use of boxes or fittings which would not necessarily conform to the letter of the law as exemplified by the standards listed in subsections **(1)** and **(2),** but would compare favorably under test with a box sized as is required for the conductor and conduit.

A pull or junction box used with 2-in. (50.8-mm) conduit and conductors No. 4 AWG or larger must be 16 in. (406.4 mm) long if straight pulls are made and must be 12 in. (305 mm) long if angle pulls are to be made. If we have a 2-in. (305-mm) conduit and we are installing eight No. 4 RHH conductors, all pull or junction boxes would have to conform to these measurements. If, however, we are installing five No. 4 RHH conductors, a smaller box would be acceptable, provided it has been tested for and is marked with this number.

Figure 370-32 shows how the rules of Sec. 370-28(a) apply to conduit bodies. **Important:** The Exception given in Sec. 370-28(a) (2) establishes the minimum dimension of L2 for angle runs, but this Exception only applies to conduit bodies which have the removable cover opposite one of the entries, such as a Type LB body. Types LR, LL, and LF do not qualify under that Exception, and for such conduit bodies the dimension L2 would have to be at least equal to the dimension L1 (that is, 6 times raceway diameter).

Figure 370-33 shows the racking of cable required by part **(b)** of this section.

Figure 370-34 shows another consideration in sizing a pull box for angle conduit layouts. A pull box is to be installed to make a right-angle turn in a group of conduits consisting of two 3-in. (76-mm), two 2½-in. (64-mm), and four 2-in. (50.8-mm) conduits.

Subparagraph **(2)** of Sec. 370-28(a) gives two methods for computing the box dimensions, and both must be met.

First method:

$$
\begin{aligned}
6 \times 3 \text{ in.} &= 18 \text{ in.} \\
1 \times 3 \text{ in.} &= \ \ 3 \\
2 \times 2\tfrac{1}{2} \text{ in.} &= \ \ 5 \\
4 \times 2 \text{ in.} &= \ \ \underline{8} \\
\text{Total} &= 34 \text{ in.}
\end{aligned}
$$

Second method:

Assuming that the conduits are to leave the box in the same order in which they enter, the arrangement is shown in Fig. 370-34 and the distance A between the ends of the two conduits must be not less than 6 × 2 in. = 12 in. It can be assumed that this measurement is to be made between the centers of the two conduits. By calculation, or by laying out the corner of the box, it is found that the distance C should be about 8½ in. (203 mm).

The distance B should be not less than 30½ in. (774.7 mm) approximately, as determined by applying practical data for the spacing between centers of conduits,

$$30\tfrac{1}{2} \text{ in.} + 8\tfrac{1}{2} \text{ in.} = 39 \text{ in.}$$

STRAIGHT RUN

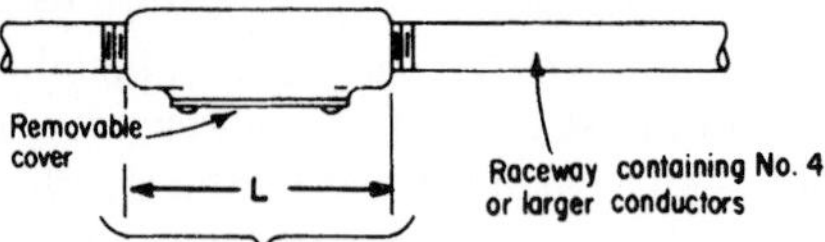

Type C conduit body must have length L equal to 8 times diameter of the raceway

Examples

If four No. 4 THHN are used in 1-in. conduit, conduit body must be at least 8 in. long.

If four 500 kcmil XHHW are used in 3-in. conduit, conduit body must be at least 24 in. long.

ANGLE RUN

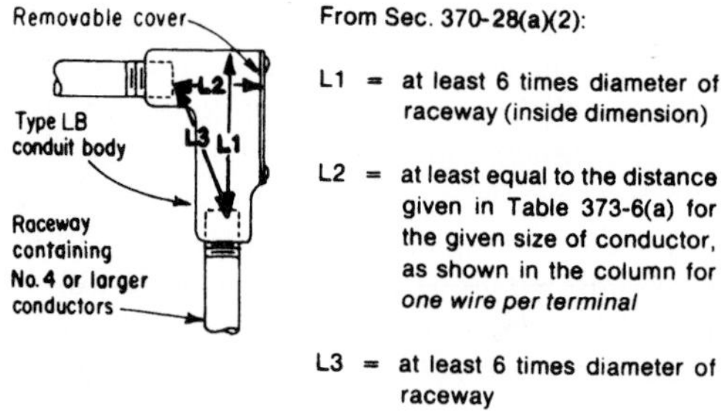

Removable cover
Type LB conduit body
Raceway containing No. 4 or larger conductors
L1
L2
L3
From Sec. 370-28(a)(2):
L1 = at least 6 times diameter of raceway (inside dimension)
L2 = at least equal to the distance given in Table 373-6(a) for the given size of conductor, as shown in the column for *one wire per terminal*
L3 = at least 6 times diameter of raceway

Examples

If four No. 4 THW conductors are used in 1¼-in. conduit, minimum dimensions would be calculated as follows:

L1 = 6 × 1¼ in. = 7.5 in.
L2 = 2 in., from Table 373-6(a) for one No. 4 conductor per terminal
L3 = 6 × 1¼ in. = 7.5 in.

If four 500 kcmil THW conductors are used in 3½-in. conduit, minimum dimensions would be:

L1 = 6 × 3½ in. = 21 in.
L2 = 6 in., from Table 373-6(a)
L3 = 6 × 3½ in. = 21 in.

Fig. 370-32. Conduit bodies must be sized as pull boxes under these conditions (Sec. 370-28.)

In this case the box dimensions are governed by the second method. The largest dimension computed by either of the two methods is of course the one to be used. Of course, if conduit positions for conduits carrying the same cables are transposed—as in Fig. 370-30—then box size can be minimized.

The most practical method of determining the proper size of a pull box is to sketch the box layout with its contained conductors on a paper.

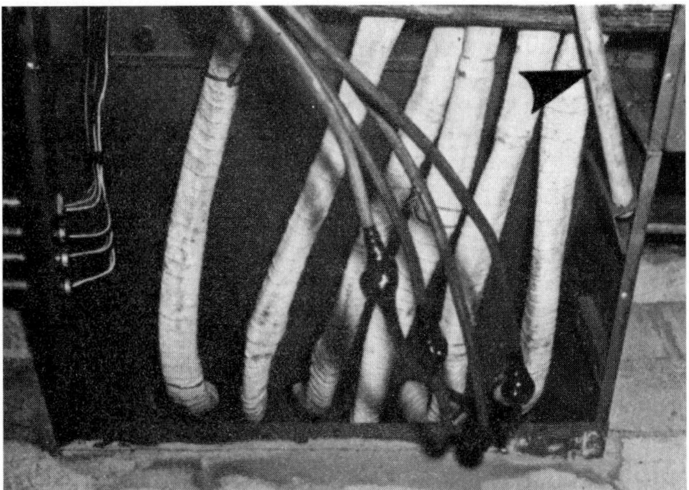

Fig. 370-33. If a pullbox has *any* dimension over 6 ft (1.83-m), the conductors within it must be supported by suitable racking (arrow) or cabling, as shown here for arc-proofed bundles of feeder conductors, to keep the weight of the many conductors off the sheet metal cover that attaches to the bottom of the box. (Sec. 370-28.)

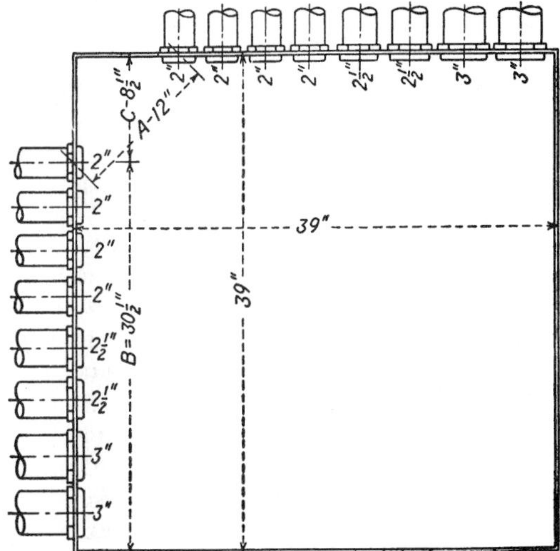

Fig. 370-34. Distance between conduits carrying same cables has great impact on overall box size. (Sec. 370-28.)

Section 370-28 applies particularly to the pull boxes commonly placed above distribution switchboards and which are often, and with good reason, termed *tangle boxes.* In such boxes, all conductors of each circuit should be cabled together by serving them with twine so as to form a self-supporting assembly that can be formed into shape, or the conductors should be supported in an orderly manner on racks, as required by part **(b)** of Sec. 370-28. The conductors should not rest directly on any metalwork inside the box, and insulating bushings should be provided wherever required by Sec. 373-6(c).

For example, the box illustrated in Fig. 370-34 could be approximately 5 in. (126.6 mm) deep and accommodate one horizontal row of conduits. By making it twice as deep, two horizontal rows or twice the number of conduits could be installed.

Insulating racks are usually placed between conductor layers, and space must be allowed for them.

370-40. Metal Boxes, Conduit Bodies, and Fittings. This section through Sec. 370-44 covers construction of boxes. UL data on application of boxes supplement this **Code** data as follows:

1. Cable clamps in outlet boxes are marked to indicate the one or more types of cables that are suitable for use with that clamp.
2. Box clamps have been tested for securing only one cable per clamp, except that multiple-section clamps may secure one cable under each section of the clamp, with each cable entering the box through a separate KO.

Part **(c)** of this section covers the pull boxes regulated by Sec. 370-28. UL data on such boxes are important and must be related to the **Code** rules. Listed pull and junction boxes may be sheet metal, cast metal, or nonmetallic, and all of these have a volume greater than 100 cu in. (1,640 cu mm). Because listed boxes of this type are available, the intent of **NE Code** Sec. 110-2 and the clear regulations of OSHA on equipment acceptability demand that only listed pull and junction boxes be used. To use a pullbox or junction box that is not listed is a violation of those regulations. Boxes marked "Raintight" or "Rainproof" are tested under a condition simulating exposure to beating rain. "Raintight" means water will not enter the box. "Rainproof" means that exposure to beating rain will not interfere with proper operation of the apparatus within the enclosure. Use of a box with either designation must satisfy **NE Code** Sec. 370-15, which notes that boxes in wet locations (such as outdoors where exposed to rain or indoors where exposed to water spray) must prevent moisture from entering *or* accumulating within the box. That is, water *may* enter the box if it does not accumulate in the box, where the box is drained. A box that is raintight or rainproof may satisfy that rule. Be sure, though, that any equipment installed in a box labeled "rainproof" is mounted within the location restrictions marked in the box.

In part **(d)** of Sec. 370-40, *connection for grounding conductor is required in metal boxes used with nonmetallic raceway or cable.* This rule is intended to assure a suitable means within a metal box to connect the equipment grounding conductor that is required to be used with such wiring methods to provide equipment grounding. This is required here for metal boxes that are "designed for use" with the nonmetallic systems. Without this rule, use of metal boxes often results in drilling holes in the box to take a nut-and-bolt connection of the grounding conductor. Such unauthorized holes in an enclosure void listing of the box and diminish its concrete-tight or liquidtight integrity.

Note that the "means" for connection of the equipment grounding conductor must be provided "in" each metal box designed for use with a nonmetallic wiring method. Use of a grounding clip (a G-clip) on the edge of the box does not appear to satisfy that wording. Instead, a grounding lug or tapped hole that is part of the box must provide effective grounding connection to the box.

370-71. Size of Pull and Junction Boxes. Figure 370-35 shows the rules on sizing of pull boxes for high-voltage circuits.

STRAIGHT PULLS

L

Box

All covers for boxes enclosing circuits over 600 volts must be permanently marked "DANGER: HIGH VOLTAGE KEEP OUT" on the outside in block-type letters at least 1/2 in. high

L - not less than 48 times the outside diameter, over sheath, of the largest shielded or lead-covered *conductor* or *cable* entering the box, *OR* not less than 32 times the outside diameter of the largest nonshielded conductor or cable.

NOTE: The box length must be 48 times the conductor or cable diameter, *not the conduit* diameter.

ANGLE PULLS

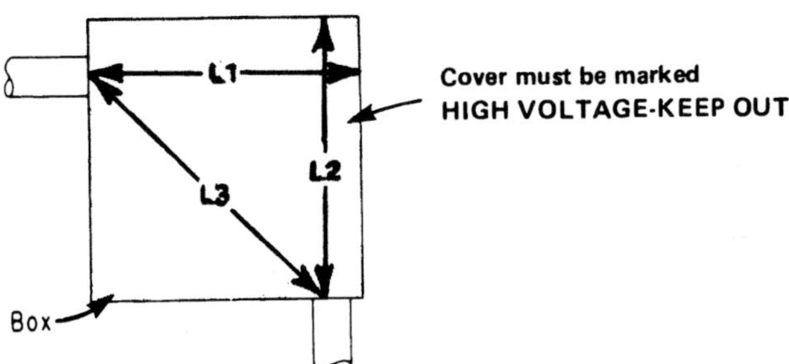

L1, L2, L3—not less than 36 times the outside diameter, over sheath, of the largest *conductor* or *cable*

Fig. 370-35. Minimum dimensions are set for high-voltage pull and junction boxes. [Sec. 370-71(a).]

370-72. Construction and Installation Requirements. Part (e) requires that covers of pull and junction boxes for systems operating at over 600 V must be marked with readily visible lettering at least ½ in. (12.7-mm) high, warning "DANGER HIGH VOLTAGE KEEP OUT."

All required warning signs must be properly worded to include the command "KEEP OUT." While certain sections of the Code, such as this one, as well as Secs. 230-203 and 110-34(c), clearly require the inclusion of the command "KEEP OUT," other sections do not (e.g., Sec. 710-43). Be aware that courts have held that warning signs that fail to include some sort of instruction or command with respect to an appropriate action that must be taken are inadequate and constitute negligence on the part of the individual posting the sign. Always include some phrase that will tell the individual what to do about the condition or hazard that exists.

ARTICLE 373. CABINETS, CUTOUT BOXES, AND METER SOCKET ENCLOSURES

373-1. Scope. Cabinets and cutout boxes, according to the definitions in Art. 100, must have doors and are thus distinguished from large boxes with covers consisting of plates attached with screws or bolts. Article 373 applies to all boxes used to enclose operating apparatus, i.e., apparatus having moving parts or requiring inspection or attention, such as panelboards, cutouts, switches, circuit breakers, control apparatus, and meter socket enclosures.

373-3. Position in Wall. Figure 373-1 shows how the ¼-in. (6.35-mm) setback relates to cabinets installed in noncombustible walls.

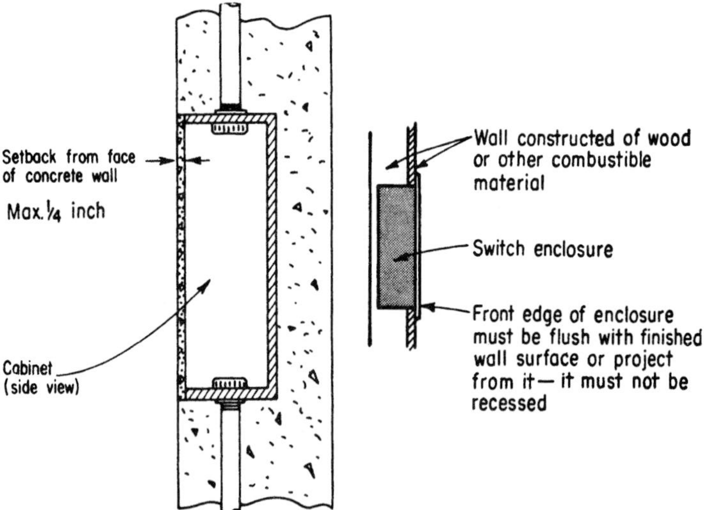

Fig. 373-1. In masonry wall, cabinet does not have to be flush with wall surface—as it does in wood wall. (Sec. 373-3.)

373-5. Conductors Entering Cabinets or Cutout Boxes. Part **(c)** makes clear that all cables used with cabinets or cutout boxes must be attached to the enclosure. NM cable, for instance, does not have to be connected by clamp or connector device to a single gang nonmetallic outlet box as in Sec. 370-17(c), *but must always* be connected to KOs in panelboard enclosures and other cabinets (Fig. 373-2).

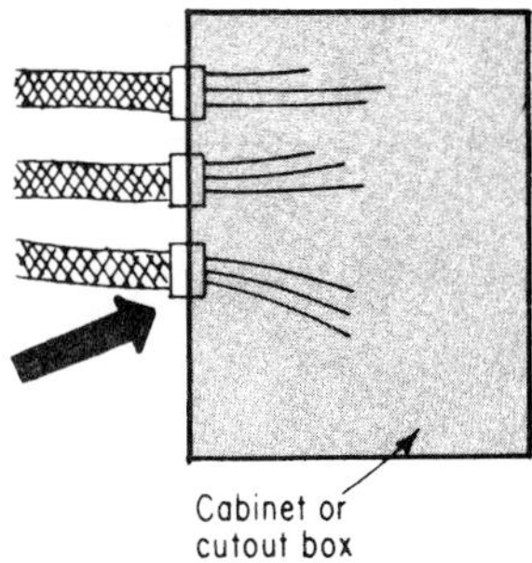
Any cable (BX, NM, etc.) must be secured to any cabinet or cutout box, whether metal or nonmetallic
Cabinet or cutout box

Fig. 373-2. All cables must be secured to all cabinets or cutout boxes. (Sec. 373-5.)

373-6. Deflection of Conductors. Parts **(a)** and **(b)** cover a basic Code rule that is referenced in a number of Code articles to assure safety and effective conductor application by providing enough space to bend conductors within enclosures.

A basic concept of evaluating adequate space for bending conductors at terminals of equipment installed in cabinets is presented in this section. The matter of bending space for conductors at terminals is divided into two different configurations, as follows:

1. The conductor does not enter (or leave) the enclosure through the wall opposite its terminals. This would be any case where the conductor passes through a wall of the enclosure at right angles to the wall opposite the terminal lugs to which the conductor is connected or at the opposite end of the enclosure. In all such cases, the bend at the terminals is a single-angle bend (90° bend), and the conductor then passes out of the bending space. It is also called an "L" bend, as shown at the top of Fig. 373-3. For bends of that type, the distance from the terminal lugs to the wall opposite the lugs must conform to Table 373-6(a), which requires lesser distances than those of Table 373-6(b) because single bends are more easily made in conductors.
2. The conductor enters (or leaves) the enclosure through the wall opposite its terminals. This is a more difficult condition because the conductor must make an offset or double bend to go from the terminal and then align with the raceway or cable entrance. This is also called an "S" or a "Z" bend because of its configuration, as shown at the top left of Fig. 373-3. For such bends, Table 373-6(b) specifies a greater distance from the end of the lug to the opposite wall to accommodate the two 45° bends, which are made difficult by the short lateral space between lugs and the stiffness of conductors (especially with the plastic insulations in cold weather).

Table 373-6(b) provides increased bending space to accommodate use of factory-installed connectors that are not of the lay-in or removable type and to allow use of field-installed terminals that are not designated by the manufacturer as part of the equipment marking. Exception No. 1 to part **(b)(1)** is shown in the bottom drawing of Fig. 373-3.

Note: For providing Code-required bending space at terminals for enclosed switches or individually enclosed circuit breakers, refer to Sec. 380-18. For

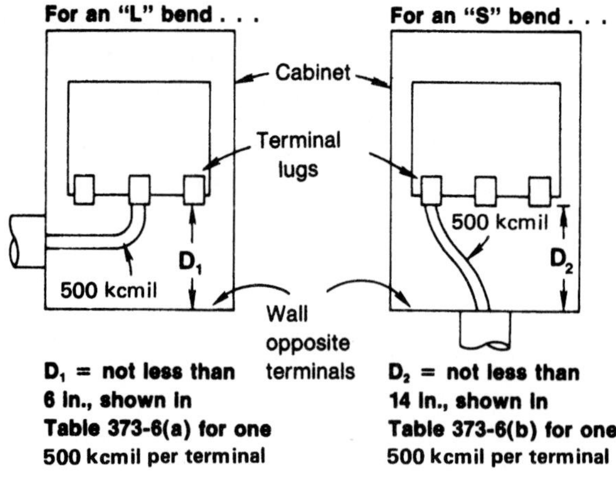

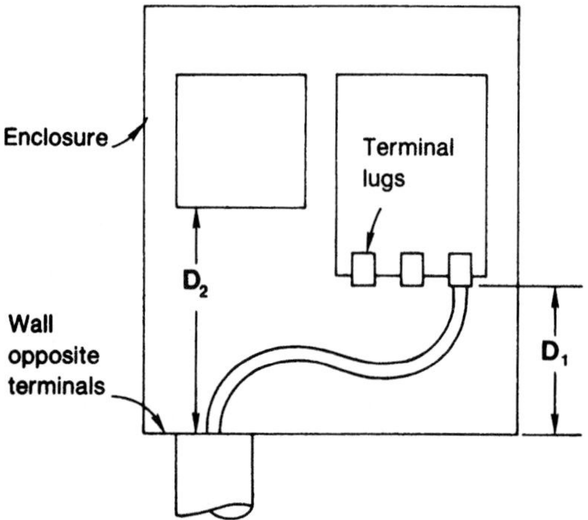

EVEN THOUGH CONDUCTOR LEAVES ENCLOSURE THROUGH WALL OPPOSITE LUGS, D_1 MAY BE SIZED FROM TABLE 373-6(a) PROVIDED THAT D_2 CONFORMS TO TABLE 373-6(b)

Fig. 373-3. These clearances are minimums that must be observed. (Sec. 373-6.)

conductor bending space at panelboard terminals, refer to Sec. 384-35. In Fig. 373-3, the clearances shown are determined from Table 373-6(a) or Table 373-6(b), under the column for one wire per terminal. For multiple-conductor circuit makeups, the clearance at terminals and in side gutters has to be greater, as shown under two, three, four, etc., wires per terminal.

Exception No. 2 of part **(b)(1)** covers application of conductors entering or leaving a meter-socket enclosure, and was based on a study of 100- and 200-A meter sockets.

Paragraph **(c)** applies to all conductors of size No. 4 or larger entering a cabinet or box from rigid metal conduit, flexible metal conduit, electrical metallic tubing, etc. To protect the conductors from cutting or abrasion a smoothly rounded insulating surface is required. While many fittings are provided with insulated sleeves or linings, it is also possible to use a separate insulating lining or sleeve to meet the requirements of the Code. Figure 373-4 shows use of a

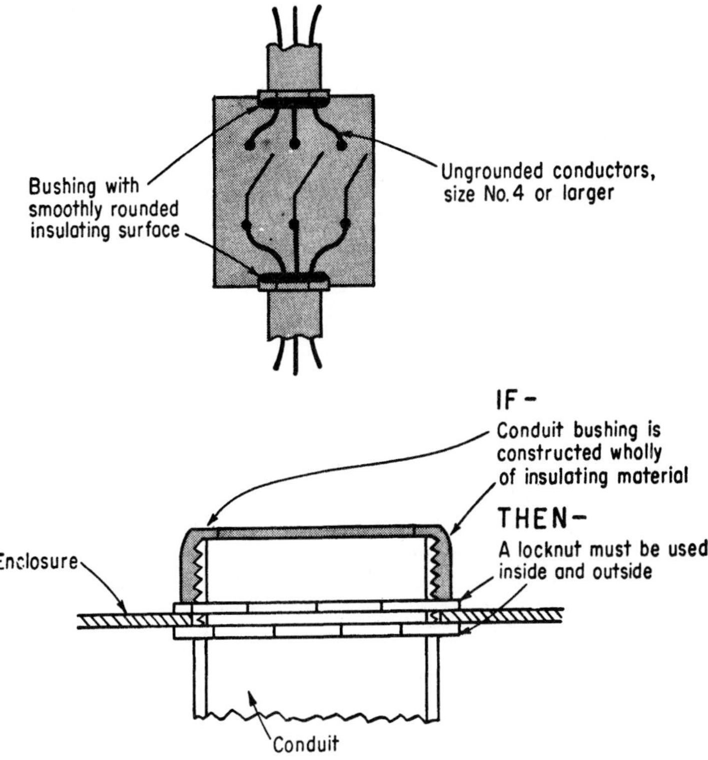

Fig. 373-4. An insulated-throat bushing or other protection must be used at enclosure openings. (Sec. 373-6.)

bushing with an insulated edge or a completely nonmetallic bushing to satisfy this rule. Figure 373-5 shows an approved sleeve which may be used to separate the conductors from the raceway fitting, which may be installed after the conductors are already installed and connected.

In the Exception to part **(c)** an insulated throat is not required for conductor protection on enclosure threaded hubs or bosses that have a rounded or flared entry surface. This is recognition of a long-standing reality—that there is no need for protective insulating material around the interior opening of integral hubs and bosses on equipment enclosures. Insulated-throat bushings and connectors are needed only for entries through KOs in sheet-metal enclosures.

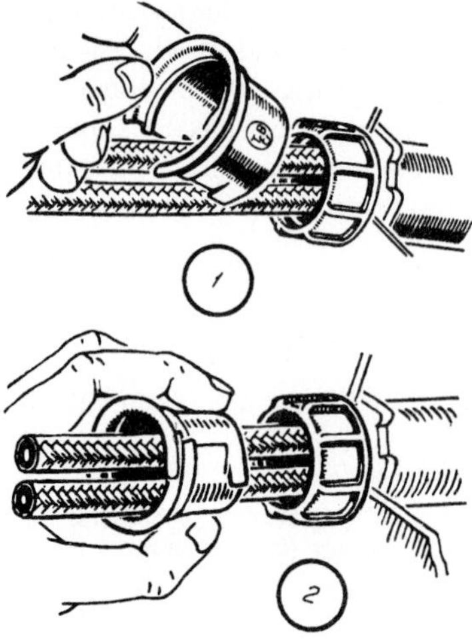

Fig. 373-5. Slip-over nonmetallic sleeve may be used to cover metal bushing throat. (Sec. 373-6.)

The last paragraph of part **(c)** prohibits use of a plastic or phenolic bushing ("wholly of insulating material") as a device for securing conduit to an enclosure wall. On a KO, there must be a metal locknut outside and a metal locknut inside to provide tight clamping to the enclosure wall, with the nonmetallic bushing put on after the inside locknut. An EMT or conduit connector must also be secured in position by a metal locknut and not by a nonmetallic bushing.

Part **(c)** also requires that any insulating bushing or insulating material used to protect conductors from abrasion must have a temperature rating at least equal to the temperature rating of the conductors.

373-8. Enclosures for Switches or Overcurrent Devices. The basic rule here is a follow-up to the rule of Sec. 373-7.

Most enclosures for switches and/or overcurrent devices have been designed to accommodate only those conductors intended to be connected to terminals within such enclosures. And in designing such equipment it would be virtually impossible for manufacturers to anticipate various types of "foreign" circuits, feed-through circuits, or numerous splices or taps.

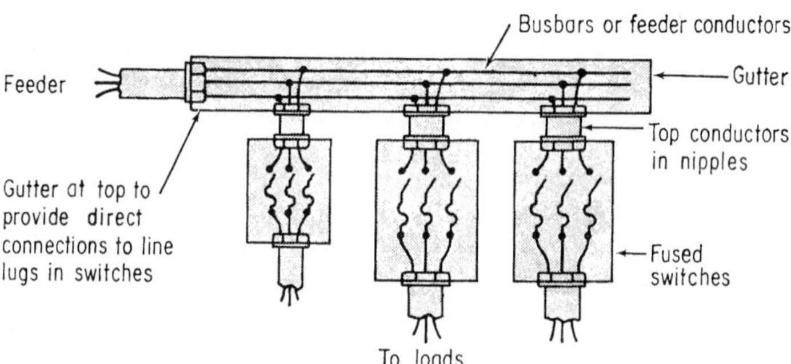

Fig. 373-6. Feeder taps in auxiliary gutter keep feeder cables and tap connectors out of switch enclosures. (Sec. 373-8.)

The rule here states enclosures for switches, CBs, panelboards, or other operating equipment must not be used as junction boxes, troughs, or raceways for conductors feeding through or tapping off, unless designs suitable for the purpose are employed to provide adequate space. This rule affects installations in which a number of branch circuits or subfeeder circuits are to be tapped from feeder conductors in an auxiliary gutter, using fused switches to provide disconnect and overcurrent protection for the branch or subfeeder circuits. It also applies to feeder taps in panelboard cabinets.

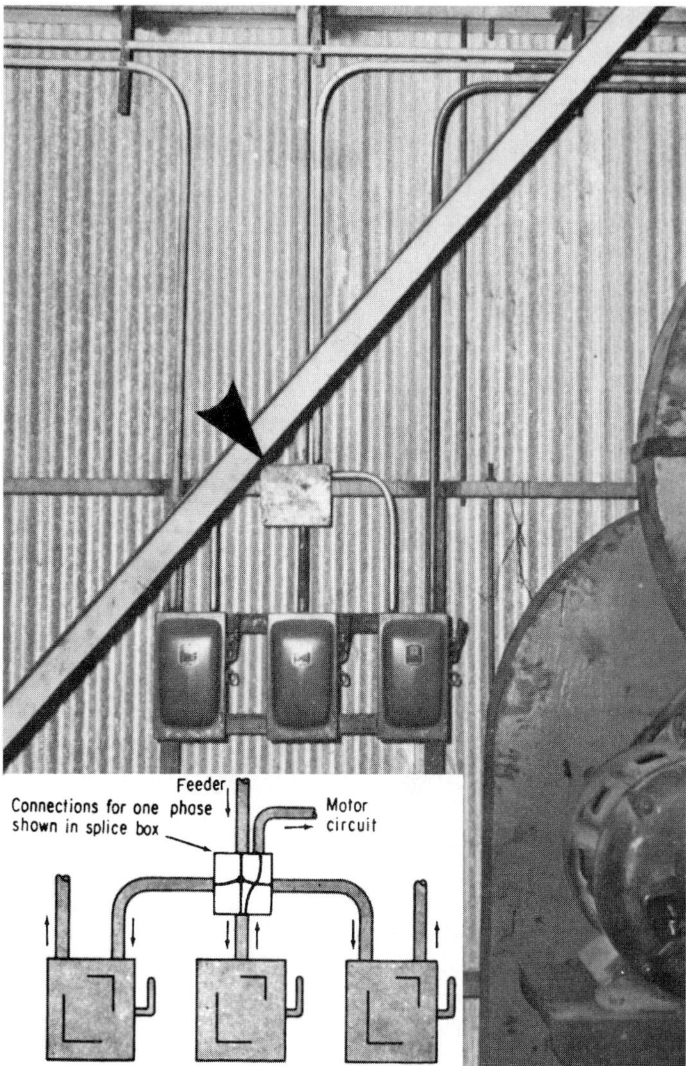

Fig. 373-7. Junction box (arrow) is used for tapping feeder conductors to supply individual motor branch circuits—as shown in inset diagram. (Sec. 373-8.)

In general, the most satisfactory way to connect various enclosures together is through the use of properly sized auxiliary gutters (Fig. 373-6) or junction boxes. Figure 373-7 shows a hookup of three motor disconnects, using a junction box to make the feeder taps. Following this concept, enclosures for switches and/or overcurrent devices will not be overcrowded.

There are cases where large enclosures for switches and/or overcurrent devices will accommodate additional conductors and this is generally where the 40 percent (conductor space) and 75 percent (splices or taps) at one cross section would apply. An example would be control circuits tapped off or extending through 200-A or larger fusible switches or CB enclosures. The csa within such enclosures is the *free gutter wiring space* intended for conductors.

The Exception to this rule is shown in Fig. 373-8 and applied as follows:

example: If an enclosure has a gutter space of 3 by 3 in., the csa would be 9 sq in. Thus, the total conductor fill (use Table 5, Chap. 9) at any cross section (including conductors) could not exceed 6.75 sq in. (9 × 0.75).

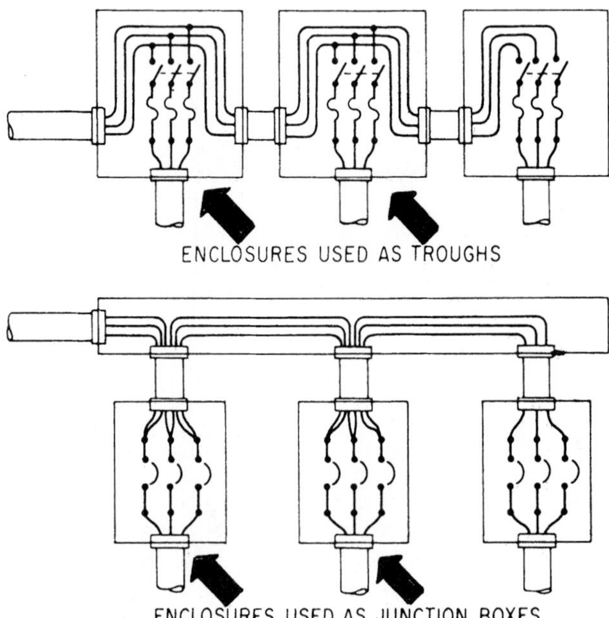

Fig. 373-8. These hookups are permitted where space in enclosure gutters satisfies Exception to basic rule. (Sec. 373-8.)

In the case of large conductors, a splice other than a wire-to-wire "C" or "tube" splice would not be acceptable if the conductors at the cross section are near a 40 percent fill, because this would leave only a 35 percent space for the splice. Most splices for larger conductors with split-bolt connectors or similar types are usually twice the size of the conductors being spliced. Accordingly, where larger conductors are to be spliced within enclosures, the total con-

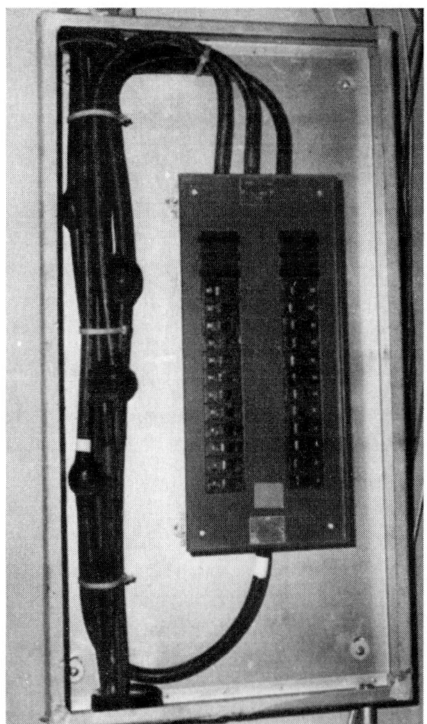

Fig. 373-9. The Exception to Sec. 373-8 permits feeding through and tapping off in cabinets for panelboards on feeder risers, where the side gutter is specially oversized for the application. (Sec. 373-8.)

ductor fill should not exceed *20 percent* to allow for any bulky splice at a cross section.

Figure 373-9 shows an example of feeder taps made in panelboard side gutter where the cabinet is provided with adequate space for the large feeder conductors and for the bulk of the tap devices with their insulating tape wrap.

ARTICLE 374. AUXILIARY GUTTERS

374-1. Use. The Code recognizes both nonmetallic and metallic auxiliary gutters just as it does wireways. Auxiliary gutters are sheet-metal troughs in which conductors are laid in place after the gutter has been installed, i.e., nonmetallic gutter is made from a plastic-like material. Auxiliary gutters are used as parts of complete assemblies of apparatus such as switchboards, distribution centers, and control equipment, as shown in Fig. 374-1. But auxiliary gutters may not

Fig. 374-1. Typical applications of auxiliary gutters provide the necessary space to make taps, splices, and other conductor connections involved where a number of switches or CBs are fed by a feeder (top) or for multiple-circuit routing, as at top of a motor control center (right) shown with a ground bus in gutter (arrow). (Sec. 374-1.)

contain equipment even though it looks like surface metal raceway (Art. 352), which may contain devices and equipment (Fig. 374-2).

374-2. Extension Beyond Equipment. Auxiliary gutters are not intended to be a type of general raceway and are not permitted to extend more than 30 ft (9.14 m) beyond the equipment which they supplement, except in elevator work. Where an extension beyond 30 ft (9.14 m) is necessary, Art. 362 for wireways must be complied with. The label of Underwriters Laboratories Inc. on each length of trough bears the legend "Wireways or Auxiliary Gutters," which indicates that they may be identical troughs but are distinguished one from the other only by their use. See comments following Sec. 362-1 in this handbook.

374-5. Number of Conductors. The rules on permitted conductor fill for auxiliary gutters are basically the same as those for wireways. Refer to Sec. 362-5. Note that Exception No. 3 permits more than 30 general circuit wires; but where over 30 wires are installed, the correction factors specified in Note 8 to Tables 310-16 through 310-19 must be applied.

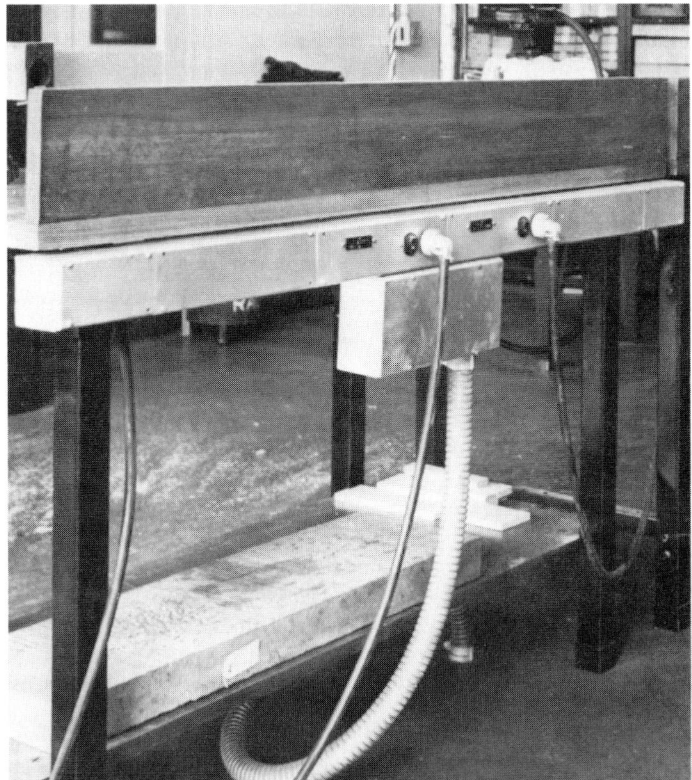

Fig. 374-2. Surface metal raceway may be used with accessory circuit breakers and/or receptacles in cover plates—but auxiliary gutters may not be used like this. (Sec. 374-1.)

No limit is placed on the size of conductors that may be installed in an auxiliary gutter.

The csa of rubber-covered and thermoplastic-covered conductors given in Table 5, Chap. 9, must be used in computing the size of gutters required to contain a given combination of such conductors.

Figure 374-3 shows a typical gutter application where the conductor fill must be calculated to determine the acceptable csa of the gutter. There are several factors involved in sizing auxiliary gutters that often lead to selecting the wrong size. The two main factors are how conductors enter the gutter and the contained conductors at any cross section. The minimum required width of a gutter is determined by the csa occupied by the conductors and splices and the space necessary for bending conductors entering or leaving the gutter. The total csa occupied by the conductors at any cross section of the gutter must not be greater than 20 percent of the gutter interior csa at that point (Sec. 374-5). The total csa occupied by the mass of conductors and splices at any cross section of the gutter must not be greater than 75 percent of the gutter interior csa at that point [Sec. 374-8(a)].

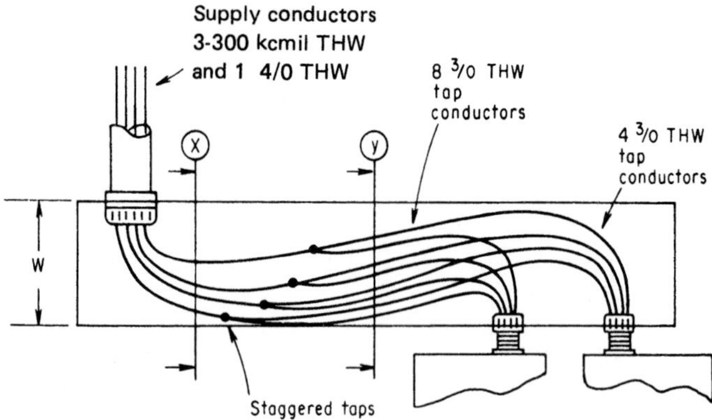

Fig. 374-3. Minimum acceptable gutter cross section and depth must be calculated. (Sec. 374-5.)

In the gutter installation shown, assume that the staggering of the splices has been done to minimize the area taken up at any cross section—to keep the mass of splices from all adding up at the same cross section. The greatest conductor concentration is therefore either at section x, where there are three 300-kcmil and one 4/0 THW conductors, or at section y, where there are eight 3/0 THW conductors. To determine at which of these two cross sections the fill is greater, apply the appropriate csa's of THW conductors as given in Table 5, Chap. 9:

1. The total conductor csa at section x is 3 × 0.5581 sq in. plus 1 × 0.3904, or 2.0647 sq in.
2. The total conductor csa at section y is 8 × 0.3288 sq in. or 2.6304 sq in.

Section y is, therefore, the determining consideration. Because that fill of 2.6304 sq in. can at most be 20 percent of the gutter csa, the total gutter area must be at least 5 times this conductor fill area, or 13.152 sq in.

Assuming the gutter has a square cross section (all sides of equal width) and the sides have an integral number of inches, the nearest square value would be 16 sq in., indicating a 4- by 4-in. (102- by 102-mm) gutter, and that would be suitable if the 300-kcmil conductors entered the end of the gutter instead of the top. But because those conductors are deflected entering and leaving the gutter, the first two columns of Table 373-6(a) must also be applied to determine whether the width of 4 in. (102 mm) affords sufficient space for bending the conductors. That consideration is required by Sec. 374-9(d). The worst condition (largest conductors) is where the supply conductors enter; therefore the 300-kcmil cable will determine the required space.

Table 373-6(a) shows that a circuit of one 300 kcmil per phase leg (or wire per terminal) requires a bending space at least 5 in. (127 mm) deep (in the direction of the entry of the 300-kcmil conductors), calling for a standard 6- by 6-in. (152- by 152-mm) gutter for this application.

In Fig. 374-3, if the 300-kcmil conductors entered at the left-hand end of the gutter instead of at the top, Sec. 374-9(d) would require Table 373-6(a) to be applied only to the deflection of the No. 3/0 conductors. The table shows, under one wire per terminal, a minimum depth of 4 in. (102 mm) is required. In that case, a 4- by 4-in. (102- by 102-mm) gutter would satisfy.

374-8. Splices and Taps. Part **(a)** is discussed above, under Sec. 374-5.

Part **(b)** covers cases where bare busbar conductors are used in gutters. The insulation might be cut by resting on the sharp edge of the bar or the bar might become hot enough to damage the insulation. When taps are made to bare conductors in a gutter, care should be taken so as to place and form the wires in such a manner that they will remain permanently separated from the bare bars.

Part **(c)** requires that identification be provided wherever it is not clearly evident what apparatus is supplied by the tap. Thus if a single set of tap conductors are carried through a short length of conduit from a gutter to a switch and the conduit is in plain view, the tap is fully identified and needs no special marking; but if two or more sets of taps are carried in a single conduit to two or more different pieces of apparatus, each tap should be identified by some marking such as a small tag secured to each wire.

ARTICLE 380. SWITCHES

380-1. Scope. Note that all the provisions of this article that cover switches *also* apply to circuit breakers, which are operated exactly as a switch whenever they are manually moved to the ON or OFF position.

380-2. Switch Connections. The rule of part **(a)** is shown in Fig. 380-1. Keeping both the supply and return conductors in the same raceway or cable minimizes inductive heating, as described under Sec. 300-20(a).

The rule of part **(b)** is illustrated in Fig. 380-2. The "ACCEPTABLE" three-pole switch satisfies Exception No. 1. Opening only the grounded wire of a 2-wire circuit would leave all devices that are connected to the circuit alive and at a voltage to ground equal to the voltage between wires on the mains. In case of an accidental ground on the grounded wire, the circuit would not be controlled by the single-pole switch.

In Fig. 380-3, the load consists of lamps connected between the neutral and the two outer wires and is not balanced. Opening the neutral while the other wires are connected would cause the voltages to become unbalanced and might burn out all lamps on the more lightly loaded side.

Except for Sec. 514-5, which requires a switch in a grounded neutral for a circuit to a pump at a gas station, the neutral does not need to be switched. But, where a grounded neutral or grounded phase leg is switched, it must never be by a single-pole switch or single-pole CB, even if the CBs are provided with handle ties.

A switch may be arranged to open the grounded conductor if the switch simultaneously opens all the other conductors of the circuit.

380-3. Enclosure. Figure 380-4 shows the basic rule of this section and Sec. 380-4. This rule also requires adequate wire bending space at terminals and in side gutters of switch enclosures. In this section and in other sections applying to wiring space around other types of equipment it is a mandatory **Code**

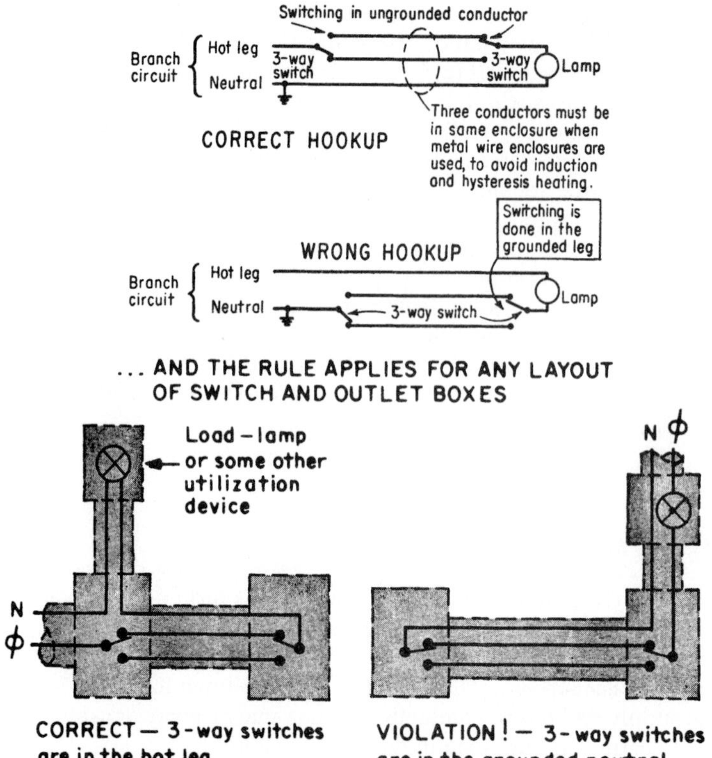
Switching in ungrounded conductor
Branch circuit
Hot leg
Neutral
3-way switch
3-way switch
Lamp
CORRECT HOOKUP
Three conductors must be in same enclosure when metal wire enclosures are used, to avoid induction and hysteresis heating.
Switching is done in the grounded leg
WRONG HOOKUP
Branch circuit
Hot leg
Neutral
3-way switch
Lamp
... AND THE RULE APPLIES FOR ANY LAYOUT OF SWITCH AND OUTLET BOXES
Load – lamp or some other utilization device
N
ϕ
N ϕ
CORRECT — 3-way switches are in the hot leg (ungrounded leg) of the circuit to the load.
VIOLATION! — 3-way switches are in the grounded neutral leg to the load.
NOTE: Wiring between switches – in the armor of BX or in metal raceway – must have all three conductors within the single cable or raceway.

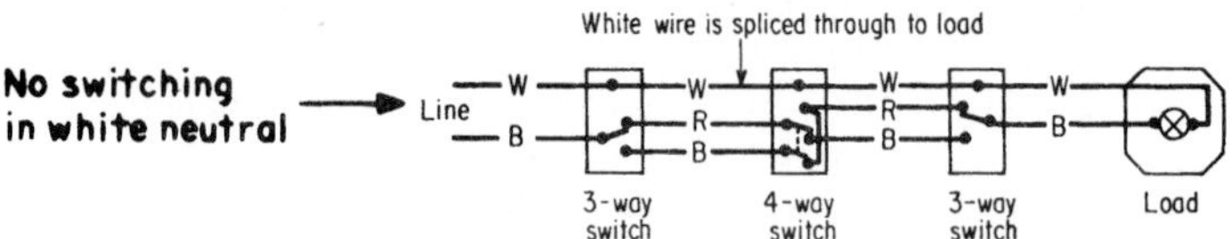
White wire is spliced through to load
No switching in white neutral
Line
W
B
R
3-way switch
4-way switch
3-way switch
Load

Fig. 380-1. All three-way and four-way switches must be placed in the hot conductor to the load. (Sec. 380-2.)

requirement that wire bending space and side gutter wiring space conform to the requirements of Table 373-6(a) for side gutters and to Table 373-6(b) for wire bending space at the line and load terminals, as described under Sec. 380-18. Those tables establish the minimum distance from wire terminals to enclosure surface or from the sides of equipment to enclosure side based on the size of conductors being used, as shown in Fig. 380-5.

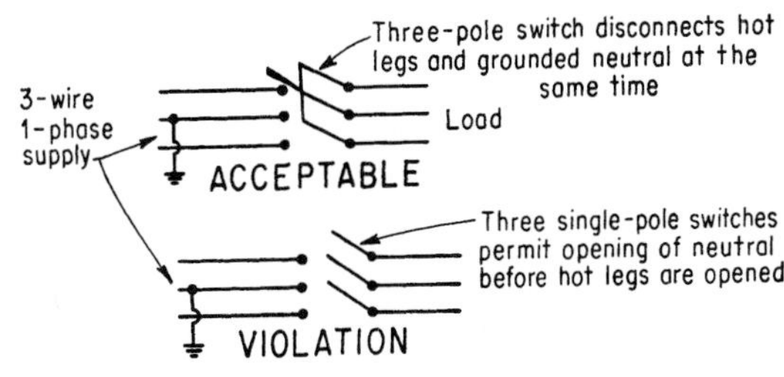
Three-pole switch disconnects hot legs and grounded neutral at the same time
3-wire 1-phase supply
Load
ACCEPTABLE
Three single-pole switches permit opening of neutral before hot legs are opened
VIOLATION

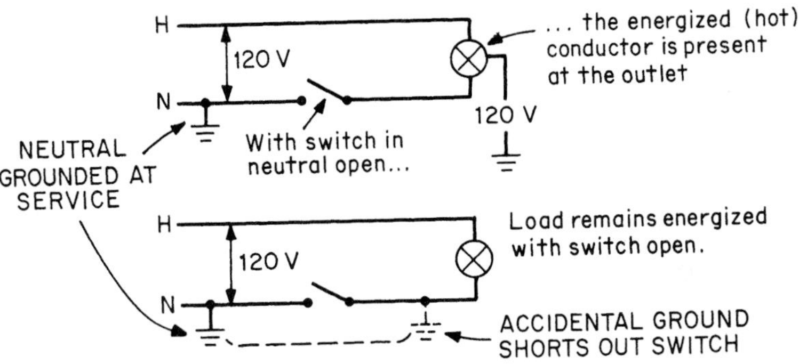
H
120 V
N
... the energized (hot) conductor is present at the outlet
120 V
With switch in neutral open...
NEUTRAL GROUNDED AT SERVICE
H
120 V
N
Load remains energized with switch open.
ACCIDENTAL GROUND SHORTS OUT SWITCH

Fig. 380-2. A single-pole switch must not be used in a grounded circuit conductor. (Sec. 380-2.)

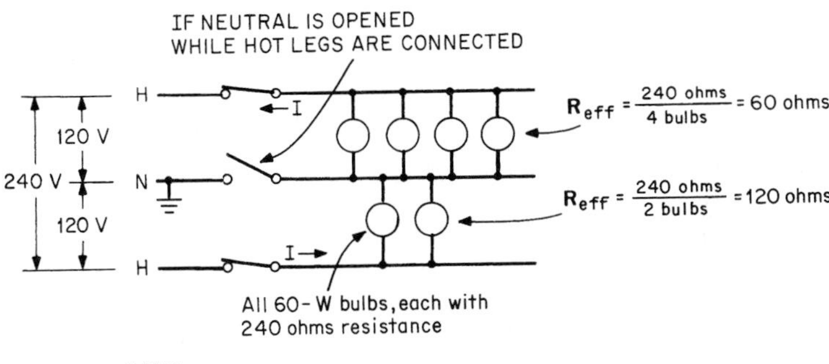
IF NEUTRAL IS OPENED WHILE HOT LEGS ARE CONNECTED
H
I
120 V
240 V
N
120 V
H
I
R eff = 240 ohms / 4 bulbs = 60 ohms
R eff = 240 ohms / 2 bulbs = 120 ohms
All 60-W bulbs, each with 240 ohms resistance
Current = 240 V / 60 + 120 = 1.33 A
Voltage on top bulbs = 1.33 X 60 ohms = 80 V
Voltage on bottom bulbs = 1.33 X 120 ohms = 160 V

Fig. 380-3. A single-pole switch in neutral can cause damaging load unbalance if opened. (Sec. 380-2.)

This whole concern for adequate wiring space in all kinds of equipment enclosures reflects a repeated theme in many Code sections as well as in Art. 110 on general installation methods. One of the most commonly heard complaints from constructors and installers in the field concerns the inadequacy of wiring space at equipment terminals. Section 380-3 is designed to assure sufficient space for the necessary conductors run into and through switch enclosures.

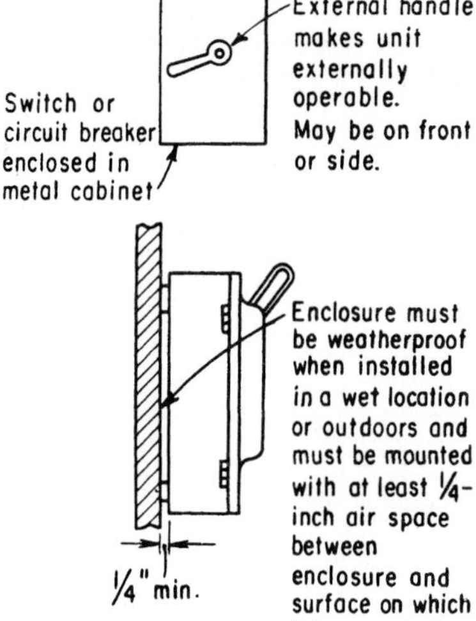
External handle
makes unit
externally
operable.
May be on front
or side.
Switch or
circuit breaker
enclosed in
metal cabinet
Enclosure must
be weatherproof
when installed
in a wet location
or outdoors and
must be mounted
with at least ¼-
inch air space
between
enclosure and
surface on which
it is mounted.
¼" min.

Fig. 380-4. Switch and CB enclosures must be suitable. (Sec. 380-3.)

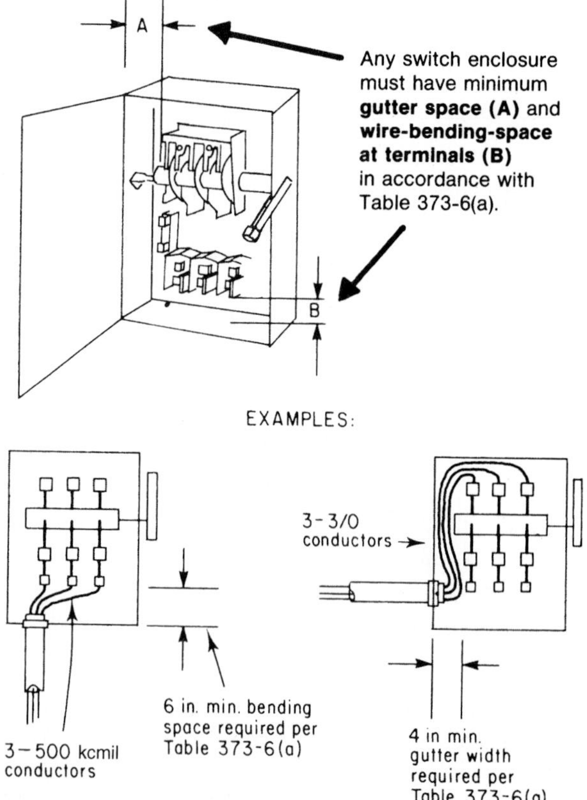
A
Any switch enclosure
must have minimum
gutter space (A) and
wire-bending-space
at terminals (B)
in accordance with
Table 373-6(a).
B
EXAMPLES:
3-3/0
conductors
3-500 kcmil
conductors
6 in. min. bending
space required per
Table 373-6(a)
4 in min.
gutter width
required per
Table 373-6(a)

Fig. 380-5. Terminating and gutter space in switch enclosures must be measured. (Sec. 380-3.)

380-4. Wet Locations. Refer to Fig. 380-4 and discussion under Sec. 373-2.

380-5. Time Switches, Flashers, and Similar Devices. Any automatic switching device should be enclosed in a metal box unless it is a part of a switchboard or control panel which is located as required for live-front switchboards. Such devices must not present exposed energized parts, except under very limited conditions where they are accessible only to qualified persons.

380-6. Position of Knife Switches (Enclosed and Open Types). The NE Code requires that knife switches be so mounted that gravity will tend to open them rather than close them (Fig. 380-6). But the Code recognizes use of an upside-down or

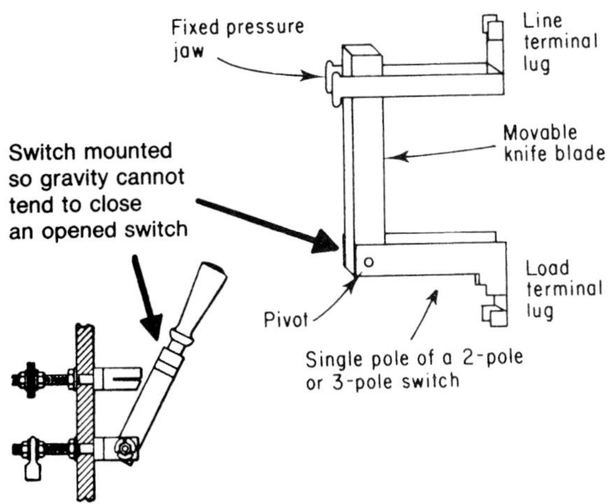

Fig. 380-6. Movable knife blade of a knife switch must be pivoted at its bottom. (Sec. 380-6.)

reverse-mounted knife switch where provision is made on the switch to prevent gravity from actually closing the switch contacts. This permission is given in recognition of the much broader use of underground distribution, with the intent of providing a switch with its line terminals fed from the bottom and its load terminals connected at the top (Fig. 380-7). With such a configuration, an upside-down knife switch provides the necessary locations of such terminals, that is, "line" at bottom and "load" at top. However, use of any knife switch in the reverse or upside-down position is contingent upon the switch being approved for such use, which virtually means UL-listed for that application and also upon the switch being equipped with a locking device that will prevent gravity from closing the switch. The same type of operation is permitted for double-throw knife switches.

As required by part **(c)**, knife-switch blades must be "dead" in the open position, except where a warning sign is used (Fig. 380-8). In a number of electrical system hookups—UPS systems, transformer secondary ties, and emergency generator layouts—electrical backfeed can be set up in such a way as to make the load terminals, blades, and fuses of a switch energized when the switch is

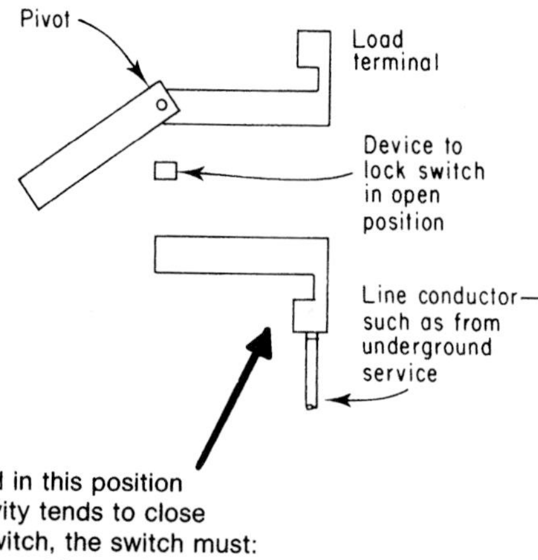
Pivot
Load terminal
Device to lock switch in open position
Line conductor—such as from underground service
To be used in this position where gravity tends to close an open switch, the switch must:
1. Be approved for such use, and
2. Be equipped with a locking device to hold switch open.

Fig. 380-7. This type of knife-switch operation is permitted. (Sec. 380-6.)

in the OFF or open position. Where that might happen, the Exception to this section says a permanent sign must be prominently placed at or near the switch to warn of the danger. The sign must read, "Warning—load side of switch may be energized by backfeed."

This potential hazard has long been recognized for high-voltage systems (over 600 V), and Sec. 710-24(o)(2) covers the matter. This Exception to the basic rule that the load side of the switch be de-energized when the switch is open applies the same concept to systems operating up to 600 V.

380-7. Indicating. Switches and circuit breakers in individual enclosures must be marked to clearly show ON and OFF positions, and vertically operated switches and CBs must be ON when in the up position. This is basically a repetition of the rule that has been in Sec. 240-81 for circuit breakers used in switchboards and panelboards.

380-8. Accessibility and Grouping. The rule of part **(a)** of this section, along with the Exceptions, is shown in Fig. 380-9. Exception No. 1 cross-references with Sec. 364-12.

Part **(b)** of this section applies where 277-V switches, mounted in a common box (such as two- or three-ganged), control 277-V loads, with the voltage between exposed line terminals of *adjacent* switches in the common box being 480 V. If the adjacent switches have exposed live terminals, anyone changing one of the switches without disconnecting the circuit at the panel could contact 480 V, as shown in Fig. 380-10. The rule of this section requires permanent barriers between adjacent switches located in the same box where the voltage between such switches exceeds 300 V and terminals are exposed.

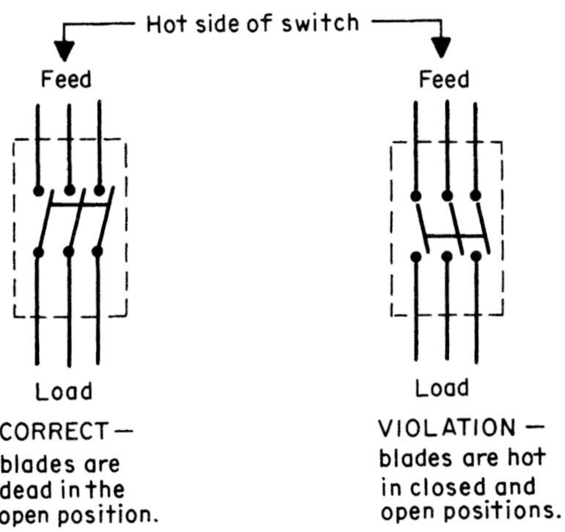

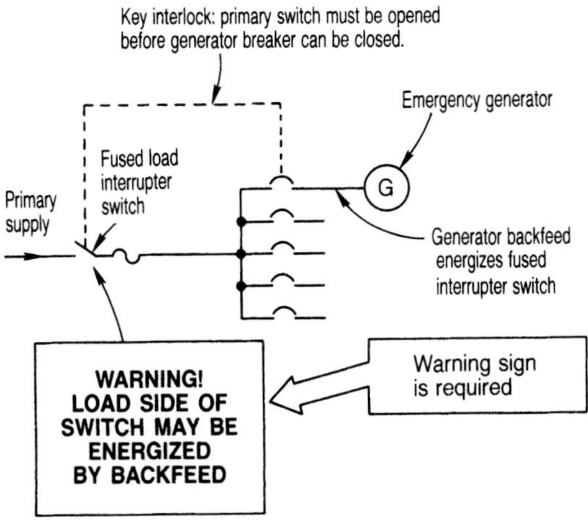

Fig. 380-8. Supply conductors must connect to "LINE" terminals of switch, but backfeed is permitted if carefully marked. (Sec. 380-6.)

If screwless terminal switches (with no exposed live parts) are used, it would *not* be a violation if any number of such switches are ganged in a common box. Where screwless switches are mounted side by side in a two-gang box, it would seem to satisfy the intent (and literal text) of Sec. 380-8 because the switches would be "arranged" to prevent exposure to 480 V. Of course, the hookup shown would be acceptable if a separate single-gang box and plate are used for each switch, or a common wire from only one phase (A, B, or C) supplies all the three switches in the three-gang box.

380-9. Faceplates for Flush-Mounted Snap Switches. Figure 380-11 shows the basic rule of the first sentence of this section. Note that the recessed metal box

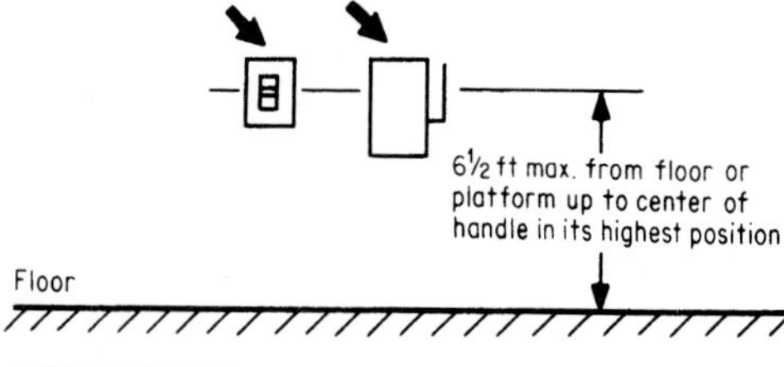

EXCEPTIONS

1. Fused switch or CB may be up on busway . . .

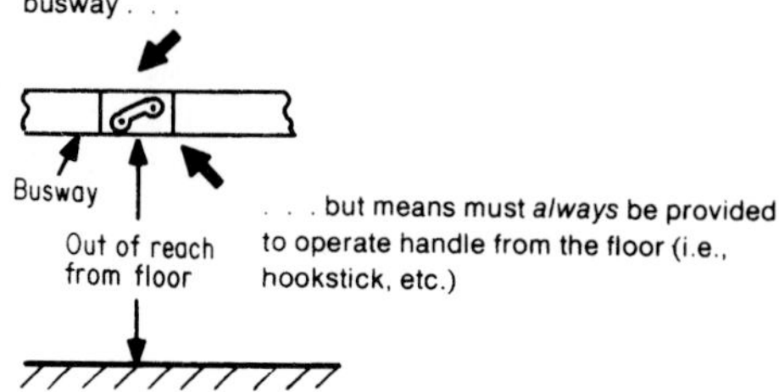

. . . but means must *always* be provided to operate handle from the floor (i.e., hookstick, etc.)

2. Switch adjacent to motor, appliance, or other equipment it supplies, at high mounting, but accessible by portable ladder or similar means

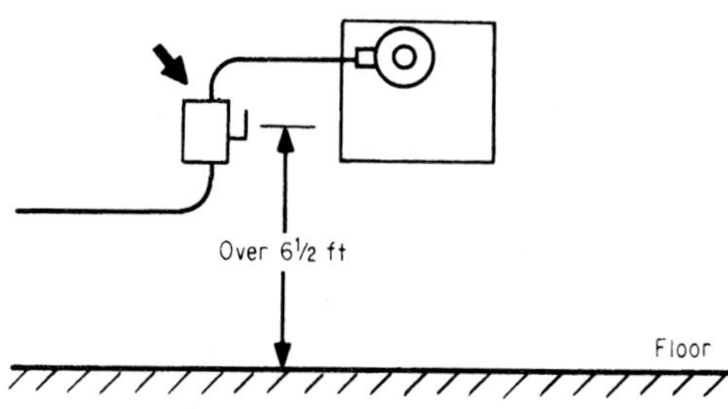

3. Hookstick-operable isolating switches are permitted at heights over 6½ ft

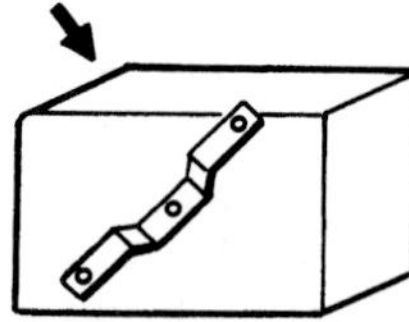

Fig. 380-9. All switches and circuit breakers used as switches must be capable of being operated by a person from a readily accessible place. (Sec. 380-8.)

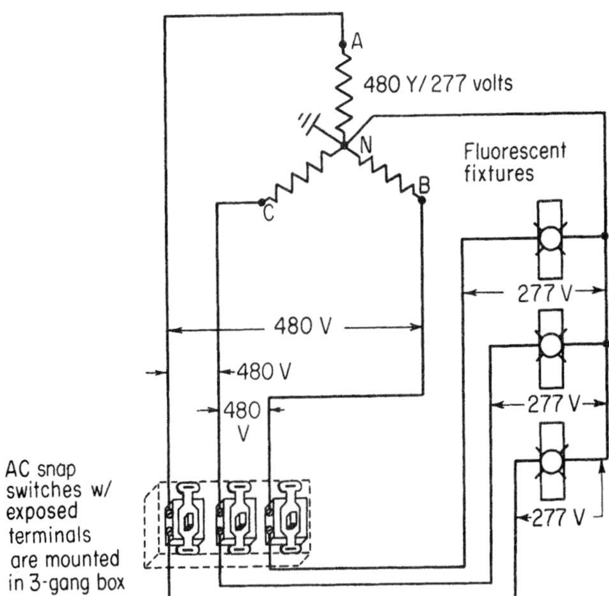

Fig. 380-10. This is a violation if barriers are not used between switches in the box. (Sec. 380-8.)

is not grounded and that is acceptable because the box is not exposed to contact. Section 250-42 would apply to an exposed box. This rule conforms to the spirit of Sec. 410-18(a), which requires any metal faceplate (metal faceplates come under "exposed conductive parts of . . . equipment") to be grounded when it is attached to a box fed by a wiring system that contains an equipment grounding conductor. That is discussed also under Sec. 370-25(a).

The wording of this rule of Sec. 380-9 requires the nonmetallic faceplate on an ungrounded metal box only when the faceplate is "within reach of conducting floors," etc. But Sec. 410-18(b) is often taken as requiring a nonmetallic faceplate on every box that does not contain an equipment grounding "means" (metal raceway, metal cable armor, ground wire in NM cable)—whether or *not* it is "within reach of conducting" or grounded parts. Of course, ungrounded metal boxes are not generally encountered in new work.

The last sentence of this section requires that faceplates be installed to cover the wall opening completely to assure that the box behind the faceplate is properly covered and to prevent any openings that could afford penetration to energized parts.

380-10. Mounting of Snap Switches. The purpose of paragraph **(b)** is to prevent "loose switches" where openings around *recessed* boxes provide no means of seating the switch mounting yoke against the box "ears" properly. It also permits the maximum projection of switch handles through the installed switch plate. The cooperation of other crafts, such as dry-wall installers, will be required to satisfy this rule.

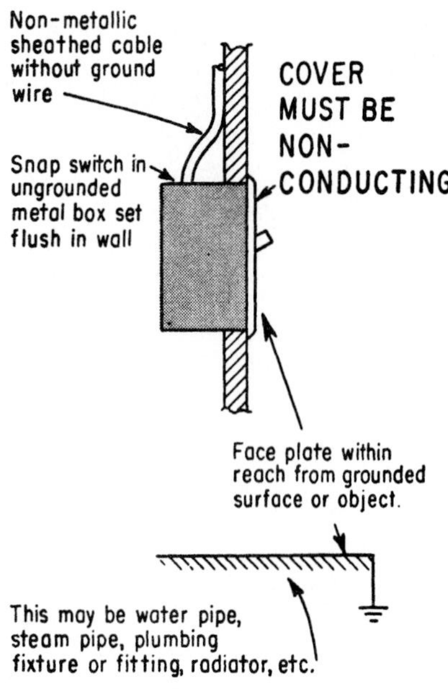

Fig. 380-11. Nonmetallic faceplate eliminates shock hazard. (Sec. 380-9.)

380-11. Circuit Breakers as Switches. Molded-case CBs are intended to be mounted on a vertical surface in an upright position or on their side. Use in any other position requires evaluation for such use. ON and OFF legends on CBs and switches are not intended to be mounted upside down.

380-12. Grounding of Enclosures. This section calls for the indicated equipment to be grounded in accordance with Art. 250. Sec. 250-42 requires *all* exposed metal parts (including enclosures) of fixed equipment to be grounded under any of the conditions described. And any switch or CB enclosure that is fed by metal raceway or metal-covered cable must be grounded. Additionally, provisions must be made when nonmetallic enclosures are used with metallic raceways and cables to assure grounding continuity between all interconnected raceways, cables, and any equipment within the enclosure.

380-13. Knife Switches. UL data on ratings and application correspond to the Code data. Specific UL rules are as follows:

Knife (WIOV)

This listing covers open type knife switches either with or without fuseholders for plug fuses or for 0-600 amp cartridge fuses having no current interrupting rating included in their marking; switches having individual bases designed for either front or rear wiring connection; and switch parts without bases designed for mounting on switchboards and panelboards. Switches may be single- or multiple-pole, and with or without quick-break or auxiliary contacts, except where such contacts are specifically required.

Knife switches without fuseholders (unfused) have been tested to determine their acceptability for continuous operation at their marked rated load.

Knife switches are provided with studs or terminal pads to which listed pressure wire connectors can be factory or field installed to accommodate field wiring.

"Knife switches are marked with a short-circuit current withstand rating."

Standard voltage ratings for knife switches are: 125, 250, 250 dc-500 ac, 600. Unless otherwise indicated, the rating includes both alternating and direct current.

Standard current ratings for knife switches are: 30, 60, 100, 200, 400, 600, 800, 1200, 1600, 2000, 2500, 3000, 4000, 5000, 6000.

"Switches with knife blade action but with external operating handles are covered as enclosed switches, dead-front switches, service equipment or fused power circuit devices."

The basic standard used to investigate products in this category is UL363, "Knife Switches". (See Fig. 380-12.)

380-14. Rating and Use of Snap Switches. For a noninductive load not including any tungsten-filament lamps, a snap switch is merely required to have an ampere rating at least equal to the ampere rating of the load it controls. Electrically heated appliances are about the only common examples of such loads.

For the control of loads consisting of tungsten lamps alone, or tungsten lamps combined with any other noninductive load, snap switches should be "T" rated, or for alternating current circuits, a general-use AC snap switch should be used.

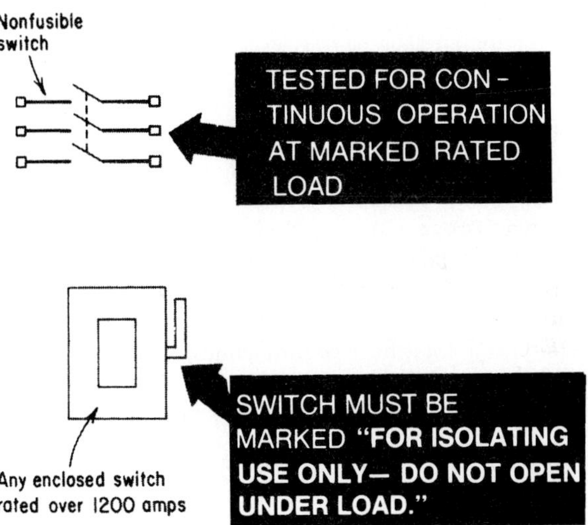

Fig. 380-12. UL data must be correlated to the Code rules. (Sec. 380-13.)

The term *snap switch* as used here and elsewhere in the Code is intended to include, in general, the common types of flush and surface-mounted switches used for the control of lighting equipment and small appliances and the switches used to control branch circuits on lighting panelboards. These switches are now usually of the tumbler or toggle type but can be the rotary-

Fig. 380-13. Top wire-bending space contains offset ("S") bends and must have dimensions from Table 373-6(b). Bottom wiring space has single ("L") bends—but must conform to Table 373-6(b), not Table 373-6(a). That varies from Section 373-6(b)(1), which permits terminal space from Table 373-6(a) when wires go out the side of the enclosure.

snap or pushbutton type. The term is not applied to CBs or to switches of the type that are commonly known as safety switches or *knife switches.* See definition of "switches" in Art. 100.

380-18. Wire Bending Space. At terminals of individually enclosed switches and circuit breakers, the spacing from lugs to the opposite wall *must* be at least equal to that of Table 373-6(b) for the given size and number of conductors per lug. The larger spacing of that table, rather than the smaller spacing of Table 373-6(a), must be used regardless of how conductors enter or leave the enclosure—on the sides or opposite terminals.

Fig. 380-13 shows the rule on wire bending space in switch or CB enclosures.

ARTICLE 384. SWITCHBOARDS AND PANELBOARDS

384-3. Support and Arrangement of Busbars and Conductors. Part **(a)** notes that only those conductors intended for termination in a vertical section of a switchboard may be run within that section, other than required inner connections and/or control wiring. This rule was intended to prevent repetition of the many cases on record of damage to switchboards having been caused by termination failures in one section being transmitted to other parts of the switchboard. In order to comply with this requirement, it will be necessary in some cases to provide auxiliary gutters. The basic concept behind the rule is that any load conductors originating at the load terminals of switches or breakers in a switchboard must be carried vertically, up or down, so that they leave the switchboard from that vertical section. Such conductors may not be carried horizontally to or through any other vertical section of the switchboard, except as indicated in the sketch at the top of Fig. 384-1. Because of field conditions, in which the installer did not know the location of protective devices in a switchboard at the time of conduit installations, the Exception modifies the requirement that conductors within a vertical section of a switchboard terminate in that section. Conductors may pass horizontally from one vertical section to another *provided* that the conductors are isolated from switchboard busbars by some kind of barrier.

The last sentence of part **(a)** requires that all service switchboards have a barrier installed within the switchboard to isolate the service busbars and the service terminals from the remainder of the switchboard as shown in Fig. 384-1. Because it is usually impossible to kill the circuit feeding a service switchboard, it has become very common practice for mechanics to work on switchboards with the service bus energized. The hazard associated with this has caused concern and is the reason for this addition to the Code.

Switchboard manufacturers in many parts of the country have been supplying switchboards with these barriers in place; this Code rule aims at making such protection for personnel a standard requirement. With a barrier of this type installed in a service switchboard mechanics working on feeder devices for other sections of the switchboard will not be exposed to accidental or surprise contact with the energized parts of the service equipment itself.

Part **(c)** requires a bonding jumper in a switchboard or panelboard used for service equipment to connect the grounded neutral or grounded phase leg to the equipment grounding conductor (the metal frame or enclosure of the equipment). UL data apply to this rule:

1. Switchboard sections or interiors are optionally intended for use either as a feeder distribution switchboard or as a service switchboard. For service use, a switchboard must be marked "Suitable for use as service equipment."
2. Some switchboard sections or interiors include neutral busbars factory-bonded to the switchboard enclosure. **Such switchboards are marked "Suitable *only* for use as service equipment" and may *not* be used as sub-distribution switchboards (Fig. 384-2).** A bonded neutral bus in a service switchboard may also serve as an equipment grounding busbar.
3. UL-listed unit substations have the secondary neutral bonded to the enclosure and have provision on the neutral for connection of a grounding conductor, as shown in Fig. 384-3. A terminal is also provided on the enclosure near the line terminals for use with an equipment grounding conductor run from the enclosure of primary equipment feeding the unit sub to the enclosure of the unit sub. Connection of such an equipment grounding conductor provides proper bonding together of equipment enclosures where the primary feed to the unit sub is direct-buried underground or is run in nonmetallic conduit without a metal conduit connection in the primary feed.

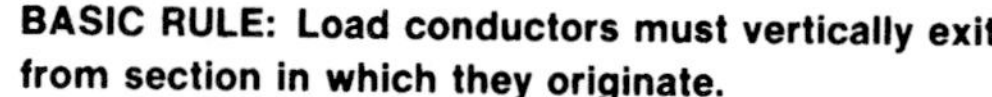

BASIC RULE: Load conductors must vertically exit from section in which they originate.

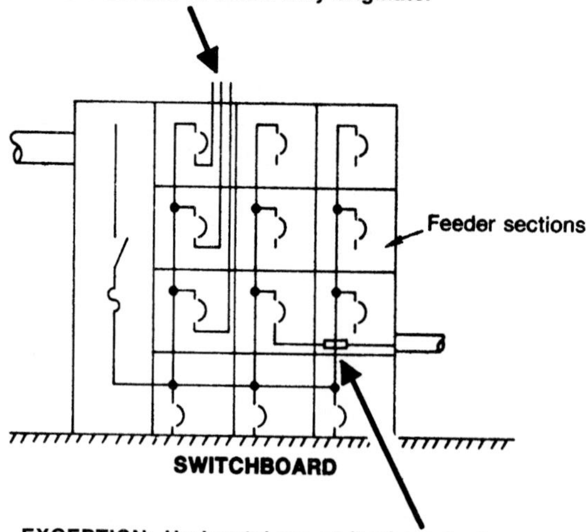

All conductor terminals within a switchboard must be used only for connection to conductors that leave the switchboard vertically from the same switchboard section in which the terminals are located—with the conductors run out the top or out of the bottom of the switchboard.

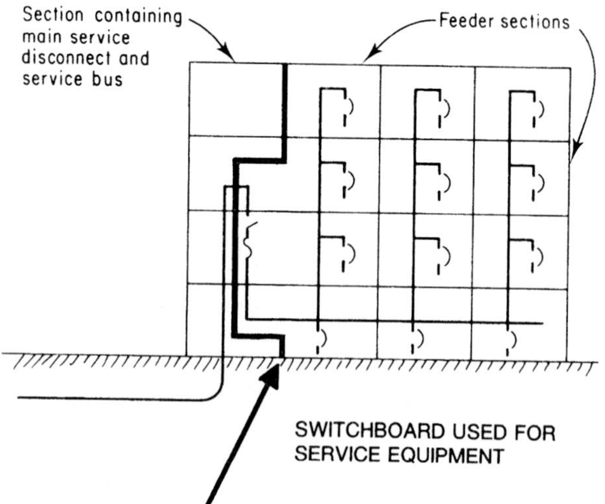

Fig. 384-1. These are basic rules on switchboard wiring. (Sec. 384-3.)

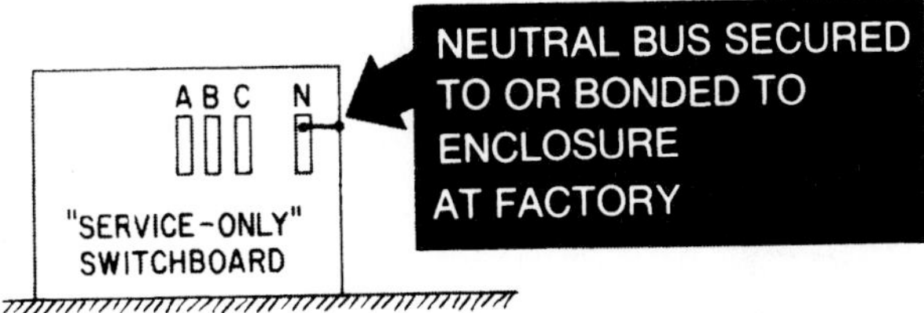

Fig. 384-2. Bonded neutral bus limits switchboard to service applications. (Sec. 384-3.)

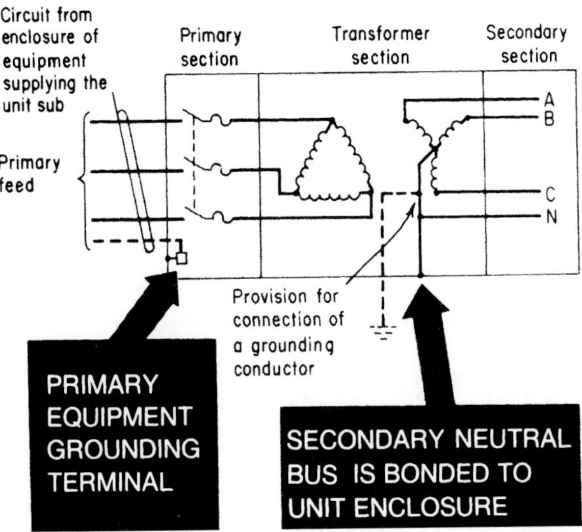

Fig. 384-3. Unit subs have bonded neutral in secondary switchboards. (Sec. 384-3.)

4. Unless marked otherwise (with both the size and temperature rating of wire to be used), the termination provisions on switchboards are for 60°C wire from No. 14 to No. 1 and 75°C for No. 1/0 and larger wires.

The rule of part **(e)** is shown in Fig. 384-4 and correlated to the rule of part **(f)**. On a 3-phase, 4-wire delta-connected system (the so-called "red-leg" delta, with the midpoint of one phase grounded), the phase busbar or conductor having the higher voltage to ground must be marked, and the higher leg to ground must be phase B, as required by part **(f)**. Without identification of the higher voltage leg, an installer connecting 120-V loads (lamps, motor starter coils, appliances) to the panelboard shown in the diagram might accidentally connect the loads from the high leg to neutral, exposing the loads to burnout with 208 V across such loads.

Part **(f)** requires a fixed arrangement (or phase sequence) of busbars in panels or switchboards. The installer must observe this sequence in hooking up such equipment and must therefore know the phase sequence (or rotation) of the

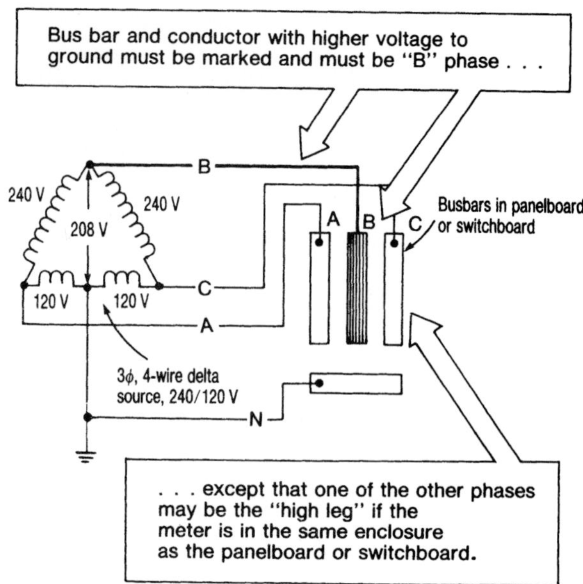
Bus bar and conductor with higher voltage to ground must be marked and must be "B" phase . . .
B
240 V
240 V
208 V
120 V
120 V
C
A
A
B
C
Busbars in panelboard or switchboard
3φ, 4-wire delta source, 240/120 V
N
. . . except that one of the other phases may be the "high leg" if the meter is in the same enclosure as the panelboard or switchboard.

Fig. 384-4. Safety requires "high leg" identification on 4-wire delta systems. (Sec. 384-3.)

feeder or service conductors. This new rule has the effect of requiring basic phase identification at the service entrance and consistent conformity to that identification and sequence throughout the whole system (Fig. 384-5).

Difficulty has been encountered with the rule of part **(f)**, requiring the high leg to be the B phase, because utility company rules may call for the high leg to be the C phase and the right-hand terminal in a meter socket—rather than the middle terminal. As shown in Fig. 384-6, the utility phase rotation can be converted to a Code phase rotation by applying the concept that phase rotation is relative, not absolute. If the utility C phase is designated as the NEC B phase, then the other phase legs are identified for NEC purposes as shown. The phase rotation C-A-B is the same as A-B-C, with voltage alternations such that wave B follows wave A by 120°, wave C follows wave B by 120°, wave A follows wave

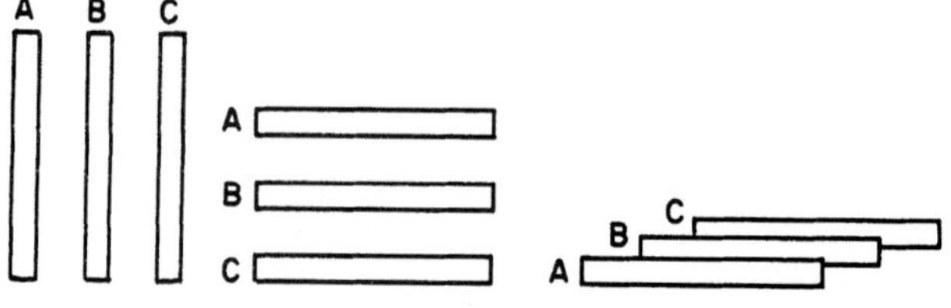
Three-phase buses must be arranged as A, B, C . . .
A
B
C
A
B
C
A
B
C
. . . as viewed from the front of the switchboard or panelboard

Fig. 384-5. Phase sequence in panelboards and switchboards must be fixed. (Sec. 384-3.)

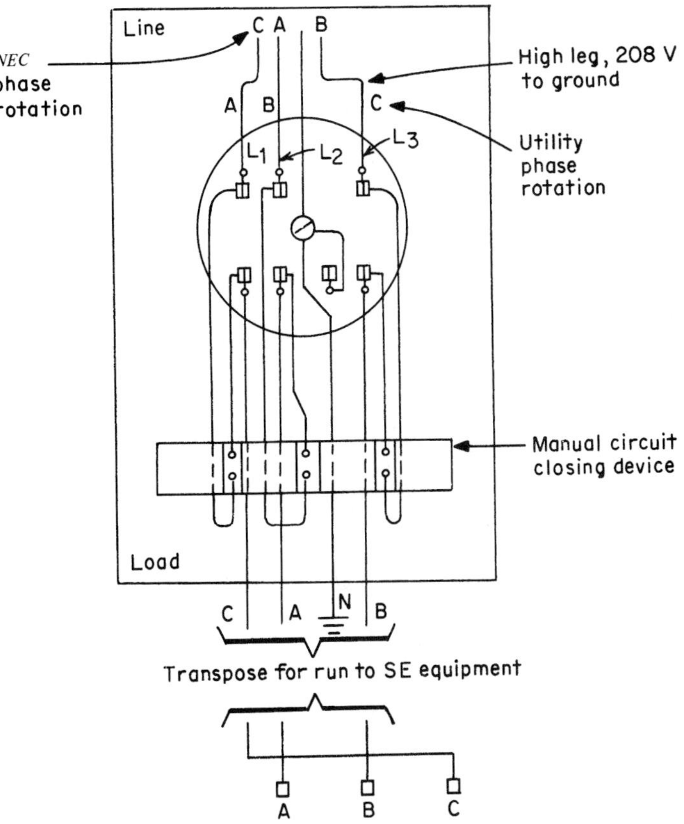

Fig. 384-6. Utility "C" phase becomes NEC "B" phase for high-leg identification. (Sec. 384-3.)

C by 120°, etc. With the phase legs identified as at the bottom of the drawing, each is carried to the appropriately designated phase lug (A-B-C, left to right) at the panelboard shown in Fig. 384-4.

The Exception to part **(f)** will permit the high leg (the one with 208 V to ground) to be other than the "B" phase (such as the "C" phase) where the meter is within the same enclosure as the switchboard or panelboard and the phase configuration of the utility supply system requires other than the "B" phase for the high leg at the meter.

The concept behind this Exception is to permit the same phase identification (such as "C" phase at 208 V) for the metering equipment and the busbars *within the switchboard or panelboard.* And it is the intent of the Code panel that the different service phase identification (such as "C" phase as the high leg) will apply to the entire switchboard or panelboard and that no transposition of phases "B" and "C" is needed within the switchboard or panelboard. However, beyond the service switchboard or panelboard, the basic rule of Sec. 384-3(f) must be observed to have the "B" phase (the middle busbar) as the high (208-V) leg, requiring phase transposition on the load side of the service switchboard or panelboard.

Part **(g)** refers to the need for specific clearances in top and side gutters in both panelboards and switchboards and makes it mandatory that wire bending space at terminals and gutter spaces must afford the room required in Sec. 373-6. This is a repeated requirement throughout the **Code** and is aimed at assuring safe termination of conductors as well as adequate space in the side gutters of panelboards and switchboards for installing the line and load conductors in such equipment. This concern for adequate wire bending space and gutter space is particularly important because of the very large size cables and conductors so commonly used today in panelboards and switchboards. Sharp turns to provide connection to terminal lugs do present possible damage to the conductor and do create strain and twisting force on the terminals themselves. Both of those objections can be eliminated by providing adequate wiring space.

384-4. Installation. Pipes, ducts, etc., must be kept out of the way of circuits from panelboards and switchboards. This rule is aimed at ensuring clean, unobstructed space for proper installation of switchboards and panelboards, along with the connecting wiring methods used with such equipment.

The argument for this **Code** rule is based on the following:

> Sections 450-47 and 710-9 are the only areas within **NE Code** prohibiting foreign piping (water pipes) in areas containing electrical equipment. With the advent of the large office and apartment-house complexes, it has become more economical to purchase primary voltage power, feed through the switchgear, step down to utilization voltage, and distribute throughout the complex. We have seen hospitals and building complexes wherein chilled-water pipes, steam pipes, cold-water pipes, sanitary cleanouts and other piping pass directly over the building's secondary or primary switchgear. In addition, some architects still utilize the electrical closets as a chase for other than electrical conduit. This rule will aid the inspection authority having jurisdiction in performing its function and assure a safer installation.

The wording of this rule has created much confusion among electrical people as to its intent and correct application in everyday electrical work. And, it seems that one can develop a complete understanding of this rule only by repeated readings.

On first reading, there are certain observations about the rule that can be made clearly and without question:

1. Although the rule is aimed at eliminating the undesirable effects of water or other liquids running down onto electrical equipment and entering and contacting live parts—which should always be avoided both indoors and outdoors—the wording of the first sentence of Sec. 384-4 limits the requirement to equipment within the scope of Art. 384 and to motor control centers. Individual switches and CBs and all other equipment are not subject to the rule—although the same concern for protection against liquid penetrations ought to be applied to all such other equipment.
2. The designated electrical equipment covered by the rule (switchboards, panelboards, etc.) does *not* have to be installed in rooms dedicated exclusively to such equipment, although it may be. This rule applies only to the area *above* the equipment, for the width and depth of the equipment.

Part **(a)** (for indoor installations) of this rule very clearly defines the "zones" for electrical equipment to include any open space *above* the equipment. Dedicated clear space above switchboards and panelboards must extend to the structural ceiling above the space *but* is *not* required to extend more than 25 ft (7.62 m) from the floor in high-bay locations. This has the effect of permitting water piping, sanitary drain lines, and similar piping for liquids to be located above switchboards or panelboards *if* such piping is at least 25 ft (7.62 m) above the floor. The permission for switchboards and panelboards to be installed below liquid piping over 25 ft (7.62 m) above the floor must be carefully considered. The object is to keep foreign piping (chilled-water pipes, steam pipes, cold-water pipes, and other piping) from passing directly over electrical equipment and thereby eliminate the problem of destructive arcing and violent burndown that can be caused by water leaking from the piping down onto the equipment (Fig. 384-7). The second-to-last sentence in Part **(a)(1)** clarifies the phrase "structural ceiling:" "A dropped, suspended, or similar ceiling that does not add strength to the building structure is not a structural ceiling."

Note Carefully: It is *not* a requirement of this rule that "foreign" piping, ducts, etc., must always be excluded from "electrical rooms." Because the rule can be satisfied completely by keeping the "foreign" piping, ducts, etc., out of the "space" dedicated to the equipment, the rule, literally, *permits* such "foreign" piping, ducts, etc., to be installed in electrical rooms, mechanical rooms, electrical closets, and similar enclosed spaces. The second sentence of Sec. 384-2 can be fully satisfied if the "foreign" piping, ducts, etc., are not "installed in" and do not "enter or pass through" the "space" that is "dedicated" to the equipment. It is all right for such piping to be in an electrical room as long as it is not in the "zones" dedicated to the equipment—meaning the space taken up by the equipment, the space above the equipment [up to 25 ft (7.62 m)] and, of course, the clean and unobstructed working space required around the equipment by NEC Sec. 110-16 (and even Sec. 110-34). (Fig. 384-7.)

From the wording of the argument for this Code rule, as quoted above, it seems clear that the object is to keep "foreign" piping (chilled-water pipes, steam pipes, cold-water pipes, and other piping) from passing "directly over" electrical equipment and thereby eliminate the problems caused by water leaking from the piping down onto the equipment.

Sprinkler piping, which is intended to provide fire suppression in the event of electrical ignition or arcing fault, would not be foreign to the electrical equipment and would not be objectionable to the Code rule. (See the second FPN at end of the section.) Another confirmation of Code acceptance of sprinkler protection for electrical equipment (which means sprinkler piping within electrical equipment and even *directly over* electrical equipment) is very specifically verified by Sec. 450-47, which states, "Any pipe or duct system foreign to the electrical installation shall not enter or pass through a transformer vault. Piping or other facilities *provided for vault fire protection* or for transformer cooling shall not be considered foreign to the electrical installation."

It is certainly effective design practice to locate the sprinkler piping in overhead space so it is not directly over electrical equipment, thereby minimizing the chance of water leaking down into the equipment. Layouts of piping can be made to assure effective fire suppression by water from the sprinkler heads

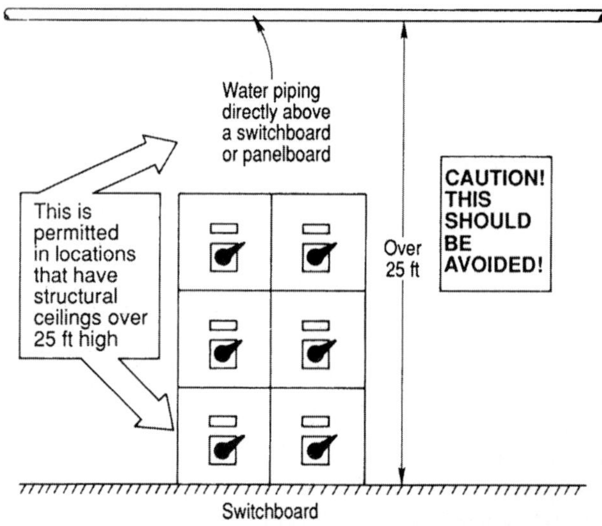
Water piping
directly above
a switchboard
or panelboard
This is
permitted
in locations
that have
structural
ceilings over
25 ft high
Over
25 ft
CAUTION!
THIS
SHOULD
BE
AVOIDED!
Switchboard

Fig. 384-7. Water pipes and other "foreign" piping must not be located less than 25 ft (7.62 m) above switchboard. (Sec. 384-4.)

when needed, without exposing equipment to shorts and ground faults that can be caused by accidental water leaks from the piping.

Caution: Section 710-9 contains a similar rule on fire-suppression means at high-voltage switching and control assemblies.

The Exceptions to this section identify types of equipment which are not subject to the rules of this section.

384-5. Location of Switchboards. Live-front switchboards, as in Fig. 384-8, must always be applied with cautious regard for the conditions stated in this rule.

384-7. Location Relative to Easily Ignitible Material. A combustible floor under a switchboard must be protected against fire hazard, as noted in Fig. 384-9.

384-8. Clearances. Although it has long been a Code rule that a clearance of at least 3 ft (914 mm) be provided from the top of a switchboard to a "combustible" ceiling above, the opening phase of part **(a)** excludes totally enclosed switchboards from this rule. The original rule requiring a 3-ft (914-mm) clear-

Fig. 384-8. A switchboard with "*any*" exposed live parts is limited to use in "permanently dry" locations, accessible only to qualified persons. (Sec. 384-5.)

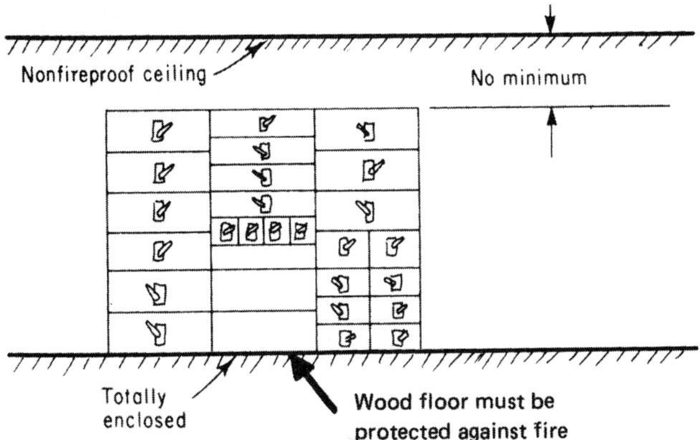

Fig. 384-9. No minimum top clearance is required to nonfireproof ceiling above enclosed switchboard. (Sec. 384-8.)

ance was based on open-type switchboards and did not envision totally enclosed switchboards. The sheet-metal top of such switchboards provides sufficient protection against heat transfer to nonfireproof ceilings. As a result of this Exception, there now is no minimum clearance required above totally enclosed switchboards, as shown in Fig. 384-9.

As covered in part **(b)**, accessibility and working space are very necessary to avoid possible shock hazards and to provide easy access for maintenance, repair, operation, and housekeeping—as required by Sec. 110-16. It is preferable to increase the minimum space behind a switchboard where space will permit.

384-10. Clearance for Conductors Entering Bus Enclosures. Figure 384-10 shows the rules of this section which is aimed at eliminating high conduit stubups

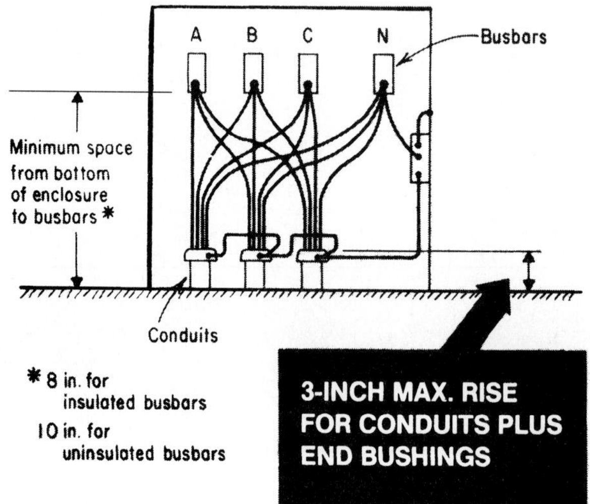

Fig. 384-10. Conduit stubups must have safe clearance from busbars. (Sec. 384-10.)

under equipment containing busbars to prevent contact or dangerous proximity between conduit stubups and the busbars. On this matter, UL says that "the acceptability of conduit stubs serving unit sections with respect to wiring space and spacing from live parts can be determined only by the local inspection authorities at the final installation."

384-13. General. The first sentence here establishes the minimum acceptable rating of any panelboard.

All panelboards—lighting and power—are required by this section to have a rating (the ampere capacity of the busbars) not less than the NE Code minimum feeder conductor capacity for the entire load served by the panel. That is, the panel busbars must have a nameplate ampere rating at least equal to the required ampere capacity of the conductors which feed the panel (Fig. 384-11). A panel may have a busbar current rating greater than the current rating of its

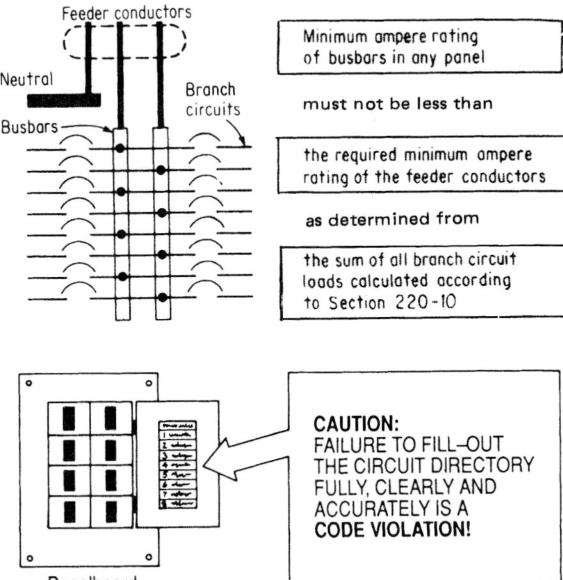

Fig. 384-11. Rating of panelboard bus must at least match required feeder ampacity and circuits must be identified. (Sec. 384-13.)

feeder but must never have a current rating lower than that required for its feeder. [Although Sec. 220-10(b) notes that a feeder for a continuous load must be rated at least 125 percent of the load current, it is not clear that the panel busbars would have to be rated for more than 100 percent of the load current.]

Although selection of a panelboard is based first on the number of circuits which it must serve, it must be assured that the busbars in a panelboard for any application have at least the Code-minimum circuits.

With respect to panelboards, marking may appear on the individual terminals, but terminals can often be changed in the field, and wiring space and the means of mounting the terminals may not be suitable. Therefore, panelboards should be marked independently of the marking on the terminals to identify the terminals and switch or CB units which may be used with aluminum wire. If all terminals are suitable for use with aluminum conductors as well as with copper conductors, the panelboard will be marked "use copper or aluminum wire." A panelboard marked "use copper wire only" indicates that wiring space or other factors make the panelboard unsuitable for any aluminum conductors.

The last sentence of this section specifically requires full and legible marking of a panelboard's *circuit directory* to show the loads and functions of each circuit originating in the panel. The FPN also refers to Sec. 110-22, where the general rule requires thorough identification of the loads fed by all circuit breakers and switches (Fig. 384-11).

384-14. Lighting and Appliance Branch-Circuit Panelboard. This definition is intended to describe the types of panelboards to which the requirements in Secs. 384-15 and 384-16(a) are applied.

Even though a panelboard may be used largely for other than lighting purposes, it is to be judged under the requirements for lighting and appliance branch-circuit panelboards if it conforms to the specific conditions stated in the definition.

Watch out for this definition! There are many panel makeups that supply no lighting and appear to be power panels or distribution panels, yet they are technically lighting and appliance panels in accordance with the above definition and must have protection for the busbars. Figure 384-12 shows an example

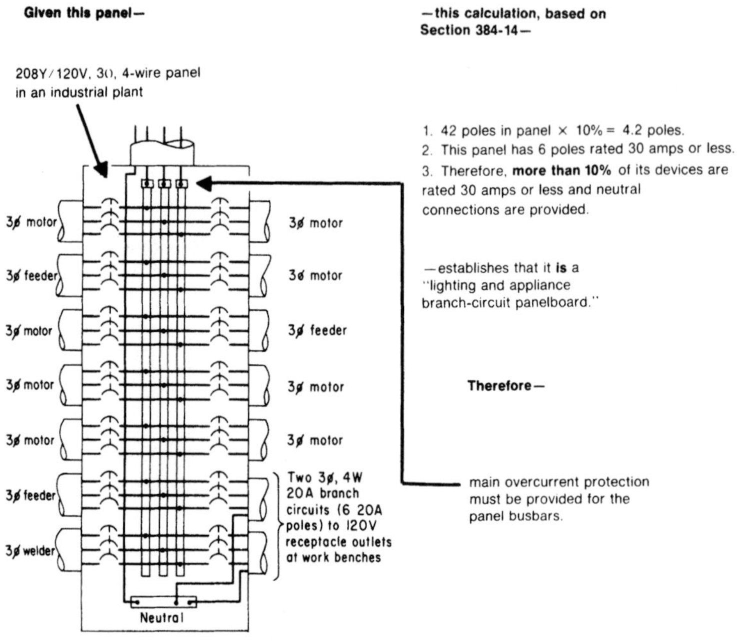

Fig. 384-12. Definition of a "lighting and appliance" panelboard hinges on a specific calculation. (Sec. 384-14.)

of how it is determined whether a panelboard *is* or *is not* "a lighting and appliance branch-circuit panelboard." The determination is important because it indicates whether or not main protection is required for any particular panel, to satisfy Sec. 384-16.

Figure 384-13 shows panels that do not need main protection because they are not lighting and appliance panels, which are the only types of panels required by Sec. 384-16 to have main protection. Just as it is strange to identify a panel that supplies no lighting as a lighting panel, as in Fig. 384-12, it is also strange that some panels that supply *only* lighting, as in Fig. 384-13, are technically *not* lighting panels. Because of the definition of Sec. 384-14, if the protective devices in a panel are *all* rated over 30 A *or* if there are *no* neutral

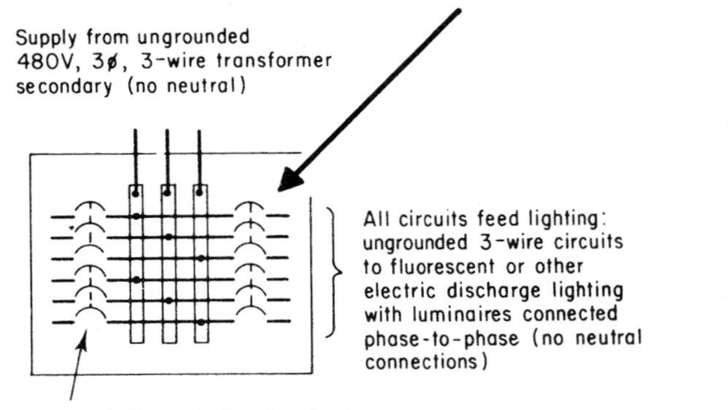
Example: No neutral connections provided
MAIN PROTECTION IS NOT REQUIRED
Supply from ungrounded
480V, 3ø, 3-wire transformer
secondary (no neutral)
All circuits feed lighting:
ungrounded 3-wire circuits
to fluorescent or other
electric discharge lighting
with luminaires connected
phase-to-phase (no neutral
connections)
3-pole CB for each 3-wire circuit per
Section 240-20(b) rated not over 20A
for fluorescent fixtures, up to 50A for other
electric discharge lighting (Sections 210-23
and 210-21 (a)
NOTE: Fusible equipment is also permissible for such
applications.

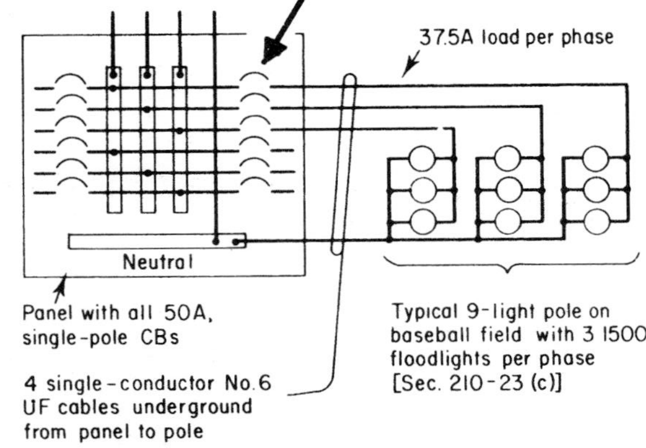
Example: OC devices rated over 30 amps
MAIN PROTECTION IS NOT REQUIRED
This is not a lighting and appliance
panelboard as defined in Section 384-14
37.5A load per phase
Neutral
Panel with all 50A,
single-pole CBs
4 single-conductor No.6
UF cables underground
from panel to pole
Typical 9-light pole on
baseball field with 3 1500W
floodlights per phase
[Sec. 210-23 (c)]

Fig. 384-13. Some panels that supply only lighting are technically not "lighting and appliance panels." (Sec. 384-14.)

connections provided in the panel, then the panel is not a lighting and appliance panel, and it does not require main overcurrent protection.

384-15. Number of Overcurrent Devices on One Panelboard. Figure 384-14 illustrates a panelboard with a 200-A main which provides for the insertion of class

Fig. 384-14. Slots for push-in CB units have different configurations to limit the total number of poles to no more than 42. This is a CTL panelboard (or loadcenter). (Sec. 384-15.)

CTL overcurrent devices. The top stab receivers are of an F-slot configuration. Each F slot will receive only one breaker pole. The remainder of the slots are of an E configuration which will receive two breaker poles per slot. Thus there is provision for installing not more than 42 overcurrent devices, which does not include the main CB. This panelboard may also be supplied without main overcurrent protection where overcurrent protection is supplied elsewhere, such as at the supply end of the feeder to the panel.

Class CTL is the Underwriters Laboratories Inc. designation for the Code requirement for circuit limitation within a lighting and appliance branch-circuit panelboard. It means "circuit-limiting."

384-16. Overcurrent Protection. Rules in this section of the NE Code concern the protection of "lighting and appliance branch-circuit panelboards." In general, lighting and appliance branch-circuit panels must be individually protected on the supply side by not more than two main CBs or two sets of fuses having a combined rating not greater than that of the panelboard, as shown in Fig. 384-15. Individual protection is not required when a lighting and appliance branch-circuit panelboard is connected to a feeder which has overcurrent protection not greater than that of the panelboard (case 1 in Fig. 384-16), as noted in Exception No. 1.

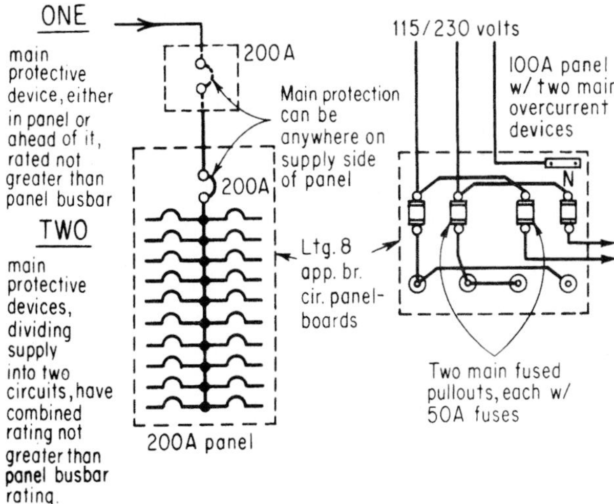

Fig. 384-15. "Main protection" may consist of one or two CBs or sets of fuses. (Sec. 384-16.)

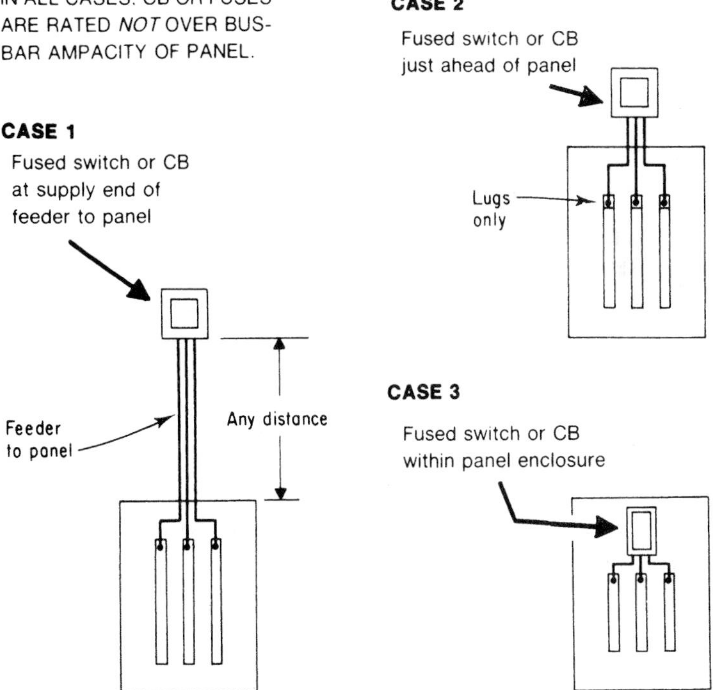

Fig. 384-16. Main panel protection may be located at any one of these locations. (Sec. 384-16.)

Because of the wording of the definition in Sec. 384-14, it is vitally important to evaluate a panel carefully to determine if main protection is required.

Where a number of panels are tapped from a single feeder protected at a current rating higher than that of the busbars in any of the panels, the main protection may be installed as a separate device just ahead of the panel or as a device within the panel feeding the busbars (case 2 and case 3 in Fig. 384-16). The main protection would normally be a CB or fused switch of the number of poles corresponding to the number of busbars in the panel.

Figure 384-17 shows other variations on the same protection requirements. As shown in the drawing with 400-A panels, it is often more economical to order the panels with busbar capacity higher than required for the load on the panels so the panel bus rating matches the feeder protection, thereby eliminating the need for panel main protection.

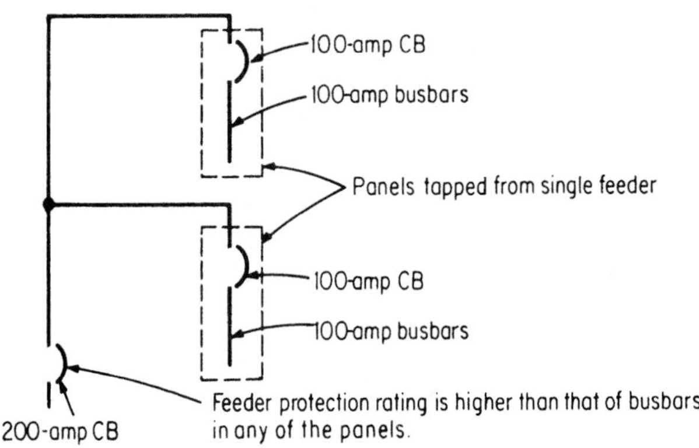

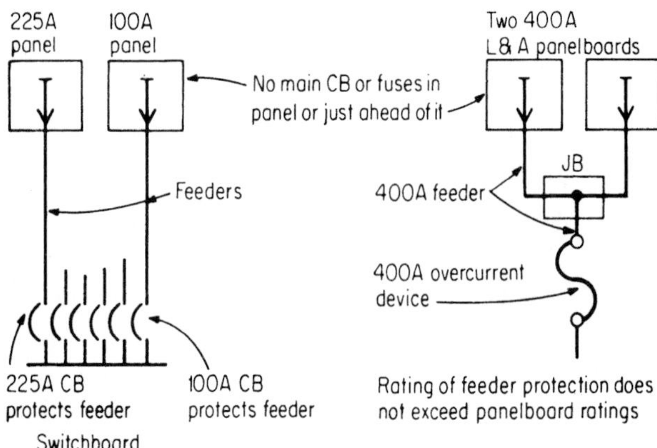

Fig. 384-17. Panel protection may be provided in a variety of ways. (Sec. 384-16.)

Although part **(a)** of this section does spell out those general requirements for main protection of lighting and appliance branch-circuit panelboards, Exception No. 2 of that section notes that a panelboard used as residential service equipment for an "existing" installation may have up to *six* main protective devices [as permitted by Sec. 230-90(a), Exception No. 3]. Such usage is limited to "individual residential occupancy"—such as a private house or an apartment in multifamily dwellings where the panel *is* truly service equipment and not a subpanel fed from service equipment in the basement or from a meter bank on the load side of a building's service.

But it must be noted that the phrase "for existing installations" makes it a violation to use a service panelboard (loadcenter) with more than two main service disconnect-and-protection devices for a residential occupancy in a new building. See Fig. 384-18. That applies to one-family houses and to "dwelling units"

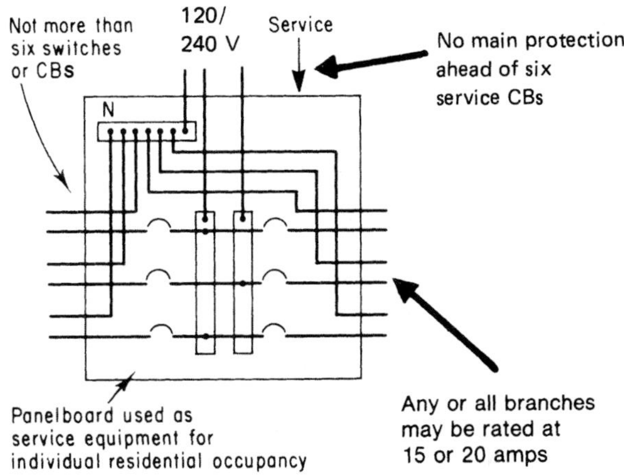
Not more than six switches or CBs
120/ 240 V
Service
No main protection ahead of six service CBs
N
Panelboard used as service equipment for individual residential occupancy
Any or all branches may be rated at 15 or 20 amps
This application is prohibited for new installations, which may have no more than two main protective devices.

Fig. 384-18. Exception No. 2 eliminates need for main in residential service panel, *but only* where such a panel is installed as a replacement in an existing installation. (Sec. 384-16.)

in apartment houses, condominiums, and the like. Use of, say, a split-bus panelboard with four or six main protective devices (such as 2-pole CBs) in the busbar section that is fed by the service-entrance conductors is limited to use only for a service panel in "an individual residential occupancy" in an existing building, such as service modernization. See Fig. 384-19.

Exception No. 2 brings the rule on overcurrent protection for residential service panels into agreement with the basic rule of Sec. 384-16(a) as it applies to all lighting and appliance panelboards in new construction.

The rules of Sec. 384-16(a) have been altered in past editions of the NE Code as follows:

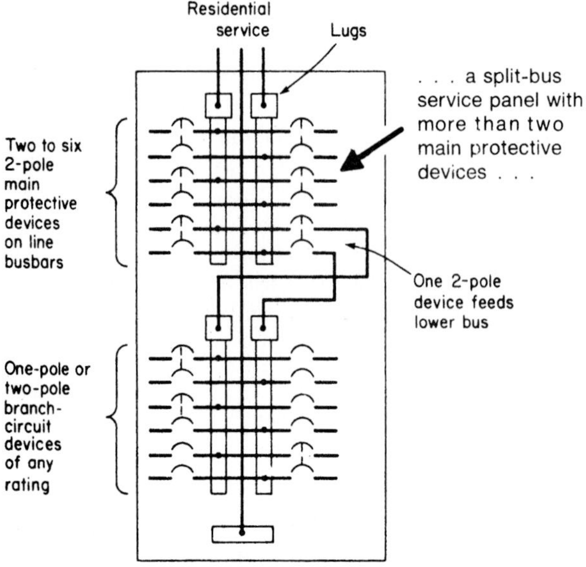

Fig. 384-19. Use of a split-bus loadcenter with more than two main overcurrent devices is permitted for residential service equipment *only* in "existing installations." (Sec. 384-16.)

1. In the 1975 NE Code, Exception No. 2 of Sec. 384-16(a) permitted up to six main protective devices in a residential service panel for *either* new work or additions. But, the rule prohibited use of any 15- or 20-A CBs or fuses without main overcurrent protection ahead of them. Under that rule, a split-bus panel with up to six main CBs or sets of fuses could be used for a residential service, but any 15- or 20-A protective devices had to be installed in the panel bus section fed by the service-entrance conductors. The rule prohibiting 15- and 20-A protective devices without a main ahead of them ruled out use of small residential panels that had only six single- or 2-pole devices rated 15 or 20 A, with no main ahead. Such panels had been used for a long time for service to one-family houses that required only two small appliance circuits and a few lighting circuits where the dwelling unit had no electric water heating, space heating, or major appliances.
2. In the 1978 NE Code, Exception No. 2 was changed to permit use of 15- or 20-A protective devices without need for a main ahead of them, such as in the main busbar section of a split-bus panel or in small six-circuit panels for service in not-all-electric houses or apartments.
3. Now, in the present NE Code, for any new job, any panelboard—for service or otherwise, in any type of occupancy—may have no more than two main protective devices if the panel is a lighting and appliance panel (as deter-

mined from Sec. 384-14). Residential service panels do qualify as lighting and appliance panelboards.

Panelboards used for service equipment are required by NE Code Sec. 230-70 to be marked as "suitable for use as service equipment," and panelboards are so marked.

The rule of part **(b)** of this section is covered in Fig. 384-20. Any panel, a lighting panel or a power panel, which contains snap switches (and CBs are not snap switches) rated 30 A or less must have overcurrent protection and not in excess of 200 A. Panels which are not lighting and appliance panels and do not

Fig. 384-20. Any panelboard containing snap switches rated 30 A or less must have main or feeder protection rated not over 200 A. (Sec. 384-16.)

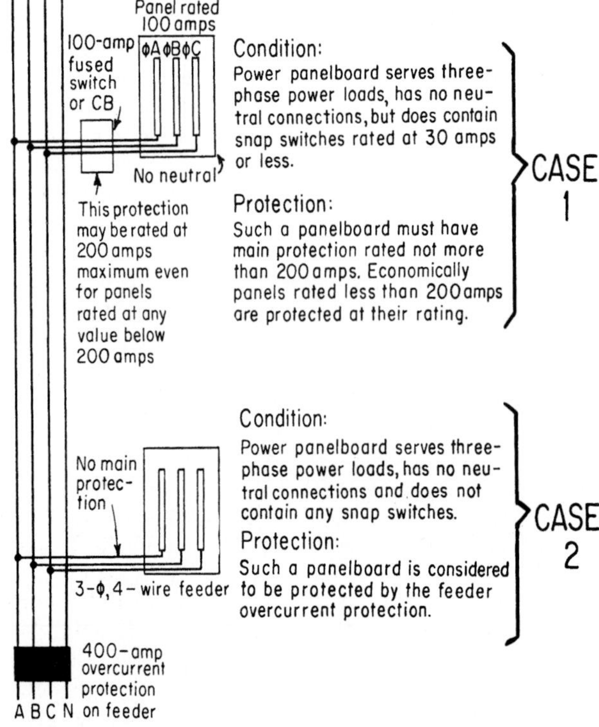

Fig. 384-21. Power panels have very limited requirements for protection. (Sec. 384-16.)

contain snap switches rated 30 A or less do not have to be equipped with main protection and may be tapped from any size of feeder. Figure 384-21 shows these two examples of overcurrent protection requirements for panelboards.

Part **(c)** applies to *any* overcurrent device in a panel and is a similar but stricter rule than those of Sec. 210-22 or Sec. 220-2, as shown in Fig. 384-22.

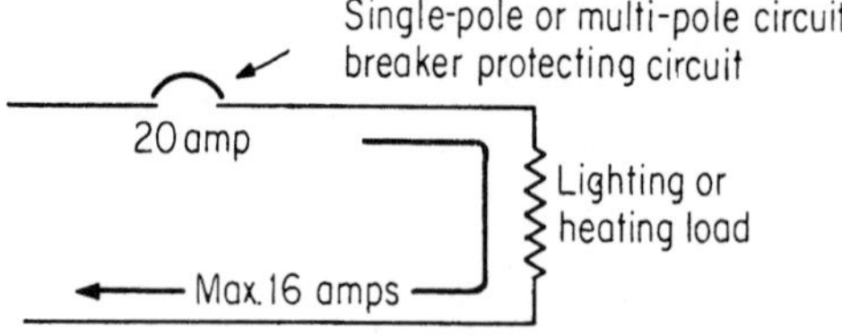

Total load on circuit must not exceed 80% of circuit rating when load is continuous (operates for 3 hours or more) - such as store lighting. CB must not be loaded over 80% of its rating.

Fig. 384-22. This applies for all fuses *and* breakers in a panelboard. (Sec. 384-16.)

The only exception to the rule is for overcurrent-device assemblies that are "approved" (which means UL listed) for continuous loading at 100 percent of their current rating. Refer to Sec. 220-10(b) on feeder protection.

Part **(d)** of this section applies to a panelboard fed from a transformer. The rule requires that overcurrent protection for such a panel, as required in **(a)** and **(b)** of the same section, must be located on the secondary side of the transformer. An exception is made for a panel fed by a 2-wire, single-phase transformer secondary. Such a panel may be protected by a primary device. This concept of prohibiting use of panel protection on the primary side of a transformer feeding the panel is consistent with the rules covered under Sec. 240-3(i). Refer to the discussion there.

Part **(e)** is a rule prohibiting the installation of any 3-phase disconnect or 3-phase overcurrent device in a single-phase panelboard. It is now required that any 3-pole disconnect or 3-phase protective device supplied by the bus within a panelboard may be used only in a 3-phase panelboard. The effect of this new rule is to outlaw the so-called delta breaker, which was a special 3-pole CB with terminal layouts designed to be used in a single-phase panel fed by a 3-phase, 4-wire, 120/240-V delta supply where the loads served by the panel were predominantly single-phase, but where a single 3-phase motor or 3-phase feeder was needed and could readily be supplied from this type of delta breaker. The delta CB plugged into the space of three single-pole breakers, with high leg of delta feeding directly through one pole of the common-trip assembly. Unit was used to protect motor branch circuit or feeder to 3-phase panel, rated up to 100 A.

Use of delta breakers has been found hazardous. When a delta breaker is used in a single-phase panel and the main disconnect for the single-phase panel is opened, there is still the high hot leg supplying the delta breaker. This has caused confusion to personnel who were surprised to find the energized conductor and were exposed to shock hazards.

As noted in part **(e)**, a plug-in circuit breaker that is connected for "backfeed" (with the plug-in stabs being the load-side of the CB) must be mechanically secured in its installed position. This rule requires all plug-in-type protective devices (CB or fusible) and/or main lug assemblies, in panelboards, to have some mechanical means that secure them in position, requiring more than just a pull to remove the devices. This is intended to eliminate the hazard of exposed, energized plug-in stabs of a device that is readily dislodged from its plug-in or connected position (Fig. 384-23).

384-17. Panelboards in Damp or Wet Locations. UL data supplement the Code rules:

Enclosed panelboards marked "Raintight" will not permit entry of water when exposed to a beating rain. Enclosed panelboards marked "Rainproof" will not permit a beating rain to interfere with successful operation of the apparatus within the enclosure but may permit entry of water.

But note this carefully: NE Code Secs. 384-17 and 373-2 require that panelboard enclosures in "*damp or wet*" locations must be placed or equipped to "prevent moisture or water from *entering and accumulating* within" the enclosure, and there must be at least a ¼-in. (6.35-mm) air space between the enclosure and the wall or surface on which the enclosure is mounted. When they are installed exposed outdoors or in other wet locations, the NE Code requires that

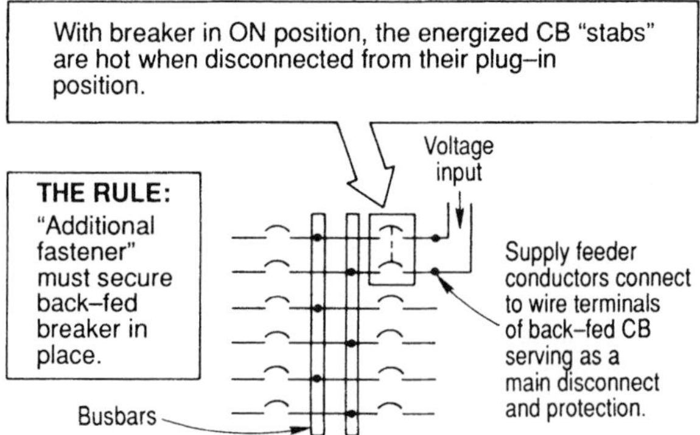

Fig. 384-23. This is an important safety rule for plug-in breakers. (Sec. 384.16.)

panelboard enclosures must be weatherproof. The NE Code definition of "weatherproof" is similar to the NE Code and UL definitions of "rainproof." Yet, NE Code Sec. 373-2 requires exclusion of water entry—which clearly demands a "raintight" enclosure for outdoor, exposed panelboards (and not "rainproof"). These same considerations apply to other cabinets or enclosures used outdoors.

384-18. Enclosure. UL data cover these considerations:

1. Panelboards labeled as "Enclosed Panelboards" have been established as having adequate wiring space in the enclosure.
2. Unless a panelboard is marked otherwise, the wiring space in the assembly and the current-carrying capacity are based on use of 60°C wire in sizes No. 14 up to No. 1 or the use of 75°C wires for sizes No. 1/0 and larger. This limitation on use is covered in the general data of the UL Green Book. If wires of higher than 60 or 75°C rating are used, such wires must be used at ampacities not greater than those given in NE Code Table 310-16 for wires rated 60 or 75°C.

384-19. Relative Arrangement of Switches and Fuses. For service equipment, switches are permitted on either the supply side or the load side of the fuses. In all other cases if the panelboards are accessible to other than qualified persons, Sec. 240-40 requires that the switches shall be on the supply side so that when replacing fuses, all danger of shock or short circuit can be eliminated by opening the switch.

384-20. Grounding of Panelboards. The effect of this rule is to *require* a panelboard to be equipped with a terminal bar for connecting all equipment grounding conductors run with the circuits connected in the panel. Such a bar must be one made by the manufacturer of the panel and must be installed in the panel in the position and in the manner specified by the panel manufacturer—to assure its compliance with UL rules, as well as the NEC. The terminal bar for connecting equipment grounding conductors may be an inherent part of a panelboard, or terminal bar kits may be obtained for simple installation in any panelboard. Homemade or improvised grounding terminal bars are contrary to the intent of this Code section.

The terminal bar that is provided for connection of equipment grounding conductors must be bonded to the cabinet *and* frame of a metal panelboard enclosure. If such a panel enclosure is nonmetallic, the equipment grounding terminal bar must be connected to the equipment grounding conductor of the feeder supplying the panel.

Equipment grounding conductors must not be connected to terminals of a neutral bar—unless the neutral bar is identified for that purpose and is in a panel where Art. 250 requires or permits bonding and grounding of the system neutral (or grounded) conductor, such as at a service panel or a panel fed from another building (Sec. 250-24).

Figure 384-24 shows some details of grounding at panelboards. There have been many field problems relative to terminating grounding conductors in panelboards where nonmetallic wiring methods have been involved. The rule here requires an "equipment grounding terminal bar" in such panels so that these grounding conductors can be properly terminated and bonded to the panel.

In other than service equipment, the grounding conductor terminal bar must not be connected to the neutral bar (that is, the neutral bar must not be bonded to the panel enclosure). Refer to Secs. 250-24, 250-50, 250-53, and 250-61. In a service panel, with the neutral bonded to the enclosure, equipment grounding conductors may, as noted above, be connected to the bonded neutral terminal bar (or block). But the neutral bar must *not* be used for equipment grounding conductors on the load side of the service, except that a bonded neutral block in a subpanel of an outbuilding that is fed from another building is an acceptable grounding terminal bar. Where a panel is used to supply loads in a building, with the panel fed from another building, the equipment ground bus in such a panel must be bonded to the grounded conductor (e.g., the neutral) of the feeder to the building, if the grounded conductor is grounded at the building by a grounding electrode conductor run to a grounding electrode (Fig. 384-25).

The Exception to this section allows an isolated ground conductor run with the circuit conductors to pass through the panelboard without being connected to the panelboard grounding terminal bar, in order to provide for the reduction of electrical noise (electromagnetic interference) on the grounding circuit as provided for in Sec. 250-74 Exception No. 4.

In order to maintain the isolation of the grounding wire necessary for a low-noise ground, the grounding wire must be connected directly to the grounding terminal bar in the service-entrance equipment if the service equipment is within the same building. To do this it may be necessary for the grounding wires to pass through one or more panelboards, but the grounding conductor *must not* leave the building in which the isolated ground is installed. Of course, such isolated grounding conductors may be spliced together by use of a terminal block installed in the panel but insulated from conductive contact with the metal enclosure of the panel. A "quiet ground" keeps grounding conductors apart from and independent of the metal raceways and enclosures (Fig. 384-26).

Sensitive electronic equipment utilized in hospitals, laboratories, and similar locations may malfunction because of electrical noise (electromagnetic interference) present in the electrical supply. This effect can be reduced by the

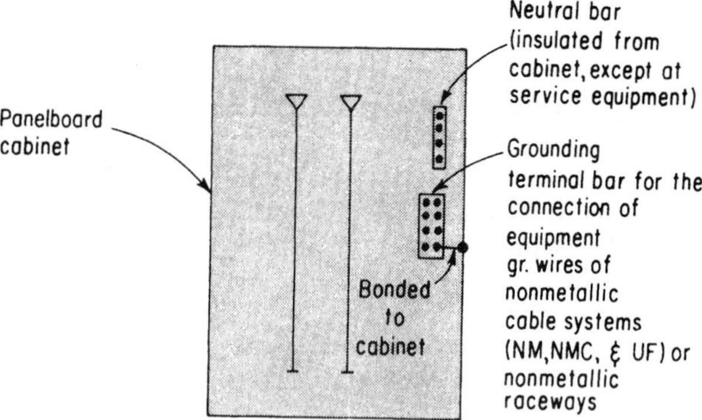

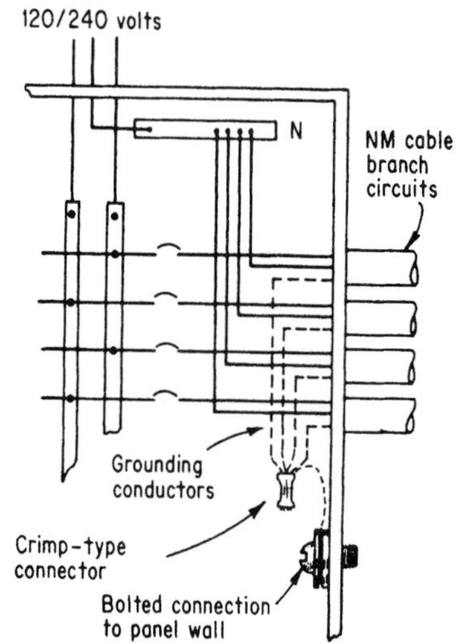

Fig. 384-24. Grounding in panelboards must use listed components. (Sec. 384-20.)

proper use of an isolated grounding wire which connects directly to the service-entrance panel grounding terminal bar. Such systems are being used in increasing numbers where computers are in use.

384-35. Wire Bending Space in Panelboards. This section of the Code correlates wire bending space at terminals in panelboards to the basic concepts of Sec. 373-6(b), as follows:

- The basic rule requires the wire bending space at the top *and* bottom of a panel to satisfy the distances called for in Table 373-6(b), regardless of position of conduit entries. And the side wiring gutters may have a width in

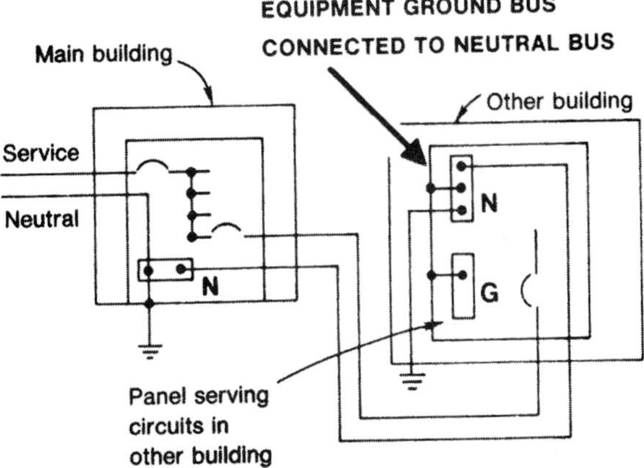

Fig. 384-25. A grounding terminal bar must be bonded to the neutral conductor in a subpanel, which is not "service equipment," when a grounding electrode is used at an out-building to ground the neutral in the subpanel, as may be required by the rule of Sec. 250-24.

Fig. 384-26. "Quiet ground" terminal block for equipment grounding conductors provides for carrying isolated grounding conductors from circuits back to service bonded neutral, with single grounding conductor connecting terminal bar back to service. Terminal block is insulated from metal panel enclosure. Check with manufacturer to assure that isolated bar is approved. (Sec. 384-20.)

accordance with the lesser distances of Table 373-6(a), based on the largest conductor to be terminated in that space. Exceptions are:

1. For a "lighting and appliance branch-circuit panelboard rated 225 A or less," *either* the top *or* the bottom bending space may conform to Table 373-6(a)—but the other space at top or bottom (whichever is the terminal-lug space) must comply with Table 373-6(b).

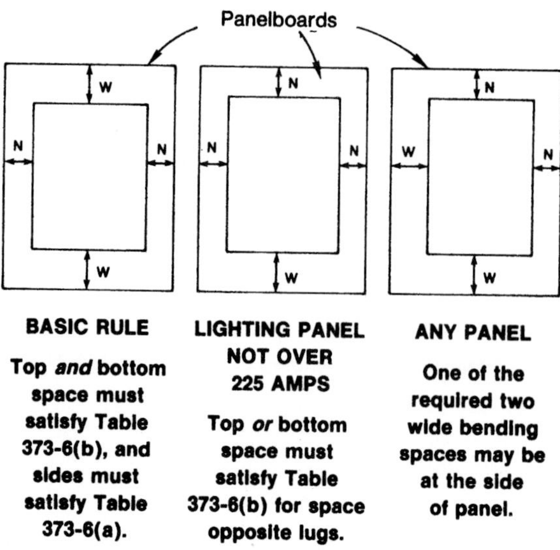

N = NARROW SPACE, TABLE 373-6(a)
W = WIDE SPACE, TABLE 373-6(b)
Panelboards
W
N
N
W
N
N
N
W
N
W
N
W
BASIC RULE
Top *and* bottom space must satisfy Table 373-6(b), and sides must satisfy Table 373-6(a).
LIGHTING PANEL NOT OVER 225 AMPS
Top *or* bottom space must satisfy Table 373-6(b) for space opposite lugs.
ANY PANEL
One of the required two wide bending spaces may be at the side of panel.

Fig. 384-27. This summarizes the requirements on wire-bending space at terminals where the feeder conductors supply a panelboard. (Sec. 384-35.)

2. For any panelboard, *either* the top *or* bottom bending space may conform to Table 373-6(a), provided that at least one of the side wiring terminal spaces satisfies Table 373-6(b), based on the largest conductor terminated in that space.
3. Depth of the wire bending space at the top and bottom of a panel enclosure may be as given in Table 373-6(a) rather than Table 373-6(b)—which is a deeper space requirement; but this may be done only where the panel is designed and constructed for a 90° bend (an "L" bend) of the conductors in the panel space and the panelboard wiring diagram is marked to show and describe the acceptable conditions of hookup.
4. This permits minimum space [Table 373-6(a)] where no conductors are terminated, in either the top or bottom of a panelboard.

Figure 384-27 shows the basic rule and Exceptions No. 1 and No. 2. Of course, the rules of Sec. 373-6(b) must be fully satisfied in the choice of location for the wire bending space. And the size of the largest conductor determines the minimum required space for all applications of the tables.

Chapter Four

ARTICLE 400. FLEXIBLE CORDS AND CABLES

400-3. Suitability. This rule requires that any application of flexible cord or cable may require use of "hard usage" cord (such as SJ cord) or "extra hard usage" cord (such as S or SO cord) if the cord is used where it is exposed to abrasion or dragging or repetitive flexing and/or pulling, depending upon severity of use. As noted in Table 400-4, cords for portable heaters must be one of those types when used in damp places. Determination of the need for a particular cable on the basis of use severity is subjective. Table 400-4 also indicates the types of portable cable—that is, for data processing and elevator circuits—and conditions under which each type is suitable, as for hazardous or nonhazardous locations.

Other data on suitability are given in the UL Green Book, as follows:

Flexible Cord (ZJCZ)

Flexible cords are constructed as described in, and Listed for use in accordance with Article 400 of the National Electrical Code. All conductors are stranded copper.

Voltage rating:

Types XT and CXT (20-24 AWG); TS and TST (27 AWG) are rated 125 volts.

Type TPT (27 AWG) is rated 250 volts.

Types C (14-10 AWG), PD (14-10 AWG), S, SO, SOO, ST, STO, STOO, SE, SEO and SEOO are rated 600 volts.

Types CXT (18 AWG), C (18-16 AWG), PD (18-16 AWG), and all other types are rated 300 volts.

Conductor size:

The conductor size ranges are specified in the National Electrical Code with the following exceptions:

Types XT and CXT, 24-18 AWG; SEOO, 18-2 AWG; HSJOO, 18-12 AWG; HSOO, 14-12 AWG.

Temperature rating:

Types XT, CXT, C, PD, SP-1, -2, -3, SRD, E, EO, EN, ET, ETLB, ETP and ETT are rated 60 C.

Type SRDT is rated 60 or 90 C.

Type SRDE is rated 90 or 105 C.

Types SVE, SVEO, SVEOO, SJE, SJEO, SJEOO, SE, SEO and SEOO are rated 90 or 105 C.

Types HPN and HPD are rated 90 or 105 C.

Types HS, HSO, HSJ and HSJO have a core temperature rating of 90 or 105 C and a jacket temperature of 60, 75, 90 or 105 C.

All other cord types are rated 60, 75, 90 or 105 C. Cords having a temperature rating higher than 60 C have the temperature rating printed on the outer surface of the cord. If the cord is rated 60 C, no temperature rating appears.

Additional Markings:

"Water Resistant"—indicates the cord is suitable for immersion in water.

"For Mobile Home Use" or "For Recreational Vehicle Use," or "For Mobile Home and Recreational Vehicle Use," followed by current rating in amperes—indicates suitability for use in mobile homes or recreational vehicles.

"Outdoor" or "W-A"—indicates suitability for use outdoors. The minimum temperature rating for these cords is –35 C unless otherwise marked on the cord.

"VW-1"—indicates the cord complies with a vertical flame test.

"—50 C"—indicates a cord which complies with a bend test (not a suppleness test) at —50 C.

Cords which have been evaluated for leakage currents between the circuit conductor and the grounding conductor and between the circuit conductor and the outer surface of the jacket may have the values so marked on the cable jacket.

Additional cord types or characteristics not covered by the National Electrical Code:

Type XT twisted or braidless parallel assembly of two conductors intended for use in decorative lighting strings.

Type CXT twisted assembly, two conductors intended for use in decorative lighting strings.

Type HPN, parallel heater cord is oil resistant.

Types HSJOO (hard usage) and HSOO (extra hard usage) jacketed heater cords with oil resistant individual conductors as well as oil resistant jackets.

Shaver Cord rated 125 volts, 27 and 20 AWG sizes, no type designation.

Standard-

The basic standard used to investigate products in this category is UL62, "Flexible Cord and Fixture Wire".

Listing Mark-

The Listing Mark of Underwriters Laboratories Inc. in suitable footage denominations on the attached tag, the reel, or the smallest unit container in which the product is packaged is the only method provided by UL to identify these products manufactured under its Listing and Follow-Up Service. The Listing Mark for these products includes the name and/or symbol of Underwriters Laboratories Inc. (as illustrated in the Introduction of this Directory) together with the word "Listed", a control number, and the following product name: "Flexible Cord".

400-5. Ampacities for Flexible Cords and Cables. A three-conductor cord set is permitted by Sec. 250-60(a) to be used with one conductor serving as *both* the neutral conductor *and* the equipment grounding conductor, with the frame of the range or dryer grounded by connection to the neutral. The last sentence of this rule points out that the common neutral-grounding conductor does not count as a current-carrying conductor, thereby making the 3-wire cord suitable for use at the higher ampacity shown under column B (fourth from the left) in

Table 400-5A—which is for cord with not more than two wires.

400-7. Uses Permitted. Figure 400-1 shows accepted uses for flexible cord. Flexible cord may be used for lighting fixtures under **(a)(2).** Refer to Sec. 410-14 for limitations on use with electric-discharge lighting fixtures and Sec. 410-30(b) and (c) for fixtures that require aiming or adjusting after installation and conditional use with other electric discharge lighting.

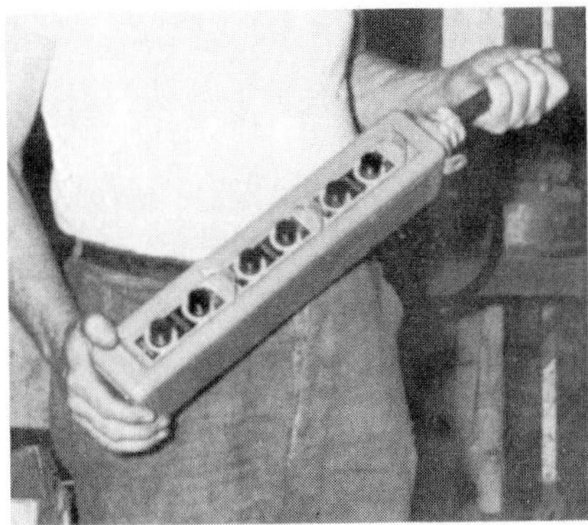

Fig. 400-1. Permitted uses for flexible cable and cord include pendant pushbutton station for crane and hoist controls (left), and connection of portable lamps. (Sec. 400-7.)

Part **(b)** states that *if* flexible cord is used to connect portable lamps or appliances, stationary equipment to facilitate frequent interchange, or fixed or stationary appliances to facilitate removal or disconnection for maintenance or repair, the cord "shall be equipped with an *attachment plug* and shall be energized from an approved *receptacle outlet.*"

It should be noted that the cords referred to under this section are the cords attached to the appliance and not extension cords supplementing or extending the regular supply cords. The use of an extension cord would represent a conflict with the requirements of the Code in that it would serve as a substitute for a receptacle to be located near the appliance.

Extension cords are intended for temporary use with portable appliances, tools, and similar equipment which are not normally used at one specific location.

But bus-drop cable may be used to feed down to machines in factories. Such cable is UL-listed for that application in accordance with Sec. 400-7 (Fig. 400-2).

400-8. Uses Not Permitted. Although Sec. 400-7 says that flexible cord may be used for "wiring of fixtures," that is a simple, general, broad recognition that may be used by any electrical inspector to accept almost any specific assembly of cord supply to a lighting fixture. It is the kind of rule that actually requires individual inspectors to spell out their own design and installation details. And it ties into the general rule of the second sentence in Sec. 90-4 which makes the inspector the final judge of Code compliance on all questions about NE Code rules.

Fig. 400-2. Bus-drop power cables are flexible cables listed by UL for feeding power down from plug-in fusible switches on busway to supply machines. Cables here have connector bodies on their ends for machine cord caps to plug into. (Sec. 400-7.)

But Sec. 400-8(1) says that flexible cord must *not* be used "as a substitute for fixed wiring." That rule could be strictly enforced to require all lighting fixtures to be supplied by fixed wiring methods—approved, Code-recognized cables like NM or BX or by a standard raceway method (EMT, rigid, flex, etc.). The rule does create a conflict with Sec. 400-7(a) by raising the question, Is there ever a case where a lighting fixture could *not* be fed by a fixed wiring method? Certainly, any fixture that might be supplied by a cord connection from a junction box to the fixture could just as easily be fed by conductors in flexible metal conduit or in liquidtight flexible metal conduit—both of which conduit-and-wire connections are considered "fixed wiring" methods. If there are no cases where a fixture could not be fed by such a fixed wiring connection, then every use of flexible cord to supply a fixture is "a substitute for fixed wiring." The relationship between Secs. 400-7(a)(2), 400-8(1), 410-14, 410-30(b), and 410-30(c) must be carefully evaluated to assure ready compliance with Code rules—particularly since cord connection of lighting fixtures has been used so long and so successfully for both indoor and outdoor applications.

Whether any use of flexible cord is a violation of Sec. 400-8(1) must be related to the rule of Sec. 400-7. If a use of flexible cord does not conform to one of the permitted uses in Sec. 400-7, it becomes a violation of this section.

Figure 400-3 shows one of a number of twin floodlight units that were installed outdoors for lighting of the facade of a building. The use of cord from a junction box to a stab-in-the-ground twin lampholder assembly does not comply with "(2) wiring of fixtures" in Sec. 400-7 because it does not satisfy Secs. 410-14 or 410-30(b) or (c), which regulate use of cord for fixtures, as noted under Sec. 400-7 above. And the application does not comply with the other permitted uses in Sec. 400-7. Because floodlights could have been installed in lampholders that thread into hubs on a weatherproof box, use of the cord is an evasion of a fixed or permanent connection technique that would totally avoid the potential shock hazard of cord pull-out or breakage. Mounting the floodlights on the box would still allow adjustment. This use of cord is a substitute for fixed wiring and is a violation.

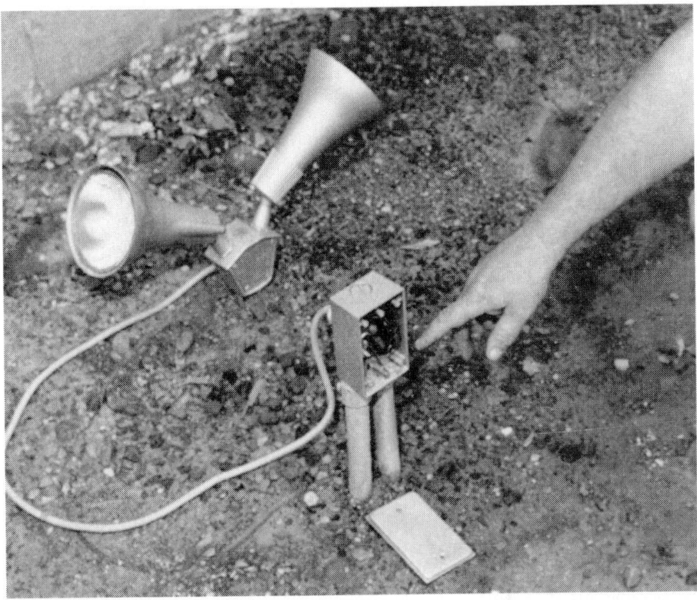

Fig. 400-3. This use of flexible cord to supply an outdoor lampholder assembly can readily be described as a "substitute for fixed wiring"—which is a prohibited use of cord. Here, the lampholders could have been attached to one or more threaded openings on an outlet box. (Sec. 400-8.)

Flexible cords and cables may not be installed in raceways, except where the NEC specifically recognizes such uses. Part **(6)** clarifies such use of flexible cords and cables, limiting their use in raceways to applications described or inferred in Sec. 400-7 and other Code rules—such as Sec. 550-5(g) for sleeving of a mobile home power-supply cord, Sec. 551-46(a)(2) on the same technique for recreational vehicles, Sec. 645-5(b)(3) for computer-room connecting cables, and Sec. 680-20(b) on the flexible cord run in conduit for a wet-niche lighting fixture, and similar limited applications.

The Exception to this rule permits a flexible cord to have connection to a building surface for a "tension take-up device," as shown in Fig. 400-4. That is an Exception to part **(4)** of Sec. 400-8. Connection of a tension take-up device to support the slack in a run of flexible cord is an exception to the rule prohibiting flexible cord and cable from being "attached to the building surfaces." This

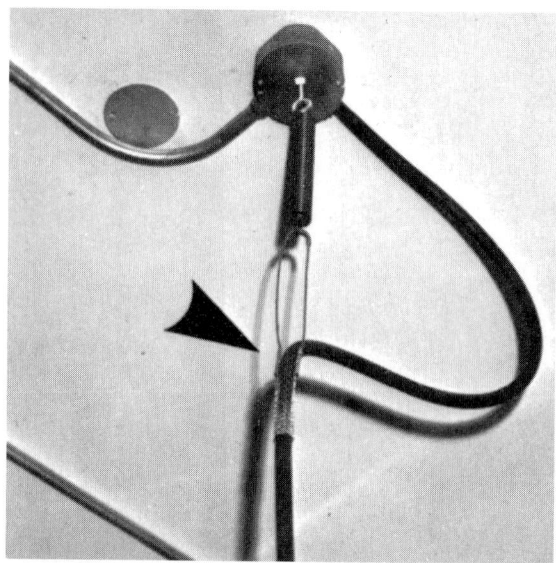

Fig. 400-4. Strain-relief for flexible cord must protect cable jacketing from damage at box connectors and protect wire terminations from pull-out. Spring-loaded come-along support supports cable against weight on bottom end of pendant and also provides up-and-down movement of cable end. (Sec. 400-10.)

is a widely used technique for supporting horizontal lengths of cord that supply equipment that has some movement or travel. Travel from the cord to the tension take-up device is limited to a maximum of 6 ft (1.83 m).

400-10. Pull at Joints and Terminals. Figure 400-5 shows methods of strain relief for cords. The "Underwriters' knot" has been used for many years and is a good method for taking the strain from the socket terminals where lamp cord is used for the pendant, through the hole in the lampholder or switch device. For reinforced cords and junior hard-service cords, sockets with cord grips such as shown in Fig. 400-4 provide an effective means of relieving the terminals of all strain. Figure 400-4 shows a support technique that comes under "other approved means."

400-11. In Show Windows and Show Cases. Because of the flammable material nearly always present in show windows, great care should be taken to ensure that only approved types of cords are used and that they are maintained in good condition.

ARTICLE 402. FIXTURE WIRES

402-5. Allowable Ampacities for Fixture Wires. Note that Table 402-5 gives the ampacity for each size of fixture wire *regardless of the type of insulation used*

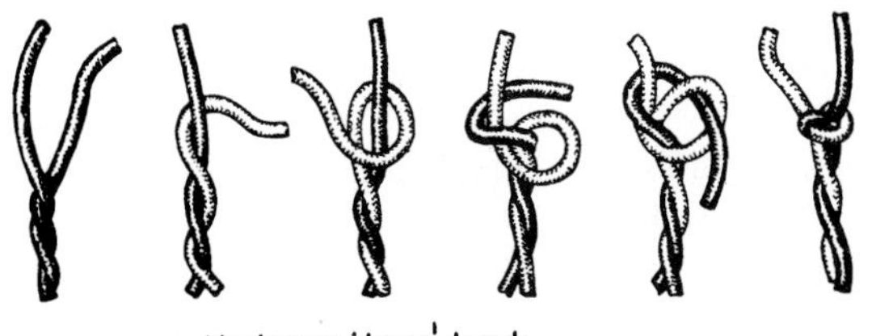

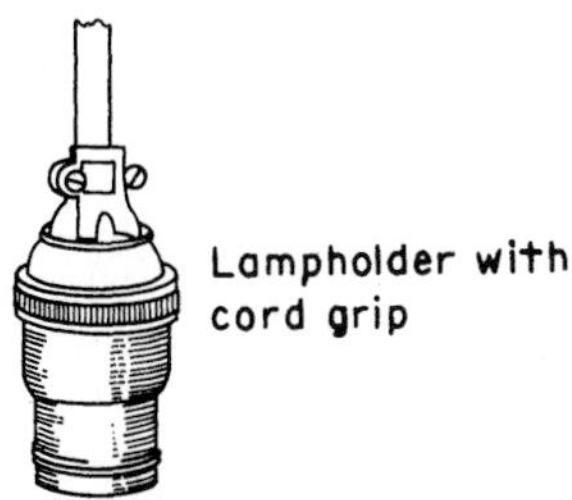

Fig. 400-5. Strain-relief must be provided at cord connections to devices. (Sec. 400-10.)

on the wire. For instance, a No. 18 fixture wire is rated for 6 A whether it is Type TFN or PF or any other type.

402-7. Number of Conductors in Conduit or Tubing. The maximum number of any size and type of fixture wire permitted in a given size and type of conduit is selected from the same tables as the ones used for determining conduit fill for building wire (THW, THHN, etc.). This must be carefully observed, especially when using fixture wires for Class 1 remote-control, signaling, or power-limited circuits, as permitted and regulated by Secs. 725-27 and 725-28.

402-10. Uses Permitted. Fixture wires may be used for internal wiring of lighting fixtures and other utilization devices. They may also be used for connecting lighting fixtures to the junction box of the branch circuit—such as by a flex whip to satisfy Sec. 410-67(c) (Fig. 402-1).

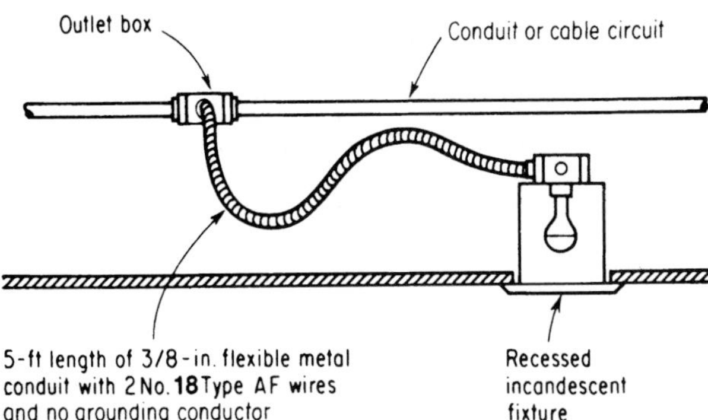

Fig. 402-1. Fixture wires may connect fixtures to branch-circuit wires. (Sec. 402-10.)

402-11. Uses Not Permitted. With the exception of their use for remote-control, signaling, or power-limited circuits, fixture wires are not to be used as general-purpose branch-circuit wires. An example of the use permitted by Sec. 725-16 would be, say, No. 18 fixture wires run as remote-control wires in a raceway from a motor starter to a remote push-button station, where the 6-A rating of the wire is adequate for the operating current of the coil in the starter.

402-12. Overcurrent Protection. This rule refers to Sec. 240-4, which permits No. 18 and No. 16 fixture wire of any type to be protected by a 15- or 20-A fuse or CB. That covers use of No. 18 or No. 16 in fixture "whips" on 15- or 20-A branch circuits and use for remote-control, signaling, or power-limited circuits.

ARTICLE 410. LIGHTING FIXTURES, LAMPHOLDERS, LAMPS, AND RECEPTACLES

410-4. Fixtures in Specific Locations. Part (a) covers the kind of installations shown in Fig. 410-1. At left, the lighting fixture on the covered vehicle-loading dock is in a damp location and must be marked "SUITABLE FOR DAMP LOCATIONS" or marked "SUITABLE FOR WET LOCATIONS." At right, the lighting fixtures at a vehicle-washing area are in a wet location and must be marked "SUITABLE FOR WET LOCATIONS"—unless the fixtures are so high mounted or otherwise protected so that there is no chance of water being played on them.

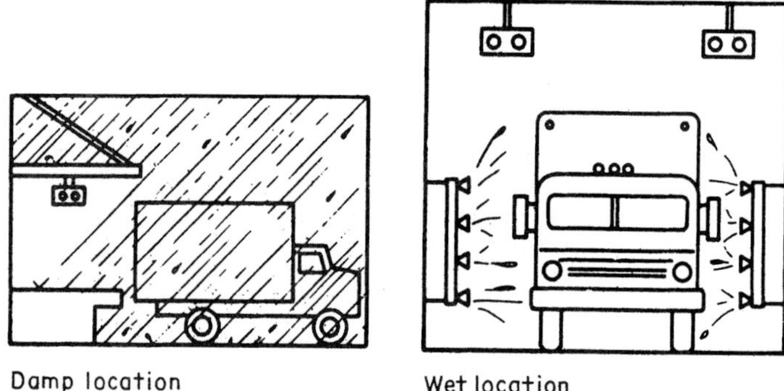

Fig. 410-1. Fixtures must be marked as suitable for their place of application. (Sec. 410-4.)

An enclosed and gasketed fixture would fulfill the requirement that water shall be prevented from entering the fixture, although under some conditions water vapor might enter and a small amount of water might accumulate in the bottom of the globe.

Fixtures in the form of post lanterns, fixtures for use on service-station islands, and fixtures which are marked to indicate that they are intended for outdoor use have been investigated for outdoor installation.

An example of fixtures in "damp" locations would be those installed under canopies of stores in shopping centers where they would be protected against exposure to rain but would be subject to outside temperature variation and corresponding high humidity and condensation. Thus the internal parts of the fixture need to be of nonhygroscopic materials which will not absorb moisture and which will function under conditions of high humidity.

The UL listing of "Fixtures and Fittings" notes:

> These fixtures are incandescent-lamp and electric discharge lamp types (including show-window and showcase type) designed for installation in ordinary locations. Unless marked "Suitable for damp locations" or "Suitable for wet locations," in combination with the Listing Mark, fixtures are only suitable for dry locations.

Part **(c)** recognizes use of lighting fixtures in commercial and industrial ducts and hoods for removing smoke or grease-laden vapors from ranges and other cooking devices. The rule spells out the conditions for using fixtures and their associated wiring in all types of nonresidential cooking hoods. The requirement that such a lighting fixture be "approved for the purpose" may be taken as "listed" by UL for such use.

Part **(d)** covers the use of chandeliers, swag lamps, and pendants over bathtubs, which could pose a potential hazard. Although there is no **Code** prohibition on use of fixtures over tubs and although a bathroom is not technically a damp or wet location, there is considerable concern over exposing persons in water to possible contact with energized parts. Where installed over a tub, a hanging fixture or pendant must be at least 8 ft (2.44 m) above "the top of the tub." In addition, hanging fixtures must be excluded from a zone 3 ft (914 mm) horizontally and 8 ft (2.44 m) vertically from the top of a bathtub rim. This defines the volume of space from which a chandelier-type lighting fixture is excluded above and around a bathtub. This excludes the entire fixture and its cord or chain suspension. Such application is difficult, if not impossible, in most interiors.

410-5. Fixtures Near Combustible Material. Figure 410-2 shows this rule. Much concern has been expressed by electrical inspectors because of instances where the fixture temperature has done damage to wires in outlet boxes and even to nonmetallic boxes themselves. Underwriters Laboratories tests and evaluates fixtures for such heating, with much useful data given in their *Electrical Construction Materials Directory.*

410-6. Fixtures Over Combustible Material. This refers to pendants and fixed lighting equipment, not to portable lamps. Where the lamp cannot be located out of reach, the requirement can be met by equipping the lamp with a guard.

410-8. Fixtures in Clothes Closets. The intent is to prevent lamps from coming in contact with cartons or boxes stored on shelves and clothing hung in the closet, which would, of course, constitute a fire hazard.

Use of lighting fixtures in clothes closets is covered by this section, with specific rules based on the given definition of "storage space" in a closet and the isometric drawing that illustrates the locations and dimensions of regulated space within the closet.

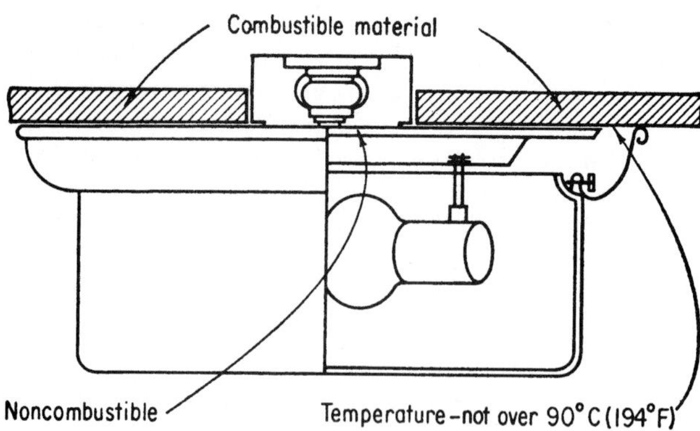

Fig. 410-2. Fixtures must not pose threat of heat to combustible materials. (Sec. 410-5.)

This section presents rules that can be divided into three categories as follows:

1. Part **(a)** is a definition of storage space, and it fully identifies those spaces that *must* be taken to be storage space for purposes of positioning lighting fixtures, regardless of whether or not any of the space is eventually used for storage of clothes, hats, shoes, etc.
2. Parts **(b)** and **(c)** describe the kind of lighting fixtures that may be used and the kinds that are prohibited from use in closets.
3. Part **(d)** then tells how the various types of lighting fixtures have to be installed in relation to defined storage space.

The Code book diagram, Fig. 410-3, clearly shows the closet space from which the various lighting fixtures in **(d)** *must* be spaced. Note that the storage space above the level of the clothes-hanging rod must be as deep as the width of the board used for the shelf if the width of board is greater than 12 in. (305 mm). If the actual width of the shelf is less than 12 in. (305 mm), or even if there is no shelf, the defined space must be taken to be at least 12 in. (305 mm) deep on all sides of the closet and must extend vertically from the level of the clothes-hanging rod up to the ceiling.

The 24-in. (610-mm) -deep storage space that extends up from the floor must be considered to exist up to at least 6 ft (1.83 m) above the closet floor *or* up to the level of the clothes-hanging rod if it is installed at a level over 6 ft (1.83 m) above the floor.

Part **(b)** requires that any lighting fixture in a closet must be a UL-listed fixture. The fixture may be surface-mounted or recessed, incandescent or fluorescent. If the fixture is incandescent, the lamp must be "completely enclosed," but if the fixture is fluorescent, the lamp may be exposed or enclosed.

Part **(c)** prohibits use of any incandescent fixture that has a nonenclosed lamp. Pendant fixtures are also prohibited. *And* it should be especially noted that this rule prohibits use of surface-mounted lampholders—such as the widely used porcelain lampholder.

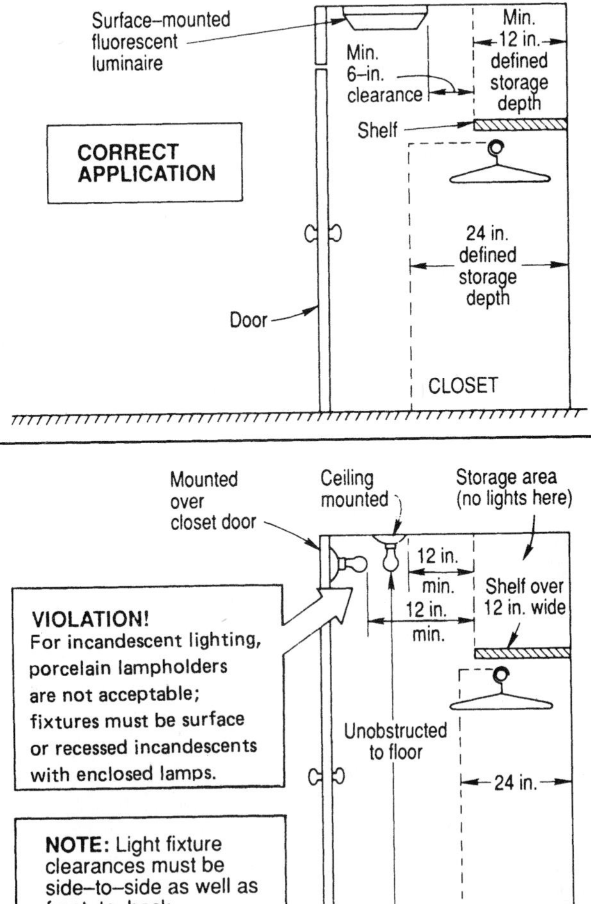

Fig. 410-3. These clearances apply to lighting fixtures in closets. (Sec. 410-8.)

Part **(d)** notes that a surface-mounted lighting fixture may be mounted on the closet wall space above the closet doorway or on the closet ceiling. For surface-mounted fixtures, an incandescent fixture must have at least 12 in. (610 mm) of clearance from "the nearest point of a storage area" and a fluorescent fixture must have at least a 6-in. (152 mm) clearance from storage space.

Recessed fixtures may be mounted in the closet wall above the door or in the ceiling. Either an incandescent or fluorescent recessed type of fixture must have at least a 6-in. (152-mm) clearance from any storage space. Figure 410-3 shows the clearances for fixtures.

For small clothes closets proper lighting may be achieved by locating fixtures on the outside ceiling in front of the closet door—especially in hallways where such fixtures can serve a dual function. Flush recessed fixtures with a solid lens are considered outside of the closet because the lamp is recessed behind the wall or ceiling line.

410-9. Space for Cove Lighting. Adequate space also improves ventilation, which is equally important for such equipment.

410.11. Temperature Limit of Conductors in Outlet Boxes. Fixtures equipped with incandescent lamps may cause the temperature in the outlet boxes to become excessively high. The remedy is to use fixtures of improved design, or in some special cases to use circuit conductors having insulation that will withstand the high temperature.

The first sentence of this rule is related to the rule of Sec. 410-5. Figure 410-4 shows how a fixture may be "so installed" that the branch-circuit wires are not subjected to excessive temperature. That hookup relates to Sec. 410-67(c) for recessed fixtures, which requires a 4- to 6-ft (1.22- to 1.83-mm) length of flex with high-temperature wire (say, Type AF) to connect hot fixture junction point (150°C) to the lower-rated branch-circuit wires.

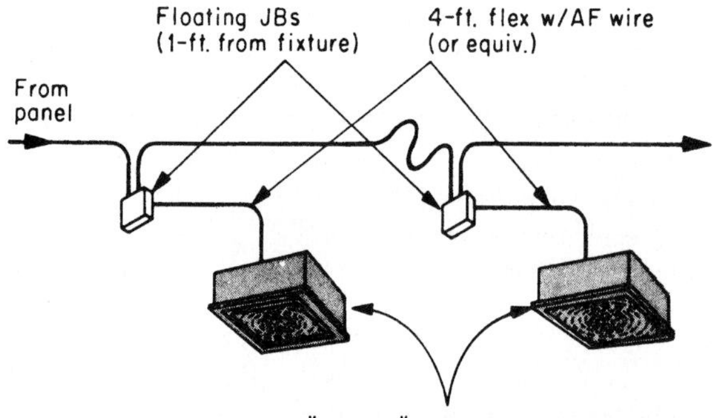

Fig. 410-4. Flex "whip" may be used to keep 60 or 75°C wire away from 150°C terminal space in light fixtures. (Sec. 410-11.)

The second sentence of this section applies to "prewired" recessed incandescent fixtures, which have been designed to permit 60°C supply conductors to be run into an outlet box attached to the fixture. Such fixtures have been listed by UL on the basis of the heat contribution by the supply conductors at *not more* than the *maximum* permitted lamp load of the fixture. Some fixtures have been investigated and listed by UL for "feed-through" circuit wiring. Accordingly, this rule requires careful use of prewired fixtures and, where necessary, the rule calls for use of fixtures which can be connected at the *start* of the circuit as well as the *end* of a circuit without the need of "throw-away" JBs that are required for "unwired" recessed incandescent fixtures.

For quite some time this problem has created considerable controversy in the field because many inspectors have been enforcing the "feed-through" concept on the basis of an Underwriters Laboratories Inc. ruling which states:

With the exception of fluorescent-lamp fixtures, recessed fixtures are marked with the required minimum temperature rating of wiring supplying the fixture. Unless marked "Maximum of___No.___AWG branch-circuit conductors suitable for at least___°C (___°F) permitted in junction box," no allowance has been made for any heat contributed by branch-circuit conductors which pass through, or supply and pass through, an outlet box or other splice compartment which is part of the fixture.

The effect of that UL limitation is this: Some prewired fixtures (with attached outlet box) are suitable only for connecting the 60°C branch-circuit wires to the fixture *and* any prewired fixture for feeding the branch circuit through its outlet box must be marked to allow such use—as shown in Fig. 410-5. To use a branch circuit to feed a number of prewired fixtures that are listed for only one set of 60°C supply wires, those prewired fixtures must be connected with a flex whip from each to a separate junction box, just as if they were "unwired" fixtures as shown in Fig. 410-4. The top hookup in Fig. 410-5 may be rectified by using two 60°C wires run in a 4-ft (1.22-mm) flex length to a separate outlet box mounted at least 1 ft (305 mm) away from the fixture.

410-12. Outlet Boxes to Be Covered. This rule is similar to that of Sec. 370-25. The canopy may serve as the box cover, but if the ceiling or wall finish is of

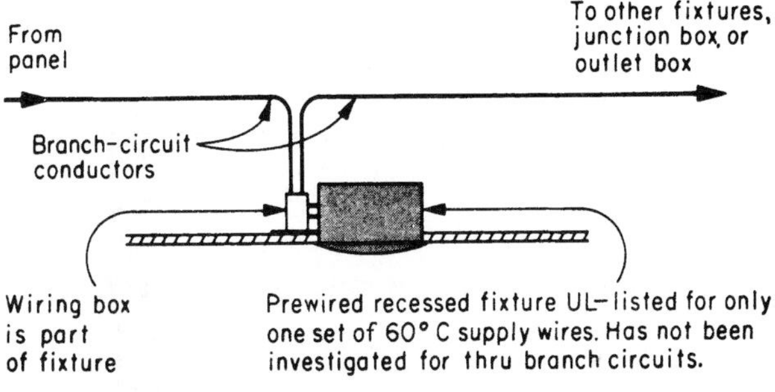

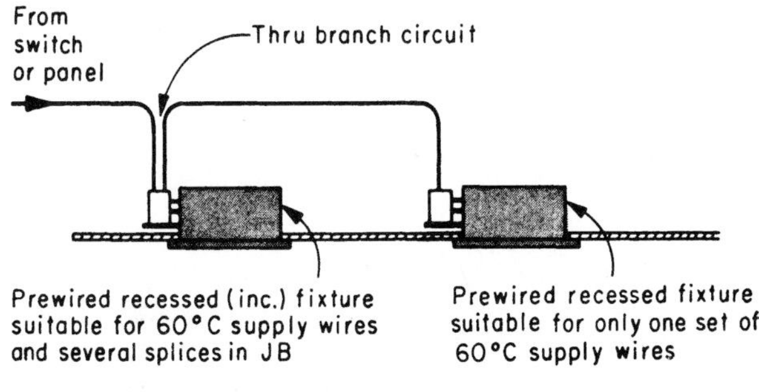

Fig. 410-5. Care must be exercised in hooking up "prewired" *types* of fixtures. (Sec. 410-11.)

combustible material, the canopy and box must form a complete enclosure. The chief purpose of this is to require that no open space be left between the canopy and the box edge if the finish is wood or fibrous or any similar material.

410-13. Covering of Combustible Material at Outlet Boxes. See comments under Sec. 370-25.

410-14. Connection of Electric-Discharge Lighting Fixtures. As Sec. 410-14 is worded, the rules presented apply to *both indoor* and *outdoor* applications of electric-discharge fixtures. The rules cover general lighting in commercial and industrial interiors as well as all kinds of outdoor floodlighting and area lighting. Part **(a)** of this section covers *only* connection of electric-discharge luminaires "where . . . supported independently of the outlet box." Chain-hung fixtures, fixtures mounted on columns, poles, structures, or buildings, and any other fixture that is not "supported" by the outlet box that provides the branch-circuit conductors to feed the fixture are covered by part **(a).** In the Exception to part **(a),** cord-fed swag-type lighting fixtures with chain suspension from a ceiling hook independent of the outlet box are recognized.

The basic rule requires a fixed or permanent wiring method to be used for supply to all "electric-discharge lighting fixtures," which includes all fixtures containing mercury-vapor, fluorescent, metal-halide, high-pressure sodium, or low-pressure sodium lamps. BUT *incandescent* luminaires are *not* covered by Sec. 410-14. As a result, incandescent luminaires using cord connection are regulated only by Secs. 400-7(a)(2), 400-8(1), and 410-30(b).

410-15. Supports. Figure 410-6 shows a 7-lb (3.17 kg) fixture shade that is 17 in. (431.4 mm) in diameter and supported by the screw-shell of the lamp and holder, on both counts violating the rule of part **(a).**

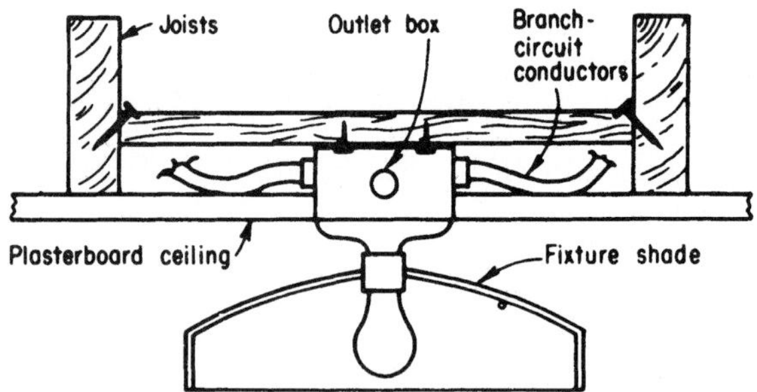

Fig. 410-6. Fixture shade assembly may be supported from a screw-shell lampholder if it is not too heavy or too big in diameter. (Sec. 410-15.)

In part **(b),** the rules cover use of metal poles for supporting lighting fixtures. A metal pole supporting a lighting fixture must have a readily accessible handhole [minimum 2 in. by 4 in. (50.8 by 102 mm)] to provide access to the wiring within the pole or its base. A grounding terminal for grounding the metal pole must be provided and be accessible through the handhole. Any metal raceway supplying the pole from underground must be bonded to the pole with an equipment grounding conductor. And conductors run up within metal poles used as raceway must have vertical supports as required by Sec. 300-19 (Fig. 410-7).

As noted in the Exception to part **(b)(2)**, a metal pole supporting a fixture does not have to have an access handhole at its base if the pole is not over 8 ft (2.44 m) in height (such as a common post light) and the enclosed wiring is accessible at the fixture end. This Exception excludes the typical short [not over 8 ft (2.44 m) high] post light from the need for a wiring access handhole at its base. That handhole is important for higher poles used on commercial and industrial properties and does add safety. But it is unnecessary for a post light,

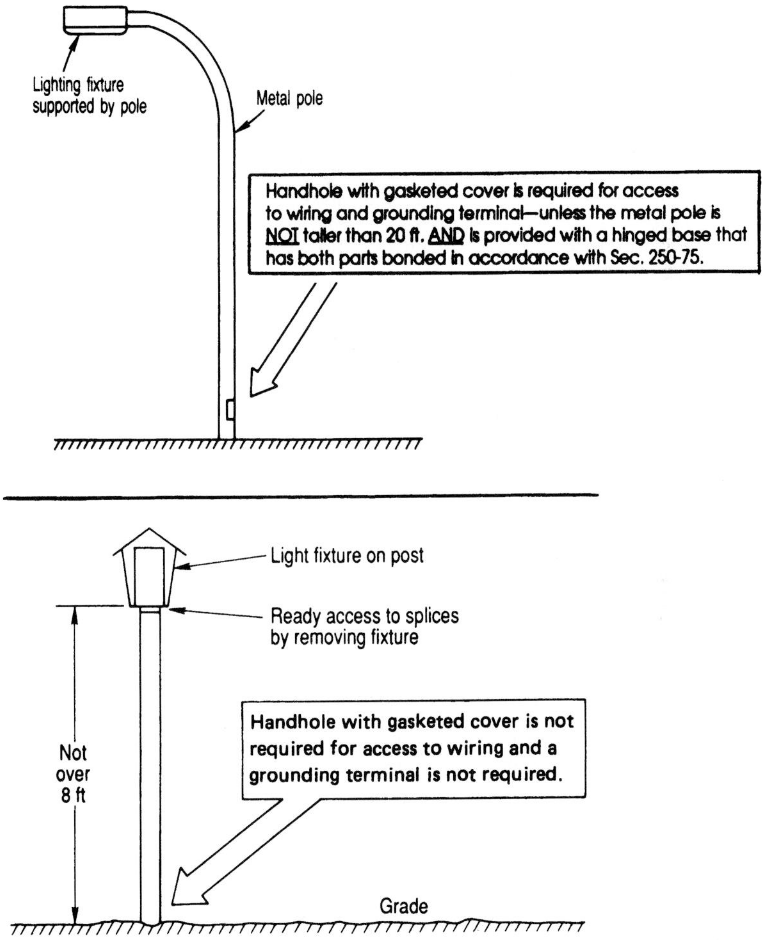

Fig. 410-7. Metal pole must provide access handhole and internal terminal for connecting grounding wire to metal of pole. (Sec. 410-15.)

and this exception allows omission of the handhole where the wiring runs "without splice or pull point" to a fixture mounted on a metal pole not over 8 ft (2.44 m) high and where splices of the fixture wires to the branch circuit supply conductors are accessible by removal of the fixture.

410-16. Means of Support. A lighting fixture may be supported by attachment to an outlet box that is securely mounted in position (see Sec. 370-23), or a fixture may be rigidly and securely attached or fastened to the surface on which it is mounted or it may be supported by embedment in concrete or masonry. As shown in Fig. 410-8, heavy fixtures must have better support than the outlet box.

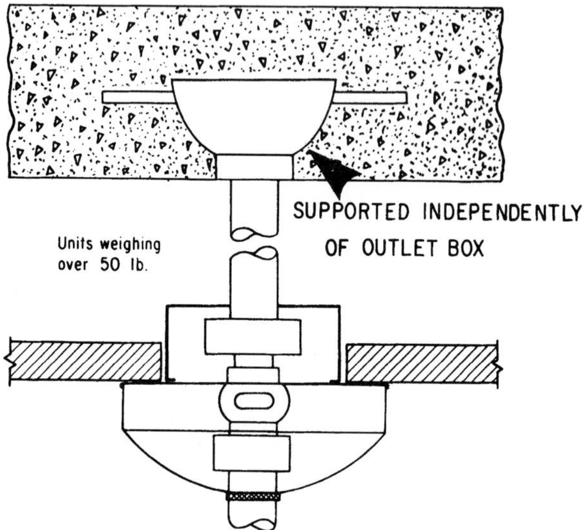

Fig. 410-8. Any fixture over 50 lb (22.7 kg) must be supported from the structure or some other means than the outlet box. (Sec. 410-16.)

Various techniques are used for mounting luminaires independently of the outlet box, depending somewhat on the total weight of the individual luminaires. In general, pipe or rods are usually used to attach the luminaires to the building structure, and the electrical circuit is made by using flex between the luminaire and the outlet box concealed in the ceiling cavity. If provision is made for lowering the luminaire, by means of winch or otherwise, provision must also be made for disconnecting the electrical circuit.

The most common method of supporting fixtures is by means of fixture bars or straps bolted to the outlet boxes, as shown in Fig. 410-9. A fixture weighing over 50 lb (22.7 kg) can be supported on a hanger such as is shown for boxes under Sec. 370-27 for a tile-arch ceiling. Care should be taken to see that the pipe used in the construction of the hanger is of such size that the threads will have ample strength to support the weight.

Any luminaire may be attached to an outlet box where the box will provide adequate support, but, as noted in Sec. 410-15, units which weigh more than 6

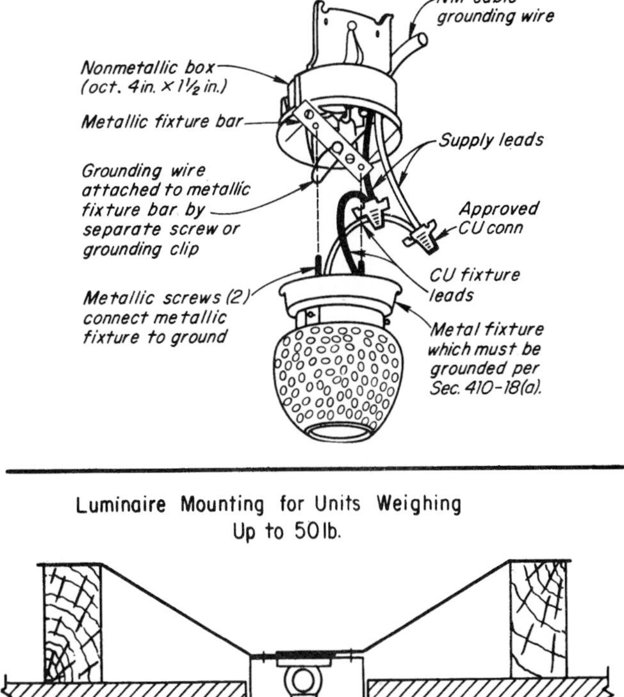

Fig. 410-9. Fixtures must be supported by approved methods. (Sec. 410-16.)

lb (2.72 kg) or exceed 16 in. (406 mm) in any dimension, "shall not be supported by the screw-shell of a lampholder."

A normal method of securing an outlet box in place is to use strap iron attached to back of outlet box and fastened to studs, lathing channels, steel beams, etc., nearby. Lightweight units are sometimes attached to outlet box by means of screws through luminaire canopy which thread into outlet box ears, or flanges, tapped for this purpose. For heavier luminaires, fixture studs, hickeys, tripods, or crowfeet are normally used.

Part **(c)** covers the support of fixtures installed in suspended ceilings. The Code rule wording was based on the following:

SUBSTANTIATION: The Uniform Building Code requires that suspended ceilings be adequately supported. This is usually in the form of an iron wire support attached to the structural ceiling members and the other end of the wire attached to the suspended ceiling frame members. The lighting fixtures are then laid in the openings and secured only by light metal clips. There have been numerous accidents occur when these metal clips have been dislodged causing fixtures to fall to the floor. There have been several instances, where fixtures are installed in end-to-end rows, when one fixture becomes dislodged from construction vibration causing the entire row to also fall to the floor.

There is also the danger of fixtures being shaken loose by seismic disturbances—Los Angeles, Oroville and Santa Rosa areas, to mention a few locations.

Having these fixtures attached to the framing members also becomes a severe problem to firemen. When the ceiling area becomes involved in a fire or enough heat generated from the fire, the framing members distort and cause the fixtures to fall through the openings.

The last sentence of part **(c)** does clearly recognize support of fixtures by means of "clips" that are listed for use with a particular framing member and type of fixture.

Part **(h)** correlates to Sec. 225-26, which prohibits trees from being used to support conductor spans, by specifically permitting outdoor lighting fixtures and their boxes and support means to be mounted on trees.

410-18. Exposed Fixture Parts. The wording of this section, in referring to "lighting fixtures and equipment," does not make clear that the rules are intended to also apply to faceplates used on snap switches, receptacles, and other devices used in outlet boxes. This section went through extensive revisions and even relocation with Art. 410 since the 1968 NEC when, as Sec. 410-95, it referred to "ungrounded metal lighting fixtures, lampholders and face plates." But the intent of the rules was not changed.

Part **(a)** says metal faceplates must be grounded if the box to which they are attached is fed by a wiring system with an equipment grounding means—metal raceway, metal cable armor, or a ground wire in NM cable. That applies to metal and nonmetallic boxes with a metal faceplate attached.

Part **(b)** says if the wiring system to the box does not contain a grounding means, any faceplate must be nonmetallic. That is a stricter rule than the similar rule in the first sentence of Sec. 380-9.

Other discussions on faceplates are given under Secs. 370-25 and 380-9. A note in Sec. 370-25 refers to Sec. 410-56(d) for faceplates.

Of course, the rules here also apply to lighting fixtures—of the metallic type in part **(a)** and nonmetallic fixtures as required by part **(b)**.

Relating Sec. 250-42(a) to the rule of Sec. 380-9, the effect is to require metal faceplates on switches and receptacles used in such places as kitchens or bathrooms to be grounded if they are within the specified distances of grounded surfaces. This rule requires covers of flush snap switches that are mounted in ungrounded metal boxes and located within reach of conducting floors or other conducting surfaces to be made of nonconducting, noncombustible material. But the way the rule now stands in Sec. 410-18, it simply requires metal faceplates to be grounded if the wiring method permits it.

When a metal faceplate used on a switch mounted in a nonmetallic box has to be grounded, a simple, effective way to do it is to use a switch that has a grounding terminal (a green hex-head screw) attached to the metal mounting yoke. When the grounding conductor in the cable is connected to that screw, the mounting yoke is grounded and so is the metal faceplate that is attached by screw to the mounting yoke.

410-20. Equipment Grounding Conductor Attachment. This rule requires lighting fixtures to have some terminal or other connection for an equipment grounding conductor when such fixtures have exposed metal parts and are supplied by NM cable or nonmetallic raceway, which must carry an equipment grounding conductor for grounding metal parts of the fixture (Fig. 410-10). For fixtures supplied by metal raceway, proper connection of the raceway to the metal of the fixture provides an equipment ground return path through the raceway, in accordance with Sec. 250-91(b).

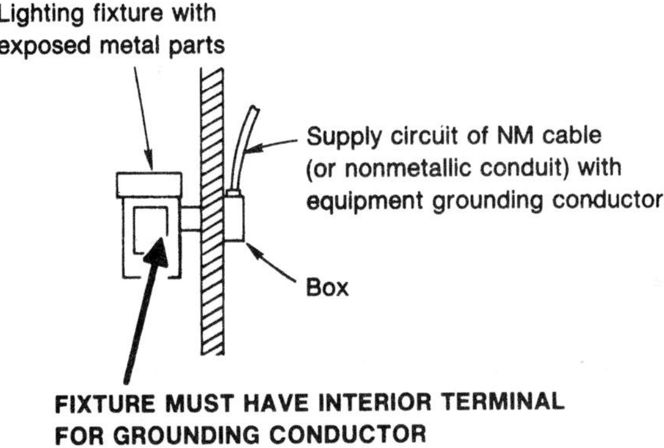

Fig. 410-10. Exposed metal parts of lighting fixtures must incorporate some suitable means for connecting the equipment grounding conductor of a nonmetallic-enclosed supply circuit. (Sec. 410-20.)

410-23. Polarization of Fixtures. This method of wiring fixtures is required in order to ensure that the screw shells of sockets will be connected to the grounded circuit wire.

410-28. Protection of Conductors and Insulation. As noted in part **(e)**, a lighting fixture fed by a conduit stem suspended from a threaded swivel-type conduit body must be supplied by stranded not solid wires run through the conduit stem—because the swivel fitting permits movement of the conductors.

410-29. Cord-Connected Showcases. Figure 410-11 shows an arrangement of cord-supplied illuminated showcases in a store. The details of this layout are lettered and involve the following rules:

1. The first showcase is supplied by flexible cord plugged into grounding-type receptacle rated 20 A. And that is permitted.
2. Flexible cord feeds the second showcase; the cord is spliced in JBs in each showcase. That is a clear violation of the requirement that in such connections, separable locking-type (twist-type) cord connectors must be used and spliced cord connections would be a violation.
3. Cord is No. 14 AWG, hard-service type. That is a violation because the cord conductors must be No. 12, the size of the branch-circuit conductors for the 20-A circuit, as required in part **(a)**.

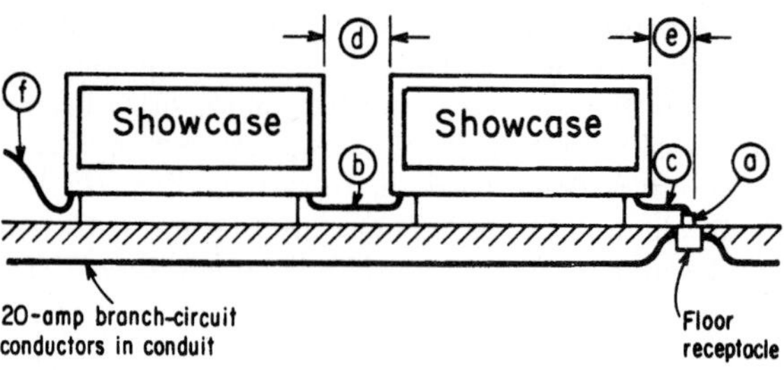

Fig. 410-11. Hookup of lighted showcases must satisfy a number of rules. (Sec. 410-29.)

4. Showcases are separated by 2 in. (50.8 mm). That is O.K. but it is the maximum permitted separation, as noted in part **(c)**.
5. The first case is 14 in. (355.8 mm) from the supply receptacle. No good! The maximum permitted distance is 12 in. (305 mm).
6. The second showcase feeds a spotlight. Violation! Part **(d)** says no other equipment may be connected to showcases.

410-30. Cord-Connected Lampholders and Fixtures. Part **(b)** recognizes the use of a fixed-cord connection for energy supply to lighting fixtures that require aiming or adjustment after installation (Fig. 410-12). Use of a cord supply to lighting fixtures has been a recurring controversial issue, although Sec. 400-7 has permitted cord supply to lighting fixtures for a long time. Sec. 410-30(c) has required that electric-discharge lighting fixtures, if suitable for supply by cord, must make use of plug-and-receptacle connection of the fixture to the supply circuit. The rule here in part **(b)** permits floodlights—such as those used for outdoor and indoor areas for sporting events, for traffic control, or for area lighting—to have a fixed-

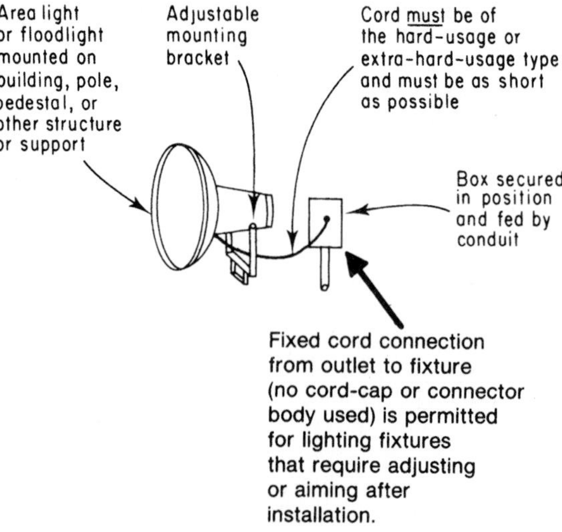

Fig. 410-12. Cord connection—either fixed cord or cord with plug cap—is permitted for adjustable fixtures. (Sec. 410-30.)

cord connection from a bushed-hole cover of the branch-circuit outlet box to the wiring connection compartment in the lighting fixture itself. This rule gives adequate recognition to the type of cord connection that has long been used on floodlights, spotlights, and other fixtures used for area lighting applications.

It should be noted this rule in part **(b)** permits fixed-cord connection but does *not* prohibit use of plug connection to a receptacle or to a connector body. If plug connection is used, however, it might be required that the rules of part **(c)** be satisfied—requiring the receptacle to be mounted directly above the fixture.

In part **(c)**(1) covering electric-discharge fixtures, the rule permits cord-equipped fixtures to be located *directly below* the supply outlet, provided that the cord is continuously visible throughout its entire length outside the fixture (i.e., sight unobstructed by lift-out ceiling panels, etc.) and that it is terminated in a grounding-type cap or busway plug. Except for those fixtures that are UL-listed for cord-and-plug connection as part of listed modular wiring systems for use in suspended ceiling spaces, cord connection may not be used for fixtures installed in lift-out ceilings (Fig. 410-13).

The phrase "directly below the outlet box" has been ruled by inspectors to mean underneath the box and not just at a lower level. The phrase is intended

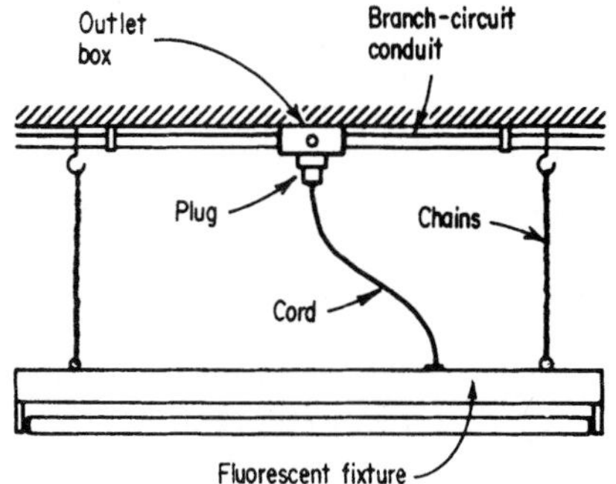

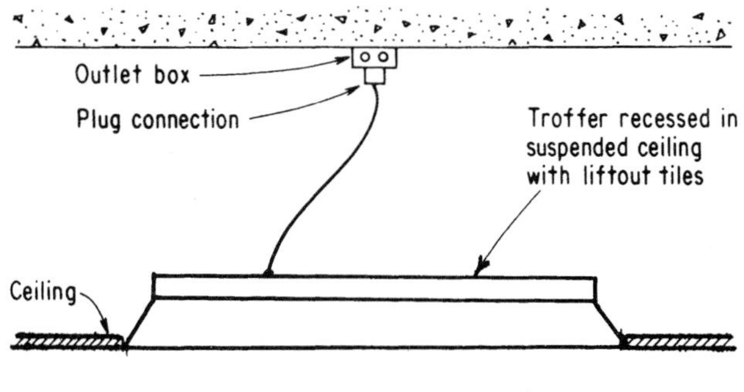

Fig. 410-13. Cord-and-plug fixture supply is O.K. only if cord is "continuously visible." (Sec. 410-30.)

to prevent cases where the fixture is connected by a cord that runs horizontally as well as vertically to an outlet box that is mounted higher than the fixture but off to the side so that the box is not directly above the fixture.

That rule raises a number of questions:

- If an electric-discharge fixture is not mounted "directly below the outlet box," is it a violation to use cord connection from the box to the fixture? The answer seems to be Yes.
- Is it a violation any time an electric-discharge fixture is supplied by a flexible cord with fixed cable connectors at both ends of the cord? Yes, except as permitted in Sec. 410-30(b) for fixtures that require aiming or adjustment.
- If outdoor floodlights—mounted on the ground, or on a building, or pole, or crossarms, or standards, or towers—do not comply with the precise conditions of hookup presented in the first and second sentences of Sec. 410-30(c), is it clearly mandatory that a "fixed wiring method" (and not flexible cord) must be used? No, as covered in Sec. 410-30(b).

But, difficulty in interpretation does arise because connection of "electric-discharge lighting fixtures" is strictly regulated and connection of incandescent fixtures is virtually ignored.

As noted in the Exception to Sec. 410-30(c)(1), there is limited permission to supply an electric-discharge fixture with a flexible cord that is hard-wired. But, the fixture must be *listed* for such installation.

Part **(c)**(2) of this section permits electric-discharge lighting with mogul-base screw-shell lampholders (such as mercury-vapor or metal-halide units) to be supplied by branch circuits up to 50 A with the use of receptacles and caps of lesser ampere rating if such devices are rated not less than 125 percent of the fixture full-load current.

Paragraph (3) at the end of part **(c)** expands coverage of the use of wiring methods suitable for supplying electric-discharge lighting fixtures. Though no longer specified here, such fixtures are permitted to be supplied from busways as described in Sec. 364-12. This paragraph provides for cord connection of a lighting fixture where the cord is equipped with a connector body at its lower end for insertion into a flanged inlet recessed in the lighting fixture housing (Fig. 410-14). This method of cord supply to electric-discharge lighting fixtures presents an alternative to the other method recognized in this section using a cord from the fixture with a plug cap on the other end of the cord for insertion into a receptacle mounted in a box directly above the fixture. The use of a connector body and flanged inlet supply affords greater ease in maintenance of the fixture, since maintenance people can disconnect the fixture at the lower end of the cord to remove it for cleaning or repair.

410-31. Fixtures as Raceways. This Code rule has long stated basically that fixtures shall not be used as a raceway for circuit conductors (Fig. 410-15). Exception Nos. 1 and 2 have permitted variations from that rule.

Exception No. 1 permits fixtures to be used for circuit conductors if the fixtures are approved for use as a raceway (i.e., UL-listed and marked for general use as a raceway for conductors other than the circuit supplying the fixture).

Exception No. 2 permits limited use of fixture wiring compartments to carry through the circuit that supplies the fixtures, provided that the fixtures are

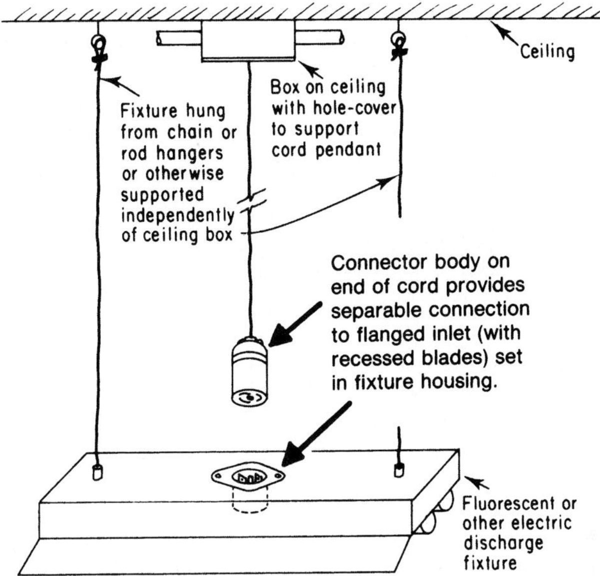

Fig. 410-14. This is the second method recognized for cord supply to lighting fixtures. (Sec. 410-30.)

designed for end-to-end assembly to form a continuous raceway or the fixtures are connected together by recognized wiring methods (such as rigid conduit and EMT). Most self-contained fluorescent luminaire units now available are designed for end-to-end assembly. Each luminaire contains a metal body, or housing, which serves as the structural member of the luminaire and provides a housing for the ballast, wiring, etc., which is of sufficient size to permit running the branch-circuit wiring through the unit. Each luminaire is then tied to the branch circuit by means of a single tap. When a fixture is specifically approved as a raceway, any number of branch-circuit conductors may be installed within the capacity of the raceway. When housings are approved as raceways, luminaires carry an Underwriters Laboratories label which states "Fixtures Suitable for Use as Raceway." Any type of circuit may be run through the fixture (Fig. 410-16).

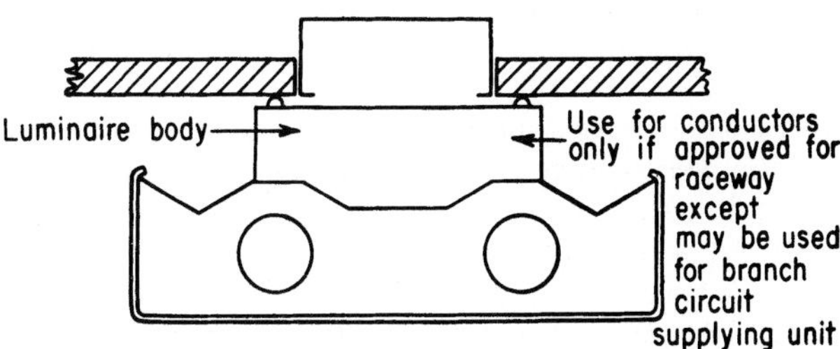

Fig. 410-15. Use of wiring through lighting fixtures is clearly limited. (Sec. 410-31.)

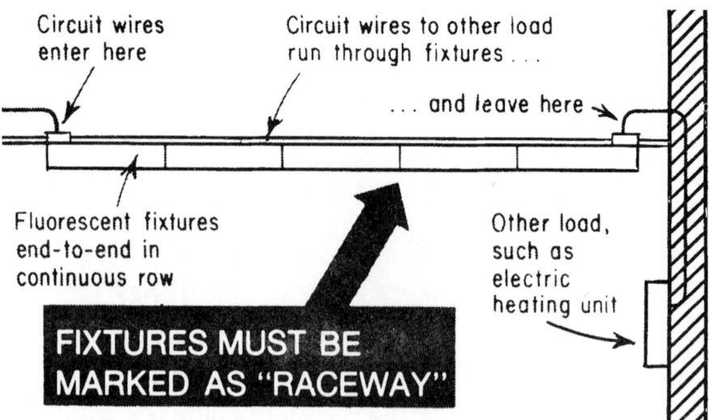

Fig. 410-16. Only fixtures approved as "Raceway" may be used for carrying-through circuit wires that supply any load other than the fixtures. (Sec. 410-31.)

It should be noted that in Exception No. 2, the permitted fixture layouts may carry only conductors of either a 2-wire or a multiwire branch circuit where the wires of the branch circuit supply only the lighting fixtures through which the circuit conductors are run. Thus, it is permissible to use a 3-phase, 4-wire branch circuit through fixtures so connected, with the total number of fixtures connected from all the phase legs to the neutral, that is, with the fixture load divided among the three phase legs. But Exception No. 2 limits such use to a single 2-wire or multiwire branch circuit.

Exception No. 3 permits one additional 2-wire branch circuit to be run through such fixtures (connected end-to-end or connected by recognized wiring methods) in addition to the 2-wire or multiwire branch circuit recognized by Exception No. 2, and this additional 2-wire branch circuit may be used only to supply one or more of the connected fixtures throughout the total fixture run supplied by the other branch circuit run through the raceway (Fig. 410-17). This was added to permit separate control of some of the fixtures fed by the additional branch circuit, providing the opportunity to turn off some of the fixtures for energy conservation during the night or other times when they are not needed.

It should be noted that UL rules tie into the above **Code** rules: Fixtures that are suitable for use as raceway—i.e., for carrying circuit wires other than the wires supplying the fixtures—must be so marked and must show the number, size, and type of conductors permitted. **NE Code** Sec. 410-31 correlates with this UL rule.

The last sentence of this section regulates use of wires run through or within fixtures where the wire would be exposed to possible contact with the ballast which has a hot-spot surface temperature of 90°C. Thus, such wires must be rated at least 90°C—which is the temperature at which the wire will operate when carrying its rated current in an ambient not over 30°C (which is 86°F). Note that Type THW wire is permitted for this use in fixtures. Although Type THW is listed as a basic 75°C wire in Table 310-13, the table does show it as a 90°C-rated wire for use in fixtures in accordance with Sec. 410-31.

This may now be done:

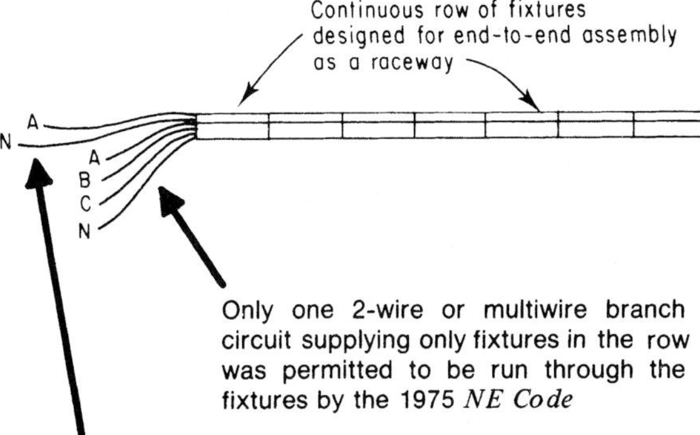
Continuous row of fixtures designed for end-to-end assembly as a raceway
A
N
A
B
C
N
Only one 2-wire or multiwire branch circuit supplying only fixtures in the row was permitted to be run through the fixtures by the 1975 *NE Code*

Now the *Code* rule accepts one more circuit—only a **2-wire** circuit—run through the row. But this additional circuit **must** supply one or more of the fixtures in the row, such as night lighting by, say, every fifth fixture, to enable the others to be turned off for energy conservation.

And the same permission applies in this case—

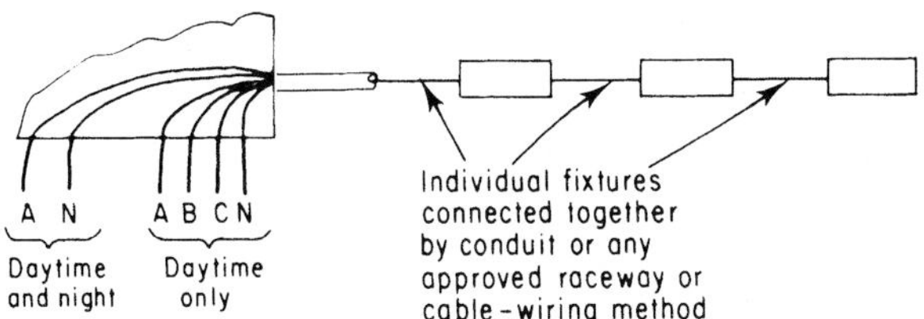
A N
A B C N
Daytime and night
Daytime only
Individual fixtures connected together by conduit or any approved raceway or cable-wiring method

Fig. 410-17. Expanded use of fixtures as raceways provides better control for conservation. (Sec. 410-31.)

The question often arises, May Type AF (150°C) fixture wire be used for circuiting through end-to-end connected continuous-row fluorescent fixtures? Sec. 402-10 permits "fixture wires" to be installed "in lighting fixtures" where they will not be subject to bending or flexing in normal use. That would seem to approve Type AF through the fixtures connected in a row. But, Sec. 402-11 prohibits fixture wires used as branch-circuit conductors. The conductors installed in the fixture "ballast compartment" are referred to as "branch-circuit conductors" in Sec. 410-31, because they feed directly from the branch circuit and are tapped at each fixture to feed each fixture. The branch circuit extends from the point where it is protected by a CB or fuse to the last point where it feeds to the final outlet, device, apparatus, equipment, fixtures, etc. Under these conditions, it seems clear that the wiring must be that approved for branch circuits. Type AF and any other fixture wires are not approved for branch-circuit wiring.

To satisfy the last paragraph of Sec. 410-31, conductors such as Type RHH, Type THHN, or Type THW must be installed. Type AF is definitely not permitted to be installed as branch-circuit conductors. But check with the local inspector if there are any doubts.

410-35. Fixture Rating. This section specifically requires that any fixture be suitably marked to indicate the need for supply wires rated higher than 90°C to withstand the heat generated in the fixture. Such marking must be prominently made on the fixture itself and also on the shipping carton in which the fixture is enclosed (Fig. 410-18). And UL rules note: Fixtures marked for use in commercial or industrial occupancies must *not* be used in residential occupancies, because the fixtures have maintenance features beyond the capabilities of ordinary householders or involve voltages higher than that permitted by the NE Code for residences.

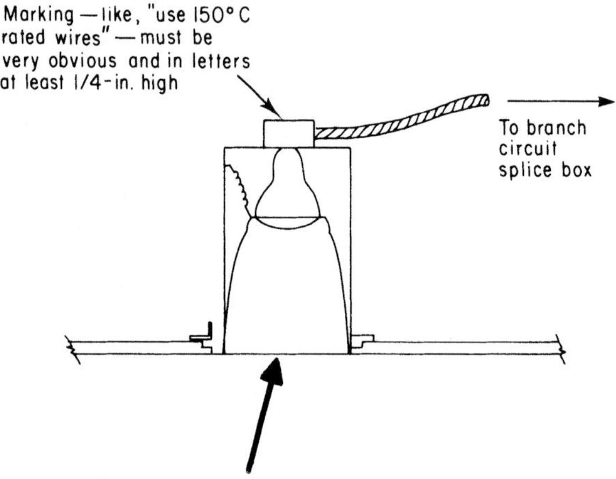
Marking — like, "use 150° C rated wires" — must be very obvious and in letters at least 1/4-in. high
To branch circuit splice box
If heat developed in fixture wiring box requires wire rated over 90C — the required temperature rating of supply wire must be marked in the fixture and on shipping carton.

Fig. 410-18. Where high-temperature wire is needed, fixture must be marked. (Sec. 410-35.)

410-42. Portable Lamps. Part **(a)** requires portable lamps (table lamps and floor lamps) to be wired with flexible cord approved for the purpose and to be equipped with polarized- or grounding-type attachment plugs. Two-prong nonpolarized attachment plugs are no longer permitted. Polarized-type plug caps permit a single orientation of the plug for insertion in the receptacle outlet. Such polarizing of the plug will provide for connecting the grounded conductor of the circuit to the screw shell of the lampholder in the lamp.

In part **(b)**, four specific rules are given on the use of portable handlamps, as shown in Fig. 410-19. The requirements of part **(a)** calling for polarized or grounding type attachment plugs also are made applicable to portable handlamps.

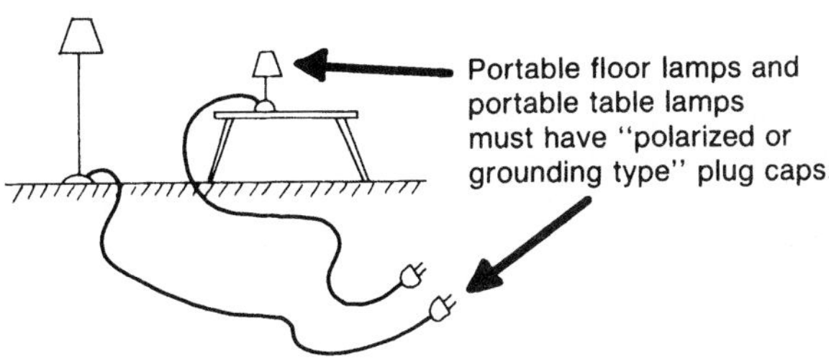
Portable lamps
Portable floor lamps and portable table lamps must have "polarized or grounding type" plug caps.

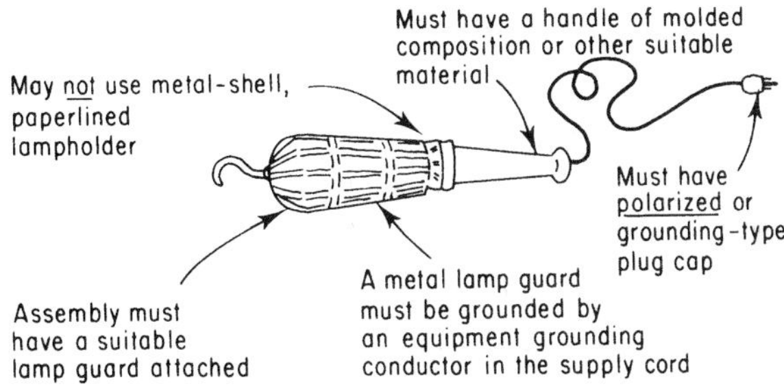
Portable hand lamps
Must have a handle of molded composition or other suitable material
May not use metal-shell, paperlined lampholder
Must have polarized or grounding-type plug cap
A metal lamp guard must be grounded by an equipment grounding conductor in the supply cord
Assembly must have a suitable lamp guard attached

Fig. 410-19. NEC rules aim at greater safety in use of portable lamps and portable handlamps. (Sec. 410-42.)

410-47. Screw-Shell Type. This warns against the previously common practice of installing screw-shell lampholders with screw-plug adapters in baseboards and walls for the connecting of cord-connected appliances and lighting equipment and thereby exposing live parts to contact by persons when the adapters were moved from place to place. See Sec. 410-56(a).

410-48. Double-Pole Switched Lampholders. On a circuit having one wire grounded, the grounded wire must always be connected to the screw shell of the socket, and sockets having a single-pole switching mechanism may be used. (See Sec. 410-52.) On a 2-wire circuit tapped from the outside (ungrounded) wires of a 3-wire or 4-wire system, if sockets having switching mechanisms are used, these must be double-pole so that they will disconnect both of the ungrounded wires.

410-54. Electric-Discharge Lamp Auxiliary Equipment. Part **(b)** requires that a switch controlling the supply to electric-discharge-lamp auxiliary equipment must simultaneously disconnect both hot conductors of a 2-wire ungrounded circuit. This is required to prevent an energized screw shell where only one circuit wire is disconnected—which would be a hazard during relamping.

410-56. Rating and Type. Part **(b)** requires that receptacles rated 20 A or less for direct aluminum connections be CO/ALR type. Part **(d)** is intended to prevent short circuits when attachment plugs (caps) are inserted in receptacles mounted with metal faceplates—in which case, the metal of the plate could short (or bridge) the blades of the plug cap if the faceplate is not set back from the receptacle face. The rule requires the "faces" of receptacles to project at least 0.015 in. (381 micrometers) through the faceplate opening when the faceplate is metallic. And it is necessary to assure a solid backing for receptacles so that attachment plugs can be inserted without difficulty. The requirement for receptacle faces to project at least 0.015 in. (381 micrometers) from installed metal faceplates will also prevent faults caused by countless existing attachment plugs with exposed bare terminal screws. The design requirements for attachment plugs and connectors in part **(e)** should prevent such faults at metal plates, but the problem of existing attachment plugs in this regard will be around for many years.

Part **(c)** covers the so-called isolated-ground receptacles that are intended for use in accordance with Sec. 250-74, Exception No. 4. Such receptacles may *only* be used as indicated by Sec. 250-74, Exception No. 4.

It should be noted that isolated-ground receptacles must be identified by an orange triangle on the face. Although the 1993 and previous **Codes** recognized an overall orange color as a means of identifying an isolated-ground receptacle, such identification is no longer permitted.

With receptacle faces and faceplates installed according to Sec. 410-56**(d)**, attachment plugs can be fully inserted into receptacles and will provide a better contact. The cooperation of other crafts, such as plasterers or dry-wall applicators, will be needed to satisfy the requirements.

As required by part **(e)**, receptacles must be securely mounted in recessed boxes and must assure effective ground continuity. The last two sentences in this section require a receptacle to be tightly screwed to the box in which it is mounted—whether the box is set back from the finished wall surface [not over ¼ in. (6.35 mm)], is flush with the surface, or projects from the wall surface. This rule is the same as the existing rule in Sec. 380-10(b), which covers snap switches in recessed boxes.

410-57. Receptacles in Damp or Wet Locations. The definition of "location" in Art. 100 describes places that fall into either the "wet" or "damp" category. Any receptacle used in a damp location—such as an open or screened-in porch with a roof or overhang above it—may not be equipped with a conventional receptacle cover plate. It must be provided with a cover that, although not UL-listed as "weatherproof," will make the receptacle(s) weatherproof when the cover or covers are in place. The type of cover plate that has a thread-on metal cap held captive by a short metal chain would be acceptable for damp locations but not wet locations (Fig. 410-20). And any other plate-and-cover assembly that is not listed as weatherproof and does not satisfy the conditions of part **(b)** of this section may be used in a damp location provided it covers the receptacle when not in use. The type of receptacle cover that has horizontally opening hinged flaps (doors) to cover the receptacles may be used in a damp but not wet location if the flaps are not self-closing, i.e., if the flaps can stay open (Fig. 410-21). Of course, any cover plate that is listed for weatherproof use may also be used in damp locations.

For wet locations, part **(b)** receptacles must be used with either of two types of cover assemblies:

1. For a receptacle outdoors or in any other wet location where a plug-connected load is normally left connected to the receptacle—such as for

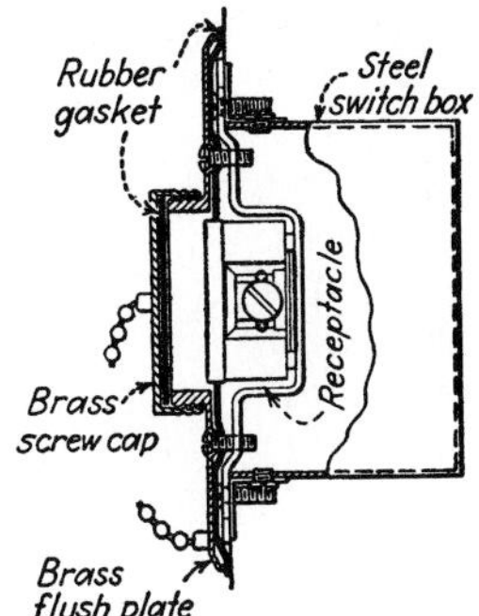

Fig. 410-20. Chain-held screw-cap cover is suitable for damp, but not wet, locations. (Sec. 410-57.)

outdoor landscape lighting or for constant supply to an appliance or other load—only a fully weatherproof, listed cover plate may be used. Such an assembly maintains weatherproof protection of the receptacle at all times—either with the plug out or the plug in. Figure 410-22 shows the way in which a vertically lifting cover shields the receptacle against driving rain (coming at an angle). Such assemblies are made with one cover for a single receptacle or two covers for duplex receptacles. Other cover assemblies use a vertically lifting "canopy" that protects either a single or duplex receptacle.

2. For outdoor receptacles used solely for occasional connection of portable tools or appliances (lawnmowers, hedge-trimmers, etc.), it is permissible, under the Exception, to use a cover that only provides protection against weather when the cover is closed (but not when a plug is inserted). But such a cover, whether installed for vertical or horizontal movement of the cover, must have spring-loaded *self-closing* covers or gravity-close for vertical-lift covers.

Part **(d)** pertains to flush-mounted boxes in which receptacles are installed in wet locations, and part **(e)** requires an elevation of outdoor receptacles to prevent accumulation of water.

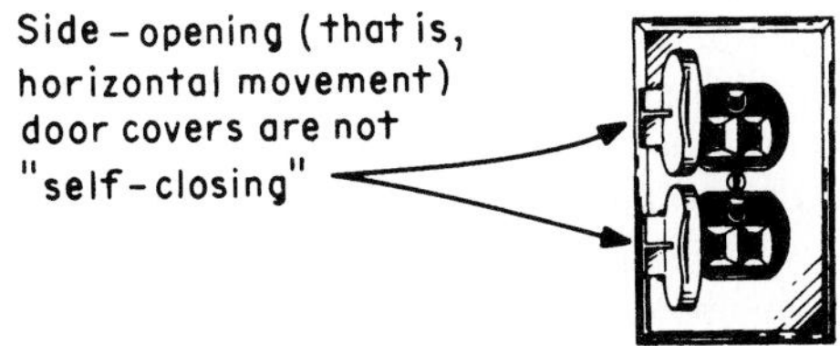

Note: If door covers were self-closing type, such a cover assembly could be used outdoors to supply a portable tool or equipment that is not left connected.

Fig. 410-21. Cover assembly with stay-open doors may be used in damp but *not* in wet locations. (Sec. 410-57.)

Receptacle is kept
weatherproof with or
without plug inserted.

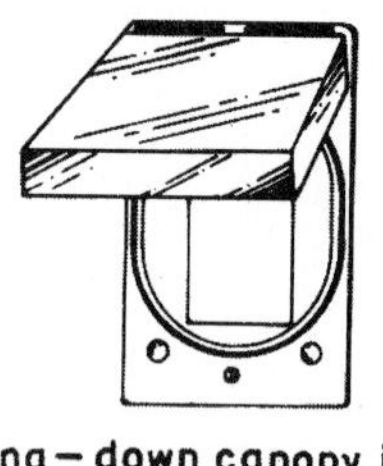
Swing-down canopy is
self-closing

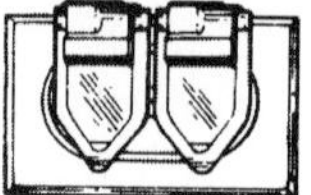
Duplex cover

Fig. 410-22. UL-listed weatherproof covers protect receptacles at all times. (Sec. 410-57.)

410-58. Grounding-type Receptacles, Adapters, Cord Connectors, and Attachment Plugs. Paragraph **(3)** of part **(b)** requires a "rigid" terminal for equipment grounding connection in grounding adapters for insertion into nongrounding receptacles (Fig. 410-23). Adapters with pigtail leads are not acceptable to the rule.

Use of grounding adapters to convert a nongrounding receptacle for connection of a three-prong grounding plug cap involves a number of NE Code and UL regulations. When a grounding receptacle connection is required for a cord-connected appliance or tool, a nongrounding-type receptacle should be replaced with a grounding type, as required by Sec. 210-7(d). And the branch circuit or branch-circuit raceway must include or provide a grounding conductor to which the receptacle ground terminal must be connected.

Plug-in grounding adapters for converting nongrounding (two-slot) receptacles to grounding type are listed by UL under "Attachment Plugs" in the *Electrical Construction Materials Directory* (Green Book). But the NE Code makes no reference to the use of such devices. Section 410-58(b) does describe the construction of adapters and thus implies that their use is permitted.

The problem with grounding adapters having a green grounding pigtail is that such adapters used on nongrounding receptacles present the inherent risk that they will be used without assuring integrity of the grounding path. It is very easy for anyone to plug such an adapter into a nongrounding receptacle and connect the spade lug in the green pigtail under the screw that secures the faceplate on the receptacle. But experience has shown a number of conditions that might exist and prevent effective grounding of the ground terminal of the adapter. The nongrounding receptacle may be fed by NM cable without a grounding conductor (or even by 2-wire knob-and-tube wiring). If the non-

***NE Code* AND UL VIOLATION!** The *NE Code* prohibits use of the pigtail grounding adapters that have been available and widely used for many years. Such devices also do not satisfy UL construction standards.

THIS IS UL AND *NE Code* RECOGNIZED. Grounding adapter has rigid tab with spade end for connecting grounding terminal of the adapter to metal screw contacting the grounded metal yoke that mounts receptacle to the grounded metal box of the outlet (or to an equipment grounding conductor in NM cable used with a nonmetallic outlet box). Different-width blades on adapter polarize it for insertion in only one way into the polarized blade-openings on the receptacle.

Fig. 410-23. This is the NEC and UL position on grounding adapters. (Sec. 410-58.)

grounding receptacle fed by NM cable is in a nonmetallic box, the screw that secures the faceplate is not grounded, even if the cable includes a grounding conductor.

Another potential hazard of the pigtail adapter is shown in Fig. 410-24. A spade lug on the end of the green pigtail wire might accidentally contact a hot terminal through the blade opening of the other receptacle on a duplex outlet. Or a spade lug could touch the hot blade of a plug cap that is not fully inserted into lower receptacle.

Figure 410-25 shows a very important rule from part **(e)** of Sec. 410-58.

410-64. General. Underwriters Laboratories rules comment on use of fixtures installed in hung ceilings, as follows:

Fixtures marked "Suitable for Use With Suspended Ceilings" have not been tested for use in ceiling spaces containing other heat sources such as steam pipes, hot-water pipes, or heating ducts.

Air-handling fixtures must be used fully in accordance with the conditions marked on the fixtures. When used in fire-rated ceilings, such fixtures must be related to the "Design Information Section" in the *Fire Resistance Index,* published by UL.

410-65. Temperature. Heat is a major problem in lighting system design, and with the trend to the use of recessed luminaires and equipment, the problem is increased. In the case of luminaires using incandescent lamps, the problem is primarily the prevention of concentrated spots of heat coming into contact with the building structure. In the case of fluorescent luminaires, the major

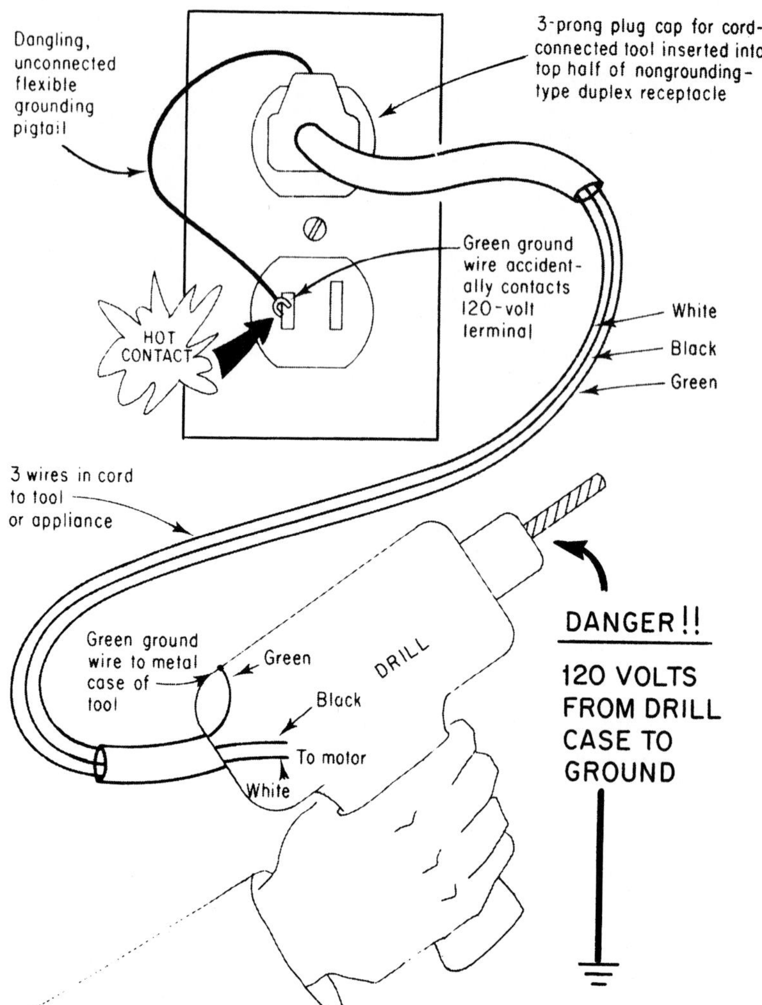

Fig. 410-24. Pigtail adapter can present shock hazard to personnel. (Sec. 410-58.)

heat problem is related to the ballast, which can build up severely high temperatures when not properly ventilated, or designed for cooler operation through adequate radiation and convection. These are problems which must be solved (1) through proper luminaire design, and (2) through proper installation methods and techniques (Fig. 410-26).

The rule of part **(c)** requiring thermal protection of recessed incandescent fixtures is intended to prevent fires that have been caused by overlamping or misuse of insulating materials. (See Fig. 410-27.) As noted in Exception No. 2, a nonthermal-protected recessed incandescent fixture may be used in direct contact with thermal insulation if it is listed as suitable by design for performance

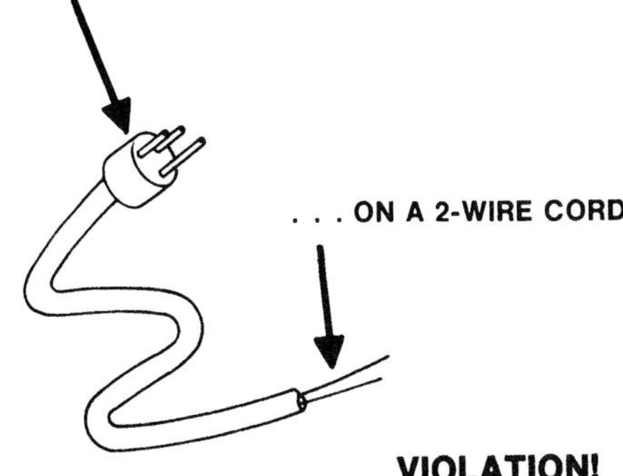
3-PRONG GROUNDING PLUG CAP . . .
. . . ON A 2-WIRE CORD
VIOLATION!

Fig. 410-25. A 3-wire cord must be used when connecting a grounding-type plug-cap. (Sec. 410-58.)

equal to thermally protected fixtures and is so identified. There are presently available recessed incandescent fixtures of such design as to prevent overheating even when installed in contact with thermal insulation. Such fixtures cannot be overlamped or mislamped to cause excessive temperature and are listed and marked for such application.

410-66. Clearance and Installation. When recessed fixtures are used with thermal insulation in the recessed space, thermal insulation must have a clearance of at least 3 in. (76 mm) on the side of the fixture and at least 3 in. (76 mm) at the top of fixture and shall be so arranged that heat is not trapped in this space. Free circulation of air must be provided with this 3-in. (76-mm) spacing. If, however, the fixture is approved for installation with thermal insulation on closer spacing, it may be so used (Fig. 410-28). In parts **(a)** and **(b)**, the Exception recognizes recessed incandescent fixtures in contact with thermal insula-

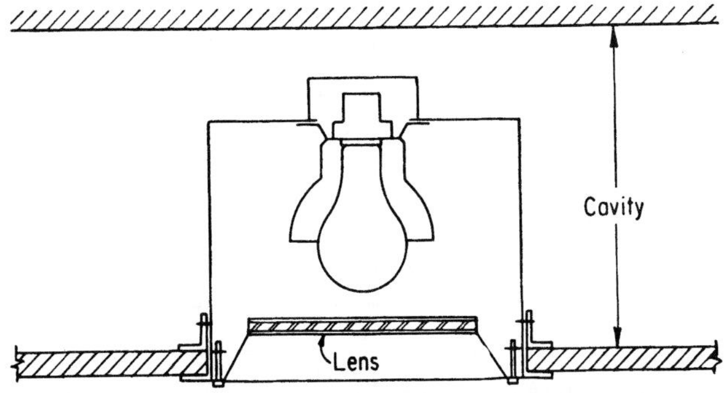
Cavity
Lens
Luminaires shall not subject adjacent combustible material to a temperature in excess of 90°C (194° F).

Fig. 410-26. Recessed fixtures must not threaten combustion of building materials. (Sec. 410-65.)

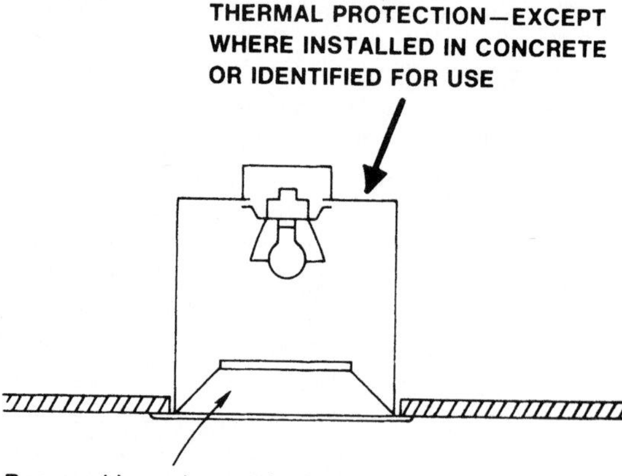

Fig. 410-27. Recessed incandescent fixtures that incorporate thermal protection must be "identified as thermally protected," but a fixture without thermal protection may be used if listed and identified for use in contact with insulation. (Sec. 410-65.)

tion or combustible material. Recessed incandescent fixtures must be listed and identified for use in contact with thermal insulation.

In the past, thermal insulation has been installed in direct contact with recessed fixtures not approved for that use and has caused overheating in fixtures with resulting failures and fires. Obviously, the installer of thermal insu-

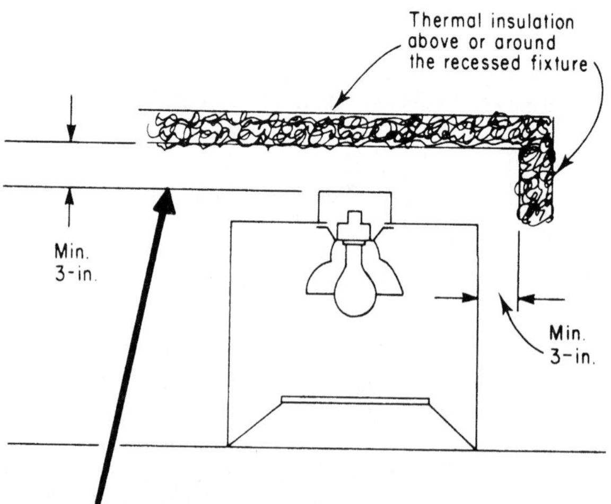

Fig. 410-28. Clearance of recessed fixture from thermal insulation is required unless fixture is UL-listed for use in direct contact with insulation. (Sec. 410-66.)

lation will have to be educated on this subject, because electrical installers have little control over how the insulation will be applied.

Figure 410-29 shows an application that involves the ½-in. (12.7-mm) clearance covered in the first sentence of this section. It shows two 40-W fluorescent strips installed in a residential kitchen ceiling. The ceiling has been furred down on all sides of the 4- by 4-ft (1.22- by 1.22-m) fixture area as shown. The question arises: Is this a "recessed" installation according to the **Code**? How small or large must such an enclosed space be to be considered a recess? The **Code** does not mention "recessed installations" but refers to "recessed fixtures." The installation as shown is basically a field-fabricated recessed fixture. With only two 40-W lamps in this space, it is not likely there will be much of a heat problem. But sufficient information on the total construction of the cavity would be needed for an inspector to make an evaluation. The temperature limitations of Secs. 410-5 and 410-65 must be observed, and wiring must be in accordance with Sec. 410-67.

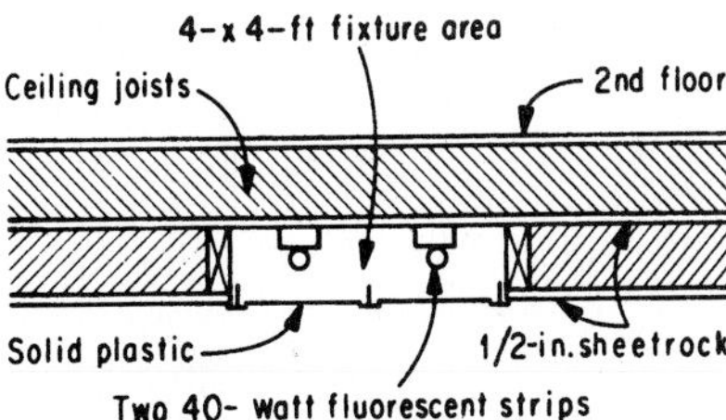

Fig. 410-29. Clearances on custom recessed lighting applications must be evaluated carefully. (Sec. 410-66.)

The next question is: Must a ½-in. (12.7-mm) clearance be maintained between the fixtures and the Sheetrock? Sheetrock is fire-rated by UL but is not fireproof. It is likely that inspectors would require the ½-in. (12.7-mm) spacing between the fixtures and the Sheetrock because the paperboard surfaces of the Sheetrock are combustible. The local inspector would have the final say on this concern.

410-67. Wiring. Supply wiring to recessed fixtures may be the branch-circuit wires, if their 60, 75, or 90°C temperature rating at least matches the temperature that will exist in the fixture splice compartment under operating conditions. Typically fluorescent fixtures do not have very high operating temperatures where the branch-circuit wires splice to the fixture wires and branch-circuit conductors may be run right into the fixture. But, incandescent fixtures develop much higher localized heat because all the wattage is concentrated in a much smaller bulb, thereby requiring higher-temperature wire where the branch-circuit splices to the fixture leads.

The rules of part **(b)** and part **(c)** of this section are related to the details discussed under Sec. 410-11. Figure 410-30 shows branch-circuit wires rated at 75°C coming out of ceiling boxes (left) and then connected directly to fluores-

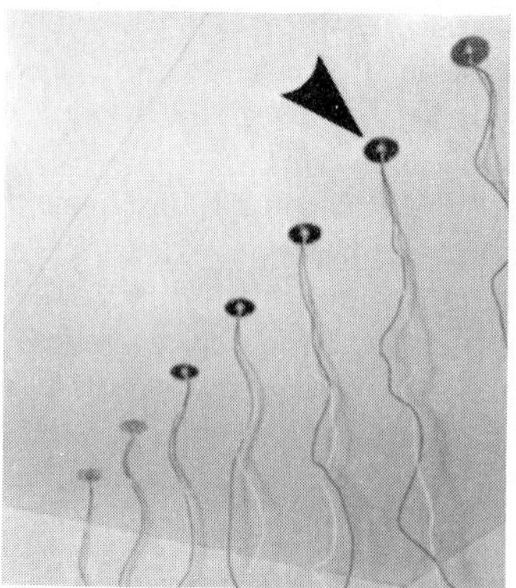

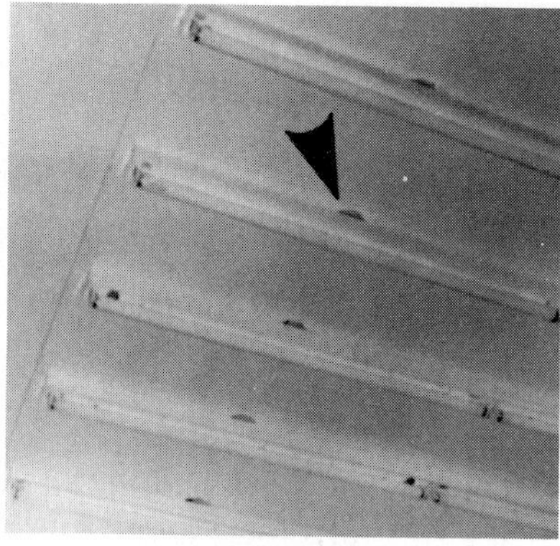

Fig. 410-30. Branch-circuit wires rated at 75°C are brought out of ceiling boxes (left) and then connected to fixture leads within the relatively cool wiring compartment of the fluorescent units. (Sec. 410-67.)

cent strip units mounted over the boxes and attached to the ceiling (right). The 75°C THW branch-circuit wires are rated at 90°C for use within the fluorescent units (Sec. 410-31, last sentence).

Figure 410-31 shows 60°C branch-circuit wiring run directly to integral junction boxes of "prewired" incandescent fixtures. The box protects the branch-circuit wires from the heat generated within the fixtures. Note these fixtures are used for supply and feed-through of the branch-circuit wires and must be UL-listed for that application (refer to Sec. 410-11).

Figure 410-32 shows an application of part **(c)** of this **Code** section, where a fixture wiring compartment operates so hot that the temperature exceeds that of the branch-circuit-rated value and high-temperature fixture wires must be run to the unit. The circuit outlet box supplying the high-wattage incandescent fixture must be mounted not less than 1 ft (305 mm) away from the fixture. The flex whip may be ⅜-in. flex for the number and type of fixture wires as specified in Table 350-3 (Sec. 350-3). If the branch-circuit supplying the fixture is protected at 20 A or 15 A, No. 18 fixture wires may be used for fixture loads up to 6 A or No. 16 fixture wire may be used for loads up to 8 A (Sec. 402-5), and the metal flex may serve as the equipment grounding conductor [Sec. 250-91(b), Exception No. 1]. The flex may not be less than 4 ft (1.22 m) long but not more than 6 ft (1.83 m) long. The fixture wires could be Type AF for conditions requiring a 150°C rating or could be another type of adequate temperature rating for the fixture's marked temperature, selected from Table 402-3. For flex, refer also to Sec. 350-5.

The 4- to 6-ft (1.22- to 1.83-m) fixture "whip" may be Type AC or Type MC cable, with conductors of adequate temperature rating for the temperature of the fixture wiring compartment. This is an alternative to using a 4- to 6-ft (1.22- to 1.83-m) length of flexible raceway to enclose the tap conductors of the whip. (Fig. 410-32.)

Recessed fixtures are, in all cases, marked with the required minimum-temperature rating of wiring supplying the fixture. See Sec. 410-35(a), the last sentence. This marking does not allow for any heat contributed by a branch cir-

Fig. 410-31. Prewired recessed incandescent fixtures may have 60°C branch-circuit wiring run directly into their junction boxes. (Sec. 410-67.)

cuit passing through the fixture enclosure or through a splice compartment (outlet box or otherwise) that is part of the fixture construction. An insulation with a temperature rating higher than that indicated on the fixture may be required for such branch-circuit conductors. See Sec. 410-11.

The requirements in part **(c)** are special provisions that apply to recessed fixtures and take precedence over the general requirements. The tap conductors (usually in flexible conduit) connecting an unwired recessed fixture to the outlet box must be in metal or nonmetallic raceway (such as electrical nonmetallic conduit or nonmetallic liquidtight flex) at least 4 ft (1.22 m) in length and not over 6 ft (1.83 m). The box is required to be at least 1 ft (610 mm) away from the fixture and the flexible conduit may be looped to use up the excess length (see Sec. 350-18, Exception No. 3). This rule does not apply to "prewired" fixtures designed for connection to 60°C supply wires.

The purpose of this requirement is to allow the heat to dissipate so that heat from the fixture will not cause an excessive temperature in the outlet box and

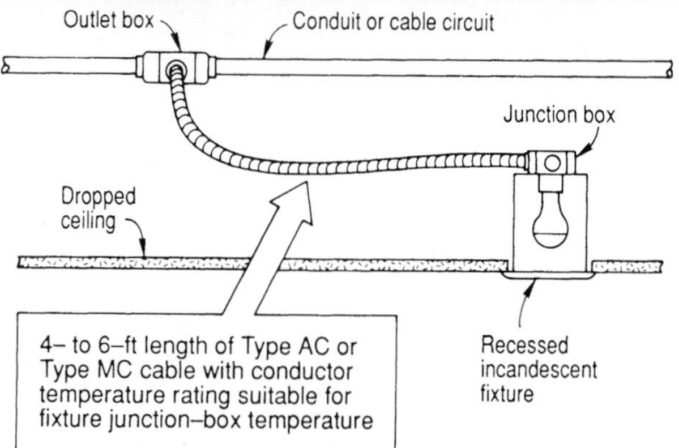

This is an alternative to tap conductors run in metal flex or liquidtight metal flex.

Fig. 410-32. For fixture with circuit-connection compartment operating very hot, high-temperature wires must be run in flex (other metal raceway or Type AC or MC metal armored cable) between 4 and 6 ft. (1.22 and 1.83 m) from an outlet box at least 12 in. (610 mm) away. (Sec. 410-67.)

thus overheat the branch-circuit conductors which could be of the general-use type limited to 60°C or 75°C temperatures.

410-73. General. Paragraph **(e)** pertains only to fluorescent lamp ballasts used indoors. The protection called for must be a part of the ballast. Underwriters Laboratories Inc. has made an extensive investigation of various types of protective devices for use within such ballasts, and ballasts found to meet UL requirements for these applications are listed and marked as "Class P." The protective devices are thermal trip devices or thermal fuses, which are responsive to abnormal heat developed within the ballast because of a fault in components such as autotransformers, capacitors, reactors, etc.

Simple reactance-type ballasts are used with preheat-type fluorescent lamp circuits for lamps rated less than 30 W. Also, a manual (momentary-contact) or automatic-type starter is used to start the lamp. The simple reactor-type ballast supplies one lamp only, has no autotransformer or capacitor, and is exempted from the protection rule of part **(e)**.

The thermal protection required for ballasts of fluorescent fixtures installed indoors must be within the ballast. Previous wording permitted the interpretation that the supplementary protection for the ballast could be in the fixture and not necessarily within the ballast.

Part **(f)** requires that recessed HID fixtures used *either* indoors or outdoors must have thermal protection. The fixture—and not just the ballast—must be thermally protected and so identified. All *indoor, recessed* fixtures using mercury-vapor, metal-halide, or sodium lamps must be thermally protected. If remote ballasts are used with such fixtures, the ballasts must also be thermally protected.

410-75. Open-Circuit Voltages Exceeding 300 Volts. As required by UL, fixtures which are intended for use in other than dwelling occupancies are so marked. This usually indicates that the fixture has maintenance features which are considered to be beyond the capabilities of the ordinary householder, or involves voltages in excess of those permitted by the National Electrical Code for dwelling occupancies.

410-76. Fixture Mounting. Underwriters Laboratories presents data which apply to part **(b)** of this section:

1. Fluorescent fixtures suitable for mounting on combustible low-density cellulose fiberboard ceilings which have been evaluated for use with thermal insulation above the ceiling and which have been investigated for mounting directly on combustible low-density cellulose fiberboard ceilings are marked "Suitable for Surface Mounting on Combustible Low-Density Cellulose Fiberboard" (Fig. 410-33).

   If a fluorescent fixture is not marked to show it is approved for surface mounting on combustible low-density cellulose fiberboard, it must be mounted with at least a 1½-in. (38-mm) space between it and such a ceiling.
2. Fluorescent fixtures that may not be used directly against *any* ceiling are marked "For Suspended Mounting Only. Minimum Distance From Ceiling Six Inches."
3. Surface-mounting incandescent and HID fixtures (mercury-vapor, metal-halide, etc.) are suitable for use on any ceiling including low-density

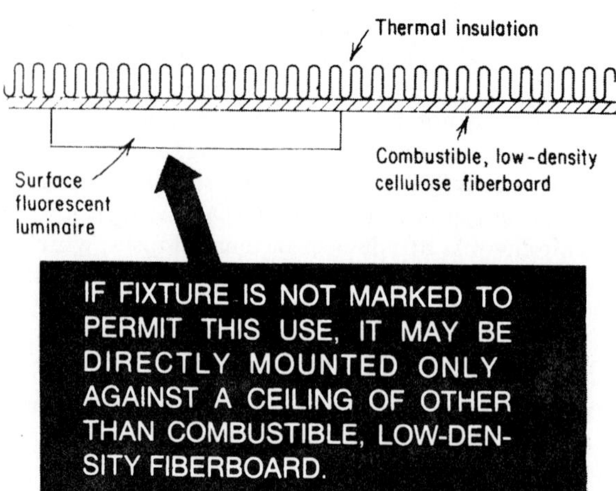
Thermal insulation
Surface fluorescent luminaire
Combustible, low-density cellulose fiberboard
IF FIXTURE IS NOT MARKED TO PERMIT THIS USE, IT MAY BE DIRECTLY MOUNTED ONLY AGAINST A CEILING OF OTHER THAN COMBUSTIBLE, LOW-DENSITY FIBERBOARD.

Fig. 410-33. UL lists fluorescent fixtures for surface mounting on low-density cellulose fiberboard ceilings with insulation above. (Sec. 410-76.)

fiberboard, without marking—but, except for incandescent fixtures marked "Type I.C.," such fixtures must never be used where there is thermal insulation above the ceiling over the fixture. Type I.C. (insulated ceiling) incandescent fixtures may be mounted on a ceiling with insulation above.

4. If surface-mounted or suspended fixtures require supply circuit wires rated over 60°C, they will be marked to show the required minimum temperature rating of the circuit wires. And where such fixtures are marked to require wires rated over 60°C, the marking does not allow for any heat added by a branch circuit passing through the fixture. In such cases, it is up to the installer to determine the need for using wires of a temperature rating higher than indicated.

The note after this section describes "combustible low-density cellulose fiberboard." Material meeting these requirements is listed in Underwriters Laboratories Inc. *Building Materials Directory* and in addition to other pertinent information includes the following: "This material has been found to comply with the flame spread requirements stipulated in Sec. 410-76 of the National Electrical Code as described therein."

410-77. Equipment Not Integral with Fixture. Part (c) covers interconnection of "paired" lighting fixtures, with a ballast supplying a lamp or lamps in both, using up to a 25-ft (7.62-m) length of ⅜-in. flexible metal conduit to enclose the circuit conductors. Lighting manufacturers have been making UL-listed pairs of lighting fixtures for use with each other. Because some inspection agencies have questioned use of this equipment, this section describes Code-conforming use of such equipment to thereby resolve field controversy.

410-78. Autotransformers. This rule ties in with the rules of Sec. 210-6 on voltage of branch circuits to lighting fixtures. On neutral-grounded wye systems (such as 120/208 or 277/480) incandescent, fluorescent, m halide, high-pressure sodium, and low-pressure sodium equ nected from phase to neutral on the circuits. If fluorescent fixtures are to be connected phase to phase, some Code author autotransformer-type ballasts cannot be used when they rai more than 300 V, because, they contend, the reference to "a gr in this rule of Sec. 410-78 calls for connection to a circuit grounded wire and a hot wire (Fig. 410-34). On phase-to-phase co would require use of 2-winding (electrically isolating) ballast trans wording of Sec. 410-78 does, however, lend itself to interpretation t necessary for the supply *system* to the ballast to be *grounded*—thus the two hot legs of a 208- or 480-V circuit to supply an autotransform the hot legs are derived from a neutral-grounded "system." But Se which calls for a "grounded conductor" to be common to the primary ondary of an autotransformer that supplies a branch circuit, can become plicating factor. Use of a 2-winding (isolating) ballast is clearly acceptab avoids all confusion.

410-80. General. These sections apply to interior neon-tube lighting, ligh with long fluorescent tubes requiring more than 1,000 V, and cold-cath fluorescent-lamp installation arranged to operate with several tubes in seri As noted in part **(b)**, electric-discharge lighting equipment with open-circu voltage exceeding 1,000 V shall not be installed in dwelling occupancies.

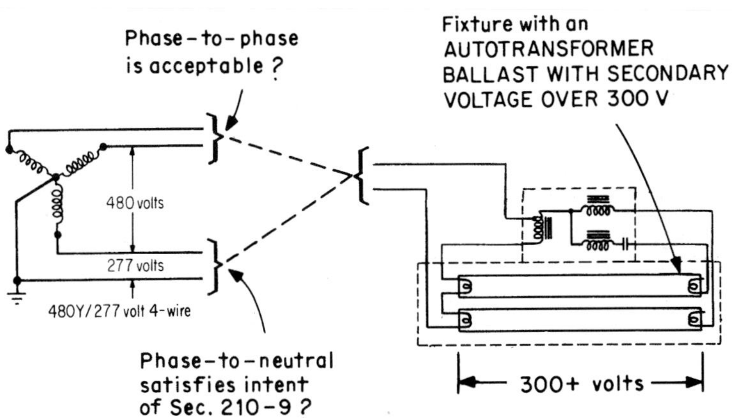

Fig. 410-34. Intent of Sec. 210-9 raises questions about connection of autotransformer ballasts. (Sec. 410-78.)

410-81. Control. When any part of the equipment is being serviced, the primary circuit should be opened and the servicers should have assurance that the disconnecting means will not be closed without their knowledge.

410-87. Transformer Loading. See comments following Sec. 600-32.

410-88. Wiring Method—Secondary Conductors. This type of cable is not included in the table in Chap. 3 listing various types of insulated conductors, but ıerwriters Laboratories Inc. have standards for such cables. The following ɔrmation is an excerpt from the Underwriters Laboratories Inc. *Electrical nstruction Materials Directory:*

> Gas tube sign and ignition cable is classified as Type GTO-5 (5,000 volts), GTO-10 (10,000 volts), or GTO-15 (15,000 volts), and is labeled in sizes Nos. 18-10 AWG copper and Nos. 12-10 AWG aluminum and copper-clad aluminum. This material is intended for use with gas tube signs, oil burners, and inside lighting.
>
> "L"-used as a suffix in combination with any of the preceding type letter designations indicates that an outer covering of lead has been applied.
>
> The label of Underwriters' Laboratories, Inc., . . . on the product is the only method provided by Underwriters' Laboratories, Inc., to identify Gas Tube Sign and Ignition Cable which has been produced under the Label Service.

410-100. Definition. This describes the very popular and widely used track lighting made by a number of manufacturers and used so commonly today in residential and commercial interiors (and even exteriors). In Secs. 410-101 through 410-105, extensive rules cover applications of track lighting mounted on ceilings or walls. Specifically, these rules cover the installation and support of lighting track used to support and supply power to lighting fixtures designed to be attached to the track at any point along the track length (Fig. 410-35).

410-102. Track Load. The 1996 Code clarifies the application of the rules given in this section. (See Fig. 410-36.) In providing minimum required capacity in feeders, a load of 150 VA must be allowed for each 2 ft (610 mm) or fraction thereof of lighting track. That amount of load capacity must be provided in feeders and service conductors (see Fig. 410-37) in both residential and nonresidential installations and would have to be added in addition to the general lighting load in voltamperes per square foot from NEC Table 220-3(b).

ARTICLE 411. LIGHTING SYSTEMS OPERATING AT 30 VOLTS OR LESS

411.1 Scope. This article, which is new to the 1996 NEC, covers the installation of so-called low-voltage lighting. Generally, such lighting must satisfy the rules of Chaps. 1 to 4, except those for swimming pools and those other installations covered by Art 680.

ARTICLE 422. APPLIANCES

422-1. Scope. See the definition for "appliance," Art. 100. For purposes of the Code, the definition for an appliance indicates that it is utilization equipment other than industrial and generally means small equipment such as may be used in a dwelling or office (clothes washer, clothes dryer, air conditioner, food mixer, coffee maker, etc.). See also the definition for "utilization equipment" in Art. 100.

422-4. Branch-Circuit Sizing. Part **(a)** states that the amp rating of an individual branch circuit to a single appliance must not be less than the marked ampere rating of the appliance.

Fig. 410-35. Part **R** of Art. 410 contains rules on installation, application, fastening, and circuit loading for standard and heavy-duty track lighting. (Secs. 410-101 through 410-105.)

422-5. Branch-Circuit Overcurrent Protection. The second sentence presents a rule based on the fact that some appliances are marked to indicate the maximum permitted rating of a protective device (fuse or CB) for the branch circuit supplying that appliance.

422-7. Central Heating Equipment. This rule requires a dedicated (individual) branch circuit to supply the electrical needs of "Central Heating Equipment" other than fixed electric space heating. This requires a separate circuit for the electrical ignition, control, fan(s), and circulating pump(s) of gas- and oil-fired central heating plants. The Exception notes that auxiliary equipment "such as a pump, valve, humidifier, or electrostatic air cleaner directly associated with the heating equipment" *may* be connected to the same branch circuit (or another branch circuit).

The purpose of this rule is to prevent loss of heating when its circuit is opened due to a fault in a lamp or other appliance that is connected on the same circuit with the heater as was permitted in the 1987 NEC and previous editions. All heating equipment should be supplied by one or more individual branch circuits that supply nothing but the heater and its auxiliary equipment.

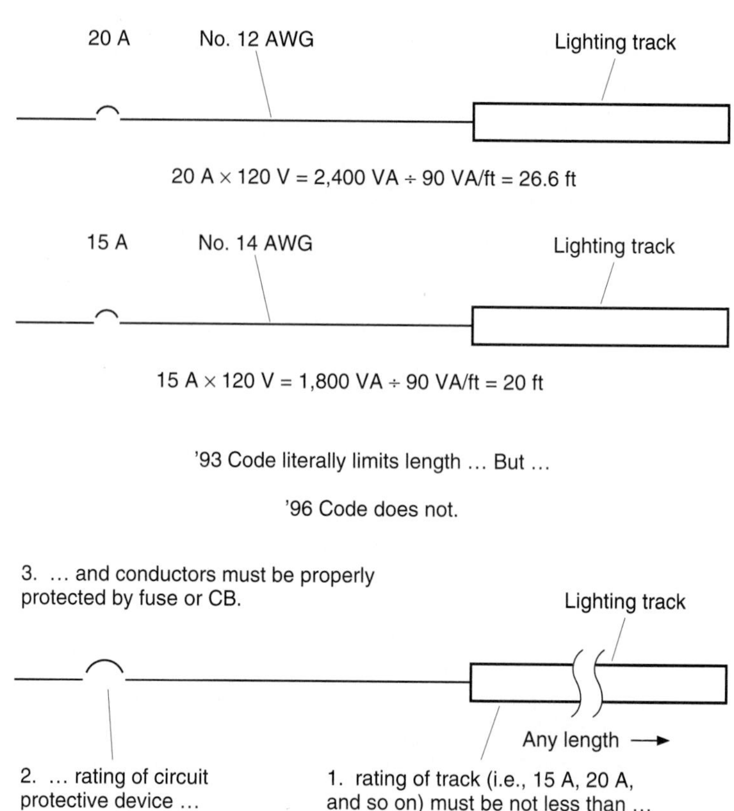
20 A
No. 12 AWG
Lighting track
20 A × 120 V = 2,400 VA ÷ 90 VA/ft = 26.6 ft
15 A
No. 14 AWG
Lighting track
15 A × 120 V = 1,800 VA ÷ 90 VA/ft = 20 ft
'93 Code literally limits length ... But ...
'96 Code does not.
3. ... and conductors must be properly protected by fuse or CB.
Lighting track
Any length
2. ... rating of circuit protective device ...
1. rating of track (i.e., 15 A, 20 A, and so on) must be not less than ...

Fig. 410-36 A single-circuit lighting track must be taken as a load of 150 VA for each 2 ft (610 mm) or fraction thereof, divided among the number of circuits. For a 2-circuit lighting track, each 2-ft (610-mm) length is a 75-VA load for each circuit. For a 3-circuit track, each 2-ft (610-mm) length is a 50 VA load for each circuit. (See Fig. 410-37.)

422-8. Flexible Cords. Figure 422-1 shows application of rules on the hookup of kitchen garbage disposers. The hookup of dishwashers and trash compactors is the same, except that the cord must be 3 to 4 ft (914 mm to 1.22 m) long, instead of 1½ to 3 ft (457 to 914 mm) long.

As required by part **(d)(3)**, a portable high-pressure spray washing machine is required to have a "factory-installed" GFCI. *And* this factory-installed device *must* be "an integral part" of the plug cap or in the cord itself no more than 12 in. (305 mm) from the plug cap. The three Exceptions recognize double-insulated high-pressure spray washers, those rated over 250 V, and 3-phase operated high-pressure spray washers *without* integral GFCI protection. And the double-insulated spray washer would also be required to be permanently marked to indicate that it must be connected to a GFCI-receptacle.

422-12. Signals for Heated Appliances. The standard form of signal is a red light so connected that the lamp remains lighted as long as the appliance is con-

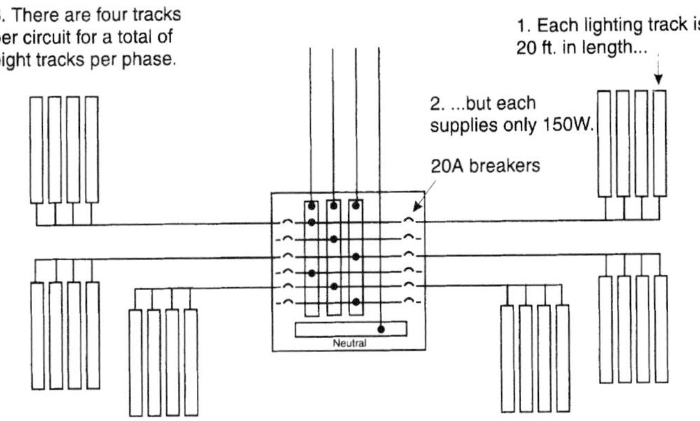

1. According to Sec. 410-102, the feeder must have capacity equal to 75VA/ft (150VA/2ft). Therefore, each phase conductor must have capacity for:
8 tracks x 20ft/track x 75 VA/ft = minimum VA capacity
12,000VA = minimum VA capacity
To find, ampere value divide by 120V.
12,000VA/120V = 100A

2. Due to the continuous nature of the load, the minimum rating of the overcurrent protective device must be at least 125A (100A x 1.25).

3. And, if the load is electric discharge lighting, an additional 80% derating is needed for the "number of conductors" (Note 8), which means.....

4. A No. 1/0, THHN or THW copper must be used.

Fig. 410-37 A single-circuit lighting track must be taken as a load of 150 VA for each 2 ft (610 mm) or fraction thereof, divided among the number of circuits. For a 2-circuit lighting track, each 2-ft (610-mm) length is a 75-VA load for each circuit. For a 3-circuit track, each 2-ft (610-mm) length is a 50 VA load for each circuit. (See Fig. 410-36.)

nected to the circuit. No signal lamp is required if the appliance is equipped with a thermostatic switch which automatically opens the circuit after the appliance has been heated to a certain temperature.

422-14. Water Heaters. Part **(b)** ties into Sec. 422-4(a) Exception No. 2 to require 120-gal (454.2 L) water heaters or any water heater of lesser capacity to be fed by branch-circuit conductors that have an ampacity and protective device rating not less than 1.25 times the marked ampacity of the water heater (Fig. 422-2). Or, to put it another way, the amp rating of the water heater must not exceed 80 percent of the amp rating of the branch circuit. And that rule must be related to Sec. 422-28(e). The only case where the water-heater current may

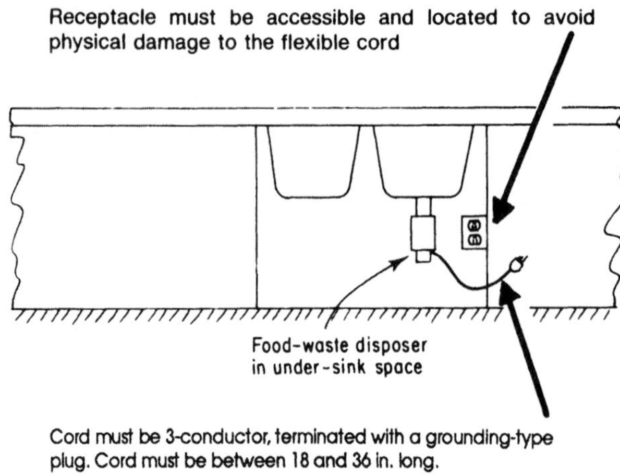
Receptacle must be accessible and located to avoid physical damage to the flexible cord
Food-waste disposer in under-sink space
Cord must be 3-conductor, terminated with a grounding-type plug. Cord must be between 18 and 36 in. long.
NOTE: Double-insulated disposers do not have to be grounded.

Fig. 422-1. Code rules aim at effective grounding for kitchen garbage disposers. (Sec. 422-8.)

load the circuit to 100 percent is where the circuit protective device is listed for continuous operation at 100 percent of its rating. But there are no standard protective devices of that type available at the circuit ratings required for water-heater loads.

422-15. Infrared Lamp Industrial Heating Appliances. So-called infrared lamps are tungsten-filament incandescent lamps, similar to lamps used for lighting except that they are designed for operation with the filaments at a lower temperature, resulting, for a given wattage, in more heat radiation and less light output, and also in a much longer lamp life. In a typical infrared heating oven for industrial use, the lampholders are mounted on panels which are hinged so that the axis of each lamp is at an angle of about 45° from the surface of the panel, to ensure that all sides of an object passing through the oven will receive a uniform amount of heat radiation.

422-18. Support of Ceiling Fans. A ceiling fan up to 35 lb (15.88 kg) may be hung from an outlet box identified for such use, and the box must be properly supported. Heavier ceiling fans must be supported independently of the box.

422-21. Disconnection of Permanently Connected Appliances. In part **(a)** lower-rated appliances may use the "branch-circuit overcurrent device" as their disconnect means and such a device could be a plug-fuse or a CB. Part **(b)** for higher-rated appliances does not permit use of a plug-fuse as the disconnect but requires a definite switch-action device—a switch or CB. A permanently connected appliance rated over 300 VA or over ⅛ hp may use its branch circuit switch or CB as the required disconnecting means *if* the switch or CB is within sight from the appliance or it can be locked in the "open" position. Note that this section applies *only* to permanently connected appliances—i.e., those

Fig. 422-2. Any fixed storage water heater with capacity of 120 gal (454.2 L) or less must be treated as a "continuous duty load" that does not load the circuit to more than 80 percent of its capacity. (Sec. 422-14.)

with fixed-wiring connection (so-called hard wired) and not cord-and-plug connection.

In part **(a)**, the overcurrent device for the circuit to the appliance is not required to be accessible to the user. But the switch or CB in part **(b)** must be "readily accessible" to the user. See definition of "readily accessible" in Art. 100.

Sections 422-27 and 422-25 relate to the rules of this section.

422-22. Disconnection of Cord- and Plug-Connected Appliances. Examples of the application of this section for disconnecting means for appliances are found in the installation of household electric ranges and clothes dryers. The purpose of these requirements is to provide that for every such appliance there will be some means for opening the circuit to the appliance when it is to be serviced or repaired or when it is to be removed.

In part **(b)**, household electric ranges may be supplied by cord-and-plug connection to a range receptacle located at the rear base of the range. The rule permits such a plug and receptacle to serve as the disconnecting means for the range if the connection is accessible from the front by removal of a drawer.

This rule refers to "electric" ranges but the concept also applies to gas ranges. For instance, there have been 115-V receptacle outlets installed behind gas ranges in mobile homes. Such a receptacle is only used as an outlet for the oven light and clock on the range and is not accessible after the range is installed. In order to disconnect the attachment plug or plugs from the receptacle, the range gas supply pipe has to be disconnected, the frame of the range disconnected from its floor fastening, and then the range moved in order to reach the receptacle outlet where the cords are plugged in (Fig. 422-3). Is such an installation in conformity with the intent of the Code? The answer seems surely to be No. The inaccessibility of the receptacle would be objectionable.

422-25. Unit Switch(es) as Disconnecting Means. As shown in Fig. 422-4, the ON-OFF switch on an appliance, such as a cooking unit in a commercial establishment, is permitted by part **(d)** to serve as the required disconnect means if the user of the appliance has ready access to the branch-circuit switch or CB. Note that the wording does not recognize simply "the branch-circuit overcurrent device," and a plug-fuse in the circuit to the appliance would not, therefore, be acceptable as the additional means for disconnection.

422-27. Disconnecting Means for Motor-Driven Appliances. The basic rule requires that a switch or CB serving as the disconnecting means for a permanently connected motor-driven appliance of more than ⅛ hp must be within sight from the motor controller, as shown in Fig. 422-5 for the power unit of a built-in vacuum cleaner system. As permitted by Sec. 430-109 Exception No. 2, general-use AC snap switches may be used as the disconnect for motors rated up to 2 hp, not over 300 V, provided that the motor full load is not greater than 80 percent of the ampere rating of the switch.

Under the Exception to this rule, the branch-circuit switch or CB serving as the other disconnect required by Secs. 422-25(a), (b), (c), or (d) is permitted to be out of sight from the motor controller of an appliance that is equipped with a unit switch that has a marked OFF position and opens all ungrounded supply conductors to the appliance, as shown in Fig. 422-6.

422-28. Overcurrent Protection. The Exception to part **(a)** requires application of overload protection provisions of Secs. 430-32 and 430-33 to motors of motor-operated appliances. And overload protection for sealed hermetic compressor motors must satisfy Secs. 440-52 through 440-55. But motors that are not continuous-duty motors—and most appliances are intermittent, short-time, or varying-duty types of motor loads—do not require running overload protection. The branch-circuit protective device may perform that function for such motors. See Sec. 430-33.

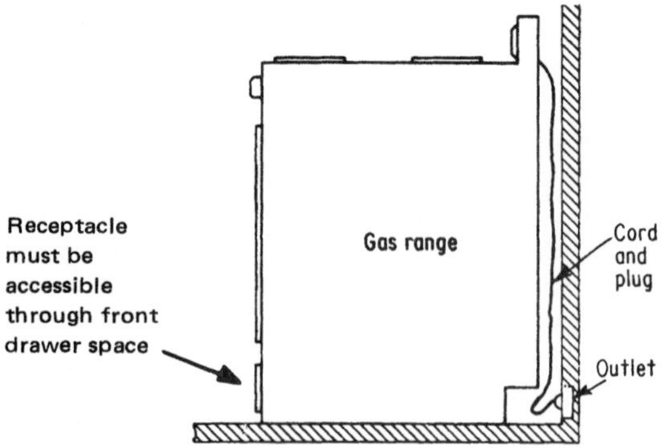

Fig. 422-3. Rule on receptacle behind "electric" ranges could be applied to gas ranges. (Sec. 422-22.)

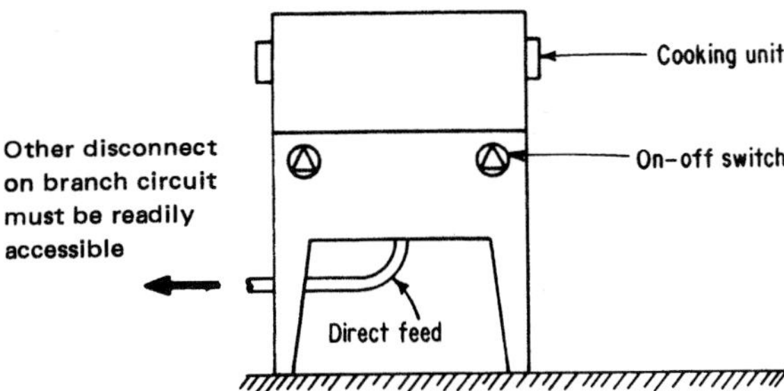

Fig. 422-4. An ON-OFF switch on an appliance must be supplemented by an additional disconnect means. (Sec. 422-25.)

Fig. 422-5. Toggle switch in outlet box on this central vacuum cleaner serves as the disconnecting means within sight from the motor controller installed in the top of the unit. (Sec. 422-27.)

Part (e) sets limits on the maximum permitted branch-circuit protection for an individual branch circuit supplying a single nonmotor-operated appliance. This rule is aimed at providing overcurrent protection for appliances that would not be adequately protected if too large a branch-circuit protective device were used ahead of them on their supply branch circuits.

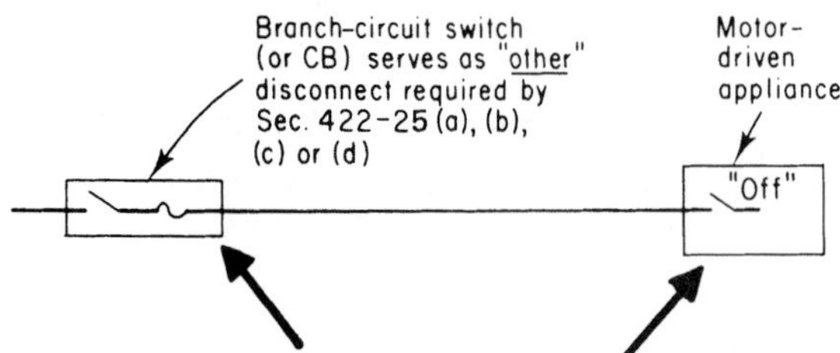
Branch-circuit switch
(or CB) serves as "other"
disconnect required by
Sec. 422-25 (a), (b),
(c) or (d)
Motor-
driven
appliance
"Off"
If appliance has internal unit switch with "off" position
that disconnects all ungrounded conductors, then other
disconnect may be "out of sight."

Fig. 422-6. Disconnect for motor-driven appliance may be "out of sight from the motor controller." (Sec. 422-27.)

The basic rule says that if a maximum rating or branch-circuit protective device is marked on the appliance, that value must be observed in selecting the branch-circuit fuse or CB. If, however, a particular appliance is *not* marked to show a maximum rating of branch-circuit protection, the values specified in this section must be used as follows:

For an appliance drawing more than 13.3 A—the branch-circuit fuse or CB must not be rated more than 150 percent (1½ times) the full-load current rating of the appliance. For an appliance drawing up to 13.3 A—the branch-circuit fuse or CB must not be rated over 20 A.

Those values were chosen to limit the maximum rating of protection to not over 150 percent of the appliance rating to afford greater protection to the appliance, as well as to provide for continuous-load appliances (that operate for 3 hr or more continuously). In the latter case, Sec 210-22(c) requires that a continuous load not exceed 80 percent of the branch-circuit fuse or CB; that is, the protective device is rated not less than 125 percent of the continuous-load current.

An appliance drawing 13.3 A must have protection of not over 150 percent of that value (1.5 × 13.3 = 20 A), which sets 20 A as the maximum protection. And 80 percent of a 20 A rating is 16 A—so that the 13.3-A load does not exceed that value for a continuous load. For appliances drawing more than 13.3 A, the exception says that if 150 percent of the appliance current rating does not correspond to a standard rating of protective device (from Sec. 240-6), the next standard rating of protective device above that value may be used even though it is rated more than 150 percent of the appliance current rating.

Electric water heaters are typical nonmotor-operated appliances rated over 13.3 A, and Sec. 422-14(b) says such water heaters of 120-gal (454.2 L) capacity or less must have a branch-circuit overcurrent device rated not less than 125 percent of the unit's current rating. And Sec. 422-28(e) says the same branch-circuit device must not be rated more than 150 percent of the unit's current rating.

Part **(f)** of this section covers electric heating appliances using resistance-type heating elements. It requires that, where the elements are rated more than 48 A, the heating elements must be subdivided. Each subdivided load shall not exceed 48 A and shall be protected at not more than 60 A.

The rules of this section are generally similar to the rules contained in Sec. 424-22 for fixed electric space heating using duct heaters as part of heating, ventilating, and air-conditioning systems above suspended ceilings. But Exception No. 2 applies to commercial kitchen and cooking appliances using sheath-type heating elements. This Exception permits such heating elements to be subdivided into circuits not exceeding 120 A and protected at not more than 150 A under the conditions specified.

Exception No. 3 of this same section permits a similar subdivision into 120-A loads protected at not more than 150 A for elements of water heaters and steam boilers employing resistance-type immersion electric heating elements contained in an ASME rated and stamped vessel.

ARTICLE 424. FIXED ELECTRIC SPACE-HEATING EQUIPMENT

424-3. Branch Circuits. The basic rule limits fixed electric space heating equipment to use on 15-, 20-, or 30-A circuits, *if the circuit has more than one outlet.* The Exception applies only to fixed infrared equipment on industrial and commercial premises, permitting use of 40- and 50-A circuits for multioutlet circuits to *fixed* space heaters.

The 125 percent requirement in paragraph **(b)** means that branch circuits for electric space heating equipment cannot be loaded to more than 80 percent of the branch-circuit rating. Even though electric heating is thermostatically controlled and is a cycling load, it must be taken as a continuous load for sizing branch circuits.

The three parts of this rule are as follows:

1. The branch-circuit wires must have an ampacity not less than 1.25 times the total amp load of the equipment (heater current plus motor current).
2. The overcurrent protective device must have an amp rating not less than that calculated for the branch-circuit wires.
3. If necessary, the rating or setting of the branch-circuit overcurrent device may be sized according to part **(b)** of Sec. 240-3. That is, if the ampacity of the selected branch-circuit wires does not correspond to a standard rating or setting of protective device (Sec. 240-6), the next higher standard rating or setting of protective device may be used. This accommodates those applications of large unit heaters, where often there is not agreement in ampacity of wires and ratings of standard fuses or CBs.

Figure 424-1 shows an example of branch-circuit sizing for an electric heat unit.

Many line thermostats and contactors are approved for 100 percent load, and derating of such devices is not required.

Section 220-15 covers sizing of feeders for electric space heating loads. The computed load of a feeder supplying such equipment shall be the total connected load on all branch circuits, with an exception left up to the authority enforcing the Code which allows permission for feeder conductors to be of a capacity less than 100 percent, provided the conductors are of sufficient capacity for the load serving units operating on duty-cycle, intermittently, or from all units not operating at one time. The second exception to the rule states that Exception No. 1 does not apply if the optional method in Sec. 220-30 is the method used for calculating the load for a single-family dwelling or individual apartment of a multifamily dwelling.

424-9. General. For instance, heating cable designed for use in ceilings may not be used in concrete floors and vice versa.

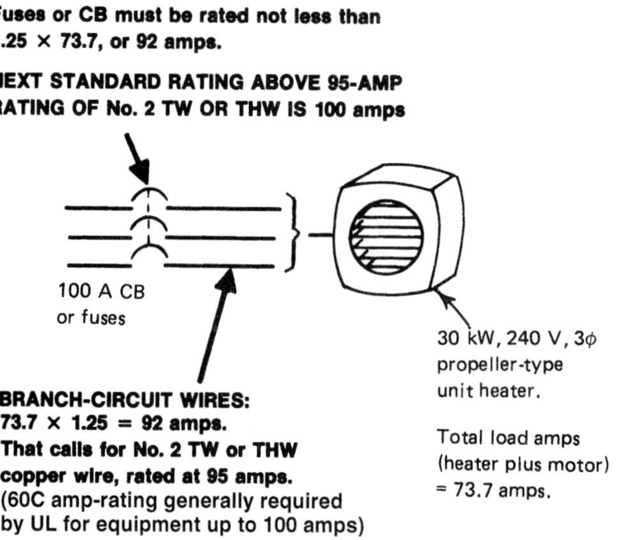
Fuses or CB must be rated not less than
1.25 × 73.7, or 92 amps.
NEXT STANDARD RATING ABOVE 95-AMP
RATING OF No. 2 TW OR THW IS 100 amps
100 A CB
or fuses
30 kW, 240 V, 3ϕ
propeller-type
unit heater.
BRANCH-CIRCUIT WIRES:
73.7 × 1.25 = 92 amps.
That calls for No. 2 TW or THW
copper wire, rated at 95 amps.
(60C amp-rating generally required
by UL for equipment up to 100 amps)
Total load amps
(heater plus motor)
= 73.7 amps.

Fig. 424-1. Branch-circuit conductors and overcurrent protection must be rated not less than 125 percent of heater unit nameplate current. (Sec. 424-3.)

This rule ties into that of Sec. 210-52(a). The fine-print note says that UL-listed baseboard heaters must be installed in accordance with their instructions, which *may* prohibit installation below outlets. This applies to all occupancies—residential, commercial, institutional, and industrial.

In buildings warmed by baseboard heaters, the question arises: How can wall receptacles be provided to satisfy the requirement of NE Code Sec. 210-52(a) that no point along the floor line be more than 6 ft (1.83 m) from an outlet? Should the receptacles be installed in the wall above the heaters? The answer is that receptacles should not be placed above the heaters if they are marked to prohibit that. Fires have been attributed to cords being draped across heaters. Continued exposure to heat causes the cord insulation to become brittle, leading to possible short circuits or ground faults.

Article 210 does not specifically prohibit installation of receptacles over baseboard heaters, and Art. 424 does not specifically prohibit installation of baseboard heaters under receptacles. However, Sec. 110-3(b) says:

> **Installation and Use.** Listed or labeled equipment shall be used, installed, or both, in accordance with any instructions included in the listing or labeling.

Underwriters Laboratories now requires in Standard 1042, *Electric Baseboard Heating Equipment,* a warning that a heater is not to be located below an electrical convenience receptacle. A similar statement is included in the UL *Electrical Appliance and Utilization Equipment Directory* (Orange Book):

> To reduce the likelihood of cords contacting the heater, the heater should not be located beneath electrical receptacles.

This instruction and the provisions of Sec. 110-3(b) make it very clear that installation of electric baseboard heaters under receptacles may be a **Code** violation.

Receptacles can be provided in electrically heated (baseboard-type) occupancies as required by Sec. 210-50(b) by making use of the receptacle accessories made available by baseboard heater manufacturers. These units are designed to be mounted at the end of a baseboard section or between two sections.

424-14. Grounding. The basic rule on grounding applies to all electric space heating equipment. The wiring method supplying any fixed electric space heating equipment must provide a means for grounding all exposed metal parts of such equipment. That must be suitable metal raceway (rigid metal conduit, IMC, EMT) or metal cable armor (BX), or an equipment grounding conductor in NM cable, or nonmetallic conduit.

424-19. Disconnecting Means. The basic rule requires disconnecting means for the heater, motor controller(s), plus supplementary overcurrent protective devices for all fixed electric space heating equipment.

Part **(a)** of this section applies to heating equipment provided with supplementary overcurrent protection (such as fuses or CBs) to protect the subdivided resistance heaters used in duct heating, as required by Sec. 424-22. The basic rules are shown in Fig. 424-2. The disconnect in that drawing must comply with the following:

1. The disconnect must be "within sight from" the supplementary overcurrent panel. If circuit breakers are used as the supplementary overcurrent protective devices for the subdivided electric heating loads [Sec. 424-22(b)], the circuit breaker may constitute the disconnect required by this section. This rule has to be correlated to the other rules of part **(a)** and **(c).**
2. The disconnect must also be within sight from the motor controller(s) and the heater, or it must be capable of being locked in the open position.
3. The single disconnect may serve as disconnect for all equipment.

Part **(b)** applies to heating equipment *without* supplementary overcurrent protection.

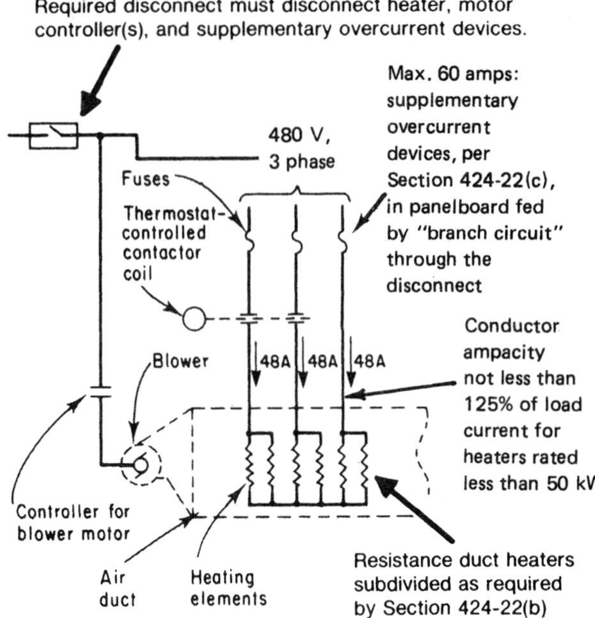

Fig. 424-2. Rules on disconnects for heating equipment demand careful study for HVAC systems with duct heaters and supplementary overcurrent protective devices. (Sec. 424-19.)

Care must be taken to evaluate each of the specific requirements in this Code section to actual job details involved with electric heating installations.

Figure 424-3 shows two conditions that relate to the rules of part **(b)** of this section. For the two-family house, the service panel is located in a rear areaway and is accessible to both occupants. Circuit breakers in the panel constitute suitable means of disconnect for heaters in both apartments—*but* only if they are lockable. If the house uses baseboard heaters without motors in them, lockable breakers would be acceptable as the disconnects in accordance with the rule of **(b)(1).** For the four-family house (a "multifamily dwelling"), the service panels accessible to all tenants are grouped in the hallway under the first-floor stairway in this four-family occupancy. If lockable, the switches in the panels may constitute suitable means of disconnect for heaters in all apartments, under the rule of **(b)(1).** It should be noted that if the CBs are *not* capable of being locked "open," then another disconnect that *is* lockable *or* "within sight" of the heater must be provided. But note that plug-fuses (without switches) on branch circuits to the heaters would not satisfy **(b)(1).**

WATCH OUT!

The rules of part **(b)** were revised in the 1978 NEC. Previous Code rules permitted "the branch-circuit overcurrent device"—which could be a plug-fuse instead of a CB—to serve as the disconnect for electric heaters up to 300 VA or ⅛ hp. Part **(b)(1)** refers only to "branch-circuit switch or circuit breaker."

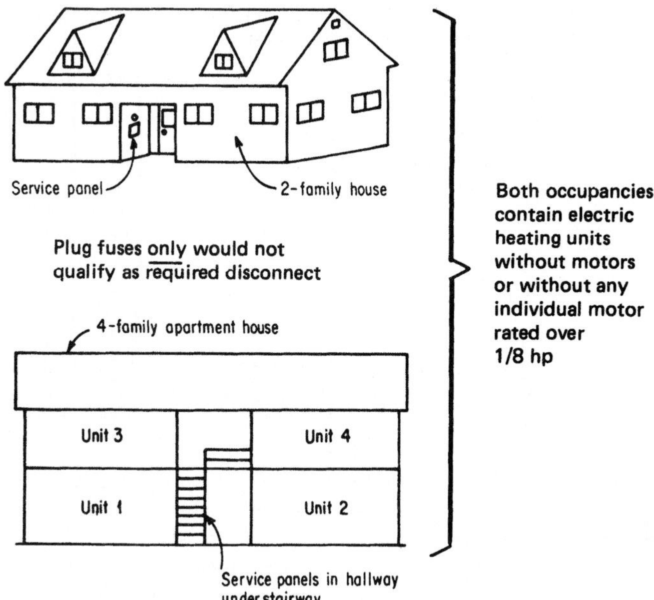
Service panel
2-family house
Plug fuses only would not qualify as required disconnect
4-family apartment house
Unit 3
Unit 4
Unit 1
Unit 2
Service panels in hallway under stairway
Both occupancies contain electric heating units without motors or without any individual motor rated over 1/8 hp

Fig. 424-3. For nonmotored electric heating units, a branch circuit switch or CB that is "within sight" from the heater or lockable may serve as disconnect. (Sec. 424-19.)

Note that **(b)(2)** requires a disconnect "within sight from" a motor controller serving a motor-operated space heater with a motor over ⅛ hp; with the limited Exception referred to part **(a)(2)**c.

Part **(c)** permits the "ON-OFF" switch on the heating unit to be used as the disconnecting means where "other means for disconnection" are provided, depending on the occupany.

424-20. Thermostatically Controlled Switching Devices. Figure 424-4 shows a hookup of duct heaters that are controlled by a magnetic contactor that responds to a thermostatic switch in its coil circuit. The subdivided heater load of 48 A per leg satisfies the rule of Sec. 424-22(b), *but* the contactor does not constitute a combination controller-and-disconnect means because it does not open all the ungrounded conductors of the circuit. And because the heater contains the 60-A supplementary overcurrent protection, a disconnect ahead of the fuses would be required by Sec. 424-19(a) even if the contactor did open all three ungrounded conductors of the circuit.

424-22. Overcurrent Protection. Heating equipment employing resistance-type heating elements rated more than 48 A must have the heating elements subdivided, with each subdivided circuit loaded to not more than 48 A and protected at not more than 60 A. And each subdivided load must not exceed 80 percent of the rating of the protective device, to satisfy Sec. 424-3(b). Such a 60-A circuit could be classed as an individual branch circuit supplying a "single" outlet that actually consists of all the heater elements interconnected. By con-

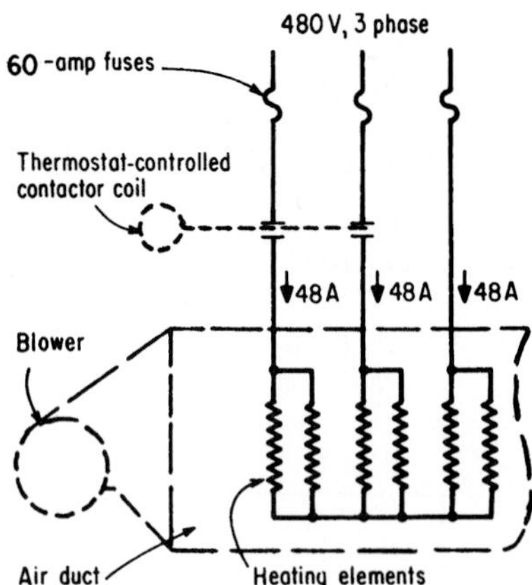

Fig. 424-4. This contactor may not serve as both controller and disconnect. (Sec. 424-20.)

sidering it as an "individual branch circuit" there is no conflict with Sec. 424-3(a), which sets a maximum rating of 30 A for a multioutlet circuit. The resistance-type heating elements on the market are not single heating elements in the 48-A size. They are made up of smaller wattage units into a single piece of equipment. The **Code** rule states that this single piece of equipment made up of smaller units must not draw more than 48 A and must be protected at not more than 60 A. Thus a heater of this type is limited to 48 A for each subdivided circuit. The subdivision is usually made by the manufacturer in the heater enclosure or housing.

Part **(b)** requires that a resistance load less than 48 A have protection rated not less than 125 percent of heater current. The sentence at the end of this paragraph covers overcurrent protection sizing for subdivided resistance-heating-element loads that are less than 48 A.

Figure 424-5 shows an example of subdivision of heater elements in a heat pump with three 5-kW strip heaters in it. At 230 V, each 5-kW strip is a load of about 22 A. Two of them in parallel would be 44 A and that is not in excess of the 48-A maximum set by part **(b)** of this section. The three heaters would be a load of 66 A in parallel. There are a number of ways the total load might be supplied, but the **Code** rules limit the actual permitted types of hookup:

Section 424-3 states that an individual branch circuit may supply any load, but that permission is qualified by part **(b)** of this section, which requires resistance heating loads of more than 48 A to be subdivided so that no subdivided heater load will exceed 48 A, protected at not more than 60 A. As shown at A, one possible way to hook up the heaters is to use two strips on one circuit for a connected load of 2 × 22, or 44 A, and the other on one circuit with a connected load of 22 A. The two heaters would require a minimum ampacity for overcurrent protection of 44 × 1.25, or 55 A, calling for a 60-A overcurrent device. For the other circuit, 22 × 1.25 = 27.5 A, requiring a 30-A fuse or breaker. Both circuits would thus be within the limits of a 48-A connected load and 60-A protection. As shown at B, it would also be acceptable to use a 30-A, 2-wire, 230-V circuit to each heater. But use of a single circuit sized at 1.25 × 66 A (82.5 A)

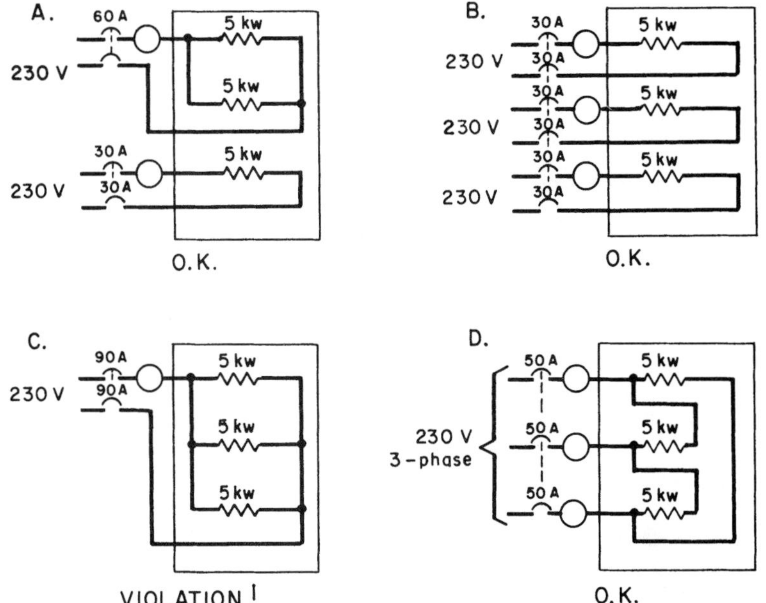

Fig. 424-5. Heater units must be limited to 48-A load with protection not over 60 A. (Sec. 424-22.)

and protected by 90-A fuses or breaker would clearly violate the Code rule of 60-A maximum protection, as at C.

Another possibility which would give a better balance to the connected load would be to feed the load with a 3-phase, 3-wire, 230-V circuit, if available. For such a circuit, the loading would be:

$$\frac{15{,}000}{1.732 \times 230} = 38 \text{ A}$$

The minimum rating would be 38 × 1.25, or 47.5 A, which calls for 50-A overcurrent protection, as at D.

It should be noted that the rule of part **(b)** in this section applies to *any type* of space heating equipment that utilizes resistance-type heating elements. The rule applies to duct heaters (as in Fig. 424-4), to the strip heaters in Fig. 424-5, and to heating elements in furnaces.

The purpose of paragraph **(c)** of this section is to require the heating manufacturer to furnish the necessary overcurrent protective devices where subdivided loads are required.

Main conductors supplying overcurrent protective devices for subdivided loads are considered as branch circuits to avoid controversies about applying the 125 percent requirement in Sec. 424-3(b) to branch circuits *only.* It is not the intent, however, to deny the use of the *feeder tap* rules in Sec. 240-21 for these *main* conductors.

Paragraph **(e)** requires that the conductors used for the subdivided electric resistance-heat circuits specified in Sec. 424-22(c) must have an amp rating not less than 100 percent of the rating or setting of the overcurrent protective device protecting the subdivided circuit(s) (Fig. 424-6). Exception is made for heaters rated 50 kW or more where under the conditions specified it is permissible for the conductors to have an ampacity not less than the load of the respective subdivided circuits, rather than 100 percent of the rating of the protective devices protecting the subdivided circuits.

The wording of part **(e)** clarifies the need for field-wired conductors rated not less than 125 percent of the load-current rating, which must not exceed 48 A. The rating of supplementary overcurrent protection must protect these circuit wires at their ampacity, although the next higher standard rating of protection may be used where the ampacity of the circuit wires does not correspond to a standard protective device rating.

For instance, if the subdivided resistance-heating load is 43 A per phase leg, the branch-circuit conductors must have an ampacity at least equal to 1.25 × 43, or 53.8 A. That would call for No. 6 TW with an ampacity of 55 A. The next standard rating of protection is 60 A, and that size protection may be used, although 60 A is the maximum size permitted for the subdivided load. The Exception is shown in the bottom part of Fig. 424-6.

424-35. Marking of Heating Cables. Note that there is a color-code for voltage identification of nonheating leads on heating cables to minimize the chance for use on a circuit of excessive voltage.

424-36. Clearances of Wiring in Ceilings. Figure 424-7 shows the details of this rule. The wire at a is O.K. because it is not less than 2 in. (50.8 mm) above the ceiling, but it must be treated as operating at a 50°C ambient. The same is true of the wire at c, because it is within the insulation. The Correction Factors table below Code Table 310-16 shows that TW wire (60°C-rated wire) must be derated to 58 percent (0.58) of its normal table ampacity when operating in an ambient of 41 to 50°C.

424-37. Location of Branch-Circuit and Feeder Wiring in Exterior Walls. For branch-circuit wires supplying electric heating cables, this rule no longer requires, as it once did, that branch-circuit wiring "be located outside the thermal insulation." The rule refers to both branch-circuit and feeder wiring in exterior walls and simply references all the wiring rules of Art. 300 and Sec. 310-10, which says in part **(3)** that "Thermal insulation which covers or surrounds conductors will affect the rate of heat dissipation," thereby requiring derating of conductor ampacity.

424-38. Area Restrictions. Figure 424-8 shows installations of heating cables and their relation to Code rules.

Heating cables shall not be installed under or over walls or partitions which extend to the ceiling except that "single runs of cable shall be permitted to pass over partitions where they are embedded." The intent here is to avoid repeated crossings of cable over (or under) partitions, since radiation from these sections would be restricted or the cable would be unnecessarily exposed to possible physical damage. While the Code specifically speaks of partitions, the same reasoning would apply to arches, exposed ceiling beams, etc.

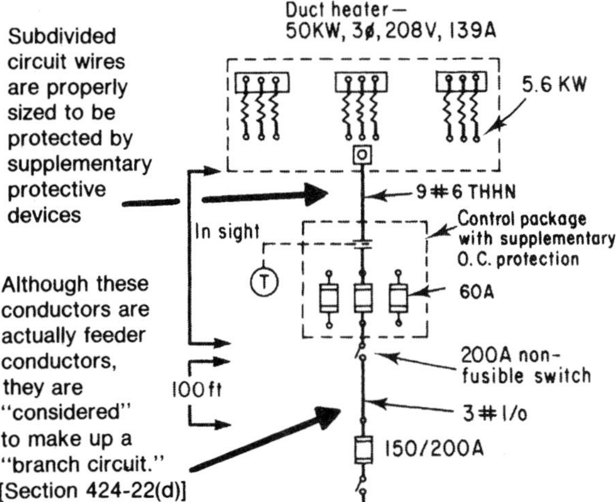
Duct heater—
50KW, 3ø, 208V, 139A
Subdivided circuit wires are properly sized to be protected by supplementary protective devices
5.6 KW
9 #6 THHN
In sight
Control package with supplementary O. C. protection
T
60A
Although these conductors are actually feeder conductors, they are "considered" to make up a "branch circuit." [Section 424-22(d)]
200A non-fusible switch
100 ft
3 # 1/0
150/200A

NOTE: There are exceptions for certain equipment.

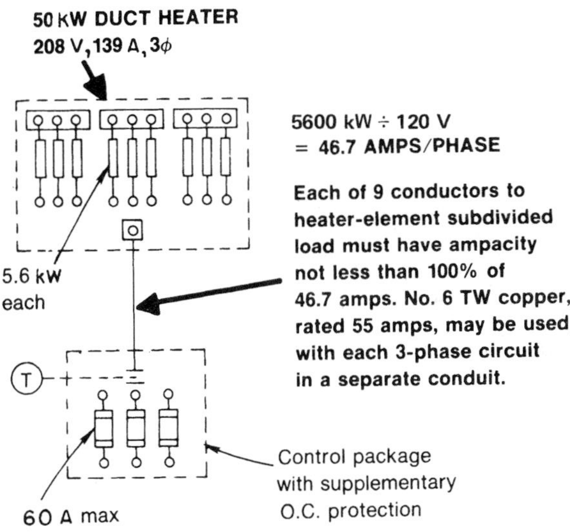
50 KW DUCT HEATER
208 V, 139 A, 3ϕ
5600 kW ÷ 120 V
= 46.7 AMPS/PHASE
Each of 9 conductors to heater-element subdivided load must have ampacity not less than 100% of 46.7 amps. No. 6 TW copper, rated 55 amps, may be used, with each 3-phase circuit in a separate conduit.
5.6 kW each
T
Control package with supplementary O.C. protection
60 A max

NOTE: Conditions a, b,and c of the Exception must be satisfied.

Fig. 424-6. Conductors for subdivided heater circuits must fully match overcurrent device rating. (Sec. 424-22.)

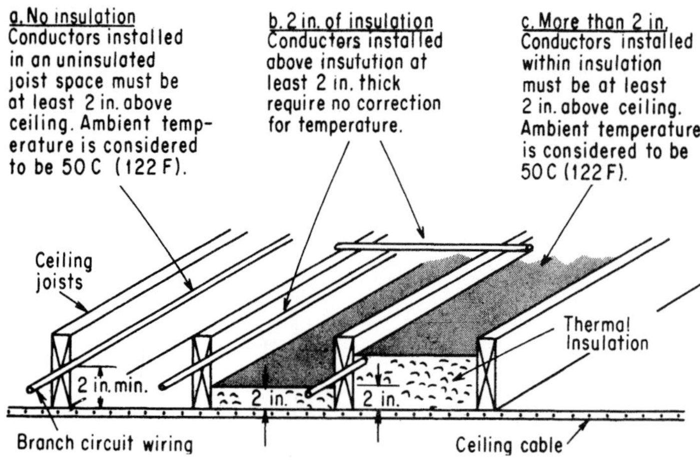

Fig. 424-7. Wiring above a heated ceiling may require derating because of heat accumulation. (Sec. 424-36.)

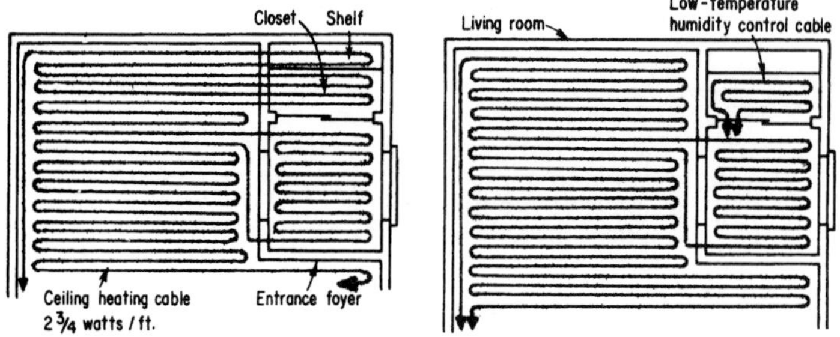

VIOLATION – cable extends beyond room and is installed in the closet. Cable in foyer is O. K.

O. K. – this cable is permitted in closet. Single cable runs that are embedded may cross partition.

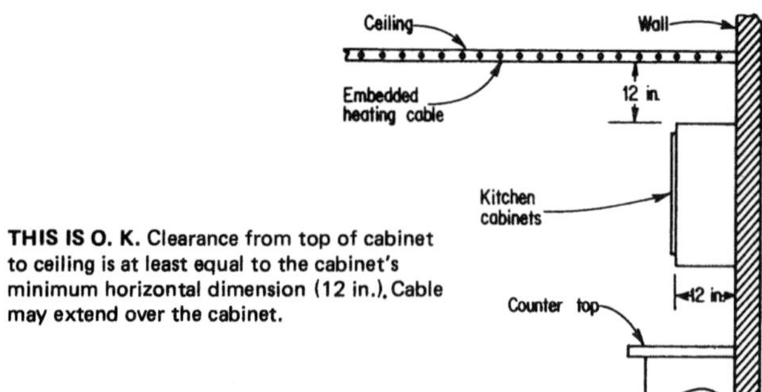

THIS IS O. K. Clearance from top of cabinet to ceiling is at least equal to the cabinet's minimum horizontal dimension (12 in.). Cable may extend over the cabinet.

Fig. 424-8. Layouts of heating cables must generally be confined to individual rooms or areas. (Sec. 424-38.)

However, there are times when a small ceiling area (such as over a dressing room or entryway) is separated from a larger room by such an arch or beam, yet it is impractical to install a separate heating cable and control. The Exception was intended as a solution to this problem. A typical floor plan of such a situation is shown in Fig. 424-9 with two methods of getting the heating cable past the partition or beam. In the upper drawing the cable is brought up into the attic space, through a porcelain tube, and back down through the gypsum board. Plaster is then forced into the tube and puddled over the exposed cable and tube in the attic. This should be the same plaster or joint cement that is used between the two layers of gypsum board.

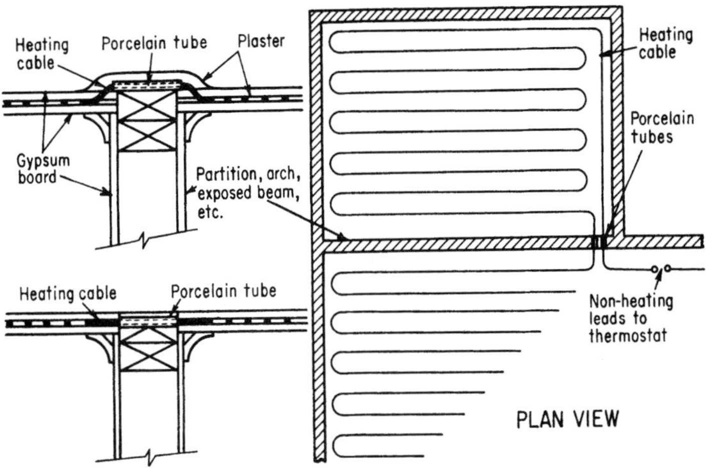

Fig. 424-9. Exception to rule permits single runs across partitions. (Sec. 424-38.)

In the lower drawing a hole is drilled through the top plate of the partition (or beam) and a porcelain tube pressed into the hole. Plaster is packed into the tube after the cable has been passed through. In both cases, the plaster serves to conduct heat away from the cable, avoiding hot spots and possible burnouts.

424-39. Clearance from Other Objects and Openings. Figure 424-10 shows application of the specified clearance distances for different conditions.

424-41. Installation of Heating Cables on Dry Board, in Plaster, and on Concrete Ceilings. All heating cables must observe these application methods. Figure 424-11 shows the rules of paragraph **(b)** and paragraph **(f)**.

Figure 424-12 shows the rule of paragraph **(d)** and refers to the rule of Sec. 424-43(c) at the outlet box.

Heating cable installed in plaster or between two layers of gypsum board must be kept clear of ceiling fixtures and side walls. In drywall construction, cable must be embedded in mastic or plaster. Without it, dead air space between the cable runs acts as a heat reservoir, increasing the possibility of cable burnouts (Fig. 424-13). Cable movement caused by the expansion and

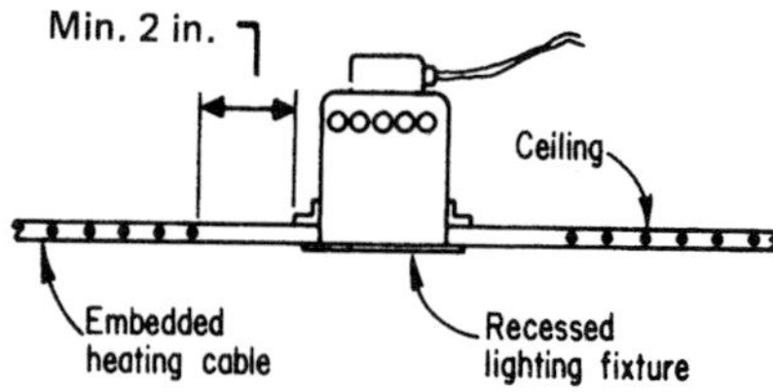

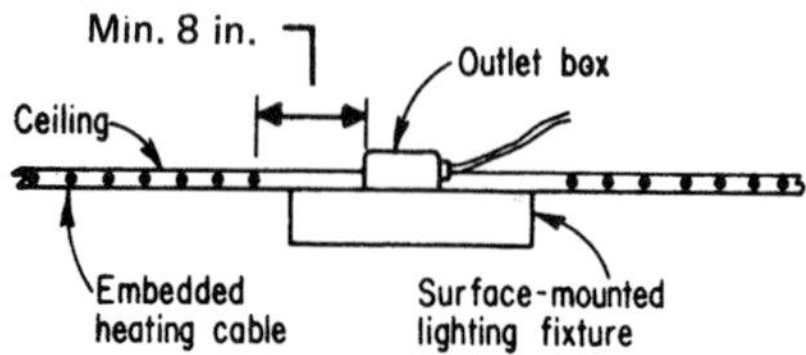

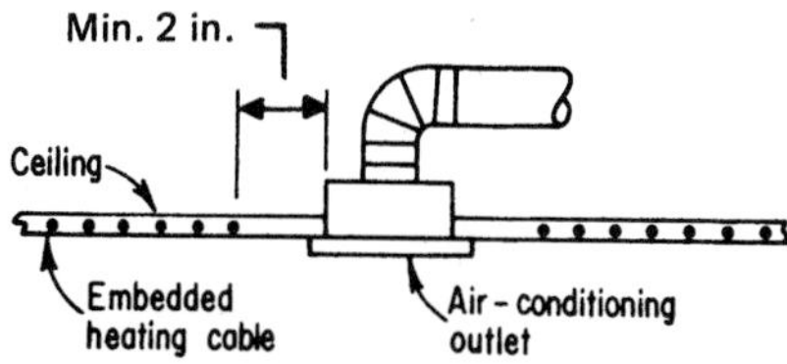

Fig. 424-10. Heating cables and panels must be kept clear of equipment. (Sec. 424-39.)

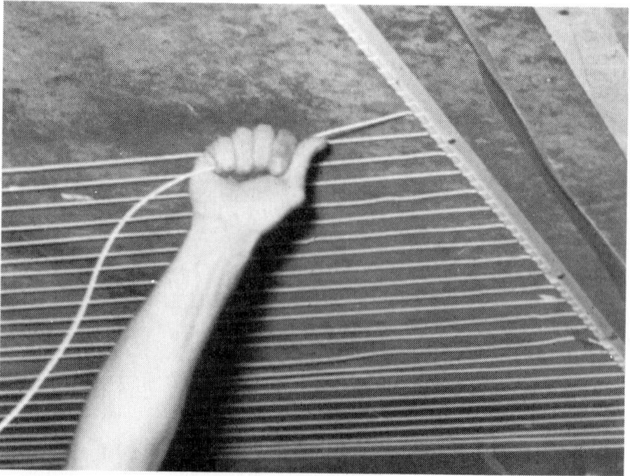

Fig. 424-11. Heating cable not over 2¾ W/ft (305 mm) must have at least 1½-in. (38 mm) spacing between adjacent conductors. (Sec. 424-41.)

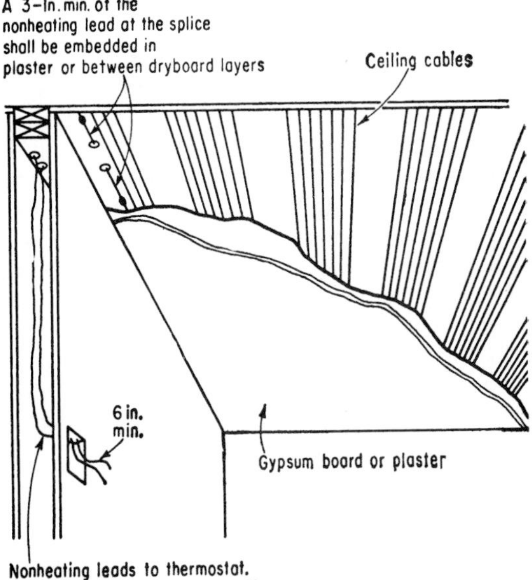

Fig. 424-12. Ends of nonheating leads must be embedded in ceiling material. (Sec. 424-41.)

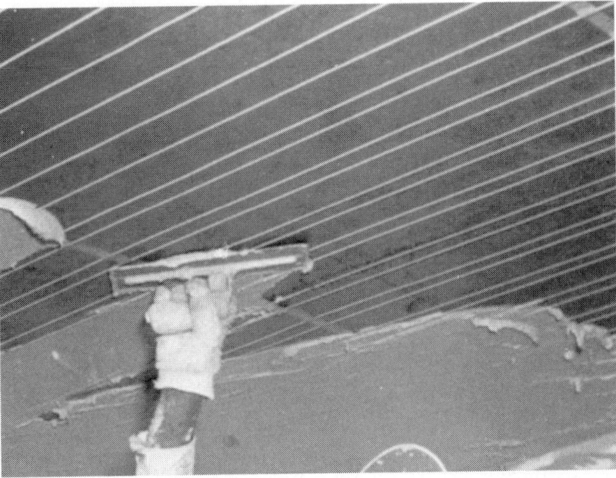

Fig. 424-13. In a drywall ceiling, the heating cable must be covered with a thermally conductive mastic before second course of gypsum board is applied to the ceiling, over the heating cable. (Sec. 424-41.)

contraction accompanying temperature changes is also prevented. Laboratory tests on cable used without mastic have found properly spaced adjacent cable runs actually making contact with each other, producing a hot spot and subsequent burnout. The plaster and sand mixture normally used as a mastic is a good conductor of heat and thus accelerates the dissipation of heat from the entire circumference of the cable. In addition, it improves the conductance from the cable to the gypsum board where the cable does not make direct contact. Even in the most careful installations small irregularities in construction and material prevent perfect contact between the cable and both layers of gypsum board throughout the entire cable length. In no case should an insulating plaster be used. Thickness of the plaster coat should be just sufficient to cover the cable. Installations have been made, unfortunately, with plaster thickness as great as ¾ in. (19 mm). Nails are not capable of supporting the resulting excessive weight, and such ceilings have collapsed. Figure 424-14 shows rules that apply to installation of heating cables in drywall ceilings.

Where the cable is to be embedded between two layers of gypsum board ("drywall" construction), after the cable is stapled to the layer of gypsum lath, it is covered with noninsulating plaster or gypsum cement, and a finishing layer of gypsum board (Sheetrock) is nailed in place covering the cable and plaster. To make sure that nails driven to secure this gypsum board to the ceiling joists do not penetrate the cable, a clear space at least 2½ in. (64 mm) wide must be left between adjacent cable runs immediately beneath each joist. That is, while adjacent cable runs must in general be at least 1½ in. (38 mm) apart, the spacing beneath joists must be increased to at least 2½ in. (64 mm). This means, of course, that the cable must be run parallel to the joists, as in part **(i)**.

Part **(j)** requires that heating cables in ceilings must cross joists only at the ends of the room, except where necessary to satisfy manufacturer's instructions. If manufacturer's instructions advise the installer to keep the cable away from ceiling penetrations and light fixtures, the Exception here will permit crossing joists at other than the ends of the room.

424-43. Installation of Nonheating Leads of Cables. Part **(c)** of this rule prohibits cutting off any of the length of heating leads that are provided by the manufacturers on the ends of heating cable. Any excess length of such leads must be secured to the ceiling and embedded in plaster or other approved material.

424-44. Installation of Cables in Concrete or Poured Masonry Floors. Details of these rules are shown in Fig. 424-15.

Paragraph **(c)** requires cables to be secured in place by nonmetallic frames or spreaders or other approved means. Metallic supports such as those commercially available for use in roadways or sidewalks are not to be used in floor space heating installations. Lumber is often used, although a more common method is to staple the cable directly to the base concrete after it has set about 4 hr. It was not the intention that this Code paragraph prohibit the use of metal staples. The object was to reduce the possibility of short circuits because of continuous metallic conducting materials spanning several adjacent cable runs.

Paragraph **(d)** requires spacing between the heating cable and other metallic bodies embedded in the floor. The intent is to reduce the possibility of contact

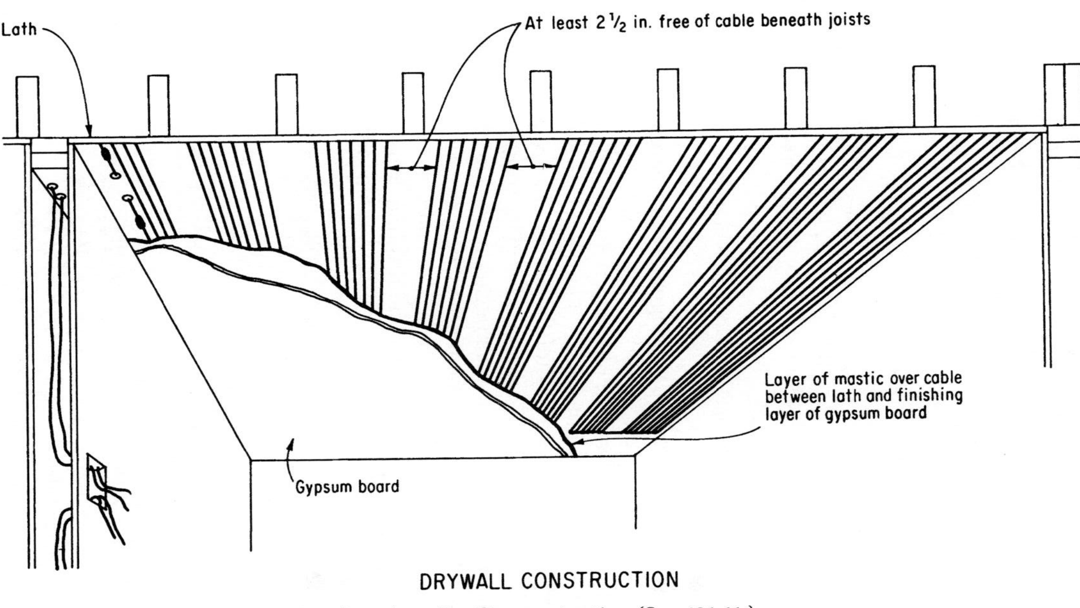

Fig. 424-14. Additional rules apply only to drywall ceiling construction. (Sec. 424-41.)

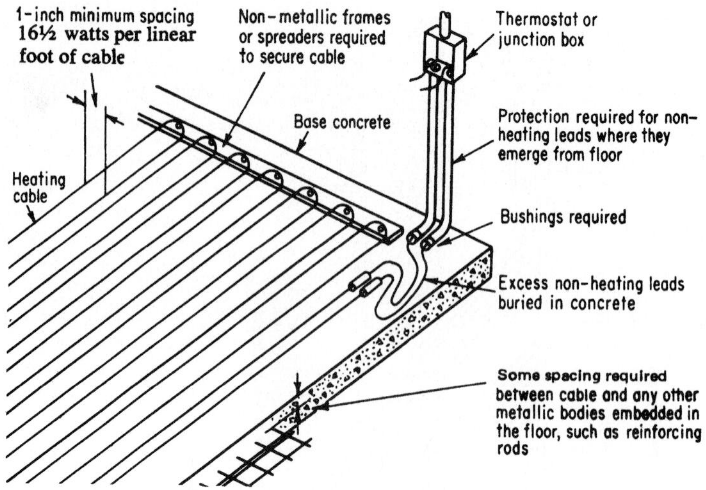

Fig. 424-15. Specific rules apply to heating cable in concrete floors. (Sec. 424-44.)

between the cable and such conducting materials as reinforcing mesh, water pipes, and air ducts (Fig. 424-16).

Paragraph **(e)** requires leads to be protected where they leave the floor, and paragraph **(f)** adds that bushings shall be used where the leads emerge in the floor slab. These provisions refer to the nonheating leads which connect the branch circuit homerun to the heating cable. The splices connecting the nonheating leads to the cable are always buried in the concrete. About 6 in. (152 mm) of leads is left available in the junction box; any remaining length of nonheating leads is buried in the concrete. The conductors should not be shortened. This applies even though the nonheating leads are Type UF cable. Section 339-3(b)(8) does prohibit the use of UF embedded in poured cement, concrete, or aggregate; however, an Exception to Sec. 339-3(a) is noted for nonheating leads of UF cable because these heating cable assemblies are tested by UL and listed as suitable for such use.

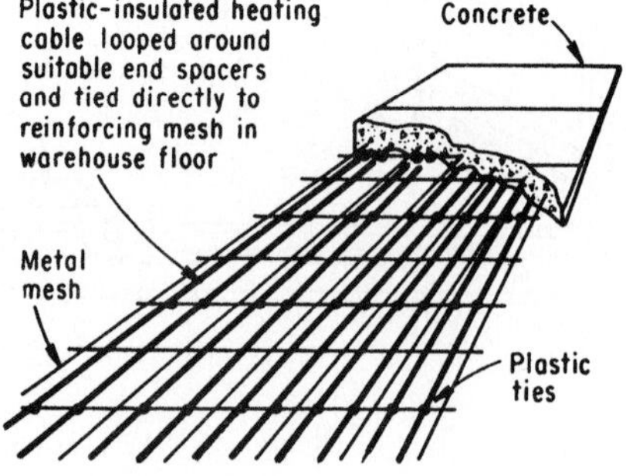

Fig. 424-16. This would be a violation of paragraph **(d)**. (Sec. 424-44.)

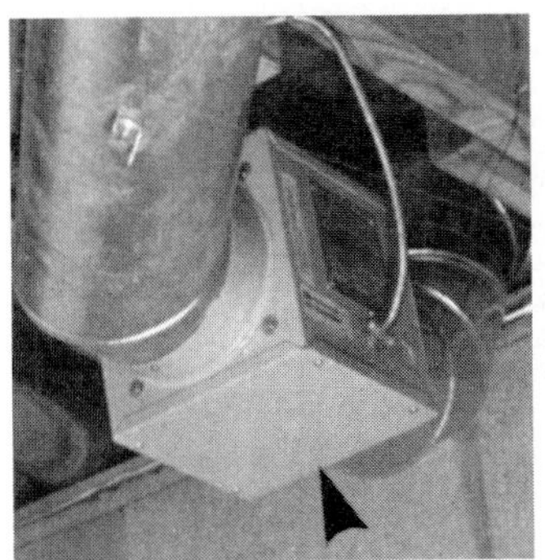

Fig. 424-17. Electric heaters designed and installed to heat air flowing through the ducts of forced-air systems are covered by Secs. 424-57 through 424-66. (Sec. 424-57.)

424-57. General. The rules of part **F** of Art. 424 apply to heater units that are mounted in air-duct systems, as shown in Fig. 424-17.

424-70. Scope. In applying Code rules, care must be taken to distinguish between "resistance-type" boilers and "electrode-type" boilers (Fig. 424-18).

424-72. Overcurrent Protection. Heating elements of resistance-type electric boilers must be arranged into load groups not exceeding the values specified in paragraph **(a)** or **(b)**. Figure 424-19 shows a 360-kW electric boiler used to heat a large school.

Part **(e)** requires that the ampacity of conductors used between the heater and the supplementary overcurrent protective devices for the subdivided heating

Fig. 424-18. Electric boilers with resistance-type heating elements are regulated by Secs. 424-70 through 424-75. (Sec. 424-70.)

Fig. 424-19. Electric boiler contains subdivided heating-element circuits, totalling 360 kW. Fuses protect the subdivided circuit loads. (Sec. 424-72.)

circuits within such boilers must not be less than 125 percent of the load served. Again, however, an Exception is added for heaters rated 50 kW or more under certain given conditions.

424-90. Scope. The rules of Secs. 424-91 through 424-98 apply to radiant heating panels and heating panel sets. Rules on electric radiant heating panels are separated from the rules on heating cables. The NEC at one time did cover both heating cables and heating panels within the same set of rules. Panels are generically different from cables and require separate specific rules on installation and layout details essential to safety.

424-95. Location of Branch-Circuit and Feeder Wiring in Walls. When electric heating panels are mounted on interior walls of buildings, any wiring within the walls behind the heating panel is considered to be operating in an ambient of 40°C rather than the normal 30°C for which conductors are rated. Because of this, the ampacity of such conductors in wall space behind electric heating panels must be reduced in accordance with the correction factors given as part of Tables 310-16 and 310-18 [Sec. 424-95(b)] (Fig. 424-20).

424-97. Nonheating Leads. Excess length of the nonheating leads of heating panels may be cut to the particular length needed to facilitate connection to the branch-circuit wires (Fig. 424-21).

ARTICLE 426. FIXED OUTDOOR ELECTRIC DE-ICING AND SNOW-MELTING EQUIPMENT

426-1. Scope. In addition to covering the longtime standard methods and equipment for electric de-icing and snow melting, this article gives detailed coverage of "skin-effect heating"—a system for utilizing the alternating-current phenomenon of skin effect, which derives its name from the tendency of AC

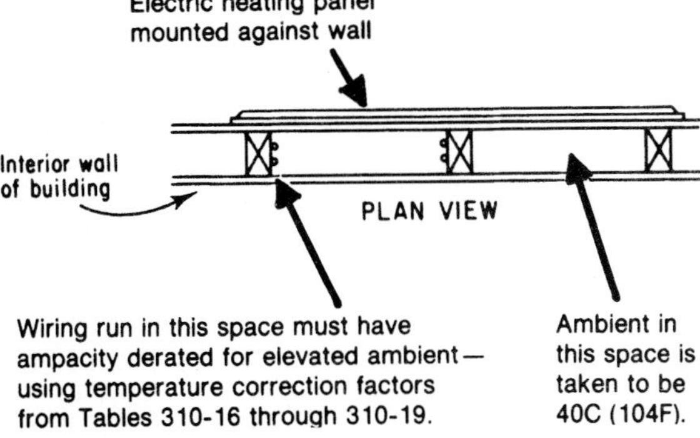

Fig. 424-20. Wiring in walls behind heating panels must have ampacity corrected for more than 30°C. (Sec. 424-95.)

current to flow on the outside (the skin) of a conductor. This action is produced by electromagnetic induction, which has an increasing opposition to current flow from the outside to the core of the cross section of a conductor.

In the layout of the article, separate coverage is given to "Resistance Heating," "Impedance Heating," and "Skin-Effect Heating."

426-4. Branch-Circuit Sizing. This basic rule requires that where electric de-icing or snow-melting equipment is connected to a branch circuit, the rating of the branch circuit overcurrent protection and the branch-circuit conductors must be not less than 125 percent of the total load of the heaters (Fig. 426-1).

Both the circuit conductors and the overcurrent protection must be rated so the load is not over 80 percent of their amp rating. When the branch-circuit conductors are selected to have ampacity of at least 125 percent of the amp

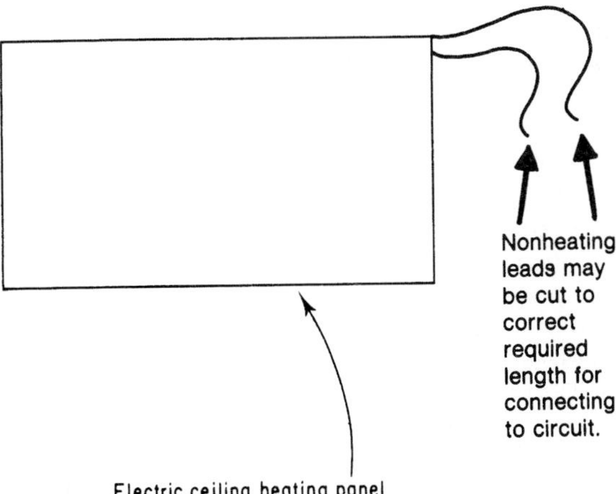

Fig. 424-21. The rule permits cutting of nonheating leads for "panels." (Sec. 424-97.)

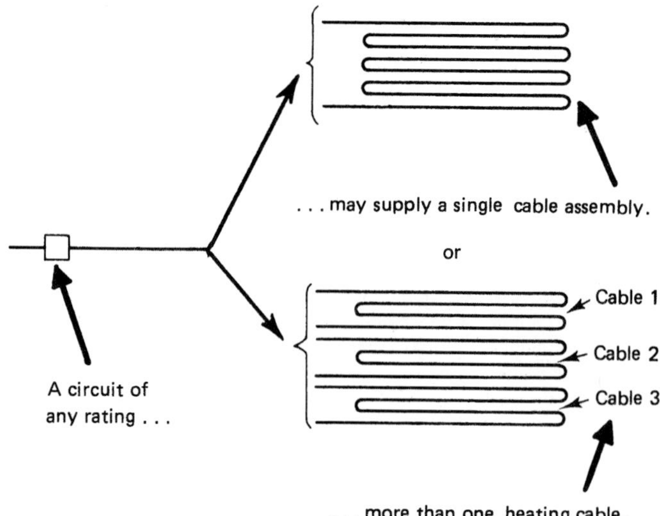

Fig. 426-1. Electric de-icing or snow melting equipment may be fed by a circuit of any rating. (Sec. 426-4.)

value of connected snow-melting and/or de-icing load equipment, it is permissible to go to the next higher amp value of overcurrent protection where the ampacity of the conductors does not correspond to a standard rating of protective device from Sec. 240-6.

426-11. Use. Whether used in concrete, blacktop, or other building material, any de-icing or snow-melting cable, panel, mat, or other assembly must be properly recognized (as by UL) for installation in the particular material.

Application data from the UL's *Electrical Appliance and Utilization Equipment Directory* include the following:

De-icing and Snow-Melting Equipment (KOBQ)

This category covers fixed outdoor electric de-icing and snow melting systems for use in accordance with Article 426 of the National Electrical Code. The equipment is provided with means for permanent wiring connections, except that equipment rated 20A or less and 150V ac or less to ground may be of cord-and-plug connected construction.

To supplement the general requirements in the National Electrical Code, the manufacturer is required to provide with the units or mats, specific installation instructions concerning any limitations of the installation and/or use of the equipment. The instructions for mats or cable units intended for burial in concrete specifically indicate that the slab must be a double pour (poured in two parts) if that is the only acceptable means of installation. If such a limitation is not specifically mentioned, either a single or double pour may be used.

For Listing of pipe heating cable see "Pipe Heating Cables" (KQUF).

The basic requirements for products in this category are contained in the Subject 1588 Outline of Investigation.

The listing Mark of Underwriters Laboratories Inc. on the product is the only method provided by UL to identify products manufactured under its Listing and Follow-Up Service. The Listing Mark for these products includes the name and/or symbol

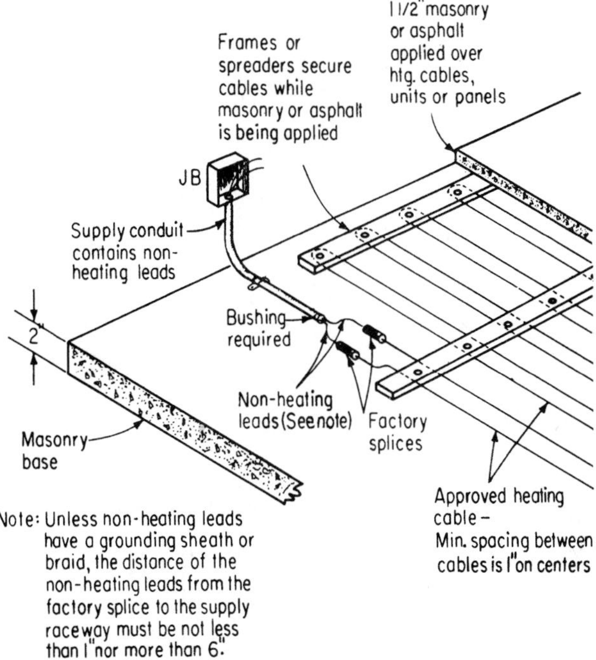
1 1/2" masonry or asphalt applied over htg. cables, units or panels
Frames or spreaders secure cables while masonry or asphalt is being applied
JB
Supply conduit contains non-heating leads
2"
Bushing required
Non-heating leads (See note)
Factory splices
Masonry base
Approved heating cable – Min. spacing between cables is 1" on centers
Note: Unless non-heating leads have a grounding sheath or braid, the distance of the non-heating leads from the factory splice to the supply raceway must be not less than 1" nor more than 6".

Fig. 426-2. Detailed rules cover installation of heating and nonheating conductors. (Secs. 426-20 and 426-22.)

of Underwriters Laboratories Inc. (as illustrated in the Introduction of this Directory) together with the word "Listed", a control number, and the following product name: "De-icing and Snow-Melting Equipment".

426-13. Identification. This rule requires the presence of a de-icing or snow-melting installation to be effectively indicated to assure safety and to prevent disruption.

After a snow-melting installation is made, some type of caution sign or other marking must be posted "where clearly visible" to make it evident to anyone that electric snow-melting equipment is present.

426-23. Installation of Nonheating Leads for Exposed Equipment. Part **(a)** of this rule permits cutting (shortening) of the nonheating leads provided the required marking on the leads (catalog numbers, volts, watts) is retained. That is not permitted for space heating cables, which may not be cut [Sec. 424-43(c)]. The marking on the nonheating leads of snow-melting cable must be within 3 in. (76 mm) of *each* end of the lead. The wording appears to permit cutting the nonheating leads back to the marking closest to the connection to the heating cable—which would mean a length of 3 in. (76 mm) plus that needed for the marking is all that would be required. Or a length could be cut out between the two markings, provided that any splicing that would necessitate is made in boxes, as specified in Sec. 426-24(b).

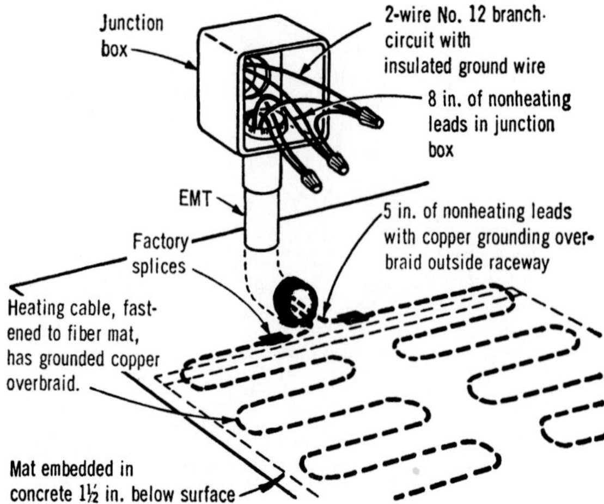

COMPLIES WITH RULES – Nonheating leads with copper grounding braid may have any length embedded in concrete.

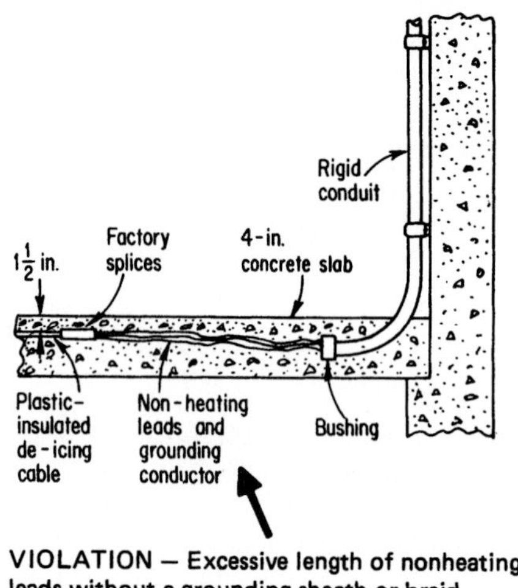

VIOLATION – Excessive length of nonheating leads without a grounding sheath or braid.

Fig. 426-3. Installation of nonheating leads must observe all the rules. (Sec. 426-22.)

426-50. Disconnecting Means. For outdoor de-icing and snow-melting equipment, the CB or fusible switch for the branch circuit to the equipment is adequate disconnecting means as long as it is readily accessible to the user. The cord-plug of plug-connected equipment rated up to 20 A, and 150 V may serve as the disconnect device.

ARTICLE 427. FIXED ELECTRIC HEATING EQUIPMENT FOR PIPELINES AND VESSELS

427-22. Equipment Protection. Electric heating equipment applied to pipelines or vessels must be supplied by a branch circuit with ground fault protection *if* the heating equipment does not have a metal outer covering.

Although the title of this section is "Equipment Protection," the major concern is to prevent shock hazard to personnel, which might occur when metal pipe or vessels are energized by a ground fault in electric heating cable or other heating equipment that is not enclosed by a grounded outer metallic enclosure. The "ground-fault protection" referred to in this rule may be a GFCI.

427-23. Metal Covering. This rule requires an overall grounded metal jacket on any heating cables intended to be installed on pipelines or vessels. Additionally, for heating panels, only the side that is *not* in contact with the pipe or vessel must have a grounded metal covering. The outer metal covering serves as an equipment ground-return path for fault current in the event of failure of the insulation on the heating conductors. Proper grounding will trip open the faulted circuit and prevent the type of fires that have been reported—as well as providing greater personnel safety.

ARTICLE 430. MOTORS, MOTOR CIRCUITS, AND CONTROLLERS

430-1. Scope. Two articles in the National Electrical Code are directed specifically to motor applications:

1. Article 430 of the NE Code covers application and installation of motor circuits and motor control hookups—including conductors, short-circuit and ground-fault protection, starters, disconnects, and running overload protection. Article 430 also covers adjustable speed drives in addition to motors, motor circuits, and controllers. Specific Code rules are given throughout this Article.
2. Article 440, covering "Air-Conditioning and Refrigerating Equipment," contains provisions for such motor-driven equipment and for branch circuits and controllers for the equipment, taking into account the special considerations involved with sealed (hermetic-type) motor compressors, in which the motor operates under the cooling effect of the refrigeration.

Diagram 430-1 in the NEC shows how various parts of Art. 430 cover the particular equipment categories that are involved in motor circuits. That Code diagram can be restructured, as shown in Fig. 430-1 in this handbook, to present the six basic elements which the Code requires the designer to account for in any motor circuit. Although these elements are shown separately here, there are certain cases where the Code will permit a single device to serve more than one function. For instance, in some cases, one switch can serve as both disconnecting means and controller.

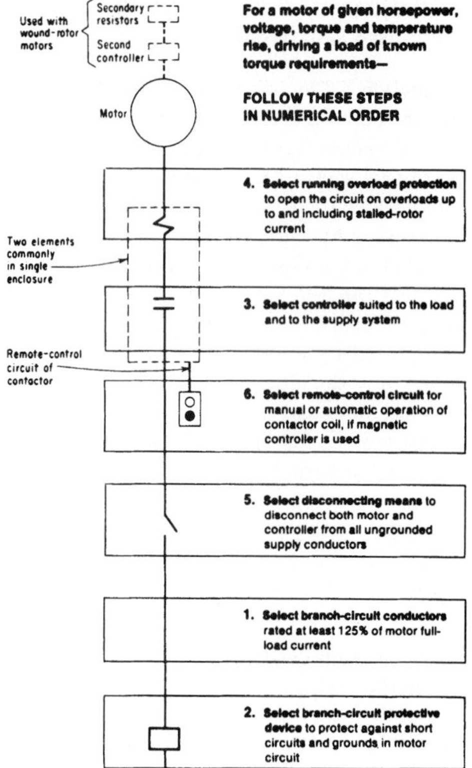

Fig. 430-1. Code rules cover these considerations. (Sec. 430-1.)

In other cases, short-circuit protection and overload protection can be combined in a single CB or set of fuses.

Throughout this article, all references to "running overcurrent devices" or simply "overcurrent devices" have been changed to "overload devices." And references to motor "running overcurrent protection" have been changed to "overload protection." This has been done to correlate with the definition of "overload," given in Art. 100, which refers to "Operation of equipment in excess of normal, full-load rating. . . ." A fault, such as a short circuit or ground fault, is not an overload. "Overload" for motors means current due to overload, up to and including locked-rotor current—or failure to start, which is the same level of current.

430-2. Adjustable Speed Drive Systems. The elements of this rule are shown in Fig. 430-2.

430-3. Part-Winding Motors. A part-winding starter is an automatic type of starter for use with squirrel-cage motors which have two separate, parallel windings on the stator. It can be used with the commonly used 220/400-V (dual-voltage) motors when they are used at the lower voltage, with the two

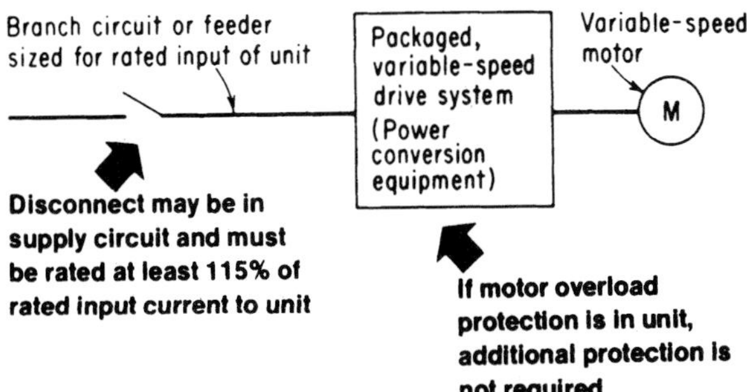

Fig. 430-2. Circuit to packaged drive systems is sized for rating of unit. (Sec. 430-2.)

windings operating in parallel. Single-voltage motors and 440- and 550-V motors must be ordered as specials if part-winding starting is to be used. The starter contains two magnetic contactors, each of which is rated for half the motor horsepower and is used to supply one winding.

430-6. Ampacity and Motor Rating Determination. For general motor applications (excluding applications of torque motors and sealed hermetic-type refrigeration compressor motors), whenever the current rating of a motor is used to determine the current-carrying capacity of conductors, switches, fuses, or CBs, the values given in Tables 430-147, 430-148, 430-149, and 430-150 must be used instead of the actual motor nameplate current rating. However, selection of separate motor-running overload protection *must* be based on the actual motor nameplate current rating.

As noted in part **(b)**, for any torque motor, the rated nameplate current is the locked-rotor current of the motor, and that value must be used in calculating the motor branch-circuit short-circuit and ground-fault protection in accordance with Sec. 430-52(b). The branch circuit for a torque motor must have its conductors and equipment protected at the motor nameplate rating (locked-rotor current) by selecting a fuse or CB in accordance with Sec. 240-3(f). The rule also requires that conductor ampacity and the setting or rating of the overload protective device be based on this value.

For shaded-pole motors, permanent-split-capacitor motors, and AC adjustable voltage motors, the other rules apply.

430-7. Marking on Motors and Multimotor Equipment. This section covers markings that manufacturers are required to put on the equipment. Code Table 430-7(b) can be used to calculate the locked-rotor current of a motor, where that value of current is related to selection of overload protection or short-circuit protection. A typical example would be selection of an instantaneous-trip CB as short-circuit protection of a motor branch circuit. The locked-rotor current of the motor represents the current value above which the breaker (and not the running overload device) must open the circuit. This is described in Sec. 430-52.

430-9. Terminals. As required by part **(b)**, unless marked otherwise, copper conductors must be used with motor controllers, and screw-type terminals of control-circuit devices must be torqued. Part **(b)** is intended to assure that only copper wires are used for wiring of motor starters, other controllers, and control-circuit devices because such equipment is designed, tested, and listed for use with copper only. If, however, equipment is tested and listed for aluminum wires, that must be indicated on the equipment, and use of aluminum wires is acceptable.

Part **(c)** requires that a torque screwdriver be used to tighten screw terminals of control-circuit devices used with No. 14 or smaller copper wires. Such terminals must be tightened to a value of 7 lb-in. (0.79 N-m)—unless marked for a different torque value. It should be noted that, although the rule literally mandates a "minimum" torque value, torque values are exact values. That is, equipment terminals should be torqued to the designated value and should *not* be overtorqued. (See top of Fig. 430-3.)

430-10. Wiring Space in Enclosures. As noted in part **(a)**, standard types of enclosures for motor controllers provide space that is sufficient only for the branch-circuit conductors entering and leaving the enclosure and any control-circuit conductors that may be required. No additional conductors should be brought into the enclosure. Section 430-12 provides a comprehensive set of rules and tables for motor terminal housings to solve the complaint by installers that motor terminal housings are too small to make satisfactory connections.

Part **(b)** on "Wire Bending Space in Enclosures" is based on Table 430-10(b), which shows the minimum distance from the end of the lug or connector to the wall of the enclosure or to the barrier opposite the lug, for each size of conductor and for one or two wires per terminal lug (Fig. 430-3). But a provision notes that the minimum wire bending space in a motor control center must conform to the requirements of Sec. 373-6(b).

A further rule applies to use of terminal lugs other than those supplied in the controller by its manufacturer. Such substitute terminal lugs or connectors must be of a type identified by the manufacturer for use with the controller, and use of such devices must *not* reduce the minimum wire bending space. That rule would apply, say, where mechanical set-screw lugs are replaced with crimp-on lugs (as for better connection of aluminum conductors). As with switches, CBs, and other equipment, the controller should be marked on its label to indicate acceptability of field changing of the lugs and to specify what type and catalog number of replacement terminal may be used, along with designation of the correct crimping tool and compression die.

430-11. Protection Against Liquids. Excessive moisture, steam, dripping oil, etc., on the exposed current-carrying parts of a motor may cause an insulation breakdown which in turn may be the cause of a fire.

430-12. Motor Terminal Housings. In part **(e)** the Code rule requires some provision for connecting an equipment grounding conductor at the terminal box where the branch circuit supplies a motor. The grounding connection may be either a "wire-to-wire" connection or a "fixed terminal" connection, and the ground terminal provision may be either inside the junction box—for connection of an equipment grounding conductor run with the circuit wires within the supply raceway—or outside the junction box—for connection of an "equipment bonding jumper" on the outside of a length of flexible metallic conduit or

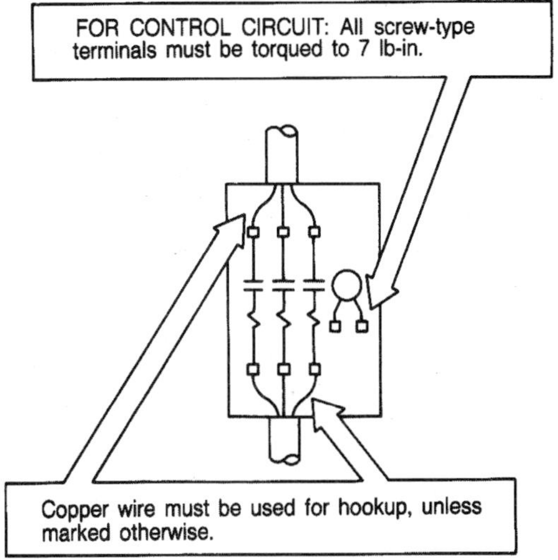
FOR CONTROL CIRCUIT: All screw-type terminals must be torqued to 7 lb-in.
Copper wire must be used for hookup, unless marked otherwise.

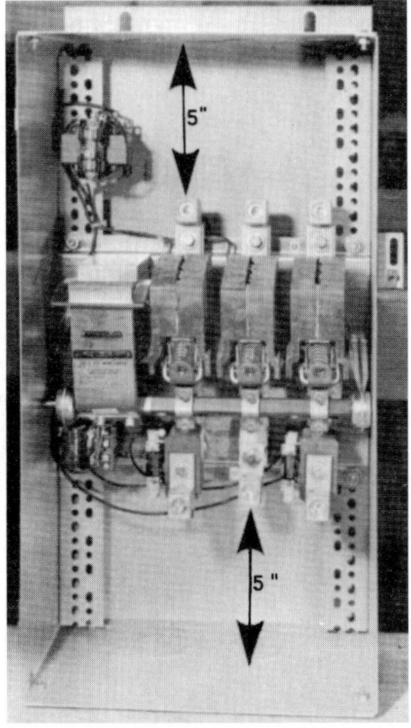
5"
5 "

Fig. 430-3. For this size 5 motor starter, use of No. 1/0 THW for the line and load conductors would require a minimum gutter height of 5 in. (127 mm), from Table 430-10(b), to provide Code-acceptable wire bending space. (Sec. 430-9.)

liquidtight flex, either of which is so commonly used. As required by Secs. 350-14 and 351-9, a bonding jumper is required for even short lengths of flex [up to 6 ft (1.83 m)] when the wires within the flex are protected at their origin by fuses or CBs rated over 20 A; and liquidtight flex over 1¼ in. (31.8 m) in size must have a bonding jumper for the typical length [up to 6 ft (1.83 m)] used with motor connections. See Sec. 430-145 and Fig. 430-71.

This rule permits either an inside or an outside connection of the bonding jumper to correlate with Sec. 250-79(e), which permits the bonding jumper [up to 6 ft (1.83 m) long] to be run either inside or outside the flex.

The Exception to this rule eliminates the need for providing "a separate means for motor grounding" at the junction box where a motor is part of "factory-wired equipment" in which the grounding of the motor is already provided by some other conductive connection that is an element of the overall assembly.

430-13. Bushing. Refer also to Sec. 373-6(c).

430-16. Exposure to Dust Accumulations. The conditions described in this section could make the location a Class II, Division 2 location; the types of motors required are specified in Art. 502.

430-17. Highest Rated or Smallest Rated Motor. Note that the current rating, not the horsepower rating, determines the "highest rated" motor where **Code** rules refer to such. See Sec. 430-62.

430-22. Single Motor. The basic rule of part **(a)** says that the conductors supplying a single-speed motor used for continuous duty must have a current-carrying capacity of not less than 125 percent of the motor full-load current rating, so that under full-load conditions the motor must not load the conductors to more than 80 percent of their ampacity. For a multispeed motor, selection of branch-circuit conductors on the supply side of the controller must be based on the highest full-load current rating shown on the motor nameplate.

Figure 430-4 shows the sizing of branch-circuit conductors to four different motors fed from a panel. (Sizing is also shown for branch-circuit protection and running overload protection, as discussed in Secs. 430-34 and 430-52. Refer to Table 430-150 for motor full-load currents and Table 430-152 for maximum ratings of fuses.) Figure 430-4 is based on the following:

1. Full-load current for each motor is taken from Table 430-150.
2. Running overload protection is sized on the basis that nameplate values of motor full-load currents are the same as values from Table 430-150. If nameplate and table values are not the same, OL (overload) protection is sized according to nameplate.
3. Conductor sizes shown are for copper. Use the amp values given and Table 310-16 to select correct size of aluminum conductors.

It is important to note that this rule establishes minimum conductor ratings based on temperature rise only and does not take into account voltage drop or power loss in the conductors. Such considerations frequently require increasing the size of branch-circuit conductors.

Exception No. 1 in part **(a)** includes requirements for sizing individual branch-circuit wires serving motors used for short-time, intermittent, periodic, or other varying duty. In such cases, frequency of starting and duration of oper-

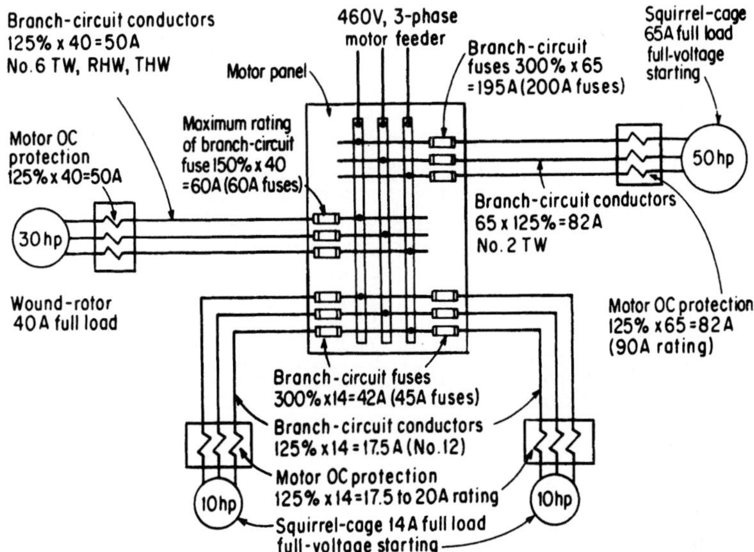

Fig. 430-4. Circuit conductors are sized at 1.25 times motor current. (Sec. 430-22.)

ating cycles impose varying heat loads on conductors. Conductor sizing, therefore, varies with the application. But, it should be noted that the last sentence of Sec. 430-33 says any motor is considered to be for continuous duty unless the nature of the apparatus that it drives is such that the motor cannot operate continuously with load under any condition of use.

When a motor is used for one of the classes of service listed in Table 430-22(a), Exception, the necessary ampacity of the branch-circuit conductors depends upon the class of service and upon the rating of the motor. A motor having a 5-min rating is designed to deliver its rated horsepower during periods of approximately 5 min each, with cooling intervals between the operating periods. The branch-circuit conductors have the advantage of the same cooling intervals and hence can safely be smaller than for a motor of the same horsepower but having a 60-min rating.

In the case of elevator motors, the many considerations involved in determining the smallest permissible size of the branch-circuit conductors make this a complex problem, and it is always the safest plan to be guided by the recommendations of the manufacturer of the equipment. This applies also to feeders supplying two or more elevator motors and to circuits supplying noncontinuous-duty motors used for driving some other machines.

430-23. Wound-Rotor Secondary. The full-load secondary current of a wound-rotor or slip-ring motor must be obtained from the motor nameplate or from the manufacturer. The starting, or starting and speed-regulating, portion of the controller for a wound-rotor motor usually consists of two parts—a dial-type or drum controller and a resistor bank. These two parts must, in many cases, be assembled and connected by the installer, as in Fig. 430-5.

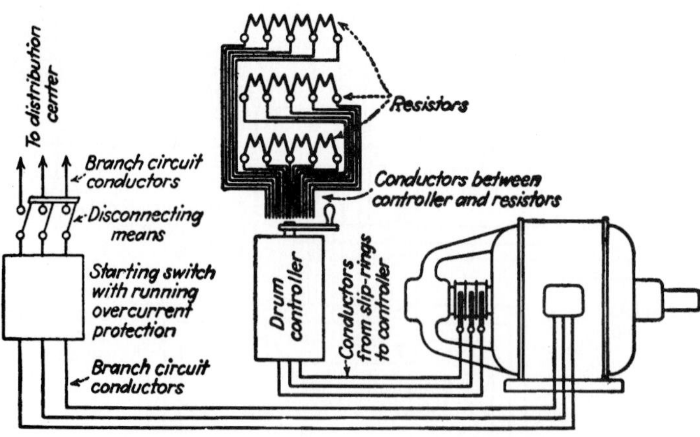

Fig. 430-5. Wound-rotor motor may be used with rotary drum switch for speed control. (Sec. 430-23.)

The conductors from the slip rings on the motor to the controller are in circuit continuously while the motor is running and hence, for a continuous-duty motor, must be large enough to carry the secondary current of the motor continuously.

If the controller is used for starting only and is not used for regulating the speed of the motor, the conductors between the dial or drum and the resistors are in use only during the starting period and are cut out of the circuit as soon as the motor has come up to full speed. These conductors may therefore be of a smaller size than would be needed for continuous duty.

If the controller is to be used for speed regulation of the motor, some part of the resistance may be left in circuit continuously and the conductors between the dial or drum and the resistors must be large enough to carry the continuous load without overheating. In Table 430-23(c) the term *continuous duty* applies to this condition.

Conductors connecting the secondary of a wound-rotor induction motor to the controller must have a carrying capacity at least equal to 125 percent of the motor's full-load secondary current if the motor is used for continuous duty. If the motor is used for less than continuous duty, the conductors must have capacity not less than the percentage of full-load secondary nameplate current given in Table 430-22(a) Exception. Conductors from the controller of a wound-rotor induction motor to its starting resistors must have an ampacity in accordance with Table 430-23(c), as shown in Fig. 430-6 for a magnetic starter used for reduced inrush on starting but not for speed control.

430-24. Several Motors or a Motor(s) and Other Load(s). Conductors supplying two or more motors (such as feeder conductors to a motor control center, to a panel supplying a number of motors, or to a gutter with several branch circuits tapped off) must have a current rating not less than 125 percent of the full-load current rating of the largest motor supplied plus the sum of the full-load current ratings of the other motors supplied.

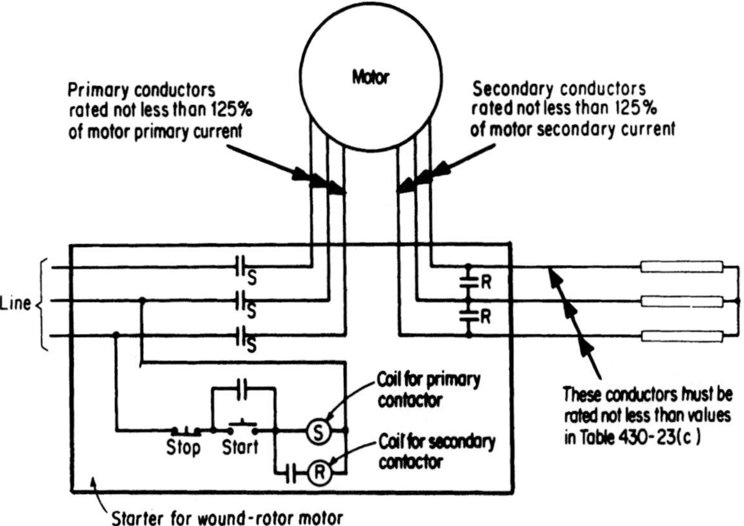

Fig. 430-6. Rules cover conductor sizing for wound-rotor motors without speed control. (Sec. 430-23.)

Figure 430-7 shows an example of sizing feeder conductors for a load of four motors, selecting the conductors on the basis of ampacities given in Table 310-16 and using conductors with a 60 or 75°C insulating rating—or using 90°C-rated conductors at the ampacities of 75°C. UL rules generally prohibit use of 90°C conductors at the 90°C ampacities shown in **Code** Table 310-16. (Refer to Sec. 310-15.)

For the overcurrent protection of feeder conductors of the minimum size permitted by this section, the highest permissible rating or setting of the protective device is specified in Sec. 430-62. Where a feeder protective device of higher rating or setting is used because two or more motors must be started simultaneously, the size of the feeder conductors shall be increased correspondingly.

These requirements and those of Sec. 430-62 for the overcurrent protection of power feeders are based upon the principle that a power feeder should be of such size that it will have an ampacity equal to that required for the starting current of the largest motor supplied by the feeder, plus the full-load running currents of all other motors supplied by the feeder. Except under the unusual condition where two or more motors may be started simultaneously, the heaviest load that a power feeder will ever be required to carry is the load under the condition where the largest motor is started at a time when all the other motors supplied by the feeder are running and delivering their full-rated horsepower.

Where other loads are also supplied, conductor sizing is determined as follows:

1. The current-carrying capacity of feeder conductors supplying a single motor plus other loads must include capacity at least equal to 125 percent of the full-load current of the motor.

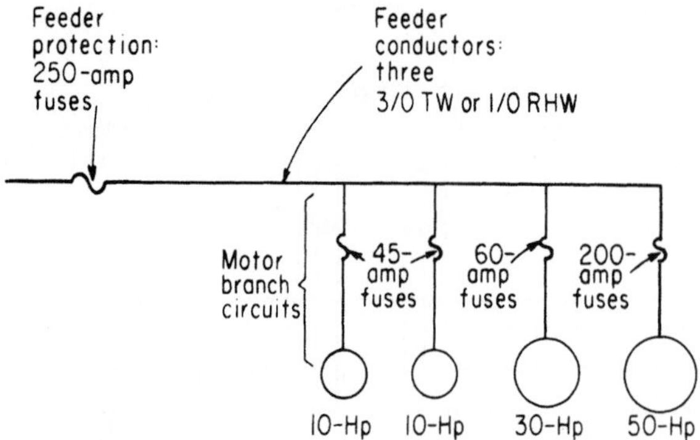

The four motors supplied by the 3-phase, 440-volt, 60-cycle feeder, which are not marked with a code letter (see Table 430-152), are as follows:

1 50-hp squirrel-cage induction motor (full-voltage starting)
1 30-hp wound-rotor induction motor
2 10-hp squirrel-cage induction motors (full-voltage starting).

Step 1. Branch-circuit loads

From Table 430-150, the motors have full-load current ratings as follows:

50-hp motor—65 amps
30-hp motor—40 amps
10-hp motor—14 amps

Step 2. Conductors

The feeder conductors must have a carrying capacity as follows (see Section 430-24):

1.25 × 65 = 81 amps
81 + 40 + (2 × 14) = 149 amps

The feeder conductors must be at least No. 3/0 TW, 1/0 THW or 1/0 RHH or THHN (copper).

Fig. 430-7. Feeder conductors are sized for the total motor load. (Sec. 430-24.)

2. The current-carrying capacity of feeder conductors supplying a motor load and a lighting and/or appliance load must be sufficient to handle the lighting and/or appliance load as determined from the procedure for calculating size of lighting feeders, plus the motor load as determined from the previous paragraphs.

The Code permits inspectors to authorize use of demand factors for motor feeders—based on reduced heating of conductors supplying motors operating intermittently or on duty-cycle or motors not operating together. Where necessary this should be checked to make sure that the authority enforcing the Code deems the conditions and operating characteristics suitable for reduced-capacity feeders, as noted in Sec. 430-26.

For computing the minimum allowable conductor size for a combination lighting and power feeder, the required ampacity for the lighting load is to be determined according to the rules for feeders carrying lighting (or lighting and appliance) loads only. Where the motor load consists of one motor only, the required ampacity for this load is the capacity for the motor branch circuit, or 125 percent of the full-load motor current, as specified in Sec. 430-22. Where the motor load consists of two or more motors, or a motor(s) and other loads, the required ampacity for the motor load is the capacity computed according to Sec. 430-24.

Figure 430-8 shows a typical installation for which calculation of required feeder ampacity is as follows:

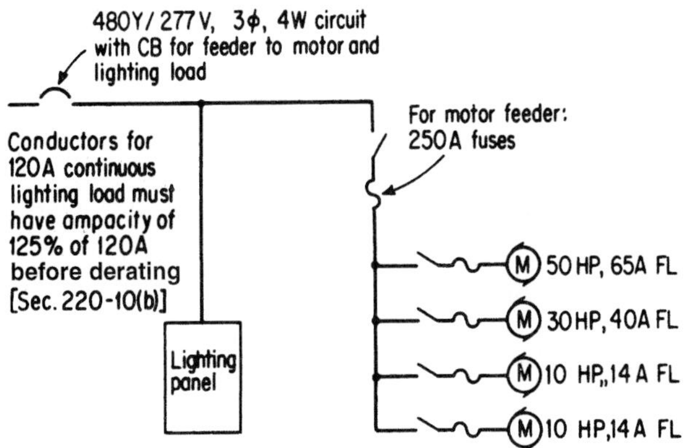

Fig. 430-8. Other load must be properly combined with motor load. (Sec. 430-24.)

Step 1. Total Load

Section 430-24 says that conductors supplying a lighting load and a motor load must have capacity for both loads, as follows:

$$\begin{aligned}\text{Motor load} &= 65\text{ A} + 40\text{ A} + 14\text{ A} + 14\text{ A} + (0.25 \times 65\text{ A}) = 149\text{ A per phase}\\ \text{Lighting load} &= 120\text{ A per phase} \times 1.25 = 150\text{ A}\\ \text{Total load} &= 149 + 150 = 299\text{ A per phase leg}\end{aligned}$$

Step 2. Conductors

Table 310-16 shows that a load of 299 A can be served by the following copper conductors:

500-kcmil TW
350-kcmil THW, RHH, XHHW, or THHN

Table 310-16 shows that this same load can be served by the following aluminum or copper-clad aluminum conductors:

700-kcmil TW
500-kcmil THW, RHH, XHHW, or THHN

430-26. Feeder Demand Factor. A demand factor of less than 100 percent may be applied in the case of some industrial plants where the nature of the work is such that there is never a time when all the motors are operating at one time. But the inspector must be satisfied with any application of a demand factor.

Sizing of motor feeders (and mains supplying combination power and lighting loads) may be done on the basis of maximum demand current, calculated as follows:

$$\text{Running current} = (1.25 \times I_f) + (DF \times I_t)$$

where I_f = full-load current of largest motor
DF = demand factor as permitted by Sec. 430-26
I_t = sum of full-load currents of all motors except largest

But modern design dictates use of the maximum-demand starting current in sizing conductors for improved voltage stability on the feeder. This current is calculated as follows:

$$\text{Starting current} = I_s + (DF \times I_t)$$

where I_s = average starting current of largest motor (Use the percent of motor full-load current given for fuses in Table 430-152.)

430-28. Feeder Taps. This Code rule is an adaptation of Sec. 240-21(b) and (c), covering use of 10- and 25-ft (3.05- and 7.62-m) feeder taps with no overcurrent protection at the point where the smaller conductors connect to the higher-ampacity feeder conductors. The adaptation establishes that the tap conductors must have an ampacity as required by Secs. 430-22, 430-24, or 430-25.

In applying condition (1), the conductor may have an ampacity less than one-tenth that of the feeder conductors but must be limited to not more than 10 ft (3.05 m) in length and be enclosed within a controller or raceway.

If conductors equal in size to the conductors of a feeder are connected to the feeder, as in condition (3), no fuses or other overcurrent protection are needed at the point where the tap is made, since the tap conductors will be protected by the fuses or CB protecting the feeder.

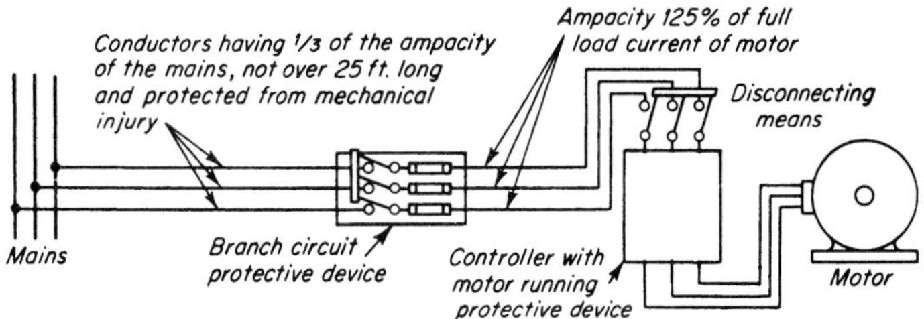

Note: Branch-circuit fuses (or CB) may be rated higher than ampacity of the tap conductors.

Fig. 430-9. Feeder tap may be sized as provided by part (c) of Sec. 240-21. (Sec. 430-28.)

The more important circuit arrangement permitted by the above rule is shown in Fig. 430-9. Instead of placing the fuses or other branch-circuit protective device at the point where the connections are made to the feeder, conductors having at least one-third the ampacity of the feeder are tapped solidly to the feeder and may be run a distance not exceeding 25 ft (7.62 m) to the branch-circuit protective device. From this point on to the motor-running protective device and thence to the motor, conductors are run having the standard ampacity, i.e., 125 percent of the full-load motor current, as specified in Sec. 430-22. If the tap conductors shown did not have an ampacity at least equal to one-third of that of the feeder conductors, then the tap conductors must not be over 10 ft (3.05 m) long.

Note that this rule actually modifies the requirements of Sec. 240-21 for taps to motor loads. Section 240-21(b)(2)b literally calls for 10-ft (3.05-m) tap conductors to have ampacity at least equal to the rating or setting of the fuses or CB (whichever is used) at the load end of the tap. And such protection may be rated up to 4 times motor full-load current. But condition (1) of this Code section requires sizing of the 10-ft (3.05-m) tap conductors to be at least one-tenth the rating of the overcurrent device protecting the feeder from which the tap conductors are supplied. And condition (2) does *not* require a 25-ft (7.62-m) tap to terminate in a protective device rated to protect the conductors at their ampacity (Fig. 430-10).

example: A 15-hp 230-V 3-phase motor with autotransformer starter is to be supplied by a tap made to a 250-kcmil feeder. All conductors are to be Type THW.

The feeder has an ampacity of 255 A; one-third of 255 A equals 85 A. Therefore the tap cannot be smaller than No. 4, which has an ampacity of 85 A for 75°C ratings.

The full-load current of the motor is 40 A and, according to part **D** of Art. 430, assuming that the motor is not marked with a Code letter, the branch-circuit fuses should be rated at not more than 300 percent of 40 A, or 120 A, which calls for 125-A fuses (Sec. 430-52) or less. With the motor-running protection set at 50 A (125 percent × 40 A), the tap conductors are well protected from overload.

The conductors tapped solidly to the feeder must never be smaller than the size of branch-circuit conductors required by Sec. 430-22.

The Exception in this rule notes that a branch-circuit or subfeeder tap up to 100 ft (30.5 m) long may be made from a feeder to supply motor loads. The spe-

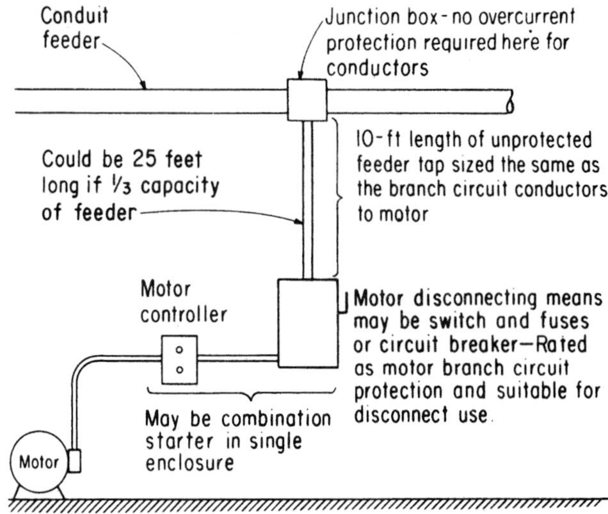

Fig. 430-10. Tap conductors may terminate in protective device rated above their ampacity. (Sec. 430-28.)

cific conditions are given for making a tap that is over 25 ft (7.62 m) long and up to 100 ft (30.5 m) long—where no protection is provided at the point of tap from the feeder conductors. This is a motor-circuit adaptation of the 100-ft (30.5-m) tap permission, which is fully described under Sec. 240-21(e).

430-29. Constant-Voltage DC Motors—Power Resistors. These rules cover sizing of conductors from a DC motor controller to separate resistors for power accelerating and dynamic braking. This section, with its table of conductor ampacity percentages, assures proper application of DC constant-potential motor controls and power resistors.

430-31. General. Detailed requirements for the installation of fire pumps are not included in the *National Electrical Code*, but this is covered in NFPA 20.

As intended by Sec. 430-52, the motor branch-circuit protective device provides short-circuit protection for the circuit conductors. In order to carry the starting current of the motor, this device must commonly have a rating or setting so high that it cannot protect the motor against overload.

For a squirrel-cage induction motor, overload protection must be of the inverse time type with a setting of not over 20 sec at 600 percent of the motor full-load current. It is the intent that the fire-pump motor be permitted to run under any condition of loading and not be automatically disconnected by an overcurrent protection device.

NFPA 20 requires a CB instead of a fuse as the short-circuit protection for the branch circuit and also requires an unfused isolating switch ahead of the CB.

Except where time-delay fuses provide both running overload protection and short-circuit protection as described in Sec. 430-55, in practically all cases where motor-running overload protection is provided the motor controller consists of two parts: (1) a switch or contactor to control the circuit to the motor and (2) the motor-running protective device. Most of the protective devices make use of a heater coil, usually consisting of a few turns of high-resistance metal, though the heater may be of other form.

430-32. Continuous-Duty Motors. The Code makes specific requirements on motor running overload protection intended to protect the elements of the branch circuit—the motor itself, the motor control apparatus, and the branch-circuit conductors—against excessive heating due to motor overloads. Overload protection may be provided by fuses, CBs, or specific overload devices like OL relays.

Overload is considered to be operating overload up to and including stalled-rotor current. When overload persists for a sufficient length of time, it will cause damage or dangerous overheating of the apparatus. Overload does not include fault current due to shorts or grounds.

Typical overload devices include:

1. Heaters in series with line conductors acting upon thermal bimetallic overload relays.
2. Overload devices using resistance or induction heaters and operating on the solder-ratchet principle (Fig. 430-11).
3. Magnetic relays with adjustable instantaneous setting or adjustable time-delay setting.

Of course, the provisions for overload protection are integrated in the enclosure of the controller.

Overload protective devices of the straight thermal type are available with varying tripping and time-delay characteristics. In such devices, the heater coils are made in many sizes and are interchangeable to permit use of the required heater sizes to provide running protection for different motor full-load current ratings. In some units, the heater coil can be adjusted to exact current values. Individual covers are used on the heating elements in some starters to isolate the relay from possible effect on its operation because of the temperature of surrounding air.

In general, it is required that every motor shall be provided with a running protective device that will open the circuit on any current exceeding prescribed percentages of the full-load motor current, the percentage depending upon the type of motor. The running protective device is intended primarily to protect the windings of the motor; but by providing that the circuit conductors shall have an ampacity not less than 125 percent of the full-load motor current, it is obvious that these conductors are reasonably protected by the running protective device against any overcurrent caused by an overload on the motor.

Part **(a)** covers application for motors of more than 1 hp. If such a motor is used for continuous duty, running overload protection must be provided. This may be an external overcurrent device actuated by the motor running current and set to open at not more than 125 percent of the motor full-load current for motors marked with a service factor of not less than 1.15 and for motors with a temperature rise not over 40°C. See examples in Fig. 430-4. Sealed (hermetic-type) refrigeration compressor motors must be protected against overload and failure to start, as specified in Sec. 440-52. The overload device must be rated or

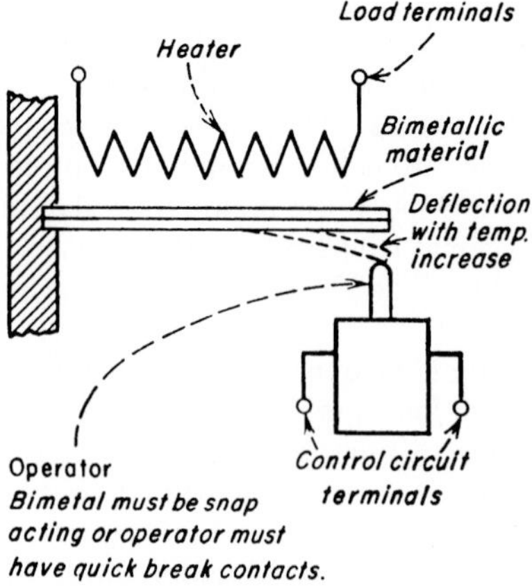

BIMETALLIC TYPE

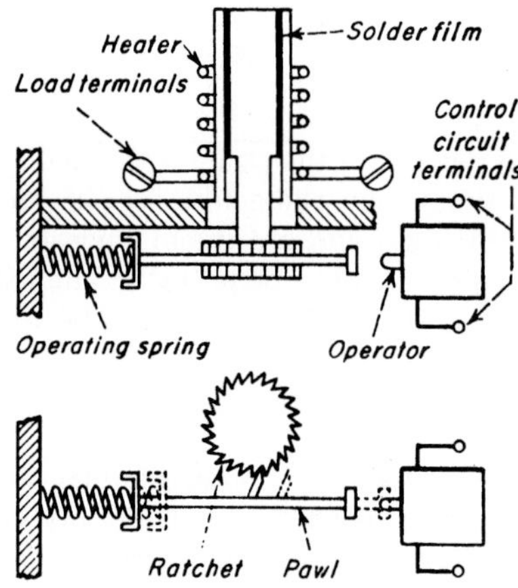

SOLDER-RATCHET TYPE

NOTE: For a manual starter, the contacts shown are the main load-current contacts of the switch—connected in series with the heater coil.

Fig. 430-11. Overload relay devices are made in various operating types. (Sec. 430-32.)

set to trip at not more than 115 percent of the motor full-load current for all other motors, such as motors with a 1.0 service factor or a 55°C rise (Fig. 430-12).

The term *rating,* or *setting,* as here used means the current at which the device will open the circuit if this current continues for a considerable length of time.

Note: Refer to Sec. 460-9, which discusses the need to correct the sizing of running overload protection when power-factor capacitors are installed on the load side of the controller.

A motor having a temperature rise of 40°C when operated continuously at full load can carry a 25 percent overload for some time without injury to the motor. Other types of motors, such as enclosed types, do not have so high an overload capacity and the running protective device should therefore open the circuit on a prolonged overload which causes the motor to draw 115 percent of its rated full-load current.

Basic **Code** requirements are concerned with the rating or setting of overcurrent devices separate from motors. However, the **Code** permits the use of integral protection. Paragraph **(2)** of part **(a)** covers use of running overload protective devices within the motor assembly rather than in the motor starter. A protective device integral with the motor as used for the protection of motors is shown in Fig. 430-13. This device is placed inside the motor frame and is connected in series with the motor winding. It contains a bimetallic disk carrying two contacts, through which the circuit is normally closed. If the motor is overloaded and its temperature is raised to a certain limiting value, the disk snaps to the "open" position and opens the circuit. The device also includes a heating coil in series with the motor windings which causes the disk to become heated more rapidly in case of a sudden heavy overload.

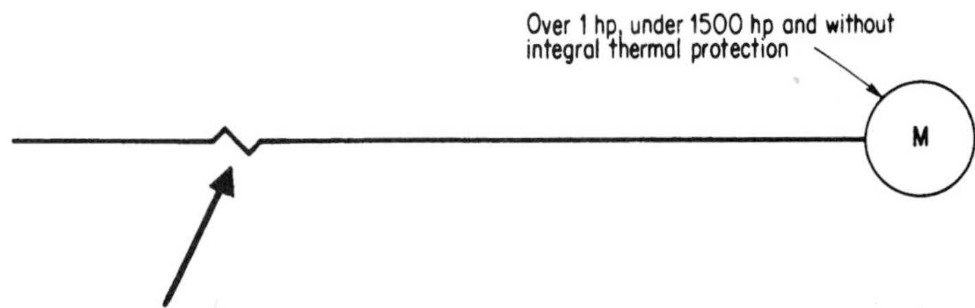

Separate overcurrent device, responsive to motor current, rated or selected to trip at no more than the following percentage of the motor full-load current rating:

Motors with marked service factor not less than 1.15 .... 125%
Motors with marked temperature rise not over 40C 125%
Sealed (hermetic type) motor compressors
Using overload relays 140%
Using other devices 125%
All other motors 115%

Each winding of a multispeed motor must be considered separately. This value may be modified as permitted by Section 430-34.

Fig. 430-12. Specific rules apply to continuous-duty motors rated over 1 hp. (Sec. 430-32.)

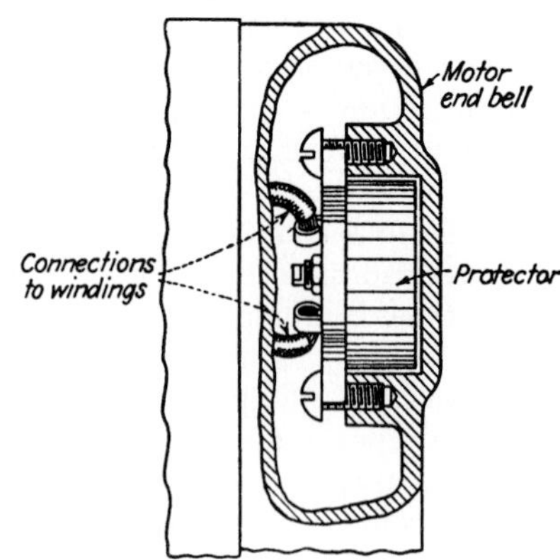

Fig. 430-13. Running overload protection may be built into the motor. (Sec. 430-32.)

Where the circuit-interrupting device is separate from the motor and is actuated by a device integral with the motor, the two devices must be so designed and connected that any accidental opening of the control circuit will stop the motor, otherwise the motor would be left operating without any overcurrent protection.

There is special need for running protection on an automatically started motor because, if the motor is stalled when the starter operates, the motor will probably burn out if it has no running protection.

Part **(b)** of this section applies to smaller motors. Motors of 1 hp or less which are not permanently installed and are manually started are considered protected against overload by the branch-circuit protection if the motor is within sight from the starter (Fig. 430-14). Running overload devices are not required in such cases. A distance of over 50 ft (15.24 m) is considered out of sight.

It should be noted that any motor of 1 hp or less which is not portable, is not manually started, and/or is not within sight from its starter location must have specific running overload protection. Automatically started motors of 1 hp or less must be protected against running overload in the same way as motors rated over 1 hp—as noted in part **(c)**. That is, a separate or integral overload device must be used.

There are alternatives to the specific overload protection rules of parts **(a)** and **(c)**. Under certain conditions, no specific running overload protection need be used: The motor is considered to be properly protected if it is part of an approved assembly which does not normally subject the motor to overloads and which has controls to protect against stalled rotor. Or if the impedance of the motor windings is sufficient to prevent overheating due to failure to start, the branch-circuit protection is considered adequate.

430-33. Intermittent and Similar Duty. A motor used for a condition of service which is inherently short-time, intermittent, periodic, or varying duty does not require protection by overload relays, fuses, or other devices required by Sec. 430-32, but, instead, is considered as protected against overcurrent by the branch-circuit overcurrent device (CB or fuses rated in accordance with Sec. 430-52). Motors are considered to be for continuous duty unless the motor is completely incapable of operating continuously with load under any condition of use.

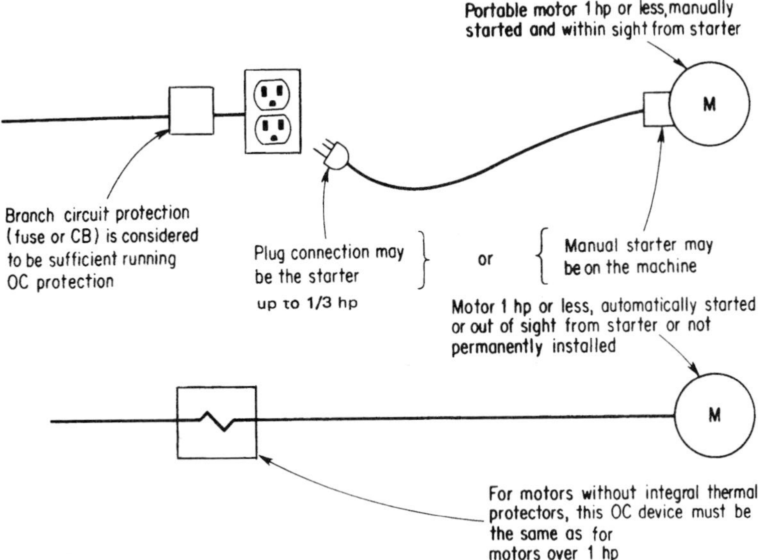
Portable motor 1 hp or less, manually started and within sight from starter
M
Branch circuit protection (fuse or CB) is considered to be sufficient running OC protection
Plug connection may be the starter
up to 1/3 hp
or
Manual starter may be on the machine
Motor 1 hp or less, automatically started or out of sight from starter or not permanently installed
M
For motors without integral thermal protectors, this OC device must be the same as for motors over 1 hp

Fig. 430-14. The rules for automatic-start motors are different. (Sec. 430-32.)

430-34. Selection of Overload Relay. This rule sets the absolute maximum permitted rating of an overload relay where values are higher than the 125 or 115 percent trip ratings of Secs. 430-32(a)(1) and (c). Motors with a marked service factor not less than 1.15 and 40°C-rise motors may, if necessary to enable the motor to start or carry its load, be protected by overload relays with trip settings up to 140 percent of motor full-load current. Motors with a 1.0 service factor and motors with a temperature rise over 40°C (such as 55°C-rise motors) must have their relay trip setting at not over 130 percent of motor full-load current.

BUT WATCH OUT! The maximum settings of 140 or 130 percent apply only to OL relays, such as used in motor starters.

Fuses or CBs may be used for running overload protection but may not be rated or set up to the 140 or 130 percent values. Fuses and breakers must have a maximum rating as shown in Secs. 430-32(a) and (c). If the value determined as indicated there does not correspond to a standard rating of fuse or CB, the next smaller size must be used. A rating of 125 percent of full-load current is the absolute maximum for fuses or breakers.

430-35. Shunting During Starting Period. As covered in part **(a)** for motors that are *not* automatically started, where fuses are used as the motor-running protection, they may be cut out of the circuit during the starting period. This leaves the motor protected only by the branch-circuit fuses, but the rating of these fuses will always be well within the 400 percent limit specified in the rule. If the branch-circuit fuses are omitted, as allowed by the rule in Sec. 430-53(d), it is not permitted to use a starter that cuts out the motor fuses during the starting period unless the protection of the feeder is within the limits set by this rule.

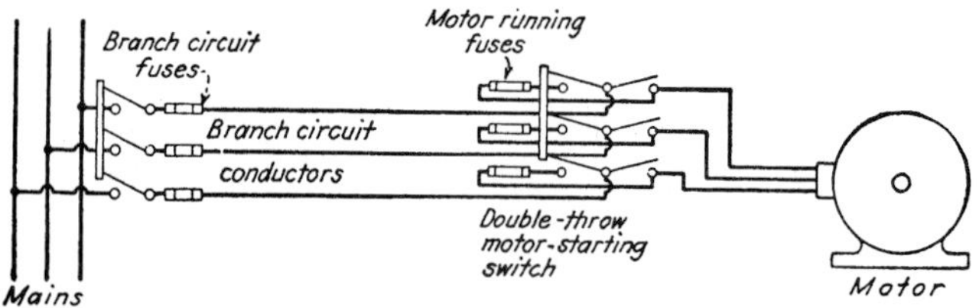

Fig. 430-15. Motor OL fuses may be shunted out for starting. (Sec. 430-35.)

As shown in Fig. 430-15, a double-throw switch is arranged for across-the-line starting. The switch is thrown to the right to start the motor, thus cutting the running fuses out of the circuit. The switch must be so made that it cannot be left in the starting position.

In the Exception to part **(b)**, conditions are given for shunting out overload protection of a motor that is automatically started. In previous Code editions, any motor that was automatically started was not permitted to have its overload protection shunted or cut out during the starting period. This Exception now accommodates those motor-and-load applications that have a long accelerating time and would otherwise require an overload device with such a long trip time that the motor would not be protected if it stalled while running.

430-36. Fuses—In Which Conductor. This rule is listed in Sec. 240-22 as Exception No. 2 to the rule that prohibits use of an overcurrent device in an intentionally grounded conductor. When fuses are used for protection of service, feeder, or branch-circuit conductors, a fuse must never be used in a grounded conductor, such as the grounded leg of a 3-phase, 3-wire corner-grounded delta system. But, if fuses are used for OL protection for a 3-phase motor connected on such a system, a fuse must be used in all three phase legs—EVEN THE GROUNDED LEG. Figure 430-16 shows two conditions of such fuse application for OL protection for a motor.

430-37. Devices Other than Fuses—In Which Conductor. Complete data on the number and location of overcurrent devices are given in Code Table 430-37.

Table 430-37 requires three running overload devices (trip coils, relays, thermal cutouts, etc.) for all 3-phase motors unless protected by other approved means, such as specifically designed embedded detectors with or without supplementary external protective devices.

Figure 430-17 points up this requirement.

If fuses are used as the running protective device, Sec. 430-36 requires a fuse in each ungrounded conductor. If the protective device consists of an automatically operated contactor or CB, the device must open a sufficient number of conductors to stop the current flow to the motor and must be equipped with the number of overcurrent units specified in Table 430-37.

430-42. Motors on General-Purpose Branch Circuits. Refer to Fig. 430-19, Type 3.

Branch circuits supplying lamps are usually 115-V single-phase circuits, and on such circuits the effect of subparagraphs **(a)** and **(b)** is that any motor larger than 6 A must be provided with a starter that is approved for group operation.

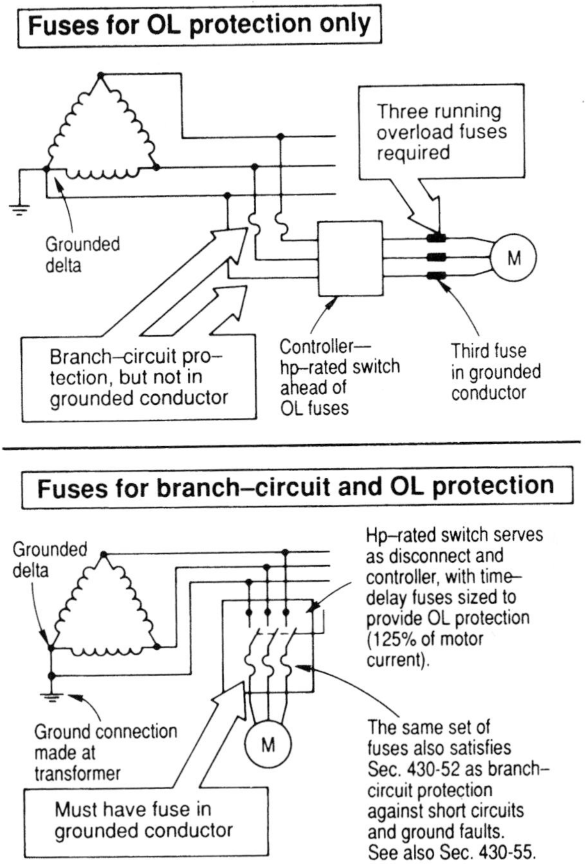

Fig. 430-16. A fuse for OL protection *must* be used in each phase leg of circuit. (Sec. 430-36.)

It is provided in Sec. 210-24 that receptacles on a 20-A branch circuit may have a rating of 20 A, and in such case subparagraph **(c)** requires that any motor or motor-driven appliance connected through a plug and receptacle must have running overcurrent protection. If the motor rating exceeds 1 hp or 6 A, the protective device must be permanently attached to the motor and subparagraph **(b)** must be complied with.

The requirements of Sec. 430-32 for the running overcurrent protection of motors must be complied with in all cases, regardless of the type of branch circuit by which the motor is supplied and regardless of the number of motors connected to the circuit.

430-43. Automatic Restarting. As noted in the comments to Sec. 430-32, an integral motor-running protective device may be of the type which will automatically restart, or it may be so constructed that after tripping out it must be closed by means of a reset button.

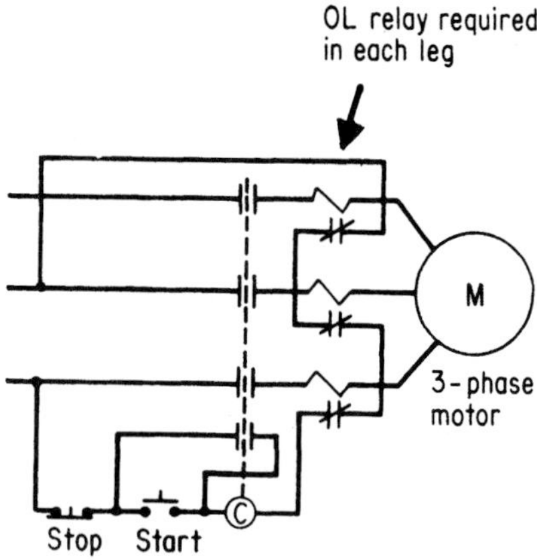

Fig. 430-17. Three OL units are required for 3-phase motors. (Sec. 430-37.)

430-44. Orderly Shutdown. Although the NE Code has all those requirements on use of running overload protection of motors, this section recognizes that there are cases when automatic opening of a motor circuit due to overload may be objectionable from a safety standpoint. In recognition of the needs of many industrial applications the rule here permits alternatives to automatic opening of a circuit in the event of overload. This permission for elimination of overload protection is similar to the permission given in Sec. 240-12 to eliminate overload protection when automatic opening of the circuit on an overload would constitute a more serious hazard than the overload itself. However, it is necessary that the circuit be provided with a motor overload sensing device conforming to the Code requirement on overload protection to indicate by means of a supervised alarm the presence of the overload (Fig. 430-18). Overload indication instead of automatic opening will alert personnel to the objectionable

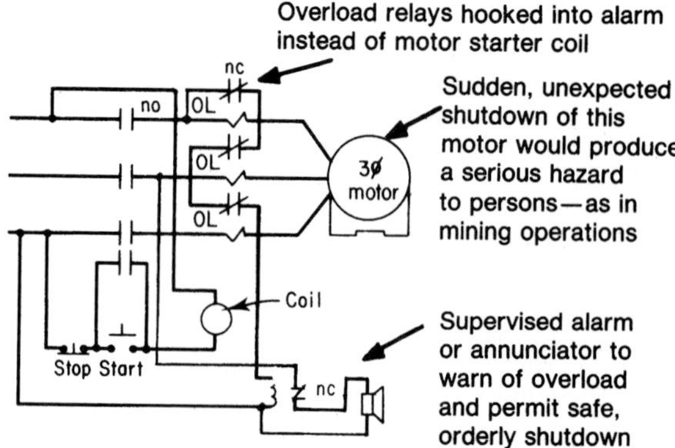

Fig. 430-18. This type of hookup may be used to warn of, but not open, an overload. (Sec. 430-44.)

condition and will permit corrective action, either immediately or at some more convenient time, for an orderly shutdown to resolve the difficulty. But, as is required in Sec. 240-12, short-circuit protection on the motor branch circuit must be provided to take care of those high-level ground faults and short circuits that would be more serious in their hazardous implications than simple overload.

Note: Section 445-4 also has an Exception that permits this same use of an alarm instead of overcurrent protection where it is better to have a generator fail than stop operating.

430-51. General. This section indicates the coverage of part **D**, which requires "Motor Branch-Circuit Short-Circuit and Ground-Fault Protection." Although the phrase "ground-fault protection" is used in several of the sections of part **D**, it should be noted that it refers to the protection against ground fault that is provided by the fuses or CB that are used to provide short-circuit protection. The single CB or set of fuses is referred to as a "short-circuit and ground-fault protective device." The rule is *not* intended to require the type of ground-fault protective hookup required by Sec. 230-95 on service disconnects (such as a zero-sequence transformer and relay hookup).

Motor branch circuits are commonly laid out in a number of ways. With respect to branch-circuit protection location and type, the layouts shown in Fig. 430-19 are as follows:

Type 1

An individual branch circuit leads to each motor from a distribution center. This type of layout can be used under any conditions and is the one most commonly used.

Type 2

A feeder or subfeeder with branch circuits tapped on at convenient points. This is the same as Type 1 except that the branch-circuit overcurrent protective devices are mounted individually at the points where taps are made to the subfeeder, instead of being assembled at one location in the form of a branch-circuit distribution center. Under certain conditions, the branch-circuit protective devices may be located at any point not more than 25 ft (7.62 m) distant from the point where the branch circuit is tapped to the feeder.

Type 3

Small motors, lamps, and appliances may be supplied by a 15- or 20-A circuit as described in Art. 210. Motors connected to these circuits must be provided with running overcurrent protective devices in most cases. See Sec. 430-42.

Figure 430-20 shows the typical elements of a motor branch circuit in their relation to branch-circuit protection, so that the protection is effective for the circuit conductors, the control and disconnect means, and the motor. Motor controllers provide protection for the motors they control against all ordinary

Type 1

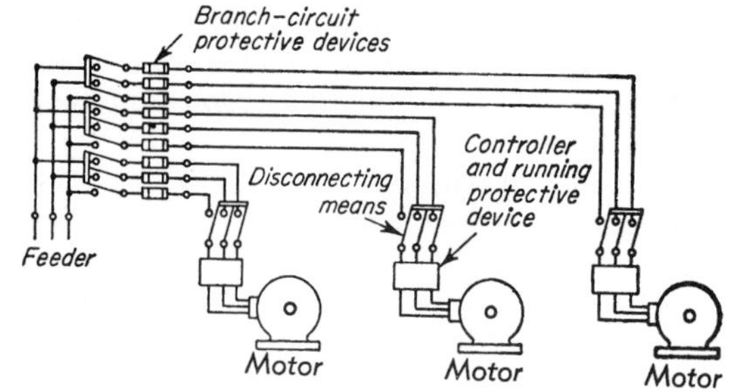

Type 2

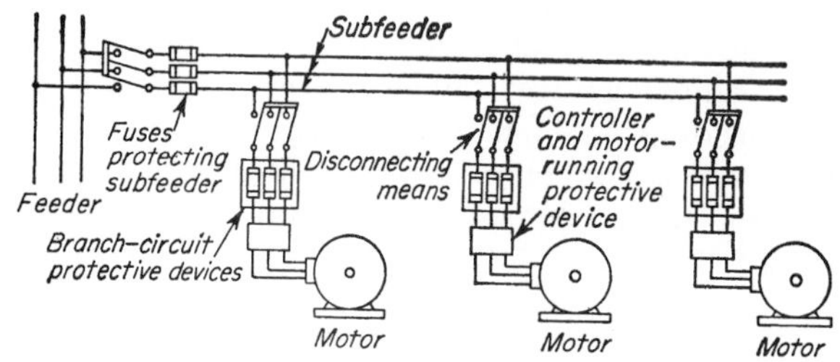

Type 3

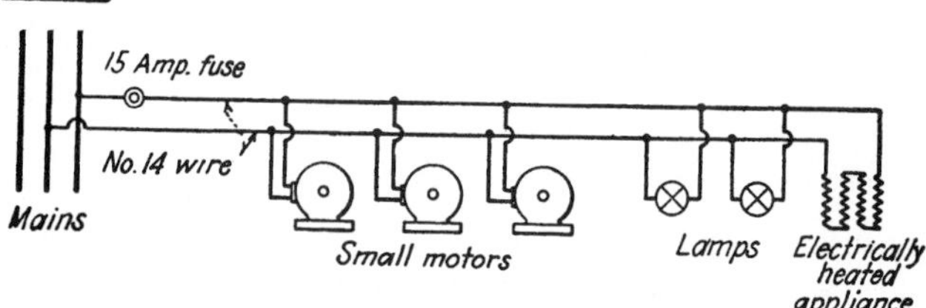

Fig. 430-19. Motor branch-circuit protection is used in various types of layouts. (Sec. 430-51.)

overloads but are not intended to open short circuits. Fuses, CBs, or motor short-circuit protectors used as the branch-circuit protective device will open short circuits and therefore provide short-circuit protection for both the motor and the running protective device. Where a motor is supplied by an individual branch circuit, having branch-circuit protection, the circuit protective devices may be either fuses or a CB and the rating or setting of these devices must not exceed the values specified in Sec. 430-52. In Fig. 430-20, the fuses or CB at the panelboard must carry the starting current of the motor, and in order to carry this current the fuse rating or CB setting may be rated up to 300 or 400 percent of the running current of the motor, depending on the size and type of motor. It is evident that to install motor circuit conductors having an ampacity up to that percent of the motor full-load current would be unnecessary.

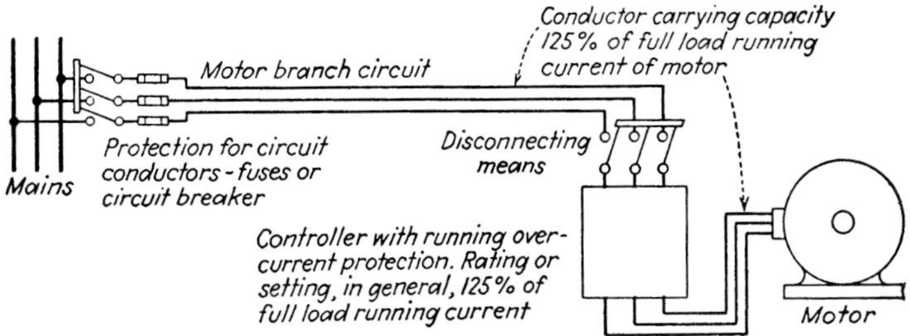

Fig. 430-20. Branch-circuit protection is on the line side of other components. (Sec. 430-51.)

There are three possible causes of excess current in the conductors between the panelboard and the motor controller, viz., a short circuit between two of these conductors, a ground on one conductor that forms a short circuit, and an overload on the motor. A short circuit would draw so heavy a current that the fuses or breaker at the panelboard would immediately open the circuit, even though the rating or setting is in excess of the conductor ampacity. Any excess current due to an overload on the motor must pass through the protective device at the motor controller, causing this device to open the circuit. Therefore with circuit conductors having an ampacity equal to 125 percent of the motor-running current and with the motor-protective device set to operate at near the same current, the conductors are reasonably protected.

430-52. Rating or Setting for Individual Motor Circuit. The Code requires that branch-circuit protection for motor circuits must protect the circuit conductors, the control apparatus, and the motor itself against overcurrent due to short circuits or ground (Secs. 430-51 through 430-58).

The first, and obviously necessary, rule is that the branch-circuit protective device for an individual branch circuit to a motor must be capable of carrying the starting current of the motor without opening the circuit. Then the Code proceeds to place maximum values on the ratings or settings of such overcurrent devices. It says that such devices must not be rated in excess of the values given in Table 430-152.

In case the values for branch-circuit protective devices determined by Table 430-152 do not correspond to the standard sizes or ratings of fuses, nonadjustable CBs, or thermal devices, or possible settings of adjustable CBs adequate to carry the load, the next higher size, rating, or setting may be used.

Under exceptionally severe starting conditions where the nature of the load is such that an unusually long time is required for the motor to accelerate to full speed, the fuse or CB rating or setting recommended in Table 430-152 may not be high enough to allow the motor to start. It is desirable to keep the branch-circuit protection at as low a rating as possible, but in unusual cases, it is permissible to use a higher rating or setting. Where absolutely necessary in order to permit motor starting, the device may be rated at other maximum values, as follows:

1. The rating of a fuse that is *not* a dual-element time-delay fuse and is rated not over 600 A may be increased above the Code table value but must never exceed 400 percent of the full-load current.

2. The rating of a time-delay (dual-element) fuse may be increased but must never exceed 225 percent of full-load current.
3. The setting of an instantaneous trip CB that is part of a *listed* combination starter (which contains a magnetic short-circuit trip element, without time delay, and independent overload device) may be increased but never over 1,300 percent of the motor full-load current, unless supplying a Design E motor, in which the setting may be increased to not more than 1,700 percent of motor full-load current.
4. The rating of an inverse time CB (a typical thermal-magnetic CB with a time-delay and instantaneous trip characteristic) may be increased but must not exceed 400 percent for full-load currents of 100 A or less and must not exceed 300 percent for currents over 100 A.
5. A fuse rated 601 to 6,000 A may be increased but must not exceed 300 percent of full-load current.
6. Torque motors must be protected at the motor nameplate current rating, and if a standard overcurrent device is not made in that rating, the next higher standard rating of protective device may be used.

The rules of this section establish maximum values for branch-circuit protection, setting the limit of safe applications. However, use of smaller sizes of branch-circuit protective devices is obviously permitted by the **Code** and does offer opportunities for substantial economies in selection of CBs, fuses and the switches used with them, panelboards, etc. In any application, it is only necessary that the branch-circuit device which is smaller than the maximum permitted rating must have sufficient time delay in its operation to permit the motor starting current to flow without opening the circuit.

But a CB for branch-circuit protection must have a continuous current rating of not less than 115 percent of the motor full-load current, as required by Sec. 430-58.

Where maximum ratings for the branch-circuit protection are shown in the manufacturer's heater table for use with a marked controller or are otherwise marked with the equipment, they must not be exceeded even though higher values are indicated in **Code** Table 430-152 and in the other rules of this section. That requirement is in the last sentence of this **Code** rule and is also specified in UL regulations which regulate the exposure of motor controllers to short-circuit currents to protect internal components, such as overload relays and contacts, from damage or destruction. Those rules state:

Motor controllers incorporating thermal cutouts, thermal overload relays, or other devices for motor-running overcurrent protection are considered to be suitably protected against overcurrent due to short circuits or grounds by motor branch circuit, short circuit and ground-fault protective devices selected in accordance with the **National Electrical Code** and any additional information marked on the product. Motor controllers may specify that protection is to be provided by fuses or by an inverse time circuit breaker. If there is no marking of protective device type, controllers are considered suitably protected by either type of device. Motor controllers may specify a maximum rating of protective device. If not marked with a rating, the controllers are considered suitably protected by a protective device of the maximum rating permitted by the **National Electrical Code**.

Unless otherwise marked, motor controllers incorporating thermal cutouts or overload relays are considered suitable for use on circuits having available fault currents not greater than [refer to Fig. 430-21]:

Horsepower rating	*RMS symmetrical amperes*
1 or less	1,000
1½ to 50	5,000
51 to 200	10,000
201 to 400	18,000
401 to 600	30,000
601 to 900	42,000
901 to 1,600	85,000

Typical application of the basic rule of Sec. 430-52 on short-circuit protection for motor circuits is shown in Fig. 430-4. Overcurrent (branch-circuit) protection (from Table 430-152 and Sec. 430-52) using nontime-delay fuses is calculated as follows:

1. The 50-hp squirrel-cage motor must be protected at not more than 200 A (65 A × 300 percent).
2. The 30-hp wound-rotor motor must be protected at not more than 60 A (40 A × 150 percent).
3. Each 10-hp motor must be protected at not more than 45 A (14 × 300 percent).

As shown in **Code** Table 430-152, if thermal-magnetic CBs were used, instead of the fuses, for branch-circuit protection, the maximum ratings that are permitted by the basic rule are:

1. For the 50-hp motor—65 A × 250 percent, or 162.5 A, with the next higher standard CB rating of 175 A permitted but *ONLY* if a 150-A CB, the next lower rating, is not capable of carrying the motor's starting current.

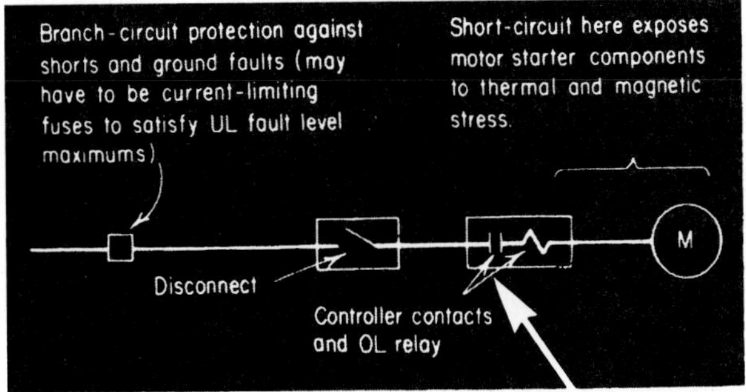

Fig. 430-21. UL specifies maximum short-circuit withstand ratings for controllers. (Sec. 430-52.)

2. For the 30-hp wound-rotor motor—40 A × 150 percent, or 60 A, calling for a 60-A CB.
3. For each 10-hp motor—14 A × 250 percent, or 35 A, calling for a 35-A CB.

Instantaneous Trip CBs

The NE Code recognizes the use of an instantaneous trip CB (without time delay) for short-circuit protection of motor circuits. Such breakers—also called "magnetic-only" breakers—may be used only if they are adjustable and if combined with motor starters in combination assemblies. An instantaneous-trip CB or a motor short-circuit protector (MSCP) may be used *only* as part of a "listed" (such as by UL) combination motor controller. A combination motor starter using an instantaneous trip breaker must have running overload protection in each conductor (Fig. 430-22). Such a combination starter offers use of a smaller CB than would be possible if a standard thermal-magnetic CB were used. And the smaller CB offers faster operation for greater protection against grounds and short circuits—in addition to offering greater economy.

A combination motor starter, as shown in Fig. 430-22, is based on the characteristics of the instantaneous trip CB, which is covered by the third percent column from the left in Code Table 430-152. Molded-case CBs with only magnetic instantaneous trip elements in them are available in almost all sizes. Use of such a device requires careful accounting for the absence of overload protection in the CB, up to the short-circuit trip setting. Such a CB is designed for use as shown in Fig. 430-22. The circuit conductors are sized for at least 125 percent of motor current. The thermal overload relays in the starter protect the entire circuit and all equipment against operating overloads up to and including stalled rotor current. They are commonly set at 125 percent of motor current. In such a

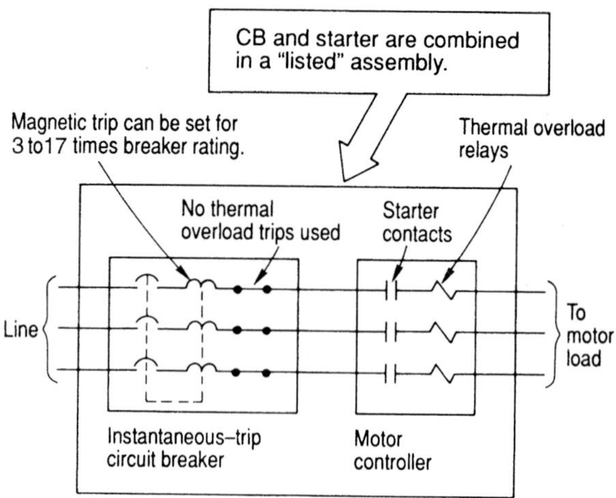

Fig. 430-22. Section 430-52 accepts use of a magnetic-only circuit breaker if it is part of a "listed" assembly of a combination starter. (Sec. 430-52.)

circuit, a CB with an adjustable magnetic trip element can be set to take over the interrupting task at currents above stalled rotor and up to the short-circuit duty of the supply system at that point of installation. The magnetic trip in a typical unit might be adjustable from 3 to 17 times the breaker current rating; i.e., a 100-A trip can be adjusted to trip anywhere between 300 and 1,700 A. Thus the CB serves as motor circuit disconnect and short-circuit protection.

Selection of such a listed assembly with an instantaneous-only CB is based on choosing a nominal CB size with a current rating at least equal to 115 percent of the motor full-load current to carry the motor current and to qualify under Secs. 430-58 and 430-110(a) as a disconnect means. Then the adjustable magnetic trip is set to provide the short-circuit protection—the value of current at which instantaneous circuit opening takes place, which should be just above the starting current of the motor involved—using a multiplier of something like 1.5 on locked-rotor current to account for asymmetry in starting current. Asymmetry can occur when the circuit to the motor is closed at that point on the alternating voltage wave where the inrush starting current is going through the negative maximum value of its alternating wave. That is the same concept as asymmetry in the initiation of a short-circuit current. Where supplying Design E motors, a greater inrush can be anticipated on start-up. As a result, higher initial and maximum settings are recognized.

Listed equipment using an instantaneous CB type is available with very simple instructions by the manufacturer to make proper selection and adjustment of the instantaneous-trip CB combination starter a quick, easy matter. The following describes the concept behind the application of listed combination starters with instantaneous-only CBs.

Given: A 30-hp, 230-V, 3-phase, squirrel-cage motor marked with the code letter M, indicating that the motor has a locked-rotor current of 10 to 11.19 kVA per horsepower, from **Code** Table 430-7(b). A full-voltage controller is combined with the CB, with running overload protection in the controller to protect the motor within its heating damage curve on overload in a listed unit.

Required: Select the maximum setting and minimum rating for the CB which will provide short-circuit protection and will qualify as the motor circuit disconnect means.

Solution: The motor has a full-load current of 80 A (**Code** Table 430-150). A CB suitable for use as disconnect must have a current rating at least 115 percent of 80 A. As covered in Sec. 430-52(c)(3), for instantaneous-trip CBs, the initial setting from Table 430-152 would be limited to 800 percent of the 80-A full-load current. The maximum setting—for other than the high-efficiency Design E motors—is 1,300 percent. For Design E motors, the initial setting may be 1,100 percent of motor full-load current with a maximum setting of 1,700 percent of the motor full-load current.

It should be noted that settings above 800 or 1,100 percent of the motor's full-load current are only permitted if nuisance-tripping occurs on starting *or* if evaluation of the motor's starting characteristics and the time-current trip curve of the breaker indicates that a greater setting is needed. Although not completely clear, the trip-value established through the "engineering evaluation" should be considered as the maximum setting.

Because the use of a magnetic-only CB does not protect against low-level grounds and shorts in the circuit conductors on the line side of the starter running overload relays, the NE Code rule permits such application only where the CB and starter are part of a *listed* combination starter in a single enclosure.

MSCPs

A motor short-circuit protector, as referred to in the second paragraph of Sec. 430-52, is a fuselike device designed for use only in its own type of fusible-switch combination motor starter. The combination offers short-circuit protection, running overload protection, disconnect means, and motor control—all with assured coordination between the short-circuit interrupter (the motor short-circuit protector) and the running OL devices. It involves the simplest method of selection of the correct MSCP for a given motor circuit. This packaged assembly is a third type of combination motor starter—added to the conventional fusible-switch and CB types.

The NE Code recognizes motor short-circuit protectors in Secs. 430-40 and 430-52 provided the combination is a "listed" assembly. This means a combination starter equipped with motor short-circuit protectors and listed by Underwriters Laboratories Inc., or another nationally recognized testing lab, as a package called an MSCP starter.

430-53. Several Motors or Loads on One Branch Circuit. A single branch circuit may be used to supply two or more motors as follows:

Part **(a):** Two or more motors, each rated not more than 1 hp and each drawing not over 6 A full-load current, may be used on a branch circuit protected at not more than 20 A at 125 V or less, or 15 A at 600 V or less. And the rating of the branch-circuit protective device marked on any of the controllers must not be exceeded. That is also a UL requirement.

Individual running overload protection is necessary in such circuits, unless: the motor is not permanently installed, is manually started, and is within sight from the controller location; or the motor has sufficient winding impedance to prevent overheating due to stalled rotor current; or the motor is part of an approved assembly that does not subject the motor to overloads and that incorporates protection for the motor against stalled rotor; or the motor cannot operate continuously under load.

Part **(b):** Two or more motors of any rating, each having individual running overload protection, may be connected to a branch circuit which is protected by a short-circuit protective device selected in accordance with the maximum rating or setting of a device which could protect an individual circuit to the motor of the smallest rating. This may be done only where it can be determined that the branch-circuit device so selected will not open under the most severe normal conditions of service which might be encountered.

This permission of part **(b)** offers wide application of more than one motor on a single circuit, particularly in the use of small integral-horsepower motors installed on 440-V, 3-phase systems. This application primarily concerns use of small integral-horsepower 3-phase motors as used in 208-, 220-, and 440-V industrial and commercial systems. Only such 3-phase motors have full-load operating currents low enough to permit more than one motor on circuits fed from 15-A protective devices.

There are a number of ways of connecting several motors on a single branch circuit, as follows:

In case I, Fig. 430-23, using a three-pole CB for branch-circuit protective device, application is made in accordance with part **(b)** as follows:

1. The full-load current for each motor is taken from NE Code Table 430-150 [as required by Sec. 430-6(a)].
2. Choosing to use a CB instead of fuses for branch-circuit protection, the rating of the branch-circuit protective device, 15-A, does not exceed the maximum value of short-circuit protection required by Sec. 430-52 and Table 430-152 for the smallest motor of the group—which is the 1½-hp motor. Although 15 A is greater than the maximum value of 250 percent times motor full-load current (2.5 × 2.6 A = 6.5 A) set by Table 430-152 (under the column "Inverse Time Breaker" opposite "polyphase squirrel-cage" motors), the 15-A breaker is the "next higher size, rating, or setting" for a standard CB—as permitted in Sec. 430-52. A 15-A CB is the smallest standard rating recognized by Sec. 240-6.
3. The total load of motor currents is:

$$4.8 \text{ A} + 3.4 \text{ A} + 2.6 \text{ A} = 10.8 \text{ A}$$

This is well within the 15-A CB rating, which has sufficient time delay in its operation to permit starting of any one of these motors with the other two already operating. Torque characteristics of the loads on starting are

CASE I—USING A CIRCUIT BREAKER FOR PROTECTION

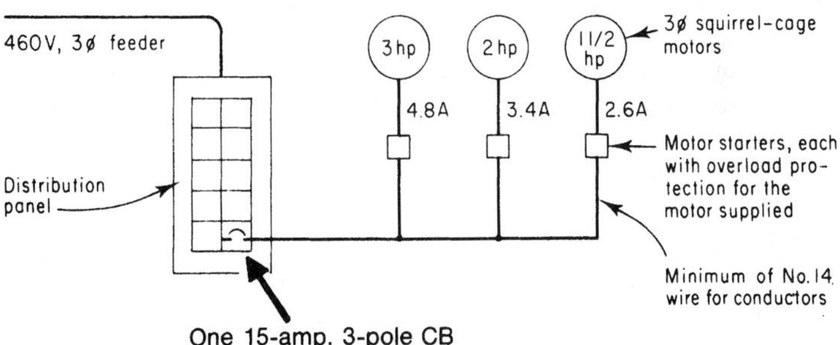

HERE IS THE KEY: A 15-amp, 3-pole CB is used, based on Section 430-52 and Table 430-152. This is the "next higher size" of standard protective device above 250% × 2.6 amps (the required rating for the smallest motor of the group). The 15-amp CB makes this application possible, because the 15-amp CB is the smallest standard rating of CB and is suitable as the branch-circuit protective device for the 1½-hp motor.

Fig. 430-23. Three integral-horsepower motors may be supplied by this circuit makeup. (Sec. 430-53.)

not high. It was therefore determined that the CB will not open under the most severe normal service.
4. Each motor has individual running overload protection in its starter.
5. The branch-circuit conductors are sized in accordance with Sec. 430-24:

$$4.8\text{ A} + 3.4\text{ A} + 2.6\text{ A} + (25\text{ percent of }4.8\text{ A}) = 12\text{ A}$$

Conductors must have an ampacity at least equal to 12 A. No. 14 THW, TW, RHW, RHH, THHN, or XHHW conductors will fully satisfy this application.

In case II, Fig. 430-24, a similar hookup is used to supply three motors—also with a CB for branch-circuit protection.

1. Section 430-53(b) requires branch-circuit protection to be not higher than the maximum amps set by Sec. 430-52 for the lowest rated of the motors.
2. From Sec. 430-52 and Table 430-152, that maximum protection rating for a CB is 250 percent × 1 A (the lowest rated motor), or 2.5 A. But, 2.5 A is not a "standard rating" of CB from Sec. 240-6; and the third paragraph of Sec. 430-52 permits use of the "next higher size, rating, or setting" of standard protective device.
3. Because 15 A is the lowest standard rating of CB, it is the "next higher" device rating above 2.5 A and satisfies Code rules on the rating of the branch-circuit protection.

The applications shown in cases I and II permit use of several motors up to circuit capacity, based on Secs. 430-24 and 430-53(b) and on starting torque characteristics, operating duty cycles of the motors and their loads, and the time

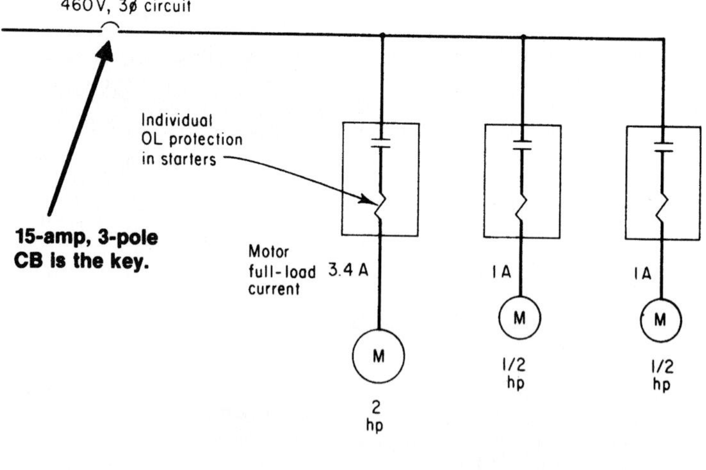

Fig. 430-24. Fractional-horsepower and integral-horsepower motors may be supplied by the same circuit. (Sec. 430-53.)

delay of the CB. Such applications greatly reduce the number of CB poles, the number of panels, and the amount of wire used in the total system. One limitation, however, is placed on this practice in the last sentence of Sec. 430-52, as noted previously. Where more than one fractional- or small-integral-horsepower motor is used on a single branch circuit of 15-A rating in accordance with NE Code Sec. 430-53(a) or (b), care must be taken to observe all markings on controllers that indicate a maximum rating of short-circuit protection ahead of the controller (Fig. 430-25).

In case III, Fig. 430-26, the same three motors shown in case II would be subject to different hookup to comply with the rules of Sec. 430-53(b) when fuses, instead of a CB, are used for branch-circuit protection, as follows:

1. To comply with Sec. 430-53(b), fuses used as branch-circuit protection must have a rating not in excess of the value permitted by Sec. 430-52 and Table 430-152 for the smallest motor of the group—one of the ½-hp motors.
2. Table 430-152 shows that the maximum permitted rating of nontime-delay type fuses is 300 percent of full-load current for 3-phase squirrel-cage motors. Applying that to one of the ½-hp motors gives a maximum fuse rating of

$$300 \text{ percent} \times 1 \text{ A} = 3 \text{ A}$$

3. BUT, there is no permission for the fuses to be rated higher than 3 A—BECAUSE 3 A IS A "STANDARD" RATING OF FUSE (but not a standard rating of CB). Section 240-6 considers fuses rated at 1, 3, 6, and 10 A to be "standard" ratings.

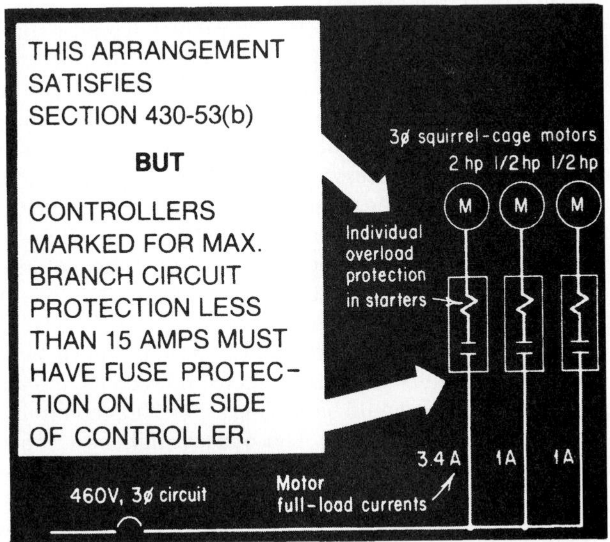

Fig. 430-25. Branch-circuit protection must not exceed marked maximum value. (Sec. 430-53.)

BUT, WATCH OUT!!!

CASE III—USING FUSES FOR CIRCUIT PROTECTION

Interpretation of *NE Code* rules of Section 430-53(b) in conjunction with the "standard" ratings of fuses in Section 240-6 may require different circuit makeup when fuses are used to protect the branch circuit to several motors.

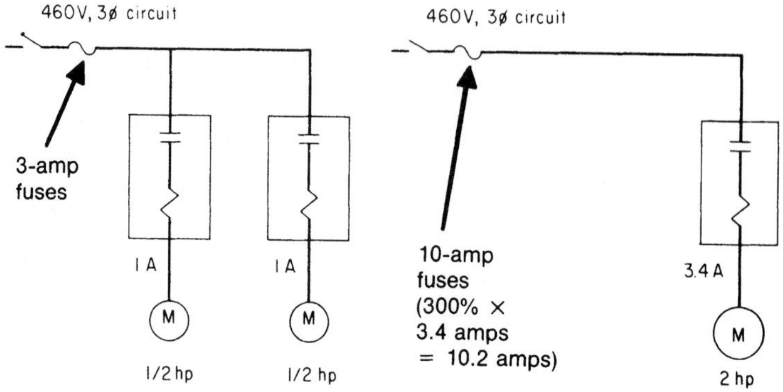

Fig. 430-26. Fuse protection may require different circuiting for several motors. (Sec. 430-53.)

4. The maximum branch-circuit fuse permitted by Sec. 430-53(b) for a ½-hp motor is 3 A.
5. The two ½-hp motors may be fed from a single branch circuit with three 3-A fuses in a three-pole switch.
6. Following the same Code rules, the 2-hp motor would require fuse protection rated not over 10 A (300 percent × 3.4 A = 10.2 A).

Note: Because the standard fuse ratings below 15 A place fuses in a different relationship to the applicable Code rules, it will require interpretation of the Code rules to resolve the question of acceptable application in case II versus case III. Interpretation will be necessary to determine if CBs are excluded as circuit protection in these cases where use of fuses, in accordance with the precise wording of the Code, provides lower rated protection than CBs—when applying the rule of the third paragraph of Sec. 430-52. And if the motors of case I are fed from a circuit protected by fuses, the literal effect of the Code rules would require different circuiting for those motors.

Figure 430-27 shows one way of combining cases II and III to satisfy Secs. 430-53(b), 430-52, and 240-6; but the 15-A CB would then technically be feeder protection, because the fuses would be serving as the "branch-circuit protective devices" as required by Sec. 430-53(b). Those fuses might be acceptable in each starter, without a disconnect switch, in accordance with Sec. 240-40—which allows use of cartridge fuses at any voltage without an individual disconnect for each set of fuses, provided only qualified persons have access to the fuses. But, Sec. 430-112 would have to be satisfied to use the single CB as a disconnect for the group of motors. And part **(b)** of that Exception recognizes one

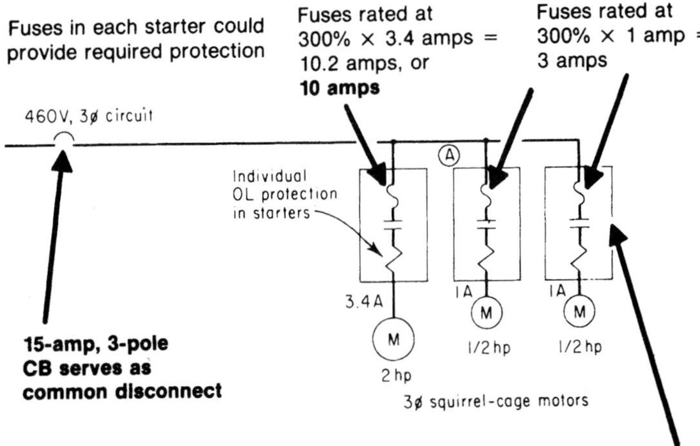

Fig. 430-27. Multimotor circuit may be acceptable with fused starters. (Sec. 430-53.)

common disconnect in accordance with Sec. 430-53(a) but not 430-53(b). Certainly, the use of a fusible-switch-type combination starter for each motor would fully satisfy all rules.

Figure 430-28 shows another hookup that might be required to supply the three motors of Fig. 430-23.

Figure 430-29 shows another hookup of several motors on one branch circuit—an actual job installation which was based on application of Sec. 430-53(b). The installation was studied as follows:

Problem: A factory has 100 1½-hp, 3-phase motors, with individual motor starters incorporating overcurrent protection, rated for 460 V. Provide circuits.

Solution: Prior to 1965, the NE Code would not permit several integral-horsepower motors on one branch circuit fed from a three-pole CB in a panel. Each of the 100 motors would have had to have its own individual 3-phase circuit fed from a 15-A, 3-pole CB in a panel. As a result, a total of 300 CB poles would have been required calling for seven panels of 42 circuits each plus a smaller panel (or special panels of greater numbers than 42 poles per panel).

Under the present Code, depending upon the starting torque characteristics and operating duty of the motors and their loads, with each motor rated for 2.6 A, three or four motors could be connected on each 3-phase, 15-A circuit—greatly reducing the number of panelboards and overcurrent devices and the amount of wire involved in the total system. Time delay of CB influences number of motors on each circuit.

BUT, an extremely important point that must be strictly observed is the requirement that the rating of branch-circuit protection must not exceed any maximum value that might be marked on the starters used with the motors.

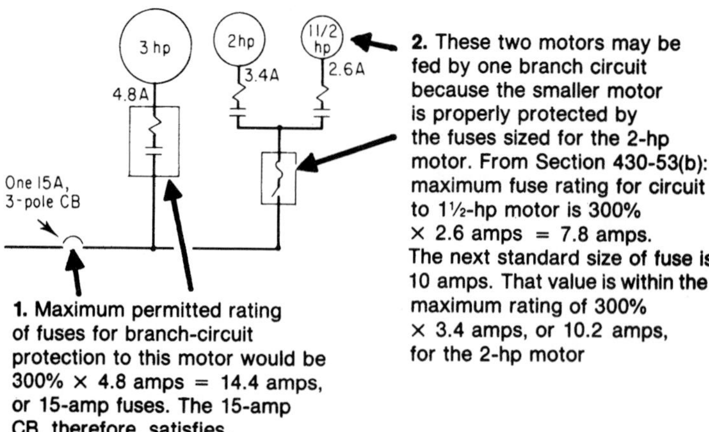
3 hp
2hp
11/2 hp
4.8A
3.4A
2.6A
One 15A, 3-pole CB
2. These two motors may be fed by one branch circuit because the smaller motor is properly protected by the fuses sized for the 2-hp motor. From Section 430-53(b): maximum fuse rating for circuit to 1½-hp motor is 300% × 2.6 amps = 7.8 amps. The next standard size of fuse is 10 amps. That value is within the maximum rating of 300% × 3.4 amps, or 10.2 amps, for the 2-hp motor
1. Maximum permitted rating of fuses for branch-circuit protection to this motor would be 300% × 4.8 amps = 14.4 amps, or 15-amp fuses. The 15-amp CB, therefore, satisfies.

Fig. 430-28. This hookup might be required to satisfy literal Code wording. (Sec. 430-53.)

Part **(c)**: In selecting the wording for part **(c)**, it was the intent of the Code-making panel to clarify the intent that several motors should not be connected to one branch circuit unless careful engineering is exercised by qualified persons to determine that all components of the branch circuit are selected and specified to meet the present requirements and to function together. The intent is to allow:

a. Completely factory-assembled equipment, or

b. A factory-assembled unit with a separate branch-circuit short-circuit and ground-fault protective device of a type and rating specified, or

c. Separately mounted components which are listed for use together and are specified for such use together by manufacturer's instructions and/or nameplate markings.

It is not the intent to change requirements for supplemental overcurrent protection such as in Sec. 422-28(f) or 424-22(c).

The change will inform the user that no interchange of components should be made without negating the manufacturer's warranty and listing by an approved laboratory.

Two or more motors of any rating may be connected to one branch circuit if each motor has running overload protection, if the overload devices and controllers are approved for group installation, and if the branch-circuit fuse or time-delay CB rating is in accordance with Sec. 430-52 for the largest motor plus the sum of the full-load current ratings of the other motors (Fig. 430-30).

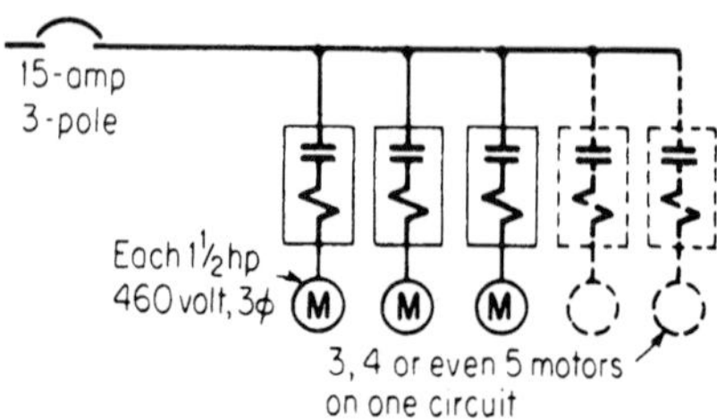
15-amp
3-pole
Each 1½hp
460 volt, 3ϕ
M
M
M
3, 4 or even 5 motors on one circuit

Fig. 430-29. Multimotor circuits offer economical supply to small integral-horsepower motors. (Sec. 430-53.)

The branch-circuit fuses or CB must not be larger than the rating or setting of short-circuit protection permitted by Sec. 430-52 for the smallest motor of the group, unless the thermal device is approved for group installation with a given maximum size of fuse or time-delay CB for short-circuit protective device. (See Sec. 430-40.) Underwriters Laboratories notes that motor controllers for group installation are marked with a maximum rating of *fuse* required to suitably protect the controller. Section 430-53(c)(2), however, calls for a group installation controller to be marked for the rating of fuse or CB ahead of it.

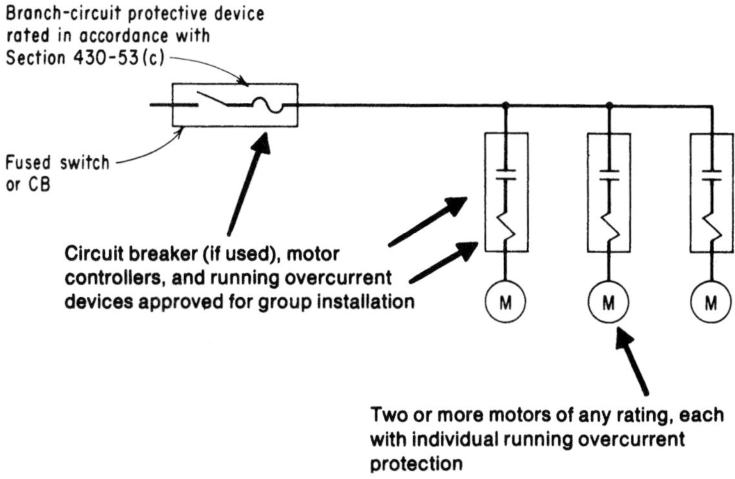

Fig. 430-30. Motors of any horsepower rating require circuit equipment for group installation. (Sec. 430-53.)

Part **(d)**: For installations of groups of motors as covered in part **(c)** above, tap conductors run from the branch-circuit conductors to supply individual motors must be sized properly. Such tap conductors would, of course, be acceptable where they are the same size as the branch-circuit conductors themselves. However, tap conductors to a single motor may be smaller than the main branch-circuit conductors provided that: they have an ampacity at least one-third that of the branch-circuit conductors, their ampacity is not less than 125 percent of the motor full-load current, they are not over 25 ft. (7.62 m) long, and they are in a raceway or are otherwise protected from physical damage (Fig. 430-31).

The principle applied here is that, since the conductors are short and protected from physical damage, it is unlikely that trouble will occur in the run between the mains and the motor protection which will cause the conductors to be overloaded, except some accident resulting in an actual short circuit. A short circuit will blow the fuses or trip the CB protecting the mains. An overload on the conductors caused by overloading the motor or trouble in the motor itself will cause the motor protective device to operate and so protect the conductors.

Condition 1

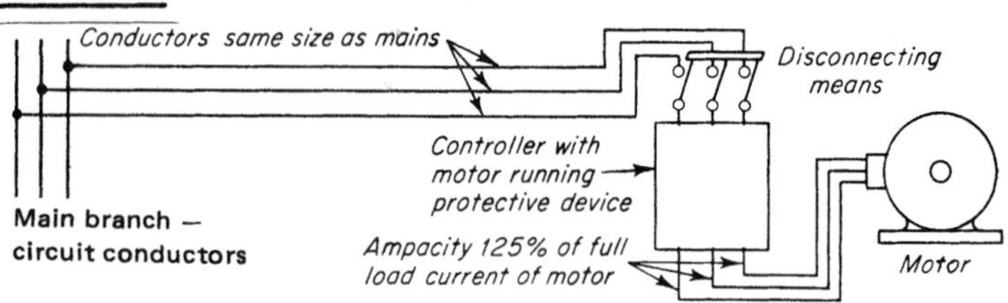

Condition 2

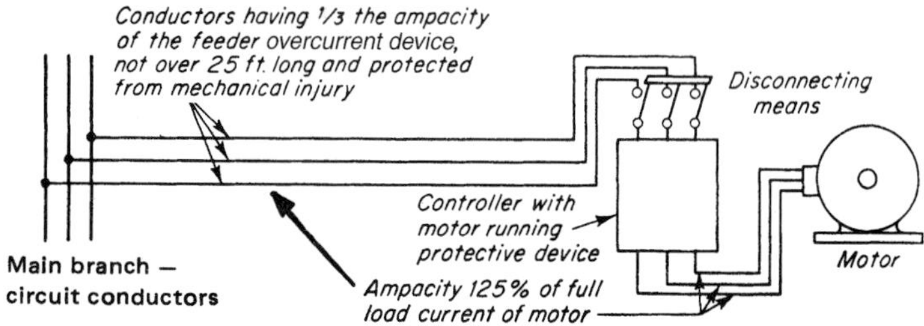

Fig. 430-31. Overcurrent protection not required for taps to single motors of a group. (Sec. 430-53.)

430-55. Combined Overcurrent Protection. A CB or set of fuses may provide both short-circuit protection and running overload protection for a motor circuit. For instance, a CB or dual-element time-delay fuse sized at not over 125 percent of motor full-load current (Sec. 430-32) for a 40°C-rise continuous-duty motor would be acceptable protection for the branch circuit and the motor against shorts, ground faults, and operating overloads on the motor. See the bottom of Fig. 430-16 for a typical fuse application.

Figure 430-32 shows a CB used to fulfill four **Code** requirements simultaneously. For the continuous-duty, 40°C-rise motor shown, the CB may provide running overload protection if it is rated not over 125 percent of the motor's full-load running current. Therefore, 28 A × 1.25 = 35 A, which satisfies Sec. 430-32(a). Because the rating of the thermal-magnetic CB is not over 250 percent times the full-load current (from Table 430-152), the 35-A CB satisfies Secs. 430-52 and 430-58 as short-circuit and ground-fault protection. The CB may serve both those functions, as noted in Sec. 430-55. The CB may serve as the motor controller, as permitted by Sec. 430-83 Exception No. 2. The CB also satisfies as the required disconnect means in accordance with Sec. 430-111 and has the rating "of at least 115 percent of the full-load current rating of the motor," as required by Sec. 430-110(a). And because it satisfies Sec. 430-110(a) on the disconnect minimum rating, it therefore satisfies Sec. 430-58, which sets the same minimum rating for a CB used as branch-circuit protection.

430-56. Branch-Circuit Protective Devices—In Which Conductor. Motor branch circuits are to be protected in the same way as other circuits with regard to the

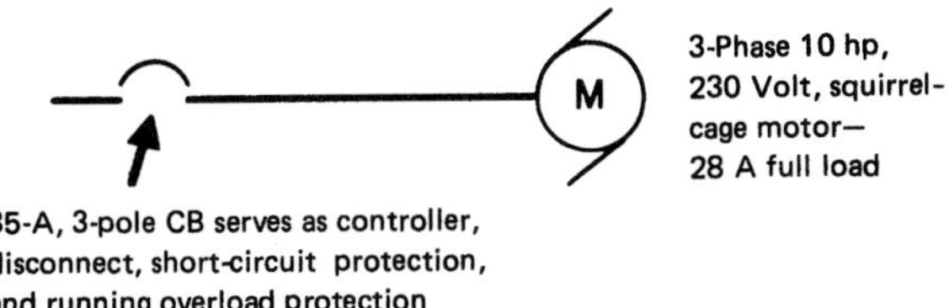

Fig. 430-32. Overcurrent functions may be combined in a single CB or set of fuses. (Sec. 430-55.)

number of fuses and the number of poles and overcurrent units of CBs. If fuses are used, a fuse is required in each ungrounded conductor. If a CB is used, there must be an overcurrent unit in each ungrounded conductor.

430-57. Size of Fuseholder. The basic rule of this section covers sizing of fuseholders for standard nontime-delay fuses used as motor branch-circuit protection. The Exception recognizes that time-delay fuses permit use of smaller switches and lower-rated fuseholders.

A fusible switch can take either standard NE Code fuses or time-delay fuses—up to the rating of the switch. Because a given size of time-delay fuse can hold on the starting current of a motor larger than that which could be used with a standard fuse of the same rating, fusible switches are given two horsepower ratings—one for use with standard fuses, the other for use with time-delay fuses. For example, a 3-pole, 30-A, 240-V fused switch has a rating of 3 hp for a 3-phase motor if standard fuses without time-delay characteristics are used. If time-delay fuses are used, the rating is raised to 7½ hp.

Consider a 7½-hp, 230-V, 3-phase motor (full-voltage starting, without code letters, or with code letters F to V), with a full-load current of 22 A. NE Code Table 430-152 shows that such a motor may be protected by nontime-delay fuses with a maximum rating equal to 300 percent of the full-load current (66 A), or time-delay fuses with a maximum rating equal to 175 percent of the full-load current (38.5 A).

If standard, nontime-delay fuses were used, the maximum size permitted would be 70 A (the next standard size larger than 66 A). From the table, this would require a 100-A, 15-hp switch, which would have fuseholders that could accommodate the fuses, as required by the basic rule. Or, a 60-A, 7½-hp switch might be used with standard fuses rated 60 A maximum. But such a switch would be required by the basic rule to have fuseholders that could accommodate 70-A fuses. Because such a fuse has knife-blade terminals instead of end ferrules and is larger than a 60-A fuse, fuseholders in the 60-A switch could be held in conflict with the Code rule even though the level of protection would be better with 60-A fuses in the 60-A switch. Wording of the rule is not clear. But cost, labor, and space savings would be realized using a 30-A, 7½-hp switch with 30-A time-delay fuses, with no worry about nuisance blowing of the fuses on motor starting current, and that would be acceptable under the Exception.

430-58. Rating of Circuit Breaker. This rule sets a maximum and minimum rating for a CB as branch-circuit protection. Refer to Sec. 430-55.

In the case of a CB having an adjustable trip point, this rule refers to the capacity of the CB to carry current without overheating and has nothing to do with the setting of the breaker. The breaker most commonly used as a motor branch-circuit protective device is the nonadjustable CB (see Sec. 240-6), and any breaker of this type having a rating in conformity with the requirements of Sec. 430-52 will have an ampacity considerably in excess of 115 percent of the full-load motor current.

430-62. Rating or Setting—Motor Load. Overcurrent protection for a feeder to several motors must have a rating or setting not greater than the largest rating or setting of the branch-circuit protective device for any motor of the group plus the sum of the full-load currents of the other motors supplied by the feeder.

The second paragraph notes that there are cases where two or more motors fed by a feeder will have the same rating as the branch-circuit device. And that can happen where the motors are of the same or different horsepower ratings. It is possible for motors of different horsepower ratings to have the same rating as the branch-circuit protective device, depending upon the type of motor and the type of protective device. If two or more motors in the group are of different horsepower rating but the rating or setting of the branch-circuit protective device is the same for both motors, then one of the protective devices should be considered as the largest for the calculation of feeder overcurrent protection.

And because Table 430-152 recognizes many different ratings of branch-circuit protective devices (based on use of fuses or CBs and depending upon the particular type of motor), it is possible for two motors of equal horsepower rating to have widely different ratings of branch-circuit protection. If, for instance, a 25-hp motor was protected by nontime-delay fuses, Table 430-152 gives 300 percent of the full-load motor current as the maximum rating or setting of the branch-circuit device. Thus, 250-A fuses would be used for a motor that had a 78-A full-load rating. But another motor of the same horsepower and even of the same type, if protected by time-delay fuses, must use fuses rated at only 175 percent of 78 A, which would be 150-A fuses, as shown in Fig. 430-33. If the two 25-hp motors were of different types, one being a wound-rotor motor, it would still be necessary to base selection of the feeder protection on the largest rating or setting of a branch-circuit protective device, regardless of the horsepower rating of the motor.

Figure 430-34 shows a typical motor feeder calculation, as follows:

The four motors supplied by the 3-phase, 440-V, 60-cycle feeder, which are not marked with a code letter (see Table 430-152), are as follows:

- One 50-hp squirrel-cage induction motor (full-voltage starting)
- One 30-hp wound-rotor induction motor
- Two 10-hp squirrel-cage induction motors (full-voltage starting)

Step 1. Branch-Circuit Loads

From Table 430-150, the motors have full-load current ratings as follows:

50-hp motor—65 A
30-hp motor—40 A
10-hp motor—14 A

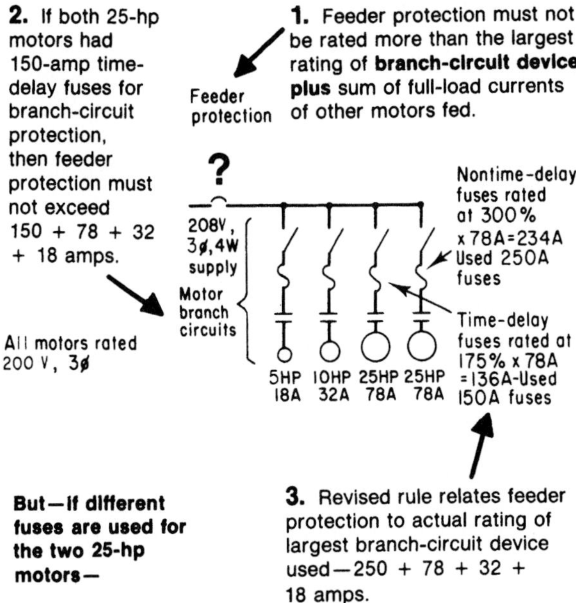
2. If both 25-hp motors had 150-amp time-delay fuses for branch-circuit protection, then feeder protection must not exceed 150 + 78 + 32 + 18 amps.
1. Feeder protection must not be rated more than the largest rating of branch-circuit device plus sum of full-load currents of other motors fed.
Feeder protection
?
208V, 3ø,4W supply
Motor branch circuits
All motors rated 200 V, 3ø
Nontime-delay fuses rated at 300% x 78A=234A Used 250A fuses
Time-delay fuses rated at 175% x 78A =136A-Used 150A fuses
5HP 18A
10HP 32A
25HP 78A
25HP 78A
But—If different fuses are used for the two 25-hp motors—
3. Revised rule relates feeder protection to actual rating of largest branch-circuit device used—250 + 78 + 32 + 18 amps.

Fig. 430-33. Feeder protection is based on largest branch-circuit protection, not on motor horsepower ratings. (Sec. 430-62.)

Step 2. Conductors

The feeder conductors must have a carrying capacity as follows (see Sec. 430-24):

$$1.25 \times 65 = 81 \text{ A}$$
$$81 + 40 + (2 \times 14) = 149 \text{ A}$$

The feeder conductors must be at least No. 3/0 TW, 1/0 THW, or 1/0 RHH or THHN (copper).

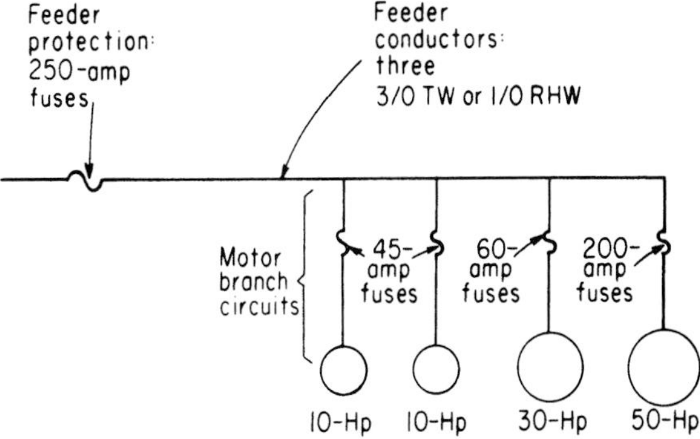
Feeder protection: 250-amp fuses
Feeder conductors: three 3/0 TW or 1/0 RHW
Motor branch circuits
45-amp fuses
60-amp fuses
200-amp fuses
10-Hp
10-Hp
30-Hp
50-Hp

Fig. 430-34. Rating of feeder protection is based on branch protection and motor currents. (Sec. 430-62.)

Step 3. Branch-Circuit Protection

Overcurrent (branch-circuit) protection (from Table 430-152 and Sec. 430-52) using nontime-delay fuses is as follows:

1. The 50-hp motor must be protected at not more than 200 A (65 A × 300 percent).
2. The 30-hp motor must be protected at not more than 60 A (40 A × 150 percent).
3. Each 10-hp motor must be protected at not more than 45 A (14 × 300 percent).

Step 4. Feeder Protection

As covered in Sec. 430-62, the maximum rating or setting for the overcurrent device protecting such a feeder must not be greater than the largest rating or setting of branch-circuit protective device for one of the motors of the group plus the sum of the full-load currents of the other motors. From the above, then, the maximum allowable size of feeder fuses is 200 + 40 + 14 + 14 = 268 A.

This calls for a maximum standard rating of 250 A for the motor feeder fuses, which is the nearest standard fuse rating that does not exceed the maximum permitted value of 268 A.

The Exception to part **(a)** addresses those installations where instantaneous-trip CBs and/or MSCPs are used as short-circuit and ground-fault protection for the largest motor supplied by the feeder to be protected. Under certain conditions those devices may set or rated to trip at 13 times or, for Design E motors, even as high as 17 times the motor full-load current. To prevent the feeder conductors from being underprotected, this Exception requires that the rating or setting of the feeder protective device be based on the type of device used. That is, if one-time fuses are used for feeder protection, the rating of those fuses must be based on the rating of fuse that would be permitted to protect the motor-branch-circuit if a fuse were used as branch-circuit protection instead of an instantaneous-trip CB or MSCP. For example, consider a 460-V, 3-phase, 100-hp motor, which draws approximately 100 A. If an instantaneous-trip breaker were used for branch-circuit short-circuit and ground-fault protection, it could be rated at 1,300 A (13 × 100 A).

When establishing the rating of protection for the feeder supplying a motor branch-circuit so protected, if one-time fuses are used the value of current that is summed with the ratings of the other type branch-circuit protective devices must be no more than that which would be permitted if one-time fuses were used as branch-circuit protection instead of an instantaneous-trip CB or MSCP. In this example, Table 430-152 would permit a one-time fuse to be 300 percent of the motor full-load current. If the branch-circuit in question is protected by one-time fuses, they would be rated at 300 A (3 × 100 A). And 300 A, *NOT* 1,300 A, would be used to calculate the maximum rating of one-time fuses that are permitted to protect the feeder conductors.

Note: There is no provision in Sec. 430-62, which permits the use of "the next higher size, rating, or setting" of the protective device, for a motor feeder when the calculated maximum rating does not correspond to a standard size of device.

According to part **(b)** of this section, in large-capacity installations where extra feeder capacity is provided for load growth or future changes, the feeder overcurrent protection may be calculated on the basis of the rated current-carrying capacity of the feeder conductors. In some cases, such as where two or more motors on a feeder may be started simultaneously, feeder conductors may have to be larger than usually required for feeders to several motors.

In selecting the size of a feeder overcurrent protective device, the NE Code calculation is concerned with establishing a maximum value for the fuse or CB. If a lower value of protection is suitable, it may be used.

430-63. Rating or Setting—Power and Light Loads. Protection for a feeder to both motor loads and a lighting and/or appliance load must be rated on the basis of both of these loads. The rating or setting of the overcurrent device must be sufficient to carry the lighting and/or appliance load plus the rating or setting of the motor branch-circuit protective device if only one motor is supplied, or plus the highest rating or setting of branch-circuit protective device for any one motor plus the sum of the full-load currents of the other motors, if more than one motor is supplied.

Figure 430-35 presents basic NE Code calculations for arriving at minimum requirements on wire sizes and overcurrent protection for a combination power and lighting load as follows:

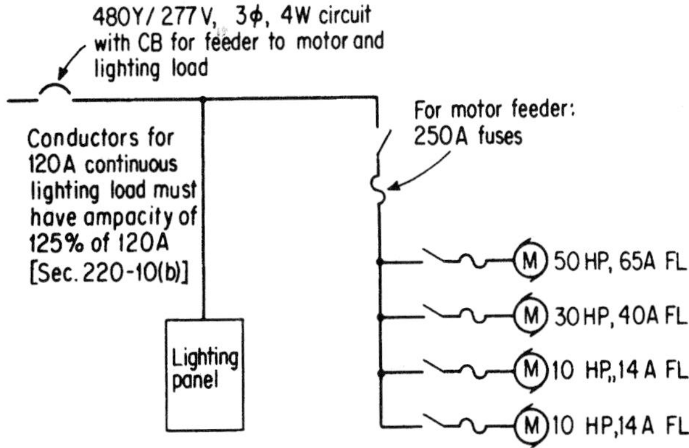

Fig. 430-35. Feeder protection for combination load must properly add both loads. (Sec. 430-63.)

Step 1. Total Load

Section 430-25(a) says that conductors supplying a lighting load and a motor must have capacity for both loads, as follows:

$$\text{Motor load} = 65\text{ A} + 40\text{ A} + 14\text{ A} + 14\text{ A} + (0.25 \times 65\text{ A}) = 149\text{ A per phase}$$
$$\text{Lighting load} = 120\text{ A per phase} \times 1.25 = 150\text{ A}$$
$$\text{Total load} = 149 + 150 = 299\text{ A per phase leg}$$

Step 2. Conductors

Table 310-16 shows that a load of 299 A can be served by the following copper conductors:

500-kcmil TW
350-kcmil THW

Table 310-16 shows that this same load can be served by the following aluminum or copper-clad aluminum conductors:

700-kcmil TW
500-kcmil THW, RHH, or THHN

Step 3. Protective Devices

Section 430-63 says, in effect, that the protective device for a feeder supplying a combined motor load and lighting load may have a rating not greater than the sum of the maximum rating of the motor feeder protective device and the lighting load, as follows:

1. Motor feeder protective device = rating or setting of the largest branch-circuit device for any motor of the group being served plus the sum of the full-load currents of the other motors:

   200 A (50-hp motor) + 40 + 14 + 14 = 268 A maximum

   This calls for a maximum standard rating of 250 A for the motor feeder fuses, which is the nearest standard fuse rating that does not exceed the maximum permitted value of 268 A.
2. Lighting load = 120 A × 1.25 = 150 A

   Rating of CB for combined load = 268 + 150 = 418 A maximum

   This calls for a 400-A CB, the nearest standard rating that does not exceed the 418-A maximum.

Again: There is no provision in Sec. 430-63 which permits the use of "the next higher size, rating, or setting" of the protective device for a motor feeder when the calculated maximum rating does not correspond to a standard size of device.

Such considerations as voltage drop, I^2R loss, spare capacity, lamp dimming on motor starting, etc., would have to be made to arrive at actual sizes to use for the job. But, the circuiting as shown would be safe—although maybe not efficient or effective for the particular job requirements.

430-71. General. Figure 430-36 shows the "motor control circuit" part of a motor branch circuit, as defined in second paragraph of this section. A control circuit, as discussed here, is any circuit which has as its load device the oper-

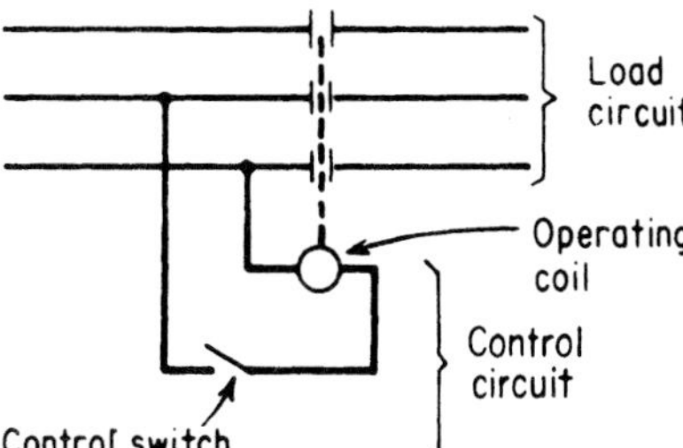

Fig. 430-36. A control circuit governs the operating coil that switches the load circuit. (Sec. 430-71.)

ating coil of a magnetic motor starter, a magnetic contactor, or a relay. Strictly speaking, it is a circuit which exercises control over one or more other circuits. And these other circuits controlled by the control circuit may themselves be control circuits or they may be "load" circuits—carrying utilization current to a lighting, heating, power, or signal device.

The elements of a control circuit include all the equipment and devices concerned with the function of the circuit: conductors, raceway, contactor operating coil, source of energy supply to the circuit, overcurrent protective devices, and all switching devices which govern energization of the operating coil.

The NE Code covers application of control circuits in Art. 725 and in Secs. 240-3 and 430-71 through 430-74. Design and installation of control circuits are basically divided into three classes (in Art. 725) according to the energy available in the circuit. Class 2 and 3 control circuits have low energy-handling capabilities; and any circuit, to qualify as a class 2 or 3 control circuit, must have its open-circuit voltage and overcurrent protection limited to conditions given in Sec. 725-31.

Most control circuits for magnetic starters and contactors could not qualify as class 2 or 3 circuits because of the relatively high energy required for operating coils. And any control circuit rated over 150 V (such as 220- or 440-V coil circuits) can never qualify, regardless of energy.

Class 1 control circuits include all operating coil circuits for magnetic starters which do not meet the requirements for class 2 or 3 circuits. Class 1 circuits must be wired in accordance with Secs. 725-21 to 725-29.

430-72. Overcurrent Protection. Part **(a)** tells the basic idea behind protection of the operating coil circuit of a magnetic motor starter, as distinguished from a manual (mechanically operated) starter:

1. Section 430-72 covers motor control circuits that are derived within a motor starter from the power circuit which connects to the line terminals of the starter. The rule here refers to such a control circuit as one "tapped from the load side" of the fuses or circuit breaker that provides branch-circuit protection for the conductors which supply the starter. See the top of Fig. 430-37.
2. The control circuit that is tapped from the line terminals within a starter is *not* a branch circuit itself.
3. Depending on other conditions set in Sec. 430-72, the conductors of the control circuit will be considered as protected by *either* the branch-circuit

protective device ahead of the starter or the supplementary protection (usually fuses) installed in the starter enclosure.

4. Any motor control circuit that is not tapped from the line terminals within a starter must be protected against overcurrent in accordance with Sec. 725-23. Such control circuits would be those that are derived from a panelboard or a control transformer—as where, say, 120-V circuits are derived external to the starters and are typically run to provide lower-voltage control for 230-, 460-, or 575-V motors. See the bottom of Fig. 430-37.

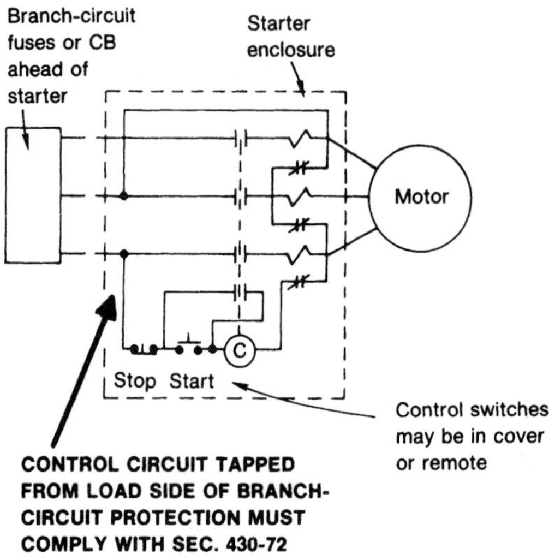
Branch-circuit fuses or CB ahead of starter
Starter enclosure
Motor
C
Stop Start
Control switches may be in cover or remote
CONTROL CIRCUIT TAPPED FROM LOAD SIDE OF BRANCH-CIRCUIT PROTECTION MUST COMPLY WITH SEC. 430-72

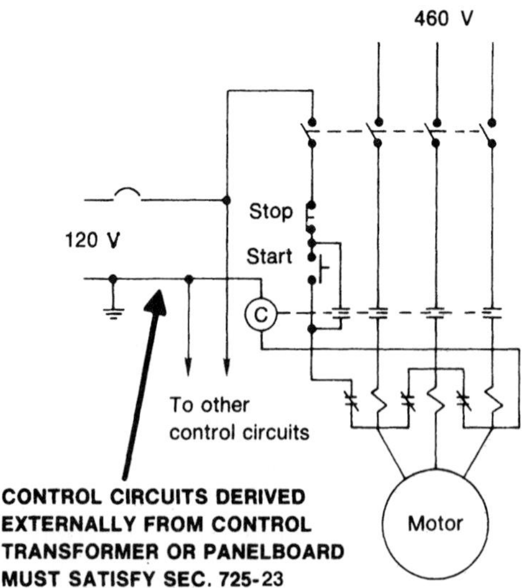
460 V
Stop
Start
120 V
C
To other control circuits
Motor
CONTROL CIRCUITS DERIVED EXTERNALLY FROM CONTROL TRANSFORMER OR PANELBOARD MUST SATISFY SEC. 725-23

Fig. 430-37. Source of power supply to the control circuit determines which Code section applies to the coil circuit. (Sec. 430-72.)

Part **(b)** applies to overcurrent protection of conductors used to make up the control circuits of magnetic motor starters. Such overcurrent protection must be sized in accordance with the amp values shown in Table 430-72(b). And where that table makes reference to amp values specified in Tables 310-16 through 310-19, as applicable, it does *not* specify that Note 8 of those tables must be observed by derating conductor ampacity where more than three current-carrying conductors are used in a conduit. Previously, the rule in part **(b)** of this section specifically recognized the use of control-circuit wires in raceway "without derating factors." Section 725-28, however, does require class 1 remote-control wires to have their ampacity derated in accordance with Note 8, based on the number of conductors, when the conductors "carry continuous loads" in excess of 10 percent of each conductor's ampacity. The application shown in Fig. 430-38 is, therefore, open to controversy.

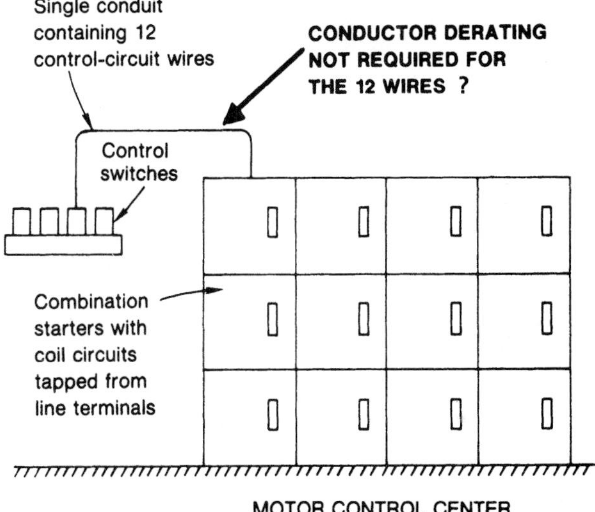

Fig. 430-38. Derating of control-wire ampacity is *not* specifically required when more than three conductors are run within the same raceway. (Sec. 430-72.)

The basic rule of part **(b)** requires coil-circuit conductors to have overcurrent protection rated in accordance with the maximum values given in column A of Table 430-72(b). That table shows 7 A as the maximum rating of protection for No. 18 copper wire and 10 A for No. 16 wire and refers to Table 310-16 for larger wires—15 A for No. 14 copper, 20 A for No. 12, etc. The Exceptions to the basic rule cover conditions under which other ratings of protection may be used, as follows:

Exception No. 1 covers protection of control wires for magnetic starters that have their START-STOP buttons in the cover of the starter enclosure.

In Exception No. 1, the value of branch-circuit protection must be compared to the ampacity of the control-circuit wires that are factory-installed in the starter and connected to the START-STOP buttons in the cover. If the rating of the branch-circuit fuse or CB does not exceed the value of the current shown in column B of Table 430-72(b) for the particular size of either copper or aluminum wire used to wire the coil circuit within the starter, then other protection is not required to be installed within the starter (Fig. 430-39). If the rating of branch-circuit protection *does* exceed the value shown in column B for the size of coil-circuit wire, then separate protection must be provided within the starter, and it must be rated not greater than the value shown for that size of wire in column A of Table 430-72(b). For instance, if the internal coil circuit of a starter is wired with No. 16 copper wire and the branch-circuit device supplying the starter is rated over the 40-A value shown for No. 16 copper wire in column B of Table 430-72(b), then protection must be provided in the starter for the No. 16 wire and the protective device(s) must be rated not over the 10-A value shown for No. 16 copper wire in column A of Table 430-72(b).

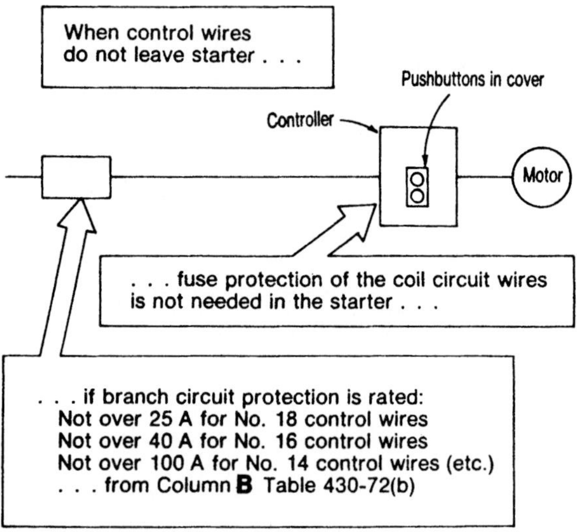

Fig. 430-39. This is the rule of Exception No. 1 to part **(b)** of Sec. 430-72.

Because most starters are the smaller ones using Nos. 18 and 16 wires for their coil circuits, Exception No. 1 and its reference to column B are particularly applicable to those wire sizes. For No. 16 control wires, branch-circuit protection rated up to 40 A would eliminate any need for a separate control-circuit fuse in the starter. And for No. 18 control wires, separate coil-circuit protection is not needed for a starter with branch-circuit protection rated not over 25 A. For Nos. 14, 12, and 10 copper control wires, maximum protective-device ratings are given in column B as 100, 120, and 160 A, respectively. For conductors larger than No. 10, the protection may be rated up to 400 percent of (or 4 times) the free-air ampacity of the size of conductor from Table 310-17.

Exception No. 2 covers protection of control wires that run from a starter to a remote-control device (pushbutton station, float switch, limit switch, etc.).

Such control wires may be protected by the branch-circuit protective device—without need for separate protection within the starter—if the branch-circuit device has a rating not over the value shown for the particular size of copper or aluminum control wire in column C of Table 430-72(b) (Fig. 430-40). Note that the maximum ratings of 7 A for No. 18 and 10 A for No. 16 require that *fuse* protection at those ratings must always be used to protect those sizes of control-circuit wires connected to motor starters supplied by CB branch-circuit protection, because 15 A is the lowest available standard rating of CB. But branch-circuit fuses of 7- or 10-A rating could eliminate the need for protection in the starter where No. 18 or No. 16 control wires are used. Figure 430-41 shows an application that was permitted for many years under previous wording of the Code rule but is now contrary to the letter and intent of the rule.

For any size of control wire, if the branch-circuit protection ahead of the starter has a rating greater than the value shown in column C of Table 430-72(b), then the control wire must be protected by a device(s) rated not over the amp value shown for that size of wire in column A of Table 430-72(b). For instance,

If branch-circuit protection ahead of starter is rated not over the ampere value shown in column C of Table 430-72(b) for the size of control-circuit wire used . . .
P
P
Motor
Coil
Controller enclosure
Remote-control device
Stop
Start
. . . then separate protection is *not* required within the starter to protect the control wires.
BUT, if the rating of branch-circuit protection exceeds the value in Column C for the size of control wire used, separate protective devices rated at the ampere value shown in Column A of Table 430-72(b) for the size of control wire used must be installed within the starter at points "P" to protect each ungrounded control wire (Sec. 240-20).

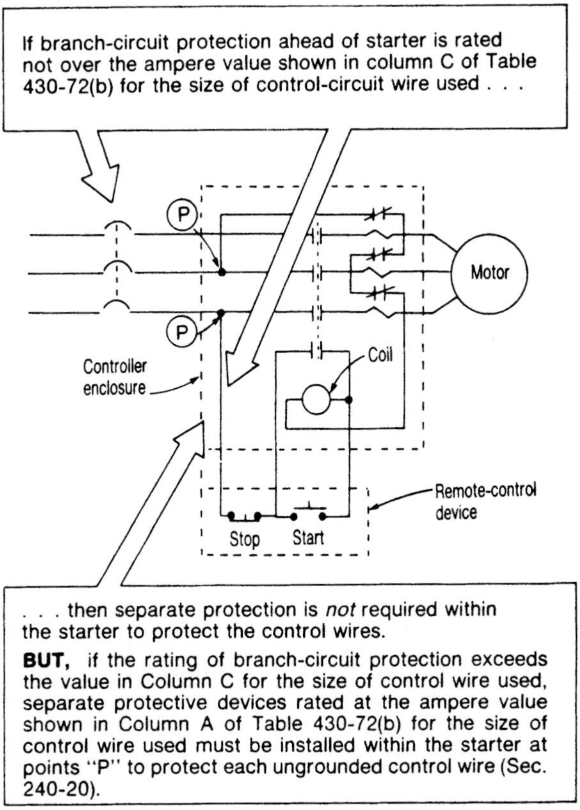

Fig. 430-40. This is covered by Exception No. 2 to part **(b)** of Sec. 430-72.

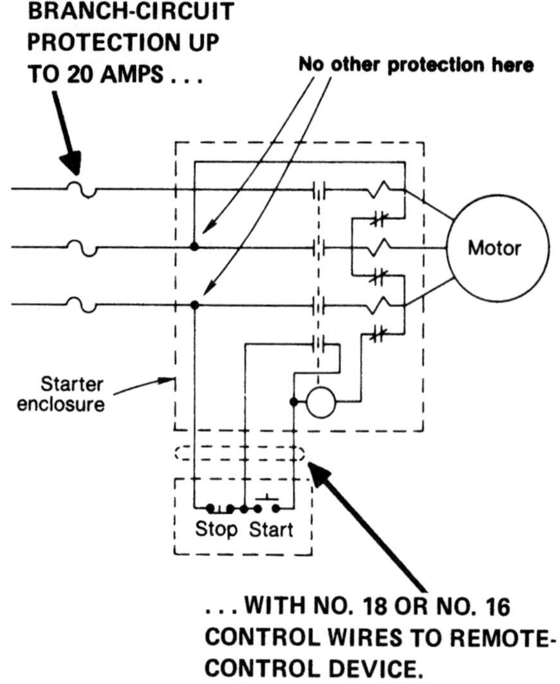

Fig. 430-41. This was permitted by previous NEC editions but is now a violation of Exception No. 2 of part **(b).** (Sec. 430-72.)

if No. 14 copper wire is used for the control circuit from a starter to a remote pushbutton station and the branch-circuit protection ahead of the starter is rated at 40 A, then the branch-circuit device is not over the value of 45 A shown in column C, and separate control protection is not required within the starter. But if the branch-circuit protection were, say, 100 A, then No. 14 control wire would have to be protected at 15 A because column A shows that No. 14 must have maximum protection rating from Note 1—which refers to Table 310-16 where No. 14 wire in conduit is shown, by the footnote, to require protection at 15 A.

It should be noted that column A gives the values to be used for overcurrent protection placed within the starter to protect control-circuit wires in any case where the rating of branch-circuit protection exceeds the value shown in either column B (for starters with no external control wires) or column C (for control wires run from a starter to a remote pilot control device).

Exception No. 3 permits protection on the primary side of a control transformer to protect the transformer in accordance with Sec. 450-3 and the secondary conductors in accordance with the amp value shown in Table 430-72(b) for the particular size of the control wires fed by the secondary. This use is limited to transformers with 2-wire secondaries (Fig. 430-42). Because Sec. 430-72(a) notes that the rules of Sec. 430-72 apply to control circuits tapped from the motor branch circuit, the rule of Exception No. 3 must be taken as applying

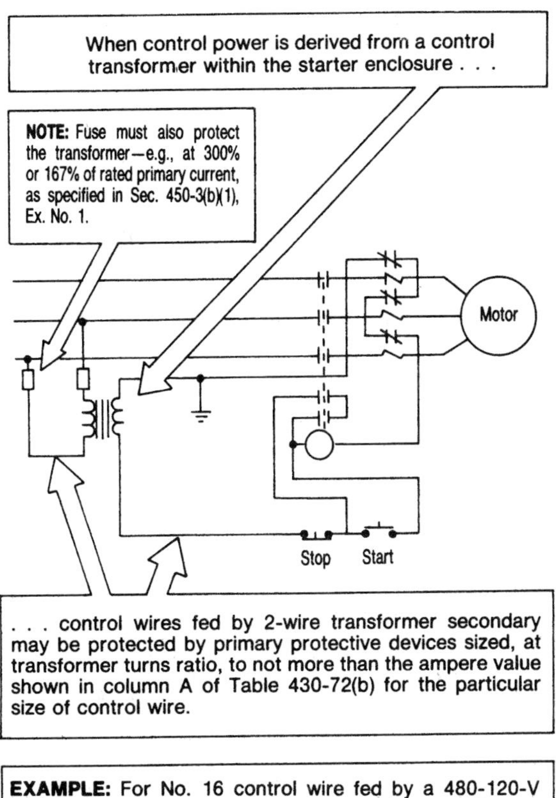
When control power is derived from a control transformer within the starter enclosure . . .
NOTE: Fuse must also protect the transformer—e.g., at 300% or 167% of rated primary current, as specified in Sec. 450-3(b)(1), Ex. No. 1.
Motor
Stop
Start
. . . control wires fed by 2-wire transformer secondary may be protected by primary protective devices sized, at transformer turns ratio, to not more than the ampere value shown in column A of Table 430-72(b) for the particular size of control wire.
EXAMPLE: For No. 16 control wire fed by a 480-120-V transformer, fuses must not be rated over 10 A × (120/480), or 2.5 A.

Fig. 430-42. Exception No. 3 to part **(b)** of Sec. 430-72 permits the secondary wires of the coil circuit to be protected by primary-side overcurrent protection. (Sec. 430-72.)

to a control transformer installed within the starter enclosure—although the general application may be used for any transformer because it conforms to Sec. 240-3, Exception No. 5, and to Sec. 450-3.

Exception No. 4 eliminates any need for control-circuit protection where opening of the circuit would be objectionable, as for a fire-pump motor or other essential or safety-related operation.

Part **(c)** covers the use of control transformers and requires protection on the primary side. And, again, it must be taken to apply specifically to such transformers used in motor control equipment enclosures. The basic rule calls for each control transformer to be protected in accordance with Sec. 450-3 (usually by a primary-side protective device rated not over 125 or 167 percent of primary current), as shown in Fig. 430-42. But, exceptions are given.

Exception No. 1 eliminates any need for protection of any control transformer rated less than 50 VA, provided it is part of the starter and within its enclosure.

Exception No. 2 permits a control transformer with a rated primary current of *less* than 2 A to be protected at up to 500 percent of rated primary current by a protective device in each ungrounded conductor of the supply circuit to the transformer primary, as shown in Fig. 430-43.

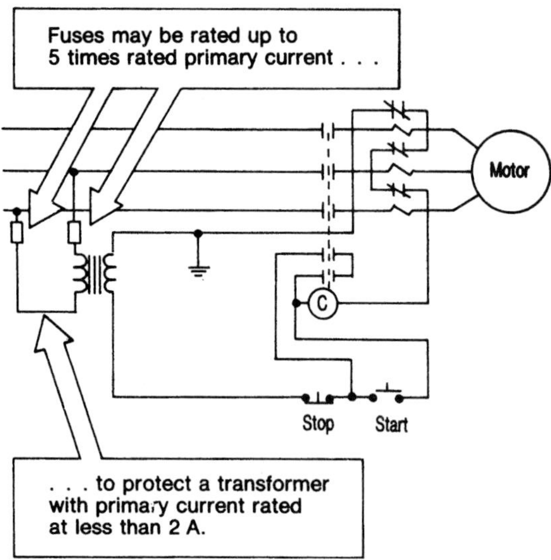

Fig. 430-43. A control circuit fed by a transformer within the starter enclosure may have overcurrent protection in the primary rated up to 500 percent of the rated primary current of a small transformer. (Sec. 430-72.)

In the majority of magnetic motor controllers and contactors, the voltage of the operating coil is the voltage provided between two of the conductors supplying the load, or one conductor and the neutral. Conventional starters are factory wired with coils of the same voltage rating as the phase voltage to the motor. However, there are many cases in which it is desirable or necessary to use control circuits and devices of lower voltage rating than the motor. Such could be the case with high-voltage (over 600 V) controllers, for instance, in which it is necessary to provide a source of low voltage for practical operation of magnetic coils. And even in many cases of motor controllers and contactors for use under 600 V, safety requirements dictate the use of control circuits of lower voltage than the load circuit.

Although contactor coils and pilot devices are available and effectively used for motor controllers with up to 600-V control circuits, such practice has been prohibited in applications where atmospheric and other working conditions make it dangerous for operating personnel to use control circuits of such voltage. And certain OSHA regulations require 120- or 240-V coil circuits for the 460-V motors. In such cases, control transformers are used to step the voltage down to permit the use of lower-voltage coil circuits.

430-73. Mechanical Protection of Conductor. The condition under which physical protection of the control circuit conductor becomes necessary is where damage to the conductors would constitute either a fire or an accident hazard. Damage to the control circuit conductors resulting in short-circuiting two or more of the conductors or breaking one of the conductors would either cause the device to operate or render it inoperative, and in some cases either condition would constitute a hazard either to persons or to property; hence, in such cases the conductors should be installed in rigid or other metal conduit. On the other hand, damage to the conductors of the low-voltage control circuit of a domestic oil burner or automatic stoker does not constitute a hazard, because the boiler or furnace is equipped with an automatic safety control.

The second paragraph of this section focuses on the hazard of accidental starting of a motor. Figure 430-44 shows an example of a control circuit installation that should be carefully designed and is required by the second sentence of Sec. 430-73 to be observed for any control circuit which has one leg grounded. Whenever the coil is fed from a circuit made up of a hot conductor and a grounded conductor (as when the coil is fed from a panelboard or separate control transformer, instead of from the supply conductors to the motor), care must be taken to place the push-button station or other switching control device in the hot leg to the coil and not in the grounded leg to the coil. By switching in the hot leg, the starting of the motor by accidental ground fault can be effectively eliminated.

Combinations of ground faults can develop to short the pilot starting device—push-button, limit switch, pressure switch, etc.—accidentally starting the motor even though the pilot device is in the OFF position. And because many remote-control circuits are long, possible faults have many points at which they might occur. Insulation breakdowns, contact shorts due to accumu-

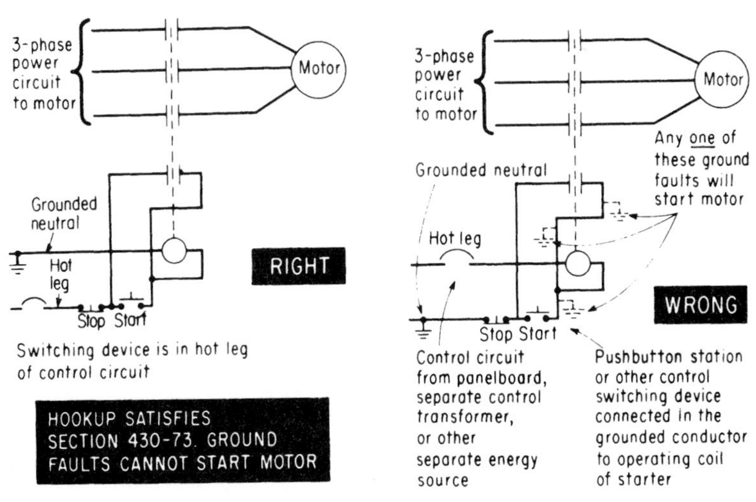

Fig. 430-44. Control hookup must prevent accidental starting. (Sec. 430-73.)

lation of foreign matter or moisture, and grounds to conduit are common fault conditions responsible for accidental operation of motor controllers.

Although not specifically covered by Code rules, there are many types of ground-fault conditions that affect motor starting and should be avoided.

As shown in Fig. 430-45, any magnetic motor controller used on a 3-phase, 3-wire ungrounded system always presents the possibility of accidental starting of the motor. If, for instance, an undetected ground fault exists on one phase of the 3-phase system—even if this system ground fault is a long distance from the controller—a second ground fault in the remote-control circuit for the operating coil of the starter can start the motor.

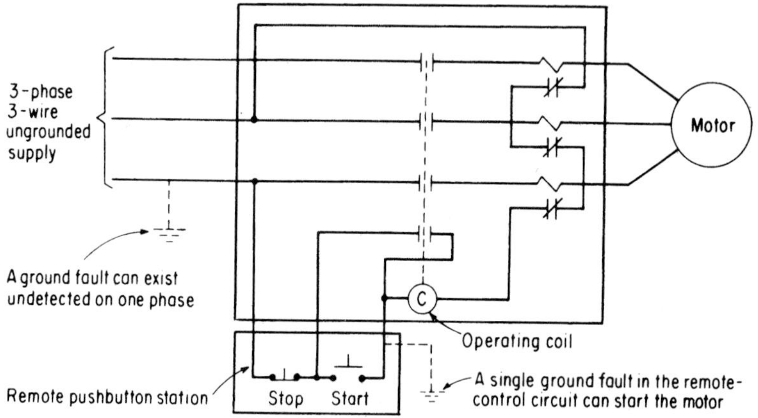

Fig. 430-45. Accidental motor starting can be hazardous and contrary to Code rule. (Sec. 430-73.)

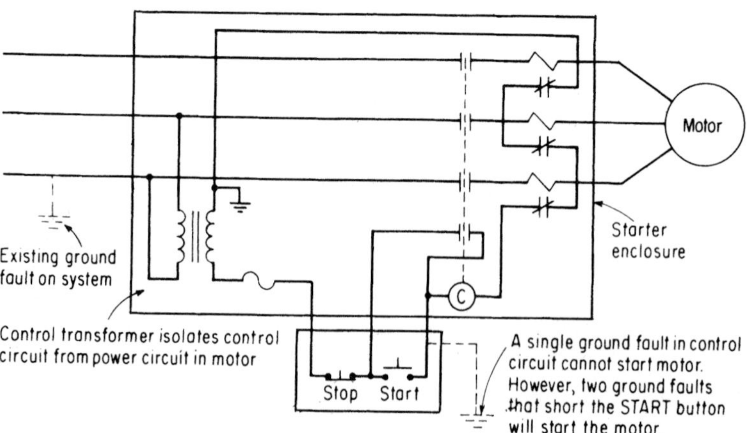

Fig. 430-46. Control transformer can isolate control circuit from accidental starting. (Sec. 430-73.)

Figure 430-46 shows the use of a control transformer to isolate the control circuit from responding to the combination of ground faults shown in Fig. 430-45. This transformer may be a one-to-one isolating transformer, with the same primary and secondary voltage, or the transformer can step the motor circuit voltage down to a lower level for the control circuit.

In the hookup shown in Fig. 430-47, a 2-pole START button is used in conjunction with two sets of holding contacts in the motor starter. This hookup protects against accidental starting of the motor under the fault conditions shown in Fig. 430-45. The hookup also protects against accidental starting due to two ground faults in the control circuit simply shorting out the START button and energizing the operating coil. This could happen in the circuit of Fig. 430-45 or the circuit of Fig. 430-46.

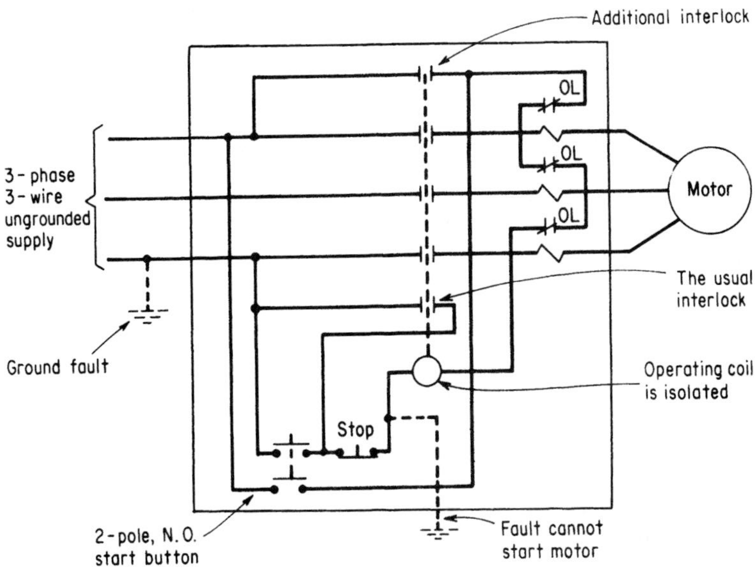

Fig. 430-47. Use of 2-pole start button can prevent accidental starting. (Sec. 430-73.)

Another type of motor control circuit fault can produce a current path through the coil of a closed contactor to hold it closed regardless of the operation of the pilot device for opening the coil circuit. Again this can be done by a combination of ground faults which short the STOP device. Failure to open can do serious damage to motors in some applications and can be a hazard to personnel. The operating characteristics of contactor coils contribute to the possible failure of a controller to respond to the opening of the STOP contacts. It takes about 85 percent of rated coil voltage to operate the armature associated with the coil; but it takes only about 50 percent of the rated value to enable the coil to hold the contactor closed once it is closed. Under such conditions, even partial grounds and shorts on control contact assemblies can produce paths for sufficient current flow to cause shorting of the stop position of pilot devices. And faults can short-out running overload relays, eliminating overcurrent protection of the motor, its associated control equipment, and conductors.

Figure 430-48 is a modification of the circuit of Fig. 430-47, using a 2-pole START button and a 2-pole STOP button—protecting against both accidental starting and accidental failure to stop when the STOP button is pressed. Both effects of ground faults are eliminated.

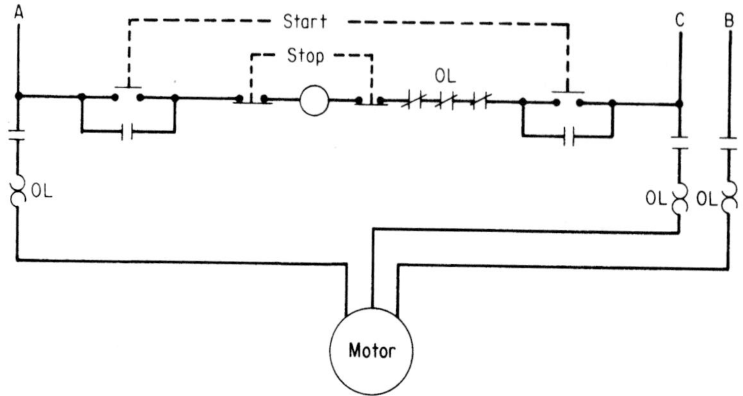

Fig. 430-48. This circuit prevents accidental starting and assures stopping. (Sec. 430-73.)

430-74. Disconnection. The control circuit of a remote-control motor controller shall always be so connected that it will be cut off when the disconnecting means is opened, unless a separate disconnecting means is provided for the control circuit.

When the control circuit of a motor starter is tapped from the line terminals of the starter—in which case it is fed at line-to-line voltage of the circuit to the motor itself—opening of the required disconnect means ahead of the starter de-energizes the control circuit from its source of supply, as shown in Fig. 430-49. But, where voltage supply to the coil circuit is derived from outside the starter enclosure (as from a panelboard or from a separate control transformer), provision must be made to assure that the control circuit is capable of being de-energized to permit safe maintenance of the starter. In such cases, the required power-circuit disconnect ahead of the starter can open the power circuit to the starter's line terminals; but, unless some provision is made to open the externally derived control circuit voltage supply, a maintenance worker could be exposed to the unexpected shock hazard of the energized control circuit within the starter.

The disconnect for control voltage supply could be an extra pole or auxiliary contact in the switch or CB used as the main power disconnect ahead of the starter, as shown in Fig. 430-50. Or the control disconnect could be a separate switch (like a toggle switch), provided this separate switch is installed "immediately adjacent" to the power disconnect—so it is clear to maintenance people that *both* disconnects must be opened to kill *all* energized circuits within

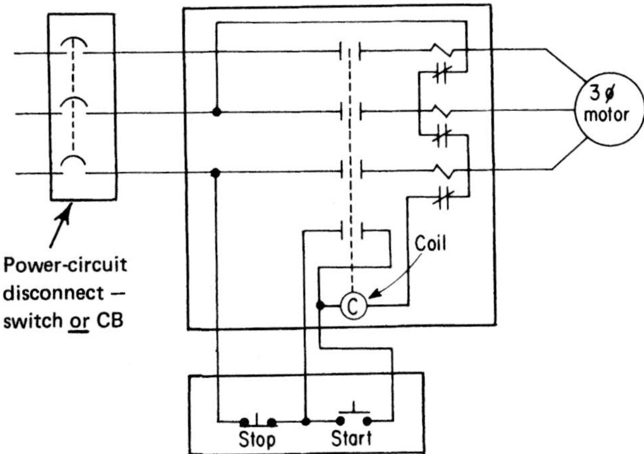

Fig. 430-49. Disconnect ahead of starter opens supply to line-voltage coil circuit. (Sec. 430-74.)

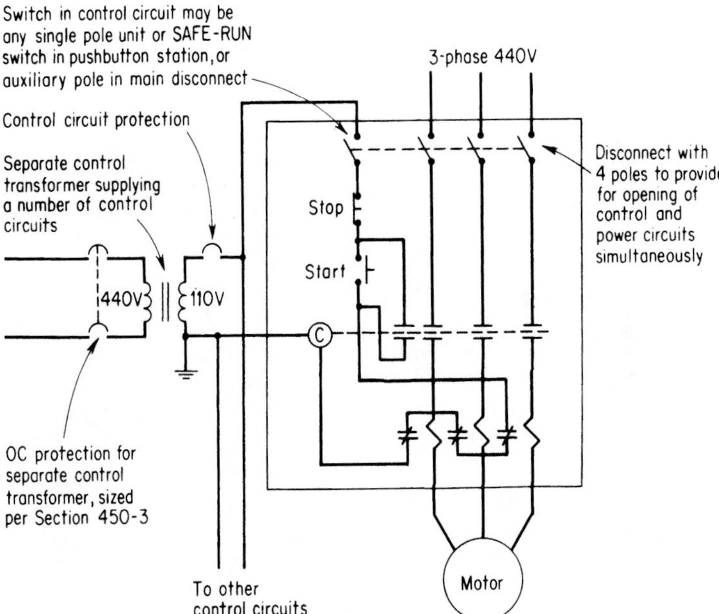

Fig. 430-50. Control disconnect means must supplement power-circuit disconnect. (Sec. 430-74.)

the starter. Control circuits operating contactor coils, etc., within controllers present a shock hazard if they are allowed to remain energized when the disconnect is in the OFF position. Therefore, the control circuit either must be designed in such a way that it is disconnected from the source of supply by the controller disconnecting means or must be equipped with a separate disconnect immediately adjacent to the controller disconnect for opening of both disconnects. [For grounding of the control transformer secondary in Fig. 430-50, refer to Sec. 250-5(b), Exception No. 3.]

Exception No. 1 of part **(a)** is aimed at industrial-type motor control hookups which involve extensive interlocking of control circuits for multimotor process operations or machine sequences. In recognition of the unusual and complex control conditions that exist in many industrial applications—particularly process industries and manufacturing facilities—Exception No. 1 alters the basic rule that disconnecting means for control circuits must be located "immediately adjacent one to each other" (Fig. 430-51). When a piece of motor control equipment has more than 12 motor control conductors associated with it, remote locating of the disconnect means is permitted under the conditions given in Exception No. 1. As shown in Fig. 430-52, this permission is applicable only where qualified persons have access to the live parts and sufficient warning signs are used on the equipment to locate and identify the various disconnects associated with the control circuit conductors.

Fig. 430-51. Industrial control layouts with more than 12 control circuit conductors for interlocking of controllers and operating stations (arrow) do not require control disconnects to be "immediately adjacent" to power disconnects. (Sec. 430-74.)

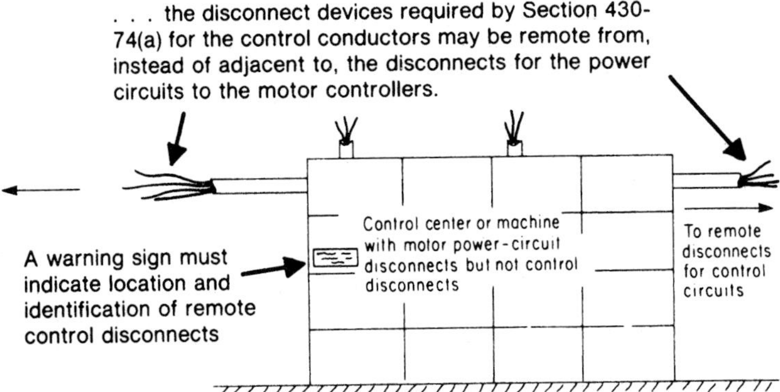
Where an assembly of motor control equipment or a machine or process layout has **more than 12** control conductors coming into it and requiring disconnect means . . .
. . . the disconnect devices required by Section 430-74(a) for the control conductors may be remote from, instead of adjacent to, the disconnects for the power circuits to the motor controllers.
Control center or machine with motor power-circuit disconnects but not control disconnects
To remote disconnects for control circuits
A warning sign must indicate location and identification of remote control disconnects

Fig. 430-52. For extensively interlocked control circuits, control disconnects do not have to be adjacent to power disconnects. (Sec. 430-74.)

Exception No. 2 presents another instance in which control circuit disconnects may be mounted other than immediately adjacent to each other. It notes that where the opening of one or more motor control circuit disconnects might result in hazard to personnel or property, remote mounting may be used where the conditions specified in Exception No. 1 exist, i.e., that access is limited to qualified persons and that a warning sign is located on the outside of the equipment to indicate the location and the identification of each remote control circuit disconnect.

The requirement of part **(b)** of this section is shown in Fig. 430-53. When a control transformer is in the starter enclosure, the power disconnect means is on the line side and can de-energize the transformer control circuit. Grounding of the control circuit is not always necessary, as noted in Exception No. 3 of Sec. 250-5(b). Overcurrent protection must be provided for the control circuit when a control circuit transformer is used, as covered in Sec. 430-72(b). Such protection may be on the primary or secondary side of the transformer, as described. In Sec. 450-1, Exception No. 2 notes that the rules of Art. 450 do not apply to "dry-type transformers that constitute a component part of other apparatus. . . ." A control transformer supplied as a factory-installed component in a starter would therefore be exempt from the rules of Sec. 450-3(b), covering overcurrent protection for transformers, but would have to comply with Sec. 430-72(c).

430-81. General. As used in Art. 430, the term "controller" includes any switch or device normally used to start and stop a motor, in addition to motor starters and controllers as such. As noted, the branch-circuit fuse or CBs are considered an acceptable control device for stationary motors not over ⅛ hp where the motor has sufficient winding impedance to prevent damage to the motor with its rotor continuously at standstill. And a plug and receptacle connection may serve as the controller for portable motors up to ⅓ hp.

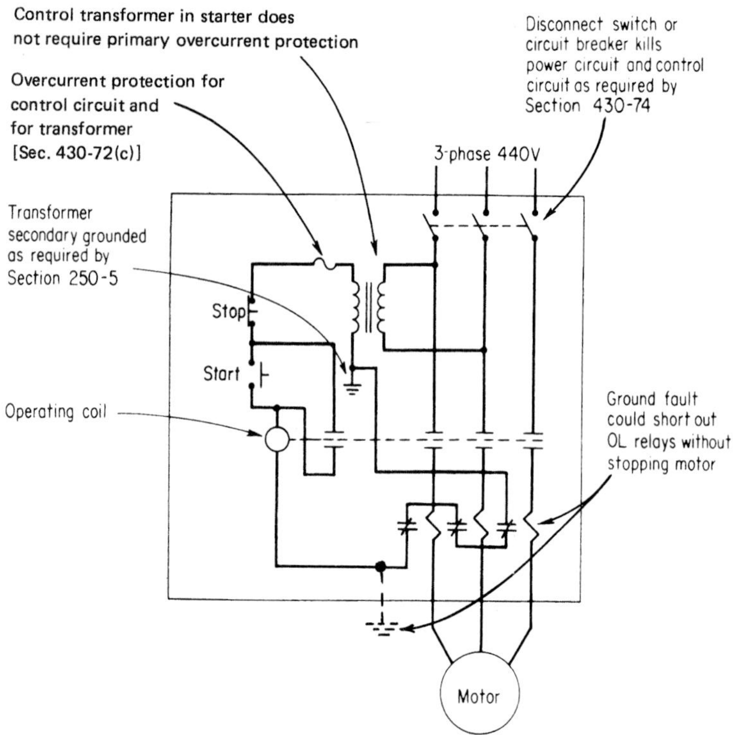
Control transformer in starter does not require primary overcurrent protection
Overcurrent protection for control circuit and for transformer [Sec. 430-72(c)]
Disconnect switch or circuit breaker kills power circuit and control circuit as required by Section 430-74
3-phase 440V
Transformer secondary grounded as required by Section 250-5
Stop
Start
Operating coil
Ground fault could short out OL relays without stopping motor
Motor

Fig. 430-53. Control transformer in starter must be on load side of disconnect. (Sec. 430-74.)

As described in the definition here, a "controller" is a device that starts and stops a motor by "making and breaking the motor circuit current"—that is, the power current flow to the motor windings. A pushbutton station, a limit switch, a float switch, or any other pilot control device that "carries the electric signals directing the performance of the controller" (see the definition of "Motor Control Circuit" in Sec. 430-71) is not the controller where such a device is used to carry only the current to the operating coil of a magnetic motor controller. For purposes of **Code** application, the contactor mechanism is the motor "controller."

430-82. Controller Design. Every controller must be capable of starting and stopping the motor which it controls, must be able to interrupt the stalled-rotor current of the motor, and must have a horsepower rating not lower than the rating of the motor, except as permitted by Sec. 430-83.

430-83. Ratings. (a) Horsepower Rating at the Application Voltage. Figure 430-54 shows the basic requirements on rating of a controller. Although the basic rule calls for a horsepower-rated switch or a horsepower-rated motor starter, there are exceptions as noted in Sec. 430-81 and as follows:

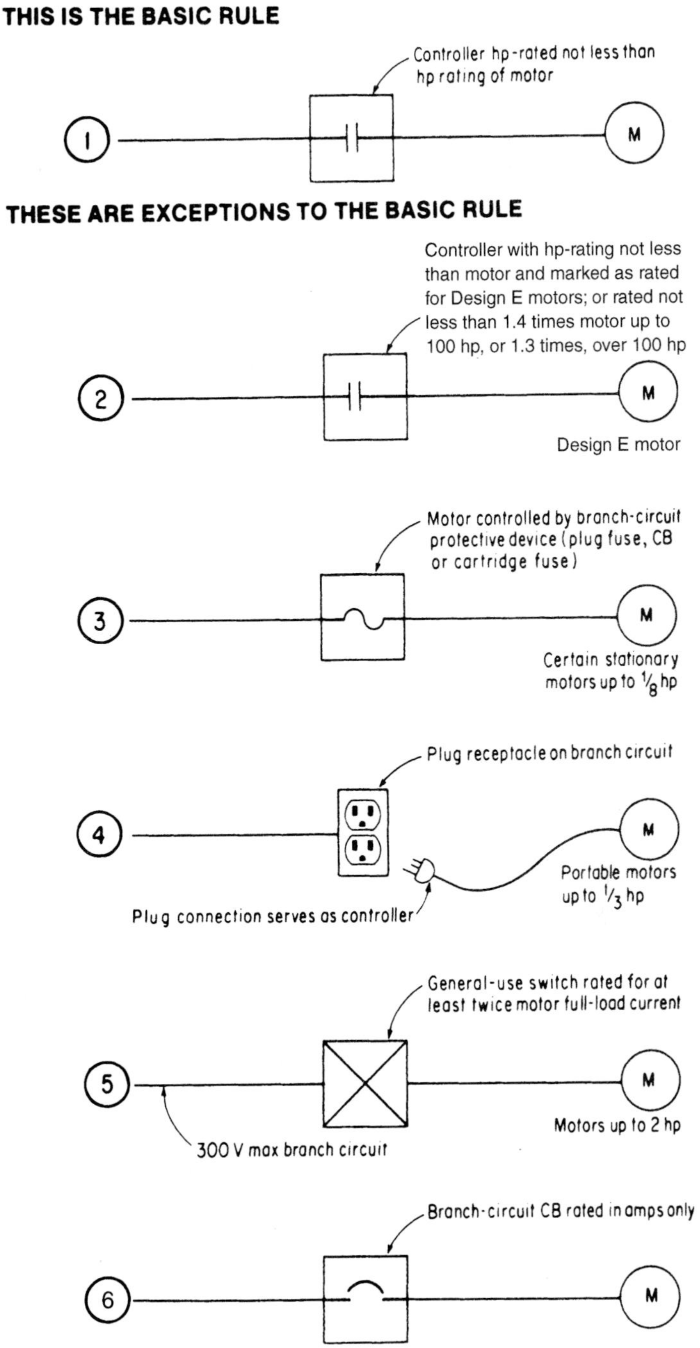

THIS IS THE BASIC RULE
Controller hp-rated not less than hp rating of motor
1
M
THESE ARE EXCEPTIONS TO THE BASIC RULE
Controller with hp-rating not less than motor and marked as rated for Design E motors; or rated not less than 1.4 times motor up to 100 hp, or 1.3 times, over 100 hp
2
M
Design E motor
Motor controlled by branch-circuit protective device (plug fuse, CB or cartridge fuse)
3
M
Certain stationary motors up to 1/8 hp
Plug receptacle on branch circuit
4
M
Portable motors up to 1/3 hp
Plug connection serves as controller
General-use switch rated for at least twice motor full-load current
5
M
Motors up to 2 hp
300 V max branch circuit
Branch-circuit CB rated in amps only
6
M

Fig. 430-54. Controller must be a horsepower-rated switch or CB—but other devices may satisfy. (Sec. 430-83.)

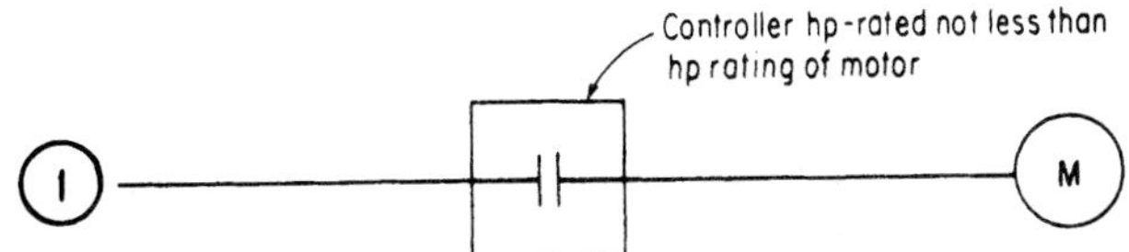

- The wording of Exception No. 1 calls for controllers supplying the high-efficiency Design E motors to be marked—by the manufacturer—as suitable for use with Design E motors and to have an hp rating at least equal to that of the motor. Alternately, it would be acceptable to use a standard controller with a rating at least 1.4 times the horsepower rating of the motor, up to 100 hp, and 1.3 times the horsepower rating of motors over 100 hp.
- A general-use switch rated at not less than twice the full-load motor current may be used as the controller for stationary motors up to 2 hp, rated 300 V or less. On AC circuits, a general-use snap switch suitable only for use on AC may be used to control a motor having a full-load current rating not over 80 percent of the ampere rating of the switch.
- A branch-circuit CB, rated in amperes only, may be used as a controller. If the same CB is used as controller and to provide overload protection for the motor circuit, it must be rated in accordance with Sec. 430-32.

In the UL's *Electrical Construction Materials Directory,* data are presented on use of switches in motor circuits, as follows:

1. Enclosed switches with horsepower ratings in addition to current ratings may be used for motor circuits as well as for general-purpose circuits. Enclosed switches with ampere-only ratings are intended for general use but may also be used for motor circuits (as controllers and/or disconnects) as permitted by NE Code Sec. 430-83 (Exception No. 1), Sec. 430-109 (Exceptions No. 2, 3, and 4), and Sec. 430-111.
2. A switch that is marked "MOTOR CIRCUIT SWITCH" is intended for use *only* in motor circuits.
3. For switches with dual-horsepower ratings, the higher horsepower rating is based on the use of time-delay fuses in the switch fuseholders to hold in on the inrush current of the higher-horsepower-rated motor.
4. Although Sec. 430-83 permits use of horsepower-rated switches as controllers and UL lists horsepower-rated switches up to 500 hp, UL does state in its Green Book that "enclosed switches rated higher than 100 hp are restricted to use as motor disconnect means and are not for use as motor controllers." But a horsepower-rated switch up to 100 hp may be used as both a controller and disconnect if it breaks all ungrounded legs to the motor, as covered in Sec. 430-111.

Figure 430-55 covers two of those points.

For selection of a controller for a sealed (hermetic-type) refrigeration compressor motor, refer to Sec. 440-41.

430-84. Need Not Open All Conductors. It is interesting to note that the NE Code says that a controller need not open all conductors to a motor, except when the controller serves also as the required disconnecting means. For instance, a 2-pole starter of correct horsepower rating could be used for a 3-phase motor if running overload protection is provided in all three circuit legs by devices separate from the starter, such as by dual-element, time-delay fuses which are sized to provide running overload protection as well as short-circuit protection for the motor branch circuit. The controller must interrupt only enough conductors to be able to start and stop the motor.

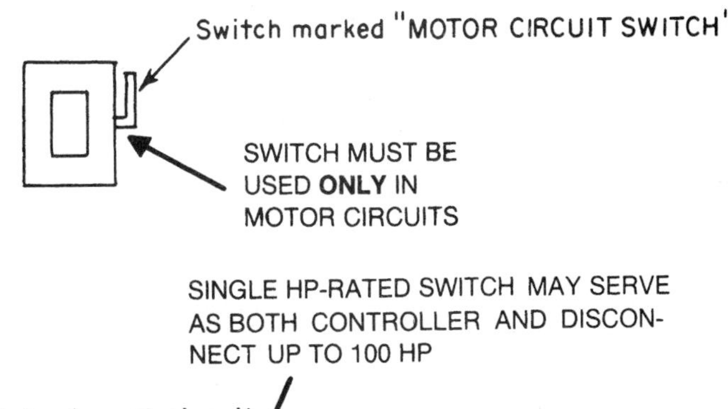
Switch marked "MOTOR CIRCUIT SWITCH"
SWITCH MUST BE USED ONLY IN MOTOR CIRCUITS
SINGLE HP-RATED SWITCH MAY SERVE AS BOTH CONTROLLER AND DISCONNECT UP TO 100 HP

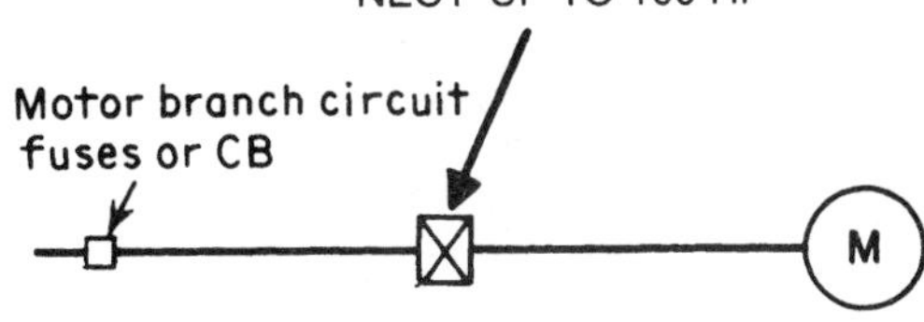
Motor branch circuit fuses or CB
M

Fig. 430-55. UL rules limit Code applications. (Sec. 430-83.)

However, when the controller is a manual (nonmagnetic) starter or is a manually operated switch or CB (as permitted by the Code), the controller itself also may serve as the disconnect means if it opens all ungrounded conductors to the motor, as covered in Sec. 430-111. This eliminates the need for another switch or CB to serve as the disconnecting means. But, it should be noted that only a manually operated switch or CB may serve such a dual function. A magnetic starter cannot also serve as the disconnecting means even if it does open all ungrounded conductors to the motor.

Figure 430-56 shows typical applications in which the controller does not have to open all conductors but a separate disconnect switch or CB is required ahead of the controller. In the drawing, the word *ungrounded* refers to the condition that none of the circuit conductors is grounded. These may be the ungrounded conductors of grounded systems.

Generally, one conductor of a 115-V circuit is grounded, and on such a circuit a single-pole controller may be used connected in the ungrounded conductor, or a 2-pole controller is permitted if both poles are opened together. In a 230-V circuit there is usually no grounded conductor, but if one conductor is grounded, Sec. 430-85 permits a 2-pole controller.

430-85. In Grounded Conductors. This rule permits a 3-pole switch, CB, or motor starter to be used in a 3-phase motor circuit derived from a 3-phase, 3-wire, corner-grounded delta system—with the grounded phase leg switched along with the hot legs, as in Sec. 430-36.

430-87. Number of Motors Served by Each Controller. Generally, an individual motor controller is required for each motor. However, for motors rated not over 600 V, a single controller rated at not less than the sum of the horsepower ratings of all the motors of the group may be used with a group of motors if any one of the conditions specified is met. Where a single controller is used for

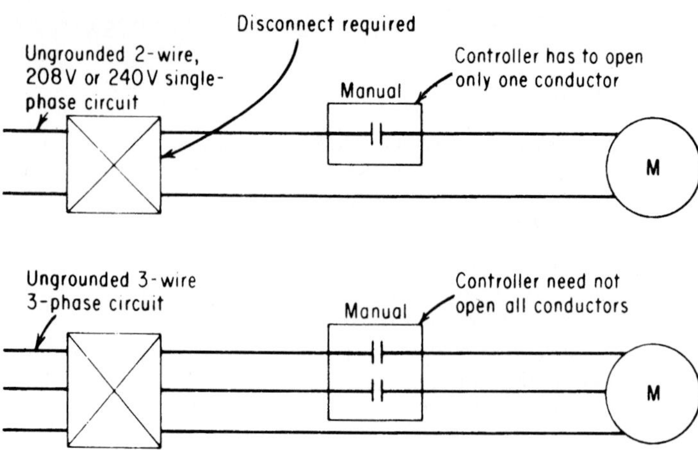

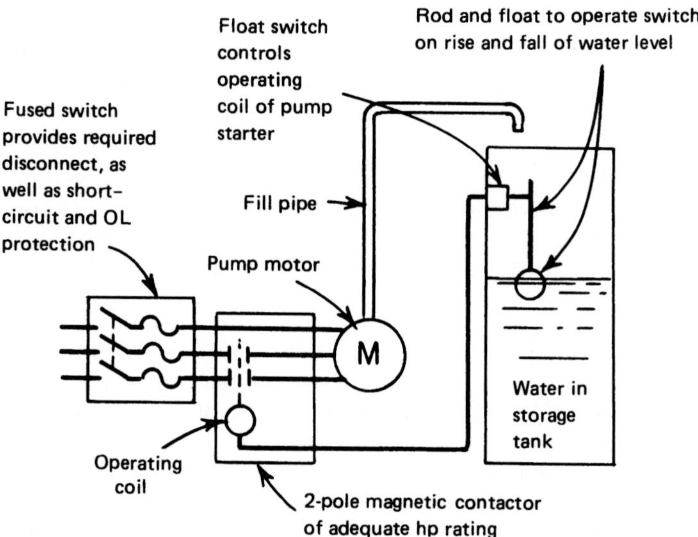

Fig. 430-56. "Controller" does not have to break *all* legs of motor supply circuit. (Sec. 430-84.)

more than one motor connected on a single branch circuit as permitted under condition *b,* it should be noted that the reference is to part (**a**) of Sec. 430-53. That use of a single controller applies only to cases involving motors of 1 hp or less and does not apply for several motors used on a single branch circuit in accordance with parts (**b**) and (**c**) of Sec. 430-53—unless the several motors satisfy conditions *a* or *c* of this section.

See Sec. 430-112, where the same conditions are set for a single disconnect means to serve a group of motors.

430-88. Adjustable-Speed Motors. Field weakening is quite commonly used as a method of controlling the speed of DC motors. If such a motor were started under a weakened field, the starting current would be excessive unless the motor is specially designed for starting in this manner.

430-89. Speed Limitation. A common example of a separately excited DC motor is found in a typical speed control system that is widely used for electric elevators, hoists, and other applications where smooth control of speed from standstill to full speed is necessary. In Fig. 430-57, G_1 and G_2 are two generators having their armatures mounted on a shaft which is driven by a motor, not shown in the diagram. M is a motor driving the elevator drum or other machine. The fields of generator G_1 and motor M are excited by G_1. By adjusting the rheostat R, the voltage generated by G_2 is varied, and this in turn varies the speed of motor M. It is evident that if the field circuit of motor M should be accidentally opened while the motor is lightly loaded, the motor would reach an excessive speed. In many applications of this system the motor is always loaded and no speed-limiting device is required.

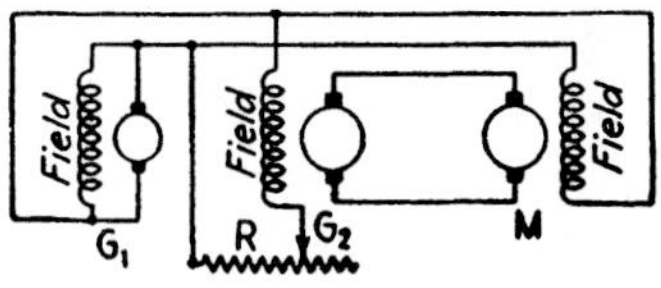

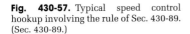

Fig. 430-57. Typical speed control hookup involving the rule of Sec. 430-89. (Sec. 430-89.)

The speed of a series motor depends upon its load and will become excessive at no load or very light loads. Traction motors are commonly series motors, but such a motor is geared to the drive wheels of the car or locomotive and hence is always loaded.

Where a motor generator, consisting of a motor driving a compound-wound DC generator, is operated in parallel with a similar machine or is used to charge a storage battery, if the motor circuit is accidentally opened while the generator is still connected to the DC buses or battery, the generator will be driven as a motor and its speed may become dangerously high. A synchronous converter operating under similar conditions may also reach an excessive speed if the AC supply is accidentally cut off.

A safeguard against overspeed is provided by a centrifugal device on the shaft of the machine, arranged to close (or open) a contact at a predetermined speed, thus tripping a CB which cuts the machine off from the current supply.

430-90. Combination Fuseholder and Switch as Controller. The use of a fusible switch as a motor controller with fuses as motor-running protective devices is practicable when time-delay types of fuses are used. The rating of the fuses must not exceed 125 percent, or in some cases 115 percent, of the full-load motor current, and nontime-delay fuses of this rating would, in most cases, be blown by the starting current drawn by the motor, particularly where the motor turns on and off frequently. (See Sec. 430-35.)

It may be found that a switch having the required horsepower rating is not provided with fuse terminals of the size required to accommodate the branch-circuit fuses. For example, assume a 7½-hp 230-V 3-phase motor started at full-line voltage. A switch used as the disconnecting means for this motor must be rated at not less than 7½ hp, but this would probably be a 60-A switch and therefore, if fusible, would be equipped with terminals to receive 35- to 60-A fuses. Section 430-90 provides that fuse terminals must be installed that will receive fuses of 70-A rating. In such case a switch of the next higher rating must be provided, unless time-delay fuses are used.

430-91. Motor Controller Enclosure Types. This section and table cover selection criteria—but no mandatory rule—on types of motor controller enclosures. This section gives selection data, with characteristics tabulated, for application of the various NEMA types of motor controller enclosures for use in specific non-hazardous locations. It can be argued that the data in this section are made mandatory by the general rules of Secs. 110-3 and 110-11—both of which require equipment to be suitable for its environment.

UL and NEMA have developed a new Type 5 motor controller that is shown in NEC Table 430-91. This is an enclosure for indoor use only and protects against settling airborne dust, falling dirt, and dripping noncorrosive liquids. This type of enclosure is suited to use for motor control centers in industrial environments.

430-102. Location. Along with Sec. 430-101, this section specifically requires that a disconnecting means—basically, a motor-circuit switch rated in horsepower, or a CB—be provided in each motor circuit. Figure 430-58 shows the basic rule on "in-sight" location of the disconnect means. This applies always for all motor circuits rated up to 600 V—even if an "out-of-sight" disconnect can be locked in the open position.

Because the basic rule here requires a disconnecting means to be within sight from the "controller location," the question arises, Is the magnetic contactor the controller or is the pushbutton station the controller? The NEC makes clear that the contactor of a magnetic motor starter *is* the controller for the motor, *not* the pushbuttons that actuate the coil of the contactor. The NEC establishes that identification by the definition of "controller" in Art. 100 and by the definition of a "motor control circuit" in Sec. 430-71, as follows:

> *Controller:* A device or group of devices that serves to govern, in some predetermined manner, the electric power delivered to the apparatus to which it is connected.
>
> *Motor control circuit:* The circuit of a control apparatus or system that carries the electric signals directing the performance of the controller, but does not carry the main power current.

In a magnetic motor starter hookup, it is the contactor that actually governs the electric power delivered to the motor to which it is connected. The motor connects to the contactor and *not* to the pushbuttons, which are in the control circuit that carries the electric signals directing the performance of *the controller* (that is, the contactor). The pushbuttons do *not* carry "the main power current," which is "delivered" to the motor by the contactor and which is, therefore, "the controller." It is well established that the intent of the Code rule, as well as the letter of the rule, is to designate the *contactor* and *not* the pushbutton station as the "controller," and the disconnect must be within sight from

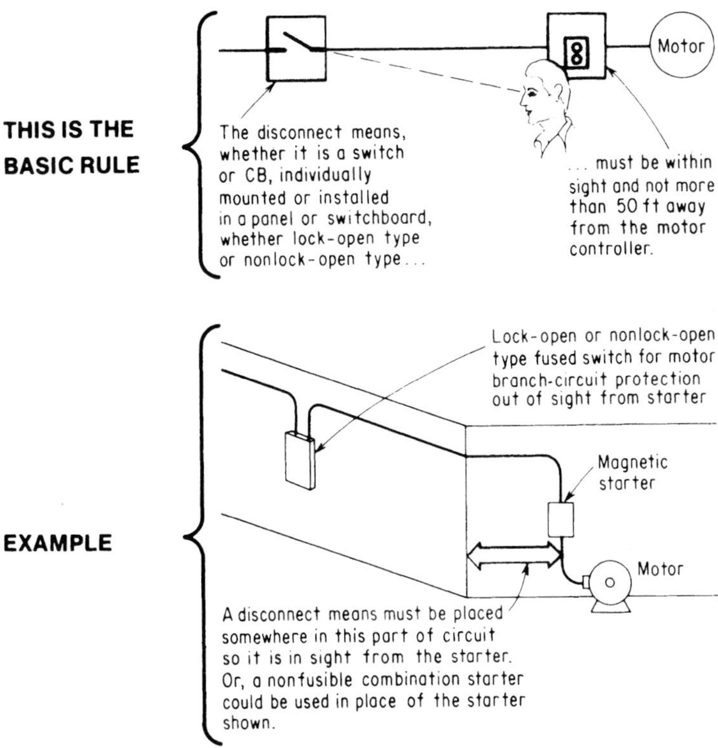
THIS IS THE BASIC RULE
Motor
The disconnect means, whether it is a switch or CB, individually mounted or installed in a panel or switchboard, whether lock-open type or nonlock-open type...
... must be within sight and not more than 50 ft away from the motor controller.
EXAMPLE
Lock-open or nonlock-open type fused switch for motor branch-circuit protection out of sight from starter
Magnetic starter
Motor
A disconnect means must be placed somewhere in this part of circuit so it is in sight from the starter. Or, a nonfusible combination starter could be used in place of the starter shown.

Fig. 430-58. The required disconnect must be visible from the controller. (Sec. 430-102.)

it and not from a pushbutton station or some other remotely located pilot control device that connects into the contactor.

There are two exceptions to this basic Code rule requiring a disconnect switch or CB to be located in sight from the controller:

Exception No. 1 permits the disconnect for a high-voltage (over 600 V) motor to be out of sight from the controller location, as shown in Fig. 430-59. But, such use of a lock-open type switch as an out-of-sight disconnect for a motor circuit rated 600 V or less is a clear Code violation.

Exception No. 2 is aimed at permitting practical, realistic disconnect means for industrial applications of large and complex machinery utilizing a number of motors to power the various interrelated parts of the machine. The Exception recognizes that a single common disconnect for a number of controllers (as permitted by part *a* of the Exception of Sec. 430-112) is often impossible to be installed "within sight" of all the controllers even though the controllers are "adjacent one to each other." On much industrial process equipment, the components of the overall structure obstruct the view of many controllers. Exception No. 2 permits the single disconnect to be technically out of sight from some or even all the controllers if the disconnect is simply "adjacent" to them—i.e., nearby on the equipment structure, as shown in Fig. 430-60.

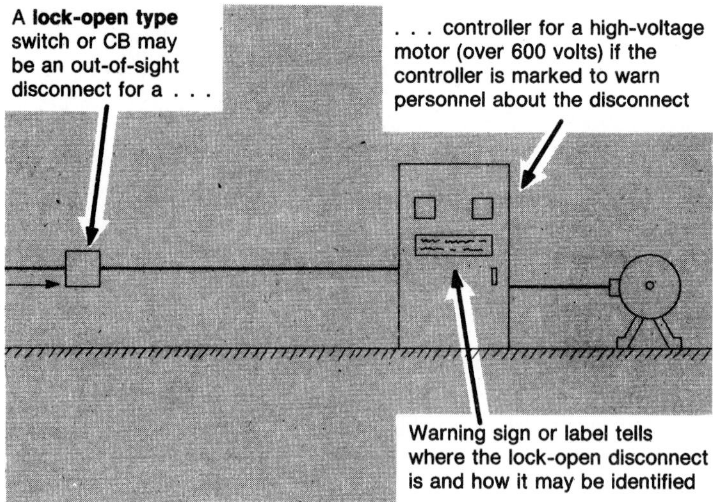
A lock-open type switch or CB may be an out-of-sight disconnect for a . . .
. . . controller for a high-voltage motor (over 600 volts) if the controller is marked to warn personnel about the disconnect
Warning sign or label tells where the lock-open disconnect is and how it may be identified

Fig. 430-59. An out-of-sight disconnect may be used for a high-voltage motor. (Sec. 430-102.)

Part (**b**) basically requires a disconnect means (switch or CB) to be within sight and not more than 50 ft (15.24 m) away from "the motor location and the driven machinery location." But the Exception to that basic requirement says that a disconnect does not have to be within sight from the motor and its load *if* the required disconnect ahead of the motor controller is capable of being locked in the open position.

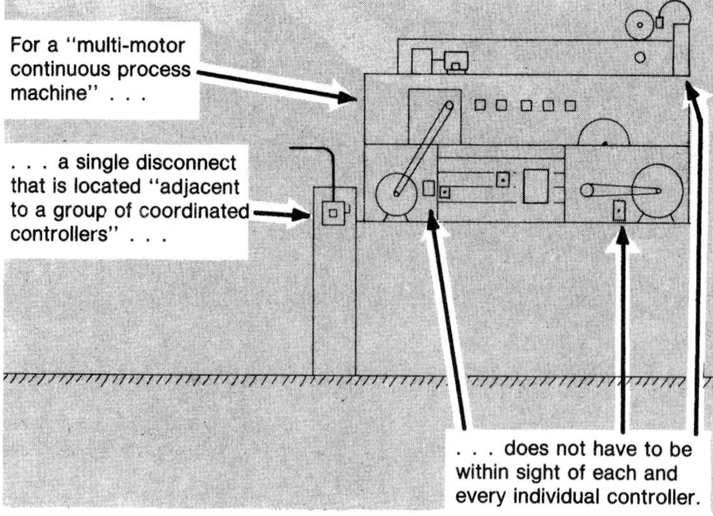
For a "multi-motor continuous process machine" . . .
. . . a single disconnect that is located "adjacent to a group of coordinated controllers" . . .
. . . does not have to be within sight of each and every individual controller.

Fig. 430-60. For multimotor machines, the disconnect may be "adjacent" to controller. (Sec. 430-102.)

According to the basic rule of part (**b**), a manually operable switch, which will provide disconnection of the motor from its power supply conductors, must be placed within sight from the motor location. And this switch *may not* be a switch in the control circuit of a magnetic starter. (The NE Code at one time permitted a switch in the coil circuit of the starter installed within sight of the motor. Such a condition is NOT acceptable to the present Code.)

These requirements are shown in Fig. 430-61. Specific layouts of the two conditions are shown in Fig. 430-62. (*Note:* Code provisions shown in these drawings are minimum safety requirements. Additional use of disconnects, with and without lock-open means, may be made necessary or desirable by job conditions.)

The intent of the Exception to part (**b**) is to permit maintenance workers to lock the disconnecting means ahead of the controller in the open position and keep the key in their possession so that the circuit cannot be energized while

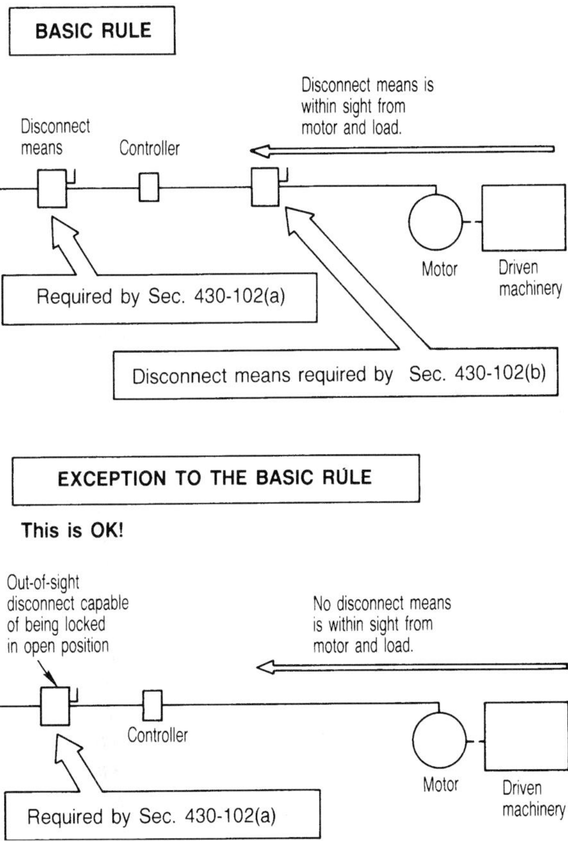

Fig. 430-61. Disconnect means must be within sight from the motor and its driven load, unless out-of-sight disconnect can be locked open. [Sec. 430-102(b).]

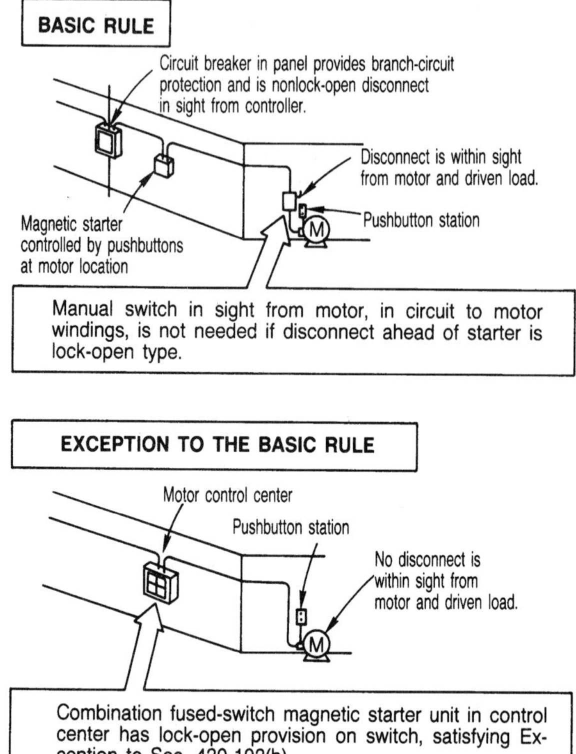

Fig. 430-62. Here's an example of the rules, showing physical layout. [Sec. 430-102(b).]

they are working on it. Over the years, many questions and controversies have arisen over the concept of a lock-open-type disconnect that satisfies the intent of this Exception. The various considerations may be addressed as follows:

Question: Does the rule specify "how" or "by what means" a disconnect must be "locked in the open position"?

Answer: The rule says nothing at all about that and does not actually require "locking" in the open position. It simply stipulates a condition that must be "capable" of being achieved—that is, some provision must assure that the disconnect is simply "capable of being locked in the open position."

Question: Does the NEC make it permissible to use the lock on a panelboard door or other enclosure door as the means of making the disconnect capable of being locked in the open position?

Answer: The very clear and straightforward working of the Code rule cannot be construed to prohibit use of such a lock to render the disconnect capable of being locked open. Panelboards made by all of the many manufacturers are provided with key-operated locks on their doors and are UL-certified with such locking provision. There is no Code rule that prohibits locking of enclosure doors. In fact, in the part (**a**) of Sec. 620-71 covering elevators and moving walks, the NEC specifically recognizes use of "cabinets with doors or removable panels capable of being locked in the closed position." And where a number of motors must all operate together, as in "Integrated Electrical Systems" (NEC Art. 685), it is more realistic and convenient—and therefore more contributive to effective safety—to be able to use a single lock to lock out *all* the disconnects that must be kept open when *any one* is open. It can be dangerous to lock out a single motor disconnect of a group of integrated motors and leave the others exposed to being turned on during maintenance operations.

Although the literal wording of the Exception of part (**b**) can be fully satisfied either by a lock on the door of a panel containing motor disconnects or by a locking hasp at the handle for each individual disconnect, there are differences between the two methods. The use of individual locking hasps provides greater ease, speed, and convenience of maintenance by limiting lockout to only the disconnect hasps on which locks are placed, keeping other disconnects available. However, the use of a lock on the panel door is really much more conducive to personnel safety than the use of lock hasps for the individual disconnects (switches or breakers) in a panel, for the following reasons:

1. Provision of lock hasps for each circuit disconnect affords no assurance that the disconnect is "capable" of being locked out—that is, that a person who might want to lock open a circuit carries a padlock to accomplish the actual locking. It is totally unrealistic to believe that a significant percentage of maintenance and operating personnel in commercial and industrial buildings carry padlocks in their pockets. Even in the rare case where workers do own a padlock, they commonly do not have it with them, have misplaced or lost it, or find it uncomfortable to have the heavy weight in their pockets. Safety cannot be based on fantasy or fiction. And especially in a panel containing two or more disconnects that must *all* be locked in the open position to assure safe maintenance—as for a conveyor system or other multimotor, integrated machine operations—it is totally ridiculous to expect that each person who may have to work on the motors will carry padlocks to use on all the individual disconnects.
2. Where lock-open provision is made by means of a lock installed in the panel door instead of the use of individual lock hasps, it is certain that the lock is always available to be used. Although the use of individual hasps, as described above, does not in itself make the disconnect "capable of being locked in the open position," the use of a panel with a built-in lock more closely follows the Code rule because the presence of a "lock" is guaranteed. The lock exists and cannot be removed, and the opportunity for real, effective safety is afforded simply by giving a key to the lock to everyone who needs to lock open the disconnects. People will more readily carry a key than a padlock.

The pushbutton station in Fig. 430-62, Exception, operates only the holding coil in the magnetic starter. The magnetic starter "controls" the current to the motor; for example, the control wires to a pushbutton station could become shorted after the motor is in operation, and pushing the STOP button would not release the holding coil in the magnetic starter and the motor would continue to run. This is the reason that a disconnecting means is required to be installed within sight from the motor and its load or a lock-open switch installed ahead of the controller. In this case, operating the disconnecting means will open the supply to the controller and shut off the motor.

430-103. Operation. This rule actually defines the meaning of "disconnecting means."

In order that necessary periodic inspection and servicing of motors and their controllers may be done with safety, the Code requires that a switch, CB, or other device shall be provided for this purpose. Because the disconnecting means must disconnect the controller as well as the motor, it must be a separate device and cannot be a part of the controller, although it could be mounted on the same panel or enclosed in the same box with the controller. The disconnect must be installed ahead of the controller. And note that the disconnect must open only the "ungrounded" conductors of a motor circuit.

In case the motor controller fails to open the circuit if the motor is stalled, or under other conditions of heavy overload, the disconnecting means can be used to open the circuit. It is therefore required that a switch used as the disconnecting means shall be capable of interrupting very heavy current.

430-105. Grounded Conductors. Although Sec. 430-103 requires a disconnect means only for the ungrounded conductors of a motor circuit, if a motor circuit includes a grounded conductor, one pole of the disconnect *may* switch the grounded conductor provided all poles of the disconnect operate together—as in a multipole switch or CB. For instance, a 120-V, 2-wire circuit with one of its conductors grounded only requires a single-pole disconnect switch, but a 2-pole switch *could* be used, with one pole switching the grounded leg.

430-107. Readily Accessible. Although a motor circuit may be provided with more than one disconnect means in series ahead of the controller—such as one at the panel where the motor circuit originates and one at the controller location—*only one* of the disconnects is required to be "readily accessible," as follows:

> *Readily accessible:* Capable of being reached quickly for operation, renewal, or inspection, without requiring those to whom ready access is requisite to climb over or remove obstacles or to resort to portable ladders, chairs, etc. (See "Accessible.")

The disconnecting means must be reached without climbing over anything, without removing crates or equipment or other obstacles, and without requiring the use of portable ladders.

Note carefully: A disconnect that has to be "readily accessible" must be so only for "those to whom ready access is requisite"—which clearly and intentionally allows for making equipment *not* readily accessible to other than authorized persons, such as by providing a lock on the door, with the key possessed by or available to those who require ready access.

Because the definition of "readily accessible" contains a last phrase that says "See 'Accessible'," logic dictates that the installation must also satisfy the definition of "Accessible." And the wording of the definition clearly establishes that there is no Code violation in putting the disconnect means in a room or area under lock and key to make it accessible only to authorized persons.

The definition reads:

> *Accessible:* (As applied to Equipment.) Admitting close approach because not guarded by locked doors, elevation, or other effective means. (See "Readily Accessible.")

Again note carefully: That definition does not say that a door to an electrical room is prohibited from being locked. In fact, the wording of the definition, by referring to "locked doors," actually presumes the existence and, therefore, the acceptability of "locked doors" in electrical systems. The only requirement implied by the wording is that locked doors, where used, must not "guard" against access—that is, disposition of the key to the lock must be such that those requiring access to the room are not positively excluded. The rule is satisfied if the key is available to provide access to authorized persons.

In reference to the definition of "Accessible," the critical word is "guarded." The definition is *not* intended to mean that equipment *cannot* be "behind" locked doors or that equipment *cannot* be mounted up high where it *can* be reached with a portable ladder. To make equipment "not accessible," a door lock or high mounting must be such that it positively "guards" against access. Equipment behind a locked door for which a key is not possessed by or available to persons who require access to the equipment is *not* "accessible." A common example of that latter condition occurs in multitenant buildings where a disconnect for the tenant of one occupancy unit is located behind the locked door of another tenant's occupancy unit from which the first tenant is effectively and legally excluded. And even that application *is* Code-acceptable if the disconnect is *not required* by the NEC to be "readily accessible."

Equipment may be fully "accessible" even though installed behind a locked door or at an elevated height. Equipment that is high-mounted but can be reached with a ladder that is fixed in place or a portable ladder *is* "accessible" (although the equipment would not be "readily accessible" if a portable ladder had to be used to reach it). Similarly, equipment behind a locked door *is* "accessible" to anyone who possesses a key to the lock or to a person who is authorized to obtain and use the key to open the locked door. In such cases, conditions do *not* "guard" against access.

Refer to the definitions of "Accessible" and "Readily Accessible" in Art. 100 of this book.

430-109. Type. In a motor branch circuit, every switch or CB in the circuit, from where the circuit is tapped from the feeder to the motor itself, must satisfy the requirements on type and rating of disconnect means. A CB switching device with no automatic trip operation, a so-called molded-case switch, may be used as a motor disconnect instead of a conventional CB or a horsepower-rated switch. Such a device either must be rated for the horsepower of the motor it is used with or must have an amp rating at least equal to 115 percent of that of the motor with which it is used. Figure 430-63 covers the basic rules on types of disconnect means.

For a motor larger than 2 hp but not larger than 100 hp, and not portable, a motor-circuit switch or a CB must be used as the disconnecting means (Fig. 430-64).

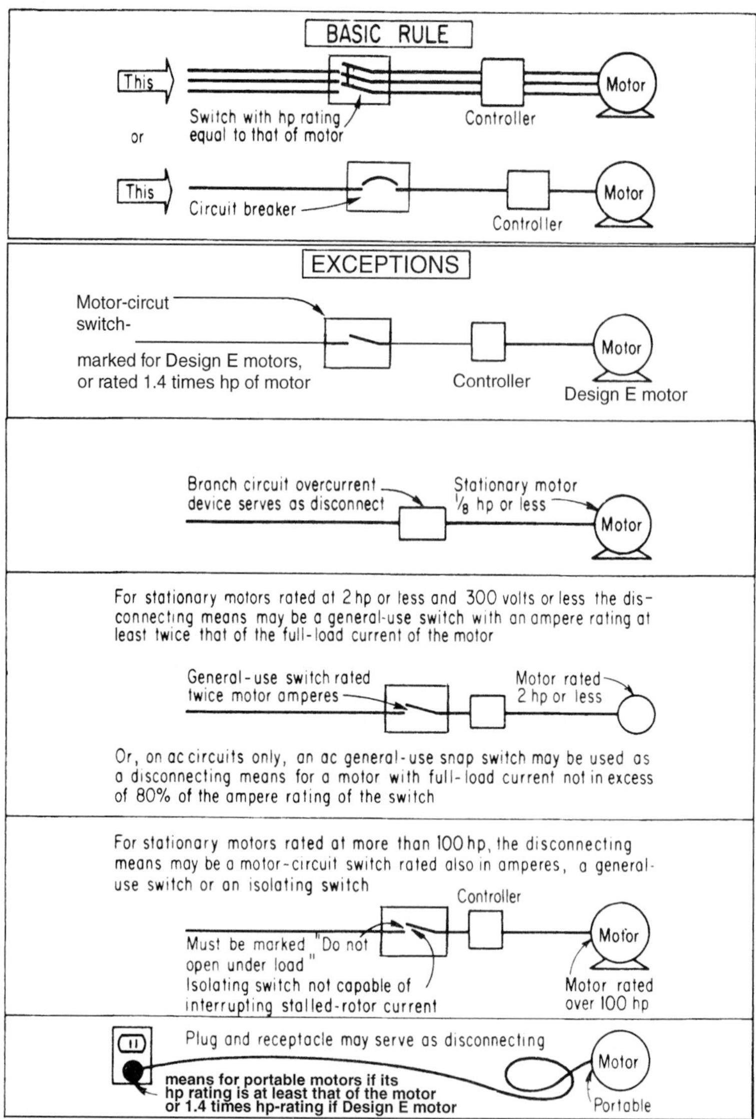

Fig. 430-63. One of these disconnects must be used for a motor branch circuit. (Sec. 430-109.)

A motor-circuit switch is a horsepower-rated switch. If in addition to the disconnecting means there is any other switch in the motor circuit and it is at all likely that this switch might be opened in case of trouble, this switch must have the interrupting capacity required for a switch intended for use as the disconnecting means.

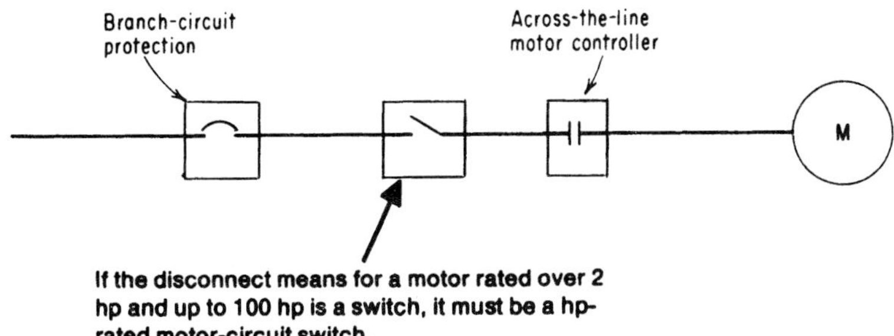

Fig. 430-64. From 2 to 100 hp, a disconnect *switch* must be horsepower-rated. (Sec. 430-109.)

Exception No. 5 to Sec. 430-109 sets the maximum horsepower rating required for motor-circuit switches at 100 hp. Higher-rated switches are now available and will provide additional safety. The first sentence of this section makes a basic requirement that the disconnecting means for a motor and its controller be a motor-circuit switch rated in horsepower. For motors rated up to 500 hp, this is readily complied with, inasmuch as the UL lists motor-circuit switches up to 500 hp and the manufacturers mark switches to conform. But for motors rated over 100 hp, the **Code** does not require that the disconnect have a horsepower rating. It makes an exception to the basic rule and permits the use of an ampere-rated switch or isolation switch, provided the switch has a carrying capacity of at least 115 percent of the nameplate current rating of the motor [Sec. 430-110(a)]. And UL notes that horsepower-rated switches over 100 hp *must not* be used as motor controllers. And Exception No. 5 notes that isolation switches for motors over 100 hp must be plainly marked "Do not operate under load," if the switch is not rated for safely interrupting the locked-rotor current of the motor. Figure 430-65 shows an example of disconnect switch application for a motor rated over 100 hp.

example Provide a disconnect for a 125-hp, 3-phase, 460-V motor. Use a nonfusible switch, inasmuch as short-circuit protection is provided at the supply end of the branch circuit.

The full-load running current of the motor is 156 A, from **NEC** Table 430-150. A suitable disconnect must have a continuous carrying capacity of 156 × 1.15, or 179 A, as required by Sec. 430-110(a).

This calls for a 200-A, 3-pole switch rated for 480 V. The switch may be a general-use switch, a current- and horsepower-marked motor-circuit switch, or an isolation switch. A 200-A, 3-pole, 480-V motor-circuit switch would be marked with a rating of 50 hp, but the horsepower rating is of no concern in this application because the switch does not have to be horsepower-rated for motors larger than 100 hp.

If the 50-hp switch were of the heavy-duty type, it would have an interrupting rating of 10 × 65 A (the full-load current of a 460-V, 50-hp motor), or 650 A. But the locked-rotor current of the 125-hp motor might run over 900 A. In such a case, the switch is required by Exception No. 4 to be marked "Do not operate under load."

If a fusible switch had to be provided for the above motor to provide both disconnect and short-circuit protection, the size of the switch would be determined by the size and type of fuses used. Using a fuse rating of 250 percent of motor current (which does not exceed the 300 percent maximum in Table 430-152) for standard fuses, the application

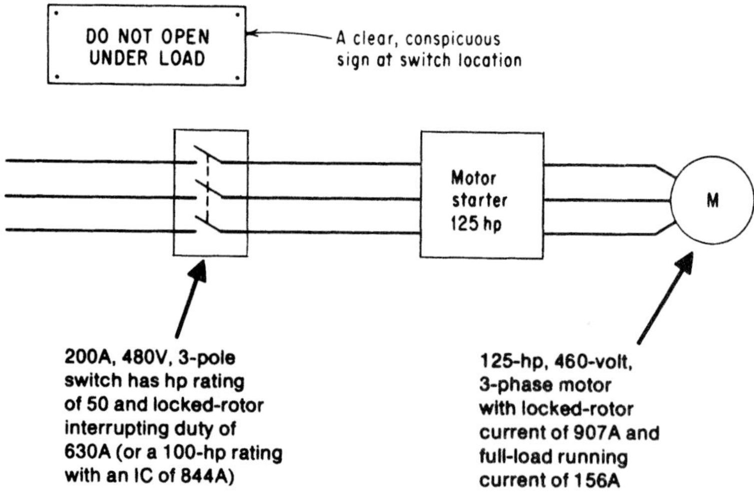

Fig. 430-65. Above 100 hp, a switch does not have to be horsepower-rated. (Sec. 430-109.)

would call for 400-A fuses in a 400-A switch. This switch would certainly qualify as the motor disconnect. However, if time-delay fuses are used, a 200-A switch would be large enough to take the time-delay fuses and could be used as the disconnect (because it is rated at 115 percent of motor current).

In the foregoing, the 400-A switch might have an interrupting rating high enough to handle the locked-rotor current of the motor. Or the 200-A switch might be of the CB-mechanism type or some other heavy current construction that has an interrupting rating up to 12 times the rated load current of the switch itself. In either of these cases, there would be no need for marking "Do not operate under load."

Up to 100 hp, a switch which satisfies the Code on rating for use as a motor controller may also provide the required disconnect means—the two functions being performed by the one switch—provided it opens all ungrounded conductors to the motor, is protected by an overcurrent device (which may be the branch-circuit protection or may be fuses in the switch itself), and is a manually operated air-break switch or an oil switch not rated over 600 V or 100 A—as permitted by Sec. 430-111.

430-110. Ampere Rating and Interrupting Capacity. An ampere-rated switch or a CB must be rated at least equal to 115 percent of a motor's full-load current if the switch or CB is the disconnect means for the motor.

When two or more motors are served by a single disconnect means, as permitted by Sec. 430-112, or where one or more motors plus a nonmotor load (such as electric heater load) make use of a single common disconnect, part (**c**) must be used in sizing the disconnect.

430-111. Switch or Circuit Breaker as Both Controller and Disconnecting Means. As described under Sec. 430-84, a manual switch—capable of starting and stopping a given motor, capable of interrupting the stalled-rotor current of the motor, and having the same horsepower rating as the motor—may serve the functions of controller and disconnecting means in many motor circuits, if the switch opens all ungrounded conductors to the motor. That is also true of a manual motor starter. A single manually operated CB may also serve as controller and disconnect (Figs. 430-66 and 430-67). However, in the case of an autotransformer type of controller, the controller itself, even if manual, may not also serve as the disconnecting means. Such controllers must be provided with a separate means for disconnecting controller and motor.

Although this **Code** section permits a single horsepower-rated switch to be used as both the controller and the disconnect means of a motor circuit, UL rules note that "enclosed switches rated higher than 100 hp are restricted to use as motor disconnecting means and are not for use as motor controllers."

The acceptability of a single switch for both the controller and the disconnecting means is based on the single switch satisfying the **Code** requirements for a controller and for a disconnect. It finds application where general-use switches or horsepower-rated switches are used, as permitted by the **Code**, in conjunction with time-delay fuses which are rated low enough to provide both running overload protection and branch-circuit (short-circuit) protection. In such cases, a single fused switch may serve a total of four functions: (1) con-

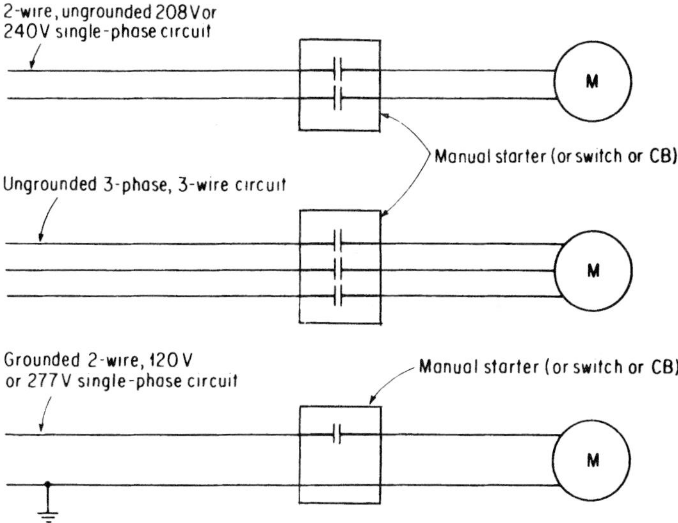

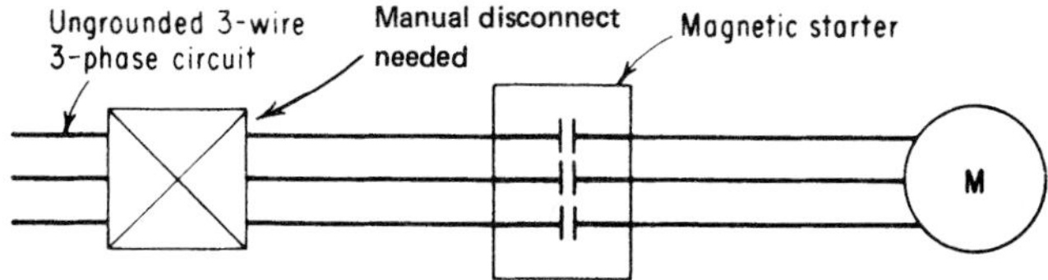

Fig. 430-66. A manual switch or CB may serve as both controller and disconnect means. (Sec. 430-111.)

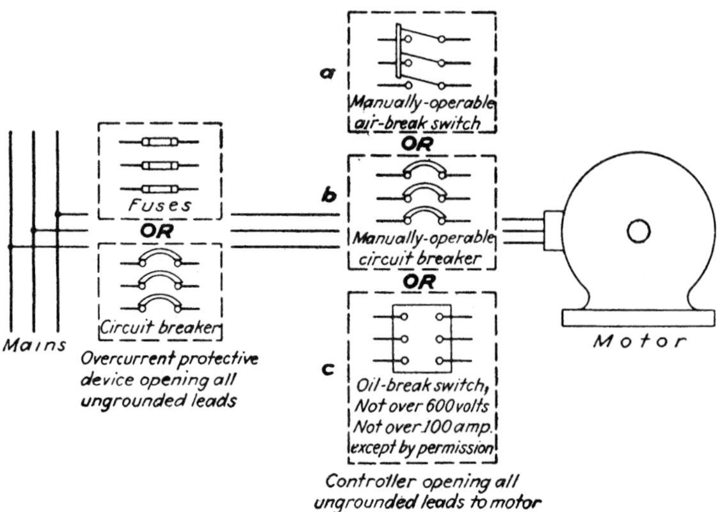

Fig. 430-67. Use of a single controller disconnect is limited. (Sec. 430-111.)

troller, (2) disconnect, (3) branch-circuit protection, and (4) running overload protection. And it is possible for a single CB to also serve these four functions.

For sealed refrigeration compressors, Sec. 440-12 gives the procedure for determining the disconnect rating, based on nameplate rated-load current or branch-circuit selection current, whichever is greater, and locked-rotor current of the motor-compressor.

430-112. Motors Served by Single Disconnecting Means. In general, each individual motor must be provided with a separate disconnecting means. However, a single disconnect sometimes may serve a group of motors under the conditions specified, which are the same as in Sec. 430-87. Such a disconnect must have a rating sufficient to handle a single load equal to the sum of the horsepower ratings or current ratings.

Exception a

In Sec. 610-31 it is required that the main collector wires of a traveling crane shall be controlled by a switch located within sight of the wires and readily operable from the floor or ground. This switch would serve as the disconnecting means for the motors on the crane. When repair or maintenance work is to be done on the electrical equipment of the crane, it is safer to cut off the current from all this equipment by opening one switch, rather than to use a separate switch for each motor. Also, in the case of a machine tool driven by two or more motors, a single disconnecting means for the group of motors is more serviceable than an individual switch for each motor, because repair and maintenance work can be done with greater safety when the entire electrical equipment is "dead."

Exception *b*

Such groups may consist of motors having full-load currents not exceeding 6 A each, with circuit fuses not exceeding 20 A at 125 V or less, or 15 A at 600 V or less. Because the expense of providing an individual disconnecting means for each motor is not always warranted for motors of such small size, and also because the entire group of small motors could probably be shut down for servicing without causing inconvenience, a single disconnecting means for the entire group is permitted.

Exception c

"Within sight" should be interpreted as meaning so located that there will always be an unobstructed view of the disconnecting switch from the motor, and Sec. 430-102 limits the distance in this case between the disconnecting means and any motor to a maximum of 50 ft (15.24 m).

These conditions are the same as those under which the use of a single controller is permitted for a group of motors. (See Sec. 430-87.) The use of a single disconnecting means for two or more motors is quite common, but in the majority of cases the most practicable arrangement is to provide an individual controller for each motor.

If a switch is used as the disconnecting means, it must be of the type and rating required by Sec. 430-109 for a single motor having a horsepower rating equal to the sum of the horsepower ratings of all the motors it controls. Thus, for six 5-hp motors the disconnecting means should be a motor-circuit switch rated at not less than 30 hp. If the total of the horsepower ratings is over 2 hp, a horsepower-rated switch must be used.

430-113. Energy from More than One Source. The basic rule of this section, which is similar to that of Sec. 430-74, requires a disconnecting means to be provided from each source of electrical energy input to equipment with more than one circuit supplying power to it, such as the hookup shown in Fig. 430-68, where two switches or a single 5-pole switch could be used. And each source is permitted to have a separate disconnecting means. This **Code** rule is aimed at the need for adequate disconnects for safety in complex industrial layouts. But an exception to the **Code** rule states that where a motor receives electrical energy from more than one source (such as a synchronous motor receiving both alternating current and direct current energy input), the disconnecting means for the main power supply to the motor shall *not* be required to be immediately adjacent to the motor—provided that the controller disconnecting means, which is the disconnect ahead of the motor starter in the main power circuit, is capable of being locked in the open position. If, for instance, the motor control disconnect can be locked in the open position, it may be remote; but the disconnect for the other energy input circuit would have to be adjacent to the machine itself, as indicated in Fig. 430-69.

430-124. Size of Conductors. For motors rated over 600 V, the circuit conductors to the motor are selected to have a current rating equal to or greater than the trip setting of the running overload protective device for the motor.

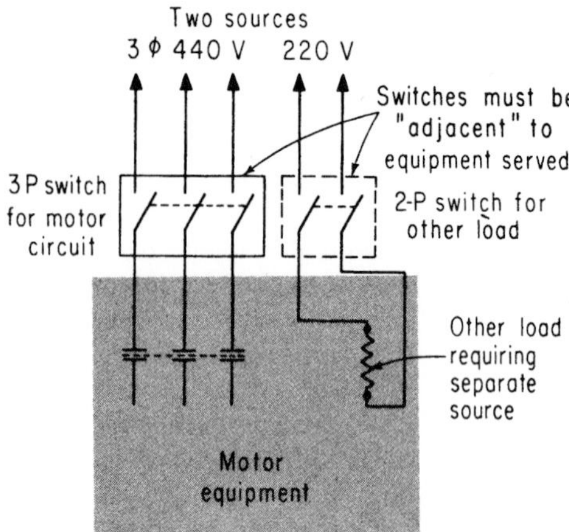

Fig. 430-68. A disconnect must be used for each power input to motorized equipment. (Sec. 430-113.)

430-125. Motor Circuit Overcurrent Protection. Overload protection must protect the motor and other circuit components against overload currents up to and including locked-rotor current of the motor. A CB or fuses must be used for protection against ground faults or short circuits in the motor circuit.

430-142. Stationary Motors. Usually stationary motors are supplied by wiring in a metal raceway or metal-clad cable. The motor frames of such motors must be grounded, the raceway or cable armor being attached to the frame and serving as the grounding conductor. [See Sec. 250-91(b).]

Any motor in a wet location constitutes a serious hazard to persons and should be grounded unless it is so located or guarded that it is out of reach. *All*

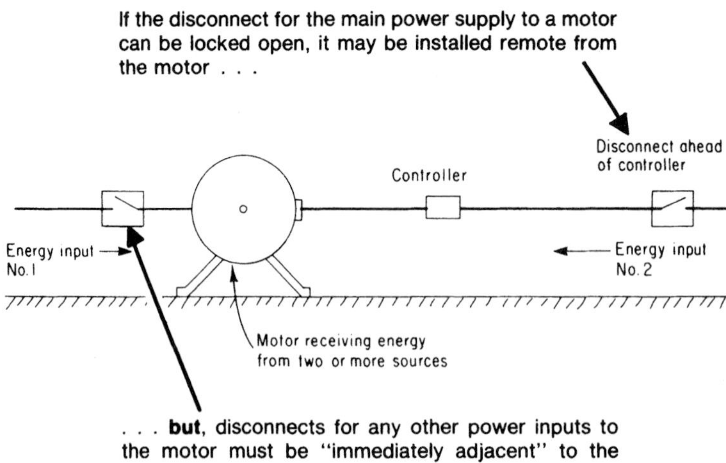

Fig. 430-69. An exception is made for disconnects for multiple power sources. (Sec. 430-113.)

water pump motors, including those in the submersible-type pump, must be grounded, regardless of location, to comply with Sec. 250-43(k).

430-145. Method of Grounding. Good practice requires in nearly all cases that the wiring to motors which are not portable shall, at the motor, be installed in rigid or flexible metal conduit, electrical metallic tubing, or metal-clad cable and that such motors should be equipped with terminal housings. The method of connecting the conduit to the motor where some flexibility is necessary is shown in Fig. 430-70. The motor circuit is installed in rigid conduit and a short length of liquidtight flexible metal conduit is provided between the end of the rigid conduit and the terminal housing on the motor. But because the size of flex is over 1¼-in. (31.8 mm), a separate equipment grounding conductor (or bonding jumper) must be used within or outside the flex as noted in Secs. 351-9 and 250-91(b), Exception No. 2. Refer to Sec. 430-12(e), which requires provision of a suitable termination for an equipment grounding conductor at every motor terminal housing, as shown in Fig. 430-71.

This section permits the use of fixed motors without terminal housings. If a motor has no terminal housing, the branch-circuit conductors must be brought to a junction box not over 6 ft (1.83 m) from the motor. Between the junction box and the motor, the specified provisions apply.

According to Sec. 300-16, the conduit, tubing, or metal-clad cable must terminate close to the motor in a fitting having a separable bushed hole for each

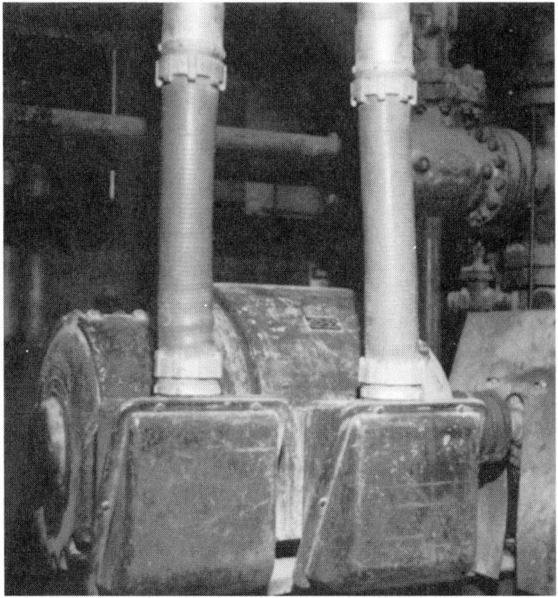

Fig. 430-70. Liquidtight flex provides flexible connection from rigid conduit supply to motor terminals but does require a separate equipment grounding conductor run within the flex with the circuit conductors or a separate external bonding jumper from the rigid metal conduit to the metal terminal box for each of the two runs. (Sec. 430-145.)

Fig. 430-71. Motor terminal housings must include some lug or terminal for connecting an equipment grounding conductor that may be run inside the raceway with the circuit wires or may be run as a bonding jumper around a length of flex or liquidtight flex, as commonly used for vibration-free motor connections. The terminal box here must have internal provision for connecting the equipment grounding conductor, required for this short length of liquidtight flex, that is larger than 1¼-in. (31.8 mm) in size. The static grounding connection shown here (arrow) on the box does not satisfy Secs. 351-9 and 250-79(e) as a bonding jumper for the flex, and it does not satisfy Sec. 250-57(b) as an equipment ground for an AC motor. (Sec. 430-145.)

wire. The method of making the connection to the motor is not specified; presumably, it is the intention that the wire brought out from the terminal fitting shall be connected to binding posts on the motor or spliced to the motor leads. The conduit, tubing, or cable must be rigidly secured to the frame of the motor.

ARTICLE 440. AIR-CONDITIONING AND REFRIGERATING EQUIPMENT

440-3. Other Articles. Article 440 is patterned after Art. 430, and many of its rules, such as on disconnecting means, controllers, conductor sizes, and group installations, are identical or quite similar to those in Art. 430. This article con-

tains provisions for such motor-driven equipment and for branch circuits and controllers for the equipment, taking into account the special considerations involved with sealed (hermetic-type) motor-compressors, in which the motor operates under the cooling effect of the refrigeration.

It must be noted that the rules of Art. 440 are *in addition to* or are *amendments to* the rules given in Art. 430 for motors in general. The basic rules of Art. 430 also apply to A/C (air-conditioning) and refrigerating equipment unless exceptions are indicated in Art. 440.

Article 440 further clarifies the application of NE Code rules to air-conditioning equipment and refrigeration equipment as follows:

1. A/C and refrigerating equipment which does not incorporate a sealed (hermetic-type) motor-compressor must satisfy the rules of Art. 422 (Appliances), Art. 424 (Space Heating Equipment), or Art. 430 (Conventional Motors)—whichever apply. For instance, where refrigeration compressors are driven by conventional motors, the motors and controls are subject to Art. 430, not Art. 440. Furnaces with air-conditioning evaporator coils installed must satisfy Art. 424. Other equipment in which the motor is not a sealed compressor and which must be covered by Arts. 422, 424, or 430 includes fan-coil units, remote forced air-cooled condensers, remote commercial refrigerators, and similar equipment.
2. Room air conditioners are covered in part **G** of Art. 440 (Secs. 440-60 through 440-64), but must also comply with the rules of Art. 422.
3. Household refrigerators and freezers, drinking-water coolers, and beverage dispensers are considered by the Code to be appliances, and their application must comply with Art. 422 and must also satisfy Art. 440, because such devices contain sealed motor-compressors.

Air-conditioning equipment (other than small room units and large custom installations) is manufactured in the form of packaged units having all necessary components mounted in one or more enclosures designed for floor mounting, for recessing into walls, for mounting in attics or ceiling plenums, for locating outdoors, etc. Figure 440-1 shows the difference between room A/C units (such as window units) and the larger so-called packaged units or central air conditioners. Room units consist of a complete refrigeration system in a unit enclosure intended for mounting in windows or in the wall of the building, with ratings up to 250 V, single phase. Unitary assemblies may be console type for individual room use rated up to 250 V single phase or central cooling units rated up to 600 V for commercial or domestic applications. This type may consist of one or more factory-made sections. If it is made up of two or more sections, each section is designed for field interconnection with one or more matched sections to make the complete assembly. Dual-section systems consist of separate packaged sections installed remote from each other and interconnected by refrigerant tubing, either with the compressor within the outdoor section or within the indoor section.

Electrical wiring in and to units varies with the manufacturer, and the extent to which the electrical contractor need be concerned with fuse and CB calculations depends upon the manner in which the units' motors are fed and the type of distribution system to which they are to be connected. A packaged unit is

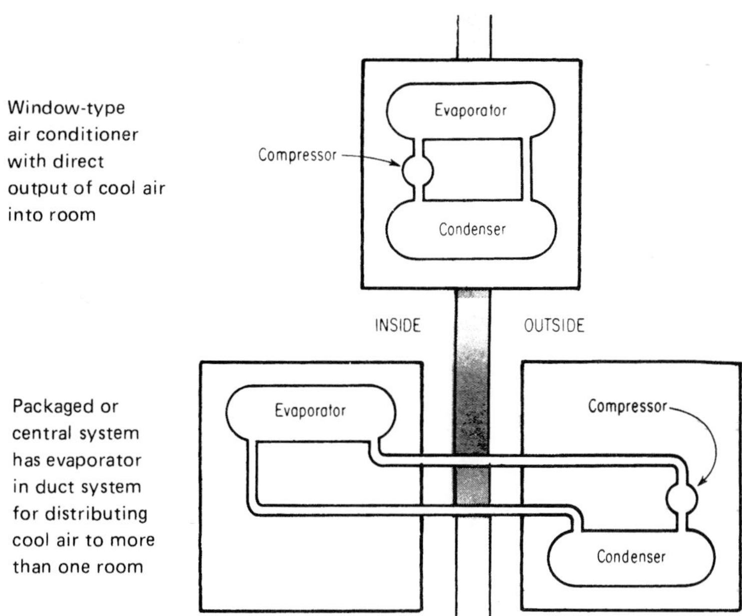

Fig. 440-1. Code rules differentiate between unit room conditioners and central systems. (Sec. 440-3.)

treated as a group of motors. This is different from the approach used with a plug-in room air conditioner, which is treated as an individual single-motor load of amp rating as marked on the nameplate.

440-4. Marking on Hermetic Refrigerant Motor-Compressors and Equipment. Important in the application of hermetic refrigerant motor-compressors are the terms "rated-load current" and "branch-circuit selection current." Definitions of these terms are given in Sec. 440-2. When the equipment is marked with the branch-circuit selection current, this greatly simplifies the sizing of motor branch-circuit conductors, disconnecting means, controllers, and overcurrent devices for circuit conductors and motors.

As noted in the definition for branch-circuit selection current (Sec. 440-2), the value of branch-circuit selection current will always be *equal to* or greater than the market rated-load current. This advises installers that for some A/C equipment that is not required to have a "branch-circuit selection current," the value of rated-load current will appear on the equipment nameplate; and that same value of current will also appear in the nameplate space reserved for branch-circuit selection current. In such cases, the branch-circuit selection current appears to be equal to the rated-load current.

440-5. Marking on Controllers. Note that a controller may be marked with "full-load and locked-rotor current (or horsepower) rating." That possibility of two methods of marking requires careful application of the rules in Sec. 440-41 on selecting the correct rating of controller for motor-compressors.

440-6. Ampacity and Rating. Selection of the rating of branch-circuit conductors, controller, disconnect means, short-circuit (and ground-fault) protection, and running overload protection is *not* made the same for hermetic motor-compressors as for general-purpose motors. In sizing those components, the "rated-load current" marked on the equipment and/or the compressor must be used in the calculations covered in other rules of this article. That value of current must always be used, instead of full-load currents from Code Tables 430-148 to 430-150, which are used for sizing circuit elements for nonhermetic motors. And if a "branch-circuit selection current" is marked on equipment, that value must be used instead of rated-load current.

440-12. Rating and Interrupting Capacity. Note that the rules here are qualifications that apply to the rules of Secs. 430-109 and 430-110 on disconnects for general-purpose motors.

A disconnecting means for a hermetic motor, as covered in part (**a**)(**2**), must be a motor-circuit switch rated in horsepower or a CB—as required by Sec. 430-109.

If a CB is used, it must have an amp rating not less than 115 percent of the nameplate "rated-load current" or the "branch-circuit selection current"—whichever is greater.

But, if a horsepower-rated switch is to be selected, the process is slightly involved for hermetic motors marked with locked-rotor current and rated-load current or rated-load current plus branch-circuit selection current—but *not* marked with horsepower. In such a case, determination of the equivalent horsepower rating of the hermetic motor must be made using the locked-rotor current and either the rated-load current or the branch-circuit selection current—whichever is greater—based on Code Tables 430-148, 430-149, or 430-150 for rated-load current or branch-circuit selection current and Table 430-151 for locked-rotor current, as follows:

For example, a 3-phase, 460-V hermetic motor rated at 11-A branch-circuit selection and 60-A locked-rotor is to be supplied with a disconnect switch rated in horsepower. The first step in determining the equivalent horsepower rating of that motor is to refer to Code Table 430-150. This table lists 7½ hp as the required size for a 460-V, 11-A motor. To ensure adequate interrupting capacity, Code Table 430-151 is used. For a 60-A locked-rotor current, this table also shows 7½ hp as the equivalent horsepower rating for any locked-rotor current over 45 to 66 A for a 400-V motor. Use of both tables in this manner thus establishes a 7½ hp disconnect as adequate for the given motor in both respects. Had the two ratings as obtained from the two tables been different, the higher rating would have been chosen.

Figure 440-2 shows an example of disconnect sizing for a horsepower-rated switch when a hermetic motor is used, in accordance with Sec. 430-53(c), along with fan motors on a single circuit, as covered in part (**b**) and in Sec. 440-33. Fan motors are usually wired to start slightly ahead of the compressor-motor through use of interlock contacts or a time-delay relay. In some units, however, all motors start simultaneously and that is covered by part (**b**) of this section in sizing the horsepower-rated disconnect switch. Where this is the case, the starting load will be treated like a single motor to the disconnect switch, and the

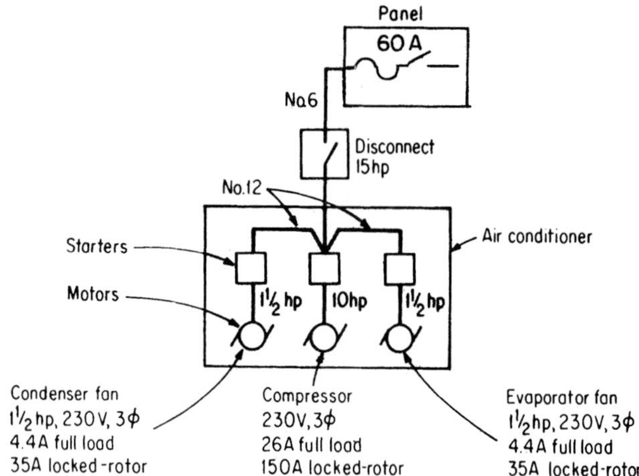

Fig. 440-2. Disconnect for multiple motors is sized from rated-load or branch-circuit selection currents and locked-rotor currents. (Sec. 440-12.)

sum of the locked-rotor currents of all motors should be used with Code Table 430-151 to determine the horsepower rating of the disconnect. The disconnect normally must handle the sum of the rated-load or branch-circuit selection currents; hence the rating as checked against Code Table 430-150 will be on the basis of the sum of the higher of those currents for all the motors. Code Table 430-150, using the full-load total of 34.8 A (4.4 A + 26 A + 4.4 A) in this example, indicates a 15-hp disconnect. Code Table 430-151, assuming simultaneous starting of all three motors and using the total locked-rotor current of 220 A (35 A + 150 A + 35 A), also shows 15 hp as the required size.

If motors do not start simultaneously, the compressor locked-rotor current (150 A) used with Code Table 430-151 gives a 10-hp rating. However, the higher of the two horsepower ratings must be used; hence the running currents impose the more severe requirements and dictate use of a 15-hp switch. See data under Sec. 440-22.

As required by part (**d**) of this section, all disconnects in a branch circuit to a refrigerant motor-compressor must have the required amp or horsepower rating and interrupting rating. This provides for motor-compressor circuits the same conditions that Sec. 430-108 requires for other motor branch circuits.

440-14. Location. Section 440-13 recognizes use of a cord-plug and receptacle as the disconnect for such cord-connected equipment as a room or window air conditioner. But this section (440-14) applies to fixed-wired equipment—such as central systems or units with fixed circuit connection. For conditioners with fixed wiring connection to their supply circuits, the rule poses a problem. If the branch-circuit breaker or switch which is to provide disconnect means is located in a panel that is out of sight [or more than 50 ft (15.24 m) away] from the unit conditioner, another breaker or switch must be provided at the equipment. If the panel breaker or switch does not satisfy the rule here, a separate disconnect means would have to be added in sight from the conditioner as shown in Fig. 440-3. This is also true if the service switch is installed as shown in Fig. 440-4.

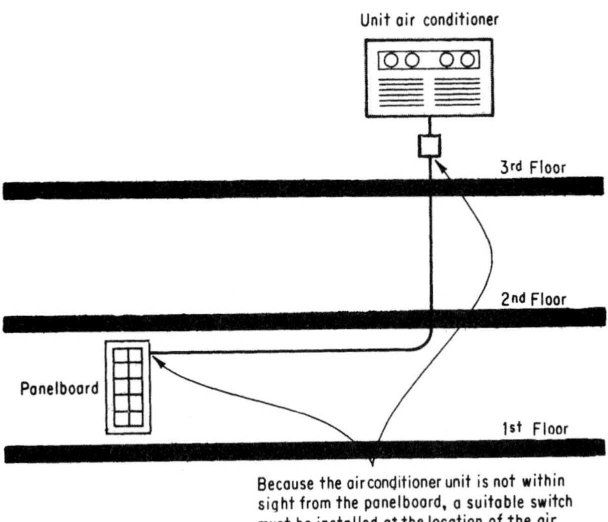
Unit air conditioner
3rd Floor
2nd Floor
Panelboard
1st Floor
Because the airconditioner unit is not within sight from the panelboard, a suitable switch must be installed at the location of the air conditioner unit.

Fig. 440-3. For any fixed-wire A/C equipment, disconnect must be "within sight." (Sec. 440-14.)

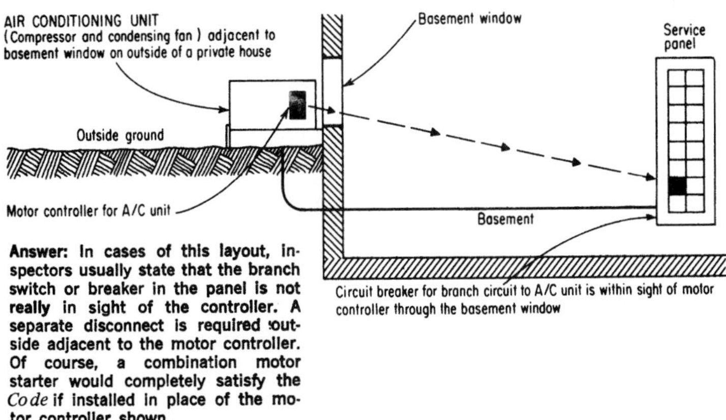
AIR CONDITIONING UNIT (Compressor and condensing fan) adjacent to basement window on outside of a private house
Basement window
Service panel
Outside ground
Motor controller for A/C unit
Basement
Answer: In cases of this layout, inspectors usually state that the branch switch or breaker in the panel is not really in sight of the controller. A separate disconnect is required outside adjacent to the motor controller. Of course, a combination motor starter would completely satisfy the Code if installed in place of the motor controller shown.
Circuit breaker for branch circuit to A/C unit is within sight of motor controller through the basement window

Fig. 440-4. "Within sight" disconnect must also be "readily accessible" at the equipment. (Sec. 440-14.)

As stated in the second sentence, the required disconnect means for air-conditioning or refrigeration equipment may be installed on or within the equipment enclosure. That is recognized as an equivalent of the basic rule that the disconnect must be readily accessible and within sight [visible and not over 50 ft (15.24 m) away] from the A/C or refrigeration equipment. Such equipment is being manufactured now with the disconnect incorporated as part of the assembly.

440-21. General. Part **C** of this article covers details of branch-circuit makeup for A/C and refrigeration equipment; Sec. 440-3(a) says that the provisions of Art. 430 apply to A/C and refrigeration equipment for any considerations that are not covered in Art. 440. Thus, because Art. 440 does *not* cover feeder sizing and feeder overcurrent protection for A/C and refrigeration equipment, it is necessary to use applicable sections from Art. 430. Section 430-24 covers sizing of feeder conductors for standard motor loads and for A/C and refrigeration loads. Sections 430-62 and 430-63 cover rating of overcurrent protection for feeders to both standard motors and A/C and refrigeration equipment. That fact is noted by a new phrase added to the end of Sec. 430-62(a).

440-22. Application and Selection. Part (**a**) of this section is illustrated in Fig. 440-5, where a separate circuit is run to the compressor and to each fan motor of a packaged assembly, containing a compressor with 26-A rated-load current and fan motors rated at 4.4 A full-load each. The compressor protection is sized at 1.75 × 26 A (175 percent of rated-load current), or 45.5 A—calling for 45- or 50-A fuses.

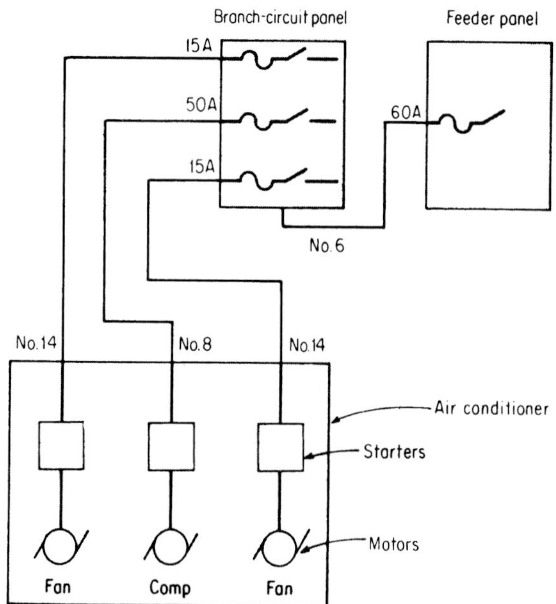

Fig. 440-5. A separate circuit may be run to each motor of A/C assembly. (Sec. 440-22.)

Although the two maximum values of 175 and 225 percent are placed on the rating of the branch-circuit fuse or CB, the last sentence of part (**a**) specifies that the branch-circuit protection is *not* required to be rated less than 15 A.

Sizing of branch-circuit protection for a single branch circuit to the same three motors is permitted by Sec. 430-53(c) as well as by Sec. 440-22(b) and is shown in Fig. 440-12. That layout is a specific example of the general rules covered in Sec. 440-22(b) (1), which ties the rules of Sec. 430-53(c) and (d) into the rules of Sec. 440-22(b), as shown in Fig. 440-6. Such application is based on certain factors, as covered in the UL *Electrical Appliance and Utilization Directory,* listed under "Air Conditioners, Central Cooling," as follows:

> The proper method of electrical installation (number of branch circuits, disconnects, etc.) is shown on the wiring diagram and/or marking required to be attached to the air conditioner.
>
> In air conditioners employing two or more motors or a motor(s) and other loads operating from a single supply circuit, the motor running overcurrent protective devices (including thermal protectors for motors) and other factory-installed motor circuit components and wiring are investigated on the basis of compliance with the motor-branch-circuit short circuit and ground fault protection requirements of Sec. 430-53(c) of the 1984 Edition of the **National Electrical Code**. Such multimotor and combination load equipment is to be connected only to a circuit protected by fuses or a circuit breaker with a rating which does not exceed the value marked on the data plate. This marked protective device rating is the maximum for which the equipment has been investigated and found acceptable. Where the marking specifies fuses, or

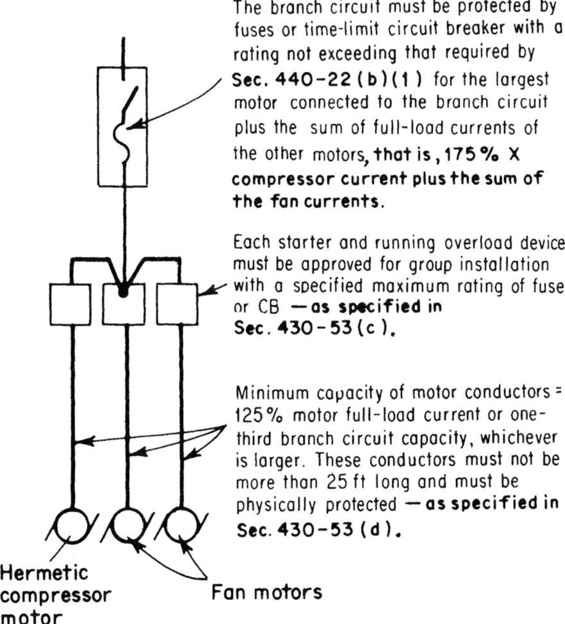
The branch circuit must be protected by fuses or time-limit circuit breaker with a rating not exceeding that required by Sec. 440-22 (b)(1) for the largest motor connected to the branch circuit plus the sum of full-load currents of the other motors, that is, 175 % X compressor current plus the sum of the fan currents.
Each starter and running overload device must be approved for group installation with a specified maximum rating of fuse or CB —as specified in Sec. 430-53 (c).
Minimum capacity of motor conductors = 125% motor full-load current or one-third branch circuit capacity, whichever is larger. These conductors must not be more than 25 ft long and must be physically protected —as specified in Sec. 430-53 (d).
Hermetic compressor motor
Fan motors

Fig. 440-6. Single multimotor branch circuit must conform to several rules. (Sec. 440-22.)

"HACR Type" circuit breakers, the circuit is intended to be protected by the type of protective device specified.

The electrical contractor and inspector charged with wiring and approving such an installation can be sure that Code requirements have been met—provided that the branch-circuit protection as specified on the unit is not exceeded and the wiring and equipment is as indicated on the wiring diagram. Provision is made in such a unit for direct connection to the branch-circuit conductors; motors are wired internally by the manufacturer.

Units are sometimes encountered in which the manufacturer has wired separate fuse cutouts for the fan motors inside the enclosure to avoid meeting the requirements of Sec. 430-53(c) for group fusing as shown in Fig. 440-7. The cutouts are normally fed from the line terminals of the compressor starter.

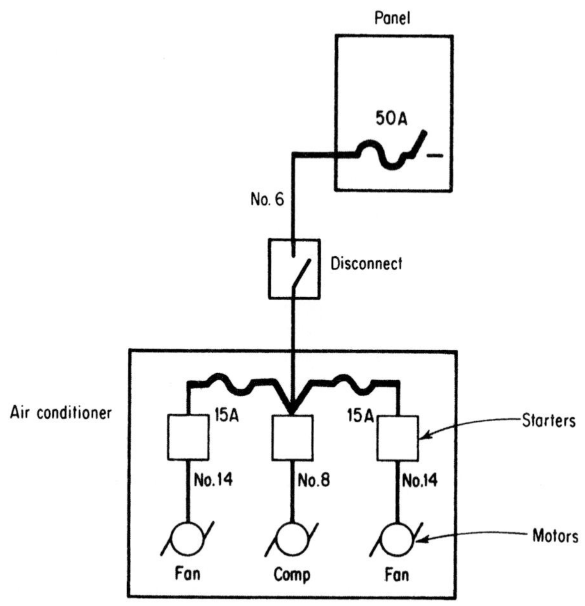

Main fuses: (175% X 26) = 45.5 amps (50-amp fuses)
Fan fuses: (300% X 4.4) = 13.2 amps (15-amp fuses)
Fan conductors: (125% X 4.4) = 5.5 amps (No.14)

Fig. 440-7. Fan circuits are sometimes fused in multimotor assemblies. (Sec. 440-22.)

Starter and disconnect sizes are the same as in Fig. 440-2, but starters and their overcurrent protection no longer need be approved for group fusing, and wiring inside the unit need not conform to Sec. 430-53(c). Fan motors may now be wired with No. 14 wire and protected with 15-A fuses. The supply circuit, feeding the same motors, will again be No. 6.

Since the fan motors are not subject to group fusing requirements, they will not restrict the maximum value of the main fuses. However, these fuses provide the only short-circuit protection for the compressor starter and conductors. Unless the compressor starter is approved for group fusing at a higher fuse rating, the fuses must not exceed 175 percent of the compressor full-load rating, or 45.5 A, calling for 50-A fuses. If needed to permit effective starting of all the motors, the fuses at the panel could be increased to 225 percent, or 60-A fuses.

Figure 440-8 shows still another hookup for supplying a multimotor assembly. Units are wired with fuse blocks for all motors, as shown. The compressor motor would be fused at 50 A, with remaining fuses and conductors as above.

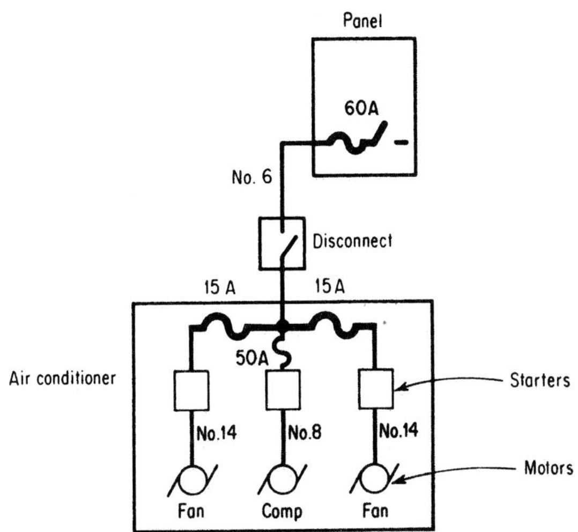

Fig. 440-8. Each motor may have individual short-circuit protection. (Sec. 440-22.)

Main fuses no longer are subject to restriction by motor starters and are sized as feeder protection per Sec. 440-33 as shown by the diagram. This results in 60-A fuses, the differential between the two being too small to matter. The need for time-delay fuses is indicated, using perhaps 30-A fuses for the compressor and 50-A fuses for the main.

The foregoing examples refer exclusively to fused disconnects and fused switches to maintain the continuity of calculations. However, Sec. 430-109 also recognizes the CB as a disconnecting means.

Figure 440-9 shows an arrangement which includes, in addition to the branch-circuit panel, a feeder panel for distribution to other units. The breakers in the branch-circuit panel serve as branch-circuit protection as well as the disconnecting means, and their ratings are computed from Secs. 430-52 and 440-22. **Code** Tables 430-150 and 430-151 would not be involved, since breakers are not rated in horsepower. Ratings of CBs in the branch-circuit panel are com-

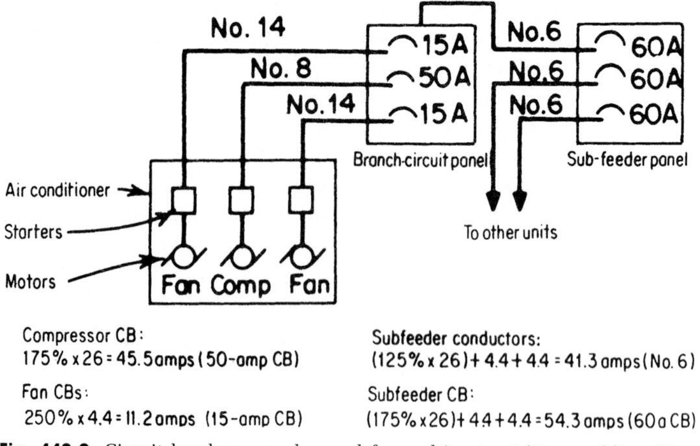

Fig. 440-9. Circuit breakers may be used for multimotor A/C assemblies. (Sec. 440-22.)

puted at 175 percent of motor current for the hermetic motor and at 250 percent of full-load current for the fan motors, to satisfy Table 430-152. Breakers in subfeeder panel are rated using Secs. 430-62 and 440-33.

Part (**c**) points out that data on a manufacturer's heater table take precedence over the maximum ratings set by Sec. 440-22(a) or (b).

440-32. Single Motor-Compressor. Branch-circuit conductors supplying a motor in a packaged unit are not sized in the same manner as other motor loads (Sec. 430-22). Instead of using the full-load current from **Code** Tables 430-148 to 430-150, the *marked* rated-load current or the *marked* branch-circuit selection current must be used in determining minimum required conductor ampacity. Note that branch-circuit selection current must be used where it is given.

Examples are shown in the typical circuits shown in Figs. 440-2 and 440-6.

440-33. Motor-Compressor(s) With or Without Additional Motor Loads. Where more than one motor is connected to the same feeder or branch circuit, calculation of conductor sizes must provide ampere capacity at least equal to the sum of the nameplate rated-load currents or branch-circuit selection currents (using the higher of those values in all cases) plus 25 percent of the current (either rated-load or branch-circuit selection current for a hermetic motor or **NEC** table current for standard motor) of the largest motor of the group. Examples are shown in Figs. 440-2 and 440-6.

In Fig. 440-6, the question arises as to whether the No. 6 conductors feeding the unit may be decreased to No. 8 inside the unit to feed the compressor motor in the absence of fuses for this motor at the point of reduction. The status of the main feed to the unit—whether it should be considered a feeder or a branch circuit—is in doubt, since it is a branch circuit as far as the compressor motor is concerned and a feeder in that it also supplied the two fused fan circuits.

Considered solely as a branch circuit to the compressor, these conductors normally would be No. 8 to handle the 26-A compressor motor full-load current, protected at not more than 175 percent or 50-A fuses. Therefore, since

50-A fuses (or less) will actually be used for the main feed, they constitute proper protection for No. 8 conductors and their use should be permitted. The existence of No. 6 conductors over part of the circuit adds to its capacity and safety rather than detracting from it.

It is particularly important to keep in mind when selecting conductor sizes that the nameplate current ratings of air-conditioning motors are not constant maximum values during operation. Ratings are established and tested under standard conditions of temperature and humidity. Operation under weather conditions more severe than those at which the ratings are established will result in a greater running current, which can approach the maximum value permitted by the overcurrent device. Operating voltage less than the limits specified on the motor nameplate also contributes to higher full-load current values, even under standard conditions. Conductor capacity should be sufficient to handle these higher currents. Motor feeders are sized according to Sec. 440-33. Since overload protection may permit motors to run continuously overloaded (up to 140 percent full load), feeders must be sized to handle such overload. By basing calculations on the largest motor of the group, the extra capacity thus provided will normally be enough to handle any unforeseen overload on the smaller motors involved with enough diversity existing in any normal group of motors to make consistent overloads on all motors at one time unlikely. However, a group of air-conditioning compressor motors all of the same size on a single feeder have a common function—reducing the ambient temperature. Except for slight possible variations, weather conditions affect each conditioner to the same degree and at the same time. Therefore, if one unit is operating at an overload, it is likely that the rest are also.

440-35. Multimotor and Combination-Load Equipment. This rule ties into the data required by UL to be marked on such equipment. Refer to the UL data quoted in Sec. 440-22.

440-41. Rating. The basic rule calls for a compressor controller to have a full-load current rating and a locked-rotor current rating not less than the compressor nameplate rated-load current or branch-circuit selection current (whichever is greater) and locked-rotor current. But, as noted for the disconnect under Sec. 440-12, for sealed (hermetic-type) refrigeration compressor motors, selection of the size of controller is slightly more involved than it is for standard applications. Because of their low-temperature operating conditions, hermetic motors can handle heavier loads than general-purpose motors of equivalent size and rotor-stator construction. And because the capabilities of such motors cannot be accurately defined in terms of horsepower, they are rated in terms of full-load current and locked-rotor current for polyphase motors and larger single-phase motors. Accordingly, selection of controller size is different than in the case of a general-purpose motor where horsepower ratings must be matched, because controllers marked in horsepower only must be carefully related to hermetic motors that are *not* marked in horsepower.

For controllers rated in horsepower, selection of the size required for a particular hermetic motor can be made after the nameplate rated-load current, or branch-circuit selection current, whichever is greater, and locked-rotor current of the motor have been converted to an equivalent horsepower rating. To get this equivalent horsepower rating, which is the required size of controller, the tables in Art. 430 must be used. First, the nameplate full-load current at the operating voltage of the motor is located in **Code** Tables 430-148, 430-149, or 430-150 and the horsepower rating which corresponds to it is noted. Then the nameplate locked-rotor current of the motor is found in **Code** Table 430-151, and again the corresponding horsepower is noted. In all tables, if the exact value of current is not listed, the next higher value should be used to obtain an equivalent horsepower, by reading horizontally to the horsepower column at the left side of those tables. If the two horsepower ratings obtained in this way are not the same, the larger value is taken as the required size of controller.

A typical example follows:

Given: A 230-V, 3-phase, squirrel-cage induction motor in a compressor has a nameplate rated-load current of 25.8 A and a nameplate locked-rotor current of 90 A.

Procedure: From **Code** Table 430-150, 28 A is the next higher current to the nameplate current of 25.8 under the column for 230-V motors and the corresponding horsepower rating for such a motor is 10 hp.

From **Code** Table 430-151, Art. 430, a locked-rotor current rating of 90 A for a 230-V, 3-phase motor requires a controller rated at 5 hp. The two values of horsepower obtained are not the same, so the higher rating is selected as the acceptable unit for the conditions. A 10-hp motor controller must be used.

Some controllers may be rated not in horsepower but in full-load current and locked-rotor current. For use with a hermetic motor, such a controller must simply have current ratings equal to or greater than the nameplate rated-load current and locked-rotor current of the motor.

440-52. Application and Selection. The basic rule of part (**a**) calls for a running overload relay set to trip at not more than 140 percent of the rated-load current of a motor-compressor. If a fuse or time-delay CB is used to provide overload protection, it must be rated not over 125 percent of the compressor rated-load current. Note that those are absolute maximum values of overload protection and no permission is given to go to "the next higher standard rating" of protection where 1.4 or 1.25 times motor current does not yield an amp value that exactly corresponds to a standard rating of a relay or of a fuse or CB.

Running overload protective devices for a motor are necessary to protect the motor, its associated controls, and the branch-circuit conductors against heat damage due to excessive motor currents. High currents may be caused by the motor being overloaded for a considerable period of time, by consistently low or unbalanced line voltage, by single-phasing of a polyphase motor, or by the motor stalling or failing to start.

Damage may occur more quickly to a hermetic motor which stalls or fails to start than to a conventional open-type motor. Due to the presence of the cool refrigerant atmosphere under normal conditions, a hermetic motor is permitted to operate at a rated current which is closer to the locked-rotor current than is the same rated current of an open-type motor of the same nominal horsepower rating. The curves of Fig. 440-10 show the typical relation between locked-rotor and full-loaded currents of small open-type and hermetic motors. Because a

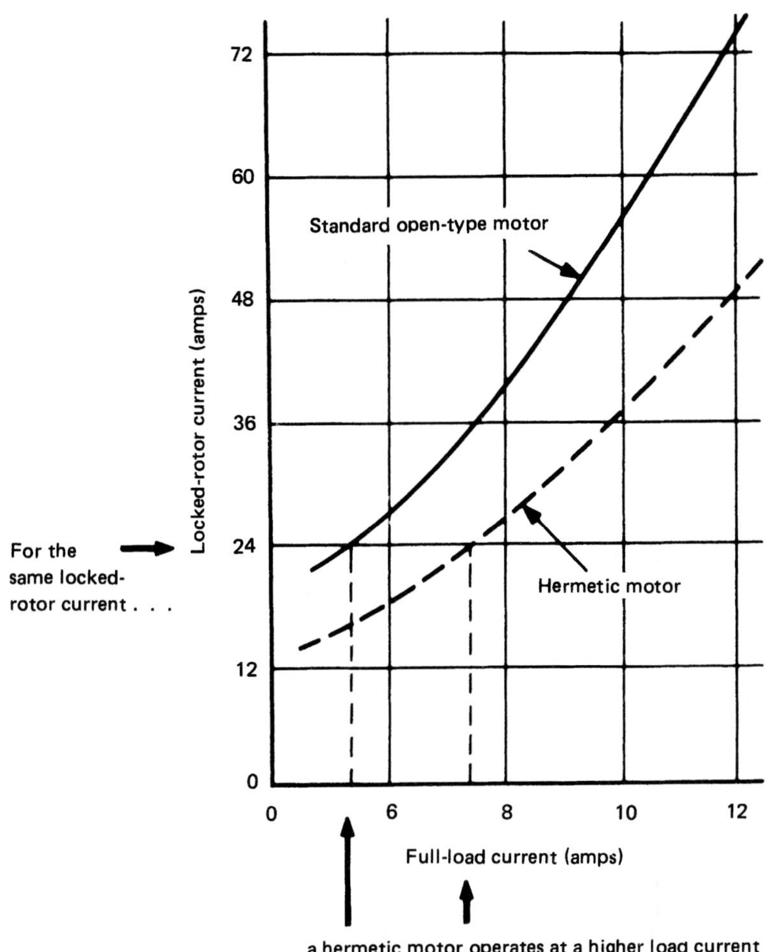

Fig. 440-10. Hermetic motors operate at full-load currents closer to locked-rotor currents. (Sec. 440-52.)

hermetic motor operates within the refrigerant atmosphere, it is constantly cooled by that atmosphere. As a result, a given size of motor may be operated at a higher current than it could be if it were used as an open, general-purpose motor without the refrigerant cycle to remove heat from the windings. In effect, a hermetic motor is operated overloaded because the cooling cycle prevents overheating. For instance, a 5-hp open motor can be loaded as if it were a 7½-hp motor when it is cooled by the refrigerant. The full-load operating current of such a motor is higher than the normal current drawn by a 5-hp load and is, therefore, closer to the value of locked-rotor current, which is the same no matter how the motor is used.

When the rotor of a hermetic motor is slowed down because of overload or is at a standstill, there is not sufficient circulation of the refrigerant to carry away the heat; and heat builds up in the windings. Special quick-acting thermal and hydraulic-magnetic devices have been developed to reduce the time required to disconnect the hermetic motor from the line before damage occurs when an overload condition develops.

Room air conditioners and packaged unit compressors are normally required to incorporate running overload protection which will restrict the heat rise to definite maximum safe temperatures in case of locked-rotor conditions. Room conditioners normally use inherent protectors built into the compressor housing which respond to the temperature of the housing. Larger units also often use inherent protection in addition to quick-acting overload heaters installed in the motor starter, which respond only to current. These protective methods are covered in paragraphs (**2**) and (**4**) of Sec. 440-52(a).

The electrical installer will normally be concerned with the running overcurrent protection of a hermetic motor only when it becomes necessary to replace the existing devices supplied with the equipment. For this purpose, compressor manufacturers' warranties explicitly specify catalog numbers of replacements which are to be used to ensure proper operation of the equipment.

440-60. General. These rules on room air-conditioning units recognize that such units are basically appliances, are low-capacity electrical loads, and may be supplied either by an individual branch circuit to a unit conditioner or by connection to a branch circuit that also supplies lighting and/or other appliances. For all Code discussion purposes, an air-conditioning unit of the window, console, or through-the-wall type is classified as a "fixed appliance" —which is described in Art. 100 as "fastened or otherwise secured at a specific location." Such an appliance may be cord-connected or it may be fixed-wired (so-called "permanently connected").

Section 210-23 of Art. 210 on "Branch Circuits" must also be applied in cases where a unit room air conditioner is connected to a branch circuit supplying lighting or other appliance load.

When a unit air conditioner is connected to a circuit supplying lighting and/or one or more appliances that are not motor loads, the rules of Art. 210 must be observed:

1. Section 210-22(a) says that "where a circuit supplies only air-conditioning and/or refrigerating equipment, Article 440 shall apply."
2. For plug connection of the A/C unit, Sec. 210-7(a) says that receptacles installed on 15- and 20-A branch circuits must be of the grounding type and must have their grounding terminal effectively connected to a grounding conductor or grounded raceway or metal cable armor.
3. On 15- and 20-A branch circuits, the total rating of a unit air conditioner ("utilization equipment fastened in place") must not exceed 50 percent of the branch-circuit rating when lighting units or portable appliances are also supplied [Sec. 210-23(a)]. It was on the basis of that rule that the 7½-A air conditioner was developed. Being 50 percent of a 15-A branch circuit, such units are acceptable for connection to a receptacle on a 15- or 20-A circuit that supplies lighting and receptacle outlets.

4. A branch circuit larger than 20 A may *not* be used to supply a unit conditioner plus a lighting load. Circuits rated 25, 30, 40, or 50 A may be used to supply fixed lighting or appliances—but not both types of loads.

440-61. Grounding. Air-conditioner units that are connected by permanent wiring must be grounded in accordance with the basic rules of Sec. 250-42 covering equipment that is "fastened in place or connected by permanent wiring." Section 250-45 covers grounding of cord-and-plug-connected air conditioners by means of an equipment grounding conductor run within the supply cord for each such unit.

The nameplate marking of a room air conditioner shall be used in determining the branch-circuit requirements, and each unit shall be considered as a single motor unless the nameplate is otherwise marked. If the nameplate is marked to indicate two or more motors, Secs. 430-53 and 440-22(b) (1) must be satisfied, covering the use of several motors on one branch circuit.

440-62. Branch-Circuit Requirements. Even though a room air conditioner contains more than one motor (usually the hermetic compressor motor and the fan motor), this rule notes that for a cord-and-plug-connected air conditioner the entire unit assembly may be treated as a single-motor load under the conditions given.

Examples of the rule of part (**b**) are shown in Fig. 440-11. The total marked rating of any cord-and-plug-connected air-conditioning unit must *not* exceed 80 percent of the rating of a branch circuit which does not supply lighting units or other appliances, for units rated up to 40 A, 250 V, single phase.

As noted under Sec. 440-60, Sec. 210-22(a) seems to say that only Art. 440 and not Art. 210 applies when the circuit supplies only a motor-operated load. But, since Arts. 430 and 440 do not "rate" branch circuits—either on the basis of the size of the short-circuit protective device or the size of the conductors—a question arises about the meaning of the phrase "80 percent of the rating of a branch circuit." Does that mean 80 percent of the rating of the fuse or CB? The answer is: It means 80 percent of the rating of the protective device, which rating is not more than the amp rating of the circuit wire. The circuit as described here is taken to be a circuit with "rating" as given in Art. 210 and covered by part (**4**) of Sec. 440-62(a).

As part (**c**) of this section notes, the total marked rating of air-conditioning equipment must *not* exceed 50 percent of the rating of a branch circuit which *also* supplies lighting or other appliances. And Secs. 210-22(a) and 210-23 must be observed. From the rule, we can see that the Code permits air-conditioning units to be plugged into existing circuits which supply lighting loads or other appliances. By the provisions of this section, such a conditioner must not draw more than 7½ A full load (nameplate rating) when connected to a 15-A circuit and not more than 10 A when connected to a 20-A circuit. In addition, Sec. 210-22(a) requires that the branch-circuit capacity must not be less than 125 percent of the air-conditioner load plus the sum of the other loads. The existing load on the circuit (lights or other appliances) must be low enough so that the total load on the circuit after the addition of 125 percent of the ampere load of the air conditioner is not greater than 15 A in the case of the 15-A circuit or greater than 20 A on a 20-A circuit. This is in accordance with Sec. 210-22(a) which restricts the total loading on such a circuit.

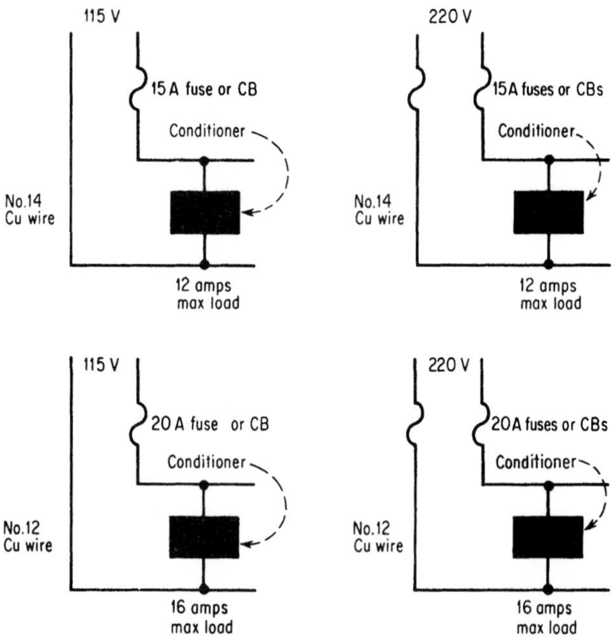

Fig. 440-11. Room air conditioners must not load an individual branch circuit over 80 percent of rating. (Sec. 440-62.)

Assuming that a 7½-A conditioner is connected to such a 15-A existing circuit, it would mean that the circuit before the addition of the air-conditioner load of 7.5 × 1.25, or 10 A, could have been loaded to no more than 5 A, as shown in Fig. 440-12.

A problem exists in connecting two or more conditioners to the same circuit. Compressor and fan motors and their controls, when installed in the same enclosure and fed by one circuit, are approved by UL for group installation when tested as a unit appliance. However, an air conditioner's component parts carry no general group-fusing approval which would permit the several separate conditioners to operate on the same circuit in accordance with Sec. 430-53(c). To connect more than one conditioner to the same branch circuit, the provisions of either Secs. 430-53(a) or 430-53(b) must be fulfilled, treating each cord-connected conditioner as a single-motor load.

According to Sec. 430-53(a), which applies only to motors rated not over 6 A, two 115-V, 6-A conditioners could be used on a 15-A circuit, three 5-A conditioners could be used on a 20-A circuit which supplies no other load, and two 220-V, 6-A units could be operated on a 15-A circuit as shown in Fig. 440-13. But it could be argued that the maximum load in any such application may be

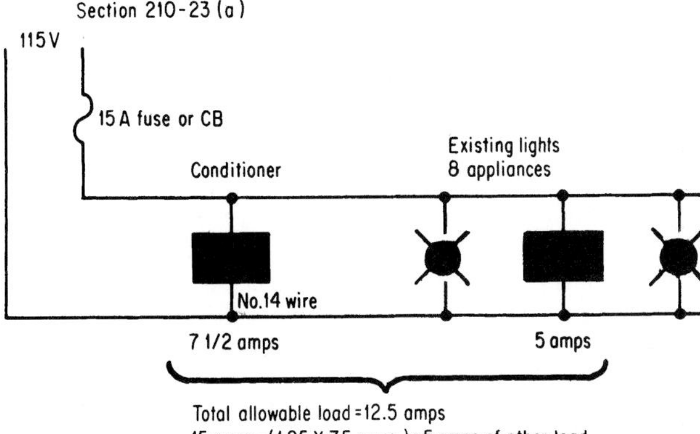

Fig. 440-12. Room air conditioner must not exceed 50 percent of circuit rating if other loads are supplied. (Sec. 440-62.)

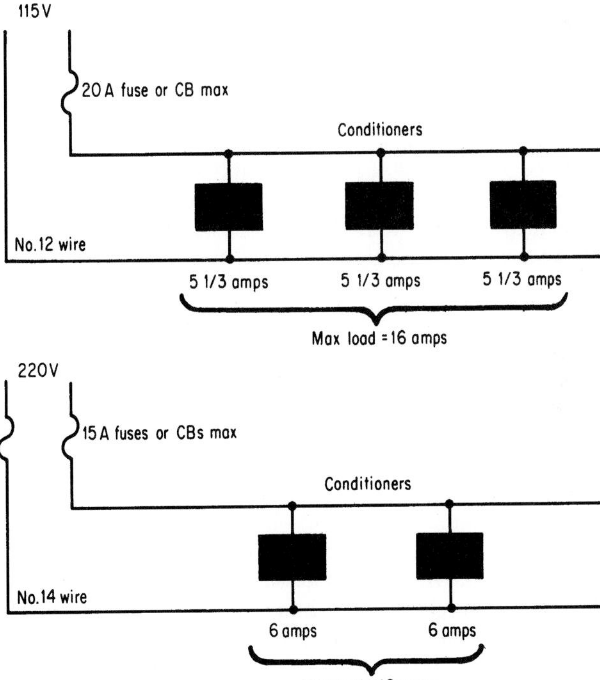

Fig. 440-13. Rules limit use of two or more room conditioners on single circuit. (Sec. 440-62.)

calculated at 125 percent times the current of largest air conditioner plus the sum of load currents of the additional air conditioners, with that total current being permitted right up to the rating of the circuit.

However, most conditioners sold today exceed 6-A full-load current. As a result, the application of two or more units as permitted by Sec. 430-53(a) is limited. But Sec. 430-53(b) does offer considerable opportunity for using more than one air conditioner on a single circuit. Figure 440-14 shows two examples of such application, which can be used if the branch-circuit protective device will not open under the most severe normal conditions which might be encountered. Although that usage is a complex connection among several Code rules and requires clearance with inspection authorities, it can provide very substantial economies.

Many local codes avoid the complications of connecting conditioners to existing circuits and connecting more than one conditioner to the same circuit by requiring a separate branch circuit for each conditioner. Multiple installations involving many room conditioners such as are frequently encountered in hotels, offices, etc., require careful planning to meet Code requirements and yet minimize expensive branch-circuit lengths.

WATCH OUT!

The NE Code refers to motor-operated appliances and/or to room air conditioners in Arts. 210, 422, and 440. Great care must be exercised in correlating the various Code rules in these different articles to assure effective compliance with the letter and spirit of Code meaning. There is much crossover in terminology and references, making it difficult to tell whether a room air conditioner should be treated as an appliance circuit load or a motor load. However, a step-by-step approach to the problem which keeps in mind the intent of these provisions can resolve confusing points. Since the manufacturer is required to supply the motor-running overcurrent protection, no problems should arise concerning these devices. For larger units connected permanently to the distribution system, these can be treated directly as hermetic motor loads, using the provisions of Art. 440.

It may be assumed that a window or through-the-wall unit will operate satisfactorily on a standard fuse of the same rating as its attachment plug cap if there is no marking to the contrary on the unit. In any event, a time-delay fuse of the same or smaller rating could be substituted. If CBs are used for branch-circuit protection, a 15-A breaker will normally hold the starting current if a standard 15-A fuse will, since such breakers have inherent time delay. If the unit is marked to require a 15-A time-delay fuse and a 15-A breaker will not hold the starting current, few inspectors will object to the use of a 20-A breaker, since Art. 430 permits such a procedure for motor loads.

Normally, starting problems are not severe with these units, since the low inertia of present-day motor-compressor combinations permits them to reach full speed within a few cycles. Such a rapid drop in starting current is usually well within the time permitted by the trip or rupture characteristics of the breaker or fuse.

Similarly, the question of wire size may be resolved by application of Arts. 210, 422, or 440. Rarely do room conditioners even as large as 2 tons take

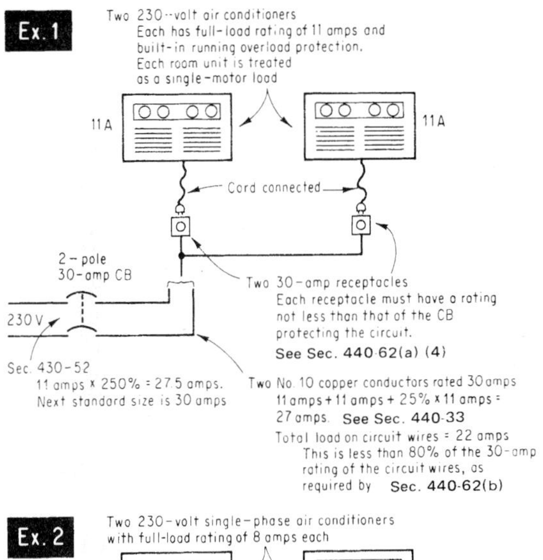
Ex. 1
Two 230-volt air conditioners
Each has full-load rating of 11 amps and built-in running overload protection.
Each room unit is treated as a single-motor load
11 A
11 A
Cord connected
2-pole 30-amp CB
230 V
Two 30-amp receptacles
Each receptacle must have a rating not less than that of the CB protecting the circuit.
See Sec. 440-62(a) (4)
Sec. 430-52
11 amps x 250% = 27.5 amps.
Next standard size is 30 amps
Two No. 10 copper conductors rated 30 amps
11 amps + 11 amps + 25% x 11 amps = 27 amps. See Sec. 440-33
Total load on circuit wires = 22 amps
This is less than 80% of the 30-amp rating of the circuit wires, as required by Sec. 440-62(b)

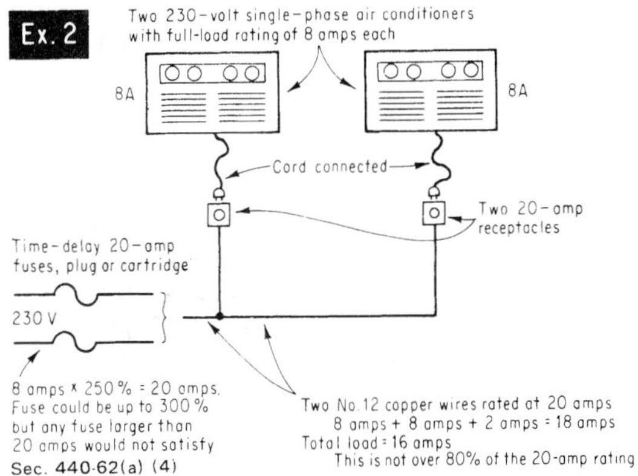
Ex. 2
Two 230-volt single-phase air conditioners with full-load rating of 8 amps each
8A
8A
Cord connected
Two 20-amp receptacles
Time-delay 20-amp fuses, plug or cartridge
230 V
8 amps x 250% = 20 amps.
Fuse could be up to 300% but any fuse larger than 20 amps would not satisfy Sec. 440-62(a) (4)
Two No. 12 copper wires rated at 20 amps
8 amps + 8 amps + 2 amps = 18 amps
Total load = 16 amps
This is not over 80% of the 20-amp rating

NOTE: Fuse or CB sizing may be required to conform to Sec. 440-22 (b), with a maximum rating of 175% times load current of one conditioner plus the current of the other conditioner. Or the 175% value itself may be held as the maximum rating of branch-current protection.

Fig. 440-14. These hookups have been accepted as conforming to rules of Arts. 440 and 430. (Sec. 440-62.)

more than 13-A running current; hence No. 12 copper or No. 10 aluminum conductors are more than sufficient. In addition, many local codes prohibit use of conductors smaller than No. 12. In localities where No. 14 wire may be used, provisions of Sec. 440-62(b), restricting the loading to 80 percent of the circuit rating, must determine the wire size, where "rating" is interpreted as referring to the conductor carrying capacity. If Art. 440 is used to determine the wire size, the 125 percent requirement of Sec. 440-32 gives the same result.

Figure 440-15 shows one feeder of an installation involving many room conditioners which practically eliminates branch-circuit wiring and will serve to illustrate the complications of circuit calculations for a multiple-unit installation. Total running current of each unit is 12 A as shown; hence No. 14 copper wire could be used for branch-circuit conductors, protected by a 15-A fuse—either standard or time-delay. However, if the appropriate conductors of a 4-wire, 3-phase feeder were routed to the location of each conditioner and a combination fuseholder and receptacle installed as shown, the only existing branch-circuit conductors would be the jumpers between the feeder, the fuseholder, and the receptacles. These jumpers, then, could be No. 14 wire. Assuming that the fuse-receptacle unit is mounted directly on or in close proximity with the junction box in which the tap to the feeder is made, the No. 14 wire is justified from the fuse to the feeder since it is not over 10 ft (3.05 m) long and is sufficient for the load supplied (Sec. 240-21, Exception No. 2). Since both motors usually start simultaneously, the total unit current is used to compute feeder conductor size and protection: 125 percent times 12 plus 24 is 39 A, permitting No. 8 conductors. However, this is practically the limit of the circuit's capacity; there is no provision for overload, and voltage drop is very likely to be a factor at the end of the feeder. Therefore No. 6 conductors should be used.

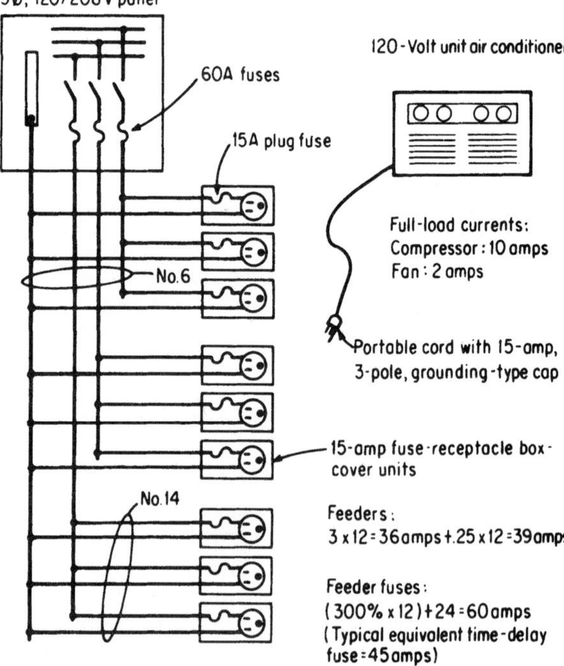

Fig. 440-15. This type of circuiting was used for air conditioners in a hotel modernization project. (Sec. 440-62.)

Feeder protection is calculated on the basis of 300 percent times 12 plus 24, or 60-A fuses. Substitution of time-delay fuses for this 39-A feeder load would likely permit 45-A fuses.

Important: The rules of Sec. 440-62 apply only to cord-and-plug-connected room air conditioners. A unit room air conditioner that has a *fixed* (not cord and plug) connection to its supply must be treated as a group of several individual motors and protected in accordance with Secs. 430-53 and 440-22(b), covering several motors on one branch.

440-63. Disconnecting Means. A disconnect is required for every unit air conditioner. An attachment plug and receptacle or a separable connector may serve as the disconnecting means (Fig. 440-16).

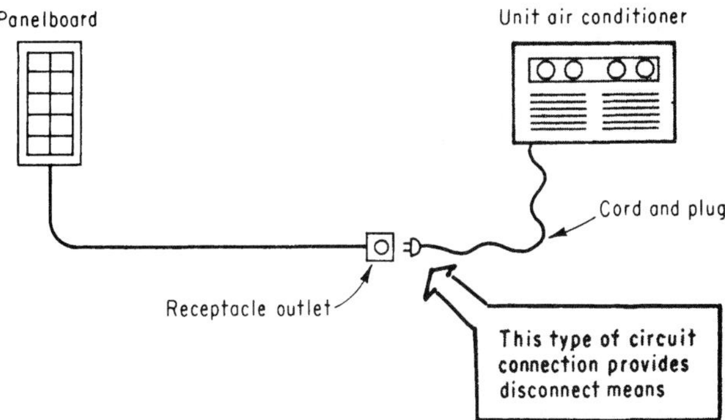

Fig. 440-16. Plug-and-receptacle serves as required disconnect means. (Sec. 440-63.)

If a fixed connection is made to an A/C unit from the branch-circuit wiring system (i.e., not a plug-in connection to a receptacle), consideration must be given to a means of disconnect, as required in Secs. 422-21 and 422-25 for appliances:

- For unit air conditioners in any type of occupancy, the branch-circuit switch or CB may, where readily accessible to the user of the appliance, serve as the disconnecting means. Figure 440-17 shows this, but the switch or CB is permitted to be out of sight by the Exception to Sec. 422-27 when the A/C unit has an internal OFF switch—which all units do have. And Sec. 422-27 requires the disconnect means for a motor-driven appliance to be within sight from the air-conditioner unit.

  Because air conditioners have unit switches within them, the disconnect provisions of Sec. 422-25 may be applied. The internal unit switch with a marked OFF position that opens all ungrounded conductors may serve as the disconnect and is considered within sight as required by Sec. 422-26 in any case where there is another disconnect means as follows:
- In multifamily (more than two) dwellings, the other disconnect means must be within the apartment where the conditioner is installed or on the same floor as the apartment.

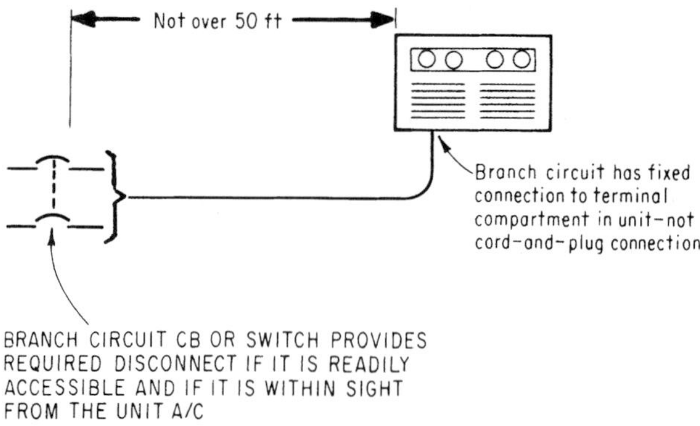

Fig. 440-17. Branch circuit CB or switch may serve as disconnect. (Sec. 440-63.)

- In two-family dwellings, the other disconnect may be outside the apartment in which the appliance is installed. It may be the service disconnect.
- In single-family dwellings, the service disconnect may serve as the other disconnect means—whether the branch circuit to the conditioner is fed from plug fuses or from a breaker or switch (Fig. 440-18).

ARTICLE 445. GENERATORS

445-4. Overcurrent Protection. Alternating-current generators can be so designed that on excessive overload the voltage falls off sufficiently to limit the current and power output to values that will not damage the generator during a

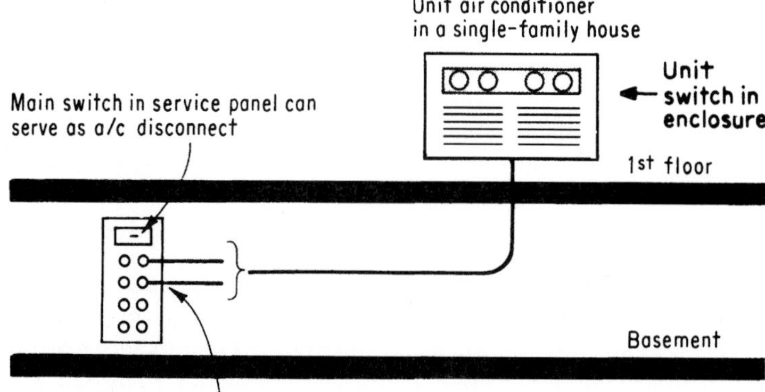

Fig. 440-18. Service disconnect may be the "other disconnect" for A/C unit in private house. (Sec. 440-63.)

short period of time. Whether or not automatic overcurrent protection of a generator should be omitted in any particular case is a question that can best be answered by the manufacturer of the generator. It is common practice to operate an exciter without overcurrent protection, rather than risk the shutdown of the main generator due to accidental opening of the exciter fuse or CB.

Figure 445-1 shows the connections of a 2-wire DC generator with a single-pole protective device. If the machine is operated in multiple with one or more other generators, and so has an equalizer lead connected to the positive terminal, the current may divide at the positive terminal, part passing through the series field and positive lead and part passing through the equalizer lead. The entire current generated passes through the negative lead; therefore the fuse or CB, or at least the operating coil of a CB, must be placed in the negative lead. The protective device should not open the shunt-field circuit, because if this circuit were opened with the field at full strength, a very high voltage would be induced which might break down the insulation of the field winding.

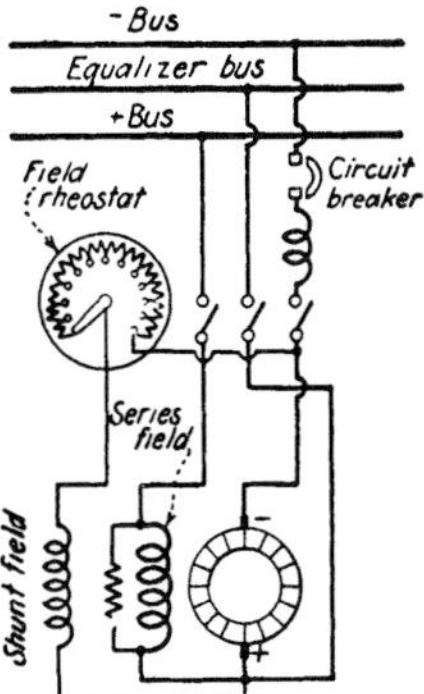

Fig. 445-1. With this connection, a single-pole CB can protect a 2-wire DC generator. (Sec. 445-4.)

Paragraph (**c**) is intended to apply particularly to generators used in electrolytic work. Where such a generator forms part of a motor-generator set, no fuse or CB is necessary in the generator leads if the motor-running protective device will open when the generator delivers 150 percent of its rated full-load current.

In paragraph (**d**), use of a balancer set to obtain a 3-wire system from a 2-wire main generator is covered, as shown in Fig. 445-2. Each of the two generators used as a balancer set carries approximately one-half the unbalanced load; hence these two machines are always much smaller than the main generator. In case of an excessive unbalance of the load, the balancer set might be overloaded while there is no overload on the main generator. This condition may be guarded against by installing a double-pole CB with one pole connected in each lead of the main generator and with the operating coil properly designed to be connected in the neutral of the 3-wire system. In Fig. 445-2, the CB is arranged so as to be operated by either one of the coils A in the leads from the main generator or by coil B in the neutral lead from the balancer set.

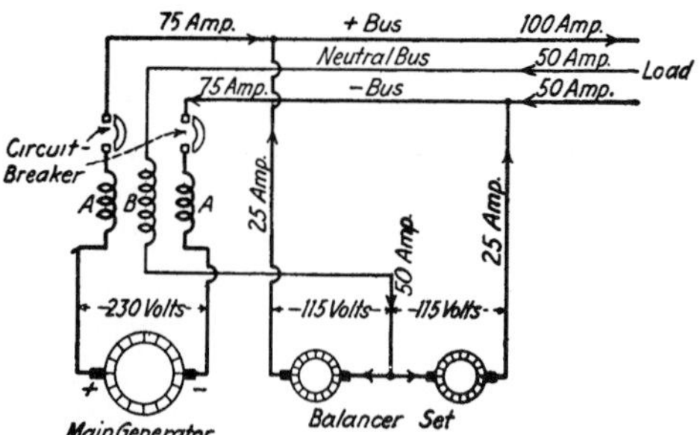

Fig. 445-2. A balancer set supplies the unbalanced neutral current of a 3-wire system, with each generator carrying 25 of the 50-A unbalance. (Sec. 445-4.)

445-5. Ampacity of Conductors. Two sentences at the end of this section clarify sizing of circuit conductors connecting a generator to the control and protective device(s) it serves:

1. The neutral of the generator feeder may have its size reduced from the minimum capacity required for the phase legs. As with any feeder or service circuit, the neutral has to have only enough ampacity for the unbalanced load it will handle—as covered by Sec. 220-22.
2. When a generator neutral is not grounded at its terminals, as is permitted by Sec. 250-5(d), the neutral conductor from the generator must be sized not only for its unbalanced load, as required in Sec. 220-22, but also for carrying ground-fault current. For a generator feeder neutral to be adequately sized as an equipment grounding conductor, to effectively carry enough current to operate overcurrent devices in a grounded system, Sec. 250-23(b) requires that the neutral be not smaller than 12½ percent of the cross-sectional area of the largest phase leg of the generator feeder. (See Fig. 445-3.)

445-6. Protection of Live Parts. As a general rule, no generator should be "accessible to unqualified persons." If necessary to place a generator operating at over 50 V to ground in a location where it is so exposed, the commutator or collector rings, brushes, and any exposed terminals should be provided with guards which will prevent any accidental contact with these live parts.

ARTICLE 450. TRANSFORMERS AND TRANSFORMER VAULTS

450-1. Scope. The Exceptions indicate those transformer applications that are not subject to the rules of Art. 450. The Exceptions shown in Fig. 450-1 are as follows:

Exception No. 2 excludes any dry-type transformer that is a component part of manufactured equipment, provided that the transformer complies with the

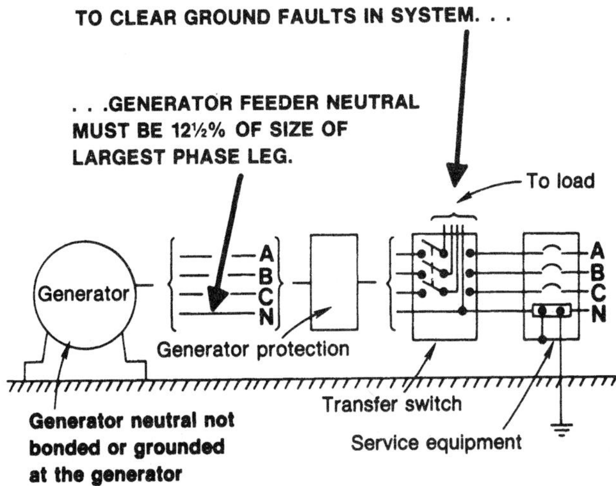

Fig. 445-3. Neutral must have adequate capacity for generator that is not a "separately derived" system source (that is, does not have its neutral bonded and grounded). (Sec. 445-5.)

requirements for such equipment. Those requirements include UL standards on the construction of the particular equipment. This exclusion applies, for instance, to control transformers within a motor starter or within a motor control center. However, although such transformers do not have to be protected in accordance with Sec. 450-3(b), such control transformer circuits must have their control conductors protected as described under Sec. 430-72(b). But a separate control transformer—one that is external to other equipment and is not an integral part of any other piece of equipment—must conform to the protection rules of Sec. 450-3 and other rules in Art. 450.

Exception No. 6 points out that ballasts for electric-discharge lighting (although they *are* transformers—either autotransformers or separate-winding, magnetically coupled types) are treated as lighting accessories rather than transformers.

Exception No. 8 notes that liquid-filled or dry-type transformers used for research, development, or testing are exempt from the requirements of Art. 450 provided that effective arrangements are made to safeguard any persons from contacting energized terminals or conductors. Again, in the interest of the unusual conditions that frequently prevail in industrial occupancy, this rule recognizes that transformers used for research, development, or testing are commonly under the sole control of entirely competent individuals and exempts such special applications from the normal rules that apply to general-purpose transformers used for distribution within buildings and for energy supply to utilization equipment, controls, signals, communications, and the like. (See Fig. 450-2.)

UL listing The *Electrical Construction Materials Directory* of the UL lists "Transformers—Power." To satisfy NE Code and OSHA regulations, as well as local code rules on acceptability of equipment, any transformers of the types

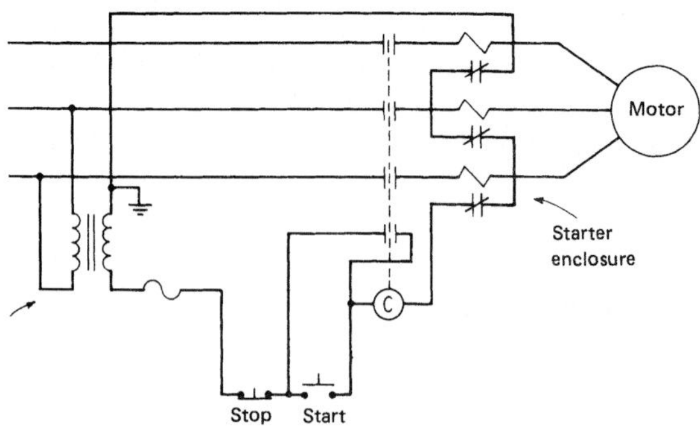

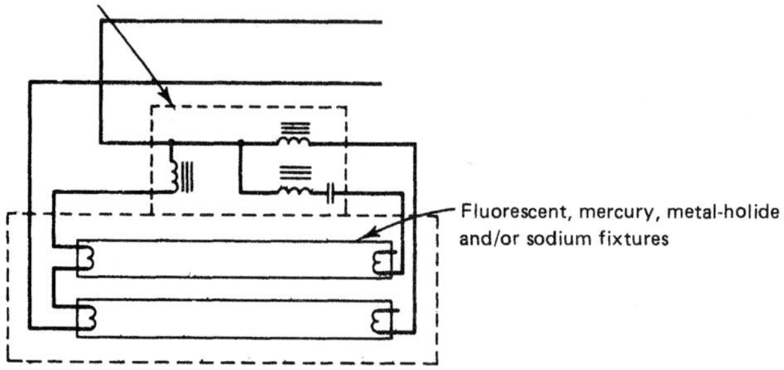

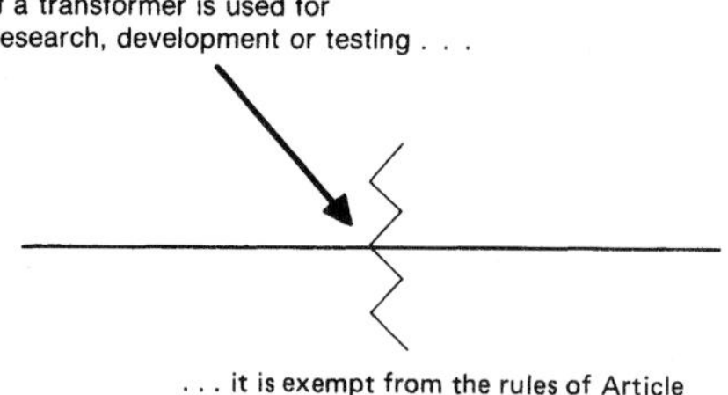

Fig. 450-1. These transformer applications are exempt from the rules of Art. 450. (Sec. 450-1.)

and sizes covered by UL listing must be so listed. Use of an unlisted transformer of a type and size covered by UL listing would certainly be considered a violation of the spirit of NE Code Sec. 110-2.

UL listing covers "air-cooled" types rated up to 500 kVA for single-phase transformers and up to 1,500 kVA for 3-phase units (all up to 600-V rating).

450-2. Definitions. A "transformer" is an individual transformer, single or polyphase, identified by a single nameplate, unless otherwise indicated in this

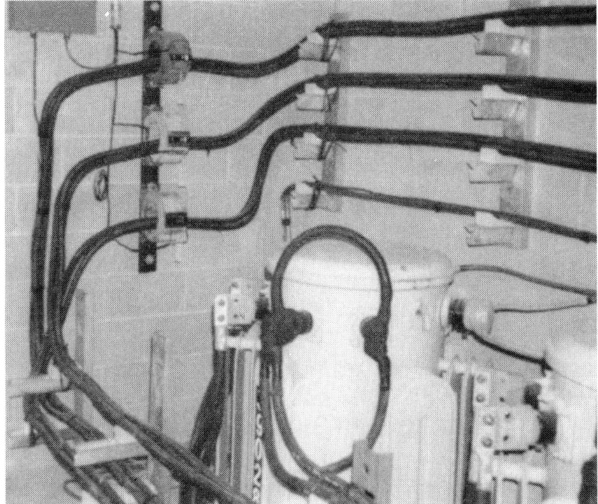

Fig. 450-2. Transformers that are set up in a laboratory to derive power for purposes of testing other equipment or powering an experiment are exempt from the rules of Art. 450 provided care is taken to protect personnel from any hazards due to exposed energized parts. (Sec. 450-1.)

article. Three single-phase transformers connected for a 3-phase transformation must be taken as three transformers, not one. This definition helps to clarify the contents of some of the rules of Art. 450.

450-3. Overcurrent Protection. This section covers overcurrent protection in great detail, and other Code rules (Secs. 240-21, 240-40, and 384-16 in particular) usually get involved in transformer applications. Although there is no rule on disconnects, use of required overcurrent protection results in the presence of a fused switch or CB that may serve as disconnecting means.

It should be understood that the overcurrent protection required by this section is for transformers *only.* Such overcurrent protection will not necessarily protect the primary or secondary conductors or equipment connected on the secondary side of the transformer. Using overcurrent protection to the maximum values permitted by these rules would require much larger conductors than the full-load current rating of the transformer (other than permitted in the 25-ft (7.62-m) tap rule in Sec. 240-21(d). Accordingly, to avoid using oversized conductors, overcurrent devices should be selected at about 110 to 125 percent of the transformer full-load current rating. And when using such smaller overcurrent protection, devices should be of the time-delay type (on the primary side) to compensate for inrush currents which reach 8 to 10 times the full-load primary current of the transformer for about $\frac{1}{10}$ sec when energized initially.

In approaching a transformer installation it is best to use a one-line diagram, such as shown in the accompanying sketches. Then by applying the tap rules in Sec. 240-21 proper protection of the conductors and equipment, which are part of the system, will be achieved. See comments following Sec. 240-21.

Section 230-207 and Sec. 240-3(i) are the only Code rules that consider properly sized primary overcurrent devices to protect the secondary conductors without secondary protection and no limit to the length of secondary conductors. The strict requirements in Sec. 230-207 apply where the transformers are in a *vault,* the primary load-interrupter switch is manually operable from outside the vault, and large secondary conductors are provided to achieve reflected protection through the transformer to the primary overcurrent protection. *It is important to note that in all other cases primary circuit protection for transformer protection is not acceptable as suitable protection for the secondary circuit conductors*—even if the secondary conductors have an ampacity equal to the ampacity of the primary conductors times the primary/secondary voltage ratio. See Sec. 240-3(i).

On 3- and 4-wire transformer secondaries, it is possible that an unbalanced load may greatly exceed the secondary conductor ampacity, which was selected assuming balanced conditions. Because of this, the NE Code does not permit the protection of secondary conductors by overcurrent devices operating through a transformer from the primary of a transformer having a 3-wire or 4-wire secondary. For other than 2-wire to 2-wire transformers, protection of secondary conductors has to be provided completely separately from any primary-side protection. Section 384-16(d) states that required main protection for a lighting panel on the secondary side of a transformer must be located on the secondary side. However, Sec. 240-3(i) recognizes such primary protection of the secondary of single-phase, 2-wire to 2-wire transformers if the primary OC protection complies with Sec. 450-3 and does not exceed the value determined by multiplying the secondary conductor ampacity by the secondary-to-primary transformer voltage ratio. The lengths of the primary or secondary conductors are not limited by this Exception.

In designing transformer circuits, the rules of Sec. 450-3 can be coordinated with Sec. 240-21(d), which provides special rules for tap conductors used with transformers. This rule would be used mainly where the primary OC devices are rated according to Sec. 450-3(b)(2), or where the combined primary and secondary feeder lengths from the primary OC device to the secondary tap exceed 10 ft (3.05 m).

Where secondary feeder taps do *not* exceed 10 ft (3.05 m) in length, the requirements of Sec. 240-21(b) could apply as in the case of any other feeder tap. In applying the tap rules in Sec. 240-21(b) and (d) the requirements of Sec. 450-3 on transformer overcurrent protection must always be satisfied.

Part (**a**) of this section sets rules for overcurrent protection of any transformer (dry-type or liquid-filled) rated over 600 V. Protection may be provided either by a protective device of specified rating on the transformer primary or by a combination of protective devices of specified ratings on both the primary and secondary. Figure 450-3 shows the basic rules of such overcurrent protection. The fact that E-rated fuses used for high-voltage circuits are given melting times at 200 percent of their continuous-current rating explains why this Code rule used to set 150 percent of primary current as the maximum fuse rating but permits CBs up to 300 percent. In effect, the 150 percent for fuses times 2 (200 percent) becomes 300 percent—the maximum value allowed for a CB. Now such fuses may be rated up to 250 percent (instead of 150 percent).

WHERE QUALIFIED PERSONS MONITOR AND SERVICE THE TRANSFORMER INSTALLATION:

Either primary-only protection . . .

On the primary side, either at the transformer or at the supply end of the primary circuit, by fuses rated at not more than 250% of rated primary current

Primary
Secondary

On the primary side, either at the transformer or at the supply end of the primary circuit, by a circuit breaker rated at not more than 300% of rated primary current

Primary
Secondary

NOTE: Where the indicated percentage of primary current does not correspond to a standard fuse rating or CB setting, the next higher size is permitted.

. . . or primary-and-secondary protection

On the primary side of a feeder overcurrent device sized from Table 450-3(a)(2)b, provided that the transformer is equipped with a coordinated thermal overload protection or has a secondary overcurrent device sized from Table 450-3(a)(2)b

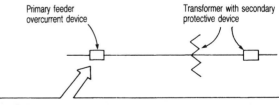

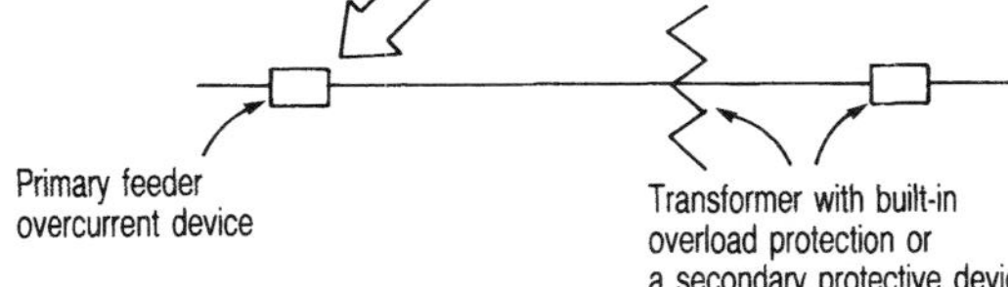

WHERE TRANSFORMER IS NOT MONITORED AND SERVICED BY QUALIFIED PERSONS:

Primary feeder overcurrent device
Transformer with secondary protective device

On the primary side by a feeder overcurrent device sized from Table 450-3(a)(1), and a secondary overcurrent device sized from Table 450-3(a)(1)

Fig. 450-3. High-voltage transformers (rated over 600 V, dry or oil- or askarel-filled) must be protected in one of these ways. (Sec. 450-3.)

- Part (**a**) (**1**) says that *any* high-voltage transformer must have *both* primary and secondary protection based on Table 450-3(a) (1) for maximum ratings of the primary and secondary fuses or circuit breakers.
- Part (**a**) (**2**) gives two alternative ways of protecting high-voltage transformers where "conditions of maintenance and supervision assure that only qualified persons will monitor and service the transformer." The two alternatives are as follows:
  1. Primary-protection-only may be used, with fuses set at not over 250 percent of the primary current or circuit breakers set at not over 300 percent of primary current. And if that calculation results in a fuse or CB rating that does not correspond to a standard rating or setting, the next higher standard rating or setting may be used.
  2. Primary and secondary protection based on Table 450-3(a) (2)b is the alternative.

These rules resulted from concerted industry action to provide better transformer protection. As stated in the "substantiation" for Sec. 450-3(a), "It is felt that this approach will aid in reducing the number of transformer failures due to overload, as well as maintaining the flexibility of design and operation by industry and the more-complex commercial establishments."

Part (**b**) of this section covers all transformers—oil-filled, high-fire-point liquid-insulated, and dry-type—rated up to 600 V. The step-by-step approach to such protection is as follows:

1. For any transformer rated 600 V or less (i.e., the rating of neither the primary nor the secondary winding is over 600 V), the basic overcurrent protection may be provided just on the primary side [Sec. 450-3(b) (1)] or may be a combination of protection on *both* the primary and secondary sides [Sec. 450-3(b) (2)].

   If a transformer is to be protected by means of a CB or set of fuses only on the primary side of the transformer, the basic arrangement is as shown in Fig. 450-4.

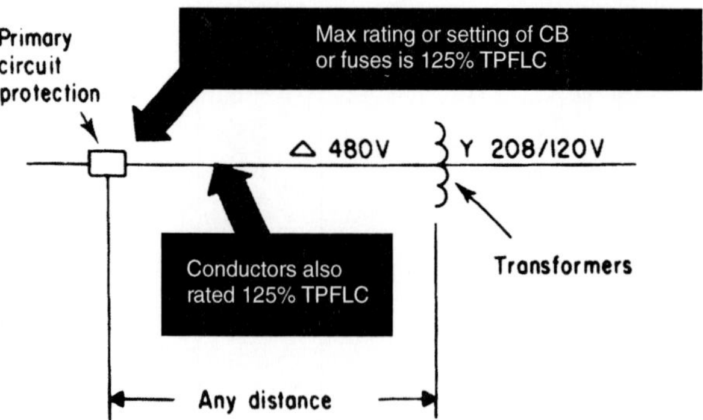

Fig. 450-4. This is the basic rule on primary-side transformer protection. (Sec. 450-3.)

In that layout, a CB or a set of fuses rated not over 125 percent of the transformer rated primary full-load current provides all the overcurrent protection required by the NE Code for the transformer. This overcurrent protection is in the feeder circuit to the transformer and is logically placed at the supply end of the feeder so the same overcurrent device may also provide the overcurrent protection required for the primary feeder conductors. There is no limit on the distance between primary protection and the transformer. When the correct maximum rating for transformer protection is selected and installed at any point on the supply side of the transformer (either near or far from the transformer), then feeder circuit conductors must be sized so that the CB or fuses selected will provide the proper protection as required for the conductors. The ampacity of the feeder conductors must be at least equal to the amp rating of the CB or fuses unless Sec. 240-3(b) is satisfied. That is, when the rating of the overcurrent protection selected is not more than 125 percent of rated primary current, the primary feeder conductor may have an ampacity such that the overcurrent device is the next higher standard rating.

The rules set down for protection of a 600-V transformer by a CB or set of fuses in its primary circuit are given in Fig. 450-5 for transformers with rated primary current of 9 A or more. Note that "the next higher standard" rating of protection may be used, if needed. Figure 450-6 shows the *absolute* maximum values of protection for smaller transformers. When using

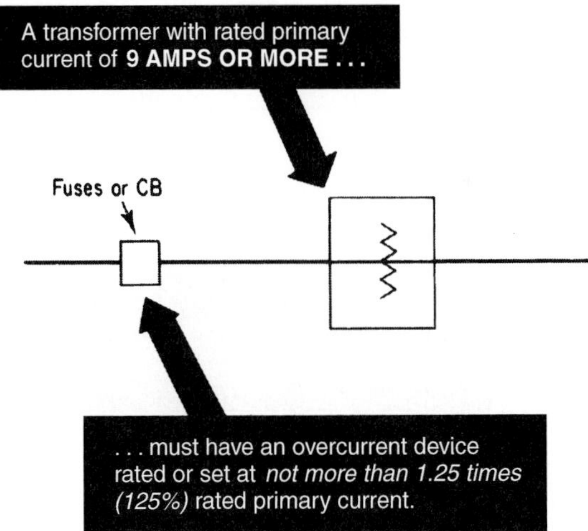

NOTE: **Where 1.25 times primary current does not correspond to a standard rating of protective device, the next higher standard rating from Section 240-6 is permitted.**

Fig. 450-5. Protection sizing for larger transformers is 125 percent of primary current. (Sec. 450-3.)

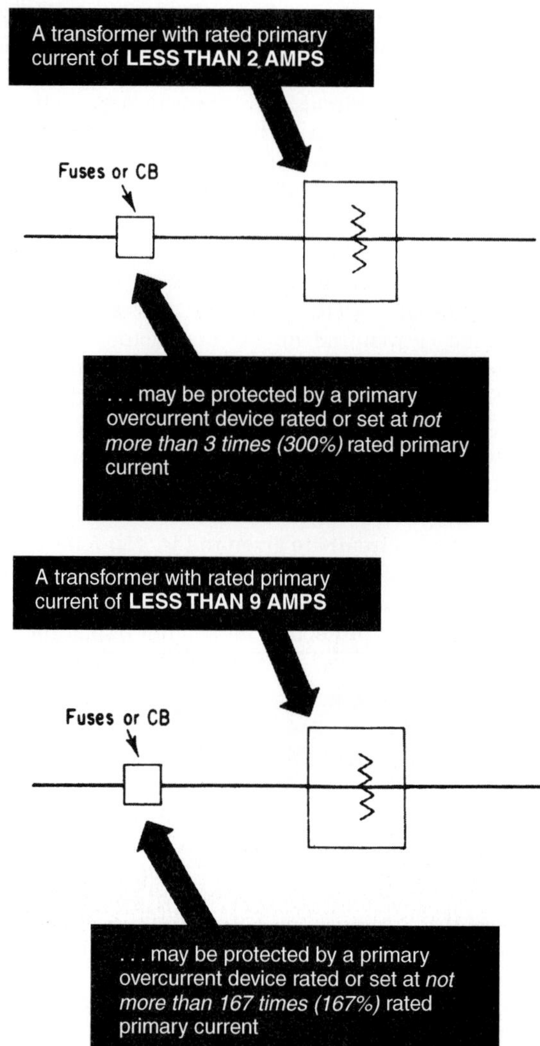

Fig. 450-6. Higher-percent protection is permitted for smaller transformers. (Sec. 450-3.)

the 1.67 or 3 times factor, if the resultant current value is not exactly equal to a standard rating of fuse or CB, then the next *lower* standard rated fuse or CB must be used.

When the rules of Sec. 450-3(b) (1) are observed, the transformer itself is properly protected and the primary feeder conductors, if sized to correspond, may be provided with the protection required by Sec. 240-3. But all considerations on the secondary side of the transformer then have to be separately and independently evaluated. When a transformer is provided with primary-side overcurrent protection, a whole range of design and installation possibilities are available for secondary arrangement that satisfies the Code. The basic approach is to provide required overcurrent protection for the secondary circuit conductors right at the transformer—such as by a fused switch or CB attached to the transformer enclosure, as shown in Fig. 450-7. Or 10- or 25-ft (3.05- or 7.62-m) taps may be made, as covered in Sec. 240-21.

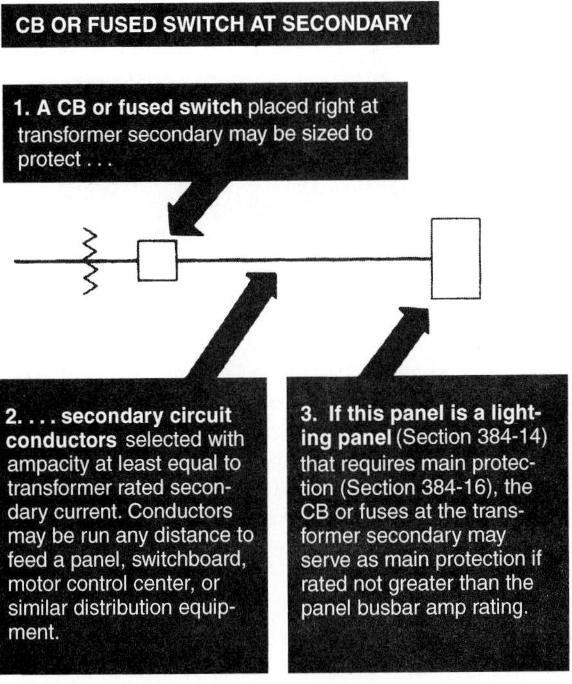

CB OR FUSED SWITCH AT SECONDARY
1. A CB or fused switch placed right at transformer secondary may be sized to protect . . .
2. . . . secondary circuit conductors selected with ampacity at least equal to transformer rated secondary current. Conductors may be run any distance to feed a panel, switchboard, motor control center, or similar distribution equipment.
3. If this panel is a lighting panel (Section 384-14) that requires main protection (Section 384-16), the CB or fuses at the transformer secondary may serve as main protection if rated not greater than the panel busbar amp rating.

Fig. 450-7. Protection of secondary circuit must be independent of primary-side transformer protection. (Sec. 450-3.)

2. Another acceptable way to protect a 600-V transformer is described in Sec. 450-3(b) (2). In this method, the transformer primary may be fed from a circuit which has overcurrent protection (and circuit conductors) rated up to 250 percent (instead of 125 percent, as above) of rated primary current—*but,* in such cases, there must be a protective device on the secondary side of the transformer, and that device must be rated or set at not more than 125 percent of the transformer's rated secondary current (Fig. 450-8). This secondary protective device must be located right at the transformer secondary terminals or not more than the length of a 10- or 25-ft (3.05- or 7.62-m) tap away from the transformer, and the rules of Sec. 240-21 on tap conductors must be fully satisfied.

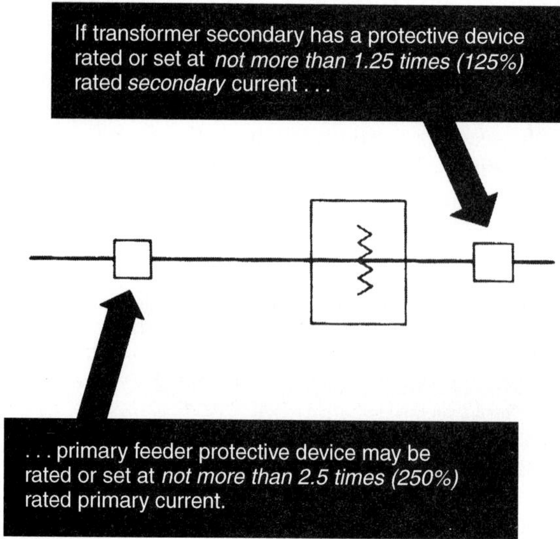
If transformer secondary has a protective device rated or set at not more than 1.25 times (125%) rated secondary current . . .
. . . primary feeder protective device may be rated or set at not more than 2.5 times (250%) rated primary current.

Fig. 450-8. Secondary protection permits higher-rated primary protective device. (Sec. 450-3.)

The secondary protective device covered by Sec. 450-3(b) (2) may readily be incorporated as part of other required provisions on the secondary side of the transformer, such as protection for a secondary feeder from the transformer to a panel or switchboard or motor control center fed from the switchboard. And a single secondary protective device rated not over 125 percent of secondary current may serve as a required panelboard main as well as the required transformer secondary protection as shown at the bottom of Fig. 450-9.

The use of a transformer circuit with primary protection rated up to 250 percent of rated primary current offers an opportunity to avoid situations where a particular set of primary fuses or CB rated at only 125 percent would cause nuisance tripping or opening of the circuit on transformer inrush current. But the use of a 250 percent rated primary protection has a more common and widely applicable advantage in making it possible to feed two or more transformers from the same primary feeder. The number of transformers that might be used in any case would depend on the amount of continuous load on all the transformers. But in all such cases, the primary protection must be rated not more than 250 percent of any one transformer, if they are all the same size, or 250 percent of the smallest transformer, if they are of different sizes. And for each transformer fed, there must be a set of fuses or CB on the secondary side rated at not more than 125 percent of rated secondary current, as shown in Fig. 450-10.

Figure 450-11 shows an example of application of 250 percent primary protection to a feeder supplying three transformers (such as at the bottom of Fig. 450-10). The example shows how the rules of Sec. 450-3(b) must be carefully related to Sec. 240-21 and other Code rules:

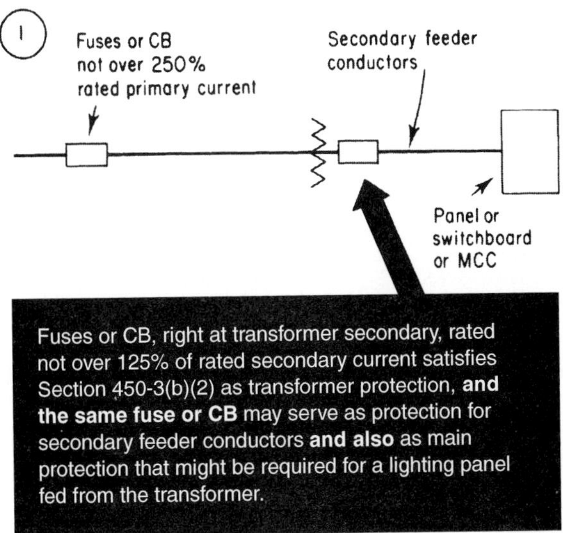
1
Fuses or CB
not over 250%
rated primary current
Secondary feeder
conductors
Panel or
switchboard
or MCC
Fuses or CB, right at transformer secondary, rated not over 125% of rated secondary current satisfies Section 450-3(b)(2) as transformer protection, **and the same fuse or CB** may serve as protection for secondary feeder conductors **and also** as main protection that might be required for a lighting panel fed from the transformer.

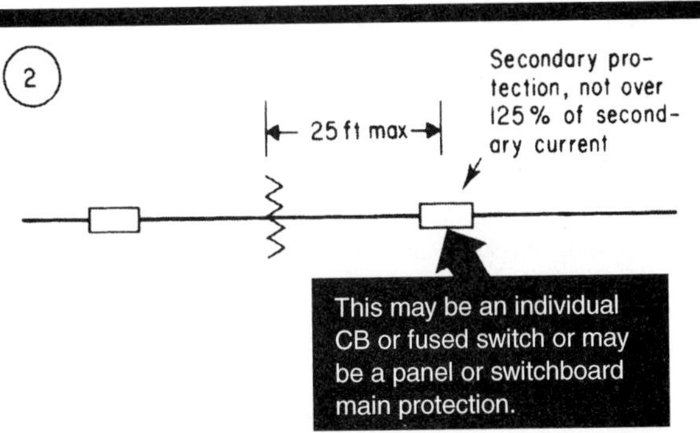
2
25 ft max
Secondary pro-
tection, not over
125% of second-
ary current
This may be an individual CB or fused switch or may be a panel or switchboard main protection.

NOTE: In both cases, primary and secondary feeder conductors must be sized to be properly protected by the fuses or CB in both the primary and secondary circuit, which will give them more than adequate ampacity for the transformer full-load current.

Fig. 450-9. With 250 percent primary protection, secondary protection may be located like this. (Sec. 450-3.)

Part (**d**) of Sec. 240-21 of the NE Code is a rule that covers use of a 25-ft (7.62-m) *unprotected* tap from feeder conductors, with a transformer inserted in the 25-ft (7.62-m) tap. This rule does not eliminate the need for secondary protection—it makes a special condition for placement of the secondary protective device. It is a restatement of part (d) as applied to a tap containing a transformer and applies to both single-phase and 3-phase transformer feeder taps.

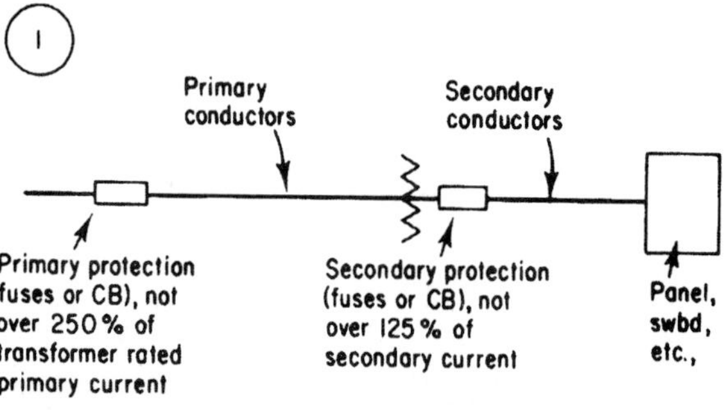

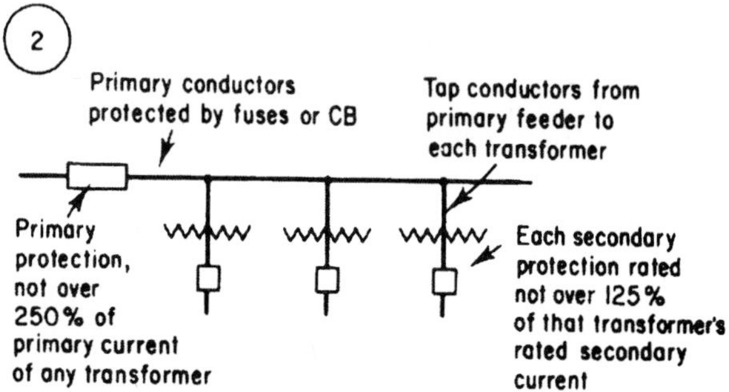

NOTE: Each set of conductors from primary feeder to each transformer may be same size as primary feeder conductors **OR** may be smaller than primary conductors if sized in accordance with Section 240-21(d)—which permits a 25-ft. tap from a primary feeder to be made up of both primary and secondary tap conductors. The 25-ft tap may have any part of its length on the primary or secondary but must be longer than 25 ft. and must terminate in a single CB or set of fuses.

Fig. 450-10. With primary 250 percent protection, primary circuit may vary. (Sec. 450-3.)

Figure 450-11 shows a feeder supplying three 45-kVA transformers, each transformer being fed as part of a 25-ft (7.62-m) feeder tap that conforms to part (d) of Sec. 240-21.

Although each transformer has a rated primary current of 54 A at full load, the demand load on each transformer primary was calculated to be 41 A, based on secondary loading. No. 1 THW copper feeder conductors were considered adequate for the total noncontinuous demand load of 3 × 41 A, or 123 A. A step-by-step analysis of this system follows. Refer to circled letters in the figure:

A. The primary circuit conductors are No. 6 TW rated at 55 A, which gives them "an ampacity at least ⅓ that of the *conductors or overcurrent protec-*

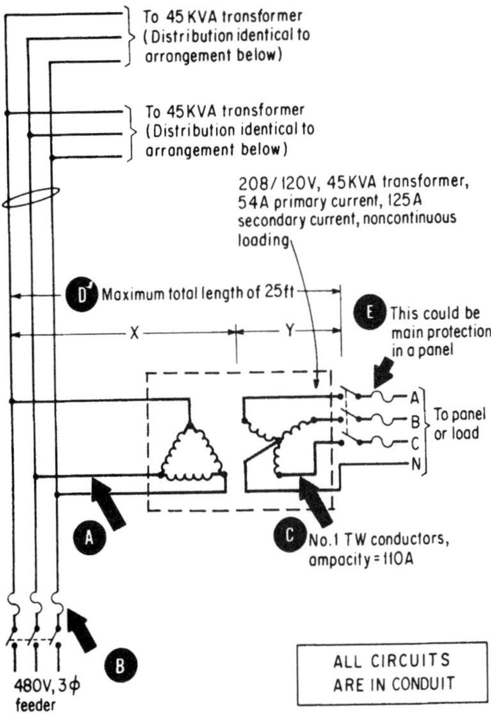

Fig. 450-11. Code rules must be tied together. (Sec. 450-3.)

tion from which they are tapped . . . ,” because these conductors are tapped from the feeder conductors protected at 125 A. No. 6 TW is O.K. for the 41-A primary current.

B. The 125-A fuses in the feeder switch properly protect the No. 1 THW feeder conductors, which are rated at 130 A.

C. The conductors supplied by the transformer secondary must have “an ampacity that, when multiplied by the ratio of the secondary-to-primary voltage, is at least ⅓ the ampacity of the conductors *or* overcurrent protection from which the primary conductors are tapped . . ." The ratio of secondary-to-primary voltage of the transformer is

$$\frac{208 \text{ V}}{480 \text{ V}} = 0.433$$

Note that phase-to-phase voltage must be used to determine this ratio.

Then, for the secondary conductors, Sec. 240-21(d) says that

$$\text{Minimum conductor ampacity} \times 0.433 = \tfrac{1}{3} \times 125 \text{ A}$$
$$\tfrac{1}{3} \times 125 = 41.67$$

Then, minimum conductor ampacity equals

$$\frac{41.67}{0.433} = 96 \text{ A}$$

The No. 1 TW secondary conductors, rated at 110 A, are above the 96-A minimum and are, therefore, satisfactory.

D. The total length of the unprotected 25-ft (7.62-m) tap—i.e., the primary conductor length *plus* the secondary conductor length ($x + y$) for any circuit leg—must not be greater than 25 ft (7.62 m).

E. The secondary tap conductors from the transformer must terminate in a single CB or set of fuses that will limit the load on those conductors to their rated ampacity from Table 310-16. Note that there is no Exception given to that requirement and the "next higher standard device rating" may not be used if the conductor ampacity does not correspond to the rating of a standard device.

The overcurrent protection required at E, at the load end of the 25-ft (7.62-m) tap conductors, must not be rated more than the ampacity of the No. 1 TW conductors.

$$\text{Max. rating of fuses or CB at E} = 110 \text{ A}$$

But a 100-A main would satisfy the 96-A secondary load. *(Note: The overcurrent protective device required at E could be the main protective device required for a lighting and appliance panel fed from the transformer.)*

WATCH OUT FOR THIS TRAP!!

Although the foregoing calculation shows how unprotected taps may be made from feeder conductors by satisfying the rules of Sec. 240-21(d), the rules of Sec. 240-21 are all concerned with PROTECTION OF CONDUCTORS ONLY. Consideration must now be made of *transformer* protection, as follows:

1. Note that Sec. 240-21 makes no reference to *transformer* protection. But Sec. 450-3 calls for protection of transformers, and there is no exception made for the conditions of part (d) to Sec. 240-21.

2. It is clear from Sec. 450-3(b) (1) that the transformer shown in Fig. 450-11 is *not* protected by a primary-side overcurrent device rated not more than 125 percent of primary current (54 A), because 1.25 × 54 A = 68 A, maximum.
3. But Sec. 450-3(b) (2) does offer a way to provide required protection. The 110-A protection at E *is* secondary protection *rated not over* 125 percent of rated secondary current (1.25 × 125 A secondary current = 156 A). With that secondary protection, a primary feeder overcurrent device rated not more than 250 percent of rated primary current will satisfy Sec. 450-3(b) (2). That would call for fuses in the feeder switch (or a CB) (at B in the diagram) rated not over

$$250 \text{ percent} \times 54 \text{ A primary current} = 135 \text{ A}$$

But, the fuses in the feeder switch are rated at 125 A—which are not in excess of 250 percent of transformer primary current and, therefore, satisfy Sec. 450-3(b) (2).

In addition to the two basic methods described above for protecting transformers, Sec. 450-3(b) (2) also provides for protection with a built-in thermal overload protection, as shown in Fig. 450-12.

450-4. Autotransformers 600 Volts, Nominal, or Less. The rules of this section cover connections and overcurrent protection for autotransformers. The rule calling for overcurrent protection rated at not more than 125 percent of the rated input current for any autotransformer with a rated current of 9 A or more

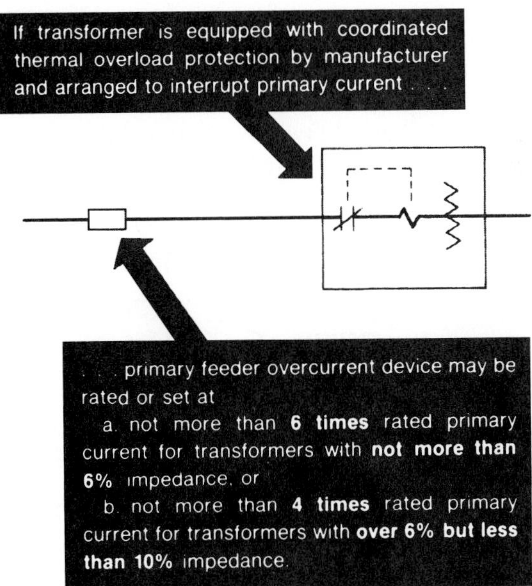

Fig. 450-12. Built-in protection is another technique for transformers. (Sec. 450-3.)

and not more than 167 percent of the rated current for smaller autotransformers is the same as that in Sec. 450-3(b) for 2-winding transformers.

450-5. Grounding Autotransformers. An existing, ungrounded, 480-V system derived from a delta transformer hookup can be converted to grounded operation in two basic ways:

First, one of the three phase legs of the 480-V delta can be intentionally connected to a grounding electrode conductor that is then run to a suitable grounding electrode. Such grounding would give the two ungrounded phases (A and B) a voltage of 480 V to ground. The system would then operate as a grounded system, so that a ground fault (phase-to-conduit or other enclosure) on the secondary can cause fault-current flow that opens a circuit protective device to clear the faulted circuit.

But corner grounding of a delta system does not give the lowest possible phase-to-ground voltage. In fact, the voltage to ground of a corner-grounded delta system is the same as it is for an ungrounded delta system because voltage to ground for ungrounded circuits is defined as the greatest voltage between the given conductor and any other conductor of the circuit. Thus, the voltage to ground for an ungrounded delta system is the maximum voltage between any two conductors, on the assumption that an accidental ground on any one phase puts the other two phases at full line-to-line voltage aboveground.

In recognition of increasing emphasis on the safety of grounded systems over ungrounded systems, Sec. 450-5 covers the use of zig-zag grounding autotransformers to convert 3-phase, 3-wire, ungrounded delta systems to grounded wye systems. Such grounding of a 480-V delta system, therefore, lowers the voltage to ground from 480 V (when ungrounded) to 277 V (the phase-to-grounded-neutral voltage) when converted to a wye system (Fig. 450-13).

A zig-zag grounding autotransformer gets its name from the angular phase differences among the six windings that are divided among the three legs of the transformer's laminated magnetic core assembly. The actual hookup of the six windings is an interconnection of two wye configurations, with specific polarities and locations for each winding. Just as a wye or delta transformer hookup

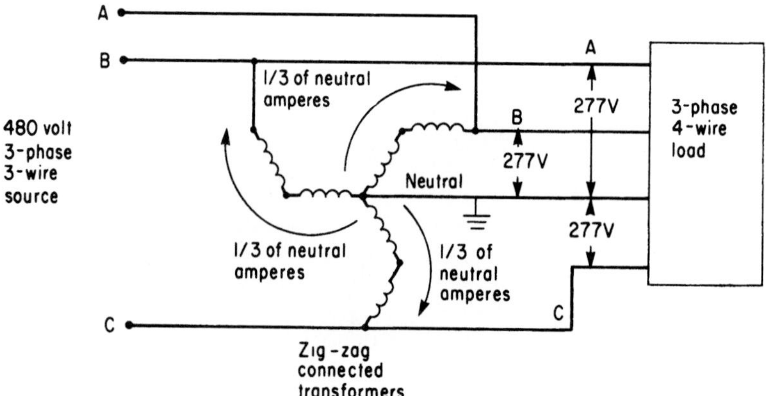

Fig. 450-13. Zig-zag transformer changes voltage to ground from 480 to 277 V. (Sec. 450-5.)

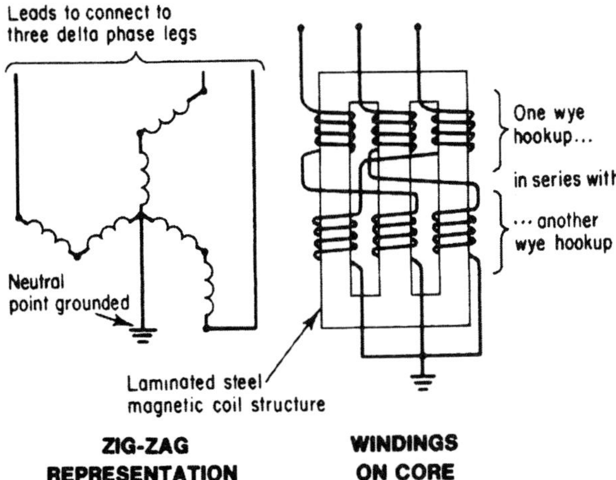

Fig. 450-14. Windings of zig-zag transformer provide for flow of fault or neutral current. (Sec. 450-5.)

has a graphic representation that looks like the letter "Y" or the Greek letter "delta," so a zig-zag grounding autotransformer is represented as two wye hookups with pairs of windings in series but phase-displaced, as in Fig. 450-14.

With no ground fault on any leg of the 3-phase system, current flow in the transformer windings is balanced, because equal impedances are connected across each pair of phase legs. The net impedance of the transformer under balanced conditions is very high, so that only a low level of magnetizing current flows through the windings. But when a ground fault develops on one leg of the 3-phase system, the transformer windings become a very low impedance in the fault path, permitting a large value of fault current to flow and operate the circuit protective device—just as it would on a conventional grounded-neutral wye system, as shown in Fig. 450-15.

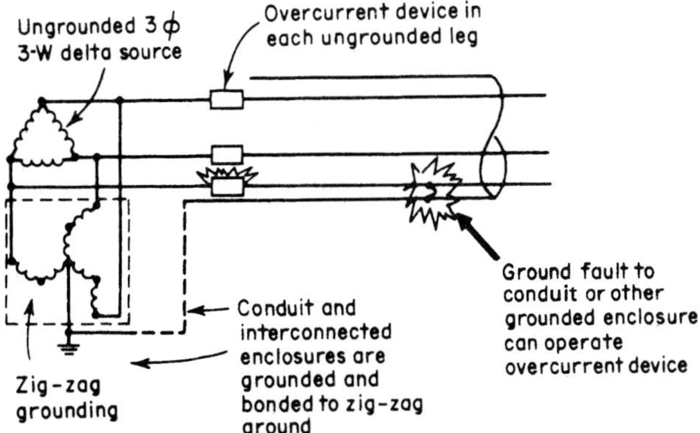

Fig. 450-15. Zig-zag transformer converts ungrounded system to grounded operation. (Sec. 450-5.)

Because the kilovoltampere rating of a grounding autotransformer is based on short-time fault current, selection of such transformers is much different from sizing a conventional 2-winding transformer for supplying a load. Careful consultation with a manufacturer's sales engineer should precede any decisions about the use of these transformers.

Section 450-5 of the **NE Code** points out that a grounding autotransformer may be used to provide a neutral reference for grounding purposes *or* for the purpose of converting a 3-phase, 3-wire delta system into a 3-phase, 4-wire grounded wye system. In the latter case, a neutral conductor can be taken from the transformer to supply loads connected phase-to-neutral—such as 277-V loads on a 480-V delta system that is converted to a 480Y/277-V system.

Section 450-5 requires such transformers to have a continuous rating and a continuous neutral current rating. The phase current in a grounding autotransformer is one-third the neutral current, as shown in Fig. 450-13.

Part (**a**) (**2**) of this section requires use of a 3-pole CB, rated at 125 percent of the transformer phase current. The requirement for "common-trip" in the overcurrent device excludes conventional use of fuses in a switch as the required overcurrent protection. A three-pole CB prevents single-phase opening of the circuit.

450-6. Secondary Ties. In industrial plants having very heavy power loads it is usually economical to install a number of large transformers at various locations within each building, the transformers being supplied by primary feeders operating at voltages up to 13,800 V. One of the secondary systems that may be used in such cases is the network system.

The term *network system* as commonly used is applied to any secondary distribution system in which the secondaries of two or more transformers at different locations are connected together by secondary ties. Two network layouts of unit subs are shown in Fig. 450-16. In the spot network, two or three transformers in one location or "spot" are connected to a common secondary bus and divide the load. Upon primary or transformer fault, the secondary is isolated from the faulted section by automatic operation of the network protector, providing a high order of supply continuity in the event of faults. The general form of the network system is similar except that widely separated individual substations are used with associated network protectors and tie circuits run between the secondary bus sections. The system provides for interchange of power to accommodate unequal loading on the transformers. Limiters protect the ties. The purpose of the system is to equalize the loading of the transformers, to reduce voltage drop, and to ensure continuity of service. The use of this system introduces certain complications, and, to ensure successful operation, the system must be designed by an experienced electrical engineer.

The provisions of Sec. 450-3 govern the protection in the primary. Refer to Fig. 450-16. The network protector consists of a CB and a reverse-power relay. The protector is necessary because without this device, if a fault develops in the transformer, or, in some cases, in the primary feeder, power will be fed back to the fault from the other transformers through the secondary ties. The relay is set to trip the breaker on a reverse-power current not greater than the rated secondary current of the transformer. This breaker is not arranged to be tripped by an overload on the secondary of the transformer.

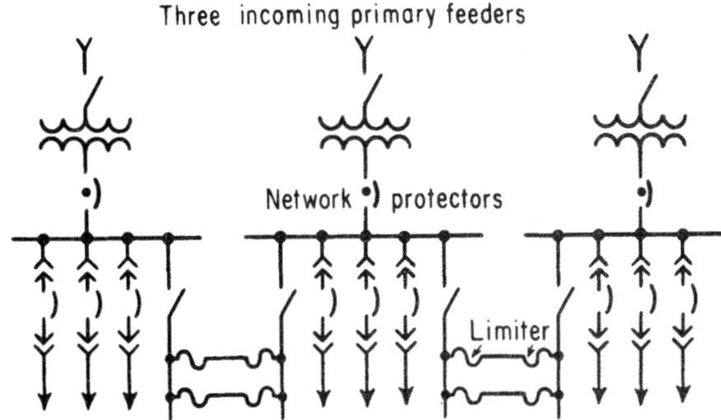

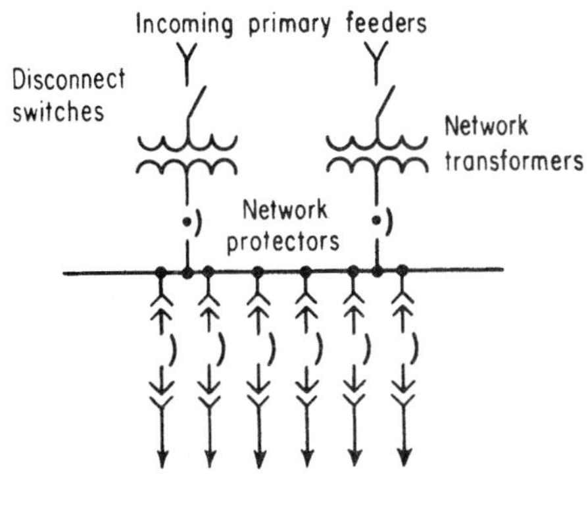

Fig. 450-16. These are the two basic types of "network" systems. (Sec. 450-6.)

Section 450-6(a) (3) provides that:

1. Where two or more conductors are installed in parallel, an individual protective device is provided at each end of each conductor.
2. The protective device (fusible link or CB) does not provide overload protection, but provides short-circuit protection only.

In the case of a short circuit, the protective device must open the circuit before the conductor reaches a temperature that would injure its insulation. The principles involved are that the entire system is so designed that the tie conductors will never be continuously overloaded in normal operation—hence protection against overloads of less severity than short circuits is not necessary—and that the protective devices should not open the circuit and thus cause an interruption of service on load peaks of such short duration that the conductors do not become overheated.

A limiter is a special type of fuse having a very high interrupting capacity. Figure 450-17 is a cross-sectional view of one type of limiter. The cable lug, the fusible section, and the extension for connection to the bus are all made in one piece from a length of copper tubing, and the enclosing case is also copper. A typical device of this type is rated to interrupt a current of 50,000 A without perceptible noise and without the escape of flame or gases from the case.

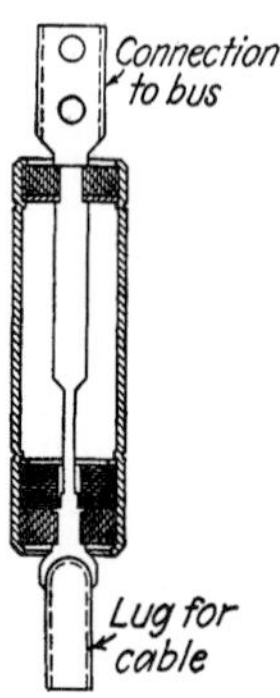

Fig. 450-17. A "limiter" is a cable connection device containing a fusible element. (Sec. 450-6.)

Figure 450-18 is a single-line diagram of a simple 3-phase industrial-plant network system. The primary feeders may operate at any standard voltage up to 13,800 V, and the secondary voltage would commonly be 480 V. The rating of the transformers used in such a system would usually be within the range of 300 to 1,000 kVA. The diagram shows two primary feeders, both of which are carried to each transformer so that by means of a double-throw switch each transformer can be connected to either feeder. Each feeder would be large enough to carry the entire load. It is assumed that the feeders are protected in accordance with Sec. 450-3(a)(2) so that no primary overcurrent devices are required at the transformers. The secondary ties consist of two conductors in multiple per phase and it will be noted that these conductors form a closed loop. Switches are provided so that any section of the loop, including the limiters protecting that section, can be isolated in case repairs or replacements should be necessary.

450-7. Parallel Operation. To operate satisfactorily in parallel, transformers should have the same percentage impedance and the same ratio of reactance to resistance. Information on these characteristics should be obtained from the manufacturer of the transformers.

450-8. Guarding. Figure 450-19 summarizes these rules. Refer to Sec. 110-17 on guarding of live parts. Safety to personnel is always important, particularly where a transformer is to operate with live parts. To protect against accidental contact with such components, isolate the unit or units in a room or place accessible only to qualified personnel and guard live parts, such as with a railing. When elevation is used for safeguarding live parts, consult Secs. 110-34(e) and 110-17.

As noted in part (**c**), switches and other equipment operating at up to 600 V and serving only circuits within a transformer enclosure may be installed

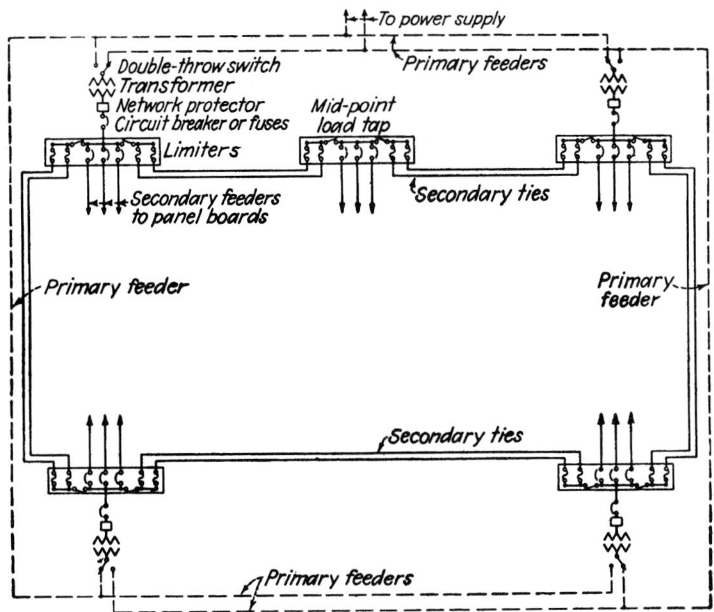

Fig. 450-18. A typical industrial plant network distribution system. (Sec. 450-6.)

within the enclosure *if* it is accessible to qualified persons only. This is intended to be part of the requirement that exposed, energized parts must be properly guarded.

450-9. Ventilation. As noted in the last paragraph, ventilation grills or slots in the sides or back of transformer enclosures must have adequate clearance from walls and objects to ensure free and substantial air flow through the case. Because of the tight quarters in today's electrical rooms and the tendency to give as little space as possible to electrical equipment, this is a critically important installation requirement to ensure safety and trouble-free operation of enclosed transformers. The substantiation for this proposal stated the following:

> Today's widespread use of dry type power transformers indoors has resulted in the common practice of their being installed directly up against walls completely blocking the rear vents. As inspectors, we frequently wind up trying to find out if the transformer installation instructions are anywhere around so we can see what clearances the manufacturer has specified.
>
> Clearly, this is not the best system. Furnaces, for example, commonly have required clearances marked on the nameplate. This proposal will let people know that clearances are required. Secondly, it will let us know what these clearances are.

(See Fig. 450-20, top.)

450-13. Location. Accessibility is an important location feature of transformer installation. The NE Code generally requires a transformer (whether liquid-filled or dry-type) to be readily accessible to qualified personnel for inspection and maintenance (Fig. 450-20). That is, it must be capable of being reached

1. Transformers must be protected against physical damage.

2. Exposed live parts must be protected against accidental contact by putting the transformer in a room or place accessible only to qualified personnel **or** by keeping live parts above the floor in accordance with Table 110-34(e).

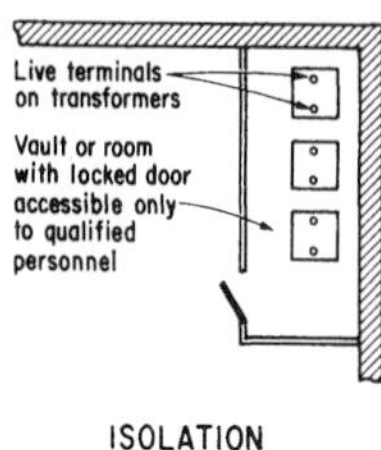

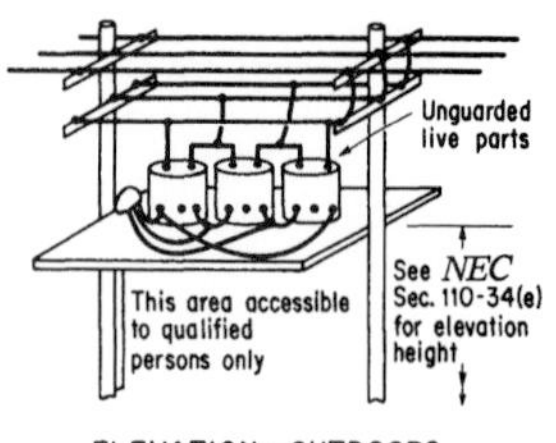

3. Signs or other visible markings must be used on equipment or structure to indicate the operating voltage of exposed live parts

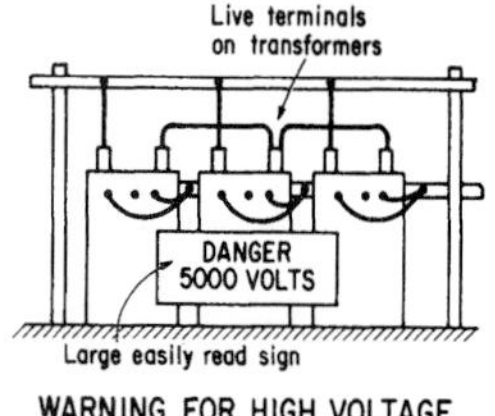

Fig. 450-19. Transformer installations must be effectively guarded. (Sec. 450-8.)

quickly for operation, repair, or inspection without requiring use of a portable ladder to get at it, and it must not be necessary to climb over or remove obstacles to reach it. A transformer may be mounted on a platform or balcony, but there must be fixed stairs or a fixed ladder for access to the transformer (Fig. 450-21). But Exceptions are made:

Exception No. 1 permits dry-type transformers rated 600 V or less to be located "in the *open* on walls, columns, or structures"—without the need to be readily accessible. A transformer suspended from the ceiling or hung on a wall—in which cases a ladder would be required to reach them because they are over 6½ ft (1.98 m) above the floor (see Sec. 380-8)—would be O.K., as shown in Fig. 450-22. And Exception No. 2 permits dry-type transformers up to 600 V, 50 kVA, to be installed in fire-resistant hollow spaces of buildings not permanently closed in by structure, provided the transformer is designed to have adequate ventilation for such installation. (See Fig. 450-23.)

Note in Exception No. 1 that a transformer *may* be *not* readily accessible only if it is located in the *open*. The words "in the open" do not readily and surely relate to such words as "concealed" or "exposed." But it is reasonable to conclude that "in the *open*" would be difficult to equate with "above a suspended ceiling." The latter location is in a generally smaller, confined space in which the transformers must be readily accessible; Exception No. 1 must be taken as a condition in which a transformer *does* not have to be readily accessible—i.e.,

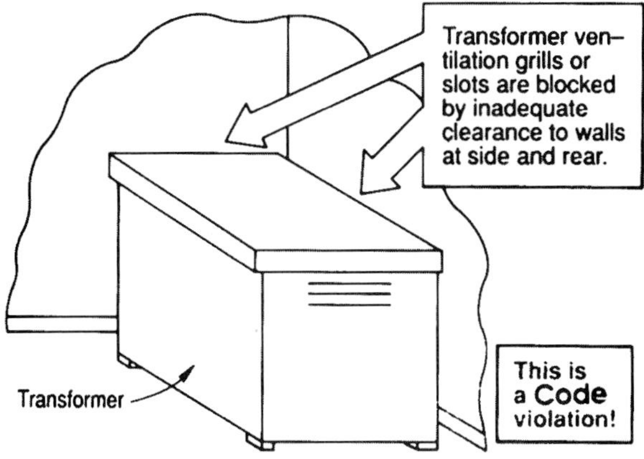
Transformer ven-
tilation grills or
slots are blocked
by inadequate
clearance to walls
at side and rear.
Transformer
This is
a Code
violation!

DANGER

Fig. 450-20. Basic rule calls for every transformer to have adequate ventilation and ready, easy, direct access for inspection or maintenance. (Secs. 450-9 and 450-13.)

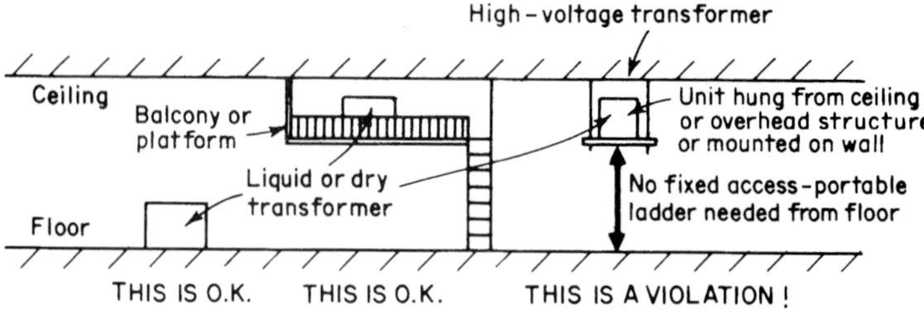

Fig. 450-21. Transformers must be "readily accessible" without need for portable ladder to reach them. (Sec. 450-13.)

Fig. 450-22. Transformer mounted on wall (or suspended from ceiling) would be considered *not* readily accessible because a ladder would be needed to reach it. But, because it is "in the open," such use conforms to Exception No. 1. (Sec. 450-13.)

Fig. 450-23. Recessed mounting of dry-type transformers is permitted within "fire-resistant hollow spaces," as in hospitals, schools, and other commercial or institutional buildings. (Sec. 450-13.)

that it may be mounted up high where a portable ladder would be needed to get at it. But the words "in the open" could logically be taken to prohibit use of the transformer above a suspended ceiling (Fig. 450-24).

It should be noted that Exception No. 1 to this rule, which permits high mounting that makes transformers *not* "readily accessible," applies to "dry-type transformers" only. As a result, high mounting of oil-filled transformers—as shown in Fig. 450-25—is not covered by that Exception and could be considered a violation of the *basic* rule of this section. However, Sec. 450-27 on oil-filled transformers installed outdoors actually contains an Exception to the rule of Sec. 450-13. The FPN following Sec. 450-27 points out that additional information can be found in the *National Electrical Safety Code* (*not* the NE Code). That code covers use of such transformers as shown in Fig. 450-25.

450-21. Dry-type Transformers Installed Indoors. This rule differentiates between dry-type transformers based on kVA rating. All dry-type transformers rated at 112½ kVA or less, at up to 35 kV, must be installed so that a minimum clear-

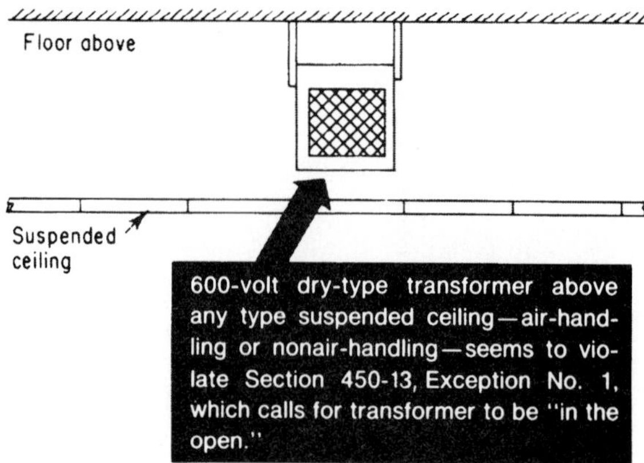
Floor above
Suspended ceiling
600-volt dry-type transformer above any type suspended ceiling—air-handling or nonair-handling—seems to violate Section 450-13, Exception No. 1, which calls for transformer to be "in the open."

Fig. 450-24. Watch out for transformers above suspended ceilings. (Sec. 450-13.)

ance of 12 in. (305 mm) is provided between the transformer and any combustible material.

Exception No. 1 recognizes the use of fire-resistant heat-insulating barriers instead of space separation for transformers rated not over 112½ kVA. *But* be aware that clearances required to ensure proper ventilation of the transformer *must* be provided to satisfy Sec. 450-9.

Exception No. 2 permits those transformers rated 600 V or less to be installed closer than the 12-in. (305-mm) minimum, but, consideration must also be given to the requirements of Sec. 450-9. For units rated over 112½ kVA, the rule basically calls for such transformers to be installed in vaults, with two Exceptions provided. Figure 450-26 shows the rules of this section. Related application recommendations are as follows:

- Select a place that has the driest and cleanest air possible for installation of open-ventilated units. Avoid exposure to dripping or splashing water or other wet conditions. Outdoor application requires a suitable housing. Try to find locations where transformers will not be damaged by floodwater in case of a storm, a plugged drain, or a backed-up sewer.
- Temperature in the installation area must be normal, or the transformer may have to be derated. Modern standard, ventilated, dry-type transformers are designed to provide rated kilovoltampere output at rated voltage when the maximum ambient temperature of the cooling air is 40°C and the average ambient temperature of the cooling air over any 24-hr period does not exceed 30°C. At higher or lower ambients, transformer loading can be adjusted by the following relationships:
  1. For each degree Celsius that average ambient temperature exceeds 30°C, the maximum load on the transformer must be reduced by 1 percent of rated kilovoltamperes.
  2. For each degree Celsius that average ambient temperature is less than 30°C, the maximum load on the transformer may be increased by 0.67 percent of rated kilovoltamperes.

Fig. 450-25. High mounting of oil-filled transformers would require use of a portable ladder for access to the units. But Sec. 450-27 covers such units installed on poles or structures. (Sec. 450-13.)

Depending on the type of insulation used, transformer insulation life will be cut approximately in half for every 10°C that the ambient temperature exceeds the normal rated value—or doubled for every 10°C below rated levels. Estimates assume continuous operation at full load. With modern insulations this rule is actually conservative for ambient temperature below normal operating temperatures and optimistic above it.

For proper cooling, dry-type transformers depend upon circulation of clean air—free from dust, dirt, or corrosive elements. Filtered air is preferable and may be mandatory in some cases of extreme air pollution. In any case, it can reduce maintenance.

In restricted spaces—small basement mechanical rooms, etc.—ventilation must be carefully checked to assure proper transformer operating temperature. The usual requirement is for 100 cfm of air movement for each kilowatt of transformer loss. Areas of inlet and outlet vent openings should be at least 1 net sq ft (0.093 sq m) per 100 kVA of rated transformer capacity.

TRANSFORMERS RATED 112½ KVA OR LESS

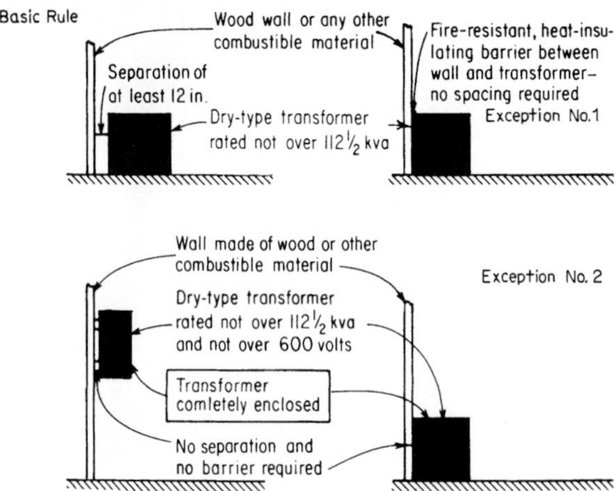

TRANSFORMERS RATED OVER 112½ KVA

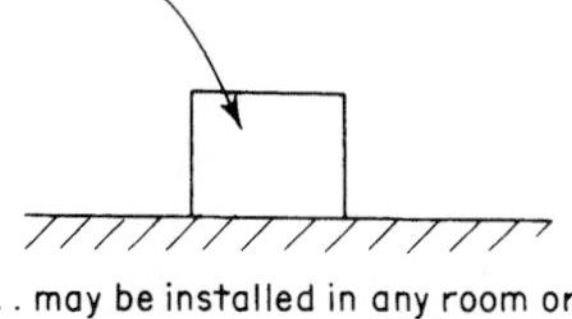

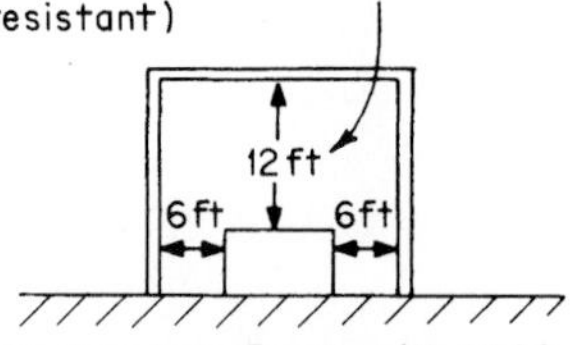

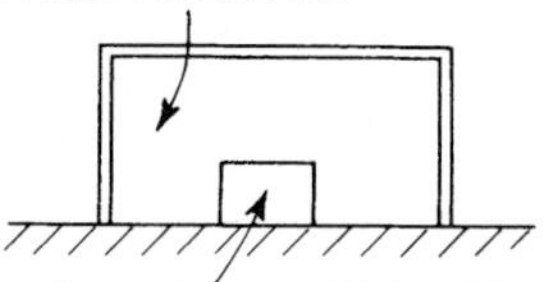

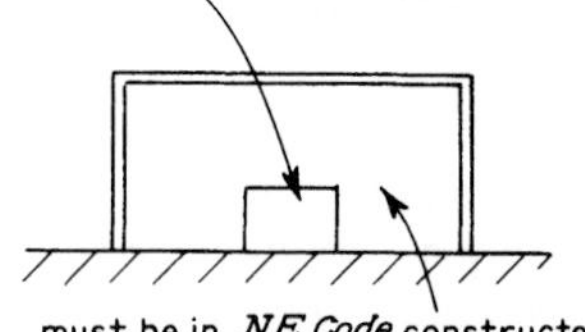

Fig. 450-26. Construction of dry-type transformer affects indoor installation rules. (Sec. 450-21.)

Height of vault, location of openings, and transformer loading affect ventilation. One manufacturer calls for the areas of the inlet and outlet openings to be not less than 60 sq ft per 1,000 kVA when the transformer is operating under full load and is located in a restricted space. And a distance of 1 ft (305 mm) should be provided on all sides of dry-type transformers as well as between adjacent units.

Freestanding, floor-mounted units with metal grilles at the bottom must be set up off the floor a sufficient distance to provide the intended ventilation draft up through their housings.

The installation location must not expose the transformer housing to damage by normal movement of persons, trucks, or equipment. Ventilation openings should not be exposed to vandalism or accidental or mischievous poking of rubbish, sticks, or rods into the windings. Adequate protection must be provided against possible entry of small birds or animals.

450-22. Dry-type Transformers Installed Outdoors. A transformer that sustains an internal fault which causes arcing and/or fire presents the same hazard to adjacent combustible material whether it is installed indoors or outdoors. For that reason, a clearance of at least 12 in. (305 mm) is required between any dry-type transformer rated over 112½ kVA and combustible materials of buildings.

In the Exception, the clearance of 12 in. (305 mm) from combustible building materials is *not* required for outdoor dry-type transformers that have an 80°C rise (or higher) rating and are completely enclosed except for ventilation openings. The same consideration given for an 80°C rise transformer is made outdoors as it is indoors, in Exception No. 2 of Sec. 450-21(b).

450-23. Less-Flammable Liquid-Insulated Transformers. Section 450-23 covers the liquid-filled transformers designed to replace askarel-insulated transformers. Because oil-filled transformers used indoors require a transformer vault, the "less flammable" (also called "high fire point") insulated transformer offers an alternative to the oil-filled transformers, without the need for a vault. This Code section permits installation of these "high fire point liquid-insulated" transformers indoors or outdoors. Over 35 kV, such a transformer must be in a vault.

The rules of this section recognize that these various high fire point liquid-insulated dielectrics are less flammable than the mineral oil used in oil-filled transformers but not as fire-resistant as askarel. Because these askarel substitutes will burn to some degree, Code rules are aimed at minimizing any fire hazards:

1. Less-flammable liquid dielectrics used in transformers must, first of all, be listed—that is, tested and certified by a testing laboratory or organization and shown in a published listing as suitable for application. "Less-flammable" liquids for transformer insulation are defined as having "a fire point of not less than 300°C."
2. Transformers containing the high fire point dielectrics may be used *without a vault* but only within noncombustible buildings (brick, concrete, etc.) and then only in rooms or areas that do not contain combustible materials. A "Type I" or "Type II" building is a building of noncombustible construction, as defined in the fine-print note, and there must be no combustible materials stored in the area where the transformer is installed.
3. The entire installation must satisfy all conditions of use, as described in the "listing" of the liquid.
4. A liquid-confinement area must be provided around such transformers that are not in a vault, because tests indicate these liquids are not completely nonpropagating—that is, if they are ignited, the flame will be propagated

along the liquid. A propagating liquid must be confined to a given area to confine the flame of its burning (Fig. 450-27). The liquid-confinement area (a curb or dike around the transformer) must be of sufficient dimensions to contain the entire volume of liquid in the transformer.

A less-flammable liquid-insulated transformer installed in such a way that all of conditions 2 and 3 above are not satisfied must be *either*

1. Provided with an automatic fire extinguishing system and a liquid-confinement area, or
2. Installed in a Code-specified transformer vault (part **C** of Art. 450), without need for a liquid-confinement area.

Less-flammable liquid-insulated transformers rated over 35 kV and installed indoors must be enclosed in a Code-constructed transformer vault. All less-flammable liquid-insulated transformers installed outdoors may be attached to or adjacent to or on the roof of Type I or Type II buildings. Such installation at other than Type I or Type II buildings, where adjacent to combustible material, fire escapes, or door or window openings, must be guarded by fire barriers, space separation, and compliance with instructions for using the particular liquid.

Because these rules are general in nature and lend themselves to a variety of interpretations, application of these requirements may depend heavily on consultation with inspection authorities.

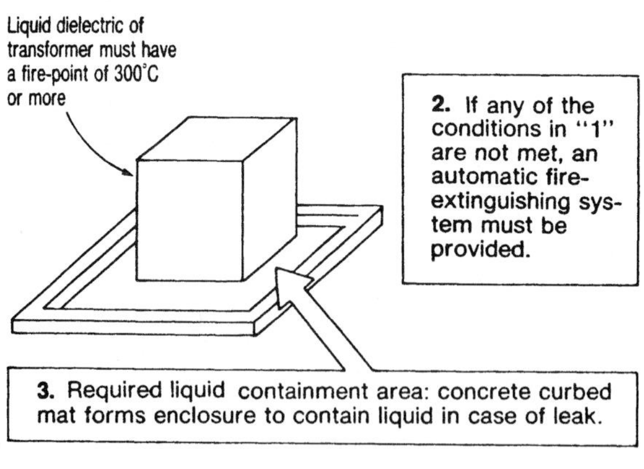

1. A vault is not required for a transformer installed in a noncombustible building, with no combustible materials stored near it, provided any restrictions given in the listing of the liquid dielectric are satisfied.
Liquid dielectric of transformer must have a fire-point of 300°C or more
2. If any of the conditions in "1" are not met, an automatic fire-extinguishing system must be provided.
3. Required liquid containment area: concrete curbed mat forms enclosure to contain liquid in case of leak.
NOTE: If any of the above conditions are not satisfied, the transformer must be installed in a *Code*-constructed transformer vault.

Fig. 450-27. Transformers containing askarel-substitute liquids must satisfy specific installation requirements. (Sec. 450-23.)

Although askarel-filled transformers up to 35 kV were used for many years for indoor applications because they do not require a transformer vault, there has been a sharp, abrupt discontinuance of their use over recent years. Growth in the ratings, characteristics, and availability of dry-type high-voltage transformers has accounted for a major part in the reduction of askarel units. But another factor that led to rejection of askarel transformers in recent years is the environmental objections to the askarel liquid itself.

A major component of any askarel fluid is polychlorinated biphenyl (PCB), a chemical compound designated as a harmful environmental pollutant because it is nonbiodegradable and cannot be readily disposed of. Thus, although the askarels are excellent coolants where freedom from flammability is important, environmental objections to the sale, use, and disposal of PCBs have eliminated new applications of askarel transformers and stimulated a search for a nontoxic, environmentally acceptable substitute.

Proper handling and disposal of askarel is important for units still in use. A regulation of the EPA (Environmental Protection Agency), No. 311, required that all PCB spills of 1 lb or more must be reported. Failure to report a spill is a criminal offense punishable by a $10,000 fine and/or one-year imprisonment. Both the EPA and OSHA have objected to use of askarels. A manufacturer of askarel has established a program for disposal of spent or contaminated PCB fluid using an incinerator that completely destroys the fluid by burning it at over 2,000°F.

Non-PCB dielectric coolant fluids for use in small- and medium-sized power transformers as a safe alternate to askarels are available and transformers using these new high fire point dielectric coolants have been widely used.

Extensive data from tests on available askarel substitutes show that they provide a high degree of safety. Such fluids do have **NE Code** and OSHA recognition. Responsibility for proper clearances with insurance underwriters, government regulating agencies, and local code authorities rests with the user or purchaser of the fluid in new or refilled transformers. Underwriters Laboratories does not test or list liquid-filled equipment. Both UL and Factory Mutual Research Laboratory have been involved in providing a classification service of flammability. The EPA has commented favorably on such high fire point fluids.

Few physical changes to transformers are necessary when using the new fluid dielectrics. However, load ratings on existing units may be reduced about 10 percent because of the difference in fluid viscosity and heat conductivity compared with askarel. The high fire point fluids cost about twice as much as askarel. Purchase price of a new transformer filled with the fluid (such as a typical 1,000-kVA loadcenter unit) is about 10 to 15 percent more than an askarel-filled unit. But the economics vary for different fluids and must be carefully evaluated.

High fire point liquid-insulated transformers require no special maintenance procedures. The liquids exhibit good dielectric properties over a wide range of temperatures and voltage stress levels, and they have acceptable arc-quenching capabilities. They have a high degree of thermal stability and a high resistance to thermal oxidation that enables them to maintain their insulating and other functional properties for extended periods of time at high temperatures.

Since silicone liquids will ignite at 750°F (350°C), they are not classed as fire-resistant. However, if the heat source is removed or fluid temperature drops below 750°F, burning will stop. The silicone fluids are thus self-extinguishing.

450-24. Nonflammable Fluid-Insulated Transformers. This section permits indoor and outdoor use of transformers that utilize a noncombustible fluid dielectric, which is one that does not have a flash point or fire point and is not flammable in air (Fig. 450-28). As an alternative to askarel-insulated transformers, these transformers offer high BIL ratings and other features of operation similar to high-fire-point dielectric-insulated transformers—without concern for flammability. Such transformers do not, therefore, have the restrictions that are set down in Sec. 450-23 for the high-fire-point-liquid transformers.

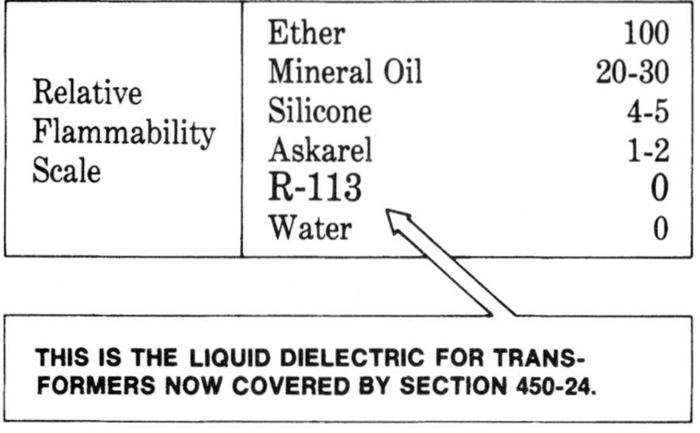

Relative Flammability Scale
Ether 100
Mineral Oil 20-30
Silicone 4-5
Askarel 1-2
R-113 0
Water 0
THIS IS THE LIQUID DIELECTRIC FOR TRANSFORMERS NOW COVERED BY SECTION 450-24.

Chart is based on a UL-defined flammability scale as tested by ASTM.

Fig. 450-28. Comparison of the relative flammabilities of liquid dielectrics compared with ether. (Sec. 450-24.)

Nonflammable fluid-insulated transformers installed indoors must have a liquid-confinement area and a pressure-relief vent. In addition, such transformers must be equipped to absorb gases generated by arcing inside the tank, or the pressure-relief vent must be connected to a flue or duct to carry the gases to "an environmentally safe area." Units rated over 35 kV must be installed in a vault when used indoors.

450-25. Askarel-Insulated Transformers Installed Indoors. Although askarel transformers are being phased out, the Code rule says such transformers installed indoors must conform to the following:

1. Units rated over 25 kVA must be equipped with a pressure-relief vent.
2. Where installed in a poorly ventilated place, they must be furnished with a means for absorbing any gases generated by arcing inside the case, or the pressure-relief vent must be connected to a chimney or flue which will carry such gases outside the building (Fig. 450-29).
3. Units rated over 35,000 V must be installed in a vault.

450-26. Oil-Insulated Transformers Installed Indoors. The basic rule is illustrated in Fig. 450-30. Oil-insulated transformers installed indoors must be installed in

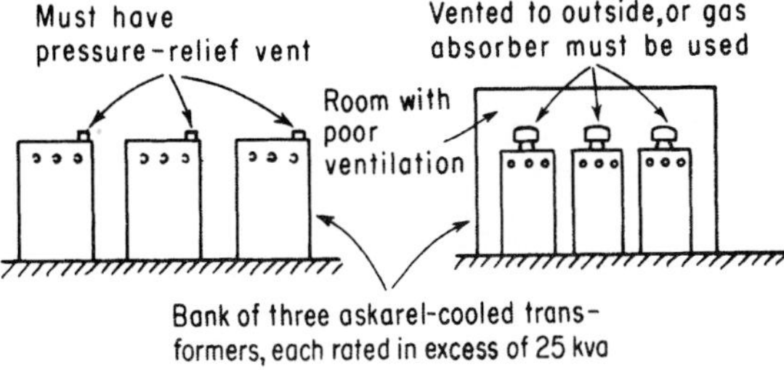

Fig. 450-29. Code rules still cover askarel transformers. (Sec. 450-25.)

a vault constructed according to Code specs, but the Exceptions note general and specific conditions under which a vault is not necessary. The most commonly applied Exceptions are as follows:

1. A hookup of one or more units rated not over 112½ kVA may be used in a vault constructed of reinforced concrete not less than 4 in. (102 mm) thick.
2. Units installed in detached buildings used only for providing electric service do not require a Code-constructed vault if no fire hazard is created and the interior is accessible only to qualified persons.

450-27. Oil-Insulated Transformers Installed Outdoors. Figure 450-31 shows how physical locations of building openings must be evaluated with respect to potential fire hazard from leaking transformer oil.

450-28. Modification of Transformers. Askarel transformers that are drained and refilled with another liquid dielectric must be identified as such and must satisfy all rules of its retrofilled status. This rule is intended to maintain safety in all cases where askarel transformers are modified to eliminate PCB hazards. Marking must show the new condition of the unit and must not create code violations. For instance, an indoor askarel transformer that is drained and refilled with oil may require construction of a vault, which is required for oil-filled transformers as specified in Sec. 450-26.

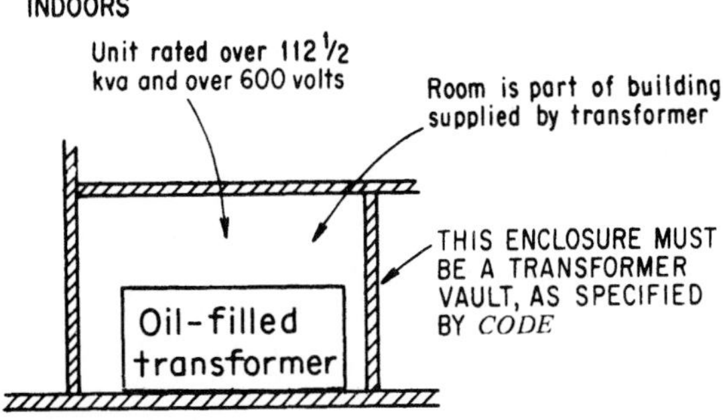

Fig. 450-30. Oil-filled transformers generally require installation in a "vault." (Sec. 450-26.)

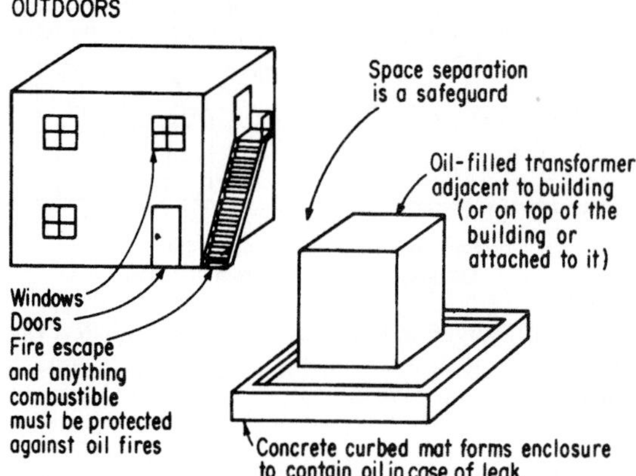

Fig. 450-31. Precautions must be taken for outdoor oil transformers. (Sec. 450-27.)

450-41. Location. Ideally, a transformer vault should have direct ventilating openings (grilles or louvers through the walls) to outdoor space. Use of ducts or flues for ventilating is not necessarily a Code violation, but they should be avoided wherever possible.

450-42. Walls, Roof, and Floor. Basic mandatory construction details are established for an NEC-type transformer vault, as required for oil-filled transformers and for all transformers operating at over 35,000 V. The purpose of a transformer vault is to isolate the transformers and other apparatus. It is important that the door as well as the remainder of the enclosure be of proper construction and that a substantial lock be provided. Details required for any vault are shown in Fig. 450-32 and include the following details:

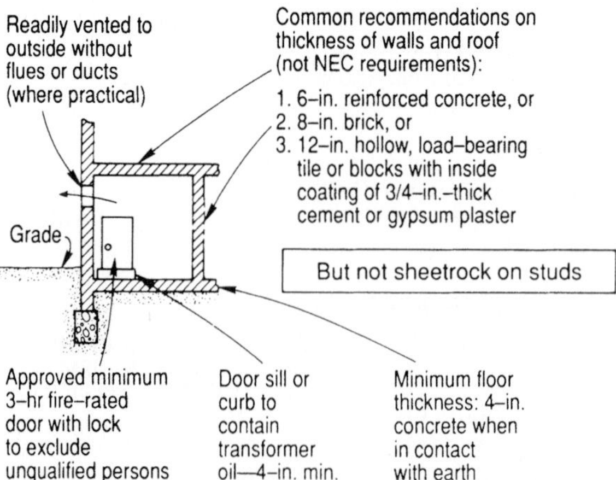

Fig. 450-32. Transformer vault must assure containment of possible fire. (Sec. 450-42.)

1. Walls and roofs of vaults shall be constructed of reinforced concrete, brick, load-bearing tile, concrete block, or other fire-resistive constructions with adequate strength and a fire resistance of 3 hr according to ASTM Standard E119-75. "Stud and wall board construction" may not be used for walls, roof, or other surfaces of a transformer vault. Although the rule here does *not* flatly mandate "concrete" or "masonry" construction, that is essentially the objective of the wording. The substantiation for this rule said the following:

> The only guidance provided in Sec. 450-42 as to the type of material to be used in vault construction is in a fine print note which states that "six-in. (152-mm) thick reinforced concrete is a typical 3-hour construction." This, of course, is only advisory. Sheet rock can be so installed as to have a 3-hour fire rating, however, it should not be considered as being suitable for this type of installation. It would not have adequate structural strength in case of oil fire. It is not too difficult to break through such a wall, either intentionally or unintentionally.
>
> Your attention is also called to Sec. 230-6(3), which states that conductors installed in a transformer vault shall be considered outside of a building. The major thrust of Sec. 230-6 has been that the conductors are to be masonry-encased.

2. A vault must have a concrete floor not less than 4 in. (102 mm) thick when in contact with the earth. When the vault is constructed with space below it, the floor must have adequate structural strength and a minimum fire resistance of 3 hr. Six-in. (152-mm) -thick reinforced concrete is a typical 3-hr-rated construction.
3. Building walls and floors that meet the above requirements may serve for the floor, roof, and/or walls of the vault.

An exception to the basic regulations establishing the construction standards for transformer fireproof vaults notes that the transformer-vault fire rating may be reduced where the transformers are protected with automatic sprinkler, water spray, or carbon dioxide. The usual construction standards for transformer vaults [such as 6-in. (152-mm) -thick reinforced concrete] provide a minimum fire-resistance rating of 3 hr. Where automatic sprinkler, water spray, or carbon dioxide is used, a construction rating of only 1 hr will be permitted.

450-43. Doorways. Each doorway must be of 3-hr fire rating as defined in the *Standard for the Installation of Fire Doors and Windows* (NFPA No. 80-1983). The Code-enforcing authority may also require such a door for doorways leading from the vault to the outdoors, in addition to any doorways into adjoining space in the building.

As required in part (c), vault doors must swing *out* and must be equipped with "panic bars" or other opening means that requires only "simple pressure." This is intended to provide the greater safety of a push-open-rather than a rotating-knob-type of door release. As noted in the substantiation:

> Conventional rotating door knob hardware is used on transformer vault doors due to lack of specific wording in the paragraph as presently written. The *National Electrical Safety Code* is believed to be very specific, or has been formally interpreted to be, requiring "panic type" door hardware. In an electrical flash or arc an electrical worker may lose the use of hands for twisting a conventional door knob.

450-45. Ventilation Openings. This rule sets the size and arrangement of vent openings in a vault where such ventilation is required by ANSI C57.12.00-1987—"General Requirements for Liquid-Immersed Distribution, Power, and Regulating Transformers," as noted in Sec. 450-9. Figure 450-33 shows the openings as regulated by part (c) of this section. One or more openings may be used, but if a single vent opening is used, it must be in or near the roof of the vault—and not near the floor.

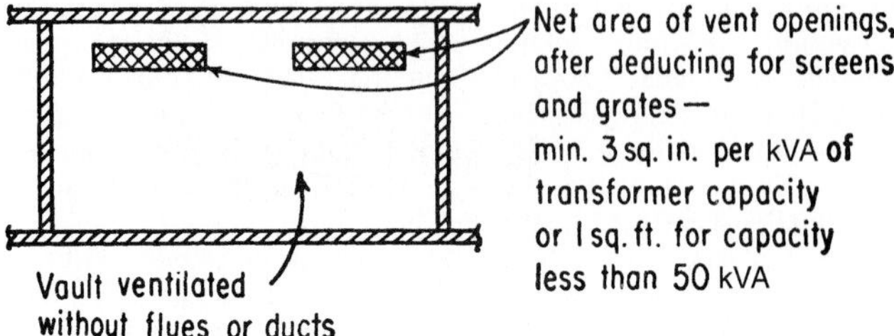
Net area of vent openings, after deducting for screens and grates —
min. 3 sq. in. per kVA of transformer capacity or 1 sq. ft. for capacity less than 50 kVA
Vault ventilated without flues or ducts

Fig. 450-33. Vault vent opening(s) may be in or near the roof—or near floor level also, if one is at or near roof level. (Sec. 450-45.)

ARTICLE 455. PHASE CONVERTERS

Refer to Fig. 455-1 which shows details of Secs. 455-2, 455-4, 455-6, 455-7, and 455-8.

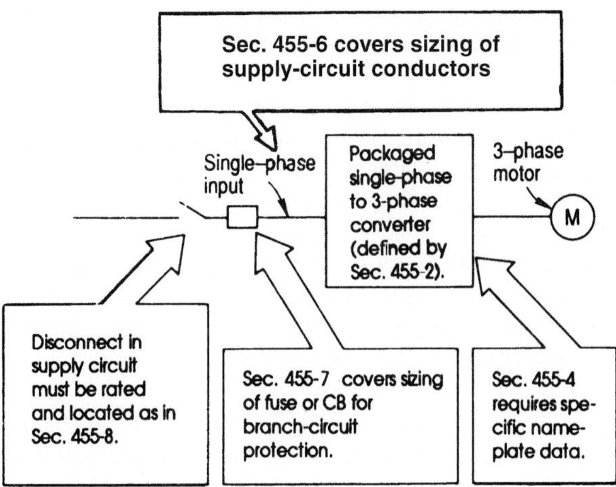
A wide range of Code rules cover single-to-3-phase converters for motors and other 3-phase loads in new Art 455.
Sec. 455-6 covers sizing of supply-circuit conductors
Single-phase input
Packaged single-phase to 3-phase converter (defined by Sec. 455-2).
3-phase motor
M
Disconnect in supply circuit must be rated and located as in Sec. 455-8.
Sec. 455-7 covers sizing of fuse or CB for branch-circuit protection.
Sec. 455-4 requires specific name-plate data.

Fig. 455-1. Code rules on phase converters cover these considerations. (Art. 455.)

ARTICLE 460. CAPACITORS

460-1. Scope. The sections in this article apply chiefly to capacitors used for the power-factor correction of electric-power installations in industrial plants and for correcting the power factors of individual motors (Fig. 460-1). These provisions apply only to capacitors used for surge protection where such capacitors are not component parts of other apparatus.

Fig. 460-1. A typical power-factor correction capacitor bank is this 300-kvar bank of twelve 25-kvar, 480-V capacitor units installed in a steel enclosure in an outdoor industrial substation. (Sec. 460-1.)

In an industrial plant using induction motors, the power factor may be considerably less than 100 percent, particularly when all or part of the motors operate most of the time at much less than their full load. The lagging current can be counteracted and the power factor improved by installing capacitors across the line. By raising the power factor, for the same actual power delivered the current is decreased in the generator, transformers, and lines, up to the point where the capacitor is connected.

Figure 460-2 shows a capacitor assembly connection to the main power circuit of a small industrial plant, consisting of capacitors connected in a 3-phase hookup and rated at 90 kVA for a 460-V system. An externally operable switch mounted on the wall is used as the disconnecting means and the discharge device required by Sec. 460-6 consists of two high-impedance coils inside the switch enclosure which consume only a small amount of power, but, having a

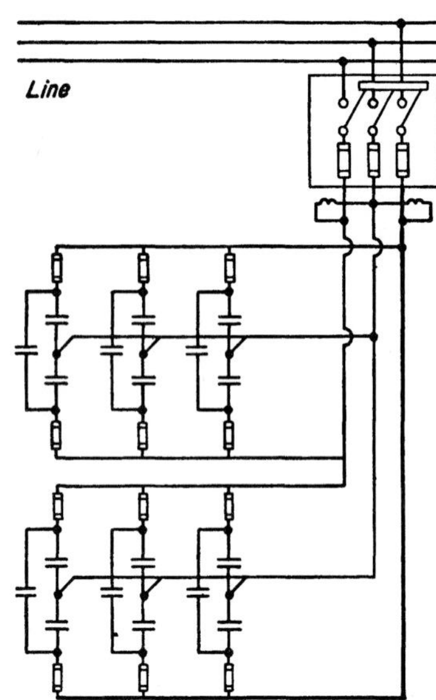

Fig. 460-2. Six internally delta-connected capacitors form a 3-phase capacitor bank. (Sec. 460-1.)

comparatively low DC resistance, permit the charge to drain off rapidly after the capacitor assembly has been disconnected from the line.

460-6. Discharge of Stored Energy. If no means were provided for draining off the charge stored in a capacitor after it is disconnected from the line, a severe shock might be received by a person servicing the equipment or the equipment might be damaged by a short circuit. If a capacitor is permanently connected to the windings of a motor, as in Fig. 460-3, the stored charge will drain off rapidly through the windings when the circuit is opened. Reactors or resistors used as discharge devices must either be permanently connected across the terminals of the capacitor (such as within the capacitor housing) or a device must be provided that will automatically connect the discharge devices when the capacitor is disconnected from the source of supply. Most available types of capacitors have discharge resistors built into their cases. When capacitors are not equipped with discharge resistors, a discharge circuit must be provided.

Figure 460-3 shows a capacitor used to correct the power factor of a single motor. The capacitor may be connected to the motor circuit between the starter and the motor or may be connected between the disconnecting means and the starter, as indicated by the dotted lines in the diagram. If connected as shown by the dotted lines, an overcurrent device must be provided in these leads, as required by Sec. 460-8(b). The capacitor is shown as having discharge devices consisting of resistors.

In previous NEC editions, Sec. 460-7 covered selection of the size of power-factor capacitors. That section is no longer in the NEC.

Power capacitors, in most applications, are installed to raise the system power factor, which results in increased circuit or system current-carrying capacity, reduced power losses, and lower reactive power charges (most utility

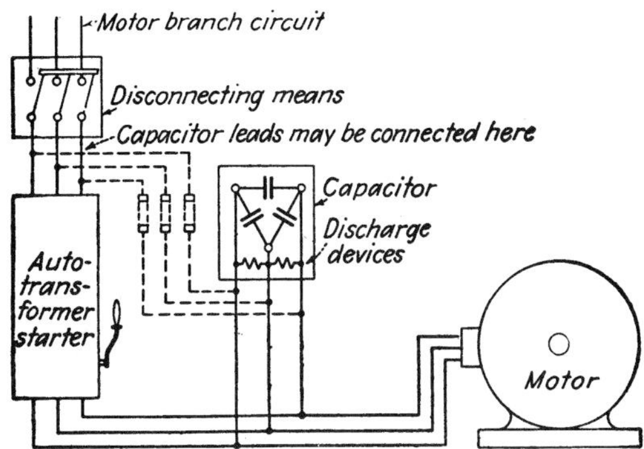

Fig. 460-3. Capacitor voltage must be discharged when circuit is opened. (Sec. 460-6.)

companies include a power-factor penalty clause in their industrial billing). Also, additional benefits derived as a result of a power capacitor installation are reduced voltage drop and increased voltage stability. Figure 460-4 presents basic data on calculating size of capacitors for power-factor correction. However, manufacturers provide tables and graphs to help select the capacitor for a given motor load.

The former rule of Sec. 460-7 limited power-factor correction to unity (100 percent, or 1.0) when there is no load on the motor. That will result in a power factor of 95 percent or better when the motor is fully loaded. The old rule recognized the use of capacitors sized *either* for the value that will produce 100 percent power factor of the circuit when the motor is running at no load *or* for a value equal to 50 percent of the kilovoltampere rating of the motor input for motors up to 50 hp, 600 V (Fig. 460-5).

A number of comments made in the discussions that led to that rule are informative and important:

1. Setting the capacitor rating at 50 percent of the kilovoltampere rating of the motor input does not afford, in most cases, an appreciable gain in power factor over that achieved by setting the capacitor rating to limit the motor no-load power factor to unity. In two examples, 7.5-hp, 1,200-rpm and 10-hp, 900-rpm ratings, the motor no-load kvar is more than 50 percent of the kilovoltampere rating of the motor input. In these cases, which are typical of ones having inherently low motor power factor and where power-factor improvement would be most beneficial, the italicized "Exception" affords no potential for additional gain in power-factor improvement. In other examples, 40-hp, 1,200-rpm and 200-hp, 1,800-rpm ratings, the corrective kvar value to raise the motor no-load power factor to unity yields better power-factor improvement up through 75 percent load than attained by corrective kvar equal 50 percent of the kilovoltampere rating of the motor input. For loads above 75 percent, the converse is true; but there is no notable overall difference in power-factor improvement between the two approaches.

Power-factor capacitors can be connected across electric lines to neutralize the effect of lagging power-factor loads, thereby reducing the current drawn for a given kilowatt load. In a distribution system, small capacitor units may be connected at the individual loads or the total capacitor kilovolt-amperes may be grouped at one point and connected to the main. Although the total kvar of capacitors is the same, the use of small capacitors at the individual loads reduces current all the way from the loads back to the source and thereby has greater PF corrective effect than the one big unit on the main, which reduces current only from its point of installation back to the source.

Calculating Size of Capacitor:

Assume it is desired to improve the power factor a given amount by the addition of capacitors to the circuit.

Then $kvar_R = kw \times (\tan \theta_1 - \tan \theta_2)$

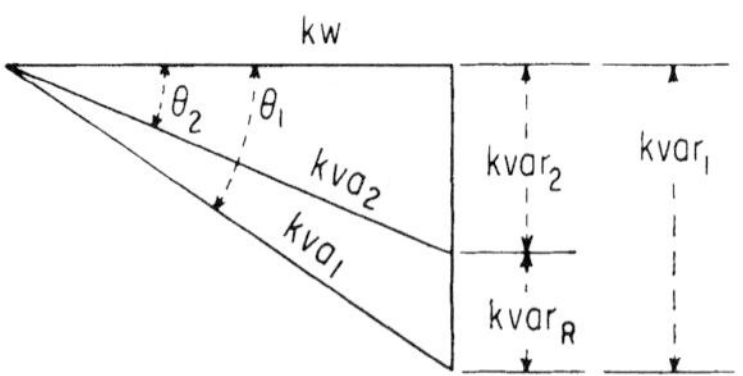

where $kvar_R$ = rating of required capacitor
$kvar_1$ = reactive kilovolt-amperes at original PF
$kvar_2$ = reactive kilovolt-amperes at improved PF
θ_1 = original phase angle
θ_2 = phase angle at improved PF
kw = load at which original PF was determined.

NOTE: The phase angles θ_1 and θ_2 can be determined from a table of trigonometric functions using the following relationships:
θ_1 = The angle which has its cosine equal to the decimal value of the original power factor (e.g., 0.70 for 70% PF; 0.65 for 65%; etc.)
θ_2 = The angle which has its cosine equal to the decimal value of the improved power factor.

Fig. 460-4. Capacitors reduce circuit current by supplying the magnetizing current to motors. (Sec. 460-6.)

2. Most motor-associated capacitors are related to low-voltage, 5- to 50-hp, 1,800- and 1,200-rpm, across-the-line-start motors. In this range, the no-load rule for determination of maximum capacitor kvar restricts capacitors to less than 50 percent of horsepower.

Noticeable economies can be made by applying larger capacitors to such motors. This has been done for years, with excellent results and no field trouble, as has been attested by several engineers whose views have been made known to the Code panel.

The larger motors should certainly have capacitor kvars limited by stringent rules, but this should not interfere with the field-proved use of augmented capacitor sizing of the integral-horsepower motors most commonly used in industrial, commercial, and institutional plants.

The no-load power factor of a motor is a design constant of the motor and may be obtained from the manufacturer of the motor—or it may be measured or

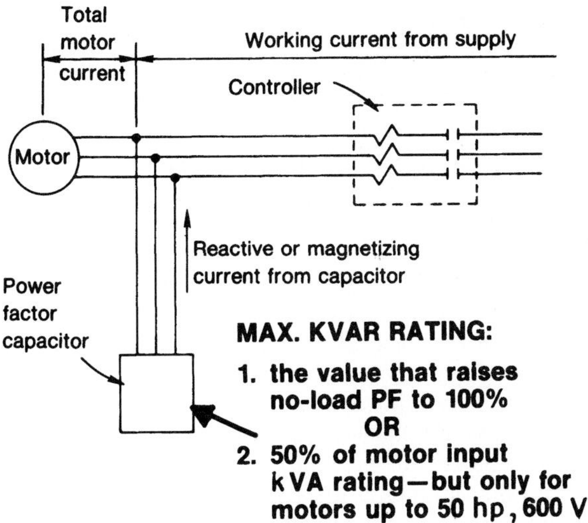

Fig. 460-5. Maximum rating of capacitor kvar may be determined by either of two approaches. (Sec. 460-6.)

calculated. In Fig. 460-4, using the known no-load PF (power factor) of a motor, the kilowatts can be calculated from kW = PF × kVA_1. Then $kvar_1$ (the required rating of PF capacity to raise the no-load value of PF to 100 percent) equals the square root of $(kVA_1)^2 - (kW)^2$, where kVA_1 is calculated from circuit voltage and current measured with a clamp-on ammeter.

A handy rule-of-thumb method for determining the kilovar rating of a capacitor required to provide optimum power-factor correction for a given motor is:

1. With no load on the motor, measure the no-load kilovoltampere. That can be determined by using a clamp-on ammeter to measure the amount of current drawn by the motor under no-load condition and then using a voltmeter to get the phase-to-phase voltage of the motor circuit. Then, for a 3-phase motor, the kilovoltampere input to the motor is derived from the formula:

   $$\text{Input kVA} = [\text{phase-to-phase voltage} \times \text{line current} \times 1.732] + 1{,}000$$

2. Because the power factor of an unloaded motor is very low—say, about 10 percent—the kilovoltampere vector for the original PF condition as shown in Fig. 460-4 is lagging the kilowatt vector by an angle that is approaching 90°. That results from the working current being small (only the resistance of the windings) while the reactive (magnetizing) current is at its normal and very much larger value. In that condition, the reactive current causes the kilovar vector to be almost the same length as the kilovoltampere vector—so close in fact that it is generally safe to take the kilovoltampere input value as the required kilovar rating of capacitor needed to correct to 100 percent PF at no-load, which will result in a 95 to 98 percent PF at full load.

3. Then select a capacitor assembly that has a kilovar rating as close as possible to—but *not* in excess of—the calculated value of input kilovoltampere of the motor. This method may be used on rewound motors or on other motors where it is not possible to make a better determination of needed capacitor kilovar.

Capacitors of the type used for PF correction of motors are commonly rated in kilovoltamperes, or the rating may be in "kilovars," meaning "reactive kilovoltamperes," abbreviated kvar. The capacitors are usually designed for connection to a 3-phase system and constructed as a unit with three leads brought out.

Corrective measures for improving the power factor may be designed into motor branch circuits. Generally, the most effective location for installation of individual power-factor-correction capacitors is as close to the inductive load as possible. This provides maximum correction from the capacitor back to the source of power. At individual motor locations, power-factor-correcting capacitors offer improved voltage regulation. As shown in Fig. 460-6, power-factor capacitors installed at terminals of motors provide maximum relief from reactive currents, reducing the required current-carrying capacities of conductors from their point of application all the way back to the supply system. Figure 460-7 shows a typical example. Such application also eliminates extra switching devices, since each capacitor can be switched with the motor it serves.

Capacitors also may be installed as a group or bank at some central point, such as a switchboard, loadcenter, busway, or outdoor substation. Usually this method serves only to reduce the utility company penalty charges; however, in many instances, installation costs also will be lower.

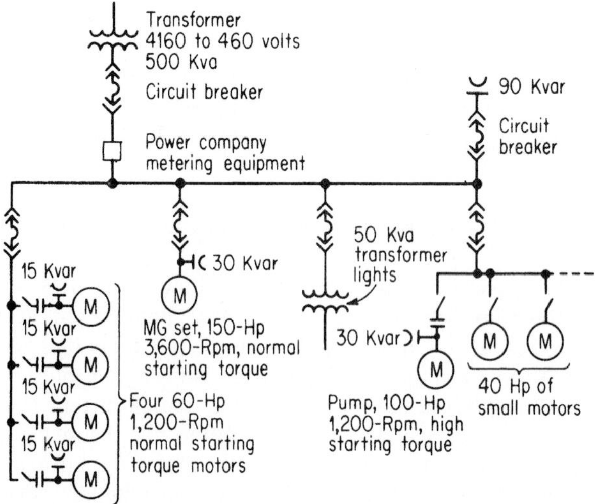

Fig. 460-6. PF capacitors at individual motors offer maximum corrective advantages. (Sec. 460-6.)

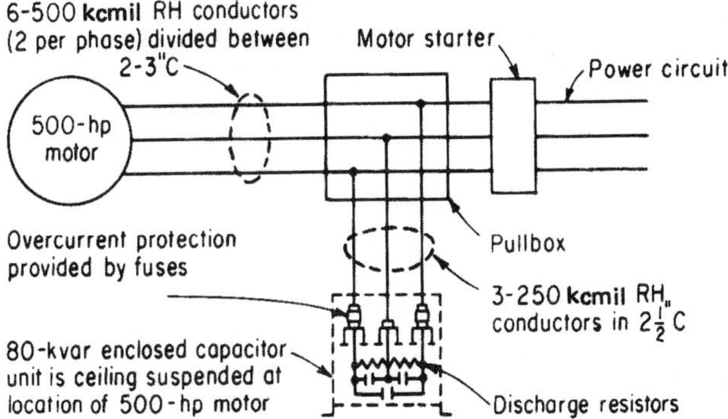

Fig. 460-7. Typical capacitor installation connected on the load side of a motor starter for a 500-hp motor consists of ceiling-suspended enclosed capacitors (arrow) that are rated at 80 kvar, connected through the pull box to the circuit conductors for the motor. Diagram shows how conductors were added to the equipment shown in the photo. (Sec. 460-6.)

When motors are small, numerous, and operated intermittently, it is often economically more desirable to install required capacitor kvar at the motor control center.

Capacitor installations may consist of an individual unit connected as close as possible to the inductive load (at the terminals of a motor, etc.) or of a bank of many units connected in multiple across a main feeder. Units are available in specific kvar and voltage ratings. Standard low-voltage capacitor units are rated from about 0.5 to 25 kvar at voltages from 216 to 600 V. For high-voltage applications, standard ratings are 15, 25, 50, and 100 kvar. Available in single-,

2-, or 3-phase configurations, power capacitors may be supplied either unfused or equipped with current-limiting or high-capacity fuses (single-phase units are furnished with one fuse; 3-phase capacitors usually have two fuses). On low-voltage units, fuses may be mounted on the capacitor bushings inside the terminal compartment.

Use of capacitor power-factor application is generally not acceptable for *any* motor application involving repetitive switching of the motor load, as in plugging, jogging, rapid reversals, reclosings, etc., because of the severe overvoltages and overtorques that are generated in such motor applications when capacitors of the permitted rating are connected on the load side of the motor starter. The objectionable effects can lead to premature failure of motor insulation.

460-8. Conductors. Part (**a**) of this section covers sizing of circuit conductors. The current corresponding to the kilovoltampere rating of a capacitor is computed in the same manner as for a motor or other load having the same rating in kilovoltamperes. If a capacitor assembly used at 460 V has a rating of 90 kVA, the current rating is 90,000/(460 × 1.73) = 113 A. The minimum required ampacity of the conductors would be 1.35 × 113 A, or 153 A.

The manufacturing standards for capacitors for power-factor correction call for a rating tolerance of "–0, +15 percent," meaning that the actual rating in kilovoltamperes is never below the nominal rating and may be as much as 15 percent higher. Thus, a capacitor having a nameplate rating of 100 kVA might actually draw a current corresponding to 115 kVA. The current drawn by a capacitor varies directly with the line voltage, so that, if the line voltage is higher than the rated voltage, the current will be correspondingly increased. Also, any variation of the line voltage from a pure sine wave form will cause a capacitor to draw an increased current. It is for these reasons that the conductors leading to a capacitor are required to have an ampacity not less than 135 percent of the rated current of the capacitor.

example Given the kvar rating of capacitors to be installed for a motor, determining the correct capacitor conductor size is relatively simple. The rule here requires that the ampacity of the capacitor conductors be not less than one-third the ampacity of the motor circuit conductors and not less than 135 percent of the capacitor rated current. The capacitor nameplate will give rated kvar, voltage, and current. It is then a simple matter of multiplying rated current by 1.35 to obtain the ampacity value of the conductor to be installed and selecting the size of conductor required to carry that value of current, from Table 310-16. Then check that the ampacity is not less than one-third the ampacity of the motor circuit conductors.

For a motor rated 100 hp, 460 V, 121 A full-load current, a 25-kvar capacitor would correct power factor to between 0.95 and 0.98 at full load. The nameplate on the capacitor indicates that the capacitor is rated 460 V, 31 A. Then 31 × 1.35 = 42 A. From **Code** Table 310-16, a No. 6 TW or THW conductor rated to carry 55 A would do the job. (No. 8 THW rated at 45 A would most likely be considered not acceptable because UL generally calls for use of 60°C wires in circuits up to 100 A.) The motor circuit conductors are found to be 2/0 THW, with an ampacity of 175 A. Since ⅓ × 175 = 58 A, the No. 6 THW, with an ampacity of 65 A, should be used.

If these conductors are connected to the load terminals of the motor controller, the overload protection heaters may have to be changed (or if the OL is adjustable, its setting may have to be reduced), because the capacitor will cause a reduction in line current and adjustment of relay setting is required by Sec. 460-9.

Although part (**b**) of the rule requires overcurrent protection (fuses or a CB) in each ungrounded conductor connecting a capacitor assembly to a circuit, the Exception considers the motor-running overload relay in a starter to be adequate protection for the conductors when they are connected to the motor circuit on the load side of the starter. Where separate overcurrent protection is provided, as required for line-side connection, the device must simply be rated "as low as practicable." When a capacitor is thrown on the line, it may momentarily draw an excess current. A rating or setting of 250 percent of the capacitor current rating will provide short-circuit protection. Being a fixed load, a capacitor does not need overload protection such as is necessary for a motor.

Most power capacitors are factory equipped with fuses which provide protection in case of an internal short circuit. These fuses are usually rated from 165 to 250 percent of the rated kilovar current to allow for maximum operating conditions and momentary current surges. When installed on the load side of a motor starter, as noted above, capacitors do not require additional fusing. However, for bank installations, separate fuses are required.

Part (**c**) of the rule requires a disconnecting means for all the ungrounded conductors connecting a capacitor assembly to the circuit—but a disconnect is not needed when the capacitor is connected on the load side of a starter with overload protection. The disconnect must be rated at least equal to 1.35 times the rated current of the capacitor.

Note that part (**c**)(**2**) requires a multipole switching device for the disconnect. This is a direct reversal from the former statement that "The disconnecting means shall *not* be required to open all ungrounded conductors simultaneously." This change was made because of the inherent danger of single-pole switching of low-voltage capacitors. Normal switching or closing on faults may cause arcs or splattering of molten metal.

Two accepted methods of wiring capacitors are illustrated in Fig. 460-8. Diagram A shows the method of connection at a central location, such as at a power center or on a busway feeder. In such an installation, the Code rule requires an overcurrent device in each ungrounded conductor, a separate disconnecting means, and a discharge resistor (usually furnished with capacitors). The current rating of both the capacitor disconnect switch and the conductors supplying the capacitor must be not less than 135 percent of the rated current of the capacitor. In B, the capacitor is connected directly to motor terminals. Installation on the load side of the motor starter eliminates the need for separate overcurrent protection and separate disconnecting means. However, motor-running overcurrent protection must take into account the lower running current of the motor, as required by Sec. 460-9.

460-9. Rating or Setting of Motor Overload Device. When a power-factor capacitor is connected to a motor circuit at the motor—i.e., on the load or motor side of the motor controller—the reactive current drawn by the motor is provided by the capacitor and, as a result, the total current flowing in the motor circuit up to the capacitor is reduced to a value below the normal full-load current of the motor. With that hookup, the total motor full-load current flows only over the

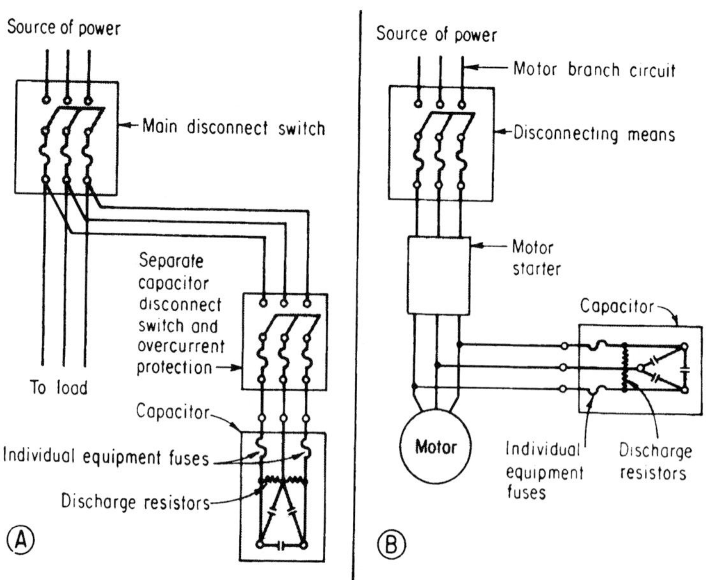

Fig. 460-8. PF capacitor assembly may be connected on the line or load side of a starter. (Sec. 460-8.)

conductors from the capacitor connection to the motor and the entire motor circuit up to that connection carries only the so-called "working current" or "resistive current." That is shown in the top part of Fig. 460-9.

Under the condition shown, it is obvious that setting the overload relay in the starter for 125 percent of the motor nameplate full-load current (as required by Sec. 430-32) would actually be an excessive setting for real protection of the motor, because considerably less than full-load current is flowing through the starter. The rule of this section clearly requires that the rating of a motor overload protective device connected on the line side of a power-factor correction capacitor must be based on 125 percent (or other percentage from Sec. 430-32) times the circuit current produced by the improved power-factor—rather than the motor full-load current (see Fig. 460-9).

The 25-kvar capacitor used on the 100-hp, 460-V motor in the example in Sec. 460-8 will reduce the motor line current by about 9 percent. Section 430-32(a) (also Secs. 430-34 and 460-9) requires that the running overload protection be sized not more than 125 percent of motor full-load current produced with the capacitor. If the OL protection heaters were originally sized at 125 percent of the motor full-load current (1.25×121), they would have been sized at 151 A. With the motor current reduced by 9 percent (0.09×121, or 11 A), the motor full-load current with the capacitor installed would be $121 - 11$, or 110 A. Since 125 percent of 110 A is 137.5 A, the heaters must be changed to a size not larger than 137.5 A.

If the capacitor conductors could be connected on the line side of the heaters, the heaters would not have to be reduced in size, since the reduction of line current occurs only from the source back to the point of the capacitor connec-

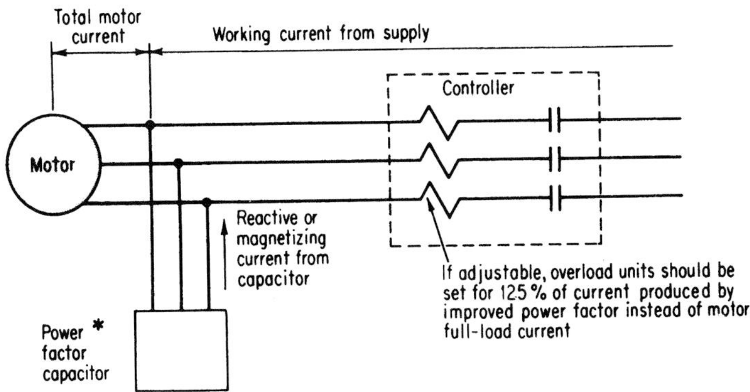

TOTAL MOTOR CURRENT = VECTOR SUM OF REACTIVE AND WORKING CURRENTS

EXAMPLE: Motor with 70% power factor has full-load current of 143 amps. Capacitor corrects to 100% PF.

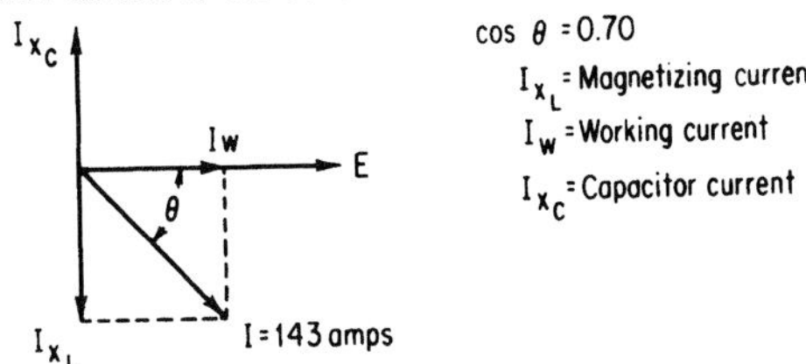

I_{X_C} cancels I_{X_L} leaving only working current to be supplied from circuit. Working current = 143 x cos θ = 143 x 0.70 = 100

OVERLOAD RELAYS SHOULD BE SET FOR 125% OF 100 AMPS

* The rating of such capacitors should not exceed the value required to raise the no-load power factor of the motor to unity. Capacitors of these maximum ratings usually result in a full-load power factor of 95 to 98 percent.

Fig. 460-9. Motor overload protection must be sized for the current at improved PF. (Sec. 460-9.)

tion. Conductor connections at this location are extremely difficult to make because of the lack of space and the large size of the connecting lugs. Controller load terminals are furnished with connectors that will accept an additional conductor, or they can be easily modified to permit a dependable connection.

460-10. Grounding. The metal case of a capacitor is suitably grounded by locknut and bushing connections of grounded metal nipples or raceways carrying the conductors connecting the capacitor into a motor circuit or feeder.

ARTICLE 470. RESISTORS AND REACTORS

470-1. Scope. Except when installed in connection with switchboards or control panels that are so located that they are suitably guarded from physical damage and accidental contact with live parts, resistors should always be completely enclosed in properly ventilated metal boxes.

Large reactors are commonly connected in series with the main leads of large generators or the supply conductors from high-capacity network systems to assist in limiting the current delivered on short circuit. Small reactors are used with lightning arresters, the object here being to offer a high impedance to the passage of a high-frequency lightning discharge and so to aid in directing the discharge to ground. Another type of reactor, having an iron core and closely resembling a transformer, is used as a remote-control dimmer for stage lighting. Reactors as well as resistors are sources of heat and should therefore be mounted in the same manner as resistors.

ARTICLE 480. STORAGE BATTERIES

480-1. Scope. Storage cells are of two general types: the so-called lead-acid type, in which the positive plates consist of lead grids having openings filled with a semisolid component, commonly lead peroxide, and the negative plates are covered with sponge lead, the plates being immersed in dilute sulfuric acid; and the alkali type, in which the active materials are nickel peroxide for the positive plate and iron oxide for the negative plate, and the electrolyte is chiefly potassium hydroxide (Fig. 480-1).

480-2. Definitions. "Stationary installations of storage batteries" provide an independent source of power for emergency lighting, switchgear control, engine-generator set starting, signal and communications systems, laboratory

Fig. 480-1. Article 480 applies only to "stationary installations of storage batteries"—whether they are used for supply to lighting, generator cranking, switchgear control, or in UPS (Uninterruptible Power Supply) systems. (Sec. 480-1.)

power, and similar applications. They are an essential component of UPS systems. This Code article does not cover batteries used to supply the motive power for electric vehicles.

The most commonly used battery is the lead-acid type—either lead-antimony or lead-calcium. Nickel-cadmium batteries offer a variety of special features that, in many instances, offset their higher initial cost. Other types include silver-zinc, silver-cadmium, and mercury batteries.

The lead-antimony battery is readily available at a moderate price, has a high efficiency (85 percent to 90 percent), is comparatively small, and has a relatively long life if operated and maintained properly under normal conditions. Voltage output is about 2 V per cell; ratings range to about 1,000 amp-hr (based on an 8-hr discharge rate).

Lead-calcium batteries offer features similar to the lead-antimony type, and they require less maintenance. They do not require an "equalizing" charge (application of an overvoltage for a period of time to assure that all cells in a battery bank will produce the same voltage). For this reason, they are often selected for use in UPS systems.

This type of cell can usually be operated for a year or more without needing water, depending on the frequency and degree of discharge. Sealed or maintenance-free batteries of this type never need water. Voltage output is 2 V per cell, with ratings up to about 200 amp-hr (8-hr rate).

Nickel-cadmium batteries are particularly useful for application in temperature extremes. They are reputed to have been successfully operated at temperatures from −40°F to +163°F. They have a very high short-time current capability and are well suited to such applications as engine starting and UPS operation. Initial cost is higher than lead-acid types; however, they offer long life (25 to 30 years), reliability, and small size per unit. Voltage is about 1.2 V per cell.

480-3. Wiring and Equipment Supplied from Batteries. As indicated in Fig. 480-2, whatever kinds of circuits and loads a battery bank serves, all rules of the NE Code covering operation at that voltage must be applied to the wiring and equipment.

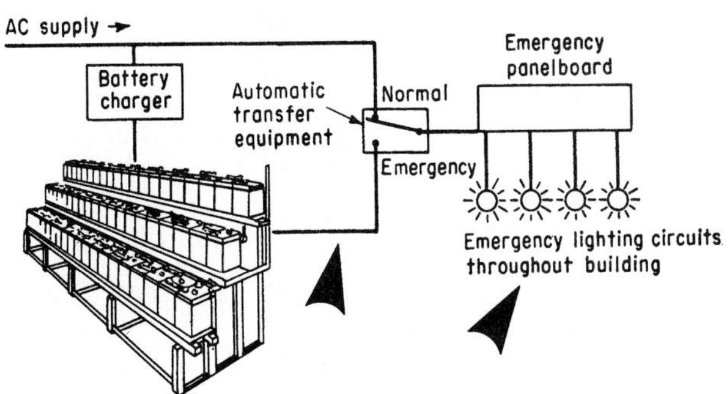

Fig. 480-2. Applicable Code rules must be observed for load circuits fed by batteries. (Sec. 480-3.)

480-6. Insulation of Batteries of Over 250 Volts. Racks of adjacent batteries must be so placed as to have a minimum of 2 in. (50.8 mm) of air space between any pair of exposed, live battery terminals of opposite polarity.

480-8. Battery Locations. Although specific "battery rooms" or enclosures are no longer required for installation of any batteries (not since the 1971 NE Code), part (**a**) does require ventilation at battery locations. A specific "battery room" was previously required for open-tank or open-jar batteries, but such units are no longer made or in use.

The overcharging of a battery can result in the breaking down of the electrolyte into gases that, if permitted to accumulate in the room, may result in an explosive mixture. Overcharging indicates problems with the charging equipment requiring correction. Proper ventilation will resolve this explosive mixture, assuring that the location is not a hazardous location subject to Art. 501.

Because the fumes given off by a storage battery are very corrosive, the type of wiring must be such that it will withstand the corrosive action, and special precautions are necessary as to the type of insulation used, as well as protection of all metalwork. It is stated by the respective manufacturers that conduit made of aluminum or Everdur (silicon-bronze) is well suited to withstand the corrosive effects of the fumes in battery rooms. If steel conduit is used, it is recommended that the conduit be zinc-coated and that it be kept well painted with asphaltum paint.

Batteries of the lead-acid type sometimes throw off a fine spray of the dilute acid which fills the air around the cells; hence steel conduit or tubing should not be brought close to any cell.

There are no special requirements on the type of fixtures or other electrical equipment used in the battery room. Proper ventilation of the room will prevent explosions. See Secs. 300-6 and 410-4(b).

Chapter Five

ARTICLE 500. HAZARDOUS (CLASSIFIED) LOCATIONS

500-1. Scope—Articles 500 Through 505. In the heading of this article, the word "classified" makes clear that hazardous locations are those which have been "classified" as hazardous by the inspection authority. Hazardous locations in plants and other industrial complexes are involved with a wide variety of flammable gases and vapors and ignitible dusts—all of which have widely different flash points, ignition temperatures, and flammable limits. And these explosive or flammable substances are processed and handled under a wide range of operating conditions. In such places, fire or explosion could result in loss of lives, facilities, and/or production.

Note: One of the many changes to the 1996 NEC was the inclusion of a new Art. 505, "Class I Zone 0, 1, and 2 Locations." Subsequent to the issuing of the 1996 NEC, the inclusion of Art. 505 was appealed. It may, or may not, be accepted. Use of those rules may, or may not, be permitted depending on the disposition of the appeal. Check with the local inspection authority to see if Art. 505 *is,* or *is not,* in force.

500-2. Location and General Requirements. Classification of hazardous areas must be approached very carefully, based on experience and a detailed understanding of electrical usage in the various kinds of locations. After study and analysis—and consultation with inspection authorities or other experts in such work—hazardous areas may be identified and delineated diagrammatically by defining the limits and degree of the hazards involved. In all cases, classification must be carefully based on the type of gas involved, whether the vapors are heavier or lighter than air, and similar factors peculiar to the particular hazardous substance.

Locations used for pyrophoric gases, which ignite spontaneously on contact with air, are exempted from designation as Class I hazardous locations. Electrical equipment approved for classified locations is not needed for places where pryophoric gases are handled.

Classification takes into account that all sources of hazards—gas, vapor, dust, fibers—have different ignition temperatures and produce different pressures when exploding. Electrical equipment must, therefore, be constructed and installed in such a way as to be safe when used in the presence of particular explosive mixtures. The source of hazard must be evaluated in terms of those characteristics that are involved with explosion or fire, as follows:

Diesel Oil and Heating Oil

Questions often arise about the need for hazardous location wiring for electrical equipment installed in areas containing diesel fuel oil or heating oil. National Fire Codes (NFC), Vol. 3, classifies diesel fuel oil as a Class II liquid having a flash point at or above 100°F (37.8°C) and below 140°F (60°C). Chapter 6 of the same NFC Vol. 3 on bulk plants, paragraph 6-5-.1, states in part that in areas where Class II or Class III liquids are stored or handled, the electrical equipment *may* be installed in accordance with the provisions of the NEC for ordinary (i.e., nonhazardous) locations. Diesel fuel oil is classified as a Class II liquid and does not come under requirements for hazardous (classified) locations. With this type of liquid, explosionproof wiring methods *are not* required and the wiring methods listed in Chaps. 1 through 4 of the NEC may be used. The NFC does, however, caution that, if any Class II flammable liquid is heated, it may be necessary that Class I Group D wiring methods be used. In some geographic areas with hot climates, local regulations do require diesel fuel areas to be treated as hazardous areas because of high ambient temperatures. Temperatures in the Southwest often exceed 115°F (143°C), especially in closed, nonventilated areas.

The *flash point* of a liquid is the minimum temperature at which the liquid will give off sufficient vapor to form an ignitible mixture with air near the surface of the liquid or within the vessel used. (This characteristic is not applicable to gases.)

The *ignition temperature* of a substance is the lowest temperature which will initiate explosion or cause self-sustained combustion of the substance.

Explosive limits: When flammable gases or vapors mix with air or oxygen, there is a minimum concentration of the gas or vapor below which propagation of flame does not occur upon contact with a source of ignition. There is also a maximum concentration above which propagation does not occur. These boundary-line mixtures are known as the lower and upper explosive (or flammable) limits and usually are expressed in terms of the percentage of gas or vapor in air, by volume. (See NFPA Bulletin No. 325M.)

Vapor density is the weight of a volume of pure vapor or gas (with no air present) compared to the weight of an equal volume of dry air at the same temperature and pressure.

Section 500-2 recognizes use of "intrinsically safe" equipment in hazardous locations and exempts such equipment from the rules of Arts. 500 through 517. Intrinsic safety is obtained by restricting the energy available in an electrical system to much less than that required for the ignition of flammable atmospheres such as gases and vapors that exist in processing industries. Intrinsically safe systems operate at low voltage (e.g., 24 V) and are designed safe, regardless of short circuits, grounding, overvoltage, equipment damage, or component failure. But such equipment must be "approved," which requires careful attention to UL listing and application data from the *Hazardous Location Equipment Directory* of UL.

Any applications of "intrinsically safe apparatus" in Class I, II, or III locations must satisfy the rules of Art. 504 covering such apparatus. In this section the reference to new Art. 504 notes that the "provisions of Article 501 through 503, 505, and 510 through 516 shall *not* be considered applicable" to intrinsically safe apparatus and wiring "except as required by Article 504."

Intrinsically safe circuits and equipment for use in Division 1 locations must be carefully applied. It is up to the designer and/or the installer to be sure that the energy level available in such equipment is below the level that could ignite the particular hazardous atmosphere. That must be assured for both normal and abnormal conditions of the equipment. Testing of an intrinsically safe system by UL is based on a maximum distance of 5,000 ft (1,525 m) between the equipment installed in the nonhazardous or Division 2 location and the equipment installed in the Division 1 location.

Wiring of intrinsically safe circuits must be run in separate raceways or otherwise separated from circuits for all other equipment to prevent imposing excessive current or voltage on the intrinsically safe circuits because of fault contact with the other circuits.

A note in this section: Maximum effort should be made to keep as much electrical equipment as possible out of the hazardous areas—particularly minimizing installation of arcing, sparking, and high-temperature devices in hazardous locations. It is generally economically and operationally better to keep certain electrical equipment out of hazardous areas. Figure 500-1 shows an example. There the drive shaft of the motor is extended through a packing gland in one of the enclosing walls. To prevent the accumulation of flammable vapors or gas within the motor room, it should be ventilated effectively by clean air or kept under a slight positive air pressure. A gas detector giving a visual and/or audible alarm would be an additional desirable safety feature.

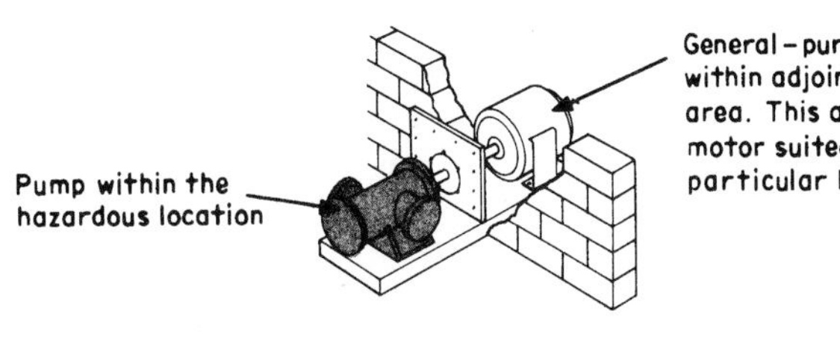
General-purpose motor within adjoining nonhazardous area. This avoids need for motor suited to use in the particular hazardous location.
Pump within the hazardous location
DETAIL

Fig. 500-1. Keeping electrical equipment in nonhazardous area eliminates costly hazardous types. (Sec. 500-2.)

Positive-pressure ventilation is also cited as a means of reducing the level of hazard in areas where explosive or flammable substances are or might be present. Air-pressurized building interiors can provide safe operation without explosionproof equipment. When explosionproof equipment is not justified financially, pressurized building interiors can alter the need. For example, if a motor control center is to be located in a building in a Class I hazardous location, the entire room may be pressurized. But construction must comply with certain specific requirements:

1. The building area or room must be kept as airtight as possible.
2. The interior space must be kept under slight overpressure by adequate positive-pressure ventilation from a source of clean air.
3. The pressurizing fans should be connected to an emergency supply circuit.
4. Ventilating louvers must be located near ground level to achieve effective air flow within the pressurized room.
5. Safeguards against ventilation failure must be provided.

Purging of electrical raceways and enclosures is another means of reducing the degree of hazard (Fig. 500-2). But that requires both the manufacturer and

Fig. 500-2. A seal fitting (upper arrow) is used for equipment enclosure. But nitrogen purging of the conduit system was also applied at this installation at a space-rocket launch-pad. Equipment was specified to be listed for Class I, Division 1, Group B where exposed to hydrogen and for Group D where the equipment was in a rocket-fuel atmosphere. Where motors, panelboards, enclosures, etc., were not available in the proper Group rating, nitrogen purging and pressurization were used in addition to sealing, to add another measure of safety. The valve (lower arrow) was used in the cover of each conduit body to provide continuous bleedoff of the nitrogen to maintain a pressure [2 in. (50.8 mm) of water] within the conduit, enabling the steady flow of nitrogen to keep the conduit free of any explosive mixture. (Sec. 500-2.)

the user of purged equipment to ensure the integrity of the system. Prepackaged purge controls for both Division 1 and Division 2 locations are on the market.

The purging medium—such as inert gas, like nitrogen, or clean air—must be essentially free from dust and liquids. The normal ambient air of an industrial interior is usually not satisfactory. And because the purge supply can contain only trace amounts of flammable vapors or gases, the compressor intake must be in a nonhazardous area. The compressor intake line should not pass through a hazardous atmosphere. If it does, it must be made of a noncombustible material, be protected from damage and corrosion, and must prevent hazardous vapors from being drawn into the compressor.

500-3. Special Precaution. Because of the inherently higher level of danger, design and installation of electrical circuits and systems in hazardous locations must be done in particularly strict compliance with the instructions given in product standards. Although NE Code Sec. 110-3(b) requires all product applications to conform to the conditions and limitations specified in the directories issued by third-party testing labs (UL, Factory Mutual, ETL, etc.), the correlation of hazardous location electrical equipment is much more thoroughly dictated than that in nonhazardous areas.

In addition to the rules given in Arts. 500 through 504, an alternative method of classification is explained and permitted by the new Art. 505. Application of the classifications indicated by Art. 505—i.e., zone 0, 1, or 2—is *only* permitted under the supervision of a "Registered Professional Engineer." That wording prohibits anyone *but* a P.E. from using that Code article.

1. Section 500-3 requires that construction and installation of equipment in all hazardous areas "will ensure safe performance under conditions of proper use and maintenance." A note urges designers, installers, inspectors, and maintenance personnel to "exercise more than ordinary care" for hazardous location work. Parts **(a)** and **(b)** designate the various groups of hazardous locations. Explanatory material describes the nature of various hazardous atmospheres.
2. Paragraph **(c)** of this section requires that all equipment in hazardous locations be approved not only for the class of location (such as Class I, II, or III) but also for the particular type of hazardous atmosphere (such as Group A, B, C, or D for locations involving gases or vapors, or Group E, F, or G if the atmosphere involves combustible or flammable dusts). The Code section describes the specific atmospheres that correspond to those letter designations.

   An important regulation is given in the third paragraph of Sec. 500-3(a), which permits use of "general-purpose equipment" or "equipment in general-purpose enclosures" in Division 2 conditions of Class I, II, or III locations. That rule permits equipment that is *not* listed for hazardous locations but is listed for general use—BUT such use is acceptable only where a Code rule specifically mentions such application. For instance, Sec. 501-4(b) does say that boxes and fittings in Class I, Division 2 locations do *not* have to be explosionproof type; i.e., sheetmetal boxes could be used. But controversy arises over that third paragraph of Sec. 500-3(c) because general-use enclosures are permitted only where the equipment does not pose a threat of ignition "under *normal* operating conditions." Division 2 locations, however, *are* those where the hazardous atmosphere is not present under normal operating conditions. So the last phrase of the Code rule seems to be superfluous, unless the phrase "normal operating conditions" is meant to apply to the equipment itself instead of the surrounding atmosphere. Equipment that is operating normally might not pose a threat of ignition, but a ground fault or short in the equipment—which is *not* a "normal operating condition"—could ignite a combustible atmosphere that might exist at a Division 2 location. The Code rule here is obscure.

3. In addition to being "approved" for the class and group of the hazardous area where it is installed, equipment is required by paragraph **(d)** of Sec. 500-3 to be "marked" with that data, along with its operating temperature when used in an ambient not over 40°C. Table 500-3(b) gives identification numbers that are used on equipment nameplates to show the operating temperature for which the equipment is approved.

Position of OSHA

With respect to equipment approval, great care must be taken to understand clearly the rules on hazardous locations equipment as covered in the electrical standards of the Occupational Safety and Health Administration (OSHA) of the U.S. Department of Labor. Those rules constitute federal law on this matter.

In the present OSHA standard, Sec. 1910.307 Hazardous (Classified) Locations is listed as one of the totally retroactive sections that apply to all electrical systems—both new ones and old ones, no matter when they were installed. On the matter of acceptability or approval of equipment used in hazardous locations, Sec. 1910.307(b) says:

> Equipment, wiring methods, and installations of equipment in classified locations shall be intrinsically safe, approved for the hazardous (classified) location, or safe for the hazardous (classified) location. Requirements for each of these options are as follows:
>
> **(1) Intrinsically safe.** Equipment and associated wiring approved as intrinsically safe shall be permitted in any hazardous (classified) location for which it is approved.
>
> **(2) Approved for the hazardous (classified) location.** [i]Equipment shall be approved not only for the class of location but also for the ignitible or combustible properties of the specific gas, vapor, dust, or fiber that will be present.
>
> Note—NFPA 70, the National Electrical Code lists or defines hazardous gases, vapors, and dusts by "Groups" characterized by their ignitible or combustible properties.
>
> [ii] Equipment shall be marked to show the class, group, and operating temperature or temperature range, based on operation in a 40°C ambient, for which it is approved. The temperature marking shall not exceed the ignition temperature of the specific gas or vapor to be encountered. However, the following provisions modify this marking requirement for specific equipment:
>
> (A) Equipment of the non-heat-producing type, such as junction boxes, conduit, and fittings, and equipment of the heat-producing type having a maximum temperature not more than 100°C (212°F), need not have a marked operating temperature or temperature range.
>
> (B) Fixed lighting fixtures marked for use in Class I, Div. 2 locations only, need not be marked to indicate the group.

(C) Fixed general-purpose equipment in Class I locations, other than lighting fixtures, which is acceptable for use in Class I, Div. 2 locations need not be marked with the class, group, division, or operating temperatures.

(D) Fixed dust-tight equipment, other than lighting fixtures, which is acceptable for use in Class II, Div. 2 and Class III locations need not be marked with the class, group, division, or operating temperature.

(3) Safe for the hazardous (classified) location. Equipment which is safe for the location shall be of a type and design which the employer demonstrates will provide protection from the hazards arising from the combustibility and flammability of vapors, liquids, gases, dusts, or fibers.

Those regulations apply to *all* electrical installations in hazardous locations—both new and existing systems. Note that there are actually *three* alternative ways for equipment to be acceptable for use in hazardous locations. A piece of equipment is acceptable if it is "Intrinsically safe" *or* "Approved for the hazardous (classified) location" *or* "Safe for the hazardous (classified) location." A piece of equipment would only have to satisfy *one* of the three conditions described in the rule.

The three alternative conditions can be verified as follows:

"*Intrinsically safe*" equipment must be "approved" as such by UL, Factory Mutual Corporation, or some nationally recognized testing laboratory. In the UL *Hazardous Location Equipment Directory,* listings of various product categories indicate which equipment is intrinsically safe and the class and group of hazardous locations for which they are approved. Such equipment is evaluated and listed in accordance with Standard UL913, Intrinsically Safe Apparatus and Associated Apparatus for Use in Class I, II and III, Division 1, Hazardous Locations. Another source of information on the subject is Installation of Intrinsically Safe Instrument Systems in Class I Hazardous Locations (ANSI/ISA RP12.6-1987).

"*Approved for the hazardous (classified) location*" equipment is basically "listed" and "labeled" equipment certified by a nationally recognized testing laboratory—such as equipment covered in the UL *Hazardous Location Equipment Directory.*

"*Safe for the hazardous (classified) location*" equipment is a category that becomes much more difficult to identify by firm, specific criteria. Just how does an "employer demonstrate" that equipment is safe for a hazardous location application? Although the OSHA rules themselves do not address that question anywhere, a note following Sec. 1910.307(b)(ii)(3) in the OSHA standard as published in the *Federal Register* does state that "guidelines" for making such a judgment are contained in Chap. 5 of the current National Electrical Code (NFPA 70). However, these guidelines are not the only means of complying with the standard. Any equipment or installation shown by the employer to provide protection from the hazards involved will be acceptable. This performance-oriented approach will allow the employer maximum flexibility in providing safety for employees. Obviously, that is a very general and vague explanation that ultimately leaves determination of acceptability completely up to the OSHA compliance officer.

A clear effect of OSHA regulations is to require "listed," "labeled," "accepted," and/or "certified" equipment to be used whenever available. If any electrical system component is "of a kind" that *any* nationally recognized testing lab "accepts, certifies, lists, labels or determines to be safe," then that component *must* be so designated in order to be acceptable for use under OSHA regulations. Every electrical designer and installer must exercise great care in evaluating any and all equipment and products used in hazardous electrical work to assure compliance with OSHA rules requiring certification by a nationally recognized testing lab.

In UL's *Hazardous Location Equipment Directory* (Red Book), limitations and application conditions are first set down for all equipment in general, as follows:

1. When equipment is listed and marked to show that it has been tested and is recognized for use in one or more of the Code-designated groups of hazardous locations, such marking indicates that such equipment is suitable for use in *either* a Division 1 or Division 2 location of the particular class of hazardous location, even though no reference is made to the "division." Such equipment is, of course, also acceptable if used in a nonhazardous location. Figure 500-3 shows a typical nameplate used on such equipment as required by parts **(c)** and **(d)** of Sec. 500-3, showing suitability as "explosion-proof" (Class I) and "dust-tight" (Class II).

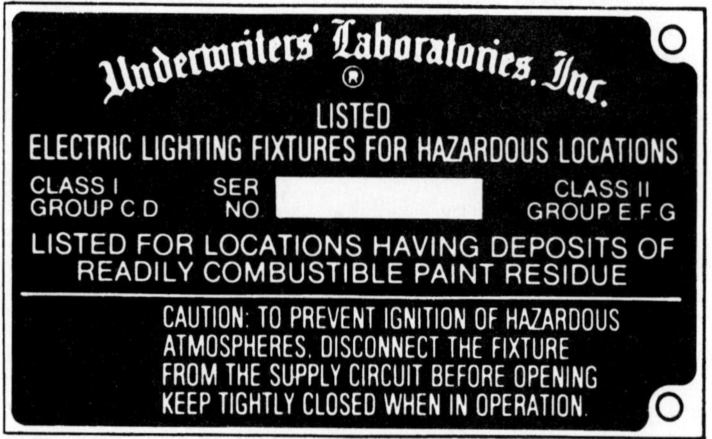

Fig. 500-3. Typical UL part of equipment nameplate describes "Approval for Class and Properties." (Sec. 500-3.)

2. BUT, equipment that is marked "Division 2" or "Div. 2" is suitable for use in only such a division and may *not* be used in a Division 1 location. However, a piece of equipment may have other marking to indicate its acceptability for other specific uses. Figure 500-4 shows a nameplate that is an addition to the nameplate in Fig. 500-3, noting that the same fixture is "Suitable for Wet Locations." Section 410-4 of the NE Code requires such marking on "all fixtures installed in wet locations. . . ." Care must be taken to distinguish between different parts of nameplates to precisely determine what is third-party certification.

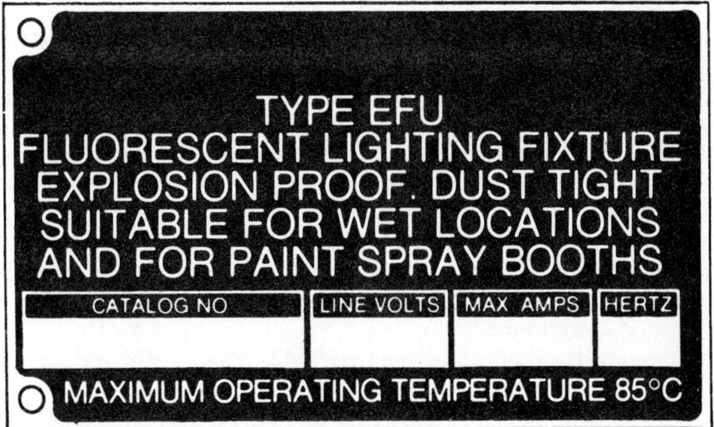

Fig. 500-4. Additional data on a nameplate may show other acceptability—as for "wet locations." (Sec. 500-3.)

3. Equipment that is listed and marked for "Class I" locations (explosion-proof) may be used for "Class II" locations if it is dusttight to exclude combustible dusts and if its external operating temperature is not at or above the ignition level of the particular dust that might accumulate on it. Obviously, these characteristics must be carefully established before Class I equipment is used in a Class II location.
4. Equipment listed for Class II, Group G (for flour, starch, or grain dust)—as used in a grain elevator—is also generally suitable for use in Class III locations, where combustible lint or flyings are present. The Exception noted is for fan-cooled type motors which might have their air passages choked or clogged by large amounts of the lint or flyings.
5. Because hazardous location equipment is critically dependent upon proper operating temperature, a UL note warns that the ampere or wattage marking on power-consuming equipment is based on the equipment being supplied with voltage exactly equal to the rated voltage value. Voltage higher or lower than the rated value will produce other than rated amps or watts, with the possibility that the heating effect of the current within the equipment will be greater than normal. Higher than normal current will be produced by overvoltage to resistive loads and by undervoltage to induction motors. Because of this, the actual circuit voltage, rather than the nameplate value, must be used when calculating the required ampacity of branch-circuit conductors, rating or setting of overcurrent protection, rating of disconnect, etc.—all to assure adequate sizing and avoid overheating.
6. Hazardous location equipment is tested and listed for use at normal atmospheric pressure in an ambient temperature not over 40°C (104°F), unless indicated otherwise. Use of equipment under higher-than-normal pressure, in oxygen-enriched atmospheres, or at higher ambient temperatures

can be dangerous. Such abnormal conditions may increase the chance of igniting the hazardous atmosphere and may increase the pressure of explosion within equipment.

7. Openings or modifications must not be made in explosionproof or dust-ignition-proof equipment, because any such field alterations would void the integrity and tested safety of the equipment. Field alteration of listed products for nonhazardous application is also generally prohibited.
8. All bolts as well as all threaded parts of enclosures must be tightly made up.
9. Indoor hazardous location equipment that is exposed to severe corrosive conditions must be listed as suitable for those conditions as well as for the hazardous conditions.

The requirements of Sec. 500-3 and **Code** Table 500-3(d) provide the means of properly identifying and classifying equipment for use in hazardous locations. The identification numbers in **Code** Table 500-3(d) pertain to temperature-range classifications as used by Underwriters Laboratories Inc., in UL *Hazardous Location Standards.*

While the **Code** rules for Class I locations do not differ for different kinds of gas or vapor contained in the atmosphere, it is to be noted that it is necessary to select equipment designed for use in the particular atmospheric group to be encountered. This is necessary for the reason that explosive mixtures of the different groups have different flash points and explosion pressures. It is also necessary because the ignition temperatures vary with the groups of explosive mixtures.

Underwriters Laboratories Inc. lists fittings and equipment as suitable for use in all groups of Class I, although the listings for Groups A and B are not as complete as those for Groups C and D.

In Class II locations the **Code**, in a few cases, differentiates between the different kinds of dust, particularly dusts which are electrically conductive and those which are not conductive. Here again, as in Class I locations, care must be used to determine that the equipment selected is suitable for use where a particular kind of dust is present.

In addition to the use of more than ordinary care in selecting equipment for use in hazardous locations, special attention should be given to installation and maintenance details in order that the installations will be permanently free from electrical hazards. In making subsequent additions or changes, the high standards that were applied during the original installation must always be maintained.

For a more thorough knowledge of specific hazardous areas and equipment selection and location it is essential to obtain copies of the various NFPA and ANSI standards referenced in Arts. 500 through 517.

500-5. Class I Locations. In each of the three classes of hazardous locations discussed in Secs. 500-5, 500-6, and 500-7, the **Code** recognizes varying degrees of hazard; hence under each class two divisions are defined. In the installation rules that follow, the requirements for Division 1 of each class are more rigid than the requirements for Division 2.

Briefly, the hazards in the three classes of locations are due to the following causes:

Class I, highly flammable gases or vapors
Class II, combustible dust
Class III, combustible fibers or flyings

The classifications are easily understood, and, if a given location is to be classed as hazardous, it should not be difficult to determine in which of the three classes it belongs. However, it is obviously impossible to make rules that will in every case determine positively whether the location is or is not hazardous. Considerable common sense and good judgment must be exercised in determining whether the location under consideration should be considered as hazardous or likely to become hazardous because of a change in the processes carried on, and if so, what portion of the premises should be classed as coming under Division I and what part may safely be considered as being in Division 2.

ARTICLE 501. CLASS I LOCATIONS

501-1. General. This section generally requires compliance with Chaps. 1 through 4 of the NEC, *except* where Art. 501 calls for something different. Where Art. 501 requires a specific piece of equipment, or the use of a specific installation technique, the rule of Art. 501 must be satisfied. Of course, as noted in Exception No. 2 to this section, if the Class I location is wired in accordance with Art. 505, "Zone 0, 1, and 2 Locations," Art. 501 does *not* apply. The more common Class I locations are those where some process is carried on involving the use of a highly volatile and flammable liquid, such as gasoline, petroleum naphtha, benzene, diethyl ether, or acetone, or flammable gases.

In any Class I location, an explosive mixture of air and flammable gas or vapor may be present which can be caused to explode by an arc or spark. To avoid the danger of explosions all electrical apparatus which may create arcs or sparks should if possible be kept out of the rooms where the hazardous atmosphere exists, or, if this is not possible, such apparatus must be "of types approved for use in explosive atmospheres."

All equipment such as switches, CBs, or motors must have some movable operating part projecting through the enclosing case, and any such part, for example the operating lever of a switch or the shaft of a motor, must have sufficient clearance so that it will work freely; hence the equipment cannot be hermetically sealed. Also, the necessity for subsequent opening of the enclosures for servicing makes hermetic sealing impracticable. Furthermore, the enclosure of the equipment must be entered by a run of conduit, and it is practically impossible to make conduit joints absolutely air- and gastight. Due to slight changes in temperature, the conduit system and the apparatus enclosures "breathe"; that is, any flammable gas in the room may gradually find its way inside the conduit and enclosures and form an explosive mixture with air. Under this condition, when an arc occurs inside the enclosure an explosion may take place.

When the gas and air mixture explodes inside the enclosing case, the burning mixture must be confined entirely within the enclosure, so as to prevent the ignition of flammable gases in the room. In the first place it is necessary that the enclosing case be so constructed that it will have sufficient strength to withstand the high pressure generated by an internal explosion. The pressure in pounds per square inch produced by the explosion of a given gas-and-air mixture has been quite definitely determined, and the enclosure can be designed accordingly.

Since the enclosures for apparatus cannot be made absolutely tight, when an internal explosion occurs, some of the burning gas will be forced out through any openings that exist. It has been found that the flame will not be carried out through an opening that is quite long in proportion to its width. This principle is applied in the design of so-called explosionproof enclosures for apparatus by providing a wide flange at the joint between the body and the cover of the enclosure and grinding these flanges to a definitely determined fit. In this case the flanges are so ground that when the cover is in place the clearance between the two surfaces will at no point exceed 0.0015 in. (38.1 micrometers). Thus, if an explosion occurs within the enclosure, in order to escape from the enclosure the burning gas must travel a considerable distance through an opening not more than 0.0015 (38.1 micrometers) in. wide.

The basic construction characteristics of equipment for Class I hazardous locations are detailed in various sections of this article and in standards of testing laboratories. Application of the products hinges on understanding those details:

An explosionproof enclosure for Class I locations is capable of withstanding an explosion of a specified gas or vapor which may occur within it and of preventing the ignition of the specified gas or vapor surrounding the enclosure by sparks, flashes, or explosions of the gas or vapor within. Explosionproof equipment must provide three things: (1) strength, (2) joints which will not permit flame or hot gases to escape, and (3) cool operation, to prevent ignition of surrounding atmosphere.

UL requires that explosionproof enclosures must withstand a hydrostatic test of 4 times the maximum explosion pressure developed inside the enclosure. Explosionproof enclosures are not vapor- or gastight, and it is simply assumed that any hazardous gases in the ambient atmosphere will enter them either through normal breathing or when maintenance is performed on the enclosed equipment.

When an explosion occurs inside a rectangular explosionproof enclosure, the resulting force exerts pressure in all directions. The enclosure must be designed with sufficient strength to withstand these forces and avoid rupture (Fig. 501-1).

The energy generated by an explosion within an enclosure must be permitted to dissipate through the joints of the enclosure under controlled conditions. There are two generally recognized joint designs intended to provide this control—threaded and flat:

1. Threaded construction of covers and other removable parts that have five full threads engaged produces a safe, flame-arresting, pressure-relieving joint. When an explosion occurs within a threaded enclosure, the flame and hot gases create an internal pressure against the cover, thus locking

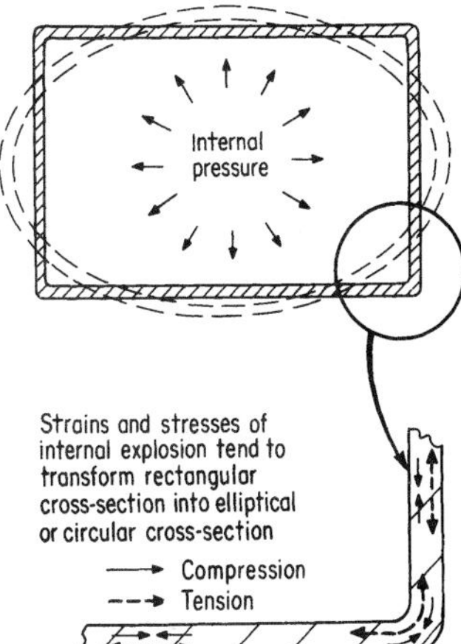
Internal
pressure
Strains and stresses of
internal explosion tend to
transform rectangular
cross-section into elliptical
or circular cross-section
Compression
Tension

Fig. 501-1. An explosion creates strains and stresses in cross section of enclosure walls. (Sec. 501-1.)

the threads and forcing the gases out through the path between the threaded surfaces. When the gases reach the outside hazardous atmosphere, they have been cooled by the heat-sink effect of the mass of metal, down to a point below the ignition temperature of the outside atmosphere, as shown in Fig. 501-2.

2. A flat joint is constructed by accurate grinding or machining of the mating surfaces of the cover and the body. This flat joint works in a manner similar to the threaded joint. The two surfaces are bolted closely together, and as flame and hot gases are forced through the narrow opening, they are cooled by the mass of the metal enclosure, so that only cool gas enters the

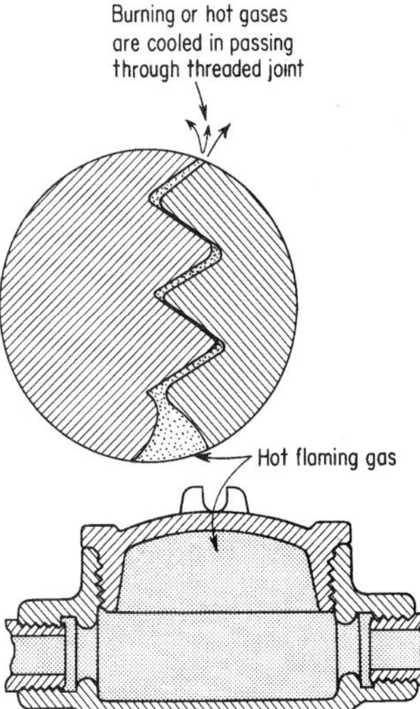
Burning or hot gases
are cooled in passing
through threaded joint
Hot flaming gas

Fig. 501-2. Threaded joints cool the heated gas as it escapes from enclosure under pressure. (Sec. 501-1.)

hazardous atmosphere. Figure 501-3 shows the flat joint and a variation on it called a "rabbet" joint. Care must be taken to assure that all cover screws are tight and that no particles of dirt or other foreign matter get in between the cover and the body. Even a small particle could prevent tight closing and might allow the joint to pass flame.

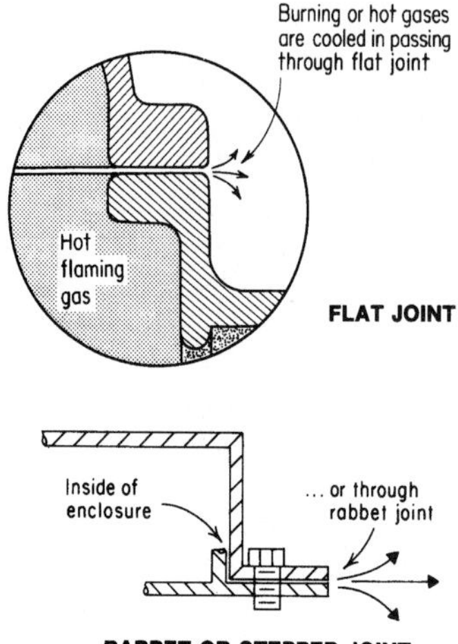

Fig. 501-3. Ground surfaces of flat or rabbet joint provide release and cooling of internal gases. (Sec. 501-1.)

UL standards on explosionproof enclosures contain rules on "Grease for Joint Surfaces": "Paint or a sealing material shall not be applied to the contacting surfaces of a joint. A suitable corrosion inhibitor (grease) such as petrolatum, soap-thickened mineral oils, or nondrying slushing compound may be applied to the metal joint surfaces before assembly. The grease shall be of a type that does not harden because of aging, does not contain an evaporating solvent, and does not cause corrosion of the joint surfaces."

501-4. Wiring Methods. Threaded steel intermediate metal conduit has been added to the wiring methods suitable for use in Class I, Division 1 locations. This permission, plus recognition by other sections of the Code, gives IMC full recognition as a general-purpose raceway equivalent in application to rigid metal conduit. Type MI cable is the only cable assembly that is permitted in Class I, Division 1 locations (Fig. 501-4).

The term "approved for the location" in paragraph **(a)** means that approval is to be based on the performance of a fitting or equipment when subjected to a specific atmosphere. As applied to rigid metal conduit, to be explosionproof, threaded joints must be used at couplings, and for connection to fittings the threads must be cleanly cut, five full threads must be engaged, and each joint must be made up tight. Conduit elbows and short-radius capped elbows provide for 90° bends in conduit but only where wires may be guided when being pulled into the conduit, to prevent damaging the conductors by pulling them

Fig. 501-4. Type MI cable is recognized for use in Class I, Division 1 locations, provided that the termination fittings (arrow) are listed as suitable for hazardous location use. (Sec. 501-4.)

around the sharp turn in the elbow. Figure 501-5 shows two types of fittings used in hazardous locations to facilitate pull-in of conductors that have stiff or heavy-wall insulation. The capped elbow is especially suited to use in tight quarters.

All fittings, such as outlet, junction, and switch boxes, and all enclosures for apparatus, should have threaded hubs to receive the conduit and must be explosionproof. Explosionproof junction boxes are available in a wide variety of types (Fig. 501-6). Box covers may have threaded connections with the boxes, or the cover may be attached with machine screws, in which case a carefully ground flanged joint is required.

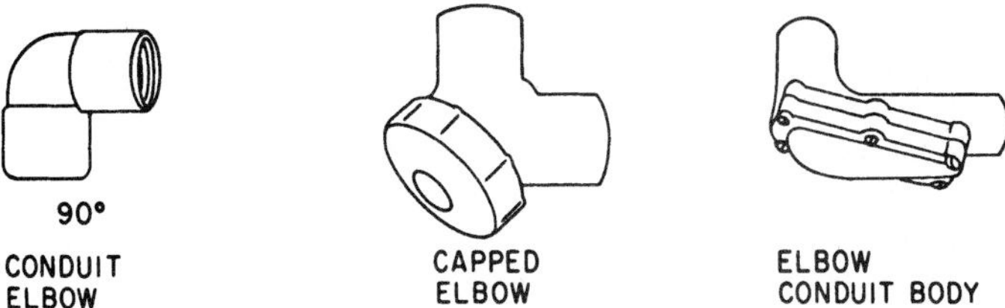

Fig. 501-5. Conduit elbows and similar fittings may be used where wires may be guided into conduit. (Sec. 501-4.)

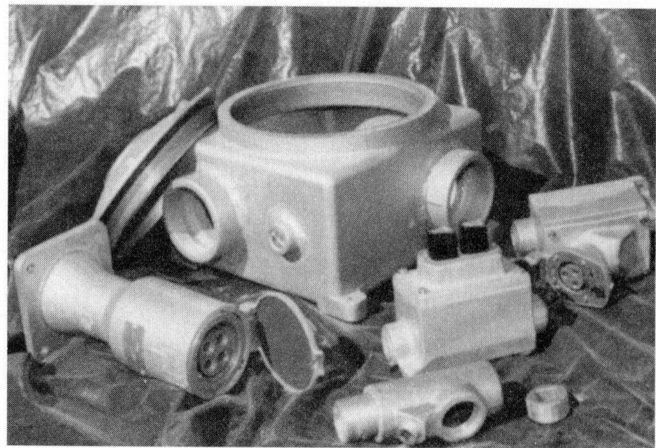

Fig. 501-6. A wide assortment of boxes, conduit bodies, fittings, and other enclosures are made in explosionproof designs listed for use in Class I locations. (Sec. 501-4.)

A flexible, explosionproof fitting, suitable for use in Class I hazardous locations, is shown in Fig. 501-7. The flexible portion consists of a tube of bronze having deeply corrugated walls and reinforced by a braid of fine bronze wires. A heavy threaded fitting is securely joined to each end of the flexible tube, and a fibrous tubular lining, similar to "circular loom," is provided in order to prevent abrasion of the enclosed conductors that might result from long-continued vibration. The complete assembly is obtainable in various lengths up to a maximum of 3 ft (914 mm).

Flexible connection fittings that are recognized by NE Code Sec. 501-4(a) for use in Class I, Division 1 locations are intended by UL and the NE Code to be used where it is necessary to provide flexible connections in threaded rigid

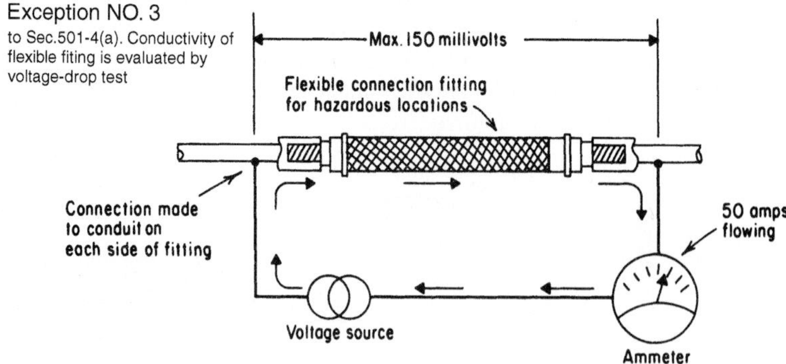

Fig. 501-7. Exception No. 3 to Sec. 501-4(a).

conduit systems—as at motor terminals. Use of such flexible fittings must observe the minimum inside radius of bend for which the fitting has been tested. Those data are provided with the fitting.

Note: The UL warns that acceptability of the use of flexible connection fittings must be cleared with local inspection authorities. In general, use of such flexible fittings should be avoided wherever possible and should be limited to situations where use of threaded rigid conduit is completely ruled out by the needs or conditions of the application.

At the end of Sec. 501-4(a), there are three exceptions. The first permits limited use of rigid nonmetallic conduit for underground installations where buried at least 2 ft (610 mm) below grade *and* encased in at least 2 in. (50.8 mm) of concrete. In addition, transition to either rigid metal or intermediate metal conduit must be made. The last 2 ft (610 mm) of the overall run must be in threaded rigid metal conduit or threaded IMC. Remember: the 2-ft (610-mm) burial depth is measured to the top of the concrete encasement. The last condition is that an equipment grounding conductor—bare, covered, or insulated—must be run to ground all noncurrent-carrying metal parts. The size of the equipment grounding conductor must be sized as required by Sec. 250-95.

Exception No. 2 recognizes the use of a specialized Type MC cable with a full-sized grounding conductor. As described in this Exception, a continuous, aluminum sheath—not spiral wound—Type MC cable is permitted for use in Class I, Division 1 locations. Such Type MC cable must be listed for Class I, Division 1 locations, provided with an overall "polymeric" jacket, provided with a full-size grounding conductor, used with fittings listed for Division 1-listed Type MC, and installed at an industrial establishment where qualified people service the equipment. The intent of the phrase "separate grounding conductors" is to require *one* equipment grounding conductor in addition to the aluminum sheath, which should *not* be relied upon as the sole means of grounding.

Where flexible connection fittings *are* used, the corrugated metal inner wall and the metal braid construction of the fitting provide equipment grounding continuity between the end connectors and the fitting. The UL test for conductivity through a flexible fitting is shown in Fig. 501-7. Although Sec. 501-16(b) requires either an internal or external bonding jumper to be used with standard flexible metal conduit in Class I, Division 2 locations, that rule does not apply to listed flexible fittings.

At the end of part **(a),** the exception to the normal wiring methods required in Class I, Division 1 locations indicates that Sec. 501-11 permits flexible cord in Class I, Division 1 locations for portable lighting equipment and portable utilization equipment. This Exception eliminates a long-time conflict between the clear and direct rules of Sec. 501-4(a) on wiring methods and the limited use of portable cord as an alternative to the wiring methods used for fixed wiring. For this, refer to Sec. 516-3 covering places where flammable materials are used for spraying, dipping, and coating—in which applications flexible cord might be used.

In Class I, Division 2 locations explosionproof outlet boxes are not required at lighting outlets or at junction boxes containing no arcing device. However, where conduit is used, it should enter the box through threaded openings as shown in Fig. 501-6, or if locknut-bushing attachment is used, a bonding jumper and/or fittings must be provided between the box and conduits, as required in Sec. 501-16(b).

As noted in part **(b)** of this section, flexible connections permitted in Class I, Division 2 locations may consist of flexible conduit (Greenfield) with approved fittings, and such fittings are not required to be specifically approved for Class I locations. It should be noted that a separate grounding conductor is necessary to bond across such flexible connections, as required in Sec. 501-16(b).

Ordinary knockout-type boxes may be installed in such locations, but Sec. 501-16(a) rules out the use of locknuts and bushings for bonding purposes, and the requirement specifies either bonding jumpers or other approved means (such as bonding locknuts on knockouts without any concentric or eccentric rings left in the wall of the enclosure) to assure adequate grounding from the hazardous area to the point of grounding at the services.

Cord connectors for connecting extra-hard-service type of flexible cord to devices in hazardous locations must be carefully applied. Section 501-4(b) permits extra-hard-usage flexible cord in Division 2 locations. But Sec. 502-4 permits its use in Division 1 and Division 2 areas of Class II locations. Section 503-3(a) (2) and (b) permits cords in Class III, Division 1 and Division 2 locations. Listed cord connectors are recognized for use in Class I, Groups A, B, C, and D or Class II, Group G locations—using Types S, SO, ST, or STO multiconductor, extra-hard-usage cord *with* a grounding conductor.

IMPORTANT! Section 501-4(b) adds power-limited tray cable (Type PLTC) to the list of wiring methods permitted in Class I, Division 2 locations, in accordance with the provisions of Art. 725 covering remote-control, signaling, and low-energy circuits. Similarly, instrumentation tray cable (Type ITC) is also permitted in Class I, Division 2 locations, as covered in Art. 727. And, the last paragraph of Sec. 501-4(b) makes clear that high-voltage circuits (i.e., circuits over 600 V) may employ the wiring methods covered in the first part of Sec. 501-4(b) and, where protected from physical damage, may be made up using metallic-shielded, high-voltage cable in cable trays when installed in accordance with Art. 318. And Art. 326 dictates that such cable must be Type MV cable.

Armored cable (Type AC) has been removed as an acceptable wiring method in Class I, Division 2 locations. Armored cable has a construction that is considered to be similar to flexible metal conduit, which is permitted in these locations. It seems clear that removal of the reference to "armored cable" is to exclude the use of Type AC (so-called BX), which is the NEC designation for "armored cable." Type MC (so-called metalclad cable within the NEC) has long been acceptable.

Figure 501-8 shows some applications of wiring methods that are covered by the rules of Sec. 501-4. At the top, use of standard flex (Greenfield) in a Division 1 area violates part **(a)** of this section. At the center, use of aluminum-sheathed cable (Type ALS) is O.K. in a Division 2 area. Even though Type ALS is no longer mentioned in part **(b)**, that type of cable is now covered by Art. 334 and is considered as one form of Type MC cable, which is mentioned in part **(b)** as acceptable in Division 2 areas. At the bottom, use of Type MC is O.K. in a Division 2 location.

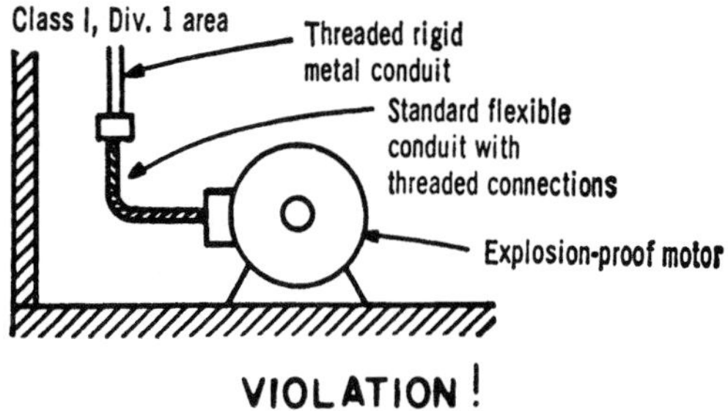

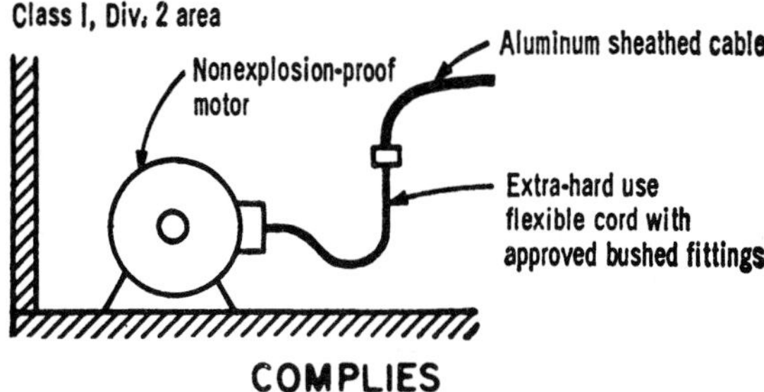

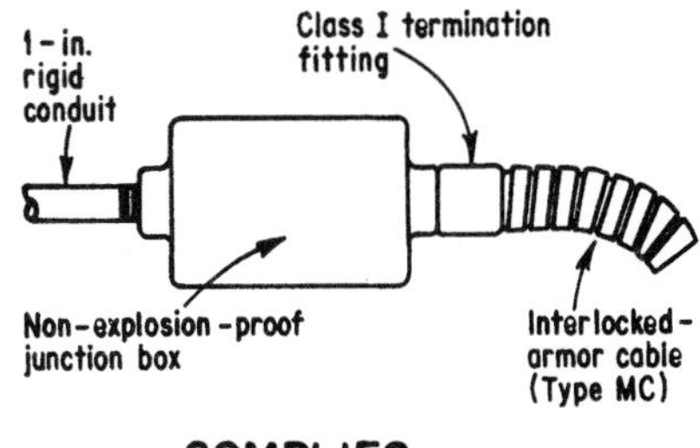

Fig. 501-8. Wiring methods in Class I locations are clearly regulated. (Sec. 501-4.)

501-5. Sealing and Drainage. The proper sealing of conduits in Class I locations is an important matter. In Class I, Division 2 locations, each piece of apparatus that produces arcs or sparks, such as a motor controller, switch, or receptacle, should be isolated from all other apparatus by sealing within the conduit so that an explosion in one enclosure cannot be communicated through the conduit to any other enclosure. Whether used in an enclosure or in conduit, seals are necessary to prevent gases, vapors, or flames from being propagated into an enclosure or conduit run and to confine an explosion that might occur within an enclosure.

The note after the first paragraph points out that seal fittings properly installed are not normally capable of preventing the passage of liquids, gases, or vapors if there is a continuous pressure differential across the seal. However, as indicated, seals may be specifically designed and tested for preventing such passage. This explanation, along with the wording in such rules as Sec. 501-5(a)(4), makes clear that seals will only "minimize," not "prevent," passage of gases or vapors through the seal.

When an explosion takes place within an enclosure because of arc ignition of gas or vapor that has entered the enclosure, flames and hot gases could travel rapidly through unsealed conduits, and the resultant buildup of pressure could exceed the strength of conduit, wireways, or enclosures, causing explosive rupture. *Pressure piling* is the name given to the action that takes place when an explosion occurs inside an enclosure because of flammable gas within the enclosure being ignited by a spark or overheated wiring. When this happens, and there are no seals in the conduits connecting to the enclosure, exploding gas will compress the entire atmosphere within the conduit system and flames or heat will ignite compressed gas some distance down the conduit and cause another, more-powerful explosion. The pressure and succeeding explosions are repeated through the system of raceways and enclosures, with each succeeding explosion increasing in intensity. To prevent such occurrences, it is mandatory that seal-off fittings be used in certain enclosures or conduit runs to block and confine potentially hazardous vapors.

The necessary sealing may be accomplished by inserting in the conduit runs special sealing fittings, as shown in Fig. 501-9, or provision may be made for sealing in the enclosure for the apparatus. An explosionproof motor is made with the leads sealed where they pass from the terminal housing to the interior of the motor, and no other seal is needed where a conduit terminates at the motor, except that if the conduit is 2 in. (50.8 mm) or larger in size, a seal must be provided not more than 18 in. (457 mm) from the motor terminal housing.

Class I, Division 1

Part **(a)** of this section covers mandatory use of seals in Class I, Division 1 locations:

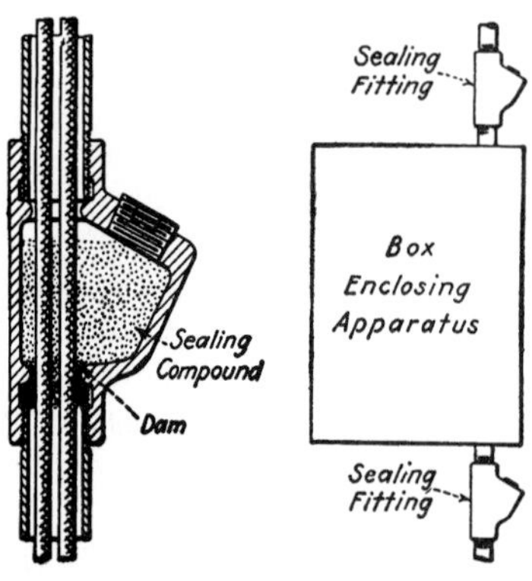

Fig. 501-9. Seal fitting is filled with compound to prevent passage of flame or vapor through the conduit. (Sec. 501-5.)

1. A seal is required in each and every conduit (regardless of the size of the conduit) entering (or leaving) an enclosure that contains one or more switches, CBs, fuses, relays, resistors, or any other device that is capable of producing an arc or spark that could cause ignition of gas or vapor within the enclosure or any device that might operate hot enough to cause ignition. In each such conduit, a seal fitting must be placed never more than 18 in. (457 mm) from such enclosure. The **Code** rule has eliminated the phrase that said seals had to be installed "as close as practicable" to the enclosure—leaving the remainder of the requirement, that the seal must not be more than 18 in. (457 mm) from the enclosure, intact. As shown in Fig. 501-10, a conduit seal fitting is installed in the top conduit and one of the bottom conduits—close to the enclosure of the arcing device. But a seal is not used in the conduit to the pushbutton because that is a factory-sealed device and that seal is not over 18 in. (457 mm) from the starter. That complies with the intent of the **Code** rule, as well as the rule of Sec. 501-5(c) which recognizes "approved integral means for sealing"—as in the pushbutton. Figure 501-11 shows a seal fitting as close as possible to a box housing a receptacle.

   The Exception to part **(a)(1)** notes that a seal is not required in a conduit entering an enclosure for a switching device in which the arcing or sparking contacts are internally sealed against the entrance of ignitible gases or vapors. Such conduit is applied to a condition similar to a conduit connection to an explosionproof junction box—that is, any gas that enters the enclosure will contact only wiring terminals and is not exposed to arcs. But because a conduit seal is always required for any conduit of 2-in. (50.8-mm) size or larger that enters a junction box or terminal housing, this Exception permits elimination of the conduit seal *only* for conduits of 1½-in. (38-mm) size or smaller that enter an enclosure for switching devices with sealed or inaccessible contacts—such as mercury-tube switches, as shown at the bottom of Fig. 501-10. Any switch, CB, or contactor with its contacts in a hermetically sealed chamber or immersed in oil might be applied under this Exception.

   Recognition of sealed-contact devices without a separate conduit seal is similar to recognition of "an approved integral means for sealing" [Sec. 501-5(c)(1)], which is a seal provision that is manufactured directly into some enclosures for Class I equipment.

   But questions have always risen about the acceptability of boxes or fittings between the sealing fittings and the enclosure being sealed. This section identifies the devices that may be used between the seal and the enclosure. Explosionproof unions, couplings, reducers, elbows, capped elbows, and conduit bodies similar to L, T, and cross type shall be the only enclosures or fittings permitted between the sealing fittings and the enclosure. A reducing bushing (a "reducer") may be connected at a conduit entry to an explosionproof enclosure so that the bushing is connected between the seal and the enclosure. Because a reducer is commonly used to provide a conduit bushing in the wall of an enclosure and does not pose a threat to the integrity of the seal function, "reducers" have been added

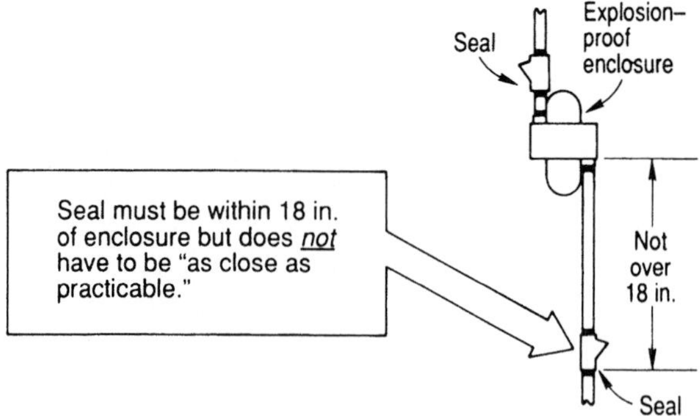
Seal
Explosion–proof enclosure
Seal must be within 18 in. of enclosure but does *not* have to be "as close as practicable."
Not over 18 in.
Seal

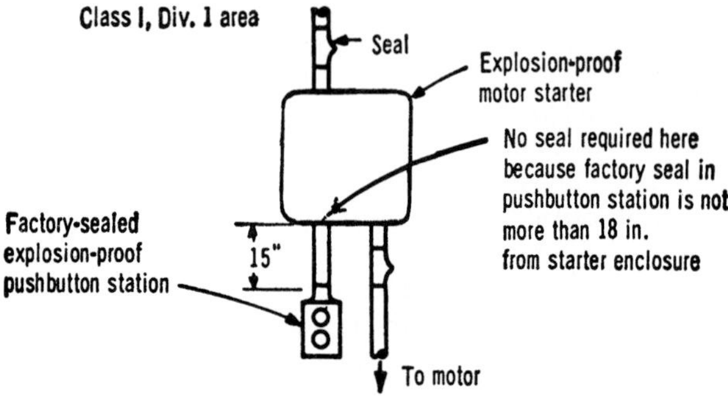
Class I, Div. 1 area
Seal
Explosion-proof motor starter
No seal required here because factory seal in pushbutton station is not more than 18 in. from starter enclosure
Factory-sealed explosion-proof pushbutton station
15"
To motor

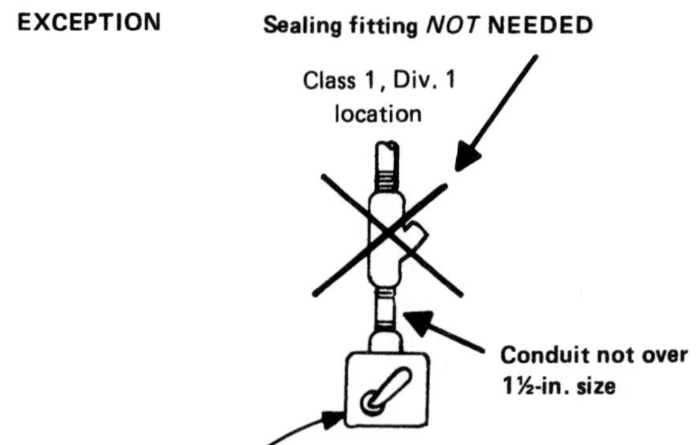
EXCEPTION
Sealing fitting *NOT* NEEDED
Class 1, Div. 1 location
Conduit not over 1½-in. size
Explosion-proof enclosure for tumbler switch with mercury-tube contacts.

Fig. 501-10. Seal should be in each conduit within 18 in. (457 mm) of the sealed enclosure, center, but may not be needed at all, bottom. (Sec. 501-5.)

Fig. 501-11. Seal fitting is as close as possible to enclosure, providing maximum effectiveness. (Sec. 501-5.)

to the list of devices permitted to be used between a seal and the enclosure it supplies. The rule goes on to note that any conduit body used in that position must be of a size not larger than the trade size of the conduit with which it is used. This clearly rules out the use of a box or any similar large-volume enclosure between the seal fitting and the enclosure being sealed, as shown in Fig. 501-12.

The fittings listed as acceptable for use between the seal and the enclosure were selected on the basis that their internal volume was sufficiently

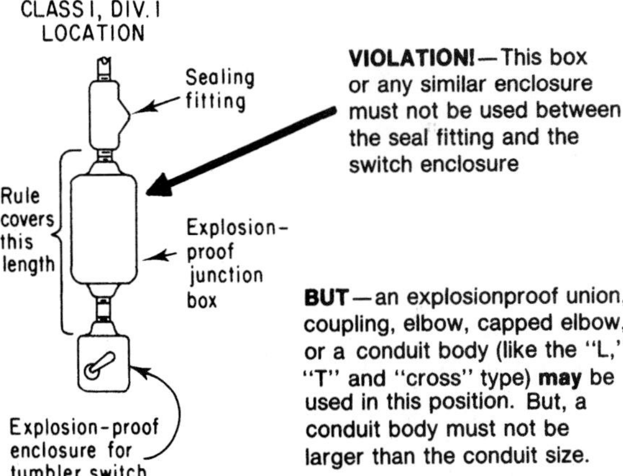

Fig. 501-12. Any type of junction "box" may not be used between seal and enclosure. (Sec. 501-5.)

small as to prevent the accumulation of any dangerous volume of gas or vapor. Acceptability was based on limiting the volume of gas or vapor that may accumulate between the seal and the enclosure being sealed. It was on this basis also that conduit bodies are prohibited from being of a larger size than the conduit with which they are used. If they were permitted, they would present the opportunity for accumulation of a larger volume of gas or vapor, which is considered objectionable.

Figure 501-13 shows an interesting variation on the above concern for use of a box between a seal and an enclosure. *No splices are permitted within seal-off fittings,* according to the UL *Hazardous Location Equipment Directory.* The illustration shows a round-box type of seal fitting that is used for pulling power and control wires. Such a fitting takes a large, round, threaded cover equipped with a pouring spout. The cover, shown removed, is readily unscrewed to provide maximum unobstructed access to the fitting interior, which facilitates damming either one or both conduit hubs. When the cover is replaced, it can be rotated so the spout points up to permit compound fill. This fitting can be used to seal conduit regardless of its direction or run.

Fig. 501-13. Seal fitting of the round box type is used to seal the conduit run into the bottom of an explosionproof starter enclosure, with flexible fitting connection to the Class I, Division 1, Group D motor below and a watertight Class I cord connector control cable. (Sec. 501-5.)

2. A seal is required in any conduit run of 2-in. (50.8-mm) size or larger, where such a conduit enters an "enclosure or fitting" that is required to be explosionproof and houses terminals, splices, or taps, as shown in Fig. 501-14. Note in such cases, however, that the rule does not call for the seal to be "as close as practicable" to the enclosure or fitting, as required for a housing of an arcing or sparking device. Here, it simply requires that the seal be not over 18 in. (457 mm) away from the enclosure.

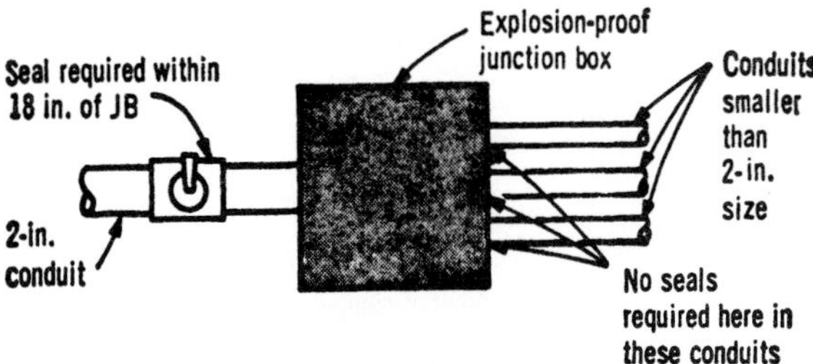

Fig. 501-14. Seals for conduits to junction boxes or fittings are required only for conduits of 2-in. (50.8-mm) size and larger. (Sec. 501-5.)

Another example of seal application in accordance with parts **(1)** and **(2)** of Sec. 501-5(a) is shown in Fig. 501-15.

The third part of Sec. 501-5(a) covers use of a single seal to provide the required seal for a conduit connecting two enclosures. Where two such pieces of apparatus are connected by a run of conduit not over 3 ft (914 mm) long, a single seal in this run is considered satisfactory if located at the center of the run. Figure 501-16 shows this rule. Although the wording is not detailed, the reference to "within 18 in. (457 mm) of such enclosure" must be understood to be 18 in. (457 mm) measured along the conduit, to avoid the misapplication of

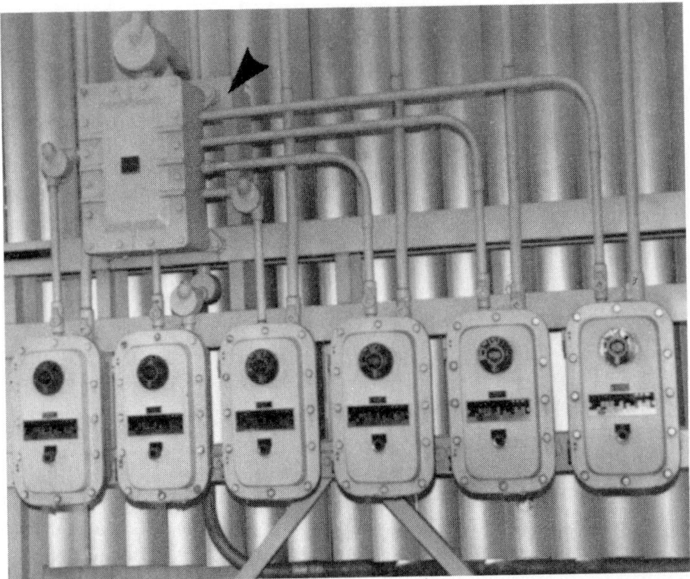

Fig. 501-15. Seal fittings are very close to points where conduits enter the motor starter enclosures. Seals are not required for the conduits entering the junction box (arrow) because they are not 2-in. (50.8-mm) or larger size—although seals may be used there. (Sec. 501-5.)

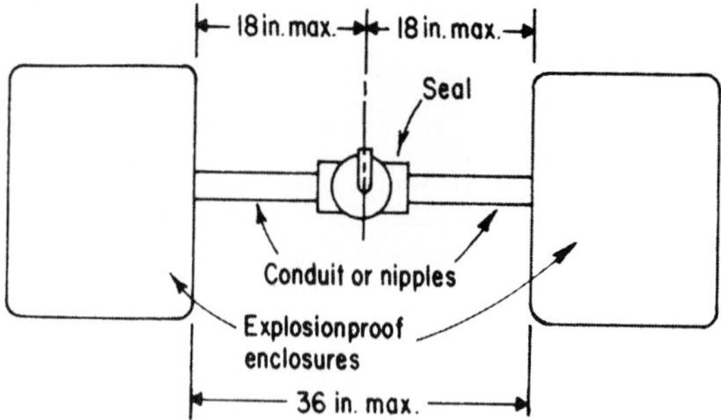

Fig. 501-16. A single seal serves two conduit entries into separate enclosures. (Sec. 501-5.)

the rule shown in Fig. 501-17. The single seal at A is not over 18 in. (457 mm) from the CB enclosure and is literally not over 18 in. (457 mm) from the starter [it is 10 in. (254 mm) from the starter]. But that use of a single seal for the two enclosures violates the Code intent that the 18 in. (457 mm) in each case *must* be measured along the conduit.

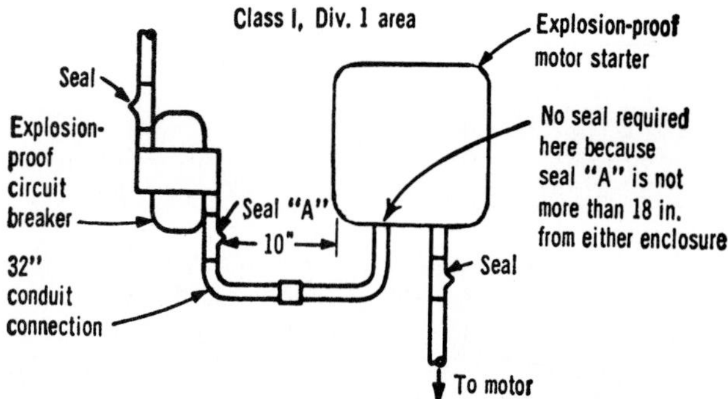

Fig. 501-17. Although this complies with the rule literally, it is a violation of the intent. (Sec. 501-5.)

Part **(4)** requires a seal in each and every conduit that leaves the Class I, Division 1 location—whether it passes into a Division 2 location or into a nonhazardous location. This required seal may be installed on either side of the boundary *and* within 10 ft (3.05 m) of the boundary (Fig. 501-18). There must be no union, coupling, box, or fitting between the sealing fitting and the point where the conduit leaves the hazardous location. The rule does specify a maximum distance that must be observed between the sealing fitting and the boundary. And that distance is 10 ft (3.05 m). The purpose of this sealing is twofold: (1) The conduit usually terminates in some enclosure in the Division 2 or nonhazardous area containing an arc-producing device, such as a switch or fuse. If

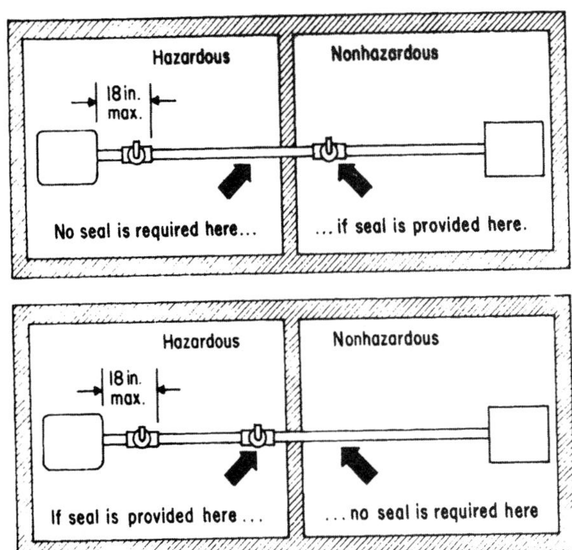

Fig. 501-18. Conduit must be provided with seal fitting where it crosses boundary. (Sec. 501-5.)

not sealed, the conduit and apparatus enclosure are likely to become filled with an explosive mixture and the ignition of this mixture may cause local damage in the Division 2 or nonhazardous location. (2) An explosion or ignition of the mixture in the conduit in the nonhazardous area would probably travel back through the conduit to the hazardous area and might cause an explosion there if because of some defective fitting or poor workmanship the installation is not completely explosionproof. If the conduit is unbroken (no union, coupling, etc.) between an enclosure seal and the point where the conduit leaves the hazardous area, an additional seal is not required at the boundary. Figure 501-19 shows two violations where conduit leaves the hazardous area.

The Exception to Sec. 501-5(a)(4) covers the case where a metal conduit system passes from a nonhazardous area, runs through a Class I, Division 1 hazardous area, and then returns to a nonhazardous area. Such a run is permitted to pass through the hazardous area without the need for a seal fitting at either of the boundaries where it enters and leaves the hazardous area. But, the wording of this Exception requires that such conduits, in order to be acceptable, must contain no union, coupling, box, or fitting in any part of the conduit run extending 12 in. (305 mm) into each of the nonhazardous areas involved (Fig. 501-20).

In the 1975 NE Code, the Exception merely referred to "unbroken rigid-metal conduits" that pass through the hazardous area. The effect of the present wording is to clarify the meaning of the word "unbroken."

An identical Exception is made to Sec. 501-5(b)(2) covering the use of a metal conduit passing through a Class I, Division 2 location. It has the same statement

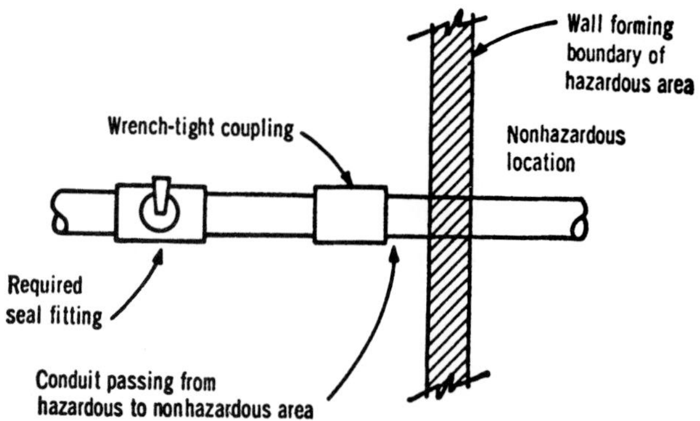

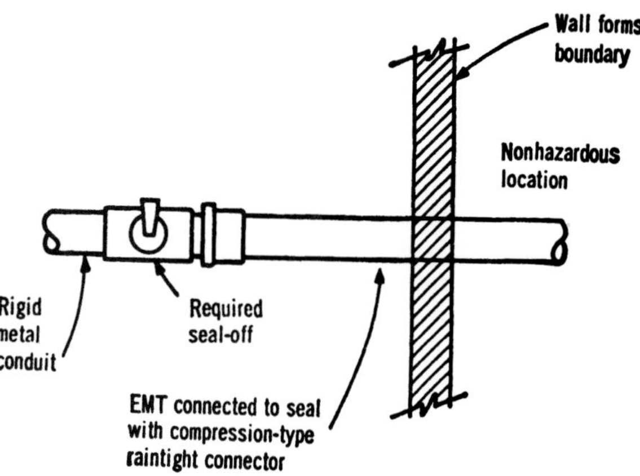

Fig. 501-19. These violate rules on seal fittings where conduit crosses boundary. (Sec. 501-5.)

prohibiting unions, couplings, boxes, or fittings that are spelled out in Exception No. 1 to Sec. 501-5(a)(4).

For some installations of conduits crossing boundaries, straightforward application of NE Code rules on conduit sealing is difficult. Most of these cases involve determination as to what constitutes the boundaries of a hazardous area; the Code provides no definition. The inspection authority should be consulted in cases not specifically covered by the Code. Because there are no provisions in the Code that *prohibit* the use of seals, "*if in doubt, seal*" would be a safe practice to follow.

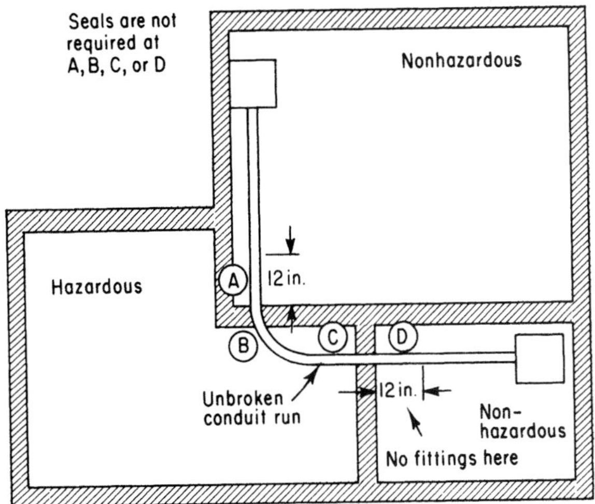

Conduit length in hazardous area must not contain any union, coupling, box, or other fitting within the hazardous area and for 12 in. beyond each boundary.

Fig. 501-20. Seals may be eliminated for conduit passing "completely through" hazardous area. (Sec. 501-5.)

At the top of Fig. 501-21, the conduit run is sealed within 18 in. (457 mm) of an explosionproof enclosure, as shown below, and extends into a concrete floor slab, emerging in a nonhazardous area. It is not clear what constitutes the boundary of the Class I, Division 1 area. Must a seal be placed at A or might it instead be placed at B in the nonhazardous area? An NE Code Official Interpretation, pertaining to a hospital operating room, ruled that the entire concrete slab through which the conduit traveled constituted the boundary of the hazardous area, and that the seal could be placed either at A, where the conduit leaves the hazardous area, or at B, where it enters the nonhazardous area. But some authorities may require seals at A and B. With a seal at A and not at B, a heavier-than-air gas or liquid (like gasoline) might penetrate a crack in the floor, enter the conduit through a coupling, and pass into the enclosure in the nonhazardous area. Or, a seal at B but not at A might not prevent vapor in the conduit from entering the nonhazardous area through a coupling in the concrete and then through a crack in that floor. That kind of gas passage has occurred.

At the bottom of Fig. 501-21, the conduit run is not in the floor slab, but in the ground below the slab. Now what constitutes the boundary? Can the seal still be placed either at point A or at point B? Code rules applying to gasoline stations and aircraft hangars may be used as a guide. The real question is whether the ground beneath the slab is a hazardous or nonhazardous location. Section 514-8 defines dispensing and service-station wiring and equipment, any portion of which is below the surface of a hazardous area, as a Class I, Division 1 location. Also, Sec. 513-3 requires that the sealing rules of Sec. 501-

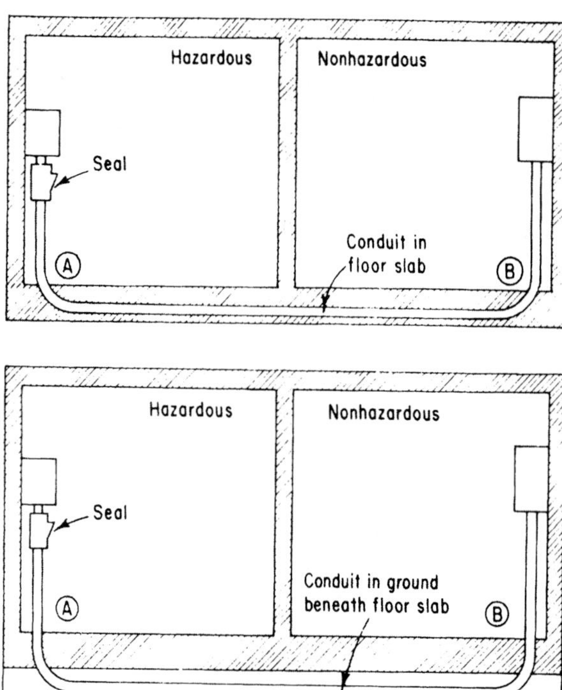

Fig. 501-21. Conduit in floor-slab boundary may require seals at both "A" and "B." (Sec. 501-5.)

5(a)(4) and 501-5(b)(2) be applied to horizontal as well as vertical boundaries of defined hazardous areas in aircraft hangars. And the last sentence of Sec. 513-7 says that raceways in or beneath a floor slab are considered as being in the hazardous location above the floor. This is also stated in Sec. 511-3.

A safe conclusion is that, unless specifically defined to the contrary, the ground beneath a hazardous area is an extension of that hazardous area. Or like the concrete floor in the example, the concrete *and* the ground beneath the hazardous area through which the conduit passes can be considered to be the "boundary" when there is any question of boundary. And the boundary itself is considered part of the hazardous location. That means that the conduit does not leave the hazardous area until it emerges at B. The seal should be placed there and an argument can be made that use of two seals—at A and B—would be better practice in both drawings in Fig. 501-21.

It should also be noted that Sec. 514-8 states that the underground Class I, Division 1 location beneath a pump island of a gas station extend at least to the point where underground conduit emerges from the ground. That concept can be logically applied to the drawings in Fig. 501-21.

Figure 501-22 is a wiring layout for a Class I, Division 1 location. The wiring is all rigid metal conduit with threaded joints. All fittings and equipment are

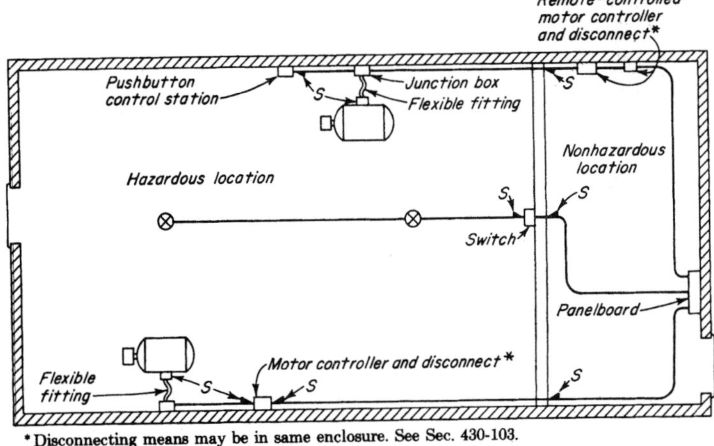

Fig. 501-22. Required seals are shown in points marked "S." (Sec. 501-5.)

explosionproof; this includes the motors, the motor controller for motor No. 1 (lower part of drawing), the pushbutton control station for motor No. 2 (upper part of drawing), and all outlet and junction boxes. The panelboard and controller for motor No. 2 are placed outside the hazardous area and hence need not be explosionproof.

Each of the three runs of conduit from the panelboard is sealed just outside the hazardous area. A sealing fitting is provided in the conduit on each side of the controller for motor No. 1 (lower part of drawing). The leads are sealed where they pass through the frame of the motor into the terminal housing, and no other seal is needed at this point provided that the conduit and flexible fitting enclosing the leads to the motor are smaller than 2 in. (50.8 mm). The pushbutton control station for motor No. 2 (upper part of drawing) is considered an arc-producing device, unless the contacts are oil-immersed as described in Sec. 501-6(b)(1), but, because they are not, the conduit is sealed where it terminates at this device.

A seal is provided on each side of the switch controlling the lighting fixture. One of these seals is in the nonhazardous room and that single seal serves as both the seal for the arcing device and the seal for conduit crossing the boundary. The lighting fixtures are hung on rigid conduit stems threaded to the covers of explosionproof boxes on the ceiling.

About seal fittings In using seal fittings in conduits in hazardous locations, application data of the UL must be observed, as follows:

- Conduit seal-off fittings to comply with NE Code Sec. 501-5 or 502-5 must be used *only* with the sealing compound that is supplied with the fitting and specified by the fitting manufacturer in the instructions furnished with the fitting.
- Seal-off fittings are listed for sealing listed conductors in conduit, where the conductors are thermoplastic insulated, rubber-covered, or lead-covered.

- Any instructions supplied with a seal-off fitting must be carefully observed with respect to limitation on the mounting position (e.g., vertical only) or location (e.g., elbow seal). Figure 501-23 shows a variety of available seal fittings. Sealing fittings are designed for vertical orientation only, for optional vertical or horizontal positioning, or as combination elbow seals. Others are compatible with conduits installed at any angle, since covers can be rotated until sealing spouts point upward.

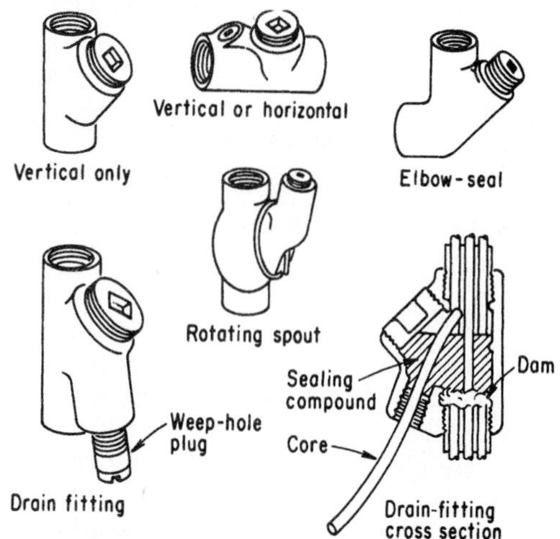

Fig. 501-23. A variety of seal fittings are suited to different applications. (Sec. 501-5.)

Because conduits are installed vertically, horizontally, and at angles and require ells, tees, and offsets, the fittings used for sealing differ in construction features, orientation, and method of sealing.

Sealing fittings intended solely for vertical orientation have threaded, upward-slanting ports slightly larger than conduit hub openings to permit asbestos-fiber dams to be tamped into fitting bases. The dam prevents the fluid-sealing compound from running down into the conduit before the seal has solidified.

A second type of fitting is designed either for vertical or horizontal positioning. These units are identified by two seal-chamber plugs that can be removed to facilitate tamping dam fibers into both conduit hubs when the device is aligned horizontally. The compound is poured into the chamber through the larger of the two ports. The ports are then replugged, and the plugs tightened flush with their collars. When these fittings are oriented vertically, however, only lower conduit hubs need be dammed.

A third type of seal which can be oriented in any position is shown in the center of Fig. 501-23 and is described in Fig. 501-13. That same fitting may be used as a drain-type seal when its spout is turned down.

Elbow seals (as at upper right of Fig. 501-23) are double-duty devices that are practical either when horizontal conduits must elbow-down to connect with an enclosure's top (as indicated), or when vertical conduits must turn to enter explosionproof enclosures horizontally. In either case, sealant application openings must slant upward.

Another fitting, designed for drainage purposes, is installed only in vertical runs of conduits. Where conduit is run overhead and is brought down vertically to an enclosure for apparatus, any condensation of moisture in the vertical run would be trapped by the seal above the apparatus enclosure. The lower part of Fig. 501-23 shows a sealing fitting designed to provide drainage for a vertical conduit run. Any water coming down from above runs over the surface of the sealing compound and down to an explosionproof drain, through which the water is automatically drained off. These fittings permit passage of condensation while also blocking the passage of explosive pressures or flames. They are equipped with plugs containing minute weep-holes that can either be opened and closed periodically as need develops or allow continuous drainage.

Drain-type seal fittings must be oriented so that compound-application ports remain above the lower downward-slanting drainage plugs. To install the seal and drain, both ports are unplugged and the lower conduit hub is dammed. The drainage plug-hole is then closed temporarily by a washer through which a rubber core is inserted. This core protrudes up into the upper part of the sealing chamber, although it must be guided so as not to remain in contact with any of the conductors. Sealing compound then is poured into the chamber through the upper access port, which is replugged and screwed tight.

After the compound has initially set (but has not yet had time to permanently harden), the washer is removed and the rubber core is pulled down and out. This creates a clear drainage canal which extends from above the seal down into the drainage weep-hole. A drainage plug is then screwed into the threaded hole, with not less than five full threads engaged to fulfill the requirement for an explosionproof joint.

Class I, Division 2

Because Division 2 locations are of a lower degree of hazard than Division 1, the requirements for sealing are somewhat less demanding, as follows:

1. Where the rules of other sections in Art. 501 require an explosionproof enclosure for equipment in a Division 2 location, all conduits connecting to any such enclosure must be sealed exactly the same as if it were in a Division 1 enclosure. And the conduit, nipple, or any fitting between the seal and the enclosure being sealed must be approved for use in Class I, Division 1 locations—as specified in Sec. 501-4(a).

   As shown in Fig. 501-24, where a nonexplosionproof enclosure is permitted by other sections to be used in a Division 2 location, a seal is *not* required in any size of conduit. Note that in Division 2 locations, there is *not* always the need to seal conduits 2 in. (50.8 mm) and larger, as required in Division 1 locations, by Sec. 501-5(a)(2).

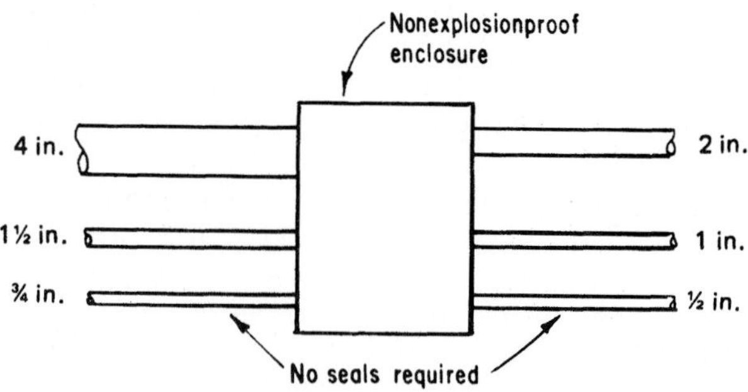

Fig. 501-24. Seals are not required for conduits connected to nonexplosionproof enclosures that are permitted in Division 2 locations. (Sec. 501-5.)

2. Any and every conduit run passing from a Class I, Division 2 area into a nonhazardous area must be sealed in the same manner as described above for conduit passing from a Division 1 area into a Division 2 or nonhazardous area (Fig. 501-25). Rigid metal conduit or IMC must be used between the seal and the point where the conduit passes through the boundary.

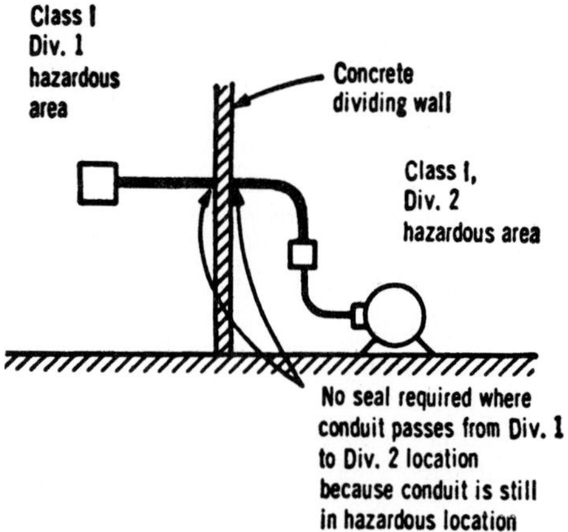

Fig. 501-25. This is a violation; a seal *is* required. (Sec. 501-5.)

Exception No. 1 to part **(b)(2)** is worded the same as the Exception in part **(a)(4)** for Class I, Division 1 locations. Exception No. 1 covers the case where a metal conduit system passes from a nonhazardous area, runs through a Class I, Division 2 hazardous area, and then returns to a nonhazardous area. Such a run is permitted to pass through the hazardous area without the need for a seal fitting at either of the boundaries where it enters and leaves the hazardous area, provided that the conduit in the hazardous area does not contain unions, couplings, boxes, or fittings. In a Class I, Division 2 location, the same prohibition against unions, couplings, etc., is applicable, and the method in Fig. 501-26 is not acceptable if seals are omitted at the boundary crossings. A seal would not be needed at A, B, C, or D, if the conduit passes through the Class I, Division 2 location without any coupling or other fittings in the conduit.

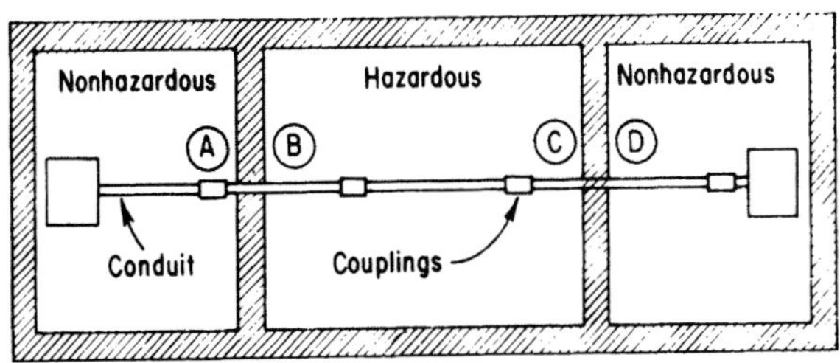

Fig. 501-26. Couplings in conduit through Division 2 location would require seals at the boundaries. (Sec. 501-5.)

Exception No. 2 to part **(b)(2)** addresses installations where the conduit passing from the Class I, Division 2 location into an unclassified location does not enter any enclosures that produce an arc or spark. Such an installation does *not* require a conduit seal provided it is also installed outdoors, or if the conduit system is installed in a single room, it may be installed indoors. But, in no case shall the unsealed conduit be connected to any enclosure that contains a source of ignition.

Exception No. 3 recognizes that where an enclosure or room is unclassified because it is pressurized to prevent accumulation of an explosive concentration, there is no need to seal the raceway where it leaves the Division 2 location and enters the pressurized spaces because the press will prevent gas from entering that unclassified area.

Exception No. 4 eliminates sealing for those portions of the conduit system that satisfy *all* the conditions given in parts **(a)** to **(e)**.

Part **(c)** of Sec. 501-5 sets regulations about the kind of seals that must be used where seals are required by foregoing rules. Part **(1)** calls for an integral seal within the enclosure itself or use of a separate seal fitting in each conduit connecting to the enclosure, as described above. The use of factory-sealed devices eliminates the need for field sealing and generally is less expensive to install. In fittings of this type, the arcing device is enclosed in a chamber, with the leads or connections brought out to a splicing chamber. No external sealing fitting is required (Fig. 501-27).

Where a seal fitting is used in the conduit, it must be *explosionproof.* The sealing compound must develop enough mechanical strength as it hardens to withstand the forces of explosions. Seals used only to prevent condensation accumulation do not have to be explosionproof; a vaportight seal is sufficient for that purpose.

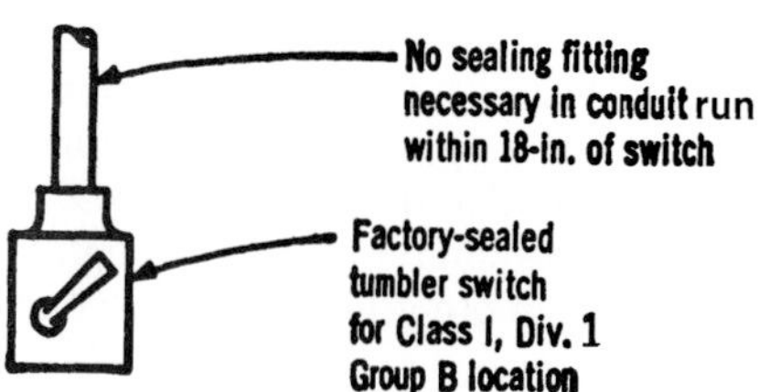

Fig. 501-27. Seal in conduit is not needed where enclosure has built-in seal. (Sec. 501-5.)

A seal must be *vaportight* to stop gases and vapors. To do that, the sealing compound must adhere to the fitting and to the conductors. It must expand as it hardens to close all voids without producing objectionable mechanical stresses in the fitting.

Liquid or condensed vapor may present a problem in Class I locations. Where such is the case, joints and conduit systems must be arranged to minimize entrance of liquid. Periodic draining may be necessary, which necessitates the inclusion of a means for draining in the original design of the motor (Sec. 501-5).

Installation instructions furnished by the manufacturer must be carefully followed. The seal fitting must be carefully packed with fibrous damming material, which packs more tightly and effectively around conductors when it is dampened, and then filled with the compound supplied with the fitting to a depth at least equal to the inside diameter of the conduit and never less than ⅝ in. (16 mm) deep, as required in part **(3)** and in the UL standard on seal fittings.

Part **(2)** of Sec. 501-5(c) covers the compound used in seals (Fig. 501-28). The sealing compound used must be one which has a melting point of not less than 93°C (200°F) and is not affected by the liquid or gas which causes the location to be hazardous. Most of the insulating compounds commonly used in cable splices and potheads are soluble in gasoline and lacquer solvents and hence are unsuitable for sealing conduits in locations where these liquids are used. A mixture of litharge and glycerin is insoluble in nearly all liquids and gases found in Class I locations and meets all other requirements, though this mixture is open to the objection that it becomes very hard and is difficult to remove if the wires must be pulled out. No sealing compounds are listed by Underwriters Laboratories Inc. as suitable for this use except in connection with the explosionproof fittings of specific manufacturers.

Part **(4)** prohibits splices or taps in seal fittings.

Part **(5)** recognizes use of listed Class I assemblies that have a built-in seal between a compartment housing devices that may cause arcs or sparks and a separate compartment for splicing or taps. Conduit connection to the splice or tap compartment requires a seal fitting only in conduit of 2-in. (50.8-mm) size or larger, as specified for junction boxes in Sec. 501-5(a)(2).

Part **(d)** limits the fill in any seal fitting to *not* more than 25 percent of its cross section. A fitting is to be evaluated on the basis of a conduit with the same trade size and conductor fill in that fitting must be no greater than 25 percent of the cross section, unless the fitting is specifically marked for greater fill.

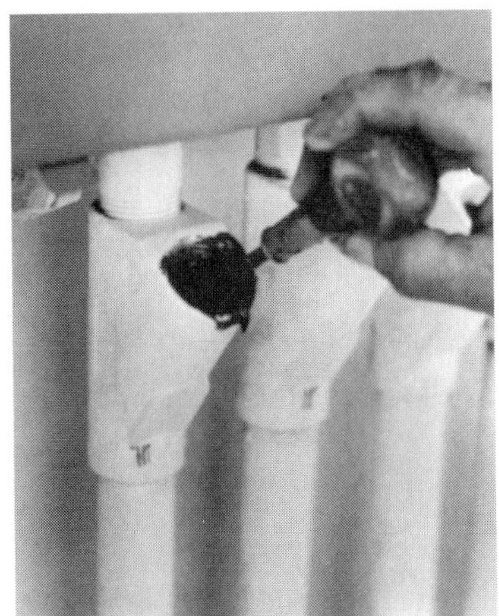

Fig. 501-28. Conduit seal fitting must be carefully packed with fibrous damming material, which packs more tightly and effectively around conductors when it is dampened and then filled with the compound supplied with the fitting to a depth at least equal to the inside diameter of the conduit and never less than ⅝ in. (16 mm) deep (UL standard). This type of seal is for vertical mounting only. (Sec. 501-5.)

Part **(d)** covers seals for cables in conduit and for Types MI and MC cables—the only cables permitted by Sec. 501-4(a) and Exception No. 1 to be used in Class I, Division 1 locations. Part **(e)** covers sealing of any of the cables permitted by Sec. 501-4(b).

In part **(f)(3),** a secondary seal fitting must be installed to prevent flammable or combustible fluids from entering the conduit system after a failure of the primary "single compression seal, diaphragm, or tube." In this section a "compression-type seal" is recognized as a method of sealing between the process fluid and the electrical enclosure. Previous wording of this rule required both draining (or venting) to detect primary seal failure and installation of a second "approved seal" in the conduit system *only* when the primary seal against process fluid entry into electrical conduit was either a single diaphragm or tube. But the same secondary draining and sealing is now also required where a *compression seal* is used.

501-6. Switches, Circuit Breakers, Motor Controllers, and Fuses. Part **(a)** requires explosionproof enclosures (listed for Class I locations, which means suitable for use in Division 1 areas), although purged enclosures are not ruled out. Explosionproof equipment is required in Class I areas only. And, any of the equipment covered by Sec. 501-6(a) including its enclosure must be listed as an assembly. In addition, the equipment must operate at a temperature low enough that it will not ignite the atmosphere around it.

Division 1 Electrical equipment described must be explosionproof and specifically designed for the specific class and group.

Division 2 Equipment selected must be explosionproof only in certain cases. Purged and pressurized enclosures are recognized as an alternative to explosionproof types, as noted in Sec. 501-3(a). General-purpose enclosures may be used only if all arcing parts are immersed under oil or enclosed within a chamber that is hermetically sealed against the entrance of gases or vapors. In addition, the surface temperature of the apparatus in the general-purpose enclosure should not exceed 80 percent of the ignition temperature of the hazardous substances involved. If those conditions are not satisfied, the enclosure must be explosionproof.

Important: UL standards on explosionproof enclosures contain rules on "Grease for Joint Surfaces":

> Paint or a sealing material shall not be applied to the contacting surfaces of a joint. A suitable corrosion inhibitor (grease) such as petrolatum, soap-thickened mineral oils, or nondrying slushing compound may be applied to the metal joint surfaces before assembly. The grease shall be of a type that does not harden because of aging, does not contain an evaporating solvent, and does not cause corrosion of the joint surfaces.

Figure 501-29 shows two explosionproof panelboards. Each panelboard consists of an assembly of branch-circuit CBs, each pair of CBs being enclosed in a cast-metal explosionproof housing. Access to the CBs and to the wiring compartment is through handholes with threaded covers, and threaded hubs are provided for the conduits. Individual CBs and motor starters are also shown. "Panelboards" for light and power are limited in the UL Red Book. Listed panelboards for Class I and Class II hazardous locations are for "lighting and *low-capacity* power distribution." *High-capacity* panelboards (like 1,200-A floor-standing panels) and switchboards must be kept out of hazardous locations wherever possible. Enclosure requirements and details are generally the same as those described under "General rules" and for CBs and boxes. Typical hazardous-location panelboards are shown at the right in Fig. 501-30.

Fig. 501-29. Explosionproof panelboards (arrows) are assemblies made up of circuit-breaker housings coupled to wiring enclosures. Large explosionproof CB enclosure at center feeds the panelboards. Explosionproof motor controllers are at lower left. (Sec. 501-6.)

Fig. 501-30. Explosionproof panelboards (at right) are combined with separate enclosures for motor starters on a rack, to make up a "modular assembly" which is listed by UL under "Industrial control equipment"—which may be assembled either at the factory or in the field. (Sec. 501-6.)

"Industrial control equipment" is a broad category in the UL Red Book, covering "control panels and assemblies" and "motor controllers." Control panels and assemblies include both enclosures and the components within them—such as motor controllers, pushbuttons, pilot lights, receptacles (Fig. 501-31). Either a single enclosure or a group of interconnected enclosures may be used for mounting the components. Where a number of interconnected enclosures are included in an assembly, it is called a "modular assembly" and may be assembled either at the factory or in the field. An example of that is shown in Fig. 501-30.

Components are provided with the enclosures, to be installed either at the factory or in the plant. Wiring between components of modular assemblies is to be field installed. Conduit seal-off fittings must be used in accordance with Sec. 501-5.

As noted at the end of part **(b)(1)**, in a Class I, Division 2 location, a general-purpose enclosure may be used for a circuit breaker, a motor controller, or switch *if* "interruption of current occurs within a factory sealed explosionproof chamber." This is a condition under which a general-purpose enclosure may be used for a current-interrupting device—instead of requiring a Class I, Division 1 enclosure.

A snap switch in an explosionproof enclosure is shown in Fig. 501-32. If a snap switch has an internal factory seal between the switch contacts and its supply wiring connection in its enclosure, it will be so identified by a marking on it. Such switches do not require a seal fitting in a conduit entering the enclosure. The integral seal satisfies Secs. 501-5(a)(1) and 501-5(c)(1).

Fig. 501-31. Combination motor starter is typical explosionproof control unit for Class I locations. Note the drain-type seal fittings that provide for draining the conduits of any accumulated condensation or other water. (Sec. 501-6.)

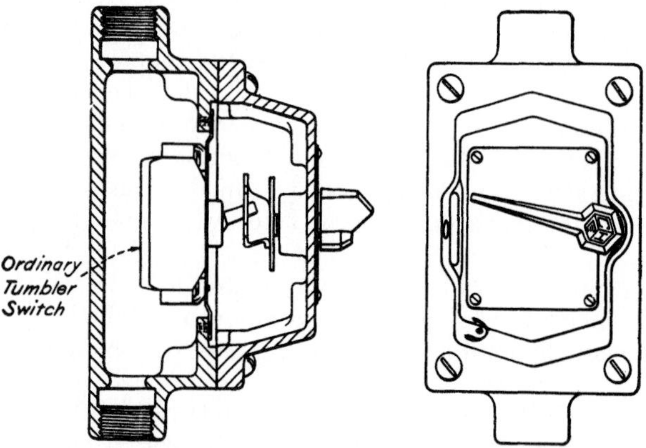

Fig. 501-32. Explosionproof enclosure suits snap switch to use in hazardous location. (Sec. 501-6.)

The last sentence of part **(4)** indicates that solid-state switches—such as those commonly used in PLCs and other static switches—are exempt from being installed in explosionproof enclosures.

As noted in part **(b)(2),** isolating switches in Class I, Division 2 locations may be used in general-purpose enclosures either with or without fuses in the enclosure. In previous Code editions, this rule recognized a general-purpose enclosure only for an isolating switch that did *not* contain fuses. But because the fuse in such a switch is for short-circuit protection and neither the switch nor the fuse operates as a "normal" interrupting device, fuses are permitted in isolating switches in general-purpose enclosures.

With reference to subparagraph **(b)(4)**, it is assumed that fuses will very seldom blow, or CBs will very seldom open, when used to protect feeders or branch circuits that supply only lamps in fixed positions. In Division 2 locations the conditions are not normally hazardous but may sometimes become so. There is very little probability that one of the overcurrent devices will operate at the same time that the hazardous conditions exist; hence it is not considered necessary to require that these overcurrent devices be in explosionproof enclosures.

Part **(b)(5)** requires that fuses in lighting fixtures must *not* discharge flame on operation. This rule covers the type of fuse that may be used for internal protection of individual lighting fixtures in Class I, Division 2 locations.

UL Red Book Data

As described under general regulations in the UL's *Electrical Construction Materials Directory* (Green Book) and repeated in Sec. 110-14(c) and in the Red Book of UL for individual types of switching and control devices, the wiring space and current-carrying capacity of CBs and other equipment used in hazardous locations are based on the use of 60°C wire connected to the breaker terminals for circuits wired with No. 14 to No. 1 conductors, and on the use of 75°C wire connected to the terminals for circuits wired with conductors of No. 1/0 or larger.

Although the reference clearly bases the listing on wires of those maximum-temperature ratings, wires of higher-temperature ratings may be used to take advantage of the smaller conduit sizes needed for thin-wall-insulated wires (such as 90°C THHN and XHHW), along with the greater ampacities of higher-temperature wires, which offer an advantage when conductors are derated. *But,* the load on any such wire must not exceed the ampacity of the same size of 60°C- or 75°C-rated wire. The same rules also apply to switches used in hazardous locations.

Terminal lugs on switches, CBs, etc., are suitable for use with copper conductors, as noted in the UL Red Book. Lugs on equipment are commonly marked "AL-CU" or "CU-AL," indicating that the terminal is suitable for use with either copper or aluminum conductors. But, such marking on the lug itself is not sufficient evidence of suitability for use with aluminum conductors. UL requires that equipment found to be suitable for use with either copper or aluminum conductors must be marked to indicate such use on the label or wiring diagram of the equipment—completely independent of a marking, like "AL-CU," on the lugs themselves. A typical CB, for instance, would have lugs marked "AL-CU" but also must have a notation on the label or nameplate to specify "Lugs suitable for copper or aluminum conductors"—if aluminum conductors are to be used with the breaker.

A hazardous-location *enclosure* for equipment that is to be fed by aluminum conductors must have a marking on it to indicate that it is permissible to use aluminum conductors with the switch or CB that is to be mounted within the enclosure. All enclosures for CBs are marked to indicate what labeled CBs are acceptable for use within the enclosures. Only the breakers specified may be used in an enclosure.

501-7. Control Transformers and Resistors. The term *control transformer* is commonly applied to a small dry-type transformer used to supply the control circuits of one or more motors, stepping down the voltage of a 480-V power circuit to 120 V.

Part **(a)** requires either explosionproof or purged and pressurized enclosures for Division 1 locations—the same as required for meters and instruments in Sec. 501-3(a).

501-8. Motors and Generators. Four different types of motor applications are recognized for use in a Class I, Division 1 location. The first is a motor approved for Class I, Division 1 locations—such as an explosionproof motor. The totally enclosed, fan-cooled motor (referred to as a "TEFC" motor) is recognized and listed by UL for use in explosive atmospheres. A motor of a type approved for use in explosive atmospheres of the totally enclosed, fan-cooled type is shown in Fig. 501-33. The main frame and end housings are made with sufficient strength to withstand internal pressures due to ignition of a combustible mixture inside the motor. Wide metal-to-metal joints are provided between the frame and housings. Circulation of the air is maintained inside the inner enclosure by fan blades on each end of the rotor. At the left side of the sectional view a fan is shown in the space between the inner and outer housings. This fan draws in air through a screen and drives it across the surface of the stator punchings and out through openings at the drive end of the motor. The motors described in **(2)** and **(3)** of part **(a)** of this section are the only ones available for Class I, Groups A and B locations and for medium-voltage, high-horsepower applications. Cost of ducts and ventilating systems limits their application in other areas.

The UL Red Book lists motors for Class I, Groups C and D locations. To date, UL lists no motors for Groups A and B; therefore, where such conditions are encountered, motors must be located outside the hazardous area or must conform to the alternate arrangements and conditions of Sec. 501-8(a). Air or inert-gas purging are recognized as alternate methods. Motors suitable for Groups C and D, Class I locations are designated as explosionproof.

In part **(b)**, the rule relaxes the requirements for Division 2 areas somewhat. In Class I, Division 2 locations open or nonexplosionproof enclosed motors may be used if they have no brushes, switching mechanisms, or integral resistance devices. However, motors with any sparking or high-temperature devices must be approved for Class I, Division 1 areas—as described above.

"Motors and Generators, Rebuilt" May Be Listed

A number of years ago, a procedure was established to provide third-party certification of rewound or rebuilt motors in hazardous locations. Refer to the UL Red Book on hazardous-location equipment for motor-repair centers authorized to provide certified repairs.

501-9. Lighting Fixtures. In these locations, part **(a)** requires that each fixture be *approved* for the Class I, Division 1 location and marked to show the maximum

Fig. 501-33. A motor approved for use in the explosive atmosphere of a Class I, Division 1 location is the basic one of the four types recognized for such locations. (Sec. 501-8.)

wattage permitted for the lamps in the fixture (Fig. 501-34). Reference to the listings in the UL Red Book shows many manufacturers listed for Class I fixtures for use in various Groups of atmospheres. Listings range from "Class I, Group C" to "Class I, Groups A, B, C, and D." For application of a lighting fixture in a particular Group, it is simply a matter of assuring that a manufacturer's fixture is listed for Class I and the Group. The designation "Class I" indicates the fixture is suitable for Division 1, except where the listing contains the phrase "Division 2 only" following the "Class I" reference or following the "Class I" plus Group references.

But in a Class I, Division 1 location, all fixtures must be listed and marked for such use by UL or other national product testing lab. That is necessary to sat-

Fig. 501-34. Fluorescent luminaire, fed by Type MI cable, is listed and marked as an explosionproof unit for use in a Class I, Division 1 location. (Sec. 501-9.)

isfy OSHA's definition of the word "approved" as it appears in Sec. 501-9(a) (1). And electrical inspectors invariably give the same meaning to the word "approved"—if a test-lab-listed product of the same generic type is a violation of Sec. 110-2.

Clearly, given the nature of the hazards involved, extra care should be exercised *and* the installation should be held to a higher standard. And that higher standard should include the exclusive use of equipment that is listed for the Class and Division in which the equipment is to be installed and used (see Art. 505 for zone classifications).

Part **(a)(3)** permits support of a suspended fixture on rigid metal conduit, IMC, or an explosionproof flexible connection fitting. Such application is acceptable provided the fixture stem is no longer than 12 in. (305 mm), unless lateral bracing is provided. Where bracing *is* needed, it must be provided not more than 12 in. (305 mm) from the fixture end of the stem. As indicated, where flexibility is needed, a flexible fitting, no longer than 12 in. (305 mm) is permitted.

Lighting Fixtures Require Careful Application

The UL data on hazardous-location lighting fixtures are given under the heading "Fixtures and Fittings" in the Red Book of UL. A lighting fixture recognized for use in Division 1 hazardous locations will be marked "Electric Lighting Fixture for Hazardous Locations" and will show the one or more

Groups of hazardous atmospheres for which it is suited. If a fixture is not recognized for Division 1 locations but is limited to Division 2 installations, it will be marked "Electric Lighting Fixture for Division 2 Hazardous Locations." Other UL data are as follows:

- Class I, Division 1 fixtures with *external* surface temperatures over 100°C will have the operating temperature marked on the fixture.
- Fixtures for Class I, Division 1 and Division 2 locations are designed to operate without igniting the atmosphere of the one or more Groups for which the fixture is listed. A Class I, Division 1 fixture (explosionproof) has its lamp chamber sealed off from the terminal compartment for the supply conductors. All modern explosionproof lighting fixtures are designed by the manufacturer to be factory-sealed, eliminating the need for seal fittings immediately adjacent to the fixtures.
- Any fixture subject to breakage must be equipped with a guard.
- A fixture with one or more germicidal lamps must have a warning to assure that its method of installation does not present a chance of injurious radiation to any person.
- Fixtures for wet locations and those suitable for use where residue of combustible paint will accumulate on them are marked to indicate such recognition.

Class I, Division 2 locations In these locations, the selection of a suitable fixture becomes a little more involved and has caused problems in the field. Correlation between the requirements of Sec. 501-9 and the application data and listings of the UL must be carefully established.

Watch Out! Controversy!

Section 501-9(b) (2) does *not* say that a fixture in a Division 2 location must "be approved for the Class I, Division 2 location"—unlike Sec. 501-9(a) (1), which requires a fixture "approved" for the specific location. Instead, Sec. 501-9(b)(2) gives a description of the type of fixture that would be acceptable, citing a number of requirements:

1. The fixture must be protected from physical damage by suitable guards or "by location"—which can be taken to mean that mounting it high or otherwise out of the way of any object that might strike or hit it eliminates the need for a guard.
2. If falling sparks or hot metal from the fixture could possibly ignite local accumulation of the hazardous atmosphere, then an enclosure or other protective means must be used to eliminate that hazard.
3. Where lamps used with the fixture may, under normal conditions, reach surface temperature above 80 percent of the atmosphere's ignition temperature, then either of two conditions must be satisfied—(a) A fixture approved for Class I, Division 1 location must be used, or (b) the fixture must be of a type that has been tested and found incapable of igniting the gas or vapor if the ignition temperature is not exceeded.

Figure 501-35 shows a Class I, Division 1 fixture that could be used in a Division 2 location. But it is not necessary to use a Division 1 fixture and have questions develop.

Fig. 501-35. In a Class I, Division 2 location, a lighting fixture listed for Class I, Division 1 would satisfy—such as this factory-sealed mercury-vapor luminaire in a Class I, Group D location. Otherwise, a fixture listed for Class I, Division 2 must be used. (Sec. 501-9.)

For many years, in Class I, Division 2 locations simply an enclosed- and gasketed-type fixture was the usual choice. The fixture does not need to be explosionproof but must have a gasketed globe. The primary requirement is that any surface, including the lamp, must operate at less than 80 percent of the ignition temperature of the gas or vapor that may be present. The effect of Sec. 501-9(b) (2) is to recognize the use of general-purpose lighting fixtures if the conditions of fixture-operating temperature and atmosphere-ignition temperature are correlated as required or if the fixture has been "tested" to verify its safety. At best, the described task of determining the suitability of a general-purpose fixture for use in the hazardous location by relating its lamp-operating temperature to 80 percent of the atmosphere-ignition temperature could be difficult for any electrical designer and/or installer. It also seems highly unlikely that any of them would have the facilities or the experience to perform the testing described in the last part of the **Code** rule and make a sound judgment on the suitability, even though the manufacturer provides the necessary temperature data on the fixture. It was the intention of the authors of that last part of the **Code** rule that the testing mentioned be done by a "qualified testing agency" (such as UL, Factory Mutual, ETL). And as a result of such testing, fixtures for Class I, Division 2 locations would be approved and listed on the same basis that fixtures are certified for Class I, Division 1 locations.

Class I, Division 2 fixtures are listed in two ways in the Red Book. Some are listed as "Class I, Division 2 only," without reference to Group or Groups. Others are listed with an indication of the Groups for which they are listed—for instance, "Class I, Groups A, B, C, and D, Division 2 only." Great care must be used in evaluating the detail of these listing designations. If such a fixture is not marked otherwise, the temperature of the fixture is lower than the ignition temperature of any of the atmospheres for which it is listed. Where a Class I Group designation is not mentioned, the fixture must not be used where its marked operating temperature is above the ignition temperature of the hazardous atmosphere. Class I, Division 2 fixtures with *internal* parts operating over 100°C will be marked to show the actual operating temperature of internal parts.

Based on the foregoing, precise enforcement of the **NE Code** and OSHA insistence on the maximum use of third-party certified products would seem to suggest the following approach:

1. In Class I, Division 1 locations, only fixtures listed by a nationally recognized test lab may be used.
2. And because fixtures are listed for Class I, Division 2 locations (in the UL Red Book), any fixture in a Class I, Division 2 location *must be listed* for that application (or, of course, a Class I, Division 1 fixture could be used). Consistent with OSHA's rationale on the matter of listing, if a third-party certified product is available (that is, Class I, Division 2 fixtures), then use of a nonlisted fixture in a Class I, Division 2 location would be a clear violation. Based on that analysis, it seems that the **Code**-rule certification of a general-purpose fixture with lamp-temperature-not-over-80-percent-of-ignition-temperature would be abrogated.

Recessed Fixtures

Recessed fixtures of both incandescent and electric-discharge lamp types are listed in the UL Red Book and are suitable only for dry locations—unless marked "SUITABLE FOR DAMP LOCATIONS" or "SUITABLE FOR WET LOCATIONS." Other rules are:

- Each fixture is marked to show the minimum temperature rating of conductors used to supply the fixture. Care must be taken to observe all such markings on these fixtures with respect to the number, size, and temperature rating of wires permitted in junction boxes or splice compartments that are part of such fixtures. Generally, no allowance is made in such boxes or compartments for heat produced by current to other loads that may be fed by taps or splices in the fixture supply wires within the JB or splice compartment. Allowance is made only for the I^2R heat input of the current to the fixture itself. If the fixture is recognized for carrying through other conductors to other loads, the fixture will be marked to cover the permitted conditions of wiring.
- Fixtures are listed to assure safe application in both Class I, Division 1 and Class I, Division 2 locations.
- In every Class I, Division 1 fixture, the wiring compartment for supply circuit connections is internally sealed from the lamp chambers.
- Fixtures that may be used as raceways for carrying through circuit conductors other than those supplying the fixtures are marked "Suitable for Use as Raceway" and show the number, size, and type of wires permitted.
- Fixtures are marked to show suitability for installation in concrete and some may be used *only* in concrete.
- Fixtures are marked when they may be used *only* with fire-resistive building construction.

- Some fixtures are marked to show acceptable use *only* in Class I, Division 2 locations.

Portable Lighting

Although the basic rule of part **(b)(1)** is that portable lighting in a Class I, Division 2 location must be listed for Class I, Division 1 locations. Portable lighting equipment in Class I, Division 2 locations does not have to be approved for Class I, Division 1 use if it is mounted on a movable stand and is cord-connected. This is a relaxation of the rule in the first sentence of part **(b)(1)**. The rule recognizes the common need for temporary lighting for maintenance work in Class I, Division 2 locations, where handlamps would not be adequate.

501-10. Utilization Equipment. This section covers devices that utilize electrical energy—other than lighting fixtures and motors or motor-operated equipment. Electric heaters in Class I, Division 1 locations must be listed for such application. The UL Red Book lists convection-type heaters under "Heaters," for Groups C and D. Industrial and laboratory heaters are also listed—heat tracing systems, hot plates, paint heaters, and steam-heated ovens.

According to Sec. 501-10(b)(1)a., Exception No. 2, in a Class I, Division 2 location, an electrically heated utilization equipment may be used *if* some current-limiting means is provided to prevent heater temperature from exceeding 80 percent of the ignition temperature of the gas or vapor. Such current limitation may be part of the control equipment for the heater to ensure safe operation by preventing dangerously high temperatures.

501-11. Flexible Cords, Class I, Divisions 1 and 2. Although Sec. 501-4(a) does not mention flexible cord as an approved method of wiring in Division 1 locations, this rule does permit such cord for connection of portable equipment. The Exception, however, refers back to Sec. 501-3(b)(6), which permits cord and plug connection of process control instruments in Division 2 locations—to facilitate replacement of such units, which are not portable equipment. And Sec. 501-4(b) covers use of cord in Division 2 locations.

An explosionproof handlamp, listed for use in Class I locations, is an example of portable equipment covered by this rule, which requires that a three-conductor cord be used and that the device be provided with a terminal for the third, or grounding conductor, which serves to ground the exposed metal parts. Such handlamps are listed under "Portable Lighting Units" in the UL Red Book, which notes that flexible cords should be used only where absolutely necessary as an alternative to threaded rigid conduit hookups. Cords, plugs, and receptacles must be protected from moisture, dirt, and foreign materials. Frequent inspection and maintenance are critically important. Consultation with inspection authorities is always recommended where plug and receptacle applications are considered.

As noted in the next to last paragraph of this section, flexible cord may supply submersible pumps in Class I locations. This Code section recognizes use of flexible cord with such pumps because they are designated as "portable utilization equipment." Commentary in favor of this Code rule stated:

Many authorities in the wastewater field (sewage, stormwater, etc.) in recognizing especially the contribution to maintenance of submersible pumps, have stipulated that in any wet-well installation they be easily removed without the need for personnel to enter or dewater the wet-well. This is provided by all member companies in the industry by means of guide-rail remote guidance system and a simple automatic discharge connection system which allows indexing and a tight connection (or removal) to be automatically accomplished between the pump discharge flange and the effluent piping flange.

In order to maintain the workability of the system and the intent of the specifying Authorities, it is imperative that flexible cord or cable be used between the place where the service enters the wet-well from the pump control (gas-tight conduit seal or—in the case of Class I, Division 1 locations—explosion-proof junction or splice box hard wired to the pump cable with a suitable compression cable-entry) and the pump cable-entry assembly.

This will allow the pump to be lifted from the wet-well through the opened cover in the access frame in the ground-level slab by its chain or wire-rope without personnel entering or dewatering the wet-well.

In the case of Class I, Division 1 locations, the pump is either explosion-proof and suitable for the installation (there are none Approved at the present time) or redundant low-level shut-off sensing is provided which guarantees the uppermost portion of a standard submersible pump is always submerged. These approaches are specified and accepted by the Administrative Code for the State of Wisconsin and the Department of Industrial Safety for the State of California for some time.

Extra hard-usage cord and cable of the S, SO, ST, STO, W, G, PCG, etc., classes have been used for many years in submersible wastewater handling with a perfect safety record in classified locations.

501-12. Receptacles and Attachment Plugs, Class I, Divisions 1 and 2. The basic rule calls for receptacles and plug caps to be approved for Class I locations, which suits them to use in either Division 1 or Division 2 locations. The Exception notes that cord connection of process control instruments in Class I, Division 2 locations does not require devices approved for Class I locations. General-purpose receptacles may be used as outlined in Sec. 501-3(b) (6).

Figure 501-36 shows a 3-pole 30-A receptacle and the attachment plug which is so designed as to seal the arc when the circuit is broken, and therefore is suitable for use without a switch. The circuit conductors are brought into the base or body through rigid conduit screwed into a tapped opening and are spliced to pigtail leads from the receptacle. The receptacle housing is then attached to the base, the joint being made at wide flanges ground to a suitable fit. All necessary sealing is provided in the device itself, and no additional sealing is required when it is installed. The plug is designed to receive a three-conductor cord for a 2-wire circuit or a four-conductor cord for a 3-wire, 3-phase circuit, and is provided with a clamping device to relieve the terminals from any strain. The extra conductor is used to ground the equipment supplied.

UL data are as follows:

- Class I receptacles for Division 1 or Division 2 locations are equipped with boxes for threaded metal conduit connection, and a factory seal is provided between the receptacle and its box.
- Receptacles for Class I, Division 2 only may be used with general-purpose enclosures for supply connections, with factory sealing of conductors in

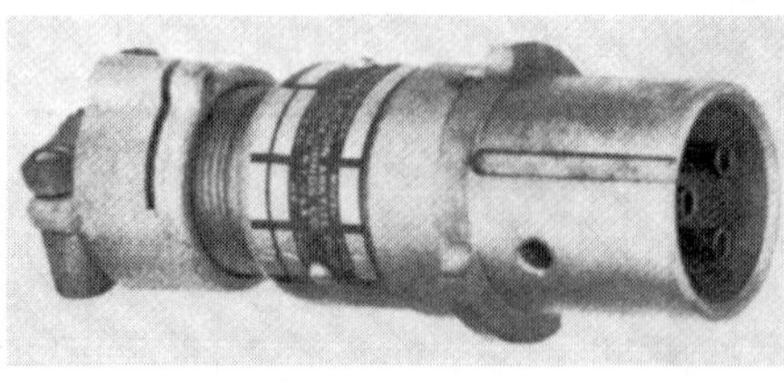

Fig. 501-36. Receptacle and plug must generally be explosionproof type for Divisions 1 and 2. (Sec. 501-12.)

the receptacle. The plugs for such receptacles are suitable for Class I, Division 1 locations.

- Frequent inspection is recommended for flexible cords, receptacles, and plugs, with replacement whenever necessary.
- For Class I, interlocked CBs and plugs are made for receptacles so that the plug cannot be removed from the receptacle when the CB is closed and the CB cannot be closed when the plug is not in the receptacle (Fig. 501-37).

Mechanical-interlock construction requires that the plug be fully inserted into the receptacle and rotated to operate an enclosed switch or CB that energizes the receptacle. The plug cannot be withdrawn until the switch or breaker has first de-energized the circuit.

The delayed-action type of plug and receptacle has a mechanism within the receptacle that prevents complete withdrawal of the plug until after electrical connection has been broken, permitting any arcs or sparks to be quenched inside the arcing chamber. And insertion of the plug seals the arcing chamber before electrical connection is made. Threaded conduit connection to the CB compartment is provided. The plug is for Type S flexible cord with an equipment grounding conductor.

501-13. Conductor Insulation, Class I, Divisions 1 and 2. Because of economics and greater ease in handling, nylon-jacketed Type THHN-THWN wire, suitable for use where exposed to gasoline, has in most cases replaced lead-covered conductors.

An excerpt from Underwriters Laboratories Inc. *Electrical Construction Materials Directory* states as follows:

> Wires, Thermoplastic.
> Gasoline Resistant TW—Indicates a TW conductor with a jacket of extruded nylon suitable for use in wet locations, and for exposure to mineral oil, and to liquid gasoline and gasoline vapors at ordinary ambient temperature. It is identified by tag marking

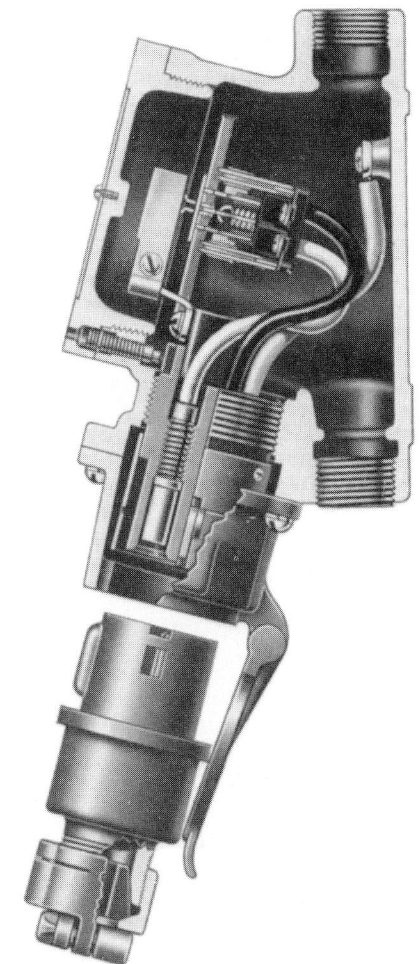

Fig. 501-37. A receptacle with a plug interlocked with a circuit breaker is an explosionproof assembly with operating safety features. (Sec. 501-12.)

and by printing on the insulation or nylon jacket with the designation "Type TW Gasoline and Oil Resistant I."

Also listed for the above use is "Gasoline Resistant THWN" with the designation "Type THWN Gasoline and Oil Resistant II."

It should be noted that other thermoplastic wires may be suitable for exposure to mineral oil; but with the exception of those marked "Gasoline and Oil Resistant," reference to mineral oil does not include gasoline or similar light-petroleum solvents.

The conductor itself must bear the marking legend designating its use as suitable for gasoline exposure; such designation on the tag alone is not sufficient.

501-14. Signaling, Alarm, Remote-Control, and Communication Systems. Nearly all signaling, remote-control, and communication equipment involves make-

or-break contacts; hence in Division 1 locations all devices must be explosionproof, and the wiring must comply with the requirements for light and power wiring in such locations, including seals.

Figure 501-38 shows a telephone having the operating mechanism mounted in an explosionproof housing. Similar equipment may be obtained for operating horns or sirens. Figure 501-39 shows fire-alarm hookups at a distillery. Alcohol is generally categorized as creating a Class I, Group G location by **Code** Table 500-2. A vapor-laden atmosphere with an alcoholic content ranging from 3.5 to 19 percent could become flammable or explode at an ignition temperature of approximately 80°F under certain conditions of air pressure and humidity.

Fig. 501-38. Explosionproof telephones are made and listed for Class I, Groups B, C, and D, and must be connected with the necessary seal fittings required in conduits to enclosures housing arcing or sparking devices. [Sec. 501-14(b) (4).]

Referring to subparagraph **(b)**, covering Division 2 locations, it would usually be the more simple method to use explosionproof devices, rather than devices having contacts immersed in oil or devices in hermetically sealed enclosures, though mercury switches, which are hermetically sealed, may be used for some purposes. Of course, reference to Sec. 501-3(a) recognizes the use of purged enclosures as an alternative method.

The UL Red Book lists "Telephones" as follows:

- Telephones, sound-powered telephones, and communications equipment and systems are listed for Class I and Class II use in Division 1 locations and are explosionproof equipment. Such equipment complies with Secs. 501-14(a) and (b).
- Intrinsically safe sound-powered telephones are also listed for Class I, Division 1, Group D and may be used in both Divisions 1 and 2 locations in accordance with Secs. 501-14(a) and (b).

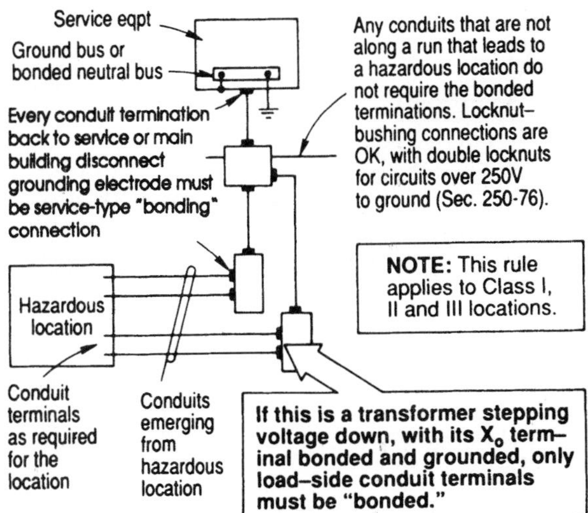

Fig. 501-40. Bonding of raceways and equipment must be made back to service ground. (Sec. 501-16.)

According to the Exception for part **(b)** for Class I, Division 2 areas, liquidtight flexible metal conduit, in lengths not over 6 ft (1.83 m), may be used *without* a bonding "jumper" (an internal or external equipment grounding conductor) to enclose conductors protected at not more than 10 A. This permission is given only for circuits to a load that is "not a power utilization load." The same use of liquidtight flex is permitted for Class II and Class III locations—in Secs. 502-16(b) and 503-16(b).

ARTICLE 502. CLASS II LOCATIONS

502-1. General. Referring to Sec. 500-5, the hazards in Class II locations are due to the presence of combustible dust. These locations are subdivided into three groups, as follows:

- Group E, atmospheres containing metal dust
- Group F, atmospheres containing carbon black, coal dust, or coke dust
- Group G, atmospheres containing grain dust, such as in grain elevators

It is important to note that some equipment that is suitable for Class II, Group G, is not suitable for Class II, Groups E and F.

Any one of four hazards, or a combination of two or more, may exist in a Class II location: (1) an explosive mixture of air and dust, (2) the collection of conductive dust on and around live parts, (3) overheating of equipment because deposits of dust interfere with the normal radiation of heat, and (4) the possible ignition of deposits of dust by arcs or sparks.

A large number of processes which may produce combustible dusts are listed in Sec. 500-6. Most of the equipment listed as suitable for Class I locations is

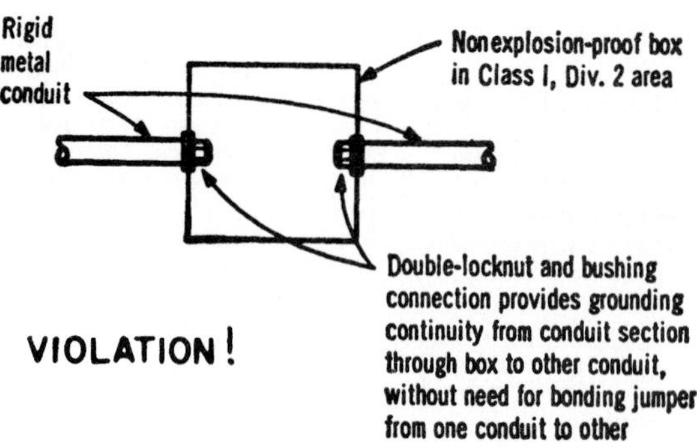
Rigid metal conduit
Nonexplosion-proof box in Class I, Div. 2 area
VIOLATION!
Double-locknut and bushing connection provides grounding continuity from conduit section through box to other conduit, without need for bonding jumper from one conduit to other

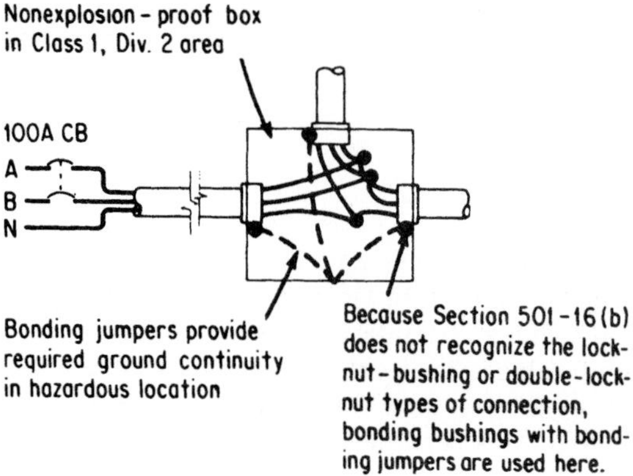
Nonexplosion - proof box in Class 1, Div. 2 area
100A CB
A
B
N
Bonding jumpers provide required ground continuity in hazardous location
Because Section 501-16(b) does not recognize the lock-nut-bushing or double-lock-nut types of connection, bonding bushings with bonding jumpers are used here.

Fig. 501-41. Bonding bushings with jumpers comply with grounding rule. (Sec. 501-16.)

also dusttight, but it should not be taken for granted that all explosionproof equipment is suitable for use in Class II locations. Some explosionproof equipment may reach too high a temperature if blanketed by a heavy deposit of dust. Grain dust will ignite at a temperature below that of many of the flammable vapors.

Location of service equipment, switchboards, and panelboards in a separate room away from the dusty atmosphere is always preferable.

In Class II locations, with the presence of combustible dust, UL standards call for a type of construction designed to preclude dust and to operate at specified limited temperatures. Dust-ignition-proof equipment is generally more economical to use in Class II areas; however, explosionproof devices are often used if such devices are approved for Class II areas and for the particular Group involved.

Dust-ignition-proof equipment is enclosed in a manner so as to exclude ignitible amounts of dusts or amounts which might affect equipment perfor-

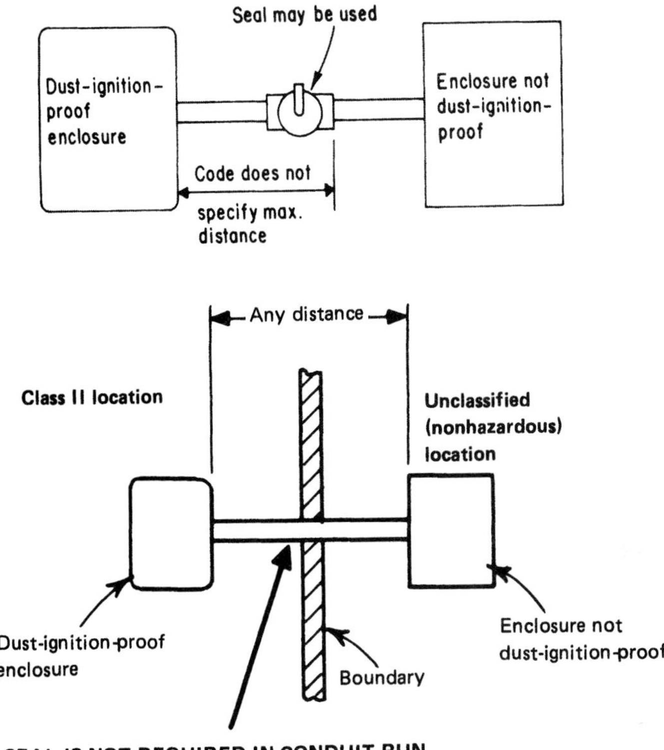

Fig. 502-2. A seal fitting must be used in short [less than 10-ft (3.05-m)] connections, as shown at top, but is not required, as shown at bottom.

502-8. Motors and Generators. In Class II, Division 1 locations, motors must be dust-ignition-proof (approved for Class II, Division 1) or totally enclosed with positive-pressure ventilation. The same motors may be used also in Class II, Division 2 areas; however, if dust accumulation is very slight, either a standard, open-type motor (without arcing or sparking parts), a self-cleaning textile motor, or a squirrel-cage motor may be used.

Part **(a)(2)** refers to a totally enclosed pipe-ventilated motor. A motor of this type is cooled by clean air forced through a pipe by a fan or blower. Such a motor has an intake opening, where air is delivered to the motor through the pipe from the blower. The exhaust opening is on the opposite side, and this should be connected to a pipe terminating outside the building, so that dust will not collect inside the motor while it is not running.

For Class II, Division 2 locations, part **(b)** permits use of totally enclosed, nonventilated motors and totally enclosed, fan-cooled (TEFC) motors in addition to totally enclosed, pipe-ventilated and dust-ignition-proof motors. In addition, a totally enclosed water-air-cooled motor would be permitted. This rule eliminates any interpretation that only a labeled motor (dust-ignition-proof) is acceptable for Class II, Division 2 locations. Experience has shown the other motors to be entirely safe and effective for use in such locations.

Motors of the common totally enclosed type without special provision for cooling may be used in Division 2 locations, but to deliver the same horsepower, a plain totally enclosed motor must be considerably larger and heavier than a motor of the open type or an enclosed fan-cooled or pipe-ventilated motor.

The UL Red Book lists motors for Class II, Divisions 1 and 2, Groups E, F, and G locations.

502-9. Ventilating Piping. In locations where dust or flying material will collect on or in motors to such an extent as to interfere with their ventilation or cooling, enclosed motors which will not overheat under the prevailing conditions must be used. It may be necessary to require the use of an enclosed pipe-ventilated motor or to locate the motor in a separate dusttight room, properly ventilated with clean air (Sec. 430-16).

The reference to ventilation is clarified in this section. Vent pipes for rotating electrical machinery must be of metal not lighter than No. 24 MSG gauge, or equally substantial. They must lead to a source of clean air outside of buildings, be screened to prevent entry of small animals or birds, and be protected against damage and corrosion.

In Class II, Division 1 locations, vent pipes must be dusttight. In Division 2 locations, they must be tight to prevent entrance of appreciable quantities of dust and to prevent escape of sparks, flame, or burning material.

Typical conditions where these requirements may apply include processing machinery or enclosed conveyors where dust may escape only under abnormal conditions, or storage areas where handling of bags or sacks may result in small quantities of dust in the air.

502-10. Utilization Equipment. As noted in part **(b)(1)**, Exception, dusttight metal-enclosed radiant heating panels may be used in Class II, Division 2 locations even though they are *not* approved for Class II locations. The Exception to this section will permit such electric-heat panels *provided* that surface temperature limitations of Sec. 500-3(d) are satisfied. Heaters of that type are available with low surface temperature.

502-11. Lighting Fixtures. Fixtures used in Class II, Division 1 areas must prevent the entry of the hazardous dust and should prevent the accumulation of dust on the fixture body. In Division 1 locations, all fixtures must be approved for that location. Where metal dusts are present, lighting fixtures must be specifically approved for use in Group E atmospheres. Fixtures are listed by Underwriters Laboratories Inc. as suitable for use in all three of the locations classed as Groups E, F, and G, for Divisions 1 and 2.

The purpose of the latter part of subparagraph **(a)(3)** is to specify the type of cord to be used for wiring a chain-suspended fixture. It is not the intention to permit a fixture to be suspended by means of a cord pendant or drop cord.

The only special requirements for lighting fixtures in Class II, Division 2 locations are that the lamp must be enclosed in a suitable glass globe and that a guard must be provided unless the fixture is so located that it will not be exposed to physical damage. The enclosing globe should be tight enough so that it will practically exclude dust, although dusttight construction is not called for. For Class II, Group G, Division 2 areas, fixtures normally are the enclosed and gasketed type. However, in addition, such fixtures must not have an exposed surface temperature exceeding 165°C.

There are portable handlamps approved for use in any Class II, Group G location, i.e., where the hazards are due to grain dust.

The UL Red Book notes that a Class II fixture for Divisions 1 and 2 is tested for dust tightness and safe operation in the dust atmosphere for which it is listed. A note points out the importance of effective maintenance—regular cleaning—to prevent buildup of combustible dust on such equipment.

502-12. Flexible Cords, Class II, Divisions 1 and 2. Figure 502-3 shows flexible cord used in a Division 2 area of a grain elevator (Class II, Group G).

Fig. 502-3. Flexible cords and listed connectors are used in a grain elevator that handles combustible grain dust. (Sec. 502-12.)

Section 502-4 permits its use in Division 1 and Division 2 areas of Class II locations. Listed cord connectors are recognized for use in Class II, Group G locations—using Types S, SO, ST, or STO multiconductor, extra-hard-usage cord *with* a grounding conductor. Cord connectors for connecting extra-hard-service type of flexible cord to devices in hazardous locations must be carefully applied.

The UL Red Book notes, under "Receptacles with Plugs," that Type S flexible cord should be frequently inspected and replaced when necessary.

502-13. Receptacles and Attachment Plugs. Class II receptacles listed as approved for Division 1 locations are equipped with boxes for threaded metal conduit connection, and a factory seal is provided between the receptacle and its box. Only receptacles and plugs listed for Class II locations are permitted in Division 1 areas.

Receptacles for Class II, Division 2 locations do not have to be approved for Class II but must satisfy the connection method described.

Frequent inspection is recommended for flexible cords, receptacles, and plugs, with replacement whenever necessary.

As shown in Sec. 501-12 for Class I locations, for Class II locations, interlocked CBs and plugs are made for receptacles so that the plug cannot be removed from the receptacle when the CB is closed and the CB cannot be closed when the plug is not in the receptacle.

502-16. Grounding, Class II, Divisions 1 and 2. The requirements of this section are the same as those of Sec. 501-16. Refer to the discussion and illustrations in that section.

502-17. Surge Protection, Class II, Divisions 1 and 2. A common application of this surge protection is found in grain-handling facilities (grain elevators) in localities where severe lightning storms are prevalent. Assuming a building supplied through a bank of transformers located a short distance from the building, the recommendations are, in general, as shown in the single-line diagram in Fig. 502-4. The surge-protective equipment consists of primary lightning arresters at the transformers and surge-protective capacitors connected to the supply side of the service equipment. The lightning arrester ground and the secondary system ground should be solidly connected together. All grounds should be bonded together and to the service conduit and to all boxes enclosing the service equipment, metering equipment, and capacitors.

Complete information on methods of providing surge protection may be obtained from the Mill Mutual Fire Prevention Bureau, 2 North Riverside Plaza, Chicago, IL 60606.

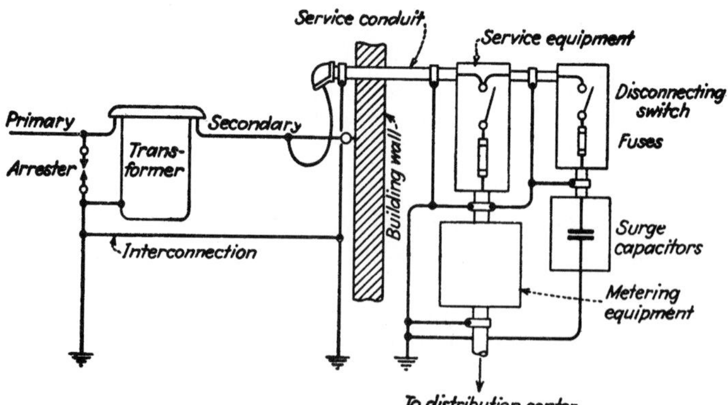

Fig. 502-4. Surge protection is often used to protect Class II systems against lightning.

ARTICLE 503. CLASS III LOCATIONS

503-1. General. The small fibers of cotton that are carried everywhere by air currents in some parts of cotton mills and the wood shavings and sawdust that collect around planers in woodworking plants are common examples of the combustible flyings or fibers that cause the hazards in Class III, Division 1 locations. A cotton warehouse is a common example of a Class III, Division 2 location.

503-3. Wiring Methods. Rigid nonmetallic conduit, EMT, and Type MC (metal-clad cable) are permitted for Class I and Class II, Division 1 locations—in addition to rigid metal conduit, IMC, Type MI cable, Type SNM cable, and dusttight wireways. Type MC (Art. 334) includes interlocked armor cable, corrugated metal armor, smooth aluminum-sheathed cable (Type ALS), and smooth copper-sheathed cable (Type CS).

Fittings and boxes must be dusttight, whether or not they contain taps, joints, or terminal connections. As part **(a)(2)** notes about flexible connections, it is necessary to use dusttight flexible connectors, liquidtight flexible metal conduit, or extra-hard-usage flexible cord that complies with Sec. 503-10. That clear, simple rule calls for an equipment ground wire in flexible cord.

Part **(b)** requires the same wiring methods for Division 2 as for Division 1. As indicated in Fig. 503-1, there are no seal requirements in Class III locations.

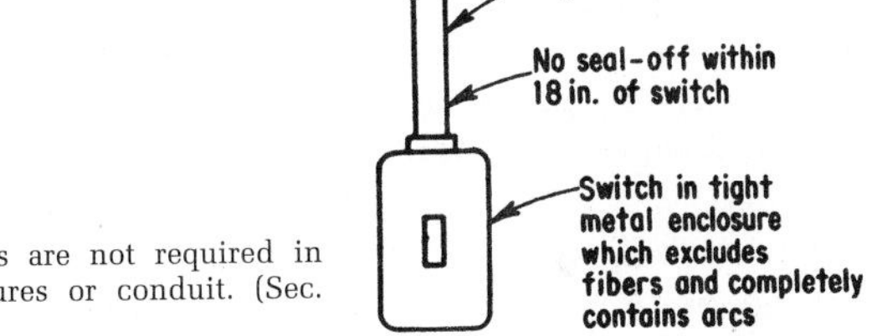

Fig. 503-1. Seals are not required in Class III enclosures or conduit. (Sec. 503-3.)

503-4. Switches, Circuit Breakers, Motor Controllers, and Fuses, Class III, Divisions 1 and 2. Equipment suitable for Class III locations must function at full rating without developing surface temperatures high enough to cause excessive dehydration or gradual carbonization of accumulated fibers or flyings. These devices have the same surface temperature limitations as Class II equipment, and the construction is similar.

Enclosures for equipment in a Class III location must be dusttight—that is, provided with telescoping or close-fitting covers or other effective means which prevent the escape of sparks or burning material and have no openings through which sparks or burning material might escape, or through which adjacent combustible material might be ignited.

503-6. Motors and Generators, Class III, Divisions 1 and 2. UL lists no Class III motors as such; however, totally enclosed nonventilated motors and the so-called lint-free or self-cleaning textile squirrel-cage motors are commonly used. The latter may be acceptable to the local inspection authority if only moderate amounts of flyings are likely to accumulate on or near the motor, which must be readily accessible for routine cleaning and maintenance. Or the motor may be a squirrel-cage motor, or a standard open-type machine having any arcing or heating devices enclosed within a tight metal housing without ventilating or other openings.

503-9. Lighting Fixtures, Class III, Divisions 1 and 2. In Class III, Divisions 1 and 2 areas, lighting fixtures must minimize the entrance of fibers and flyings and prevent the escape of sparks or hot metal. And again, the surface temperature of the unit must be limited to 165°C. Available fixtures are third-party certified (by a national test lab) for Class III locations, Divisions 1 and 2. In the past, enclosed and gasketed types of fixtures, of the type that was used in Class I, Division 2 areas, have been acceptable as suitable for use in this application. But because there are listed Class III fixtures available, inspection agencies and OSHA might insist on use of *only* listed fixtures in such applications—to be consistent with the trend to third-party certification.

503-13. Electric Cranes, Hoists, and Similar Equipment, Class III, Divisions 1 and 2. A crane operating in a Class III location and having rolling or sliding collectors making contact with bare conductors introduces two hazards:

1. Any arcing between a collector and a conductor rail or wire may ignite flyings of combustible fibers that have collected on or near to the bare conductor. This danger may be guarded against by proper alignment of the bare conductor and by using a collector of such form that contact is always maintained, and by the use of guards or barriers which will confine the hot particles of metal that may be thrown off when an arc is formed.
2. Dust and flyings collecting on the insulating supports of the bare conductors may form a conducting path between the conductors or from one conductor to ground and permit enough current to flow to ignite the fibers. This condition is much more likely to exist if moisture is present. Operation on a system having no grounded conductor makes it somewhat less likely that a fire will be started by a current flowing to ground. A recording ground detector will show when the insulation resistance is being lowered by an accumulation of dust and flyings on the insulators, and a relay actuated by excessively low insulation resistance and arranged to trip a CB provides automatic disconnection of the bare conductors when the conditions become dangerous.

ARTICLE 504. INTRINSICALLY SAFE SYSTEMS

504-1. Scope This NEC Article covers design, layout, and installation of electrical equipment and systems that are not capable of releasing sufficient electrical or thermal energy to ignite flammable or combustible atmospheres. All equipment and apparatus that is "intrinsically safe" must be "approved"—which can be understood to mean evaluated by a testing laboratory such as UL. Reference to the NEC definition of "approved" and Sec. 110-2 and its fine-print note (FPN) dictate the most rigorous and objective determination that such equipment is "approved."

The substantiation for adoption of this **Code** Article noted the following:

SUBSTANTIATION: Installation personnel need specific requirements for the installation of intrinsically safe equipment that are clear and concise. The authority having jurisdiction needs requirements that are enforceable.

Currently, **NEC** Sec. 500-2 exempts intrinsically safe equipment and associated wiring from the requirements of Arts. 500 through 517. The requirements of Art. 500 are applicable to all hazardous (classified) location equipment and intrinsic safety installation should not be expected from them.

The applications of intrinsically safe apparatus and associated apparatus used to be limited to a few specific applications of process control equipment. The need for optimizing process control for quality, economy, and environmental regulations has resulted in a great expansion of the process control industry. In recent years, the use of computerized process control equipment, process logic controllers (PLCs), and similar equipment has grown dramatically. Pneumatic controls are no longer as acceptable as they were years ago and discrete instruments that used to switch 110 V to control processes have also become unsatisfactory.

With the current technology of microelectronics, there is now a vast array of sensors, process transmitters, actuators, etc., that were not possible a decade ago. These devices work with low power levels in the 24-V, 20-mA range and are ideally suited to the application of intrinsic safety techniques. Installation is more cost-effective than other techniques such as purging or use of explosionproof enclosures and conduits. More and more instrument manufacturers are obtaining intrinsic safety approval of their products from UL and FM. One approval agency has indicated an average growth rate of almost 30 percent per year over the last 10 years.

When new requirements are placed in the **NEC**, they get tremendous coverage by the trade journals, seminars, etc. People in the trade become educated and are able to apply the new requirements in practice. Placing installation requirements for intrinsically safe installations in the **NEC** will further the practice more than any other single action. Installers and inspectors will have a common basis for proper installation, and final acceptance of installations will be achieved with much less difficulty.

ARTICLE 505. CLASS I, ZONE 0, 1, AND 2 LOCATIONS

This completely new **Code** article is intended as an alternate method for classifying Class I hazardous locations. As indicated by the NOTICE that appears immediately after the title, this entire article is the subject of an appeal, which is seeking to have Art. 505 deleted. There is every indication that this article will withstand the appeal and remain in the **Code**. However, the only way one can be certain that Art. 505 *did* survive the appeal is to check with the authority having jurisdiction.

The most important prerequisite for application of the Zone classifications in Class I hazardous locations is that the designation of the degree of hazard, as well as the selection of equipment and wiring methods, must be performed under the "supervision of a qualified Registered Professional Engineer." Curiously enough, that requirement is not included anywhere in Art 505. It is given in Sec. 500-3, "Special Precautions." The wording used in Sec. 500-3, unfortunately, is not as clear as it could be. The term "supervision" infers that an engineer—and *only* an engineer—can make final decisions on where the various Zones begin and end, as well as the methods and equipment used.

The reference to a "qualified Registered Professional Engineer" also raises questions. It seems as if the local inspector must decide who is, and who is not, "qualified." In addition, because the literal wording does *not* limit the type of engineering to electrical engineering, the application of this article could be performed under the direction of, say, a Registered Professional Chemical Engineer or Mechanical Engineer, provided either one was deemed "qualified" by the local electrical inspector.

505-5. Groupings and Classifications. Art. 505 uses different designations and groupings for the various hazardous gases than does Art. 500. Section 505-5 gives the group designation and spells out which gases are included in that group.

The familar method given in Art. 500 has four gas groups, A, B, C, and D. But, there are only three designations for gas groups in Art. 505: IIC, IIB, and IIA. Basically stated, Art 500 groups can be equated to Art. 505 groups as follows:

Art. 500	Art. 505
Groups A and B	IIC
Group C	IIB
Group D	IIA

505-10. Listing and Marking. As is the case with the Division-classification system given in Art. 500, the Zone-classification system permits the use of equipment suited for a more hazardous location within a lesser hazardous location. Any equipment listed for Zone 0 may be used in Zone 1 or 2 locations provided the equipment is listed for the same gas group. And, any equipment listed for Zone 1 may be used in Zone 2 locations of the same gas group.

Table 505-10(b) in the **Code** gives the temperatures that correspond to the "T" designation assigned to a piece of equipment. The major difference between this table and Table 500-3(d), which applies to the Divison system of classification, is the number of designation. The table for the Zone system only has six different temperature ratings: T1 through T6, which range from 85 to 450°C. The Division-system temperature table has 14, which also range from 85 to 450°C, but with eight other intermediate values between T1 and T6.

505-15. Wiring Methods. This section identifies those wiring methods that may be used in the diferent Zones. As covered in part **(a)**, for Zone 0, only nonincendive circuits or intrinsically safe circuits may be installed in Zone 0 locations. The intrinsically safe circuits may be installed as permitted by Art. 504, and either intrinsically safe circuits or nonincendive circuits may be installed in threaded rigid metal conducit, threaded IMC, or Type MI cable. Regardless of which method is selected, the conduit or cable and all fittings and enclosures, etc., must be listed for use in Zone 0 locations.

Additionally, fiber optic cable of the nonconductive type, which, as stated in Sec. 770-4, contains no metallic components, may be installed in Zone 0 locations. This system excludes any other equipment or wiring from being installed within the Zone 0 classified area.

The rule of part **(b)** indicates the acceptable wiring methods for Zone 1 locations. Any of the methods recognized in part **(a)** of this section for Zone 0 locations may be used. And any of those wiring methods permitted in Class I, Division 1 locations—Sec. 501-4(a)—would also be suitable for use within a Zone 1 location.

The last part recognizes the use of any wiring method permitted by either **(a)** or **(b)** in this section, as well as any of the wiring methods permitted by Sec. 501-4(a) and (b), which cover wiring methods for Division 1 and 2 locations.

It should be noted that all sealing requirements spelled out in Sec. 501-5 must be satisfied where the Zone system of classification is applied.

505-20. Equipment. The equipment permitted within any of the hazardous locations classified according to the Zone system described in Art. 505 must comply with Sec. 505-20.

As required by Sec. 505-20(a), in Zone 0 locations, equipment specifically *listed* for use Zone 0 locations, and *only* equipment *listed* for use in Zone 0, may be used.

According to the basic rule in Sec. 505-20(b), any equipment that is listed for use in Zone 1 locations, or, as covered in the exception, any equipment "approved" for use in Division 1 or Zone 0 locations within the same gas group and with "similar temperature marking" may be used. Notice that the temperature marking does not have to correspond directly to the maximum permitted. It simply must be "similar." That wording recognizes the use of equipment with temperature markings lower than the maximum acceptable temperature permitted, but not greater than that.

ARTICLE 511. COMMERCIAL GARAGES, REPAIR AND STORAGE

511-2. Locations. At one time (1971 NEC), this section considered parking garages as hazardous locations. Now, the rule no longer requires that this article apply to locations in which more than three cars, trucks, or other gas vehicles are or may be stored. Specifically, parking garages used simply for parking or storage of gasoline-powered vehicles are not classified as hazardous areas, no matter how many vehicles are present. But, such parking areas in enclosed buildings must be adequately ventilated.

Below-grade areas occupied for repairing, or communicating areas located below a repair garage, shall be continuously ventilated by a mechanical ventilating system having positive means for exhausting indoor air at a rate of not less than 0.75 cfm/sq ft of floor area. An approved means shall be provided for introducing an equal amount of outdoor air.

Operations involving open flame or electric arcs, including fusion, gas, and electric welding, shall be restricted to areas specifically provided for such purposes.

All enclosed, basement, and underground parking structures shall be continuously ventilated by a mechanical system capable of providing a minimum of six air changes per hour.

Heating equipment may be installed in motor vehicle repair or parking areas where there is no dispensing or transferring of Class I or II flammable liquids (as defined in the Flammable and Combustible Liquids Code, NFPA No. 30-1984) or liquefied petroleum gas, provided the bottom of the combustion chamber is not less than 18 in. (457 mm) above the floor, the heating equipment is protected from physical damage by vehicles, and continuous mechanical ventilation is provided at the rate of 0.75 cfm/sq ft of floor area. The heating system and the ventilation system shall be suitably interlocked to ensure operation of the ventilation system when the heating system is in operation.

Approved suspended unit heaters may be used provided they are located not less than 8 ft (2.44 m) above the floor and are installed in accordance with the conditions of their approval.

The question often arises, Does diesel fuel come within the classification of volatile flammable liquids, thereby requiring application of Art. 511 to places used exclusively for repair of diesel-powered vehicles? The third paragraph under Sec. 514-1 reads: "Where the authority having jurisdiction can satisfactorily determine that flammable liquids having a flash point below 38°C (100°F), such as gasoline, will not be handled, such location shall not be required to be classified."

The NFPA Inspection Manual, under identification of flammable liquids, says, "Minimum flash points for fuel oils of various grades are: No. 1 and No. 2, 100°F; No. 4, 110; No. 5, over 130; No. 6, 150 or higher. Actual flash points are commonly higher and are required to be higher by some state laws. No. 1 fuel is often sold as kerosene, range oil or coal oil."

Diesel fuel is a Class 3 flammable liquid, having flash points above 70°F (21°C). One listing of flash points of flammable liquids showed no diesel fuel below 120°F (49°C). Therefore, a diesel-fuel installation may be classified as a nonhazardous area and wired as such, unless it can be firmly established that the particular fuel has a flash point under 38°C (100°F). But, of course, the authority enforcing the **Code** is the one responsible for classifying such areas as nonhazardous.

511-3. Class I Locations. In part **(a)**, the wording considers any floor area to be a Class I, Division 2 hazardous location up to 18 in. (457 mm) above the floor. Any wiring within this space must be suitable for Class I, Division 2 locations.

Figure 511-1 shows the basic rule of parts **(a)** and **(b)**. Note that part **(b)** no longer refers to "floor below *grade*," but simply covers "pit or depression below *floor* level." Previous wording of these two parts resulted in the classification as Class I, Division 2 of the total shaded areas in Fig. 511-2, where vehicular servicing is done on the below-grade floor, the first floor, and the second floor. For each floor at or above grade, part **(a)** used to define the entire area up to a level of 18 in. (457 mm) above the floor as a Class I, Division 2 location. And part **(b)** stated that below-grade areas up to a level of 18 in. (457 mm) above the bottom of outside doors or other openings that are at or above grade level must be considered Class I, Division 2 locations, which is the lower shaded area in Fig. 511-2. Now part **(b)** simply requires any pit or depression below floor level to be considered as Class I, Division 2 area.

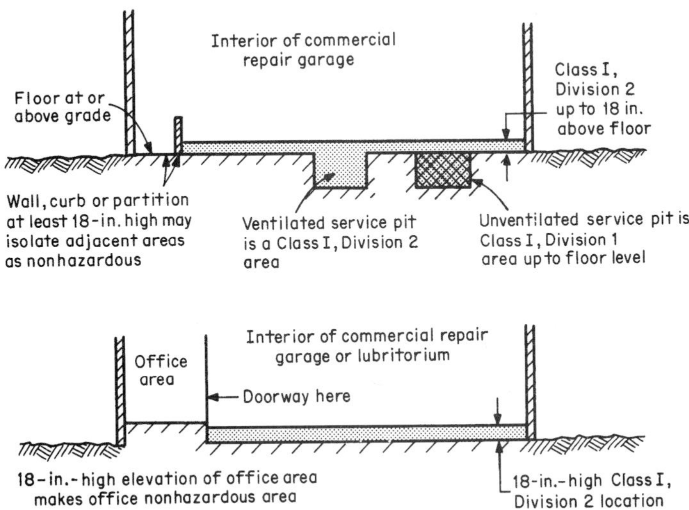

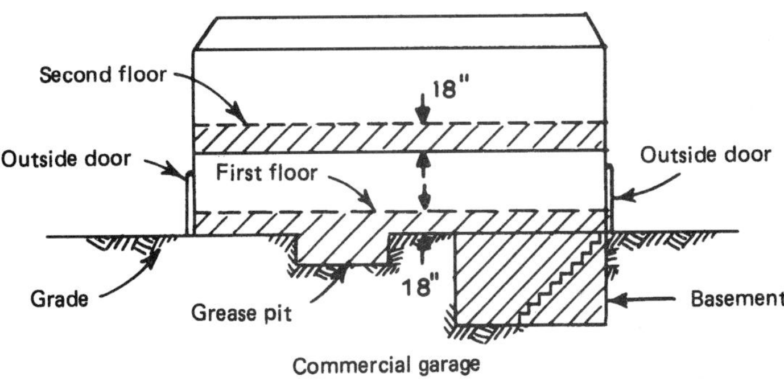

Fig. 511-1. Hazardous areas must be carefully established. (Sec. 511-3.)

Fig. 511-2. Previous wording of Code rule identified shaded areas as Class I, Division 2 locations. (Sec. 511-3.)

Under the new wording, for floors below grade, such as the basement in the drawing, the enforcing authority may judge that the hazardous location extends up to a level of only 18 in. (457 mm) above the floor. And that means that any wiring and equipment installed in any of these defined hazardous areas must be approved for Class I, Division 2 locations. Above these hazardous areas, Sec. 511-6 applies.

Part **(d)** allows the authority having jurisdiction to classify areas adjacent to hazardous locations as nonhazardous if proper ventilation, air pressure differentials, or physical separation are provided in a specific garage installation.

Equipment located in a suitable room or enclosure provided for the purpose or in a showroom separated from the garage proper by a partition which is reasonably tight up to 18 in. (457 mm) above the floor need not conform to the requirements of this section.

In all garages within the scope of this chapter, because of the possible presence of gasoline vapor near the floor, any equipment which in its normal operation may cause arcs or sparks, if less than 18 in. (457 mm) above the floor, is considered as in a hazardous location. It is seldom necessary to make use of devices having exposed live parts, but where this is unavoidable, even though the device is 18 in. (457 mm) above the floor, any such device should be well guarded.

511-6. Wiring in Spaces Above Class I Locations. The rules here apply to the lubritorium areas in service stations and *any* other space above the defined hazardous locations.

511-7. Equipment Above Class I Locations. Part **(a)** notes that equipment that may produce arcs or sparks and is within 12 ft (3.66 m) of the floor above hazardous areas must be enclosed or provided with guards to prevent hot particles from falling into the hazardous area, but lamps, lampholders, and receptacles are excluded. Standard receptacles are O.K. Lighting fixtures that are within 12 ft (3.66 m) of the floor over hazardous areas, over traffic lanes, or otherwise exposed to physical damage, must be totally enclosed, as required in part **(b)**.

511-9. Electric Vehicle Charging. The requirements for battery-charging cables and connectors are similar to the requirements for outlets for the connection of portable appliances, except that when hanging free the battery-charging cables and connectors may hang within 6 in. (152 mm) from the floor. The common form is a plug which is inserted into a receptacle on the vehicle, and, since the prongs are "alive," they must be covered by a protecting hood.

511-10. Ground-Fault Circuit-Interrupter Protection for Personnel. This rule is shown in Fig. 511-3.

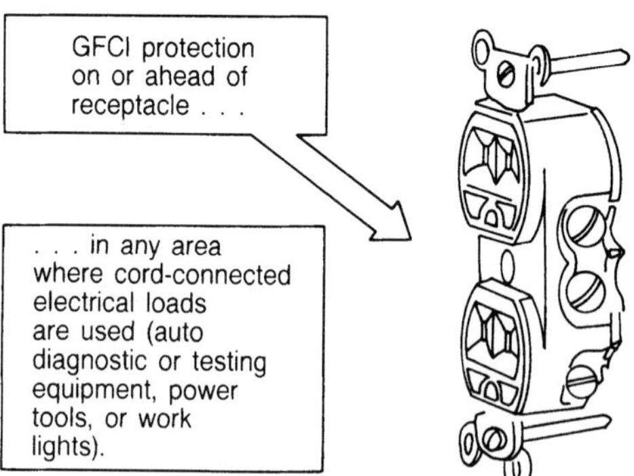
GFCI protection on or ahead of receptacle . . .
. . . in any area where cord-connected electrical loads are used (auto diagnostic or testing equipment, power tools, or work lights).

Fig. 511-3. GFCI protection is required for all 125-V, single-phase, 15- and 20-A receptacles where electrical auto-testing equipment, electrical hand tools, and portable lighting are used. (Sec. 511-10.)

ARTICLE 513. AIRCRAFT HANGARS

513-2. Classification of Locations. Figure 513-1 shows the details of hangar classifications. The entire floor area up to 18 in. (457 mm) above the floor, and adjacent areas not suitably cut off from the hazardous area or not elevated at least 18 in. (457 mm) above it, are classified as Class I, Division 2 locations. Pits below the hangar floor are classified as Class I, Division 1. Within 5 ft (1.52 m) horizontally from aircraft power plants, fuel tanks, or structures containing fuel, the Class I, Division 2 location extends to a level that is 5 ft (1.52 m) above the upper surface of wings and engine enclosures.

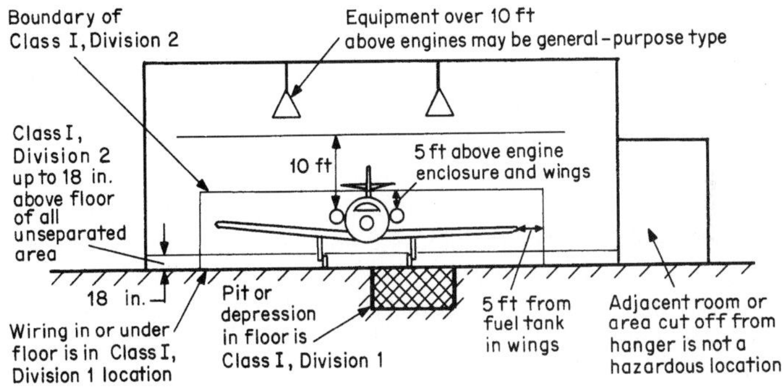

Fig. 513-1. Boundaries of hazardous areas are clearly defined. (Sec. 513-2.)

513-5. Equipment Not Within Class I Locations. Fixtures and other equipment that produce arcs or sparks may not be general-use types but are required to be totally enclosed or constructed to prevent escape of sparks or hot metal particles if less than 10 ft (3.05 m) above aircraft wings and engine enclosures, as indicated in Fig. 513-1.

ARTICLE 514. GASOLINE DISPENSING AND SERVICE STATIONS

514-1. Definition. As noted under Sec. 511-1, there is a question about application of the rules of Arts. 511 and 514 to areas used for service of vehicles using diesel fuel and to dispensing pumps and areas for diesel fuel. Fuel with a flash point above 38°C (100°F) may be ruled to be *not* "a volatile flammable liquid" to which the regulations of Arts. 511 and 514 are addressed.

Note that vehicle repair rooms or areas and lubritoriums at gas stations must comply with Art. 511.

514-2. Class I Locations. As noted, Table 514-2 delineates and classifies the various areas at dispensing pumps and service stations. This table brings the Code rules into agreement with data given in NFPA 30 S/C, General Storage and Handling of Flammable and Combustible Liquids.

Fig. 1 in the NEC book, in Art. 514, gives a visual representation of the dimensions of Class I locations—both Division 1 and Division 2—as described in the text of Table 514-2. In the wording of Table 514-2, the space within the dispensing-pump enclosure is a Class I, Division 1 location as described in ANSI/UL 87, "Power Operated Devices for Petroleum Products." This Code rule has become less specific because the previous specific dimensions—"within a dispenser enclosure up to 4 ft (1.22 m) vertically above the base except that space defined as Division 2"—has been removed.

Around the outside of a dispenser pump housing, the Division 2 location extends 18 in. (457 mm) horizontally in all directions, from grade up to the height of the pump enclosure or up to the height of "that portion of the dispenser enclosure containing liquid-handling components" (Fig. 514-1).

In the table, there is no direct statement that the ground under the Division 1 and Division 2 locations at a dispenser island is a Division 1 location. However, that seems to be the intent of the rule given for "Pits" under "Dispensing Units," where it refers to "space below grade level," where "any part [of that grade level] . . . is within the Division 1 or 2 location." But the phrase "which shall extend at least to the point of emergence above ground" is no longer part of the rule in the table, but is made mandatory at the end of the first paragraph of Sec. 514-8.

Outdoor areas within 20 ft (6.1 m) of a pump are considered a Class I, Division 2 location and must be wired accordingly. Such a hazardous area extends 18 in. (457 mm) above grade. If a building with a below-grade basement is within this hazardous area, gasoline fumes could enter the building if there were any windows within the 18-in. (457-mm) -high classified space, thereby making the basement a Class I, Division 2 area. This condition could be eliminated by "suitably cutting off the building from the hazardous area" by installing an 18-in. (457-mm) -high concrete curb between the service station and the residential property or enclosing the window openings up to that height.

Table 514-2 sets a 20-ft (6.1-m) -diameter, 18-in. (457-mm) -high, Division 2 area around each fill-pipe for the underground gasoline tanks at a gas station—as shown in Fig. 514-2.

As noted above, any wiring or equipment that is installed beneath any part of a Class I, Division 1 or Division 2 location is classified as being within a Class I, Division 1 location to the point where the wiring method is brought up out of the ground. That means the Division 1 area may extend far beyond the limits of the 20-ft (6.1-m) -radius perimeter around the dispensing pumps. The Division 1 area extends to the point where the conduit comes up to the supply panelboard or runs up to a lighting standard or a sign, as shown under Sec. 514-6.

As covered in Table 514-2, under "Dispensing Device, Overhead Type," where the dispensing unit and/or its hose and nozzle are suspended from overhead, the space within the dispenser enclosure and "all electrical equipment

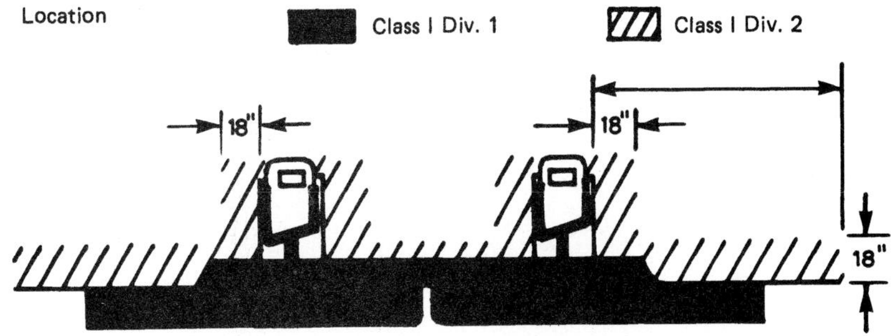

Division 1 space is *within* the dispenser, within any pit or box *below* the dispenser, and within the ground below the Class I, Division 2 location.

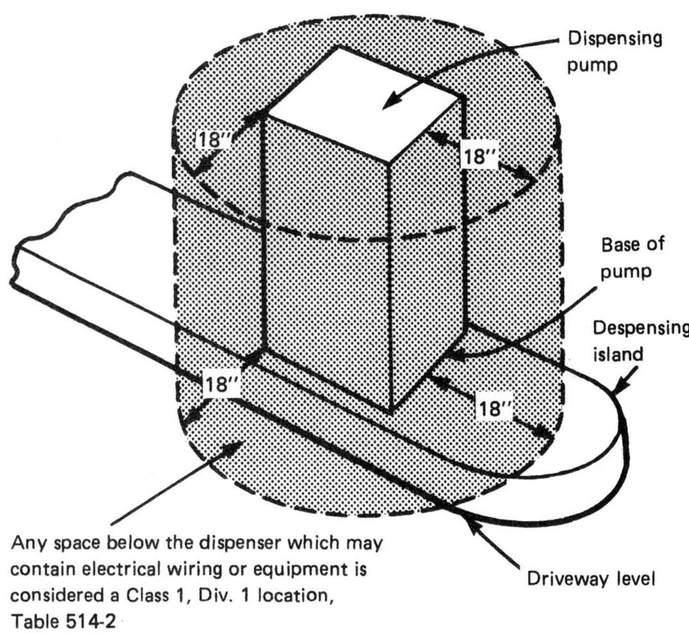

Fig. 514-1. The shaded space around the *outside* of the pump enclosure is a Class I, Division 2 hazardous location. (Sec. 514-2.)

integral with the dispensing hoze or nozzle" are classified as Class I, Division 1 areas. The space extending 18 in. (457 mm) horizontally in all directions from the enclosure and extending down to grade level is classified as a Class I, Division 2 area. The horizontal area, for 18 in. (457 mm) above grade and extending 20 ft (6.1 m) measured from those points vertically below the outer edges of an overhead dispenser enclosure, is also classified as a Class I, Division 2 area. All

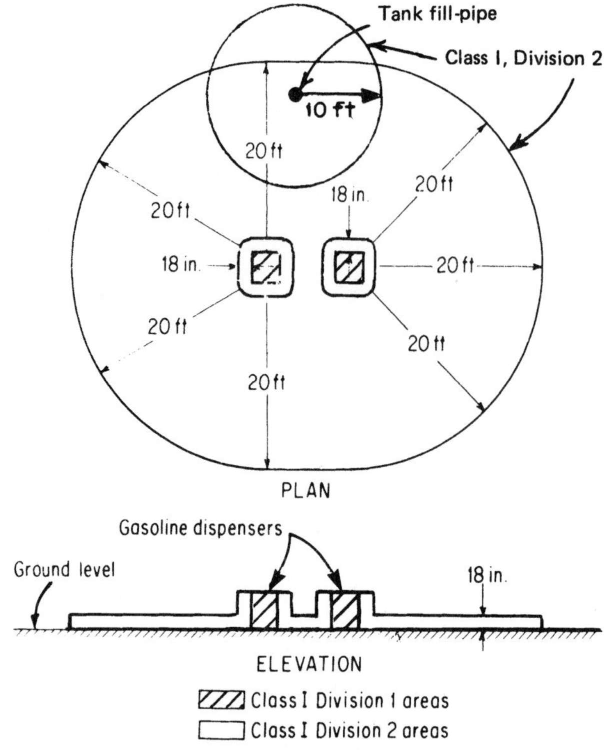

Fig. 514-2. Class I, Division 2 area extends around pumps and tank fill-pipes. (Sec. 514-2.)

equipment integral with an overhead dispensing hose or nozzle must be suitable for a Class I, Division 1 hazardous area.

Table 514-2 notes the hazardous space around any vent pipe for an underground tank at a gas station is simply a 3-ft (914-mm) -radius *sphere* (a ball-like volume; Class I, Division 1) and a 5-ft (1.52-mm) -radius *sphere* (Class I, Division 2) around the top opening of a pipe that discharges upward, but the hazardous space around a pipe opening that does not discharge up includes the sphere described *plus* a cylinder of space from that sphere down to the ground. The space beyond the 5-ft (1.52-m) radius from tank vents that discharge upward and spaces beyond unpierced walls and areas below grade that lie beneath tank vents are not classified as hazardous.

Table 514-2 designates a pit or depression below grade in a lubritorium as a Class I, Division 1 location if it is within an unventilated space, but does allow for the possibility of classification as Division 2, as permitted in Sec. 511-3(b) for a pit in a repair garage where ventilation exists. Refer to Fig. 511-1 on classification of repair area.

514-5. Circuit Disconnects. When the electrical equipment of a pump is being serviced or repaired, it is very important that there be no "hot" wire or wires inside the pump. Since it is always possible that the polarity of the circuit wires may have been accidentally reversed at the panelboard, control switches or CBs must open all conductors, including the neutral.

To satisfy this Code rule, a special panel application is commonly used in gas stations. Figure 514-3 shows how the hookup is accomplished using a gas-

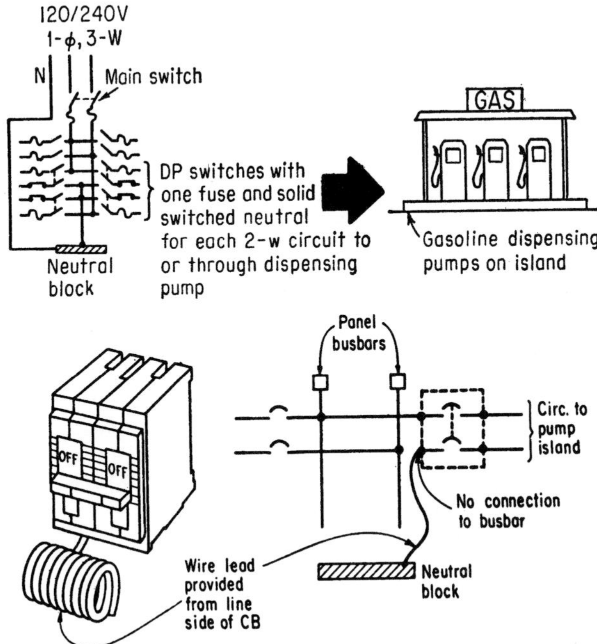

Fig. 514-3. "Gas-station" switches or CBs provide neutral disconnect. (Sec. 514-5.)

station-type panelboard, which has its bussing arranged to permit hookup of standard solid-neutral circuits in addition to the switch-neutral circuits required. Another way of supplying such switched-neutral circuits is with CB-type panelboards for which there are standard accessory breaker units, which have a trip element in the ungrounded conductor and only a switching mechanism in the other pole of the common-trip breaker, as shown. Either 2- or 3-pole units may be used for 2-wire or 3-wire circuits, rated 15, 20, or 30 A. Use of single-pole circuit breakers with handle ties would be a Code violation. No electrical connection is made to the panel busbar by the plug-in grip on the neutral breaker unit. A wire lead connects line side of neutral breaker to neutral block in panel, or two clamp terminals are used for neutral.

514-6. Sealing. Every conduit connecting to a dispenser pump must have a seal in it, as shown in Fig. 514-4. Conduits connecting to gas pumps are commonly connected through an explosionproof junction box that is set in the pump island, as shown in Fig. 514-5. This box is approved as raintight and provided with integral sealing wells. All the conduits connecting to the box are sealed without need for separate individual sealing fittings. Additional individual seals are required where the conduits enter the pump cavity as shown. And, of course, a seal must be used in each conduit that leaves the hazardous area—such as in the conduit that feeds each lighting standard, with no fitting or coupling between the seal in the base of each standard and the boundary at

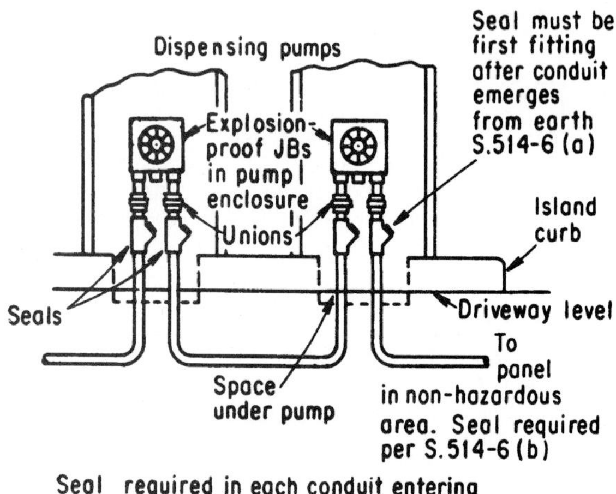

Fig. 514-4. Seal fitting must be used for every conduit at dispenser. (Sec. 514-6.)

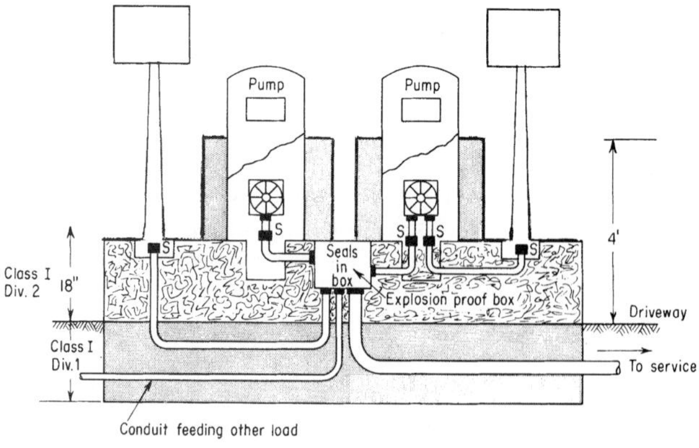

Fig. 514-5. Seals are required in conduits to pumps and to lighting fixtures or signs at points marked "S." (Sec. 514-6.)

the 18-in. (457-mm) height where the circuit crosses into nonhazardous areas. And the conduit at bottom right, which extends back to the panelboard, must also be sealed where it comes up out of the earth at the panelboard location.

The lighting fixtures in Fig. 514-5 must satisfy Sec. 511-7(b), which refers to "fixed lighting" that may be exposed to physical damage—such as impact by a vehicle. If the fixtures are not at least 12 ft (3.66 m) above the ground, they must be totally enclosed or constructed to prevent escape of sparks or hot metal. Section 514-4 specifies that.

In Fig. 514-6, four seals are shown. Normally panelboards are located in a nonhazardous location so that a seal is shown where the conduit is emerging from underground, which Sec. 514-8 identifies as the point where the conduit is leaving the hazardous location.

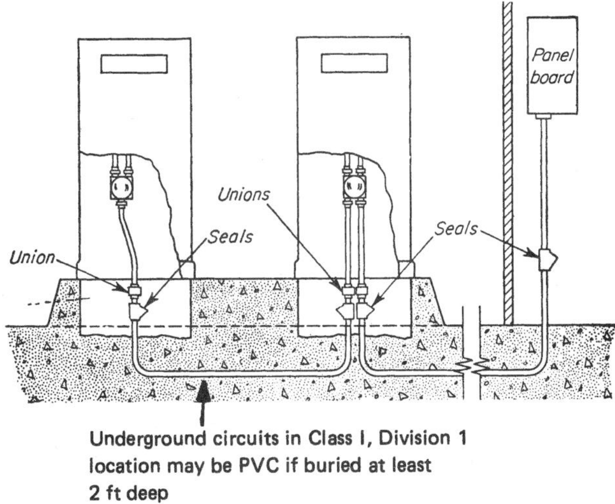
Panel board
Unions
Seals
Seals
Union
Underground circuits in Class I, Division 1 location may be PVC if buried at least 2 ft deep

Fig. 514-6. Conduit from pump island must be sealed at panelboard location. (Sec. 514-6.)

514-8. Underground Wiring. Note that Exception No. 2 of this rule permits rigid nonmetallic conduit for circuits buried at least 2 ft (610 mm) deep in the earth, even though the conduit in the earth may be in a Class I, Division 1 area under a Division 1 or Division 2 area, as noted in the second sentence of this rule. Section 347-3(a) normally excludes rigid nonmetallic conduit from hazardous locations but specifically cites Sec. 514-8 as an exception, as shown at the bottom of Fig. 514-6.

Where rigid nonmetallic conduit is buried at least 2 ft (610 mm) in the ground, as permitted for underground wiring at a gas station, a length of *threaded* rigid metal conduit or *threaded* IMC—at least 2 ft (610 mm) long—must be used at the end of the nonmetallic conduit where it turns up from the 2-ft (610-mm) burial depth. This clarifies that the entire length of the nonmetallic conduit must be down at least 2 ft (610 mm) (Fig. 514-7).

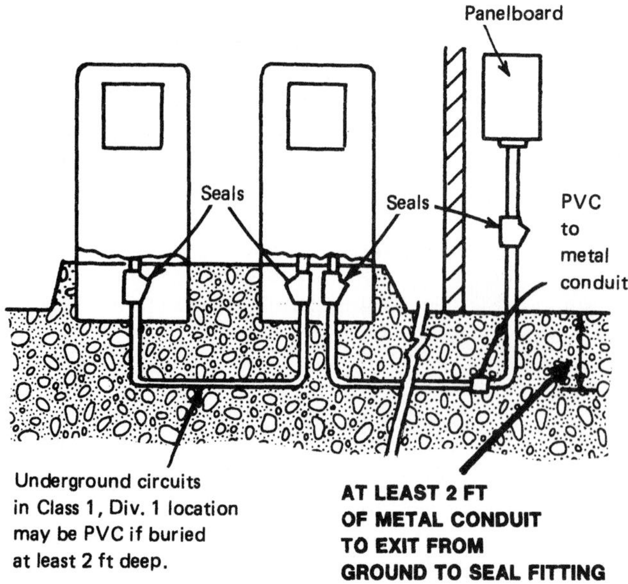

Fig. 514-7. The nonmetallic conduit permitted in the underground Class I, Division 2 location must never come above the required 2-ft burial depth. (Sec. 514-8.)

The phrase in this rule also requires the 2-ft (610-mm) length of metal conduit to be used on the end of the nonmetallic conduit where the conduit run does *not* turn up, but passes horizontally into the nonhazardous area of a basement. In that case or where the conduit turns up, a length of metal conduit is needed to provide for installation of a seal-off fitting because the conduit is, in effect, emerging from the Class I, Division 1 location below ground; and a seal is required at the crossing of the boundary between the classified location and the nonhazardous location in, say, the office or other general area of a gas station.

514.16. Grounding. Because of the danger at gas stations, grounding is very important and the rule here calls for thorough grounding.

ARTICLE 515. BULK STORAGE PLANTS

515-1. Definition. A flammable liquid is said to become volatile when the ambient temperature is equal to, or greater than, its flash point. Typical flash points are gasoline, −45°F; kerosene, 100°F (38°C); diesel oil, 100°F (38°C). Thus, the status of gasoline is definitely established as a volatile flammable liquid regardless of geographical location of the storage facility. Other liquids may change from one state to another depending upon the relation of the ambient temperature to their respective flash points. (See Fire Hazard Properties of Flammable Liquids, Gases and Volatile Solids, NFPA No. 325M.)

515-2. Class I Locations. This section sets its rules in a table indicating the extent of Division 1 and Division 2 locations at various equipment locations. As with the table in Sec. 514-2, this table brings the Code rules into agreement with NFPA No. 30 S/C, General Storage and Handling of Flammable and Combustible Liquids. Figure 515-1 shows rules from Table 515-2, covered under "Pumps, Bleeders, Withdrawal Fittings, Meters and Similar Devices, Indoors."

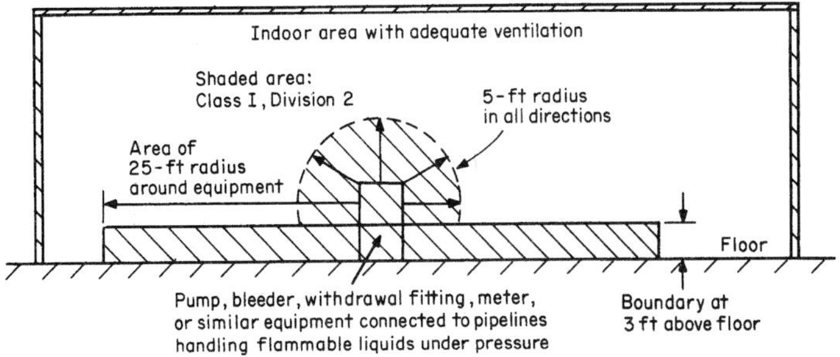

Fig. 515-1. Space around indoor equipment is a Class I, Division 2 location if adequately ventilated; or a Class I, Division 1 location, if not. (Sec. 515-2.)

For outdoor use of the same kinds of equipment the 5-ft (1.52-m) radius is reduced to 3 ft (914 mm), the 25-ft (7.62-m) radius is reduced to 10 ft (3.05 m), and the 3-ft (914-mm) -high level is reduced to 18 in. (457 mm). And where *transfer* of gasoline or similar liquid is done outdoors or in a ventilated indoor place, the space around the vent or fill opening becomes a Class I, Division 1 location for 3 ft (914 mm) in all directions from the opening and a Division 2 location out to 5 ft (1.52 m) from the opening.

The hazardous area around a volatile flammable liquid outdoor storage tank ("Tank-Aboveground" in Table 515-2) extends 10 ft (3.05 m) horizontally beyond the periphery of the tank (Fig. 515-2). Code designation is Class I, Division 2, and wiring installations within this range must conform to Code rules for this category. Space around the vent is a Division 1 location.

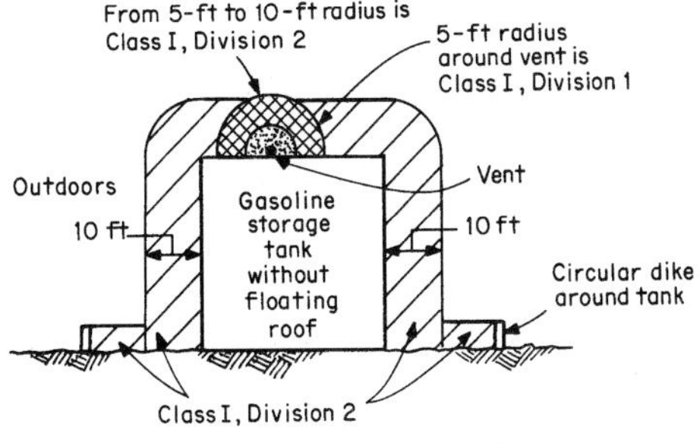

Fig. 515-2. (Sec. 515-2.)

515-5. Underground Wiring. This rule is somewhat similar to that of Sec. 514-8. However, in this article no underground space is designated as a Class I, Division 1 location—as it is in Table 514-2. Therefore, use of nonmetallic conduit or approved cable buried at least 2 ft (610 mm) underground is *not* made in a hazardous location as it is under Sec. 514-2. But the reference that is made to Sec. 515-5 in Sec. 347-3(a) implies that underground wiring at a bulk-storage plant may be in a hazardous location.

ARTICLE 516. SPRAY APPLICATION, DIPPING, AND COATING PROCESSES

516-1. Scope. Note that this article applies to "locations" used for finishing processes—which means open spraying areas as well as enclosed or semienclosed "booths."

The safety of life and property from fire or explosion in the spray application of flammable paints and finishes and combustible powders depends upon the extent, arrangement, maintenance, and operation of the process.

An analysis of actual experience in industry demonstrates that the largest fire losses and greatest fire frequency have occurred where good practice standards were not observed.

516-2. Classification of Locations. Two different types of hazardous conditions are present in a paint-spraying operation: the spray and its vapor which create explosive mixtures in the air and combustible mists, dusts, or deposits. And each must be treated separately.

In part **(a)**, the interior of every spray booth **(1)** and some area spraying **(3)** are Class I or Class II, Division 1 locations depending on whether the atmosphere contains vapors or dusts. When spray operations are not contained within a booth, there is greatly reduced control of flammable atmosphere, and the area of hazard is increased considerably. This is shown in Fig. 516-1, which is a typ-

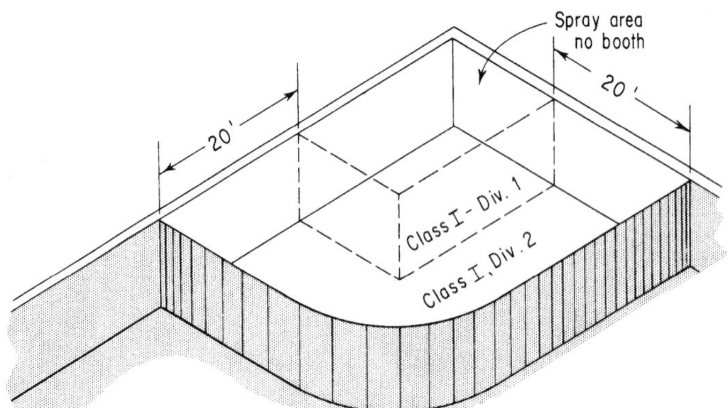

Fig. 516-1. Open spraying involves Division 1 and Division 2 locations, for Class I or Class II conditions. (Sec. 516-2.)

ical specific application of the concept shown in Code Fig. 1 of Sec. 516-2(b)(1). A Class I or Class II, Division 1 area exists at the actual spraying operation plus a Division 2 area extending 20 ft (6.1 m) horizontally in any direction from the actual spraying area and for 10 ft (3.05 m) up. Although only one corner of the room is used for spraying, the entire room area inside the 20-ft (6.1-m) line around the spraying is classified as Class I or Class II, Division 2 and must be wired accordingly. And the Division 2 area extends 10 ft (3.05 m) above the spray operation.

The NFPA "Standard for Spray Finishing" (No. 33) notes that the inspection department having jurisdiction may, for any specific installation, determine the extent of the hazardous "spraying area."

The interior of any enclosed coating or dipping process must be considered a Class I or Class II Division 1 location. A Division 2 location exists within 3 ft (914 mm) in all directions from any opening in such an enclosure.

516-3. Wiring and Equipment in Class I Locations. In part **(a)**, spray operations that constitute a hazardous location solely on the basis of the presence of flammable vapors (and no paint or finish residues) contain wiring and equipment for Class I locations, as specified in Art. 501.

Part **(b)** places tighter restrictions on spray booths or areas where readily ignitible deposits are present in addition to flammable vapors. In general, electrical equipment is not permitted inside any spray booth, in the exhaust duct from a spray booth, in the entrained air of an exhaust system from a spraying operation, or in the direct path of spray, unless such equipment is specifically approved for both readily ignitible deposits and flammable vapor.

Only rigid metal conduit, IMC, and Type MI cable and threaded boxes or fittings containing no taps, splices, or terminal connections may be installed in such locations.

However, for that part of the hazardous area where the fixtures or equipment may not be subject to readily ignitible deposits or residues, fixtures and equipment approved for Class I, Division 1 locations may be installed. The authority having jurisdiction may decide that because of adequate positive-pressure ventilation the possibility of the hazard referred to in paragraph **(b)** has been eliminated.

Sufficient lighting for operations, booth cleaning, and repair should be provided at the time of equipment installation in order to avoid the unjustified use of "temporary" or "emergency" electric lamps connected to ordinary extension cords. A satisfactory and practical method of lighting is the use of ¼-in. (6.35-mm) -thick wired or tempered glass panel in the top or sides of spray booths with electrical light fixtures outside the booth, not in the direct path of the spray. Part **(c)** covers lighting fixtures that illuminate the spray operation through "windows" in the top or walls of a spray booth. Any such fixture used on the outside of the booth must be approved for a Class I, Division 2 location when used in any part of the top or sides of a booth that is within the Division 2 locations as shown in Figs. 2 and 3 in Sec. 516-2 of the Code.

Automobile undercoating spray operations in garages, conducted in areas having adequate natural or mechanical ventilation, may be exempt from the requirements pertaining to spray-finishing operations, when using undercoating materials not more hazardous than kerosene (as listed by Underwriters'

Laboratories in respect to fire hazard rating 30–40) or undercoating materials using only solvents listed as having a flash point in excess of 38°C (100°F). There should be no open flames or other sources of ignition within 20 ft (6.1 m) while such operations are conducted.

ARTICLE 517. HEALTH CARE FACILITIES

517-1. Scope. Although the NEC is *not* intended as a design specification, the requirements for "electrical construction and installation" given in parts **B** and **C** of Art. 517, essentially present design requirements. For example, the rule in Sec. 517-30 mandating segregation of emergency and normal circuits necessitates an additional raceway system with all attendant components. Similarly, the rules for equipment ground-fault protection on the "next level of feeder disconnecting means down stream" as given in Sec. 517-17 present a design requirement. Therefore, in reality this article covers the *design and installation* of electric circuits and equipment in hospitals, nursing homes, residential custodial care facilities, mobile health care units, and doctors' and dentists' offices. *But,* this article does not cover "performance, maintenance, and testing" of electrical equipment in such facilities. Such considerations are covered in other industry standards—such as NFPA 99.

Any specific type of health care location—such as a doctor's office or a dental office—must comply with Code rules whether the location is a sole occupancy itself or is part of a larger facility (like a hospital containing other types of health care locations) or is within a school, office building, etc.

Veterinary facilities are not subject to the requirements of Art. 517.

517-3. Definitions.

Patient Vicinity:

This term provides a definite value for limiting the area—horizontally and vertically—in which special grounding requirements are to be observed in patient care areas.

Psychiatric Hospital:

This is a facility used around the clock to provide only psychiatric care for not less than four resident patients.

Selected Receptacles:

This phrase designates specific receptacles that will provide power to appliances used for patient care emergencies. A dissenting vote in the panel acceptance of this definition noted that the wording would allow task receptacles of any kind to be supplied by the emergency system, even receptacles as unimportant as those for floor cleaners—which is contrary to the basic concept that the emergency system is intended to supply only extremely limited loads.

517-10. Applicability. All the Code rules on "Wiring and Protection" apply to the entire wiring system in hospitals and to "patient-care areas" of clinics, medical and dental offices, outpatient facilities and doctor examining rooms or treatment rooms in nursing homes and residential care facilities. The basic rules of part **B** apply to *all* health care facilities except those areas covered by Exceptions No. 1 and 2.

Exception No. 1 exempts those areas of a health care facility that are *not* intended for examining or treating patients. Areas that are dedicated to other purposes—business offices, corridors, waiting rooms, restrooms, etc.—need not be wired as indicated in part **B.**

The wording used in Exception No. 2 is intended to exclude those health care facilities where patient rooms are used "exclusively" as sleeping quarters. The last sentence further reinforces the idea that such rooms are only permitted to be exempt where there is no intention to ever use the rooms as a treatment area.

517-13. Grounding of Receptacles and Fixed Electrical Equipment. In patient care areas of *all* health care facilities, nonmetallic wiring methods are excluded. Electrical nonmetallic tubing, rigid nonmetallic conduit, and nonmetallic sheathed cable (Romex) may not be used in any part of a hospital or in patient care areas of clinics, medical and dental offices, outpatient facilities, nursing homes, or residential custodial-care facilities. (See Exceptions No. 1 and 2 of Sec. 517-10.)

Part **(a)** requires the use of a separate, insulated equipment grounding conductor run with the branch-circuit conductors in a metal raceway or metal-clad cable from a panelboard to any receptacle or metal surface of fixed electrical equipment operating over 100 V in all health care facilities. But a separate grounding conductor is not required in a feeder conduit to such a panel. For feeders, the metal conduit is a satisfactory grounding conductor, as recognized generally in Sec. 250-91(b). But for all branch circuits to "receptacles and all . . . fixed equipment . . ." in "areas used for patient care," neither metal conduit, jumpers with box clips (G-clips), nor a receptacle with self-grounding screw terminals (Sec. 250-74, Exception No. 2) may be used alone without the grounding wire run with the branch-circuit wires (Fig. 517-1), which must be in a metal raceway or in Type MC, Type MI, or Type AC cable (so-called BX). But, those metal-clad cables must have an "outer metal armor or sheath of cable" that "is identified as an acceptable grounding return path." Type AC cable, Type MI cable, and Type MC cable with a smooth or corrugated continuous metal sheath all satisfy that grounding requirement. Type MC with a spiral-wrap metal sheath with two grounding conductors—one of which is insulated copper—has been a source of controversy.

This controversy centers on the fact that Sec. 250-91(b)(8) specifically recognizes the spiral metal sheath in conjunction with the internal equipment grounding conductor as a type of equipment ground. And, it certainly seems that where an additional grounding conductor is provided and at least one of the two is insulated copper, the requirement for redundant grounding—which is the objective here—has certainly been satisfied.

The present wording is intended to prohibit the use of Type MC with two grounding conductors; however, some inspectors feel the wording used does *not* do so. "Identified" is defined in Art. 100 as "Recognizable as suitable." And, those inspectors feel that the specific mention of Type MC in conjunction

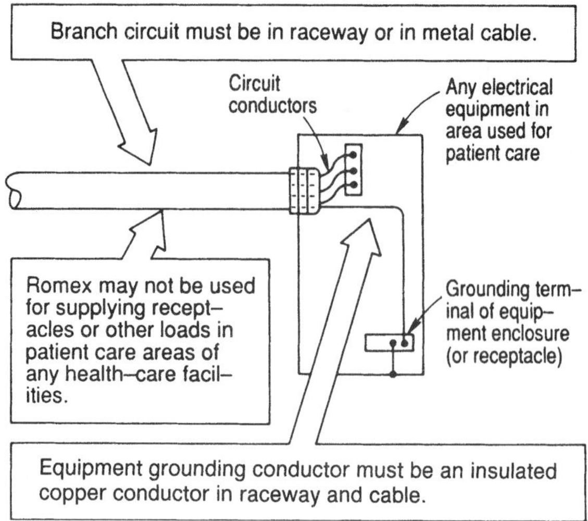

Fig. 517-1. Grounding conductor run with branch circuit in metal raceway or cable must ground receptacles and equipment. (Sec. 517-13.)

with its grounding *is* a ground return by Sec. 250-91(b)(8), which essentially makes Type MC with one ground "recognizable"—and there "identified"—as one return path and one additional insulated copper conductor within a Type MC cable. Therefore, it satisfies the rule. Check with the inspection authority before using Type MC with two ground wires because not all are accepting such installation.

It should be noted that the phrase requiring an insulated equipment grounding conductor within these type cables (which appeared in Exception No. 1 to Sec. 517-11 in the 1987 NEC) has been deleted. But it is the intent of the Code Making Panel to require an insulated copper equipment grounding conductor, sized in accordance with Sec. 250-95, to be run in any such cable assembly in a patient care area in every health care facility.

The ground terminal of receptacles must be grounded to an equipment grounding conductor run in a *metal raceway or metal-covered cables.* Either metal raceway or metal cable must be used for circuits in patient care areas of hospitals, clinics, medical and dental offices, outpatient facilities, nursing homes, and residential custodial care facilities—*always* with an insulated *copper* equipment grounding conductor included in the raceway or cable.

This rule applies to "areas used for patient care"—which, in hospitals, covers patient bedrooms and any other rooms, corridors, or areas where patients are treated, like therapy areas or EKG areas. But, for other than hospitals, Exception No. 1 of Sec. 517-10 excludes waiting rooms, admitting rooms, solariums, recreation areas, as well as business offices and other places used solely by medical personnel or where a patient might be present but would not be treated.

Exception No. 2 to this rule clarifies the use of metal faceplates on wall switches or receptacles without actually connecting an "insulated copper conductor" to each faceplate. They are acceptable as grounded simply by screw connection to a grounded box or grounded mounting strap of a grounded wiring device.

Exception No. 3 excludes lighting fixtures from the rule for grounding by an insulated copper conductor, provided the fixture is mounted "more than 7½ ft (2.29 m) above the floor." Although fixtures so located do *not* need an insulated grounding conductor, they must be fed by a conduit or cable that satisfies part **(b)** of this section.

Part **(b)** of this section emphasizes that a redundant metallic grounding path is required in patient care areas. Part **(b)** requires that *all* branch circuits supplying patient care areas must be run in a metal-enclosed wiring method—rigid metal conduit, IMC, EMT, or MI, MC, or AC cable—to provide a redundant metallic grounding path in parallel with the insulated copper ground wire required by part **(a)** in the wiring method. This rule emphasizes the need for high reliability of the ground-fault current return path as major protection against electrical shock.

517-14. Panelboard Bonding. Normal and essential electrical system panelboards serving either the same general care or critical care patient location must have their equipment grounding terminal bars bonded together with an insulated, continuous copper bonding jumper not smaller than No. 10 AWG. Although required to be "continuous," the wording of this rule recognizes terminating this conductor at ground buses and terminals as satisfying the requirement for a "continuous" conductor.

517-16. Receptacles with Insulated Grounding Terminals. Insulated-ground receptacles must be clearly and externally identified. This rule applies to those receptacles that have their grounding terminals insulated from the metal of the box and conduit.

517-17. Ground-Fault Protection. At least one additional level of ground-fault protection is required for health care facilities where ground-fault protection is used on service equipment (see Sec. 230-95). Where the installation of ground-fault protection is made on the normal service disconnecting means, each feeder must be provided with similar protective means. This requirement is intended to prevent a catastrophic outage. By applying appropriate selectivity at each level, the ground fault can be limited to a single feeder, and thereby service may be maintained to the balance of the health care facility.

As shown in Fig. 517-2, with a GFP (ground-fault protection) hookup on the service, a GFP hookup must be put on each feeder derived from the service.

The second paragraph is aimed at ensuring that essential systems are *not* isolated when the additional level of feeder GFP is actuated. This requirement previously appeared as a FPN but now is included in the rule itself. Consequently, it is no longer advisory, but rather mandatory. Installation of GFP on the load-side of an emergency transfer switch, at the output of a generator, or on *any* system rated for voltages other than 480Y/277 V or 600Y/347 V is expressly prohibited. And part **(b)** requires that selection of the tripping time of the main GFP be such that each feeder GFP will operate to open a ground fault on the

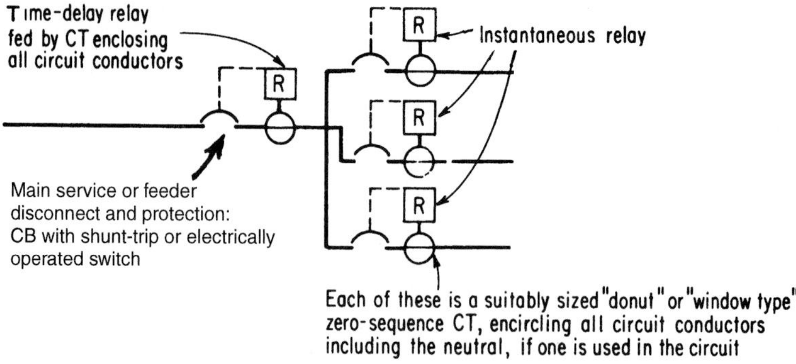

Fig. 517-2. GFP on the service requires GFP on main feeders also. (Sec. 517-17.)

feeder, without opening the service GFP. And a time interval of not less than 0.1 sec (i.e., the time of six cycles) must be provided between the feeder GFP trip and the service GFP trip. As shown, if the feeder GFP relays are set for instantaneous operation, the relay on the service GFP must have at least a 0.1-sec time delay. A zone-selective GFP system with a feedback lock-out signal to an instantaneous relay on the service could satisfy the rule for selectivity.

517-18. General Care Areas. Two circuits must supply each bed used for inpatient care. But two branch circuits are not required for each patient bed in nursing homes, outpatient facilities, clinics, medical offices, limited care facilities, and the like. "Psychiatric, substance abuse, and rehabilitation hospitals" are also exempted from the branch-circuit and receptacle requirements for general care patient bed locations.

As noted in part **(b)**, receptacles at patient bed locations in "General Care Areas" must be "listed hospital grade" and "so identified." The minimum of six required receptacles at each such bed location may be single or duplex types, or a combination of the two. (Three duplex receptacles provides a total of six receptacles.) *All* receptacles at *all* patient bed locations—general care and critical care—must be "listed hospital grade" devices.

As noted in part **(c)**, only tamper-resistant receptacles are permitted in pediatric locations. This rule requires that all 15- or 20-A, 125-V receptacles in pediatric locations be "tamper resistant." Although the **Code** does not contain a definition for that word and the UL *Electrical Construction Materials Directory* does not refer to "tamper-resistant" receptacles, the last phrase of part **(c)** gives an indication of its meaning. It seems clear that a tamper-resistant receptacle is one that would make it extremely difficult, if not impossible, to insert a pin, paper clip, or similar small metal object into a slot on the receptacle and make contact with an energized part. Obviously, the concern here is to protect infants or children from shock hazard as a result of playful or inadvertent tampering with the receptacle.

The Exception recognizes the use of covers instead of "tamper-resistant" receptacles. It is no longer required that protection against tampering be provided by design of the receptacle and not by attachment of an accessory device, such as a plastic plug, which must be removed to use the receptacle. Receptacles are available with rotating slot covers or internal contact mechanisms, both of which make it necessary to use a cord plug cap to gain access to energized parts. But, caps that limit "improper access" may *now* be used.

517-19. Critical Care Areas. Patient bed locations in general care areas (Sec. 517-18) must be supplied by six single or three duplex receptacles, just as critical care area patient beds must be provided with at least six receptacles (single or duplex devices totaling six points for connecting a cord plug cap). The two or more branch circuits to each critical care area patient bed location must include one or more from the emergency system and one or more from the normal system (Fig. 517-3). In both cases, at least two branch circuits must supply these receptacles. In the case of general care areas, additional receptacles serving other patient locations may be served by these branch circuits, but in the case of critical care areas at least one of these branch circuits is required to be an individual branch having no other receptacles on it except those of a single bed location.

As covered in part **(b)**, "hospital grade" receptacles must be used at patient bed locations in "Critical Care Areas." Six single or three duplex receptacles (or any combination totaling six receptacles) that are UL-listed as "Hospital Grade" devices must be used at each patient bed location and must be so identified at each patient bed location in "Critical Care" patient areas. The best point to call a reference grounding point is the grounding bus in the distribution panel, which is the transition connection point between the branch-circuit grounding wires and the feeder grounding system.

The meaning of part **(c)** is that a "patient equipment grounding point" (defined as a "point for redundant grounding of electric appliances") is *not* mandatory. Such a grounding point may be used, if desired, and connected as described in the rule. In its original application, years ago, the "patient equipment grounding point" was a special jack used to make a grounding connection to the metal enclosures of electrical medical equipment because, at the time, many power cords did not contain a grounding conductor. In addition, this was a carryover from procedures for operating rooms, where all metal surfaces had to be grounded to minimize static charge buildup. With today's universal use of good 3-wire (grounding type) power cords and plugs or double-insulated equipment, there is no real justification for requiring this patient equipment grounding jack. Each three-contact receptacle, in effect, becomes a patient equipment grounding point.

When a patient equipment grounding point is used in a patient vicinity, it must be grounded to the ground terminal of *all* grounding-type receptacles in the patient vicinity by means of a minimum No. 10 copper conductor looped to all of the receptacles or by individual No. 10 conductors run from the patient grounding point to each receptacle.

Regardless of what additional methods are employed, in order to keep potential differences within the required limits, equipotential grounding is essential to the electrical safety of critical care areas. Some of the earliest equipotential grounding installations consisted of copper busbars run around the walls of patient rooms to which furniture and equipment were attached by means of

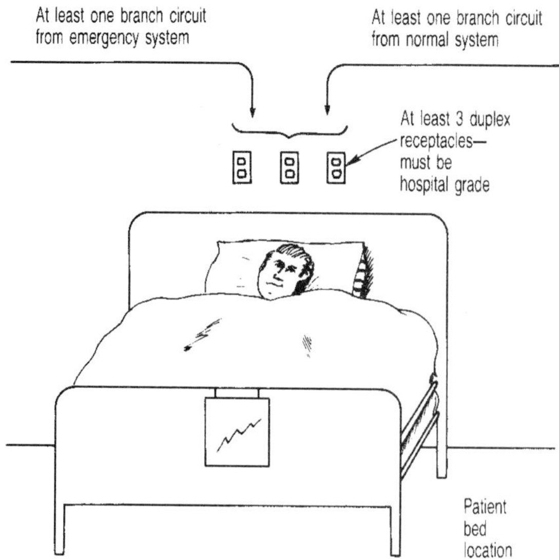

Sec. 517-19. Both emergency system and normal system must supply branch circuits to critical-care bed location.

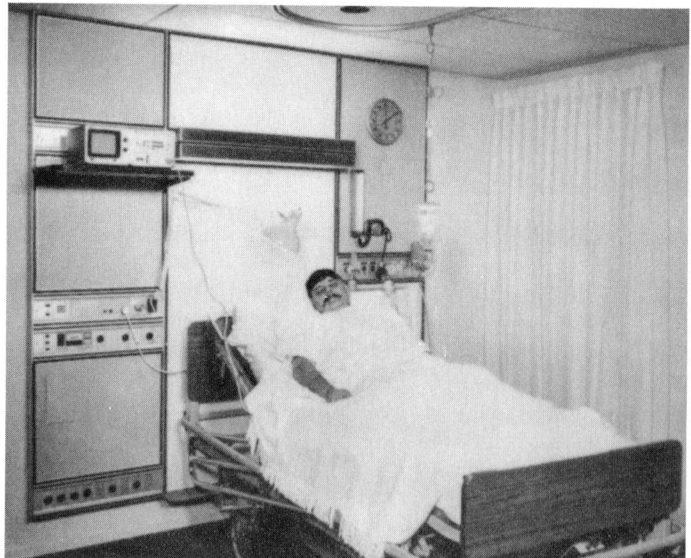

Fig. 517-3. Multipurpose patient-care modules may incorporate a variety of the circuit, receptacle, and grounding requirements for patient care areas. In addition to a patient equipment grounding point and room bonding point, such a preassembled unit might include facilities for communication, patient monitoring, lighting, and lines for air, water, and medical gas. (Sec. 517-19.)

grounding jumpers. Based on experience obtained through these early installations as well as the refinements produced by the NFPA Committee on Hospitals, the **National Electrical Code** now contains the requirements which correlate with the pertinent NFPA standards on the subject. At the same time these new requirements also permit the achievement of the desired end with a minimum of expenditure in labor and materials.

Objection has been raised to the concept of grounding every piece of exposed metal in sight. Doing this may actually increase the hazard. Because a shock occurs when a person touches two surfaces with a voltage difference between the surfaces, the fewer surfaces that are deliberately grounded the better. Thus, a door frame or window frame that is not likely to become energized is not required to be grounded. It was not good safety engineering to propose that metal furniture in a patient's room be deliberately grounded. Figure 517-4 shows the type of grounding and building "points" that past editions of the **Code** regulated, along with typical grounding hookups in older hospitals.

As required by part **(d)** of this section, a bonding-type connection is required for *feeder* metal conduit or metal cable (Type MC or MI) terminations—using a bonding bushing plus a *copper* bonding jumper from a lug on the bushing to the ground bus in the panelboard fed by the conduit (see illustrations in Sec. 250-72). Bonding locknuts or bonding bushings may be used on clean knockouts (all punched, concentric, or eccentric rings removed), and bonding may be provided by threaded connection to hubs or bosses on panel- or switchboard enclosures. This is required for feeder conduits but not for branch-circuit conduits. And it seems clear that bonded terminals are required at both ends of each and every feeder to a panelboard that serves the critical care area (Fig. 517-5).

Part **(e)** makes use of an isolated power system for critical care areas a completely optional technique, simply noting that such systems are "permitted" to be used if the design engineer or the hospital-client wants it. That approach ties in with the deletion of the maximum potential difference of 100 mV "under conditions of line-to-ground fault" in a critical care area—which in past **Codes** made the isolated power system mandatory. Grounding of this optional power system must satisfy part **(f)**.

Part **(g)** covers receptacles that are intended for use by specific pieces of equipment. Those receptacles intended for special purposes—e.g., "mobile x-ray equipment"—must have their equipment grounding conductor "extended" to the reference point within the branch-circuit panelboard that supplys the patient area.

517-20. Wet Locations. As covered by this section, locations intended for ground-fault protection are limited to patient care areas. So, even though the governing body of the hospital may wish to extend this form of protection to such areas as laundries, boiler rooms, and kitchens, the fact that these are not considered patient care areas does not make GFCI (ground-fault circuit interrupter) protection mandatory in such locations. The designer and/or hospital authorities must designate such "wet locations." Locations which are intended for protection would include hydrotherapy, dialysis facilities, selected wet laboratories, and special-purpose rooms where wet conditions prevail.

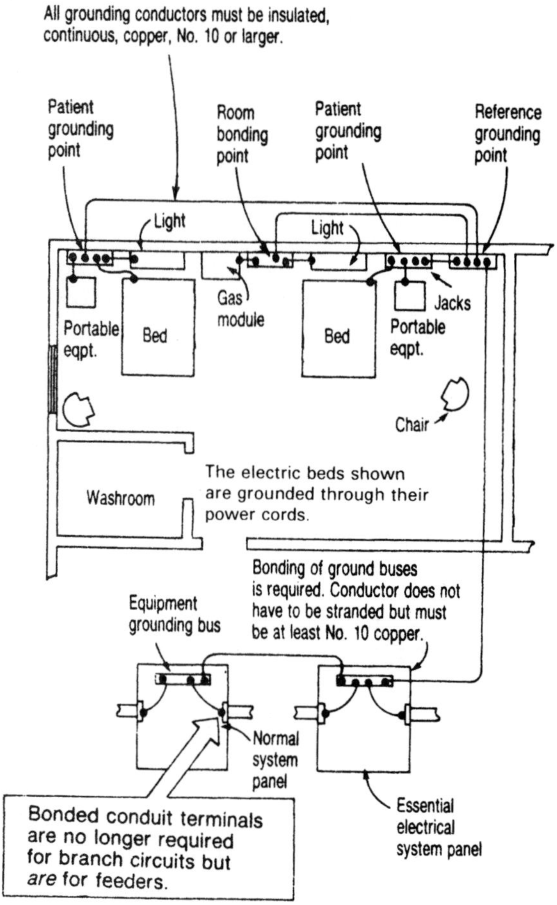

Fig. 517-4. The grounding techniques shown here were regulated by past editions of the NEC but are now optional at the patient bed locations. [Sec. 517-19(c).]

All receptacles (of any rating) *and fixed electrical equipment* at a wet location must have GFCI protection, where power interruption "can be tolerated." Otherwise, they must be fed from an isolated power supply. The only Exception eliminates the need for GFCI protection or supply from isolated power for "listed, fixed" therapeutic and diagnostic equipment.

517-21. Ground-Fault Circuit-Interrupter Protection for Personnel. This rule correlates with the rules in Secs. 210-8(b)(1) and 210-7(d)(2). Those two Code sections deal with GFCI protection of 15- and 20-A, 125-V receptacles installed in new bathrooms and replacements made in existing bathrooms in occupancies other than dwellings. The intent of Sec. 517-21 is to exempt receptacles installed in critical care areas—whether new installations or where replace-

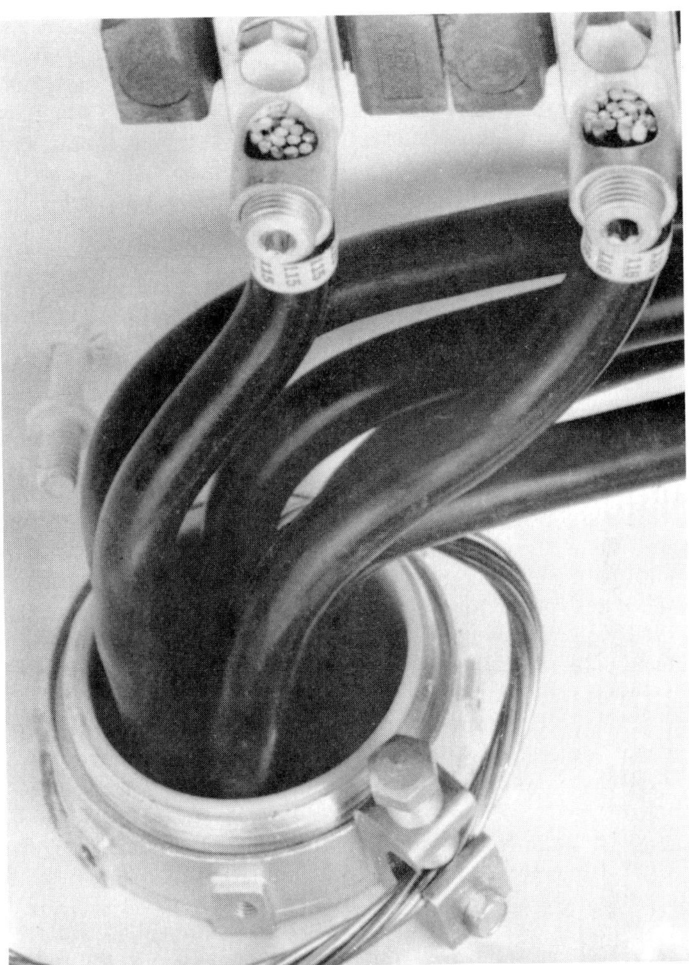

Fig. 517-5. Feeders to branch-circuit panelboards for critical care areas must be bonded. Figure 517-4 covers rules on grounding and bonding for the patient vicinity of a critical care area, with feeder conduit bonding as in photo above. [Sec. 517-19(d).]

ment is made in an existing installation—from being GFCI-protected where a toilet and basin is within the same room. Literally, the rules of Secs. 210-8(b)(1) and 210-7(d)(2) would mandate GFCI protection for any 15- and 20-A, 125-V receptacles installed in such a location. *But,* for any "general care" patient bed location where the toilet and basin are in the room, and for bathrooms separate from the care area—critical or general—GFCI protection *is* required for 15- and 20-A, 125-V receptacles.

517-25. Scope. Essential electrical systems are covered for hospitals, clinics, medical and dental offices, outpatient facilities, nursing homes, residential custodial care facilities, and other health care facilities for patient care.

517-30. Essential Electrical Systems for Hospitals. This section makes clear that essential electrical systems which are to be installed in hospitals must observe the rules of part **(c)**. Essential electrical systems in hospitals are subdivided into the emergency system (consisting of the life safety and critical branches) and the equipment system, whereas essential electrical systems for nursing homes and the like are the branches shown in Code diagrams 517-41(A) and (B) for the emergency system. It should be noted that the critical branch in hospitals comprises different equipment and connections than does the critical system in nursing homes.

In Sec. 517-30(c)(3), although wiring of the emergency system is required to be installed in a "metal raceway," Exception No. 3 permits Schedule 80 rigid nonmetallic conduit for such circuits. Only the Schedule 80 version of rigid PVC conduit may be used without concrete encasement. Exception No. 4 says Schedule 40 PVC conduit and electrical nonmetallic tubing is permitted for other than branch circuits in patient care areas—where encased in no less than 2 in. (50.8 mm) of concrete. *But* note that although such a raceway may be used for emergency circuits, it is *not permitted* for such circuits that are "branch circuits serving patient care areas."

Exception No. 5 allows limited use of Type MI cable instead of metal raceways to supply branch circuits in patient care areas provided the metal armor permits redundant grounding in accordance with Sec. 517-13. And Exception No. 6 formally recognizes flexible metal conduit and cable assemblies in prefabbed "headwalls" and where flexibility is needed.

517-31. Emergency System. Code diagrams 517-30(A), (B), and (C) in the Code book clarify interconnections and transfer switches required. Handbook Table 517-1 summarizes the loads supplied by the hospital emergency system, which must restore electrical supply to the loads within 10 sec of loss of normal supply.

517-41. Essential Electrical Systems. In part **(b)**, a single transfer switch may be used for the entire essential electrical system instead of using a separate transfer switch for each branch, as shown in Code diagram 517-41(C). One transfer switch may supply one or more branches of the essential electrical system in a nursing home or residential custodial care facility where the essential electrical system has a maximum demand of 150 kVA. Separate transfer switches are required only if dictated by load considerations. For small facilities, the essential electrical system generally consists of the life safety branch and the critical branch. For larger systems the critical branch is divided into three separate branches for patients, heating, and sump pumps and alarms. Code diagrams 517-41(A) and 517-41(B) illustrate typical installations.

517-42. Automatic Connection to Life Safety Branch. Part **(a)** describes the switching arrangements for night transfer of corridor lighting. The rule is intended to assure that some lighting will always be provided in the corridor regardless of the mode of operation.

517-43. Connection to Critical Branch. This section details the loads requiring transfer from normal source to the alternate power source. In part **(b)(2)**, elevator operation must not trap passengers between floors. This rule, which was only a recommendation in the 1978 Code, is now mandatory.

517-60. Anesthetizing Location Classification. Figure 517-6 shows the classified hazardous locations of part **(a)**.

Note that part **(b)** requires designation by the hospital administration that a particular location (operating room, anesthesia room, etc.) is nonhazardous. Section 5-2 of NFPA 56A requires signs *prohibiting* the use of flammable anesthetics.

Hazardous locations rules are separated from those for other-than-hazardous locations. A third category is also designated—"above-hazardous locations."

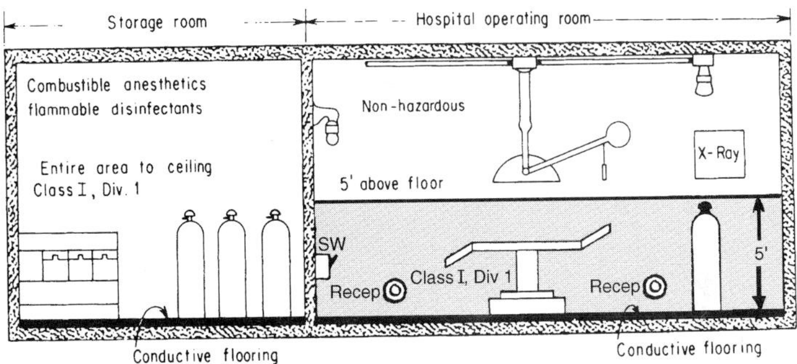

Fig. 517-6. Two types of hazardous locations must be identified. (Sec. 517-60.)

Table 517.1 A Hospital Emergency System Must Serve These Loads (Sec. 517.31)

Life Safety Branch	Critical Branch
Lighting and receptacles for: —Means of egress illumination —Exit and directional signs —Alarm systems •Manual fire stations •Sprinklers •Fire & smoke detection —Alarms for nonflammable medical gas —Communications for emergency use —Generator—set location	—Isolating transformers in anesthetizing locations —Task illumination and selected receptacles in: •Nurseries •Medication preparation •Pharmacy •Acute nursing •Psychiatric beds (no receptacles) •Nurses stations •Ward treatment rooms •Surgery and obstetrics •Angiographic labs •Cardiac catheter labs •Coronary care •Delivery •Dialysis •Emergency •Human physiology labs •Intensive care •Operating rooms •Post-operative •Recovery rooms

The Code covers wiring and equipment in three relations to anesthetizing locations: Sec. 517-61(a) (within-hazardous), Sec. 517-61(b) (above-hazardous), and Sec. 517-61(c) (other-than-hazardous).

517-61. Wiring and Equipment. Part **(a)** calls for explosionproof wiring methods, in general, for such locations.

Part **(a)(4)** notes an extension of the hazardous boundary. Section 517-60(a)(1) defines the area of a flammable anesthetizing location as a Class I, Division 1 location from the floor to a point 5 ft (1.52 m) above the floor. The question then arises, Is the seal required in the upper conduit entering the switch box on the wall of a hospital operating room as shown in Fig. 517-7? The box is partly below and partly above the 5-ft (1.52-m) level.

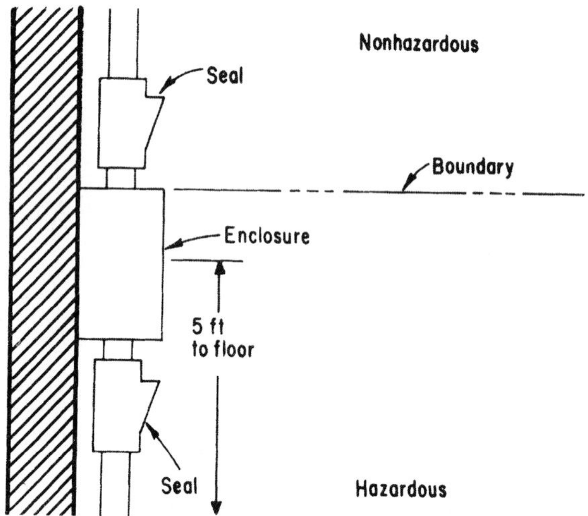

Fig. 517-7. Boundary of Class I, Division 1 location may be extended. (Sec. 517-61.)

Part **(a)(4)** states that, if a box or fitting is partially, but not entirely, beneath the 5-ft (1.52-m) level, the boundary of the Class I, Division 1 area is considered to extend to the *top* of the box or fitting. Therefore, the box or fitting is entirely within the hazardous area, and a seal is required in conduit entering the enclosure from either above or below, as shown in Fig. 517-7.

If the box or fitting is entirely *above* the 5-ft (1.52-m) level, a seal would not be required at the box or fitting, but conduit running to the box from the hazardous area would have to be sealed at the boundary, on the hazardous-location side of the box. If the box shown were recessed in the wall instead of surface-mounted, some means would have to be provided to make the seals accessible [Sec. 501-5(c) (1)], such as removable blank covers at the locations of the seals.

Part **(a)(5)** calls for explosionproof receptacles and plugs within hazardous locations described in Sec. 517-60(a).

In part **(b)(1)** of this section, rigid metal conduit, electrical metallic tubing (EMT), and intermediate metal conduit (IMC) or Type MI or MC cable that has a continuous "gas/vaportight" sheath are permitted in "above-hazardous anesthetizing locations." In "other-than-hazardous anesthetizing locations" [Sec. 517-61(c)(1)], wiring may be in rigid metal conduit, EMT, or IMC—or may be Type MC, Type AC, or MI cable. Again, the armor on the Type MI or MC cable must be "identified" as a ground return path. But rigid nonmetallic conduit may not be used for wiring in anesthetizing locations that use nonflammable anesthetics. And it may not be used in or above a hazardous anesthetizing location. Many of the requirements for safety in the above-hazardous anesthetizing locations are not included for other-than-hazardous locations.

Hospital-Grade Receptacles

The NE Code generally requires that "hospital-grade" receptacles (the UL-listed "Green-dot" wiring devices) be used at patient bed locations in health care facilities. However, it should be noted that Secs. 517-61(b)(5) and 517-61(c)(2) of the Code also require use of "receptacles and attachment plugs" that are "listed for hospital use" *above* "hazardous" anesthetizing locations and *in* "other-than-hazardous" anesthetizing locations. As described, those rules require that all 2-pole, 3-wire grounding-type receptacles and plugs for single-phase 120-, 208-, or 240-V AC service must be marked "Hospital Only" or "Hospital Grade" and have a a green dot on the face of each receptacle (Fig. 517-8). The relation between the phrase "listed for hospital use" and the phrase "Hospital Grade" is explained in the UL *Electrical Construction Materials Directory* (the UL Green Book), under the heading "Attachment Plug Receptacles and Plugs." It says:

> Receptacles listed for hospital use in other than hazardous locations in accordance with Article 517 of the National Electrical Code are identified (1) by the marking "Hospital Only" or (2) by the marking "Hospital Grade" and a green dot on the receptacle. The green dot is on the face of the receptacle where visible after installation.

Of course, in the defined hazardous areas of flammable anesthetizing locations [Secs. 517-60 and 517-61(a)(5)], receptacles must be explosionproof type, listed for Class I, Division 1 areas.

Underwriters' Laboratories Inc. devised a special series of tests for wiring devices intended for hospital use. These tests are substantially more abusive than those performed on general-purpose devices and are designed to ensure the reliability of the grounding connection in particular, when used in the hospital environment. Hospital-grade receptacles have stability and construction in excess of standard specifications and can stand up to abuse and hard usage. Devices which pass this test are listed as "Hospital Grade" and are identified with these words and a green dot, both of which are visible after installation. UL listings include 15- and 20-A, 125-V grounding, nonlocking-type plugs, receptacles, and connectors. This class of device is acceptable for use in any nonhazardous anesthetizing location. As noted in the Exception to Sec. 517-

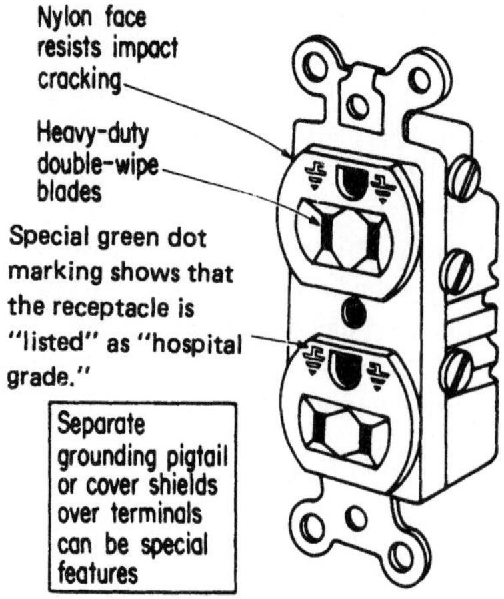
Nylon face resists impact cracking
Heavy-duty double-wipe blades
Special green dot marking shows that the receptacle is "listed" as "hospital grade."
Separate grounding pigtail or cover shields over terminals can be special features

Hospital-grade plugs and receptacles are required *only* for anesthetizing locations. (Sec. 517-61.)

Sec. 517-18(c). Receptacles in pediatric and psychiatric locations must have "construction" that prevents improper contact

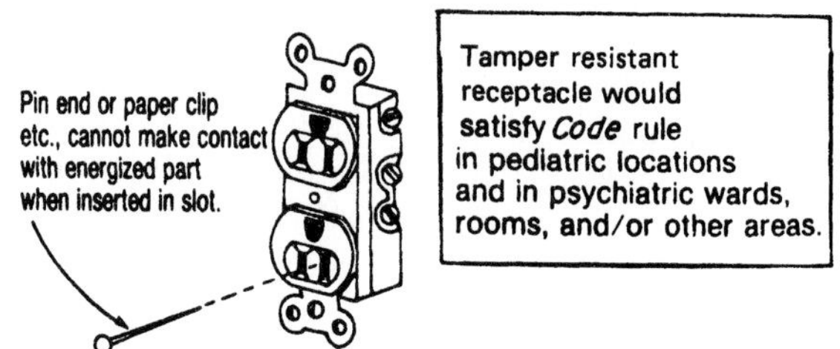
Pin end or paper clip etc., cannot make contact with energized part when inserted in slot.
Tamper resistant receptacle would satisfy *Code* rule in pediatric locations and in psychiatric wards, rooms, and/or other areas.

Fig. 517-8. Hospital-grade plugs and receptacles are required *only* for anesthetizing locations. (Sec. 517-61.)

61(b)(2), receptacles above a hazardous location do not have to be totally enclosed and may be standard, available hospital-grade receptacles of the type that would be used to satisfy the rule of Sec. 517-61(b)(5).

As covered in part **(c)**(1), in anesthetizing locations that are not hazardous (no flammable agents used), Type AC cable is recognized along with Types MI and MC cable and rigid metal conduit, IMC, and EMT—but any such cable or raceway must contain an insulated-copper equipment grounding conductor and its outer jacket must be an approved grounding conductor. Note that although the rule here does not specify a "copper" conductor, because an anesthetizing location is a patient care area, the copper wire is required by Sec. 517-13.

517-63. Grounded Power Systems in Anesthetizing Locations. Part **(a)** calls for a general-purpose lighting circuit, fed from the normal grounded service, to be installed in each operating room. And the Exception recognizes feed from an emergency generator or other emergency service that is separate from the

Fig. 517-9. Altough the NE Code requires use of hospital-grade receptacles at patient bed locations and in "anesthetizing" locations, the ruggedness and high degree of connection reliability strongly recommend their use for such critical applications as plug-connection of respiratory or life-sustaining equipment.

source of the hospital's "Emergency System," as defined in Sec. 517-2. Figure 517-10 shows a layout where such an emergency supply (at lower right) may be the source of supply to the general-purpose lighting circuit in the operating room because it is a separate supply from that to the Emergency System (at left).

In part **(f)**, an isolated power system (ungrounded operation with a line isolation monitor) is required only in an anesthetizing location with flammable anesthetics. As the rule has appeared in recent editions of the NEC, isolated power systems have been required for locations with both flammable and nonflammable anesthetics. But in 1984, a revision was made in Sec. 517-104(a)(1) to change the phrase an "anesthetizing" to a "flammable anesthetizing" in this section. As a result, an anesthetizing location that does not use flammable anesthetics will not require use of an isolated power system. Section 517-63(f) now makes the requirement for an isolated power system applicable only for anesthetizing locations where "flammable" anesthetics are used. For those cases where the anesthetizing location is used solely for nonflammable anesthetizing materials, an isolated power system is not mandatory in the NEC (although NFPS 99 would require it for wet anesthetizing locations).

Figure 517-11 shows the application of a completely packaged transformer loadcenter to provide power for the ungrounded, isolated circuits in hospital operating suites.

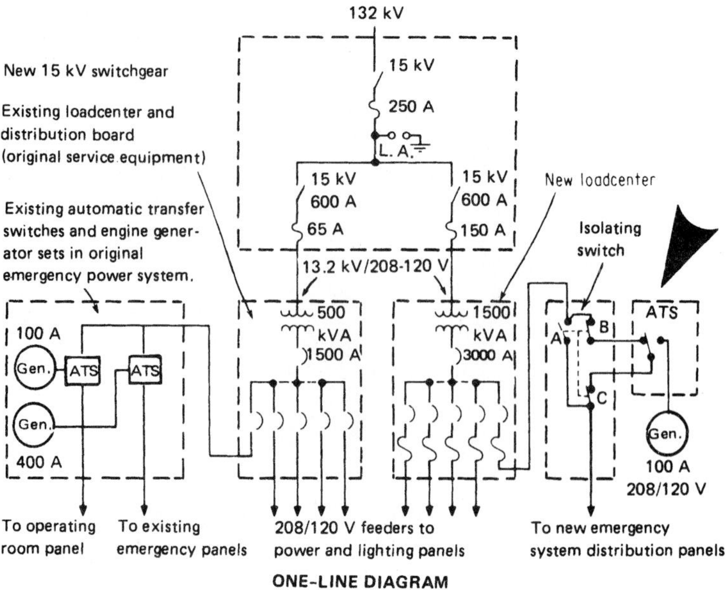

Fig. 517-10. General-purpose lighting circuit must be fed from normal service or separate alternate source. (Sec. 517-63.)

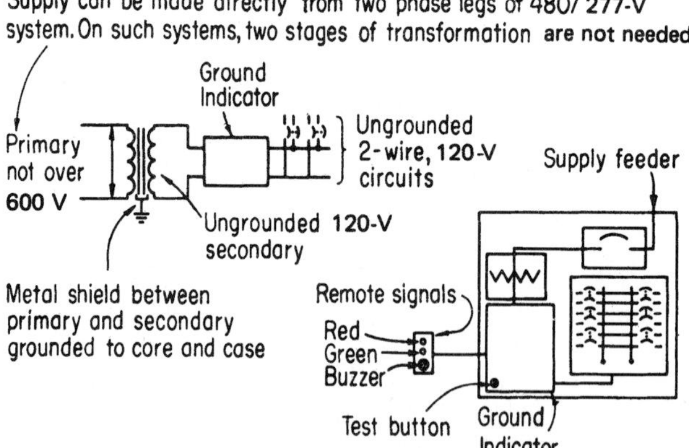

Fig. 517-11. Isolated power supply is required for circuits only in flammable anesthetizing locations. (Sec. 517-63.)

517-64. Low-Voltage Equipment and Instruments. Specific details are given for use of low-voltage equipment in an anesthetizing location. Figure 517-12 shows some of the rules. Section 517-160(a)(2) limits isolating transformers to operation with primary at not over 600 V.

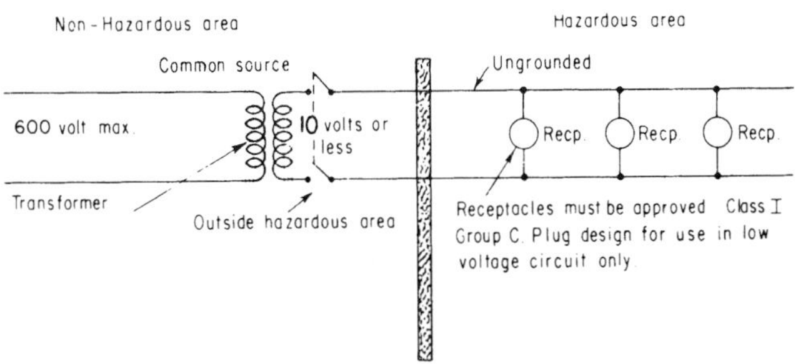

Fig. 517-12. Low-voltage circuits in anesthetizing locations must operate at 10 V or less or be otherwise approved. (Sec. 517-64.)

517-160. Isolated Power Systems. Each isolated power circuit must be controlled by a switch that has a disconnecting pole in each isolated circuit conductor to simultaneously disconnect all power. That is in part **(a)(1)**.

As covered in part **(a)(2)**, any transformer used to obtain the ungrounded circuits must have its primary rated for not more than 600 V between conductors and must have proper overcurrent protection. This **Code** rule used to limit the transformer primary to 300 V between conductors, which often required two stages of voltage transformation to comply with the rule, as shown in Fig. 517-13. That diagram shows the circuit makeup used in a hospital to derive the 120-V ungrounded circuits, with transformation down from 480 to 240 V and then to 120 V. The ungrounded secondary system must be equipped with an

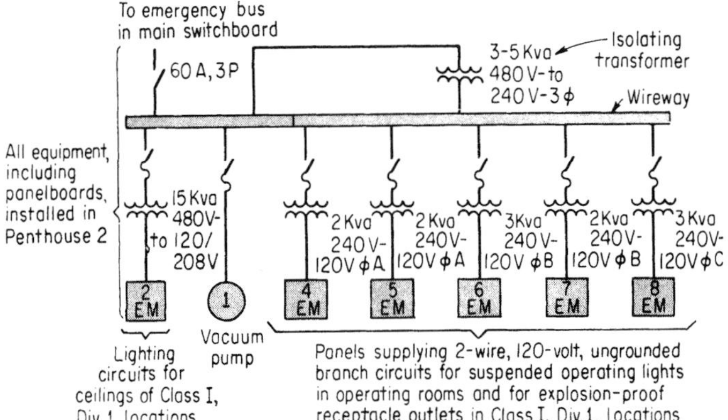

Fig. 517-13. Two-stage transformation was often needed for isolated circuits but is no longer necessary. (Sec. 517-160.)

approved ground contact indicator—to give a visual and audible warning if a ground fault develops in the ungrounded system.

Isolating transformers must be installed out of the hazardous area [part **(a)(3)**]. The ground indicator and its signals must also be installed out of the hazardous area. In an anesthetizing location, the hazardous area extends to a height of 5 ft (1.52 m) above the floor.

Fixed lighting fixtures above the hazardous area in an anesthetizing location, other than the surgical luminaire, and certain x-ray equipment may be supplied by conventional grounded branch circuits [Sec. 517-63(b) and Sec. 517-63(c)].

Part **(a)(5)** requires isolated circuit conductors to be identified by brown and orange colors.

Part **(b)(1)** details a line isolation monitor and clarifies line isolation monitor alarm values, specifying 5.0 mA as the lower limit of alarm for total hazard current. Figure 517-14 shows the basic concept behind detection and signal of a ground fault. The diagram shows major circuit components of a typical ground

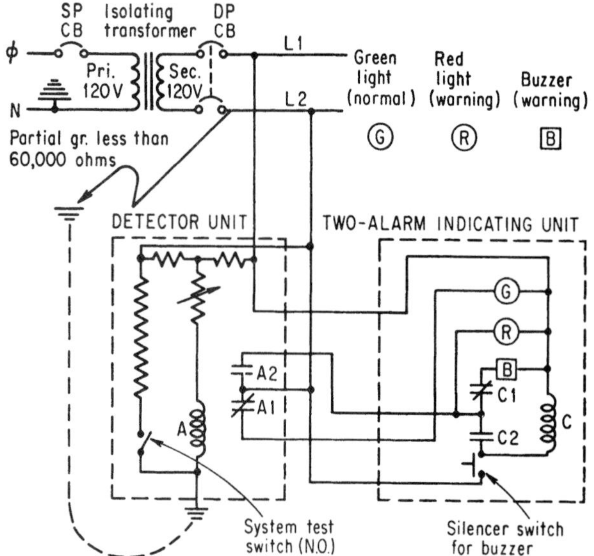

Fig. 517-14. Detection and alarm on ground fault is required for isolated power systems. (Sec. 517-160.)

detector/alarm system. Partial ground energizes current-relay A, opening contact A2 (energizing red light and warning buzzer). Pressing the momentary-contact silencer switch energizes coil C, opening contact C1 (disconnecting buzzer), and closing holding contact C2. When ground is cleared, contacts resume the position shown in drawing.

The purpose of such a ground indicator is to provide warning of the danger of shock hazard and the possibility of a fault in the system due to accidental grounding of more than one conductor. If one conductor of an isolated system becomes grounded at one point, normal protective devices (fuses or CBs) will not operate because there is no return path and, therefore, no flow of short-circuit fault current. However, if an accidental ground subsequently develops on the other conductor, a short circuit will occur with possible disastrous consequences, such as ignition of ether vapors by arc or a lethal shock to personnel.

ARTICLE 518. PLACES OF ASSEMBLY

518-1. Scope. This article covers places of assembly which used to be covered along with theaters in Art. 520. This article does not apply to theaters. It covers any single indoor space (a whole building or part of a building) designed or intended for use by 100 or more persons for assembly purposes. That includes dining rooms, meeting rooms, entertainment areas (other than with a stage or platform or projection booth), lecture halls, bowling alleys, places of worship, dance halls, exhibition halls, museums, gymnasiums, armories, group rooms, funeral parlor chapels, skating rinks, pool rooms, transportation terminals, court rooms, sports arenas, and stadiums. A school classroom for less than 100 persons is not subject to this article. See Sec. 518-2.

The clear differentiation given in this section points out that any such building or structure or part of a building that contains a projection booth or stage platform or even just an area that may, on occasion, be used for presenting theatrical or musical productions—whether the stage or platform is fixed or portable—must comply with the rules of Art. 520, as if it were a theater, and not Art. 518. A restaurant, say, that has a piano player for entertainment on Saturday night, could readily be classed as a theater and subject to Art. 520.

The question often arises, Does Art. 518 apply to supermarkets and department-store types of occupancies because such places are regularly crowded with far more than 100 persons? Although occupancies of those types have capacity to hold more than 100 persons at any given time, Art. 518 is not generally applicable.

Article 518 directs attention "to a building or portion of a building" that would be used for the purposes outlined; therefore, you would have to determine how the occupancy is used.

A supermarket generally would not have a public assembly area. However, a department store could incorporate a community room for shows and similar audience functions. This room would be subject to Art. 518. The main areas of supermarkets and department stores, unlike theaters and assembly halls, have many aisles and exits that could be used in case of emergency evacuation of the building. It is these characteristics that permit conventional wiring methods to be accepted.

A proposal was once made to include supermarkets and department stores as "places of assembly," but it was rejected.

Note that places of assembly covered by this article must be for 100 or more people. No rules are given for determining the number of people, but a note refers questions of population to local building codes or to the NFPA Life Safety Code. Of course, the number of seats is an index of capacity, and other reasonable indications must be observed.

The following information is found in NFPA No. 101, Life Safety Code, for determining occupant load in places of assembly.

Occupant Load. The occupant load permitted in any assembly building, structure, or portion thereof shall be determined by dividing the net floor area or space assigned to that use by the square feet per occupant as follows:

(a) An assembly area of concentrated use without fixed seats such as an auditorium, church, chapel, dance floor, and lodge room—7 square feet (0.651 sq m) per person.

(b) An assembly area of less concentrated use such as a conference room, dining room, drinking establishment, exhibit room, gymnasium, or lounge—15 square feet (1.4 sq m) per person.

(c) Standing room or waiting room—3 square feet per person.

The occupant load of an area having fixed seats shall be determined by the number of fixed seats installed. Required aisle space serving the fixed seats shall not be used to increase the occupant load.

The occupant load permitted in a building or portion thereof may be increased above that specified in "Occupant Load" if the necessary aisles and exits are provided subject to the approval of the authority having jurisdiction. An approved aisle, exit and/or seating diagram may be required by the authority having jurisdiction to substantiate an increase in occupant load.

518-4. Wiring Methods. The basic rule says that fixed wiring must be in a metal raceway, nonmetallic raceways encased in *not less than 2 in. (50.8 mm) of concrete,* Types ALS, MI, or MC cable. The first Exception says that nonmetallic-sheathed cable, BX (Type AC cable), electrical nonmetallic tubing (ENT), and rigid nonmetallic conduit may be used in building areas that are *not* required by the local building code to be of fire-rated construction. Note that use of those methods no longer relates to the number of persons that the place holds, which was once in this rule. Another Exception permits the use of other wiring methods for sound systems, communication circuits, Class 2 and 3 remote-control and signal circuits, and fire-alarm circuits.

ARTICLE 520. THEATERS, AUDIENCE AREAS OF MOTION PICTURE AND TELEVISION STUDIOS, AND SIMILAR LOCATIONS

520-1. Scope. Where only a part of a building is used as a theater or similar location, these special requirements apply only to that part and do not necessarily apply to the entire building. A common example is a school building in which there is an auditorium used for dramatic or other performances. All special requirements of this chapter would apply to the auditorium, stage, dressing rooms, and main corridors leading to the auditorium but not to other parts of the building that do not pertain to the use of the auditorium for performances or entertainment.

520-5. Wiring Methods. Building laws usually require theaters and motion picture houses to be of fireproof construction; hence practical considerations limit the types of concealed wiring for light and power chiefly to raceways. Only Type MI or MC cables may be used. Cables were long ago found unsuitable for circuits in theaters because they do not readily offer increase in the size of conductors for load growth. Many instances of overfusing dictated the value of raceways, which do permit replacement of larger conductors for safely handling load growth.

Much of the stage lighting in a modern theater is provided by floodlights and projectors mounted in the ceiling or on the balcony front. In order that the projectors may be adjustable in position, they may be connected by plugs and short cords to suitable receptacles or "pockets."

There are three Exceptions to the basic rule. Portable loads—switchboards, lighting, etc.—may be supplied by cords or cables, and recording, communications, remote control signaling and fire alarm circuits may be installed using other wiring methods.

520-23. Control and Overcurrent Protection of Receptacle Circuits. The term "auditorium receptacles" should be understood as including all receptacles, wherever they may be located, that are intended for the connection of stage lighting equipment. Circuits to such receptacles must of necessity be controlled at the same location as other stage lighting circuits.

520-24. Metal Hood. Because of the large amount of flammable material always present on a stage, and because of the crowded space, a stage switchboard must have no live parts on the front, and the back must be so guarded as to keep unauthorized persons away from the space in back of the board and the wall, with a door at one end of the enclosure.

The more important stage switchboards are commonly of the remote-control type. Pilot switches mounted on the stage board control the operation of contactors installed in any convenient location where space is available, usually below the stage. The contactors in turn control the lighting circuits.

The stage switchboard is usually built into a recess in the proscenium wall, as shown in the plan view, Fig. 520-1. After passing through the switches and dimmers, many of the main circuits must be subdivided into branch circuits so that no branch circuit will be loaded to more than 20 A. Where the board is of the remote-control type, the branch-circuit fuses are often mounted on the same panels as the contactors. Where a direct-control type of board is used, and sometimes where the board is remotely controlled, the branch-circuit fuses are mounted on special panelboards known as *magazine panels,* which are installed in the space back of the switchboard, usually in the location of the junction box shown in Fig. 520-2.

520-25. Dimmers. Figure 520-2 shows typical connections of two branch circuits arranged for control by one switch and one dimmer plate or section. The single-pole switch on the stage switchboard is connected to one of the outside buses, and from this switch a wire runs to a short bus on the magazine panel. The magazine panel is similar to an ordinary panelboard, except that it contains no switches and the circuits are divided into many sections, each section having its own separate buses. One terminal of the dimmer plate, or variable

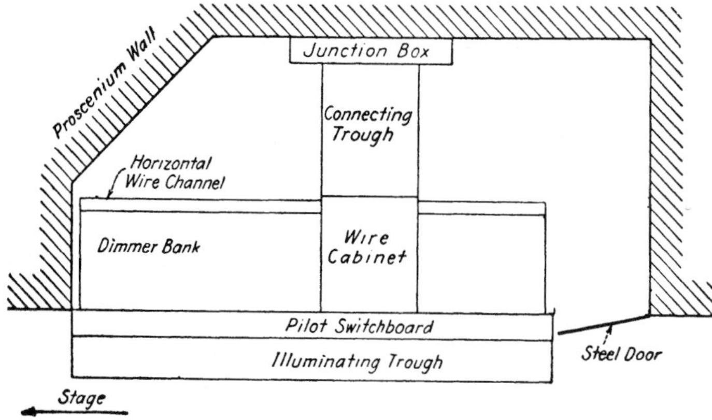

Fig. 520-1. Stage switchboards must be circuited to provide highly flexible usage. (Sec. 520-24.)

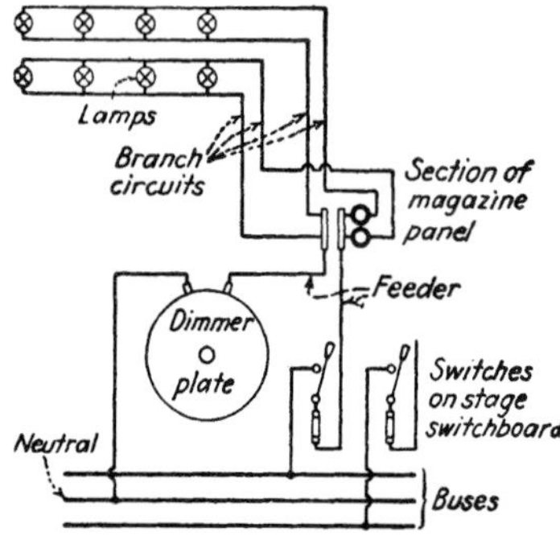

Fig. 520-2. Branch circuits of lighting may be controlled by single dimmer in grounded or ungrounded conductor. (Sec. 520-25.)

resistor, is connected to the neutral bus at the switchboard, and from the other terminal of the dimmer a wire runs to the neutral bus on the magazine panel. This neutral bus must be well insulated from ground and must be separate from other neutral buses on the panel; otherwise the dimmer would be shunted and would fail to control the brightness of the lamps.

While the dimmer is permanently connected to the neutral of the wiring system, this neutral is presumed to be thoroughly grounded and hence the dimmer is dead. A dimmer in the grounded neutral does not require overcurrent protection, as noted in part **(a)**.

Figure 520-3 shows an autotransformer used as a dimmer. By changing the position of the movable contact, any desired voltage may be supplied to the lamps, from full-line voltage to a voltage so low that the lamps are "black out." As compared with a resistance-type dimmer, a dimmer of this type has the advantages that it operates at a much higher efficiency, generates very little heat, and, within its maximum rating, the dimming effect is not dependent upon the wattage of the load it controls.

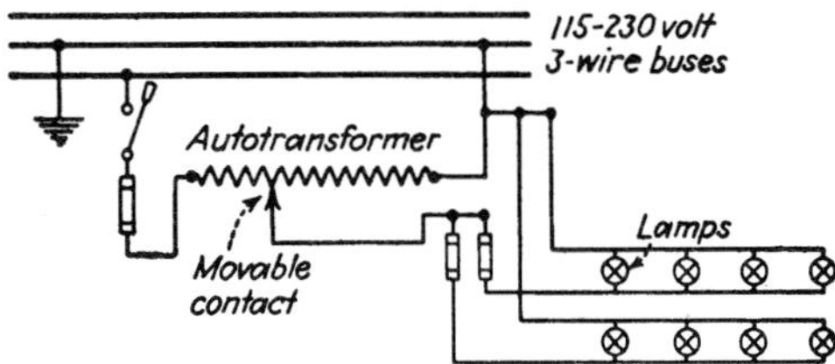

Fig. 520-3. Autotransformer dimmer must have grounded leg common to primary and secondary. (Sec. 520-25.)

520-43. Footlights. A footlight of the disappearing type might produce so high a temperature as to be a serious fire hazard if the lamps should be left burning after the footlight is closed. Part **(c)** calls for automatic disconnect when the lights disappear.

There is no restriction on the number of lamps that may be supplied by one branch circuit. The lamp wattage supplied by one circuit should be such that the current will be slightly less than 20 A.

Individual outlets as described in part **(b)** are seldom used for footlights, as such construction would be much more expensive than the standard trough type.

A modern type of footlight is shown in Fig. 520-4. The wiring is carried in a sheet-iron wire channel in the face of which lamp receptacles are mounted. Each lamp is provided with an individual reflector and glass color screen or

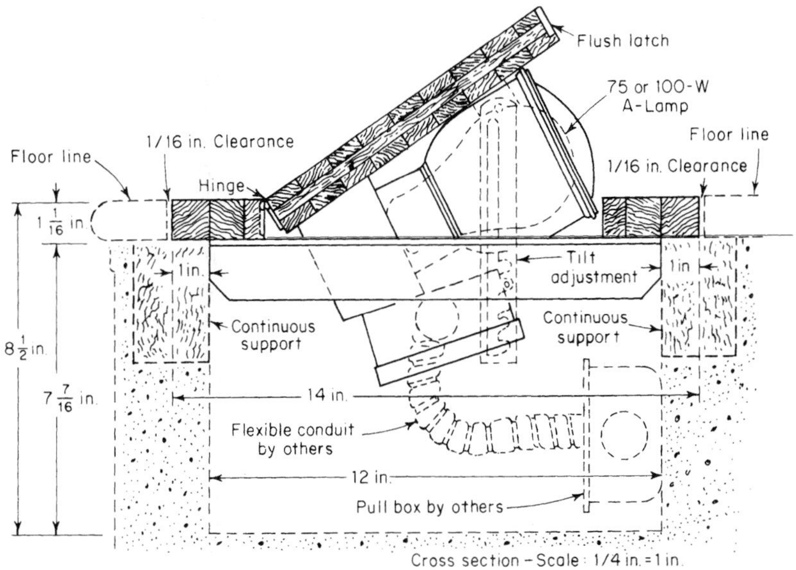

Fig. 520-4. Footlights must be automatically de-energized when the flush latch is closed down. (Sec. 520-43.)

"roundel." The circuit wires are usually brought to the wire channel in rigid conduit. In the other type of footlight, still used to some extent, the lamps are placed vertically or nearly so, and an extension of one side of the wire channel is shaped so as to form a reflector to direct the light toward the stage.

520-44. Borders and Proscenium Sidelights. Figure 520-5 is a cross section showing the construction of a border light over the stage. This particular type is intended for the use of 200-W lamps. An individual reflector is provided for each lamp so as to secure the highest possible efficiency of light utilization. A

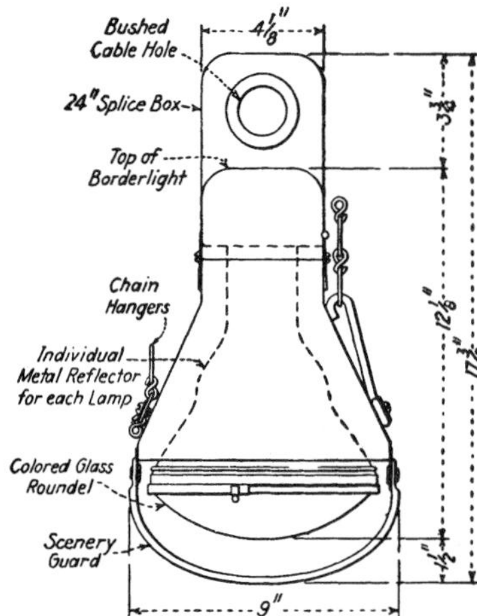

Fig. 520-5. Border lights must comply with NEC construction rules. (Sec. 520-44.)

glass roundel is fitted to each reflector; these may be obtained in any desired color, commonly white, red, and blue for three-color equipment and white, red, blue, and amber for four-color equipment. A splice box is provided on top of the housing for enclosing the connections between the border-light cable and the wiring of the border. From this splice box, the wires are carried to the lamp sockets in a trough extending the entire length of the border.

Border lights are usually hung on steel cables so that their height may be adjusted and so that they may be lowered to the stage for cleaning and replacing lamps and color screens; hence the circuit conductors supplying the lamps must be carried to the border through a flexible cable. The individual conductors of the cable may be of No. 14, though No. 12 is more commonly used.

520-49. Smoke Ventilator Control. A normally closed-circuit device has the inherent safety feature that in case the control circuit is accidentally opened by the blowing of a fuse, or in any other way, the device immediately operates to open the flue dampers.

520-53. Construction and Feeders. In part **(f)**, specs cover conductors within portable stage dimmer switchboards. A broad detailed rule covers the temperature ratings of conductors permitted in dimmer boards, based on the type of dimmer used. The rule recognizes the difference in temperature of dimmers and permits lower-rated conductors for solid-state dimmers.

520-65. Festoons. "Lanterns or similar devices" are very likely to be made of paper or other flammable material, and the lamps should be prevented from coming in contact with such material.

520-72. Lamp Guards. Lamps in dressing rooms should be provided with guards that cannot easily be removed to prevent them from coming in contact with flammable material.

ARTICLE 525. CARNIVALS, CIRCUSES, FAIRS, AND SIMILAR EVENTS

525-1. Scope. Although previously considered subject to the rules of Art 305, "Temporary Wiring," the inclusion of this new article makes clear that wiring for such installations is regulated by this article. However, unless otherwise modified by Art 525, the requirements given in Chaps. 1 through 4 apply.

For example, Sec. 525-18 specifically excludes installations for carnivals, etc., from the need for GFCI protection as spelled out in Sec. 305-6. That section requires all 125-V, single-phase, 15- and 20-A, receptacles, whether part of the permanent installation or not, to be provided with GFCI protection where used to supply cord- and plug-connected portable tools on a jobsite. Although that requirement is waived for those receptacles used for the carnival itself, it seems as if any 15- or 20-A, 125-V receptacle that is used during the setup or assembly of the rides and attractions for the supply of portable tools *would* be covered by the rules of Art 305.

ARTICLE 530. MOTION PICTURE AND TELEVISION STUDIOS AND SIMILAR LOCATIONS

530-1. Scope. Article 520 covers theaters used for TV, motion picture, or live presentations where the building or part of a building includes an assembly area for the audience. Article 530, however, applies to TV or motion picture studios where film or TV cameras are used to record programs and to the other areas of similar application—but where the facility does not include an audience area.

The term "motion picture studio" is commonly used as meaning a large space, sometimes 100 acres or more in extent, enclosed by walls or fences within which are several "stages," a number of spaces for outdoor setups, warehouses, storage sheds, separate buildings used as dressing rooms, a large substation, a restaurant, and other necessary buildings. The so-called stages are large buildings containing numerous temporary and semipermanent setups for both indoor and outdoor views.

The Code rules for motion picture studios are intended to apply only to those locations where special hazards exist. Such special hazards are confined to the buildings in which films are handled or stored, the stages, and the outdoor spaces where flammable temporary structures and equipment are used. Some of these special hazards are due to the presence of a considerable quantity of highly flammable film; otherwise, the conditions are much the same as on a theater stage and, in general, the same rules should be observed as in the case of theater stages.

530-11. Permanent Wiring. The 1975 NEC used the word "metal"—between "approved" and "raceways," in the first sentence. Because the word "metal" no longer appears in the rule, rigid nonmetallic conduit is, therefore, acceptable for use in motion picture and TV studios.

In the Exception, Class 2 and 3 remote control or signaling circuits and power-limited fire-protective signaling circuits are exempted from the basic rule requiring permanent wiring to be in raceways or Type MC or MI cable. Those circuits along with communications and sound recording and reproducing circuits are exempt from the basic rule.

ARTICLE 540. MOTION PICTURE PROJECTORS

540-1. Scope. According to the definition of hazardous locations in Art. 500, a motion picture booth is not classed as a hazardous location, even though the film is highly flammable. The film is not volatile at ordinary temperatures and hence no flammable gases are present, and the wiring installation need not be explosionproof but should be made with special care to guard against fire hazards.

540-2. Professional Projector. Figure 540-1 shows a professional movie projector, which is subject to lengthier and stricter requirements than those of nonprofessional projectors.

540-10. Motion Picture Projection Room Required. Professional projectors must be installed in a projector booth, which does not have to be treated like a hazardous (classified) location.

Figure 540-2 shows the arrangement of the apparatus and wiring in the projection room of a large modern motion picture theater. This room, or booth, contains three motion picture projectors P, one stereopticon or "effect machine" L, and two spot machines S.

The light source in each of the six machines is an arc lamp operated on DC. The DC supply is obtained from two motor-generator sets which are installed in the basement in order to avoid any possible interference with the sound-reproducing apparatus. The two motor generators are remotely controlled from the generator panel in the projection room. From each generator a feeder consisting of two 500-kcmil cables is carried to the DC panelboard in the projection room.

From the DC panelboard to each picture machine and to each of the two spot machines a branch circuit is provided consisting of two No. 2/0 cables. One of these conductors leads directly to the machine; the other side of the circuit is led through the auxiliary gutter to the bank of resistors in the rheostat room and from its rheostat to the machine. The resistors are provided with short-circuiting switches so that the total resistance in series with each arc may be preadjusted to any desired value.

Fig. 540-1. Professional projector. But note that Art. 540 applies to *both* professional and nonprofessional movie projectors. The article is divided into part **C** on professional equipment and part **D** on nonprofessional units. (Sec. 540-2.)

Two circuits consisting of No. 1 conductors are carried to the stereopticon or "effect machine," since this machine contains two arc lamps.

The conduit leading to each machine is brought up through the floor.

It is provided in Sec. 540-13 that the wires to the projector outlet shall not be smaller than No. 8, but in every case the maximum current drawn by the lamp should be ascertained and conductors should be installed of sufficient size to carry this current. In this case, when suitably adjusted for the large pictures, the arc in each projector takes a current of nearly 150 A.

In addition to the main outlet for supplying the arc, four other outlets are installed at each projector machine location for auxiliary circuits.

Outlets F are for foot switches which control the shutters in front of the lenses for changeover from one projector to another.

Outlets G are for a No. 8 grounding conductor which is connected to the frame of each projector and to the water-piping system.

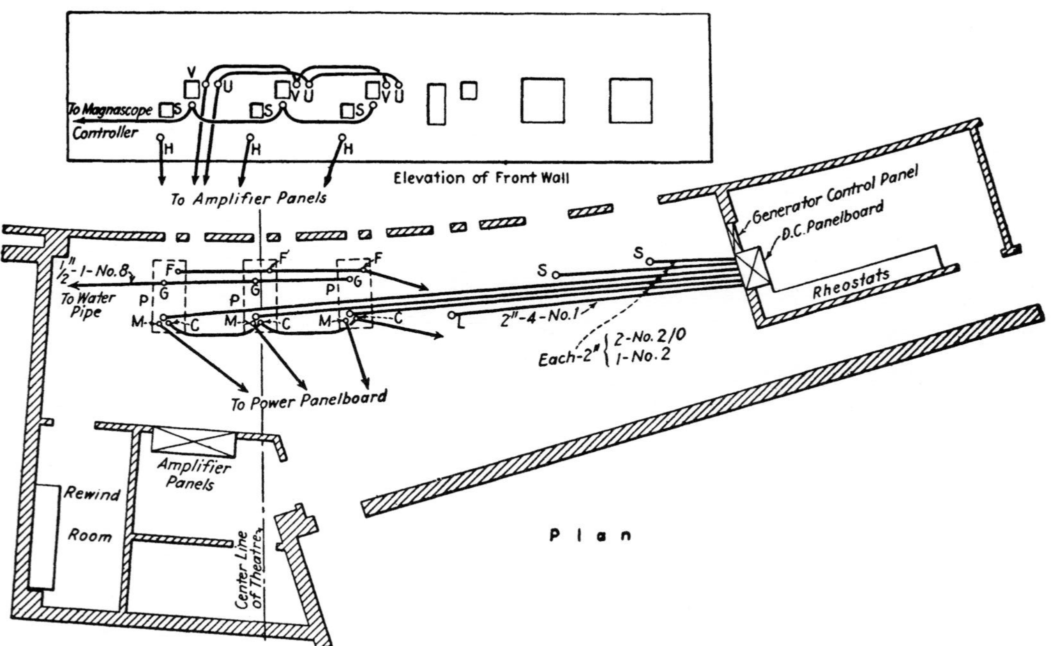

Fig. 540-2. Code rules cover many electrical details in a motion-picture projection room (or "booth"). (Sec. 540-10.)

From outlets C a circuit is brought up to each machine for a small incandescent lamp inside the lamp house and a lamp to illuminate the turntable. Outlets M are for power circuits to the motors used to operate the projector machines.

Ventilation is provided by two exhaust fans and two duct systems, one exhausting from the ceiling of the projection room and one connected to the arc-lamp housing of each machine. (See Fig. 540-1.)

A separate room is provided for rewinding films, but since this room opens only into the projection room, it may be considered that the rewinding is performed in the projection room.

540-11. Location of Associated Electrical Equipment. All necessary equipment *may* be located in a projector booth, but equipment which is not necessary in the normal operation of the motion picture projectors, stage-lighting projectors, and control of the auditorium lighting and stage curtain *must* be located elsewhere. Equipment such as service equipment and panelboards for the control and projection of circuits for signs, outside lighting, and lighting in the lobby and box office must not be located in the booth.

ARTICLE 547. AGRICULTURAL BUILDINGS

547-1. Scope. Any agricultural building without the environments covered in parts **(a)** and **(b)** must be wired in accordance with all other **Code** rules that apply to general building interiors. Article 547 covers *only* agricultural buildings with the dust, water, and/or corrosive conditions described in parts **(a)** and **(b)**.

547-4. Wiring Methods. The wording leaves much of the determination of acceptability up to the inspection authority. But Type NMC cable (nonmetallic, corrosion-resistant—so-called barn wiring cable), UF, SNM, or copper SE are specifically recognized for these buildings. PVC conduit and other nonmetallic or protected products would be suitable for the wet and corrosive conditions that prevail. The rule accepts wiring for Class II hazardous locations, as well as open wiring on insulators (Art. 320).

Note that boxes and fittings must be both dust- and watertight. Flexible connections must use dusttight flex or liquidtight flex or cord. Also note that *nonmetallic* boxes, fittings, etc., are exempt from the provisions of Sec. 300-6(c). If such components and cables are made from a *metallic* material, then the ¼-in. (6.35-mm) clearance called for in Sec. 300-6(c) would apply.

547-5. Switches, Circuit Breakers, Controllers, and Fuses. In this part, the description of the type of enclosure required corresponds to the following NEMA designations on enclosures:

Type 4. Watertight and dusttight For use indoors and outdoors. Protect against splashing water, seepage of water, falling or hose-directed water, and severe external condensation. They are sleet-resistant but not sleet- (ice) proof.

Type 4X. Watertight, dusttight, and corrosion-resistant These have the same provisions as Type 4 enclosures, but in addition are corrosion-resistant.

The rule of this section seems to clearly call for NEMA 4X enclosures (Fig. 547-1).

Stainless steel NEMA Type 4X enclosures are used in areas which may be regularly hosed down or are otherwise very wet, and where serious corrosion problems exist. Typical enclosures are made from 14-gauge stainless steel, with an oil-resistant neoprene door gasket.

NEMA Type 4X

Fig. 547-1. The Code rule seems to make use of this type of enclosure mandatory in agricultural buildings. (Sec. 547-4.)

Epoxy powdered resin coated NEMA Type 4X enclosures are designed to house electrical controls, terminals, and instruments in areas which may be regularly hosed down or are otherwise very wet. These enclosures are also designed for use in areas where serious corrosion problems exist. They are suitable for use outdoors, or in dairies, packing plants, and similar installations. These enclosures are made from 14-gauge steel. All seams are continuously welded with no holes or knockouts. A rolled lip is provided around all sides of the enclosure opening. This lip increases strength and keeps dirt and liquids from dropping into the enclosure while the door is open.

547-8. Grounding, Bonding, and Equipotential Plane. As noted in Exception No. 1 to part **(a)**, panelboard bonding and grounding is not required for a building that houses livestock if an equipment grounding conductor is run as part of a feeder to the panel. This rule is essentially the same as that given in Sec. 250-24(a), Exception No. 2, and permits the main panelboard in the livestock building to be fed and connected as if it were a subpanel in the building where the feeder originates, without the neutral block being bonded to the ground block and panel enclosure. The objective is to separate the neutral conductor and the equipment grounding conductor so that any voltage drop on the neutral as a result of neutral current flow will not be transferred to metal equipment enclosures that might be contacted by cows, hogs, or other livestock—which are very sensitive to even low voltage-to-ground potentials.

The equipment grounding conductor run to the panel must be the same size as the hot legs of the feeder, must be connected to the equipment ground bus in the panel, and must be covered or insulated copper from Sec. 250-24(a), Exception No. 2, if it is run underground. In addition, a grounding electrode must be connected to the ground bus in the panelboard (Fig. 547-2).

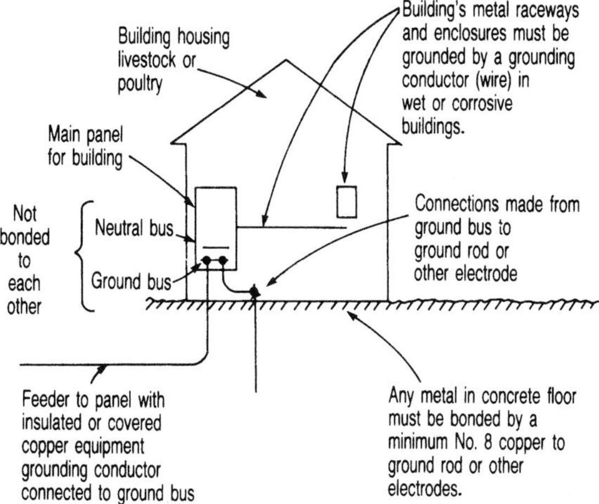

Fig. 547-2. Specific rules cover use of a panelboard in a building housing livestock or poultry. (Sec. 547-8.)

As permitted by Exception No. 2 of part **(a)**, a "listed" *impedance device* may be used to ground interior metal piping in livestock buildings to prevent exposure of animals and humans to neutral-to-earth voltage. This specifically permits a technique that will overcome grounding problems that have caused loss of milk output from dairy cows and other objectionable conditions.

As required by parts **(b)** and **(c)**, reinforcing metal in a concrete floor of an animal confinement area must be bonded to the building's grounding electrode system by a No. 8 copper wire; and in wet or corrosive locations, metal enclosures in the building must be grounded by a copper equipment grounding conductor connected back to the ground bus in the building panel. The concern here is to minimize the possibility of shock hazard and unwanted voltages by assuring the integrity of the entire grounding system. The last sentence of part **(c)** calls for any underground equipment grounding wires to be insulated or covered.

Part **(d)** requires that the uninsulated metal frame of a water-pump motor *must* be grounded. A submersible pump must have any metal well casing bonded to the equipment grounding conductor of the pump or to the ground bus in the panel supplying the pump.

ARTICLE 550. MOBILE HOMES, MANUFACTURED HOMES, AND MOBILE HOME PARKS

550-1. Scope. The provisions of this article cover the electric conductors and equipment installed within or on mobile homes, and also the conductors that connect mobile homes to a supply of electricity. But the service equipment which is located "adjacent" to the mobile home is not covered in Art. 550, and all applicable Code rules on such service equipment—as in Arts. 230 and 250—must be observed.

550-2. Definitions. A *double-wide mobile home* is manufactured in two sections, each being approximately 12 by 60 ft (3.66 by 18.28 m). Each section is mounted on a chassis, with one side in each section open. The sections are moved to location and joined together to make a 24- by 60-ft (7.32- by 18.28-m) complete unit. Because the double-wide will most likely be placed on a foundation, the enforcing authority will usually classify it as a prefabricated structure, because it does not meet the definition of "mobile home." *But* Art. 550 would still apply.

The definition for "manufactured home" clarifies when the service equipment may be hung on the dwelling. The service may *never* be installed on, or in, a mobile home. And, generally, it may not be installed in, or on, a manufactured home. But, where a manufactured home is *not* equipped with a power cord, it may be supplied by a service that is installed adjacent to or even on or in the manufactured home, provided Exception No. 1 to Sec. 550-23(a) is satisfied. Notice that the last sentence in the definition makes all rules for "mobile homes" applicable to "manufactured homes" as well.

550-4. General Requirements. Many so-called mobile homes do not have their main service-entrance equipment located *adjacent* to the mobile home, as required by Sec. 550-5. In some, the service equipment is mounted on the outside of the mobile home, and in others it is mounted inside. That is commonly the case with mobile homes that have had the wheels removed and are on permanent foundations. Some such mobile homes are used as living units, some as business offices, coin-operated laundries, and for many other purposes.

When a mobile home is altered by removing its wheels and installing a permanent foundation, it is no longer mobile and does not satisfy the definition of "mobile home" given in Sec. 550-2. Many inspection authorities treat such installation as a prefabricated building or structure and apply the rules of Art. 545 instead of Art. 550. The requirement for a prefabricated structure is that such buildings must satisfy Code requirements the same as a building being built on the site. As with any constructed-on-site building, a prefabricated building can have the service equipment inside or outside.

550-5. Power Supply. Part **(a)** requires that the mobile home service equipment be located adjacent to the mobile home and not mounted in or on the mobile home. It further specifies that the power supply to the mobile home shall be a feeder circuit consisting of not more than *one* 50-A rated approved mobile home supply cord, or that feeder circuit could be a permanently installed circuit of fixed wiring. (Fig. 550-1.)

Part **(b)** covers use of a cord instead of permanent wiring. The power-supply cord to a mobile home is actually a feeder, and must be treated as such in applying Code rules. The service equipment must be located adjacent to the mobile home and could be either a fused or breaker type in an appropriate enclosure or enclosures, with not over 50-A overcurrent protection for the supply cord (or 40 A, as in the Exception). The equipment must be approved service-entrance equipment with an appropriate receptacle for the supply cord, installed to meet Code rules the same as any installation of service equipment. The panel or pan-

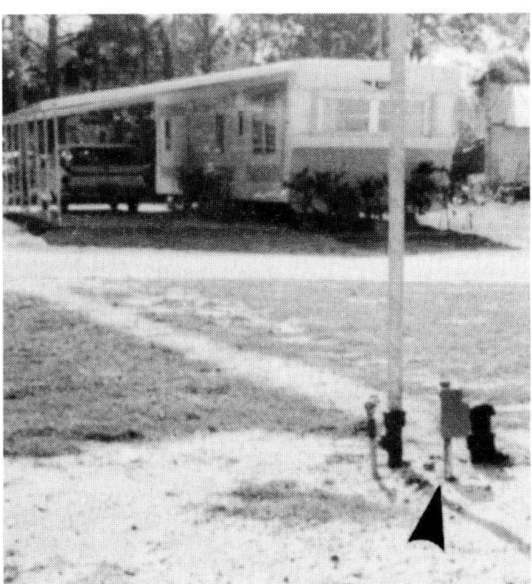

Fig. 550-1. Service equipment for a mobile home lot consists of disconnect, overcurrent protection, and receptacle for connecting *one* 50-A (or 40-A) power-supply cord from a mobile home parked adjacent to the service equipment. (Sec. 550-5.)

els in the home are feeder panels and are *never* to be used as service-entrance equipment according to Secs. 550-5 and 550-6(a). This means that the neutral is isolated from the enclosure and the equipment grounding goes to a separate bus for that purpose only. As a result, there must be an equipment grounding conductor run from the service-entrance equipment to the panel or panels in the home. This is true whether there is cord connection or permanent wiring.

In the 1975 NEC, part **(j)** required special permission to permit two or three 50-A power-supply cords to a mobile home. Most mobile homes parks were not equipped to handle more than one such power-supply cord for each mobile home lot. But in this Code edition, Sec. 550-5(a) does *not* permit mobile home parks to be wired with more than one 50-A receptacle at each mobile home lot.

In some areas mobile homes are permanently connected as permitted in paragraph **(i)**. Accordingly, local requirements must be checked in regard to the approved method of installing feeder assemblies where a mobile home has a calculated load over 50 A. In many such cases, a raceway is stubbed to the underside of a mobile home from the distribution panelboard. It is optional as to whether the feeder conductors are installed in the raceway by the mobile home manufacturer or by field installers. When installed, four continuous, insulated, color-coded conductors, as indicated, are required. The feeder conductors may be spliced in a suitable junction box, but in no case within the raceway proper.

550-6. Disconnecting Means and Branch-Circuit Protective Equipment. As shown in Fig. 550-2, the required disconnect for a mobile home may be the main in the panelboard supplying the branch circuits for the unit. Details of this section must be observed by the mobile home builder.

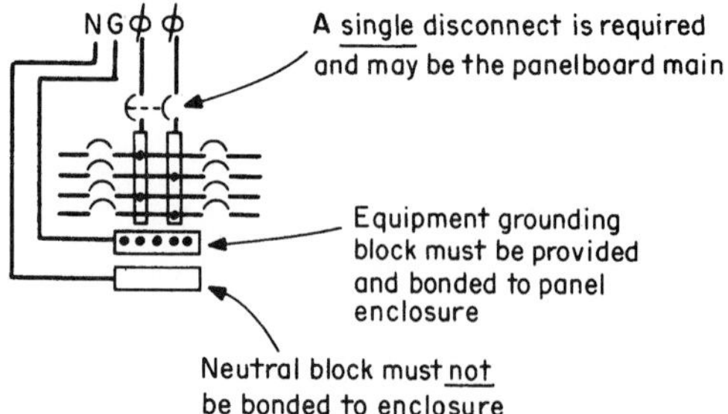

Fig. 550-2. A "distribution panelboard," not "service panelboard" may be used *in* mobile home. (Sec. 550-6.)

550-7. Branch Circuits. The manufacturer of the mobile home must assure this minimum circuiting.

550-10. Wiring Methods and Materials. In the Exception to part **(j)**, the smaller-dimensional box mentioned would usually be a box designed for a special switch or receptacle, or a combination box and wiring device. Such combinations can be properly evaluated and tested with a limited number of conductors and connections and a specific lay of conductors to ensure adequate wiring space in the spirit of the first paragraph in Sec. 370-6.

550-11. Grounding. The white (neutral) conductor is required to be run from the "insulated busbar" in the mobile home panel to the service-entrance equipment, where it is connected to the terminal at the point of connection to the grounding electrode conductor.

The green-colored conductor is required to be run from the "panel grounding bus" in the mobile home to the service-entrance equipment, where it is connected to the neutral conductor at the point of connection to the grounding electrode conductor.

The requirements provide that the grounded (white) conductor and the grounding (green) conductor be kept separate within the mobile home structure in order to secure the maximum protection against electric-shock hazard if the supply neutral conductor should become open.

A common point of discussion among electrical authorities and electricians is whether or not the green-colored grounding conductor in the supply cord should be connected to the grounded circuit conductor (neutral) outside the mobile home, say at the location of the service equipment. The grounding conductor in the supply cord or the grounding conductor in the power supply to a mobile home is always required to be connected to the grounded circuit conductor (neutral) outside the mobile home on the supply side of the service disconnecting means, but *not* in a junction box under the mobile home or at any other point on the *load side* of the service equipment (pedestal).

550-21. Distribution System. The mobile home park supply is limited to nominal 120/240-V, single-phase, 3-wire to accommodate appliances rated at nominal 240 V or a combination nominal voltage of 120/240 V. Accordingly, a 3-wire 120/208-V supply, derived from a 4-wire 208Y/120-V supply, would not be acceptable.

While the demand factor for a single mobile home lot is computed at 16,000 VA, it should be noted that Sec. 550-23(b) requires the feeder circuit conductors extending to each mobile home lot to be not less than 100 A.

550-23. Mobile Home Service Equipment. The service equipment disconnect means for a mobile home must be mounted with the bottom of its enclosure at least 2 ft (610 mm) above the ground, because some very low mounted disconnects are subject to flooding and are difficult to operate. But the disconnect must not be higher than 6½ ft (1.98 m) above the ground or platform (Fig. 550-3).

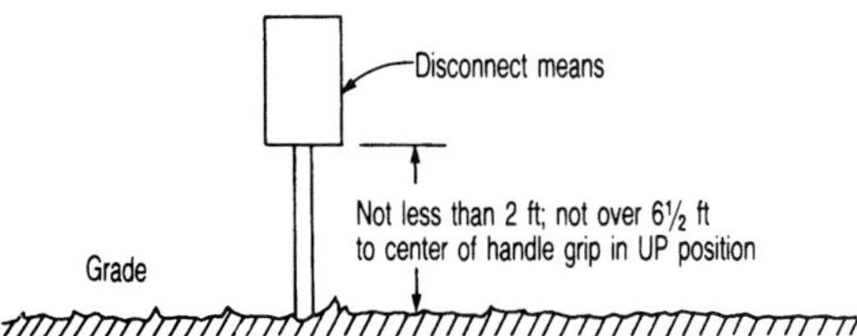

Fig. 550-3. Mobile home service disconnect must comply with new minimum mounting height rule. (Sec. 550-23.)

ARTICLE 551. RECREATIONAL VEHICLES AND RECREATIONAL VEHICLE PARKS

551-1. Scope. Some states have laws that require factory inspection of recreational vehicles by state inspectors. Such laws closely follow NFPA No. 501C, Standard for Recreational Vehicles. This standard contains electrical requirements in accordance with part **A** of Art. 551. It also contains requirements for plumbing and heating systems.

551-10. Low-Voltage Systems. This section concerns 12-V systems for running and signal lights similar to those in conventional automobile systems. Also, many recreational vehicles use 12-V systems for interior lighting or other small loads. The 12-V system is derived from an on-board battery or through a transfer switch from a 120/12-V transformer often equipped with a full-wave rectifier.

551-20. Combination Electrical Systems. As explained in the last Exception of part **(b)**, "momentary" operated electric appliances do not affect converter sizing. This Exception excludes from calculation of the required converter rating any appliance that operates only momentarily (by a momentary contact switch) and cannot have its switch left in the closed position. Such appliances draw current for only momentary periods and do not have to be counted as "load" in sizing the converter rating.

551-42. Branch Circuits Required. This rule coordinates the rules on branch circuits to those of Sec. 551-45 on distribution panelboard.

ARTICLE 553. FLOATING BUILDINGS

553-1. Scope. This article covers the electrical system in a building—either residential (dwelling unit) or nonresidential—that floats on water, is moored in a permanent location, and has its electrical system supplied from a supply system on land. The rules apply to any floating building and are not limited only to floating "dwelling units."

553-4. Location of Service Equipment. The service-disconnect means and protection for a floating building must not be mounted on the unit. This assures the ability to disconnect the supply conductors to the floating building in an emergency, such as in a storm, in the event that it is necessary to move the unit quickly (Fig. 553-1).

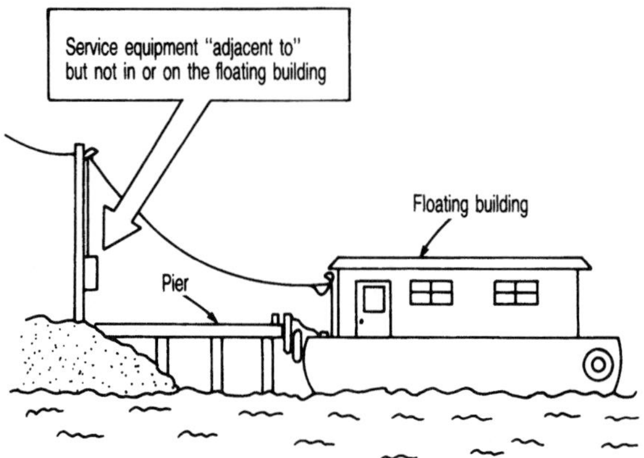

Fig. 553-1. Service equipment for a floating building must be on the dock, pier, or wharf.

553-8. General Requirements. A green-colored, insulated equipment grounding conductor must be used in a feeder to the main panel of a floating building. This positive equipment grounding conductor must be run to the panel from an equipment grounding terminal (or bonded neutral bus) in the building's service equipment on land.

ARTICLE 555. MARINAS AND BOATYARDS

555-1. Scope. This article covers both fixed and floating piers, wharfs, and docks—as in boat basins or marinas. In Fig. 555-1, branch circuits and feeder cables run from panelboards in the electrical shed at the left, down, and underground into a fabricated cable space running the length of the pier shown at the right, supplying shore-power receptacle pedestals (arrows) along both sides of the pier.

555-3. Receptacles. Figure 555-2 shows typical configurations of locking- and grounding-type receptacles and attachment plugs used in marinas and boatyards. A complete chart of these devices can be obtained from the National Electrical Manufacturers Association or wiring-device manufacturers. Locking-

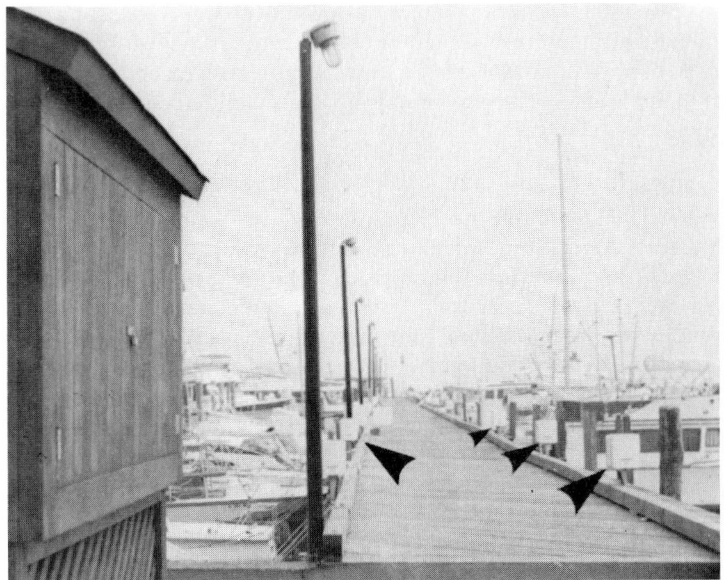

Fig. 555-1. This is one part of a 406-boat marina where shore power is supplied to moored boats from receptacle power pedestals (arrows) supplied by cables run under the pier from a panelboard in the shed at the left (Sec. 555-1.)

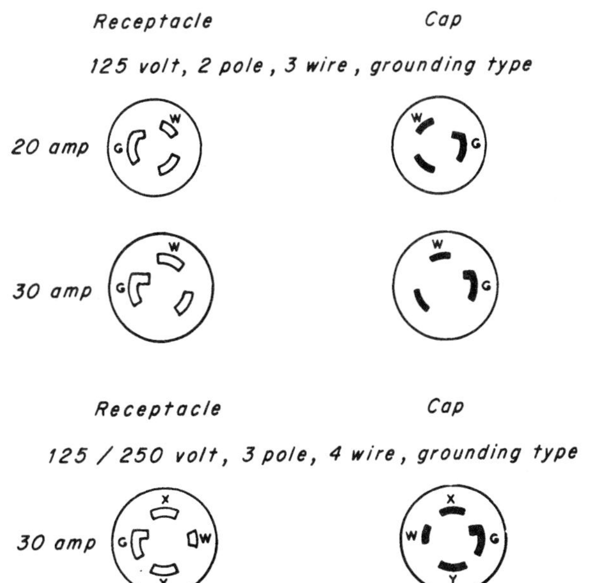

Fig. 555-2. These types of connections provide shore power for boats. (Sec. 555-3.)

type receptacles and caps are required to provide proper contact and assurance that attachment plugs will not fall out easily and disconnect on-board equipment such as bilge pumps or refrigerators. Shore-power receptacles for boats up to 20 ft (6.1 m) long must be rated at least 20 A, and for boats longer than 20 ft (6.1 m) must be rated at least 30 A.

555-4. Branch Circuits. Each single receptacle must be installed on an individual or multiwire branch circuit, with only the one receptacle on the circuit. As shown in Fig. 555-3, a receptacle pedestal unit (two mounted back to back at each location) contains receptacles providing plug-in power to boats at their berths along the pier, with CB protection and control in each housing. As required by NEC Sec. 553-3, each receptacle must be rated not less than 20 A and must be a single locking- and grounding-type receptacle. There is no requirement for ground-fault circuit interruption on these receptacles. (However, at a marina, any 15- or 20-A, 120-V receptacles that are *not* used for shore power to boats must be provided with GFCI protection.)

Also, as required by NEC Sec. 555-4, each individual receptacle in the pedestal unit is supplied by a separate branch circuit of the voltage and current rating that corresponds to the receptacle rating. At each pedestal location a separate bare, stranded No. 6 copper conductor (arrow) is available as a static grounding conductor bonded to all pedestals and lighting fixtures. The inset in Fig. 555-3 shows one receptacle wiring arrangement. Hookup details at pedestal units vary with voltage ratings, current ratings, and phase configuration of power required by different sizes of boats—from small motorboats up to 100-ft (30.48-m) yachts.

555-6. Wiring Methods. The rules here present various options that are available for the circuiting to the loads at marinas and boatyards. This section recognizes any wiring method "identified" for use in wet locations. Examples of wiring methods that are recognized by the NEC for use in wet locations are

1. Rigid nonmetallic conduit.
2. Type MI cable.
3. Type UF cable.
4. Corrosion-resistant rigid metal conduit—which is taken to mean either rigid aluminum conduit or *galvanized* rigid steel conduit. The use of the words "corrosion-resistant" is not intended to require a plastic jacket on galvanized rigid steel conduit, although such a jacket does provide significantly better resistance to natural corrosion, such as rusting.
5. Galvanized IMC.
6. Type MC (metal-clad) cable.

In the design and construction of a marina it is usually necessary to compare the material and labor costs involved in each of those methods. Emphasis is generally placed on long, reliable life of the wiring system—with high resistance to corrosion as well as high mechanical strength to withstand impact and to accommodate some flexing in the circuit runs. The need for great flexibility in running the circuits under the pier and coming up to receptacle pedestals and lighting poles is often extremely important in routing the circuits over, around, and below the many obstructions commonly built into pier construction. And that concern for flexibility in routing can weigh heavily as a labor cost if a rigid conduit system is used.

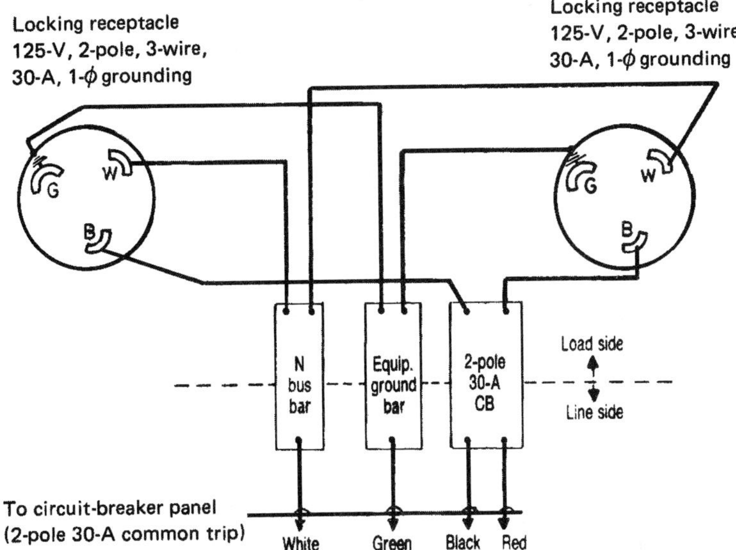

Fig. 555-3. Receptacle providing shore power to each boat is contained in a "power pedestal" (called a "power outlet" in Sec. 555-4) and is a locking and grounding type. (Sec. 555-4.)

Any of the recognized types of cable offer the material-labor advantage of a preassembled, highly flexible "raceway and conductor" makeup that is pretested and especially suited to the bends, offsets, and saddles in the circuit routing at piers, as shown in Fig. 555-4. Cable with a metal armor can offer a completely sealed sheath over the conductors, impervious to fluids and water. For added protection for the metal jacket against oils and other corrosive agents, the cable assembly can have an overall PVC jacket.

Fig. 555-4. PVC-jacketed armored cable—with a continuous, corrugated aluminum armor that is completely impervious to any moisture and water and resistant to corrosive agents—is used for branch circuits and feeders under this marina pier. (Sec. 555-6.)

555-7. Grounding. The purpose is to require an insulated equipment grounding wire that will ensure a grounding circuit of high integrity. Because of the corrosive influences around marinas and boatyards, metal raceways and boxes are not permitted to serve as equipment grounding conductors.

555-8. Wiring Over and Under Navigable Water. There are some federal and local agencies that have specific control over navigable waterways. Accordingly, any proposed installations over or under such waterways should be cleared with the appropriate authorities.

555-9. Gasoline Dispensing Stations—Hazardous (Classified) Locations. Figure 555-5 shows the rules that define the need for hazardous location wiring at a marina. A fuel-dispensing area at the end of the pier consists of gasoline and diesel fuel pumps at the pier edge, with a shack for service personnel at the right (arrow). A panel installed in the shack supplies lighting and receptacles

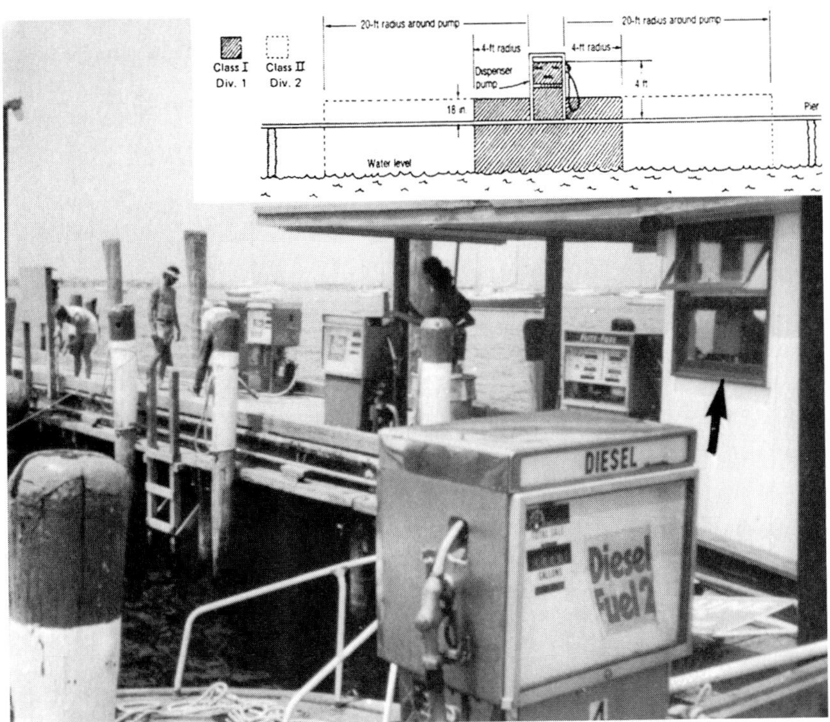

Fig. 555-5. Gasoline pumping areas at a marina must utilize Class I wiring and equipment within the specific classified boundaries. (Sec. 555-9.)

in the shack, as well as the outdoor sign and pumps in the fuel dispensers. Electrical connections from the dispenser pumps tie into the panel. The inset shows the limits and classifications of hazardous locations around each gasoline-dispensing pump. As indicated, some of the space is classified as Class I, Division 1, and other space as Class I, Division 2—both Group D, gasoline.

Type MC cable is suited for use in the Division 2 spaces, but only threaded metal conduit or Type MI cable is suited for the Division 1 spaces.

Chapter Six

ARTICLE 600. ELECTRIC SIGNS AND OUTLINE LIGHTING

600-1. Scope. The 1996 Code represents the first time that this article has been extensively revised and updated in over 40 years. The evolution of technology has essentially made the rules and requirements given in the 1993 and previous Codes irrelevant, if not obsolete. This rewrite, which was championed by the National Electric Sign Association, addresses the latest techniques and clarifies long-time application. The latter is accomplished by incorporating standard industry terminology to distinguish which light sources are subject to a given regulation. The former is accomplished by inclusion of presently recognized industry practice and equipment.

In the case of signs that are constructed at a shop or factory and sent out complete and ready for erection, the inspection department must require listing and installation in conformance with the listing. In the case of outline lighting and signs that are constructed at the location where they are installed—the so-called skeleton tubing covered by part **B**—the inspection department must make a detailed inspection to make sure that all requirements of this article are complied with. In some cities, inspection departments inspect signs in local shops.

600-2. Definitions. This section contains a number of definitions that are critical to the proper application of the Code rules for electric signs and outline lighting.

600-5. Branch Circuits. In part **(a)**, no limit is placed on the number of outlets that may be connected on one circuit for a sign or for outline lighting, except that the total load shall not exceed 20 A where "lamps, ballasts, and transformers, or combinations" of these loads are supplied. A 30-A maximum is established for circuits supplying *only* transformers for electric discharge lighting. Where in normal operation the load will continue for 3 hr or more, the load shall not exceed 80 percent of the branch-circuit rating. See Sec. 210-22(c).

Part **(b)** requires a sign circuit. The 20-A branch circuit for sign and/or outline lighting for commercial occupancies with ground-floor pedestrian entry may supply one or more outlets for the purpose but not any other loads. The intent is that one dedicated 20-A circuit supply one or more outlets.

As noted in part **(c)**, a 1,200-VA load must be allowed for the required sign circuit. Part **(b)** of this section requires that a sign outlet be installed for every ground-level store—even if an outdoor electric sign is not actually installed or planned. That also applies to a whole commercial building with ground floor accessible to pedestrians. Part **(c)** follows up on that and requires that this outlet(s) and its required 20-A branch circuit be taken as a minimum load of 1,200 VA (Fig. 600-2).

600-6. Disconnects. Figure 600-1 depicts the disconnecting means that shall be within sight of the sign, outline lighting, or remote controller. The term "within sight" is clearly defined in Art. 100, and it is well understood that it means the

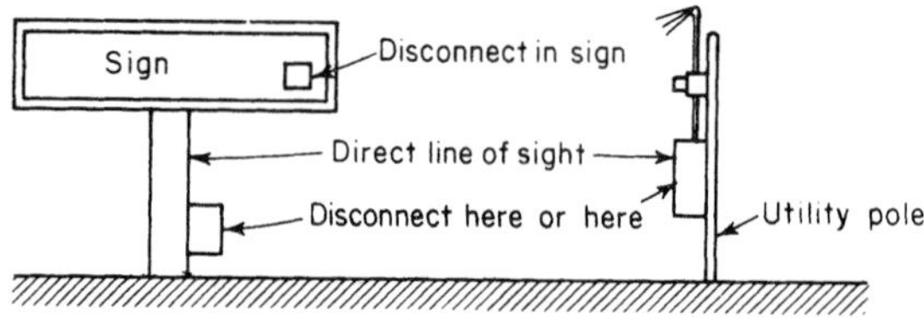

Fig. 600-1. An "in-sight" disconnect may be *in* the sign or visible from the sign. (Sec. 600-2.)

Fig. 600-2. Commercial buildings must have outdoor sign outlet. (Sec. 600-5.)

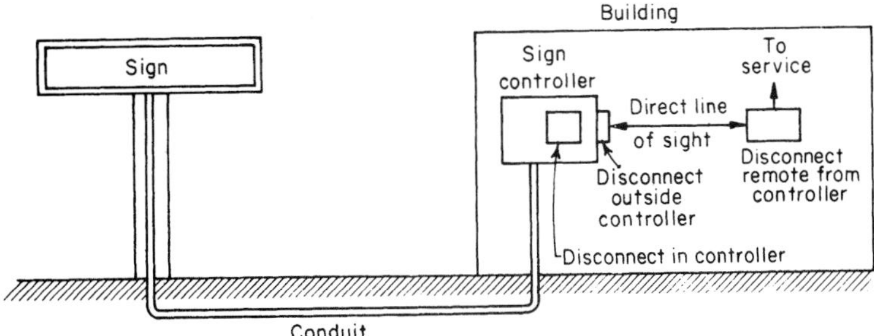

Fig. 600-3. Controller disconnect location may vary but disconnect must be lock-open type. (Sec. 600-6.)

same thing as the term "in sight from," which specifies that it shall be visible and not more than 50 ft (15.24 m) distant from the other.

Figure 600-3 illustrates the conditions recognized by the Exception—which allows the disconnecting means to be located within sight of the controller where the signs are operated by electronic or electromechanical controllers located external to the sign.

With respect to part **(b)**, any switching device controlling the primary of a transformer that supplies a luminous gas tube operates under unusually severe conditions. In order to avoid rapid deterioration of the switch or flasher due to arcing at the contacts, the device must be a general-use AC snap switch or have a current rating of at least twice the current rating of the transformer it controls.

600-10. Portable or Mobile Signs. This rule applies to outdoor portable or mobile signs that are plug-connected. The rule calls for "factory-installed" GFCI protection in or within 12 in. (305 mm) of the plug cap at the end of the supply cord from the sign to protect personnel from potential shock hazards. Documentation for the need for this new rule cited six accidents—three deaths and three shocks—due to ground faults in such outdoor signs that were plug-connected but in which there was no grounding connection or a failed grounding connection.

600.21. Ballasts, Transformers, and Electronic Power Supplies. The transformers used to supply luminous gas tubes are, in general, constant-current devices and, up to a certain limit, the voltage delivered by the transformer increases as the impedance of the load increases. The impedance of the tube increases as the length increases and is higher for a tube of small diameter than for one of larger diameter. Hence a transformer should be selected which is designed to deliver the proper current and voltage for the tube. If the tube is too long or of too small a diameter, the voltage of the transformer may rise to too high a value.

ARTICLE 604. MANUFACTURED WIRING SYSTEMS

604-1. Scope. This article covers modular prefab wiring systems for ceiling spaces.

Prior to the 1981 NE Code, so-called modular wiring systems—those highly engineered, ultra-flexible, plug-in branch-circuit systems for supplying and controlling lighting fixtures in suspended ceilings—encountered opposition in the field. Although these systems have been UL-listed and made of UL-listed raceway, cable, and connector components, inspection agencies adopted a negative attitude toward such systems on the basis that they could not be squared with the NE Code. It was a serious contradiction when UL and the NE Code, both sturdy cornerstones of electrical safety, were viewed to be in conflict.

But now the NE Code has Art. 604 to recognize the various types of modular wiring systems that provide plug-in connections to lighting fixtures, switches, and receptacles in all kinds of commercial and institutional interiors that use suspended ceilings made of lift-out ceiling tiles.

These manufactured wiring systems were logically dictated by a variety of needs in electrical systems for commercial-institutional occupancies. In the interest of giving the public a better way at a better price, a number of manufacturers developed basic wiring systems to provide plug-and-receptacle interconnection of branch-circuit wires to lighting fixtures in suspended-ceiling spaces. Such systems afford ready connection between the hard-wired circuit homerun and cables and/or ducts that form a grid- or treelike layout of circuiting to supply incandescent, fluorescent, or HID luminaries in the ceiling.

Acknowledged advantages of modular wiring systems are numerous and significant:

- Factory-prewired raceways and cables provide highly flexible and accessible plug-in connection to multicircuit runs of 120- and/or 227-V conductors.
- Drastic reductions can be made in conventional pipe-and-wire hookups of individual circuits, which are costly and inflexible.
- Plug receptacles afford a multiplicity of connection points for fixtures to satisfy needs for specific types and locations of lighting units to serve any initial layout of desks or other work stations while still offering unlimited, easy, and extremely economical changes or additions of fixtures for any future rearrangements of office landscaping or activities.
- Systems may also supply switches and/or convenience receptacles in walls or partitions, with readily altered switching provisions to provide energy conservation through effective ON-OFF control of any revised lighting layout.
- Work on the systems has been covered by agreement between the IBEW and associated trades.
- Such systems have a potential for a tax advantage of accelerated depreciation as office equipment rather than real estate.

604-4. Uses Permitted. Modular systems may be used in air-handling ceilings. Equipment may be used in the specific applications and environments for which it is listed by UL.

604-6. Construction. Prewired plug-in connections may be BX or MC cable or metal flex. A minimum of No. 12 copper equipment grounding conductor (insulated or bare) is always required in each cable or flex length—even though flex itself is otherwise permitted to be used by the NE Code without an equipment grounding conductor in lengths not over 6 ft (1.83 m), provided the wires within it are protected at not over 20 A, and BX cable in other uses is recognized by the Code and by UL for equipment grounding through its armor and enclosed aluminum bonding wire (Fig. 604-1).

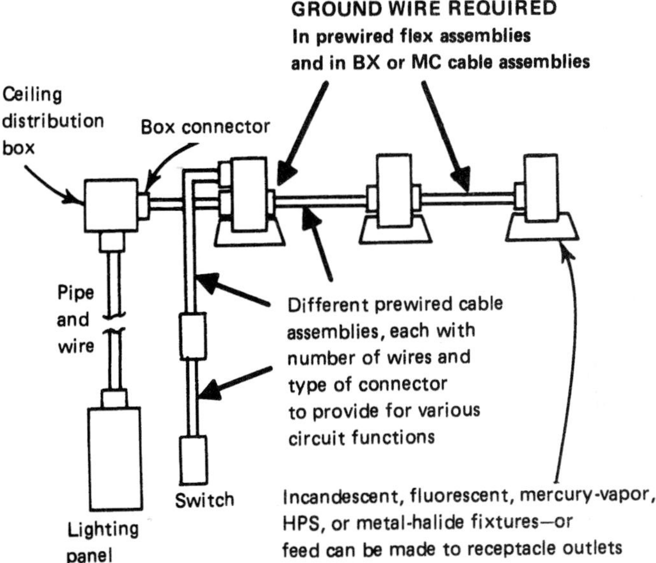

Fig. 604-1. Modular wiring systems are now fully recognized by the Code. (Sec. 604-6.)

ARTICLE 605. OFFICE FURNISHINGS

605-1. Scope. This Code article covers electrical equipment that is part of manufactured partitions used for subdividing office space, as shown in Fig. 605-1.

The substantiation given by the office furniture association that submitted this Code article stated:

> This proposal concerns itself with wiring systems as provided by members of our industry, with office furniture systems that are now being used extensively in offices throughout the United States. Although not exclusively office furniture systems are primarily used in areas referred to as "open plan" or "landscape" office layouts.
>
> Within our industry, office systems furniture has grown in popularity to a great extent over the past several years. Today the sales of this type of furniture are well over $800 million dollars annually and growing. Due to energy conservation requirements users have demanded the inclusion of task and ambient lighting with this type of furniture. Current industry estimates show that approximately 80% of all office furniture systems sold contain electrical power. When such power is provided by manufacturers within our association, safety is foremost in their consideration, and all wiring systems have been or are in the process of being submitted to and listed by Underwriters Laboratories Inc.

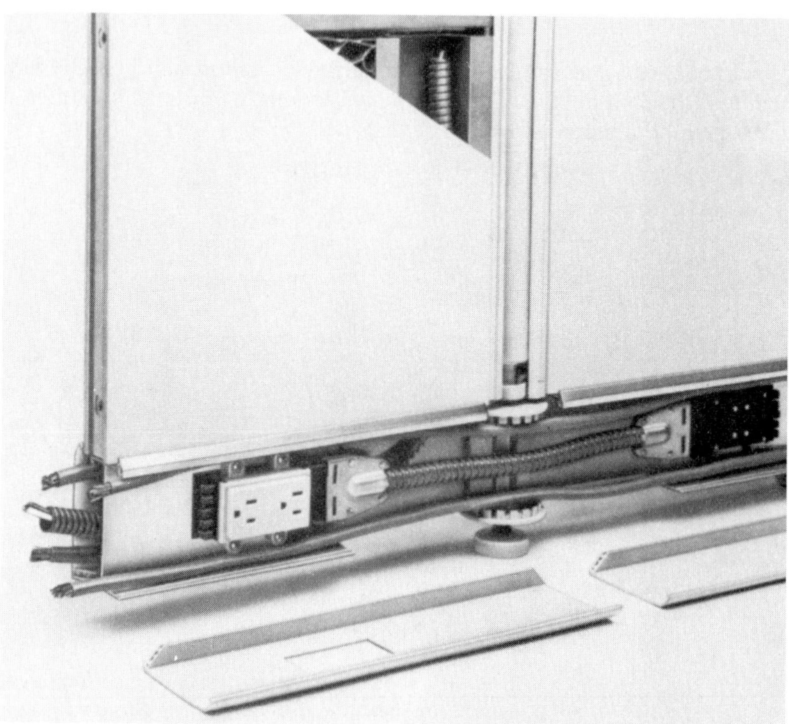

Fig. 605-1. This article covers electrical wiring and electrical components within or attached to manufactured partitions, desks, cabinets, and other equipment that constitute "Office Furnishings." Photo at top shows interior wiring in base of partitions, to supply lighting fixtures and receptacle outlets—as shown at arrows in bottom photo of a typical electrified office work station.

Our industry is very proud of its concern for product safety and performance and the good record that is currently enjoyed. Our purpose in submitting the enclosed proposal to the National Fire Protection Association is to establish a category within the National Electrical Code that deals specifically with products made within our industry that contain wiring systems and to provide in writing the standard of quality that must be adhered to by those making such systems.

605-4. Partition Interconnections. Wired partitions may be interconnected by a cord and plug. The basic rule calls for interconnection of partitions by a "flexible assembly identified for use with wired partitions."

605-8. Free-Standing-type Partitions, Cord- and Plug-Connected. A partition or group of connected partitions that is supplied by cord- and plug-connection to the building electrical system must not be wired with multiwire circuits (all wiring must be 2-wire circuits) and not more than thirteen 15-A, 125-V receptacles may be used.

ARTICLE 610. CRANES AND HOISTS

610-11. Wiring Method. In general, the wiring on a crane or a hoist should be rigid-conduit work or electrical metallic tubing. Short lengths of flexible conduit or metal-clad cable may be used for connections to motors, brake magnets, or other devices where a rigid connection is impracticable because the devices are subject to some movement with respect to the bases to which they are attached. In outdoor or wet locations liquidtight flexible metal conduit should be used for flexible connections.

610-21. Installation of Contact Conductors. Part **(f)** permits use of the track as one of the circuit conductors. In some cases, particularly where a monorail crane or conveyor is used for handling light loads, for the sake of convenience and simplicity it may be desirable to use the track as one conductor of a 3-phase system. Where this arrangement is used, the power must be supplied through a transformer or bank of transformers so that there will be no electrical connection between the primary power supply and the crane circuit, as in Fig. 610-1. The secondary voltage would usually be 220 V, and the primary of the transformer would usually be connected to the power-distribution system of the building or plant. The leg connected to the track must be grounded at the transformer only, except as permitted in Sec. 610-21(f)(4).

610-32. Disconnecting Means for Cranes and Monorail Hoists. This disconnect is an emergency device provided for use in case trouble develops in any of the electrical equipment on the crane or monorail hoist, or to permit maintenance work to be done safely.

610-33. Rating of Disconnecting Means. It is possible that all the motors on a crane might be in operation at one time, but this condition would continue for only a very short while. A switch or CB having a current rating not less than 50 percent of the sum of full-load current rating of all the motors will have ample capacity.

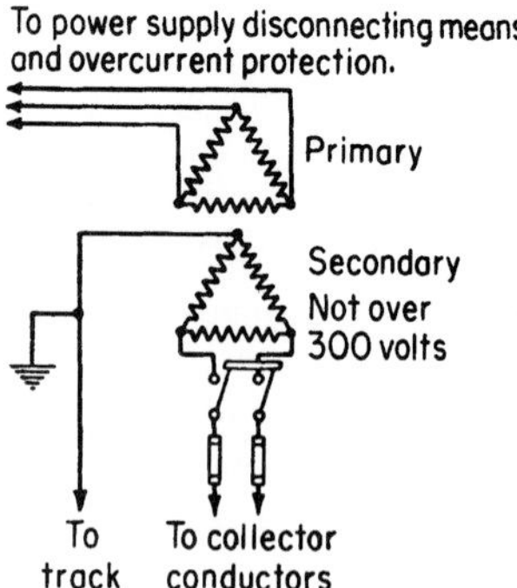

Fig. 610-1. Isolating transformer is used to power track of crane or conveyor. (Sec. 610-21.)

ARTICLE 620. ELEVATORS, DUMBWAITERS, ESCALATORS, MOVING WALKS, WHEELCHAIR LIFTS, AND STAIRWAY CHAIR LIFTS

620-1. Scope. These provisions may also be considered as applying to console lifts, equipment for raising and lowering or rotating portions of theater stages, and all similar equipment.

620-11. Insulation of Conductors. A distinction is made here between the conductors carrying the power current and the smaller wires of operating circuits, such as wires connected to the magnet coils of contactors. The operating current passing through the magnet coils may be quite small, and a small current leaking through damp slow-burning insulation where two insulated wires are in contact might be sufficient to operate a contactor.

620-12. Minimum Size of Conductors. Code Tables 310-16 to 310-19 do not include the ampacity for No. 20 AWG copper conductors. However, it is generally considered that No. 20 conductors up to two conductors in cable or cord may safely carry 3 A.

Because of wider use of advanced semiconductor computer equipment, use of wire smaller than No. 20 is permitted for other than traveling cables by part **(b)**, with No. 24 the new minimum.

The development of elevator control equipment, which has been taking place for many years, has resulted in the design and use of equipment including electronic unit contactors requiring very much smaller currents (milliamperes) for their operation.

620-36. Different Systems in One Raceway or Traveling Cable. It would be difficult, if not practically impossible, to keep the wires of each system completely isolated from the wires of every other system in the case of elevator control and signal circuits. Hence such wires may be run in the same conduits and cables if all wires are insulated for the highest voltage used and if all live parts of apparatus are insulated from ground for the highest voltage, provided that the signal system is an integral part of the elevator wiring system.

ARTICLE 630. ELECTRIC WELDERS

630-1. Scope. There are two general types of electric welding: arc welding and resistance welding. In arc welding, an arc is drawn between the metal parts to be joined together and a metal electrode (a wire or rod), and metal from the electrode is deposited on the joint. In resistance welding, the metal parts to be joined are pressed tightly together between the two electrodes, and a heavy current is passed through the electrodes and the plates or other parts to be welded. The electrodes make contact on a small area—thus the current passes through a small cross section of metal having a high resistance—and sufficient heat is generated to raise the parts to be welded to a welding temperature.

In arc welding with AC, an individual transformer is used for each operator; i.e., a transformer supplies current for one arc only. When DC is used, there is usually an individual generator for each operator, though there are also "multi-operator" arc-welding generators.

630-11. Ampacity of Supply Conductors. The term *transformer arc welder* is commonly used in the trade and hence is used in the **Code**, though the equipment might more properly be described as an *arc-welding transformer.* Reference should be made here to the FPN following Sec. 630-31 where the term *duty cycle* is explained.

It is evident that the load on each transformer is intermittent. Where several transformers are supplied by one feeder, the intermittent loading will cause much less heating of the feeder conductors than would result from a continuous load equal to the sum of the full-load current ratings of all the transformers. The ampacity of the feeder conductors may therefore be reduced if the feeder supplies three or more transformers.

630-12. Overcurrent Protection. Arc-welding transformers are so designed that as the secondary current increases, the secondary voltage decreases. This characteristic of the transformer greatly reduces the fluctuation of the load on the transformer as the length of the arc, and consequently the secondary current, is varied by the operator.

The rating or setting of the overcurrent devices specified in this section provides short-circuit protection. It has been stated that with the electrode "frozen" to the work the primary current will in most cases rise to about 170 percent of the current rating of the transformer. This condition represents the heaviest overload that can occur, and of course this condition would never be allowed to continue for more than a very short time.

630-31. Ampacity of Supply Conductors. Subparagraph **(a)(1)** applies where a resistance welder is intended for a variety of different operations, such as for welding plates of different thicknesses or for welding different metals. In this case the branch-circuit conductors must have an ampacity sufficient for the heaviest demand that may be made upon them. Because the loading is intermittent, the ampacity need not be as high as the rated primary current. A value of 70 percent is specified for any type of welding machine which is fed automatically. For a manually operated welder, the duty cycle will always be lower and a conductor ampacity of 50 percent of the rated primary current is considered sufficient.

example 1 A spot welder supplied by a 60-Hz system makes 400 welds per hour, and in making each weld, current flows during 15 cycles.

The number of cycles per hour is 60 × 60 × 60 = 216,000 cycles.

During 1 hr, the time during which the welder is loaded, measured in cycles, is 400 × 15 = 6,000 cycles.

The duty cycle is therefore (6,000/216,000) × 100 = 2.8 percent.

example 2 A seam welder operates 2 cycles "on" and 2 cycles "off," or in every 4 cycles the welder is loaded during 2 cycles.

The duty cycle is therefore ²⁄₄ × 100 = 50 percent.

Transformers for resistance welders are commonly provided with taps by means of which the secondary voltage, and consequently the secondary current, can be adjusted. The rated primary current is the current in the primary when the taps are adjusted for maximum secondary current.

When a resistance welder is set up for a specific operation, the transformer taps are adjusted to provide the exact heat desired for the weld; then in order to apply subparagraph **(a)(2)** the actual primary current must be measured. A special type of ammeter is required for this measurement because the current impulses are of very short duration, often a small fraction of a second. The duty cycle is controlled by the adjustment of the controller for the welder.

The procedure in determining conductor sizes for an installation consisting of a feeder and two or more branch circuits to supply resistance welders is first to compute the required ampacity for each branch circuit. Then the required feeder ampacity is 100 percent of the highest ampacity required for any one of the branch circuits, plus 60 percent of the sum of the ampacities of all the other branch circuits.

Some resistance welders are rated as high as 1,000 kVA and may momentarily draw loads of 2,000 kVA or even more. Voltage drop must be held within rather close limits to ensure satisfactory operation.

630-32. Overcurrent Protection. In this case, as in the case of the overcurrent protection of arc-welding transformers (Sec. 630-12), the conductors are protected against short circuits. The conductors of motor branch circuits are protected against short circuits by the branch-circuit overcurrent devices and depend upon the motor-running protective devices for overload protection. Although the resistance welder is not equipped with any device similar to the motor-running protective device, satisfactory operation of the welder is a safeguard against overloading of the conductors. Overheating of the circuit could result only from so operating the welder that either the welds would be imperfect, or parts of the control equipment would be damaged, or both.

ARTICLE 640. SOUND-RECORDING AND SIMILAR EQUIPMENT

640-1. Scope. Centralized distribution systems consist of one or more disc or tape recorders and/or radio receivers, the audio-frequency output of which is distributed to a number of reproducers or loudspeakers.

A public-address system includes one or more microphones, an amplifier, and any desired number of reproducers or speakers. A common use of such a system is to render the voice of a speaker clearly audible in all parts of a large assembly room.

640-2. Application of Other Articles. In general, the power-supply wiring from the building light or power service to the special equipment named in Sec. 640-1, and between any parts of this equipment, should be installed as required for light and power systems of the same voltage. Certain variations from the standard requirements are permitted by the following sections. For radio and television receiving equipment, the requirements of Art. 810 apply except as otherwise permitted here.

Part **(b)** covers wiring to loudspeakers and microphones and signal wires between equipment components—tape recorder or record player to amplifier, etc. As shown in Fig. 640-1, amplifier output wiring to loudspeakers handles energy limited by the power (wattage) of the amplifier and must conform to the rules of Art. 725. As shown in **Code** Table 725-31(a), the voltage and current rating of a signal circuit will establish it as either Class 2 or 3 signal circuit. Amplifier output circuits rated not over 70 V, with open-circuit voltage not over 100 V, may use Class 3 wiring as set forth in **Code** Table 725-31(a) of Art. 725.

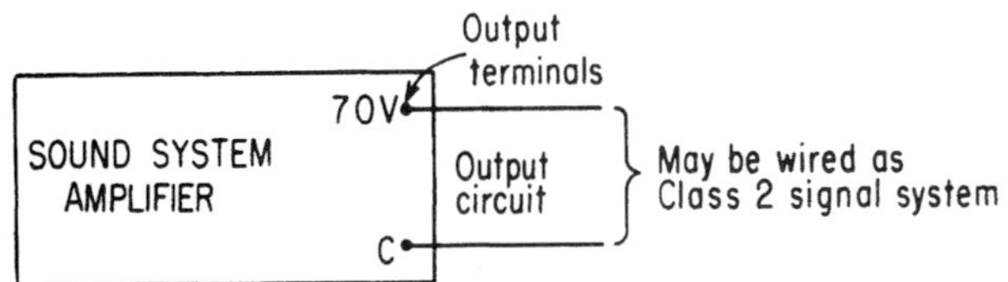

Fig. 640-1. Sound-system speaker wiring may be either Class 2 or 3 signal system. (Sec. 640-2.)

Article 725 of the **Code** covers, among other things, signal circuits. A signal circuit is defined as any electrical circuit which supplies energy to a device—like a loudspeaker or an amplifier—that gives a recognizable signal.

640-4. Wireways and Auxiliary Gutters. Wireways and auxiliary gutters may be used with conductor occupancy up to 75 percent of cross-section area (instead of 20 percent as for power and light wires) and may be used in concealed places where run in straight lines between wiring boxes (Fig. 640-2).

Wireway for sound system

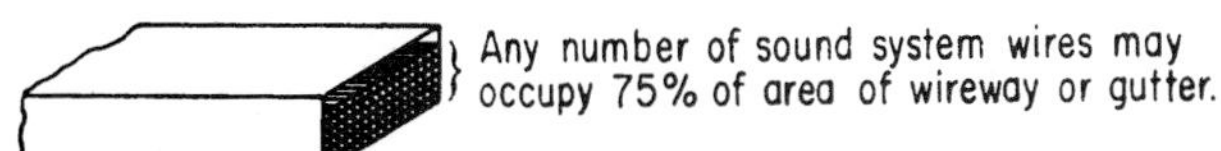

Fig. 640-2. Signal-circuit wires may fill wireway more fully than power wires. (Sec. 640-4.)

640-6. Grouping of Conductors. In this class of work, the wires of different systems are in many cases closely associated in the apparatus itself; therefore little could be gained by separating them elsewhere.

The input leads to a motor-generator set or to a rotary converter would commonly be 115- or 230-V power circuits. These wires are not a part of the sound-recording or reproducing system and should be kept entirely separate from all wires of the sound system.

640-10. Circuit Overcurrent Protection. Although use of solid-state electronic equipment has been steadily replacing vacuum-tube constructions, these rules apply to tube-type components. The overcurrent protection described here is actually involved with the internal circuiting of electronic-tube equipment. Other protection for external signal circuits must comply with Secs. 725-35 and 725-36 on protection of Class 2 and 3 circuits.

As mentioned in part **(a)**, a 20 amp-hr battery is capable of delivering a heavy enough current to heat a No. 14 or smaller wire to a dangerously high temperature, and overcurrent protection is therefore quite necessary. A storage B battery might be capable of delivering enough current to overheat some part of the equipment. Several different positive connections may be made to the battery in order to obtain different voltages, and each such lead must be provided with overcurrent protection.

ARTICLE 645. INFORMATION TECHNOLOGY EQUIPMENT

645-1. Scope. This article applies only to a Information Technology Equipment room. Because this article covers all of the designated "equipment," "wiring," and "grounding" that is contained in a room, there is no question about the mandatory application of these rules to such "rooms." But the specific nature of that word *room* in the Sec. 645-1 statement of "Scope" leaves doubt and uncertainty about electronic computer/data processing equipment and systems that are *not* installed in a dedicated room. For such applications, the safest conclusion would be to simply follow Art. 645 rules as closely as possible, especially because all of the rest of the **Code** applies.

The original proposal for Art. 645 (1968) also contained other material that gives insight into the meaning of the concepts, as follows:

> **Definitions.** In addition to those included in Article 100, the following definitions are applicable to this Article.
>
> **(a) Console.** Unit containing main operative controls of the system.
>
> **(b) Data Processing System.** Any electronic digital or analog computer, along with all peripheral, support, memory, programming or other directly associated equipment, records, storage and activities. The most common types of electronic computer systems are of the digital computer type and are usually classed as Electronic Data Processing Machines (EDPM), Automatic Data Processing Machines (ADPM), and/or Integrated Data Processing Systems.
>
> **(c) Interconnecting Cables.** Signal and power cables for operation and control of the system.

(d) Raised Floor. Platform on which machines are installed for housing interconnecting cables, and at times as a means of supplying conditioned air to various units.

(a) Wiring Under Raised Floors. Data processing equipment, if connected to the power supply system by means of an approved computer cable or flexible cord and attachment plug cap, or cord set assembly located under a raised floor construction, shall be acceptable provided that:

(1) Existing combustible, structural floors shall be covered with an insulating noncombustible material before the raised floor is installed.

(2) The supporting members for the raised floor shall be of concrete, steel, aluminum or other noncombustible material.

(3) The decking for the raised floor shall be one of the following: (a) Concrete, steel, aluminum or other noncombustible material; (b) Pressure impregnated, fire retardant treated lumber having a flame spread rating of 25 or less. (See NFPA Method of Test Surface Burning Characteristics of Building Materials, No. 255); (c) Wood or similar core material which is encased on the top and bottom with sheet, cast or extruded metal with all openings or cut edges covered with metal or plastic clips or grommets so that none of the core is exposed, and has an assembly flame spread rating of 25 or less. (See NFPA Method of Test of Surface Burning Characteristics of Building Materials, No. 255).

(4) Access sections or panels shall be provided in the raised floor so that all space beneath is accessible (by the use of simple hand tools, available on the site, if necessary) so that any area or space beneath the floor can be exposed in not over one and one-half minutes.

(5) The underfloor area, if ventilated, shall be used for air handling associated with the data processing equipment only.

(6) The power supply computer cable or flexible cord assembly shall be not longer than 15 feet.

Grounding

(a) All external noncurrent-carrying metallic parts of a data processing system liable to become energized or liable to have a potential above ground, shall be bonded and grounded in accordance with one of the methods outlined in Sections 250-51 to 250-59, and Sections 250-71 to 250-79 inclusive.

(b) One or more individual grounding conductors shall be used to bond the separate units of a data processing system, and the conduit and other metallic raceways associated with the system, and to connect those parts to a common ground point, which, in turn, is connected to the grounded building structure or other recognized system ground.

(c) The grounding conductor required by Section 645-6(b) shall be copper or other corrosion resistant material, stranded or solid, insulated or bare with an ampacity at least as great as the supply wiring connecting that unit or component to the source of supply, and installed so that it will not be subjected to physical damage.

NOTE: The method of grounding is critical to the operation of a data processing system. The design of the system shall determine the grounding method to be used.

Most data processing systems are designed to have a single point for the grounding of the frames and exposed metallic parts of each unit. This point provides both safety ground and logic or DC ground.

Metal conduit, metal clad cable or other metallic raceway systems in most cases do not provide a sufficient ground for the data processing system. If they are used in the supply wiring installation to a system, it may be necessary to connect those portions of the metallic wiring system separately to the common ground point, and isolate them from the frame of the equipment being served so that equipment ground is obtained solely by the separate conductor required by 645-7(b,c).

645-2. Special Requirements for Information Technology Equipment Room. Six conditions are described under which the rules of Art. 645 may be applied to data processing equipment:

1. One *or* two "grouped and identified" disconnects must be provided to open the supply to *all* electronic equipment and *all* HVAC equipment in the computer room, with this disconnect (or these disconnects) controlled at the "principal exit doors" of the room.
2. A dedicated HVAC system must be used for the room, or strictly limited use may be made of an HVAC system that "serves other occupancies."
3. Only "listed" (such as by UL) electronic equipment may be installed.
4. The computer "room" must be "occupied only by those personnel needed for" operating and maintaining the computer/data processing equipment.
5. The "room" must have *complete* fire-rated separation from "other occupancies."
6. All applicable building codes are satisfied.

It is very clear from the wording that if any of those conditions is *not* met, the entire computer/data processing installation is *not* subject to the rules of Art. 645. But such an installation would be subject to *all* other applicable rules of the NEC and could even be made subject to Art. 645 on an optional basis.

645-5. Supply Circuits and Interconnecting Cables. Part **(a)** limits every branch circuit supplying data processing units to a maximum load of not over 80 percent of the conductor ampacity (which is an ampacity of 1.25 times the total connected load).

Part **(b)** covers use of computer or data processing cables and flexible cords. As shown in Fig. 645-1 (under a raised floor), part **(c)** permits data processing units to be "interconnected" by flexible connections that are "approved as a part" of the system.

Part **(d)** permits a variety of wiring methods under a raised floor serving a data processing system: metal surface raceway with metal cover, metal wireway (as shown in Fig. 645-2), liquidtight flexible conduit, rigid metal conduit, EMT, flexible metal conduit, IMC, Type MI cable, Type AC cable (commonly known as "BX"), and Type MC cable (Fig. 645-2).

Although Sec. 352-1 prohibits use of a metal surface raceway where it would be concealed, Exception No. 2 of that rule recognizes its use under raised flooring for data processing by referencing Art. 645. It should be noted that, in addition to a surface metal raceway, a metal wireway may be used under a raised floor. Section 645-5(d)(2) does recognize use of "wireway" under raised floors. Section 362-2 accepts wireway "*only* for exposed work," but it may be used under raised floors and above suspended ceilings of lift-out tiles because of the definition of "exposed." In addition, Sec. 300-22(c) does recognize metal wireway above a suspended ceiling space used for air-handling purposes.

Part **(d)(4)** requires openings in raised floors to provide abrasion protection for cables passing up through the floor and requires that openings be only as large as needed and made in such a way as to "minimize the entrance of debris beneath the floor."

Part **(e)** adds important information by stating clearly that any cable, boxes, connectors, receptacles, or other components that are "listed as part of, or for,

Fig. 645-1. Connection of data-processing units to their supply circuits and interconnection between units (power supply, memory storage, etc.) may be made only with cables or cord-sets specifically approved as parts of the data processing system. (Sec. 645-5.)

information technology equipment" are *not* required to be secured in place, *but,* any cable or equipment that is *not* "listed as part of" the computer equipment *must* be secured in accordance with all **Code** rules covering them.

645-10. Disconnecting Means. As shown in Fig. 645-3, a master means of disconnect (which could be one or more switches or breakers) must provide disconnect for all computer equipment, ventilation, and air-conditioning (A/C) in the data processing (DP) room.

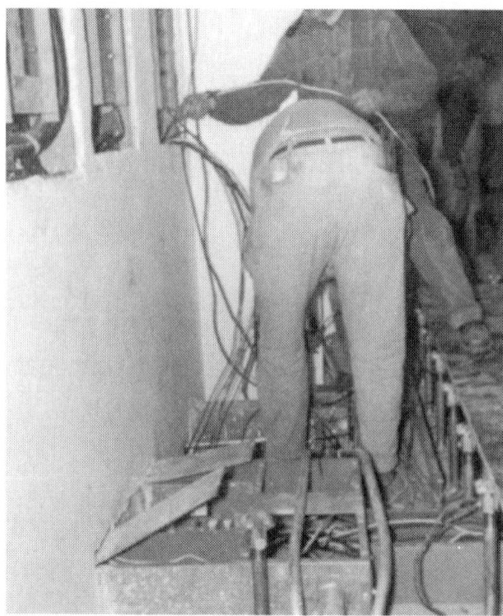

Fig. 645-2. Branch circuits from a panelboard to data processing receptacle outlets must be in a metal-clad raceway system or use Type MI, Type AC, or Type MC cable. (Sec. 645-5.)

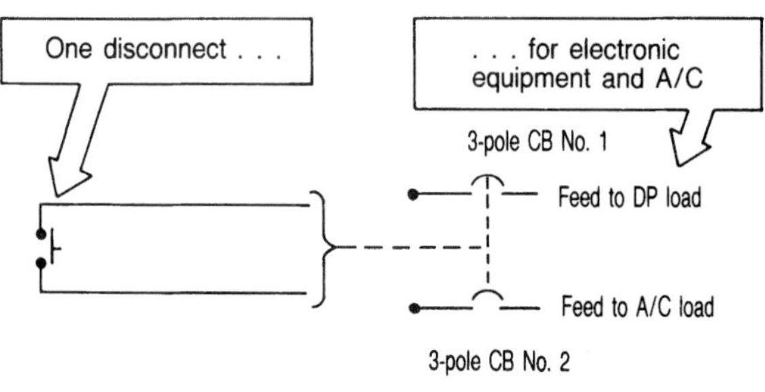

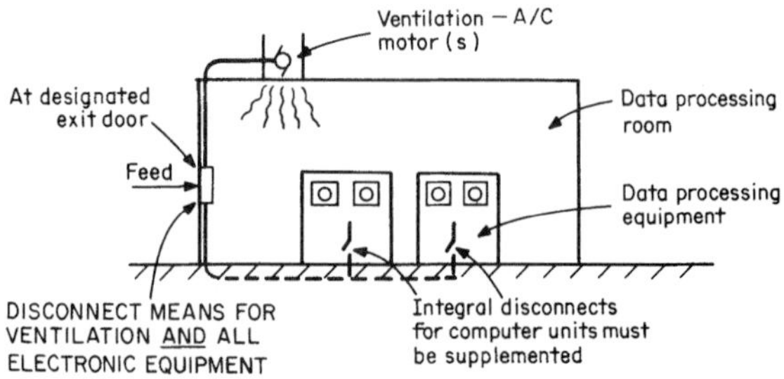

Fig. 645-3 Data processing room must have arrangements like those shown above. (Sec. 645-10.)

The disconnects called for in this rule are required to shut down the DP system and its dedicated HVAC and to close all required fire/smoke dampers under emergency conditions, such as fire in the equipment or in the room. For that reason, the rule further requires that the disconnect for the electronic equipment and "a similar" disconnect for A/C (which could be the same control switch or a separate one) must be grouped and identified and must be "controlled" from locations that are readily accessible to the computer operator(s) or DP manager. And then the rule specifies that these one or more emergency disconnects must be installed at all "principal" exit doors—any doors that occupants of the room might use when leaving the DP room under emergency conditions. The concept is that operators would find it easy to operate the one or more control switches as they exited the room through the doorway. Figure 645-4 shows two control switches—one in the control circuit of the A/C system and the other a shunt-trip pushbutton in the CB of the feeder to the DP branch circuits—with a collar guard to prevent unintentional operation.

Although the present wording of this rule readily accepts the use of a single disconnect device (pushbutton) that will actuate one or more magnetic contactors that switch the feeder or feeders supplying the branch circuits for the computer equipment and the circuits to the A/C equipment, the wording also recognizes the use of separate disconnect control switches for electronic equipment and A/C. Control of the branch circuits to electronic equipment may be provided by a contactor in the feeder to the transformer primary of a computer power center, as shown in Fig. 645-5.

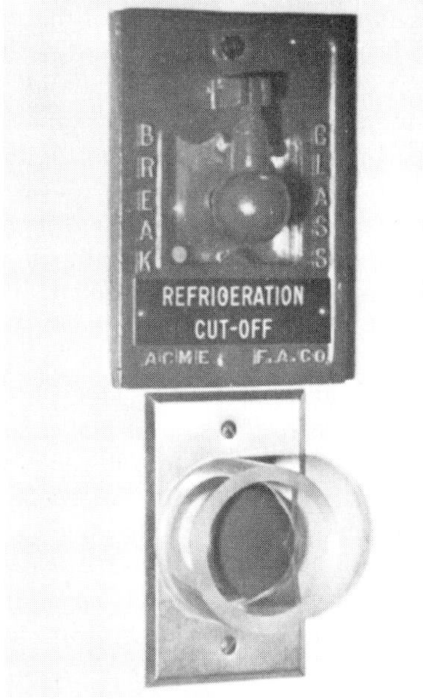

Fig. 645-4. Adjacent to the door of a DP room, a break-glass station (at top) provides emergency cutoff of the A/C system in the DP room; and a mushroom-head pushbutton—with an extended collar guard that requires definite, intentional pushing action—energizes a shunt-trip coil in the feeder circuit breaker supplying branch circuits for the electronic DP equipment. (Sec. 645-10.)

Fig. 645-5. Rapidly accelerating application of DP equipment in special DP rooms with wiring under a raised floor of structural tiles places great emphasis on Art. 645. "Computer power centers" (arrow) are complete assemblies for the supply of branch circuits to DP equipment, with control, monitoring, and alarm functions. (Sec. 645-10.)

A single means used to control the disconnecting means for *both* the electronic equipment and the air-conditioning system offers maximum safety. In the event of a fire emergency, having two separate disconnecting means (or their remote operators) at the principal exit doors will require the operator to act twice and thus increase the hazard that only the electronic equipment or the air-conditioning system will be shut down. If only the electronic equipment is disconnected, a smoldering fire will become intensified by the air-conditioning system force-ventilating the origin of combustion. Similarly, if only the air-conditioning system is disconnected, either a fire within the electronic equipment will become intensified (since the electric energy source is still present), or the electronic equipment could become dangerously overheated due to the lack of air conditioning in this area.

Wording of this rule requires means to disconnect the "dedicated" A/C "system serving the room." If the DP room has A/C from the ceiling for personnel comfort and A/C through the raised floor space for cooling of the DP equipment, both A/C systems would have to be disconnected. There have been rulings that *only* the A/C serving the raised floor space must be shut down to minimize fire spread within the DP equipment and that the general room A/C, which is tied into the whole building A/C system, does not have to be interrupted. In other cases, it has been ruled that the general room A/C must be shut down, while the floor space A/C may be left operating to facilitate the dispersion of fire-suppressant and extinguishing materials within the enclosures of DP equipment that is on fire. Because of the possibility of various specific interpretations of very general rules, this whole matter has become extremely controversial.

The Exception to this rule waives the need for disconnect means in any "Integrated Electrical Systems" (Art. 685), where orderly shutdown is necessary to ensure safety to personnel and property. In such cases, the entire matter of type of disconnects, their layout, and their operation is left to the designer of the specific installation.

645-11. Uninterruptible Power Supplies (UPS). This rule requires disconnects for "supply and output circuits" of any UPS "within" the computer room. The UPS disconnecting means *must* satisfy Sec. 645-10, and it *must* "disconnect the battery from its load." The wording of this rule leaves questions about a disconnecting means for a UPS installed *outside* the computer room (Fig. 645-6).

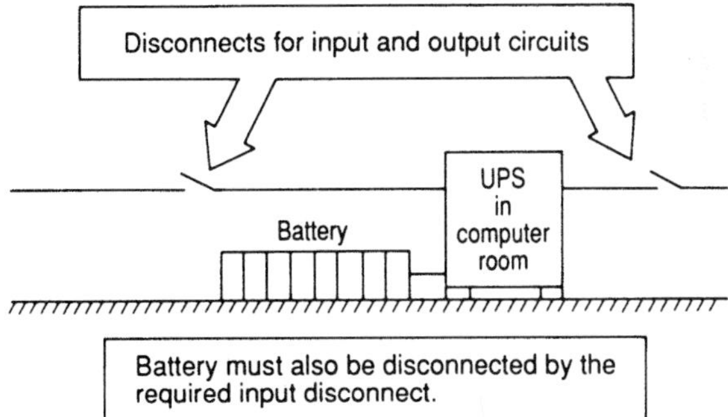

Fig. 645-6. Considerable interpretation latitude is inherent in the rule that requires "grouped and identified" switches or circuit breakers "at principal exit doors" to control an uninterruptible power supply. (Sec. 645-11.)

645-15. Grounding. Data processing equipment must *either* be grounded in full compliance with Art. 250 or it *must* be "double insulated." But contradiction is expressed in two specific rules:

1. Any power system "derived within listed" computer equipment that supplies the computer systems through "receptacles or cable assemblies supplied as part of this equipment" must *not* be considered to be a separately derived system [Sec. 250-5(d)] and may *not* be grounded in accordance with Sec. 250-26 covering such systems.
2. *All* exposed noncurrent-carrying metal parts of an electronic computer/data processing system "*shall* be grounded."

The rule in this section—exempting computer power equipment from the requirements for a separately derived system—is in conflict with the substantiation given of the revision of the definition of premises wiring in Art. 100 of the 1990 NEC. It also appears to conflict with part **(d)** of Sec. 250-21. Members of the

Code Making Panel for Art. 645 made strong objections to the wording and ideas covered here. It seems certain that controversy and confusion will continue.

ARTICLE 650. PIPE ORGANS

650-4. Grounding. Organ control systems are usually supplied from a motor-generator set consisting of a 115- or 230-V motor driving a generator that operates at about 10 V. Neither the generator windings nor the control wires are necessarily insulated for the motor voltage. Assume that the frames of the two machines are electrically connected together by being mounted on the same base and that the frames are not grounded. If a wire of the motor winding becomes grounded to the frame of the motor, the frames of both machines may be raised to a potential of 115 or 230 V above ground, and this voltage may break down the insulation of the generator winding or of the circuit wiring. If the generator is insulated from the motor, or if both frames are well grounded, this trouble cannot occur.

650-5. Conductors. In part **(d)**, the wires of the cable are normally all of the same polarity and hence need not be heavily insulated from one another. The full voltage of the control system exists between the wires in the cable and the common return wire; therefore the common wire must be reasonably well insulated from the cable wires.

650-6. Installation of Conductors. A 30-V system involves very little fire hazard, and the cable may be run in any manner desired; but for protection against injury and convenience in making repairs, the cable should preferably be installed in a metal raceway.

650-7. Overcurrent Protection. The "main supply conductors" extend from the generator to a convenient point at which one conductor is connected through 6-A fuses to as many circuits as may be necessary, while the other main conductor is connected to the common return.

ARTICLE 660. X-RAY EQUIPMENT

660-1. Scope. An x-ray tube of the hot-cathode type, as now commonly used, is a two-element vacuum tube in which a tungsten filament serves as the cathode. Current is supplied to the filament at low voltage. In most cases unidirectional pulsating voltage is applied between the cathode and the anode. The applied voltage is measured or described in terms of the peak voltage, which may be anywhere within the range of 10,000 to 1,000,000 V, or even more. The current flowing in the high-voltage circuit may be as low as 5 mA or may be as much as 1 A, depending upon the desired intensity of radiation. The high voltage is obtained by means of a transformer, usually operating at 230-V primary, and usually is made unidirectional by means of two-element rectifying vacuum tubes, though in some cases an alternating current is applied to the x-ray tube. The x-rays are radiations of an extremely high frequency (or short wavelength) which are the strongest in a plane at right angles to the electron stream passing between the cathode and the anode in the tube.

As used by physicians and dentists, x-rays have three applications: *fluoroscopy,* where a picture or shadow is thrown upon a screen of specially prepared glass by rays passing through some part of the patient's body; *radiography,* which is similar to fluoroscopy except that the picture is thrown upon a photographic film instead of a screen; and *therapy,* in which use is made of the effects of the rays upon the tissues of the human body.

660-24. Independent Control. In radiography it is important that the exposure be accurately timed, and for this purpose a switch is used which can be set to open the circuit automatically in any desired time after the circuit has been closed.

660-35. General. A power transformer supplying electrical systems is usually supplied at a high primary voltage; hence in case of a breakdown of the insulation on the primary winding, a large amount of energy can be delivered to the transformer. An askarel-filled x-ray transformer involves much less fire hazard because the primary voltage is low, and it is therefore not required that such transformers be placed in vaults of fire-resistant construction.

660-47. General. This section definitely requires that all new x-ray equipment shall be so constructed that all high-voltage parts, except leads to the x-ray tube, are in grounded metal enclosures, unless the equipment is in a separate room or enclosure and the circuit to the primary of the transformer is automatically opened by unlocking the door to the enclosure. Conductors leading to the x-ray tube are heavily insulated.

ARTICLE 665. INDUCTION AND DIELECTRIC HEATING EQUIPMENT

665-1. Scope. Induction and dielectric heating are systems wherein a workpiece is heated by means of a rapidly alternating magnetic or electric field.

665-2. Definitions

Dielectric Heating:

In contrast, dielectric heating is used to heat materials that are nonconductors, such as wood, plastic, textiles, rubber, etc., for such purposes as drying, gluing, curing, and baking. It uses frequencies from 1 to 200 MHz, especially those from 1 to 50 MHz. Vacuum-tube generators are used exclusively to supply dielectric heating power, with outputs ranging from a few hundred watts to several hundred kilowatts.

Whereas induction heating uses a varying magnetic field, dielectric heating employs a varying electric field. This is done by placing the material to be heated between a pair of metal plates, called electrodes, in the output circuit of the generator. When high-frequency voltage is applied to the electrodes, a rapidly alternating electric field is set up between them, passing through the material to be heated. Because of the electrical charges within the molecules of this material, the field causes the molecules to vibrate in proportion to its frequency. This internal molecular action generates the heat used for dielectric heating.

Induction Heating:

Induction heating is used to heat materials that are good electrical conductors, for such purposes as soldering, brazing, hardening, and annealing. Induction heating, in general, involves frequencies ranging from 3 to about 500 kHz, and power outputs from a few hundred watts to several thousand kilowatts. In general, motor-generator sets are used for frequencies up to about 30 kHz; spark-gap converters, from 20 to 400 kHz; and vacuum-tube generators, from 100 to 500 kHz. Isolated special jobs may use frequencies as high as 60 to 80 MHz. Motor-generator sets normally supply power for heating large masses for melting, forging, deep hardening, and the joining of heavy pieces, whereas spark-gap and vacuum-tube generators find their best applications in the joining of smaller pieces and shallow case hardening, with vacuum-tube generators also being used where special high heat concentrations are required.

To heat a workpiece by induction heating, it is placed in a work coil consisting of one or more turns, which is the output circuit of the generator (Fig. 665-1). The high-frequency current which flows through this coil sets up a rapidly alternating magnetic field within it. By inducing a voltage in the workpiece, this field causes a current flow in the piece to be heated. As the current flows through the resistance of the workpiece, it generates heat (I^2R loss) in the piece itself. It is this heat that is utilized in induction heating.

665-22. Access to Internal Equipment. This section allows the manufacturer the option of using interlocked doors or detachable panels. Where panels are used

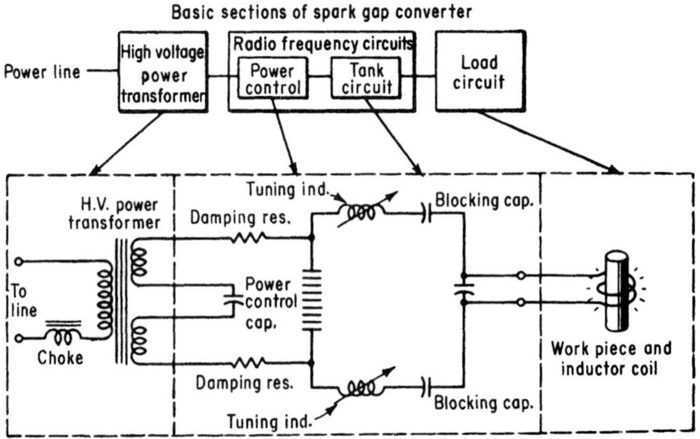

Fig. 665-1. A "generator" circuit supplies the "work coil" of an induction heater. (Sec. 665-2.)

and are not intended as normal access points, they shall be fastened with bolts or screws of sufficient number to discourage removal. They should not be held in place with any type of speed fastener.

665-25. Work Applicator Shielding. See the discussion under Sec. 665-44. This section is intended primarily to apply to dielectric heating installations where it is absolutely essential that the electrodes and associated tuning or matching devices are properly shielded.

665-26. Grounding and Bonding.

Bonding

At radio frequencies, and especially at dielectric-heating frequencies (1 to 200 MHz), it is very possible for differences in radio-frequency potential to exist between the equipment proper and other surrounding metal objects or other units of the complete installation. These potentials exist because of stray currents flowing between units of the equipment or to ground. Bonding is therefore essential, and such bonding must take the form of very wide copper or aluminum straps between units and to other surrounding metal objects such as conveyors, presses, etc. The most satisfactory bond is provided by placing all units of the equipment on a flooring or base consisting of copper or aluminum sheet, thoroughly joined where necessary by soldering, welding, or adequate bolting. Such bonding reduces the radio-frequency resistance and reactance between units to a minimum, and any stray circulating currents flowing through this bonding will not cause sufficient voltage drop to become dangerous.

Shielding

Shielding at dielectric-heating frequencies is a necessity to provide operator protection from the high radio-frequency potentials involved, and also to prevent possible interference with radio communication systems. Shielding is accomplished by totally enclosing all work circuit components with copper sheet, copper screening, or aluminum sheet.

665-44. Output Circuit.

RF Lines

When it is necessary to transmit the high-frequency output of a generator any distance to the work applicator, a radio-frequency line is generally used. This usually consists of a conductor totally enclosed in a grounded metal housing. This central conductor is commonly supported by insulators, mounted in the grounded housing, and periodically spaced along its length. Such a line, rectangular in cross section, may even be used to connect two induction generators to the load.

While contact with high-voltage radio frequencies may cause severe burns, contact with high-voltage DC could be fatal. Therefore, it is imperative that generator output (directly, capacitively, or inductively coupled) be effectively grounded with respect to DC so that, should generator failure place high-voltage DC in the tank oscillating circuit, there will still be no danger to the operator. This grounding is generally internal in vacuum-tube generators. In all types of induction generators, one side of the work coil should usually be externally grounded.

In general, all high-voltage connections to the primary of a current transformer should be enclosed. The primary concern is the operator's safety. Examples would be interlocked cages around small dielectric electrodes and interlocking safety doors.

On induction heating jobs, it is not always practical to completely house the work coil and obtain efficient production operation. In these cases, precautions should be taken to minimize the chance of operator contact with the coil.

665-61. Ampacity of Supply Conductors. Quite often where several pieces of equipment are operated in a single plant, it is possible to conserve on power-line requirements by taking into account the load or use factor of each piece of equipment. The time cycles of operation on various machines may be staggered to allow a minimum of current to be taken from the line. In such cases the Code requires sufficient capacity to carry all full-load currents from those machines which will operate simultaneously, plus the stand-by requirements of all other units.

665-67. Keying. Radio-frequency generators are often turned on and off by applying a blocking bias to the grid circuit of the oscillator tube, for the purpose of obtaining fast, accurate control of power. If this keying circuit does not completely block the tube oscillations, high-frequency power will appear at the work applicator, even though the operator thinks it has been turned off. However, if this residual output voltage is limited to a value of 100-V peak, the operator will be protected from any serious burns.

665-68. Remote Control. In part (a), if interlocking were not provided, there would be a definite danger to an operator at the remote-control station. It might then be possible, if the operator had turned off the power and was doing some work in contact with a work coil, for someone else to apply power from another point, seriously injuring the operator.

ARTICLE 668. ELECTROLYTIC CELLS

668-1. Scope. This was a new article in the 1978 NE Code and provides effective coverage of basic electrical safety in electrolytic cell rooms.

The presentation of these requirements in the "Proposed Amendments for the 1978 National Electrical Code" was accompanied by a commentary from the technical subcommittee that developed them. Significant background information from that commentary is as follows:

> In the operation and maintenance of electrolytic cell lines, however, workmen may be involved in situations requiring safeguards not provided by existing articles of the NEC. For example, it is sometimes found that in the matter of exposed conductors or surfaces it is the man or his workplace which has to be insulated rather than the conductor. Work practices and rules such as are included in IEEE Trial Use Standard 463-

1974 pertinent to such specific situations have been developed which offer the same degree of safety provided by the traditional philosophy of the NEC.

As a corollary to this concept, overheating of conductors, overloading of motors, leakage currents and the like may be required in cell lines to maintain process safety and continuity.

Proposed Article 668 introduces such concepts as these as have been proven in practice for electrolytic cell operation.

668-2. Definitions. The subcommittee noted:

An electrolytic cell line and its DC process power supply circuit, both within a cell line working zone, comprise a single functional unit and as such can be treated in an analogous fashion to any other individual machine supplied from a single source. Although such an installation may cover acres of floor space, may have a load current in excess of 400,000 amperes DC or a circuit voltage in excess of 1,000 volts DC, it is operated as a single unit. At this point, the traditional NEC concepts of branch circuits, feeders, services, overload, grounding, disconnecting means are meaningless, even as such terms lose their identity on the load side of a large motor terminal fitting or on the load side of the terminals of a commercial refrigerator.

It is important to understand that the cell line process current passes through each cell in a series connection and that the load current in each cell is not capable of being subdivided in the same fashion as is required, for example, in the heating circuit of a resistance-type electric furnace by Section 424-72(a).

668-3. Other Articles. Electrical equipment and applications that are not within the space envelope of the "cell line working zone," as dimensioned in Sec. 668-10, must comply with all the other regulations of the NEC covering such work.

668-13. Disconnecting Means. As shown in Fig. 668-1, each DC power supply to a single cell line must be capable of being disconnected. And the disconnecting means may be a removable link in the busbars of the cell line.

668-20. Portable Electrical Equipment. This section and the rules of Secs. 668-21, 668-30, 668-31, and 668-32 cover installation and operating requirements

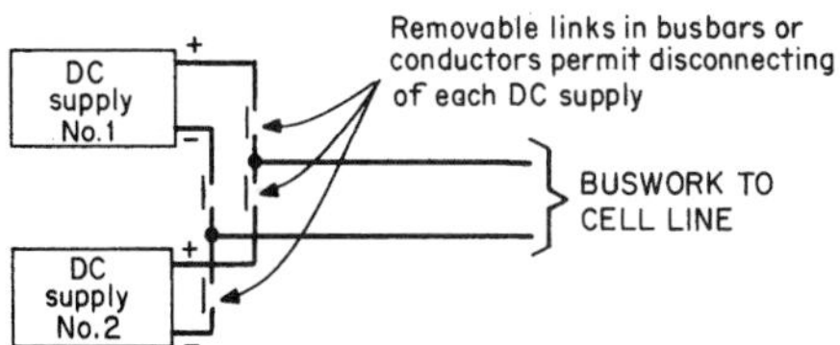

Fig. 668-1. Removal of busbar sections may provide disconnect of each supply. (Sec. 668-13.)

for cells with exposed live conductors or surfaces. These rules are necessary for the conditions as noted by the subcommittee:

In some electrolytic cell systems, the terminal voltage of the cell line process power supply can be appreciable. The voltage to ground of exposed live parts from one end of a cell line to the other is variable between the limits of the terminal voltage. Hence, operating and maintenance personnel and their tools are required to be insulated from ground.

ARTICLE 670. INDUSTRIAL MACHINERY

670-2. Definitions.

Industrial Machinery (Machines):

It should be noted that these provisions do not apply to woodworking machines or to any other type of motor-driven machine which is not included in this definition of machine tools. The provisions do not apply to any machine or tool which is not normally used in a fixed location and can be carried from place to place by hand.

670-4. Supply Conductors and Overcurrent Protection. For the disconnect required by part **(b)**, NFPA No. 79 states: "The center of the grip of the operating handle of the disconnecting means when in its highest position, shall not be more than 6½ feet above the floor. The operating handle shall be so arranged that it may be locked in the 'Off' position."

ARTICLE 680. SWIMMING POOLS, FOUNTAINS, AND SIMILAR INSTALLATIONS

680-1. Scope. Electrification of swimming, wading, therapeutic, and decorative pools, along with fountains, hot tubs, spas, and hydromassage bathtubs, has been the subject of extensive design and Code development over recent years. Details on circuit design and equipment layout are covered in NE Code Art. 680. Careful reference to this article should be made in connection with any design work on pools, fountains, etc.

Because "therapeutic" pools are covered by the Code, hot tubs and spas have been added to coverage. A note tells what is meant when any Code rule refers to "pool" and what is meant to be covered under the term "fountain." This is critically important to correct application of the rules of Art. 680.

Research work conducted by Underwriters Laboratories Inc. and others indicated that an electric shock could be received in two different ways. One of these involved the existence in the water of an electrical potential with respect to ground, and the other involved the existence of a potential gradient in the water itself.

A person standing in the pool and touching the energized enclosure of faulty equipment located at poolside would be subject to a severe electrical shock because of the good ground which his or her body would establish through the water and pool to earth. Accordingly, the provisions of this article specify construction and installation that can minimize hazards in and adjacent to pools and fountains.

Very Important: Therapeutic pools in a health care facility are not exempt from this article. Therapeutic pools in hospitals are subject to all applicable rules in Art. 680.

680-4. Definitions. These definitions are important to correct, effective application of Code rules of Art. 680. Figure 680-1 shows a typical dry-niche swim-

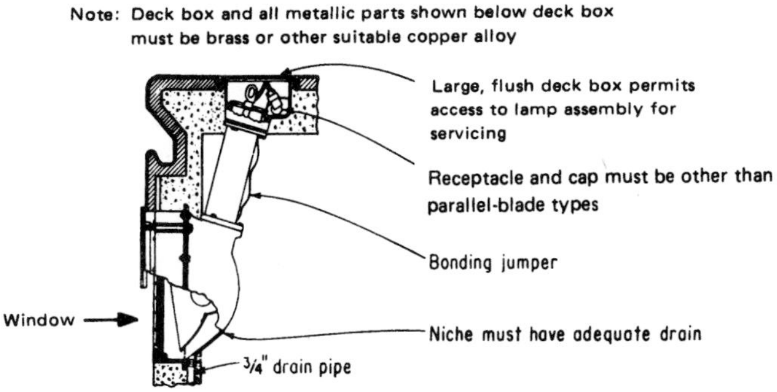

Fig. 680-1. Dry-niche fixture lights underwater area through glass "window." (Sec. 680-4.)

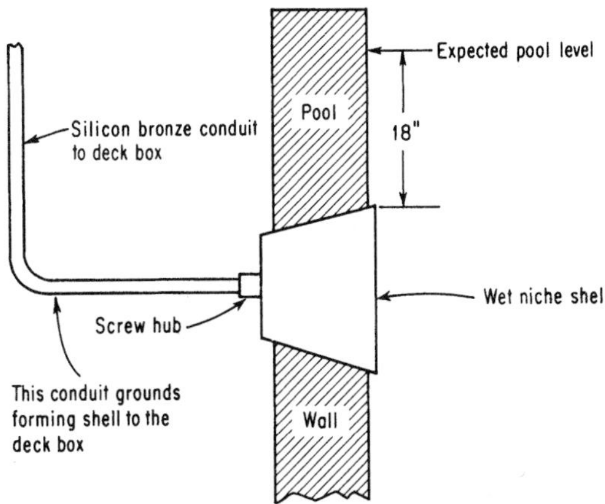

Fig. 680-2. Forming shell is a support for the lamp assembly of a wet-niche fixture. (Sec. 680-4.)

ming pool lighting fixture. Figure 680-2 shows a forming shell for a wet-niche lighting fixture.

The definition of "cord- and-plug-connected lighting assembly" covers a lighting fixture of all-plastic construction for use in the wall of a spa, hot tub, or storable pool. This type of fixture operates from a cord- and plug-connected transformer, and it does not require a metal niche around the fixture.

A "hydromassage bathtub" is a "whirlpool" bath for an individual bather—which is smaller than a hydromassage pool (spa or hot tub)—but is covered by Secs. 680-70 and 680-71.

The definition of "no-niche lighting fixture" covers a fixture for installation above or below the water without any niche. This definition correlates to Sec. 680-20(d), which describes the installation of such fixtures.

Because there are differences in the requirements for "permanently installed" pools and "storable" pools, there has been some confusion in the past as to just what a "storable" pool is. A "storable swimming or wading pool" must be

. . . so constructed that it may be readily disassembled for storage and reassembled to its original integrity.

The definition of "storable swimming or wading pool" defines such a pool by dimensions—one with wall height not over 3½ ft (1.07 m) and no dimension over 18 ft (5.49 m). The introduction of the limiting dimensions now serves to differentiate storable pools from permanently installed pools.

Figure 680-3 shows a "wet-niche lighting fixture."

680-5. Transformers and Ground-Fault Circuit-Interrupters. A swimming pool transformer must be in a weatherproof enclosure to suit it to outdoor use, and a grounded metallic shielding between the primary and secondary winding prevents a primary to secondary short that would connect primary voltage (120 V) to the 12-V secondary circuit—thereby creating a hazardous condition (Fig. 680-4).

Part **(b)** describes general rules on ground-fault circuit-interrupters (GFCI) that are required to be used by other rules of this article. Additional protection may be accomplished, even where not required, by the use of a GFCI. Since the ground-fault interrupter operates on the principle of line-to-ground leaks or breakdowns, it senses, at low levels of magnitude and duration, any fault currents to ground caused by accidental contact with energized parts of electrical equipment. Because the ground-fault interrupter operates at a fraction of the

Fig. 680-3. Wet-niche lighting fixture consists of forming shell set in pool wall with cord-connected lamp-and-lens assembly that attaches to the forming shell, with cord coiled within the shell housing. (Sec. 680-4.)

Fig. 680-4. Transformers for low-voltage swimming pool lighting are listed by UL under "Swimming Pool Transformers" and such listing is used by inspectors as evidence that the unit is "approved for the purpose." (Sec. 680-5.)

current required to trip 15-A CB, its presence is mandatory under following Code rules and is generally very desirable.

Part (c) calls for keeping wires on the load side of a ground-fault interrupter independent of other wiring. Figure 680-5 shows a hookup that might be considered a violation because of the presence of "other conductors"—the wires in the JB that taps to the floodlight JB. But Exception No. 1 qualifies that rule, to permit GFCI-protected conductors to be used in a panelboard enclosure with conductors not protected by GFCI. In the past when a GFCI was used in a panelboard to supply swimming pool circuits, it was necessary to use supplementary insulation (such as nonmetallic sleeving or tubing) on the GFCI conductors in the panelboard gutter to protect these conductors against excessive leakage because of capacitive coupling to the other conductors in the gutter. Excessive leakage was considered a problem because it could cause unwanted circuit opening due to the sensitivity of GFCI, especially when used in the highly conductive conditions that exist at wet locations, in such applications as swimming pools.

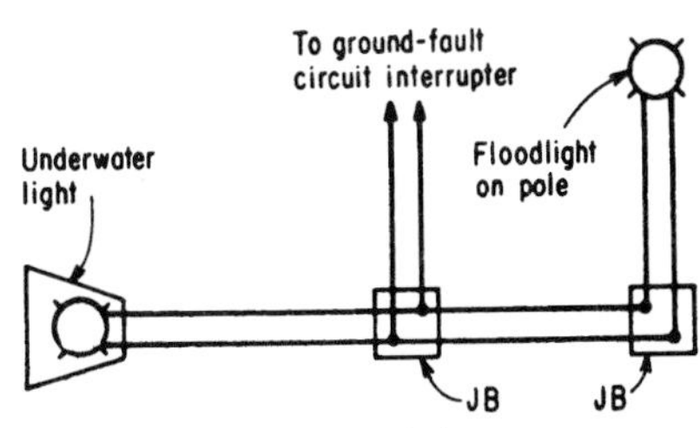

Fig. 680-5. GFCI conductors are in junction box with other than conductors for underwater light. (Sec. 680-5.)

Exception No. 1 no longer specifies a need for insulating sleeving on the GFCI circuit conductors in panelboard gutters. It was concluded that such insulation did not offer sufficient protection against the problem of leakage (Fig. 680-6).

Although conductors on the load side of a GFCI are prohibited from running in raceways or other enclosures containing "other" conductors, Exception No. 3 permits GFCI load conductors to run in an enclosure that contains other GFCI load-side conductors. Conductors having equal protection may be used in the same raceway or enclosure.

680-6. Receptacles, Lighting Fixtures, Lighting Outlets, Switching Devices, and Ceiling Fans. The basic rule here prohibits receptacles within 10 ft (3.05 m) from the pool edge, and part **(a)(3)** calls for GFCIs to protect all 120-V receptacles located between 10 and 20 ft (3.05 and 6.1 m) of the inside walls of indoor and outdoor pools. But the Exception to part **(a)(1)** permits the installation of a receptacle for a swimming pool recirculating pump less than 10 ft (3.05 m) but not closer than 5 ft (1.52 m) from the inside wall of the pool. Normally, receptacles are prohibited from installation anywhere within the 10-ft (3.05-m) boundary around the edge of the pool. However, because swimming pool pump motors are commonly cord-connected to permit their removal during cold weather in areas where freezing may damage them, this Exception applies to a receptacle for the pump motor. Such a receptacle must be a single receptacle of the locking and grounding type and must have GFCI protection for any receptacle fed at 120 or 240 V for supply to the cord- and plug-connected pump. A 240-V receptacle for a 240-V pump motor, as well as a 120-V pump, does require GFCI protection. Of course, on a residential property, all outdoor recep-

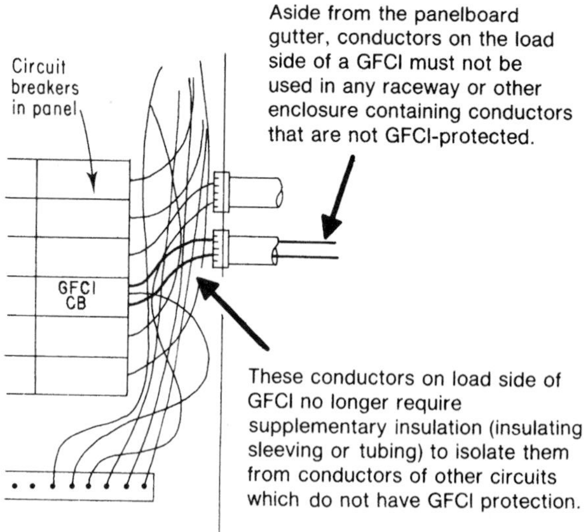

Aside from the panelboard gutter, conductors on the load side of a GFCI must not be used in any raceway or other enclosure containing conductors that are not GFCI-protected.
Circuit breakers in panel
GFCI CB
These conductors on load side of GFCI no longer require supplementary insulation (insulating sleeving or tubing) to isolate them from conductors of other circuits which do not have GFCI protection.

Fig. 680-6. GFCI conductors must be protected against leakage from capacitive coupling to other conductors. (Sec. 680-5.)

tacles beyond the 20-ft (6.1-m) band around the pool must also have GFCI protection, as required by Sec. 210-8(a)(3). But for any property that does not conform to the definition of "dwelling unit" (Art. 100), GFCI protection is not required for outdoor receptacles more than 20 ft (6.1 m) away from the pool's edge.

Part **(a)(2)** requires at least one 120-V receptacle to be "located" within the 10- to 20-ft (3.05- to 6.1-m) band around the pool for *any* permanent pool at a dwelling unit (e.g., a one-family house). The word "located" was put in to replace the word "installed," because there is no need to install such a receptacle if there is already one located within that area around the pool. The rule at one time required this receptacle only where a pool is installed at an "existing dwelling"—which could mean that a receptacle did *not* have to be installed in the 10- to 20-ft (3.05- to 6.1-m) band if the pool was being installed at the same time as the house; that is, the house was not "existing." The rule requires a minimum of one 120-V receptacle at *every* pool installed at a dwelling unit. This rule ensures that a receptacle will be available at the pool location to provide for the use of cord-connected equipment. It was found that the absence of such a requirement resulted in excessive use of long extension cords to make power available for appliances and devices used at pool areas.

Part **(a)(3)** requires that all receptacles within 20 ft [6.1 m; it used to be 15 ft (4.57 m)] of the inside wall of the pool must be protected by a GFCI. Of course, if the pool is on the property of a private home, *all* outdoor receptacles with direct grade-level access must have GFCI protection—at any distance from the pool.

Figure 680-7 summarizes the rules with respect to receptacles. Note that the **Code** wording does not distinguish between "indoor" or "outdoor" receptacles. The present wording appears to apply to all pools and refers simply to "receptacles on the *property*. . . ."

The fine-print note after part **(a)(3)** says measurement of the prescribed distances of a receptacle from a pool is made over an unobstructed route from the receptacle to the pool, with hinged or sliding doors, windows and walls, floors, and ceilings considered to be "effective permanent barriers." If a receptacle is physically only, say, 3 ft (914 mm) from the edge of the pool but a hinged or sliding door is between the pool edge and the receptacle, then the distance from the receptacle to the pool is considered to be infinite, and the receptacle is thus more than 20 ft (6.1 m) from the inside wall of the pool and does not require GFCI protection (Fig. 680-7, bottom).

Figure 680-8 summarizes graphically the rules set forth in part **(b)** of this section.

The reference to "existing lighting fixtures" in Exception No. 1 to part **(b)(1)** must be understood to refer to lighting fixtures that are already in place on a building or structure or pole at the time construction of the pool begins. Where a pool is installed close to, say, a home or country club building, lighting fixtures attached to the already existing structure may fall within the shaded area for a band of space 5 ft (1.52 m) wide, extending from 5 ft (1.52 m) above the water level to 12 ft (3.66 m) above water level all around the perimeter of the pool. The requirement for a lighting outlet so located to be provided with GFCI

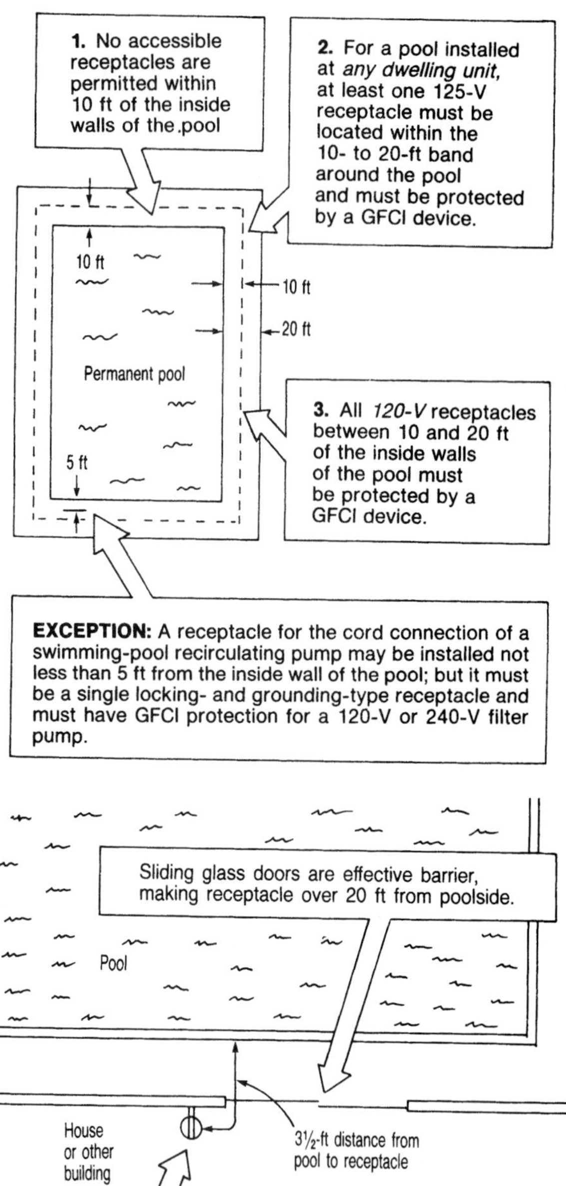
1. No accessible receptacles are permitted within 10 ft of the inside walls of the pool
2. For a pool installed at any dwelling unit, at least one 125-V receptacle must be located within the 10- to 20-ft band around the pool and must be protected by a GFCI device.
10 ft
10 ft
20 ft
Permanent pool
3. All 120-V receptacles between 10 and 20 ft of the inside walls of the pool must be protected by a GFCI device.
5 ft
EXCEPTION: A receptacle for the cord connection of a swimming-pool recirculating pump may be installed not less than 5 ft from the inside wall of the pool; but it must be a single locking- and grounding-type receptacle and must have GFCI protection for a 120-V or 240-V filter pump.
Sliding glass doors are effective barrier, making receptacle over 20 ft from poolside.
Pool
House or other building
3½-ft distance from pool to receptacle
Receptacle does not need GFCI protection

Sec. 680-6(a)(3), Fine Print Note. Doors or windows block a receptacle from connection of appliances that might be hazardous at poolside.

Fig. 680-7. Rules cover all receptacles within 20 ft (6.1 m) of the pool's edge. (Sec. 680-6.)

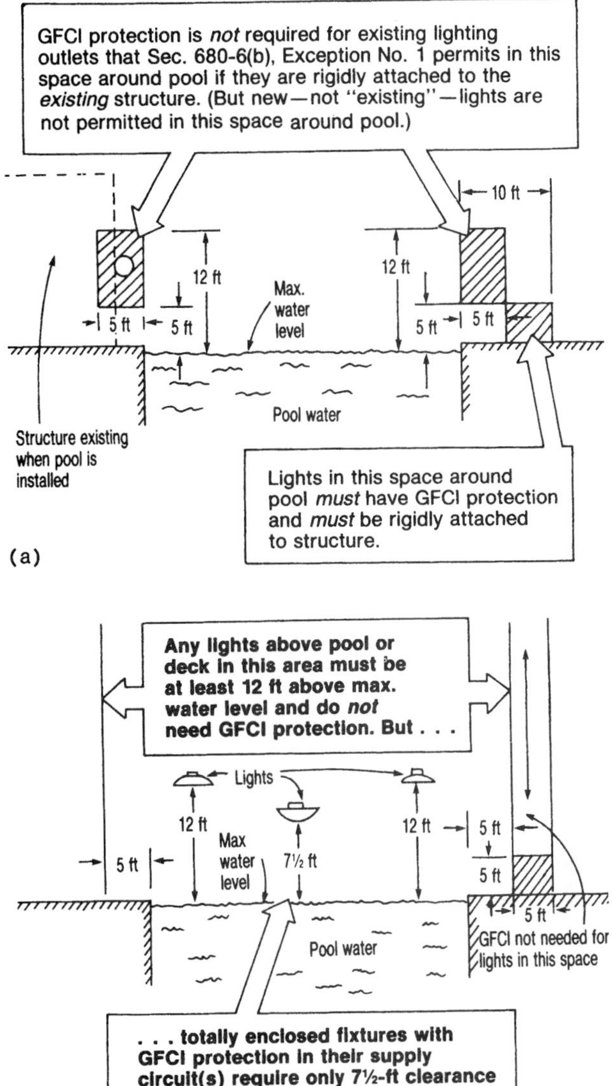
GFCI protection is *not* required for existing lighting outlets that Sec. 680-6(b), Exception No. 1 permits in this space around pool if they are rigidly attached to the *existing* structure. (But new—not "existing"—lights are not permitted in this space around pool.)
10 ft
12 ft
5 ft
5 ft
Max. water level
12 ft
5 ft
5 ft
Pool water
Structure existing when pool is installed
Lights in this space around pool *must* have GFCI protection and *must* be rigidly attached to structure.
(a)
Any lights above pool or deck in this area must be at least 12 ft above max. water level and do *not* need GFCI protection. But . . .
Lights
12 ft
Max water level
7½ ft
12 ft
5 ft
5 ft
5 ft
5 ft
Pool water
GFCI not needed for lights in this space
. . . totally enclosed fixtures with GFCI protection in their supply circuit(s) require only 7½-ft clearance
(b)

Fig. 680-8. For lighting fixtures and switching devices, installed locations are governed by space bands around pool perimeter. (Sec. 680-6.)

protection has been removed from this section on the basis that such protection is a negligible safety factor. However, new lighting fixtures may not be installed in that space band around the pool.

Under the conditions given in Exception No. 2, lighting fixtures may be installed less than 12 ft (3.66 m) above the water of indoor pools. Lighting fixtures that are totally enclosed and supplied by a circuit with GFCI protection

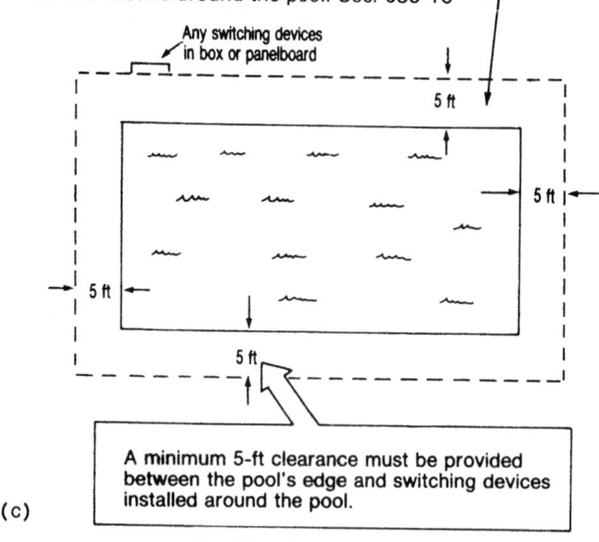
If at all possible, wiring not associated with pool equipment must be kept out of the ground under a pool and under a 5-ft band around the pool. Sec. 680-10
Any switching devices in box or panelboard
5 ft
5 ft
5 ft
5 ft
A minimum 5-ft clearance must be provided between the pool's edge and switching devices installed around the pool.
(c)

Fig. 680-8. (*Continued*)

may be installed where there is at least 7½ ft (2.29 m) of clearance between the maximum water level and the lowest part of the fixture. Note that this Exception covers *only indoor* pools.

Figure 680-9 applies the foregoing rules to lighting at an enclosed pool.

Part **(c)** requires that switching devices must be at least 5 ft (1.52 m) from the pool's edge or be guarded [Fig. 680-8(c)]. To eliminate possible shock hazard to persons in the water of a pool, all switching devices—toggle switches, CBs, safety switches, time switches, contactors, relays, etc.—must be at least 5 ft (1.52 m) back from the edge of the pool, or they must be behind a wall or barrier that will prevent a person in the pool from contacting them.

680-7. Cord- and Plug-Connected Equipment. The 3-ft (914-mm) cord limitation mentioned in this rule would not apply to swimming pool filter pumps used with storable pools under part **C** of Art. 680, because these pumps are considered as portable instead of *fixed or stationary*. See the comments following Sec. 680-30.

680-8. Overhead Conductor Clearances. The general rule states that service drops and open overhead wiring must not be installed above a swimming pool or surrounding area extending 10 ft (3.05 m) horizontally from the pool edge, or diving structure, observation stands, towers, or platforms. But, the Exceptions exempt only utility company lines from the rule, provided the designated clearances are satisfied.

Item C to the table of Exception No. 1 and the diagram clarify the horizontal dimensions around the pool to which the clearances of the table apply for utility lines over a pool area (Fig. 680-10, top). The dimension C, measured hori-

Fig. 680-9. The lighting fixtures over the pool and for 5 ft (1.52 m) back from the edge must be at least 12 ft (3.66 m) above the maximum water level if their supply circuit is without GFCI protection. If GFCI protection is provided by a GFCI-type circuit breaker in the supply circuit(s) to these totally enclosed fixtures, their mounting height may be reduced to a minimum of only 7½ ft (2.29 m) clearance above water level for an indoor (but not an outdoor) pool. The fixtures at right side do not require GFCI protection because they are over 5 ft (1.52 m) back from pool edge and are over 5 ft (1.52 m) above the water level. Refer to Fig. 680-8. (Sec. 680-6.)

zontally around a pool and its diving structure, establishes the area above which utility lines (and *only utility* lines) are permitted, provided the clearance dimensions of A or B in the table are observed. The dimensions A and B do *not* extend to the ground as radii, and the dimension C is the sole ruling factor on the "horizontal limit" of the area above which the clearances of A and B apply.

As the basic rule and Exception No. 1 are worded—and the table and diagram specify—the clearances of A and B must be observed for utility lines above the water and above that area at least 10 ft (3.05 m) back from the edge of the pool, all around the pool. But the horizontal distance would have to be greater if any part of the diving structure extended back farther than 10 ft (3.05 m) from the pool's edge. If, say, the diving structure extended back 14 ft (4.27 m) from the edge, then the overhead line clearances of A and B would be required above the area that extends 14 ft (4.27 m) back from the pool edge, not just 10 ft (3.05 m) back (Fig. 680-10, bottom).

Part C of the table says that the horizontal limit of the area over which the required vertical clearances apply extends to the "outer edge of the structures listed in (1) and (2) above." That wording clearly excludes item 3 of the basic rule (observation stands) from the need to extend the horizontal limit over 10 ft (3.05 m), as shown at the bottom of Fig. 680-10.

Exception No. 2 in this section provides guidance on use of telephone-company overhead lines and community antenna system cables above swimming pools. Although the first sentence of Sec. 680-8 generally prohibits

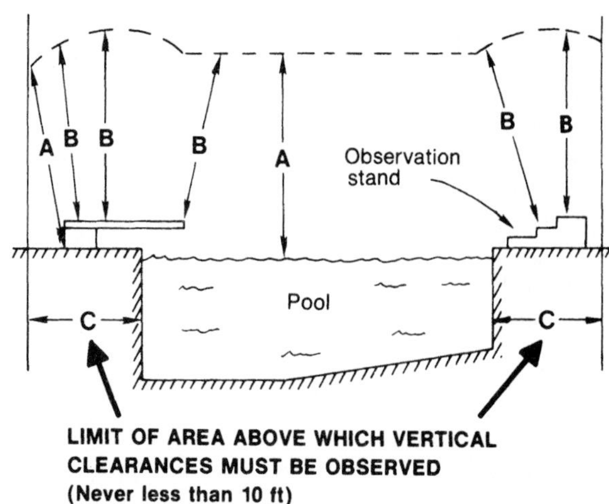

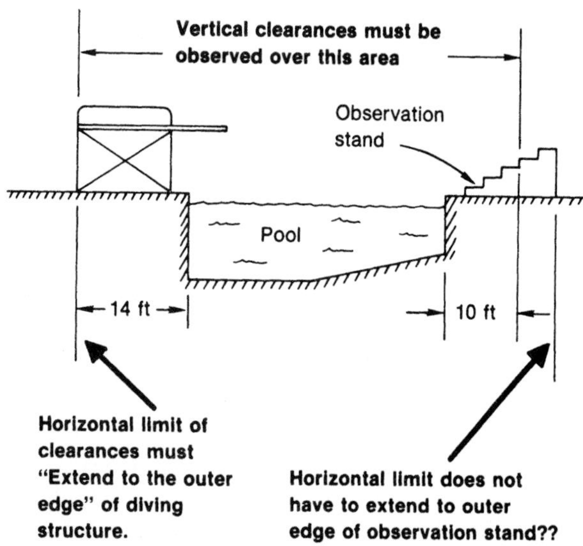

Fig. 680-10. Clearances from Table of Exception No. 1 apply as indicated in these diagrams. (Sec. 680-8.)

"service-drop or other open overhead wiring" above pools, it was never the intent that the rules of this section apply to telephone lines. The general concept of this Exception is to specifically permit such lines above pools *provided that* such conductors and their supporting messengers have a clearance of *not less* than 10 ft (3.05 m) above the pool and above diving structures and observation stands, towers, or platforms.

680-9. Electric Pool Water Heaters. A swimming pool heater requires branch-circuit conductor ampacity and rating of the CB or fuses at least equal to 125 percent of the nameplate load current. An electrically powered swimming pool heater is considered to be a continuous load and is therefore made subject to the same requirements given in Sec. 424-3(b) for fixed electric space heating equipment.

680-10. Underground Wiring Location. This section is aimed at eliminating the hazard that underground wiring can present under fault conditions that create high potential fields in the earth and in the deck adjacent to a pool. Aside from the electric circuits associated with pool equipment, underground wiring must not be run within the ground closer than 5 ft (1.52 m) from the sides of a pool. When inadequate space requires that extraneous underground circuits be run within the ground under the 5-ft (1.52-m) horizontal band around the pool, such wiring is permitted provided that (1) any such circuits are in rigid metal conduit, IMC, or rigid nonmetallic conduit, (2) the raceways are galvanized steel or otherwise provided with corrosion resistance, (3) the raceways are suitable for the location (by complying with underground application data from the UL's *Electrical Construction Materials Directory*), and (4) the burial depths of the raceways conform to the table of burial depths, given in this section.

680-20. Underwater Lighting Fixtures. In part **(a)** of this section, the wording must be followed carefully to avoid confusion on the intent.

Part **(1)** starts by requiring that any underwater lighting fixture must be of such design as to assure freedom from electric shock hazard when it is in use and must provide that protection without a GFCI. *But,* a GFCI *is* required for all line-voltage fixtures (any operating over 15 V, such as a 120-V fixture) to provide protection against shock hazard during relamping. A GFCI is not required for low-voltage swimming pool lights (12-V units).

In the 1971 **NE Code**, the rule recognized the use of line-voltage, self-grounding fixtures (listed for use in pools) to prevent shock hazard—without need for a GFCI. But, although such fixtures satisfy the **Code** as safe in operation, the rule now calls for safety during relamping and requires that a GFCI must be installed in the circuit to any fixture operating at more than 15 V to prevent shock hazard during relamping.

The UL presents certain essential data on use of GFCI devices, which must be factored into application of such devices—as follows:

> A ground-fault circuit-interrupter is a device whose function is to interrupt the electric circuit to the load when a fault current to ground exceeds some pre-determined value that is less than that required to operate the overcurrent protective device of the circuit.
>
> A ground-fault circuit-interrupter is intended to be used only in a circuit that has a solidly grounded conductor.
>
> A Class A ground-fault circuit-interrupter trips when the current to ground has a value in the range of 4 through 6 milliamperes. A Class A ground-fault circuit-interrupter is suitable for use in branch and feeder circuits, including swimming pool circuits. However, swimming pool circuits installed before local adoption of ANSI C1-1965 **National Electrical Code** may include sufficient leakage current to cause a Class A ground-fault circuit-interrupter to trip.

A Class B ground-fault circuit-interrupter trips when the current to ground exceeds 20 milliamperes. This product is suitable for use with underwater swimming pool lighting fixtures only.

A ground-fault circuit-interrupter of the enclosed type that has not been found suitable for use where it will be exposed to rain, is so marked.

The last sentence of part **(a)(1)** requires that *only* an "approved" lighting fixture be used—which means only a fixture listed by UL or other test lab. UL data on listed fixtures must be carefully observed:

These fixtures are for installation in or on the walls of swimming pools not less than 18 in. below the normal water level as measured to the top of the lens opening.

"Wet-Niche" fixtures are intended for installation in a metallic fixture housing (forming shell) mounted in or on the side of a swimming pool wall where the fixture will be completely surrounded by pool water. Such fixtures are provided with a factory-installed, permanently-attached flexible cord that extends at least 12 ft outside the fixture enclosure to permit the fixture to be removed from the forming shell and lifted to the pool deck for servicing without lowering the pool water level or disconnecting the fixture from the branch circuit wiring. Fixtures with longer cords are available for installations in which the junction box or splice enclosure is so located that a 12-foot-long cord will not permit its removal from the forming shell and placement on the pool deck for servicing. To avoid possible cord damage, cord length in excess of that necessary for servicing should be trimmed rather than stored in the forming shell. The trimming must be at the supply end. Each fixture is marked to indicate the proper housing or housings with which it is to be used, and the fixture housing is marked to indicate the fixture or fixtures with which it is to be used.

"Dry-Niche" fixtures are intended for permanent installation in the wall of the pool, having provisions for threaded conduit entries, being designed for servicing from the rear in a passageway or tunnel behind the pool wall, or from the deck surrounding the pool.

Fixtures which are suitable for use only in fresh water pools are marked "Fresh Water Only." Fixtures which are suitable for use in either fresh or salt water pools are marked "Salt Water" or "Salt or Fresh Water."

Fixtures which have been investigated for operation only in contact with water are marked "Submerse Before Lighting," or the equivalent, and such marking is visible after installation of the fixtures.

Part **(2)** sets 150 V as the maximum permitted for a pool lighting fixture—which means that the usual 120-V listed fixtures are acceptable.

Part **(3)** repeats the UL limitation on mounting distance of a fixture below water level. When installed, the top edge of the fixture must be at least 18 in. (457 mm) below the *normal* level of the pool water (Fig. 680-11). This 18-in. (457-mm) rule was adopted to keep the fixture away from a person's "chest area," because this is the vital area of the body concerning electric shocks in swimming pools. Keeping the top of the fixture 18 in. (457 mm) below the normal water level avoids a swimmer's chest area when he or she is hanging on to the edge of the pool while in the water. *But,* as the Exception notes: An underwater lighting fixture may be used at less than 18 in. (457 mm) below the water surface if it is a unit that is identified for use at a depth of not less than 4 in. (102 mm).

Part **(4)** presents an interesting requirement on the use of wet-niche lighting fixtures. The rule here requires that some type of cutoff or other inherent means

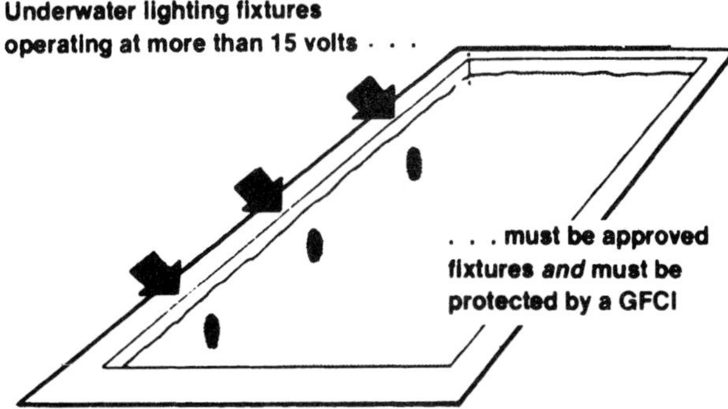

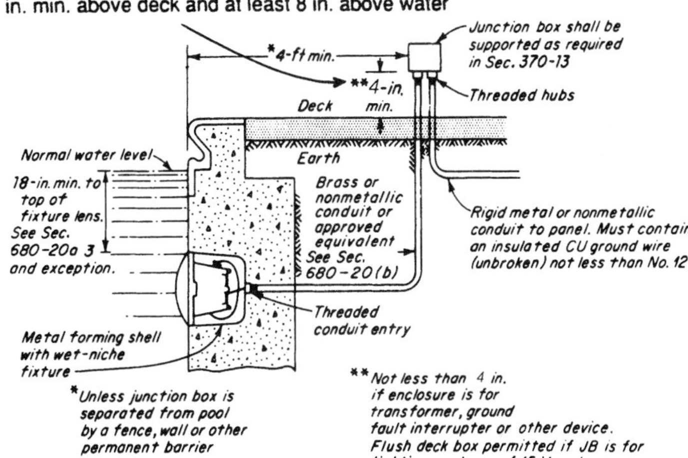

Fig. 680-11. Mounting of lighting fixture and circuit components must observe all Code rules and their specific dimensions. (Sec. 680-20.)

be provided to protect against overheating of wet-niche fixtures that are not submerged but are types that depend on submersion in water for their safe operation. Note that UL rules quoted above require some fixtures to be marked "Submerse Before Lighting." Manufacturers of such fixtures should incorporate this protection—such as in the form of a bi-metal switch similar to those used in motor end-bells for motor overload protection.

Part **(b)** details the use of wet-niche fixtures. A wet-niche underwater lighting assembly consists of two parts—a forming shell, which is a metal structure designed to support a wet-niche fixture in the pool wall, and a lighting fixture, which usually consists of a lamp within a housing furnished with a waterproof flexible cord and a sealed lens that is removable for relamping.

Part **(b)(1)** requires that the conduit between the forming shell and junction box or transformer enclosure must be: (1) rigid metal conduit or IMC and made of brass or other approved corrosion-resistant metal, or (2) rigid nonmetallic conduit with a No. 8 *insulated* copper conductor installed in the conduit and connected to the junction box or transformer enclosure and to the forming shell enclosures. The No. 8 insulated copper wire must be stranded. Each enclosure—the forming shell as well as the box—must contain approved grounding terminals.

Figure 680-12 shows a typical connection from a forming shell to a transformer enclosure supplying the 12-V lamp in the fixture. If a 120-V fixture is used, the conduit from the shell terminates in a junction box, as shown in Fig. 680-11. In the drawing of Fig. 680-12, from the forming shell, a length of 1-in. (25.4-mm) PVC conduit extends directly to a 120/12-V transformer mounted on the back wall of a planter adjoining the pool [observing the 4-ft (1.22 m) back and 8-in. (203-mm) -high provisions of Sec. 680-21(b)(4)]. Where the nonmetallic conduit stubs up out of the planter soil, a metallic LB connects the conduit to the transformer. The required No. 8 conductor in the PVC conduit is terminated at the grounding bar in the transformer enclosure and on the *inside* terminal of an inside/outside grounding/bonding terminal on the forming shell. The external bonding lug provides for connecting the forming shell to the common bonding grid, as required by Sec. 680-22(a) and (b). The No. 8 in the PVC conduit bonds the forming shell up to the transformer enclosure. Note that this No. 8 conductor is not needed if metal conduit connects the shell to the transformer enclosure. One of the No. 12 conductors in the supply circuit is an equipment grounding conductor that runs back to the panelboard grounding block and thereby grounds the No. 8 and the metal fittings and transformer enclosure.

Part **(b)(1)** also requires that the inside forming shell termination of the No. 8 be covered with, or encapsulated in, a UL-listed potting compound. Experience has shown that corrosion occurs when connections are exposed to pool water. Epoxies are available to achieve this protection; however, some inspection agencies do accept a waterproof, permanently pliable silicone caulk compound.

Note that the illustrated assembly includes three noncurrent-carrying conductors: (1) a No. 8 *bonding* conductor connecting the forming shell to the bonding grid; (2) a No. 8 insulated conductor in PVC conduit between the forming shell and the transformer enclosure; and (3) a *grounding* conductor in the fixture flexible supply cord.

Part **(b)(2)** requires sealing of the fixture cord end and terminals within the wet-niche to prevent water from entering the fixture. And grounding terminations must also be protected by potting compounds.

Part **(b)(3)** states that an underwater lighting fixture must be secured and grounded to the forming shell by a positive locking device which will assure a low-resistance contact and which will require a tool to remove the fixture from the forming shell. This provides added assurance that fixtures will remain grounded because, in the case of wet-niche fixtures, the metal forming shell provides a bond between the raceway (or No. 8 conductor in PVC) connected to the forming shell and the noncurrent-carrying metal parts of the fixture.

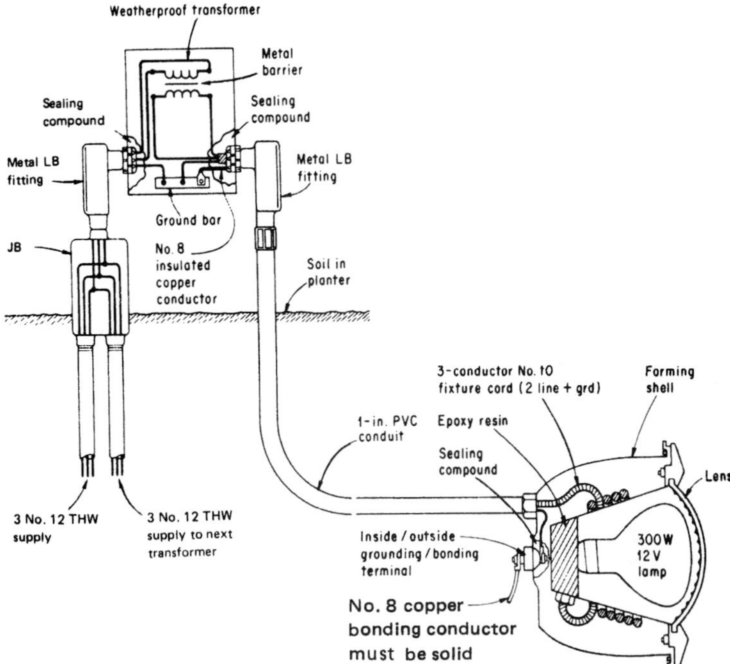

Fig. 680-12. Grounding and bonding is required in a typical hookup of low-voltage wet-niche fixture with PVC conduit. (Sec. 680-20.)

Part **(c)** permits use of an approved dry-niche lighting fixture that may be installed outside the walls of the pool in closed recesses which are adequately drained and accessible for maintenance. For a dry-niche fixture, a "deck box"—set in the concrete deck around the pool—may be used and fed by metal (rigid or IMC) or nonmetallic conduit from the service equipment or from a panelboard. Where the circuit conductors to the fixture are run on or within a building, the Exception to the rule permits the conductors to be enclosed in EMT—but rigid metal or IMC or rigid nonmetallic conduit must be used outdoors when not on a building. And such a deck box does not have to be 4 in. (102 mm) up and 4 ft (1.22 m) back from the pool edge, as required for a junction box for a wet-niche fixture. (See Fig. 680-1.)

Some approved dry-niche fixtures are provided with an integral flush deck box used to change lamps. Such fixtures have a drain connection at the bottom of the fixture to prevent accumulation of water or moisture.

680-21. Junction Boxes and Enclosures for Transformers or Ground-Fault Circuit-Interrupters. Part **(a)** covers junction boxes that *connect to a conduit that extends directly to a pool-lighting forming shell,* such as shown in Fig. 680-11. The junction box must be of corrosion-resistant material provided with threaded hubs for the connections of conduit.

For line-voltage (120 V) pool fixtures the so-called deck box (set in the concrete deck around the pool) is no longer permissible (except where approved dry-niche fixtures include flush boxes as part of an approved assembly), because the deck box, which was installed flush in the concrete adjacent to the pool, was the major source of failure of branch-circuit, grounding, and fixture conductors due to water accumulation within them. The rule of part **(4)** states that these junction boxes must be located not less than 4 in. (102 mm) above the ground level or above the pool deck, and not less than 8 in. (203 mm) above the maximum pool water level (whichever provides the greatest elevation), and not less than 4 ft (1.22 m) back from the pool perimeter.

The wording of part **(4)** does make clear that the elevated junction box could be less than 4 ft (1.22 m) from the pool's edge if a fence or wall is constructed around the pool, with the box on the side of the wall away from the pool, isolating the box from contact by a person in the pool. Or the box could be within 4 ft (1.22 m) of the edge if the box is in a permanent nonconductive barrier.

Important: The Exception to part **(a)** permits flush deck boxes where underwater lighting systems are 15 V or less if approved potting compound is used in the deck boxes and the deck boxes are located 4 ft (1.22 m) from the edge of the pool. In Fig. 680-13, a deck box for a 12-V fixture could be used in the deck but the use of the box less than 4 ft (1.22 m) from the pool's edge might be considered a violation of *b* of the Exception, which does not recognize the fence along the pool in the same way as in part **(a)(4)**. That is, the fence is not mentioned in the Exception as sufficient isolation of the box from the pool—although the installation certainly does comply with the basic concept in part **(a)(4)**.

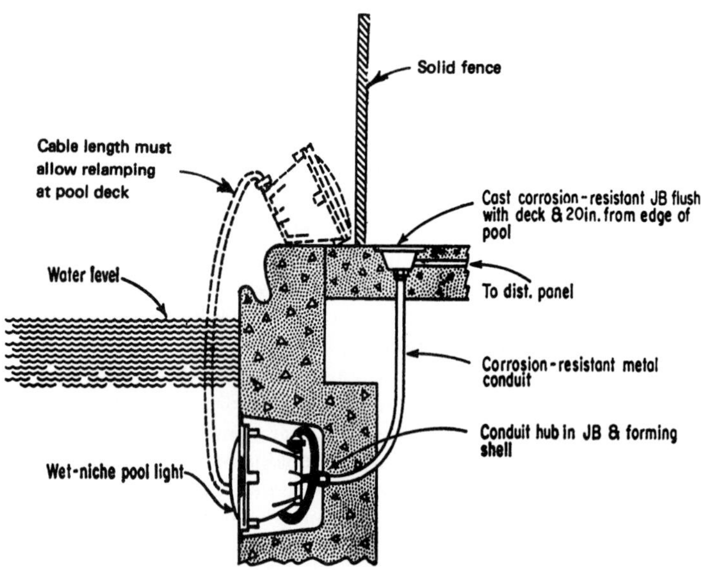

Fig. 680-13. The fence here permits the box to be closer than 4 ft (1.22 m) from pool's edge. (Sec. 680-21.)

Part **(b)** covers installation of enclosures for 12-V lighting transformers and for GFCIs that are required for line-voltage fixtures. Such enclosures may be installed indoors or at the pool location. If a ground-fault interrupter is utilized at a pool, its enclosure must be located not less than 4 ft (1.22 m) from the perimeter of the pool, unless separated by a permanent means and must be elevated not less than 8 in. (203 mm), measured from the inside bottom of the box down to the pool deck or maximum water level, whichever provides higher mounting. These rules cover installation of transformer or GFCI enclosures that connect to a conduit that "extends directly" to a forming shell.

Part **(b)(1)** requires any such enclosure connected to a conduit that extends directly to an underwater pool-light forming shell to have threaded hubs or bosses. An enclosure of cast construction with raised, threaded hubs or with threaded openings in the enclosure wall would satisfy that rule. But because approved swimming pool transformers are available *only* in sheet metal enclosures with knockouts, some type of threaded hub fitting must be provided in the field and connected to the knockout. The intent of this rule on threaded raceway connections is to provide a high degree of bonding and grounding of the underwater fixture.

Connection to a transformer enclosure has been accepted by inspectors when made up with locknuts. The intent of the section, however, would not be considered satisfied even though the transformer enclosure can be well grounded and bonded by connections at the grounding bar within the enclosure. In most instances, the grounding bar must be added by the contractor.

Part **(b)** also requires that transformer or GFCI enclosures must be provided with an approved seal (such as duct seal) at conduit connections to prevent circulation of air between the conduit and the enclosure; must have electrical continuity between every connected metal conduit and the grounding terminals by means of copper, brass, or other approved corrosion-resistant metal that is integral with the enclosures; must be located not less than 4 ft (1.22 m) from the inside walls of the pool (unless separated by a solid fence, wall, or other permanent barrier); and must be located not less than 8 in. (203 mm) from the ground level, pool deck, or maximum pool water level, whichever provides the greatest elevation. This distance is measured from the inside bottom of the enclosure. (See Fig. 680-12.)

Note that part **(b)(3)** intends to assure a grounding path *from the enclosure and its grounding terminals* to any metal conduit. The section specifically states "metal conduit." Where PVC conduit is used, the provision is not applicable, and the No. 8 ground wire in the PVC bonds to the forming shell. However, the section requires electrical continuity between an enclosure and "*every* connected metal conduit." The conduit feeding the transformer primary does not seem to be involved with that rule because the concern is with the grounding path between the transformer or GFCI enclosure and the forming shell and because the No. 12 equipment grounding conductor in the primary supply will carry any current from a fault originating within the transformer enclosure. Local **Code** authorities should be consulted on the point.

The phrase "integral with the enclosures" is meant to cover a situation where the enclosure is nonmetallic. In this case, electrical continuity between the metal conduits and the grounding terminals must be provided by one of the metals specified, and this "jumper" must be permanently attached to the nonmetallic box so that it is "integral."

In Fig. 680-12, the transformer enclosure is being used as a junction box to an underwater light, with the equipment grounding conductors terminated at the grounding bar and carried through. However, the primary purpose of this enclosure is to house the transformer. Thus parts **(b)**, **(c)**, and **(d)** of Sec. 680-21 would apply. Section 680-21(a) would apply to boxes connected directly to underwater lights and is intended to cover situations where splices, terminations, or pulling of conductors might be required.

Figure 680-12 also involves Sec. 680-21(b)(1), which requires that the enclosure be equipped with provisions for threaded conduit entries, and Sec. 680-21(b)(3), which requires that the enclosure be provided with electrical continuity between every connected metal conduit and the grounding terminals by means of copper, brass, or other approved corrosion-resistant metal that is integral with the enclosures. The intent of those rules is to assure maximum safety with a high degree of bonding and grounding. An enclosure housing a transformer with conduit connection directly to an underwater light could be equipped with raised hubs for conduit connections, be watertight, and if nonmetallic, be provided with a permanently attached bonding jumper between all metal conduits to provide the required electrical continuity.

Part **(c)** of this section warns against creating a tripping hazard or exposing enclosures to damage where they are elevated as required. It is also important to remember that these junction boxes must be afforded additional protection against damage if located on the walkway around the pool. For protection against impact, they may be installed under a diving board or adjacent to a permanent structure such as a lamp post or service pole.

Part **(d)** must be carefully satisfied and part **(e)** calls for strain relief to be added to the flexible cord of a wet-niche lighting fixture at the termination of the cord within a junction box, transformer enclosure, or a GFCI. If this device is not supplied with the fixture, the contractor must provide it.

680-22. Bonding. At the beginning of this section, the fine-print note clarifies that the No. 8 pool bonding wire does not have to be taken back to the panelboard. When all the required bonding connections are made at a pool, the entire interconnected hookup will be grounded by the "equipment grounding conductor" that is required to be run to the filter pump and to the lighting junction boxes and is connected to the No. 8 bonding conductor at the pump and in the boxes. In a pool without underwater lighting, the equipment grounding conductor run with a pump-motor circuit will be the sole grounding connection for the bonded parts—and that is all that is required.

The No. 8 bonding conductor does not have to be run to a panelboard, service equipment ground block, or grounding electrode.

Part **(a)** spells out in detail the pool components that must be bonded together. In general, all metal parts that are within 5 ft (1.52 m) of the inside walls of the pool and are not separated by a permanent barrier and all metal parts of electrical equipment associated with the pool water circulating system must be bonded together. That usually includes forming shells of underwater lights, ladders, rails, fill spouts, drains, reinforcing bars, transformer enclosures supplying underwater lights, and equipment in the pump room (Fig. 680-14). The bonding grid must always include metal parts of a pool cover mechanism, including the housing of a drive motor for a power-operated pool cover. The objective of "bonding" all metal together and then "grounding" the interconnected metal components is to bring everything within touch to the same electrical potential—earth potential. This eliminates shock hazard from any stray currents that may be induced in or conducted to the metal from outside the pool environment or from faults in any of the pool electrical equipment that, for one reason or another, are not cleared by the circuit protective devices.

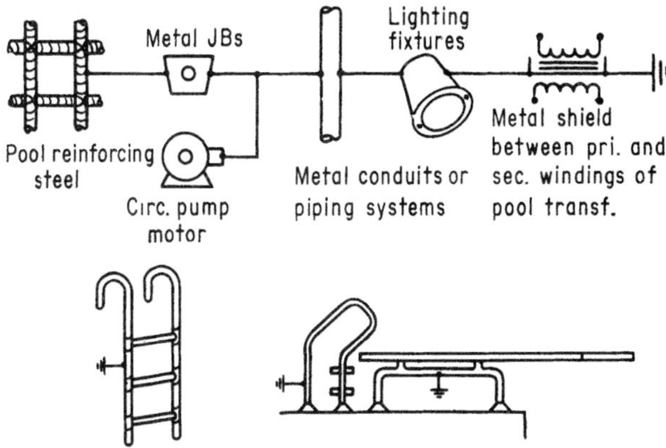

Fig. 680-14. All of these metallic, non-current-carrying parts of a pool installation must be "bonded together." (Sec. 680-22.)

In addition to specifying that underwater lighting fixtures and lighting fixture housings shall be grounded, the rule also requires that all metallic conduit and piping, reinforcing steel and other noncurrent-carrying metal components, located within 5 ft (1.52 m) of a pool, must likewise be bonded together and grounded (Fig. 680-15). Part **(a)(6)** states that bonding of metal parts with No. 8 solid copper wire is required within 5 ft (1.52 m) horizontally of the inside wall of a pool, *but* bonding is *not* required for parts over 12 ft (3.66 m) above the pool area. Any metal piping, raceways, structural parts, or any other metal that is more than 12 ft (3.66 m) above the maximum pool water level or above observation stands, towers, platforms, or diving structures does not require connection to the No. 8 solid bonding grid or conductor.

These references are all-inclusive. For example, conduit and piping may relate to power circuitry, intercom or telephone wiring, supply and return water, or to gas lines serving nonelectric heaters. Reinforcing steel refers to that which is installed in deckslabs and walkways, as well as to pool structures which are poured in place, cast in forms, or "gunnited." Other noncurrent-carrying components include metal parts of ladders, diving boards, platforms

Fig. 680-15. No. 8 insulated bonding conductor connects to clamps on both sides of coupling in brass conduit from forming shell to transformer housing. Section 680-22(a) (5) requires bonding of all "metal conduit" within 5 ft (1.52 m) of pool edge. This bonding connection, although literally required, is not always required by inspection authorities because the conduit connects to bonded enclosures at both ends. (Sec. 680-22.)

and supports, scuppers, strainers, filters, pump and transformer housings, etc. All these items must be bonded together with an insulated or bare solid copper conductor not smaller than No. 8 and connected to a common electrode.

Exception No. 1 recognizes that the usual steel tie wires provide suitable bonding for the individual bars of the reinforcing steel and no special type of clamps or welding is required. The structural reinforcing steel may be used as a common bonding grid [part **(b)** of this section] where connections of the No. 8 bonding conductors are made to the steel rods by suitable clamps. Such connections must be used to bond metal parts to the reinforcing steel grid. Usually, the center-line rebars are bonded together and bonded to the No. 8 bonding conductor at several points.

The Code does not require each individual reinforcing bar to be bonded. It recognizes that the steel tie wires used to secure the rebars together where they cross each other provide the required bonding of the individual rods. Tests conducted over a period of several years by the NE Code Technical Subcommittee on Swimming Pools have shown the resistance of the path from one end to the other through the structural steel to remain at less than 0.001 ohm. The Code thus states in Exception No. 1 that clamping or welding these rods at their intersection will not be required.

If the pool is of metal construction and suitably welded or bolted together, only one bonding connection need be made to the pool.

The rule does not require individual sections of such pools to be bonded. However, the overlapping ends of each section to be bolted must not be painted. If they are, the paint must be removed completely to restore conductivity. In addition, resistance tests should be made across each bolted section after assembly to assure low resistance. These sections normally are fastened together by corrosion-resistant bolts at least ⅜ in. (9.5 mm) in diameter, and such an installation satisfies the bonding objectives (Fig. 680-16). But, electrical parts of such a pool must be tied into that common bonding grid by No. 8 bonding conductors.

Figure 680-17 shows how a metal junction box is bonded into the required bonding grid by No. 8 conductor run in conduit to reinforcing steel. This bonding of metal enclosures within 5 ft (1.52 m) of pool edge is required by part **(a)(6).** When the conduit to the forming shell is PVC, the No. 8 bonding wire required in the PVC by Sec. 680-20(b)(1) bonds the junction box to the forming shell, which is bonded to the rebars. In such cases, some inspectors would not require the No. 8 shown at the bottom of the box.

Important: Exception No. 3 should be carefully noted. It excludes relatively small parts—like bolts, clamps braces, etc.—from the need to be bonded.

Part **(b)** of Sec. 680-22 describes how the bonding must be achieved. Part **(b)** requires all the parts specified in part **(a)** to be connected by means of a No. 8 solid copper conductor (insulated, covered or bare) to a *common bonding grid.* Brass, copper, or copper-alloy connectors or clamps must be used in the common bonding grid, because field reports noted failures of other connectors due to corrosion. This grid could be pool structural reinforcing steel, a metal pool wall, or a solid copper conductor not smaller than No. 8. The idea of connecting all parts to a common grid accomplishes more reliably the objective of equipotential interconnection. Loosening of a connection at one of the parts would not disconnect the bonded parts into two unconnected groups. But it should be noted that the rule does *not* require the common bonding grid to be *continuous,* although the word "grid" does convey the idea of a loop, as in Fig. 680-18.

Figure 680-19 shows how the reinforcing steel of a concrete pool is used as a common grid for connecting all parts together. In that drawing, the steel reinforcing rods, tied together with steel tie wires at intersections, are used as a common bonding grid to bond together pool equipment. Equipment shown here is required to be bonded. In addition, any metal parts (lighting standards, pipes, etc.) within 5 ft (1.52 m) of the inside walls of the pool and not separated from the pool by a permanent barrier must be bonded. All connections made must be in accordance with Sec. 250-113, that is, with proper connectors, lugs, etc.

As shown in Fig. 680-19, when all the required bonding connections are made, the entire interconnected hookup will be grounded by the "equipment grounding conductor" that is required to be run to the filter pump and to the lighting junction boxes and is connected to the No. 8 bonding conductor at the pump and in the boxes. In a pool without underwater lighting, the equipment grounding conductor run with the pump-motor circuit will be the sole grounding connection for the bonded parts—and that is all that is required.

NOTE: THE NO. 8 BONDING CONDUCTOR DOES *NOT* HAVE TO BE RUN TO A PANELBOARD OR SERVICE EQUIPMENT GROUND BLOCK. Because water-circulating equipment is always used and such equipment is required to be both bonded and grounded, the enclosure of the motor terminal connections provides for bringing the bonding and grounding conductors together. The **Code** does not contain a specific provision requiring the bonding network to be

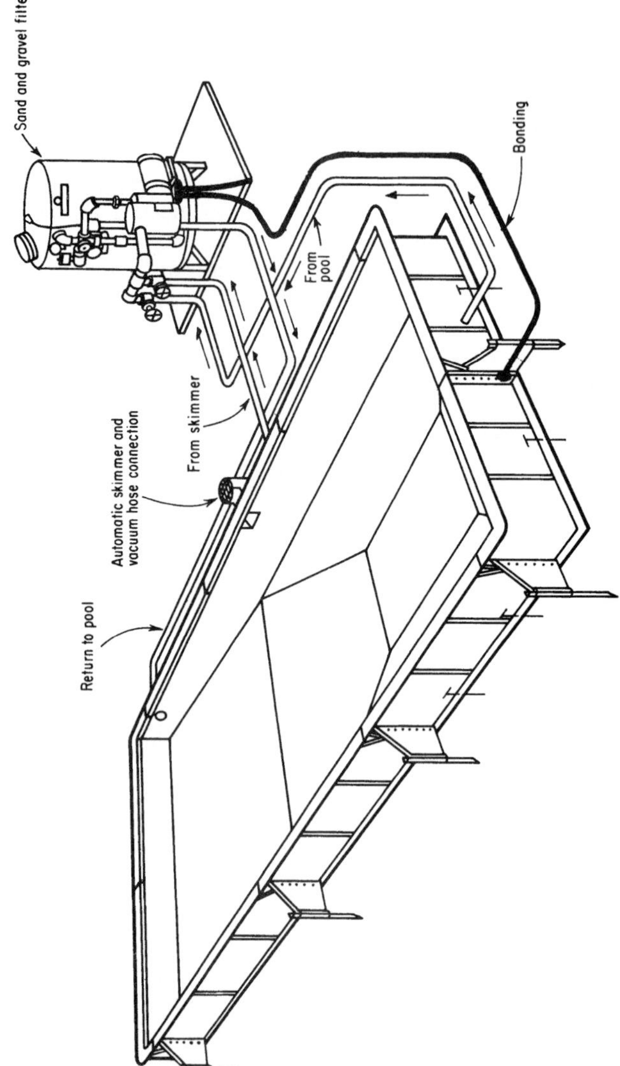

Fig. 680-16. Walls of bolted or welded metal pool may be common bonding grid for *nonelectrical* parts only. (Sec. 680-22.)

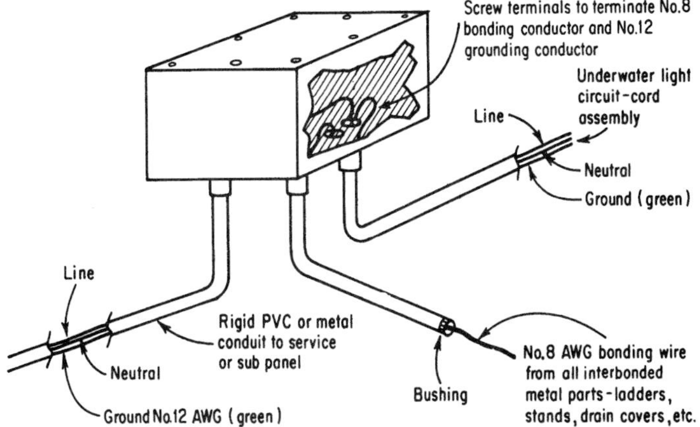

Fig. 680-17. Elevated metal junction box within 5 ft (1.52 m) of pool edge must be bonded. (Sec. 680-22.)

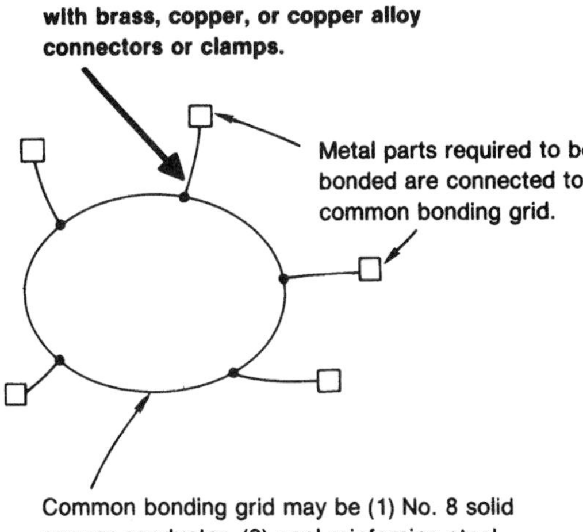

Fig. 680-18. All designated parts must be connected to a "common bonding grid." (Sec. 680-22.)

grounded. Where wet-niche underwater fixtures are used, the metal conduit between the *bonded* forming shell and the *grounded and bonded* junction box accomplishes another connection between the No. 8 bonded hookup and the equipment grounding conductor. Or the No. 8 wire with the PVC conduit to the forming shell will make the connection.

In part **(b)**, the rule requires that the bonding conductor be of *solid copper* not smaller than No. 8 AWG. It is not required to be insulated. If it is insulated,

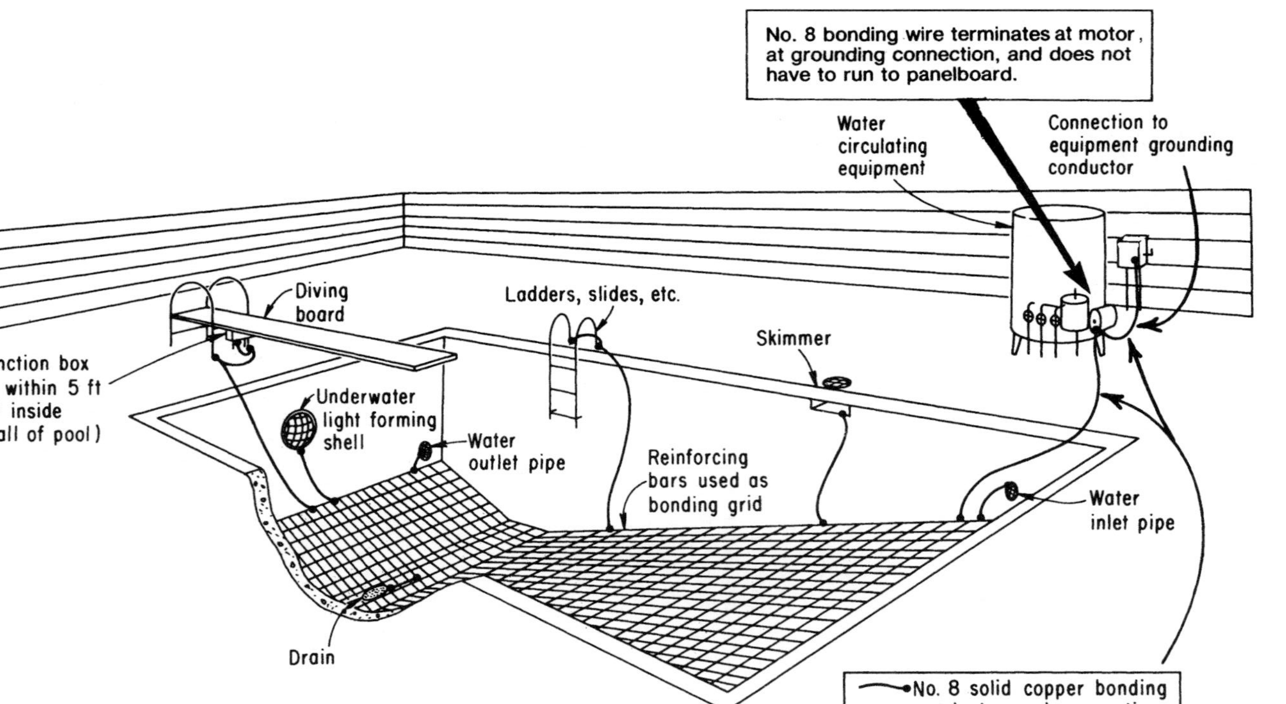

Fig. 680-19. A No. 8 bonding jumper ties each of the indicated parts to the rebar grid, completing the bonding. (Sec. 680-22.)

inspectors might require green color coding at any permanently exposed termination, if a rigid interpretation is put on Sec. 250-57(b), Exception No. 1 and if the "bonding" conductor is considered to be an "equipment grounding" conductor. Section 250-79 on "bonding jumpers" does not specify color of insulation or covering. Because the **Code** rules do not cover the matter of color of bonding jumpers, inspectors commonly do not require green color, permitting black or other colors. The bonding conductor is usually installed underground and under the pool deck, except where it extends into the pump room. At all visible termination points, the conductor can be wrapped with green tape for identification, similar to the permission in *c* of Exception No. 1 of Sec. 250-57(b) which applies to equipment grounding conductors larger than No. 6 (Fig. 680-20).

Conflict has arisen in the past over use of this No. 8 bonding conductor because Sec. 310-3 requires No. 8 and larger conductors to be *stranded* where installed in raceways. That rule had the effect of limiting use of *solid* No. 8 conductors, and their manufacture seemed to be curtailed. Section 310-3 has Exception No. 2 which references Sec. 680-22(b) and exempts the swimming pool No. 8 bonding conductor from the need to be stranded. If solid No. 8 copper cannot be obtained, the local inspection agency must be consulted.

To comply with the requirement that all the bonded parts "shall" be connected to "a common bonding grid," it is common practice to connect metal parts of diving boards, slides, and ladders as well as the drain cover, skimmers, and water-circulating equipment to steel reinforcing in the concrete bottom or walls of the pool. These connections are made and inspected prior to the pouring of concrete, of course. The No. 8 copper bonding conductor may, in some

Fig. 680-20. Water-fill pipe and metal housings associated with the water-circulation system are "bonded" with an insulated No. 8 solid copper conductor, as required. (Sec. 680-22.)

instances, have to be connected to aluminum ladders, rails, or junction boxes (Fig. 680-21). Care must be taken to use a connector suitable for copper to aluminum connections. On some jobs, the ladders, spouts, and forming shells are made of stainless steel, and drains are cast bronze. High-quality, red brass compression connectors can be used, and connections at the iron rebars made with silicon bronze ground clamps.

Part (c) points out that water heaters rated over 50 A must have parts of the unit bonded to the other bonded metal parts by a No. 8 conductor and other

Fig. 680-21. Fittings for pool ladders have bonding strips attached to them for connection of the No. 8 bonding conductor. When the supports are set in the concrete deck, the bonding connections tie them all together. The No. 8 bonding connections are shown (arrow) and a protective coating is painted on the connectors to protect against corrosion. (Sec. 680-22.)

parts grounded by connection to the equipment grounding conductor of the circuit supplying the heater. And the parts to be grounded or bonded will be designated by instructions with the heater.

680-23. Underwater Audio Equipment. This section treats connection of loudspeakers for underwater audio output in the same way as a wet-niche pool lighting fixture. Wording and rules are almost identical to that in Sec. 680-20(b). Connection from the speaker forming shell is made to a junction box installed as set forth in Sec. 680-21(a) for a lighting fixture.

680-24. Grounding. This section lists equipment that must be grounded: wet-, no- and dry-niche underwater lighting fixtures, all electric equipment within 5 ft (1.52 m) of the inside walls of the pool, all electric equipment associated with the recirculating system of the pool, junction boxes and transformer enclosures, GFCIs, and panelboards supplying any electric equipment associated with the pool.

After metal parts have been properly "bonded" to comply with the rules of Sec. 680-22, then "grounding" must be provided to satisfy the many specific rules of Sec. 680-25.

The prime objective of grounding is to provide both connection to the grounding electrode at the service and also to assure a low-impedance path for fault currents to flow back to the grounded neutral to permit proper operation of overcurrent devices. Grounding brings all metallic parts to ground potential, reducing shock hazard.

680-25. Methods of Grounding. The basic difference between *bonding* and *grounding* becomes evident when the provisions of this section are considered. Essentially, this section calls for grounding with an equipment grounding conductor run in conduit along with the supply conductors. The bonding conductor required by Sec. 680-22 does not have to be in conduit. Also, all equipment grounding conductors must terminate at an *equipment grounding terminal.* Bonding conductors may be connected directly to enclosures.

This section provides rules on methods of grounding metal junction boxes and transformer enclosures, panelboards, underwater lighting fixtures, cord-connected equipment, and other equipment. In general, metal raceways are not depended upon to provide the grounding path for swimming pool electrical systems. This is to make sure that the grounding path is maintained even though the conduit metallic current path might open because of corrosion. This can be a problem in pump houses, for example, where the conduit may be exposed to chlorine or acids. For the same reason, the equipment grounding conductor is required to be insulated.

Section 680-25(b) covers grounding of wet-niche lighting fixtures and grounding required for junction boxes, transformer enclosures, and any other enclosures in the supply circuit to a wet-niche fixture and the field-wiring chamber of a dry-niche lighting fixture. Grounding of all those pieces of equipment must be made by an equipment grounding conductor run with the circuit conductors back to the equipment grounding terminal block of the panel supplying the equipment.

Paragraph **(b)(1)** describes sizing of an equipment grounding conductor that runs from an elevated junction box back to the branch-circuit panel to provide grounding of a wet-niche lighting fixture that is fed from the box. Such an equipment grounding conductor must be sized from **Code** Table 250-95, based on the rating of the overcurrent device that protects the circuit supplying the lighting fixture, and may never be smaller than No. 12 insulated copper wire (Fig. 680-22).

Although the basic rule of **(b)(1)** calls for the lighting circuit and its equipment grounding conductor to be installed in rigid metal conduit, IMC, or rigid nonmetallic conduit where it is run outdoors, overhead or underground, any part of such a circuit that is installed on or within a building *may* be run in EMT instead of the other raceways, as noted in Exception No. 1, because it is a better-protected condition.

Exception No. 2 requires careful sizing of an equipment grounding conductor from the secondary side of a low-voltage transformer for 12-V pool lights to the junction box. That conductor *must* be sized from **Code** Table 250-95 on the

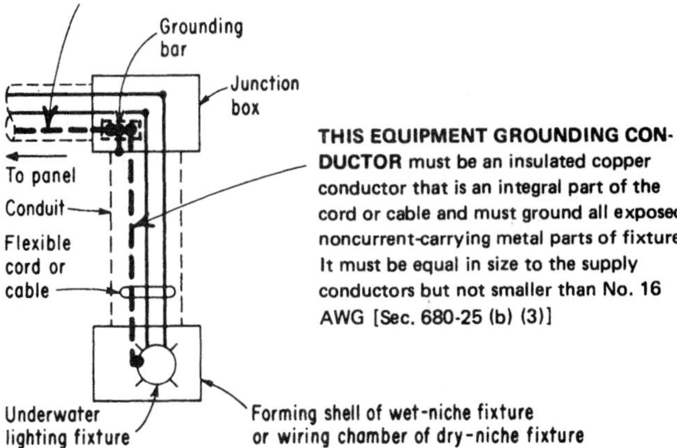
THIS EQUIPMENT GROUNDING CONDUCTOR must be without joint or splice [Sec. 680-25(b)(2)] and must be installed in IMC or rigid metal or rigid nonmetallic conduit with circuit conductors. It must be sized per Table 250-95 but not smaller than No. 12 AWG. It must be an insulated, copper conductor [Sec. 680-25 (b) (1)]
Grounding bar
Junction box
To panel
Conduit
Flexible cord or cable
THIS EQUIPMENT GROUNDING CONDUCTOR must be an insulated copper conductor that is an integral part of the cord or cable and must ground all exposed noncurrent-carrying metal parts of fixture. It must be equal in size to the supply conductors but not smaller than No. 16 AWG [Sec. 680-25 (b) (3)]
Underwater lighting fixture
Forming shell of wet-niche fixture or wiring chamber of dry-niche fixture

Fig. 680-22. Wet-niche fixture must be grounded back to its supply panelboard. (Sec. 680-25.)

basis of the overcurrent-device-rating of the circuit supplying the primary of the transformer. It is not necessary to use a larger grounding conductor on the secondary even though the secondary circuit conductors are larger than the primary conductors because of the 10-to-1 current step-up (120 ÷ 12 V) (Fig. 680-23). A 300-W, 12-V fixture would require No. 10 copper secondary supply conductors (300 W ÷ 12 V = 25 A), but needs only a No. 12 copper grounding conductor (Fig. 680-23).

The rule of part **(b)(2)** on grounding of the junction box for the lighting fixture is also described at the top left of Fig. 680-22. Note that the size of this equipment grounding conductor must be based on the rating of the overcurrent protection for the circuit to the lighting fixture—using Code Table 250-95, but not smaller than No. 12 [as specified in part **(b)(1)** of this section]. No. 12 would be required for a 15- or 20-A branch circuit.

Exception No. 1 of part **(b)(2)** is shown in Fig. 680-24, where connection of the equipment grounding conductor to the terminal blocks is an *acceptable*

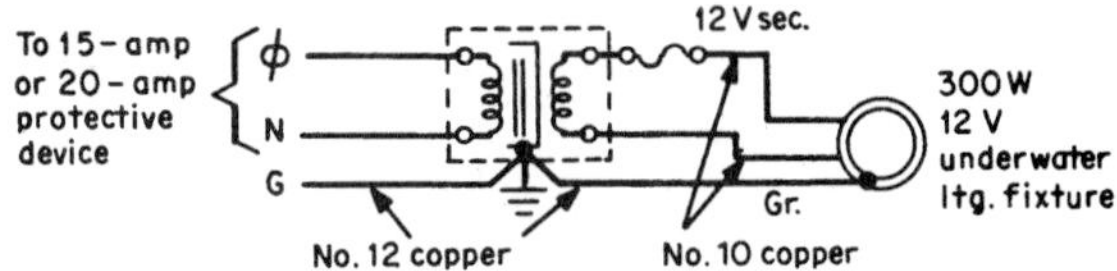
To 15-amp or 20-amp protective device
ϕ
N
G
12 V sec.
Gr.
300 W 12 V underwater ltg. fixture
No. 12 copper
No. 10 copper

Fig. 680-23. Equipment grounding conductor does not have to be larger on transformer secondary side. (Sec. 680-25.)

joint in the conductor as an Exception to the last sentence of part **(b)(2)**, which calls for the conductor to be "without joint or splice."

Exception No. 2 is illustrated in Fig. 680-25, showing that the grounding conductor requires joints when a circuit from a panelboard to the lighting fixture(s) feeds through an enclosure for a low-voltage transformer, a GFCI, a time switch, or a manual snap switch.

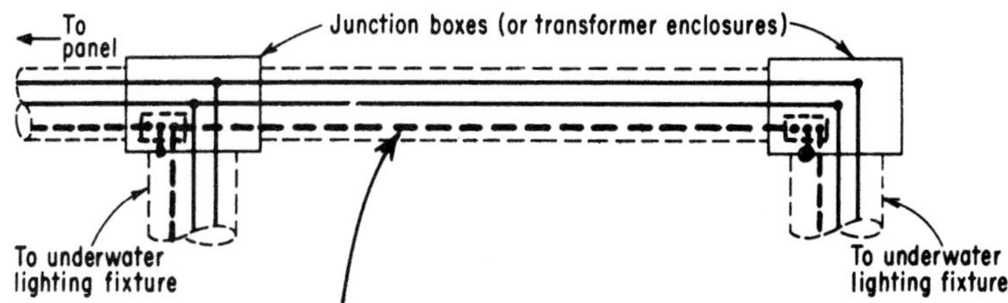
To panel
Junction boxes (or transformer enclosures)
To underwater lighting fixture
To underwater lighting fixture

THIS EQUIPMENT GROUNDING CONDUCTOR must be terminated on approved grounding terminals in both enclosures. It must be copper, insulated, sized per Table 250-95, but not smaller than No. 12 AWG. It must be installed with the circuit conductors in IMC or rigid metal or rigid nonmetallic conduit

Fig. 680-24. Grounding conductor between junction boxes may have terminal "joints" where same circuit supplies more than one pool light. (Sec. 680-25.)

Part **(b)(3)** requires that the cord supplying a wet- or no-niche lighting fixture must be provided with an equipment grounding conductor in the cord from the forming shell to the junction box, as described in the right-hand text in Fig. 680-22.

In part **(c)**, the rule specifically requires an equipment grounding conductor for "pool-associated motors." This is actually a specific requirement that is a

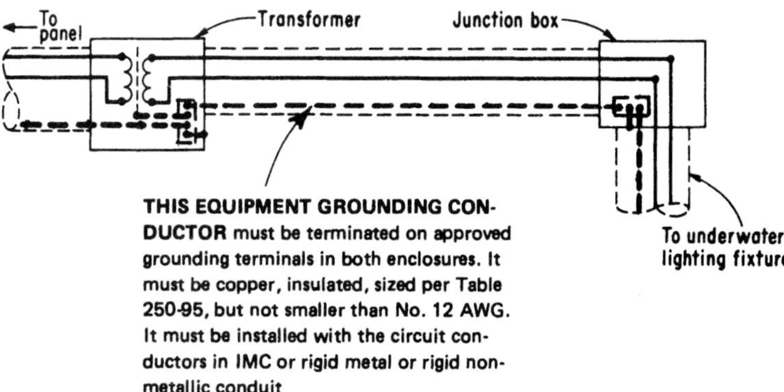
To panel
Transformer
Junction box
THIS EQUIPMENT GROUNDING CONDUCTOR must be terminated on approved grounding terminals in both enclosures. It must be copper, insulated, sized per Table 250-95, but not smaller than No. 12 AWG. It must be installed with the circuit conductors in IMC or rigid metal or rigid nonmetallic conduit
To underwater lighting fixture

Fig. 680-25. Proper terminals must be used where grounding conductor runs through enclosures. (Sec. 680-25.)

follow-up to the general rule, in part **(a)** of this section, that grounding must be provided for "motors" at pools. The rule here requires that a circuit to a pool filter pump—or any other "pool-associated motor"—*must* be run in rigid metal conduit (steel or aluminum), in intermediate metal conduit (so-called IMC), in rigid nonmetallic conduit (such as Schedule 40 or Schedule 80 PVC conduit), or in Type MC cable that is "listed" for the application. And, for all such circuits to pool motors, a separate equipment grounding conductor of the proper size must be run in the raceway or cable with the branch-circuit conductors. The equipment grounding conductor must be sized from NEC Table 250-95, based on the rating of the overcurrent protective device (the fuse or CB) protecting the branch-circuit wires. It must *never* be smaller than No. 12 and must *always* be *copper* and *insulated,* colored green for its entire length in sizes up to No. 6.

Note: THIS RULE CLEARLY ELIMINATES PAST CONFUSION AND DISPUTED Code PRACTICE. IT REQUIRES ONE OF THE THREE RACEWAYS OR TYPE MC CABLE TO FEED A FILTER PUMP AND DOES *NOT* PERMIT USE OF TYPE UF OR TYPE USE CABLE FOR THE PUMP CIRCUIT, as shown in Fig. 680-26.

The first two Exceptions to part **(c)** offer only limited use of EMT or liquidtight flexible conduit as part of the circuit to a pool motor. EMT may be used as part of the circuit where the raceway is within or on a building—but may *not* be used overhead or underground outdoors. Any part of the circuit outdoors (not on a building) *must* be one of the three rigid conduits described in the

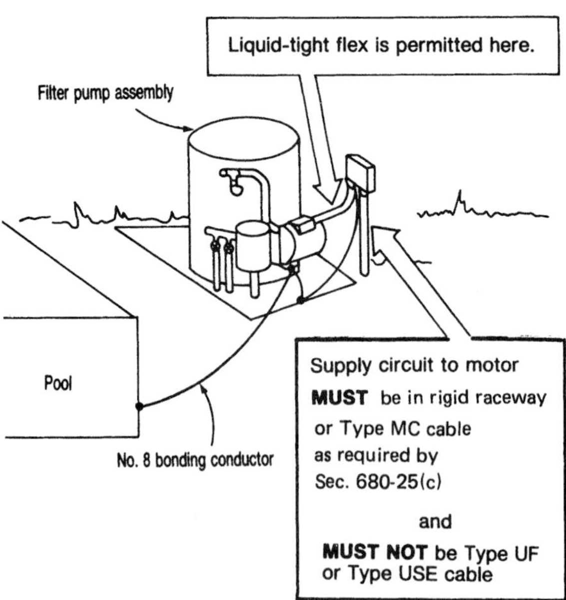

Fig. 680-26. A very clear rule is now applied to wiring method required for pump. (Sec. 680-25.)

basic rule. Exception No. 2 recognizes the use of liquidtight flexible conduit as the supply raceway for a pool motor where flexibility is needed. Such flex would have to be a metallic type, unless it is nonmetallic flex UL listed for outdoor use, as permitted by Sec. 351-23(a)(3). *But in all cases* where EMT or liquidtight flex is used as permitted by the Exceptions, a separate equipment grounding conductor must be used, as described above, within the raceway.

As permitted by Exception No. 3, wiring to a pool-associated motor may be NM cable or any of the NEC wiring methods for that part of the circuit that is run in the interior of a one-family dwelling unit. *But,* it should be noted that Exception No. 3 specifies that interior wiring of a one-family dwelling may be part of the circuit to a filter pump *only* if the interior circuit has at least a No. 12 insulated or covered ground wire. That would recognize Romex with a No. 12 ground wire, but BX (Type AC) with its No. 16 aluminum bonding strip would have to contain an additional insulated grounding conductor that is at least a No. 12. This Exception recognizes that wiring within a building is under better protection for the reliability of the equipment grounding conductor.

Exception No. 4 of this section permits use of flexible cord for cord- and plug-connected equipment, as covered in Sec. 680-7.

Part **(d)** requires an equipment grounding conductor between the service equipment and a panelboard that supplies circuits to pool electrical equipment, as shown in Fig. 680-27. The basic intent is to require an insulated equipment grounding conductor, sized in accordance with Code Table 250-95, but not smaller than No. 12 AWG, and run back to the equipment grounding terminal in service equipment. Note that this conductor does not have to be copper and may be aluminum.

The equipment grounding conductor required by part **(d)** of this section must be sized from Code Table 250-95 in accordance with the rating of the protective

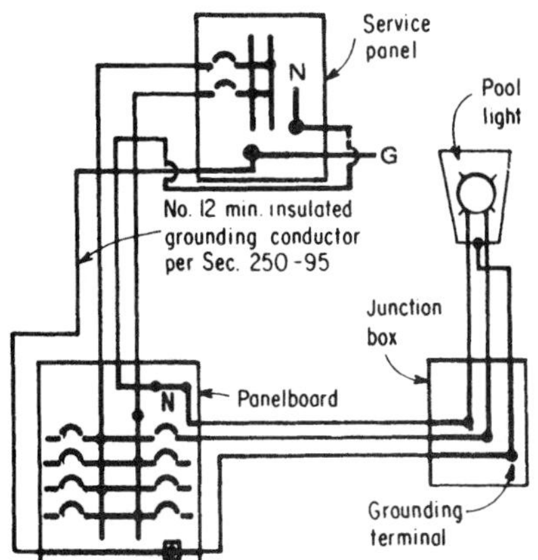

Fig. 680-27. Grounding conductor must connect subpanel ground block to equipment grounding terminal of service equipment. (Sec. 680-25.)

device for the circuit that is involved—*but* No. 12 is the minimum size. From Code Table 250-95, a 20-A fuse or CB protecting a branch circuit to pool lighting, as in part **(b)(1)**, would require a No. 12 copper grounding wire with the circuit. But if 40-A fuses or a 40-A CB protected a feeder from the service equipment to a subpanel serving the pool, then Code Table 250-95 would require a No. 10 size of copper conductor for the grounding.

The equipment grounding conductor must be installed along with the feeder conductors to the panel in rigid metal conduit, intermediate metal conduit, or rigid nonmetallic conduit. AND ALL OF THESE GROUNDING CONDUCTORS *MUST* BE GREEN IN COLOR FOR THEIR ENTIRE LENGTHS IF THEY ARE UP TO NO. 6 IN SIZE. LARGER THAN NO. 6 MAY BE OF OTHER COLOR IF MARKED GREEN AT TERMINALS [see Sec. 250-57(b)].

Exception No. 1 to part **(d)** is shown in Fig. 680-28. When a nonmetallic cable assembly is used between panels A and B in Fig. 680-29, a No. 12 or larger insulated conductor must be available. The circuit between panel B and deck boxes for lights must be in a conduit. In the drawing, a nonmetallic 4-wire cable may be used between the two panelboards. Two conductors can be used for the hot-leg conductors, the third used for the neutral, and the last insulated conductor used for the grounding of pool equipment. The grounding conductor in the cable assembly may be either "insulated or covered." If it is covered or insulated, it must be finished in green color or green with a yellow stripe [Sec. 250-57(b)].

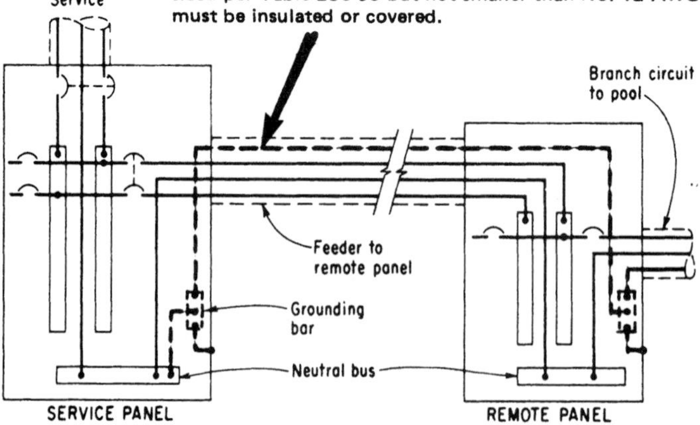
THIS EQUIPMENT GROUNDING CONDUCTOR must be sized per Table 250-95 [Sec. 680-25(d)]. It must be insulated, not smaller than No. 12 AWG, and installed with circuit conductors in IMC or rigid metal or rigid nonmetallic conduit
EXCEPTION: If pool equipment is fed from an existing remote panel, the feeder to the remote panel may be in flexible metal conduit or a cable. The equipment grounding conductor must be sized per Table 250-95 but not smaller than No. 12 AWG, and must be insulated or covered.
Service
Branch circuit to pool
Feeder to remote panel
Grounding bar
Neutral bus
SERVICE PANEL
REMOTE PANEL

Fig. 680-28. Equipment grounding conductor may be in cable between panelboards. (Sec. 680-25.)

Note: Exception No. 1 is applicable *only* where the subpanel, as shown in Fig. 680-28, is an "existing" panel. With that wording, if a new subpanel were installed in the garage or at the pool location, it would have to be fed from the service equipment by a feeder in rigid metal conduit, IMC, or rigid nonmetallic conduit—with a separate grounding conductor in whatever type of conduit is used. In Exception No. 1, a "remote panel" is one which is "not part of the service equipment."

The integrity of this ground-return path is all the more important now that ground-fault circuit protection is required for outdoor residential outlets, and for circuits supplying electrical equipment used with storable pools. Special attention should be paid to avoiding any connection between grounding terminals and the neutral (except at the service entrance). The grounding terminal block *must* be connected to the neutral bus in the *service panel* but *not in any remote panel* unless the provisions of Sec. 250-24 are carefully considered and GFCI breakers are not used in the service panel to feed any remote panel. Bonding the neutral to ground in a subpanel can make ground-fault protection in the service panel inoperative.

Exception No. 2 permits use of EMT for the circuits covered by Sec. 680-25(d) where such circuits are under the better-protected conditions when installed on or inside buildings (Fig. 680-29).

Part **(f)** of this section applies to electrical equipment other than the underwater lighting fixture (and its related equipment—the junction box, transformer enclosure or other enclosure in the supply circuit to a wet-niche lighting fixture, and the field-wiring chamber of a dry-niche fixture) and other than motors [covered by part **(c)**] and other than panelboards [covered by part **(d)**]. It simply recognizes that grounding of all such "other" equipment may be done by any of the Code-recognized equipment grounding means covered in Sec. 250-57 or 250-59. That means the metal frame of any electrical equipment *other than* the types specifically covered in parts **(b)** through **(e)** of Sec. 680-25 may be grounded by an equipment grounding conductor in a raceway or in a recognized cable assembly. Type UF or USE cable, with an equipment grounding conductor, is therefore an alternative to the use of a raceway for "other" equip-

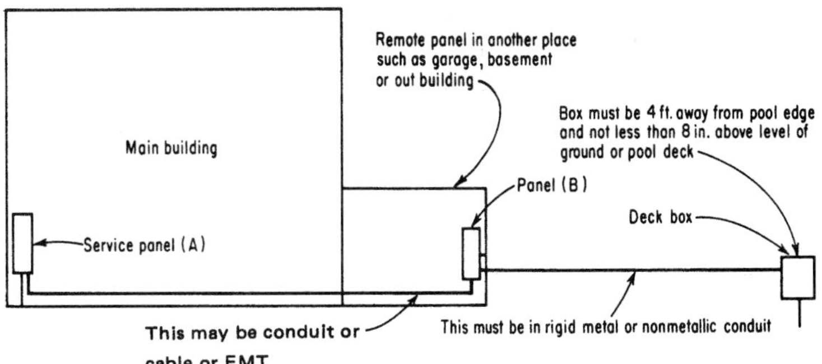

Fig. 680-29. Subpanel is commonly used to supply circuits for pool electrical equipment. (Sec. 680-25.)

ment. And if a metal raceway is used for such a circuit, the raceway itself—rigid, IMC, or even EMT—would satisfy Secs. 250-57, 250-59, and 250-91(b), and a separate equipment grounding conductor would not have to be run in the raceway. The requirement for an insulated, copper equipment grounding conductor installed in rigid metal conduit, IMC, or rigid nonmetallic conduit—as specified in parts **(b)(1)** and **(c)**—applies only for equipment used in circuits to pool lighting fixtures and pool-associated motors.

680-27. Deck Area Heating. These rules cover safe application of unit and radiant heaters. Such units must be securely installed, must be kept at least 5 ft (1.52 m) back from the edge of the pool, and must not be installed over a pool. Radiant heating cables are prohibited from use embedded in the concrete deck.

680-30. Pumps. There are portable filter pumps listed by Underwriters Laboratories Inc., and they comply with Sec. 680-30.

680-31. Ground-Fault Circuit-Interrupters Required. Ground-fault circuit-interrupters must be installed so that all wiring used with storable pools will be protected. For GFCIs see comments following Sec. 210-8.

680-41. Indoor Installations. Part **(a)** requires that at least *one* 15- or 20-A, 125-V convenience receptacle must be installed at a spa or hot tub—not closer than 5 ft (1.52 m) from the inside wall of the unit and not more than 10 ft (3.05 m) away from it. This is intended to prevent the hazards of extension cords that might otherwise be used to operate radios, TVs, etc. (Fig. 680-30).

As required by part **(2)** of this section, this receptacle and any others within 10 ft (3.05 m) of the spa or tub must be GFCI protected.

Exception No. 2 of part **(b)** permits lighting fixtures of the designated types to be used above a spa or hot tub at any mounting height [less than 7½ ft (2.29 m)]. Recessed and surface lighting fixtures recognized for use over a spa or hot tub now only have to be suitable for "damp" locations, not for "wet" locations.

680-70. Protection. Any hydromassage bathtub and its associated electrical equipment must be supplied from a circuit protected by a GFCI.

680-71. Other Electric Equipment. Hydromassage bathtubs are not subject to the requirements for spas and hot tubs—as they were in the 1984 NEC. Receptacles

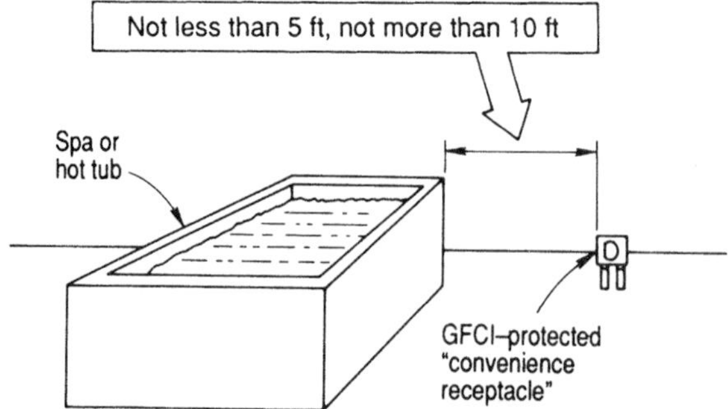

Fig. 680-30. At least one 15- or 20-A general-purpose receptacle must be installed at a spa or hot tub. [Sec. 680-41(a).]

do not have to be at least 5 ft (1.52 m) from the tub's inside wall, and lighting fixtures do not have to be mounted at least 7½ ft (2.29 m) above the tub's water level.

ARTICLE 690. SOLAR PHOTOVOLTAIC SYSTEMS

690-1. Scope. This complete, detailed article covers this developing technology for direct conversion of the sun's light into electric power.

In 1983, work was completed on the world's largest facility converting sunlight directly to electricity. ARCO Solar, Inc., and Southern California Edison Company are building the facility on Edison property near the community of Hesperia in San Bernardino County. Edison will meter and purchase electricity generated by the facility, which will be designed, built, owned, and operated by ARCO Solar, an Atlantic Richfield subsidiary.

The ARCO Solar photovoltaic system, rated at 6 MW at peak power, was installed on 20 acres adjacent to an existing Edison station.

Mounting the photovoltaic panels on approximately 100 double-axis trackers orients the panels toward the sun throughout the day, taking into account seasonal changes of the sun's position. These computer-controlled trackers, developed by ARCO Power Systems, increase the average daily power output of the panels, thus lowering the average cost of electricity.

This is an operating and not a research project. Technology was used that had not previously been demonstrated commercially on this scale. These technical advances included the combination of double-axis trackers with mass-produced photovoltaic modules, use of large-scale inverters to convert the DC electricity generated by the panels into AC current, and the introduction of a module that achieves the highest efficiency in converting sunlight to electricity of any mass-produced photovoltaic module.

Federal and state tax credits, applicable to photovoltaic installations, enabled the project to be built. Improved technology and further cost reductions over the next several years could make such large systems economic in many foreign countries where electricity is more expensive than in the United States. ARCO Solar also believes that photovoltaic technology eventually will be applied economically in large-scale systems in the United States without solar tax credits.

Chapter Seven

ARTICLE 700. EMERGENCY SYSTEMS

700-1. Scope. Note that all the regulations of this article apply to the designated "circuits, systems, and equipment"—ONLY WHEN THE SYSTEMS OR CIRCUITS ARE *REQUIRED BY LAW* AND CLASSIFIED AS EMERGENCY PROVISIONS BY FEDERAL, STATE, MUNICIPAL, OR OTHER CODE OR BY A GOVERNMENTAL AUTHORITY. THE NE CODE, ITSELF, DOES *NOT* REQUIRE EMERGENCY LIGHT, POWER, OR EXIT SIGNS.

The effect of the first paragraph of this section is to exclude from all these rules any emergency circuits, systems, or equipment that are installed on a premises but are not legally mandated for the premises. Of course, any emergency provisions that are provided at the option of the designer (or the client) must necessarily conform to all other NE Code regulations that apply to the work.

The placement or location of exit lights is not a function of the National Electrical Code but is covered in the Life Safety Code, NFPA No. 101 (formerly Building Exits Code). But, where exit lights are required by law, the NEC considers them to be parts of the emergency system. The NEC indicates how the installation will be made, not where the emergency lighting is required, except as specified in part **C** of Art. 517 for essential electrical systems in health care facilities.

OSHA regulations on exit signs are presented in Subpart E of the *Occupational Safety and Health Standards* (Part 1910). These requirements on location and lighting of exit signs apply to all places of employment in all new buildings and also in all existing buildings.

Prior to OSHA, there was no universal requirement that all buildings or all places of employment have exit signs. The National Electrical Code does not require them. And although the NFPA Life Safety Code does cover rules on emergency lighting and exit signs, that code was enforced where state or local government bodies required it—that is, in some areas and for specific types of occupancies. As a result, there are existing occupancies which do not have the exit signs now required by federal law.

Section 1910.37 of OSHA makes clear that every building must have a means of egress—a continuous, unobstructed way for occupants to get out of a building in case of fire or other emergency, consisting of horizontal and vertical ways, as required. Egress from all parts of a building or structure must be provided at all times the building is occupied. The law says:

> (5) Every exit shall be clearly visible or the route to reach it shall be conspicuously indicated in such a manner that every occupant of every building or structure who is physically and mentally capable will readily know the direction of escape from any point, and each path of escape, in its entirety, shall be so arranged or marked that the way to a place of safety outside is unmistakable. Any doorway or passageway not constituting an exit or way to reach an exit, but of such a character as to be subject to being mistaken for an exit, shall be so arranged or marked as to minimize its possible confusion with an exit and the resultant danger of persons endeavoring to escape from fire finding themselves trapped in a dead-end space, such as a cellar or storeroom, from which there is no other way out.

Then it says:

> (q) *Exit marking.* (1) Exits shall be marked by a readily visible sign. Access to exits shall be marked by readily visible signs in all cases where the exit or way to reach it is not immediately visible to the occupants.
>
> (2) Any door, passage, or stairway which is neither an exit nor a way of exit access, and which is so located or arranged as to be likely to be mistaken for an exit, shall be identified by a sign reading "Not an Exit" or similar designation, or shall be identified by a sign indicating its actual character, such as "To Basement," "Storeroom," "Linen Closet," or the like.
>
> (3) Every required sign designating an exit or way of exit access shall be so located and of such size, color, and design as to be readily visible. No decorations, furnishings, or equipment which impair visibility of an exit sign shall be permitted, nor shall there be any brightly illuminated sign (for other than exit purposes), display, or object in or near the line of vision to the required exit sign of such a character as to so detract attention from the exit sign that it may not be noticed.
>
> (4) Every exit sign shall be distinctive in color and shall provide contrast with decorations, interior finish, or other signs.
>
> (5) A sign reading "Exit", or similar designation, with an arrow indicating the directions, shall be placed in every location where the direction of travel to reach the nearest exit is not immediately apparent.
>
> (6) Every exit sign shall be suitably illuminated by a reliable light source giving a value of not less than 5 foot-candles on the illuminated surface. Artificial lights giving illumination to exit signs other than the internally illuminated types shall have screens, discs, or lenses of not less than 25 square inches area made of translucent material to show red or other specified designating color on the side of the approach.
>
> (7) Each internally illuminated exit sign shall be provided in all occupancies where reduction of normal illumination is permitted.
>
> (8) Every exit sign shall have the word "Exit" in plainly legible letters not less than 6 inches high, with the principal strokes of letters not less than three-fourths-inch wide.

A lot of discussion has been generated by that last rule. Note that an exit sign does not have to be internally illuminated, although it may be. And this lighting is required on "every exit sign" which means signs over exit doors and exit signs indicating direction of travel.

The phrase "reliable light source" raises questions as to its meaning. Just what is reliable? Does this mean that the light source must operate if the utility supply to a building fails? And is there a difference in required application in new buildings versus existing buildings?

Because the OSHA regulations on exit signs do not require emergency power for lighting of such signs, the light units that illuminate exit signs may be supplied from regular (nonemergency) circuits. OSHA does not make it mandatory to supply such circuits from a tap ahead of the service main or from batteries or an emergency generator. And this applies to new buildings as well as existing buildings. As far as OSHA is concerned, Art. 700 of the NE Code on emergency systems does not apply to circuits for exit sign lighting. Of course, OSHA does not object to the extra reliability such arrangements give. But the NEC does classify exit lights as emergency equipment (Fig. 700-1).

Fig. 700-1. Exit lights and wall-hanging battery-pack emergency lighting units are covered by Art. 700 whenever such equipment or provisions for emergency application are legally required by governmental authorities. (Sec. 700-1).

An emergency lighting system in a theater or other place of public assemblage includes exit signs, the chief purpose of which is to indicate the location of the exits, and lighting equipment commonly called "emergency lights," the purpose of which is to provide sufficient illumination in the auditorium, corridors, lobbies, passageways, stairways, and fire escapes to enable persons to leave the building safely.

These details, as well as the various classes of buildings in which emergency lighting is required, are left to be determined by state or municipal codes, and where such codes are in effect, the following provisions apply.

700-3. Equipment Approval. This rule has the effect of requiring use of only emergency equipment that is listed for such application by UL or another testing laboratory.

Under the heading of "Emergency Lighting and Power Equipment," the UL's *Electrical Construction Materials Directory* states:

> This listing covers battery-powered emergency lighting and power equipment, for use in ordinary indoor locations in accordance with Article 700 of the National Electrical Code. The lighting circuit ratings do not exceed 250 volts for tungsten lamps or 277 volts ac for electric discharge lamps. Other ratings may be included (motor loads, inductive loads, resistance loads, etc.) to 600 volts. This listing covers unit equipment, automatic battery charging and control equipment, inverters, central station battery systems, distribution panels, exit lights, and remote lamp assemblies, but not lighting fixtures. The investigation of emergency equipment includes the determination of their suitability of transferring operation from normal supply circuit to an immediately available emergency supply circuit.

700-4. Tests and Maintenance. Part **(a)** of Sec. 700-4 calls for inspection of an operational test at the time of installation *and* at periodic intervals thereafter. The frequency of these witnessed tests is to be determined by the local inspection authority [part **(b)**]. It is not clear if any documentation of such tests is required. Nor is there any indication of who should retain such documentation. In the absence of specific guidance from the NEC, the local inspector will have to be consulted with regard to any paperwork that may be required.

Emergency systems must be tested "during maximum anticipated load conditions." Testing under less than full-load conditions can be misleading and is not a true test.

700-5. Capacity. It is extremely important that the supply source be of adequate capacity. There are two main reasons for adequate capacity:

1. It is important that power be available for the necessary supply to exit lights, emergency and egress lighting, as well as to operate such equipment as required for elevators and other equipment connected to the emergency system.
2. In such occupancies as hospitals, there may be a need for an emergency supply for lighting in hospital operating rooms, and also for such equipment as inhalators, iron lungs, and incubators.

It is also essential to safety and effective, long-time successful operation of emergency system equipment that it be rated to sustain the maximum available short circuit that the supply circuit could deliver at the terminals of the equipment.

Part **(b)** of this section permits one generator to be used as a single power source to supply emergency loads, essential (legally required) standby loads, and optional standby loads when control arrangements for *selective* load pickup and load shedding are provided to ensure that adequate power is available first for emergency loads, then legally required standby loads, and finally optional standby loads. (See Fig. 700-2 (top) and Arts. 701 and 702.)

An on-site generator may be used for peak load shaving—in addition to its use for supplying emergency, legally required standby, and optional standby loads. The title of part **(b)** refers to "Peak Load Shaving," which is recognized as a func-

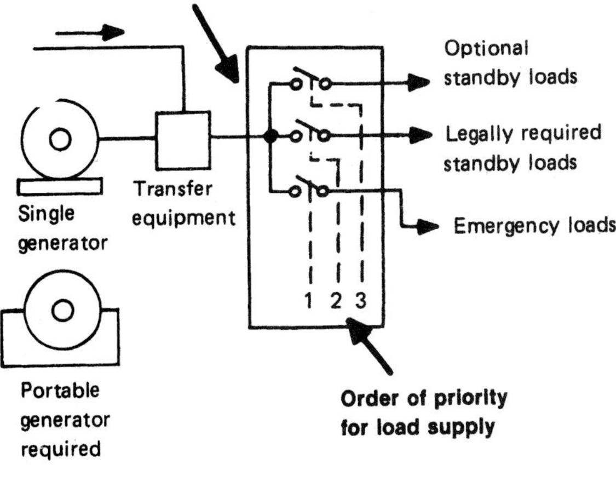

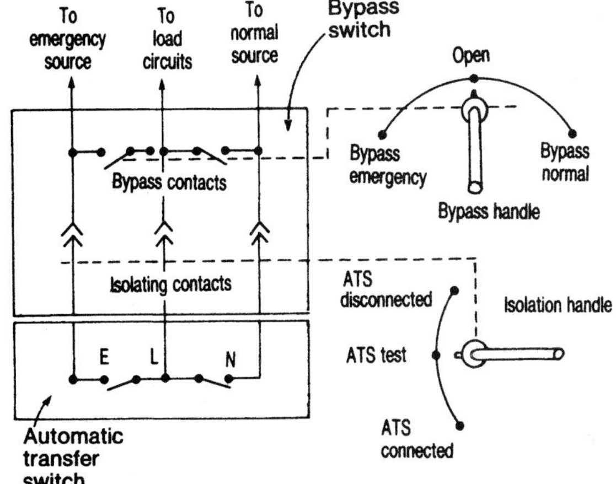

Fig. 700-2. Special load-control must be used where a single generator supplies emergency and standby loads (top). And a bypass switch may be used to isolate an automatic transfer switch. (Secs. 700-5 and 700-6.)

tion of the generator in the last sentence of the paragraph. For peak shaving use, however, the generator must be equipped with the selective load-control equipment described in the paragraph to assure the order of priority for various purposes. The second-to-last sentence essentially permits the peak load shaving operation to satisfy the testing and maintenance requirements of Sec. 700-4.

Another important requirement of this section specifies that a portable or temporary generator, or other alternate source of power, must be provided for any time that the emergency generator is out of service for major maintenance or repair.

700-6. Transfer Equipment. Any switch or other control device that transfers emergency loads from the normal power source of a system to the emergency power supply *must* operate automatically on loss of the normal supply. Transfer equipment must also be automatic for legally required standby systems, as covered in Sec. 701-7, but a manual transfer switch may be used for switching loads from the normal to an optional standby power source, as in Sec. 702-6.

As described in the last paragraph, a bypass switch is recognized on an automatic transfer switch (ATS) to provide for repair or maintenance of the ATS (Fig. 700-2, bottom). Because hospitals and many industrial systems contain transfer switches that cannot be shut down, some means is required to isolate a transfer switch for routine maintenance or for some specific task like contact replacement. When a bypass switch is used, inadvertent parallel operation must not be possible.

700-7. Signals. In order to be effective, the signal devices should be located in some room where an attendant is on duty. Lamps may readily be used as signals to indicate the position of an automatic switching device. An audible signal in any place of public assemblage should not be so located or of such a character that it will cause a general alarm.

The standard signal equipment furnished by a typical battery manufacturer with their 60-cell battery for emergency lighting includes an indicating lamp which is lighted when the charger is operating at the high rate, and a voltmeter marked in three colored sections indicates (1) that the battery is not being charged, or is discharging into the emergency system, (2) that the battery is being trickle charged, or (3) that the battery is being charged at the high rate. This last indication duplicates the indication given by the lamp.

Part **(d)** of this section requires the use of a hookup for ground-fault indication on the output of a grounded-wye 480/277-V generator. The rule, however, makes the ground-fault indication signal mandatory only "where practicable." Section 700-26 states that emergency systems do not require ground-fault protection of equipment. Yet, when the load is served by the emergency source, the possibility of a ground fault is no less than when the load is served by the utility source. Should a ground fault occur within the emergency system during a power failure, it is debatable whether essential loads should be immediately disconnected from the emergency power supply. However, for reasons of safety and to minimize the possibility of fire and equipment damage, at least an alarm should indicate if a hazardous ground-fault condition exists so appropriate corrective steps will be initiated. When ground-fault indication is used on the generator output, the fault sensor must be located within or on the line side of the generator main disconnect. And the rule also makes it mandatory to have specific instructions for dealing with ground-fault conditions. (Refer to Sec. 700-26 and Fig. 700-16.)

700-8. Signs. A sign at the service entrance is required to designate the type and location of emergency power source(s).

700-9. Wiring, Emergency System. Part **(a)** requires that all boxes and enclosures for emergency circuits must be readily identified as parts of the emergency system. Labels, signs, or some other *permanent* marking must be used on all enclosures containing emergency circuits to "readily" identify them as components of an emergency system. The "boxes and enclosures" that must be marked include enclosures for transfer switches, generators, and power panels. *All* boxes and enclosures for emergency circuits must be painted red, marked with red labels saying "EMERGENCY CIRCUITS," or marked in some other manner clearly identifiable to electricians or maintenance personnel (Fig. 700-3, top).

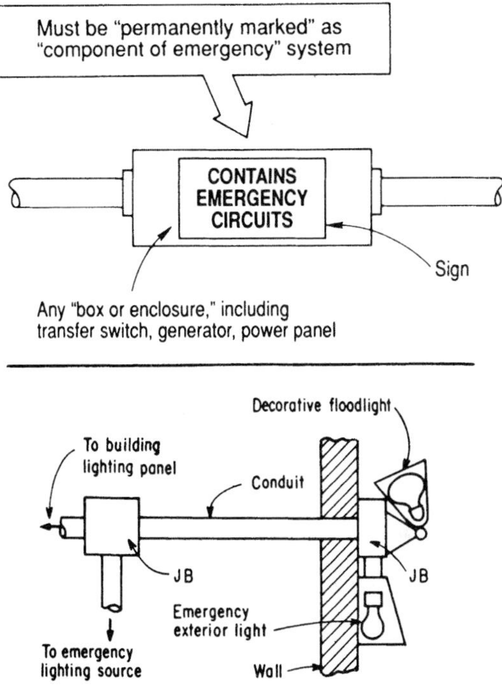

Fig. 700-3. Emergency wiring may not be run in enclosures with wiring for nonemergency circuits, and emergency enclosures must be marked. (Sec. 700-9.)

The bottom of Fig. 700-3 shows a clear violation of the rule of part **(b)** because the wiring to the emergency light (or to an exit light, which is classed as an emergency light), is run in the same raceway and boxes as the wiring to the decorative floodlight.

This section requires that the wiring for emergency systems be kept entirely independent of the regular wiring used for lighting, and it thus needs to be in separate raceways, cables, and boxes. This requirement is to ensure that where faults may occur on the regular wiring, they will not affect the emergency system wiring, as it will be in a separate enclosure.

Exception No. 1 for transfer switches is intended to permit normal supply conductors to be brought into the transfer-switch enclosure and that these conductors would be the only ones within the transfer-switch enclosure which were not part of the emergency system. Exception Nos. 2 and 3 permit two sources supplying emergency or exit lighting to enter the fixture and its common junction box.

Exception No. 4 specifically permits more than one emergency circuit within a raceway, etc., *provided* both circuits are supplied from the same source. By inference, two emergency circuits supplied from different sources may *not* occupy a common enclosure, unless of course there are two emergency supplies to a single piece of equipment as covered in Exception Nos. 1, 2, 3, and 5. Exception No. 5 also allows a "common junction box" on other equipment to contain both emergency and nonemergency circuits.

700-12. General Requirements. This section lists and describes the types of emergency supply systems that are acceptable—with one or more of such systems required where emergency supply is mandated by law. It specifies that the normal-to-emergency transfer must not exceed 10 sec. This is a change from previous wording that required emergency supply to be "immediately available" on loss of normal supply.

The last paragraph presents important information on protection of emergency power sources for certain occupancies. Such protection may be provided by (1) automatic sprinkler or (2) enclosing the equipment in a room with 1-hr fire-rated walls.

Part **(a)** recognizes storage batteries for an emergency source. A storage battery for emergency power must maintain *voltage to the load* at not less than 87½ percent of the normal rated value. This was changed from "87½ percent of system voltage" (that is, battery voltage) because the concern is to keep the voltage to the lamps at 87½ percent. The "electronics" between the battery and the lamps maintains the required "load voltage," and the battery voltage is not in itself the major concern.

Part **(b)** covers use of engine-generator sets for emergency supply as an alternate to utility supply. Engine-driven generators (diesel, gasoline, or gas) are commonly used to provide an alternate source of emergency or standby power when normal utility power fails. Gas-turbine generators also are used.

The first step in selecting an on-site generator is to consider applicable requirements of the National Electrical Code, which differ depending on whether the generating set is to function as an emergency system, a standby power system, or as a power source in a health care facility, such as a hospital.

For example, an *internal-combustion*-type engine-generator set selected for use under Art. 700 must be provided with *automatic starting and automatic load transfer,* with enough on-site fuel to power the full demand load for at least 2 hr (Fig. 700-4). If a standby power system selected under the regulations of Art. 701 is *legally* required, it must be provided with enough on-site fuel to power the full demand operation of the load for *not less than 2 hr.*

In part **(b)(3)**, the engine driving an emergency generator must not be dependent on a public water supply for its cooling. That means a roof tank or other on-site water supply must be used and its pumps connected to the emergency source (Fig. 700-5).

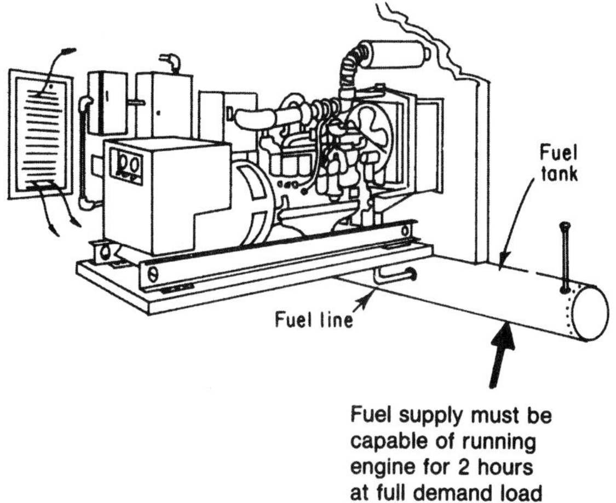

Fig. 700-4. Generator must have automatic start and adequate fuel supply. (Sec. 700-12.)

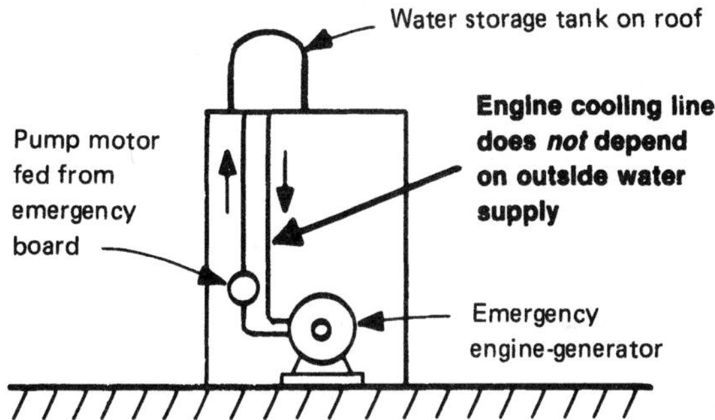

Fig. 700-5. Engine cooling for an emergency generator set must be assured for continuous operation of the generator. (Sec. 700-12.)

An Exception in part **(b)(3)** permits use of a utility gas supply to the engine of an emergency generator—at the discretion of the local inspector—where simultaneous outage of both electric power and gas supply is highly unlikely.

As required by part **(b)(4)** a battery used with a generator set must have adequate capacity whether it cranks the generator for starting or is simply used for control and signal power for another means (such as compressed air) for starting the generator.

Part **(b)(5)** of this section requires another power supply to pick up an emergency load in not over 10 sec where the main generator cannot come up to power output in 10 sec.

Part **(c)** allows an uninterruptible power supply to supply emergency power provided battery capacity satisfies part **(a)**, and if a generator is also used—which is generally the case—it must satisfy part **(b)**.

As recognized by part **(d)**, two separate services brought to different locations in the building are always preferable, and these services should at least receive their supply from separate transformers where this is practicable. In some localities, municipal ordinances require either two services from independent sources of supply, or auxiliary supply for emergency lighting from a storage battery, or a generator driven by a steam turbine, internal-combustion engine, or other prime mover. Figure 700-6 shows two different forms of the separate-service type of emergency supply. The method at bottom makes use of two sources of emergency input.

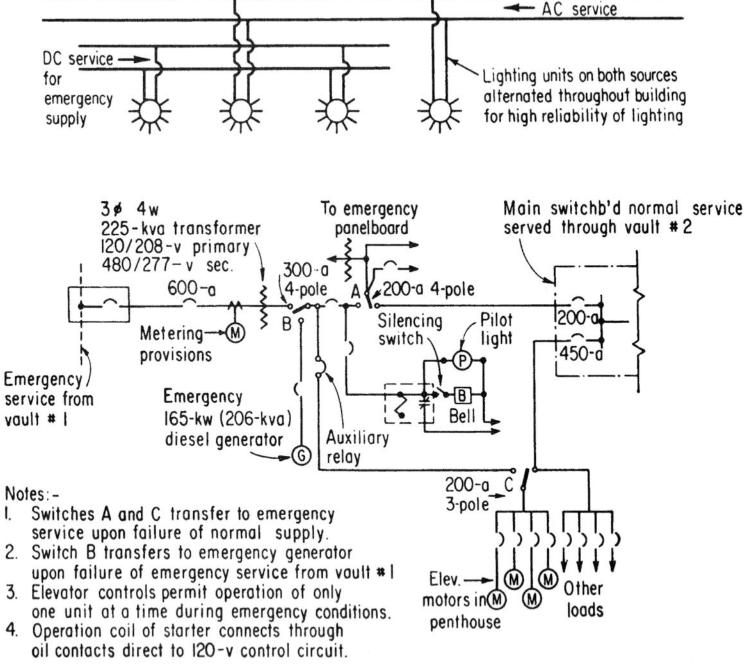

Fig. 700-6. Dual-service emergency provisions can take many different forms. (Sec. 700-12.)

The method shown in Fig. 700-7 was recognized in the 1993 and previous Codes as a means of providing emergency power. However, this permission in the NEC was at odds with the NFPA's *Life Safety Code*. In an effort to harmonize the two standards, this permission, formerly covered in part **(e)**, was deleted in the 1996 Code. In that diagram, the tap ahead of the main could supply the emergency panel directly, without need for the transfer switch.

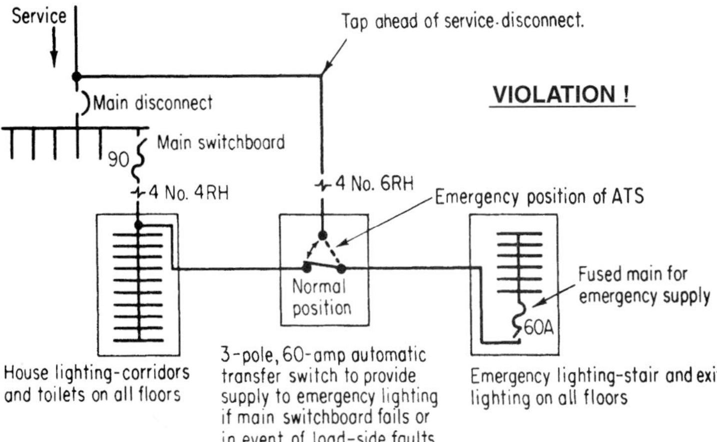

Fig. 700-7. A tap ahead of the service main protects only against internal failures and is no longer recognized. (Sec. 700-12.)

Part **(e)** covers typical wall-hanging battery-pack emergency lighting units, as shown in Fig. 700-8. The 1971 NEC accepted only connection of emergency light units by means of fixed wiring. Part **(e)** now recognizes permanent wiring connection *or* cord-and-plug connection to a receptacle.

Even though the unit equipment is allowed to be hooked up with flexible cord and plug-connections, it is still necessary that the unit equipment be permanently fixed in place.

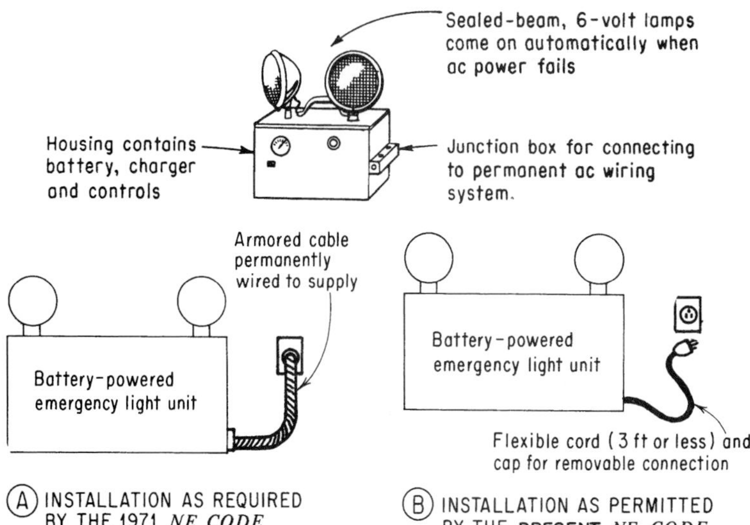

Fig. 700-8. Unit emergency lights may serve as required source of emergency supply. (Sec. 700-12.)

Individual unit equipment provides emergency illumination only for the area in which it is installed; therefore, it is not necessary to carry a circuit back to the service equipment to feed the unit. This section clearly indicates that the branch circuit feeding the normal lighting in the area to be served is the same circuit that should supply the unit equipment.

In part **(e)** the intent of the 87½ percent value is to assure proper *lighting output* from lamps supplied by unit equipment. It is generally considered acceptable to design equipment that will produce acceptable lighting levels for the required 1½ hr, even though the 87½ percent rating of the battery would not be maintained during this period. The objective is adequate light output to permit safe egress from buildings in emergencies. Hence, the unit equipment shall supply and maintain not less than 60 percent of the initial emergency illumination for a period of at least 1½ hr.

As shown at the top of Fig. 700-9, a battery-pack emergency unit must not be connected on the load side of a local wall switch that controls the supply to the unit or the receptacle into which the emergency unit is plugged. Such an arrangement exposes the emergency unit to accidental energization of the lamps and draining of the battery supply. But, as permitted by the Exception, if a panelboard supplies at least three normal lighting circuits for a given area,

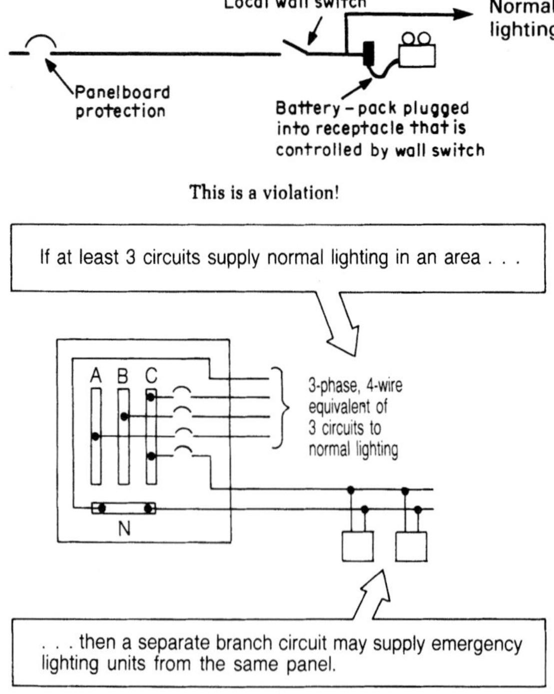

Fig. 700-9. Circuiting of battery-pack emergency lighting may be on normal lighting circuit or separate circuit. [Sec. 700-12(f).]

emergency battery-pack lighting units for that area may be connected to a separate branch circuit from the panel, with lock-on provision for that circuit. This is an Exception to the rule of the previous paragraph, which says that unit equipment must be connected on a branch circuit supplying normal lighting in the area. By allowing unit equipment on a separate branch circuit from the same panel, the unit equipment will sense loss of power to the panel and activate. The advantage of such application from a design standpoint is that there is no need to observe the rule that the unit equipment must be "connected ahead of any local switches," which applies when unit equipment is connected on normal lighting circuits.

700-15. Loads on Emergency Branch Circuits. Figure 700-10 shows a clear violation of this rule, because appliances are excluded from emergency circuits.

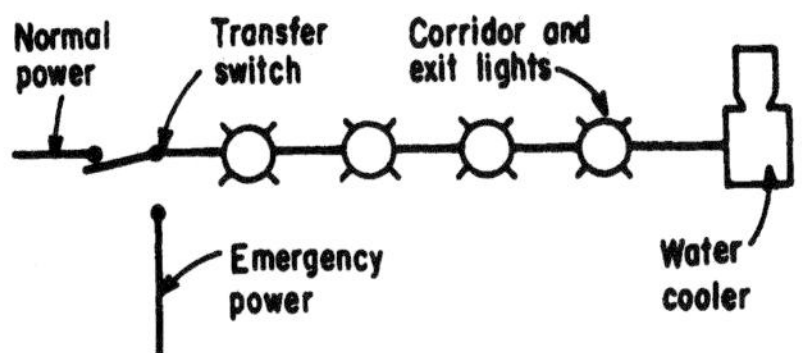

Fig. 700-10. The water cooler may not be on an emergency circuit. (Sec. 700-15.)

700-16. Emergency Illumination. Note that all exit lights are designated as part of the emergency lighting, and, as such, their circuiting must conform to Sec. 700-17.

Where HID lighting is the *sole* source of *normal* illumination, the emergency light system must continue to operate for a sufficient time after return of normal power to enable the HID lighting to come up to brightness. This rule in the third paragraph is intended to prevent the condition that return of normal power and the disconnect of the emergency lighting leave the building in darkness because of the inherent, normal time delay in the light output upon energizing HID lamps. The Exception after that rule permits "alternative means" to keep emergency lighting on.

700-17. Circuits for Emergency Lighting. Figure 700-11 shows the basic rule on transfer of emergency lighting from the normal source to the emergency source. If a single emergency system is installed, a transfer switch shall be provided which, in case of failure of the source of supply on which the system is operating, will automatically transfer the emergency system to the other source. Where the two sources of supply are two services, the single emergency system may normally operate on either source, as in Fig. 700-12. Where the two sources of supply are one service and a storage battery, or one service and a generator set, as in Fig. 700-11, the single emergency system would, as a general rule, be operated normally on the service, using the battery or generator only as a reserve in case of failure of the service. Figure 700-13 shows an emergency hookup that has two separate supplies tied into the emergency lighting system.

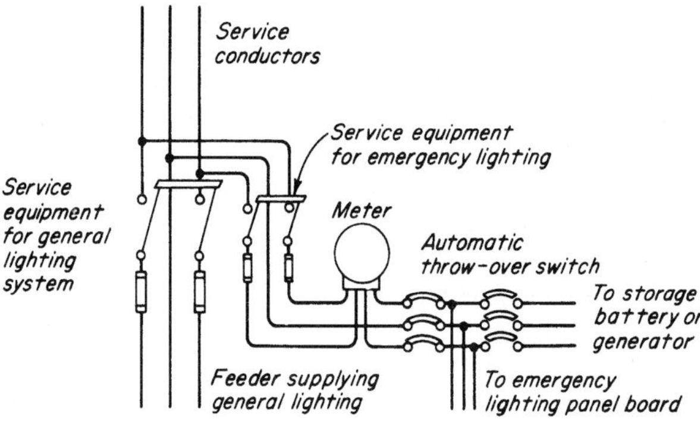

Fig. 700-11. Emergency lighting is automatically switched from normal service to the battery or generator. (Sec. 700-17.)

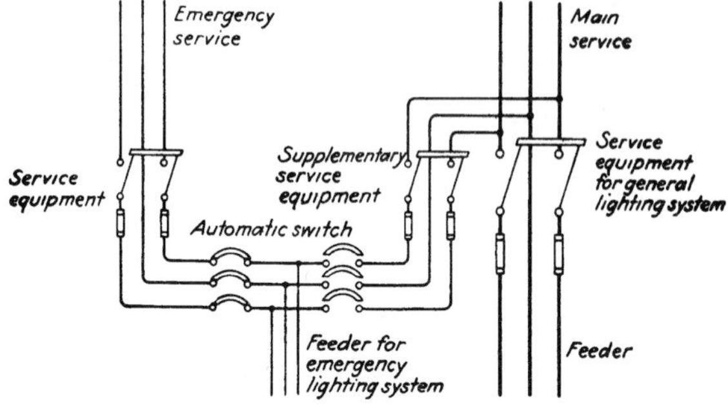

Fig. 700-12. Emergency lighting may be supplied from an emergency service. (Sec. 700-17.)

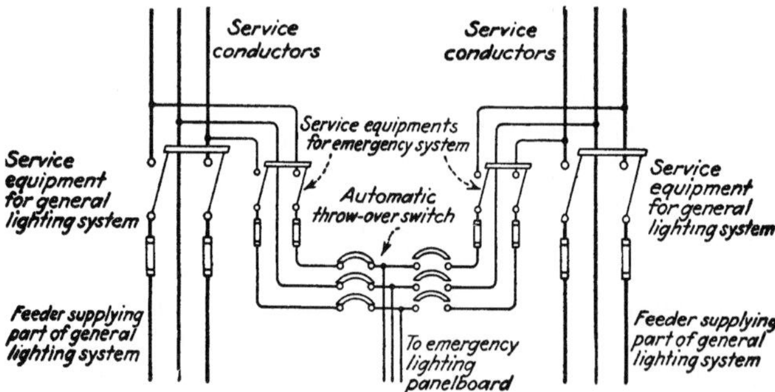

Fig. 700-13. A single emergency lighting system may be fed from two services. (Sec. 700-17.)

Part **(2)** of this section provides for use of two or more "separate and complete" emergency systems. If two emergency lighting systems are installed, each system shall operate on a separate source of supply, as where each emergency disconnect in Fig. 700-12 feeds a separate, independent emergency lighting system. Either both systems shall be kept in operation, or switches shall be provided which will automatically place either system in operation upon failure of the other system.

700-20. Switch Requirements. Figure 700-14 shows a violation of the last sentence of this rule, where use of three- and four-way switches is prohibited.

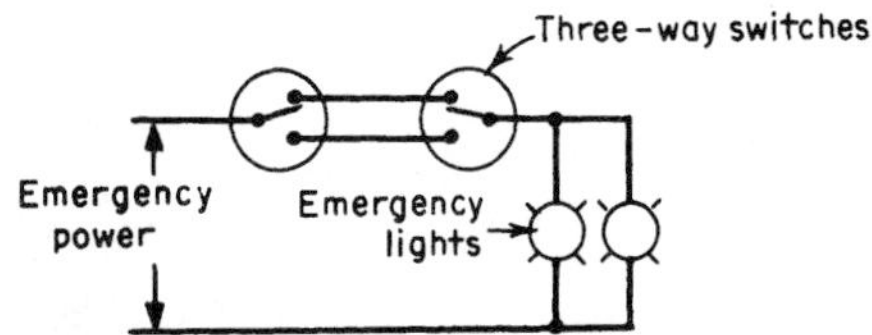
These switches are a code violation in this circuit...
Three-way switches
Emergency power
Emergency lights

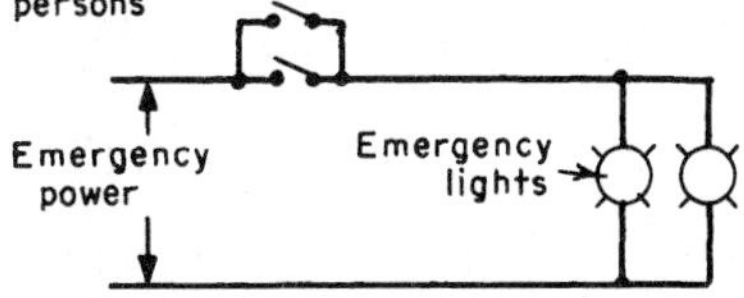
... BUT this use of parallel switches is acceptable if one of the switches is accessible only to authorized persons
Emergency power
Emergency lights

Fig. 700-14. Watch out for switches in emergency lighting circuits. (Sec. 700-20.)

700-21. Switch Location. The sole switch for emergency lighting control in a theater may not be placed in a projection booth (Fig. 700-15). The rule prohibits use of an emergency control switch in the motion-picture projection booth or on stage. In the 1981 NEC, the rule that used to be in Sec. 700-19(b)—which prohibited use of an emergency control switch in the motion-picture projection booth or on stage—was deleted. The comment that accompanied the proposal to remove this rule stated:

> Since the place in a theater where there is most likely to be someone present at all times is the projection booth, the emergency lighting switch should be in the projection booth. There was general agreement that this was proper and is conventionally being done. It should be pointed out that the typical motion-picture theater has no one present on the stage to operate the emergency lighting.

In the present NE Code, the old rule prohibiting the sole control switch in the projection booth or on the stage is reinstated. The comment that accompanied the proposal to again prohibit installation of an emergency lighting switch in a projection booth or on a stage made the following points:

> The primary control of the life safety emergency system must be restored to those responsible for the management of the building and not rest with those primarily concerned with the performance. These systems are the exit lighting, aisle lighting, and alarm systems that are "first on-last off" during the operation of these occupancies.
>
> Overzealous and unfamiliar production staff persons have been known to disconnect the life safety systems when given the cue to darken the auditorium. Many traveling companies bring their own equipment and do not bother to familiarize themselves with fixed controls nor do they wish to assume responsibility. Motion-picture projectionists

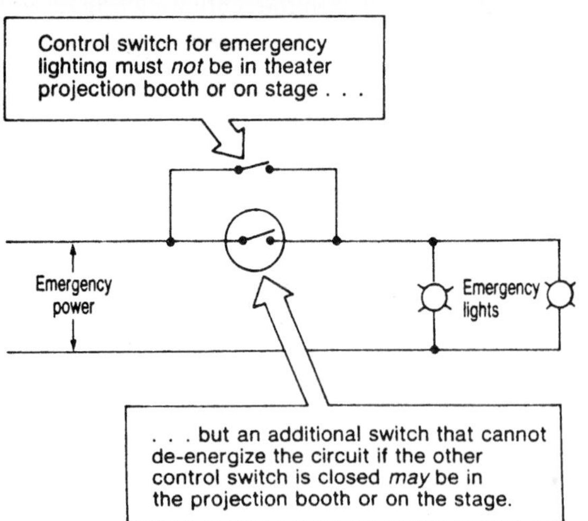
Control switch for emergency lighting must *not* be in theater projection booth or on stage . . .
Emergency power
Emergency lights
. . . but an additional switch that cannot de-energize the circuit if the other control switch is closed *may* be in the projection booth or on the stage.

Fig. 700-15. The sole switch that controls emergency lighting in a theater may *not* be installed in a projection booth or on the stage, with the Exception shown. (Sec. 700-21.)

in booths are concerned with their equipment, film, and presenting the show. Their function is further compounded by the multi-auditorium theaters of today and the hazards of dealing with xenon lamps in modern projection equipment, leaving little time to monitor the auditorium and render decision regarding the emergency system.

Figure 700-15 shows the Exception to the prohibition against an emergency lighting control switch in a projection booth or on stage. An emergency lighting switch may be used in a projection booth or on a stage if it can energize such lighting but cannot deenergize the lighting if another switch located elsewhere is in the closed (ON) position.

700-26. Ground-Fault Protection of Equipment. Emergency generator disconnect does not require ground-fault protection. This rule clarifies the relationship between Sec. 230-95, calling for GFP on service disconnects, and the disconnect means for emergency generators. To afford the highest reliability and continuity for an emergency power supply, GFP is not *required* on any 1,000-A or larger disconnect for an emergency generator, although it is permissible to use it if desired (Fig. 700-16).

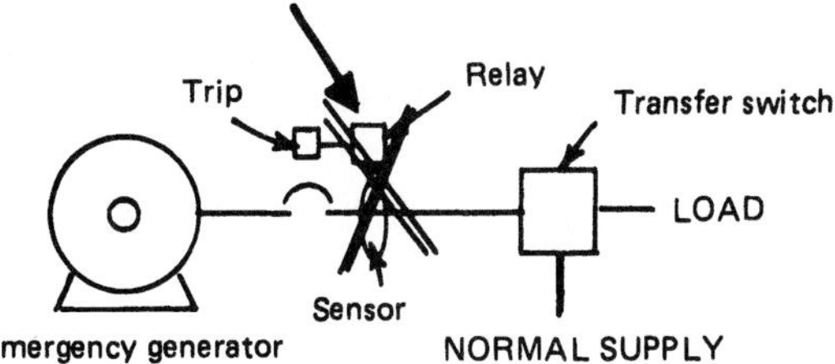

. . . But, Sec. 700-7 (d) REQUIRES ALARM ON FAULT.

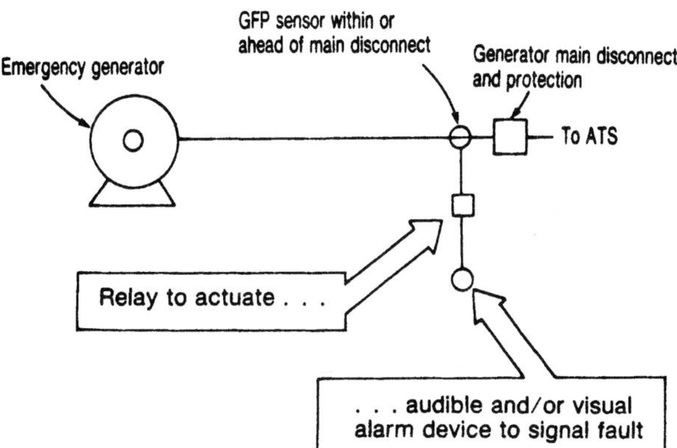

Fig. 700-16. Ground-fault protection is optional for any 480/277-V grounded-wye generator disconnect rated 1,000 A or more, but ground-fault "indication" is required. (Sec. 700-26.)

ARTICLE 701. LEGALLY REQUIRED STANDBY SYSTEMS

701-1. Scope. This article covers *standby* power systems that are required by law. Legally required standby power systems are those systems required and so classed as legally required standby by municipal, state, federal, or other codes or by any government agency having jurisdiction. These systems are intended to supply power automatically to selected loads (other than those classed as emergency systems) in the event of failure of the normal source.

Legally required standby power systems are typically installed to serve such loads as heating and refrigeration systems, communication systems, ventilation and smoke-removal systems, sewerage disposal, lighting, and industrial processes that, when stopped during any power outage, could create hazards or hamper rescue or firefighting operations.

This article covers the circuits and equipment for such systems that are permanently installed in their entirety, including power source.

ARTICLE 702. OPTIONAL STANDBY SYSTEMS

702-1. Scope. Life safety is *not* the purpose of optional standby systems. Optional standby systems are intended to protect private business or property where life safety does not depend on the performance of the system. Optional standby systems (other than those classed as emergency or legally required standby systems) are intended to supply on-site generated power to selected loads, either automatically or manually.

Optional standby systems are typically installed to provide an alternate source of electric power for such facilities as industrial and commercial buildings, farms, and residences to serve such loads as heating and refrigeration systems, data processing and communications systems, and industrial processes that, when stopped during any power outage, could cause discomfort, serious interruption of the process, or damage to the product or process.

Because of the constant expansion in electrical applications in all kinds of buildings, the use of standby power sources is growing at a constantly accelerating rate. Continuity of service has become increasingly important with the widespread development of computers and intricate, automatic production processes. More thought is being given to and more money is being spent on the provision of on-site power sources to back up or supplement purchased utility power to ensure the needed continuity as well as provide for public safety in the event of utility failure.

The intent of Arts. 701 and 702 is to recognize and regulate use of any permanently installed standby system that is not considered to be an "Emergency System" as covered in Art. 517 or 700 of the National Electrical Code, or "Essential Electrical Systems for Health Care Facilities" as covered in NFPA Pamphlet No. 99.

The most fundamental application of standby power is the portable alternator used for residential standby power where electric utility supply is not sufficiently reliable or is subject to frequent outages (Fig. 702-1).

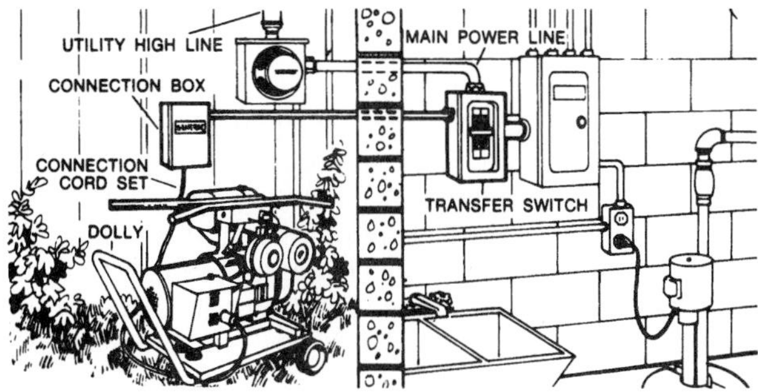

Fig. 702-1. Standby generator with manual transfer is common residential application. (Sec. 702-1.)

Note that an optional "standby" power load may be manually or automatically transferred from the normal supply to the "standby" generator. Any generator used as an "emergency" source or legally required standby system must always have provision for automatic transfer of the load from "normal" supply to the emergency generator. But automatic transfer must be used if a standby power system is required by law to be installed.

702-5. Capacity and Rating. Any standby power source and system must be capable of fully serving its demand load. This can generally be satisfied relatively easily for generator loads. But the task can be complex for uninterruptible power supply UPS systems. The UPS is an all-solid-state power conversion system designed to protect computers and other critical loads from blackouts, brownouts, and transients. It is usually connected in the feeder supplying the load, with bypass provisions to permit the load to be fed directly. Figure 702-2

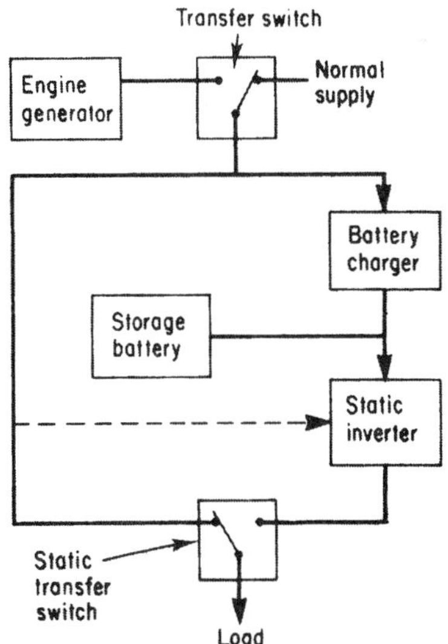

Fig. 702-2. UPS system is a common standby power system. (Sec. 702-5.)

shows a typical basic layout for a UPS system. Such a system utilizes a variety of power sources to assure continuous power. Circuits are shown for normal power operation. If normal power fails, the static switch transfers the load to the inverter within ¼ cycle. The battery is the power source while the engine-generator is being started. When the generator is running properly, it is brought onto the line through its transfer switch, and the static switch transfers the load to the generator supply. The dashed line indicates the synchronizing signal, which maintains phase and frequency of inverter output.

Calculation of the required capacity of a UPS system must be carefully made. Data processing installations normally require medium-to-large 3-phase UPS systems ranging from 37.5 to over 2,000 kVA. Some of the typical ratings available from UPS manufacturers are 37.5 kVA/30 kW, 75 kVA/67.5 kW, 125 kVA/112.5 kW, 200 kVA/180 kW, 300 kVA/270 kW, 400 kVA/360 kW, and 500 kVA/450 kW. Larger systems are configured by paralleling two or more of these standard-size single modules.

The necessary rating is chosen based on the size of the critical load. If a power profile itemizing the power requirements is not available from the computer manufacturer, the load may be measured using a kilowattmeter and a power-factor meter. Since UPS modules are both kVA- (apparent power) and kW- (real power) limited, a system should be specified with both a kVA and kW rating. The required kVA rating is obtained by dividing the actual load kW by the actual load power factor. For example, an actual 170-kW load with an actual 0.85 power factor would require a 200-kVA-rated UPS. The system should be specified as 200 kVA/170 kW, and the standard 200 kVA/180 kW UPS module should be selected for the application.

ARTICLE 705. INTERCONNECTED ELECTRIC POWER PRODUCTION SOURCES

705-1. Scope. This NEC article covers interconnection of electric power sources, including utility supply, emergency systems, standby systems, on-site generator supply, and solar photovoltaic systems.

The substantiation for addition of this Code article stated the following:

> There is a need for an article within the NEC that deals with on-site power production that is interconnected as well as operating in parallel with an electric utility source. This is not the same as emergency generators, which by definition are for an emergency condition when the normal power is not available, nor is it the same as either legally required or optional standby generators not intended to be continuously operated nor interconnected to the utility grid.
>
> The on-site power-production aspect needs to be viewed as possibly ranging from a singular low-wattage, backyard wind machine up to an industrial application that can possibly have a dozen machines all interconnected.
>
> The similar yet even more obvious need exists for multiple power-production sources that are interconnected within the same site or structure: there must be the ability to separate the sources without jeopardizing either the equipment, personnel, or other sources that are allowed to continue operating.

ARTICLE 710. OVER 600 VOLTS, NOMINAL, GENERAL

710-1. Scope. An important consideration in selection and application of conductors for systems operating over 600 V is careful correlation of NE Code regulations and the data made available by third-party testing laboratories. Underwriters Laboratories and Electrical Testing Labs have been deeply involved in listing cables for use in high-voltage systems. The expanded testing and listing of high-voltage conductors and cable assemblies have increased the need for the electrical designer and installer to be particularly thorough and careful in establishing full and effective compliance with all applicable codes and standards.

The term *high voltage* is commonly used to refer to any circuit operating at voltage above 600 V, phase to phase. It should be noted, however, that circuits from 601 V up to 35 kV are frequently called "medium-voltage" circuits and the term "high voltage" also is used for circuits operating above 35 kV. As far as the NEC is concerned, any voltage in excess of 600 V is "high voltage" (Sec. 710-2).

The NE Code contains very extensive rules on all aspects of high-voltage work. These rules must be evaluated and studied carefully. Check all questions about interpretations of Code rules with the authority having jurisdiction for Code enforcement.

710-3. Other Articles. Article 710 covers general requirements on all circuits operating at more than 600 V between conductors. Specific requirements on high-voltage application are covered within other Code articles on services, motors and controllers, transformers, capacitors, outside wiring, and other specific categories of equipment.

710-4. Wiring Methods. In the past, high-voltage circuits used for commercial and industrial feeders, both outdoors and indoors, most commonly operated at voltages up to 15,000 V (15 kV). But today, higher voltages (26 and 35 kV) can offer economy for extremely large installations. Typical circuits today operate at 4,160/2,400 and 13,200/7,600 V—both 3-phase, 4-wire wye hookups.

Modern high-voltage circuits for buildings include: overhead bare or covered conductors, installed with space between the conductors which are supported by insulators at the top of wood poles or metal tower structures; overhead aerial cable assemblies of insulated conductors entwined together, supported on building walls or on poles or metal structures; insulated conductors installed in metal or nonmetallic conduits or ducts run underground, either directly buried in the earth or encased in a concrete envelope under the ground; insulated conductors in conduit run within buildings; multiple-conductor cable assemblies (such as nonmetallic jacketed cables, lead-sheathed cable, or interlocked armor cable), installed in conduit or on cable racks or trays or other types of supports. Another wiring method gaining wide acceptance consists of plastic conduit containing factory-installed conductors, affording a readily used direct-earth-burial cable assembly for underground circuits but still permitting removal of the cable for repair or replacement.

(a). Conductors aboveground must be in rigid metal conduit, in intermediate metal conduit (IMC), in rigid nonmetallic conduit, in cable trays, in a cable bus, in other suitable raceways (check inspector for suitability of EMT), or as open runs of metal-armored cable suitable for the use and purpose. The NE Code now equates IMC to rigid metal conduit for indoor and outdoor (including underground) applications for high-voltage circuits.

The phrase "other identified raceway," as used in part (a), seems to mean raceways listed by a testing laboratory and installed in accordance with any instructions given with the listing. If it is meant to accept all raceways covered under the NE Code definition for "raceway," then it includes EMT, flexible metal conduit, wireways, and rigid nonmetallic conduit (Fig. 710-1). But that should be checked with the local inspector having jurisdiction.

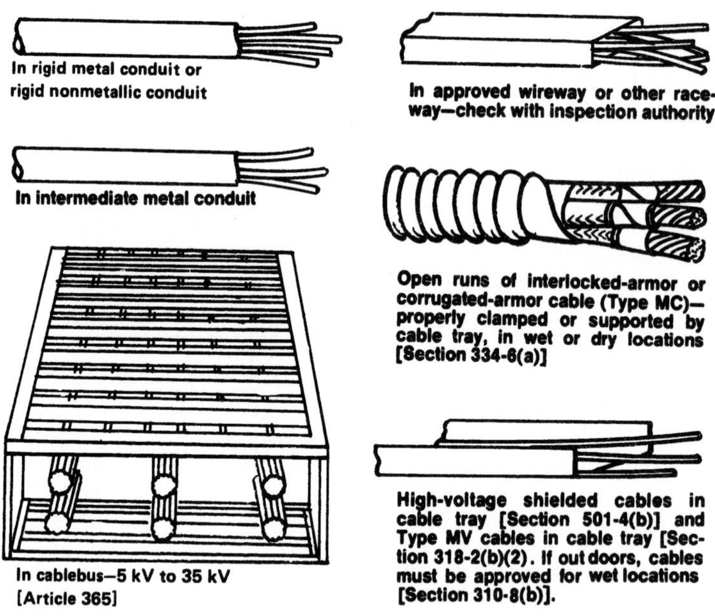
In rigid metal conduit or
rigid nonmetallic conduit
In approved wireway or other race-
way—check with inspection authority
In intermediate metal conduit
Open runs of interlocked-armor or
corrugated-armor cable (Type MC)—
properly clamped or supported by
cable tray, in wet or dry locations
[Section 334-6(a)]
In cablebus—5 kV to 35 kV
[Article 365]
High-voltage shielded cables in
cable tray [Section 501-4(b)] and
Type MV cables in cable tray [Sec-
tion 318-2(b)(2). If outdoors, cables
must be approved for wet locations
[Section 310-8(b)].

Fig. 710-1. A variety of wiring methods may be used for aboveground high-voltage circuits. (Sec. 710-4.)

Rigid nonmetallic conduit has been added to the list of raceways acceptable for running high-voltage circuits aboveground—*and the PVC conduit does not have to be encased in concrete.* The previous requirement in Sec. 347-2(b), which required concrete encasement of rigid PVC conduit aboveground if it carried circuits operating at over 600 V, is no longer in the Code.

Directly buried nonmetallic conduit carrying high-voltage conductors does not have to be concrete-encased if it is a type approved for use without concrete encasement. If concrete encasement is required, it will be indicated on the UL label and in the listing. Sections 347-2(c) and 710-4(b) permit direct burial rigid nonmetallic conduit (without concrete encasement) for high-voltage circuits.

In locations accessible to qualified persons only, open runs of Type MV cable, bare conductors, and busbars may be used (Secs. 710-32 and 710-33). In locations accessible to qualified persons only there are no restrictions on the types of wiring that may be used. The types more commonly employed are open wiring on insulators with conductors either bare or insulated, and rigid metal conduit or nonmetallic rigid conduit containing lead-covered cable.

(b). Underground conductors may be installed in "raceways identified for the use" or in approved direct burial cable assemblies.

Table 710-4(b) makes clear that underground circuits may be installed in rigid metal conduit, in intermediate metal conduit, or in rigid nonmetallic conduit. Rigid metal conduit or IMC does not have to be concrete-encased, but it may be, of course. Direct burial nonmetallic conduit must be an approved (UL listed and labeled) type, specifically recognized for use without concrete encasement. If rigid nonmetallic conduit is approved for use only with concrete encasement, at least 2 in. (50.8 mm) of concrete must enclose the conduit. All applications of the various types of nonmetallic conduit must conform to the data made available by UL in the *Electrical Construction Materials Directory.*

In Table 710-4(b), Exceptions No. 2 and No. 3 are modifications to the burial-depth table for high-voltage circuits and provide more specific guidance. Exception No. 2 permits the table burial depth for direct buried cable or rigid nonmetallic conduit to be reduced by 6 in. (152 mm) for each 2 in. (50.8 mm) of concrete "placed in the trench over the underground installation." Note that the reduction in burial depth is not allowed for concrete that is above the raceway or cable, up at grade level, such as a roadway, walkway, or patio. The phrase "placed in the trench" indicates that such concrete must be specifically placed in the trench to provide the protection. And both rigid metal conduit and intermediate metal conduit are excluded from this reduction in burial depth because their 6-in. (152-mm) burial depth would become zero. That is consistent with Table 300-5 for underground wiring methods up to 600 V.

Exception No. 3 eliminates the need for any burial depth in earth for "conduits or other raceways" that are run under a building or exterior concrete slab not less than 4 in. (102 mm) thick and extending at least 6 in. (152 mm) "beyond the underground installation"—that is, overlapping the raceway by 6 in. (152 mm) on each side. Note that in this Exception the 4-in. (102-mm) thick concrete *may* be up at grade level in the form of a slab or patio or similar concrete area not subject to vehicular traffic.

The last sentence in Exception No. 3 requires a "warning ribbon or other effective means" such as a flag or stake indicating the presence of high-voltage circuits.

Fig. 710-2 covers burial of underground high-voltage circuits and Exceptions 2 and 3.

Part **(b)** notes that unshielded cable (i.e., cable without electrostatic shielding on the insulation) must be installed in rigid metal conduit, in IMC, or in rigid nonmetallic conduit encased in not less than 3 in. (76 mm) of concrete. The effect of this rule is that unshielded, or nonshielded, cables may not be used directly buried in the earth.

By reference to Sec. 310-7, nonshielded cables may be directly buried up to a rating of 2,000-V, except that metal-encased, nonshielded conductors (as in Type MC or lead-jacketed cables) may be used in ratings up to 5,000 V. But *all direct burial cables must be "identified" for such use,* which really means listed by a qualified testing agency.

DIRECT-BURIED CABLES

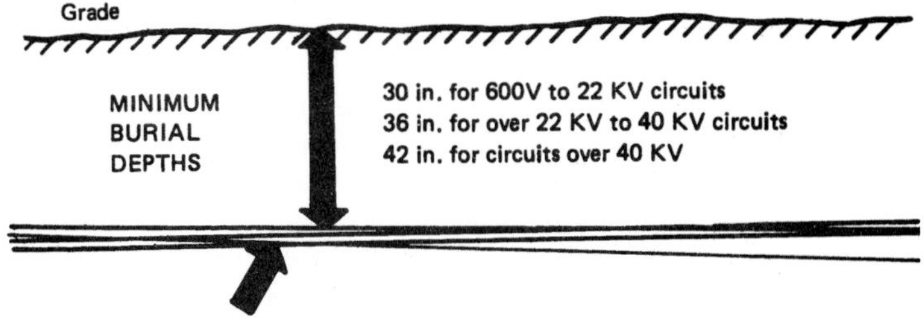
Grade
MINIMUM
BURIAL
DEPTHS
30 in. for 600V to 22 KV circuits
36 in. for over 22 KV to 40 KV circuits
42 in. for circuits over 40 KV
Direct-buried cables must be concentric-neutral or drain-wire shielded type or with a conducting sheath of equivalent ampacity
NOTE: Unshielded cables are not acceptable directly buried!

IN RIGID METAL CONDUIT OR INTERMEDIATE METAL CONDUIT (IMC)

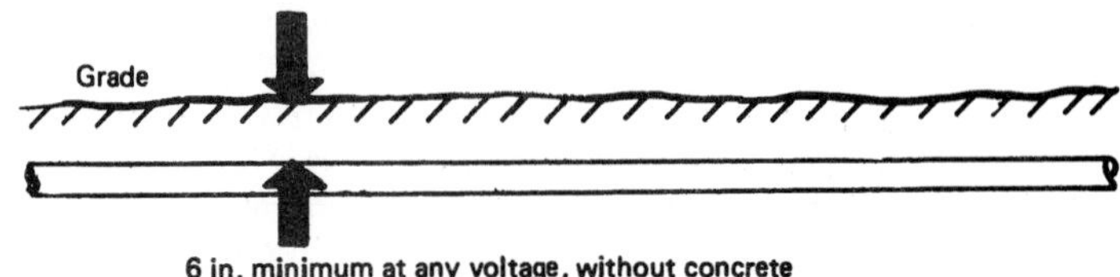
Grade
6 in. minimum at any voltage, without concrete encasement. See Note 2 below.

IN RIGID NONMETALLIC CONDUIT

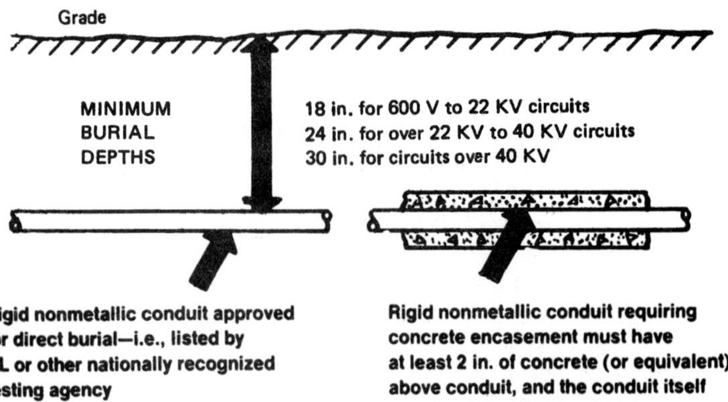
Grade
MINIMUM
BURIAL
DEPTHS
18 in. for 600 V to 22 KV circuits
24 in. for over 22 KV to 40 KV circuits
30 in. for circuits over 40 KV
Rigid nonmetallic conduit approved for direct burial—i.e., listed by UL or other nationally recognized testing agency
Rigid nonmetallic conduit requiring concrete encasement must have at least 2 in. of concrete (or equivalent) above conduit, and the conduit itself must be at the depths shown.

Fig. 710-2. Direct burial high-voltage cables must be of correct type, at specified depth. (Sec. 710-4.)

For underground use, all nonshielded cables were required by previous Code editions to be installed in rigid metal conduit, in IMC, or in rigid nonmetallic conduit with a 3-in. (76-mm) -thick concrete encasement. Now Type MC cable and lead-sheathed cable—with nonshielded conductors in either assembly—may be directly buried, as permitted by Sec. 334-3(5).

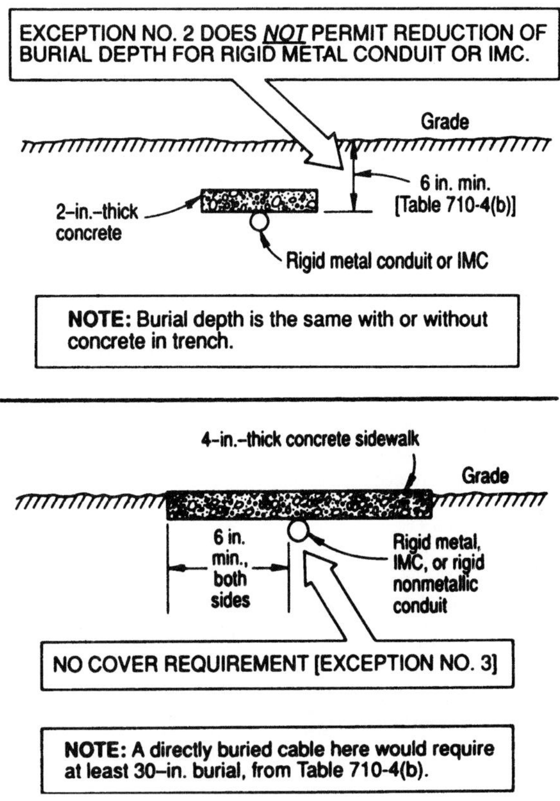

Fig. 710-2. *(Continued)*

As indicated above, when nonshielded cable (of the nonmetallic-jacket type) is used *underground* in rigid nonmetallic conduit, the conduit must have a 3-in. (76-mm) -thick concrete encasement. But if the same nonshielded cable is used in rigid nonmetallic conduit *aboveground,* concrete encasement is *not* required.

Figure 710-2 demonstrates uses of direct burial high-voltage cables in relation to the Code rules. Figure 710-3 covers underground high-voltage circuits in raceways.

Figure 710-4 covers the basic rules of Table 710-4(b), subject to the considerations noted in the rules and Exceptions, as follows:

1. Burial depths shown in diagrams may be reduced 6 in. (152 mm) for each 2 in. (50.8 mm) of concrete or equivalent above the conductors.
2. Where run under a building or slab that is at least 4 in. (102 mm) thick and extends 6 in. (152 mm) in each direction beyond the cable or raceway.
3. Areas of heavy traffic (public roads, commercial parking areas, etc.) must have minimum burial depth of 24 in. (610 mm) for any wiring method.
4. Lesser depths are permitted where wiring rises for termination.
5. Airport runways may have cables buried not less than 18 in. (457 mm) deep, without raceway or concrete encasement.

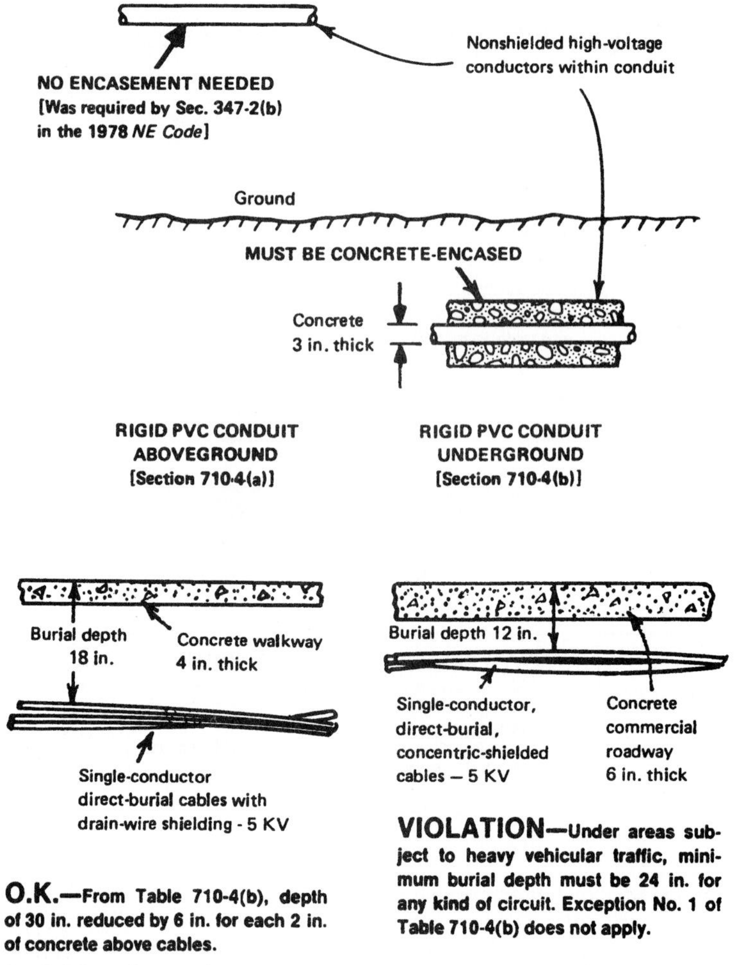

Fig. 710-3. Underground raceway circuits may vary widely in acceptable conditions of use. (Sec. 710-4.)

6. Conduits installed in solid rock may be buried at lesser depths than shown in diagrams when covered by at least 2 in. (50.8 mm) of concrete that extends to the rock surface.

As noted in part **(b)(2)** splices or taps are permitted in a trench without a box—but only if approved methods and materials are used—that is, "listed" splice kits. Taps and splices must be watertight and protected from mechanical injury. For shielded cables, the shielding must be continuous across the splice or tap.

Figure 710-5 shows a permanent straight splice for joining one end of a cable off a reel to the start of a cable off another reel, or for repairing a cable that is cut through accidentally by a backhoe or other tool digging into the ground.

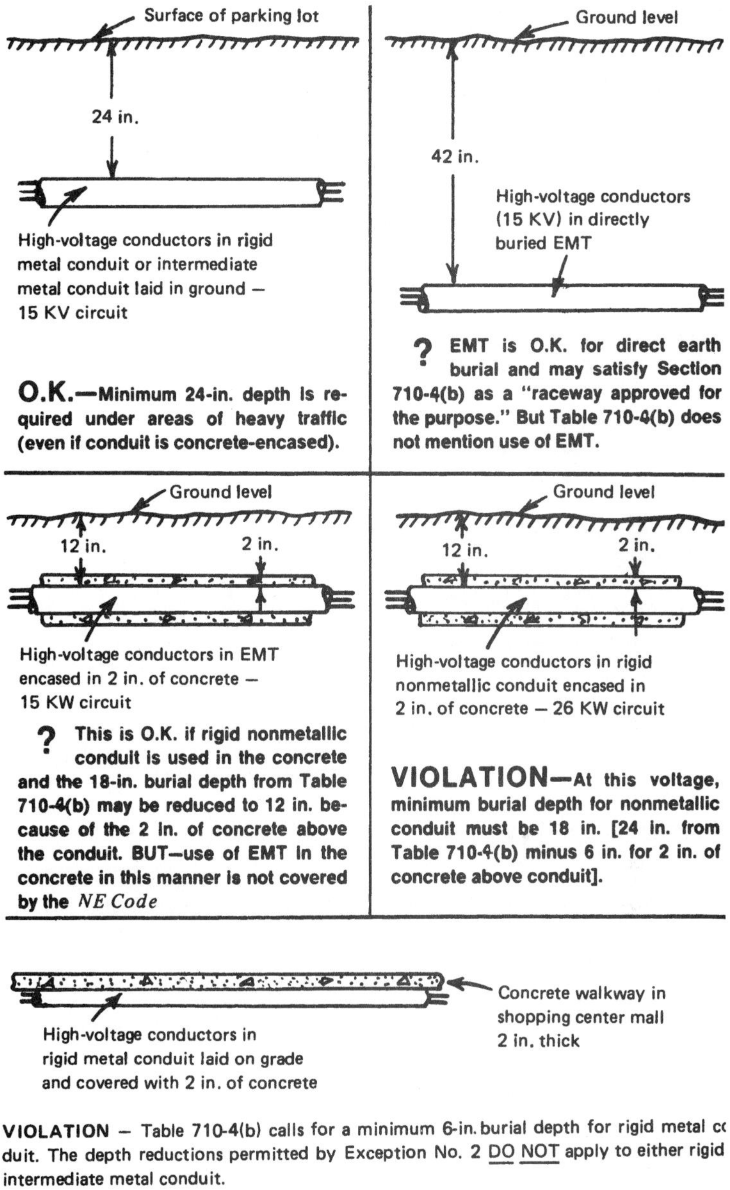

Surface of parking lot
24 in.
High-voltage conductors in rigid metal conduit or intermediate metal conduit laid in ground — 15 KV circuit
O.K.—Minimum 24-in. depth is required under areas of heavy traffic (even if conduit is concrete-encased).
Ground level
42 in.
High-voltage conductors (15 KV) in directly buried EMT
? EMT is O.K. for direct earth burial and may satisfy Section 710-4(b) as a "raceway approved for the purpose." But Table 710-4(b) does not mention use of EMT.
Ground level
12 in.
2 in.
High-voltage conductors in EMT encased in 2 in. of concrete — 15 KW circuit
? This is O.K. if rigid nonmetallic conduit is used in the concrete and the 18-in. burial depth from Table 710-4(b) may be reduced to 12 in. because of the 2 in. of concrete above the conduit. BUT—use of EMT in the concrete in this manner is not covered by the NE Code
Ground level
12 in.
2 in.
High-voltage conductors in rigid nonmetallic conduit encased in 2 in. of concrete — 26 KW circuit
VIOLATION—At this voltage, minimum burial depth for nonmetallic conduit must be 18 in. [24 in. from Table 710-4(b) minus 6 in. for 2 in. of concrete above conduit].
Concrete walkway in shopping center mall 2 in. thick
High-voltage conductors in rigid metal conduit laid on grade and covered with 2 in. of concrete
VIOLATION — Table 710-4(b) calls for a minimum 6-in. burial depth for rigid metal cc duit. The depth reductions permitted by Exception No. 2 DO NOT apply to either rigid intermediate metal conduit.

Fig. 710-4. Underground high-voltage circuits must observe burial depths. (Sec. 710-4.)

"T" and "Y" splices are made with similar techniques. Disconnectable splice devices provide watertight plug-and-receptacle assembly for all types of shielded cables and are fully submersible.

Figure 710-6 covers the rules of parts **(b)(3)** and **(b)(4)**. Figure 710-7 covers part **(b)(1)**.

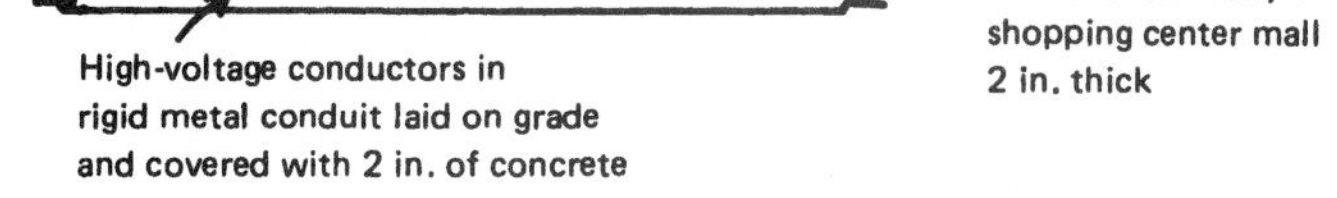

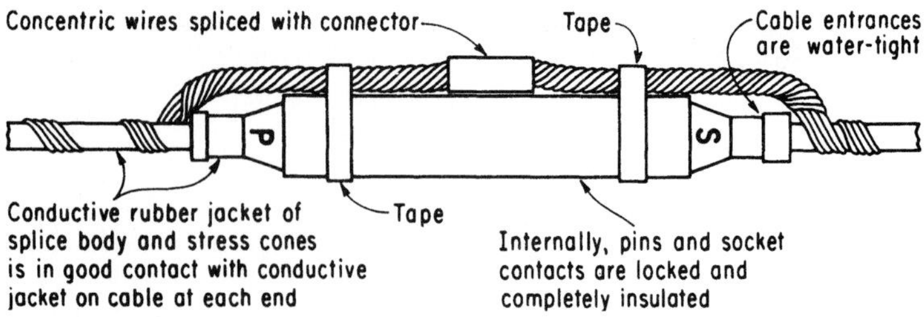

Concentric wires spliced with connector

Cable entrances are water-tight

Conductive rubber jacket of splice body and stress cones is in good contact with conductive jacket on cable at each end

Internally, pins and socket contacts are locked and completely insulated

Fig. 710-5. Splice may be made in direct-burial cable if suitable materials are used. (Sec. 710-4.)

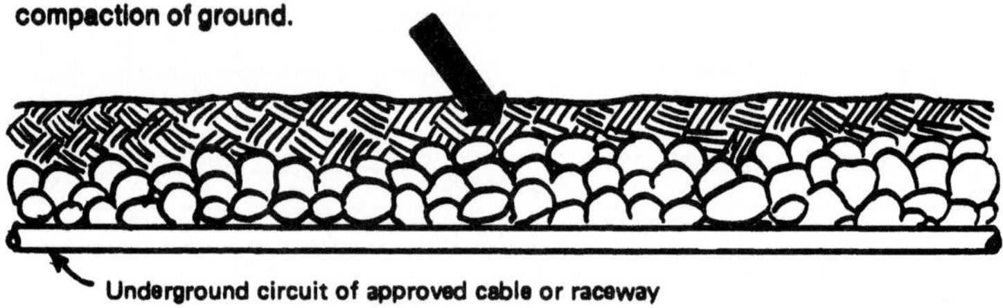

BACKFILL MUST NOT DAMAGE DUCTS, CABLES OR RACEWAYS
Backfill of heavy rocks or sharp or corrosive materials must not be used if it may cause damage or prevent adequate compaction of ground.
Underground circuit of approved cable or raceway
Protection shall be provided to prevent physical damage to the raceway or cable in the form of granular or selected material or suitable sleeves.

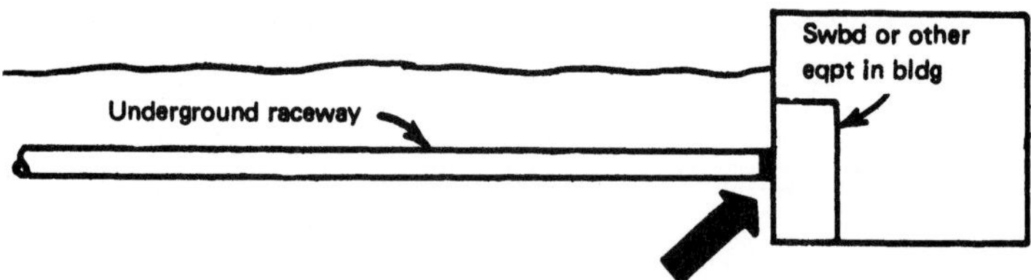

Swbd or other eqpt in bldg

Where raceway enters from an underground system, the end in the building must be sealed with suitable compound to prevent entry of moisture or gases; or it must be arranged to prevent moisture from contacting live parts.

Fig. 710-6. Circuits must be protected and sealed where they enter equipment. (Sec. 710-4.)

710-6. Insulation Shielding. One of the basic decisions to make in selecting high-voltage conductors is whether or not electrostatic insulation shielding is required on the cable. In the 1971 NE Code (and previous editions), Table 710-5 sets forth an elaborate variety of conditions under which solid dielectric insulated conductors had to be shielded or were permitted to be unshielded. That

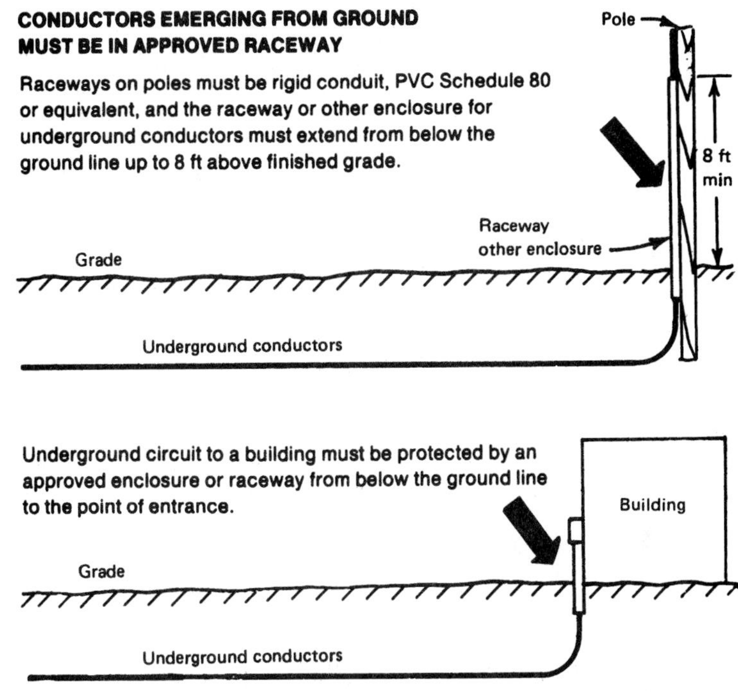

CONDUCTORS EMERGING FROM GROUND
MUST BE IN APPROVED RACEWAY
Raceways on poles must be rigid conduit, PVC Schedule 80 or equivalent, and the raceway or other enclosure for underground conductors must extend from below the ground line up to 8 ft above finished grade.
Pole
8 ft min
Raceway other enclosure
Grade
Underground conductors
Underground circuit to a building must be protected by an approved enclosure or raceway from below the ground line to the point of entrance.
Building
Grade
Underground conductors

Fig. 710-7. Direct burial cables must be protected aboveground. (Sec. 710-4.)

table set the same basic shielding requirements as recommended by the IPCEA. BUT, the NE Code *no longer contains* that table and now takes a different approach to mandatory shielding. The basic requirements on electrostatic shielding of high-voltage conductors are presented in Sec. 310-6 and are explained there.

This section sets forth very general rules on terminating shielded high-voltage conductors. The metallic shielding or any other conducting or semi-conducting static shielding components on shielded cable must be stripped back to a safe distance according to the circuit voltage—at all terminations of the shielding. At such points, stress reduction must be provided by such methods as the use of potheads, terminators, stress cones, or similar devices.

The wording of this regulation makes clear that the need for shield termination using stress cones or similar terminating devices applies to semiconducting insulation shielding as well as to metallic-wire insulation shielding systems.

A stress cone is a field-installed device or a field-assembled buildup of insulating tape and shielding braid which must be made at a terminal of high-voltage shielded cable, whether a pothead is used or not. A stress-relief cone is required to relieve the electrical stress concentration in cable insulation directly under the end of cable shielding. Some cable constructions contain stress-control components that afford the cable sufficient stress relief without the need for stress-relief cones. If a cable contains inherent stress-relief components in its construction, that would satisfy Sec. 710-6 as doing the work of a stress cone. As a result, separate stress cones would not have to be installed at the ends of such cable. Or, heat-shrinkable-tubing terminations may be used with stress-control material that provides the needed relief of electrostatic stress.

At a cable terminal, the shielding must be cut back some distance from the end of the conductor to prevent any arcing-over from the hot conductor to the grounded shield. When the shield is cut back, a stress is produced in the insulation. By providing a flare out of the shield, i.e., by extending the shield a short distance in the shape of a cone, the stress is relieved, as shown in Fig. 710-8.

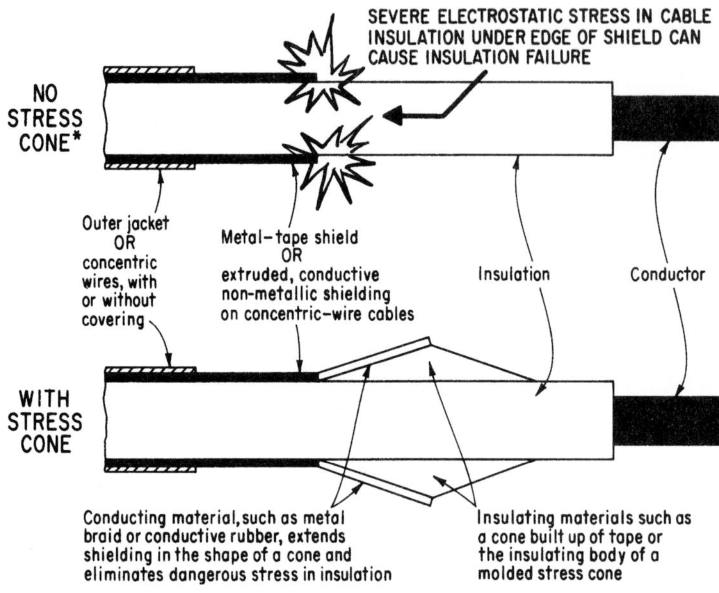

Fig. 710-8. Here is how a stress cone protects insulation at cable ends. (Sec. 710-6.)

Stress cones provide that protection against insulation failure at the terminals of shielded high-voltage cables. Manufacturers provide special preformed stress cones (Fig. 710-9) and kits for preparing cable terminals with stress cones for cables operating at specified levels of high voltage (Fig. 710-10). A wide assortment of stress-relief terminators are made for all the high-voltage cable assemblies used today.

Metallic shielding tape must be grounded, as required by Secs. 710-6 and 300-5(b), which refers to "metallic shielding"—as in Fig. 710-11. The shield on shielded cables must be grounded at one end at least. It is better to ground the shield at two or more points. Grounding of the shield at all terminals and splices will keep the entire length of the shield at about ground potential for the safest, most effective operation of the cable. Cable with improperly or ineffectively grounded shielding can present more hazards than unshielded cable.

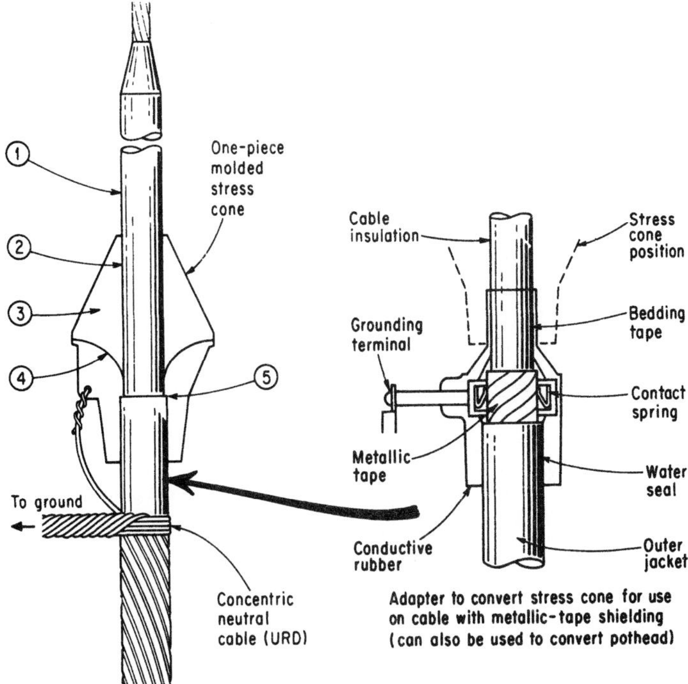

APPLICATION:
Cable shield is cut back about 12 in. Then, using silicone lubricant, the cable insulation surface and the inner bore of the stress cone are lubricated. The stress cone is simply pushed down over the cable end until it bottoms on the cable shield. After cable is prepared, termination takes about 30 seconds.

REFER TO DIAGRAM:
1. Cable insulation with shielding cut back
2. Tight fit between insulation and bore of stress cone
3. Insulating rubber
4. Stress relief provided by conductive rubber flaring away from insulation along bond between insulating rubber and conductive rubber
5. Conductive rubber of cone tightly fit to conductive cable shield

Fig. 710-9. Typical preformed stress cone is readily applied on cables up to 35-kV indoors. (Sec. 710-6.)

710-8. Moisture or Mechanical Protection for Metal-Sheathed Cables. A "pothead" is one specific form of stress-reduction means referred to in Sec. 710-6 and has long been a common means of protecting insulation against moisture or mechanical injury where conductors emerge from a metal sheath (Fig. 710-12). Such protection for metal-sheathed cables (such as lead-covered, paper-insulated cables) is required by this section. A pothead is a cable terminal

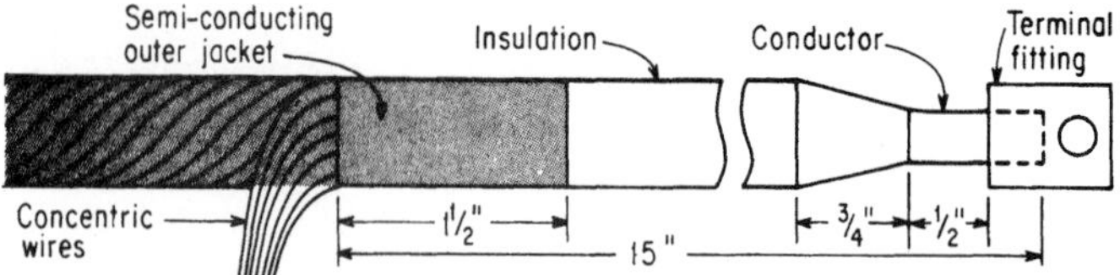

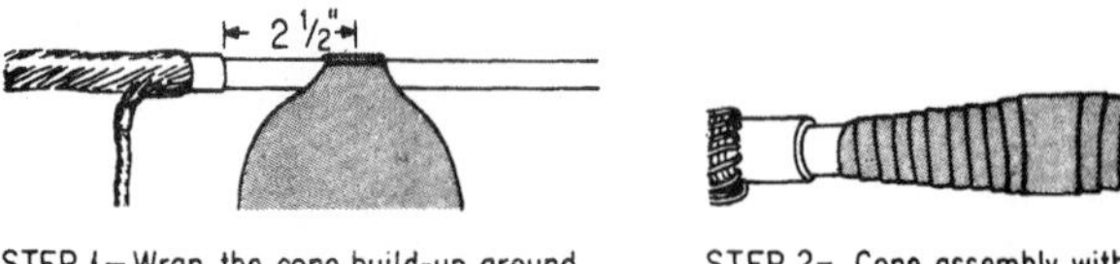

THIS IS CABLE PREPARED FOR PENNANT STRESS CONE

$2\frac{1}{2}$"
STEP 1 – Wrap the cone build-up around the insulation at a given place

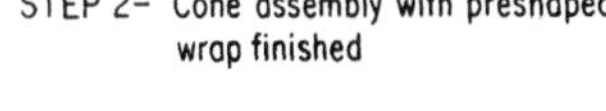

STEP 2 – Cone assembly with preshaped wrap finished

STEP 3 – Semi-conducting tape contacts semi-conducting cable jacket and extends up to peak of cone

STEP 5 – Insulate terminal fitting area at end of cable and (for outdoor use) cover entire assembly with silicone tape or use potheads.

Fig. 710-10. "Pennant" method is one of a variety of job-site termination buildups. (Sec. 710-6.)

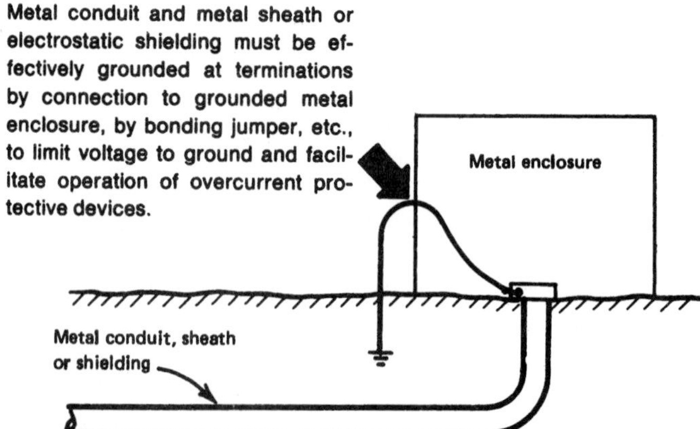

Metal conduit and metal sheath or electrostatic shielding must be effectively grounded at terminations by connection to grounded metal enclosure, by bonding jumper, etc., to limit voltage to ground and facilitate operation of overcurrent protective devices.
Metal enclosure
Metal conduit, sheath or shielding

Fig. 710-11. Metallic shielding must be grounded for all high-voltage conductors—under- or aboveground. (Sec. 710-6.)

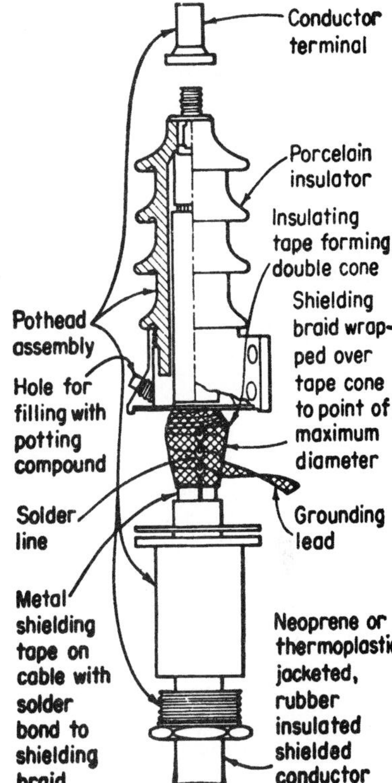

Fig. 710-12. Typical single-conductor pothead protects metallic- or nonmetallic-jacketed cable. (Sec. 710-8.)

which provides sealing to the sheath of the cable for making a moisture-proof connection between the wires within the cable and those outside.

When metal-jacketed high-voltage cables are terminated outdoors exposed to the weather, a pothead is commonly used to protect the insulation of conductors against moisture or mechanical injury where conductors emerge from a metal sheath, as shown in Fig. 710-13.

On use of potheads:

1. Paper-insulated cables must be terminated in potheads. This requirement also extends to such cables operated at under 600 V.
2. Varnished-cambric-insulated cables should be terminated in potheads but may be terminated with taped connections in dry locations.
3. Rubber-insulated cables are commonly terminated in potheads in locations where moisture protection is critical but may be terminated without potheads in accordance with manufacturer's instructions.
4. Although many modern high-voltage cables can be terminated without potheads, many engineers consider potheads the best terminations for high-voltage cable.
5. The use of potheads offers a number of advantages:
   a. Seals cable ends against moisture that would damage the insulation.

Fig. 710-13. A pothead is used on the end of each paper-insulated, lead-covered cable to protect against entry of moisture, with a wiped lead joint at the terminal. (Sec. 710-8.)

b. Provides a compartment for surrounding the termination with insulating compound to increase strength of electrical insulation.
c. Seals cable ends against loss of insulating oils.
d. Provides engineered support of connections.

710-9. Protection of Service Equipment, Metal-Enclosed Power Switchgear, and Industrial Control Assemblies. The basic rule of the first sentence in this section excludes "pipes or ducts foreign to the electrical installation" from the "vicinity of the service equipment, metal-enclosed power switchgear, or industrial control assemblies." Then, addressing the case where foreign piping is unavoidably close to the designated electrical equipment, the next sentence calls for "protection" (such as a hood or shield above such equipment) to prevent damage to the equipment by "leaks or breaks in such foreign systems."

Piping for supplying a fire protection medium for the electrical equipment is not considered to be "foreign" and may be installed at the high-voltage gear. The reason given for that sentence was to prevent the first sentence from being "interpreted to mean that no sprinklers should be installed." Fire suppression at such locations may use water sprinklers or protection systems of dry chemicals and/or gases specifically designed to extinguish fires in the equipment without jeopardizing the equipment. Water is sometimes found to be objectionable; leaks in piping or malfunction of a sprinkler head could reduce the switchgear integrity by exposing it to a flashover and thereby initiate a fire.

710-20. Overcurrent Protection. Figure 710-14 shows the basic rule for high-voltage circuit protection. Refer to Sec. 230-208 for high-voltage service conductors, to Sec. 240-100 on feeders, and Sec. 240-101 on branch circuits. The

OVERCURRENT DEVICE FOR EACH UNGROUNDED CONDUCTOR . . .

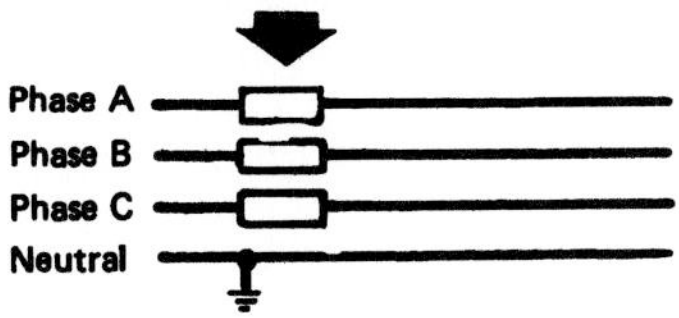

. . . must be either . . .

a CIRCUIT BREAKER with three overcurrent relays operated from three current transformers [see Exceptions in Section 710-20(a)]

OR

a FUSE connected in series with each ungrounded conductor.

Fig. 710-14. Basic rule calls for overcurrent protection. (Sec. 710-20.)

specified ratings of protection for high-voltage conductors are presented in those sections.

710-21. Circuit-Interrupting Devices. High-voltage power CBs provide load switching, short-circuit protection, electrical operation, adjustable time delays of trip characteristics for selectively coordinated protection schemes, quick reclosing after tripping, and various protective hookups such as differential relay protection of transformers. There are oil-type, oil-less (or air-magnetic), and vacuum-break CBs. The air-magnetic CB is the common type for indoor applications in systems up to 15 kV and higher. Oil CBs are sometimes used for indoor and outdoor high-voltage service equipment where they provide economical disconnect and protection on the primary of a transformer.

Modern high-voltage CB equipment meets all the needs of control and protection for electrical systems from the simplest to the most complex and sophisticated. In particular, its use for selectively coordinated protection of services, feeders, and branch circuits is unique. In current ratings up to 3,000 A, CB gear has the very high interrupting ratings required for today's high-capacity systems. Available in "metal-clad" assemblies, all live parts are completely enclosed within grounded metal enclosures for maximum safety. For applications exposed to lightning strikes or other transient overcurrents, CB equipment offers quick reclosing after operation. Drawout construction of the CB units provides ease of maintenance and ready testing of breakers. CB gear offers unlimited arrangements of source and load circuits and is suited to a variety of AC or DC control power sources. Accessory devices are available for special functions.

Figure 710-15 covers the basic rules of this section on use of CBs. Figure 710-16 shows an oil CB, with the line-side isolating switch required by Sec. 230-204(a).

Part **(b)** covers power fuses, which are available in current-limiting and noncurrent-limiting types. (See Fig. 710-17 for an example of the uses to which power fuses may be put.) The current-limiting types offer reduction of thermal

A. Indoor installations of circuit breakers must consist of metal-enclosed units or fire-resistant, cell-mounted units, except that open-mounted CBs may be used in places accessible to qualified persons only.

B. All CBs must be rated for short-circuit duty at point of application.

C. Circuit breakers controlling oil-filled transformers *must* either be located outside the transformer vault *or* be capable of being operated from outside the vault.

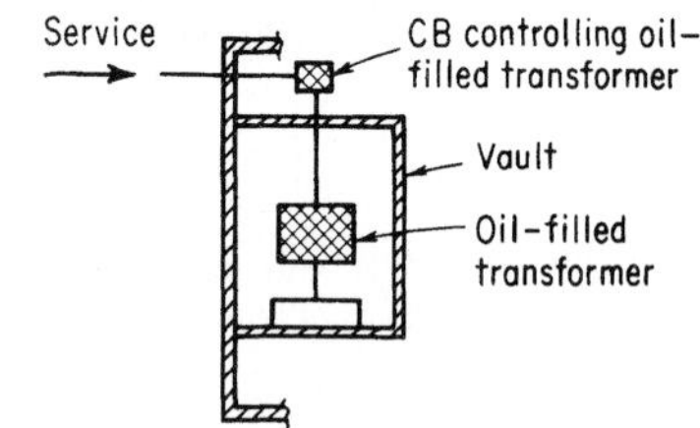

D. Oil CBs must be arranged or located so that adjacent readily combustible structures or materials are safeguarded in an approved manner. Adequate space separation, fire-resistant barriers or enclosures, trenches containing sufficient coarse crushed stone, and properly drained oil enclosures such as dikes or basins are recognized as suitable safeguards.

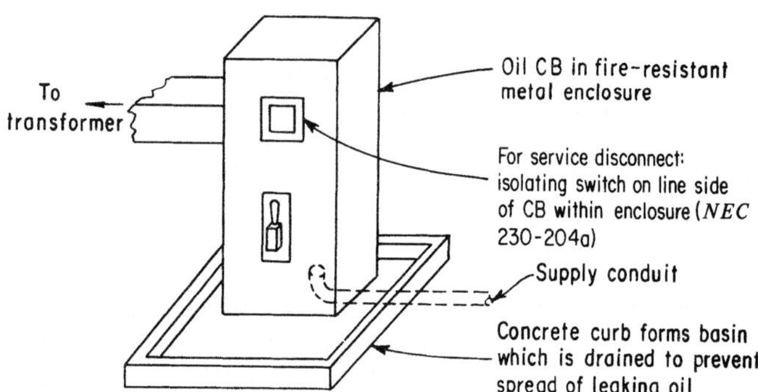

Fig. 710-15. Detailed rules regulate use of high-voltage circuit breakers. (Sec. 710-21.)

and magnetic stresses on fault by reducing the energy let-through. They are constructed with a silver-sand internal element, similar to 600-V current-limiting fuses. Such fuses generally have higher interrupting ratings at some voltages, but their continuous current-carrying ratings are limited.

Noncurrent-limiting types of power fuses are made in two types of operating characteristics: expulsion type and nonexpulsion type. The expulsion fuse gets its name from the fact that it expels hot gases when it operates. Such fuses should not be used indoors without a "snuffer" or other protector to contain the exhaust, because there is a hazard presented by the expelled gases. At the end of part **(b)(1)**, the rule says that vented expulsion-type power fuses used indoors, underground, or in metal enclosures must be "identified for the use." Vented power fuses are not safe for operation in confined space—unless specif-

Fig. 710-16. Oil circuit breaker for high-voltage application has a disconnecting switch on its supply side to isolate the line terminals of the breaker. (Sec. 710-21.)

ically tested and "identified" for such use. Part **(b)(5)** of this section requires that fuses expelling flame in operation must be designed or arranged to prevent hazard to persons or property. The boric-acid fuse with a condenser or other protection against arcing and gas expulsion is a typical nonexpulsion, noncurrent-limiting fuse (Fig. 710-18).

Parts **(b)(6)** and **(b)(7)** cover very important safeguards for the use of fuses and fuseholders:

In coordinating power fuses, care must be taken to account for ambient temperature adjustment factors, because time-current curves are based on an ambient of 25°C. Adjustment also must be made for preheating of fuses due to load current to assure effective coordination of fuses with each other and/or with CBs. Manufacturer's curves of adjustment factors for ambient temperature and fuse preloading are available.

Figure 710-19 shows the time-current characteristic for an R-rated, current-limiting (silver-sand) fuse designed for 2,400- and 4,800-V motor applications. Such fuses must be selected to coordinate with the motor controller overload protection, with the controller clearing overloads up to 10 times motor current and the fuse taking over for faster opening of higher currents up to the interrupting rating of the fuse. The amp rating of R-rated fuses is given in values

Fig. 710-17. Typical power fuses are used in load-interrupter switchgear, which is an alternative to circuit-breaker gear for control and protection in indoor high-voltage systems. (Sec. 710-21.)

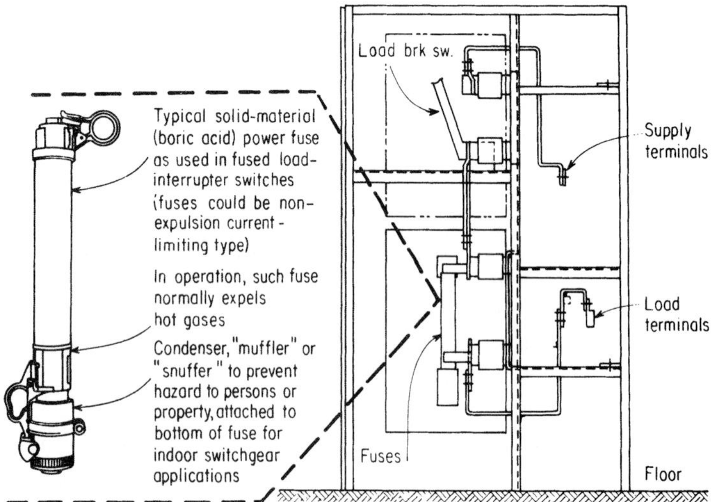

Fig. 710-18. Boric-acid fuse uses a device to protect against flame expulsion. (Sec. 710-21.)

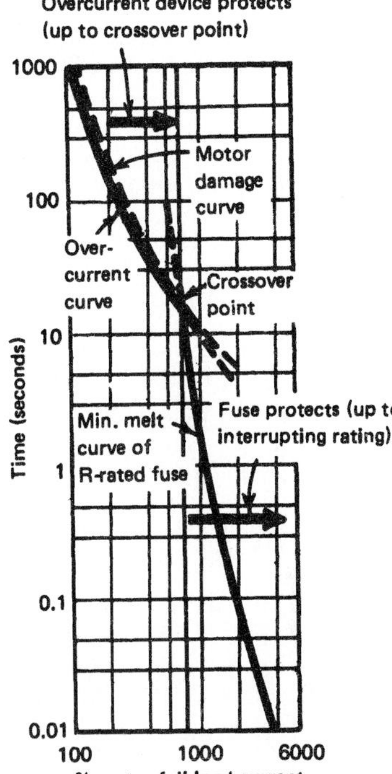

Fig. 710-19. "R-rated" fuses are used in motor starters for 2,400- and 4,800-V motors. (Sec. 710-21.)

such as 2R or 12R or 24R. If the number preceding the R is multiplied by 100, the value obtained is the ampere level at which the fuse will blow in 20 sec. Thus, the rating designation is *not* continuous current but is based on the operating characteristics of the R-rated fuse. Continuous current rating of such fuses is given by the manufacturer at some value of ambient temperature.

Fused cutouts for high-voltage circuits, as shown in Fig. 710-20, are available for both indoor and outdoor application, as regulated by part **(c)** of this section. Pull-type fuse cutouts are used outdoors on pole-line crossarms or indoors in electric rooms where accessible only to qualified persons, as shown in Fig. 710-21. Such fused cutouts are acceptable for use as an isolating switch, as permitted by Sec. 710-22.

Part **(d)** of this section covers oil-filled cutouts. In addition to air CBs, oil CBs, and fused load-interrupter switches, another device frequently used for control of high-voltage circuits is the oil-filled cutout. Compared to breakers and fused switchgear, oil-filled cutouts are inexpensive devices that provide economical switching and, where desired, overload and short-circuit protection for primary voltage circuits.

The oil-filled cutout is a completely enclosed, single-pole assembly with a fusible or nonfusible element immersed in the oil-filled tank that makes up the major part of the unit, and with two terminals on the outside of the housing. Figure 710-22 shows the basic construction of a typical cutout with a listing of

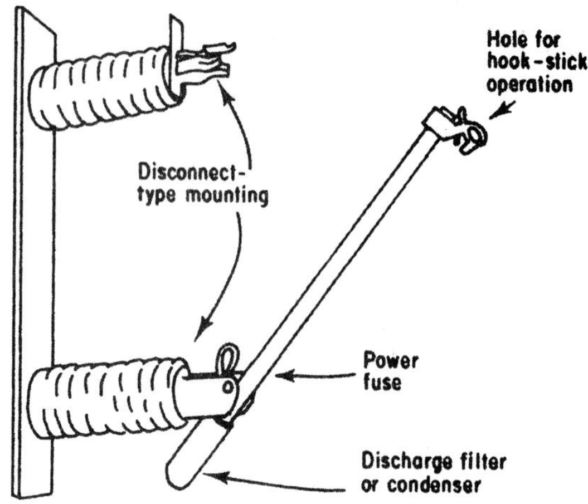

Fig. 710-20. Expanded rules cover use of distribution cutouts. (Sec. 710-21.)

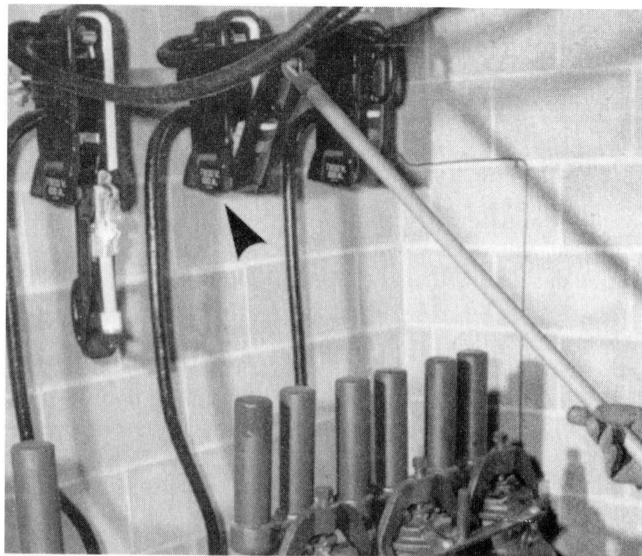

Fig. 710-21. Distribution cutouts are single-pole, fused, protective, and disconnect devices that are hook-stick-operable. Note voltage and current rating on case of each cutout (arrow), as required by part **(c)(5)** of Sec. 710-21. (Sec. 710-21.)

available entrance fittings for the terminals to suit them to various cable and job requirements. The circuit is broken or closed safely and rapidly by the internal switching mechanism. The switch mechanism is made up of a rotating element that, in the closed position, bridges two internal contacts—each contact connecting to one of the outside terminals. The rotating element is completely

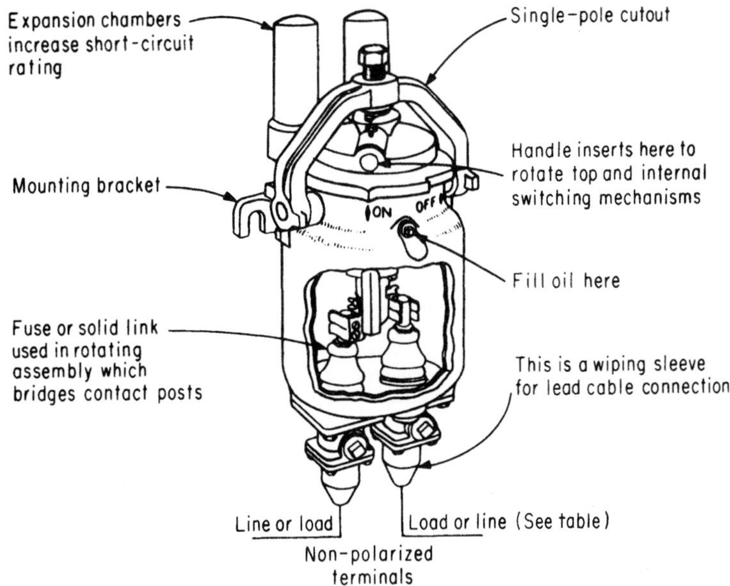

Fig. 710-22. Oil-filled cutout is a fused or unfused, single-pole disconnect device. (Sec. 710-21.)

insulated from the external case and from the external handle that operates the element. The rotating element may be simply a shorting blade when the cutout is used as an unfused switch. When the cutout is to be used as a fused switching unit, the rotating bridging element is fitted with a fuse. Operation of an oil-filled cutout is controlled at the top end of the shaft extending out through the top of the housing.

As a single-pole switching device, the oil-filled cutout is not polarized—i.e., either terminal may be a line or load terminal. This is a result of the symmetrical construction of the switching element and suits the device to use in circuit sectionalizing or as a tie device in layouts involving two or more primary supply circuits. Note that these **Code** rules on oil-filled cutouts are different from those in part **(c)** on distribution cutouts.

With an oil-filled cutout, the switching of load current or the breaking of fault current is confined within a sturdy metal housing. Operation is made safe and quiet by confining arcs and current rupture forces within the enclosure. This operating characteristic of the cutouts especially suits them to use where there are explosive gases or flammable dusts, where complete submersion is possible, where severe atmospheric conditions exist, or where exposure of live electrical parts might be hazardous.

Oil-filled (sometimes called "oil-fuse") cutouts are made in three sizes based on continuous current—100, 200, and 300 A, up to 15 kV. In one line there are three basic types. *Pole-type cutouts* are equipped with rubber-covered leads from the terminals for use in open wiring. Pothead-type cutouts have a cable lead from one terminal for open wiring and a sleeve on the other for connecting a lead or rubber-covered cable from an underground circuit. *Subway-type cutouts* are for underground vaults and manholes, particularly where submersion might occur, and are equipped with a sleeve on each terminal for rubber- or lead-sheathed cable. Figure 710-23 lists the various types of terminal connections that are available on oil-filled cutouts.

Application	Cable cover	No. of conduits	Entrance type	Method of sealing
Indoor	Rubber or neoprene	Single	Porcelain	Tape Always tape this connection!
		Multiple	Stud bushing	
Indoor or outdoor	Rubber, lead or neoprene	Single	Stuffing box	Compression fittings
	Polyethylene	Single		Compression fittings and tube seal
	Lead	Single	Wiping sleeve	Solder wipe

Fig. 710-23. Terminals on oil-filled cutouts must be matched to application and cable type. (Sec. 710-21.)

For multiphase circuits, two or three single-phase cutout units can be group-mounted with a gang-operating mechanism for simultaneous operation. Figure 710-24 shows three-gang assemblies. For pole mounting, linkage and a long handle are available for operating cutouts from the ground. Or cutouts can be flange-mounted on a terminal box, as shown in Fig. 710-25, where the three-gang assembly was added to a high-voltage switchgear on a modernization job.

Because oil-filled cutouts provide load-break capability and overcurrent protection, they may be used for industrial and commercial service equipment, for switching outdoor lighting of sports fields and shopping centers, for transformer load centers, for primary-voltage motor circuits, or for use in vaults and manholes of underground systems.

In 100- and 200-A ratings, oil-filled cutouts are available in combination with current-limiting power fuses in double-compartment indoor or outdoor enclosures. These fused oil interrupter switches provide moderate load-break and high fault-current interrupting capability in an economical package.

Part **(e)** of this section recognizes the use of so-called load-interrupter switches used in high-voltage systems. In parts **(1)** to **(6)**, a wealth of specific data provides guidance to design engineers, electrical installers, and electrical inspectors on the proper installation, operation, and maintenance of high-voltage interrupter switches—with particular emphasis on safety to operators and maintenance personnel.

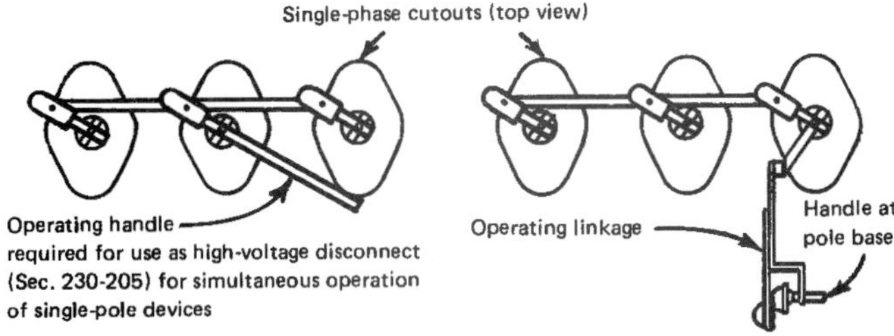

Fig. 710-24. Oil-filled cutouts can be assembled as a 3-pole device for 3-phase circuits. (Sec. 710-21.)

Switching for modern high-voltage electrical systems can be provided by a number of different equipment installations. For any particular case, the best arrangement depends on several factors: the point of application—either for outside or inside distribution or as service equipment; the voltage; the type of distribution system—radial, loop, selective, network; conditions—accessibility, type of actual layout of the equipment; job atmosphere; use; future system expansion; and economic considerations.

Types of switches used in high-voltage applications include:

1. Enclosed air-break load-interrupter switchgear with or without power fuses
2. Oil-filled cutouts (fused or unfused)
3. Oil-immersed-type disconnect switches

Modern load-interrupter switchgear in metal safety enclosures finds wide application in high-voltage distribution systems, in combination with modern power fuses (Fig. 710-26). Section 230-208 of the NE Code covers use of air load-interrupter switches, with fuses, for disconnect and overcurrent protection of

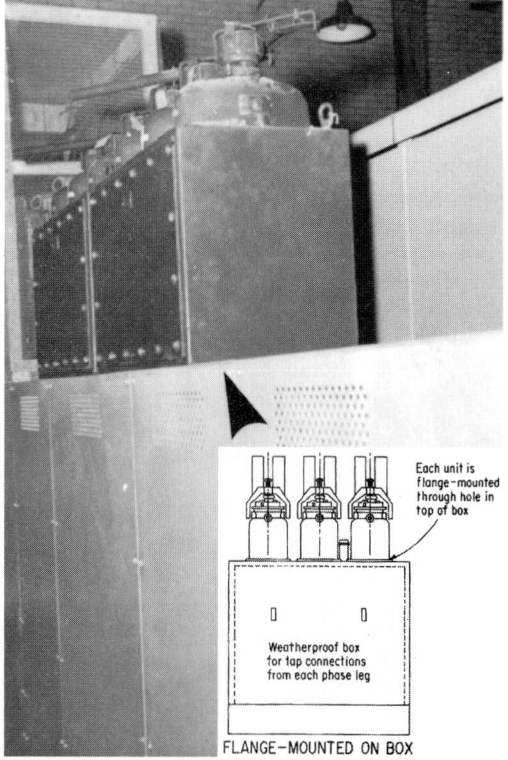

Fig. 710-25. Gang-operated 3-pole assembly of oil-filled cutouts provided addition of a new high-voltage circuit on a modernization project, but location of the units was questioned because part **(d)(7)** imposes a 5-ft (1.52-m) maximum mounting height. (Sec. 710-21.)

high-voltage service-entrance conductors. Part **(e)** of Sec. 710-21 covers use of fused air load-interrupter switches for high-voltage feeder in distribution systems.

Metal-enclosed fused load interrupters offer a fully effective alternative to use of power CBs, with substantial economies, in 5- and 15-kV distribution systems for commercial, institutional, and industrial buildings. Typical applications for such switchgear parallel those of power CBs and include the following:

1. *In switching centers*—Switchgear is set up for control and protection of individual primary feeders to transformer loadcenters.
2. *In substation primaries*—Load-interrupter switchgear is used for transformer switching and protection in the primary sides of substations.
3. *In substation secondaries*—Here the switchgear is used as a switching center closely coupled to a high-voltage transformer secondary.
4. *In service entrances*—This is a single-unit application of a switchgear bay for service-entrance disconnect and protection in a primary supply line.

Fig. 710-26. Load-interrupter switchgear is generally used with fuses to provide protection as well as load-break switching for high-voltage circuits. The fuses must be rated to provide complete protection for the load interrupter on closing, carrying, or interrupting current—up to the assigned maximum short-circuit rating. (Sec. 710-21.)

Fused load-interrupter switchgear, typically rated up to 1,200 A, can match the ratings and required performance capabilities of power CBs for a large percentage of applications in which either might be used.

Fuse-interrupter switches for high-voltage circuits are available with manual or power operation—including types with spring-powered, over-center mechanisms for manual operation or motor-driven, stored-energy operators. Available in indoor and outdoor housings, assemblies can be equipped with a variety of accessory devices, including key-interlocks for coordinating switch operation with remote devices such as transformer secondary breakers.

Vacuum switchgear, with their contacts operating in a vacuum "bottle" that is enclosed in a compact cylindrical assembly, has gained wide acceptance as load interrupters for high-voltage switching and sectionalizing. Available in 200- and 600-A ratings for use at 15.5, 27, and 38 kV, this switching equipment is suited to full-load interruption and is rated for 15,000- or 20,000-A short-circuit current under momentary and make and latch operations. BIL ratings are 95, 125, or 150 kV.

Vacuum switch assemblies, with a variety of accessories, including stored-energy operators and electric motor operator for remote control, are suited to all indoor and outdoor switching operations—including submersible operation for underground systems. The units offer fireproof and explosionproof operation, with virtually maintenance-free life for its rated 5,000 load interruptions. Units are available in standard 2-, 3-, and 4-way configurations, along with automatic transfer options. Accessory CTs and relays can be used with stored-energy operators to apply vacuum switches for fault-interrupting duty.

710-22. Isolating Means. Air-break or oil-immersed switches of any type may be used to provide the isolating functions described in this section. Distribution cutouts or oil-filled cutouts are also used as isolating switches.

Oil-immersed disconnect switches are used for load control and for sectionalizing of primary-voltage underground-distribution systems for large commercial and industrial layouts (Fig. 710-27). Designed for high-power handling—such as

Fig. 710-27. Oil switches are commonly used for isolating equipment and circuits for sectionalizing and for transfer from preferred to emergency supply. (Sec. 710-22.)

400 A up to 34 kV—this type of switch can be located at transformer loadcenter primaries or at other strategic points in high-voltage circuits to provide a wide variety of sectionalizing arrangements to provide alternate feeds for essential load circuits.

Oil-immersed disconnect switches are available for as many as five switch positions and ground positions to ground the feeder or test-ground positions for grounding or testing. Ground positions are used in such switches to connect circuits to ground while they are being worked on to assure safety to personnel.

Oil switches for load-break applications up to 15 kV are available for either manual or electrically powered switching for all types of circuits. When electrical operation is used, the switch functions as a high-voltage magnetic contactor.

Switches intended only for isolating duty must be interlocked with other devices to prevent opening of the isolating switch under load, or the isolating switch must be provided with an obvious sign warning against opening the switch under load.

710-24. Metal-Enclosed Power Switchgear and Industrial Control Assemblies. Where the previous sections presented regulations on the individual switching and protective devices, this section covers enclosure and interconnection of such unit devices into overall assemblies. Basically, the rules of this section are aimed at the manufacturers and assemblers of such equipment.

Use of all high-voltage switching and control equipment must be carefully checked against information given with certification of the equipment by a test laboratory—such as data given by UL in their Green Book. Typical data are as follows:

Unit substations listed by UL have the secondary neutral bonded to the enclosure and have provision on the neutral for connection of a grounding conductor. A terminal is also provided on the enclosure near the line terminals for use with an equipment grounding conductor run from the enclosure of primary equipment feeding the unit sub to the enclosure of the unit sub. Connection of such an equipment grounding conductor provides proper bonding together of equipment enclosures where the primary feed to the unit sub is direct-buried underground or is run in nonmetallic conduit without a metal conduit connection in the primary feed (Fig. 710-28).

The rule of part **(o)** is particularly aimed at the designers and installers of equipment, rather than at the manufacturer. Part **(o)(2)** emphasizes that careful

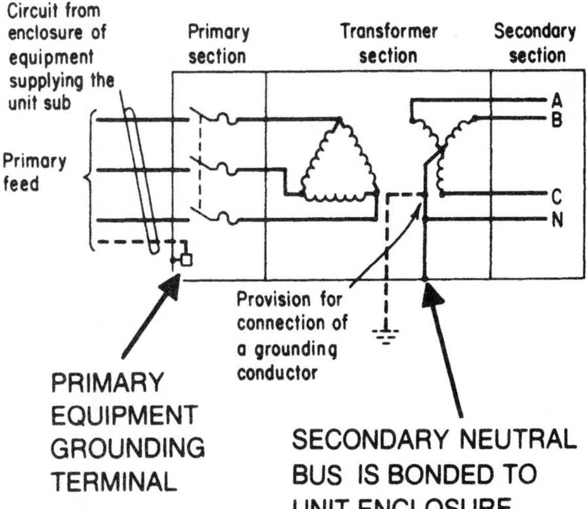

Fig. 710-28. NEC rules on equipment construction are supported by UL data. (Sec. 710-24.)

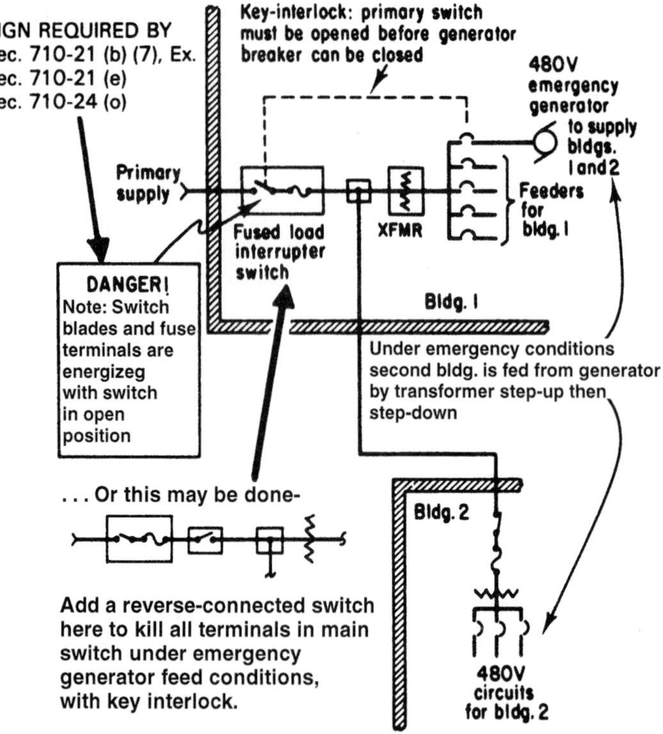

Fig. 710-29. Feedback in high-voltage hookups can be hazardous. (Sec. 710-24.)

layout and application of switching components of all types is important. Figure 710-29 shows the kind of condition that can be extremely hazardous in high-voltage layouts where there is the chance of a secondary to primary feedback—such as the intentional one shown to provide emergency power to essential circuits in Building 2. Under emergency conditions, the main fused interrupter is opened and the secondary CB for the generator is closed, feeding power to the 480-V switchboard in Building 1 and then feeding through two transformers to supply power to the 480-V circuits in Building 2. This hookup makes the load side of the main interrupter switch alive, presenting the hazard of electrocution to any personnel who might go into the switch thinking that it is dead because it is open. A second switch can eliminate this difficulty, if applied with interlocks.

ARTICLE 720. CIRCUITS AND EQUIPMENT OPERATING AT LESS THAN 50 VOLTS

720-1. Scope. This article covers low-voltage applications that are not power-limited circuits as defined in Sec. 725-3 and are not remote-control or signal circuits. Determination that Art. 720 applies to any circuit or equipment must be carefully made on the basis of the particular load being supplied and circuit conditions. This article covers circuits operating at more than 30 V but not over 50 V—such as 32-V circuiting.

720-4. Conductors. The minimum No. 12 conductor size, rather than No. 14 as permitted for standard power and light wiring, is aimed at the higher current required for a given wattage load at low voltage. For instance, at 32 V, the current corresponding to a given wattage is 3.6 times the current for the same wattage at 115 V. It should also be noted that for a given load in watts and a given size of wire and circuit length, the voltage drop in percentage is about 13 times as great at 32 V as at 115 V.

720-5. Lampholders. Where medium-base sockets are used, there is no good reason for using any but those having a 660-W rating. The ampere ratings of candelabra and intermediate base sockets would permit the use of 25-W lamps at 32 V, but it is not considered safe to allow the installation of these low-wattage sockets on circuits operating at 50 V or less.

Fixtures, regardless of supply voltage, would need to meet the requirements given in Art. 410. If for outdoor use, they would need to comply with Sec. 410-4.

720-10. Grounding. If any circuit connected to the system is carried overhead from one building to another, a system ground must be installed. The grounding conductor should be connected to one of the buses at the switchboard or generator and battery control panel. The requirements of Art. 250 should be followed in general. If water piping connected to a fairly extensive system of street mains is not available, a local water-piping system may be used as the grounding electrode; but unless the piping system is carried for some distance at a depth where it will always be in moist earth, supplementary driven rod or pipe electrodes should also be employed. Refer to Sec. 250-81.

ARTICLE 725. CLASS 1, CLASS 2, AND CLASS 3 REMOTE-CONTROL, SIGNALING, AND POWER-LIMITED CIRCUITS

725-1. Scope. Article 725 of the Code covers power-limited circuits and remote-control and signal circuits. A signal circuit is defined as any electrical circuit which supplies energy to an appliance or device that gives a visual and/or audible signal. Such circuits include those for doorbells, buzzers, code-calling systems, signal lights, annunciators, fire or smoke detection, fire or burglar alarm, and other detection indication or alarm devices.

A "remote-control" circuit is any circuit which has as its load device the operating coil of a magnetic motor starter, a magnetic contactor, or a relay. Strictly speaking, it is a circuit which exercises control over one or more other circuits. And these other circuits controlled by the control circuit may themselves be control circuits or they may be "load" circuits—carrying utilization current to a lighting, heating, power, or signal device. Figure 725-1 clarifies the distinction between control circuits and load circuits.

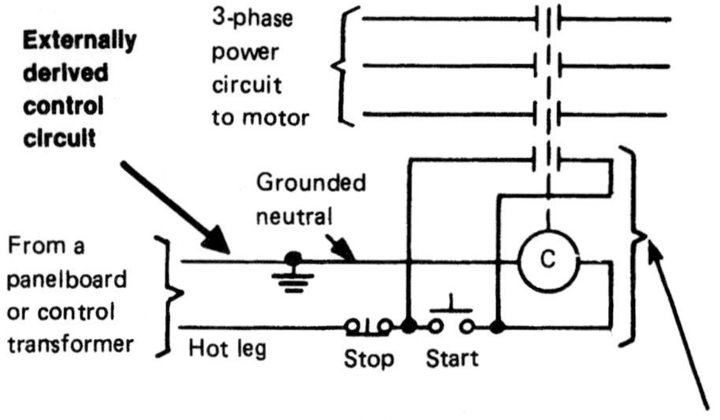

ARTICLE 725 APPLIES

Fig. 725-1. A control circuit governs the operating coil or some other element, to switch the load circuit. (Sec. 725-1.)

The elements of a control circuit include all the equipment and devices concerned with the function of the circuit: conductors, raceway, contactor operating coil, source of energy supply to the circuit, overcurrent protective devices, and all switching devices which govern energization of the operating coil. Typical control circuits include the operating-coil circuit of magnetic motor starters (NEC Secs. 430-71 and 430-72), magnetic contactors (as used for switching lighting, heating, and power loads), and relays. Control circuits include wiring between solid-state control devices as well as between magnetically actuated components. Low-voltage relay switching of lighting and power loads is also classified as remote-control wiring (Fig. 725-2).

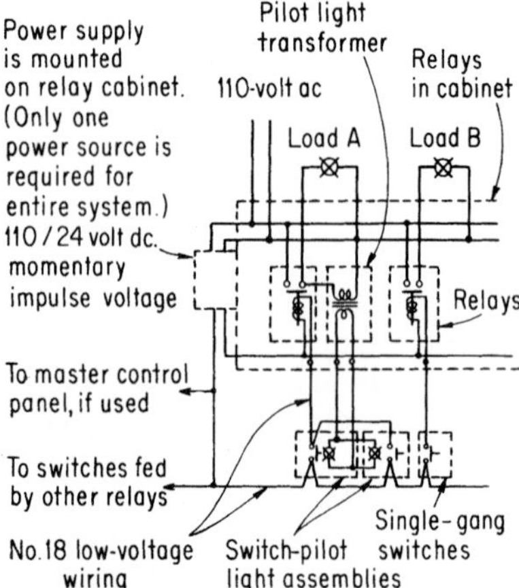

Fig. 725-2. Low-voltage switching involves typical "remote-control circuits." (Sec. 725-1.)

Power-limited circuits are circuits used for functions other than signaling or remote-control—but in which the source of the energy supply is limited in its power (volts times amps) to specified maximum levels. Low-voltage lighting, using 12-V lamps in fixtures fed from 120/12-V transformers, is a typical "power-limited circuit" application.

725-2. Definitions. The provisions of this section divide all signaling and remote-control systems into three classes.

Class 1 includes all signaling and remote-control systems which do not have the special current limitations of Class 2 and Class 3 systems.

Class 2 and Class 3 systems are those systems in which the current is limited to certain specified low values by fuses or CBs, and by supply through transformers which will deliver only very small currents on short circuit, or by other means which are considered satisfactory. The current values depend upon the voltage at which the system operates and range from 5 mA up, as shown in Table 725-31(a) and (b). All Class 2 and 3 circuits must have a power source with the power-limiting characteristics assigned in the tables in addition to the overcurrent device.

725-3. Locations and Other Articles. All the applications under this article must observe the specified sections that also rule on use of general power and light wiring.

Part **(a)** prohibits any installation of remote-control, signaling, or power-limited wiring in such a way that there is an appreciable reduction in the fire rating of floors, walls, or ceilings.

Part **(b)** requires that circuits covered by this article must be run in metal raceway or metal cable assembly when used in an air handling ceiling—as required by Sec. 300-22. The Exception permits Class 2 and/or 3 circuits (as defined in Sec. 725-2) to be used without a metal raceway or metal cable cover in ducts, plenums, or ceiling spaces used for environmental air *provided that the conductors are "listed" as Type CL2P or CL3P, as required in Sec. 725-53(a).* Because of the definition of the word "listed" (see Art. 100, "Definitions"), this rule would require that any such nonmetallic assembly of conductors for those circuits must be specifically listed in the UL's *Electrical Construction Materials Directory* (or with similar third-party certification) as having the specified characteristics for use without a metal raceway or covering in ducts, plenums, and air handling ceilings.

Note: For any such application, check that the conductors of the circuit are definitely listed by UL or others.

Part **(e)** notes that Art. 725 does not apply to control circuits tapped from line terminals in motor starters. As described under Sec. 430-72, the control circuit for the operating coil in a magnetic motor starter where the coil voltage supply is tapped from the line terminals of the starter is regulated by the rules of Sec. 430-72 and not by the remote-control rules of Art. 725. Where a control circuit for the operating coil of one or more magnetic motor starters is derived from a separate control transformer, one that is not fed from a motor branch circuit, the control circuit(s) and all components are covered by the rules of Art. 725. In the same way, a control circuit that is taken from a panelboard for power supply to the operating coil of one or more motor starters is also covered by Art. 725 and not by the rules of Sec. 430-72. (See Fig. 725-1.)

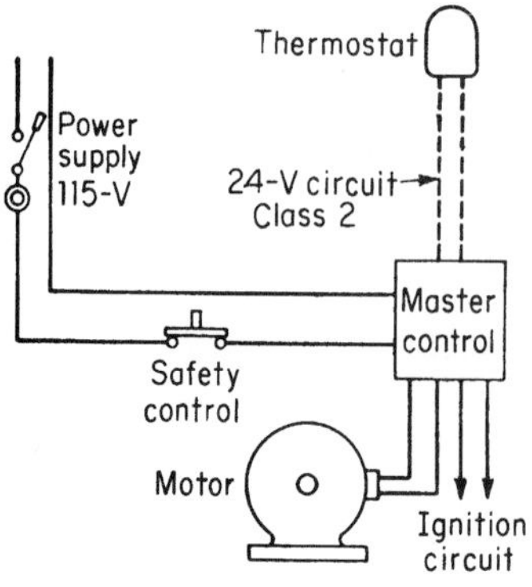

Fig. 725-3. Remote-control circuit *must* be Class 1 if failure would create a hazard. (Sec. 725-8.)

725-8. Safety-Control Equipment. The application of this rule is illustrated by Fig. 725-3, which is a simplified diagram of a common type of automatic control for a domestic oil burner. Assuming a steam boiler, the safety control is a switch that opens automatically when the steam pressure reaches a predetermined value and, preferably, also opens if the water level is allowed to fall too low. The master control includes a transformer of the current-limiting type which supplies the thermostat circuit at a voltage of 24 V. When the thermostat contacts close, a relay closes the circuits to the motor and to the ignition transformer.

Failure of the safety control or ignition to operate would introduce a direct hazard; hence, the circuits to this equipment are Class 1. The thermostat circuit fulfills all requirements of a Class 2 circuit and can be short-circuited or broken without introducing any hazard. The wiring of this circuit can therefore be done with any type of wire or cable that is sufficiently protected from physical damage to ensure serviceability.

725-21. Class 1 Circuit Classifications and Power Source Requirements. Class 1 systems may operate at any voltage not exceeding 600 V. They are, in many cases, merely extensions of light and power systems, and, with a few exceptions, are subject to all the installation rules for light and power systems.

Part **(a)** requires that Class 1 power-limited circuits must have energy limitation on the power source that supplies them. And such circuits may be supplied from either a transformer or another type of power supply—such as a generator, batteries, or manufactured power supply. Note that a Class 1 power-limited circuit must be supplied at *not over* 30 V, 1,000 VA.

Part **(b)**, however, permits Class 1 remote-control or signaling circuits to operate at up to 600 V, and no limitation is placed on the power rating of the source to such circuits (Fig. 725-4).

The most common example of a Class 1 remote-control system is the circuit wiring and devices used for the operation of a magnetically operated motor controller. The term *remote-control switch* is used in various **Code** references to designate a switch or contactor used for the remote control of a feeder or branch circuit, with the operating-coil circuit as a Class 1 remote-control circuit.

The signaling systems which are included in Class 1 operate at 115 V with 20-A overcurrent protection, although they are not necessarily limited to this

POWER-LIMITED CIRCUITS

Properly rated maximum overcurrent protection

Supply

Load

Open-circuit voltage not in excess of 30 volts

REMOTE-CONTROL OR SIGNALING CIRCUITS

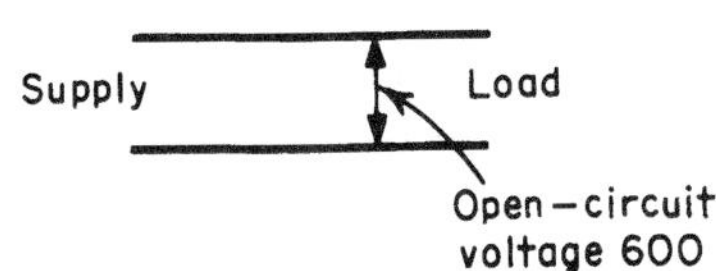

Fig. 725-4. Class 1 circuits are divided into two maximum voltages. (Sec. 725-21.)

voltage and current. Some of the signaling systems which may be so operated include electric clocks, bank alarm systems, and factory call systems. An example of a lower voltage Class 1 signaling system is a nurses' call system, as used in hospitals. Such systems commonly operate at not over 25 V.

Most control circuits for magnetic starters and contactors could not qualify as Class 2 or Class 3 circuits because of the relatively high energy required for operating coils. And any control circuit rated over 150 V (such as 220- or 440-V coil circuits) can never qualify, regardless of energy.

Class 1 control circuits include all operating-coil circuits for magnetic starters or contactors which do not meet the requirements for Class 2 or 3 circuits. Class 1 circuits must be wired in accordance with Secs. 725-11 to 725-20.

725-23. Class 1 Circuit Overcurrent Protection. In general, conductors for any Class 1 remote-control, signaling, or power-limited circuit must be protected against overcurrent. No. 14 and larger wires must generally be protected at their ampacities from Table 310-16. In the second sentence of this rule, an important statement indicates that it is *not* necessary to take any ampere derating—for either elevated ambient temperature or for more than three wires in a conduit or cable. The wires may simply be used and protected at the ampacity values given in the table.

Important. The statement "Derating factors *shall not* be applied" is a strange prohibition, which can be disregarded without violating this or any **Code** rule. And there seems to be a conflict between that statement and the requirement in Sec. 725-28 that Class I conductors must be derated under some conditions.

It is important to note that Nos. 18 and 16 control or signal-circuit conductors must always be protected at 7 or 10 A, respectively. Those smaller sizes of wire may be used for control and signal circuits supplying coils, relays, or signal devices that are current loads up to 10 A, as recognized by the first sentence of Sec. 725-27(a).

The rule of Exception No. 2 is the same as that of Sec. 240-3, Exception No. 2, but applies to the case where the 2-wire transformer secondary supplies a control circuit to one or more operating coils in motor starters or magnetic contactors. A properly sized circuit breaker or fuses may be used at the supply end of the circuit that feeds the transformer primary and may provide overcurrent protection for the primary conductors, for the transformer itself, and for conductors of the control circuit which is run from the transformer secondary to supply power to motor starters or other control equipment, as follows:

1. The primary-side protection must not be rated greater than that required by Sec. 450-3(b)(1) for transformers rated up to 600 V. For a transformer rated 9 A or more, the rating of the primary CB or fuses must not be greater than 125 percent of (1.25 times) the rated transformer primary current. And if 1.25 times rated primary current does not yield a value exactly the same as a standard rating of fuse or CB, the next-higher-rated standard protective device may be used. Where the transformer-rated primary current is less than 9 A—as it would be for all the usual control transformers rated 5,000 VA and stepping 480 V down to 120 V—the maximum permitted rating of primary protection must not exceed 167 percent of (1.67 times) the rated primary current. For a transformer with a primary rated less than 2 A, the primary protection must never exceed 300 percent of (3 times) the rated primary current. With most control transformers—with primary ratings well below 10 A—fuse protection will be required on the primary because the smallest standard CB rating is 15 A, and that will generally exceed the maximum values of primary protection permitted by Sec. 450-3(b)(1) (Fig. 725-5).
2. Primary protection must not exceed the amp rating of the primary circuit conductors. And when protection is sized for the transformer, as described above, No. 14 copper primary conductors will be protected well within their 15-A rating.
3. Secondary conductors for the control circuit can then be selected to have an ampacity at least equal to the rating of primary protection times the primary-to-secondary transformer voltage ratio. Of course, larger conductors may be used if needed to keep voltage drop within limits.

In Fig. 725-6, covering use of a magnetic contactor, No. 14 and larger remote-control conductors may be properly protected by the feeder or branch-circuit overcurrent devices if the devices are rated or set at not more than 300 percent of (3 times) the ampacity of the control conductors. If the branch-circuit overcurrent devices were rated or set at more than 300 percent of the rating of the control conductors, the control conductors would have to be protected by separate protective devices located within the contactor enclosure at the point where the conductor to be protected receives its supply.

This is covered by Exception No. 3 of this section, which applies to the remote-control circuit that energizes the operating coil of a magnetic contactor, as distinguished from a magnetic motor starter. Although it is true that a magnetic starter is a magnetic contactor with the addition of running overload

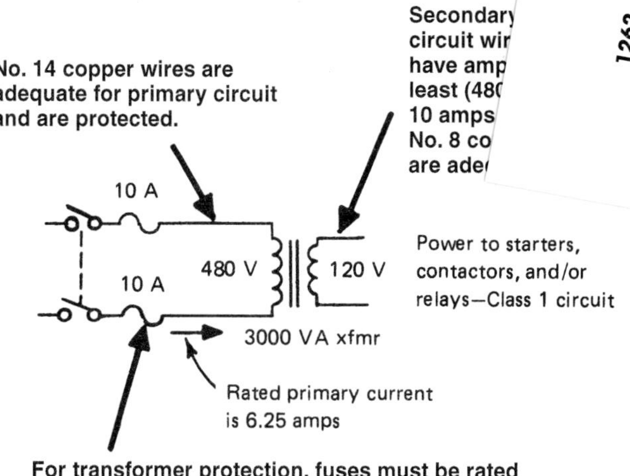
No. 14 copper wires are adequate for primary circuit and are protected.
Secondary circuit wir have amp least (48 10 amps No. 8 co are ade
10 A
10 A
480 V
120 V
Power to starters, contactors, and/or relays—Class 1 circuit
3000 VA xfmr
Rated primary current is 6.25 amps
For transformer protection, fuses must be rated not over 1.67 × 6.25, or 10 amps [Sec. 450-3(b)(1), Ex. No. 1]
1262

Fig. 725-5. A separate control transformer supplying a number of coil circuits for motor starters or magnetic contactors must have primary protection that protects the secondary control conductors as well as the transformer. (Sec. 725-23.)

relays, Exception No. 3 covers only the coil circuit of a magnetic contactor. That applies to control wires for magnetic contactors used for control of lighting or heating loads, but not motor loads. Section 430-72 covers that requirement for motor-control circuits.

In Fig. 725-6, for instance, 45-A fuses at A in the feeder or branch circuit ahead of the contactor would be adequate protection if No. 14 wire, with its ampacity of 15 A (reduced from 20 A in Table 310-16 by the footnote to that table), were used for the remote-control circuit, because 45 A is not more than 300 percent of (3 times) the 15-A ampacity. Larger fuses ahead of the contactor would require overcurrent protection in the hot leg of the control circuit, at B, rated not over 15 A. (See footnote to Table 310-16.)

It should be noted that the overcurrent protection is required for the control conductors and not for the operating coil. Because of this, the size of control conductors can be selected to allow application without separate overcurrent protection. When overcurrent protection is added in the enclosure, its rating must be such that it conforms to the first paragraph of this rule.

725-25. Class 1 Circuit Wiring Methods. In general, wiring of Class 1 signal systems must be the same as power and light wiring, using any of the cable or raceway wiring methods that are Code-recognized for general-purpose wiring. The two Exceptions refer to the details of wiring permitted by Secs. 725-26, 725-27, and 725-28.

725-26. Conductors of Different Circuits in Same Cable, Enclosure, or Raceway. Any number and any type of Class 1 circuit conductors—for remote control, for signaling, and/or for power-limited circuits—may be installed in the same con-

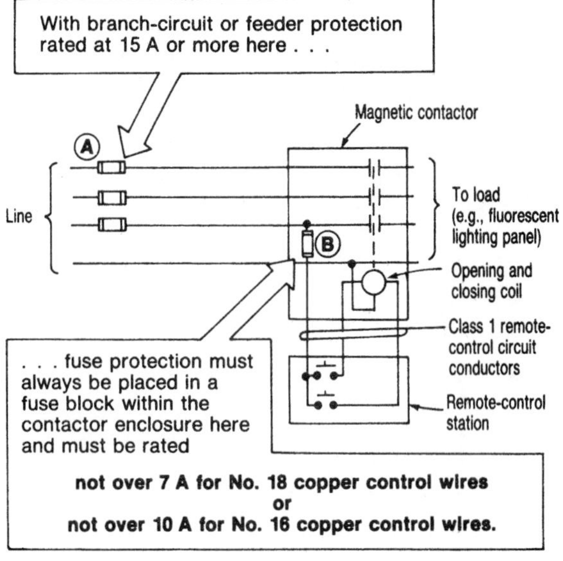
With branch-circuit or feeder protection rated at 15 A or more here . . .
Magnetic contactor
A
Line
To load (e.g., fluorescent lighting panel)
B
Opening and closing coil
Class 1 remote-control circuit conductors
Remote-control station
. . . fuse protection must always be placed in a fuse block within the contactor enclosure here and must be rated
not over 7 A for No. 18 copper control wires
or
not over 10 A for No. 16 copper control wires.

NOTE: If No. 14 or larger control-circuit wires are used, Exception No. 3 of this Section permits omission of separate protection in the control circuit when the rating of the branch-circuit or feeder protection does not exceed three times the ampacity of the particular size of control wire from Table 310-16.

EXAMPLE: 30-A fuses at "A" would be adequate protection if No. 14 wire, rated at 15 A, is used for the remote-control circuit, because 30 A is *less than* 3 × 15 A. If fuses at "A" were rated over 45 A, 15-A protection would be required at "B" for No. 14 wire.

Fig. 725-6. Protection of coil circuit of a magnetic contactor is similar to that of a starter. (Sec. 725-23.)

duit, raceway, box, or other enclosure—*if* all conductors are insulated for the maximum voltage at which any of the conductors operates.

Class 1 circuit wires (starter coil-circuit wires, signal wires, and power-limited circuits) may be run in raceways by themselves in accordance with the first sentence of this section. A given conduit, for instance, may carry one or several sets of Class 1 circuit wires. And Sec. 725-25 says use of Class 1 wires must conform to the same basic rules from NE Code Chap. 3 that apply to standard power and light wiring.

But, it should be noted that two specific sections of the NE Code cover the use of Class 1 circuit conductors in the same raceway, cable, or enclosure containing circuit wires carrying power to a lighting load, a heating load, or to a motor load (Fig. 725-7). Section 300-3 covers the general use of "conductors of differ-

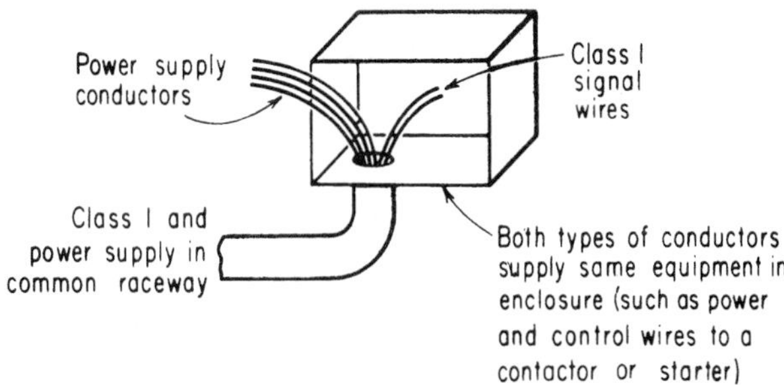

Fig. 725-7. This is permitted by Secs. 300-3(a) and 725-26. (Sec. 725-26.)

ent systems" in raceways as well as in cable assemblies and in equipment wiring enclosures (i.e., cabinets, housings, starter enclosures, junction boxes, etc.). But Class 1 circuit wires are also regulated by this section, which strictly limits use of Class 1 wires in the same box and/or raceway with power wires. Figure 725-8 shows a clear violation, if the annunciator has no relationship to the motor load.

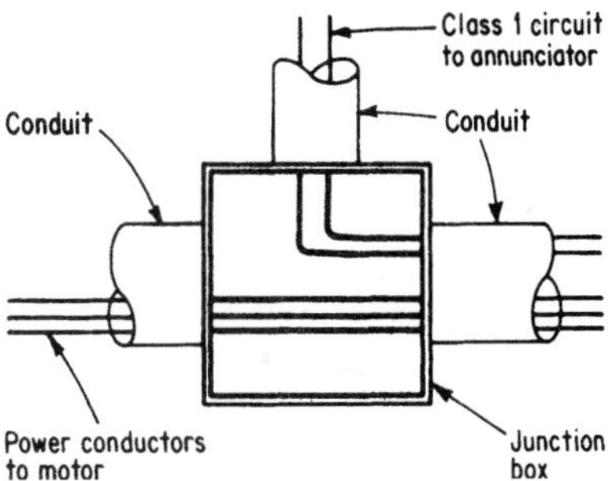

Fig. 725-8. Class 1 wires must *not* be used in raceways with "unrelated" wires. (Sec. 725-26.)

Note that this section permits Class 1 circuit wires to be installed in the same raceway or enclosure as "power supply" conductors *only* if the Class 1 wires and the power wires are "functionally associated" with each other. That would be the case where the power conductors to a motor are run in the same conduit along with the Class 1 circuit wires of the magnetic motor starter used to control or to start or stop the motor. Refer to the commentary in Sec. 300-3.

The same permission would apply to the hookup of a magnetic contactor controlling a lighting or heating load, as shown in Fig. 725-9. There, the circuit wires for the Class 1 remote-control run to the pushbutton station may be run in the same conduit carrying the wires supplying the lighting fixtures. A typical application would have the magnetic contactor adjacent to a panelboard, with the control and power wires run in the same raceway to a box at some point where it is convenient to bring the control wires down to the control switch and carry the power wires to the lighting fixtures being controlled. The

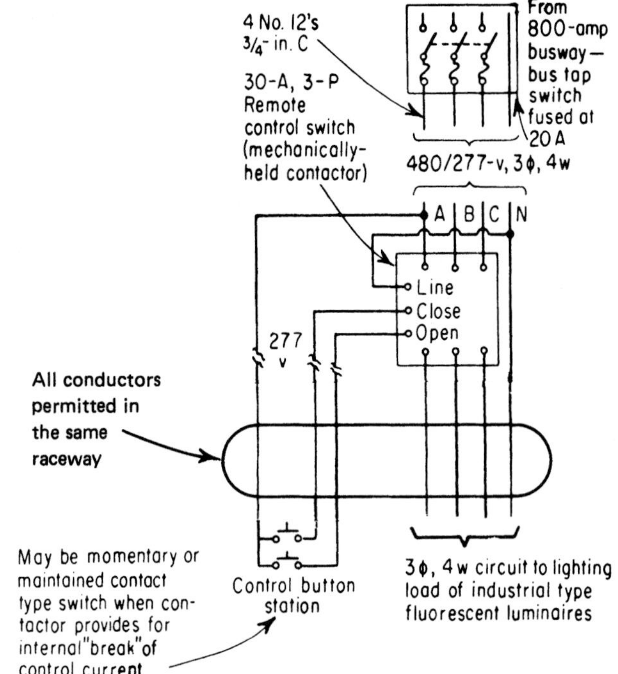

Fig. 725-9. Class 1 wires and power wires may be used in same raceway for "functionally associated" equipment. (Sec. 725-26.)

contactor can be located at the approximate center of its lighting load to keep circuit wiring as short as possible for minimum voltage drop and the control wires are then carried to one or more control points. In such a layout, the control and power wires are definitely "functionally associated" because the control wires provide the ON-OFF function for the lighting. *But,* other control or power wires are prohibited from being in the same conduit, boxes, or enclosures with the single set of associated Class 1 and power wires.

Note: Exception No. 1 in this section permits power and control wires for more than one motor in a common raceway. Factory- or field-assembled control centers may group power and Class 1 control conductors that are not "functionally associated." This Exception recognizes the use of listed motor control centers that have power and control wiring in the same wireway or gutter space supplying motors that are *not* "functionally associated." The basic rule generally prohibits that condition when hooking up motor circuits.

Exception No. 2 says that Class 1 circuit conductors and unassociated power-supply conductors are permitted in a manhole if either of them is in metal-enclosed cable or Type UF cable *or* if effective separation is provided between the Class 1 conductors and the power conductors. This new Exception covers the conditions under which Class 1 conductors and unrelated power conductors may be used in the same enclosure (a manhole).

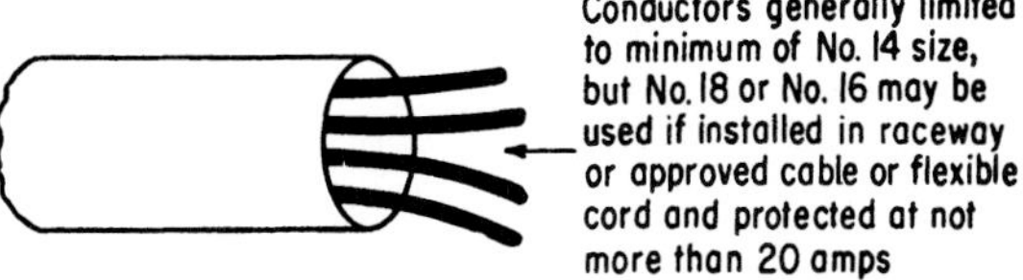

Conductors generally limited
to minimum of No. 14 size,
but No. 18 or No. 16 may be
used if installed in raceway
or approved cable or flexible
cord and protected at not
more than 20 amps

EXAMPLE :

If branch-circuit protection is rated at 15 A or 20 A . . .

3ϕ motor

Coil

Starter enclosure

Stop Start

. . . the Class 1 remote-control wires may be No. 18 or No. 16 fixture wire or No. 14 building wire, depending upon the ampere load of the starter coil.

Fig. 725-10. No. 16 or No. 18 fixture wire may be used for Class 1 circuits. (Sec. 725-27.)

725-27. Class 1 Circuit Conductors. Figure 725-10 shows this basic rule, which accepts use of building wire to a minimum No. 14 size. *But,* No. 16 or 18 fixture wires of the types specified in part **(b)** *may* be used for running starter coil circuits, signal circuits, and any other Class 1 circuits. Of course, use of No. 16 or 18 fixture wire for Class 1 circuits depends upon such conductors having sufficient ampacity for the current drawn by the contactor or relay operating coil or by whatever control device is involved. Wires larger than No. 16 must be building types (TW, THW, THHN, etc.). Class 1 circuits may not use fixture wires larger than No. 16. And ampacity of any Class 1 circuit wires larger than No. 16 must have that value shown in Table 310-16.

Note: Section 402-5 shows that the ampacity of any No. 18 fixture wire is 6 A and the ampacity of any No. 16 fixture wire is 8 A. Any No. 18 or 16 fixture wire or No. 14 building wire used for a Class 1 circuit is considered adequately protected by a fuse or CB rated not over 20 A. See Secs. 725-23 and 240-4.

725-28. Number of Conductors in Cable Trays and Raceway, and Derating. The number of Class 1 remote-control, signal, and/or power-limited circuit conductors in a conduit must be determined from Tables 1 through 5 in Chap. 9 of the NE Code.

When more than three Class 1 circuit conductors are used in a raceway, ampacity derating of Note 8, Table 310-16, applies only if the conductors carry continuous loads in excess of 10 percent of the ampacity of each conductor (Fig. 725-11). This rule is aimed at relieving the need to derate conductors that

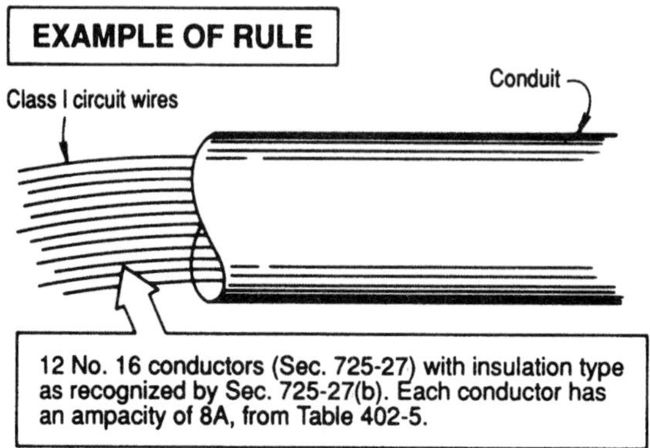
EXAMPLE OF RULE
Conduit
Class I circuit wires
12 No. 16 conductors (Sec. 725-27) with insulation type as recognized by Sec. 725-27(b). Each conductor has an ampacity of 8A, from Table 402-5.

Derating of conductor ampacity (Note 8, Table 310-16) is required (70% of 8A) only if these conductors carry "continuous loads" in excess of 10% of 8A, or 0.8A. If each conductor carries no more than 0.8A, no derating is required.

With 12 wires at 0.8A, the total heating effect is equivalent to 12 x 0.8, or a total of 9.6A divided among the 12 conductors. If the conductors carry different load currents but their sum does not exceed 9.6A, it would be reasonable to eliminate derating. But even if loading is greater than 9.6A and derating is required, using the 70% factor from Note 8 for 10 to 24 conductors gives each No. 16 wire here an ampacity of 0.7 x 8A or 5.6A—which is the amount of current each conductor is rated to carry continuously.

Fig. 725-11. Determining conductor ampacity for more than three Class I circuit wires in a raceway requires careful evaluation of conditions and the Code rule. (Sec. 725-28.)

are usually carrying very low values of current (like up to 2 A for most coil circuits of contactors and motor starters). The wording does not spell out whether all the conductors are carrying continuous current or if only some of them are. Actual application of this rule can get involved, depending upon the number of conductors and the type of devices they supply. And it is important to note that there is a direct contradiction of this requirement for derating in the flat statement in Sec. 725-12 that "Derating factors shall not be applied" to Class I circuit wires that are No. 14 or larger.

The same concept of "10 percent of the ampacity of each conductor" has been applied in part **(b)(1)** and **(2)** of Sec. 725-28.

When power conductors and Class 1 circuit conductors are used in a single conduit or EMT run (as permitted by Sec. 725-26), the derating factors of Note 8 must be applied as follows:

1. Note 8 must be applied to all conductors in the conduit when the remote-control conductors carry continuous loads in excess of 10 percent of each conductor's ampacity and the total number of conductors (remote-control and power wires) is more than three. For example, in Fig. 725-12, the conduit size must be selected according to the number and sizes of the wires. Because two of the control wires to the pushbutton and the power wires to the motor will carry a continuous load that is usually less than 10 percent of conductor ampacity, a derating factor of 80 percent (from Note 8) does not have to be applied.

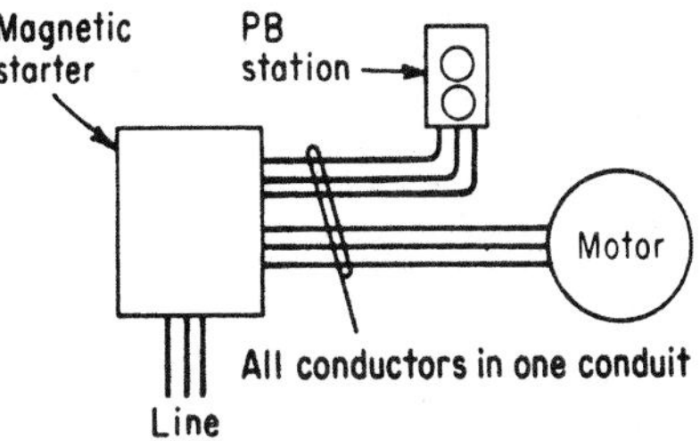

Fig. 725-12. Derating of conductor ampacity is usually not required for this circuit makeup. (Sec. 725-28.)

2. Note 8 must be applied only to the power wires when the remote-control wires do not carry continuous load and when the number of power wires is more than three. In Fig. 725-13, no derating at all is applied because the control wires do not carry continuous current (only for the instant of switching operation), and there are only three power wires.

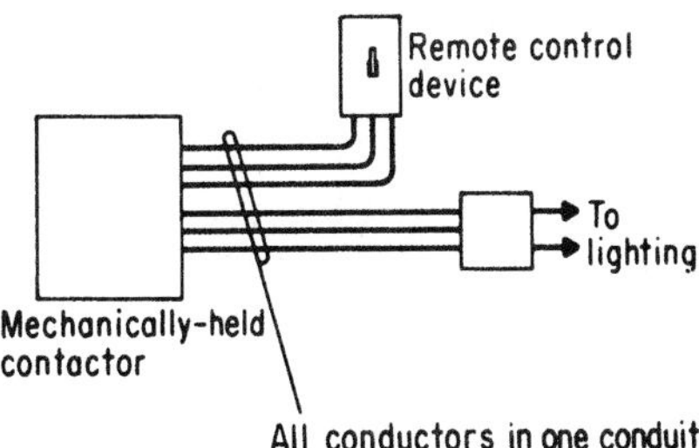

Fig. 725-13. Derating is not required here if Class 1 conductors do not carry continuous load. (Sec. 725-28.)

Those rules of part **(b)** have created controversy. It usually starts with the question, If a conduit from a starter carries the three power wires of a motor circuit and also contains three control wires run from the starter to a pushbutton station, is it necessary to derate any of the conductor ampacities?

Answer: Section 725-28(b) covers this. (Read that rule several times.) If the starter is the usual magnetically held type of contactor, the two control wires to the STOP button at the pushbutton station will carry the holding current to the coil as long as the starter is closed. Section 725-28(b)(1) says that all conductors in the raceway must be derated in ampacity if the total number of conductors (power wires plus control wires) is more than three—*but only if the Class 1 circuit conductors carry continuous loads* of more than 10 percent of conductor ampacities.

The **Code** definition for *continuous load* is: "A load where the maximum current is expected to continue for three hours or more." On that basis, derating of all the conductors (power and control wires) would be required if the starter is left closed for 3 hr or more. It may be argued, however, that because the holding current of the coil is so low (only an amp or less in most cases), the load on the control wires (for instance, No. 14) is nowhere near the ampacity of those wires and does not constitute a "maximum current" in the meaning of the definition for continuous load; therefore, derating is not required for any of the conductors—as noted in Sec. 725-28(b)(2). But it may also be argued that because the three power wires in the conduit are rated at 125 percent of motor current and protected at that value by the OL relays, such conductors could be continuously subjected to maximum operating temperature by motor overload. Under such a condition, any additional current—even 1 A in the control wires—could produce heat that would push the temperature over the limit for the particular insulation.

Because motors are not generally even fully loaded and because continuous operation of motors at 25 percent overload is extremely unlikely, it would seem unreasonable to require derating of conductors for the usual conduit with three power wires and three control wires from a magnetic starter.

Conflict: All this consideration is further confused by a conflict between part **(b)(1)** of Sec. 725-28 and Exception No. 1 to Note 8 of Table 310-16/19, which states that derating factors apply *only* to the number of "power and lighting conductors." Part **(b)(1)** says derating applies to "all conductors"—Class 1 wires as well as the power and light wires.

The certain answer to the puzzle is this: Evaluate each installation on its own conditions and circumstances. Table 310-16 gives maximum continuous current ratings with not more than three conductors in a conduit. If for any control circuit layout, there is indication that the condition of maximum temperature is likely for the power wires, then all the wires must be derated if control wires in the same conduit would be adding any heat at all—even the I^2R of very low coil current.

And because Sec. 725-28, requiring derating under some conditions, is in part **B Class 1 Circuits** of Art. 725, it should be noted that the second sentence of Sec. 725-23 (also in part **B**) flatly says "Derating factors shall not be applied"—and the words "shall not" makes it strangely mandatory.

We should always keep in mind:

The Code *is a guide to safe electrical practice; it is not a desi*
Its rules must be analyzed and applied with knowledge, care, a

725-41. Power Sources for Class 2 and Class 3 Circuits. Part of the of this article relocated specific data regarding maximum curre ratings for Class 2 and 3 circuits to Tables 11a and 11b in Chap installation is concerned, the marking on any listed piece of equip ... will be the determining factor as to how a power source is classified. And, the installation must satisfy all rules related to that classification.

725-51. Wiring Methods on Supply Side of the Class 2 or Class 3 Power Source. Conductors and equipment on the line side of devices supplying Class 2 systems must conform to rules for general power and light wiring.

725-52. Wiring Methods and Materials on Load Side of the Class 2 or Class 3 Power Source. This section requires that conductors be installed in accordance with Sec. 725-54 and 725-61. The present wording of that rule and its Exceptions clarifies many past misunderstandings about the intent of the rule. Previous Code editions only mentioned raceways, porcelain tubes, or "loom" as a means of separating Class 2 circuits, such as "bell wiring," from conductors of light and power systems where such systems were closer together than 2 in. (50.8 mm). And on that basis, some inspectors required such bell wiring or similar Class 2 wires to have a 2-in. (50.8-mm) clearance from any type of cable (NM, UF, AC, etc.) that contained conductors for power or lighting circuits. The old rule was also commonly applied to prohibit bell wires and NM cables in the same bored holes through studs, etc. With the present wording the 2-in. (50.8-mm) clearance from Class 2 wiring applies only to "open" light, power, and Class 1 circuit conductors. Power and light circuits or Class 1 circuits that are in raceway or cable do not require 2-in. (50.8-mm) separation from Class 2 and/or 3 circuits (Fig. 725-14).

Class 2 or 3 conductors must be insulated and must be separated at least 2 in. (50.8 mm) from open power and light conductors, unless the Class 2 or 3 conductors are enclosed in a continuous and firmly fixed nonconductor.

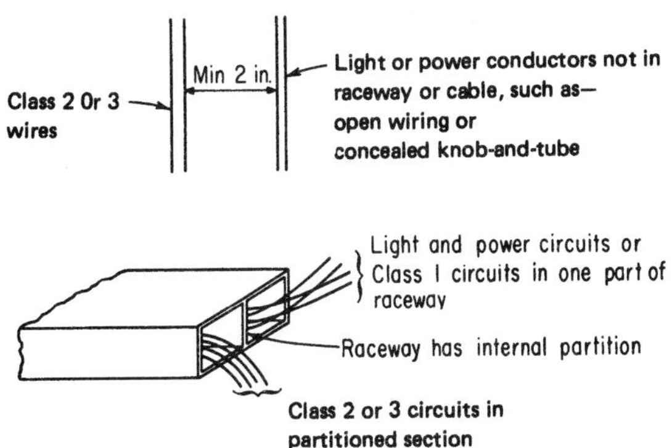

Fig. 725-14. Class 2 or 3 wiring must be separated from open-wiring for power and light. (Sec. 725-52.)

Part **(a)(1)** of Sec. 725-54 says that Class 2 or 3 conductors must not be used in any raceway, compartment, outlet box, or similar fitting with light and power conductors or with Class 1 signal or control conductors, unless the conductors of the different systems are separated by a partition. But this does not apply to wires in outlet boxes or similar fittings or devices, where power-supply conductors have to be brought in to supply power to the signal equipment to which the other conductors in the enclosure are connected.

In hoistways, conductors must be installed in rigid conduit, IMC, or EMT, except as provided for elevators in Art. 620.

In part **(a)(3)**, Class 2 or 3 circuit conductors in *shafts* must be separated not less than 2 in. (50.8 mm) from conductors for light, power, Class 1, or nonpower-limited fire-protective signaling circuits in the shaftway.

The rule of part **(b)(3)** requires separation of Class 2 and 3 circuits, *unless* the Class 2 wires have insulation that is at least equivalent to that required for Class 3 wires.

725-71. Listing and Marking of Class 2, Class 3, and Type PLTC Cables. In this section extensive requirements are presented on fire resistance and other characteristics of UL-listed Class 2 and 3 conductors and PLTC (power-limited tray cable). This is very detailed information on the variety of cables commonly used today for signal, remote control, and alarm circuits. These sections cover single- and multiconductor cables of Class 2 and 3 circuits installed "within buildings," in "vertical runs," and in "ducts, plenums, and other airhandling spaces."

Here, distinction must be made between Class 2 and 3 wiring. Although any type of insulation is permitted for the conductors of Class 2 systems, in order to ensure continuity of service, a type of insulation should be selected which is suitable for the conditions, such as the voltage to be employed and possible exposure to moisture.

There is a marked distinction between power-limited circuits supplied by limited power sources with overcurrent protection and those without; but within the same source category, little difference exists in the power limitation.

The significant distinction between Class 2 and 3 circuits is the character and magnitude of the voltages. The classifications of Class 2 circuits recognize their acceptability from both fire and shock hazard. Class 3 circuits recognize fire hazard only; hence the reason for more restrictive conductor and insulation requirements.

Low-voltage relay switching is a common application of Class 2 remote-control circuits regulated by Art. 725. Low-voltage relay switching is commonly used where remote control or control from a number of spread-out points is required for each of a number of small 120- or 277-V lighting or heating loads. In this type of control, contacts operated by low-voltage relay coils are used to open and close the hot conductor supplying the one or more luminaires or load devices controlled by the relay. The relay is generally a 3-wire, mechanically held, ON-OFF type, energized from a step-down control transformer.

In some cases, all the relays may be mounted in an enclosure near the panelboard supplying the branch circuits which the relays switch, with a single transformer mounted there to supply the low voltage. Where a single panelboard serves a large number of lighting branch circuits over a very large area—such as large office areas in commercial buildings—a number of relays associated with each section of the overall area may be group-mounted in an enclosure in that area.

Figure 725-15 shows 24-V control of 277-V fixtures, with constantly illuminated switchplates alongside doorways to define interior exit routes and prac-

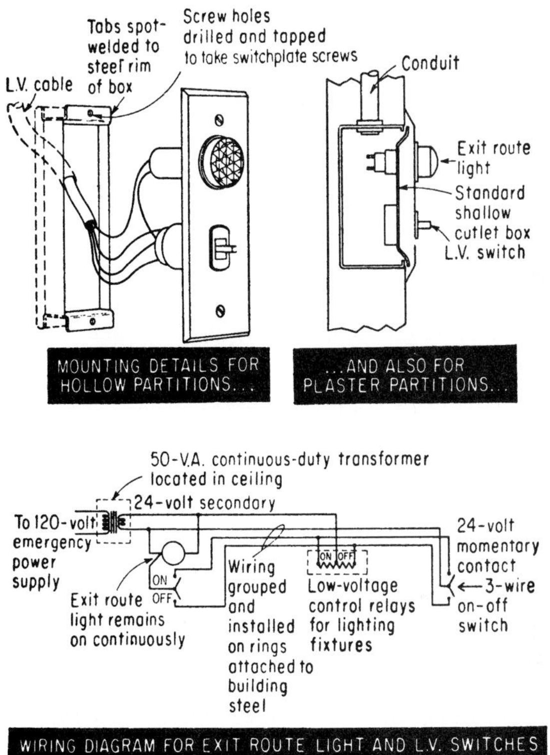

Fig. 725-15. Low-voltage relay switching is a typical application of Class 2 conductors. (Sec. 725-71.)

tical hollow-partition clip-in switch boxes in interior labs of a medical research center. Control relays are in compact boxes atop luminaires. Relays are connected to switches and to 50-VA continuous-duty 120/24-V transformers by Class 2 remote-control circuits routed through overhead nonair-handling ceiling plenums and supported by insulator rings attached to fixture hangers by spring clips. Wiring of the Class 2 circuits may be done with multiwire low-voltage cable, such as 3- or 4-conductor thermostat cable. Interior-area route-indicating lights and general lighting switches are mounted together in thin boxes set in partitions.

Low-voltage lighting—using recessed and/or surface-mounted fixtures containing 12-V, 20-W incandescent reflector lamps—is another application of Class 2 wiring for the power-limited circuits supplying the lamps from a transformer. In such lighting systems, which have gained popularity for interior decorative and architectural purposes, the supply transformers are available in three ratings—20 VA for one lamp, 60 VA for three lamps, and 180 VA for nine lamps.

ARTICLE 760. FIRE ALARM SYSTEMS

760-1. Scope. As indicated by FPN No. 1, the various elements associated with the fire alarm system are now all covered in a single NFPA Standard—NFPA 72. The overall article here has been revised to consolidate the various requirements to facilitate comprehension as well to correlate with changes made in Art. 725.

760-2. Location and Other Articles. The Exception to part **(b)** involves use of plastic-jacketed cables in air-handling ceilings. The same considerations are involved here as described under Sec. 725-2.

ARTICLE 770. OPTICAL FIBER CABLES AND RACEWAYS

770-1. Scope. In the Technical Committee Reports on development of the 1984 NE Code, the following substantiation was given for adding Art. 770 to the Code:

> Fiber optic technology should be included in the Code to permit its orderly development and usage for communications, signaling and control circuits in lieu of metallic conductors. It is reasonable for an optical fiber cable to be installed in electrical raceway and enclosures along with associated electrical conductors. An example of such an application is an optical fiber control circuit for electrically noisy equipment. Since optical fibers are not affected by electrical noise, one could, if permitted by the Code, run the optical fiber cable in the same raceway with the power wiring. A further example is the use of optical fiber communications cable. One would expect to place this cable in a common raceway along with ordinary metallic conductor telephone cable. However, if the Code is not changed to recognize optical fiber technology, a separate conduit system may be demanded by some local authorities.
>
> The proposed article divides optical fiber cables into three types: nonconductive, conductive, and hybrid. Obviously the nonconductive types cannot be accidentally energized when placed in raceway so it is proposed that they be permitted in a raceway with conductors for electric light, power, or Class 1 circuits operating at less than 600 volts only where the functions of optical fiber cables and electrical conductors are associated. Since the conductive optical fiber cables have a potential for inadvertent energizing of metallic strength members and metallic vapor barriers, it is proposed that these cables be permitted to share raceway with low voltage wiring systems only, and the conductive members of these cables must be grounded. Grounding (or isolation) is also proposed for entrance cables in a manner consistent with the Code requirements for ordinary communications cable.

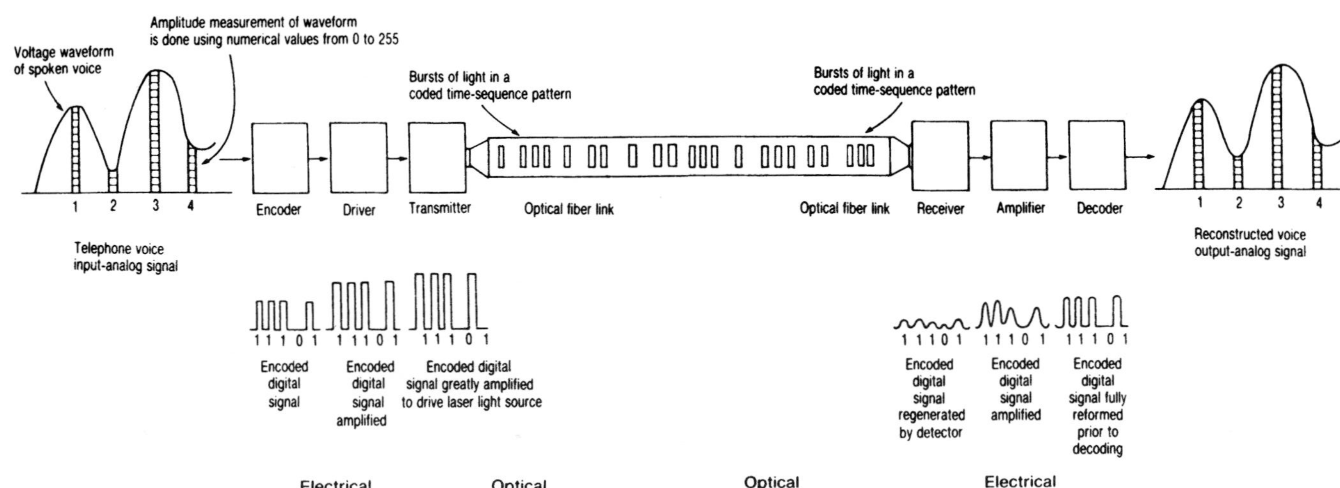

Fig. 770-1. The transmission of a signal (left to right) along an optical fiber link (a cable) at center. At the left, an electrical signal is converted to light pulses that are sent through the fiber cable by a laser diode (a light signal generator), and then, at the right, the light pulses are received and reconstructed into the original electrical signal from the left. (Sec. 770-1.)

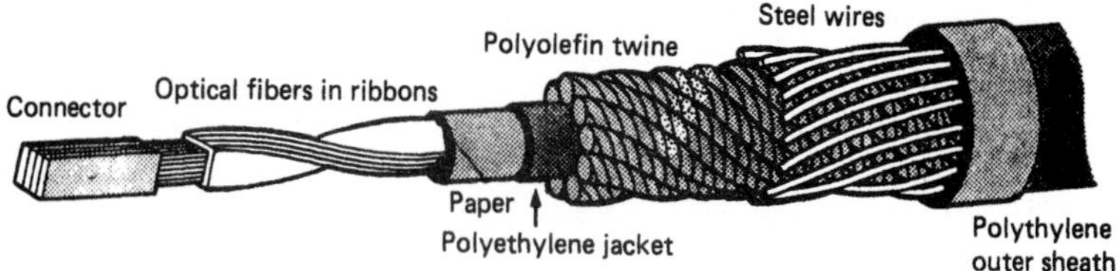

Fig. 770-2. This fiber-optic cable, which is used in a mile-long telephone communication line, is classified as a "conductive" type of optical fiber cable because of the steel wires used to provide an outer mechanical sheath over the fiber assembly. (Sec. 770-4.)

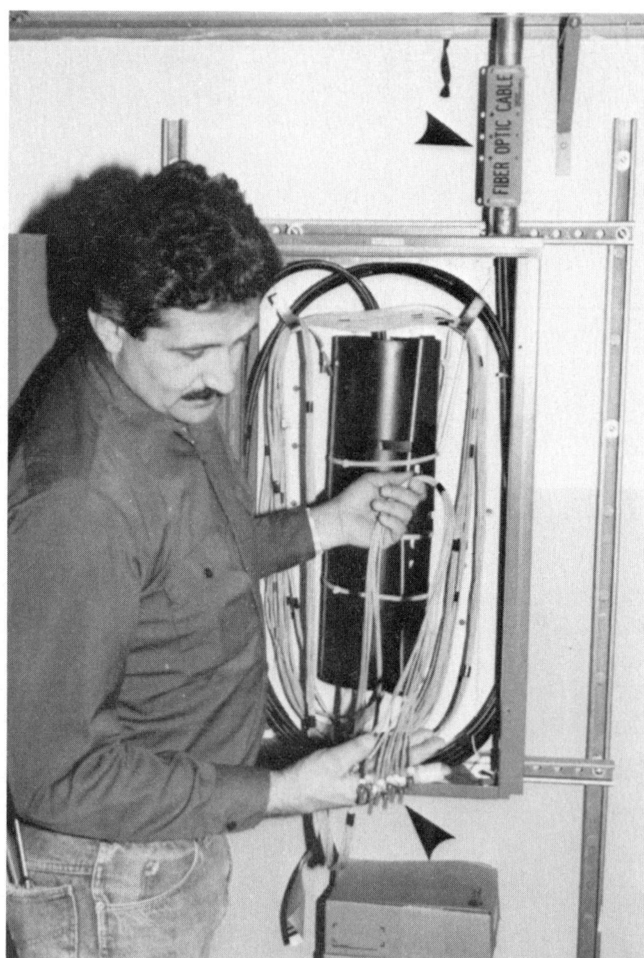

Fig. 770-3. FO conductors are used as data communication links between five different buildings in which a banking firm has branch locations in a large city. The overall circuiting includes the runs within the buildings (connecting to computers, video equipment, telephones, and telecommunications equip-

The proposed article deals with fire properties of optical fiber cables in a manner identical with other low voltage wiring.

The scope statement, in order to be consistent with the purpose of the Code, which is the "practical safeguarding of persons and property from hazards arising from the use of electricity," limits the coverage of the proposed article to joint installations of electrical cable and optical fiber cable.

As part of the high-technology revolution in industrial and commercial building operations, the use of light pulses transmitted along optical fiber

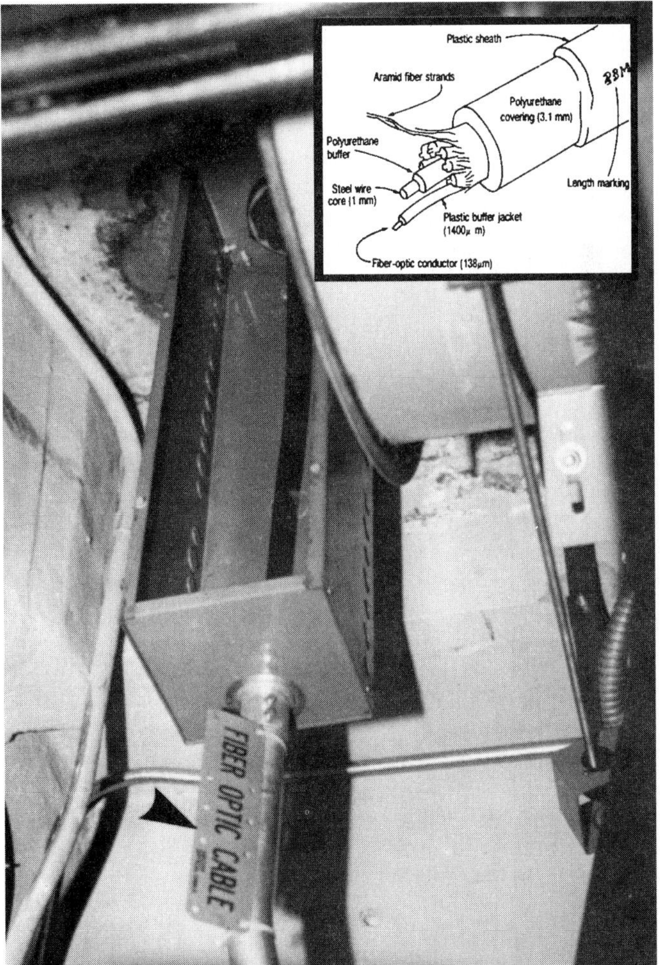

ment) and over 5 mi. of underground 30-conductor FO cable with a metal armor sheath—run under the city streets in ducts. A NEMA 3R splice box in the telecommunications room of one of the buildings, with two 10-conductor cables entering at top (arrow) and the individual FO conductors terminated in special connectors (bottom arrow). (Right) A pullbox in a 1½-in. (38 mm) EMT run carrying the 10-conductor cables, with a bright orange tag identifying the FO conduit (inset shows FO cable). (Sec. 770-4.)

cables has become an alternative method to electrical pulses on metal conductors for control, signals, and communications. The technology of fiber optics has grown dramatically over recent years as a result of great strides in development of the fiber cables and associated equipment that converts electrical pulses to light pulses, and vice versa. For high rates of data transmission involved in data processing and computer control of machines and processes, optical fiber cables far outperform metallic conductors carrying electrical currents—all at a small fraction of the cost of metallic-conductor circuiting.

NEC Art. 770, "Optical Fiber Cables and Raceways," covers the use of such cables in association with conventional metallic-conductor circuits. Nonconductive optical fiber cables are permitted to be installed in the same raceway and enclosures as metallic-conductor circuits where the functions of the two different types of cables are associated with the same equipment, operation, or process.

Figure 770-1 shows operation of a fiber-optic (FO) link in the transmission of a telephone signal.

770-4. Types. Figure 770-2 shows an optical fiber cable that is a "conductive" type, as covered in the descriptions of the three types of cables in this section. Figure 770-3 shows details of a typical FO cable installation.

ARTICLE 780. CLOSED-LOOP AND PROGRAMMED POWER DISTRIBUTION

780-1. Scope. This NEC article covers electrical distribution systems that are controlled by signaling between energy-controlling equipment and energy-utilization equipment.

The purpose of this Code article is described in the substantiation given with the proposal to include this in the NEC:

> Article 780 is included to permit the orderly development and usage of Closed-Loop Power Distribution systems. This method of power distribution offers advantages in safety and new features over the present open-loop method.
>
> With closed-loop control, any electrical circuit or appliance is activated only after proper identification during a power startup sequence, and operation is continued only so long as an acknowledgement signal is received from the equipment indicating normal operation. The control channel may also be used by the energy-consuming equipment to communicate operational status as well as to call for adjustments in the characteristics of the power being fed to the circuit.
>
> *SUBSTANTIATION:* Closed-Loop Power Distribution Systems should be permitted because they have the following safety advantages:
>
> a. Power feed to a branch circuit is initiated only when the energy-utilization equipment requests it by a characteristic electrical signal.
>
> b. Power feed to a branch circuit continues only on continuous presence of a nominal operation signal from the energy-utilization equipment.
>
> c. Power consumption is continuously monitored by the energy-control equipment, and power feed to the branch circuit is interrupted at any time that power consumption falls outside the range being requested by the energy-utilization equipment.
>
> d. Low-voltage DC power that does not present a shock hazard is provided to appliances that can utilize it.

Chapter Eight

ARTICLE 800. COMMUNICATIONS CIRCUITS

800-1. Scope. The sections of this chapter apply basically to those systems which are connected to a central station and operate as parts of a central-station system.

The paragraph titled "Code Arrangement" of the "Introduction to the Code" (Sec. 90-3) states that Chap. 8, which includes Art. 800, "Communication Circuits," is independent of the preceding chapters except as they are specifically referred to.

800-52. Installation of Communications Wires, Cables, and Equipment. It should be noted that the requirements of Secs. 800-50 and 800-51, covering the listing and marking of the various types of communications wires and cables, are related to the rules of Sec. 800-52 and give authorities enforcing the Code the tools whereby they can judge whether the types of cables used in communication circuits substantially contribute to the hazards during fire conditions where fire fighters must of necessity be subjected to these products of combustion. Statistics show that most people die from smoke and products of combustion and not from the heat of fires. Therefore, it is imperative that electrical materials which contribute products of combustion be held to a minimum. The requirements give electrical inspectors and fire marshals criteria whereby they can judge the hazards which conductors contribute to fire problems.

See discussion of the Exception to Sec. 725-2(b).

ARTICLE 810. RADIO AND TELEVISION EQUIPMENT

810-13. Avoidance of Contacts with Conductors of Other Systems. For service drops and conductors on the exteriors of buildings, the requirements for insulating covering and methods of installation depend upon the likelihood of crosses occurring between signal conductors and light or power conductors. Where communication wires are run on poles in streets, it is assumed that they are exposed to contact with other wires. But where the overhead wires are run in an alley or from building to building and kept away from streets, lighter insulation is permitted.

Where a communication system is connected to a distribution system that is entirely underground except within the block in which the building is located, and any overhead wires in alleys or attached to buildings are not likely to become crossed with light or power wires, nearly all restrictions as to insulating covering and methods of installation are eliminated.

810-14. Splices. The antenna may unavoidably be so located that in case of a break in the wire, it may come in contact with electric light or power wires. For this reason, the wire should be of sufficient size to have considerable mechanical strength, and the joints should be as reliable as the wire. Joints will have sufficient mechanical strength if properly made with the standard double-tube connectors used in telephone and telegraph work.

810-19. Electric Supply Circuits Used in Lieu of Antenna—Receiving Stations. The device referred to usually consists of a small fixed capacitor connected between one wire of the lighting circuit and the antenna terminal of the receiving set. As most receiving sets are arranged, a breakdown in this capacitor would result in a short circuit to ground through the antenna coil of the set, and the capacitor should therefore be one that is designed for operation at 300 V or higher in which mica is used as the dielectric so that it will have a high factor of safety.

810-20. Antenna Discharge Units—Receiving Stations. Where the lead-in is enclosed in a continuous metallic shield, i.e., is run in rigid conduit or electrical metallic tubing, or consists of a lead-covered conductor or pair of conductors, and the metallic enclosure is well grounded, a lightning discharge will usually jump from the lead-in conductor to the metallic shield, because this path to ground offers a much lower impedance than the path through the antenna coil of the receiving set. A lightning arrester is therefore not required where the lead-in is so shielded.

810-21. Grounding Conductors—Receiving Stations. In order to avoid potential differences between various masses of metal, in or on buildings, and lead-in conductors, the metal portions of antenna masts should never be grounded to soil pipes, soil vent pipes, metal gutters, downspouts, etc. In other words, grounding must be done in accordance with Art. 250, and it is required to use the same grounding electrode for the grounding of masts as for the electrical system in the building.

810-54. Clearance on Building. The creepage distance is the distance from the conductor to the building measured on the surface of the supporting insulator. The air gap is the distance measured straight across from the conductor to the building.

810-57. Antenna Discharge Units—Transmitting Stations. A transmitting station should be protected against lightning, either by an arrester or by a switch that connects the lead-in to ground and is kept closed at all times when the station is not in operation.

Chapter Nine

A. TABLES

A major revision of the tables covering conduit fill was completed for the 1996 Code. Table 1, which used to differentiate between lead-covered and all other conductors, no longer does so. As a result, the table has been abbreviated. Now, it shows a 53 percent fill for one wire, a 31 percent fill for two wires, and for more than two wires, it shows 40 percent.

In previous editions of the NEC, Tables 2, 3A, 3B, and 3C in Chap. 9 were used to determine the maximum number of fixture wires (Table 2) and building wires (3A, 3B, and 3C) permitted in *any* conduit *or* tubing—without regard for the actual type of conduit *or* tubing—provided all conductors were of the same size and insulation type. This procedure, however, ignored the fact that the internal dimensions of the various conduits and tubing recognized by Chap. 3 were *not* the same. As a result the area inside the various raceways was different and, therefore, the permissible number of conductors should also be different.

The tables that used to apply for determining the maximum number of conductors in *either* conduit or tubing where the conductors are all of the same size and type of insulation—Tables 2, 3A, 3B, and 3C—have been deleted. As of the 1996 NEC, where a raceway contains conductors of all the same size and insulation type, fill must be in accordance with the table that covers the specific raceway or tubing—e.g., Table C8 for rigid metal conduit or Table C5 for liquidtight flexible nonmetallic conduit—that is being used.

In addition to the replacement of Tables 2, 3A, 3B, and 3C with the tables in the new Appendix C, Tables 4, 5, 5A, and 5B have also been radically revised to correct errors that previously existed. Table 4 was revised to permit greater fill for those raceways with greater internal area. And Tables 5, 5A, and 5B were revised to correct inaccuracies regarding the dimensions of various conductor types. The revised tables still appear in Chap. 9.

Table 1 does not apply where conduit sleeves are used to protect various types of cables from physical damage. As indicated in Note 2, it applies only to "complete systems." And Note 1 directs the reader to the new Appendix C for conduit fill where all conductors are the same size and insulation type.

While Note 3 mentions bare (as well as insulated) equipment grounding or bonding conductors, Note 8 to **Code** Table 1 applies to all forms of *bare* conductors (equipment grounding conductors and neutral or grounded conductors). Where any bare conductors are used in conduit or tubing, the dimensions given in Table 8 may be used. Since *all* wires utilize space in raceways, they must be counted in calculating raceway sizes whether the conductors are insulated or bare.

In regard to Note 5, there are conductors (particularly high-voltage types) that do not have dimensions listed in Chap. 9. Conduit sizes for such conductors may be determined by computing the cross-sectional area of each conductor as follows:

$$D^2 \times 0.7854 = \text{cross-sectional area}$$

where D = outside diameter of conductor, including insulation. Then the proper conduit size can be determined by applying Tables 1 and 4 for the appropriate number of conductors.

example Three single-conductor, 5-kV cables are to be installed in conduit. The outside diameter (D) of each conductor is 0.750 in. Then $0.750^2 \times 0.7854 \times 3 = 1.3253$ sq in. From Tables 1 and 4 (40 percent fill) a 2-in. rigid metal conduit would be permitted. **Code** Tables in Appendix C are based on Table 1 allowable percentage fills, and have been provided for the sake of convenience. Although revisions have been made to eliminate conflict, in any calculation, Table 1 is the table to be used where any conflict may occur.

Table 1 is also used for computing conduit sizes where various sizes of conductors or conductor types are to be used in the same conduit.

An example of Note 4 would be to determine how many No. 14 Type TW conductors would be permitted in a ½-in. rigid metal conduit. For three or more such conductors, Table 1 permits a 40 percent fill. From Table 4, 40 percent of the internal cross-sectional area of a ½-in. rigid metal conduit is 0.125 sq in. From Table 5 the cross-sectional area of a No. 14 type TW conductor is 0.0139 sq in. Thus 0.125/0.0139 = 8.9 or 9 such conductors would be permitted in a ½-in. conduit. Where the decimal is less than 0.8 (such as 0.7), the decimal would be dropped and the whole number would be the maximum number of equally sized conductors permitted; e.g., 8.7 would be 8 conductors (see Note 7).

The following is an example for computing a conduit size for various conductor sizes:

Number	Wire Size and Type	Table 5 Cross-sectional Area (ea.)	Subtotal Cross-sectional Area
3	No. 10 TW	0.0243	0.0729
3	No. 12 TW	0.0181	0.0543
3	No. 6 TW	0.0726	0.2178
		Total cross-sectional area	0.345

Table 1 permits a 40 percent fill for three or more conductors. Following the 40 percent column in Table 4, 1-in. rigid metal conduit would be required for these nine conductors, which have a combined cross-sectional area of 0.345 sq in.

Table 5A gives the maximum number of compact conductors permitted in trade sizes of conduit or tubing. Conductors with "compact-strand" construction have the cross-section areas of their strands shaped as trapezoids to provide tight "nesting" of the strands when they are twisted together. Such construction eliminates the air voids that occur when individual strands of circular cross section are twisted together and results in a smaller overall diameter of the total bundle of strands. Thus a 600-kcmil compact-strand assembly has an overall cross-section area of a conventional 500 kcmil with circular strands.

Note 6 of part **A. Tables** in Chap. 9 recognizes the fact that compact-strand conductors have increased conduit fill because of their overall smaller area. To cover such compact-strand conductors, Table 5A is in the NEC.

Table 10 gives "Expansion Characteristics of PVC Rigid Nonmetallic Conduit." This is an excellent source of important engineering data that ensures proper design, layout, and installation of rigid nonmetallic conduit runs. The substantiation for this proposed Code change stated as follows:

> Section 347-9 alerts an inspector to the requirement for expansion joints for rigid nonmetallic conduit. Now, the inspector must rely on manufacturer's literature, which is generally not available at the jobsite, for expansion characteristics. Generally, the expansion coupling will indicate an expansion/contraction range. With this chart included in Chap. 9, the inspector can determine the proper number of expansion couplings for a conduit run in an accurate, timely fashion.

And the same benefit will accrue to design engineers, electrical contractors, and operating personnel.

B. EXAMPLES

In Chap. 9, part **B,** the NE Code offers sample calculations using Code rules to determine load currents for several types of occupancies. It has been our experience that much confusion and many questions arise in applying the provisions of Arts. 220 and 230 to computing branch-circuit, feeder, and service loads in single-family residences. To clarify the brief calculation data given in the Code book, the following presents an expanded discussion of the steps covered in Examples 1(a) and 1(b). This will explore in detail NE Code examples of sizing the services and circuits for single-family dwellings (including individual apartments in multifamily dwellings), with and without air-conditioning units.

Two general procedures spelled out at the beginning of part **B,** Chap. 9, involve voltage values to be used and the method of handling fractions of an ampere in the calculations.

To standardize calculations, part **B** of Chap. 9 in previous NEC editions had specified that nominal voltages of 230 and 115 V are to be used in computing the ampere load on a conductor. [Dividing these voltages into the watts load will produce higher current values than would 240 and 120 V, thus resulting in larger (safer) conductor sizes.]

Now, however, the paragraph on "voltage" in part **B** has designated voltage values as "120, 240/120, 240, and 208Y/120." This correlates with Sec. 220-2, which further designates those nominal values to be used with other than single-phase and 3-phase 120-V systems. Although it says that those values "shall be used in computing the ampere load on the conductor," there would be no Code violation in making all ampere calculations with a voltage of 115 or multiples of 115 (or even lower values). Use of the lowest possible voltage value that might be encountered assures the greater adequacy of higher-ampere values when kilowatt or kilovoltampere ratings are divided by voltage to arrive at current values. The higher current values would dictate use of larger conductor sizes, with greater adequacy and safety. The use of values higher than 120 V (or multiples of it) would be a violation and would result in lower ampere values.

Where a particular calculation produces a current value involving a fraction of an ampere, the fraction may be dropped if it is 0.4 or less. Presumably, a value such as 20.7 A should be continued to be used as 20.7. We have chosen here to round off such values as the next higher whole number, in this case 21 A. Again, this is on the safe side. There are occasions, however, when current values must be added together. In such cases, it is on the safe side to retain fractions less than 0.5, since several fractions added together can result in the next whole ampere.

It is assumed that the loads in the following examples are properly balanced on the system. If they are not properly balanced on the system, additional feeder capacity may be required.

Example No. 1(a). One-Family Dwelling. The basic dwelling we will consider here is assumed to have a total usable floor area of 1,500 sq ft. As indicated in Sec. 220-3(b), when load is determined on a voltamperes/square feet basis, those areas not used as normal living quarters are excluded from the area calculation. Open porches, garages, unfinished basements and attics, and unused areas are not counted as part of the house area. Area calculation is made using the *outside* dimensions of the "building, apartment, or other area involved."

example A two-story dwelling 30 by 25 ft. First and second floors 30 × 25 ft × 2 = 1,500 sq ft. The "floor" area is computed from the "outside" dimension of the building and multiplied by the number of floors. [Section 220-3(b).]

Cooking will be done using a 12-kW electric range. And the house has a 5.5-kW 240-V electric dryer. The kilowatt ratings are taken as kilovoltampere values.

The steps taken in arriving at the branch-circuit and feeder loads follow.

General Lighting Circuits

Because lighting usage in dwelling occupancies is a random, noncontinuous application—with no control over sizes and types of light bulbs—the Code simply requires that a minimum amount of branch-circuit capacity be provided. Based on experience, the minimum required branch-circuit capacity will accommodate a fairly heavy and extensive use of general-purpose lighting fixtures in a home. Of course, if a given dwelling is provided with an unusually heavy amount of built-in indoor and outdoor lighting, then the provision of specific branch circuits for the loads is the best design approach.

Calculation of the **Code** minimum required branch-circuit capacity for general lighting is done also to determine the required capacity in feeders and service-entrance conductors.

In Sec. 220-3(b), the **NE Code** requires a minimum unit load in voltamperes/square feet for general lighting in the various types of occupancies listed in Table 220-3(b). For a dwelling occupancy (other than a hotel), circuit capacity for general lighting must be not less than 3 VA/sq ft times the square-foot area of usable living space. For the dwelling in this example, *minimum capacity for general lighting* would be

$$1{,}500 \text{ sq ft} \times 3 \text{ VA/sq ft, or } 4{,}500 \text{ VA}$$

When the total load capacity of branch circuits for general lighting is known, it is a simple matter to determine how many lighting circuits are needed. By dividing the total load by 120 V, the total current capacity of circuits is determined:

$$\frac{4{,}500 \text{ VA}}{120 \text{ V}} = 37.5 \text{ A}$$

Then, using either 15- or 20-A, 2-wire, 120-V circuits (and not dropping the major fraction of an ampere),

$$\frac{3.75 \text{ A}}{15 \text{ A}} = 2.5$$

which means three 15-A circuits.

—OR—

$$\frac{37.5 \text{ A}}{20 \text{ A}} = 1.87$$

which means two 20-A circuits.

Small-Appliance Circuits

The next step is to provide for at least 20-A, 2-wire circuits to supply *only* those designated receptacle outlets in the kitchen, pantry, breakfast room, dining room, family room, and similar areas. Section 220-4(b) requires a minimum of *two* such small-appliance circuits.

In addition, Sec. 220-4(c) requires at least one 20-A, 2-wire appliance circuit for the receptacle outlet required by Sec. 210-52(f) at the laundry location.

Range Circuit

A branch circuit for the 12-kW range is selected in accordance with Note 4 of Table 220-19, which says that the branch circuit load for a range may be selected from the table itself. Under the heading "Number of Appliances," read across from 1. The maximum demand to be used in sizing the range circuit for a 12-kW range is shown under the heading "Maximum Demand" to be not less than 8 kW. The minimum rating of the range-circuit ungrounded conductors will thus be

$$\frac{8{,}000 \text{ VA}}{240 \text{ V}} = 33.33 \text{ or } 33 \text{ A}$$

Table 310-16 shows that the minimum size of copper conductors that may be used is No. 8 (TW—40 A, THW—50 A, XHHW or THHN—55 A). A 40-A circuit rating is also designated in Sec. 210-19(b) as the minimum size of conductor for any range rated 8¾ kW or more. And the UL regulation calls for 60°C circuit conductors for sizes No. 14 to No. 1—or use of 75 or 90°C conductors at the ampacity of the corresponding size of 60°C conductor.

The overload protection for this circuit of No. 8 TW conductors would be 40-A *fuses or a* 40-A *circuit breaker.* THW or THHN or XHHW conductors must be used as if they are TW (60°C) wires and should not be protected at their higher ampacities.

Although the two hot legs of the 240/120-V, 3-wire circuit must be not smaller than No. 8, Exception No. 2 of Sec. 210-19(b) permits the neutral conductor to be smaller, but it specifies that it must have an ampacity not less than 70 percent of the branch-circuit rating (the rating of the protective device) and may never be smaller than No. 10.

For the range circuit in this example, the neutral may be rated

$$70 \text{ percent} \times 40 \text{ A (the rating of the CB of fuses)} = 28 \text{ A}$$

This calls for a *No.* 10 *neutral.*

If THHN or XHHW conductors are used for the hot legs, they must be used as 40-A wires, which calls for the No. 10 neutral. The No. 10 neutral would be acceptable for *any* of the conductor insulations (Fig. 1).

The branch circuit for the dryer is sized from Sec. 220-18 at 5,000 W (VA) or the nameplate rating, whichever is greater. The nameplate value of 5,500 VA (5.5 kW) is used here:

$$\frac{5{,}500 \text{ VA}}{240 \text{ V}} = 22.9 \text{ or } 23 \text{ A}$$

That calls for No. 10 wires.

Service Conductors

After calculating the required circuits for all the loads in the dwelling, the next step is to determine the minimum required size of service-entrance conductors to supply the entire connected load.

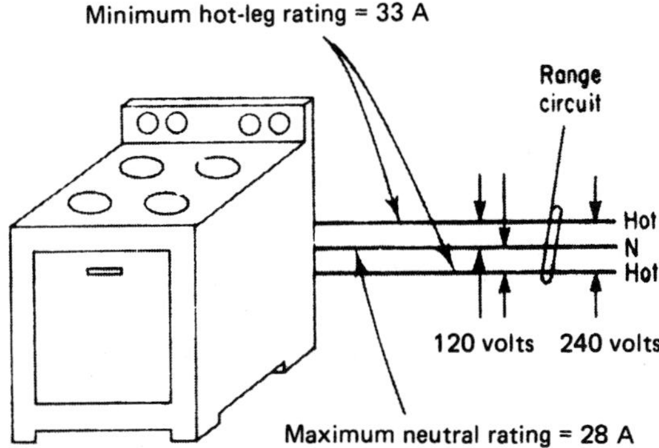

Fig. 1. Neutral for range is only 70 percent of the rating of the circuit protective device. (Chap. 9.)

The NE Code procedure is the same as sizing feeder conductors for the entire load—as set forth in Sec. 220-10. Basically, the service "feeder" capacity must be not less than the sum of the loads on the branch circuits for the different applications.

The *general lighting load* is subject to demand factors from Table 220-11, which takes into account the fact that simultaneous operation of all branch-circuit loads, or even a large part of them, is highly unlikely. Thus, feeder capacity does not have to equal the connected load.

Section 220-16(a) and (b) permits the *small appliance loads* to be added to the general lighting load before applying the demand factor from Table 220-11.

General lighting (Three 15-A or two 20-A circuits)	4,500 VA
Kitchen appliance load (two circuits) [1,500 VA/circuit, Sec. 220-16(a)]	3,000 VA
Laundry load [One circuit, Sec. 220-16(b)]	1,500 VA
Total	9,000 VA

Then the demand factors are applied:

3,000 VA at 100 percent	3,000 VA
6,000 (9,000 – 3,000) VA at 35 percent	2,100 VA
Basic feeder load	5,100 VA

The feeder demand load for the 12-kW range must be added to the 5,100 VA. As stated in Sec. 220-19, the range feeder demand load is selected from Code Table 220-19. In this case, it is 8 kW (column A, one appliance).

Basic load	5,100 VA
Range feeder capacity	8,000 VA
Dryer load	5,500 VA
Total feeder load	18,600 VA

The minimum required ampacity of the ungrounded service-entrance conductors of a 240/120-V, 3-wire, single-phase service is readily found by dividing the *total feeder load* by 240 V:

$$\frac{18{,}600\text{ VA}}{240\text{ V}} = 77.5 \text{ or } 78\text{ A}$$

Although a load current of that rating could be readily supplied by No. 3 copper TW, THW, THHN, or XHHW conductors, there is an important provision of Sec. 230-42(b) (2) that comes into play here. Where the initial computed total feeder load is 10 kW or more for a single-family dwelling, the ungrounded conductors of a service feeder must be rated at least 100 A; i.e., **the minimum capacity of the service-entrance hot legs = 100 A** (Fig. 2).

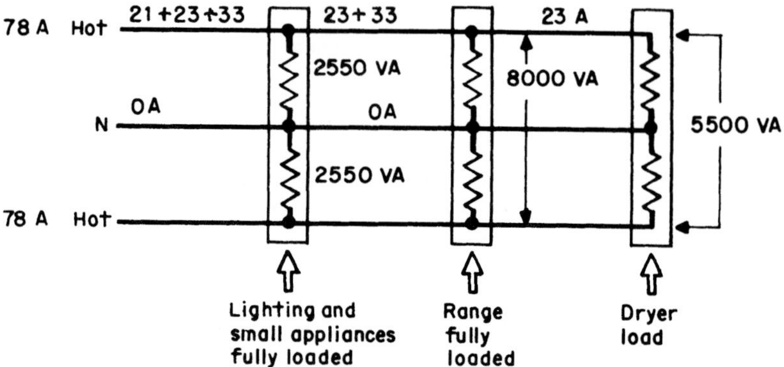

Fig. 2. Service hot legs are based on total demand load. (Chap. 9.)

That would call for a minimum of one of the following:

- No. 1 TW copper conductors (110 A)
- No. 1 THW copper conductors (130 A)
- No. 1 THHN copper conductors (150 A)
- No. 1 XHHW copper conductors (150 A)
- No. 1/0 TW, THW, THHN, or XHHW aluminum conductors (100 to 135 A)

IMPORTANT NOTE: It is a basic UL requirement that, unless otherwise marked, listed switches, CBs, and panelboards be used *only* with 60°C conductors in conductor sizes from No. 14 up to No. 1. The terminals on switches and breakers used as service equipment require the use of TW wire, in this case; or if higher temperature wires are used (THW, THHN, XHHW), they must be used at the ampacity of the corresponding size of TW wire. It would violate that UL rule to use No. 2 or 3 copper THW wire at its 115- or 110-A rating for the service conductors, because TW wire in those sizes does not have a 100-A rating and would operate too hot with a 100-A load on it. Of course, in the example here, the calculated demand load current is only 78 A, and the 100-A rating of service-entrance conductors is dictated by the Code on the basis of experience with load additions over the life of the system. But when and if demand load does come up to 100 A, effective operation of a 100-A service CB or fused switch or panelboard would then require that the thermal condition at terminals not exceed that produced when the 100-A demand load is supplied by No. 1 TW (60°C) conductors, rated 110 A. In any situations that pose difficulty in correlating NE Code rules and UL limiting conditions, it is in the best interests of real, long-time economy and reliable life of equipment to resolve all choices of conductor selection in favor of larger size and greater adequacy. Certainly, experience indicates that far too many services have to be increased in capacity shortly after installation simply because initial design was skimpy or based on squeezing as much as possible out of "apparent" ratings of equipment. Although virtually all new overcurrent devices, panelboards, etc., are now marked for use with 75°C insulated conductors at their full 75°C ampacity, where performing renovations, remodeling, or adding onto existing systems, verify the temperature limitations of all terminations. Remember, the conductor's ampacity must be correlated with the equipment terminations at both ends of the conductor.

Service Neutral

Because a neutral conductor of a 3-wire, single-phase service carries only the unbalance (or difference in) current of the two hot legs, the NE Code does permit the very realistic reduction of neutral conductor size from the size of the service hot legs.

Section 220-22, which covers sizing of any feeder neutral, also applies to the neutral of a service. The neutral must have an ampacity at least equal to the maximum unbalance of loads connected from the two hot legs to the neutral. This load is taken to be "the maximum connected load between the neutral and any one ungrounded conductor" (hot leg). On a service of the type considered here, it is assumed that all the 120-V loads will be divided between the two hot legs, with one half connected from one hot leg to neutral and the other half connected from the other hot leg to neutral. Section 220-4(d) requires loads to be evenly proportioned among the branch circuits to assure optimum sharing of load by the hot legs of the service (Fig. 3). Thus, in the example here, half of the 5,100-W basic demand load for the 120-V loads can be considered connected from either hot leg to neutral, and that would require ampacity in the neutral of

$$\frac{5{,}100\text{ W}}{2} \div 120\text{ V} = 21.25\text{ A}$$

Then, in addition, the neutral must have capacity for some unbalance in the 3-wire, 240/120-V circuit to the 12-kW range and in the 3-wire circuit to the dryer. Section 220-22 notes that the feeder neutral load (i.e., the required ampere capacity in the neutral service-entrance conductor) may be taken to be "70 percent of the load on the ungrounded conductors." This recognizes that a circuit to a range or dryer usually has some load connected from each hot leg to neutral, requiring the neutral to carry *only* the unbalance. From the above calculation of the current load on the hot legs of the range branch circuit (using an 8,000-VA demand load for the 8,000-W range), it was found that the load was 33 (actually 33.33) A. Then, as shown in Fig. 4, the feeder neutral ampacity for the range load must be not less than

$$70\text{ percent} \times 33.33\text{ A} = 23.33\text{ A}$$

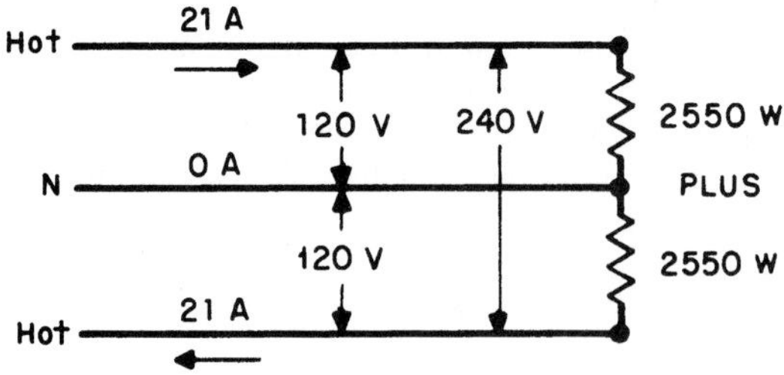

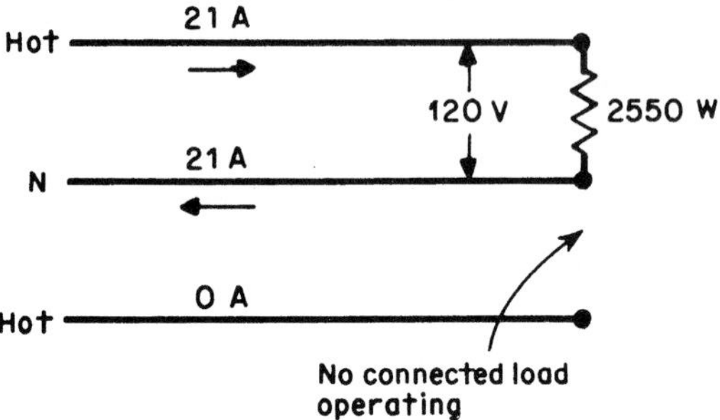

Fig. 3. Half of the total 120-V load is considered to be from hot leg to neutral. (Chap. 9.)

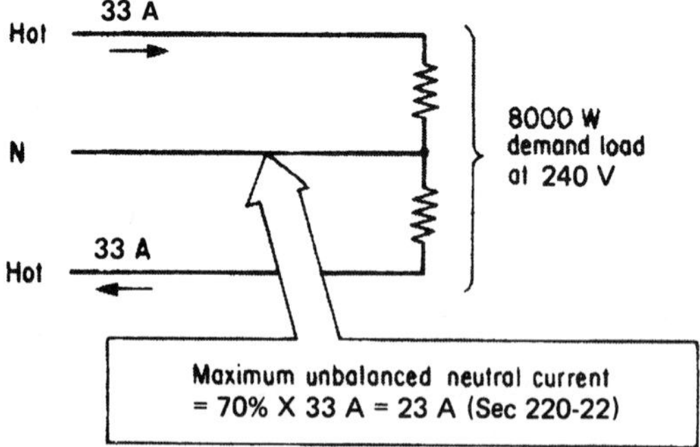

Fig. 4. Range neutral load is taken at 70 percent of hot leg current. (Chap. 9.)

And the feeder neutral ampacity for the dryer load must be at least

$$70 \text{ percent} \times 22.9 \text{ A} = 16.03 \text{ A}$$

Adding the three values of minimum neutral conductor ampacity for the different loads, the total minimum required neutral conductor ampacity is determined:

Neutral load, lights and appliances	21.25 A
Neutral load, range circuit	23.33 A
Neutral load, dryer circuit	16.03 A
Total neutral load	60.61 A

Since 0.61 is greater than 0.5, the total neutral load should be rounded off from 60.61 A to 61 A. From the above, it is evident that some judgment must be exercised in dropping fractions. If the contributions to the neutral load had been rounded off to 21, 23, and 16 A, respectively, the total would have been 60 instead of 61. We chose to stay with the larger value since, if it ultimately makes any difference at all, it will be in the safe direction.

Of course, as with so many other calculations, there is more than one way to arrive at the same result. For instance, the total lighting and appliance load can be added to the total range demand load and dryer demand load, with both of the latter modified by a 70 percent neutral demand factor, and the grand total divided by 240 V to get the required amp capacity of the service neutral:

Lighting and small appliance load	5,100 VA
Range load (8,000 VA × 70 percent)	5,600 VA
Dryer load (5,500 VA × 70 percent)	3,850 VA
Total	14,550 W

Then, since that is the total load that forms the basis for feeder unbalance to determine the required neutral capacity, there are two ways to determine the neutral load.

1. Under conditions of maximum possible unbalance, half of that total divided by 120 V yields the result:

$$\frac{14{,}550 \text{ VA}}{2} \div 120 \text{ V} = 60.6 \text{ A or } 61 \text{ A}$$

—OR—

2. Simply take the total load and divide it by 240 V, giving the same result:

$$\frac{14{,}550 \text{ VA}}{240 \text{ V}} = 60.6 \text{ A or } 61 \text{ A}$$

Selection of the neutral conductor can then be made from Table 310-16 as required for either copper or aluminum conductor in a raceway or service cable, with 60°C insulation or higher-temperature conductors of the same size as required for a 60°C conductor as may be required or permitted [see Sec. 110-14(c)]. The above calculations dictate the following conductor choices for this example:

Copper: Not less than No. 4 (TW, THW, THHN, XHHW)
Aluminum: Not less than No. 3 (TW, THW, THHN, XHHW)

Single-Family Dwelling with Air-Conditioning Units

Example No. 1(b) in Chap. 9 of the NE Code deals with the same dwelling as the previous example (1,500 sq ft of living space with a 12-kW range and a 5.5-kW dryer) plus one 6-A 230-V room air-conditioning unit, one 12-A, 120-V room air-conditioning unit, one 10-A, 115-V dishwasher, and one 8-A, 115-V disposal unit. The approach to the calculations is the same as in the preceding example, with simple addition of the extra loads. The amp loadings on the two hot legs and the neutral break down as shown in the following table:

Line A (One Hot Leg), amps	Neutral Leg, amps	Line B (Other Hot Leg), amps	Load
78	61	78	From first example
6	—	6	One 230-V A/C
12	12	10	One 115-V A/C and one 115-V dishwasher
—	8	8	One 115-V disposal
3	3	3	25% largest motor load
99	84	105	Total

Because the unit air conditioners are cord and plug-connected units, each may be treated as a single motor load of the A/C unit's nameplate rating. Room air conditioners are considered to be fixed appliances and must be added into required service-entrance conductor capacity as required by Sec. 220-17. As indicated in the listing above, the 6-A, 230-V A/C unit places a 6-A load on each service hot leg but has no effect on neutral capacity because it has no neutral supply conductor. The three 115-V loads are divided as shown on the two service hot legs—the 115-V A/C unit connected from line A to neutral and the other two 115-V units connected from line B to neutral. Maximum unbalance of those loads would exist when the 12-A A/C unit on line A is off and the 10- and 8-A loads on line B are running. The current drawn by the neutral under that condition would be 18 A to match the 18 A (10 + 8) drawn by the 10-A dishwasher and the 8-A disposal unit. Thus, 18 A of load capacity must be added to the 61-A neutral load originally calculated.

Section 430-24, on sizing of conductors supplying several motors, requires that such conductors have an ampacity equal to the sum of the full-load current ratings of all the motors plus 25 percent of the highest-rated motor in the group. Because the 12-A-rated 120-V unit has the highest current rating of the motors, a value of 3 A (25 percent × 12 A) is added to line A hot leg and 25 percent × 10, or 2.5 A, is taken as an addition of 3 to line B hot leg. This is not required, but is surely desired. A load of 3 A is also added to the neutral because each 120-V motor load is connected hot-leg-to-neutral, and the neutral is part of the motor circuit and thus is a conductor "supplying several motors."

Note that the calculated demand load here (99 A for line A and 105 A for line B) is above the minimum 100-A required rating of service conductors from Sec. 230-42(b)(2). The service hot-leg conductors may be No. 1 TW copper conductors, with their 110-A ampacity or No. 2/0 TW aluminum conductors, which are rated at 115 A.

The neutral conductor for the service conductors to this house will have to be rated for not less than 61 A + 18 A + 3 A, or 82 A. Such a neutral conductor would be:

No. 3 copper TW, THW, THHN or XHHW; or

No. 1 aluminum TW, THW, THHN, or XHHW.

Index

Boldface numbers refer to article numbers, *not* page numbers.

ABOUT THE AUTHORS

Joseph F. McPartland is the nation's foremost expert on electrical system design and construction. He is the former publisher of *Electrical Design and Installation* magazine. Known throughout the industry as "Mr. Electrical Construction," Mr. McPartland is the author or coauthor of 28 books on electrical design, electrical construction methods, electrical equipment, and the National Electrical Code®, including the popular *Handbook of Practical Electrical Design* and *McGraw-Hill's Handbook of Electrical Construction Calculations.*

Brian J. McPartland is an electrical consultant and editor of *Electrical Contractors: Design and Installation Update,* a bimonthly technical report. He was chief editor at "edi" (*Electrical Design and Installation*) magazine and an assistant editor at *EC&M* magazine. Both he and Joseph F. McPartland are co-authors of *McGraw-Hill's National Electrical Code® Yearbook Supplements, McGraw-Hill's Handbook of Electrical Construction Calculations,* and *McGraw-Hill's Handbook of Practical Electrical Design.*